Visit the Ricklefs and Miller Web Companion at
www.whfreeman.com/biology

ECOLOGY

ECOLOGY

Fourth Edition

Robert E. Ricklefs
University of Missouri at St. Louis

Gary L. Miller
University of Mississippi

W. H. FREEMAN AND COMPANY
NEW YORK

Executive Editor: Sara Tenney
Senior Editor/Development Editor: Deborah Allen
Editoral Assistant: Jessica Olshen
Project Editor: Diane Maass
Cover and Text Designer: Diana Blume
Illustration Coordinator: Lou Capaldo
Illustration: Fine Line Illustrations
Production Coordinator: Paul W. Rohloff
Composition: Progressive Information Technologies
Manufacturing: RR Donnelley and Sons Company

Cover: Jacqueline Bishop, "From the Vine to the Vein," 1992. The Roger Houston Ogden Collection. Photo, Arthur Roger Gallery.

Library of Congress Cataloging-in Publication Data
Ricklefs, Robert E.
Ecology/Robert E. Ricklefs and Gary L. Miller.—4th ed.
p. cm.
Includes bibliographical references (p.) and index.
ISBN 0-7167-2829-X
1. Ecology. I. Miller, Gary L. (Gary Leon), 1954-
II. Title.
QH541.R53 1999
577—dc21 99-18604
CIP

Printed in the United States of America

First printing 1999

To
Leon David Miller,
teacher

ABOUT THE AUTHORS

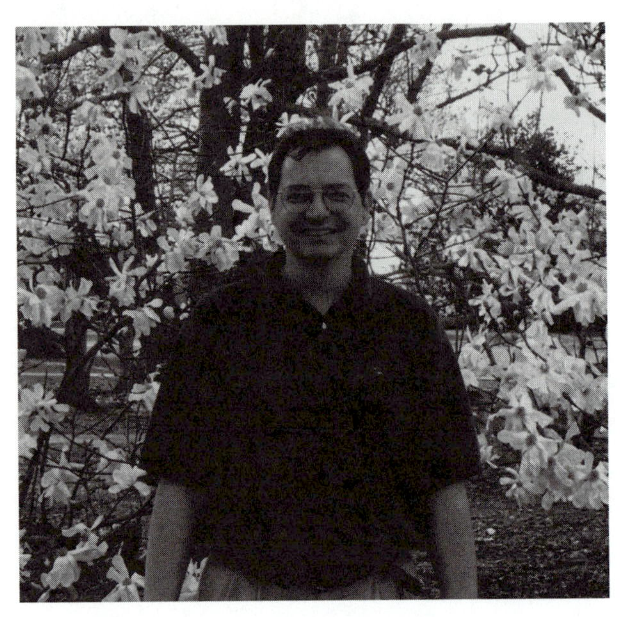

Robert E. Ricklefs (left) is Curators' Professor of Biology at the University of Missouri—St. Louis, where he joined the faculty in 1996 after 27 years in the Biology Department at the University of Pennsylvania. Professor Ricklefs is a native of California and holds an undergraduate degree from Stanford University and a Ph.D. from the University of Pennsylvania. His interests include the energetics of reproduction in birds, evolutionary differentiation of life histories, biogeography, and the historical development of biological communities, including the generation and maintenance of large-scale patterns of biodiversity. His research has taken him to a wide variety of habitats from the lowland tropics to seabird islands in Antarctica. Professor Ricklefs is a Fellow of the American Association for the Advancement of Science and the Academy of Arts and Sciences. He is the author of *The Economy of Nature* and co-author of the Scientific American Library volume *Aging: A Natural History,* both published by W. H. Freeman and Company.

Gary L. Miller (right) is Professor and Chair of the Department of Biology at the University of Mississippi, where he joined the faculty in 1989. Professor Miller grew up in the Shenandoah Valley of Virginia. He graduated with a B.S. in biology with a minor in anthropology from The College of William and Mary in 1976, and obtained his Ph.D. in Biological Sciences with an emphasis on statistics from Mississippi State University in 1982. His research focuses on the evolution of courtship behavior in wolf spiders. He has published in the areas of community and population ecology, theoretical ecology, and animal behavior. He is currently Associate Editor of the *Journal of Arachnology* and past editor of the Aquatic Ecology Section of the *Bulletin of the Ecological Society of America.* At the University of Mississippi he teaches courses in theoretical ecology and biostatistics and, prior to becoming chair, taught general ecology for six years. This is his first book published with W. H. Freeman and Company.

BRIEF TABLE OF CONTENTS

DETAILED TABLE OF CONTENTS

ANALYTICAL MODELS OF ECOLOGY

Allometric equation: $Y = aX^b$ 49

Y = physiological process, anatomical feature, or ecological or behavioral characteristic
X = measure of body size
a = constant of proportionality
b = allometric constant

Poisson distribution: $P(x) = M^x e^{-M}/x!$ 278

$P(x)$ = the probability of the value x
M = mean number of individuals per unit time or space
$x!$ = the factorial of x

Effective population size: $N_e = 4[(1/N_{males}) + (1/N_{females})]^{-1}$ 289

N_e = effective population size
N_{males} = effective population size for males
$N_{females}$ = effective population size for females
$N_{males} = [(N_m K_m - 1)]/[K_m + V_m/K_m) - 1]$
$N_{females} = [(N_f K_f - 1)]/[K_f + V_f/K_f) - 1]$
N_m and N_f are number of breeding males and females respectively
K_m and K_f are the average number of offspring produced by males and females respectively
V_m and V_f represent the variance in the number of offspring produced by each sex

Survivorship: $l_x = \Pi s_i$ 291

l_x = the probability that a newborn individual will be alive at age x
s_i = the probability of survival between ages x and $x + 1$

Exponential population growth: $N_t = N_0 e^{rt}$ 299

N_t = population density at time t
N_0 = initial population size ($t = 0$)
r = exponential growth rate

Geometric population growth: $N(t) = N(0)\lambda^t$ 302

$N(t)$ = population density at time t
$N(0)$ = initial population density
λ = geometric growth rate.

Relationship between exponential and geometric population growth rates: $r = \ln \lambda$. 302

Growth rate of age-structured population with overlapping generations: $dN/dt = N\int r_x c_x \, dx$ 303

dN/dt = rate of growth of the population
r_x = exponential growth rate for age x
c_x = proportion of individuals in the population that are at age x

Growth in age-structured population with discrete generations: $N(t + 1) = \Sigma n_x(t)s_x + \Sigma n_x(t)s_x b_{x+1}$ 304

$N(t + 1)$ = number of individuals in the population at time $t + 1$
$n_x(t)$ = number of individuals in age class x at time t
s_x = proportion of individuals in age class x that survive to the next age class
b_{x+1} = number of newborn contributed to the population by age class $x + 1$

The first part of the right side of the equation is the survival term, the second part is the birth term. The expression may be simplified to $\Sigma\, n_x(t)s_x(1 + b_{x+1})$

Lewis-Leslie matrix model of population growth: $A^t n(0) = n(t)$ 306

A^t = Lewis-Leslie matrix containing age specific survival birth rates
n(0) = a vector containing the number of individuals in each age class at time $t = 0$
n(t) = a vector containing the number of individuals in each age class at time $t = t + 1$

Characteristic equation of a population (Euler's equation): $1 = \Sigma e^{-rx} l_x b_x$ (for populations with exponential growth); $1 = \Sigma \lambda^{-x} l_x b_x$ (for populations with geometric growth) 309

r = exponential growth rate
λ = geometric growth rate
l_x = survivorship to age x
b_x = age-specific birth rate

Logistic equation of growth of population i: $dN_i/dt = r_i N_i(1 - N_i/K_i)$ 315

dN/dt = rate of population growth
N_i = population size
K_i = carrying capacity of population i
r_i = intrinsic growth rate of population i

Rate of change of proportion of occupied patches in a metapopulation: $dp/dt = mp(1 - p) - ep$ 331

dp/dt = rate of change in the proportion of occupied patches, p
m = rate of colonization of patches
p = proportion of total number of patches that are occupied
e = rate of patch extinction

General form of the population recruitment function: $N(t + 1) = f[N(t)]$ 351

$N(t + 1)$ = population size at time $t + 1$
$f[N(t)]$ = some function f of the population size at time t
$t = 0, 1, 2, \ldots$ (discrete-time model)

Ricker recruitment model: $N(t + 1) = N(t)e^{r(1-N(t)/K)}$ 353

$N(t + 1)$ = population size at time $t + 1$
$N(t)$ = population size at time t
K = carrying capacity
r = exponential growth
$t = 0, 1, 2, \ldots$ (discrete-time model)

Probability of extinction of a population near equilibrium: $p_0(t) = [bt/(1 + bt)]^N$ 365

$p_0(t)$ = probability of extinction at time t
b = birth rate
N = population size

Time to extinction of a population: $T = 2/Vc((K^c - 1)/c) - \ln K)$ 366

T = time to extinction (also known as persistence time)
V = variance in the intrinsic rate of increase, r
K = carrying capacity
$c = 2r/V - 1$

The Monod equation expressing population growth rate in terms of resource level: $1/C\, dC/dt = qR/(k + R)$ 394

$1/C\, dC/dt$ = per capita growth rate
C = size of the consumer population
q = growth rate of the consumer population in the absence of crowding
k = amount of resource at which the growth rate of the population is exactly one-half the maximum growth rate q
R = level of the resource

Lotka-Volterra model of the competitive effect of population j on population i: $dN_i/dt = r_i N_i(1 - N_i/K_i - a_{ij}N_j/K_i)$ 408

dN_i/dt = rate of growth of population i
r_i = intrinsic rate of increase of population i
N_i and N_j = size of population i and j respectively
K_i and K_j = carrying capacity of populations i and j respectively
a_{ij} = coefficient of competition; the effect of an individual of species j on the exponential growth rate of the population of species i.

(The complementary equation for the competitive effect of population i on population j is: $dN_j/dt = r_j N_j(1 - N_j/K_j - a_{ji}N_i/K_j)$)

General Lotka-Volterra equations for the rate of growth of a predator and its prey: $dH/dt = f(H,P)$ (prey population); $dP/dt = g(H, P)$ (predator population) 450

dH/dt and dP/dt = rates of growth of prey and predatory populations respectively
$f(H, P)$ = some function f of the population sizes of prey (H) and predator (P)
$g(H, P)$ = some function g of the population sizes of prey (H) and predator (P)

Lotka-Volterra model of the rate of growth of prey population in presence of predator: $dH/dt = rH - pHP$ 451

dH/dt = rate of growth of the prey population
r = intrinsic rate of population growth of the prey
H = size of the prey population
P = size of the predator population
p = proportion of encounters that result in a kill

Lotka-Volterra model of the rate of growth of a predator population: $dP/dt = apHP - dP$ 451

dP/dt = rate of growth of the predator population
H = size of the prey population
P = size of the predator population
p = proportion of encounters that result in a kill
a = efficiency with which prey is converted to predator reproduction

Nicholson-Bailey equations of parasitoid-host interactions: $H(t+1) = bH(t)[e^{-aP(t)}]$ **(host);** $P(t+1) = cH(t)[1 - e^{-aP(t)}]$ **(parasitoid)** 457

$H(t+1)$ = number of hosts in the next generation $(t+1)$
$P(t+1)$ = number of parasitoids in the next generation $(t+1)$
$H(t)$ and $P(t)$ = density of hosts and parasitoids respectively
a = search efficiency of the parasitoid
c = number of parasitoid offspring resulting from an attack of a host
b = per capita birth rate of hosts

Holling disk equation: $E = aHT/(1 + aHT_h)$ 459

E = number of encounters
H = density of prey
T_h = handling time per prey
T = total handling time
a = efficiency of searching

Lotka-Volterra model of predator-prey patch dynamics: $dH/dt = aH - bHM - cH$ **(prey);** $dP/dt = bHP - dP$ **(predator)** 476

dH/d and dP/dt = rate of increase of prey and predator populations respectively
H and P = size of prey and predator populations respectively
a = rate at which dispersing prey establish new colonies
b = rate at which predators invade prey patches
c = rate at which arbitrary patches of prey go extinct
d = rate at which arbitrary patches of predators go extinct

Infection model of pathogen population dynamics: $dx/dt = b(x + y + z) - \beta xy - dx$ **(susceptible portion of the population);** $dy/dt = \beta xy - (D + \gamma)y$ **(infected portion of the population);** $dz/dt = \gamma y - dx$ **(portion of the population that has recovered from the infection)** 498

dx/dt, dy/dt and dz/dt = the rate of change in the susceptible, infected, and recovered portions of the population respectively.
x, y, and z = the proportion of the population that is susceptible, infected, and recovered respectively
β = the transmission coefficient
d = constant rate of mortality of susceptible individuals
D = constant rate of mortality for the infected group
γ = recovery rate

Reproductive rate of infection: $R(x) = \beta x/(D + \gamma)$ 500

$R(x)$ = reproductive rate of infection; the average of the number of individuals infected in the lifetime of the infected individuals in a population
β = the transmission coefficient
D = constant rate of mortality for the infected group
γ = recovery rate

Species-area relationship: $S = cA^z$ **(linear form:** $\log S = \log c + z \log A$**)** 548

S = number of species
A = area
c and z = constants fitted to the data

May's community stability model: $b(SC)^{1/2} < 1$ 556

b = average strength of species interactions
S = number of species
C = food web connectance

Markov process of state transitions: $PN(t) = N(t + 1)$ 578

$N(t)$ = vector of states at time t
$N(t+1)$ = vector of states at time $t + 1$
P = matrix of transition probabilities

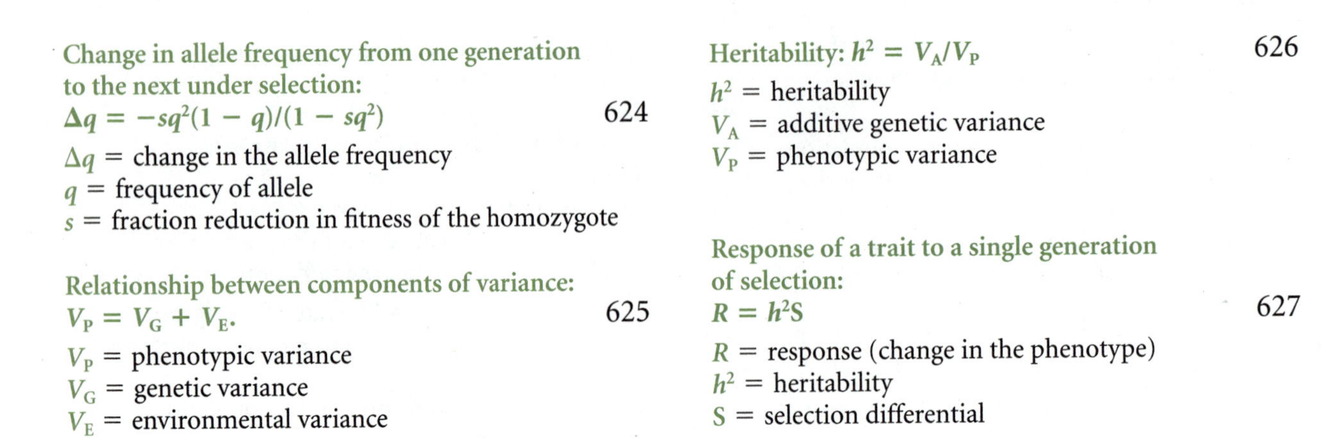

Change in allele frequency from one generation to the next under selection:
$\Delta q = -sq^2(1 - q)/(1 - sq^2)$ 624
Δq = change in the allele frequency
q = frequency of allele
s = fraction reduction in fitness of the homozygote

Relationship between components of variance:
$V_P = V_G + V_E$. 625
V_P = phenotypic variance
V_G = genetic variance
V_E = environmental variance

Heritability: $h^2 = V_A/V_P$ 626
h^2 = heritability
V_A = additive genetic variance
V_P = phenotypic variance

Response of a trait to a single generation of selection:
$R = h^2S$ 627
R = response (change in the phenotype)
h^2 = heritability
S = selection differential

MODELS AND TECHNIQUES OF ECOLOGY

Many different techniques and approaches are used in ecological study. Here you will find some of the most common and most important listed in the order that they first appear in the text.

HUMAN INTERFACES OF ECOLOGY

All the principles discussed in this book pertain to the human condition. Following is a list of some ideas from each chapter that will serve as a guide to the application of ecology to human activities.

PREFACE

Few fields of study have more direct relevance to the human condition than the field of ecology, the study of the interactions of organisms with their physical and biological environment. The human population continues to grow and with it the demand for food and the pressure of human activities on the natural systems that sustain life on earth. The increasing globalization of our economy and social and political structures has resulted in both intentional and accidental introductions of organisms, including pests and diseases, to all corners of the earth—ecological globalization on a grand scale. Industry, agriculture, and forestry, transform the energy of the natural world for human needs and in the process affect the systems from which the energy is derived. The long-term effects of our interactions with nature are uncertain.

It has been nearly a decade since the publication of the third edition of *Ecology.* This period has seen an increasing awareness of the importance of understanding the dynamics of the natural world. In response, we have seen a proliferation of ecological journals and organizations that focus on the application of ecological principles to conservation and restoration. Equally significant has been a shift in the place of ecology in undergraduate biology curricula and in programs of study for nonscience majors. Ecology is increasingly taught early in the undergraduate curriculum, and is now often required in basic courses at the freshman and sophomore levels. Juniors, seniors, and graduate students are now offered a greater variety of advanced ecology

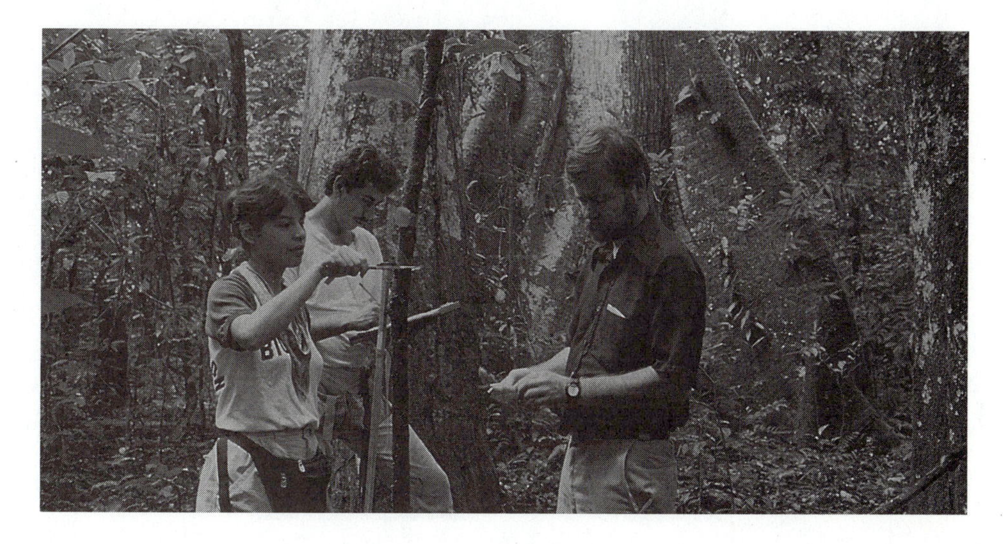

courses than ever before. Many premedical curricula now include courses emphasizing environmental interactions, parasite ecology, and the evolutionary dynamics of disease. Much of this shift in emphasis has been in response to student interest in courses that will give them an understanding of the natural world that they have come to see imperiled.

The new emphasis in ecological studies has greatly heartened us as we have prepared this fourth edition of *Ecology.* More than ever, we believe that there is a great need for a comprehensive survey of the field of ecology, with a historical perspective, an evolutionary foundation, and an unwavering emphasis on contemporary theory, which we believe are strengths of previous editions of *Ecology.* As with past editions, a major goal of this revision has been to reflect current trends and approaches in the field of ecology. We also believe that a comprehensive treatment of the field need not be inaccessible. Thus, our second goal was to substantially enhance the pedagogical features of the book through a new two-column design, several key organizational changes, and new pedagogical approaches.

Ecological themes and new emphases

As with previous editions of *Ecology* the essential principles of evolution and adaptation, energy flow, population and community interactions, and behavioral ecology form the foundation of our narrative. We retain the view that the responses of individual organisms to conditions of the physical and biotic environment form the basic information of ecological study. To this foundation, we have added new emphasis in several areas.

Methods of ecological study
Interactions of humans with the natural world
Importance of spatial scale in ecology
Global element cycles
Biodiversity
Conservation biology
Plant ecology
Coevolution and mutualism

Organization

Those of you who are familiar with previous editions of *Ecology* will recognize the overall organization. However, a number of important changes have made the book easier to use and help to articulate the relationships among the various subfields of ecology. Part 1 has been reduced to two introductory chapters that explain why the study of ecology is important and how ecological study is conducted. As before, we also use this section to emphasize the importance of evolution in the natural world.

The relationship between individual organisms and their physical and biotic environment is covered in Part 2. The emphasis on the physical and chemical features of the environment in this part of the book leads naturally to the discussion of the dynamics of energy and element cycling presented in Part 3. This early placement of ecosystems ecology, which is an innovation of earlier editions of *Ecology,* allows us to reinforce at an early point ideas about human interactions with the environment that are introduced in Part 2. The placement of Part 4 (population ecology) and Part 5 (population interactions) conform to the traditional organizational hierarchy of ecology texts. We have included new chapters on metapopulations (Chapter 17) and extinction, conservation, and restoration (Chapter 19) in Part 4. In Part 5 the order of presentation has been changed so that competition is discussed before predation, herbivory, parasitism, and mutualism. Treating herbivory and predation in separate chapters reflects a greater emphasis compared to previous editions on the interaction of plants and their consumers.

In this edition, the chapters on community ecology have been moved from the very end of the book to become Part 6, following directly the two parts dealing with populations. We have eliminated the chapter on the ecological niche and incorporated the most relevant material from that chapter into a new chapter focusing on species diversity. We now end the book with Part 7, which contains chapters on evolutionary ecology. The extensive treatment of evolutionary ecology is unique to *Ecology.* We believe that this section is significant both because it supports our emphasis on evolutionary theory and, perhaps more important, because it promotes the unification of ecology with behavior, genetics, and evolution. However, we understand that this material is often not covered in general ecology courses and, thus, the placement of it at the end may make the text easier to use.

Pedagogical enhancements

Much of ecological theory is complex and abstract. Thus, besides striving for a clear narrative, we have added a number of pedagogical devices that we believe will help students organize their thinking about ecology.

Among these are:

Opening questions to guide the study of each chapter and reinforce the importance of inquiry in scientific discovery
A detailed table of contents to provide a comprehensive summary of the book
Specialized tables of contents to help students find material quickly
Numbered chapter subheadings
Secondary headings to mark transitions and emphasize major themes
Chapter summaries
Chapter exercises including problems and thought questions where appropriate
Boldface key words that are defined in the narrative and included in the glossary
Cross-references between chapters to connect topics

Ecology was the first general textbook in the field to include a rigorous and comprehensive introduction to the mathematical theory of ecology. We continue to believe very strongly in the value of mathematical models in ecological study and, thus, we have retained the mathematical emphasis of the book. Because we also believe that ecological theory must be accessible to the beginning ecology student, we have explained the development of mathematical theory as thoroughly and clearly as possible. In most cases, numerical examples have been included in the narrative to emphasize major highlights of theory. Exercises are used in some cases to help develop a deeper, intuitive appreciation for models and quantitative relationships. With few exceptions, readers who can perform basic algebra, understand exponents, logarithms, and inequalities, and who are proficient at constructing and interpreting graphs, will find little difficulty with the mathematics presented here.

This revision has been the result of a collaboration between the senior author, who wrote the first three editions of this text, and the junior author, who has had the major responsibility for the fourth edition. We have enjoyed working together on the major features of the revision, and we have also benefited wonderfully from the efforts of the editorial and production staff at W. H. Freeman and many of our colleagues, who have offered information and insights and read countless drafts of chapters. Our greatest gratification will come if this book helps students learn about nature, appreciate the grave state of our relationship to the natural environment, and see how we can find remedies to environmental problems through a better understanding of ecological principles.

Acknowledgments

We were extremely fortunate to have had the opportunity to work with two fine editors at W. H. Freeman. Deborah Allen nurtured the book through its early development, organizing the reviews, editing the first revisions, and developing the basis of the new design. Her creativity, scientific knowledge, good humor, and constant encouragement moved the project forward at a crucial stage. Sara Tenney brought the project to a successful conclusion with great care and professionalism. Her insights regarding both the scientific content and pedagogical elements resulted in significant improvements. Norma Roche copyedited the manuscript with great skill and precision, improving it immeasurably. Diane Maass, Kathy Bendo, Jessica Olshen, Diana Blume, Lou Capaldo, and Paul Rohloff at W. H. Freeman were of enormous help.

Ken Klemow and Jack Grubaugh read two revisions of the entire manuscript, providing detailed comments on both style and content. This is a service for which we are most appreciative. Many fellow ecologists spent considerable time reading parts of the manuscript. Among these are: S. Allison, A. Blaustein, R. Brugam, R. Burke, G. Capelli, F. Chew, S. Dobson, N. Dronen, P. Ewald, D. Fong, C. Freeman, M. Fulton, S. Heard, V. Hutchison, T. Miller, P. Mulholland, H. Neufeld, M. Pigliucci, G. Polis, S. Reilly, D. Reznik, T. Schoener, W. Sousa, E. Temeles, and D. Wise. We discussed

many aspects of the book with S. Threlkeld, R. Holberton, S. Brewer, C. Ochs, G. Stratton, W. Garrison, and P. Miller, all of the ecology and evolution group at the University of Mississippi. B. Baca, J. Chambers, W. Houston, K. Overstreet, K. Rhew, S. Samples, and M. Villegas provided a thorough review of Part 3 as part of C. Och's graduate course in biogeochemistry. UM undergraduate students B. Myles and K. Macy served as research assistants and readers, and their work is much appreciated. L. Irving, A. Houston, A. Williams, and R. Landreth provided logistical support. S. Lewis and P. O'Neill provided encouragement at crucial times. We are most grateful to Martha Swan and the staff of the Interlibrary Loan Office of the J. D. Williams Library at the University of Mississippi for their tireless work on our behalf, and to Eileen Hebets at the University of Arizona and Bob Suter at Vassar College for providing us with so many photos. W. Miller was a most welcome source of encouragement. GLM is especially appreciative of the unfailing support of Mary Stuckey, who, having authored six scholarly books, knew precisely when to offer advice and encouragement and exactly when to just laugh it off. Thanks, Mary.

PART 1
INTRODUCTION

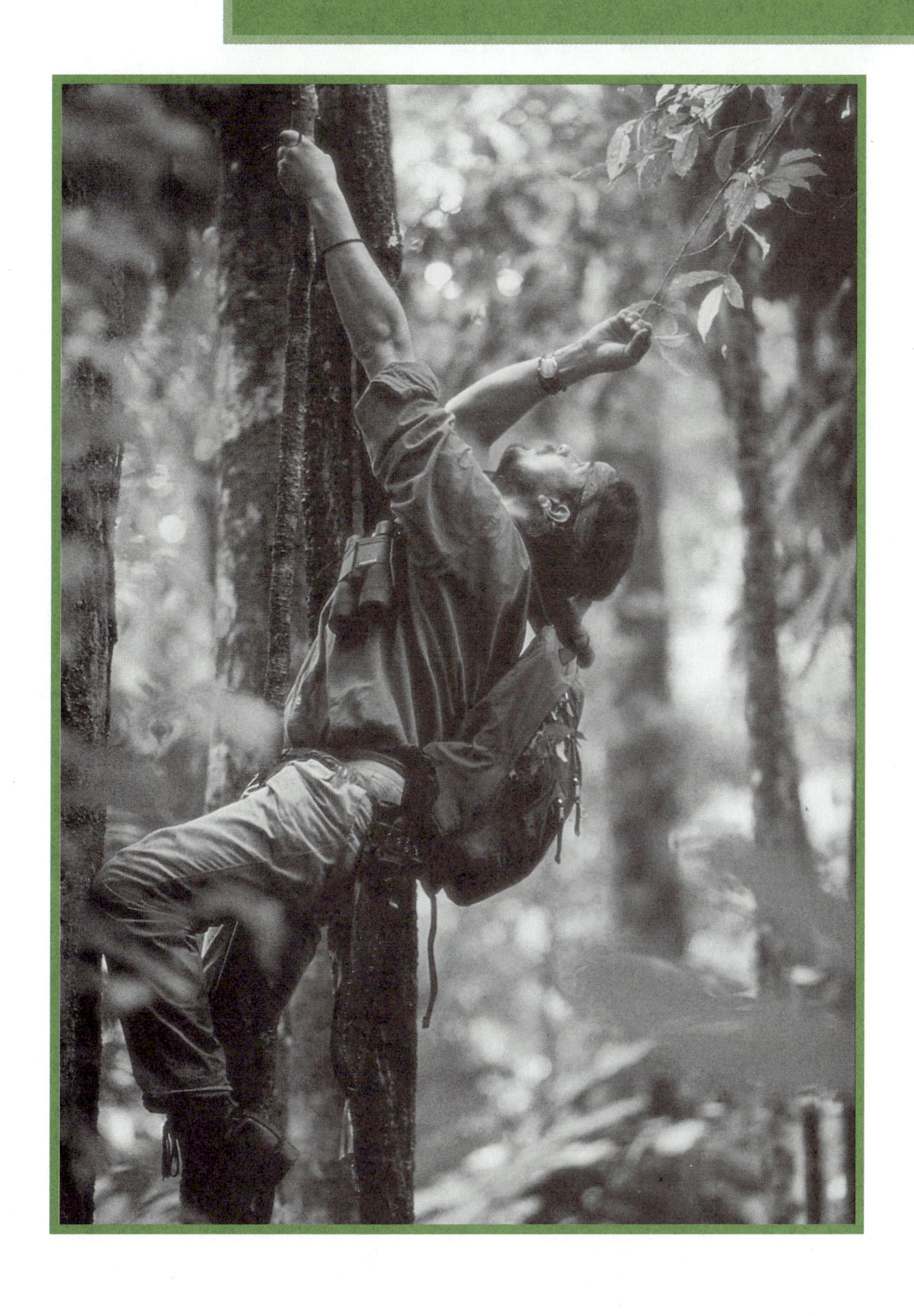

All living things are both dependent on the natural world for their fundamental energy needs and agents of change in the natural systems in which they live. The rapid expansion of the human population—and with it the globalization of economic, social, and political systems—represents the strongest force of change in the natural world today. To appreciate the effects of this change on the systems that sustain life on earth, we must first understand how the natural world works, and that understanding is embodied in the principles of ecology. In Part 1 we will explore the approaches and techniques that ecologists use to study the natural world and develop the principles of the theory of evolution by natural selection, the foundation of all biological study.

The natural world is at once orderly and beautifully and bewilderingly complex. Ecologists are drawn to their studies by simple curiosity, which is manifest in endless questions about form and function. Such questions are the engines of discovery and the basis of experimentation. In Chapter 1, we will examine how the order of the natural world is revealed to us through the questions that we ask. We will explore how our perception influences and limits our understanding of nature. We will see that the beauty of the natural world is also an inspiration for ecological study.

Simple observations and questions help ecologists—and other scientists—search for answers to their questions through experimentation and systematic sampling. Theory guides these activities by providing alternative ideas about the nature of the process. These ideas are then shared through the publication of books and scientific articles. We will explore the process of scientific discovery in Chapter 2 by examining how ecologists are addressing an ecological question of current importance.

CHAPTER 1

The Order of the Natural World

GUIDING QUESTIONS

- How do ecologists use questions to reveal patterns in nature?
- What are the types of questions that ecologists ask?
- How is the natural world both diverse and complex?
- How is nature both dynamic and self-replenishing?
- How predictable are patterns in nature?
- How does evolution by natural selection occur?
- How do perception and point of view influence our understanding of nature?
- What is natural history and how is it used to study ecology?
- How have human activities affected the natural processes of Lake Victoria, Africa?

Countless varieties of animals, plants, and microorganisms occupy the earth. To nourish themselves, reproduce, and maintain proper body conditions, living things must interact with other organisms and respond to the physical conditions of the place in which they live. Each organism possesses a set of morphological, physiological, and behavioral features that are characteristic of the kind of organisms to which it belongs and that reflect its ancestry. We refer to the set of all living things, the physical conditions under which they live, and the interactions among organisms and between organisms and the physical world as the **natural world,** or, alternatively, as **nature.** To understand the way nature works is the business of modern **ecology** (from the Greek *oikos,* meaning house, the immediate environment of man). Ecology is the study of the relations of organisms to one another and to their surroundings.

An appreciation of the principles of ecology is essential to an understanding of the human condition. Humans are no less dependent upon the natural world than an oak tree, or an earthworm, or the pet cat sleeping contentedly on your best sweater. While the conveniences of the modern world may dull our awareness of this dependency, they do not relieve us of our physiological needs for oxygen, water, and nourishment, which must be obtained ultimately from the natural world. Like all living things, we meet these needs through interactions with other components of nature, and these interactions have effects that alter the very natural world on which we depend. Our potential to change nature is unquestionably vast and already demonstrated. To understand the short- and long-term effects of this change will require a deep knowledge of the underlying dynamics of the natural world—that is, a knowledge of the principles of ecology.

Like other organisms, humans depend on and affect the natural world. But the human interaction with the natural world is distinct from other aspects of nature in that it has an aesthetic component. The inspiration for much of art, music, and literature emerges from our interpretation of the mysteries and beauty of nature. The work of art displayed on the cover of this book is an interpretation of nature that, though representative of fundamental ecological relationships, takes liberties with reality in order to stimulate, persuade, and entertain. This aesthetic awareness of nature derives from a curiosity about and appreciation of nature that comes from close observation, an activity that can be extremely satisfying and enjoyable. Like the artist, the ecologist savors the sheer pleasures of the sights, smells, sounds, and textures of the natural world.

The ecologist seeks to understand the order of the natural world. What is that order? Can the patterns and connections among living things and between organisms and the physical environment be described? Can these patterns and connections be predicted? How do the processes of the natural world affect humans and other organisms? These are some of the questions of ecology. The news of the day focuses our attention on the importance of these questions. During the preparation of this book, the world experienced the warmest summer on record (1998), with temperatures in some parts of the southwestern United States exceeding 100°F for over 20 consecutive days. The year brought record rainfall to southern California, a downpour caused by El Niño. Starvation caused by drought again ravaged central Africa, and fires consumed vast areas of Indonesia. Fish populations are on the decline worldwide, and downward trends in migratory bird populations continue. Reports that populations of frogs and salamanders are rapidly declining throughout the world raise questions about the viability of life. Meanwhile, the human population continues to grow. These changes are phenomena of the natural world and thus come under the purview of the science of ecology.

Our goals are to illuminate the details of the complex interrelationships between living things and the environment in which they live, to describe the patterns in those relationships, and to explain the underlying mechanisms responsible for those patterns. Along the way, we will show how ecologists, like other scientists, employ observation, experimentation, and mathematical modeling to uncover the mysteries of the natural world.

1.1 We can observe patterns in the natural world.

As you will see in the many examples of ecological studies presented throughout this book, the advancement of ecological knowledge requires the evaluation of hypotheses through sampling and experimentation. But, more often than not, ecologists begin their studies with the rather more simple activities of observing and asking questions. To ecologists, making observations and asking questions about what they see in the natural world surrounding them are essential and enjoyable first steps in revealing patterns and interconnections—an order, if you will, in nature. From these initial observations and this discovery of order, hypotheses about the mechanisms at work in the natural world are formulated and, eventually, the details of those mechanisms are unraveled through experimentation and more observation. In this chapter we will examine how ecologists observe the natural world, and we will describe for you some fundamental features of the order of nature.

Ecological Questions

Ecologists conduct their work by asking questions. Some questions are simple inquiries on the state or condition of an organism, interaction, or ecological system. (We use the word "system" in the first two chapters to identify any complex set of ecological interactions. We shall identify specific types of ecological systems in later chapters.) "What are the different kinds of trees in the forest?" "What does the raccoon eat?" and "What are the times of year when the plant produces fruit?" are examples of such questions. For most ecologists, these simple questions represent a kind of background curiosity that does not necessarily require extensive study. For the experienced forest ecologist who can easily identify many different kinds of trees on sight and who has spent many years studying forest dynamics, the question "What are the different kinds of trees in the forest?" is asked and answered nearly simultaneously as she mentally characterizes the forest during a walk through it. However, such basic questions are important when we encounter unknown places and processes. At one time, that forest ecologist was not so experienced in forest dynamics, and her curiosity may have focused almost entirely on very simple "what" questions. "What" questions are the engines of initial discovery.

Ecologists often ask questions about the nature of an underlying mechanisms or functions. The question "How do the leaves of oak trees turn color in the fall?" inquires about the physiological mechanisms of the leaf cells in which leaf pigments are contained. "How does the female wolf spider select a male with which to mate?" is a question that delves into the mechanisms of communication between male and female spiders (Figure 1-1). The evaluation of questions about mechanisms make up a substantial part of ecological work because an understanding of functional relationships is essential to an understanding of the natural world.

Knowing how a particular system functions does not explain why it functions the way that it does. An ultimate understanding of the natural world requires that we sort out potential alternative functions. The

FIGURE 1-1 Male wolf spider in courtship. The spider waves its front legs, which contain bristles of dark hairs, to attract the female. (*Courtesy of G. E. Stratton.*)

question "Why do oak leaves turn color in the fall?" inquires about the relative advantages of at least two alternative phenomena. On the one hand, the leaves could change color and subsequently die and fall from the tree, the pattern that we observe. On the other hand, they could remain green and live through the winter. The evaluation of such alternatives reveals the evolutionary basis for a phenomenon or condition. As we shall see shortly, ecological processes are based in evolution, in which relative advantage has to do with the ability to survive and leave offspring.

Questioning Nature: A Walk Through a Rain Forest

Evolution shapes all interactions among living things and thus provides the foundation for our understanding of the natural world. The principles of ecology can be fully understood only in the context of evolutionary theory. We shall show in this chapter that the root evolutionary principles arose from careful observations of nature by early naturalists such as Charles Darwin and Alfred Wallace. Throughout the book you will see how the details of evolutionary mechanisms are revealed in field and laboratory experiments as performed by ecologists all over the world (see Color Insert " Ecologists at Work").

To the casual observer, the natural world may seem chaotic and without pattern or organization. The placement of one tree with respect to another in a woodlot, the circuitous flight of a passing butterfly, or the cacophony of night sounds that fills the forest in springtime may seem random and without explanation. But to the naturalist, the natural world is full of shape and interconnections. The placement of the magnificently engineered orb web by the garden spider reflects a decision by the spider based on an assessment of possible web locations (Figure 1-2a). The chorus of springtime frog calls represents a carefully orchestrated symphony of competing males, each striving to gain advantage over other males for access to a female (Figure 1-2b). The seemingly random distribution of grasses, shrubs, and sapling trees in an abandoned agricultural field represents a snapshot of a complex, continuous, and organized process of change among the plant populations in the field (Figure 1-3a). These observations represent patterns and predictable phenomena governed by physical and biological processes. To discover pattern in the natural world is the first step in ecological study, a step that requires few implements. You need only slip on an old pair of boots and strike off down a forest path, asking questions as you go (Figure 1-3b).

What better place to begin our search for patterns in nature than in the tropical rain forest, where life is luxurious and diverse (Figure 1-4)? Nowhere else is one so acutely aware of nature (see Color Insert "A Walk Through the Rain Forest"). At night, especially, the multitude of active creatures makes the life of the greatest cities seem paltry by comparison. In the Panamanian rain forest, the delicate first light of a November day is shattered by the cries of a nearby troop

(a)

(b)

FIGURE 1-2 (a) Orb web of a spider. (b) Singing male frog.(*a, photograph by G. L. Miller; b, courtesy of H. Carl Gerhardt.*)

(a)

(b)

FIGURE 1-3 (a) Agricultural fields. (b) Ecology students using a seine to collect small fish in a stream. (*a, courtesy of M. Greenstone; b, courtesy of C. Britson.*)

of howler monkeys. The howling subsides, but it is presently answered by that of another troop farther away, and then another and another. We can easily locate the nearby troops of monkeys, and so, we would guess, can the howlers themselves. Are we witnessing a morning proclamation of territory by each group?

It is now light enough to begin to wend our way along the narrow forest trails, being careful not to brush against the long, sharp spines that viciously ornament the trunk of the black palm. We walk around the giant buttresses that spread out from the bases of trees over the forest floor (Figure 1-5). There is a faint snapping sound, and our eyes quickly search for some small animal moving in the bushes. Another snap. A small object catapults in front of us and strikes the ground nearby. Finally we locate the source—a small bush whose fruits have opened and are shooting their seeds over the forest floor. As the fruit, which resembles a three-sided pea pod, dries out, the edges of the pod come together, squeezing the seeds with greater and greater force until the stalk of the seed finally breaks and the seed is shot explosively from the pod (Figure 1-6). How does this plant generate the force necessary to launch its seeds?

The drab browns and greens of the inner rain forest are occasionally broken by a flash of iridescent butterfly wings. A blue morpho butterfly streaks by, just out of reach. How striking it is that these butterflies should be so conspicuous (Figure 1-7), so different from the moths that are attracted at night to the lights around buildings, which possess every conceivable device for remaining unobtrusive. The brown and gray colors of the moths look like dead leaves and bark. Many conceal their legs, normally a sure betrayal of any animal, beneath their wings, and others benefit from legs modified to resemble bark or lichens (see below). One species of moth characteristically protrudes one leg or another from underneath its regular outline to break up the symmetry so characteristic of animal forms. Others have wings that, although intact, often have the broken and contorted appearance of dead leaves. The hind edge of the wing of another species bears the unmistakable picture of a rolled-up leaf, complete with its shadow. How is it that these

FIGURE 1-4 Lush tropical rain forest of Panama. (*Courtesy of R. B. Suter.*)

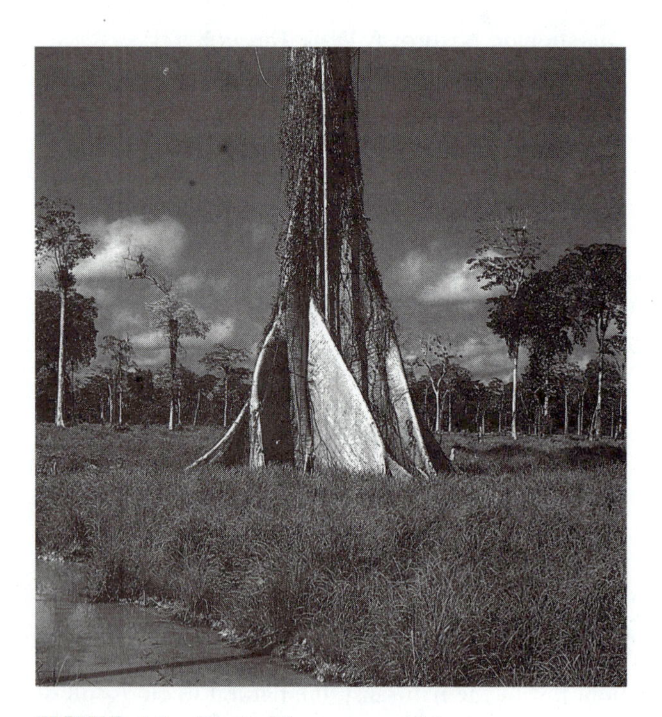

FIGURE 1-5 Tropical forest tree with buttresses. (*Courtesy of M. L. Crump and P. Feinsinger.*)

FIGURE 1-6 Ballistically dispersing seeds.

FIGURE 1-7 Temperate and tropical butterflies often have striking patterns and colors.

moths have come to resemble a leaf so closely? Why is it that other kinds of butterflies and moths are so conspicuous?

Our thoughts are interrupted as we pass, as though through a door, from the shadowy forest into a large clearing, where the sun momentarily blinds us. A recent treefall caused this break in the forest canopy. The giant did not exit gracefully, but took with it at least a dozen other trees directly in its path or bound to it by vines. The clearing provides our first glimpse of the sky in several hours. A long, graceful ribbon of hawks silently glides southward many hundreds of feet above. Their flight seems effortless—none of the birds moves its wings. The narrow ribbon is caught by a small updraft, sending it spiraling upward in wide, lazy circles until it breaks away again, seemingly impatient to be gone. Are these the same birds that we saw earlier in the year, gliding along the ridges of the Appalachians in Pennsylvania? What has sent them south? How far will they go? What cue will tell them spring is returning in the north and that it is time to make the return journey?

By the time our own journey through the forest comes to an end, our eyes and ears will have experienced more than our minds can sort out in a day. The questions we have asked about howler monkeys, ballistically dispersing seeds, the coloration of butterflies and moths, and the flight of hawks are windows into the patterns of nature. We are tired after the long walk, and grateful for a shower and a good meal to prepare us for an evening of conversation, of recounting our observations, and of asking many questions far into the night.

Of course, not all of us have the opportunity to walk through a tropical rain forest. But a hike through a nearby woodland or field, if undertaken with a sense of adventure and curiosity, will reveal just as many patterns and just as many surprises.

1.2 The natural world is diverse, complex, and interconnected.

The naturalist stands in awe of the diversity of forms in nature and the intricate interactions of these forms with one another and with their environments. In our walk through the tropical rain forest, we passed several hundred varieties of trees, and yet to untrained eyes they all seemed to be alike and to be "doing" about the same thing. They were using the energy of sunlight assimilated by their green leaves to convert carbon dioxide from the air and water and minerals from the soil into the organic molecules that make up their structure. Why should there be so many kinds of plants when one could seemingly perform the same functions as all the others?

There are hundreds of species of butterflies and moths in the forest, and thousands of species of other insects. Their appearance varies so much that one wonders if there are as many different environments within one forest as there are kinds of insects. In spite of how different the butterflies and moths look to us (and, presumably, to their natural predators), they are remarkably similar in their feeding habits. As adults, they all have long tubular snouts, which normally lie curled beneath their heads, but which can extend to the depths of tubular flowers for feeding (Figure 1-8). Adult butterflies and moths, as well as many other insects, birds, and bats, perform a vital function in the forest by carrying pollen from flower to flower, thus ensuring that the plants will set seed. How do they decide which flowers to visit? As larvae (caterpillars), they all eat green vegetation (Figure 1-9). Most species are picky eaters and can be found on only one or a few species of plants. How do the adults know which are suitable food plants upon which to lay their eggs? Why are these organisms specialists, feeding on a single or very few types of food, occupying places with very specific qualities, or laying eggs in only one or

FIGURE 1-8 Head of a Panamanian butterfly showing the tubular snout. The snout is uncoiled and inserted into flowers for feeding. (*Courtesy of R. B. Suter.*)

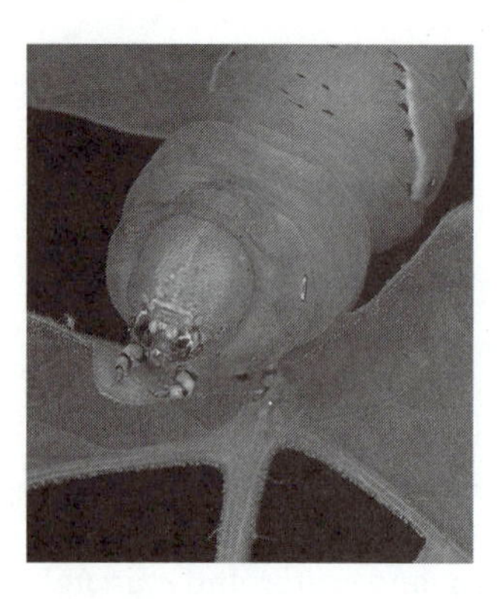

FIGURE 1-9 Caterpillar eating vegetation. (*Courtesy of J. Burgett.*)

FIGURE 1-11 Lichens. (*Photograph by R. E. Ricklefs.*)

two preferred places? A few species are generalists, feeding widely from the forest's smorgasbord and having little preference for where they live or where they lay their eggs. Why do they develop this lifestyle? And why does the adult silk moth refrain from eating altogether and die within a few days of emerging from the pupa?

Species are dependent upon one another for their survival. The delicately colored harlequin beetle, for example, carries on its back a small community of mites and pseudoscorpions that feed upon the mites (Figure 1-10a,b). The mites perform a beneficial function for the beetle by scouring fungi from its delicate membranous hind wings (fungi seem to grow on almost everything in the Tropics). The pseudoscorpions take advantage of this ready food supply, but the mites are at least partly protected when they hide in the numerous small pits that dot the forewings of the beetle. Examples of other organisms living in intimate, mutually interdependent relationships abound. Lichens are a combination of an alga and a fungus (Figure 1-11). The alga contributes carbohydrates made by photosynthesis; the fungus contributes water and minerals obtained from the rock or tree trunk on which the lichen lives. Much of the digestion that occurs in the guts of animals is carried out by specialized bacteria and protozoa living there. Plants are aided in the uptake of minerals from the soil by specialized fungi intimately associated with their roots. Birds and mammals, including some bats, eat the fruit of trees and later deposit the seeds when they defecate. How do species come to be interdependent on one another? Are all such relationships beneficial to both species in an interacting pair?

(a)

(b)

FIGURE 1-10 (a) Harlequin beetle of Central America. (b) Pseudoscorpions catching a ride. (*Courtesy of D. Zeh and J. Zeh.*)

1.3 The natural world is dynamic, but it is also stable and self-replenishing.

We have seen how our questions and observations may lead us to an appreciation of pattern and complexity in the natural world. If we inquire even more closely and observe even more keenly, we shall also discover that there is a constant tension between change and equilibrium in nature. The forest near your neighborhood may appear to change little from year to year. But the forest community actually undergoes constant turnover, with the removal of individuals by death and their replacement by birth. Just as the general shape and size of your body remains relatively constant while the cells and molecules within it are continuously replaced during your lifetime, the forest remains a forest, even though its components are constantly changing.

Populations of organisms are also continuously replaced. An insect may lay thousands of eggs each year, and some marine organisms shed millions of eggs into the water—more than necessary to compensate for losses of individuals from their populations. Yet in spite of the tremendous potential for growth, various checks and balances keep the size of most populations within rather narrow limits.

All systems suffer disturbances (from weather, fire, treefalls—even a cow pat creates a major disturbance for some organisms) from which they are continually recovering (Figure 1-12). Patches of disturbance may range from a few millimeters, as when an earthworm eats its way through the soil, to large portions of the earth, as when global weather patterns shift. Much variation is imposed by the physical environment—climate, ocean currents, landscape—but much is also generated by the biological world. Many organisms create disturbances for others as they forage and move about. The dynamics of population interactions can establish cycles of population change that send ripples throughout the biological community. Variation must be considered a part of the equilibrium of the natural world.

Renewal and replenishment are also characteristics of the natural world. The dead bodies of organisms and the wastes of biological processes do not pile up. They are broken down, and their component parts are recycled by the community. The dead leaves rustling under our feet cover the decomposing remains of other leaves in the soil beneath them. Soil organisms transform their elegant shapes into an amorphous mass of decaying and decomposing plant tissue called humus, finally reducing them to the mineral elements from which they were once, in part, synthesized.

FIGURE 1-12 A grass fire is an example of a natural disturbance. (*Courtesy of J. S. Brewer, Jr.*)

1.4 The natural world is organized by physical and biological processes.

We recognize patterns, organization, and interrelationships in the complexity of our surroundings because of the predictability of their elements. We can anticipate nature in various ways, but most efficiently by the generalization of past experience. For example, experience with the movement of objects thrown in the air enables us to predict their trajectories accurately. A good outfielder can predict where a baseball will land long before it begins to drop, thanks to having fielded thousands of fly balls.

Only if nature is predictable can we respond properly to it. Birds live in the woods; fish inhabit the sea. Without ever having been in a particular forest, we would expect to find birds rather than fish there simply on the basis of past experience with birds and fish and forests and seas. By experiencing the unnaturalness of surrealism and dreams, we realize how completely our minds are bound up in the various patterns we recognize in nature. As ecologist G. Evelyn Hutchinson once pointed out, "[if] we imagine ourselves encountering in the middle of a desert a rock crystal carving of a sewing machine associated with a dead fish to which postage stamps are stuck, we may suspect that we have entered a region of the imagination in which ordinary concepts have become completely disordered."

Patterns have two sources of predictability. One relies upon observation, and the other upon understanding the mechanisms that produce the pattern. In the first case, predictions are made by extrapolating previous observations to new but similar situations. We do this when we predict the flight path of a ball. But by applying the laws of motion, we can predict the trajectory of the ball without any previous experience with the phenomenon, knowing only its initial speed and direction. Similarly, by understanding the principles governing the form and functioning of animals, we know that we are not likely to find an organism resembling a fish living in a forest, even though the forest clearly supports such diverse creatures as trees, birds, fungi, worms, beetles, and mosses.

In science, empirically observed patterns often precede discovery of the causative principles that produce those patterns. After detailed observations, the German astronomer Johannes Kepler discovered that the time required for a planet to revolve around the sun is inversely related to its distance from the sun. Only later did the English physicist Isaac Newton formulate the laws of motion whose predictive powers are so great that they made possible the detection of unseen planets on the basis of their gravitational effects on the motion of some of the known planets. Alfred Wegener did not understand the mechanism of continental drift when, in 1915, he proposed, based on his observations of the geologic and geographic relationships of the continents, that the major landmasses slowly drift over the surface of the earth. Plausible explanations of how this might occur have been proposed and validated only recently. The same is true in the biological sciences. Early naturalists classified organisms into a regular hierarchy of species and other taxonomic groupings based on similarities. But only after Charles Darwin proposed his theory of evolution was the basis for those patterns understood.

1.5 Patterns in nature are understood in terms of evolution by natural selection.

One of the keenest observers of nature was Charles Darwin (1809–1882) (Figure 1-13). After finding himself unsuited for a career in medicine, Darwin received training in theology at Cambridge University. But his desire for a life in the clergy was overshadowed by his deep passion for nature and its mysteries. In 1831, shortly after receiving his degree, Darwin joined the ship's company of H.M.S. *Beagle* as naturalist for a five-year voyage to prepare charts for the British navy. During the voyage, he was able to explore and collect specimens from a vast array of different habitats throughout the world. It was his pondering of the patterns he observed during his voyage and during subsequent studies conducted after he returned to England that led Darwin to formulate a theory of evolution by natural selection. His book, *On the Origin of Species by Means of Natural Selection* (1859), is widely regarded as the seminal work in evolutionary theory. Darwin's theory of evolution, which was revolutionary in his time, now forms the basis for all contemporary biology and ecology.

FIGURE 1-13 Charles Darwin. (*Courtesy of the American Museum of Natural History.*)

Evolution and pattern in the natural world

When Darwin departed on his round-the-world voyage, questions were emerging regarding two important features of the natural world: the age of the earth, and the nature of change in the different types, or species, of animals and plants. Most scientists and intellectuals believed the earth to be quite young—too young, in fact, to have allowed time for any species to have changed since the earth's creation. Traditional views held that each species was immutable—that is, unchangeable once it had appeared on earth.

But observers of nature were curious about inconsistencies between generally accepted views about the natural world and their own observations. In the early 1800s, the geologist Charles Lyell proposed that the modern physical features of the earth, such as mountains and valleys, were the result of processes of geologic change that occurred very slowly over very long periods of time. Lyell's idea suggested that the earth was extremely old. The discovery of more and more fossil organisms that bore little or no resemblance to modern forms seemed to support this idea. To some, the fossil record also drew into question the idea of immutability of species. The notion that species might undergo change can be traced to the Renaissance, and some consideration had been given to the mechanism of such change prior to Darwin. One theory of species change was proposed in 1809 by the French naturalist Jean Baptiste de Lamarck in his book *Philosophie Zoologique.* Lamarck's idea was that species change through some "inner drive" to improve themselves. Whatever improvements individuals achieved during their lifetime were then passed on to their young. The classic example of this idea, which we call Lamarckian evolution, is the explanation for the length of giraffe necks. In changing from eating grass to eating leaves on trees, the explanation goes, ancestral giraffes continually "improved" themselves by stretching their necks to reach higher and higher into the trees. Each improvement was passed on to their offspring, and eventually, long-necked giraffes appeared. Although Lamarckian evolution is now discredited, it was accepted by many scientists of Darwin's time, and it represents the origin of modern ideas of species change. For Darwin, the notion that species might undergo change was reinforced by his understanding of domestic plant and animal breeding, in which individuals having specific desirable characteristics were actively *selected* by the breeder to reproduce.

In our imaginary walk through a tropical rain forest, we emphasized the great variety of organisms in the natural world. It was his extensive observations of this variety, against a background of emerging questions about nature, that stimulated Darwin to question how such diversity could arise. On the Galápagos Islands, a small group of islands lying near the equator about 900 miles off the west coast of South America (in Chapter 14 we will discuss some work of contemporary ecologists on the Galápagos Islands), Darwin observed a variety of small, rather drab finches that, although overall quite similar to one another in size and general form, were distinct enough to be identifiable as different species. He also observed that each island in the Galápagos group possessed one or more unique types of finches having a slightly different beak morphology from finches from other islands. Perhaps most intriguing of all, he noticed that the finches on the Galápagos Islands appeared to be very similar to species that inhabited mainland South America to the west. How did the variety of finches arise on the Galápagos Islands?

After nearly twenty years of pondering his observations of nature, and at the urging of a friend, Darwin joined another biologist, Alfred Russel Wallace, in advancing the idea of evolution by natural selection. The components of the idea are quite simple. Organisms have a great capacity to reproduce, and the offspring of two parents most often do not resemble one another exactly. So, in a population of any species, there will be great variety in form and function. Not all rabbits will be exactly the same size, or have the same sprint speed, or have exactly the same metabolic rate. Resources such as food, space, water, light, and the availability of mates are limited, and thus not every individual will survive and reproduce. Some characteristics or traits give individuals a better chance of obtaining resources and surviving to reproduce than other traits do. The individuals born with such favorable traits are more likely to survive. In other words, there is **natural selection** for favorable traits.

We now know that the variations that we see among individuals in a population have a genetic basis, and because of this, specific traits may be inherited. Those individuals that possess favorable traits are more likely to reproduce and pass those traits to the next generation. This differential reproduction leads to a change in the genetic composition of the population. For example, consider a population in which each organism has either trait A or trait B. If trait A is a favorable trait, one that gives the bearer an advantage in, say, gathering food or finding shelter, then more individuals with trait A are likely to reproduce than individuals with trait B. Trait A is referred to as an **adaptation,** a genetically determined characteristic that enhances the ability of an individual to cope with its environment. Individuals with trait A are selected for, whereas individuals with trait B are selected against. The next generation will therefore have more individuals with trait A, and consequently, the genetic composition of the population will change.

This simple example brings up a number of important aspects of basic evolutionary theory. Natural selection works only on heritable variation, that is, traits having a genetic basis. Traits that are acquired during your lifetime, such as the ability to hit a baseball, or to write well, or to operate a car, are not passed along to your offspring (although genetically based characteristics that predispose you to excel at these activities might be passed on). The "inner force" of Lamarck does not drive evolution. Second, the process of natural selection operates at the level of the individual, but evolutionary change is a population phenomenon. In the example above, an individual with trait A may be favored in its environment and reproduce, but with respect to the two traits, the individual does not change. An individual will always have either trait A or trait B. What changes is the proportion of the population having trait A—the population has evolved.

There are two common mischaracterizations of evolutionary theory that must be mentioned. First, natural selection is often portrayed in popular writing as an overt and aggressive struggle for survival among individuals. What really matters, we are led to believe, is strength, agility, and the ability to win a fight. In reality, as the simple example above demonstrates, the process of natural selection is considerably more benign. Individuals enter the world possessing a suite of features that adapt them more or less to their environment. The real struggle is in obtaining food and surviving to reproduce. Second, evolutionary change does not proceed toward some ideal form or being, nor does it necessarily lead to greater complexity. Selection simply favors traits that give organisms an advantage in their environment, whether those traits represent elaborations or simplifications of the current form. Successive evolutionary change results in a lineage of types from ancestor to modern form, each type giving way to one better adapted to the changing environment. At one time, each type represented the most "modern" or "best" form, that is, the form best adapted to the environment. The evolutionary step from one type to the next is not a step up a ladder of increasing value. Rather, it is a step along a long pathway of change.

Mechanisms of evolution

Since Darwin proposed the theory of evolution, a vast literature has emerged that supports the precepts of the theory and increases our understanding of the mechanisms of evolution. One of the most striking cases of evolution in action is that of industrial melanism in the peppered moth of England. Early in the nineteenth century, occasional dark, or **melanistic,** specimens of the common peppered moth (*Biston betularia*) were collected (Figure 1-14). Over the next 100 years, this dark form, referred to as *carbonaria,* became increasingly common in forests near heavily industrialized regions of England, which is why the phenomenon is often referred to as **industrial melanism.** In places without factories and other heavy industry, the lighter, salt-and-pepper form of the moth prevailed. This phenomenon aroused considerable interest among geneticists, who showed by cross-mating light and dark forms that melanism is an inherited trait determined by a single dominant gene. Because the melanistic trait is an inherited characteristic, its spread reflected genetic changes (evolution) in the population.

Peppered moths inhabit dense woods and rest on tree trunks during the day. Where melanistic individuals had become common, the environment must somehow have been altered so as to give dark forms a survival advantage over light forms. It seemed reasonable to suppose that natural selection had led to the replacement of typical light individuals with *carbonaria* individuals. To test this hypothesis, the English biologist H. B. D. Kettlewell captured a sample of moths of both forms, marked each individual with a dot of cellulose, and then released them back into the woods (Kettlewell 1955, 1956, 1959). The mark was placed on the underside of the wing so that it would not attract the attention of predators to a moth resting on a tree trunk. Some days later, Kettlewell captured more moths by attracting them to a mercury vapor lamp in the center of the woods, or to caged virgin females at the edge of the woods. (Only males could be used in the study because females are attracted neither to lights nor to virgin females.) This type of study is referred to as a mark-recapture study, and we shall show in Chapter 14 that it can be used to estimate population size.

In one such mark-recapture experiment, Kettlewell marked and released 201 typicals (the light form) and 601 melanics in a wooded area near industrial Birmingham. The results indicated that a greater percentage of the dark form survived over the course of the experiment (Table 1-1). A similar experiment in a nonindustrial area revealed higher survival by the typical salt-and-pepper form of the moth.

The specific agent of selection was easily identified. Kettlewell reasoned that in industrial areas, pollution had darkened the trunks of trees so much that typical moths stood out against them and were readily found by predators. Any aberrant dark forms were

(a)

(b)

FIGURE 1-14 (a) The melanic form of the peppered moth. (b) The typical salt-and-pepper form. (*Courtesy of H. B. D. Kettlewell.*)

TABLE 1-1 Results of Kettlewell's mark-recapture experiment

	Typicals	Melanics
Number of moths released	201	601
Number of moths recaptured	34	205
Percentage recaptured	16	34

better camouflaged against darkened tree trunks, and their coloration made them more likely to survive (see Figure 1-14). Eventually, differential survival of dark and light forms would lead to changes in their relative frequency in the population. To test this idea, Kettlewell placed equal numbers of light and dark forms on tree trunks in polluted and unpolluted woods and watched them carefully at some distance from behind a blind. (A blind is a tentlike structure intended to conceal observers from their subjects; it is more often called a hide in England.) He quickly discovered that several species of birds regularly searched tree trunks for moths and other insects, and that these birds more readily found a moth that contrasted with its background than one that resembled the bark it clung to (Table 1-2). These data were consistent with the results of the mark-recapture experiments. Together they clearly demonstrated the operation of natural selection, which over a long period resulted in genetic changes in populations of the peppered moth in polluted areas.

One of the most gratifying aspects of the peppered moth story is that, with the advent of smoke control programs and the return of forests to a cleaner state, frequencies of melanistic moths have decreased. In the area around the industrial center of Kirby in northwestern England, for example, the *carbonaria* form decreased from more than 90% of the population to about 30% over a period of 20 years (Figure 1-15).

TABLE 1-2 Bird predation on typical (light) and melanic (dark) forms of the peppered moth

	Individuals taken by birds	
	Typicals	Melanics
Unpolluted woods	26	164
Polluted woods	43	15

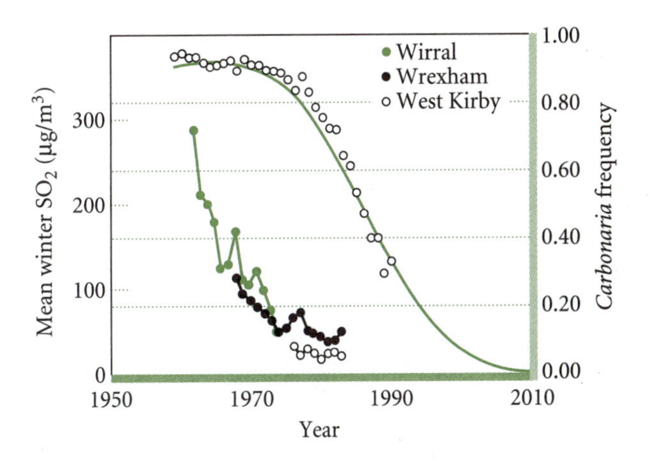

FIGURE 1-15 Changes in frequency of the melanistic *carbonaria* form of the peppered moth since the beginning of pollution control programs in England in the 1950s. The index of pollution is the winter level of sulfur dioxide, where 1 = 300 micrograms per cubic meter of air. Sulfur dioxide directly affects lichens growing on tree trunks, against which moths rest by day. The lag in evolutionary response to changes in air pollution levels reflects the time required for forests to return to a more natural (unpolluted) state, as well as a low initial frequency of the recessive allele for typical coloration. (*After Clarke et al. 1985, Mani and Majerus 1993.*)

1.6 Perception influences and limits our understanding of nature.

Because of our size and mobility and the pace of our lives, we are sensitive to particular scales of variation in time and space. Our perception of distance differs from that of a tiny collembolan, which finds a jungle in a pinch of soil, or a small water flea (Figure 1-16), which perceives a drop of pond water as we would the entire pond. For us, a few decades are a lifetime. But our lives are instants to a redwood tree and eternities to microorganisms.

Our limited perception of nature can encumber our ability to understand natural phenomena. For years, most ecologists were trained and worked in Europe and temperate North America, where seasonal fluctuations in temperature are a major aspect of the environment. When these naturalists visited the

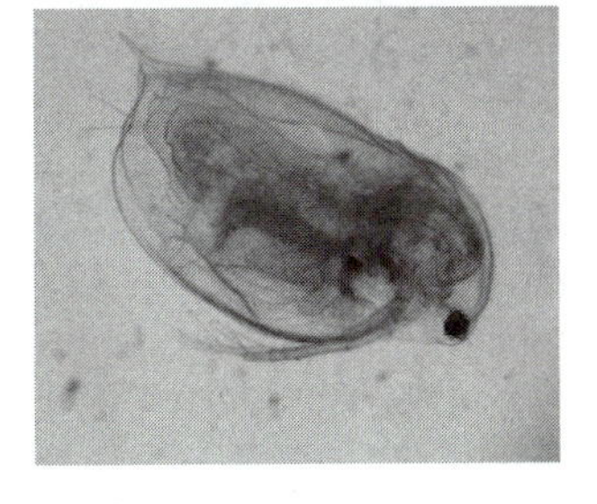

FIGURE 1-16 The water flea *Daphnia magna,* a small aquatic animal just visible to the naked eye. (*Courtesy of M. Moore.*)

Tropics, they were impressed by the constant year-round temperature, and they assumed that biological communities would be less variable in tropical regions than in temperate regions. Only relatively recently have ecologists bothered to count individuals of tropical species over long periods. Surprisingly, they have found that populations undergo marked seasonal fluctuations in the Tropics, and additionally, that they can vary considerably from year to year. Temperature defines the seasons in a temperate climate. But the patterns are different in the Tropics, where the seasons are marked by wet and dry periods, and where rainfall is notoriously unpredictable at many times of the year.

Our interpretation of the natural world is also affected by the way size determines pertinent scales of time and space. A flea can jump a hundred times its length. We immediately react by translating that distance to a familiar scale—comparing flea lengths to human lengths. The comparable human jump would clear two football fields. How incredibly strong fleas are! But the flea does not actually perform athletic miracles. As size changes, the relationships between distance, power, and time change accordingly. Consider the wingbeats of flying organisms. Try to move your arms up and down as rapidly as possible. How fast can you do it? Two or three times per second? Most large birds can flap their wings between 2 and 20 times per second, and some small insects can do so up to 500 times per second. This may seem amazing, but as the size scale changes, so does the relevant time scale. You can demonstrate this very simply for yourself. Tie a weight to a string and start it swinging like a pendulum. As you shorten the length of the string, you will notice that the weight swings back and forth much faster. In fact, the rate of swinging varies inversely with the length of the string. Halve the length of the pendulum, and its swinging frequency doubles. The wings of most small insects are less than a centimeter long, so we should not be surprised that an insect can beat its wings more than a hundred times faster than we can flap our arms.

1.7 Natural history forms the basis of ecological inquiry.

Let us imagine ourselves transported from the lush Panamanian rain forest northward to the great Mojave Desert of southern California. The desert is normally a forsaken place with little rain, searing summer heat, and chilling winter cold, conditions so forbidding that, except for a few struggling plants, it appears to be nearly devoid of life for most of the year. But the desert's silence and apparent sterility are occasionally broken during the milder days in winter, and it is in the early morning of one such day in January that we arrive. As the morning sun warms the desert, there is suddenly life everywhere. Swarms of insects and other creatures appear on the sand, scurrying and flying about in a frenzied blur, taking advantage of the few hours of moderate temperatures that this winter morning offers to move out of burrows and other hiding places and to briefly go about the business of feeding and finding mates. Our attention is drawn to what appear to be red dots, thousands of them, moving about in the sand, and we drop to our knees to get a closer look. We are lucky to be accompanied on this day by ecologists Lloyd Tevis and Irwin Newell, who are familiar with the desert life. They inform us that we are seeing *Dinothrombium pondorae,* the giant red velvet mite, whose scientific name, meaning "terrible clot," comes from its resemblance to a blood clot (Figure 1-17). Tevis and Newell know quite a bit about this creature, because in the 1960s their curiosity led them to ask many questions about it (Tevis and Newell 1962). The story they tell us is fascinating. It encompasses the basic information about the giant red velvet mite, about where it lives, what it eats, and its reproductive behavior. What Tevis and Newell offer for the mite is a **natural history,** the basis of all ecological study.

Tevis and Newell began by asking whether the emergence behavior of the mites was related in any way to the physical conditions of the desert. They found that the mites spend most of the year in burrows dug in the sand. They also found that the particular conditions that favor the emergence of the mites occur infrequently in the Mojave Desert. During four years of observations—patience is a quality of all good naturalists—adults appeared above ground only ten times, always during the cooler months of December, January, or February, when they can tolerate the temperatures on the desert's surface. From their observations, Tevis and Newell could predict that an emergence would occur on the first sunny day after a rain of more than three-tenths of an inch, provided that air temperatures were moderate.

FIGURE 1-17 Giant red velvet mite.

On the day of our visit, the mites came out of their burrows between 9:00 and 10:00 A.M., and by late morning we could see thousands of mites scurrying across the desert sands in all directions. At midday, between 11:30 and 12:30, the mites dug back into the sand, where they would wait until the following year before emerging again.

These "terrible clots" do not leave their burrows simply to terrorize unsuspecting desert travelers. During its 2- or 3-hour stay above ground each year, a mite must perform two important functions: feeding and mating. On the same day that the mites emerge, large swarms of termites appear flying over the desert sand, their own emergence presumably triggered by the same physical factors that cause the mites to leave their burrows. It is upon these termites that the mites feed. A flying termite cannot, of course, be caught by the earthbound mites; a mite must locate its prey after the termite drops to the ground and sheds its wings, but before it burrows into the sand. All this happens very quickly, giving the mites about an hour to find their prey.

Because the mites are solitary in their burrows, they must mate during their brief period above ground each year. The courtship of the giant red velvet mite is similar to that of spiders and other relatives. The males walk nervously around and over a feeding female, tapping and stroking her, and cover the sand around her with loosely spun webs. Males court feeding females for two reasons. First, the females have ravenous appetites (they have not eaten for a year, after all) and would just as likely devour a male mite as a termite. Second, and perhaps more important, females can produce eggs only if they have had a meal. Thus, by mating with a feeding female, the male guarantees that his efforts to reproduce will not be wasted.

About midday, after the mites have fed and mated, they congregate in troughs on the windward sides of sand dunes, where surface temperatures and the size of the sand particles are just right (less than half a millimeter in diameter), and reenter the sand almost simultaneously, as if urged on by the same unseen hand. The mites continue to dig their new burrows on the first day until the coolness of the late winter afternoon slows their activity. Burrowing continues on subsequent days when the sand becomes warm enough, until the burrows are completed. During the rest of the year, the adult mite spends its time moving up and down in its burrow to follow the movements of its preferred temperature zone as the surface of the sand heats and cools each day.

Females lay eggs during the early spring. The eggs soon hatch, and the young mites crawl to the surface of the desert to search for hosts, usually grasshoppers, to which they attach themselves. While it is growing, the young mite remains with its host, obtaining its nourishment from the host's body fluids. When it is fully grown, the young mite drops off its unwilling host and seeks a suitable spot to dig its own burrow in the sand, thus renewing the life cycle of the giant red velvet mite.

Natural history observations are not limited to the details of the daily lives of organisms. They also include observations of interactions among organisms in populations and communities, and of the intricate interplay among the physical and biological components of complex ecosystems. Our persistent questioning of nature, and our experimental investigations of it, enable us to expand natural history, enriching it until it becomes a magnificent story, one no less exciting than the best mystery novel.

Natural history provides the basic theme for every chapter of this book, from the most descriptive to the most theoretical. Each story that we present will emphasize a different suite of ecological principles.

1.8 The order of nature is affected by human activity.

We emphasized in the beginning of this chapter that in order to gain a full understanding of the natural world, we must come to appreciate the significance of human activities in nature. Human activities such as agriculture have real ecological effects that alter natural processes, processes that sustain life on earth. It is not our intention in this book to focus strictly on the effects that humans have on nature, but we will make frequent reference to such effects. We wish to emphasize the importance of this aspect of ecology, however, by ending this chapter with an examination of a case in which human activities, though undertaken with the best of intentions, dramatically altered an ecological system.

The highest rate of human population growth on earth (2–3.5% per year) prevails in sub-Saharan Africa, a vast region where human survival has long been linked closely to ecosystem health because of its dependence on subsistence-level farming and fishing. The region is subject to severe drought and plagued by overexploitation of natural resources. In a well-intentioned effort to provide additional food and exports for people living in East Africa, the Nile perch (*Lates niloticus*) was introduced into Lake Victoria beginning in 1954. More than 30 million people depend on Lake Victoria for all or part of their food (Kaufman 1992), but because simple ecological principles were ignored, the lake's entire fishery became jeopardized (Barel et al. 1985, Kaufman 1992, Goldschmidt et al. 1993).

Until the introduction of the Nile perch, Lake Victoria supported a wide variety of species of cichlids and other fishes. Cichlids (a common name referring to the fish family Cichlidae) are freshwater fishes that are found predominantly in the lakes and rivers of Central and South America and Africa, where they are often collected and sold for the aquarium trade. In most places where they are found, they have undergone extensive ecological diversification through evolution and are represented by many species—about 300 species in Lake Victoria alone—each of which has adopted relatively specialized feeding habits and habitat preferences (Figure 1-18). This process of diversification through evolution is referred to as **adaptive radiation.**

Cichlids are herbivores that feed on plant material and detritus (fresh or partially decomposed plant and animal material). The Nile perch (Figure 1-19) is a large predator that feeds upon other fish—the smaller cichlids, in this instance. When one organism eats another, not all of the energy contained in the consumed organism is available to the organism doing the consuming. Some energy is lost because parts of the consumed plant or animal are not ingested or not digestible, and some energy is lost because of inefficiencies in metabolism. Because of this, predatory fish like the Nile perch cannot be harvested at as high a rate as their herbivorous prey, the cichlids. Furthermore, the perch was alien to Lake Victoria, and the evolution of the local cichlids did not include behaviors to escape predation. Inevitably, the perch annihilated the cichlid populations, destroying the native fishery and all but eliminating its own food, and thereby in turn hastening its own demise as an exploitable fish. To be sure, the native fishery was already precariously overexploited owing to growth of the local human population and the recent application of advanced fishing technology, but the appropriate solution to these problems would have been better management of the cichlids, not the introduction of an efficient predator upon them.

FIGURE 1-18 African cichlid fish. (*Photo ©Ken Lucas.*)

FIGURE 1-19 Nile perch. (*Courtesy of L. Kaufman.*)

Still other results of the introduction of the alien fish turn the story into a tragic comedy of errors. The flesh of the Nile perch is not particularly liked by the people living near the shores of Lake Victoria. They prefer the texture and flavor of the native fishes. Moreover, the oily meat of the Nile perch must be preserved by smoking rather than by sun-drying, so local forests are being cut rapidly for firewood. Because the larger Nile perch requires larger and more elaborate nets, local subsistence fishermen cannot compete with more prosperous outsiders who are equipped for commercial fishing. The tragic consequence is that much of the tonnage of fish taken from Lake Victoria today is unavailable to the local small-scale fisherman who does not have the sophisticated gear required to catch the Nile perch. Furthermore, there is concern about the stability of the Nile perch population itself. Since the 1980s, a number of unexplained large-scale die-offs of Nile perch have occurred in Lake Victoria (Kaufman 1992).

The consequences of perturbing natural ecosystems by introducing exotic species are often far-reaching and unpredictable. The devastation of the Lake Victoria cichlid fishery is not the only repercussion of the introduction of the Nile perch. Considerable evidence suggests that the fundamental physical and biological properties of the lake, such as oxygen distribution, turbidity, productivity (the rate of photosynthesis), and shoreline plant community structure, have deteriorated dramatically since the introduction of the Nile perch. Of course, some of these changes in the characteristics of Lake Victoria resulted from the steady increase in nutrient inputs from the growing human population surrounding the lake, causing an increase in algal productivity, a process known as

eutrophication. As a group, the native cichlids of Lake Victoria consumed a wide variety of algal types and occupied many different habitats, features that probably resulted in their ability to recycle nutrients and thus forestall eutrophication in the face of increased human inputs. The Nile perch, by removing the cichlids from the lake, circumvented this natural recycling system, thus speeding the eutrophication process (Kaufman 1992, Goldschmidt et al. 1993).

The lesson to be drawn from the Nile perch experience is simple: Humans are an integral part of the ecology of the Lake Victoria area. Traditional local fishing had been sustained for thousands of years until the pressure of population growth and the perception of an opportunity for an export fishery led to an ecologically unsound decision and to an economic and social disaster in that region of East Africa.

In the next chapter, we shall examine in more detail the methods that ecologists use to study situations such as that of Lake Victoria and other ecological systems. We shall see that all ecological study is based on the identification of patterns in nature and on understanding those patterns in terms of evolutionary processes.

SUMMARY

1. The activities and phenomena that we observe in nature represent patterns. Our questions about nature represent the beginning of ecological inquiry. Ecologists ask questions that inquire about the condition or state of a system, about function or mechanisms, and about the evolutionary basis of a pattern.

2. A vast array of different kinds of organisms inhabit the natural world. This diversity may manifest itself as subtle variations on a basic theme, as in the bewildering variety of butterflies and moths, all with a similar body form and feeding structure.

3. The natural world is characterized by complex interactions whereby different species are dependent upon one another.

4. Death and change are characteristics of the natural world, but materials and populations are replenished by cycles of reproduction and nutrient recycling.

5. Physical and biological processes underlie the patterns that we observe in nature. An understanding of these processes enables us to predict and explain patterns.

6. Patterns in the natural world are best understood as resulting from evolution by natural selection, as originally proposed by Charles Darwin. Natural selection refers to the differential survival of organisms that possess certain favorable traits called adaptations. The process is based on the fact that features vary among individuals, and some variants are more likely to reproduce and leave offspring than others. Industrial melanism in the peppered moth of England is an example of natural selection.

7. Our interpretation of nature is limited both by our sensory acuity and by the point of view that we adopt.

8. Ecologists understand nature by asking questions, discovering connections, and developing from them a detailed story of a natural phenomenon. This information is called natural history, and it forms the basis of all ecological study.

9. The effect of the introduced predatory Nile perch on the ecology of the native cichlid fishes of Lake Victoria, in East Africa, is an example of how a lack of knowledge about natural history can lead to ecological disaster.

EXERCISES

1. Humans depend on the natural world for their survival. To illustrate this, consider the different types of food that you had for lunch yesterday. List what organism the food represents, where on earth the organism lives, what type of environment it is likely to be found in, and whether it is something that is cultivated by humans or found in the wild.

2. Consider the piece of art on the cover of this book, which represents the artist's interpretation of the natural world. What natural interactions are represented in the work? Explain how the artist has taken liberties with the order of nature in order to inform and entertain.

3. Find a quiet patch of forest, or desert, or grassland in which to spend an hour or two. During that time, write ten questions that begin with the word "how" and five questions that begin with the word "why." Using this book or any other source, provide an answer to one of the "how" questions and one of the "why" questions.

4. During your visit to nature, write down all of the interactions of nature that result from your presence there. Think about how the animals around you respond to your presence or about how your movements affect the plants in the area. Also, note how the components of the natural world affect you.

5. Part of our responsibility as scientists is to interpret nature for those not trained in scientific principles. Provide a brief explanation of the importance and beauty of the diversity and complexity of the natural world to be presented to a town council or a corporate board of directors.

6. Write a brief description of evolution by natural selection that uses neither the term "evolution" nor "selection." Your explanation may take the form of an example.

7. Using observations made over a period of days or weeks, develop a natural history of a pet cat or dog, a houseplant, or some other organism that you can observe regularly in its habitat. The natural history should include a description of its significant interactions with other organisms, such as you and your friends.

CHAPTER 2

Discovering the Order of Nature

GUIDING QUESTIONS

- How are hypotheses used in ecology?
- How are theoretical models used in ecology?
- What is the importance of statistics in the study of ecology?
- What types of questions may be answered using sampling studies?
- What are the important features of an experiment?
- How have experiments been used to investigate amphibian population declines?
- Why is understanding the evolutionary history of organisms important to an understanding of their ecology?
- What are the steps that ecologists follow in publishing the results of their studies?
- What is the relationship between ecology and the practice of agriculture?
- What are the various levels of complexity of ecological study?
- What is the difference between the science of ecology and the practice of environmentalism?

We begin our study of the natural world by recognizing the patterns and connections in nature, by posing simple questions about those patterns, and by describing the natural history of what we see (Chapter 1). Our exploration of nature is sustained and extended by more energetic and deliberate strategies of inquiry, ones that are framed within the context of evolutionary theory, driven by systematic observation and experimentation, and supported by mathematics, statistics, and a knowledge of the physical forces of nature. We are challenged in these inquiries by the great variation in nature, which introduces uncertainty into our inferences about it, and by the sometimes profound logistical problems of working in the natural world. As scientists, we are obligated to disseminate our findings and to join in the public discussion about the importance and possible applications of new discoveries. In this chapter we will explain the practical features of ecological work. That is, we will show how ecologists expand on their simple questions, their knowledge of natural history, and their understanding of evolutionary processes to perform the work of ecological study.

The process of ecological inquiry involves the following fundamental steps: (1) the generation of questions through observation and exploration, (2) the development of ideas about the answers to those questions, (3) the evaluation of competing possible answers through systematic observation and experimentation, and, using the results of those observations and experiments, (4) the drawing of an inference about the natural world. In the first part of this chapter we will describe these steps in some detail, emphasizing how they are influenced by natural variation. In the last part of the chapter we will discuss some of the practical problems and challenges of ecological research and explore how the knowledge derived from ecological studies is used.

2.1 Questions about nature are extended and refined into hypotheses and theory.

When we suggest an answer to a particular question that we have posed, we are expressing an idea about how something works. For example, ecologists who first asked why the populations of cichlid fishes in Lake Victoria were in decline (see Chapter 1) might have naturally suggested that the Nile perch was having a detrimental effect on those populations. The idea that cichlid decline was caused by Nile perch predation is called a **hypothesis.** An idea that we identify for specific consideration at a particular time is called the **null hypothesis.** Any one of the other possible ideas about the nature of the process is called an **alternative hypothesis.** Alternative hypotheses about why cichlid populations declined in Lake Victoria might include the introduction of a new fish parasite, dramatic changes in the aquatic vegetation on which the cichlids feed, or dramatic changes in the water chemistry of the lake caused by increased agricultural runoff.

When expressed in language, hypotheses take the form of declarative sentences, not questions. For example, "The introduction of Nile perch caused the decline of cichlid fishes in Lake Victoria" is a hypothesis, whereas the question "Did the introduction of Nile perch into Lake Victoria cause the decline in cichlid populations?" is not. But hypotheses are most often more precisely stated in the form of a mathematical expression about the relationship between some measured features of the natural world. These features are called **variables** because they may take on any one of a number of values. The number of blue morpho butterflies that we observed in our imaginary walk through the rain forest, the number of Nile perch hatched in Lake Victoria each year, and the time of year of emergence of the giant red velvet mite are all examples of variables.

The variability and complexity of the natural world can make the development of null hypotheses difficult. Often no single idea emerges as a good suggestion for how the system works. Ecologists have recognized for a long time that it is the interconnections among different components of nature that often give a natural system its distinctive characteristics. As we shall see throughout this book, these interconnections are manifested in the transfer of materials, transformations of energy, feeding relationships, social interactions, and many other processes. Our first simple questions may not lead directly to simple hypotheses. Because of this, many hypotheses in ecology emerge first through the development of theoretical models that explain the interactions of the essential variables of complex systems.

Theoretical ecological models are verbal, algebraic, or graphic constructions that explicitly identify the relationship among the variables of a system. A simple hypothesis may be considered a theoretical model in its simplest form. For example, the null hypothesis "Male frogs sing on warm evenings in the spring following rain" sets forth the relationship between a variable representing a particular behavior of the frog (the timing of the male courtship song) and variables that represent environmental conditions (temperature and precipitation).

Most ecological theories are presented as algebraic expressions or equations. The power of such expressions lies in the explicit nature of their predictions of the dynamics of one component of a system with respect to other components of that system. Mathematical expressions represent step-by-step instructions for how one variable changes with respect to another variable. A very simple example is the equation for a straight line, $y = mx + b$, where y and x are variables and m and b are constants. So long as you know m (the slope of the line) and b (the y-intercept), the value of y is known for any value of x. Thus, changes in y track changes in x. This simple model shows unambiguously how y and x are related when m and b are known.

Mathematical expressions cannot portray the precise dynamics of ecological processes, however, because of the presence of variation in the natural world. Mathematical expressions are only *models* of nature. Thus, in the case of two variables that are hypothesized to have a positive linear relationship ($y = mx + b$), this relationship might be *generally* true, but it is unlikely that it is *exactly* true. This is illustrated in Figure 2-1, in which a scattering of points is shown around a straight line. The line—the ecological model, if you will—is our hypothesis about the relationship between the variables x and y. The dots represent actual values of x and y that we have measured in nature. The variables do appear to have a linear relationship, but we cannot draw one straight line that will pass simultaneously through all of the dots. The importance of the model is that it accurately represents the *essence* of the relationship.

The utility of using mathematical expressions (we might call them expressions of theory) as statements of hypotheses can be demonstrated with a further examination of the equation for a straight line. If x and y are variables of ecological interest, the equation for a straight line would represent a hypothesis about the

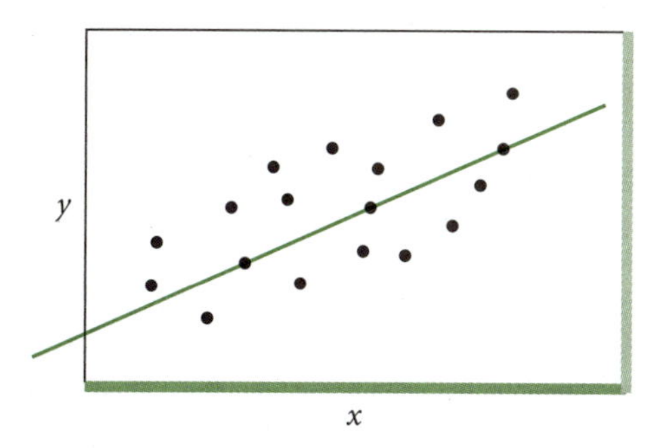

FIGURE 2-1 A straight line is a mathematical model that describes the exact relationship between two variables, x and y. If x and y are two factors of the natural world, one—say, y—may tend to increase as the other—x—increases, but this relationship will not be perfect, as indicated by the points scattered about the line. In such a case the straight line serves as a model for the relationship between the two factors.

dynamics of the two variables. If an ecologist suggests that the relationship is $y = mx + b$, he is suggesting only that the variables y and x are linearly related. The actual configuration of the line—that is, where the line crosses the y-axis (b) and whether it has a negative or positive slope (m)—is left unspecified. However, if the ecologist suggests that the relationship is $y = -1.5x + 2.0$, he has suggested not only that there is a linear relationship, but, in addition, that it is a very specific negative linear relationship. Mathematical expressions like this one, and many others that we shall encounter in this book, provide straightforward and unambiguous avenues for the successive refinement of hypotheses about nature.

Another quality of mathematical ecological theory, which is difficult to demonstrate with a simple relationship like the straight line, has to do with the heuristic nature of mathematics itself. To put it another way, explorations of the algebraic relationships among variables may reveal suggested patterns in the system—that is, further hypotheses—that are not readily apparent. To appreciate this aspect of mathematical theory, suppose that you are an ecologist interested in the pattern by which a foraging animal obtains reward, in the form of energy, from the single type of prey that it eats. Based on sampling studies and other observations, you believe that the following variables are important: (1) the amount of time that the foraging animal spends searching for prey, which you denote T_s for time$_{\text{searching}}$, (2) the frequency with which the foraging animal encounters prey in the environment, denoted λ, (3) the amount of time it takes for the animal to subdue, kill, and eat each prey item, denoted h for handling time, and (4) the amount of energy that is contained in each prey item, which you call E_p. Your gut feeling is that the most important of these is the total amount of time spent searching. Before you attempt an experiment, you spend some time think-ing about the possible relationships among these variables—that is, you develop some theory. Your pondering leads you to propose the following relationships:

> The number of prey encountered in a period of search time is $T_s\lambda$.
>
> The total amount of energy E that the foraging animal gets during that time is $(T_s\lambda)E_p$.

The total amount of time spent handling prey is $(T_s\lambda)h$; thus the total amount of time required for the animal to get some amount of energy E is the total search time plus the total handling time, $T_s + (T_s\lambda)h$, which you call T.

Using these ideas, you suggest that the rate at which the animal obtains reward during foraging is the amount of energy obtained, E, divided by the total amount of time, T, spent, or

$$\frac{(T_s\lambda)E_p}{T_s + (T_s\lambda)h}.$$

When you look carefully at this expression, you realize that it may be simplified to

$$\frac{\lambda E_p}{1 + \lambda h}.$$

That is, the total search time, which you believed was important, does not appear in this form of the equation. What is important, apparently, is the rate at which prey are encountered in the environment, λ, and the time it takes for the foraging animal to kill and eat them, h. To be sure, this is a very simplified treatment of a complex body of theory that is discussed in much more detail in Chapter 31. But it shows how the simple act of performing algebra on simple expressions of relationships among variables can reveal important features of natural processes.

In many places in this book, we begin our discussion of a set of ecological principles in much the same way that we have presented the hypothetical situation above. We do this in order to emphasize the importance of theoretical thinking in ecology, and in science in general. You might use the theoretical work above as the basis of hypotheses to explore using experiments and sampling studies. Or you might choose to extend your theoretical treatment to include more than one prey, or to constrain the activities of the forager in some way. In this way, you might find an expression that suggests a general trend or fundamental process, one that organizes your thinking about a broad suite of patterns. Once such a body of theory is developed, you can set about verifying its assumptions and predictions using experiments.

2.2 Inferences about the natural world include some uncertainty.

We have emphasized the role that variation plays in the process of evolution by natural selection (see Chapter 1). The presence of variation in nature also directly affects our ability to discover the order of nature. The questions that ecologists ask are stimulated by observations of individual events, but they are meant to apply to the entire set of such observations. We see a honeybee visit the flowers of a particular plant species, and we wonder how many flowers the bee visits each day (Figure 2-2). (The number of flowers visited per day is a variable.) Our interest is not restricted to the individual honeybee that we are watching at the moment. Rather, we are interested in all honeybees that visit that species of plant in the area. Natural variation imposes a fundamental constraint on our understanding of such a question. The area in which we observe the bee may contain thousands of bees, and each may differ in the number of flowers it visits during the day. In the face of this variation, we refine our question and ask, "What is the *average* number of flowers visited by a bee in one day?" If we could count the number of flowers that each bee in the area visits in a day, sum those values, and divide that quantity by the total number of bees present, we would obtain the true average number of daily flower visits per bee per day. It might be possible to obtain an estimate of the true average number of visits by observing a subset of all of the bees in the area, called a sample, but because there are so many bees, the determination of the true average number of daily flower visits is practically impossible.

FIGURE 2-2 The average number of times that bees visit flowers in a field may be obtained by a diligent researcher for a sample of bees. That number is called a statistic, and it estimates the unknown true average number of visits, called the parameter, which, for practical reasons, cannot be obtained. Statistical analysis is used to determine the likelihood that the sample average is a good estimate of the parameter. (*Courtesy of R. B. Suter.*)

This simple example illustrates an important feature of all scientific inquiry. This idea may be reinforced by imagining the following situation. You and a friend sit facing each other on opposites sides of a table, on top of which is an opaque screen that shields all but your faces. Your friend has in front of him an object the size and shape of which is unknown to you because it is blocked by the screen. Your job is to determine the configuration of the object by asking your friend questions. Your friend must answer only the exact question that you ask. Thus, if the object happens to be a plastic cube, and you ask "Is the object square?" your friend would respond by saying "No." because a square is a two-dimensional shape and thus describes only one side of the object. You question your friend for an hour, making careful notes and drawings as you proceed. With careful questioning, you learn a great deal about the object, such as its general size, color, and the material from which it is constructed, and after a while, a number of possible configurations emerge. Finally, even though some of the information that you have obtained is not consistent with your conclusion, you believe that you have developed a good description of your friend's object and, excited at the prospect of finding out if you are correct, you ask to see it. At this point, your friend quickly places the object in a paper bag and exits the room, leaving you with no confirmation of whether your description is correct. In a way, this is the principal challenge that scientists face. If we question nature carefully and systematically, we may learn a great deal about it, but we may never be absolutely certain that our conclusion is correct.

Because of this uncertainty, scientists rely on probabilistic models of inquiry embodied in the science of **statistics,** the study and analysis of quantitative data. Statistical approaches provide the means of understanding a process with some specified level of uncertainty. For example, you would not be so disappointed by your friend's sudden exit from the room if you were, say, 90% sure that your idea of the shape of the object was correct. It is this type of certainty—not perfect, but substantial—that statistical analysis of ecological data provides. We will explain certain aspects of statistical analysis in our examination of ecological studies throughout this book. For now, let us mention a few basic ideas.

Basic Statistics

In statistics, an unknown true characteristic of a system, such as the true average number of flowers visited per bee per day, or the exact size and shape of the object hidden from your sight by your friend, is called a **parameter.** The parameters of ecological systems are usually unknowable for practical reasons. For example, as we mentioned earlier, while it might be technically possible to observe the number of flower visits in a day for every honeybee in an area, the amount of energy required to obtain such information would be prohibitive. To be sure, some ecological systems are small enough, or sufficiently isolated, that all of the measurements of a particular variable may be obtained. For example, populations of endangered plants and animals may include just a handful of individuals. When the value of a parameter may be obtained by direct observation or measurement, then there is no need to apply statistical analysis.

The goal of scientific research is either to estimate the parameter of interest or to use an experimental procedure to evaluate a hypothesis about the parameter. A parameter is estimated by a **statistic,** which is some value or index calculated from a **sample** of measurements of the variable that is obtained from the entire group of possible measurements of that variable. The group of all the possible measurements of a variable is called the **statistical population.** A statistical population differs from a biological population: the former refers to the total set of *observations* of a variable, the latter to a group of individuals of the same species in the same area. All possible observations of the number of daily visits to flowers by bees in some prescribed area constitute a statistical population. The average of all of the daily numbers of flower visits is a parameter. Parameters are characteristics of statistical populations. If we were able to obtain observations of the daily flower visits of, say, fifty bees, those fifty observations would represent a sample from the statistical population of all possible daily flower visits. The average number of daily flower visits for those fifty bees is a statistic that estimates the true parametric average. A statistic is a characteristic of a sample.

Both estimation and hypothesis testing rely on knowledge of the underlying probabilities of the different values of the variable of interest. In order to evaluate a sample of observations and the statistics calculated from that sample, we must assume some underlying probability distribution of the variable. Many variables of natural systems approximate what is called a **normal distribution** (the familiar bell-shaped curve; Figure 2-3). The average of such a distribution is the parametric mean, denoted μ (Greek letters are used to denote parameters, while Roman letters are used for statistics). In a true normal distribution, one-half of the values of the variable will fall below the mean, and one-half above. A few values will be quite a bit smaller or considerably larger than the mean, and many values will be close to the mean, differing from it just a little on the high side or the low side. The shape of the normal distribution is determined by the way in which observations are spread about the mean; this is measured by a parameter called the **variance,** denoted σ^2, and calculated as $\Sigma(x_i - \mu)^2/N$, where N is the total number of observations in the statistical population ($i = 1 \ldots N$). We can estimate the parameters μ and σ^2 by obtaining a sample of measurements of the variable and calculating the statistic $\bar{x}$, the sample mean, which estimates μ, and the statistic s^2, which estimates σ^2. The sample variance, s^2, is calculated as $s^2 = \Sigma(x_i - \bar{x})^2/n - 1$. Dividing by $n-1$, a quantity called the degrees of freedom, is necessary in order to make s^2 an unbiased estimator of σ^2. If an ecologist can reasonably assume that the variable has a normal distribution, and if the sample size, n, is large enough, the sample statistics $\bar{x}$ and s^2 are good estimators of the mean and variance of the variable.

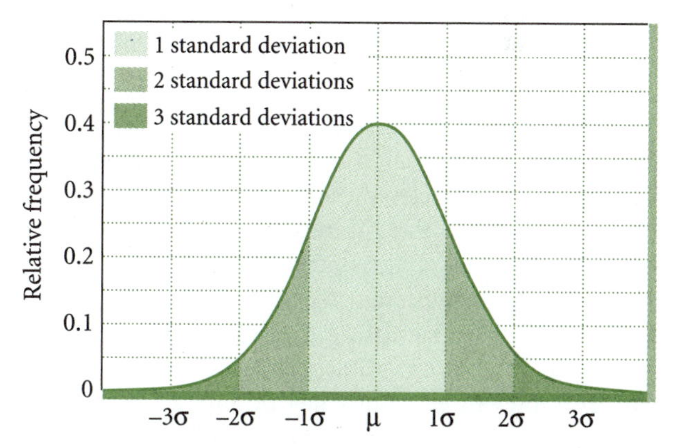

FIGURE 2-3 Many natural variables have a bell-shaped probability distribution called the normal distribution. A normal variable has a true average value μ (parametric mean). The values of the variable fall in a symmetrical pattern around μ with one-half lower and one-half higher than that value. The extent to which values are spread around μ is measured by a parameter called the variance, σ^2, which is calculated as $\Sigma(x_i - \mu)^2/N$, where N is the total number of observations in the statistical population and x_i are the i individual values of the variable. The square root of the variance is the standard deviation of the distribution. About 68% of the observations will fall within one standard deviation of the mean.

You will notice that while the mean is given in the same units as the variable, the variance is given in units squared. That is, if the variable is measured in meters, then the average will be in meters, but the variance will be in meters squared (m^2). Another measure of the spread about the mean of a normal distribution is the square root of the variance, $\sqrt{s^2}$, called the **standard deviation,** s. The standard deviation is given in the same units as the variable. In a normal distribution, about 68% of all observations of the variable fall within one standard deviation of the mean, about 95% fall within two standard deviations, and 99% fall within three standard deviations (see Figure 2-3). A measure of the precision of a statistic is called the **standard error,** denoted SE, and it is calculated as $SE = s/\sqrt{n}$. **Precision** refers to how close repeated measurements of the same quantity are to one another. A small standard error means that repeated calculations of the statistic will yield values close to one another. The size of the standard error depends on the sample size, n. Larger samples will have smaller standard errors for a given variation in the variable. Whether or not the sample mean is near the true mean is a question of **accuracy.** The accuracy of a measurement is how close that measurement comes to the actual value of the variable being measured.

2.3 Ecologists employ sampling studies to estimate ecological parameters.

In order to learn about basic quantitative features of the natural world, such as the number of individuals of a particular species present in an area, the age of those individuals, the timing of an ecological event such as flowering or the onset of courtship behavior, or the number of different types of organisms present, ecologists employ **sampling studies.** Sampling studies involve a set of programmed observations of nature. For example, an ecologist may be interested in what time of year a particular species of lake-dwelling freshwater fish spawns (produces eggs). The presence of a large number of fish larvae (immature forms) in the water is evidence of recent spawning (Figure 2-4). Thus, if the ecologist sampled the lake for larval fish periodically, say every week, during the year, she could estimate the time of spawning by observing a peak number of larval fish in the samples.

Sampling studies often include the measurement of a large number of different variables that are later subjected to **correlation analysis.** The correlation between two variables is the way in which they vary together. If two variables are positively correlated, then high values of one of the variables are associated with

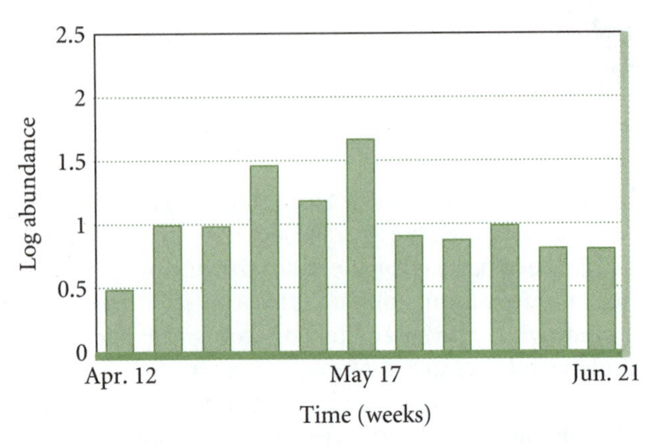

FIGURE 2-4 Abundances of larval fish over time (log scale) obtained from a sampling study conducted in the Tallahatchie River, Mississippi. The peak represents a spawning event, which occurs in late April when the water temperatures become warm enough to allow the eggs to develop into larvae.(*From Turner et al. 1994.*)

high values of the other. For example, the timing of spawning of fishes is often positively correlated with water temperature. In the sampling study mentioned above, more larval fish will be observed as the temperature of the water increases in the spring. The correlation of one variable with another does not imply that the two variables have any direct effect on each other. What triggers spawning may have nothing to do with rising temperature. The correlation between spawning and temperature may be incidental, or it may be that the two are related in some complex way through the actions of another variable that was not measured.

Sometimes, large numbers of variables are measured, and all of the intercorrelations among them are subjected to a simultaneous analysis using what are referred to as **multivariate statistics.** Such procedures analyze groups of variables all at one time, usually with the goal of finding patterns in the groups that can be represented as linear combinations or some other index of the variables. These procedures are important in community ecology where multiple species interact.

Long-term Ecological Monitoring

Ecological processes often take place relatively slowly and can be understood only if they are observed over a period of years or even decades. Thus ecologists sometimes employ **long-term monitoring studies** of individual species or entire ecological systems. These long-term studies involve obtaining samples year in and year out, often for decades. Much of our current understanding of the dynamics of the fishes of Lake Victoria is the result of long-term sampling of the fishes in the lake. Because of the enormous cost and labor requirements of such studies, they are most often conducted by consortia of individuals and educational and government institutions. In many cases, monitoring studies are conducted at research stations (Figure 2-5) that are situated in particularly vulnerable parts of the earth, or in places where human intervention is not yet significant and where the features of pristine natural processes can be examined. Such studies have proved to be enormously important in understanding the extent and nature of global climate change.

The need for and importance of long-term ecological monitoring studies is clearly demonstrated by the current crisis in amphibian (frogs, toads, and salamanders) populations around the world. Since the early 1980s scientists have noticed a precipitous decline in the size of some amphibian populations. In many places, populations that formerly contained thousands of frogs have dwindled to several hundred individuals, and in a number of places populations have disappeared altogether in just a few years. The golden toad (*Bufo periglenes*) of Costa Rica (Figure 2-6) provides a dramatic example of such a decline. The toad lives in the Monteverde Cloud Forest Preserve of Costa Rica, a protected area that appears safe from direct human intervention. Like many frogs and toads, adult golden toads gather in great numbers in temporary ponds in the spring, where they mate and lay masses of eggs. Such a breeding episode at the principal breeding site at Brillante in 1987 involved over 1,500 toads (Crump et al. 1992). But in the years 1988 and 1989, only a single toad appeared at that site, and no toads were observed during the 1990–1992 breeding seasons. Intensive searching during those years turned up only a few individuals at other breeding sites. Similar precipitous declines have been observed for other species in other parts of the world. A study of six populations of leopard frogs (*Rana pipiens*) in Colorado between 1973 and 1982 revealed that by 1981, reproductive failure and extinction had occurred in all six populations (Corn and Fogleman 1984).

FIGURE 2-5 The Arctic Tundra Long Term Ecological Research Site at Toolik Lake, Alaska. This research station is located in a very remote part of northern Alaska near the Brooks Range. (*Courtesy of R. L. Holberton.*)

FIGURE 2-6 The golden toad (*Bufo periglenes*) of the Monteverde Cloud Forest Preserve of Costa Rica. Populations of the toad experienced a dramatic and unexplained decline between 1987 and 1989. (*Courtesy of M. L. Crump.*)

Unfortunately, long-term studies of amphibian populations are not common, and it is therefore sometimes difficult to determine whether a population decline represents a disastrous loss of a species or a natural low point in a population cycle (see Chapter 18). For this reason, in 1991, a worldwide network of scientists and others interested in conservation began monitoring amphibian populations. This organization, which is sponsored by the Open University of the United Kingdom, is called the Declining Amphibian Populations Task Force (DAPTF). Its goal is to determine the extent of amphibian population decline and to promote collaboration in research to determine the causes of that decline. The monitoring efforts of DAPTF, along with the careful experimental studies discussed below, have given ecologists reason to believe that the recent decline in amphibian populations is the result of climate change associated with human activities.

One of the most important long-term monitoring efforts is the Long-Term Ecological Research program begun in 1980 by the U.S. National Science Foundation. The purpose of this program is to establish long-term research sites, referred to as LTER Sites, in sensitive and important ecosystems throughout the United States. These sites serve as the focal point for collaborative research and monitoring of ecological and climatological information. Today the program includes twenty LTER sites (Table 2-1) involving the work of over 900 ecologists from various educational and governmental institutions. The program has expanded to include two sites in the Antarctic. In recognition of the importance of the ecology of the urban environment, two urban LTER sites (Phoenix and Baltimore) were established in 1997. In 1993 the

TABLE 2-1 List of Long-Term Ecological Research (LTER) sites currently funded by the National Science Foundation

Site
H. J. Andrews Experimental Forest, Blue River, Oregon
Arctic Tundra, Toolik Lake, Brooks Range, Alaska*
Baltimore Urban LTER, Baltimore, Maryland
Bonanza Creek Experimental Forest, Fairbanks, Alaska
Central Arizona-Phoenix Urban LTER, Phoenix, Arizona
Cedar Creek Natural History Area, Minneapolis, Minnesota
Coweeta Hydrologic Laboratory, Otto, North Carolina
Harvard Forest, Petersham, Massachusetts
Hubbard Brook Experimental Forest, West Thornton, New Hampshire
Jornada Basin, Las Cruces, New Mexico
W. K. Kellogg Biological Station, Hickory Corners, Michigan
Konza Prairie Research Natural Area, Manhattan, Kansas
Luquillo Experimental Forest, near San Juan, Puerto Rico
McMurdo Dry Valleys, McMurdo Station, Antarctica
Niwot Ridge/Green Lakes Valley, near Boulder, Colorado
North Temperate Lakes, near Boulder Junction, Wisconsin
Palmer Station, Antarctica
Sevilleta National Wildlife Refuge, near Albuquerque, New Mexico
Shortgrass Steppe, Nunn, Colorado
Virginia Coast Reserve, near Oyster, Virginia
LTER Network Office, Albuquerque, New Mexico

*See Figure 2-5.

International LTER Network was established to coordinate the activities of ecologists throughout the world who study the ecological processes that take place over large time and spatial scales. One of the most important features of the LTER program is the development of a database of basic information about the ecological factors and climate conditions at LTER sites. These data, which are available for all scientists to use, represent a history of ecological change that will help ecologists understand the way the natural world works.

2.4 Ecologists use experiments to study causation in nature.

Careful observation and systematic long-term sampling can provide ecologists with essential information about the way nature works. For instance, after repeatedly observing, in tropical regions, hawks that were banded in the hills of Pennsylvania, ecologists feel comfortable accepting the idea that hawks migrate. The ecologist interested in the spawning time of lake fish will conclude to her satisfaction that spawning occurs in late March if her sampling studies show that, year after year, the numbers of larval fish increase dramatically at that time of the year. But sampling studies do not yield information about the *causes* of hawk migration or the fish spawning cycle. Information about how one component of a natural system directly affects another is what is most relevant to an understanding of the natural world. In order to obtain information on causes, one must conduct experiments.

An **experiment** is an activity whereby natural processes are allowed to proceed under conditions that are controlled by the experimenter. In some cases, the experimenter exercises a great deal of control by recreating a portion of nature, either in the laboratory or outdoors in small containers (Figure 2-7), where all or most of the important variables are under his control. In other cases, he may control selected parts of the system in the field while allowing the rest to operate under the local conditions. The goal of an experiment is to make an inference about a null hypothesis, that is, to determine whether it is more likely that the

FIGURE 2-7 Experimental facility at the University of Mississippi Field Station. Experimental treatments are applied to sets of individual tanks. Since all of the tanks use the same water sources and are exposed to the same environmental conditions, they differ only in the treatment applied by the researcher. (*Courtesy of S. T. Threlkeld.*)

null hypothesis is true or false. If the conclusion is that the null hypothesis is false, then some alternative hypothesis must be true. However, most experiments cannot distinguish between competing alternative hypotheses, and so, in such a case, more experimentation is required.

Just as with the estimation of a parameter through sampling, we can usually never determine conclusively whether or not a null hypothesis is actually true. Thus, there is some uncertainty inherent in any experimental procedure. Several types of errors may be made in the evaluation of a null hypothesis. To illustrate, take the case of American jurisprudence in which a person is accused of a committing a crime. The accused person is presumed innocent of the crime. That is, the null hypothesis is "the person is not guilty of the crime." Think of the trial as an experiment to determine whether the null hypothesis is true or false. After hearing the evidence, the jury renders a verdict that is either "guilty" or "not guilty." But does the jury's verdict actually determine the guilt or innocence of the accused? It does not, because only the accused knows, assuming there are no witnesses, whether or not he or she is guilty. Suppose that the accused did not commit the crime—in other words, that the null hypothesis "the accused is not guilty" is actually true. If the jury comes back from their deliberations with a verdict of "guilty," they have convicted an innocent person and have clearly made a mistake. They have rejected a true null hypothesis, which, in the terminology of statistics, is a **type I error** (Figure 2-8). If the jury finds that the accused is not guilty, they have taken the correct action. Now, suppose the null hypothesis "the accused is not guilty" is actually false—that is, the person did commit the crime. If the jury comes back with a verdict of not guilty, they have made another kind of mistake, called a **type II error,** which is the acceptance of a false null hypothesis. Properly designed experiments control for the level of type I and type II error. Thus, an experimenter may not know for sure whether he has rejected a true null hypothesis, but he will know what the probability is that such a mistake has been made.

Decision about Null Hypothesis	Null Hypothesis: True	Null Hypothesis: False
Reject	type I error	correct decision
Accept	correct decision	type II error

FIGURE 2-8 The two types of experimental error. When the null hypothesis is true (a fact that is unknown to the researcher), a type I error is made if the hypothesis is rejected. When the null hypothesis is false, a type II error is made if the hypothesis is accepted. The probability of type I and type II error is controlled by the experimental design.

It is important to appreciate that the presence of uncertainty and the possibility of making an error do not invalidate scientific information or hamper progress toward our understanding of the natural world. Uncertainty is a part of all fields of scholarly inquiry. A political scientist can never be absolutely certain that she knows the opinion of a population regarding a specific issue. An art historian can never be absolutely certain of his conclusion that a recently discovered unsigned painting was actually painted by the early-twentieth-century French painter Henri Rousseau. Both proceed with confidence based on the information that they obtain and with the full knowledge that there is a possibility that their conclusions may be incorrect. The experimental approach of science quantifies and minimizes the uncertainty inherent in inquiry. Experiments proceed at some predetermined level of probability of committing a type I or type II error.

The way in which an ecological experiment is conducted will depend upon the particular null hypotheses of interest as well as many other factors. The size and physical characteristics of the phenomenon of interest will affect the type of experiment that can be conducted. For example, an ecologist interested in the way in which human-made chemicals such as chlorofluorocarbons (CFCs), which are used as propellants in aerosol spray products, affect the earth's protective ozone (O_3) layer would have a difficult time conducting an experiment in the high atmosphere (though some have been done). Nevertheless, the ecologist can learn a great deal about the chemistry of interactions of CFCs and ozone by performing laboratory experiments in which chemical reactions can be controlled and more easily observed. Experiments must also accommodate the peculiarities and unique features of the organisms involved. This requires that the experimenter fully understand the natural history of the organisms of interest, an understanding that comes from the careful observation and questioning of nature (as we have seen in Chapter 1). Thus, an ecologist conducting an experiment on how the females of a night-breeding spider select a male with which to mate will want the experimental conditions to include low light levels. An experiment designed to determine the relationship between soil pH and plant growth must be conducted within a range of pH that allows the growth of the plant of interest.

2.5 Experimental studies involve the application of treatments and the observation of responses.

For a number of reasons, considering some recent experimental approaches to expanding our understanding of the alarming and rapid decline of many frog populations worldwide is a particularly instructive way to explore the role and techniques of experimentation in ecology. First, like most ecological processes, the sizes of amphibian populations are affected by many interacting factors. The importance of the various factors and their interactions with one another may differ from one geographic location to the next, or from one species to the next. The experimental protocols used to address the problem must take this complexity into consideration.

Second, many of the null hypotheses regarding the decline of amphibian populations focus on interactions of humans with the environment that have been suggested as detrimental to the global environment. Amphibians have life cycles that link them very closely to both aquatic and terrestrial habitats. They lay their eggs in the water, and most species spend their entire lives near wetland areas. Larvae and juveniles often live completely in the water, breathing with gills. Adults most often live in moist terrestrial areas, where they obtain much of their oxygen through the skin, which must be kept moist for that purpose. Some ecologists have suggested that for these reasons, amphibians are more sensitive than other organisms to environmental change brought about by human activity, and that the declines in amphibian populations forecast dramatic changes in life on earth resulting from human-induced global climate change.

Finally, because experimental work toward an understanding of amphibian decline is relatively recent, owing to our recent realization of the scope of the problem, competing null hypotheses are still being sorted out, and scientists are engaged in a lively debate about the importance of various factors. By examining the experimental work on amphibian decline, you can obtain a feel for the excitement of early experimental discovery.

The reason for the apparent decline in amphibian populations is currently unknown, but it is probably the result of one or a combination of the following factors: (1) the presence of human-produced toxic residues from pesticides and herbicides, (2) increasing amounts of ultraviolet light of a type very harmful to biological systems, resulting from damage to the earth's ozone layer, (3) human land use, which has destroyed much of the wetland habitat required by amphibians, and (4) the activity of fungi and parasites that may have reached higher than usual population levels (Blaustein and Wake 1990, Carey 1993, Blaustein et al. 1994, 1995, Licht and Grant 1997). Let us focus on one of these factors, ultraviolet radiation.

The solar radiation incident upon the outer atmosphere of the earth is composed of a spectrum of wavelengths of light having different energy levels. Those wavelengths of light between 400 and 700 nanometers (nm; one-billionth of a meter, or 10^{-9}) are visible to humans (the visible spectrum). Visible light passes through the earth's atmosphere, where it is absorbed by the earth's land surface, vegetation, and waters. Ultraviolet radiation, with wavelengths shorter than the visible spectrum, has high energy. Ultraviolet radiation can be divided into three forms based on wavelength: Ultraviolet A (UV-A), with wavelengths between 320 and 400 nm, UV-B, with wavelengths between 280 and 320 nm, and UV-C, sometimes called far-UV, with wavelengths less than 280 nm. Of these forms of ultraviolet radiation, UV-C is the most damaging to biological systems. However, most of that form is absorbed by atmospheric gases and does not reach the earth's surface. UV-B is also disruptive to living forms; its effect is most obviously demonstrated by the sunburn that it causes when you go exposed at the beach. Overexposure to UV-B radiation has also been linked to cancer and immune system suppression (Black and Chan 1977, Jagger 1985). UV-B acts by disrupting the structure of the DNA molecule. Such DNA damage may be repaired in a number of ways by a cellular mechanism involving the enzyme photolyase and, interestingly, by exposure to UV-A radiation. UV-A radiation appears to have little harmful effect on biological systems.

The stratosphere (the upper part of the earth's atmosphere, between about 10 and 50 kilometers [km] above the surface of the earth) contains a natural filter for UV-B radiation in the form of a thin layer of ozone (O_3). Ozone is formed when ultraviolet radiation causes the dissociation of oxygen (O_2) molecules in the upper atmosphere, releasing highly reactive individual atoms of oxygen ($O_2 \rightarrow O + O$). Some of these O atoms unite to form O_2, and others join with O_2 molecules to form ozone ($O + O_2 \rightarrow O_3$). The effect of this reaction is the absorption of a good portion of the ultraviolet radiation that is incident upon the upper atmosphere of the earth.

Since the early 1970s, levels of stratospheric ozone have declined markedly over some portions of the earth, particularly in the Antarctic region, which contains one of the world's most productive marine ecosystems. Such "ozone holes" are now apparent in temperate regions as well. This degradation of the earth's natural ultraviolet filter has resulted almost entirely from the release into the atmosphere of human-made chlorofluorocarbons (CFCs), which have been widely used as refrigerants and to pressurize spray cans. In the stratosphere, chlorine atoms from CFCs react with ozone to form chlorine monoxide (ClO) and oxygen ($Cl + O_3 \rightarrow ClO + O_2$). Chlorine monoxide molecules may then react with oxygen atoms to release more chlorine atoms ($ClO + O \rightarrow Cl + O_2$), which are involved in further dissociation of ozone.

Andrew Blaustein of Oregon State University and his coworkers have conducted a number of field experiments to determine how UV-B affects amphibian populations (Blaustein et al. 1994, 1995). In their studies of frog populations, Blaustein and his group focused on how UV-B radiation affects the hatching success of eggs. Many species of frogs and salamanders deposit their eggs in masses in shallow parts of ponds and lakes or in temporary bodies of water. Unlike the adults, which are highly mobile and can avoid direct exposure to sunlight, the eggs develop where they are deposited, often in full sunlight. If UV-B radiation reduces the hatching success of the eggs, the population will decline because fewer individuals will grow to reproductive age. The variable of interest in Blaustein's experiments was thus the number of eggs surviving to hatch. This variable is called the **dependent variable** (or sometimes the **response variable**) of the experiment because it is the variable for which a change resulting from the experimental manipulation is observed. The null hypothesis of the experiment was that "hatching success (dependent variable) is unaffected by the level of UV-B radiation." A rejection of this hypothesis would mean that some alternative is true. As we mentioned before, if the hypothesis is rejected, there is a chance that a type I error has been committed. In the experimental design used by Blaustein, the probability of such an error was held at 5%, acceptably small for most ecological studies.

The questions that underlie these experimental studies are questions of mechanism (see Chapter 1). The goal of the experiments was to determine the effect of one factor, UV-B light level, on another, hatching success. That is, the experimenters wanted to know if there is evidence that high UV-B levels cause poor hatching success. As we shall see below, Blaustein and his group also asked a number of evolutionary questions of this system.

A number of other factors, such as water temperature and the presence of predators of eggs, may affect the hatching success of frog eggs. Thus, in order to evaluate the effect of UV-B radiation on hatching success, Blaustein had to design an experiment that reduced or eliminated the influence of these other factors. His group also had to devise a way to expose eggs to different levels of UV-B light. The various levels of UV-B radiation exposure are the **treatments** of the experiment. Treatments are variables that come under the control of the investigator, and are called **independent variables.** Experiments may have more than one independent variable. Usually, one treatment is the normal condition, along with any peculiar aspects of the experimental procedure. This treatment is called the **control.** For example, in an experiment involving the injection of different amounts of a drug into animals, the doses of the drug would be the treatments. One treatment would include none of the drug, but would include an injection of a benign liquid such as saline solution. Thus, animals receiving the control (no-drug) treatment would be treated the same as the others: they would receive an injection, just like the animals receiving the other treatments. Blaustein controlled the amount of UV-B to which frog eggs were exposed by employing three different types of filters. In one treatment, UV-B was completely blocked using Mylar. In another, an acetate filter that transmitted about 80% of the incident UV-B radiation was used, and in a third—the control—no filter was used at all.

The natural history of the frogs that Blaustein studied presented a number of experimental design challenges. His group studied three species of frogs that occur around several lakes in the Cascade Mountains of Oregon: *Hyla regilla, Rana cascadae,* and *Bufo boreas.* (The genera *Hyla* [tree frogs], *Rana* [bullfrogs], and *Bufo* [toads] represent three common genera of frogs.) These frogs lay their eggs in masses near the shore of the lake. As we mentioned, many factors other than the incidence of UV-B, such as water temperature and predation, may affect the mortality of eggs. These factors are likely to differ in their effects on egg hatching success between lakes and among different locations within a single lake. In order to evaluate the effect of UV-B on hatching success, these other factors had to be eliminated or controlled in some way. To accomplish this, the experimenters constructed 28 cm × 38 cm × 7 cm enclosures that could be placed in the water to hold groups of eggs at locations determined by the experimenter. A screen door on each container allowed lake water to circulate into and out of the container. Sets of twelve containers were placed linearly along the shore at two places in the lake (a total of twenty-four enclosures per lake; Figure 2-9). Each container was positioned in water 5 to 10 cm deep. The experimenters then collected newly laid eggs (less than 24 hours old) of a particular frog species from various places in the lake and placed 150 eggs in each of the enclosures. Using these containers, all of the eggs could be held in roughly the same conditions during the experiment. It was not necessary to control the water temperature or other aspects of water chemistry since all the eggs were held in the same lake environment.

Although the experimenters expected that all the enclosures would have similar conditions, there was

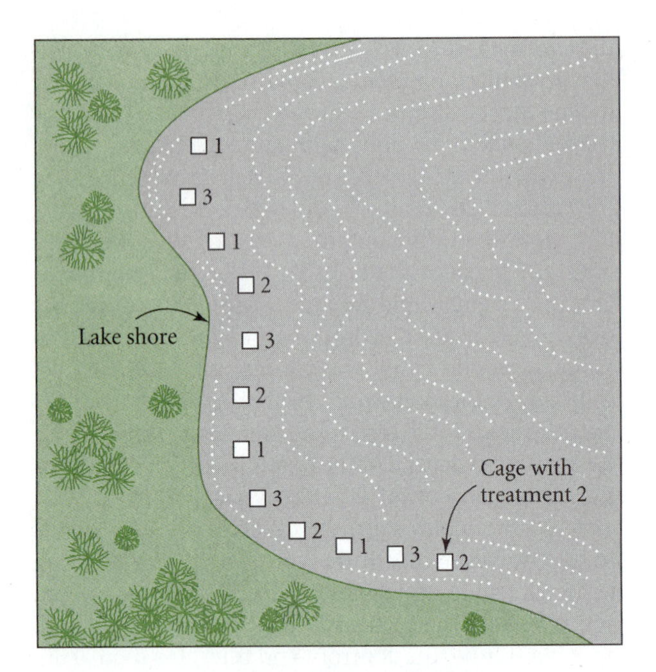

FIGURE 2-9 Schematic diagram of the placement of cages in Blaustein's experiment to determine the importance of exposure to UV-B light in frog egg hatching success. The cages were placed along the lakeshore, each in water 5 to 10 cm deep. Three treatments (1 = UV-B blocked; 2 = 80% transmission allowed; 3 = complete transmission allowed [control]) were assigned to the cages at random. Four cages received each treatment. (*Based on Blaustein et al. 1994.*)

no way to control completely for small changes from one enclosure location to the next. To diminish the effect of such changes, the three treatments (UV-B blocked, 80% transmission, and complete transmission) were randomly assigned to each of the twelve enclosures (see Figure 2-9). Since there were twelve enclosures at each site and three treatments, each of the treatments was assigned to four of the enclosures. The four enclosures receiving the same treatment are called **replicates** of the experiment. Replication is essential in experiments because it is the only way to determine the extent to which a treatment has a variable effect on the dependent variable. Once the experiment was constructed, Blaustein and his coworkers checked each enclosure until all of the eggs had either hatched or died. Thus, at the end of the experiment, there would be one value of the dependent variable for each enclosure. They expressed this value as the proportion of the eggs in the enclosure that hatched.

The results of one of Blaustein's experiments are displayed in Figure 2-10. Each bar in the figure represents the average of the replicates of a treatment. The line protruding from the bar is the standard error of the mean (see above). The standard error is a measure of how precise the measurement of hatching success is. If the standard error lines of two bars overlap, then the measurement is not precise enough to say that there is a difference between the two treatments represented by those bars. If the error lines do not overlap, then we can say with a certainty corresponding to [1—(the probability of making a type I error)] that the treatments have different effects. In these results, we see that *Hyla regilla* eggs appear to be insensitive to different levels of UV-B, whereas the eggs of both *Rana cascadae* and *Bufo boreas* have a much higher hatching success when UV-B is filtered out completely. The results also show that there appears to be little difference in this pattern among the different lakes in which the experiment was conducted.

Blaustein and his colleagues (1995) conducted a similar experiment on the northwestern salamander (*Ambystoma gracile*) and showed that its eggs were very sensitive to UV-B. But a more recent experiment by P. Corn (1998) on *Bufo boreas* using an experimental design similar to Blaustein's failed to show a sensitivity to UV-B radiation. Such seemingly contradictory results are common in science, particularly in areas where knowledge is just emerging, as in the case of amphibian population decline. Corn's study raises interesting and important questions about the effects of genetic variation among populations of frogs and the effects of a pathogenic fungus that has been sug-

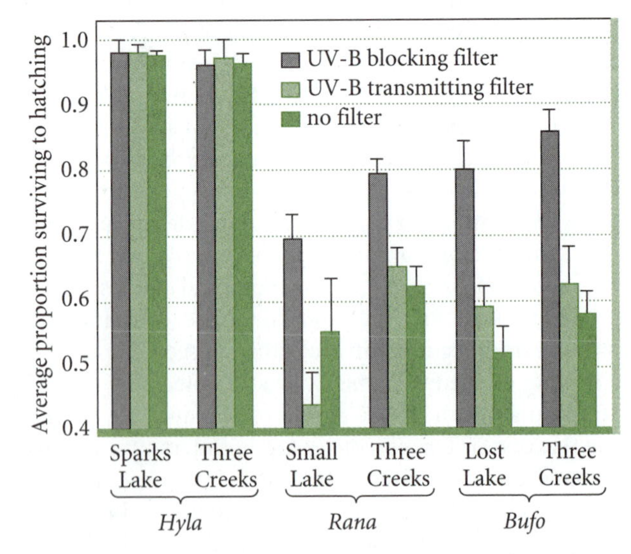

FIGURE 2-10 Results of Blaustein's study to determine the effects of UV-B light on the hatching success of frog eggs. The experiment was conducted in three different lakes with three different species of frogs (*Hyla regilla, Rana cascadae,* and *Bufo boreas*). Each bar represents the mean proportion of eggs in cages surviving to hatching for a particular treatment. The line that protrudes from the top of each bar shows the standard error ($s/\sqrt{n}$), a measure of precision. The results indicate that *Hyla regilla* is not sensitive to different levels of UV-B light, whereas the other two species do display some sensitivity. (*From Blaustein et al. 1994.*)

gested as a contributor to frog population declines. The differences in the outcomes of these studies stimulate further research and increase the rate at which we learn about the natural world.

These experiments provide important information about the possible causes of frog population declines, but we are left wondering about how UV-B disrupts hatching in frog eggs and why it has an effect on some species and not others. Blaustein and his colleagues were interested in this aspect of the problem as well, because observations by many ecologists have shown that some populations of amphibians have declined dramatically and others have not. This observation raises a number of evolutionary questions, and led Blaustein to suggest that since UV-B incidence is not uniform around the globe, amphibians have evolved in different UV-B environments. In places where the natural incidence of UV-B is high, natural selection will have favored a greater capability for DNA repair of UV-B damage, whereas populations evolving in areas with low UV-B incidence might not be expected to have such well-tuned repair mechanisms. An indication of the level of DNA repair capability is the amount of the enzyme photolyase in the cells. Photolyase, along with exposure to UV-A radiation, promotes DNA repair. To test the null hypothesis that photolyase levels are the same in all species of frogs, Blaustein performed assays of the enzyme in a variety of different amphibian species. He found that *R. cascadae* and *B. boreas* had very low levels of the enzyme, whereas the levels in *Hyla* species were much higher.

2.6 Historical analysis provides information about evolutionary change.

The life of every organism is influenced by its interactions with the environment and with the other organisms with which it shares its time and space. As we emphasized in Chapter 1, an organism brings to these interactions a set of genetically based physical, physiological, and behavioral features that shape their outcome. Individuals share characteristics with living members of their species, though it is rare that any two individuals are identical. Individuals also share features with their ancestors, and with other taxa with which they share ancestors. Thus, your pet cat bears a striking resemblance to African lions, Asian tigers, and North American bobcats, to which it is distantly related. The pattern of evolutionary relationships among species or other taxonomic groups is called a **phylogeny.** Ecologists appreciate that an understanding of the phylogeny of a species is essential to a deep understanding of its ecological relationships.

Phylogenetic relationships are important in understanding the nature of adaptation, as demonstrated in Figure 2-11. The figure shows the evolutionary relationships among five species, all of which have a common ancestor. This diagram is called a **cladogram.** Each species has a feature that may occur in state 1, the ancestral state, or in state 0. An ecologist observes that species A and E have state 1 and the other species have state 0. If the ecologist were not aware of the evolutionary relationships among the five species, she might conclude that natural selection has favored state 1 in species A and E and, thus, that those species have been subjected to similar selection pressures with respect to that trait. She might arrive at the same conclusion for species B, C, and D. However, an understanding of the phylogeny of the species leads to a different conclusion. Consider the relationship between species A and B. Species A is a direct descendant of the ancestor and has retained the ancestral state, 1. Species B is descended from the ancestor, but it does not have the ancestral trait; rather, it shows a new trait, called a **derived trait,** state 0. This means that an evolutionary event has occurred between the ancestor and species B, denoted by the solid box. Species C and D, which share a common ancestor with B, also have the derived state, 0. A second evolutionary event has occurred between species D and species E that changed the trait back to state 1, the ancestral condition. This information shows that the state 1 of species A and the state 1 of species E are derivative of different processes. In the case of species A, the trait is simply a holdover from the ancestral condition. In the case of species E, it is the result of two evolutionary events.

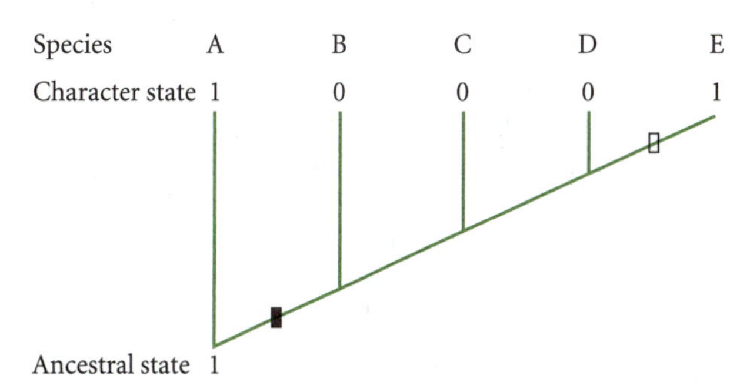

FIGURE 2-11 A hypothetical cladogram showing how the features of five living species (A through E) reflect evolutionary history. The five species each have a feature or character that may have one of two states, 0 or 1, with the ancestral state being 1. Species A and species E have the ancestral state, whereas species B, C, and D have state 0. The state of B, C, and D resulted from some evolutionary event (solid black box) between species A and B. Another event (open box) between species D and E changed the state back to the ancestral state. Knowledge of the evolutionary relationships among these species will help ecologists determine the utility of the character in each.

Thus, a direct comparison of the utility of the feature between species A and species E is inappropriate. More and more, this type of analysis is being applied to ecological problems.

We have presented an overview of the strategies that ecologists use to explore nature. In the remaining sections of this chapter we will turn our attention to some of the practical problems that ecologists face in doing their work, how the results of ecological studies are disseminated, and how ecological knowledge may be applied. We will also identify some of the different fields of ecological study.

2.7 Ecological study poses many practical and logistical challenges.

There are enormous logistical challenges to overcome in conducting ecological experiments and sampling studies. In many cases, the phenomenon of interest takes place in a remote or inaccessible location, such as the very top of the tropical forest canopy, a remote area of the desert, or the depths of a body of water. Many organisms are too small for us to see without the aid of a microscope, and some, like the beetle-riding pseudoscorpions that we mentioned in Chapter 1 (see Figure 1-10b), can be observed only with great difficulty. Many animals move quickly or operate over a great range, making detailed observations of all parts of their natural history difficult. Thus, the observation, sampling, and monitoring of the natural world is limited as much by ingenuity, creativity, and courage as by simple curiosity.

One of the most interesting and exciting challenges for ecologists has been working in the high canopy of tropical rain forests. From our vantage point on the rain forest floor, we are unable to see the activities in the forest canopy, which lies well over 30 meters above our heads. The logistical problems associated with obtaining access to the tops of the tallest tropical trees, and with moving about in the canopy in order to collect and observe organisms, are enormous. Nevertheless, a number of adventurous souls have succeeded in gaining access to these unexplored reaches of the natural world. One project in French Guiana involved laying an enormous rubber raft on the very tops of the dense canopy trees (Halle 1990). The raft was a hexagonal affair constructed of large inflatable tubes and equipped with a wire mesh floor (Figure 2-12). The raft was floated over the canopy by a hot-air dirigible and gently lowered onto the tops of the trees, where it sat swaying as the trees moved in the wind. Scientists lived on the exposed raft for weeks, obtaining samples and taking measurements of the canopy life. Another system for accessing the rain forest canopy was constructed at the U.S. Smithsonian Tropical Research Station in Panama. A crane, whose 40 m tall tower extends above the top of the canopy, was constructed there in the forest. A 50 m long movable boom extends outward from the crane, at the end of which is attached a small gondola big enough to hold several people. The gondola can be raised and lowered, allowing for sampling at various heights in the canopy, and the entire boom can be rotated, giving scientists enormous access to the canopy. These are just two examples of the extraordinary efforts to which ecologists go to study the natural world.

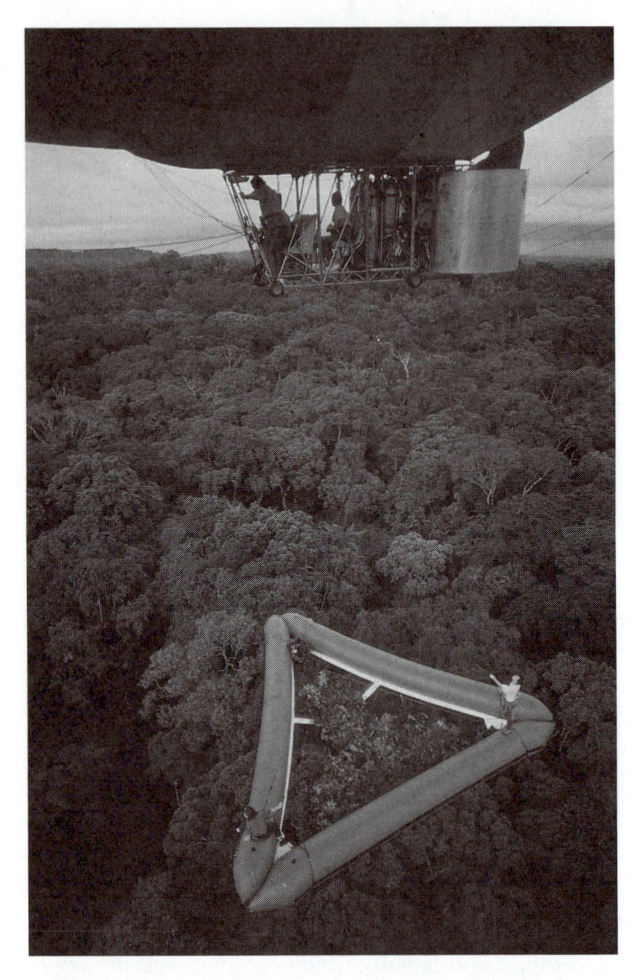

FIGURE 2-12 Canopy raft atop a tropical rain forest. The raft was lifted above the trees and lowered into place using a hot-air dirigible. (*Courtesy of Setsumasa.*)

Funding Ecological Research

Like all scientific work, ecological research costs money. Ecologists often have to travel to remote locations, buy materials for experiments, pay workers to help them run sampling or long-term monitoring programs or experiments, and pay for the publication of their work. A number of governmental agencies hire ecologists to study ecological systems that have some importance to the nation's well-being. For example, the U.S. Department of Agriculture, the

Environmental Protection Agency, and even the military services employ ecologists to conduct experiments on crop systems, diseases, the fate and action of human-made chemicals, and the effects of global change, to name just a few areas. In these cases, research is supported by part of the agency budget. Many ecologists hold positions in universities and colleges, where their duties include teaching undergraduates, training graduate students, working on university committees, and conducting research. Most of these scientists rely on grants and contracts from private industry or governmental agencies to supplement the funding provided by the college or university for their research. Such grants are usually awarded on a competitive basis based on the strength of a detailed research proposal submitted by the scientist. Among the agencies of the U.S. government that fund such work are the National Science Foundation, the National Institutes of Health, the Environmental Protection Agency, and various divisions of the Department of Agriculture.

2.8 Ecologists share their ideas through publication.

As we learned in our imaginary walk through the rain forest in Chapter 1, we begin our study of nature by asking questions and carefully describing what we see. In the very early stages of an ecological study, we expand these more or less casual observations by determining what is already known about the questions of interest. This requires that we review the published literature, talk with other ecologists who have considered the question, and participate in scientific meetings, where new ideas and results are presented.

Like other scholars, ecologists publish the results of their studies in scientific journals and books. Most ecological journals are published by scientific organizations or societies whose members share common goals, approaches, or geography. For example, the Ecological Society of America, the largest ecological organization, with a membership exceeding 7,500, publishes the journals *Ecology, Ecological Monographs,* and *Ecological Applications.* You will find papers from each of these journals and from hundreds of other journals listed in the bibliography of this book.

When an ecologist completes a study and is ready to draw some conclusions from his work, he prepares a manuscript that summarizes the methods used, the specific results obtained, and the principal conclusions that may be drawn from the results. The manuscript also includes an explanation of the importance of the question and a review of the work that others have done on the system he has studied. Near the end of the paper, some suggestions about what experiments might be conducted in the future to advance the understanding of the system are provided. The manuscript is then submitted to the editor of an ecological journal, who asks several scientists, often members of an editorial board, who are familiar with the system and techniques of the study to offer an opinion about the strengths and weaknesses of the work and its importance in increasing knowledge about the subject. Using the input from these reviewers, the editor may decide to publish the paper, reject it, or ask the author to revise it in light of the reviewers' comments and resubmit it for further consideration. Papers that appear in print as a result of such a process are referred to collectively as the primary literature.

When an ecologist or group of ecologists has studied a particular system intensively, they may elect to make their work available in the form of a book. Darwin's *On the Origin of Species,* which was mentioned in Chapter 1, is an example of such a work. Books may take the form of a comprehensive and unified narrative, or they may be composed of a series of separate contributions by the leading workers in the field.

Because there is growing public interest in and concern about nature and the effects of human activities on the environment, more and more ecologists are electing to provide summaries of their work in articles for popular magazines or newspapers, popular books, or, in some cases, television documentaries. And, of course, ecologists produce textbooks like this one and other materials to be used in teaching. Information published in this form usually omits the fine details of the underlying studies and focuses instead on the importance of the work and the impact of the findings on the human condition.

It is through the dissemination of research findings in print that knowledge is advanced. Publication creates a basically continuous discussion about an intriguing problem or system. In recent years, the scope and timeliness of these discussions has been significantly enhanced by use of the Internet, where scientific discourse can proceed more or less continuously. Several ecological journals are now published on the Internet, and many others provide abstracts and tables of contents of recent issues. A number of excellent literature databases are available, and most university and college libraries subscribe to one or more of these.

2.9 Agriculture and resource management are based on ecological principles.

Human activities such as agriculture and aquaculture are unalterably dependent on the ecological relationships of the natural world (Figure 2-13). The energy that you obtained from last evening's juicy quarter-pound hamburger came ultimately from the sun through the photosynthetic activities of the grass eaten by the steer whose ground and cooked skeletal muscles were placed between the halves of the sesame-seed bun. The tuna that you had for lunch was a member of a large population of tuna, the size of which is dependent on birth and death rates, ecological processes that depend on resource availability and biological interactions.

Modern ecology includes the study not only of pristine, undisturbed parts of the natural world, but also of croplands, replanted forests, and managed fish and game populations. In the former, the focus is on the discovery of how nature works in the absence of human intervention. In the latter, the goal is to apply ecological principles to sustain or increase yield. The yield of wheat or corn or soybeans is determined by a myriad of factors, including the physiological responses of the crop plants to environmental conditions, which are influenced by global climate change; the ecology of diseases and herbivores (plant eaters) that prey on those crops; the nature of soil processes; and the local hydrological cycle, all of which may be influenced by the ecological processes happening in adjacent fields and forests and by the activities of humans. The fundamental approach to the study of managed ecosystems is the same as that employed in the study of amphibian declines discussed above: questions, detailed observations, sampling, and experimentation. Indeed, gaining an understanding of the dynamics of natural systems, such as amphibian populations, leads to direct applications in agriculture. Many fisheries' stocks, for example, have experienced rapid declines on the order of those observed in amphibian populations (Myers et al. 1997). It is important to appreciate that the focus of ecological study is toward an understanding of how systems of living organisms operate. Such knowledge, if it is carefully developed through detailed sampling and experimentation, is often very general in nature, and thus applicable to managed systems such as agricultural crops or game.

FIGURE 2-13 A field of soybeans used in experimental studies of the effects of spiders on pest control. Crops like soybeans are planted in long rows, and are thus called row crops. (*Courtesy of A. Rypstra.*)

2.10 Ecological study may be organized around levels of increasing complexity.

The way in which an ecological study proceeds very much depends on how broadly or how narrowly the ecologist focuses her observations of nature. For example, one's interest in photosynthesis might be in the biochemical mechanisms of the process, requiring careful laboratory experiments, or in the rate of photosynthesis of an entire forest, requiring a field study of some type. There are widely accepted levels of resolution of ecological study that help to organize one's thoughts about ecology.

Ecological studies may focus on how individual organisms interact with their environment. Investigations of how organisms are adapted to respond to temperature, maintain proper water and salt balance, balance levels of oxygen and carbon dioxide, or deal with other factors of their physical environment are examples of such a focus. This approach is often referred to as **physiological ecology** or **ecophysiology** (Chapters 4–6). Studies of physiological ecology play an important role in agriculture since crop yield is very much dependent on the performance of individual plants. Ecophysiology is also an important focus of conservation studies. For example, one avenue of inquiry into the reason that many migratory bird species are in decline focuses on how changes in the environment affect the physiological mechanisms that prepare birds for long-distance migration (Wingfield 1990).

Individual organisms must obtain energy either through photosynthesis or by consuming other organisms. Associated with these energy transformations are the movements of materials within and between organisms and between organisms and the physical environment. Some ecologists are interested in the dynamics of energy transformations and material transfers among large assemblages of organisms and the physical environment occupied by those organisms, an interacting entity called an **ecosystem.** This is the subfield of ecology called **ecosystems ecology** (Part 3, Chapters 9–13). Contemporary ecosystems ecologists are very interested in how human activities affect the global cycles of materials. Studies about what happens to the excess carbon that is released into the atmosphere by human activity and the fate and action of human-made chemicals in the natural world are within the purview of ecosystems ecologists.

The study of the ecology of groups of organisms takes place at various levels. Organisms of the same species living in the same place and time are said to constitute an ecological **population.** We may consider the dynamics of a single population of snails, long-leaf pine trees, whales, wombats, or any other type of living thing (Part 4, Chapters 14–19), or we may focus on how two populations—for example, that of a predator and its prey, or a parasite and its host—interact with each other (Part 5, Chapters 20–25). Evolutionary change occurs at the level of the population. Population theory is directly relevant to the management of fish and game populations, forestry, and agriculture. It is also fundamental to our understanding of the dynamics of disease.

Wherever we look in nature we see coexisting organisms. Each species of organism is part of a population. The populations of the many different organisms that we see in a particular place are tied to one another by feeding relationships and other interactions. These suites of interacting populations are called **ecological communities,** and the study of them is **community ecology** (Part 6, Chapters 26–29). Community ecology is distinguished from ecosystems ecology in that the focus of community studies is principally on how biotic interactions such as predation, herbivory, and competition influence the numbers and distributions of organisms, rather than on energy transfer and material movement among individuals and their biotic and abiotic (nonliving) environment. Community ecology has particular relevance in our understanding of the nature of biological diversity.

Questions related to how animals choose mates, determine the sex of their offspring, forage for food, and live in groups, or how plants attract pollinators, disperse seeds, or allocate resources between growth and reproduction, are examples of the types of questions that are asked in the subfield of ecology referred to as **evolutionary ecology** (Part 7, Chapters 30–34). Evolutionary ecologists are interested in how form and function adapt organisms to their environment. The precepts of evolutionary ecology are particularly important to gaining an appreciation of the long-term responses of nature to natural and human-induced changes.

A number of fields of ecological study require the synthesis of several subfields of ecology, as well as an approach that takes into consideration how ecological processes operate on large spatial scales. **Landscape ecology** focuses on how ecological processes operate against a background of great heterogeneity of habitat, which is evident in most geographic regions. The field of **conservation biology** blends the concepts of genetics with population and community ecology, often taking a landscape approach, in the consideration of questions about the maintenance of biodiversity and the preservation of endangered species. More and more ecologists are accepting as their responsibility not only the discovery, explanation, and conservation of nature, but also the application of their knowledge to reestablishing the integrity of natural systems that have been damaged by human activity. This new and exciting field of ecology is called **restoration ecology.**

Evolutionary theory provides the underpinning of all ecological study. The mechanisms of evolutionary change are revealed in the science of genetics, and the avenues of expression of genetic makeup are found in the sciences of physiology, endocrinology, and behavior. The study of ecology involves all of these.

It is important to distinguish the field of ecology from a number of other related areas of scientific inquiry. Ecology is the study of the interrelationships among organisms and their environment. **Environmental science** focuses on understanding the specific impacts of humans on the environment. The study of environmental science integrates the principles of ecology, chemistry, physics, and other sciences with economics, politics, and ethics. Some environmental scientists study the fate and action of human-made substances, such as pesticides and detergents, in the natural world. This avenue of scientific inquiry is often referred to as **ecotoxicology.** While ecotoxicologists are ultimately interested in the way in which human-made substances affect human health, they often use other animals, such as fish or small invertebrates, as models for the action of the particular substance under study.

Finally, ecology must be distinguished from a number of other endeavors that are often confused with it. **Environmentalism, conservationism,** and **preservationism** are social and political movements, not fields of scientific inquiry. In their most constructive and responsible form, these movements seek to educate the public about human-induced environmental problems and to effect changes that will alleviate such problems. Responsible environmentalism is based on a deep understanding of ecological principles, but it is not a subfield of the science of ecology. Neither does ecology encompass well-intentioned public beautification and cleanup activities such as roadside trash pickups and city tree planting drives. Almost everyone applauds such civic responsibility, but such activities are not science, and they do not increase our understanding of the natural world—though they do make it much more pleasant to observe (Figure 2-14).

The subfields of ecological study provide convenient ways to think about the various approaches in

FIGURE 2-14 Example of a place where the activities of environmentalists are much needed. Cleaning up such areas serves an important function in reducing the effects of careless human activity on the natural world. (*Courtesy of W. J. Garrison.*)

ecology, but in many cases, individual ecologists conduct work that crosses the boundaries of these subdisciplines. The complexity of nature, in complement with the natural curiosity of most ecologists, often encourages broad approaches. The population ecologist who sets out to understand the dynamics of a plant species may find that those dynamics depend on the physiological features of the plant (physiological ecology), its relationship with other species in the area (community ecology), and the behavior of a suite of herbivores that feed on it (behavioral ecology). His studies may lead to ideas about the relationship among the important variables of his system that he develops into a body of theory, which may form the hypotheses for the experiments of other ecologists. And, in time, he may write about his work in order to communicate its significance to other ecologists and to the public in general. Ecological study is an integrative science, one that requires great innovation, breadth, and curiosity. The rewards of such work are great, as we hope that you will come to believe.

SUMMARY

1. Ideas about how things work are called hypotheses. The specific hypothesis of interest at a particular time is called the null hypothesis, and all other possible explanations are called alternative hypotheses. Hypotheses may be posed in language or as mathematical expressions, called theoretical ecological models, that show the relationship among measured features of nature, or variables. Ecological theory provides a way to explore the complex relationships among variables before experiments are conducted.

2. Any inference made about nature is accompanied by a degree of uncertainty owing to variation in the natural world. Ecologists use statistical methods in order to manage the amount of uncertainty in their investigations.

3. In order to obtain basic information on ecological variables, ecologists employ sampling studies, in which they make systematic observations of nature. In order to understand the dynamics of processes that take place over long periods of time, ecologists employ long-term monitoring programs, which are sampling programs that extend over many years.

4. In order to uncover causal relationships in nature, ecologists must conduct experiments. An experiment allows for some component of the natural world to operate under conditions that are controlled by the experimenter. Experiments are direct tests of one or more null hypotheses. Since the truth of a null hypothesis is unknown, experimenters may make two major types of errors: rejecting a true null hypothesis, called a type I error, or accepting a false null hypothesis, called a type II error. Experimental procedures control for the probability of making these types of errors.

5. Some experiments to determine the causes of amphibian population declines have focused on the effects of certain wavelengths of ultraviolet (UV) light that are known to cause damage to biological tissues. Human activity has depleted the earth's ozone layer, which serves as a natural filter of UV light, leading to the null hypothesis that increased UV light damages frog eggs and reduces the number of eggs that survive to hatching. Experimenters in Oregon exposed frog eggs to three levels of damaging UV-B light. The different light levels were the treatments of the experiment. They found that, in two of three frog species studied, increased levels of UV-B reduced hatching success, the response variable of the experiment.

6. Phylogeny is the pattern of evolutionary relationships among species or other taxa. Ecologists use this information to determine the nature of the adaptations of living species.

7. Ecological study requires considerable resources and poses many logistical problems. Ecologists must study in remote areas that are often inhospitable. Funding must be obtained to support ecological research.

8. Ecologists publish the results of their work in ecological journals, which are sponsored by ecological organizations

such as the Ecological Society of America, or in books. Publications that are first reviewed by other scientists constitute the primary literature.

9. Agriculture and resource management are based on ecological principles. The application of ecology in these areas is intended to increase yields and to sustain the environment for future production.

10. The science of ecology is divided into a number of subfields, including physiological, ecosystems, population, community, evolutionary, landscape, conservation, and restoration ecology. Environmental science is the study of human impacts on the environment. Environmentalism, conservationism, and preservationism are social or political movements, not branches of ecology.

EXERCISES

1. Refer to Exercise 3 of Chapter 1, in which you developed questions about nature. Develop a null hypothesis for each of the questions you asked.

2. Referring to the natural history that you developed in Exercise 7 in Chapter 1, develop a simple mathematical model that shows the hypothetical relationship between two components of the organism's natural history. (Hint: By "component" we mean some identifiable activity of the organism, some physical condition of the environment, or some behavior of another interacting organism.)

3. Using what you have learned so far about the nature of uncertainty in scientific inquiry, how would you respond to the assertion that scientific ideas are always suspect because scientists can never know the truth?

4. Develop an experiment to test one of the null hypotheses that you set forth in Exercise 1 above.

5. Estimate how much it would cost to set up and perform the experiment that you designed in Exercise 4. Take into consideration the cost of materials and labor, the cost of purchasing or fabricating specialized equipment, and, of course, the cost of your time.

6. Using the World Wide Web, develop a list of scientific organizations that emphasize ecological research. (Hint: Many such organizations focus on one particular group of animals, plants, or microorganisms. For example, the membership of the American Arachnological Society includes ecologists who are interested in spiders.)

PART 2

ORGANISMS IN PHYSICAL ENVIRONMENTS

We can easily distinguish the two great realms of the natural world: the living and the nonliving. But these two realms do not exist in isolation from one another. Organisms depend on the physical environment. They also affect the physical world and thereby the conditions of their environment. The changes in global climate resulting from human technology and land-use patterns represents a dramatic example of how organisms affect the physical environment. Here in Part 2 we shall examine the nature of these interactions. We will also introduce some ideas about the interaction of organisms with other organisms.

In Chapter 3 we will show how the biological and physical worlds differ, focusing on how organisms maintain improbably high levels of energy in the face of physical forces that tend to dissipate that energy. We shall discuss the nature of the flux of energy and materials between living things and the physical environment and adaptations for the control of this flux. Water is critical for life, and in Chapter 4 we will describe the special features of water that promote life on earth and examine in some detail how organisms manage water and solute balance in their bodies. We will also describe the movement of water from the soil to the tops of trees.

Most biological transformations of energy are based on the chemistry of carbon and oxygen. The transformations of inorganic carbon to organic carbon in photosynthesis represents the key energetic link in the ecosystem. The rate of photosynthesis is affected by temperature as well as by the intensity and quality of light. In Chapter 5 we will focus on the ecological factors affecting photosynthesis and the adaptations of organisms for different regimes of temperature and light. The features of the physical environment change through time and vary from one place in the environment to another, often corresponding to daily or seasonal cycles. As we will see in Chapter 6, organisms must respond to these changes in order to survive.

Living things must not only respond to the conditions of the physical environment, they must also interact with other organisms. Although the responses of organisms to the physical environment promote the convergence of form and function, we will see that biological interactions promote diversification. In Chapter 7 we will examine the nature of the adaptations of organisms to their biological environment.

Part 2 ends with Chapter 8, which describes the patterns of climate and topography that occur on earth. These patterns are the result of the dynamics of the earth's rotation about the sun interacting with regional and local topographic and geologic features. The climate and topographic features of the earth affect the distributions of plants and animals. Based on their unique features of vegetation and climate, ecologists have identified a number of biomes around the world.

CHAPTER 3

Life and the Physical Environment

GUIDING QUESTIONS

- How do living things depend on the physical world for survival?
- What are some ways in which living things change the physical environment?
- What are some ways in which humans affect the physical environment?
- What are the properties of living things that distinguish them from nonliving things?
- How do organisms control the flux of materials and energy between their internal and external environments?
- How do form and function change allometrically with body size?
- How do adaptations reflect compromise in living things?

We often see the living and the nonliving as opposites: biological versus physical and chemical, organic versus inorganic, biotic versus abiotic, animate versus inanimate. While these two great realms of the natural world are almost always distinguishable and separable, they do not exist in isolation from each other. Life depends upon the physical world. To stay alive, organisms must continually exchange materials and energy with the physical environment. Living beings, in turn, affect the physical world: soils, the atmosphere, lakes and oceans, and many sedimentary rocks owe their properties in part to the activities of plants and animals. Indeed, the earth's oxygen-rich atmosphere owes its character to the oxygen produced by photosynthetic organisms over the past 3.5 billion years. The biological and physical transformations that interconnect the realms of the living and the nonliving define the global ecosystem. An understanding of these relationships is the first step in appreciating how nature works.

3.1 The biological and physical worlds are interdependent.

The ultimate source of energy for most life on earth is sunlight. Pigments in the green tissues of plants and algae absorb light and capture its energy; that radiant energy is converted to chemical energy through the manufacture of carbohydrates from the simple inorganic compounds carbon dioxide and water. This process is called **photosynthesis**—literally, "putting together with light." Energy locked in the chemical bonds of sugars and other organic compounds is used by plants, by animals that eat plants, by animals that eat other animals that eat plants, and so on. The energy contained in your hamburger lunch is just a few steps away from the energy transformation of photosynthesis.

The physical environment both provides the raw materials of life and sets the conditions in which life exists. Organisms ultimately receive their energy from sunlight. They also depend on the physical environment for basic elements such as oxygen and nitrogen, important molecules such as water and carbon dioxide, and other essential materials such as vitamins and minerals. Organisms must also tolerate the extremes of temperature, moisture, salinity, and other physical factors of their surroundings. The heat and dryness of deserts excludes most life forms, just as the bitter cold of polar regions turns back all but the most hardy. The form and function of plants and animals must obey the rules of the physical world. The fact that water has a higher viscosity than air requires that fish be streamlined according to restrictive hydrodynamic design principles if they are to be both efficient and swift. The fact that oxygen has limited solubility in air and water places upper bounds on the metabolic rates of animals and microorganisms.

While organisms depend totally upon the physical world, they also affect the physical world, sometimes in a profound manner. The composition of the earth's atmosphere, the condition and quality of soil, the formation of local topography, the local climatic conditions, and the extent of desertification are not conditions that are imposed on living things by an unyielding physical world. Rather, they are the result of an intricate interplay among living and nonliving components of the environment. That interplay is characterized by continual biological innovation of physical and chemical transformations through the process of evolution, allowing organisms to function better within the constraints of the physical environment. From the very beginning of life on earth, the characteristics of the physical environment have been modified by the activities of living things. We shall see that humans are having a profound effect on the physical conditions of the world and thus on life on the planet.

The Role of Organisms in the Formation of the Earth's Atmosphere

The earth began with no atmosphere and no oceans. Today, the earth's atmosphere contains a stable concentration of about 20% oxygen, a concentration that is neither too oxygen-rich (a concentration greater than 25% would allow even very moist things to burn) or too oxygen-poor (less than 15% would preclude fire) (Lovelock 1979). The story of the formation of the earth's atmosphere is unquestionably one of the most fascinating and important examples of how organisms affect their physical environment. Of course, organisms weren't around at the very beginning of the earth's history. The first components of the atmosphere, such as water, carbon, carbon dioxide, chlorine, nitrogen, and sulfur, arrived on earth in meteors or were released from the earth's crust by volcanic action (Schlesinger 1991). The cooling of the earth's crust caused water condensation, and eventually oceans were formed, about 3.8 billion years ago. Although it is difficult to know for sure, it is thought that by about 3.5 billion years ago, an atmosphere composed largely of reduced gases such as methane (CH_4) or carbon dioxide (CO_2), ammonia (NH_3) or molecular nitrogen (N_2), water vapor (H_2O), and hydrogen (H_2) had developed (Berkner and Marshall 1965, Chapman and Schopf 1983, Schlesinger 1991), and in the earth's oceans, very simple organic molecules had been assembled. This early environment held no oxygen. This was fortuitous, since free oxygen would have quickly broken down the first simple organic molecules that were the necessary precursors of life. Within this environment, the first very simple living organisms developed, and the dramatic changes that would lead to a fundamentally different type of atmosphere began.

Perhaps the most important innovation of living things in the history of life on earth was the development of photosynthesis around 3.5 billion years ago. The earliest photosynthetic reactions probably involved the oxidation of reduced gases such as hydrogen sulfide (H_2S). **Oxidation** is a process whereby one or more electrons are removed from an atom or a molecule. The addition of electrons is called **reduction.** Oxidation of H_2S on the early earth may have been carried out by sulfur bacteria in areas with abundant volcanic activity, such as in the following reaction:

$$2H_2S + CO_2 \longrightarrow CH_2O + 2S + H_2O.$$

Because H_2S was scarce, photochemical processes that employed the splitting of water (H_2O), a very abundant substance at the time, were favored, with the consequent evolution of the familiar photosynthetic reaction

$$H_2O + CO_2 \longrightarrow CH_2O + O_2.$$

The availability of oxygen gas created by photosynthetic reactions dramatically changed the nature of the earth's atmosphere, though not immediately. For over a billion years, most of the oxygen released by photosynthesis was immediately used in oxidation reactions. First, O_2 reacted with ferrous iron (Fe^{2+}) in the oceans, causing most of the oceanic iron to oxidize and form deposits of iron ore. But between 2.0 billion and 1.5 billion years ago, the amount of O_2 released by photosynthesis exceeded that used in oceanic and atmospheric oxidation reactions, and oxygen began to accumulate in the atmosphere (Figure 3-1). Thus the earth's atmosphere was changed from a predominantly reducing environment to an oxidizing one, thanks to the activities of its living organisms.

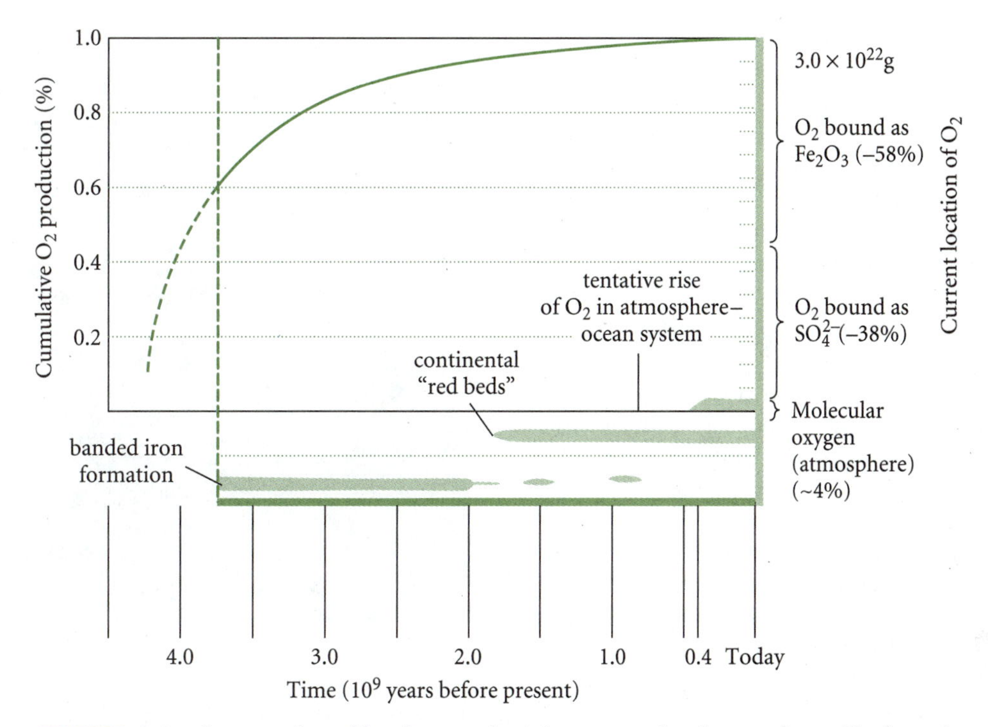

FIGURE 3-1 Oxygen released by photosynthesis has accumulated on earth steadily through geologic time, but the occurrence of molecular oxygen in the atmosphere is a relatively recent event. Oxygen released by photosynthesis prior to about 2.0 billion years before present was immediately bound with ferrous iron to form iron ore deposits in the earth's crust. After that, until about 4 million years before present, oxygen went into oxidation reactions with reduced metals in the oceans, forming deposits of SO_4^{2-} and Fe_2O_3 in continental areas called "red beds." Oxygen began to appear in the atmosphere about 400 million years ago. Today, about 58% of the earth's oxygen is bound as iron ore and about 38% is found in the red beds. The rest, about 4%, is found in the atmosphere. (*From Schlesinger 1991, modified from Schidlowski 1983.*)

The reducing atmosphere of the earliest earth promoted the formation of simple organic molecules and, eventually, the more complex molecules that constitute living things. Simple combinations of molecules, conformed in some cases as spherical structures that anticipate complex cells, appeared on earth billions of years ago. Eukaryotic organisms are thought to have first appeared around 1.3 billion years ago, at a time when the oxygen concentration of the earth's atmosphere was about 1% of what it is today (Chapman and Schopf 1983). The evolution of membrane-bound chloroplasts packaging photosynthetic pigments provided for more efficient photosynthesis, and thus the release of more oxygen into the atmosphere. Important biochemical reactions arose within this increasingly oxygen-rich environment. Among these were the nitrogen transformations that occur in some bacteria, whereby a series of oxidation reactions transforms ammonia (NH_3) into nitrate (NO_2^-) and nitrite (NO_3^-). As we shall see in Chapter 11, these reactions release energy, which is used to drive other biochemical processes.

Some Ways in Which Plants and Microorganisms Affect the Physical Environment

The activities of organisms produced an atmosphere on earth capable of sustaining life as we know it. The effect of plants and microorganisms on the physical conditions of barren, lifeless areas today provides another example of the transforming power of living things. Volcanic eruptions, erosion, the retreat of glaciers (a process called deglaciation), and the accumulation of waste heaps from human industrial activity may create or expose areas that are essentially devoid of life. Initially, such places may be lacking soil and nutrients, presenting a harsh physical environment that is apparently unsuitable for life. But life is not long denied access to such locations. Plants and microorganisms play an equally influential role in the development of soil from rock, sand dunes, or other barren areas (Olson 1958, Shure and Ragsdale 1977; see Chapter 28). On rock outcrops, plant roots invade tiny crevices and pulverize the rock as they grow and expand. The "rotting" of plant detritus by bacteria and fungi produces organic acids, which dissolve minerals out of the rock, thereby weakening its crystalline structure and speeding its weathering. Fragments of detritus eventually alter the physical structure of the resulting soil. Certain bacteria and cyanobacteria (once called blue-green algae) are responsible for the biological fixation of nitrogen from the atmosphere, thus providing a link between the abiotic and biotic worlds (Broda 1975, Schlesinger 1991).

Plants and microorganisms play a part in controlling moisture and temperature conditions both locally and globally. They help maintain soils and thus influence the way in which water cycles through the environment. And they influence the content of the atmosphere and thus the climate. It is often easiest to appreciate the way in which plants and microorganisms influence the physical environment when they are removed, as is often the case when humans alter the natural world for agriculture or to procure natural products. A vivid demonstration of what happens when these functions of living organisms are disrupted occurred in the so-called Dust Bowl of the American Midwest during the 1930s. The Dust Bowl region is normally dry and windy, but the native perennial grasses, which lived for several or many years, had roots extensive enough to hold the soil in place. When the prairies were brought under the plow in the late 1800s, annual crops, which have less extensive root systems, replaced the perennial grasses. A series of dry years reduced crop growth and turned the soil surface to fine dust. The result is illustrated in Figure 3-2.

(a)

(b)

FIGURE 3-2 The Dust Bowl of the midwestern United States. Wind erosion begins when soils are plowed but too little rain falls to support crop growth. (a) A winter wheat crop failure in Finney County, Kansas, in March 1954 has resulted in soil blowing. (b) A dust storm, composed of windblown soil particles, approaches Springfield, Colorado, in May 1937, during the height of the Dust Bowl tragedy. (*Courtesy of the U.S. Soil Conservation Service.*)

The importance of the activities of plants and microorganisms is also revealed when forests are cut indiscriminately, an activity that is increasing at an alarming rate worldwide. People cut or burn forests to obtain firewood and wood for export or to convert the land for cultivation and grazing livestock. When forests are removed, so are the important functions that they perform in the natural world. Forests normally provide shade and reduce water evaporation and surface runoff. Moreover, vegetation absorbs sunlight more efficiently than does bare ground. Because it also has a humidifying effect on the atmosphere, vegetation alters local heat budgets and fosters precipitation, thereby affecting local weather. Thus deforestation has substantially altered environmental conditions and climate in many locations around the world. In northern Africa, the once extensive forests of Algeria are now nearly depleted. The result has been a local decrease in rainfall, with a subsequent reduction in mountain snow cover, more frequent and extreme droughts, shortages of drinking water, and massive soil erosion (Zaimeche 1994). In vast regions of Africa just south of the Sahara Desert, the removal of most native vegetation due to overgrazing and the collection of wood for fuel greatly intensified the devastating drought of the 1980s (Figure 3-3).

Plants and microorganisms also influence the way in which water cycles through the environment. Rain does not accumulate where it falls on land. If it did, the state of New York would be under 60 meters of water within a human lifetime. Some of the water flows over the soil or through the underlying earth to rivers, lakes, and, eventually, the ocean. The remainder escapes by evaporation from the ground surface and through transpiration by vegetation. In places such as the eastern United States, where the leaves of deciduous trees have about four times the surface area of the ground underneath, vegetation provides the major pathway for water escaping from the soil to the atmosphere. When a forest is cut, much of the water that would have evaporated from its leaves flows instead into rivers. Without provision for extensive replanting, deforestation causes flooding, increased erosion, downstream silt deposition, and removal of mineral nutrients from the denuded soil. In the humid Tropics, where vegetation holds most of the nutrients in the system, deforestation can greatly decrease soil fertility, with tragic consequences.

The gaseous content of the earth's atmosphere is influenced by the activities of plants and microorganisms. The plants of natural wetlands such as peat bogs and marshes, agricultural crops, and cultivated wetland plants such as rice add carbon to the earth's

FIGURE 3-3 Overgrazing by sheep has obliterated most of the vegetation in this area of northern Kenya. (*Courtesy of the World Wildlife Fund/UCN.*)

atmosphere in the form of carbon dioxide. The microorganisms associated with these plants produce methane (CH_4). Although, in comparison to carbon dioxide, methane is a relatively minor constituent of the world's total carbon, its input into the atmosphere is increasing faster than that of carbon dioxide (Schlesinger 1991). This increase is cause for some concern, inasmuch as methane is 20 times more efficient than carbon dioxide as a greenhouse gas (see Chapter 11; Lashof and Ahuja 1990). The global increase in methane production is in part attributable to inputs from cultivated plants, particularly microbial activity associated with rice paddies (Schlesinger 1991, Ehrlich et al. 1994).

Some Ways in Which Animals Affect the Physical Environment

Animals, too, may affect the conditions of the physical environment, though their effects are not as dramatic as those of plants and microorganisms. Animal activities such as burrowing, trampling, and defecating play a part in the development and alteration of soil. The activities of earthworms may cause changes in the structure and nutrient composition of soils (Lee 1985) as well as in the rate at which decomposition occurs in the leaf litter of forest and agricultural soils (Wolters and Schaefer 1993). The earth-moving activities of burrowing rodents, which involve much trampling, and their deposition of urine and feces modify the vegetation near the burrow entrance. Rodent tunnels can affect the structure and species composition of the overlying plant community (Tilman 1983, Koide et al. 1987, Swihart 1991). The large mounds created by pocket gophers (Figure 3-4) are a striking example of the influence of burrowing activity on local topography (Cox 1984, Cox and Allen 1987). The activities of aquatic animals also affect their physical environment. It has been suggested, for example, that the abundance and population dynamics of small floating algae and crustaceans (collectively called plankton) contribute to the temperature dynamics of freshwater lakes. Of course, most of us are familiar with the way in which beaver dams alter the local flow of water.

How Humans Affect the Natural World

One animal species possesses more potential to change the physical characteristics of the natural world than any other: human beings (*Homo sapiens*). The Dust Bowl, discussed above, and the effect of the introduction of the Nile perch into Lake Victoria, Africa (see Chapter 1) are examples of how human activity may significantly alter the environment. It is important, we believe, to examine the effects of our own species in a bit more detail.

Human technology, which has increased agricultural productivity and raised life expectancy in some parts of the world through advances in medicine and nutrition, comes with a cost. The processes and byproducts of technology redistribute natural materials in ways that can cause dramatic changes in the physical conditions of the earth. Perhaps the best example

FIGURE 3-4 Pocket gopher mounds near San Diego, California. Mounds form around the gophers' burrow entrances as excavated dirt accumulates. The gophers locate new entrances on established mounds because the lower areas between mounds are flooded during spring periods of high rainfall. (*Courtesy of G. W. Cox; from Cox 1984.*)

is the redistribution of large amounts of carbon, an activity that appears to have already caused an increase in average global temperatures. Carbon occurs on earth in various forms, but the total amount present is more or less fixed. Carbon dioxide, CO_2, is an extremely important form of carbon because, through photosynthesis, the carbon atoms in carbon dioxide are assembled into organic compounds, from which energy is released in the process of respiration. Carbon dioxide, as well as other gases in the atmosphere, also play an important role in maintaining the earth's environment within a relatively narrow range of temperatures. This is a property of the natural world that is being changed by human activity.

The visible light and ultraviolet radiation that pass through the earth's atmosphere are absorbed by its land surface, vegetation, and waters. This energy may then be reradiated as infrared radiation, low-energy radiation having wavelengths between 700 and 1,000 nm. Carbon dioxide and other **greenhouse gases** in the atmosphere (along with water vapor and clouds) absorb that infrared radiation, so that heat is retained. Thus the earth's atmosphere works very much like a greenhouse, in which the glass roof admits sunlight but retains reradiated long-wave radiation coming from the inside of the building, warming its interior. We call this phenomenon the **greenhouse effect.** This natural greenhouse maintains the earth's temperatures within the limits necessary for physiological function, making life as we know it possible. Human activities, particularly the burning of fossil fuels, have vastly increased the concentration of carbon dioxide and other greenhouse gases in the atmosphere, thereby increasing the efficiency of the earth's greenhouse and warming the earth's atmosphere, a process that is now well documented (Robert and MacArthur 1994, Vitousek 1994). Current estimates suggest that even a moderate increase in the average global temperature could result in a significant redistribution of arable land, the inundation of some coastal biological communities, and a significant reduction in biodiversity owing to the redistribution of vegetation types.

3.2 Life has unique properties not shared by physical systems.

Living things are typically very distinct from inanimate objects. They may move, reproduce, metabolize, and maintain internal conditions that are different from the external environment. Nevertheless, living beings must function within constraints set by physical laws. The burning of gasoline in an automobile engine is a chemical process; transmission of power from the cylinder to the wheels is a mechanical process. The metabolism of carbohydrates in a living organism and the movement of its appendages follow the same rules. Like internal combustion engines, organisms transform energy to perform work. Both the engine and the organism obey the laws of thermodynamics. Neither can create or destroy energy, but they can transform it from one form to another—for example, chemical to mechanical (first law of thermodynamics). And, in both the engine and the organism, some energy is lost during transformation (second law of thermodynamics). The biological world is therefore not an alternative to the physical world, but an extension of it.

While biological systems operate on the same principles as physical systems, there is an important difference. In the physical world, energy transformations tend to minimize differences in energy level throughout the system, always following paths of least resistance. But in biological systems, the transformation of energy keeps an organism out of equilibrium with the physical forces of gravity, heat flow, diffusion, and chemical reaction. A living organism remains distinct from the physical world, whether it is pursuing prey, producing seeds, or maintaining basic body functions.

In a sense, an organism's use of energy is the secret of life. A boulder rolling down a steep slope releases energy during its descent, but it performs no useful work. The source of the energy—in this case, gravity—is external, and as soon as the boulder comes to rest in the valley below, it is once more in equilibrium with the forces in its physical environment. A bird in flight, on the other hand, constantly expends energy to maintain itself aloft against the pull of gravity. The bird's source of energy—the food it has assimilated—is internal, and the bird uses that energy to perform useful work, such as the pursuit of prey or migration.

The ability to act against external physical forces is the one common property of all living things, the source of animation that distinguishes the living from the nonliving. Bird flight supremely expresses this property, but plants just as surely perform work to counter physical forces when they absorb soil minerals into their roots and synthesize the highly complex carbohydrates and proteins that make up their structures.

3.3 Living organisms can increase their energy levels by thermodynamically improbable transformations.

The energy level of a chemical or physical system decreases with time as the system loses energy to the surroundings; in other words, the system spontaneously changes from a higher to a lower energy state. The burning of a piece of paper—a simple chemical system—releases energy in the form of light and heat, and the products of this oxidation, carbon dioxide and water, contain less energy than the reactants, oxygen and carbohydrates. The energy in a swinging pendulum—a simple physical system—is periodically transformed between the **kinetic energy** (energy of motion) of the weight moving at the bottom of its swing and the **potential energy** (energy of position) stored at its highest point. In a frictionless environment, a pendulum would continue to swing forever without loss of energy. But in an atmosphere, it sets molecules of oxygen and nitrogen in motion and thereby transfers some of its energy to its surroundings. As it loses energy, the pendulum swings through smaller and smaller arcs. In this way, energy initially residing in the pendulum becomes more evenly distributed throughout the larger system.

If we could perceive energy density as light, organisms would appear to us as beacons against the dim background of the physical world. Animals and plants represent immense concentrations of energy—energy derived from the brightest light in their surroundings, the sun. Physical and chemical processes work to balance out the high amount of energy in the organism and the low amount in the organism's environment. In other words, organisms tend to lose energy to their surroundings; organisms perform work to slow this process and maintain their own integrity. By way of analogy, imagine yourself as a high mound of sand piled steeply on a flat landscape. Little avalanches of grains tumble down from your sides, the wind blows other grains away, and rain erodes your stature and carries your substance off in milky rivulets. To maintain your prominence, you continually scoop up nearby sand and pile it on top of your head. You may even reach across to another pile, where sand is easier to get, to maintain or perhaps add to your own substance. If you are clever, you build walls at your base to help retain your sand. In many ways living forms elaborate this theme, with the sand representing their energy and substance.

Living organisms continuously expend energy stored in the form of chemical bonds of organic molecules. These molecules are oxidized in chemical reactions in the cell—a process referred to as cellular respiration—to produce high-energy intermediates such as ATP, which are used to drive biological processes. The energy expended to maintain life must be balanced by energy inputs if the organism is to continue living. Thus the organism must procure organic molecules—either by the assimilation of food or, in the case of plants, through the process of photosynthesis—to fuel cellular respiration. Further, in order to grow and reproduce, organisms need energy above and beyond that required for the maintenance of basic life processes. These energy needs must be met in the face of physical forces that work to dissipate energy. When energy is transformed in chemical reactions such as metabolism, some of that energy is lost as heat. Because of the inefficiency of these transformations, all assimilated energy is eventually dissipated to the physical system. Organisms can only slow this process. In the end, physical laws prevail, an energy balance between the organism and its environment is established, and the organism dies.

3.4 Organisms can control the flux of energy and material between their internal and external environments.

Throughout its life, an organism works both to gain energy and to forestall the loss of energy resulting from the chemical and physical forces of nature. The movement of energy into and out of the organism is an example of an important process in ecology referred to as **flux.** Ecological systems are characterized by the flux, or movement, of energy and materials between the system and the environment surrounding the system. The flux of water in your body involves the loss of water through evaporation from the skin and by urination and the gain of water by drinking. The flux of water in the forest of a small mountain valley involves the loss of water by runoff and stream flow and the gain of water by precipitation. In both of these examples, the amount of water held in the living system—your body or the forest—is greater than that in the environment, yet there is still a loss and a gain of water.

This example illustrates an important characteristic of flux that may be demonstrated by thinking about a simple physical system such as a uniform (solid) sphere of metal in space. Completely lifeless, the sphere passively intercepts light emanating from the sun, or from some more distant star, and reradiates energy into the black depths of space (Figure 3-5). When the sphere absorbs light, its temperature rises as molecules are caused to move more rapidly; the energy in the light is transformed to heat energy. Energy is assimilated, transformed, and lost from the system. The hotter an object, the more rapidly it loses energy to its surroundings, and so the sphere radiates more and more energy to empty space. Eventually, the sphere heats up so much that it loses energy as rapidly as it absorbs it, and at this point the temperature of the sphere comes into equilibrium with its environment. This equilibrium is a **thermodynamic steady state,** in which thermal characteristics of the system,

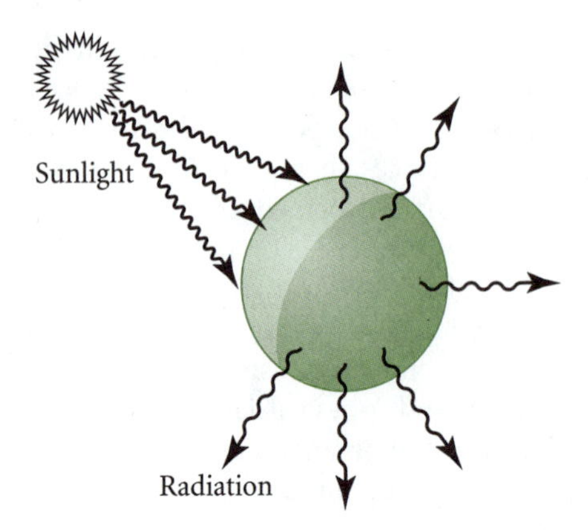

FIGURE 3-5 Thermal model of a solid sphere, which receives energy as it absorbs light and loses energy by radiation to space. When the sphere heats up to the point that it loses energy as rapidly as it gains it, it is at a thermodynamic steady state.

the most important of which is the equilibrium temperature, remain constant, but the system sustains continual loss and gain, or flux, of energy. The sphere remains hotter than its environment—that is, it contains more energy—even though the loss and gain of energy continue. Such is the case with the flux of energy and materials in living systems. So long as inputs are maintained, the amount of material or energy contained in the system may remain higher than in the surroundings, even though there may be continual loss and gain. Thus, for example, your body contains much more water than the air that surrounds you, yet there is a flux of water between you and your environment.

The flux of heat energy through an inanimate sphere in space is a result of the sphere's position relative to the source of heat energy, a situation upon which the sphere has no influence. Unlike the sphere, organisms can exercise some control over the flux of energy through their bodies. For example, when you restrict the loss of heat from your body in winter by wearing warm clothes, a coat and a hat, you have changed the flux of energy between your body and the environment. To reduce heat gain, an animal might simply move from the sunlight into the shade. All of us have engaged in this behavior on a hot summer day. Indeed, the responses of plants and animals to the many factors that will be considered in other chapters of this book can be understood, at least in part, in terms of how energy and materials move across their surfaces.

The flux of energy or materials across the surface of an organism can be described in general terms as

$$\text{flux} = \text{gradient} \times \text{conductance} \times \text{surface area}.$$

The **gradient** is the difference between the level or concentration of energy or material inside the organism and that in its surroundings. **Conductance** is the ease with which the energy or material crosses the surface barrier. Organisms can control flux by altering any of the three components of the above equation.

A mammal's thick winter fur reduces the conductance of heat (thermal conductance) from its body to its surroundings. The hard, waxy cuticles covering the exoskeletons of insects decrease the conductance of water across the body's surface. Because pale colors reflect more light than dark colors do, they enable desert organisms to reduce radiational heat gain. Biological surfaces may also actively transport materials either from the inside out or from the outside in.

Organisms can manipulate their surface area in many ways. They may effectively reduce it by sealing off some portion of the surface, thus restricting flux to smaller regions, which may be highly specialized. For example, the exterior surface of a leaf is protected from water loss by a waxy cuticle not unlike the surface of an insect. Gas exchange between the interior of the leaf and the environment takes place through stomates, tiny holes whose size can be controlled precisely in accordance with concentrations of water within the leaf. Surface area can also be increased by elaborate folding, as in the external gills of salamander larvae or in the interior of the lungs of mammals.

The gradient between an organism and its surroundings depends, of course, on the internal and external environments. The internal environment is dictated in part by the ranges of body temperature, ion concentrations, and other conditions suitable for life processes. Moreover, biochemical transformations often exaggerate gradients and thereby maintain high fluxes. For example, as our tissues consume oxygen, our internal oxygen concentrations are reduced, enhancing the outside-inside gradient and encouraging oxygen flow across the lung surface. Conversely, metabolism produces carbon dioxide, and as its concentration builds up in the tissues and the blood, its outward flow increases.

The gradient between internal and external environments may also change as organisms move through a heterogeneous environment. Shady and sunny spots, for example, present dramatically different radiative environments to an organism that must regulate its temperature. When humidity, ion concentrations, pH, soil nutrients, dissolved oxygen, prey abundance, presence of disease organisms, and other factors vary over distances that are small compared with the individual's daily range of movement, choice of environment has a direct bearing on the magnitude of flux.

Finally, we must distinguish between passive (physical) and active (biological) flux. **Passive flux** is the natural thermodynamic tendency of materials or energy to move from areas of high concentration to areas of low concentration. Passive fluxes across a surface occur in direct proportion to the gradient across the surface. Given time, and in the absence of other processes, passive flux continually reduces a gradient until the gradient, and the flux, drops to zero. Thus, in the absence of energy input, a hot sphere gradually cools to the temperature of its surroundings.

Active flux allows biological systems to accumulate substances against a physical gradient (as, for example, when we were sand piles accumulating sand into unstable piles of self) and thus requires the expenditure of energy. Organisms interact with their environment at their surfaces, and as we've shown, for a specific gradient and a particular conductance, an increase in surface area will increase the flux of materials or energy. The surfaces of living things are extremely complex biological structures with capacities for excluding or accepting materials at different rates, depending on the internal environment, the biochemical structures of the surface boundary, and the physical conditions. It is these processes that form the basis for active flux.

3.5 Form and function change allometrically with body size.

Large organisms relate differently to their environment than do small ones because many physical and physiological processes vary out of proportion to size (Calder 1984, Schmidt-Nielsen 1990). This phenomenon is referred to as **allometry.** In ecology, relationships between rates of processes and dimensions of objects are often described by the **allometric equation,**

$$Y = aX^b,$$

where Y represents a physiological process, such as metabolic rate or heart rate, an anatomic feature such as brain mass or seed size, or an ecological or behavioral characteristic such as parental investment in offspring. The variable X represents some measure of size, such as body mass, body length, or basal area. The constant a, the **constant of proportionality,** reflects the magnitude of the relationship, whereas b, the **allometric constant,** measures the rate of change of X with Y. When $b = 1$, Y is directly proportional to X ($Y = aX$). When $b > 1$, Y increases proportionately more rapidly than X, and the ratio Y/X increases with larger X. When $b < 1$, Y increases proportionately less rapidly than X, and consequently the ratio Y/X decreases with larger X. The allometric equation is a power function—that is, the independent variable X is raised to the power b—and is thus nonlinear for values of b other than 1. For convenience, the relationship between Y and X is often transformed to the logarithmic form

$$\log(Y) = \log(a) + b\log(X),$$

which is an equation describing a straight line with Y-intercept $\log(a)$ and slope b.

It is important to appreciate that employing the logarithmic form does not change the nonlinear relationship of X and Y. It just changes the scale on which X and Y are considered. That is, the relationship between some measure of body size X and physiological process Y is still nonlinear, even though the relationship between $\log(X)$ and $\log(Y)$ is linear. This type of transformation is common in mathematical models. As an analogy, think of a map of the world that is represented on a piece of paper. The world is a sphere, and thus in order to represent the various countries in two dimensions, their shape and relative positions must be altered. This representation does not change the spherical nature of the globe or the actual relationship of one place on the globe to another. It is just a convenience that is adopted in order to help visualize the various locations on the globe.

Allometry arises from simple physical and geometric considerations. Two three-dimensional objects that are geometrically similar (same shape), such as two spheres or two cubes, are said to be **isometric.** Isometric objects have two important properties: their surface area varies with the square of the linear dimension, and their volume varies with the cube of the linear dimension. Thus, the volume, V, of a sphere increases in proportion to the cube of its diameter, d

$$V = ad^3,$$

where a is a constant, but the surface area, S, increases in proportion to only the square of the diameter

$$S = ad^2.$$

We may represent the surface area as

$$S = a(d^3)^{2/3}.$$

Substituting $V = ad^3$ into the above equation yields

$$S = aV^{0.67}.$$

Hence the relationship between surface area and volume has an allometric constant of 2/3.

If organisms are conceptualized as spheres, larger organisms have relatively smaller surfaces compared with the mass of their bodies. A sphere of radius r has a surface area of $4\pi r^2$ and a volume of $(4\pi r^3)/3$; the

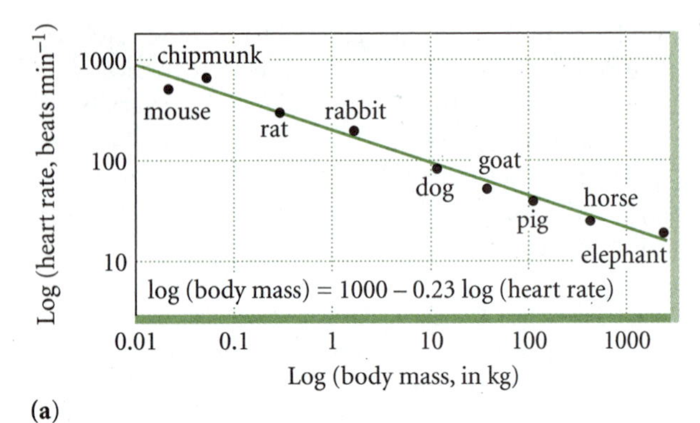

(a)

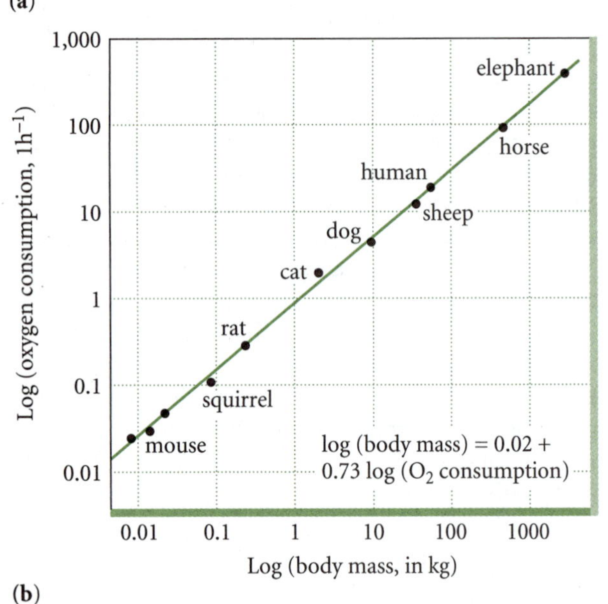

(b)

FIGURE 3-6 (a) Allometric relationship between heart rate and body mass in a variety of mammals, ranging in size from mouse to elephant. (b) Allometric relationship between resting metabolic rate (RMR), measured as the amount of oxygen consumed per hour, and body mass in a variety of mammals. (*Data from Altman and Dittmer, 1964.*)

ratio of surface to volume is $(4\pi r^2)/(4\pi r^3)/3 = 3/r$. Thus, a sphere with a diameter of 1 millimeter, comparable in size to a small water flea, has 1,000 times as much surface area per unit of volume as a bear-sized sphere 1 meter in diameter. The smaller animal will benefit from its high surface-to-volume ratio when it comes to surface-related phenomena such as oxygen uptake. But because of its small size, heat and water loss may pose severe problems.

Figure 3-6a shows the relationship between heart rate and body mass for mammals ranging over many orders of magnitude of size. The equation for the line that best fits the points has an allometric constant $b = -0.23$, reflecting the slower heart rates of larger species. In contrast, the resting metabolic rate (RMR) of mammals increases with size, but less rapidly than mass itself; the allometric constant of the relationship between RMR and body mass is 0.73 (Figure 3-6b). Again, we emphasize that the underlying relationship between RMR and size is nonlinear. We choose to use a linear transformation of that relationship for convenience.

An enormous array of physiological, anatomic, and behavioral characteristics have been shown to vary allometrically with size. In plants, seed production and plant architecture are often scaled to plant size. Among animals, traits as diverse as the size of the individual egg mass, the structure of the feeding and reproductive apparatus, parental investment, and susceptibility to toxicants display allometric relationships to body size. Table 3-1 gives examples of allometric relationships that are well documented in fishes.

Because allometry represents an interplay among specific traits that are subject to natural selection, we can think of allometric relationships as having adaptive significance. Such **adaptive allometry** has been hypothesized with respect to the seeds of plants

TABLE 3-1 Processes and characteristics known to vary allometrically with body size in fishes

Process or trait	Size dimension	Range or typical value of b
Gill area	Body weight	0.8
Food consumed over 24-hour period	Body weight	1.1–0.40
Standard metabolism	Body weight	<1.0
Body weight	Body length	Variable
Growth rate	Body weight	−0.4
Number of eggs per batch	Body length	1.0–5.0
Instantaneous daily mortality rate	Body weight	<0.0

(*From Wootton 1990*)

that rely on the ingestion of fruit by birds for dispersal of their seeds. It might be assumed that, in such plants, elongate fruits would be easier for birds to swallow, and that fruits thus should be progressively more elongate as fruit diameter increases. Such an allometry of fruit length and diameter has been shown in the Neotropical tree *Ocotea tenera* (Lauraceae), which is a major food source for some birds (Mazer and Wheelwright 1993).

3.6 Compromise dominates the adaptations of life forms.

Compromise is a consistent theme in the relationship of organisms to their environments. Terrestrial organisms cannot reduce water loss without also reducing their access to oxygen or, in the case of plants, carbon dioxide in the atmosphere. The same thick coat of fur that promotes the conservation of body heat in cold surroundings prevents the dissipation of excess heat in a warm environment. Modifications of the legs and feet of horses that enable them to run swiftly also produce a built-in stiffness that makes the limbs useless for scratching and swatting flies. Of course, horses have found ways around this problem. They have long tails to swish flies off their hindquarters, and as for scratching, horses love nothing more than rolling in the dust.

Still, every adaptation has it costs. No organism has unlimited time, resources, or body tissue. What it allocates to one function, it must take from another—nothing is free. In the absence of any benefit, even small costs become apparent. The eye, so important to humans, is useless to cave-dwelling fish that live in total darkness. The cost of producing eyes and their associated muscles and nerves is apparently so great that in the course of evolution, the eyes of many species of cave organisms have been reduced to tiny, rudimentary structures (Figure 3-7).

Although the examples in this chapter have concentrated on physical factors in the struggle for existence, organisms must also contend with biological aspects of their environment: predators, prey, parasites, and even collaborators. These factors, too, impose certain requirements of structure and function for successful living, and they also create conflicts of allocation. Time taken to watch for predators is time taken from feeding. Carbohydrates that a plant devotes to spines as defense against herbivores are carbohydrates that cannot be packaged in seeds. Evolutionarily, the bargain that each kind of organism strikes among these conflicting needs is molded by the pressures it faces in its environment.

FIGURE 3-7 The northern cavefish (*Amblyopsis spelaea*), an inhabitant of subterranean streams in Indiana and Kentucky, has rudimentary, nonfunctional eyes. (*Courtesy of T. C. Barr; from Barr 1968.*)

SUMMARY

1. Organisms receive water and other essential materials from the physical world, and in the process, they make changes in the physical environment. The oxygen-rich atmosphere of the earth resulted from the earliest photosynthetic activities of organisms. Organisms alter soil composition and chemistry, modify the movement of water in the environment, and influence climate.

2. Unlike inanimate objects, living things move, reproduce, metabolize, and maintain internal conditions often very different from those in the surrounding environment.

3. Living things are unique in their ability to accumulate large amounts of energy for their own use. Organisms must work to maintain these high energy concentrations against physical forces that tend to dissipate energy.

4. The flux of materials and energy across the surface of an organism is related to the surface area of the organism, the gradient of the material from the inside to the outside of the organism, and the ease with which the material is conducted through the skin or covering of the organism in the following way: flux = gradient × conductance × surface area. The surfaces of living things are dynamic structures, having an important function in managing the flux of materials and energy.

5. Many factors bearing on the ecology of organisms scale disproportionately with respect to overall body size. Such relationships are described by the slope (allometric constant) of the line relating the logarithm of a measurement of some structure or function to the logarithm of body size. Surface area scales to the 0.67 power of body size when shape is held constant; hence larger organisms have proportionately smaller surface-to-volume ratios. This and other allometric relationships have important consequences for organisms of different sizes.

6. Characteristics of organisms that are well suited for dealing with one aspect of the physical environment may be ill suited for dealing with another component of the environment. Thus, an organism's features reflect compromise between competing environmental constraints.

EXERCISES

1. If all life ceased to exist on the earth at this instant, what would the physical conditions of the earth be? How different would they be 100 years from now? Consider such changes as temperature and atmospheric composition. How different would these conditions be from those on the moon? On Mars?

2. List four or five activities that you engaged in today that have an effect on the physical environment in which you live, and explain what the effect is. (Hint: Sometimes an effect is several steps removed from the activity that causes it.)

3. Design an experiment or describe a series of observations that would provide an estimate of the thermodynamic properties of a house mouse.

4. Design a life form that would operate efficiently in an environment with one-third the gravity of the earth.

CHAPTER 4

Water and Solute Balance

GUIDING QUESTIONS

- How do covalent and hydrogen bonds give water its unique properties?
- What are the different ecological challenges of organisms that are hyperosmotic and hypo-osmotic to their environments?
- What osmoregulatory challenges do freshwater, marine, and terrestrial organisms face?
- What are some of the ways that fluid-feeding organisms maintain water and solute balance?
- How is the challenge of nitrogen excretion met in terrestrial and aquatic organisms?
- What are the factors that determine how much water is held in the soil?
- How does water move from the soil to the uppermost parts of plants?
- What are the sources and functions of six important inorganic compounds?

Water is perhaps the most prominent and essential feature of the natural world. Much of our understanding of how the natural world works at all levels, from individual organisms to ecosystems, depends on an appreciation of the properties of water and of the processes that occur in or are facilitated by water. All organisms, whether they dwell in oceans, lakes, or rivers, or on land, are composed mostly of water. Many compounds dissolve in water, and others become suspended in water and move with it. Ultimately, the nutrients required to fuel the metabolism of an organism arrive via its internal aqueous environment. Water is also often required for the removal of toxic materials produced during metabolism. Perhaps the most important challenge of living organisms is the maintenance of the proper volume and concentration of this aqueous environment.

In this chapter, we examine the properties of water and the challenges that individual organisms face in managing the flux of water and materials between their bodies and the environment. We shall see in later chapters that water also plays an important role at levels of ecological organization above that of the individual organism. The abundance of water limits the distributions of plants and animals. The movement of materials in water represent an essential dynamic component of many ecosystems, both terrestrial and aquatic. On a global scale, climate patterns are, to a large extent, determined by the movement of water.

4.1 The properties of water make it favorable to life.

The simplicity of the water molecule (H_2O), composed of two atoms of the simplest element, hydrogen, and one atom of oxygen, corresponds to the elegance and utility of its features—features that are essential to life. Here we discuss the chemical, physical, and thermal characteristics of water. You will see later in the chapter how each of these properties is important to life.

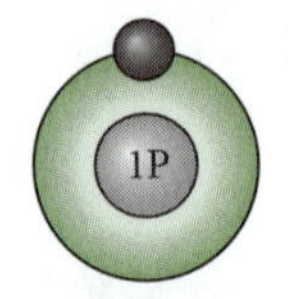

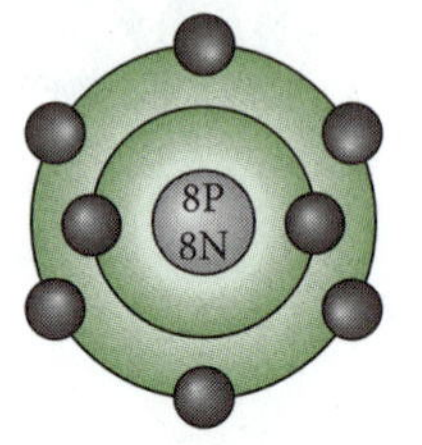

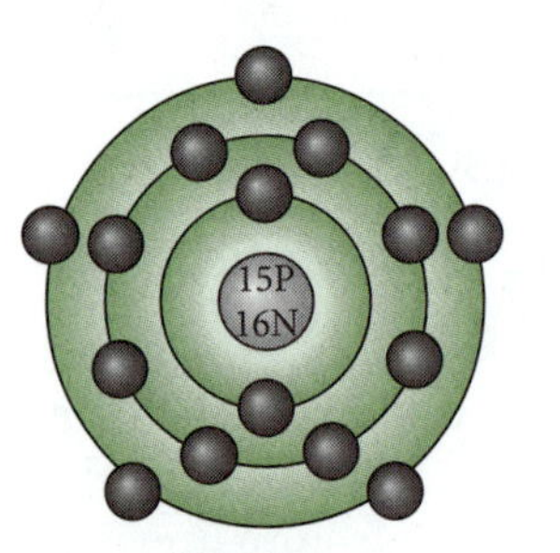

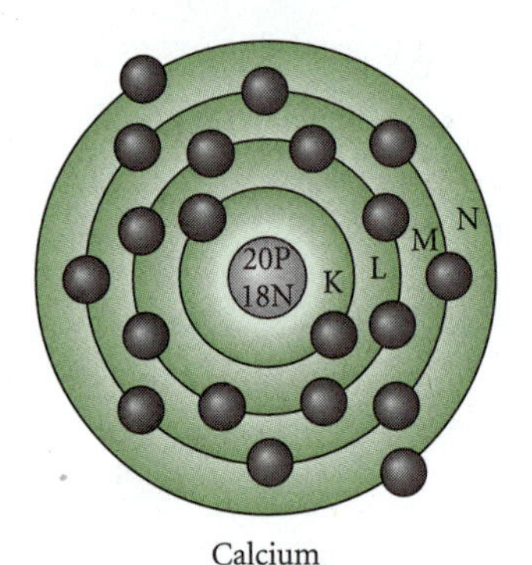

FIGURE 4-1 Atoms of hydrogen (H), oxygen (O), phosphorus (P), and calcium (Ca), showing the numbers and positions of electrons in the K, L, M, and N energy levels. For simplicity, electrons are shown to occur in fixed circular orbits around the atomic nucleus.

Chemical Properties of Water

The unique properties of water can be explained by the characteristics of hydrogen and oxygen atoms and by the nature of the bonds between the two hydrogen atoms and the one oxygen atom that form each molecule of water (H_2O). Atoms are composed of an atomic nucleus, which contains positively charged protons and neutrons having no charge, as well as negatively charged electrons, which are in constant motion around the nucleus. Electrons occur at different distances from the nucleus, depending on their energy state: the higher the energy level, the greater the distance from the nucleus. Since electrons are in constant motion, we think of them as occurring at some most probable distance from the nucleus corresponding to their energy level. These energy levels (sometimes called shells) occur in discrete steps, which are given the designations K, L, M, N, and so on, where K is the energy level nearest the nucleus. Each energy level can hold only a specific number of electrons. Two electrons may reside in the K level, 6 in the L level, 18 in the M level, and 32 in the N level.

Figure 4-1 shows atoms of hydrogen, oxygen, phosphorus, and calcium. In the figure, the electrons at the different energy levels are drawn in circular orbits around the nucleus. A more realistic representation of a hydrogen atom is shown in Figure 4-2a, in which the position of the electron is shown as a cloud, indicating that its exact location at a particular time is unknown. The hydrogen electron is shown to occur somewhere in a more or less spherical area at the K energy level. This spherical cloud is referred to as an **orbital,** the volume within which an electron is found about 90% of the time. The two spherical orbitals at the K energy level, one corresponding to each of the two electrons that may reside at that energy level, are designated 1*s* (notice that only one electron occurs at the K level in the hydrogen atom). But not all orbitals are spherical in shape. For example, the oxygen atom has eight electrons, two at the K energy level and six at the L level. The two electrons at the K level have 1*s* orbitals, which are spherical. Two of the oxygen electrons at the L level have spherical orbitals, called 2*s* orbitals, but the other four electrons have dumbbell-shaped orbitals, designated 2*p* (Figure 4-2b).

Atoms form **covalent** bonds by sharing electrons. For example, the hydrogen atom has a single electron at the K energy level (and that electron has a 1*s* orbital), but it can accommodate two electrons there. Two hydrogen atoms could fill their K energy levels by reciprocally sharing electrons, as shown by this simple formula, in which each dot represents an electron:

$$\mathrm{H\cdot + H\cdot \longrightarrow H{:}H.}$$

Such a covalent bond results in molecular hydrogen, H_2. Now consider covalent bonds between the two hydrogen atoms and one oxygen atom that form a water molecule. Oxygen has eight electrons, two at the K energy level and six at the L level. The L level of oxygen can accommodate two additional electrons, which can be supplied by two hydrogen atoms, each having a single electron. By sharing electrons, the K energy levels of the two hydrogen atoms and the L energy level of the oxygen atom can all be filled:

$$\mathrm{H + H + O \longrightarrow H{:}\,{:}O{:}\,{:}H.}$$

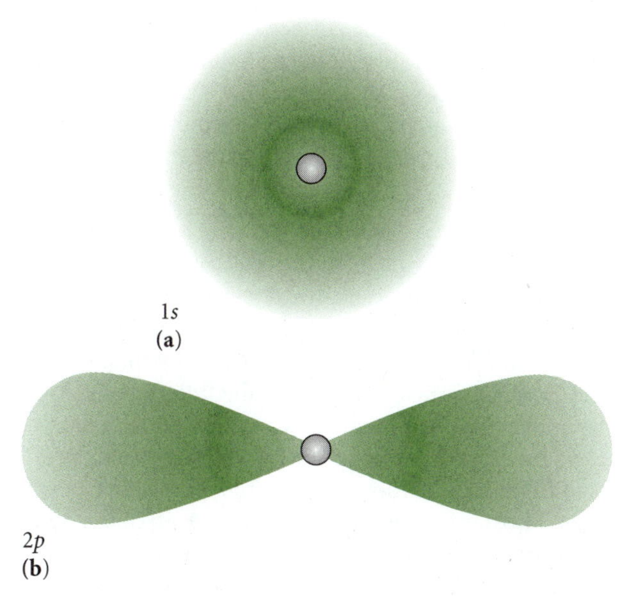

FIGURE 4-2 (a) The spherical 1*s* orbital of the hydrogen atom. The orbital is drawn as a cloud to emphasize that the position of the electron at a particular time is unknown. The orbital is the volume in which the electron is found 90% of the time. (b) The dumbbell-shaped 2*p* orbital of the L energy level of the oxygen atom.

No two elements exert the same attractive force on electrons. Thus, in a covalent bond between two atoms, electrons may spend more time around one of the members of the bonded pair—the one with the stronger attraction for electrons—than the other. This arrangement is called a **polar covalent** bond. In a water molecule, electrons are more strongly attracted to oxygen than to hydrogen, and so more negative charge is associated with the oxygen atom than with the hydrogen atom. If the three atoms of the water molecule were arranged in a linear fashion—H: :O: :H, as we have drawn them in the formula above—the electric charge would be distributed symmetrically across the molecule. But when oxygen covalently bonds with two hydrogen atoms to form water, the 2*s* and 2*p* orbitals in the L energy level of the oxygen atom are deformed in such a way that the molecule adopts an arrangement whereby the hydrogen atoms are forced to one side of the oxygen atom, forming a molecule with a tetrahedral shape (Figure 4-3). This can be drawn as:

H: :O $^-$
:
:
$^+$H

In this arrangement, the oxygen end of the molecule has a negative charge, and the end with the hydrogen atoms has a positive charge, giving the entire molecule a polarity.

Water as a Solvent

Because of the polarity of the water molecule, water has an immense capacity to dissolve inorganic compounds, making them accessible to living systems and providing a medium within which they can react to form new compounds. Many molecules are composed of electrically charged atoms or groups of atoms called **ions,** which are held together by the attraction of opposite charges. For example, common table salt, sodium chloride (NaCl), is made up of a positively charged sodium atom (Na^+) and a negatively charged chlorine atom (Cl^-). When salt is placed in water, however, the attraction of the water molecules for the charged sodium and chlorine atoms of the salt molecule is greater than the bonds that hold the molecule of salt together, that the salt molecule readily dissociates into its component ions—another way of saying that the salt dissolves.

Many different kinds of compounds dissolve in water, and thus lakes, streams, and oceans contain a variety of cations, which are positively charged, and anions, which are negatively charged. Among the most important cations are calcium (Ca^{2+}), potassium (K^+), and ammonia (NH_4^+). Among the most important anions are the nitrate ion (NO_3^-), the phosphate ion (PO_4^{3-}), the carbonate ion (CO_3^{2-}), and the bicarbonate ion (HCO_3^-). The availability of these ions

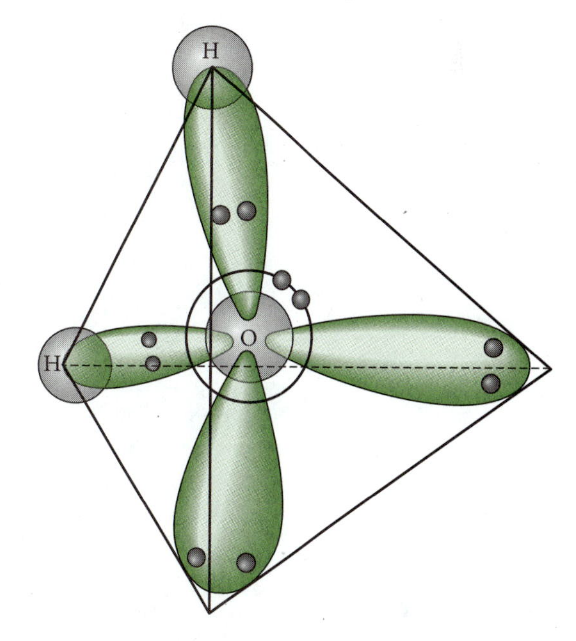

FIGURE 4-3 The tetrahedral shape of the water molecule results from the deformation of the orbitals during covalent bonding of two hydrogen atoms and one oxygen atom. This characteristic shape, along with the tendency of electrons to spend more time near the oxygen atom, gives the water molecule polarity, with a negative charge at the oxygen end and a positive charge at the hydrogen end.

is extremely important to organisms, as we shall see in Chapters 11 and 12.

Hydrogen Bonds Between Water Molecules

Since water molecules are polar, they are attracted to one another. This relatively weak attraction is referred to as a **hydrogen bond.** The arrangement of the atoms in the water molecule allows each molecule to form hydrogen bonds with four other molecules (Figure 4-4). If you imagine each of the outer four water molecules in Figure 4-4 forming hydrogen bonds with three others, and, in turn, each of those three forming bonds with three others, you can appreciate that the volume of water in a cup of water, or a swimming pool, or a lake is, in fact, a more or less continuous chemical entity. To be sure, hydrogen bonds are weak, and thus they continually break and reform. Therefore, the volume of water is not solid, but is rather the fluid you are so familiar with. The attraction of water molecules to one another is called **cohesion,** and at the surface of a body of water, this cohesion gives water its characteristic **surface tension.**

Water molecules are attracted to surfaces having a charge. This property, called **adhesion,** explains why some surfaces get wet. The cohesive and adhesive properties of water explain the phenomenon of **capillary action,** the movement of water in a small tube against the force of gravity. In a small tube, the water molecules adhere to the surface of the tube, and molecules not near the tube surface are pulled along by cohesive forces. As we shall see below, this phenomenon is responsible in part for the movement of water in plants and within the tiny spaces between soil particles.

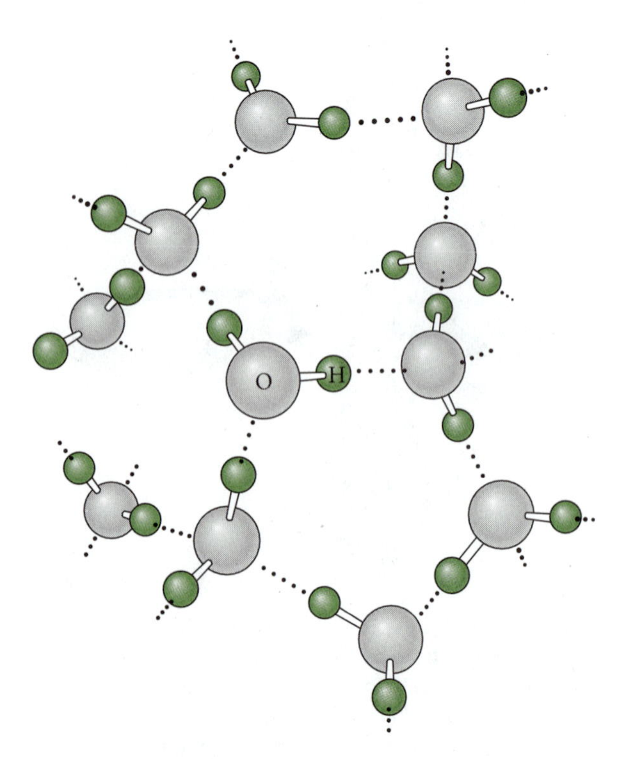

FIGURE 4-4 Hydrogen bonding between water molecules. Each molecule can form hydrogen bonds with four other molecules.

Ice

When water is cooled, the motion of the molecules decreases, and the weak hydrogen bonds are not as readily broken. At the point of ice formation (0° C), water molecules are rigidly connected by their hydrogen bonds. Because of their tetrahedral shape, the water molecules form a lattice structure, which contains a great deal of open space (see Figure 4-4). This property of water is rather serendipitous, because whereas most substances become more dense at colder temperatures, water becomes less dense as it cools below 4° C. Water also expands and becomes even less dense upon freezing. As a consequence, ice floats (Figure 4-5), which not only makes ice skating possible, but also prevents the bottoms of lakes and oceans from freezing, enabling aquatic plants and animals to find refuge there in winter.

Thermal Properties of Water

Many of the thermal properties of water are favorable to life (Table 4-1). For example, one must add or remove a large amount of heat energy to change the temperature of water. In addition, water conducts heat rapidly. Because of these two properties, referred to respectively as the **specific heat** and the **thermal conductivity** of water, the temperatures of organisms and aquatic environments tend to remain relatively constant and homogeneous. Water also resists change of state between solid (ice), liquid, and gaseous (water vapor) phases. Over 500 times as much energy must be added to evaporate a quantity of water (the **heat of vaporization**) as is needed to raise its temperature by 1° C. Freezing requires the removal of 80 times as much heat (the **heat of melting**) as is

FIGURE 4-5 The lattice structure of the ice in this antarctic iceberg contains many open spaces. The formation of these open spaces makes it less dense than liquid water (see Figure 4-4).

TABLE 4-1 Thermal properties of water

Property	Definition	Quantity
Specific heat	The quantity of heat energy required to raise the temperature of 1 g of water 1° C	1 calorie (cal) or 4.2 joules (J)
Heat of melting	The quantity of heat energy that must be added to ice to melt 1 g of water at 0° C	80 cal or 335 J
Heat of vaporization	The quantity of heat energy that must be added to evaporate 1 g of water	597 cal or 2,498 J at 0° C; 536 cal or 2,243 J at 100° C
Thermal conductivity	The flux of heat through a 1 cm^2 cross section at a gradient of 1° C cm^{-1}	(units are J cm^{-1} s^{-1} ° C^{-1}): 0.0055 at 0° C; 0.0060 at 20° C; 0.0063 at 40° C; 0.022 for ice at 0° C
Density	Mass per unit of volume	Water at 30° C = 0.99565 g cm^{-3} 20° C = 0.99821 10° C = 0.99970 4° C = 0.99997 (maximum density) 0° C = 0.99984 Ice at 0° C = 0.917

needed to lower the temperature of the same quantity of water by 1° C.

4.2 Substances dissolved in the aqueous environment pose osmotic challenges for organisms.

Organisms obtain nutrients from the soil, the water, or their food. Often, these substances are much more concentrated in the tissues of the organism than in its surroundings, and organisms must therefore assimilate them against the prevailing gradient. But organisms also must exclude from their bodies many abundant substances in the environment that are metabolically useless, or even toxic at high concentrations. When a surface permits the flux of desirable substances, it can keep out others only by selective permeability (if wanted and unwanted substances differ greatly in size or electric charge) or by actively pumping unwanted substances out across the surface.

Following purely physical laws, ions would diffuse across the surfaces of organisms from regions of high concentration to regions of low concentration, and would thereby tend to equalize their concentrations. Water also moves across permeable membranes toward regions of high ion concentration (that is, low water concentration), which also tends to equalize concentrations of dissolved substances on both sides of the membrane. This process is called **osmosis.** The tendency of a solution to attract water is known as its **osmotic pressure.**

Pressure may be represented in a number of different units. The pressure exerted by the atmosphere at sea level is about 14.7 pounds per square inch, which is designated as 1 atmosphere (atm). In the International System of Units, 1 atm=101,325 pascals (Pa) or about 0.1 megapascals (MPa) (the prefix *mega-* represents a factor of 100). Sometimes pressure is given in units called bars. One bar is equivalent to 0.987 atm, and thus roughly 0.1 MPa.

The osmotic pressure of a solution depends on its concentration. The higher the concentration, the greater the attraction of water to the solution, and the greater the osmotic pressure. A molar concentration of a substance in solution (1 mole of the substance per liter) exerts an osmotic pressure of about 2.1 MPa. The osmotic pressure of seawater is about 1.2 MPa, and that of fresh water is practically zero. The body fluids of vertebrate animals, which have an osmotic pressure 30–40% that of seawater (0.3–0.5 MPa), occupy an intermediate position. Plant cells attract water because the cytoplasm is more concentrated than the outside of the cell. It is this osmotic pressure that gives plant cells the rigidity called **turgor.** When water is plentiful in the environment, the plant cell attracts water, causing the cell membrane to push against the relatively rigid cell wall, thus giving the cell some rigidity. Turgor helps keep the plant erect. A loss of

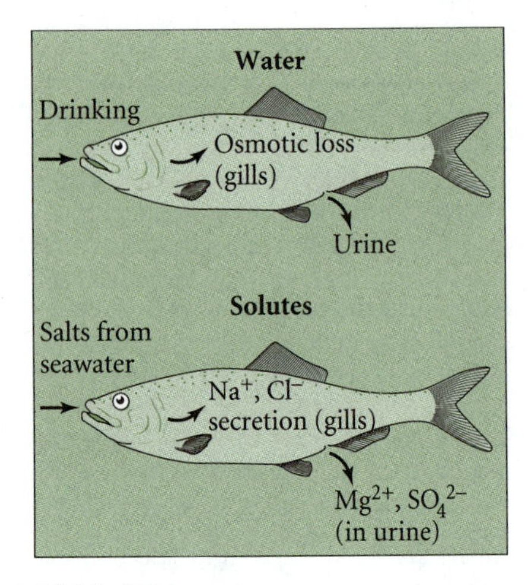

(a) Marine

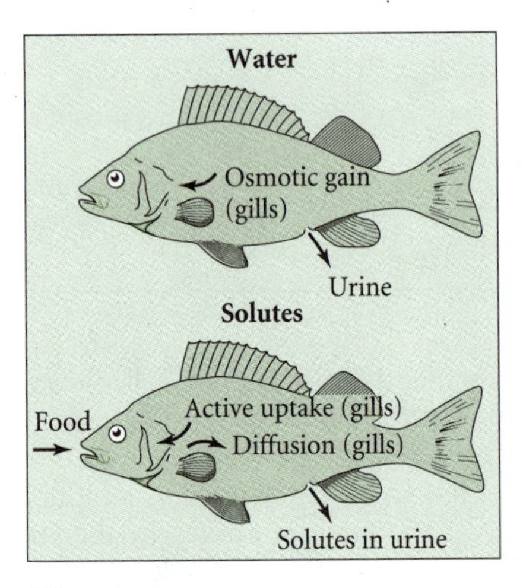

(b) Freshwater

FIGURE 4-6 Pathways of exchange of water and solutes by (a) marine fish, whose body fluids are hypo-osmotic (contain lower solute concentrations than the surrounding water), and (b) freshwater fish, whose body fluids are hyperosmotic. The gills and kidneys actively exclude or retain solutes to maintain salt balance. Marine fish must drink to acquire water. *(After Schmidt-Nielson, 1975.)*

cell turgor—known as **wilting**—may occur when water is in short supply.

The tissues of freshwater fish have higher concentrations of salts than the surrounding water. Such organisms are **hyperosmotic** and tend to gain water from, and lose solutes to, their surroundings. Marine fish, which have lower concentrations of salts than the surrounding seawater, are **hypo-osmotic.** They tend to gain solutes and lose water, and must drink to acquire water (Figure 4-6). As we shall see below, fish address these osmotic challenges by using active transport mechanisms to pump ions in one direction or the other across various body surfaces (skin, kidney tubules, and gills), expending considerable energy in the process.

4.3 Salt and water balance go hand in hand.

Certain environments pose special challenges with respect to the maintenance of solute and water balance. An organism must expend energy to maintain an ionic imbalance between itself and the surrounding environment against the physical forces of diffusion and osmosis. This process, referred to as **osmoregulation,** often is accomplished by organs specialized for salt retention or excretion. As we shall see, ion retention is particularly important to terrestrial and freshwater organisms.

Osmoregulation in Aquatic Habitats

Freshwater fish are hyperosmotic, continually gaining water by osmosis through the mouth and gills, which are the most permeable tissue exposed to the surrounding environment. The skin is relatively impermeable. To counter this influx, the fish continually eliminate water in the urine. But, of course, if they did not also selectively retain dissolved ions, they would soon become lifeless bags of water. A fish's kidneys retain salts by actively removing ions from the urine and feeding them back into the bloodstream. In addition, its gills can selectively absorb ions from the surrounding water and secrete them into the bloodstream (Schmidt-Nielsen 1997).

Marine fish are surrounded by water with a salt concentration higher than that of their bodies. As a result, they tend to lose water to the surrounding seawater and must drink seawater to replace this loss. The salt that comes in with the water and with their food, as well as that which diffuses in across body surfaces, must be excreted, at great metabolic cost, from the gills and kidneys.

Among marine vertebrates, salt excretion is not limited to fishes. Oceanic birds and reptiles ingest more salt in their food than the kidneys can excrete. Many species have additional regulatory organs known as **salt glands,** usually located near the eyes or nasal passages, which help to rid the body of excess salt while conserving water (Peaker and Linzell 1975).

Some sharks and rays have achieved a rather elegant solution to the problem of osmoregulation. Sharks retain urea [CO(NH)]—a common nitrogenous waste product of metabolism in vertebrates—in the bloodstream instead of excreting it in urine. The urea raises the ion concentration of the blood—and thus its osmotic pressure—to the level of seawater without having to increase the concentration of sodium chloride. The high levels of urea in the blood effectively cancel the tendency of water to leave by osmosis, and there is consequently no net movement of water across a shark's surfaces. This strategy makes it much easier to regulate the flux of ions such as sodium because the shark does not have to drink salt-laden water to replace water lost by osmosis. Urea does tend to destabilize proteins, but this problem is overcome by high blood levels of another compound, trimethylamine oxide [$NO(CH_3)_3$] (Yancey and Somero 1980, Yancey et al. 1982). The fact that the few freshwater sharks and rays do not accumulate urea in their blood (Thorson et al. 1967) emphasizes the osmoregulatory role of urea in marine species.

Aquatic environments with salt concentrations greater than that of seawater occur in some land-locked basins, particularly in arid zones where evaporation is great. Great Salt Lake (20% salt) in Utah and the Dead Sea (23% salt), lying between Israel and Jordan, are examples of such evaporation basins. The osmotic pressure of such environments would suck the water from most animals and plants, but a few aquatic creatures, such as brine shrimp (*Artemia*), can survive in salt water concentrated to the point of crystallization (300 g per liter, or 30%). Brine shrimp excrete salt at a prodigious rate to maintain their body fluids at a lesser concentration than their surroundings (Croghan 1958, Gilles 1975).

The small copepod *Tigriopus* (Figure 4-7) lives in pools high in the splash zone along rocky coasts. These pools receive fresh seawater infrequently and, as the water evaporates, their salt concentration rises to high levels. Unlike *Artemia, Tigriopus* solves its water loss problem by increasing the osmotic pressure of its body fluids. It accomplishes this by synthesizing certain amino acids abundantly (Burton and Feldman 1982). These small molecules increase the osmotic pressure of the body to match that of the habitat without the deleterious physiological effects of high levels of salt (Yancey et al. 1982).

Osmoregulation in Terrestrial Habitats

Water and salt balance are intimately related in terrestrial as well as in aquatic organisms. Terrestrial animals acquire mineral ions in the water they drink and the food they eat. Animals, especially meat eaters, obtain salts in their food in excess of their requirements. Where water is abundant, such animals can drink large quantities of water and produce large quantities of urine, which carries salt from the body. Where water is scarce, however, animals must produce concentrated urine to conserve water. And so, as one would expect, desert animals have champion kidneys (Schmidt-Nielsen 1964, Amanova 1994). For example, humans can concentrate salt ions in their urine to about four times the level in blood plasma, but the desert-dwelling kangaroo rat's kidneys produce urine with a salt concentration as high as 14 times that of its blood (Schmidt-Nielsen 1964). The Australian hopping mouse, another desert-adapted species, produces urine with a salt concentration 25 times that of its blood (MacMillen and Lee 1967).

The accumulation of urea in body fluids as a way of decreasing osmotic pressure is not restricted to marine sharks. Some amphibians adapted to living in deserts remain in underground burrows for up to a year or more, waiting for rain. As time passes, the soil around the burrow becomes depleted of water and its salt concentration increases, creating a situation in which the osmotic pressure of the soil is great relative to the body fluids of the amphibian. Like sharks and rays, spadefoot toads (*Scaphiopus*) and desert salamanders (*Ambystoma tigrinum*) retain large quantities of urea in their body fluids in order to prevent or slow down the loss of water across the skin during dormant periods in their burrows (Shoemaker et al. 1992).

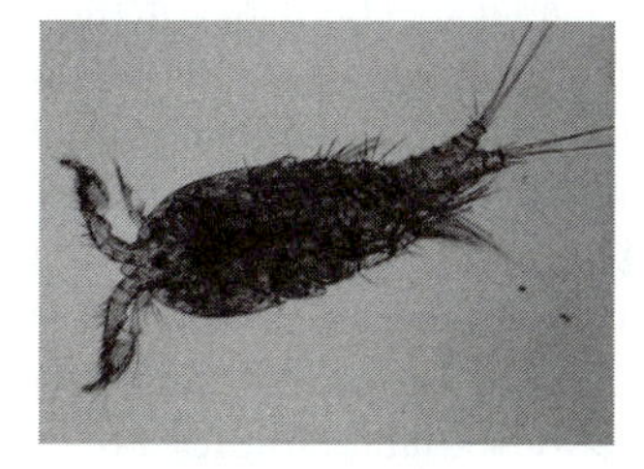

FIGURE 4-7 The copepod *Tigriopus californicus.* Adults are 1–2 mm long. This small organism lives in salty pools along rocky coasts. It synthesizes amino acids to increase the osmotic potential of its body fluids to prevent water loss. (*Courtesy of R. S. Burton.*)

Osmoregulation in Terrestrial and Aquatic Plants

Some plants can tolerate habitats with high salt concentrations. These **halophytes** are confronted with the problem of losing water to an environment having a greater osmotic pressure than their own bodies; thus, they face dehydration of their tissues. Many halophytes solve this problem by producing **compatible solutes,** compounds such as amino acids, some polyols, and some methylated ammonium compounds, which reside in the cytoplasm in high concentrations, thereby increasing its osmotic pressure, but that have no harmful effect on enzyme systems in the cell. When Sabry and coworkers (1995) exposed seedlings of spring wheat (*Triticum aestivum*) to either drought stress, by withholding water for 3 days, or to salinity stress, by watering with a concentrated sodium chloride solution, accumulations of glycine betaine, proline, and the amino acids asparagine, glutamine, and valine were observed in the plant tissues.

Mangrove plants grow in marine sediments that are inundated daily by high tides. (A recent review of mangrove ecology is edited by Jaccarini and Martens, 1992.) Not only does this habitat present a high salt load, but the high osmotic pressure surrounding the roots makes it more difficult for them to take up water. To counter this problem, many mangrove plants contain high levels of compatible solutes such as proline, sorbitol, and glycine-betaine in their roots and leaves, which increase their osmotic pressure. In addition, salt glands secrete salt to the exterior surface of the leaves (Figure 4-8). The roots of many species exclude salts, apparently by means of semipermeable membranes that do not allow the salts to enter. We know that active transport is not involved, as it is in the salt glands of the leaves, because neither cooling nor metabolic inhibitors diminish salt exclusion by the roots. Mangrove plants further reduce their salt loads by decreasing the transpiration of water from their leaves. Because many of these adaptations resemble those of plants from arid environments, where water is scarce, the mangrove habitat has been referred to as an **osmotic desert.**

The precise mechanism of control of osmotic pressure in plants is uncertain (Hellebust 1976, Morgan 1990). The observation that the concentration of abscisic acid, a plant hormone, often increases when plants are found in saline soils (Morgan 1990, Skriver and Mundy 1990) suggests an important role for hormones in regulating stress responses in plants. Abscisic acid is thought to turn on genes that produce the protein osmotin, which is known to provide some protection against salt stress in some plants (Skriver and Mundy 1990). It has been recently suggested that the response of plants to stress is a centralized, hormonally controlled process that involves integrated changes in nutrient uptake, water balance, and photosynthesis (Chapin 1991).

(a)

(b)

FIGURE 4-8 (a) The roots of mangrove vegetation are immersed in salt water. Some species excrete excess salt from their roots. (b) Specialized glands in the leaves of the button mangrove (*Conocarpus erecta*) excrete salt, which precipitates on their outer surfaces.

4.4 High fluid turnover is an osmoregulatory adaptation of fluid-feeding animals.

Some animals consume large quantities of water as a consequence of feeding on sap, nectar, or blood. Among the insects there is a wide variety of species with piercing or sucking mouthparts that penetrate plant and animal tissues in order to extract fluids. Most notable are the sucking mouthparts of aphids, bugs (Hemiptera), fleas, and mosquitoes. Some insects, such as the tetse fly, consume pooled blood from a wound (Chapman 1982, Hadley 1994). In nearly all such cases, fluid uptake is enhanced by some form of pumping action in the mouth or esophagus.

Fluid-feeding insects face two major problems: the elimination of excess water after feeding and the conservation of ions. Morphological adaptations, such as abdominal membrane folds and a reduction in the rigidity of the cuticle—the noncellular waxy covering of arthropods—that allow for a significant enlargement of the abdomen during fluid feeding are common in fluid-feeding insects (Hadley 1994). Water is eliminated, often rapidly, from the anus, or transferred into the hemolymph (body fluids) for processing through the Malpighian tubules (the excretory organs of many arthropods). In some ticks, fluid is passed from the gut into the hemolymph and then voided from the body through specialized salivary glands. An unusual mechanism for the elimination of excess water is employed by the tephritid fly *Rhagoletis pomonella* after obtaining large amounts of fluid from ingesting the sap of walnuts (Hendrichs et al. 1992, Hadley 1994). In a behavior called "bubbling," the engorged fly produces a droplet of water from the tip of its proboscis. Water is eliminated by evaporation from the bubble, and valuable solutes are retained inside the bubble and eventually taken back into the body.

The retention of ions is critical in nectar feeders because the solute content of the diet is well below that of the body. The nectar diet of the carpenter bee (*Xylocopa capitata*) provides an excess of water with an extremely low sodium and potassium concentration (Nicolson 1990). The problem is magnified by the bee's extremely high metabolic rate, which results in substantial metabolic water production. Carpenter bees produce a very dilute urine in order to eliminate the excess water, but nearly all sodium and potassium is recycled in the gut.

The liquid diet of hummingbirds presents osmoregulatory challenges similar to those of nectar-feeding insects. Studies of Anna's hummingbird (*Calypte anna*) show that it meets this challenge in much the same way as do freshwater fishes, with high urinary water loss and rapid water turnover time (Beuchat et al. 1979, Powers and Nagy 1988, Beuchat et al. 1990). The total body water content of a 4.5 g Anna's hummingbird is about 2.8 ml, about 63% of its body mass. Measurements of water intake and urinary water loss show that an Anna's hummingbird may drink as much as 15 ml per day and urinate as much as 14 ml per day. Water is also lost in feces and, to a lesser extent, through evaporation. The time it takes for an Anna's hummingbird to completely turn over a volume of water equal to its own body mass has been estimated at 15 hours, which is much faster than most other vertebrates of similar mass, including many freshwater fish species (Calder 1984, Beuchat et al. 1990).

Sodium deficiencies in some habitats may force animals to obtain salt directly from mineral sources—salt licks (Hebert and Cowan 1971, Botkin et al. 1973, Weeks and Kirkpatrick 1976)—or to develop mechanisms for concentrating salt from low-salt sources. One of the most remarkable examples of the latter strategy is the puddling behavior observed in many species of butterflies and moths, in which adults, usually males, gather at a water source to drink. Smedley and Eisner (1995) showed how the moth *Gluphisia septentrionis* concentrates sodium ions from the huge quantities of water that it drinks and then quickly expels in the form of an anal jet (Figure 4-9). The mouthparts of this species are specialized for rapid fluid intake, and specializations of the hindgut allow the moth to absorb sodium ions from the water. Smedley and Eisner observed that males drinking at natural puddles produced an average of 18.4 anal ejections each minute. The amount of fluid expelled during each ejection was found to be around 8.2 μl, which is equivalent to about 12% of the moth's body mass. Some males may remain at a puddle drinking and ejecting jets of water for over an hour. In one such case, the male passed 38.4 ml of fluid.

4.5 Excretion of nitrogenous wastes presents terrestrial animals with special problems.

Most carnivores, regardless of whether they eat crustaceans, fish, insects, or mammals, consume excess nitrogen, as well as excess salts, in their diets. This nitrogen, ingested as a component of their prey's proteins and nucleic acids, must be eliminated from the body when these compounds are metabolized (Figure 4-10) (Campbell 1970, Schmidt-Nielsen 1983). Animals lack the biochemical mechanisms possessed by some microorganisms for producing molecular nitrogen (N_2), and consequently, they cannot dispose of nitrogen as a gas that would escape from the blood through the lungs or gills. Many oxidized, inorganic forms of nitrogen—nitrates, for one—are highly poisonous and cannot be produced in quantity without toxic effects. Most aquatic organisms excrete nitrogen as the simple metabolic by-product ammonia (NH_3). Although ammonia is mildly poisonous to tissues, aquatic organisms can eliminate it rapidly in a copious, dilute urine before it reaches a dangerous concentration in the body.

Terrestrial animals cannot afford to use large quantities of water to excrete nitrogen. To circumvent this problem, they produce protein metabolites that are less toxic than ammonia and therefore can be con-

FIGURE 4-9 Jet of water expelled from the anus of the moth *Gluphisia septentrionis* engaged in puddling behavior. The moth obtains sodium from water that it imbibes in large quantities and periodically expels from the anus with great force. (*Courtesy of S. R. Smedley and T. Eisner; from Smedley and Eisner 1995.*)

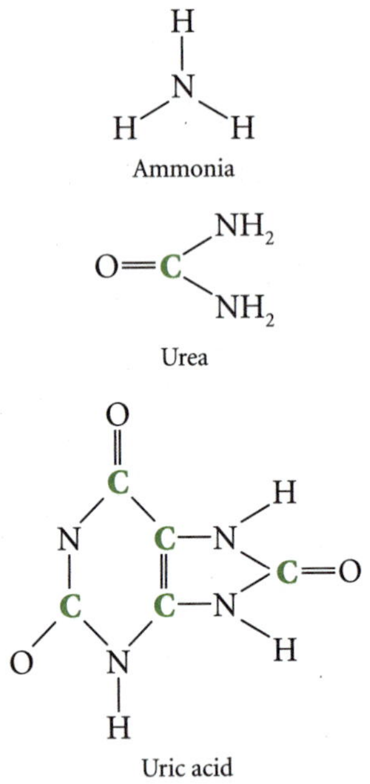

Ammonia

Urea

Uric acid

FIGURE 4-10 The chemical structures of common nitrogenous waste products. Ammonia lacks carbon and is toxic to living tissues. Urea and uric acid are less harmful, but the considerable energy contained in reduced carbon represents a loss to the organism.

centrated in blood and urine without dangerous effects (Hochachka and Somero 1973). In mammals, this waste product is urea [$CO(NH_2)_2$], the same substance produced and retained by sharks and rays to achieve osmotic balance in marine environments (Mommsen and Walsh 1989). Because urea dissolves in water, it can be excreted in the urine with some urinary water loss; the amount of water loss depends upon the concentrating power of the kidneys.

Birds and reptiles excrete nitrogen in the form of a double-ringed compound, uric acid ($C_5H_4N_4O_3$). Uric acid production is an adaptation for nitrogen excretion by the embryos of birds and reptiles (Needham 1931). Because water cannot cross the shell, nitrogenous waste produced by embryos must be retained within the egg. This waste is sequestered harmlessly as uric acid crystals in special membranes external to the embryo, which are discarded at hatching (Clark and Fisher 1957). Uric acid crystallizes out of solution and can therefore be greatly concentrated in the urine, enhancing water conservation by birds and reptiles, particularly in desert environments. Uric acid is also the most common nitrogenous waste product of terrestrial arthropods (Hadley 1994). Although water is saved by excreting urea or uric acid, there is also a cost, the energy lost in the organic carbon used to form these compounds. For each atom of nitrogen excreted, 0.5 and 1.25 atoms of organic carbon are lost in urea and uric acid respectively.

4.6 The ability of soil to retain water is related to the size of soil particles.

Most terrestrial plants obtain the water they need from the soil. The amount of water that soil holds, and its availability to plants, vary according to the physical structure of the soil particles (Brady 1974). Soil consists of grains of clay, silt, and sand, as well as particles of dead and decomposing organic matter, called **detritus.** Grains of clay (particles smaller than 0.002 mm), produced by the weathering of minerals in certain kinds of bedrock, are the smallest; grains of sand (particles larger than 0.05 mm), derived from quartz crystals that remain after minerals more susceptible to weathering dissolve out of the rock, are often the largest; silt particles are intermediate in size. Collectively, these particles make up the **soil skeleton.** As the name implies, the soil skeleton is a stable component that influences the physical structure of the soil and its water-holding ability, but does not play a major role in its chemical transformations.

Water Potential

The concept of water potential is central to the dynamics of water, plants, and soil. The **water potential** of a system is the free energy of water in the system. Like osmotic pressure, water potential is measured in units of pressure (MPa). One way to grasp the idea of water potential is to think of the process of diffusion. Just as solutes diffuse from areas of high solute concentration to areas of low solute concentration, water diffuses in response to changes in water potential, moving from areas of high water potential to areas of low water potential. Pure water has a higher water potential than a solution. Water moves down a gradient of water potential. Thus, a living cell surrounded by pure water will gain water through osmosis, as we have discussed. The cell has a lower water potential than the surrounding water. By convention, the water potential of pure water at atmospheric pressure is zero. Thus, the water potential of a solution will be negative (for example, -3.2 MPa is a lower water potential than -3.0 MPa).

There are a number of components of water potential (Figure 4-11), one of which we have just revealed. The water potential of a cell in pure water is determined by the solute concentration in the cell. The more concentrated the solute, the lower (more negative) the water potential. This component of water potential is called the osmotic or solute component. Water potential may also be affected by pressure. If the pressure on a system of pure water is increased above atmospheric pressure, the water potential will increase (it will become positive). The water potential

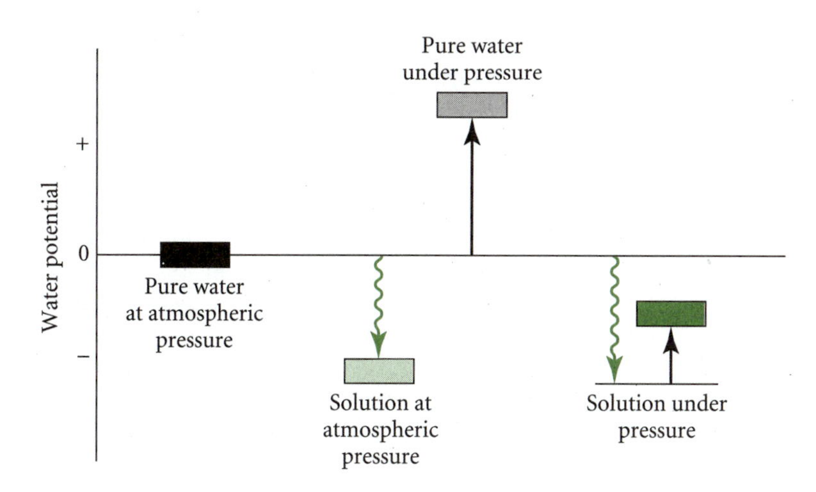

FIGURE 4-11 Effects of solute concentration and pressure on water potential. The water potential of pure water is zero. The addition of solutes (wavy green arrows) lowers the water potential to a point below pure water (negative pressure units). The application of pressure (black arrows) increases the water potential of both pure water and solutions. (Note that pressure can be negative when tension is applied, a situation not shown in this figure.) (*After Salsbury and Ross 1992.*)

of solutions is also increased when pressure is increased. The water potential, then, may be thought of as the sum of the **osmotic component** and the **pressure component:**

Water potential = osmotic component
+ pressure component

The osmotic component of the water potential is always negative because any solution has a water potential less than that of pure water. The pressure component may be positive or negative. The latter occurs when there is a pull or suction, called **tension,** applied to the system. Water potential will be positive if the pressure applied to the system overcomes the decrease in water potential caused by the presence of solutions. The relationship between the osmotic and the pressure components of water potential has an important implication in plant cells. As a cell takes in water, its internal pressure increases because of the presence of the rigid cell wall. This pressure gives the cell turgor. The increase in pressure also raises the water potential of the cell slightly.

A comparison of the water potentials of the soil, the atmosphere, and various parts of plants helps us understand how water moves from the soil to the upper portions of a plant, a topic that we discuss more fully in section 4.7 below. Typically, water potential is lowest (most negative) in the atmosphere and highest in the soil. Leaves, stems, and other parts of a plant usually have water potentials that are higher than that of the atmosphere but lower than that of the soil. This difference creates a downward water potential gradient from the soil to the leaves and out to the atmosphere, which helps move water through the plant.

The water potential of a plant is affected mainly by the concentration of solutes in and tension (negative pressure) applied to some parts of the plant. Water potential in the soil, at least for parts of the soil that are above the water table, is usually not affected as much by pressure components. But another factor, the attraction of water molecules to particles, does have an influence.

Water in the Soil

Plants obtain water from the soil. The availability of water in the soil is only partly determined by the amount of water present. It is also determined by how strongly the water is held in the soil. If you have houseplants, you are already aware of some obvious features of soil. When you saturate the soil in the pot containing your plant, some of the water that you poured in at the top ends up in the trough in which the pot sits, or, if you have really poured it on, on your floor. This water moves downward in the soil by the force of gravity. The soil in the pot cannot hold all of the water that you pour into it, though some of the water does remain in the soil. The amount of water held in the soil against gravity is called the **field capacity** of the soil. The water that runs through the soil and ends up in the trough—or, in nature, the water that moves downward until it becomes part of the groundwater—is called **gravitational water.** How is water held in the soil?

Water is held in the soil by capillary action resulting from adhesion of the water to the elements of the soil skeleton. This capillary action holds water in the soil with a force equivalent to a pressure of about 0.1 atmosphere (atm), or 0.01 MPa. Water that is drawn to soil particles with a force of less than 0.01 MPa—that is, water in the interstices between large soil particles, generally more than 0.005 mm from their surfaces—becomes gravitational water and joins the groundwater in the crevices of the bedrock below. Water within the spaces between soil particles is held by cohesion, which generates pressures of about −0.015 MPa. Water in large pores in the soil, at a distance from the surfaces of soil particles, is held with pressures less than −0.015 MPa, and usually drains through the soil under the pull of gravity (note that "less than −0.015 MPa" means any number that is more negative, that is, a larger negative value). Imagine a particle of silt with a diameter of 0.01 mm enlarged to the size of this page (×25,000); the film of water held at field capacity by forces of capillary attraction would be as thick as half the width of the page.

The tendency of soil to hold additional water molecules may be referred to as its **matric potential.** The matric potential is equal to the average strength with which the least tightly held water molecules are held. The matric potential of a soil contributes to its water potential. If the soil is very dry, the matric potential may be extremely low (e.g., −200 MPa or lower). This will lower the water potential of the soil, thereby making the water potential gradient from soil to plant weaker.

As soil water is depleted, the remainder is held by increasingly strong forces, on average, because a greater proportion of the water lies close to the surfaces of soil particles. The relationship between water content and water potential is shown in Figure 4-12 for a typical soil with a more or less even distribution of soil particle sizes, from clay (up to 0.002 mm) through silt (0.002–0.05 mm) to sand (0.05–2.0 mm). Such soils are called **loams.** When saturated, a loam holds about 45 grams of water per 100 grams of dry soil; this is a soil moisture content of 45%. The field capacity of a loam is about 32%. Thus, when a loam is saturated, about 13% of the water will be lost to groundwater by gravity. The **wilting coefficient** of a loam—the minimum water content of the soil at which plants can no longer obtain water—is about 7% soil moisture. The difference between the field capacity and the wilting coefficient, about 25% in this case, measures the water available to plants. Of course, plants obtain water most readily when the soil moisture is close to the field capacity.

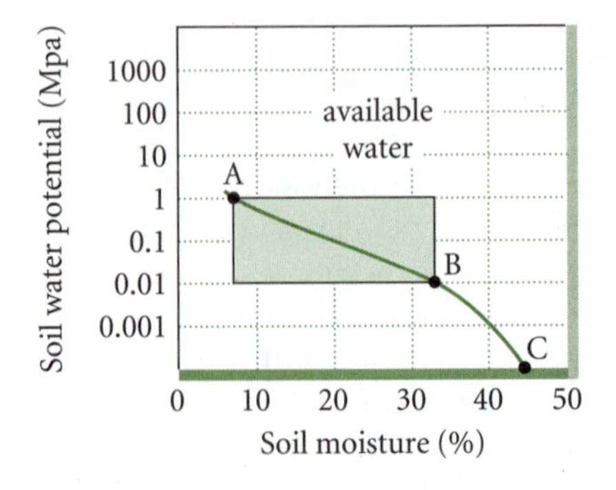

FIGURE 4-12 Relationship between the water content of a loam and the average force of attraction of the water to soil particles (water potential). The difference between the soil water content at field capacity (B, 0.1 atm) and the wilting coefficient (A, 15 atm) is the water available to plants. Point C is the saturation capacity of the soil. (*After Brady 1974.*)

In soils with predominantly smaller particles, the soil skeleton has a relatively large surface area; such soils hold a larger amount of water at both the wilting coefficient and the field capacity, and a correspondingly larger proportion of soil water is held by forces greater than −0.015 MPa. Soils with predominantly larger skeletal particles have less surface area and larger interstices between particles. More of the soil water is held loosely and is thus available to plants, but such soils have lower field capacities due to higher drainage. Plants can obtain the most water from soils having a variety of particle sizes between sand and clay.

4.7 The movement of water from soil to plant to atmosphere depends on transpiration and the cohesive properties of water.

In order to live, terrestrial plants must obtain water from the soil and transport it to all areas of the plant where cellular function is ongoing. This transport mechanism must include a way to get water from the soil into the roots, and a means to move that water from the roots to the topmost parts of the plant, all against the force of gravity and the friction exerted by the structures through which the water moves. When we recall that the tallest trees may exceed 100 meters, we can appreciate the challenge of water movement in plants.

Xylem

Vascular plants are those in which the internal transport systems are composed of tissues that form structures through which fluids move. Xylem (from the Greek *xylon,* meaning "wood") is the tissue primarily responsible for the transport of water in plants. Two types of xylem cells, **tracheids** and **vessel elements,** are recognized based on anatomical differences (Figure 4-13). At maturity, both types of cells die, and the cellular constituents disintegrate, leaving a hollow tube. These dead cells are arranged end to end in the stems of the plant. Tracheids are elongate cells with tapered ends and very thick cell walls. Thus, they provide support for the plant as well as an avenue of fluid transport. Vessel elements are thought to be evolutionarily more recent than tracheids because they occur only in angiosperms (flowering plants). In general, they are shorter and wider than tracheids. Vessel

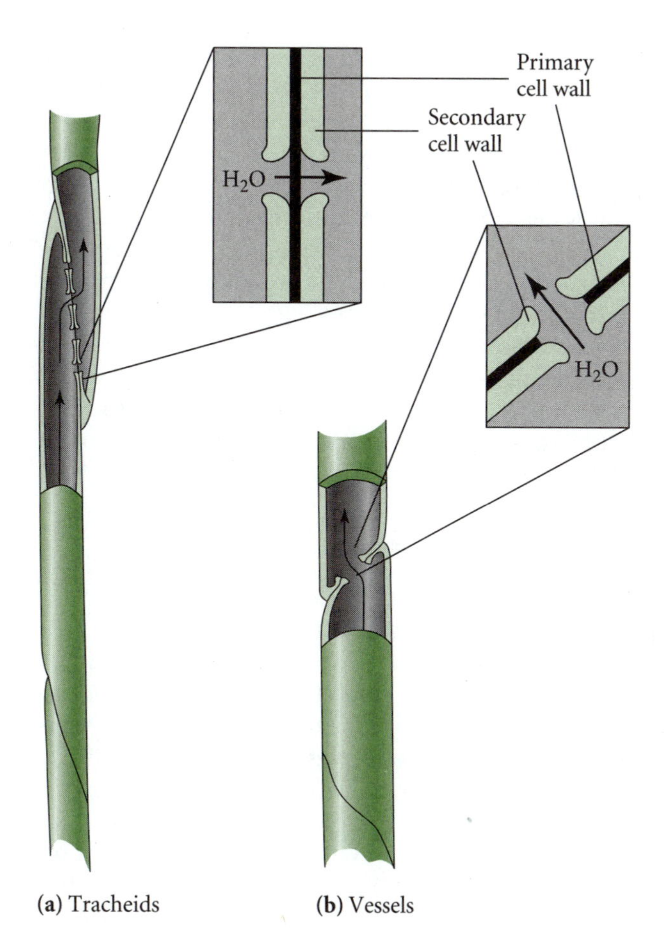

FIGURE 4-13 Two types of xylem: (a) tracheids and (b) vessels. One difference between the two is the structure of the pits. In tracheids, the pit opening is covered by a thin extension of the cell wall. No covering is found in the pits of vessel elements. Because of this and the presence of more perforations in the ends of vessel elements, water moves more freely in vessels.

elements are aligned end to end to form structures called **vessels,** some of which may contain millions of vessel elements and be over 10 meters in length. Both tracheids and vessel elements contain numerous structures called **pits,** which are openings in the cell wall through which material moves from one cell to the next. The pits of tracheids and vessel elements are somewhat different in structure, and this explains why fluids move through vessels more easily than through tracheids. In tracheids, the pit opening is covered with a thin layer of tissue that is part of the cell wall. This layer is selectively permeable to water and dissolved materials. The cell wall covering of the pit of tracheids represents a resistance to the movement of fluids that is not present in vessel elements, whose pit openings have no cell wall covering. Vessel elements also tend to be highly perforated at the ends where they abut other vessel elements, and water can move freely through these perforations.

Water Movement in Plants

The xylem tissue forms a conduit for water movement from the roots of the plant to the stems and leaves. But moving water through this conduit is not a simple matter. Consider the problems that must be overcome. The xylem tissue itself presents a problem in that any movement of fluid within it must overcome the frictional forces of the inside of the tracheid cells and vessel elements. Further, xylem is composed of cells arranged end to end, and the passage of water is affected by movement from one cell to the next. Resistance to water movement occurs wherever one cell meets another because water must pass through a pit or through perforations in the cell wall. The force of gravity, which tends to push water downward in the xylem, must also be overcome, a formidable challenge for the tallest trees.

What is the magnitude of the forces that must be overcome to move water to the top of a tall tree? Consider a long thin glass tube filled with water and sealed at one end (Figure 4-14). If the open end is immersed in a pool of water, gravity will push down on the water in the tube, but some of the water will remain in the tube. This is because the weight of the atmosphere above the surface of the water in the pool forces water up the tube. At sea level, the atmosphere exerts a pressure of about 0.1 MPa (1.0 atm) on the surface of the pool, which is enough to keep a column of water a bit over 10 meters high in the tube. To push the column of water in the tube to a height of 100 meters, which would be close to the height of the tallest trees, would require 1.0 MPa (10 atm) or more of pressure. Of course, plants are much more complex than a glass

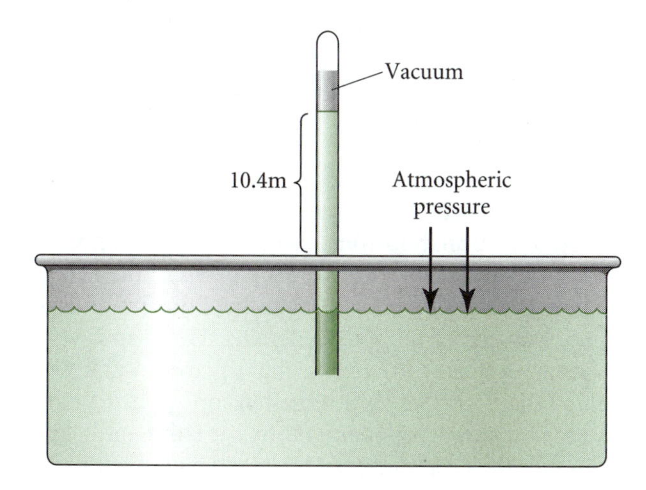

FIGURE 4-14 At sea level, the pressure of the atmosphere above a pool of water will exert enough pressure (1 atm) to hold a column of water in a thin tube about 10 meters high against the force of gravity. A pressure of 10 atmospheres would be required to sustain a column of water 100 meters high, the height attained by the tallest trees.

tube. Already we have pointed out that the movement of water in xylem must overcome frictional forces created by the walls of the tracheids and vessel elements and as well as resistance to fluid movement created by the presence of pits. These features of the plant's internal transport system suggest that even greater force is required to move water in plants. How is this force overcome?

The answer to this question is not completely understood, but the most plausible explanation involves the process of transpiration and the cohesive properties of water. The xylem tissue of the plant creates a continuous column of water from root to leaf. In order to enter the roots, water must overcome the force of gravity pushing down on this water column. We know that the process of water intake by the roots can create considerable osmotic pressure, called **root pressure,** in some plants. But this pressure is not enough to force water from the roots to the highest parts of a tree.

Water is lost from the leaves through the process of transpiration. Leaves are composed principally of parenchyma tissue, which contains relatively unspecialized cells that are loosely packed, thereby creating air spaces in the leaf. Water vapor is lost through the stomates of the leaf due to evaporation within these moist air spaces of the leaf tissue. As water vapor is moved from these spaces through the stomates and out of the leaf, it is replaced by water from the leaf cells surrounding those spaces, thereby increasing the osmotic pressure of those cells. Since leaf cells are typically very thin-walled, water can move from one cell to the next in response to an osmotic gradient. Thus, cells with high osmotic pressure draw water from neighboring cells. If we imagine a row of cells lining an air space in which one end of the row is near the stomate, we can see that evaporation and subsequent loss of water in the air space could set up an osmotic gradient in the cells, which could extend through the leaf tissue all the way to the xylem of the stem of the leaf. Thus transpiration exerts a pulling force, or tension, on water in the leaves—a force sometimes referred to as **transpirational pull**—and this tension is felt in the xylem tissue. It is this tension in the xylem that makes the xylem tissue water potential more negative than that of the soil. Generally, xylem sap is very dilute, and thus it has only a slightly lower water potential than pure water. But the tension applied to the sap makes its water potential more negative and helps maintain the decreasing water potential gradient from soil to atmosphere. The cohesive property of water, and the fact that the xylem tissue creates a continuous column of water from root to leaves, explain the movement of water. When tension is applied to the top of the column by transpiration, the entire column is pulled upward.

If we return to the simple example of a glass tube in a pool of water, we can see how transpirational pull substitutes for the application of atmospheric pressure. In order to move water up the tube, we must exert pressure on the surface of the water, thereby pushing water from the pool into the tube. Alternatively, we can remove the seal at the top of the tube and apply suction (tension) to draw the water from the pool into the tube, as one does with a straw and a soft drink. Mechanical devices would allow us to exert considerable pressure on the column of water and raise it to a great height. Likewise, the pull associated with transpiration from the leaves provides some of the force of water movement.

Soil Water and Wilting

Plants obtain water from the soil by osmosis, so the ability of their roots to take up water depends on their osmotic pressure. Osmotic pressure, in turn, is a function of the concentration of dissolved molecules (including sugars) and ions within the root cells. By manipulating the osmotic pressure of their root cells, plants can alter their ability to remove water from the soil. Plants growing in deserts and in salty environments can increase the water potential of their roots to as much as −6.0 MPa by increasing the concentration of amino acids, carbohydrates, or organic acids in their root cells. They pay a metabolic price, however, to maintain such high concentrations of dissolved substances.

Plant roots easily take up the water that clings loosely to soil particles by adhesion, but close to the surfaces of soil particles, stronger forces between water molecules and the particles themselves hold water tightly in the soil with potentials of −5.0 MPa or more. Plants first tap the soil water that is held with the least potential and is easiest to extract. As soils dry out, the water remaining is held with greater and greater force, eventually causing drought stress and wilting. As a general rule, plants from moist environments cannot obtain water held with pressures more negative than −1.5 MPa, the permanent wilting coefficient of the soil (Slatyer 1967, Meidner and Sheriff 1976). Once plants under drought stress have taken up all the water in the soil held by forces weaker than −1.5 MPa, they can no longer obtain water, and they wilt, even though water may remain in the soil.

4.8 Life requires inorganic nutrients.

The bodies of organisms are not composed solely of carbon, oxygen, and hydrogen. Other elements, such as nitrogen, phosphorus, sulfur, potassium, calcium,

TABLE 4-2 Major mineral nutrients required by organisms, with some of their primary functions

Element	Function
Nitrogen (N)	Structural component of proteins and nucleic acids
Phosphorus (P)	Structural component of nucleic acids, phospholipids, and bone
Sulfur (S)	Structural component of many proteins
Potassium (K)	Major solute in animal cells
Calcium (Ca)	Structural component of bone and of material between woody plant cells; regulator of cell permeability
Magnesium (Mg)	Structural component of chlorophyll; involved in function of many enzymes
Iron (Fe)	Structural component of hemoglobin and many enzymes
Sodium (Na)	Major solute in extracellular fluids of animals

magnesium, and iron, are important as well. These elements are called **mineral nutrients,** and their primary functions are summarized in Table 4-2. Thus, organisms must assimilate a wide variety of chemical elements. Many other elements, such as boron, selenium, zinc, copper, and molybdenum, sometimes referred to as **micronutrients,** are known to be required in smaller quantities.

Plants acquire mineral nutrients from water. They obtain nitrogen in the form of ammonium ions (NH_4^+) or nitrate ions (NO_3^-), phosphorus in the form of phosphate ions (PO_4^{3-}), calcium and potassium in the form of their elemental ions (Ca^{2+}, K^+), and so on. The availability of each of these elements varies with its chemical form in the soil and with the temperature, acidity, and presence of other ions in the soil water. Some elements are scarcer than others (Table 4-3). Phosphorus in particular often limits plant production, especially in acidic or alkaline soil, because it forms insoluble complexes with certain cations (see Chapter 11). Table 4-3 also shows that some elements, particularly calcium, potassium, and nitrogen, are taken up in amounts much higher than their concentrations in the soil.

Active Uptake of Scarce Soil Nutrients

Plants acquire mineral nutrients—nitrogen, phosphorus, potassium, calcium, and others—from dissolved forms of these elements in the soil water (Haynes and Goh 1978, Chapin 1980). When an element is abundant and highly mobile in the soil solution, as in the case of calcium and magnesium ions (Ca^{2+} and Mg^{2+}), its uptake is limited primarily by the absorptive capacity of the roots (Nye 1977). Plants compensate for low levels of a limiting nutrient in the soil by active uptake of that nutrient and by increasing the extent of the root system, that is, the absorptive surface of the roots. In laboratory experiments, barley and beet roots passively took up phosphorus by diffusion when its concentration in the water surrounding the root was greater than 0.2–0.5 millimolar (mM) (Barber 1972). Under these conditions, concentrations of phosphorus in root tissues and in the water conducted to the leaves were less than that in the soil solution. At external concentrations lower than 0.2–0.5 mM, roots actively transported phosphorus across their surfaces and concentrated the element within the root cortex. This active absorption requires the expenditure of energy as the root tissue moves ions against a concentration gradient.

Plants may also respond to decreased soil nutrient availability by increasing root growth at the expense of shoot growth. This strategy brings the nutrient requirements of the plant into line with the availability by reducing the nutrient demand created by the leaves, by increasing the absorptive surface area of the root system, and by growing roots into new areas of soil whose scarce minerals may not have been depleted (Chapin 1980). Such modifications cannot completely compensate for decreasing soil fertility, however, and both the concentration and total amount of a limiting nutrient in a plant vary in direct relationship to the availability of the nutrient in the soil.

Crop plants and wild species growing on fertile soils have a great capacity to absorb nutrients across their root surfaces, and their growth rates vary in response to variation in soil nutrient levels. Species adapted to nutrient-poor soils are much more conservative in increasing their rate of growth (Chapin 1980). They cope with low nutrient availability by allocating a large proportion of their biomass to roots; by establishing symbiotic relationships with fungi, which enhance mineral absorption; and by growing slowly and retaining leaves for long periods, thereby reducing nutrient demand. Such species typically cannot respond to artificially increased nutrient levels by increasing their growth rate. In response to a nutrient flush, their roots

TABLE 4-3 Typical concentrations of elements in soils and annual uptake by plants

Element	Soil content (weight %)*	Annual plant uptake (kg ha^{-1} yr^{-1})	Soil content/annual plant uptake (years)†
Silicon (Si)	33	20	21,000
Aluminum (Al)	7	0.5	180,000
Iron (Fe)	4	1	52,000
Calcium (Ca)	1	50	260
Potassium (K)	1	30	430
Sodium (Na)	0.7	2	4,600
Magnesium (Mg)	0.6	4	2,000
Titanium (Ti)	0.5	0.08	62,000
Nitrogen (N)	0.1	30	40
Phosphorus (P)	0.08	7	150
Manganese (Mn)	0.08	1	1,000
Sulfur (S)	0.05	2	320
Fluorine (F)	0.02	0.01	26,000
Chlorine (Cl)	0.01	0.06	220
Zinc (Zn)	0.005	0.01	6,500
Copper (Cu)	0.002	0.006	4,200
Boron (B)	0.001	0.03	400
Molybdenum (Mo)	0.0003	0.0003	13,000
Selenium (Se)	0.0000001	0.0003	40

**Carbon, oxygen, hydrogen, and some additional trace elements make up the remaining percentage.*

†Soil content (g m^{-2}) divided by annual uptake (g m^{-2} yr^{-1}) yields a ratio, the time to soil depletion in the absence of replenishment, in years.

(From Bohn et al. 1979.)

absorb more nutrients than the plant requires and store them for subsequent utilization when the nutrient status of the soil decreases (Grime and Hunt 1975, Grime 1979).

Dissolved Minerals in Water

All natural waters contain some dissolved substances. Although nearly pure, rainwater acquires some dissolved minerals from dust particles and droplets of ocean spray in the atmosphere (Ingham 1950, Likens et al. 1977). Most lakes and rivers contain 0.01–0.02% dissolved minerals and roughly 0.05–0.02% of the average salt concentration of the oceans (3.4%), in which salts and other minerals have accumulated over the millennia (Hutchinson 1957).

The dissolved minerals found in fresh water and in salt water differ in proportions as well as quantities due to the different rates of solution and solubilities of substances (Table 4-4). Seawater contains high concentrations of sodium and chlorine, with significant amounts of magnesium and sulfate. Fresh water contains a more even distribution of diverse ions, but Ca^{2+} is usually the most abundant cation, and carbonate (CO_4^{2-}) and sulfate (SO_4^{2-}) are the most abundant anions. Few compounds reach their maximum solubilities in fresh water. Their concentrations reflect the composition and rates of solution of materials in the rock and soil with which the water comes in contact. Limestone consists primarily of calcium carbonate, which dissolves quickly; thus, water in limestone areas contains abundant calcium ions, making it "hard." Granite contains such minerals as quartz and feldspar, which do not contain calcium and which dissolve slowly; water flowing through granitic areas contains few dissolved substances and is "soft."

The oceans are like large stills in that they concentrate minerals as pure water evaporates from the

TABLE 4-4 Percentage composition of dissolved minerals in rivers (fresh water), in seawater, and in the blood plasma and cells of organisms (frogs)

Mineral ion	Delaware River	Rio Grande	Seawater	Frog plasma	Frog cells
Sodium (Na^+)	6.7	14.8	30.4	35.4	1.3
Potassium (K^+)	1.5	0.9	1.1	1.3	77.7
Calcium (Ca^{2+})	17.5	13.7	1.2	1.2	3.1
Magnesium (Mg^{2+})	4.8	3.0	3.7	0.4	5.3
Chlorine (Cl^-)	4.2	21.7	55.2	39.0	0.8
Sulfate (SO_4^{2-})	17.5	30.1	7.7	—	—
Carbonate (CO_3^{2-})	33.0	11.6	0.4	22.7	11.7

***Note:** The percentages of the negatively charged ions (anions) exceed those of the positively charged ions (cations) because, ion for ion, anions are much heavier; the numbers of positive and negative ions are approximately equal. The sums of all columns do not equal 100 because not all dissolved substances are included.*

(Data from Reid 1961, Gordon 1968.)

surface and nutrient-laden water arrives via streams and rivers. Here the concentrations of some minerals, particularly calcium carbonate, are limited by their maximum solubilities. Calcium carbonate dissolves only to the extent of 0.000014 grams per gram (ca. 1 cm^3) of water. Its concentration in the oceans reached this level eons ago, and the excess Ca^{2+} entering oceans each year from streams and rivers precipitates to form limestone sediments. At the other extreme, the solubility of NaCl (0.36 g per g of water) far exceeds its concentration in seawater; most of the NaCl washing into ocean basins remains dissolved.

SUMMARY

1. Water is the basic medium of life. The chemical, physical, and thermal characteristics of water result from the nature of the covalent bonds between hydrogen and oxygen and the hydrogen bonds between water molecules.

2. The tendency of solutions to attract water is called osmotic pressure. In aquatic habitats, differences in the concentrations of dissolved salts establish osmotic gradients between the organism and its surroundings. Hyperosmotic organisms, which have greater salt concentrations than their environment, tend to lose salt and gain water; hypo-osmotic organisms gain salt and lose water. Organisms achieve osmotic balance by altering the solute concentration of their body fluids and by actively pumping salts across membranes.

3. Organisms maintain the proper ionic balance between themselves and the surrounding environment through a process referred to as osmoregulation. Freshwater organisms are hyperosmotic and face the problem of ridding themselves of excess water while simultaneously retaining solutes. Marine organisms are hypo-osmotic, and thus tend to lose water to the surrounding seawater.

4. Animals that sustain themselves on fluid diets face the problem of eliminating excess fluid and conserving ions, similar to the problem of freshwater organisms. A number of adaptations, including high fluid turnover, have arisen in fluid-feeding organisms to meet the challenges of high fluid intake.

5. Meat-eating animals take in excess nitrogen, the oxidized, inorganic forms of which are toxic in high quantities. Aquatic animals excrete nitrogen as ammonia (NH_3), a process that requires a considerable amount of water, and thus is not available to terrestrial animals, which must conserve water. Mammals produce the waste product urea ($CO[NH_2]_2$), which dissolves in water and is excreted in the urine. Birds and reptiles produce uric acid ($C_5H_4N_4O_3$), which may be greatly concentrated in the urine. Both adaptations allow for the conservation of water.

6. The water potential of a system is the free energy of water in the system measured in units of pressure. The water potential of pure water is zero, and that of solutions less than zero. Pressure can increase the water potential both of pure water and solutions. Water tends to move from areas of high water potential to areas of low water potential. Atmospheric water potential is very low, whereas soil water potential is high. Plant tissue usually has a water potential between that of the soil and that of the atmosphere.

7. Most terrestrial plants obtain water from the soil. Grains of clay, silt, and sand and particles of decomposing organic matter called detritus make up the soil skeleton. Water is held in the soil by attraction to soil particles. A volume of soil having relatively more small soil particles (clay) than large particles (sand) will hold more water than the same volume of soil composed mostly of large particles because there will be more surface area to attract water molecules. The amount of water held in the soil against gravity is called the field capacity.

8. Water moves in vascular plants through an internal transport system called xylem, a system of hollow dead cells that allows for the presence of a continuous column of water from the roots to the leaves of the plant. Transpiration in the leaves exerts tension on the water column in the xylem and, because of the cohesive nature of water, the entire column is pulled upward.

9. In addition to carbon, oxygen, and hydrogen, living things require mineral nutrients such as nitrogen, phosphorus, sulfur, potassium, calcium, magnesium, and iron. Plants acquire these nutrients from water. The expenditure of energy (active uptake) may be required to obtain scarce nutrients.

EXERCISES

1. Considering the properties of water, how might the oceans and other large bodies of water affect the climate near them?

2. Suppose that in order to osmoregulate, you had to consciously monitor various environmental and body conditions and purposely take action to maintain the proper water balance in your body. What factors and conditions would you have to continuously monitor, and what actions would you instruct your body to take in order to maintain the proper water balance? Would you monitor different parameters and issue different instructions on summer and winter days? (Hint: Think about the functions of various organ systems in your body.)

3. Compare the osmoregulation challenges and mechanisms of humans with those of a freshwater fish.

CHAPTER 5

Energy and Heat

GUIDING QUESTIONS

- What are the chemical equations of photosynthesis and respiration?
- What are the different challenges faced by aquatic and terrestrial plants with respect to carbon availability?
- How is light affected by the earth's atmosphere?
- How does water limit the availability of light for photosynthesis?
- How do C_4 and CAM photosynthesis increase water use efficiency?
- What kinds of adaptations are required for living at extremely high and extremely low temperatures?
- What are the four avenues of heat transfer between organisms and their environment?
- How do some organisms tolerate wide ranges of environmental conditions?
- How does temperature affect the conservation of water in organisms?

As we emphasized in Chapter 2, living things accumulate energy. The energetic link between the physical and the biological worlds is photosynthesis, in which light energy from the sun is transformed into potential or chemical energy in the bonds of organic molecules. Through the process of respiration, animals and plants transform the energy of these organic molecules into energy of a form usable in the physiological processes of life. The energetic transformations of photosynthesis and respiration are dependent on the conditions of the physical environment. Photosynthesis requires light of the proper wavelength and ready sources of water and carbon. Oxygen is needed for respiration. A proper water balance and temperature regime must be maintained in order for these transformations to proceed.

In this chapter, we shall explain the constraints placed on energy transformations by conditions of the physical environment. Our presentation here is divided into two parts. The first part of the chapter deals with the challenges faced by aquatic and terrestrial organisms in obtaining a proper balance of light, carbon dioxide, and oxygen for photosynthesis and respiration. Adaptations of photosynthesis for the conservation of water are discussed in detail. The range of adaptations for life at different temperatures and the mechanisms by which organisms interact with the thermal environment are the focus of the second portion of the chapter.

5.1 Most biological energy transformations are based on the chemistry of carbon and oxygen.

Plants and animals consist of many elements joined together into the organic molecules that form the structure of the individual. Such organic compounds also supply the energy needed to maintain the organism. This energy is present in the form of chemical bonds between atoms and molecules. These energy-rich bonds arise from chemical changes in the atoms of various elements. These chemical changes are the fundamental transactions of the natural world. Understanding them is the first step in gaining an appreciation of nature.

Photosynthesis and Respiration

All living systems are driven energetically by the chemical reduction of carbon, which is accomplished when electrons are added to the carbon atom. The typical oxidized form of carbon is carbon dioxide (CO_2), which is found as a gas in the atmosphere and dissolved in water. During photosynthesis, plants reduce the carbon atom in CO_2. Reduced carbon, in turn, is used to form organic molecules, such as the carbohydrate glucose ($C_6H_{12}O_6$), within which its energy level is greatly increased. With few exceptions, the added energy comes from light. Photosynthesis may be summarized by the following formula showing the reduction of six moles of CO_2 to form a mole of sugar:

$$6CO_2 + 6H_2O \longrightarrow 6CO_2 + 12H + 6O \longrightarrow (CH_2O)_6 + 6O_2.$$

To release this stored energy for other purposes, both plants and animals undo the results of photosynthesis by oxidizing the organic carbon in the sugar back to CO_2 through the process of respiration. This transformation releases energy, a portion of which organisms harness as ATP. Respiration may be summarized by the following equation showing the transformation of a mole of sugar:

$$(CH_2O)_6 + 6O_2 \xrightarrow[\text{Heat}]{36\,ADP + 36P \;\to\; 36\,ATP} 6CO_2 + 6H_2O$$

Photosynthesis and respiration involve the complementary reduction and oxidation of carbon and oxygen (Figure 5-1). Oxygen's common oxidized state is molecular oxygen (O_2), which occurs as a gas in the atmosphere and dissolved in water. In a reduced state, oxygen readily forms water molecules (H_2O). Thus, as carbon is reduced during photosynthesis, oxygen is oxidized from its form in water to its molecular form. In essence, the electrons in water are transferred to carbon. During respiration, those electrons are shifted back to oxygen, which is inhaled or absorbed once again, producing water. Why, then, does the coupling of an oxidation reaction to a reduction reaction result in a net release of energy? Because the reduction of oxygen to water is thermodynamically more favorable (requires less energy input) than the reduction of carbon to organic molecules. Thus, the oxidation of organic carbon releases more energy than the reduction of oxygen requires. (This is why oxygen is such a good oxidizer.)

Oxygen and Carbon Dioxide in Aquatic and Terrestrial Habitats

Though we often focus on the photosynthetic function of plants, they also respire. However, plants reduce more carbon than they oxidize (otherwise they would not grow), and they therefore require an external source of carbon. The only practical source of inorganic carbon, CO_2, has an extremely low concentration in the atmosphere (about 0.03%, or a partial pressure at sea level of 0.2 torr [1 torr = 0.0013 atm]). As a result, gradients of carbon concentration between the atmosphere and the interior of plant cells are not very large—certainly much smaller than gradients of water vapor pressure between the plant and the surrounding atmosphere. This creates special problems for water conservation by plants, especially in arid environments. Whenever stomates of the plant leaf open to take in CO_2, water is lost from the leaf. Plants may lose up to 500 grams of water for every gram of carbon assimilated.

Carbon availability poses less of a problem for aquatic plants than for terrestrial plants because CO_2 has a high solubility in water. When CO_2 dissolves, some of the molecules react with water to form

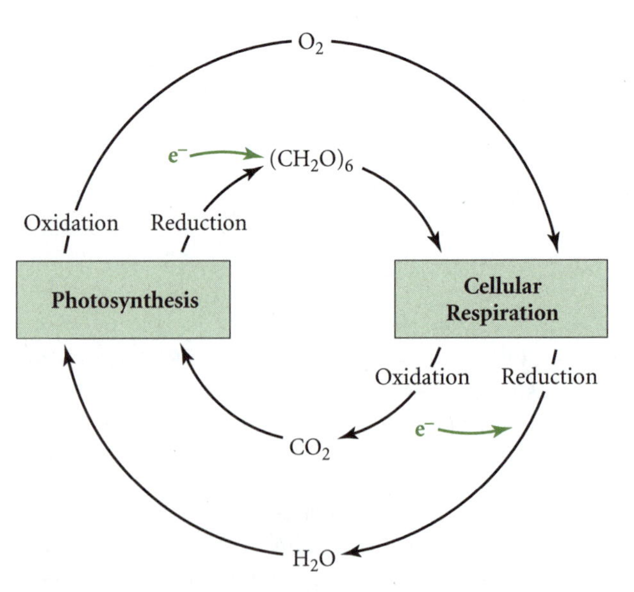

FIGURE 5-1 In photosynthesis, CO_2 is reduced to form organic compounds (CH_2O) and H_2O is oxidized to form molecular oxygen (O_2), thus in essence transferring the electrons in water to carbon. The electrons shift back to oxygen in cellular respiration. The reduction of O_2 to H_2O is thermodynamically more efficient than the reduction of carbon to organic molecules. Thus, energy is released during the oxidation of organic molecules in respiration.

carbonic acid (H_2CO_3) and associated compounds, which provide a reservoir of inorganic carbon. The concentration of CO_2 in the atmosphere is about 0.0003 cm^3 cm^{-3} (cm^3 of carbon dioxide gas per cm^3 of water vapor); its solubility in fresh water under ideal conditions is nearly the same. In turn, carbonic acid molecules dissociate into bicarbonate (HCO_3^-) and carbonate (CO_3^{2-}) ions. Within the range of acidity of most natural waters (pH 6–9), bicarbonate ions are the most common form of carbon; however, more carbonate is produced when water is acidic (see Chapter 11) (Stumm and Morgan 1981, Wetzel 1983). Bicarbonate ions dissolve readily in water. As a result, seawater normally contains concentrations of bicarbonate ions equivalent to 0.03–0.06 cm^3 cm^{-3}, over 100 times the concentration of dissolved carbon dioxide gas (Nicol 1967), making it relatively easy for aquatic plants to take up carbon from the water.

Once inside the plant cell, bicarbonate ions can be used directly as a source of carbon for photosynthesis, although at only 10–40% of the efficiency of utilizing CO_2. As CO_2 is depleted, however, the bicarbonate ions also produce CO_2 at the site of photosynthesis. Bicarbonate ions and CO_2 exist in a chemical equilibrium, which represents the balance achieved between H^+ and HCO_3^-, on the one hand, and CO_2 and H_2O on the other. We can represent this equilibrium as

$$H^+ + HCO_3 \longrightarrow CO_2 + H_2O.$$

As CO_2 is used in photosynthesis, some of the bicarbonate ions reassociate with hydrogen ions to replenish the CO_2 supply.

Unfortunately, the rate of diffusion of CO_2 through unstirred water is about 10,000 times slower than it is in air, and bicarbonate ions diffuse even more slowly. Every surface of an aquatic plant, alga, or microbe is covered by a **boundary layer,** an area of unstirred water, which may range from as little as 10 micrometers (μm) thick for single-celled algae in turbulent waters to 500 μm for large aquatic plants in stagnant waters (Figure 5-2). Thus, in spite of the high concentration of bicarbonate ions in the water surrounding these organisms, photosynthesis may nonetheless be limited by a diffusion barrier created by the boundary layer.

Just as the availability of carbon dioxide poses difficulties for plants in terrestrial environments, the availability of oxygen limits animals in aquatic habitats. Compared with its concentration of 0.2 cm^3 cm^{-3} in the atmosphere, the maximum solubility of oxygen in water (at 0° C in fresh water) is 0.01 cm^3 cm^{-3}, only 14 parts per million (ppm) by weight. The solubility of oxygen decreases as water temperature increases. Furthermore, below the limit of light penetration in deep bodies of water and in waterlogged

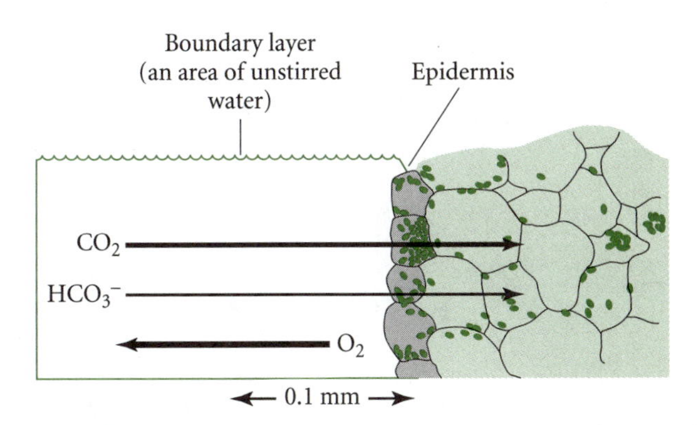

FIGURE 5-2 Diagrammatic cross section of a leaf of the aquatic plant *Vallisneria spiralis,* showing the thickness of the boundary layer in still water relative to the sizes of the cells. The widths of the arrows are proportional to the rates of diffusion of carbon dioxide (CO_2), bicarbonate ions (HCO_3^-), and oxygen (O_2) through the boundary layer. The larger bicarbonate ions diffuse more slowly through the boundary layer. (*After Prins and Elzenga 1989.*)

sediments and soils, where aquatic plants are absent, no oxygen can be produced by photosynthesis. Therefore, as animals and microbes use oxygen to metabolize organic materials, their habitats may become severely depleted of dissolved oxygen.

An adaptation that enhances the uptake of dissolved oxygen from water is **countercurrent circulation,** an arrangement seen in the structure of gills whereby water and blood flow in opposite directions (Figure 5-3). In a countercurrent system, as blood picks up oxygen from the water flowing past the gills, it comes into contact with water having progressively greater oxygen concentrations because the water has flowed past a progressively shorter distance of the gill lamella (Figure 5-4). With this arrangement, the oxygen concentration in the blood plasma can approach that of the surrounding water. If blood and water were to flow through the gills in the same direction (concurrent flow), an equilibrium oxygen concentration would quickly be established, with equal, intermediate levels in blood and water. The countercurrent system keeps the blood and water out of equilibrium and maintains a constant gradient across which oxygen can flow.

Habitats that are devoid of oxygen, such as deep layers of water in lakes and mucky sediments in marshes, are referred to as **anaerobic** or **anoxic** habitats. The waterlogged soils of swamps pose problems for terrestrial plants, whose roots need oxygen for respiration. In these habitats, many plants contain specialized vascular tissues, called **aerenchyma,** that conduct air directly from the atmosphere to the roots. The roots of cypress trees and many mangroves grow vertical extensions that project above the anoxic soil

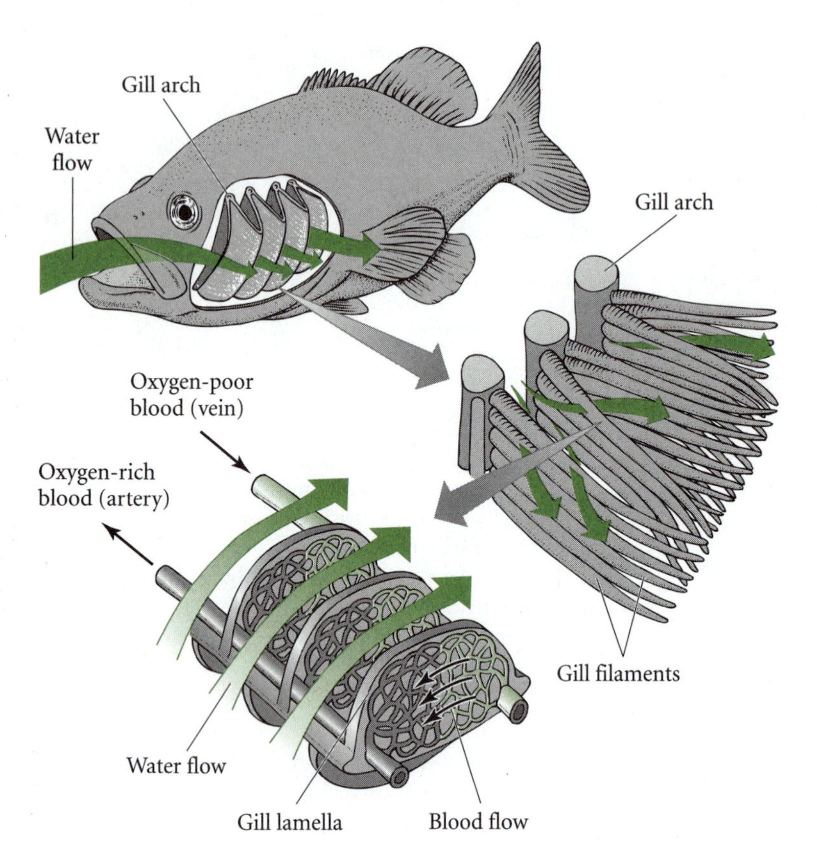

FIGURE 5-3 A fish's gill consists of several gill arches, each of which carries two rows of filaments. The filaments bear thin lamellae oriented in the direction of the flow of water through the gill. Within the lamellae, blood flows in a direction counter to the movement of water past the surface. (*See Figure 5-4; From Randall 1968.*)

and conduct oxygen directly from the atmosphere to the roots (Figure 5-5).

5.2 Light energy is the ultimate driving force of life processes.

Light from the sun is the primary source of energy for the biosphere. An exception occurs in deep-sea hydrothermal vent ecosystems, where symbiotic bacteria residing in the guts of pogonophores (a specialized phylum which includes vent-inhabiting worms) possess the capability of fixing carbon using energy derived from the oxidation of hydrogen sulfide gas. Green plants, algae, and some bacteria absorb light from the sun and assimilate its energy by photosynthesis. But not all light striking the earth's surface is useful in photosynthesis. Rainbows and prisms show that light consists of a spectrum of wavelengths that we perceive as different colors. Recall from our discussion of amphibian population declines in Chapter 2 that wavelengths of light are generally expressed in nanometers (nm; one-billionth of a meter, or 10^{-9} m). The visible spectrum extends between wavelengths of about 400 nm (violet) and 700 nm (red), and it is these wavelengths of light that are suitable for photosynthesis. Wavelengths shorter than 400 nm make up the **ultraviolet** part of the spectrum, and we refer to wavelengths longer than 700 nm as **infrared.** The energy content of light varies with wavelength and hence with color; short-wavelength blue light has a higher energy level than longer-wavelength red light.

The light that reaches the upper part of the earth's atmosphere from the sun extends far beyond the visible range: through the ultraviolet region toward the short-wavelength, high-energy X-rays at one end of the spectrum, and through the infrared region to extremely long-wavelength, low-energy radiation such as radio waves at the other end of the spectrum (Figure 5-6; Gates 1980). Because of its high energy level, ultraviolet light can damage exposed cells and tissues. It is the presence of enhanced levels of ultraviolet B (UV-B) radiation, resulting from depletion of the ozone layer of the earth's atmosphere, that is suspected as a major cause of the worldwide decline in amphibian populations.

Photochemical conversion of light energy to chemical energy by plants occurs primarily within the portion of the solar spectrum at the earth's surface containing the greatest amount of energy. The pattern of light absorption depends on the nature of the absorbing substance. Water only weakly absorbs light whose characteristic wavelengths fall in the visible region of the spectrum; as a result, a glass of water appears colorless. Dyes and pigments are strong absorbers of certain wavelengths in the visible region

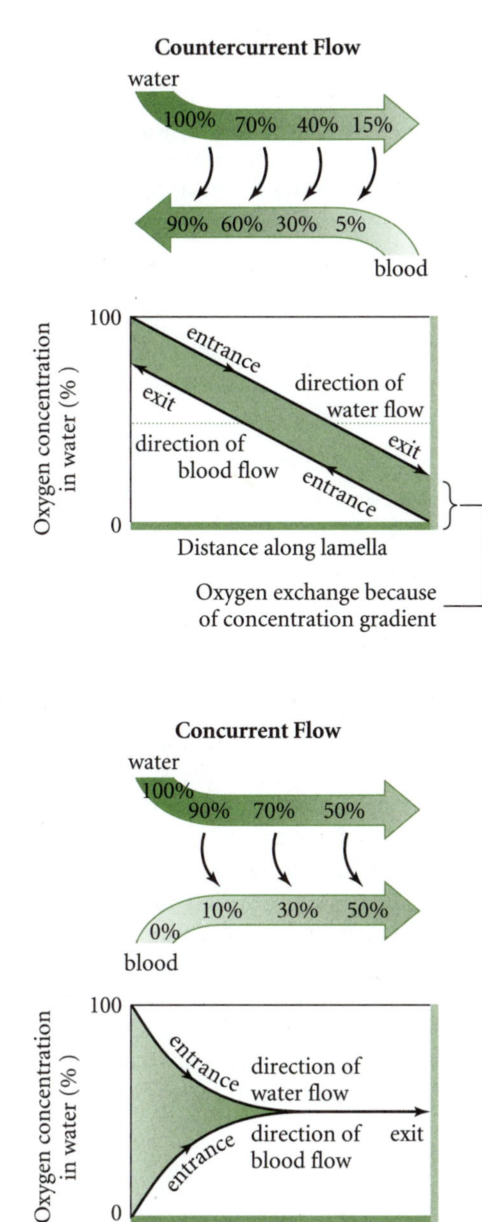

FIGURE 5-4 Changes in oxygen concentration of blood and water in countercurrent and concurrent systems. The more dense the shading, the higher the concentration of oxygen. In a countercurrent system, water with a high oxygen concentration enters the gill lamella (from the top left of the graph). Blood returning to the gills flows in the opposite direction (from the bottom right of the graph) through vessels that lie very close to the lamella. Since the returning blood is low in oxygen, a concentration gradient exists between the water and the blood, and oxygen diffuses into the blood vessels. The concentration gradient is maintained because water with a high oxygen concentration and blood with a low oxygen concentration are always in proximity. In a concurrent system, in which water and blood move in the same direction, the oxygen concentration gradient cannot be maintained. (*After Schmidt-Nielsen 1975.*)

and reflect or transmit light of definite colors. Plant leaves contain several kinds of pigments, particularly **chlorophylls** (green) and **carotenoids** (yellow), that absorb light and harness its energy (Figure 5-7). Carotenoids, which give carrots their orange color, absorb primarily blue and green light and reflect light in the yellow and orange regions of the spectrum. Chlorophyll absorbs red and violet light while reflecting green and blue. When chlorophyll and carotenoids occur together in a plant, green light is reflected most, hence leaves appear green to us.

5.3 The attenuation of light in water limits photosynthesis in aquatic environments.

Water absorbs or scatters enough light to limit the depth at which photosynthesis is possible in aquatic environments. The transparency of a glass of water is deceptive. In pure seawater, the energy content of light in the visible part of the spectrum diminishes to 50% of its surface value within 10 meters, and to less than 7% within 100 meters. Furthermore, water absorbs longer wavelengths more strongly than shorter ones; virtually all infrared radiation disappears within the topmost meter of water (Weisskopf 1968). Short waves of light (violet and blue) tend to be scattered by water molecules and thus also do not penetrate deeply. As a consequence, green light tends to predominate with increasing depth. The photosynthetic pigments of aquatic plants are adapted to this spectral shift. Plants growing near the surface of the oceans, such as the green alga *Ulva* (sea lettuce), contain pigments resembling those of terrestrial plants and best absorb blue and red light. The deep-water red alga *Porphyra* has additional pigments that enable it to utilize green light more effectively (Figure 5-8).

Because photosynthesis requires light, the depth at which algae can exist in the oceans and lakes is limited by the penetration of light. Algae are limited to a fairly narrow zone close to the surface where photosynthesis exceeds respiration. This range of depths is called the **euphotic zone.** The lower limit of the euphotic zone, where photosynthesis just balances respiration, is called the **compensation point** (Figure 5-9). It may be defined by either depth or light level. If algae sink below the compensation point or are carried below it by currents, and do not soon return to the surface on upwelling currents, they die, because they are unable to convert energy by photosynthesis.

In some exceptionally clear ocean and lake waters, the compensation point may lie 100 meters below the surface. But this is a rare condition. In productive waters (i.e., having high rates of photosynthesis)

FIGURE 5-5 The knees of these bald cypress trees in a freshwater swamp in South Carolina conduct air from the atmosphere to roots growing in waterlogged, anaerobic sediments. Other marsh and swamp plants have air-conducting tissue (aerenchyma) in their stems. (*Courtesy of the U. S. Forest Service.*)

containing dense phytoplankton, or in waters turbid with suspended silt particles, the euphotic zone may be as shallow as 1 meter. In some polluted rivers, little light penetrates beyond a few centimeters.

5.4 C_4 and CAM photosynthesis increase water use efficiency.

Plants require a constant supply of water, primarily to offset losses incurred when the stomates are open. As we saw above, because the concentration of CO_2 in the atmosphere is so low, the concentration gradient for water loss from the leaf is several orders of magnitude greater than that for CO_2 assimilation into the leaf. And because the vapor pressure of water increases with temperature, the problem of water loss is magnified in hot environments.

Heat-adapted and drought-adapted plants exhibit modifications of anatomy and physiology that reduce transpiration across their surfaces, reduce heat loads, and enable the plants to tolerate high temperatures (Hadley 1970, Berry and Bjorkman 1980). When plants absorb sunlight, they heat up. Overheating can be minimized by an increased surface area for heat dissipation and by protection of surfaces from direct

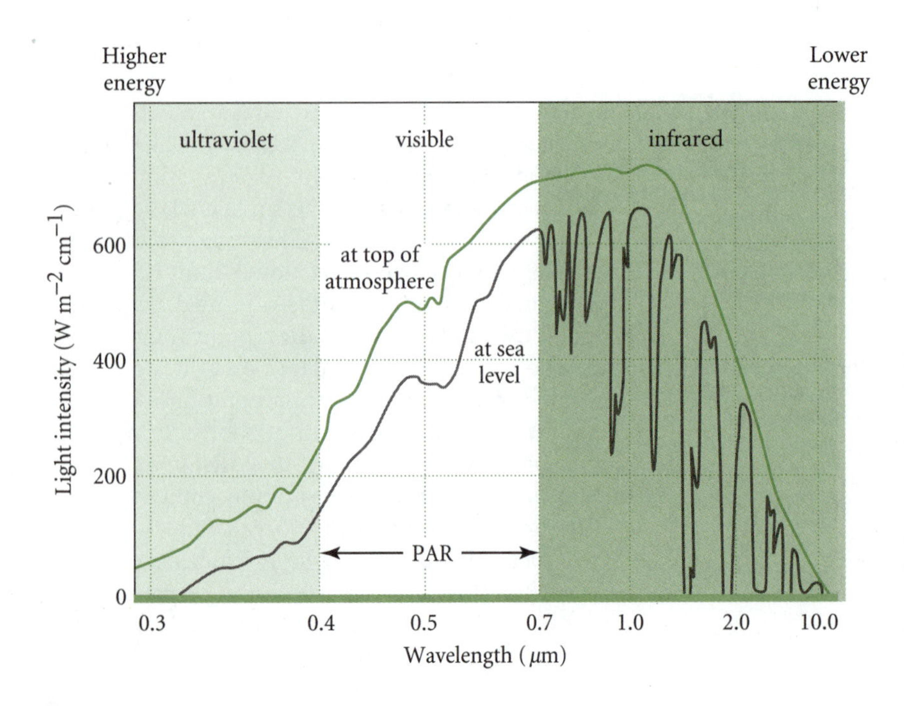

FIGURE 5-6 The spectral distribution of direct sunlight above the earth's atmosphere and at sea level. PAR indicates the photosynthetically active region of the spectrum. Ozone in the upper atmosphere absorbs light in the ultraviolet region of the spectrum, and water vapor and carbon dioxide absorb light in the infrared region. (*After Gates 1980.*)

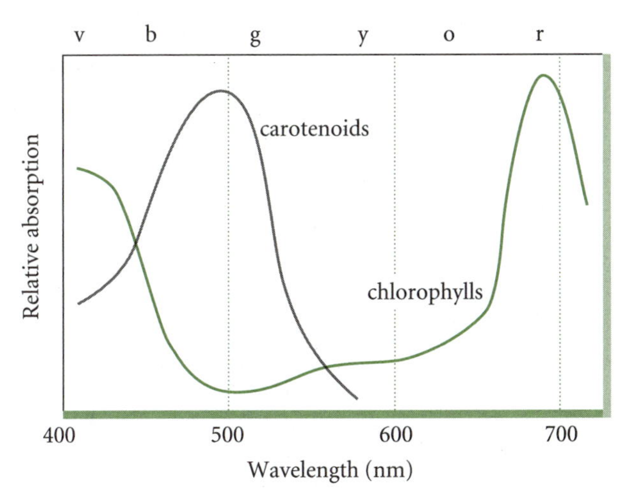

FIGURE 5-7 Absorption of light of different wavelengths by the two groups of plant pigments—chlorophylls (gray line) and carotenoids (green line)—that capture light energy for photosynthesis. The colors of the spectrum are violet (v), blue (b), green (g), yellow (y), orange (o) and red (r). (*After Emerson and Lewis 1942.*)

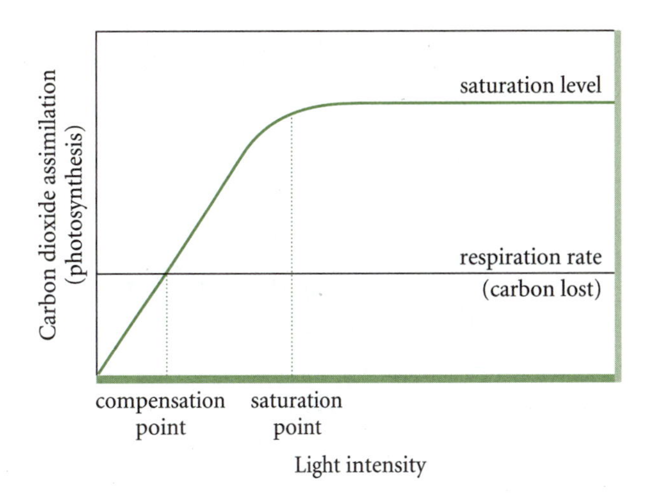

FIGURE 5-9 Relationship between CO_2 assimilation (photosynthesis) and light intensity in water, illustrating the compensation point, at which photosynthesis balances respiration, and the saturation point, beyond which additional light does not increase photosynthesis. (*After Barbour et al. 1980.*)

sunlight with dense hairs and spines (Ehleringer 1984; Figure 5-10). Spines and hairs also produce a still boundary layer of air that traps moisture and reduces evaporation. Because the thick boundary layers retard heat loss as well, hairs are prevalent in cool, arid environments, but occur less frequently in hot deserts. Transpiration can be further reduced by having thick, waxy cuticles impervious to water and by placement of the stomates in deep pits, often themselves filled with hairs (Figure 5-11). Even the process of photosynthesis has been modified in some plants to conserve water, as we shall see.

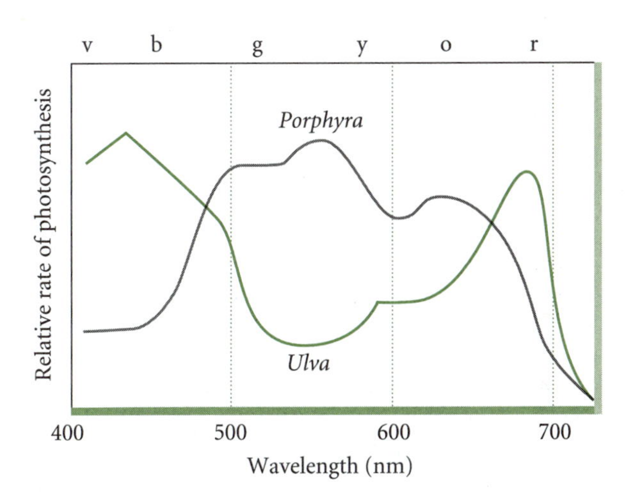

FIGURE 5-8 Relative rates of photosynthesis by the green alga *Ulva* (sea lettuce) and the deep-water red alga *Porphyra* as a function of the color of light. (*After Haxo and Blinks 1950.*)

C_3 Photosynthesis

Most plant species in mesic environments (environments with adequate water) assimilate the carbon in CO_2 into an organic molecule in a single biochemical step of the pathway known as the **Calvin-Benson cycle** (Figure 5-12). This step may be represented as

$$CO_2 + \text{RuBP} \longrightarrow 2\ \text{PGA},$$

where RuBP (ribulose bisphosphate) is a five-carbon organic compound and PGA (phosphoglycerate) is a three-carbon compound. Because the immediate

(a) **(b)**

FIGURE 5-10 Cross section (a) and surface view (b) of the leaf of the desert perennial herb *Enceliopsis argophylla,* showing protective hairs. (*Courtesy of J. R. Ehleringer; from Ehleringer 1984.*)

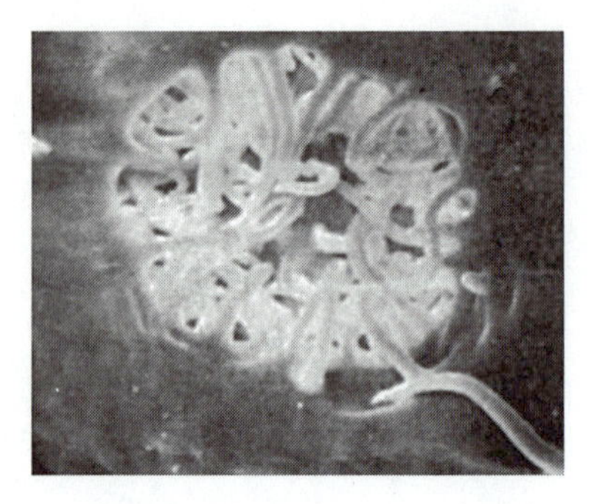

FIGURE 5-11 In oleander, a drought-resistant plant, the stomates lie deep within hair-filled pits on the leaf's undersurface (magnified about 500 times). The hairs reduce water loss by lowering air movement and trapping water. (*Courtesy of M. V. Pathasarathy.*)

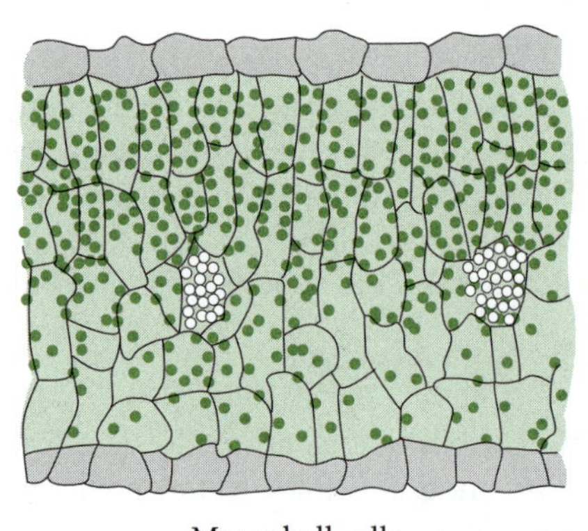

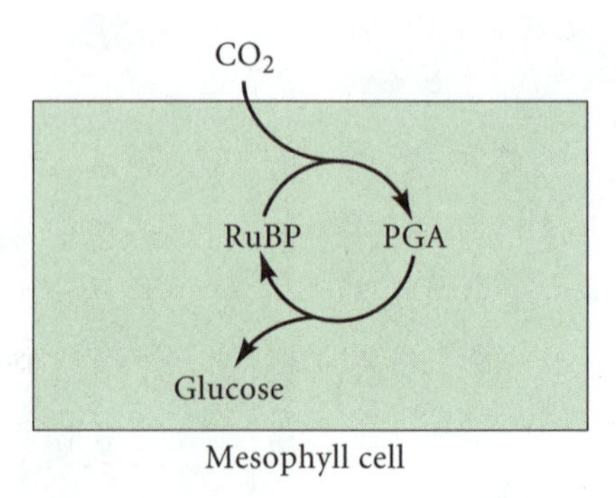

FIGURE 5-12 (a) Cross section of a leaf of a C_3 plant, illustrating the dispersion of chloroplasts throughout the mesophyll. (b) The major steps of the Calvin-Benson cycle of CO_2 assimilation. The assimilation of carbon involves the formation of the three-carbon compound PGA (phosphoglycerate); thus the mechanism is called C_3 photosynthesis. (*Courtesy of J. Ehleringer.*)

product of carbon assimilation is a three-carbon compound, this mechanism is referred to as **C_3 photosynthesis.** Several biochemical steps beyond the production of PGA, one molecule of RuBP is regenerated to complete the Calvin-Benson cycle, while one carbon atom is made available for the synthesis of glucose. Inasmuch as glucose is a six-carbon compound, each molecule of glucose produced requires six turns of the Calvin-Benson cycle.

The enzyme responsible for the assimilation of carbon is ribulose bisphosphate carboxylase/oxygenase (Rubisco), which has a very low affinity for carbon dioxide. As a result, plants assimilate carbon very inefficiently. To achieve high rates of assimilation, cells must be packed with large amounts of Rubisco. An additional problem with this enzyme is that it facilitates the oxidation of RuBP in the presence of high oxygen and low carbon dioxide concentrations, especially at elevated leaf temperatures. Oxidation of RuBP undoes what Rubisco accomplishes when it reduces carbon, thus making photosynthesis inefficient. In fact, carbon assimilation tends to be self-inhibiting unless CO_2 is maintained at a high level in the cell. This can be accomplished only by opening the stomates to allow gas to enter the leaf. And that, of course, leads to water loss.

C_4 and CAM Photosynthesis

Many plants, such as corn and crabgrass, that live in hot climates exhibit a modification of C_3 photosynthesis. This modification involves an additional step in the assimilation of CO_2 and the spatial separation of the initial assimilation step from the Calvin-Benson cycle pathway within the leaf. It is called **C_4 photosynthesis** because the assimilation of CO_2 initially results in a four-carbon compound:

$$CO_2 + PEP \longrightarrow OAA,$$

where PEP (phosphoenol pyruvate) is a three-carbon compound and OAA (oxaloacetic acid) is a four-carbon product (Hatch and Slack 1966, Bjorkman and Berry 1973, Edwards and Walker 1983). This assimilatory reaction is catalyzed by PEP carboxylase, which, in contrast to Rubisco, has a high affinity for CO_2. Assimilation occurs in the mesophyll cells of the leaf, but in most C_4 plants, photosynthesis, including the Calvin-Benson cycle, occurs in specialized cells surrounding the leaf veins, called **bundle sheath cells** (Figure 5-13). OAA, or another four-carbon derivative such as malate, diffuses into the bundle sheath cells, where it is metabolized to produce CO_2 plus pyruvate, a three-carbon compound. The pyruvate moves back into the mesophyll cells, where enzymes convert it to PEP to complete the carbon assimilation cycle. The CO_2 released by the metabolism of OAA in the bundle

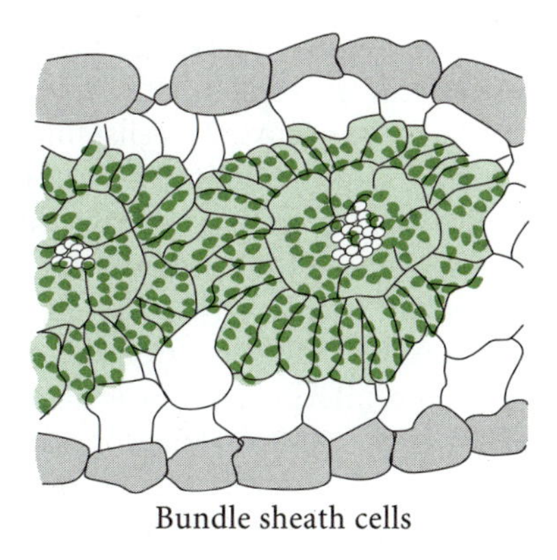

(a)

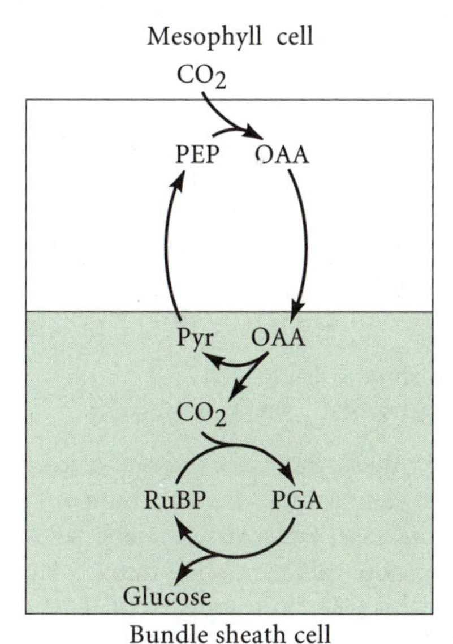

(b)

FIGURE 5-13 (a) Cross section of a leaf of a C_4 plant, illustrating the concentration of chloroplasts found in bundle sheath cells. (b) The initial assimilation step in C_4 photosynthesis involves the transformation of CO_2 into the four-carbon compound OAA (thus the name C_4 photosynthesis). This step moves CO_2 from a mesophyll cell to a bundle sheath cell, where it undergoes transformation in the Calvin-Benson cycle (see Figure 5-12). (*Courtesy of J. Ehleringer.*)

sheath cells enters the Calvin-Benson cycle, as it does in C_3 plants.

C_4 photosynthesis confers an advantage because CO_2 can be concentrated within the bundle sheath cells at a level, determined by the concentration of OAA or malate, that far exceeds its equilibrium level established by diffusion from the atmosphere. At this higher concentration, the Calvin-Benson cycle operates more efficiently. Further, C_4 photosynthesis can bind CO_2 at a lower concentration in the cell, thereby allowing the plant to increase stomatal resistance and reduce water loss.

C_4 plants segregate the initial carbon assimilation step and the Calvin-Benson cycle spatially in different tissues. Certain succulent plants in desert environments utilize the same biochemical pathways as C_4 plants, but segregate assimilation and the Calvin-Benson cycle temporally, between day and night. Because this arrangement was first recognized in plants of the family Crassulaceae (stonecrop family; *Sedum* is one example), and because it entails the storage of certain four-carbon organic acids (malic acid and OAA), it is referred to as **crassulacean acid metabolism,** or **CAM** (Kluge and Ting 1978, Osmond 1978).

CAM plants open their stomates for gas exchange during the cool desert night, at which time water loss is minimal. CAM plants initially assimilate CO_2 in the form of four-carbon OAA and malic acid, which the leaf tissues store in high concentrations (Figure 5-14). During the day, the stomates close, and the stored organic acids are gradually recycled to release CO_2 to the Calvin-Benson cycle. Assimilation of CO_2 and regeneration of PEP are regulated by enzymes having different temperature optima. CAM photosynthesis results in extremely high water use efficiencies and enables some types of plants to exist in habitats too hot and dry for other, more conventional species.

CAM and C_4 Photosynthesis in Aquatic Plants

Submerged aquatic plants face problems entirely different from those confronting terrestrial plants. Water loss is not a problem, but many of these plants

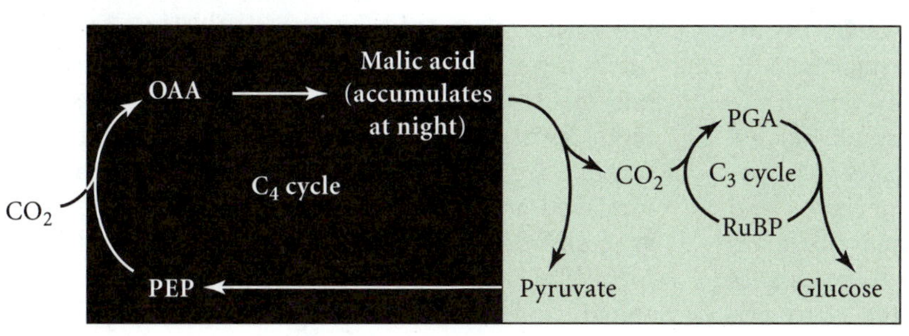

FIGURE 5-14 Diagram of the photosynthetic pathway in CAM plants. CO_2 is taken into the leaf at night when it is cool and water loss is minimal. CO_2 is released to the Calvin-Benson cycle during the day, when stomates are closed. (*After Harbourne 1982.*)

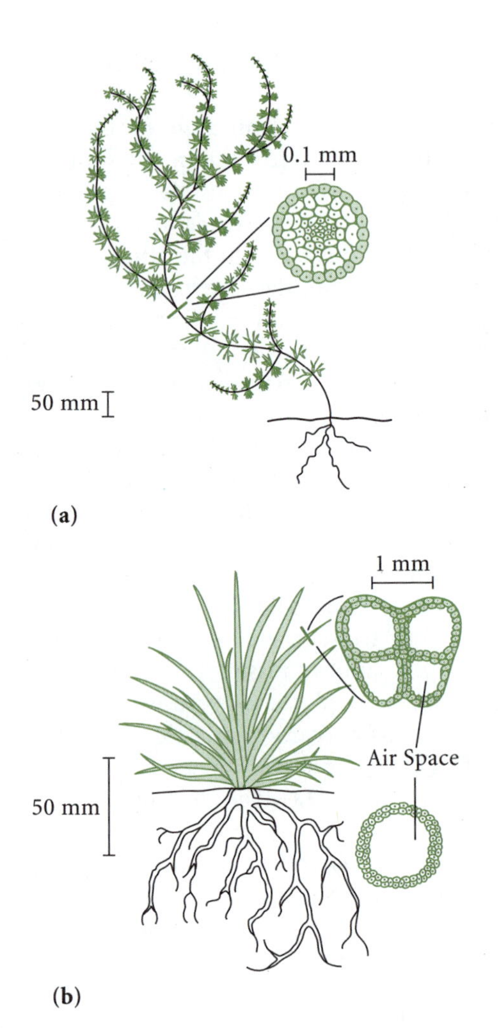

FIGURE 5-15 Structures of two submerged aquatic plants. (a) *Myriophyllum* is found in water with high nutrient concentrations. (b) *Isoetes* is characteristic of lakes and ponds with lower nutrient concentrations. Note the air-filled chambers, which allow internal gas exchange between the leaves and the roots. (*After Keeley and Mooney 1991.*)

nonetheless exhibit CAM and C_4 photosynthesis. *Myriophyllum* inhabits nutrient-rich ponds and lakes (Figure 5-15a). Photosynthetic rates are high, and the plant grows rapidly. One consequence of these adaptations is that the oxygen produced by photosynthesis builds to high levels within the leaf cells, which might limit further photosynthesis. Plants such as *Myriophyllum* circumvent this problem by using the C_4 mechanism, which elevates the level of CO_2 within the leaf cells and helps to maintain high rates of photosynthesis.

Isoetes is a primitive aquatic plant distantly related to the ferns. Unlike *Myriophyllum,* it grows in nutrient-poor ponds and lakes. It obtains nutrients and CO_2 primarily from sediments. Accordingly, its leaves have low surface-to-volume ratios and thick cuticles to prevent the loss of ions and CO_2 to the surrounding water (Figure 5-15b). Its root system is highly developed, and the entire interior of the plant, from the roots to the leaves, is filled with large air spaces. Most photosynthesis occurs in cells lining the air spaces in the leaves. In the sediments, carbon dioxide is produced by the respiration of bacteria and animals, which also deplete the oxygen in the surrounding water. The plant takes up CO_2 from the sediments through the roots. CO_2 diffuses from the roots to the leaves through the interior air spaces, and is stored during the night by CAM assimilation. Oxygen produced by photosynthesis during the day diffuses to the roots, where it is used by the respiring root cells.

Stylites, a relative of *Isoetes,* grows in peat deposits in seasonal bogs at over 4,000 m elevation in the Andes of South America. The leaves of *Stylites* have a thick cuticle that is essentially impermeable to CO_2 and water vapor; this cuticle prevents water loss to the thin air during dry weather. Carbon dioxide is obtained from the decomposing peat through the roots, just as it is in aquatic relatives. In both cases, the leaves can have impermeable outer surfaces because they exchange gases internally with the roots and then with the root environment.

The Evolutionary Trade-Off Between C_3 and C_4 Photosynthesis

C_4 photosynthesis increases the rate of carbon assimilation in hot environments with abundant solar radiation, and increases water use efficiency in dry habitats. Why, then, don't all plants use the C_4 pathway? The answer to this question appears to have three parts. First, C_4 plants must expend energy to regenerate PEP from pyruvate, so C_4 photosynthesis is energetically less efficient than C_3 photosynthesis. Second, at temperatures below 25°C, Rubisco operates relatively efficiently because it has less tendency to behave as an oxidase. Hence the advantage of the C_4 pathway diminishes in cool environments, particularly when the availability of light, rather than that of CO_2, limits photosynthesis (Figure 5-16a). Maximum photosynthesis is achieved by C_4 plants at about 45°C, close to the maximum tolerable temperature for plants, and by C_3 plants at between 20° and 30°C. As a result, C_4 plants predominate in hot climates and C_3 plants in cool climates (Teeri and Stowe 1976, Pearcy and Ehleringer 1984, Ehleringer et al. 1991). Where they occur together, C_4 plants often remain green during hot, dry summers—conditions that cause C_3 species to go dormant. Third, the ability of C_4 plants to concentrate CO_2 in the bundle sheath cells is greatly diminished at high levels of CO_2 (Figure 5-16b). It is generally thought that photosynthesis evolved when high levels of CO_2 were present in the early atmosphere, and thus, even though hot and arid climates were common for much of the history of the earth,

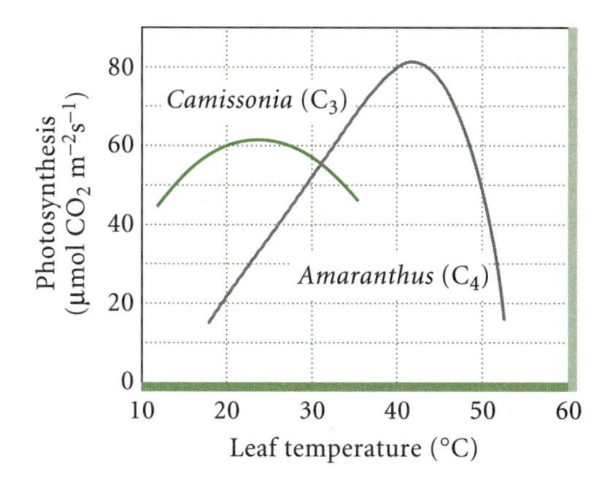

(a)

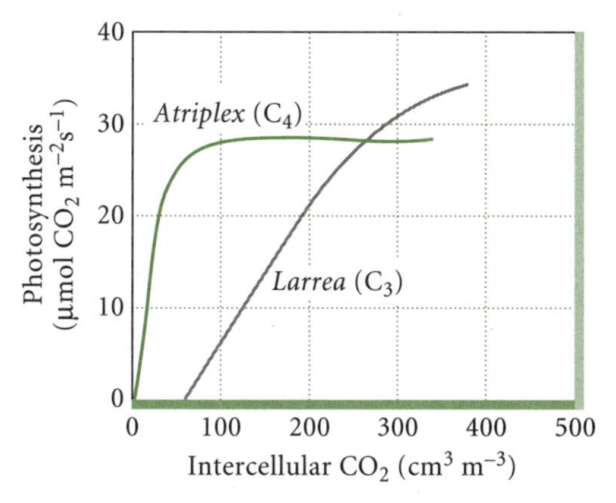

(b)

FIGURE 5-16 (a) Relationship of photosynthesis (measured in micromoles of CO_2 assimilated per square meter of leaf area per second) to leaf temperature in the winter-active C_3 desert herb *Camissonia claviformis* and the summer-active C_4 desert herb *Amaranthus palmeri.* The C_3 plant is more efficient at temperatures below 30°C. (b) Relationship of photosynthesis to intercellular CO_2 concentration in two desert shrubs, *Atriplex hymenelytra* (C_4) and *Larrea divaricata* (C_3), under high light intensities and leaf temperatures of 30°C. Under these conditions, the C_4 plant utilizes CO_2 more efficiently at low concentrations. (*After Pearcy and Ehleringer 1984.*)

there existed no evolutionary advantage for C_4 photosynthesis until atmospheric levels of CO_2 dropped to their current levels in the Paleozoic era (Ehleringer et al. 1991).

Photosynthetic Mechanisms and Global Climate Change

Much of the sunlight that passes through the atmosphere is absorbed by the earth's land surface, vegetation, and waters. This energy may be reradiated as infrared radiation, low-energy radiation having wavelengths of between 700 and 1,000 nm. Carbon dioxide and other so-called **greenhouse gases** in the atmosphere (along with water vapor and clouds) absorb this infrared radiation, trapping heat near the earth's surface. Thus, the earth's atmosphere works very much like a greenhouse, in which the glass roof admits sunlight but retains long-wave radiation coming from the inside of the building, thus warming the interior. We call this phenomenon the **greenhouse effect.** The natural greenhouse maintains the earth's temperature within the limits necessary for physiological function, making life as we know it possible.

Human activities—particularly the burning of fossil fuels in automobile engines—have vastly increased the concentration of CO_2 and other greenhouse gases in the atmosphere (see Chapter 11), thereby increasing the efficiency of the earth's greenhouse and promoting an enhanced warming of the earth's atmosphere, a process that is now well documented (Vitousek 1994, Robert and MacArthur 1994). The summer before the publication of this book was the warmest summer on record.

Our input of CO_2 into the earth's atmosphere affects the physical environment of plants in at least two important ways. First, of course, the amount of CO_2 available to plants for photosynthesis is increased. Second, because CO_2 enhances the greenhouse effect, the temperature of the plants' environment may rise. Over the long term, changes in atmospheric CO_2 concentrations may affect the distributions of C_3 and C_4 plants differently because of their different affinities for the gas. Elevated CO_2 levels would be likely to increase the efficiency of C_3 plants, whereas C_4 plants would have an advantage under low CO_2 levels. Paleoecological studies (studies of ancient ecology using fossils and other means of studying the past, such as carbon dating) show that C_4 plants expanded on the early earth during times of low CO_2 availability in the atmosphere, such as during the end of the Miocene (approximately 25 million years before present), when there was a great expansion of C_4 grasslands (Ehleringer et al. 1991).

Photosynthesis increases with temperature in both C_3 and C_4 plants, but as we have seen (see Figure 5-16a), this increase is not without bound. Each plant species has an optimum photosynthetic temperature, and this often corresponds to the temperature of greatest plant growth. Thus, changes in the temperature of the environment may make growing conditions unfavorable for the plants in that environment. The distributions of both C_3 and C_4 plants are determined in part by temperature conditions.

Because both the distributions of plants and their rates of growth are related to the amount of CO_2 in the atmosphere and to the temperature, which is related to the CO_2 level, global climate change may have important effects on world agriculture (Adams et al. 1990, Parry 1990, Parry et. al. 1990, Rosenzweig and Parry 1994). Moreover, the effects of these changes may be different for C_3 and C_4 plants. Many crop species, such as wheat, rice, beans, and many vegetables, are C_3 plants. Such plants may be expected to thrive in high-CO_2 environments if temperatures do not rise appreciably. The most important plants to the global economy are maize, millet, sugarcane, and sorghum, all of which are C_4 plants. These plants are not as responsive to high levels of CO_2. As temperatures increase with rising CO_2, some crop plants will no longer be profitably grown in certain regions. Under the most commonly used models of global climate change, global temperature increases will have negative effects on both C_3 and C_4 plants unless the higher levels of CO_2 in the atmosphere increase plant growth. But even if higher CO_2 levels results in greater photosynthetic efficiency, C_3 plants will fare better than C_4 plants in a high-CO_2, high-temperature environment (Rosenzweig and Parry 1993).

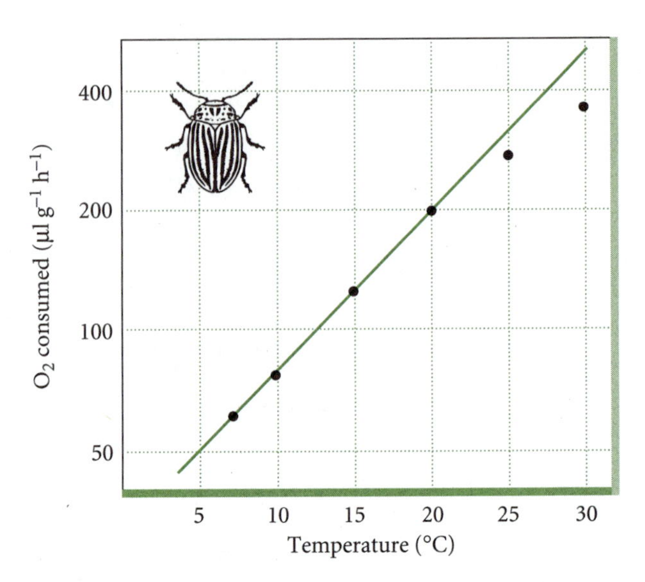

FIGURE 5-17 The rate of oxygen consumption (log scale) of the Colorado potato beetle as a function of temperature. The rise in oxygen consumption is exponential, increasing by a factor of 2.5 for each 10°C. This factor of increase is called the Q_{10} of the process. (*After Marzusch 1952.*)

5.5 Life processes occur within a narrow range of temperatures.

Life processes as we know them are restricted to the temperatures at which water is liquid: 0°–100° C at the earth's surface. Temperature has several opposing effects on life processes. Heat increases the kinetic energy of molecules and thereby accelerates chemical reactions. The rates of biological processes commonly increase between two and four times for each 10°C rise in temperature throughout the physiological range (Schmidt-Nielsen 1983, Hochachka and Somero 1973, 1984). This factor of increase is called the $\mathbf{Q_{10}}$ of a process, and it is estimated by the relationship between the rate of a physiological process, plotted on a logarithmic scale, and temperature (Figure 5-17). Also, enzymes and other proteins become less stable and may not function properly or retain their structures at high temperatures. In addition, the heat energy in a cell influences the conformation of proteins, which are balanced between the natural kinetic motions induced by heat and the forces of chemical attraction between different parts of the molecule. For example, the physical properties of the fat molecules in cell membranes and those that many animals accumulate as a reserve of food energy depend on temperature. When cold, fats become stiff (picture the fat on a piece of meat taken from the refrigerator); when warm, they become fluid. For most organisms, the range of suitable temperatures is somewhere between 0°C and 45°C.

Life at Extremely High Temperatures

Relatively few organisms can survive body temperatures above 45°C. Photosynthetic cyanobacteria tolerate temperatures as hot as 75°C (Brock 1970, 1985, Brock and Darland 1970). Some bacteria, referred to as **thermophilic bacteria,** occur in hot springs close to the boiling point of water (Figure 5-18). Thermophilic bacteria are also found near hydrothermal vents located in the deep oceans in places where the earth's crustal plates join one another. These vents

FIGURE 5-18 A hot spring in Yellowstone National Park. Even though the temperature of the water approaches the boiling point, some thermophilic bacteria thrive in this environment. (*Photo by R. Ricklefs.*)

often lie more than 3,000 meters below the surface, where the water pressure may exceed 40 MPa. Because of the intense pressure at these sites, the boiling point of water is much higher than at sea level, and thus water stays liquid at a much higher temperature. Recent studies suggest that a remarkable variety of microorganisms inhabit these extremely hot deep-sea habitats (Straube et al. 1990, Deming and Baross 1993), and that eukaryotic organisms are common in the very warm sediments surrounding the vents (Juniper et al. 1992). The properties that permit existence at high temperatures are not well understood. Compared with most bacteria, the proteins of thermophilic bacteria have subtly different proportions of amino acids, which increase the intramolecular forces that hold the proteins together and allows the structure of these proteins to remain stable at temperatures up to 97°C (Singleton and Amelunxen 1973, Hochachka and Somero 1984).

Some plants, most notably desert plants of the agave and cactus families, are adapted to withstand temperatures in excess of 50°C (Nobel 1988). In arid and semiarid environments, plant surfaces—in particular, leaves, which are exposed to extremely high incident solar radiation—rapidly heat up to well above the environmental temperature. This heating increases the rate of evaporation of water from the plant to the point at which it may become heat-stressed. Plants exposed to high temperatures often have adaptations that reduce their exposure to radiation. Rolling leaves, as in many grasses, the development of waxy surfaces, the presence of leaf hairs that reflect sunlight, or an increase in leaf thickness with decreased surface area, are common adaptations.

Many plants are especially sensitive to high temperatures during early development or other high-growth periods. This vulnerability has enormous economic implications in agricultural ecosystems, where extremely hot (or cold) temperatures at critical times during the growing season may damage crops and reduce yields. Crops such as soybeans, cotton, tomatoes, rice, sorghum, and cowpeas have been shown to be particularly sensitive to high temperatures during early development. In most varieties of these species, seed or pod development is reduced with exposure to high temperatures.

Life at Extremely Low Temperatures

Although temperatures on the earth rarely exceed 50°C, except in hot springs and at the soil surface in deserts, temperatures below the freezing point of water are common over large portions of the earth's surface. When living cells freeze, the crystal structure of ice disrupts most life processes and may damage delicate cell structures, leading rapidly to death. Many kinds of organisms successfully cope with freezing temperatures by activating mechanisms that enable them either to resist freezing or to tolerate its effects (Somme 1964, Baust 1973).

The freezing point of water may be depressed by dissolved substances that interfere with the formation of ice. For example, the freezing point of seawater, which contains about 3.5% dissolved salts, is −1.9°C. The blood and body tissues of most vertebrates contain less than half the salt content of seawater and thus may freeze at a higher temperature than the freezing point of the ocean. Saltier blood could enable vertebrates to live in polar seas, but protein structure and function are too sensitive to salt concentration to make this a practical solution. Many organisms reduce the freezing point of their body fluids with large quantities (up to 30% in some terrestrial invertebrates) of glycerol and glycoproteins, which act like antifreeze; their presence in the blood and tissues allows antarctic fish, for example, to remain active in seawater that is colder than the normal freezing point of the blood of fish in temperate or tropical seas (DeVries 1980, 1982).

Supercooling provides a second solution to the problem of freezing. Under certain circumstances, fluids can fall below the freezing point without ice crystals forming. Ice generally forms around an object, called a seed, which can be a small ice crystal or some other particle. If seeds are absent, pure water can be cooled to more than 20°C below its melting point without freezing. Thus, if an organism can exclude seeds from its body fluids, it can survive very cold temperatures. Supercooling has been recorded to −10°C in reptiles (Packard et al. 1997) and −18°C in invertebrates, particularly insects and spiders (e.g., Baust and Morrissey 1975, Catley 1992, Bayram and Luff 1993, Storey et al. 1993, Danks et al 1994, Lee et al. 1994, Van der Merwe et al. 1997). In insects, the supercooling point may be raised by the presence of bacteria or fungi that promote the formation of ice crystals (Lee et al. 1993, Lee et al. 1994). When such bacteria or fungi are ingested by the insect or applied topically, significant overwinter mortality may occur. It has been suggested that some species of pest insects could be controlled by applying ice-promoting bacteria and fungi to overwintering stages (Lee et al. 1993). Supercooling is notably absent in some groups of animals, such as amphibians, because the water-permeable skin of these animals cannot prevent the entry of small ice crystals or other particles that might act as seeds (Pinder et al. 1992). The woody tissues of plants may be supercooled to −15°C, and some tissues, such as the buds of conifers, may be supercooled to nearly −40°C (Burke et al. 1976, George et al. 1982, Hopkins 1995). Supercooling in plants is accomplished in much the same way as it is in animals:

tissue fluids lack structures around which ice crystals may form.

Finally, some plants and animals can tolerate the freezing of most or all of the water in their bodies. Such organisms employ mechanisms to restrict ice formation to the spaces between cells rather than within them. Because salts are excluded from ice and are therefore concentrated in the liquid water within cells, freezing-tolerant organisms must cope with extremely high salt levels in their tissues during winter.

Among vertebrate animals, only a limited number of frogs (four species), a salamander, and early developmental stages of the midland painted turtle (*Chrysemys picta marginata*) have been found to possess mechanisms for withstanding the freezing of body fluids (Pinder et al. 1992, Packard et al. 1997). In these species, extracellular fluids are permitted to freeze at temperatures below the freezing point of water (around $-10°C$). The freezing of the interstitial fluids removes water from solution, thereby increasing the salt concentration of the remaining fluids, causing them to be hypertonic to the interior of the cell and resulting in a net movement of water out of the cell and a net movement of solutes into the cell. As this process continues, the solute concentration in the dehydrating cell increases until its freezing point is extremely low (this is akin to lowering the freezing point of water by adding salt to it). Restricting freezing to the extracellular spaces prevents delicate cell organelles from being damaged by ice crystals as well as lowers the freezing point of the fluids inside the cell. However, the cell must still function, at least at some minimal level, in the face of its drastically elevated solute concentration if the animal is to survive freezing.

Plants that live at extremely high latitudes or altitudes are exposed to temperatures low enough to freeze some of their tissues. Some plants, such as the larch (*Laryx dahurica*), which must withstand Siberian winters in which temperatures may drop to nearly $-70°C$, tolerate the freezing of their tissues. Often in such species, tissues such as seeds, which are largely dehydrated, do not freeze. As in amphibians and other freezing-tolerant animals, ice formation in these plants is restricted to areas outside the cells, thereby preventing cellular damage by ice crystal formation (Hopkins 1995).

5.6 Radiation, conduction, and convection define the thermal environments of terrestrial organisms.

Much of the solar radiation absorbed by objects is converted to heat. The earth is warmed during the day and cools at night. As the days lengthen and the sun rises higher in the sky toward summer, the surroundings become warmer as more heat is added each day than is lost. The warmth of the sun heats the atmosphere, causing it to expand, thus creating wind. Light absorbed by water provides the major source of heat for evaporation.

Heat is the total kinetic energy of the molecules in a system. The temperature of a system is the average kinetic energy of its molecules. As we discussed in Chapter 3, each object and each organism continually exchanges heat with its environment. When the temperature of the environment exceeds that of the organism, the organism gains heat and becomes warmer. When the environment is cooler, the organism loses heat and cools. The heat budget of an organism includes avenues of heat gain and avenues of heat loss (Figure 5-19). When temperature reaches an equilibrium, heat gains equal losses. When gains exceed losses, energy is stored or accumulated in the body, and the organism's temperature rises. When losses exceed gains, its temperature drops.

Radiation is the absorption or emission of electromagnetic energy. Sources of radiation in the environment are the sun, the sky (scattered light), and the landscape, including vegetation. At night, although we cannot see the infrared radiation, objects that have warmed up in the sunlight reradiate their stored heat to colder parts of the environment and, eventually, to space. The bodies of organisms, especially warm-blooded birds and mammals, often are the brightest objects in the night (Figure 5-20). Because radiation increases with the fourth power of thermodynamic temperature (K), we radiate tremendous quantities of energy to the clear, black night sky. We can also receive radiation from atmospheric water vapor and from vegetation, which balances much of our radiation loss.

Conduction is the transfer of kinetic energy between substances in contact or from one part of a substance to another part. The thermal conductance (k) of a substance is expressed in watts (joules per second), normalized by the cross-sectional area (cm^2), the inverse of the distance traversed (cm^{-1}), and the temperature gradient (°C). Thus, the units of conductance are $W\ cm^{-2}\ cm\ °C^{-1}$, or $W\ cm^{-1}\ °C^{-1}$. Substances differ greatly in their ability to conduct heat. No heat is conducted to a vacuum ($k = 0$). Water ($k = 0.006$) conducts heat better than does air ($k = 0.00026$), owing to its greater density. Some metals, such as silver ($k = 4.3$) and copper ($k = 4.0$), conduct heat very rapidly. The rate of conductance between two objects, or between the inside and outside of an organism, depends on the insulative properties of the surface (the resistance to heat transfer, a function of k, and surface thickness), surface area, and temperature gradient. An organism may either gain or

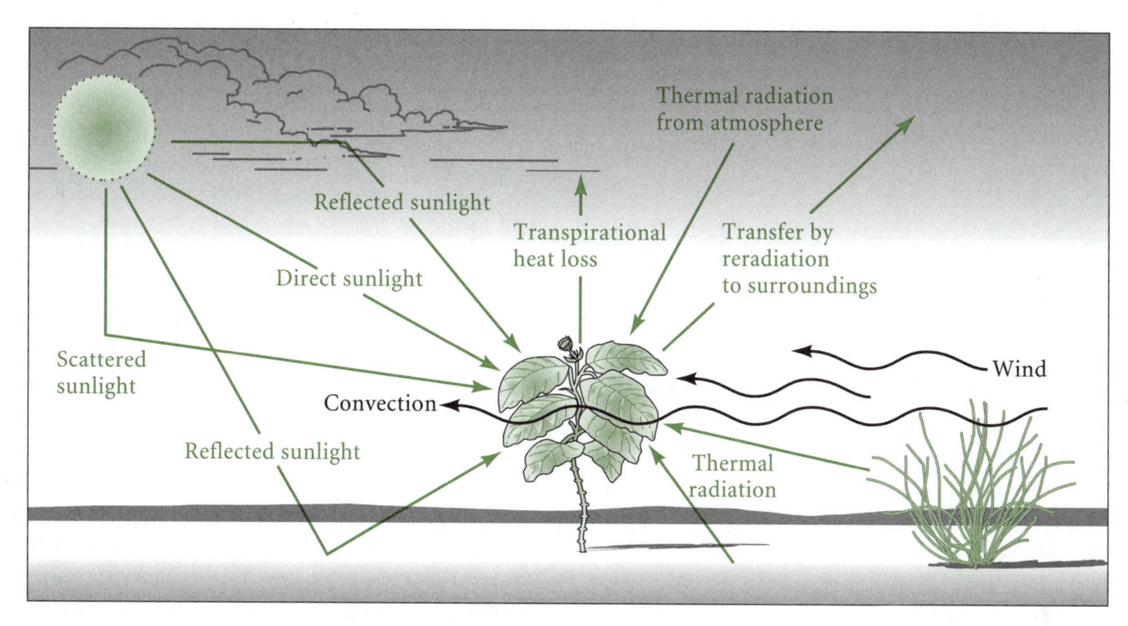

FIGURE 5-19 Pathways of heat exchange between a plant and its environment by radiation, convection, and transpiration. (*After Gates 1980.*)

lose heat depending on its temperature relative to that of the environment.

Convection is the movement of liquids and gases of different temperatures, particularly over surfaces across which heat is transferred by conduction. Often liquids and gases move in swirls or eddies. Air conducts heat poorly. In still air, a boundary layer of air forms over a surface. A warm body tends to warm this boundary layer to its own temperature, effectively insulating itself against heat loss. A current of air flowing past a surface tends to disrupt the boundary layer and increase the rate of heat exchange by conduction. This convection of heat away from the body surface is the basis of the "wind-chill factor" heard on the evening weather report. On a cold day, air movement makes it feel the way it would on an even colder windless day. For example, a wind blowing 32 km per hour at an air temperature of −7°C has the cooling power of still air at −23°C.

FIGURE 5-20 Thermal images of Canada geese in an open meadow on a cool morning. It is clear that the geese lose more heat across their necks and legs than from their well-insulated bodies. (*Courtesy of R. Boonstra; from Boonstra et al. 1995.*)

Evaporation of water requires heat. The evaporation of 1 gram of water from the body surface removes 2.43 kj of heat at 30°C. As plants and animals exchange gases with the environment, some water evaporates from their respiratory surfaces. In plants, the evaporation of water from the surface of a leaf is referred to as **transpiration.** The rate of evaporative or transpirational heat loss depends on the amount of water exposed on the surface of the organism, the relative temperatures of the surface and the air, and the **vapor pressure** of the atmosphere. Vapor pressure is a measure of the capacity of the atmosphere to hold water. When vapor pressure is expressed in atmospheres, it represents the fractional weight of water vapor in saturated air. Thus, at 30°C, the vapor pressure of the atmosphere is 0.042 atm, meaning that the air can hold 4.2% water by weight. When the temperature of air saturated with water drops from 30° to 20°C, its capacity to hold water decreases from 4.2% to 2.3%, and the difference—almost 2%—condenses to form clouds or precipitation.

Like heat, moisture can be trapped in the boundary layer of air that forms around bodies. Convection tends to disrupt boundary layers and therefore increases evaporative as well as conductive heat loss. Because warm air holds more water than cold air (51 g m^{-3} at 40°C, 17.3 g m^{-3} at 20°C, and 4.8 g m^{-3} at 0°C), it has greater potential for evaporating water than does cold air. When water is plentiful in hot climates, animals evaporate water from their skin and respiratory surfaces to cool themselves. For warm-blooded animals in cold climates, evaporation can become an unavoidable problem as cold air containing little water is warmed in contact with the body surface. We see the evidence of such water loss on winter days when water evaporated from the warm surfaces of our lungs condenses as our breath mingles with the cold atmosphere.

5.7 Adaptations match the temperature optima of organisms to the temperature of the environment.

Unlike consumable resources, such as carbon dioxide, oxygen, and soil nutrients, conditions such as temperature and salt concentration influence organisms through their effects on the rates at which physical and biochemical processes proceed, and through their effects on the structures of such biologically important molecules as proteins and lipids (Hochachka and Somero 1973, 1984). As a result of these interactions, each organism generally has a narrow range of conditions to which it is best suited, which define its **optimum.** The optimum is subject to natural selection, which acts on variations in the properties of enzymes and lipids, in the structures of cells and tissues, and in the form of the body to enable the organism to function well under the particular conditions of its environment. The breadth of conditions that organisms can tolerate is also quite variable. Organisms with wide tolerance ranges (physiological generalists, so to speak) are referred to as **eurytypic;** those with narrow tolerance ranges (physiological specialists) are **stenotypic.** Shifts in optimum environmental conditions may be seen by comparing the effects of salt concentration on the activities of various enzymes found in bacteria adapted to low salt concentrations and in **halophilic** species that live in concentrated brine (Figure 5-21). Note that although the optimum salt concentration for halophylic *Halobacterium salinarium* is very high, not all of its individual enzymes have high salt optima. They are merely higher than those of species not adapted to high salt concentrations.

As a rule, between the freezing point of water (0°C) and the upper temperature limit for most life forms (40°–50°C), higher temperatures quicken the pace of life by increasing the kinetic energy of the organism and its surroundings. Temperature affects such physical processes as diffusion and evaporation

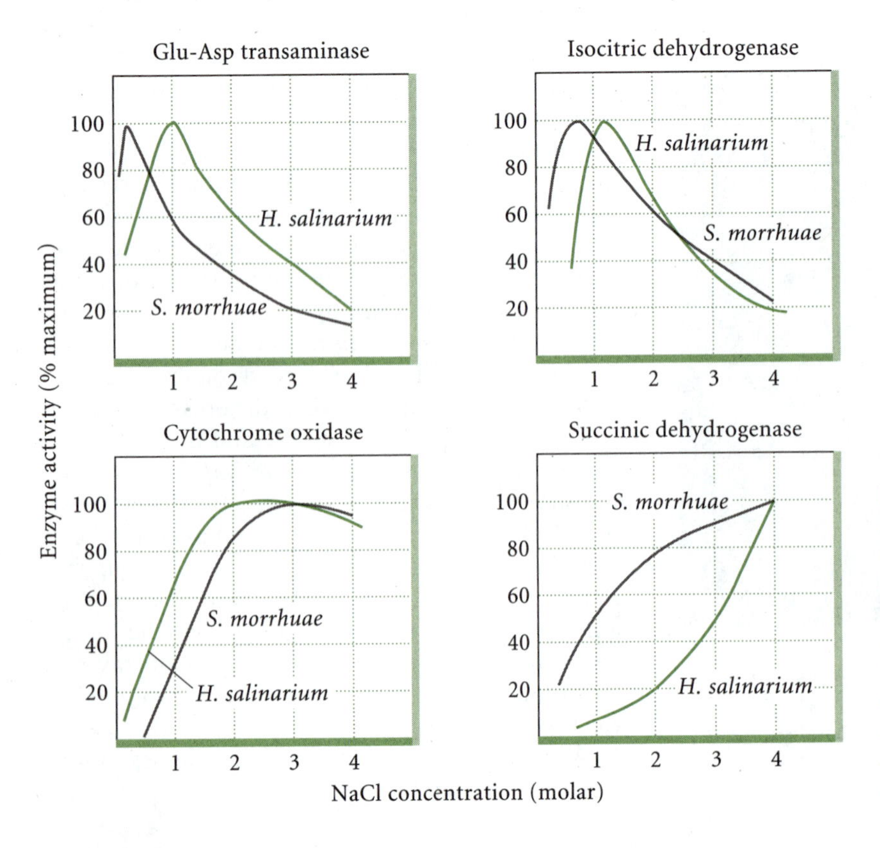

FIGURE 5-21 Activity levels of selected enzymes found in the halophilic bacterium *Halobacterium salinarium* (green curves) at various salt concentrations, compared with those of the same enzymes in *S. morrhuae* (gray curves), a bacterium that cannot tolerate high salt concentrations. The concentration of salt in seawater is about 0.6 molar. (*After Larsen 1962.*)

as well as biochemical reactions. For example, increased temperature speeds the diffusion of gases through the shell of a bird's egg or of mineral ions through the groundwater. Temperature also influences the shapes of enzymes and other proteins because heat energy imparted to such molecules can break the weak forces holding the structure together. Each enzyme functions best over a narrow range of temperatures determined by the amino acid sequences of its protein chain (the primary structure of the protein) and the way in which they affect the bonds that hold the protein in precise configurations. These sequences are, of course, subject to evolutionary modifications.

Temperature adaptation is strikingly revealed by comparing organisms taken from environments with different characteristic temperatures. When we examine the performances of such organisms over a range of temperatures, each typically exhibits its maximum activity at a temperature corresponding to that of its normal environment. Many fish in the freezing oceans surrounding Antarctica swim just as actively as fish living among tropical coral reefs. A graph of the relationship between oxygen consumption and water temperature shows that the metabolic rates of cold-water fish are adjusted so that their activity level in cold water is on a par with that of warm-water fish at the higher temperatures characteristic of their usual environments (Figure 5-22). Put a tropical fish in cold water, however, and it becomes sluggish and soon dies; conversely, antarctic fish cannot tolerate temperatures warmer than 5°–10°C.

How can fish from cold environments be as active metabolically as fish from the Tropics? Metabolism consists of a series of biochemical transformations, most of which are catalyzed by enzymes. Because a given transformation occurs more rapidly at high temperatures than at low temperatures, the compensation observed in cold-adapted organisms must involve either a quantitative increase in the amount of substrate or in the amount of enzyme that catalyzes each step, or a qualitative change in the enzyme itself (Somero 1978). Many enzymes occur in slightly different forms whose characteristics reflect the substitution of amino acids in the protein chain. Those forms are termed **isoenzymes,** or **isozymes.** Their differences usually result from small changes in the gene that encodes the structure of the protein. Not only do the structures of enzymes sometimes differ among species and among populations of the same species, but more than one form of an isozyme may occur within the same population or even the same individual (different variants of the enzyme may be inherited from the father and from the mother).

Laboratory studies of the function of isolated enzymes have shown, in many cases, that the different isozymes have different catalytic properties when tested over ranges of temperature, pH, salt concentration, and availability of substrate. One measure of catalytic ability is the facility with which the enzyme binds with its substrate. This is usually expressed as the **Michaelis-Menten constant** (K_m), which is the concentration of the substrate at which the velocity of the reaction is one-half of its maximum. Optimally adapted enzymes are thought to have K_ms in the range of substrate concentrations normally encountered within the tissues (Graves et al. 1983). The rate of the catalyzed reaction is thus very sensitive to changes in substrate concentration (Figure 5-23). K_m values for lactate dehydrogenase (LDH) isozymes in three species of barracudas (*Sphyraena*) illustrate compensation: within the range of temperatures normally

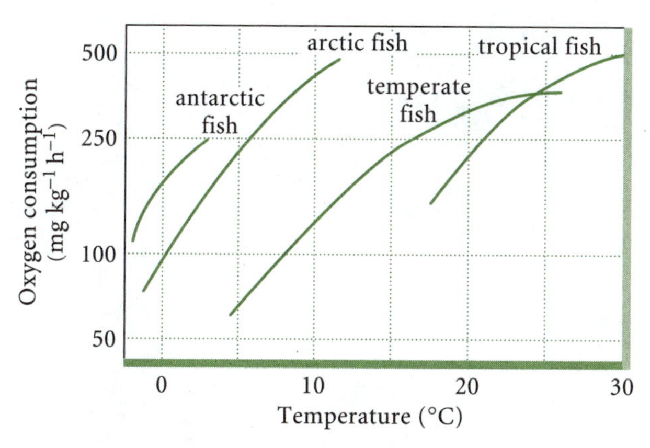

FIGURE 5-22 Temperature compensation in the metabolic rates (represented by rates of oxygen consumption) of fish from different thermal environments. Although metabolism in each increases with temperature, fish from different thermal environments have similar metabolic rates at the prevailing temperature of their native habitats. (*After Hochachka and Somero 1984.*)

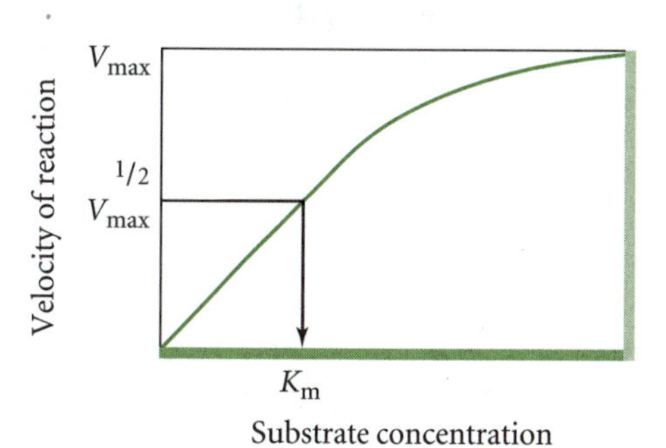

FIGURE 5-23 The velocity of an enzymatic reaction as a function of substrate concentration. The reaction rate levels off when the enzyme is saturated (V_{max}). The Michaelis-Menten constant (K_m) is the concentration of the substrate at which the reaction proceeds at half its maximum rate.

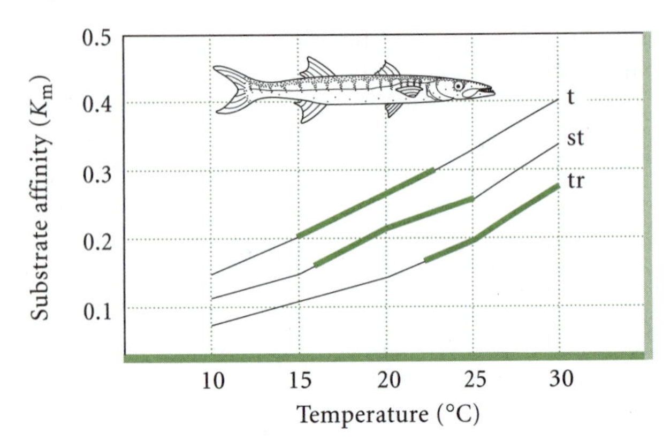

FIGURE 5-24 Relationship between K_m and temperature for the LDH enzymes of three species of barracudas (*Sphyraena*) from temperate (t), subtropical (st), and tropical (tr) waters. Because of temperature compensation, the K_ms of the three species are nearly identical within the normal temperature ranges of their environments. (*After Hochachka and Somero 1984.*)

encountered by each of the species, the K_m values of their particular LDHs are similar (Figure 5-24).

This picture of metabolic compensation is greatly simplified, of course. Adapting to changes in the environment requires a complete retuning of metabolic pathways, which may involve changes in enzyme structure or concentration or even the use of alternative metabolic pathways. The evolution of metabolic systems is not fully understood; it is certainly beyond the scope of this discussion. Clearly, however, plants and animals employ a wide range of structural and functional adaptations that optimize the fit between themselves and their environments (Hochachka and Somero 1984, Prosser 1986).

5.8 For terrestrial organisms, conservation of water becomes more difficult with increasing temperatures.

Avoiding heat stress is critical to an individual's survival. Mammals and birds generally maintain their body temperatures only 6°C, or thereabouts, below the upper lethal maximum temperature—the temperature at which it is too hot for physiological processes to proceed. In cool environments, plants and animals dissipate excess heat, generated by activity or absorbed from the sun, by conduction and radiation to their surroundings. But when air temperature approaches or exceeds body temperature, individuals can dissipate heat only by evaporating water from their skin and respiratory surfaces. In deserts, the scarcity of water makes evaporative heat loss a costly mechanism. In hot, humid climates, the water vapor pressure of the air slows evaporation from body surfaces and reduces the effectiveness of evaporative heat loss. In either case, to balance heat budgets in hot environments, organisms are often forced to reduce heat-generating activity, use cool microclimates, or undertake seasonal migrations to cooler regions (Figure 5-25). Many desert plants orient their leaves to avoid the direct rays of the sun (Ehleringer and Forseth 1980, Schulze et al. 1987), or they may shed their leaves and become inactive during periods of combined heat and water stress.

Even in the absence of heat stress, hot, dry environments rob the body of water simply because the vapor pressure of water at body surfaces increases with temperature, roughly doubling with each 10°C rise. Where water is scarce, one often sees adaptations to reduce water loss. Among mammals, the kangaroo rat is well adapted to life in a nearly waterless environment owing primarily to a number of highly developed water-conserving mechanisms (Schmidt-Nielsen and Schmidt-Nielsen 1952, 1953). For example, the large intestine of the kangaroo rat's digestive tract resorbs water from waste material so efficiently as to produce virtually dry feces.

In most animals, the respiratory passages are a major avenue of water loss. Kangaroo rats recover much of the water that evaporates from the lungs by condensation in enlarged nasal passages (Schmidt-Nielsen et al. 1970). When the kangaroo rat inhales dry air, the moisture in its nasal passages evaporates, cooling the nose and saturating the inhaled air with water. When moist air is exhaled from the lungs, much of its water condenses on the cool nasal surfaces. By alternating condensation with evaporation

FIGURE 5-25 A jackrabbit seeking refuge from the hot sun of southern Arizona in the shade of a mesquite tree. The large ears and long legs of desert-inhabiting jackrabbits effectively radiate heat when the environment is cooler than the body. (*Courtesy of the U.S. Fish and Wildlife Service.*)

during breathing, the kangaroo rat minimizes its respiratory water loss.

In extremely hot environments where the **ambient** (surrounding) temperature exceeds that of the body, organisms can cool themselves only by evaporation. So unless water is freely available, individuals must reduce their activity or seek cool microenvironments to balance both their heat and water budgets. The kangaroo rat does this in part by restricting its feeding activity to the cooler nighttime periods, and spending the daylight hours below ground in relatively cool, humid burrows. In sharp contrast, ground squirrels remain active during the day. They conserve water by restricting evaporative cooling. As a consequence, their body temperatures rise when they are above ground and exposed to the hot sun. But before their body temperatures become dangerously high, they return to their cool burrows, where they dissipate their heat load by conduction and radiation, rather than by evaporation. In this manner, ground squirrels extend their activity into the heat of the day and pay a relatively small price in water loss (Hudson 1962).

SUMMARY

1. Energy transformations in living systems are based on the chemistry of carbon and oxygen. In photosynthesis, carbon is reduced and organic molecules are produced, thereby transforming light energy into the potential energy of chemical bonds. In respiration, organic molecules are oxidized, releasing energy in the form of ATP and heat.

2. The CO_2 gradient between the atmosphere and the bodies of terrestrial plants is smaller than the gradient of water vapor pressure. Thus, water conservation is a problem for plants in arid environments. Aquatic plants are not as limited by CO_2 availability as are terrestrial plants because CO_2 has a high solubility in water. Some dissolved CO_2 forms carbonic acid (H_2CO_3), which dissociates into bicarbonate (HCO_3^-) and carbonate (CO_3^{2-}) ions in many natural waters. The carbon in these ions can be used directly for photosynthesis.

3. Wavelengths of light in the visible spectrum, between 400 nm (violet) and 700 nm (red), are suitable for photosynthesis. Plant leaves contain pigments such as chlorophylls (green) and carotenoids (yellow) that absorb this light and harness its energy.

4. The absorption and scattering of light by water limits the depth at which photosynthesis can occur in aquatic environments. Water absorbs longer wavelengths more strongly than shorter ones; thus green light tends to predominate with increasing depth. Deep-water algae have pigments that absorb green light. Algae near the surface, where short wavelengths (violet and blue) predominate, have pigments that absorb those wavelengths. Photosynthesis occurs only to the depth of light penetration, in an area usually close to the surface, called the euphotic zone. The lower limit of the euphotic zone, where photosynthesis just balances respiration, is called the compensation point.

5. During photosynthesis, plants assimilate carbon through the C_3 pathway (Calvin-Benson cycle), catalyzed by the enzyme Rubisco. This enzyme has a low affinity for CO_2 and brings about oxidation at high temperatures, resulting in low efficiency. Plants adapted to high temperatures (C_4 plants) interpose a more efficient carbon assimilation step, which is spatially separated from the C_3 reactions in the leaf. In desert environments, some plants utilize CAM photosynthesis to separate carbon assimilation and the Calvin-Benson cycle reactions temporally, between nighttime and daytime. Both C_4 and CAM increase water use efficiency at high temperatures, but are energetically less efficient than C_3 photosynthesis, which prevails in cool, moist environments.

6. Temperature affects the rate of chemical reactions, the stability of enzymes, and the conformation of proteins. Most organisms operate between 0° C and 45°C. A few organisms, such as the thermophilic bacteria found in hot springs, survive at temperatures above 45°C. Some desert plants of the agave and cactus families are adapted to withstand temperatures above 50°C by reducing their exposure to radiation. A larger number of organisms can survive very low temperatures (below freezing) through mechanisms that enable them either to resist freezing or to tolerate the effects of freezing.

7. There are four avenues of heat transfer between organisms and their environment. Radiation is the absorption or emission of electromagnetic energy. Conduction is the transfer of kinetic energy of heat between substances in contact or from one part of a substance to another. Convection is the movement of liquids or gases of different temperatures over surfaces across which heat is transferred by conduction. Water vapor carried off in the process of evaporation represents a heat loss.

8. Although most physical and biological processes are accelerated at higher temperatures, temperature optima of metabolic processes may be adjusted to match the characteristic temperature of the environment by altering the structure and quantity of key enzymes.

9. Water stress increases with temperature. In dry environments, animals seek cool microclimates, and plants increase the stomatal resistance of their leaves. Such responses uniformly reduce productivity in return for enhanced survival.

EXERCISES

1. Photosynthesis takes place in cell organelles called chloroplasts, and cellular respiration is carried out in mitochondria. Why do all eukaryotic cells have mitochondria, but only some plant cells have chloroplasts?

2. When a fluid (gas or liquid) is in contact with a surface, a boundary layer—an area near the surface where the fluid does not flow—is formed. The boundary layer on the surface of aquatic plants limits diffusion of carbon dioxide from the water to the plant. Think of some other boundary layers in ecology and comment on what their significance might be. Explain the relationship between the boundary layer and convection and discuss adaptations of plants and animals that affect that relationship.

3. Consider a shallow lake that contains a population of fish and a community of photosynthesizing plankton. In one winter, the lake is covered by a layer of very clear ice. In the next winter, after the ice is formed, a foot of snow is deposited on top of the ice. Contrast the conditions of the lake and the fate of the organisms in the lake in the two winters.

CHAPTER 6

Response to Variation in the Environment

GUIDING QUESTIONS

- How are feedback loops used in the regulation of homeostasis?
- What is the relationship between ambient temperature and the energetic costs of temperature regulation in homeotherms?
- What are the differences in the mechanisms of thermoregulation in homeotherms and poikilotherms?
- What are the regulatory challenges of desert organisms, and how are those challenges met?
- What is the activity space of an animal?
- How do regulatory, acclimatory, and developmental responses differ?
- What are the advantages of migration, storage, and dormancy?
- What is the importance of the developmental response of the water striders of Europe?
- What distinguishes proximate and ultimate cues for predicting environmental change?
- What are ecotypes?

In Chapters 4 and 5 we described the features of the physical environment and discussed how organisms maintain themselves out of equilibrium with that environment with respect to water, salts, chemical energy, and heat. We emphasized that maintaining this state of imbalance in which organisms uphold gradients in concentrations of energy and materials between their bodies and their surroundings requires the expenditure of energy. Complicating the relationship of every organism to its environment is the fact that change pervades its surroundings—daily periods of light and dark, the annual cycle of the seasons, frequent unpredictable turns of climate. The survival of each individual depends on its ability to cope with variation in the environment. Humans can be aware of their own responses to change. When we step from a warm room into the outdoors on a cold day, our shivering generates heat. Our responses to environmental change maintain our internal conditions at optimum levels for proper functioning. But what determines the best internal condition? At what rate should the individual function? How can one respond to environmental change most effectively?

Responses to environmental change must be analyzed, like a problem in economics, in terms of costs and benefits. When the environment cools, shivering burns glucose to create extra heat, thus warming the body. The longer-term benefit is that it increases the probability of survival of the organism. But shivering also requires energy to produce body heat, which in turn may deplete fat reserves and render life more precarious in the face of a sudden food shortage. The organism may reduce the cost of such temperature regulation by lowering the regulated temperature, just as we turn down the thermostat to save fuel. But turning down the fire of an organism's life also reduces its rate of activity, and hence its food-gathering and predator-avoiding abilities.

When we consider the many factors affecting the costs and benefits of a particular response, the optimum becomes a subtle concept. In this chapter we shall explore some facets of response to environmental change in order to understand why different organisms regulate their internal conditions at different levels, or not at all, and why they employ different means of response to environmental change.

6.1 Homeostasis depends upon negative feedback.

Homeostasis is the ability of an individual to maintain constant internal conditions in the face of a varying external environment. All organisms exhibit homeostasis to some degree with respect to some environmental conditions, although the occurrence and effectiveness of homeostatic mechanisms varies.

All homeostasis exhibits the properties of a **negative feedback** system, exemplified by the working of a thermostat. When a room becomes too hot, a temperature-sensitive switch turns off the heater; when the temperature drops too low, the switch turns the heater on. When we walk from a dark room into bright sunlight, the pupils of our eyes rapidly contract, restricting the amount of light entering the eye. A sudden exposure to heat brings on sweating, which increases evaporative heat loss from the skin and helps to maintain body temperature at its normal level.

Such responses are forms of negative feedback: when a system deviates from its norm, or desired state, internal response mechanisms act to restore that state. The system may be changed from its norm either by external influences—that is, environmental conditions—or by internal influences resulting from changes in the physiological state of the organism. For example, activity is accompanied by increased metabolic production of heat, which must be dissipated by sweating or some other physical mechanism. In the blood, the concentration of oxygen decreases and that of carbon dioxide increases; these changes are countered by an increase in the rate of breathing, which increases gas exchange between the blood and air in the lungs.

The essential elements of such negative feedback systems are (1) a mechanism that senses the internal condition of the organism; (2) a means of comparing the actual internal condition with the desired condition, or **set point;** and (3) an effector apparatus that alters the organism's internal condition in the direction of the preferred condition (Figure 6-1). In some cases, a negative feedback system exists within a single cell, or is even a property of a single molecule. For example, as metabolic rate increases, oxygen concentrations in blood and tissues decrease; at the same time, carbon dioxide and hydrogen ion concentrations (from the formation of carbonic acid) increase. The structure of hemoglobin molecules in our red blood cells responds to increasing hydrogen ion concentrations in such a way that the molecules hold oxygen less tightly and release it more readily to the blood plasma and tissues. Thus, hemoglobin has a mechanism built into its structure for increasing the delivery of oxygen to tissues where it is needed to support metabolism.

In other cases, many parts of the body are involved in a single negative feedback system. For example, mammals measure their body temperatures by taking the temperature of the blood flowing through the hypothalamus of the brain. In response, appropriate signals are sent out to various parts of the body by the nervous system, which stimulates the muscles to shiver or the skin to produce sweat, and by hormones secreted into the blood, which may increase the metabolic heat production of the visceral organs. Some control over heat exchange is retained at the local level, however. When an area of skin is exposed to cold, the blood vessels contract and restrict the flow of blood—and heat—to the surface. Responses to temperature changes also include behaviors: many small birds and mammals huddle together in cold weather;

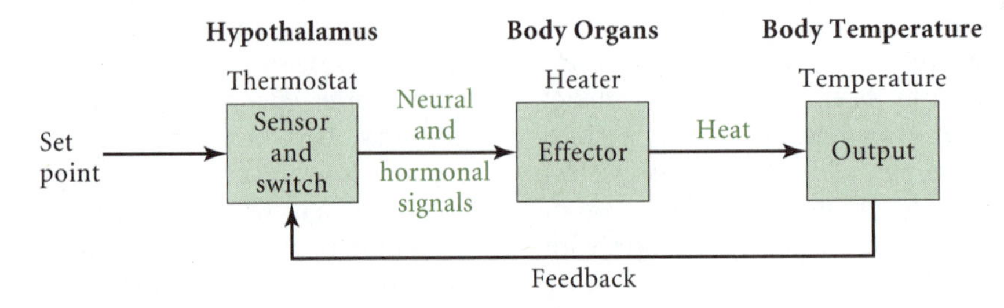

FIGURE 6-1 The major features of a negative feedback system for the regulation of body temperature. The thermostat compares body condition with a set point; when the two differ, it signals the effector organs to bring body condition back into line with the set point. In the case of body temperature regulation, the thermostat resides in the hypothalamus of the brain.

dogs curl up in a ball to reduce exposed surface area; we humans put on sweaters and coats.

Organisms fall into two broad categories with respect to their responses to variation in the physical environment. **Regulators** are organisms that maintain constant internal environments, whether we refer to temperature, pH, or solutes. **Conformers** allow their internal environments to follow external changes. Few organisms fit either ideal in all respects. Frogs regulate the salt concentration of their blood, but conform to the external temperature. Even "warm-blooded" animals conform partially to the ambient temperature; in cold weather our hands, feet, noses, and ears—our exposed extremities—become noticeably cool.

6.2 The regulation of body temperature is an important aspect of homeostasis in animals.

Temperature is one of the most important determinants of biological function because the rates of most biochemical processes increase rapidly with increasing temperature. For example, the Q_{10}—the factor of increase corresponding to an increase in temperature of 10°C (see Chapter 5)—for the metabolic rate of the salamander *Plethodon jordani* is 2.4 $\mu l\ O_2\ h^{-1}$ in the temperature range 10°–20°C (Feder et al. 1984). Most organisms live in places where the environmental temperature varies. Often this variation is temporal, as in the cooling of the night air or the lowering of temperatures in winter. Temperatures may vary spatially as well, as exemplified by the difference in temperature between shady and sunny places. In this section we discuss adaptations to variation in environmental temperature.

Most mammals and birds maintain a constant body temperature somewhere between 36°C and 41°C, even though the temperature of their surroundings may vary from −50°C to +50°C. Such regulation, referred to as **homeothermy** (the Greek root *homo* means "same"), creates constant temperature (homeothermic) conditions within the cells, under which biochemical processes can proceed most efficiently. In addition, these high temperatures speed reactions and support high rates of activity. In contrast, the internal environments of "cold-blooded," or **poikilothermic,** organisms, such as amphibians and insects, conform to the external temperature (the root *poikilo* means "varying"). Of course, these organisms cannot function at either high or low temperature extremes, so they can be active only within a narrow part of the range of environmental conditions under which mammals and birds thrive.

Many of the so-called "cold-blooded" animals, including reptiles and insects, adjust their heat balance behaviorally simply by moving into or out of shade or by changing their orientation with respect to the sun. Because their source of heat lies outside the body, biologists refer to them as **ectotherms** ("external heat"); "warm-blooded" animals (birds and mammals) are referred to as **endotherms** ("internal heat"). However, the distinction between ectotherms and endotherms is not sharp. Some endotherms are able to lower their body temperatures at times when food is not plentiful or during rest periods. This strategy is referred to as **heterothermy** (Hutchison and Dupré 1992). For example, insectivorous (insect-eating) bats, which, of course, are endotherms, may find it difficult to maintain high body temperatures when it is cold because their ectothermic prey (insects) are not active. This problem may be overcome by facultatively lowering their body temperature while roosting. Similarly, some ectothermic animals may increase their internal body temperature physiologically for short periods of time—a strategy referred to as **facultative endothermy.** Table 6-1 summarizes these forms of temperature regulation.

TABLE 6-1 Terms associated with body temperature homeostasis

Term	Definition
Homeothermy	Maintenance of a constant body temperature, usually warmer than that of the environment (hence "warm-blooded")*
Endothermy	Use of elevated metabolism in response to body cooling to maintain homeothermy
Ectothermy	Reliance on external sources of heat (solar radiation, conduction of heat from warm surfaces) to maintain an elevated body temperature
Poikilothermy	Failure to regulate body temperature; hence, conformance to environmental temperature ("cold-blooded")†
Heterothermy	Facultative reduction in body temperature by an endothermic animal
Facultative endothermy	Increase in body temperature of an ectothermic animal by means of some physiological process

**Note that homeothermic organisms may have to dissipate heat, usually by evaporative cooling, when environmental temperature exceeds preferred body temperature.*

†Although called "cold-blooded," poikilotherms may become very warm at high environmental temperatures.

Energetic Costs of Body Temperature Regulation in Endotherms

How is body temperature regulated? Most homeotherms have a sensitive thermostat in the brain (Hammel 1968, Calder and King 1974, Heller et al. 1978). This thermostat responds to changes in the temperature of the blood by secreting hormones into the bloodstream to slow down or accelerate the generation of heat in the body tissue. In addition, most homeotherms partly regulate body temperature by altering gain or loss of heat from the environment. For example, humans put on heavy clothes in cold weather and avoid standing in the sun when it is hot; birds fluff up their feathers to provide greater insulation against the cold.

To maintain internal conditions significantly different from the external environment requires energy and work. The maintenance of constant body temperatures by birds and mammals in cold environments exemplifies the metabolic cost of homeostasis. As air temperatures decrease, the gradient between the internal and external environments increases, and the body surface loses heat proportionately more rapidly. An animal that maintains its body temperature at 40°C loses heat twice as fast at an ambient temperature of 20°C (a gradient of 20°C) as at an ambient temperature of 30°C (a gradient of 10°C). The principle that heat loss varies in direct proportion to the gradient between body and ambient temperature is called **Newton's law of cooling.** To maintain a constant body temperature, an organism must replace lost heat by generating heat metabolically. Thus, the rate of metabolism required to maintain body temperature increases in direct proportion to the difference between body and ambient temperature, all other things being equal. For example, in the moustached bat *Pteronotus davyi,* a small cave-dwelling species of Central and South America, oxygen consumption (ml O_2 g^{-1} h^{-1}) increases at a rate of about 0.40 for each degree drop in ambient temperature. Temperatures below 15°C are lethal to these bats (Figure 6-2).

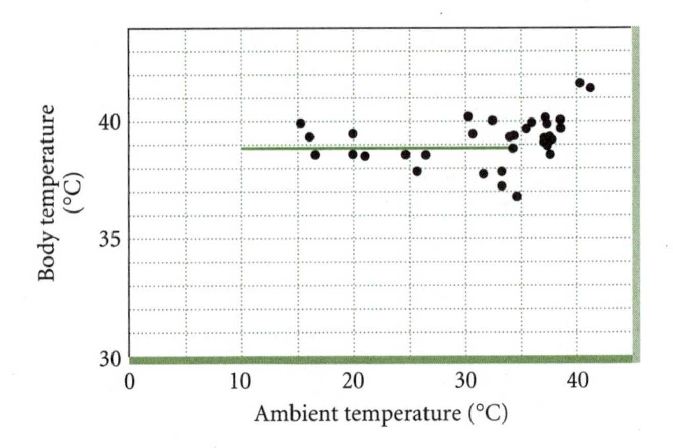

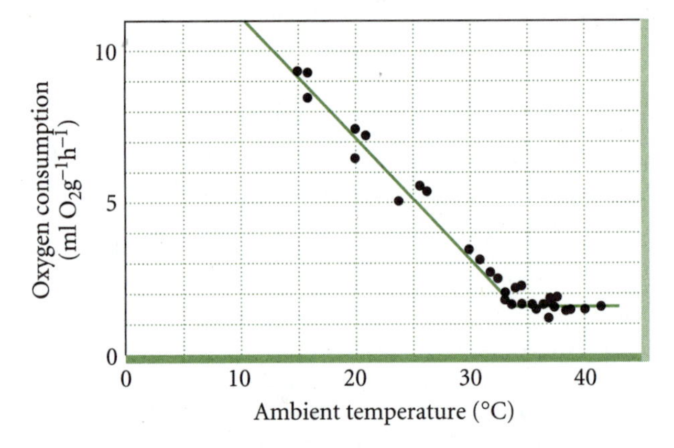

FIGURE 6-2 Relationship between the ambient temperature and the body temperature and rate of oxygen consumption of the moustached bat *Pteronotus davyi* of Venezuela. Below about 33°C, the bat must increase its oxygen consumption proportionally with decreases in ambient temperature in order to maintain its body temperature at a constant 38°C. Temperatures below about 15° C are lethal. (*After Bonaccorso et al. 1992.*)

If the only purpose of metabolism were temperature regulation, an organism would require no metabolic heat production when body temperature equaled ambient temperature, at which point no heat would flow between it and its surroundings. But organisms release energy unrelated to temperature regulation to support such functions as heartbeat, breathing, muscle tone, and kidney function, regardless of the ambient temperature. The metabolic rate of an organism that is resting quietly and without food in its digestive tract (postabsorptive condition) is called the **basal** or **resting metabolic rate** (BMR or RMR). BMR represents the lowest level of energy use under normal conditions. At this metabolic rate, the individual produces sufficient heat to maintain its body temperature as long as ambient temperatures remain above a certain point, referred to as the **lower critical temperature** (T_{lc}) (Figure 6-3). Below the lower critical temperature, the individual must increase its metabolism to balance increased heat loss. The lower critical temperature depends on BMR and thermal conductance (the rate at which heat passes through the skin). As body size increases, the basal metabolic rate increases more rapidly (allometric slope 0.75; see Chapter 3) than body surface area (0.67), and thicker fur and feathers tend to reduce thermal conductance. As a result, T_{lc} decreases with increasing size, from about 30°C in sparrow-sized birds to below 0°C in penguins and other large species (Calder and King 1974).

An organism's ability to sustain a high body temperature while exposed to very low ambient temperatures is limited over the short term by its physiological capacity to generate heat, and over the long term by

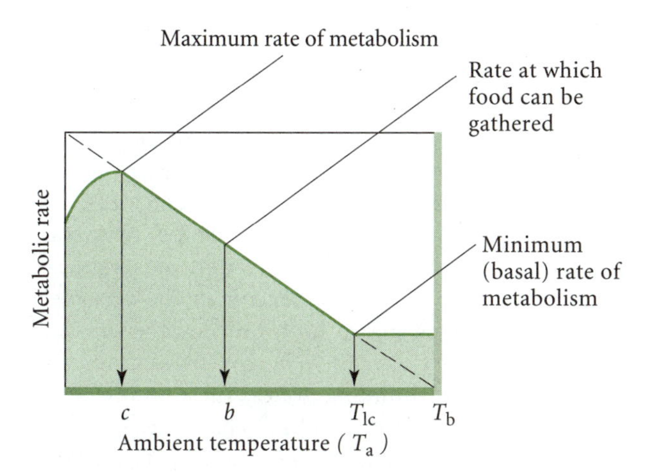

FIGURE 6-3 Relationship between metabolic rate and ambient temperature for a homeothermic bird or mammal whose body temperature is normally maintained at T_b. T_{lc} is the lower critical temperature, below which metabolism must increase to maintain body temperature. Point *c* is the lower critical physiological temperature—the lowest temperature at which an organism can sustain metabolic energy production for temperature regulation. Point *b* is the lower critical ecological temperature—the lowest temperature at which an organism can maintain itself indefinitely in its natural environment, limited by its ability to procure food.

its ability to gather food to satisfy the metabolic requirements of generating heat. The maximum rate at which an organism can perform work generally is no more than 10 to 15 times its BMR (Kleiber 1961, King 1974). Over the course of a day, few organisms expend energy at a rate exceeding four times BMR (Drent and Daan 1980). When the environment becomes so cold that heat loss exceeds the organism's ability to produce heat (point *c* in Figure 6-3), body temperature begins to drop, a condition fatal to most homeotherms.

The lowest temperatures that homeotherms can survive for long periods often depend on their ability to gather food rather than their ability to assimilate and metabolize energy. Animals can quite literally starve to death at low temperatures because they metabolize food energy more rapidly than they can gather it. In temperate and arctic climates, the coldest temperatures are often associated with storms and snow accumulation, which make it difficult to gather food, thereby taxing both physiological and ecological sources of heat energy. If the temperature drops below point *b* in Figure 6-3, the organism will be unable to produce enough heat both to gather food and to maintain body temperature. This temperature is called the **lower critical ecological temperature.** If the temperature of the environment drops below point *c* in Figure 6-3, the organism will be unable to sustain metabolic energy production for body temperature regulation, and will die. Point *c* is referred to as the **lower critical physiological temperature.** Between points *c* and *b*, the organism can survive, but only for relatively brief periods and with a negative energy balance. Above point *b*, the energy balance is positive, and not only can the organism survive indefinitely, but it can also engage in energy-consuming activities besides foraging.

Lowering Energetic Costs

When homeostasis costs more than the individual can afford, certain economy measures are available. For example, an individual may lower the regulated temperature of portions of its body, thereby reducing the temperature difference between environment and body. Because the legs and feet of birds are not feathered, they would be a major avenue of heat loss in cold regions if they were not kept at a lower temperature than the rest of the body (Figure 6-4). Gulls accomplish this by means of a countercurrent heat exchanger (see Chapter 5) in which warm blood in the arteries leading to the feet cools as it passes close to the veins that return cold blood to the body (Scholander 1955, Scholander and Schevill 1955). In this way, heat is transferred from arterial to venous blood and transported back into the body rather than being lost to the environment.

Some organisms lower their body temperature at certain times of day in order to reduce the difference

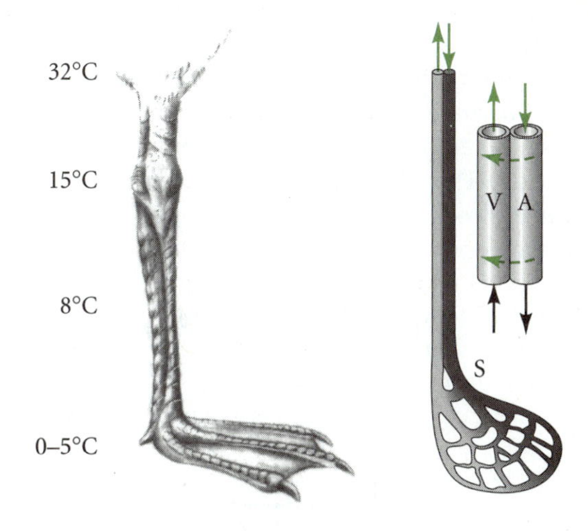

FIGURE 6-4 Skin temperatures of the leg and foot of a gull standing on ice. The anatomical arrangement of blood vessels that permits countercurrent heat exchange between arterial blood (A) and venous blood (V) is diagramed at right. Arrows indicate the direction of blood flow; dashed arrows indicate heat transfer. A shunt at point S allows the gull to constrict the blood vessels in its feet, thereby reducing blood flow and heat loss even further, without having to increase its blood pressure. (*After Irving 1966, Schmidt-Nielsen 1983.*)

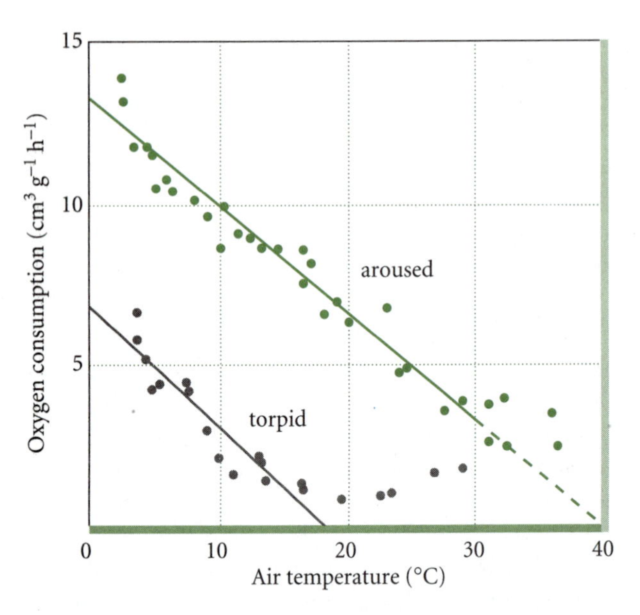

FIGURE 6-5 Relationship between energy metabolism and air temperature in the West Indian hummingbird *Eulampis jugularis* during periods of torpor and normal arousal. The bird is endothermic in each case, but regulates its temperature at different set points. (*After Hainsworth and Wolf 1970.*)

between their temperature and that of the environment. Because of their small size, hummingbirds have a large surface area relative to their weight and consequently lose heat rapidly compared with their ability to produce it. As a result, hummingbirds require very high metabolic rates to maintain their at-rest body temperatures near 40°C. Species that inhabit cool climates would risk starving to death overnight if they did not become **torpid,** assuming a condition of lowered body temperature and inactivity resembling hibernation. The West Indian hummingbird, *Eulampis jugularis,* drops its temperature to between 18°C and 20°C when resting at night. It does not cease to regulate its body temperature, but merely changes the setting on its thermostat to reduce the difference between ambient and body temperature (Figure 6-5) (Hainsworth and Wolf 1970, Hainsworth et al. 1977).

Some species of mammals and birds spend a significant amount of time at the surface or beneath the surface of water. Seabirds, such as diving petrels, puffins, and penguins, forage for fish and other sea creatures at considerable depths. Some temperate mammals, such as beavers, otters, and muskrats, are principally aquatic animals. Because of the extremely low resistance of water to heat flow and the substantial difference between their body temperatures and those of the surrounding water, these animals are faced with a formidable thermal challenge. In water, body heat is rapidly lost from exposed surfaces. Morphological and physiological adaptations to prevent or reduce heat loss, such as a thick pelage (fur coat) and lots of body fat, are common in such animals. The exquisite adaptations of marine mammals such as whales and dolphins provide sufficient protection from body heat loss to allow for an obligate aquatic existence. But many aquatic birds and mammals are unable to remain active in water for indefinite periods of time because of loss of body heat. The activities of these animals are determined, in part, by thermal constraints. For example, the foraging pattern of the North American beaver (*Castor canadensis*) in cold water is composed of short foraging bouts in the water, during which the body temperature drops rapidly because of heat dissipation to the water, alternating with lengthy warm-up periods in the lodge (MacArthur and Dyck 1990) (Figure 6-6). The desert antelope ground squirrel (*Ammospermophilus leucurus*) is faced with the opposite problem: it heats up during foraging bouts on the hot desert surface and must periodically return to its burrow to cool down (Hainsworth 1995). Alternating locations in this way to maintain body temperature is referred to as **shuttling,** and such behavior is common in animals. As we shall see in the next section, shuttling behavior may be employed to regulate salt and water balance as well.

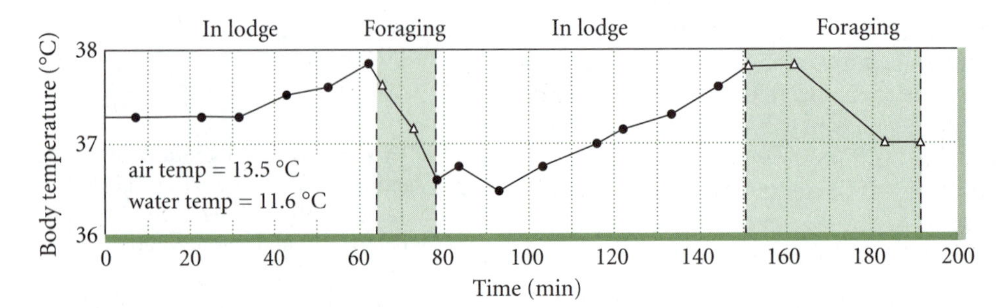

FIGURE 6-6 Body temperature changes associated with the foraging activity of the beaver *Castor canadensis.* Body temperature drops during foraging bouts in the cold water, forcing the beaver to return to the lodge periodically to warm up. (*After MacArthur and Dyck 1990.*)

Thermal Ecology of Ectotherms

Nearly every aspect of the ecology and behavior of ectothermic vertebrates (fishes, amphibians, and reptiles) is in some way influenced by the requirements of the regulation of body temperature. Because these organisms cannot generate heat unless they are very active, many adjust their heat balance behaviorally by simply moving into or out of the shade, or by orienting the body with respect to the sun. As we mentioned earlier, some ectothermic vertebrates also possess physiological mechanisms for regulating body temperature to some degree (e.g., Holland and Sibert 1994, Sinervo and Dunlap 1995).

Most fish display a very low rate of metabolic heat production when compared with homeotherms or terrestrial poikilotherms, and this, coupled with the very high heat capacity of water, means that they lose heat rapidly to their aquatic environment. For the most part, fish have very limited thermal regulation abilities (Moyle and Cech 1988). Some species, particularly warm-water species, can make small adjustments in body temperature behaviorally by selecting thermally favorable portions of the water column (Snucins and Gunn 1994, Schurmann and Christiansen 1994). Digestion in fish is strongly affected by water temperature, and thus fish who have recently completed a foraging bout may seek out warm water. Physiological thermoregulation in fish is restricted to just a few species. The albacore tuna (*Thunnus alalunga*) employs a countercurrent exchange system to maintain a core body temperature considerably higher than the surrounding water. Blood vessels that transport warm blood from the core of the fish's body lie close to vessels bringing cold blood from the gills. Heat is transferred from the outflowing vessels into the inflowing vessels, thereby retaining the heat in the core of the body (Carey et al. 1971). Recent studies of physiological thermoregulation in the bigeye tuna (*Thunnus obesus*) show that the countercurrent system gives these fish the ability to alter body temperature rapidly in response to changes in ambient temperature (Holland and Sibert 1994).

Amphibians rely principally on behavioral thermoregulation (Hutchison and Dupré 1992). The aquatic larvae of frogs and salamanders may display patterns of vertical migration and aggregation in small ponds in response to daily changes in the thermal environment (Figure 6-7). Preferences for thermally favorable microhabitats are common in amphibians. In adult amphibians, thermoregulation may be complicated by the need to exchange gases across the skin. Cutaneous respiration, which accounts for a significant amount of the respiratory gas exchange in many species of amphibians, and is the only means of respiration in some (e.g., the lungless salamanders of the family Plethodontidae), requires that the skin remain moist. This requirement may limit the extent to which an individual can raise its body temperature above the ambient temperature by basking in sunlight, since evaporation from the moist skin will tend

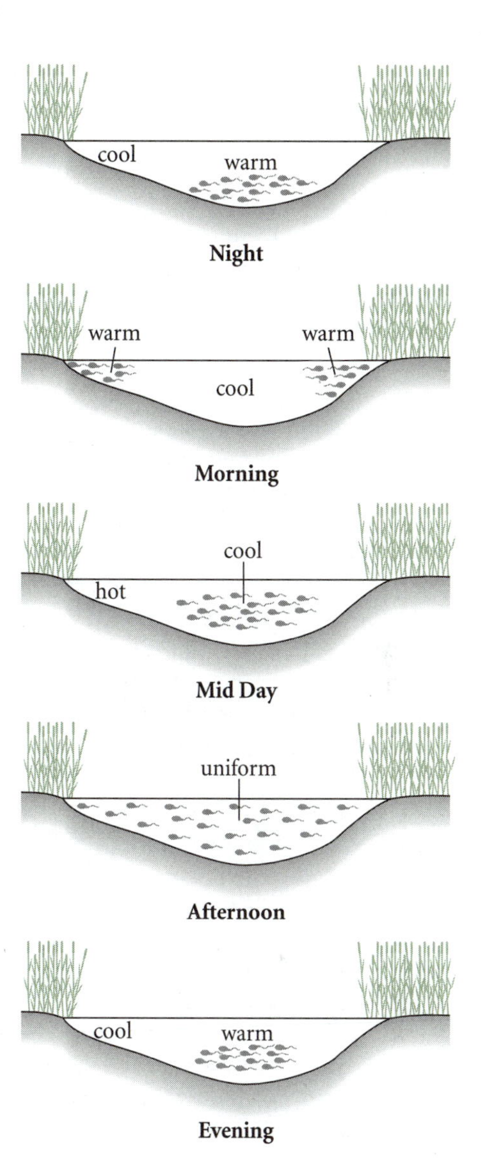

FIGURE 6-7 Changes in the spatial distribution of tadpoles of the frog *Rana boylii* associated with daily temperature changes in a pond in Del Norte County, California, in June. In the evening, the tadpoles move to the warmest part of the pond and remain there through the night. As the water is warmed by the morning sun, aggregations form in the shallow water until midday, when those areas become too hot and there is movement toward the cooler deep water. Tadpoles may be equally distributed throughout the pond during the period in the afternoon when the water temperature is equal throughout the pond. (*From Brattstrom 1962.*)

to lower body temperature. Also, the need to prevent the desiccation of the skin places limits on the amount of exposure to sunlight that can be tolerated. For these reasons, basking is uncommon in amphibians, and when it is observed, individuals are not able to raise their body temperatures as far above the ambient temperature as are most reptiles (Hutchison and Dupré 1992). Physiological mechanisms for thermoregulation in amphibians appear to be rather limited.

For many people, the image of a large snake basking on a rock in the hot sun, or of an alert lizard stationed atop a sunny fence post, exemplifies the life of reptiles. Such behaviors, which are principally related to thermoregulation, have been widely studied (e.g., Bogert 1949, McGinnis and Dickson 1967, Heatwole 1976, Stevenson 1985, Huey et al. 1989, Hertz 1992, Adolph and Porter 1993). It is clear that reptiles have an enormous capacity to maintain their body temperatures, within a narrow range, at levels well above the temperature of the surrounding air by adjusting their location and position in the environment (Cowles and Bogart 1944, Hammel et al. 1967, Huey 1974). When cold, horned lizards (*Phrynosoma*) increase the profile of their bodies exposed to the sun by lying flat against the ground. When hot, they decrease their exposure by standing erect upon their legs (Heath 1965). By lying flat against the ground, horned lizards also gain heat by conduction from the sun-warmed surface. Garter snakes (*Thamnophis elegans*) choose retreats that lie under rocks having a thickness of 20–30 cm because the thermal characteristics of such sites are more favorable than potential retreats beneath rocks of greater or lesser thickness (Huey et al. 1989). The lizard *Anolis cristatellus* modifies the amount of time it spends basking with elevation, basking longer at higher elevations (Hertz 1992). Extreme environmental temperatures may limit the amount of time available for reproduction and development, or prevent survival over the winter, and thus constrain the geographic distribution of snakes (Peterson et al. 1993). Behaviors that require locomotion, such as prey capture, predator avoidance, and habitat selection, are dependent on body temperature, as are physiological processes such as digestion, metabolism, and the periodic shedding of the outer skin, referred to as **ecdysis.**

The physiological control of body temperature may be better developed in reptiles than once thought. Studies of the diamond python (*Morelia spilota*), the northern water snake (*Nerodia sipedon*) and several species of garter snakes (*Thamnophis*) have revealed that in these species, body temperature rises significantly after eating, an example of facultative endothermy (Slip and Shine 1988, Gibson et al. 1989, Lutterschmidt and Reinert 1990). In some cases, the higher body temperature may persist for 30 or more hours, during which time the snake may function in microhabitats that are cooler than would be possible if it had not just eaten (Peterson et al. 1993).

6.3 The level of regulation balances costs and benefits.

Organisms sometimes regulate their internal environments over moderate ranges of external conditions, but conform under extremes. This strategy is called **partial regulation.** The phenomena of heterothermy and facultative endothermy demonstrate the phenomenon of partial regulation. Only birds and mammals are true endotherms in that they generate metabolic heat to regulate body temperature. But pythons maintain high body temperatures while incubating eggs (Hutchison et al. 1966, Vinegar et al. 1970, and see review by Gans and Pough 1982). As we mentioned, some large fish, such as the tuna, use a countercurrent arrangement of blood vessels to maintain temperatures up to 40°C in the center of their muscle masses (Carey et al. 1971). Swordfish employ specialized metabolic heaters, derived from muscle tissue, to keep their brains hot (Carey 1982, Block 1987). Large moths and bees often require a preflight warm-up period during which the flight muscles shiver to generate heat (Heinrich 1979, Heinrich and Bartholomew 1971, Stone 1994). Even among plants, temperature regulation based upon metabolic heat production has been discovered in the floral structures of philodendron and skunk cabbage (Nagy et al. 1972, Knutson 1974).

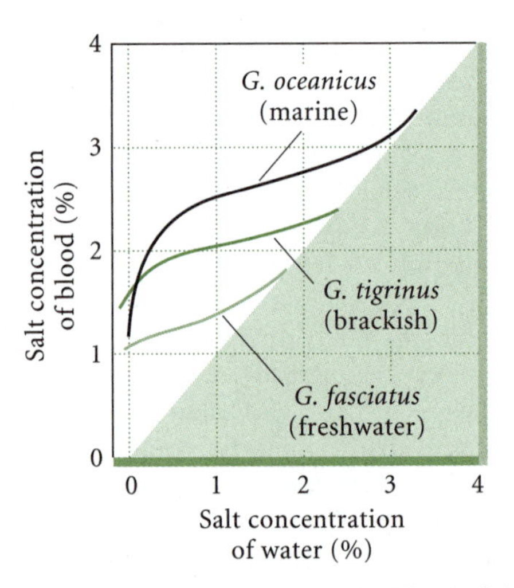

FIGURE 6-8 Salt concentrations in the blood of three gammarid crustaceans from different habitats as a function of the salt concentration of their external environment. The normal salt concentration of seawater is 3.5%. (*After Prosser and Brown 1961.*)

Partial regulation is also demonstrated by the regulation of salt and water balance in organisms. For example, small aquatic amphipods of the genus *Gammarus* regulate the salt concentrations of their body fluids when placed in water with a salt concentration less than that of their blood, but not in water with a higher salt concentration (Figure 6-8). The freshwater species *Gammarus fasciatus* maintains the salt concentration of its blood at a lower level than the saltwater species *G. oceanicus,* and thus begins to conform to salt solutions at a lower concentration. In their natural habitat, however, neither the freshwater species nor the saltwater species encounters salt more concentrated than that in its blood. Animals that inhabit salt lakes and brine pools, however, actively keep the salt concentration of the body below that of the surrounding water. The brine shrimp *Artemia* maintains an internal salt concentration below 3% even when placed in a 30% salt solution (Croghan 1958).

Whereas most fully aquatic invertebrates cannot maintain the salt concentration of their blood below that of the surrounding water, land crabs and many intertidal invertebrates that are periodically exposed to air do possess this capability (Prosser 1973). The ghost crab (*Ocypode quadrata*) forages actively during the daytime when the temperature of the beach may exceed 40°C, a situation that imposes extremely high evaporative water loss. During periodic visits to their damp burrows, ghost crabs rehydrate themselves by sucking fresh water from the interstitial spaces of the sandy soil with the aid of specialized tufts of hairs situated near the branchial chamber. Some of the water is taken directly into the gill chambers, and some is moved forward to the mouth. Ghost crabs eat mostly invertebrates, a nitrogen-and ion-rich food source. Extreme flexibility of ion conservation in the urine and the ability to excrete NH_3^- and NH_4^+-loaded urine have evolved as mechanisms to balance ion concentrations in the body fluids (Figure 6-9) (Wolcott 1984, 1992, Hadley 1994).

Constraints on the Evolution of Homeothermy

Clearly, organisms other than birds and mammals are physiologically capable of generating heat to maintain elevated body temperatures, and many do so under certain conditions. Why, then, is the distribution of endothermy throughout the animal and plant kingdoms so limited? Part of the answer certainly lies in a consideration of body size. Birds and mammals are relatively large as animals go. As body size increases, volume increases relatively more rapidly than surface area. This means that the amount of surface across which heat leaves the body is smaller per unit of volume for larger animals than for smaller ones. The practical implication is that larger animals have relatively smaller amounts of surface over which they must exercise control of heat exchange, and thus are able to achieve a greater degree of precision in body temperature regulation.

Although body size may explain why mammals are endotherms and insects generally are ectotherms (the large moths that exhibit preflight warm-up behavior approach the size of small mammals), large fish and reptiles also generally have not made the shift to homeothermy. For most fish, the low availability of oxygen in the aquatic environment and the high rate of heat loss owing to the high thermal conductance of water preclude the high metabolic rates necessary for temperature regulation in all but the most active species. The metabolic rates of resting birds and mammals may be 10 times as high as those of fish, amphibians, and reptiles of similar size (Schmidt-Nielsen 1983).

Why have reptiles not evolved temperature regulation? After all, the body sizes of contemporary

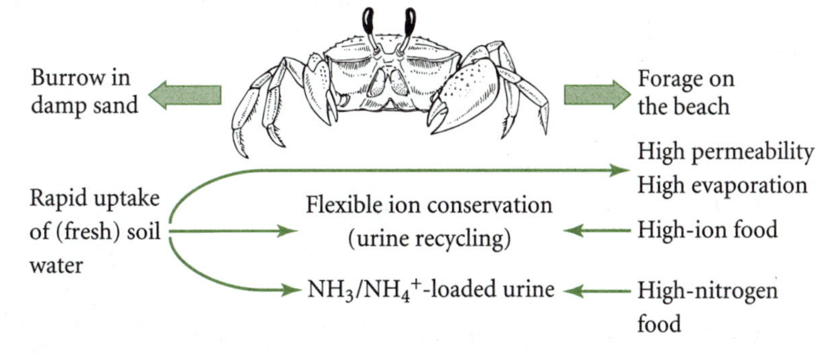

FIGURE 6-9 Strategy for maintenance of water and salt balance in the ghost crab (*Ocypode quadrata*). The crab experiences high evaporative water loss as it forages during the daytime on the open beach, where temperatures may exceed 40°C. Rehydration occurs during periodic visits to a damp burrow, where the crab draws fresh water from the soil with the aid of specialized hairs. The invertebrate food of the ghost crab is high in ions and nitrogen, which are eliminated in the urine during the time the crab is in the burrow. (*After Wolcott 1992.*)

reptiles are at least comparable to those of small mammals, and they breathe air. The major innovation leading to endothermy appears to have been the insulation provided by fur and feathers. Birds and mammals evolved from reptile lineages, but did not displace their ancestors completely, probably owing the to advantages of ectothermy under certain environmental conditions. Reptiles lack the high activity levels of endothermic animals, but their low energy requirements are well suited to highly seasonal and erratic food supplies (Pough 1980).

Advantages of High Body Temperatures

Most homeotherms maintain their body temperature considerably above that of the surrounding air. Logic tells us that to minimize the energetic cost of thermoregulation, temperature should be regulated close to the average ambient temperature. Why do birds and mammals maintain such high body temperatures, irrespective of whether they live in the Tropics or close to the poles (McNab 1966)? The advantages of a high body temperature are several. First, an increase in temperature raises the level of sustained activity and may increase alertness, both of which contribute to the capture of food and avoidance of predators. Second, an elevated body temperature reduces the need for frequent use of evaporative cooling to dissipate body heat; that is, when body temperature is well above the ambient temperature, metabolic heat may be lost by radiation or convection alone. For terrestrial organisms, the potential savings in water may partly justify the energetic cost of elevated temperatures. Third, a large gradient in temperature between body and environment may allow more rapid and precise adjustment of heat loss in response to changes in the rate of activity than is possible with a small gradient.

These considerations illuminate one facet of the optimization of physiological functions. Clearly, however, organisms have developed a variety of physiological mechanisms, morphological devices, and behavioral ploys to lessen the tension in their relationship to the physical environment. The adaptations of desert birds and mammals to their stressful environment illustrate this versatility particularly well.

6.4 Temperature regulation and water balance in hot deserts require diverse homeostatic adaptations.

Organisms inhabiting desert environments face severe survival challenges presented by the interrelated problems of water loss and thermal stress. The high temperatures that are characteristic of most (but not all) deserts promote water loss in an already nearly waterless environment. Thus, in order for desert organisms to survive, precious body water must be conserved, excessive overheating must be avoided, and excess heat must be dissipated. And all this must be done while the organism goes about the business of feeding, avoiding predators, finding a mate, and reproducing—all activities that produce heat and exacerbate the water loss problem.

Desert Birds

Among vertebrates, birds are perhaps the most successful inhabitants of deserts. They remain active in the heat long after other animals have sought refuge. Their success derives from their low excretory water loss (recall from Chapter 4 that birds excrete nitrogen as crystallized uric acid) and from feeding on insects, from which they obtain some water. Even some seed-eating species can persist without water in the desert, provided they avoid full sun and shade temperatures above 35°C.

The behavior of the cactus wren, a desert insectivore (Figure 6-10), shows that it too must respect the physiological demands of its environment. In cold air, wrens lose 2 to 3 cm^3 of water per day in the air they exhale. Water loss increases rapidly at ambient temperatures above 30°–35°C, up to over 20 cm^3 per day at 45°C; active birds may use water at five times that rate to dissipate their heat load. The wren's body contains only about 25 cm^3 of water.

In the cool temperatures of early morning, cactus wrens forage throughout most of the environment, actively searching for food among the foliage and on the ground (Ricklefs and Hainsworth 1968). As the day brings warmer temperatures, they select cooler parts of their habitat, particularly the shade of small trees and large shrubs, always managing to avoid feeding where the temperature of the microhabitat exceeds 35°C (Figure 6-11). When the minimum temperature in the environment rises above 35°C, the wrens become less active; they even feed their young less frequently during hot periods.

FIGURE 6-10 The cactus wren (*Campylorhynchus brunneicapillus*) at its nest. (*Courtesy of R. B. Suter.*)

FIGURE 6-11 Microhabitat use by cactus wrens in southeastern Arizona during the course of a day in late spring. Microhabitats vary in their degree of thermal stress between exposed ground (a) and the deep shade of trees (e). Wrens distribute their activity among all microhabitats in the cool hours of early morning (7:00 A.M.)(a–e), but restrict their activity to cool shade (e) during the hottest part of the day (2:30 P.M.), when other microhabitats are above 40°C. (*From Ricklefs and Hainsworth 1968.*)

Many desert birds build enclosed nests or place their nests in holes in the stems of large cacti, where the young are protected from the sun and from extremes of temperature. Cactus wrens build an untidy nest, a bulky and somewhat haphazard ball of grass, with a side entrance into the cavity of a cactus stem (see Figure 6-10). Once a pair of wrens have built their nest, they cannot change its position or orientation. For a month and a half, from the laying of the first egg until the young leave the nest, the nest must provide a suitable environment day and night, in hot and cool weather. Cactus wrens usually rear several broods of young during the breeding period of March through September in southern Arizona. They build their early nests so that the entrances face away from the direction of the cold winds of early spring; during the hot summer months, nests are oriented to face prevailing afternoon breezes, which circulate air through the nest and facilitate heat loss (Ricklefs and Hainsworth 1969; Figure 6-12). This strategy makes a difference! Nests oriented properly for the season are successful 82% of the time, whereas only 45% of nests facing in the wrong direction produce successful broods (Austin 1974).

The reuse of fecal and excretory water is common in desert organisms. Cactus wrens normally remove the fecal sacs of their young in order to keep the nest sanitary. However, in the hottest part of the breeding season, the sacs may be left in the nest, where evaporative water loss from them presumably cools the air in the nest and increases humidity, thereby reducing respiratory water loss. In the desert iguana (*Dipsosaurus dorsalis*) secretions from salt glands, located in the orbits of the eyes, run down to small pits at the entrance of the nasal passages, where the water evaporates into the inhaled air, thereby reducing respiratory water loss (Dunson 1976). Some storks defecate on their legs during periods of hot weather, thus benefiting from the evaporative cooling of their fecal water. Adult roadrunners eat the fecal sacs of their young; the kidneys of adults apparently have greater urine-concentrating abilities than the kidneys of the young, so adults can extract water from the ingested fecal sacs (Calder 1968).

Desert Arthropods

Some desert arthropods regulate body temperature in much the same way as the cactus wren, by shuttling among microhabitats with different temperature characteristics. Desert locusts (*Schistocerca gregaria*)

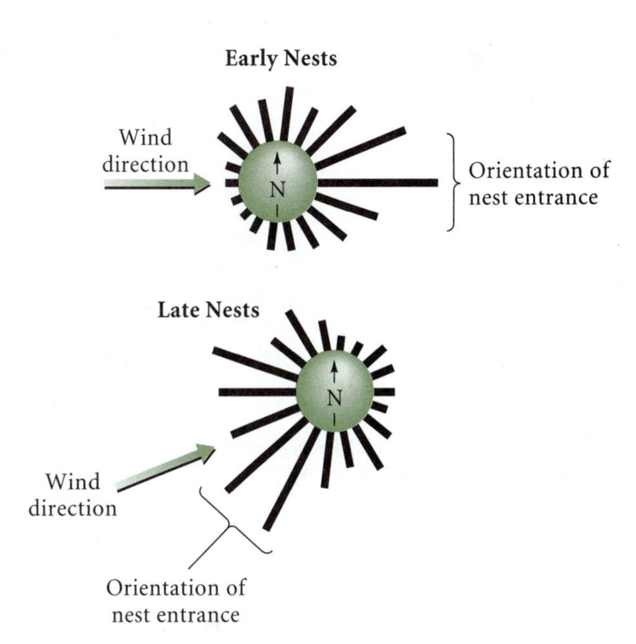

FIGURE 6-12 Orientation of the entrances of nests of the verdin (*Auriparus flaviceps*) during the early (cool) and late (hot) part of the breeding season in Nevada and Arizona. Each bar shows the relative concentration of nest entrances that are oriented in that direction. The orientation of cactus wren nest entrances is similar. (*Data courtesy of G. T. Austin.*)

aggregate in the shade of bushes and shrubs when the ambient temperature becomes extremely high. Tenebrionid beetles, species of which occur in many of the world's deserts, are known to shuttle between shade and sun to avoid overheating (Cloudsley-Thompson 1991). But attaining a thermally suitable microhabitat is not the only behavioral mechanism employed to regulate body temperature by desert arthropods. Body posture and rapid movements are also used. The flightless desert grasshopper *Taeniopoda eques* of the Chihuahuan desert spends the coldest (midnight) and hottest (midday) parts of each 24-hour period in roosts in bushes, descending to the ground in early morning and late afternoon when the ambient temperature is more moderate. During this cycle of daily movement, a grasshopper may adopt one of four body positions, depending on whether it needs to heat up or cool down (Figure 6-13). By positioning the long axis of its body perpendicular to the sun's rays—called flanking—or by crouching against a warm substrate, such as the desert sand or a rock, the grasshopper can warm up. By positioning itself in the shade of a stem—referred to as stem-shading—or stilting its body above a hot substrate, the grasshopper can cool down. The frequencies of these four behaviors correspond to changes in air temperature during the day. Flanking is the most common posture adopted during the early morning, when air temperatures are 12°–14°C. Shading and stilting are most commonly observed around noon, when the temperature may reach 32°C (Whitman 1987, 1988).

FIGURE 6-13 Thermoregulatory postures of the desert grasshopper *Taeniopoda eques:* (a) flanking, (b) stilting, (c) crouching, (d) stem-shading. Flanking and crouching warm the grasshopper; stilting and stem-shading promote cooling. (*From Whitman 1987; courtesy of D. Whitman.*)

Some insects can reduce their body temperatures by sprinting rapidly across the desert floor or by flying. The desert tenebrionid beetle *Onymacris plana* sprints among patches of vegetation during the hottest part of the day, thereby cooling itself through convection. Cooling may be accomplished in a similar manner by flying in the relatively cooler air above the hot desert surface. The digger wasp (Sphecidae) spends most of its day hovering a few meters above the ground, dipping to the desert floor occasionally to capture an insect, and quickly descending to its burrow with its catch (Amos 1959, Cloudsley-Thompson 1991).

6.5 Variation in the environment occurs at different temporal and spatial scales.

The natural world varies in time and space. We perceive temporal variations in our environment as the alternation of day and night and the seasonal cycles of temperature and precipitation. Superimposed on these cycles are irregular and unpredictable variations. Winter weather is generally cold and wet, but the weather at any particular time cannot be predicted much in advance; it varies perceptibly over intervals of a few hours or days as cold fronts and other atmospheric phenomena pass through. Some irregularities in conditions, such as the alternation of series of especially wet and dry years, occur over longer periods. Some events of great ecological consequence, such as fires and tornadoes, strike a particular place only at very long intervals.

Each type of variation in the environment has a characteristic dimension or scale. Variation between night and day has a dimension of 24 hours; seasonal variation has a dimension of 365 days. Waves pound a rocky shore at intervals of seconds; winter storms bringing rain or snow may follow one another at intervals of days or weeks; hurricanes may strike a particular coast at intervals of decades. In general, the more extreme the condition, the lower its frequency, and the longer the interval between events.

Both the severity and the frequency of events are relative measures, depending on the organism that experiences them. Fire may touch a tree many times within its life span but skip dozens of generations of an insect population. As we shall see, how organisms and populations respond to change in their environments depends on the scale of temporal variation. The dimensions of ecological processes may also be intrinsic properties of the systems themselves. For example, in pine woodlands, the probability of a destructive fire increases with time since the last such event. As litter and other fuels accumulate over time, they increase the probability of another fire, producing a characteristic fire cycle for a particular habitat. Similarly, the rapid spread of contagious disease through a population often depends on the accumulation of young, nonimmunized individuals following the last epidemic.

The environment also differs from place to place. Variations in climate, topography, and soil type cause large-scale heterogeneity (across meters to hundreds of kilometers). At smaller spatial scales, much heterogeneity is generated by the structures of plants, by the activities of animals, and even by the particle structures of soil. A particular dimension of spatial variation may be important to one organism but not to another. The difference between the top and the underside of a leaf is important to an aphid, but not to a moose, which happily eats the whole leaf, aphid and all.

Spatial heterogeneity combined with speed of movement determines how frequently a moving individual encounters a new environment. That is, spatial variation is perceived as temporal variation by an animal traveling through the environment. For a plant, the dimension of spatial variation determines the variety of conditions that its roots encounter in the soil and the variety of habitats in which its offspring germinate, which depends on how far its pollen and seeds travel.

Animals and plants live in spatially varied environments in which the particular size and arrangement of habitat patches have important repercussions for the activities of individuals. At any given moment, a cactus wren may feed within any of several parts of the habitat, each with different thermal characteristics and food supplies. These distinct microhabitats shift continuously throughout the daily cycle of solar radiation and temperature, and seasonally as both the physical environment and populations of suitable prey organisms vary. The wren must adjust behaviorally and physiologically to these ever-changing conditions. Spatial variation also affects the growth and regulation of populations and the interactions between species, as we shall see in Parts 4, 5, and 7 of this book.

6.6 An organism's choice of patches defines its activity space.

Each organism functions best within a limited range of conditions, which is called its **activity space.** Animals actively move among places in the environment that provide different conditions as the environment changes temporally. These places in the environment having different conditions may be referred to informally as patches. (The word *patch* has a more specific meaning in population ecology, as we shall see in Part 4.) Conditions within each spatially defined patch vary over diurnal and seasonal cycles; by moving among patches, animals can remain within ranges of conditions close to their optima. Although plants cannot uproot themselves and move, most regulate their activity according to the suitability of conditions at a particular time. Simply by closing their stomates, for example, plants can shut themselves off from some unfavorable conditions.

Plant Responses to Temporal Environmental Change

Plants have no choice about where they grow: they survive in suitable sites, elsewhere they die. Thus, once a seed has germinated and the plant has taken root, its principal responses are to the temporal rather than the spatial variation in its environment. To survive, it must regulate its activity according to the suitability of conditions in the place where it stands—conditions that change daily and seasonally.

However, spatial variation in the environment is important in determining which seeds germinate and which do not. For proper germination, seeds require quite specific combinations of light, temperature, and moisture—conditions that vary even among closely related species (Harper et al. 1965). Irregularities in the surfaces of natural soils provide the variety of conditions needed to allow the germination of many species. British plant ecologist John Harper and his coworkers dramatized these differences between species by creating an artificially heterogeneous soil environment in which they sowed three species of plantains, common lawn and roadside weeds of the genus *Plantago*. The three species responded differently to modifications in the environment produced by slight depressions in the soil, by squares of glass placed on the soil surface, and by vertical walls of glass or wood (Figure 6-14). Relatively few seeds germinated on the smooth surfaces of soil that had not been disturbed experimentally.

The Activity Space of Animals

Unlike plants, animals have some choice about where they live. Their choices may encompass major categories of habitat, such as forest, grassland, or marsh, within which an individual may spend all its time. Or they may allow the individual to follow the most suitable conditions within a habitat, as we have seen the cactus wren do. As conditions within different patches change, the activity space of an animal also changes. Some animals may make seasonal movements among major categories of habitat in response to seasonal changes in environmental conditions, such as with migration. While in a particular habitat, individuals

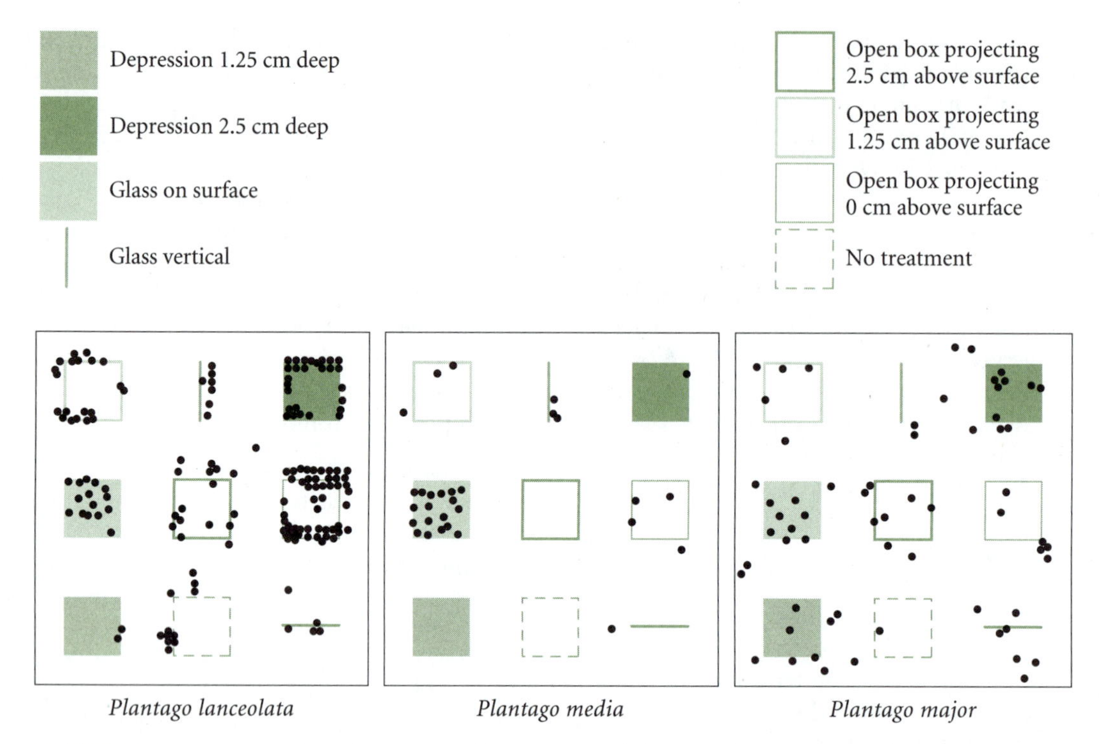

FIGURE 6-14 Germination of seedlings of three species of plantains (genus *Plantago*) with respect to artificially produced variation in the soil surface. The activity spaces for seed germination include different soil surface conditions for each species. (*After Harper et al. 1965.*)

may adjust their position and their behavior in response to diurnal cycles of temperature and other environmental conditions in such a way that the location and vigor of their activity tracks environmental change. There are many examples of such tracking in animals, but some of the best are found among ectothermic animals that must withstand the severe conditions of arid environments.

The diurnal behavioral cycle of animals is often geared toward the varying temperature conditions of habitat patches. While surface temperatures in deserts may fluctuate over 40°C in a 24-hour period, the temperatures in the burrows of some desert scorpions and spiders may change only a few degrees over the same period of time. The surface activity times of such animals correspond to times when their body temperature optima can be maintained (Cloudsley-Thompson 1991).

The diurnal behavioral cycle of lizards is also geared to the varying temperatures of habitat patches (Heatwole 1970, Gans and Pough 1982). Although few lizards generate heat metabolically for temperature regulation, they do take advantage of solar radiation and warm surfaces to maintain their body temperatures within the optimum range. At night, these sources of heat are not available, and the lizard's body temperature gradually drops to that of the surrounding air. The mallee dragon (*Amphibolurus fordi*), an agamid lizard of Australia, is fully active only when its body temperature lies between 33°C and 39°C (Cogger 1974). In the early morning, when its body temperature is below 25°C and it still moves sluggishly, the mallee dragon basks within large clumps of grass of the genus *Triodia*, within which it finds protection from predators (Figure 6-15a). In fact, the dragon depends so much upon these grass clumps that it occurs only where *Triodia* grows. When the dragon's temperature rises above 25°C, it leaves the *Triodia* clump and basks in the sunshine nearby, with its head and body in direct contact with the ground surface, from which it absorbs additional heat. When its body temperature enters the range for normal activity (33°–39°C), the mallee dragon ventures farther from *Triodia* clumps to forage, with its head and body raised above the ground as it moves. When its body temperature exceeds 39°C, it moves less rapidly and seeks the shade of small *Triodia* clumps; above 41°C, it reenters large *Triodia* clumps, at whose centers it finds cooler temperatures and deeper shade. It may also pant to dissipate heat by evaporation. If heat stress is not avoided, the lizard loses locomotor ability above 44°C, and will die if its body temperature exceeds 46°C.

On a typical summer day, during which air temperature varies from about 23°C at dawn to 34°C at midday, the mallee dragon does not begin to forage until about 8:30 A.M. By 11:30, the habitat has become too hot for normal activity, and most individuals seek

(a)

(b)

(c)

FIGURE 6-15 The mallee dragon (*Amphibolurus fordi*) at different times during its activity cycle: (a) early morning, basking in *Triodia* grass clump; (b) midmorning, basking on ground (note body flattened against surface to increase exposed profile and contact with warm soil); (c) normal foraging attitude. (*Courtesy of H. Cogger; from Cogger 1974.*)

shade and become inactive. By 2:30 P.M., the habitat has cooled off enough for the dragons to resume foraging, but by 6:00, it has cooled so much that they must retreat back into *Triodia* clumps, where their bodies rapidly cool. Individuals remaining in the open after this time of day are sluggish and easily caught by endothermic predators.

The desert iguana (*Dipsosaurus dorsalis*) of the southwestern United States faces a more severe environment, with greater annual fluctuations, than the mallee dragon in Australia (Beckman et al. 1973, DeWitt 1967, Porter et al. 1973). Shade temperatures can reach 45°C in summer and plunge below freezing in winter. During mid-July, the thermal environment courses so rapidly between extremes that the desert iguana can be normally active within its preferred body temperature range of 39°–43°C for only about 45 minutes in mid-morning and a similar period in the early evening (Figure 6-16a). During the remainder of the day, it seeks the shade of plants or the coolness of its burrow, where the temperature rarely rises above the preferred range. The desert iguana spends the night in its burrow, where it is safe from predators; at dawn, the burrow is warmer than the desert surface, and so the early morning warm-up period is correspondingly brief.

Whereas in summer the desert iguana restricts its activity to two brief bouts separated by inactivity through midday to avoid heat stress, it finds more favorable temperatures for activity in spring (Figure 6-16b). The thermal environment in May does not exceed the iguana's preferred range, and individuals forage actively above ground from 9:00 A.M. to 5:00 P.M., only occasionally seeking the cool shade of plants. Winter cold restricts the iguana to brief periods of activity in the middle of the day, when body temperature rises to the point that individuals can come above ground and forage. Between early December and the end of February, most days are so cold that the desert iguana cannot venture from its burrow.

For a lizard, patches of different environmental temperatures vary in both time and space. Its response to this environment, particularly its movement between patches, allows it to exploit this mosaic of conditions to best advantage. In lizards, the effect of a change in activity space can have consequences

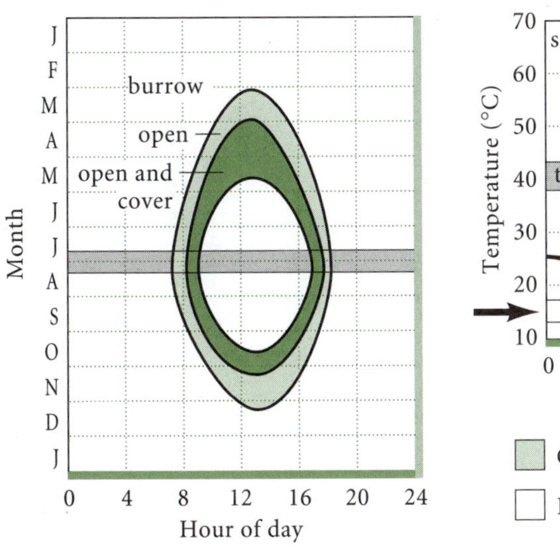

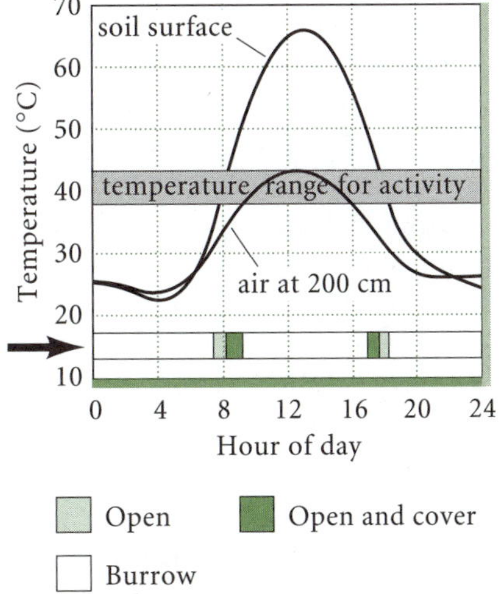

FIGURE 6-16 Seasonal activity space of the desert iguana (*Dipsosaurus dorsalis*) in southern California. (a) The activity budget for July 15 is shown with the time course of environmental temperature. (b) The daily activity budget for an entire seasonal cycle. (*After Beckman et al. 1973.*)

beyond those daily activities such as foraging. Life history characteristics such as annual survival and fecundity have been shown to be closely related to activity space in lizards (Adolph and Porter 1993).

Not all desert organisms have activity spaces that are as tightly synchronized with the environmental conditions as those of the lizards discussed above. The diurnal activity patterns of the desert locust *Schistocerca gregaria* have been shown to be extremely variable. The availability of suitable feeding and oviposition sites is unpredictable in the habitats of the locust, and the ability to quickly adjust foraging and reproductive activity, even in the face of unfavorable temperature conditions, is critically important. In the case of the locust, the activity space is determined more by the biotic environment (e.g, abundance of food and reproductive sites) than by physical conditions (Cloudsley-Thompson 1991).

6.7 Homeostatic responses vary in their time courses.

The time course of a response to changing conditions must be substantially shorter than the period of environmental change. Otherwise today's form and function may reflect yesterday's conditions. Responses to environmental change fall naturally into three general categories: **regulatory, acclimatory,** and **developmental** responses. Regulatory responses are accomplished most rapidly, developmental responses most slowly.

Regulatory responses include both changes in the rates of physiological processes and changes in behavior. The mallee dragon's shade-seeking behavior under heat stress is an example of a regulatory response. These responses do not require modification of existing morphology or biochemical pathways. Acclimatory responses involve more substantial changes, such as thickening of the fur in winter, an increase in the number of red cells in the blood at high altitude, or the production of enzymes with different temperature optima. These changes may be thought of as shifts in the ranges of the regulatory responses of the individual. Both regulatory and acclimatory responses are reversible, as they must be to follow the ups and downs of the environment. When the environment changes more slowly, a given set of conditions may persist during the adult life span of an individual, and the individual may alter its development to produce the phenotype most suitable to the prevailing conditions. Such developmental responses are slow and generally are not reversible.

Acclimation

Organisms exposed to warmer or colder temperatures than normally experienced may undergo reversible changes in morphology and physiology referred to as **acclimation.** Such changes take days to weeks, so acclimation is a strategy restricted to seasonal and other persistent variations in conditions. During the cold winter months, many birds don a heavier plumage, providing greater insulation, than they wear during the hot summer months. These species replace their body feathers in spring and fall; each plumage is suited to the typical conditions of the environment encountered between each molt. The willow ptarmigan, a ground-feeding arctic bird, trades its lightweight, brown summer plumage in the fall for a thick, white winter plumage, which provides both insulation and camouflage against a background of snow. With this increased insulation, ptarmigans expend less energy to maintain their body temperatures during the winter (Figure 6-17). The seasonal change in plumage thickness effectively shifts the regulatory response range to match the prevalent temperature range of the season, so that the ptarmigan needs a similar metabolic rate to maintain its body temperature when the surroundings are −40°C in the winter and −10°C in the summer (a conceivable temperature in its arctic home). Although winter-acclimated individuals would seem well adapted for both winter and summer climates when at rest, summer activity combined with a winter plumage would quickly produce heat prostration. Adjusting insulation to enhance heat conservation in winter and to facilitate heat dissipation in summer maintains a constant body temperature at the least possible cost.

Many temperature-conforming animals and plants also acclimate to seasonal changes in their environments (Schmidt-Nielsen 1983, Hochachka and Somero 1984). By employing a variety of enzymes and

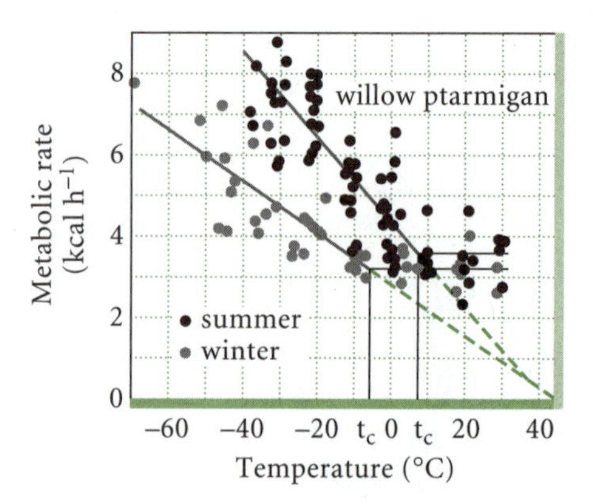

FIGURE 6-17 Metabolic responses of willow ptarmigan acclimated to summer and winter temperatures. Winter-acclimated birds have thicker plumage, providing better insulation than that of summer-acclimated birds. Hence their metabolic rates are lower at any given temperature, and their lower critical temperatures (see Figure 6-3) are also lower. (*After West 1972.*)

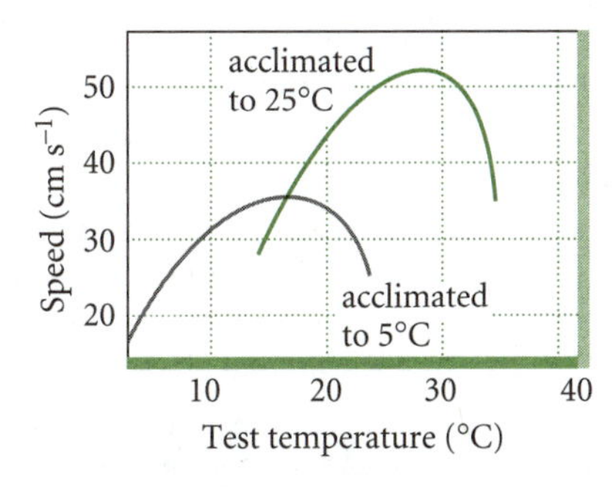

FIGURE 6-18 Swimming speed of goldfish as a function of temperature. Fish acclimated at 25°C display higher swimming speeds (top speed just over 50 cm s^{-1}) than those acclimated at 5°C (top speed around 35 cm s^{-1}). (*After Fry and Hart 1948.*)

other biochemical systems with different temperature optima, ectothermic animals can adjust their tolerance ranges to prevalent environmental conditions. The relationship of the swimming speed of fish to water temperature shows at once the capabilities and limitations of acclimation (Fry and Hart 1948). Goldfish swim most rapidly when acclimated to 25°C and placed in water between 25° and 30°C, conditions that closely resemble their natural habitat (Figure 6-18). Lowering the acclimation temperature to 5°C increases the swimming speed at 15°C, but reduces it at 25°C. Increased tolerance of one extreme often brings reduced tolerance of the other.

The mechanisms of acclimation, and the extent to which acclimation can compensate for changes in the environment, are not equivalent among all organisms. For example, when compared with fish and other ectothermic animals, amphibians appear to have less capacity for acclimatory change in the amount or efficiency of muscle fiber, less ability to manufacture isozymes having greater efficiency in the cold, and less ability to alter enzyme function (Rome et al. 1992).

In plants, acclimation of photosynthetic rate to temperature shows that the capacity for acclimation is often linked to the range of temperatures experienced in the natural environment (Figure 6-19). *Atriplex glabriuscula,* a species of saltbush native to cool coastal regions of California, does not increase its photosynthetic rate at high temperatures when acclimated at an unnaturally high temperature of 40°C; plants acclimated to the 16°C seen in the natural habitat are uniformly more productive at all temperatures. Likewise, the thermophilic (heat-loving) species *Tidestromia oblongifolia* cannot acclimate to cool temperatures. A third plant species that inhabits interior deserts, but is photosynthetically active during the cool winters as well as the hot summers (*Larrea divaricata*), shows the classic shift in temperature optima characteristic of thermal acclimation. The basis for

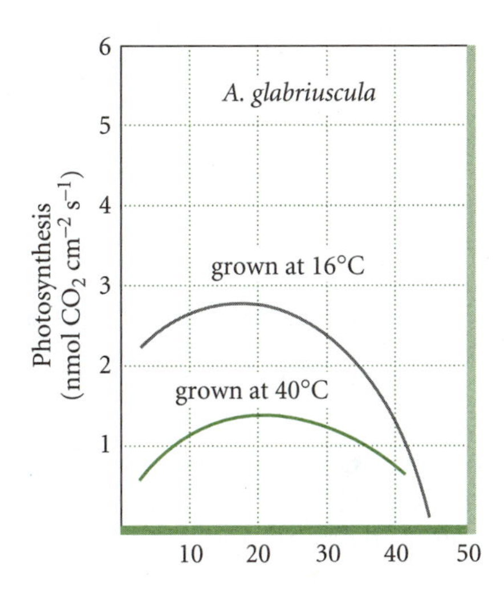

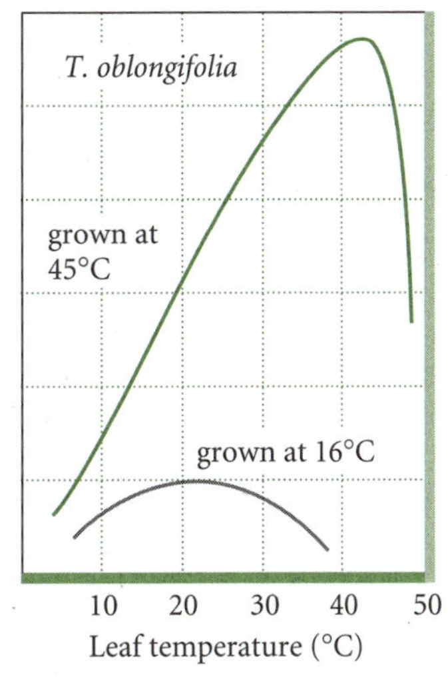

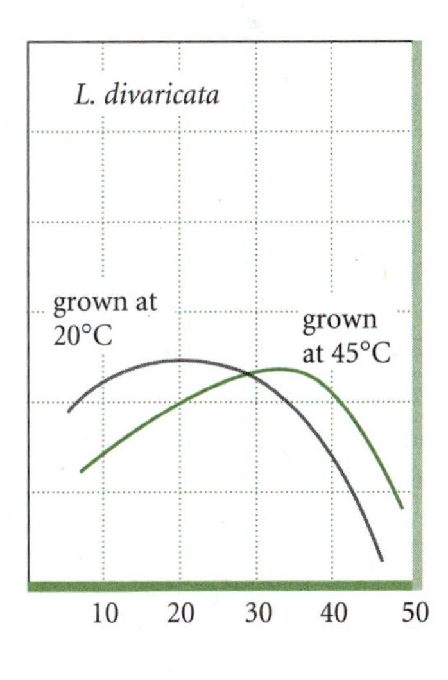

FIGURE 6-19 Light-saturated photosynthetic rate as a function of leaf temperature in three species of plants (genera *Atriplex, Tidestromia,* and *Larrea*) grown under moderate and hot temperatures. The highest point of each curve indicates the temperature at which the maximum photosynthetic rate occurs. The three plant species respond differently to being grown under two temperature conditions. Individuals of *A. glabriuscula* grown at moderate temperatures display a higher photosynthetic rate for nearly all temperatures than those grown at high temperatures. Individuals of the heat-loving species *T. oblongifolia* show an opposite response, with plants grown at high temperatures having greater photosynthetic rates. In *L. divaricata,* plants grown at high temperatures have roughly the same maximum photosynthetic rate (about 2.4 nmol CO_2 cm^{-2} s^{-1}) as those grown at moderate temperatures. However, the temperature at which the maximum rate occurs increases from about 20°C in the plants grown at moderate temperatures to about 35°C in the plants grown at high temperatures. (*From Hochachka and Somero 1984; after Bjorkman et al. 1980.*)

this acclimation appears to be associated with changes in the viscosity of membranes directly related to photosynthetic pathways (Raison et al. 1980). That some plant species lack this capability suggests that mechanisms to acclimate photosynthetic rate to temperature entail a cost to individuals, and that these mechanisms have been dispensed with when plants experience only narrow ranges of temperatures.

Acclimation of photosynthesis may occur in response to environmental factors other than temperature, such as the quality and quantity of light, CO_2 and water availability, and the availability of nutrients (Anderson et al. 1995, Walters and Horton 1995, Terashima and Hikosaka 1995). The mechanisms of acclimation of photosynthesis to light appear to be complex, involving chlorophyll molecules, accessory pigments, and associated electron acceptors organized in the thylakoid membrane of the chloroplast (Anderson et al. 1995). The acclimation of photosynthesis to light highlights the importance of the modular construction of plants. In a plant, photosynthesis takes place in each leaf, and each leaf is subject to different amounts of incident radiation, depending on its position in the canopy or its location with respect to neighboring plants that may provide shade. Conceivably, acclimation to local light levels could occur in each leaf. Thus, from the perspective of the entire plant, achieving acclimation of photosynthesis to light levels or light quality requires a mechanism that integrates the acclimation responses of individual leaves to their local light conditions. This integration is apparently achieved in two ways. First, gradients of light in the leaf canopy may be reduced by gradual changes in leaf distribution and individual leaf inclination from the top of the canopy to the bottom. The light-scattering properties of leaves may also be different depending on the position of the leaf in the canopy. Second, it is known that the maximum photosynthetic rate of a photosystem is related to the leaf nitrogen content. Photosynthetic rates may be adjusted by redistributing nitrogen within the chloroplast, and thus local acclimation responses may be different for leaves having the same leaf nitrogen content (Percy and Sims 1994, Terashima and Hikosaka 1995). It has been suggested that these two mechanisms operate to optimize photosynthetic output for the entire plant (Terashima and Hikosaka 1995).

Developmental Responses

Developmental responses, in which growth and differentiation are sensitive to environmental variation, are conspicuous in plants and in animals with several generations per year. Developmental responses are generally not reversible; once fixed during development, they remain unchanged for the remainder of the organism's life. Developmental responses cannot accommodate short-term environmental changes owing to their long response times and their irreversibility; as a general rule, therefore, plants and animals exhibit developmental flexibility only in environments with persistent variation in conditions experienced by individuals. When environmental changes occur slowly compared with the life span of an organism, as do changes in seasonal conditions for short-lived animals, developmental responses are often appropriate. Also, for plant species whose seeds may settle in many different kinds of habitats, the strategy of developmental flexibility makes good sense.

Light intensity is one of the most important influences on the course of development in plants. Photoreceptors, such as phytochrome, a molecule composed of two identical polypeptide chains, absorb light, which, through a variety of complex mechanisms collectively known as **photomorphogenesis,** influences seed germination, seedling growth, plant development, and flowering. The rapidity with which a seed germinates, the rate at which the seedling grows, the plant's shape and size, and the success with which it reproduces—all of which collectively represent the plant's response to environmental conditions—are dependent to some extent on the light regime. Seed germination in many temperate herbaceous plants is promoted by light (Baskin and Baskin 1988). Loblolly pine (*Pinus taeda*) seedlings grown in shade have smaller root systems and more foliage than seedlings grown in full sunlight (Bormann 1958). Because the shaded environment taxes the plant's water economy less, shade-grown seedlings can allocate more of their production to stem and needles (65% vs. 48% for sun-grown seedlings). Sun-grown seedlings must develop more extensive root systems to obtain sufficient water. The larger proportion of foliage in the shade-grown seedlings results in a higher rate of photosynthesis per plant under given light conditions, particularly under low light intensities (Figure 6-20).

A more complicated example of a developmental response involves wing development in water striders, which are freshwater insects of the genus *Gerris* (Brinkhurst 1959, Vepsalainen 1973, 1974a, 1974b, 1974c). European water strider species fall into four categories of wing length depending on where they live, a situation referred to as **alary polymorphism** (Table 6-2). At one extreme, species inhabiting large, permanent lakes have short wings, or none at all, and do not disperse between lakes. At the other extreme, species living in temporary ponds usually have long, functional wings and disperse to find suitable sites for breeding each year. Between these extremes, species

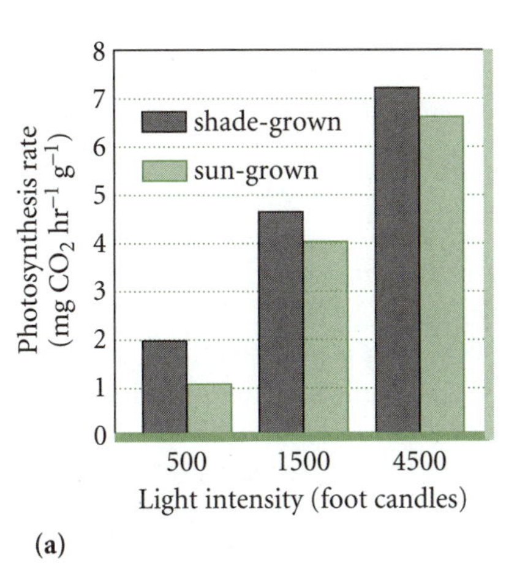

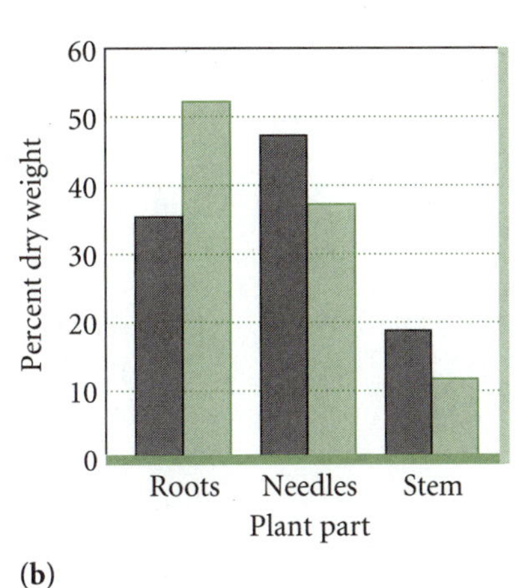

FIGURE 6-20 (a) Rates of photosynthesis and (b) distribution of dry matter between roots, needles, and stems in loblolly pine (*Pinus taeda*) seedlings grown in shade and in full sunlight. Shade-grown plants (gray bars) have a higher rate of photosynthesis at each of three light intensities and allocate less energy to roots than do sun-grown plants (green bars). (*Data from Borman 1958.*)

characteristic of small ponds, which are more or less persistent from year to year but tend to dry up during the summer, frequently have both long-winged and short-winged forms.

The life cycle of most *Gerris* species in central Europe, England, and southern Scandinavia includes two generations per year. The first (summer) generation hatches during the spring, reproduces during the summer, and then dies. The second (winter) generation, hatched from eggs laid by females of the summer generation, develops to the adult stage during late summer, then overwinters before breeding in early spring the following year. In species that inhabit seasonal ponds, the summer generation is dimorphic, having both long-winged and short-winged forms (Figure 6-21). All the individuals in the winter generation have long wings and can fly. They leave the pond in late summer and move into nearby woodlands for the winter. In the spring, they return to the ponds to lay eggs.

Dimorphism in the summer generation reflects two extreme strategies. The long-winged forms can fly to other habitats if their pond dries up, especially when this happens early in the season. The short-winged forms depend on the persistence of the pond. What do water striders gain by producing short wings rather than functional ones? The advantage of having short wings may be higher fecundity, because nutrients that would have been directed to wings and flight muscles can be converted into eggs (Roff 1994).

Because all of the winter generation have long wings, seasonal dimorphism must be a developmental response. K. Vepsalainen (1971) found that wing length is determined primarily by day length. In southern Finland, when day length increases continually during the larval period and exceeds 19 hours during the last nymphal (immature) stage prior to adulthood, individuals grow short wings. When day length begins to decrease before the end of the last nymphal stage (as it does when larval development extends beyond June 21, the summer solstice), long-winged adults are produced. Thus summer generation individuals become short-winged or long-winged depending upon hatching date and rate of nymphal development. The switch between long and short wings is also influenced by temperature (Vepsalainen

TABLE 6-2 **Wing lengths of water strider species (*Gerris*) inhabiting bodies of fresh water with different levels of permanence and predictability**

Habitat	Characteristic wing length	Characteristic determination mechanism
Permanent	Short	Genetic
Fairly persistent, but unpredictable	Both short and long	Genetic dimorphism
Seasonal	Seasonally dimorphic (summer generation)	Developmental switch
Very unpredictable	Long	Genetic

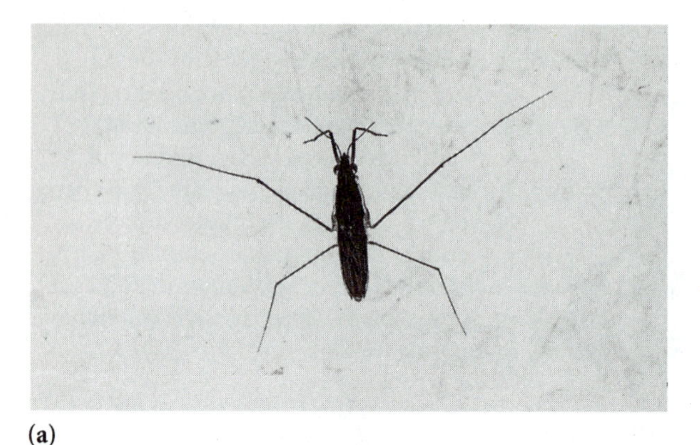

(a)

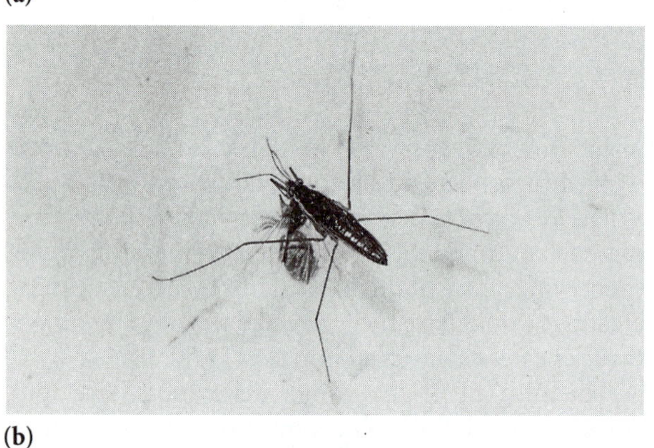

(b)

FIGURE 6-21 Alary polymorphism in the water strider *Gerris odontogaster:* (a) long-winged form and (b) short-winged form. (*Courtesy of K. Vepsalainen.*)

1974c); high temperatures, which cause ponds to dry quickly, favor the development of long-winged forms. The developmental response described by Vepsalainen for wing length in water striders is only one of a large number of similar responses found in various insects with short generations (for example, see Fraser Rowell 1970, Lamb and Pointing 1972, Lees 1966, Steffan 1973, Young 1965, Partridge et al. 1994, Legaspi and O'Neil 1994).

6.8 Migration, storage, and dormancy allow organisms to tolerate seasonally unsuitable conditions.

In many parts of the world, freezing temperatures, extreme drought, low light, and other adverse conditions prevent animals and plants from continually pursuing their usual activities. Under such conditions, organisms resort to a number of extreme responses. These include moving to another region where conditions are more suitable, relying on resources stored during bountiful seasons for use during lean periods, or becoming inactive.

Migration

Many animals, particularly among those that fly or swim, undertake extensive migrations (Baker 1978, Gathreaux 1981, Gwinner 1990, Alerstam 1990, 1991, Berthold 1993). The arctic tern (*Sterna paradisaea*) probably holds the record for long-distance migration with a yearly round trip of 30,000 kilometers between its North Atlantic breeding grounds and its antarctic wintering grounds. Each fall, hundreds of species of land birds leave temperate and arctic North America, Europe, and Asia for more equatorial climes in anticipation of cold winter weather and dwindling supplies of their invertebrate food (Keast and Morton 1980). In East Africa, many of the large ungulates, such as the wildebeest, migrate long distances, following the geographic pattern of seasonal rainfall and fresh vegetation (Maddock 1979). A few insects, such as the monarch butterfly, perform impressive migratory movements each year, and some others undertake local movements (Dingle 1978). Some marine organisms undertake large-scale migrations to reach spawning grounds, to follow a food supply, or to keep within suitable temperature ranges for development. The migration of salmon from the ocean to their spawning grounds at the headwaters of rivers and the reverse migration of adult freshwater eels to their breeding grounds in the Sargasso Sea are striking examples (Jones 1968, Gross et al. 1988, Kimura et al. 1994). Many whales undertake seasonal migrations between feeding and breeding areas (for example, see Jones et al. 1984, Melnikov and Bobkov 1994).

Some populations exhibit irregular or sporadic movements that are tied to food scarcity during particular years, rather than to seasonal conditions (Svardson 1957). The occasional failure of cone crops in the coniferous forests of Canada and in the mountains of the western United States forces large numbers of birds that rely on seeds to move to lower elevations or latitudes. Birds of prey that normally feed on rodents disperse widely when their prey populations decline sharply (Gross 1947, Snyder 1947). Even insects are subject to irregular movements. Outbreaks of migratory locusts, spreading out from areas of high local density where food has been depleted, can reach immense proportions and cause extensive crop damage over wide areas (Gunn 1960, Waloff 1966; Figure 6-22). This behavior is a developmental response to population density (Uvarov 1961). When the locusts grow up in sparse populations, they become solitary and sedentary as adults. In dense populations, frequent contact with others stimulates young individuals to develop a gregarious, highly mobile behavior, often leading to mass emigration following local depletion of food resources.

FIGURE 6-22 A dense swarm of migratory locusts in Somalia, Africa, in 1962. Swarming is thought to be a developmental response to locally high population densities. (*Courtesy of the U. S. Department of Agriculture.*)

Storage

Although homeostasis and migration help maintain function in the face of a changing physical environment, environmental changes often plunge organisms from feast into famine. When an environment marginally supports life, even small fluctuations in food or water supply can be critical. To prevent disaster in such circumstances, many plants and animals store resources during periods of abundance for use in times of scarcity. During infrequent rainy periods, desert cacti store water in their succulent stems. Plants growing on infertile soils absorb more nutrients than they require in times of abundance and use them when soil nutrients are depleted. Many temperate and arctic animals accumulate body fat during mild weather in winter as a reserve of energy for periods when snow and ice make food sources inaccessible. Some winter-active mammals (beavers and squirrels) and birds (acorn woodpeckers and jays) cache food supplies underground or under the bark of trees for later retrieval (Ritter 1938, Swanberg 1951, Andersson and Krebs 1978, Tomback 1980, Sherry 1984). In habitats that frequently burn, such as the chaparral of southern California, perennial plants store food reserves in fire-resistant root crowns, which sprout and sent up new shoots shortly after a fire has passed (Figure 6-23).

Dormancy

The environment sometimes becomes so extremely cold, dry, or nutrient-poor that animals and plants can no longer function normally. Under such circumstances, those incapable of migration may enter physiologically dormant states. Many mammals hibernate because they cannot find food, not because they are physiologically unable to cope with the physical environment. Dormancy in plants refers to the situation in which roots, buds, seeds, or leaves stop growing and reduce their metabolism in anticipation of extreme intolerable conditions of cold or drought. Many tropical and subtropical trees shed their leaves during

(a)

(b)

FIGURE 6-23 Root-crown sprouting by chamise (*Adenostoma fasciculatum*) following a fire in the chaparral habitat of southern California. (a) May 4, 1939, 6 months after the burn. (b) July 16, 1940, showing extensive regeneration. (*Courtesy of the U.S. Forest Service.*)

seasonal periods of drought; temperate and arctic broad-leaved trees shed theirs in the fall because they cannot obtain from frozen soil the moisture needed to maintain them (Vegis 1964). The breaking of dormancy is often dependent on the plant's experiencing extreme conditions such as prolonged freezing (Hopkins 1995). Otherwise, a plant might emerge from the dormant state before the season of extreme conditions had passed.

For many small invertebrates and cold-blooded vertebrates, freezing temperatures directly curtail activity and lead to dormancy. In rare instances, organisms become nearly completely dehydrated and cease metabolic activity altogether, a condition known as **anhydrobiosis.** The larvae of the chironomid fly *Polypedilum vanderplanki,* an inhabitant of exposed rock pools, become completely dehydrated when their pool dries up. When the rains come, the larvae rehydrate within hours to resume their development (Hochachka and Somero 1984). In most species, conditions requiring dormancy are anticipated by a series of physiological changes (for example, production of antifreezes, dehydration, fat storage) that prepare the organism for a complete shutdown (Prosser 1973, Mrosovsky 1976, Clutter 1978, Gregory 1982, Lyman 1982).

Prior to winter, insects enter a resting state known as **diapause,** in which water is chemically bound or reduced in quantity to prevent freezing and metabolism drops to near zero (Hochachka and Somero 1984, Lee et al. 1987). In summer diapause, the bodies of drought-resistant insects either dry out to tolerate desiccation or are covered by an impermeable outer covering to prevent drying. Plant seeds and the spores of bacteria and fungi have similar dormancy mechanisms (for example, see Koller 1969).

6.9 Proximate cues enable organisms to anticipate predictable environmental change.

What stimulus indicates to birds wintering in the Tropics that spring is approaching in northern forests, or forces salmon to leave the seas and migrate upstream to their spawning grounds? How do aquatic invertebrates in the Arctic sense that if they delay entering diapause, a quick freeze may catch them unprepared for winter? J. R. Baker (1938) made the important distinction between **proximate factors,** which are aspects of the environment, such as day length, that an organism uses as a cue for behavior, and **ultimate factors,** which are features of the environment, such as the amount of food available, that bear directly upon the well-being of the organism. Virtually all plants and animals sense the length of the day (photoperiod) as a proximate factor that indicates season, and many can distinguish periods of lengthening and shortening days (Bunning 1967, Lofts 1970, Vince-Prue 1975, Murton and Westwood 1977, Beck 1980). In plants, the timing of germination, flowering, and leaf abscission are all controlled by day length.

Similar organisms may differ strikingly in their responses to photoperiod in different locations. Under controlled cycles of light and dark, southern populations (30° N) of side oats grama grass flowered when day length was 13 hours, whereas more northerly populations (47° N) flowered only when the light period exceeded 16 hours each day (Olmsted 1944). The longer period of light suppressed flowering in the southern populations. At 45° N in Michigan, populations of small freshwater crustaceans known as water fleas (*Daphnia*) form diapausing broods at photoperiods of 12 hours (mid-September) or less (Stross and Hill 1965). In Alaska, at 71° N, related species enter diapause when the light period decreases to fewer than 20 hours per day, which corresponds to mid-August (Stross 1969). Warm temperatures and low population densities tend to shorten the day length necessary for diapause (and hence delay the inception of diapause in the fall), suggesting that these factors portend more favorable environmental conditions for *Daphnia.*

When day length does not accurately predict sporadically changing conditions, animals and plants must take their cues more directly from changes in ultimate factors in the environment. Annual cycles in equatorial regions, where day length is nearly constant, follow upon seasonal cycles of rainfall and their effects on humidity and vegetation. In such highly unpredictable environments as deserts, many organisms adopt a conservative strategy of readiness during the entire period in which sporadic rains are likely to occur. In some desert birds, photoperiod stimulates the development of reproductive organs to a point just short of breeding. The gonads maintain this state of readiness throughout the time when rains might occur, but rainfall itself finally kicks off the completion of their physiological development and the initiation of breeding (Marshall and Disney 1957, Immelmann 1971).

6.10 Ecotypic differentiation reflects adaptation to local conditions.

About 75 years ago, the Swedish botanist Gote Turesson collected seeds of several species of plants that occurred in a wide variety of habitats and grew them in his garden. He found that even when grown under identical conditions, many of the plants exhibited different forms depending upon their habitats of origin. Turesson (1922) called these forms **ecotypes,** a name that persists to the present, and suggested that ecotypes represent genetically differentiated strains of a population, each restricted to specific habitats. Because Turesson grew these plants under identical conditions, he realized that the differences between the ecotypes must have a genetic basis, and that they must have resulted from evolutionary differentiation within the species according to habitat.

Botanists have long recognized that individuals of a species grown in different habitats may exhibit various forms corresponding to the conditions under which they are grown. In many cases, these differences result from developmental responses. But experiments such as Turesson's have revealed genetic adaptations to local conditions. The phenotype of an individual is fixed, but phenotypes vary among individuals from place to place. In the hawkweed *Hieracium umbellatum,* for example, woodland plants generally have an erect habit; those from sandy fields are prostrate; and those from sand dunes are intermediate in form. Leaves of the woodland ecotype are broadest, those of the dunes ecotype are narrowest, and those of sandy fields are intermediate. Plants from sandy fields are covered with fine hairs, a trait the others lack.

Jens Clausen and coworkers (1948) conducted similar experiments on a species of yarrow, *Achillea millefolium,* in California. *Achillea,* a member of the sunflower family, grows in a wide variety of habitats ranging from sea level to more than 3,000 meters elevation. Clausen collected seeds from plants at various points along the altitude gradient and planted them all at Stanford, California, near sea level. Although the plants were grown under identical conditions for several generations, individuals from montane populations retained their distinctively small size and low seed production (Figure 6-24), thereby demonstrating ecotype differentiation within the population. Such region and habitat differences in adaptations undoubtedly broaden the ecological tolerance ranges of many species by dividing them into smaller subpopulations, each differently adapted to consistent local environmental conditions (Kruckenberg 1951, McMillan 1959, Mooney and Billings 1961, Hiesey and Milner 1965, Bjorkman 1968, Antonovics 1971, McNaughton 1973, Eickmeier et al. 1975, Coyne et al. 1983, Schemske 1984, Silander 1985, Loik and Nobel 1993). Such ecotypic differentiation may influence the response of plants to climate change in different regions. Norton et al. (1995) found that five ecotypes of the plant *Arabidopsis thaliana* increased in biomass when exposed to elevated CO_2 levels in the atmosphere, but that the level of increase was different among the ecotypes.

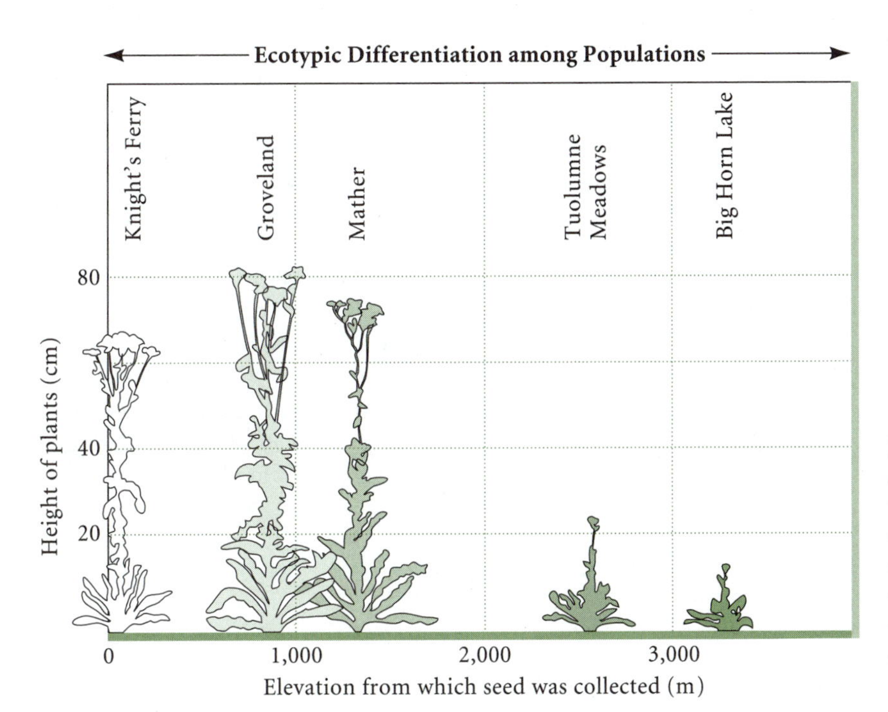

FIGURE 6-24 Ecotypic differentiation in populations of the yarrow *Achillea millefolium,* demonstrated by raising plants from different elevations under identical conditions in the same garden. Seeds obtained from high altitudes produced shorter plants than those from lower elevations. (*After Clausen et al. 1948.*)

SUMMARY

1. Maintenance of constant internal conditions, called homeostasis, depends upon negative feedback responses. Organisms sense changes in their internal environments and respond to return those conditions toward their optimum.

2. Organisms that maintain constant internal environments in response to variation in the physical environment are called regulators. Those that allow their internal environments to follow external changes are called conformers.

3. The maintenance of body temperature within a narrow range, as in birds and mammals, is called homeothermy. In poikilothermy, the internal body temperature conforms to the ambient temperature. Organisms that use the heat produced by metabolism to regulate body temperature are referred to as endotherms. Some endotherms, such as some bats, may facultatively lower their body temperatures, a phenomenon known as heterothermy. Organisms that rely on external sources of heat to warm their bodies are called ectotherms. Some ectotherms, such as some snakes and fish, have physiological mechanisms to raise their body temperatures in some circumstances, a process referred to as facultative endothermy.

4. The magnitude of the body-environment gradient and the constancy of the internal conditions that organisms maintain balance the costs of homeostasis (maintaining the physiological apparatus and sustaining the gradient) against the benefits of closely regulated internal conditions.

5. A major component of homeostasis is the selection of microhabitats that minimize the body-environment gradient. This strategy is illustrated by the temperature-dependent foraging behavior of desert birds and their construction of nests so as to ensure that the environments of their chicks remain stable.

6. Conditions of the physical environment have different temporal and spatial scales. The changes in temperature, moisture, and light that occur with the alternation of day and night and those that occur with the change of seasons are examples of regular temporal variations. Interacting with these changes are less predictable variations caused by fires, tornadoes, hurricanes, and drought. Spatial variation in the environment is determined by the size and arrangement of suitable habitat patches and by the speed with which the organism moves through the environment.

7. The activity space of an individual is the set of conditions in which it functions. The activity space depends on the suitability of conditions and availability of resources in each patch at each time. This phenomenon is illustrated by the influence of changing thermal conditions on habitat utilization by desert lizards. Since plants cannot move, they regulate their activity primarily in response to temporal rather than spatial variation in conditions.

8. Behavioral selection of habitats cannot fully compensate for environmental change, and organisms must bring a variety of homeostatic responses into play. The most rapid of these are regulatory responses, which involve changes in rates of processes and the attitude of the organism, but not in its morphological structure or biochemical pathways. Shivering in response to cold stress and contraction of the pupil in response to bright light are examples. Regulatory responses are rapid and reversible.

9. Acclimatory responses involve reversible changes in structure (for example, fur thickness) or biochemical pathways (changes in the amounts of enzymes and their products), and require longer periods (usually days to weeks) than regulatory responses. Acclimation plays a prominent role in response to seasonal change by long-lived organisms.

10. Developmental responses express the interaction between an organism and its environment during the growth period. Different environments lead to characteristic structures and appearances that generally are not reversible. Responses to sun versus shade by plant seedlings and variation in the length of wings in water striders depending on the permanence of their habitat are examples of developmental responses.

11. When conditions exceed their ranges of tolerance, organisms may migrate elsewhere, rely on materials stored during periods of abundance, or enter inactive states (for example, torpor, hibernation, diapause).

12. To respond appropriately to environmental change, the individual must anticipate changes in conditions. Organisms rely on proximate cues, such as day length, to predict changes in ultimate factors, such as food supply and temperature, that directly affect their well-being.

13. When populations become subdivided in habitats with different conditions, each subpopulation may evolve genetic adaptations to the local environment. Such differentiated forms are called ecotypes. Well-known examples are found among plants, which, because of their low mobility, exhibit evolutionary isolation of subpopulations on small geographic scales.

EXERCISES

1. All animals have adaptations for regulating their body temperature. Construct a chart that compares these adaptations in insects, amphibians, reptiles, fish, and mammals.

2. Construct an activity space diagram (see Figure 6-16) for your activities during the past week.

3. We discussed ecotypes of plant species. Are there animal ecotypes?

CHAPTER 7

Biological Factors in the Environment

GUIDING QUESTIONS

- How do evolutionary responses to the biotic and abiotic environments differ?
- What are some of the adaptations of predators for capturing and eating prey?
- What are some of the ways in which prey avoid detection by predators?
- What is the advantage of polymorphism?
- How do insects use chemicals to protect themselves from predators?
- What are some of the ways in which plants avoid herbivores?
- How do herbivores circumvent the protective devices of plants?
- What are some characteristic adaptations of parasite life cycles?

We saw in Chapters 4, 5, and 6 how the form and function of plants and animals evolve in response to physical factors of the environment—referred to as the **abiotic** environment—so as to match the organism to its surroundings. These adaptations arise through the natural selection of those individuals genetically best suited to the environment at a particular time and place. The winners of the evolutionary contest are those that leave the most progeny that themselves reproduce (see Chapter 1). Were the individual's accommodation to the physical environment the only consideration for survival, we would expect life to be much less complex and variable. Presumably, each set of physical conditions selects for organisms with the combinations of anatomic, physiological, and behavioral traits that best suit them to local conditions. Once this ideal has been achieved, and providing the environment does not change, the engines of evolution, to the degree that they are fueled by the exigencies of physical factors, should slow to a halt. Furthermore, when organisms experience similar conditions, selection by physical factors would seem to promote uniformity of appearance among them. Indeed, plants and animals do arrive at common forms dictated by factors in the physical environment. For example, in arid habitats around the world, many different types of plants grow small, leathery leaves as adaptations to reduce water loss. But what, then, accounts for the enormous variety of living things?

To answer this question, we must appreciate that an organism's environment is composed of much more than a set of physical conditions. There is another dimension—an evolving, interactive, complex, and extremely variable dimension—to which organisms must respond. This **biotic** environment is composed of all of the living things, including plants, animals, microbes, fungi, and protists, with which an individual shares its space and time. Organisms take up space, consume limited resources, and in some cases, represent a direct danger to other organisms through processes such as predation, herbivory, parasitism, and disease. The morphological and physiological characteristics of a plant are, in part, adaptations to the range of local physical conditions, but to ensure its survival, those features of body plan and function must also protect it from herbivorous insects and plant diseases. We saw in Chapter 1 that the giant red velvet mite of the Mojave Desert of southern California spends most of the year in a burrow, emerging only in the winter months, when it can tolerate the temperatures on the desert surface. But it is no coincidence that the emergence of the mite coincides with that of its favored prey, termites.

In this chapter, we present examples of the challenges that organisms face in adapting to the biotic environment, and we describe some of the solutions that have evolved. Just as with our discussion of how plants, animals, and microorganisms contend with the abiotic environment, we will focus on individual adaptations of form and function and on the costs and benefits of those adaptations. Our discussion will center on three of the most important ecological interactions, those between predator and prey, herbivore and plant, and parasite and host. In Part 4 of the book we will reconsider these interactions, along with a fourth one—competition—focusing on how their effects are manifested in populations of individuals.

7.1 Biotic factors promote diversification while abiotic factors promote convergence.

The markings that allow moths to blend into their resting places during the day clearly allow them to escape the notice of most predators. By their insistent colors and fragrances, flowers draw our attention as well as the notice of insects and birds that carry pollen from one flower to the next and effect fertilization. The agents that have selected these adaptations are biological, and their effects differ from those of physical factors in two ways. First, biological factors elicit **interactive traits:** a predator shapes its prey's adaptations for escape, but its own adaptations for pursuit and capture are just as surely shaped by the prey. Second, biological factors tend to diversify adaptations, rather than promoting **convergence**—resemblance among organisms of different taxonomic groups resulting from adaptations to similar environments. Organisms are specialists with respect to biological factors in their environments; each pursues a different combination of prey, strives to avoid a different combination of predators and disease organisms, and engages in cooperative arrangements with a unique set of pollinators, seed dispersers, or gut flora.

The biotic environment promotes diversification not only because of the many different kinds of organisms present, but also because of the nearly limitless set of local conditions created by the interactions of these organisms with one another and with the physical environment. The physical environment within a habitat offers a variety of places—called **microenvironments** or **microhabitats**—each presenting a distinct set of conditions to the individual. (Harper's studies of seed germination in *Plantago* that we presented in Chapter 6 exemplify the importance of microhabitats.) The presence and activities of plants and animals change the local conditions for other organisms. For example, large trees may reduce the light available to vegetation below them. So, the number of microhabitats increases with the presence of plants and animals. This variety allows for some specialization with respect to physical factors, but very little, really, compared with the diverse adaptations of organisms to biological factors in their environments. But because each population of animal, plant, or microbe has a separate evolutionary history and future, and because the interaction of each pair of organisms creates its own special "biological microenvironment," the possibilities for diversification are limited only by the number of kinds of organisms.

7.2 Adaptations of carnivores demonstrate the importance of the biotic environment as an agent of natural selection.

Predators capture and consume other organisms to obtain nourishment. One need only watch a house cat stalking a mouse to appreciate how superbly evolution has endowed predators for this task. Having slept away the day on your living room sofa, the cat, excluded from the house in the evening—perhaps rudely—moves about the darkened lawn with remarkable stealth. Using eyes adapted for night vision, it spots the mouse, and aided by the loose articulation of its front legs with the shoulder girdle and its powerful hind legs, it springs for the capture, subduing the mouse with retractable claws and dispatching it with several quick bites of its sharp teeth. To be sure, not all predation events are as dramatic as those exhibited by wild and domestic cats. The adaptations of the blue whale, which eats small, shrimplike krill and small fish by the millions, or the water filtration systems used by oysters, clams, and mussels are just as exquisite and just as exemplary of the complex evolutionary interplay between predator and prey.

Predators that stalk and subdue prey one at a time face multiple challenges (Endler 1991). Like all organisms, prey species occur in areas where the physical and biological conditions are suitable for their growth and reproduction, and within those areas, they expose themselves to predators in daily and monthly rhythms that correspond to their activity space. Predators must position themselves in locations where prey occur, and once there, they must be capable of detecting individual prey and determining the profitability of eating that prey. An individual prey must be captured, and the predator is faced with the challenge of circumventing the early warning system of the prey to maneuver within attack range. And, of course, the attack must succeed, the prey must be subdued and consumed—often in the presence of other hungry predators who might find it easier to steal prey from another predator than to hunt down and kill prey for themselves. To accomplish all this, the predator must have the proper sensory mechanisms to locate prey. It must be large enough to overpower the prey, or it must be otherwise suitably equipped to subdue the prey. It must be capable of ingesting the prey, and its gut must be adapted for extracting nourishment from the prey's flesh. In evolutionary terms, all of this must be accomplished against a background of change in the adaptations of the prey to escape predation. What are some of the prominent adaptations of predators for capturing prey?

Size and Predator-Prey Allometry

In general—the krill-feeding whale mentioned above notwithstanding—large predators take large prey (Figure 7-1). But predators experience limits in the size of prey that they can capture and consume. In general, as the size of prey in relation to predator increases, prey become more difficult to capture, and predators become more specialized for pursuing and subduing their prey. Beyond a critical size ratio, predators lack sufficient strength and swiftness to capture potential prey. Lions will attack animals their own size or a little larger, but they are no match for fully grown elephants (Schaller 1972). Of course, predators may cooperate to kill prey that are larger than what any individual predator can subdue.

The ratio of predator body size to prey body size is, to some extent, determined by the immediate foraging decisions of the predator. Most predators prefer to take prey of the specific size that provides the most energetic gain for the least energetic cost (highest profitability). But most predators also have the ability to take prey of a variety of different sizes. In some situations, such as when prey are scarce, predators may opt to take extremely large prey, even though the cost of subduing the prey, and the risk of being injured during the process, is high. Thus, within the limits of prey size, the relationship between the body size of the predator and that of the prey is very much influenced by prey abundance and predator nutritional state. This is discussed in more detail in Chapter 31, in which we describe foraging theory.

The relationship between organisms and their environment may change with the size of the organism because some physical and biological processes scale nonproportionally with respect to size (see Chapter 3). That is, shape and physiological function do not change isometrically (geometrically) with body size. Such allometric relationships also play an important role in the morphological relationship between predator and prey. Consider how the force of the jaw muscles of a predator, as measured by the cross-sectional area of the muscle mass, might change as the body mass of the predator changes. If muscle force were isometric to body size, then it would scale as (body mass)$^{2/3}$, which would mean that larger predators would have relatively less muscle force at their disposal than smaller predators, thereby effectively diminishing the advantage of size in predation. Thus, equivalence of predatory function among predators of different sizes usually implies the presence of allometry.

FIGURE 7-1 With their powerful legs,and jaws, lions can subdue prey somewhat larger than themselves. But because they cannot maintain speed over long distances, successful hunting relies on stealth and surprise.

FIGURE 7-2 The appearance of flowers to the human eye (*left*) and to eyes that are sensitive to ultraviolet light (*right*): (a) marsh marigolds; (b) five species of yellow-petaled Compositae from central Florida. (*Courtesy of T. Eisner; from Eisner et al. 1969.*)

Adaptations for Locating, Capturing, and Subduing Prey

Many adaptations have arisen among predators for the capture of prey. We are familiar with the speed and agility of the big cats, the ferocity and razor-sharp teeth of sharks, the springlike power and injectable venom of pit vipers, and the controlled crash of the gannet as it plunges into the water to grab a fish just beneath the surface. Some species, such as chimpanzees and other primates, wolves, hyenas, lions, and army ants, hunt cooperatively to run down and subdue prey larger than themselves (Rettenmeyer 1963, Mech 1970, Kruuk 1972, Boesch 1994).

To locate and capture food, predators possess sensory capabilities commensurate with their habitats, feeding tactics, and their prey's ability to avoid detection (Gordon 1968, Prosser 1973, Lythgoe 1979, Dusenbery 1992). We ourselves primarily use vision to locate food, particularly as it is now displayed on the shelves of supermarkets. Yet our vision is pitiable compared with that of hawks and falcons (Fox et al. 1976), and most insects can perceive ultraviolet light invisible to us (Figure 7-2). Insects also can detect rapid movement, such as that of wings beating 300 times per second; we see such movement as only a blur. Many animals have mechanisms to accommodate to changes in light intensity that occur during the day (Dusenbery 1992). For example, nocturnal animals such as cats possess slit-shaped pupils that allow them to restrict light from entering their eyes better than the round pupils of day-active animals (like humans) can (Lythgoe 1979). The anatomic position of the eyes of a predator, like a cat, and a prey animal, such as a squirrel, is illustrative of the different selective pressures brought to bear on predator and prey. The forward-directed eyes of the cat provide overlapping fields of view and binocular vision. Thus, the cat has great depth of field, a distinct advantage in spotting and capturing prey. But the cat has a rather limited field of view; that is, it must move its head to see to the side. The eyes of the squirrel, on the other hand, are on the sides of the head, providing an increased field of view to the side and rear of the animal (Dusenbery 1992).

Among the more unusual sensory organs of predators are the pit organs of pit vipers, a group that includes the rattlesnake (Figure 7-3). The pit organs, located on each side of the head in front of the eyes, detect the infrared (heat) radiation given off by the warm bodies of potential prey—a sort of "seeing in the dark" (Grinnell 1968). Pit vipers are so sensitive to

FIGURE 7-3 Head of western rattlesnake (*Crotalus veridis*), showing the location of the infrared-sensitive pit organ between, and slightly lower than, the eye and the nostril. (*Courtesy of R. B. Suter.*)

infrared radiation that they can detect a small rodent several feet away in less than a second. Moreover, because the pits are directionally sensitive, vipers can locate warm objects precisely enough to strike them.

Some sensory adaptations are nothing short of exotic. For example, some species of "electric fish" produce continuous discharges of electricity from specialized muscle organs, creating a weak electric field around them. Nearby objects distort the field, and these changes are picked up by receptors on the surface of the electric fish (Machin and Lissman 1960). Some species use electric signals to communicate between individuals. The specialized electric ray *Torpedo* uses powerful electric currents, up to 50 volts at several amperes, to defend itself and to kill prey (Keynes and Martins-Ferreira 1953). As one might expect, the production and sensation of electric fields are most highly developed in fish inhabiting murky waters. In other habitats with poor visibility, bottom-dwelling species such as catfish use elongated fins and barbels around the mouth as sensitive touch and taste receptors.

In contrast to the magnificent senses of many predators, others perceive their surroundings only dimly and rely upon chance to bump into prey. (Of course, their prey must be equally oblivious for this tactic to work.) But even such "blind" predators adopt searching patterns to increase their chances of encountering prey. For example, the predatory larvae of the ladybird beetle feed on mites and aphids that infest the leaves of certain plants, and they must physically contact their prey to recognize them (Figure 7-4). Their movements on the leaves are not oriented toward the prey, but neither are they random. The veins and rims of leaves make up less than 15% of the leaf surface, yet the larvae spend most of their time searching on these areas—which they recognize by touch—where almost 90% of aphids are distributed (Dixon 1959).

FIGURE 7-4 Adult and larval ladybird beetle (family Coccinellidae) feeding on aphids in a laboratory culture. Note the hairs on the veins of the leaf, which deter the aphids from penetrating the plant and sucking its juices. (*Courtesy of the U.S. Department of Agriculture.*)

Some predators construct traps or snares in order to capture prey. The most obvious and, arguably, the most beautiful such devices are the webs of spiders, which may take on one of a nearly limitless variety of forms depending on the habitat arrangement and preferred prey type. Spiders of the family Araneidae build the familiar orb webs that are composed of a long thread of sticky silk laid in a spiral pattern on a frame of nonsticky threads. Orb webs are specialized for capturing aerial prey. Many spiders build webs close to the ground or nestled within the branches of vegetation, where they may snare walking prey or where low-flying insects may be knocked out of the air into a tangle of webbing. Burrowing and funnel web spiders forage from the mouths of tunnels that they construct in the ground, the webbing deposited around the mouth of the burrow serving as an early warning system of approaching pedestrian prey. With the exception of the orb webs, which to even the most casual observer appear highly organized, spider webs may seem haphazardly arranged. This apparent disorganization is misleading. The web, like the teeth of the lion or the electric field of the ray, represents a solution to the selective pressures of the biotic environment. Indeed, the architecture of spider webs is so attuned to the biotic environment that arachnologists can often identify the genus or species of a particular spider (and sometimes the relative age of the spider) simply by examining the web.

Adaptations for Feeding in Predators

For the predator to make use of the prey that it has captured and killed, it must get the prey into its gut, where it can be digested and where the nutrients contained in the flesh of the prey may be assimilated. For some animals, such as snakes, the process of capturing and subduing prey is more or less continuous with the process of ingestion. The snake partially subdues the prey with a bite or by wrapping, and then finishes it off as it swallows the prey whole. Some species of snakes compensate for their lack of grasping appendages with distensible jaws that enable them to swallow large prey whole (Figure 7-5). Many fish predators, such as trout and pike, expend little energy in subduing prey, inasmuch as it is sucked into the mouth and then swallowed in a single powerful gulp. Diving birds often eat large fish, but must swallow them whole because their hind legs are specialized for swimming and diving rather than for grasping and dismantling prey. In these animals, ingestion begins nearly immediately after the attack.

For many other predators, the process of killing the prey is only the beginning of the foraging event.

FIGURE 7-5 The jaw articulation of some snakes, such as this garter snake shown here consuming a whole fish, allows them to open their mouths wide enough to eat relatively large prey whole. (*Courtesy of R. B. Suter.*)

These animals must dismember the prey and take it into the gut bit by bit. Many predators use their forelegs to tear their food into small morsels. Among birds, for example, the hawks, eagles, owls, and parrots use their powerful, sharp-clawed feet and hooked beaks for this purpose. Spiders are not equipped to chew prey or ingest it whole or in large parts. Instead, enzymes from the gut are pumped through the mouth onto the dead prey, where they liquefy the prey's tissue and begin the digestion process prior to ingestion. The spider feeds on the partially digested liquid prey.

7.3 Prey defend themselves by avoidance, deception, and protective morphology.

The evolutionary interplay between predators and prey is not one-sided. The predator is very much a part—an extremely important part—of the prey's biotic environment. Innovations by predators present new evolutionary challenges for their prey. Just as those predators that are a bit more observant, a bit quicker, or a bit more efficient will be at a selective advantage, those prey that avoid detection by the predator, or foil an attack and escape, or which, through some arrangement of morphology or behavior, confuse the predator, are more likely to survive and reproduce. The innovations for avoiding predation are as exquisite and beautiful as those that have arisen to promote predation.

Avoiding Detection by Predators

One way to avoid predation is to occupy microhabitats that are not frequented by predators, or to adopt activity periods that do not overlap extensively with those of potential predators. Prey species may be adapted to seek inconspicuous retreats during periods when they are not active or to function at times or under circumstances that disadvantage the predator in some way. For example, prey might avoid predators that are adapted to hunting during the daytime by being active at night.

Predators often do not take prey in proportion to their availability in the environment. Rather, they preferentially consume the most common or the most obvious prey, a strategy referred to as **apostatic prey selection** (Figure 7-6). By engaging in such behavior, the predator may decrease the abundance of one prey type—the current favored type—to the point at which that prey is less abundant or less obvious than a second prey type. When this happens, the predator may switch to the second prey type. This predation strategy tends to increase the advantage to the prey of being rare, since predators often switch to more numerous prey long before a rare prey species becomes impossible to find (Endler 1991). Some prey species may exist in more than one distinct form. Such **polymorphisms** may bestow an advantage on the prey by making it appear to the predator to be more rare than it actually is. A predator faced with choosing from among apparently distinct forms, each of which is relatively uncommon, may switch to a more common prey type, thereby reducing the predation pressure on the polymorphic species.

Prey whose population sizes, habitat preferences, or activity periods result routinely in encounters with

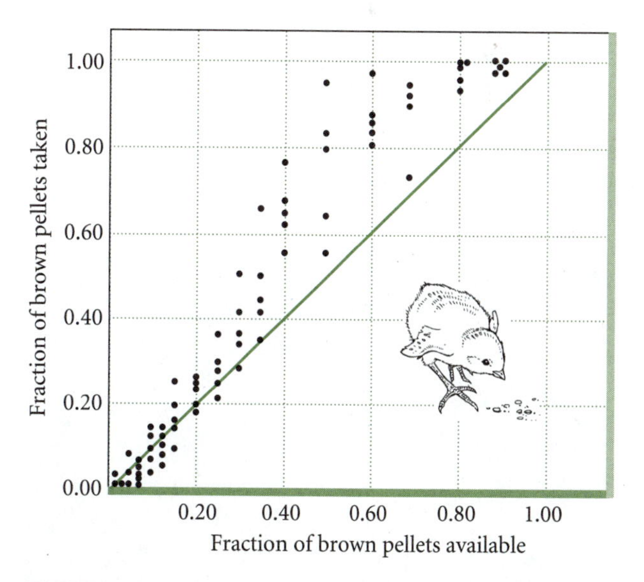

FIGURE 7-6 Apostatic prey selection. In a series of trials, different proportions of brown and green food pellets were presented to chicks. Each dot indicates the proportion of brown pellets taken during one trial. Dots that fall on the line represent trials in which the number of brown pellets taken equaled the number available. When brown pellets were rare in the environment, the chicks took them roughly in proportion to their availability. However, when brown pellets were more common than green pellets, brown pellets were preferred. (*From Endler 1991; redrawn from Fullick and Greenwood 1979.*)

predators face the challenge of remaining undetected during such encounters. Such prey may have evolved characteristics of morphology and behavior that render them undetectable, or nearly so, by predators. Such **crypsis** may be achieved by matching the color and pattern of the background upon which they rest. Various animals resemble inedible objects: sticks, leaves, flower parts, even bird droppings. The elaborate adaptations that have arisen to conceal their heads, antennae, and legs underscore the importance of these cues to predators. In the stick-mimicking phasmids (stick insects) and leaf-mimicking katydids, the legs are often concealed in the resting position either by being folded back upon themselves or upon the body, or by being protruded in a stiff, unnatural fashion (Robinson 1969). The dead-leaf-mimicking mantis *Acanthops* partially conceals its head under its folded front legs (Figure 7-7). Asymmetry is also a good cover for animals, but it is difficult to achieve. The leaf-mimicking moth *Hyperchiria nausica* produces an asymmetrical appearance by folding one forewing over the other (Figure 7-8). Moths sometimes rest with a leg protruding on one side, but not the other, or with the abdomen twisted to one side to break their symmetry. The behavior of cryptic organisms must correspond to their appearance. A leaf-mimicking insect resting on bark, or a stick insect moving rapidly along a branch, would not be likely to fool many predators.

Caterpillars, the plant-feeding larvae of moths and butterflies, are subject to intense predation pressure by predators that rely heavily on vision, such as birds and some invertebrates, particularly spiders (Heinrich 1993, Stamp and Wilkens 1993). It is not surprising, therefore, that some of the most extraordinary adaptations for avoiding detection have arisen among these animals. Among the most remarkable are those of the "stem" caterpillars (family Geometridae, which also includes inchworms), which gain protection from bird predators by resembling, in exquisite detail, the stems of the plants on which they feed (Figure 7-9). The behaviors of these caterpillars are carefully choreographed to ensure that potential predators see only a "stem" and not a caterpillar. Leaf-eating caterpillars face a special problem with respect to detection by visually oriented predators: they change the size and shape of the item that they are feeding on—the leaf—in such a way as to make their location obvious. Insect-feeding black-capped chickadees (*Parus atricapillus*), when offered a choice of damaged and undamaged leaves in experimental situations, quickly learned to fly toward damaged leaves, where caterpillars resided (Heinrich and Collins

FIGURE 7-8 The moth *Hyperchiria nausica* partly disguises its symmetry by folding one wing over the other.

FIGURE 7-7 This Central American mantis of the genus *Acanthops* resembles a dead, curled-up leaf and thus escapes the notice of most predators.

(a)

(b)

FIGURE 7-9 Two species of stem-mimicking geometrid caterpillars: (a) a species of *Betula* mimicking a birch stem. (b) *Populus tremuloides* mimicking quaking aspen. (*Courtesy of B. Heinrich.*)

1983). To reduce this risk, caterpillars often consume only a portion of the leaf on which they are feeding before moving to another leaf.

Advertisement and Mimicry

Crypsis is a strategy of palatable animals. Others have rejected crypsis and taken a bolder approach to antipredator defense: they produce noxious chemicals, or accumulate them from food plants, and advertise the fact with conspicuous color patterns. Species having such conspicuous warning colors are called **aposematic** forms. A number of advantages to being conspicuous in these circumstances have been suggested (Guilford 1990). Predators learn more easily to associate bad-tasting prey with conspicuous color patterns, such as the black and orange stripes of the monarch butterfly, than with less obvious color patterns (Brower 1969, Roper and Redstone 1987). It is also possible that predators may have a generalized instinctive aversion to certain conspicuous patterns (Smith 1975, 1977, Guilford 1990).

Caterpillars include not only some of the most superb masters of disguise in the natural world, but also some of its most conspicuous inhabitants. Many caterpillars possess stinging hairs or spines, regurgitate noxious chemicals when touched, or sequester toxic substances from the leaves that they eat, making themselves unpalatable to predators. To advertise their distasteful nature, most such species are endowed with bright red, yellow, and black warning colors. Besides the possible innate aversion to such colors mentioned above, it is possible that these colors ensure that predators learn quickly which caterpillars are unpalatable (Bowers 1993). Many aposematically colored caterpillars live in aggregations that may increase the effectiveness of their warning coloration.

Unpalatable animals that display warning coloration often serve as models for palatable species that evolve to mimic their color patterns (Wickler 1968, Rettenmeyer 1970, Edmunds 1974, Turner 1984). Some potential prey may even resemble their predators to avoid predation (Figure 7-10; Mather and Roitberg 1987, Greene et al. 1987). Such relationships are collectively called **Batesian mimicry,** named for the nineteenth-century English naturalist Henry Bates. In his journeys to the Amazon region of South America, Bates found numerous cases of palatable insects that had forsaken the cryptic patterns of their close relatives and had come to resemble brightly colored, unpalatable species. Experimental studies have subsequently demonstrated that Batesian mimicry does confer advantage to the mimic. Brower and Brower (1962) showed that toads that were fed live bees thereafter avoided the palatable drone fly, which mimics bees. When toads were fed only dead bees from which the stingers had been removed, they relished the drone fly mimics.

FIGURE 7-10 The wing markings of the tephritid fly *Rhagoletis pomonella* closely resemble the forelegs and pedipalps of jumping spiders (Salticidae), which are common predators of the fly. (*Courtesy of the U.S. Department of Agriculture.*)

Another type of mimicry, called **Müllerian mimicry** after its discoverer, occurs among unpalatable species that come to resemble one another. Many species form Müllerian mimicry complexes in which each participant is both model and mimic. When a single pattern of warning coloration is adopted by several unpalatable species, avoidance learning by predators is more efficient because a predator's bad experience with one species confers protection on all the other members of the mimicry complex as well. For example, most of the bumblebees and wasps that co-occur in Rocky Mountain meadows share a pattern of blue and yellow stripes. In the Tropics, dozens of species of unpalatable butterflies, many of them distantly related, share patterns of black and orange "tiger stripes" or black, red, and yellow coloration patterns. The brightly colored orange and black viceroy butterfly (*Limenitis archippus*) was once thought to be a palatable Batesian mimic of the monarch butterfly (*Danaus plexippus*) (Brower 1958), which is distasteful to blue jays, one of the monarch's common predators. It is now known to be unpalatable to blue jays as well, and thus a Müllerian mimic of the monarch (Ritland 1991, 1995).

Aggressive Mimicry

Mimicry has arisen not just as an adaptation of prey to avoid predation. Some predators use mimicry to lure prey into their clutches. Mature male fireflies of the genus *Photuris* fly about at night, emitting flashes of light from their bioluminescent abdomens in a pattern that is specific to their species. Females, who position themselves on the ground or in low vegetation, respond with flashes of their own in a pattern recognizable to males of their own species. A male thus signaled approaches the female and, if all goes well, copulation ensues. But females of some species of *Photuris* (and several other genera) also emit the signal patterns of other species. Males of these species move toward them as they would approach a female of their own species—only to be captured and eaten by the mimicking female (Lloyd 1975, 1981). Such **aggressive mimicry** has also been observed in the bolas spiders (*Mastophora* spp.) of Central America. From its position in the vegetation, a bolas spider extends a sticky ball of silk into the space beneath it. Insects that fly into the ball adhere to it and are consumed by the spider. Eberhard (1977) has shown that bolas spiders produce a chemical that mimics the female sex attractant pheromones of certain moth species, thereby luring them to the sticky ball.

Protection and Escape

Protective prey defenses rarely involve physical combat because few prey can match their predators in combat, and predators carefully avoid those that can. Instead, seemingly defenseless organisms may produce foul-smelling or stinging chemical secretions to dissuade predators (Eisner and Meinwald 1966). Whip scorpions, bombardier beetles, and skunks direct sprays of noxious liquids at threatening animals. Similarly, many plants and animals contain toxic substances that make them inedible (Whittaker and Feeny 1971).

When predators discover cryptic organisms, they may be confronted with a variety of second-line defenses, including startle-and-bluff displays and various attack-and-escape mechanisms (Blest 1964). The green caterpillar of the hawkmoth *Leucorampha omatus* normally assumes a cryptic position. When disturbed, however, it puffs up its head and thorax, looking for all the world like the head of a small poisonous snake, complete with a false pair of large, shiny eyes; the caterpillar consummates this display by weaving back and forth while hissing like a serpent (Figure 7-11). The eyespots displayed by many moths and other insects when disturbed frighten their predators because they resemble the eyes of large birds of prey (Blest 1957) (Figure 7-12).

Chemicals are involved in nearly every aspect of the ecology of insects (Robertson et al. 1995). Chemicals are used for species recognition, for orientation during flight and terrestrial movements, to locate and identify host plants and animals, in the maintenance of social organization, and for protection against predators (Bell and Cardé 1984). **Alarm pheromones** are chemicals exuded by an injured insect and detected by others in its group, causing a defensive or aggressive response by those individuals. They are found in most species of ants, termites, and many other insects (Bradshaw and Howse 1984, Howse 1984). In response to injury or the detection of a predator, aphids release an alarm pheromone from the tips of a pair of specialized extensions of the exoskeleton referred to as cornicles. The pheromone, which is highly volatile, is quickly detected by other aphids within about 1–3 cm from the release point (an area large enough to contain many aphids), causing them to jump or fall from the host plant to avoid predation (Nault and Phelan 1984). Aphids under attack by a predator may also exude a sticky substance (triglycerides) from the cornicles that may foul the mouthparts of an attacking predator (Figure 7-13).

Alarm pheromones are utilized by vertebrate animals as well as invertebrates. An interesting example of this phenomenon, which demonstrates the dynamic nature of adaptations to features of the biotic environment, is the alarm pheromone of the fathead minnow (*Pimephales promelas*). The alarm chemical is not released by the minnows themselves; rather, it is released in the feces of the minnow's predator, the

FIGURE 7-11 Snake display by the caterpillar of the hawkmoth *Leucorampha omatus*. (*From Robinson 1969.*)

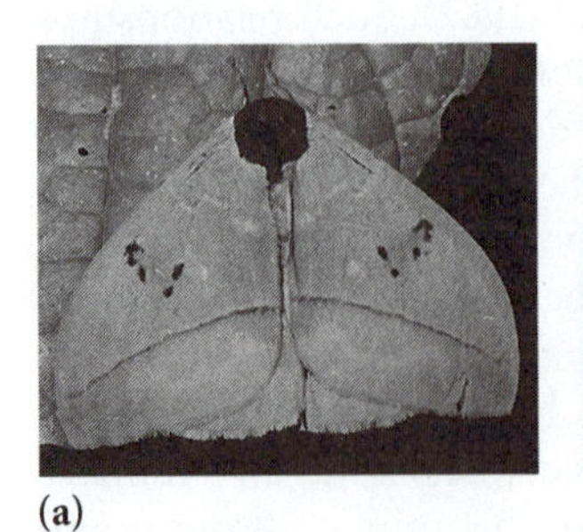

(a)

(b)

FIGURE 7-12 The eyespot display of an automerid moth from Panama: (a) normal resting attitude; (b) reaction when touched.

FIGURE 7-13 Sticky droplets exuded from the cornicles of a pea aphid (*Acyrthosiphon pisum*) being attacked by a predatory nabid. The mouthparts of the nabid are fouled by the sticky substance. (*Courtesy of L. R. Nault.*)

northern pike (*Esox lucius*), after the pike has eaten a minnow. Fathead minnows detect the chemical in the water wherever the pike defecates, and swim away. At first, this may seem like a big disadvantage for the pike. However, recent experiments have demonstrated that when pike eat fathead minnows, they deposit their feces in areas where they do not forage, thereby reducing the chance that their prey will detect their presence (Brown et al. 1996).

Since no predator is perfectly stealthy in its approach to prey, adaptations to detect and elude an attacking predator have arisen. Some of the most interesting of these adaptations are seen in insects that avoid aerial attacks by echolocating bats. Insect-eating bats emit ultrasonic calls whose echoes are reflected by stationary objects, allowing the bat to avoid colliding with trees, buildings, and other structures. The echoes also bounce off the bodies of flying insects, which the bats capture in flight. Anyone who has watched bats fly about in the early evening will appreciate their agility and the apparent efficiency with which they pluck flying insects from the air. But the insects, it seems, are not just clay pigeons. Many insects can detect the ultrasonic signals of bats and steer away from them (May 1991). Indeed, experimental studies of bat-insect interactions have shown that many insects perform elaborate evasive maneuvers in response to bat signals (Miller and Olesen 1979, Moiseff et al. 1978, Yager et al. 1990a,b).

Slow-moving animals, such as the porcupine and armadillo, are protected with spines or armored body coverings. Such defenses require time, energy, and materials, which may be limited in supply. The cost of such antipredator adaptations indicates the major influence of predation on fitness, and the strength of predators as agents of natural selection.

While organisms that are extremely small compared with their predators—those captured by filter feeders, for example—exhibit few adaptations to avoid being caught, larger quarry may either hide, fight, or flee. The circumstances of the predator-prey relationship determine which strategy, or combination of strategies, minimizes the chance of being caught. Grassland offers no hiding places for large ungulates (hoofed mammals), and so escape depends on early detection of predators and on swiftness. Other prey possess the wherewithal to make themselves hard to swallow. By grabbing its own tail or that of a would-be snake predator, the lizard *Gerrhonotus multicarinatus* prevents its ingestion (Arnold 1993).

The appropriate line of defense sometimes depends on features of the environment other than predators. Large-bodied moths at low elevations in the Tropics rarely exhibit protective displays such as eyespots or flashy coloration. During the day, the air temperature rises to within 6°C of the optimum working temperature of the flight muscles (Heinrich and Bartholomew 1971), and moths can escape from predators by flight. At higher elevations (1,500 meters), moths require a lengthy preflight warm-up period because air temperatures may be 15°C lower than the working temperature of the flight muscles. Because the cold precludes escape, a large proportion of moth species living at these elevations do have special protective displays (Blest 1963).

7.4 Herbivores must overcome the unique defensive features of plants.

Fully one-fourth of the species of nonmicroscopic animals on earth are herbivores, obtaining their nourishment by eating or sucking the juices from plants (Bernays and Chapman 1994). The importance of herbivores as the principal energetic link between primary production (photosynthesis) and carnivores and animal parasites is fully appreciated by ecologists. But the intricacies of the evolutionary interplay between herbivores and plants, and the consequences of those interactions for plant and herbivore populations and communities, and for the population dynamics of carnivores and parasites, are only now emerging. The adaptations of predators to their prey may serve as a model in understanding the evolutionary processes involved in the development of specific adaptations of plant-eating organisms. As with meat-eating animals, the most successful herbivores are those having the most efficient configuration of mouthparts, feeding behavior, gut physiology, and the myriad of other morphological, physiological, and behavioral characteristics needed for feeding on the preferred plant.

And like prey animals, plants are subject to selective pressures of their own, making the biotic environment of the herbivore as richly diverse and dynamic as that of the predator. But the challenges presented by the biotic environment of plant-eating organisms are, in many important ways, very different from the challenges faced by predators. These differences are primarily due to the sedentary nature and the enormous physiological and morphological plasticity of plants.

Locating and Selecting Plants to Eat

One need only watch a horse grazing in a field, or examine the pattern of tree bark removal in the forest around a beaver pond, to understand that herbivores do not feed indiscriminately on the plants that they encounter. The horse makes its way through the field, carefully selecting young, tender shoots of grass and avoiding thistles and other less palatable plants. Beavers show preferences for specific species or even ages of trees. In the case of the horse and the beaver, finding the preferred plant food involves active searching within a local area. The migrations of large ungulates are often associated with regional patterns of plant food availability. Wildebeests of the plains of the southern Serengeti spend the months of December through April—the wet season—feeding on short grass. Beginning in May, when the grassy plains dry up, large herds of wildebeests move northward to areas near Lake Victoria where grass is still available (Dingle 1980).

Phytophagous (plant-eating) insects represent the most diverse collection of herbivores on earth—and the most intensively studied, owing to their importance as pests in agroecosystems. In some species of insects, such as moths, butterflies, flies, and some beetles, the larvae are herbivores, whereas the adults are nectar or fluid feeders, predators, or in some cases, unable to feed at all, spending their brief adult life in the acts of reproduction and dispersal. Typically, the eggs from which herbivorous larvae develop are placed by the female on the host plant, which will provide food for the developing larvae. Some insects lay their eggs on plants other than the preferred host plant, and the larvae must locate and move to a suitable feeding location or face death by starvation. The adaptations of larval and adult phytophagous insects for locating food are varied, and may involve detection of and orientation toward chemicals that are produced by the plant, or an attraction to specific shapes, sizes, or colors that are characteristic of the preferred host plant (Figure 7-14; Bernays and Chapman 1994).

Adaptations for Eating Plants

Once an herbivore locates a suitable host plant, it must consume some of the plant and assimilate its nutrients. Large herbivores, especially those that eat grass, have teeth with large grinding surfaces to break down tough, fibrous plant materials. In comparison, the teeth of carnivores have cutting and biting surfaces that both immobilize the prey in the mouth and cut it into pieces small enough to swallow (Figure 7-15). Even seemingly simple differences in dentition reflect important ecological differences. The upper and lower incisors of horses, for example, are strongly opposed so that they can cut the fibrous stems of grasses. Other ungulates, such as cows, sheep, and

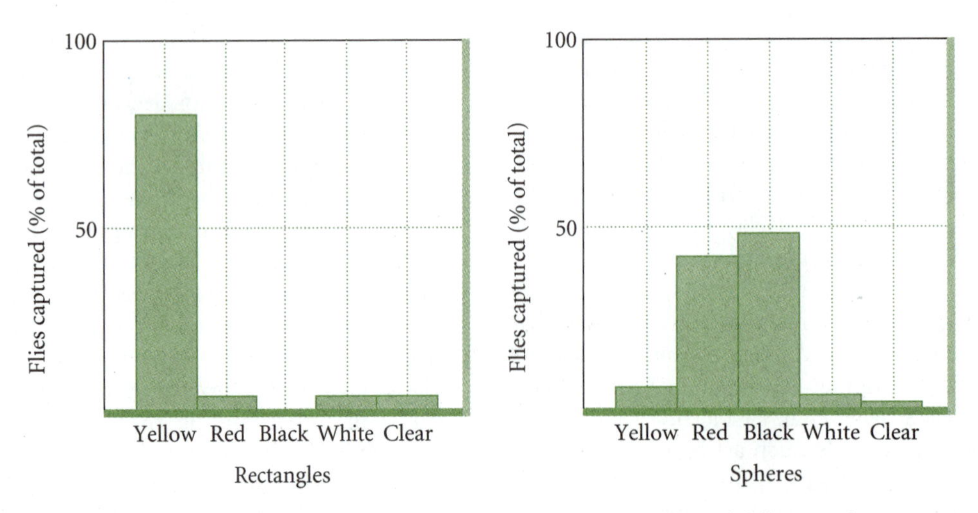

FIGURE 7-14 Rectangular (30 × 40 cm) and spherical (7.5 cm in diameter) plates of different colors were coated with an adhesive and positioned near an apple tree to trap apple maggot flies (*Rhagoletis pomonella*). Each bar represents the percentage of the total number of flies captured that were captured by that trap. The flies appear to prefer yellow rectangles and red or black spheres, indicating an interaction of form and color in microhabitat selection. (*From Bernays and Chapman 1994; after Prokopy 1968.*)

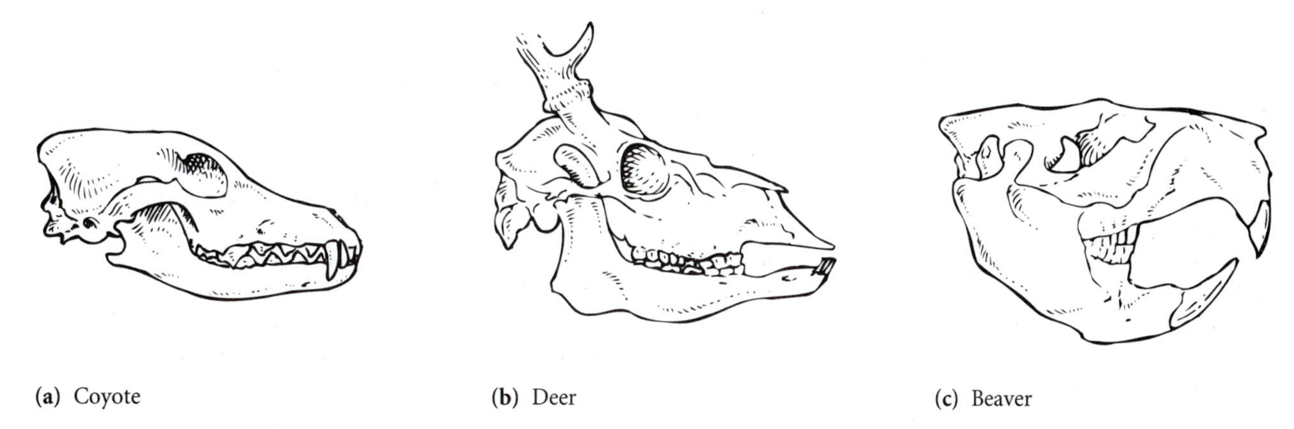

FIGURE 7-15 Skulls of three mammals, illustrating adaptations of the jaws and teeth to different diets. (a) Coyote, with daggerlike canine teeth and knifelike premolars for securing prey and tearing flesh. (b) Fallow deer, with well-developed, flat-surfaced molars and premolars for grinding plant materials; note the absence of canines and upper incisors. The lower incisors are used to secure vegetation against the upper jaw; the deer then rips the leaves from the plant. (c) Beaver, with greatly enlarged chisel-like incisors that are used to gnaw on wood. (*After Vaughn 1986.*)

deer, lack upper incisors; their lower teeth press against the upper jaw at an angle for gripping and pulling plant material (Gwynne and Bell 1968). Herbivorous fish, which feed on or near the bottom where plants and algae grow, may be broadly classified as **browsers** or **grazers** depending on whether they bite off pieces of the macroalgae or other material they consume (browsers) or scrape or suck plant material from some surface (grazers). Both browsers and grazers possess shorter snouts than midwater-feeding fish, and are endowed with teeth that are adapted for cropping and scraping (Horn 1992). The terms *browser* and *grazer* are also used to describe large terrestrial vertebrates. Browsers eat leaves from woody plants, and grazers eat grasses and other herbs.

Phytophagous insects may be categorized as chewers, miners and borers, leaf rollers, or plant suckers (Weis and Berenbaum 1989). Many species, including beetles (order Coleoptera), bees, wasps, and ants (order Hymenoptera), and grasshoppers (order Orthoptera), as well as a species of stick insect (order Phasmida) and the larval forms—caterpillars—of moths and butterflies (order Lepidoptera) and flies (order Diptera), chew plant material prior to digesting it. The head region of these insects is endowed with several sets of paired hardened appendages, referred to, from the most anterior to the most posterior, as the mandibles, maxilla, and labium. These structures allow the insect to rip up plant material and to masticate it prior to ingestion. The feeding apparatus also constrains food selection. Juvenile grasshoppers (*Chorthippus parallelus*) select host plants having a relatively thin leaf cross section because thicker leaves are too big for their narrow intermandibular gap (Figure 7-16). Adult grasshoppers have larger mouths and can eat thicker leaves (Bernays and Chapman 1970a,b).

The larval stages of some insects, and some small adult beetles, derive nutrition from the inner tissues of the host plant upon which the adult laid her eggs. These **endophytophagous** species chew their way into the interior of the plant, where they "mine" the plant tissue and consume it, creating obvious tunnels in the leaf tissue that are easily observed by the alert student

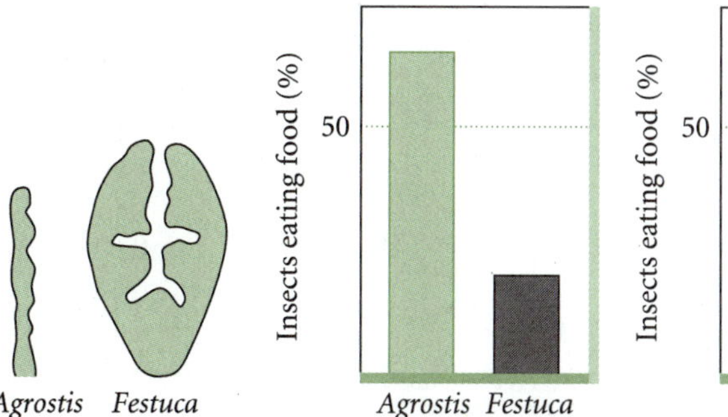

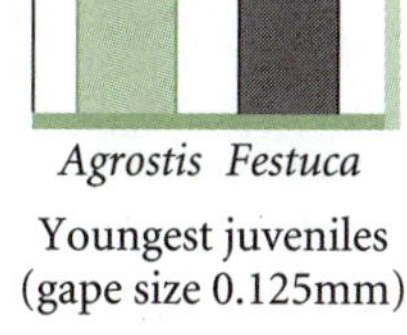

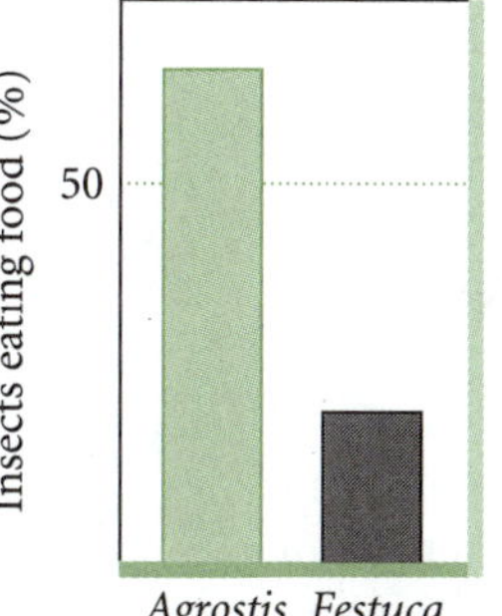

FIGURE 7-16 The type of food selected by the grasshopper *Chorthippus parallelus* is dependent on its developmental stage. The youngest grasshoppers have small mouthparts and are unable to feed on thick-leaved plants such as *Festuca*, preferring instead to consume the thin leaves of *Agrostis*. The oldest juveniles have bigger mouthparts and are thus able to consume thick *Festuca* leaves. (*After Bernays and Chapman 1994.*)

of natural history. Most often the entire larval development of leaf-mining insects is completed within one leaf.

Leaf-rolling caterpillars use silk to form a leaf wrapping around themselves, then consume the leaf from inside the roll. Generally, rollers consume several leaves during larval development (Stamp and Wilkens 1993). Highly modified mouthparts configured in a tubelike structure have arisen among some other groups of insects, most notably the leafhoppers (order Hemiptera). These structures are used to pierce the outer tissues of plants for the purpose of sucking liquids from the plant tissue.

Nutritional Ecology of Herbivory

Like animals, plants contain a wide variety of chemical constituents, such as proteins and amino acids, carbohydrates, lipids, and other substances, that are potentially nutritious to herbivores. But plants differ substantially from animals in the manner in which these substances are distributed within the body of the organism. Unlike captured prey that is nearly entirely consumed by the predator, only parts of a particular plant are selected for consumption by herbivores. This is due to variation in nutritional quality among the parts of the plant, or to the herbivores' inability to gain access to certain parts of the plant. The reproductive structures and leaves are generally higher in protein content than the stems and roots, and are likely to provide a better nutritional source to herbivores (Berneys and Chapman 1994). Whatever the nutritional value of the roots, they are unavailable to a caterpillar feeding above ground. Moreover, plants contain a variety of metabolic by-products, which may be harmful to herbivores, and volatile chemicals, which may be attractants to herbivores. Plant tissues also contain structural molecules, such as cellulose and lignin, that are difficult to digest.

Because vegetation is more difficult to digest than the high-protein diets of carnivores, the digestive tracts of herbivorous animals are often greatly elongated. In addition, many contain saclike offshoots (caecae of rabbits; rumens of cows) that function like fermentation vats, housing bacteria and protozoa that aid in digestion (Swenson 1977). With a larger volume of intestines, herbivores can keep meals in the digestive tract longer and digest them more completely. Balancing this seemingly workable adaptation, herbivores must carry quantities of undigested food in their bellies, adding weight and reducing mobility.

The nutritional quality and digestibility of plant foods is critical to herbivores (Moran and Hamilton 1980, Scriber and Slansky 1981, Scriber 1984). Because young mammals have a high protein requirement for growth, the reproductive success of grazing and browsing mammals depends upon the protein content of their food. Herbivores usually select plant food according to its nutrient content (Gwynne and Bell 1968). Young leaves and flowers are chosen frequently because of their low cellulose content; fruits and seeds are particularly nutritious compared with leaves, stems, and buds because of their higher nitrogen, fat, and sugar content (Short 1971). Comparing saplings of 46 species of canopy-forming trees on Barro Colorado Island, Panama, Coley (1983) analyzed rates of grazing on leaves with respect to their nutritional value and physical structure. Coley found that increased cellulose content and leaf toughness significantly reduced herbivory (signified by the high negative correlation coefficient, r), but that the presence of hairs on the leaf surface and high levels of water and nitrogen seemed to encourage grazing (high positive correlation; Table 7-1).

TABLE 7-1 Correlations between rates of herbivory and defenses of leaves of saplings among 46 species of trees in lowland rain forest habitat in central Panama

Leaf attribute	Range of values	Correlation coefficient (r)
Chemical		
Total phenols (% dry mass)	1.7–22.6	−0.10
Cellulose (% dry mass)	10.2–30.4	−0.47**
Lignin (% dry mass)	3.3–20.8	−0.23
Physical		
Toughness (Newtons)	2.5–11.6	−0.52
Undersurface hairs (number mm^{-2})	0–18	0.64**
Nutritional		
Water (%)	49–82	0.51**
Nitrogen (% dry mass)	1.7–3.1	0.29*

*Statistical significance: *$P < 0.05$; **$P < 0.01$.*

(From Coley 1983.)

Many plants use chemicals to reduce the availability of their proteins to herbivores. For example, **tannins** sequestered in vacuoles in the leaves of oaks and other plants combine with leaf proteins and digestive enzymes in an herbivore's gut, inhibiting protein digestion (Feeny 1969, Hagerman and Butler 1991). As a consequence, tannins slow the growth of caterpillars and other herbivores, an effect that reduces the quality of tannin-laden hosts as food plants. In experimental studies of the Neotropical shrub *Psychotria horizontalis* in Panama, Sagers and Coley (1995) found that plants with high tannin concentrations and tough leaves suffered less herbivory than plants with lower tannin concentration. It is important to appreciate, however, that there is a cost to producing tannins. Sagers and Coley found that plants that invested heavily in the production of tannins had a reduced growth rate when compared with plants that produced fewer tannins. The implication of this finding is that tannin production is of value to the plant only when the plant is exposed to high levels of herbivory. We discuss the dynamics of such ecological trade-offs in more detail in Part 7.

Circumventing Plant Defenses

Plants possess a plethora of morphological and chemical adaptations for combating herbivory; these plant defenses constitute a major part of the biotic environment of the herbivore. We should not be surprised, then, that adaptations have arisen among herbivores for circumventing or avoiding these defense mechanisms. Countermeasures may include the active avoidance of unpalatable plants or of potentially harmful parts of plants, the removal of toxic or unpalatable structures prior to feeding, the biochemical neutralization of toxic substances in the gut of the herbivore, or habituation or sensitization to the plant's chemical defenses.

The chemical and morphological features of a plant may give potential herbivores an early warning and lead to avoidance of that plant, or provide information about which parts of the plant are edible. As we have seen, phytophagous insects may be repelled or attracted to a potential host plant by chemicals exuded by the plant or by the morphological features of the plant. Once a host plant is selected, however, the herbivore is faced with a decision about where on the plant it may safely feed. The presence of surface odors, urticating hairs (detachable hairs that attach to and cause discomfort to the herbivore), areas with waxy or impenetrable surfaces, and even the shape and size of the particular plant surface being considered as a feeding site will determine which part of the plant is consumed (Dussourd 1993, Berneys and Chapman 1994). Caterpillars possess an extensive array of behaviors to avoid or deactivate plant structural defenses. The larvae of some species of nymphalid moths simply spin silk to bridge sticky or prickly hairs on the leaf surface. Noctuid moth caterpillars often bite off and remove spines and hairs on the ribs of leaves before beginning to feed; by doing so, they avoid being entrapped or injured. Other caterpillars are known to sever the leaf veins of the host plant, thus stopping the flow of harmful substances to the feeding area (Dussourd 1993).

Strategies for circumventing plant defenses, like nearly everything else in ecology, are scale dependent. The distance between urticating hairs on a leaf surface, which is on the order of millimeters, has no bearing on whether a browsing deer will be successful in munching on that leaf. Those same hairs, however, may represent an impenetrable forest for a foraging caterpillar. In phytophagous insects and other small herbivores, strategies for the avoidance or circumvention of plant defenses may change as the individual grows. Very young caterpillars, for example, may have mouthparts small enough to enable feeding between reservoirs of toxic material in the leaf, whereas the mouthparts of older caterpillars would be too large to avoid tissues with concentrated toxins (Reavey 1993).

7.5 Plants use structural and chemical defenses against herbivores.

Even though plants cannot uproot and flee from an approaching herbivore, or, through the use of physical force, throw off a hungry grasshopper or push away a ravenous moose, they are by no means defenseless. As we shall see below, formidable structural and biochemical barriers to herbivory have evolved in plants. We caution against using our understanding of predator-prey adaptations as a strict model for the relationships between herbivores and their plant food. Plants are very different from animal prey, and as a consequence, so is the phenomenon of herbivory different from predation. In terms of individual survival, plants enjoy a distinct advantage over prey that are hunted, killed, and eaten by predators, in that they are often not fully consumed, and the damage inflicted by an herbivore may not fully disrupt the life-sustaining functions of a plant. Successful predation leads to the death of the prey; not so herbivory. Additionally, herbivores, particularly phytophagous insects, tend toward greater specialization on their host plants than do predators toward their prey. Thus, even slight changes in the chemical structure of a plant result in rapid and specific alterations in herbivore physiology.

This strong reciprocal selection between two (or more) organisms having a close ecological relationship is called coevolution, a topic that we cover in detail in Chapter 25. Coevolutionary relationships between herbivores and their host plants are often fine-tuned. For this reason, in our discussion of chemical defenses below, we discuss the adaptations of plant and herbivore together.

Defenses of Structure and Nutritional Quality

Plants that are difficult to handle and ingest may be less attractive to herbivores. Morphological adaptations, such as waxy coatings and the presence of spines and hairs, may act as deterrents to herbivory (Figure 7-17). The production of tannins, discussed in the previous section, is a defensive mechanism that makes plant tissue difficult to digest. Plants may also lower their nutritive value to herbivores by allocating growth to tissues that contain few of the nutrients needed by the herbivore. Herbivores may overcome this strategy by increasing their rate of feeding or by feeding longer on the plant (Augner 1995).

Like prey, plants can use deception as protection against being consumed. For example, one species of Australian mistletoe (*Dendrophthoe shirleyi*), which lives as a parasite of other plants, produces leaves that are nearly indistinguishable from those of one of the three species of host plants that it inhabits, all of which are inedible for most herbivores. Through crypsis, then, the mistletoe is not detected by potential herbivores as a separate, edible plant (Weis and Berenbaum 1989).

Biochemical Defenses

Much of the conflict between herbivores and plants is waged on biochemical battlegrounds delineated by the toxic properties of **secondary compounds.** Secondary compounds are produced and sequestered for defense by many species of plants (Fraenkel 1959, 1969, Beck 1965, Whittaker and Feeny 1971, Robbins and Moen 1975, Levin 1976, Seigler and Price 1976, Rosenthal and Janzen 1979, Harborne 1982). These substances can be divided into three major classes of chemical structures: nitrogen compounds ultimately derived from amino acids, terpenoids, and phenolics (Table 7-2). Among the nitrogen-based substances are lignin, a highly condensed polymer that resists digestion; alkaloids, such as morphine (derived from poppies), and atropine and nicotine (from various members of the tomato family, Solanaceae). This class also includes nonprotein amino acids, such as L-canavanine, and cyanogenic glycosides, which produce cyanide (HCN). Terpenoids include essential oils, latex, and plant resins. Many simple phenolics have antimicrobial properties.

Whereas tannins exhibit a generalized reaction with proteins of all types, many secondary compounds interfere with specific metabolic pathways or physiological processes of herbivores. However, because the sites of action of such substances are localized biochemically, herbivores may counter their toxic effects by modifying their own physiology and biochemistry. Such **detoxification** may involve one or several biochemical steps, including oxidation, reduction, or hydrolysis of the toxic substance, or its conjugation with another compound.

Several early studies of the chemical give-and-take between plants and herbivores focused upon the larvae of bruchid beetles, many types of which infest seeds of legumes (pea family). Adult bruchids lay their eggs on developing seed pods. The larvae then hatch and burrow into the seeds, which they consume as they grow. Their host plants have developed a variety of defenses via natural selection (Janzen 1969, Center and Johnson 1974), including the evolution of tiny seeds. Each larva feeds on only one seed. To pupate successfully and metamorphose into an adult, the

(a)

(b)

FIGURE 7-17 Spines protect the stems and leaves of many plants: (a) an agave (century plant) from Baja, California; (b) a *Parkinsonia* (bean family) from the Galápagos Islands.

TABLE 7-2 Secondary plant compounds involved in plant-herbivore interactions

Class	Approximate number of chemical structures	Distribution	Physiological activity
Nitrogen compounds			
Alkaloids	5,500	Widely in angiosperms, especially in roots, leaves, and fruits	Many toxic and bitter-tasting
Amines	100	Widely in angiosperms, often in flowers	Many repellent-smelling, some hallucinogenic
Amino acids (nonprotein)	400	Especially in seeds of legumes, but relatively widespread	Many toxic
Cyanogenic glycosides	30	Sporadic, especially in fruits and leaves	Poisonous (as HCN)
Glucosinolates	75	Cruciferae and 10 other families	Acrid and bitter
Terpenoids			
Monoterpenes	1,000	Widely, in essential oils	Pleasant-smelling
Sesquiterpene lactones	600	Mainly in Compositae, but increasingly found in other angiosperms	Some bitter and toxic, also allergenic
Diterpenoids	1,000	Widely, especially in latex and plant resins	Some toxic
Saponins	500	In over 70 plant species	Hemolyze blood cells
Limonoids	100	Mainly in Rutaceae, Meliaceae, and Simaroubaceae	Bitter-tasting
Curcurbitacins	50	Mainly in Cucurbitaceae	Bitter-tasting and toxic
Cardenolides	150	Especially common in Apocynaceae, Asclepiadaceae, and Scrophularaceae	
Phenolics			
Simple phenols	200	Universal in leaves, often in other tissues as well	Antimicrobial
Other			
Polyacetylenes	650	Mainly in Compositae and Umbelliferae	Some toxic

(From J. B. Harborne 1982)

larva must attain a certain size, ultimately limited by the amount of food in the seed. The small seeds of some species of legumes contain too little food to support the growth of even a single bruchid larva (Janzen 1969).

Most legume seeds also contain substances referred to generally as protease inhibitors, which inhibit proteolytic enzymes produced in the herbivore's digestive organs. Protease inhibitors are found in a wide variety of plants (Casaretto and Corcuera 1995).

```
H      H   H H H   O
 \     |   | | |  //
  N-C-N-O-C-C-C-C
 /  ||     | | |  \
H   N      H H N   OH
    |         / \
    H        H   H
```

Canavanine

```
H      H H H H H   O
 \     | | | | |  //
  N-C-N-C-C-C-C-C
 /  ||   | | | |  \
H   N    H H H N   OH
    |         / \
    H        H   H
```

Arginine

FIGURE 7-18 The toxic nonprotein amino acid L-canavanine and its protein amino acid analog arginine. The shaded area highlights the difference between the two molecules. (*From Harborne 1982.*)

While these toxins provide an effective biochemical defense against most insects, many bruchid beetles have metabolic pathways that either bypass or are insensitive to them (Applebaum 1964, Applebaum et al. 1965). Among legume species, however, soybeans stand out as being resistant to attack even by most bruchid species. When bruchids lay their eggs on soybeans, the larvae die soon after burrowing beneath the seed coat; chemicals isolated from soybeans have been shown to inhibit the development of bruchid larvae in experimental situations.

Seeds of the tropical leguminous tree *Dioclea megacarpa* contain 13% by dry weight of L-canavanine, a nonprotein amino acid that is toxic to most insects because it interferes with the incorporation into proteins of the amino acid arginine, which it closely resembles (Figure 7-18). One species of bruchid, *Caryedes brasiliensis,* possesses enzymes that discriminate between L-canavanine and arginine during protein formation, and additional enzymes that degrade L-canavanine to forms that can be utilized as a source of nitrogen (Rosenthal et al. 1976).

Tobacco (*Nicotiana tabacum*) is notorious because of its ability to produce nicotine. In nature, nicotine is a powerful insecticide because it disrupts the normal functioning of the nervous system by preventing the transmission of impulses from nerve to nerve. The tobacco hornworm (the larval stage of the moth *Manduca sexta;* Figure 7-19) can tolerate nicotine concentrations in its food far in excess of those that would kill other insects. The hornworm has circumvented the plant's defense by excluding nicotine from the nerve at the cell membrane; in other species of moths, nicotine readily diffuses into nerve cells (Yang and Guthrie 1969). Resistance to nicotine enables *M. sexta* to feed on tobacco, but some species of *Nicotiana* produce other alkaloid toxins that the tobacco hornworm cannot tolerate. When tobacco hornworms were grown on forty-four species of *Nicotiana* in greenhouse experiments, the larvae grew normally on twenty-five species, but were retarded or stopped completely on the others. In addition, fifteen of the species caused moderate to severe mortality in the hornworm (Parr and Thurston 1968). These results emphasize the degree of specialization that can develop in plant-herbivore interactions.

Where herbivory is more intense, plants have developed more varied and concentrated toxins (see, for example, Dolinger et al. 1973). In turn, where plant defenses are strong, adaptations of herbivores to detoxify poisonous substances proliferate. This war between plants and herbivores promotes biochemical specialization of herbivores on certain restricted groups of plants with similar toxins. Associations of plants and herbivores in groups based on plant chemistry and structure have been referred to as **plant defense guilds** (Atsatt and O'Dowd 1976).

FIGURE 7-19 Larva of the tobacco hornworm *Manduca sexta* (Sphingidae).

Plant defenses may be induced by herbivore damage in much the same way that foreign proteins induce an immune response in vertebrate animals. Alkaloids, phenolics, N-oxidases, and proanthocyanins, all of which are linked to antiherbivore defenses, were found to increase dramatically in many plants following defoliation by herbivores (or the clipping of leaves by investigators). Other studies have shown that plant defenses induced by herbivory can substantially reduce subsequent herbivory (Bryant and Kuropat 1980, Haukioja 1980, Karban and Carey 1984, Fowler and MacGarvin 1986; Figure 7-20). Fungal pathogens and plant diseases may also induce a response in plants, including programmed cell death at the site of the disease (Mittler and Lam 1996). This inducibility suggests that some chemical defenses are too costly for plants to maintain economically under light grazing pressure. Several studies have shown trade-offs between production of defensive chemicals and growth. In addition, where soils are low in the nutrients required for the production of some defensive chemicals, the costs of defense are relatively high. Undoubtedly, the offensive biochemical tactics of herbivores are also expensive.

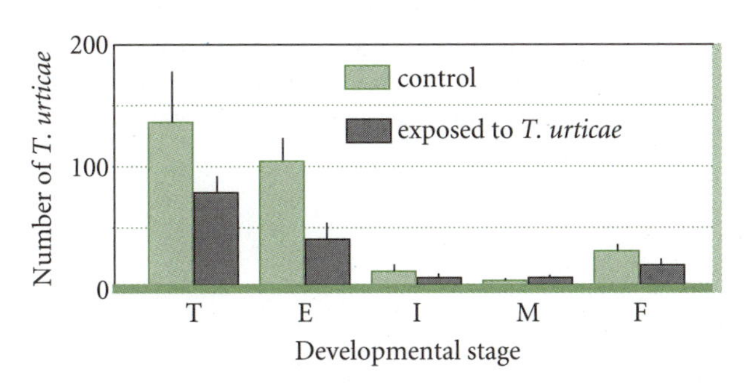

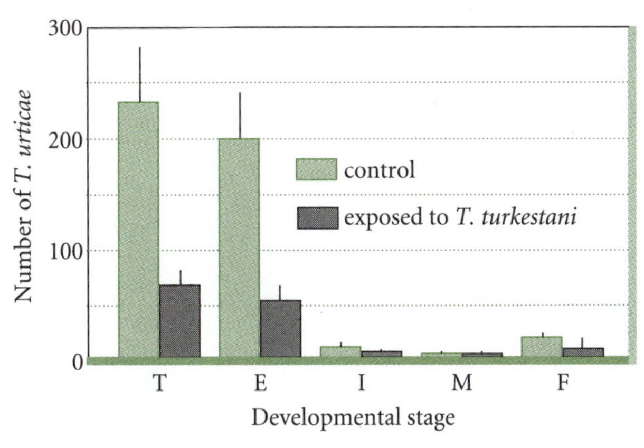

FIGURE 7-20 Mean number of the mite *Tetranychus urticae* found on cotton plants previously exposed to mites and on controls with no previous exposure. The effect is similar whether the initial exposure is to (a) *T. urticae* or to the related mite species (b) *T. turkestani*. (T = total population, E = eggs, I = immatures, M = adult males, F = adult females.) (*From Karban and Carey 1984.*)

7.6 Host specificity and complex life cycles characterize many parasites.

Another major type of biotic interaction is that between parasites and their hosts. **Parasites** are organisms that feed on another organism (the host), usually without killing it. Unlike the predator-prey and plant-herbivore systems discussed above, host-parasite interactions are typically prolonged affairs.

Parasites usually are much smaller than their hosts. **Ectoparasites** live on the surface of their host; **endoparasites** live inside the host's body. Endoparasites that cause diseases, such as dysentery or malaria, may be referred to as **microparasites.** Larger, non-disease-causing parasites are often referred to as **macroparasites** (May and Anderson 1979). Macroparasites include both endoparasites, such as the pinworms that sometimes infect the intestinal tracts of humans, and ectoparasites, such as ticks and lice.

Both ectoparasites and endoparasites demonstrate characteristic adaptations to their way of life. First, because parasites generally live inside, or in close association with, a larger organism, they expend little effort to maintain their own internal environments. Thus, endoparasites often have little more than food-processing and egg-producing capacities. Most sensory organs and complex nervous and circulatory systems have been lost. Second, parasites must disperse through the hostile outside environment between hosts. Many accomplish this through complicated life cycles, one or more stages of which can cope with the external environment, and by producing huge numbers of offspring. These principles apply equally to parasitic plants such as mistletoes (Kuijt 1969, Lamont 1983).

Parasite Life Cycles

Ascaris, an intestinal roundworm (endoparasite) that parasitizes humans and many other vertebrates, has a relatively simple life cycle. A female *Ascaris* may lay tens of thousands of eggs per day, which pass out of the host's body in the feces. Where sanitation is poor, or where human excrement is applied to farmland as fertilizer, the eggs may be inadvertently ingested by a new host. The egg is the only stage during the life cycle of *Ascaris* that occurs outside the host, and it is well protected by a sturdy, impermeable outer covering. The parasite relies on the host to consume these eggs and complete the life cycle.

Many parasites have distinctive life stages, each of which occupies a different host. For example, the life cycle of the protozoan parasite *Plasmodium,* which causes malaria, involves two hosts. The sexual phase of the *Plasmodium* life cycle takes place in a mosquito.

The mosquito is the **primary host**—the host in which the reproductive stage of the parasite occurs. The nonreproductive stages of *Plasmodium* occur in a human or some other mammal, bird, or reptile. These hosts are called **intermediate hosts** (Mattingly 1969, Figure 7-21). When an infected mosquito bites a human, cells called sporozoites are injected into the bloodstream with the mosquito's saliva. Eventually, the sporozoites enter red blood cells and feed upon hemoglobin. When the sporozoite becomes large enough, it undergoes a series of divisions (asexual reproduction), and the daughter cells break out of the red blood cell. Each daughter cell enters a new red blood cell, grows, and repeats the cycle, which takes about 48 hours. When the infection has built up to a high level, the emergence of daughter cells corresponds to periods of high fever in the malaria victim. After several of these cycles, some of the daughter cells change into sexual forms. If these sexual cells are swallowed by a mosquito along with a meal of blood, they are transformed into eggs and sperm, and fertilization (sexual reproduction) takes place. The fertilized egg then divides and produces sporozoites, which work their way into the salivary glands of the mosquito, from which they may enter a new intermediate host, thereby completing the life cycle.

The life cycle of *Schistosoma*, a trematode worm (blood fluke) that commonly infects humans and other mammals in tropical regions, is similar to that of *Plasmodium*. Humans are the primary host. Male-female pairs of adult worms live in the blood vessels that line the human intestine or the bladder, depending on the species of *Schistosoma*. The eggs pass out of the host's body in the feces or urine. When the eggs are deposited in water, they develop into a free-swimming larval form, the miracidium, which burrows into a freshwater snail—the intermediate host—within 24 hours. The miracidium produces cells that eventually develop into free-swimming cercariae, which leave the snail and can penetrate the skin of the next primary host in the cycle. A snail infected with one miracidium may liberate from 500 to 2,000 cercariae per day over a period of a month (McClelland 1965). Once inside the body of a human or other primary host, the cercariae travel a circuitous route through blood vessels until they become lodged in an appropriate place, where they metamorphose into adult worms (Jordan and Webbe 1969).

Many ectoparasites serve as vectors or carriers of disease organisms and other parasites. Among the most important of these vectors are ticks, which transmit more infectious diseases of humans, including Lyme disease and tick-borne encephalitis, than any other arthropods (Sonenshine 1991). Ticks also serve as the principal vectors of a wide variety of livestock and wildlife diseases. Indeed, ticks are known to serve as vectors for protozoan, fungal, bacterial, and viral pathogens, and there is some evidence that even

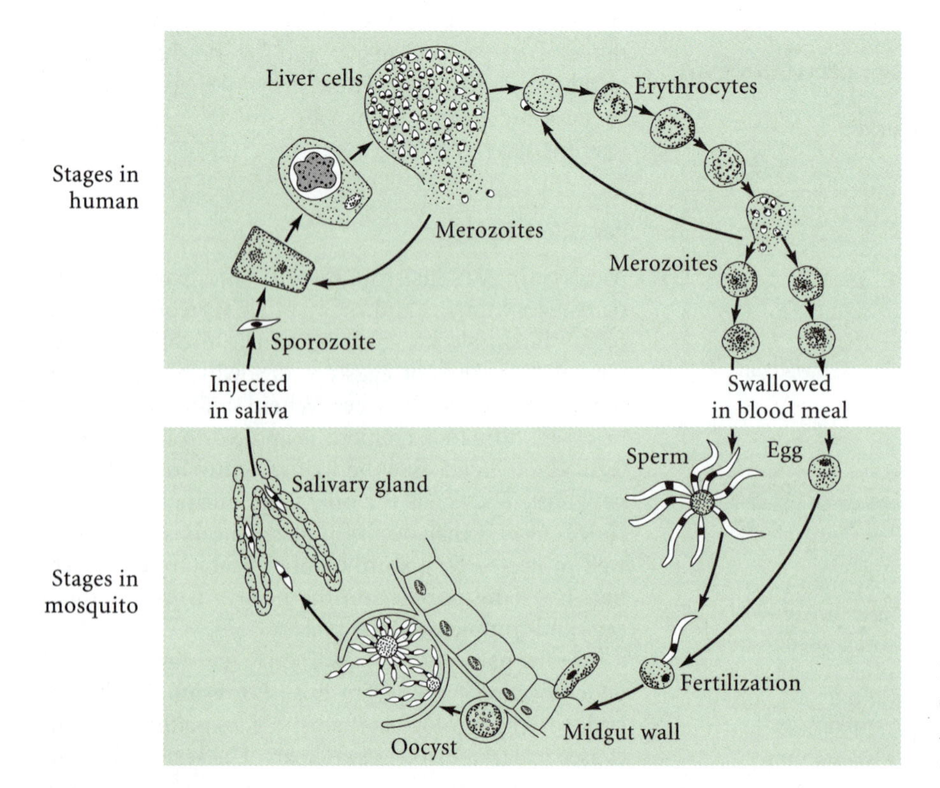

FIGURE 7-21 Stages in the life cycle of the malaria parasite *Plasmodium*. (*After Buchsbaum 1948, Sleigh 1973.*)

nematodes may be transmitted by ticks (Sonenshine 1991). Ticks are marvelously adapted for their obligate blood-sucking lifestyle, and are found nearly everywhere on earth where there are animals with blood. As a group, their life cycles involve an enormous variety of host animals. Unlike most blood-sucking organisms, ticks have very long life spans, often years, and can accommodate an enormous blood volume per feeding—up to 4 ml per feeding for some species. Most ticks develop from eggs through one larval stage and three or more nymphal stages prior to becoming adults. In a three-host life cycle (Figure 7-22), eggs are typically deposited in the environment, such as in leaf litter, grass, or the bedding of a host, and develop there into the first larva, which may overwinter before feeding. That larva eventually attaches to a host and takes a blood meal, feeding until it is engorged, a process that may take several days. Once the tick is engorged, it drops off of the first host and undergoes development through one or more nymphal stages before finding a second host, which may or may not be a different species than the first host. It takes another blood meal and, again, drops off, this time to develop either into a mature male or a mature female. Mating takes place on the third host, after which the females drop off to produce eggs; the males remain and feed again while they wait for other females (Sonenshine 1991).

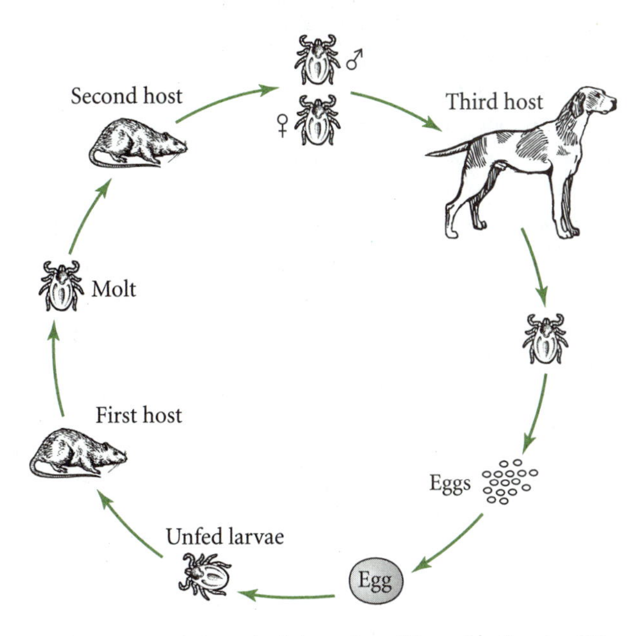

FIGURE 7-22 A typical three-host life cycle of an ixodid tick. Ticks take a blood meal on each intermediate host, then drop off to undergo development in the environment. Mating takes place on the third host, after which females drop off and lay eggs in the environment. (*After Sonenshine 1991.*)

Host Specificity

Parasites usually are host-specific, each species being restricted to one host species, or to a few closely related ones, in each stage of its life cycle. Furthermore, most parasite-host relationships have evolved a fine balance; parasites rarely impair the health of their host. Parasitic organisms so virulent that they kill their hosts also die. As a general rule, therefore, selection favors parasites that do not kill the host—provided that different family lines of the parasite do not compete in the same host (see Ewald 1983). This delicate balance depends in part upon the adaptation of the parasite to its host environment. Diseases or other stresses can change the internal environment of the host and upset the normal host-parasite balance, often leading to the death of the host (Latham 1975). The famous plagues of medieval Europe are thought to have followed periods of widespread famine, which generally increased susceptibility to all disease (Cartwright and Biddis 1972).

Responses to Host Physiology

The balance between parasite and host also depends on the immune responses and other defenses of the host. Parasites introduced to human populations that have not developed mechanisms of immunity can increase in number very rapidly, with disastrous results. Time and time again, foreigners have introduced parasites that caused no disease to them into susceptible native populations, in which the parasites became virulent and widespread. One example of this phenomenon is the devastating effects that diseases introduced by Europeans had on Native Americans.

The mutual accommodation of parasites and their hosts raises many questions (Jackson et al. 1969, 1970). How, for example, does an intestinal worm avoid being digested? How do organisms that invade the bloodstream of vertebrates avoid being destroyed by immune mechanisms, which usually stimulate the formation of antibodies that specifically attack foreign proteins?

Successful parasites have found several ways to circumvent their hosts' immune mechanisms (Bloom 1979). Some microscopic disease organisms produce chemical factors that suppress the immune system of the host (Schwab 1975). Others have surface proteins that mimic host antigens and thus escape notice by the immune system (Damien 1964). Some schistosomes are known to elicit an immune response when they enter the host, but do not succumb to antibody attack because they coat themselves with proteins of the host before antibodies become numerous (Smithers et al., 1969). As a consequence, other parasites that subsequently infect the host face a barrage of antibodies stimulated by the earlier entrance of the now-entrenched parasite individuals. Schad (1966) suggested that parasites could use the immune response of their hosts in this way to exclude competing parasites. When this response affects closely related species of parasites, it is known as **cross-resistance** (Cohen 1973, Kazacos and Thorson 1975). For example, most of the predominantly human forms of schistosomes disobey the common rule and are extremely virulent. However, in a person who has previously been infected by other schistosome species—some of which have little effect on humans—the effect of the parasite is moderated considerably (Lewert 1970).

Parasitoids

Among the insects, particularly in the orders Hymenoptera (bees, wasps, and ants) and Diptera (flies), are some species, referred to as **parasitoids,** whose life cycles include features of both predation and parasitism. In a typical parasitoid life cycle, after mating with a male, an adult female goes in search of a host, which will be used as a food source by her offspring. She deposits eggs on the outside of the host in the case of an **ectoparasitoid,** or inside it in the case of an **endoparasitoid,** the latter being accomplished with the aid of a long ovipositor. The female must locate a host, subdue it if necessary, and sometimes, carry it to a suitable location. She possesses adaptations similar to those of predators. The egg, once deposited on or inside of the host, develops into a larva, which uses the host as a food supply. In the case of endoparasitoids, the larva literally eats its way out of the host, destroying it in the process. The study of parasitoids occupies a central place in the development of ecological theory, and in later chapters, we will discuss the details of a number of parasitoid life histories (Godfray 1994).

SUMMARY

1. The environment of organisms includes both biotic and abiotic components. The biotic component of the environment consists of interactions among organisms, such as predation, herbivory, and parasitism. Unlike the abiotic environment, the biotic environment evolves and diversifies.

2. The evolutionary responses of organisms to the abiotic environment promote convergence of form and function, while responses to the biotic environment promote diversification. The reason for this is that biotic factors interact with one another; innovations to escape predation or herbivory represent new selective pressures for the predator or the herbivore, and thus lead to evolutionary change.

3. Predators must be adapted to find prey, choose from among individual prey, capture and kill the prey, and then eat the prey. Large size relative to the prey, sharp teeth, acute vision, and speed are among the adaptations of carnivores.

4. Organisms may avoid predation by avoiding detection, by chemical, structural, and behavioral defenses, and by escape behaviors. Crypsis and warning coloration are examples of contrasting defenses of edible and unpalatable organisms. Some palatable organisms mimic unpalatable ones as a means of escaping predation.

5. Insects use chemicals as a means of protection from predators. Some chemicals are noxious to predators and repel them or interrupt an attack; other chemicals, such as alarm pheromones, warn other insects in the group of a predator attack.

6. Herbivores are very selective with respect to both the plant species they prefer and the particular parts of the plant that they will eat. Like predators, herbivores must have adaptations for finding a host plant, selecting an individual plant and that part of the plant that is palatable, detoxifying or avoiding the secondary chemicals produced by the plant, and digesting plant material.

7. Phytophagous insects represent one of the largest groups of herbivores. Among the most prevalent of these are the caterpillars, the larval stages of moths and butterflies.

8. Herbivores may circumvent plant defenses by avoiding harmful plants or harmful portions of plants, by neutralizing plant defenses directly, as by removing spines, or by feeding in areas away from reservoirs of toxic substances.

9. Plants resist the attacks of herbivores in a number of ways, including the production of harmful or toxic secondary compounds and the presence of spines, hairs, or waxy coverings.

10. Host-parasite relations represent a specialized type of interaction characterized by complex life cycles of the parasite, made necessary by the difficulty of locating and infecting new hosts. In addition, parasite and host often evolve a delicate balance because the parasite's well-being may depend upon the survival of its host.

EXERCISES

1. Think of ways in which the biotic environment of humans has influenced human evolution. What are some of the important organisms in the biotic environment of humans?

2. Parasitologists have long held that coadaptation of parasite and host leads to a situation in which the parasite has only a moderate negative effect on the host. That is, a parasite that is highly virulent in a host is thought to be an evolutionarily recent parasite of that host. Ecologists have questioned this assumption. Using what you know about evolution, natural selection, and parasite life cycles, develop an argument that counters this conventional view.

3. A limited number of organisms are adapted to living entirely in caves. Among these are a handful of cave salamanders. Why are there no cave-adapted frogs?

4. Evolution in response to the biotic environment results in diversification. Does this mean that organisms are improved by evolution? That is, are modern organisms better than their ancestors of 100 million years ago?

CHAPTER 8

Climate, Topography, and the Diversity of the Natural World

GUIDING QUESTIONS

- Why is precipitation greater in the Tropics than in arctic regions?
- How does the position of the sun influence global climate patterns?
- How does the temperature in the waters of a temperate lake change with season?
- What are the effects of El Niño on ocean currents of the southern Pacific and on climate patterns throughout the world?
- How are global climate patterns and local conditions related?
- What is the relationship among evapotranspiration, photosynthesis, and temperature in deserts and temperate deciduous forests?
- What is the importance of scale in the consideration of plant distributions?
- What factors are important in determining the local distribution of plants?
- What are the important biomes of the world, and where are they located?

No single type of animal or plant or fungus or bacterium or protist can tolerate all the conditions found on the earth. Each thrives only within relatively narrow ranges of temperature, precipitation, soil conditions, salinity, and other physical factors (see Chapters 4–6). Biotic factors also influence the ability of populations to maintain themselves under these conditions (see Chapter 7). The preferences and tolerances of each species differ from those of every other species.

Because all species are adapted to their surroundings by modifications of form and function, differences in physical conditions from place to place often reveal themselves as differences in the forms of plants and animals. Variation in physical conditions derives from two major classes of factors: climate and topography. For most terrestrial organisms, climate includes the characteristic temperature and precipitation patterns of a region. Climate interacts with topography and other features of the land to create local variation. For aquatic organisms, temperature and salinity serve as climatic factors, and the land underlying oceans, streams, and lakes creates topographic diversity to which aquatic organisms respond by way of their adaptations and the distributions of their populations.

Clues to the origin of diversity in the biological world can be found in the study of variation in the physical world. The surface of the earth, its waters, and the atmosphere above it behave as a giant heat machine, obeying the laws of thermodynamics. Climate is determined by absorption of the energy in sunlight and its redistribution over the globe. As the earth's surface varies from bare rock to forested soil, open ocean, and frozen lake, its ability to absorb sunlight varies as well, creating differential heating and cooling. The heat energy absorbed by the earth is eventually radiated back into space, but not before undergoing further transformations that perform the work of evaporating water and contributing to the circulation of the atmosphere and oceans. All these factors have created a great variety of physical conditions, which in turn have fostered the diversification of ecosystems. In this chapter, we shall describe these patterns of climate and topography.

The activities of organisms may change the physical environment, as we emphasized in Chapter 3. There is great concern about how the industrial and agricultural activities of humans are affecting the natural climate control mechanisms of the earth. Humans may change climatic conditions by altering the flux of energy and materials in the natural world. We shall discuss those interactions in Part 3 of this book when we talk about how energy and materials move about in the natural world.

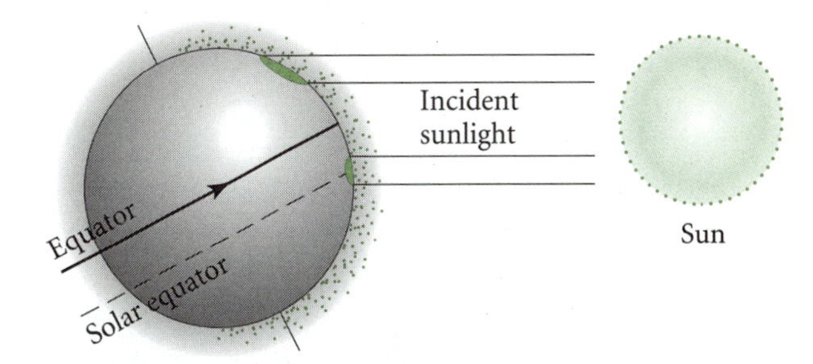

FIGURE 8-1 The warming effect of the sun is greatest at the solar equator because the sun's rays are closer to the perpendicular there and shine directly down on the surface of the earth during the middle of the day. At higher latitudes, sunlight strikes the earth's surface at a lower angle, and is therefore spread over a greater area; it must also pass through more of the earth's atmosphere.

8.1 Variation in solar radiation with latitude creates major global patterns in temperature and rainfall.

The earth's climate tends to be cold and dry toward the poles and hot and moist toward the equator. Although this oversimplification has many exceptions, climate nonetheless does exhibit broadly defined global patterns. The primary cause of this global variation in climate is the greater intensity of sunlight at the equator than at higher latitudes. This is a simple consequence of the angle of the sun relative to the surface of the earth at different latitudes (Figure 8-1). The sun warms the atmosphere, oceans, and land most when it is directly overhead. A beam of sunlight is spread over a greater area when the sun approaches the horizon, and it also travels a longer path through the atmosphere, where much of its energy is either reflected or absorbed by the atmosphere and reradiated into space as heat. The sun's highest position each day varies from directly overhead in the Tropics to near the horizon in polar regions; hence the warming effect of the sun diminishes from the equator to the poles.

Air Currents and Precipitation

Warming air expands, becomes less dense, and thus tends to rise. As air heats up, its ability to hold water vapor increases, and evaporation quickens; the rate of evaporation from a wet surface nearly doubles with each 10°C rise in temperature. The mass of air that rises in the Tropics under the warming sun eventually spreads to the north and south in the upper layers of the atmosphere. It is replaced from below by surface-level air from subtropical latitudes. The rising tropical air mass cools as it radiates heat back into space. By the time this air mass has extended to about 30° north and south of the **solar equator** (the parallel of latitude that lies directly under the sun), it has become dense enough to sink back to the earth's surface, completing a cycling of air within the atmosphere (Figure 8-2a). This type of circulation pattern is called a **Hadley cell.** One Hadley cell forms around the earth immediately to the north of the equator and another to the south. The sinking air of these tropical Hadley cells at about 30° north and south of the equator drives secondary Hadley cells in temperate regions, which circulate in the opposite direction. The circulation of these Hadley cells causes air to rise at about 60° north and south of the equator, which leads to the formation of polar Hadley cells. All this circulation of air is driven by the differential heating of the atmosphere with respect to latitude (Graedel and Crutzen 1995).

The region within which surface currents of air from the northern and southern subtropics meet in the equatorial region and begin to rise under the warming influence of the sun is referred to as the **intertropical convergence.** As moisture-laden tropical air rises and begins to cool, the moisture condenses to form clouds and precipitation. Thus, the Tropics are wet not because there is more water at tropical latitudes than elsewhere, but because water cycles more rapidly through the tropical atmosphere. The heating effect of the sun causes water to evaporate and warmed air masses to rise. As warm, moist air rises, it cools. Eventually, the moisture condenses, forming clouds. Moisture droplets in clouds bump into each other and aggregate, eventually forming droplets too large to be supported by the rising air; that moisture falls as precipitation (Figure 8-2b).

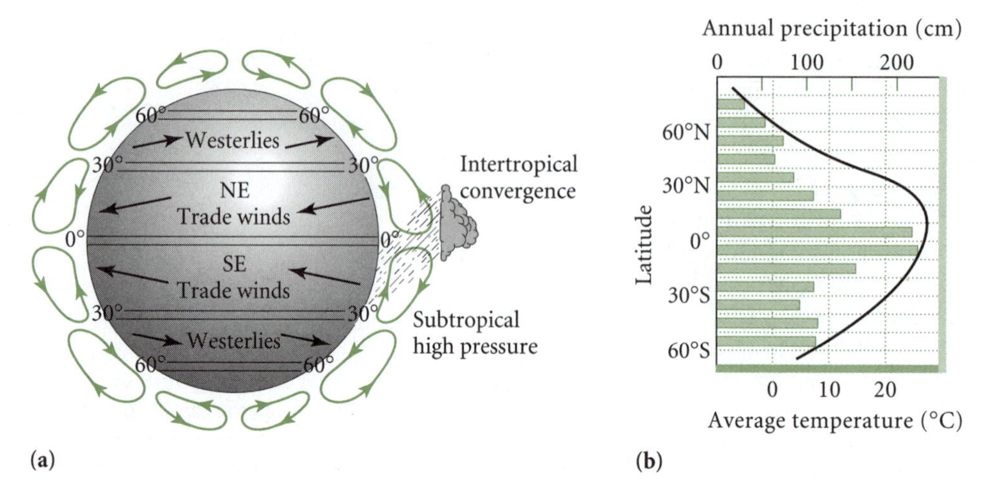

FIGURE 8-2 (a) Differential warming of the earth's surface creates Hadley cells of air circulation. Warm, moist air rises in the Tropics, resulting in abundant rainfall. Cool, dry air descends to the surface at subtropical latitudes. The intertropical convergence is the latitudinal belt at the solar equator within which surface winds converge from the north and south. (b) Average annual precipitation (green bars) and temperature (black curve) are shown for 10° latitudinal belts within continental landmasses. Because the chart presents averages from many localities, it obscures the great variation within each latitudinal belt. (*Data from Clayton 1944.*)

The air mass moving high in the atmosphere to the north and south, away from the intertropical convergence, has already lost much of its water to precipitation in the Tropics. As it sinks and begins to warm at subtropical latitudes, its capacity to evaporate and hold water increases. As the air mass descends to ground level in the subtropics and spreads to the north and south, it draws moisture from the land, creating zones of arid climate centered at latitudes of about 30° north and south of the equator (Figure 8-3). All the great deserts of the world—the Arabian, Sahara, Kalahari, and Namib of Africa, the Atacama of South America, the Mojave, Sonoran, and Chihuahuan of North America, and the Australian—fall within these belts.

The positions of the continental landmasses exert a secondary effect on the global pattern of precipitation. At a given latitude, rain falls more plentifully in the Southern Hemisphere because oceans and lakes cover a greater proportion of its surface (81%, com-

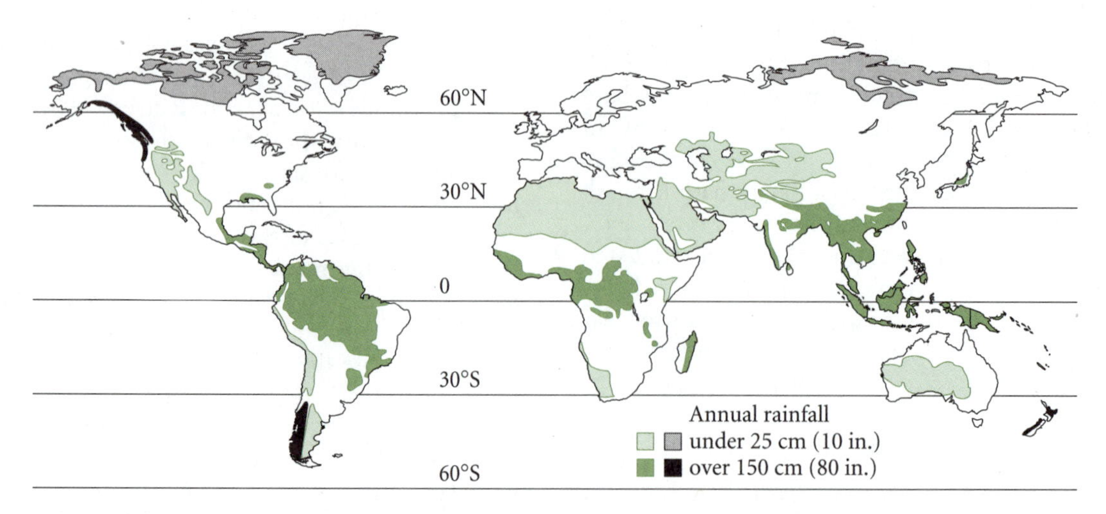

FIGURE 8-3 Distribution of the major deserts (regions with less than 25 cm [10 inches] annual precipitation) and wet areas (with more than 150 cm [80 inches] annual precipitation) of the world. Dark-colored areas are tropical rain forests; light-colored areas are subtropical deserts; black areas in western North America, Chile, and New Zealand are temperate rain forests; and gray areas are arctic deserts. (*After Espenshade 1971.*)

pared with 61% of the Northern Hemisphere: List 1966). Water evaporates more readily from exposed water surfaces than from soil and vegetation.

Energy from the sun drives the winds. The rotation of the earth deflects the surface flows in the Hadley cells to the west in the Tropics and to the east in the middle latitudes. The resulting wind patterns are called the **trade winds** and the **westerlies,** respectively (see Figure 8-2). Such air currents help to distribute water vapor through the atmosphere. Indeed, wind patterns interact with topographic features to create precipitation. Mountains force air upward, causing it to cool and lose its moisture as precipitation on the windward side of a mountain range. As the air descends the leeward slopes of the mountains and travels across the lowlands beyond, it picks up moisture and creates arid environments called **rain shadows** (Figure 8-4). The Great Basin deserts of the western United States and the Gobi Desert of Asia lie in the rain shadows of extensive mountain ranges.

The interior of a continent is usually drier than its coasts simply because the interior is farther removed from the major site of water evaporation, the surface of the ocean. Furthermore, coastal (maritime) climates vary less than interior (continental) climates because the great heat storage capacity of the oceans reduces temperature fluctuations. For example, the hottest and coldest mean monthly temperatures near the Pacific coast of the United States at Portland, Oregon, differ by 16°C. Farther inland, this range increases to 18°C at Spokane, Washington, 26°C at Helena, Montana, and 33°C at Bismarck, North Dakota.

Ocean Environments

Ocean currents play a major role in moving heat over the surface of the earth and thereby influence the climates of the continents. In large ocean basins, cold water tends to move toward the Tropics along the western coasts of the continents, and warm water tends to move toward temperate latitudes along the eastern coasts of the continents (Figure 8-5). The cold Peru Current moving north from the Antarctic Ocean along the coasts of Chile and Peru creates cool, dry environments along the west coast of South America right to the equator. Conversely, the warm Gulf Stream, emanating from the Gulf of Mexico, carries a mild climate far to the north into western Europe and the British Isles.

The physical conditions of the oceans themselves are as complex as those of the atmosphere. Variation in those conditions is caused by winds, which propel the major surface currents of the oceans, and by the underlying topography of the ocean basin. In addition, deeper currents are established by differences in the density of ocean water caused by variations in temperature and salinity.

Any upward movement of ocean water is referred to as an **upwelling.** Vertical upwelling currents occur wherever surface currents diverge, as in the western tropical Pacific Ocean. As surface currents move apart, they tend to draw water upward from deeper layers. Strong upwelling zones are also established on the western coasts of continents where surface currents flow toward the equator. A curious consequence of the rotation of the earth is the deflection of surface currents away from the continental margins, which is aided by the trade winds. As water moves away from the continents, it is replaced by water from greater depths. Because this water tends to be nutrient-rich, upwelling zones are often regions of very high biological productivity. The most famous of these support the rich fisheries of the Benguela Current along the western coast of southern Africa and the Peru Current along the western coast of South America.

Cold temperatures and high salinity at the ocean surface can result in vertical currents because cold, salty water is dense and tends to sink. These conditions are most pronounced around the continent of Antarctica. When water freezes each winter, salt is excluded from the crystalline ice. This causes an increase in the salinity of the water immediately underneath. This dense water descends and forms the Antarctic

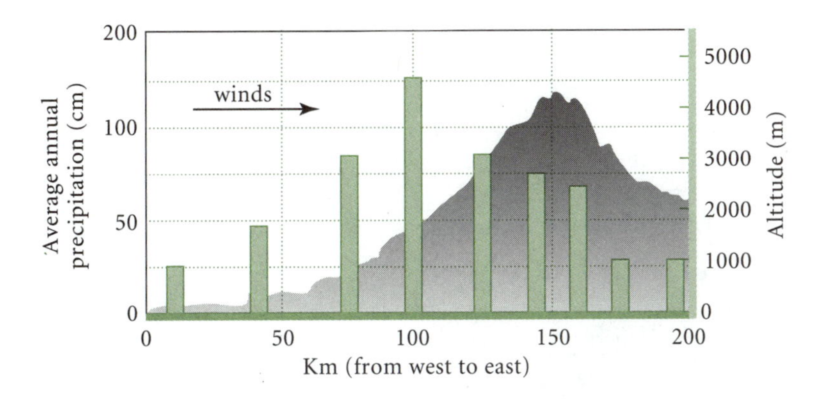

FIGURE 8-4 The influence of the Sierra Nevada mountain range on local precipitation causes a rain shadow to the east. Winds come predominantly from the west across the Central Valley of California. As moisture-laden air is deflected upward by the mountains, it cools, and its moisture condenses, resulting in heavy precipitation on the western slope of the mountains. As the air rushes down the eastern slope, it warms and begins to pick up moisture, creating arid conditions in the Great Basin (*After Pianka 1988.*)

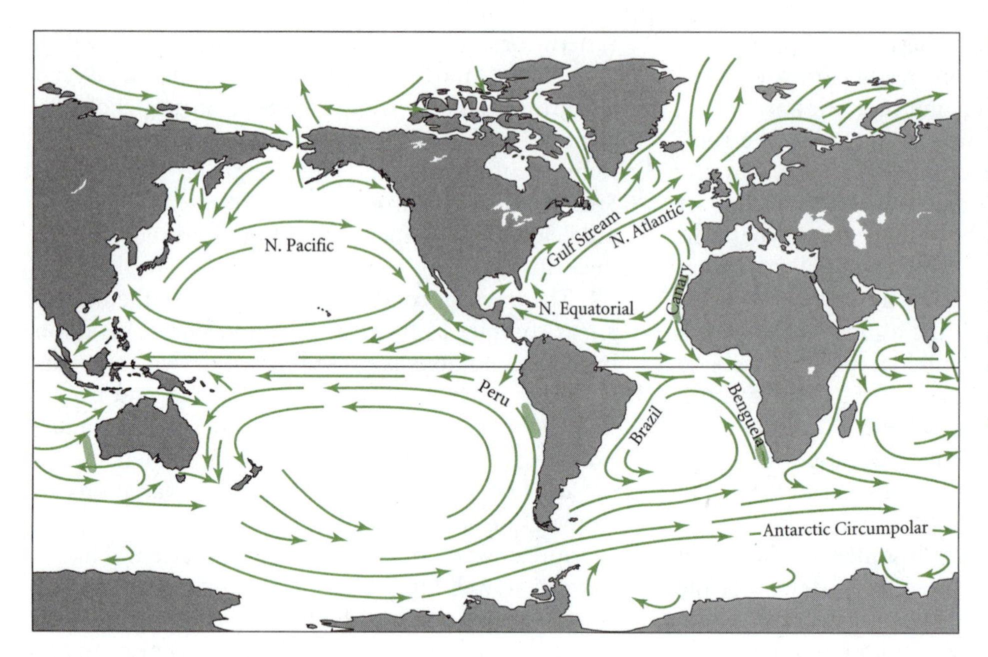

FIGURE 8-5 Major surface currents of the world's oceans. Water movement generally proceeds clockwise in the Northern Hemisphere and counterclockwise in the Southern Hemisphere. Zones of strong upwelling are indicated by green shading. The Peru Current of the eastern Pacific is often called the Humboldt Current. (*After Schlesinger 1991; from Knauss 1978.*)

Bottom Water, which spreads northward over the ocean basins. Close to the antarctic continent, it is replaced at the surface by the so-called North Atlantic Deep Water, which has been slowly moving south at great depth for hundreds of years. This water reaches the surface rather devoid of oxygen, but rich enough in accumulated nutrients to support tremendously high productivity. Just beyond this belt of upwelling around Antarctica is a region of downwelling caused by the increasing density of water as it warms toward 4°C. Because this pattern of downwelling and upwelling is driven by temperature and salinity differences in surface waters, it is referred to as **thermohaline circulation.**

Upwelling zones are extremely important to marine ecosystems. When organisms living in the surface waters die, their remains sink to the bottom, taking with them all of the organic material in their bodies. Because of this, surface waters tend to be depleted of nutrients. Deep water rising to the surface brings this organic material back to the surface and usually creates a region of high productivity. Just as mountains force air currents upward, irregularities in the ocean bottom, especially around islands and archipelagoes, force water upward. Tropical coral reefs are productive in part because the islands around which they develop deflect deeper water to the surface. The Galápagos archipelago, which lies on the equator 1,000 km off the west coast of South America, has a particularly complex marine environment because its many islands intercept several ocean currents (Figure 8-6). To the north of the island group is deep ocean, warm at

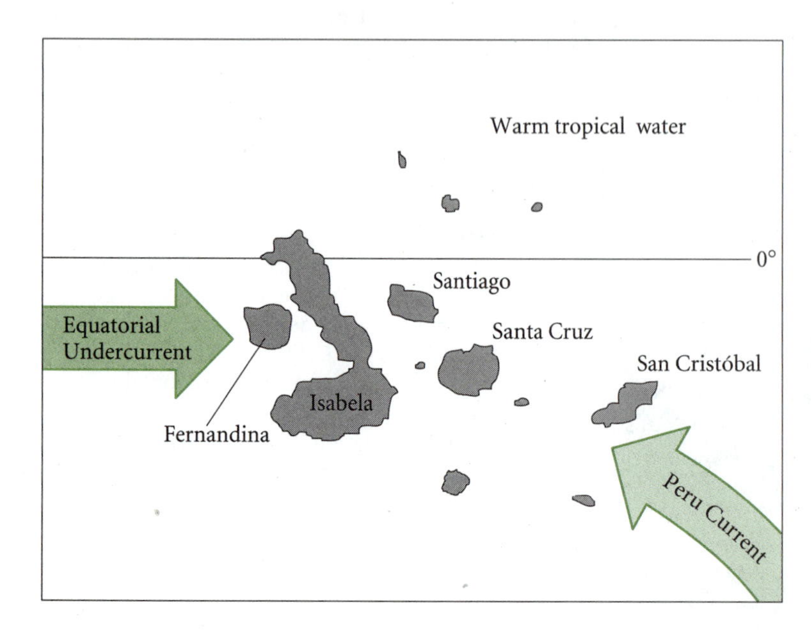

FIGURE 8-6 Map of the major ocean currents in the region of the Galápagos archipelago. The primary influences are the deep, warm tropical waters of the north, the cold Peru Current, which approaches the islands from the southeast, and the cold, nutrient-laden Equatorial Undercurrent, which is forced to the surface as it encounters the western islands in the group. (*After Houvenaghel 1984*).

the surface and unproductive. From the coast of Peru to the southeast comes the Peru Current, cold and laden with nutrients received from coastal upwelling. These surface waters flow past the Galápagos to the west, bringing nutrient-rich waters to the southernmost islands in the group. From the west, the strong Equatorial Undercurrent at mid-depth is forced to the surface by the westernmost islands of the archipelago, creating a zone exceptionally rich in ocean life and home to the only species of penguin living within the Tropics.

8.2 The seasons bring predictable changes in the environment.

Patterns of climate change are as important to biological systems as the average temperature and precipitation (Wolfe 1979). Periodic cycles in climate follow astronomical cycles: the rotation of the earth upon its axis causes daily periodicity; the revolution of the moon around the earth creates lunar cycles in the amplitude of the tides; and the revolution of the earth around the sun brings seasonal change.

Seasonal Changes in Solar Radiation, Temperature, and Rainfall

The equator is tilted slightly with respect to the path the earth follows in its orbit around the sun (see Figure 8-1). As a result, the Northern Hemisphere receives more solar energy than the Southern Hemisphere during the northern summer, and less during the northern winter. The seasonal change in temperature increases with distance from the equator (Figure 8-7). At high latitudes in the Northern Hemisphere, mean monthly temperatures vary by an average of 30°C, with extremes of more than 50°C annually; the mean temperatures of the warmest and coldest months in the Tropics differ by as little as 2–3°C.

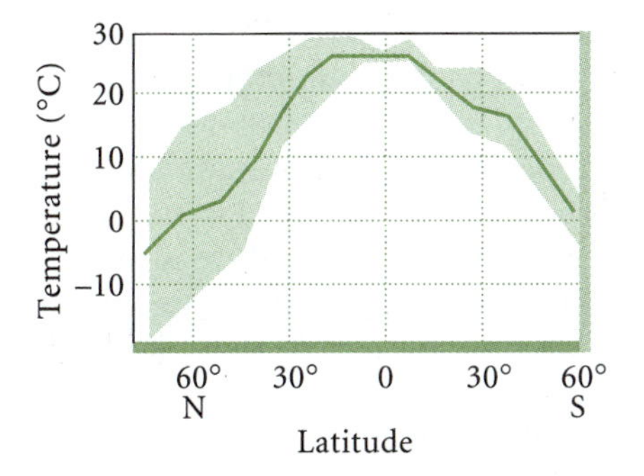

FIGURE 8-7 Annual range of mean monthly temperatures (shaded area) as a function of latitude. The mean annual temperature is indicated by the line. The width of the shaded area corresponds to the extent of seasonal fluctuation in temperature. The most variation occurs at high northern latitudes, the least at the equator. (*Data from Clayton and Clayton 1947.*)

Latitudinal patterns in rainfall seasonality result in part from the seasonal northward and southward movement of the solar equator. This movement of the intertropical convergence results in two seasons of heavy precipitation at the equator and a single wet season alternating with a pronounced dry season at the edges of the Tropics. Seasonality of rainfall is most pronounced in broad latitudinal belts lying about 23° north and south of the equator. As the seasons change, these regions alternately come under the influence of the solar equator, bringing heavy rains, and of subtropical high-pressure belts, bringing clear skies (Figure 8-8).

Panama, at 9° N, lies within the wet Tropics, but even there the seasonal movement of the solar equator profoundly influences the climate. The major tropical belt of high rainfall remains south of Panama during most of the northern winter, but it lies directly overhead during the northern summer. Hence the winter is dry and windy, the summer humid and rainy. Panama's climate is wetter on the northern (Caribbean) side of the isthmus—the direction from which the prevailing winds come—than on the southern (Pacific) side; mountains intercept moisture coming from the Caribbean side of the isthmus and produce a rain shadow. The Pacific lowlands are so dry during the winter months that most trees lose their leaves. Bare branches in tinder-dry forests contrast sharply with the wet, lush, more typically tropical forest seen during the wet season (Figure 8-9).

Farther to the north, at 30° N, in the Chihuahuan Desert of central Mexico, rainfall comes only during the summer, when the solar equator reaches its most northward limit (Figure 8-10a). During the rest of the year this region falls within the dry subtropical high-pressure belt. Summer rainfall extends north into the Sonoran Desert of southern Arizona and New Mexico (Figure 8-10b). This area also receives moisture during the winter from the Pacific Ocean, carried by the southwesterly winds emanating from the subtropical high-pressure belt farther south. Southern California lies beyond the summer rainfall belt and has a winter rainfall, summer drought climate (Figure 8-10c), often referred to as a **Mediterranean climate** because the Mediterranean region of Europe has the same seasonal pattern of temperature and rainfall. Mediterranean climates are also found in western South Africa, Chile, and Western Australia, all lying along the western sides of continents at about the same latitude north or south of the equator.

Seasonal Changes in Oceans

The sun warms the seas just as it does the continents and the atmosphere, but the ocean's great mass of

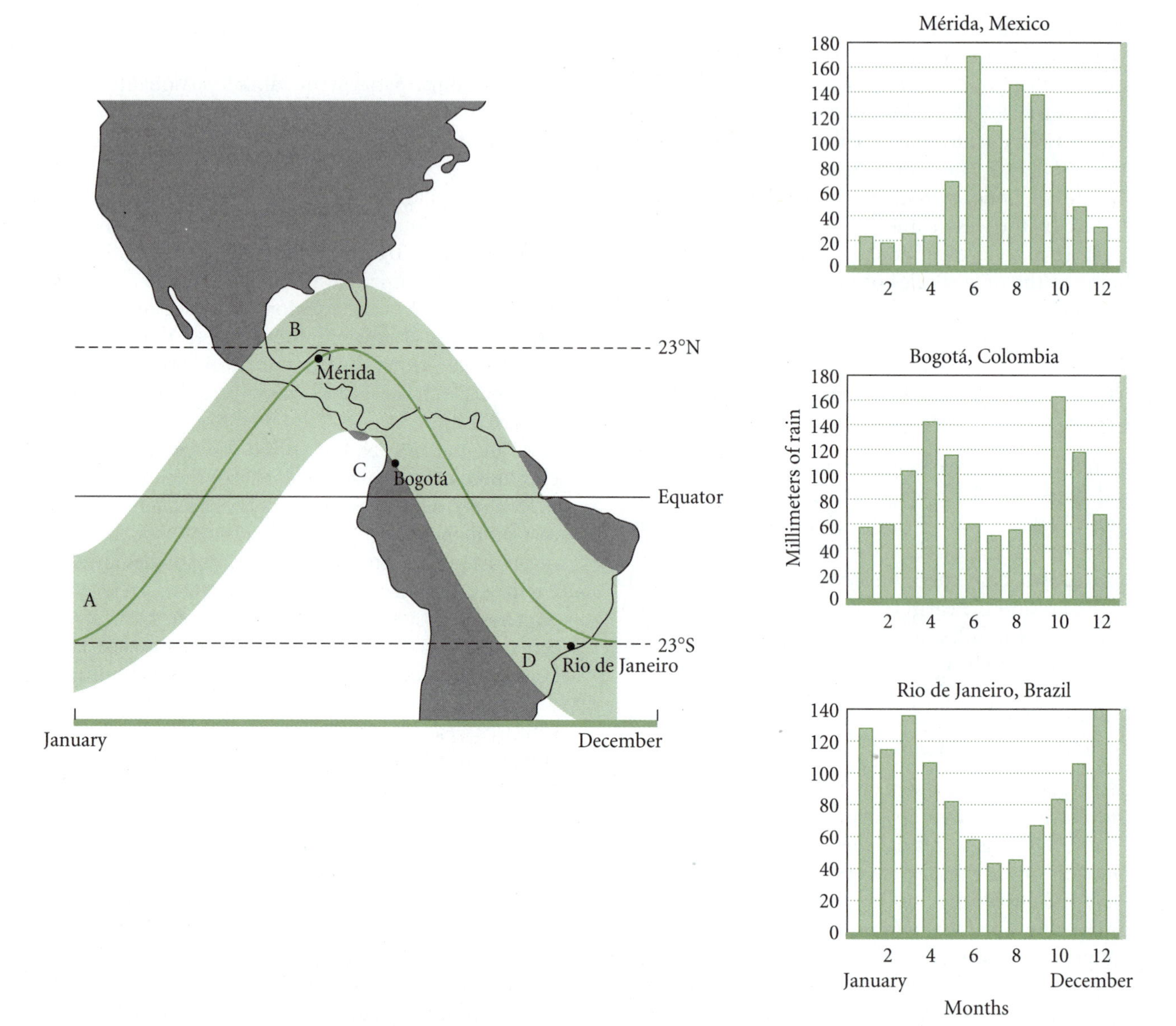

FIGURE 8-8 The dark line represents the intertropical convergence (see Figure 8-2), which moves seasonally within the shaded area (A). When the intertropical convergence is centered at Mérida, Mexico, that location experiences heavy rain (B), whereas Bogotá, Colombia, experiences a dry season (C) and Rio de Janeiro (D) an intermediate condition.

water acts as a heat sink to dampen daily and seasonal fluctuations in temperature. Where ocean temperature does change seasonally, it reflects seasonal movements of water masses of different temperatures more often than it does local heating and cooling. During the Panamanian dry season, roughly January to April, steady trade winds blowing in a southwesterly direction create strong upwelling currents in the Pacific Ocean along the southern and western coasts of Central America. During these upwelling periods, winds blow warm surface water away from the coast, where cooler water moves upward from deeper regions to replace it. As a result, the annual fluctuation in seawater temperature is three times higher on the Pacific coast of Panama than on the Caribbean coast.

Seasonal Changes in Temperate Lakes

Small temperate zone lakes respond quickly to the changing seasons (Figure 8-11). In winter, a typical lake has an inverted **temperature profile;** that is, the coldest water (0°C) lies at the surface, just beneath the ice. (Because the density of water increases between freezing and 4°C, the warmer water within this range sinks, and the temperature increases to as much as 4°C toward the bottom of the lake.) In early spring, the sun warms the lake surface gradually. But until the surface temperature exceeds 4°C, the sun-warmed surface water tends to sink into the cooler layers immediately below. This vertical mixing distributes heat throughout the water column from the

FIGURE 8-9 Many trees on the Pacific slope of Panama shed their leaves during the dry season, which lasts from January through April. (*Photograph by M. A. Guerra; courtesy of the Smithsonian Tropical Research Institute.*)

surface to the bottom, resulting in a uniform temperature profile. Winds cause deep vertical movement of water in early spring, a phenomenon called **spring overturn,** bringing nutrients to the surface from the bottom sediments and bringing oxygen from the surface to the depths.

Later in spring and early summer, as the sun rises higher each day and the air above the lake warms, surface layers of water gain heat faster than deeper layers, Now, at temperatures exceeding 4°C, the warmer and less dense surface water literally floats on the cooler, denser water below, a condition known as **stratification.** Stratification creates a zone of rapid temperature change at intermediate depth, called the **thermocline.** Once the thermocline is well established, water does not mix across it. The depth of the thermocline varies with local winds and with the depth and turbidity of the lake. It may occur anywhere between 5 and 20 meters below the surface; lakes less than 5 meters deep usually lack stratification.

The thermocline demarcates an upper layer of warm water called the **epilimnion** and a deep layer of cold water called the **hypolimnion.** Most of the primary production of the lake occurs in the epilimnion, where sunlight is intense. Photosynthesis supplements mixing of oxygen from air, which is aided by wind, at the lake surface to keep the epilimnion well aerated and thus suitable for animal life, but growing plants in the **euphotic zone**—that part of the water column receiving sufficient light for photosynthesis—often deplete dissolved mineral nutrients and thereby curtail their own production. The hypolimnion is cut off from the surface of the lake, and its animals and bacteria, remaining mostly below the euphotic zone, deplete the water of oxygen, creating anaerobic conditions.

In the fall, the surface layers of the lake cool more rapidly than the deeper layers and, becoming heavier than the underlying water, begin to sink. This second period of vertical mixing, called **fall overturn,** persists into late fall, until the temperature at the lake surface drops below 4°C and winter stratification resumes. Fall overturn causes greater vertical mixing of water than spring overturn because temperature differences in the lake during summer stratification exceed those during winter stratification. Fall overturn speeds the movement of oxygen to deep waters and rushes nutrients to the surface. Where the hypolimnion becomes fairly warm in midsummer, deep vertical mixing may take place in late summer, when temperatures remain favorable for plant growth. The infusion of nutrients into surface waters at this time often causes a burst of phytoplankton population growth, resulting in a fall bloom. In deep, cold lakes, vertical mixing does not penetrate to all depths until late fall or early winter, when water temperatures are too cold to support plant growth.

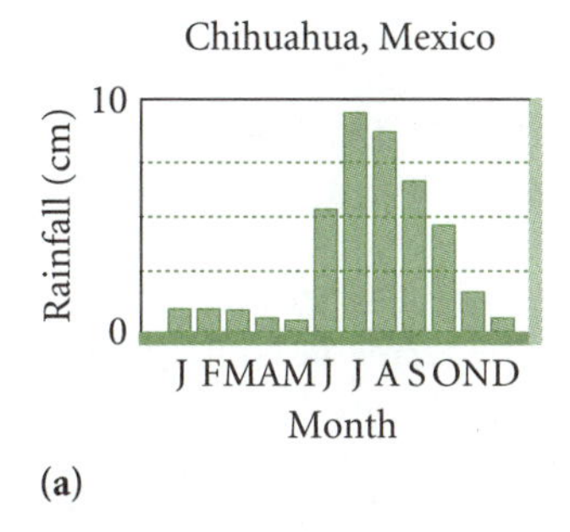

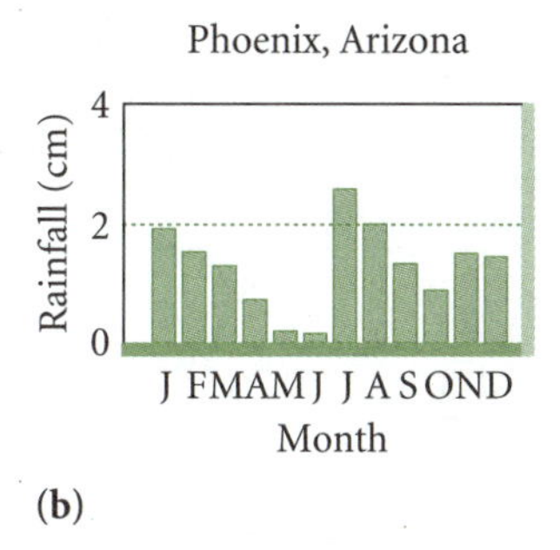

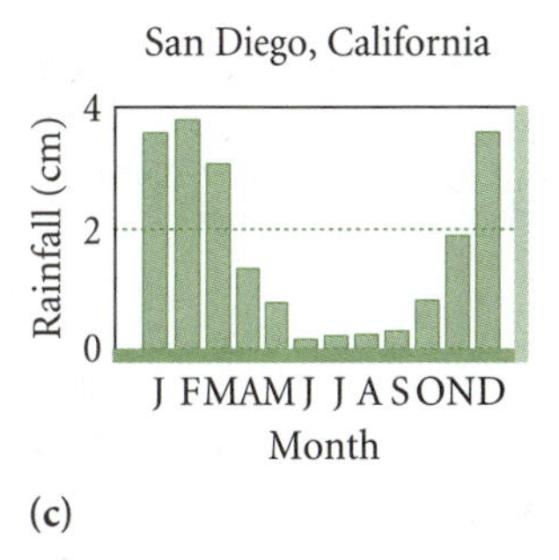

FIGURE 8-10 Seasonal occurrence of rainfall at three localities in western North America. (a) The summer rainy season of the Chihuahuan Desert in central Mexico. (b) The combined climate pattern of the Sonoran Desert. (c) The winter rain and summer drought pattern of the Pacific coast (Mediterranean climate). (*Data from Clayton 1944.*)

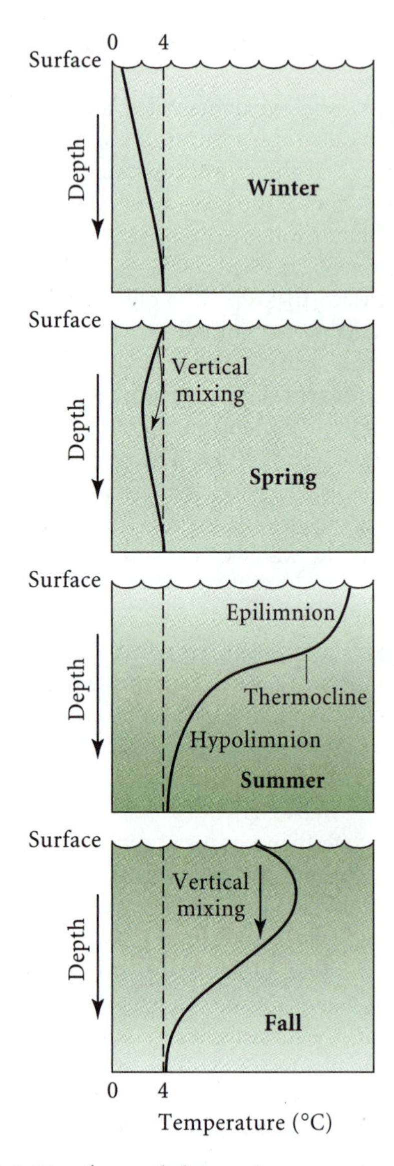

FIGURE 8-11 Seasonal changes in water temperature in a temperate lake. Because the density of water is temperature dependent (densest at about 4°C), thermal stratification may occur during the coldest and warmest parts of the year. In winter, dense water is overlain by a sheet of ice. In summer, warm water (> 4°C) overlays cooler, denser water. The density difference between the top portion of the lake (epilimnion) and the bottom portion (hypolimnion) may be so great as to prevent mixing of the two areas. When surface waters cool down in the fall, or warm up in the spring, the lake water becomes a uniform temperature, and mixing of the top and bottom waters may occur.

8.3 Irregular fluctuations in the environment are superimposed on periodic cycles.

Everyone knows that weather is difficult to predict. We often remark that a year was particularly dry or cold compared with others. The flooding in recent years in the Mississippi Valley and along many rivers in Europe drives home the capriciousness of nature. Most aspects of climate seem unpredictable. Rainfall varies most where it is sparsest: in deserts, and, in wetter localities, during the driest season. Year-to-year variation in temperature on a particular date is greatest where temperature fluctuates most during the year. The most extreme conditions occur infrequently, but they may affect organisms disproportionately.

The Peruvian fishing industry, as well as some of the world's largest seabird colonies (Figure 8-12), thrives on the abundant fish in the anchovy-rich waters of the Peru Current, a mass of cold water that flows up the western coast of South America and finally veers offshore at Ecuador, toward the Galápagos archipelago. North of this, warm tropical inshore waters prevail along the coast. Each year a warm countercurrent known as **El Niño** ("little boy" in Spanish, referring to the infant Jesus because the phenomenon appears about Christmastime) moves down the coast of Peru. Some El Niños are strong enough to force the cold Peru Current offshore, taking with it the food supply of millions of birds as well as the livelihoods of local fishermen (Glynn 1988).

During years between strong El Niño events, a steady wind blows across the equatorial central Pacific Ocean from an area of high atmospheric pressure centered over Tahiti to an area of low pressure centered over Darwin, Australia. The difference in atmospheric pressure between the western Pacific and the eastern Pacific normally fluctuates somewhat between years in a see-saw pattern, with high pressure in the west corresponding to low pressure in the eastern regions. This fluctuation of pressure is called the **Southern Oscillation** (SO). Occasionally, the SO will bring extremely high pressures in the western Pacific, resulting in a reversal of the trade winds and, in some instances, the normally westward-flowing ocean currents. Such conditions, coupled with a strong El Niño, cause warm water to pile up along the western coast of South America and greatly weaken or even halt upwelling there. The simultaneous occurrence of an unusually strong El Niño and unusually high atmospheric pressures in the western Pacific is called an **El Niño-Southern Oscillation** (ENSO) **event** or, sometimes, a **warm event.** Such events not only affect the Peruvian fishing industry, but have important widespread climatic effects. It is important to appreciate that the term "El Niño," which was coined by Peruvian fishermen to describe an annual change in environmental conditions, is used by many to refer to all the global events that occur in an ENSO year. Thus, "El Niño" is often used synonymously with "ENSO" or "warm event" when discussing global changes, even though it has a more restrictive meaning.

FIGURE 8-12 Nesting colony of Peruvian boobies on an island off the coast of Peru. This dense population depends on the anchovy stocks in the nutrient-rich Peru Current. *(Photograph by R. C. Murphy; courtesy of the Department of Library Services, American Museum of Natural History.)*

Historical records of atmospheric pressure at sea level, direction of surface winds, sea surface temperature, and air temperature just above the surface of the ocean are combined using statistical techniques to calculate the **multivariate ENSO index** (MEI) (Wolter 1987, Wolter and Timlin 1993). MEI values above 1 indicate warm sea temperatures and westerly wind directions, indicating an ENSO event. MEI values below 0 indicate periods when ocean temperatures are unusually cold. MEI data recorded from 1950 until the present (Figure 8-13a) shows the occurrence of significant ENSO events in 1957–1958, 1965–1966, 1972–1973, 1982–1983, and 1986–1987. An unusually long ENSO event began in 1991 and extended until 1995.

The climatic and oceanographic effects of an ENSO event extend over much of the world, affecting ecosystems in such distant areas as India, South Africa, Brazil, and western Canada. When the El Niño countercurrent warms the water of the eastern Pacific, rainfall in western North America increases and temperatures moderate, leading to warmer winters. The warming of the waters of the eastern Pacific is associated with a cooling of the waters of the western Pacific, which results in a reduction in the intensity of the monsoon rainfalls in that region (Schlesinger 1991). The ENSO event in 1982–1983, one of the largest in recorded history, disrupted fisheries and destroyed kelp beds in California, caused reproductive failure of seabirds in the central Pacific Ocean, and resulted in widespread mortality of coral in Panama. Precipitation was also dramatically affected in many terrestrial ecosystems, notably in the deserts of northern Chile, normally one of the driest places on earth, which received the first recorded rainfall in over a century. At the time of the writing of this book (1997–1998) the world is in the grip of the largest ENSO event ever recorded. A comparison of the MEI for the seven largest ENSO events, including the one in progress, shows that in the late summer of 1997 the current event peaked at the highest level of the 1982–1983 event (Figure 8-13b). Of course, we can't wait for the current warm event to play itself out before proceeding with the publication of this book, so we cannot compare it with the 1982–1983 event. But we know from weather reports that the effects of the event are being felt in the form of extremely dry conditions in Australia, resulting in frequent and intense fires, and extremely wet conditions along the west coast of the United States and Mexico, where rains have been torrential at times.

The lengthy period of warm ocean temperatures between 1990 and 1995 and the occurrence of more frequent ENSO events since 1975 (see Figure 8-13a) have led some scientists to question whether there has been a fundamental change in the periodicity of the ENSO and, more importantly, whether such a change is related to global climate change. Trenberth and Hoar (1996) studied sea temperature data from the 1991–1995 event to determine whether the event was unexpectedly long, based on information from 113 years of ENSO observations. They concluded that, indeed, a change in the pattern of ENSO events has occurred. They speculate that this change may be caused by a general warming of the earth as a result of the enhancement of the earth's greenhouse (see Chapter 3).

Just as fishermen noticed that the waters off Peru sometimes become unusually warm, they also noticed that those waters occasionally become unusually cold. They referred to this phenomenon as **La Niña** ("little girl" in Spanish). Unusually strong La Niña conditions can also have widespread effects on environmental conditions. A **cold event** is one associated with unusually cool sea surface temperatures in the eastern

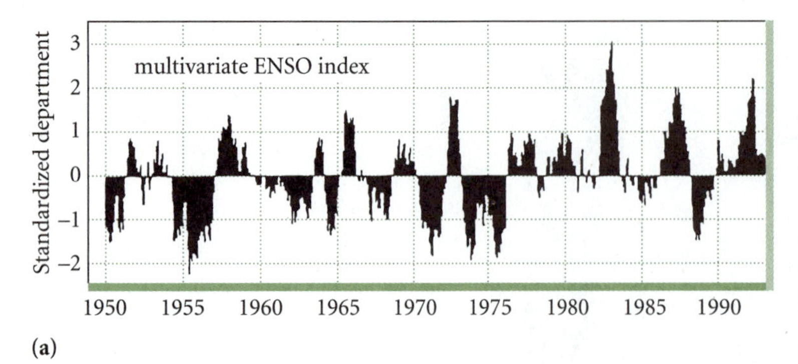

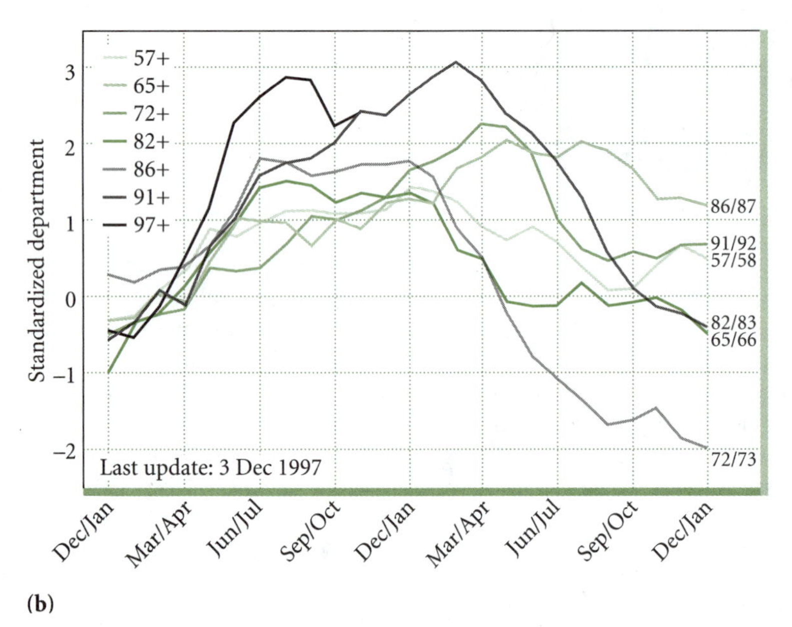

FIGURE 8-13 (a) Multivariate ENSO index (MEI) for the years between 1950 and 1996. The MEI includes information about wind currents, ocean and air temperatures, and atmospheric pressure. Values above 0 indicate times of warm ocean water temperatures. Six significant ENSO events (MEI > 1) are indicated. The most recent ENSO (1997–1998) is not shown. (b) Comparison of MEI indexes for seven ENSO events. The 1997–1998 event (black line) was very strong, as shown by its peak near that of the 1982–1983 event, which was the largest previous event. (*Data from NOAA-CIRES Climate Diagnostics Center, University of Colorado at Boulder.*)

Pacific (a strong La Niña) along with unusually high atmospheric pressures in the eastern Pacific associated with the Southern Oscillation. During such events, cold sea surface temperatures extend well to the west in the equatorial Pacific. The global climate effects of a cold event tend to be opposite of those of an ENSO event ("El Niño"). For example, whereas winter temperatures are warmer than usual in the southwestern and southeastern United States during an ENSO event, those areas would experience cooler temperatures during a cold event.

8.4 Local topographic and geologic features produce additional variation in global climate patterns.

Climate patterns are large-scale phenomena. In the United States, we talk about the sunny South or the desert Southwest. Our appreciation of the differences in average snowfall between Mississippi and Maine gives us an intuitive understanding that the two areas will have different associations of plants and animals, and that both areas will be very different in that respect from a western state such as Utah. But were we to travel within Mississippi, Maine, or Utah, we would still observe a variety of patterns. Hilltops would possess different types of vegetation than would valleys, and riverbanks would differ from upland areas. Variation in topography and geology can create variation in the environment within regions of uniform climate.

Topography

In hilly areas, the slope of the land and its exposure to the sun influence the temperature and moisture content of the soil. Soils on steep slopes drain well, often causing moisture stress for plants at the same time that the soils of nearby lowlands are saturated with water. In arid regions, stream bottomlands and seasonally dry riverbeds may support well-developed **riparian** forests—located along the banks of a river, stream, or lake—that accentuate the bleakness of the surrounding desert. In the Northern Hemisphere, south-facing slopes directly face the sun, whose warmth and drying power limit vegetation to shrubby, **xeric** (drought-resistant) forms. The adjacent north-facing slopes remain relatively cool and wet and harbor **mesic** (moisture-requiring) vegetation (Figure 8-14).

Elevation

Air temperature decreases with altitude by about 6°C for each 1,000-meter increase in elevation, depending on the region. This decrease in temperature, which is caused by the expansion of air with lower atmospheric pressures at higher altitudes, is referred to as **adiabatic cooling.** In north temperate latitudes, a 6°C drop in temperature corresponds to the temperature change over an 800-kilometer increase in latitude. Even in the Tropics, if one climbs high enough, one eventually encounters freezing temperatures and perpetual snow. Where the temperature at sea level is 30°C, freezing temperatures are reached at about 5,000 meters, the approximate altitude of the snow line on tropical mountains. In many respects, the climate and vegetation of **alpine,** or high-altitude, areas resemble those of sea-level localities at higher latitudes (Billings and Mooney 1968). But, despite their similarities, alpine environments usually vary less from season to season than their low-elevation counterparts at higher latitudes. Temperatures in tropical montane environments remain nearly constant and frost-free over the year, allowing many tropical plants and animals to live in the cool environments found there.

As you move from the base of a mountain toward the summit, you will observe changes in the plant communities with elevation. These more or less distinct belts of vegetation were called **life zones** by the nineteenth-century naturalist C. H. Merriam (1894). Merriam's scheme of classification included five broad zones in the mountains of the southwestern United States, which he named, from low to high elevation (or from north to south), Lower Sonoran, Upper Sonoran, Transition, Canadian (or Hudsonian), and Alpine (or Arctic-Alpine).

FIGURE 8-14 The influence of exposure to the sun on the vegetation of a series of mountain ridges near Aspen, Colorado. The north-facing (left-facing) slopes are cool and moist, permitting the development of spruce forest. Shrubby, drought-resistant vegetation grows on the south-facing slopes.

At low elevations in the Southwest, one encounters a cactus and desert shrub association characteristic of the Sonoran Desert of northern Mexico and southern Arizona (Figure 8-15). In the woodlands along streambeds, the plants and animals have a distinctly tropical flavor. Many hummingbirds and flycatchers, ring-tailed cats, jaguars, and peccaries make their only temperate zone appearances in this area. In the Alpine zone, 2,500 meters higher, one finds a landscape resembling the tundra of northern Canada and Alaska. Thus, by climbing 2,500 meters, one experiences changes in climate and vegetation that would require a journey to the north of 2,000 kilometers or more at sea level.

Geology and Soil

Local variation in the bedrock underlying a region promotes the differentiation of soil types and enhances biotic heterogeneity. In the northern Appalachian Mountains and in mountains near the Pacific coast of the United States, outcrops of **serpentine,** a kind of igneous rock, weather to form soils containing so much magnesium that plant species characteristic of the surrounding soil types cannot grow (Walker 1954, Whittaker 1954, Proctor and Woodell 1975). Serpentine **barrens,** as they are called, usually support little more than a sparse covering of grasses and herbs, many of which are distinct **endemics**—species found nowhere else—that have evolved a high tolerance for magnesium (Figure 8-16). Depending on the composition of the bedrock and the rate of weathering, granite, shale, and sandstone also can support a barren type of vegetation. The extensive pine barrens of southern New Jersey, where mature trees attain no more than waist height in some areas, occur on a large outcrop of sand, which produces a dry, acid, infertile soil (McPhee 1968, McCormick 1970, Forman 1979).

8.5 Environmental conditions are influenced by the balance between rainfall and evapotranspiration.

Precipitation falling on a landscape returns to the atmosphere in two ways. First, some water is directly evaporated from the soil, from open water (lakes and oceans), and from the surfaces of objects—especially plants—wet from recent rain. A second pathway is via transpiration, the process by which plants evaporate from their leaves the water taken up from the soil

FIGURE 8-15 Vegetation at different elevations in the mountains of southeastern Arizona. The Lower Sonoran zone supports mostly saguaro cactus, small desert trees such as paloverde and mesquite, numerous annual and perennial herbs, and small succulent cacti. Agave, ocotillo, and grasses are conspicuous elements of the Upper Sonoran zone, with oaks appearing toward its upper edge. Large trees predominate at higher elevations: ponderosa pine in the Transition zone, spruce and fir in the Hudsonian zone. These gradually give way to bushes, willows, herbs, and lichens in the Alpine zone above the timberline. *(Courtesy of the U. S. Soil Conservation Service, the U. S. Forest Service, W. J. Smith, and R. H. Whittaker; from Whittaker and Niering 1965.)*

by their roots (see Chapter 4). The combination of transpiration and evaporation is referred to as **evapotranspiration.** It is a process that is very much dependent on temperature. Evaporation and transpiration nearly double with each 10°C rise in temperature, other things being equal, although the character of the soil and vegetation cover also influences water loss.

In natural environments, evapotranspiration is sometimes limited by the availability of water in the soil. **Potential evapotranspiration** (PE) is the amount of water that would be drawn from the soil if moisture were unlimited and the vegetation cover were 100%, as in some tropical areas. Potential evapotranspiration can be calculated from temperature and precipitation

FIGURE 8-16 A small serpentine barren in eastern Pennsylvania. The soils surrounding the barren support oak-hickory-beech forest.

(Ward 1967, Schlesinger 1991). Because potential evapotranspiration increases with temperature, temperature and water stress go hand in hand. Regions at high latitudes receiving 25 to 50 cm of precipitation each year have more favorable water budgets for plant production than tropical regions with similar levels of precipitation and higher temperatures.

The actual evapotranspiration in an area is related to the amount of soil water, the leaf area of the local vegetation, and the temperature. The rate of photosynthesis is proportional to evapotranspiration (Figure 8-17; Aber and Melillo 1991). In deserts, where precipitation and soil water are limited, most of the evapotranspiration occurs during the brief period after winter or early spring rains, when plant foliage is greatest and before the hottest part of the year. In grasslands, the relatively cool winter conditions, which reduce plant growth, and the heavy summer rains mean that plant photosynthetic activity and, consequently, evapotranspiration coincide with relatively high temperatures. Evapotranspiration in temperate

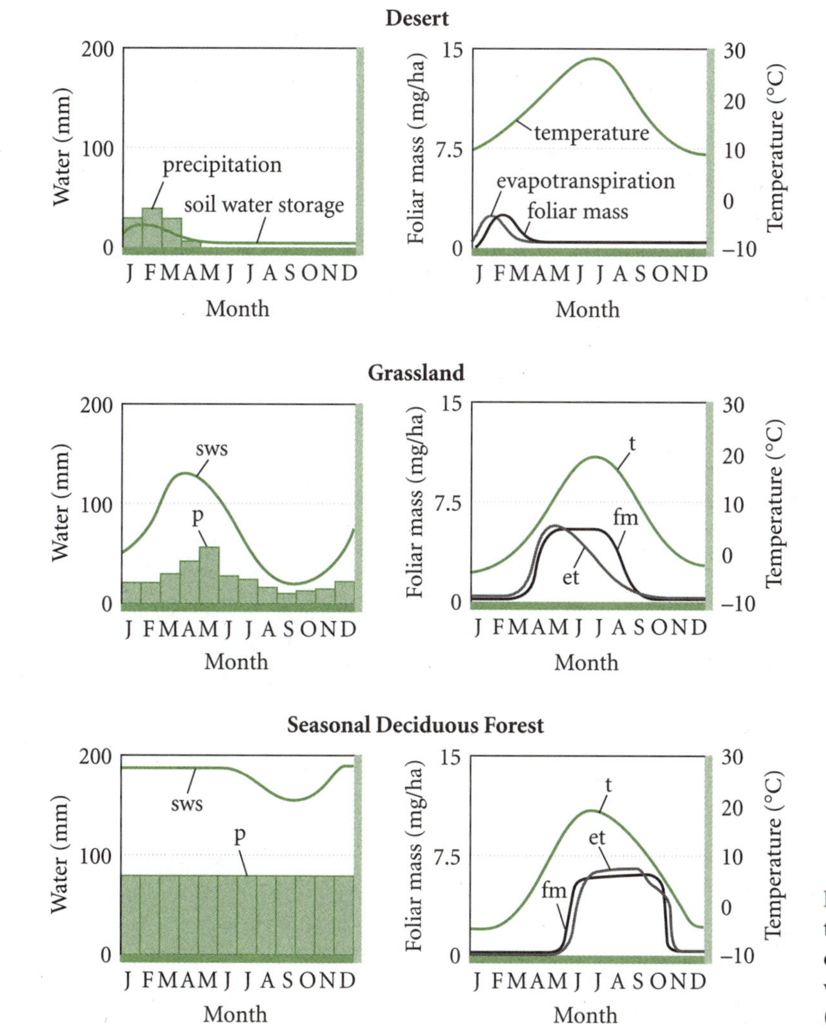

FIGURE 8-17 Relationship among temperature (t), evapotranspiration (et), mass of the foliage (fm), precipitation (p), and soil water storage (sws) for three different habitats. (*After Aber and Melillo 1991.*)

deciduous forests, where precipitation and soil water content are high, is constrained more by temperature than by the availability of water. Photosynthetic activity is reduced or halted by leaf fall during the winter months. Spring refoliation brings high leaf mass (and high leaf surface area), high photosynthetic activity, and, consequently, increased evapotranspiration (see Figure 8-17; Aber and Melillo 1991).

8.6 The distributions of plants are related to climate patterns, topography, and soil.

The geographic range of a plant species is that area within which overall climate patterns provide rainfall and temperature regimes suitable for its growth and reproduction. But not every patch of ground within the range of a plant species is suitable for the survival of individuals of that species. Variations in topography may affect local temperature and hydrologic conditions, and soil conditions may vary owing in part to differences in the underlying geologic processes that drive the formation of soils, as we saw above. Broad climate patterns interact with local topography and soil conditions to define the area over which physical conditions are suitable for a particular plant species.

Plant Distributions in Relation to Climate Patterns

The range of the sugar maple, a common forest tree in the northeastern United States and southern Canada, is limited by cold winter temperatures to the north, hot summer temperatures to the south, summer drought to the west, and the ocean to the east (Figure 8-18a). Attempts to grow sugar maples outside their normal range have shown that they cannot tolerate average monthly high temperatures above 24°–27°C or below about −18°C (Fowells 1965). The western limit of the sugar maple, determined by dryness, coincides with the western limit of forest vegetation in general. Because temperature and rainfall interact to determine the availability of moisture, sugar maples tolerate lower annual precipitation at the northern edge of their range (about 50 cm) than at the southern edge (about 100 cm). Within its geographic range, sugar maple is more abundant in northern forests, where it sometimes forms single-species stands, than in the more diverse forests to the south. Sugar maple occurs most frequently on moist, slightly acid soils.

Differences in the distributions of the sugar maple and other tree-sized species of maples—black, red, and silver—suggest differences in ecological tolerances (Figure 8-18a–d). Where their ranges overlap, maples exhibit distinct preferences for local environmental conditions created by differences in soil and

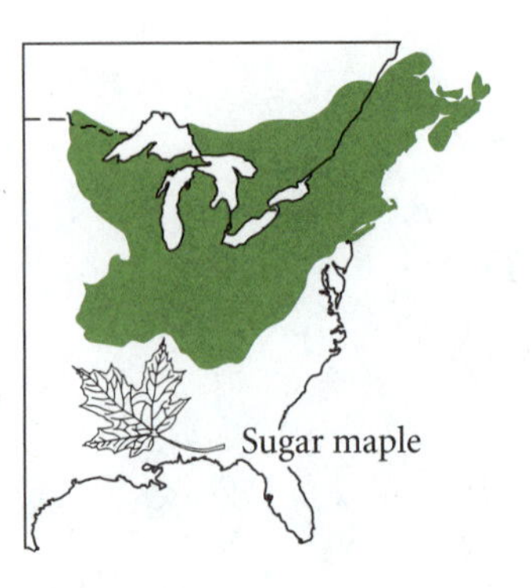

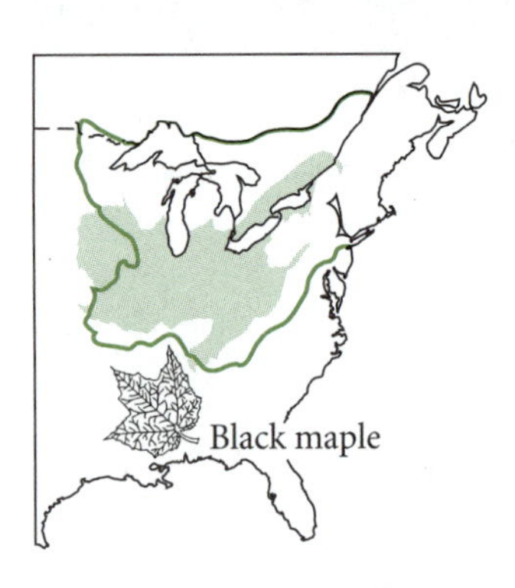

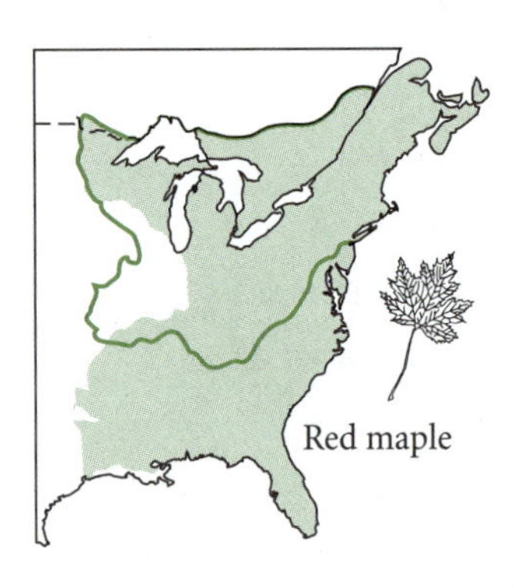

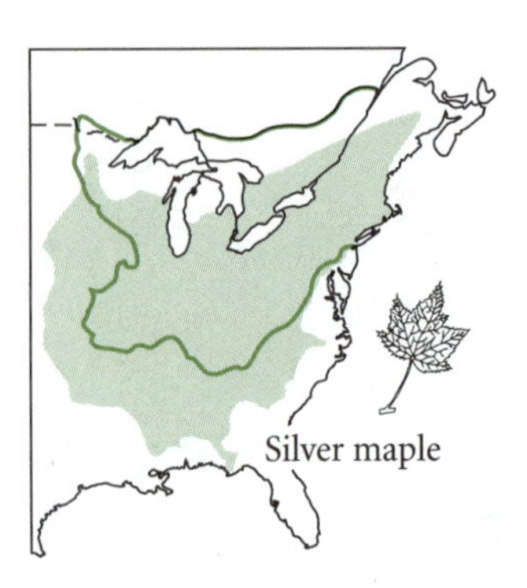

FIGURE 8-18 The ranges of sugar, black, red, and silver maples in eastern North America. The range of the sugar maple is outlined on each map to show the area of overlap. Where the ranges of these species overlap, each shows preferences for different conditions of soil type and moisture. (*After Fowells 1965.*)

topography. Black maple frequently occurs with the closely related sugar maple, but prefers drier, better-drained soils whose higher calcium content renders them less acidic. Red maple is peculiar in its ability to tolerate either wet, swampy conditions or dry, poorly developed soils. Silver maple is widely distributed, but prefers the moist, well-drained soils of the Ohio and Mississippi river basins.

Local Plant Distributions

Locally, plant distributions are influenced by several factors, including elevation, slope, exposure, and underlying bedrock. These factors vary most in mountainous regions, and therefore ecologists frequently turn to mountain habitats to study plant distribution.

Along the coast of northern California, mountains create conditions for a variety of plant communities, ranging from dry coastal chaparral to tall forests of Douglas fir and redwood (Waring and Major 1964). When localities are ranked on scales of available moisture, the distribution of each species among the localities exhibits a distinct optimum (Figure 8-19). The coast redwood dominates the central portion of the moisture gradient and frequently forms pure stands. Cedar, Douglas fir, and two broad-leaved evergreen species with small, thick leaves—manzanita and madrone—occur at the drier end of the moisture gradient. Three deciduous species—alder, big-leaf maple, and black cottonwood—occupy the wetter end.

Change in one environmental condition usually brings about changes in others. Increasing soil moisture alters the availability of nutrients. Variation in the amount and source of organic matter in the soil creates parallel gradients of acidity, soil moisture, and available nitrogen. Such factors often interact in complex ways to determine the distributions of plants.

Figure 8-20 relates the distributions of some forest-floor shrubs, seedlings, and herbs in the woodlands of eastern Indiana to levels of organic matter and calcium in the soil (Beals and Cope 1964). These soils contain between 2% and 8% organic matter and between 2% and 6% exchangeable calcium. Within the range of soil conditions in these woodlands, each species shows different preferences. Black cherry seedlings occur only within a narrow range of calcium, but tolerate variation in the percentage of organic matter. Bloodroot is narrowly restricted by the percentage of organic matter in the soil, but is insensitive to variation in calcium. The distributions of yellow violets and cream violets extend more broadly over levels of organic matter and calcium in the soil, but the two species do not overlap. Cream violets prefer relatively higher calcium and lower organic matter content than do yellow violets; where one occurs, the other usually does not.

8.7 The adaptations of plants and animals match the conditions within their environments.

As we have seen in Chapters 4, 5, and 6, the adaptations of an organism cannot easily be separated from the environment in which it lives. Insect larvae from stagnant aquatic environments in ditches and sloughs can survive longer without oxygen than can related species from well-aerated streams and rivers; species of marine snails that occur high in the intertidal zone, where they are frequently exposed to air, tolerate desiccation better than do species from lower levels. These are examples of **specializations** that suit organisms to particular, restricted ranges of environmental conditions.

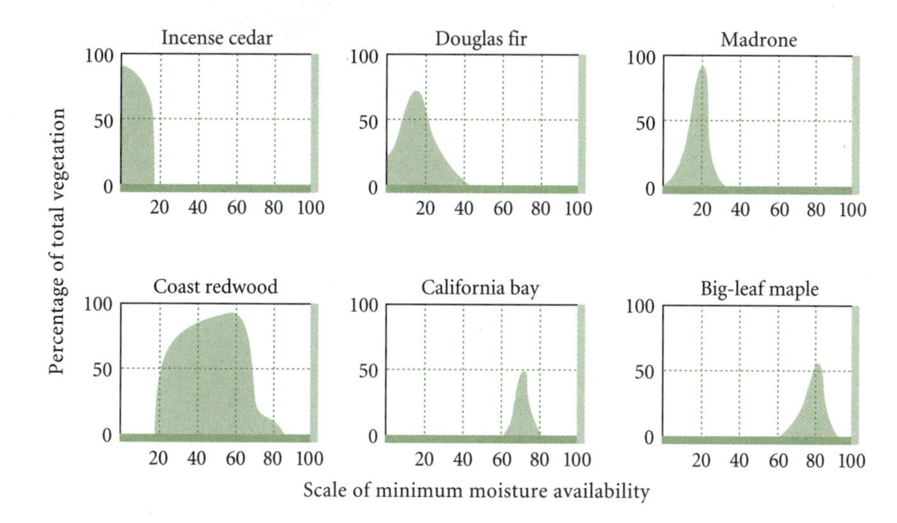

FIGURE 8-19 The distribution of tree species along a gradient of minimum available soil moisture in the northern coastal region of California. (*After Waring and Major 1964.*)

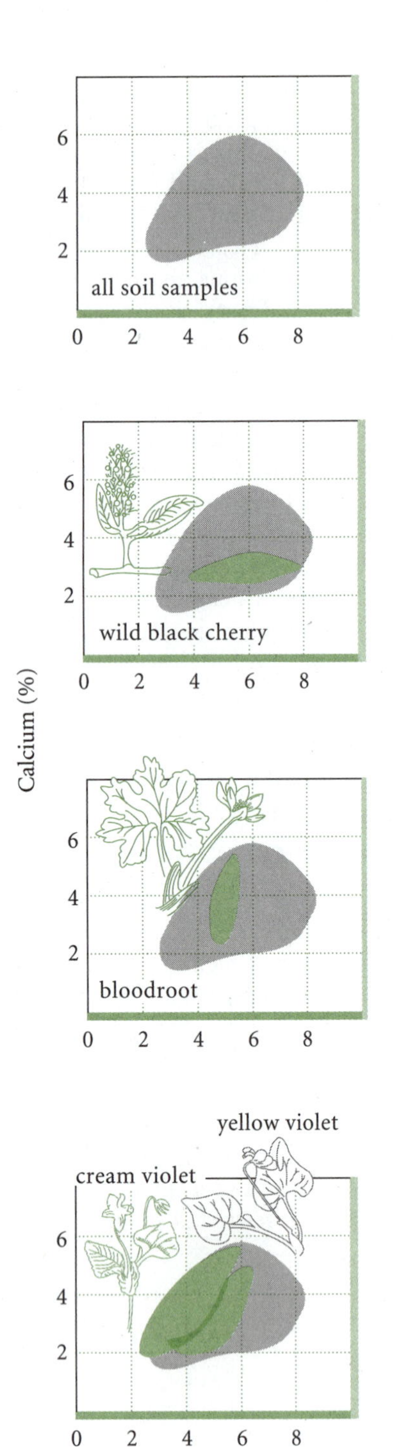

FIGURE 8-20 The occurrence of four forest-floor plants with respect to the calcium and organic matter content of the soil in woodlands of eastern Indiana. The gray shaded area represents the relationship between calcium and organic matter for all soil samples taken. The green shading shows the relationship between these two factors for the four species examined. (*After Baels and Cope 1964.*)

Environment, Form, and Function in Plants

Compare the leaves of deciduous forest trees with those of desert species. The former are typically broad and thin, providing a large surface area for light absorption and—unavoidably, therefore—for water loss. Desert trees have small, finely divided leaves—or sometimes none at all (Figure 8-21). Leaves heat up in the desert sun. Structures lose heat by convection most rapidly at their edges, where wind currents disrupt the insulating boundary layers of still air. The more edges, the cooler the leaf, and the lower the

FIGURE 8-21 Leaves of some desert plants from Arizona. (a) Mesquite (*Prosopis*) leaves are subdivided into numerous small leaflets, which facilitate the dissipation of heat when exposed to sunlight. (b) The paloverde (*Cercidium*) carries this adaptation even further; its leaves are tiny, and the thick stems, which contain chlorophyll, are responsible for much of the plant's photosynthesis (hence the name paloverde, which is Spanish for "green stick). (c) Unlike most desert plants, limberbush (*Jatropha*) has broad, succulent leaves, which it produces for only a few weeks, during the summer rainy season in the Sonoran Desert.

water loss (Vogel 1970). Smaller size means that a larger portion of each leaf is given over to its edge. Even on a single plant, leaves exposed to full sun may be differently shaped to dissipate heat and conserve water better than shade leaves (Figure 8-22).

Coastal sage and chaparral plant communities in southern California demonstrate divergent courses of adaptation to deal with the challenge of a limited water supply (Mooney and Dunn 1970, Harrison et al. 1971). Chaparral plants generally occur at higher elevations than coastal sage plants, and thus experience cooler and moister conditions. During the prolonged summer drought, soils have greater water deficits in the coastal sage habitat. Coastal sage plants typically have shallow roots and small, delicate leaves, which most species shed during the summer (Figure 8-23). Chaparral species have deep roots that often extend through tiny cracks and fissures far into the bedrock; their leaves are typically thick and have a waxy outer covering (cuticle) that reduces water loss. The leaves of chaparral plants usually persist through the drought period.

The influence of leaf morphology on photosynthetic rate parallels its influence on transpiration (Table 8-1). The leaves of coastal sage species, with their numerous stomates, are designed for rapid gas exchange with the surrounding air. This means that they lose water rapidly, but they can also assimilate carbon rapidly from the atmosphere when water is available in the soil to replace water lost by

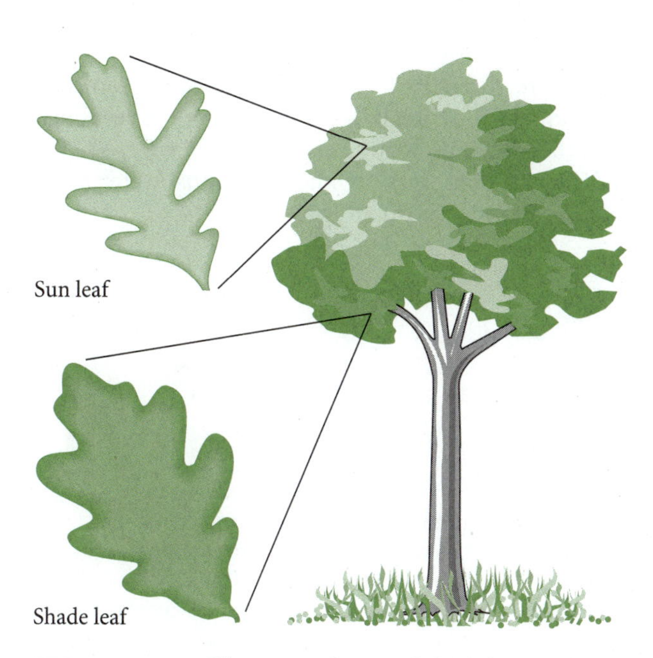

FIGURE 8-22 Silhouettes of sun and shade leaves of white oak. The sun leaves have more edge per unit of surface area and therefore dissipate heat more rapidly. *(After Vogel 1970.)*

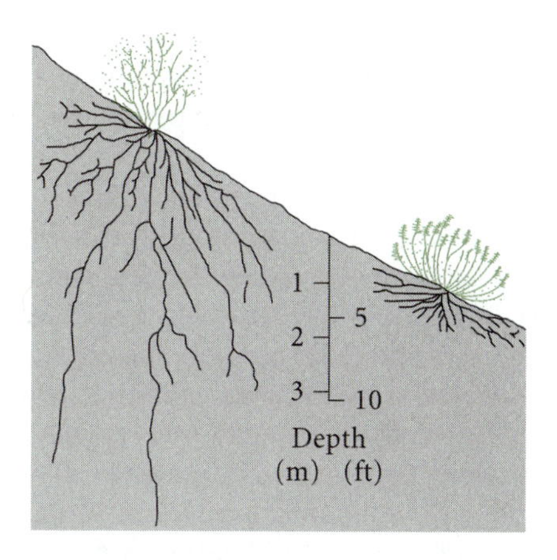

FIGURE 8-23 Profiles of the root systems of chamise (*Adenostoma fasciculatum*), at left a chaparral species and at right black sage (*Salvia mellifera*), a member of the coastal sage community. The two species have different adaptations to their limited water supply. The extensive root systems of chaparral species allow them to obtain water from deep fissures in the bedrock. Chaparral leaves are small and thick to prevent water loss, and they are usually retained during the summer drought. The shallow root system of the sage requires it to drop its leaves during drought to minimize water loss. *(After Hellmers et al. 1955.)*

TABLE 8-1 Characteristics of chaparral and coastal sage vegetation in southern California

	Vegetation type	
Characteristic	Chaparral	Coastal sage
Roots	Deep	Shallow
Leaves	Evergreen	Summer deciduous
Average leaf duration (months)	12	6
Average leaf size (cm^2)	12.6	4.5
Leaf weight (g dry weight dm^{-2})	1.8	1.0
Maximum transpiration (g H_2O dm^{-2} h^{-1})	0.34	0.94
Maximum photosynthetic rate (mg C dm^{-2} h^{-1})	3.9	8.3
Relative annual CO_2 fixation	49.8	46.8

(Data from Harrison et al., 1971; Mooney and Dunn, 1970.)

transpiration. The relationship between transpiration and carbon assimilation can be demonstrated readily in the laboratory. In one set of experiments, Harrison and coworkers (1971) clipped leaves from plants and placed them in a chamber within which they could monitor transpiration and photosynthesis. Both functions declined as the leaves dried out and their stomates closed to prevent further water loss. Rates of both photosynthesis and transpiration for coastal species such as the black sage (a member of the mint family) were high initially, but decreased rapidly (Figure 8-24). However, even when the stomates were fully closed and carbon assimilation had ceased, the leaves continued to lose water across their thinly protected outer surfaces. Photosynthetic rates for chaparral species such as the toyon (a member of the rose family) were at most only one-fourth to one-third those of the black sage, but the leaves resisted desiccation better and continued to be active under drying conditions for longer periods. The outer surfaces of the leaves of the toyon have a thick, waxy cuticle to minimize water losses when the stomates are fully closed.

Coastal sage and chaparral plants are differently specialized. Black sage is active only during the rainy season of winter and early spring. Its leaves are designed for high rates of photosynthesis and high rates of growth, but they are dropped and the plant becomes dormant as soon as water becomes scarce in the soil. Black sage is thus specialized for the transient moist conditions of the Mediterranean-climate winter. Toyon and other chaparral species make use of the more limited water that lies deeper in the soil, but nonetheless persists through a longer part of the year; they cannot utilize the winter water bonanza in the upper layers of the soil as efficiently as coastal sage species.

Where chaparral and coastal sage species grow together near the overlapping edges of each other's ranges, they exploit different parts of the environment: deep, perennial sources of water versus shallow, ephemeral sources of water. In spite of these differences and their corresponding adaptations of leaf morphology and drought response, the two groups of species are equally productive at intermediate levels of water availability. In drier habitats, the prolonged seasonal absence of deep water tips the balance in favor of the deciduous coastal sage vegetation. Increasing availability of deep water at higher elevations favors the evergreen chaparral vegetation.

These examples drive home the point that plant growth form is closely related to the physical conditions of the environment. With respect to terrestrial plants, those with large growth forms are often competitively superior to those with smaller growth forms, but they require moister soils. Thus, we should not be surprised that the availability of water is the single predominating factor determining the character and distribution of terrestrial biomes. Because temperature influences moisture stress and moisture availability, it also makes an important contribution.

Factors other than the availability of water are also important in determining plant growth forms. One of these is fire, whose influence is greatest where moisture availability is intermediate in level and highly seasonal. Deserts and moist forests burn infrequently: deserts rarely accumulate enough plant debris to fuel a fire, and moist forests rarely dry out enough to be highly flammable. Grassland and shrub biomes have the combination of abundant fuel and seasonal drought that makes fire a frequent visitor. In these areas, fire is a predominating factor to which all organisms must be adapted and, indeed, for which many are specialized: for some species, fire is necessary for germination of seeds and growth of seedlings. Another factor is conditions in the soil, referred to as **edaphic** factors, such as nutrient status, water availability, and toxic mineral content.

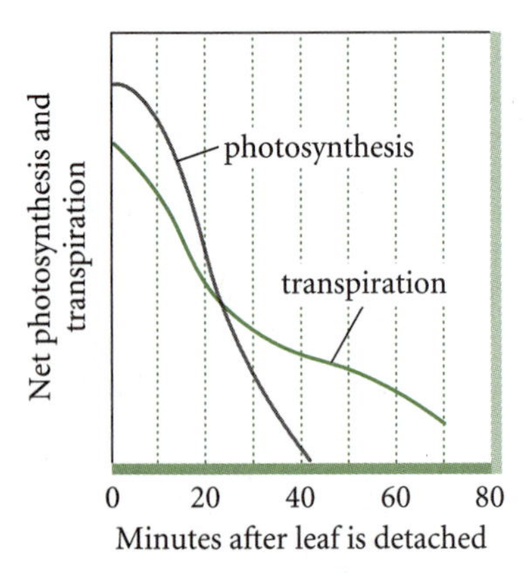

(a) Black sage

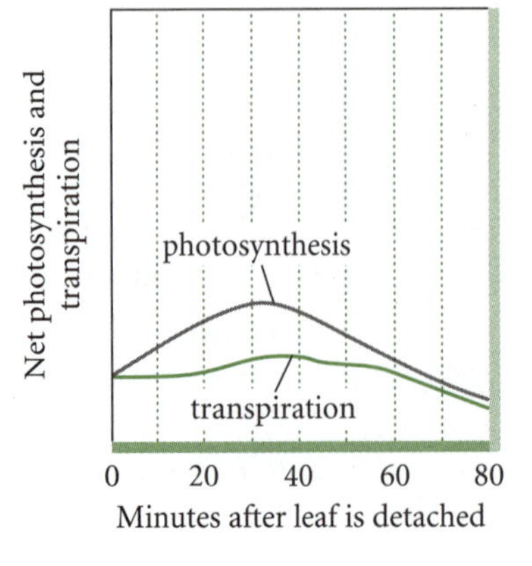

(b) Toyon

FIGURE 8-24 Time courses for photosynthesis and transpiration under standard laboratory drying conditions for (a) black sage (*Salvia mellifera*), a coastal sage species, and (b) toyon (*Heteromeles arbutifolia*), a chaparral species. Note that transpiration continues well after photosynthesis has been shut off; hence leaf dormancy is an ineffective long-term solution to drought. (*After Harrison et al. 1971.*)

Environment, Form, and Function in Animals

In general, the life forms of animals are less sensitive to climate than are those of plants. This does not mean that animals are not adapted to physical conditions and are not specialized for narrow ranges of these conditions, only that such adaptations usually are not expressed as major differences in life form. Birds of deserts and birds of forests have the same basic body plan; so do mammals, reptiles, amphibians, and insects.

Why the different responses of plants and animals? A facile answer—which cannot fully explain the plant-animal dichotomy—is that because animals are mobile, they can seek out favorable microclimates no matter where they are. That is to say, the difference between the environment in Arizona and the environment in Illinois is not as great for animals as it is for plants, which are firmly rooted in place and must therefore tolerate all extremes of conditions. Three other factors seem important, however. First, plants have very high surface-to-volume ratios, both above and below ground level, and are therefore much more sensitive than animals to changes in, and extremes of, environmental conditions. Second, because of the modular construction of plants, most cells of the plant "body" are more or less independent, and they must survive extremes of environmental conditions on their own, with little help from other parts of the body—with the conspicuous exception of the delivery of water from the roots. The animal body has many specialized organs, such as the kidneys, lungs, lymphoid system, and muscles, that support other tissues, particularly by ameliorating the cellular environment. Third, all plants play basically the same ecological role: they draw water and minerals through their root systems and fix carbon by means of photosynthesis in their leaves. As a result, their entire existence is dominated by water balance and the problems of delivering an adequate water supply to the aboveground portion of the plant organism. Because of the modular design of plants, the only effective ways of manipulating water balance are through the design of the water delivery systems and the leaves, and through the overall size of the plant and the ratio of its aboveground to belowground parts. As we shall see below, these are precisely the characteristics of life form that define the major terrestrial biomes of the world.

Variations in animal form, such as those between worms, insects, and vertebrates, reflect the different roles that animals play within the natural world. Thus, differences in animal form are associated more with different ecological roles, such as soil dwelling, seed gathering, or predation, than with differences in the physical environment from one place to the next. Also, animals can compensate for differences in life form with behavior. For example, in the deserts of Arizona, seeds are gathered from the soil surface by rodents, birds, and ants; other rodents, birds, and ants gather seeds in the forests of North Carolina and in the savannas of East Africa. In spite of their different body plans, these animals' ecological roles are the same. It is important to remember that in spite of their similarity, plants can also occupy different positions within the same ecosystem. Tropical forests exhibit the greatest diversity of plant growth forms, which include trees (both deciduous and evergreen in some localities), vines (both woody and herbaceous), epiphytes, understory treelets and shrubs, and herbaceous plants on the forest floor. Each of these growth forms lives in a very different environment with respect to light and moisture stress.

8.8 Classifications of plant associations based on plant form correspond closely to climate.

Natural history, and later ecology, grew out of classification schemes by which animals and plants were given names based upon their similarities. European botanists had described most local plant species by the end of the nineteenth century; they then began to develop systems of classification for entire communities of plants. They based most of these schemes on structural characteristics such as height of vegetation, leaf or needle structure, and deciduousness, and on which plant growth form predominated. Because these properties or specializations adapt plants to the physical environment in which they live, vegetation type and climate correspond closely.

The earliest classifications of vegetation described the most important plants of each community (see Shimwell 1971, Mueller-Dombois and Ellenberg 1974). These classifications included a complete **floristic analysis** of the community, including the names of the species as well as a description of plant forms, regardless of species. Floristic analysis proved useful in Europe, where botanists knew all the species and where minor differences between communities involved the replacement of species by similar ones with slightly different ecological requirements. But floristic analysis proved unworkable on a global scale. Biogeographic barriers restrict the distributions of individual species, rendering such floristic comparisons ecologically meaningless. Forests in Europe and the United States, shrublands in California and Australia, grasslands in Africa and South America, while structurally similar, have few species in common.

Early in the twentieth century, the Danish botanist Christen Raunkiaer classified plants according to the position of their buds (regenerating parts), and

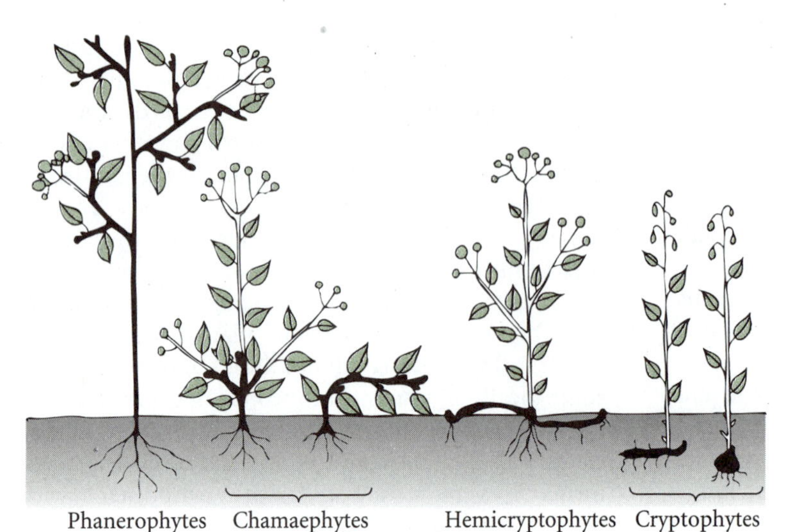

FIGURE 8-25 Diagrammatic representation of Raunkiaer's life forms. The lightly shaded parts of the plant die back during unfavorable seasons, while the solid black portions persist and give rise to the following year's growth. Proceeding from left to right, the buds are progressively better protected. Therophytes, whose persistent parts are seeds, are not illustrated. (*After Raunkiaer 1937.*)

found that the occurrence of his major categories corresponded closely to climatic conditions (Raunkiaer 1934). He distinguished five principal life forms (Figure 8-25). **Phanerophytes** (from the Greek *phaneros,* "visible") carry their buds on the tips of branches, exposed to extremes of climate. Most trees and large shrubs are phanerophytes. As one might expect, this plant form predominates in moist, warm environments where buds require little protection. **Chamaephytes** (from the Greek *chamai,* "on the ground," "dwarf") comprise small shrubs and herbs that grow close to the ground (prostrate life forms). Proximity to the soil protects the buds, and in winter, snow cover often protects the buds from extreme cold. Chamaephytes occur most frequently in cool, dry climates. **Hemicryptophytes** (from the Greek *kryptos,* "hidden") persist through the extreme environmental conditions of the winter months by dying back to ground level, where the regenerating bud is protected by soil and withered leaves. This growth form is characteristic of cold, moist zones. **Cryptophytes** are further protected from freezing and desiccation because their buds are completely buried beneath the soil. The bulbs of irises and daffodils are the regenerating buds of cryptophyte plants. Like hemicryptophytes, cryptophytes are found in cold, moist climates. **Therophytes** (from the Greek *theros,* "summer") die during the unfavorable season of the year and do not have persistent buds. Therophytes are regenerated solely by seeds, which resist extreme cold and drought. The therophyte form includes most annual plants and occurs most abundantly in deserts and grasslands. Raunkiaer's life forms are correlated with climate. Tropical and subtropical areas contain primarily phanerophytes. Temperate and arctic regions contain mostly hemicryptophytes. Therophytes predominate in desert regions.

8.9 Global life zones may be differentiated by the relationship between temperature and precipitation.

Most of us intuitively categorize the natural world around us. We can easily distinguish forest from desert by the numbers and types of plants that occur in each, even though we may not know the names of the plants, or care to contemplate the patterns of precipitation, temperature, topography, and soil conditions that form the basis of the different plant communities. Classifications of life zones are extremely useful because they provide us with a more or less unambiguous way of mentally organizing the natural world on a large scale. When applied on a global scale, they divide the natural world into a relatively small number of categories that may be distinguished by the relationship of temperature and precipitation or by the dominant plant form.

The botanist L. R. Holdridge (1967) proposed a classification of the world's plant communities based solely on climate (Figure 8-26). Holdridge considered temperature and rainfall to prevail over other environmental factors in determining what type of vegetation is found in a certain area, although he recognized that soils and exposure may strongly influence plant communities within each climate zone.

Holdridge's scheme, while not widely used, classifies climates according to the biological effects of temperature and rainfall on vegetation. Temperature and rainfall are seen as interacting to define humidity provinces separated by critical ratios of potential evapotranspiration to precipitation. Humidity provinces relate temperature and rainfall to the water relations of plants. Holdridge's formula indicates, for ex-

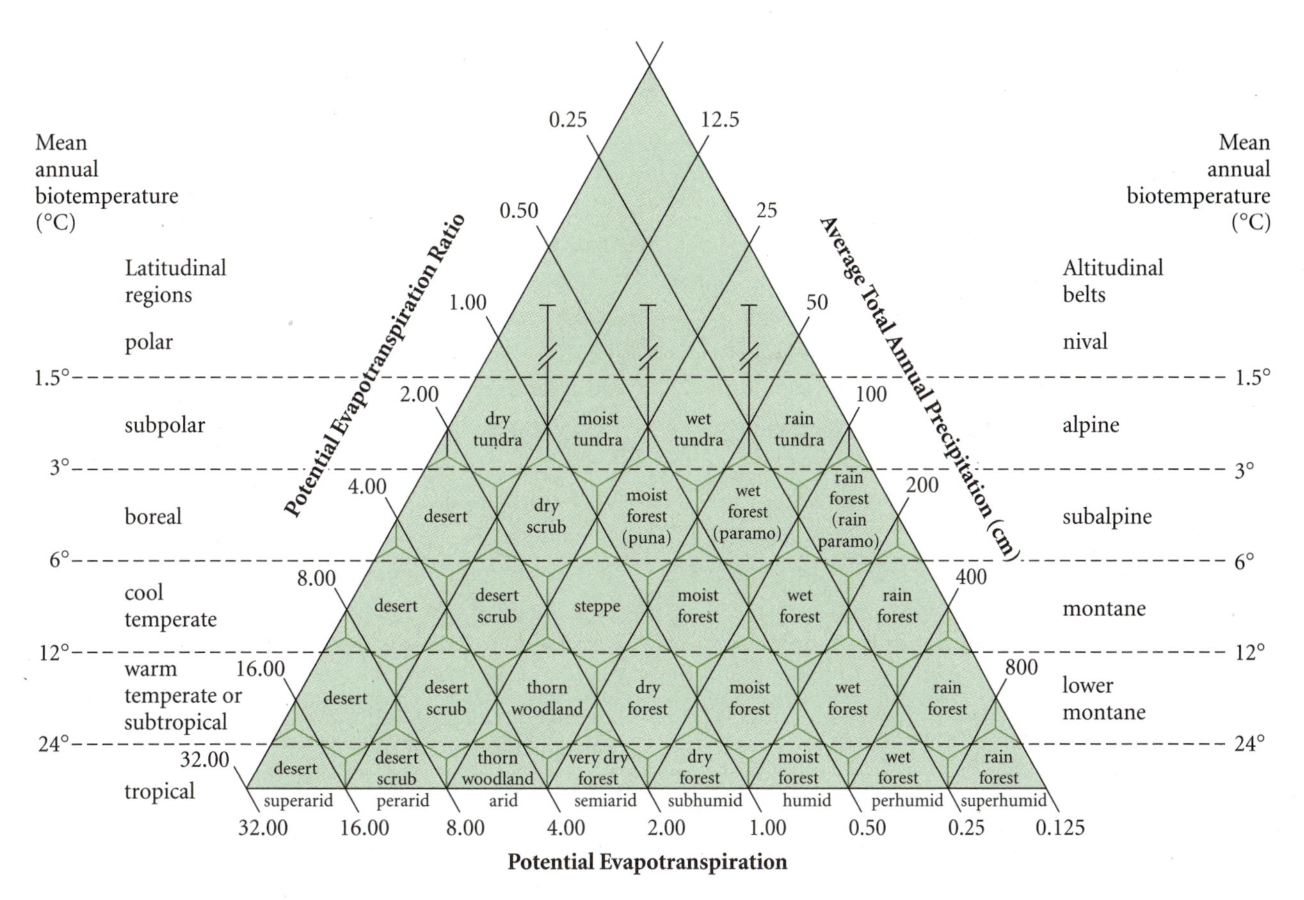

FIGURE 8-26 The Holdridge scheme for the classification of plant communities. Mean annual biotemperature (MAB) is calculated from monthly mean temperatures after converting means below freezing to 0°C. The potential evapotranspiration ratio is the potential evapotranspiration divided by the precipitation; the ratio increases from humid to arid regions. (*After Holdridge 1967.*)

ample, that moisture is equally available to plants in wet tundra, with an annual precipitation of 25 cm and an average temperature near freezing, and wet tropical forest, with 400 cm annual precipitation and an average temperature of 27°C.

One of the most widely adopted classification schemes is the climate zone system of the German ecologist Heinrich Walter. This system, which has nine major divisions, is based on the annual course of temperature and precipitation. The important attributes of the climate and the characteristic vegetation in each of these zones is presented in Table 8-2. The boundaries of the climate zones correspond to conditions of moisture and cold stress that are particularly important determinants of plant form. For example, within the Tropics, tropical climates are distinguished from equatorial climates by periods of water stress during a pronounced dry season. Subtropical climate zones are perpetually water-stressed. The typical vegetation of these climate zones is evergreen rain forest, deciduous forest or savanna, and desert scrub, respectively.

Another approach is that of plant ecologist R. H. Whittaker (1975), who combined several structural classifications of plant communities and Holdridge's idea of humidity provinces into a single scheme, which he transposed onto a graph of temperature and rainfall (Figure 8-27). Within the tropical and subtropical realms, with mean temperatures between 20°C and 30°C, vegetation types grade from true rain forest, which is wet throughout the year, to desert. Intermediate climates support seasonal forests, in which some or all trees lose their leaves during the dry season, and short, dry forests or scrublands with many thorn trees. Plant communities in temperate areas follow the pattern of tropical communities, with the same vegetation types distinguishable in both. In colder climates, however, precipitation varies so little from one locality to another that vegetation types are poorly differentiated on the basis of climate. Where mean annual temperatures are below −5°C, Whittaker lumps all plant associations into one type: tundra.

Toward the drier end of the rainfall spectrum within each temperature range, fire plays a distinct role in shaping the form of plant communities (Borchert 1950, Daubenmire 1968b; see Part 6). For example, in the African savannas and Midwestern

TABLE 8-2 H. Walter's classification of the climate zones of the world

Climate zone	Corresponding vegetation
I **Equatorial** Always moist and lacking temperature seasonality	Evergreen tropical rain forest
II **Tropical** Summer rainy season and cooler "winter" dry season	Seasonal forest, scrub, or savanna
III **Subtropical** Highly seasonal, arid climate	Desert vegetation with considerable exposed surface
IV **Mediterranean** Winter rainy season and summer drought	Sclerophyllous (drought-adapted), frost-sensitive shrublands and woodlands
V **Warm temperate** Occasional frost, often with summer rainfall maximum	Temperate evergreen forest, somewhat frost-sensitive
VI **Nemoral** Moderate climate with winter freezing	Frost-resistant, deciduous, temperate forest
VII **Continental** Arid, with warm or hot summers and cold winters	Grasslands and temperate deserts
VIII **Boreal** Cold temperate with cool summers and long winters	Evergreen, frost-hardy needle-leaved forest (taiga)
IX **Polar** Very short, cool summers and long, very cold winters	Low, evergreen vegetation, without trees, growing over permanently frozen soils

American prairies, frequent fires kill the seedlings of trees and prevent the establishment of tall forests, for which favorable conditions otherwise exist. Burning favors perennial grasses with extensive root systems that can survive fire. After an area has burned over, grass roots send up fresh shoots and quickly revegetate the surface. In the absence of frequent fires, tree seedlings can become established and eventually shade out prairie vegetation.

As in all classification schemes, exceptions appear frequently. Boundaries between vegetation types are at best fuzzy. Moreover, not all plants respond to climate in the same way. For example, some species of Australian eucalyptus trees form forests under climatic conditions that support only shrubland or grassland on other continents. Finally, plant communities reflect factors other than temperature and rainfall. Topography, soils, fire, and seasonal

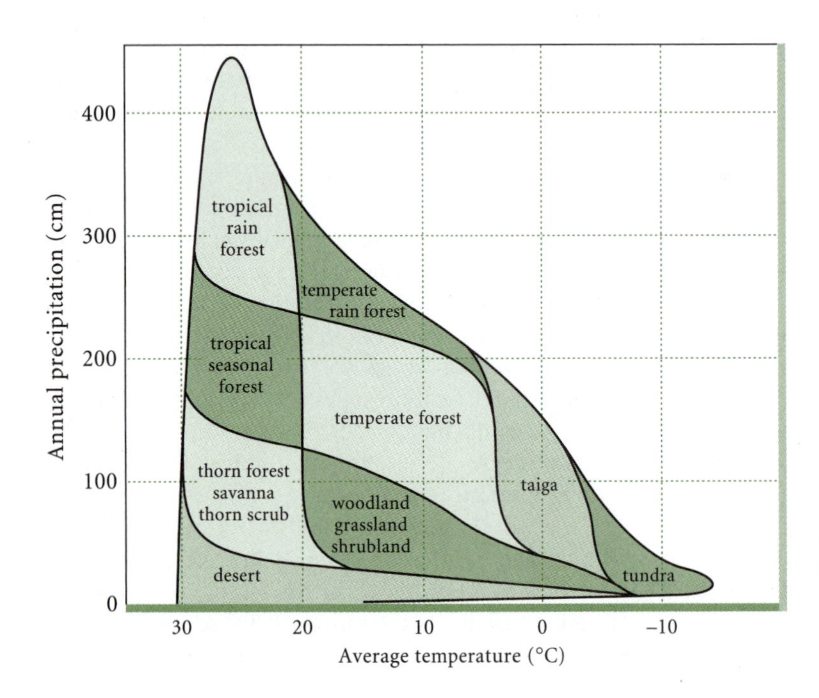

FIGURE 8-27 Whittaker's classification of vegetation types superimposed upon the range of terrestrial climates. In climates intermediate between those of forested and desert regions, fire, soil, and climate seasonality determine whether woodland, grassland, or shrubland develops. (*From Whittaker 1975.*)

variation in climate all leave their mark, further emphasizing the adaptation of life forms to the diversity of environments on the earth.

8.10 The biome concept organizes large-scale variation in the natural world.

Although no two places are inhabited by exactly the same assemblage of species, biological communities can be grouped into categories based on the dominant plant form, which gives the community its overall character. These categories are referred to as **biomes.** Important terrestrial biomes of North America are tundra, boreal forest, temperate deciduous forest, temperate evergreen forest, shrubland, grassland, and desert (Figure 8-28). As one would expect, these biomes show a close correspondence to the major climate zones of North America. Although each biome is immediately recognizable by its distinctive vegetation, it is important to realize that different systems of classification make coarser or finer distinctions among biomes, and that the characteristics of one biome usually intergrade gradually into the next. The following brief overview of the major biomes emphasizes the distinguishing features of the physical environment and how these are reflected in the form of the dominant plants.

Terrestrial Biomes

Because most readers of this book live in the so-called temperate zone, it is a good place to start. Temperate climates are characterized by average annual temperatures in the range of 5°–50° C at low elevations. Such climates are distributed between approximately 30° N and 45° N in North America and between 40° N and 60° N in Europe, which is warmed by the Gulf Stream current. Frost is an important factor throughout the temperate zone, perhaps even a defining character distinguishing it from subtropical and tropical climates. Within the temperate zone, biomes are differentiated primarily by total amounts and seasonal patterns of precipitation, although the length of the frost-free season, which is referred to as the **growing season,** and the severity of frost are also important. We distinguish seven terrestrial biomes of the temperate zone.

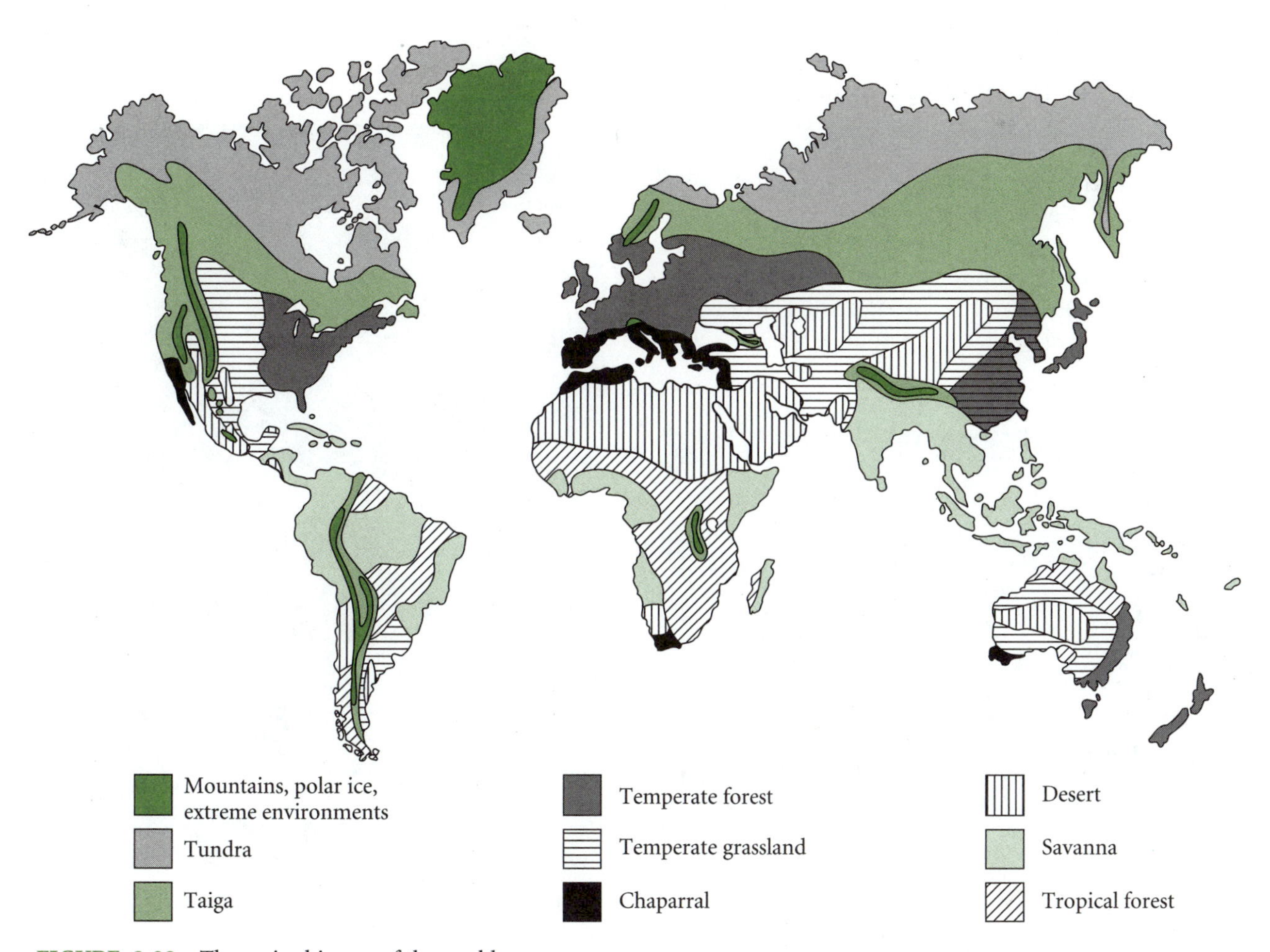

FIGURE 8-28 The major biomes of the world.

The **deciduous forest biome** (Figure 8-29) is found in North America principally in the eastern part of the United States and southern Canada, but also occurs widely in Europe and eastern Asia. It is poorly developed in the Southern Hemisphere (New Zealand and southern Chile) because of the milder winter temperatures at moderate latitudes. The length of the growing season varies from 130 days at higher latitudes to 180 days as lower latitudes. Precipitation usually exceeds the potential evapotranspiration; as a result, water tends to move downward through the soils and to drain from the landscape as groundwater and as aboveground streams and rivers. The vegetation is dominated by deciduous trees, predominantly oak, maple, beech, birch, and hickory, often with a subcanopy layer of small trees and shrubs. Herbaceous plants complete their growth and flower early in spring, before the trees have fully leafed out, blocking sun from the forest floor.

FIGURE 8-30 Temperate needle-leaved biome. Jeffrey pine forest of Inyo National Forest, California. (*Courtesy of the U. S. Forest Service.*)

The **temperate needle-leaved biome** (Figure 8-30) is dominated by pines and exists under conditions of water and nutrient stress, often on sandy soils. The most important of these formations in North America are the pine forests of the coastal plains of the Atlantic and Gulf states, the jack pine forests of the northern parts of the Great Lakes states and central Canada, and the montane pine forests of the American West. The low availability of nutrients and water favors evergreen, needle-leaved trees, which resist desiccation and give up nutrients slowly because they retain their needles for several years. Because soils tend to be dry, fires are frequent, and most species are able to resist fire damage.

The **temperate rain forest biome** occurs near the Pacific coast in the northwestern United States and British Columbia, and also in southern Chile, New Zealand, and Tasmania. Mild winters, heavy winter rains, and summer fog create conditions that support extremely tall evergreen forests. In North America, these forests are dominated toward the south by coast redwood and toward the north by Douglas fir (Figure 8-31). Trees are typically 60–70 meters high and may

FIGURE 8-29 Deciduous forest biome. This stand of hardwoods in Indiana is dominated by white oak and has a well-developed understory of small shrubs. (*Courtesy of the U. S. Forest Service.*)

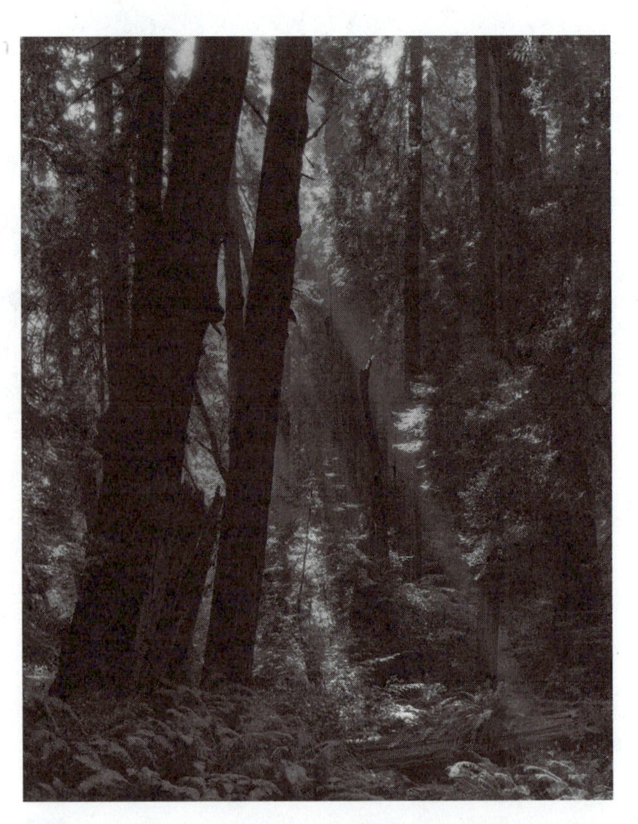

FIGURE 8-31 Temperate rain forest biome. Redwood trees growing in a temperate rain forest in northern California. (*Courtesy of the U. S. Forest Service.*)

grow to over 100 meters. It is not well understood why these sites are dominated by needle-leaved trees, but the fossil record shows that these plant formations are very old, and that they are mere remnants of forests that were vastly more extensive during the Mesozoic era, as recently as 70 million years ago. In contrast to rain forests in the Tropics, the species diversity of temperate rain forests typically is very low.

The **temperate grassland biome** (Figure 8-32) develops where rainfall is between 30 and 85 cm per year, depending on the average temperature. Summers are hot and wet; winters are cold. The growing season increases from north to south from about 120 to 300 days. North American grassland biomes are often called **prairies.** Extensive grasslands are also found in central Asia, where they are called **steppes.** Because annual precipitation is low, organic detritus does not decompose rapidly, and the soils are rich in organic matter. Because of their low acidity, the soils are not heavily leached and tend also to be rich in nutrients. The vegetation is dominated by grasses, which grow to over 2 meters in the moister parts of the grassland biome and to less than 0.2 meters in more arid regions. There are also abundant nongrass herbaceous species, which are called **forbs.** Fire is a dominant influence in grasslands, particularly where the habitat dries out during the late summer. Most grassland species have fire-resistant underground stems, or **rhizomes,** from which shoots resprout, or have fire-resistant seeds. Indeed, some areas of grassland receive sufficient rainfall to support forests, yet fires kill invading trees. Much of this biome has been converted to agriculture.

Where precipitation ranges between 25 and 50 cm per year, and the winters are cold and the summers hot, grasslands grade into the **temperate shrubland biome** (Figure 8-33). The shrubland biome covers most of the Great Basin of the western United States. In the northern part of the region, sagebrush is the dominant plant, whereas toward the south and on somewhat moister soils, widely spaced juniper and piñon pine trees predominate, forming open woodlands of less than 10 meters in height with sparse coverings of grass. In these shrublands, potential evapotranspiration exceeds precipitation during most of the year, and so soils are dry and little water percolates through them to form streams and rivers. Fire occurs infrequently in shrublands because the habitat produces little fuel. However, because of the low productivity of this biome, grazing can exert strong pressure on the vegetation and may even favor the persistence of shrubs, which are not good forage. Indeed, many dry grasslands in the western United States and elsewhere in the world have been converted to shrublands by overgrazing.

The **Mediterranean woodland biome** (Figure 8-34) is distributed at 30°–40° latitude north and south of the equator—somewhat higher in Europe—

FIGURE 8-32 Temperate grassland biome. Prairie in Pottawatomie County, Kansas. (*Courtesy of the U. S. Soil Conservation Service.*)

FIGURE 8-33 Temperate shrubland biome. Zion National Park, Utah. (*Courtesy of the U. S. Soil Conservation Service.*)

FIGURE 8-34 Mediterranean woodland biome. Chaparral on the Southern California Coast. (*Courtesy of the U. S. Forest Service.*)

on the western sides of continental landmasses. Representatives of this biome include southern Europe and southern California in the Northern Hemisphere, and central Chile, the Cape region of South Africa, and southwestern Australia in the Southern Hemisphere. Mediterranean climates are characterized by mild temperatures, winter rain, and summer drought. These climates support thick, evergreen, shrubby vegetation 1–3 meters in height, with deep roots and drought-resistant foliage. Mediterranean-climate plants, which typically have small, durable leaves, are known as **sclerophyllous** (hard-leaved) vegetation. Fires are frequent in Mediterranean biomes, and most plants have either fire-resistant seeds or root crowns that resprout soon after a fire.

What people call deserts varies tremendously, showing the danger in using everyday terms to name biomes. Many people refer to the dry areas of central Asia as deserts—the Mongolian Desert and the Gobi Desert are names familiar to most of us—but the climate and vegetation of these "deserts" differ utterly from those of the arid areas located within the subtropical belts of high pressure that girdle the earth. The Gobi Desert falls within Walter's continental climate zone (see Table 8-2), characterized by low precipitation and cold winters. This climate, which is similar to that of the Great Basin and the high western plains of North America, produces dry grasslands—steppes—and shrublands. Where precipitation dwindles to near zero, the vegetation dwindles accordingly, leaving a landscape that is more rock and sand than vegetation.

The **subtropical desert biome** develops at latitudes of 20°–30° north and south of the equator in areas with very sparse rainfall (less than 25 cm per year) and generally long growing seasons (Figure 8-35). Because of the low rainfall, the soils of subtropical deserts are shallow, virtually devoid of organic matter, and neutral in pH. Impermeable hardpans of calcium carbonate often develop at the limits of water penetration—depths of a meter or less. Whereas sagebrush dominates Great Basin "deserts," creosote bush takes its place in the subtropical deserts of the Americas. Wetter sites support a profusion of succulent cacti, shrubs, and small trees, such as mesquite and paloverde. Most subtropical deserts receive summer rainfall, during which many herbaceous plants sprout from dormant seeds and quickly grow and reproduce before the soils dry out again. Many of the plants in subtropical deserts are not frost-tolerant. Species diversity is usually much higher than it is in temperate arid lands.

FIGURE 8-35 Subtropical desert biome. The Sonoran Desert of Arizona and northern New Mexico. (*Courtesy of the U. S. Park Service.*)

Boreal and Polar Biomes

Three biomes are characteristic of the high latitudes of the Northern Hemisphere and of areas of high elevation in temperate and tropical regions. The **boreal forest biome** (also known as **taiga;** Figure 8-36) stretches in a broad belt centered at about 50° N in North

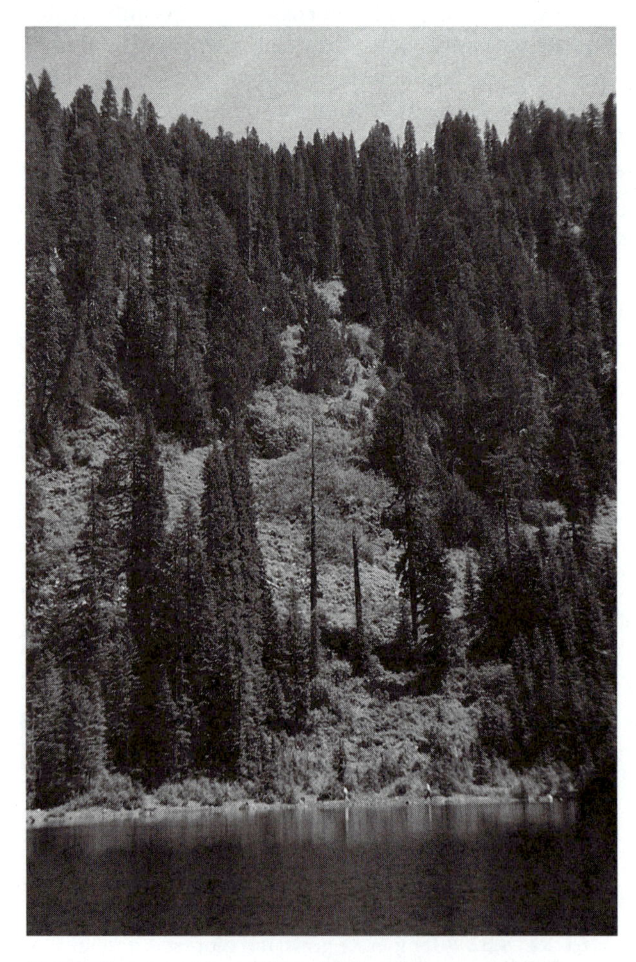

FIGURE 8-36 Boreal forest biome. (*Courtesy of R. B. Suter.*)

America and about 60° N in Europe and Asia. The average annual temperature is below 5°C, and winters are severe. Precipitation is in the range of 40–100 cm, and because evaporation is low, soils are moist throughout most of the growing season. The vegetation consists of vast, dense stands of evergreen needle-leaved trees, mostly spruce and fir, that grow to be 10–20 meters high. Because of the low temperatures, leaf litter decomposes very slowly and accumulates at the soil surface. When the needles do eventually decompose, however, they produce high levels of organic acids, and so the soils are acid and generally of low fertility. Growing seasons are rarely as long as 100 days, and more often half that. The vegetation is extremely frost-tolerant, as temperatures may reach −60°C during the winter. Species diversity is very low.

The **tundra biome** (Figure 8-37) lies north of the boreal forest in the polar climate zone. It is a treeless expanse underlain by permanently frozen soil, or **permafrost.** The soils thaw to a depth of 0.5–1 meter during the brief summer growing season. Precipitation is generally less than 60 cm, and often much less, but in low-lying areas where drainage is prevented by the permafrost, soils may remain saturated with water throughout most of the growing season. Soils tend to be acid because of their high organic matter content, and they are very low in nutrients. In this nutrient-poor environment, plants hold their foliage for years. Most plants are dwarf, prostrate, woody shrubs—that is, chamaephytes—that grow low to the ground to gain protection under the winter blanket of snow and ice. Anything protruding above the surface of the snow is sheared off by blowing ice crystals. For most of the year, the tundra is an exceedingly harsh environment, but during the 24-hour-long summer days, the tundra biome exuberantly testifies to the remarkable adaptability of life.

At high elevations in temperate climate zones, and even within the Tropics, one finds vegetation resembling that of the arctic tundra and containing some of the same species, or their close relatives. This life zone is called the **alpine tundra biome.** These areas occur above the treeline (Figure 8-38), most broadly in the Rocky Mountains of North America and, especially, on the Tibetan Plateau of central Asia. In spite of their similarities, alpine and arctic tundra have important differences. Alpine tundra generally has warmer and longer growing seasons, higher precipitation, less severe winters, greater productivity, better-drained soils, and higher species diversity than arctic tundra. Also, alpine tundra does not contain permafrost. Still, as in the high-latitude tundra, it is the harsh winter conditions that ultimately limit the growth of trees.

Equatorial and Tropical Biomes

In regions of the world within 20° north and south of the equator, temperatures vary more throughout the day than average monthly temperatures vary through the year. Average temperatures at sea level generally exceed 20°C. Environments within tropical latitudes are distinguished by the seasonal course of rainfall, which creates a continuous gradient of vegetation from wet, seasonal rain forests, to seasonal forests, to scrub, savanna, and desert. Frost is not a factor in tropical biomes, even at high elevations, and tropical plants and animals generally do not tolerate freezing. Three biomes are typically distinguished within these equatorial and tropical climate zones.

FIGURE 8-37 Tundra biome. Alaskan tundra, showing the characteristic polygonal patterns in the ground surface created by freezing and thawing of the surface layers of the soil. (*Courtesy of the U. S. Soil Conservation Service.*)

FIGURE 8-38 Wind-driven ice has stripped bark and branches from these trees near the forest-alpine tundra border (timberline) in the Rocky Mountains of Colorado. (*Courtesy of the U. S. Forest Service.*)

The **tropical rain forest biome** develops in climates that are always warm and that receive at least 200 cm of precipitation though the year, with not less than 10 cm during any one month. These conditions exist in three important regions within the Tropics: the Amazon and Orinoco basins of South America, with additional areas in Central America and along the Atlantic coast of Brazil, constitute the American rain forest; the area from southernmost West Africa and extending eastward through the Congo River basin constitutes the African rain forest; and the Indo-Malayan rain forest covers parts of Southeast Asia (Vietnam, Thailand, and the Malaysian Peninsula), the islands between Asia and Australia, including the Philippines, Borneo, and New Guinea, and the Queensland coast of Australia.

The tropical rain forest climate often exhibits two peaks of rainfall centered around the equinoxes, corresponding to the periods when the intertropical convergence overlies the equatorial region. Rain forest soils are typically old and deeply weathered. Because they are relatively devoid of humus and clay, they take on the reddish color of aluminum and iron oxides and have poor ability to retain nutrients. In spite of the low nutrient status of the soils, rain forest vegetation is dominated by a continuous canopy of tall evergreen trees rising to 30–40 meters, with occasional **emergent trees** rising above the canopy to heights of 55 meters. Because water stress on emergents is great, owing to their height and exposure, they are often deciduous, even in an evergreen rain forest. Tropical rain forests have understory tree, shrub, and herb layers, but these are usually quite sparse because so little light penetrates the canopy. Climbing **lianas,** or woody vines, and **epiphytes,** plants that grow on the branches of other plants and are not rooted in the soil (also called air plants), are prominent in the forest canopy itself (Figure 8-39). Species diversity is higher than anywhere else on earth.

The productivity of the rain forest biome is greater than that of any other terrestrial biome, and its standing biomass exceeds all others except the temperate rain forest. Because of the continuously high temperatures and abundant moisture, plant litter decomposes quickly, and the nutrients released are immediately taken up by the vegetation. This rapid nutrient cycling supports the high productivity of the rain forest, but also makes the rain forest ecosystem extremely vulnerable to disturbance. When tropical rain forests are cut, many of the nutrients are carted off in logs or go up in smoke. The vulnerable soils erode rapidly and fill the streams with silt. In many cases, the environment degrades rapidly and the landscape becomes unproductive.

The **tropical seasonal forest biome** (Figure 8-40) develops within the Tropics beyond 10° north and south of the equator. These climates often exhibit a pronounced dry season, corresponding to winter at higher latitudes. Seasonal forests in the Tropics have a preponderance of deciduous trees that shed their leaves during the season of water stress. Increasingly longer and more severe dry seasons generally result in forests with lower stature and more thorny vegetation that protects leaves from grazing, leading to thorn scrub and finally true desert under the extremely dry conditions that occur in rain shadows of mountain ranges or along coasts with cold ocean currents running alongside.

Savanna may be defined as grassland with scattered trees, and it typifies large areas of the dry

FIGURE 8-39 Tropical rain forest biome. Lianas and epiphytes drape the trees in this lowland tropical rain forest in Panama. (*Courtesy of W. J. Smith.*)

FIGURE 8-41 Savanna biome. Drought-adapted trees are interspersed with grassland in the Samburu district of Kenya. (*Courtesy of the U. S. Department of Agriculture and the Soil Conservation Service.*)

Tropics, especially in Africa. The **tropical savanna biome** (Figure 8-41) has an average rainfall of 90–150 cm per year, but the driest three or four months receive less than 5 cm each. Fire and grazing play important roles in maintaining the character of the savanna biome, particularly in wetter regions, as grasses can persist better than other forms of vegetation under these influences. Often when grazing and fire are controlled within a savanna habitat, seasonal forest begins to develop. It is possible that vast areas of African savanna owe their character to the influence of human activity, including burning, over many millennia.

8.11 Classifications of aquatic ecosystems are based on physical characteristics.

The biome concept was developed for terrestrial ecosystems, and biomes are distinguished principally by the growth form of their dominant vegetation.

FIGURE 8-40 Tropical seasonal forest biome. (*Courtesy of P. Feinsinger and M. L. Crump.*)

Throughout most of the development of ecology as a science, terrestrial and aquatic ecologists have generated concepts and descriptive terms for ecological systems independently. As a consequence, aquatic "biomes" do not exist in the sense in which the term is applied to terrestrial ecosystems. Indeed, employing a vegetation concept would be impossible in aquatic systems because the primary producers in many aquatic systems are single-celled algae, which do not form "vegetation" with a characteristic structure. As a result, classifications of aquatic systems have been based primarily on physical characteristics: salinity, water movement, depth, and so on.

The major kinds of aquatic environments are streams, lakes, estuaries, and oceans. **Streams** form wherever precipitation exceeds evaporation and excess water drains from the surface of the land. Within small streams, ecologists distinguish areas of **riffles,** where water runs rapidly over a rocky substrate, and **pools,** which are deeper stretches of slowly moving water. Water is well oxygenated in riffles; pools tend to accumulate silt and organic matter. Production in small streams is often dominated by **allochthonous** material (from the Greek *chthonos,* "of the earth," and *allos,* "other")—that is, organic material, such as leaves, that enters the aquatic system from the outside. Streams grow with distance as they join together to form rivers. The larger a river, the more of its production is home-grown, or **autochthonous.** A **river continuum** concept has grown up around the continuous change between the headwaters and the mouth of a river drainage. As one moves downstream, the water becomes warmer, more slowly flowing, and richer in nutrients, and ecosystems are more complex and generally more productive. **Fluvial** systems, as river systems are called, are also distinguished by the fact that material, including animals and plants, is continually moved downstream by currents. Inasmuch as fluvial systems exist in steady states, this so-called **downstream drift** is balanced by the active movement of animals upstream, by the productivity of the upstream portions of the system, and by the input of materials from outside.

Lakes form in any kind of depression. For the most part, such bodies of water are the products of glaciation, which leaves behind gouged-out basins and blocks of ice buried in glacial deposits, which eventually melt and form lakes. Lakes also form in geologically active regions, such as the Rift Valley of Africa, where the vertical shifting of blocks of the earth's crust creates basins within which water accumulates. Broad river valleys, such as those of the Mississippi and Amazon, may contain oxbow lakes, which are broad bends of the former river cut off by shifts in the main channel. An entire lake could be considered a biome, but lakes are usually subdivided into regions, each of which has its own character. The **littoral** zone is the shallow zone around the edge of a lake within which one finds rooted vegetation, such as water lilies and pickerel weed (Figure 8-42). The open water beyond the littoral zone is the **limnetic** zone, where primary production is accomplished by floating single-celled algae, or **phytoplankton.** Lakes may also be subdivided vertically on the basis of light penetration and the formation of thermally stratified layers of water (see Section 8.2). The sediments at the bottoms of lakes and ponds form a special **benthic** habitat for burrowing animals and microorganisms.

Estuaries are special environments found at the mouths of rivers, especially where the outflow is partially enclosed by landforms or barrier islands. The unique character of estuaries derives from the mixture of fresh and salt water, within which the larvae of many species of marine organisms grow in great profusion. In addition, the nutrients carried in by rivers and the rapid exchange between surface waters and sediments contribute to the extremely high biological productivity of estuaries. Because estuaries tend to be shallow areas within which sediments are deposited, they are often edged by extensive tidal marshes characterized by **emergent vegetation** (plants rooted in water whose upper parts emerge above the water line). Indeed, the marshes that surround many estuaries are among the most productive habitats on earth, owing to a combination of high nutrient levels and freedom from water stress. These marshes then contribute abundant additional organic matter to the estuarine ecosystem, which in turn supports abundant populations of estuarine and marine species.

The largest portion of the surface of the earth is covered by **oceans.** Beneath the surface of the water lies an immensely complex realm harboring a great variety of ecological conditions and ecosystems. Variation in marine systems results from temperature,

FIGURE 8-42 Cattails growing in the littoral zone of a shallow lake in New York State.

depth, current, substrate, and, at the edge of the seas, tides. Many marine ecologists have recognized several zones differentiated by depth. The **littoral** zone (compare with the littoral zone of lakes) extends between the highest and lowest tidal levels and, to a varying extent, depending on position within the intertidal range, is exposed periodically to air (Figure 8-43). The rapid changes in ecological conditions within the intertidal range often create sharp zonation of organisms according to their ability to tolerate the stresses of terrestrial conditions. Beyond the range of the lowest tidal level, the **neritic** zone extends to depths of about 200 meters, which corresponds to the edge of the continental shelf. It is often a region of high productivity because the sunlit surface layers of water are not far removed from the regeneration of nutrients in the sediments below. Even strong waves can move suspended materials from depths of 100–200 meters to the surface. Beyond the neritic zone—looking outward from the shore—the seafloor drops rapidly to the great depths of the **oceanic** zone, thousands of meters below. Here, production usually is strictly limited by the low availability of nutrients. Both the neritic and the oceanic zones may be subdivided vertically into a superficial **photic,** or euphotic, zone, in which there is sufficient light for photosynthesis, and an **aphotic** zone, without light, in which organisms depend mostly on organic material raining down from above.

Whereas the open ocean has been compared to a desert, coral reefs are like tropical rain forests, both in the richness of their biological production and in the diversity of their inhabitants (Figure 8-44). Reef-building corals occur in shallow waters of warm oceans, usually where water temperatures remain above 20° C year-round. Many coral reefs develop

FIGURE 8-43 The littoral zone, which is exposed to air twice each day, may support prolific growth of algae and a variety of animals, as in this area of the New Brunswick coast of Canada.

FIGURE 8-44 The high productivity of coral reefs in warm tropical waters provides abundant food for a diverse biological community. (*Courtesy of P. J. Tzimoulis, American Littoral Society.*)

around volcanoes, which are widely distributed throughout the western Pacific and Indian oceans. The volcanoes themselves may gradually disappear through erosion or subsidence under their own weight, but as long as the rate of coral growth exceeds the rate of subsidence, the reef continues to build. Eventually, all that may be left is a ring of coral—an **atoll**—outlining the position of the former volcanic island. The high production of the reef is fed by nutrients eroding from the encircled volcano and by deep-water currents forced upward by the profile of the island. Corals are doubly productive because they contain symbiotic photosynthetic algae within their tissues, which generate the carbohydrate energy base for their phenomenal rates of growth.

The unique qualities that characterize each type of biome or aquatic system are manifested in every aspect of ecosystem structure and function. The most direct way to evaluate these attributes is to measure the flux of energy through and the cycling of nutrients within an ecosystem. These aspects of ecological structure and function, and how they differ among the terrestrial biomes and aquatic ecosystems, are the subject of the next part of this book.

SUMMARY

1. Global patterns of temperature and rainfall are determined by the local interception of solar radiation and the redistribution of heat energy by winds and ocean currents. Warm tropical air rises at the equator and moves to the north and south, cooling and becoming more dense as it does so. At about 30° north and south of the equator, the cool, dense air sinks and moves toward the equator as surface air currents, which meet at the intertropical convergence. This circulation pattern is called a Hadley cell.

2. Ocean currents affect climate by moving heat over the surface of the earth. Cold-water ocean currents tend to move toward the Tropics along the western coasts of continents, and warm-water currents tend to move toward temperate regions along the eastern coasts of continents.

3. Seasonality is caused by the annual progression of the sun's path northward and southward and the latitudinal movement of associated belts of wind and precipitation. Seasonal changes in water temperatures cause changes in water movement patterns in the oceans.

4. Seasonal warming and cooling profoundly changes the characteristics of temperate zone lakes. During summer, such lakes are stratified, with a warm surface layer (epilimnion) separated from a cold bottom layer (hypolimnion) by a sharp thermocline. In spring and fall, the profile of temperature with depth becomes more uniform, allowing vertical mixing.

5. Periodic variations in climate, such as those resulting from El Niño-Southern Oscillation (ENSO) events, may cause major disruptions of biological communities on global scales. Recent large-scale ENSO events occurred in 1982–1983 and 1997–1998.

6. Topography and geology superimpose local variation in environmental conditions on more general climate patterns. Mountains intercept rainfall, creating rain shadows in their lees. Conditions at higher altitudes resemble conditions at higher latitudes. Soil characteristics reflect the quality of the underlying bedrock and sometimes foster specialized floras, such as those of serpentine barrens.

7. The amount of evapotranspiration is related to temperature, soil water storage, and foliage mass. An increase in photosynthesis increases evapotranspiration. Regional differences in evapotranspiration are evident.

8. The geographic distributions of plants are determined primarily by climate, whereas their local distributions within regions vary according to topography and soils.

9. Climate profoundly affects the adaptations of plants and animals. Each climate region has characteristic vegetation forms differing in general habit, leaf morphology, and pattern of growth.

10. Recognizing that plant form is directly related to climate through adaptation, Holdridge and Whittaker characterized major regions of vegetation on the basis of local temperature and precipitation. All such classification schemes emphasize the interaction between temperature and water availability, and further recognize the modifying effects of seasonality, soils, and fire.

11. Global life zones may be delineated based on the dominant vegetation. Such areas are referred to as biomes. Biomes are not cohesive biological units, but they provide an effective way of organizing the natural world on a global scale.

12. The area of the globe between 30° N and 45° N in North America and 40° N and 60° N in Europe is referred as the temperate zone. Temperate zone biomes include deciduous forest, temperate needle-leaved forest, temperate rain forest, temperate grassland, temperate shrubland, Mediterranean woodland, and subtropical desert.

13. Boreal forest, tundra, and alpine tundra are biomes characteristic of high latitudes. Tropical rain forest, tropical seasonal forest, and tropical savanna are biomes found within 20° of the equator.

14. The biome concept, based on vegetation features, was developed for terrestrial systems. Aquatic systems are classified based primarily on physical characteristics such as salinity, water movement, and water depth. The major types of aquatic systems are streams, lakes, estuaries, and oceans.

EXERCISES

1. Write a brief description of the seasonal weather patterns and climatic conditions in your area. Include in your description details about temperature and precipitation, and comment on significant unusual events that sometimes occur (e.g., tornadoes, hurricanes, flooding). Describe how the prominent topographic features of your area influence weather patterns.

2. Describe seasonal climatic conditions that would result in a freshwater lake having only one period of turnover each year. Find a place on earth where such conditions would occur.

3. You find yourself standing by Hudson Bay just north of Churchill, Manitoba. What biome are you in? Using a globe or a world map, locate a town or a city in each of the biomes mentioned in this chapter.

PART 3

ENERGY AND MATERIALS IN THE ECOSYSTEM

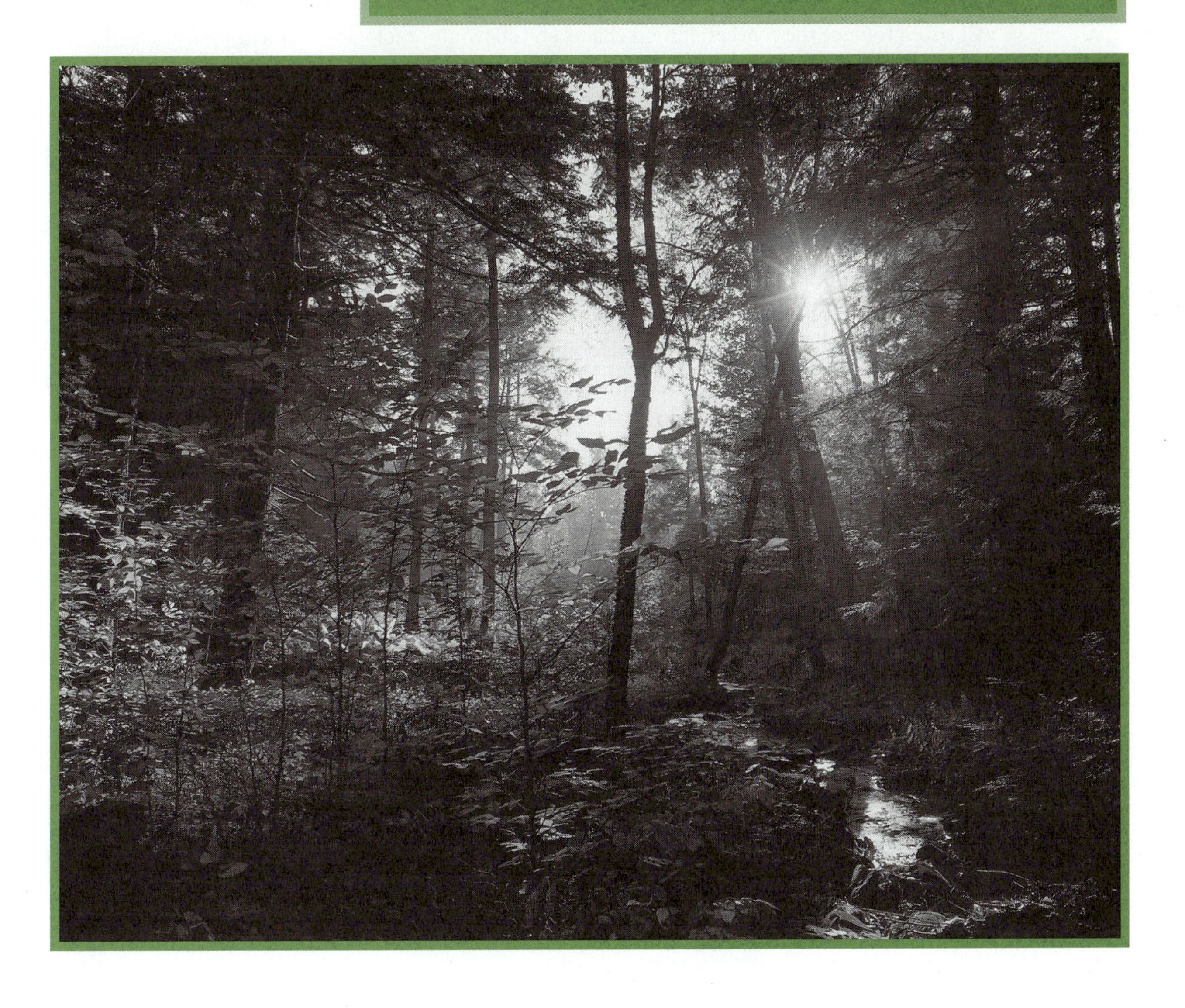

The energy of life ultimately flows from the nonliving physical environment through the process of photosynthesis. The flux of energy through the ecosystem is accomplished through the feeding relationships of organisms. Energetic transformations involve the movement of elements like carbon, nitrogen, and phosphorus through the ecosystem. In Part 3 we will explain the processes by which energy and materials move through ecosystems, processes involving interactions between the physical and biological environments.

The development of the concept of the ecosystem represents an important part of the early thinking in ecology, as we will explain in Chapter 9. Chapter 10 will focus on how energy is transformed in the natural world. We shall see that the rate of photosynthesis is affected by different factors in terrestrial and aquatic ecosystems. We will also introduce the concept of food chains and show how ecological efficiencies in the food chain influence the rate of energy flow in the ecosystem.

The flux of energy through ecosystems is accompanied by the movement of materials, which occurs on both global and local scales. In Chapter 11 we will describe the global cycles of water, carbon, nitrogen, phosphorus, and other important materials. We shall explore the potential consequences of human-induced alterations of these cycles. Many scarce elements must be regenerated locally through decomposition and mineralization. These regenerative processes, which depend on fungi and microorganisms, may limit the productivity of the ecosystem, as we show in Chapter 12.

The productivity of ecosystems is determined by the availability of sunlight and nutrients and is shaped by trophic interactions within the ecosystem. Ecologists use experimentation and theoretical modeling to explore the dynamics of ecosystem regulation. In Chapter 13, we will discuss the dynamics of such regulation.

CHAPTER 9

The Ecosystem Concept

GUIDING QUESTIONS

- How did F. E. Clements and H. A. Gleason differ in their views about the nature of biological communities?
- How did Charles Elton envision biological communities?
- How is energy transformed in ecosystems?
- What were A. J. Lotka's ideas regarding the regulation of ecosystem function?
- What is the organismal view of communities and ecosystems?
- What are the important goals of landscape ecology?

In Chapter 2 we defined an ecosystem as all the interacting parts of the physical and biological world. The branch of ecology dealing with the study of such systems is called **ecosystems ecology,** which is the focus of this part of the book. We begin in this chapter with a brief overview of the history of the development of the ideas that form the foundation of ecosystems ecology and, indeed, much of the rest of ecology. We believe that this historical perspective is important because much of modern ecological thinking grew from the insights of early naturalists who were interested in the interaction of organisms and the physical environment. Many of the ideas presented in the early part of this chapter have been discredited, and we shall point these out as we go along. Nevertheless, as with the development and growth of any body of knowledge, it was the debate—sometimes acrimonious—about these ideas that gave shape and direction to contemporary ecology.

9.1 Much of modern ecology is based on two concepts that emerged from the observations of twentieth-century naturalists.

The foundation of our unified picture of nature rests on the detailed observations of naturalists during the early part of the twentieth century. From these observations emerged two new concepts of the natural world. The first was the realization that species of plants and animals form natural associations, each with distinctive members. For example, along rocky coastlines one expects to find rockweed, barnacles, starfish, and limpets, while in a woodlot one expects to see a mixture of oaks, maples, ferns, squirrels, earthworms, and perching birds. Just as morphological data had allowed systematists to assign species to a hierarchy of taxonomic groups, detailed studies of the ecological distributions of plants led ecologists to think that they could classify biological communities in a parallel manner (Shimwell 1971). Second came the realization that organisms are linked both directly and indirectly by means of their feeding relationships. Humans have appreciated since their beginnings that an organism may both prey upon and be preyed upon by others, but the idea that these feeding relationships linked species into a functional unit was a novel one at the turn of the twentieth century.

Now, at the dawn of a new century, the concepts of natural assemblages and feeding interactions, which so strongly shaped the science of ecology in its formative years, are still paramount to our understanding of how the natural world works. But a new, broader view of ecological relationships is emerging in the face of the dramatic changes in global vegetation and climate patterns that have resulted from the activities of humans. There is a new appreciation of the importance of scale, both as a characteristic of ecosystems and as a determinant of ecosystem processes; of the dynamics of how different ecosystems, such as, for example, a lake and the surrounding forest, interact at their boundaries; and of how patterns of species abundance and distribution relate to ecosystem function. More and more, ecologists view ecological processes in the context of interacting ecosystems within vast landscapes.

As ecologists came to appreciate that organisms occur in natural assemblages, they began to ask questions about the nature of these assemblages. The debate over one question in particular dominated ecological thought in the early twentieth century. That question was whether groups of animals and plants, through their interactions with one another, possessed some ability to function as a coordinated unit—as an organism, if you will.

Before we proceed, it is important to point out that, because of the historical approach that we have adopted, the word *community* is used in the next several sections of this chapter (Sections 9.2 through 9.4) rather than the word *ecosystem* to describe natural assemblages. Ecosystem thinking grew from early community ecology, and thus early ecologists did not yet have the ecosystem concept at their disposal. In Part 6 of the book, we shall discuss contemporary community ecology in great detail.

9.2 The analogy of the organism was applied to biological communities by F. E. Clements and rejected by H. A. Gleason and A. G. Tansley.

From its inception, thinking about systems of interacting populations has been divided over a major issue: whether the whole exceeds the sum of its parts. On the one hand, we may believe that the system as a whole has attributes that cannot be understood in terms of the workings of its component parts, just as organisms have functions that cannot be assessed by looking at their individual parts. On the other hand, the system may be viewed merely as a collection of independently functioning populations.

Some ecologists believe that there is determinism at the level of the multispecies system analogous to the determinism imparted at the organismal level by natural selection. In other words, no matter how much one knows about the anatomy of elephants, one cannot explain the special relationship between a whole elephant and its surroundings. Is it also true, then, that all possible knowledge of grass and crickets and mice and lizards cannot unlock the secrets of a prairie?

Ecologists have frequently compared associations of species living together (biological communities) to organisms. Functional similarities between communities and organisms—between primary production and feeding, predation and metabolism, species and organs, the regular succession of stages from fallow field to mature forest and the development of the individual—are obvious. The influential American plant ecologist F. E. Clements (1916, 1936) extrapolated these similarities into a concept of mature biological communities as discrete vegetation types, referred to as **climaxes.** Each climax occurs in a particular region defined by climate and soil, and has a characteristic sequence of developmental stages, called the **sere,** leading from bare or cleared ground to the mature climax state (see Chapter 28). Clements viewed plant communities as **superorganisms,** that is, entities that function much like an individual animal or plant. He had little need to question the analogy between organism and superorganism because his work was primarily descriptive and the organism concept adequately embraced what he saw in nature.

A contemporary of Clements was the American plant ecologist H. A. Gleason. Gleason believed that the local community, far from being a distinct unit like an organism, was merely a fortuitous association of species whose adaptations enabled them to live in a particular place (Gleason 1926, 1939). While recognizing that species do interact (all animals must eat!), Gleason argued that the presence or absence of any one species is independent of all others. In his view, we may define an association for convenience, but it does not represent a natural unit, and it has no functional significance beyond the roles played by each of its members.

The English plant ecologist A. G. Tansley (1935) also rejected the superorganism notion of Clements, preferring to regard the animals and plants in associations, together with the physical factors of their surroundings, simply as systems.

> The more fundamental conception is, as it seems to me, the whole *system* (in the sense of physics), including not only the organism-complex, but also the whole complex of physical factors forming what we call the environment of the biome—the habitat factors in the widest sense. Though the organisms may claim our primary interest, when we are trying to think fundamentally we cannot separate them from their special environment, with which they form one physical system.

Tansley called this integration of organisms and the physical world they share the *ecosystem.*

Despite the writings of Gleason and Tansley, because of Clements' dominant personality, his superorganism idea persisted as a major paradigm in ecology for some time. However, more detailed studies of plant community dynamics by R. H. Whittaker and others in the 1960s supported Gleason's view that species assemblage patterns can be understood by focusing on individual species. Contemporary ecologists do not accept the superorganism view of natural assemblages, though they do recognize that functions that are not attributable to individual species may derive from interactions among species.

9.3 Charles Elton described communities in terms of feeding relationships.

Clements and Gleason were concerned with the species composition of communities. By the mid-1920s, however, other ecologists had begun to consider functional patterns within communities. Foremost among the proponents of this viewpoint was the English ecologist Charles Elton. During his student days at Oxford, Elton accompanied an ecological expedition to Bear Island in the North Atlantic Ocean, where, in collaboration with the botanist V. S. Summerhayes, he worked out the feeding relationships among the inhabitants of a simple tundra community (Figure 9-1).

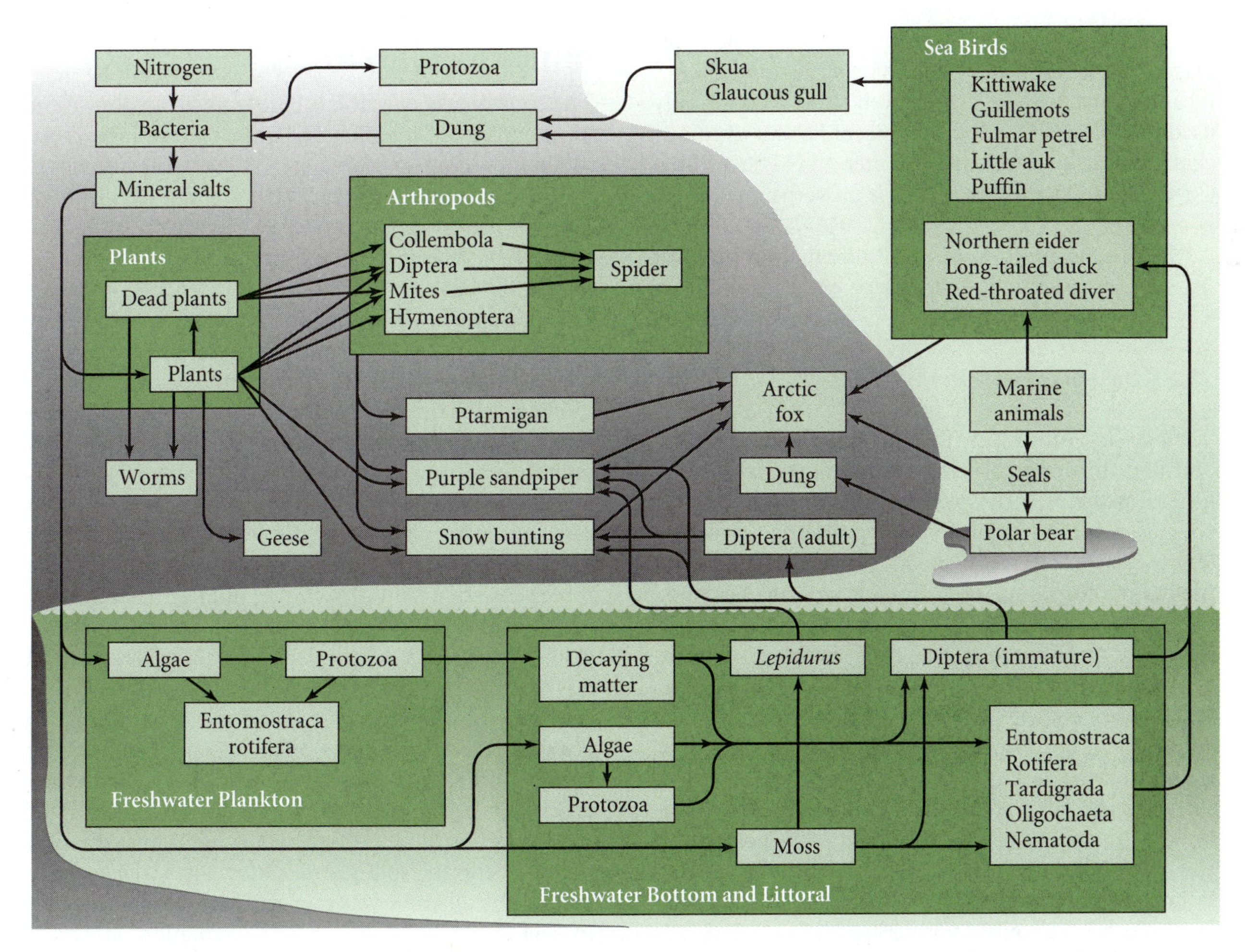

FIGURE 9-1 Summerhayes and Elton's 1923 depiction of the feeding relationships among animals on Bear Island, near Spitsbergen. (*From Elton 1927.*)

By the time Elton was twenty-six, he had developed a new concept of communities organized by the feeding relationships within them. In his book *Animal Ecology* (1927), which was to become a landmark for modern ecology, he wrote:

> Food is the burning question in animal society, and the whole structure and activities of the community are dependent upon questions of food-supply . . . animals have to depend ultimately upon plants for their supplies of energy, since plants alone are able to turn raw sunlight and chemicals into a form edible to animals. Consequently herbivores are the basic class in animal society. . . . The herbivores are usually preyed upon by carnivores, which get the energy of the sunlight at third-hand, and these again may be preyed upon by other carnivores, and so on, until we reach an animal which has no enemies, and which forms, as it were, a terminus on this food-cycle. There are, in fact, chains of animals linked together by food, and all dependent in the long run upon plants. We refer to these as "food-chains," and to all the food-chains in a community as the "food-cycle."

Over the years, the term "food-cycle" has been replaced by the term "food web." The term "food chain," meaning one part of a food web, is still commonly used in ecology. Elton's basic concept has survived unchanged. He prefaced his Chapter V ("The Animal Community") with three Chinese proverbs:

> The large fish eat the small fish; the small fish eat the water insects; the water insects eat plants and mud.
>
> Large fowl cannot eat small grain.
>
> One hill cannot shelter two tigers.

The first proverb is Elton's "food-chain." The second and third are cornerstones of another of Elton's general principles, the **pyramid of numbers.** As one goes up the food chain, one also ascends a more or less regular progression of body sizes because most predators consume prey somewhat smaller than themselves. Progressively larger animals require progressively more space to find food; hence their numbers are lower. Elton noted that an oak wood harbors "vast numbers of small herbivorous insects like aphids, a large number of spiders and carnivorous ground beetles, a fair number of small warblers [insectivorous birds], and only one or two hawks. Similarly in a small pond, the numbers of protozoa may run into millions, those of *Daphnia* and *Cyclops* into hundreds of thousands, while there will be far fewer beetle larvae, and only a very few small fish."

Elton explained this pyramid of numbers, large at the base of the food chain and small at the top, by using arguments from population biology and the scaling of biological functions to body size:

> The small herbivorous animals which form the key-industries in the community are able to increase at a very high rate (chiefly by virtue of their small size), and are therefore able to provide a large margin of numbers over and above that which would be necessary to maintain their population in the absence of enemies. This margin supports a set of carnivores, which are larger in size and fewer in numbers. These carnivores in turn can only provide a still smaller margin, owing to their large size which makes them increase more slowly, and to their smaller numbers.

And so on to the tiger on the hill.

During the 1930s, the idea of the community as an association of interacting species became more and more the focus of ecological thinking, but was far from universally accepted.

9.4 A. J. Lotka espoused a thermodynamic view of ecosystems.

The controversy over the superorganism analogy went unnoticed by A. J. Lotka, a chemist by training, whose different vision of biological systems still influences us through the legacy of his book *The Elements of Physical Biology,* published in 1925. Lotka was the first to treat populations and communities as thermodynamic systems. In principle, he said, each system can be represented by a set of equations that govern transformations of mass among its components. Such transformations include the assimilation of carbon dioxide into organic carbon compounds by green plants and the consumption of plants by herbivores and of animals by carnivores.

Lotka believed that the size of a system and the rate of transformations within it were determined according to certain thermodynamic principles. In the same sense that heavy machines and fast machines require more fuel to operate than do their lighter and slower counterparts, and efficient machines require less fuel than inefficient ones, the energy transformations of ecosystems grow in direct relation to their size (roughly the total masses of their constituent organisms), productivity (rate of transformations), and inefficiency. Lotka regarded the ecosystem as a part of the world machine responsible for the transformation of the energy of sunlight reaching the surface of the earth. Not all the energy enters biological pathways; in fact, most of it drives the circulation of winds and ocean currents and the evaporation of water. But the portion that plants and algae assimilate by

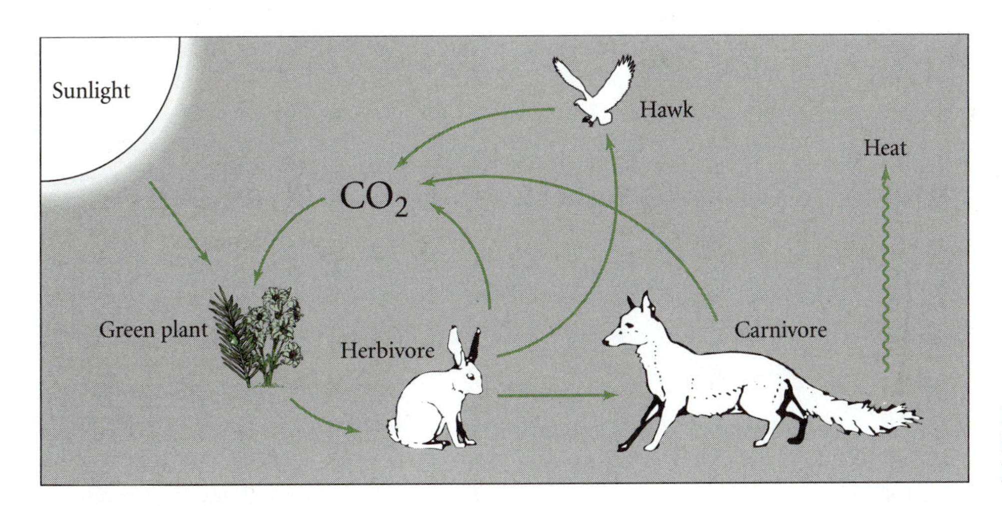

FIGURE 9-2 Lotka's idea of the mill-wheel of life. (*After Lotka 1925.*)

photosynthesis drives a part of the world machine—what Lotka referred to, by way of analogy, as the "mill-wheel of life" (Figure 9-2).

When Lotka published his book, most ecologists were preoccupied with the problem of species associations and missed the implications of a thermodynamic characterization of the natural world (Kingsland 1985). Tansley never referred to *The Elements of Physical Biology,* even though Lotka had provided a mechanical analogy—and thus a tangible model—for his concept of the ecosystem.

9.5 Raymond Lindeman developed the trophic-dynamic concept of the ecosystem.

Lotka's idea of the ecosystem as an energy-transforming system was brought to the attention of many ecologists for the first time in a paper published in 1942 by Raymond Lindeman, a young aquatic ecologist from the University of Minnesota. The full story behind this historic paper has been recounted by Robert Cook (1977): the journal *Ecology* first rejected the manuscript on the advice of reviewers who felt the treatment was too theoretical, but strong advocacy by Yale ecologist G. E. Hutchinson finally led to its publication. Lindeman's framework for understanding ecological succession based on sound thermodynamic principles made a deep impression on Hutchinson. Lindeman adopted Tansley's notion of the ecosystem as the fundamental unit in ecology and Elton's concept of the food web, including inorganic nutrients at the base, as the most useful expression of ecosystem structure (Figure 9-3).

Lindeman's food chain consisted of steps—primary producers (organisms that photosynthesize), herbivores, carnivores—that he referred to as **trophic levels.** But rather than seeking regularity in a trophic pyramid of numbers, as Elton had, he visualized a pyramid of energy transformation. He argued that less energy is available to each higher trophic level owing to the work performed and to the inefficiency of biological energy transformations at the trophic level below. Thus, of the light energy impinging on a lake (Λ_0), plants use only a fraction (Λ_1)—the primary production of the system. Herbivores assimilate less

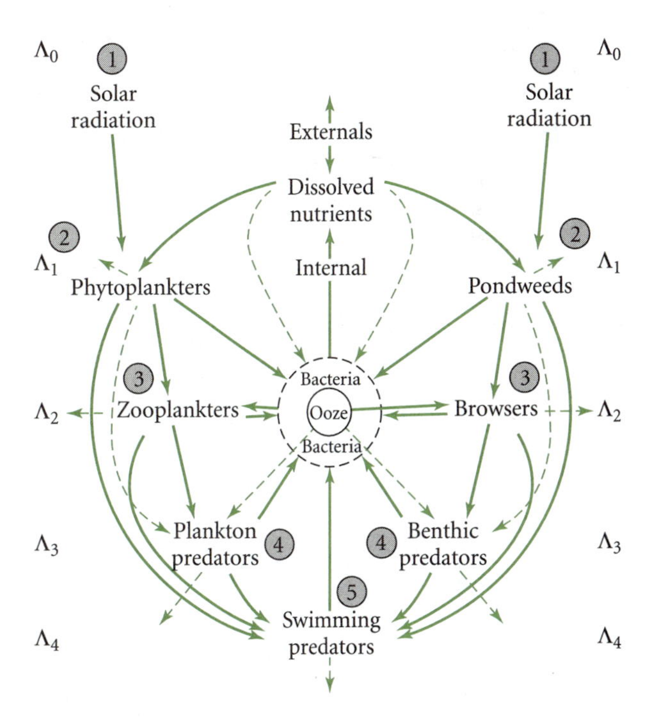

FIGURE 9-3 Lindeman's diagram of the generalized "food-cycle" relationships within a temperate zone lake. Energy enters the system through (1) photosynthesis by organisms such as (2) phytoplankters and pondweeds. These are consumed by (3) zooplankters and browsers that are eaten by (4) plankton and benthic predators, which, in turn, are eaten by (5) swimming predators. Organisms at all trophic levels die and become part of the lake ooze, where the action of bacteria releases nutrients back into the water. (*After Lindeman 1942.*)

energy (Λ_2) than plants do because plants utilize some of their own production to maintain themselves before herbivores consume them. The ratio of production on one trophic level to that on the level below it (for example, Λ_2/Λ_1) is the biological efficiency of that link in the food chain.

By the 1950s, the ecosystem concept had fully pervaded ecological thinking and provided the foundations for a new branch of ecology. University of Michigan ecologist Francis C. Evans (1956) summarized the essential points of this concept in a brief essay:

> In its fundamental aspects, an ecosystem involves the circulation, transformation, and accumulation of energy and matter through the medium of living things and their activities. Photosynthesis, decomposition, herbivory, predation, parasitism, and other symbiotic activities are among the principal biological processes responsible for the transport and storage of materials and energy, and the interactions of the organisms engaged in these activities provide the pathways of distribution. The food-chain is an example of such a pathway. . . . The ecologist, then, is primarily concerned with the quantities of matter and energy that pass through a given ecosystem and with the rates at which they do so.

Thus, the cycling of matter and the associated flux of energy through an ecosystem provided a basis for characterizing its structure and function. Currencies of energy and the masses of elements such as carbon allowed direct comparison of plants, animals, microbes, and abiotic sources of energy and elements in the ecosystem. Taxonomic lists of species and the numbers of individuals in populations gave way to measurements of energy assimilation and energetic efficiencies in this new thermodynamic concept of the ecosystem.

9.6 A. J. Lotka described the regulation of ecosystem function in terms of the ecological relationships of the component populations.

In his writing, Evans expressed a mechanical notion of the function of the ecosystem and its regulation. Each population was an individual machine, and the work it performed was added to that performed by all the others to yield the work performed by the ecosystem as a whole. In Evans's view—probably the prevalent one in the 1950s—energy transformation by each population depended on factors in its environment that directly affected the activities of individuals in the population.

Ecosystems are further characterized by a multiplicity of regulatory mechanisms that depend on the number of organisms present and the ways in which those organisms interact with one another. By limiting the numbers of organisms and influencing their physiology and behavior, the quantities and rates of movement of both matter and energy can be controlled. Processes of growth and reproduction, agencies of mortality (physical as well as biological), patterns of animal movement into and out of the ecosystem, and habits of adaptive significance are among the most important kinds of regulatory mechanisms.

Lotka (1925) developed a mathematical model of such relationships. His model was composed of a set of equations, one for each population in the community. Each equation described the rate of change in the number of individuals in the population, or the energy or biomass equivalents of those individuals, in terms of physical and biological factors in the environment, including the species it fed upon and which fed upon it. To understand Lotka's model, consider a system having three populations, or "components" as Lotka called them, denoted X_1, X_2, and X_3. Further, suppose that there are three factors of the physical environment, denoted P_1, P_2, and P_3, that affect these populations—that is, these factors contribute to either an increase or a decrease in the number of individuals in the populations. Lotka suggested that the rate of change (increase or decrease) in the size of any one of the three populations is dependent on the interactions among all three populations and all three physical factors. Thus, the rate at which one of the populations—say, X_1—increases or decreases is dependent on the interactions of individuals in population X_1 with those in X_2 and X_3 and the effects of physical factors P_1, P_2, and P_3 on all three populations. We denote the rate of change of population X_1 as dX_1/dt, and a general equation for the rate of change of X_1 is given by

$$\frac{dX_1}{dt} = F_1(X_1, X_2, X_3; P_1, P_2, P_3).$$

This equation says that the rate of change in X_1 with respect to time (t) is equal to some function, denoted F_1, of six parameters (X_1, X_2, X_3; P_1, P_2, P_3). This type of equation, which shows the rate of change of one component as a function of one or more other components, is called a differential equation. Lotka did not suggest precisely how the populations and physical factors interact—that is, the exact form of F_1—the way in which the various components are added to, subtracted from, or multiplied with one another. We will discover some possible functions of the components later in this book.

According to Lotka's model, there would be similar functions for X_2 and X_3. These are

$$\frac{dX_2}{dt} = F_2(X_1, X_2, X_3; P_1, P_2, P_3).$$

$$\frac{dX_3}{dt} = F_3(X_1, X_2, X_3; P_1, P_2, P_3).$$

Thus, the complete three-population model would include all three equations (dX_1/dt, dX_2/dt, and dX_3/dt). The system may be summarized and generalized with the following single equation:

$$\frac{dX_i}{dt} = F_i(X_1, X_2, \ldots, X_n; P_1, P_2, \ldots, P_n),$$

where $i = 1 \ldots n$. When Xs are expressed in terms of energy, the sum of the X_is is the total energy content of the system at a given time, and the sum of the dX_i/dt values is the overall rate of change in the structure of the system. Energy transformations are dictated by the individual terms of the function F, which include all the increments of gain and loss to X.

In Raymond Lindeman's symbolism, Λ_n is the total energy content of trophic level n, and the rate of change in Λ_n ($d\Lambda_n/dt$) is the sum of the gains (λ_n) and losses (λ'_n) of energy from that trophic level; hence,

$$d\Lambda_n/dt = \lambda_n + \lambda'_n.$$

Here, the form of the function $F_n(\lambda_n, \lambda'_n)$ is known. That is, the two components are added together ($\lambda_n + \lambda'_n$). The component λ_n represents the rate at which energy enters the nth trophic level from the preceding, $n - 1$, level. λ_n is positive. The component λ'_n represents the rate at which energy leaves the nth trophic level, and thus is negative. The quantity λ_n is very interesting from an ecological point of view because it represents the productivity of the nth trophic level.

According to Lotka's equations, the dynamics of a single link in the food web of the ecosystem—for example, between components (species, physical factors) i and j, are defined by terms in F_i and F_j. The dynamic properties of the entire system are governed by all the terms of the individual functions F_i. The whole is equal to, and can be understood in terms of, the sum of its parts.

9.7 Eugene P. Odum popularized the study of ecosystem energetics.

With a clear conceptual framework for the ecosystem and a currency of energy to describe its structure, ecologists began to measure energy flow and the cycling of nutrients in ecosystems. One of the strongest proponents of this approach has been Eugene P. Odum of the University of Georgia, whose text *Fundamentals of Ecology*, first published in 1953, influenced a generation of ecologists.

Odum depicted ecosystems as simple energy flow diagrams (Figure 9-4). For any one trophic level, such a diagram consists of a box representing the biomass (or its energy equivalent) at any given time and pathways through the box representing the flow of energy. The energy flow within and through a single trophic level is shown in Figure 9-4a, in which the size of the shaded box is proportional to the total amount of biomass at the trophic level. Energy enters the trophic level when food is ingested by the organisms at that level. Some of the ingested energy is simply not used and is lost to the trophic level; the rest is assimilated. Some of the assimilated energy is required for respiration or is stored in the form of fat. The rest is used for growth or is egested. The amount of energy or biomass that is available for the next trophic level is referred to as production. Odum's diagram may represent the dynamics of an entire trophic level or of a

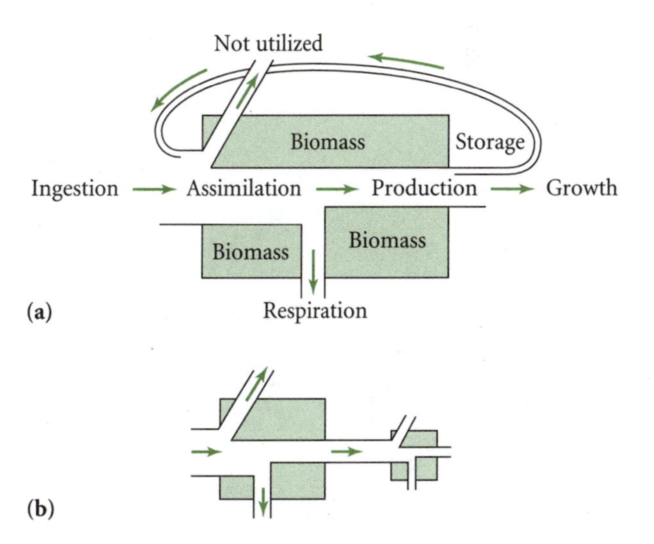

FIGURE 9-4 E. P. Odum's "universal" model of ecological energy flow. (a) Energy flow in one trophic level. (b) A link between two trophic levels in a food web. Some of the energy that enters the trophic level represented by the large box on the left is lost or used for respiration and thus is not available to the next trophic level, which is represented by the smaller box on the right. The size of each box represents the total biomass in that trophic level. These diagrams could depict the interactions of individual organisms as well as those of trophic levels, in which case each individual would be depicted as a separate box. (*After Odum 1968.*)

single individual within the trophic level. That is, the diagram summarizes the energy inputs and outputs of a particular individual or of all of the individuals at a trophic level.

Within a food web there are a number of trophic levels linked by feeding relationships. Odum showed this in his model by linking boxes (Figure 9-4b). Because energy is lost at each trophic level, successive boxes are drawn smaller because they will contain, overall, less biomass. For example, a box representing the predators in a food web will be smaller than the box representing the prey, which are at the trophic level below predators.

Energy flow diagrams have been elaborated to include the cycling of mineral elements (Figure 9-5). Energy flows in one direction through the system. Absorbed light is either lost as heat or transformed into chemical energy through the process of photosynthesis (gross primary production), which is carried out by autotrophic organisms. Autotrophs use some energy for respiration, but much of the rest is put into growth and thus is available to the heterotrophs at the next trophic level (net primary production). Unlike energy, which ultimately comes from sunlight and leaves the ecosystem as heat, nutrients are typically regenerated and retained within the system, as shown by the colored circle in Figure 9-5. Points of nutrient input and output exist, and often, nutrients pool in a particular component of the ecosystem. But the predominant dynamic is the physical transfer of nutrients among the various trophic levels of the system.

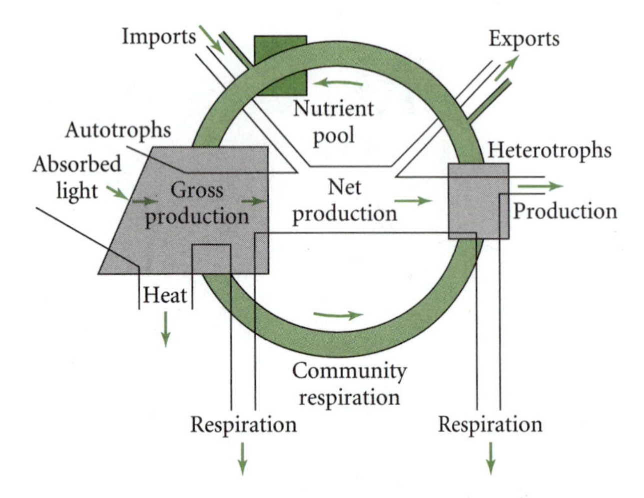

FIGURE 9-5 E. P. Odum's flow diagram of an ecosystem, showing the one-way flow of energy and the recycling of materials. Gross primary production is the total amount of energy absorbed by the system. Imports and exports to the system include organisms that move in or out or debris that may be blown in or out by wind or carried by water. (*After Odum 1960.*)

In the development of ecosystem studies, the cycling of elements has assumed a standing nearly equal to that of the flow of energy. One reason for its prominence is that the amounts of elements and their movement between ecosystem components can provide a convenient index to the flow of energy, which is difficult to measure directly. Carbon, in particular, bears a close relationship to energy content because of its intimate association with the assimilation of energy by photosynthesis. A second reason for the prominence of nutrient cycling is the fact that levels of certain nutrients regulate primary production. In most deserts, plant growth reflects the amount of water available rather than sunlight or minerals in the soil. By contrast, the open oceans are "deserts" in terms of their scarce nutrients. Understanding how elements cycle among ecosystem components is crucial to understanding the regulation of ecosystem structure and function.

9.8 Watershed studies emphasized the flux of elements and energy within and among ecosystems

By the early 1960s, the precepts of modern ecosystems ecology were well established. It was known that energy flows through a medium of trophic interactions and biogeochemcial transformations of nutrients and materials. And it was clear that an understanding of the structure and function of an ecosystem, regardless of whether the ecosystem is viewed as an organism or as a simple sum of its parts, is fundamentally dependent on knowledge of the details of the transformations within the ecosystem and the fluxes of energy and materials across the ecosystem boundary. But precious few ecosystems had been studied in detail, and none had been examined for any appreciable length of time. Moreover, experimental approaches to the study of ecosystems ecology were in their infancy (Likens and Bormann 1985, Likens 1992).

The decade of the 1960s saw the beginning of detailed studies of the hydrology, nutrient cycling, and materials and energy flux of watersheds. A watershed is the valley or basin surrounding a stream or lake from which water drains into that stream or lake. The most intensive watershed studies were initiated in the Hubbard Brook watershed in New Hampshire (Likens 1985), the Walker Branch watershed in eastern Tennessee (Johnson and VanHook 1989), and the Coweeta watershed in eastern North Carolina (Swank and Crossley 1988). (The Coweeta research area was actually established in 1931, but intensive ecosystems studies began there only in 1968.) These studies, which continue today, have as their basis the concept of **mass balance,** the large-scale movement of materials. Individual watersheds are viewed as distinct ecosystems having definable boundaries and including both abiotic and biotic components among which nutrients cycle (Figure 9-6). Inputs and outputs represent connections of the watershed ecosystem with the rest of the biosphere. Materials are transported into or out of the ecosystem by meteorologic or geologic processes or by the movement of organisms. The hydrologic features of the watershed, such as precipitation, surface runoff, and groundwater flow, serve to integrate the cycling of nutrients within the ecosystem and the flux of materials into and out of the ecosystem. The overarching goal of these studies is to determine the mass of particular nutrients in the various components of the ecosystem in relation to inputs and outputs of those materials and in response to experimental changes in the watershed, such as logging. Changes in other parts of the biosphere affect what enters the watershed. For example, pollution produced elsewhere may enter the watershed in precipitation. Outputs from the watershed are inputs into other ecosystems and thus have an effect on their function. The long-term monitoring of inputs and outputs of watersheds has provided essential information about how global environmental change is manifested in particular ecosystems.

Watershed studies led to a proliferation of published papers in the 1960s and 1970s detailing the energy and materials fluxes in many different ecosystems. By the early 1980s, a vast literature describing the functional details of most of the world's prominent ecosystems was available (Strayer 1991). Serious attempts to synthesize and generalize this vast amount of information emerged in the 1970s and 1980s, including studies that investigated bacterial production and other functional attributes of freshwater and marine ecosystems (Nixon 1980, Cole et al. 1988), primary productivity (Lieth 1975), the role of herbivory in primary production (McNaughton, et al. 1989) forest productivity (O'Neill and DeAngelis 1981; DeAngelis et al. 1981), and global patterns of nutrient cycling (Post et al. 1982, 1985).

In 1990, prominent ecosystems ecologists gathered at the Institute of Ecosystem Studies in Millbrook, New York, to discuss what advances had been made in the comparative analysis of ecosystems, to present case studies using the comparative approach, and to develop new comparative theory (Cole

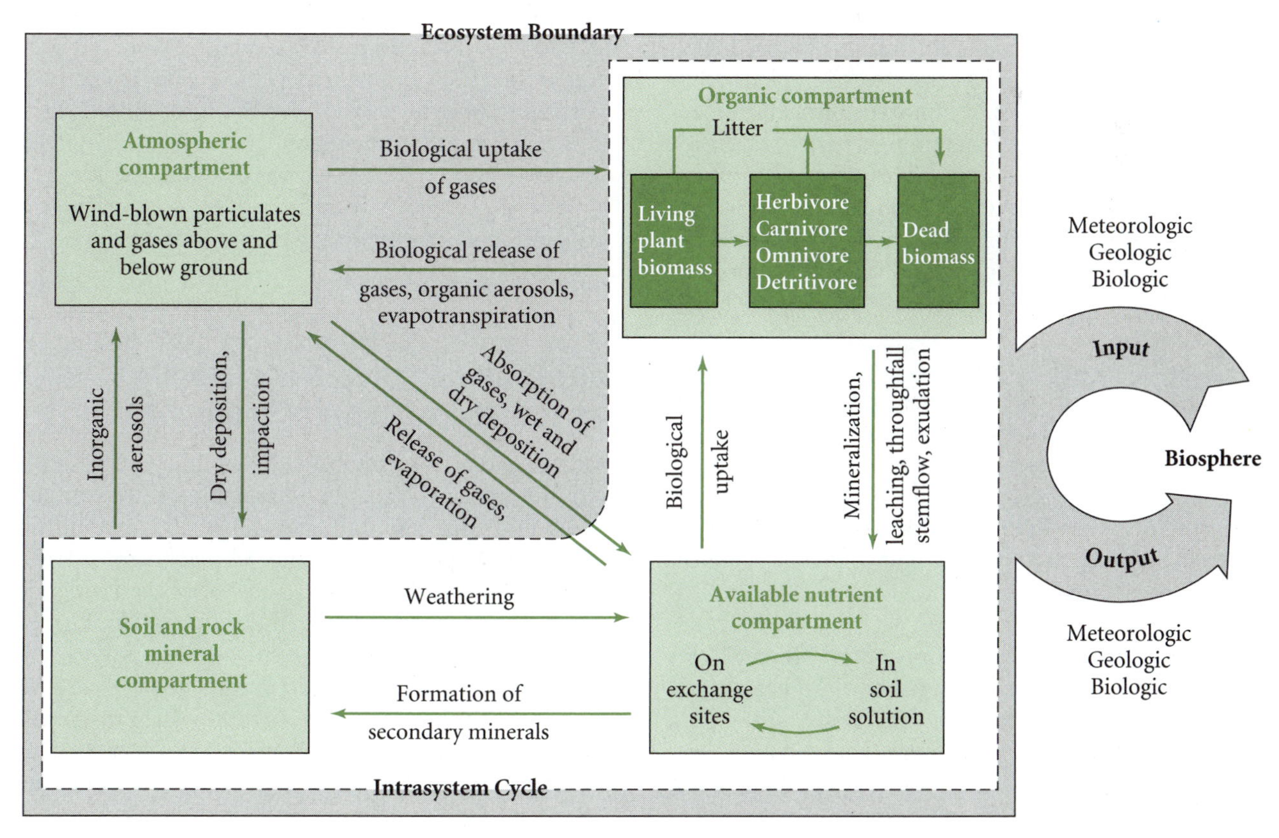

FIGURE 9-6 A model of the nutrient dynamics of a watershed ecosystem based on studies at Hubbard Brook. The model depicts the inputs and outputs of an ecosystem having relatively defined boundaries. Such ecosystems are connected to other ecosystems by the flux across their boundary. (*Redrawn from Likens 1985; after Bormann and Likens 1967, Likens and Barmann 1972, Likens et al. 1977.*)

et al. 1991). Detailed analyses of individual ecosystems, particularly those, such as the arctic tundra, that are unique or thought to be highly sensitive to global environmental change, will continue to provide the information base for ecosystems ecology. In addition, broad comparisons of ecosystem structure and function hold the promise of revealing new insights into how the natural world works.

9.9 Landscape ecology considers the effect of spatial scale on ecosystem function.

The intensity of global environmental change, which is fueled by alterations of human land use patterns, industrial activity, and agriculture, has compelled ecologists to take a global view of ecosystem structure and function. The comparative studies spawned by the work at Hubbard Brook and other experimental watersheds are rooted in the need to identify patterns of ecosystem structure and function of sufficient scale to reflect global environmental change. In adopting this broader view, ecologists have come to appreciate that the relative sizes and shapes of ecosystems, along with their positions in relation to other natural systems, are characteristics of ecosystems as well as effectors of ecosystem function. This realization has fueled an intense interest in how ecosystems interact within vast tracts of land (tens of hundreds of hectares), referred to as **landscapes,** containing a variety of different and distinct ecosystems. This way of thinking about ecosystem structure and function is part of the field of **landscape ecology,** the focus of which M. G. Turner (1989) of the Oak Ridge National Laboratory at Oak Ridge, Tennessee, has clearly characterized:

> Landscape ecology emphasizes broad spatial scales and the ecological effects of the spatial patterning of ecosystems. Specifically, it considers (a) the development and dynamics of spatial heterogeneity, (b) interactions and exchanges across heterogeneous landscapes, (c) the influences of spatial heterogeneity on biotic and abiotic processes, and (d) the management of spatial heterogeneity.

Two aspects of the landscape approach merit emphasis. First, in viewing ecosystems within landscapes, the importance of ecosystem size (scale), shape, and position in the landscape is more than just an operational consideration. It is now understood that ecosystem structure, function, and response to perturbations are dependent in many important ways on the physical characteristics and position of the ecosystem. Whether a watershed faces to the east or to the west (position) will influence nutrient cycling and other processes. The distance between two woodlots, their sizes relative to each other, the extent of the boundary they have in common, and the nature of the intervening landscape may determine whether a population of birds persists in one or both woodlots. The hydrologic processes that connect forests to lakes determine the nature of energy and materials fluxes in both. Turner emphasizes this point when she writes, "Elucidating the relationship between landscape pattern and ecological processes is a primary goal of ecological research on landscapes" (Turner 1989).

The importance of spatial scale and position emphasizes a second point regarding landscape ecology that turns on the question of how, precisely, qualitatively different ecosystems—such as, say, a forest and a lake—interact: What happens at the ecosystem boundaries—called **ecotones**—within landscapes? This question becomes compelling when one appreciates that ecotones are very much influenced by human activities, such as agriculture and forestry, and that ecotone processes are scale dependent. Changes in land use patterns will result in changes in the extent and configuration of ecotones and, inevitably, in alterations in patterns of biological diversity, energy flow, and materials flux (Hansen and di Castri 1992).

By adopting a landscape perspective, ecologists can address such questions as how large a forest patch should be in order to maintain a population of a particular species, or how much of a land corridor is needed between two forest patches in order to maintain gene flow between populations of endangered species in those patches. As we shall see, the ideas of landscape ecology have been applied to concepts in both population ecology and community ecology.

9.10 Both tactical and strategic approaches have been applied to the study of ecosystems.

Many ecologists feel that understanding an ecosystem requires an analysis of the interactions of all or most of its individual components—that is, the adoption of a tactical approach. Others feel that the properties of ecosystems transcend the behavior of their individual components and can be understood by looking at properties of the entire system—that is, by using a strategic approach. The tactical approach seeks to mimic the fine details of the natural system to the extent possible, whereas the strategic approach seeks to represent the essence of the system by examining the behavior of selected, important components of it.

The tactical approach to the study of ecosystems is called **systems ecology.** Systems ecology relies heavily on mathematical analysis to handle the complexity of natural systems. The goal of the systems approach is to provide a model of the ecosystem capable of mimicking its dynamic properties. Lotka's models, which we described above, are tactical models. Disturbances, such as pollution, grazing, fire, introduced pests, hunting, and severe climatic conditions, alter Lotka's system of equations dX_i/dt for certain species by changing the conditions of the environment (P) or the abundances of interacting species (X). By their effect on X_i, such disturbances indirectly affect other species or components (X_j) to which i is linked in the system. A detailed mathematical model that faithfully depicts the interactions among all components could assess the effect of a stress or other disturbance on the structure and function of the ecosystem as a whole (Watt 1968, Van Dyne 1969), but the development of such models has proven difficult.

Systems ecologists understand ecosystems by taking them apart and seeing how each piece functions in the whole. Their philosophy asserts that just as one cannot understand a house without detailed study of the blueprints and direct nail-by-nail observation of its construction, one cannot understand an organism without a map of its genes and step-by-step knowledge of its development and functioning.

Other ecologists believe that the essential workings of ecological systems may be comprehended in the absence of a detailed examination of their parts. An analogy or two will help explain this view. An understanding of an organism's function in its particular environment differs from an understanding of the biochemistry of its cell processes or the control of gene expression during its development. Darwin did not have a detailed, or even correct, knowledge of genetics, and yet he was able to appreciate the mechanism of evolution and the purpose of adaptation. An architect designs the structure of a house with a purpose in mind. The detailed arrangements of the bricks and plumbing are merely the means to achieve this design; they do not offer insight into the purpose of the house as a complete, integrated structure intended for human habitation.

E. P. Odum may be credited with the early development of the strategic approach to ecosystem study. Odum was interested in how communities change through time, a process called succession, which we shall examine in some detail in Chapter 28. Put very simply, he suggested that the evolution of the species that make up an ecosystem in response to one another results in increased stability of the system. That is, older, more developed communities should be more stable (Odum 1969). He wrote:

> [Ecological succession] culminates in a stabilized ecosystem in which maximum biomass (or high information content) and symbiotic function between organisms are maintained per unit of available energy flow. In a word, the "strategy" of succession as a short-term process is basically the same as the "strategy" of long-term evolutionary development of the biosphere—namely increased control of, or homeostasis with, the physical environment in the sense of achieving maximum protection from its perturbations.

To Odum, "information content" was related mainly to the number of different species in the community, that is, to biological diversity (Margalef 1963, Odum 1988). It is the increase in information, which comes from increasing ecosystem complexity, that increases ecosystem resilience and drives ecosystem development. Thus, ecosystems may be understood strategically by understanding their development.

One of the difficulties with the strategic approach is that it requires that many processes be represented by an index or an average. In large systems, this may mean that some parts of the system are under- or overrepresented in the strategic model. G. M. Weinberg (1975) suggested that ecosystems pose conceptual difficulties because they fall in the range of "middle-number systems" that cannot be treated appropriately either by differential equations, as Lotka proposed, or by statistical approaches, whereby we may characterize the properties of a collection of components by calculating means and variances (see Chapter 2). Simple physical systems with few components, such as the gravitational attraction among the nine planets of our solar system or contrived laboratory microcosms with few species, are "small-number simple systems." "Large-number simple systems" consist of immense numbers of similar items—a quantity of gas consisting of billions and billions of atoms, for example. The temperature and pressure of the gas depend on the motion and collisions of individual atoms, but averages of these properties adequately describe the whole system.

According to Weinberg, whole ecosystems are too complex to model as small-number systems of differential equations. Furthermore, their parts differ so much in function that simple averages, even within trophic levels, would obscure essential features of ecosystem structure. Thus, the ecosystem might be what Weinberg calls a "middle-number system"—its parts are too numerous to describe fully, and too few and diverse to average meaningfully. We hasten to add, however, that specific, and important, components of whole ecosystems, such as primary production and some nutrient cycles, may not be too complex to model as small-number systems.

One way to overcome the middle-number "problem" is to view individual organisms, rather than trophic levels or complex processes such as net primary production and nutrient cycles, as the basic unit of the ecosystem. In so-called **individual-based**

models, the ecosystem may be viewed as approaching a large-number system in which organism-level processes, such as growth, reproduction, or nutrient uptake, can be reasonably averaged (as is done routinely in population studies) and described statistically (Huston et al. 1988). Computers make scaling such models up to the ecosystem level possible. Such individual-based models have been used to study interactions within biological communities (Botkin et al. 1972, Shugart and West 1977, Huston and Smith 1987) as well as ecosystems (Aber et al. 1982, Pastor and Post 1986).

SUMMARY

1. An ecosystem is composed of organisms and the physical environments they inhabit. Although this concept is of central importance to present-day ecology, its development was slow and tentative.

2. An early theme in ecology was the dichotomy between one concept of systems as having properties that result from the interactions of their components, and a second concept of systems as analogous to organisms, expressing properties of organization and regulation that cannot be deduced from their parts. Early in the twentieth century, F. E. Clements favored the second, organismal viewpoint in his writings about plant associations, while H. A. Gleason felt that the properties of these associations were only the sums of the properties of their parts. In 1935, A. G. Tansley coined the term *ecosystem* to include organisms and all the abiotic factors in their habitat.

3. Charles Elton described communities in terms of feeding relationships and emphasized the pyramid of numbers as a dominant organizing principle in community structure.

4. A. J. Lotka, in 1925, provided a thermodynamic perspective on ecosystem function, whereby the movements and transformations of mass and energy conform to thermodynamic laws. He also showed how the behavior of whole systems could be described mathematically in terms of the interactions of their components.

5. Raymond Lindeman, in 1942, popularized the idea of the ecosystem as an energy-transforming system, providing a formal notation for the energy flux among trophic levels and for ecological efficiency.

6. The study of ecosystem energetics dominated ecology during the 1950s and 1960s, largely due to the influence of Eugene P. Odum, who championed energy as a common currency for describing ecosystem structure and function.

7. The watershed studies at Hubbard Brook, Walker Branch, and Coweeta provided a foundation for the experimental study of ecosystem function and ecosystem interactions and gave rise to a large number of detailed studies of ecosystems in the 1960s, 1970s, and 1980s. Watershed studies emphasize the flux of elements and energy within ecosystems and between ecosystems and other parts of the biosphere.

8. The interactions of ecosystems within vast landscapes are dependent in large measure on the sizes and relative positions of the ecosystems. This type of thinking forms the basis of contemporary landscape ecology.

9. Ecosystems may be studied using either a tactical or a strategic approach. Systems ecology, based upon Lotka's idea that ecosystems can be represented by a set of equations describing the dynamic interrelationships of their components, is an example of the tactical approach. Systems ecologists model ecosystems in ever-increasing complexity with the objective of understanding the dynamic behavior of the whole system from the behavior of mathematical analogies. Odum viewed communities strategically by relating their properties to the number of species in the community. Some ecologists have attempted to cope with the complexity of ecosystems by using individual organisms as the basic unit of an ecosystem. Such individual-based models may be treated as large-number systems.

EXERCISES

1. Construct a time line of the development of ideas related to the concept of the ecosystem. Using other sources, include important world historical events not related to science. Extend the time line backward to include Darwin and his contemporaries.

2. Ecological interactions at the boundary between ecosystems—ecotones—are of great interest to ecologists. Why are such areas of importance? Design an experiment or series of observations that uses a tactical approach to determine the characteristics and function of an ecotone in the area where you live. (Hint: This would be a good time to pull on your boots and head to the field.)

3. How might the ecosystem concept be applied in agriculture? In conservation biology?

4. Consider Lindeman's equation $d\Lambda_n/dt = \lambda_n + \lambda'_n$, where Λ_n is the total energy content of a trophic level, λ_n is the energy gain, and λ'_n is the energy loss at the trophic level. Suppose that a trophic level in a simple ecosystem has an energy content of $\Lambda_1 = 100{,}000$ units of energy, and that the energy gain at every step in the ecosystem is $\lambda_n = 0.10\ \Lambda_{n-1}$. If the next trophic level contains 1,000 units of energy, what is the value of λ'_n?

CHAPTER 10

Energy Flow in Ecosystems

GUIDING QUESTIONS

- What is the relationship among gross production, net production, and net ecosystem production?
- How is primary production measured in different habitats?
- How can the effect of light intensity on the rate of photosynthesis in terrestrial plants be demonstrated experimentally?
- How does productivity in terrestrial and aquatic ecosystems compare?
- What is the importance of the concept of ecological efficiency in terms of production, exploitation, and assimilation?
- What is the relationship between biomass turnover time and net productivity?
- How does energy flow from primary producer to predator to decomposer?

As we emphasized in the previous chapter, one of the fundamental concepts upon which modern ecology rests is that organisms interact with one another, directly or indirectly, through feeding relationships, which we call trophic interactions. Trophic interactions involve biochemical transformations of energy and the transfer of energy from one individual to the next through the process of consumption. Materials move within ecosystems, and the pathways of such movements are closely associated with the flow of energy (see Chapter 11).

In Chapter 3 we drew a distinction between living and nonliving things based on differences in the dynamics of the energy flux of each. While both inanimate objects and living things lose energy according to the laws of thermodynamics, the many biochemical and morphological adaptations of living things work to forestall the loss of energy, giving them improbably high levels of energy when compared with similarly sized nonliving objects. Organisms dissipate energy in two ways. First, they perform work on the system in which they live. (For example, as an animal runs, it transfers kinetic energy to the atmosphere and to the ground.) Second, organisms lose energy in the form of heat at each biochemical step in the transformations required for movement, biosynthesis, secretion, and cell maintenance because these transformations are inefficient. In spite of these losses of energy by individual organisms, the ecosystem as a whole is not depleted of energy because plants and algae assimilate energy from the sun. Therefore, there is a constant input of energy into the ecosystem. Plants use some of this energy, stored in the chemical bonds of carbohydrates, for their metabolic needs. But herbivores also take a portion of this energy, and thus start some of the energy assimilated by plants on its way up the food chain.

In this chapter, we shall see how the flux of energy through the ecosystem depends on the rates at which plants assimilate energy, the rates of consumption at each trophic level, and the energetic efficiencies of transforming food into biomass. A logical place to start is at the beginning of the food chain, with plant production.

10.1 Plants assimilate energy by photosynthesis.

Photosynthesis is the process by which plants and algae capture light energy and transform it into the energy of chemical bonds in carbohydrates (see Chapter 3). Glucose and other organic compounds (starches and oils, for example) may be stored and their energy later released in respiratory metabolic pathways. Photosynthesis chemically unites two common inorganic compounds, carbon dioxide (CO_2) and water (H_2O), to form glucose ($C_6H_{12}O_6$), with the release of oxygen (O_2). The overall stoichiometry (chemical balance) of photosynthesis is

$$6\,CO_2 + 6\,H_2O \longrightarrow C_6H_{12}O_6 + 6\,O_2.$$

Photosynthesis transforms an atom of carbon from an oxidized state in CO_2 to a reduced state in carbohydrates (see Chapter 5). Reactions in which the reactants have less energy than the products are referred to as endergonic reactions. The reduction of CO_2 is endergonic because the reactants, CO_2 and H_2O, have less energy than the product, $C_6H_{12}O_6$. Thus, photosynthesis requires energy, which is provided by visible light. For each gram of carbon assimilated, a plant gains 39 kilojoules (kJ) of energy. But because of inefficiencies in the biochemical steps of photosynthesis, no more than 34%, and usually much less, of the light energy absorbed by photosynthetic pigments eventually appears in carbohydrate molecules (Rabinowitch and Govindjee 1969).

Photosynthesis supplies the carbohydrate building blocks and the energy the plant needs to synthesize tissues and grow. Rearranged and joined together, glucose molecules can be used to build fats, oils, and cellulose. Combined with nitrogen, phosphorus, sulfur, and magnesium, simple carbohydrates derived ultimately from glucose produce an array of proteins, nucleic acids, pigments, and secondary metabolites such as alkaloids. Plants cannot grow unless they have all these basic building materials (Clarkson and Hanson 1980). Chlorophyll, for example, contains an atom of magnesium; even though all other necessary elements might be present in abundance, a plant lacking magnesium cannot produce chlorophyll, and thus cannot grow.

Plants build and maintain their tissues by complex, energy-requiring biochemical transformations. Because plants utilize much of the energy assimilated by photosynthesis to supply these needs, their tissues always contain substantially less energy than the total assimilated. Therefore, ecologists must distinguish two measures of assimilated energy: gross production, the total energy assimilated by photosynthesis, and net production, the energy accumulated in plant biomass—that is, plant growth and reproduction. Because plants are the first link in the food web, ecologists refer to these measures respectively as **gross primary production** and **net primary production.** The difference between them is the energy of respiration, that energy utilized by the plant for maintenance and biosynthesis (Figure 10-1). By *production,* ecologists mean the accumulation of energy or biomass, not the creation of energy. Often, we are interested in the rate at which energy is accumulated, and this is referred to as **productivity.** Thus, **primary productivity** is the rate at which energy, usually measured in terms of biomass, is accumulated in plants by the process of photosynthesis.

Terrestrial plants are rooted in soil, which contains microorganisms and other organisms that decompose dead plant and animal material. The energy used by these soil organisms is unavailable to consumers of plants. Because of the intimate relationship between terrestrial plants and the soil, ecologists often

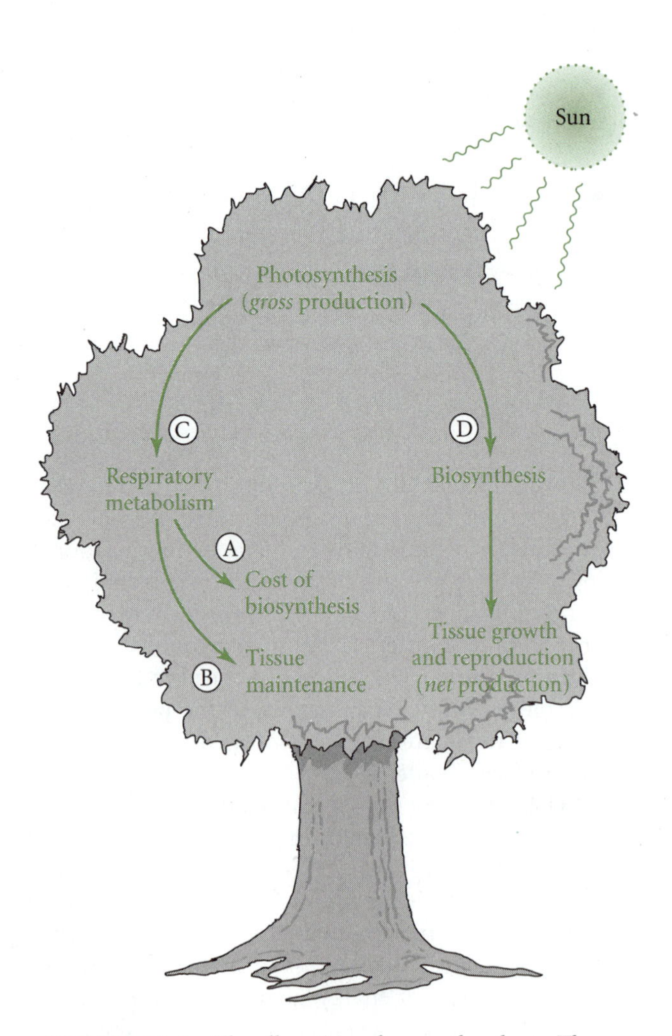

FIGURE 10-1 The allocation of energy by plants. The metabolic costs of biosynthesis (A) and tissue maintenance (B) represent energy lost by respiration (C) and thus unavailable to higher trophic levels (D).

consider the two as a unit when measuring net primary production. The **net ecosystem production** (NEP) is the difference between gross primary production and the energetic costs of both the plants and the soil organisms:

$$\text{NEP} = \text{gross primary production} - \text{respiration of plants} - \text{respiration of soil organisms.}$$

This measure is used as an index to whether the plants and the soils of an ecosystem constitute an energy source (NEP positive) or an energy sink (NEP negative).

10.2 Methods of measuring plant production vary with habitat and growth form.

Net primary production is an important characteristic of ecosystems because it is a measure of the amount of energy available for living things other than primary producers. Thus, ecologists have invested considerable effort in developing appropriate ways to measure it (Table 10-1). Primary production involves fluxes of carbon dioxide, oxygen, minerals, and water on the one hand and the accumulation of plant biomass on the other. The rates of any of these should provide an index to the overall rate of plant production. In practice, however, the appropriate measure is dependent on the form and habitat of the producers. Whereas production may be quite easily estimated for large terrestrial or aquatic plants by harvesting them and measuring the accumulation of biomass, such an approach may be impractical for small phytoplankton. The measurement of minute changes in oxygen concentration is considerably easier to accomplish in aquatic environments, where oxygen concentrations are low, than in terrestrial ones. Different measures may be required in order to estimate gross and net production, or the production of an entire system and that of a part of a single plant. Following are some common techniques for measuring primary production.

Energy, Biomass, or Carbon Content

Net production can be expressed conveniently as grams of carbon assimilated, dry weight of plant tissues, or their energy equivalents. Dry weight, obtained simply by drying the plant material, is preferred because the amount of water in plant tissues fluctuates widely. Ecologists use such indices interchangeably because they have found a high degree of correlation among them. The energy content of an organic compound depends primarily on its carbon content (assimilated at a cost of 39 kJ per gram) and the energy added or subtracted during various transformations. For example, glucose contains 40% carbon by mass, and should therefore contain an energy equivalent of $0.40 \times 39 = 15.6$ kJ g^{-1}. The observed value is about 17.6 kJ g^{-1}, as determined by burning samples in devices called **bomb calorimeters** (Paine 1971).

A calorimeter has a small chamber in which the sample is burned, in a fast chemical version of cellular respiration, to oxidize it completely to carbon dioxide and water. Oxygen is forced under high pressure into the chamber to ensure complete combustion. A water jacket surrounds the chamber and absorbs the heat produced. The increase in temperature of a known amount of water in the jacket provides a direct

TABLE 10-1 Common methods used to measure primary production

Method	Description	Habitat
Calorimetry	Sample of plant tissue is burned in bomb calorimeter. Energy content of organic compounds is calculated from heat energy released during burning.	Terrestrial
Harvesting	Samples of plant tissue are obtained, dried, and weighed. Dry weight of plant biomass is used as measure of production.	Terrestrial; large aquatic plants
CO_2 flux	CO_2 uptake of plant or plant part is measured in a closed container.	Terrestrial and aquatic
O_2 flux	Production of O_2 by plants or algae is measured in a closed container.	Usually aquatic
Chlorophyll concentration	Chlorophyll is extracted and its concentration measured. Concentration is compared with the rate of assimilation of carbon per gram of chlorophyll for the particular plant or alga under study.	Usually aquatic

estimate (4.2 kJ °C^{-1} kg^{-1}) of the heat energy released by combustion. Measured in this way, the amounts of energy released in complete oxidation average approximately 17.6 kJ g^{-1} for carbohydrates, 23.8 kJ g^{-1} for proteins, and 39.7 kJ g^{-1} for fats (King and Farner 1961, Kleiber 1961). In other words, fats are the most energy-rich organic molecules, while carbohydrates are the least energy-rich. Thus, by measuring the relative proportions of these substances in plant tissues, one can estimate the energy content of those tissues.

In terrestrial ecosystems, ecologists usually estimate plant production by the annual increase in plant biomass, although measurements of CO_2 flux also are practical (Newbould 1967, Milner and Hughes 1968). In areas of seasonal production, annual growth is determined by cutting, drying, and weighing the plants at the end of the growing season, a technique sometimes referred to as **harvesting** (Odum 1960). Harvesting measures only the **annual aboveground net productivity** (AANP), the most common basis for comparing terrestrial communities.

Carbon Dioxide Flux

Because the atmosphere contains so little carbon dioxide (0.03%), uptake by plants can measurably reduce its concentration in an enclosed chamber within a short period. Ecologists have used this principle to measure production in terrestrial habitats, beginning with the classic study of H. N. Transeau (1926) on field corn. The most convenient application of this method is to enclose samples of vegetation, usually whole herbaceous plants or branches of trees, in a clear chamber that allows light to penetrate for photosynthesis, and to measure the change in the concentration of CO_2 (or O_2) in air passed through the chamber. Modern ecologists measure the absorbance of infrared light by the airstream, inasmuch as CO_2 strongly absorbs light in that part of the spectrum (Mooney and Billings 1961). CO_2 uptake per gram of dry weight, or per square centimeter of surface area of leaves in the chamber, can then be extrapolated to the entire tree or forest.

Carbon dioxide flux during the light period of the day includes both assimilation (uptake) and respiration (output), and thus measures net production. Respiration can be estimated separately by carbon dioxide production during the night, when photosynthesis shuts down. Botkin and coworkers (1970) measured CO_2 uptake (grams per square meter of ground area per day) by the three most common species of trees (two oaks and a pine) in the Brookhaven National Laboratory forest over the course of a single growing season. During daylight hours the plants assimilated 4,336 g m^{-2} d^{-1}. Over 24 hours the value was reduced by nighttime respiration to 3,702 g m^{-2} d^{-1}, a difference of 635 g m^{-2} d^{-1}. Because the average chemical composition of plant carbohydrate is $C_6H_{12}O_6$, one gram of CO_2 assimilated is equivalent to 0.614 g of carbohydrate produced. Using this ratio, Botkin and his colleagues estimated the net dry matter production of the forest to be $3{,}702 \times 0.614 = 2{,}273$ g m^{-2} d^{-1}, which agreed well with measurements using the harvesting technique in the same forest (Whittaker and Woodwell 1969).

Radioactive Isotope Techniques

The radioactive isotope carbon 14 (^{14}C) provides a useful variation on the gas exchange method of measuring productivity. When one adds a known amount of ^{14}C-carbon dioxide to an airtight chamber containing a plant, the plant assimilates the radioactive carbon atoms in the same proportion as they occur in the air inside the chamber. Thus, one may calculate the rate of carbon fixation by dividing the amount of ^{14}C found in the plant by the proportion of ^{14}C in the chamber at the beginning of the experiment. For example, if a plant assimilated 10 milligrams of ^{14}C in an hour, and the proportion in the chamber was 0.05, we could calculate that the plant assimilated carbon at a rate of 200 mg h^{-1} (10 divided by 0.05).

Oxygen Flux in Aquatic Systems

Although the production of large aquatic plants, such as kelps, can be estimated by harvesting (Penfound 1956, Mann 1973), the small size and rapid turnover of phytoplankton preclude that method as a general approach to aquatic production. Whereas high levels of atmospheric oxygen preclude using the production of oxygen by photosynthesis as a measure in terrestrial habitats, the low natural concentration of oxygen dissolved in water makes the measurement of small changes in oxygen concentration practical in most aquatic systems (Strickland 1960, Strickland and Parsons 1968).

To measure production, samples of water containing phytoplankton are suspended in pairs of sealed bottles at desired depths beneath the surface of a lake or the ocean; one of each pair (a "light bottle") is clear and allows sunlight to enter; the other (a "dark bottle") is opaque. In the light bottles, photosynthesis and respiration occur together, and part of the oxygen produced by the first process is consumed by the second. In the dark bottles, respiration consumes oxygen without its being replenished by photosynthesis. Thus, to estimate gross production, one adds the change in oxygen concentration in the light bottle (photosynthesis minus respiration) to the amount consumed in the dark bottle.

In unproductive waters, such as those of deep lakes and the open ocean, changes in oxygen concentration due to photosynthesis and respiration are small compared with the amount dissolved in the water, and light and dark bottle measurements are not practical. The measurement of ^{14}C uptake is used to estimate production under such conditions. The principle of the ^{14}C method is the same in aquatic systems as in terrestrial ones, except that the isotope is usually provided in the form of bicarbonate ion (HCO_3^-).

Chlorophyll Concentrations

Another method for estimating plant production in aquatic habitats is based on the idea that the concentration of chlorophyll sets an upper limit on the rate of photosynthesis at high light intensities. J. H. Ryther and C. S. Yentsch (1957), who advocated the application of this method, determined that marine algae assimilate a maximum of 3.7 grams of carbon per gram of chlorophyll per hour (with variation between 2.1 and 5.7 grams over several studies). With experimentally determined relationships between light intensity and rates of photosynthesis by known concentrations of phytoplankton (Ryther 1956, Steele 1962), as well as measurements of chlorophyll concentrations and light penetration through the water column, one can estimate total production per unit of surface area. What the chlorophyll method lacks in precision it makes up in speed and simplicity. It has been used in surveys of production over large areas. In one such study, samples were taken at 1 kilometer intervals between Portland, Maine, and Yarmouth, Nova Scotia, a distance of about 300 km, revealing considerable heterogeneity in production (Huntley and Boyd 1984) (Figure 10-2).

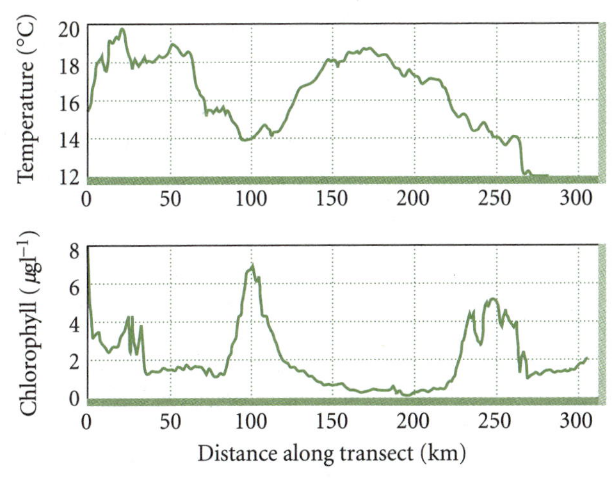

FIGURE 10-2 Temperatures and chlorophyll concentrations at a depth of 4 meters along a transect between Portland, Maine, and Yarmouth, Nova Scotia, during the summer of 1979. (*From Huntley and Boyd 1984.*)

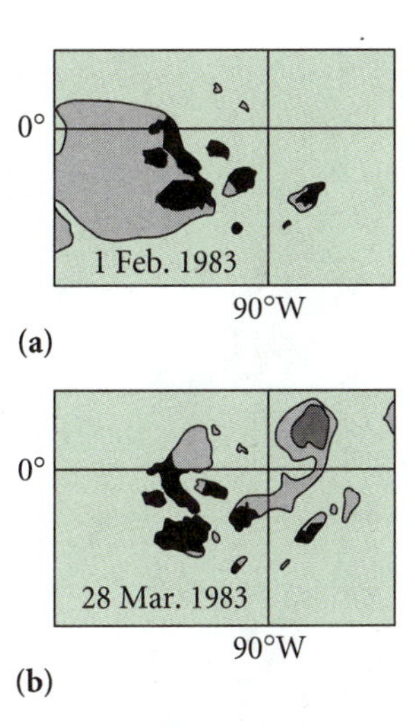

FIGURE 10-3 Phytoplankton pigment concentrations around the Galápagos Islands obtained from satellite imagery on February 1, 1983 (a) and March 28, 1983 (b). The light shading indicates 1–3 mg chlorophyll m^{-3}; the dark shading, 3–10 mg chlorophyll m^{-3}. In general, concentrations throughout the region were less than 0.2 mg m^{-3}. Note the shift in ocean current direction from westerly to northeasterly with the development of El Niño conditions during March 1983. (*After Feldman 1984.*)

Instrumentation aboard orbiting satellites can also be used to estimate production by measuring the reflectance of the earth at specific locations. In the visible and near-infrared portion of the solar spectrum, bare ground has a lower reflectance than the leaves of vegetation. Peterson and coworkers (Peterson et al. 1987) have shown that the near-infrared reflectance of a coniferous forest of the Pacific Northwest is correlated highly with the leaf area index ($m^2\ m^{-3}$), which is known to be directly related to primary production. In aquatic environments, chlorophyll concentrations may be estimated by measuring phytoplankton pigment concentrations, which can be obtained from satellite images of surface waters (Figure 10-3).

Satellite imagery provides a way to map and follow changes in productivity over vast areas. The potential of satellite imagery for the observation and analysis of ecosystem structure and function is immense. In addition to measuring patterns of primary production, the technique can be used to measure land use, erosion, vegetation distribution, ocean hydrology, ice flow, forest dynamics, and many other features of the landscape (Quattrochi and Pelletier 1991, Schlesinger 1991). Satellite images showing phytoplankton concentrations and water temperature and flow are shown in Figure 10-4.

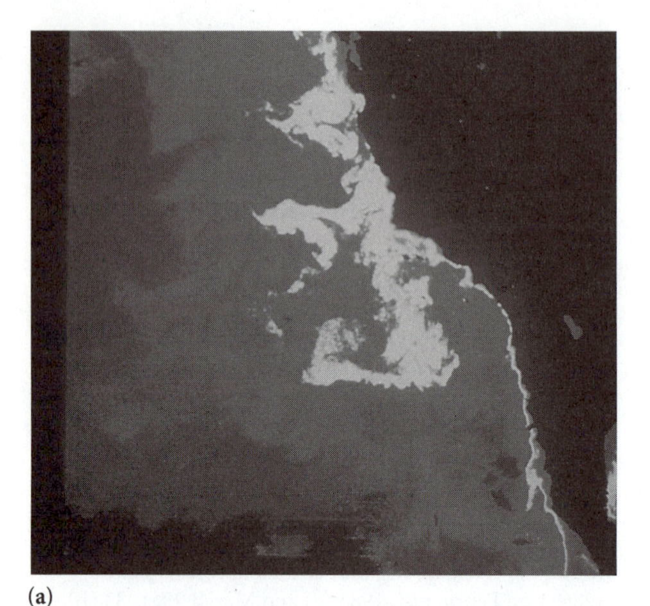

(a)

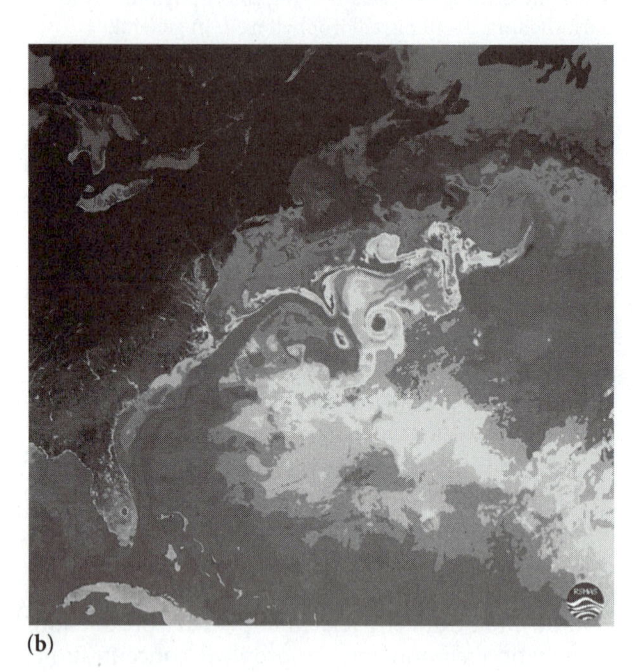

(b)

FIGURE 10-4 Many kinds of information about escosystem structure and function can be obtained from satellite images. (a) This image of the California coast from June 15, 1981 shows concentrations of phytoplankton (light areas) near the coast and in large eddy currents that extend out from the coast. (b) This image of the North Atlantic Ocean was taken during the first week of June 1984. Here, warm water is indicated by lighter shades and cold water by darker shades. The image shows that the Gulf Stream not only transports considerable heat to the vicinity of northern Europe, but also conveys a tropical community of marine organisms far to the north. (*a, courtesy of J. A. McGowan, from McGowan et al. 1986; b, courtesy of O. Brown, R. Evans, and M. Carle, University of Miami Rosentiel School of Marine and Atmospheric Science.*)

10.3 The rate of photosynthesis varies in relation to light, temperature, and availability of water and nutrients.

In terrestrial plants, the rate of photosynthesis is influenced by the intensity of the incident solar radiation, the temperature, the ability of the plant to tolerate water loss, and the availability of nutrients.

Light Intensity and Temperature

The rate of photosynthesis varies in direct proportion to light at low intensities, usually less than one-fourth that of full sunlight (Ryther 1956, Kramer 1958, Mooney and Billings 1961, Loach 1967). Brighter light saturates the photosynthetic pigments, and, as light intensity increases, the rate of photosynthesis increases more slowly or levels off (Berry 1975). In many plants, extremely bright light impairs photosynthesis because photosynthetic reactions are deactivated. Moreover, in the presence of bright light, the enzymes responsible for CO_2 assimilation also enhance the oxidation of carbohydrates to CO_2 and water, a process referred to as **photorespiration** (Ogren 1984, Downes and Hesketh 1968).

The response of photosynthesis to light intensity has two reference points. The first, called the **compensation point,** is the level of light intensity at which photosynthesis and respiration balance each other. Above the compensation point, the energy balance of a plant is positive; below it, its energy balance is negative. The second reference point is the **saturation point,** above which the rate of photosynthesis no longer responds to increasing light intensity. Among terrestrial plants, the compensation points of species that normally grow in full sunlight (approximately 500 watts per square meter annually) occur between 1 and 2 W m^{-2}. The saturation points of such species usually are reached between 30 and 40 W m^{-2}, less than a tenth of the energy level of bright, direct sunlight.

Like most other physiological processes, photosynthesis proceeds most rapidly within a narrow range of temperatures. The optimum temperature varies with environment, from about 16° C in many temperate species to as high as 38° C in tropical species. Optimum temperature also varies with light intensity in some species, such as the alpine heath *Loiseleuria* of Austria (Figure 10-5). Net production depends on the rate of respiration as well as that of photosynthesis, and respiration generally increases with increasing leaf temperature (Kramer 1958).

Photosynthetic efficiency is the percentage of the energy in incident radiation converted to net primary production during the growing season. It provides a useful index to rates of primary production under natural conditions. Where water and nutrients do not

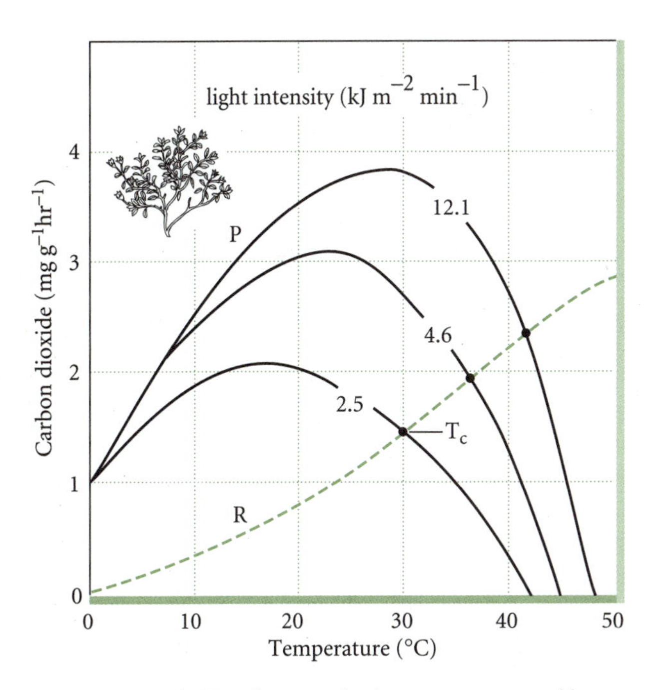

FIGURE 10-5 Net photosynthetic rate, as measured by uptake of carbon dioxide (mg g^{-1} h^{-1}) as a function of leaf temperature, at three light intensities (solid lines denoted by P) in the heath *Loiseleuria*. The dashed line shows the level of respiration, which increases with temperature. A compensation point exists wherever the dashed line crosses a solid line (denoted T_c for the lowest light level) (*After Larcher et al. 1975.*)

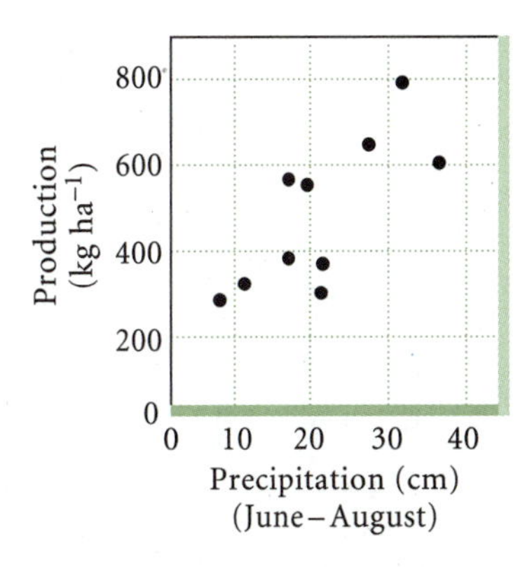

FIGURE 10-6 The relationship between annual production and summer precipitation on perennial grassland in southern Arizona. (*After Cable 1975.*)

severely limit plant production, photosynthetic efficiency varies between 1% and 2%.

What happens to the remaining 98–99% of the energy? Leaves and other surfaces reflect anywhere from one-fourth to three-fourths of the incident energy from the sun. Molecules other than photosynthetically active pigments absorb most of the remainder, which is converted to heat and either radiated or conducted across the leaf surface or dissipated by the evaporation of water from the leaf in the process of transpiration. For example, in a study of an oak forest, annual incident photosynthetically active radiation was 1.9×10^6 kJ m^{-2} (an average of 60 W m^{-2}). Of this, 56% was absorbed by leaves. Annual transpiration was approximately 0.5×10^6 kJ m^{-2}. At 2.24 kJ g^{-1} of water evaporated (the heat of vaporization), transpiration accounted for 1.1×10^6 kJ m^{-2}, which was approximately the total light energy absorbed (Loucks 1977).

Transpiration Efficiency

The tiny openings, called stomates, through which leaves exchange carbon dioxide and oxygen with the atmosphere also allow the passage of water vapor (transpiration). As the moisture content of soil decreases, plants obtain water with increasing difficulty, as we discussed in some detail in Chapter 4. As soil moisture approaches the wilting point, leaves close their stomates to reduce water loss; this prevents uptake of CO_2, and photosynthesis slows to a standstill. Consequently, the rate of photosynthesis depends on a plant's ability to tolerate water loss, the availability of soil moisture, and the influence of air temperature and solar radiation on the rate of transpiration (Kramer 1969; Fisher and Turner 1978).

Agronomists have devised the measure of **transpiration efficiency,** also called **water use efficiency,** which is the number of grams of dry matter produced (net production) per kilogram of water transpired. In most plants, transpiration efficiencies are less than 2 grams of production per kilogram of water transpired, but they may be as high as 4 g kg^{-1} in drought-tolerant crops (Odum 1971). Because transpiration efficiency varies within narrow limits across a wide variety of plants, production is directly related to water availability in the environment (Webb et al. 1978). For example, the annual production of perennial grasses in southern Arizona varies in direct relation to summer rainfall (Figure 10-6).

Nutrients and Plant Production

Nutrients stimulate plant growth in most habitats. When nitrogen and phosphorus fertilizers were applied singly and in combination to chaparral habitats in southern California, most species responded to the application of nitrogen, but not to that of phosphorus, with increased production (McMaster et al. 1982) (Figure 10-7). This result suggests that production in most chaparral species is limited by the availability of nitrogen. However, the growth of California lilac bushes (*Ceanothus greggii*), which harbor nitrogen-fixing bacteria in their root systems, responded to the application of phosphorus, but not to that of nitrogen. The productivity of annual plants (forbs and grasses) in the same habitat increased when nitrogen was applied, but was depressed somewhat with the

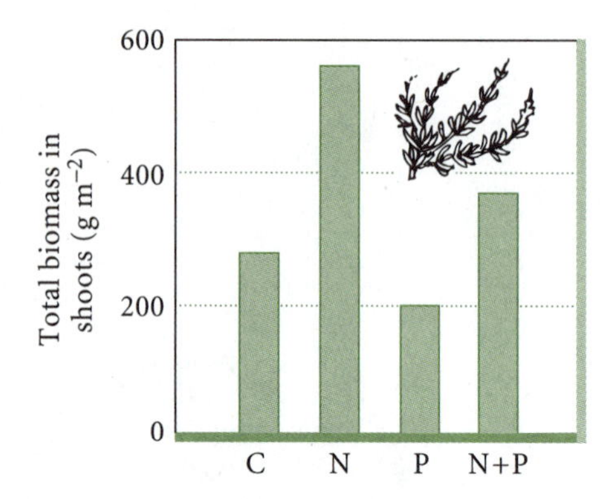

FIGURE 10-7 Response of the chaparral shrub *Adenostoma* to nitrogen (N) and phosphorus (P) fertilization. (C = control.) (*After McMaster, et al. 1982.*)

application of phosphorus alone. When, however, equal amounts of the two elements were applied together, however, production soared. Evidently, the plants could take advantage of increased phosphorus only in the presence of high levels of nitrogen. We will discuss the different roles of nitrogen and phosphorus in plant production more thoroughly in Chapter 13.

10.4 The productivity of terrestrial and aquatic ecosystems is limited by different ecological factors.

Productivity—the rate of production—varies greatly from one ecosystem to another due to a variety of ecological factors. The highest productivity on earth is found in the humid Tropics due to the favorable combination of intense sunlight, warm temperatures, and abundant rainfall. Conversely, low winter temperatures and long winter nights curtail production in temperate and arctic ecosystems. Within a given latitude belt, where light and temperature do not vary appreciably from one locality to the next, net production varies with annual precipitation. For example, Pandey and Singh (1992) found that net primary production in a dry tropical savanna ranged from 329 to 741 g m^2 yr^{-1}, with higher values associated with higher rainfall. Webb and coworkers (1978) have shown that above a certain threshold of water availability, production increases by 0.4 grams per kilogram of water in hot deserts and by 1.1 g kg^{-1} in short-grass prairies and cold deserts. Thus, a given

TABLE 10-2 Average net primary production and related dimensions of the earth's major ecosystem types

Ecosystem	Net primary production ($g\ m^{-2}\ yr^{-1}$)	Biomass ($kg\ m^{-2}$)	Chlorophyll ($g\ m^{-2}$)	Leaf surface area ($m^2\ m^{-2}$)	Biomass accumulation ratio (yr)
Terrestrial					
Tropical forest	1,800	42	2.8	7	23
Temperate forest	1,250	32	2.6	8	26
Boreal forest	800	20	3.0	12	25
Shrubland	600	6	1.6	4	10
Savanna	700	4	1.5	4	6
Temperate grassland	500	1.5	1.3	4	3
Tundra and alpine	140	0.6	0.5	2	4
Desert scrub	70	0.7	0.5	1	10
Cultivated land	650	1	1.5	4	1.5
Swamp and marsh	2,500	15	3.0	7	6
Aquatic					
Open ocean	125	0.003	0.03	—	0.02
Continental shelf	360	0.01	0.2	—	0.03
Algal beds and reefs	2,000	2	2.0	—	1.00
Estuaries	1,800	1	1.0	—	0.56
Lakes and streams	500	0.02	0.2	—	0.04

(*From Whittaker and Likens 1973.*)

amount of water supports almost three times as much plant production in cold climates as in hot climates. Forest ecosystems, whose transpiration efficiencies are on the order of 0.9–1.8 g kg^{-1}, are relatively insensitive to variation in water availability, but do not develop unless precipitation exceeds at least 50 cm annually in the cooler regions of the United States and perhaps 100 cm in the warmer eastern and southeastern areas.

Ecologists R. Whittaker and G. Likens (1973b) estimated the net primary production for representative terrestrial and aquatic ecosystems (Table 10-2). Their values were based on the results of many studies employing a wide variety of techniques, but what they lack in strict comparability does not override the general patterns they reveal. Whittaker and Likens's summary shows that production of terrestrial vegetation is greatest in the wet Tropics and least in tundra and desert habitats. Swamp and marsh ecosystems, which occupy the interface between terrestrial and aquatic habitats, can be more productive than tropical forests because of the continuous abundance of water and the rapid regeneration of nutrients in the mucky sediments surrounding the plant roots.

The open ocean is a virtual desert, where scarcity of mineral nutrients limits productivity to a tenth or less that of temperate forests (see Table 10-2). Upwelling zones, where nutrients are brought to the surface from deeper waters, and continental shelf areas, where exchange between shallow, nutrient-rich bottom sediments and surface waters occurs, support greater production. In shallow estuaries, coral reefs, and coastal algal beds, production approaches that of adjacent terrestrial habitats (Bunt 1973, Mann 1973, Barnes and Mann 1980, McLusky 1981). Primary production in freshwater habitats compares favorably with that in marine habitats, being greatest in rivers, shallow lakes, and ponds and least in clear streams and deep lakes. In general, phosphorus limits production in freshwater systems; nitrogen limits production in marine systems.

10.5 Ecological efficiencies characterize the movement of energy along the food chain.

Plants manufacture their own "food" from raw inorganic materials. Hence they are referred to as **autotrophs** (literally, "self-nourishers"). Animals and most microorganisms, which obtain their energy and most of their nutrients by eating plants, animals, or their dead remains, are called **heterotrophs** (literally, "nourished from others"). The dual roles of living organisms as food producers and food consumers give the ecosystem a trophic structure, determined by feeding relationships, through which energy flows and nutrients cycle. Most ecosystems contain a myriad of interconnected trophic interactions, which, taken as a whole, are referred as a **food web.** The food chain from grass to caterpillar to sparrow to snake to hawk delineates one particular path of energy through the food web. With each link in a food chain, a great deal of energy is dissipated before it can be consumed by organisms feeding at the next higher trophic level. All the grass in Africa piled together would dwarf a mound of all the grasshoppers, gazelles, zebras, wildebeests, and other animals that eat grass. But that mound of herbivores would be overwhelming beside the pitiful heap of all the lions, hyenas, and other carnivores that feed on them.

FIGURE 10-8 An ecological pyramid. The breadth of each bar represents the net productivity of each trophic level in the ecosystem. For this particular system, ecological efficiencies are 20%, 15%, and 10% between trophic levels, but these values vary widely among communities and ecosystems.

As Lindeman pointed out in 1942, the amount of energy reaching each trophic level depends on the net primary production at the base of the food chain and on the efficiencies with which animals convert food energy into their own biomass energy through growth and reproduction at each higher trophic level. Of the light energy assimilated by photosynthesis, plants use between 15% and 70% for their own maintenance, thereby making that portion unavailable to consumers. Herbivores and carnivores are more active than plants and expend correspondingly more of their assimilated energy on maintenance. As a result, the productivity of each trophic level is typically only 5–20% that of the level below it (Figure 10-8). Ecologists refer to the percentage of energy transferred from one trophic level to the next as the **ecological efficiency** or, synonymously, the **food chain efficiency,** between those two levels.

10.6 The individual link in the food chain is the basic unit of trophic structure.

Once consumed, food energy follows a variety of paths through the organism (Figure 10-9). Regardless of the source of its food, what the organism digests and absorbs constitutes its **assimilated energy,** which supports maintenance, builds tissues, or may be excreted in the form of unusable metabolic by-products. Animals excrete another, usually smaller, portion of their assimilated energy as nitrogen-containing organic wastes, primarily ammonia, urea, or uric acid, produced when the diet contains an excess of nitrogen: this is called the **excreted energy.** The portion of the assimilated energy used to fulfill metabolic needs, most of which escapes the organism as heat, makes up the **respired energy.** The assimilated energy retained by the organism becomes available for the synthesis of new biomass (production) through growth and reproduction, which animals feeding at the next higher trophic level may then consume. Many components of food resist digestion and assimilation: chitin in insect exoskeletons, hair, feathers, cartilage, bone, and some constituents of the skin such as keratin in animal foods, and cellulose and lignin in plant foods. These substances may be defecated or regurgitated, and the energy they contain is referred to as **egested energy.** Egested energy is not wasted, however. Some organisms are specialized for digestion of these recalcitrant materials. Much of the egested material becomes part of the soil organic matter, along with the bodies of dead plants and animals. These materials are collectively called **detritus.** In the soil, they enter a second, **detritus-based food chain.**

10.7 Assimilation and production efficiencies determine ecological efficiency.

Ecological efficiency is the product of the efficiencies with which organisms exploit their food resources and convert them into biomass: the exploitation, assimilation, and net production efficiencies (Table 10-3). The **exploitation efficiency** is the efficiency with which the

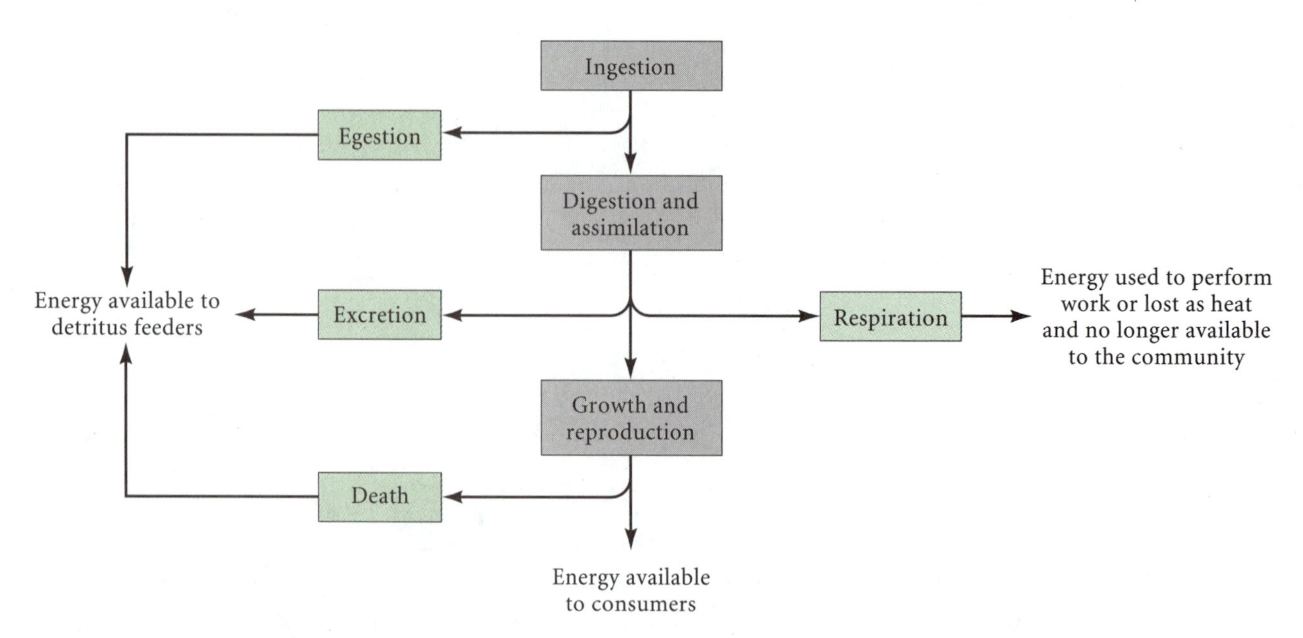

FIGURE 10-9 Allocation of energy within one link of a food chain. Energy is ingested, digested, and used for growth and reproduction. Energy is lost through egestion, excretion, and, ultimately, by death.

TABLE 10-3 Definitions of several energetic efficiencies

$$\text{Exploitation efficiency} = \frac{\text{Ingestion of food}}{\text{Prey production}}$$

$$\text{Assimilation efficiency} = \frac{\text{Assimilation}}{\text{Ingestion}}$$

$$\text{Net production efficiency} = \frac{\text{Production (growth and reproduction)}}{\text{Assimilation}}$$

$$\text{Gross production efficiency} = \text{Assimilation efficiency} \times \text{Net production efficiency} = \frac{\text{Production}}{\text{Ingestion}}$$

$$\text{Ecological efficiency} = \text{Exploitation efficiency} \times \text{Assimilation efficiency} \times \text{Net production efficiency} = \frac{\text{Consumer production}}{\text{Prey production}}$$

biological production of an entire trophic level is consumed. For a particular ecosystem, the exploitation efficiency of the prey would be the proportion of the total prey biomass that is eaten, or exploited, by predators. Because most biological production is consumed by one organism or another, the exploitation efficiency approaches 100%. But only part of what is consumed is assimilated by the consumer. The **assimilation efficiency** is the proportion of the consumed energy that is assimilated. The efficiency with which assimilated energy is incorporated into growth, storage, and reproduction is referred to as the **net production efficiency.**

For plants, net production efficiency is defined as the ratio of net to gross production. This index has been found to vary between 30% and 85%, depending on habitat and growth form. Rapidly growing plants in temperate zones, whether trees, old-field herbs, or crop species, have uniformly high net production efficiencies (75–85%). Similar vegetation types in the Tropics exhibit lower net production efficiencies, perhaps 40–60%. As we would expect, respiration increases relative to photosynthesis at low latitudes.

The nutritional value of plant foods depends upon the amount of cellulose, lignin, and other indigestible materials present (Grodzinski and Wunder 1975). Herbivores assimilate as much as 80% of the energy in seeds, and 60–70% of that in young vegetation (Chew and Chew 1970). Most grazers and browsers (elephants, cattle, grasshoppers) assimilate 30–40% of the energy in their food. Millipedes, which eat decaying wood composed mostly of cellulose and lignin (and the microorganisms that occur in decaying wood), assimilate only 15% (O'Neill 1968).

The effect of food quality on assimilation can be seen in the efficiencies with which a single animal retains energy from different portions of its diet. In studies with mountain hares (*Lepus timidus*), Pehrson (1983) determined that the efficiency of energy assimilation on a diet of small willow twigs was 39%, of which 5% was lost through urinary nitrogen excretion. Assimilation efficiencies were lower for larger twigs (31%), presumably because of their thick, less digestible bark, and for birch twigs (23–35%), on which the hares could not maintain a constant weight. By measuring the fiber content (cellulose and lignin) of the hares' food and feces, Pehrson found that the digestibility of fiber was between 15% and 25%. Assimilation of nutritional components of the diet, estimated by similar input-output measurements, varied between 9% for phosphorus and 81% for magnesium (Figure 10-10).

Food of animal origin is more easily digested than food of plant origin; assimilation efficiencies of predatory species vary between 60% and 90%. Vertebrate prey are digested more efficiently than insect prey because the indigestible exoskeletons of insects constitute a larger proportion of the body than the hair, feathers, and scales of vertebrates. Assimilation efficiencies of insectivores vary between 70% and 80%, whereas those of most carnivores are about 90%. Maintenance, movement, and heat production require energy that animals otherwise could use for growth and reproduction. Active warm-blooded animals exhibit low net production efficiencies: birds less than

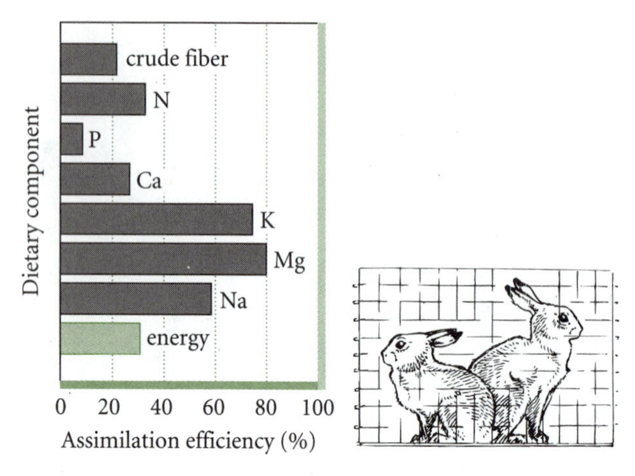

FIGURE 10-10 Proportions of dietary components assimilated by caged mountain hares (*Lepus timidus*) during the winter. (*After Pehrson 1983.*)

1%, and small animals with high reproductive rates up to 6%. More sedentary cold-blooded animals, particularly aquatic species, channel as much as 75% of their assimilated energy into growth and reproduction. The **gross production efficiency,** which is the product of the assimilation efficiency and net production efficiency, represents the overall efficiency of biomass production within a trophic level. The gross production efficiencies of warm-blooded terrestrial animals rarely exceed 5%, and those of some birds and large mammals fall below 1%. For insects, gross production efficiencies lie within the range of 5% to 15%, and for some aquatic animals they exceed 30% (Figure 10-11).

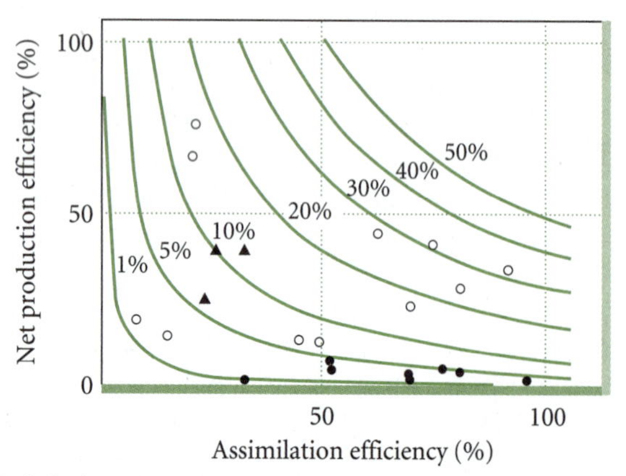

FIGURE 10-11 Relationships between assimilation efficiency and net production efficiency for a variety of animals. Gross production efficiencies are indicated by the curved lines on the graph. Homeotherms are warm-blooded animals; poikilotherms have variable body temperatures. (*From Ricklefs 1979.*)

10.8 Terrestrial ecosystems are dominated by detritus-based food chains.

Terrestrial plants, especially woody species, allocate much of their production to structures that are difficult to ingest, let alone digest. As a result, most terrestrial plant production ends up as dead organic matter or detritus. In forests, for example, it is estimated that 90% or more of the plant biomass produced becomes detritus. Detritus is consumed by a host of organisms, including many kinds of worms, snails, insects, mites, bacteria, and fungi. The larger of these organisms (worms, snails, etc.) assimilate only about 30–45% of the energy available in leaf litter, and even less from wood, but they speed the decay of litter because they macerate plant detritus in their digestive tracts, and the finer particles in their egested waste expose new surfaces to microbial feeding. The mechanical breakdown of detritus by large detritivores increases the overall consumption of that material. In an experiment in which large detritivores were excluded from detritus material, the actions of bacteria and fungi resulted in the processing of only about 25% of the material (Figure 10-12).

The partitioning between herbivory and detritus feeding establishes two food chains in terrestrial communities. The first originates with relatively large animals feeding on leafy vegetation, fruits, and seeds; the second originates with relatively small animals and microorganisms consuming detritus in the litter and soil layer. These separate food chains sometimes mingle considerably at higher trophic levels, but the energy of

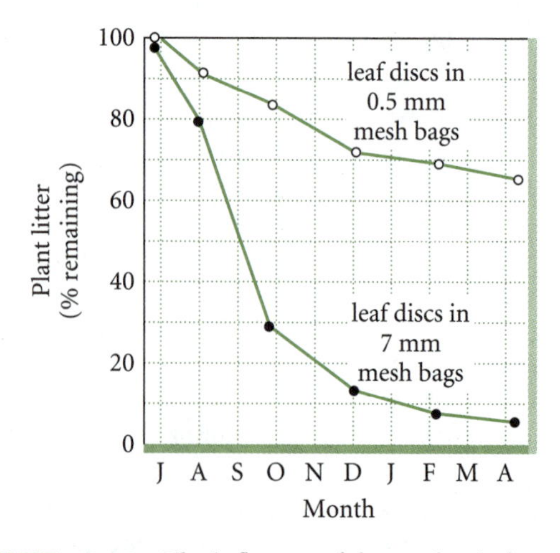

FIGURE 10-12 The influence of the mechanical breakdown of plant litter by large detritus feeders on its overall consumption. Leaf discs (small cutouts of leaves) were enclosed in mesh bags with either large (7 mm) or small (0.5 mm) openings. The small openings admitted bacteria, fungi, and small arthropods, but excluded such larger detritivores as earthworms and millipedes. (*After Edwards and Heath 1963; in Phillipson 1966.*)

detritus tends to move into the food chain much more slowly than the energy consumed by herbivores.

The relative importance of predatory-based and detritus-based food chains also varies greatly among communities. Predators predominate in plankton-dominated aquatic communities, detritus feeders in terrestrial communities. The proportion of net production that enters herbivore-predator strands of the food web depends on the relative allocation of plant tissue between structural and supportive functions on the one hand and growth and photosynthetic functions on the other. Herbivores consume 1.5–2.5% of the net production of temperate deciduous forests, and 12% of that in old-field habitats. Herbivore consumption in aquatic communities is usually much higher, as much as 60–99% in some plankton communities (Wiegert and Owen 1971). But recent work suggests that detritus-based food chains may also be important in some aquatic systems (Wetzel 1995).

10.9 How long does energy take to flow through the ecosystem?

Food chain efficiencies indicate the amount of energy that eventually reaches each trophic level of the ecosystem. The rate of transfer of energy, or inversely, its **residence time** in each trophic level, provides a second index to the energy dynamics of the ecosystem. For a given rate of production, the residence time, also called the **transit time,** of energy in the community and the storage of energy in living biomass and detritus are directly related: the longer the residence time, the greater the accumulation of energy.

The average residence time in a particular link of the food chain is equal to the energy stored in biomass divided by the rate at which energy is converted into biomass—the net productivity:

$$\text{Residence time (yr)} = \frac{\text{Energy stored in biomass (kJ m}^{-2})}{\text{Net productivity (kJ m}^{-2}\text{ yr}^{-1})}.$$

We can also calculate the residence time defined by this equation in terms of mass (g m^{-2}) rather than energy, in which case it expresses the **biomass accumulation ratio.** Wet tropical forests produce an average of 1,800 g m^{-2} yr^{-1} and have an average living biomass of 42,000 g m^{-2} (Whittaker and Likens 1973a). Thus, the biomass accumulation ratio for wet tropical forests is 23 years (42,000/1,800). Biomass accumulation ratios for representative ecosystems may be more than 20 years in forested terrestrial environments and less than 20 days in aquatic plankton-based ecosystems (Table 10-4). These figures are only averages for the system as a whole. Some energy remains in the system longer; much of it disappears more quickly. For example, leaf eaters and root feeders consume much of the energy assimilated by forest trees during the year of its production—some of it within days of assimilation by the plant (Lawton 1994).

TABLE 10-4 Average transit time of energy in living plant biomass (biomass/net primary production) for representative ecosystems

Ecosystem	Net primary production (g m^{-2} yr^{-1})	Biomass (g m^{-2})	Transit time (yrs)
Tropical rain forest	2,000	45,000	22.5
Temperate deciduous forest	1,200	30,000	25.0
Boreal forest	800	20,000	25.0
Temperate grassland	500	1,500	3.0
Desert scrub	70	700	10.0
Swamp and marsh	2,500	15,000	6.0
Lake and stream	500	20	0.04*
Algal beds and reefs	2,000	2,000	1.0
Open ocean	125	3	0.024†

** 15 days.*

† 9 days.

(Data from Whittaker and Likens 1973.)

The figures in Table 10-4 also underestimate the average residence time of energy in plant biomass because they do not include the accumulation of dead organic matter in the litter. The residence time of energy in accumulated litter can be determined by the equation:

$$\text{Residence time (yr)} = \frac{\text{Litter accumulation } (\text{g m}^{-2})}{\text{Rate of litterfall } (\text{g m}^{-2}\,\text{yr}^{-1})}.$$

In forests, the average litter biomass turnover time varies from 3 months in the wet Tropics to 1 to 2 years in dry and montane tropical habitats, 4 to 16 years in the southeastern United States, and more than 100 years in temperate mountains and boreal regions (Olson 1963). Warm temperatures and an abundance of moisture in lowland tropical regions create optimum conditions for rapid decomposition.

10.10 Energy transfer and accumulation describe the structure and function of ecosystems.

The flux of energy and its efficiency of transfer summarize certain aspects of the structure of an ecosystem: the number of trophic levels, the relative importance of detritus, herbivore, and predatory feeding, the steady-state values for biomass and accumulated detritus, and the turnover rates of organic matter in the community. The importance of these measures was argued by Lindeman (1942), who constructed the first energy budget for an entire biological community—that of Cedar Bog Lake in Minnesota. The proliferation of energy flow studies during the 1950s and 1960s more clearly reflected energy's value as a universal currency—a common denominator to which all populations and their acts of consumption could be reduced.

The overall energy budget of an ecosystem reflects the balance between income and expenditure, just as it does in a bank account. The ecosystem gains energy through photosynthetic assimilation of light by green plants and the transport of organic matter into the system from outside, as when a leaf is blown into a stream. As we saw in Chapter 9, inputs of the latter type are referred to as allochthonous inputs; local production is referred to as autochthonous input.

G. W. Minshall (1978) surveyed the relative amounts of autochthonous and allochthonous inputs in aquatic systems. He suggested that large rivers, lakes, and most marine ecosystems are dominated by autochthonous production due to their resident populations of phytoplankton and large aquatic plants, whereas allochthonous inputs are most important in small streams and springs under the closed canopies of forests due to imports of plant debris falling onto these small bodies of water. While this generalization is widely accepted, recent studies suggest that in some lakes and rivers allochthonous inputs may be more important than previously imagined. The animal biomass of large rivers that periodically inundate their floodplains may be sustained principally by the floodplain production (an allochthonous source) (Junk et al. 1989), a situation that may also be true for smaller streams (Meyer 1990). A survey of twenty lakes in southern Quebec (del Giorgio and Peters 1994) showed that, in all cases, the amount of carbon respired by the plankton community exceeded the amount fixed by phytoplankton photosynthesis. The researchers suggested that the shortfall in respiratory carbon needs was made up by allochthonous dissolved organic carbon.

Lindeman constructed his Cedar Bog Lake energy budget from measurements of the harvestable net production at each of three trophic levels (all carnivores were lumped together) and from laboratory determinations of respiration and assimilation efficiencies. The animals and plants collected at the end of the growing season constituted the net production of the trophic levels to which each species was assigned (Table 10-5).

Lindeman estimated the energy dissipated by respiration from ratios of respiratory metabolism to production measured in the laboratory: 0.33 for aquatic plants, 0.63 for herbivores, and 1.4 for the more active carnivores in the lake. He calculated the gross production of carnivores as the sum of their harvestable production (54 kJ $\text{m}^{-2}\,\text{yr}^{-1}$) and respiration (54 × 1.4 =

TABLE 10-5 Initial measurements for Lindeman's calculation of energy flow for Cedar Bog Lake, Minnesota

Trophic level	Harvestable production ($\text{kJ m}^{-2}\,\text{yr}^{-1}$)	Harvestable production (W m^{-2})
Primary producers (green plants)	2,944	0.0934
Primary consumers (herbivores)	293	0.0093
Secondary consumers (carnivores)	54	0.0017

TABLE 10-6 An energy flow model for Cedar Bog Lake, Minnesota

Energy production or removal	Energy ($kcal\ m^{-2}\ yr^{-1}$)		
	Primary producers	Primary consumers	Secondary consumers
Harvestable production*	704	70	13
Respiration	234	44	18
Removal by consumers			
Assimilated	148	31	0
Unassimilated	28	3	0
Gross production (totals)	1,114	148	31

**Does not include net production removed by consumers. Actual net production, including removal by consumers, was 879 kcal m^{-2} yr^{-1} for primary producers, 104 kcal m^{-2} yr^{-1} for primary consumers, and 13 kcal m^{-2} yr^{-1} for secondary consumers.*

(From Lindeman 1942.)

76 kJ m^{-2} yr^{-1}), a total of 130 kJ m^{-2} yr^{-1} (Table 10-6). He then estimated the gross production of primary consumers as the sum of their harvestable production, respiration (production × 0.63), and the consumption of primary consumers by secondary consumers. In making the last calculation, Lindeman assumed that the assimilation efficiencies of secondary consumers were 90%. By backtracking another step in this manner, he calculated the production of primary producers as well.

Lindeman's energy budget is somewhat startling in that organisms at the next higher trophic level failed to consume 83% of the net primary production and 70% of the net secondary production. This surplus production is transported out of the system by sedimentation; the lake is filling with layers of organic detritus.

Even with sedimentation, the overall ecological efficiency of energy transfer between trophic levels in Cedar Bog Lake was about 12%. After comparing similar analyses for five aquatic communities, Kozlovski (1968) concluded that (1) assimilation efficiency increases at higher trophic levels; (2) net production efficiency decreases at higher trophic levels; (3) gross production efficiency also decreases; and (4) ecological efficiency (assimilation or gross production at level n divided by that at level $n - 1$) remains constant between trophic levels at about 10%.

10.11 The length of food chains is limited by ecological efficiencies.

Kozlovski's 10% generalization is not a fixed law of ecological thermodynamics. In Silver Springs, Florida, Odum (1957) measured ecological efficiencies of 17% between the producer and herbivore levels but only 5% between herbivores and secondary consumers. Ecological efficiencies are usually lower in terrestrial habitats, and a useful rule of thumb states that the top carnivores in terrestrial communities can feed no higher than the third trophic level on average, whereas aquatic carnivores may feed as high as four or five levels (Fenchel 1988). This is not to say that there can be no more than three links in a terrestrial food chain; a tiny fraction of the total energy may travel through a dozen links before it is dissipated by respiration. Such high trophic levels usually do not, however, contain enough energy to fully support a single predator population.

We can estimate the average length of food chains in a community from net primary production, average ecological efficiency, and average energy flux of a top predator population. The energy available [$E(n)$] to a predator at a given trophic level n (plants being level 1) is equal to the product of the net primary production (NPP) and the intervening ecological efficiencies (Eff). Thus

$$E(n) = (\mathrm{NPP})(\mathrm{Eff}^{n-1}),$$

where Eff is the geometric mean of the efficiencies of transfer between each level. The geometric mean is the $(n - 1)$th root of the product of the efficiencies:

$$\sqrt[n-1]{(f_1 f_2 \ldots f_{n-1})}.$$

Using the fact that $\log(a)(b)^n = \log(a) + n\log(b)$, we can obtain the following intermediate expression from the equation for $E(n)$ above:

$$\log E(n) = \log(\mathrm{NPP}) + (n - 1)\log(\mathrm{Eff}).$$

This expression may then easily be solved for n, the number of trophic levels, in terms of mean ecological efficiency and net primary production:

$$n = 1 + \frac{\log[E(n)] - \log(\mathrm{NPP})}{\log(\mathrm{Eff})}.$$

TABLE 10-7 Average number of trophic levels in various ecosystem types calculated from primary production, consumer energy flux, and ecological efficiencies

Community	Net primary production (kcal m^{-2} yr^{-1})	Predator ingestion (kcal m^{-2} yr^{-1})	Ecological efficiency (%)	Number of trophic levels
Open ocean	500	0.1	25	7.1
Coastal marine	8,000	10.0	20	5.1
Temperate grassland	2,000	1.0	10	4.3
Tropical forest	8,000	10.0	5	3.2

Note: These values are approximations based on many studies.

Using this equation and some rough estimates for the values on its right-hand side, we can calculate the average number of trophic levels to be about 7 for marine plankton-based ecosystems, 5 for inshore aquatic communities, 4 for grasslands, and 3 for wet tropical forests (Table 10-7). These estimates should be taken with a grain of salt, to be sure, but they do indicate the general size of the pyramid of energy built upon a base of primary production within an ecosystem.

SUMMARY

1. The ecosystem is a giant thermodynamic machine that continually dissipates energy in the form of heat. This energy initially enters the biological realm of the ecosystem by photosynthesis and plant production, which is the ultimate source of energy for all animals and other nonphotosynthetic organisms.

2. Gross primary production is the total energy fixed by photosynthesis. Net primary production is the accumulation of energy in plant biomass; hence, it is the difference between gross production and plant respiration. Net ecosystem production is the difference between the gross primary production and the energetic costs of plants and soil organisms combined.

3. Primary production can be measured by one or some combination of a variety of methods: calorimetry, harvesting, measurements of gas exchange (carbon dioxide in terrestrial habitats, oxygen in aquatic habitats), assimilation of radioactive carbon (^{14}C), and production indices based on chlorophyll content.

4. The rate of photosynthesis varies in direct relation to light intensity up to the saturation point (usually 30–40 W m^{-2}), above which it levels off or decreases (bright sunlight has an intensity of approximately 500 W m^{-2}). The compensation point, above which photosynthesis exceeds respiration, occurs at light intensities of 1–2 W m^{-2} in terrestrial vegetation. The efficiency of photosynthesis (gross production/total incident light energy) is 1–2% in most habitats.

5. Because plants lose water in direct proportion to the amount of carbon dioxide they assimilate, plant production in dry environments is limited by, and varies in direct proportion to, availability of water. Transpiration efficiency, the ratio of net production (grams of dry matter) to water transpired (kilograms), is typically 1 to 2 grams, and rarely as high as 4 grams in drought-adapted species.

6. The movement of energy and materials through the food chain can be characterized by efficiencies of assimilation (assimilation/digestion) and of net production (production/assimilation). Ingested material that resists digestion and assimilation contains energy that is egested and becomes part of detritus-based food chains.

7. Assimilation efficiency depends on the quality of the diet, particularly the amount of digestion-resistant structural material it contains (cellulose, lignin, chitin, keratin), and varies from about 15% to 90%. Net production efficiencies are lowest—1% to 5%—in animals whose costs of maintenance and activity are greatest, especially endothermic vertebrates, in contrast to the values of 15% to 45% typical of invertebrates.

8. Gross production efficiencies (production/ingestion) vary between about 5% and 20% in most studies.

9. The average residence time of energy or biomass in a single link of the food chain is the ratio of biomass to the rate of net production. Residence times for primary production vary from 20 years in some forests to 20 days or less in aquatic plankton-based communities.

10. Considerations of energy flux and ecological efficiency suggest that the highest trophic level at which a consumer population can be maintained ranges from the fourth level in terrestrial food chains to the seventh level in plankton-based communities of the open ocean.

EXERCISES

1. Respiration among the plants in a small ecosystem requires 9,201 g of $C_6H_{12}O_6$ each day. How much CO_2 is required for photosynthesis in the ecosystem just to meet this demand?

2. Besides cellulose and lignin, a plant contains 5% carbohydrate, 8% protein, and 2% fat. What is the energy content of 1 g of the plant tissue?

3. Plants in a shaded area receive 100 W m^{-2} of incident sunlight annually. Assuming a photosynthetic efficiency of 1.5% and a reflectance of 26%, construct a bar diagram that shows the amount of the incident light (W m^{-2}) that (1) is used in photosynthesis, (2) is reflected, (3) is lost in transpiration (assume that transpiration is the only avenue of energy loss). What is the amount of energy (kJ g^{-1}) associated with each of these categories?

4. Consider the following allocation of energy within a trophic level of a food web (in arbitrary energy units): growth and reproduction = 600, assimilation = 720, ingestion = 1,000, prey production = 10,000. Calculate (1) exploitation efficiency, (2) assimilation efficiency, (3) net production efficiency, (4) gross production efficiency, and (5) ecological efficiency for the system (see Table 10.3). Based on the values of ecological efficiency given in Table 10-7, which type of ecosystem does this example best represent?

5. The productivity of a trophic level is 1,000 kJ m^{-2} yr^{-1}, and the residence time is 2 years. How many years will it take to accumulate 10,000 kJ m^{-2} in energy stored in biomass?

6. The geometric mean of ecological efficiencies for a particular ecosystem is 0.264, and NPP = 2,000 kcal m^{-2} yr^{-1}. How many trophic levels will the ecosystem have? If the geometric mean of the ecological efficiencies is 0.40, how many trophic levels will there be in the system?

CHAPTER 11

Pathways of Elements in Ecosystems

GUIDING QUESTIONS

- How do energy and materials flow through an ecosystem?
- How is energy flux related to carbon oxidation and reduction reactions?
- How is energy flux related to the global hydrologic cycle?
- What happens to the excess carbon that human activities have added to the carbon cycle?
- How might the oceans serve as a sink for carbon dioxide?
- What is the relationship between energy flow and the nitrogen cycle?
- How might the carbon and nitrogen cycles interact in the arctic tundra in response to increased atmospheric levels of carbon dioxide?
- How do anthropogenic changes in the sulfur cycle affect the problem of acid rain?
- What roles do microorganisms play in the carbon and nitrogen cycles?

We need only recall the transformations of carbon in photosynthesis and respiration to appreciate that the flux of energy in ecosystems is dependent on the movement and availability of elements and materials. Energy flows via biochemical transformations involving carbon, nitrogen, phosphorus, and many other constituents of the biosphere. The thermodynamic inefficiency of these transformations makes energy transient in the ecosystem, requiring a constant input of light energy from the sun. Nutrients, unlike energy, are retained within the ecosystem, where they are continually recycled among organisms and between organisms and the physical environment. Most of these nutrients originate in rocks of the earth's crust, but because they are released very slowly, any ecosystem that has high rates of organic production requires that materials assimilated by one organism be reused by others.

The biochemical transformations of an element determine the unique paths that it will take through the ecosystem. Elements cycle on at least two important spatial scales: local and global. Locally, within an ecosystem, processes of **mineralization**—the transformation of elements from organic to inorganic forms—and mobilization in the soil or the sediments affect the availability of nutrients to primary producers. Trophic interactions move materials among subcomponents of the ecosystem; litterfall and death return material to the soil, where the process of recycling begins. Globally, element cycles are characterized by the transport of materials over long distances between large reservoirs or pools. For instance, carbon is transported from the terrestrial compartment of the biosphere (one pool of carbon) to the atmosphere (another pool of carbon) by the process of respiration. Nitrogen may be removed from terrestrial ecosystems by runoff and river flow and carried to the oceans. In this chapter, we will turn our attention primarily to the global pathways of elements, focusing in particular on how modern element cycles are balanced. In Chapter 12 we will take up the issue of how nutrients are regenerated in the local environment.

The dynamics of global element cycles are of considerable importance in ecology. As we shall see, human activities have upset the delicate balance of a number of important elements such as carbon, nitrogen, phosphorus, and sulfur. In most cases, the imbalance results from the release into the biosphere of heretofore sequestered element pools. These new inputs alter global ecosystem function and may affect many of the natural climate patterns that we discussed in Chapter 8.

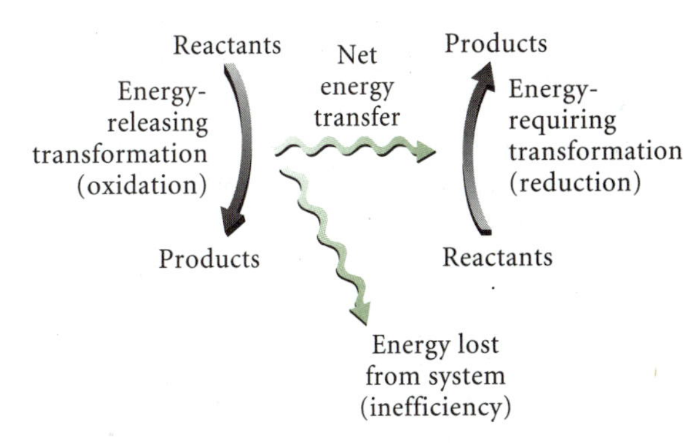

FIGURE 11-1 The coupling of energy-releasing and energy-requiring transformations is the basis of energy flow in the ecosystem. Some of the energy released in oxidation is used in transformations requiring energy. Some energy is lost as heat.

11.1 The movement of many elements parallels energy flow through the community.

Transformations that result in the production of organic forms of a particular element are often called **assimilatory** processes. Photosynthesis, by which inorganic carbon (carbon dioxide) is reduced to the organic carbon of carbohydrates, is the most obvious assimilatory transformation of an element, although others involving nitrogen and sulfur, for example, are equally important. In the overall cycling of carbon, photosynthesis is balanced by respiration, a complementary **dissimilatory** process that involves the oxidation of organic carbon, with the accompanying release of energy and the return of carbon to its available inorganic form.

Most energy transformations are associated with biochemical oxidation and reduction of carbon, oxygen, nitrogen, phosphorus, and sulfur. In each case, an energy-releasing transformation is coupled with an energy-requiring transformation, so that energy is transferred from the reactants in the first to the products in the second (Figure 11-1). When more energy is released by the first transformation than is required by the second, the unutilized energy is lost as heat—hence the thermodynamic inefficiency of life processes. A typical coupling between transformations might involve the oxidation of carbon in a carbohydrate (perhaps glucose), which releases energy, and the reduction of nitrate-nitrogen to amino-nitrogen, which requires energy. These particular transformations are not directly linked; many intermediate steps of the type portrayed in Figure 11-1 are required for the transfer of energy between them.

The initial input of energy into the ecosystem is accomplished by an assimilatory transformation—the reduction of carbon—in which the source of energy is light rather than a coupled dissimilatory process. A portion of that energy is lost with each subsequent transformation up the food chain (Figure 11-2). Some of these transformations involve the assimilation of other elements required for growth and reproduction; most involve biochemical transformations required for maintaining the cellular environment and for movement. The cycling of elements between biotic and abiotic compartments of the ecosystem is related to energy flow by the coupling of the dissimilatory part of one cycle to the assimilatory part of another.

Not all transformations of elements in the ecosystem are biological, nor do all involve the net assimilation or release of useful quantities of energy. Many chemical reactions take place in the air, soil, and water. Some of these, like the physical weathering of bedrock, release certain elements (potassium,

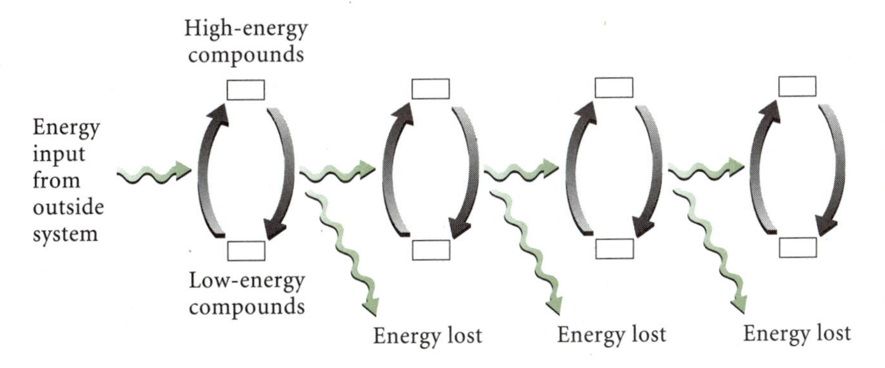

FIGURE 11-2 Energy is input into the ecosystem by the reduction of carbon, an assimilatory reaction. As energy flows through the ecosystem, elements alternate between assimilatory and dissimilatory transformations, with heat lost at each step.

phosphorus, and calcium, for example) to the ecosystem. Other physical and chemical processes, such as the sedimentation of calcium carbonate in the oceans, remove elements from ecosystem circulation to rocks in the earth's crust, where they can remain locked up for eons. The mining activities of humans release large quantities of phosphorus and other elements from these pools into the global ecosystem.

11.2 Elements cycle among compartments in the ecosystem.

Each form of an element can be thought of as occupying a separate compartment or pool in the ecosystem. Sometimes compartments are separated in space, like different rooms in a house, but often they are not. Biochemical transformations are fluxes of the elements among the compartments; that is, movement between rooms (Figure 11-3). Carbon, for example, occurs in the form of carbon dioxide both in the atmosphere (the atmospheric pool) and dissolved in water (the oceanic and freshwater pools), and as organic carbon in the plants and soil of the terrestrial realm (the terrestrial pool). Photosynthesis moves carbon from the atmospheric pool into the compartment containing organic forms of carbon (assimilation); respiration returns it to the atmosphere (dissimilation). The organic carbon compartment has many subcompartments: animals, plants, microorganisms, and detritus. Herbivory, predation, and detritus feeding move carbon between these subcompartments.

Some transformations involve changes in energy state. Photosynthesis adds energy to carbon in the form of chemical bonds, which may be thought of as lifting the element to the second floor of a house (raising its potential energy). In descending the respiration "staircase," carbon releases this stored chemical energy, which can then be used for other purposes.

Both organic and inorganic forms of elements may be removed from rapid circulation within the ecosystem to pools that are not readily accessible to transforming agents. For example, coal, oil, and peat contain vast quantities of carbon. Although this carbon remains within the ecosystem, it is not readily available for cycling. Sedimentation removes inorganic (oxidized) carbon from ecosystem circulation primarily by deposition as calcium carbonate, which forms thick layers of limestone over large areas presently or formerly covered by seas. These forms of carbon are returned to the cycling pools only by the slow geological processes of uplift and erosion. Of course, by burning fossil fuels, humans have greatly accelerated the rate at which stored organic carbon is returned to the active pool.

Sinks and Sources

Compartment models like those discussed here are tactical models. That is, in constructing such models, the goal is to represent the transformations of materials among the various compartments as accurately as possible. The two important attributes of these models are the sizes of the element pools and the rates of flux of materials among the compartments, which are related to one another in an important way. The size of a pool is the amount of the element of interest in that compartment. Thus, we might measure pool size in units of, for example, g/ha or

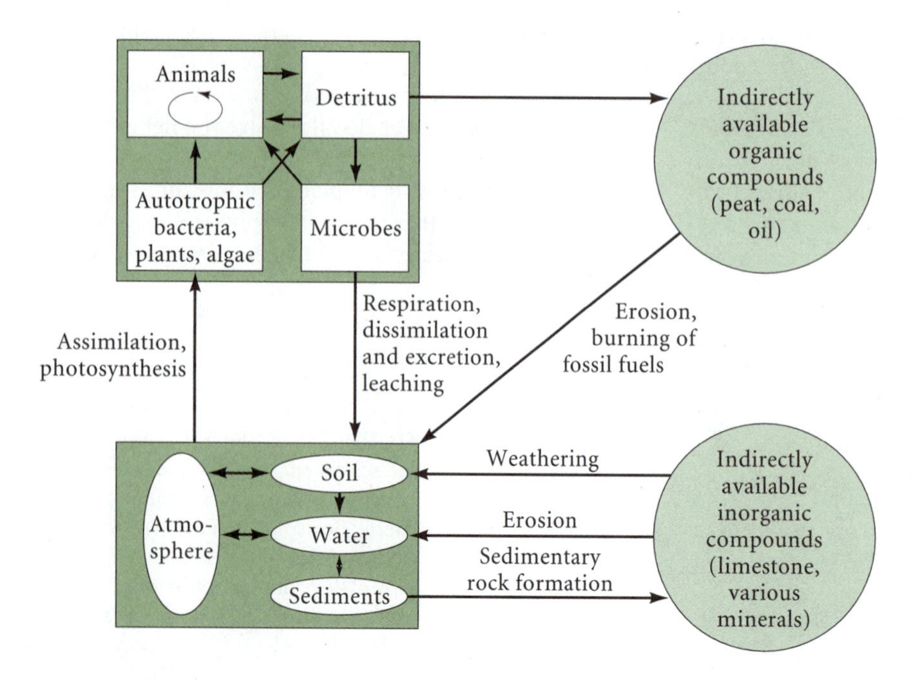

FIGURE 11-3 A generalized compartment model of the ecosystem. Elements reside in various compartments or pools. Movement within and among the compartments is accomplished by biochemical transformations.

mg/ha. Fluxes are measurements of rate and thus may be represented as g/yr or some other appropriate rate. If we compare the rate of input of an element to a pool to the rate of output from that pool, we can determine whether the pool is increasing in size, in which case it is often referred to as a **sink** for that element, or decreasing in size, in which case we say that it is a **source** for that element. The source-sink concept has other applications in ecology, as we shall see in Chapter 17.

11.3 The water cycle provides a physical model for element cycling in the ecosystem.

Although water is chemically involved in photosynthesis, most water flux through the ecosystem occurs by the physical processes of evaporation, transpiration, and precipitation (Figure 11-4). Light energy absorbed by water performs the work of evaporation. The condensation of atmospheric water vapor to form clouds releases the potential energy in that water vapor as heat. Thus, evaporation and condensation resemble photosynthesis and respiration thermodynamically, and they provide an instructive model for the cycling of nutrients within the ecosystem more generally.

Most of the water that circulates within the biosphere today originated through volcanic outpourings of steam from deep within the earth. The total available water at the earth's surface amounts to about 1.4 billion cubic kilometers, or $1{,}400{,}000 \times 10^{18}$ g. More than 97% of all available water resides in the oceans. Other reservoirs of available water are ice caps, glaciers ($29{,}000 \times 10^{18}$ g), underground aquifers ($8{,}000 \times 10^{18}$ g), lakes and rivers (100×10^{18} g), soil moisture (100×10^{18} g), water vapor in the atmosphere (13×10^{18} g), and all the water in living organisms (1×10^{18} g).

Over land surfaces, precipitation (111×10^{18} g yr^{-1}, which is 22% of the global total) exceeds evaporation and transpiration (71×10^{18} g yr^{-1}; 16% of the global total). Over the oceans, evaporation exceeds precipitation by a similar amount. Much of the water that evaporates from the surface of the oceans is carried by winds to the continents, where it is captured as precipitation by the land. This net flow of atmospheric water vapor from oceans to land (40×10^{18} g yr^{-1}) is balanced by runoff from the land back into the ocean basins.

We can calculate the energy that drives the global hydrologic cycle by multiplying the total weight of water evaporated (456×10^{18} g yr^{-1}) by the energy required to evaporate 1 g of water (2.24 kJ). The product, approximately 10^{21} kJ yr^{-1}, represents about one-quarter of the total energy of the sun's radiation striking the earth. Evaporation, not precipitation, determines the rate of movement of water through the ecosystem. The absorption of radiant energy by liquid water couples an energy source to the hydrologic cycle. Evaporation and precipitation are closely linked because the atmosphere has a limited capacity to hold water vapor; any increase in the evaporation of water into the atmosphere creates an excess of vapor and causes an equal increase in precipitation.

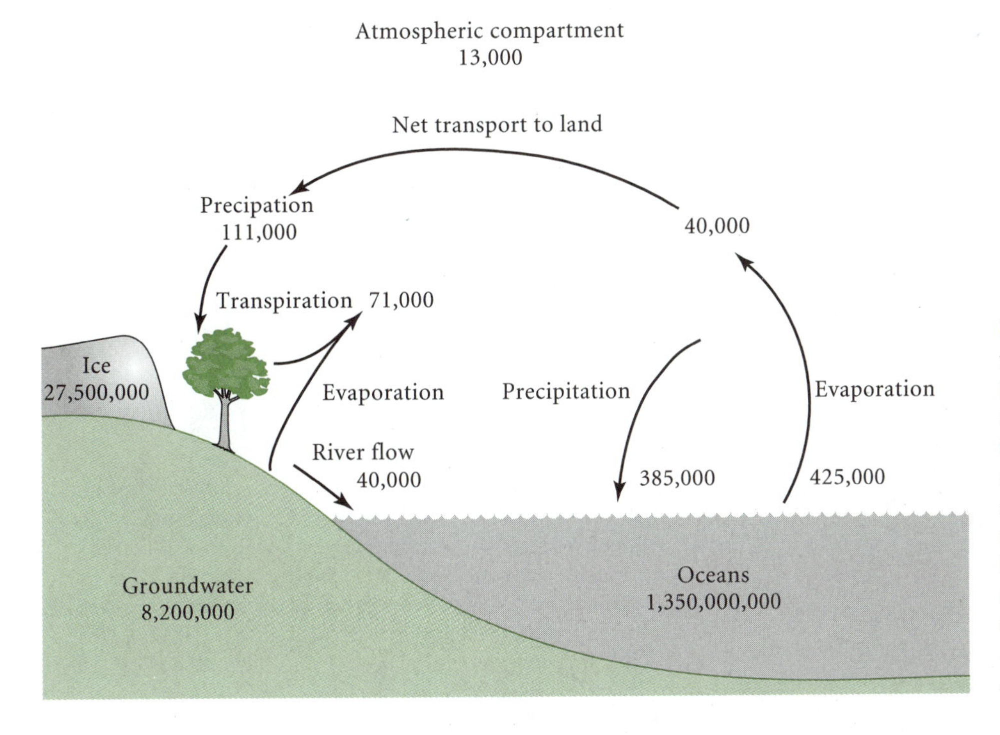

FIGURE 11-4 The global water cycle in the biosphere. Pools are expressed in units of volume (km^3), and transfer values are expressed as km^3/yr. Water that is sequestered in the rocks of the earth's core is not shown here. Most water is contained in the oceans. *(Redrawn from Schlesinger 1991; after Speidel and Agnew 1982.)*

The water vapor in the atmosphere at any one time (the pool size) corresponds to an average of 2.5 cm (1 in) of water spread evenly over the surface of the earth. An average of 65 cm (26 in) of rain or snow falls each year (the water flux), which is 26 times the average amount of water vapor. Thus, the steady-state content of water in the atmospheric compartment replaces itself 26 times each year on average. (Conversely, water has an average residence time in the atmosphere of $\frac{1}{26}$ of a year, or 2 weeks.) The soils, rivers, lakes, and oceans contain more than 100,000 times the amount of water in the atmosphere. Fluxes through both pools are the same, however, because evaporation balances precipitation. Thus the average residence time of water in liquid form at the earth's surface (about 2,800 years) is about 100,000 times longer than its average residence time in the atmosphere.

Human activities may affect the global water cycle and, thus, the local availability of water. Deforestation, agricultural practices, the alteration of wetland ecosystems near coastal areas, and the rerouting or damming of streams are some of the activities that may affect rates of evaporation, transpiration, and precipitation on a broad scale and, thus, alter the global water cycle. As we shall see below, inputs of materials such as CO_2 into the environment may change temperature regimes on earth and further alter the global water cycle.

11.4 The oxidation-reduction (redox) potential of a system indicates its energy level.

Just as the energy transformation of evaporation drives the water cycle, chemical energy transformations propel the cycling of elements through the biosphere. The energy potential of a chemical transformation is measured by the **redox potential** of a reaction, which reflects the capacity of an atom to accept electrons—that is, to become reduced (see Chapter 4). Redox potentials are expressed as volts (electric potential). A high value indicates that an atom accepts electrons readily and, therefore, that the reaction has the potential to oxidize some other substance. Lower values indicate more powerful reducing potential.

Molecular oxygen (O_2) is an example of a strong oxidizer—an electron acceptor. In its organic form, carbon is a strong reducer—a donor of electrons. Giving and taking electrons are opposite sides of the same coin; each substance may serve as an oxidizer or a reducer. But because strong oxidizers hold onto electrons tightly, they are always weak reducers, and vice versa. To put it another way, a strong oxidizer, like molecular oxygen, can always oxidize (be reduced by) a weaker oxidizer, like organic carbon.

Redox Reactions

Oxidation-reduction reactions are referred to as **redox reactions** (Figure 11-5). They can be characterized as a pair of half-reactions, one for the reduction (electron-accepting) step and the other for the oxidation (electron-donating) step. For example, when molecular oxygen acts as an oxidizer, each oxygen atom accepts two electrons:

$$O_2 + 4e^- = 2O^{2-}.$$

In this reduced form, oxygen readily combines with any of a variety of positive ions, such as H^+, giving

$$O_2 + 4e^- + 4H^+ = 2H_2O,$$

or C^{4+}, giving

$$O_2 + 4e^- + C^{4+} = CO_2.$$

These half-reactions include only the reduction. The electrons must come from an oxidation half-reaction; for example,

$$CH_2O = C^{4+} + H_2O + 4e^-,$$

in which case the carbon atom donates electrons (it is oxidized) and the electrons are available to reduce an oxidizer. The overall reaction combining the last two half-reactions has the form

$$\underset{\text{reduced C}}{CH_2O} + \underset{\text{oxidized O}}{O_2} = \underset{\text{oxidized C}}{CO_2} + \underset{\text{reduced O}}{H_2O}$$

which is the equation for respiration. On the left side of this equation, carbon, C, in CH_2O is in a reduced form, and O in O_2, is in an oxidized form. On the right side of the equation, C in CO_2 is in an oxidized form and O in water is in a reduced form. Electrons are transferred from C to O in the process.

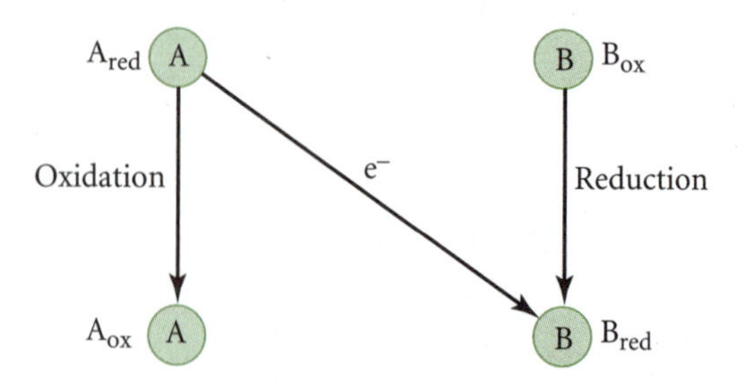

FIGURE 11-5 Oxidation and reduction. When A is in the reduced state (A_{red}), it may act as a reducer of B in the oxidized state (B_{ox}) by donating electrons to B. When A donates electrons, it changes from the reduced state to the oxidized state (A_{ox}). When B accepts electrons, it changes from the oxidized state to the reduced state (B_{red}). If A has a higher redox potential than B (A_{red} higher than B_{red}), then energy is released in the reaction.

The redox potentials, or the relative strengths, of oxidizers and reducers are measured by their electric potentials (Eh), expressed in volts (V) (Table 11-1). An electric current represents the movement of electrons; a chemical battery generates an electric potential by a pair of redox reactions, one at each electrode. Because oxygen has a high redox potential (0.81 volts at pH 7 and 25° C), as we mentioned before, it is an excellent oxidizer. So are nitrate (NO_3^-), ferric iron (Fe^{3+}), and, to a lesser extent, sulfate (SO_4^{2-}). Near the bottom of the redox scale are carbon dioxide (CO_2) and molecular nitrogen (N_2). The electric potentials of the half-reactions reducing CO_2 carbon (C^{4+}) to organic carbon (C^0, −0.43 V) and N_2 to organic nitrogen (ammonia) (N^{3-}, 0.28 V) are unfavorable thermodynamically.

Any oxidation reaction having a high redox potential coupled with a reduction reaction having a low redox potential can proceed with the release of energy. Thus the oxidation of carbon by oxygen (O_2) releases energy because oxygen is a stronger oxidizer than CO_2. For the reaction to proceed in the opposite direction (for example, fixation of carbon by photosynthesis), energy must be added.

Oxygen and carbon are not the only electron acceptors and donors of importance to biological reactions. Nitrate is a powerful oxidizer; hydrogen sulfide (H_2S) and ammonium (NH_3) are good reducers.

TABLE 11-1 Redox potentials of selected half-reactions at 25° C and pH 7

Reaction	Eh (V)
$O_2 + 4H^+ + 4e^- = 2H_2O$	0.81
$NO_3^- + 6H^+ + 6e^- = \frac{1}{2}N_2 + 3H_2O$	0.75
$NO_3^- + 2H^+ + e^- = NO_2^- + H_2O$	0.42
$NO_3^- + 10H^+ + 8e^- = NH_4^+ + 3H_2O$	0.36
$Fe^{3+} + e^- = Fe^{2+}$	0.36
$NO_2^- + 8H^+ + 6e^- = NH_4^+ + 2H_2O$	0.34
$CH_3OH + 2H^+ + 2e^- = CH_4 + H_2O$	0.17
$CH_2O + 2H^+ + 2e^- = CH_3OH$	−0.18
$SO_4^{2-} + 8H^+ + 6e^- = S + 4H_20$	−0.20
$SO_4^{2-} + 10H^+ + 8e^- = H_2S + 4H_2O$	−0.21
$CO_2 + 8H^+ + 8e^- = CH_4 + 2H_2O$	−0.24
$N_2 + 8H^+ + 6e^- = 2NH_4^+$	−0.28
$H^+ + e^- = \frac{1}{2}H_2$	−0.41
$CO_2 + 4H^+ + 4e^- = \frac{1}{6}C_6H_{12}O_6 + H_2O$	−0.43
$CO_2 + 4H^+ + 4e^- = CH_2O + H_2O$	−0.48
$Fe^{2+} + 2e^- = Fe$	−0.85

(From Stumm and Morgan 1981.)

Indeed, the synthesis and metabolism of many organic compounds, including amino acids, involve assimilatory and dissimilatory redox reactions of nitrogen and sulfur. In general, oxygen is the oxidizer of choice, owing to its ubiquity and to the fact that its common reduced form, H_2O, is innocuous. In contrast, the reduction of nitrate produces nitrite, NO_2^-, which most organisms can tolerate only in minute quantities. Oxidizers such as nitrate and ferric iron can nonetheless become important in soils and aquatic sediments when oxygen has been depleted (anaerobic conditions), as we shall see below.

When all other oxidizers have been exhausted (extremely reduced conditions), hydrogen ions can be used as an electron acceptor:

$$2H^+ + 2e^- = H_2.$$

This reaction is, however, thermodynamically very unfavorable (redox potential = −0.41 V) and therefore requires considerable energy. Organic carbon also can be an electron acceptor when other oxidizers are absent:

$$CH_2O + 2H^+ + 4e^- = CH_4 + O^{2-}.$$

This reaction occurs in anaerobic fermentation in oxygen-depleted sediments and, notably, in the rumens of ungulates, which lack inorganic electron acceptors (Wolin 1979). Methane, or natural gas (CH_4), is one of the end products of this set of reactions.

The hydrogen molecule (H_2) can be a powerful electron donor, by the half-reaction $H_2 = 2H^+ + 2e^-$, depending on the hydrogen ion concentration of the environment, or pH. At lower pH, where hydrogen ions are more abundant (more acid conditions), hydrogen gives up electrons less readily, and the electric potential of the redox reaction increases (that is, hydrogen is a poorer reducer). As pH increases, the supply of hydrogen ions (H^+) decreases, the equilibrium of the reaction $H_2 = 2H^+ + 2e^-$ shifts to the right, and electric potentials decrease.

Characterization of Environmental Redox

With respect to oxidation-reduction reactions, an environment may be characterized by pH and electric potential (Eh). The first measures the availability of hydrogen ions, the second the availability of electrons. Because hydrogen and oxygen are so abundant, their reactions limit naturally occurring values of pH and Eh. For a given pH, electric potential has a practical upper bound set by the reaction $O_2 + 2e^- = 2O^{2-}$ (Eh = 0.81 V at pH 7) and a lower bound set by the reaction $2H^+ + 2e^- = H_2$ (Eh = −0.41 V). When a powerful oxidizer raises Eh above 0.81 V, it oxidizes the O_2 in water, liberating oxygen (O_2^-), until the oxidizer is fully consumed (reduced) and Eh drops to

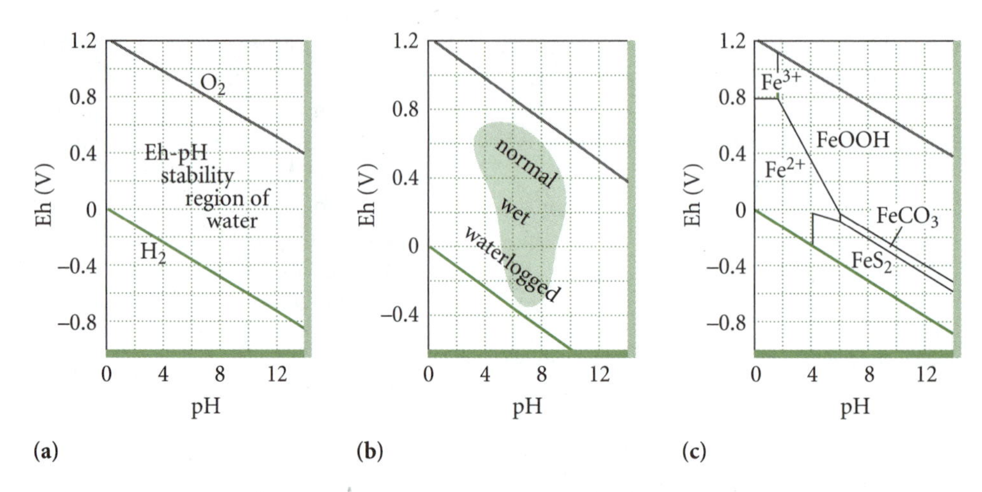

FIGURE 11-6 Distribution of pH and electric potentials of soils and the ion compounds associated with these conditions. (a) The stability region of water sets limits on soil conditions. (b) Measured values of pH and Eh in soils fall within the shaded region; waterlogged soils tend to have lower (more reduced) electric potentials. (c) Stability regions for various iron ions and compounds. Ferric iron (Fe^{3+}) occurs under oxidizing conditions and ferrous iron (Fe^{2+}) occurs under reducing conditions. At low pH (acidic conditions), both forms occur as ions. At high pH, ions of iron form hydroxides (FeOOH), carbonates ($FeCO_3$), or sulfides (FeS_2), which are not soluble in water and, thus, may precipitate from the soil. (*After Bohn et al. 1979.*)

that of O_2. When a powerful reducer decreases Eh below −0.41 V, it reduces hydrogen ions to molecular hydrogen (H_2) until the reducer is fully oxidized. In general, high values of Eh occur in well-oxygenated soils; lower values occur in waterlogged soils and the deep waters and sediments of lakes and estuaries, into which oxygen diffuses from the atmosphere much more slowly than it is consumed by the respiration of organisms (Figure 11-6).

Each element that undergoes redox reactions assumes a predominant form that may vary depending upon the Eh and pH of the environment. For example, iron occurs in its oxidized ferric form (Fe^{3+}) under oxidizing conditions, but in its reduced ferrous form (Fe^{2+}) under reducing conditions. Under acidic conditions, both forms tend to occur as dissociated ions, but as the hydrogen ion concentration decreases, they more readily form insoluble compounds: hydroxides (e.g., FeOOH), carbonates (e.g., $FeCO_3$), and sulfides (e.g., FeS_2) (Figure 11-6c). Conditions of pH and Eh have important consequences for the cycling of many elements.

11.5 The modern carbon cycle includes a missing sink of carbon.

The assimilatory and dissimilatory redox reactions of carbon in photosynthesis and respiration are the major energy-transforming reactions of life. Approximately 10^{17} grams (10^{11} metric tons) of carbon enters into such reactions worldwide each year. Because of this, the cycling of carbon is closely associated with energy flow in ecosystems, and as we shall see, the cycles of other important elements, such as nitrogen, are often closely linked to the carbon cycle. Gaining a complete understanding of the global carbon cycle, then, presents one of the most important challenges of modern ecology.

The modern carbon cycle is something new under the sun. Carbon pools that were previously excluded from the cycle—stored in coal, oil, and natural gas—have been released by human activities. There is now little doubt that these new inputs, particularly the increase in atmospheric CO_2, are changing global ecosystems and the global climate. Our challenge is to understand the responses of the atmosphere, the oceans, and terrestrial ecosystems to elevated carbon inputs.

The Modern Carbon Cycle

The earth's carbon resides in four great compartments: the oceans, the atmosphere, the terrestrial biomass, and fossil deposits. Soils and land plants are subunits of the terrestrial compartment that are often considered separately when accounting for global carbon flux. As with the cycles of other materials, the carbon cycle depends both on the relative amounts of carbon residing in each pool and on the rates of flux among the pools. Underlying the large-scale and relatively rapid transfers of carbon among pools are numerous, slower, transfers among the subcompartments of each pool (Schlesinger 1991).

The oceans contain the largest carbon pool ($38{,}000 \times 10^{15}$ g), followed by soil and land plants, the two major subcomponents of the terrestrial system, which account jointly for approximately 2,060 × 10^{15} g (560 × 10^{15} g for land plants and 1,500 × 10^{15} g for soils) (Figure 11-7). Atmospheric carbon, primarily in the form of carbon dioxide and methane, accounts for less than 1% of global carbon (720 × 10^{15} g). By way of comparison, this amount represents 35% of the amount found in the terrestrial system and only a fraction of the ocean carbon pool. The total amount of carbon bound up in fossil fuels is estimated to be 4,000 × 10^{15} g, roughly twice the amount found in the terrestrial carbon pool (Post et al. 1990).

Land plants take in more carbon from the atmosphere in the process of primary production (120 × 10^{15} g C yr^{-1}) than they return via respiration (60 × 10^{15} g C yr^{-1}). However, terrestrial-atmospheric carbon exchange is roughly balanced in the long term. Plants transfer the products of primary production to the soil in the form of litter and dead organic material at a rate of about 62 × 10^{15} g C yr^{-1}, and much of that is returned to the atmosphere (60 × 10^{15} g C yr^{-1}) by respiration and decomposition in the soils. Human destruction of vegetation, primarily by fires related to the clearing of land for agriculture, adds carbon to the atmosphere (2 × 10^{15} g C yr^{-1}), as does the combustion of fossil fuels (5 × 10^{15} g C yr^{-1}). These inputs are chiefly responsible for the imbalance in the modern global carbon cycle (see below). The link between the terrestrial carbon pool and the oceans is primarily via river flow, which transfers about 0.4 × 10^{15} g C yr^{-1} to the oceans.

Globally, the oceans take in more carbon than they release into the atmosphere, and are thus considered to be carbon sinks (see Figure 11-7). The dynamics of the flux of CO_2 at the air-sea interface is controlled almost entirely by physical and biological processes within the vast deeper waters of the ocean. Winds at the ocean's surface ensure a rapid flux of CO_2 across the surface waters, creating a situation in which the partial pressure of CO_2 in the air is nearly in equilibrium with that just below the ocean's surface (Post et al. 1990). Processes that move the products of primary production from the surface waters into deeper waters, referred to as **biological pumping,** lower the CO_2 concentration of the surface waters.

The oceans are vast and heterogeneous with respect to circulation patterns, depth, and temperature profiles. Thus, it is not surprising that there is considerable global variation in the flux of carbon at the air-sea interface. At high latitudes, the extremely low temperatures and a circulation profile that is characterized by a downwelling of cold, dense surface water appear to provide an avenue for the entry of large quantities of carbon from the atmosphere (CO_2 is roughly twice as soluble in water at 0° C as it is at 20° C). In these areas, such as the North Atlantic,

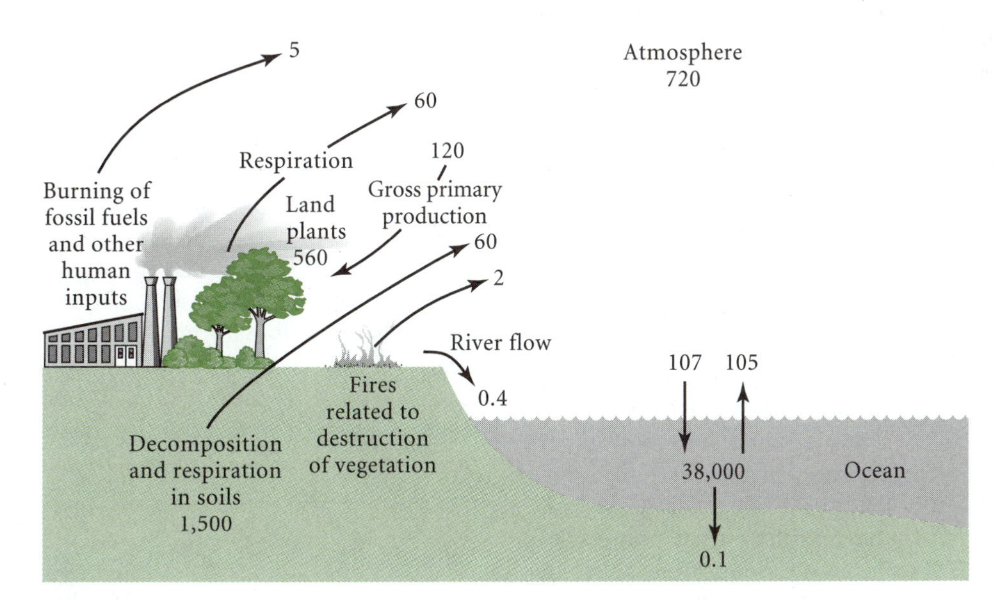

FIGURE 11-7 The global carbon cycle. Pools are presented in billions of metric tons (10^{15} g); fluxes are given in 10^{15} g C/yr. Land plants take in about twice as much carbon via primary production as they return in respiration. The input of carbon to the atmosphere from soil decomposition and respiration equals the input via plant respiration, making the flux between the terrestrial compartments (land plants + soil) and the atmosphere balance the input from the atmosphere to land plants. Anthropogenic (human-made) inputs from automobile emissions, industrial processes, and fires used for deforestation provide additional carbon inputs that throw the modern carbon cycle out of balance. A small amount of carbon is lost to the seafloor by burial. (*Redrawn from Schlesinger 1991.*)

atmospheric CO_2 entering the surface waters is quickly transported downward, thus maintaining a low CO_2 concentration in the surface waters.

The terrestrial portion of the carbon cycle is also enormously variable owing to the vast differences in the rates of primary production in different types of terrestrial habitats (see Table 10-2). For example, tundra and desert scrub habitats have rates of productivity much less than those of temperate or tropical forests (Whittaker and Likens 1973b). Moreover, the rate of decomposition, which accounts for about one-half of the carbon output from terrestrial ecosystems, is generally lower in cold soils than in warm soils. Thus, turnover rates are much lower at high latitudes (Post et al. 1990, Shaver et al. 1991).

Carbon in Fresh Waters

Although lakes account for less than 2% (approximately 2.5×10^6 km^2) of the earth's surface, and thus represent only a fraction of the earth's store of water (Wetzel 1983), they are extremely important both as a source of water for humans and, more important from a global ecosystem perspective, as critical components of some of the earth's most sensitive habitats. Phytoplankton, algae, and macrophytes (large aquatic plants) account for most of the primary production in lakes. The relative contribution of each to the total carbon budget of a lake is highly variable, owing to differences among lakes in depth profiles, temperature regimes, and nutrient inputs. As with oceans, carbon enters lake waters at the surface, where the relative pressures of CO_2 determine the rate of flux. However, unlike the open oceans, most lakes sustain relatively high rates of allochthonous (external) inputs (see Chapter 8) of nutrients and organic material via streams and runoff. Additionally, primary production in freshwater lakes is usually higher than in oceans, and lake sediments may store carbon. The carbon balance in a lake is dependent upon these various fluxes.

The oceans take in more carbon than they release; they represent a carbon sink. Is the same true of bodies of fresh water? The answer appears to be no. In a study of the partial pressures of CO_2 of over 1,800 lakes from around the world, Jonathan Cole and coworkers (1994) found that the partial pressure of CO_2 exceeded that of the overlying atmosphere in most lakes studied, and that the average partial pressure of lake CO_2 was three times that of the atmosphere (Figure 11-8). Globally, lakes appear to be sources of carbon.

Whether bodies of fresh water serve as sinks or sources of carbon may be especially important in ecosystems that are expected to be particularly sensitive to global climate change. Intake of carbon by the vegetation of the arctic tundra, for example, slightly exceeds the rate of carbon return to the atmosphere by plant and soil respiration and decomposition, creating a terrestrial carbon sink that accumulates carbon at the rate of $0.1-0.3 \times 10^{15}$ g C yr^{-1} (Kling et al. 1991, 1992). Arctic lakes appear to be supersaturated with CO_2 and, thus, may serve as sources of carbon to the arctic ecosystem. An increase in temperature in the arctic region resulting from global warming would increase both the rate of primary production and the rates of respiration and decomposition (Shaver et al. 1992), but whether this would lead to a net output of CO_2, which would exacerbate climate change, or a net intake of CO_2, which would tend to ameliorate the warming trend, is unclear. The interactions between the terrestrial arctic sink and the aquatic arctic source of CO_2 will determine the carbon balance in the face of global warming.

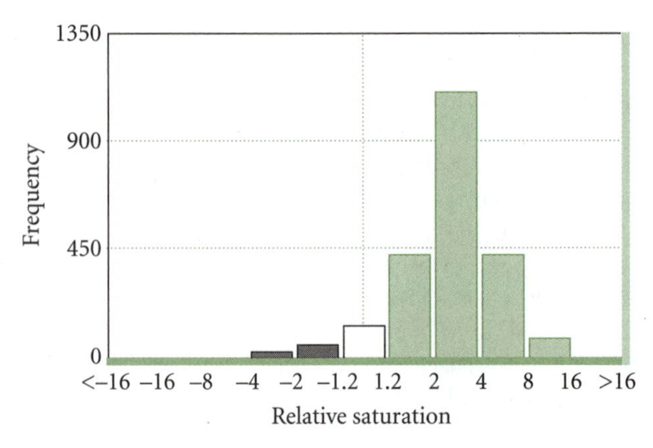

FIGURE 11-8 Relative CO_2 saturation in the surface waters of lakes from around the world. The height of each bar represents the number of the 2,395 samples taken from 69 lakes having a particular relative saturation. Relative saturation for lakes that are supersaturated with CO_2 is calculated as P_{CO_2} (water)/P_{CO_2} (air) (green bars; positive saturation values). Relative saturation for lakes that are undersaturated with CO_2 is calculated as P_{CO_2} (air)/P_{CO_2} (water) (gray bar; negative values). The white bar represents samples near equilibrium. In most lakes, the partial pressure of CO_2 exceeds that of the atmosphere. These data suggest that globally, lakes are a source of carbon. (*From Cole et al. 1994.*)

Balancing the Global Carbon Cycle

Before modern humans arrived on the scene, the carbon cycle involved exchanges between the oceans, the atmosphere, and terrestrial biomes; the carbon locked in the fossil compartment was not involved in the cycle in any appreciable way. And, until modern humans appeared, the relative sizes of the three active carbon pools remained unchanged. The burning of fossil fuels and the destruction of terrestrial biomes by humans, which typically leads to the release of large quantities of carbon in fires, have resulted in both new inputs of carbon into the global ecosystem and a significant reduction in the size of the terrestrial compartment of the carbon cycle. Carbon dioxide is released into the atmosphere as a result of the burning of fossil fuels at a rate of about 5×10^{15} g yr^{-1}, and as a result of deforestation at a rate of roughly 2×10^{15} g yr^{-1} (see Figure 11-7). Where does this additional carbon go?

The return of carbon to the fossil compartment occurs too slowly to be relevant to the rate of current global climate change. Thus, we must look for the excess carbon in one of the three active carbon pools: the atmosphere, the oceans, or the terrestrial ecosystem. Current estimates of the rate of increase of atmospheric carbon suggest that about 45% of the carbon released from the burning of fossil fuels such as gasoline winds up in the atmospheric compartment each year (3.2×10^{15} g C yr^{-1}) (Table 11-2). Approximately 28% appears to enter the oceanic carbon pool each year via gas exchange at the surface (2.0×10^{15} g C yr^{-1}) (Schlesinger 1991, Siegenthaler and Sarmiento 1993). A small amount of the carbon released by the destruction of vegetation is transported by river flow and runoff into the oceans in the form of organic material. But the fate of the rest of the carbon released by human activity from the fossil pool each year (1.8×10^{15} g C yr^{-1}) is unclear. There is, in the modern carbon cycle, a missing carbon sink.

The ocean's ability to dissolve CO_2 is related in part to the dynamics of biogenic carbonates in the water. When carbon dioxide dissolves in water, it forms carbonic acid,

$$CO_2 + H_2O \rightleftharpoons H_2CO_3,$$

which readily dissociates into bicarbonate and carbonate ions,

$$HCO_3 \rightleftharpoons H^+ + HCO_3^-,$$
$$HCO_3^- \rightleftharpoons H^+ + CO_3^2.$$

At low pH, abundant hydrogen ions in the environment drive these reactions to the left, creating more carbonic acid (Figure 11-9). When calcium is present, it also equilibrates with the carbonate and bicarbonate ions,

$$CaCO_3 \rightleftharpoons Ca^{2+} + CO_3^{2-}.$$

Calcium carbonate ($CaCO_3$) has low solubility under most conditions and readily precipitates out of the water column. In oceans, the amount of $CaCO_3$ is affected locally by the activities of organisms. The overall equilibrium of the carbonate system is $CaCo_3$ (insoluble) + H_2O + $CO_2 \rightleftharpoons Ca(HCO_3)$ (soluble).

When CO_2 is removed from the surface waters by photosynthesis, the equilibrium shifts to the left, resulting in the formation and precipitation of calcium carbonate.

Calcium carbonate deposits are found in the world's oceans, but only in shallow waters, because there tends to be more CO_2 there. Carbon dioxide dissolves more readily in the deep ocean because of the high pressures and low temperatures there. The lack of primary production at great depths, which would remove CO_2, coupled with inputs from the respiration of deep-sea organisms and decomposition of organic material transported below by biological pumping, supersaturates (relative to the atmosphere) deep waters with CO_2, moving the equilibrium above to the right, thereby undersaturating the deep oceans with calcium carbonate (Schlesinger 1991).

TABLE 11-2 Fate of anthropogenic CO_2 in the global carbon budget

Compartment or process	Rate $\times 10^{15}$ g C yr^{-1}
Source of anthropogenic CO_2	
Fossil fuel combustion	5.0
Deforestation and changing land use patterns	2.0
Uptake of anthropogenic CO_2	
Atmospheric accumulation	3.2
Uptake by oceans	2.0
Net balance (sources − uptake)	−1.8

(Data from Schlesinger 1991, Siegenthaler and Sarmiento 1993.)

Land Plants as the "Missing Sink"

Of the 7.0×10^{15} g of anthropogenic CO_2 input into the global carbon cycle each year, 3.2×10^{15} g accumulates in the atmosphere, and the ocean serves as a sink for 2.0×10^{15} g (see Figure 11-7 and Table 11-2). What happens to the remaining 1.8×10^{15} g that is released each year by human activity? Ecologists

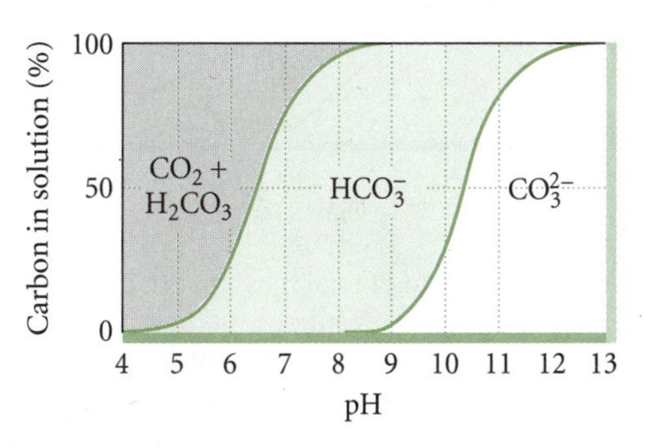

FIGURE 11-9 Proportion of carbon that occurs as carbonic acid, bicarbonate, and carbonate in solution as a function of pH.

have focused their attention on the terrestrial system in their search for an explanation for this "missing sink." In many plants, particularly those with C_3 photosynthesis (see Chapter 5), photosynthetic activity increases under conditions of elevated CO_2 concentrations in the atmosphere. Such an increase in production might account for the missing CO_2 if it were accompanied by an increase in the amount of carbon storage in plant tissues. As we mentioned earlier, some terrestrial ecosystems, notably the arctic tundra, store large quantities of carbon and thus serve as carbon sinks. But our current understanding of the fate of CO_2 in the terrestrial ecosystem suggests that this may be too simple an explanation.

The response of plants to elevated CO_2 may be transient and is most certainly constrained by the availability of nutrients (Grodzinski 1992). In a study of how seedling ash, birch, maple, and oak trees—common species of the north temperate forest of the United States—grew in response to elevated CO_2 levels in greenhouses, F. A. Bazzaz and colleagues at Harvard University (1993) showed that, for some species, high atmospheric CO_2 levels led to seedlings having a greater average plant biomass than seedlings grown at ambient CO_2 levels (Figure 11-10). But the rate of gain in mass declined substantially over the 3-year period of the study and was dependent on the light level and the amount of nitrogen available. Additionally, Bazzaz and coworkers found that some species showed more enhanced growth with respect to elevated CO_2 levels in the atmosphere than others. These results indicate that the sequestration of anthropogenic carbon in the terrestrial ecosystem, if it occurs, is likely to be highly complex and variable on both a local and global scale.

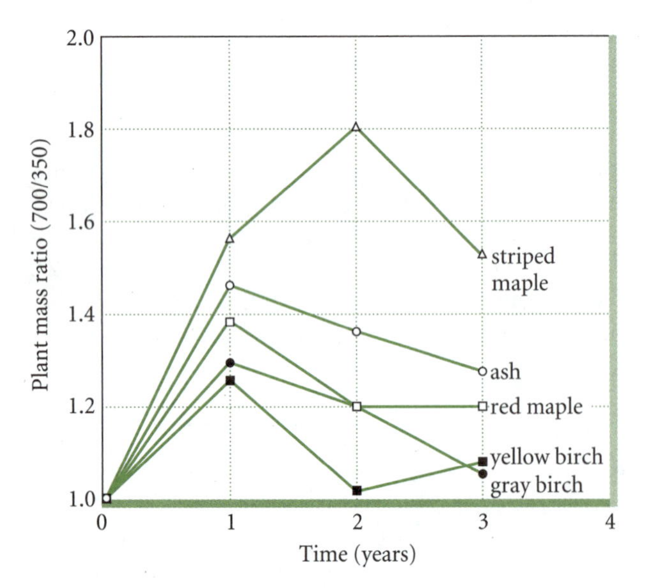

FIGURE 11-10 Ratio of total masses of five species of seedling trees grown in a greenhouse at high (700 $\mu l \, l^{-1}$) and at low (350 $\mu l \, l^{-1}$) concentrations of CO_2. Trees grown in a high-CO_2 environment increased in biomass relative to those grown in the low-CO_2 environment. The difference in the biomass between the two experimental groups decreased after the first year for all species except striped maple. (*From Bazzaz et al. 1993.*)

The mechanism of sequestration of carbon in plants is complicated not only by variation in the response of plant species to elevated CO_2, as demonstrated by the work of Bazzaz et al., but also by the complex interrelationship between climate change and the global carbon cycle. Increases in atmospheric CO_2 owing to human activities, if continued unabated, will lead to increases in global temperatures. Photosynthetic rates, which are enhanced, at least in the short term, by increases in the availability of CO_2, may be further stimulated by increased temperatures. But the rates of respiration and decomposition may also increase, resulting in the release of more CO_2. Moreover, warmer climates may mean greater cloud cover, thereby reducing the amount of incident solar radiation and providing a countervailing force to increased photosynthetic rates. Thus, it is not clear whether the response of the global ecosystem to enhanced carbon will be compensatory or will result in a progressive destabilization of the ecosystem.

The Role of Soils in the Global Carbon Cycle

In our description of the major carbon pools, we referred to terrestrial plants and the soils in which they are rooted collectively as the terrestrial compartment. However, this simplification should not suggest that the two subcompartments of terrestrial ecosystems are uninteresting in their own right. As we shall see in the next chapter, local nutrient regeneration on land depends largely on soil processes. Soil dynamics also affect global nutrient cycling, and ecologists are particularly interested in the role that soils may play in the global carbon cycle.

The principal source of organic matter in soils is dead plant material, often referred to as litterfall or leaf litter. The rate at which organic matter accumulates in developing soils varies widely among the terrestrial ecosystems of the world, and depends to a large extent on the origin of the soil and the plant community in the area. It is estimated that the total amount of organic carbon sequestered in soils is about $1{,}456 \times 10^9$ metric tons, with relatively wet ecosystems such as swamps and marshes containing more than dryer areas (Table 11-3). The decomposition of this organic material, which occurs most rapidly at the soil surface, results in the release of CO_2 into the spaces between soil particles and, eventually, into the atmosphere. Respiration by plant roots and by microbes and other soil organisms also

TABLE 11-3 Distribution of soil organic matter for different types of ecosystems

Ecosystem type	Mean soil organic matter (kg C m^{-2})	World area (ha × 10^8)	Total world soil organic carbon (mt C × 10^9)	Total world litter (mt C × 10^9)
Tropical forest	10.4	24.5	255	3.6
Temperate forest	11.8	12	142	14.5
Boreal forest	14.9	12	179	24.0
Woodland and shrubland	6.9	8.5	59	2.4
Tropical savanna	3.7	15	56	1.5
Temperate grassland	19.2	9	173	1.8
Tundra and alpine	21.6	8	173	4.0
Desert scrub	5.6	18	101	0.2
Extreme desert, rock, and ice	0.1	24	3	0.02
Cultivated	12.7	14	178	0.7
Swamp and marsh	68.6	2	137	2.5
Totals	147	1,456	55.2	

(*From Schlesinger 1991.*)

releases CO_2. How might this CO_2 flux be affected by human activities such as agriculture and by climate change?

The question is complicated by the complexity of soils themselves. Fresh organic material at the soil surface may undergo rapid decomposition, but some organic material becomes complexed with clay particles and thus becomes difficult for microbes to process. Cultivation causes a rapid decline in soil organic matter (Figure 11-11). Some of the organic matter is lost in runoff, but a large portion is released into the atmosphere as CO_2 (Schlesinger 1991). The effect of climate change—in particular, global warming—on the release of CO_2 from soils is still an open question.

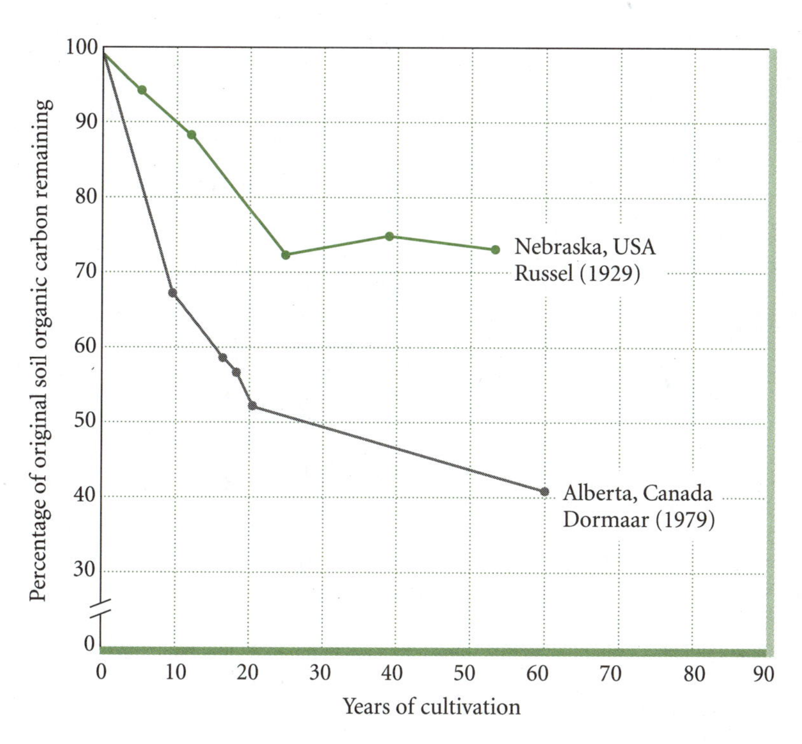

FIGURE 11-11 The effect of cultivation on the amount of organic carbon (presented as a percentage of the original amount) in soils. Cultivation reduces the amount of organic carbon that was originally available in the soil. (*From Schlesinger 1991.*)

11.6 Nitrogen assumes many oxidation states in its paths through the ecosystem.

Nitrogen is an essential component of the enzymes that guide and regulate the assimilatory and dissimilatory reactions of carbon and the other elements. Because of its essential role in the biochemical processes of living things, nitrogen limits the rate of production in terrestrial and ocean ecosystems. Most of the nitrogen on earth occurs in gaseous form (N_2) in the atmosphere. In contrast to CO_2, which is taken in directly by photosynthesizing plants, the strong triple bond of atmospheric nitrogen makes it inaccessible to most organism. Nitrogen becomes accessible to organisms thanks to biological transformations of gaseous nitrogen into forms such as nitrate (NO_3^-) and ammonium (NH_3). Nitrogen in these forms is referred to as "fixed" nitrogen, which implies that it is in a form that may enter biological processes. (We may also say that carbon is "fixed" when the carbon atoms from CO_2 molecules are assembled into $C_6H_{12}O_6$ molecules during photosynthesis.) As we shall see, the nitrogen cycle is characterized by numerous transformations among the oxidation states of nitrogen.

The Nitrogen Cycle

Like carbon, nitrogen exists in a number of interacting pools represented by different inorganic and organic forms (Figure 11-12). Nitrogen is the most common atmospheric gas, representing approximately 78% (3.8×10^{21} g total mass) of the constituents of air. (Oxygen and carbon dioxide make up 21% [1.2×10^{21} g] and 0.03% [0.06×10^{21} g] of the atmosphere, respectively.) The amount of nitrogen held in terrestrial plants and soil is relatively small (98×10^{15} g) compared with that in the atmosphere. The ocean receives nitrogen from rivers, precipitation, and via biochemical transformations of gaseous nitrogen.

Molecular nitrogen enters the biological pathways of the nitrogen cycle primarily through assimilation by certain microorganisms, particularly bacteria, a process referred to as **nitrogen fixation** (Stacey et al. 1992, Young 1992). It is estimated that in excess of $1{,}200 \times 10^{12}$ g N yr^{-1} is required for primary production by land plants. The best estimate of the amount of nitrogen provided by biological nitrogen fixation is on the order of 12% of that amount, or about 140×10^{12} g yr^{-1}. The rest of the nitrogen is provided by recycling and decomposition within ecosystems. Nitrogen may also be fixed by lightning. Such abiotic nitrogen fixation may deliver as much as 20×10^{12} g N yr^{-1} to terrestrial ecosystems and substantially more to the oceans (Schlesinger 1991).

Once in the biological realm, the cycling of nitrogen is much more complicated than that of carbon because of its numerous oxidation states (Figure 11-13). Beginning arbitrarily with reduced (organic) nitrogen, the first step in the nitrogen cycle is **ammonification:** the hydrolysis of protein and oxidation of amino acids, which results in the production of ammonia (NH_3). This transformation is carried out by all organisms.

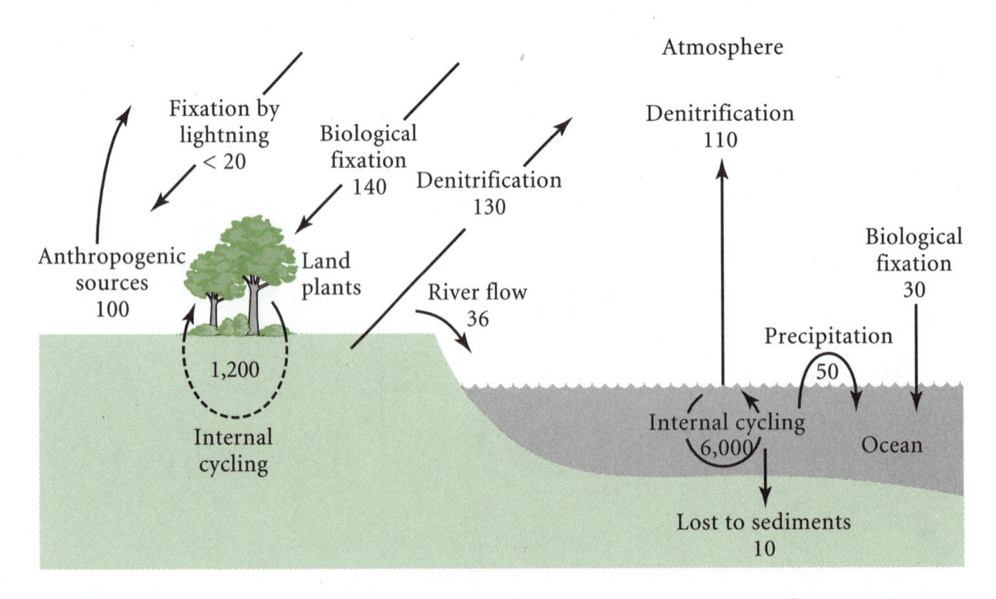

FIGURE 11-12 The global nitrogen cycle. Pools are given in 10^{12} g N; fluxes are given in 10^{12} g N/yr. The total amount of nitrogen in the atmosphere is 3.8×10^{21} g, considerably larger than that in terrestrial plants and soils (98×10^{15}). Oceans receive nitrogen from river flow, precipitation, and biological fixation and lose nitrogen by denitrification and in losses to sediments. Terrestrial ecosystems gain usable nitrogen by biological fixation. Both ocean and terrestrial ecosystems recycle nitrogen internally. The use of nitrogen fertilizers and emissions from internal combustion engines constitute the principal anthropogenic inputs of nitrogen to the global cycle. (*Redrawn from Schlesinger 1991 from data from Söderlund and Rosswall 1982)*

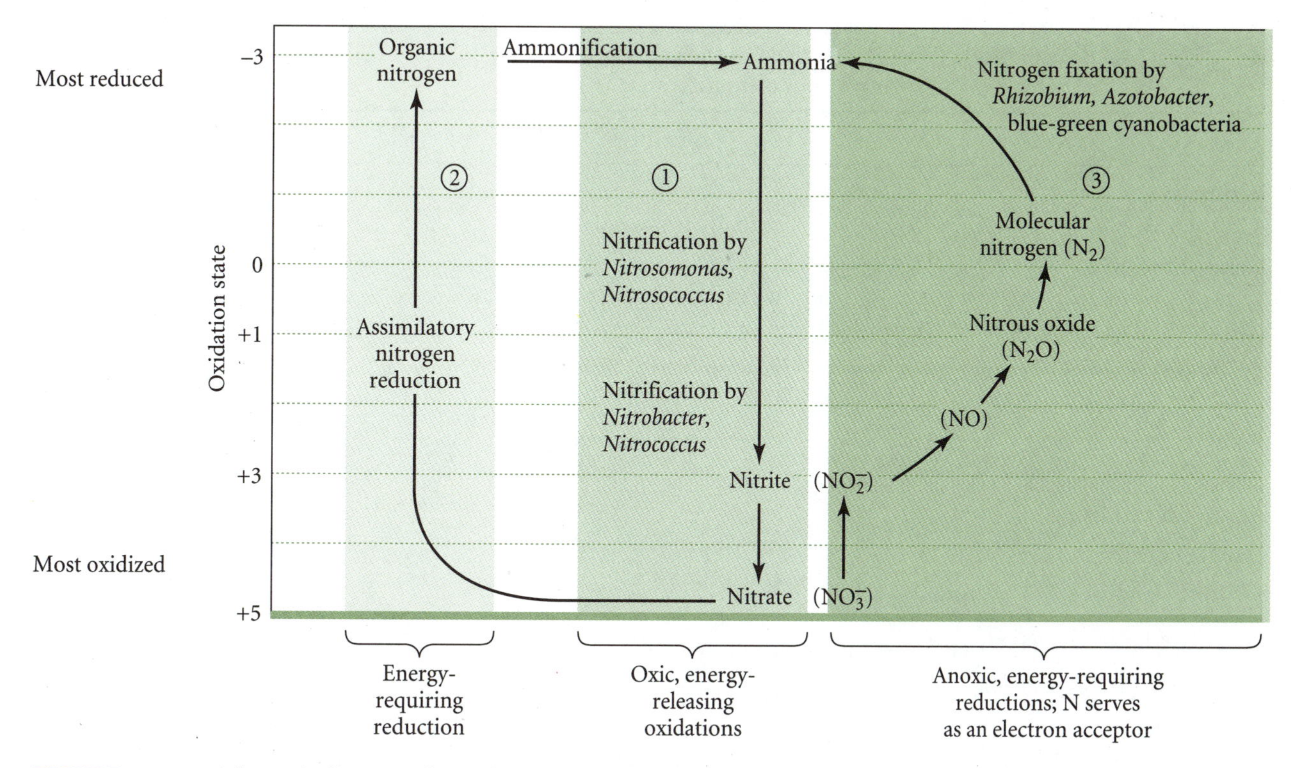

FIGURE 11-13 Schematic diagram of transformations and oxidation states of compounds in the nitrogen cycle. Ammonification produces ammonia, which undergoes nitrification by bacteria to form nitrite (1). Further oxidation of nitrite yields nitrate, which undergoes assimilatory nitrogen reduction to produce organic forms of nitrogen (2). Physical reactions of nitrite result in the formation of molecular nitrogen, thus removing nitrogen from the biological system (3).

During the initial breakdown of amino acids, although some carbon is oxidized, releasing energy (Table 11-4), the energy potential of the nitrogen atom does not change from its oxidation state of −3 (N^{3-}).

Nitrification involves the oxidation of nitrogen, first from ammonia to nitrite:

$$NH_3 \longrightarrow NO_2^- (N^{3-} \longrightarrow N^{3+}),$$

then from nitrite to nitrate:

$$NO_2^- \longrightarrow NO_3^- (N^{3+} \longrightarrow N^{5+}),$$

during which the nitrogen atom releases much of its potential chemical energy (Brown et al. 1974, Fochte and Verstraete 1977). Both of these steps are carried out only by specialized bacteria: $NH_3 \rightarrow NO_2^-$ by *Nitrosomonas* in the soil and by *Nitrosococcus* in marine systems; $NO_2^- \rightarrow NO_3^-$ by *Nitrobacter* in the soil and *Nitrococcus* in the seas.

Because nitrification steps are oxidations, they require the presence of oxygen to act as an electron acceptor. In waterlogged, anoxic soils and sediments and in oxygen-depleted bottom waters, nitrate and nitrite can act as electron acceptors (oxidizers), and the nitrification reactions reverse (Delwiche and Bryan 1976):

$$NO_3 \longrightarrow NO_2 \longrightarrow NO(N^{5+} \longrightarrow N^{3+} \longrightarrow N^{2+}).$$

This **denitrification** occurs in soil when the redox potential is less than 0.2 V, and is accomplished by such bacteria as *Pseudomonas denitrificans.* Denitrification also occurs in anaerobic waters of some deep oceans. Additional physical reactions can result in the production of molecular nitrogen; that is,

$$NO_3^- \longrightarrow NO_2^- \longrightarrow NO \longrightarrow N_2O \longrightarrow N_2,$$

with the consequent loss of nitrogen from biological circulation. Denitrification may be one of the major causes of the low availability of nitrogen in marine systems. When organic remains of plants and animals sink to the depths of the oceans, their oxidation by bacteria in deep waters and bottom sediments often is accomplished anaerobically, using nitrate as an oxidizer. This results in the conversion of nitrate and nitrite to the dissolved gases NO and N_2, which cannot be used by algae.

The Energetics of Nitrogen Fixation

Denitrification is balanced in terrestrial and aquatic systems by nitrogen fixation (Postgate and Hill 1979, Subba Rao 1980). This assimilatory reduction of nitrogen,

$$N^0 \longrightarrow N^{3-},$$

TABLE 11-4 Biochemical processes involved in the ecological cycling of nitrogen and several other elements

Process	Organism	Yield (kJ/mol)
Respiration		
$C_6H_{12}O_6 + 6O_2 \longrightarrow 6CO_2 + 6H_2O$	Virtually universal	2,870
Denitrification		
$C_6H_{12}O_6 + 6KNO_3 \longrightarrow 6CO_2 + 3H_2O + 6KOH + 3N_2O$	*Pseudomonas denitrificans*	2,280
$5C_6H_{12}O_6 + 24KNO_3 \longrightarrow 30CO_2 + 18H_2O + 24KOH + 12N_2$	*Pseudomonas denitrificans*	2,385
$5S + 6KNO_3 + 2CaCO_3 \longrightarrow 3K_2SO_4 + 2CO_2 + 3N_2$	Anaerobic sulfur bacteria	
Ammonification		
$C_2H_5NO_2 + 1\frac{1}{2}O_2 \longrightarrow 2CO_2 + H_2O + NH_3$	Many bacteria; most plants and animals	736
Nitrification		
$NH_3 + 1\frac{1}{2}O_2 \longrightarrow HNO_2 + H_2O$	*Nitrosomonas*	276
$KNO_2 + \frac{1}{2}O_2 \longrightarrow KNO_3$	*Nitrobacter*	73
Nitrogen fixation		
$2N_2 + 3H_2 \longrightarrow 2NH_3$	Some cyanobacteria, *Azotobacter*	−616
Oxidation of sulfur		
$2H_2S + O2 \longrightarrow S_2 + 2H_2O$		335
$S_2 + 3O_2 + 2H_2O \longrightarrow 2H_2SO_4$		1,004
Oxidation of iron		
$Fe^{2+} \longrightarrow Fe^{3+}$		48

Note: $C_6H_{12}O_6$ *= glucose;* CO_2 *= carbon dioxide;* $C_2H_5NO_2$ *= glycine (an amino acid);* $CaSO_4$ *= calcium sulfate;* $CaCO_3$ *= calcium carbonate;* HNO_2 *= nitrous acid;* H_2S *= hydrogen sulfide;* H_2SO_4 *= sulfuric acid;* KNO_2 *= potassium nitrite;* KNO_3 *= potassium nitrate; KOH = potassium hydroxide;* NH_3 *= ammonia;* N_2O *= nitrous oxide; S = sulfur.*

(From Delwiche 1970, Rheinheimer 1980.)

is accomplished by bacteria such as *Azotobacter,* which is a free-living species, and *Rhizobium,* which occurs in symbiotic association with the roots of some legumes (members of the pea family) and other plants (Figure 11-14) as well as by cyanobacteria (Quispel 1974, Stewart 1975, Shanmugam et al. 1978). The enzyme responsible for nitrogen fixation—nitrogenase—consists of two protein subunits, one containing one atom each of iron and molybdenum and the other containing an atom of iron. The enzyme is extremely sensitive to oxygen and works efficiently only under extremely low oxygen concentrations. This explains why *Azotobacter* bacteria, living freely in the soil, exhibit only a small fraction of the nitrogen-fixing capacity of *Rhizobium* bacteria, which are sequestered in the relatively anoxic cores of root nodules.

Nitrogen fixation requires energy (see Table 11-4). It has been estimated that between 8 and 12 g of glucose ($C_6H_{12}O_6$) is required to fix 1 gram of nitrogen (Gutschick 1981). Nitrogen-fixing microorganisms obtain the energy and reducing power (organic carbon) they need to reduce N_2 to NH_3 by oxidizing sugars or other organic compounds. Free-living bacteria must obtain these resources by metabolizing organic detritus in the soil, sediments, or water column. More abundant supplies of energy are available to bacteria that enter into symbiotic relationships with plants, which provide them with photosynthate. The best known of these associations is that of root nodules of leguminous plants (peas, alfalfa, and their relatives) with the nitrogen-fixing bacterium *Rhizobium* (Quispel 1974, Broughton 1983, Bottomley 1992). The nodules are specialized structures of the root cortex whose development is stimulated by *Rhizobium* (see Figure 11-14). The nodules, having abundant photosynthate and low oxygen concentrations, provide an optimum environment for nitrogen fixation (Werner 1992). Nitrogen-fixing symbioses are by no means limited to legumes, however,

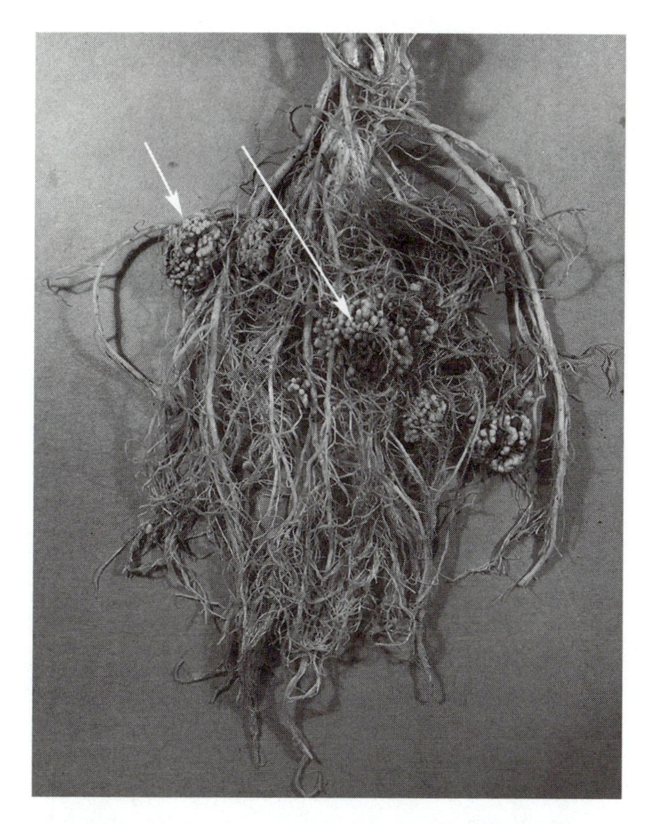

FIGURE 11-14 The root system of an Austrian winter pea plant, showing the clusters of nodules that harbor symbiotic nitrogen-fixing bacteria. The arrows point to two clusters of nodules. (*Courtesy of the U. S. Soil Conservation Service; from Grant and Long 1981.*)

and such associations with bacteria or blue-green algae have been identified in other terrestrial plants (Becking 1992), some marine algae (Wiebe 1975), lichens (Milbank and Kershaw 1969), and shipworms (Carpenter and Culliney 1975).

Balancing the Global Nitrogen Budget

Denitrification, and the subsequent production of molecular nitrogen by physical reactions, puts slightly more nitrogen into the global ecosystem than total nitrogen fixation, pushing the current global nitrogen cycle, like that of carbon, out of equilibrium. For the most part, denitrification produces N_2, but small amounts of NO and N_2O (nitrous oxide) are also released into the atmosphere. This is particularly true in tropical rain forests, which contribute between 2.4 and 7.4×10^{12} g N_2O yr^{-1} through both nitrification and denitrification processes, principally in the soils, or about one-half of the biogenic N_2O produced each year (McElroy and Wofsy 1986, Matson and Vitousek 1990). But the widespread use of nitrogen fertilizers and the emissions of internal combustion engines add in excess of 90×10^{12} g N yr^{-1} in the form of NO and NO_2 to the terrestrial environment (Schlesinger 1991). Some ecologists have estimated that the amount of fixed nitrogen deposited into the environment will increase 25% in the next 25 years because of human activities (Galloway et al. 1994). Most of this excess nitrogen is lost from terrestrial ecosystems by denitrification; some is transported into the oceans by river flow.

The fate of N_2O is of particular importance because of its exceptional efficiency as a greenhouse gas (see Chapter 3)—one molecule of N_2O contributes 200-fold more to the earth's greenhouse than a molecule of CO_2 (Schlesinger 1991)—and because the only known sink of N_2O is the stratosphere, where it is destroyed in a reaction with ozone to produce NO. Current estimates suggest that the amount of N_2O in the atmosphere increases by about 3% each year. The total N_2O input from all sources is approximately 11.1×10^{12} g yr^{-1}. About 95% of this input (10.5×10^{12} g yr^{-1}) is destroyed in the stratosphere. The rest accumulates in the atmosphere and enhances the greenhouse effect. But the atmospheric increase in the amount of N_2O is thought to be about 3.0×10^{12} g yr^{-1}, leaving about 2.4×10^{12} g yr^{-1} in atmospheric N_2O for which a source is unknown. The search for N_2O inputs continues, with a focus on agricultural soils, where fertilization may increase N_2O production.

Nitrogen in the forms of nitric oxide (NO) and nitrogen dioxide (NO_2), along with sulfur dioxide (SO_2), are the chief industrial products that contribute to the phenomenon of **acid rain.** These compounds react with other atmospheric constituents, particularly hydroxyl radicals (OH^-), to form nitric acid (HNO_3) and sulfuric acid (H_2SO_4), both of which are soluble in water. These compounds lower the pH of the rainwater that brings them to the surface of the earth. When the pH of rainwater is below 5.0, it is called acid rain (Schindler 1988). Forest and freshwater environments over large parts of eastern North America and Europe have been damaged as a result of acid rain.

11.7 Phosphorus cycling is closely linked with pH in the soil and with trophic interactions in aquatic environments.

Ecologists have intensively studied the role of phosphorus in ecosystems because living organisms require phosphorus, which is a major constituent of nucleic acids, cell membranes, energy transfer systems, bones, and teeth. Low levels of phosphorus limit plant productivity in many freshwater aquatic habitats, but the influx of phosphorus into many rivers and lakes, in the form of sewage and runoff from fertilized agricultural lands, artificially stimulates production in aquatic habitats, with undesirable consequences.

The Global Phosphorus Cycle

Unlike carbon, nitrogen, and, as we shall see, sulfur, phosphorus does not occur in any appreciable amount in the atmosphere; thus, the global phosphorus cycle involves only soil and aquatic compartments (Figure 11-15). Phosphorus occurs in the environment in several forms. Particulate organic phosphate is that phosphorus incorporated into the cells of living organisms or dead organic material. Phosphorus also occurs in two inorganic forms: orthophosphate (PO_4^{3-}), a soluble form readily assimilated by organisms, and colloidal phosphate, another soluble form of phosphate that is not available to organisms. In addition, under certain pH conditions (see below), phosphorus undergoes reactions with soil minerals, particularly iron and aluminum, to form compounds that precipitate out of the soil and become unavailable to living organisms.

The total amount of phosphorus in the various forms in the soil and ocean pools is quite large, but the amount available for use by organisms in most environments is extremely small in comparison. The principal global flux of phosphorus is the movement of about 21×10^{12} g P yr^{-1} from the terrestrial pool to the oceans by river flow (see Figure 11-15), most of which (about 20×10^{12} g P yr^{-1}) is in the form of particulate phosphate in dead organic matter. Phosphate fertilizers used for agriculture, which are obtained by mining phosphate-rich rock, are added to the soil compartment at a rate of about 14×10^{12} g yr^{-1}; most of this input is carried into the oceans by runoff and subsequent river transport (Schlesinger 1991, Caraco 1993).

Phosphorus Cycling in the Soil

Because the soluble form of phosphorus usable by organisms is in such limited supply in most environments, its availability depends on well-developed recycling processes. In soils, phosphorus recycling involves both biological and geochemical transformations through which soluble phosphate is made available for uptake by soil biota. In developing soils, soluble phosphate is released primarily by weathering of the mineral apatite [$Ca_5(PO_4)_3OH$] and may be assimilated immediately by organisms, becoming part of the particulate phosphate soil pool. Inorganic (soluble) phosphate (PO_4^{3-}) is released into the soil waters by the process of mineralization of dead organic material, in which CO_2 and nutrients in inorganic form are released through decomposition.

Much of the soluble phosphate mobilized by the weathering of apatite undergoes reactions with other minerals in the soil such as iron and aluminum, a process that is very much controlled by the pH of the soil environment. In general, at low pH (acid conditions), phosphorus binds tightly to clay particles in the soil and forms relatively insoluble compounds with ferric iron [strengite, $Fe(OH)_2H_2PO_4$] and aluminum [variscite, $Al(OH)_2H_2PO_4$]. At high pH, other insoluble compounds are formed; for example, with calcium [hydroxyapatite, $Ca_{10}(PO_4)_6(OH)_2$]. These relationships can be portrayed in a solubility diagram

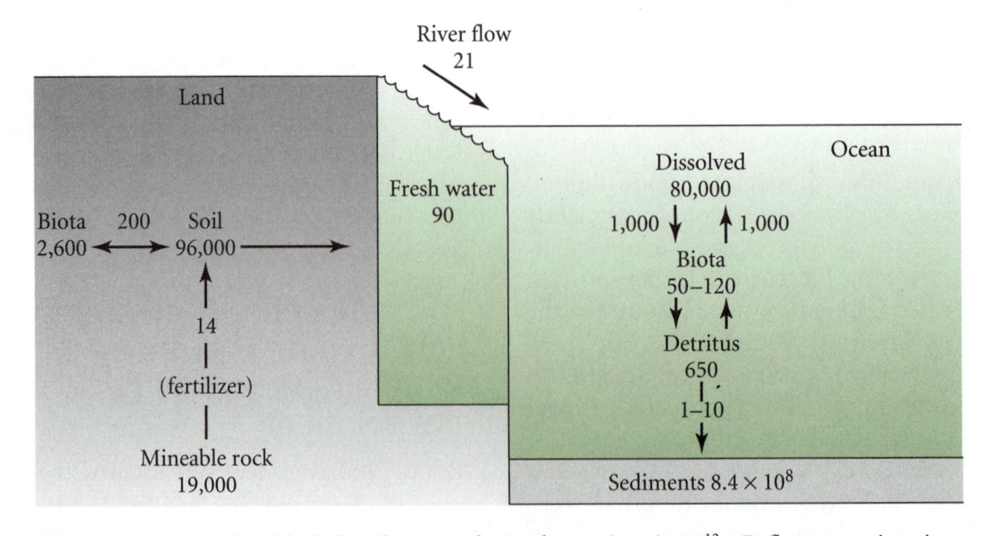

FIGURE 11-15 The global phosphorus cycle. Pools are given in 10^{12} g P; fluxes are given in 10^{12} g P/yr. While large amounts of phosphorus reside in the ocean and in the terrestrial ecosystem, most of it is not accessible to living organisms. The largest flux of phosphorus is movement from the terrestrial compartment to the ocean. (*Redrawn from Schlesinger 1991; based on data from Graham and Duce 1979, Maybeck 1982, and Richey 1983.*)

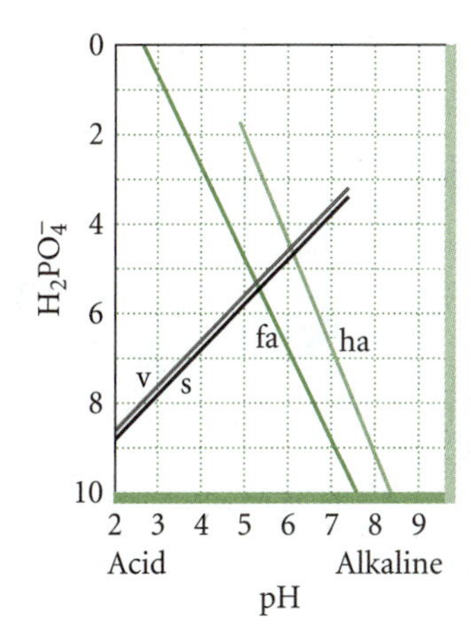

FIGURE 11-16 The solubility of phosphate ions in equilibrium with several insoluble forms of phosphorus as a function of pH of the soil. v = variscite; s = strengite; fa = fluoroapatite; ha = hydroxyapatite. (*After Lindsay and Moreno 1960.*)

(Figure 11-16), in which the concentration of dissolved $H_2PO_4^-$ in equilibrium with insoluble compounds is shown as a fraction of the acidity of the soil. At low pH and in the presence of ferric iron or aluminum (which are present in virtually all soils, sediments, and waters), the equilibrium is driven to a very low level of dissolved $H_2PO_4^-$. Similarly, in alkaline soils in the presence of calcium, the equilibrium concentration is also very low. When both ferric iron or aluminum and calcium are present, the highest concentration of dissolved phosphate—that is, the greatest availability of phosphorus—occurs at a pH of between 6 and 7.

Phosphorus in Aquatic Systems

Most of the phosphorus found in freshwater lakes (about 98%) is held as organic or particulate phosphorus within the plankton (this accounts for the heavy input of particulate phosphate in the river flow to the oceans). The rest is dissolved in the water as orthophosphate or suspended as colloidal phosphate. The rate of phosphorus recycling is very much dependent on trophic interactions in freshwater lakes. Beginning with bacteria and phytoplankton, organic phosphate moves through the aquatic food chain by consumption, first by protozoa and zooplankton, then by invertebrates and fish. Soluble phosphate is returned to the water column by phytoplankton and zooplankton excretion, bacterial decomposition, and mineralization. Some phosphate is lost to the sediments during detrital decomposition or in reactions with secondary minerals. Overlying these trophic interactions in temperate lakes are the processes of stratification and seasonal turnover. In strongly stratified lakes, phosphorus may become extremely limited in the upper waters as phytoplankton die and sink. In such situations, the amount of primary production may depend entirely on the amount of allochthonous phosphorus input (Schindler 1978).

11.8 Sulfur takes part in many redox reactions.

Sulfur is required by organisms as a component of the amino acids cysteine and methionine, but the importance of sulfur in the ecosystem goes far beyond this. Like nitrogen, sulfur has many oxidation states. It follows complex chemical pathways and affects the cycling of other elements.

Sulfur Redox Reactions

The most oxidized form of sulfur is sulfate (SO_4^{2-}; that is, oxidation state S^{6+}); the most reduced forms are sulfide (S^{2-}) and the organic (thiol) form of sulfur (also S^{2-}). Under aerobic conditions, assimilatory sulfur reduction ($SO_4^{2-} \rightarrow$ organic S) is balanced by the oxidation of organic sulfur back to sulfate, either directly or with SO_3^{2-} (sulfite, S^{4+}) as an intermediate step (Figure 11-17). This oxidation is accomplished by most animals when they excrete excess dietary organic sulfur and by microorganisms when they decompose plant and animal detritus.

Under anaerobic conditions (Eh < 0, dependent upon pH), sulfate may function as an oxidizer (see Table 11-1). In sediments, the bacteria *Desulfovibrio* and *Desulfomonas* couple dissimilatory sulfate reduction ($SO_4^{2-} \rightarrow S^{2-}$) to the oxidation of organic carbon in order to make energy available (Le Gall and Postgate 1973). The reduced S^{2-} may then be used as a reducer ($S^{2-} \rightarrow S^0 + 2e^-$) by photoautotrophic bacteria, which assimilate carbon by pathways analogous to photosynthesis in green plants. In these reactions, sulfur takes the place of the oxygen atom in water as an electron donor. Elemental sulfur (S^0) accumulates unless the sediments are exposed to aeration or oxygenated water, at which point the sulfur may be further oxidized to SO_3^{2-} and SO_4^{2-}.

The fate of S^{2-} produced under oxic conditions depends on the availability of positive ions. Frequently hydrogen sulfide is formed; it escapes from shallow sediments and moist soils as a gas having the familiar smell of rotten eggs. Because oxic conditions generally favor the reduction of ferric iron (Fe^{3+}) to ferrous iron (Fe^{2+}), the presence of iron in sediments leads to the formation of iron sulfide (FeS). Sulfides are commonly associated with coal and oil. When these materials are exposed to the atmosphere in mine wastes or

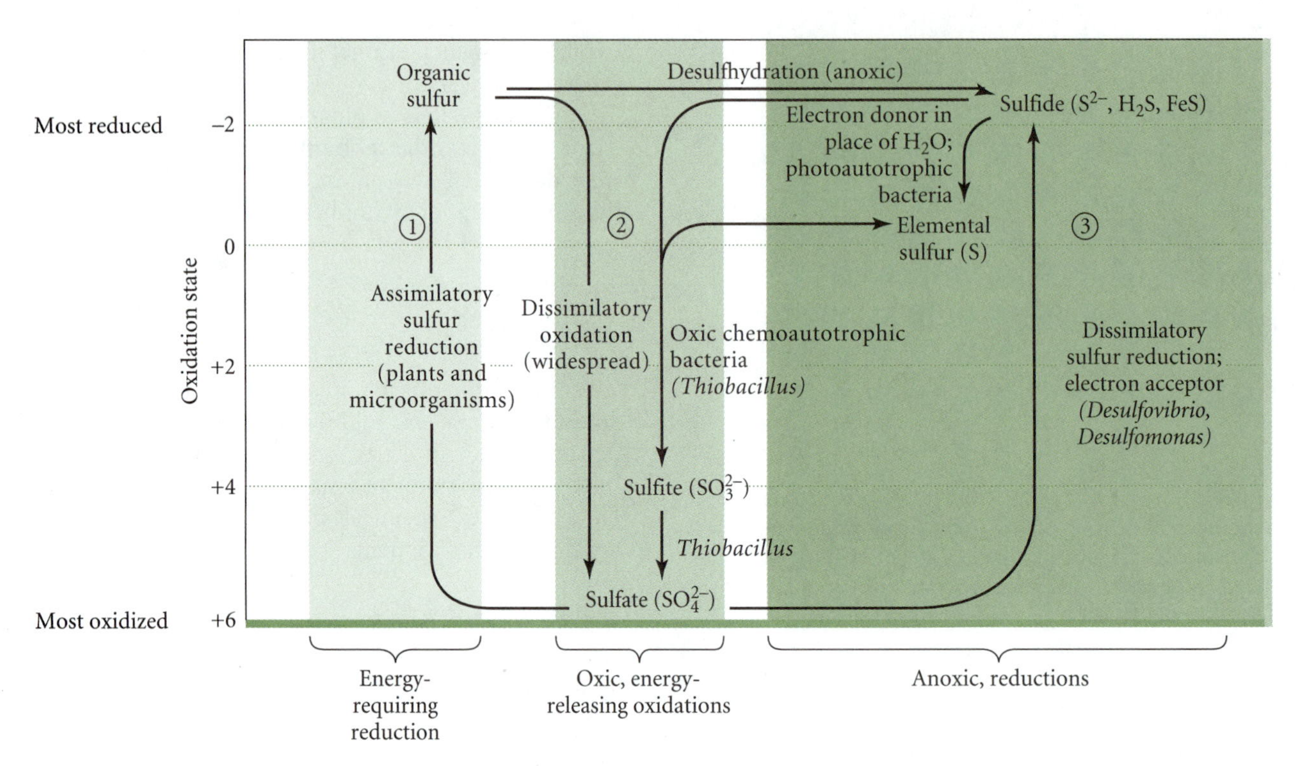

FIGURE 11-17 Schematic diagram of the transformations and oxidation states of compounds in the sulfur cycle. Assimilatory sulfur reduction is accomplished by plants and microorganisms (1). This is balanced by dissimilatory oxidation by animals (2). Under some conditions, bacteria couple dissimilatory sulfate reduction with the oxidation of organic carbon to release energy (3).

burned for energy, the reduced sulfur is oxidized, with the help of *Thiobacillus* bacteria in mine wastes, and the oxidized forms combine with water to form the sulfuric acid of acid mine drainage and acid rain.

The Global Sulfur Cycle

The sulfur cycle is one of the least understood of the global element cycles (Schlesinger 1991). Sulfur gas does not reside very long in the atmosphere because it is quickly oxidized to SO_4 and redeposited, either in dry form or in precipitation. Thus, the atmosphere contains relatively little sulfur at any time. However, the flux of sulfur through the atmosphere is large because of the presence of substantial inputs. There are at least four major inputs of sulfur into the atmosphere from land (Figure 11-18). Volcanic activity, which contributes sulfur primarily in the form of H_2SO_4, accounts for 10×10^{12} g yr^{-1}. Roughly twice that amount enters the atmosphere in soil dust, which contains sulfur in the form of substances such as gypsum ($CaSO^4H_2O$). Both volcanic activity and dust storms are episodic, resulting in considerable year-to-year variation in the level of sulfur input into the atmosphere. Human industrial activity adds large amounts of sulfur (93×10^{12} g yr^{-1}) to the atmosphere, principally in the form of SO_2, much of which is returned to the earth near the area where it was produced. The activity of sulfur bacteria in soils releases 22×10^{12} g yr^{-1} into the atmosphere in the form of biogenic gases such as H_2S. Approximately 84×10^{12} g yr^{-1} is redeposited on land via precipitation or in dry form, and a small amount is imported from the oceans (20×10^{12} g yr^{-1}).

Sulfur enters the atmosphere from the oceans in three ways. Deep-sea hydrothermal vents release about 10×10^{12} g yr^{-1}. An additional 43×10^{12} g yr^{-1} enters in the form of biogenic gas, primarily in the form of dimethylsulfide [$(CH_3)_2S$]. Seawater whipped by winds forms aerosols containing SO_4 (144×10^{12} g yr^{-1}). Current estimates of inputs of sulfur to the oceans exceed estimates of the flow of sulfur from the oceans. Much of the sulfur released into the atmosphere by the oceans is redeposited in the oceans. A small amount is transported to land by winds. Losses from the ocean system, however, are far smaller than inputs to the oceans from the atmosphere (258×10^{12} g yr^{-1}) and from the terrestrial system in the form of river water containing SO_4 (213×10^{12} g yr^{-1}).

As with the carbon and nitrogen cycles, human activities contribute a substantial amount of sulfur to the biosphere each year. Globally, anthropogenic inputs of sulfur account for over half of the sulfur input into the atmosphere from the terrestrial compartment; in industrialized areas, they may contribute

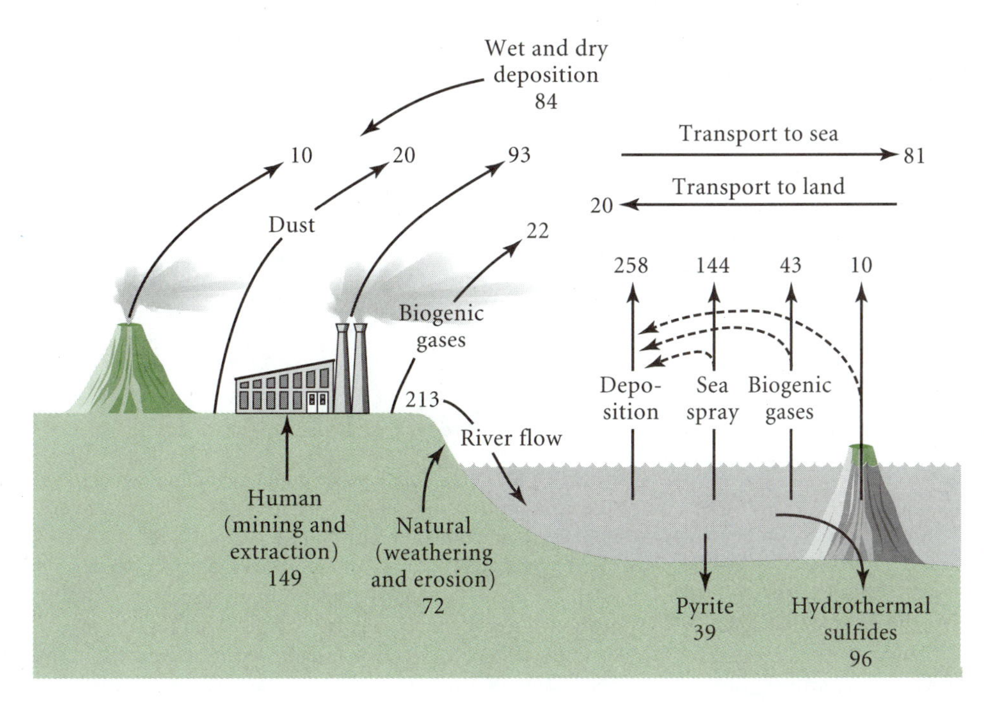

FIGURE 11-18 The global sulfur cycle. Sulfur enters the atmosphere primarily through volcanic activity and deep-sea venting; the release of sulfur by bacteria, particularly in marshes and wetlands; and industrial emissions. Additional sulfur enters the atmosphere in soil dust and through the production of aerosols at the surface of the oceans. In the atmosphere, sulfur oxidizes readily and is returned to the earth via precipitation or dry deposition.

90% of the sulfur input into the air (Schindler 1988). Most of the input is in the form of SO_2, which is highly reactive in the atmosphere and which, in the presence of water, may be converted to SO_4^{2-} or HSO_3 by the following reactions:

$$SO_2 \longrightarrow SO_3 \longrightarrow H_2SO_4 \rightleftharpoons H^+ + SO_4^{2-},$$
$$SO_2 \longrightarrow H^+ + HSO_3.$$

The deposition of SO_4^{2-}, HSO_3, and the H^+ ions lowers the pH of precipitation, causing acid rain.

11.9 Element cycles interact in complex ways in ecosystems.

One might think that carbon, nitrogen, phosphorus, sulfur, and the other elements cycle among their respective global pools more or less independently of one another, each subject only to the special biogeochemical transformations peculiar to its origin, chemistry, and biological function. In reality, such an impression is too simplistic to account for the complexity of ecosystem function. Element cycles are linked at many different levels. The assimilatory and dissimilatory reactions of carbon in photosynthesis and respiration link the carbon and oxygen cycles. The rate of primary production in ocean waters may be influenced by the nitrogen-to-phosphorus ratio of the phytoplankton, thereby linking the carbon cycle with those of nitrogen and phosphorus (Howarth 1988). The availability of phosphorus in freshwater systems is related in part to the indirect influence of nitrate on iron and oxygen cycles in the bottom sediments (Caraco 1993).

Because of these linkages, amendments to the global carbon, nitrogen, or phosphorus budgets by human activities will change the biogeochemical cycling of these elements in complex and often unpredictable ways. Adding more CO_2 to the atmosphere, in particular, may have far-reaching consequences, owing both to the potential increase in rates of photosynthesis and to global warming and possible concomitant reductions in light intensity and soil moisture. At the physiological level, plant responses to elevated CO_2 levels and to increases in temperature interact strongly with each other, as well as with the levels of other nutrients. For example, the response of temperate forest trees to high levels of CO_2 is constrained by the availability of nitrogen (see above). An understanding of the consequences of human-induced changes in global nutrient cycles requires an understanding of how those cycles interact. Our understanding of these interactions is far from complete, and will thus be the subject of much ecological research in the coming decades.

A complete discussion of element cycle interactions is beyond the scope of this book. To demonstrate the importance of such interactions, we will examine how interactions of carbon and nitrogen provide some possible explanations of the response of a particularly sensitive environment to global warming.

Response of the Arctic Tundra to Elevated CO_2

G. R. Shaver, of the Marine Biological Laboratory at Woods Hole, and his colleagues (1991, 1992) have found that the arctic tundra is an excellent model ecosystem for the study of the interactions of nutrient cycles in response to global warming. Because of the presence of permafrost (see Chapter 8), a permanently frozen soil layer lying 10–30 cm below the surface, soil processes occur near the surface and groundwater activity is limited, making the tundra simpler than other terrestrial ecosystems. Moreover, soil types and, consequently, vegetation types tend to be easily distinguished and sharply divided, making the study of the transport of materials among the various compartments of the ecosystem more straightforward.

The tundra is thought to be more sensitive to global climate change than most other ecosystems on earth. Huge amounts of carbon are stored in the soils (see Table 11-3) and in the parts of the tundra vegetation below ground; the cold temperatures of the high latitudes retard decomposition and respiration, thereby restraining the release of this carbon into the atmosphere. Presumably, an increase in temperature would act as a countervailing force on the delicate balance of this system by increasing respiration and decomposition, thus releasing CO_2 and exacerbating climate change. But it is unclear whether increased photosynthetic rates in tundra plants resulting from the higher atmospheric CO_2 levels would counterbalance an increase in atmospheric CO_2.

A number of factors in addition to the increase in atmospheric CO_2 would come into play. For example, global warming might bring greater cloudiness because of changes in the hydrologic cycle. Such a change would lower light intensity and reduce the rate of photosynthesis. Also, photosynthesis requires nutrients such as nitrogen and phosphorus. The cycles of these nutrients may be affected by global warming as well. The role of high-latitude freshwater ecosystems must also be considered. Thus, the question of whether the tundra can be expected to be a sink or a source of carbon is unresolved.

Careful experimental studies of the arctic tundra indicate that the response of the tundra ecosystem to global warming will depend upon the interaction between the carbon and the nitrogen cycles there (Chapin et al. 1980, Bazzaz 1990, Shaver et al. 1992). Shaver and his coworkers (1992) have proposed a conceptual model of how carbon:nitrogen (C:N) ratios could shape the response of the tundra to increases in atmospheric CO_2 (Figure 11-19). Carbon and nitrogen occur in both of the two main subcompartments of the tundra system: plants and soil organic matter. Carbon is assimilated by photosynthesis in plants and transferred in organic form (in litterfall) to the soil, where respiration and decomposition return CO_2 to the atmosphere. This process also results in the mineralization of nitrogen held in the soil organic matter, releasing it into the dissolved inorganic nitrogen pool, where it may be taken up by plants. But, as we

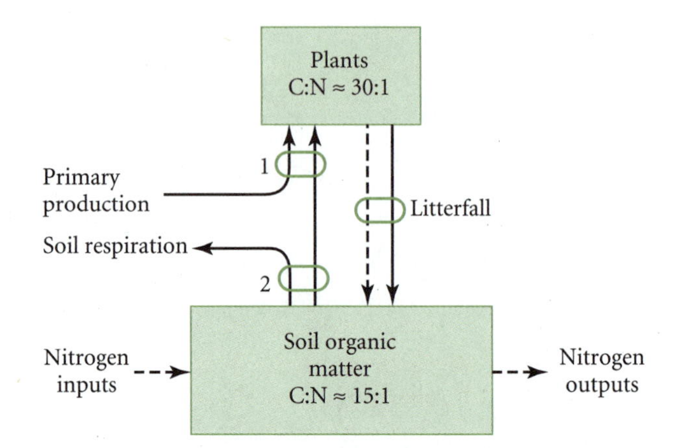

FIGURE 11-19 Conceptual model of the interaction of carbon and nitrogen in the arctic tundra. The tundra has two main compartments: plants and soil organic matter. The cycles of carbon and nitrogen are coupled. Carbon enters the ecosystem via primary production. The sequestration of carbon as plant biomass requires nitrogen from the soil compartment. Carbon leaves the system in soil respiration. Warmer temperatures may increase primary production, thereby increasing carbon input, and soil respiration, thereby increasing carbon output. The extent to which production may be increased is constrained by the availability of nitrogen. Thus, whether the tundra is a sink or a source of carbon may be determined by the availability of nitrogen. (*After Shaver et al. 1992.*)

mentioned above, decomposition occurs slowly in the cold tundra. Microbial nitrogen fixation and atmospheric inputs are also extremely low in tundra ecosystems. Thus, the vast majority of the nitrogen in the tundra resides in soil organic matter (Chapin et al. 1980). Generally, the soil has a lower C:N ratio than does the plant compartment, and the C:N ratio is more or less fixed for a particular vegetation type over the long term.

These characteristics suggested to Shaver and his coworkers a set of possible effects of warmer temperatures and increased atmospheric levels of CO_2 on the tundra ecosystem. Their model focuses on the connection between the carbon and nitrogen cycles in the tundra. If warmer temperatures increase the uptake of CO_2 by plants (carbon input) more than the release of CO_2 by soil respiration (carbon output), then the tundra will be a carbon sink. If the release of CO_2 from soil respiration is greater than the uptake via photosynthesis, the tundra will be a carbon source. The determining factors are the availability of nitrogen and the ratio of carbon to nitrogen. Because the vast majority of nitrogen in the tundra is held in the soil organic matter, the availability of nitrogen to plants depends on the rate of mineralization of nitrogen in the soil. If nitrogen mineralization does not increase with increasing temperatures, then plant production will be limited, outputs from increased soil respiration will exceed input via photosynthesis, and the tundra would become a carbon source. On the other hand, if warmer temperatures enhance nitrogen mineralization, thereby providing plants with additional nitrogen from the soil compartment, primary production will store carbon more rapidly than it is released into the atmosphere by soil processes, making the tundra a carbon sink (Shaver et al. 1992). The entire system is constrained by the differences in C:N ratios between the plant and soil compartments. Because soils typically have a lower C:N ratio than plants, the availability of nitrogen greatly increases carbon storage in plants. The C:N ratios given in Figure 11-19 are typical for moist tundra. Trees typically have C:N ratios of 200:1, suggesting that small increases in nitrogen availability would result in considerable carbon storage.

To be sure, Shaver's model is a simplification. The increase in the availability of nitrogen resulting from higher rates of mineralization might be counterbalanced by a loss of nitrogen resulting from greater leaching from soils. Other nutrients, such as phosphorus, are likely to have a strong interaction with the carbon cycle in some ecosystems, particularly aquatic ones.

11.10 Microorganisms assume special roles in element cycles.

Many of the transformations discussed in this chapter are accomplished mainly or entirely by microorganisms (Coleman et al. 1983). In fact, if it were not for the activities of bacteria and fungi, many element cycles would be drastically altered and the productivity of the ecosystem much reduced. For example, wood is broken down for the most part only by microorganisms and, especially, fungi (Kirk et al. 1977, Crawford and Crawford 1980). In their absence, organic carbon would accumulate as cellulose and lignin, rather than passing through detritus-based food chains producing CO_2. Over long periods, primary production would drop as levels of CO_2 in the atmosphere decreased. Further, without the capacity of some microbes to utilize sulfur and iron as electron acceptors, little decomposition would occur in anoxic organic sediments. The resulting accumulation would further reduce the amount of inorganic carbon in the ecosystem.

The origin of forms of nitrogen that can be assimilated by plants depends mostly on nitrogen fixation by microorganisms, although some useful nitrogen is also produced by lightning discharges in the atmosphere. Without nitrogen-fixing bacteria, however, denitrification under anaerobic conditions would slowly deplete ecosystems of useful nitrogen and reduce biological productivity. Under aerobic conditions, plants and bacteria compete for ammonia in the soil and water column. Because plants and algae can assimilate ammonia as easily as, and at less energetic expense than, nitrate, the activities of nitrifying bacteria may reduce the primary productivity of terrestrial and aquatic plants by forcing them to allocate more of their energy to nitrogen assimilation. Nonetheless, without microorganisms, much organic detritus would never decompose, and its minerals would not be released to support plant production.

Many of the transformations carried out by microorganisms, such as the metabolism of sugars and other organic molecules, are accomplished in similar ways by plants and animals. The bacteria and cyanobacteria ("blue-green algae") are distinguished physiologically primarily by the ability of many species to metabolize substrates under anoxic conditions and to use substrates other than organic carbon as energy sources. The requirements of fungi for inorganic substrates as sources of nitrogen, phosphorus, potassium, sulfur, and other elements are similar to those of plants (Griffin 1981). What distinguishes the fungi from the bacteria is their greater ability to break down such complex polysaccharides as cellulose and lignin, which make up a large proportion of terrestrial plant litter (Grant and Long 1981).

Every organism needs, above all, a source of carbon for building organic structures and sources of energy to fuel the life processes. As we saw in Chapter 10, organisms can be distinguished according to their source of carbon. Heterotrophs obtain carbon in reduced (organic) form by consuming other organisms or organic detritus. All animals and fungi, and many bacteria, are heterotrophs. Autotrophs assimilate carbon as carbon dioxide and expend energy to reduce it to an organic form. **Photoautotrophs** utilize sunlight as their source of energy for photosynthesis. These include all green plants, algae, and the cyanobacteria, which use H_2O as an electron donor ($CO^{2-} \rightarrow O^0 + 2e^-$) and are aerobic (Wolk 1973), and the purple and green bacteria, which have light-absorbing pigments different from those green plants, use H_2S organic compounds as electron donors, and are anaerobic. **Chemoautotrophs** obtain their energy through the oxidation of inorganic compounds.

Organisms, then, may obtain energy from three sources: sunlight (photoautotrophs), the oxidation of organic compounds (heterotrophs), and the oxidation of inorganic compounds (chemoautotrophs). Oxidation of glucose (represented as $CH_2O + O_2 \rightarrow CO_2 + H_2O$) incorporates an oxidation reaction that releases energy (in this case, $C^0 \rightarrow C^{4+} + 4e^-$) and an electron-accepting (reduction) reaction ($2O^0 + 4e^- \rightarrow$

$2O^{2-}$) (see Section 11.4). Heterotrophs all use C^0 as a reduced substrate for oxidation. Under aerobic conditions, O_2 is used as an electron acceptor. But many bacteria (sulfate-reducing bacteria, or methanogens) are either facultatively or obligately capable of utilizing SO_4^{2-}, Fe^{3+}, or CO_2 as electron acceptors under anaerobic conditions (Zehnder 1988, Fenchel and Finlay 1995).

Chemoautotrophs all use CO_2 as a carbon source, but obtain energy for its reduction by the aerobic oxidation of inorganic substrates, such as methane (for example, *Methanosomonas, Methylomonas*), hydrogen (*Hydrogenomonas, Micrococcus*), ammonia (nitrifying bacteria, *Nitrosomonas, Nitrosococcus*), nitrite (nitrifying bacteria, *Nitrobacter, Nitrococcus*), hydrogen sulfide, sulfur, and sulfite (*Thiobacillus*), and ferrous iron salts (*Ferrobacillus, Gallionella*). The chemoautotrophs are almost exclusively bacteria, which apparently are the only kind of organism that is biochemically specialized enough to make efficient use of inorganic substrates and efficiently dispose of the products of chemoautotrophic metabolism.

The special role of microorganisms in ecosystem function is illustrated nicely by the highly productive communities of marine organisms that develop around deep-sea hydrothermal vents (Grassle 1985, 1996, Janinasch and Mottl 1985). These miniature ecosystems were first discovered in deep waters off the Galápagos archipelago in 1977 and have since been found more widely in ocean basins around the world. The most conspicuous members of the community are giant white-shelled clams and tube worms (pogonophorans) that grow to 3 meters long, but numerous crustaceans, annelids, mollusks, and fishes also cluster at great densities around hydrothermal vents. The high productivity of vent communities contrasts strikingly with the desertlike appearance of the surrounding ocean floor. As you might suspect, it is based on the unique qualities of the water issuing from the vents themselves, which is hot and loaded with a reduced form of sulfur, hydrogen sulfide (H_2S). Where vent water and seawater mix, conditions are ideal for chemoautotrophic sulfur bacteria. These bacteria use the oxygen in seawater to oxidize the hydrogen sulfide in vent water as a source of energy for assimilatory reduction of the inorganic carbon and nitrogen in seawater. All the other members of the vent community feed on the bacteria, which thus form the base of the food chain. The pogonophoran worms have gone so far as to house symbiotic colonies of the bacteria within a specialized organ, the trophosome, trading a protected environment for a share of the carbohydrate and organic nitrogen produced by the bacteria.

In this chapter we have examined the cycling of several important elements from the standpoint of their chemical and biochemical reactions. Elements are cycled through the ecosystem primarily because the metabolic activities of organisms result in chemical transformations of those elements. The kinds of transformations that predominate depend on the physical and chemical conditions of the system. Each type of habitat presents a different chemical environment, particularly with respect to the presence or absence of oxygen and possible sources of energy. It stands to reason, therefore, that patterns of element cycling should differ greatly among habitats and ecosystems. In the next chapter, we shall contrast element cycling in aquatic habitats and in terrestrial habitats by focusing on how some of the unique physical features of each of these environments affect the chemical and biochemical transformations involved in organic production and recycling of elements.

SUMMARY

1. Unlike energy, nutrients are retained within the ecosystem and are cycled between its abiotic and biotic components. The paths of elements through the ecosystem depend upon chemical and biological transformations, which are functions of the chemical characteristics of each element and the ways in which organisms use it.

2. The movement of energy through the ecosystem parallels the paths of several elements, particularly carbon, whose transformations either require or release energy.

3. The cycling of each element may be thought of as movement between compartments of the ecosystem, the major compartments being living organisms, organic detritus, immediately available inorganic forms, and unavailable organic and inorganic forms, generally in sediments. The size of an element pool is represented by the amount of the element in that pool. If the rate of input of an element to a compartment is larger than the rate of output, then the compartment is referred to as a sink. If the rate of output is greater than the rate of input, the compartment is referred to as a source.

4. The water cycle provides a physical analogy for element cycling. Energy is required to transform water from its liquid phase to atmospheric vapor. Upon condensation and precipitation, that energy is released as heat.

5. Energy transformations in biological systems occur primarily during oxidation-reduction (redox) reactions.

An oxidizer is a substance that readily accepts electrons (O_2, NO_3^-); a reducer is one that readily donates electrons. Upon being reduced, oxidizers gain energy; upon being oxidized, reducers release energy. Elements in organic compounds tend to be in reduced forms; hence, biological assimilation of many elements (carbon, nitrogen, sulfur) requires energy.

6. All organisms require organic carbon as the primary substance of life. It also is the major source of energy for most animals and microorganisms. Carbon shuttles between living forms and the carbon dioxide compartment of the ecosystem by way of photosynthesis and respiration.

7. The oceans serve as a carbon sink, taking in more carbon than they release. The movement of the products of primary production from ocean surface waters to deeper waters is referred to as biological pumping, a process that lowers CO_2 concentrations in the surface waters. In contrast, freshwater lakes appear to constitute a source of carbon for the atmosphere.

8. The global carbon cycle is out of balance owing to the release of large amounts of carbon from the fossil pool by human activities. Oceans are thought to absorb much of this excess carbon. The idea that terrestrial plants increase their rate of biomass accumulation under high atmospheric CO_2 concentrations and thereby serve as a carbon sink is under active investigation.

9. Nitrogen has many oxidation states and consequently follows many pathways through the ecosystem. Quantitatively, most of the flux follows the cycle nitrate → organic nitrogen → ammonia → nitrite → nitrate. The last two steps, referred to as nitrification, are accomplished by certain bacteria under oxic conditions. Under anoxic conditions in soils and sediments, nitrate replaces oxygen as an electron acceptor (denitrification) and the reactions reverse: nitrate → nitrite → (eventually) molecular nitrogen (N_2). This loss of nitrogen from the biotic compartments is balanced by nitrogen fixation by certain microorganisms.

10. Like the carbon cycle, the nitrogen cycle is slightly out of balance owing to human activities. Widespread use of nitrogen fertilizers and emissions from internal combustion engines add significant quantities of nitrogen to the global ecosystem each year. The concentration of nitrous oxide, an exceptionally efficient greenhouse gas, in the atmosphere is increasing. The source of about 95% of the yearly input of nitrous oxide is known.

11. Phosphorus occurs primarily in the terrestrial and aquatic pools in the form of particulate phosphate, inorganic or orthophosphate, or colloidal phosphate.

12. Phosphorus does not change its oxidation state. It is assimilated by plants in the form of phosphate (PO_4^{3-}), whose availability is largely determined by the acidity and oxidation level of the soil or water. In aquatic systems, the rate of phosphorus cycling is dependent on trophic interactions. Phosphorus may become extremely limited in the surface waters of stratified temperate lakes because of downward transport of phosphorus contained in dead plankton.

13. Sulfur is an important redox element in anoxic habitats, where it may serve as an electron acceptor in the form of sulfate (SO_4^{2-}) or as an electron donor (for photoautotrophic bacteria) in the forms of elemental sulfur (S^0) and sulfide (S^{2-}).

14. Element cycles interact in complex ways. As an example, the balance between carbon uptake related to primary production and the rate of nitrogen mineralization in the soil will determine whether the arctic tundra will function as a carbon sink or a carbon source in the face of global warming.

15. Many elemental transformations, particularly under anaerobic conditions, are accomplished by biochemically specialized microorganisms (bacteria, cyanobacteria). The activities of these organisms therefore assume important roles in the cycling of elements through the ecosystem.

EXERCISES

1. The assimilatory-dissimilatory coupling of photosynthesis and respiration involves both energy flow (Chapter 10) and the movement of carbon (Chapter 11). If an ecosystem has a gross production of 1,000 kJ m^{-2} yr^{-1}, how much carbon m^{-2} yr^{-1} in the form of CO_2 is moved from the atmosphere into photosynthesizing plants? (Hint: Review Section 10.2 and Exercise 10-1.)

2. Selenium, a material that occurs in aquatic sediments and is potentially toxic to aquatic organisms, exists in oxidation states −2, 0, +4, and +6. When it exists in the +4 and +6 states, it becomes soluble in water and toxic to organisms. When it exists in states −2 and 0, selenium is insoluble in water and is less toxic. Would oxidizing or reducing sediments favor the reduction of exposure of aquatic organisms to selenium?

3. Consider one compartment of the carbon cycle (for example, terrestrial plants or the atmosphere). Consulting Figure 11-7, write a simple set of equations that models the fluxes into and out of the compartment. Try constructing such models for the dynamics of two compartments in the cycle.

4. It has been estimated that land plants fix 140×10^{12} g N yr^{-1}. Assuming that it takes 10 g of $C_6H_{12}O_6$ to fix 1 g of nitrogen (the actual value is between 8 and 12), how much energy (kJ) is represented in the $C_6H_{12}O_6$ required for this nitrogen fixation? How many grams of CO_2 must be fixed by photosynthesis in order to fix this much nitrogen?

5. Write a brief article for your local newspaper that explains the possible responses of the arctic tundra to elevated levels of CO_2. In your article, address the question of how changes in the tundra might affect ecological processes in the area where the reader of your article lives.

6. In Section 11.10, we write, "In fact, if it were not for the activities of bacteria and fungi, many element cycles would be drastically altered and productivity of the ecosystem much reduced." Choose any two of the element cycles that we have discussed and describe how the cycles would be altered in the absence of microorganisms.

CHAPTER 12

Nutrient Regeneration in Terrestrial and Aquatic Ecosystems

GUIDING QUESTIONS

- What factors determine the characteristics of soils?
- What are the major soil horizons, and what is the content and origin of each?
- How do precipitation and the activities of soil biota influence the weathering of granite?
- How does acid precipitation affect soil cation mobility and plant growth?
- How do podsolization and laterization differ, and under what conditions does each predominate?
- How would the cation budget of a small watershed change if most of the trees were removed?
- What is the role of fungi in the cycling of nutrients in the soil?
- How do nutrient cycling processes differ in a wet tropical forest and in a temperate forest?
- How does the nutrient regeneration process in shallow-water aquatic sediments differ from that in temperate forest soils?
- What is the difference between the regeneration of nitrogen and that of phosphorus in aquatic sediments?
- How does the seasonal stratification pattern of some lakes influence nutrient availability in those lakes?
- What is a microbial loop, and how does it alter the classic phytoplankton-based food chain?
- What factors influence the process of nutrient spiraling in streams?
- Why are estuaries and marshes among the most productive ecosystems on earth?

In the last chapter we saw how elements move among the various ecosystem compartments through chemical and biochemical transformations. Carbon is transferred from the atmosphere into the terrestrial and aquatic pools by the process of photosynthesis and returned to the atmospheric pool by respiration. Through the biochemical processes of nitrogen fixation, organisms move nitrogen from the atmosphere into the biotic realm. The global cycle of a particular element represents the gross movement of that element among the compartments of the world's ecosystems.

But elements and materials do not simply collect in one compartment or pool and await transport to the next. Rather, they are used by biota within the compartment. In fact, the utilization of elements and materials within most ecosystem compartments is so rapid, and the demand for materials so high, that the demand cannot be met by the processes that move materials between global compartments. We know, for example, that phosphorus is an essential material for life, but that biologically usable forms of phosphorus are very limited in most environments. Forms of phosphorus that can be utilized by organisms must be regenerated within the ecosystem in order to maintain high productivity. This is true of most nutrients, and it is these processes of regeneration that are the subject of this chapter.

There are significant differences in the process of regeneration between terrestrial and aquatic ecosystems, both freshwater and marine. Whereas the chemical and biochemical transformations involved in the two types of ecosystems are basically the same, the material bases for nutrient regeneration differ. In terrestrial habitats, most elements cycle through detritus at the soil surface, where plant roots have ready access to nutrients. In aquatic habitats, particularly in lakes and oceans, sediments are the ultimate source of regenerated nutrients; these sediments are often far removed from the sites of primary production.

12.1 Regenerative processes in terrestrial ecosystems occur in the soil.

Soils provide the substrate within which plant roots grow and many animals burrow. The characteristics of soil determine its ability to hold water and to make available the minerals required for plant growth. Thus, its variation provides a key to understanding the distribution of plant species and the productivity of biological communities.

While soil is difficult to define, it can be described as the generally porous material that overlies unaltered rock at the surface of the earth. It includes minerals derived from the parent rock, altered minerals formed by weathering, organic material contributed by plants, air and water within the pores of the soil, living roots of plants, microorganisms, and the larger worms and arthropods that make the soil their home. Five factors largely determine the characteristics of soils: climate, parent material (underlying rock), vegetation, local topography, and to some extent, age (Jenny 1941, 1980, Brady 1974, Wardenaar and Sevink 1992). Also, the activities of organisms within the soil affect the arrangement of particles and the size and degree of pores in the soil (Oades 1993) as well as its chemical characteristics (Gonzalez-Prieto and Carballas 1995).

Soils are dynamic, changing from the moment they begin to develop on newly exposed rock material. But even after soils achieve stable properties, they remain in a constant state of flux. Groundwater removes some material; other material enters the soil from vegetation, in precipitation, as dust from above, or by the weathering of rock from below. Where little rain falls, the parent material weathers slowly and plant production adds little organic detritus to the soil. Thus arid regions typically have shallow soils, with bedrock lying close to the surface (Figure 12-1). Soils may not form at all where weathered material and detritus erode as rapidly as they form. Soil development also stops short on alluvial deposits, where fresh layers of silt deposited each year by floodwaters bury weathered material. At the other extreme, weathering proceeds

FIGURE 12-1 Profile of a poorly developed soil in Logan County, Kansas, illustrating shallow soil depth and absence of soil zonation. (*Courtesy of the U.S. Soil Conservation Service.*)

rapidly in parts of the humid Tropics, where chemical alteration of parent material may extend to depths of 100 meters (Bunting 1967, Eyre 1968). Most soils of temperate zones are intermediate in depth, extending to about 1 meter, as a rough average.

Where recent road development or excavation exposes soil in cross section, one often notices distinct layers, called **horizons.** These horizons have been described with complex and sometimes conflicting terminology by soil scientists (Buol et al. 1973, Jenny 1980). A generalized, and somewhat simplified, soil profile has five major divisions: O, A, E, B, and C horizons. The nonsoil layer below these horizons is referred to as the R layer (Table 12-1, Figure 12-2). The O horizon is further subdivided into two layers. The very top soil layer, called the O_i horizon, is composed of **litter,** the dead materials such as leaves, needles, or twigs that have recently arrived at the soil surface and which have started to decompose. Beneath the litter is a layer called the O_a layer, which is composed mostly of **humus,** fine particles of partially decomposed plant and animal material. The A horizon lies just below the O_a layer. The upper part of the A horizon may contain a considerable amount of organic material and may be difficult to distinguish from the O_a layer. Most organic material in the soil is restricted to the O_i and O_a horizons, and the upper part of the A horizon.

When water percolates through the soil, it causes **eluviation** (sometimes called **leaching**) of minerals from the top horizons. These materials may accumulate in lower regions of the soil, a process called **illuviation.** Because of the accumulation of minerals in lower regions of the soil, the horizons below the A horizon, and sometimes the lower part of the A horizon, are called mineral horizons. The most extensive eluviation occurs in the E horizon, or in soils that do not have an E horizon, in the lower part of the A horizon. The E horizon is the region of the soil profile where most plant roots are found. Much of the material that is leached from the E horizon is deposited in the next layer, the B horizon, which is often characterized by a dark band of redeposited materials such as clay minerals and oxides of aluminum and iron. There is usually very little organic material in this layer. The C horizon lies just above the underlying rock—the R horizon—which is part of the parent material of the soil. The C horizon is often light-colored and heavily leached of much of its calcium. The area nearest the parent material is loosely aggregated, calcium-rich material. As we shall see below, weathering of the parent rock contributes to the composition of the C horizon.

TABLE 12-1 **Soil horizons and their predominant characteristics, arrayed in descending order from the surface of the soil**

Horizon	Predominant characteristics
O_i	Composed of material such as leaves, needles, or twigs that are newly arrived at the soil surface and just beginning to decompose.
O_a	Rich in humus, consisting of partly decomposed material mixed with mineral soil.
A	Contains organic material, particularly in the upper part. May be difficult to distinguish from the O_e horizon. Some eluviation in the lower part.
E	Most plant roots are found in this horizon. The site of most eluviation.
B	A horizon containing little organic material, whose chemical composition resembles that of the underlying rock. Clay minerals and oxides of aluminum and iron leached out of the overlying E horizon are sometimes deposited here (illuviation).
C	Primarily weakly weathered material, similar to the parent rock. Calcium and magnesium carbonates accumulate in this layer, especially in dry regions, sometimes forming hard, impenetrable layers.
R	Underlying rock.

12.2 Weathering is the physical and chemical breakdown of rock near the earth's surface.

Soil develops initially from the weathering of one of three types of parent rock found in the earth's crust. **Igneous rock** is formed when the hot magma of the earth's mantle rises to near the surface and cools. Igneous rock is composed of **primary minerals,** products of the original rock formation, such as olivine and plagioclase. **Sedimentary rock** is formed when deposits of materials in lakes and oceans accumulate and merge over thousands of years, and are pushed to the surface by geologic activity. Sedimentary rocks, such as limestone and shale are composed of **secondary minerals,** which are the products of weathering. When either igneous rock or sedimentary rock is subjected to the intense heat and pressure of the earth's core, the minerals in those rocks melt, forming new material, called **metamorphic rock.**

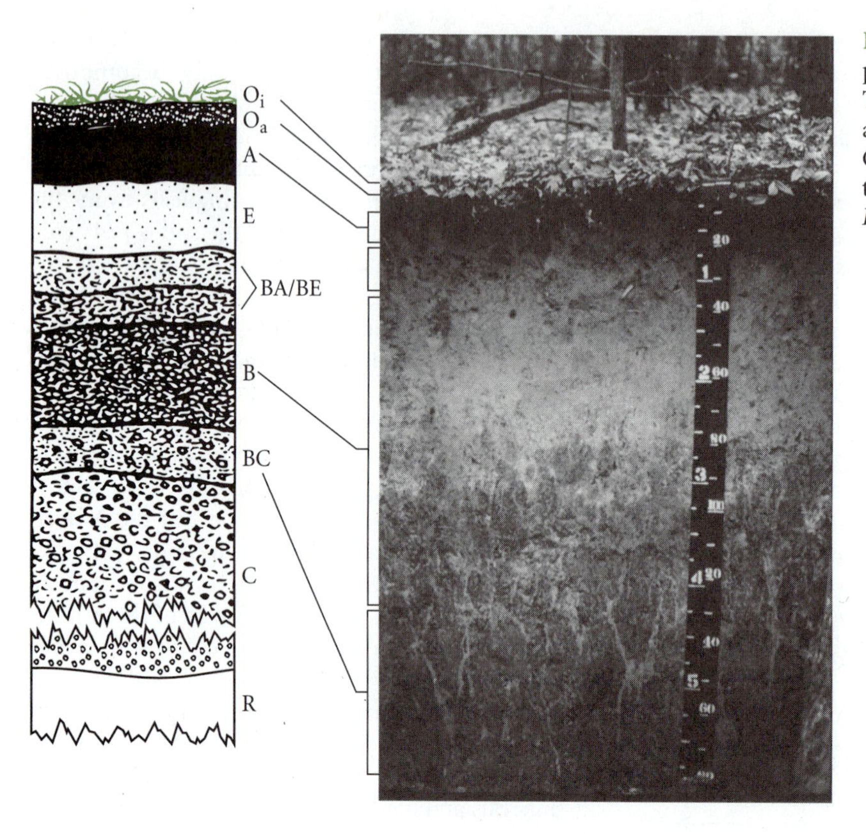

FIGURE 12-2 (a) Diagram of soil profile, showing the major soil horizons. There are five major horizons: O, A, E, B, and C. (b) A soil profile from Crockett County, Tennessee, showing the horizons that occur there. *(Photograph courtesy of F. E. Rhoton; from Rhoton et al. 1996.)*

Weathering of Granite: An Example of Chemical Weathering

Weathering is the process whereby parent rock is broken down by physical and chemical processes. Scouring by wind, repeated freezing and thawing of water in rock crevices, and the physical actions of roots are the chief modes of **mechanical weathering.** Initial **chemical weathering** of the rock occurs when water dissolves some of its more soluble constituent minerals, especially sodium chloride (NaCl) and calcium sulfate ($CaSO_4$). Other materials, particularly the oxides of titanium, aluminum, iron, and silicon, dissolve less readily.

The weathering of granite exemplifies some basic processes of chemical weathering and soil formation. Granite, an igneous rock, forms when less dense molten material from deep within the earth rises to the surface, cools, and crystallizes. The minerals making up the grainy texture of granite—feldspar, mica, and quartz—consist of various combinations of oxides of aluminum, iron, silicon, magnesium, calcium, and potassium, along with other, less abundant compounds. The key to weathering is the displacement of elements in these minerals (notably calcium, magnesium, sodium, and potassium) by hydrogen ions, followed by the reorganization of the remaining oxides into new minerals. For example, in the weathering of bedrock in the Hubbard Brook Experimental Forest of New Hampshire, virtually all the calcium is displaced and leached from the developing soil, as is 16–24% of the sodium, magnesium, and potassium. In contrast, only 2% of the aluminum and silicon is lost (Likens et al. 1977).

Feldspar ($KAlSi_3O_8$), which consists of aluminosilicates of potassium, weathers rapidly owing to the displacement of potassium (K) by hydrogen ions:

$$KAlSi_3O_8 + 4H_2O + 4H^+ \longrightarrow K^+ + Al^{3+} + 3Si(OH)_4$$

(Bohn et al. 1979). Silicon hydroxide ($Si[OH]_4$) is soluble and subject to leaching under these conditions, but new insoluble materials, particularly clay particles, normally form, and most of the aluminum and silicon stay in the soil:

$$2Al^{3+} + 2Si(OH)_4 + H_2O \longrightarrow 5H^+ + Al_2Si_2O_5(OH)_4$$

The resulting mineral is called kaolinite. Along with other clay minerals, it helps retain leachable ions (for example, Ca^{2+}, K^+) in the soil.

Mica grains consist of aluminosilicates of potassium, magnesium, and iron. As in feldspar, the potassium and magnesium are displaced readily during weathering, and the remaining iron, aluminum, and silicon form various kinds of clay particles. Quartz, a form of silica (SiO_2), is relatively insoluble, and therefore

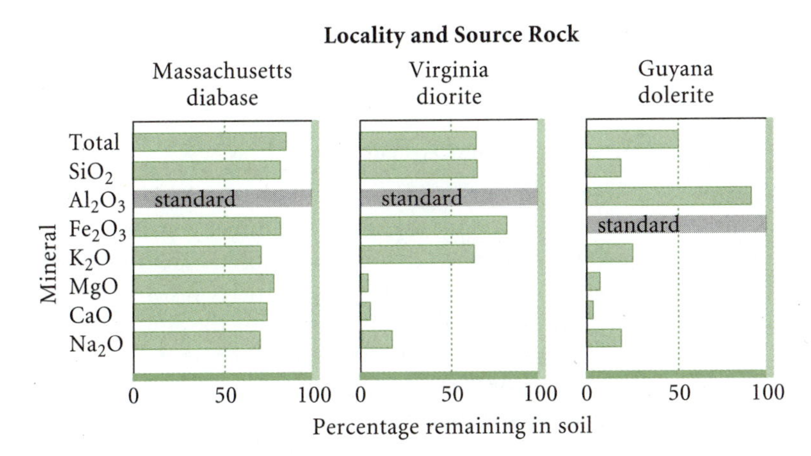

FIGURE 12-3 Differential removal of minerals from granitic rocks as a result of weathering in different climates. Values are compared with those for either aluminum or iron oxides (standards = 100%), which are assumed to be the most stable components of the mineral soil. More minerals are removed from the Virginia and Guyana soils because conditions are warmer and wetter in those places than in Massachusetts. (*After Russell 1961.*)

remains more or less unaltered in the soil as grains of sand. Changes in chemical composition as granite weathers from rock to soil in different climates show that weathering is most severe under tropical conditions of high temperature and rainfall (Figure 12-3).

The Influence of Precipitation and Biota on Chemical Weathering

Regardless of the chemical nature of rock, an important factor in the initial weathering of parent material is the presence of hydrogen ions (H^+) in the water that percolates to the bedrock. Hydrogen ions may come from precipitation and from processes that occur within the soil. Hydrogen ions appear as part of the carbonation reaction of rainwater:

$$H_2O + CO_2 \rightleftharpoons H_2CO_3 \rightleftharpoons H^+ + HCO_3^-,$$

a component of which is carbonic acid (H_2CO_3). All precipitation contains dissolved carbon dioxide, some of which dissociates to H^+ and HCO_3^-, as shown above. In regions not affected by acid precipitation (see below), concentrations of hydrogen ions in rainwater are between 1 and 2×10^{-5} moles per liter, equivalent to a pH of about 5.1 (Likens et al. 1987, Galloway et al. 1995).

Carbon dioxide is also released into the soil by decomposition and by the respiration of soil organisms and underground plant parts, activities that drive the carbonic reaction to the right and enhance chemical weathering. Nitrification of ammonia acidifies the soil by the reaction

$$NH_4^+ + 2O_2 \longrightarrow 2H^+ + H_2O + NO_3^-,$$

and also contributes hydrogen ions to the soil. In the Hubbard Brook Experimental Forest, these processes account for about 50% of the hydrogen ions needed for weathering of bedrock. In general, high concentrations of soil CO_2 tend to occur in areas of high plant growth. These areas also have high rates of weathering (Johnson et al. 1977, Brook et al. 1983, Schlesinger 1991).

12.3 The leaching of nutrient cations is determined by the clay and humus content of the soil.

Plants obtain mineral nutrients from the soil in the form of dissolved ions, which we call nutrient cations. Recall from our discussion of the solute properties of water (see Chapter 4) that positively charged ions, such as Na^+, are called cations; negatively charged ones, such as Cl^-, are called anions. Because ions are soluble in water, those not immediately taken up by plants or fungi may leach out of the soil profile if they are not strongly attracted to stable soil particles. Such particles occur in the form of associations of clay and humus. These associations are called **micelles.** Micelles have a net negative charge on their surface and thus attract cations in the soil, thereby reducing cation mobility (Figure 12-4). Due to this cation-binding property of micelles, the potential long-term fertility of soil, determined by its capacity for storing nutrients, depends largely on its clay content.

The negative charges of micelles arise by two processes. First, during the formation of clay particles by crystallization, an atom of magnesium might substitute in the crystal lattice for one of aluminum, or one of

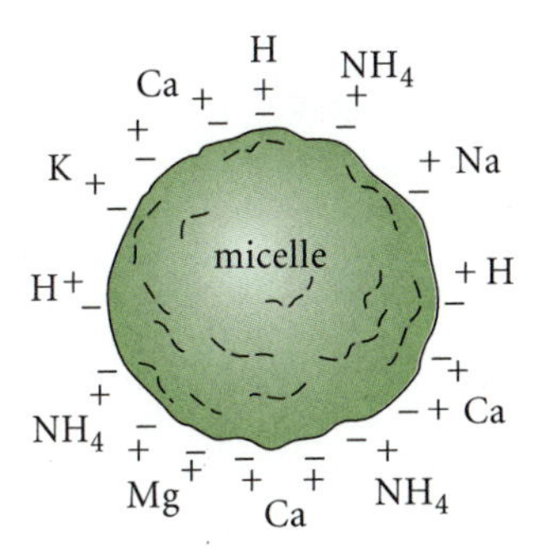

FIGURE 12-4 Schematic representation of a clay or humus particle (micelle) with hydrogen ions and mineral ions attracted by negative charges at its surface. (*After Eyre 1968.*)

aluminum for one of silicon. In each of these substitutions, a less positively charged atom replaces a more positively charged atom (that is, Mg^{2+} for Al^{3+}, or Al^{3+} for Si^{4+}). The result is a net negative charge on the clay particle. The second process is the ionization of several types of functional groups; for example, the hydroxide ($—OH$) in clay and the carboxyl ($—COOH$), amine ($—NH_2$), or phenolic ($—C_6H_4OH$) groups in organic compounds. The dissociation of such groups (for example, $—COOH \rightleftharpoons COO^- + H^+$) exposes negative charges on the particle that can then hold other ions in the soil. Charges resulting from the substitution of metal atoms in clay particles are called **permanent charges** because they are insensitive to the concentration of hydrogen ions (pH) in the soil. In contrast, the dissociation of functional groups varies inversely with hydrogen ion concentration and is called **pH-dependent charge.** The total negative charge (expressed as milliequivalents/100 g) measures the ability of the soil to hold positive ions and is referred to as the **cation exchange capacity** (CEC).

Some ions cling more strongly to micelles than others: in order of decreasing tenacity, hydrogen (H^+), calcium (Ca^{2+}), magnesium (Mg^{2+}), potassium (K^+), and sodium (Na^+). Hydrogen ions thus tend to displace calcium and all other cations in the soil. The bonds between soil particles and such ions as K^+ and Ca^{2+} are relatively weak, so they constantly break and reform. When a potassium ion (K^+) dissociates from a micelle, it may be replaced by any other cation dissolved in the water close by.

The immediate availability of ions to plants depends largely on the pH of the soil, which is determined by the availability of hydrogen ions (H^+). If cations were not added to or removed from the soil, their relative proportions in association with micelles would assume a steady state. But carbonic acid in rainwater and organic acids produced by the decomposition of organic matter continually add hydrogen ions to the upper layers of the soil; these readily displace other cations, which are then washed out of the soil and into the groundwater. The influx of hydrogen ions in water percolating through the soil largely initiates the mobility of ions and the differentiation of layers in the soil profile.

Anions important to plant nutrition, such as nitrate, phosphate, and sulfate, can be held onto clay particles by means of ion "bridges." These bridges form under acid conditions by the association of an additional hydrogen ion with a functional group on the micelle; for example, $—OH + H^+ \rightarrow —OH_2^+$, which makes possible $—OH_2^+ \cdots NO_3^-$. However, the strength with which the negatively charged NO_3^- is held to the negatively charged micelle is much less than the attraction of NH_4^+ to the micelle. This accounts for the greater mobility of NO_3^- in the soil and, consequently, its greater availability to plants.

Breakdown and Availability of Clay

Acid conditions affect not only the availability of nutrient cations, but also the structure of clay itself. Under mild, temperate conditions of temperature and rainfall, sand grains and clay particles resist weathering and form stable components of the soil skeleton. In acid soils, however, clay particles break down in the A horizon of the soil profile, and their soluble ions are transported downward and deposited in lower horizons. This process, known as **podsolization,** reduces the ion exchange capacity and, therefore, the fertility of the upper layers of the soil.

Acid soils occur primarily in cold regions where coniferous trees dominate the forests. The slow decomposition of abundant plant litter produces organic acids. Rainfall usually exceeds evaporation in such regions, creating moist conditions in which water continually moves downward through the soil profile. The result is that little clay-forming material is transported upward from the weathered bedrock below. In North America, podsolization advances furthest under spruce and fir forests in New England and the Great Lakes region and across a wide belt of southern and western Canada.

The abundance of clay particles is also affected by weathering. The warm, wet climates of many tropical regions weather the soil to great depths. A conspicuous feature of weathering under these conditions is the breakdown of clay particles, which results in the leaching of silica from the soil and leaves oxides of iron and aluminum to predominate in the soil profile. This process is called **laterization,** and the iron and aluminum oxides give lateritic soils (oxisols) their characteristic reddish coloration. Even though the rapid decomposition of organic material in tropical soils contributes an abundance of hydrogen ions, these are quickly neutralized by the bases formed by the breakdown of clay minerals; consequently, oxisols usually are not very acidic. Laterization is enhanced in certain soils that develop on parent material deficient in quartz (SiO_2) but rich in iron and magnesium (basalt, for example); these soils contain little clay to begin with because they lack silicon. Regardless of the parent material, weathering reaches deepest and laterization proceeds furthest on low-lying soils, such as those of the Amazon basin, where highly weathered surface layers are not eroded away and the soil profiles are very old.

One of the consequences of laterization in many parts of the Tropics is that the capacity of the soil to hold nutrients is very poor. Without clay and humus, CEC can be low, in which case mineral nutrients are readily leached out of the soil. Where soils are weathered deeply, new minerals formed by the decomposition of the parent material are simply too far from the surface layers of the soil to contribute to soil fertility. Besides, heavy rainfall keeps water moving down through the soil profile, preventing the upward movement of nutrients. In general, the deeper the ultimate source of nutrients in the unaltered bedrock, the poorer the surface layers. Rich soils do, however, develop in many tropical regions, particularly in mountainous areas where erosion continually removes the nutrient-depleted surface layers of the soil, and in volcanic areas where the parent material of ash and lava is often rich in nutrients such as potassium.

Acid Rain and Forest Decline

In the 1980s ecologists became concerned about population declines in a number of forest tree species in Europe and North America. The Norway spruce (*Picea abies*) in Germany and the red spruce (*Picea rubens*) in North America were two species that appeared to have undergone dramatic declines. The decline in these and other forest tree species coincided with growing evidence of the effects of acid precipitation on soil and aquatic processes. Ecologists began to question whether acid precipitation was the primary cause of forest decline (Pitelka and Raynal 1989).

Studies of the effects of acid precipitation on forest ecosystems have focused on the availability of cations (Ca^{2+}, Mg^{2+}, K^+, Na^+) under conditions of low pH (Johnson et al. 1988, Schulze 1989, McLaughlin et al. 1993, Likens et al. 1996). Recall that hydrogen ions (H^+) displace other cations in the soil, releasing those cations to the soil water, where they may be lost from the ecosystem. In addition, the presence of strong anions such as SO_4^{2-} and NO_3^-, both constituents of acid rain, may pull cations from the soil (Reuss and Johnson 1986). In some forest ecosystems, these processes, along with cation uptake by the forest trees, cause a depletion of cations, particularly Ca^{2+} and Mg^{2+}, from the soil. Such nutrient depletion could result in slowed growth or tree death.

Mass balance studies, which measure the components of rainwater and stream outflow (Figure 12-5), have demonstrated that acid precipitation alters the cation exchange capacity of forest soils. Studies of chestnut oak and yellow poplar forests in the Walker Branch watershed revealed that high levels of H^+ input into the ecosystem in the form of acid precipitation resulted in the net export of K^+, Ca^{2+}, and Mg^{2+} from the ecosystem (Johnson and Henderson 1989).

Measurements of the chemistry of precipitation (input) and stream water (output) have been made nearly continuously since 1963 in the Hubbard Brook Experimental Forest. The pH of rain and snow in the area ranged between 4.05 and 4.30 from 1963 to 1994. Over this period of time large amounts of Ca^{2+} and Mg^{2+} were lost from the soil (Likens et al. 1996). While nutrient cation depletion most certainly has an effect on tree growth and forest dynamics, the role played by acid rain is unclear. Cation depletion can occur in the absence of acid precipitation via downward movement of soil water during high rainfall. Soil acidification may lead to aluminum toxicity in some cases, and inputs of other toxicants, such as heavy metals, may play a role in forest ecosystem change in some locations. Moreover, most forest tree species exhibit natural variation in abundance resulting from pest outbreaks, variation in the availability of

FIGURE 12-5 This stream gauge at the lower end of a watershed at the Coweeta Hydrological Laboratory, North Carolina, is part of a mass balance study. The V-shaped notch is engineered so that the flow of water through the weir can be estimated from the water level in the basin. Analyses of water samples taken from this outflow give measurements of watershed nutrient loss. (*Courtesy of the U.S. Forest Service.*)

nutrients, and adverse weather conditions such as unusual frost conditions or drought (Pitelka and Raynal 1989). Experimental approaches are required to determine whether and how cation depletion affects tree growth.

D. W. Johnson and S. B. McLaughlin and their colleagues at the Oak Ridge National Laboratory in Tennessee have investigated the effects of soil nutrient levels on forest tree growth and physiology. They found that, between 1972 and 1982, the soil nutrient content of Ca^{2+} and Mg^{2+} in a number of study plots in the Walker Branch watershed in Tennessee was significantly reduced as a result of plant uptake coupled with leaching promoted by acid precipitation (Johnson et al. 1988). Their data suggested a reduction in forest fertility as a result of this nutrient deficiency. In an experimental study of the effects of acid deposition on photosynthesis and respiration, McLaughlin and colleagues (1993) subjected red spruce (*Picea rubens*) seedlings to four different soil treatments: Ca^{2+} added to the soil, Mg^{2+} added to the soil, both Ca^{2+} and Mg^{2+} added, and no amendments to the soil. Each of these treatments was given under watering conditions simulating acid precipitation (pH 3.0) and normal precipitation (pH 5.0). The ratio of photosynthesis (P) to respiration (R) was used as an index of production by the seedlings. Seedlings grown under low pH conditions showed a substantial decrease in both photosynthesis and respiration (measured as gas exchange) in all soil treatments (Figure 12-6). However, the decrease in the unamended soil was substantially greater than in the soils enriched with additional nutrients. This decrease was attributed to the substantial loss of calcium from the soil under acid conditions.

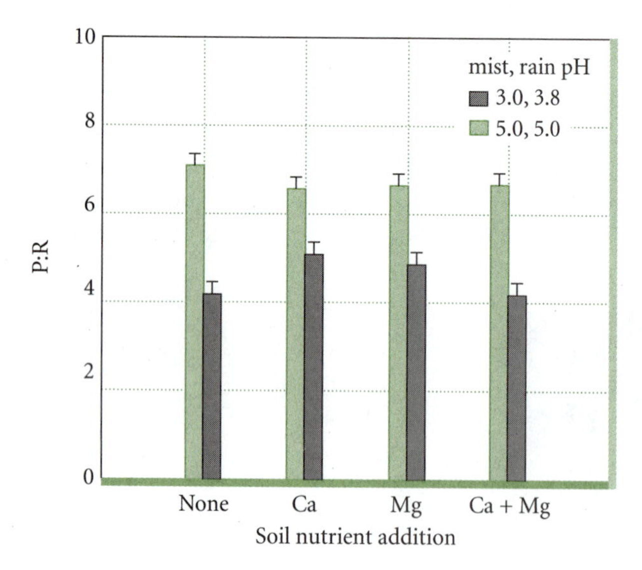

FIGURE 12-6 Effects of acid mist and acid rain on the photosynthesis : respiration (P : R) ratio of red spruce (*Picea rubens*) seedlings in a controlled nutrient addition experiment. Seedlings were grown in soils amended with calcium (Ca), magnesium (Mg), both calcium and magnesium (Ca + Mg), or in native soil with no cations added (None). Each treatment was subjected to conditions of low pH (gray bars) or high pH (green bars) rain and mist. A reduction in the P : R ratio at low pH, indicating reduced growth, was observed in all treatments, with the unamended soil showing the greatest reduction. (*From McLaughlin et al. 1993.*)

12.4 Most nutrients in terrestrial ecosystems cycle through detritus.

Plants assimilate elements from the soil far more rapidly than they are generated by weathering of the parent material. Important nutrients such as nitrogen, phosphorus, and sulfur are poorly represented in parent material. Igneous rocks, such as granite and basalt, contain no nitrogen, and only 0.3% of phosphate (P_2O_5) and 0.1% of sulfate (SO_3) by mass (Bohn et al. 1979). Hence, weathering adds little of these nutrients to the soil. Inputs of these important elements from precipitation, except in areas subjected to acid rain, are also small. Even nitrogen fixation adds relatively small amounts of usable nitrogen to the soil. Plant production therefore depends on the rapid regeneration of nutrients from detritus.

Nutrient Cycling in Detritus

Dead or partially decomposed organic matter, referred to as organic detritus, occurs everywhere. It is most conspicuous in terrestrial ecosystems, where the resistance of woody plant parts to herbivory results in the accumulation of abundant plant remains (Harmon et al. 1986, Vogt et al. 1986). Regardless of the habitat, however, this reservoir of nutrients is regenerated by the activities of a wide variety of worms, snails, insects, mites, bacteria, and fungi that consume detritus as their primary source of carbon and energy (Griffin 1972, Harley 1972, Petersen and Cummins 1974, Anderson and Sedell 1979, Coleman et al. 1983).

Of all the earth's detritus-based communities, the organisms that consume the litter of leaves and branches on the forest floor are probably the best known (Witkamp 1966, Minderman 1968, Dickinson and Pugh 1974, Hayes 1979). The breakdown of leaf litter occurs in four ways: (1) leaching of soluble minerals and small organic compounds by water, (2) consumption by large detritus-feeding organisms (millipedes, earthworms, wood lice, and other invertebrates); (3) further decomposition by fungi, and (4) eventual mineralization of phosphorus, nitrogen, and sulfur by bacteria. Between 10% and 30% of the substances in newly fallen leaves, including salts, sugars, and amino acids, dissolves in cold water. Leaching rapidly removes most of these from the litter (Daubenmire and Prusso 1963), leaving behind complex carbohydrates and other organic compounds. Although large detritus feeders assimilate no more than 30–45% of the energy available in leaf litter, and even less from wood, they nonetheless speed the decay of litter because they macerate leaves in their digestive tracts, and the finer particles in their egested wastes expose new surfaces to microbial feeding (Hartenstein 1986).

FIGURE 12-8 Shelf fungi speed the decomposition of a fallen log. The brackets are fruiting structures produced by the fungal hyphae—together called the mycelium—that grow throughout the interior of the log, slowly digesting its structure. (*Courtesy of the U.S. National Park Service.*)

The Role of Fungi in Nutrient Regeneration

The rigidity and shape of the cells of leaves, wood, and other plant parts is maintained by a complex extracellular matrix composed of carbohydrates and proteins, which is organized into primary and secondary cell walls. The primary cell wall surrounding young living plant cells, such as those found in leaves and young stems, is composed primarily of **cellulose,** which is a long unbranched chain of glucose molecules. Cellulose molecules are linked together in groups of 50 or 60 molecules to form microfibrils, which give the plant tissue considerable strength while maintaining a degree of flexibility (Figure 12-7). The secondary cell wall found in mature cells that are no longer growing, as in wood, may constitute as much as a third of the structure's dry weight. It is composed primarily of **lignin,** which is made up of polymers of phenolic alcohols. Lignin provides considerable rigidity to plant tissue, and it is one of the most important structural compounds in nature. The structural complexity of cellulose and lignin presents a formidable challenge to the degradation of plant material that falls to the forest floor or that is deposited in an aquatic ecosystem. The fungi are highly adapted to meet that challenge.

Most fungi consist of a network, or **mycelium,** of **hyphae,** which are threadlike elements made up of cells connected end to end. Fungal hyphae can penetrate the woody cells of plant litter that bacteria cannot reach. The mushrooms and shelf fungi that we see in the forest are merely fruiting structures produced by the mass of the fungal organism deep within the litter or wood (Figure 12-8). Fungi secrete enzymes into the litter or wood and absorb the simple sugar and amino acid breakdown products of this extracellular digestion. Bacteria are also capable of extracellular digestion, but unlike fungi, most are unable to digest cellulose or, especially, lignin. That task is left mostly to the fungi (along with a few bacteria, protozoans, and snails). Cellulose digestion by fungi begins with hydrolysis of polysaccharides (long chains of sugar subunits) into simple sugars, which can then be absorbed into the hyphae. The breakdown of lignin apparently is initiated by an oxidation reaction that cleaves the phenolic ring structure (Kirk et al. 1977).

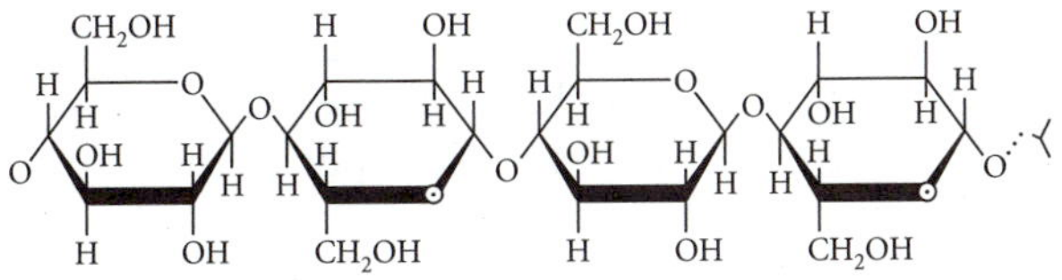

Some species of fungi form a mutually beneficial association with the roots of woody plants (Smith 1980). This association, which is called a **mycorrhiza** (literally, "fungus root"), enhances the plant's ability to extract mineral nutrients from the soil. Although many forms of mycorrhizae are recognized, they are classified as **endomycorrhizae** when the fungus penetrates the root tissue and **ectomycorrhizae** when the fungus grows as a sheath over the root surface (Figure 12-9). Ectomycorrhizae are the most common form found in the Tropics (Janos 1980).

The commonest type of endomycorrhiza is the **vesicular-arbuscular** mycorrhiza, so called because of the structures it develops within the host tissue. These fungi do not grow freely in the soil, but rather infect plant roots from spores left behind by dead, previously infected roots. (Presumably, spores blow into virgin soils along with other dust.) The fungi that form vesicular-arbuscular mycorrhizae derive their carbon exclusively from the plant root. Their mineral nutrients are derived from the soil, into which the fungi send long hyphae. Apparently, they can use sources of phosphorus not available to plants, such as the highly insoluble "rock phosphorus" $Ca_3(PO_4)_2$, which dissolves only under acid conditions. Without the help of mycorrhizae, plants can use only the more soluble forms $CaHPO_4$ and KH_2PO_4. Over and above their unique ability to secrete hydrogen ions and organic

FIGURE 12-7 The chemical structure of cellulose.

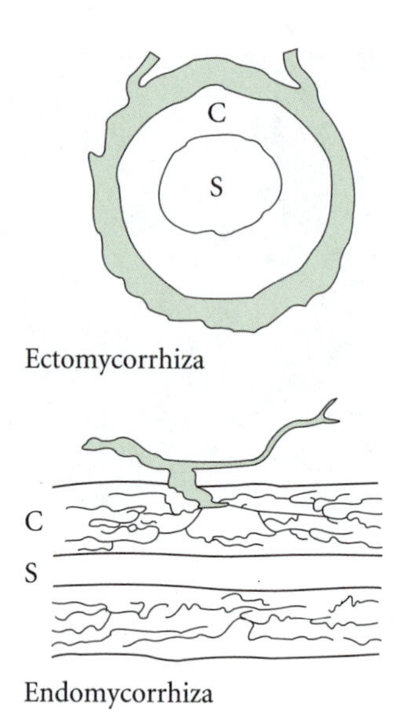

FIGURE 12-9 Generalized structure of ectomycorrhizae and endomycorrhizae. C = root cortex; S = vascular tissue stele. (*After Grant and Long 1981.*)

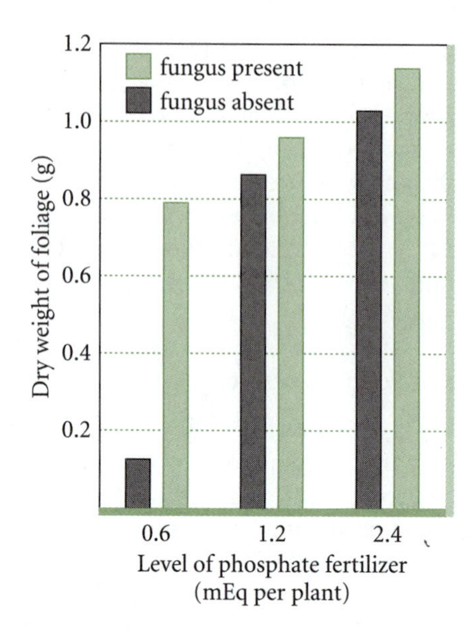

FIGURE 12-10 The effects of phosphate fertilizer and inoculation with the mycorrhizal fungus *Endogone macrocarpa* on the growth of tomato plants (*Lycopersicon esculentum*). The mycorrhizae stimulated growth under low phosphate concentrations, but not under high phosphate concentrations, at which point other nutrients become limiting. (*After Harley and Smith 1983.*)

acids, fungi may be aided in this task when their hyphae grow in association with phosphate-solubilizing bacteria. The biochemical mechanisms by which phosphate and other nutrients are translocated by the fungus from the soil to the plant are largely unknown, although recent studies of the molecular genetics of mycorrhizal fungi have identified DNA encoding sequences for a phosphate transporter protein (Harrison and Van Buren 1995). The relationship between the fungus and the plant is beneficial for both and is thus an example of a mutualistic relationship (see Chapter 25).

Fungi are nearly ubiquitous in soil, and thus the role of mycorrhizae has been examined in many different types of terrestrial ecosystems (Carroll and Wickler 1992). Such studies reveal that many factors influence the extent to which mycorrhizal relationships occur and the level of benefit gained by the participants. Mycorrhizae are particularly important to plant species growing in sterile or infertile soils (Bowen 1973) and in some deserts (Cui and Nobel 1992). In an experiment with *Pinus strobus* seedlings, for example, ectomycorrhizae increased the uptake of nitrogen, phosphorus, and potassium per unit of root mass by two to three times and greatly improved the growth of the plants (Bowen 1973). Mycorrhizae promote plant growth the most in soils that are relatively depleted of nutrients, as shown by the experiments with tomatoes summarized in Figure 12-10 (Harley and Smith 1983). The growth of nursery-grown giant redwood trees (*Sequoiadendron giganteum*) is increased dramatically when mycorrhizal fungi are present in the roots (Molina 1994).

Mycorrhizae play an important role in the energy budgets of the ecosystems in which they occur because they can sequester as much as 15% of total net primary production (Vogt et al. 1982), and this energy is not available for plant growth. For example, in a study of mycorrhizal associations in a prairie grass, Dhillion and Anderson (1993) found that, after a prairie community was burned—a relatively common and natural phenomenon—plants in the burned area were less efficient in their nutrient uptake and use than plants in an unburned area. They attributed this lowered efficiency in part to the presence of vesicular arbuscular mycorrhizae, which used energy to support their own biomass at the expense of the plant. However, global estimates of the extent of such drains on primary production are currently unavailable (Schlesinger 1991).

12.5 Nutrients are regenerated more rapidly in tropical forests than in temperate forests.

Nutrient cycling differs in tropical and temperate ecosystems because of the effects of climate on weathering, soil properties, and the decomposition of detritus. Tropical soils tend to be deeply weathered and low in clay content, which results in a poor ability to retain nutrients (low cation exchange capacity; see above). The high productivity of tropical forests is supported by rapid regeneration of nutrients from detritus under the warm, humid conditions, rapid uptake of nutrients by plants from the top layers of the soil, and efficient retention of nutrients by plants (Vitousek and Stanford 1986, Grubb 1995). Ecologists have concluded that in very old tropical forests, whose soils are the most weathered and nutrient-poor, most of the nutrients occur in the living biomass, and nutrients are regenerated and assimilated very rapidly. This appears not to be the case for younger forests occurring on more nutrient-rich soils (Grubb 1995). This pattern has important implications for tropical agriculture and conservation (Jordan 1985).

Nutrient cycling in tropical ecosystems may be strongly pulsed in correspondence with rainy and dry seasons (Barrios and Herrera 1994, Lodge et al. 1994). For example, tropical forests in Venezuela are flooded during the wet season (May through December), causing considerable stress as the floodwaters create anaerobic conditions in the soil. Barrios and Herrera (1994) have suggested that nitrogen inputs to the forest via mineralization and plant uptake of nitrogen are synchronized during the dry season, when the forest is under less stress. Such synchronization of nutrient regeneration processes and plant nutrient utilization may be typical of tropical ecosystems (Lodge et al. 1994). We hasten to note, however, that tropical regions are far from homogeneous. In comparing tropical and temperate ecosystems, it is necessary to stipulate the precise environmental conditions of each (Lugo and Brown 1991).

Planting crops such as corn on clear-cut land in the Tropics has had disastrous consequences (Figure 12-11). The practice of cutting and burning native vegetation releases many mineral nutrients into the soil. But while these nutrients may support a year or two of crop growth, they are readily leached from the soil because there is no natural tropical vegetation to assimilate them. As a result, soil fertility declines rapidly. Furthermore, as the exposed soil dries, upward movement of water draws iron and aluminum oxides to the surface, where they form a concretelike substance called **laterite.** Surface runoff over the impenetrable laterite accelerates erosion, further depleting nutrients and choking streams with sediment.

The ecological lesson taught by such experiences is that vegetation is critical to the development and maintenance of soil fertility. Even in temperate zones, retention of vegetation is important in retention of soil nutrients. Clear-cutting of small watersheds in the Hubbard Brook Forest increased stream flow several times, owing to the removal of transpiring leaf surfaces; losses of cations increased 3 to 20 times over comparable undisturbed systems (Figure 12-12; Likens et al. 1977). The nitrogen budgets of the cutover watersheds sustained the most striking change. Plants assimilate available soil nitrogen so rapidly that undisturbed

(a)

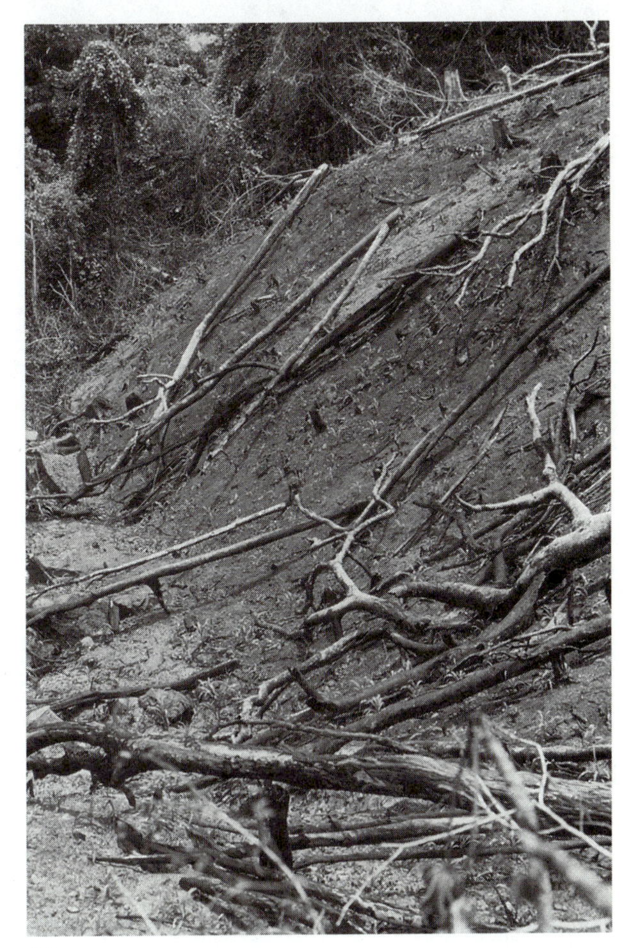

(b)

FIGURE 12-11 (a) An area of about two hectares on a steep slope in Costa Rica that has been cut and burned for shifting agriculture. One can see small corn plants emerging among the debris in the closeup (b). This practice exposes the soil to erosion and promotes leaching of nutrients and laterization. Such clearings produce crops for two or three years at most; decades of forest regeneration are required to restore the fertility of the soil. (*Courtesy of D. H. Janzen.*)

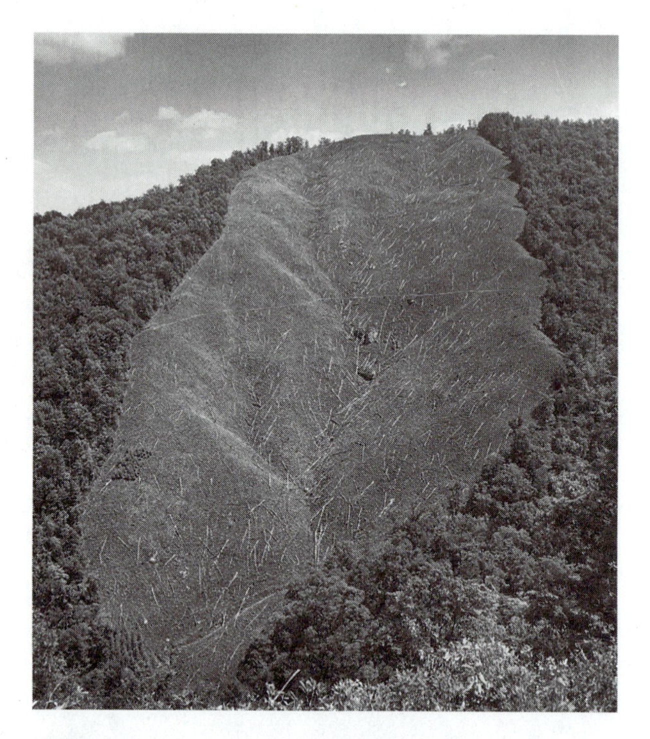

FIGURE 12-12 A clear-cut watershed at the Coweeta Hydrological Laboratory, North Carolina, employed in studies of evapotranspiration and runoff in forest ecosystems. Clear-cuts such as this increase stream flow because of the loss of leaf surfaces, the site of much transpiration. (*Courtesy of the U.S. Forest Service.*)

forests gained nitrogen at the rate of 1–3 kg ha^{-1} yr^{-1}. In the clear-cut watersheds, the net loss of nitrogen as nitrate soared to 54 kg ha^{-1} yr^{-1}, a value comparable to the annual assimilation of nitrogen by vegetation, and several times the precipitation input (7 kg ha^{-1} h^{-1}). The loss of nitrate resulted from nitrification of organic nitrogen at the normal annual rate by soil microorganisms without simultaneous rapid uptake by plants.

Comparative studies of temperate and tropical forests shed further light on their differences. Litter on the forest floor constitutes an average of about 20% of the total biomass of vegetation and detritus in temperate coniferous forests, 5% in temperate hardwood forests, and only 1–2% in tropical rain forests (Ovington 1965). The ratio of litter to the biomass of living leaves is between 5 and 10 in temperate forests, but less than 1 in tropical forests. Of the total organic carbon in the ecosystem, more than 50% occurs in the soil and litter in northern forests, but less than 25% in tropical rain forests (Kira and Shidei 1967). Clearly, dead organic material decomposes rapidly in the Tropics and does not form a substantial nutrient reservoir, as it does in temperate regions.

Fewer data are available on the relative proportions of nutrients in the soil and in the living vegetation. The distributions of potassium, phosphorus, and nitrogen in a temperate and a tropical forest with similar living biomass are compared in Table 12-2. The accumulation of nutrients in vegetation, on a weight for weight basis, is somewhat greater in the tropical forest. For example, the total dry weight of living vegetation in the Belgian ash-oak forest exceeds that of the tropical deciduous forest in Ghana by 14%, but the accumulation of the three elements per gram of dried vegetation is 32–38% lower in the temperate forest. Also, the ratio of each element in the soil to its level in the biomass is much lower in the Tropics; more than 90% of the phosphorus in the tropical ecosystem resides in the living biomass.

C. F. Jordan and R. Herrera (1981) recognized the general nutrient poverty of many tropical soils, but they also distinguished between nutrient-rich

TABLE 12-2 **Distribution of mineral nutrients in the soil and living biomass of a tropical and a temperate forest ecosystem**

Forest (locality)	Biomass (T ha^{-1})	NUTRIENTS (kg ha^{-1}) Potassium	Phosphorus	Nitrogen
Ash and oak (Belgium)	380			
Living vegetation		624	95	1,260
Soil		767	2,200	14,000
Ratio of soil to biomass		1.2	23.1	11.1
Tropical deciduous (Ghana)	333			
Living vegetation		808	124	1,794
Soil		649	13	4,587
Ratio of soil to biomass		0.8	0.1	2.0

(*Data from Greenland and Kowal 1960, Ovington 1965, Duvigneaud and Denayer-de-Smet 1970.*)

TABLE 12-3 Standing crops and fluxes of dry biomass and calcium in a nutrient-rich and a nutrient-poor tropical rain forest

Characteristic	Eutrophic*	Oligotrophic*
Soil		
Exchangeable calcium† (kg ha^{-1})	1,900	306
Standing crop		
Living vegetation (T ha^{-1})	263	298
Production		
Living vegetation (g m^{-2} yr^{-1})	1,033	1,012
Calcium in standing crop		
Living vegetation (kg ha^{-1})	760	529
Calcium flux		
Precipitation (kg ha^{-1} yr^{-1})	21.8	16.0
Subsurface runoff (kg ha^{-1} yr^{-1})	43.1	13.2
Net change (kg ha^{-1} yr^{-1})	−21.3	+2.8

**Eutrophic: montane tropical rain forest, Puerto Rico; oligotrophic: Amazonian rain forest, Venezuela.*

†Soil calcium measured to a depth of 40 cm.

(From Jordan and Herrera 1981.)

and nutrient-poor soils within the Tropics. The former, which they called **eutrophic** (literally, "well-nourished") soils, develop in geologically active areas where natural erosion is high and soils are relatively young. With the bedrock closer to the surface, weathering adds nutrients more rapidly and soils retain nutrients more effectively. In the Western Hemisphere, such eutrophic soils occur widely in the Andes Mountains, in Central America, and in the Caribbean. By contrast, **oligotrophic** (nutrient-poor) soils develop in old, geologically stable areas, particularly on sandy alluvial deposits (as in much of the Amazon basin), where intense weathering removes clay and reduces nutrient retention. The terms *eutrophic* and *oligotrophic* are also used to describe nutrient-rich and nutrient-poor aquatic ecosystems, as we shall see below.

Comparison of an oligotrophic forest (Amazon basin) with a eutrophic forest (Puerto Rico) illustrates that although production and nutrient flux are similar in the two forests, the distributions of nutrients among ecosystem compartments differ (Table 12-3). Moreover, nutrients are held more tightly in the oligotrophic forest; that is, loss of calcium from the eutrophic forest is equivalent to about half the annual flux through vegetation, whereas the oligotrophic forest appears to gain calcium through precipitation input.

Especially in nutrient-poor areas, nutrient retention by vegetation is the key to the productivity of tropical ecosystems. This is accomplished by a very dense mat of roots (and associated fungi), which lie close to the surface (Stark and Jordan 1978) and even extend up the trunks of trees to intercept nutrients washing down from the canopy (Sanford 1987). Data from Africa revealed that between 68% and 85% of the root biomass of forests is concentrated within the top 25–30 centimeters of the soil (Greenland and Kowal 1960). Studies using radioactively labeled compounds showed that nutrients regenerated by the leaching and decomposition of detritus are intercepted by the root mat before they can penetrate into the mineral soil and be washed out of the system (Luse 1970, Odum 1970).

Thus far, we have focused on nutrient regeneration in terrestrial ecosystems and, in particular, the activities of microbes and fungi in the soil. Let us now turn to aquatic ecosystems. We shall begin with a discussion of the role of sediments in nutrient cycling.

12.6 The sediments of shallow waters play an important role in nutrient regeneration.

Many ecology students experience the pleasure, though some would say torment, of mucking about in the shallow margins of a muddy pond or a small stream. The aquatic sediments underfoot (or inside your boot) are much more than a slimy impediment to your examination of the exciting world of aquatic plants and animals. Like terrestrial detritus, they represent the site of important nutrient regeneration processes.

Most of the lakes of the world are relatively shallow, with shoreline areas, or **littoral zones,** characterized by emergent and floating vegetation and open-water **limnetic zones** populated by free-floating plankton. Primary production occurs in both regions of such lakes. In deep bodies of water, light does not penetrate all the way to the bottom. The **photic zone** is that portion of the water column where there is light. The bottoms of streams, lakes, and oceans are referred to as **benthic zones** (Figure 12-13). Benthic sediments are composed primarily of partially decomposed organic matter, mineral particles, particularly clays and carbonates, and an inorganic component that includes such things as the silicate remains of diatoms (Wetzel 1983). These sediments provide a medium in which rooted emergent and submerged vegetation, bacteria, fungi, and numerous types of burrowing or bottom-dwelling microfauna and macrofauna live. The extent and complexity of benthic sediments, and the diversity of the life within those sediments, tends to decrease with water depth, with the very deepest oceans—the **profundal** zone—having relatively less sediment structure and biotic diversity than regions closer to the photic zone.

Organic matter may be transported to the sediments in a number of ways. Emergent vegetation in the littoral zones dies and accumulates at the bottom. Plankton in the open waters also die and sink to the sediments, though the contribution of plankton biomass to shallow sediments is usually small relative to other sources. Most of the organic matter in rivers and streams originates in the terrestrial realm. This material, which undergoes degradation as it moves along the river course, is eventually deposited in oceans or lakes, and may account for a large portion of lake carbon. Lakes may themselves be recipients of substantial amounts of organic material from the surrounding terrestrial ecosystem. When organic material reaches the sediments, decomposition and mineralization occur primarily through the actions of bacteria and fungi.

The importance of decomposition in the sediments to the productivity of the water above them depends, in large measure, on the proximity of the sediments to the photic zone. In shallow, productive (eutrophic) lakes having extensive littoral vegetation, detrital recycling processes may dominate. In deep, unproductive (oligotrophic) lakes and open oceans, where littoral vegetation contributes little to overall production and where the sediments lie at vast distances from the photic zone, sediment mineralization processes are less important. A map of productivity in the oceans (Figure 12-14) shows that the rate of carbon fixation is greatest in shallow seas, both in the Tropics (for example, the Coral Sea and the waters surrounding Indonesia) and at high latitudes (the Baltic Sea, the Sea of Japan), and in zones of upwelling. The latter occur along the western coasts of Africa and the Americas, where winds blow surface waters away from shore, thus establishing a vertical current to replace them (see Chapter 8) (Boje and Tomczak 1978).

Nutrient Regeneration Processes in Organic Sediments

Organic sediments of rivers, lakes, and oceans are comparable to terrestrial detritus in that dead organic matter accumulates there and undergoes decomposition. But there are important differences between aquatic sediments and terrestrial detritus. Except in the littoral zones of shallow lakes, where primary producers like rooted vegetation directly connect with the sediments, sediments generally lie some distance from the source of organic production. As a consequence, nutrients released back into the water column may require additional processing or transport in order to reenter the planktonic food web. Another difference is that decomposition of terrestrial detritus is accomplished, for the most part, aerobically, hence relatively rapidly. Aquatic sediments often become anoxic, greatly slowing most biochemical transformations, a situation that is exacerbated by the low temperatures of most benthic regions.

Nutrients are regenerated from aquatic sediments in a number of ways. Rooted aquatic macrophytes may assimilate large amounts of nutrients directly from littoral sediments in much the same way that terrestrial plants take up nutrients from the soil (Barko and Smart 1981, Lodge et al. 1988). Most organic matter in aquatic sediments is transformed into dissolved material or gaseous compounds by

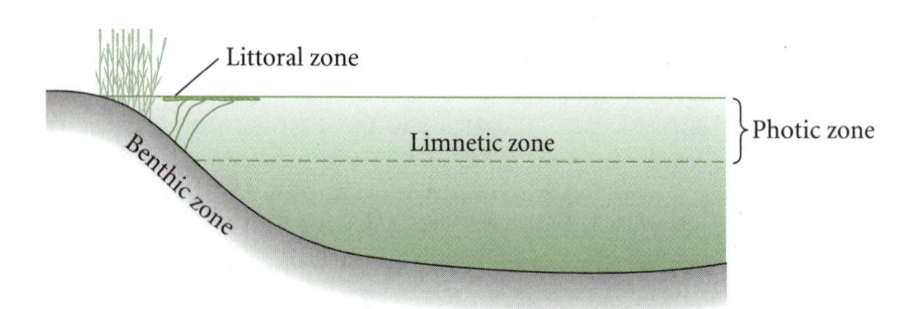

FIGURE 12-13 Diagram of a lake ecosystem, showing the major regions.

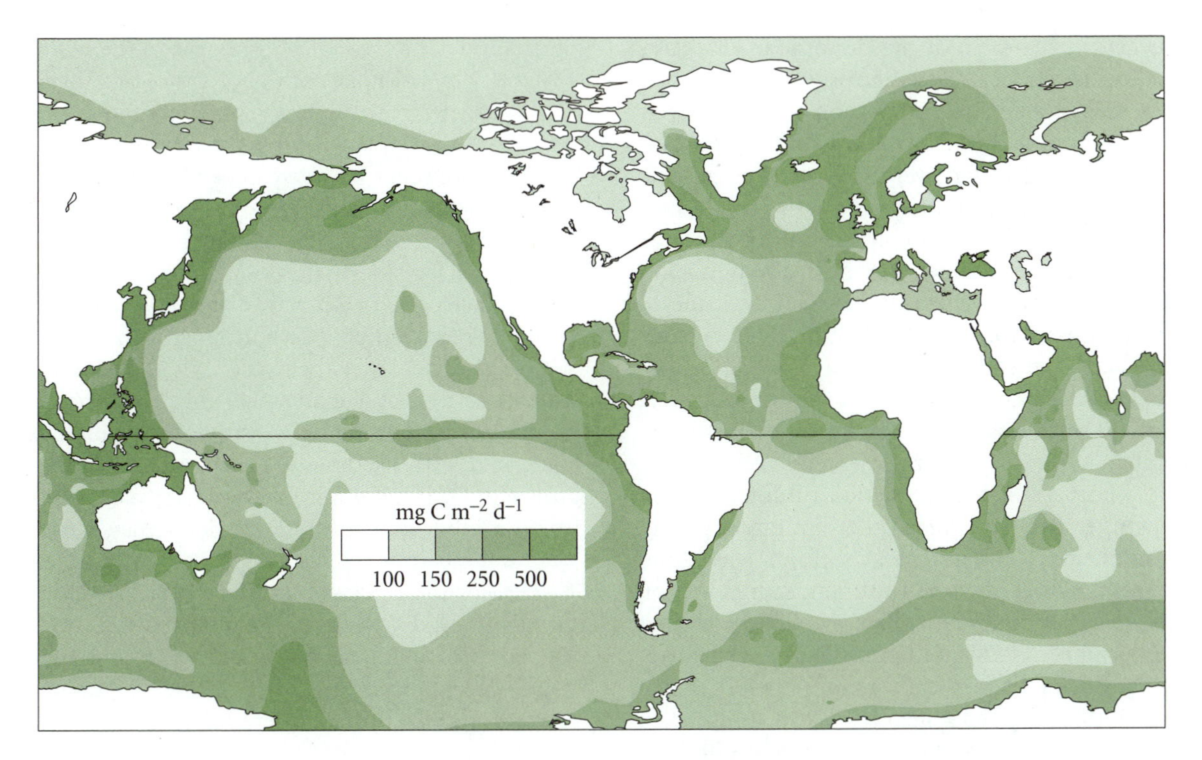

FIGURE 12-14 Primary productivity in the world's oceans, in milligrams of carbon fixed per square meter per day. Productivity is greatest on the continental shelves and in regions of upwelling on the west coasts of Africa and South and Central America. (*After Barnes and Mann 1980.*)

microbial action that takes place primarily under anaerobic conditions. Acid-forming bacteria hydrolyze complex carbohydrates, proteins, and lipids into amino acids, simple sugars, and fatty acids. Those products then undergo fermentation, producing organic acids and alcohols that are utilized by methanogenic bacteria to form either methane (CH_4) or, through the process of denitrification, carbon dioxide and molecular nitrogen (Wetzel 1983).

A number of processes slow the rate of nutrient recycling in aquatic sediments. In order to reach the sediments, organic material must sink (be sedimented) by the force of gravity. The material most rapidly sedimented is large material not dissolved in the water. Microbial degradation of such material, once it reaches the sediments, is slow. Additionally, the degradation of cellulose from littoral vegetation produces acid fermentation products that lower the pH of the sediments and inhibit bacterial metabolism (Wetzel 1983).

Nitrogen and Phosphorus Dynamics in Aquatic Sediments

In the nitrogen pathway, organic nitrogen first undergoes ammonification (hydrolysis of protein and oxidation of amino acids) and then, under aerobic conditions, nitrification from ammonia to nitrite (NO_2^-) and then to nitrate (NO_3^-) (see Figure 11-13). Ammonification is accomplished in the cells of all organisms, whereas nitrification is a process unique to nitrifying bacteria, a number of types of which occur in fresh-water and ocean sediments. Under anaerobic conditions, nitrite and nitrate become oxidizers (electron acceptors), and the nitrification process reverses, leading eventually to the production of N_2 gas by the process of denitrification. Because of the anoxic conditions of most sediments, denitrification can account for substantial losses of fixed nitrogen from lake ecosystems (Lijklema 1994). In Lake Mendota, Wisconsin, 13% of nitrogen losses (approximately 28,100 kg N yr^{-1}) was the result of denitrification (Brezonik and Lee 1968). In general, denitrification is greatest when there is high organic nitrogen input to the system coupled with high temperatures, poor mixing of the water column, and consequently, low oxygen availability (Lijklema 1994).

Aquatic ecosystems can be highly productive only where there is strong interchange between the sediments and the surface, at least on occasion. Nitrogen budgets measured by C. F. Liao and D. R. S. Lean (1978) in the Bay of Quinte, Lake Huron, Ontario, illustrate the relative magnitudes of assimilation and regeneration. The studies were conducted within columns of water in "limnocorrals": enclosures, triangular or circular in cross section, made of sheets of plastic suspended by floats at the surface and entrenched in sediment at the bottom (in this case, 4 m below the surface). Such enclosures allow studies of the fluxes of elements by addition of radioactively labeled compounds, but they may disrupt normal patterns of mixing of water layers; thus the vertical profiles of production and nutrient cycling within limnocorrals may differ from those in adjacent open water.

In the Bay of Quinte, ammonia accounted for about 65% of the regenerated nitrogen available to plants, and nitrate for the other 35%, both early (June 5) and late (September 4) in the growing season (Table 12-4). But although levels of nitrogen were similar on the two sample dates, gross production in September was nearly seven times the level in June, probably because of a combination of differences in water temperature and some other limiting nutrient. Short-term uptake of nitrogen by plants was about one-tenth the level of carbon fixation. Sedimentation of nitrogen in sinking particulate matter was 14% of uptake in June and 28% of uptake in September. Accordingly, during the September sample period, when the total nitrogen in the system (NH_4^+, NO_3^-, and particulate) was 586 g l^{-1} and sedimentation was 36 g l^{-1} d^{-1}, physical removal of nitrogen from the system could deplete the resource quickly. But sedimentation of particulate nitrogen was approximately balanced throughout the season by return of ammonia, because total nitrogen in the water column varied little relative to internal cycling.

In many lakes, iron and aluminum bind with phosphorus, resulting in the precipitation of iron and aluminum complexes of phosphorus to the sediments (Wetzel 1983). Unlike nitrogen, the sediment flux of which is determined by microbial physiology, the rate of exchange of phosphorus between the sediments and the lake water is largely dependent on the chemical dynamics of the sediments and the water. In particular, the redox potential at the sediment-water interface determines the rate of interchange of phosphorus across the sediment-water boundary. In the low redox environment of bottom sediments and waters immediately over them, the shift of such redox elements as iron and manganese from oxidized to reduced forms greatly increases their solubility. In particular, as ferric iron (Fe^{3+}) is reduced to ferrous iron (Fe^{2+}), insoluble iron-phosphate complexes become solubilized, and both elements may move into the water column. This was first demonstrated in C. H. Mortimer's classic studies (1941, 1942) on the exchange of dissolved substances between the mud and water of lakes, and has been reconfirmed in many more recent studies (for example, Holdren and Armstrong 1980, Riley and Prepas 1984).

Effects of Thermal Stratification and Vertical Mixing on Lake Nutrient Dynamics

Seasonal changes create cycles of temperature stratification in many lakes and some temperate oceans (see Figure 8-11). In summer, the warm waters of the top layer, the epilimnion, and the cooler waters at the lake bottom, the hypolimnion, are prevented from mixing.

TABLE 12-4 Estimates of release and uptake of nitrogen (μg N l^{-1} d^{-1}) in limnocorrals in the Bay of Quinte, Ontario

Characteristic	June 5, 1974	September 4, 1974
Concentration (μg N l^{-1} d^{-1})		
Ammonia (NH_4^-)	120	176
Nitrate (NO_3^-)	84	72
Primary production (μg C l^{-1} d^{-1})		
Gross	185	1,281
Net	−139	+807
Uptake (μg N l^{-1} d^{-1})		
Ammonia	20	117
Nitrate	4.5	4.5
Nitrogen fixation	1.2	2.7
Total	26	124
Release (μg N l^{-1} d^{-1})		
Zooplankton grazing	9.2	27
Sedimentation	3.7	63
Total	13	90

(From Liao and Lean 1978.)

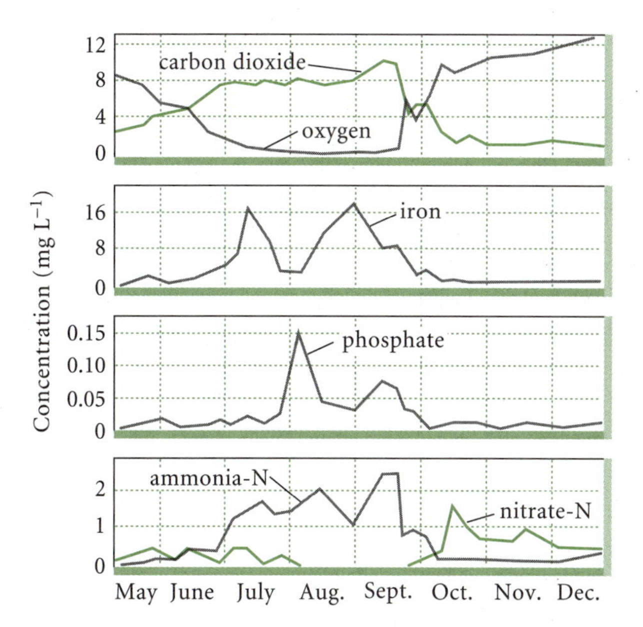

FIGURE 12-15 Seasonal course of water chemistry of the hypolimnion of Esthwaite Water, England, showing the increase in phosphorus and iron compounds during the period of summer anoxia resulting from reduction processes in the sediments and at the sediment-water boundary. (*From Hutchinson 1957b.*)

As the summer progresses, nutrients are assimilated by organisms in the photic zone of the epilimnion and then removed by sedimentation to the hypolimnion, thereby decreasing nutrient supplies in the photic zone and reducing production. A uniform temperature profile arises in the fall as air temperatures fall and epilimnetic waters cool. Surface winds during this time provide sufficient energy to mix the top and bottom waters (fall overturn), and nutrients are recycled from the hypolimnion. A weaker thermal stratification exists in the winter, when production is reduced by lower temperatures and, in some cases, occlusion of light by ice cover. Warming in the spring starts another round of mixing (spring overturn), which often results in a bloom of production (spring bloom).

Changes in the water chemistry of the hypolimnion of an English lake, Esthwaite Water, during the course of a single season show the effects of anoxic conditions on iron and phosphate dynamics (Figure 12-15). After summer stratification, dissolved oxygen at the deepest level of the lake decreased gradually, while dissolved carbon dioxide increased. The water became anoxic by early July, and remained so until the end of stratification and the onset of vertical mixing in late September. During the period of anoxia, levels of ferrous iron, phosphate, and ammonia increased dramatically as they were released by reduction processes in the sediments and at the sediment-water boundary. The return of oxidizing conditions in the fall completely altered the chemistry of the hypolimnion, initially because of the replacement of bottom with surface water, but ultimately because oxidized forms of several redox elements formed insoluble compounds, which precipitated out of the water column. Nitrogen was a conspicuous exception: under oxic conditions, nitrifying bacteria convert ammonia to nitrate, which generally remains in solution.

Life in the anoxic world can often be surprising, as in the case of the white sulfur bacterium *Thioploca*, which forms dense, thick mats on the surface of sediments at depths of 40–280 m off the coast of Peru and Chile (Fossing et al. 1995) (Figure 12-16). The habitat of *Thioploca* occurs within an upwelling zone that supports a rich fishery, but the water chemistry there is unusual in that there is little oxygen in the deep waters, but relatively high concentrations of nitrate. The sediments themselves are anoxic, but other bacteria oxidize organic carbon in a reaction coupled with the reduction of sulfate to hydrogen sulfide. *Thioploca* is chemoautotrophic and gains energy by oxidizing the hydrogen sulfide back to sulfate, using nitrate as an oxidizer ($NO_3^- \rightarrow N_2$). The only problem is that the nitrate it needs is in the water column above the sediments. Filaments of *Thioploca* cells live within tubular sheaths, which extend from the water-sediment interface 5–10 cm into the sediment. The bacterial filaments slowly glide up and down in these sheaths, shuttling between the water and sediment environments. At the top of the sheath, the bacterium takes up nitrate, which it stores at high concentrations in a large vacuole in its center. Armed with a supply of oxidizer, *Thioploca* slides down the sheath into the sediment, where it converts hydrogen sulfide to elemental sulfur, forming globules of pure sulfur in its cytoplasm, and then further oxidizes the sulfur to sulfate. When its nitrate supply runs low, it slides back up to the sediment surface to restock.

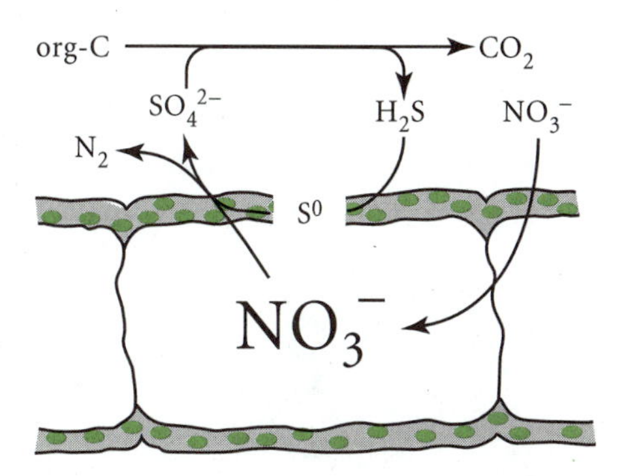

FIGURE 12-16 A schematic diagram of a *Thioploca* bacterium, showing its biochemical transformations involving sulfur and nitrogen. The center of the cell is a large vacuole within which nitrate is stored at high concentrations. Globules of pure elemental sulfur embedded in the cytoplasm are represented as colored dots. (*From Fossing et al. 1995.*)

12.7 Microbial processes dominate food webs and nutrient cycling in unproductive open waters.

Primary production in the open waters of large lakes and oceans, areas referred to as the **pelagic** zone, is the result of the photosynthetic activity of small free-floating plankton. Plankton can be conveniently categorized by size: those cells less than 2 μm are referred to as **picoplankton,** those measuring between 2 and 20 μm are called **nanoplankton,** those between 20 and 200 μm are called **microplankton,** and plankton larger than 200 μm are referred to as **mesoplankton** (Sieburth 1979). Each of these size categories may contain both photosynthesizing **phytoplankton** and heterotrophic organisms referred to as **phagoplankton** that assimilate dissolved organic material from the water. Some plankton graze on smaller plankton; these are referred to as **zooplankton.**

Historically, limnologists have attributed most of the production in pelagic waters to nanoplankton and microplankton species, which are consumed by zooplankton, which are, in turn, consumed by small invertebrates and fish in the classic plankton-based aquatic food chain (Figure 12-17). The application of epifluorescence microscopy in the 1960s, however, led to the discovery that the open waters of many lakes and oceans contain vast numbers of both autotrophic and heterotrophic picoplankton, the latter made up largely of bacteria (Stockner and Porter 1987, Fenchel 1988). The immense diversity of aquatic bacteria has been confirmed more recently with the use of ribosomal RNA sequencing techniques (Höfle 1990). Subsequent investigations revealed that as much as 70% of the carbon fixation in oligotrophic systems is the result of the activities of algal picoplankton, and that up to 50% of the carbon fixed by phytoplankton is released into the water as dissolved organic matter (DOM) and is thus not retained as producer biomass (Azam et al. 1983, Fenchel 1988). Over one-half of this "lost" organic material is assimilated by heterotrophic picoplankton (bacteria) in marine systems (Fenchel 1988), and considerably more is assimilated in some freshwater systems (Currie and Kalff 1984a,b). However, most zooplankton are unable to capture and consume these very small picoplankton, creating what would appear to be a missing link in the food web of open-water systems. To account for this missing connection, F. Azam and colleagues at Scripps Institution of Oceanography hypothesized that heterotrophic picoplankton-sized bacteria are grazed by small, nanoplankton-sized, heterotrophic flagellates and ciliated protozoans, which are in turn eaten by microzooplankton that are large enough to be utilized by macrozooplankton (Azam et al. 1983). In essence, they suggested that there is a **microbial loop** in the pelagic food chain (see Figure 12-17) that lengthens the chain. This microbial loop does exist, and is important in oceans (Fenchel 1988), freshwater lakes (Currie and Kalff 1984a,b, Stockner and Porter 1987), and streams and rivers (Edwards et al. 1990, Meyer 1990).

The availability of nutrients interacts with the size distribution of the plankton producers to shape the way in which nutrients are recycled in the plankton-based food web of pelagic waters and to determine, in large measure, whether the food web is dominated by the microbial loop. Nutrient uptake by plankton involves the active transport of dissolved materials across

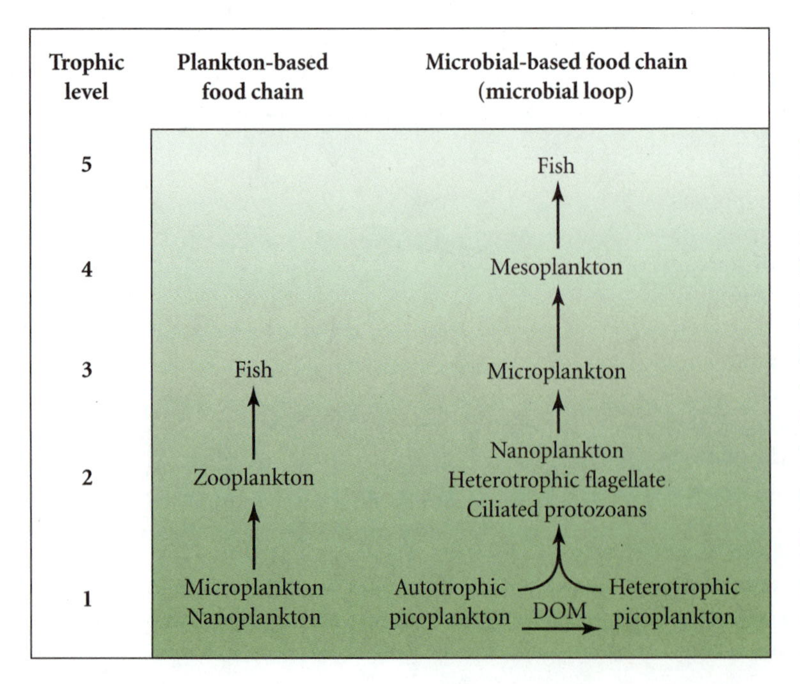

FIGURE 12-17 Comparison of the plankton-based food chain and the microbial-based food chain. Microplankton and nanoplankton constitute the lowest trophic level of the plankton-based food chain. The microbial-based food chain is longer owing to the presence of very small autotrophic and heterotrophic picoplankton. Much of the carbon fixed by autotrophic picoplankton is released as dissolved organic matter (DOM) and consumed by heterotrophic picoplankton (mostly bacteria), thus forming a microbial loop at the bottom of the food chain. The presence of the picoplankton producers gives the microbial-based food chain more trophic levels than the plankton-based food chain. (*After Stockner and Shortreed 1988, Stockner and Porter 1988.*)

the cell membrane (Fenchel 1988, Chróst 1990). In nutrient-rich environments, the amount of nutrients available to the inside of the cell is limited by the efficiency of the transport mechanism rather than by the concentration gradient of the material in the water. Relatively large cells, such as microplankton or mesoplankton, with small surface-to-volume ratios suffer no particular disadvantage in competition for resources with smaller picoplankton or nanoplankton when nutrients are abundant. However, in nutrient-poor environments, such the open waters of oceans or large, deep lakes, smaller producers enjoy a competitive advantage owing to the reduced physiological demands of their smaller cell volume and their relatively large surface area. This advantage accounts for the importance of the microbial loop in open oceans, where nutrients are extremely limited (Platt et al. 1983, Fenchel 1988).

Bacterial and autotrophic picoplankton do not sink readily from the epilimnion. Moreover, bacteria sequester important nutrients such as phosphorus and nitrogen. Thus, nutrients and organic matter tend to remain near the surface (Currie and Kalff 1984b, Stockner and Porter 1987). The implication of these features of the microbial loop is that nutrient recycling in the epilimnion is more dependent on trophic interactions—the grazing of heterotrophic bacteria by nanozooplankton and microzooplankton—than on the microbial degradation processes observed in soils and aquatic sediments. Indeed, it has been suggested that carbon and nutrient flux are more closely related in the pelagic zone than in terrestrial ecosystems or other aquatic ecosystems (Azam et al. 1983).

12.8 Nutrient regeneration is strongly influenced by the movement of water in streams and rivers.

The physical dynamics of moving water exercise a profound influence on the processes by which energy flows and nutrients cycle in stream and river ecosystems. Organic material is transported, sometimes rapidly, downstream. Nutrients that are regenerated at one location may be assimilated at another location some distance away. Rates of water flow determine the flux of material through a particular section of a stream and thus, in part, constrain productivity in that portion of the stream. Moving water modifies aquatic habitats, shapes the river basin, and serves as the medium of interaction between the river and the surrounding floodplains. The life cycles of the river biota reflect the unidirectional movement patterns of the medium of their environment.

Organic Material in Rivers and Streams

Nutrient dynamics in rivers and streams depend on the flux of organic material across the river-land boundary and through particular reaches of the river itself. Thus, the balance between the amount of organic material contributed to the stream ecosystem by primary production within the stream and the amount contributed by terrestrial inputs and upstream sources is extremely important (Allen 1995). For the most part, streams and rivers are thought to be heterotrophic, having a much larger allochthonous than autochthonous input of organic material. In a study of a small blackwater stream, for example, Edwards and Meyer (1987) found that in-stream gross primary production accounted for only about 6% (0.25 kg C m^{-2} yr^{-1}) of the annual carbon budget in a square meter of the stream; most of the organic material came from the surrounding floodplain (44%; 1.76 kg C m^{-2} yr^{-1}) or from upstream sources (approximately 50%; 2.0 kg C m^{-2} yr^{-1}). But the relative importance of primary production varies with the relationship of the stream to the surrounding terrestrial habitats and with the size of the stream. Forest streams, which run beneath the forest canopy, are subject to considerably more litterfall and other external inputs than are grassland or desert streams. Primary production in the latter may contribute more material to the carbon budget than do external sources. Large rivers may support abundant plankton and macrophyte communities having high productivity (Minshall 1978, Allen 1995).

Allochthonous organic material, particularly that from terrestrial sources, such as litterfall, enters a stream in large chunks that are broken down initially by the physical action of the moving water or the activities of grazing invertebrates. Material greater than 1 mm in a linear dimension is referred to as **coarse particulate organic matter** (CPOM). Further processing of CPOM while in transit downstream produces material whose particle sizes range from 1 mm to 0.5 μm, referred to as **fine particulate organic matter** (FPOM). FPOM may also be imported as soil organic matter. **Dissolved organic matter** (DOM) arises from the breakdown of FPOM or, more importantly, as a product of in-stream primary production. The fates of the various types of organic matter in the stream food web are depicted in Figure 12-18.

Nutrient Spiraling

We may think of an atom of nitrogen or phosphorus or some other nutrient as undergoing repeated cycles of assimilation and regeneration within a single hectare of a forest or within a well-defined region of the epilimnion of an oligotrophic lake. But in streams and rivers, nutrients are on the move. Applied to the cycle of assimilation, food web transfer, and release is

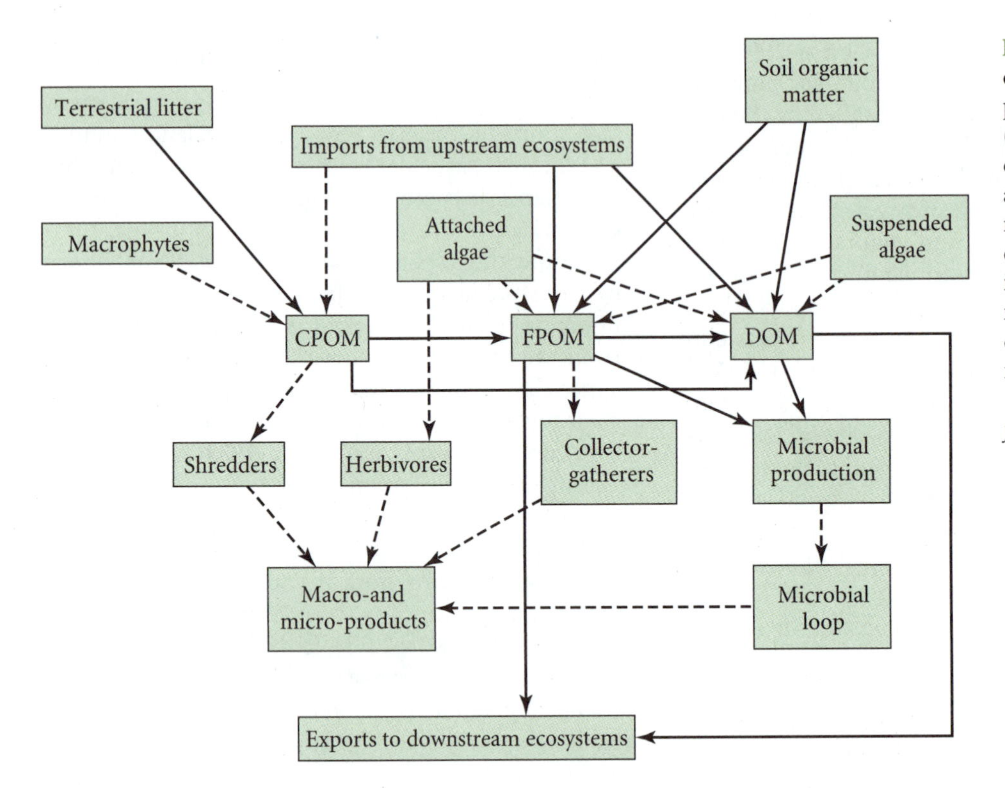

FIGURE 12-18 The origins and fates of coarse particulate organic matter (CPOM), fine particulate organic matter (FPOM), and dissolved organic matter (DOM) in a stream ecosystem. Dashed lines indicate transfers that are not as strong or as common as those represented by solid lines. (*From Allen 1995; modified from Wetzel 1983.*)

the pervading unidirectional force of the stream current. Nutrients are continuously displaced downstream as they undergo chemical transformations from one phase of their cycle to another. We conceptualize this pattern of movement as a **nutrient spiral** (Webster and Patten 1979), an idea that integrates the unique features of flowing water with nutrient regeneration processes.

The essential features of nutrient spiraling are represented in Figure 12-19. Nutrients may exist either as particulate organic matter that is part of the stream biota (represented by a crayfish in Figure 12-19) or as dissolved inorganic forms in the water column. Nutrient regeneration occurs when nutrients are released from the biotic compartment into the water column. Nutrient assimilation occurs when

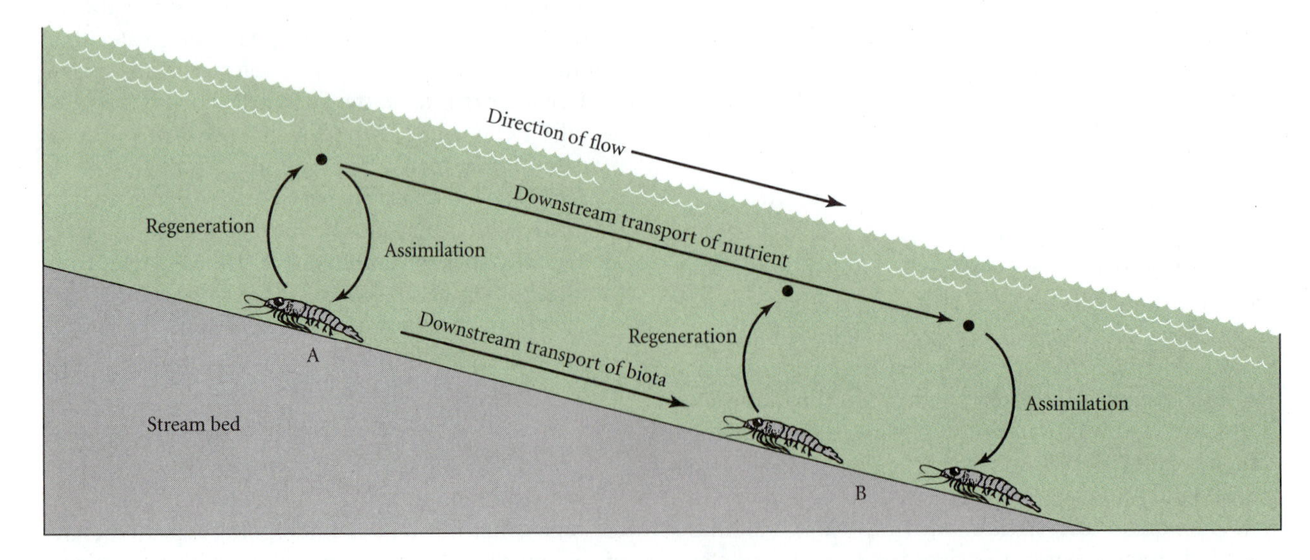

FIGURE 12-19 Nutrients exist in streams either as particulate organic matter that is part of the stream biota (represented by the crayfish) or in dissolved inorganic form in the water column. Nutrient regeneration occurs when nutrients are released from the biotic compartment into the water column. Nutrient assimilation is the uptake of nutrients by the biota. The flow of the water in the stream displaces nutrients. Thus, nutrients regenerated at point A may be assimilated downstream at point B. The movement of animals in the stream may also displace nutrients. (*After Newbold et al. 1982.*)

nutrients are taken from the water column by the biotic constituents of the stream ecosystem. A nutrient atom is said to have completed one cycle when it has passed through both the water column and biotic compartments of the system. We may think of a stream as being at a steady state with respect to nutrient cycling when the rate of regeneration equals the rate of assimilation. If there were no stream flow, we could imagine one cycle occurring more or less in the same location. But nutrients move, both while in the biotic compartment and while in the water column. Thus, nutrients regenerated at one location will travel downstream some distance before being assimilated. Downstream transport is generally slower for nutrients that are part of the biota, owing to the ability of some organisms to maintain their position or move against the current. The distance that nutrients travel will depend on the rate of flow of the water and on the rates of regeneration and assimilation. Assimilation is particularly important since it is a measure of the demand for nutrients. The distance traveled by nutrients will be shorter in streams with high assimilation rates.

The simple conceptual model of nutrient spiraling presented in Figure 12-19 underrepresents the complexity of nutrient cycling. In most streams, many biotic compartments—CPOM, FPOM, and organisms with different feeding strategies (e.g., grazers, shredders, collectors)—are probably components of a complete nutrient cycle. The likelihood that a single atom will pass through every compartment during each cycle is small, and thus a complete representation of nutrient dynamics would have to account for such probabilities (Newbold et al. 1982, 1983).

Because phosphorus is an essential nutrient, and because it limits primary production in many natural waters (Elser et al. 1990), its spiraling dynamics in streams and rivers have been examined intensively (e.g., Newbold et al. 1983, Mulholland et al. 1985). To understand how phosphorus moves in a small stream, J. D. Newbold and coworkers at the Oak Ridge National Laboratory released ^{32}P (in the form of $^{32}PO_4$), a radioactive isotope of phosphorus, into a small section of a stream in the Walker Branch watershed of Tennessee. They then measured the concentrations of the ^{32}P tracer in the water and in the biotic compartments of a section of the stream below the release point (Newbold et al. 1983). Their results showed that in Walker Branch, each atom of phosphorus completed one cycle in about 18 days, and during that time, traveled approximately 190 m downstream (spiraling length). By sampling the amount of ^{32}P in the water at downstream locations and comparing those measurements with ^{32}P uptake in the various biotic compartments of the stream, they were able to estimate the flux of phosphorus among the compartments of the ecosystem (Table 12-5). CPOM accounted for most of the transfer of phosphorus from the water column; transfers among the biotic compartments were small. Of the 190 m spiraling length of phosphorus, 165 m was spent in the water compartment.

12.9 Estuaries and marshes may provide net inputs of energy and nutrients to marine ecosystems.

Shallow estuaries—semi-enclosed coastal regions subject to both freshwater inputs from rivers and tidal inputs from the sea—are among the most productive ecosystems on earth (McClusky 1981). Examples include Chesapeake Bay and the Mississippi River Delta region. Salt marshes, which are intertidal areas with

TABLE 12-5 Selected ^{32}P exchange fluxes among compartments of a Walker Branch stream ecosystem

From	To	Flux ($mg\ m^{-2}\ s^{-1} \times 10^5$)
Water	All biotic compartments	1.5
Water	CPOM	2.7
Water	FPOM	1.6
Water	Aufwuchs*	0.3
Particulate organic matter	Consumers	0.1
Consumers	Predators	0.04

(Data from Newbold et al. 1983.)

"Aufwuchs" includes organisms that form a coating on stones and other submerged objects (periphyton) and associated heterotrophic microbes.

FIGURE 12-20 Salt marshes are a common feature of protected bays along most temperate coasts.

emergent vegetation (Figure 12-20), exhibit similarly high productivity (Odum 1988). Marshes can be found in many places along the Atlantic and Gulf of Mexico coasts of North America. The high production of these ecosystems results from rapid and local regeneration of nutrients and external loading in the form of nutrients brought into the system by rivers. The effects of high production in estuaries and coastal marshes extend to marine ecosystems in many areas through net exports of organic matter. A Georgia salt marsh exports nearly 10% of its gross primary production and almost half of its net primary production to surrounding marine ecosystems in the form of organisms, particulate detritus, and dissolved organic matter carried out with the tides (Teal 1962; Figure 12-21). Because of their high productivity and structural complexity, coastal marshes and estuaries are important feeding areas for the larvae and immature stages of many fish and invertebrates that later complete their life cycles in the sea.

The high production of coastal ecosystems reflects their high nutrient levels. Because such a large fraction of the production is carried out to sea, exported nutrients must be replaced by imports. Studies of the nitrogen budget of Great Sippewissett Marsh on Cape Cod have revealed nutrient inputs to the marsh through precipitation (minor), groundwater flow from surrounding terrestrial ecosystems, and local fixation of atmospheric nitrogen. These inputs approximately balance losses through denitrification, local accumulation of sediments, and exports in tidal water (Table 12-6). A high rate of denitrification,

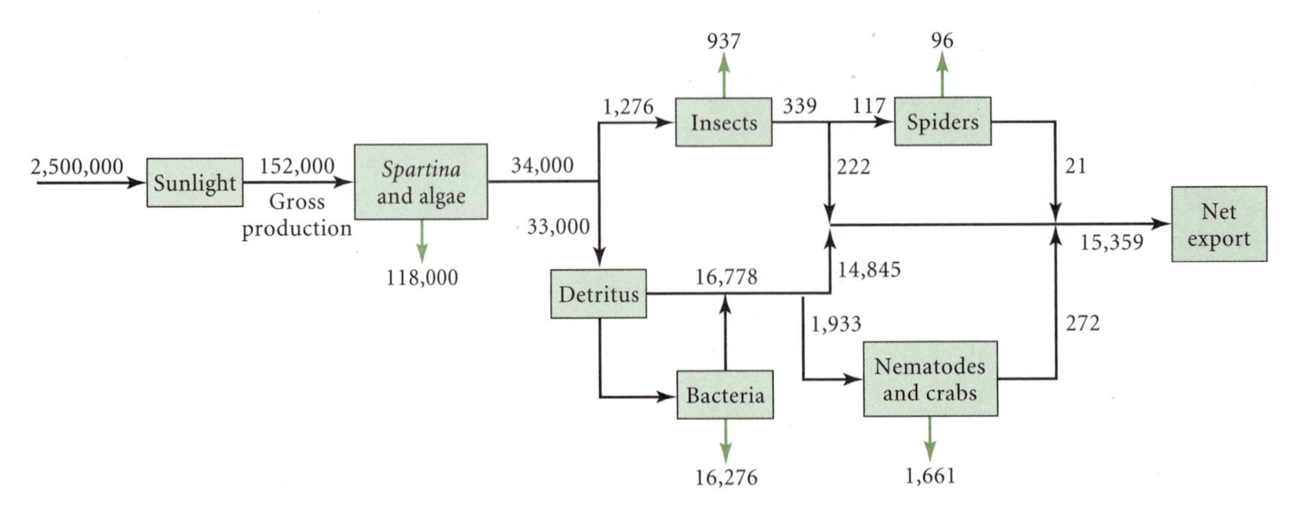

FIGURE 12-21 Energy flow diagram for a Georgia salt marsh. Fluxes are in kJ m^{-2} yr^{-1}. The colored arrowheads represent respired energy lost from the system. The net export of energy from the system represents energy that goes to the oceans. The rest of the energy in the system is contained in the plants and animals that remain in the marsh. (*From McClusky 1981; after Teal 1962.*)

TABLE 12-6 Nitrogen budget for Great Sippewissett marsh ($kg\ N\ yr^{-1}$)

	Input	Output	Net exchange
Precipitation			
Nitrate	110		
Ammonia	70		
Dissolved organic matter	190		
Particulate organic matter	15		
Total	390		390
Groundwater flow			
Nitrate	2,920		
Ammonia	460		
Dissolved organic matter	2,710		
Total	6,120		6,120
Nitrogen fixation	3,280		3,280
Tidal water exchange			
Nitrate	390	1,210	
Nitrite	150	170	
Ammonia	2,620	3,540	
Dissolved organic matter	16,300	18,500	
Particulate organic matter	6,740	8,200	
Total	26,200	31,600	−5,350
Denitrification		6,940	−6,940
Sedimentation		1,295	−1,295
Totals	35,990	39,835	−3,845

(From Valiela and Teal 1979.)

which occurs primarily in the anoxic sediments at the bottom of the creek draining the marsh, underscores the role of chemoautotrophic and photoautotrophic bacteria in salt marsh metabolism. The rich organic sediments are mostly anoxic just below their surfaces owing to high rates of microbial decomposition of organic matter. As a result, oxidations based on denitrification ($N^{5+} \rightarrow N^{0}$) and sulfate reduction are important for the regeneration of nutrients (Howarth 1984).

As we have seen, the basic chemical and biochemical transformations of the element cycles are uniquely molded by each type of terrestrial and aquatic ecosystem according to the physical and chemical conditions created in those environments. The paths of elements discussed in this chapter describe patterns of nutrient cycling. Next we shall examine the regulation of these nutrients through their cycles; hence, the overall productivity of the ecosystem.

SUMMARY

1. Nutrient cycles in terrestrial and aquatic ecosystems are based upon similar chemical and biochemical reactions, but differ according to the physical conditions of the habitat.

2. Nutrient regeneration in terrestrial ecosystems takes place in the soil. The characteristics of soil reflect the influences of the bedrock below and the climate and vegetation above. The influences of the factors that form soils are reflected in the presence of well-defined soil horizons. The influence of biotic and climate factors is most important in the formation of the soil horizons near the surface; weathering is the most important factor in the formation of the deep horizons.

3. Soil develops initially through the mechanical and chemical weathering of parent rock, which results in the

breakdown of native minerals such as feldspar, mica, and quartz and their reformation into clay particles, which mix with organic detritus entering the soil from the surface. The concentration of hydrogen ions (H^+) influences weathering. Two sources of hydrogen ions are the carbonation reaction of rainwater ($H_2O + CO_2 \rightleftharpoons H^+ + HCO_3^- \rightleftharpoons H_2CO_3$) and soil nitrification ($NH_4^+ + 2O_2 \rightarrow 2H^+ + H_2O + NO_3^-$).

4. The clay and humus content of soil determines its ability to retain nutrients required by plants. Clay and humus particles (micelles) have negative charges on their surfaces that attract cations (Ca^{2+}, K^+, NH_4^+) directly and anions (PO_4^{3-}, NO_3^-) indirectly under acid conditions. Acid soils have more hydrogen ions and fewer mineral cations than neutral soils. Hydrogen ions tend to displace other cations on micelles and thereby reduce soil fertility. The ability of soil to hold positive ions is called its cation exchange capacity. Acid rain can increase cation mobility in soils and lead to nutrient depletion. In acid temperate zone (podsolized) soils and deeply weathered (laterized) tropical soils, clay particles break down and the fertility of the soil is much reduced.

5. Nutrients are regenerated from litter by leaching of soluble substances, consumption by large detritus feeders (millipedes, earthworms), further decomposition by fungi to break down cellulose and lignin, and eventual mineralization of phosphorus, organic nitrogen, and sulfur by bacteria.

6. In many lowland tropical habitats, the soils are deeply weathered and retain nutrients poorly. In such habitats, regeneration and assimilation of nutrients is rapid, and most of the nutrients, especially phosphorus, occur in the living vegetation. If such soils are clear-cut for agriculture, they soon lose their fertility as a consequence of nutrients being removed along with the native vegetation.

7. The productivity of shallow-water aquatic ecosystems is maintained by the transport of nutrients from bottom sediments. Nitrogen is mobilized by microbial activity; phosphorus regeneration depends on the redox environment of the sediments. The vertical mixing of seasonally stratified lakes and upwelling in oceans replenish the upper waters with nutrients.

8. One of two food chains predominates in open waters. In nutrient-rich waters, relatively large ($>20\ \mu m$) microplankton and mesoplankton are the principal primary producers; these organisms are consumed by zooplankton, which in turn are eaten by fish and invertebrates. Where nutrients are limited, as in open oceans and large, deep lakes, very small ($<20\ \mu m$) picoplankton are the predominant primary producers. These organisms are too small for zooplankton to eat, and thus several additional trophic steps, referred to as the microbial loop, occur in the food chain.

9. Water currents in streams and rivers displace nutrients downstream as those nutrients move from one compartment of the nutrient cycle to another. This movement of cycling nutrients is called a nutrient spiral. The distance traveled during one cycle by an atom of a nutrient is related to the flux of the nutrient in the water and biotic compartments of the ecosystem, the rate at which nutrients are assimilated by the biota, the rate of respiration, and the rate of water flow.

10. Shallow-water communities, particularly estuaries and salt marshes, are extremely productive owing to rapid and local regeneration of nutrients and the external loading of additional nutrients from nearby terrestrial habitats. Studies of nutrient budgets indicate that marshes and estuaries are major exporters of both organic carbon and mineral nutrients to surrounding marine systems, and hence are an indispensable component of marine production in some areas.

EXERCISES

1. The phenomenon of cation exchange capacity was discussed in this chapter. Ecologists also recognize the phenomenon of anion exchange capacity. What soil characteristics might affect such a process?

2. Tropical forests are luxuriant and diverse, but the soils on which these forests grow are usually very poor in nutrients. How do tropical forests maintain such high biomass in the face of limited soil nutrients? What are the implications of this for the management, preservation, and restoration of tropical forests?

3. Develop a simple analytical model that depicts the dynamics of nutrient spiraling shown in Figure 12-19.

CHAPTER 13

Regulation of Ecosystem Function

GUIDING QUESTIONS

- What are the various methods that ecologists use to study ecosystem structure and function?
- Under what conditions do phosphorus and nitrogen limit production in freshwater and marine ecosystems, and how do their roles differ?
- How does the physical transport of nutrients regulate production in terrestrial and aquatic environments?
- What are the features of a three-compartment ecosystem model?
- What is the difference between bottom-up and top-down ecosystem regulation, and how do these two processes interact?

In previous chapters, we have noted that ecosystems have both structure and function. The trophic interrelationships, numbers of different species present, patterns of element cycling, development of soils, and the myriad other biotic and abiotic processes that characterize a forest, stream, or lake constitute the structure of an ecosystem (Odum 1962). We distinguish one ecosystem from another, at least superficially, based on such structural characteristics. Forests are dominated by trees and soils, lakes by open water and muddy sediments, rivers and streams by flowing water. But what of ecosystem function? What does an ecosystem do? And, more importantly, what regulates its activity? These questions provide the focus of this chapter.

The rate of production and the fluxes of energy and nutrients among the compartments of an ecosystem constitute the function of the ecosystem (Odum 1962). Function is integrated with structure and, thus, is dependent on it. As we have seen, the most productive ecosystems are tropical rain forests, coral reefs, and estuaries, where favorable combinations of high temperatures, abundant water, and intense sunlight promote rapid photosynthesis and assimilation. Plant growth depends on the regeneration of nutrients through biological processes, which themselves are sensitive to temperature, moisture, and other conditions. How, then, can we untangle the factors that influence the rate of primary production? What generalizations can be made about the regulation of ecosystem function? In this chapter, we provide an overview of the generalizations that have come from studies of ecosystem function. In the first part of the chapter we will focus on the role of nutrients as ecosystem regulators, providing as part of that discussion a taste of mathematical ecology. In the last part of the chapter we will turn our attention to the influence of organisms and the interaction between biotic and abiotic factors in the regulation of ecosystem function. But first, let us take a look at the tools ecologists use to study ecosystems.

13.1 Ecologists seek to understand ecosystem regulation through experimentation, comparative studies, and mathematical modeling.

Discerning how ecosystem function is regulated would be relatively simple if, for example, there were a single resource necessary for production that could be used up; whatever controlled the supply of that resource would control ecosystem function, the way an accelerator pedal controls the speed of a car by providing more or less gasoline to the engine. But even under the most favorable circumstances, plants assimilate only 1–2% of the light energy striking the earth's surface. Not all the water entering the system leaves by transpiration from plant surfaces. Furthermore, plants do not exhaust nutrient elements from the soil. It is not surprising that ecologists have yet to resolve the control of productivity in most ecosystems.

Three general research methodologies are applied to the study of ecosystem regulation: experimental manipulations in the field and in the laboratory, comparative ecosystem studies, and mathematical modeling. The type and resolution of information yielded by each of these approaches, and their applicability to specific questions, vary.

Field experiments generally involve the addition of materials, such as nutrients, to an ecosystem or to some portion of an ecosystem that has been isolated for the experiment (recall the limnocorral experiments in Chapter 12) and the subsequent observation of one or more response variables of the ecosystem. Such experiments are expensive and difficult to carry out, and because of this, are often limited to the exploration of just a handful of the many factors involved in regulating production. Laboratory experiments, while more easily controlled than field experiments, introduce problems of scale. For example, does the response of a seedling in a pot to the addition of a nutrient reflect the response of a natural field full of such seedlings? **Comparative studies** seek patterns of ecosystem function by comparing the features of similar or disparate ecosystems from around the world or from a particular region. These studies suffer from lack of resolution and problems with the interpretation of data generated by different sampling techniques. **Mathematical models,** because they allow the inclusion of virtually an unlimited number of factors, have been used extensively by ecosystems ecologists. Mathematical models are very useful because they require the explicit representation of the interrelationships of various compartments of the ecosystem, and thus the act of constructing the model sets forth many testable hypotheses about the ecosystem being modeled. But, except for the simplest models, which may be of limited usefulness, mathematical models often remain unsupported by empirical studies. Because of the relative utilities of the methods and because of the complexity of ecosystems, ecologists increasingly employ all three methods in a single study.

Nutrient addition experiments are the most common type of experimental ecosystem study. Such studies have been most widely used in freshwater lake and stream ecosystems (Newbold et al. 1983, Elser et al. 1990, Peterson et al. 1993). The typical protocol involves the introduction of a specific element or suite of elements into an ecosystem and a subsequent assay for the uptake of that nutrient in the various compartments of the ecosystem. In some cases, radiolabeled isotopes such as ^{32}P are released and tracked (see Chapter 12). Nutrient enrichment experiments may be performed in mesocosms or laboratory aquaria in which the response of a particular organism or assemblage of organisms to the addition is measured (see Figure 2-7). Studies such as this in which the response of a focal organism or group of organisms is measured are called **bioassay experiments.** These studies provide information on the rates of nutrient uptake or energy flow between specific ecosystem compartments.

Experimental studies may also include manipulations of biota in the field either by the exclusion or by the addition of particular grazing or predatory animals (e.g., Power et al. 1988, Carpenter et al. 1987, Lamberti et al. 1987). Large-scale manipulations may be conducted in watersheds, where the effects of logging and other land management practices within the confines of a stream basin can be measured by comparing the mass balance of nutrients in the basin (see Chapter 12). The goal of these field manipulations is to determine the effect of specific trophic relationships on the production of the entire system. The hypotheses and conclusions of large-scale field experiments are often supported by data from long-term monitoring studies. For example, the amounts of certain cations have been monitored in precipitation and in stream water at the Hubbard Brook Experimental Forest in New York since 1963 (Likens and Bormann 1995). We discussed the need for such long-term data in Chapter 2, where we described the National Science Foundation Long-Term Ecological Research program.

Ecologists may learn a great deal about the function of ecosystems by comparing the results from studies conducted on different ecosystems representing gradients of environmental, nutrient, or trophic conditions (Heal and Grime 1991, Peters et al. 1991, Vitousek and Matson 1991). These comparative studies seek to identify patterns of biotic or abiotic conditions that explain similarities or differences among ecosystems. Large-scale comparisons of both terrestrial and aquatic ecosystems have been conducted (Cole et al. 1991).

A third approach to the study of ecosystem regulation is to model ecosystem function by studying the behavior of mathematical or electrical analogues of systems in response to the alteration of symbolic or electrical equivalents of resources and conditions. This approach, called systems modeling, requires detailed knowledge of all the processes critical to regulating ecosystem function (which could be interpreted as everything that happens in the system!), as well as validation of each part of the model by field observation and experiment.

Using these techniques, ecologists attempt to understand the complex interactions that form the basis of ecosystem function. Our discussion of ecosystem function begins with an examination of the role of nutrients in primary production in aquatic ecosystems.

13.2 The regulatory roles of nitrogen and phosphorus differ in freshwater and marine ecosystems.

Phytoplankton account for much of the primary production in aquatic ecosystems (see Chapter 12). The ratio of the amount of carbon to nitrogen to phosphorus in the tissues of algae is approximately 106:16:1, roughly proportional to the atomic weights of the carbon, nitrogen, and phosphorus atoms. This ratio is called the **Redfield ratio** (Redfield et al. 1963). Thus, phytoplankton are assumed to assimilate nitrogen and phosphorus from the environment in a ratio of roughly 16:1. The N:P ratio in the environment, then, is an indicator of the relative availability of nitrogen and phosphorus to phytoplankton. Values of N:P less than 16:1 suggest that nitrogen is limiting, whereas N:P ratios greater than 16:1 indicate that there is less phosphorus per unit of nitrogen, and thus, that phosphorus is limiting. In general, the Redfield ratio for freshwater lakes is greater than 16:1, and phosphorus is considered the limiting nutrient in such ecosystems (Edmondson 1970, Schindler 1978, Howarth 1988). The production in most stream ecosystems is also thought to be limited by phosphorus (Allen 1995), although streams in arid and semiarid regions appear to be exceptions (Grimm and Fisher 1986). Most estuaries and open-water marine ecosystems have very low N:P ratios and are therefore nitrogen limited (Howarth 1988).

Phosphorus Limitation in Fresh Water

The importance of phosphorus as a limiting nutrient in lake ecosystems was established by a number of classic whole-lake fertilization studies conducted in lakes on the Canadian Shield in the late 1970s (Schindler 1974, 1977). In these studies, small lakes fertilized with phosphorus showed a dramatic increase in productivity, whereas the addition of nitrogen or carbon had no effect (Figure 13-1). The increased production resulting from phosphorus addition is referred to as **eutrophication.** Much of the interest in phosphorus as a limiting nutrient grew from an appreciation of the importance of eutrophication in lake ecosystem function. The release of raw sewage into rivers and lakes creates what is called **biological oxygen demand** (BOD) due to the oxidative breakdown of the detritus by microorganisms. Inorganic nutrients such as phosphorus stimulate the production of organic detritus, adding to the BOD. In its worst manifestations, this type of pollution can deplete the surface water of its oxygen, leading to the suffocation of fish and other obligately aerobic organisms.

Redfield ratios in freshwater streams are generally high, and thus phosphorus is limiting in many types of rivers and streams (Allen 1995). In each of the four years between 1983 and 1986, B. J. Peterson and coworkers (Peterson et al. 1993) added phosphorus (in the form of PO_4, phosphoric acid) to the Kuparuk River, a pristine clearwater tundra river in northern Alaska. During the initial two summers of the study, algal biomass, primary productivity, and chlorophyll a accumulation increased in the section of the river below the site of phosphorus introduction (Figure 13-2).

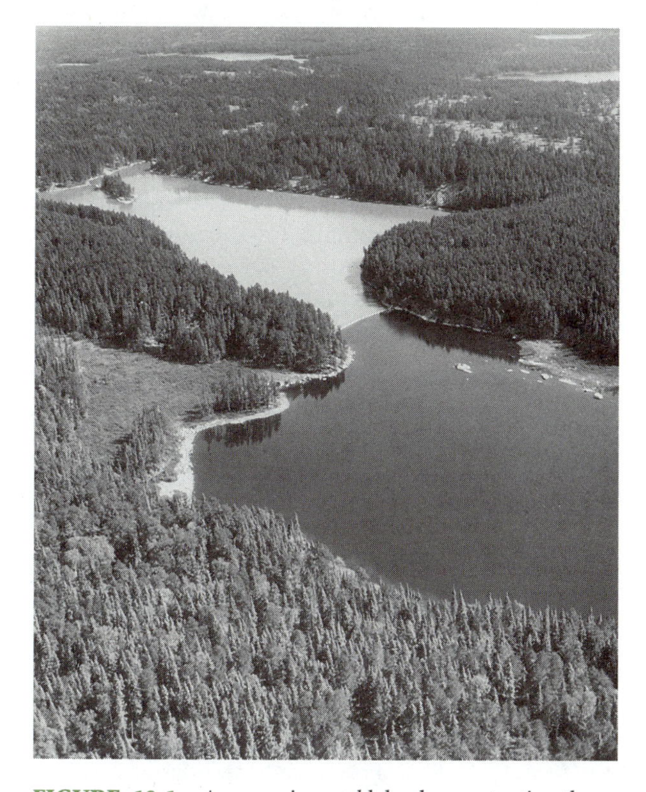

FIGURE 13-1 An experimental lake demonstrating the crucial role of phosphorus in eutrophication. The near basin, fertilized with carbon (in sucrose) and nitrogen (in nitrates), exhibited no change in organic production. The far basin, separated from the first by a plastic curtain, received phosphate in addition to carbon and nitrogen, and was covered by a heavy bloom of blue-green algae within 2 months. (*Courtesy of D. W. Schindler; from Schindler 1974.*)

Bioassay experiments in which photosynthetic algae were incubated on river rocks in 10-liter Plexiglas chambers held either in the dark or in the light showed that the photosynthetic rate (CO_2 uptake, measured as $\mu g\ C\ cm^{-2}\ h^{-1}$) of algae incubated in water amended with phosphorus was greater than that of algae incubated in unamended river water (control).

However, the effect of adding phosphorus to the Kuparuk River was far from simple. With respect to nutrient dynamics, the addition of phosphorus resulted in a decline in the supply of nitrogen ions (NH_4^+ and NO_3^-) in the fertilized portions of the stream. Whereas the N:P ratios of the control portions of the stream (above the phosphorus release site) were usually greater than the Redfield ratio, ranging between 13:1 and 78:1, those in the treatment portions of the stream ranged from 4.4:1 to 7.6:1, low enough to suggest that nitrogen was limiting. Apparently, the increased production induced by the additional phosphorus resulted in increased nitrogen uptake and thus a reduction of nitrogen in the water, resulting in low Redfield ratios. Biotic interactions in the river were also affected by the introduction of phosphorus. In the last two years of the experiment, apparently in response to greater overall production, the growth rates of both fish and invertebrates increased, as did the population sizes of some aquatic insects (Peterson et al. 1993).

Recent comparative studies suggest that the role of phosphorus as the principal nutrient regulator of phytoplankton production in lakes should be reexamined (Elser et al. 1990). A survey of enrichment bioassay experiments in which chlorophyll, plankton biomass, or carbon uptake were measured after the addition of nitrogen, phosphorus, or a combination of nitrogen and phosphorus to lake water samples revealed that, in most cases, the addition of nitrogen and phosphorus together resulted in more significant increases in production than the addition of either of the nutrients alone. Whole-lake fertilization experiments support the proposition that both phosphorus and nitrogen may limit production in many lakes (Elser et al. 1990). A mechanism for nitrogen and phosphorus limitation is suggested in our discussion of marine systems below.

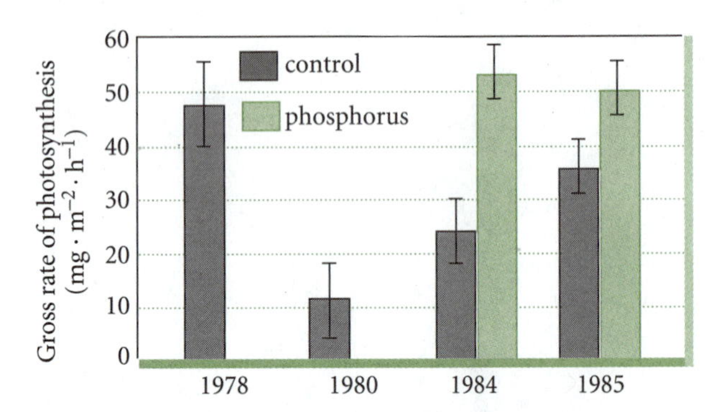

FIGURE 13-2 Rate of photosynthesis (measured as change in O_2 level, $mg\ m^{-2}\ h^{-1}$) of rock-borne primary producers in the Kuparuk River, Alaska, for two years (1978 and 1980) before and two years (1984 and 1985) after experimental additions of phosphorus. Gray bars in 1984 and 1985 are control measurements taken in the unfertilized portion of the river. The rate of photosynthesis in the fertilized section of the river was higher than that in the control (unfertilized) part of the stream in the years following fertilization. (*From Peterson et al. 1993.*)

Nitrogen Limitation in Marine Ecosystems

Two lines of evidence suggest that nitrogen is the most important nutrient regulator of production in marine ecosystems. The first comes from bioassay experiments in which nitrogen, phosphorus, or combinations of nitrogen and phosphorus were added to samples of seawater in which natural assemblages of plankton or monocultures of a single species of algae were allowed to grow. Typically, when such studies are conducted on water from estuaries and other marine ecosystems, measurements of plankton biomass or the amount of radiolabeled carbon (C^{14}) uptake indicate that adding nitrogen alone or adding nitrogen along with phosphorus promotes algal growth or production. Such experiments usually indicate that phosphorus alone has little effect on growth or production (Figure 13-3; Howarth 1988).

Observations of the amounts of inorganic nitrogen and phosphorus dissolved in the waters of a variety of estuaries provide a second line of evidence supporting the idea that nitrogen limits production in marine ecosystems. Boynton and coworkers (1982) compared the amounts of inorganic nitrogen and phosphorus in twenty-seven estuaries to determine whether the environmental N:P ratio approached the Redfield ratio. They found that the ratio of dissolved inorganic nitrogen to dissolved inorganic phosphorus (DIN:DIP) was well below 16:1 in twenty-two of the ecosystems, suggesting that nitrogen limitation is widespread in such ecosystems.

There are exceptions to the generalization that nitrogen limits function in marine ecosystems. In some marine ecosystems, DIN:DIP ratios are quite high, reaching over 30:1 in some cases. Moreover, because nutrient input—nutrient loading—fluctuates seasonally in most estuaries and nearshore marine ecosystems, the DIN:DIP ratio may show large seasonal variation, suggesting that nitrogen may be important at one time of the year and phosphorus at another (Howarth 1988).

Production in open-water marine ecosystems is closely related to the supply of nitrogen in the surface

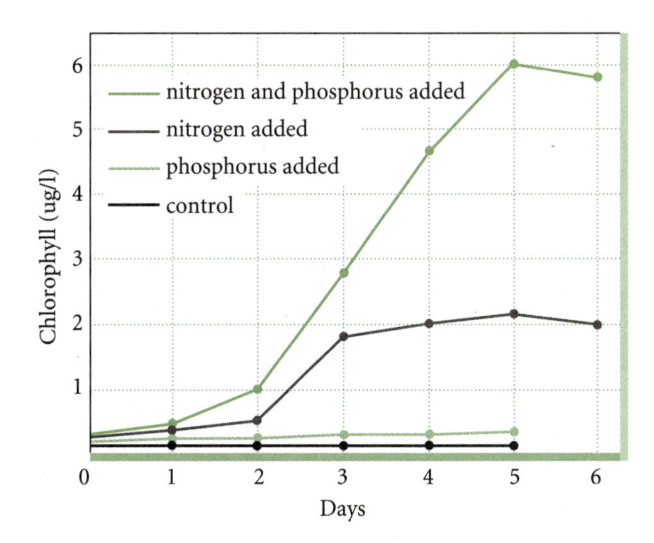

FIGURE 13-3 Results of bioassay experiments measuring chlorophyll concentrations in seawater from near Woods Hole, Massachusetts. Nutrients were added to unfiltered seawater containing the natural assemblage of plankton. Addition of phosphorus alone had little effect on production. (*After Howarth 1988; data from Vince and Valiela 1973.*)

layers of water where light penetration permits photosynthesis to occur. As a result, the highest levels of production are observed in shallow seas, where vertical mixing reaches to the bottom, and in areas of strong upwelling. However, there appear to be areas of the oceans where nitrogen and phosphorus are present in abundance, but phytoplankton concentrations and primary production are low. These conditions suggest limitation by other elements, among which iron and silicon are strong candidates. Iron is an important component of many electron transport and catalytic systems. Silicate is the primary material in the glass shells of diatoms, which are the predominant kind of phytoplankton in the oceans, and is lost from the photic zone when the organisms die and their shells fall to the bottom.

High phytoplankton densities in the Southern Ocean are clearly associated with the proximity of continental sources of nutrients: plankton blooms are concentrated in waters downcurrent of Australia and New Zealand, South America and the Antarctic Peninsula, and southern Africa (Figure 13-4). However, concentrations of nitrogen and phosphorus are high enough to sustain high phytoplankton densities throughout the entire area due to the upwelling currents established by thermohaline circulation in the oceans around Antarctica. The absence of dense phytoplankton concentrations in much of the region suggests limitation by other nutrients. In particular, the area to the west of southern South America between 40° S and 50° S appears to have too little silicon (Sullivan et al. 1993).

Over 20% of the open oceans appears to have abundant nitrogen and phosphorus, but low densities of phytoplankton. These regions are referred to as **high-nutrient low-chlorophyll** (HNLC) **areas,** and they have puzzled marine biologists for years. One hypothesis is that phytoplankton populations in these areas are kept low by zooplankton grazers, although it is unclear why this would happen in some regions of the sea and not others. In the late 1980s, J. H. Martin of the Moss Landing Marine Laboratories in California proposed that production in these areas is limited by iron (Martin and Gordon 1988, Martin et al. 1989). In well-aerated surface waters, iron complexes with other elements, including phosphorus, forms precipitates, and sediments out of the system. Inputs of iron to remote parts of the oceans come almost exclusively from windblown dust.

In an enormous experiment conducted in 1993 off the Pacific coast of South America, about 5° south

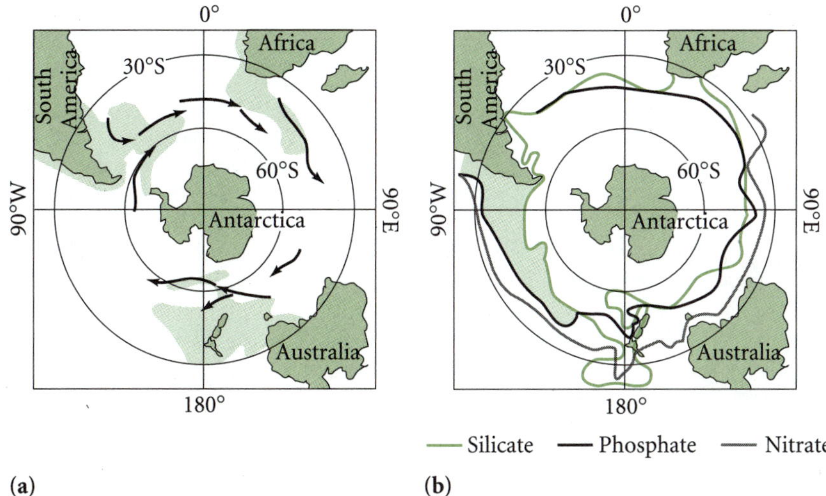

FIGURE 13-4 Maps of the Southern Ocean showing (a) areas of high phytoplankton concentration, with predominant surface current directions indicated, and (b) regions within which nutrients are sufficient for the abundant growth of phytoplankton (nitrogen > 10 μM nitrate; phosphorus > 1 μM phosphate; silicon > 5 μM silicate). The light green areas have sufficient nitrogen but insufficient silicon, suggesting silicon limitation. (*From Sullivan et al. 1993.*)

of the equator, scientists fertilized a target area by distributing 450 kg of dissolved iron over 64 km^2, increasing the concentration of iron in that area almost a hundredfold. Within a few days, phytoplankton production inside the fertilized patch, as measured by the concentration of chlorophyll, tripled. The result clearly demonstrated iron limitation in natural surface waters. The original motivation for conducting the experiment was to determine whether stimulation of marine production could quickly sequester large amounts of carbon in biomass and help to reduce the carbon dioxide concentration of the atmosphere, ameliorating global warming (see Chapter 5). In this respect, however, the experiment was a failure, probably because zooplankton populations increased along with phytoplankton populations and regenerated much of the assimilated carbon dioxide by respiration. Nonetheless, the point that production might be limited by particulate essential nutrients, heterogeneously distributed throughout the oceans, was well made.

The Importance of Sediments and Nitrogen Fixation

What determines whether nitrogen or phosphorus is the principal limiting nutrient in aquatic ecosystems? R. W. Howarth of Cornell University (1988) has suggested that biogeochemical processes in the sediments and in the water column interact with nitrogen and phosphorus inputs to determine whether nitrogen or phosphorus will regulate ecosystem function (Figure 13-5). It is presumed that phytoplankton assimilate nitrogen and phosphorus in the Redfield ratio (16:1). Nutrient loading affects the amount of nitrogen and phosphorus available for assimilation, as do the amounts of nutrients released by decomposition in the sediments. Because most estuaries and nearshore marine systems are subject to sewage influx, which increases phosphorus input, the N:P ratio of the nutrient load from the terrestrial realm in such systems is typically lower than in temperate lakes, indicating nitrogen limitation. Moreover, lake sediments often sequester phosphorus, raising the N:P ratio of sediment release to near or above the Redfield ratio while limiting the amount of phosphorus available to producers. In contrast, the sediments of many estuary systems release phosphorus readily, resulting in a low N:P.

The flux of phosphorus in the sediments, however, cannot explain the difference in the importance of nitrogen and phosphorus as limiting nutrients in marine and freshwater systems. Howarth (1988) suggested that the amount of nitrogen fixation in the two systems, interacting with the phosphorus dynamics, also plays a role. Nitrogen-fixing bacteria are more common in freshwater systems, whereas in marine systems much nitrogen is lost via nitrification and denitrification in the sediments (see Figure 13-5). The combination of higher phosphorus flux and reduced nitrogen fixation helps explain nitrogen limitation in most estuarine ecosystems.

Other mechanisms not involving the dynamics of nitrogen and independent of the redox conditions of the sediments have been proposed to explain the roles of nitrogen and phosphorus in the regulation of production in marine and freshwater systems. For example, the level of sulfate may control the release of phosphorus from sediments; the level of phosphorus mobilization is positively correlated with sulfate concentration (see Chapter 12). The sediments of saline waters typically have higher sulfate concentrations than do those of freshwater systems, and thus phosphorus mobilization in those sediments is greater. The precise mechanism by which sulfate promotes phosphorus mobilization is unclear (Caraco et al. 1989).

13.3 Physical transport processes may regulate ecosystems that are nutrient limited.

As we have seen, essential nutrients, particularly nitrogen and phosphorus, are often limited in the environment. Because of this, the balance between the rate of external input of nutrients and the rate of assimilation

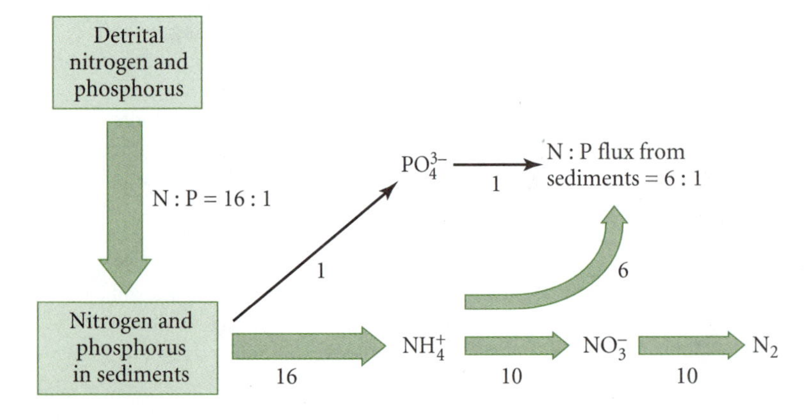

FIGURE 13-5 Nutrient regeneration dynamics of the sediments of Narragansett Bay, Rhode Island, an estuarine ecosystem. Nitrogen and phosphorus enter the sediments in approximately the Redfield ratio, but, whereas phosphorus mineralization releases PO_4^{3-} into the water column, much of the nitrogen is lost from the system through nitrification and denitrification, resulting in nitrogen limitation. Nitrogen loss may be prevented in freshwater systems owing to the presence of nitrogen-fixing bacteria. (*After Howarth 1988.*)

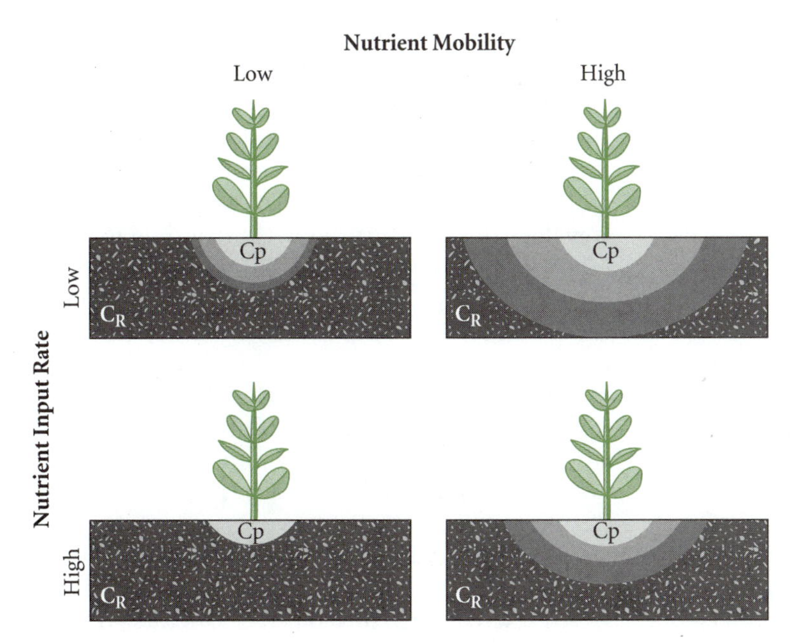

FIGURE 13-6 Hypothetical interaction of nutrient uptake by plants, nutrient mobility, and nutrient input rate on nutrient concentration gradients in terrestrial ecosystems. Darker shading indicates higher concentrations of the nutrient. When nutrient input into the area where the plant grows is low and nutrient uptake by the plant is high, low concentrations of the nutrient will exist in the immediate vicinity of the plant (*top right*). When nutrient input is high and plant uptake is low, high concentrations of the nutrient will exist near the plant (*bottom left*). Intermediate levels of nutrient availability will exist when both nutrient input and plant uptake are low (*top left*) or when both are high (*bottom right*). The overall productivity of an area will be affected by the relative abilities of plants to sequester nutrients. (*After Huston and DeAngelis 1994.*)

of nutrients by producers sets a maximum limit on production. Additionally, nutrients generally exhibit considerable spatial and temporal heterogeneity in the environment. Because primary producers possess limited mobility—this is true even of plankton, which move limited distances within the water column—the spatial distribution of nutrients also influences ecosystem function. Thus, producer productivity depends, to a large extent, on the rate of physical transport of nutrients into a specific region of the ecosystem, coupled with the ability of the producer to influence the utilization of nutrients in the region where it lives—to get its share of the resources, if you will (Huston and DeAngelis 1994).

Plants and algae assimilate regenerated nitrogen rapidly, especially in nutrient-depleted water. Under conditions of high nutrient concentrations, phytoplankton can take up more nitrogen than they need for growth. This ability to store nitrogen, called **luxury consumption,** enables them to take advantage of short-term abundances or local "pockets" of nutrients made available by discrete bursts of excretion or decomposition.

For terrestrial ecosystems, we may conceptualize the interplay between external nutrient flux and plant uptake by considering an individual plant growing in a region where it competes for nutrients with neighboring plants (Figure 13-6). The plant assimilates nutrients in the area near its roots, thereby creating a nutrient concentration gradient, with concentrations increasing with distance from the plant. The rate at which nutrients are input into the region either by external transport or by nutrient regeneration processes within the region, the rate at which nutrients move through the soil, and the rate at which the plant assimilates nutrients determine the concentration gradient. The influence of an individual plant on the soil concentration of a limiting nutrient is greatest when nutrient mobility is high and nutrient input is low. In that situation, the plant may exert a negative competitive influence on neighboring plants, thereby altering the plant community structure and overall productivity. Competitive influences are lowest when nutrient mobility is low and nutrient input is high.

The influence of physical transport on ecosystem function is particularly strong in streams and rivers, where, as we saw in Chapter 12, rates of flow and nutrient spiraling influence the flux of material and the productivity in a particular section of the stream (Newbold et al. 1982, 1983). Nutrient availability in a particular area is affected by assimilation and regeneration processes upstream. Thus, in many streams, particularly those that are nutrient limited, there is a pronounced upstream-downstream linkage in ecosystem function, whereby productivity and producer community composition reflect longitudinal changes—that is, changes along the length of the stream—in nutrient availability (Fisher et al. 1982, Mulholland and Rosemond 1992).

The influence of river flow on production may also be viewed at the scale of the entire river. The size and shape of the stream bed, in-stream fauna and flora, and the composition of the surrounding terrestrial vegetation all change dramatically from a river's source to its mouth. Gradients of physical characteristics such as light, temperature, and discharge create continuous variation in ecosystem structure. Variation in ecosystem function parallels this structural continuum. Headwater sections are typically narrow and strongly influenced by the overhanging forest canopy, which contributes large amounts of allochthonous detritus. In-stream production in these areas is usually less than in the open waters near the river mouth, where large populations of plankton are found. The conceptualization of a river as a single ecosystem having gradients of structure and function is referred to as the river continuum concept (see Chapter 8) (Vannote et al. 1980).

13.4 Production does not vary in direct proportion to nutrient cycling.

Energy provides the most generalizable currency of ecosystem structure and function, and assimilated energy ultimately drives virtually all ecosystem processes. However, the availability of energy appears to cause little of the variation observed among ecosystems. Ecologists agree that basic aspects of ecosystem structure and function respond to external factors, particularly temperature and precipitation in the case of terrestrial ecosystems. What remains to be discovered is how these factors act to regulate ecosystems.

Peter Vitousek, a plant ecologist at Stanford University, attempted to pinpoint the influence of external factors by comparing the cycling of individual elements with the flow of energy through the system as a whole (Vitousek 1982). He reasoned that the process responsible for regenerating the element whose cycling shows the strongest correlation with primary production must exercise predominant control. When a forest achieves a steady state, net aboveground primary production approximately equals litter production. Similarly, the cycling of each element approximately equals the amount of that element that falls each year in litter. Vitousek compared the total dry matter of litter produced each year (which is proportional to carbon and hence to energy content) with the amounts of several elements in the litter; the data for nitrogen and phosphorus are given in Figure 13-7. As can be seen, production more closely parallels the cycling of nitrogen than that of phosphorus (or calcium, which is not shown). Vitousek concluded that factors regulating the cycling of nitrogen predominantly control primary production in forests.

Vitousek's study also showed that production (energy flow) does not vary in direct proportion to nutrient cycling. Rather, forests with low rates of cycling exhibit relatively greater production per unit of nutrient cycled—called **nutrient use efficiency** (NUE)—than do those with high fluxes. Two factors can result in higher nutrient use efficiency: first, trees may assimilate more energy per unit of nutrient assimilated, and second, trees may retain nutrients for reuse by drawing them back into their stems before they drop their leaves. For both nitrogen and phosphorus, nutrient use efficiency clearly decreases as nutrient cycling increases, and for phosphorus, NUE in the Tropics greatly exceeds that in temperate latitudes; tropical trees evidently retain phosphorus to a greater extent than do temperate trees.

Nutrient use efficiency may reflect certain adaptations of the plant that enable it to retain nutrients or reduce requirements for production. Thus, while the relationship between nutrient cycling and production may indicate which nutrients limit ecosystem function, the picture is somewhat blurred by adaptations of plants to manage nutrient use. Furthermore, even when a limiting nutrient has been identified, one still does not know at what point the cycling of that nutrient is controlled. In the next section, we shall use some simple systems models to attempt to sort out the factors responsible for the regulation of ecosystem function. These models consist of mathematical expressions representing the internal controls that govern the movement of nutrients between compartments within the ecosystem. Various parameters of the equations incorporate the effects of external factors such as temperature, moisture, and light. We shall then use these models to determine whether variations in external factors and internal controls produce diagnostic, measurable variations in ecosystem properties from which we can infer the mechanism that regulate ecosystem function.

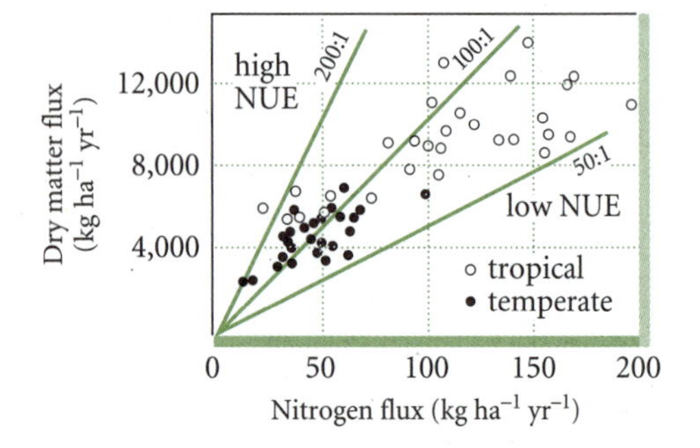

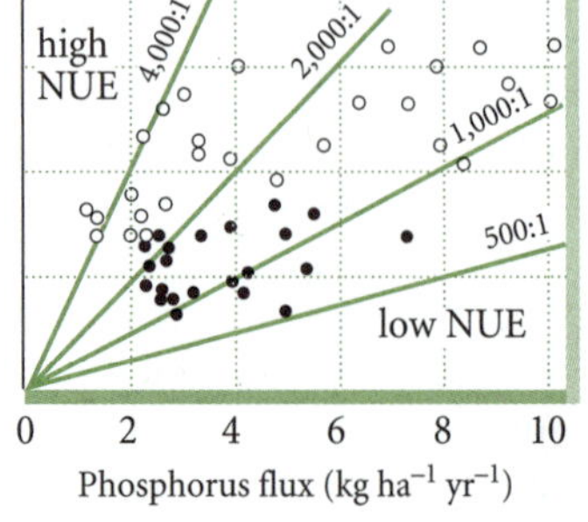

FIGURE 13-7 The relationship of dry matter flux to nitrogen and phosphorus fluxes in the litterfall of temperate and tropical forests. (NUE = nutrient use efficiency.) (*After Vitousek 1982.*)

13.5 Systems models portray ecosystem structure and function as sets of interacting transfer functions.

Mathematical ecologists attempt to capture the essence of ecosystem structure and function in sets of mathematical expressions that, when taken together, form a **systems model.** At the heart of a systems model is the idea that materials and energy move within and between the various abiotic and biotic compartments of the ecosystem (see Chapter 11). The flux between two compartments is represented by an equation referred to as a **transfer function.** Because most ecosystems are complex, with many interacting compartments, a complete representation of all of the material transfers in a system is rarely achievable, though some attempts have been made. To give you an idea of the structure of a systems model and an appreciation of the type of information about ecosystem structure and function that may be derived from such a model, we show first the dynamics of one-, two-, and three-compartment models. Then we discuss the application of these modeling techniques to aquatic and terrestrial ecosystems.

One- and Two-Compartment Systems Models

Let us begin by letting each form of a nutrient or energy within a system be represented as a distinct compartment, which we shall designate X_i for the ith form. Nitrogen, for example, could be represented as three compartments: organic nitrogen (X_1), ammonia (X_2), and nitrate (X_3). You may think of X_i as the amount of nutrient or energy held in the compartment. Each compartment has inflows and outflows, which we shall designate as J. A schematic diagram of a single compartment (X_1) with one input (J_0) and one output (J_1) is shown in Figure 13-8. One way to think of this is to imagine a bucket that receives water from a faucet above and loses water from a hole in the bottom. The amount of water, X, in the bucket is some function of the input (J_0) from the faucet and the output (J_1) from the hole at the bottom. The rate of change in the amount of water in the bucket with respect to time, denoted dX_1/dt, is equal to the difference between the input and the output, which is represented algebraically as

$$\frac{dX_1}{dt} = J_0 - J_1.$$

The amount of water in the bucket is in a steady state (that is, $dX_1/dt = 0$) when inflow equals outflow ($J_0 = J_1$). When $J_0 > J_1$, $dX_1/dt > 0$, and the amount of water in the bucket is increasing. When $J_0 < J_1$, $dX_1/dt < 0$, and the amount of water is decreasing.

FIGURE 13-8 Representation of a single compartment in an ecosystem model, with input (J_0), output (J_1), and compartment, or pool, size (X_1).

Each flux J may be a constant, or it may vary depending on the state of other factors, including the value of X and the values of external influences. We may express such variable fluxes in symbolic notation as $J = f(X,S,P)$, where f denotes that J is a function of the values inside the parentheses; S represents the state of the system (for example, the number and size of holes in the bucket, the leaf area available for photosynthesis, the population density of nitrifying bacteria); and P stands for various parameters that represent external factors (such as the flow of water from the tap, the temperature of the environment, the amount of precipitation). These external factors are sometimes referred to as **external forcing functions.** In the analogy of the bucket, the flow of water from the tap would be considered an external forcing function.

Ecologists basically agree that the fundamental aspects of ecosystem structure and function are determined by external forcing functions. Lieth (1973) summarized annual total dry matter production (g m^{-2}), average temperature, and annual precipitation for fifty-three localities distributed throughout the world. Estimates of production varied 50-fold, from 70 to over 3,500 g m^{-2}; precipitation varied 480-fold, from 94 to 4,500 mm annually; and temperature ranged from $-14.2°$ C to $27.1°$ C. Leith found that all of these factors correlate strongly with the amount of primary production.

In the bucket example, suppose that J_0 is a constant ($J_0 = k_0$), but that J_1 increases in direct proportion to the volume of water in the bucket (that is, $J_1 = k_1X_1$). As the bucket fills, the water pressure at the bottom increases the flow of water through the hole. The amount of water in the bucket (X_1) reaches a steady state when $J_0 = J_1$, and therefore $k_0 = k_1X_1$. We may rearrange this equation to show that the steady-state amount of water (X_1) equals k_0/k_1. Thus, when the value of the external factor (k_0) increases, water reaches a higher level in the bucket. Reducing the size of the hole (the variable k_1) has the same effect on X_1 (Figure 13-9), but note that the flux through the system is always controlled by k_0 alone.

We need at least two compartments to represent the internal cycling of elements within ecosystems. Realistic systems models can become much more complicated, but we shall first consider the simple case of two compartments that cycle an element

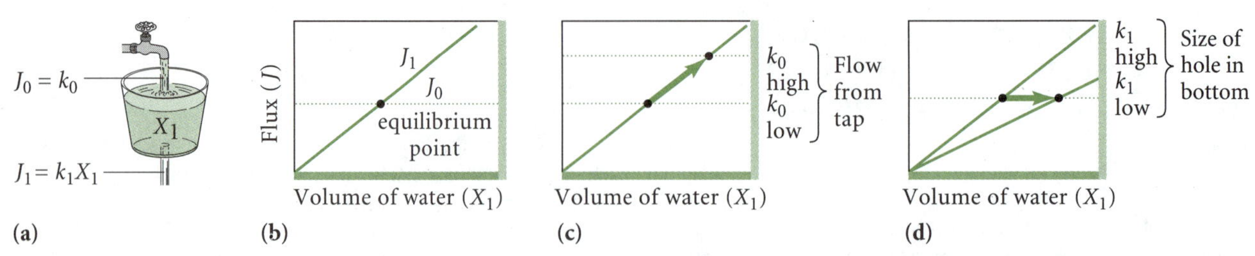

FIGURE 13-9 The effect of changing input (J_0) and output (J_1) rates on the volume of water (X_1) in a bucket. (a) Because water leaves the bucket through a hole in the bottom, the rate of output, J_1, is proportional to the water pressure at the bottom and thus to the depth and volume of water (that is, $J_1 = k_1X_1$). (b) If the input, J_0, remains at a constant value k_0 (horizontal line), the flux will increase in direct proportion to the volume of water, with an equilibrium (input = output) where $J_0 = J_1$ ($k_0 = k_1X_1$), as indicated where the lines cross. (c) If the rate of input—flow from the tap—is increased, shown by moving the horizontal line upward from a lower k_0 to a higher k_0, the equilibrium flux increases. (d) If the flow from the hole in the bottom of the bucket increases, the flux does not change. The system is controlled by the external factor (k_0).

between them (Figure 13-10). The compartment sizes are X_1 and X_2, and the fluxes are J_1 and J_2. As we have seen, the Js represent functions. These may be of zero order, in which case J is a constant ($J = k$); of first order, in which case J_1 is a function of X_1; or of second order, in which case J_1 is a function of both X_1 and X_2.

We can say little about how a particular system works without knowing the details of the flux functions. Simple models may, however, provide insight into general features of system function. The global water cycle, for example, can be thought of as having two compartments, vapor (X_1) and liquid (X_2). What we know about the change of water between vapor and liquid phases allows us to infer what the functions (f) for the fluxes (J) might look like. Precipitation (J_1) is a first-order equation depending only on the water vapor content of the atmosphere (X_1) and on various external factors, such as air temperature; the amount of water at the earth's surface (X_2) has no direct effect on precipitation. Air has a limited capacity to hold water vapor at a given temperature, and so condensation probably increases more rapidly with increasing atmospheric vapor pressure as this ceiling is approached.

Evaporation (J_2) is a second-order equation depending on the surface area of water (some function of X_2), the vapor pressure of water in the atmosphere (proportional to X_1), and external factors such as temperature and the intensity of sunlight. On a global scale, changes in the amount of water in the oceans resulting from evaporation and precipitation have little effect on the total surface area of water, so we can safely ignore X_2 in the function J_2. Accordingly, we may portray the general features of the hydrologic cycle on a graph relating J_1 and J_2 to X_1 (Figure 13-11). The model shows that the system assumes a steady state with $J_1 = J_2$ at some intermediate value of X_1; changes in external factors that affect J_1 or J_2 (change in temperature, for example) would adjust the equilibrium point for the entire system. Clearly, however, this model cannot be used to predict the local weather.

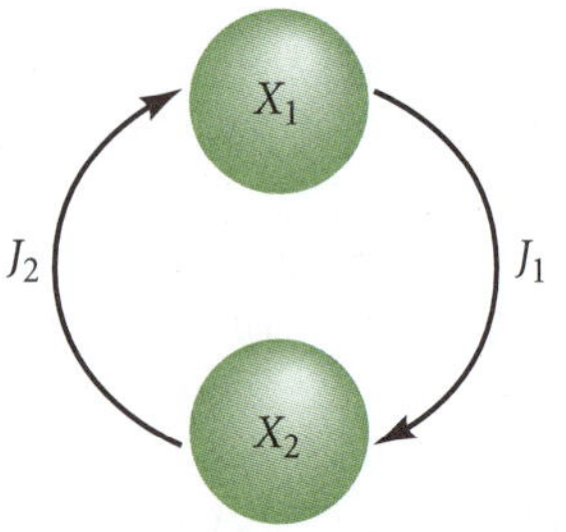

FIGURE 13-10 Diagram of a two-compartment system.

Lotka's Model of Ecosystem Function: A Three-Compartment Model

In his book *The Elements of Physical Biology* (1925), A. J. Lotka investigated the behavior of biological systems using the insights of thermodynamics and the tools of mathematical modeling borrowed from the study of chemical equilibria and other physical phenomena. In the space of a few pages, he outlined the

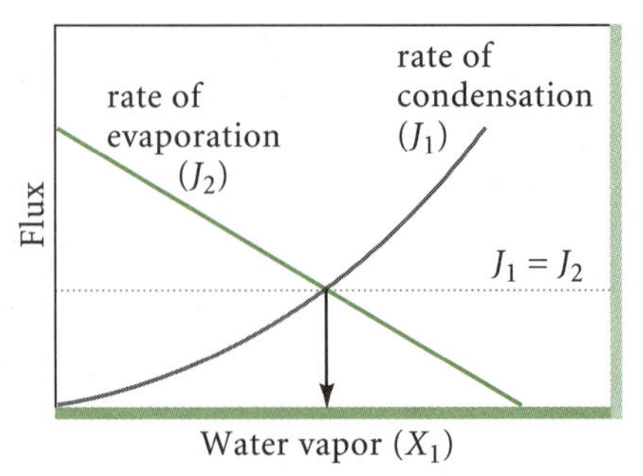

FIGURE 13-11 Rate of evaporation decreases and rate of condensation increases with increasing water vapor in the atmosphere. A steady-state level (arrow) of water vapor in the atmosphere occurs when evaporation equals condensation, at the level indicated by the dashed line.

application of systems modeling to the interpretation of nutrient cycling in ecosystems.

Lotka treated the specific case of three compartments, X_1, X_2, and X_3, through which some element or other material cycles with fluxes J_1, J_2, and J_3 (Figure 13-12). Change in any compartment—for example, dX_1/dt—equals the difference between the fluxes into and out of the compartment (J_3 and J_1 in the case of compartment X_1). To illustrate the behavior of such a system, Lotka described each flux as a first-order term $f_i(X_i)$ arbitrarily having the form g_iX_i; g_i is the rate at which material in compartment X_i is transferred to the next compartment, thus $dX_1/dt = g_3X_3 - g_1X_1$. When the system is in a steady state, all the fluxes are equal; hence $J_1 = J_2 = J_3$, or $g_1X_1 = g_2X_2 = g_3X_3$. Since $J_1 = g_1X_1$, then $X_i = J_i/g_i$. All of the *J*s are equal. Thus, the sizes of the compartments are in the relative proportions

$$X_1 : X_2 : X_3 = \frac{1}{g_1} : \frac{1}{g_2} : \frac{1}{g_3}.$$

All compartments of the system together contain a total *M* (equal to $X_1 + X_2 + X_3$) of the cycling material. The proportion of the total amount of material in any one component of the system (P_i) is the amount in that compartment divided by the total amount in all the compartments, *M*, or

$$P_i = \frac{X_i}{X_1 + X_2 + X_3} = \frac{X_i}{M}.$$

The actual amount in compartment *i* is the proportion in that compartment times the total amount in the system, or $X_i = P_iM$. Since the compartments are in the relative proportions $X_1 : X_2 : X_3 = 1/g_1 : 1/g_2 : 1/g_3$, we can write P_i as

$$P_i = \frac{X_i}{X_1 + X_2 + X_3} = \frac{1/g_i}{1/g_1 + 1/g_2 + 1/g_3}.$$

Using this expression, we can solve for the size of one of the compartments in terms of the total amount of material in the system, *M*, and the rates of exchange between components of the system, the *g*s. For example, the total amount in compartment 1 is given by

$$X_1 = \frac{M}{g_1}\left(\frac{g_1g_2g_3}{g_1 + g_2 + g_3}\right).$$

The flux of compartment 1 is $J_1 = g_1X_1$, thus

$$J_1 = g_1X_1 = M\left(\frac{g_1g_2g_3}{g_1 + g_2 + g_3}\right).$$

These equations tell us that the structure (*X*s) and function (*J*s) of the system are defined by the transfer functions (*g*s), which incorporate the external forcing functions and internal properties of the system.

Lotka further showed that differences between systems in structure and function probably derive from differences in the lowest of the transfer rates. Low transfer rates place bottlenecks in the path of material flow through a system, causing material to accumulate in the preceding compartment. Therefore, the underlying cause of variation in flux through a system should be apparent in the shifts of materials between compartments in the system. For example, if an increase in *J* were accompanied by a shift of material from compartment X_1 to compartments X_2 and X_3, we could infer that an increase in the function g_1 was responsible. Distinguishing among the roles of compartment size, internal controls, and external factors in causing this change would require additional study of the system, but research efforts would be focused by the insights of the systems model. The practical lesson to be learned is that a change in one segment of a nutrient cycle can alter the function of the entire system: a chain is only as strong as its weakest link.

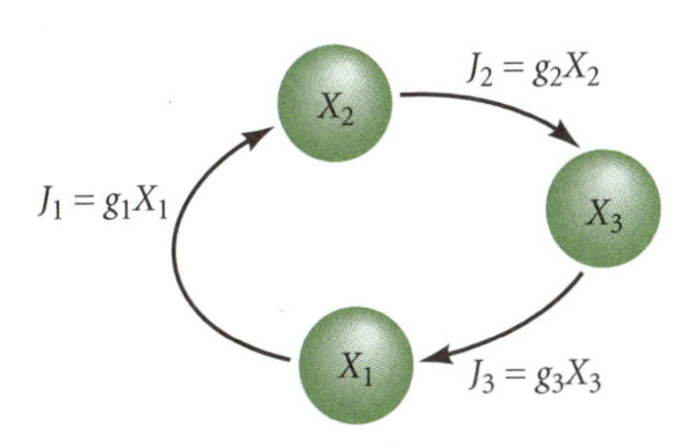

FIGURE 13-12 A diagram of Lotka's system of three compartments, in which the flux between one compartment and next is a function of the size of only the first compartment. For example, the flux between X_2 and X_3 depends on the size of X_2 ($J_2 = g_2X_2$).

13.6 A model for nutrient cycling in aquatic ecosystems incorporates nitrogen transformations in the water column.

The cycling of nitrogen within a water column by and large follows the path (ammonia → nitrate) → particulate nitrogen (organisms + detritus) → dissolved organic nitrogen (DON) → (ammonia → nitrate). Phytoplankton may assimilate either ammonia or nitrate. In a study of nitrogen transformations in the water column of the Bay of Quinte, Lake Ontario, the sizes of some compartments were found to change dramatically with the seasons (Figure 13-13). In particular, between winter and summer, particulate and dissolved organic nitrogen increased while nitrate decreased. This shift implies that the key difference between the cycling of nitrogen in summer and in winter is the rate of nitrate uptake by phytoplankton. A

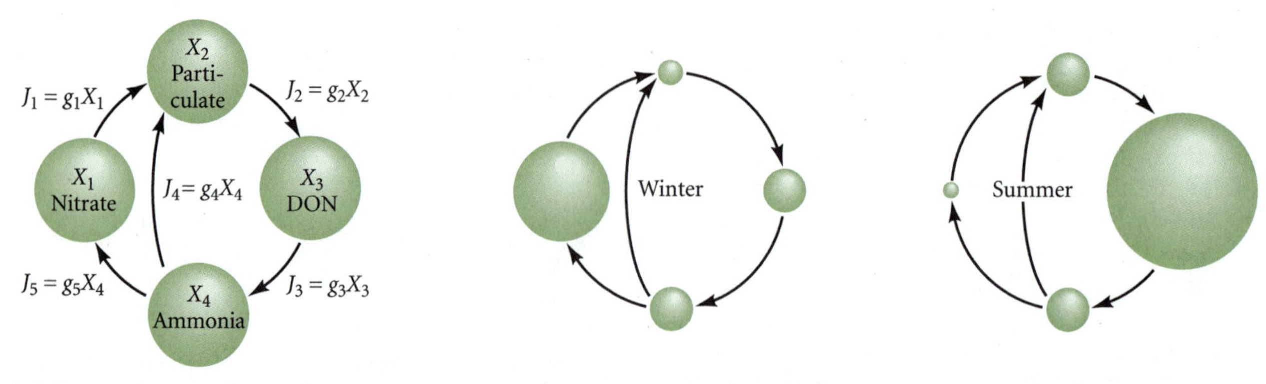

FIGURE 13-13 A first-order systems model of nitrogen cycling in the Bay of Quinte, Lake Ontario, showing the changes in compartment sizes between summer and winter. The major changes are the dramatic increase in the dissolved organic nitrogen (DON) compartment in summer, corresponding to a large decrease in the nitrate compartment.

simple first-order systems model will show how we can elaborate on this idea.

When the system diagrammed in Figure 13-13 has achieved a steady state, fluxes into and out of each compartment must be balanced: for example, $J_2 = J_3$, $J_3 = J_4 + J_5$, and so on. Hence, under steady-state conditions,

$$g_2X_2 = g_3X_3 = (g_4 + g_5)X_4 = g_1X_1 + g_4X_4.$$

From the last two quantities, we can show algebraically that $X_1/X_4 = g_5/g_1$, and by making the appropriate substitutions, we obtain the relationship

$$X_1{:}X_2{:}X_3{:}X_4 = \frac{g_5}{g_1(g_4 + g_5)}:\frac{1}{g_2}:\frac{1}{g_3}:\frac{1}{(g_4 + g_5)}.$$

Setting each X_i at its proportion of the total nitrogen content enables us to estimate relative values of the g_is during the winter and summer (Table 13-1). These estimates indicate, quite dramatically, that the ratio g_1/g_5 (the ratio of algal assimilation to nitrification) is an order of magnitude lower during the winter than it is during the summer. The sum $g_4 + g_5$ differs little between the seasons, so we may assume that g_5 is relatively constant. Evidently, it is the value of g_1 that decreases so much between summer and winter (g_1 describes the rate of assimilation of nitrate by algae). During the winter, assimilation decreases markedly compared with nitrification (g_5) and ammonification (g_3), suggesting that some factor such as light limits production. Alternatively, g_1 and g_4 (assimilation of nitrate and ammo-

TABLE 13-1 Transfer rates for nitrogen in the Bay of Quinte, Lake Ontario

Transfer rate	Description	Winter	Summer
g_2	Production of dissolved organic nitrogen by grazers, excretion, and leakage	8.0	4.1
g_3	Ammonification	4.4	2.0
$g_4 + g_5$	Assimilation of ammonium plus nitrification	4.2	4.7
$\frac{g_1}{g_5}$	Ratio of nitrate assimilation to nitrification	0.6	5.1
$\frac{g_1}{g_5}(g_4 + g_5)$		2.5	24

(Data from Liao and Lean 1978.)

nia, respectively) might both decrease in winter, in which case nitrification rate (g_5) would have to increase at the same time to keep the sum $g_4 + g_5$ constant.

The preceding model may oversimplify or even misrepresent nitrogen transformations in the water column. It was intended only to illustrate an approach. Most systems models are much more complicated, often including dozens of compartments and equations of much higher order than unity. In our simple model, for example, fluxes J_1 and J_4 almost certainly should have been represented by second-order equations involving X_2 (the populations of phytoplankton that accomplish the assimilation; no nitrogen would be assimilated in the absence of algae). More realistic equations can be developed for particular systems, but simplified models, such as this one and the one that follows, can also serve to direct inquiry into more general comparisons of function between ecosystems.

13.7 A model for nutrient cycling in terrestrial ecosystems incorporates compartments for soil, plant biomass, and detritus.

In broad comparisons among terrestrial ecosystems, measurements of net primary production of dry matter in terrestrial ecosystems vary almost 30-fold between desert shrub (70 m^{-2} yr^{-1}) and tropical rain forest (2,000 m^{-2} yr^{-1}). This variation clearly is related to climate, but ecologists have not determined where external forcing variables exert their influence within the system. That is, it is unclear which components of the ecosystem are most sensitive to temperature and moisture. A three-compartment systems model can point the way to fruitful avenues of study in this case.

Consider the cycling of nitrogen again (see Chapter 11). We shall represent an ecosystem as three compartments: mineral soil (X_1), living plant biomass (X_2), and organic detritus (X_3), with nitrogen fluxes J_1, J_2, and J_3 between them (Figure 13-14). (In this model, animals are downgraded in status to trivial hangers-on; the activities of microorganisms are implicit in the flux J_3.) J_2 (the annual dropping of leaves and other detritus) and J_3 (the mineralization of detritus by microorganisms) can be considered as first-order processes with rates g_2 and g_3 respectively. Nitrogen assimilation (J_1) must be modeled as a second-order process ($g_1X_1X_2$) because its rate depends on both nutrient availability and plant abundance. Under steady-state conditions, therefore, we obtain the relationships

$$g_1X_1X_2 = g_2X_2 = g_3X_3$$

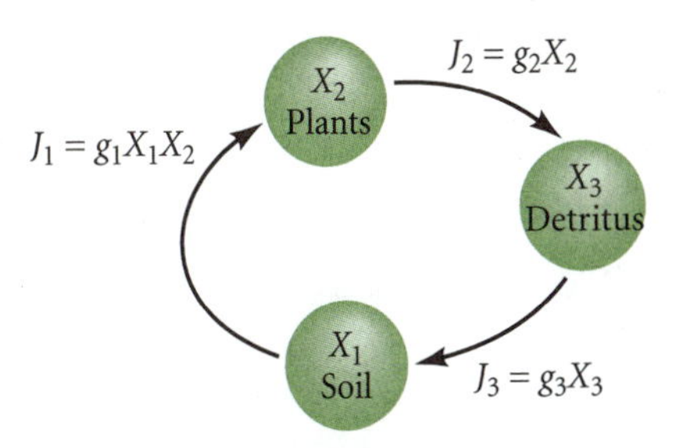

FIGURE 13-14 A second-order systems model of nitrogen cycling in a terrestrial ecosystem. Notice that the assimilation of nitrogen by plants (J_1) is a function of both the availability of nitrogen in the soil (X_1) and the amount of living plant biomass (X_2).

and, using the logic of the previous section,

$$X_1X_2:X_2:X_3 = \frac{1}{g_1}:\frac{1}{g_2}:\frac{1}{g_3}.$$

This simple model offers some surprises. For example, the level of inorganic nitrogen in the soil, mostly nitrate (X_1), is equal to the ratio of g_2 (rate of detritus production) to g_1 (rate of nitrogen assimilation), and is independent of g_3 (rate of microbial regeneration of inorganic nitrogen). To see this algebraically, set $X_1X_2 = 1/g_1$ and solve for X_1 to obtain $X_1 = 1/g_1X_2$. Substituting $1/g_2$ for X_2 in the denominator gives $X_1 = g_2/g_1$. This is not to say that nitrogen flux is independent of g_3. The equation expressing flux as a function of the rate g_1 is a bit complex, but with some algebra it can be shown to be

$$J = M\left(\frac{g_1g_3(g_1 - g_2)}{g_1(g_2 + g_3)}\right)$$

where M equals the total nitrogen in the system. Fluxes of elements are difficult to measure directly; relative values of g_i are estimated from the sizes of the compartments. Differences among systems in the relative rates of transfer between compartments can provide insights into points of internal control.

Ovington (1962) tabulated estimates of compartment sizes for various elements in forest systems. For illustration, we shall compare the cycles of nitrogen and phosphorus in a 47-year-old Scots pine (*Pinus sylvestris*) plantation in England and a 50-year-old mixed tropical forest in Ghana (Figure 13-15). The differences are striking. In England, for both nitrogen and phosphorus, values of g_1, g_2, and g_3 are of the same magnitude. For nitrogen in the Ghanian forest, g_1 and g_2 are similar, but g_3 is almost two orders of magnitude greater. Thus the regeneration of mineral nutrients apparently proceeds much more rapidly in the Tropics than in temperate areas—a conclusion that is confirmed by direct measurement of litter decomposition

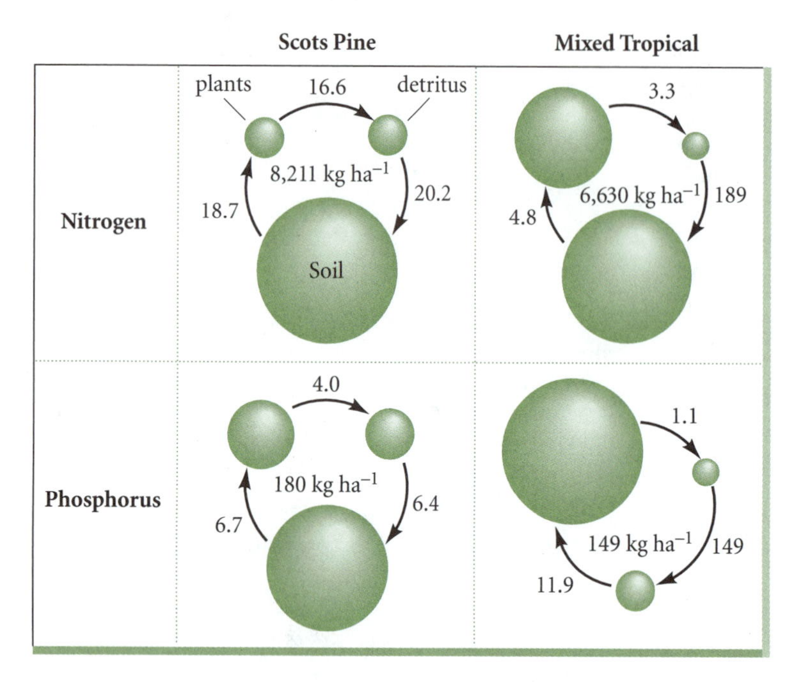

FIGURE 13-15 Compartment sizes and values of g_i for nitrogen and phosphorus, estimated from compartment sizes, in a Scots pine plantation in England and a tropical forest in Ghana. The size of each sphere is proportional to the amount of the nutrient in that compartment. The model suggests that tropical plants sequester more nitrogen and phosphorus than do temperate plants. The rates of transfer of nitrogen and phosphorus among the three compartments are of the same magnitude in England. In Ghana, the regeneration of both nitrogen and phosphorus in the detrital compartment proceeds more rapidly than in England. (*Data from Ovington 1962, Greenland and Kowal 1960.*)

(Olson 1963). External factors appear to influence ecosystem function through their effects on soil microorganisms rather than their effects on plants.

The relative transfer rates of phosphorus behave similarly to those of nitrogen, but reveal another difference between the tropical and temperate forests. In Ghana, the assimilation rate (g_1) is relatively much greater than the rate at which vegetation gives up nutrients (g_2), in contrast to the more nearly equal values in England. These values of g imply that, compared with *Pinus sylvestris,* tropical forest trees assimilate phosphorus more efficiently and hold onto it more tightly. Therefore, the systems model is consistent with the higher nutrient use efficiency observed in tropical vegetation (which, for example, withdraws phosphorus into the stem before shedding leaves).

The values estimated for the gs in Figure 13-15 are adjusted so that the fluxes (Js) are equal to 1.0 when compartments are expressed as fractions of the total. Therefore, although the gs are internally consistent, they can be compared between ecosystems only when one of the fluxes is known. The flux of either nitrogen or phosphorus in Ghana is probably no more than double that in England, and the total amounts of the two elements are similar in the two areas. Therefore, for nitrogen, both g_1 and g_2 are probably lower in the tropical locality than in the temperate locality. For phosphorus, g_1 is probably much higher in Ghana and g_2 somewhat lower. Unquestionably, however, much of the difference between temperate and tropical localities in overall flux is due to the tremendous increase in the rate of decomposition of litter.

13.8 Production may be enhanced or diminished by herbivory.

Thus far, we have focused on how production is affected by nutrient flux in the environment and how mathematical models are used to represent nutrient flux. The presumptions are that the rate of production is limited by the availability of nutrients such as phosphorus and nitrogen, which, in turn, is dependent on transfer processes among ecosystem compartments. Ecologists refer to the control of ecosystem function by nutrient flux and the conditions of the physical environment as **bottom-up control.** The term derives from our concept of the food web, in which producers are seen to occupy the lowest trophic level. "Below" the producers are the nutrient regeneration processes that make elements available for assimilation by producers. Thus, control by nutrient limitation is exerted from the "bottom." The effect of phosphorus on the ecology of Kuparuk River of Alaska (see Section 13.2) is a good example of how nutrients can affect ecosystem function from the "bottom" (Peterson et al. 1993).

But producers must be consumed, in whole or in part, in order for the potential chemical energy stored in their tissues to be utilized at higher trophic levels of the food web. This fact of nature raises the possibility of **top-down control** of ecosystem production—that is, regulation by consumption. Can consumers alter ecosystem function by simply eating producers? We will address this question by taking up two principal issues. First, in this section, we will discuss situations in which herbivores exert either a positive or a negative influence on primary production. Our examples will demonstrate that consumers can have an important regulatory effect on ecosystem function. Then, in the following section, we will turn to the issues of whether trophic interactions several steps removed from the producer level can affect production, and under what conditions bottom-up or top-down controls predominate.

Experimental studies of aquatic food webs indicate that herbivory can effectively limit production. For example, a number of studies have shown that increases in zooplankton biomass in lakes result in decreases in phytoplankton density (McCauley and Kaliff 1981, Lynch and Shapiro 1981; see also Carpenter and Kitchell 1993 for review). For the most part, these effects are negative, reducing plankton biomass and overall productivity; however, in eutrophic lakes, herbivory may have less of an effect on production (McQueen et al. 1989).

Stream rocks are often covered with a rich community of producers called **periphyton,** including cyanobacteria, diatoms, and algae. Herbivorous fish may change the community structure of such assemblages and thus alter their productivity. For example, the grazing activities of the algivorous minnows *Campostoma anomalum* and *C. oligolepis* alter the rock-borne algal community of an Ozark mountain stream (Power et al. 1988). In the Kuparuk River, increases in the abundance of grazing insects in the last two years of the study prevented the chlorophyll accumulation by phytoplankton that was observed in the first years of the study (Peterson et al. 1993).

The persistence of the dominant primary producers of grasslands, some marine intertidal areas, coral reef ecosystems, and some stream bottom communities is dependent on grazing (Lubchenco 1978, Steneck 1982, McNaughton 1985, Bazely and Jefferies 1989, Power et al. 1988). The productivity of these ecosystems is often enhanced by herbivory. Grazing of intertidal grasslands near Hudson Bay by lesser snow geese resulted in a substantial increase in net aboveground primary production in the area (Bazely and Jefferies 1989). In an exhaustive study of the grassland ecosystem of the Serengeti in Africa, McNaughton (1985) found that aboveground primary production was much higher in areas that were grazed by nomadic herds of wildebeests, zebras, and elands than in study plots from which grazers were excluded. These examples represent a coadaptation of producer and consumer whereby the vigor and growth form of individual plants, and thus the productivity of the plant community, actually benefit from continual cropping by the grazers and from the physical disturbance of the soil by the grazers as they forage. However, the balance between a grazer and its resource can be a delicate one. Moderate grazing can stimulate the growth of producers, while excessive grazing can decimate the producer community. Hence, the adage "don't kill the goose that lays the golden eggs" applies to ecological systems as well as human ones. We will have more to say about the balance between grazer and producer in Chapter 24.

13.9 Are ecosystems regulated from the top down or from the bottom up?

As we suggested in the previous section, one of the most important questions of modern ecology is whether ecosystem function is more strongly regulated from the top down or from the bottom up (Hunter and Price 1992, Power 1992, Strong 1992, Menge 1992, Carpenter and Kitchell 1993). The heart of this issue is not whether production is limited by nutrient flux or environmental conditions—it most certainly is to some degree—or whether herbivores have some regulatory effect on the producer organisms that they eat—they most certainly do. The issue is how far up bottom-up effects extend in the food web and the extent to which the effects of trophic interactions at the top of the food web are felt throughout the rest of the food web. How does the availability of phosphorus, for example, limit not only producers but also herbivores, and even predators? How do predators affect not only the populations of herbivores, but also those of producers? The most intense discussion has centered around top-down effects.

The concepts of top-down and bottom-up control bear not only on the productivity of ecosystems, but also on the structures of the populations and communities of producer and consumer organisms. Our interest here is limited to ecosystem function. In later chapters we will revisit the top-down/bottom-up concept when we explore what regulates the population dynamics and community structures of plants and animals.

Top-Down Regulation: The Trophic Cascade

Top-down regulation appears to be most important in food webs having low species diversity, few trophic levels, and in which one or a few herbivores have a strong influence on producer populations (Power 1992, Strong 1992). The algae-based food webs of lakes are a good example of such systems (Carpenter and Kitchell 1993), in which regulation may be conceptualized as a **trophic cascade** (Figure 13-16). Zooplankton herbivores—that is, those zooplankton that eat phytoplankton—play the key role in the trophic cascade model. Large zooplankton, such as *Daphnia*, consume phytoplankton of a variety of sizes, whereas small zooplankton, such as calanoid copepods and rotifers, take only small phytoplankton. Additionally, large zooplankton recycle nutrients more slowly than

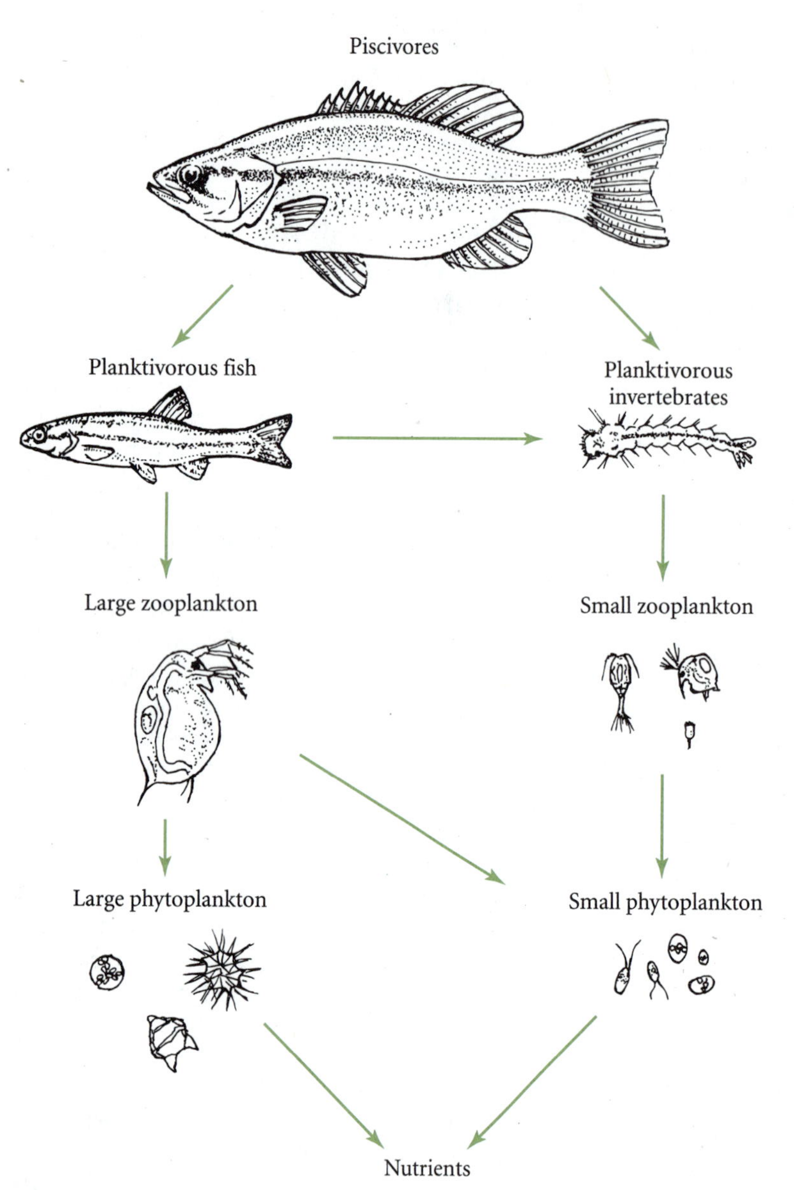

FIGURE 13-16 Diagram of a trophic cascade. Large zooplankton such *Daphnia* have slow nutrient turnover rates and feed on both large and small phytoplankton. Small zooplankton select only small phytoplankton. When *Daphnia* populations are high, production may be reduced, both because of a reduction in phytoplankton of all sizes and because of nutrient limitation owing to the sequestration of nutrients by *Daphnia.* When planktivorous fish populations are high, *Daphnia* levels drop, and production may increase. Populations of piscivorous fish may, in turn, regulate populations of planktivorous fish. (*After Kitchell and Carpenter 1993.*)

do small herbivores. Consequently, when the herbivore assemblage is dominated by large zooplankton, a wider variety of producers is consumed, fewer nutrients are available for assimilation by phytoplankton, and both phytoplankton biomass and primary production are reduced (Kitchell and Carpenter 1993).

Trophic interactions above the herbivore level cascade down through the food web (top-down effect) to regulate the herbivore populations and, consequently, the producer populations. Planktivorous fish prefer the larger zooplankton; invertebrate planktivores prefer the smaller zooplankton. When planktivorous fish are abundant, smaller zooplankton predominate, thereby increasing production (Kitchell and Carpenter 1993). Large piscivorous (fish-eating) fish may regulate planktivorous fish populations, thereby exerting additional top-down control on the system. Current estimates suggest that about half of the variation in production among lakes is the result of such top-down mechanisms, the rest being accounted for by the nutrient input and regeneration properties of the systems (Hunter and Price 1992).

Strong top-down control of production, such as that described above for plankton-based lake ecosystems, may be quite rare (Strong 1992). In some eutrophic lakes, a reduction in zooplankton as a result of fish predation has been found to have little effect on plankton biomass (McQueen et al. 1989). Marine food webs appear to be remarkably unresponsive to the addition or removal of consumers (Strong 1992). Few clear examples of trophic cascades exist in terrestrial ecosystems (Hunter and Price 1992). In communities in which there are many different types of consumers (species-rich communities), consumer feeding preferences are narrow, and thus, even at high population levels, a single consumer species may have little effect on overall production. Moreover, top-down and bottom-up processes are likely to interact with one another in complex ways (McQueen et al. 1989, Menge 1992). For example, the number of trophic levels in a food web may be limited from the bottom up by nutrient availability.

Interaction of Top-Down and Bottom-Up Mechanisms

Recent studies suggest that ecosystem function is under the simultaneous control of nutrient input and regeneration (bottom-up processes) and trophic interactions (top-down processes) (Carpenter 1988, McQueen et al. 1989, Hunter and Price 1992, Power 1992, Rosemond et al. 1993, Peterson et al. 1993). In a study of how production is regulated in stream periphyton, Rosemond and coworkers (1993) measured the productivity (measured as ^{14}C uptake, $\mu g\ cm^{-2}\ h^{-1}$) of periphyton communities in Walker Branch, Tennessee, that were subjected to or protected from grazing by snails and to which nitrogen, phosphorus, or nitrogen and phosphorus were added (Figure 13-17). Their results support the view that the periphyton community in Walker Branch is limited both from the bottom, by nutrient availability, and from the top, by snail grazing. The addition of nitrogen, phosphorus, or nitrogen and phosphorus increased productivity in all treatments. Although productivity in grazed plots was less than that in ungrazed plots for all nutrient treatments, the productivity of grazed plots was highest when nitrogen and phosphorus were added to the system. The addition of both nitrogen and phosphorus to the ungrazed plots substantially increased productivity in those plots as well.

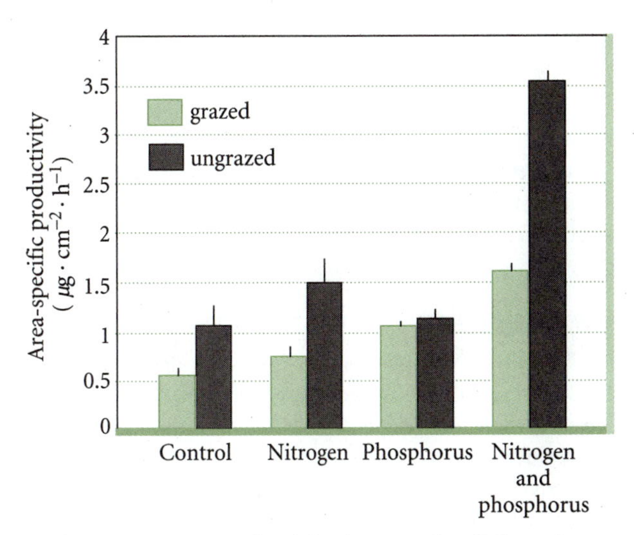

FIGURE 13-17 Productivity (measured as ^{14}C uptake, $\mu g\ cm^{-2}\ h^{-1}$) of substrate-borne producers subjected to grazing by snails under several conditions of nutrient addition. The data show the interaction between top-down control by grazing and bottom-up control by nutrient limitation. *(After Rosemond et al. 1993.)*

Energy and materials flow through ecosystems because plants, animals, and microbes must acquire energy and nutrients to grow and reproduce. As the models and discussions of this chapter have shown, variations in ecosystem function are influenced by the ways in which the activities of organisms respond to variations in the physical environment and by the particular ways in which organisms are adapted to manage their energy and nutrient resources.

SUMMARY

1. The configuration of the biological communities and the interrelationships between the biotic and abiotic compartments of an ecosystem constitute the structure of the ecosystem. Primary productivity and the flux of nutrients and energy constitute the function of an ecosystem. Ecologists use three methods to study the regulation of ecosystem function: experimentation, comparison, and mathematical modeling.

2. Limited availability of nutrients may limit production. The regulatory roles of important nutrients such as phosphorus and nitrogen differ in freshwater and marine systems. Under a wide variety of conditions, phosphorus is most limiting in freshwater ecosystems and nitrogen is most limiting in marine ecosystems. Recent studies suggest that in many cases, the availability of both phosphorus and nitrogen limits production in an interactive way.

3. Nutrients are often in short supply and are distributed heterogeneously in the environment. Thus, processes that transfer nutrients from one place to another may play a major role in ecosystem regulation. Transfer processes may act on a local or a whole-system scale.

4. Systems models based upon a particular element or upon energy consist of compartments (X) and fluxes (J) between them. When a compartment is in a steady state, input equals output. For a cycling element in a steady state, fluxes through each segment of the cycle are equal. Each flux out of a compartment is equal to a transfer rate (g) times the compartment size. Each transfer rate may be a function of the size of one or more compartments, other properties of the system, and factors external to the system.

5. A three-compartment model (soil-plant-detritus) has been developed for terrestrial ecosystems. Comparisons of compartment sizes in temperate and tropical forests indicate that nitrogen and phosphorus are regenerated from litter much more rapidly in the Tropics. Furthermore, within the Tropics, production is probably insensitive to

variation in litter decomposition rate, which is uniformly high. Such conclusions are only tentative, however, because of the simplicity of the models on which they are based.

6. Ecosystem regulation by nutrient availability or nutrient flux is referred to as bottom-up control. Regulation by trophic interactions is referred to as top-down control. The most direct top-down effects on production come from herbivores, the activities of which may enhance or diminish production.

7. In some ecosystems, consumption of herbivores by predators reduces herbivore populations, thereby reducing the grazing pressure on producers and resulting in an increase in production. Such a phenomenon is referred to as a trophic cascade and is an example of strong top-down control.

EXERCISES

1. Consider Lotka's three-compartment model of ecosystem function (Figure 13-12) in which the sizes of the three compartments are $X_1 = 100$, $X_2 = 150$, and $X_3 = 200$. If $g_1 = 2.3$, what do g_2 and g_3 have to be to bring the model into steady state? If the size of compartment X_1 increased by 7% and the other two compartments remained the same size, what would g_1 have to be for the system to remain at equilibrium?

2. The ecosystems models that we consider in this chapter are closed models. That is, nutrients do not enter or leave the system; rather, they cycle continuously among two or more compartments. Construct an open ecosystems model having three compartments in which nutrients may enter the system at at least one point and leave at at least one point. Can you think of situations in nature for which such a model would be realistic?

3. Suppose that you are studying a lake that is regulated from the top down. What would you have do to the lake to change the system of regulation to a bottom-up one? Would this be possible?

4. The ecosystem models presented here do not include an explicit spatial dimension. Revisit the nutrient spiraling model presented in Chapter 12. Develop an ecosystems model for nutrient spiraling in streams that incorporates the total distance traveled in the stream.

PART 4
POPULATION ECOLOGY

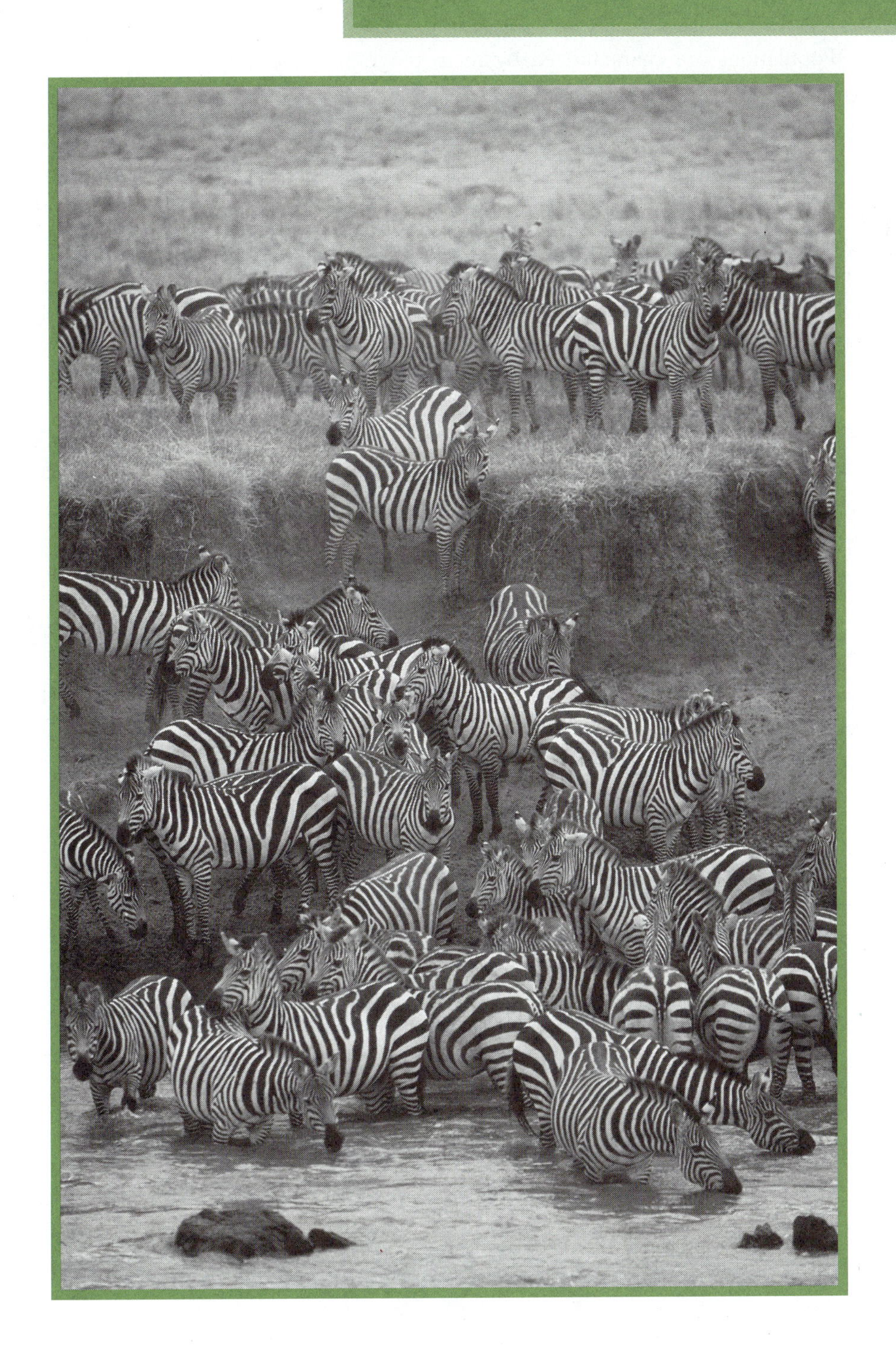

Organisms of a single type living together constitute an ecological population. Populations represent the ecological unit within which mates are found and offspring produced. And, more importantly, populations are where the manifestations of natural selection reveal themselves, where evolution occurs. Much of evolutionary and ecological theory is based on population dynamics. In this part of the book we shall explore the dynamics of single populations. The ideas and theoretical formulations developed here form the basis for most of the discussion in the rest of the book.

Populations are characterized by their geographic location, the way in which members arrange themselves in space, the genetic composition of the group, and the patterns of birth and age. These characteristics are collectively referred to as the population age structure and they will be discussed in Chapter 14. The pattern of population growth is one of its most important features. In Chapter 15, we will describe the patterns of population growth and discuss how age structure and other factors effect it. Populations tend to grow without bound unless some countervailing force acts to curb the growth. Such forces may be intrinsic to the population or external to it, as we shall see in Chapter 16.

In Chapter 17, we will focus on the effect of the arrangement of organisms in space in our discussion of metapopulations. We will see how human land use patterns, which have intensified the amount of habitat fragmentation, affect ecological processes. It has long been recognized that some populations cycle in abundance in response to environmental fluctuations or demographic factors. Understanding these fluctuations is extremely important to managing populations of game and fish and to preserving endangered species, as we shall see in Chapter 18. Much of the ecological theory regarding extinction and preservation is grounded in population ecology. In Chapter 19, we will review current ideas about extinction and discuss strategies for conserving and restoring natural populations of plants and animals.

CHAPTER 14

Population Structure

GUIDING QUESTIONS

- What are the various components of population structure?
- How do environmental factors affect geographic distribution at different scales?
- What factors influence the dispersion of individuals within a local population?
- How is the Poisson distribution used to examine patterns of dispersion in populations?
- What factors affect the density of a population?
- How is overall density maintained in a system of source and sink populations?
- How can dispersal be measured?
- How does dispersal occur in plants?
- Why are conservation biologists interested in the genetic structure of small populations?
- How are isozyme analysis, DNA fingerprinting, and RFLPs used to examine genetic variation?
- How are life tables used to characterize population structure?
- How are cohort life tables constructed?
- What is the concept of survivorship?
- What are the differences among the four classes of population models used by ecologists?

In order to understand any complex phenomenon or process, one must consider it from a number of perspectives at different resolutions or scales. For example, a full appreciation of the dynamics of a baseball team in the field requires an understanding both of the behaviors and responses of individual players and of the entire nine-member team. So it is with the natural world. We must understand not only how individual organisms respond to the conditions of the physical or biological environment, but also how the activities of individual organisms are manifested in and affect the dynamics of the groups in which they live. The next three parts of this book (Parts 4, 5, and 6) adopt the latter perspective; that is, we shall focus on how reproduction and interactions such as predation, herbivory, and parasitism affect the number and distribution of organisms in groups. We shall begin in Part 4 with a consideration of one of the most important organism groups: the population.

A **population** is composed of the individuals of a single species within a given area. A population's boundaries may be natural ones imposed by the geographic limits of suitable habitat, or they may be defined arbitrarily for the purposes of scientific study. (Take a moment to compare this definition of *population* with that of a statistical population in Chapter 2. The latter refers to a collection of measurements, not to a group of individual organisms.) In either case, a population has a **population structure,** which includes features such as the density, spacing, and movement of individuals, the proportion of individuals in different age classes, genetic variation, and the arrangement and size of areas of suitable habitat, all of which may vary in space and time. Characterizing a population's structure provides us with a snapshot of that population at one instant in time. We can understand the degree to which different parts of a population interact with or, alternatively, exhibit independence from one another by examining the distribution and movement patterns of individuals. Such knowledge may help us determine the vulnerability of populations to fragmentation, which can result from natural disturbance or human activity. Population structure also affects the dynamics of parasites and their hosts, including human diseases (see Chapter 24). By knowing the rates at which individuals of different ages produce young and the probability of death of different age classes, we can predict the pattern of growth or decline of a population, information that is of practical importance in the management of wild populations of fish and game and endangered species.

In this chapter, we shall explore several aspects of population structure. First, we shall examine the **spatial structure** of a population, the component of the population structure having to do with the geographic distribution of the species, the spatial arrangement of individuals and habitats, the size of the population, and the patterns of movement of individuals within and among populations. Second, we shall consider the collective genetic composition of the population, or its **genetic structure,** focusing in particular on the relationship between genetic variation and population size. Finally, populations are dynamic, and using life tables we shall study how the rates of births, deaths, and movements of individuals continuously change over time.

Before proceeding with our discussion of population structure, it is important to point out that for plants and other sessile organisms such as corals and fungi, we may apply some of the concepts of population ecology to seeds or other propagules as well as to individuals that have developed to be recognizable as a juvenile or adult. The most notable example is the so-called **seed pools** that reside in the soil beneath many plants. These pools are composed of seeds that have rained down from the plant above, and in many contexts they may be thought of as seed populations having specific and important spatial and genetic structure. Moreover, plants have a modular construction owing to the fact that growth occurs only at specific areas called meristems. Thus, it is possible to think of the belowground and aboveground portions of the plant as being more or less separate populations having different dynamics. While our discussion will focus primarily on populations of distinct organisms, an appreciation of this application of population theory is important, and it will be mentioned from time to time.

14.1 The geographic distributions of species and the locations of local populations are determined by ecologically suitable habitat.

Ecologists refer to the geographic area in which a species occurs as its **geographic distribution** or **geographic range.** When ecologists talk with one another, they often use the words "distribution" or "range" as shorthand for either geographic distribution or geographic range, and we will often resort to this simplification. The distribution of a species is determined by the range of ecological conditions within which individuals of that species can survive, a set of conditions that is referred to as the **ecological range.** Distributions are determined primarily by the presence or absence of suitable habitat. For example, the natural range of the sugar maple in the United States and Canada stops abruptly to the east at the Atlantic Ocean, but is limited more gradually to the west by low precipitation, to the north by cold winters, and to the south by hot summers (see Figure 8-18a).

The movement of organisms away from their place of birth or away from areas of high population concentration is referred to as dispersal, a topic that we take up in more detail in Section 14.4. Distributional limits may be imposed by barriers to long-distance dispersal movements. For example, there is undoubtedly much suitable habitat for the sugar maple throughout the world, especially in Europe and Asia, where it has been transplanted successfully, but the species evolved in North America and has not had the opportunity to colonize other areas on its own.

Barriers to dispersal often reveal themselves dramatically when introduced species expand successfully into new regions. For example, in 1890 and 1891, 160 European starlings were released in the vicinity of New York City, evidently because of someone's wish to introduce all the birds mentioned in Shakespeare's works to the New World. Within 60 years, the population had expanded to cover more than 3 million square miles and stretched from coast to coast (Figure 14-1).

Within the geographic range of a species, individuals are not equally numerous in all regions. Individuals generally live only in areas of suitable habitat in **local populations,** sometimes called **subpopulations.** A local population is that set of individuals that live within a **habitat patch,** a more or less homogeneous area of habitat containing all of the resources needed to sustain the local population. (We shall see later that a somewhat more restrictive definition of the word *patch* is sometimes employed by evolutionary ecologists when, for example, they refer to foraging patches.) Patches of habitat are separated from other

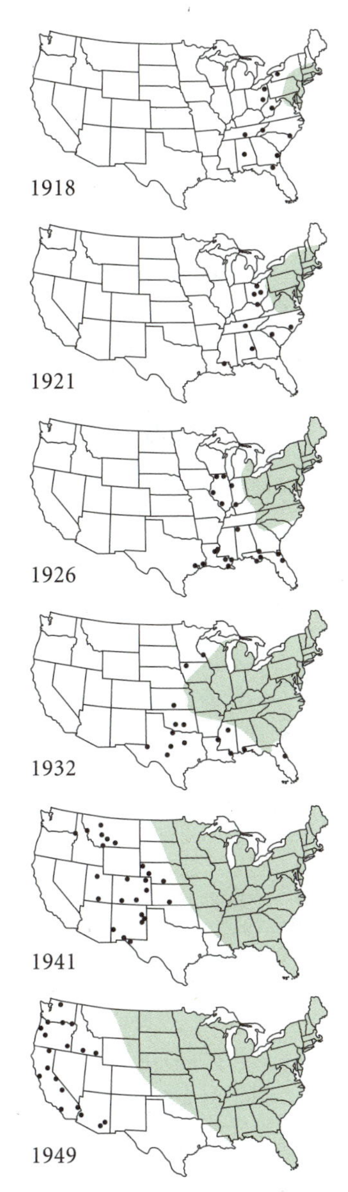

FIGURE 14-1 Western expansion of the range of the European starling (*Sturnus vulgaris*) in the United States. The green shaded areas represent the breeding range; dots indicate records of birds in preceding winters. The population now inhabits the entire country. (*After Kessel 1953.*)

patches by areas that are unsuitable for the species. The geographic range of the starling, for example, is composed of an array of local populations. During the summer months, starlings feed by probing the ground for insect larvae and other food items. Because starlings have difficulty feeding where soil is dry and hard, populations are most dense in habitat patches that are cooler and moister than other parts of the species' distribution. Starlings are most abundant in parks, golf courses, croplands, and pastures that are well irrigated (Figure 14-2). Their situation is similar to that of the sugar maple mentioned above. Sugar maples do not grow in marshes, on serpentine barrens, on newly formed sand dunes, in recently burned areas, or in a variety of other habitats that simply lie outside their range of ecological tolerance. Hence, the geographic ranges of both the sugar maple and the starling represent a patchwork of occupied and unoccupied sites, a situation that is typical of most species.

Organisms must respond to a variety of environmental factors, and these factors exert their influence on distribution at different scales. For example, cold winters may explain the northern distributional limit of sugar maples, but do not explain why sugar maples do not grow in marshes. Factors such as climate, topography, soil chemistry, and soil texture may exert progressively finer influences on the geographic range of a species, as demonstrated by the perennial shrub *Clematis fremontii* (Figure 14-3). Climate and perhaps interactions with ecologically similar species restrict *C. fremontii* to a small part of the midwestern United States. The distinctive variety of *C. fremontii* named *riehlii* occurs only in Jefferson County, Missouri. Within its geographic range, *C. fremontii* var. *riehlii* is restricted to dry, rocky soils on outcroppings of limestone. Small variations in relief and soil quality further confine the distribution of the plant within each limestone glade to sites with suitable conditions of moisture, nutrients, and soil structure. The local aggregations that occur on each of these sites consist of many more or less evenly distributed individuals.

The horned lark (*Eremophila alpestris*), a small songbird of short grasslands, occurs more widely in North America than *Clematis fremontii*, but its local distribution also reflects subtle variations in habitat. In Colorado, populations of horned larks are denser on heavily grazed rangeland than on lightly grazed land because the larks prefer open spaces with low vegetation. Even within their territories, individuals use some areas more often than others; for example, the best display sites on small shrubs are often not in the best areas for foraging (Figure 14-4).

It is important to remember that the geographic range of a species includes all of the areas its members occupy during their life cycle. Thus the distribution of salmon includes not only the rivers that are their spawning grounds, but also vast areas of the sea, where individuals grow to maturity before making the long migration back to their birthplace (Figure 14-5). Many birds make annual migrations to warm climates during the winter months. The distributions of the golden plover and blackburnian warbler, for example, lie entirely within North America during the summer, but entirely within Central and South America during the

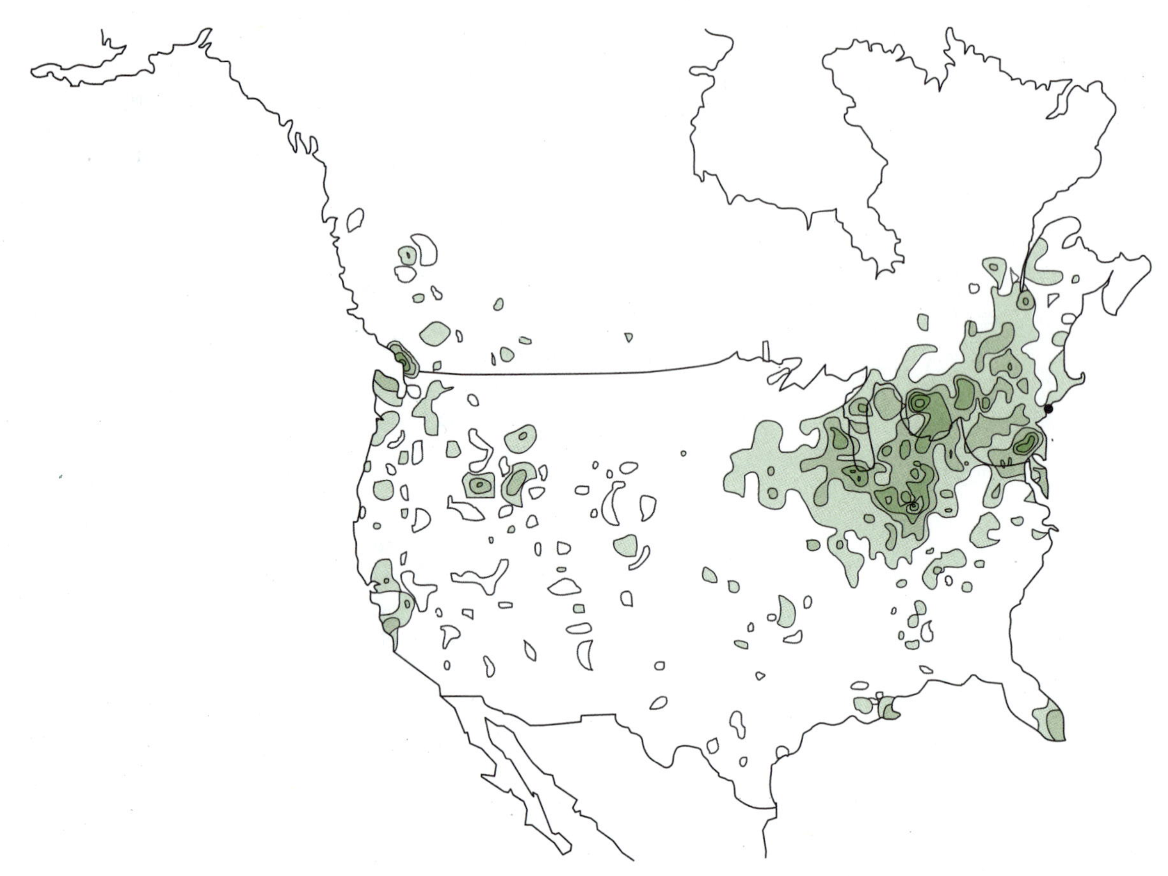

FIGURE 14-2 Relative densities, indicated by density of shading, of European starlings in North America averaged over 1966–1989, as determined by the Breeding Bird Survey. The starlings are most dense in areas of suitable habitat. (*From Maurer and Villard 1994.*)

winter (Figure 14-6). The sizes and year-to-year variations of populations of such migratory species depend on the interactions of individuals with their environments throughout the year. Thus, for many species, including a fair number of the most threatened species, ecological interactions take place on a global scale. Ecological changes in Mexico, the Amazon basin, or central Africa may affect populations of migratory species that breed in the forests of eastern North America or Europe. This presents an enormous challenge for those of us interested in the protection of such species because any effective conservation strategy requires not only a sound ecological foundation, but an innovative and inclusive diplomatic strategy as well.

14.2 The dispersion of individuals reflects habitat heterogeneity and social interactions.

The **dispersion** of individuals within a population describes their spacing with respect to one another. (Compare the term *dispersion* with the term *dispersal,* which refers to the movement of individuals, a phenomenon that was defined in Section 14.1 and will be discussed further in Section 14.4.) Patterns of dispersion range from **clumped** distributions, in which individuals are found in discrete groups, to evenly **spaced** distributions, in which each individual maintains a minimum distance between itself and its neighbors (Figure 14-7). Between these extremes, one finds **random** dispersion, in which individuals are distributed throughout a homogeneous area without regard to the presence of others.

Spaced and clumped distribution patterns derive from different processes. Even spacing—sometimes called **hyperdispersion**—most commonly arises from direct interactions between individuals. Maintenance of a minimum distance between oneself and one's nearest neighbor results in even spacing; for example, in their crowded colonies, seabirds place their nests just beyond their neighbors' reach (Figure 14-8). Plants situated too close to large neighbors often suffer from shading and root competition; as these individuals die, the spacing of individuals becomes more even.

Clumping, or **aggregation,** may result from (1) the social predisposition of individuals to form

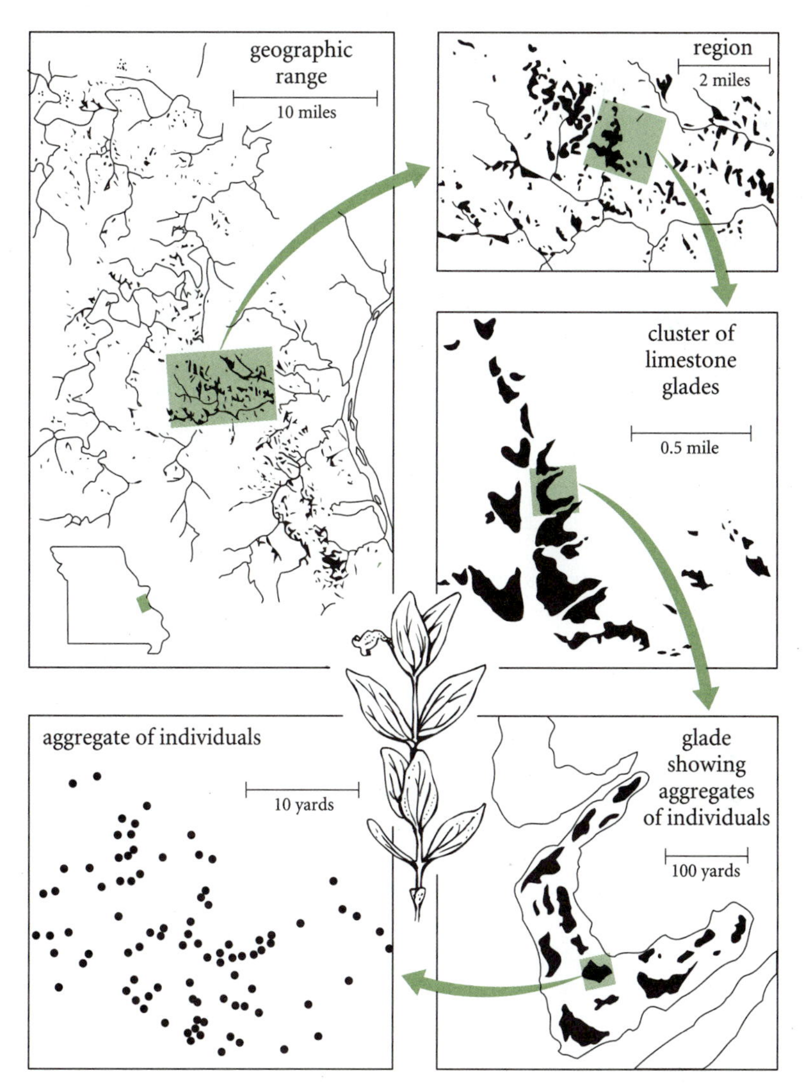

FIGURE 14-3 Hierarchy of patterns of geographic distribution of *Clematis fremontii* var. *riehlii* in Missouri. Climate and interactions with other species restrict *C. fremontii* to a particular region of the United States. Within this range, the variety *riehlii* is restricted to limestone areas with particular soil features. (*After Erickson 1945.*)

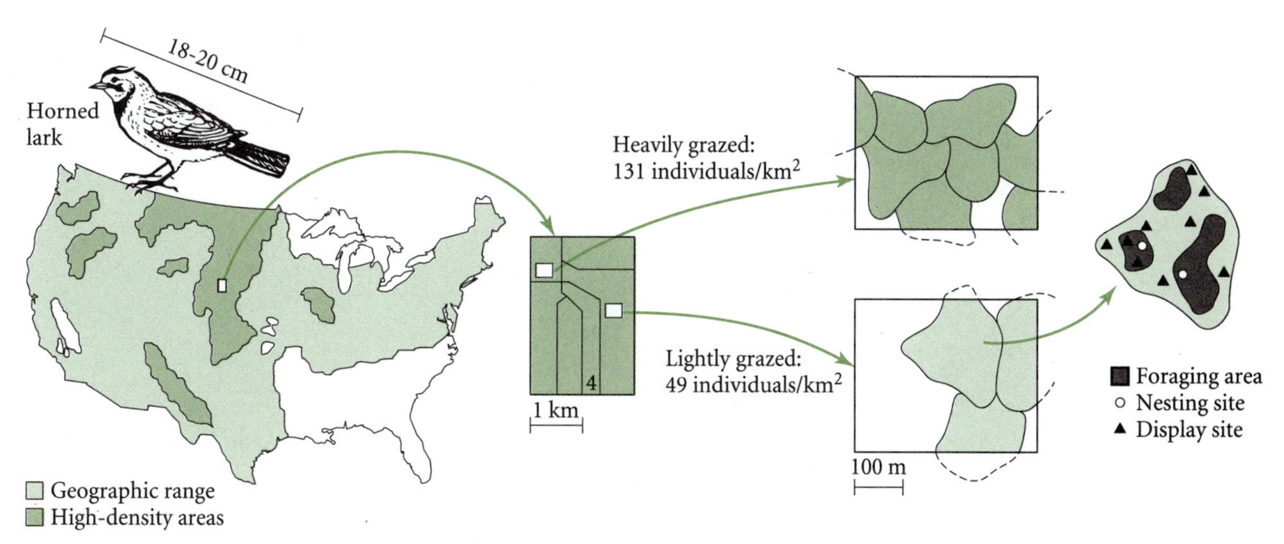

FIGURE 14-4 Hierarchy of patterns of geographic distribution of the horned lark (*Eremophilla alpestris*) during the breeding season (compare with Figure 14-3 above). (*After Wiens 1973.*)

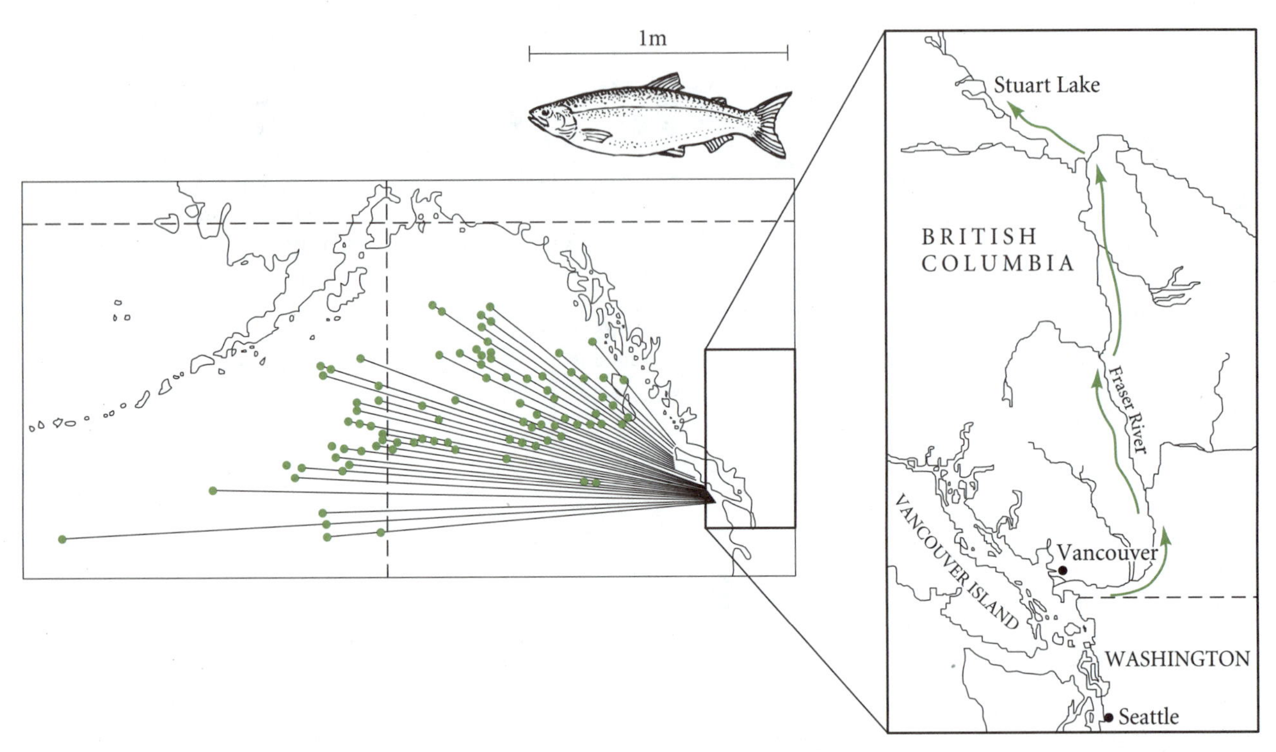

FIGURE 14-5 Individuals of the sockeye salmon (*Onchorynchus nerka*) range as adults over a vast area of the Gulf of Alaska, then migrate up rivers to their birthplaces to breed. The map shows the distribution of individual salmon tagged as adults at sea and then subsequently recaught as breeders at Stuart Lake, more than 1,000 km upriver from the mouth of the Fraser River. (*After Groot and Quinn 1967.*)

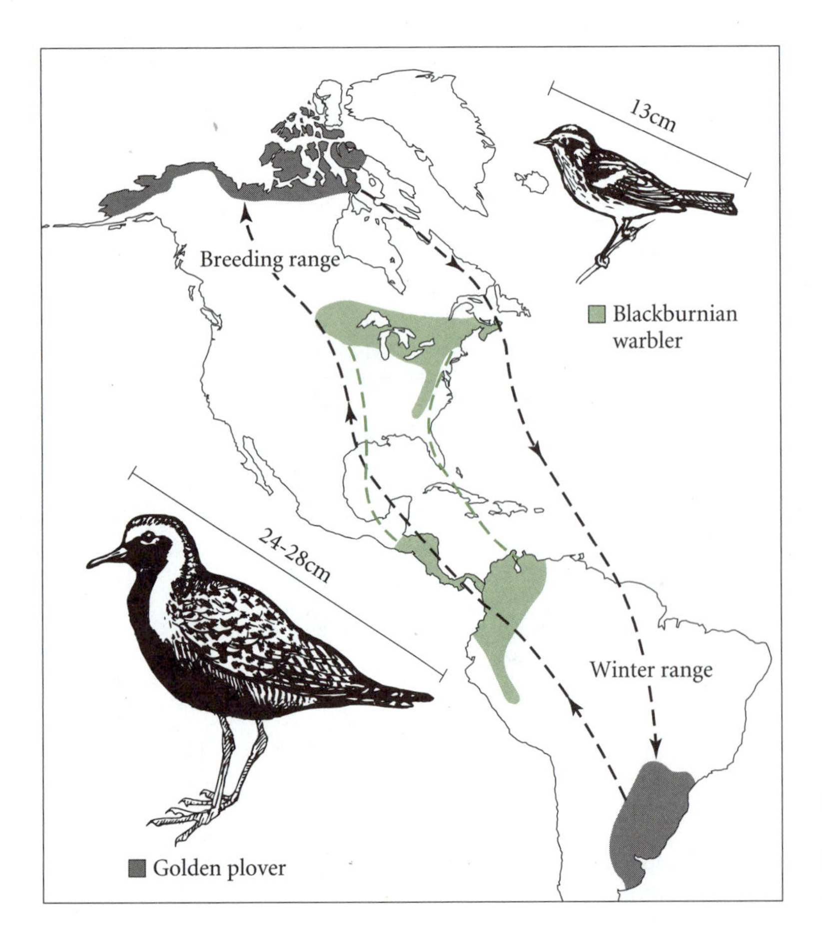

FIGURE 14-6 Breeding and wintering ranges of the golden plover (black) and blackburnian warbler (green). Migration routes are indicated by dashed lines.

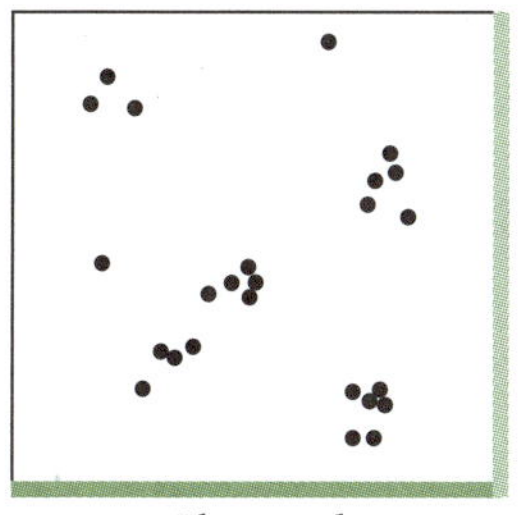

Clumped

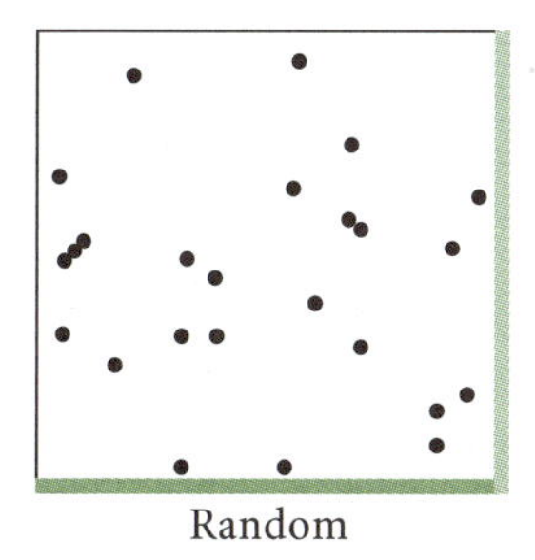

Random

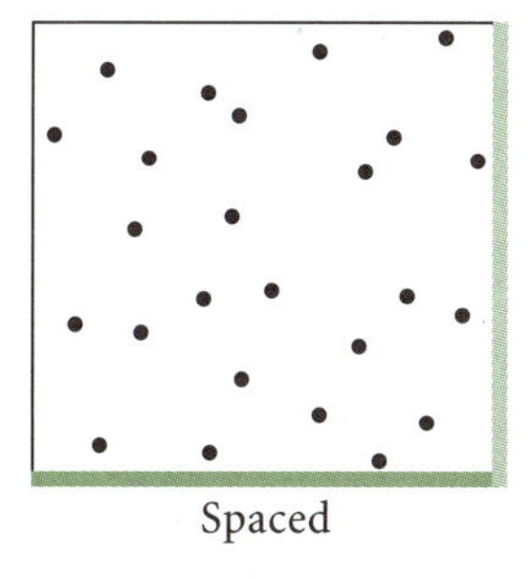

Spaced

FIGURE 14-7 Diagrammatic representation of individuals in clumped, random, and evenly spaced dispersion patterns.

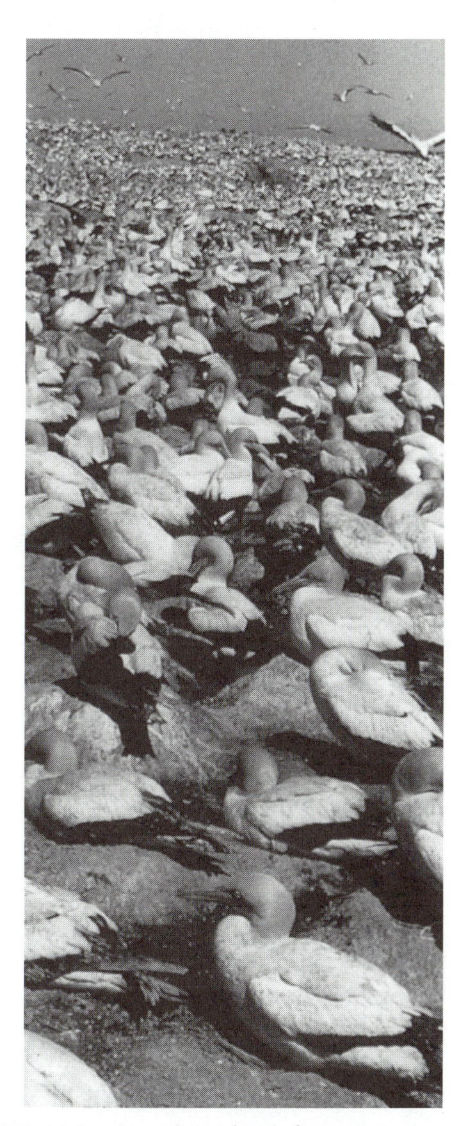

FIGURE 14-8 A nesting colony of cape gannets on an island off the coast of South Africa. The densely packed birds space their nests more or less evenly, at a distance determined by behavioral interactions between individuals. Along the entire length of the South African coast, however, seabird populations are clumped during the breeding season on a few offshore islands with suitable nesting sites.

groups, (2) clumped distributions of resources, which may be the most common cause of aggregation in most organisms, or (3) tendencies of progeny to remain in the vicinity of their parents. Birds that travel in large flocks often aggregate to find safety in numbers. Salamanders that prefer to live under logs exhibit clumped distributions corresponding to the patterns of fallen deadwood. Trees form clumps of individuals by vegetative reproduction or when seeds are dispersed through the actions of animals (Figure 14-9; see Section 14.4).

In the absence of social antagonism, which results in a spaced dispersion pattern, or mutual attraction, which leads to clumping, individuals may distribute themselves at random. In that case, the position of an individual is not influenced by the positions of other individuals in the population. Since a random dispersion pattern implies that spacing is not related to a biological process, it is often used as the model against which an observed dispersion may be compared (Poole 1974, Pielou 1977, Diggle 1989). Probability theory provides us with the tools to make such comparisons. To see how, consider the contrived situation in which organisms are dispersed within a grid of small spaces, each space being occupied by one or more individuals. If the position of each of these organisms is a random event, then we would say that the organisms are dispersed randomly on the grid. But how can we tell whether they are randomly dispersed? To put it another way, if they are randomly dispersed, how many of the grids would we expect to contain no individuals, one individual, two individuals, and so on? We know from probability theory that the probability, $P(x)$, of the number of random events x (x is a random

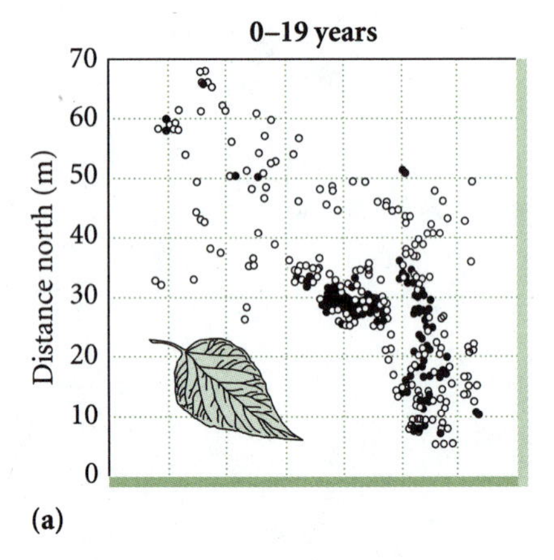

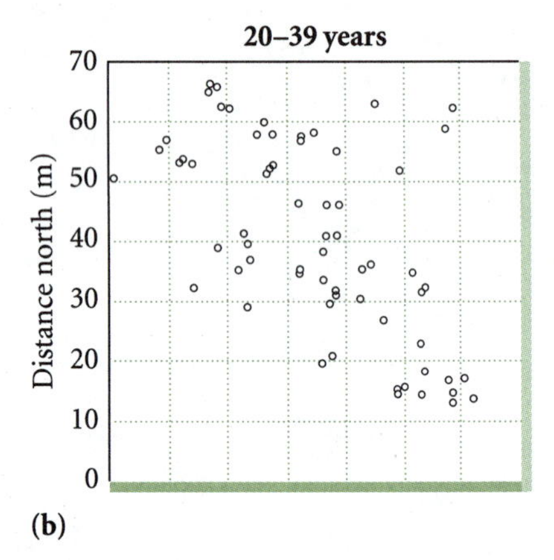

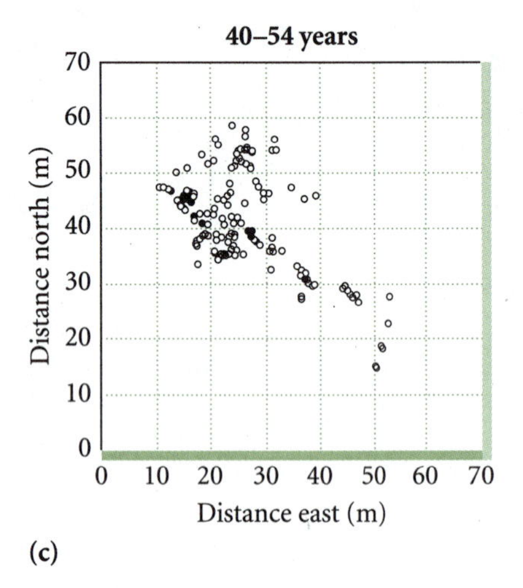

FIGURE 14-9 Map showing the distribution of stems within three age classes in a single clone of balsam poplar (*Populus balsamifera*) in Quebec, Canada. The clone reveals a highly clumped distribution of stems as well as a shift toward the southeast. Organisms that live beneath fallen logs, such as salamanders and insects, might be expected to have distributions very similar to those of the trees from which branches fall. (*After Brodie et al. 1995.*)

variable) that occur in a unit of space (or time) follows a Poisson distribution, which is one of a class of probability distributions of discrete random variables, variables that can take on only a finite number of values. The formula for the Poisson distribution is

$$P(x) = \frac{M^x e^{-M}}{x!},$$

where M is the mean number of individuals per space and $x!$ is the factorial of x (for instance, $5! = 5 \times 4 \times 3 \times 2 \times 1$; $0! = 1$ by definition). So, if on average, there are $M = 2$ individuals per space in our hypothetical grid, the probability that we would observe 10 individuals in a grid is $P(10) = 2^{10}e^{-2}/10! = 3.8 \times 10^{-5}$, a very small probability indeed. There is a much higher probability that we would observe 1 individual per grid $[P(1) = (2)(.1353)/1! = 0.2706]$ or 3 individuals per grid $[P(3) = (8)(.1353)/3! = 0.1804]$ than 10 individuals.

Let us use the Poisson distribution to analyze data from a more realistic example. Consider the number of red mites counted on each of 150 apple leaves: 70 of the leaves had no mites, 38 had 1, 17 had 2, and so on, as shown in Table 14-1. How would a random distribution of mites over the leaves appear? Suppose that each mite in the population had an equal probability of finding itself on a leaf (a probability of 1/150, or 0.067, per leaf), irrespective of the number of other mites present on the leaf. The mites will be distributed on the leaves according to the Poisson probability distribution if their distribution is a random process. In the example, 172 mites were recorded on 150 leaves, so $M = 172/150 = 1.1467$ mites per leaf. To calculate the expected number of leaves with, say, 3 mites, we substitute $M = 1.1467$ and $x = 3$ into the equation for $P(x)$; this gives a value of 0.080, or 8% of the 150 leaves (12 leaves). We see by inspection of Table 14-1 that fewer leaves than expected had 1 or 2 mites, and more leaves than expected had 0 and 4 or more mites, indicating a clumped distribution.

We may use the Poisson distribution to make a more quantitative determination of the pattern of dispersion by comparing the mean (M) of our distribution with the variance (V). The variance of a random variable is a measure of the spread about the mean of the distribution of that variable. In the example of mites on apple leaves, a low variance would mean that the number of mites per leaf is not much different from leaf to leaf. A high variance would result if the number of mites per leaf differed substantially among the leaves. An important and useful property of the Poisson distribution, which we present to you here without proof, is that the mean of the distribution is equal to the variance ($M = V$). If the mites are indeed dispersed at random, then the average number of

TABLE 14-1 Observed and expected frequencies of adult female red mites on 150 apple leaves

Mites per leaf x	Number of leaves observed $F(x)$	Poisson distribution $P(x)$	Number of leaves expected* $F(x) \times P(x)$
0	70	0.3177	47.65
1	38	0.3643	54.64
2	17	0.2089	31.33
3	10	0.0798	11.98
4	9	0.0229	3.43
5	3	0.0052	0.79
6	2	0.0010	0.15
7	1	0.0002	0.02
8 or more	0	0.0000	0.00
Total	150	1.0000	149.99

**Expected frequencies are based on the Poisson distribution.*

(From Poole 1974.)

mites per leaf will equal the variance about that average number; that is, the variance/mean ratio, V/M, will be 1. With even spacing, most leaves will have the same number of individuals, and the variance in the number of individuals per leaf will be considerably less than M (hence $V/M < 1$). With clumping, there will be many leaves with no individuals and many with a large number of individuals, and the variance will considerably exceed the mean ($V/M > 1$).

The natural history of organisms determines the way in which we analyze their dispersion patterns. Mites—at least the mites that we are considering here—live on leaves, where they make a living sucking the juices from the plant tissue. Thus, it is convenient to think of each leaf as a plot, like the individual units of our contrived grid. For some organisms, such as plants and sessile animals, it is possible to construct a grid and count the number of individuals in each unit to analyze the pattern of dispersion. Such an approach may be problematic, however, since one can never be sure whether the scale of the grid reflects the ecological processes that result in the dispersion pattern observed. A number of approaches that do not employ plots have been devised to analyze dispersion patterns. Most of these plotless sampling techniques rely on information about the distance between nearest neighbors within the population.

Clark and Evans (1954) devised an index to dispersion based on the distance between nearest neighbors within a population. When individuals are distributed at random with respect to one another, the expectation of the distance between an individual and its nearest neighbor is one-half of the square root of π divided by the density of individuals (provided, of course, that distance and area are measured in the same units). For example, at a density of 4 individuals per square meter, the expected (average) nearest-neighbor distance is $\sqrt{(\pi/4)}/2$, or 0.44 m. The ratio of the observed average nearest-neighbor distance to the expected value provides an index to dispersion. Ratios of less than 1 indicate clumping; those greater than 1 indicate even spacing. (Other nearest-neighbor methods are discussed in Pielou 1977, Warren and Batchelor 1979, and Diggle 1985.)

For many species, the pattern of dispersion reflects the arrangement of habitat patches in the environment. The apple leaves, for example, are habitat patches for the mite. But for other organisms, the pattern of dispersion may be the result of an interaction between the spatial arrangement of habitat patches and other ecological or behavioral processes. For example, in a study of the spatial structure of burrowing banner-tailed kangaroo rats (*Depodomys spectabilis*) in Arizona, Amarasekare (1994) found that, within populations, individual burrows displayed an aggregated dispersion pattern. Kangaroo rats require certain soil characteristics in order to construct their burrows, so at first glance, it might be reasonable to assume that individual kangaroo rats simply aggregate within suitable habitat patches where they can easily construct burrows. However, the aggregated dispersion in the populations that Amarasekare studied was not entirely the result of habitat patchiness. Banner-tailed kangaroo rats, it turns out, are loath to leave the place of their birth. Thus, the aggregation was partly the result of limited dispersal of individuals. Areas near aggregations of burrows with soil suitable for burrowing contained fewer burrows than would be expected if soil conditions were the primary determinant of burrow location. We shall see in Section 14.4 that the movement of individuals may also influence the pattern of dispersion, particularly among plants, in which dispersal of seeds often depends on the action of other organisms.

14.3 The estimation of population density is important to the study of population dynamics.

An important measure of a population is the number of individuals. From a management and conservation standpoint, it is important to understand the factors that cause population size to change and the processes that regulate those factors. This understanding must begin with an empirical knowledge of the numbers of individuals in populations. Total population size has two components, local density of individuals and total range of the population.

Density is defined as the number of individuals per unit of area. The density of individuals in a particular habitat depends on the intrinsic quality of that habitat for the species of concern and on the net movement of individuals into that habitat from other habitats. From the standpoint of understanding the ecological relationship of a population to its environment, local density is more revealing than total population size because it is more directly connected to local ecological interactions. As a rule, individuals are most numerous where resources are most abundant. Thus, local density provides information about the relationship of a population to its environment, and changes in density reflect changing local conditions.

Populations usually contain too many individuals, or are distributed over too large an area, to make a complete count of the population practical. When individuals are not mobile—plants and sessile marine invertebrates, for example—their densities may be estimated by counting individuals within plots of known area and extrapolating those counts to the entire population. When individuals can move between plots faster than an investigator can count the number within a plot, other methods must be employed.

One class of methods used to make density estimates of animal populations involves capturing and marking, with tags, paint, or by some other means, a sample of individuals from a population and then releasing the marked individuals back into the population. After enough time has passed to allow the marked individuals to recover from the trauma of capture and marking and to mix thoroughly with the rest of the population, a second sample is taken from the population, and the ratio of marked to unmarked individuals is noted. If one assumes that the ratio of marked to unmarked individuals in the second sample is representative of the ratio for the entire population, then an estimate of the population size can be calculated. For example, suppose we capture 20 fish from a small pond and mark them with colored fin tags. The pond contains a population of fish of size N. N is what we wish to estimate. After we release the 20 marked fish (M) into the pond, the ratio of marked to unmarked fish in the entire population is M/N. Suppose that several days later, another sample of 50 fish is captured, and 6 of those 50 fish are bearing our colored fin tags. The number of recaptures, denoted x, in the second sample is the product of the size of that sample (n) times the proportion of individuals in the population that are marked (M/N), or

$$x = \frac{nM}{N}.$$

The only variable in this simple relationship that we do not know is N, the size of the population. We can easily rearrange the equation to obtain

$$N = \frac{nM}{x},$$

providing an estimate of the population size. The population of fish in our small pond is estimated to be $N = 50(20)/6 = 167$. This method of estimating population density is one of a class of such methods called **mark-recapture** methods (Jolly 1965, 1979, Seber 1965, 1973, Overton 1969, Caughley 1977, Cormack 1979, 1981).

Because densities of populations change over time and space, no population has a single structure; one's perception of a population depends on where and when one looks. Long-term records of the population of the chinch bug (*Blissus leucopterus*) in Illinois illustrate this point (Figure 14-10). These records exist because chinch bugs damage cereal crops; the Illinois State Entomologist's Office, and later the State Natural History Survey Division, determined the importance of monitoring these populations, which they estimated from county reports of crop damage attributable to chinch bugs.

Consider the numbers involved. During 1873, when the bugs damaged crops severely over most of the state, ballpark estimates of the population indicated an average density of 1,000 chinch bugs per square meter over an area of 300,000 square kilometers, or a total of 3×10^{14} pests (300 trillion, more or less). By contrast,

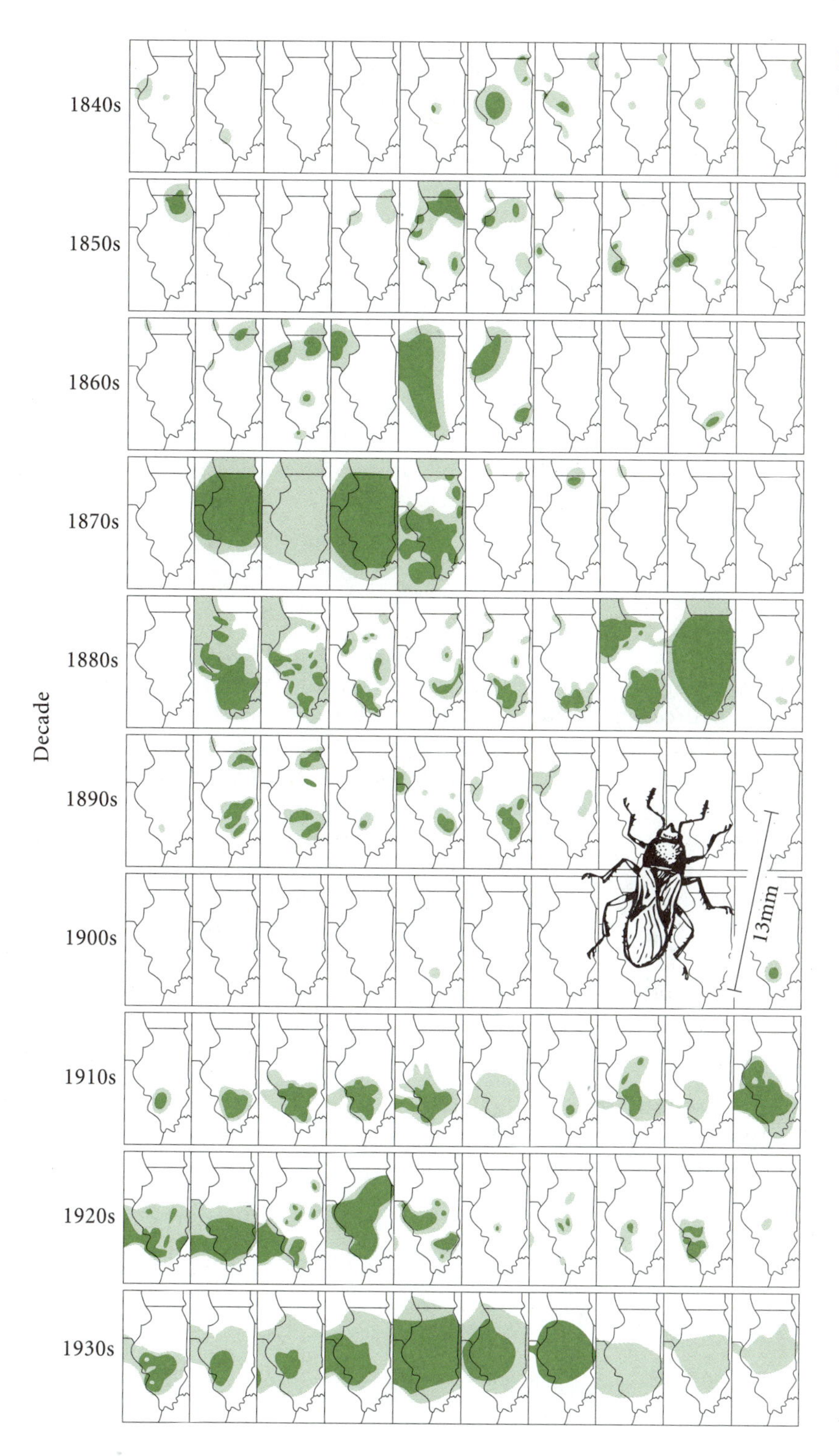

FIGURE 14-10 Distribution of crop damage caused by the chinch bug (*Blissus leucopterus*) in Illinois between 1840 and 1939. (*From Shelford and Flint 1943.*)

farmers reported hardly any damage in 1870 and 1875. Severe outbreaks were usually confined to small portions of the state, sometimes in the north, sometimes in the south. However, the spatial and temporal continuity of the population reveals itself in Figure 14-10 in the waxing and waning of infestations.

Source and Sink Populations

Local population density may be influenced by interactions with other populations because reproductive success varies among habitats. In habitats with abundant resources, individuals may produce more

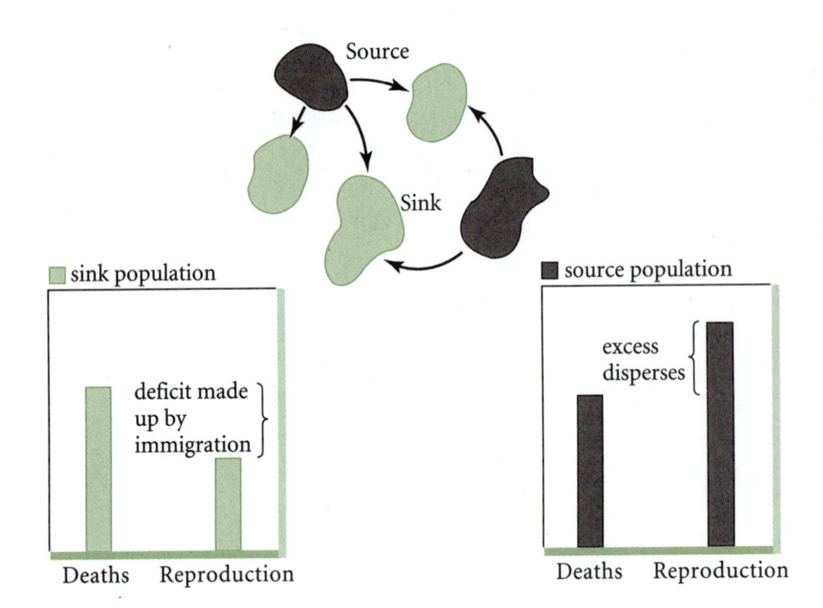

FIGURE 14-11 Diagram of the source-sink model of population structure. Populations in high-quality habitat patches produce excess offspring, which disperse to less suitable habitat patches, where immigration maintains less productive populations.

offspring than required to replace themselves, and the surplus offspring may disperse to other areas (Figure 14-11). Such populations are referred to as **source populations.** In poor habitats, populations are maintained by immigration of individuals from elsewhere, because too few offspring are produced locally to replace losses to mortality. These populations are referred to as **sink populations.** The idea of sources and sinks will be familiar to you from compartment models of global element cycles (see Chapter 11).

In southern Europe, a small songbird called the blue tit (*Parus caeruleus*) breeds in two types of habitats, one dominated by the deciduous downy oak (*Quercus pubescens*) and the other by the evergreen Holm oak (*Quercus ilex*). Comparisons of population densities and reproductive success in the two habitats suggest that deciduous oak habitat is superior for tits, and supports source populations (Blondel et al. 1993) (Table 14-2). Indeed, tit populations in deciduous oak habitats produce so many young that they would grow at almost 10% annually if individuals did not leave the population. The net rate of population decline in evergreen oak habitats would be about 13% per year in the absence of immigration. Recent genetic analyses of

TABLE 14-2 **Comparison of the density and productivity of blue tit (*Parus caeruleus*) populations from deciduous and evergreen oak woodlands**

	HABITAT	
	Deciduous oak	Evergreen oak
Breeding density (pairs/100 ha)	90	14
Laying date (average)	April 10	April 21
Clutch size	9.8	8.5
Survival to fledging	0.60	0.43
Fledglings per parent	2.9	1.8
Probable number of recruits per parent	0.59	0.37
Likelihood of death of parent	0.50	0.50
Net productivity per parent	+0.09	−0.13
Type of population	Source	Sink

(From Blondel et al. 1993.)

local populations of blue tits in southern France, which live in habitat patches separated by 10 kilometers on average, have shown that movement of young birds from deciduous oak forest to breed in evergreen oak forest (estimated at 2,000 individuals per year in the region) is about a hundred times higher than movement between patches of the same kind of habitat.

The source-sink concept has created some confusion in the ecological literature (Hanski and Gilpin 1997). In practice, it may be difficult to determine whether a population is really a sink by simply comparing the birth and death rates of two populations (see Figure 14-11). A population in which the birth rate is lower than the death rate is a candidate sink population, and immigration from a source might sustain such a population. However, immigrants would increase the density of the sink population, which could result in density-dependent interactions that actually reduce birth rates and increase mortality (Watkinson and Sutherland 1995). Also, populations may be erroneously designated as sinks simply because they are small. So, a collection of small and large local populations might be viewed uncritically as a set of sources (the large populations) and sinks (the small populations). This oversimplification arises from the appreciation that small populations often have a high risk of extinction, so that their persistence is often assumed to be the result of inputs from a source population (Hanski and Simberloff 1997).

Plant Population Density and Yield

There is a close relationship between plant population density and the amount of plant biomass that is represented in the population, a feature that is referred to as **yield.** Yield is particularly important in populations of crops and economically important tree species. The relationship may be appreciated by considering a small plot in which trees have been planted at a certain density (number/m^2). As the trees grow, each increases in biomass, and the total yield also increases. As growth continues, the trees take up more space and require more resources from the soil. Eventually, the trees will become so large that they will grow together, closing the canopy and preventing new seedlings from developing on the forest floor below. At this point the trees may be in competition with one another for resources, resulting in a slower rate of growth and a reduced increase in yield. When total yield is plotted against the density of plant populations, the yield will level off as shown in Figure 14-12. This phenomenon is called the **law of constant final yield.**

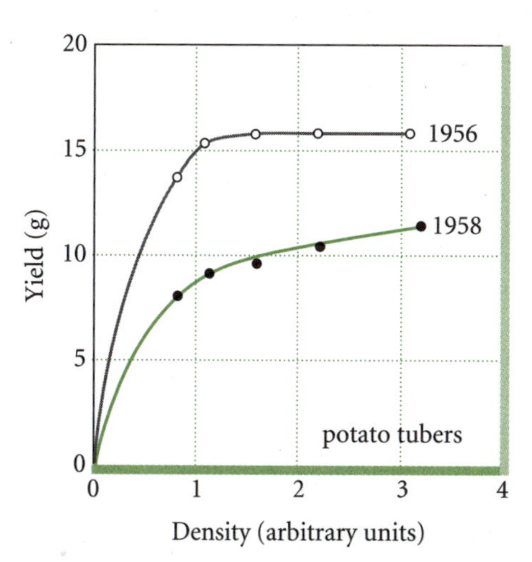

FIGURE 14-12 Density-yield curve for potato tubers from two different years. At high densities, yield may level off because of competition among plants. Above a certain density, increasing density does not increase yield. (*After Silvertown and Doust 1993, from Willey and Heath 1969.*)

14.4 The movement of individuals among populations affects population processes.

Local population changes may be due to the movement of individuals among populations. Such movement is frequently referred to as **dispersal** or, when discussed with respect to a particular population, as **emigration** (movement out of a population) or **immigration** (movement into a population) (Stenseth and Lidicker 1992). The movement of young animals from their place of birth is called **natal dispersal.**

Dispersal, particularly over long distances, is difficult to measure directly because detection of such movements requires marking and recapturing individuals. Many estimates of dispersal distances come from studies of population genetics, in which investigators wish to determine the genetic structure of a population. Indeed, the first attempts to measure dispersal in natural populations involved measuring movements away from a release point by fruit flies (*Drosophila*) that could be distinguished by a visible mutation (Dobzhansky and Wright 1943, 1947). The studies of the blue tit mentioned above are more typical of modern dispersal studies employing genetic analysis.

Mathematical Descriptions of Dispersal

Population ecologists describe dispersal with a number of mathematical indices, each of which reflects different assumptions about the structure of the population (Porter and Dooley 1993). The simplest model describes dispersal as random movement through a homogenous environment, analogous to Brownian motion in physics. With respect to a fixed point of origin, some random movements take an individual farther away and some bring it closer, although on

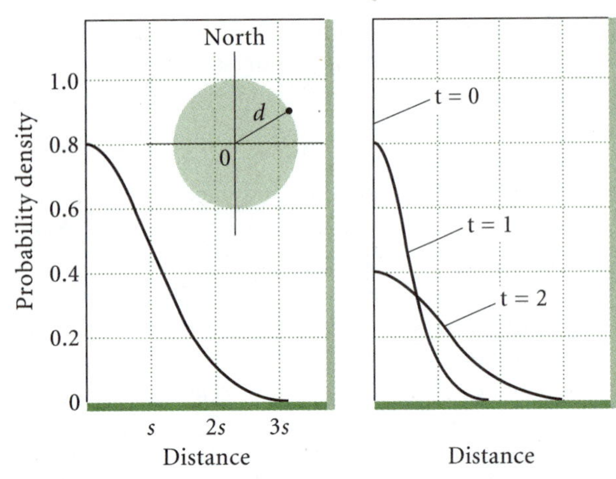

FIGURE 14-13 The normal distribution of distances of individuals from a release point based on random movement. The distribution of distances dispersed may be characterized by the standard deviation (s) of the curve (a), which increases in direct proportion to the time t since release (b). The shaded circle indicates that dispersal may be in any direction away from the point of origin (0).

average, distance tends to increase with time. In this model, the position of any individual at any moment is the sum of many random increments of distance. That is, the individual may move a distance from the origin, stop for a rest or to feed, move again, but this time toward the origin, stop, and move again, this time a great distance away from the origin, repeating the pattern many times before the measurement of its distance from the origin is made. The probability of finding an individual at a given distance from a release site is described by the **normal distribution** (Figure 14-13), a bell-shaped curve whose peak coincides with the point of origin (that is, distance = 0). The normal distribution is actually a symmetrical distribution having as many points on one side of the peak as the other. But in the case of dispersal, where there can be no movement less than 0 (one can't move negatively from the place of origin!), we simply use one half of the distribution, as shown in Figure 14-13.

Take a moment to compare the normal distribution with the Poisson distribution that we discussed in Section 14.2 with respect to dispersion. Both are distributions of a random variable. In the example of the dispersion of mites, the random variable is the number of mites per leaf, a discrete variable. The variable has a mean and a variance, which, in the case of the Poisson, are equal. With respect to dispersal, the variable of interest is distance from the origin, a continuous variable having a normal distribution.

Dispersal Distance and Neighborhood Size

The standard deviation (s) of the normal distribution provides a convenient index to dispersal distance: s_t is the standard deviation estimated over the average life span, t, of an individual. Crumpacker and Williams (1973) measured the dispersal of fruit flies (*Drosophila pseudoobscura*) in grasslands in central Colorado. Flies were marked with minute fluorescent dust particles and released; individuals were then caught in traps placed at regular intervals in eight directions of the compass at distances of up to 351 meters from the release point. After 1 day of dispersal, the estimated values of s were 139 and 171 meters at two localities, for an average standard deviation of dispersal distance of $s = 155$ m.

A felicitous property of s as a measure of dispersal is that variances (s^2) of distance add over time. Thus, when fruit flies disperse distance $s_1 = 155$ m in one day, the variance in distances after 2 days will be $2s_1^2$ ($2[155]^2 = 48{,}050$ m^2; 209.2 m), $4s_1^2$ ($4[155]^2 = 96{,}100$ m^2; 310 m) after 4 days, and ts_1^2 after t days. Accordingly, if adult fruit flies live an average of 23 days and the average dispersal distance (s) per day is 155 meters, the flies may be expected to move $23(155)^2 = 524{,}423$ m^2, or about $s_{23} = 724$ m, during their lifetimes.

Neighborhood size provides an index to the number of individuals in a population that are potentially coupled by strong interactions. The neighborhood size of a population is the number of individuals within a circle whose radius is twice the standard deviation of dispersal distance (s) within the average reproductive life span of the individual (Wright 1946). The area of a circle with radius r is $2\pi r^2$, so the area of a circle having radius $r = 2s$ is $2\pi 2s^2$, or $4\pi s^2$. In the example of the fruit flies above, a circle of area $4\pi s^2 = 4\pi 155^2$, or 301,907.8 m^2 in area. We must also have an estimate of the population density in order to employ the neighborhood size concept. For the fruit flies, the density is known to be about 0.38 flies per 100 m^2. The neighborhood size is calculated as

(density)
× (area of the circle with radius equal to two standard deviations of dispersal distance)
× (average life span).

Thus, for the fruit flies, the neighborhood size is calculated at (0.38/100 m^2)(301,907.8 m^2)(23) = 26,386.7 individuals.

Barrowclough (1980) used recaptures of birds banded as nestlings to estimate the standard deviation of dispersal distance, s, for several small songbird species in the United States. Because juvenile birds usually move much farther from their birthplaces to breed than adults move from one year to the next, Barrowclough distinguished s_{juvenile} and s_{adult} and combined them according to the rule of adding variances; that is, $s^2(t + 1) = s^2_{\text{juvenile}} + s^2_{\text{adult}}$, where t is the expected life span of an adult bird (the juvenile stage lasts about 1 year, from hatching to first breeding).

TABLE 14-3 Estimates of dispersal parameters for three populations of land snails (*Cepaea nemoralis*) in Europe

s (meters) after 1 year	Individuals per m²	Neighborhood size (number of individuals
5.5	20	7,603
9.7	2.0	2,365
10	1.4	1,759

(From Greenwood 1974.)

For eight species, $s(t + 1)$ varied between 344 and 1,681 meters, densities varied between 16 and 480 individuals per square kilometer, and neighborhood size varied between 151 and 7,679 individuals. Greenwood (1974) estimated dispersal parameters for three populations of the land snail *Cepaea nemoralis* in Europe (Table 14-3). Notice that the neighborhood sizes of these slowly dispersing snails are on the same order as those of small birds. From mark-recapture data gathered by Blair (1960) on the rusty lizard (*Sceloporus olivaceous*) near Austin, Texas, Kerster (1964) estimated $s(t)$ to be 89 meters and neighborhood size between 225 and 270 individuals.

Population Spread and Genetic Differentiation

In most studies of dispersal, the principal source of information is the distribution of distances of movement of individuals away from a point of release. Two other kinds of observations are pertinent, however. The first of these concerns the spread of introduced populations, which can occur only through the movements of individuals. The second concerns the genetic differentiation of local populations, which is hampered by movements of individuals.

The European starling spread almost 2,500 miles across the United States in 60 years, an average rate of about 40 miles per year (see Figure 14-1). This figure is much higher than Barrowclough's estimate for dispersal distance within populations of small songbirds. Adult starlings tend to nest in the same area year after year, so most dispersal is accomplished by young birds. Furthermore, the westward spread of the starling was characterized by frequent sightings of nonbreeders before breeding populations were established in an area. It is unlikely that such long-distance movements of juveniles could be detected within an established population, particularly because of the limited area of intensive population study.

Whereas a few long-distance movements may be sufficient to expand the borders of a population, they may have less effect on the dynamics of widely separated, but established, subpopulations. Given the propensity of populations for increase—a phenomenon emphasized in Chapter 15—a small number of colonizing individuals can form a large population in a relatively short period. However, those same individuals immigrating to an established population may have a negligible effect.

Genetic differentiation of populations depends far less on the movement of individuals among populations than on the forces of selection, mutation, and random change (genetic drift). Strong selection can lead to divergence between populations even in the face of considerable gene flow (Endler 1977), as we have seen in the establishment of distinct ecotypes of plants in different habitats (Chapter 6). **Gene flow** is the exchange of genetic information among populations resulting from the movement of individuals. The presence of apparently nonadaptive differences between close populations provides stronger evidence of limited gene flow. Such differences occur in populations of the small annual plant *Linanthus parryae.* In one small area in southern California, blue flowers (the normal flower color is white) are haphazardly distributed without apparent relation to environmental factors (Figure 14-14). Some samples of individuals separated by less than 1 km differed in flower color by 100%. Evidently, gene flow occurs so infrequently between these populations that frequencies of alleles for flower color diverge by genetic drift.

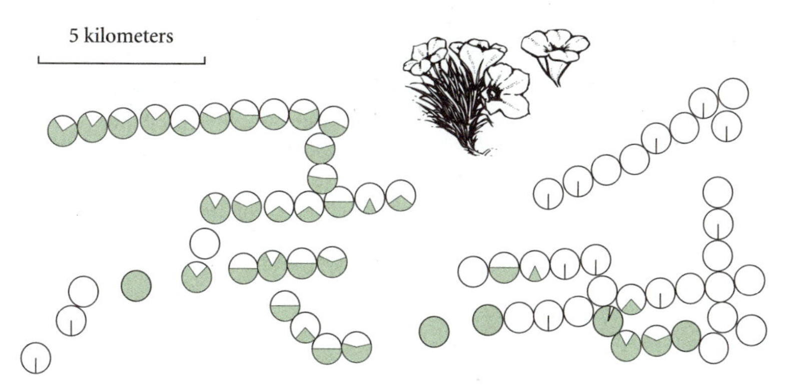

FIGURE 14-14 Frequencies of blue flowers (green portion of circle) in subpopulations of *Linanthus parryae* from a small region of southern California. Genotype frequency varies abruptly over small distances in some cases, suggesting that gene flow between these small, isolated subpopulations is minimal. *(From Epling and Dobzhansky 1942.)*

Dispersal in Plants and Sessile Animals

In plants and sessile animals, it is the **propagules**—seeds or larvae—rather than the adults that generally move from one place to another. Seeds have little or no ability to move under their own power, and larval animals, though they may dart about locally with considerable agility, do not possess sufficient mobility to move great distances under their own power. For these organisms, dispersal beyond the reach of the branches of the parent tree, or the spawning site of the adult female fish, or the web of the newly hatched spiderling requires the assistance of winds, currents, and, very often, other organisms.

The seed dispersal strategies of plants may be characterized by the dispersal agent. Some small seeds are carried great distances by wind or by currents in rivers, oceans, and other large bodies of water. Among the most interesting dispersal strategies is **ballistic seed dispersal,** whereby plants literally shoot their seeds from seed pods (see Chapter 1). The architectural features of such plants include mechanisms whereby changes in the structure of the fruit through the season lead to a conformation that allows for the rapid transfer of kinetic energy from the fruit to the seeds at the time when the seeds are to be released, resulting in an explosive projection of the seeds (van der Pijl 1969, Beer and Swaine 1977, Ohkawara and Higashi 1994, Witztum and Schulgasser 1995).

It has been estimated that animals are the principal seed dispersal agent for between 60% and 90% of tropical plant species (Frankie et al. 1974, Charles-Dominique 1993) When seeds are moved about by wind or water or shot from the plant by ballistic dispersal, the resulting pattern of seed dispersion (the spatial distribution of the seeds after dispersal) is more or less subject to the physical forces of nature. But when animals are involved, the movement patterns of the animals themselves, as well as the social interactions and behaviors of the animals, become an important determinant of the resulting seed dispersion pattern. In one study, it was found that red howler monkeys (*Alouatta seniculus*) in French Guiana were responsible for the clumped distributions of four species of plants from which they exploited ripe fruit. Seedlings of the four plant species were found to be clumped beneath the tree sleeping sites of monkey troops, where they were deposited by defecation (Julliot 1997).

The profligate destruction and fragmentation of habitats around the world by human activities have led ecologists to consider the dynamics of dispersal more thoroughly. Often, the very survival of a species depends on its ability to move about within and between small isolated areas of habitat.

14.5 The genetic structure of a population describes the amount and distribution of genetic variation.

All populations contain genetic variation, although the amount of this variation differs substantially from species to species and from place to place. The genetic structure of a population describes the distribution of this variation among individuals and among subpopulations, as well as the way in which organisms manage the consequences of genetic variation by means of mating systems. Genetic variation is important to a population because it is the basis of the population's capacity to respond to environmental change through evolution. Genetic variation is also important to individuals: variation among an individual's progeny may increase the likelihood that at least some of them will be well adapted to particular habitat patches or to changed conditions. Genetic variation is maintained primarily by mutation and by gene flow from other localities in which different genes have a selective advantage.

For the most part, our discussion of population genetic structure below assumes that the individual organisms we see are discrete reservoirs of genetic information. In a population of salamanders, for example, the genetic composition of each salamander is distinct from that of other salamanders in the population. However, for plants, fungi, corals, and a number of other organisms, such distinctions are not always so clear. The modular construction of these organisms may give rise to groups of reproducing *individuals* that arise from the same zygote and have the same genetic composition. You are no doubt familiar with plants such as strawberries that produce runners that extend outward and give rise to vertical flowering structures. Another good example of such a plant is the aspen (*Populus*) (Figure 14-15). An aspen tree that develops from a seed may produce other individuals asexually via underground horizontal roots that give rise to what appear to be other individual aspen trees. In reality, the trees are a **clone;** that is, a population of genetically identical individuals arising by asexual reproduction. Individuals produced—cloned, if you will—from the same zygote are called **genets** and, with respect to evolution, are a single genetic unit. The *individuals* within the clone, each of which arose from an asexual process and all of which share the genetic composition of the original zygote, are called **ramets.** A population of ramets, then, is a clone.

The Nature and Measurement of Genetic Variation in Populations

We have a deep appreciation for the wondrous variety of the natural world. We admire not only the diversity of

FIGURE 14-15 Clones of aspen (*Populus* spp.) trees. Each group of similarly sized trees is a clone. (*Courtesy of M. L. Crump.*)

different kinds of organisms, but also the great variation that we see among the individuals of any single type of organism. Variation abounds. No two pikas, or scrub jays, or house cats, or maple trees, or aardvarks, or, of course, humans are exactly alike. You will recall from your general biology course that this variation in **phenotype**—the outward appearance and behavior of an organism—results from the interaction of the **genotype** of the individual—all of the genetic characteristics that give rise to the morphological, physiological, and behavioral characteristics of the individual—and the environment. The unit of genetic inheritance is the **gene,** which is distinguished biochemically as a specific sequence of nucleotides of deoxyribonucleic acid, or DNA, that encodes for all or part of a particular enzyme or structural protein. In eukaryotic cells, DNA, and thus genes, reside on rodlike structures called **chromosomes,** which are usually visible only during cell division. In sexually reproducing organisms, chromosomes occur in pairs, one from each parent. The genes occur in particular positions on the chromosomes called **loci** (singular **locus**). Any gene may have two or more alternative forms, called **alleles,** which result from slight differences in the DNA sequence of the gene. Different alleles often cause slight differences in form or function. When the genes of a particular locus on each of a pair of chromosomes are identical, we say that the organism is **homozygous** with respect to that gene. When different alleles are found at a particular locus on each chromosome of a pair, the individual is **heterozygous** with respect to that locus. The amount of heterozygosity in a population is one measure of genetic variation that is commonly used by ecologists and population geneticists.

In order to understand genetic variation, the ecologist must be able to measure it. There are a number of techniques available for determining the extent of genetic variation in a population. In many cases, a particular trait that is obvious in the phenotype is inherited in Mendelian fashion, and thus the ecologist can tell whether an individual is homozygous or heterozygous for the trait. By observing many members of the population, then, he may determine the extent of heterozygosity in the population. Genetic variation may also be determined by examining the protein products of genes directly through a procedure called **electrophoresis.** This process is based on the fact that proteins produced by a heterozygote have slightly different forms, called **allozymes,** owing to the different amino acid sequences imparted by the different encoding sequences on the DNA. When tissue such as blood is placed at one end of a gel made of hydrolyzed starch, polyacrylamide, or some other suitable substance, and an electric current is passed through the gel, the gene products in the blood will move through the gel. Allozymes will travel at different rates through the gel owing to their different electrical charges, and when the gel is stained properly, will show up at different positions on the gel, indicating that the organism is heterozygous at that locus. Homozygotes will produce only a single type of protein, and this will show as a single band in the gel. The level of heterozygosity of large numbers of loci in a population may be screened in this way.

In recent years, ecologists have employed **recombinant DNA technology** to examine genetic variation by looking directly at the DNA rather than at the products of genes. One such approach takes advantage of a special class of enzymes called **restriction enzymes.** Restriction enzymes recognize a specific base sequence on the DNA molecule and cut both strands of the DNA at the site of that sequence. Different restriction enzymes may be used to cleave the DNA molecule at two places, producing a DNA fragment called a **restriction fragment.** Long sequences of DNA often vary to some degree in their nucleotide sequence. If such small differences in nucleotide sequence occur at restriction sites, the application of restriction enzymes may produce restriction fragments of slightly different lengths, a situation referred to as **restriction fragment length polymorphism** or **RFLP.** RFLPs are heritable and thus may serve as a basis for determining genetic variation or as genetic markers for studying particular genetic disorders.

A more commonly used approach for examining DNA directly is **DNA fingerprinting,** an enhancement of RFLP technology that is used to detect variation in short segments of DNA having repeating sequences of nucleotides. Such **variable number tandem repeats** (VNTR) may be found at different loci scattered about on different chromosomes. Restriction enzymes may be used to cleave out the repeating sequences from between eight and fifteen loci at one time. The lengths of the DNA fragments obtained in this way depend, of course, on the number of repeated nucleotides in the sequence, and this may vary dramatically among the loci. Indeed, there are so many VNTRs and so many different fragment lengths possible at each locus that the pattern of fragment lengths is unique to the individual, thus the term "DNA fingerprint." The extent of variation in fragment length may be used as a measure of the level of genetic variation in populations.

Influences of Population Size on Genetic Variation

Because different alleles cause slight differences in form or function, they are liable to come under the influence of natural selection, which may favor an increase of one allele in a population at the expense of others. In small populations, however, the frequencies of alleles also may change because of random variations in birth and death rates due to chance events. Such changes are referred to as **genetic drift.** Suppose that a population contains two alleles of the same gene. The rate of increase or decrease of each allele has a random component; just by chance, one may increase and the other decrease, resulting in a change in their frequencies. Even though such changes are not biased in one direction or the other, the occurrence, by chance, of a long series of changes in one direction may lead to the disappearance of one allele from the population. In this case, we say that the remaining allele becomes **fixed.** Further variation in allele frequency cannot take place, unless mutation introduces new alleles into the population.

The rate of fixation of alleles is inversely related to the size of a population. Thus, genetic variation decreases more rapidly over time in small populations than in large ones. Furthermore, a single episode of small population size, which might occur during the colonization of an island or a new habitat by a few individuals from a large parent population, can reduce genetic variation in the colonizing population. Such episodes are known as **founder events.** When founding populations consist of ten or fewer individuals, they typically contain a substantially reduced sample of the total genetic variation of the parent population. Continued existence at a low population size results in further loss of genetic variation due to genetic drift and close inbreeding. This situation is often referred to as a **population bottleneck.** Such a condition appears to have occurred in the recent past in populations of cheetahs in East Africa, which exhibit practically no genetic variation. Because of the effects of founder events and bottlenecks, the fragmentation of natural populations into small subpopulations may eventually restrict the evolutionary responsiveness of each subpopulation to the selective pressures of changing environments, thus making these small subpopulations more vulnerable to extinction. As yet, there is no agreement as to whether the cheetah's genetic uniformity poses a serious threat to its future.

Geographic Variation in the Gene Pool

As we mentioned in Chapter 6, the genotypes of individuals within a population often vary geographically. This may happen because of differences in selective factors in different parts of the population's range, or because of random changes (genetic drift, founder effects) in isolated or partially isolated subpopulations. Populations do not have to be subdivided for genetic differences to arise within them. If the difference in selective factors between two localities is strong relative to the rate of gene flow between them, then differences in allele frequency can be maintained by differential selection. This situation often results in a gradual change, or **cline,** in allele frequencies, or some phenotypic character under genetic influence, over distance. It may also give rise to ecotypes (see Chapter 6), which are varieties of species that are adapted to local conditions.

Effective and Minimum Viable Population Size

Our discussion of the genetic structure of populations suggested clearly that small populations may be in jeopardy of losing genetic variation via genetic drift, and with it, the ability to respond to environmental change. Thus, conservation biologists are keenly interested in the question of how large a population must be in order to stave off the loss of genetic diversity. In terms of the evolutionary processes involved, the question boils down to how large the population must be in order to diminish the effects of genetic drift sufficiently to maintain its genetic diversity. This question is difficult to address, however, because not every member of a population contributes to the genetic structure of the population in the same way. One sex may produce more offspring on average than the other, or the variance in the number of offspring produced by males and females may be different. Mating systems may prevent some individuals from mating with others, and there may be age-related differences in reproductive output. Because of this, simple estimates of population density do not provide enough information to evaluate the vulnerability of the population to genetic loss.

One way to overcome this difficulty is to calculate the genetic **effective population size,** denoted N_e. The effective population size is the size of an idealized population that has the same amount of genetic drift as the natural population under study. An idealized population is one in which the number of reproducing males and females is the same (1 : 1 sex ratio), random mating occurs among males and females (that is, every male has an equal probability of mating with every female), the rate of emigration and immigration is constant, and in which there is no age structure. Usually $N_e < N$, though this is not always the case.

The effective population size depends on the relative reproductive success of males and females and the variance in that success. Thus, to determine N_e, we calculate separate effective population sizes for males and females as follows:

$$N_{males} = \frac{N_m K_m - 1}{(K_m + V_m/K_m) - 1}$$

$$N_{females} = \frac{N_f K_f - 1}{(K_f + V_f/K_f) - 1},$$

where N_m and N_f are the numbers of breeding males and females respectively, K_m and K_f are the average numbers of offspring produced by males and females in their lifetime, and V_m and V_f represent the variance in the number of offspring produced by each sex.

Inspection of these equations reveals how the variance in the number of offspring produced influences the effective population size of males and females. For a given number of breeding males, N_m, producing an average number of offspring K_m, the effective population size N_{males} will decrease as the variance in offspring V_m decreases. For example, for a population with $N_m = 10$, and average number of offspring $K_m = 2$, the effective male population size, N_{males}, will be 7.6 if $V_m = 1$, but 4.7 if $V_m = 2$.

The separate values for males and females given in the equations above may be combined in the following way (here given without derivation) to calculate the overall effective population size (Lande and Barrowclough 1987):

$$N_e = 4\left[\frac{1}{N_{males}} + \frac{1}{N_{females}}\right]^{-1}.$$

Figure 14-16 shows how the number, reproductive output, and variance of reproductive output of males and females interact to determine the effective population size of a population with a total density of $N =$ 1,000. When the entire population is composed of breeding individuals (500 males and 500 females; Figure 14-16a), the effective population size is very close to the total population density. The variance in the number of offspring produced by each sex (V_m and V_f) determines the extent to which N_e is less than N. Figure 14-16b shows that as the variance of reproductive output increases ($V_m = V_f = 3$ instead of 1), the effective population size decreases. It is possible for the effective population size to be larger than the population density, as shown in Figure 14-16c, in which all individuals in the population breed and one sex (males in this case) has a higher reproductive output than the other. When the number of breeding individuals is small relative to the total number of individuals in the population, the effective population size is small (Figures 14-16d and e). Comparing Figures 14-16b and d shows that decreasing the number of breeding individuals has essentially the same effect as increasing the variance of reproduction.

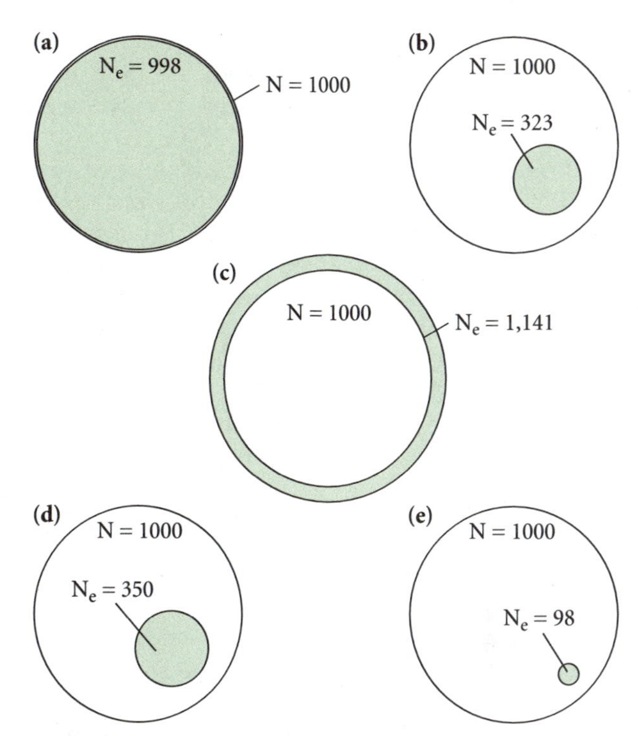

FIGURE 14-16 Calculations of effective population size, N_e, for five different hypothetical populations. $N = 1,000$ in each case. (a) The population contains an equal number of males and females, all individuals produce an average of one offspring during breeding, and the variance in the number of offspring produced is 1 ($N_m = N_f = 500$, $K_m = K_f = 1$, $V_m = V_f = 1$). In this case the effective population size is nearly equal to the total density of the population. (b) The same situation as depicted in a, except that the variance in the number of offspring produced is tripled ($N_m = N_f = 500$, $K_m = K_f = 1$, $V_m = V_f = 3$). This results in an effective population size about one-third the total density of the population. (c) The effective population size may be larger than the total density if all individuals breed and the average reproductive output of one of the sexes is high ($N_m = N_f = 500$, $K_m = 3$, $K_f = 1$, $V_m = V_f = 1$). (d) When only a portion of the population of one sex breeds and produces a small mean number of offspring, $N_e < N$ ($N_m = 100$, $N_f = 500$, $K_m = K_f = 1$, $V_m = V_f = 1$). (e) When a very small proportion of the population breeds and produces a small mean number of offspring, $N_e \ll N$ ($N_m = 50$, $N_f = 50$, $K_m = K_f = 1$, $V_m = V_f = 1$).

How large should the effective population size be in order to maintain genetic diversity in the face of genetic drift? It has been suggested that an effective population size of at least 500 is required, though it is likely that the minimum number will vary widely among populations having different sex ratios, age structures, and social arrangements (Lande and Barrowclough 1987).

Related to the concept of effective population size is that of **minimum viable population size** (MVP) (Shaffer 1981, Gilpin and Soulé 1987, Nunney and Campbell 1993). The MVP is the smallest population that can persist for a specified time, usually taken to be 1,000 years. The MVP is dependent both on the demographics of the population and the genetic diversity within the population. This concept will be discussed in more detail in Chapter 19.

14.6 Life tables summarize the survival and reproduction of individuals within populations.

Populations have a spatial structure, which includes the range of the species, the spatial arrangement of individuals within and among habitats within the range, population density, and patterns of movement of individuals. Populations also have a genetic structure, which describes the genetic variation among individuals in the population and among subpopulations, or among different areas of a large continuous population. A third aspect of population structure has to do with the rates of births and deaths and the pattern of distribution of individuals among different age classes. An appreciation of this aspect of population structure is prerequisite to our discussions of population growth and regulation (Chapters 15 and 16).

Ecologists are not the only ones interested in the way in which organisms are added to populations by birth and subtracted from populations by death. Those in the insurance and health care professions are keenly interested in such dynamics of human populations. Ecologists and business professionals use the same techniques to describe the demographic characteristics of the populations they study. Their approach, which was used first for human populations, is to construct a **life table,** which is a tabular accounting of the birth rates and probabilities of death for each age class in the population. The first life table assembled for a nonhuman species was that of Pearl and Parker (1921) for a laboratory population of the fruit fly *Drosophila melanogaster.* The life table is now an essential part of the study of natural populations. Let us begin with a description of the technical details of the construction of a life table, and then look at an example.

Elements of Life Tables

In order to construct a life table, one must first have knowledge of the **age structure** of the population; that is, the number of different age classes and the number of individuals in each age class at some time. Age is designated in a life table by the symbol x, and age-specific variables are indicated by the subscript x (the first or youngest age class is designated $x = 0$). For species in which reproduction occurs during a brief breeding season each year—a situation that is common to many species of plants and animals—each **age class** is composed of a discrete group of individuals born at approximately the same time. The individuals that make up that group are referred to as a **cohort.** The young blue jays hatched this past spring, for example, may be considered a cohort. When reproduction is continuous, as it is in the human population, each age class x is designated arbitrarily as comprising individuals between ages $x - 1/2$ and $x + 1/2$. We may still use the term *cohort* to distinguish the offspring produced during the arbitrary time interval. For example, you may have heard you and your classmates referred to as a cohort because you were all born in the same time interval. The age x is very often expressed in terms of years, but we hasten to add that this is not always the case. In short-lived species, it is more appropriate to represent age in terms of weeks, or even days.

Two features of life tables simplify our representation of natural populations, but, as is the case with any simplification, discount to some extent the complexity of interactions in populations. First, our allocation of the individuals of a population to the different age classes is done without regard to or consideration of any characteristic of the individuals except age. Thus, individual differences in size, social status, and genotype are not represented in the life table. In essence, we assume that all of the individuals within an age class are uniform with respect to all traits. Second, life tables nearly always include only females for species having distinct males and females. The reason for this is that one of the key parameters in a life table is the rate of reproduction. While males have an effect on reproduction in sexually reproducing organisms, their contribution is very difficult to measure in terms of numbers of young.

We account for birth in life tables by counting the number of female offspring produced per breeding season or age interval per female in the population, a measure that we call **fecundity.** Fecundity is designated as b_x. (Think of b for births; m [for maternity] is used in some treatments, but m is used here for mortality.) Thus, the fecundity of females of age class 4 would be designated b_4. The determination of fecundity usually requires extensive observations, which often reveal that not every female of a particular age class produces the same number of young. So, fecundity may represent average numbers of female young produced among the females in the age class. We expect that the production of young will not be the same for all age classes. Reflection on the human population supports this assumption. Very young and very old human females usually do not produce young. We emphasize again that our accounting of births includes only females.

Life tables portray the statistics of mortality in several ways. The fundamental measure is the **probability of survival** (s_x) between ages x and $x + 1$, or its alternative, the **probability of death,** m_x (where m is mortality; those who use m for fecundity often use q for mortality). Because an individual either lives or dies, $m_x = 1 - s_x$. The probabilities of survival over many age intervals are summarized by the **survivorship** to age x, l_x, which is the probability that a newborn individual will be alive at age x. Because by definition all newborns are alive at age 0, $l_0 = 1$. The proportion of newborns alive at age 1 is the probability of surviving from age 0 to age 1, hence $l_1 = s_0$. The proportion alive at age 2 is the probability of surviving both from age 0 to age 1 and from age 1 to age 2, or $l_2 = s_0 s_1$ (using the multiplicative rule of probability). By extension, $l_x = s_0 s_1 s_2 \ldots s_{x-1}$. In product notation,

$$l_x = \prod_{i=0}^{x-1} s_i.$$

(The large pi indicates the product of the terms s_i for values of i from 0 to $x - 1$.)

When the number of individuals surviving to age x is expressed on a log scale and plotted against age, one obtains a **survivorship curve** (Figure 14-17). Three hypothetical survivorship curves were introduced by Pearl (1927). Few populations display one of these curves precisely, but the curves do provide a basis for comparison of populations with different age structures and dynamics. The type I curve, for example, depicts a situation with very low mortality among younger individuals, a situation that is characteristic of some human populations. The type III curve shows an extremely high mortality rate among young individuals, a pattern typical of many fishes, marine invertebrates, and some human populations. Of course, there are many possible survivorship patterns between these extremes. Figure 14-18 shows survivorship curves for the total human population and for juvenile humans in the United States.

Because the probability of death may be considered as a rate applied over time, some life tables include the

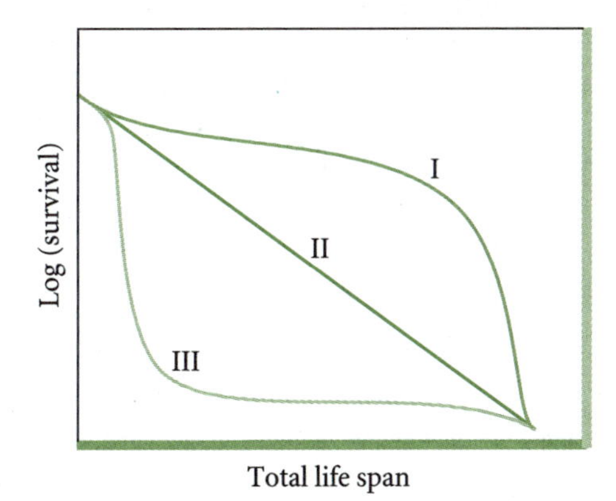

FIGURE 14-17 Depiction of three idealized survivorship curves. Populations with a type I survivorship curve have more mortality in the older age classes, whereas a considerable amount of mortality occurs among the youngest individuals in populations with type III curves.

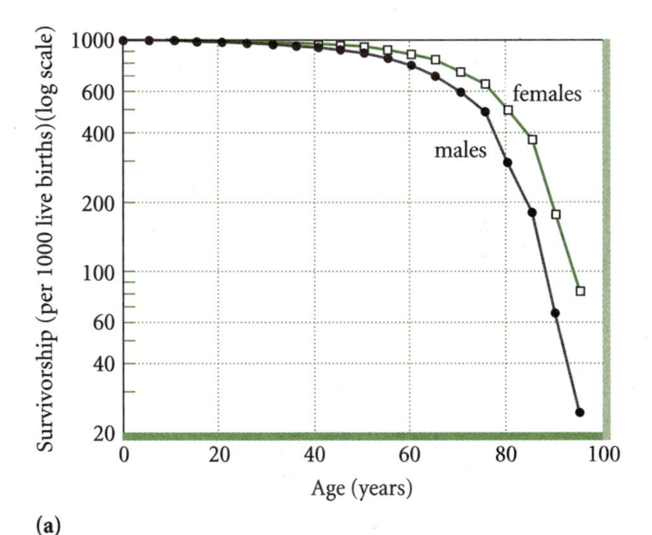

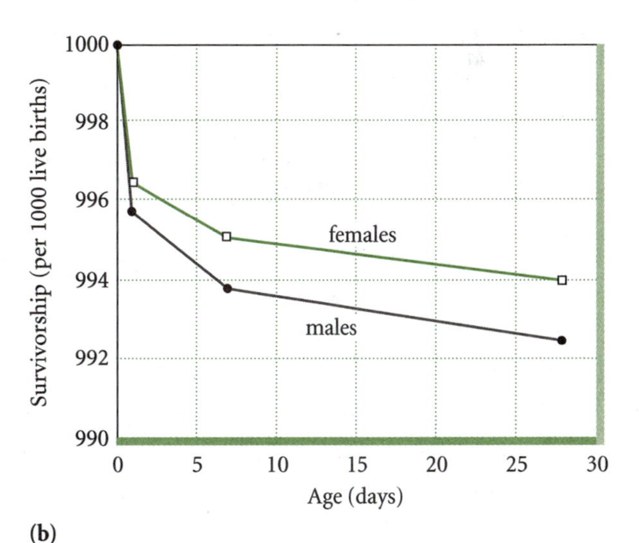

FIGURE 14-18 Survivorship curves for (a) all males and females in the United States in 1989 and (b) for infants during the first month of life in the United States in 1989. Except for the very high mortality of both males and females during the first day or two of life (b), infant mortality remains relatively stable for the first 30 days. This contributes to the low overall mortality and high survivorship of this population (type I survivorship curve; a). (*After Krebs 1994; data from the Statistical Abstracts of the United States 1991.*)

exponential mortality rate, k_x (think of k for killing power; Haldane 1949). According to the law of exponential decrease, survival over t units of time follows the expression $s(t) = e^{-kt}$ (e is a constant, 2.7182). In this equation, k represents the rate of population decline. Recall the following general relationship from algebra: if $x = a^y$, then $y = \log_a x$. Using this relationship, we can solve the equation $s(t) = e^{-kt}$ for k, the killing power. First, we obtain $-kt = \log_e s(t)$, and then, dividing through by $-t$, we get $k = -\log_e s(t)/t$. When one wishes to calculate the exponential mortality rate during each age interval, $t = 1$ and $k_x = -\log_e s_x$.

We may write a relationship between exponential mortality rate and survivorship using the following steps. Survivorship, l_x, is the product of the individual survival probabilities; that is, $l_x = s_0 s_1 s_2 \ldots s_{x-1}$. Using the algebraic relationship $\log_a(xy) = \log_a x + \log_a y$, we obtain:

$$\log_e l_x = \log_e s_0 + \log_e s_1 + \ldots + \log_e s_{x-1} = -k_0 - k_1 - \ldots - k_{x-1}.$$

In summation notation,

$$\log_e l_x = -\sum_{i=0}^{x-1} k_i.$$

One additional measure is sometimes included in the life table: the **expectation of further life** (e_x) of an individual of age x, which is derived in the following manner. Individuals that die between ages i and $i + 1$ live only to age i. These are fraction $l_i - l_{i+1}$ of all individuals and fraction $l_i - l_{i+1}/l_x$ of all individuals alive at age x. An individual dying between ages i and $i + 1$ survives $i - x$ time units beyond x. Now, the expectation of further life is simply the weighted average of these survival periods; that is,

$$e_x = \frac{1}{l_x} \sum_{i=x}^{\infty} (i - x)(l_i - l_{i+1}).$$

This is a life table parameter of particular interest to insurance companies. Clearly, if the expectation of death is high, so will be the premium for the life insurance policy issued. The life table variables are summarized in Table 14-4.

Life table information can be used to determine the growth of a population. You may already have surmised how this might be done. If we know the number of females in each age class and the age-specific fecundities, we should be able to calculate the number of young added to the youngest age class (by birth) in the next time interval. All we need to do then is apply the correct mortality factors to the various age classes, and we can estimate the population size and the age distribution at the time of the next reproductive event. Your anticipation of the details of this process notwithstanding, we shall postpone that discussion to Chapter 15, when we talk about population growth. For now, let us examine a real-life example of a life table.

TABLE 14-4 Summary of life table variables

l_x	Survival of newborn individuals to age x
b_x	Fecundity at age x (female offspring produced per breeding season or age interval)
m_x	Proportion of individuals of age x dying by age $x + 1$
s_x	Proportion of individuals of age x surviving to age $x + 1$
e_x	Expectation of further life of individuals of age x
k_x	$-\log_e s_x$, the exponential mortality rate between age x and $x + 1$

An Example of a Life Table: Darwin's Finches

One of the most comprehensive studies of a natural population was the one conducted by Peter and Rosemary Grant of Princeton University on the biology of Darwin's ground finches (*Geospiza*) of Isla Daphne Major, Galápagos. Beginning in 1975, the Grants carefully followed cohorts of a number of species of Darwin's finches, including *Geospiza scandens,* for over 15 years. This work is extraordinary not only for the length of time over which the Grants conducted their studies—a time equal to about twenty generations of Darwin's finches—and the wealth of superb ecological information that their work has produced, but also because it represents a fine example of evolution in action. Weiner (1994) has written an excellent popular account of the Grants' work in the Galápagos.

Isla Daphne Major is a small (about 40 ha) volcanic island in the Galápagos archipelago. Because the island is relatively isolated and difficult to reach by boat, the finch populations there have remained relatively undisturbed by human activities. Most important, the small size of the island allowed the Grants to observe the entire population of birds. Beginning in 1975, the Grants marked individual birds with plastic leg bands of various colors; unique color combinations identified individual birds. Two cohorts of *G. scandens,* one of 82 birds hatched in 1975 and one of 210 birds hatched in 1978, were studied until 1991 (Grant and Grant 1992). A life table for the 1978 cohort is presented in Table 14-5. Of the 210 birds alive at time 0, 91 (43.4%) survived their first year. Hence $s_0 = 0.434$, $m_0 = 1 - 0.434 = 0.566$, $l_x = 0.434$, and $k_0 = -\log_e(0.434) = 0.835$ per year. The life table shows a high rate of mortality for first-year birds followed by a generally increasing but highly variable rate

TABLE 14-5 Life table of the 1978 cohort of Darwin's ground finch *Geospiza scandens* on Isla Daphne Major, Galápagos

Age (x)*	Number alive	Survivorship (l_x)	Mortality rate (m_x)	Survival rate (s_x)	Expectation of life (e_x)	Exponential mortality (k_x)
0	(210)†	1.000	0.566	0.434	2.84	0.835
1	91	0.434	0.143	0.855	4.91	0.157
2	78	0.371	0.102	0.898	4.64	0.107
3	70	0.333	0.072	0.928	4.11	0.075
4	65	0.309	0.045	0.955	3.39	0.046
5	62	0.295	0.322	0.678	2.53	0.389
6	42	0.200	0.455	0.545	2.50	0.607
7	23	0.109	0.349	0.651	3.11	0.429
8	15	0.071	0.056	0.944	3.50	0.057
9	14	0.067	0.224	0.776	2.71	0.266
10	11	0.052	0.077	0.923	2.32	0.070
11	10	0.048	0.604	0.396	1.50	0.926
12	4	0.019	0.263	0.737	2.00	0.305
13	3	0.014	(0.004)	0.714	(1.50)	0.337
>14	(3)	(0.000)	(1.000)	0.000	. . .	. . .

Age in years

†Estimated

(After Grant and Grant 1992.)

of mortality. The high mortality of first-year birds is reflected in the low expectation for further life of individuals in that age class, about one-half that of birds who live to their second year. In general, the expectation of life declines with age.

The highly variable life table parameters for *G. scandens* reflect the high degree of environmental uncertainty on Isla Daphne Major. *Geospiza* are seed-eating birds, and thus their survival depends on seed production, which is highly correlated with rainfall. Over the 16-year period of the Grants' study, the annual rainfall on the island fluctuated from a high of approximately 1,300 mm yr^{-1} to a low of 0 mm yr^{-1}. Little or no reproductive activity took place in a few years having limited rainfall.

14.7 The estimation of survivorship in natural populations employs several sampling techniques.

In the *G. scandens* life table, survivorship is estimated by observing individuals from birth to a particular age—in this case, until death. Life tables based on this kind of information are referred to as **cohort** or **dynamic** life tables. This method is most readily applied to populations of plants, sessile animals, or populations of mobile animals located on small islands or in areas where dispersal is limited, as in the case of the finches on Isla Daphne Major. In such situations, marked individuals can be continually resampled over the course of their life span. Herein lies one of the disadvantages of this method: it can take a long time to collect the data, as Peter and Rosemary Grant well know. It is also difficult to apply to highly mobile animals.

One way to sidestep the problems of the cohort life table is to consider the survival of individuals of known age during a single time interval—say, one year. Thus, each age-specific survival value is estimated independently for each age class of the population during the same period. Life tables constructed in this way are referred to as **time-specific** or **static life tables.** A limitation of this method is that one must know the ages of individuals, so it is practical only when age can be estimated by growth rings, tooth wear, or some other reliable index. One also must assume that the sizes of the cohorts and the survivorship remain the same from year to year.

A life table may also be constructed from the distribution of the ages at death in a population. For

FIGURE 14-19 A group of Dall mountain sheep in Alaska. The size of the horns increases with age. (*Courtesy of the American Museum of Natural History.*)

example, we could obtain the age at death of each person buried in a local cemetery by examining the headstones, which typically give both the date of birth and the date of death, and, from those records, construct a distribution of ages at death. Using the total number of occupants of the cemetery, we could determine the number of individuals surviving at the beginning of each age interval, and from that, the survivorship. Of course, this method is fraught with potential biases. In particular, it assumes that an equal number of newborns forms the basis for each age class. This assumption is violated in expanding and declining populations, in which younger and older age classes, respectively, are overrepresented.

The distribution of ages at death was first used to construct a life table for a nonhuman population by Edward Deevey (1947), who worked from Olaus Murie's (1944) data on the ages at death of Dall mountain sheep in Denali (formerly Mount McKinley) National Park, Alaska. Murie estimated age at death from the size of the horns, which grow continuously during the lifetime of the sheep (Figure 14-19). In all, he found 608 skeletal remains: he judged 121 sheep to have been less than 1 year old at death; 7 between 1 and 2 years, 8 between 2 and 3 years, and so on, as shown in Table 14-6. Deevey reasoned that all 608 dead sheep must have been alive at birth; all but the 121 that died during their first year must have been alive at the age of 1 year ($608 - 121 = 487$); all but 128 (121 dying during the first year and 7 dying during the second) must have been alive at the age of 2 years ($608 - 128 = 480$); and so on. He built his life table in this fashion until the oldest individuals had died, during their fourteenth year.

Life tables may also be constructed from the age structure of a population at a particular time, but such tables suffer the same problems as tables constructed from age at death data. The number of individuals of age x alive at time i is simply the number of newborns x years in the past times the survivorship to age x. Hence survivorship can be estimated by

$$l_x = \frac{n_x(i)}{n_0(i - x)}.$$

If one is willing to assume that equal numbers of offspring are born each year, which may be an incorrect assumption, then

$$n_0(i - x) = n_0(i),$$

and

$$l_x = \frac{n_x(i)}{n_0(i)}.$$

TABLE 14-6 Life table for the Dall mountain sheep (*Ovis dalli*) constructed from the ages at death of 608 sheep in Denali National Park

Age interval (years)	Number dying during age interval	Number surviving at beginning of age interval	Number surviving as a fraction of newborns (l_x)
0–1	121	608	1.000
1–2	7	487	0.801
2–3	8	480	0.789
3–4	7	472	0.776
4–5	18	465	0.764
5–6	28	447	0.734
6–7	29	419	0.688
7–8	42	390	0.640
8–9	80	348	0.571
9–10	114	268	0.439
10–11	95	154	0.252
11–12	55	59	0.096
12–13	2	4	0.006
13–14	2	2	0.003
14–15	0	0	0.000

(Based on data of O. Murie, quoted by E. S. Deevey Jr., 1947.)

14.8 Four classes of population models are distinguished by discreteness of breeding episodes and overlap of generations.

In the next chapter we shall discuss the dynamics of populations by presenting theoretical models of population growth and population regulation. The bases of these models are population density and the birth and death characteristics of the populations. Empirically, we obtain estimates of density from direct counting or by employing mark-recapture techniques. Parameters of birth and death are obtained from the population life table. Thus, the life table provides the foundation upon which all population models are built.

The particular form of a population model depends on two aspects of the age structure of a population, which, taken together, result in four possible combinations. First, reproduction may be either discrete (that is, restricted to a narrow portion of each year) or continuous. Second, generations may overlap or not. The four combinations of these traits are summarized in Table 14-7, with examples of each type of population.

The dynamics of each type of population are modeled in a slightly different way. Those with nonoverlapping generations are the simplest because they may be described by only two variables: survival from birth to reproduction (s_0) and fecundity (b_x,

TABLE 14-7 Four types of population models

	Seasonal (discrete) reproduction	Continuous reproduction
Nonoverlapping generations	Annual plants and insects	Bacteria
Overlapping generations	Most seasonally reproducing vertebrates and higher plants	Humans Fruit flies

where x is the age at reproduction). With overlapping generations, population models become more complex because each age class contributes differently to the dynamics of the population, and each individual moves through progressively older age classes during its lifetime.

In populations with discrete breeding seasons, age classes correspond to episodes of life. When breeding is continuous, one may either divide the life span into age classes arbitrarily or treat birth and death as instantaneous rates that change continuously over time. While the latter option is attractive to theorists because such populations may be modeled by differential equations, it usually is more practical to gather and analyze data by age group. As we shall see, ecologists have used one or the other model as their purposes have required.

SUMMARY

1. Populations are composed of the individuals of a single species in a given area defined either by natural boundaries or arbitrarily for purposes of ecological study. Population structure includes the spatial arrangement of individuals (spatial structure), the pattern of genetic variation (genetic structure), and the dynamics of birth, death and age distribution.

2. Individuals of a species are found in particular areas, called the geographic distribution or the geographic range of the species, which include that range of ecological conditions, called the ecological range, within which the species can survive and reproduce. Local areas of suitable habitat support local populations, sometimes called subpopulations. The geographic distribution of most species is composed of arrays of local populations separated by areas of unsuitable habitat.

3. Dispersion describes the positions of individuals in relationship to those of other individuals in the population. Clumped distributions may result from independent aggregation of individuals in suitable habitats, spatial proximity of direct descendants, or tendencies to form social groups. Evenly spaced distributions may result from antagonistic interactions between individuals. Dispersion patterns are determined by comparing them with random dispersion using the Poisson distribution.

4. The density of a population is the number of individuals per unit of area. Density is more important than total population size because it is more directly connected to local ecological interactions. Density may be estimated either directly, by counts of individuals within plots of known size, or indirectly, by such methods as mark-recapture techniques.

Interactions among populations influence population density. Some populations produce an overabundance of offspring, some of which migrate to other populations. Such populations are called source populations. Populations that are sustained mainly by immigration from other populations are called sink populations.

5. The movement of individuals between populations is called dispersal. Dispersal is most often represented as the normal frequency distribution of distances moved by individuals. The standard deviation, s, of the normal distribution serves as an index of dispersal.

The movements of individuals may affect the genetic structure of populations if the exchange of individuals among populations introduces new genetic variation at a rate great enough to overcome the forces of natural selection, mutation, and genetic drift. Dispersal in plants and sessile animals is accomplished by seeds and larvae. The dispersal of these propagules is often aided by the movement of wind, water, and animals.

6. The distribution of genetic variation among individuals and among subpopulations constitutes the genetic structure of a population. Population size influences genetic variation. Small populations may lose genetic variation by chance in the process of genetic drift. Populations in areas colonized by a small number of individuals may have little genetic variation.

7. A life table sets out the fecundities (b_x) and probabilities of survival (s_x) of individuals by age class (x). These are the principal variables in models of the dynamics of populations. Survival rates may be estimated from the fates of individuals born at the same time (a cohort or dynamic life table), the survival of individuals of known age during a single time period (a static or time-specific life table), the age structure of a population at a particular time, or the age distribution of deaths.

8. Models of the age structure of populations may be divided into four classes depending on whether reproduction is continuous or occurs during discrete seasons, and whether generations overlap or not.

EXERCISES

1. Select a common plant in the area where you live and, by visiting various sites where the plant lives and other places where it is not found, describe in general the ecological range of the species. Using library resources, determine the geographic range of the species.

2. How might the distribution of a species be shaped by the evolutionary history of the species?

3. *Geolycosa rogersi* is a wolf spider that lives in burrows that it constructs in sandy soil. At the time of natal dispersal, young spiders (spiderlings) leave their mother's burrow and move off to construct small burrows of their own, enlarging them as they grow. An ecologist wants to know the extent to which burrows are clumped at a particular site. Explain how she would go about determining whether clumping exists. Consider all of the practical problems that the ecologist faces.

4. The Poisson distribution may be employed to determine whether events happen randomly in time. Suppose that a bird visits a number of feeding sites each day. The bird may visit a site more than once each day, but, on average, it visits each site just once. If we assume that the number of visits per site, x, is a Poisson random variable, what is the probability that the bird will (a) not visit a site during a day, and (b) visit a site twice during a day? (Hint: Let $x = 0, 1, 2 \ldots$)

5. The management and preservation of populations of endangered species often involves the conservation of genetic diversity in the population. Discuss the factors that an ecologist must consider in working toward such a goal.

6. Calculate the effective population size for each of the following two populations and discuss the results in terms of the conservation of the two populations.

Population feature	Population 1	Population 2
Density	10,000	10,000
Sex ratio (male : female)	1 : 1	1 : 2
k_m	1	1
k_f	2	1
V_m	1	1
V_f	1	1.5
Proportion of population that breed	0.50	0.25

7. Construct a life table from the following list of ages at death of the fifteen individuals buried in a small country cemetery: 5, 10, 30, 31, 36, 48, 49, 49, 50, 70, 71, 82, 83, 90, 91. Plot the survivorship curve of the population. (Hint: Use 10-year age intervals.)

CHAPTER 15

Population Growth

GUIDING QUESTIONS

- What is the difference between continuous-time and discrete-time population models?
- What are the exponential and geometric population growth equations?
- How does the rate of increase of a population depend on per capita birth and death rates?
- How are projections of population growth made for age-structured populations?
- What is a stable age distribution?
- How is the Lewis-Leslie matrix used to project the growth of a population?
- What is a stage-classified population, and how can its growth be projected?
- What is the characteristic equation of a population?

A universal characteristic of living things is that sexually mature individuals have the ability to produce one or more offspring. Thus, natural populations have the ability to grow. The capacity for population growth is enormous, a fact that is sometimes clearly demonstrated when species are introduced to new regions having a habitat suitable for them. In 1937, two male and six female ring-necked pheasants released on Protection Island, Washington, increased to 1,325 adults within 5 years, a 166-fold increase (Einarsen 1942, 1945). When domestic sheep were introduced into Tasmania, a large island off the coast of Australia, the population increased from fewer than 200,000 in 1820 to more than 2 million by 1850 (Davidson 1938).

But perhaps the most meaningful and dramatic example of the capacity of natural populations to grow is that of our own species. Although humans are not the most numerous of animal species, the proliferation of the human species is, by any accounting, a remarkable ecological event. A relatively minor component of the earth's fauna for much of its history, the human population began to expand rapidly after 1600, reaching 1 billion by the early 1800s, doubling that number by 1930, and doubling again by 1975. By the late 1990s, the global human population will have reached nearly 6 billion, representing the addition of 5 billion souls in fewer than 200 years. If current rates of population increase remain unchecked, we can expect the human population to double to 12 billion in as little as 50 years (well within the life span of most college students reading this book). The ecological consequences of the expansion of the human population cannot be ignored, and we shall discuss those consequences at various places in this book.

Ever since humankind began to understand the consequences of the rapid increase in its numbers, human population growth has been a cause for concern. This concern led to the development of mathematical techniques to predict the growth of populations—the discipline of **demography**, or the study of populations—and to intensive study of natural and laboratory populations to learn about mechanisms of population growth and regulation. As a result, we now have a general understanding of the causes of fluctuations in natural populations and the effects of crowding on birth and death rates.

Reproduction, then, provides a force of increase in all populations. But our casual observations of nature do not suggest to us that populations are everywhere exploding. We may notice year-to-year fluctuations in the numbers of turtles in our favorite fishing pond, but we do not expect that the turtle population will expand and take over the pond. Reproduction continues, but populations do not always grow. The strength of the force of reproduction depends on the number of individuals of reproductive age in the population and on other factors, such as the availability of resources and mates, that affect the ability of individuals to reproduce. Our goal in this chapter is to examine patterns of growth in natural populations. In Chapter 16, we shall show how other factors work to counteract growth and, as a result, provide some stability to population size.

15.1 Individuals may be added to populations by reproduction continuously or during discrete reproductive periods.

Population growth is accomplished principally by reproduction, although populations may also increase in size by immigration (and may decrease in size by emigration and death). Populations grow like interest-bearing bank accounts. Each time the account is compounded (that is, the interest is added to the principal), the amount bearing interest increases. As each baby in a population matures, the number of reproducing individuals similarly increases. For example, if a population initially having 1,000 individuals increases at a rate of 10% each year (one would feel fortunate to have a bank account that did that well), it will have 1,100 individuals at the end of the first year, 1,210 at the end of the second year, 1,331 at the end of the third year, and so on. The rate of increase of the population remains the same from year to year, but because that rate is applied to a larger number each year, the number of individuals added to the population increases from year to year.

In some bank accounts, the compounding of interest occurs annually. Some banks apply interest to accounts quarterly or semiannually, and it is possible, theoretically at least, to add interest to an account in a continuous fashion. In ecology, the periodicity with which offspring are produced results in important differences in the way in which population growth is conceptualized mathematically. When young are added to the population only at specific times of the year, during discrete reproductive periods, the population is said to have **geometric growth,** in which the increment of increase is proportional to the number of individuals in the population at the beginning of the breeding season. Geometric growth is the typical pattern of population growth. However, there are some organisms—humans are a good example—that have no distinct reproductive season, but instead add young at any time of the year. Such populations are augmented more or less continuously. This pattern of growth, referred to as exponential growth, is the one we shall take up first.

Exponential Population Growth

A population growing exponentially increases according to the equation

$$N_t = N_0 e^{rt},$$

where N_t is the number of individuals in the population after t units of time, N_0 is the initial population size ($t = 0$), and r is the **exponential growth rate.** The constant e is the base of the natural logarithms, having a value of approximately 2.72. Exponential growth results in a continuous curve of increase (or decrease, when the rt term is negative) whose slope varies in direct relation to the size of the population (Figure 15-1). The term e^r is the factor by which the population increases during each time unit, often written as the lowercase Greek lambda (λ). That is, when $t = 1$, $N_t = N_0 e^r$, or $N_t = N_0 \lambda$.

Of course, a population growing in this fashion would quickly climb toward infinity, something that Charles Darwin appreciated when he wrote in *On the Origin of Species*: "There is no exception to the rule that every organic being naturally increases at so high a rate, that, if not destroyed, the earth would soon be covered by the progeny of a single pair." To make his case as forcefully as possible, Darwin offered a conservative example:

> The elephant is reckoned the slowest breeder of all known animals, and I have taken some pains to estimate its probable minimum rate of natural increase; it will be safest to assume that it begins breeding when thirty years old, and goes on breeding till ninety years

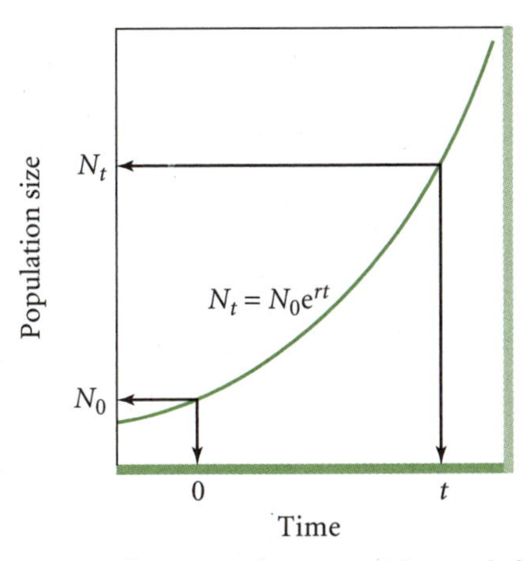

FIGURE 15-1 The curve of exponential growth for a population growing at rate r between time 0 and time t. During this period the number of individuals increases from N_0 to N_t.

> old, bringing forth six young in the interval, and surviving till one hundred years old; if this be so, after a period of from 740 to 750 years there would be nearly nineteen million elephants alive, descended from the first pair.

Let us apply the equation for exponential growth given above to Darwin's elephant problem. Although he omitted the details of his calculations, we can reasonably estimate the variables needed to solve the equation. We refer to the amount of time between the birth of a parent and the birth of the parent's offspring as the **generation time.** The generation time will be different for each individual parent, so a more useful concept is the average generation time. Using Darwin's calculations, a reasonable average generation time for elephants would be 60 years. Thus, in the equation above, the time in years from $t = 0$ to $t = 1$ is 60 years. Of the six offspring produced by one female during 60 years, three would probably be female. Hence each female would replace herself with three female progeny ($e^r = 3$) at an average generation time of 60 years ($t = 1$). Darwin's 750 years are 12.5 generations ($t = 12.5$), and we assume a starting population N_0 of 1 female (with a male, of course, to provide the passion between the sexes). Thus evaluated, the equation above is $N_{8.3} = 1 \times 3^{12.5} = 926{,}483$. Even doubled to account for the males produced, this number falls well short of Darwin's calculation.

Consider, however, that baby elephants born to a mother early in her life might mature and produce babies even while their mother is still in her reproductive years. As we shall see below, this overlapping of generations in the population effectively reduces the generation time. For elephants, a generation time of 51 years is short enough for the progeny of a single pair to swell to over 19 million in 750 years. Regardless of the details, however, the important point is simply that elephants produce more elephants at roughly an exponential rate.

Rate of Exponential Population Increase

The rate of increase of a population undergoing exponential growth at a particular instant in time—the **instantaneous rate of increase**—is the differential of the exponential equation; that is,

$$\frac{dN}{dt} = rN.$$

This equation encompasses two principles. First, the exponential growth rate (r) expresses population increase (or decrease) on a per individual basis. Second, the rate of increase (dN/dt) varies in direct proportion to the size of the population (N). In words, the equation could read: (the rate of change in population size) = (the contribution of each individual to population growth) × (the number of individuals in the population).

Per Capita Birth and Death Rates

The individual, or **per capita,** contribution to population growth is the difference between the birth rate (b) and the death rate (d) calculated on a per capita basis. Rates of birth and death are abstractions that have little meaning for the individual. An elephant dies only once, so it can't have a rate of death. Babies are produced in discrete litters separated by intervals required for gestation, not at constant rates, such as 0.05 offspring per day. But when births and deaths are averaged over a population as a whole, they take on meaning as rates of demographic events in the population. If 1,000 individuals were to produce 10,000 progeny in a year, it would be reasonable to assign a per capita birth rate of 10 per year, and to assume that a population of 1 million would produce 10 million progeny—still 10 per individual—under the same conditions. Of the groundhogs alive on their day in one year, if only half survived to February 2 of the next year, we would ascribe a death rate of 50% per year to the population, even though some of the groundhogs had died completely and others hadn't died at all.

A groundhog might become very nervous on learning that mortality rates in populations can exceed 100% annually. Of course, no individual dies twice. But the probability of death during a short interval—say, a day—may be high enough that most of the individuals in a population pass on during a period considerably shorter than a year. With a probability of death of 1% per day (0.01 day^{-1})—365% on an annual basis—99% of individuals would survive each day. Those surviving both today and tomorrow would constitute the fraction $0.99 \times 0.99 = 0.99^2 = 0.98$ of the initial population. Those surviving a full year would constitute only $0.99^{365} = 0.0255$ (2.55%) of the original population.

When death occurs at an instantaneous per capita rate $-d$, the number of individuals in a population in which there are no births, changes over time according to $N_t = N_0e^{-dt}$, and $dN/dt = -dN$. The minus sign must be included because death removes individuals from the population. A pure birth process (no deaths) adds individuals to the population at rate $dN/dt = bN$. Combining birth and death in the same equation,

$$\frac{dN}{dt} = bN - dN = (b - d)N.$$

Furthermore, because r is the difference between birth rate and death rate ($b - d$), $dN/dt = rN$ (see above).

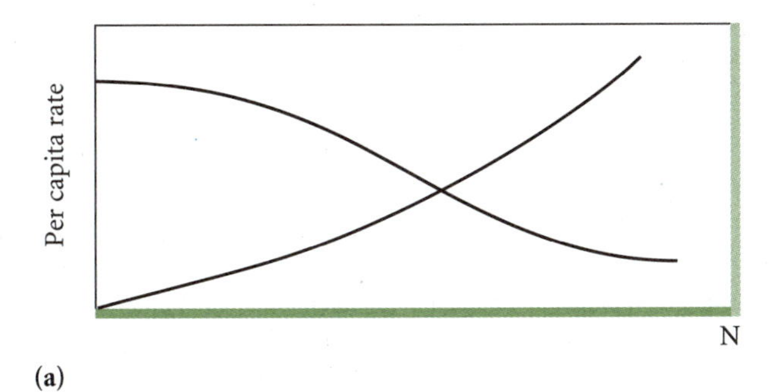

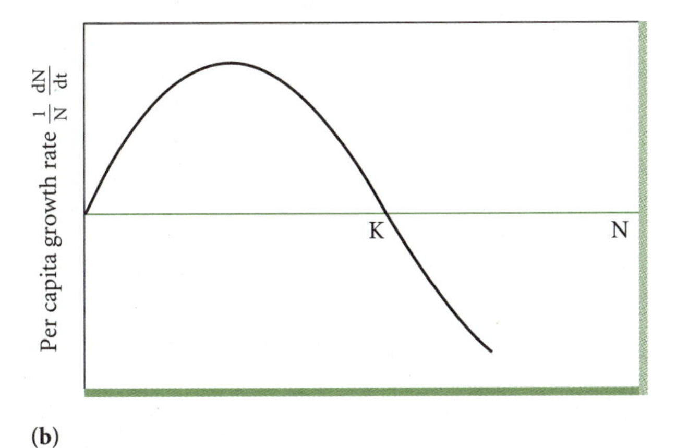

FIGURE 15-2 (a) Hypothetical instantaneous per capita birth $[b(N)]$ and death $[d(N)]$ functions of population density N. (b) When the death rate is subtracted from the birth rate $[b(N) - d(N)]$ for all levels of N, the result is the rate of per capita population increase, which is positive (curve is above the horizontal axis) when $b(N) > d(N)$ and negative when $b(N) < d(N)$. When $b(N) = d(N)$, the rate of increase, $1/N\ dN/dt$, $= 0$, denoted by K. *(After Yodzis 1989.)*

When the death rate exceeds the birth rate, r is negative, and the population declines.

To see more clearly how the rate of increase of a population, dN/dt, is related to per capita birth and death rates, consider a graphic representation of $(b - d)$ in which the birth rate declines and the death rate increases as the size of the population increases (Figure 15-2). We will discuss the ecological implications of such birth and death patterns later, but for now, think of a population in which crowding influences birth rates negatively and death rates positively. By subtracting the birth and death curves in Figure 15-2a, we obtain Figure 15-2b, in which the rate of change increases rapidly at first, levels off at the point where $b - d = 0$—a point that we designate as K—and then becomes negative for $N > K$. An appreciation for how dN/dt changes with respect to N is important to understanding population regulation, population cycles, and extinction, as we shall see in Chapters 16 and 18.

15.2 Geometric equations describe discrete time growth processes.

The exponential growth model presented above is an appropriate representation of the growth of populations in which young are added to the population continuously, as, for example, in the human population. But in cases in which young are added to the population only at specific times of the year, during discrete reproductive periods, the exponential model may not be the best representation of population growth. (It should be noted here that death is more likely to be a continuous process than is birth.)

In these cases, population growth is best represented by a **geometric growth model.** Returning to the idea of populations growing like interest-bearing bank accounts, an account might be compounded annually, quarterly, daily, or perhaps continuously, according to a particular bank's practices. Periodic compounding is an example of geometric growth, with increases posted at discrete intervals. The last case, that of continuous compounding, represents exponential growth.

The human population grows continuously because babies are born and added to the population at all seasons of the year. This situation is unusual in natural populations, most of which restrict reproduction to a particular time of year. The population grows during the breeding season, and then usually declines between one breeding season and the next because of mortality and emigration (Figure 15-3). In the case of the California quail, for example, the number of individuals doubles or triples each summer as adults produce their broods of chicks, but then dwindles by nearly the same amount during the fall, winter, and spring. Within each year, the population growth rate varies tremendously with seasonal changes in the balance of birth and death processes.

Growth in populations with discrete breeding seasons is treated somewhat differently than growth

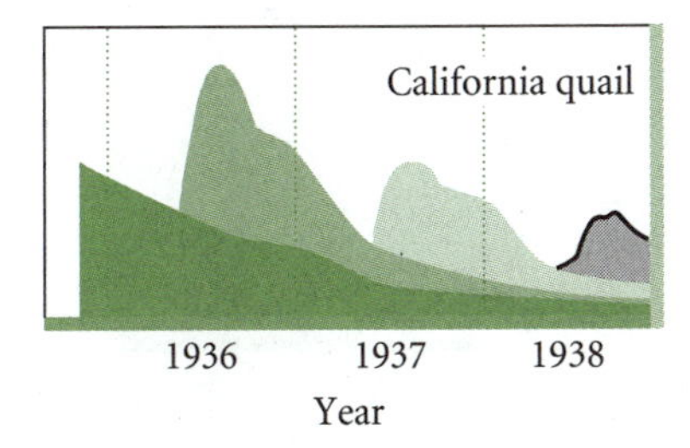

FIGURE 15-3 Growth of a population of the California quail, which has discrete (seasonal) reproduction. The curves are shaded differently to show that each curve represents the population of individuals hatched in a different year. The population declines between one breeding season and the next. *(After Emlen 1940.)*

in continuously breeding populations. The size of populations with discrete breeding seasons must be measured at a particular time of year to have any meaning. If one were interested in projecting population growth, it would be pointless to compare numbers in August, recently augmented by the chicks of the year, with numbers in May, after the winter has taken its toll. One must count individuals at the same time each year, with each count separated by the same cycle of birth and death processes.

When populations are augmented periodically, growth rate is most conveniently expressed as the ratio of the population in one year [for example, $N(1)$] to that in the preceding year [$N(0)$ in this case]. (It is customary to use subscripts to represent time increments in exponential models [e.g., N_t] and parentheses [e.g., $N(t)$] in geometric and other discrete time models.) Demographers have assigned the symbol λ to this ratio, called the **geometric growth rate.** Hence $\lambda = N(1)/N(0)$. Take a moment to compare this definition of λ with the one we presented above ($\lambda = e^r$) in our discussion of exponential population growth. In both cases, λ represents a factor of population increase.

This equation for λ can be rearranged to provide a formula for projecting the size of the population through time,

$$N(t + 1) = N(1)\lambda.$$

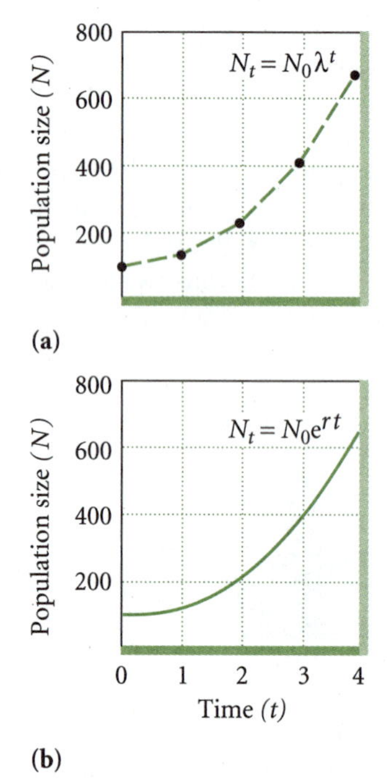

FIGURE 15-4 Increase in the number of individuals in populations undergoing (a) geometric growth and (b) exponential growth at equivalent rates ($\lambda = 1.6$, $r = 0.47$).

To project the growth of a population, the original number, $N(0)$, is multiplied by the geometric growth rate λ once for each unit of time passed. Hence,

$$N(t) = N(0)\lambda^t.$$

Note that this equation for geometric growth is identical to that for exponential growth ($N^t = N_0\ e^{rt}$) except that λ, the geometric growth rate, takes the place of e^r, the amount of exponential growth accomplished in one time period. Because of this relationship, curves depicting the two models of growth can be superimposed upon each other (Figure 15-4), and there is a direct correspondence between values of λ and r. When a population's size remains constant, $r = 0$ and $\lambda = 1$ ($r = \log_e\lambda$ and $\log_e 1 = 0$). Decreasing populations have negative exponential growth rates and geometric growth rates less than 1 (but greater than 0; a real population cannot have a negative number of individuals). Increasing populations have positive exponential growth rates and geometric growth rates greater than 1.

15.3 A population's growth rate depends upon the proportion of individuals in each age class.

In both the exponential and geometric growth models above, birth and death rates were assumed to be uniform over all members of the population. That is, we assumed that an old individual has no more chance of dying than a young individual. In order to determine the size of the population in the future under such an assumption, one need only base the calculation on the total population size, N. But when birth and death rates vary with respect to age, the contributions of younger and older individuals to population growth must be figured separately. The proportions of individuals in each age class in a population is referred to as the population's **age structure.**

Two populations with identical birth and death rates at corresponding ages, but with different age structures, will grow at different rates. A population composed wholly of prereproductive adolescents and adults too old to breed will not increase at all in the near future. While this example may be greatly exaggerated, smaller variations in age distribution can have a profound influence on population growth rate and other population processes (Charlesworth 1994).

Age Structure in Organisms with Overlapping Generations

The growth rate of an age-structured population is the sum of the growth rates owing to individuals in each age class weighted by their proportion in the population. For exponential growth,

$$\frac{dN}{dt} = N \int_{x=0}^{\infty} r_x c_x dx,$$

where r_x is the exponential growth rate attributable to age x and c_x is the proportion of individuals at age x. The utility of this formula is limited because one cannot determine birth and death rates at every age, since an individual remains a particular age only for an instant. As we mentioned earlier, demographers get around this problem by dividing populations into age classes, to which they can assign probabilities of giving birth and dying. For example, all humans who are in their twentieth year would be assigned to age class 20, even though those born in January would be nearly a year older than those born in December. In a human population, then, c_{20} would be the proportion of individuals in the population who are in their twentieth year. In effect, this practice represents continuous breeding as a discrete process. Instead of weighting the contribution of every age, the weightings correspond to groups of ages, such as all the individuals in their twentieth year.

Growth in Age-Structured Populations with Discrete Breeding Seasons

Consider a population having three age classes that produces young in a single burst each year. At the present time, t, each age class, x, contains $n_x(t)$ individuals, giving us $n_0(t)$, $n_1(t)$, and $n_2(t)$ individuals in the population. Those in the $x = 0$ age class are the youngest and those in the $x = 2$ age class are the oldest. Recall from our discussion of life tables in Chapter 14 that we designate the youngest age class $x = 0$, but we say that the youngest age class is the *first* age class. So, in this example, we will talk about the first ($x = 0$), second ($x = 1$), and third ($x = 2$) age classes. Let us consider how such a population would grow by examining what happens between the present time (t) and the next breeding time ($t + 1$).

Two processes work to determine the pattern of population growth: reproduction and maturation (Figure 15-5). Reproduction adds individuals to the first age class, and maturation moves individuals from one age class to the next (for example, from the second to the third). Reproduction takes place during a particular breeding season. In our example, there are two reproductive events, one in the present time (t) and one in the next time ($t + 1$). Assume that each individual of age class x contributes b_x newborns to the population. During a breeding season, then, the number of newborns contributed by individuals of age class x is the number of individuals in that age class multiplied by the number of newborns contributed by each individual in that age class. For the present time (t) in our example, individuals of the second age class will contribute $n_1(t)b_1$ and individuals of the third age class $n_2(t)b_2$ newborns to the population, and their total contribution will be the sum of those contributions (see Figure 15-5):

$$n_0(t) = n_1(t)b_1 + n_2(t)b_2.$$

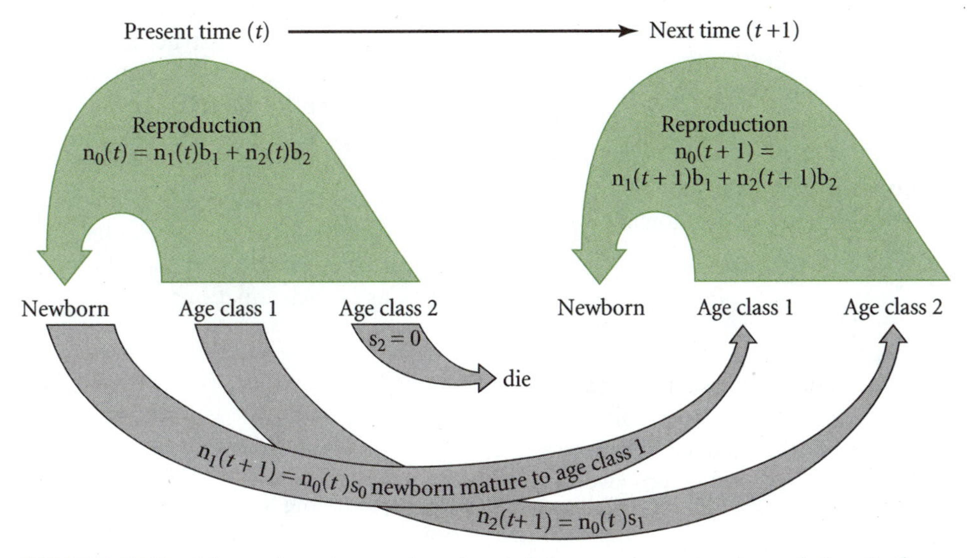

FIGURE 15-5 Schematic representation of growth in an age-structured population. In the present time (t), reproduction by age classes 1 and 2 add newborns to the population. Between the present time and the next reproductive time ($t + 1$), some of the individuals of age classes 0 and 1 mature into the next age class. All the members of age class 2 die. A second reproductive event at time ($t + 1$) adds newborns to the population. The parameter $n_x(t)$ represents the number of individuals in age class x at time t. The parameter s_x represents the proportion of age class x surviving to the next age class.

After the breeding period is concluded, there will be individuals of all three age classes in the population. Some of these individuals will mature to become members of the next age class. So, some of the individuals of age class $x = 0$ at time t will grow up to become members of age class $x = 1$ at time $t + 1$. Others will die in the time between one breeding season and the next, and the factors affecting mortality may be different in each age class. For example, it may be that survival of individuals in the first age class is extremely low, as in the case of sea turtles, in which most of the many hatchlings die during their scramble from nesting sites on the beach to the open ocean. In our example, we stipulate that all individuals in the third age class will die before the next breeding season. Let s_x be the proportion of individuals in age class x surviving from one breeding period to the next. The number of individuals of age class x that make it to the next age class, then, is the number in the next youngest age class, $x - 1$, multiplied by the proportion of individuals that survive, s_x, or $n_x(t + 1) = n_{x-1}s_{x-1}$. The specific equations for our example are:

$$n_1(t + 1) = n_0(t)s_0,$$
$$n_2(t + 1) = n_1(t)s_1,$$
$$n_3(t + 1) = n_2(t)s_2.$$

The first equation in the set gives the number of individuals surviving from the first ($x = 0$) age class to the second ($x = 1$) age class; the second equation gives the number surviving from the second to the third ($x = 2$) age class; and the third shows the number of age class 3 ($x = 2$) individuals making it to age class 4 ($x = 3$). Of course, we have already said that $s_2 = 0$, so $n_3(t + 1) = 0$; no one grows up to become a member of the fourth age class (see Figure 15-5).

When the next breeding season comes, the population will have $n_1(t + 1)$ and $n_2(t + 1)$ individuals in breeding condition, and they will contribute newborns to the $t + 1$ population in the following way:

$$n_0(t + 1) = n_1(t + 1)b_1 + n_2(t + 1)b_2,$$

which is identical to the way in which newborns were added in the previous time (t).

Some efficiency in the algebraic representation of the growth of age-structured populations can be obtained by recognizing the general relationship $n_x(t + 1) = n_{x-1}(t)s_{x-1}$. Thus, wherever we find $n_x(t + 1)$ in the equations above, we may substitute $n_{x-1}(t)s_{x-1}$. The equation showing reproduction in the time $t + 1$ can therefore be rewritten in the following way:

$$n_0(t + 1) = n_0(t)s_0b_0 + n_1(t)s_1b_1 + n_2(t)s_2b_2,$$

where a term for the youngest age class $[n_0(t)s_0b_0]$ is included for completeness (that age class would contribute no newborns because $b_0 = 0$). The advantage of this formulation is that the total size of a population can be represented as equal to the sum of its age classes: in summation notation, $N(t + 1) = \Sigma n_x(t + 1)$. Thus we can express the population in year $t + 1$ in terms of the population during year x as two sums, one of the survival terms and the other of the birth terms,

$$n(t + 1) = \sum_{x=0}^{\infty} n_x(t)s_x + \sum_{x=0}^{\infty} n_x(t)s_xb_{x+1}$$
$$= \sum_{x=0}^{\infty} n_x(t)s_x(1 + b_{x+1}). \quad (15\text{-}1)$$

The equation above allows us to project the growth of a population through time. As it stands, however, it merely formalizes the pencil-and-paper figuring that Darwin used to calculate the increase in elephants. The size of each age class must be calculated from the size of the previous age class during the previous year to determine the total population and its growth from year to year. But, as we shall see below, populations growing for long periods with constant values of b and s assume a characteristic age distribution and a constant rate of growth, which can be calculated directly from the life table.

The Lewis-Leslie Matrix Model of Population Growth

Mathematicians use the techniques of matrix algebra to organize and analyze sets of equations. These techniques have proved extremely useful in ecology as well—in particular, as a means of efficiently projecting the growth of age-structured and, as we shall see below, stage-classified populations (Caswell 1989). The **Lewis-Leslie matrix** (Lewis 1942, Leslie 1945, 1948) is an example of a simple matrix model for population projection.

Below is the system of equations for projecting population growth just presented, but written in a spatial arrangement that aligns the corresponding portions of each equation:

$$\begin{aligned} n_0(t + 1) &= n_0(t)s_0b_1 + n_1(t)s_1b_2 + n_2(t)s_2b_3 \\ n_1(t + 1) &= n_0(t)s_0 + 0 + 0 \\ n_2(t + 1) &= 0 + n_1(t)s_1 + 0 \end{aligned}$$

We may augment the zero terms in the above equations in the following way to show that the equations belong to the same set:

$$\begin{aligned} n_0(t + 1) &= n_0(t)s_0b_1 + n_1(t)s_1b_2 + n_2(t)s_2b_3 \\ n_1(t + 1) &= n_0(t)s_0 + n_1(t) \times 0 + n_2(t) \times 0 \\ n_2(t + 1) &= n_0(t) \times 0 + n_1(t)s_1 + n_2(t) \times 0 \end{aligned}$$

Matrix algebra provides a convenient way to summarize the operations depicted in these equations. To see how, we must first understand a few basic matrix terms and operations. A matrix is a rectangular array of numbers (which may be represented by letters) having r rows and c columns ($r \times c$ is said to be the order of the matrix). Some matrices have only a single row or a single column; these are called row and column vectors respectively. A vector is just a matrix with either $r = 1$ or $c = 1$. Matrices (and vectors) are denoted in print by boldface letters, with vectors often designated in lowercase. Following is a 2×2 matrix (**A**), a 1×3 row vector (**a**), and a 3×1 a column vector (**b**):

$$\mathbf{A} = \begin{bmatrix} 1 & 2 \\ 3 & 4 \end{bmatrix} \quad \mathbf{a} = [\mathrm{a} \quad \mathrm{b} \quad \mathrm{c}] \quad \mathbf{b} = \begin{bmatrix} 2 \\ 3 \\ 4 \end{bmatrix}$$

The utility of matrix algebra comes from the concise way in which operations such as addition, subtraction, multiplication, and division are represented. For example, to add or subtract two matrices, we simply add or subtract the corresponding elements of each matrix, like this:

$$\begin{bmatrix} 1 & 2 \\ 3 & 4 \end{bmatrix} + \begin{bmatrix} 6 & 7 \\ 8 & 9 \end{bmatrix} = \begin{bmatrix} 1+6 & 2+7 \\ 3+8 & 4+9 \end{bmatrix} = \begin{bmatrix} 7 & 9 \\ 11 & 13 \end{bmatrix}$$

The matrices have to be of the same order for this to work.

It is the multiplication of matrices that is most important for our purposes, and that is a bit more complicated for a number of reasons. First, in matrix algebra, the familiar property $xy = yx$ does not hold in every case. Thus, **AB** may not be equal to **BA**. To keep the order of the multiplication straight, we say that in the product **AB**, the matrix **B** is pre-multiplied by the matrix **A** (or we could say that **A** is post-multiplied by **B**). Second, for two matrices to be multiplied, the number of *columns* of the first matrix must equal the number of *rows* in the second matrix. So, the following multiplications may be performed:

$$\underset{1 \times 3}{[1 \quad 2 \quad 3]} \times \underset{3 \times 1}{\begin{bmatrix} 4 \\ 5 \\ 6 \end{bmatrix}} \qquad \underset{3 \times 1}{\begin{bmatrix} 1 \\ 2 \\ 3 \end{bmatrix}} \times \underset{3 \times 3}{\begin{bmatrix} 1 & 2 & 3 \\ 4 & 5 & 6 \\ 7 & 8 & 9 \end{bmatrix}}$$

but the following could not be:

$$\underset{3 \times 1}{\begin{bmatrix} 1 \\ 2 \\ 3 \end{bmatrix}} \times \underset{3 \times 1}{\begin{bmatrix} 4 \\ 5 \\ 6 \end{bmatrix}} \qquad \underset{1 \times 3}{[1 \quad 2 \quad 3]} \times \underset{3 \times 3}{\begin{bmatrix} 1 & 2 & 3 \\ 4 & 5 & 6 \\ 7 & 8 & 9 \end{bmatrix}}$$

The order of the matrix that results from a multiplication of two matrices is the number of columns of the first matrix and the number of rows of the second, as shown below:

$$\underset{3 \times 1}{\begin{bmatrix} 1 \\ 2 \\ 3 \end{bmatrix}} \underset{3 \times 3}{\begin{bmatrix} 1 & 2 & 3 \\ 4 & 5 & 6 \\ 7 & 8 & 9 \end{bmatrix}} = \begin{bmatrix} a \\ b \\ c \end{bmatrix}$$

Now, how do we find the elements of the 3×1 matrix resulting from the above multiplication? That is, how, precisely, is the multiplication accomplished? The multiplication results from a series of steps in which each element of the column vector is multiplied with the corresponding element of a row of the matrix and those results are summed. This operation is repeated for each of the three rows of the matrix to obtain each element of the new matrix. For example, to get the first element, a, of the matrix above, we first multiply the first element of the column vector, 1, by the first member of the first row of the matrix, 1. Then we multiply the second element of the vector, 2, by the second element of the first row of the matrix, 2, to obtain $2 \times 2 = 4$. This result is added to the first multiplication. Finally, we multiply the third, last, element of the vector by the third element of the first row of the matrix, 3, to obtain $3 \times 3 = 9$, and this is added to the results of the other two operations. Thus, the operations performed to obtain the first element of the vector that results from the multiplication are summarized as: $(1 \times 1) + (2 \times 2) + (3 \times 3) = 1 + 4 + 9 = 14$. The complete set of operations is shown below:

$$\begin{bmatrix} 1 \\ 2 \\ 3 \end{bmatrix} \begin{bmatrix} 1 & 2 & 3 \\ 4 & 5 & 6 \\ 7 & 8 & 9 \end{bmatrix} = \begin{bmatrix} (1)(1) + (2)(3) + (3)(3) \\ (1)(4) + (2)(5) + (3)(6) \\ (1)(7) + (2)(8) + (3)(9) \end{bmatrix} = \begin{bmatrix} 14 \\ 32 \\ 50 \end{bmatrix}$$

Now, let us return to the equations for the age-structured population given above. Upon examination, we see that each term of the right-hand side of the equations is multiplied by the same quantity. Thus, we may write the right-hand side of the equations as the product of a 3×1 column vector, denoted $\mathbf{n}(t)$, containing the number of individuals of each age class, and a 3×3 matrix, denoted **A**, having combinations of age-specific survival and birth rates in some positions and 0 in other positions. The result of the multiplication of vector $\mathbf{n}(t)$ and matrix **A** is a 3×1 column vector designated $\mathbf{n}(t + 1)$, shown below with dashes to indicate the three elements of the vector:

$$\begin{bmatrix} s_0b_1 & s_1b_2 & s_2b_3 \\ s_0 & 0 & 0 \\ 0 & s_1 & 0 \end{bmatrix} \times \begin{bmatrix} n_0(t) \\ n_1(t) \\ n_2(t) \end{bmatrix} = \begin{bmatrix} - \\ - \\ - \end{bmatrix}$$

$$\mathbf{A} \times \mathbf{n}(0) = \mathbf{n}(t=1) \quad (15\text{-}2)$$

If we perform the multiplication using the method described above, we can easily reconstruct the original equations. For example, to obtain the first equation in the series, multiply each element of the column vector with the corresponding elements of the first row of the matrix, and sum them to obtain

$$n_0(t)s_0b_1 + n_1(t)s_1b_2 + n_2(t)s_2b_3.$$

The square matrix, **A**, of coefficients on the left-hand side of equation (15-2) is called the Lewis-Leslie matrix. Often the terms s_0b_1, s_1b_2 . . . are replaced by f_0, f_1 . . . Each time **A** is multiplied by $\mathbf{n}(t)$, the number of individuals in the next generation, $\mathbf{n}(t+1)$ is obtained. If $\mathbf{n}(t+1)$ is then multiplied by **A**, $\mathbf{n}(t+2)$ is obtained. The same thing may be accomplished by multiplying the initial population size $\mathbf{n}(0)$ by $\mathbf{A}^t$, where t is the number of time periods into the future that the population is to be projected. Thus, the projection of the population from time t to time $t + 1$ can be expressed by the matrix multiplication

$$\mathbf{A}^t\mathbf{n}(0) = \mathbf{n}(t).$$

For a population with four age classes $x = 0, 1, 2, 3$, the terms f_x are 0.5, 2.4, 1.0, and 0. Starting with a population structure $\mathbf{n}_x(t)$ of 20, 10, 40, and 30 individuals, the matrix multiplication for the first unit of time, t to $t + 1$, is

$$\begin{bmatrix} 0.5 & 2.4 & 1.0 & 0 \\ 0.5 & 0 & 0 & 0 \\ 0 & 0.8 & 0 & 0 \\ 0 & 0 & 0.5 & 0 \end{bmatrix} \times \begin{bmatrix} 20 \\ 10 \\ 40 \\ 30 \end{bmatrix} = \begin{bmatrix} 74 \\ 10 \\ 8 \\ 20 \end{bmatrix}$$

$$\mathbf{A} \times \mathbf{n}(t) = \mathbf{n}(t=1)$$

The last column vector represents the number in each of the four age classes at time $t + 1$.

15.4 Matrix models provide a method for the analysis of growth in stage-classified and subdivided populations.

As part of their normal development, many organisms go through a series of different stages. An individual plant may occur successively as a seed, a vegetatively reproducing rosette, and a flowering plant. Salamanders and frogs have aquatic larvae and terrestrial adult stages. Individuals at different developmental stages may be found in the same habitat or, as in the case of salamanders, in different habitats. We may think of such populations as **stage-classified.** Each stage may have a different probability of producing young, dying, or moving to another location. Matrix models provide a method for projecting the growth of such populations. We will demonstrate this with a simple example (Caswell 1989).

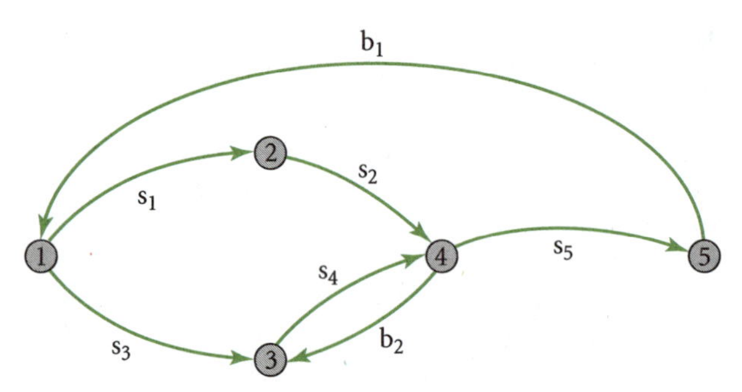

FIGURE 15-6 Life cycle graph of an organism having both vegetative and sexual reproduction. Stage 5 individuals reproduce sexually to produce stage 1 individuals. Stage 4 individuals reproduce asexually to produce stage 5 forms or sexually to produce stage 3 forms. A population of such organisms is referred to as a stage-classified population. Matrix models are used to project the growth of such populations.

Consider the situation in which an organism displays both vegetative and sexual reproduction, as depicted in the **life cycle graph** in Figure 15-6. Individuals in stage 5 reproduce sexually to produce stage 1 individuals at a rate b_1. Individuals of stage 1 may develop either into stage 2 individuals (with probability s_1) or into stage 3 individuals (s_3). Likewise, stage 2 and stage 3 individuals may grow into stage 4 individuals with probabilities s_2 and s_4 respectively. Stage 4 individuals may grow into sexually reproducing stage 5 forms (s_5), or they may reproduce vegetatively at a rate b_2 to produce more stage 3 forms. Notice that this model is most appropriately referred to as a stage-classified model rather than an age-classified model because stages 2 and 3 could be the same age.

The Lewis-Leslie matrix for such a population would be

$$\begin{bmatrix} 0 & 0 & 0 & 0 & b_1 \\ s_1 & 0 & 0 & 0 & 0 \\ s_3 & 0 & 0 & b_2 & 0 \\ 0 & s_2 & s_4 & 0 & 0 \\ 0 & 0 & 0 & s_5 & 0 \end{bmatrix}$$

Were one to obtain estimates of the number of individuals in each stage in the population, a projection of population growth could be made in the way described in Section 15.3.

15.5 A population with a fixed life table assumes a stable age distribution and grows at a constant rate.

When age-specific birth, b_x, and survival, s_x, rates remain unchanged for a sufficient length of time, a population will assume a **stable age distribution.** Under such conditions, each age class in the population grows or declines at the same rate, and so, therefore, does the total size of the population. This was first demonstrated by A. J. Lotka in 1922.

Imagine a population of 100 individuals having the characteristics depicted in Table 15-1. Because all individuals of age class 4 ($x = 3$) die, there are no individuals of age class 5 ($x = 4$) in the population; newborns have a fecundity of zero ($b_0 = 0$), as seems biologically reasonable. Applying the equations of age-specific growth [equation (15-1)] to the number of individuals in each age class at year 0, and then to the numbers at year 1, and so on gives a projection of the population through the years, as shown in Table 15-1. The geometric growth rate, $\lambda = N(t + 1)/N(t)$, is appropriate for this population since we assume discrete reproduction.

In this example, the growth rate of the population is at first very erratic, fluctuating between a low of $\lambda = 1.05$ after year 1 and a high of $\lambda = 1.69$ after year 2 before settling down to a constant value of 1.49. At this point, the population has achieved a stable age distribution; the percentage of individuals in each age class at $t = 8$ is shown in the right-hand column of Table 15-1. Even by the end of the fourth year, the population approached its stable age distribution, with proportions in each of the age classes of 62.7, 21.8, 10.0, and 5.4%. Under stable age distribution conditions, each age class grows at the same rate from year to year (Figure 15-7).

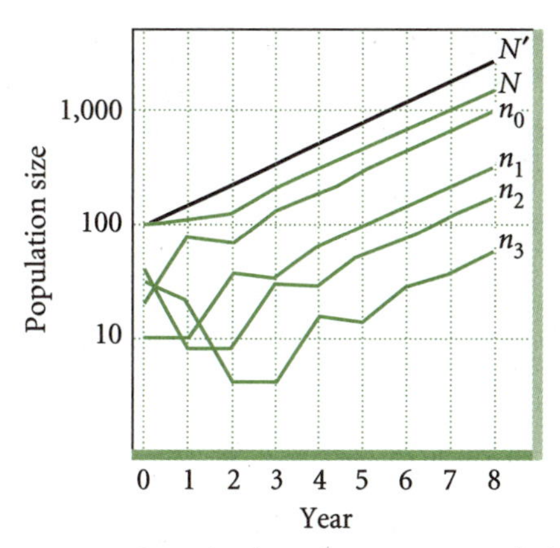

FIGURE 15-7 Growth of age classes as a population achieves its stable age distribution. Notice that each age class eventually grows at the same rate.

Had we initiated this population with a stable age distribution, its growth rate (λ) would have been constant from the beginning, as shown in the example given in Table 15-2. (The small variations in growth rate in the earlier years are due to rounding individuals to whole numbers when the population is small.) As you can see, this population grows at the same rate ($\lambda = 1.49$) as that achieved in the first example after several rounds of birth and death, and its age structure also remains unchanged.

A stable age distribution and growth rate in a population depend upon birth and survival values. Any change in the age-specific birth or mortality rates alters the stable age distribution and results in a new rate of population growth. Consider the example given in Table 15-3, in which the survival and fecundity of our imagined population are reduced. We start with a population represented by the last column of Table 15-2 ($n_0 = 1{,}491$, $n_1 = 501$, $n_2 = 270$, $n_3 = 90$). Because of the change in the survival and birth rates,

TABLE 15-1 An imaginary population with four age classes ($x = 0, 1, 2, 3$), each with survival probability s_x, birth rate b_x, and initial population size $N(0)$, projected through 9 years using equation (15-1)

Age class	s_x	b_x	Year 0	1	2	3	4	5	6	7	8	% of column 8
n_0	0.5	0	20	74	69	132	175	274	399	599	889	63.4
n_1	0.8	1	10	10	37	34	61	87	137	199	299	21.3
n_2	0.5	3	40	8	8	30	28	53	70	110	160	11.4
n_3	0.0	2	30	20	4	4	15	14	26	35	55	3.9
N			100	112	118	200	279	428	632	943	1,403	
λ			1.12	1.05	1.69	1.40	1.53	1.48	1.49	1.49		

(λ values are printed between successive year columns.)

TABLE 15-2 An imaginary population with the same properties as those in Table 15-1, but starting with a stable age distribution, projected through 9 years using equation (15-1)

	Year									% of column 8
	0	1	2	3	4	5	6	7	8	
n_0	63	94	138	207	306	453	674	1,002	1,491	(63.4)
n_1	21	31	47	69	103	153	226	337	501	(21.3)
n_2	12	17	25	38	55	82	122	181	270	(11.5)
n_3	4	6	8	12	19	27	41	61	90	(3.8)
N	100	148	218	326	483	715	1,063	1,581	2,352	
λ		1.48	1.47	1.50	1.48	1.48	1.49	1.49	1.49	

births no longer exceed deaths, and the population declines. As often happens in a declining population, the distribution of individuals shifts toward the older age classes. In this example, the shift to the new stable age distribution involves small changes and occurs quickly, after which the population achieves a growth rate of $\lambda = 0.82$. Early on, however, some age classes briefly increase (e.g., n_2 between $t = 0$ and $t = 1$) as the age distribution of the population readjusts to the new survival and fecundity values. The effects of population growth rate on age structure stand out in comparisons of human populations with stable and with growing populations (Figure 15-8). Rapid growth leads to a bottom-heavy age structure with large proportions of young individuals.

15.6 The intrinsic rate of increase of a population is determined by its life table values.

The **intrinsic rate of increase** is the exponential or geometrical rate of increase assumed by a population with a stable age distribution. Recall from Chapter 14 that life tables are a way to reveal patterns for birth and death rates in a population. The birth and death rates, and other life table measures such as survivorship, are affected by the conditions of the environment. When changing environmental conditions alter life table values, the age structure of the population continually readjusts to the new schedule of birth and death rates. In practice, therefore, populations rarely

TABLE 15-3 An imaginary population with the same properties as those in Table 15-1, projected through 9 years using equation (15-1), with a starting population size corresponding to the stable age distribution of the eighth year of Table 15-2 and with survival probabilities and birth rates reduced

			Year									% of column 8
Age class	s_x	b_x	0	1	2	3	4	5	6	7	8	
n_0	0.3	0	1,491	1,130	965	776	642	525	431	352	286	57.7
n_1	0.6	1	501	447	339	289	233	193	157	129	106	21.4
n_2	0.3	2	270	301	268	203	174	140	116	94	76	15.3
n_3	0.0	1	90	81	90	81	61	52	42	35	28	5.7
N			2,352	1,959	1,662	1,349	1,110	910	746	610	496	
λ				0.83	0.85	0.81	0.82	0.82	0.82	0.82	0.81	

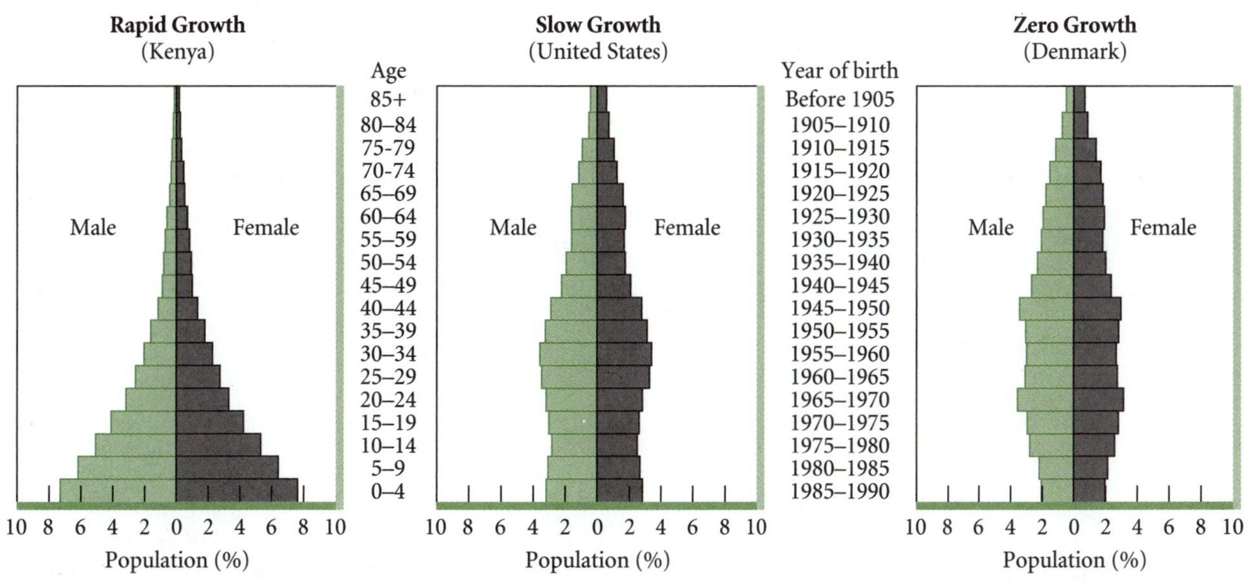

FIGURE 15-8 Age distributions for human populations in Kenya, the United States, and Denmark in 1990. The rapidly increasing population in Kenya has a bottom-heavy age distribution (many individuals in the younger age classes). Both the United States and Denmark show slower growth and thus have fewer individuals in the younger age classes. (*After Krebs 1994; from McFalls 1991.*)

achieve stable age distributions or grow at their intrinsic rates of increase. Because the growth performance of a population depends as much on past conditions, which determine its age structure, as on its present life table values, the intrinsic rate of increase is more useful than observed growth rates for assessing the effects of environmental conditions or individual attributes on population growth.

Each life table has a single intrinsic rate of increase. Lotka (1907, 1922) first showed that in a population with a stable age distribution, each age class grows at the same exponential rate, as we have seen in the examples presented above (see Figure 15-7) and as Lotka (1922) proved mathematically. This fact allowed Lotka to derive a relationship between the intrinsic rate of growth, survivorship, and the birth rate for each age class. For populations undergoing exponential growth, the relationship (given here without derivation) is

$$1 = \sum_{x=0}^{\infty} e^{-rx} l_x b_x. \qquad (15\text{-}3)$$

An equivalent expression for geometric growth is

$$1 = \sum_{x=0}^{\infty} \lambda^{-x} l_x b_x. \qquad (15\text{-}4)$$

These equations are known as **Euler's** (pronounced "oiler's") **equation,** or the **characteristic equation** of a population.

At this point it is important to recall our earlier definitions of λ. We said that, in exponential growth models, λ stood for the term e^r. So, the exponential growth equation $N_t = N_0 e^{rt}$ could be written $N_t = N_0 \lambda^t$ (see Section 15.1). Given this definition of λ, equation (15-3) can be recast as $1 = \Sigma\ \lambda^{-x} l_x b_x$, which is the same as equation (15-4). This should not surprise you, since we have shown that exponential and geometric growth curves may be superimposed (see Figure 15-4). This does not mean, however, that the λ that we inserted into equation (15-3) represents the same thing as the λ in equation (15-4). They both represent an increment of increase. But in the case of the exponential model, λ depends on the intrinsic rate of increase, r, and in the geometric model it represents the ratio of the current population size to the population size in the previous discrete time interval. This unfortunate bit of notation is somewhat confusing, but it does make it a bit easier to discuss the properties of the characteristic equation, since both equation (15-3) and (15-4) have the same analytical form.

What does the characteristic equation of a population tell us about population growth? Using the geometric form (equation [15-4]), let us write out the terms of the equation like this:

$$1 = \lambda^{-0} l_0 b_0 + \lambda^{-1} l_1 b_1 + \lambda^{-2} l_2 b_2 + \ldots + \lambda^{-k} l_k b_k. \qquad (15\text{-}5)$$

This form shows that the equation is a higher-order polynomial. We can simplify the equation a bit by recognizing that the first term is not needed, since $b_0 = 0$ (by definition, $\lambda^{-0} = 1$; anything raised to the power of zero equals one). And, we can replace the $l_x b_x$

products in each term with coefficients like a, b, c, and so forth. For example, we can let $a = l_1b_1$ in the first term, $b = l_2b_2$ in the second term and $q = l_kb_k$ in the last term. If the population has only four age classes, we would need only three coefficients, a, b, and c. We would need no coefficient for the first age class because, as we showed, it is dropped from the equation. With these simplifications, the equation becomes

$$1 = a\lambda^{-1} + b\lambda^{-2} + c\lambda^{-3} + \ldots + q\lambda^{-k}.$$

It is customary to put the above equation in a form in which the designation of the oldest age class, k, occurs as an exponent of every λ. We can do this by multiplying every term of the equation by λ^{k+1} to obtain

$$\lambda^{k+1} = a\lambda^{k+1}\lambda^{-1} + b\lambda^{k+1}\lambda^{-2} + c\lambda^{k+1}\lambda^{-3} + \ldots + q\lambda^{k+1}\lambda^{-k}.$$

Multiplying the λs in each term, which requires adding the exponents of the λs, yields the following equation:

$$\lambda^{k+1} = a\lambda^{k} + b\lambda^{k-1} + c\lambda^{k-2} + \ldots + q\lambda^{0}.$$

Rearranging to bring all terms to the same side of the equation and substituting 1 for λ^0 in the last term gives

$$\lambda^{k+1} - a\lambda^{k} - b\lambda^{k-1} - c\lambda^{k-2} - \ldots - q = 0. \quad (15\text{-}6)$$

The utility of this equation is that it embodies the survival and birth characteristics of the population (contained in the coefficients a, b, etc.) and the rate of growth of the population, λ. Birth and survival data may be obtained from a life table and the characteristic equation constructed. All we have to do then to determine the rate of growth is to find λ so that the equation above is solved. From your experience in algebra in factoring quadratic equations to determine their roots (the values of λ that satisfy the equation), you can appreciate the difficulty of trying to solve this equation when the number of age classes is greater than two or three. In practice, the roots of the equation may be found by trial and error, in which case we are grateful for one property of equation (15-6): because there is only one change of sign in the coefficients of the terms, the characteristic equation has only one real positive root, which is the intrinsic rate of increase. Microcomputer programs that perform symbolic calculus will efficiently estimate the roots of such an equation.

Estimation of the Intrinsic Rate of Increase

In most cases, ecologists are more interested in comparing the intrinsic rates of population increase among different populations than in estimating precisely the rate of growth for an individual population. Thus, finding an estimate of λ or r for Euler's equation may suffice. This can be done by choosing a value for λ or r, solving the equation, making an adjustment in the value, and solving again until, slowly, one converges on a value that solves the equation. Such a process of iteration can be made more efficient if one has a reasonable starting value for λ or r. Such a starting value was recommended by Caughley (1977), and is calculated as

$$r_a = \frac{\Sigma l_x b_x \log_e \Sigma l_x b_x}{\Sigma x l_x b_x},$$

where r_a denotes the starting value and the other elements of the equation are from the life table.

To illustrate how this equation can be used, let us examine the essential elements of the life table of the imaginary population introduced earlier (Table 15-4). We calculate r_a to be 0.38; this is equivalent to $\lambda = 1.46$, close to the observed value of about 1.48 after the population achieved a stable age distribution. When we substitute the value $r_a = 0.38$ into the right-hand terms of the characteristic equation (15-3), these sum to 1.032. This number is close to 1, and thus indicates that we have made an good initial guess. To get closer to 1, we must increase the value of r. When we evaluate the equation for $r = 0.39$ and $r = 0.40$, the terms add to 1.013 and 0.999 respectively, indicating that the root lies between these values. A linear interpolation gives an estimate of $r = 0.397$, or $\lambda = 1.487$, making the sum of the right-hand term equal to 1.000.

Effect of Environmental Fluctuation on Population Growth

The intrinsic rate of increase of a population is strictly determined by the life table values, which express the interaction between individuals and their environment. As a consequence of these interactions, the life table (and hence the intrinsic growth rate) varies with respect to the conditions of the environment. These conditions

TABLE 15-4 Life table parameters from the imaginary population described in Section 15.5

x	s_x	l_x	b_x	l_xb_x	xl_xb_x
0	0.5	1.0	0	0	0
1	0.8	0.5	1	0.5	0.5
2	0.5	0.4	3	1.2	2.4
3	0.0	0.2	2	0.4	1.2
Sum				2.1	4.1

(From Caughley 1977.)

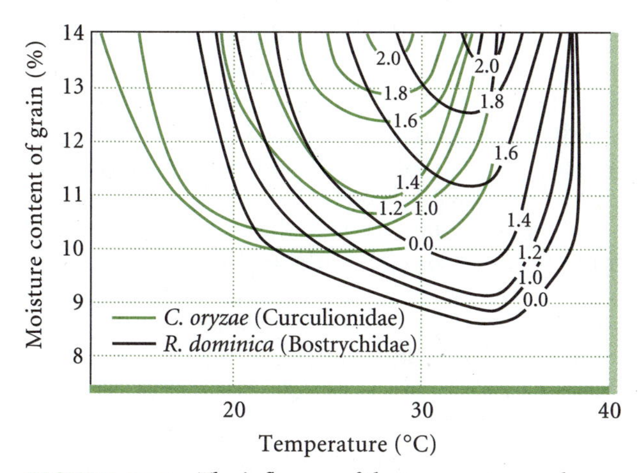

FIGURE 15-9 The influence of the temperature and moisture content of grain on the geometric rate of increase of populations of the grain beetles *Calandra oryzae* and *Rhizopertha dominica* living in wheat. Rates of increase are indicated by contour lines that describe conditions with identical values of λ. (*After Birch 1953.*)

vary spatially and temporally, creating differences in population dynamics from place to place and leading to changing population dynamics over time.

Such effects of the environment are best revealed in experimental studies in which groups of individuals from the same population are provided with different conditions or combinations of conditions. For example, L. C. Birch (1953) determined that the intrinsic rates of increase (λ) of populations of two species of grain beetles, *Calandra oryzae* and *Rhizopertha dominica*, vary over a wide range of temperature and moisture conditions (Figure 15-9). Neither species performed well at low temperatures and humidities; moreover, the optimum conditions for growth differed between the two species. *Rhizopertha* populations grew most rapidly at somewhat warmer temperatures. Of the two species, *Rhizopertha* has the more tropical distribution in nature.

Climate and other environmental conditions have been shown to be important determinants of life table values of populations in their natural settings, as one would expect. For example, the European rabbit, introduced to Australia in the nineteenth century and now widespread throughout many habitats, survives longer but produces fewer offspring per year in arid regions of Australia than in the more mesic Mediterranean climate regions (Table 15-5).

Continuous exponential growth leads, with time, to an inconceivable number. As Darwin (1872) put it, "Even slow-breeding man has doubled in twenty-five years, and at this rate, in less than a thousand years,

TABLE 15-5 Condensed life tables for European rabbits in arid and Mediterranean climates of Australia

Age (months)	Pivotal age (months)*	Survival (l_x)†	Proportion of female population	Females pregnant at any one time	Litter size (embryos)	b_x	$l_x b_x$	$x l_x b_x$
Mediterranean								
3–6	4.5	0.222	0.177	0.23	4.4	1.8	0.198	0.891
6–12	9.0	0.160	0.359	0.43	5.6	8.7	0.696	6.264
12–18	15.0	0.075	0.303	0.53	5.9	9.4	0.357	5.358
18–24	21.0	0.028	0.107	0.45	5.9	2.8	0.039	0.823
>24	37.5	0.006	0.053	0.55	6.2	1.8	0.005	0.203
Total							1.295	13.539
Arid								
3–6	4.5	0.570	0.140	0.05	3.8	0.2	0.057	0.257
6–12	9.0	0.457	0.265	0.15	4.7	1.8	0.411	3.702
12–18	15.0	0.302	0.217	0.32	4.6	3.1	0.468	7.022
18–24	21.0	0.186	0.192	0.37	4.3	3.1	0.288	6.054
>24	37.5	0.061	0.187	0.34	4.5	2.9	0.089	3.317
Total							1.313	20.352

**Midpoint of the age interval, frequently used to calculate generation time when reproduction is more or less continuous, as it is in rabbits in Australia.*

†Based on rabbits aged 0–3 months; hence $l_{1.5} = 1$.

(From Myers 1970.)

there would literally not be standing-room for his progeny." The Australian rabbit example raises the important issue of the regulation of population size. In 1859, 12 pairs of rabbits were released on a ranch in Victoria to provide sport for hunters. Within 6 years, the population had increased so rapidly that 20,000 rabbits were killed in a single hunting drive. Even by conservative estimates, the population must have increased by a factor of at least 10,000 in 6 years, an exponential rate (r) of about 1.5 per year. Yet the life tables of present-day Australian rabbit populations (see Table 15-5) suggest growth rates (r_a) of 0.30 and 0.21. Considering the difficulty of estimating survival of the young to reproduction and the statistical errors involved in such studies, these values probably do not differ significantly from $r = 0$. Furthermore, under environmental conditions so different as to produce up to twofold or threefold differences in survival rates, the intrinsic growth rates of the two populations are nearly identical. To achieve this, differences in rates of fecundity must balance the differences in survival. Where survival is higher, fecundity is lower.

How can one reconcile the initial rapid growth with the eventual stabilization of the Australian rabbit population? Either birth rates decreased, death rates increased, or both, as the population became more numerous. When there are more rabbits, there is less food for each; with fewer resources, fewer offspring can be nourished, and those offspring survive less well. Such density dependence of the life table values underlies the regulation of population size, the subject of the next chapter.

SUMMARY

1. The study of populations—demography—arose as a result of efforts to develop mathematical models that would predict the future course of human population growth. These models were quickly applied to the growth of natural populations in the 1920s and have since provided a quantitative framework for understanding population dynamics.

2. Newborn individuals may be added to the human population at any time of the year because humans do not have specific breeding periods. Thus, human population growth may be represented by a demographic model based on exponential growth—a continuous-time growth model. In most populations, newborns are added during specific breeding periods each year. Discrete-time demographic models such as the geometric growth model are used to model population growth in such cases. The exponential growth model is a special case of the discrete time model, and growth curves resulting from the two can be superimposed.

3. Continuous population growth can be described by the exponential rate of increase (r) in the expression $N_t = N_0 e^{rt}$. The factor by which a population increases in one unit of time (e^r) is the geometric growth rate of the population (λ). λ and e^r are interchangeable in population equations. The exponential growth rate is the difference between the birth and death rates averaged over individuals (per capita) in the population (that is, $r = b - d$).

4. The instantaneous rate of increase of an exponentially growing population is $dN/dt = rN$.

5. Populations with discrete breeding seasons increase geometrically by periodic increments according to the relation $N(t + 1) = N(t)\lambda$. The parameter λ is the ratio of the current population size to the population size in the previous time interval; for example, $\lambda = N(1)/N(0)$.

6. When birth and death rates vary according to the age of the individual (as summarized in the life table), one must additionally know the proportions of individuals in each age class in order to project population growth. In order to do this, the survival, s_x, and rate of birth, b_x, for each age class x must be taken into account.

7. The Lewis-Leslie matrix is a matrix of fecundities and survival probabilities that, when multiplied by a vector of initial age class sizes in an iterative fashion, will project the growth of a population having discrete breeding times.

8. A stage-classified population is one in which different developmental stages, such as egg, larva, and adult or seed, vegetatively reproducing rosette, and flowering plant, have different probabilities of surviving to the next stage and different probabilities of producing young. Matrices similar to the Lewis-Leslie matrix may be used to project the growth of such populations.

9. A population with a fixed life table assumes a stable age distribution in which the numbers in each age class, as well as the population as a whole, increase at the same exponential or geometric rate, known as the intrinsic rate of increase of the population.

10. The intrinsic rate of increase for any given life table is the positive root of a polynomial equation of the form $1 = \Sigma\lambda^{-x} l_x b_x$, called Euler's equation or the characteristic equation of the population.

11. The life table of a population may be expected to vary with the conditions of the environment and the density of the population.

EXERCISES

1. A population of $N_0 = 10$ individuals grows exponentially with $r = 1.0$. How long will it take (i.e., what will be the value of t) for the population to reach 10 times its beginning size? (Hint: Use the relationship, if $x = a^y$, then $y = \log_a x$.)

2. Project the following population through three breeding seasons. Assume that no individual survives to age class $x = 4$. Does the population achieve a stable age distribution in three seasons? (Hint: See Figure 15-5.)

Age class	b_x	s_x	N_0
0	0	0.6	0
1	0	0.6	10
2	2	0.3	100
3	1	0.0	50

3. Construct the Lewis-Leslie matrix for the population given in question 2.

CHAPTER 16

Population Regulation

GUIDING QUESTIONS

- How is population regulation depicted by the logistic equation?
- How do population densities converge on local equilibria?
- What is the difference between density-dependent and density-independent population regulation?
- How do density-dependent factors regulate animal and plant populations?
- How did H. G. Andrewartha and L. C. Birch challenge the theory of density-dependent population regulation?
- How did Frederick E. Smith use H. G. Andrewartha and L. C. Birch's data to challenge their ideas in turn?
- What is density vagueness?

Given the great capacity for populations to increase, what factors operate to curb their growth? Put another way, how can we reconcile the fact, demonstrated so dramatically by human populations, that populations have an enormous innate capacity to increase with our observation that populations do not everywhere explode? What regulates populations? These questions have a long history in ecology, arising even before the time of Darwin, and form the basis of a good deal of ecological theory and experimentation. We will begin our consideration of these questions in this chapter, but they will arise again in a number of subsequent chapters in the book.

Although the question of what regulates populations is a complex one, we may become comfortable with the range of possible answers by reflecting a bit on what we have learned about population growth. In general, in order for a population to grow, the birth rate plus the rate at which individuals immigrate into the population from other populations must be larger than the death rate plus the rate at which individuals leave the population. In simple terms, the rate of input of individuals to the population (birth and immigration) must be greater than the rate of output of individuals (death and emigration). Any factor that diminishes inputs relative to outputs may act to regulate the population. Such factors may derive from everyday interactions among organisms, such as predation, herbivory, and parasitism, or, as in the case of plants, from the dynamics of resource use by individuals. The intensity of such factors is most often related to the densities of the interacting populations. Other factors not related to how organisms interact may also regulate populations, as when a severe storm or an exceptionally cold winter causes high mortality rates in all age classes of a population. Such factors tend to have effects unrelated to the density of the population. The complexity of population regulation arises from the fact that it is often difficult to determine what particular factor is most responsible for regulation, how various regulating factors interact, and whether regulation implies some equilibrium population size.

In this chapter, we will first present formulations of the classic mathematical models of regulated population growth. While these models are too simple to mirror actual population dynamics (indeed, as we mentioned near the beginning of the book, simple analytical models reflect the essence of nature, not the fine details of it), their forms reveal much about the biological mechanisms of regulation. After that, we will look at mechanisms that actually restrict population expansion and show how passions have flared with respect to the issue of how much population density itself influences the regulation of populations.

Finally, we will briefly examine a number of ideas related to "nonequilibrium" models of population regulation. We will return to our consideration of population regulation in Part 5 of this book, where we will examine in earnest how populations may be regulated by interactions with populations of other species.

16.1 The logistic equation describes the growth of a regulated population.

In 1920, Raymond Pearl and L. J. Reed, at the Institute for Biological Research of Johns Hopkins University, published a paper in the *Proceedings of the National Academy of Sciences* titled "On the Rate of Growth of the Population of the United States since 1790 and Its Mathematical Representation." Thorough and accurate population data had been gathered even in colonial times. Indeed, the phenomenal population growth of the American colonies had greatly impressed upon Thomas Malthus how rapidly humans could multiply; this was not so evident in the more crowded European countries of his time.

Pearl and Reed wished to project the future growth of the U.S. population, which they supposed must eventually reach a limit. Data for the population to 1910, the latest census then available, had revealed a decline in the exponential rate of growth (Figure 16-1). Pearl and Reed reasoned that if this decline followed a regular pattern that could be described mathematically, it would be possible to predict the future course of the population's growth, as long as the decline in the exponential growth rate continued. They also reasoned that changes in the exponential rate of growth must be related to the size of the population rather than to time, because any time scale is arbitrary with respect to any particular population. And so, rather than using a constant value of r in the differential equation for unrestrained population growth ($dN/dt = rN$), Pearl and Reed suggested that r decreases as N increases, according to the relation

$$r = r_0\left(1 - \frac{N}{K}\right),$$

where r_0 represents the intrinsic exponential growth rate of a population when its size is very small (that is, close to 0), and K—the **carrying capacity** of the environment—represents the number of individuals that the environment can support. This equation defines a straight line with slope $-r_0/K$. (To see how to obtain the slope, multiply the right-hand side of the equation through by r_0 to obtain $r_0 - r_0N/K$. N is the independent variable, so $-r_0/K$ is the slope.) Thus, if $K = 50$ and $r_0 = 2$, the per capita population growth would decline at a rate of -0.04 as N increased. The relationship between r and the population size (N) for the population of the United States between 1790 and 1910 is shown in Figure 16-2.

If we substitute $r = r_0(1 - N/K)$ into the equation for unrestrained growth ($dN/dt = rN$), we obtain

$$\frac{dN}{dt} = r_0N\left(1 - \frac{N}{K}\right),$$

which is called the **logistic equation.** This logistic curve is hump-shaped, with $dN/dt = 0$ at two population densities, when $N = 0$ and when $K = N$ (Figure 16-3). The peak of the hump occurs when dN/dt is at its maximum value, which occurs when $N = K/2$. To see this, let $r_0 = 2$ and $K = 50$. If $N = K/2 = 25$, then, using the equation above, $dN/dt = 2(25)[1 - 25/50] = 25$,

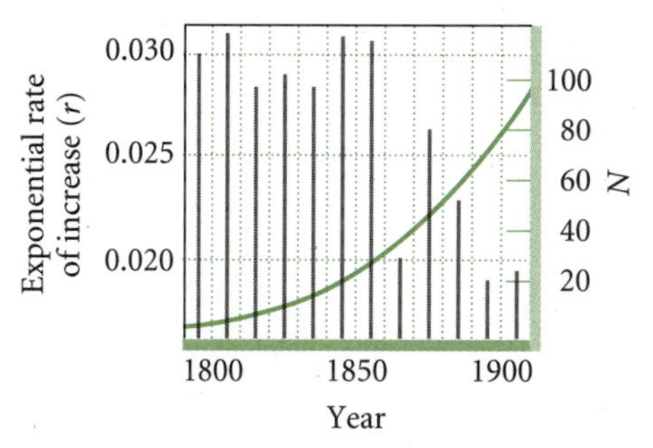

FIGURE 16-1 Increase in the population of the United States between 1790 and 1910 (curve), and the exponential rate of increase during each 10-year period (vertical bars). (*Data from Pearl and Reed 1920.*)

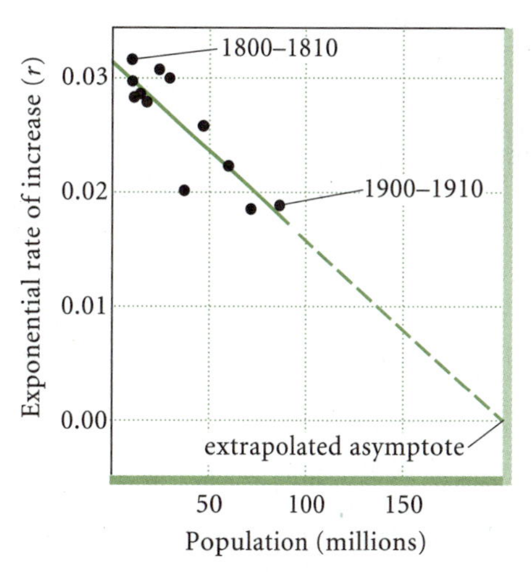

FIGURE 16-2 Exponential rates of population increase in the United States during each decade between 1790 and 1910, plotted as a function of population size during the decade (the geometric mean of the beginning and ending numbers) using the data shown in Figure 16-1. The dashed line is the extrapolation of the straight line $r = r_0(1 - N/K)$ fitted to the data.

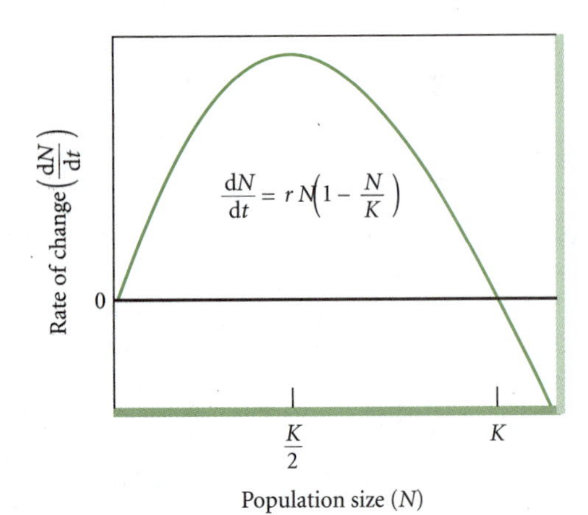

FIGURE 16-3 The logistic curve $dN/dt = rN(1 - N/K)$, where r is the population growth rate, N is the population size, and K is the carrying capacity. The population growth rate, dN/dt, is 0 when $N = 0$ or when $N = K$. The rate is at its maximum when $N = K/2$.

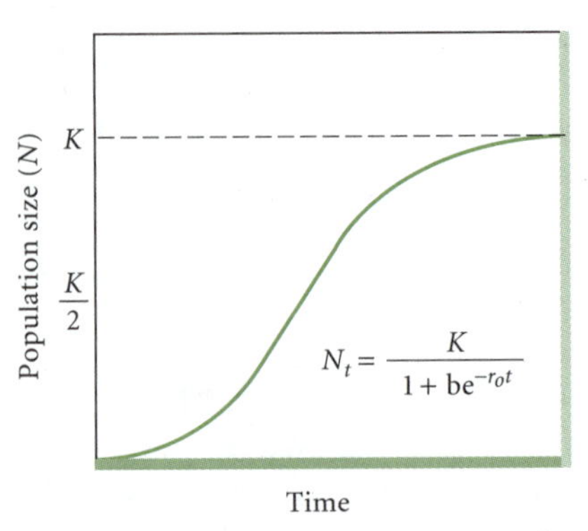

FIGURE 16-4 According to the logistic growth equation, increase in numbers over time follows an S-shaped curve that is symmetrical about the inflection point ($K/2$). That is, accelerating and decelerating phases of population growth have the same shape.

which is the maximum growth rate (the hump of the curve). Values of $N < K/2$ or $N > K/2$ yield dN/dt values less than 25. To check this, let $N = 20$. We see that $dN/dt = 2(20)[1 - 20/50] = 24$. When $N = 30$, $dN/dt = 2(30)[1 - 0.6] = 24$.

As long as population size, N, does not exceed the carrying capacity, K—that is, so long as N/K is less than 1—a population continues to increase, albeit at a slowing rate. When the value of N exceeds the value of K, the ratio N/K exceeds 1, the term in the parenthesis $(1 - N/K)$ becomes negative, and the population decreases. Because populations below K increase and those above K decrease, K is the eventual theoretical equilibrium size of a population growing according to the logistic equation. (We shall examine the implications of such an equilibrium shortly.)

The time course of population growth according to the logistic equation can be found by integrating the differential, which yields

$$N_t = \frac{K}{1 + be^{-rt}},$$

where b is a constant equal to $[K - N(0)]/N(0)$. The value of b depends arbitrarily on the size of the population at the designated time zero—1790 in the case of Pearl and Reed's data. This equation describes a sigmoid, or S-shaped, curve with time on the x-axis and the number of individuals—density—on the y-axis (Figure 16-4). This type of curve is called a **time-density curve.** The sigmoid time-density curve shows that the population grows slowly at first, then more rapidly as the number of individuals increases, and finally more slowly again, gradually approaching the equilibrium number K. Figure 16-5 shows a time-density curve for the growth of the population of the United States from 1790 to 1910.

Pearl and Reed obtained the best fit of their equation to the population data when the value of K was 197,273,000 and that of r_0 was 0.03134, which is equivalent to a population doubling time of 22 years. Recall, however, from our discussion about statistical relationships in Chapter 2 that data from the natural world rarely adhere perfectly to functional relationships, but that functional relationships may represent the trend of the data. Thus, if one feature X of nature increases linearly with another feature Y, there will not be a one-to-one relationship between each X and a corresponding Y, even though an underlying one-to-one function may adequately describe the relationship (see Figure 2-1). The "best fit" obtained by Pearl and Reed is such an underlying relationship. Thus, even though the population in 1910 was only 91,972,000, Pearl and Reed were able to extrapolate its future growth to twice the 1910 level from its earlier growth performance. Projections often prove incorrect, however, when circumstances change. The U.S. population reached 197 million between 1960 and 1970, when it was still growing vigorously. A leveling off in the

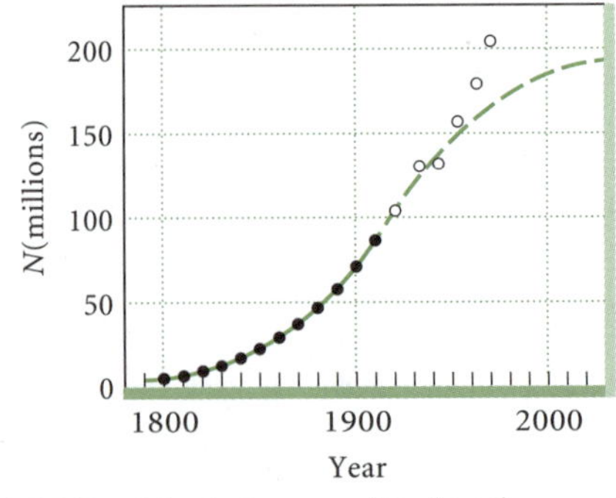

FIGURE 16-5 A logistic curve fitted to the population of the United States between 1790 and 1910 (solid black dots). Subsequent censuses (white open dots) have shown numbers above the projected population curve.

mid-200 millions can now be predicted on the basis of a much reduced birth rate in recent years, but all this could easily change.

16.2 A qualitative model of continuous population growth reveals the relationship between population density and growth rate.

The concordance of the U.S. human population data between 1790 and 1910 with the time-density function $N(t)$ of the logistic model is rather remarkable. The time-density data of few other natural populations are fitted so well with the logistic equation. Indeed, the logistic equation is generally considered far too simple a model to serve as a quantitative representation of the dynamics of specific natural populations (Yodzis 1989). The real usefulness of the model lies in its qualitative behavior; that is, populations with densities below some equilibrium K increase toward K, and populations having densities greater than K decrease toward K. Many natural populations follow this qualitative pattern.

We need not think of the rate of increase dN/dt as a specific function of N; rather, we may simply think of it as *some* function of N, $f(N)$. Beginning with this idea, and under the assumption of continuous birth and death processes, a qualitative model of population regulation can be developed that reveals a number of important characteristics of populations regulated about equilibria. P. Yodzis (1989) has presented a clear treatment of the qualitative behavior of simple continuous time population models, and we follow his presentation here.

Consider what happens when both the per capita birth rate, b, and the per capita death rate, d, are continuous functions of N and have the shapes depicted in Figure 16-6a. In this depiction, the birth rate declines as the population increases. Notice that the curves shown in Figure 16-6 are not time-density curves. Rather, they show the relationship between the rate of growth of the population (vertical axis) and the size or density of the population (horizontal axis). The death rate, which is low when the population is small, increases through some range of N and then begins to decline. Now, let us assume that $dN/dt = f(N) = b(N) - d(N)$. Subtraction of the death rate from the birth rate over the range of population sizes (as prescribed by the equation) yields Figure 16-6b, which has three equilibria [where $b(N) = d(N)$], one at K_1, one at K_2, and one at a point that we shall denote as B. For any value of N where $dN/dt > 0$, the population will increase; likewise, for any value of N where $dN/dt < 0$, the population will decrease.

To see this, consider first what happens around K_1. When $N < K_1$—say, at point a in Figure 16-6b—the rate of growth $dN/dt > 0$ (the curve is in the region above the horizontal axis). Thus the population will increase to some level greater than a during the next time period, thus moving the population toward K_1, as indicated by the rightward-pointing arrow

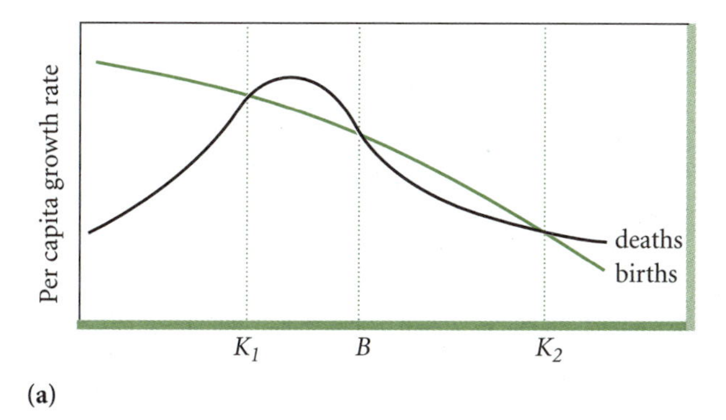

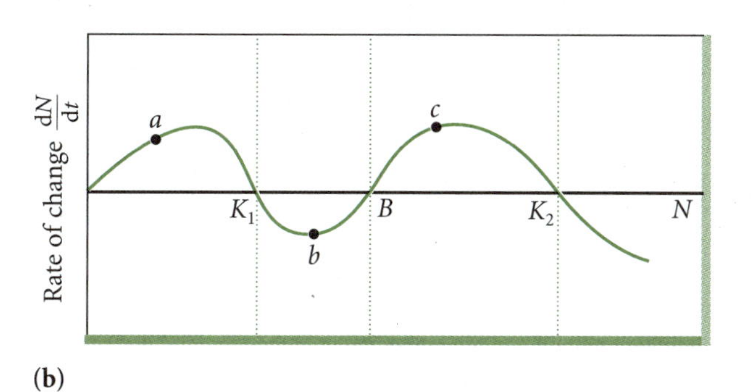

FIGURE 16-6 Qualitative behavior of a population with growth rate $dN/dt = f(N)$. (a) Birth and death rates are a function of the population density, N. In this hypothetical population, the birth rate decreases with population size, whereas the death rate is highest at intermediate population size. (b) The curve $f(N)$ is the difference between the functions of birth rate $b(N)$ and death rate $d(N)$. When the curve is below the horizontal axis, the death rate is higher than the birth rate, and the rate of population growth, dN/dt, is negative, indicating that the population will decline. For ranges of N where the curve is above the N-axis, dN/dt is positive, and the population will increase. The points K_1 and K_2 are equilibria because the population tends to move toward them from both above and below. Point B is called a breakpoint because the population tends to either increase or decrease away from it.

along the horizontal axis. If, on the other hand, the population size is at point *b*, above K_1, then $dN/dt < 0$ (the curve is in the region below the horizontal axis), and the population declines. So, the population will decrease from the level $N = b$ to some level between K_1 and *b*, thus moving the population toward K_1, as indicated by the leftward-pointing arrow along the horizontal axis. This model shows qualitatively that the behavior of the population around the point K_1 is to tend toward K_1, increasing toward it in the region below K_1 and decreasing toward it in the region above K_1. The area along the *N*-axis above and below K_1 in which *N* converges to K_1 is called the **domain of attraction** of K_1, and we may think of K_1 as a local population equilibrium. In Figure 16-6 the specific domain of attraction of K_1 is that region of the *N*-axis from 0 up to but not including point *B*. If we think of K_1 as the carrying capacity of the population, the model implies that the population density stabilizes about the carrying capacity.

Now, let us examine the behavior of this model around point *B* in Figure 16-6b. We have already seen that, if the population is of density $N = b$, it will display negative population growth, and the population will decrease toward the equilibrium point K_1. If the population is of density $N = c$, however, it will have a positive growth rate and will tend to increase away from *B* and toward another equilibrium point, K_2. That is, points between $N = B$ and $N = K_2$ are in the domain of attraction of K_2. Point *B* separates the domain of attraction of K_1 and K_2, and is called a **breakpoint.** The model we have depicted in Figure 16-6b, then, has two domains of attraction separated by the breakpoint *B*. An easy way to distinguish an equilibrium point having a domain of attraction from a breakpoint is to note the slope of the curve at the point of intersection with the *N*-axis. If the curve intersects the *N*-axis with a negative slope, the point is an equilibrium having a domain of attraction. If the slope is positive, it is a breakpoint.

Multiple domains of attraction like the ones described above help explain the observation that the application of a continuously increasing perturbation to a natural system often results in a discontinuous response. Prior to the onset of intense commercial fishing, an economically important fish species, for example, might exist at a population density, *N*, around K_2 in Figure 16-6b. If fishing pressure reduced *N* to a value between *B* and K_2, according to the model, the population would increase toward K_2 and would presumably recover from the fishing pressure. But if sufficient fish were extracted from the population to reduce the population to below the breakpoint, *B*, the population would then be under the domain of attraction of the smaller equilibrium, K_1. In this case, relieving the fishery of the fishing pressure might not result in a recovery of the population, since it would now be operating with a different domain of attraction (Yodzis 1989).

The foregoing descriptive model of population regulation and the discussion of the dynamics of the logistic model in Section 16.1 may have left you with the impression that the carrying capacity of a population is an inviolate number around which the population always fluctuates. It is important to appreciate that the equilibrium point of a population is set by the conditions of the environment and, because of this, will vary in time and space. As we shall see in Chapter 18, the amount of variation in a population depends on the magnitude of the fluctuation in the environment of that population. It is best to think of population equilibria as being reflective of environmental conditions of the moment. The population might be expected to stabilize about the equilibrium point if that point were to persist for an extended period of time. The qualitative model presented above provides a simple framework within which to envision such variation in equilibrium. One need only think of K_1 and K_2 as equilibria resulting from different environmental conditions.

16.3 Populations may be regulated by the effects of density-dependent factors.

The logistic equation has been applied successfully to describe the growth of populations in the laboratory and in natural habitats. It suggests that factors limiting growth exert stronger effects on mortality and fecundity as a population grows. But what are these factors, and how do they operate? Many things influence rates of population growth, but only **density-dependent** factors, whose effects increase with crowding, can bring a population under control. Of prime importance among these factors are limitations on food supply and places to live, as well as predators, parasites, and diseases whose effects are felt more strongly in crowded than in sparse populations. Other factors, such as temperature, precipitation, and catastrophic events, alter birth and death rates largely without regard to the number of individuals in a population. Such **density-independent** factors may influence the exponential growth rate of a population, but they do not regulate the size that the population will attain in the environment.

Density Dependence in Animals

Numerous experimental studies have revealed various mechanisms of density dependence in animals. For example, when a pair of fruit flies is confined to a bottle with a fixed supply of food, the descendants of those flies increase in number rapidly at first, but soon reach a limit. When different numbers of pairs of flies are introduced into otherwise identical culture bottles, the number of progeny raised per pair varies inversely with the density of flies in the bottle (Figure 16-7). This effect results from competition among the larvae for food, which causes high mortality in dense cultures. Adult life span also declines, but only at high densities, well above the levels that affect the survival of larvae. Juvenile stages often suffer the adverse affects of density-dependent factors more than adults.

The effects of density on the life tables of grain beetles result from fights between larvae under crowded conditions. *Rhizopertha dominica* is a tiny beetle that completes its larval development within a single grain (of wheat, for example), living off the kernel of the seed. Females lay their eggs on the surfaces of seeds. Immediately after hatching, a larva bores its way into a seed, where it commences development. Once inside the grain, the larva enjoys security as long as it is alone; a kernel of wheat cannot support more than one beetle. When two larvae meet in the same grain, they attack each other with their mandibles, sometimes fighting to the death.

The British scientist A. C. Crombie (1944, 1945) performed a number of experiments on grain beetles using the following protocol. He infested a single wheat grain with from one to eight larvae, giving them 6 hours to become established in the grain. He then transferred the infested grain to a dish with five other fresh grains and left them together for 48 hours. At the end of the experiment, he dissected all the grains to determine how many larvae had been killed and how many had migrated to a different grain. The results showed conclusively the strong effects of density on the survival and movement of larvae even within the first 2 days (Figure 16-8).

Crombie's beetles contested each grain of wheat through direct confrontation. After all, sole possession

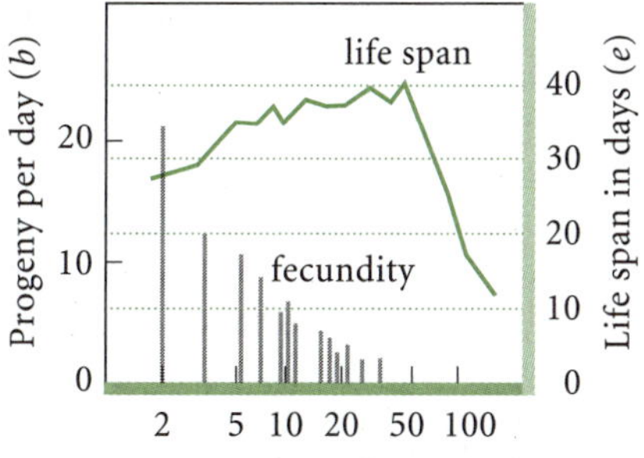

FIGURE 16-7 Density-dependent population regulation in laboratory populations of the fruit fly *Drosophila melanogaster.* The continuous line shows the influence of density on life span; the vertical bars, its influence on fecundity. (*After Pearl 1927.*)

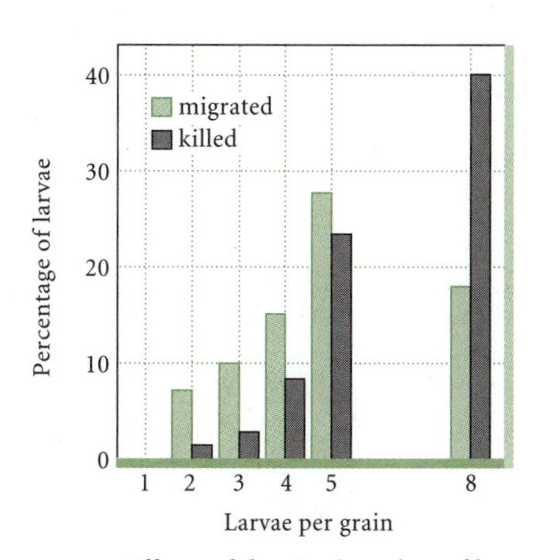

FIGURE 16-8 Effects of density (number of larvae per grain) on migration and mortality in the grain beetle *Rhizopertha dominica.* Both migration and mortality increase as density increases. (*Data from Crombie 1944.*)

of a single grain is a larva's ticket to success. In this population, density dependence came into play abruptly as the number of larvae approached the number of grains of wheat. Above that level, regardless of the number of larvae relative to wheat grains, the end result was always the same: one grain, one larva. Territorial behavior in many kinds of animals regulates population size in a density-dependent fashion, forcing young or socially subordinate individuals to leave the local population and seek space elsewhere.

Water fleas (*Daphnia pulex*) influence one another less directly by eating the same resources. Each water flea consumes millions of prey: single-celled green algae and diatoms of the plankton. Under laboratory conditions, as prey are eaten, the availability of food decreases gradually, leading to a graded response of birth and death rates to prey density. When water fleas were maintained in small beakers at densities between 1 and 32 individuals per cubic centimeter and were fed identical cultures of green algae, fecundity decreased markedly with increasing population density (Figure 16-9a). Somewhat unexpectedly, however, survival increased at densities up to 8 individuals per cm^3 before decreasing at higher densities (Figure 16-9b). Stunted body growth at densities of 8 individuals per cm^3 and above suggested that depletion of food resources between periodic feedings limited birth rates and survival; the effect was clearly density dependent. The geometric rate of population growth (λ; see Chapter 15) calculated from the water flea life tables decreased linearly with increasing density and fell below 1.0 at a density of about 20 individuals per cm^3 (Figure 16-10). Therefore, under the conditions of temperature, light, water quality, and food availability provided in the laboratory, *Daphnia* populations would attain a stable size of about 20 individuals per cm^3, regardless of the initial density of the culture.

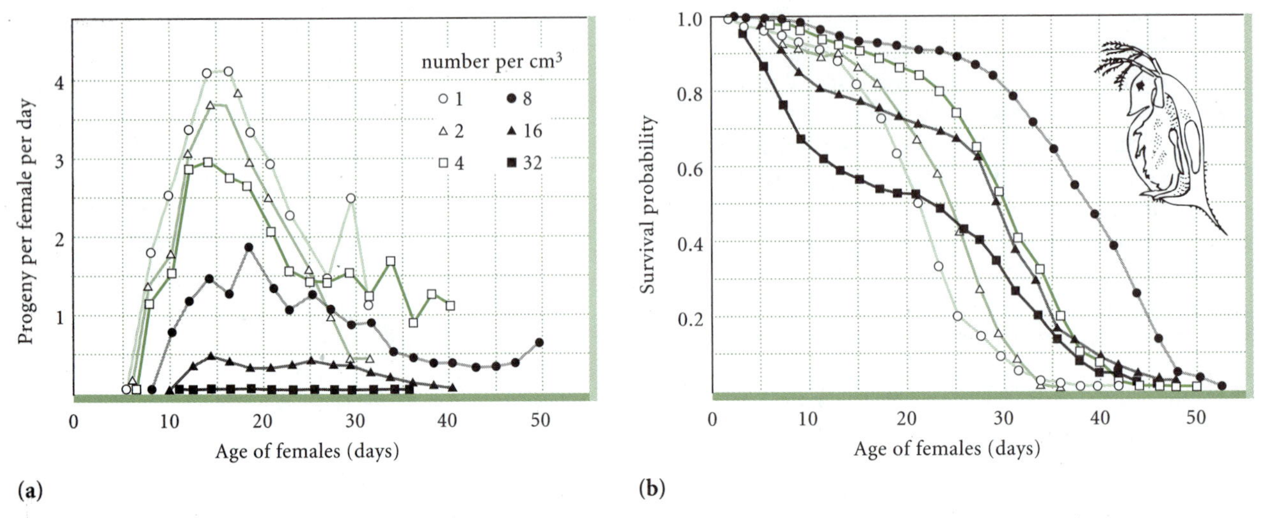

FIGURE 16-9 Fecundity (progeny per female per day) (a) and survival probability (b) in laboratory populations of *Daphnia pulex* as a function of age at different densities. Fecundity decreased as density increased. Fecundity also decreased with increasing female age at each density level. Survival was highest for densities between 1 and 8. (*From Frank et al. 1957.*)

Most studies of density dependence have been conducted in the laboratory, where factors affecting populations can be controlled experimentally. The simplicity of such systems leaves some doubt about the relevance of laboratory findings to populations in more complex natural surroundings, where physical conditions change continually and food and predation are not controlled by the experimenter. For example, winter weather and other factors have caused a population of song sparrows on Mandarte Island—a 6-ha speck of land off the coast of British Columbia—to fluctuate widely during recent years, between 4 and 72 breeding females and between 9 and 100 breeding males (Arcese and Smith 1988, Smith et al. 1991). In response to this environmentally induced variation in population size, density-dependent factors clearly limit the productivity of the population during years of high density by restricting the number of breeding males (through territoriality), reducing the number of offspring fledged by each female (through decreases in breeding season food supply), and reducing survival of juveniles in fall and winter (Figure 16-11).

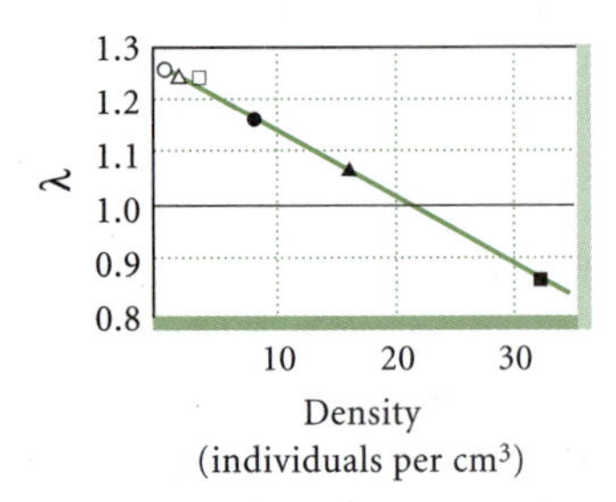

FIGURE 16-10 Values of λ, the geometric growth rate, calculated from the life table of *Daphnia pulex*. The population growth rate decreases as a function of density. (Symbols are as in Figure 16-9.) (*After Laughlin 1965.*)

Although natural variation in population size provides a method for visualizing density dependence, ideally we would like to conduct the same experiment in nature as in the laboratory—that is, to alter the density of individuals in a population while keeping everything else constant. In practice, this difficult experiment can be accomplished only with populations managed intensively for some other purpose. Game animals are sometimes maintained at altered levels by management practices, and ecologists have taken advantage of such situations to study population processes. A survey of harvested white-tailed deer (*Odocoileus virginianus*) in New York State in the 1940s provides an example of this method (Chaetum and Severinghaus 1950).

The reproduction and survival of deer depend directly on the quality of their food. Deer browse leaves, and they require large quantities of new growth with high nutritional content to maintain high growth rates and normal reproduction. In white-tailed deer in New York State, the proportion of females pregnant and the average number of embryos per pregnant female were found to be directly related to range conditions (Table 16-1). The number of corpora lutea in each ovary indicates the number of eggs ovulated, and hence the reproductive potential of a female. A difference between number of corpora lutea and number of embryos shows that poor conditions, resulting in poor nutrition of pregnant females, have caused embryo death and resorption. In the central Adirondack area, where habitat for deer was very poor, even ovulation

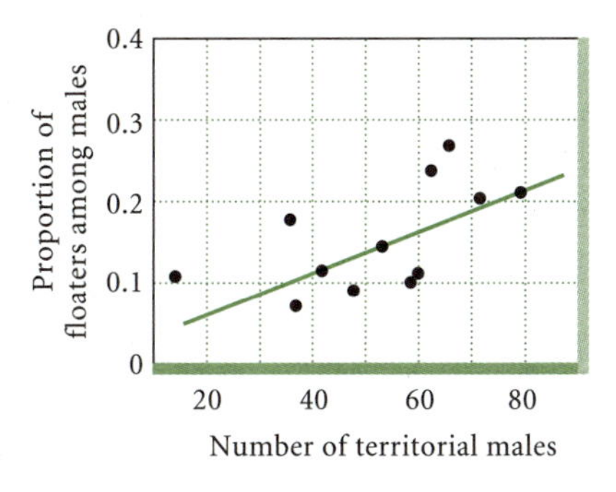

(a)

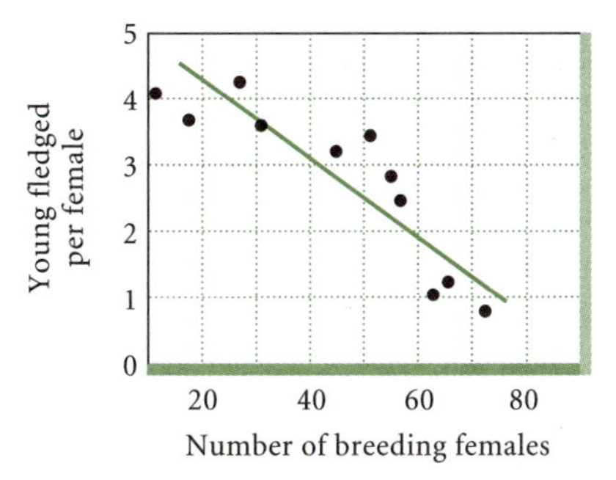

(b)

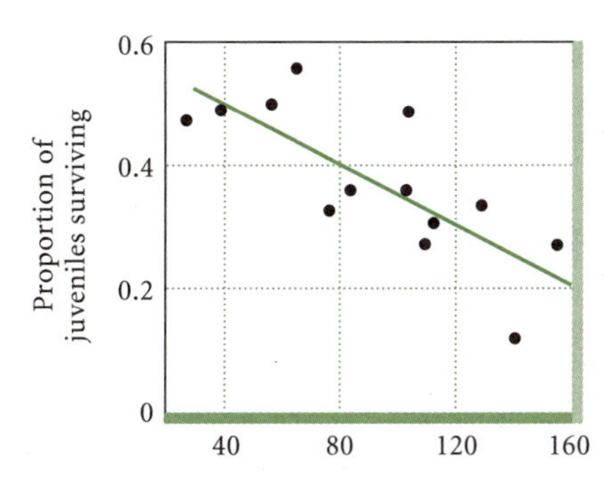

(c)

FIGURE 16-11 Density dependence in the Mandarte Island song sparrow (*Melospiza melodia*) population. With increased crowding on the small island, the proportion of males prevented from acquiring territories ("floaters") increases, and the number of fledglings produced per female, as well as the survival of those offspring through fall and winter, decreases. (*After Arcese and Smith 1988, Smith et al. 1991.*)

was greatly reduced. Range deterioration caused by overgrazing can often be reversed by selective hunting to thin dense populations. When DeBar Mountain, an area of very poor range, was opened to hunting, the population of white-tailed deer decreased, range quality recovered, and reproduction improved dramatically (Table 16-2).

Density Dependence in Plants

Like animals, plants experience increased mortality and reduced fecundity at high densities. A common response of plants to intense competition for resources is slowed growth, which has consequences for fecundity and, to a lesser extent, survival. The sizes of flax (*Linum*) plants grown to maturity at different densities reveal this flexibility (Figure 16-12). When seeds were sown sparsely at a density of 60 per square meter, the modal dry weight of individuals fell between 0.5 and 1 g, and many plants attained weights exceeding 1.5 g (Harper 1967). When seeds were sown at densities of 1,440 and 3,600 per m^2, most of the individuals weighed less than 0.5 g, and few grew to large size. Variation in size within a planting results from chance factors early in the seedling stage, particularly date of germination and quality of the site in which the seedling grows. Early germination in a favorable spot gives an individual plant an initial growth advantage over others, which increases as larger plants grow and crowd their smaller neighbors.

The flexibility of plant growth does not preclude mortality in crowded situations. When horseweed (*Erigeron canadensis*) seed was sown at a density of 100,000 per cm^2 (equivalent to about 10 seeds in the area of your thumbnail), young plants competed vigorously (Harper 1967). As the seedlings grew, many died, and the density of the surviving seedlings decreased (Figure 16-13). At the same time, however, the growth rates of surviving individual plants exceeded the rate of decline of the population, and the total weight of the planting increased. Over the entire growing season, a thousandfold increase in the average weight of each plant more than balanced the hundredfold decrease in population density.

The results of many such experiments have yielded a distinct pattern: when the logarithm of average plant weight is plotted as a function of the logarithm of density, the data points recorded during the growing season fall on a line with a slope of approximately $-3/2$ (Figure 16-14). Plant ecologists call this relationship between average plant weight and density a **self-thinning curve.** Such is the regularity of this relationship that many have referred to it as the **$-3/2$ power law.**

Density-dependent factors tend to bring populations under control and maintain their size at close to the carrying capacity set by the availability of resources and conditions in the environment. Changes in these conditions and resources continually establish new equilibrium values toward which populations grow or decline. Furthermore, catastrophic changes in the environment brought about by a sudden freeze, a violent storm, or a shift in an ocean current often

TABLE 16-1 Reproductive parameters of white-tailed deer (*Odocoileus virginianus*) in five regions of New York State, 1939–1949

Region*	Percentage of females pregnant	Embryos per female	Corpora lutea per ovary
Western (best range)	94	1.71	1.97
Catskill periphery	92	1.48	1.72
Catskill central	87	1.37	1.72
Adirondack periphery	86	1.29	1.71
Adirondack center (worst range)	79	1.06	1.11

**Arranged by decreasing suitability of range.*
(From Chaetum and Severinghaus 1950.)

TABLE 16-2 Reproductive parameters of white-tailed deer (*Odocoileus virginianus*) in the DeBar Mountain area of the Adirondack Mountains of New York State before and after hunting

	Percentage of females pregnant	Embryos per female	Corpora lutea per ovary
1939–1943 (prehunting)	57	0.71	0.60
1947 (after heavy hunting)	100	1.78	1.86

(From Chaetum and Severinghaus 1950.)

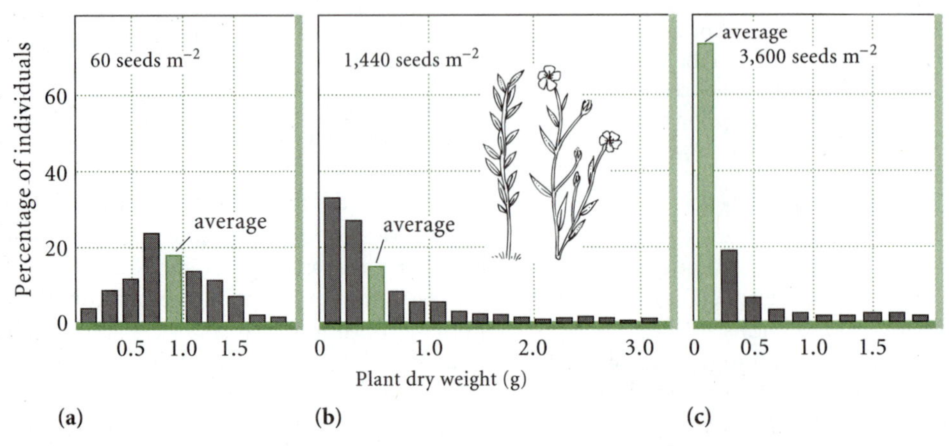

FIGURE 16-12 Distribution of dry weights of individuals in populations of flax (*Linum*) plants at a low density (a) and at two high densities (b and c). (*After Harper 1967.*)

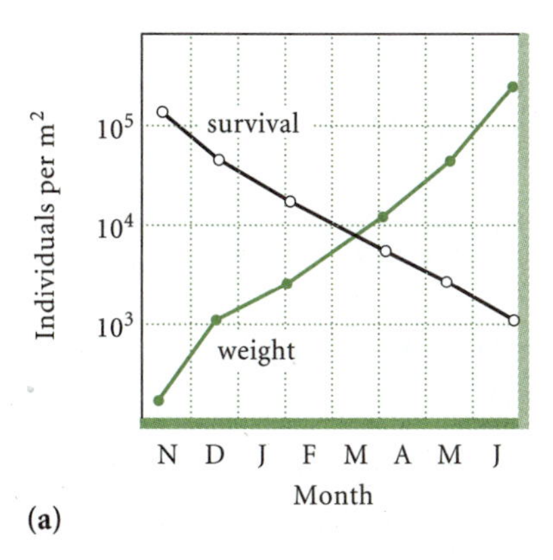

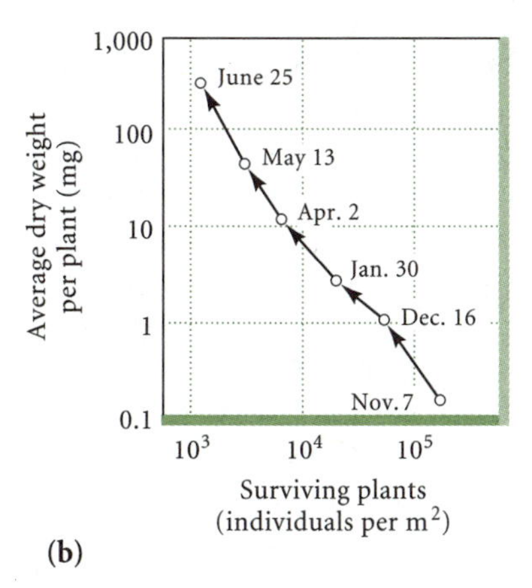

FIGURE 16-13 (a) Progressive change in plant weight and population density in an experimental planting of horseweed (*Erigeron canadensis*) sown at a density of 100,000 seeds per m^2. (b) Relationship between plant density and plant weight as the season progressed. (*After Harper 1967.*)

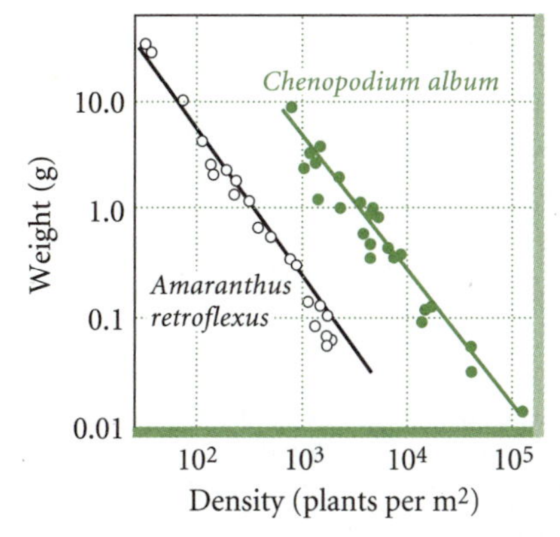

FIGURE 16-14 Changes in plant density and mean plant weight with time for plantings of *Amaranthus retroflexus* and *Chenopodium album.* The two variables are related by a slope of $-3/2$ (-1.5), which has been extremely consistent in experiments of this type. (*After Harper 1977.*)

reduce populations far below their carrying capacities and initiate periods of population recovery. Thus, although density-dependent factors regulate all populations, variations in the environment also cause all populations to fluctuate, to a greater or lesser extent, about their equilibrium sizes.

16.4 Andrewartha and Birch challenged density-dependent regulation of population size.

The year 1954 brought the publication of two important books on population biology. Each was remarkable on its own merits, but even more striking was the utter opposition of their points of view. The English ornithologist David Lack, in *The Natural Regulation of Animal Numbers,* vigorously advocated the regulation of population size by density-dependent factors, principally food, predators, and disease. His arguments and evidence added support for the theory of density dependence and the regulation of populations about stable equilibria promoted by Lotka, G. F. Gause, A. J. Nicholson, and others.

The same year, two Australian entomologists, H. G. Andrewartha and L. C. Birch, published *The Distribution and Abundance of Animals,* in which they argued that most populations, particularly those of insects and other small invertebrates, are influenced primarily by density-independent factors, and that periods of favorable environmental conditions for population growth ultimately control the size of a population:

> The numbers of animals in a natural population may be limited in three ways: (a) by shortage of material resources, such as food, places in which to make nests, etc.; (b) by inaccessibility of these material resources relative to the animals' capacity for dispersal and searching; and (c) by shortage of time when the rate of increase r is positive. Of these three ways, the first is probably the least, and the last is probably the most, important in nature. Concerning (c), the fluctuations in the value of r may be caused by weather, predators, or any other component of environment which influences the rate of increase.

Although Andrewartha and Birch marshaled abundant evidence to support their view that exponential growth rates reflect density-independent factors, they were greatly influenced in their thinking by a study done by J. Davidson and Andrewartha (1948) on populations of the tiny insect pest *Thrips imaginis* near Adelaide, South Australia. Thrips, which infest roses and other cultivated plants, undergo regular periods of increase when seasonally favorable conditions prevail, followed by rapid decline when conditions are

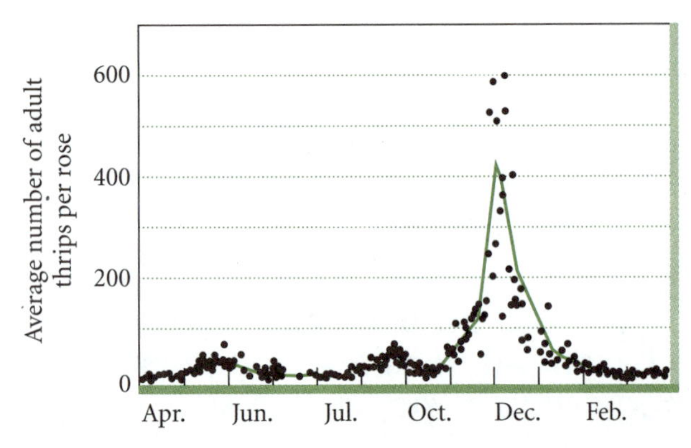

FIGURE 16-15 Number of *Thrips imaginis* per rose from April 1932 through March 1933 near Adelaide, Australia. Dots indicate daily records. The curve is a moving average for 15 days. (*After Davidson and Andrewartha 1948.*)

less suitable (Figure 16-15). Adelaide has a Mediterranean climate: winters are cool and rainy, summers hot and dry. Thrips subsist mainly on plant pollen, whose abundance varies seasonally with the production of flowers. The spring (October through December in the Southern Hemisphere) brings an ideal combination of moisture, warmth, and plant flowering. But the climate of Adelaide is sufficiently mild that some flowers are always available, and thrips remain active all year. During the winter, however, the depressing effect of cool temperatures on development rate and fecundity causes the thrips population to decline to low levels. Population growth is further checked by high mortality of immature stages: thrips take so long to mature during the winter that before many can complete their development, the flowers in which they live have withered and fallen to the ground—carrying the young thrips with them to their deaths. The warm weather of spring speeds development and increases the daily fecundity of females. Under these conditions, the thrips rapidly increase to infestation levels; their populations are not checked until the heat and dryness of the austral summer bring about a sharp rise in adult mortality.

Not only did the number of thrips mirror seasonal changes in weather, but the peak density of the population also varied with climate from year to year during the course of Davidson and Andrewartha's study. Some years were warmer or moister, others cooler or drier. Food supply apparently had little to do with the vagaries of the population because, as Andrewartha and Birch (1954) explained,

> [E]ven when the thrips are most numerous, the flowers in which they are breeding do not appear to be overcrowded except perhaps locally or temporarily: while the thrips are multiplying, the flowers increase even more rapidly. Then, when the population begins to decline, the flowers become less crowded still. . . . Considerations of this sort led to the hypothesis that the numbers achieved by the thrips during each year were determined largely by the duration of the period that was favorable for their multiplication. When this period was prolonged, the thrips would ultimately reach higher numbers; when it was briefer, the decline would set in while the numbers were still relatively low.

To determine the relationship between thrips populations and weather, Davidson and Andrewartha (1948) compared the peak population abundance in each year between 1932 and 1945 with four weather variables, subjecting the lot to a multiple regression statistical analysis. **Regression analysis** is a procedure whereby a linear equation is developed in which a group of independent, or regressor, variables (the four weather variables in this case, denoted as *X*s) predicts a dependent, or response, variable (thrips density, denoted as *Y*). In Davidson and Andrewartha's study, the dependent variable was

Y — the average of the logarithm of the number of thrips per flower over the 30 days before the population peak.

The independent, or regressor, variables were

X_1 — the effective degree-days from the first rains of the winter season to August 31, which might determine the growth of the annual plants on whose pollen the thrips would feed later in the season,

X_2 — the rainfall during September and October, the spring rains sustaining the thrips' food plants and promoting the survival of thrips pupae in the soil,

X_3 — the effective degree-days during September and October, a time when temperatures are becoming marginally adequate for thrips reproduction, and, finally,

X_4 — the same as X_1, but for the previous year, to take into account any carryover of thrips or seeds from one year to the next.

The four variables were chosen to represent conditions that could affect either the well-being of the thrips or the production of their food.

The analysis revealed that year-to-year variation in the four climate variables together predicted 78% of the year-to-year variation in the population (Figure 16-16). Individually, the warmth of the fall and winter (X_1) made the largest contribution to the size of the population, followed by spring rainfall (X_2) and the

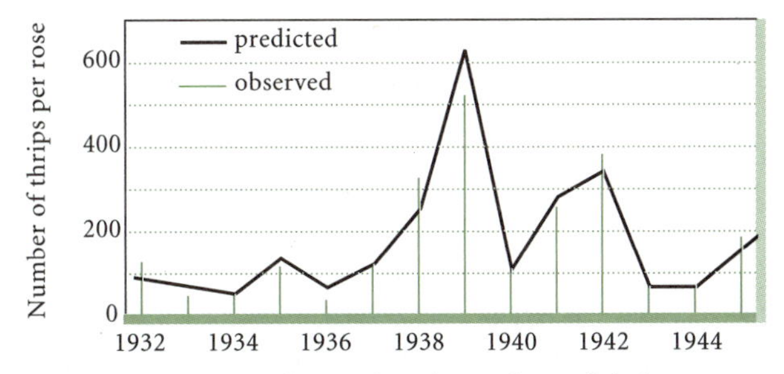

FIGURE 16-16 Observed peak numbers of thrips per rose between 1932 and 1945 (green vertical lines) compared with the predictions of a regression equation based upon four climate variables (black line). (*After Davidson and Andrewartha 1948.*)

warmth of the previous winter (X_4). Variation in spring temperature appeared not to be important.

Andrewartha and Birch made two observations concerning the thrips study. First, the thrips were at no time numerous enough to consume more than a small fraction of the food available. Second, variation in the population was accounted for satisfactorily by the physical conditions of the environment, which they presumed to affect the thrips independently of the density of their population. With 78% of the variance explained by the independent variables in their model, only 22% was left to be explained by some other effect. Their claim was that this remaining amount was so small that there was little doubt that fluctuations in the environment accounted for population size in the thrips.

How did Andrewartha and Birch address the abundant evidence of density dependence? First, they claimed, most of this evidence came from simplified and controlled laboratory populations, and had dubious application to natural populations. Second, the demonstration of density dependence does not necessarily imply regulation of the population by density-dependent factors. One must show under natural circumstances that such factors cause dense populations to decrease and allow sparse populations to increase. And what about the argument that in the absence of density-dependent factors, populations would either decrease to extinction or increase without bound? Andrewartha and Birch pointed out that thrips populations do, indeed, die out every year in many localities where climatic conditions have been particularly severe. Such areas presumably are recolonized from populations persisting in more favorable places. As to unlimited increase, the thrips population simply does not have enough time each year to reach the limits imposed by food or other resources. Each favorable season of the year—the spring—is invariably followed by an unfavorable period—the summer.

16.5 The theory of density independence created a major controversy among ecologists.

As expected, Andrewartha and Birch's challenge prompted a vigorous defense of density dependence, notably by M. E. Solomon (1957), A. J. Nicholson (1958), G. C. Varley (1963) (all entomologists), and David Lack (1966). But while these authors criticized the logic of Andrewartha and Birch's arguments, Frederick E. Smith, then at the University of Michigan, went straight to the thrips data. Smith (1961) argued that Davidson and Andrewartha (1948) had analyzed their data in such a way that they could not have detected density dependence. He pointed out that arguments about density dependence do not address the absolute size of a population at a particular time of year; rather, they relate changes in population size to the initial size of the population.

Smith assumed that thrips had an underlying exponential growth pattern (see Chapter 15). A characteristic of exponential growth is that changes in the size of N may be represented by increments of the logarithm of N: that is, $\Delta \log N(t) = \log N(t + 1) - \log N$ (given here without derivation). Using this relationship, the presence of density dependence can be determined by comparing the increase in $\log N$ from one time to the next to $\log N$ at the beginning of the period. Smith made several such comparisons based on Davidson and Andrewartha's data (for example, Figure 16-17). As you can see, ΔN is negatively related to N, as required by density dependence.

Smith posed a second argument based on the variation from year to year in thrips numbers during each month. Davidson and Andrewartha (1948) had tabulated the average numbers of thrips per rose for 81 consecutive months between April 1932 and

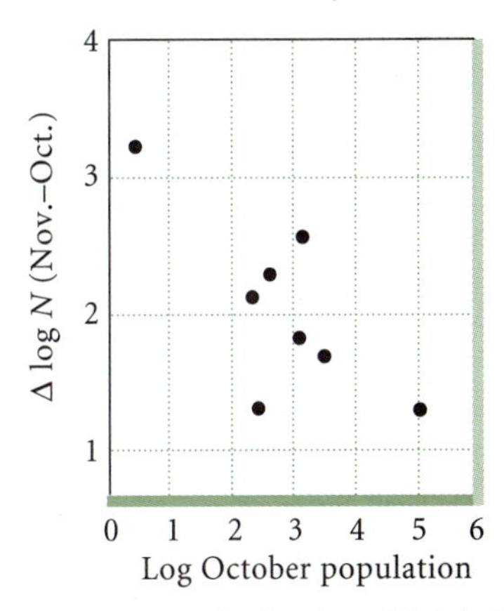

FIGURE 16-17 Increase in the size of the thrips population between October and November as a function of the size of the population in October. (*From data in Davidson and Andrewartha 1948; after Smith 1961.*)

December 1938 (Figure 16-18). From these data, Smith calculated the average and the variance for each month of the year. The variance is the sum of the squared deviations of each value from the average value for that month (see Chapter 2); hence it is a measure of variability. Smith noticed that the variance was relatively low during the fall and winter, but rose to a high level in October, the period of the most rapid population growth, after which it fell rapidly.

From his training in statistics, Smith knew that when two numbers vary independently of each other, the variance of their sum is equal to the sum of their variances; that is, var $(X + Y) =$ var (X) + var (Y). Populations grow according to the expression $\log N(t + 1) = \log N(t) + \Delta\log N(t)$. Therefore, if $\Delta\log N(t)$ is independent of N—growth is density-independent—the variance in $\log N$ must increase, on average, with each increment of population change, whether positive or negative (the variance is always positive). But the variance in $\log N$ declines between October and November (see Figure 16-18b), indicating a lack of independence between $\Delta\log N(t)$ and $\log N$.

The degree of relationship between two random variables X and Y can be described by their covariance, where cov (X,Y) is the sum of the product of the deviations in X and Y from their means. If Y tends to be below average when X is above average, as in Figure 16-18b, cov (X,Y) will contain products mostly of positive and negative deviations, and will therefore be negative. When two variables are related, var $(X + Y) =$ var (X) + var (Y) + 2cov (X,Y). Applying this to the thrips population, Smith pointed out that for the variance of $\log N$ to decrease, cov $(\log N, \Delta\log N)$ must be a larger negative number than -0.5var $(\Delta\log N)$, which implies strong density dependence.

Andrewartha was not convinced by Smith's arguments, and said so in a response printed in *Ecology* the following year (1963). Smith's reply, appended to the end of Andrewartha's note, opens with an expression of the frustrating gulf between their views: "In an effort, not to continue an argument, but to leave it as free of error as possible . . ."

The Contemporary View of Density Dependence

Uncertainties about the importance of density-dependent population regulation continue today. At the root of these uncertainties are questions about the efficacy of equilibrium models in general (Berryman 1987, Wolda 1989, Botkin 1990) and the difficulty of detecting density dependence in natural populations. So far, we have presented models of population growth and regulation that have depended on the idea of equilibrium. That is, we have assumed that populations fluctuate around some characteristic density, K, that is determined by environmental conditions and by the interactions of organisms within and between populations. Our presentation of the qualitative behavior of population growth rate as a function of birth and death rates represents the idea of equilibrium in its most reduced form: a single density, K. However, this conceptualization often does not match our observations of nature. While we draw smooth curves to represent the relationship of population density to birth and death rates and the rate of increase, our observations tell us that the real world is not so deterministic. Variation in the birth and death rates of a population, called **demographic stochasticity,** and variation in the environment, called **environmental stochasticity,** create random fluctuations, or noise, in the system that may be significant when compared with density-dependent responses, a situation referred to as **density vagueness** (Strong 1986). A population experiencing large demographic or environmental stochasticity may in fact be regulated by density-dependent processes, but because of the large random component of population fluctuation, an equilibrium density is difficult to pinpoint.

Ecologists have also been given reason to reflect on the utility of equilibrium models by the rather unsettling fact that, until recently, despite the clear evidence that natural populations are regulated, in very few cases has the mechanism of regulation been identified unambiguously (Murdoch 1994). Particularly troubling is that the available statistical tests for population regulation, all of which involve the analysis of time

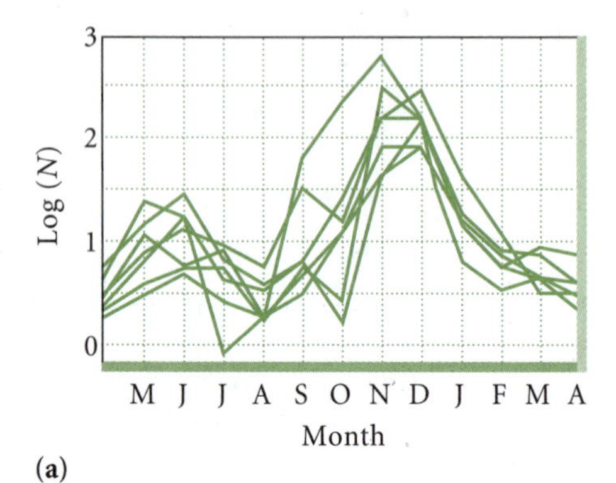

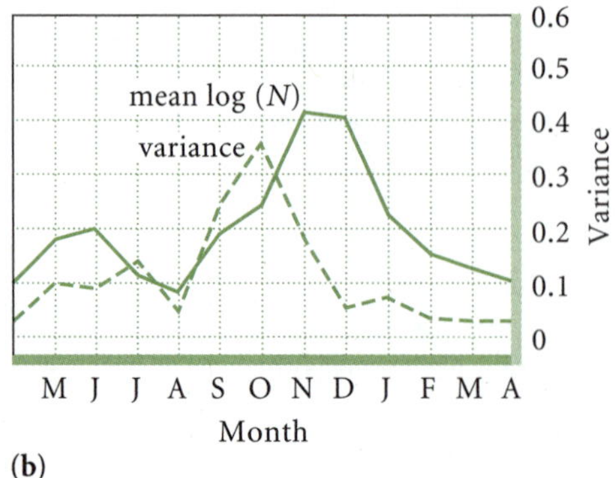

FIGURE 16-18 (a) Monthly mean numbers of thrips per rose recorded by Davidson and Andrewartha over nearly 7 years. (b) The mean and variance for each month of the year. (*From data from Davidson and Andrewartha 1948; after Smith 1961.*)

series of population densities, search for the presence of density dependence. Thus, even though it is widely appreciated that bird populations are regulated in a density-dependent manner, neither Greenwood and Baillie (1991), who analyzed population trends for 39 species of British passerine birds, or Murdoch (1994), who looked at time series density data for 59 species of birds, found density dependence in more than a few species. Newer statistical approaches provide methods for uncovering density dependence that may be obscured by trends in population dynamics (e.g., Vickery and Nudds 1991). Indeed, Smith, Lack, and other original defenders of density dependence would savor the results of Woiwod and Hanski (1992), who looked for evidence of density dependence in 5,715 time series of 447 species of moths and aphids and found strong evidence of its presence.

One of the manifestations of the new "nonequilibrium" thinking is that it has drawn into question the concept of population regulation itself. In his address to the Ecological Society of America on the occasion of being given the Robert H. MacArthur Award, W. W. Murdoch (1994) lamented, "A focus on regulation is apparently at odds with the recent emphasis on nonequilibrium dynamics. . . ." Murdoch believes that "population regulation underlies most other ecological problems of interest, such as the dynamics of disease, competition, and the structure and dynamics of communities." In his address, he developed a conceptual framework that reconciles equilibrium and nonequilibrium thinking within the context of metapopulation dynamics: "The notion of equilibrium is central to regulation, and we therefore need to inquire into the idea that populations can persist via the operation of so-called 'non-equilibrium' dynamics, in which metapopulations are key." Recent advances in metapopulation theory (see Chapter 17) in which both demographic and environmental stochasticity are explicitly included in analytical models of population dynamics (see Foley 1997), though still based on the equilibrium concept, have provided support for Murdoch's idea.

Many natural populations fluctuate, sometimes dramatically, in density. Indeed, there are animal populations that are acknowledged to be characteristically highly variable in size. One of the best studied is the population of Soay sheep (*Ovis aries*) on Hirta Island of the St. Kilda archipelago, which has been censused systematically since 1965 (Clutton-Brock et al. 1991). During that time the population has fluctuated in size between 600 and 1,600 individuals. Are such populations under density-dependent regulation? We shall see that even when we model populations using density-dependent logistic models, large fluctuations in size may occur because of intrinsic demographic characteristics of the population. But it is also possible that density-dependent factors interact with density-independent factors, such as weather, in a complex way to affect population size (Sinclair and Pech 1996).

Consider Figure 16-19a, which shows a population having a density-dependent mortality rate (the solid curve denoted d_m) and a constant (density-independent) reproductive rate (the horizontal line denoted R). Where the two lines cross is, as we have seen before, an equilibrium, K_1. Now consider a similar population that is subject to an additional mortality factor that is density-independent (Figure 16-19b),

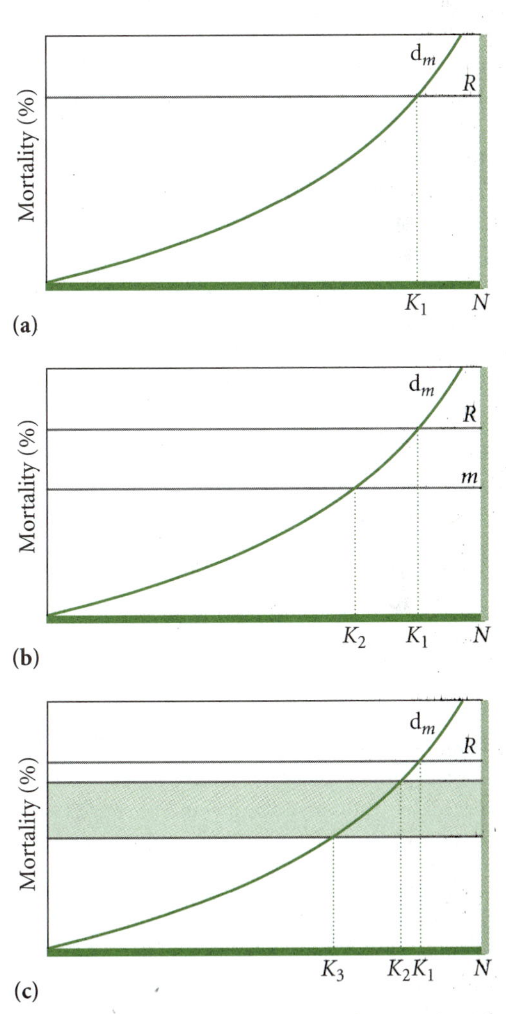

FIGURE 16-19 Possible interactions between density-dependent and density-independent mortality factors. (a) A population with a constant rate of reproduction, R, and a density-dependent mortality rate, d_m, showing an equilibrium point at K_1. (b) A similar population having both a density-dependent mortality rate and a density-independent mortality factor, m. The presence of the density-independent factor lowers the population equilibrium density to K_2. (c) A population in which the density-independent mortality factor fluctuates (within the green shaded area) in response to environmental stochasticity, creating a situation in which the equilibrium density fluctuates between K_2 and K_3. (*After Sinclair and Pech 1996.*)

represented by the horizontal line marked m. The density-independent factor effectively lowers the equilibrium density of the population from K_1 to K_2. If the intensity of the density-independent effect fluctuates in response to environmental stochasticity, then the population equilibrium will also fluctuate. This is shown in Figure 16-19c, in which the density-independent factor m falls somewhere in the shaded area of the graph, in which case the equilibrium fluctuates between K_2 and K_3. The important point to take from this example is that both density-dependent and density-independent factors may affect population dynamics.

The models of population growth and regulation that we have presented in this chapter and the previous one are based only on population density. We saw in Chapter 14 that populations possess a spatial structure that is determined by the arrangement of individuals in the area occupied by the population. In the next chapter we shall explore the effects of population spatial structure on population growth and persistence.

SUMMARY

1. An analytical model of population regulation was presented by Pearl and Reed in 1920 in an effort to understand and predict human population growth. The logistic equation of population growth is $dN/dt = r_0N(1 - N/K)$, where r_0 is the intrinsic exponential growth rate of the population when its size is small, N is the population density, and K is the carrying capacity, the number of individuals that the environment can support. The term $(1 - N/K)$ of the equation represents a density-dependent regulation factor. When $N > K$, the term is negative, and the rate of population growth is negative. When $N = K$, the rate of population growth is 0. The derivative of the logistic equation yields a time-density relationship for the population that is sigmoidal in shape.

2. The rate of growth of a population may be portrayed qualitatively as $dN/dt = f(N)$, where $f(N)$ is the difference between the birth $[b(N)]$ and death $[d(N)]$ functions of the population. Such a qualitative model reveals that populations may come under the influence of multiple equilibrium points, each with a certain domain of attraction—a range of population sizes within which the population tends toward that equilibrium. Certain points, called breakpoints, represent divisions between one equilibrium level and another. The model suggests an explanation why natural populations subjected to continuous perturbations may respond in a discontinuous fashion.

3. Populations may be regulated by factors, called density-dependent factors, whose effects increase as the density of the population increases. Density-independent factors are those whose effects are not related to the size of the population. In animals, crowding may affect survival, fecundity, and the outcome of competition for resources, among other things. Examples of density-dependent effects in plants are slowed growth and reduced seedling survival.

4. In 1954, Australian entomologists H. G. Andrewartha and L. C. Birch challenged the idea that populations are regulated by density-dependent factors, thereby setting off a controversy about the relative importance of density-dependent and density-independent regulation that continues today. Density dependence is difficult to demonstrate in nature and may be obscured by environmental variation.

5. Variation in birth and death rates, called demographic stochasticity, and variation in environmental conditions, called environmental stochasticity, may obscure the effects of density dependence, a situation called density vagueness because an equilibrium level cannot be observed. Because of this, some have called into question the whole concept of population equilibrium.

6. It is likely that both density-dependent and density-independent factors regulate population levels. Fluctuation in the intensity of density-independent factors could lead to a lower equilibrium. Year-to-year changes in density-independent factors could obscure the effects of density dependence.

EXERCISES

1. Plot the equation $r = r_0(1 - N/K)$ over the range of $N = 1 \ldots 100$ with $r_0 = 1.5$ for two different levels of carrying capacity, $K = 50$ and $K = 75$. Provide an ecological explanation of the two graphs.

2. Construct a family of three logistic curves having $r_0 = 1.5$, $r_0 = 2.0$, and $r_0 = 3.0$ respectively and $K = 25$. At what population size will the peak rate of growth occur in each of the three equations? Which population will have the highest rate of growth at the peak, and what will that rate be?

3. Sketch a qualitative population model similar to the one shown in Figure 16-6 that includes extinction. Describe the shape of the birth and death rate curves that might lead to such a population pattern.

4. Suppose that you are a wildlife biologist with responsibility for managing the white-tailed deer herds in your state. Explain why you would be interested in the debate about density dependence and density independence.

CHAPTER 17

Metapopulations

GUIDING QUESTIONS

- What is a metapopulation?
- What are the roles of extinction and colonization in metapopulations?
- What is the Levins metapopulation model, and how does it differ from the logistic model of single populations?
- What is meant by metapopulation structure?
- How do local population size and isolation affect metapopulation persistence?
- What is the rescue effect?
- How do extinction and colonization affect genetic variation in metapopulations?
- How does the metapopulation concept relate to the concept of landscape ecology?

Areas of habitat that contain the necessary resources and conditions for a population to persist are called **habitat patches** or, simply, patches. The individuals of a species that live in a habitat patch constitute a **local population** (see Chapter 14). (For simplicity, in this chapter, we will use the terms *local population* and *population* synonymously.) Because the habitat of any species is usually not distributed homogeneously in nature, a species may occur as a set of local populations, each occupying a patch separated from other patches by areas of unsuitable habitat. Often these populations interact with one another through immigration and emigration. A set of local populations occupying an array of habitat patches and connected to one another by the movement of individuals among them is called a **metapopulation,** or "population of populations."

Few ecological concepts have received more attention and generated more theoretical and empirical research in recent years than the metapopulation concept (Hanski and Gilpin 1997). It has proved to be a valuable framework for studying the population dynamics of subdivided populations (McCullough 1996), the conservation of threatened and endangered species (Schoener 1991, Lahaye et al. 1994, Hanski et al. 1995, Litvaitis and Villafuerte 1996, Thomas et al. 1996; but see Simberloff 1994 and Hanski and Simberloff 1997), the dynamics of predator-prey systems (Kareiva 1987, Taylor 1991), the biological control of pest organisms (Murdoch et al. 1984, 1985), competition among animals and plants (Bengtsson 1991, Tilman 1994), and the evolution of virulence in host-parasite interactions (May and Nowak 1994), including infectious diseases of humans (Anderson and May 1991). In this chapter we will present the basic concepts of metapopulation theory.

17.1 The metapopulation concept encompasses the dynamics of sets of interacting local populations.

The conceptual foundation of metapopulation theory, and the adoption of the term *metapopulation,* may be attributed to Richard Levins (1969, 1970). In Levins's simple model, a metapopulation of a species is thought of as a set of small populations occupying an array of similar small habitat patches situated in a matrix of uniform unsuitable habitat (Figure 17-1). At any particular time, some suitable habitat patches may contain no individuals of the species, and thus may be subject to colonization from other patches that are inhabited. Because the local populations are assumed to be small, local catastrophes and chance fluctuations in numbers of individuals have important effects on population dynamics, so there is a high probability of extinction of a local population during a particular time interval. Metapopulation dynamics, then, are represented as a

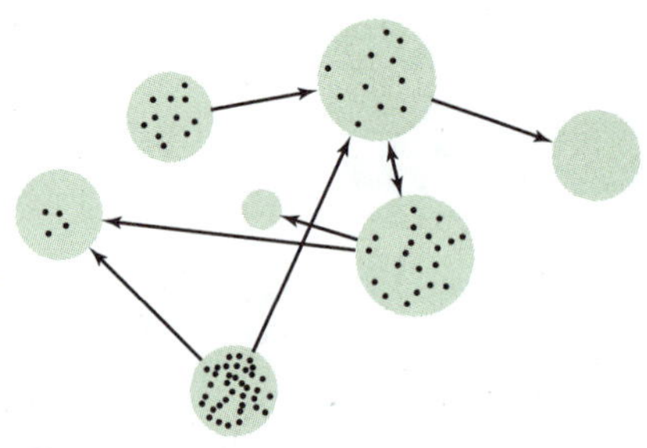

FIGURE 17-1 A metapopulation can be portrayed as a set of discrete local populations with partially independent local population dynamics. At any one time, some patches of suitable habitat may be occupied while others are not. Individuals may migrate from occupied patches to colonize unoccupied patches.

balance between extinction and colonization. We can imagine the metapopulation as a set of blinking lights, with occupied patches represented by a burning light that is extinguished if extinction occurs and turned on again if the patch is colonized. When a habitat patch becomes vacant through extinction and is then recolonized by individuals from other local populations in the area, an extinction-colonization **turnover event** has occurred. When colonization rates are lower than extinction rates, the extinction of all local populations, and thus the extinction of the metapopulation, may occur. The length of time until all populations have become extinct is referred to as the **metapopulation persistence time** (Hanski and Simberloff 1997).

The metapopulation concept accurately depicts the dynamics of many natural populations with patchy distributions (Harrison and Taylor 1997). The mistletoes, for example, are a group of plants, most species of which belong to the families Viscaceae and Loranthaceae, that live as parasites of other plants, usually trees. Each mistletoe plant obtains water and nutrients through the xylem system of the host plant. In most cases, dispersal from one host to the next takes place via birds, which eat the mistletoe fruit and defecate the seeds (which are coated with a sticky substance) onto the branch of another tree. Mistletoes may be thought of as occurring in metapopulations in which host trees are habitable patches separated from one another by areas without trees, which are thus inhospitable to the mistletoes (Overton 1994).

Other plant metapopulations may arise because of activities of other plants or animals that create a patchy environment. For example, plant species that grow best in the gaps created when trees fall in a dense forest—called **canopy gaps,**—represent metapopulations in which extinctions occur as the canopy slowly closes by regrowth. Such plants must have means of dispersing their seeds that allow new canopy gaps to be quickly colonized. In other cases, a mosaic of suitable habitats for a particular plant species may be created by the activities of animals, such as that formed by badger mounds in tallgrass prairie (Platt 1975, Platt and Weiss 1977).

The natural history of butterflies represents the essentials of basic metapopulation dynamics because many species are specialized to lay eggs on specific host plant species, which often have patchy distributions. Local populations are thus connected by the dispersal dynamics of adults, who fly about in search of patches of host plants on which to lay eggs. Butterflies have become prominent models for the study of metapopulations (Thomas and Hanski 1997). They also serve as sensitive indicators of the pressures of habitat fragmentation. Many species of butterflies, particularly those in Europe, have become endangered as a result of such fragmentation (Warren 1992, 1993). The populations of many other species of insects are also best thought of as metapopulations.

Many birds and mammals also occur in sets of local populations inhabiting patches interspersed with inhospitable areas (McCullough 1996). For example, the Lower Keys marsh rabbit (*Sylvilagus palustris hefneri*) lives in isolated patches of high marsh (above the influence of the tides) in the Lower Keys of Florida (Forys and Humphrey 1996). The Southern California spotted owl (*Strix occidentalis occidentalis*), a threatened subspecies of the endangered northern spotted owl (*Strix occidentalis caurina*), occurs in a series of isolated populations in the San Bernardino Mountains of southern California (Figure 17-2). Both the Lower Keys marsh rabbit and the California spotted owl are threatened by human development activities that tend to further fragment the landscape. In the case of the rabbit, although lower marsh areas are generally protected from development, the upper marsh is not. Thus, habitat for the rabbit is disappearing. In the case of the owl, individual subpopulations show recent declines, which are probably the result of both increased isolation of the habitat patches and natural environmental fluctuations (Lahaye et al. 1994, Gutiérrez and Harrison 1996)—factors whose effects we shall examine below.

17.2 Simple metapopulation models represent a balance between local extinction and recolonization.

The first simple model of metapopulation dynamics was presented by Levins (1969, 1970), and it serves as a basis for much of metapopulation theory (Hanski 1991, 1997). In this model, the metapopulation is

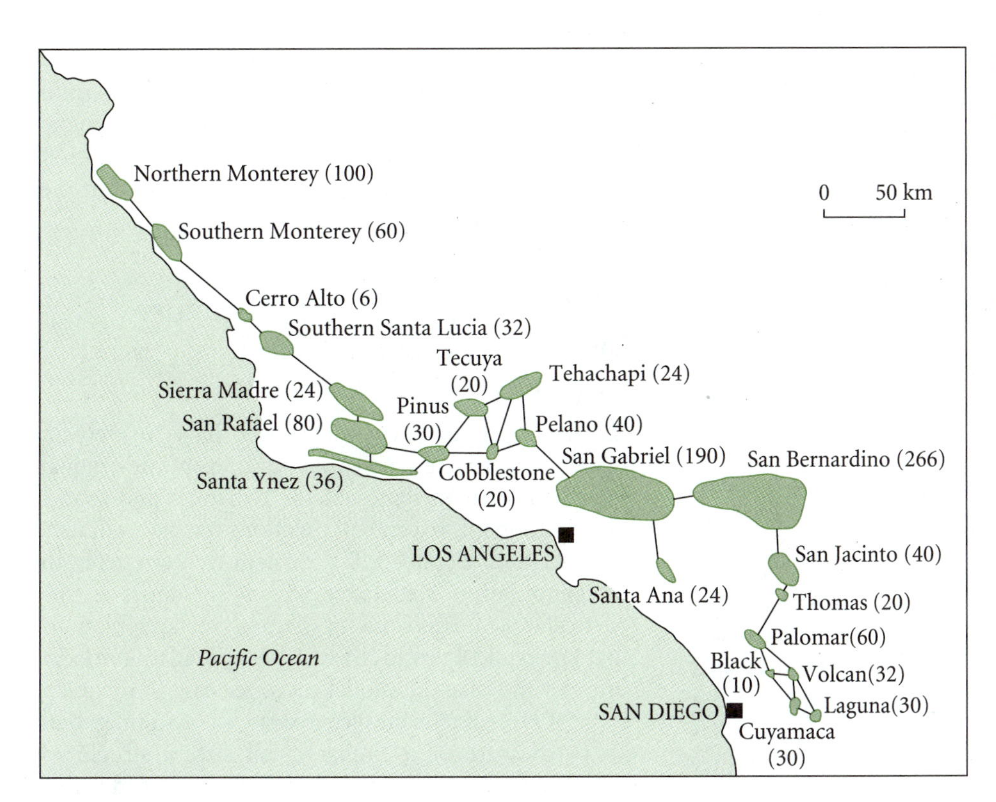

FIGURE 17-2 Metapopulation of the Southern California spotted owl in the San Bernardino Mountains. The numbers given in parentheses represent estimated carrying capacities for each patch. (*From Lahaye et al. 1994.*)

conceptualized as a group of local populations, each having a density of either 0 (extinct) or K (equilibrium density), where K is the patch carrying capacity. The carrying capacity of the patch is the number of individuals that can be supported by the resources in the patch for an indefinite period of time (the same concept of carrying capacity that was introduced in Chapter 16). The assumption that a patch has either no individuals or the carrying capacity is, of course, a simplification. At any time, some proportion, p, of the total number of patches in the metapopulation will be occupied, and the remaining fraction, $1 - p$, will be unoccupied or extinct. The rate of change of p is given by

$$\frac{dp}{dt} = mp(1 - p) - ep,$$

where m is the rate of patch colonization and e is the rate of patch extinction. When the rate of change of occupancy is 0, $dp/dt = 0$, and $\hat{p} = 1 - e/m$, the equilibrium proportion of occupied populations (Levins 1969, 1970; Hanski 1991). The main prediction of this model is that species will not persist—that is, $p < 0$—when extinction rate, e, is greater than colonization rate, m, in the metapopulation. Or, to put it another way, population persistence requires $1 < e/m$.

The Levins model has a number of simplifying assumptions. First, the processes of population growth, resulting from the dynamics of birth and death, and of population regulation, resulting from the interaction of birth and death processes with the environment, in the local populations are not considered. The metaphor of the switching on and off of small lights reflects this assumption clearly. No time is presumed to pass between the colonization of an empty patch and the turning on of the light or between the beginning of the decline of a population and the turning off of the light. Second, all the patches are assumed to be equal in area and to be equally isolated from the other patches—that is, the movement of individuals occurs with equal probability between any pair of patches. The populations within the patches are also assumed to be independent from other populations—that is, the population dynamics of one population are not affected by immigrants from another population. If populations are not independent, then extinction and colonization events may be correlated or synchronized. Such a situation becomes more likely as the rate of migration among patches increases, as we shall see below. In the Levins model, it is only when immigrants enter an unoccupied patch (colonization) that their movement has an effect on the metapopulation dynamics. Third, the extent to which individuals move from one patch to the next (immigration and emigration) is assumed to be neither extremely high nor extremely low. If movement among local populations is extremely high, it is likely that extinction will be rare, in which case the metapopulation concept is not needed to explain population dynamics. Likewise, if movement among populations is extremely low, the populations are essentially unconnected, and thus represent individual, though possibly small, populations. Finally, the Levins model does not take into consideration variation in the ease with which individuals may move from one patch to the next. Such variation might arise as a result of landscape features (see Figure 17-16). For example, inhabitants of occupied habitat patch A, located equidistant from two unoccupied patches, B and C, may colonize B more quickly than C if there is a mountain between A and C. The Levins model is referred to as a **patch metapopulation model** because the local population dynamics are ignored; all that is considered is the proportion of occupied and unoccupied patches.

Comparison of Metapopulation and Logistic Population Models

Let us take a moment to contrast the metapopulation concept with the population models that we have discussed in the previous chapters in this part of the book. Logistic population models depend upon the relationship between population density, N, and the carrying capacity, K (see Section 16.1). Metapopulation models emphasize extinction-colonization dynamics rather than the dynamics of population density. The logistic models lend themselves most naturally to questions such as "How large can the population be?" and "What factors keep the population within relatively narrow limits of size?" The metapopulation concept generates questions such as "How long will the population persist?" and "How do the landscape and the movement patterns of individuals interact to affect the persistence of the population?" Such questions arise particularly often in conservation biology, in which the metapopulation concept is proving to be most useful. Ecologists are finding that many metapopulations are the result of extensive habitat fragmentation related to human activities, and thus often contain threatened or endangered species in low numbers.

Metapopulation theory adds a new wrinkle to the debate about density-dependent regulation and the controversy over whether and in what form population equilibria exist. In our discussion of the regulation of single populations in Chapter 16, we suggested strongly that density-dependent regulation is most important. The study of metapopulations, however, can proceed relatively unencumbered by equilibrium concepts (such as carrying capacity and density dependence) because the total density of many metapopulations is very often far below the carrying capacity for a particular habitat (Hanski 1997). While the effects of density dependence are discounted somewhat in the study of metapopulations, the effects of demographic and environmental stochasticity (see Section 16.5) take on considerable importance. We shall show shortly how repeated recolonization events involving one or very few colonizers may quickly reduce genetic variation in metapopulations. A bad winter or an unusually destructive series of fires could wreak havoc in a metapopulation by destroying many small local populations or by disrupting the migration routes that interconnect patches and thus sustain the population through recolonization.

17.3 Metapopulation structure involves characteristics of patch size and density.

Enhancements of Levins's approach have focused on relaxing the very restrictive assumptions of the original model in order to represent the dynamics and spatial organization of natural populations more accurately (Hanski and Gilpin 1997). Ecologists often refer to **metapopulation structure,** which encompasses the particular characteristics of natural metapopulations that are explicitly included in a metapopulation model in order to make the model more realistic (Gyllenberg et al. 1997). For example, instead of assuming that all patches are of a similar small size, a structured metapopulation model might include a specified distribution of patch sizes. Or, the assumption that patches are equally isolated from one another might be relaxed by specifying different migration rates among local populations. Likewise, densities of local populations or population age structure might be stipulated. In practice, the stipulation of some metapopulation structure is usually required in order to lend the metapopulation concept utility.

Levins's simple metapopulation model does not specify the spatial relationship among populations either in relation to one another or in relation to their positions in the matrix of unsuitable habitat. The implication of this assumption is that all populations are equally accessible to dispersers; that is, all populations are equally connected. This type of model is called a **spatially implicit metapopulation model.** But some models relate the extent of connectedness among populations—the extent of migration—to the distances between populations. Such models, which are referred to as **spatially explicit metapopulation models,** most commonly assume an inverse relationship between the extent of exchange between two populations by migration and the distance between those populations. Some spatially explicit models even restrict the placement of populations to specific locations within an imaginary grid in the habitat matrix (Caswell and Etter 1993).

Patch Size and Density

The dynamics of metapopulations are affected not only by extinction and colonization rates, but also by the relationship of those rates to the spatial

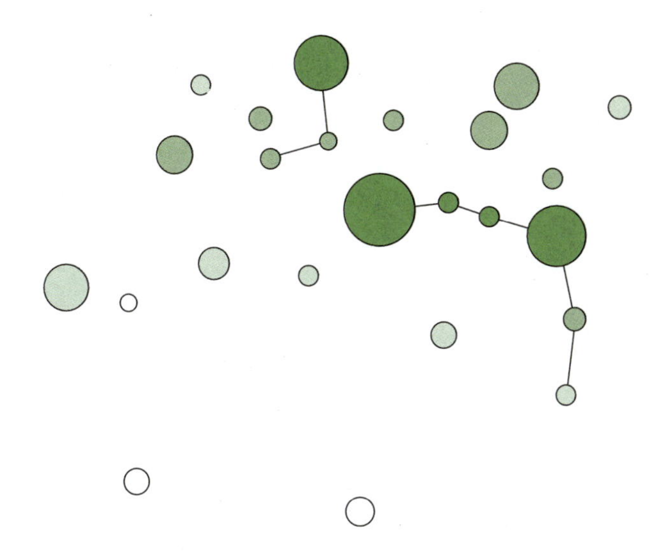

FIGURE 17-3 The pattern of probability of occupancy of habitat patches of differing size and isolation. The darker the shading, the higher the probability that the patch will be occupied. In general, large patches and patches that are close to another population have a higher probability of being occupied than those that are small and isolated. (*From Opdam 1991.*)

arrangement and density of the habitat patches (our reference to density here refers to the density of patches, not to the density of populations in patches). In order for a metapopulation to persist, the overall colonization rate must be greater than the extinction rate ($1 < e/m$). But successful colonization requires that individuals move from an occupied site to one that is not occupied, and such movement may be prevented if there is a great distance between the occupied and unoccupied sites. The distance barrier may be overcome in time if potential colonists arise continuously from an occupied site, thereby increasing the chance that one will make it to the distant unoccupied site. But in order for this to happen, the colonizing population must persist, and in general, population persistence is related to population size. Small populations suffer a higher risk of extinction than do large populations.

Figure 17-3 depicts the relationship between the pattern of patch occupancy in a metapopulation and the spatial and size relationships among patches. The larger the patch, the more likely it is to be occupied. Small, isolated patches remain unoccupied because they are too isolated from occupied patches to be colonized. Patches of similar size that are located closer to occupied patches are more likely to receive colonists. In general, where the patch density is high, the probability that a patch will be occupied is high, and that probability increases with patch size.

Studies of species under threat of extinction because of increasing habitat fragmentation provide evidence for the relationships shown in Figure 17-3. Studies of the Glanville fritillary butterfly (*Melitaea cinxia*), a species that went extinct in Finland in the 1970s but persists in other areas, represent a good example. Figure 17-4a shows the relationship between patch size and the fraction of occupied patches. Patch occupancy is much higher in areas where average patch size is large (e.g., 1.0 ha) than in areas that contain primarily small patches (<0.01 ha). Lower patch isolation, measured as a higher density of patches, results in a higher proportion of occupied patches (Figure 17-4b) (Hanski et al. 1995).

A number of different metapopulation structures have been employed (Hanski and Simberloff 1997). **Mainland-island metapopulations** are those in which a system of patches, or islands, is situated near another, larger, patch, the mainland, from which dispersers can reach all the islands. The assumption in these models is that the mainland population never goes extinct. Another type of metapopulation structure is the **source-sink metapopulation,** in which, as we described in

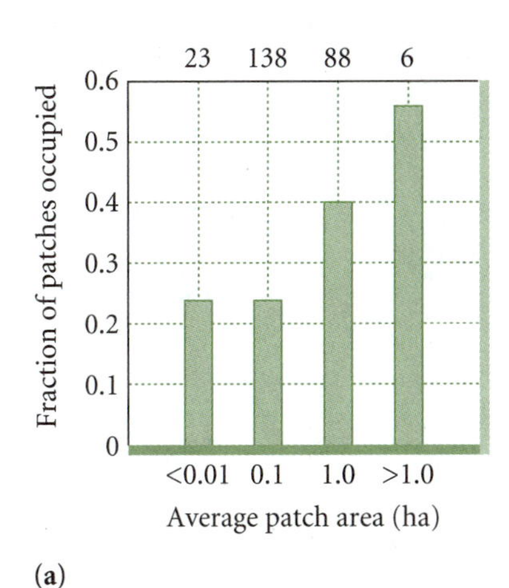

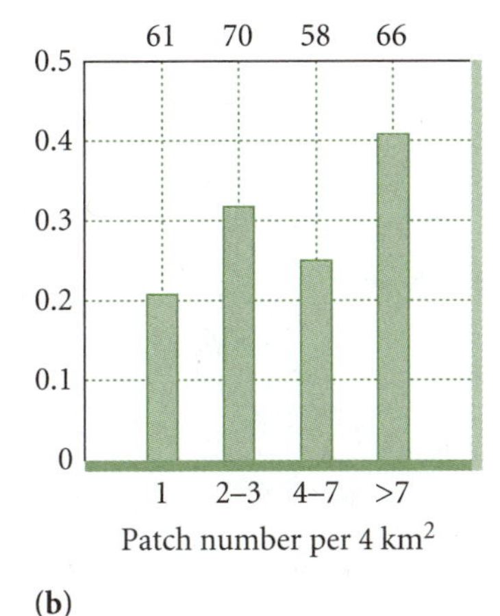

FIGURE 17-4 (a) Effect of patch area on the fraction of patches occupied in a metapopulation of the Glanville fritillary butterfly (*Melitaea cinxia*). The proportion of occupied patches was higher for large patches. (b) Effect of the density of patches. The proportion of occupied patches was highest for areas with the highest density of patches. (*From Hanski et al. 1995.*)

Chapter 14, some populations (sources) have a positive growth rate at low densities and other populations (sinks) have a negative growth rate in the absence of immigration.

Compensation of Patch Size and Density

Patch size and density (again, we are referring to the density of patches, not to the density of populations in patches) may interact in a compensatory way to affect metapopulation persistence. This interaction can be shown with a modification of the basic metapopulation model, $dp/dt = mp(1 - p) - ep$ (Hanski 1991). Suppose that the migration rate, m, is dependent on the degree to which a patch is isolated, measured as some distance, D. That is, migration to an unoccupied patch that is isolated from a colonizing patch (high D) is less likely than if the patch is in the proximity of a colonizing patch (low D). This pattern is shown in Figure 17-3, in which more isolated patches have a lower probability of occupancy. The exact relationship between D and m is not that important, so long as m declines as D increases. One possible relationship between the two is a negative exponential function,

$$m = m_0 e^{-aD},$$

where m_0 and a are parameters. The shape of this function is shown in Figure 17-5.

Now, let us suppose that the extinction rate, e, is dependent on the size of the patch, so that populations in larger patches have a lower chance of becoming extinct than do those in smaller patches. Again, this is the situation implied in Figure 17-3, in which large patches are shown to have a very high frequency of occupancy. We shall let extinction be related to patch area, A, in a negative exponential function as well,

$$e = e_0 e^{-bA},$$

where e_0 and b are parameters (see Figure 17-5).

If we substitute these two equations into the equation for the equilibrium value of p, $\hat{p} = 1 - e/m$ (see Section 17.2 above), we obtain

$$\hat{p} = 1 - \frac{e_0 e^{-bA}}{m_0 e^{-aD}},$$

which may be simplified by factoring e_0/m_0 from the right-hand term to obtain

$$\hat{p} = 1 - \left(\frac{e_0}{m_0}\right) e^{-bA+aD}.$$

This equation shows how average patch area and isolation in a metapopulation affect the equilibrium value of patch occupancy. If the term e_0/m_0 is constant, then we can see how compensation occurs by examining the term e^{-bA+aD}. For a particular set of values of a and b, any change in patch area, A, that results in a reduction in the exponent of e, since the sign on b is negative, will have to be offset by an increase in isolation, D, in order for the term to remain constant and $\hat{p}$ to be unchanged. This is easy to see if $a = b = 1$. Then, the term becomes e^{-A+D}. If $A = 1$ and $D = 1$, the term is $e^{-1+1} = e^0 = 1$. If D is increased to 3, then, in order for the term to remain 1, A will have to be 2.

Figure 17-6 provides a schematic representation of what this means for metapopulation persistence. In this figure, $a = b = 1$, $m_0 = 0.7$, and $e_0 = 0.05$; thus $e_0/m_0 = 0.071$. If patch area A is 1 ($A = 1$) and the patch is isolated by a distance of $D = 1$ distance units, as in the first example, then $\hat{p} = 1 - 0.07e^{-1+1} = 1 - 0.07 = 0.993$. In that situation, the equilibrium patch occupancy is very high because of the proximity of patches to one another. If the isolation of a patch of the same size ($A = 1$) is twice that of the first example ($D = 2$), then $\hat{p} = 1 - 0.07e^{-1+2} = 1 - (0.07)(2.718) = 0.81$, and the equilibrium occupancy declines. The decline is even greater ($\hat{p} = 0.48$) if D is increased to 3. However, if when $D = 3$, the size of the patch is increased to $A = 2$, then p is the same as for the situation in which $A = 1$ and $D = 2$ (that is, $\hat{p} = 0.81$). The greater patch area compensates for the greater isolation.

The relationship between patch size and isolation is demonstrated by the dynamics of populations of shrews (*Sorex araneus*) that occupy islands in two lakes in Finland (Figure 17-7). Larger islands are more

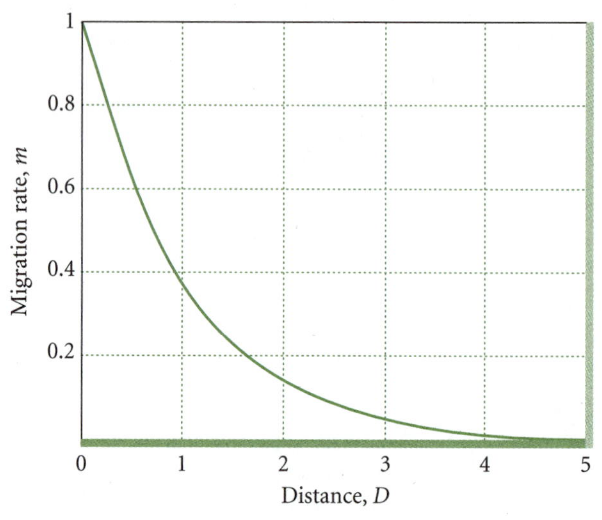

FIGURE 17-5 Negative exponential relationship between migration rate, m, and distance of patch isolation, D ($m = m_0e^{-aD}$, where m_0 and a are parameters). The relationship between extinction rate, e, and patch area, A ($e = e_0e^{-bA}$, where e_0 and b are parameters) will have the same shape.

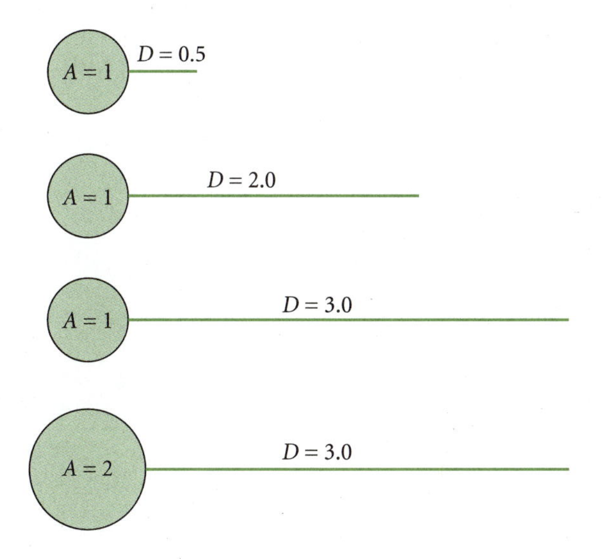

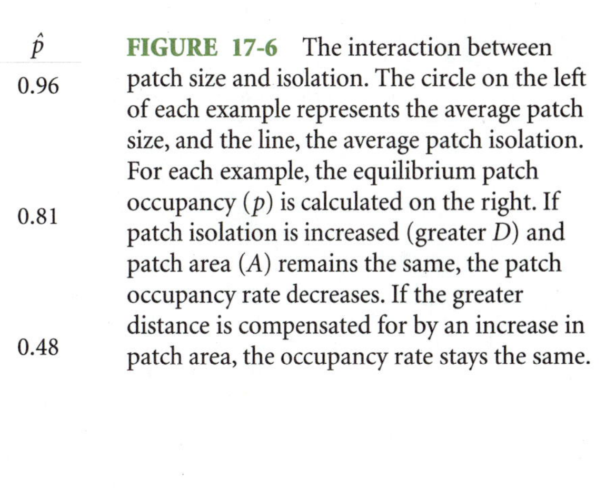

FIGURE 17-6 The interaction between patch size and isolation. The circle on the left of each example represents the average patch size, and the line, the average patch isolation. For each example, the equilibrium patch occupancy (p) is calculated on the right. If patch isolation is increased (greater D) and patch area (A) remains the same, the patch occupancy rate decreases. If the greater distance is compensated for by an increase in patch area, the occupancy rate stays the same.

likely to be occupied than smaller islands, and islands that are more than 1.5 km from an occupied island do not contain populations of shrews (Peltonen and Hanski 1991, Hanski 1991). A similar relationship exits in the metapopulation dynamics of the brown kiwi (*Apteryx australis mantelli*) of New Zealand. Kiwis are flightless birds that live in forest fragments of varying sizes separated by open areas of pasture (Potter 1990). The birds disperse from one fragment to another by walking, and generally have poor overall dispersal abilities. Potter found that fragments isolated from other fragments by more than 80 meters were generally not occupied unless they were considerably larger than the nearest fragment.

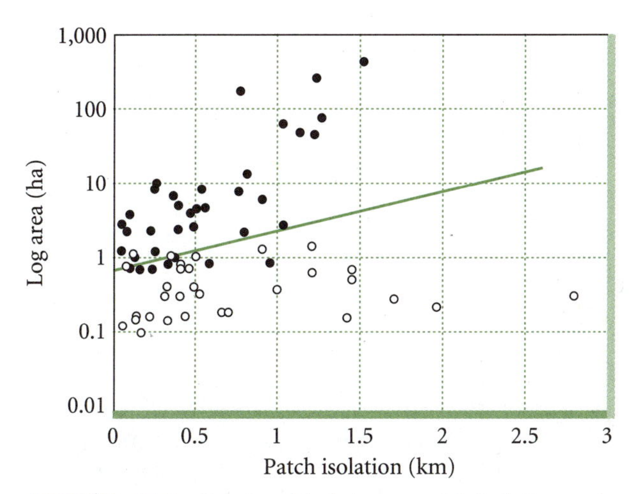

FIGURE 17-7 Relationship between patch isolation and area (shown as log [ha]) in a metapopulation of the shrew *Sorex araneus* on islands in two lakes in Finland. The solid black dots represent patches that are occupied; the open white dots, patches that are unoccupied. Large patches are more likely to be occupied, and patches that are isolated by more than 1.5 km are not occupied. (*From Hanski 1991.*)

17.4 Metapopulations may fluctuate in their level of patch occupancy.

Metapopulations are characterized by frequent extinctions of populations in patches followed by recolonization of those patches (turnover). Thus, at any point in time, some patches are unoccupied. In metapopulations, then, it is the level of patch occupancy and the rate of turnover (extinction-colonization events) that are important, rather than the density of the population in a particular patch. Studies of metapopulations reveal that patch occupancy varies through time, and that this variation may be the result of a myriad of factors, including variation in habitat quality among patches, disturbance events, demographic stochasticity within populations, and interactions with organisms at different trophic levels (e.g., herbivores or parasites).

In their lengthy study of the American pika (*Ochotona princeps*) near Bodie, California, Smith and coworkers (Smith 1974a,b, 1978, 1980, Smith and Gilpin 1997) conducted a census of patch occupancy, patch size, distance among patches, and other measures during each of four years, 1972, 1977, 1989, and 1991. They found that the percentage of patches occupied by pika in the study area varied between the four census periods, with patch occupancy dropping considerably during the last two census periods (Table 17-1). The census showed that the average patch size remained relatively stable, but that the average distance between occupied patches increased. A visual representation of the census data shows that this variation in the patch occupancy pattern resulted principally from extinctions that occurred in the southern part of the metapopulation range (Figure 17-8). The causes of the higher extinction rate in the southern region are not clear. Differences in habitat quality or in the number of predators or competitors between northern and southern regions might explain the difference. Alternatively,

TABLE 17-1 Census data for a metapopulation of American pika (*Ochotona princeps*) near Bodie, California

	Census year			
	1972	1977	1989	1991
Average size of occupied patch (perimeter, m)	96.0	90.5	96.9	85.9
Average interpatch distance (m)	101.5	110.1	102.6	193.4
Number of patches censused	78	78	77	78
Percentage of patches occupied	60.3	57.7	44.2	43.6

(Data from Smith and Gilpin 1997.)

the more numerous relatively large populations in the northern half of the range (shown as large solid squares in Figure 17-8) might serve as source populations in that area, thereby making recolonization more likely (Smith and Gilpin 1997).

Variation in patch occupancy is also observed in metapopulations of ragwort (*Senecio jacobaea*), a weedy biennial plant (2 years to first reproduction) that thrives in disturbed areas such as sand dunes and roadsides and has invaded many areas of the world.

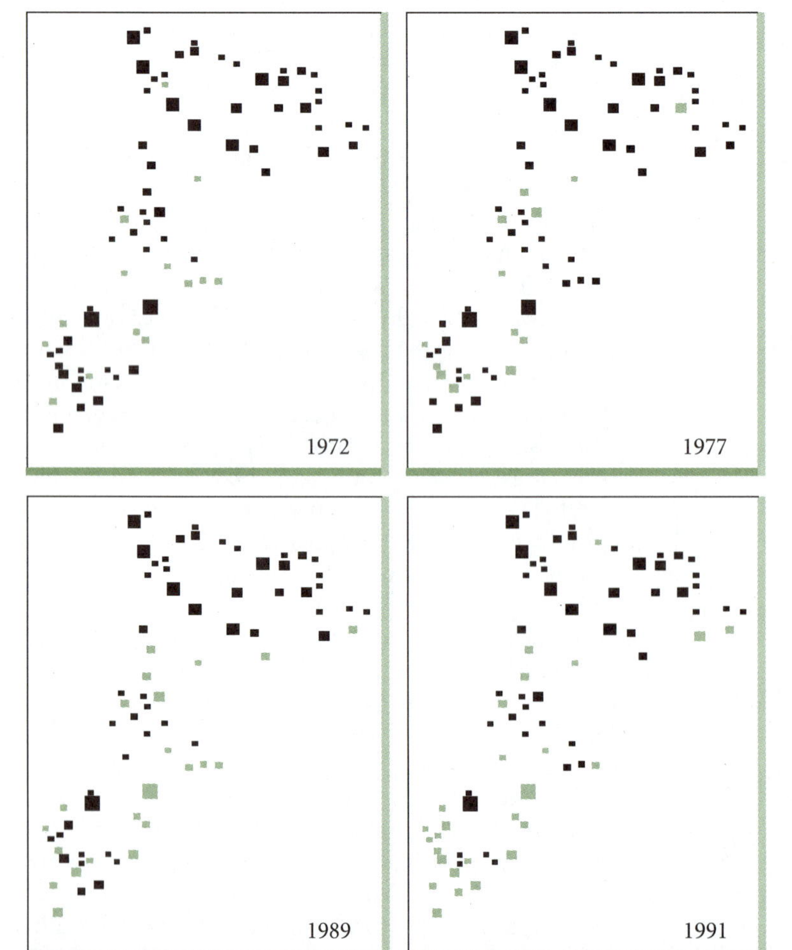

FIGURE 17-8 Occupied (black) and unoccupied (green) patches in the American pika (*Ochotona princeps*) metapopulation near Bodie, California. The size of each square is proportional to its carrying capacity. The figure shows a decrease in the proportion of occupied patches in 1989 and 1991, with more extinctions in the southern portion of the metapopulation range. (*From Smith and Gilpin 1997.*)

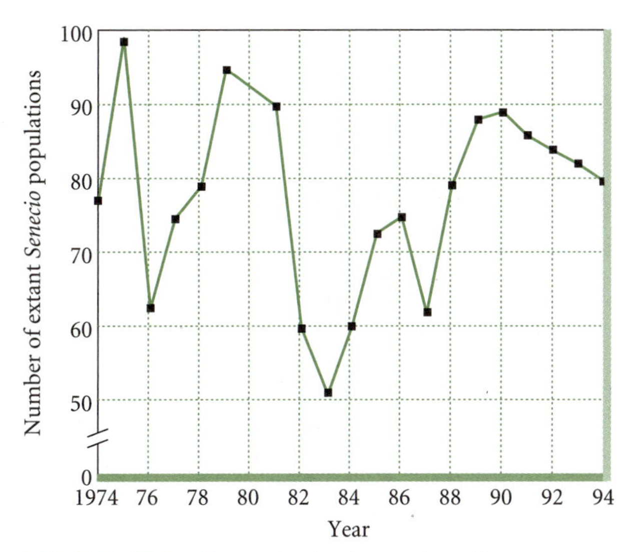

FIGURE 17-9 Fluctuation in the number of extant subpopulations in a metapopulation of ragwort (*Senecio jacobaea*) in the Netherlands over a 20-year period. (*From van der Meijden and van der Veen-van Wijk 1997.*)

Ragwort possesses an extremely high reproductive capacity: a single plant is capable of producing 20,000 or more wind-dispersed seeds. Thus, it is capable of colonizing unoccupied areas quickly. A 20-year census of 102 ragwort populations in a 6-km^2 coastal sand dune area of the Netherlands showed considerable variation in the number of extant local populations (van der Meijden and van der Veen-van Wijk 1997). Figure 17-9 shows the number of populations in each year between 1974 and 1994. Even though the ragwort metapopulation persisted during that time, the number of occupied sites fluctuated widely.

The dynamics of the ragwort metapopulation are not entirely related to the reproductive potential and dispersal abilities of the plant. Rather, they result from a complex interaction between the ragwort and two other organisms that live in close association with it. One is the cinnabar moth (*Tyria jacobaeae*), an herbivore whose larvae specialize on ragwort. The cinnabar moth has the peculiar habit of periodically completely defoliating its ragwort food, which leads to widespread starvation of moth larvae. Another player in this system is the braconid wasp *Cotesia popularis,* a parasitoid of the cinnabar moth.

17.5 Metapopulations are affected by migration and local population demographics.

We suggested above that the reason that the more isolated patches in a metapopulation are less likely to be colonized is that the greater distance represents a greater impediment to the movement of individuals. Although this is a reasonable basic relationship, other factors, such as patch quality and environmental and demographic stochasticity, affect the immigration and emigration of the individuals in a patch. Moreover, these factors are likely to have different effects in different patches, and indeed, it is clear from recent studies of butterfly metapopulations that immigration and emigration rates are patch specific (Hanski et al. 1994, Sutcliffe et al. 1997a,b). Thus, the rate at which patches are colonized has to do not only with the size of each patch and its proximity to other patches, as shown in Figure 17-3, but also with the quality of the patch, its relationship to landscape features, and the demographic features of the population occupying the patch. Let us examine a few issues related to this idea.

The Patch Population-Metapopulation Continuum

The different patterns of movement of animals in spatially structured populations lie along a continuum. Situations in which individuals move freely among habitat patches represent one end of the continuum, and situations in which there is very little movement between local populations represent the other extreme. Populations with movement patterns of the first type are sometimes called **patch populations.** Such populations display spatial patchiness, but the individuals in the population are not segregated to any great degree by the spatial structure of the habitat. When there is little or no movement between patches, the patches harbor independent populations; that is, the populations are essentially isolated from one another by the spatial structure of the habitat. Neither of these extremes represents a metapopulation. It is only when there is an intermediate level of movement among patches that the population is considered a metapopulation (Harrison 1991, 1994). The extent to which migration occurs—that is, the extent to which a population displays the characteristics of a patch population or a metapopulation—may affect the persistence of that population.

What factors affect the movement patterns of organisms in spatially structured populations? Demographic characteristics such as density and age structure, morphological features such as body size, and environmental factors such as temperature may all play a part in shaping movement patterns. Nieminen (1996) investigated the migration patterns of seven species of moths and butterflies that occupy a small group of islands in Finland. The seven species showed different overall migration rates, which were to some extent reflective of the morphology of the species. For example, migration rates were higher in moth species with long wingspans than in those with short wingspans (Figure 17-10a). Environmental factors also played a role in determining the migration rates

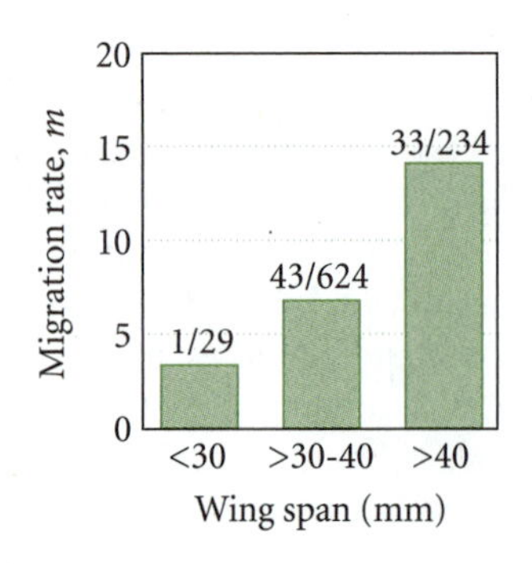

(a)

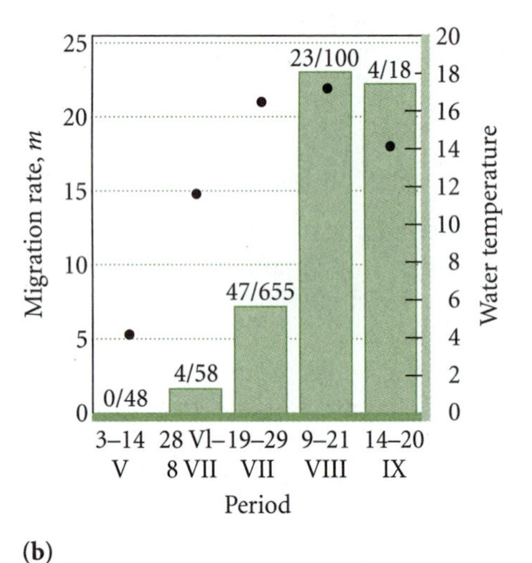

(b)

FIGURE 17-10 Factors affecting migration rates in metapopulations of moths in Finland. (a) Migration rates were greater for moth species having a greater wingspan. (b) Migration rates increased during periods (denoted by the bars) when seawater temperature (black dots) was highest. (*From Nieminen 1996.*)

of moths in the island group. Migration rates were highest during the times of year when seawater temperatures were high (Figure 17-10b). The reason for this, apparently, is that when the seawater is warm, the air temperature above the water is closer to that above the islands, and moths are more likely to move from land over the water. Nieminen's study also suggested that emigration was greatest from populations with the greatest density, though these results were inconclusive because densities of moths and butterflies are difficult to determine.

The Rescue Effect

As we have seen, one of the assumptions of simple metapopulation models is that local patches either contain a population (occupied) or do not (extinct). Structured metapopulation models relax this assumption and take into consideration the demography of individual populations. These more advanced models give us an avenue for understanding the fluctuation or stability of a local population in the context of regional processes. In particular, they afford us a conceptual framework within which to examine the question of how a population's connection to other populations through migration may affect its dynamics.

Stacey and Taper (1992) explored this question in their examination of a number of demographic variables—such as annual reproductive success and survival of juveniles and adults—in an isolated population of acorn woodpeckers (*Melanerpes formicivorus*) in the mountains of central New Mexico. From previous studies, it was known that these variables and others exhibited considerable annual variation (demographic stochasticity) as a result of local environmental stochasticity, particularly with respect to rainfall. Using simulation techniques, Stacey and Taper discovered that, if the woodpecker population was considered completely closed to immigration, it could be expected to go extinct in less than 20 years (Figure 17-11a). Extinction would occur simply because of the likelihood of at least one very poor reproductive year. Further work with the simulation model, however, showed that even low levels of migration into the population would vastly increase the persistence time of the population (Figure 17-11b). Situations like this one, in which immigration prevents a population from declining to eventual extinction, demonstrate a phenomenon called the **rescue effect.** The acorn woodpecker population is prevented from fluctuating so widely as to go extinct—it is "rescued" from extinction—by input from another population.

The rescue effect may make it difficult to determine whether or not a local population is part of a metapopulation, because migration may prevent local extinction from occurring. Extinction-recolonization events, the main process in simple metapopulations, may be replaced with patterns in which individual local populations show annual variation in size, but persist over long periods of time (Stacey et al. 1997). The rescue effect need not be unidirectional, as in strict source-sink models or mainland-island models. In metapopulations in which local population growth rates are highly variable, a particular population may serve as a source of migrants, and thus rescue another population in years when its growth rate is high and be subject to rescue by another population having a higher growth rate in years when its own growth rate is low.

Stacey et al. (1997) have identified a number of natural populations that appear to show rescue effects. Many species of butterflies, for example, are observed to occur in patches. In the case of checkerspot butterflies (*Euphydryas*) of North America, this distribution reflects the patchy distribution of the plants on

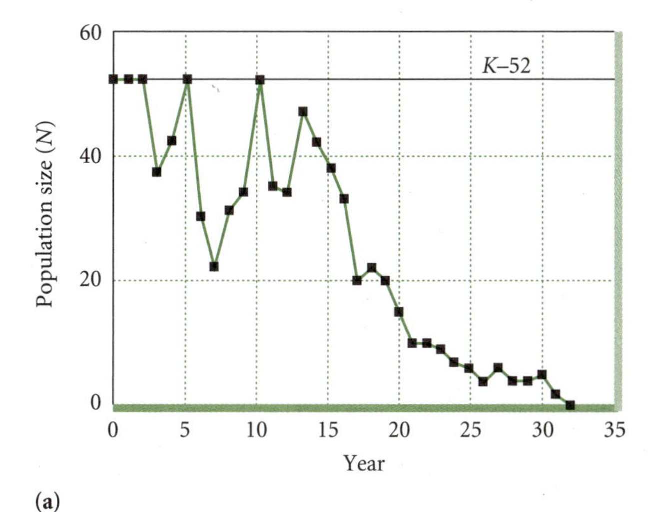

(a)

(b)

FIGURE 17-11 Results of a simulation of population size in a population of acorn woodpeckers (*Melanerpes formicivorus*) in central New Mexico. Demographic parameters from previous studies were used to construct the simulation. (a) Projected population trend if it is assumed that the population receives no migrants. (b) The number of years that the population is expected to persist at different migration rates. These figures show that even a small amount of migration will rescue the population from extinction. (*From Stacey et al. 1997.*)

which the butterflies specialize. Observations of checkerspot butterflies by a number of workers suggest that while local populations may fluctuate widely in density, the populations tend to persist. Studies of genetic diversity among populations of these butterflies suggest that relatively high levels of heterozygosity are maintained in local populations (Brussard and Vawter 1975, Debinski 1994). Taken together, these data suggest strongly that enough migration occurs among subpopulations of checkerspot butterflies to rescue local populations from extinction.

Correlated Extinction

Let us return for a moment to the idea of a continuum of movement patterns within spatially structured populations. Metapopulations are defined as those that display an intermediate level of movement among habitat patches. The demographic dynamics of a local population may affect movement. For example, owing to social interactions and resource limitation, emigration may be greater from very dense populations than from populations with low densities. Moreover, in nature, demographic dynamics may differ among local populations. Thus, immigration and emigration rates may be patch specific (Sutcliffe et al. 1997a,b). This situation will affect the persistence of the metapopulation as a whole.

Figure 17-12 depicts a simple metapopulation with just two patches, both occupied, in which the density of each local population is represented by the size of the circle denoting the patch. Suppose that density is correlated with environmental conditions; that is, that density tracks environmental stochasticity. If the density of each local population responds to environmental variation in the same manner, then they will both increase when conditions are good and decrease when conditions are poor. Further, let us assume that emigration is correlated with density, and that greater emigration from either population increases the likelihood of movement from one patch to the other. In such a situation, the metapopulation may be imperiled in times of low density because both populations will become too small to sustain movement between them (Figure 17-12a). If the populations become extinct as a result, this situation is referred to as **correlated extinction** (Harrison and Quinn 1989, Gilpin 1990).

Now, suppose that the two populations are more or less independent of each other—that is, that density is uncorrelated with environmental fluctuation—and one population may increase while the other decreases (Figure 17-12b). If that were to happen, emigration from the population with the higher density might be sufficient to sustain the smaller population via the rescue effect, and the metapopulation as a whole might persist. To be sure, two independent populations might by chance simultaneously decline in density, creating a situation like the one shown in Figure 17-12a, in which the metapopulation becomes endangered. In general, however, when the demographic dynamics of populations are largely independent of one another, the metapopulation has a better chance of persisting (Harrison and Quinn 1989, Gilpin 1990).

Such correlated dynamics have been observed in metapopulations of the ringlet butterfly (*Aphantopus hyperantus*) that inhabit the oak-ash woodlands of the

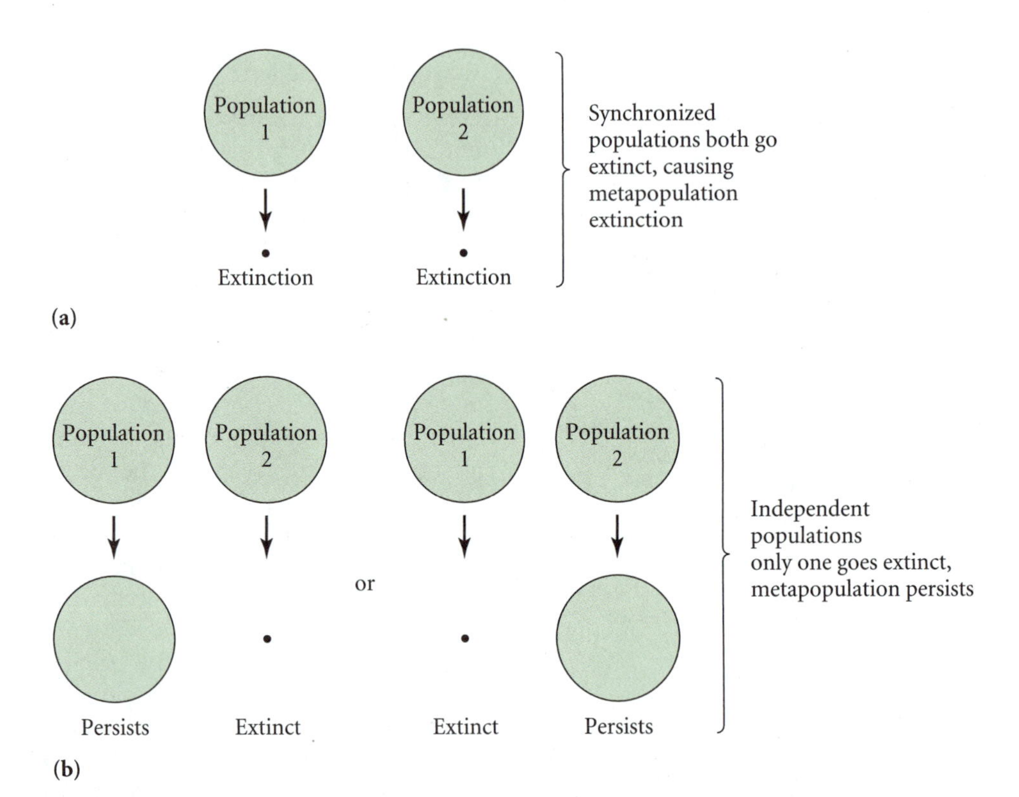

FIGURE 17-12 Schematic representation of correlated extinction in a simple metapopulation with two populations. (a) Two populations that are synchronized in their response to environmental fluctuation may simultaneously suffer a reduction in density, thereby increasing the likelihood that the metapopulation as a whole will go extinct. (b) If the demographic processes of the populations operate more or less independently of one another, then situations may arise in which one population suffers a reduction in density and the other does not. In this case, the metapopulation may persist via the rescue effect. Independent population dynamics tend to increase metapopulation persistence time.

Monks Wood National Nature Reserve in Cambridgeshire, England (Sutcliffe et al. 1997b). An extensive census of the butterflies of Monks Wood has been conducted since 1973. A 0.5-km^2 area has been divided into fourteen different sections. Each week a count of the number of individuals of all butterfly species is made in each of the sections. If no individuals of a particular species are observed in a particular section when that species had been seen in that section the year before, an extinction is declared to have occurred in the section. When a section that was unoccupied in the previous year becomes occupied, a colonization event is said to have occurred. The extinction-colonization dynamics of *A. hyperantus,* one of the species monitored in the study, show a high degree of synchronization with the environment. Figure 17-13 shows the number of sections exhibiting extinctions and colonizations over periods of 3 years. The figure shows a dramatic event in the 1977–1979 period, in which extinctions far outweighed colonizations. This period followed a severe drought that occurred in 1976. The high level of extinctions in the years following the drought indicates correlated

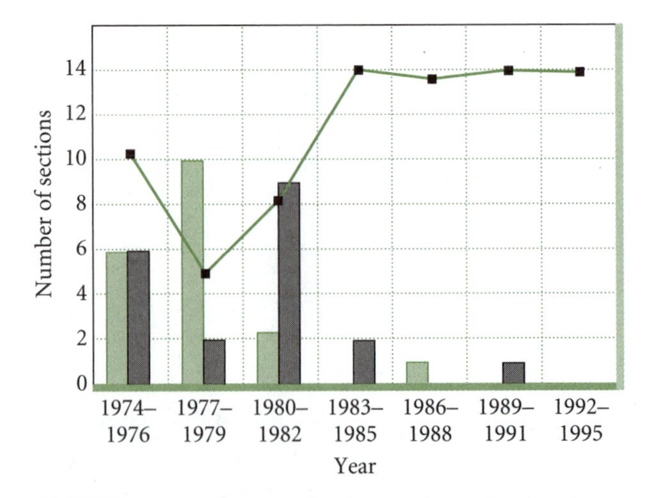

FIGURE 17-13 Pattern of extinction (green bars) and colonization (black bars) in a metapopulation of the ringlet butterfly (*Aphantopus hyperantus*) in Monks Wood National Nature Reserve in Cambridgeshire, England, between 1974 and 1991. Also shown is the average number of sections in which *A. hyperantus* was observed (colored curve). A dramatic decline occurred in 1977 following a severe drought. The populations show a high degree of correlation in their responses to the environment. (*From Sutcliffe et al. 1997.*)

extinction. Colonizations outnumbered extinctions in the period 1980–1982, indicating that the metapopulation had recovered from the drought. Such recovery may not have been possible were it not for the fact that the shady habitats the butterflies prefer were in close enough proximity to allow for immigration even in times of low population density.

17.6 The level of genetic variation in a metapopulation is determined by interaction between population size, extinction, and colonization.

The collective genetic composition of a population is known as its genetic structure. In our previous discussion of genetic structure in Chapter 14, we emphasized the importance of population size. Genetic variation is a buffer against imminent and future environmental variation, and small populations are particularly vulnerable to the loss of this variation. Random fluctuations in birth and death rates due to chance events may change allele frequencies dramatically in small populations, a situation referred to as genetic drift. Thus, the rate of fixation is higher in small populations, and they experience a more rapid decrease in genetic variation over time than do large populations.

What are the dynamics of genetic variation within metapopulations? Do the extinction and colonization events that characterize metapopulation dynamics alter the effects of population size with respect to genetic variation? For example, will the metapopulation as a whole maintain genetic variation even though genetic drift may be high in local populations? These questions are of particular importance in the conservation of endangered species, which often live in fragmented habitats or in protected areas in isolated nature preserves. They are also important to population geneticists, who have long held an interest in gene flow between populations and the processes by which genetic differentiation between populations arises, which are central to concepts of speciation.

The genetic structure of a metapopulation encompasses the entire array of local populations occupying the habitat patches, rather than a single local population. We have emphasized that a defining characteristic of metapopulations is the presence of an intermediate level of migration between patches. An array of habitat patches in which movement among patches is extensive may be considered a single population, and we would not expect to observe much genetic variation among the spatial components of such a population. Where landscape features prevent movement between habitat patches (see Figure 17-16), or where the limited dispersal capabilities of a species prevent the exchange of individuals among habitat patches, we would expect greater genetic differentiation among local populations. In metapopulations, we expect an intermediate level of genetic differentiation among local populations because there is an intermediate level of gene flow.

The extent to which genetic diversity is maintained in a metapopulation is dependent on the relationships between patch size and extinction and colonization rates. As a generalization, overall genetic variation increases as population size increases. Thus, we may expect that a metapopulation in which individuals are distributed among a number of small populations, or in which there is great variability in population size, will have less genetic variation than one in which the same number of individuals are arranged in fewer, larger populations. In general, the movement of individuals among local populations decreases genetic variation. Some migration patterns result in more genetic variation than others (Whitlock 1992). Metapopulations in which local populations have similar immigration rates are likely to display less genetic variation than those in which some local populations have very high rates of immigration and others very low rates.

To demonstrate how metapopulation dynamics can influence genetic variation, we turn to a simple simulation first performed by Gilpin (1991) and later described by Hedrick and Gilpin (1997). Consider a metapopulation that is composed of three populations residing in patches 1, 2, and 3. Each population begins the simulation with an effective population size of 500, which is equal to the actual number of individuals in the population (that is, for the purposes of the simulation, $N_e = N$; this is the situation depicted in Figure 14-16a). Let us further assume that each population has an initially high level of heterozygosity, H, indicating substantial genetic variation (Figure 17-14). Following the simple Levins metapopulation model, a population may become extinct and its patch recolonized by individuals from another population. Since colonization generally introduces very few individuals to an unoccupied patch, the genetic diversity of a newly colonized patch will be small due to the founder effect. Such a turnover event, with a consequent decrease in heterozygosity, is shown in Figure 17-14, as the population in patch 1 goes extinct around generation 18 and the patch is recolonized by individuals from the population in patch 2 in generation 20, resulting in a population in patch 1 with low H.

From this basic dynamic, Gilpin (1991) showed that if colonization is restricted to very few individuals—say, one fertilized female—certain patterns of turnover may lead to a metapopulation

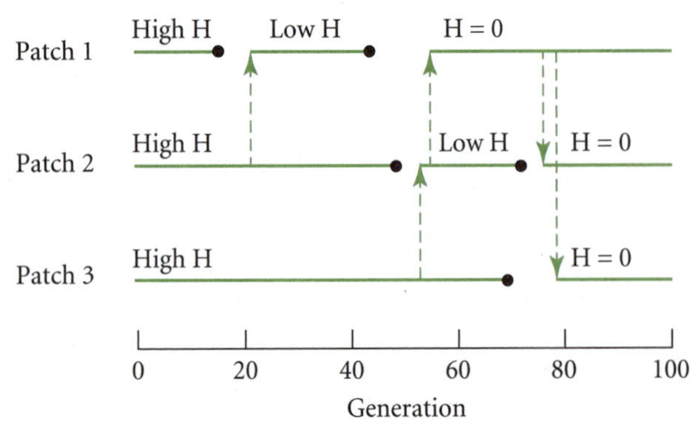

FIGURE 17-14 A metapopulation having three patches, each initially with high levels of heterozygosity (H), may lose genetic diversity by repeated recolonization of extinct patches. Extinction events are noted as breaks in the green line; recolonization with vertical arrows. The population in patch 1 goes extinct in generation 18 and is recolonized in generation 20 by individuals from patch 2. Because the recolonization involves only a few individuals, the level of heterozygosity in patch 1 is reduced. A similar event occurs in patch 2 around generation 47. A second extinction and recolonization event occurs in patch 1 between generations 40 and 60. Because of the reduced genetic variation of the colonizers from patch 2, the level of heterozygosity in patch 1 becomes 0. Subsequent recolonization events in which patch 1 (H = 0) serves as the source population result in a metapopulation having no genetic variation. (*From Hedrick and Gilpin 1997.*)

having many individuals but essentially no genetic variation. To see how this might happen, focus on the events beginning around generation 48 in Figure 17-14. The population in patch 2 goes extinct, and the patch is recolonized, in generation 51, by only two individuals from patch 3, resulting in very low H. The population in patch 1, which went extinct a second time in generation 40, is recolonized around generation 55 by a single individual from the newly colonized patch 2, rendering the genetic diversity of that population zero. Final extinction events in patches 2 and 3 result in a situation in which the entire metapopulation is without genetic variation.

To be sure, the example above is an extreme case. A number of factors will determine the level of genetic variation in a metapopulation, including (1) the maximum number of individuals that can be supported indefinitely within a patch (carrying capacity), (2) the rate of turnover in the metapopulation, (3) the number of sources of colonization in the set of populations, (4) the total number of patches in the metapopulation, and (5) the rate of gene flow between patches. Hedrick and Gilpin (1997) used computer simulations to examine the relative importance of all of these factors in determining the genetic variation in metapopulations.

17.7 The world exists as a mosaic of habitat patches called a landscape.

Uniform, homogeneous habitats extending over large areas simply do not exist. Even in the vast Amazonian rain forest of eastern Peru, satellite photographs revealed more than fifty subtly and not so subtly different types of forest, varying on scales from a few tens of meters to kilometers. Ground-level studies coordinated with the satellite images showed that distributions of forest trees responded to small variations in drainage, soil, and recent disturbance. That such observations have had a great impact on the way in which ecologists view the natural world is evidenced by the subfield of ecology called **landscape ecology** (Forman and Godron 1986, Turner 1989; see Chapter 9). On its face, landscape ecology seems very similar to the ideas of metapopulation dynamics that we have discussed in this chapter, and indeed, there are similarities, as we shall see below. But there are differences as well, and it is important that we appreciate the different origins and emphases of these two areas of inquiry in ecology.

The field of landscape ecology is not yet well defined, owing in part to the fact that its precepts are derived from such divergent fields as landscape architecture, land use policy, sociology, human geography, resource management, spatial pattern analysis, and, of course, ecology. However, four principles appear to have emerged as fundamental to landscape ecology (Wiens 1997). First, patches of habitat or of the distinct ecosystems that make up a landscape—the elements of the landscape mosaic, if you will—vary both spatially and temporally in their quality. The characteristics of the patch in which an organism finds itself may change through time, and a new patch that an organism or its offspring arrive in, though suitable, may be somewhat different than the one they just left. Second, there are boundaries between patches. The effects of boundaries on the movements of organisms and materials may be quite dramatic and may affect processes both within patches and between them. Third, and related to the influence of boundaries, the movements of organisms and materials among the elements of the landscape determine the connectivity of those elements. The degree of connectivity is likely to be extremely important in determining the effects of disturbance on the landscape. For example, a disturbance may have a wider effect in a landscape of highly connected patches. Finally, the characteristics and dynamics of a particular patch are dependent on its location with respect to the structure of the mosaic as a whole. What goes on in one patch is affected by the activities in neighboring patches.

It is the explicit way in which the scale of habitat patchiness and the spatial arrangement of habitats are considered that perhaps most distinguishes landscape ecology from other types of ecological theory. The scale of patchiness within a landscape can exert a powerful influence on populations because it affects the distribution and quality of suitable habitat. The quality of one type of habitat can be altered by the presence of different habitat types nearby. Habitat quality may be improved if other patches in the landscape provide resources such as roosting sites, nesting materials, pollinators, or water. Other kinds of neighboring habitat patches may be a serious drawback if they harbor predators or disease organisms. For example, throughout much of the eastern and midwestern United States, forest fragmentation has brought populations of forest birds into contact with the parasitic brown-headed cowbird, which lays its eggs in the nests of other species and greatly reduces the reproductive success of its hosts. Cowbirds prefer the open habitats of farms and fields, but do not hesitate to enter woodlots to seek out nests to parasitize. As a result, populations of many forest birds have decreased in areas of high landscape heterogeneity.

A recent experimental study of scallop populations inhabiting seagrass beds in North Carolina estuaries demonstrates the importance of patchiness to population processes in marine environments (Irlandi et al. 1995). Predation on juvenile scallops by fish, blue crabs, starfish, and whelks was much higher in patchy areas of seagrass than in continuous seagrass habitat, even though the immediate environment of individual scallops was the same in all cases (Figure 17-15). The increased edge per unit of habitat area in the patchy areas evidently provided increased access to the interior of the seagrass habitat by predators, just as forest edge provides access by cowbirds to birds nesting in the interior.

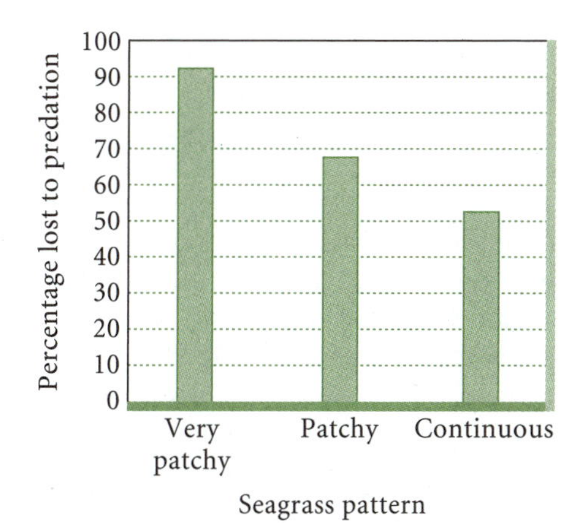

FIGURE 17-15 Rates of predation over a 4-week period on juvenile scallops placed in continuous seagrass habitat and in areas of seagrass with two levels of patchiness. (*From Irlandi et al. 1995.*)

The scallop example hints at the complexity of the interaction between the physical characteristics of the environment and the biological interactions among organisms in the environment. When the quality of habitat patches varies, individuals tend to distribute themselves over landscapes in proportion to the suitability of environments. When organisms have complete knowledge of patch quality and complete freedom of choice, we would expect them to occupy or exploit patches in direct proportion to their quality; that is, patches with the highest levels of resources should attract the most consumers. But the suitability of a habitat patch depends not only on its intrinsic characteristics, but also on the density of other individuals living in the patch. A patch becomes less attractive to newcomers as the number of individuals exploiting it increases, just as a male bird's territory becomes less attractive to a female if he already has a mate. Occupied patches are likely to have fewer remaining resources than undiscovered patches; furthermore, the presence of competing individuals may precipitate costly behavioral conflicts.

Relationship Between the Landscape and Metapopulation Concepts

You have probably already recognized some similarities between the metapopulation concept and the landscape concept. Both involve interactions among sets of habitat patches. Although a melding of the two concepts would no doubt provide new and exciting insights into the natural world, particularly with respect to conservation biology (Hanski and Simberloff 1997, Wiens 1997), the two concepts are, unfortunately, currently recognized more by their differences than by the ways in which they complement one another. The principal difference between the two ideas is in the way that the matrix of background habitat is viewed. You will recall from our description of the Levins metapopulation model that habitat patches reside in a featureless matrix of unsuitable habitat. The dynamics of the metapopulation represent a balance between extinction and colonization in the set of patches. Enhancements to the Levins model focus on how habitat patches interact and what happens within patches, not on the characteristics or dynamics of the area between patches, which is assumed to be homogeneous. In contrast, the concept of a landscape envisions a mosaic of habitats within which conditions vary continuously in space and time and between which exist boundaries that may restrict the flow of organisms and materials. There is no featureless background within which suitable habitats occur. This more realistic view of habitat arrangement is one of

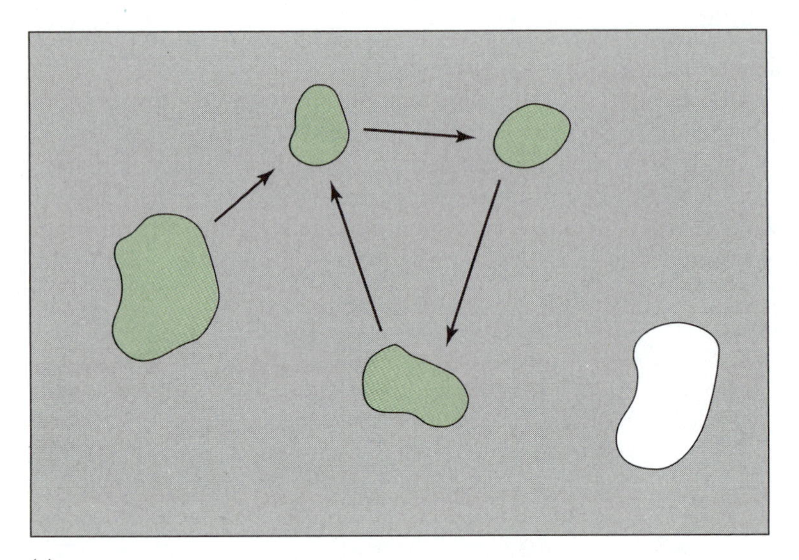

(a)

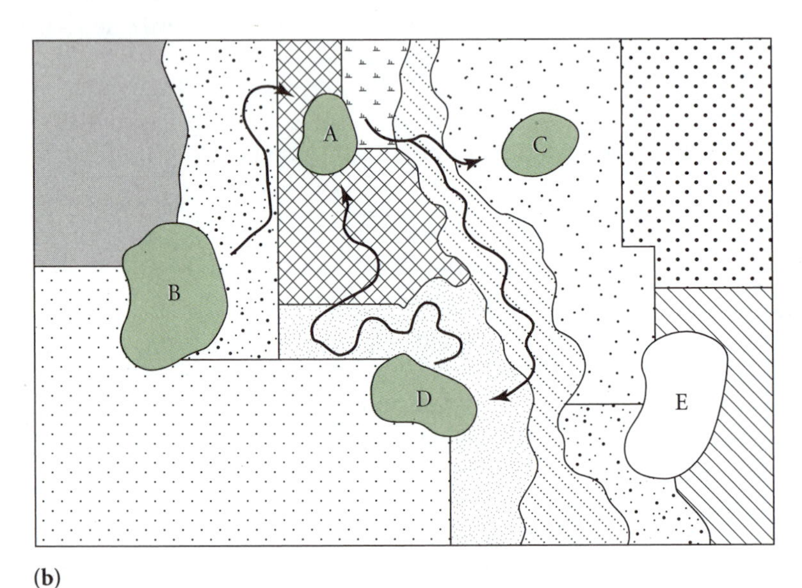

(b)

FIGURE 17-16 A schematic depiction of the principal differences between the metapopulation and landscape concepts. (a) A metapopulation with five habitat patches, four of which are occupied (green areas) and one of which is unoccupied (white, open area). Arrows represent interpatch movements. According to the metapopulation concept, habitat patches occur against a featureless background of unsuitable habitat (solid grey area). (b) The same metapopulation overlaid on a landscape composed of a mosaic of many different habitat types, which, as depicted by the arrows, affect interpatch interactions within the metapopulation.

the reasons that the development of analytical theory has proceeded more slowly in landscape ecology. Without the simplifying assumptions that characterize the metapopulation concept, the development of analytical models of landscape processes is difficult.

Wiens (1997) provides a clear representation of the differences and possible complementarity between the two concepts (Figure 17-16). Figure 17-16a represents the classic metapopulation model, showing four occupied patches connected by migration and one unoccupied patch. The background habitat does not affect the movement of individuals among the patches, one of the implicit assumptions of metapopulation models. In Figure 17-16b, the metapopulation is overlaid on a hypothetical landscape mosaic, where different stippling patterns represent patches in the mosaic having different characteristics. The figure shows how the movement of organisms among the patches of the metapopulation may be affected by the landscape. For example, movement of individuals from patch A to patch D appears to be facilitated by the presence of a corridor of habitat that stretches through the center of the landscape. Movement of individuals from patch D to patch A takes place by a different route. Would removal of the corridor through which individuals move from A to D create a situation where D became strictly a source population and A a sink? Many other questions about the dynamics of the metapopulation could be posed. For example, is patch E empty because the features of the landscape between it and the central habitat corridor block the movement of individuals? Currently, theory that predicts the interaction of these two concepts is not well developed.

SUMMARY

1. Ecologists recognize that suitable habitat for a species often occurs in patches surrounded by areas of unfavorable habitat. When populations in habitat patches—often called local populations—are connected to one another by an intermediate level of movement of individuals between them, the set of local populations is called a metapopulation, or "population of populations."

2. A local population may become extinct, thereby rendering its habitat patch unoccupied. An unoccupied patch may be colonized by individuals from occupied patches. An extinction event followed by a recolonization is called turnover. The length of time until all local populations in the metapopulation become extinct is called the metapopulation persistence time.

3. The Levins model represents the dynamics of extinction and colonization in an idealized metapopulation in which all of the patches are small and either occupied at the carrying capacity of the patch ($N = K$) or not occupied ($N = 0$). The rate of change in the proportion of occupied patches is given by $dp/dt = mp(1 - p) - ep$, where m is the rate of migration and e is the rate of extinction. Metapopulation models represent patch occupancy, whereas logistic models of single populations represent the dynamics of population density.

4. Metapopulation models are enhanced by the inclusion of metapopulation structure. Population size, patch area, the spatial arrangement of patches (patch density), and demographic properties of populations are features of metapopulation structure. In general, small, isolated patches are more likely to suffer extinctions or to remain unoccupied than larger patches that occur in the proximity of other patches. Patch size may compensate for patch isolation in a way that increases the likelihood of metapopulation persistence.

5. The level of patch occupancy in natural metapopulations may fluctuate widely owing to variation in patch quality, demographic stochasticity, or the presence of disturbance.

6. The pattern of movement of individuals affects the persistence of a metapopulation. In some cases, small populations within a metapopulation are rescued from extinction by receiving frequent immigrants from larger populations. Populations that are correlated in their response to environmental fluctuations may suffer reductions in size simultaneously, thus imperiling the entire metapopulation. Metapopulations containing populations that are more or less independent of one another may persist longer than those with populations having correlated dynamics.

7. Genetic variability in metapopulations is dependent upon population size, migration, extinction, and colonization. Small population size increases genetic variation within the metapopulation, whereas variation declines in the face of greater migration. Extinction may increase genetic differentiation among populations.

8. Landscape ecology involves the explicit study of how landscape features affect ecological processes, whereas the study of metapopulations emphasizes the dynamics of extinction and colonization. The two fields of study are merging in important ways.

EXERCISES

1. Construct a graph of the Levins metapopulation model over the entire range of p ($p = 0 \ldots 1$) for $m = 1.0$ and $e = 0.3$. Determine approximately at what level of patch occupancy the maximum rate of change, dp/dt, occurs. Compare the shape of the curve to the logistic model of population growth under regulation.

2. Consider the group of patches shown in Figure 17-1. The size of each patch is proportional to the number of individuals in the population that it contains and the rate at which emigrants leave the population. Speculate on which of the populations are more likely to go extinct and why.

3. Discuss how the metapopulation concept might be used to study the dynamics of rare species.

CHAPTER 18

Population Fluctuations and Cycles

GUIDING QUESTIONS

- What is the relationship between a population's response to environmental fluctuations and the population's growth rate?
- How are population cycles revealed in discrete time population models?
- What are recruitment functions?
- How do population oscillations and limit cycles arise?
- How do time delays cause oscillations in continuous population models?
- How have laboratory experiments demonstrated time delays in animal populations?

We often make comparisons of weather conditions from one year to the next. We might observe, "This is one of the coldest winters that I can remember," or "This summer is at least as warm as the summer I left for college." You have probably also found yourself comparing the abundances of certain plants and animals in the same way. For example, if you enjoy gardening, you might comment to your neighbor, "I have never seen so many tomato hornworms," or share your impression that "there are more Japanese beetles this year than last." Sometimes your comments reflect conditions of the natural world that affect your health, as when you mention to your friend, "My allergies tell me that pollen levels this fall are very high."

These observations of the natural world reflect real variation in the abundances of organisms. Natural populations change in size, age structure, distribution, and genetic composition over time. In some cases, there is a cycle to the change that is more or less predictable. In other cases, fluctuations occur without a set pattern. In this chapter we shall examine the principles of population change through time.

The reasons for fluctuations in population size and structure are, like most aspects of the natural world, complex. The conditions of the environment affect rates of birth and death, and thus surely have a large part to play. The interactions of individuals within the population also affect these processes and thus the size and distribution of the population as well. And, of course, populations interact with other populations through processes such as predation, herbivory, and parasitism, and these interactions may also play a large role in how the population changes through time. We will present some models that can be used to examine the dynamics of population change. These ideas focus principally on what happens within a single population. We will expand on these models in the next part of the book, where we will examine the processes of population interactions.

So far, we have presented two ways to conceptualize natural populations. One approach is based on the logistic model and turns on the idea that population density and the demographic characteristics of the population are the keys to understanding population dynamics. In the second approach—the metapopulation approach—our attention was drawn to patterns of extinction and colonization within sets of interacting populations. In the simplest metapopulation models, the local population demographic processes are ignored. In this chapter, we will focus primarily on the first approach, since it represents the historical roots of population theory.

18.1 Populations fluctuate in density.

Let us begin our discussion of population fluctuation by focusing on changes in population density. Variation in the density of a population depends on the magnitude of fluctuation in its environment and on the inherent stability of the population. After domestic sheep became established on the island of Tasmania, their population varied irregularly between 1,230,000 and 2,225,000—less than a factor of 2—over nearly a century (Figure 18-1). Much of this variation was related to changes in grazing practices, markets for wool and meat, and pasture management, which could be considered factors in the environment of the sheep industry, if not in that of the sheep themselves.

In sharp contrast, populations of small, short-lived organisms may fluctuate wildly over many orders of magnitude within short periods. Populations of the green algae and diatoms that make up the phytoplankton may soar and crash over periods of a few days or weeks (Figure 18-2). These rapid fluctuations overlie changes over longer periods that occur on, for example, a seasonal basis.

Sheep and algae differ in their degree of sensitivity to environmental change and in the response times of their populations. Because sheep are larger, they have a greater capacity for homeostasis and better resist the physiological effects of environmental change. Furthermore, because sheep live for several years, the population at any one time includes individuals born over a longer period. This tends to even out the effects of short-term fluctuations in birth rate. The lives of single-celled algae span only a few days, so populations turn over rapidly and bear the full impact of a capricious environment.

Populations of similar species living in the same place often respond to different environmental factors. For example, the densities of four species of moths, whose larvae all feed on pine needles, were found to fluctuate more or less independently in a pine forest in Germany (Varley 1949; Figure 18-3). The populations varied over three to five orders of magnitude (1,000-fold to 100,000-fold change) with irregular periods of a few years. Furthermore, the highs and lows of the four populations did not coincide closely, suggesting that even though each species fed on the same trees, their populations were regulated independently by different factors.

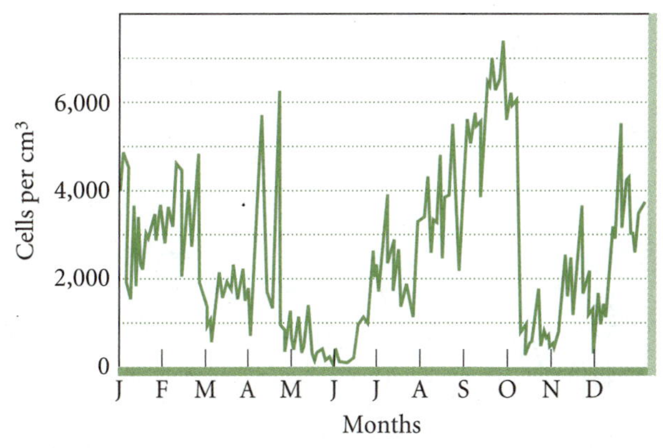

FIGURE 18-2 Variation in the density of phytoplankton in samples of water from Lake Erie during 1962. Such short-lived organisms may exhibit population fluctuations of several orders of magnitude over short periods of time. (*After Davis 1964.*)

Among the most striking population phenomena found in nature are the regular cycles of certain mammals and birds at high latitudes. For example, regular trapping of small mammals over a 26-year period in northern Finland revealed six peaks of abundance separated by intervals of 4 or 5 years (Lahti et al. 1976; Figure 18-4). This regularity is distinctly nonrandom, and the peaks and troughs exhibited by the most abundant species, the vole *Clethrionomys rufocanus*, are roughly paralleled by those of the less common species.

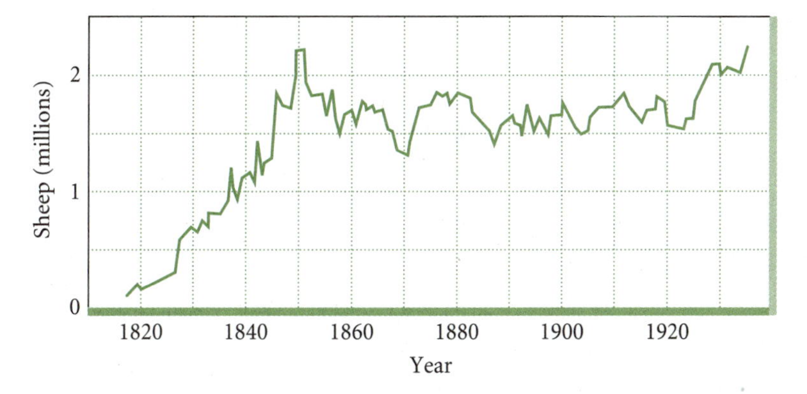

FIGURE 18-1 Number of sheep on the island of Tasmania since their introduction in the early 1800s. The sheep population grew until the mid-1950s and then leveled off. The graph shows irregular year-to-year variation in the population size, but the extent of the fluctuation is relatively small when compared with the total population size. (*After Davidson 1938.*)

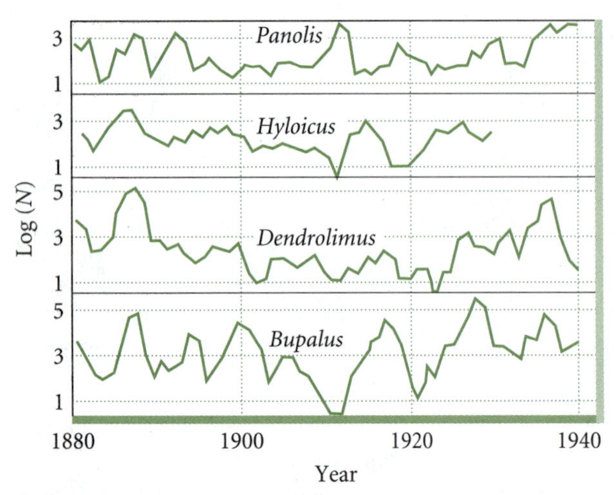

FIGURE 18-3 Fluctuations in the numbers of pupae of four species of moths (hibernating larvae in *Dendrolimus*) in a managed pine forest in Germany over sixty consecutive midwinter counts. (*After Varley 1949.*)

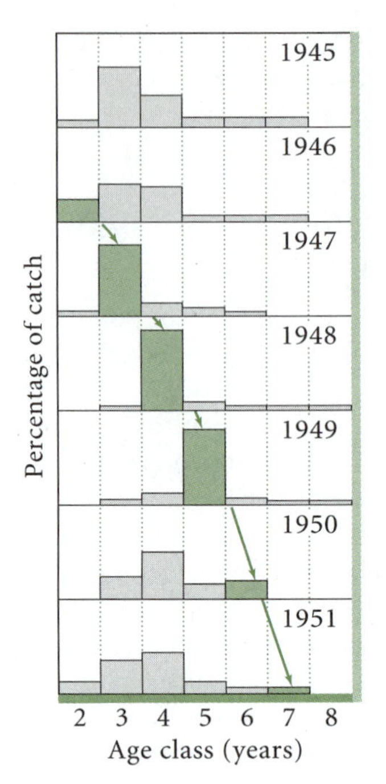

FIGURE 18-5 Age composition of samples from the commercial whitefish catch in Lake Erie between 1945 and 1951. Fish spawned in 1944 are indicated by the green bars. (*From Lawler 1965.*)

18.2 Temporal variation affects age structure in populations.

Populations may change in age structure through time—that is, the relative frequencies of individuals of each age may change. As we saw in Chapter 15, a changing age structure can affect the rate of population growth. The sizes of age classes also provide a history of population change in the past. For example, the age composition of samples from the Lake Erie commercial whitefish catch for the years 1945–1951 shows that during 1947, 1948, and 1949, most of the individuals caught belonged to the 1944 year class (Figure 18-5). Biologists estimated the ages of fish from growth rings on their scales. Their data showed that 1944 was an excellent year for spawning and recruitment, particularly compared with several years that followed.

Variation in annual recruitment is also evident in the age structure of stands of trees. The age of a tree may be estimated by counting growth rings in the woody tissue of the trunk—one ring is added each year under normal circumstances (Figure 18-6). The pattern in a virgin stand of timber surveyed near

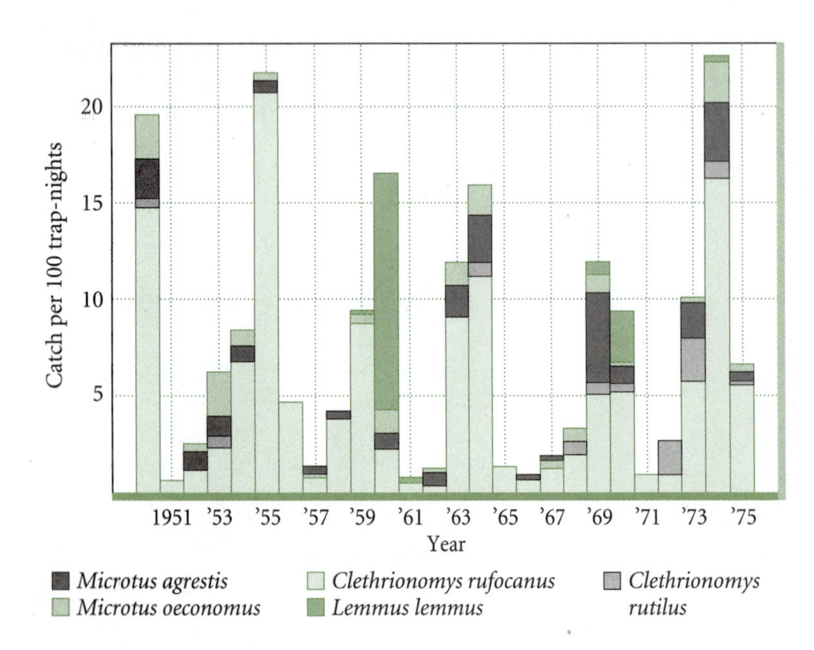

FIGURE 18-4 Numbers of small mammals trapped in an area of northern Finland from 1950 to 1975. These population cycles are dominated by variations in the numbers of the most common species, *Clethrionomys rufocanus*, but other species tended to reach peak abundances more or less in synchrony with it. (*From Brewer 1994; after Lahti et al. 1976.*)

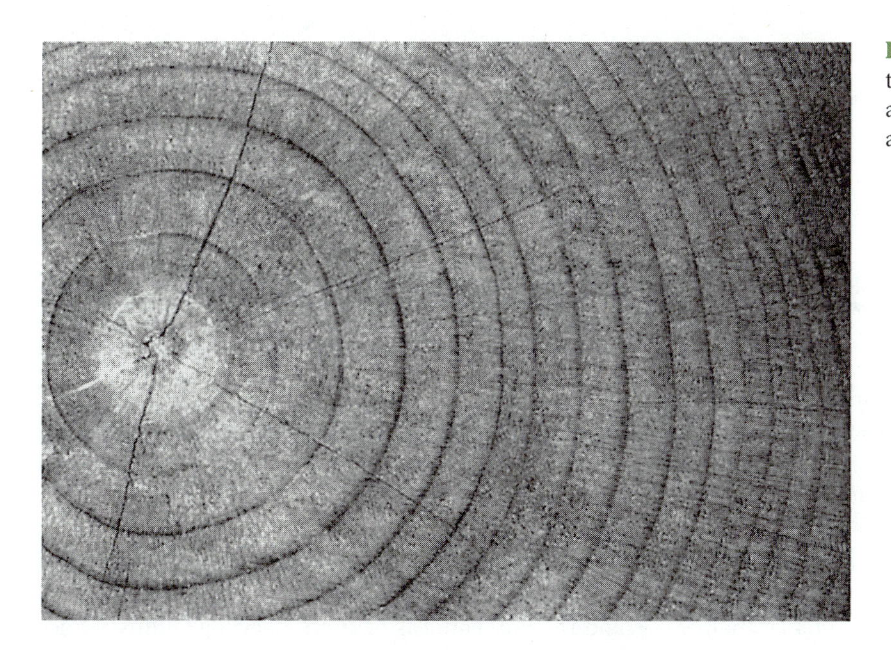

FIGURE 18-6 Cross section of the trunk of a Monterey pine, showing annual rings formed by winter (dark) and summer (light) growth.

Heart's Content, Pennsylvania, in 1928 shows that individuals of most species were recruited sporadically over the nearly 400-year span of the record (Figure 18-7). Many white pines became established between 1650 and 1710, undoubtedly following a major disturbance, possibly associated with the serious drought and fire year of 1644. Fire can open a forest enough to allow the establishment of white pine seedlings, which do not tolerate deep shade. In contrast, beech—a species whose seedlings can grow under the canopy of a closed forest—exhibited a relatively even age distribution.

One aspect of temporal variation in age structure in plants has to do with the fate of seeds in the seed pool. Seeds that fall to the ground may mature and germinate, eventually giving rise to new individual plants. They may also become dormant, maturing at a later date. To be sure, many seeds fall victim to decay or predation and are lost to the population. Thus, in a way, the seed population displays a variation in age structure through time.

18.3 Populations with high growth rates track environmental fluctuations more closely than those with low growth rates.

Fluctuations in conditions and resources continually increase and decrease the carrying capacity of the environment of each population. How a population responds to such changes through density-dependent effects depends on the intrinsic capacity of the population to increase in size, or *r*. The faster the potential rate of growth or decline of the population—that is, the higher the fecundity and the shorter the life span of individuals—the greater is its capacity to track change in its environment. Theoretical studies indicate that, as

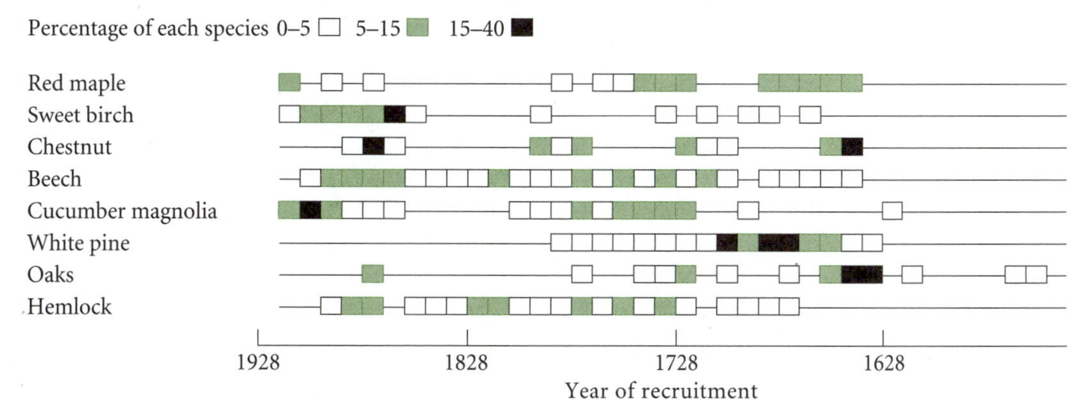

FIGURE 18-7 Age distribution of forest trees near Heart's Content, Pennsylvania, in 1928. (*After Hough and Forbes 1943.*)

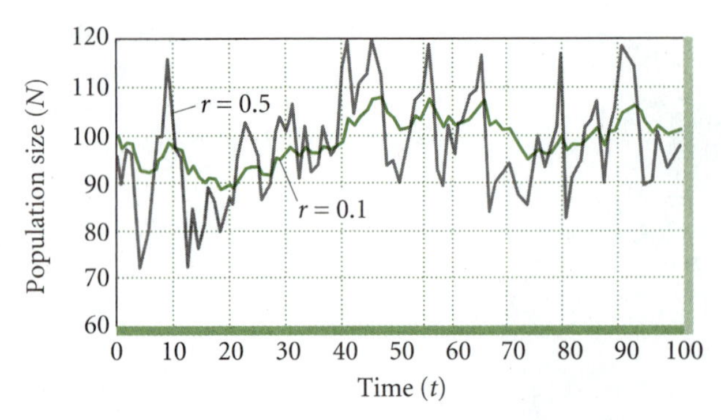

FIGURE 18-8 Population growth according to a logistic equation with random variation in carrying capacity. A population with a higher growth rate ($r = 0.5$) tracks fluctuations in carrying capacity more closely than a population with a lower growth rate ($r = 0.1$). (*From Gotelli 1995.*)

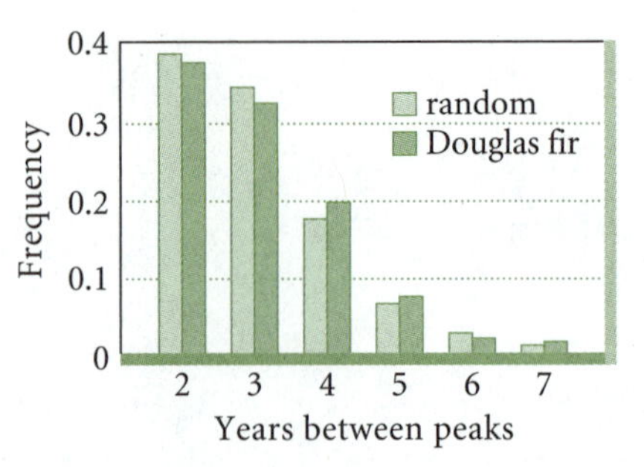

FIGURE 18-9 Frequency distribution of intervals between peaks in widths of growth rings of Douglas fir, compared with that of a series of random numbers. (*After Cole 1951.*)

a general rule, populations with r much greater than 1 will track their environments closely, quickly responding to changes in carrying capacity. Populations with r much less than 1, such as the Tasmanian sheep, tend to be sluggish and unresponsive, their numbers ironing out short-term variations in their environments. Differences in the responses of high-r and low-r populations can be demonstrated by simulations using logistic population growth equations with randomly varying carrying capacities (Figure 18-8).

Population biologists compare the responses of populations to change by their **characteristic return time** (T). T is simply the inverse of the exponential rate of growth of a population free of the effects of crowding; hence $T = 1/r$. Theoretical analyses have shown that a population will track the environment closely when its characteristic return time is less than the period of the environmental fluctuation (the time from peak to peak) divided by 2π (Nisbet and Gurney 1982). When T is much longer than the period of the pertinent environmental variation, the population varies little. Small insects have values of T much shorter than 1 year, and their populations track seasonal variation in conditions; the same variation evokes little response from populations of long-lived mammals.

18.4 Population cycles may result from intrinsic demographic processes.

Except for fluctuations associated with daily, lunar (tidal), and seasonal cycles, environmental change tends to be irregular rather than periodic. Historical records reveal that years of abundant rain or drought, extreme heat or cold, or such natural disasters as fires and hurricanes occur irregularly, perhaps even at random. Biological responses to these factors are similarly aperiodic. For example, variation in the widths of growth rings of trees is directly related to temperature and rainfall. Patterns of successive ring widths cannot be distinguished from a random series (Figure 18-9).

The sizes of many populations do, however, change with periodic frequency. Such a pattern may be seen in *Clethrionomys* in Figure 18-4. A number of small mammal cycles have become part of the lore of population ecology. The most famous are those of a number of furbearing small mammals of the Canadian boreal forest, described by Charles Elton in his paper "Periodic fluctuations in numbers of animals: Their causes and effects," published in the *British Journal of Experimental Biology* in 1924. From data held by the Hudson's Bay Company on furs brought in by trappers each year, Elton showed that the lynx and its principal prey, the snowshoe hare, display large, regular fluctuations in population size (Figure 18-10). Each cycle lasts approximately 10 years, and the cycles of the two species are highly synchronized, with peaks of lynx abundance tending to trail those of hare abundance by a year or two. Cycles of shorter duration have been observed in a number of other arctic vertebrates.

The population models developed in the 1920s and 1930s made it evident that because of their inherent dynamic properties, populations subjected to even minor, random environmental variation could be caused to cycle or oscillate in abundance. Such cycling can result from **time delays** in the response of births and deaths to changes in the environment, just as the momentum imparted to a pendulum by the acceleration of gravity carries it past the equilibrium point and causes it to swing back and forth periodically. The "momentum" imparted to a population by high birth rates at low densities or high death rates at high densities carries the population past its equilibrium when demographic responses are time-delayed.

Time delays that cause populations to oscillate in abundance when displaced from their equilibria are inherent in models based upon discrete generations.

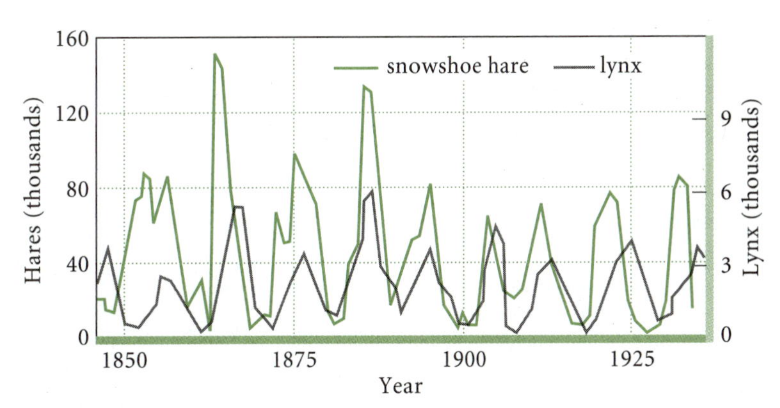

FIGURE 18-10 Population cycles of the snowshoe hare (*Lepus americanus*) and the lynx (*Lynx canadensis*) in the Hudson's Bay region of Canada, as indicated by furs brought in to the Hudson's Bay Company. (*After MacLulich 1937; photograph courtesy of U.S. Fish and Wildlife Service.*)

According to these models, such populations respond by discrete increments from one time to the next, and therefore cannot continuously readjust their growth rates as population size approaches equilibrium. As we shall see below, this can cause the population to overshoot the equilibrium, first in one direction and then the other, as population size (N) draws closer to equilibrium (K). In models of continuously growing populations, oscillations can be induced by introducing a time lag in the term for density dependence.

18.5 Population cycles are revealed in discrete time logistic population models.

In Chapter 15, we drew the distinction between discrete time and continuous time population processes. Discrete time models of population dynamics, in which newborns are added to the population during specific reproductive periods, are representative of most organisms (recall that humans are a notable exception). Here we will extend our conception of discrete time models to show how such models may reveal patterns of cycling in population density. These models will form the basis of some important discussions in future chapters regarding population interactions.

Recruitment Functions

Newborns appear in most populations at one or a few specific times of the year. Populations with discrete generations may be described as growing according to expressions of the form

$$N(t + 1) = f[N(t)],$$

where time is represented as a discrete variable, $t = 0, 1, 2, 3 \ldots$. This equation is referred to as a **difference equation** because it portrays the difference in population size between two discrete time periods. In words, the number of individuals present in the population after the next reproductive event $[N(t + 1)]$ is some function, f, of the number of individuals currently present in the population $[N(t)]$. The function $f[N(t)]$ is sometimes called a **recruitment function.** In its most general usage, the word **recruitment** means the addition of individuals to a population either by birth or by immigration. But in the context of population ecology, the term is often restricted to those individuals that are added by reproduction and that reach reproductive age. This latter meaning is the one used here. Note that the word *recruitment* does not mean that young individuals have to be convinced to join the ranks of the mature!

Recruitment functions show the number of individuals at the present time, $N(t)$, on the horizontal axis and the number of individuals at the next time, $N(t + 1) = f[N(t)]$, on the vertical axis (Figure 18-11). You will note that for any value of $N(t)$, you may obtain the quantity $N(t + 1)/N(t)$, which, you may recall, is the geometric growth rate λ (see Chapter 15). Thus, the recruitment function shows how the geometric growth rate changes as the population density changes.

Qualitative Behavior of Difference Equations

Consider the situation depicted in Figure 18-11a, in which the recruitment function $f[N(t)]$ is a monotonically increasing function. Difference equations show the relationship between population densities at different times, rather than the relationship between density and the rate of increase in density (dN/dt), which we are accustomed to seeing in continuous time population models (see Chapter 15). So, $f[N(t)]$ is always positive, and $f[N(t)] = 0$ when the population density is zero. It is very important to appreciate

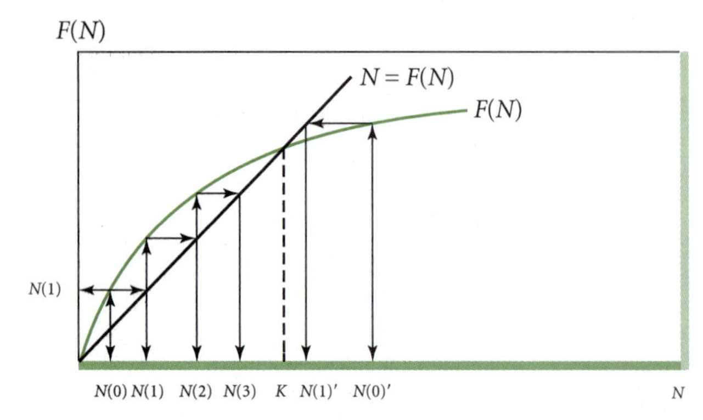

(a)

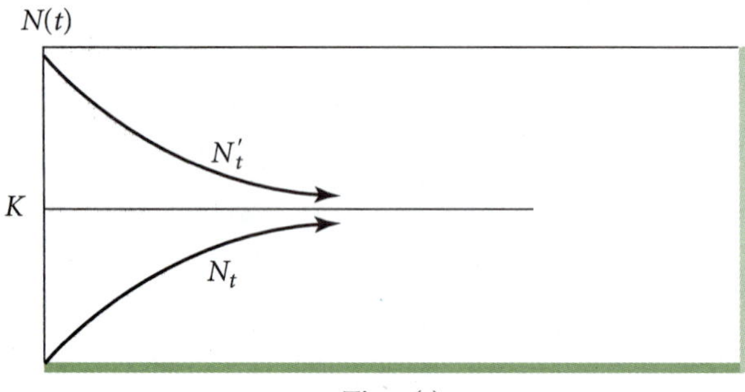

(b)

FIGURE 18-11 Qualitative behavior of difference equations. (a) The curve depicts the relationship between population density, N, and some function of N, f(N). The arrows demonstrate a graphic method of determining the population at the next time in the discrete time series $t = 0, 1, 2, 3 \ldots$ The density at $N(t + 1)$ is obtained by drawing lines from $N(t)$ on the N-axis to the curve, from there to the 45° line, and from there back to the N-axis. Repeating this procedure gives successive densities through time. Two series are presented, one beginning with $N(0) < K$, the other beginning with $N'(0) > K$. (b) The population will converge on the equilibrium level K regardless of the initial population size. The series beginning with $N(0) < K$ increases toward K; the series beginning with $N'(0) > K$ decreases toward K. (*After Yodzis 1989.*)

that time is not explicitly incorporated into the recruitment function f[$N(t)$]. f[$N(t)$] is a smooth curve corresponding to the continuous variable N (density). The notation $N(t)$ refers to a specific value of this continuous variable corresponding to some time t.

The population shown in Figure 18-11a has one equilibrium, which corresponds to the place where a straight line passing through the origin and positioned at a 45° angle crosses the recruitment function. At this point, $N(t) = \text{f}[N(t)]$, and as is our custom, we denote this point K. The 45° line represents a recruitment function for which there is no recruitment. That is, it shows that the population size at time $t + 1$ will be the same as the size of the population at time t.

We may use this line to show graphically how difference equations work. Consider a population density $N(0)$ at time $t = 0$. We may graphically determine the population size at time $N(1)$ by first drawing a line from $N(0)$ on the N-axis up to the curve f[$N(t)$]. We then draw a line from that point on the curve to the vertical axis to arrive at the next, $N(1)$, population size. The $N(1)$ value is the next value of the parental population. In order to find the $N(2)$ population, we must find the place on the horizontal axis corresponding to $N(1)$, draw a line to the curve, and then over to the vertical axis as before. But where, exactly, is $N(1)$ located on the horizontal axis? We may use the 45° line to map $N(1)$ on the vertical axis to the corresponding position on the horizontal axis. Begin at $N(1)$ on the vertical axis, draw a line to the 45° line, then drop a perpendicular line from that point to the horizontal axis to obtain the position of $N(1)$ on that axis. By continuing this process through successive time periods, the progression of population densities is revealed on the horizontal axis. The population, which began at a level below the carrying capacity, K, increases toward K. This is shown in Figure 18-11b, which depicts the population size at time t [$N(t)$] on the vertical axis and t on the horizontal axis as the line approaching K from below. The pattern with which our hypothetical population changes is dependent on the shape of the function f[$N(t)$], the recruitment function, as we shall see.

Now, consider what happens when the population begins at $N(0)'$, somewhere above the equilibrium K (see Figure 18-11a). Using the procedure described above, we see that the population decreases toward K through time (see Figure 18-11b).

We employ a graphic analysis of the difference equation here for instructive purposes. If one has the algebraic expression for the curve f[$N(t)$], then it is a simple matter to calculate population densities through time using standard microcomputer software capable of iterative calculations.

The recruitment model shown in Figure 18-11 is called the **Beverton-Holt model.** It shows a situation in which recruitment increases monotonically from zero, with the rate of increase slowing as the parental population becomes larger. The slowing of the rate of increase in recruitment is due to an assumption built into the model of a density-dependent component in the mortality rate of young. That is, the Beverton-Holt model assumes that more adults will produce more potential recruits—thus the gradual increase in the curve as the parental population, $N(t)$, increases—but that the mortality of the young will increase as the number of young increases. Thus, while young continue to be recruited into the population as the adult population increases, they are recruited at a slower rate.

The Beverton-Holt model factors in density-dependent mortality of the young, but it is also likely that there are density-dependent effects related to parental population numbers. For example, as the parental population increases, competition between maturing individuals and adults for resources might increase. Other factors, such as cannibalism of young, which is common in fishes and many other organisms, could result in pressure against recruitment at high parental populations. A recruitment model called the **Ricker model** accounts for density-dependent reduction in recruitment at high parental populations (Figure 18-12). The Ricker model is of the logistic form, and may be written as

$$f[N(t+1)] = N(t)e^{r[1-N(t)/K]}.$$

From the general form of the difference equation, we obtain

$$N(t+1) = f[N(t+1)] = N(t)e^{r[1-N(t)/K]}$$

$$N(t+1) = N(t)e^{r[1-N(t)/K]}. \qquad (18\text{-}1)$$

A plot of $N(t+1)$ against $N(t)$ following equation (18-1) may be informally described as humped, increasing through the range of $N(t)$ between 0 and some value, N_p, at which a maximum is attained, then decreasing for values of $N(t) > N_p$ (Figure 18-12; see also Figures 18-13 and 18-14). (Contrast this model with the curve for logistic population growth shown in Figure 16-3, which has a similar humped shape. The logistic model represents change in growth rate, dN/dt, with population size. The Ricker model is a difference equation that predicts the next population size from the current population size.) The location of the peak may be determined by differentiation of equation (18-1), yielding

$$\left(1 - \frac{rN(t)}{K}\right)e^{r[1-N(t)/K]}.$$

This expression is 0 when $N(t) = N_p = K/r$, and it is at that point that the maximum of equation (18-1) is found.

Figure 18-12 displays a family of five curves, all having the same equilibrium level, K, but each with a different value of r. We see from this figure that the behavior of the curve depends critically on r. When $r < 1.0$, the curve passes through K with a positive slope and peaks at $N_p > K$. For example, if $K = 10$ and $r = 0.5$, $K/r = N_p = 20$. When $r = 1.0$, the peak of the curve coincides with the equilibrium level; that is, $N_p = K$. When $r > 1.0$, the peak of the curve lies to the left of K, $N_p < K/r$, and the slope of the line passing through K is negative. The numerical result for $K = 10$ and $r = 2.0$ is $K/r = 5$.

Limit Cycles, Oscillations, and Chaos

What are the implications of the Ricker model? Let us employ the graphic approach that we used in Figure 18-11 when we introduced the concept of difference equations to consider what happens to a population whose density is close to the equilibrium level K. In Figure 18-13a the logistic recruitment curve is drawn with $r = 1.5$. Suppose that at time $t = 0$, the

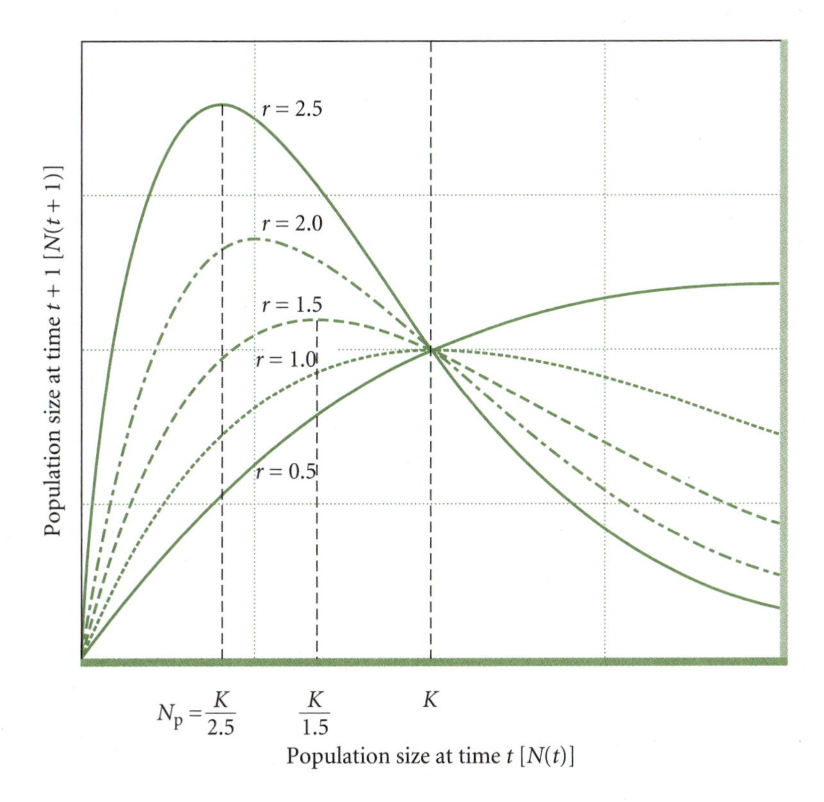

FIGURE 18-12 A family of Ricker curves of the form $N(t+1) = N(t)e^{r[1-N(t)/K]}$, each having the same equilibrium, K, and different values of r. The vertical lines at $N_p = K/2.5$ and $N_p = K/1.5$ correspond to the peaks of the curves having $r = 2.5$ and $r = 1.5$ respectively. The shape of the curve is dependent on the value of r. For $r < 1.0$, the curve passes through K with a positive slope and peaks at population density $N_p > K$. For $r = 1.0$, $N_p = K$. For $r > 1.0$, $N_p < K$, and the curve passes through K with a negative slope.

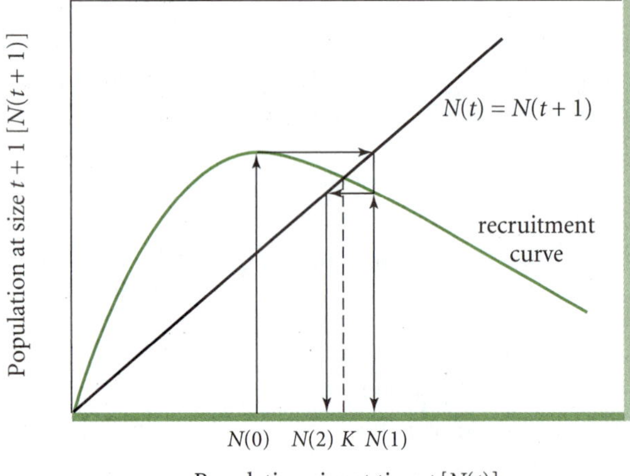

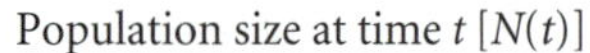

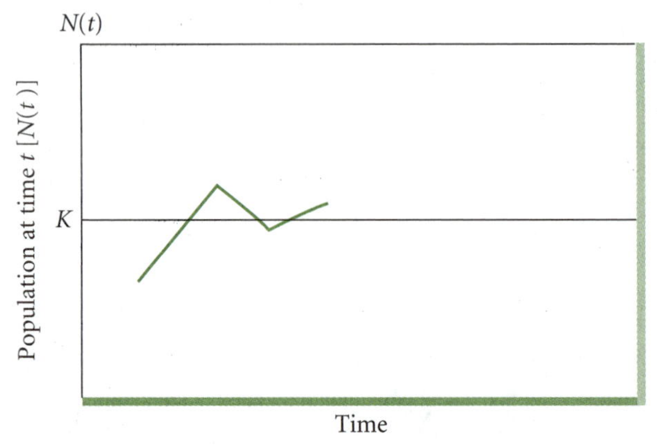

FIGURE 18-13 A discrete time logistic model $N(t + 1) = N(t)e^{r[1-N(t)/K]}$ with $r = 1.5$. (a) Using the techniques shown in Figure 18-11, arrows are drawn to show how a time series of densities in discrete time is obtained graphically. (b) A time series of densities showing damped oscillation.

population density $N(0)$ is the maximum (again, we are using parentheses to denote the discrete time variable, $t = 0, 1, 2, 3 \ldots$). To determine the population level at $t = 1$, $N(1)$, we draw lines first from position $N(0)$ on the $N(t)$-axis to the curve, then to the $N(t + 1) = N(t)$ line (45° line), then, finally, back down to the $N(t)$-axis. [Again, these values may be easily calculated by manipulation of equation (18-1) by computer.] Whereas the original population size, $N(0)$, lies below the equilibrium, K, $N(1)$ lies above K. During the first time period, the population overshoots the equilibrium. Repeating the process to obtain $N(2)$ reveals that, in the second time period, the population drops below K (undershoots), but it comes to lie closer to K than at either of the two previous times. By continuing this process through several more time periods, we can convince ourselves that eventually the population will converge on the equilibrium K by alternately overshooting and undershooting K by increasingly small amounts until $N = K$ (Figure 18-13b) This pattern of stabilization is referred to as a **damped oscillation.**

Now, consider the situation in which $r = 2.5$, shown in Figure 18-14a, where the $N(t)$ are obtained for several time periods by the procedure above. As in the previous situation, when $N(0) < K$, the population will overshoot K in the first time period and undershoot K in the second time period. But the population change between time $t = 3$ and $t = 4$ brings the population back close to the starting level, $N(0)$, thereby resulting in another overshoot of K. Indeed, this population will never converge on the equilibrium. Rather, it will adopt a stable oscillation, also called a **limit cycle** (Figure 18-14b). A sample of some of the time-density patterns that result at high r is presented in Figure 18-15.

With increasing r, the oscillations take on very complex, eventually unpredictable forms referred to

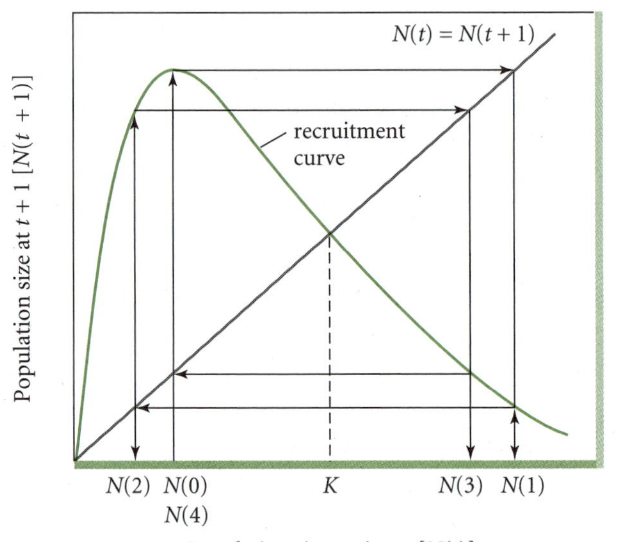

(a)

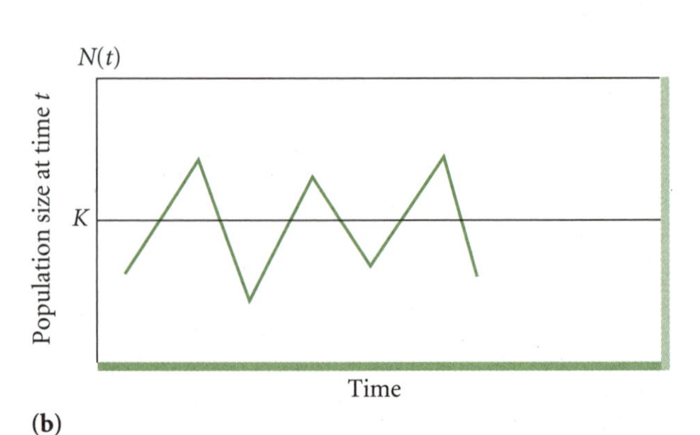

(b)

FIGURE 18-14 A discrete time logistic model $N(t + 1) = N(t)e^{r[1-N(t)/K]}$ with $r = 2.5$. (a) Using the techniques shown in Figure 18-11 and Figure 18-12, arrows are drawn to show how a time series of densities in discrete time is obtained graphically. (b) A time series of densities showing a stable oscillation or limit cycle.

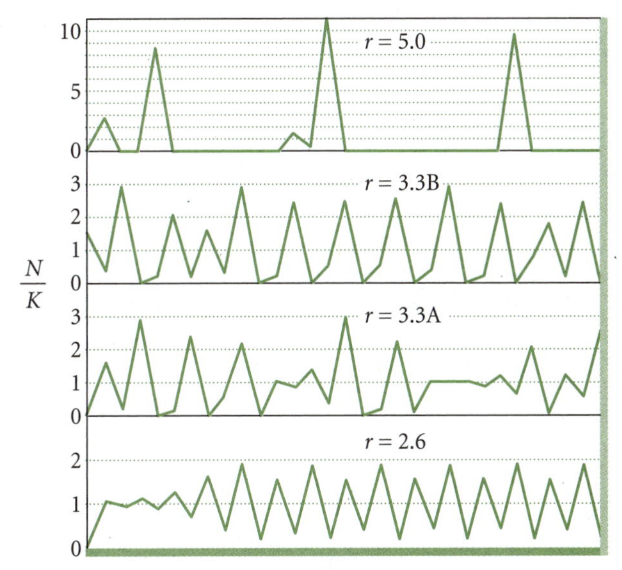

FIGURE 18-15 Patterns of change in population density through time for logistic models with $r = 5.0$ (*top*), $r = 3.3$ (*center*), and $r = 2.6$ (*bottom*). (*After Yodzis 1989; from May 1981.*)

as **chaos.** The characteristics of chaos and its presence in nature are much discussed in ecology (May 1987, Godfray and Grenfell 1993, Hastings et al. 1993, Perry et al. 1993). If the common usage of the word *chaos* is applied here, it is a misnomer. Systems that display chaotic dynamics are not randomly fluctuating, although they may appear to be so at first glance. A number of mathematical models have been developed to detect chaos dynamics using time-density data (Hastings et al. 1993). The application of these models to natural populations, while fraught with difficulties and uncertainties, has revealed what appear to be chaotic dynamics in a number of populations, including populations of Canadian lynx, several species of rodents, a few species of insects, and epidemics of some diseases such as measles (Schaffer and Kot 1986, Turchin and Taylor 1992).

18.6 Time delays cause oscillations in continuous time models.

Density dependence results when either the growth rate or the mortality rate of a population is dependent on its own density. Reproduction increases the density of a population, but the effects of the enhanced density may not be felt until the young mature. For example, the newborns of today cannot increase the growth rate of the population until they reproduce at some time in the future. Such time delays in density-dependent factors may result in oscillations in populations. This effect can be modeled by introducing a time delay term into the logistic model of population growth.

Time Delay Models of Continuous Growth

Oscillations in population density are produced in continuous time models when the response of population growth to density is time-delayed; that is, when the effect of density dependence reflects the density of the population τ time units in the past. The logistic equation may be modified to reflect this:

$$\frac{dN}{dt} = rN(t)\left[1 - \frac{N(t-\tau)}{K}\right].$$

Hutchinson (1948) pointed out that this model produces damped oscillations in N as long as the product $r\tau$ is less than $\pi/2$ (about 1.6). Below $r\tau = e^{-1}$ (0.37), the population increases or decreases monotonically without oscillation—to the equilibrium point. For $r\tau$ greater than $\pi/2$, the oscillations increase until the maximum population size reaches $N/K = e^{r\tau}$. Thus, for $r\tau = 2$, oscillations increase in amplitude until the maximum value of N is e^2 = (7.4) times K. As we indicated in our discussion of discrete time models above, population biologists refer to such stably maintained oscillations as limit cycles. Their periods—the time from peak to peak—increase from about 4 times τ to more than 5 times τ with increasing r (May 1976).

Time Delay Cycles in the Water Flea

Population cycles have been observed in many laboratory cultures of single species. Pratt's (1943) observations on the water flea *Daphnia magna* have been widely quoted, partly because the populations exhibited marked oscillations when cultured at 25°C, but strong damping at 18°C (Figure 18-16). The period at 25°C appeared to be just over 40 days for two cycles, suggesting a time delay in the density-dependent response of about 10 days. This is about the average age at which water fleas give birth at 25°C. The time lag arose in the following manner. As population density increased, reproduction decreased, to near zero when the population exceeded 50 individuals. Survival was less sensitive to density even at the highest densities, and adults lived at least 10 days. Crowding at the peak of the cycle prevented births. Then, when the population fell to densities low enough to permit reproduction, the adult population contained only senescent, nonreproducing individuals, and thus the population continued to decline. The beginning of a new cycle awaited the accumulation of young, fecund individuals. The length of the time delay was approximately the average adult life span at high densities.

At the lower temperature, the reproductive rate fell quickly with increasing density, and life span increased greatly over that at 25°C at all densities. Populations at the colder temperature apparently lacked a time delay because death was more evenly distributed

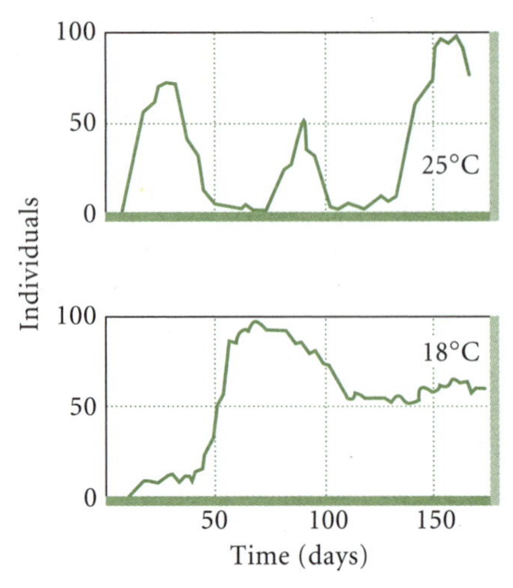

FIGURE 18-16 Growth of *Daphnia magna* populations at 25°C and at 18°C, showing the development of population cycles at the warmer temperature. (*After Pratt 1943.*)

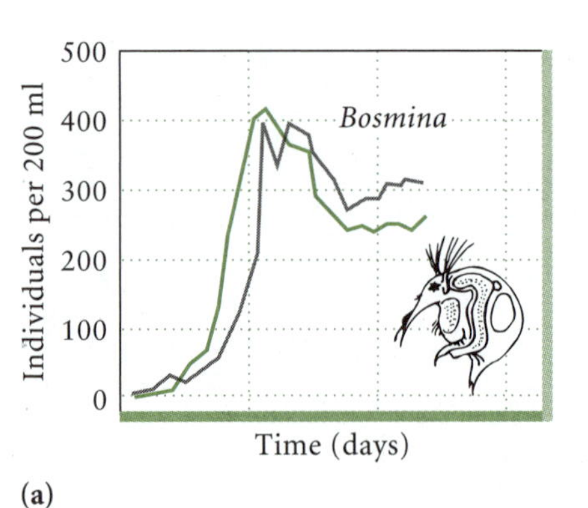

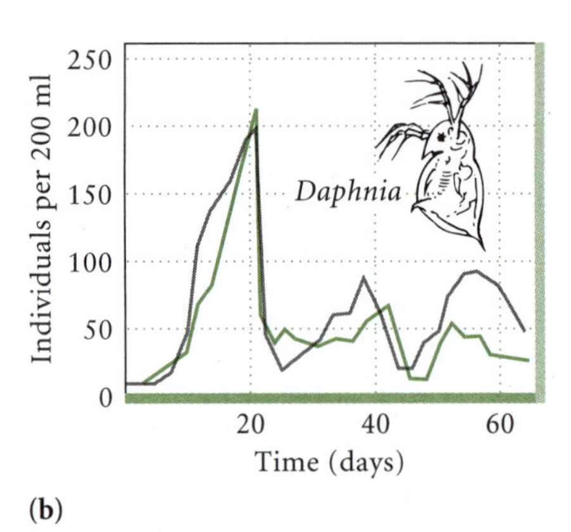

FIGURE 18-17 Densities of two populations each of (a) *Bosmina longirostris* and (b) *Daphnia galeata* under laboratory conditions. The storage of lipid droplets in *Daphnia*, which may be used as food in periods of low food availability, increases survival and introduces time delays that are manifested as limit cycles. (*From Goulden et al. 1982.*)

over ages, and some individuals gave birth at high population densities. Consequently, generations overlapped more broadly. At the higher temperature, the *Daphnia* behaved according to a discrete generation model with a built-in time delay of one generation. At the lower temperature, they behaved according to a continuous generation model, with little or no time delay.

Storage of lipid reserves by some species of water fleas reduces the sensitivity of mortality to density and therefore introduces a time delay into the population processes (Goulden and Hornig 1980). *Daphnia galeata*, a large species, stores energy in the form of lipid droplets during periods when food is abundant (that is, at low population densities), which it can then utilize when food supplies are reduced by overgrazing at high population densities. Females also pass lipid to each offspring through oil droplets in the eggs, thereby increasing the survival of young, prereproductive water fleas under poor feeding conditions (Tessier et al. 1983). The smaller water flea *Bosmina longirostris* stores little lipid, and therefore starvation increases directly in response to increases in population density. The consequences for population growth are predictable: under laboratory conditions, *Daphnia* exhibits pronounced limit cycles with a period of 15 to 20 days; *Bosmina* populations grow quickly to an equilibrium, with perhaps a single strongly damped overshoot (Figure 18-17). For *Daphnia*, Goulden et al. (1982) estimated that r was about 0.3 day^{-1}. With a cycle period of 15 to 20 days, τ must have been about 4 to 5 days, and therefore $r\tau$ was about 1.2–1.5. Because the value of $r\tau$ was somewhat less than $\pi/2$, the cycles in the *Daphnia* population should have damped out eventually.

A. J. Nicholson's Experiments with Blowfly Populations

The behavior of a population with respect to its equilibrium is sensitive to many aspects of life history that govern time delays in responses to density. Slight differences in laboratory culture conditions or the intrinsic properties of species can tip the balance between a monotonic approach to equilibrium and a limit cycle. A. J. Nicholson's (1958) experimental manipulation of time delay in laboratory cultures of the sheep blowfly *Lucilia cuprina* provide a dramatic demonstration of the relationship of time delays to population cycles. Under one set of culture conditions, Nicholson restricted blowfly larvae to 50 grams of liver per day while giving the adults unlimited food. The number of adults in the population cycled through a maximum of about 4,000 to a minimum of 0 (at which point all the individuals were either eggs or larvae) with a period of between 30 and 40 days (Figure 18-18).

In this experiment, regular fluctuations of the blowfly population were caused by a time delay in the response of fecundity and mortality to the density of adults in the laboratory cages. When adults were numerous, many eggs were laid, resulting in strong larval competition for the limited food supply. None of the larvae that hatched from eggs laid during adult population peaks survived, primarily because they did not grow large enough to pupate. Therefore, large adult populations gave rise to few adult progeny, and because adults lived less than 4 weeks, the population soon began to decline. Eventually so few eggs were laid on any particular day that most of the larvae survived, and the size of the adult population began to increase again.

Nicholson's result may be interpreted as a time-delayed logistic process, which provides a good fit to the observed oscillations with a value of $r\tau = 2.1$. This model predicts the ratio of the maximum to the minimum population to be 84, and the cycle period to be 4.54τ (May 1975b). The experiment clearly reveals that density-dependent factors did not immediately affect the mortality rate of adults as the population increased, but were felt a week or so later when the progeny were larvae. Larval mortality was not expressed in the size of the adult population until those larvae emerged as adults about 2 weeks after eggs were laid. The blowfly population resembled Pratt's *Daphnia* population at a high temperature, in which crowding created discrete, nonoverlapping generations of individuals with an inherent time delay equal to the larval development period, about 10 days. [Hassell et al. (1976) explored discrete generation models of the blowfly population; their paper should be consulted for further discussion.]

The hypothesis that the blowfly population cycles were caused by time delays could be tested directly by eliminating the time delay in the density-dependent response; that is, by making the deleterious effects of resource depletion at high densities felt immediately. Nicholson did this by adjusting the amount of food provided to adults as well as to larvae. Because adults require protein to produce eggs, by restricting the liver available to adults to 1 gram per day, Nicholson cut egg production to a level determined by the availability of liver rather than by the number of adults in the population. Under these conditions, the recruitment of new individuals into the population was determined at the egg-laying stage by the influence of

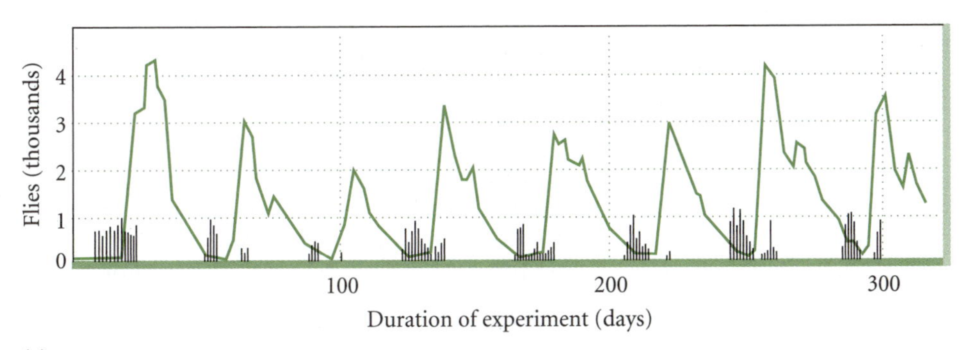

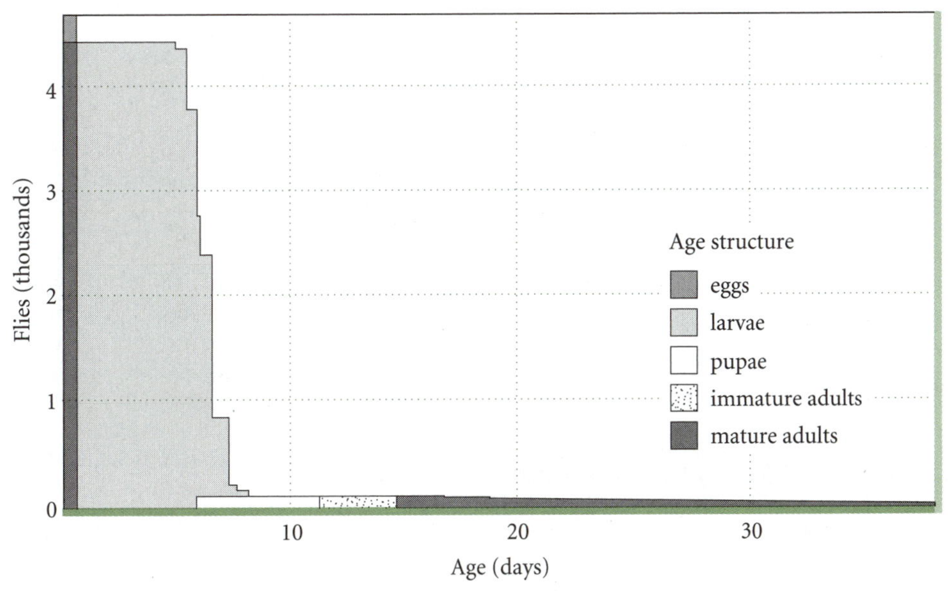

FIGURE 18-18 (a) Fluctuations in laboratory populations of the sheep blowfly *Lucilia cuprina*. Larvae were provided with 50 g of liver per day. Adults were given unlimited supplies of liver and water. The green line represents the number of adult blowflies in the population cage. The black vertical lines represent the numbers of adults that eventually emerged from eggs laid on the days indicated by the lines. (b) Average age structure of the blowfly population. (*After Nicholson 1958.*)

food supply on per capita fecundity, and most of the larvae survived. As a result, fluctuations in the population all but disappeared (Figure 18-19).

We have seen that responses of populations to density can be delayed by development time and by the storage of nutrients, both of which put off deaths to a later point in the life cycle or to a later time. Density-dependent effects on fecundity can act with little delay when eggs are produced quickly from resources accumulated over a short period. Populations controlled primarily by such factors would not be expected to exhibit marked oscillations.

Time Delay Through Maternal Effects

Time delays longer than a single generation may be introduced by the transmission of **maternal effects** to progeny through the egg or by the selection of different genotypes at different population densities. Studies of the western tent caterpillar on Vancouver Island, Canada (Wellington 1960) suggested a time delay through maternal effects. A 4-year survey included a peak year for the population, 1956, followed by a rapid decline through 1959. Tent caterpillar larvae were classified as either "active" or "sluggish" depending upon their behavior. Broods of tent caterpillars composed primarily of active larvae differed from those with a higher proportion of sluggish larvae in that they constructed more tents with a more elongate structure, foraged over greater distances along the limbs of the tree, ate more leaves, and, as a result of their greater food intake, developed more rapidly. Active caterpillars survived better than sluggish ones.

The proportion of active larvae in a particular brood was influenced by the past history of the infestation. Broods in areas where the population had been low in previous years were likely to contain more active caterpillars than broods from areas of recent high levels of infestation. The activity level of the larvae depended on the nutrition provided by the eggs from which they developed. Most adult female moths laid similar numbers of eggs, regardless of their ability to provision each individual egg; how well each female provisioned her eggs with nutrients depended on how much food she consumed as a caterpillar. When food was relatively scarce, few eggs were properly provisioned, and broods contained few active caterpillars. Thus the quality of the caterpillars in broods of a particular year depended on the availability of food to their mother when she was a caterpillar during the previous year. Furthermore, because the activity level of each caterpillar partly determined how much food it consumed, the food supply of its mother during the previous year affected the success of its own progeny during the following year. In this way, populations could exhibit a time delay of more than 1 year in their response to density or to changes in the availability of food.

Regardless of the time delay in the density-dependent response, a population at its equilibrium point will remain there until perturbed by some outside influence, whether a change in the equilibrium level (K) or a catastrophic change in population size (N). Once displaced from the equilibrium, some populations will move toward stable limit cycles, depending on the nature of the time delay and the response time. Others will return to the equilibrium directly or through damped oscillations. Cycles may be

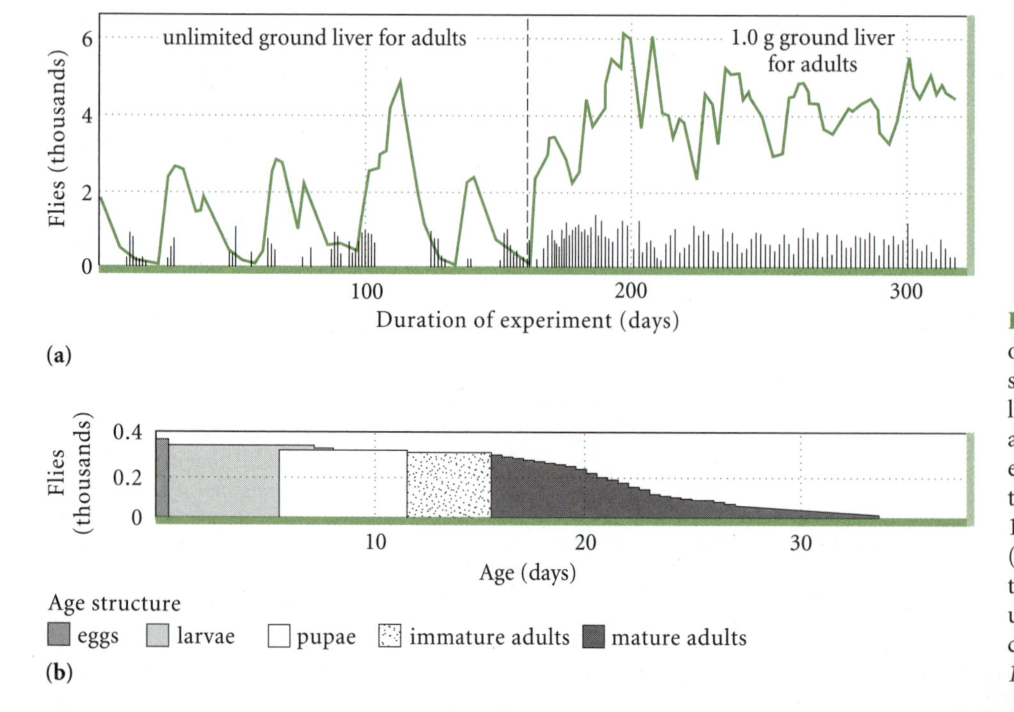

FIGURE 18-19 (a) Effect on the fluctuations of a sheep blowfly population of limiting the food supply available to adults. This experiment was similar to that depicted in Figure 18-18 in all other respects. (b) Average age structure of the blowfly population under the limited food conditions. (*After Nicholson 1958.*)

reinforced through interactions with other species—prey, predators, parasites, perhaps even competitors—with similar time constants.

The theoretical models presented in this chapter form the basis of experiments that address questions about the mechanisms and consequences of population dynamics. Fluctuation in population density may have profound effects on the persistence of natural populations. In the next chapter we shall examine situations in which natural populations are at risk, and some strategies that may be adopted to prevent premature population demise.

SUMMARY

1. Variation in the density of a population depends on the magnitude of environmental fluctuation. Some populations exhibit regular cycles of density.

2. The age structure of a population may change through time due to variation in mortality among the various age classes. Such changes may affect the rate of growth of the population.

3. The carrying capacity of a population fluctuates in response to environmental variation. The extent to which the population responds to such changes depends on the rate of population increase, r. In general, populations having high rates of growth track environmental variation more closely than do populations with low rates of growth. The characteristic return time of a population is defined as the inverse of the exponential growth rate in the absence of crowding; that is, $T = 1/r$. Populations can track environmental variation closely when the characteristic return time is less than the period of environmental variation divided by 2π.

4. Populations cycles are revealed using difference equations of the form $N(t + 1) = f[N(t)]$, where $f[N(t)]$ is a logistic recruitment function that gives the pattern with which new individuals enter the population. These discrete time models of populations with density dependence tend to oscillate when perturbed. For r between 0 and 1, population size (N) approaches equilibrium (K) monotonically. For r between 1 and 2, N undergoes damped oscillations, and eventually converges on K. For r greater than 2, oscillations in N increase in amplitude until either a stable limit cycle is achieved or the population fluctuates irregularly (chaos).

5. Continuous time models can produce cyclic population change when the density-dependent response is time-delayed. Defining the time delay as τ, such models exhibit monotonic damping when the product $r\tau$ lies between 0 and e^{-1} (0.37), damped oscillations when $r\tau$ lies between e^{-1} and $\pi/2$ (1.6), and limit cycles with a period of 4τ or more when $r\tau$ exceeds $\pi/2$.

6. Many laboratory populations of animals exhibit oscillations that arise from time delays in the response of individuals to population density. The time delays are related to the period of development from egg to adult and may be enhanced by the storage of nutrients. In laboratory populations of sheep blowflies, A. J. Nicholson experimentally removed the time delay and was able to eliminate cycles in numbers.

EXERCISES

1. List ten different plant or animal species for which you have noticed a difference in abundance from one year to the next.

2. Using the population given in question 15-2, simulate the effects of environmental fluctuations by making a random adjustment in the fecundity of age class $x = 2$, b_2, using a random numbers table in the following way. Choose a three-digit sequence in the table (for example, 348). Let the first digit indicate whether you will increase b_2 or decrease it. If the number is zero or an even integer, then increase b_2. If the number is odd, decrease b_2. So, if you choose 348, you would decrease b_2 because the first digit, 3, is odd. Let the second and third digits represent the proportion increase or decrease of b_2. For the number 348, this would mean that b_2 will be decreased by 48%. For example, if $b_2 = 2$, as in question 15-2, then for the next season use $2 - 2(0.48) = 1.04$ as the b_2 for that season. (Hint: See Figure 15-5 for a summary of the calculations.)

3. Plot the time-density graph for a population where recruitment follows the Ricker model with $K = 75$, $r = 1.2$ and letting $N(t)$ go from 0 to 100 in steps of 10. (Hint: Find the various densities using equation 18-1.)

4. Hutchinson suggested that the logistic population model with time delay, $dN/dt = rN(t)[1 - N(t - \tau)/K]$, produces damped oscillations when $r\tau$ is less than about 1.6. Show this graphically.

5. Drawing on what you learned in Chapter 17, provide a discussion of how population fluctuation would be viewed in the metapopulation context.

6. Suppose that you are a conservation biologist interested in the preservation of a population of wild turkeys. Drawing on what you have learned about population dynamics, what factors would you be most interested in knowing about your population in order to develop an effective conservation strategy? (Hint: Turkeys are game animals and thus suffer mortality via hunting.)

CHAPTER 19

Extinction, Conservation, and Restoration

GUIDING QUESTIONS

- What are the three types of extinction?
- What are the factors that may cause extinction?
- How do the activities of humans cause extinction?
- Why do small populations have a greater risk of extinction than large populations?
- How does the geographic range of a species affect its risk of extinction?
- How does environmental stochasticity affect the risk of extinction?
- How might population size and spatial arrangement buffer a population from extinction?
- What is population resilience?
- How do invading species increase the risk of extinction in island systems?
- What is the difference between conservation and restoration?
- Why is the metapopulation concept important in conservation biology?
- What is involved in a population viability analysis?
- What are some of the problems of managing genetic variation in rare and endangered populations?
- What are some of the challenges associated with captive breeding programs?
- What are some of the challenges associated with species reintroductions?

We have emphasized at various points in this book that human activities often alter and degrade the natural world. Such human-induced alterations may take the form of a number of interrelated effects, including removal of a species from parts or all of its range, degradation of whole ecosystems, changes in climate, and loss of biodiversity. It is our hope that a better understanding of how the natural world works will increase public awareness of how human activities affect nature. This has been one of our goals in writing this book. Ecologists have a lot to say about how to prevent further damage and how to conserve or restore natural systems. However, conservation and environment-friendly development are not the exclusive purview of ecologists. Indeed, public awareness and appreciation of nature must go hand in hand with social and political change, which requires the participation of all segments of society. Clearly, effective action to conserve and restore nature must be based on a thorough understanding of the principles of ecology. What ecologists bring to the discussion about humans in the natural world is knowledge about how nature works. This knowledge, integrated with an appreciation of our social and political systems, economics, and history, provides a foundation for understanding the human condition.

From previous chapters you have already developed an understanding of some of the ecological concepts essential to understanding human effects on the natural world. The features and dynamics of the physical environment, including the flow of energy and materials; the way in which evolution through natural selection shapes living systems; the specific responses of individual organisms to their environment; and the dynamics of populations represent the main themes of your study thus far. A full appreciation of how populations interact; the dynamics of communities, especially in relation to biodiversity; and the processes by which social interactions arise and are maintained in natural populations, all of which will be discussed in coming chapters, will round out your understanding of the basic principles of ecology. While the implementation of conservation and restoration practices employs concepts from all of these areas (Meffe and Carroll 1997), we have chosen to introduce some ideas about conservation and restoration here, in this section on population ecology, because of the central role that population theory plays in conservation. We will focus on other aspects of conservation and restoration in Part 6, where we will take up the issue of biodiversity in the context of our discussion of ecological communities. In this chapter, we shall discuss a number of topics and ideas of ecology that are at the interface of human activity and the natural world. Central to our discussion is the concept of extinction, a process that marks the end of the life of a species throughout its range or in some part of its distribution.

19.1 Extinction is a natural process that expresses the failure of species to adapt.

In 1810, the American ornithologist Alexander Wilson observed an immense flock of passenger pigeons in the Ohio River Valley. For days the column of birds, perhaps a mile wide, passed overhead in numbers that darkened the sky. Wilson estimated that there were more than 2 billion birds. The last passenger pigeon died in the Cincinnati Zoological Garden just over a century later. With its extinction on September 1, 1914, the passenger pigeon joined a growing list of species that have vanished from the earth. Many of those species, including the passenger pigeon, would have persisted had it not been for human activities. But the fossil record reveals that virtually all lineages have become extinct without leaving descendants. Indeed, it has been estimated that 99.9% of all species that have ever lived are now extinct (Raup 1991). Thus, the several million species of plants and animals living today are derived from a small fraction of those alive at any time in the distant past.

At least three times in the past 570 million years the earth has experienced a series of extinctions so devastating that 50% or more of the species on earth disappeared (Ward 1994). Two of these mass extinctions resulted in major shifts in the forms of life on earth. The first occurred about 245 million years ago at the end of the Paleozoic era (Permian period), at a time coinciding with great geologic upheaval associated with the movement of continental landmasses. It is estimated that about 90% of the earth's species were exterminated during this time. The second occurred about 65 million years ago, at the end of the age of the dinosaurs, the Mesozoic era (Cretaceous period). There is substantial agreement that a principal cause of this extinction was the collision of a large asteroid with the earth. Over half of all the species on earth, including the dinosaurs, went extinct during this time. A growing number of ecologists believe that a third mass extinction is now under way, one that is primarily the result of the activities of humankind, as we shall see below (Raven 1990, Soulé 1991, Wilson 1992, Myers 1997).

Extinction is part of the natural world. If we are to understand the causes of extinction, we shall have to comprehend many aspects of ecology and evolution. The questions that ecologists ask about extinction strike at the heart of population and community ecology, evolution, and life history theory. Why do species go extinct? Why do some species persist in relatively low numbers while other, more numerous species disappear? Why do some species fail to adapt quickly enough to changes in the environment? How do population and community interactions affect the probability of extinction? How does habitat fragmentation affect the rate of extinction? The answers to these questions have remained elusive owing to the characteristics of extinction itself. Extinction is merely the final event in a long sequence of subtle evolutionary and ecological processes leading to the demise of a population. Nevertheless, it is important for us to try to understand extinction, not only because it culminates the relationship between a population and its environment, but also because we humans are exercising such a strong hand in the fate of natural populations.

Types of Extinction

We have used the word *extinction* in various other places in this book, most notably in our discussion of metapopulations in Chapter 17. There, we viewed extinction as a local event, involving a single population. Indeed, when a population is removed from an area by, say, a natural disaster such as a flood, we think of that population as becoming extinct in that location. The word **extirpation** is often used for such local extinctions, though we will use the term *extinction* throughout, since the context of the discussion usually makes the scale of the extinction clear. Extinctions such as those discussed in the first part of this section involve all the individuals of a particular species.

It is useful to distinguish three types of such global extinctions. **Background extinction** reflects the fact that as ecosystems change, some species disappear and others take their places. This turnover of species, which occurs at a relatively low rate, appears to be a normal characteristic of the natural world. **Mass extinction** refers to the dying off of large numbers of species as a result of natural catastrophes, such as volcanic eruptions, hurricanes, droughts, or meteor impacts. Some disasters occur locally, others affect the entire globe, and species that happen to be in the way disappear. **Anthropogenic extinction** is extinction caused by humans. It is similar to mass extinction in the number of taxa affected and in its global dimensions and catastrophic nature. Anthropogenetic extinction differs from mass extinction, however, in that its causes are theoretically under our control.

Most information on background and mass extinction comes from the fossil record, which reveals appearances and disappearances of species through geologic time. Disappearances may occur in two ways. First, species may evolve sufficiently that individuals are no longer recognized as belonging to the same taxon as their ancestors and are given a different scientific name. True extinction has not taken place, and such instances are therefore referred to as **pseudoextinctions.** Second, a species may cease to exist, in which case its disappearance from the fossil record is a case of true extinction. The finer the resolution of the fossil record, the greater the probability of distinguishing between the two.

Where true extinction can be demonstrated, the life spans of species in the fossil record vary according to taxon, but they generally fall within the range of 1 to 10 million years. Thus, on average, the probability that a particular species will go extinct in a particular year is in the range of 1 in a million to 1 in 10 million. This is the background rate of species extinction: 10^{-6} to 10^{-7} per year. The number of species on earth at present is, conservatively, on the order of 1 to 10 million, and so the earth's biota suffers about one species extinction per year at the background rate.

Causes of Extinction

It is likely that the particular circumstances that lead to extinction are unique to each population. Local natural disturbances such as fires, volcanic eruptions, mudslides, or floods may remove a population from a particular habitat or so alter the habitat that the population can no longer survive and reproduce there. There is no way to predict this type of local extinction. If a species is already imperiled and confined to a small area because of a history of bad times, one unpredictable catastrophic event may finish it off for good. Consider the plight of the heath hen (*Tympanuchus cupido*), a species of grouse related to prairie chickens. At the time of the arrival of Europeans in North America, the heath hen was distributed throughout much of the area that became New England and south into Virginia. It was fairly common and abundant throughout its range. Hunting pressures and habitat alteration increased dramatically with the arrival of the Europeans, and by the early 1900s, the heath hen was restricted to one place, Martha's Vineyard, an island off the coast of Cape Cod, Massachusetts (Ehrlich and Ehrlich 1981). Concern about the survival of the species resulted in the establishment of a protected refuge in 1907, and the population began to increase. Then, a series of unpredictable and unfortunate events occurred. A disastrous fire during the nesting season destroyed many nests, and a subsequent population boom in goshawks, a predator of the heath hen, followed by an outbreak of disease, reduced the population to a handful of individuals by 1920. The last individual died in 1932.

We know from our study of the conditions of the physical environment that all species live within certain ranges of environmental conditions such as temperature, sunlight, oxygen concentration, and so forth (see Chapter 3). Were these conditions to shift outside the required range of a particular species in a particular location, then the existence of that species in that location would be impossible. Such changes could come about because of the activities of other organisms, including humans, because of changes in climate, or because of a large disturbance. At present, the burning of wood and fossil fuels is increasing the carbon dioxide concentration in the atmosphere, and thereby increasing the average temperature of the earth by enhancing the greenhouse effect (see Chapter 11). Such anthropogenic climate change may raise temperatures between 2°C and 6°C sometime during the twenty-first century. This is equal to the warming of the earth's climate since the last glaciation, only it is happening fifty times faster. It is likely to cause the worldwide extinction of many species, particularly plants, with narrow temperature tolerances.

Introduced organisms often wreak havoc on local, native species. An excellent example is the overexploitation of cichlid fishes of Lake Victoria, Africa, by the Nile perch (see Chapter 1), resulting in the extinction of many species of cichlids. A number of other prominent examples should be mentioned. The brown tree snake (*Boiga irregularis*), introduced from Asia, has literally eaten most of the native land birds of Guam to extinction. Most of the birds were endemics—species found nowhere else. The Hawaiian Islands have also suffered greatly from introductions of alien species, which have resulted in the extinction of a large proportion of native birds and other taxa, including land snails. The bird extinctions are probably the result of a number of different introductions. Important agents of avian mortality have been introduced, including malaria and pox virus—which would not have been a problem if the mosquito that transmits these diseases had not also been introduced to the islands. Five native bird species of Hawaii are nectar feeders that depend on the availability of various nectar-producing native plants. Introduced goats and pigs have destroyed many of the native nectar-producing plant species, resulting in the extinction of three of the species of nectarivorous birds. The American chestnut (*Castanea dentata*), which was once one of the most common hardwood trees of the eastern deciduous forests of the United States, all but disappeared in the early 1900s due to the introduction from Asia of a fungal disease called chestnut blight.

Habitat loss may cause extinction by wiping out suitable places to live. Animals of the forest will disappear when all the forest has been cut. Even when some suitable habitat remains, however, conditions within the habitat may change and cause a population to begin to decline toward extinction. As we have mentioned previously, one of the most obvious and troubling anthropogenic alterations in habitat is the destruction of forested areas, particularly in tropical regions, where biodiversity is extremely high. But many other types of habitats are rapidly declining because of human activity. Table 19-1 gives a list of some of these habitat types and where they occur.

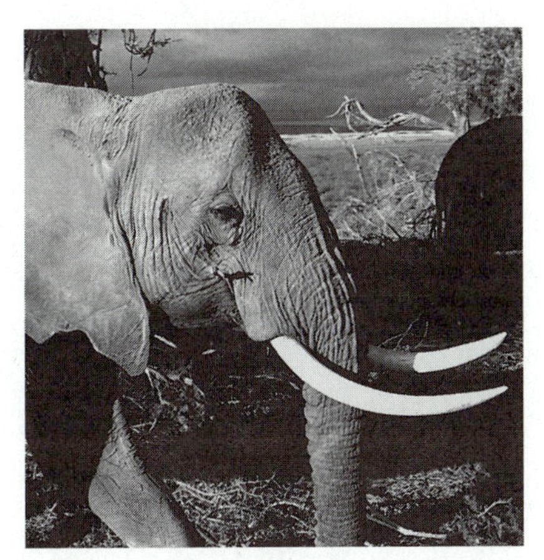

FIGURE 19-1 African elephant with large tusks. The ivory of the tusks is coveted by jewelry makers and craftspeople around the world.

Economic pressures may accelerate the natural process of extinction, as is demonstrated by the plight of the African elephant (Figure 19-1). Ivory from elephant tusks has been a popular commodity throughout the world since Europeans began their intervention in the African continent. Local population pressures in Africa and growing worldwide demand for ivory fueled by unsavory marketing strategies led to a dramatic increase in the amount of ivory taken (and elephants killed) in the 1970s and 1980s (Caughley et al. 1990). This contributed to a dramatic decline of elephant populations in the region (Figure 19-2). Other human activities have contributed to the decline of the African elephant. These magnificent animals are herbivores and, because of their great size, require vast areas of

TABLE 19-1 Major non-forest endangered habitat types

Habitat type	Location	Original extent (km^2)	Percent remaining
Tallgrass prairie	North America	1,430,000	1
Thorn scrub	Sri Lanka	19,800	25
Heathland	United Kingdom	1,432	27
Mangrove	Nigeria	24,440	50
Chaco	Paraguay	320,000	57
Fynbos	South Africa	75,000	67

(After Meffe and Carroll 1997; from Groom and Schumaker 1993; data from Nature Conservancy (UK) 1984, MacKinnon and Mackinnon 1986a,b, Mooney 1988, Redford et al. 1990, World Resources Institute 1991.)

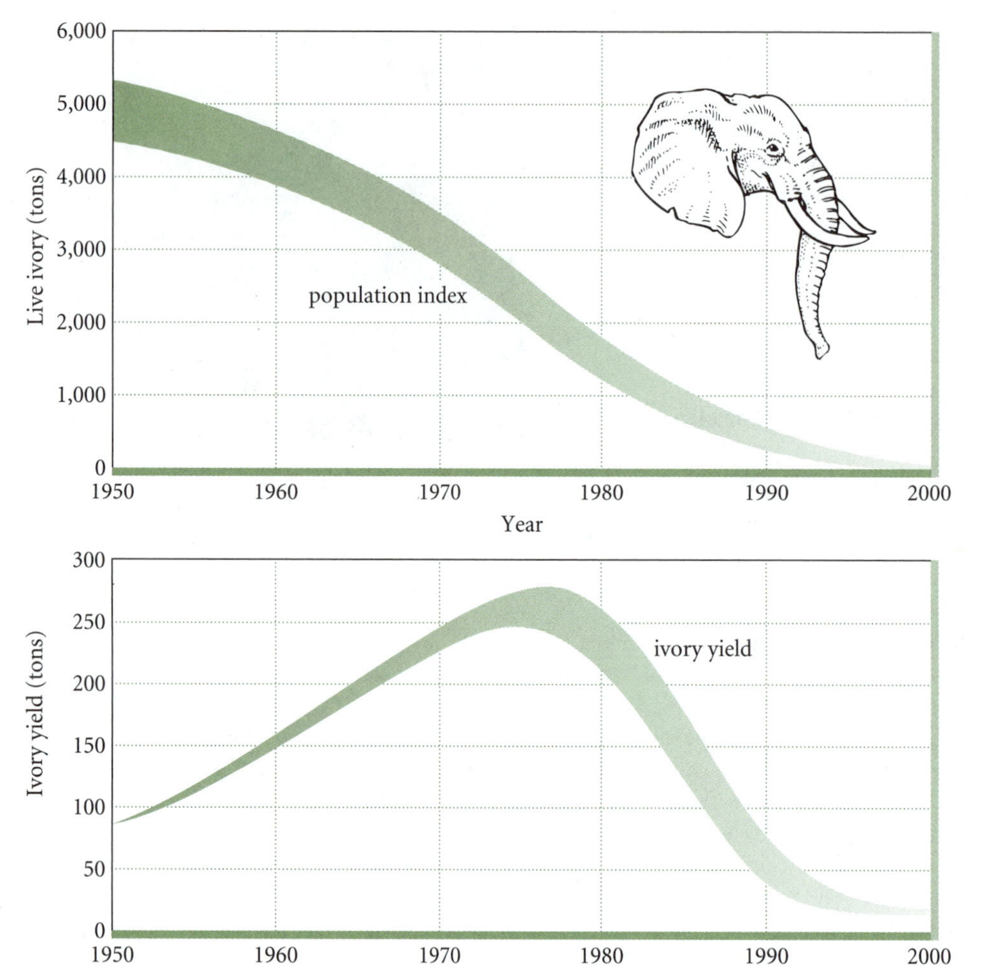

FIGURE 19-2 The amount of ivory harvested from African elephants increased dramatically in the 1970s and 1980s, contributing to the decline of elephant populations during the same time period (as measured by the amount of live ivory). *(From Krebs 1994, after Caughley et al. 1990.)*

open grassland for their survival. Human agricultural practices have reduced the amount of unaltered grassland in Africa and restricted the movement of elephants among grazing areas.

We conclude here with a reminder that anthropogenic extinction is not something that is unique to modern humankind. There is strong evidence that the activities of ancient humans in Polynesia and North and Central America contributed to the decline of species in those areas. However, technology has given modern humans considerably more destructive power than their ancestors, and few scientists doubt that the current wave of human-induced extinctions is the most dramatic of all.

19.2 The risk of extinction is affected by population size, geographic range, age structure, and spatial arrangement.

Because the loss of a species through extinction is so final, ecologists have focused their studies on factors that increase the risk of extinction. Among these factors are population size, age structure, variation in abundance, geographic range—including relative rarity and abundance—and a number of life history characteristics such as body size and life span. Ecologists are also interested in the rate at which extinctions occur in particular taxa or in specific areas or types of habitat (Lawton and May 1995). Both the risk and rate of extinction are likely to be affected by a combination of these factors (Mace 1994).

Population Size, Area, and Geographic Distribution

It is fairly widely accepted that, all else being equal, smaller populations are at greater risk of extinction than larger ones. Rosenzweig and Clark (1994) examined the data from three studies of the bird populations of the British Isles (Diamond and May 1977, Diamond 1984, Williamson 1983). Their plot of population size against probability of extinction per year shows that extinction probability decreases as population size increases (Figure 19-3). Similar relationships have been shown in spider populations on islands (Schoener and Spiller 1987, 1992), bighorn sheep (Berger 1990), and a

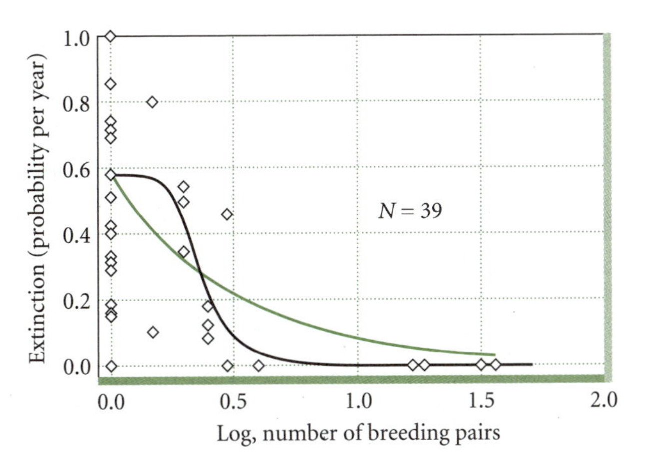

FIGURE 19-3 Probability of extinction per year as a function of population size (given as the log of the number of breeding pairs) for thirty-nine populations of birds of the British Isles. As population size increases, the probability of extinction decreases. The curves show two different mathematical models that fit the data. (*From Rosenzweig and Clark 1994; data from Diamond and May 1977, Diamond 1984, Williamson 1983.*)

number of other species for which long-term census data are available.

Small changes in birth or death rates or environmental conditions may affect small populations more strongly than large ones. Large populations are more likely to buffer such change, both because of their greater genetic variability and because they are likely to occupy a larger area and thus extend beyond the area affected by the environmental change. Thus, the risk of extinction is also influenced by the area that a population occupies. In general, the larger the area inhabited by a species, the less likely it is that extinction will occur.

Some species are widely distributed, and others are restricted to one or a few locations. Moreover, more widely distributed species are often more locally abundant than species with restricted ranges (Bock and Ricklefs 1983). Do these relationships affect the risk of extinction? Some evidence may be gleaned from the fossil record. Marine gastropods and bivalves are well represented in the fossil record because the hard shells of adults and certain parts of larval individuals are readily preserved in ancient deposits. As a consequence, the geographic distributions and longevity of many species can be estimated (Hansen 1978, 1980, Jablonski and Lutz 1980, Jablonski and Valentine 1981). When fossil bivalves and gastropods of the late Cretaceous (about 66 million years before present) from the Atlantic and Gulf coastal areas of the United States were divided into those with small (<1,000 km), medium (1,000–2,500 km), and large (>2,500 km) geographic ranges, those with small ranges were found to suffer a higher rate of extinction than those with larger ranges (Figure 19-4; Jablonski 1986).

Probability of Extinction and Persistence Time

The birth and death rates of populations depend on a variety of ecological factors, but whether a particular individual dies or successfully rears one or more progeny during a particular period is largely a matter of chance. When the annual probability of death is one-half, for example, some individuals live and, on average, an equal number die. There exists a finite probability, however, that all the individuals in such a population will die, just as ten coin tosses could all come up tails with a small but finite probability (we expect this to happen once in 1,024 trials on average).

Changes in populations owing to chance events are called **stochastic fluctuations,** and their force is felt more strongly in small populations (Shaffer 1981, Pimm et al. 1988). This becomes clear when we consider that the probability of obtaining five tails in a row with successive tosses of a coin is 1 in 32, compared with the smaller chance of 1 in more than 1,000 for obtaining twice as many tails in a row. When we visualize each individual in the population as a coin, and turning up tails as equivalent to death, we see clearly that a population of five individuals has a higher probability of extinction than one of ten individuals during the same stretch of time.

When a population is at equilibrium, the probability of extinction at a particular time, $p_0(t)$, may be given as

$$p_0(t) = \left[\frac{bt}{1 + bt}\right]^N, \qquad (19\text{-}1)$$

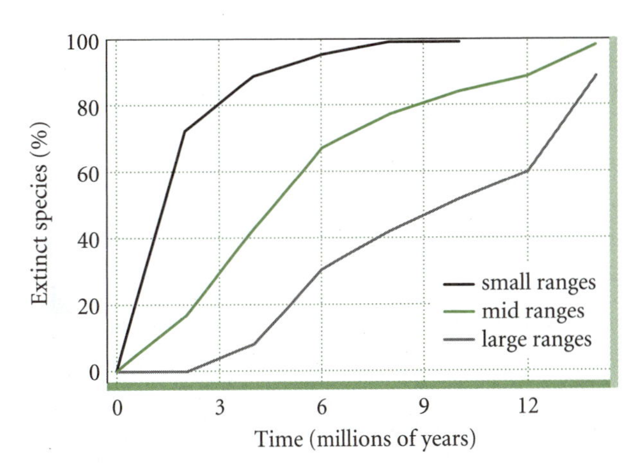

FIGURE 19-4 Percentage of species going extinct through time (millions of years) among late Cretaceous bivalves and gastropods having geographic distributions of three different sizes: small (<1,000 km), medium (1,000–2,500 km), and large (>2,500 km), measured as the linear extent of the range. The rate of extinction was highest in those species with smaller ranges. (*After Jablonski 1986.*)

where b is the birth rate and N is the population size. For a population at equilibrium, the birth rate and the death rate, d, are equal, $b = d$, and the rate of change of the population, $dN/dt = 0$ (see Figure 16-3). If we let $b = 0.5$ (thus, $d = 0.5$), a reasonable value for terrestrial vertebrates, and examine the probability of extinction for a particular time period over a range of population densities, we will see that the probability of extinction decreases as the population gets larger (Figure 19-5). The figure shows how population size and time interact to affect the persistence of a population. As time goes on, a population of a given size will have a greater chance of becoming extinct. Consider a population of size $N = 10$. The probability that the population will become extinct in 10 years is 0.162, in 50 years, 0.67, and in 100 years, 0.82. The chance that a population of size 50 will become extinct in 10 years is 1.01×10^{-4}, extremely small. The likelihood of extinction in 50 years for a population of size 50 is 0.14, and in 100 years, 0.37. These calculations assume that b and d are not density-dependent, but rather remain constant as the size of the population fluctuates stochastically from its initial value. Although this assumption is not reasonable for many populations, it probably does apply to relatively rare species.

We may also think of extinction in terms of the **time to extinction** or **persistence time,** which is generally taken to be the time that elapses between the colonization of a site and extinction. Such a concept is of enormous importance in developing strategies for the conservation of species. One model of persistence time was developed by Lande (1993). In this model, the average time to extinction, T, is given in terms of the carrying capacity, K, of the population. Additionally, the model includes a consideration of the relationship between the intrinsic rate of increase, r, of the population and the variance in that rate, which results from variation in the environment. The rather complicated formula for the relationship is

$$T = \frac{2}{Vc}\left[\left(\frac{K^c - 1}{c}\right) - \ln K\right], \qquad (19\text{-}2)$$

where $c = 2r/V - 1$ and V is the variance in the intrinsic rate of increase, r. Let us consider the parameter c first. At a given time, a population will have a specific intrinsic growth rate, r. But that rate may change over time because of responses of the population to changes in environmental conditions. If we think of r as the difference between the birth and the death rates of the population ($b - d$), it is easy to see how such variation might arise. If the death rate were to remain constant over a period of years, but for some reason the birth rate were to change each year, then r would fluctuate as well. Thus, a population may be said to have an average rate of increase, $\bar{r}$, which, like other variables, has a variance, V, associated with it. Recall that the sample variance of a variable is a measure of how widely a sample of observations of that variable deviates from the average value of that variable. The larger V, the greater the fluctuation from the average (see Chapter 2). For theoretical reasons not discussed here, the relationship $c = 2r/V - 1$ is a good way to depict the relationship between r and V for the development of the model of time to extinction given above. Figure 19-6 shows the shape of the relationship for $r = 0.2$ over a

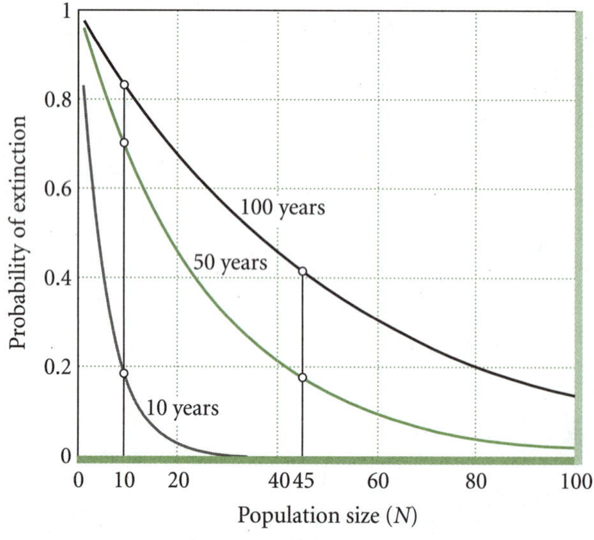

FIGURE 19-5 Change in the probability of extinction with increasing population size, N, for three different time periods: $t = 10$ years (gray line), $t = 50$ years (green line), $t = 100$ years (black line). The curves are generated using the formula $p_0(t) = [bt/(1 + bt)]^N$. For a population of size $N = 10$, the probability of going extinct in 10 years is 0.162, in 50 years, 0.67, and in 100 years, 0.82. For a population of size 45, the probability of going extinct in 10 years is nearly zero, in 50 years, 0.14, and in 100 years, 0.37.

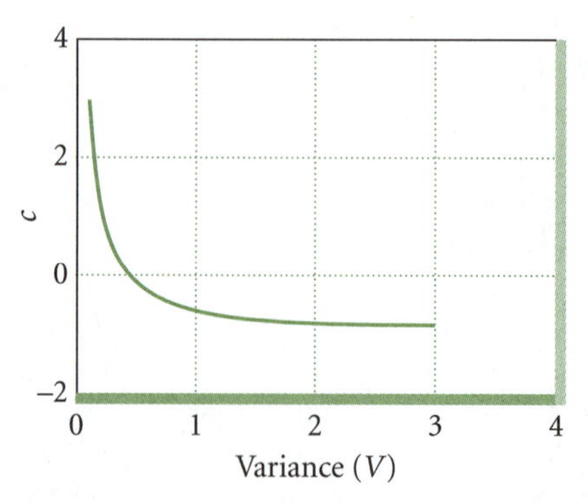

FIGURE 19-6 Relationship between the rate of population growth, r, and the variance, V, in that rate as depicted in the parameter $c = 2r/V - 1$, which is used in Lande's model of time to extinction. The parameter c decreases as the variance increases with respect to r. In this figure, $r = 0.2$.

range of values of V. As the variance increases, the value of c decreases, becoming 0 when $V = r$ (in this example, when $V = 0.2$).

Lande's equation for persistence time is plotted against a range of maximum population sizes (Ks) and for two different conditions of $\bar{r}$ and V in Figure 19-7. The upper line depicts a situation in which the average rate of increase of the population is greater than the variance of the rate of increase ($\bar{r} > V$), and the lower line shows the pattern for ($\bar{r} < V$). In both situations, the time to extinction is greater for larger equilibrium population sizes, K. The interesting prediction of the model is that environmental stochasticity, which is reflected in V, has a substantial effect that outweighs population size in some circumstances. For example, consider two populations with carrying capacities K_1 and K_2, as shown in the figure. If $\bar{r} > V$ for the first population, the one with K_1, and $\bar{r} < V$ for the second population (K_2), the first population will have a longer time to extinction, even though it is smaller, because it is subject to less environmental fluctuation—lower V relative to r—than the second.

Age and Spatial Structure

Populations of similar size are likely to differ in demographic characteristics such as age structure and sex ratio. Moreover, these features may vary from season to season. Thus, two populations of similar size may be expected to respond differently to environmental disturbance. Such differences are particularly acute in extremely small populations, in which chance can leave a population with few male offspring in a given year, or few female offspring, or, in extreme cases, no offspring at all. In such an event, which might be brought about by an intense environmental disturbance, a population may simply disappear.

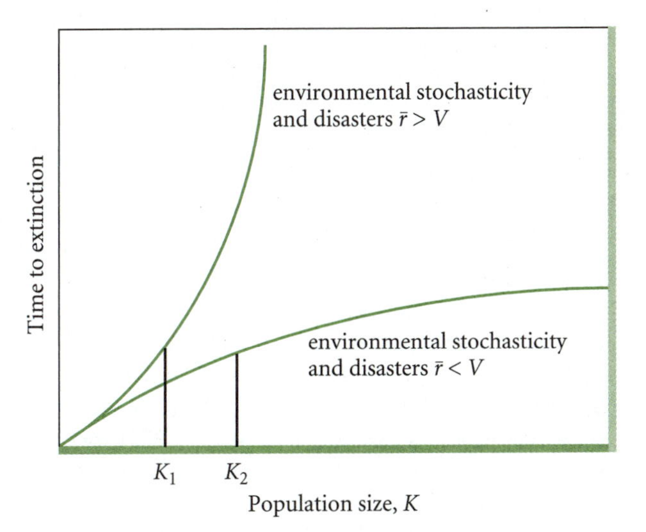

FIGURE 19-7 Lande's model of time to extinction for $\bar{r} > V$ (upper line) and $\bar{r} < V$ (lower line). A small population with carrying capacity K_1 and a low V, indicating low environmental stochasticity, may have a longer time to extinction than a large population in a highly variable environment ($\bar{r} > V$). (*After Lande 1993.*)

The spatial structure of a population can also influence the likelihood of extinction. As we have seen, many populations are subdivided into patches or local populations, each having a different size, age structure, and resilience to environmental disturbance. Such a subdivided population may persist on a regional scale even when rates of local extinction are relatively high. Extinction of subdivided populations may, however, occur when dispersal capabilities are extremely low, or when the population is so finely subdivided that each local population is very small.

It should be pointed out here that both age and spatial population structure could buffer a population from extinction. Environmental disturbances are likely not to affect all local populations in the same way. Likewise, disturbance may affect one age class more dramatically than another. The interaction of age and spatial structure also affects the probability of extinction, as the following example shows.

Shags are colonially nesting seabirds, called cormorants in North America, that inhabit coastal waters in many parts of the world. Shags forage in the ocean, where they eat fish and other small sea creatures. They are picky about their nest sites, which are located on nearly inaccessible cliff faces high above the spray line. A nest site must be big enough and flat enough to support eggs and young and sufficiently protected from the elements. Because of the limited supply of such nesting sites, males do not nest until they are at least 2 years old, and females not until they are 3 years old.

Coulson and colleagues reported an exceptional mortality event in a shag population nesting on the Farne Islands off the coast of England. In 1968, an infestation of dinoflagellates (single-celled biflagellated algae) in the oceans around the Farne Islands resulted in a substantial reduction in the fish food of the shag during the peak nesting season. Most of the adult males and females died, leaving the nests unattended and resulting in a massive decrease in reproductive output for the population. For some populations, this would have been a disaster from which recovery would have been impossible. But the age and spatial structure of the shag population provided it with a buffer to extinction. Not all colonies along the English coast were affected by the outbreak to the extent those of the Farne Islands were, and thus the potential for recolonization by immigration was great. Moreover, the age structure of the population was such that a substantial reserve of nonbreeding birds (males of age 2 or less and females of age 3 or less) were available for breeding the next year. These birds were not affected by the reduction in fish stocks because they range widely during their nonreproductive years (Coulson et al. 1968).

19.3 Body size, longevity, and population size interact to affect the risk of extinction.

We can use simple population models to help us understand the process of extinction. Think of a population at equilibrium that is "pushed" from that equilibrium by some environmental perturbation. (We note here, as we have done elsewhere in this book, that a population at equilibrium is an ideal represented by models against which we compare the dynamics of natural populations, which probably do not settle for long at any single equilibrium point; see Chapter 16.) The rate at which the population returns to its equilibrium is referred to as the **resilience** of the population. In general, if a long time is required for a return to equilibrium (long return time), then the population has low resilience. The resilience of a population is related to its age structure and spatial distribution, as we saw in the previous section. It may also be related to the life history characteristics of the population. Pimm (1991) reviews some ideas about two such characteristics that are interrelated: longevity and body size.

Consider two animal populations, one containing small-bodied and short-lived animals, the other large-bodied and long-lived animals. (Pimm suggests that we think about aphids and elephants.) What if the populations are small? Because the large animals live a long time, the likelihood that any one of them will die in, say, a year's time is small, and likewise, the likelihood that the entire population will die is small. In the absence of environmental disaster or disease, the small population of long-lived animals might be considered relatively stable. Not so for the short-lived animals. Because they are short-lived, the probability that a particular individual will die in a year's time is high, and so is the likelihood that the entire small population will be lost. We might conclude that long-lived species have greater resilience because of their lower per year mortality rates.

But the situation is not so simple. Long-lived animals usually have large bodies, and populations of large-bodied animals have longer return times—that is, lower resilience. Thus, if a population of large-bodied animals is reduced to a low level, it may remain at that level for a long time. In contrast, small-bodied animals have shorter return times and higher population resilience. A population of small-bodied animals would recover quickly from population reduction. All this suggests a complex interaction between life history characteristics and population size. In cases in which populations are small, all else being equal, large, long-lived animals may be exposed to less risk of extinction than small, short-lived species. In situations in which populations are large, small, short-lived species should be at an advantage owing to their greater resilience (Pimm 1991).

19.4 Patterns of distribution among and within islands suggest that extinction may result from a decrease in competitive ability.

The probability of extinction differs greatly among species. Some teeter perilously close to their inevitable fate and falter with the least ecological setback, while others are resilient and productive, able to withstand the perturbations of their environments. Such differences in probability of extinction can be inferred from patterns of geographic distribution and taxonomic differentiation among populations inhabiting groups of islands, such as the West Indies.

Immigrants to islands appear to be excellent competitors initially. Species that colonize islands are usually abundant and widespread on the mainland; these qualities make good colonizers. Many invaders of islands exhibit rapid population growth and expand into habitats not occupied by the parent population on the mainland (Crowell 1962, Grant 1966, MacArthur et al. 1972, Cox and Ricklefs 1977). After an immigrant population becomes established on an island, however, its competitive ability appears to wane; its distribution among habitats becomes restricted, and local population densities decrease (Ricklefs 1970, Ricklefs and Cox 1978). These trends eventually can lead to extinction. This pattern of progressive changes in the distribution and differentiation of species over time is known as the **taxon cycle** (Wilson 1961).

We can judge the relative ages of populations on islands by their patterns of geographic distribution and by differences in their appearance from the appearance of the mainland forms from which they were derived. Range maps of representative species of birds in the Lesser Antilles (Figure 19-8) demonstrate the progress of the taxon cycle. On the basis of such distribution patterns, Ricklefs and Cox (1972) assigned populations to one of four arbitrary stages (Table 19-2): expanding (I), differentiating (II), fragmenting (III), and endemic (IV). Similar patterns have been described for ants on islands of the southwestern Pacific Ocean (Wilson 1961) and for birds and some insects in the Solomon Islands (Greenslade 1968, 1969). Among birds of the West Indies, species in late stages of the taxon cycle exhibit reduced population

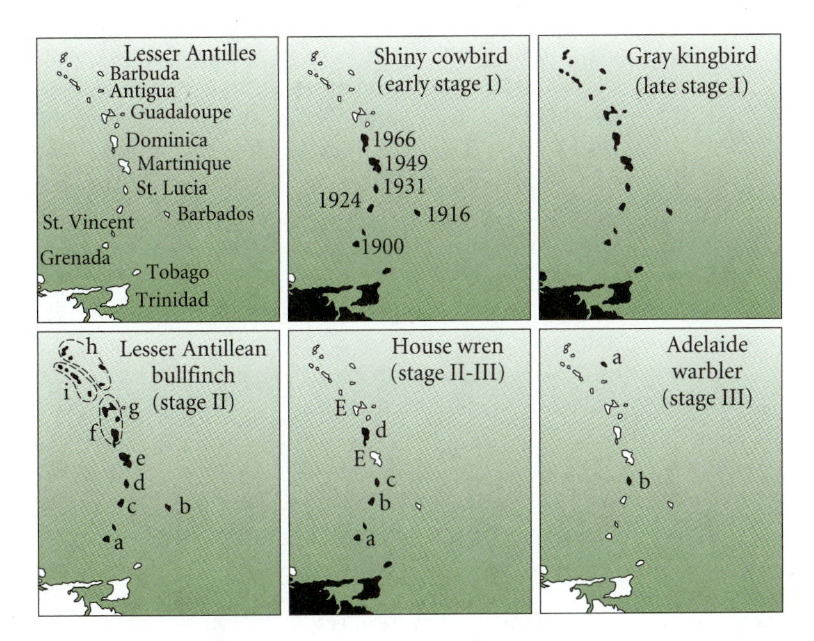

FIGURE 19-8 Distribution patterns and taxonomic differentiation of several birds in the Lesser Antilles, illustrating progressive stages of the taxon cycle. The shiny cowbird has recently expanded its range in the islands (stage I; dates of arrival are indicated): the house wren has become extinct (E) on several islands during the twentieth century (stage IV). Lowercase letters designate subspecies. (*After Ricklefs and Cox 1972.*)

densities and restriction to a narrower range of habitats, often including montane forests (Figure 19-9).

Populations become more vulnerable to extinction as the taxon cycle progresses (Table 19-3). More endemic species (stage IV) of West Indian birds have become extinct since 1850, or are currently in grave danger of extinction, than have populations of widespread species (stages I to III). The same is true of songbirds in stages III and IV in the Hawaiian Islands.

From observed patterns of distribution and taxonomic differentiation, we may infer that the time scales of ecological and evolutionary processes increase in the following order: immigration, evolutionary differentiation, extinction. Current or recent immigrants to the West Indies show little or no taxonomic differentiation, and some species have expanded their ranges in the archipelago at the rate of one island every 10 to 50 years. There are no cases of species with gaps in their distributions—which would indicate extinctions of island populations—that are not highly differentiated among islands. The fact that extinction follows the inception of visible evolutionary change on islands suggests that vulnerability to extinction may have an evolutionary component.

Ricklefs and Cox (1972) proposed that the decrease in competitive ability of island populations with age may be caused by evolutionary responses of the island's existing biota to new species. According to this scenario, immigrants are relatively free of parasites, predators, and efficiently specialized competitors when they colonize an island, so their populations increase rapidly and become widespread. Having reached this stage, however, the new immigrants constitute a larger part of the environment of other island species, which then evolve to exploit, avoid exploitation by, or outcompete them. Conceivably, a large number of existing species, when adapting to a single abundant new population, can evolve faster than the new species can

TABLE 19-2 Characteristics of distribution and taxonomic differentiation of species in the stages of the taxon cycle

Stage	Distribution among islands	Differentiation between island populations
I	Expanding or widespread	Island populations similar to one another
II	Widespread over many neighboring islands	Widespread differentiation of populations on different islands
III	Range fragmented due to extinction	Widespread differentiation
IV	Endemic to one island	

(*From Ricklefs and Cox 1972.*)

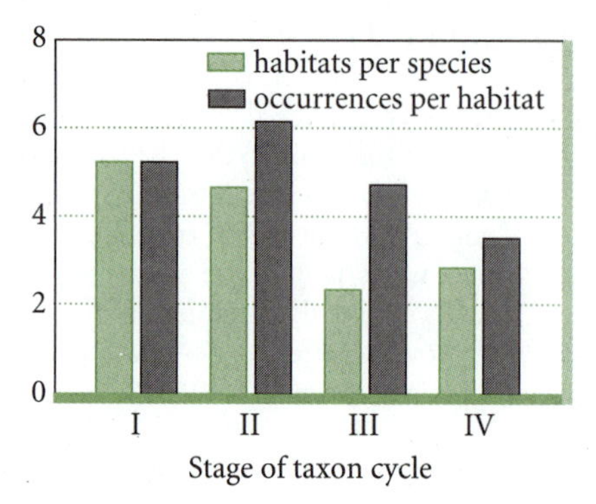

FIGURE 19-9 Relative indices of population density and ecological distribution of songbirds in the West Indies as a function of stage of taxon cycle. Figures are based on censuses in nine habitat types on Jamaica, St. Lucia, and St. Kitts. (*After Ricklefs and Cox 1978.*)

adapt to meet their evolutionary challenge. The competitive ability of the immigrants is progressively reduced by the counteradaptations of the island residents until the once-abundant new species becomes rare. Some species are eventually forced to extinction by subsequent arrivals from the mainland that are more efficient competitors. When a species becomes rare, however, other species no longer gain an evolutionary advantage by adapting to it, and evolutionary pressure upon the rare species is released. If this occurs before a species' decline has proceeded too far, the species may again increase and begin a new cycle of expansion throughout the island. This apparently has occurred many times; species distributions provide ample evidence of secondary expansions within the West Indies.

Although the scenario of the taxon cycle is speculative and has not found acceptance, even as a reasonable hypothesis, by some biologists (Pregill and Olson 1981), it is clear that islands offer promise for the study of extinction. The fact that relationships between ecological distribution and geographic distribution can be found suggests that sister populations of extinct island populations (species in stage III and IV) are themselves ecologically closer to their final demise than are populations of stage I and II species. Thus the course that leads to extinction is at least somewhat deterministic and predictable.

19.5 When conservation is no longer possible, restoration is sometimes an option.

The goal of conservation is to preserve, but sometimes degradation and decline have proceeded too far for preservation to work. In such cases, direct intervention aimed at recovery and rehabilitation is needed. In recent years, ecologists have adopted the view that, in some cases, attempts should be made to restore degraded ecological systems as close to their natural condition as possible (Jordon et al. 1987, Bowles and Whelan 1994, MacMahon 1997). From this view has grown the field of **restoration ecology.**

Restoration requires not only an extensive knowledge of life history and ecological principles, but also the ability to apply that knowledge in deliberate and well-planned steps designed to alter the trajectory of an endangered species or community (Bowles and Whelan 1994). More often than not, hard choices

TABLE 19-3 Rate of extinction of island populations of birds in the West Indies and the Hawaiian Islands as a function of stage of the taxon cycle

	Stage of cycle			
	I	II	III	IV
West Indies				
Number of recently extinct or endangered populations	0	8	12	13
Total number of island populations	428	289	229	57
Percentage extinct or endangered	0	2.8	5.2	22.8
Hawaiian Island drepanids				
Number recently extinct	2	2	9	7
Total number of island populations	23	12	12	10
Percentage extinct	8	16	75	70

(*From Ricklefs and Cox 1972; Hawaiian data from Amadon 1950.*)

must be made about which species and communities to preserve and which to let go. These choices bring ecological principles and ideals into direct conflict with economic imperatives, resulting in a cruel calculus that weighs the proximate fiscal gain of a shopping mall or housing development against the benefits of preserving a single species or community, benefits that are often difficult to appreciate in the time frame in which economic decisions are made (Brown 1994). Restoration, then, involves not only scientific work, but also organization, communication, and the necessity of working within the relevant political and social establishment.

For the most part, restoration ecology focuses on restoring whole habitats and their constituent biological communities, rather than on single populations. Nevertheless, the reintroduction of species to an area is often an important part of restoration projects. We shall discuss that aspect of restoration ecology in a number of the sections that follow.

19.6 The metapopulation concept is central to conservation biology.

Habitat fragmentation related to human development and expansion is a major reason for the decline of many endangered species. Thus one of the most daunting challenges of conservation efforts is to understand the dynamics of spatially structured populations. This is becoming more and more important as ecologists are called upon to design refuges in which endangered species are to live. The metapopulation and landscape concepts have helped shaped the approach to this challenge in recent years (Meffe and Carroll 1997a, McCullough 1996).

Much of conservation ecology has focused on the attributes of reserves in which endangered species might persist. Generally, such reserves are viewed as islands of habitat, and the questions of interest turn on how big the reserve should be, what the optimal shape is, and, if the reserve exists as a set of patches, what the spatial relationship of the patches should be. The **theory of island biogeography** (MacArthur and Wilson 1963, 1967), which we consider in detail in Chapter 29, has historically provided a theoretical basis for such thinking. The theory focuses on how species richness (number of species) is maintained in a system of islands that are subject to immigration and on which species may go extinct. Since the principal focus of island biogeography models is on species richness rather than on the dynamics of a particular population, they have proved less than satisfactory as a basis for conservation theory, and thus have given way to the metapopulation concept as the preeminent theoretical framework in conservation ecology (Hanski and Simberloff 1997, Wiens 1997). The metapopulation concept focuses attention away from species richness in isolated patches toward the dynamics of how patches are connected and on the persistence of the entire set of subpopulations. When the landscape concept is included, the focus broadens to include explicit consideration of how habitat affects the movement of individuals and other population processes (see Figure 17-16).

To be sure, idealized metapopulations, like those portrayed in the Levins model (see Chapter 17), rarely exist in nature, and therefore, the uncritical application of the metapopulation concept to the development of wildlife reserves would be inappropriate. It is not likely that a workable reserve design could be developed without consideration of the characteristics of the landscape as a whole—in particular, without an understanding of how the habitat in the intervening areas between patches affects the movement of individuals among subpopulations. Whether a species persists in a fragmented habitat will depend on the spatial relationship among habitat patches, the relative size of the patches, the reproductive potential of the species, and the dispersal ability of the species (Fahrig and Marrian 1994).

One aspect of the dynamics of wildlife reserves that has received considerable attention is the concept of habitat **corridors,** which are areas, often configured in strips or narrow lanes, that connect patches. Corridors may facilitate recolonization of patches whose subpopulations have gone extinct, and may even reduce inbreeding within the metapopulation. However, corridors present some potential difficulties (Simberloff and Cox 1987, Hanski and Simberloff 1997, Wiens 1997). Connections between patches may indeed facilitate the movement of individuals among the patches. But they may also allow the movement of predators and of disease. Moreover, the existence of a corridor does not guarantee that movement of individuals will occur. For example, the habitat within the corridor may not be suitable for the species of concern.

19.7 Recovery plans are based on the life history characteristics of the endangered species.

The Endangered Species Act of the United States requires that a recovery plan be developed for each species placed on the endangered species list. Often such plans are also developed for threatened species. Typically, these plans are prepared by teams of ecologists and others, such as representatives of industry or governmental agencies, who might have an interest in the disposition of the species or its habitat. The recovery plan includes an analysis of the predicament of the species and proposes strategies for its recovery. The plan also usually includes an analysis of the costs and benefits of the preservation of the species. The scientific foundation of the plan is based on the natural history of the species.

Population Viability Analysis

An important part of any recovery plan is an assessment of the likelihood that an endangered population can be sustained if a specific restoration plan is implemented. Such an assessment is called a **population viability analysis** (PVA) (Gilpin and Soulé 1986, Shaffer 1990). The idea of minimum viable population size (see Section 14.5) is closely associated with the concept of population viability analysis. Because PVA is a relatively new approach, it can be characterized primarily as a strategy of assessment rather than a set of specific steps. The process involves three goals. First, it is conventional to proceed toward a minimum viable population size that will have a 90% chance or better of persisting for the long term, which is usually measured in hundreds of years. Second, PVA seeks to determine what minimum viable population size would be required for the species to exist in the absence of significant intervention from managers. Finally, PVA seeks to identify restoration procedures that will maintain genetic variability in the population.

Population viability analysis includes the consideration of all information that is relevant to the history and future of the species in question. A great deal of ecological information is required. Information about the phylogenetic history and systematic relationships of the species can help determine the rate of evolutionary change in the species and the relationship of its habitat preferences and life history to those of related species. A thorough survey of the species' habitat requirements is necessary to determine how to protect the current habitat and so that alternative locations for population introductions may be found. Understanding the population characteristics of the endangered species is essential. Characteristics such as density, age distribution, rates of immigration and emigration, and birth and death rates represent fundamental information on which to base a recovery plan. In addition, some estimate of the genetic variability within the population is usually obtained in order to estimate the likelihood of fixation of genes and inbreeding problems. Finally, in most cases, it is essential to understand the basic behavioral ecology of the species. The extent of territoriality, the requirements for courtship and mating, and basic foraging behavior are examples of this type of information.

Ecological information, however, is not all that is needed. The plan will include information about the economic forces that affect the resources and habitats of the species, social and cultural attitudes about the species and the habitats within which it lives, and information about the political climate within which the recovery plan is to operate. Putting all of this information together into a recovery plan is an exercise in risk assessment requiring considerable organizational skill (Clark and Cragun 1994). Let us examine the information obtained for population viability analysis of one endangered species.

A PVA for the Woodland Caribou

The woodland caribou (*Rangifer tarandus caribou*) (Figure 19-10) once occupied the entire boreal forest zone of North America, extending into areas around the Great Lakes. The species prefers mature boreal forests, where it is thought to browse mainly on lichens, though the foraging ecology of the species is not well understood. Individual caribou lead a solitary life, spending most of their time away from others of their species, apparently in order to avoid detection by predators (Bergerud 1980). Since the early 1800s the species has disappeared from nearly all of the southern part of its range, which once included much of Michigan, Minnesota, and southern Ontario. In Canada, the species has gradually retreated northward, disappearing from much of its former range (Bergerud 1974, Gogan and Cochrane 1994). The decline of the woodland caribou is probably attributable to a combination of factors, including loss of habitat resulting from logging and fires, increased hunting, increased predation by the gray wolf (*Canis lupus*), whose population has increased in recent years, and infestations of meningeal brainworm (*Parelaphostrongylus tenuis*), a disease transmitted by the white-tailed deer

FIGURE 19-10 The woodland caribou (*Rangifer tarandus caribou*). (*Courtesy of P. J. P. Gogan.*)

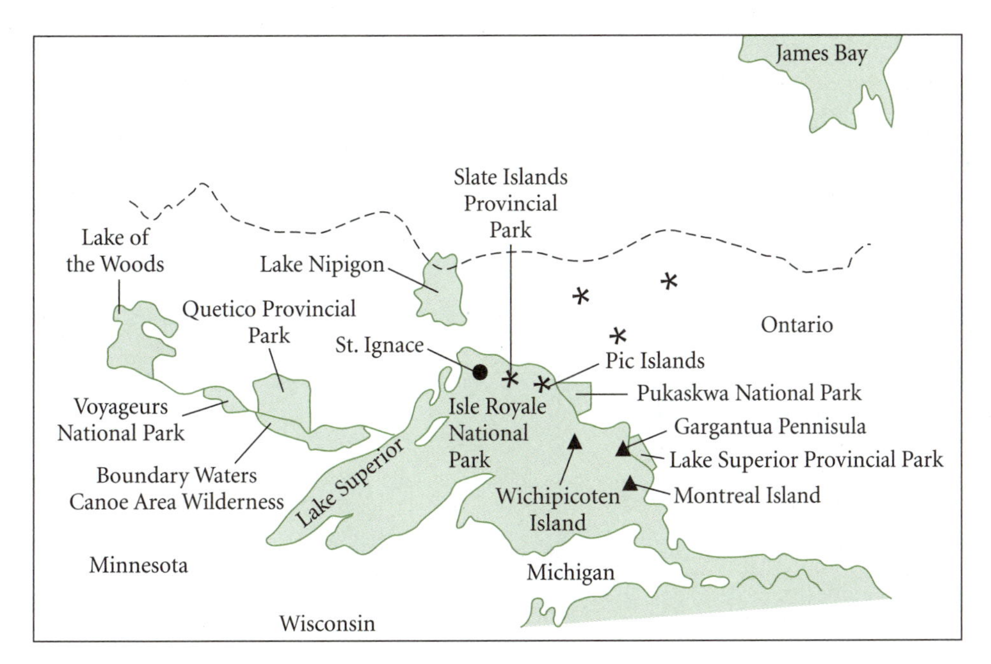

FIGURE 19-11 Distribution of the woodland caribou (*Rangifer tarandus caribou*) in southern Ontario, Canada. The dashed line indicates the southern extent of the main range of the species. The locations of remnant herds are marked with an asterisk. Herds have been successfully reintroduced in the locations marked with triangles. A reintroduction site is noted with the solid circle. (*After Gogan and Cochrane 1994.*)

(*Odocoileus virginianus*), which has expanded its range northward into areas occupied by the caribou.

Efforts are under way to restore the woodland caribou to some parts of its previous range. One effort focuses on restoring caribou populations around Lake Superior. Remnant herds occur along the northern shore of Lake Superior in southern Ontario (Figure 19-11). The species has been successfully reintroduced at three sites, all located on islands in Lake Superior where there are no white-tailed deer or predators. One other reintroduction effort failed.

Efforts are also under way to reintroduce woodland caribou in the United States, either in the Superior National Forest or at Voyageurs National Park (see Figure 19-11; Gogan et al. 1990, Gogan and Cochrane 1994). The planning and implementation of such a reintroduction is fraught with political difficulties. Federal, state, and provincial agencies from Canada and the United States would have to be involved. A number of private groups have expressed interest in and concern about the planned reintroduction. Jurisdiction over the animals, which will come from Canada and reside in the United States, is uncertain. The ecological challenge is no less daunting. How to protect introduced individuals from disease and predators, where to obtain individuals for reintroduction, the quality of the habitat, and the minimum viable population size must all be considered. An assessment flowchart for the reintroduction of woodland caribou to the Lake Superior area is presented in Figure 19-12. The planning process involves simultaneous consideration of both the political and ecological challenges. The chart fairly represents the way in which restoration planning is carried out.

19.8 Managing genetic diversity is an essential part of conservation and restoration.

Many rare and endangered species occur in small populations. What are the challenges in managing genetic diversity in the process of species recovery or restoration?

Genetics and Plant Conservation

Because of the variety of mating systems in plants, the challenge of maintaining genetic diversity in small populations, or in assembling suitable genetic variation in introduced populations, is large (Fenster and Dudash 1994, Weller 1994). Most flowering plants are hermaphroditic, producing male and female gametes either in the same flower or on different flowers on the same plant. Although a variety of adaptations have arisen in plants to prevent self-fertilization (called **selfing**), many species do undergo selfing, which may result in greatly diminished genetic variation in small populations.

One adaptation to reduce self-fertilization in seed-producing plants is **self-incompatibility,** whereby selfing is prevented by a biochemical incompatibility between the pistil and pollen (de Nettancourt 1977, Mulcahy and Mulcahy 1985). Populations with self-incompatibility are composed of a finite number of mating types, each having a different set of self-incompatibility alleles, referred to as S-alleles. Two plants with the same S-alleles are incompatible and will not produce seeds. The maintenance of genetic diversity in such a system depends on the

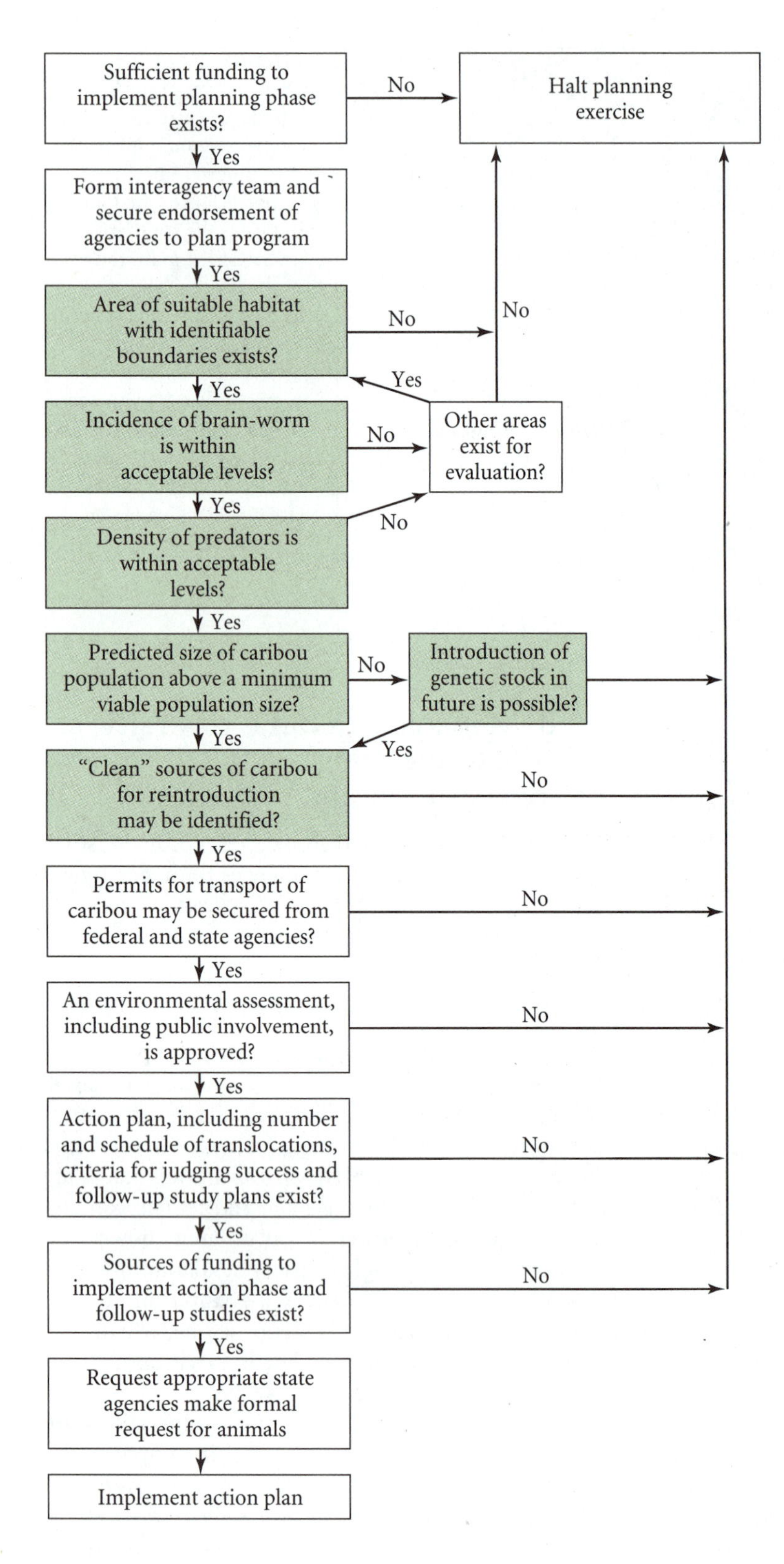

FIGURE 19-12 Flowchart of the planning process involved in the reintroduction of the woodland caribou (*Rangifer tarandus caribou*) to the Lake Superior area. The shaded boxes are part of the population viability analysis. (*From Gogan and Cochrane 1994; after Gogan et al. 1990.*)

number of different S-alleles present. In large populations, there are likely to be many compatible mating types, but genetic drift may result in the presence of relatively few mating types in small populations. For rare or endangered plants with self-incompatibility systems, this creates a severe conservation challenge, as the example of the lakeside daisy will show.

The lakeside daisy (*Hymenoxys acaulis* var. *glabra*) is a variety of the fairly widespread perennial species *Hymenoxys acaulis,* a member of the family Asteraceae, which contains such familiar plants as sunflowers, thistles, dandelions, and chrysanthemums. (Conventions of plant taxonomic nomenclature hold that the name of the variety be given by the designation var. *glabra,* which follows the genus and species.) *Hymenoxys acaulis* is distributed throughout much of western Canada through the western Great Plains of the United States and into Texas. The lakeside daisy (variety *glabra*) is found in only three small populations, two in Ontario and one in northern Ohio near Lake Erie (Figure 19-13), where it grows only in dry areas associated with dolomite and limestone or dry gravel prairies (DeMauro 1994).

DeMauro (1994) describes the recent history of the lakeside daisy populations in Illinois and Ohio. The Illinois populations had all but disappeared by the early 1970s, having been reduced to about thirty plants that were living in three small patches. No viable seeds were produced by these plants during the 1970s, and some of the plants were moved to gardens for preservation and study. In 1985, studies were initiated to compare the Illinois plants with those in three small remnant populations in Ohio. Genetic studies revealed the presence of fifteen mating types within the Illinois and Ohio populations. However, all the Illinois plants were found to be of the same mating type, and thus self-incompatible. In addition, one of the three sites examined in Ohio was found to have only a single mating type (DeMauro 1993). DeMauro suggested that the Illinois population once belonged to an extensive metapopulation in which dispersal was sufficient for local populations of a single mating type to outcross with individuals having other mating types. Once the metapopulation disappeared, the remaining local population did not set seed because there were no compatible mating types. Because the plant is capable of vegetative growth, however, the local population was able to sustain itself for a substantial period of time. It was, as DeMauro notes, already extinct before the last individual died.

The restoration plan for the lakeside daisy called for the establishment of two populations, each having a minimum viable population size of about 1,000 plants, a number determined by population viability analysis and life history study. It was estimated that this population size would buffer the plants against loss of genetic variation. Another goal of the plan was to fairly represent both the Illinois and Ohio populations. Because of the genetic structure of the populations, individuals for transplantation had to be derived from F_1 hybrid seeds of crosses between the Illinois and Ohio populations and from wild-collected seeds from Ohio and Ontario. There is cautious optimism about the outcome of the transplants. Although mortality was very high, surviving plants have flowered and continue to grow.

Captive Breeding in Animals

In the case of the woodland caribou and the lakeside daisy, viable populations, though small and

FIGURE 19-13 Distribution of populations of the rare and threatened lakeside daisy (*Hymenoxys acaulis* var. *glabra*). There are still a few remaining living individuals in the Illinois populations, but because they are of the same mating type and cannot produce seeds, their population is essentially extinct. (*From DeMauro 1994.*)

endangered, exist in part of the historic range of the species. Thus, conservation efforts can focus on the preservation of wild populations and on the transfer of individuals from those populations to other suitable places in the range. In these cases, maintaining genetic diversity depends on the genetic variation in the natural populations and on the ability to assemble new populations that mirror that diversity. Unfortunately, the natural populations of some organisms are so small and endangered that there is very little with which to begin a restoration effort. The Florida panther (*Felis concolor coryi*), for example, is now represented by fewer than fifty cats in southern Florida. The black-footed ferret, California condor, and red wolf are other such examples. Often the only hope for the preservation of such species is through captive breeding programs, in which wild animals are held and bred in order to produce offspring that may be released back into the wild (Lacy 1994). Such programs are difficult, expensive, and in some cases, impractical. Population viability analysis of the Florida panther, for example, suggests that the remaining population is too small and too inbred to sustain a recovery effort (Seal and Lacy 1989).

The success of a captive breeding program depends on the preservation of all aspects of behavior and physiology that are unique to the species. One of the most challenging goals is to maintain, or if possible, even enhance genetic variation in the population, a challenge made all the more daunting by the fact that most candidate populations are already greatly diminished and inbred. The problem of maintaining genetic diversity in a captive population is compounded both by the necessity of constructing the captive population with a small sample from an already very small population and by the circumstances of captivity itself, which introduces new selective factors that may alter the evolutionary trajectory of the captive population. Lacy (1994) offers the antelopes as a good example of the latter problem. When antelopes are startled in the wild, they turn quickly away from the disturbance and sprint in the opposite direction for some distance, then stop and reexamine the situation. This behavior is clearly an adaptation for avoiding large predators in open areas such as the plains where antelopes live. The behavior is not adaptive in the confinement of a zoo, however, where an antelope may sprint directly into a fence or wall and suffer great injury. In the wild, individuals who do not display the sprinting response will be selected heavily against because they will be more likely to fall victim to predators than their herdmates who run away. But in zoos, such individuals will have a selective advantage, and there will be a consequent evolution of the antelope herd away from the natural behavior. If reintroduced into the wild, these animals would be expected to fare poorly. Considerable effort has been expended to develop techniques to minimize such nonadaptive genetic change (e.g., Foose and Ballou 1988, Lacy 1994).

19.9 Restoration often involves the reintroduction of species.

As we have seen above, the restoration of an endangered species often involves the reintroduction of the species to parts of its former range. Augmentation of existing populations from captive breeding stocks or from extant populations at other locations in the range may also be employed.

Reintroduction programs face several problems. First, it must be determined whether the environmental conditions of the release site will sustain the released individuals. In the case of the woodland caribou (see Section 19.7), these considerations turned on the presence of predators and disease. In many cases, the factors that led to the extirpation of a species from a particular portion of its range in the first place, such as habitat fragmentation or the presence of an exotic species, may still exist, in which case some accommodation to those factors will be required. In other cases, the species may have been lost from an area as a result of a series of unique events, such as a severe drought followed by a rapid increase in a particularly effective predator (recall the plight of the heath hen in Section 19.1), and reintroduction may be more of a matter of establishing a population that exceeds the minimum viable population size. The difficulties of reintroduction have been overcome in a number of cases, as we have seen. Let us look at one program in more detail.

Reintroduction of the Swift Fox to Canadian Prairies

The swift fox (*Vulpes velox*) is a small fox that historically inhabited the prairies of the midwestern United States and southern Alberta and Saskatchewan, Canada. The fox disappeared from Canada in the early 1930s, and its range has retreated southward since that time, owing primarily to the destruction of the prairie ecosystem resulting from agriculture and development. Currently the swift fox is confined to areas in the plains of Wyoming, Nebraska, Colorado, Kansas, New Mexico, and Texas (Figure 19-14). This range lies north of the range of the closely related kit fox (*Vulpes macrotis*). Swift foxes prey on small mammals and birds, which they catch mainly at night. They themselves fall prey to coyotes (*Canis latrans*), golden eagles (*Aquila chrysaetos*), and badgers (*Taxidea taxus*).

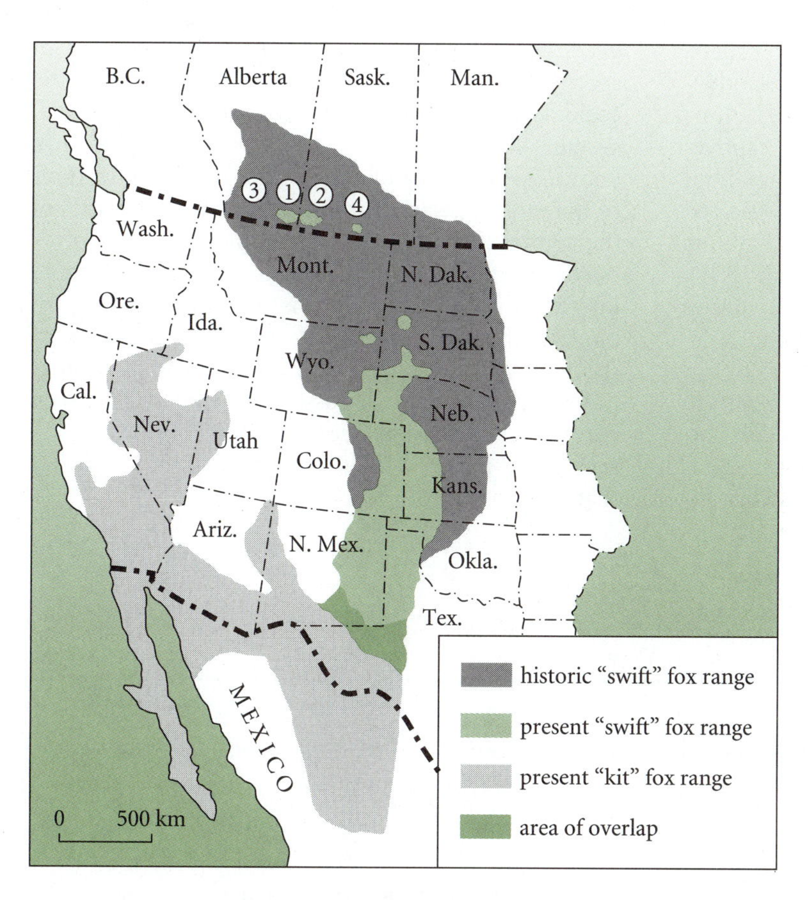

FIGURE 19-14 Range of the swift fox (*Vulpes velox*). Candidate reintroduction sites in southern Alberta and Saskatchewan are numbered 1 through 4. The closely related kit fox (*Vulpes macrotis*) lives to the south of the swift fox. (*From Carbyn et al. 1994.*)

Carbyn et al. (1994) provide an excellent description of the program developed to reintroduce swift foxes in Alberta and Saskatchewan, which we shall summarize here. An extensive population viability analysis suggested three possible reintroduction sites, and releases began in 1987. Some of the animals released were obtained from a captive breeding program, which was initiated using animals from the northernmost portion of the current range of the species in Wyoming, Colorado, and South Dakota. Other animals were taken directly from this range and transplanted to the reintroduction sites. A large number of the released animals were fitted with radio collars so that their movements could be monitored. Two types of release strategies were used. In some cases, animals were transported to the release site and placed in pens constructed in the prairie. The foxes were held in the pens for several months until they bred, after which the adults and young were released into the wild. This type of release strategy is called a soft release. In most cases, foxes were simply transported from the captive breeding area or from source populations to the reintroduction site, where they were released. This strategy is known as a hard release. There appeared to be no difference in survival between the two release methods.

The results of the reintroduction program are encouraging. Table 19-4 summarizes the survival of 155

TABLE 19-4 Survival of swift foxes (*Vulpes velox*) after soft or hard release

Release strategy	Number radio collared	Survival		
		To 6 months	To 12 months	To 24 months
Soft	45	25/45 = 55%	14/45 = 31%	6/45 = 13%
Hard	155	52/155 = 34%	20/117 = 17%	5/43 = 12%

(*From Carbyn et al. 1994.*)

hard-released foxes that were fitted with radio collars. Wild-captured individuals fared much better than those that came from captive breeding stocks. The timing of the release affected the survival of the captive-raised foxes. Coyote predation accounted for about one-third of the known predation on the released foxes. The current estimated population size of the swift fox in the area is 150–225 (Carbyn et al. 1994). To be sure, it remains to be seen whether the species will establish a permanent foothold in the area.

Here we have discussed a number of important issues of conservation and restoration that relate to population dynamics. Other approaches to conservation are important as well. It is often desirable to conserve or restore entire communities or ecosystems. Indeed, as we mentioned above in our discussion of species reintroductions, care must be taken that the ecological conditions of the reintroduction site will sustain the reintroduced population. This requires knowledge not only of the specific ecological features of the target species, but also of the characteristics of the community of plants and animals in the reintroduction area. One of the major concerns of those interested in conservation is the maintenance of biodiversity, measured most simply as numbers of species (species richness). The principles of ecological diversity will be discussed in detail in Part 6 as part of our discussion of community ecology. We will revisit the ideas of conservation and restoration in that context.

SUMMARY

1. Extinction is a natural process. The fossil record reveals that global extinction events, called mass extinctions, have occurred at least twice in the last 570 million years. Some believe that we are in the midst of another mass extinction caused by human activity. Extinction caused by humans is called anthropogenic extinction. Background extinction is the low level of turnover characteristic of most populations.

2. Extinction may be caused by large disturbances, such as volcanic eruptions, fires, or floods, by overexploitation of one species by another, and by population interactions such as predation and competition. Extinction of native species may occur when non-native plants or animals are introduced into an area. Humans may cause extinction by overexploitation or by destruction of habitat.

3. Small populations are usually at greater risk of extinction than large populations. Species having large geographic ranges are often at lower risk of extinction than species that are restricted to one or two populations.

4. The probability of extinction may be predicted using the formula $p_0(t) = [bt/(1 + bt)]^N$, where b is the birth rate, N is the population size, and t is time. The time to extinction may be calculated using the formula $T = 2/Vc[(K^c - 1/c) - \ln K])$, where $c = 2r/V - 1$, and V is the variance in the intrinsic rate of increase, r. This equation shows the relationship between environmental variation and population size. A large population subjected to great environmental stochasticity may have a shorter time to extinction than a small population in a more stable environment.

5. The resilience of a population is the time that it takes for the population to return to equilibrium after it has been "pushed" from equilibrium by an environmental perturbation. Long-lived animals with large body sizes usually have a lower population resilience than do short-lived and small-bodied animals.

6. Studies of island populations suggest that populations become more vulnerable to extinction as they progress through the taxon cycle. It has been suggested that the competitive ability of species is diminished with population age.

7. The goal of conservation biology is to preserve the habitat and genetic diversity of the natural world. Restoration ecology is an applied subfield of ecology that uses ecological principles to implement strategies to help endangered species and ecosystems to recover.

8. Because many conservation problems arise from habitat fragmentation, the concept of the metapopulation has become central to conservation biology, replacing the theory of island biogeography as the basis for conservation theory and practice.

9. Population viability analysis (PVA) is a process of determining whether a particular recovery or restoration strategy will lead to success. PVA involves the consideration of information from all aspects of the life history of the population.

10. Managing the genetic diversity of small and endangered species is particularly important for the survival of those species. Self-incompatibility in plants often results in isolated populations having only a single mating type. The restoration of such populations requires outcrossing with populations of different mating types. Captive breeding programs are used in cases in which a natural population is extremely rare and endangered.

EXERCISES

1. Suppose you are faced with the challenge of developing a plan to safeguard an endangered species that lives in an imperiled natural habitat near your home. Make a list of all of the different individuals, by profession or avocation, that you would have to involve in such a project to make it successful. How would your plans be affected by an upcoming congressional election in the district containing the endangered species? Explain how you would use ecological information about the species (information that presumably you will be able to provide after you have finished reading this book!) in meetings with groups of planners that include local business and religious leaders.

2. Write an explanation of the process of extinction, how humans affect that process, and the possible effects of anthropogenic extinction that you will present to the congressional staff of a powerful senator.

3. Equation (19-1) assumes that b, the birth rate, and, consequently, d, the death rate, are independent of population size, N. Derive a formula for the probability of extinction that provides for b to decrease as the population size increases. Plot the change in probability of extinction over some range of N. (Hint: Start with a simple linear decrease in b with N.)

4. Using equation (19-2), calculate the time to extinction for a population with the following characteristics: $r = 1.2$, $K = 120$, $V = 10$.

5. Using the library or other resources such as the World Wide Web, find information about a species for which a restoration project is planned or under way. Write a summary of the challenges involved with the conservation of that particular species and a review of the approaches that are being planned or executed.

PART 5

POPULATION INTERACTIONS

In Part 4 we explored how single populations are structured demographically, spatially, and genetically. Populations do not exist in isolation from other populations. Organisms of different species interact with one another through competition for resources, predation, and other means, and these interactions affect not only the individual participants in the interactions but their populations as well. Here in Part 5, we will examine the nature of these interactions.

The consumption of resources is an essential function of all living organisms. Resource-consumer interactions represent the direct transfer of energy from one organism to another. The availability of resources shapes the interactions between populations. The general nature of resource-consumer interactions will be discussed in Chapter 20. The competition of two or more species for limited resources will be the subject of Chapters 21 and 22. In Chapter 21 we shall see how competition affects the population growth of the competing species and how disturbance and other factors affect the outcome of competition. In Chapter 22 we shall explore some examples of competition that have been observed in nature.

One of the most dramatic species interactions is predation, which we shall explore in Chapter 23. The effect of predation is strongly negative on one participant, the prey, and strongly positive for the other, the predator. We shall see that predator-prey dynamics has important applications in pest control. In Chapter 24 we will discuss the population dynamics of plant-herbivore and host-parasite interactions, which unlike predation, do not usually result in the death of the host. The population consequences of herbivory and parasitism are the focus of this chapter. We shall also explore the population dynamics of a number of important human diseases. In Chapter 25 we will explore interactions in which both participants benefit. The adaptations involved in such interactions are some of the most extraordinary in nature.

CHAPTER 20

Resources and Consumers

GUIDING QUESTIONS

- What are the four general types of species interactions?
- What is the difference between renewable and nonrenewable resources?
- What are the three types of renewable resources?
- What is a limiting resource?
- What are the five types of consumers?
- How can population growth rate and the rate of resource supply be dynamically coupled using mathematical models?
- What makes the Monod equation a biologically reasonable way of modeling simple consumer-resource systems?
- How might an ecologist experimentally determine the critical resource level in a system with two resources?
- What are the ways in which two resources may interact to determine the growth rate of a consumer population?

Many interactions between populations involve the consumption of resources. Competitive interactions involve several populations of consumers that vie for one or more resources. In some cases, such as predation, one of the participants in the interaction is the resource (prey) and the other is the consumer (predator). Herbivores and parasites consume parts of their food plants or hosts. Mutualistic relationships, in which all participants benefit from the interaction, such as the relationship between plants and their pollinators, involve the production and consumption of a resource as part of the interaction. Often the characteristics and availability of the resource shape the interaction. In this chapter, we shall describe the various types of resources, provide an overview of the general nature of resource-consumer interactions in the context of population growth, and explain how resources may limit population growth.

Before beginning our discussion of consumer-resource dynamics, we shall summarize the various types of population interactions (Section 20.1). It is important to appreciate that these interactions have consequences at several levels. There is, of course, the effect of the interaction on the individuals involved. This is most dramatically observed in predator-prey interactions, in which one of the participants dies. But there are individual effects in all other types of interactions as well. Some competitive interactions may involve direct aggressive behavior between individuals of competing populations. Individual plants are surely affected by the loss of leaves or stems to herbivory, and hosts feel the effects of their parasites. But interactions among organisms have effects at the population level as well. Competition among the individuals of a single population may result in a reduction in the growth rate of that population. When one species is a superior competitor to another, it may eliminate its competitor from an area. Predators reduce the population size of their prey. The population-level effects of herbivory, parasitism, and mutualism are sometimes difficult to observe, as we shall see; nevertheless, such effects are no doubt important. In discussing competition and predation (Chapters 21-23), we shall focus primarily on the population consequences of the interaction. Our discussions of herbivory, parasitism, and mutualism will include more attention to the responses of individuals to the interaction, with a focus on how those individual responses may affect population factors.

Also, before proceeding, we must comment on the use of the terms *population* and *species* in this part of the book. Earlier, we defined a *population* as a group of organisms of the same species occupying a particular area defined either by natural boundaries or arbitrarily by an ecologist. A *species* is a group of actively or potentially interbreeding populations that is isolated reproductively from other such groups. We may synonymize these terms for simplicity at times, but it is important to appreciate that, with respect to mathematical models of competition and predation, we are referring to local populations, not to all populations of the species involved in the interaction. When discussing coevolutionary processes, we will use the word *species.*

20.1 Species interactions may be categorized based on the effect of the interaction on the species involved.

We may conveniently place species interactions into one of four general categories based on the effects of the interaction on the species involved (Table 20-1). One of the most obvious interactions is that in which individuals of one species consume individuals of another. These **consumer-resource interactions** have a negative effect on one of the participants and a positive effect on the other. Among the most obvious of such interactions are those of predator (the consumer) and prey (the resource; see Chapter 7), but the grazing of a horse in a field, or the sucking of blood by a mosquito, or the chewing of a leaf by a caterpillar are no less interactions of consumer and resource. Consumer-resource interactions directly link species in nature (Figure 20-1a). Predation, herbivory, and parasitism, including disease, are important consumer-resource interactions discussed in the chapters of this part of the book. (It should be mentioned that one type of consumer-resource interaction occurs *within* some populations. That interaction is called **cannibalism.**)

When interactions between individuals of two or more species have negative consequences for all the species involved, the interaction is said to be **competition.** Competition results when many species seek the same resources, and the depressing effect that each one has on the availability of the shared resources adversely affects the others. Competition may be represented schematically by expanding the model of simple consumer-resource link shown in Figure 20-1a to include more than one consumer, as shown in Figure 20-1b. Usually individuals compete indirectly through their mutual effects on the shared resources. Less frequently, when consumers can profitably defend resources, competitors may interact directly through various antagonistic behaviors. In some cases, two species that are consumed by a single consumer may appear to compete if the consumer favors one species over the other (Figure 20-1c; see Figure 21-16). We shall discuss this situation and other aspects of competition in Chapters 21 and 22.

TABLE 20-1 Categories of relationships between species

	Effects of interaction on	
Type of interaction	Species 1	Species 2
Competition	Negative (−)	Negative (−)
Consumer-resource	Positive (+) for consumer	Negative (−) for resource
Detritivore-detritus	Positive (+)	Indifferent (0)
Mutualism	Positive (+)	Positive (+)

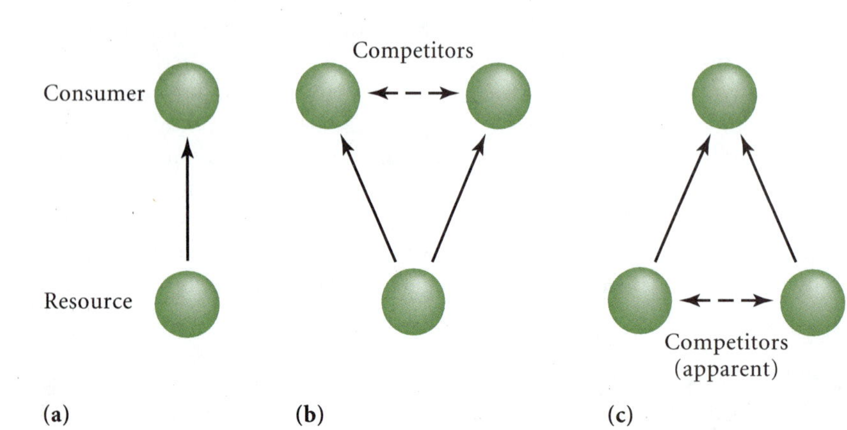

FIGURE 20-1 A schematic diagram of the relationships between species involved in (a) resource-consumer interactions, (b) competition, and (c) indirect competition. Solid arrows indicate the direction of consumption; dashed arrows indicate competitors.

Organisms that consume detritus—freshly dead or decomposing organic matter—have a special interaction with their resource. Detritus is not living, and thus the consumer cannot have an effect on its growth or behavior. This **detritivore-detritus interaction** results in a positive effect for the detritivore and no effect for the detritus.

Sometimes an interaction will result in a positive effect for all of the participants involved. Flowers provide bees with a supply of nectar, and bees carry pollen between plants and effect fertilization. Mycorrhizal fungi extract inorganic nutrients from the soil that plants can use, and plants supply their fungus partners with carbohydrates. These interactions are called **mutualisms.** In most cases, each party to a mutualism is specialized to perform a complementary function for the other. In lichens (see Figure 1-11), photosynthetic algae team up with fungi that can obtain nutrients from difficult substrates, such as bark and rock surfaces. Such intimate associations, in which the members together form a distinct entity, are referred to as **symbioses**—literally, "living together."

At the most fundamental level, species interactions involve the interactions of individual organisms. When the interaction has a positive or negative effect, as in consumer-resource interactions, it is felt by the individuals involved. The strength of the effects varies with the type of interaction. In the case of predation, the effects are clear: one of the participants is nourished by the interaction, and the other dies. In herbivory, while the plant on which an herbivore feeds may suffer a negative effect from the interaction, the effect is usually not death. The effect of competition on individuals is negative, but the strength of the effect may vary widely among individuals. The positive effects of mutualisms are enjoyed by all of the individual participants in the interactions.

The benefit or harm that individual organisms experience in their interactions with other organisms affects their evolutionary fitness. Consumers must obtain nourishment in order to produce eggs and rear young. Prey must avoid being eaten, and plants and hosts must minimize damage from herbivores and parasites, in order to live to reproduce. Competition may reduce reproductive output and growth among competitors. We will save our discussion of these individual effects for the section on evolutionary ecology at the end of the book (Part 6).

The population consequences of herbivory and parasitism are often more difficult to observe than those of predation. In the case of herbivory, a plant attacked by an herbivore often does not die, and continues to produce seeds. Thus there may be no obvious population consequence of the interaction. Nevertheless, plants do experience changes in physiology and morphology resulting from herbivory, and sometimes these changes may alter rates of reproduction and mortality, and consequently, population size. The consequences for the herbivore in an herbivore-plant interaction are likewise not as clear-cut as those experienced by predators in predator-prey interactions. We shall see, however, that population effects of these interactions can be identified and examined.

The population effects of mutualistic interactions are more difficult to discern than those of herbivore-plant and parasite-host interactions. In some mutualisms, such as the relationship between the algae and fungi that form lichens, each participant requires the presence of the other. Therefore, if for some reason one of the participant species is not present, the population of the other species is dramatically affected. However, in other cases, in which the mutualism is facultative, the effect of the relationship on the two populations is more difficult to determine, as we shall see (see Chapter 25).

20.2 What are resources and consumers?

A **resource** is a substance or an object that is required by an organism for normal maintenance, growth, and reproduction. When a resource is used, its amount is reduced. Thus, food is always a resource, even though some components of the diet may not be. For example, we consume cellulose and other forms of plant fiber in our diet, but these pass through our digestive tracts largely unused. Thus plant fibers are not themselves resources for human consumers, but the food that contains them most certainly is. Water is a resource for terrestrial plants and animals. Water is consumed, and it is critical to maintenance and growth. Furthermore, when its availability is reduced, biological processes are so severely affected as to reduce population growth.

In economics, the term *resources* takes on a broader meaning than that used here. In that usage, resources may include things, such as water, that fit our notion of resources, and other things, such as coal and oil, that do not. In the case of coal and oil, the reference is to materials that are used in human commerce in some way, not to substances that are required for the growth and maintenance of individual organisms. Referring to coal, oil, and natural minerals such as gold and copper as resources, or "natural resources," is correct in one sense because these materials accumulated in the earth as a result of lengthy (millions of years) geologic processes. But they are not ecological resources in the sense that we use the term here. (An important aside with respect to the economist's terminology for coal and oil is the misleading connotation of the term "oil production." The term suggests that the oil pumped from the depths of the earth is somehow replaced by the formation of new oil. Such is almost certainly not the case, and we suggest that the term "oil extraction" more fairly represents the process.)

Types of Resources

We may classify resources as one of two major types according to how they are affected by their consumers. Some resources, such as space, occur in fixed quantities and may be completely used up by a consumer. These are called **nonrenewable resources.** Often, nonrenewable resources are not altered by their use and thus may become available for reuse. Once occupied, space becomes unavailable, but it is "replenished" when the consumer (the occupier of the space) leaves. In contrast, **renewable resources** are constantly regenerated, or renewed. Births in a population of prey continually supply food items for predators. By continually decomposing the organic detritus in the soil, microorganisms provide a fresh supply of nitrate to plant roots.

We recognize three types of renewable resources. The first has a source external to the place where the resource is used, beyond the influence of consumers (Figure 20-2a). Sunlight strikes the surface of the earth regardless of whether plants "consume" it. Local precipitation is largely independent of the consumption of water by plants, even though transpiration plays a major role in returning water to the atmosphere. For all practical purposes, detritus rains down from the sunlit surface of the sea to the abyssal depths uninfluenced by the consumers groping there in everlasting darkness.

Renewable resources of the second type are generated near or at the place where they are consumed and are directly affected by the activities of consumers (Figure 20-2b). Most predator-prey, plant-herbivore, and parasite-host interactions depend upon resources of this type. Predation strongly influences the growth rates of prey populations, particularly when predators depress prey populations below the carrying capacity of the environment, activating density-dependent response mechanisms.

Renewable resources of the third type are also generated near or at the place where they are consumed, but resource and consumer are linked indirectly, either through other resource-consumer steps or through abiotic processes (Figure 20-2c). For example, in the nitrogen cycle of a forest, plants assimilate nitrate from the soil. Herbivores and detritivores consume plant biomass, returning large quantities of organic nitrogen compounds to the soil as detritus. These compounds are attacked by microorganisms, which release the nitrogen in a form the plants can use. Nitrogen consumption thus follows the cycle soil → plant → detritivore → microorganism → mineral → soil (Figure 20-2d). Uptake of nitrate by plants has little direct effect on its release by detritivores; similarly, consumption of detritus cannot immediately influence plant production. Clearly, however, detritivores and microorganisms do influence plant production indirectly through the rate at which they release nutrients into the soil.

Limiting Resources

Consumption reduces the availability of a resource. What is used by one organism cannot be used by another. By diminishing their resources, consumers limit their own population increase. As a population grows, its overall resource requirement grows as well, and is eventually balanced by a decreasing supply of resources available to fulfill the need. But whereas all

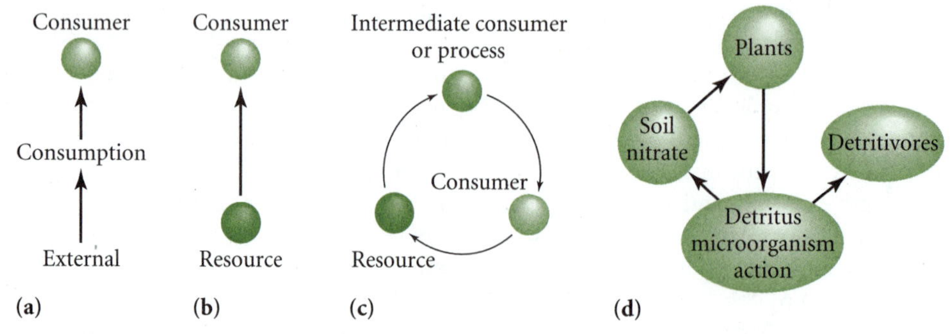

FIGURE 20-2 Three types of relationships between consumers and renewable resources. (a) A resource external to the place where it is consumed (e.g., sunlight). The supply of such a resource is out of the control of the consumer. (b) A resource generated near the place where it is consumed (e.g., prey that live in the same place as the predator). The rate of supply of such a resource may be affected by the activities of the consumer. (c) A resource that arises near where a consumer uses it, but is used first by an intermediate consumer. (d) Nitrogen cycling as an example of indirect linkage between consumer and resource. Nitrogen is available to plants only after intermediate processing by microorganisms.

resources, by definition, are reduced by their consumers, not all resources limit consumer populations. All animals require oxygen, for example, but they do not depress its level in the atmosphere even noticeably before some other resource, such as food supply, limits population growth.

The potential of a resource to limit population growth depends on its availability relative to demand. At one time, ecologists believed that populations were limited by the single resource having the greatest relative scarcity. This principle has been called **Liebig's law of the minimum,** after Justus Liebig, who expounded upon the idea in 1840. According, to this principle, each population increases until the supply of some resource, called a **limiting resource,** no longer satisfies the population's requirement for it. Although we now know that two or more resources can interact to limit population growth, Liebig's perspective first placed population regulation in the context of resource supply.

Before leaving this general discussion of resources, we should point out that there are factors required by organisms that are not resources. Temperature, for example, is not a resource. The reproductive rate of a species may increase with increases in temperature through some range, but the species does not consume temperature. This is not to imply that temperature and other nonconsumable physical and biological factors are not important, only that they must be considered in a different way from resources. Temperature, humidity, salinity, hydrogen ion concentration (pH), buoyancy, and viscosity are conditions that influence the rates of processes, and therefore the individual's ability to consume resources, but they are not themselves used and thereby transformed by the activities of organisms.

Consumption

Consumers go by many names, the most familiar of which are **predators, parasites, parasitoids, herbivores,** and **detritivores.** As we indicated earlier, from the standpoint of population interactions, some of these are useful distinctions, while others are confusing.

Each type of consumer interacts in a complex way with its principal resource: the plant, host, or prey that it consumes. Proximally, consumer-resource interactions are shaped by the densities, temporal and spatial patterns of abundance, age structures, and time delays in demographic response of both the consumer and the resource populations. And these interactions are embedded in the matrix of other interactions throughout the community. Because of this, even though we will focus primarily on interactions between single consumers and single resources in this part of the book, we proceed with caution, knowing that such a two-species perspective is too simple a representation of natural communities. Ultimately, consumer-resource dynamics are shaped by mutual evolution of consumer and resource populations. Evolution is an adaptive, genetic response to environmental change. In the case of consumers and biological resources, each is a part of the other's environment. As one changes in response to the other, it also stimulates further change in the other (see Chapter 7).

"Consumption" extends beyond the act of eating. For organisms that require a bit of space to put down roots or to hold their position against the force of wind or tide, the use of space may be considered a form of consumption. Among barnacles growing on rocks within the intertidal zone, individuals require space to grow, and larvae require space to settle and take up adult life (Figure 20-3). As the population grows, space is used up. Crowding increases adult mortality and reduces fecundity by limiting the growth of adults and recruitment of larvae. Hiding places and other safe sites constitute another kind of resource. Each area of habitat has a limited number of holes, crevices, or patches of dense cover in which an organism may escape predation or seek refuge from adverse weather (Martin 1988). As individuals occupy, or "consume," the best sites, others must settle for less favorable places; they may suffer higher mortality as a consequence.

20.3 A water flea consumer and an algal resource reveal the characteristics of a consumer-resource system.

Water fleas are small freshwater crustaceans (related to lobsters and crayfish) that are common in lakes and ponds throughout the world, often occurring in great numbers. There are over 150 species of water fleas in North America. Water fleas may inhabit open water (planktonic), bottom areas, or the surfaces of aquatic vegetation, where they feed on algae or detritus suspended in water. The creatures get their name from the way in which they dart about in the water, giving the impression of fleas hopping from one host to the next. Like other crustaceans, water fleas have a shell, or carapace, that covers the body and must be molted periodically in order for the animal to grow. The carapace is composed of two transparent structures called valves, one covering each side of the body (see Figure 1-16). One can view the appendages and the internal body parts through the transparent valves. A current of water is produced by movements of the appendages that draw water in through an opening at the anterior end of the carapace and out through an opening at the posterior end. Particles of food, such as algal cells,

FIGURE 20-3 Competition for space among barnacles on the Maine coast. (a) Above their optimum range in the intertidal zone, the barnacles are sparse, and young can settle in the bare patches. (b) Lower in the intertidal zone, dense crowding of barnacles precludes further population growth; young barnacles can settle only on older individuals. (*Courtesy of the American Museum of Natural History.*)

are filtered from this current and moved to the mouth opening by the mandibular appendages, where they are consumed.

Planktonic water fleas such as the common *Daphnia magna* make excellent model systems for the study of consumer-resource dynamics for a number of reasons. Both water fleas and their algal food are easily cultured in small aquaria in the laboratory. Under many laboratory conditions, one can reasonably assume that both the water fleas and their algal food are homogeneously distributed in the aquarium or holding vessel. This allows for experiments to proceed with some control for spatial effects, which, as we shall see below, is an important assumption in understanding filtering rates. Water fleas have relatively short life spans and reproduce readily under laboratory conditions, so it is possible to study the effects of feeding on their reproduction and population dynamics. Most water fleas can reproduce parthenogenetically; that is, females can produce eggs without fertilization by males. Eggs are produced in broods that are shed with the carapace during a molt. One female may produce several broods in a season. Thus, the researcher is afforded a direct link between consumption and reproduction in individual animals (females). Because the carapace of water fleas is transparent, experimenters can often observe various stages of egg development.

A series of experiments conducted by Porter and coworkers (1982, 1983) with *Daphnia magna* and a single algal food source, *Chlamydomonas reinhardi,* demonstrated how the amount of food in the environment affects reproduction in water fleas. Porter and coworkers established laboratory cultures under conditions of water quality, temperature, and resource densities that matched the natural lake environment of *Daphnia.* Algal concentrations were maintained between 0 and 1 million (that is, 10^6) *Chlamydomonas* cells per cubic centimeter (cm^3) of water. The investigators measured rates of water filtering and food ingestion by *Daphnia* and several population variables as a function of resource concentration. They estimated the filtering rate from the uptake of algal cells (determined by accumulation of radioactive label from cultures of *Chlamydomonas* labeled with carbon 14) and the density of cells in the culture in the following way. Assuming that the *Chlamydomonas* cells are homogeneously distributed in the water, if there are, for example, 10^5 algal cells per cm^3, then there is 10^{-5} cm^3 of water per cell. Or, to put it another way, each algal cell represents 10^{-5} cm^3 of water. Using this information, the researchers could estimate the rate at which water was filtered by simply counting the number of cells ingested. If each algal cell represents 10^{-5} cm^3 of water, and on average water fleas ingested 10^4 cells in a 60-minute trial, then 10^4 cells $h^{-1} \times 10^{-5}$ cm^3 $cell^{-1}$ (number of cells ingested times the amount of water represented by each cell), or 0.1 cubic centimeter of water, was filtered in that hour. With this approach, the researchers could measure both the filtering rate, a measure of feeding activity, and the number of cells ingested, a measure of consumption. They could also count the number of broods of eggs produced by females and number of eggs in each brood.

Porter conducted experiments to determine how the abundance of algal food affected the feeding rate

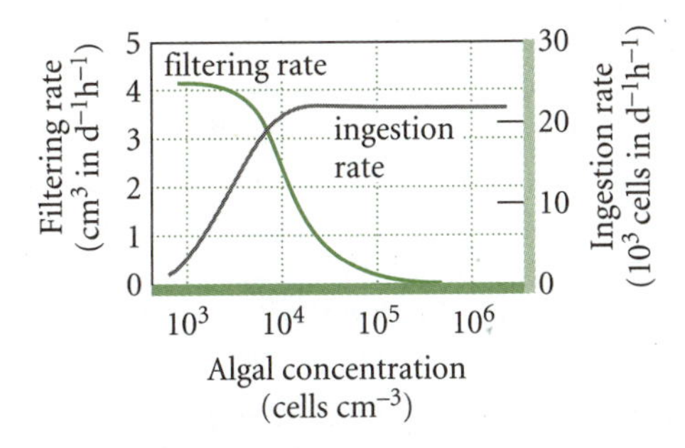

FIGURE 20-4 Filtering rate ($cm^3\ h^{-1}$) and ingestion rate (10^3 cells h^{-1}) of individual *Daphnia magna* feeding on cultures of the alga *Chlamydomonas reinhardi* at different food densities. (*From Porter et al. 1982.*)

and reproduction of *Daphnia*. Their results showed that at algal concentrations below 10^3 cells cm^{-3}, *Daphnia* ingested very few algal cells, even though their feeding effort (filtering rate) was very high (Figure 20-4). At this level of food availability they could not maintain themselves. At 10^3 cells cm^{-3}, they ingested fewer than 5,000 cells h^{-1}. The ingestion rate increased at higher concentrations, but leveled off at about 25,000 cells h^{-1} (about 7 per second) at concentrations between 10^4 and 10^5 cells cm^{-3}. Over the same range of concentrations, the apparent filtering rate decreased from 4 cells cm^{-3} h^{-1} to less than 1 cell cm^{-3} h^{-1}. Although the beat of the filtering appendages slowed from about 6 to 3.5 movements per second, this could not have accounted for the apparent drop in filtering rate. The rate of movement of the mandibular appendages also did not vary appreciably over algal densities between 10^3 and 10^6 cells cm^{-3}. Apparently, therefore, concentrations exceeding $10^{3.5}$ cells cm^{-3} provide so much food that *Chlamydomonas* cells stream by the stuffed mouths of the *Daphnia* uneaten, and ability to process rather than to capture food limits ingestion.

Although the rate at which *Daphnia* ingested resources increased up to algal concentrations between 10^3 and 10^4 cells cm^{-3}, then leveled off, resource concentration exerted a more complex influence on population processes. As the feeding rate increased between food concentrations of 10^3 and 10^4 cells cm^{-3}, the intrinsic rate of exponential increase, r, also increased, as one would expect. But as the feeding rate leveled off above concentrations of 10^4 cells cm^{-3}, r first rose to a peak at 10^5 cells cm^{-3} and then dropped at 10^6 cells cm^{-3} (Table 20-2). Between 10^4 and 10^5 cells cm^{-3}, average life span decreased somewhat, but fecundity increased greatly, more individuals survived to breed, and reproduction started earlier in life, even though ingestion rates remained constant. At high food concentrations, the efficiency of food gathering might have increased, as suggested by the decreased beat rate of the filtering appendages.

At the highest resource concentration, brood size continued to increase, but life expectancy, especially of younger water fleas, decreased markedly, and fewer individuals reproduced; thus the number of broods per female decreased markedly (Figure 20-5). Perhaps *Chlamydomonas* produces a toxic by-product at high concentrations, or the sheer numbers of prey clog the feeding apparatus of juvenile *Daphnia*. Regardless of the cause, the *Daphnia-Chlamydomonas* example highlights the complex relationship between consumers and their resources.

TABLE 20-2 **Reproductive parameters for *Daphnia magna* cohorts fed a range of *Chlamydomonas reinhardi* concentrations at 20°C**

	FOOD CONCENTRATION (CELLS CM^{-3})			
	10^3	10^4	10^5	10^6
Percentage reproducing	50	87	97	50
Eggs per brood	2.8	2.6	15.5	21.1
Broods per female	1.7	7.5	8.2	3.4
Days between broods	5.4	3.6	3.1	3.3
Age at first brood (days)	23.4	16.9	9.8	9.1
Net reproductive rate (R_0)	2.25	16.23	99.33	34.80
Exponential rate of increase (r)	0.03	0.10	0.28	0.20

(*From Porter et al. 1983.*)

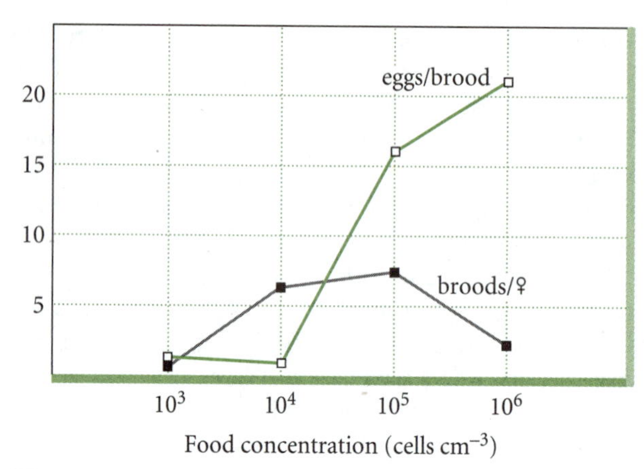

(a)

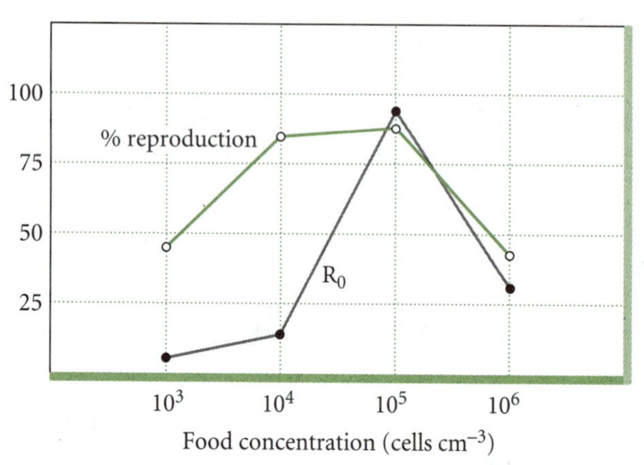

(b)

FIGURE 20-5 Graphic representation of selected data from Table 20-1 grouped according to measurement scale. (a) Increased food resulted in more eggs per brood (white squares). The number of broods per female (black squares) also increased until the highest food level, at which the number decreased because fewer females lived to that stage in the experiment. (b) The percentage of the population reproducing (white circles) showed the same pattern as broods per female because of reductions in life expectancy. The net reproductive rate (R_0; black circles) was highest at 10^5 cells cm^{-3}. (*Data from Porter et al. 1983.*)

20.4 Rates of change in resource level and consumer population growth can be dynamically coupled with logistic models.

The studies of Porter et al. show that the supply of resources may affect the number of individuals reproducing, the fecundity of individuals, and the longevity of young, and that these effects are reflected at the population level. We can integrate such experimental results into the general models of population dynamics that we described in some detail in the last part of the book.

Let us revisit the simple logistic model of population growth, $dN/dt = rN(1 - N/K)$, that we introduced in Chapter 15. As written, this equation does not say much about how resources affect the growth of a population. Indeed, the equation assumes that the environment is constant; that is, that the resources and conditions that the population experiences do not vary with respect to the size of the population. Similar assumptions hold for simple discrete time models of population growth. To be sure, the decrease in per capita population growth rate with increasing density derives from the interaction of the population with its resources, but this interaction is not made explicit in the logistic equation, nor are resources formally endowed with any particular dynamics. Once one recognizes that resources also have dynamics, it becomes necessary to formalize these dynamics and to link the resulting equations for consumers and resources by terms expressing their mutual effects. This can be done by including some biologically relevant terms in the simple logistic equation, as we show below.

Population Growth with Constant Resource Input

Let us consider the case of a renewable resource (R) whose supply is unaffected by its consumer, the situation depicted in Figure 20-2a. The rate of supply of the resource, dR/dt, reflects factors outside the consumer-resource system, and may thus be considered a constant value, k_R. Thus, we may write the resource supply dynamic as

$$\frac{dR}{dt} = k_R,$$

where R is the quantity of resource and k_R is the rate at which it is supplied.

The rate of increase of the consumer population depends both on the rate at which the resource is supplied to it and its own population density (C). We may write this algebraically as $dC/dt = f(k_R, C)$. In words, the rate of population growth of the consumer is a function of the resource supply rate, k_R (which is a constant in this example), and the consumer population size, C.

Now suppose that each individual must consume resources at some rate a just to maintain itself. (Both a and k_R are expressed in units of biomass per time, such as g/h.) A population of C individuals would then use resources at rate aC for maintenance, leaving resources supplied at rate $k_R - aC$ for reproduction. Thus, for a resource input rate of $dR/dt = k_R = 100$ g/h and a maintenance rate of $a = 0.05$ g/h, a population of $C = 1{,}000$ consumers would have 100 g/h − (0.05 g/h)1,000 = 50 g/h available for reproduction. Let us further suppose that each individual converts the resources available for reproduction into population growth (births) with an efficiency b ($0 < b < 1$). With these assumptions, we may write an equation for the population growth rate of the consumer,

$$\frac{dC}{dt} = bC(k_R - aC).$$

By factoring k_R from inside the parentheses of this equation, we obtain

$$\frac{dC}{dt} = bk_RC\left(1 - \frac{aC}{k_R}\right), \qquad (20\text{-}1)$$

which we see is in the form of the logistic equation, with $r = bk_R$ and $K = k_R/a$. According to this equation, the intrinsic growth rate of the population (r) increases as the rate of resource supply (k_R) increases (we assume that b is constant), and the carrying capacity of the environment (K) is simply the resource supply divided by the maintenance resource requirement per individual (k_R/a). Thus, we have now explicitly included consumer-resource dynamics in the logistic equation of population growth. To be sure, we have made a number of important generalizations. The resource supply rate is almost certainly not constant, and we might expect both a and b to be variable under some set of conditions.

Equation (20-1) gives us a clearer picture of the relationship between carrying capacity (the number of individuals that can be supported in an area indefinitely) and the resource level. If the individual maintenance rate, a, remains constant, an increase in the rate at which resources are supplied to the environment, k_R, means that more individuals can be supported in the environment. This is demonstrated in Figure 20-6, in which three logistic growth curves, each with a different resource supply rate and a constant maintenance rate, are shown. All three curves show the typical logistic growth pattern with the magnitude of the carrying capacity determined by the rate of resource supply.

When we discussed the logistic equation in Chapter 15, we were concerned about the conditions under which the growth rate of the population becomes negative. Using the modification of the logistic model that we have developed here, let us ask what combination of resource level, k_R, and individual maintenance rate, a, implies a negative rate of growth ($dC/dt < 0$) for the consumer. The consumer population growth rate will become negative when $bk_RC(1 - aC/k_R) < 0$, or when $1 - aC/k_R < 0$ (we obtain this by dividing both sides of the inequality by the term bk_RC). Thus, dC/dt will be negative whenever $aC/k_R > 1$ (notice that we change the direction of the inequality in order to get both sides to be positive). Any time the total amount of resource required to maintain the population (aC)

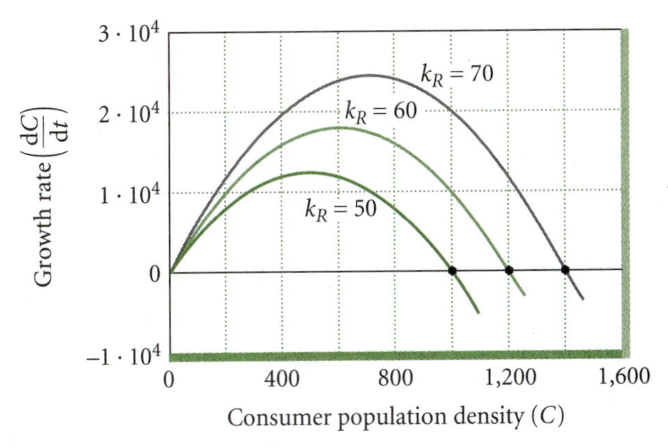

FIGURE 20-6 Logistic growth of a consumer population according to the equation $dC/dt = bk_RC(1 - aC/k_R)$, where $b = 1$ and $a = 0.05$ for three different levels of resource input rate, k_R (dark green curve, $k_R = 50$; light green curve, $k_R = 60$; gray curve, $k_R = 70$). Increasing the resource input rate elevates the curve and increases the carrying capacity (the point where the curves meet the horizontal line $dC/dt = 0$), but does not change the general logistic shape of the curve. The carrying capacity is $K = k_R/a$, which for the three values of k_R are 1,000, 1,200, and 1,400 respectively.

is greater than the resource supply rate (k_R), the growth rate will become negative. Using the current parameter values, we see that if the consumer population is at the equilibrium value of $C = 2{,}000$, then $aC/k_R = (0.05)(2{,}000)/100 = 1$. That is, the resource supply rate required for population maintenance just equals the amount supplied. If the consumer population increases to 2,500, then $aC/k_R = (0.05)(2500)/100 = 1.25$. At this consumer density, the rate of resource input required for maintenance is greater than the supply of the resource to the population, and the rate of growth will decline. You can demonstrate this in Figure 20-6 by choosing a point just to the right of the carrying capacity, K, for one of the curves and noting that the curve lies below the line $C = 0$.

Population Growth with a Density-Dependent Resource Supply Rate

Now let us expand our model by supposing that consumers gain resources in direct relation to resource abundance rather than the rate of resource supply; hence $dC/dt = f(R,C)$. That is, the growth rate is a function jointly of the amount of the resource, R, and the size of the consumer population, C. Here, we think of the resource as being a population with size R; for example, we may think of a predator (consumer population with size C) and its prey (resource population with size R). Again we shall assume that the rate of supply of the resource is independent of the consumer (although in nature, consumers deplete resources at a rate governed by both the number of consumers and the abundance of the resource—that is, for a given number of consumers, resources are consumed more rapidly when abundant than when scarce). So, $dR/dt = f(k_R, C, R)$. The following set of equations is one of many reasonable interpretations of these assumptions.

We might hypothesize that the rate of growth of the resource population is determined by the difference between the rate at which it is supplied to the environment, k_R, through reproduction or immigration and the efficiency with which it is used by the consumer, as follows:

$$\frac{dR}{dt} = k_R - gCR, \qquad (20\text{-}2)$$

where g is the efficiency of resource gathering ($0 < g < 1$).

It is worth examining the second term (gCR) of this equation in a bit more detail since such terms crop up often in population interaction models. The key feature of the term is the product of C and R, which is an algebraic way of showing how the sizes of two interacting populations relate to the level of their interaction. If either the consumer population, C, or the resource population, R, increases and the other is held constant, the product CR will increase, indicating a higher level of interaction. If there are 10 predators and 100 prey in an area, $CR = 1{,}000$. If the prey population suddenly doubles to 200, then $CR = 2{,}000$. The level of interaction between 10 predators and 200 prey is likely to be higher than that of 10 predators and 100 prey. Terms such as CR are sometimes called collision terms because they provide an index of the level of encounters, or "collisions," of individuals of interacting populations. They may also be thought of as the result of the **law of mass action** of chemistry, according to which the rate of a reaction is directly proportional to the concentration of the reacting substances. Thus, in this case, the rate of growth of the resource is proportional to the "concentration" of the consumer and the resource. As with any algebraic representation of the natural world, the CR term requires assumptions. In the case of predator and prey, we assume that there is no habitat structure and that animals move about at random, encountering one another in proportion to their respective abundance level. Such an assumption is, of course, a simplification for predator-prey systems and probably inappropriate for herbivore-plant systems, in which one of the participants, the plant, is stationary.

For the consumer, we hypothesize the following relationship:

$$\frac{dC}{dt} = bR\left(1 - \frac{aC}{R}\right). \qquad (20\text{-}3)$$

Because both C and R vary, the dynamics of the system are more complex than the model presented earlier.

Consider the features of these two equations. Equation (20-2) depicts a linear relationship between the rate of resource growth, dR/dt, and the amount of resource, R. The line has a slope $-gC$. Figure 20-7a shows how dR/dt changes for three different consumer population sizes ($C = 0$, $C = 10$, $C = 50$) when $g = 0.03$ and $k_R = 100$. When there are no consumers around ($C = 0$), the rate of change is constant, $dR/dt = k_R$, which was the situation that we discussed earlier. When $C > 0$, the term aCR acts as a discount on the input rate k_R. Increasing either the number of consumers (C) or the amount of the resource (R) will reduce the input rate. For a consumer population of size $C = 10$, dR/dt will decrease with slope $-0.03(10) = -0.3$, and $dR/dt = 0$ when $R = 333.3$. For a consumer population of size $C = 50$,

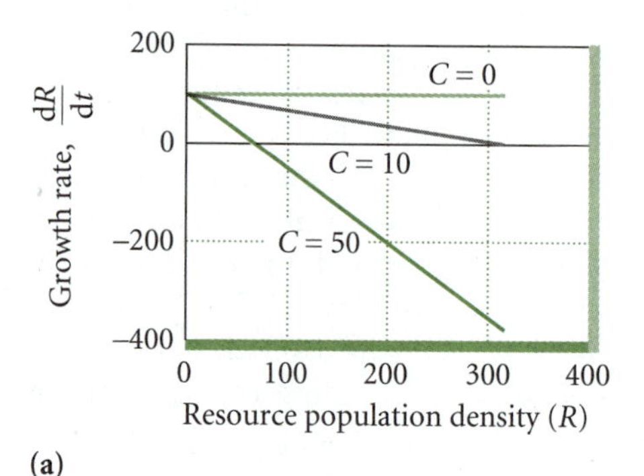

(a)

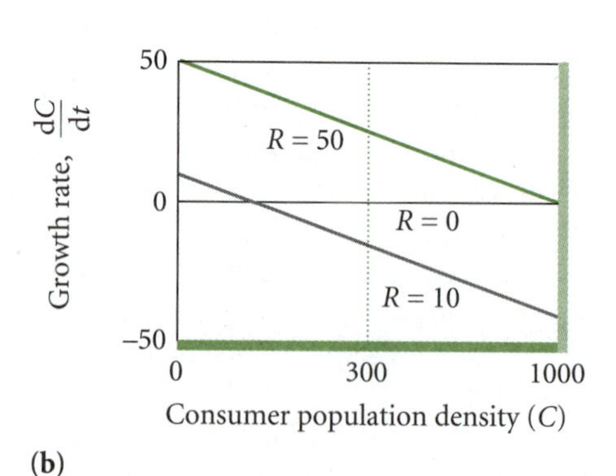

(b)

FIGURE 20-7 (a) Representations of the relationship $dR/dt = k_R - gCR$ for three different consumer population sizes (light green curve, $C = 0$; gray curve, $C = 10$; dark green curve, $C = 50$) where $k_R = 100$ and $g = 0.03$. The rate of change in the resource, dR/dt, is a straight line with slope $-gC$. Increasing C decreases the slope. Increasing R for a given C reduces the growth of the resource. (b) Representations of the relationship $dR/dt = bR(1 - aC/R)$ for three different resource levels (black curve, $R = 0$; gray curve, $R = 10$; dark green curve, $R = 50$) where $b = 1$ and $a = 0.05$. The rate of change in the consumer, dC/dt, is a straight line with slope $-ab$. Increasing C does not change the slope, but it elevates the line, thereby increasing the equilibrium value of C.

the rate of decrease of dR/dt will be $-0.03(50) = -1.5$, and the equilibrium size will be $R = 66.6$. The key point here is that, in this model, the rate of growth of the resource, dR/dt, in the presence of any number of consumers is negative, but that rate *decreases* as the level of R increases (the value -0.3 is larger than the value -1.5, and thus a change from -0.3 to -1.5 is a decrease).

At first glance, equation (20-3) looks like the familiar logistic equation. However, like equation (20-2), it is actually a straight line. This can be seen by recognizing that $dC/dt = bR(1 - aC/R)$ can be rearranged (by multiplying the term in parentheses by bR) to yield $bR - abC$, which is the equation for a straight line with slope $-ab$. The slope of equation (20-3) is constant for different levels of the resource, R. The implications of this are shown in Figure 20-7b, in which equation (20-3) is plotted for three values of R ($R = 0$, $R = 10$, $R = 50$), letting $b = 1$ and $a = 0.05$. The case $R = 0$ is trivial, since in a system of one consumer and one resource, if there is no resource, there will be no consumer. For both $R = 10$ and $R = 50$, the consumer population declines by a factor of $-ab = -0.05$. Notice that the two lines in Figure 20-7b are parallel (same slope), but the higher resource level ($R = 50$) elevates the line. So, while in both cases, the growth rate of the consumer population will decrease, when resources are high, the consumer population size at which $dC/dt = 0$ will be much higher than when resources are scarce. For $b = 1$ and $R = 10$, $dC/dt = 0 = (10)[1 - 0.05C/10]$. Solving for C gives $C = R/a = 10/0.05 = 200$. When $R = 50$, the equilibrium value is $C = 50/0.05 = 1{,}000$. The growth rate of the consumer will become negative whenever $aC/R > 1$.

To understand what these equations tell us about the consumer and resource, we must examine them near their joint equilibria, which occurs with $dR/dt = 0$ and $dC/dt = 0$. First we determine the equilibrium values for R and C by setting their respective equations equal to zero and solving for the variable. For the equation $dR/dt = 0 = k_R - gCR$, we obtain $\hat{R} = k_R/gC$. You can verify that this is the equilibrium value of R by substituting k_R/gC into equation (20-2) and solving to yield $k_R - gCk_R/gC = k_R - k_R = 0$. For the equation $dC/dt = 0 = bR(1 - aC/R)$, we solve for C to find $\hat{R} = R/a$.

The values $\hat{R}$ and $\hat{C}$ tell us only the population sizes of consumer and resource at which their respective growth rates are zero. What happens to the resource growth rate equation when the consumer is at its equilibrium value, and, conversely, what happens to the growth of the consumer when the resource is at its equilibrium value? We can determine this simply by substituting C into the equation for dR/dt and R into the equation for dC/dt. For the resource equation, the substitution yields $dR/dt = 0 = k_R - gCR = k_R - g(R/a)R$, which we simplify to $k_R - gaR^2 = 0$. Solving for R, we obtain $R = \sqrt{ak_R/g}$. Similar operations on the equation for dC/dt give $C = \sqrt{k_R/ag}$.

We emphasized at the beginning of this book that the algebraic manipulation of ecological models is a form of discovery (see Chapter 2). The work above demonstrates this. It would have been difficult to guess before going through the math that the levels of both resources and consumers varied in proportion to the square root of the rate of resource supply. Furthermore, while it makes sense that the equilibrium level of resources should decrease when resources are consumed more efficiently (that is, higher g), it may come as a surprise that more efficient consumers also assume lower equilibrium numbers. In fact, efficient consumers may eat their resources down to low levels and thereby limit their own populations.

The joint equilibrium values of $\hat{R}$ and $\hat{C}$ tell only a small part of the story of their dynamics. What happens when R and C are displaced from their joint equilibrium? Do one or both of the variables increase or decrease, and under what conditions? Do R and C return to their joint equilibrium, and if so, what path do they take? Such questions about resource-consumer systems have been dealt with in great detail with respect to the interaction between predators and their prey, which we shall consider in Chapter 23.

Before leaving this section, we wish to emphasize that the models presented here work best for those systems in which resources are individual organisms and can be accounted for by adding and subtracting discrete units. In the case of plants, resources are not individual organisms, but organic and inorganic materials obtained from the soil. Herbivores usually consume only a portion of their host plants, and thus it is often inappropriate to think of resource availability in terms of the number of plants available. Likewise, individual parasites usually do not consume an entire host, although we shall see that population models like the ones discussed above are helpful in describing the dynamics of parasites that cause disease. Let us now examine some consumer-resource models that are more appropriate for these types of systems.

20.5 The Monod equation relates population growth rate to the abundance of a single resource.

Let us take a closer look at how resources limit population growth by first examining a system with a single resource and a single consumer in detail, then expanding those ideas to include two resources. There are a number of biologically realistic properties that we wish to incorporate into our thinking. First, we wish to measure growth in the population on a per capita basis rather than for the population as a whole. That is, we want to measure individual reproductive output rather than the number of individuals added to the population as a whole. This approach is consistent with our previous use of per capita birth and death rates in our coverage of population growth and life tables (see Chapter 15). One practical reason for taking this approach is that it is often easier to make measurements on individuals than on entire populations. Second, we want the growth rate of the consumer to approach zero as the resource level approaches zero. This is saying, in effect, that the consumer does not have alternative resources on which it may rely to keep the growth rate high. This is, of course, a simplification, which we shall relax shortly. Third, we want per capita growth rate to increase as the amount of resources increases, but we do not expect that it will increase without bound, because there will be some limit to the capacity of individual consumers to use the resource. Thus, at high resource levels, we expect the per capita growth rate to slow and eventually reach some constant value beyond which increasing the level of resources will have no effect on consumer population growth. These ideas are represented graphically in Figure 20-8, which shows the relationship between consumer per capita growth, denoted $1/C\ dC/dt$, and the level of resource, R.

The French microbiologist Jacques Monod (1950) derived an equation to express the growth of a bacterial population as a function of the level of a limiting resource. His expression,

$$\frac{1dC}{Cdt} = \frac{qR}{k + R},$$

is called the **Monod equation.** The equation takes the shape shown in Figure 20-8. In this equation, q is the growth rate of the consumer population in the absence of crowding; thus it represents the maximum growth rate. The parameter q is equivalent to the intrinsic capacity of the consumer population for increase (r). The parameter k is the amount of resource at which the growth rate of the population is exactly one-half the maximum growth rate q. (Compare k in this model, which is an amount of the resource, with k_R in the previous models, which represents a rate.) For $q = 3$ and $k = 50$, the per capita consumer growth rate at a resource level of $R = 50$ is 1.5. That is, we expect 1.5 new individuals per consumer to be added to the population at those parameter values.

Some additional biological realism can be added to the Monod equation by incorporating a mortality term:

$$\frac{1dC}{Cdt} = \frac{(q - m)R}{k + R}.$$

This has the effect of lowering the overall curve (Figure 20-9). The mortality, m, is a constant that does not change with the resource level, and thus is represented as a horizontal line in Figure 20-9. The level of resources at which the mortality line and the growth curve intersect is the **critical resource level,** denoted R^*. The critical resource level is that amount of resource at which mortality balances growth. Thus, a level of $R > R^*$ indicates that there are sufficient resources for a per capita growth rate greater than 1. When $R < R^*$, the per capita growth rate is less than 1, and individuals in the population do not replace themselves. The R^* for two levels of mortality ($m = 1$ and $m = 0.5$) and a growth rate of $q = 3$ are shown in Figure 20-10. Notice that the higher the mortality rate, m, the higher the critical resource level. In the example in the figure, when the mortality rate is 1, $R^* = 50$. That is, R must be greater than 50 in order for the population to overcome the mortality effect. When $m = 0.5$, $R^* = 25$, indicating that fewer resources are required for the per capita growth rate to exceed 1.

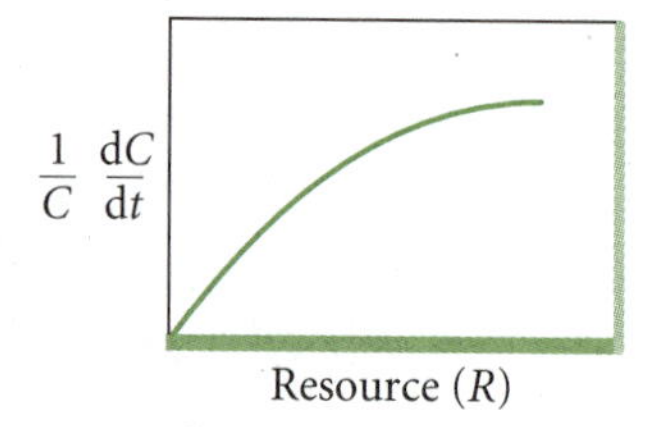

FIGURE 20-8 The per capita rate of growth of a population of consumers (C) as a function of resource availability (R). At high levels of a particular resource, R, growth rate levels off (saturation) as other resources or factors limit population growth.

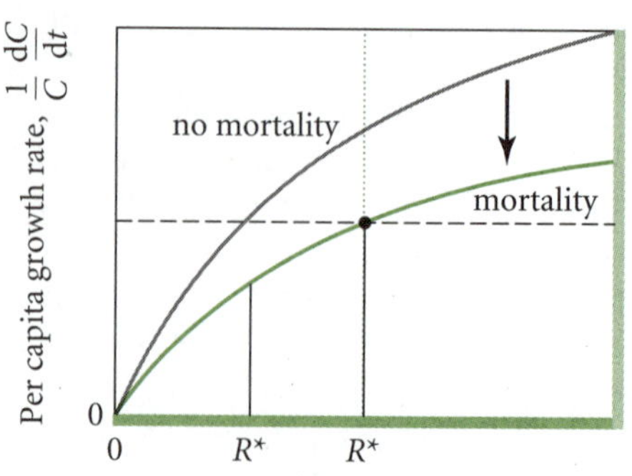

FIGURE 20-9 Graph showing how the addition of a mortality factor m decreases (arrow) the Monod equation of resource dynamics. The gray curve shows the equation without a mortality factor: $dC/Cdt = qR/(k + R)$. The green curve shows the equation with a mortality factor, m: $dC/Cdt = (q - m)R/(k + R)$. The critical resource level, R^*, is the amount of resource at which mortality balances population growth.

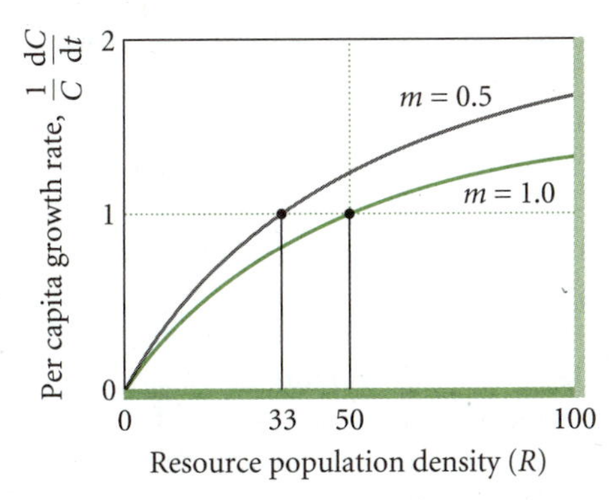

FIGURE 20-10 The effect of mortality on resource dynamics modeled with the Monod equation, $dC/Cdt = (q - m)R/(k + R)$, for $q = 3$, $k = 50$, and two levels of m (gray curve, $m = 0.5$; green curve, $m = 1.0$. The curve with the lowest value of m has a higher per capita growth rate for all values of R.

Finding R^* Experimentally

David Tilman and his colleagues (1981) determined the relationship between population growth rate and the concentration of a limiting nutrient for the diatom *Asterionella formosa* by growing it in media with varying levels of silicon (SiO_2), which diatoms require to secrete their silicate "shells." Diatom populations were established at silicate concentrations ranging from 0 to 40 micromoles (μM); population growth rates were then measured over brief periods, so that the silicate levels did not decrease substantially during the course of the experiment. Such short-term measurements provide a series of snapshots of consumer-resource dynamics: single points on the curve relating dC/Cdt to R. Because the experiments were brief, mortality was not a factor, and only the positive contribution of resources to population growth was revealed. In this particular experiment, *Asterionella* had a maximum growth rate (q) of about 0.6 d^{-1}, and the concentration of silicate at which half the maximum growth rate occurred (k) was about 9 μM (Figure 20-11). Hence one could describe the relationship between population growth and resource level (R, μM SiO_2) by the Monod equation as $dC/Cdt = 0.6R/(9 + R)$.

Tilman also established long-term cultures of *Asterionella* in a chemostat to study the course of the diatom-silicate relationship over time. A **chemostat** is a continuous-flow culture vessel to which fresh nutrient medium is added, and from which culture is removed, at a constant rate. The volume of the culture remains constant, but its contents are replaced with a time constant defined by the ratio of the flux to the volume. For example, when 100 cm^3 (0.1 liter) of culture medium is added each day to a flask containing 1 liter, the time constant of the chemostat is 0.1 d^{-1}. Growing a culture of diatoms under these conditions is equivalent to imposing a mortality rate of 0.1 d^{-1} on the population because, as one-tenth of the culture is removed each day, so are one-tenth of the diatoms growing in the culture.

In one experiment, Tilman et al. (1981) established a chemostat with a time constant of 0.11 d^{-1}. According to the relationship of diatom growth to resource level obtained in the preceding experiment ($q = 0.6$ d^{-1}, $k = 9$ μM), at a mortality rate of 0.11 d^{-1}, the Monod equation predicts that the equilibrium resource level R^* should be $0.11 \times 9/(0.6 - 0.11)$, or approximately 2 μM, which was close to the level obtained in the chemostat after the diatom population had reached an equilibrium (Figure 20-12). In general, the value of R^* depends on the time constant of the chemostat, the conditions of the experiment, and the species of diatom. For example, when Tilman et al. (1981) conducted similar chemostat experiments using *Asterionella* and a second species of diatom, *Synedra ulna*, at temperatures between 4° C and 24° C, they found that *Asterionella* could maintain positive population growth at lower resource levels than *Synedra* at temperatures below 21° C, but not at higher temperatures (Figure 20-13).

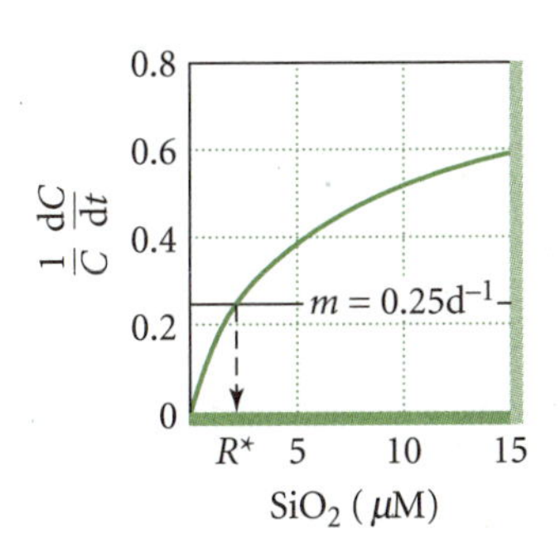

FIGURE 20-11 Relationship between per capita growth rate and concentration of silicon available to a population of the diatom *Asterionella*. (*After Tilman et al. 1981.*)

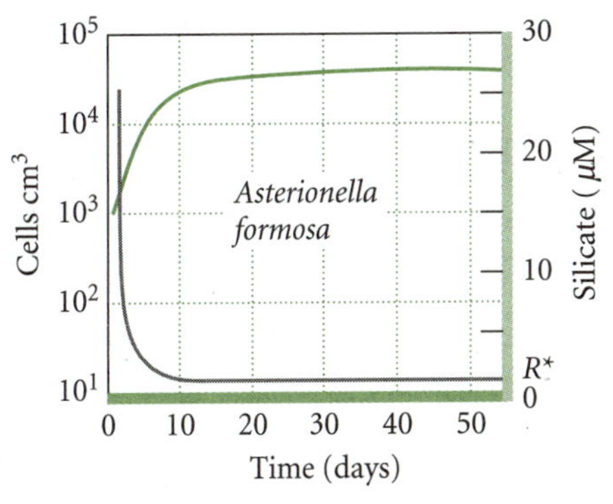

FIGURE 20-12 Growth of a diatom population (cells cm^{-3}; green line) and depletion of its silicon resource (μM; gray line) in a chemostat experiment. (*From Tilman et al. 1981.*)

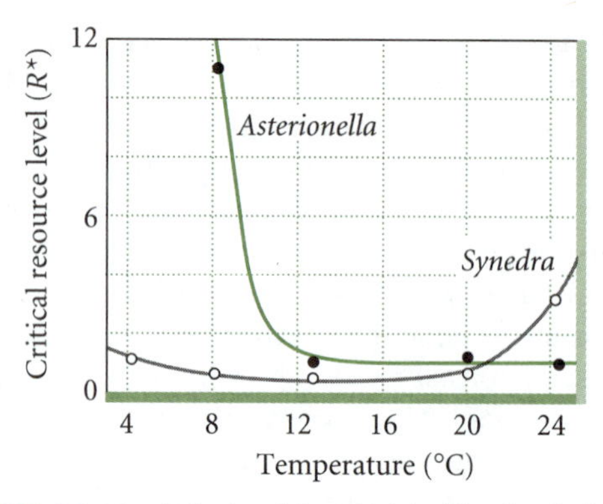

FIGURE 20-13 Relationship of critical levels of silicon for populations of the diatoms *Asterionella* and *Synedra* to temperature. *Synedra* is a superior competitor only at temperatures exceeding 20°C. (*After Tilman et al. 1981.*)

20.6 Two resources can simultaneously limit a consumer population.

Generally, organisms rely on more than a single resource. The growth of a population under a given set of conditions responds uniquely to the level of each of its resources. For example, for the diatom *Cyclotella meneghiniana* grown under silicate and phosphate limitation in a chemostat with a time constant of 0.25 d^{-1}, R^* is 0.2 μM for phosphate and 0.6 μM for silicate (Figure 20-14). According to Liebig's law of the minimum, whichever of these resources is reduced to its value of R^* first will limit the growth of the *Cyclotella* population.

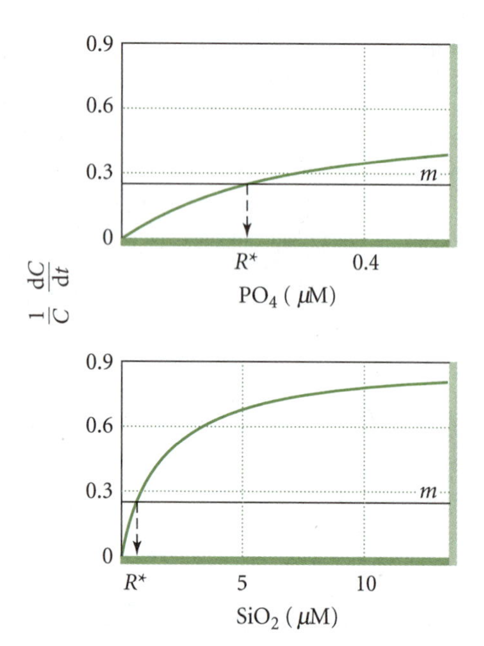

FIGURE 20-14 Experimentally determined requirements of the diatom *Cyclotella meneghiniana* for phosphorus and silicon in a chemostat with a time constant of 0.25 d^{-1}. The critical resource levels were 0.2 μM PO_4 and 0.6 μM SiO_2. (*After Tilman 1982.*)

Because silicate and phosphate resources have different dynamics, one cannot compare them on a single resource axis. Instead, one can portray the response of population growth rate to these two resources on a three-dimensional graph with the base defined as the level of silicate on one horizontal axis and the level of phosphate on the other (Figure 20-15). Accordingly, the critical resource levels for phosphate (R_1^*) and silicate (R_2^*) define a point in a two-dimensional resource space at which both of the resources are equally limiting.

Tilman (1982) recognized that the essential information in Figure 20-15 is the space defined by the two resources; that is, the "floor" of the graph (Figure 20-16). That two-dimensional region contains all possible combinations of the two resources (R_1, R_2). In the case of *Cyclotella meneghiniana,* it would represent all possible levels of silicate and phosphate, with each resource represented on the axes. If the organism needs both resources, then the growth rate will decrease if either one of the resources drops below its critical level, R^*. In Figure 20-16, the white region denotes those combinations of resource levels at which the per capita growth rate will overcome the force of mortality. The green regions include resource levels at which this is not the case. The lines delimiting these regions are called the **zero growth isoclines** because if the resource level is exactly R^*, the growth rate will be zero. A population could persist if one resource level was on the isocline so long as the other was in the white region. Of course, if both R_1^* and R_2^* lie on the isocline, the population will be at equilibrium.

Notice that a reduction of either resource below its critical level means that the population will not increase, no matter how high the level of the other resource is (Figure 20-16, point A). However, it is possible for the population to increase under a number of rather different resource conditions. For example, compare points B and C in Figure 20-16. At point B, the organism experiences very high levels of resource 2 and levels of resource 1 just above the critical level, a different resource condition than at point C, at which the relative abundances of the two are reversed. Such comparisons are problematic, of course, for resources that have very different properties, such as, for example, light level and the amount of nitrogen in the soil. Is such a case, some combinations of levels of the two resources are unrealistic. At the forest floor, for example, light levels on the order of those experienced above the canopy, though they would occur in the growth area in the simple model we have presented, would not be experienced by organisms living on the forest floor in normal circumstances.

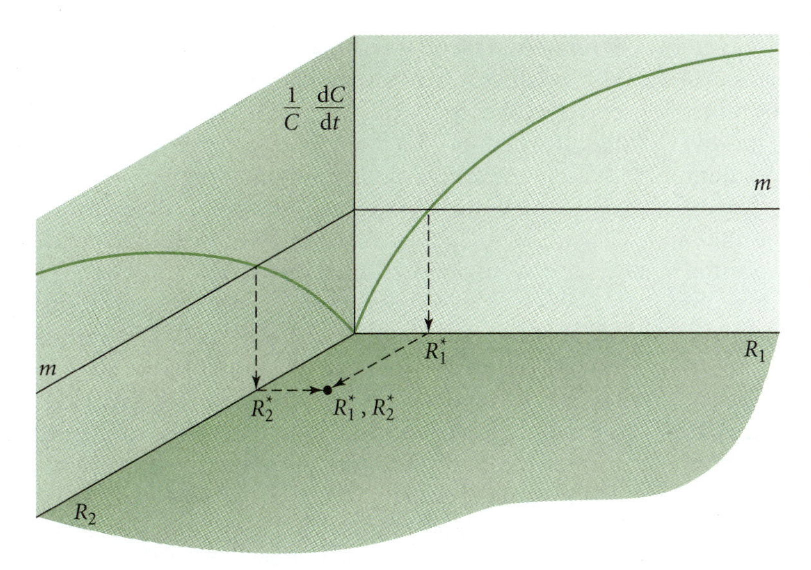

FIGURE 20-15 Determination of the joint critical level of two resources, R_1^* and R_2^*, at which both equally limit the population. For example, R_1 and R_2 might be the levels of phosphate and silicate, respectively, two important resources for diatoms. The region of the graph defined by R_1 and R_2—the "floor" of the graph—includes all possible levels of the two resources.

Figure 20-16 shows two limiting resources. We can also graphically represent a situation in which one resource is limiting and the other is not (Figure 20-17). Suppose, for example, that resource 1 in Figure 20-17 is not limiting. In that case, there will be no zero growth isocline for resource 1, since any level of that resource will do. In such a case, the growth rate will increase so long as resource 2 is above its critical level, regardless of the level of resource 1.

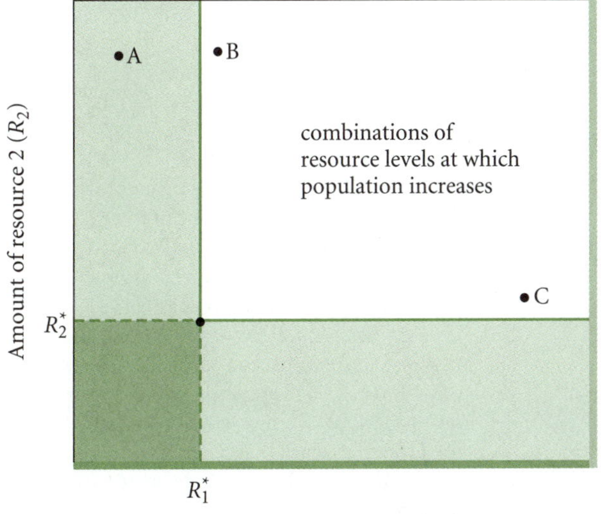

FIGURE 20-16 Two-dimensional representation of the space determined by two resources, both of which are required by the consumer. This figure represents the "floor" of Figure 20-15. The white area represents combined levels of resources 1 and 2 (including points B and C) that allow the consumer population to grow. The population will not grow at combinations of resources in the vertical light green area (e.g., point A) because there will be insufficient amounts of resource 1. The population will not grow at combinations of resources in the horizontal light green area because there will be insufficient amounts of resource 2. Insufficient amounts of both resources occur in the dark green area. Population growth is possible under a number of rather different conditions of resources, as seen by comparing resource levels at points B and C.

The simple models shown in Figures 20-16 and 20-17 assume that individuals consume each resource independently of the other. When two resources independently affect the growth of the consumer population, the consumer will reduce both the resources until one of them reaches its critical level, at which point the population will stop growing. Hence the eventual outcome of a chemostat experiment may

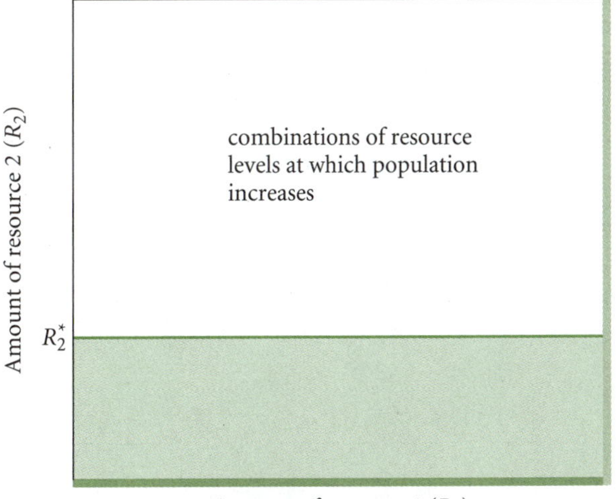

FIGURE 20-17 Two-dimensional representation of the space determined by two resources when only resource 2 is required by the consumer. At resource levels represented by the light green area, there will be insufficient amounts of resource 2, and the population will not grow. Sufficient levels of resource 2 are present in the white area. Resource 1 is sufficient at any level.

settle at any point along the lines defining the boundary of the region of consumer increase. The particular end point in any situation will depend upon the initial concentrations and relative rates of supply of the resources, but one or the other resource will limit the population in accordance with Liebig's law of the minimum. Resources that limit populations independently of one another are referred to as **essential resources.** For diatoms, silicon and phosphorus are essential resources; the consumer requires both.

In many cases, two or more resources interact to determine the growth rate of a consumer population. At least five types of resource interactions are recognized (Figure 20-18). In some cases, a reduction in the amount of one essential resource requires the organism to use more of another essential resource. Thus, for example, as R_1 approaches R_1^* from the right, the level of use of R_2 must increase. This has the effect of rounding the intersection of the isoclines (Figure 20-18b). Resources that interact in this way are called **interactive essential resources.** Such an interaction is also possible when one of the resources is essential and the other is not (see Figure 20-17). In that case, as the amount of the nonessential resource (R_1) decreases toward zero, more of the essential resource (R_2) is required, giving the R_2 isocline a curve upward (Figure 20-18c). Two resources with such characteristics are called **hemiessential resources.**

In some cases, one resource may be substituted for another. Such resources are referred to as **substitutable resources.** Substitutable resources are alternative forms of the same resource, such as the sugars glucose and fructose, which both serve as sources of carbon for bacteria. An organism may counteract a decrease in the amount of one of two substitutable resources by taking proportionally more of the other. The dynamics of this situation are depicted in Figure 20-18d. In all such cases, the level of one resource may be depressed below its R^* provided that a second substitutable resource is present.

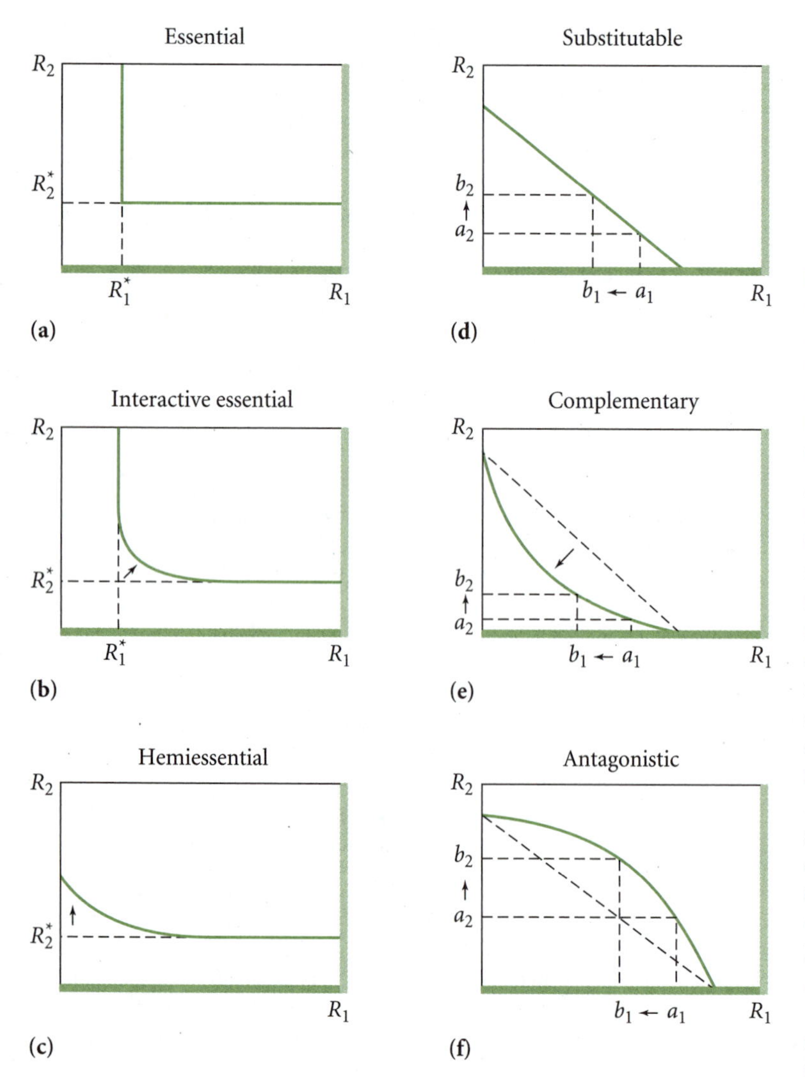

FIGURE 20-18 Graphic representation of the ways in which resources can interact to affect population growth. (a) Two independent essential resources (no interaction; same as Figure 20-16). (b) Interactive essential resources. When one of two essential resources approaches the zero growth isocline, more of the second resource is required. (c) Hemiessential resources. When the level of essential resource 1 approaches zero, additional quantities of the nonessential resource 2 are required. (d) Substitutable resources. Two essential resources may substitute for one another in a one-for-one manner. Thus, a reduction from a_1 to b_1 in the amount of resource 1 is substituted for by an equivalent increase in resource 2 (an increase from a_2 to b_2). (e) Complementary resources. A small amount of one essential resource can substitute for a relatively large amount of another essential resource. Thus, a large reduction in resource 1 (from a_1 to b_1) is complemented by a small increase in resource 2 (from a_2 to b_2). (f) Antagonistic resources. A small reduction in one essential resource (a_1 to b_1) requires the substitution of a relatively large amount of another essential resource (a_2 to b_2).

Substitutable resources replace one another one for one. When small amounts of one resource replace relatively large amounts of another resource, the resources are said to be **complementary resources.** Figure 20-18e shows a situation in which a large decrease in the level of resource 1 can be compensated for by a relatively small increase in the complementary resource 2. When substitution requires relatively large amounts of one resource to compensate for a relatively small decrease in the other, the resources are called **antagonistic resources** (Figure 20-18f).

Resource Synergism

When increases in two resources enhance the growth of a consumer population more than the sum of the two individually, the resources are said to be **synergistic** (from the Greek roots *syn,* "together," and *ergon,* "work"). Complementary resources are therefore synergistic in promoting consumer growth. Peace and Grubb (1982) provided an example of synergistic resource interaction in their study of *Impatiens parviflora,* a small herbaceous plant common in woodlands of England. Studies on other shade-tolerant forest herbs had shown that individual plants can grow in deeper shade on alkaline soils than on strongly acidic soils. For example, such plants growing at a pH of 7.5, just on the alkaline side of neutral, can tolerate light levels as low as 1–2% of full sun. When grown on soils with an acid pH of 4.5, the same plants cannot grow at light levels below 5–6%. One explanation for this difference in response is that slightly alkaline conditions solubilize mineral nutrients more readily; in acid soils, some metals become highly reactive and form insoluble complexes with phosphorus.

Peace and Grubb investigated the relationships among light, phosphorus, and nitrogen by following the growth of *Impatiens* plants under controlled laboratory conditions. They used a woodland soil of moderate fertility and near-neutral pH (6.6). They added a nitrate and phosphate fertilizer to some plants, and placed the plants under different intensities of light. In one experiment, fertilized and unfertilized plants were exposed to different levels of light from the time of seed germination until the end of the experiment at 37 days. Added light enhanced the growth of the fertilized plants more than that of the controls (Figure 20-19), demonstrating that the ability of *Impatiens* to utilize light depends on the presence of other resources. Plant growth requires both the carbon reduced by photosynthesis, as a source of energy and for structural carbohydrates, and nitrogen and phosphorus for the synthesis of proteins and amino acids.

Peace and Grubb also examined the interaction of nitrogen and phosphorus at the highest light intensities used in the previous experiment. Plants were grown in pots containing ten plants each. Plants were collected, dried, and weighed after 7 days and after 35 days. The addition of nitrogen in the absence of phosphorus had little effect on growth during the first week (Figure 20-20). Phosphorus alone enhanced growth as much as the combination of phosphorus and nitrogen. Thus, at its concentration in natural soil, nitrogen was not a limiting resource for young plants. During the subsequent month of growth, however, nitrogen and phosphorus were synergistic in promoting plant growth. The addition of either nutrient alone had no appreciable effect, whereas both together increased dry weight by about 50%.

Peace and Grubb thought that phosphorus might limit growth early on because it is relatively immobile in the soil, and plants must therefore obtain their supply from the soil immediately surrounding their roots. Being more mobile, nitrogen can enter the area of the

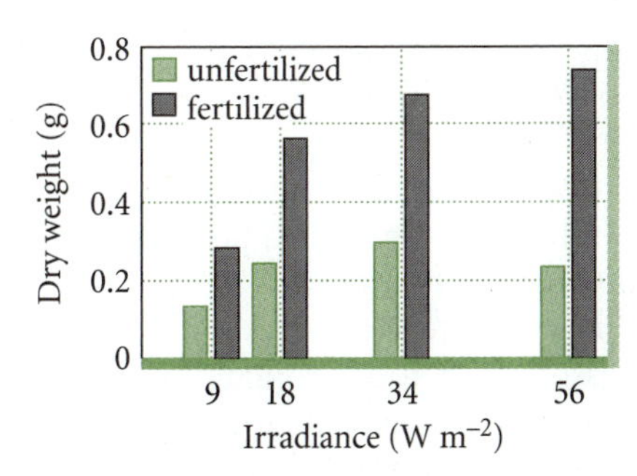

FIGURE 20-19 Joint influence of light levels and fertilizer on growth of *Impatiens.* Increasing light levels enhanced growth in fertilized plants more than in unfertilized plants. (*After Peace and Grubb 1982.*)

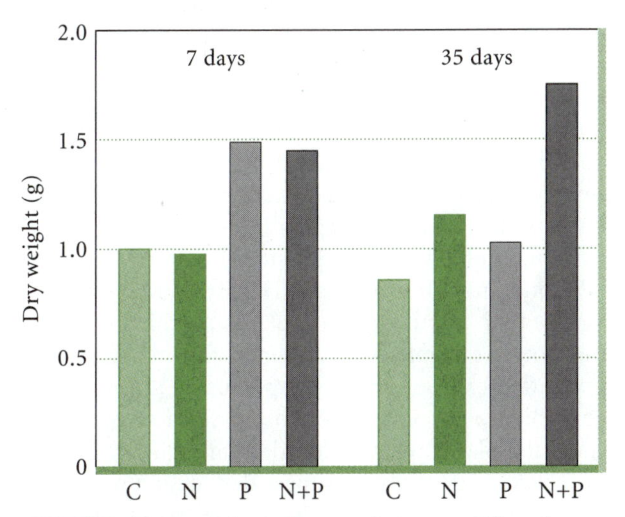

FIGURE 20-20 Joint influence of nitrogen (N) and phosphorus (P) fertilization on growth of *Impatiens.* (C, control without fertilization.) Nitrogen is not a limiting resource in young plants. Nitrogen and phosphate act synergistically to promote growth in older plants. (*After Peace and Grubb 1982.*)

roots from a greater distance. As a result, seedlings deplete local levels of phosphorus more quickly, and phosphorus thus tends to limit plant growth before nitrogen does. Eventually most of the more readily available nitrogen in the soil is taken up by the plant, and it too becomes limiting.

If this were the correct explanation, Peace and Grubb reasoned, then dense plantings would deplete soil nitrogen more quickly than plantings of one seedling per pot, and nitrogen limitation would be felt earlier in the growth period. Indeed, when *Impatiens* were planted ten to the pot instead of one, as in previous experiments, nitrogen, rather than phosphorus, proved to be the limiting resource during the first week.

20.7 Consumer-resource systems may involve more than two resources or more than one consumer.

Although one may conceive of systems dominated by a single consumer and a single resource, most species play both roles. Furthermore, species generally require more than one resource, and must share their resources with other consumers. The consequences of these interactions for population dynamics will be discussed fully in subsequent chapters, but we shall introduce them here in the form of a general class of resource-consumer interactions involving more than two species.

Consider a food chain in which a second species (an herbivore) feeds upon the first (a plant) and a third (a carnivore) feeds upon the second. In this simplest of food chains, the first species has no biological resource and the third is a consumer only. The consequences of this arrangement for the regulation of resource and consumer populations have intrigued ecologists for decades. In 1960, in a provocative and controversial paper in the *American Naturalist* titled "Community structure, population control, and competition," ecologists N. G. Hairston, F. E. Smith, and L. B. Slobodkin—then at the University of Michigan—argued that populations of plants, herbivores, and predators are regulated by different factors. They observed that cases of obvious depletion of green plants by herbivores are exceptions to the usual scenario, in which the plants remain abundant and largely intact. Moreover, cases of obvious mass destruction of plants by meteorological catastrophe are exceptional in most areas. Taken together, these two observations mean that producers are neither herbivore-limited nor catastrophe-limited, and must therefore be limited by their own exhaustion of a resource. In many areas, the limiting resource is obviously light, but in arid regions water may be the critical factor, and there are spectacular cases of limitation through the exhaustion of a critical mineral. Thus plant populations are limited by resources, but are little affected by consumers.

The authors then pointed out that there are temporary exceptions to this general lack of depletion of green plants by herbivores. Such exceptions occur when herbivores are protected either by humans or by natural events, and they indicate that herbivores are able to deplete vegetation when they become numerous enough. It therefore follows that in the usual condition populations of herbivores are not limited by their food supply.

Hairston, Smith, and Slobodkin went on to suggest that herbivore populations are limited by their consumers, the predators. Furthermore, because predators themselves have no consumers, they must be limited by their resources, the herbivores.

The three-level food chain envisioned by Hairston and his colleagues is contrasted with the simpler two-species consumer-resource system in Figure 20-21. The addition of predators reduces the herbivore population, which in turn may result in an increase in the abundance of plants. The merits of these arguments have been debated heatedly (see, for example, Ehrlich and Birch 1967, Murdoch 1966, Pimentel 1988), and will be considered in the light of theory and evidence in the following chapters. For the present, we should be aware that the behavior of complex natural systems may surprise us.

Hairston, Smith, and Slobodkin also had something to say about the interaction of species sharing the same resource (Figure 20-22). When two species share the same limiting resource, the one that can

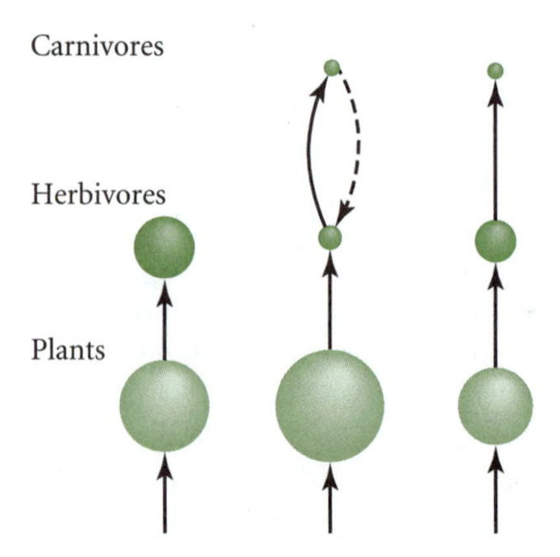

FIGURE 20-21 Food chains with and without a third (carnivore) level. In the center, carnivores depress the population of herbivores (indicated by the reduction in the size of the circle representing herbivores), allowing the plant resource populations to increase; at the right, they do not.

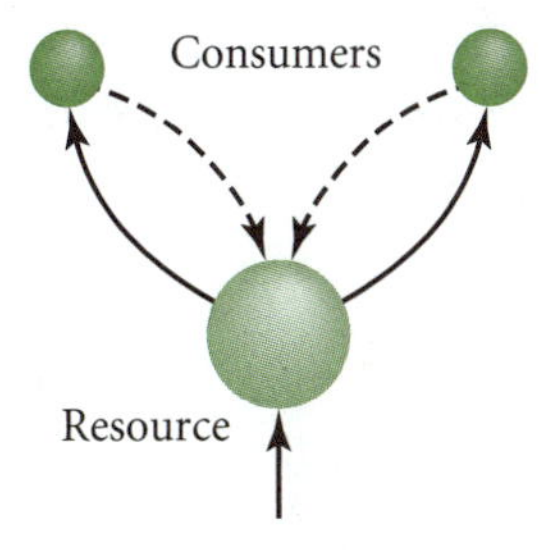

FIGURE 20-22 Food web diagram illustrating two consumers feeding upon, and depressing the level of, a single resource. Whichever consumer can continue to grow at the lower resource level will persist.

persist at a lower resource level inevitably excludes the other. When predators are superimposed on such a system, Hairston et al. suggested that because herbivore populations are predator-limited, they compete less intensely for common resources and therefore should be more likely to coexist. Here again, there are counterarguments. But the issue of the relationship between the diversity of species and consumer-resource relationships has continued to play a prominent role in ecology. As we shall see, this theme persistently recalls the question posed by G. E. Hutchinson in 1957, "Why are there so many kinds of animals?"

SUMMARY

1. Population interactions may be generally classified according to whether the participants are negatively or positively affected by the interaction. In consumer-resource interactions, such as predation, one participant benefits (consumer) and the other is harmed (resource). In competitive interactions, both participants experience negative effects. In detritivore-detritus interactions, the detritivore enjoys a positive effect, whereas there is no effect on the detritus. All participants in mutualisms experience a positive effect.

2. Population interactions most often involve the consumption of individuals of one species by individuals of another species. Such interactions affect the population dynamics of both species.

3. A resource may be defined as any factor whose increase promotes population growth and which is consumed. Thus light, food, water, mineral nutrients, and space are resources. Temperature, salinity, and other such conditions are not.

4. Some resources are nonrenewable (space); others are renewable (light, food). Three types of renewable resources are distinguished according to whether the consumer has no influence on the provisioning of the resource, a direct influence, or an indirect influence through other consumers. Of all the resources consumed, only one or a few will limit the population growth of the consumer. These are called limiting resources, and are normally those whose supply relative to demand is least. This principle is known as Liebig's law of the minimum.

5. The major types of consumers include predators, parasites, parasitoids, herbivores, and detritivores. When predation occurs, individuals of the resource population (the prey) are killed and removed from the resource population. Herbivores and parasites may consume portions of a living host, but usually do not remove the host from the resource population.

6. Experimental studies of water fleas feeding on algal cells demonstrate that consumer-resource systems are complex. Increases in algal (resource) density resulted in a longer average life span for water fleas, but a lower number of broods per female.

7. The logistic equation $dN/dt = rN(1 - N/K)$ may be modified to form a simple model of a consumer-resource system in which the resource growth rate, dR/dt, is constant and the consumer growth rate, dC/dt, is dependent on the rate of supply of the resource and the size of the consumer population. Such a model shows that an increased resource supply rate increases the carrying capacity of the consumer population.

8. The logistic equation may also be modified to model a consumer-resource system in which the rates of growth of both the consumer and resource population are related to the population densities of both. Such models show that both populations depend in a complex way on the rate of resource supply and the efficiency with which consumers use their resources.

9. The relationship between the rate of growth of a consumer population (C) and the level of its resources (R) may be described by the Monod equation, $1dC/Cdt = qR/(k + R)$, where q is the intrinsic exponential rate of increase of the consumer and k is the level of resources at which the consumer growth rate is one-half q. A constant mortality factor, m, may incorporated into the Monod equation to add biological realism. The intersection of the Monod curve with the level of mortality occurs at a resource level called the critical resource level, R^*. The critical resource level is the amount of resources at which mortality balances growth of the consumer population.

10. A graphic analysis of the use of several resources by one consumer may be used to categorize the interactions of resources in promoting consumer growth. At least five types of resource interactions are recognized: interactive essential resources, hemiessential resources, substitutable resources, complementary resources, and antagonistic resources. Laboratory and field experiments have revealed widespread synergism among resources in their effect on consumers.

11. Most species play roles as both consumer and resource. The regulation of populations depends in complex ways on the relative importance of these different roles.

EXERCISES

1. Using common names of plants, animals or microorganisms, give an example of each of the four types of species interactions described in Table 20-1 for organisms living in your area.

2. In their experiments with water fleas (*Daphnia*), Porter and coworkers estimated feeding activity by first estimating how much water each algal cell (prey) represented and then by counting radiolabled cells in the *Daphnia.* This was possible because of the filtering feeding mode of *Daphnia,* in which whole algal cells are picked out of the water. What sort of experiments and observations would you have to undertake to estimate the feeding activity of a predator such as a spider that kills prey that are often larger than itself, exudes digestive enzymes into the prey, and sucks up the half-digested prey material?

3. In a population of consumers, resources are converted to reproduction with efficiency $b = 0.7$, and each individual must consume resources at a rate of $a = 0.5$ g/h to maintain itself. Resources are supplied to the population at a constant rate of $k_R = 5$ kg/h. At what consumer population size C will its growth rate become negative under these conditions?

4. We stated that equations (20-2) and (20-3) were two of many reasonable ways to couple the rates of resource and consumer population growth in a density-dependent way. Can you think of other ways that are biologically plausible?

CHAPTER 21

Competition Theory

GUIDING QUESTIONS

- What are the different types of competition?
- How did the experimental work of Tansley, Gause, and Park contribute to the development of competition theory?
- What is the competitive exclusion principle?
- How can the logistic equation for population regulation be modified to represent interspecific competition?
- How can graphs be used to predict conditions under which competing species will coexist and under which they will not?
- Under what conditions can two species exist in competition with each other for a single limiting resource?
- How can a predator influence the outcome of competition between prey species?

Competition is any use or defense of a resource by one individual that reduces the availability of that resource to other individuals. Competition is one of the most important ways in which the activities of individuals affect the well-being of others. Competition for resources that occurs between members of the same species is called **intraspecific competition.** Intraspecific competition is very closely related to two phenomena that we discussed earlier: population regulation and evolutionary change. Within a population, competition reduces resource levels in a density-dependent manner, and thereby affects fecundity and survival. The more crowded a population, the stronger is competition between the individuals in the population. Thus, intraspecific competition underlies the regulation of population size. Evolution, as we have seen, involves changes in the representation of genotypes in a population. Such changes may arise through competition among individuals in a population that possess different forms or abilities.

When individuals of different species compete for resources, the interaction is called **interspecific competition.** Two different predatory species, for example, might compete for the same prey species. Competition between individuals of different species causes a mutually depressing effect on the populations of both. Each species contributes to the regulation of the other population as well as its own. Under conditions of intense interspecific competition, a population may be eliminated by its competitor. Because of this potential, competition is one factor that determines which species coexist within a habitat, and thus the structure of ecological communities.

The outcome of competition between two populations depends on the relative efficiencies with which individuals in the populations exploit their shared resources. When resources are scarce relative to demand for them, each act of consumption by one individual makes resources less available to others, as well as to itself. As consumption continues, resources decline to levels that no longer support the growth of the consuming population, and the population may reach a theoretical equilibrium size. When one population can continue to grow at a resource level that curtails the growth of a second population, the first will eventually replace the second. Thus, competition and its various outcomes depend on the relationships of consumers to their resources.

Several general mechanisms of competition are recognized. **Interference competition** occurs when one individual actively interferes with another individual's access to a resource. Such interference may take the form of aggressive behavior, as when two males birds compete for a good nesting territory. In Chapter 32 we shall see how interference competition affects the fitness of individuals in contests for mates.

Interference competition may be either intraspecific or interspecific. In many cases, competitive interactions are more benign. Different species may utilize the same resource, but their use of it may occur at different times of day or different seasons. Thus, while the use of the resource by the individuals of one species reduces the availability of the resource to the other, there may be no direct interaction between individuals of the species. This type of competition is called **exploitative competition.** Individuals of the same species are likely to have the same general resource use patterns, and thus exploitative competition is usually an interspecific interaction.

In developing mathematical models of population interactions, we will emphasize the interactions between pairs of populations such as two competing species or predators and prey, herbivores and food plants, and parasites and hosts. This approach is a useful way to introduce theory because it allows us to develop specific hypotheses about how the activities of one population affect another. This approach is a simplification, of course. In nature, populations certainly interact with more than one other population. Such simplifications are particularly problematic in competition theory, since in most cases species compete intraspecifically and interspecifically with any number of other species not just for one resource, but for many resources. A plant may compete with conspecifics (individuals of the same species) and heterospecifics (individuals of a different species) for water, nutrients, and sunlight. The strength of the competitive interactions within and between plant species may be different with respect to each of these resources. Thus, many species are subjected to what is referred to as **diffuse competition,** in which competitive interactions occur simultaneously across an array of resources.

Competition, like other population interactions, does not occur free of fluctuations in the conditions of the environment, or of the pressures of other biotic interactions such as predation. We shall consider how these factors affect the outcome of competitive interactions.

21.1 The emergence of competition as a central theory in ecology was slow and tentative.

Predation has obvious effects on the prey individual and the prey population. When predators are either removed from or added to a system experimentally, prey populations usually respond so dramatically as to leave little doubt about the dynamic link between the two (see Chapter 23). But competition is more subtle. When an individual of one of two competing populations eats a prey type or excludes it from the foraging area, that activity does not result directly in the death of an individual of the species with which it competes for that prey. Because competition produces less obvious effects than does predation, the development of thinking about competition and its integration into our perceptions of ecological systems have been slow and tentative.

The concept of competition has nonetheless been a part of ecology almost from its beginning. Charles Darwin, who borrowed heavily from economics for insight and analogy, fully understood its implications. Evolution, as we mentioned, is the expression of competition within populations between individuals having different genotypes (intraspecific competition). But Darwin also appreciated the implications of interspecific competition for populations. For example, in discussing the natural checks on populations in *On the Origin of Species,* he mentions "the prodigious number of plants which in our gardens can perfectly well endure our climate, but which never become naturalized, for they cannot compete with our native plants. . . ." Even so, whereas intraspecific competition was implicit in natural selection, interspecific competition left less of an impression on Darwin, preoccupied as he was with the direct effects on populations of climate, predators, and food supplies. Charles Elton's book *Animal Ecology,* the first modern, general account of ecology, published in 1927, similarly elevates these factors, especially the feeding (or trophic) relations among animals, to a preeminent position compared with competition. Elton discussed interspecific competition only in relation to ecological succession (see Chapter 28), in which the replacement of one species by a second, having similar ecological requirements, suggested the possibility of interaction between the two.

The first mathematical treatments of population processes, including competition, by A. J. Lotka, V. Volterra, and others between 1925 and 1935 created a wave of laboratory experimental work in response (Crombie 1947). Predator-prey and parasitoid-host interactions dominated early efforts, but the Russian biologist G. F. Gause (1932, 1934), who was strongly influenced by the work of Raymond Pearl and Lotka, examined competition between species in laboratory populations of yeasts and protozoans. His efforts were soon followed by work on flour beetles, fruit flies, and other organisms amenable to such experimentation. But in spite of the convincing laboratory demonstrations of interspecific competition, the process played a relatively small role in ecological thinking for decades afterward. The benchmark 1949 text, *Principles of Animal Ecology,* by W. C. Allee and his colleagues at the University of Chicago, recounted the laboratory work on competition in detail, but assigned competition a minor role in the workings of natural communities.

The role of competition in determining the population sizes of species utilizing the same resources and in governing the partitioning of resources among species was fully recognized only in the mid-1940s, when its importance was persuasively argued by the English ornithologist David Lack (1944, 1947). Lack sought to reconcile Gause's experimental result—that two similar species could not coexist in a simple laboratory environment—with the observation that natural habitats often harbor several species with similar ecological requirements. He reasoned that the process of species formation may be accompanied by sufficient divergence in ecological requirements (and hence reduced competition) between two species to permit their coexistence: "When two related bird species meet in the same region, they tend to compete, and both can persist there only if they are isolated ecologically either by habitat or food" (1947, 162). Although Elton and others were slow to accept Lack's position, by the mid-1940s the tide of opinion was turning in favor of competition. Lack's perception of competition in natural communities became firmly established (ordained, one might say) in the doctrine of modern ecology by G. E. Hutchinson's famous "concluding remarks" paper of 1957a, in which he reviewed much of the literature on the coexistence of ecologically similar species. Hutchinson concluded, at least tentatively, that "animal communities appear qualitatively to be constructed as if competition were regulating their structure. . . ." We shall see, though, that Hutchinson (1961) later concluded that competition may not be as important as other factors in some communities, an idea that has gained a strong following in recent years (Huston 1994).

21.2 The experiments of Tansley, Gause, and Park provided early experimental demonstrations of competition.

The British botanist A. G. Tansley (1917) was the first to devise an experiment to determine the existence of competition between closely related species. Tansley prefaced his report with the observation that closely related plant species occurring in the same region often grow in different habitats or on different types of soil. The observation was not new, nor was the suggestion that the ecological segregation of such species might have resulted from competition, leading to the exclusion of one species or the other from certain habitats. But no one had experimentally tested that hypothesis, or its alternative, that two species may have such different ecological requirements that each could not grow where the other flourished.

Tansley selected a pair of species of bedstraws (genus *Galium,* of the madder family, Rubiaceae), which are small, perennial, herbaceous plants. One species, *G. saxatile,* normally lives on acid, peaty soils, whereas the other, *G. sylvestre,* inhabits limestone hills and pastures. He planted the two species as seeds, both singly and together, in soils taken from areas in which each species grew. Because the seeds were planted together in a "common garden," only soil type and presence or absence of the other species were experimental treatments.

Tansley's experiments were plagued by such technical problems as poor germination and lapses in watering. His results were nonetheless quite clear. When planted singly, each of the species grew and maintained itself on both types of soil, although germination and growth were most vigorous on the soil on which the species normally grew in nature. When grown together on calcareous (limestone) soils, *G. sylvestre* plants overgrew and shaded out *G. saxatile.* The reverse occurred on the more acid, peaty soils typical of *G. saxatile* habitat. Tansley concluded that *G. saxatile* was most efficient on calcareous soils and, likewise, when grown on acid soils, *G. sylvestre* was unable to compete with *G. saxatile.*

In his brief paper reporting the results of these experiments, Tansley first put on record (1) that the presence or absence of a species could be determined by interspecific competition; (2) that the conditions of the environment affected the outcome of competition; (3) that competition might be felt very broadly (that is, from "other vegetation") throughout the community; and (4) that the present ecological segregation of species might have resulted from competition in the past. But, although Lack's and Hutchinson's concepts contain little that was not presaged by Tansley, further experimentation on competition did not come until after the publication of mathematical treatments of population interactions by Lotka, Volterra, and others.

Setting 1930, or thereabouts, as the dividing point between the earlier natural history tradition and the later population tradition in ecology, one must consider Gause's experiments (1932, 1934) on yeasts and protozoans as the first modern research on competition. Gause's protocol, essentially identical to Tansley's, was to grow two species, both singly and together, under controlled conditions. The difference in the population growth of one species in the presence and in the absence of the other was a measure of competition between them. When populations of the protozoans *Paramecium aurelia* and *P. caudatum* were established separately on the same type of nutritive medium, both populations grew rapidly to limits eventually imposed by resources. When grown together, however, only *P. aurelia* persisted (Figure 21-1). Similar experiments with fruit flies, mice, flour beetles, and annual

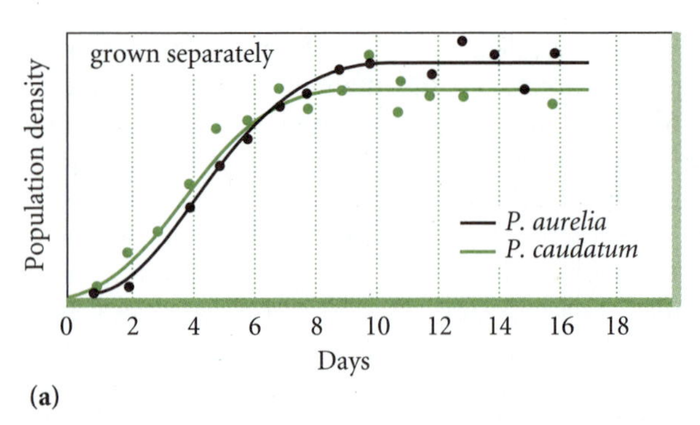

(a)

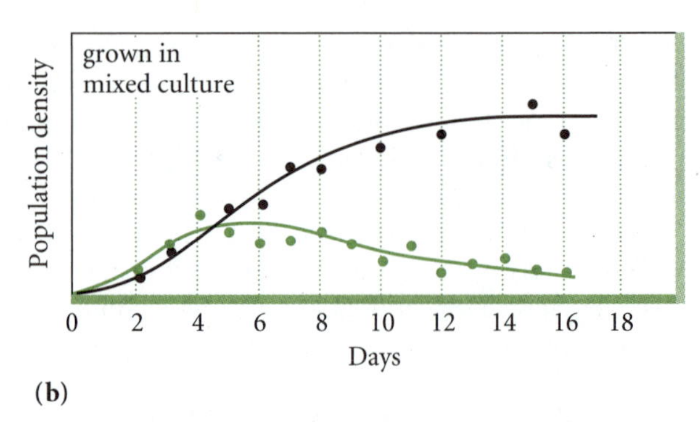

(b)

FIGURE 21-1 Increase in populations of two species of *Paramecium* when grown in separate cultures (a) and when grown together (b). Although both species thrive when grown separately, *P. caudatum* cannot survive together with *P. aurelia*. (*After Gause 1934.*)

plants have always produced the same result: one species persists and the other dies out, usually after 30 to 70 generations.

Thomas Park's work (1954, 1962) on competition between two species of flour beetles demonstrated another of Tansley's points, that the outcome of competition depends on the conditions of the environment. When Park established populations of *Tribolium castaneum* and *T. confusum* together in vials of wheat flour under cool, dry conditions, *T. castaneum* usually excluded *T. confusum*. Under warm, moist conditions, however, it was *T. confusum* that persisted (Table 21-1). The experiments showed not only that physical conditions were critical to the outcome of competition, but also that only the relative performance of the two species mattered. *T. castaneum* consistently achieved higher population densities under moist than under dry conditions when grown alone, but just as consistently was excluded under moist conditions by *T. confusum*, whose populations when grown alone were yet higher.

21.3 The competitive exclusion principle states that two species cannot coexist on a single limiting resource.

The accumulating results of laboratory experiments on competition eventually appeared so general as to warrant their elevation to the status of a principle. It has been variously named, after its principal authors, Gause's principle (Lack 1944) and the Volterra-Gause principle (Hutchinson 1957), but Garrett Hardin's (1960) **competitive exclusion principle** seems to have stuck.

Hardin traced the birth of the principle to a meeting on March 21, 1944, of the British Ecological Society, the topic of which was the ecology of closely related species. A report of the meeting published in the *Journal of Animal Ecology* (13, 176, 1944), noted that "a lively discussion . . . centered about Gause's contention (1934) that two species with similar ecology

TABLE 21-1 Competition between two species of flour beetles of the genus *Tribolium* at different temperatures and relative humidities

		Equilibrium population size*		Percentage of contests won by†	
Temperature	Humidity	*T. confusum*	*T. castaneum*	*T. confusum*	*T. castaneum*
Cool	Dry	26.0	2.6	100	0
	Moist	28.2	45.2	71	29
Moderate	Dry	29.7	18.8	87	13
	Moist	32.9	50.1	14	86
Warm	Dry	23.7	9.6	90	10
	Moist	41.2	38.3	0	100

**Number of adults, larvae, and pupae per gram of flour.*

†Based on 20–30 contests in each combination of temperature and humidity.

(Data from Park 1954, 1962.)

cannot live together in the same place. . . . Mr. Lack, Mr. Elton, and Dr. Varley supported the postulate. . . . Capt. Diver made a vigorous attack on Gause's concept, on the grounds that the mathematical and experimental approaches had been dangerously oversimplified."

The experimental results that led to the competitive exclusion principle can be summarized simply as follows: two species cannot coexist on the same limiting resource. "Limiting" is required in the definition of the principle because competition expresses itself only when consumption depresses resources and thereby limits population growth. Hardin expanded the concept, but shortened the statement: complete competitors cannot coexist.

Similar species do, however, coexist in nature. Detailed observations nearly always reveal ecological differences between such species, often based on subtle differences in habitat or diet preference. However, there are situations in which coexisting species are nearly indistinguishable in their ecological function, thereby calling into question the principle of competitive exclusion. We shall discuss this important issue below.

The principle of competitive exclusion, which states that identical species cannot exist, and the observation that coexisting species are rarely identical together prompt us to ask, how much ecological segregation is sufficient to allow coexistence? Mathematical models of competition can help us address this question by providing some understanding of the theoretical limits of the responses of competing populations. We shall develop some simple population models in the next several sections. These models, like the population regulation and consumer-resource models that we presented in earlier chapters, are grounded in the idea of population equilibrium. That is, in these models, competitive interactions among populations result in a stabilization of population growth for both populations or the elimination of one of the populations from the system. As we have emphasized before, the idea that natural forces push populations toward some fixed equilibrium point is a simplification.

21.4 Populations may be regulated by intraspecific and interspecific competition.

The above discussion of the competitive exclusion principle and our previous discussions of population growth and resource-consumer dynamics may have stimulated your thinking about the notion of population regulation. There appear to be two aspects to this important process. First, as we emphasized in our discussion of logistic population growth (see Chapter 16), populations may be regulated in a density-dependent way. The logistic population growth rate equation $dN/dt = rN(1 - N/K)$ shows how this might happen (see Figure 16-3). The closer the population size is to the carrying capacity, K, the smaller the term $(1 - N/K)$, and consequently, the smaller the right side of the equation. Thus, the rate of growth, dN/dt, depends on the density, N, of the population, assuming that r and K do not change. (We note that density dependence can also be seen in populations growing in a discrete pattern and modeled by difference equations, as we saw in Chapter 18.) If we reflect on what happens in a population when the population size, N, exceeds the carrying capacity, K, we can see how intraspecific competition comes into play in population regulation. The carrying capacity is the number of individuals in a population that the resources of the habitat can support. Thus, when $N > K$, resources are scarce relative to demand, and there will be competition for those resources. The outcome of these intraspecific competitive interactions will spell doom for those having heritable characteristics that make them less efficient at accruing resources, and will mean survival for those able to obtain enough resources to maintain their bodies and produce young. Thus, as we indicated in the introduction to this chapter, intraspecific competition and population regulation are closely related.

As the competitive exclusion principle and theories of predation, herbivory, and parasitism show, populations may also be regulated from the outside. A population may never reach a population size N that is close to its carrying capacity if it experiences strong competition with another species or if it is at risk of high levels of predation, herbivory, or parasitism. Thus, in thinking about population regulation, we must consider both the interactions of individuals within the population and the interactions of the population with other populations.

21.5 The logistic equation can be modified to incorporate interspecific competition.

Most competition theory is based upon the formulations of Lotka (1925, 1932), Volterra (1926), and Gause (1934), who used the logistic equation for population growth as their starting point. We may write the logistic equation for a single population, i, as

$$\frac{dN_i}{dt} = r_i N_i \left(1 - \frac{N_i}{K_i}\right), \tag{21-1}$$

which is the same equation that we have used before, except that all the parameters have been subscripted to show that they are specific to population i. Intraspecific competition appears as the term N_i/K_i; as N_i approaches K_i, N_i/K_i approaches 1, and the quantity $(1 - N_i/K_i)$ approaches 0. A stable equilibrium $(dN_i/dt = 0)$ is reached when $N_i = K_i$.

Now, suppose that there is a population designated j that competes with population i for resources. Population j has a density N_j and a carrying capacity K_j. The competitive pressure of population j will have a negative effect on population i and thus will lower the growth rate of population i. We may incorporate this idea into the logistic equation by subtracting a term that represents the effect of interspecific competition with species j on species i:

$$\frac{dN_i}{dt} = r_i N_i \left(1 - \frac{N_i}{K_i} - \left[\begin{matrix}\text{competitive effect} \\ \text{of } j \text{ on } i\end{matrix}\right]\right). \tag{21-2}$$

Notice that the equation now includes terms for both intraspecific competition (N_i/K_i) and interspecific competition. Both terms act as discounting terms (subtractions). We could write the equation as

$$\frac{dN_i}{dt} = r_i N_i (1 - [\text{intraspecific competition}] - [\text{competitive effect of } j \text{ on } i]).$$

In the term in parentheses on the right side of the equation, 1 is diminished by both intraspecific competition and the interspecific competitive effect of population j. It is important to appreciate that in this model, the intraspecific and interspecific effects are independent of each other. In order to show an interaction between the two effects, we would have to include a term where the product of the two was subtracted from 1.

Of course, to a theoretician, the word equations given above are not fully satisfactory. An analytical representation of the competitive effect of species j on species i must be developed. The challenge is to build an algebraic expression in terms of N_i and K_i to replace "competitive effect of j on i." That is, for simplicity, we want the equation to explain what happens to population i when it is in the presence of N_j individuals of the competing species j. One way to arrive at such a formulation is to think about what, exactly, individuals of species j do as competitors. If the two species i and j compete for food, then when an individual of species j eats, it makes some of the food unavailable for individuals of species i. If we knew how much food was removed by individuals of species j, we would have some measure of the competitive effect of the species. But we cannot assume that individuals of species i and species j consume food in exactly the same way. Thus, we do not know whether when it eats, an individual of species j takes away enough food for one individual of species i, or two, or one-half. What is needed is a conversion factor of some kind, something that will let us cast the activities of species j in *terms* of species i. Volterra provided such a factor.

Volterra incorporated interspecific competition into the logistic equation by incorporating the term $a_{ij}N_j/K_i$ into the quantity within the parentheses. Hence equation (21-2) becomes

$$\frac{dN_i}{dt} = r_i N_i \left(1 - \frac{N_i}{K_i} - \frac{a_{ij}N_j}{K_i}\right), \tag{21-3}$$

where N_j is the number of individuals of the competing species j, and a_{ij} is the **coefficient of competition;** that is, the effect of an individual of species j on the exponential growth rate of the population of species i. The multiplication of a_{ij} by N_j, the size of the competing population j, converts the population of the competing species into units of the species i on which it has an effect. Dividing the product $a_{ij}N_j$ by K_i gives the proportion of the carrying capacity of i that is taken up by equivalent units of individuals of i.

The term a_{ij} is a dimensionless constant whose value is a fraction of the effect of an individual of species i on its own population growth rate. We can appreciate the significance of a_{ij} by casting it in a simple geometric framework (Figure 21-2). Suppose that a rectangle of some area represents the carrying capacity of population i. Each individual of population i is represented by a small square. For simplicity, we shall let all the squares be the same size; that is, we assume that all members of population i are identical in every way, including the way in which they use resources. If the carrying capacity is fixed (another

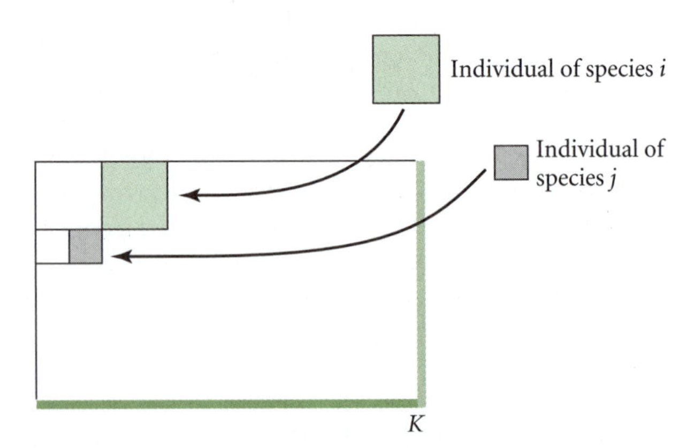

FIGURE 21-2 Schematic diagram showing the significance of competition coefficients for competition for space. Each square represents an individual of species i that inhabits the large box. The carrying capacity, K, of the box is the number of squares that will fit into the box. Individuals of a competing species, j, are represented by squares one-half the size of those representing species i. When an individual of species j enters the box, it takes up space equivalent to one-half the space taken up by an individual of species i. The space taken up by an individual of species j is the competitive effect of species j on species i, $a_{ij} = 0.5$. (*From Krebs 1994.*)

simplification), we can "fit" only a certain number, N_i, squares into the large box. The number of squares that the large box can hold is K_i, and when the box is filled up, $K_i = N_i$.

Now, suppose that individuals of the competing species *j* come in a different size, say, a square one-half the size of the square representing each individual of species *i*. Whenever an individual of species *j* is placed inside the box, it takes up the space of one-half an individual of species *i*. In this case, the coefficient of competition of species *j* with species *i*, a_{ij}, is 0.5. Conversely, one may think of the competition coefficient a_{ji} as the degree to which individuals of species *i* utilize the resources (in this case the resource is space) of individuals of species *j*. That is why $a_{ji}N_i$ is divided by K_j, the carrying capacity of species *j*. It is the degree to which individuals of species *i* usurp the resources of species *j*—expressed by the value K_j—that determines the effect of *i* on *j*'s rate of population growth.

Strictly speaking, the term N_i/K_i should have coefficient a_{ii}, but this is assumed to be 1 and left out. Although it need not be, the value of a_{ij} usually is less than 1; individuals of the same species probably compete more intensely than individuals of different species.

To summarize, we have developed an equation that shows the competitive effect of one population *j* on another population *i*. But shouldn't we also expect that population *i* will have some competitive effect on population *j*? After all, they are both competing for the same resource or set of resources. The answer to this question is, of course, yes. To show such an interaction, we need another equation:

$$\frac{dN_j}{dt} = r_j N_j \left(1 - \frac{N_j}{K_j} - \frac{a_{ji}N_i}{K_j}\right). \qquad (21\text{-}4)$$

This equation shows how intraspecific competition in population *j* and the competitive of effect of population *i* on population *j* affect the growth rate of population *j*. Thus, we see that the two species, *i* and *j*, exert effects on each other, and that the mutual relationship between them requires two equations: one for the effect of species *j* on species *i* (equation [21-3]) and a second for the effect of species *i* on species *j* (equation [21-4]).

Let us consider a numerical example in order to illustrate the mechanics of these equations. Consider the following set of parameters: $N_i = 100$, $K_i = 150$, $r_i = 0.2$, $a_{ij} = 0.5$, $N_j = 50$. The rate of growth of population *i* is $dN_i/dt = (0.2)(100)[1 - 100/150 - (0.5)(50)/150)] = 3.48$. If a_{ij} is increased from 0.5 to 0.7, corresponding to a greater competitive effect of species *j* on species *i*, and all other elements in the equation remain the same, then $dN_i/dt = 2.14$, lower than before. Whenever the sum of the intraspecific competition term and the interspecific competition term is greater than 1, the growth rate of the population will be negative. Thus, in a population near the carrying capacity, in which intraspecific competitive interactions are expected to be the strongest, the model suggests that even very low levels of interspecific competition may have a dramatic effect. For example, if we let $N_i = 145$, very close to the carrying capacity, we see that the expression inside the parentheses becomes $(1 - 0.9667 - a_{ij}N_j/K_i)$, or $(0.033 - a_{ij}N_j/K_i)$. Clearly, the term $a_{ij}N_j/K_i$ does not have to be very large to change the sign of the expression from positive to negative. Of course, very strong interspecific competition coupled with very weak intraspecific competition could also lead to negative growth. The balance between intraspecific and interspecific competition will be discussed in Chapter 24.

21.6 Equilibrium competition models reveal conditions for coexistence of two competing populations.

We may use equations (21-3) and (21-4), developed in the previous section, to explore how two competing populations can coexist. For two species to coexist, they must both reach a stable size that is greater than 0. That is, both dN_i/dt and dN_j/dt must equal 0 at some combination of positive values of N_i and N_j. Let us assume that the equations developed above for dN_i/dt and dN_j/dt represent the dynamics between two populations. What is the population size of N_i at which $dN_i/dt = 0$? To say this another way, what is the stable population size of species *i* when it is in competition with species *j* if the two species behave according to the logistic models developed above?

We can answer this question algebraically by solving equation (21-3) $[dN_i/dt = 0 = r_iN_i(1 - N_i/K_i - a_{ij}N_j/K_i)]$ for N_i, employing the following steps. First, eliminate the term r_iN_i by dividing both sides of the equation by the term to obtain $0 = 1 - N_i/K_i - a_{ij}N_j/K_i$. Rearrange the equation to obtain $1 = a_{ij}N_j/K_i + N_i/K_i$. (Notice the change of position and sign of the terms $a_{ij}N_j/K_i$ and N_i/K_i coincident with this rearrangement.) Factor $1/K_i$ from the right side of the equation and then multiply both sides by K_i to obtain $K_i = a_{ij}N_j + N_i$. Finally, isolate N_i to obtain

$$\hat{N}_i = K_i - a_{ij}N_j. \qquad (21\text{-}5)$$

Since this particular N_i represents the population size at which the population is at equilibrium, we give it the special designation $\hat{N}_i$, called "*N*-hat."

Equation (21-5) says that the equilibrium population size of species *i* is the carrying capacity of population *i* minus the number of individuals of the competing species *j* in equivalents of species *i*. That is, the effective carrying capacity of species i is reduced by the presence of the competing species j. Using the numerical parameter values from the previous section ($N_i = 100$, $K_i = 150$, $r_i = 0.2$, $a_{ij} = 0.5$, $N_j = 50$), we can calculate the equilibrium population size of species i to be $\hat{N}_i = 150 - 0.5(50) = 125$. There are 50 individuals of the competing species j present, each of which uses an amount of the contested resource equivalent to 0.5 of a single individual of species i. Thus, the competitive influence of species j in terms of species i is $0.5(50) = 25$. Even though there are 50 species j individuals, their effect on species i is equivalent to that of only 25 species i individuals.

A similar equation can be developed for species j when $dN_j/dt = 0$. That equation is

$$\hat{N}_j = K_j - a_{ji}N_i. \quad (21\text{-}6)$$

Our interest is in the conditions under which both species i and species j have populations sizes that are greater than zero. Using equations (21-5) and (21-6) for the equilibrium population sizes of each of the populations, let us explore the conditions for the coexistence of the two populations at their equilibrium sizes. Let us ask the question, what is the equilibrium population size, $\hat{N}_i$, of population i when its competitor is also at its equilibrium size, $\hat{N}_j$? We can address this question algebraically by substituting $K_j - a_{ji}N_i$ for $\hat{N}_j$ in equation (21-5) to get

$$\hat{N}_i = K_i - a_{ij}[K_j - a_{ji}N_i].$$

Our goal is to solve this equation for N_i, so we group terms having N_i in them. (Note that N_i and $\hat{N}_i$ are not different variables in this equation. $\hat{N}_i$ simply designates a specific value of the variable N.) We obtain

$$N_i - a_{ij}a_{ji}N_i = K_i - a_{ij}K_j.$$

Factoring out N_i gives $N_i(1 - a_{ij}a_{ji})$ on the left side of the equation. Dividing both sides of the equation by $(1 - a_{ij}a_{ji})$ results in an equation for the equilibrium value of population i when its competitor is also at its equilibrium size:

$$\hat{N}_i = \frac{K_i - a_{ij}K_j}{1 - a_{ij}a_{ji}}. \quad (21\text{-}7)$$

Similar reasoning leads to an equation for population j in the presence of competing population i:

$$\hat{N}_j = \frac{K_j - a_{ji}K_i}{1 - a_{ij}a_{ji}}. \quad (21\text{-}8)$$

We wish to emphasize three aspects of these equations. First, in order for both $\hat{N}_i$ and $\hat{N}_j$ to be positive, the denominators of the two equations must be positive. Thus, the relationship $a_{ij}a_{ji} < 1$ must be true. This means that the competition coefficients must be less than 1, and generally this is the case. Second, the actual values of $\hat{N}_i$ and $\hat{N}_j$ are dependent on the carrying capacities of the two populations and the competition coefficients, not on the population sizes of each. Thus, it is the relative potential sizes of the populations, rather than the actual population sizes per se, that determine whether there will be coexistence. Third, if the carrying capacity of one of the populations far exceeds that of the other, coexistence may not be possible. For example, consider the following parameter values: $K_i = 100$, $a_{ij} = 0.5$, $K_j = 500$, and $a_{ji} = 0.5$. Putting these values into equation (21-7) for $\hat{N}_i$ results in a negative numerator and a positive denominator. Indeed, the numerator of the equation for $\hat{N}_i$ will be negative when $K_i < a_{ij}K_j$ or when $K_i/a_{ij} < K_j$. The numerator of the equation for $\hat{N}_j$ will be negative when $K_j < a_{ji}K_i$ or when $K_j/a_{ji} < K_i$. Both of these conditions must be met in order for coexistence to occur.

21.7 A graphic representation illustrates the basic features of logistic competition.

We can use the algebraic expressions that we developed above to create graphs that show the dynamics of the competitive interactions between two species. Let us start by considering Volterra's equation for the growth rate of species i in the presence of N_j individuals of a competing species j (equation [21-3]): $dN_i/dt = 0 = r_iN_i(1 - N_i/K_i - a_{ij}N_j/K_i)$. The rate of growth dN_i/dt depends both on the size of the population i (N_i) and on the size of the competitor population j (N_j) (as well as r_i and K_i, which we shall consider constants for this discussion). Since there are two variables, N_i and N_j, we cannot plot the change in the growth rate in two dimensions. However, we can get an idea of the behavior of this system by examining it in two dimensions for specific values of N_j. Look first at what happens when $N_j = 0$; that is, when there are no competitors. A plot of the equation is simply the logistic curve that we presented previously (see Chapter 16) with an equilibrium at K_i (Figure 21-3, gray line). For values of $N_i < K_i$, the growth rate will be positive, and for values of $N_i > K_i$, the growth rate will be negative. The growth rate will be zero at $N_i = K_i$.

Now, what happens if there are a few competitors around? We derived equation (21-5) for just that situation. Recall that the equilibrium size of population i in the presence of N_j competitors having a competition coefficient of a_{ij} (the competitive effect of species j on species i) was shown to be $\hat{N}_i = K_i - a_{ij}N_j$. This number is just the carrying capacity discounted by the competitive effect of species j. Including this term in the equation of dN_i/dt does not change the shape of the equation. It is still a logistic curve, it just moves the equilibrium point to the left of K_i by the amount $a_{ij}N_j$. In Figure 21-3, the green line shows the new curve. Consider the following set of values for the various elements of the equation: $r_i = 2.0$, $K_i = 150$, $a_{ij} = 0.2$. If there are $N_j = 100$ individuals of the competing species, the new equilibrium for species i changes from $K_i = 150$ to $\hat{N}_i = 150 - (0.2)(100) = 130$. A similar graph can be constructed for the growth rate of the population of species j, dN_j/dt, when in the presence of the competing species i. Note that the equation for the growth rate of population i has no meaning when $N_i = 0$.

When examining the behavior of algebraic expressions, it is often easier if curved lines can be made straight. Such an adjustment can be made to the logistic equation above by considering the per capita growth rate, $1/N_i dN_i/dt$, instead of the growth rate dN_i/dt for the entire population. When this adjustment is applied to the growth equation of species i, the curves in Figure 21-3 become the straight lines shown in Figure 21-4. Now, the vertical axis shows the growth rate per individual in the population. The horizontal axis remains unchanged, still showing the population size of species i. Suppose there is a single individual of species i, and the elements of the equation have the same values listed in the paragraph above. When there are $N_j = 0$ competitors, the per capita growth rate of species i is $1/N_i dN_i/dt = (2.0)(1)[1 - 1/150 - 0] = 1.98$. When there are $N_j = 100$ competitors, the per capita growth rate is $1/N_i dN_i/dt = (2.0)(1)[1 - 1/150 - (0.2)(100/150)] = 1.72$. The presence of competitors reduces the individual growth rate. Putting the equations on a per capita basis does not change the dynamics on the horizontal axis, where $N_i = K_i - a_{ij}N_j$ gives the carrying capacity in the presence of a competitor. A similar linear relationship can be constructed for species j in the presence of competitor i.

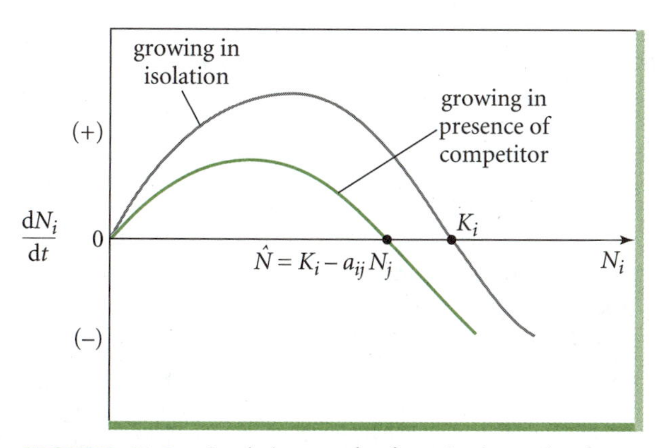

FIGURE 21-3 Logistic growth of species i growing in isolation (gray line) and in the presence of a competing species j (green line). When there is no competitor, an equilibrium is reached at the carrying capacity K_i. Competition with species j reduces the equilibrium point by $a_{ij}N_j$.

For the equation of the growth rate of population i in the presence of species j, there will be many points (N_i, N_j) for which the growth rate will be positive. We may expect that species i will survive in the presence of some level of competitive pressure from species j, but not at some other, higher, level. We can use graphs to describe the general conditions under which species i will grow in the presence of species j, and likewise, the conditions under which species j will grow in the presence of competitor i.

Consider a three-dimensional space defined as shown in Figure 21-4a. The population sizes N_i and N_j define the two-dimensional "floor" of the graph, and the per capita growth rate of population i is shown as the vertical axis. Values of $1/N_i dN_i/dt$ lie on a plane describing a single value of $1/N_i dN_i/dt$ for every combination of values N_i and N_j. In the absence of species j ($N_j = 0$), the population of species i would, according to logistic growth, increase along the N_i axis to K_i (see Figure 21-3). When the population of species i is very small (N_i close to 0, hence little intraspecific competition), $1/N_i dN_i/dt$ is positive so long as the population of the competitor j is less than K_i/a_{ij}. This intercept represents the level of population j at which its use of the resources of population i equals that of K_i individuals.

To comprehend how two species coexist, we must understand the conditions under which the populations either increase or decrease. As you can see in Figure 21-4a, the per capita growth rate for population i, $1/N_i dN_i/dt$, is positive for certain combinations of N_i and N_j (e.g., point A) and negative for others (e.g., point B). The dividing line between the two regions, which is called the equilibrium isocline ($dN_i/dt = 0$), is the intersection of two planes along the line defined by the relationship $N_i = K_i - a_{ij}N_j$, which we discussed above (equation [21-5]).

When the per capita growth rate is positive, the population will be growing, which is represented on the graph as a movement on the N_i axis away from the origin. We need only know the sign of the growth rate to determine the population trend: when the growth rate is positive, the population tends to increase (move away from the origin on the N_i axis); when the growth rate is negative, the population tends to decrease (move toward the origin on the N_i axis). Thus, we may examine the equilibrium isocline for N_i by simply looking at the "floor" of Figure 21-4a, which is a two-dimensional graph, the axes of which are the population sizes of species i and j (Figure 21-4b). Within the region that includes the origin of the graph (point A), population i increases; outside that region it decreases (point B), as indicated by the

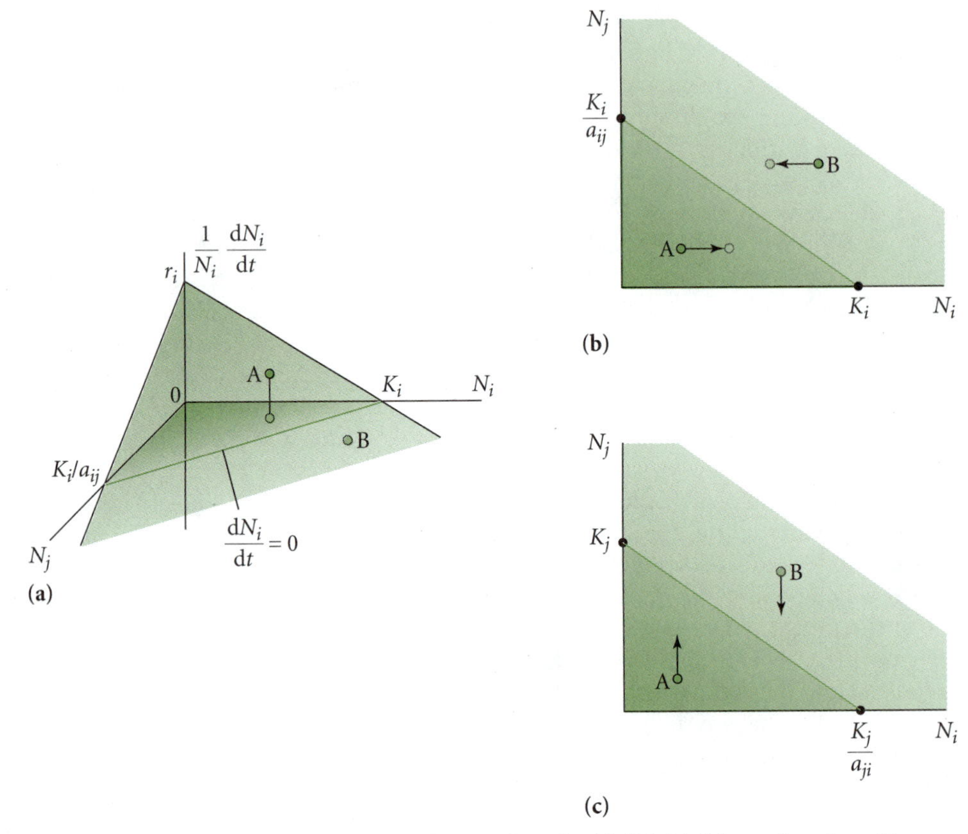

FIGURE 21-4 (a) The per capita growth rate of species i ($1/N_i dN_i/dt$) as a function of its population, N_i, and the size of a competing population (N_j). The shaded area is the plane containing all possible combinations of the population and its competitor (N_i, N_j). The line where the shaded area intersects the plane is called the equilibrium isocline, which contains all combinations (N_i, N_j) for which the growth rate is zero. Points on the plane above the isocline (e.g., point A) represent positive growth, and points below the isocline (e.g., point B) represent negative growth. (b) A simplification of the graph in (a), showing only the plane defined by the $N_i - N_j$ axes [the "floor" of graph (a)]. This view is achieved by rotating the N_j axis up into the vertical position while leaving the N_i axis in the horizontal position (the growth rate axis projects directly toward the reader). Point A is in the positive growth region of the plane, meaning that the population tends to increase, as indicated by the rightward-directed arrow. Point B is in the negative growth region of the plane, meaning that the population tends to decrease, as indicated by the leftward-directed arrow. Note that the graph in (b) shows only the growth of species i in the presence of the competitor j. (c) A separate graph is required to show the growth of species j, which is indicated along the vertical axis, in the presence of species i. Point A is in the region of positive growth for species j, as indicated by the upward-directed arrow. Point B is in the region of negative growth.

arrows. Species j has an analogous equilibrium isocline, shown in Figure 21-4c.

The behavior of species i and j together depends on the relative positions of their equilibrium isoclines. When the isocline of one species lies outside that of the second along its entire length, the first is the superior competitor, and it eliminates the second. In Figure 21-5, the isocline of species i lies outside that of species j. When both populations are small (point A), both increase, and the trajectory of the populations jointly moves upward and to the right. When both populations are large (point B), both decrease. But within the region between the isoclines of the two species, species j decreases (it is outside its equilibrium isocline) and species i increases (it is within its isocline). As a result, the joint trajectory of the populations moves downward and to the right, eventually reaching K_i on the N_i axis ($N_i = K_i$, $N_j = 0$). Thus species i eliminates species j from the system. A simulation of the time course of such an interaction is shown in Figure 21-6, which closely resembles the outcome of Gause's experiment on *Paramecium* (see Figure 21-1).

This graphic analysis of competition clearly indicates that coexistence can be achieved only when the equilibrium isoclines of species i and j cross (Figure

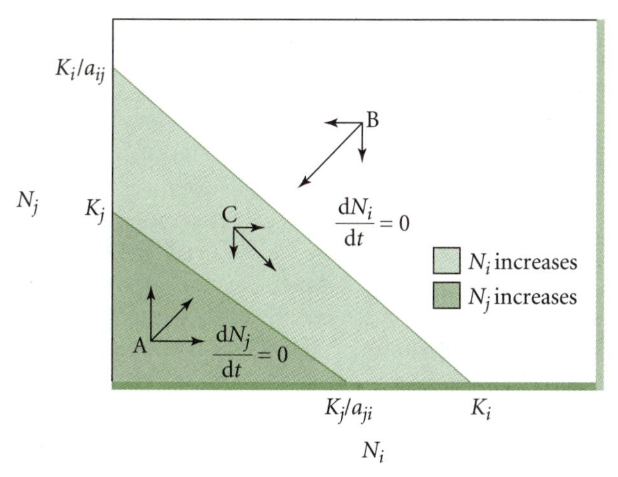

FIGURE 21-5 Graphic representation of the equilibrium conditions for two species, of which species i is the better competitor. Areas in which the populations can increase are indicated by light green for species i and dark green for species j. Both species increase in region A, both decrease in region B, and i increases and j decreases in region C.

21-7a). At the point of their intersection, corresponding to $\hat{N}_i$ and $\hat{N}_j$, the growth rates of both populations are 0. The joint population trajectories in each of the four sections of the graph defined by the isoclines in Figure 21-4 show that the joint equilibrium point is stable. Accordingly, the conditions for coexistence are the inequalities $K_i < K_j/a_{ji}$ and $K_j < K_i/a_{ij}$. By rearranging these inequalities, we obtain $a_{ij} < K_i/K_j$ and $a_{ji} < K_j/K_i$, as we have seen before. Biologically, these inequalities correspond to the situation in which each species limits itself (K_i) more than it limits the other (K_j/a_{ji}). For coexistence, intraspecific competition must exceed interspecific competition. Notice also that r_i and r_j are immaterial to the outcome of competition; they affect only its time course.

What happens when both populations are more strongly limited by interspecific competition than by intraspecific competition? This situation, shown in Figure 21-7b, has an unstable equilibrium point where the isoclines cross, away from which the joint population trajectories move. One species or the other prevails, depending on which initially holds the numerical advantage. This situation requires the unlikely circumstance of each species being superior to the other in a predominantly intraspecific environment (that is, when abundant) and inferior in a predominantly interspecific environment (that is, when rare). Conceivably, if each species inhibited the population growth of the other by the release of species-specific toxic chemicals, whichever species occurred initially in greater numbers would probably have a competitive edge. But such conditions seem unlikely in natural systems, and ecologists have paid little attention to this unstable equilibrium.

21.8 We can estimate competition coefficients from the results of competition experiments.

In his laboratory studies of competition between pests of stored grain products, A. C. Crombie (1945) estimated values of a_{ij} and a_{ji} from changes in population size when species i and j were grown together. According to the logistic interspecific competition equation (21-3), $dN_i/dt = r_iN_i(1 - N_i/K_i - a_{ij}N_j/K_i)$, the term a_{ij} may be estimated when r_i and K_i are known from the rate of change in population i. Solving the equation above for a_{ij} gives

$$a_{ij} = \frac{K_i - N_i}{N_j} - \frac{dN_i}{dt}\left(\frac{K_i}{r_iN_iN_j}\right).$$

This equation shows that the experimental challenge of estimating the competition coefficient lies in accurate measurements of the populations sizes and carrying capacities of competing species and in estimating population growth rates.

Crombie obtained values of r_i and K_i from populations of species i grown without species j. For competition between the beetle *Rhizopertha dominica* (which he called species R) and the moth *Sitotraga cerealella*

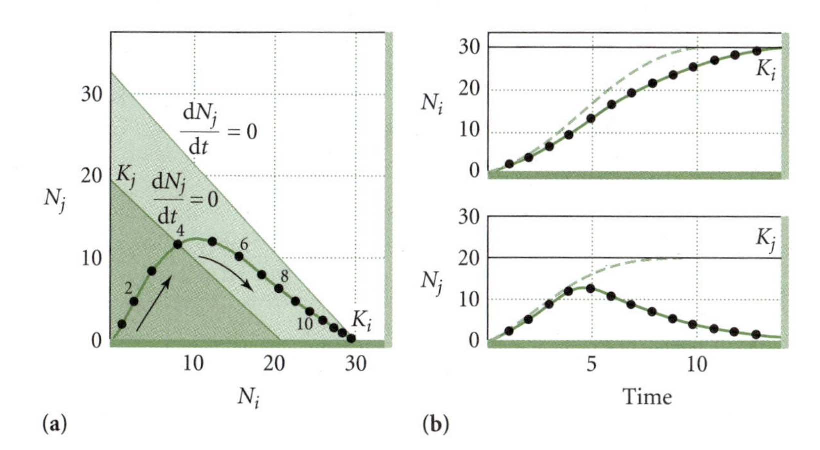

FIGURE 21-6 The course of competition between two populations portrayed (a) on a competition graph and (b) as changes in population size over time. The time intervals are indicated on the competition graph by numerals next to the sample points. Dashed lines indicate population increases when the species are grown separately.

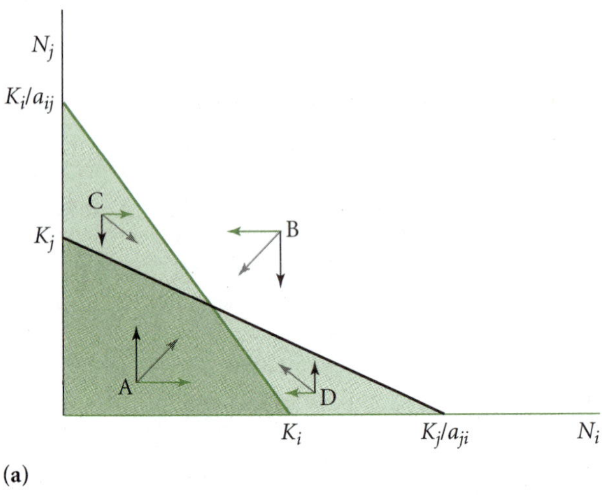

(a)

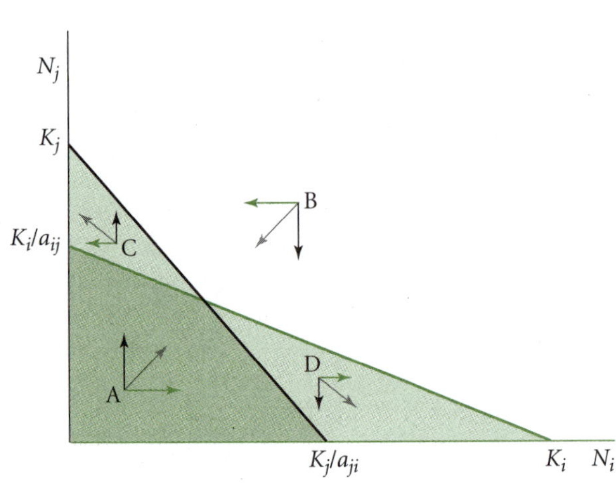

(b)

FIGURE 21-7 (a) Conditions for the stable coexistence of two competing species. The graph can be obtained from Figure 21-5 by reducing the competition coefficient of species i (the better competitor) against species j. When a_{ij} decreases, the intercept K_j/a_{ij} increases. The green and black vectors in each of the four regions created by the crossed isoclines represent the direction of population growth in that region (green corresponds to the growth of species i and black corresponds to the growth of species j). Resultant vectors are given in gray. It is evident that with this configuration of isoclines, the competing populations will tend toward a stable equilibrium, denoted by the point where the lines cross. (b) Outcome of competition between two species that are both more strongly limited by interspecific competition than by intraspecific competition (a_{ij} and a_{ji} are both large). The populations tend to diverge from the equilibrium point.

(species S), Crombie estimated $r_R = 0.05$ and $r_S = 0.10\ d^{-1}$, $K_R = 338$ and $K_S = 200$, and $a_{RS} = 1.0$ and $a_{SR} = 1.3$. Because the product $a_{RS}a_{SR}$ exceeded 1, we would not have expected the species to coexist. Furthermore, because K_R exceeded K_S/a_{SR}, *Rhizopertha* should have outcompeted *Sitotraga,* as indeed it did (Figure 21-8a). *Rhizopertha* and *Sitotraga* resemble each other ecologically in that the larvae of both species burrow into grains of wheat and feed upon the germ (Figure 21-8b). Because the larvae are aggressive toward one another, we should not be surprised that their interspecific competition coefficients are high.

Crombie also investigated competition between *Rhizopertha* and a second species of beetle, *Oryzaephilus surinamensis.* Unlike *Sitotraga,* however, the larvae of *Oryzaephilus* feed on the outside of the wheat grain. Accordingly, competition coefficients between *Rhizopertha* and *Oryzaephilus* are low (about 0.2 and 0.1), and the

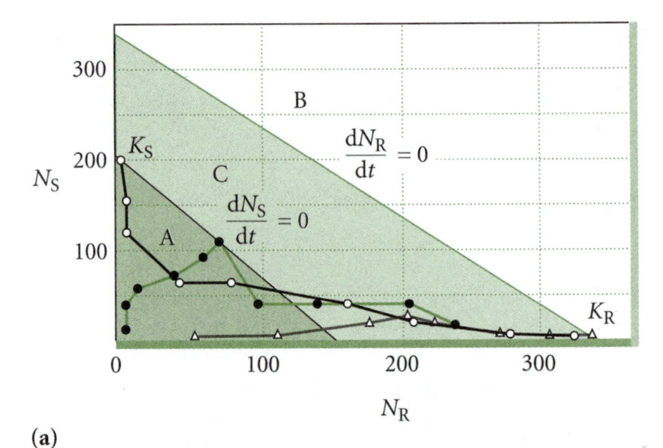

(a)

(b)

FIGURE 21-8 Population trajectories of competing grain pests, the beetle *Rhizopertha dominica* (N_R) and the moth *Sitotraga cerealella* (N_S) in renewed wheat cultures with different initial densities of the two species. The isocline for *Rhizopertha* is everywhere above the line for *Sitotraga,* suggesting that *Rhizopertha* will outcompete *Sitotraga.* The three curves indicate the population trajectories of the competing species at three different initial densities. Each point on a curve is a combination of species densities (N_i, N_j). The three regions A, B, and C correspond to the same regions in Figure 21-5. In region A both species persist. When the population *Rhizopertha* grows into region C, the population of *Sitotraga* declines, and is eventually excluded. (b) This photograph shows a similar competitive situation among rice weevils infesting grains of rice. (*a, from Crombie 1945; b, courtesy of U.S. Dept. Agriculture.*)

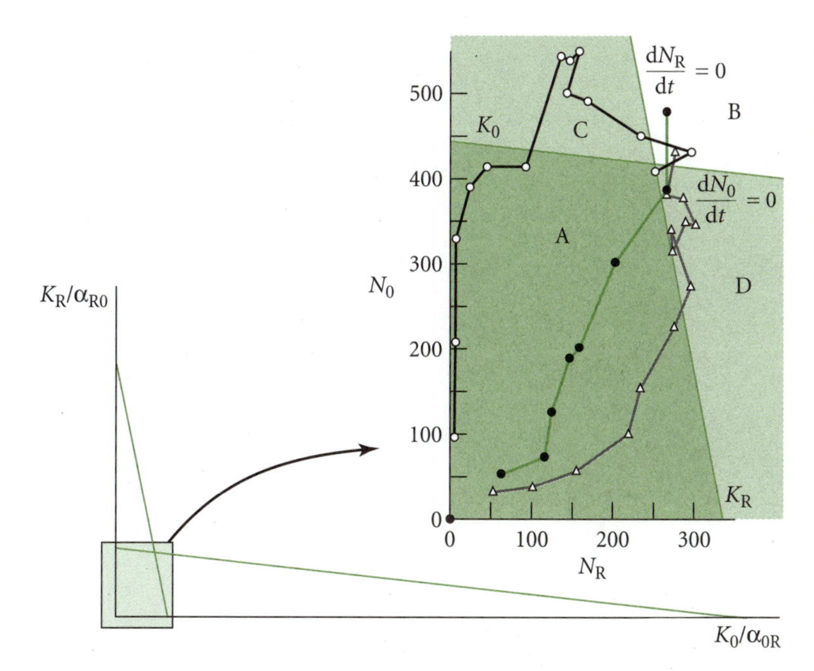

FIGURE 21-9 Population trajectories of competing beetles *Rhizopertha* (N_R) and *Oryzaephilus* (N_O) grown in renewed wheat cultures at three different initial densities of the two species (see Figure 21-8). Regions A, B, C, and D correspond to the same regions shown in Figure 21-7a. The populations tend toward the equilibrium point where the isoclines cross. (*From Crombie 1945.*)

two species coexist with little mutual inhibition (Figure 21-9).

The experimental estimation of competition coefficients has pointed to some of the oversimplifications that worried Capt. Diver. For example, when Francisco Ayala (1970) grew two species of fruit flies (*Drosophila*) together, his estimates of competition coefficients seemed to conflict with Volterra's simple model. *D. pseudoobscura,* from the western United States, and *D. serrata,* from New Guinea, coexist indefinitely in laboratory cultures at 23°C. At higher temperatures, *D. serrata* replaces *D. pseudoobscura;* at lower temperatures, *D. pseudoobscura* persists at the expense of *D. serrata.* The sizes of populations of each species maintained separately and together are summarized in Table 21-2. The competition coefficients for these interactions were estimated to be $a_{ps} = 1.49$, and similarly $a_{sp} = 3.86$. The product $a_{ps}a_{sp} = 5.75$ violates the general condition for coexistence, yet neither species outcompetes the other!

Ayala's results do not refute the general principle of competition theory developed here, but they do suggest that simple linear models, such as the logistic equation, may not adequately describe population dynamics in some situations (Law and Watkinson 1987). Ayala's result can be reconciled with the conditions for coexistence ($K_i < K_j/a_{ji}$) by bending the equilibrium isoclines into the curves shown in Figure 21-10

TABLE 21-2 **Equilibrium population sizes in competition experiments at which *Drosophila pseudoobscura* coexists with *D. serrata***

	Mathematical expression according to logistic equation	Adult population size
Species raised separately		
D. pseudoobscura	$N_p = K_p$	664
D. serrata	$N_s = K_s$	1,251
Species raised together		
D. pseudoobscura	$\hat{N}_p = K_p - a_{ps}\hat{N}_s$	252
D. serrata	$\hat{N}_s = K_s - a_{sp}\hat{N}_p$	278
		Total = 530

(*Data from Ayala 1970.*)

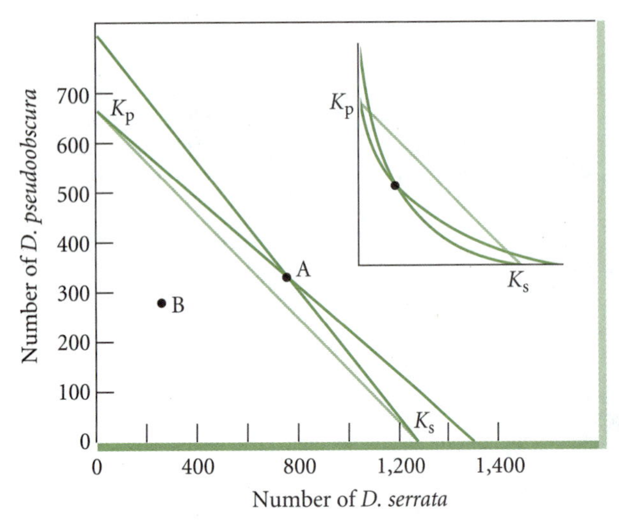

FIGURE 21-10 A comparison between the prediction of the logistic competition model for coexisting species and the outcome of an experiment with coexisting species of *Drosophila*. When coexistence occurs, the model predicts that the equilibrium population (point A) must lie outside a line connecting the two carrying capacities (K_s and K_p). Ayala's result is indicted by point B. Modification of the graphic model to accommodate point B is shown in the small graph.

(Gilpin and Justice 1972, 1973). Such bending could result when each species competes intensely with the other when rare, but less so when each is close to its own carrying capacity. Imagine a situation in which there are two resources, A and B. Each species feeds on both, but species i uses resource A much more efficiently than does species j, and specializes on it when rare; the reverse applies to species j and resource B. When i is rare and j is abundant, most of the competition between them results from their utilization of resource A, and i is the better competitor. When j is rare and i is abundant, competition centers upon resource B, and j is the better competitor (Ayala 1970, 1971; Gilpin and Ayala 1973).

21.9 Two-species competition may be represented on graphs relating population change to resource availability.

Logistic competition theory is based on the dynamics of the consumer populations involved. It does not explicitly consider changes in resources utilized by the competitors. We saw in Chapter 20 how essential resources affect population size. Two species of consumers competing for two essential resources are portrayed in Figure 21-11. Here, species i can increase at lower levels of both resources than species j. That is,

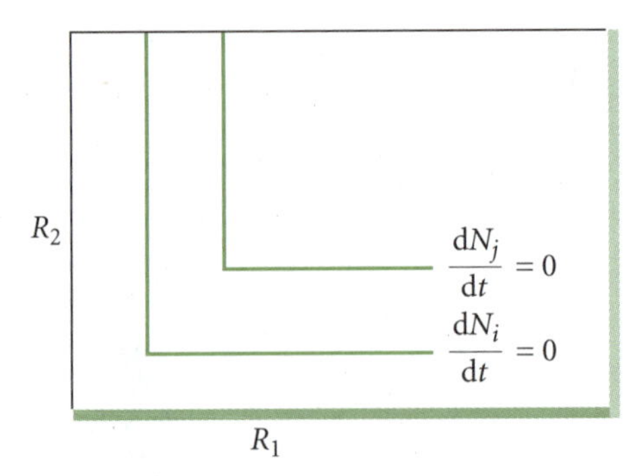

FIGURE 21-11 Portrayal of competition between two species in which one (i) can maintain population growth at lower levels of both limiting resources than can the second (j). In this case, species i wins the competition, and j is excluded from the system. (*From Tilman 1982.*)

species i requires less of both resources. In this situation, i outcompetes j.

Coexistence may occur when the equilibrium isoclines, called the **zero net growth isoclines,** or ZNGI for short, cross, as in Figure 21-12 (Tilman 1982). But two additional factors influence the nature of the joint equilibrium. The first is the rates of consumption of each of the resources by each of the species. These are indicated by the vectors C_i and C_j projecting down and to the left through the joint equilibrium. The second is the position of the supply point of the resource, which is the point in the graph representing the levels of each of the essential resources in the absence of the consumers.

Two conditions are required for stable coexistence. The first is that each species must consume relatively more of the resource that limits its growth at equilibrium. In Figure 21-12, species i is limited by resource 2 at the joint equilibrium point; because C_i indicates that species i uses resource 2 more rapidly than

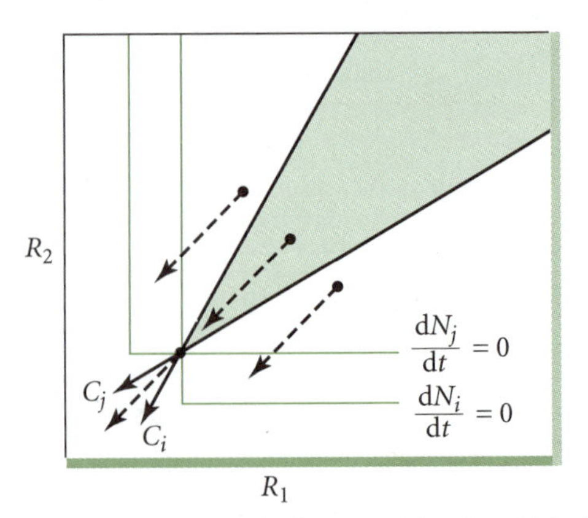

FIGURE 21-12 Portrayal of competition in which the zero net growth isoclines cross at a stable equilibrium point, shown with a dot. The point is stable because each species consumes relatively more of the resource that limits its growth at the equilibrium. (*From Tilman 1982.*)

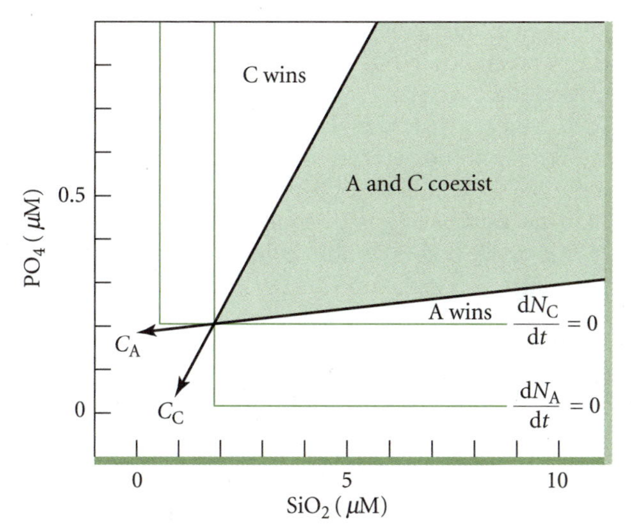

FIGURE 21-13 Observed zero net growth isoclines and consumption vectors for the diatoms *Asterionella formosa* and *Cyclotella meneghiniana* predict stable coexistence between the two species. (*After Tilman 1982.*)

resource 1 at this point, this condition is met. The second is that the resource supply point must be located in the region between the consumption vectors (shaded in Figure 21-12). Outside of this region, one species has an overwhelming advantage over the other because its limiting resource is supplied at a much greater rate than the limiting resource of the other species.

Tilman tested his model by growing two species of diatoms, *Asterionella formosa* and *Cyclotella meneghiniana,* in chemostats under controlled rates of nutrient supply. *Asterionella* requires relatively higher levels of silicon and *Cyclotella* requires relatively higher levels of phosphorus, as shown by the position of the ZNGI for each species (Figure 21-13). These facts, combined with the consumption vectors, suggest that the two species should coexist when the ratio of silicon to phosphorus is between 5.6 and 9.7, a prediction largely confirmed by the chemostat experiments.

21.10 Disturbance can influence the outcome of competition between species.

Both Tilman's model and the logistic portrayals of competition lead to a similar conclusion: coexistence is made possible by ecological segregation. For two species to persist together, each must excel at utilizing some resource that limits the other. When two species are forced to compete for the same resource in laboratory cultures, one inevitably excludes the other. However, when species are limited by different resources and cultured under heterogeneous conditions, or under ratios of resources that are limiting to both, they can co-occur without either one's gaining a competitive advantage.

The logistic competition models and the carefully controlled experiments of Tilman and others may not provide an adequate explanation for the observation that some communities of organisms contain large numbers of different species whose ecological functions—that is, the ways in which they utilize resources—are so similar as to make them nearly indistinguishable. The cichlid fishes of Lake Victoria, Africa, prior to the introduction of the Nile perch (see Chapter 1), the great diversity of plant species in many terrestrial communities, and the vast numbers of different types of planktonic organisms found in many lakes are examples of such situations. In such cases, it is often difficult to detect even the smallest difference in the way in which these coexisting species use their resources, so that they seem to violate the principle of competitive exclusion. In the case of plants, Huston (1994) and others have pointed out that all the many different types of terrestrial plants occurring in a single location are rooted in soil whose characteristics are fairly homogeneous. Thus, ecological divergence in the way in which the different species obtain water and nutrients from the soil is not likely. Yet, they coexist. Hutchinson eloquently articulated this discordance between observation and theory in his 1961 paper titled "The paradox of the plankton."

In reconciling these observations with theory, ecologists have come to understand that disturbance and fluctuating environmental conditions may prevent the populations of competing species from reaching a size at which a competitive effect is realized. Thus, whereas two species may overlap nearly completely in the manner in which they obtain resources and thus be predicted by competition theory to be unable to coexist, they may in fact coexist in nature because disturbance prevents either population from reaching a size at which it will exert a negative effect on the other. We may think of such a system as a **nonequilibrium** system, since the stable equilibrium predicted by the logistic competition models is never reached.

21.11 Consumers can influence the outcome of competition between species.

Disturbance is not the only natural force that affects the outcome of competitive interactions. Ecologists have observed since the time of Darwin that when an animal grazes a field that contains different species of plants, the grazing will affect the abundances of those plant species differently (see Chapter 22). Early experiments showed that grazing affects the competitive interactions among the plants in a grazed field. Tansley and Adamson (1925) conducted systematic and extensive experiments on the effect of rabbit grazing on the composition of British chalk grasslands. Over a 6-year period, the vegetation within rabbit-free enclosures became dominated by the grass *Zerna erecta* (now called *Bromus erectus*), a dominant competitor normally held in check by grazing. Similar studies on the effects of grazing animals on the composition of grasslands are discussed in Chapter 24. Predation too, may affect the interactions of competitors. For example, predators are known to have strong effects on competition among rodent species at high latitudes, and those predator effects influence the population cycles of the rodent species (Andersson and Erlinge 1977, Korpimäki and Norrdahl 1989, Hanski et al. 1991, Hanski and Henttonen 1996). Let us examine how predation can be incorporated into the theoretical models that we have developed in this chapter.

We can extend the Lotka-Volterra competition model to include the effects of predation on one of a pair of competing species. Suppose that, in the absence of predation, species i outcompetes species j (that is, $K_i > K_j/a_{ji}$ and $K_j < K_i/a_{ij}$; see above). Now suppose that a predator removes individuals of species i at rate m per capita. We may write the dynamics of species i as

$$\frac{dN_i}{dt} = r_i N_i \left(1 - \frac{N_i}{K_i} - \frac{a_{ij}N_j}{K_i}\right) - m_i N_i.$$

This equation is identical to equation (21-3) for competition between two species, except that dN_i/dt is now discounted not only by the competitive effects of species j, but also by predation ($m_i N_i$). The model assumes that predators take prey in direct proportion to their availability at all prey densities. As we shall demonstrate in Chapter 23, this model is an oversimplification of the way predators work in nature.

As is typical of the way we analyze such equations, we want to know what happens when $dN_i/dt = 0$. We will omit the details of the algebra here, since the procedures are the same as used before, albeit a bit messier because of the added term. Equilibria are achieved when

$$\hat{N}_i = \frac{K_i\left(1 - \frac{m_i}{r_i}\right) - a_{ij}K_j}{1 - a_{ij}a_{ji}} \tag{21-9}$$

and

$$\hat{N}_j = \frac{K_j - a_{ji}K_i\left(1 - \frac{m_i}{r_i}\right)}{1 - a_{ij}a_{ji}}. \tag{21-10}$$

These equations tell us the respective equilibrium densities of two competing populations when one of the populations, in this case population i, is exposed to predation at constant rate m_i. Coexistence of the two species means that both $\hat{N}_i$ and $\hat{N}_j$ are greater than zero.

The conditions represented in the above equations resemble the equilibria under pure competition, except that the value of K_i is discounted by the term $(1 - m_i/r_i)$, which occurs in both equations (compare these equations with equation [21-3]). In the absence of predation ($m = 0$), equilibria are determined only by the *a*s and *K*s, as in the models of pure competition (equations [21-7] and [21-8]). Thus, the dynamics of this system depend primarily on what happens with the interaction of species i and its predator; that is, on the ratio m_i/r_i. Focus your attention on what happens to species j in this system under the assumption that r_i is constant. Intuitively, we expect that if the predation pressure on species i is large, there will be fewer individuals of species i around to compete with individuals of species j, and the number of individuals of species j will increase. Consider what happens to the equation for N_j as the ratio m_i/r_i changes at values of $K_i = 200$, $K_j = 100$, $a_{ij} = 0.5$, and $a_{ji} = 0.5$ (Figure 21-14). As the ratio increases, the value of N_j increases linearly. Indeed, whereas in the pure competition models, N_j could not increase beyond $a_{ji}K_i$, in this case, N_j can be positive when it is above that point. The figure shows clearly that N_j can grow beyond $a_{ji}K_i = 0.5(200) = 100$. Figure 21-14 also shows what happens to population i when it is subject to both predation and competition. As predation pressure increases (increasing m_i/r_i), N_i decreases linearly until species i is eaten out of existence (see Exercise 8).

Gause (1935; recounted by Slobodkin 1961) explored theoretically the situation in which both species of competitors suffered predation at the same level

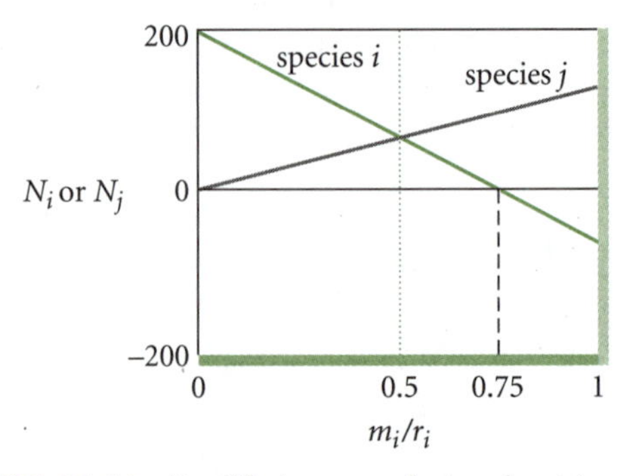

FIGURE 21-14 Equilibrium population densities (vertical axis) for two competing species i and j where species i is under pressure of predation. The horizontal axis shows a range of values of m_i/r_i where m_i is the constant rate of predation and r_i is the rate of increase of species i. As the ratio increases, the equilibrium population density of species i decreases and that of species j increases because fewer individuals of its competitor are present.

(hence nonselective predation). In some cases, such predation can promote coexistence, much like periodic disturbance, if it prevents a population having a competitive advantage from outgrowing its competitor.

21.12 Indirect interactions may lead to apparent competition.

Populations may compete for limited resources directly, as shown in the models presented in this chapter. But species may also have indirect effects on one another, and in some instances, these indirect interactions result in population dynamics that look very much like direct competition. The term **apparent competition** has been applied to situations in which two species have a negative effect on each other resulting from their interaction with a common predator species or with another potential competitor (Holt 1977, 1984, Holt and Lawton 1994, Abrams and Matsuda 1996).

Apparent competition may take a number of increasingly complex forms depending on how many predators and how many prey species are involved (Figure 21-15a). In the simplest case, prey species A may indirectly negatively affect another prey species B that is eaten by the same predator simply by benefiting the predator. That is, if the predator's reproductive rate increases with an increase in the density of prey species A—called a numerical response (see Chapter 23)—then there will be more predators around to eat prey species B. This has the same effect as if prey species A and B were in direct competition with each other. As either of their population sizes increases, it has a negative effect on the other species, but the effect is mediated indirectly through the predator. This phenomenon is often called competition for **enemy-free space,** because the prey species that can avoid predation will be benefited, or alternatively, **shared predation,** because two or more prey fall victim to a shared enemy (Holt and Lawton 1994).

More complicated situations arise when the predator takes a preferred prey, which we call the focal species, as well as many alternative prey (Figure 21-15b). In this case, the focal species may indirectly affect the other prey species, but because the effect is spread over a number of alternative prey, the effect is more diffuse. A focal species may also fall victim to more than one predator (Figure 21-15c), and each of those predators may have a number of alternative prey. It is also possible for a suite of predators to feed on both a suite of focal

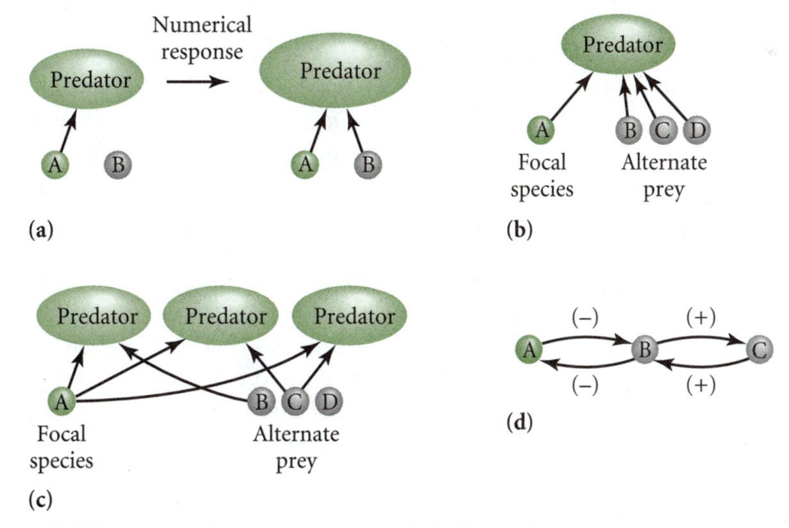

FIGURE 21-15 Possible ways in which indirect effects can cause apparent competition. (a) A predator eats both species A and B, but prefers A. If by consuming A, the predator undergoes a numerical response (an increase in predator density, shown by the larger one), then predation pressure on species B may increase. Thus species A has a negative indirect effect on species B. (b) If a predator prefers focal species A, but can consume a suite of alternative species, species A will have a diffuse apparent competitive effect on the suite of alternative species. (c) Complex indirect interactions may arise when more than one predator is involved. (d) Apparent competition may occur between three species at the same trophic level. If species A and B compete, they have a negative (−) effect on each other. Species B and C have a mutually beneficial effect on each other (+). Because A negatively affects B, the positive effect of B on C is diminished, and thus C is indirectly affected by the interaction of A and B. (Adapted from Holt and Lawton 1994.)

(preferred) prey and a group of alternative prey. We can see from these possibilities that except for the simplest cases, apparent competition may be difficult to observe in nature because its effects are so diffuse (Holt and Lawton 1994).

There is another side to the indirect interactions of two prey species with a single predator. If the consumption rate of the predator increases as the prey density increases—referred to as a functional response to the prey species (see Chapter 23)—the interaction can actually lead to a positive indirect effect. Ignoring the predator's numerical response for a moment, consider a situation in which the predator displays a strong functional response to both prey species. If the density of prey species A increases and the density of prey species B remains at some low level, species B will be relieved of predation pressure because the predator will switch from species B to the more abundant species A. Species B may also be benefited because predators become satiated from eating the plentiful species A. The release from predation pressure enjoyed by prey species B may allow its population to increase, while the increased predation pressure on species A drives the density of that species down. At some point, the population of species A will become smaller than that of species B, and the predator will focus its attention on species B, thereby relieving species A of predation pressure.

Another way that apparent competition may occur is through indirect interactions of populations at the same trophic level (Figure 21-15d). Consider three species, A, B, and C. Species A and B compete, so they have negative effects on each other. Species B and C interact in a positive way; that is, their association benefits both species, representing a mutualistic relationship (see Chapter 25). The negative effect of species A on species B will indirectly negatively affect species C because it diminishes species B, from which species C gains a benefit.

The phenomenon of shared predation may be of great importance in disease ecology, biological pest control, and conservation. Holt and Lawton (1994) present a number of examples that demonstrate this importance. Different wildlife species often fall victim to a shared disease or parasite. If one species can withstand a higher infection rate of the disease or parasite, it may serve as a reservoir for a co-occurring species that has a lower resistance to the disease. For example, white-tailed deer can survive an infestation of a meningeal worm that will kill caribou, thereby preventing wildlife managers from introducing caribou to regions where infested white-tailed deer are present (Bergerud and Mercer 1989; see Chapter 19). Predators and parasitoids that are introduced as biological pest control agents (see Chapter 23) may create serious problems if they are able to eat one or more alternative prey. Introduced to reduce the level of a pest, the predator or parasitoid may simply switch to native organisms when the pest populations drop. Ironically, it is often beneficial to introduce a biological control species that can use alternative prey. In that way, large populations of it may be maintained in the area. The problem arises when one of the alternative prey is endangered or unable to find refuge from the introduced predator.

21.13 Inferior competitors may coexist with superior competitors in metapopulations.

The spatial dynamics of competing species may affect the outcome of competition. We may appreciate this by thinking about competition between species that occur in a metapopulation—that is, species that live within an array of suitable patches of habitat that are surrounded by unsuitable area. You will recall from our discussion in Chapter 17 that one of the most important features of the metapopulation concept is that it provides a framework for the consideration of ecological phenomena at different spatial scales. A process may have a different dynamic within each of the patches and among the patches. This feature of metapopulation theory gives us another way to think about the coexistence of competing species. Competitors may coexist, or not, in an individual patch, which we may think of as local coexistence. Or they may exclude each other from patches through competitive interactions in such a way that they rarely exist in the same patch, but coexist at the landscape scale because they occupy different patches in the metapopulation. Such landscape-level coexistence can allow inferior competitors to coexist with superior competitors under some circumstances, as the following example shows.

Suppose that two species, A and B, live within a metapopulation (note that in this context we may use the word *metapopulation* even though we are talking about more than one population). Further suppose that species A is a superior competitor with species B such that whenever the two colonize the same patch, species A eliminates species B from the patch through competitive exclusion. Under these constraints, species A and species B will never coexist within an individual patch. Are there conditions under which they will coexist within the metapopulation as a whole? That is, can they coexist at the landscape level? A number of models that address this question have been developed. Let us take a look at one that is an extension of the Levins metapopulation model (see Chapter 17).

Recall the simple Levins model of metapopulation dynamics,

$$\frac{dp}{dt} = mp(1 - p) - ep,$$

showing the relationship between the rate of change, dp/dt, in the proportion of occupied patches, p, with the colonization rate, m, and the extinction rate, e. With this in mind, think about a metapopulation with two competitors, A and B, as described above. In that situation, patches may be unoccupied, having been colonized by neither species A nor B, or they may be occupied by either species A or species B. Since species A is a superior competitor with species B and eliminates it from all patches they both colonize, there will be no patches containing both species in the system. Let $x =$ the proportion of the patches in the metapopulation that are unoccupied, $y =$ the proportion that are occupied by species A, and $z =$ the proportion occupied by species B. Since these are the only three possibilities under the assumptions that we have placed on the model, $x + y + z = 1$. The two species have distinct colonization and extinction rates, which we denote m_A and m_B (colonization) and e_A and e_B (extinction).

The following formulations for the rate of change of the proportions of the three patch conditions (unoccupied, occupied by A, occupied by B) were proposed by Nee and May (1992):

$$\frac{dx}{dt} = -m_A xy + e_A y - m_B xz + e_B z$$

$$\frac{dy}{dt} = m_A y(x + z) - e_A y$$

$$\frac{dz}{dt} = m_B zy - e_B z - m_A zy.$$

The first equation represents the rate of change in the proportion of unoccupied patches. Such patches occur either because they have never been occupied or because of extinction in a patch. The two positive terms in the equation ($e_A y$ and $e_B z$) represent the extinction of the two species from patches. Since each extinction leaves an open patch, thereby increasing the proportion of unoccupied patches, z, the terms are positive. They add to the number of unoccupied patches. The two negative terms in the equation ($-m_A xy$ and $-m_B xz$) represent colonization of empty patches by the two species. Since colonization removes unoccupied patches, thereby reducing z, the terms are negative. The second equation gives the rate of change, dy/dt, of the proportion of patches occupied by species A. This equation incorporates the assumption that species A is a superior competitor with species B in the first term, $m_A y(x + z)$, which indicates that species A can colonize either empty patches or those occupied by species B. Compare the term $m_A y(x + z)$ with the term $mp(1 - p)$ of the basic Levins model. In the Levins model, the number of patches available for colonization is $1 - p$, the number of unoccupied patches. In the competition model, the number of patches available for colonization by species A is $x + z$, the number of unoccupied patches, x, plus those occupied by species B, z. The third equation gives the dynamics for the inferior competitor species B. Species B can colonize only empty patches, indicated by the term $m_B zy$. Patches occupied by species B may go extinct ($-e_B z$), or species B may be eliminated from a patch by the colonization of that patch by the superior competitor, species A ($-m_A zy$). A schematic representation of these relationships is shown in Figure 21-16.

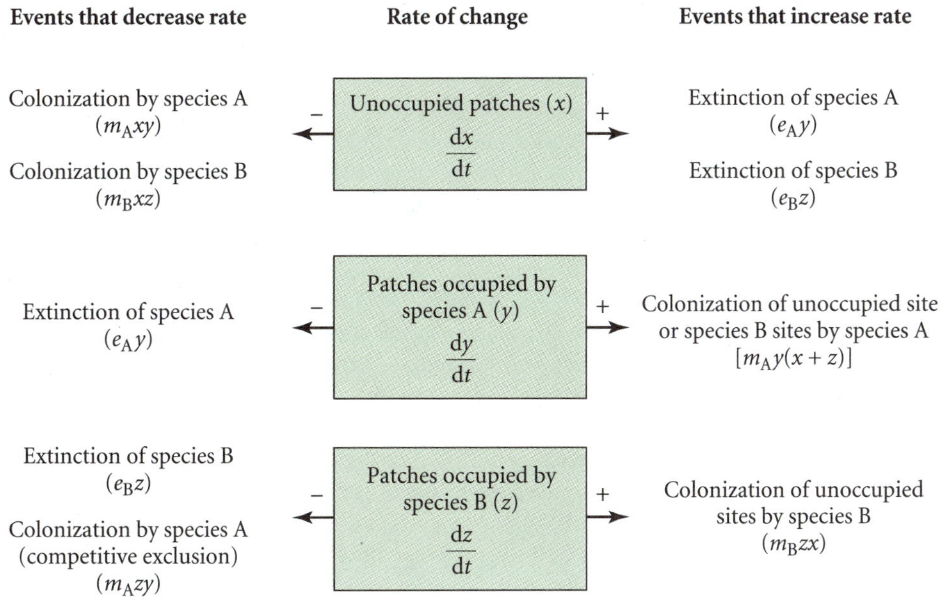

Events that decrease rate		Rate of change		Events that increase rate
Colonization by species A ($m_A xy$) Colonization by species B ($m_B xz$)	−	Unoccupied patches (x) $\frac{dx}{dt}$	+	Extinction of species A ($e_A y$) Extinction of species B ($e_B z$)
Extinction of species A ($e_A y$)	−	Patches occupied by species A (y) $\frac{dy}{dt}$	+	Colonization of unoccupied site or species B sites by species A [$m_A y(x + z)$]
Extinction of species B ($e_B z$) Colonization by species A (competitive exclusion) ($m_A zy$)	−	Patches occupied by species B (z) $\frac{dz}{dt}$	+	Colonization of unoccupied sites by species B ($m_B zx$)

FIGURE 21-16 Summary of the events that change the proportion of patches occupied by two competing species and the proportion of empty patches in a metapopulation. Species A is a superior competitor with species B, such that whenever species A colonizes a patch occupied by species B, species B is competitively excluded from the patch.

Let us return to our original question of whether two competing species, one a superior competitor with the other, can coexist in the system described above. Coexistence in this case means that some of the patches in the metapopulation are occupied by species A and others by species B. That is, we want to know whether the two species can coexist at the landscape level. You will notice that the question really has little to do with the number of unoccupied patches. We are interested only in whether there are patches occupied by species A and other patches occupied by species B. Indeed, because $x + y + z = 1$, we need only consider any two of the equations in the system above. You know by now that the way to approach such a question is to consider what happens at equilibrium; that is, when $dx/dt = 0$, $dy/dt = 0$, and $dz/dt = 0$. Such analyses, which are omitted here, have shown that the inferior competitor can coexist in a metapopulation with a superior competitor if $m_B/e_B > m_A/e_A$ (Nee and May 1992, Nee et al. 1997). That is, if the ratio of colonization to extinction for the inferior competitor is greater than that ratio for the superior competitor, the two can coexist. Again, coexistence here is viewed at the landscape level. The two species will not be found together in a patch because species A will always exclude species B locally. Notice that the inferior competitor does not have to have a higher colonization rate than the superior competitor to coexist with it.

The models presented in this chapter provide the theoretical framework within which to study competition in natural systems. However, no natural system is as simple as these models or the two-species microcosms concocted in the laboratory. Natural environments are heterogeneous over space and variable over time. Competition extends beyond pairs of species, and consumers also are consumed. In the next chapter we will provide some examples of competition in nature.

SUMMARY

1. Competition is the utilization or contesting of a resource by more than one individual consumer. When the individuals belong to the same species, their interaction is called intraspecific competition; when they belong to different species, it is called interspecific competition. Interference competition occurs when an individual actively interferes with another individual's access to a resource. When individuals use the same resource, but at different times of the day or season than other individuals, exploitative competition occurs.

2. Competition was recognized as a fundamental ecological process by Darwin, who based his theory of evolution by natural selection on intraspecific competition between individuals having different genotypes. Modern studies on competition date from the theoretical investigations of V. Volterra and A. J. Lotka during the 1920s and early 1930s. During subsequent decades, interspecific competition was a fashionable subject for laboratory investigation, but not until the 1940s and 1950s did David Lack, G. E. Hutchinson, and others recognize the potential role of interspecific competition in regulating the structure of biological communities.

3. A. G. Tansley was the first to conduct an experiment to determine the existence of competition between species. G. F. Gause demonstrated competition in laboratory experiments with protozoans. Thomas Park's experiments on flour beetles (*Tribolium*) showed that the outcome of competition depends on the conditions of the environment for any given pair of species.

4. The competitive exclusion principle holds that two species cannot coexist indefinitely on a single limiting resource.

5. Intraspecific competition is closely associated with population regulation. When a population approaches its carrying capacity, competitive interactions among the members of the population intensify, resulting in more mortality and a slower population growth rate.

6. Volterra's and Lotka's mathematical treatments of competition were based on the logistic equation of population growth, in which the term N/K represents the intensity of intraspecific competition. Competition between species i and species j is incorporated into the equation for the population growth rate of species i (dN_i/dt) by the analogous term $a_{ij}N_j/K_i$. The term a_{ij} is called the coefficient of competition, and expresses the effect of individuals of species j on the growth rate of population i. The dynamics of population i are thus described by $dN_i/dt = r_iN_i(1 - N_i/K_i - a_{ij}N_j/K_i)$.

7. The joint equilibrium of species i and j ($dN_i/dt = 0$, $dN_j/dt = 0$) is described by the equation $N_i = (K_i - a_{ij}K_j)/(1 - a_{ij}a_{ji})$, and the analogous equation for species j. In the most general terms, coexistence ($N_i > 0, N_j > 0$) requires that $a_{ij}a_{ji} < 1$.

8. The mathematical description of logistic competition may be portrayed on a graph whose axes are the sizes of populations i and j (that is, N_i and N_j). Each combination of these values describes a point at which dN_i/dt or dN_j/dt is either positive, zero, or negative. The relative positions of the equilibrium isoclines (lines describing combinations of N_i and N_j for which dN_i/dt or $dN_j/dt = 0$) determine whether two competitors will coexist, or one will exclude the other.

9. The competition coefficient, a_{ij}, can be estimated from the dynamics of competing populations or, if two populations coexist, from their equilibrium population sizes in the presence and absence of each other. These estimates depend on the assumption of linearity implicit in logistic competition.

10. Competition between species can be understood in terms of analysis of resource dynamics. The outcome of competition depends on the relative positions of the zero net growth isoclines of each species, the vectors of resource consumption, and the relative supplies of each resource in the absence of consumption (the resource supply point).

11. Logistic competition models may not provide an adequate explanation of some communities in which many different species coexist. Competitive exclusion may not occur in such communities because natural disturbance keeps all populations well below the carrying capacity, where competitive effects do not come into play.

12. Predators may influence the outcome of competitive interactions between prey species. The Lotka-Volterra models may be modified to show this by subtracting a constant rate of predation: $dN_i/dt = r_iN_i(1 - N_i/K_i - a_{ij}N_j/K_i) - m_iN_i$.

13. Two species may have a negative effect on each other even if they do not compete because of their interaction with a predator, a phenomenon called apparent competition. Apparent competition may also result from the interactions of populations within a single trophic level.

14. An inferior competitor may coexist with a superior competitor in a metapopulation so long as the ratio of its colonization rate to extinction rate is greater than the same ratio for the superior competitor.

EXERCISES

1. Write a brief historical comparison of the major ideas of ecosystems ecology as presented in Chapter 9 with the development of ideas regarding competition as presented in Section 21.1 of this chapter.

2. Suggest appropriate null and alternative hypotheses for Tansley's experiments with bedstraw and Gause's experiments with *Paramecium.* (Hint: See Chapter 2 for a discussion of hypotheses.)

3. When species coexist in nature, it is most often found that they differ somewhat in the way in which they utilize resources, thus avoiding competitive exclusion. Suppose that there are two species of wolf spiders that live in the leaf litter of the forest near your home. The species are indistinguishable in size, appear to be active at the same time of day, and take similar types of prey. Explain the kinds of observations and experiments that you would undertake to determine how the ecology of these two very similar species might differ.

4. Midway through your careful study of the two wolf spiders described in Exercise 3, your advisor asks you how you will explain the coexistence of the two species if you are unable to determine an important difference in the way in which they utilize resources. How do you answer?

5. In the development of the logistic competition models in Section 21.5, intraspecific and interspecific competition were assumed to be independent of each other. Is this a biologically reasonable assumption?

6. The population of a particular species contains $N_i = 1{,}000$ individuals. The population has the following characteristics: $K_i = 1{,}800$, $r_i = 1.2$. There are 3,000 individuals of a competing species j in the area, and the competition coefficient of the second species against the first is $a_{ij} = 0.2$. What is the growth rate of population i under the assumptions of the logistic competition model?

7. Consider two populations, i and j, each with a carrying capacity equal to 150, that are in competition with each other. Population i has a high rate of growth, $r_i = 2.0$, and population j has a low rate, $r_j = 0.5$. The competitive effect of population i on population j is $a_{ji} = 0.8$, whereas that of population j on population i is $a_{ij} = 0.2$. Show algebraically how a pattern of disturbance that keeps population i at a low level will prevent it from competitively excluding population j. (Hint: Show how dN_j/dt changes from very high to very low levels of population i.)

8. Referring to Figure 21-14 and equation (21-9), derive an algebraic expression that tells the value of m_i (predation pressure) at which population i disappears. (Hint: Start by determining the value of m_i/r_i when $N_i = 0$.)

CHAPTER 22

Competition in Nature

GUIDING QUESTIONS

- Why is competition difficult to observe in nature?
- What are some of the mechanisms of competition?
- What is allelopathy, and why is it important in agriculture?
- What can introductions of exotic species for the purpose of pest control tell us about competitive interactions among species?
- What is asymmetrical competition?
- How do researchers distinguish between the effects of interspecific and intraspecific competition?
- What are some of the mechanisms by which plants and animals compete for space?
- How do the activities of predators affect competition among their prey?

The models that we presented in the previous chapter revealed to us the conditions of population density (N) and intensity of competition (a_{ij}) under which competing populations may coexist and under which one population excludes another by competition. We also saw how predation affects the outcome of competition and how competing species coexist in metapopulations. Because competition models predict which of a group of competing populations will survive and which will disappear, they are of great interest to ecologists who wish to understand the mechanisms by which ecological communities are assembled and maintained, and the processes that give rise to biological diversity in general. Our interest in such models is heightened by results from simple laboratory experiments like those of Gause (see Section 21.2), in which the outcome of competitive interactions is clearly shown and unambiguous. Unfortunately, the relative simplicity of mathematical models belies the complexity of competition in nature, leaving ecologists with the challenge of testing the assumptions of the models with complex experiments. Questions about the importance and mechanisms of competition in nature are being intensively considered by ecologists. In particular, the role of competition as a factor affecting community structure has been the focus of intense debate since the mid-1980s (see Part 6).

The characteristics and effects of competition are murky for a number of reasons. First, we have already seen how the interactions of a predator and its prey may make it appear that the prey are competing when in fact they may not be (apparent competition; see Section 21.12). Second, it is often difficult to measure the relative strengths of intraspecific competition and interspecific competition in nature. Knowing the strengths of these two process is important because they represent two different ecological processes. Intraspecific competition is a population-level phenomenon that reflects density-dependent features of population growth. Interspecific competition is a process involving the interactions of different species. Finally, it is sometimes very difficult to identify the specific mechanisms of competition. We have already pointed out that competitive interactions may involve face-to-face contests between individuals over resources (interference competition) or indirect interactions in which individuals of competing species use a critical resource at different times, thereby reducing the availability of the resource for members of the competing species without ever facing them (exploitative competition). But the mechanism of competition may be shaped by context or mediated by special behaviors or morphological and physiological features of the competing species in such a way that the nature of the interference or the exploitation is difficult to determine.

In this chapter, we examine the evidence for competition in nature. Our discussion will show how ecologists use experiments to study the nature and importance of competition in natural settings. We shall see that the outcomes of competitive interactions are complex.

22.1 How does competition occur?

We have distinguished exploitative competition, in which individuals, by using resources, deprive others of the benefits of those resources, from interference competition, in which individuals directly inhibit access to or use of resources by other individuals, often by physical (fighting, for example) or chemical means (toxins). Schoener (1983) further subdivided competition into six categories according to its mechanisms:

1. **consumptive competition,** based on the utilization of some renewable resource
2. **preemptive competition,** based on the occupation of open space
3. **overgrowth competition,** which occurs when one individual grows upon or over another, thereby depriving the second of light, nutrient-laden water, or some other resource
4. **chemical competition,** by production of a toxin that acts at a distance after diffusing through the environment
5. **territorial competition,** the defense of space
6. **encounter competition,** involving transient interactions over a resource that may result in physical harm, loss of time or energy, or theft of food

These mechanisms of competition are defined in terms of the capabilities of organisms and the habitats in which they occur; hence their distribution among organisms and habitats is predictably heterogeneous (Table 22-1). Preemptive and overgrowth competition appear among sessile space users, primarily terrestrial plants, marine macrophytes, and animals living on hard substrates. Territorial and encounter competition occur among actively moving animals; chemical competition, among terrestrial plants (most toxins are diluted too readily in aquatic systems, but see Jackson and Buss 1975). By Schoener's count, consumptive competition is the most common mechanism, especially in terrestrial environments; preemptive and overgrowth competition predominate in marine habitats, but this may be in part because most studies have involved sessile organisms living on hard substrates.

One of Schoener's categories, chemical competition, or **allelopathy,** has received considerable attention (Whittaker and Feeny 1971, Harborne 1982, Rice 1984, Putnam and Tang 1986, Gopal and Goel 1993, Seigler 1996). Although the causing of injury (*-pathy*) to other individuals (*allelo-*) by chemical means has been reported most frequently in terrestrial plants, such interactions may also take place in aquatic plant systems (Gopal and Goel 1993). Generally, we think of plants as exuding poisons to impair the growth of other plants. However, allelopathy can be much more complex. For example, some parasites appear to exclude other, potentially competing parasites from a host by stimulating the host's immune system against them (Schad 1966, Cohen 1973). It has been suggested

TABLE 22-1 A survey of proposed mechanisms of interspecific competition in experimental field studies

	Mechanism						
Group	Consumptive	Preemptive	Overgrowth	Chemical	Territorial	Encounter	Unknown
Freshwater							
Plants	0	0	1	1	0	0	0
Animals	13	1	0	1	1	5	2
Marine							
Plants	0	6	4	1	0	0	0
Animals	9	10	6	0	7	6	0
Terrestrial							
Plants	28	3	11	7	0	1	9
Animals	21	1	0	1	11	15	6
Total	71	21	22	11	19	27	17

(Data from Schoener 1983.)

that the abundant oils in the leaves of the eucalyptus trees of Australia promote frequent fires in the leaf litter, killing the seedlings of competitors (Mutch 1970). More frequently, it is the direct effect of a toxic substance that does the damage. In an early study, A. B. Massey (1925) showed that toxic substances released into the soil by walnut trees inhibit seedling growth in other species of plants.

In shrub habitats in southern California, several species of sage of the genus *Salvia* apparently use chemicals to inhibit the growth of other vegetation (Muller 1966, 1970, Muller et al. 1968). Clumps of *Salvia* usually are surrounded by a halo of bare soil separating the sage from neighboring grassy areas (Figure 22-1). When observed over long periods, *Salvia* may expand into the grassy areas. But because sage roots extend only to the edge of the bare strip surrounding the plant and not beyond, it is unlikely that a toxin is extruded into the soil directly by the roots. However, the leaves of *Salvia* produce volatile terpenes (a class of organic compounds that includes camphor and gives foods spiced with sage part of their distinctive taste) that apparently affect nearby plants directly through the atmosphere (Muller 1966, Muller et al. 1964).

Bartholomew (1970) suggested that the halo zone around *Salvia* could be caused by grazing or by seed-eating birds and mammals, and that plant toxins, regardless of their efficacy, were not required to explain the phenomenon. In order to test this hypothesis, he placed cages in the halo zone to keep out small birds and mammals; controls were constructed in the same manner, but with one of the sides of the cage left open. After 1 year, the exclosure cages contained about 20 times as much plant biomass as the controls, and about the same amount as found in the grassy area beyond the halo. John Harper (1977, 378) interpreted these results as suggesting that "the toxin hypothesis is unnecessary to account for the observed pattern of vegetation." J. B. Harborne (1982, 215) drew a different conclusion from the same data: "The possible role of animals, especially birds and rodents, in producing the bare zones was experimentally investigated by Muller (1970) and Bartholomew (1970), but no convincing evidence that they play a causal role could be established in numerous experiments." Both are correct in that although both toxins and herbivores have demonstrable effects on grasses, decisive experiments to show the role of one or the other (or both!) in the formation of the bare zones have not been conducted.

Allelopathy is of considerable interest to ecologists who study agricultural systems. Allelopathic effects of crop plants against weeds (called heterotoxicity) might be of great benefit. But allelopathy can work against crop production in at least two ways. First, when crops are planted in the same field year after year, the buildup of secondary compounds in the soil can result in a negative allelopathic effect (called autotoxicity). When alfalfa (*Medicago sativa*) is replaced in the next year with another seeding of alfalfa, the second crop is usually poorer than the first because of autotoxicity (Miller 1996). Many other forage

(a)

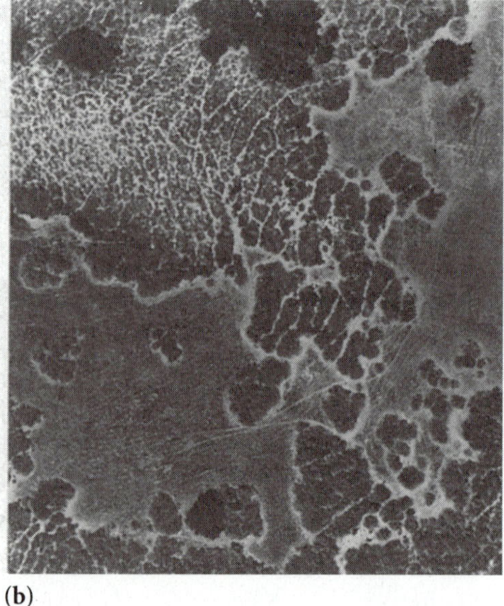

(b)

FIGURE 22-1 (a) Bare patch at edge of a clump of sage includes a 2-meter-wide strip with no plants (A-B) and a wider area of inhibited grassland (B-C) lacking wild oat and bromegrass, which are found with other species to the right of (C) in unaffected grassland. (b) Aerial view shows sage and California sagebrush invading annual grassland in the Santa Inez Valley of California. (*Courtesy of C. H. Muller; from Muller 1966.*)

crop species exhibit the same pattern. Another common problem with allelopathy in agriculture arises when crops are mixed in the same field, a technique that is sometimes used to reduce insect herbivores. Allelopathic influences can result in diminished yield in some of the crops in the mix. Because of these problems, agricultural geneticists are working actively to discover cultivars of crop plants that are resistant to the effects of their own toxic secondary compounds or those of other crops with which they may be grown.

Surveys of the competition literature have revealed a number of different suspected mechanisms of competition. But it is also entirely possible that the patterns found in some competition studies are the result of indirect interactions among species at a particular trophic level or among species who share a predator (see Section 21.12). Connell (1990) reviewed the literature in search of evidence of such apparent competition and found that out of fifty studies that claimed to show evidence of interspecific competition, a large number did not present enough evidence to determine whether indirect effects could have been at play. The difficulty comes in interpreting the results of removal or addition experiments. Consider a competition experiment involving two plant species in which one of the species is removed from experimental plots and the growth of the remaining species is measured. If the remaining species responds with increased growth, we can reasonably suspect that the two species compete. However, suppose that the removed species provided shelter for an herbivorous insect that feeds on both plant species. Then, the removal of that plant from the plot will also reduce the level of predation on the remaining plant, and that may be the cause of its increased growth. Clearly, careful design is required in competition experiments.

22.2 Which species are most likely to compete?

Darwin emphasized that competition should be most intense between closely related species or organisms. In *On the Origin of Species,* he remarked: "As species of the same genus have usually, though by no means invariably, some similarity in habits and constitution, and always in structure, the struggle will generally be more severe between species of the same genus, when they come into competition with each other, than between species of distinct genera." Darwin reasoned that similar structure indicated similar ecology, especially similar resource requirements. Most experimental studies of competition have adopted this viewpoint in the sense that the pairs of species tested often are close relatives and frequently congeners. But many resources are jointly utilized by distantly related organisms. Barnacles and mussels, as well as algae, occupy space in the intertidal zone. Both fish and aquatic birds prey on aquatic invertebrates. Krill (*Euphausia superba*), shrimplike crustaceans that abound in subantarctic waters, are fed upon by virtually every other larger type of animal, including fish, squid, diving birds, seals, and whales.

In terrestrial habitats, invertebrates in forest litter are consumed by spiders, ground beetles, salamanders, and birds. Birds also compete with lizards for many of the same prey in other habitats. One experimental study of competition for seeds between ants and rodents has attracted considerable attention (Brown and Davidson 1977, Brown et al. 1979). Seeds are a major resource for ants, rodents, and birds in desert ecosystems (Brown et al. 1979). Furthermore, these animals consume many of the same kinds of seeds, although rodents tend to take more large seeds and ants consume more of those less than a millimeter in length. Because many of the ants and rodents in deserts eat little else, the opportunity for interspecific (interphylum!) competition over seeds is certainly present.

We suggested in the previous chapter that some very diverse natural communities contain many species that are practically indistinguishable from one another in the way in which they obtain resources. Some plankton and terrestrial plant communities were offered as examples. Such systems seem to violate the principle of competitive exclusion and are not particularly well represented by equilibrium competition models (see section 21.6). Does this mean that competition is simply not important in such communities? A growing number of ecologists have suggested that, in fact, competition may be relatively rare in nature (e.g., Huston 1994), arguing that environmental fluctuation and disturbance are more important factors than competition in determining what species occur in a particular place.

22.3 Elimination of species following the introduction of competitors demonstrates the population effects of competition.

The challenge of observing competition in nature can be readily demonstrated by reconsidering the competitive exclusion principle, which states that two species cannot coexist on a single resource that is scarce relative to demand for it. If a species excludes another species from an area, then only one species, the winner of the competitive contest, remains in the area for the ecologist to observe. That is, competitive exclusion is a transient process. The evidence of exclusion having taken place is lost when the poorer competitor disappears. We can observe competitive exclusion in the laboratory because we can mix populations according to whim, as Gause did, and follow the course of their interaction. The closest natural analogy to such laboratory experiments is the accidental or intentional introduction of species by humans. For example, in agriculture, parasites or predators are often introduced into a system in order to control weeds or insect pests. If a number of such control species are introduced simultaneously, they may compete for the pest that they are introduced to control. Competitive exclusion may occur and be observed under these conditions.

An example of such a simultaneous introduction of pest control species occurred between 1947 and 1952 when the Hawaii Agriculture Department released thirty-two potential parasitoid species to combat several species of fruit pests, including the oriental fruit fly (Bess et al. 1961). Thirteen of the parasitoid species became established, but only three kinds, all of the wasp family Braconidae (Figure 22-2), proved to be important control agents of fruit flies. Populations of these three parasitoid species, all closely related members of the genus *Opius*, successively replaced one another from early 1949 to 1951, after which only *Opius oophilus* was commonly found to parasitize fruit flies (Figure 22-3). As each parasitoid population was replaced by a more successful species, the level of parasitism of fruit flies by wasps also increased, suggesting superior competitive ability.

A similar pattern of replacement involving wasps that parasitize scale insects has been more thoroughly documented in southern California (DeBach and Sundby 1963, DeBach 1966). Scale insects are pests of citrus groves, capable of causing extensive damage to the trees. As the evolution of resistance by pests reduced the effectiveness of chemical pesticides,

FIGURE 22-2 Pupae of a parasitoid braconid wasp, here emerging from a parasitized tomato hornworm larva.

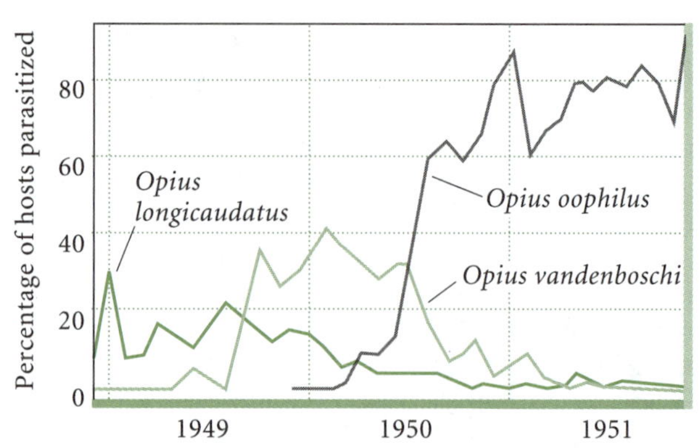

FIGURE 22-3 Successive change in predominance of three species of braconid wasps of the genus *Opius*, all of which parasitize the oriental fruit fly. (*After Bess et al. 1961.*)

agricultural biologists began to rely on insect parasites and predators (DeBach 1974). Yellow scales have infested California citrus groves since oranges and lemons were first planted there. In the late 1800s, the red scale was accidentally introduced and has replaced yellow scale almost completely, perhaps itself a case of competitive exclusion (DeBach et al. 1978).

Of the many species introduced in an effort to control citrus scale, tiny parasitic wasps of the genus *Aphytis* (from the Greek *aphyo*, "to suck") have been most successful. One species, *A. chrysomphali*, was accidentally introduced from the Mediterranean region and became established by 1900. The life cycle of *Aphytis* begins when adults lay their eggs under the scaly covering of hosts. The newly hatched wasp larva uses its mandibles to pierce the body wall of the scale and proceeds to consume the body contents. After the wasp pupates and emerges as an adult, it continues to feed on scales while producing eggs. Each female can raise twenty-five to thirty progeny under laboratory conditions, and the development period is so short (egg to adult in 14 to 18 days at 27°C [80° F]) that populations may produce eight to nine generations per year in the long growing season of southern California.

In spite of its tremendous population growth potential, *A. chrysomphali* did not effectively control scale insects, particularly not in the dry interior valleys. In 1948, a close relative from southern China, *A. lingnanensis*, was introduced as a control agent. This species increased rapidly and widely replaced *A. chrysomphali* within a decade (Figure 22-4). When both species were grown in the laboratory, *A. lingnanensis* was found to have the higher net reproductive rate, whether the two species were raised separately or together in population cages.

Although *A. lingnanensis* excluded *A. chrysomphali* throughout most of southern California, it still did not provide effective biological control of scale insects in the interior valleys because cold winter temperatures

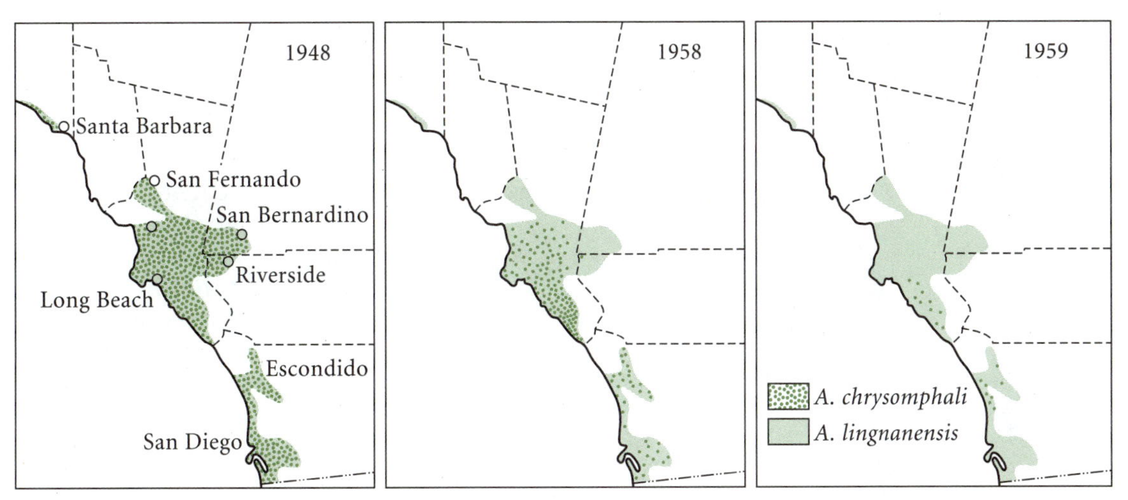

FIGURE 22-4 Successive changes in the distribution of *Aphytis chrysomphali* and *A. lingnanensis,* wasp parasites of citrus scale, in southern California. *A. lingnanensis* was first released in 1948 and rapidly replaced *A. chrysomphali* throughout the region. (*After DeBach and Sundby 1963.*)

greatly reduced parasite populations there. The larval development of both wasps slows to a standstill at temperatures below 16°C (60°F), and adults cannot tolerate temperatures below 10°C (50°F).

In 1957, a third species of wasp, *A. melinus,* was introduced from northern India and Pakistan, where temperatures range from below freezing in winter to above 40°C in summer. As hoped for, *A. melinus* spread rapidly through the interior valleys of southern California, where temperatures resemble those of the wasp's native habitat, but it did not become established in the milder coastal areas (Figure 22-5). P. DeBach and R. A. Sundby (1963) demonstrated in laboratory experiments that at a temperature of 27°C and a relative humidity of 50%, conditions resembling the typical climate of coastal areas more closely than that of the interior valleys, *A. lingnanensis* is the superior competitor.

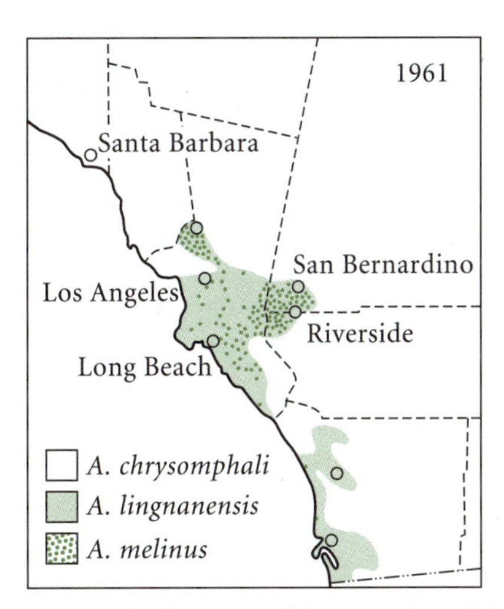

FIGURE 22-5 Distributions of three species of *Aphytis* in southern California in 1961. *A. melinus* predominates in the interior valleys, while *A. lingnanensis* is more abundant near the coast. (*After DeBach and Sundby 1963.*)

22.4 Removal, addition, and substitution experiments have been important tools in the study of plant competition.

Plants compete for resources with neighboring plants of the same species (intraspecific competition) or different species (interspecific competition). The effects of plant competition are revealed in changes in rates of plant growth, reproduction, or some aspect of plant size, shape, or physiological function, collectively referred to as performance. These features of plants vary with resource levels and therefore can provide a sensitive index to the intensity of competition. The intensity of the demand for resources by other plants of the same or different species in the neighborhood of a plant may be examined by adding or removing possible competing plants.

The depressing effect of intraspecific competition on the growth of trees has been demonstrated in experiments in which trees are removed from the forest in order to decrease tree density (forest-thinning experiments). The acceleration of growth of young long leaf pine trees in response to selective removal of trees more than 15 inches in diameter is shown in Figure 22-6. Each core of wood was obtained by boring into a tree's trunk, from the bark to the center, with a long, tubular device called an increment borer. The core of wood removed by the borer tube provides a record of annual growth without the need to cut down the entire tree. These cores showed an increased growth rate,

FIGURE 22-6 Cores of two longleaf pine trees obtained near Birmingham, Alabama, showing the effect on subsequent growth of removing large trees. (*Courtesy of the U.S. Forest Service.*)

particularly in summer (light wood), during the 18 years between the time the forest was thinned and the time the cores were taken. The implication of these results is that tree growth was somehow inhibited by competition with other trees prior to the time that some trees were removed.

Similar competitive effects have been demonstrated between other species of forest trees. One experiment in tropical forests of Surinam was initiated to determine whether the growth of commercially valuable trees could be improved by removing species of little economic importance (Schultz 1960). Foresters poisoned 70% of undesirable trees having girths greater than 30 centimeters in one experimental plot, and greater than 15 centimeters in another, leaving the desirable species untouched. The increase in girth of the desirable trees was then measured over a year in the experimental plots and in control plots that had not been selectively thinned (Figure 22-7).

Removal of the larger trees (>30 cm girth) increased the penetration of light to the forest floor by a factor of 6. The additional light greatly stimulated the growth of the trees remaining on the experimental plot. Improvement was greatest among small individuals, which had been most shaded by others before thinning; trees whose girth exceeded 100 centimeters did not grow appreciably faster. Although removal of the smaller trees (15–30 cm girth) further increased light penetration by only one-third, it led to a striking response in growth rate, particularly among the remaining large trees. The improved growth in this plot could not have been caused by increased light because many of the trees that responded were much taller than the trees that were poisoned. The added growth probably resulted, therefore, from reduced competition for belowground factors such as water or mineral nutrients in the soil, a process that we examine in more detail below.

22.5 Both aboveground and belowground competition are important in plants.

Competitive interactions among plants are complicated by the fact that the plant body essentially occupies two very different environments: the soil, where the roots are located, and the atmosphere above the soil, where the stems and leaves live. The roots of plants compete for water and nutrients, a process called **root competition,** whereas the aboveground parts compete primarily for light, in what is referred to as **shoot competition.** The relative intensities of root and shoot competition affect the overall performance of the plant, although root competition appears to be the more important of the two (Wilson 1988, Wilson and Tilman 1991).

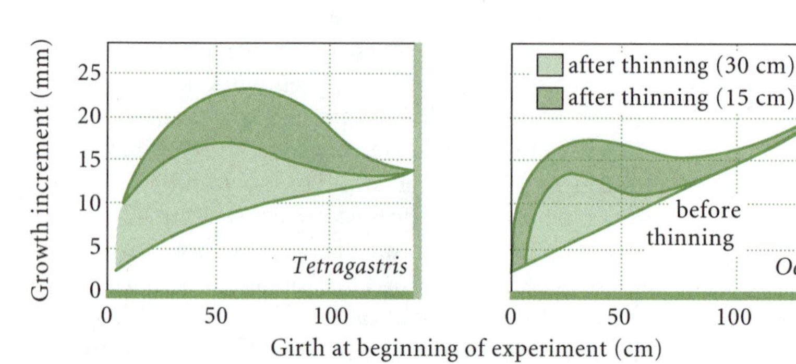

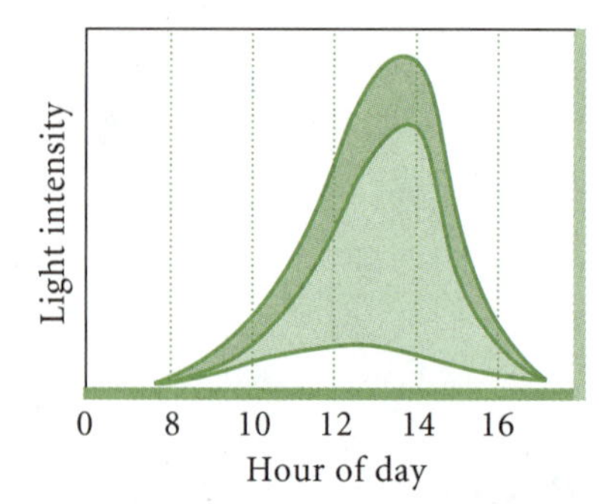

FIGURE 22-7 Effects on the girth of two species of tropical forest trees, *Ocotea* and *Tetragastris,* of removing competing trees greater than 15 centimeters or 30 centimeters in girth. The increase in light intensity that resulted from thinning is shown at the right. (*After Schulz 1960.*)

In order to determine the relative intensities of root and shoot competition, experiments must be designed that separate the two processes. Figure 22-8 shows how this could be done in a greenhouse experiment involving two potential competitors, one of which has broad leaves and the other of which has small leaves. To eliminate interspecific competition, plants of each species are simply placed in separate pots. In order to control for intraspecific competition, the experimenter would have to place the same number of plants in each pot (see Section 22.7). Root competition in the absence of shoot competition can be studied by placing individuals of the two species in the same pot, arranged in the pots in such a way that the aboveground parts of the two plants are not in close proximity. In this arrangement, the roots of the two plants compete for nutrients and water in the soil, but competition for light among the shoots is not a factor. Shoot competition in the absence of root competition can be studied either by using separate pots, but placing them close enough to each other so that the leaves of the plants touch, or by placing partitions in single pots containing both species, thus isolating the roots in the soil. The simultaneous effects of root and shoot competition may be studied by planting individuals of the competing species in such a way that the roots and the stems of the adjacent plants are in close proximity.

The results of a study of the relative strengths of root and shoot competition between white clover (*Trifolium repens*) and perennial ryegrass (*Lolium perenne*) using these techniques is presented in Figure 22-9. In this study, the effect of competition was measured in terms of yield as a percentage of the yield of the plants in the control group (no root or shoot competition). The results show that both root and shoot competition favor the ryegrass and cause a reduction in the yield of white clover (Martin and Field 1984).

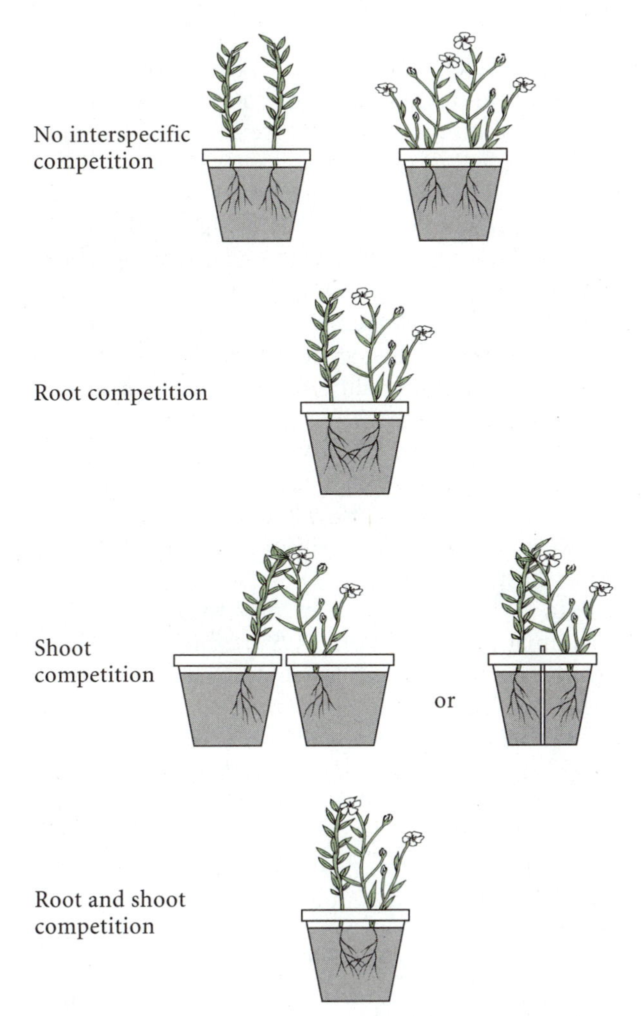

FIGURE 22-8 Experimental method for examining the relative strengths of root and shoot competition between two species of plants. Interspecific competition may be eliminated by placing plants of each species in separate pots. (Intraspecific competition is controlled by using the same density in each plot; see Figure 22-13.) Root competition in the absence of shoot competition is arranged by placing competing plants in pots in such a way that the leaves of the plants are not in close proximity to each other. Shoot competition in the absence of root competition is created by placing two pots close together or by partitioning the soil of a single pot so that the roots do not compete. Root and shoot competition together occur when the plants are placed close to one another in a single pot. (*After Silvertown and Doust 1993.*)

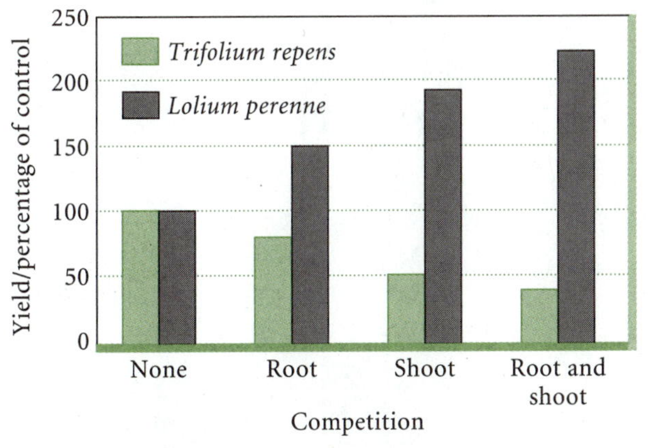

FIGURE 22-9 Results of an experiment to determine the relative effects of root and shoot competition between white clover (*Trifolium repens*) and a perennial ryegrass (*Lolium perenne*). The measure of interest is the yield as a percentage of the yield of control plants (no root or shoot competition), which is shown in the graph as 100%. The results indicate that the ryegrass wins in both root and shoot competition. (*From Martin and Field 1984.*)

22.6 The effect of competition may be different for each of the competing populations.

When two populations compete, it is possible, perhaps likely, that one of the populations will be more strongly affected by the competitive interaction than the other. This phenomenon is referred to as **asymmetrical competition.** Smith (1975) demonstrated asymmetrical competition in two species of *Desmodium,* which are small herbaceous legumes common in oak woodlands of the midwestern United States. Smith employed a addition experiment in which small individuals of each of two species, *D. glutinosum* and *D. nudiflorum,* were planted either 10 cm from a large individual of the same species (to test for the effects of intraspecific competition), 10 cm from a large individual of the other species (interspecific test), or at least 3 m from any other *Desmodium* plant (control—without competition). As an index to subsequent growth, Smith measured the total increase in length of all leaves, both old and new. The results of the experiment (Figure 22-10) showed that even when surrounded by unrelated plants, both species grew best in the absence of other *Desmodium* plants. However, the effects of interspecific competition were not the same for both species. The growth of *D. nudiflorum* was depressed more by interspecific competition than by intraspecific competition.

As we emphasized above, plants are stationary organisms, with both aboveground and belowground parts that may be affected by different competitive regimes. Root competition for water and nutrients among plants is generally symmetrical. Individual plants draw resources from their surroundings, creating what is called a zone of depletion, as described in Chapter 12. When an individual plant's zone of depletion overlaps with that of another plant, both may suffer a reduction in growth rate. Thus, the position of each plant with respect to others determines how strongly symmetrical competition affects it. Shoot competition for light is often asymmetrical in plants. Since light comes from above, taller plants have an advantage over the small plants living in the shade below.

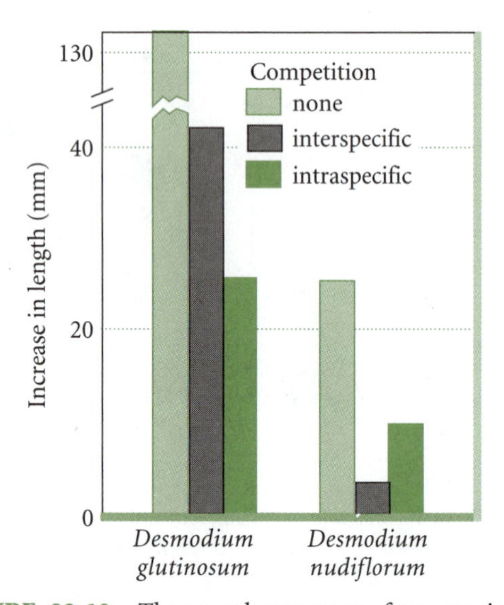

FIGURE 22-10 The growth responses of two species of *Desmodium* when planted near individuals of the same species, near individuals of the other species, and at a distance from individuals of either species. (*After Smith 1975.*)

The spatial distribution of plants affects the outcome of symmetrical aboveground and asymmetrical belowground competition. Figure 22-11a shows a random distribution of seeds in a plot. Some seeds have fallen close to neighbors and others have not. Those seedlings produced from seeds in the more crowded areas will not grow as fast as those in more open areas because of belowground competition for nutrients. Figure 22-11b shows the result of this differential growth. Not all of the seeds that are represented in (a) are found in (b) because of mortality. Aboveground asymmetrical competition for light will come into play among the plants at this point, with the larger plants outcompeting nearby small plants for light and thereby growing faster. Eventually, a few plants will dominate the plot, as shown in Figure 22-11c (Huston 1986, Huston et al., 1988, Huston 1994).

In a recent study of asymmetrical competition in animals, W. J. Resetarits (1997) examined the competitive interactions of two bottom-dwelling stream fishes, the mottled sculpin (*Cottus bairdi*) and the fantail darter (*Etheostoma flabellare*), which occur together in small streams in western Virginia. Resetarits measured the average total length and mass of samples of juveniles of the two species and then placed them in artificial streams in three different combinations for each species: alone, with juveniles of the other species, and with adults of the other species. After allowing the fish to interact in the streams for several months, he removed and counted all of the individuals of each species to obtain an estimate of survival during the period of the experiment. He also obtained another estimate of the average mass and length. The results of the experiment are shown in Figure 22-12. Juvenile sculpins showed higher survival in the presence of juvenile or adult darters than when in the streams alone, a result that is not yet completely understood. The final mass and length of the sculpins were reduced in the presence of both adult and juvenile fantail darters. Darters showed lower survival in all three treatments (alone, with adult sculpins, with juvenile sculpins) than did sculpins. Competitive effects were observed only when the darters were in the presence of juvenile sculpins, not in the presence of

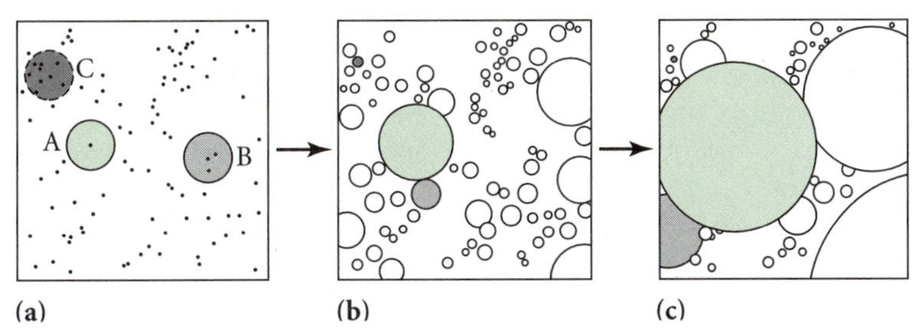

FIGURE 22-11 Effect of asymmetrical competition in a hypothetical plant community. (a) Each dot represents a seed. The circles represent the zone of depletion of the plants that will develop from three individual seeds. Plants having no (circle A), two (circle B), and many (circle C) competing individuals in their zone of depletion are shown. In general, the more competitors within a plant's zone of depletion, the slower its rate of growth. (b) Root competition is symmetrical, so in the early stages of growth, the plants with fewer neighbors in their zone of depletion will grow larger, as represented by the larger circles. (c) As the canopies of the plants spread, asymmetrical competition for light will give the larger plants an advantage. (*After Huston 1986, 1994; from Huston et al. 1988.*)

adult sculpins. Thus, the competitive effects of the two species on each other are asymmetrical.

22.7 The relative intensity of intraspecific and interspecific competition may be determined with substitution experiments.

In most studies of competition in nature, investigators assess the responses of individuals following the removal of others of the same or different species. Addition experiments usually are more difficult because their success depends on the germination of seeds, survival of transplanted seedlings, or willingness of animals to stay where they are put (a problem that is often addressed by the use of fences or walls). But both removal and addition experiments are limited in that, while they reveal interspecific competition, they do not allow a quantitative comparison of the relative magnitudes of interspecific and intraspecific competition. The problem is simply this: the response of individuals to the presence of competing species is measured at a higher total density. When

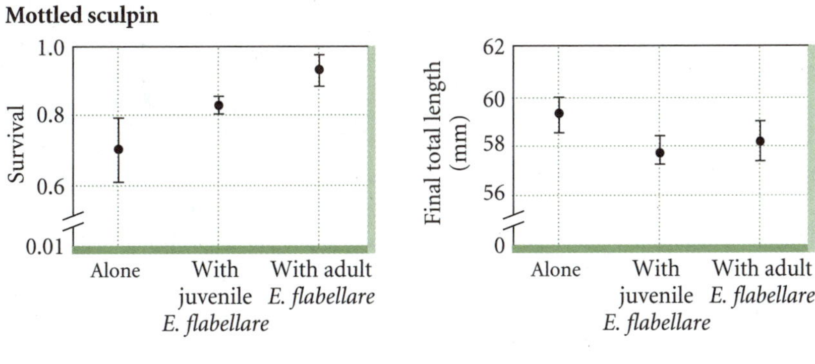

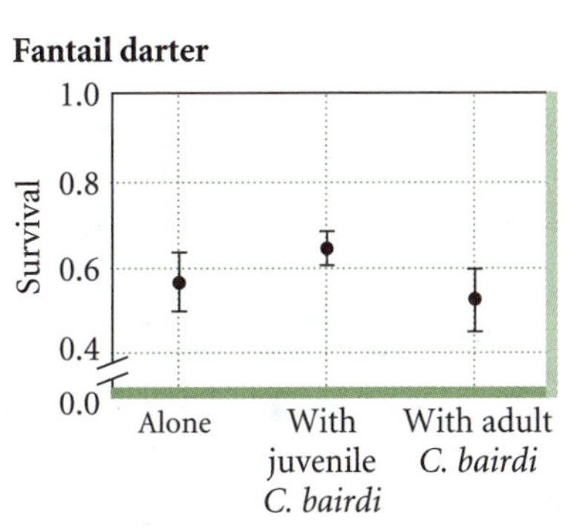

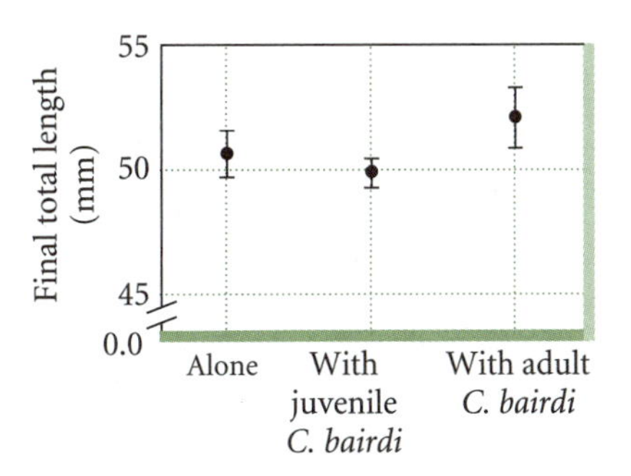

FIGURE 22-12 Results of a competition experiment between mottled sculpins (*Cottus bairdi*) and fantail darters (*Etheostoma flabellare*), showing an asymmetrical competitive response. The survival, mass, and length responses of each species in streams alone, with juveniles of the other species, and with adults of the other species are shown. Sculpins showed higher overall survival and responded to both juvenile and adult darters with reductions in length and mass. Darters showed lower survival, but suffered reduction in length and mass only in the presence of juvenile sculpins. (*From Resetarits 1997.*)

individuals of competing species are removed for the measurement of intraspecific competition, the performance of the remaining individuals under pure intraspecific conditions does not allow direct comparison with performance under interspecific conditions because the total density of individuals is reduced.

The challenge of determining the relative strengths of intraspecific and interspecific competition reveals a weakness in the logistic competition models and demonstrates why competition is sometimes difficult to observe. The logistic models predict that the growth rate of species i will be dependent on its own population density and that of the competing species adjusted by the competition coefficient: $dN_i/dt = r_iN_i(1 - N_i/K_i - a_{ij}N_j/K_i)$. In our development of this model (see Section 21.5), we pointed out that it says simply that the rate of growth of a population of species i (dN_i/dt) in the presence of a competing species j is the exponential growth rate of that species (r_iN_i) adjusted by intraspecific and interspecific competition. The term $a_{ij}N_j/K_i$ gives the strength of interspecific competition. An important characteristic of this relationship is that the effect of interspecific competition is assumed to have a linear relationship with density. For example, consider the competitive effects of species j on species i at three different levels of species j: $N_j = 10$, $N_j = 20$, and $N_j = 30$ if $a_{ij} = 0.5$. For each N_j, the competitive effect is $a_{ij}N_j/K_i$, and the respective values of this term for the three population sizes are $5/K_i$, $10/K_i$, and $15/K_i$. If you pick some value, say 100, for K_i and sketch a plot with the densities (N_j) on one axis and the competition terms ($a_{ij}N_j/K_i$) on the other, you will get a straight line. Thus, in the model, the addition or removal of individuals (a change in N_j) results in a linear change in the competitive effect ($a_{ij}N_j/K_i$). The model also assumes that competitive effects can be measured in population changes (that is, changes in N_i).

These assumptions may not accurately reflect competition in nature. We cannot assume that competitive effects are linear with density, and as we saw in the examples above, the effect of competition may be revealed as changes in individual growth rate, seed production, or survival through a part of the life cycle, rather than as a change in the population density per se.

One way in which ecologists overcome this problem is by experimentally comparing the change resulting from interspecific competition with that resulting from intraspecific competition over the same range of density in what are called **substitution** or **replacement series** experiments (De Wit et al. 1960). In such experiments, one keeps the total density of organisms constant while varying the ratio of individuals of the two species. For example, in a competition experiment with two species of plants grown in greenhouse pots, the experimenter could maintain a total density of 100 individuals per pot, but have each pot contain 30 individuals of species A and 70 individuals of species B. This would create conditions of strong interspecific competition for A and strong intraspecific competition for B. By varying the ratio of species, the degree of intraspecific and interspecific competition could be varied for each, while maintaining the total intensity of competition overall.

Marshall and Jain (1969) applied this experimental design to a study of competition in wild oats (*Avena*). Two species, *A. fatua* and *A. barbata*, which co-occur in California grasslands, were planted in pots at combined densities of 8, 16, 32, 64, 128, and 256 individuals per pot. Marshall and Jain recorded various indices of growth, survival, and reproduction in order to assess responses to density, rather than measuring changes in population density. Under the conditions of the experiment, survival to reproduction was high and independent of density. The effects of both intraspecific and interspecific competition were expressed primarily in growth and seed production. In annual plants, such as *Avena*, that mature, set seed, and die within a single season, the number of seeds produced is most relevant to long-term population change.

When Marshall and Jain sowed wild oats in single-species populations, the number of spikelets (seed heads) per pot (that is, for the total planting) increased with the number of individuals initially and then leveled off (Figure 22-13a). No matter how many seeds were planted in each pot, the number of spikelets harvested at the end of the experiment never exceeded about 500 to 600 per pot. Thus each plant produced many spikelets per plant in low-density plantings, but fewer as density and intraspecific competition increased (Figure 22-13b).

Marshall and Jain studied the effects of interspecific competition by planting the two species at ratios of 0 : 8, 1 : 7, 4 : 4, 7 : 1, and 8 : 0 at each of the six densities. The results of mixed-species plantings are usually portrayed in a **replacement series diagram,** in which the response of each species is plotted separately as a function of the planting ratio or the number of individuals of the competing species at a particular density. In the case of Marshall and Jain's experiment, the response is number of spikelets. Possible outcomes for different levels of intraspecific and interspecific competition are illustrated for a single species and a single total density in Figure 22-14. When seed production by one species is affected equally by the density of individuals of either species ($a_{ij} = a_{ii} = 1$), the effective density for each individual of species i or j is independent of the planting ratio, and the total seed output of i or j individuals increases in proportion to the

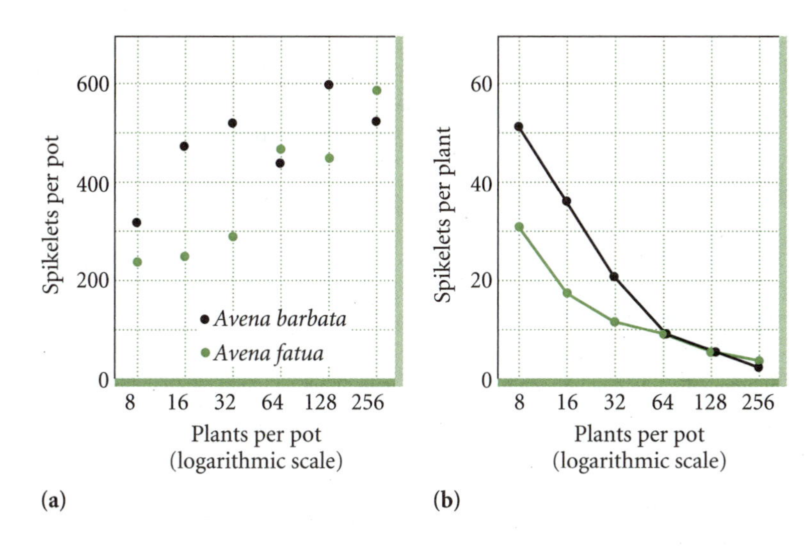

FIGURE 22-13 When pure cultures of two species of oats (*Avena*) were planted at different densities, the total number of spikelets per pot (a measure of reproductive output) increased with density (a), but the number of spikelets per plant decreased (b). (*After Marshall and Jain 1969.*)

number planted. Such a result appears as a straight diagonal line on the replacement series diagram.

When intraspecific competition exceeds interspecific competition ($a_{ii} > a_{ij}$), individuals of species *i* are more productive at lower planting ratios, and the observed output of seeds per pot lies above the diagonal. When interspecific competition exceeds intraspecific competition ($a_{ij} > a_{ii}$), production rises disproportionately as the planting ratio increases (hence as interspecific competition decreases). As a result, the output of seeds lies below the line.

The results of Marshall and Jain's mixed-species plantings at the low and high densities of 32 and 256 plants per pot are compared with those for pure populations (no interspecific competition) in Figure 22-15. The horizontal axis in this figure gives the number of conspecifics in the pot. The two panels on the left side of the graph represent a total density of 32 individuals per pot, and the two panels on the right show the results for a total density of 256 individuals per pot. At the lower density, the numbers of spikelets produced by *A. fatua* lie above the diagonal and do not differ from the numbers produced at the same density in pure populations. Thus at a density of 32 plants per pot, individuals of *A. fatua* are indifferent to the presence of individuals of *A. barbata* and behave as if subjected only to intraspecific competition. *A. barbata*, however, is influenced by competition from *A. fatua*, even at this low density. Because the numbers of spikelets produced by this species lie close to the diagonal, we can infer that competition from individuals of *A. fatua* is of the same order of magnitude as intraspecific competition (that is, $a_{fb} = 0$ and $a_{bf} = 1$). At the highest density, 256 plants per pot, *A. fatua* clearly is influenced by interspecific competition; for *A. barbata*, the effects of interspecific competition exceed those of intraspecific competition.

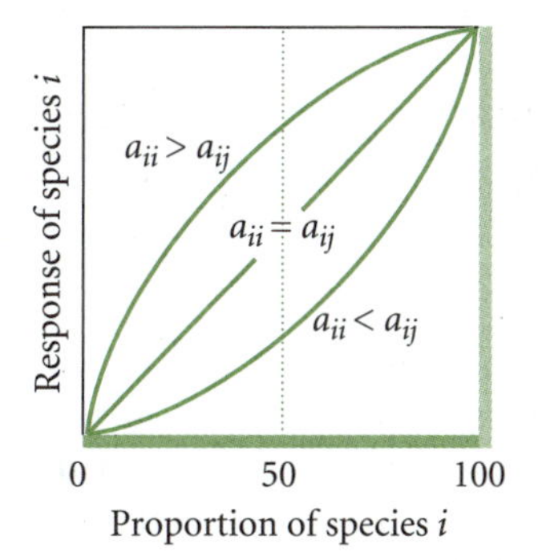

FIGURE 22-14 A replacement series diagram, in which the response of one species (i) is portrayed as a function of the proportion of that species in mixed plantings for different relationships between the coefficients of intraspecific competition (a_{ii}) and interspecific competition (a_{ij}).

Replacement series experiments may not adequately reflect the conditions of plants in nature. Yet the range and control of densities and physical conditions that can be achieved in greenhouses points up the limitations of removal experiments in nature. Whereas one can infer interspecific competition from such field experiments, it is difficult to assess its strength relative to intraspecific competition (Connell 1983) or to investigate the nature of the limiting resource. Sometimes however, the activities of humans or natural disturbances create situations in which a range of densities of competing species may be observed in the field. The introduction of exotic species to natural systems, for example, may set up a natural competition experiment. Such studies are usually of great importance since introduced species often threaten the integrity of the system into which they are introduced (Dyer and Rice 1997).

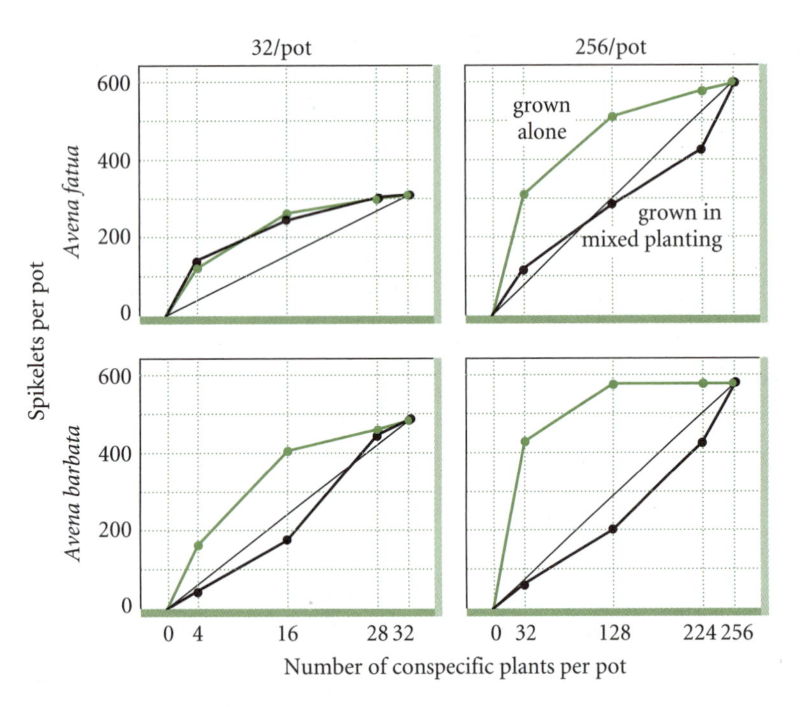

FIGURE 22-15 Production of spikelets by two species of oats grown in the absence (green lines) or presence (black lines) of interspecific competition at different densities. (*From Marshall and Jain 1969.*)

22.8 Experimental studies have revealed interspecific and intraspecific competition among species of animals.

There have been many studies of interspecific competition among animals. Such studies have been conducted both to determine the mechanism of competition and to gain a better appreciation for the processes by which natural communities are structured. Studies of intraspecific competition in animals are usually associated with questions about density-dependent population regulation. There have been relatively few studies that compare the relative importance of intraspecific and interspecific competition.

Interspecific Competition in Animals: Space

Plants occupy space. In dense plantings, roots and leaves crowd together so closely that individuals constantly vie for sunlight, water, and soil nutrients. Among animals, the space-occupying invertebrates of rocky shores most closely resemble plant systems. Among the most prominent of these are the barnacles, which may form dense, continuous populations. Barnacles feed on plankton, which they gather from the water that washes over them. And just as plant production per unit of area (or per pot) is limited by space rather than by the number of plants, the productivity of barnacles is independent of the number of spats (immatures) that settle on a bare surface, above a certain critical density.

One of the first experimental demonstrations of interspecific competition in the field resulted from the work of Connell (1961) on two species of barnacles within the intertidal zone of the rocky coast of Scotland. Connell's experiments rank among the classic experiments in ecology both because of the elegance of their execution and because of the importance of their results. Adults of *Chthamalus stellatus* normally occur on rocks higher in the intertidal zone than those of *Balanus balanoides,* the more northerly of the two species. Although the vertical distributions of newly settled larvae of the two species overlap broadly within the intertidal zone, the line between the vertical distributions of adults is sharply drawn.

Connell demonstrated that adult *Chthamalus* are restricted to the portion of the intertidal zone above *Balanus* not because of physiological tolerance limits, but rather by interspecific competition. When Connell removed *Balanus* from rock surfaces, *Chthamalus* thrived in the lower portions of the intertidal zone where they were normally absent. The two species compete directly for space. The heavier-shelled *Balanus* grow more rapidly than *Chthamalus,* and as individuals expand, the shells of *Balanus* edge underneath those of *Chthamalus* and literally pry them off the rock. *Chthamalus* can occur in the upper parts of the intertidal zone because they are more resistant to desiccation than *Balanus.* So when surfaces in the upper levels are kept free of *Chthamalus, Balanus* do not invade.

Connell's work revealed most of the principles of competition outlined by Tansley and subsequently demonstrated in laboratory populations. As in Park's studies on *Tribolium,* the outcome of the interaction depended on the environment. As in Marshall and Jain's studies on *Avena,* competition between the species was asymmetrical, with *Balanus* exerting the stronger interspecific influence except where it was limited by physical factors. Interspecific competition for space has been demonstrated in a number of other animals. For example, African fish eagles compete with other birds of prey (raptors) for foraging space in Uganda (Krueger 1997).

Interspecific Competition in Animals: Differential Aggression and Resource Use

Competition between barnacles results from physical interference rather than differential exploitation of food or other resources. But even more mobile animals may exhibit similar interference competition through occasional aggressive encounters. For example, two species of voles (small mouselike rodents of the genus *Microtus*) co-occur in some areas of the Rocky Mountain states. In western Montana, the meadow vole inhabits both dry habitats and wetter habitats surrounding ponds and watercourses, whereas the montane vole is restricted to dry habitats. When meadow voles were trapped and removed from an area of wet habitat, montane voles began to move in from surrounding dry habitats (Koplin and Hoffman 1968). Stoecker (1972) obtained the complementary result at another site in Montana. After 9 days of trapping in an area of dry habitat, which initially was occupied solely by the montane vole, meadow voles began to be caught, presumably after having moved in from moister surrounding habitats. Each of the species excludes the other by aggressive behavior or else avoids habitats where the other species is common. Both types of behavior, which occur widely among rodents, tend to sharpen the boundaries between the ecological distributions of closely related species (Grant 1972b, Heller 1971, Meredith 1977, Sheppard 1971).

Tropical hummingbirds offer another example of the relative importance of aggressive behavior, territoriality, and interference in interspecific competition. The principal challenge for hummingbirds is to meet their high energy needs using a relatively low-energy source: floral nectar. In areas rich in hummingbird species, such as Costa Rica and other places in the Neotropics, considerable interspecific variation in foraging strategy has evolved, presumably in response to intense competition for food (Tiebout 1993). The steely-vented hummingbird (*Amazilia saucerrottei*), for example, actively and aggressively defends patches of flowers using aerial combat techniques—behaviors that require considerable energy, and for which the bird is morphologically suited by having high wing loading. Living in the same area is the more mild-mannered fork-tailed emerald hummingbird (*Chlorostilbon canivetii*), a nonterritorial species that exploits undefended isolated patches. These two species compete for food asymmetrically, and it has been shown that the steely-vented hummingbird can have a strong negative effect on the population size of the fork-tailed emerald by excluding it from food sources. Steely-vented hummingbirds also compete intraspecifically at high densities (Feinsinger 1976).

Tiebout (1993) studied the mechanism of competition between these two species by comparing the behavior and energy responses over a 24-hour period of solitary birds of both species with those of conspecific and heterospecific pairs of the two species in laboratory experiments. Tiebout found that energy expenditure (J g^{-1} h^{-1}) increased when birds of either species were paired with either a conspecific or a heterospecific individual, owing to the increase in time spent flying around the food source or engaging in aggressive defense of the resource (Figure 22-16). In heterospecific pairs, the more aggressive steely-vented hummingbirds aggressively excluded fork-tailed emeralds from the food source, and thus were able to compensate for higher energy demands and maintain a constant body mass over the period of the experiment (Figure 22-17).

Although territorial defense and social aggression occur frequently within species, they are less common between species, where competition more typically occurs through exploitation of resources. Because exploitative competition expresses its effects indirectly, through differential survival and reproduction of individuals of different species, its detection may be difficult (Belovsky 1984, Kleeberger 1984, Lenski 1984). Often the best way to detect such competition is to

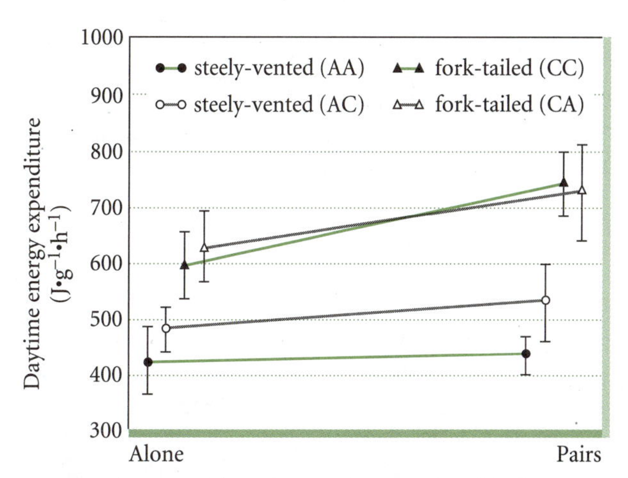

FIGURE 22-16 The amount of energy (J g^{-1} h^{-1}) expended by steely-vented hummingbirds (*Amazilia saucerrottei*) and fork-tailed emerald hummingbirds (*Chlorostilbon canivetii*), feeding on a laboratory sucrose solution either alone or paired with a conspecific (AA, CC) or with a heterospecific (AC, CA) bird. Energy expenditure increased for both species when paired with another bird, but the increase was greatest for fork-tailed emeralds. (*From Tiebout 1993.*)

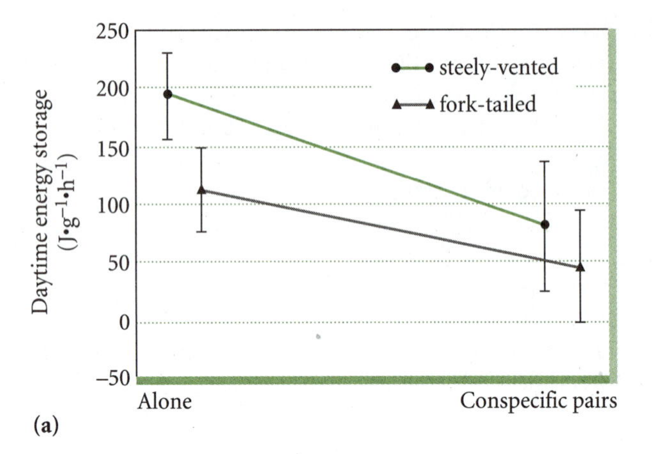

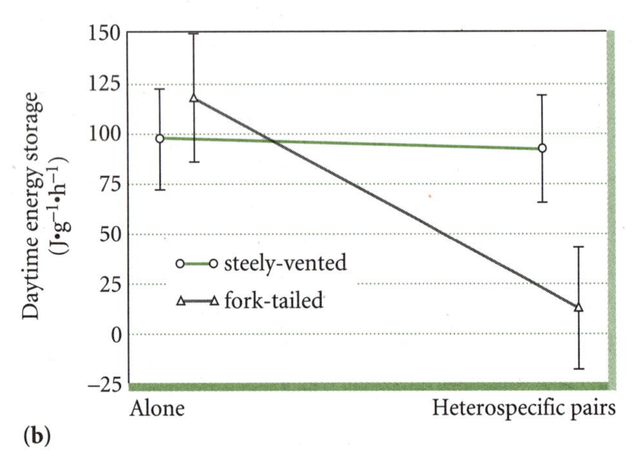

FIGURE 22-17 The amount of energy storage ($J\ g^{-1}\ h^{-1}$) during the daytime of steely-vented hummingbirds (*Amazilia saucerrottei*) and fork-tailed emerald hummingbirds (*Chlorostilbon canivetii*) feeding alone or paired with a conspecific or heterospecific bird. (a) The rate of energy storage is lower for both species when paired with a bird of their own species. (b) The aggressive steely-vented hummingbird excludes the fork-tailed emerald from the feeding site, thereby maintaining its own rate of energy storage at the expense of the fork-tailed emerald. (*From Tiebout 1993.*)

remove one of the component species and observe the response of the remaining species.

Dunham (1980) used this strategy to investigate the interaction between two species of lizards in Big Bend National Park, Texas. The canyon lizard *Sceloporus merriami* and the tree lizard *Urosaurus ornatus* both search for insect prey on exposed surfaces of large rocks. Dunham established experimental plots from which one or the other species was removed, and compared population densities, survival, feeding rates, and body growth in those plots with undisturbed control areas over 4 years from 1974 to 1977.

Population densities in the control and experimental plots are shown in Figure 22-18. Where *Sceloporus* were removed, numbers of *Urosaurus* increased over the controls during 1975 and 1976. There were no differences during 1977. In contrast, where *Urosaurus* were removed, *Sceloporus* populations did not respond. Once again, we find marked asymmetry in the effect of one species on another, with *Sceloporus* obviously being the better competitor. Fluctuations in both control and experimental populations of *Sceloporus* were closely related to rainfall (1975 and 1977 were dry years), suggesting that physical factors may more severely limit *Sceloporus* while competition may more strongly influence *Urosaurus,* a situation recalling the relative ecological positions of *Balanus* and *Chthamalus* in the intertidal zone.

Differences between experimental and control populations of *Urosaurus* additionally suggest that the intensity of competition varies with food resources. For example, feeding rates of adult male *Urosaurus* were significantly higher on experimental than on control plots during a period of low food abundance, but not during a period of high food abundance (Figure 22-19). Hence the competitive effect of a second species may be expressed only during periods of low resource availability (Wiens 1977).

The ultimate measure of competition between species is its effect on survival and fecundity. Recaptures of marked individuals indicated that *Sceloporus*

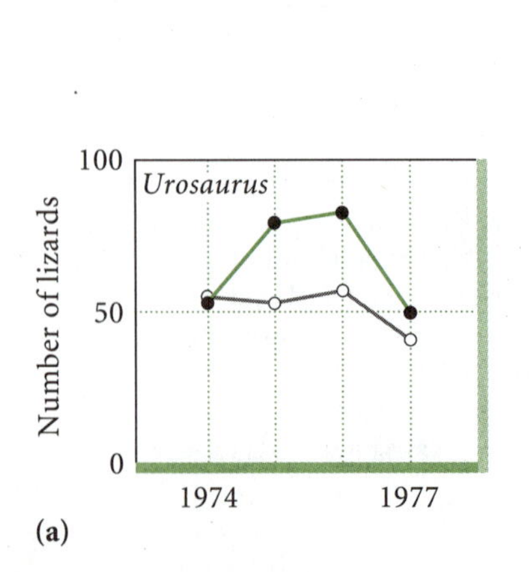

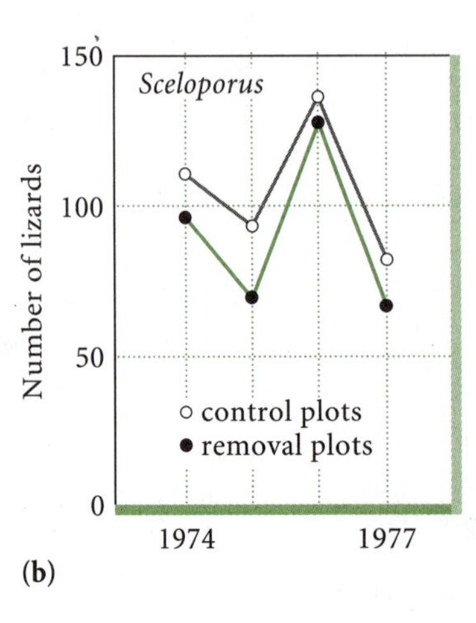

FIGURE 22-18 Populations of two species of lizards (*Urosaurus ornatus* and *Sceloporus merriami*) per hectare on plots from which one or the other of the species was removed, and on control plots where both populations remained. (a) In the absence of *Sceloporus,* the number of *Urosaurus* increased over control plots. (b) *Sceloporus* populations did not respond to the removal of *Urosaurus.* These results suggest that asymmetrical competition exists between the two lizard species. (*From Dunham 1980.*)

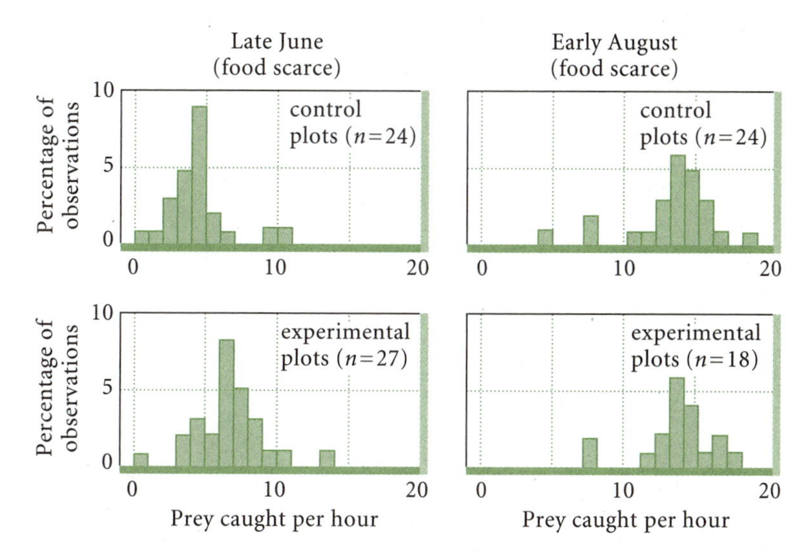

FIGURE 22-19 Feeding rates of adult male *Urosaurus ornatus* on experimental plots from which *Sceloporus merriami* were removed and on unmanipulated control plots. These results suggest that the intensity of the competition between these species is greater when food is limited. (*From Dunham 1980.*)

adults lived longer on the experimental plots, as one might expect as a result of reduced competition, but that the adult life span of *Urosaurus* decreased, perhaps owing to increased intraspecific competition associated with higher densities.

Intraspecific Competition in Animals

Intraspecific competition is no doubt an important factor in the regulation of animal populations. However, the complicating effects of animal movement and predation make it more difficult to study experimentally in animals than in plants. Simple systems, in which no predators exist, and in which resources are limited to just a few, offer the best opportunity to understand the nature of intraspecific competition in animals. The carabid cave beetles (*Neaphaenops tellkampf*) of Mammoth Cave, Kentucky, represent one such system.

Griffith and Poulson (1993) undertook a study of intraspecific competition in these remarkable beetles. Cave beetles have no significant predators or competitors, and they represent the only consumer of their single resource, the eggs of a cricketlike cave-dwelling insect (*Hadenoecus subterraneus*), which they excavate from the moist sand of the cave floor, where they are deposited by the adult crickets. The eggs represent a limiting resource at most times of the year. In addition to the simplicity of the feeding relationships, the internal environment of the cave is so nearly constant that environmental effects on the system are considered negligible.

Based on over 20 years of observations, Griffith and Poulson believed that the beetles competed with one another both by depleting the supply of eggs available to other beetles (exploitation competition) and by actively and aggressively interfering with one another's extraction of eggs from the sand (interference competition). The results of their experiments and observations supported their hypotheses. When beetles were excluded by cages from certain areas of the cave floor, the cricket egg densities there remained high in comparison with similar-sized areas where beetles were allowed to forage. Thus the foraging of beetles had a direct effect on cricket egg abundance. In addition, beetle fecundity (eggs/female) tracked the fecundity of crickets (Figure 22-20), showing that

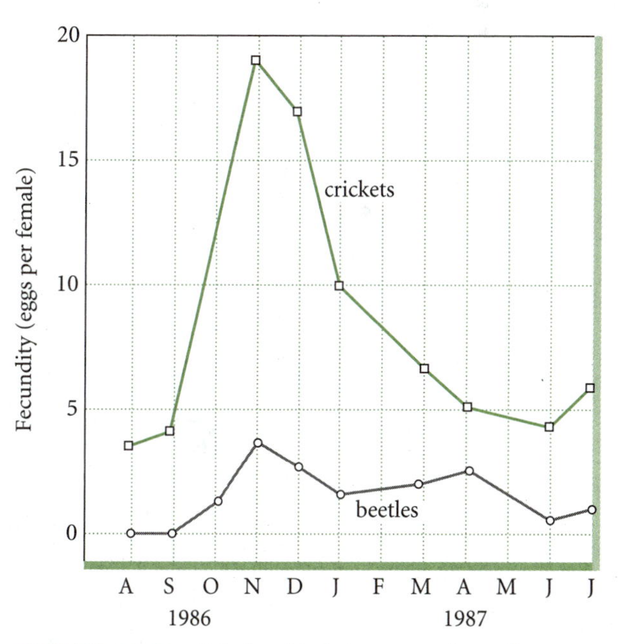

FIGURE 22-20 Relationship between the fecundity (eggs per female) of the cave beetle *Neaphaenops tellkampf* and that of the cricket *Hadenoecus subterraneus.* The beetles feed only on the eggs of the crickets. Beetle fecundity correlates closely with that of the cricket, indicating that beetle population growth is directly related to the availability of the essential limiting resource. (*From Griffith and Poulson 1993.*)

TABLE 22-2 Intraspecific competition by interference in the cave beetle *Neaphaenops tellkampf* (Carabidae) at three different densities

	Beetle numbers per bowl					
	1		2		3	
Responses	Mean	SD	Mean	SD	Mean	SD
Eggs eaten per beetle*	1.00	0.0	0.69	0.53	0.09	0.13
Hole depths (mm)†	11.9[a]	1.54	8.8[b]	0.82	5.5[c]	1.59
Holes per beetle per day	0.50	0.20	0.38	0.18	0.10	0.03
Days to find egg	5.6	3.9	7.9	4.0	10.3	4.5
% time not foraging	25		50		62	

**Significant Kruskal-Wallis.*

†Significant ANOVA. Different superscript letters indicate significant differences between means, by Tukey MSD (unplanned).

(From Griffith and Poulson 1993.)

beetle population growth is directly related to the availability of the limiting resource. In laboratory studies, Griffith and Poulson manipulated the densities of beetles in sand-filled bowls containing a constant number of cricket eggs. They discovered that as beetle density increased, the effectiveness of individual beetle foraging decreased. Beetles in crowded bowls ate fewer eggs and dug fewer and shallower holes. They also took more days to find an egg (Table 22-2). These results suggest that intraspecific competition in the cave beetle is extremely important.

22.9 Higher-order interactions between competitors may alter conditions for coexistence.

Thus far, our discussions of population interactions have focused primarily on interactions between pairs of competing species. Indeed, under the Lotka-Volterra competition models that we presented, all interactions between species are strictly additive and independent of the presence or absence of other species. That is, only the two-species interactions (*ij*, *ik*, *jk*) are considered. But we might expect that in nature, competitive interactions would include more than two species. What is known about such higher-order interactions?

Wilbur (1972) searched for multispecies interactions experimentally by raising immature stages of three species of salamanders (*Ambystoma*) at three densities (0, 32, and 64 individuals per cage) in cages that enclosed portions of natural habitat at the edge of a pond. All three species were raised alone and in combination with each and with both of the other two. Wilbur recorded survival, length of larval period, and mass at metamorphosis. He analyzed these response variables in such a way as to reveal statistical effects due to the density of the population, competition from the other two species, and any interactions involving all three species uniquely.

Wilbur found multispecies interactions to be quite important. In Figure 22-21a, the curves represent the average change in body mass for 64 *A. tremblayi* at three different densities of *A. laterale* and two different densities of *A. maculatum*. The graph shows that the body mass at metamorphosis of *A. tremblayi* is lower whenever there are *A. maculatum* present. It also shows that increasing the density of *A. laterale* has little effect on the body mass at metamorphosis of *A. tremblayi*. The response of *A. laterale* to different densities of competitors shows a more complex pattern (Figure 22-21b). The body size at metamorphosis for that species is dependent on the density of both of its competitors. When there are no *A. maculatum* in the system, the body mass declines as the density of *A. tremblayi* increases, indicating that *A. tremblayi* has a competitive effect. However, when *A. maculatum* is present, the decline in body mass of *A. laterale* occurs only at the highest density (64 individuals) of *A. tremblayi*, suggesting that the competitive effect of *A. tremblayi* on *A. laterale* is somehow affected by competition between *A. tremblayi* and *A. maculatum*. Such complex interactions are probably quite common in nature.

Neill's study (1974) of competition between four species of freshwater crustaceans—*Hyalella*, *Ceriodaphnia*, *Alonella*, and *Simocephalus*—showed directly that one species can influence the interaction between populations of two others. Neill maintained the crustaceans on diets of algae in 1,500-ml laboratory microcosms. He determined the equilibrium

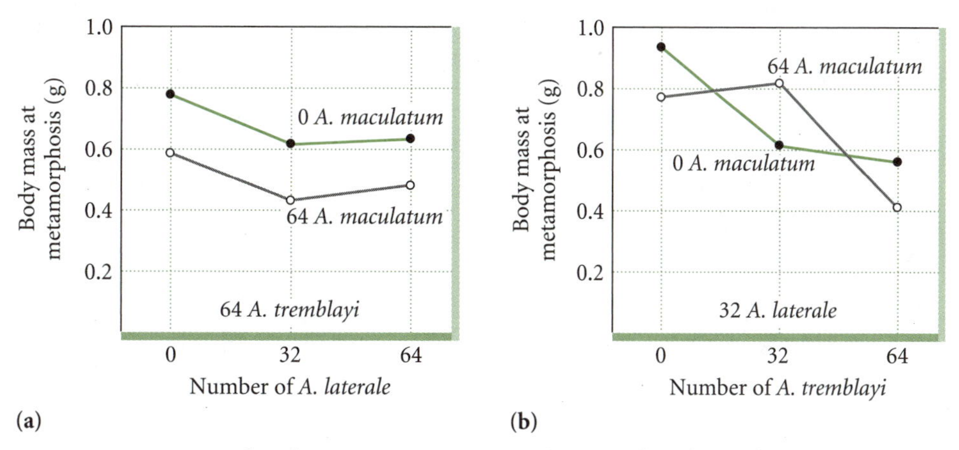

FIGURE 22-21 Results of competition experiment between the salamanders *Ambystoma tremblayi* and *A. laterale* when in the presence of another salamander, *A. maculatum.* (a) Average body mass at metamorphosis for 64 individuals of *A. tremblayi* at each of three different densities of *A. laterale* (0, 32, and 64) and two different densities of *A. maculatum* (green line, 0; gray line, 64). When *A. maculatum* is present, the body mass of *A. tremblayi* is lower at all densities of *A. laterale.* (b) Average body mass at metamorphosis for 32 individuals of *A. laterale* at each of three different densities of *A. tremblayi* (0, 32, 64) and two different densities of *A. maculatum* (green line, 0; gray line, 64). (*From Wilbur 1972.*)

densities of each species after 2 to 4 months in the presence of each of the other three species (two-species interactions), combinations of two of the other species (three-way interactions), and all three other species (the four-way interaction). His experiments revealed asymmetrical coefficients of competition in many cases. For example, *Hyalella* was influenced little by competition from any other species; its numbers did not vary (range 312-331), regardless of which other species it was raised with. In contrast, competition with *Hyalella* greatly reduced equilibrium populations of all three of the other species. So while the coefficients a_{Hi} were all close to zero, the values of a_{iH} were quite large.

Apropos of the question of multispecies interactions, the outcome of competition between some pairs of species was strikingly influenced by a third. When *Ceriodaphnia* (C; $\hat{N} = 294$) was removed from the four-species system, the population of *Alonella* (A) increased by 197 individuals ($614 - 417 = 197$). The competition coefficients were calculated as $a_{AC} = 0.67$ in the presence of *Simocephalus* (S) and *Hyallela* (H) and $a_{AS} = 0.29$ in the presence of C and H. If competition between *Ceriodaphnia* and *Alonella* were independent of competition between *Simocephalus* and *Alonella,* the increase in *Alonella* when both other species were absent would equal the sum of the increases when *Ceriodaphnia* or *Simocephalus* was removed individually; that is, $197 + 14 = 211$. Neill found, however, that removal of both resulted in an increase in *Alonella* of 322 individuals ($739 - 417$). So a_{AC} and a_{AS} were not strictly additive, or linear.

When a relationship $Y = f(X)$ is linear, it may be described by $Y = a + bX$; when it is not linear, its description requires higher-order and multispecies terms ($X^2, X^3, XZ, \ldots$). The outcome of such interactions cannot be predicted by the strictly linear Lotka-Volterra models.

By comparing a_{AC} and a_{AS} in the presence of *Hyallela* (+H) and in its absence (−H), we may judge whether a third species can affect the interaction between two others. Without *Hyalella,* $a_{AC(-H)} = 1.47$, and $a_{AS(-H)} = -0.072$, which probably does not differ significantly from 0. These values do, however, differ from $a_{AC(+H)} = 0.67$ and $a_{AS(+H)} = 0.29$. Hence three-way interactions appear to be important in this system, and it would have been unsound to extrapolate simple competition coefficients obtained in two-species competition to a prediction of the behavior of the more complicated four-species system.

The work presented in the foregoing discussion proceeded under the assumption that responses to density within and between species are linear, a basic assumption of Lotka-Volterra models. That such an assumption oversimplifies nature is widely accepted. Emlen (1984) emphasized this when he suggested that "it is probably optimistic folly to hope for additivity in competition intensity." Pomerantz (1981) and Thomas and Pomerantz (1981) have suggested that apparent multispecies and higher-order interactions in complex systems could result simply from nonlinearities in the response of population growth to density. We will leave the details of nonlinear models to other authors.

22.10 Consumers can affect the outcome of competitive interactions among resource populations.

We shall see in Chapter 24 that grazing can maintain a high diversity of plants in grasslands (Harper 1969, 1977). In the absence of grazers, dominant competitors grow rapidly and exclude others. Similar results have been obtained from experiments on marine algal communities under the pressure of grazing by limpets, snails, and urchins (Lubchenco 1978, Paine and Vadas 1969, Witman 1987). These studies indicate that consumers have a strong hand in shaping the outcome of competition.

The classic study is Paine's work (1966, 1974) on the exposed rocky coast of the state of Washington. Within the intertidal zone, several species of barnacles, gooseneck barnacles, mussels, limpets, and chitons (a kind of grazing mollusk) dominate the habitat; these are preyed upon by the starfish *Pisaster* (Figure 22-22). A study plot, 8 meters in length and 2 meters in vertical extent, was kept free of starfish by physically removing them; an adjacent control area was left undisturbed. Following the removal of the starfish, the

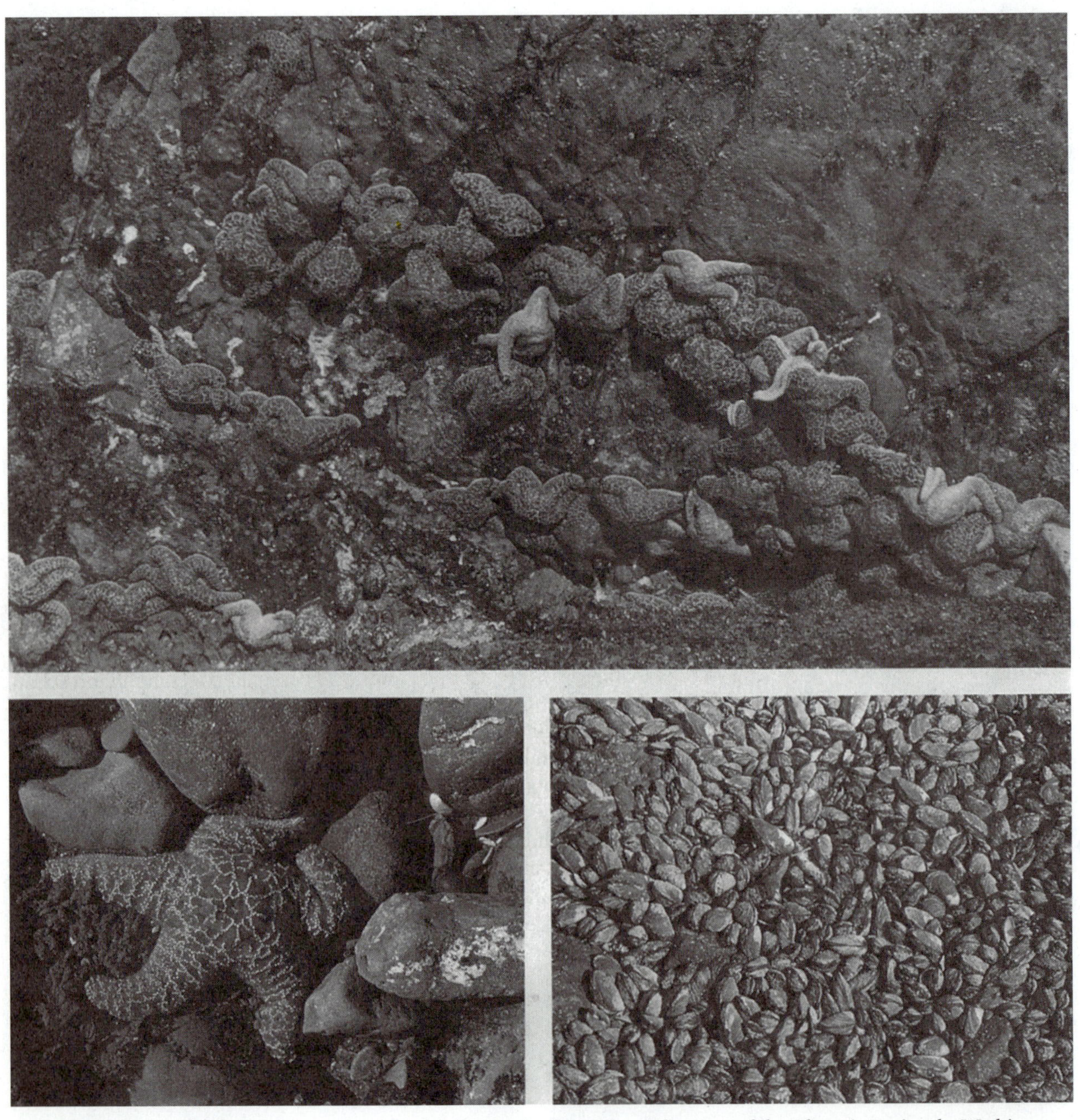

FIGURE 22-22 (*Above*) A congregation of starfish (*Pisaster*) at low tide on the coast of the Olympic Peninsula, Washington. This starfish (*lower left*) is an important predator on mussels (*lower right*).

number of prey species in the experimental plot decreased rapidly, from 15 at the beginning of the study to 8 at the end. This decline in species number occurred when populations of barnacles and mussels increased and crowded out many of the other species. Paine concluded that starfish predation of barnacles and mussels kept those populations low enough to prevent them from crowding out other species, thereby maintaining a higher diversity of prey species in the area. The crown-of-thorns starfish (*Acanthaster*) similarly enhances the diversity of coral reefs near the Pacific coast of Central America by voraciously consuming a species of coral, *Pocillopora,* that would otherwise crowd out many other species (Figure 22-23) (Porter 1972b, 1974).

In experiments with communities established in artificial ponds, P. Morin (1981) showed that predatory salamanders can reverse the outcome of competition among frog and toad tadpoles. Morin seeded each of the ponds with 200 hatchlings of the spadefoot toad (*Scaphiopus holbrooki*), 300 of the spring peeper (*Hyla crucifer*), and 300 of the southern toad (*Bufo terrestris*). To replicates of the ponds he added either 0, 2, 4, or 8 predatory broken-striped newts (*Notophthalmus viridescens*). In the absence of newt predation, *Scaphiopus* tadpoles grew rapidly, survived well, and dominated the ponds, along with smaller numbers of *Bufo; Hyla* tadpoles were all but eliminated (Figure 22-24). *Notophthalmus* apparently prefer toad tadpoles, and as Morin increased the number of predators in the ponds, the survival of both *Scaphiopus* and *Bufo* decreased markedly. With fewer toads per pond, levels of food increased, and the survival and growth of *Hyla* tadpoles improved immensely, as did the growth of surviving *Scaphiopus* and *Bufo* tadpoles.

Sources of mortality other than predation may also enhance coexistence (Witman 1987, Huston 1994). Particularly on rough coasts, physical disturbance by wave action and battering by floating ice and logs haphazardly knock individuals or whole patches of individuals off rocks. The newly opened space becomes available for colonization, often by species that do not compete well. Although these are eventually

FIGURE 22-23 A crown-of-thorns starfish consuming a coral head in Panama. (*Courtesy of J. W. Porter; from Porter 1972b.*)

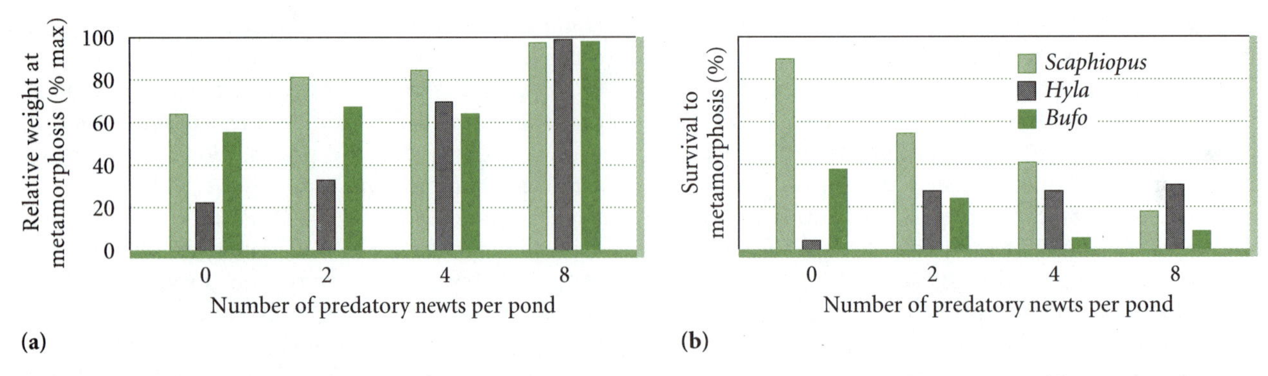

FIGURE 22-24 Effect of predators on (a) weight at metamorphosis and (b) survival in three species of frog and toad tadpoles raised in artificial ponds. (*From Morin 1981.*)

excluded from a particular patch as it becomes overgrown by superior competitors, populations of the poorer competitors may be maintained within a larger area when new patches are continually made available for colonization.

22.11 Competition may lead to evolutionary divergence of competitors.

In Chapter 7 we emphasized the importance of biological interactions as selective forces in nature. Does competition represent such a force? If so, we should find evidence in nature that competitors have partly molded one other's adaptations. We may search for such evidence in the ways in which species utilize the resources in their environment.

Related species that live together in the same place may differ in the ways in which they use the environment (using different food resources, for example). These differences, which are manifested in their morphology and behavior, may arise in one of two ways. First, the behaviors and morphological characteristics that distinguish one species from another may be the result of selection pressures resulting from the interactions among the species—interactions such as competition. An alternative explanation for such differences is that each of the species became adapted to different resources in different places, and when their populations subsequently overlapped as a result of range extensions, these ecological differences remained (see Section 26.6).

One way to distinguish between these alternatives is by looking at the ecology of a species in an area where the competitor is absent. When two species coexist within the same geographic area, they are said to be **sympatric;** when their distributions do not overlap, they are said to be **allopatric.** The terms *sympatry* and *allopatry* also can be applied to different parts of the ranges of species and partially overlapping distributions. Suppose that species 1 occurs in areas A and B, and species 2 occurs in areas B and C. The populations of the two species in area B are sympatric, and the population of species 1 in area A is allopatric with the population of species 2 in area C. If areas A, B, and C all have similar environmental conditions and habitats, and if competition causes divergence, we would expect the sympatric populations of species 1 and 2 in area B to differ more from each other than the allopatric populations of those species in areas A and C (Figure 22-25). This phenomenon is called **character displacement.** Ecologists disagree on the prevalence of character displacement in nature (Arthur 1982, Taper and Case 1985).

A number of examples do seem to fit the pattern of character displacement, one of which involves the ground finches (*Geospiza*) of the Galápagos Islands (see Chapter 14). On islands with more than one finch species, the finches usually have beaks of different sizes, indicating different ranges of preferred food size. For example, on Marchena Island and Pinta Island, the beak size ranges of the three resident species of ground finches do not overlap (Figure 22-26). On

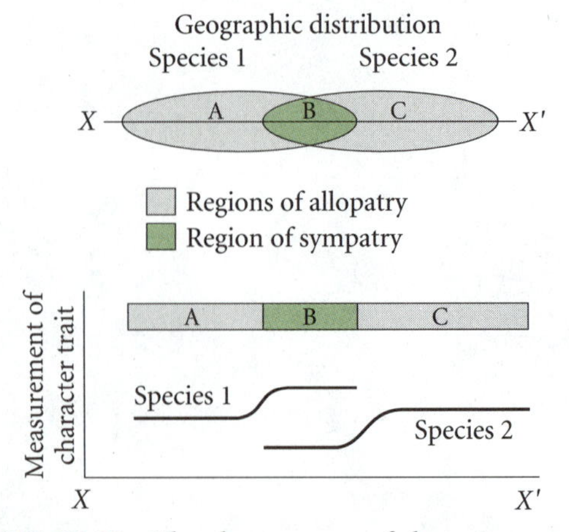

FIGURE 22-25 The phenomenon of character displacement, in which character traits of two closely related species differ more where they occur in sympatry than in allopatric portions of their geographic ranges.

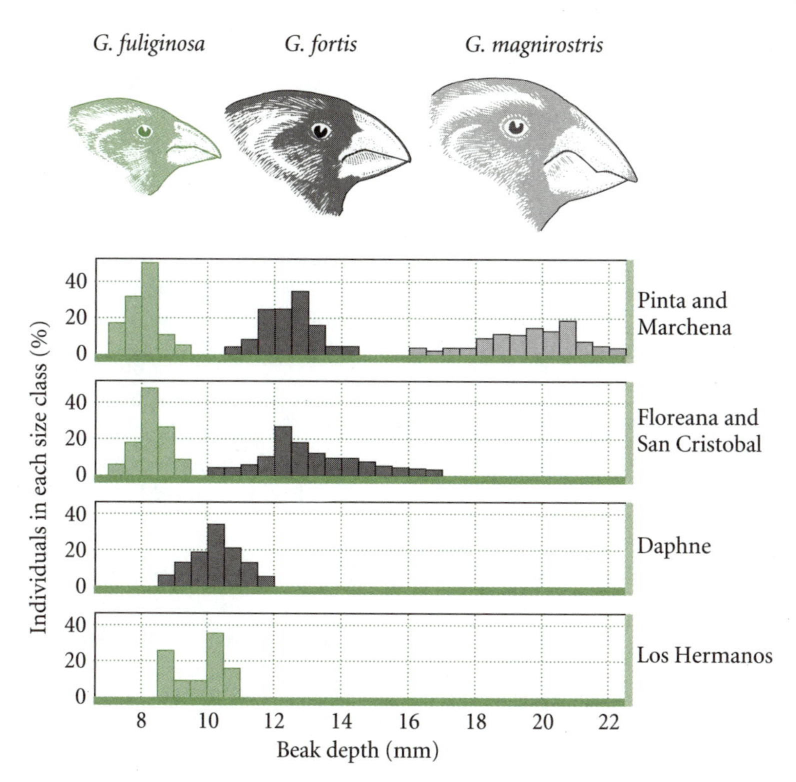

FIGURE 22-26 Proportions of individuals with beaks of different sizes in populations of ground finches (*Geospiza*) on several of the Galápagos Islands. This pattern is a possible example of character displacement. (*After Lack 1947.*)

Floreana and San Cristobal, the two resident species, *G. fuliginosa* and *G. fortis,* have beaks of different sizes. On Daphne Island, however, where *G. fortis* occurs alone, its beak is intermediate in size between those of the two species on Floreana and San Cristobal. On Los Hermanos Island, *G. fuliginosa* occurs alone, and its beak is intermediate in size.

The Galápagos ground finches illustrate the diversifying influence of competition because of the chance distribution of these species on small islands within the archipelago: some islands have two or three species and some only one. In many other cases it is difficult to know whether differences between two species arose because of competition between them or because the species evolved in response to selection by other environmental factors in different places, and then retained their differences when their populations reestablished contact. In most cases, certainly, genetic differences that lead to speciation occur in allopatry. So why not differences that allow two species to avoid strong competition? In either case, coexistence depends on some degree of ecological difference between species, whether it is achieved in allopatry or as an evolutionary consequence of competition in sympatry. In Chapter 25 we shall examine the evolutionary responses of interacting populations more closely.

In this chapter, we have seen that competition occurs in nature and may have dramatic effects on the populations of the competing species. We have also seen that it is a complex process that affects populations differently and is sometimes difficult to detect. We shall revisit these ideas about competition when we discuss the factors that regulate biological diversity in communities in Chapter 29.

SUMMARY

1. Competition is sometimes difficult to observe in nature, and thus careful experiments are required to determine whether it is occurring. Since the early 1960s, ecologists have conducted numerous field studies designed to reveal the influence of competition on the sizes of natural populations.

2. Among the mechanisms by which competition may occur are consumptive, preemptive, overgrowth, chemical, territorial, and encounter competition. The first is usually classified as exploitative competition and the last three as interference competition. Preemptive and overgrowth competition are based upon utilization of space and renewable resources, respectively, but involve close contact between competing individuals. Chemical competition, or allelopathy, has received considerable attention. Allelopathy occurs when one individual uses chemicals to cause harm to another, a phenomenon most widely studied in terrestrial plants.

3. In pest control programs, the disappearance of one introduced parasite following the introduction of a second, similar species provides indirect evidence of competition.

4. Increased growth rates of some species of trees following the removal of other species provide direct evidence of competition for light, water, and nutrients among plants. Transplant experiments with smaller species of plants, creating varying conditions of intraspecific and interspecific competition, often show striking effects of interspecific competition.

5. Competition between plants involves competition between belowground parts, called root competition, and competition between aboveground parts, referred to as shoot competition.

6. The effects of competition may be different for each of the competing populations. This phenomenon is referred to as asymmetrical competition. In plants, root competition is usually symmetrical, whereas shoot competition is asymmetrical.

7. A common method of studying plant competition is the substitution experiment, developed by C. T. de Wit, in which the ratios of two species of plants are varied, but their total density is held constant. The results are portrayed on replacement series diagrams, in which one can visualize the relative strengths of intraspecific and interspecific competition. Experiments with oats (*Avena*) and other plants have demonstrated strong asymmetry in interspecific competition.

8. Removal experiments involving intertidal invertebrates have demonstrated strong competition among such space-filling animals as barnacles, mussels, and encrusting sponges. Competitive exclusion is accomplished by direct physical interaction.

9. Exploitation competition is most convincingly demonstrated in studies that show appropriate changes in resource levels accompanying the demographic response of one species after removal of a competitor.

10. Complex competitive interactions occur in situations in which more than two species compete for resources. In such situations, the outcome of competition between a particular pair of competitors may be different in the presence of a third competitor than when the third competitor is not present.

11. When suites of competing species are preyed upon by a common predator, they may coexist because the activities of the predator keep the population levels of the strongest competitors below the level at which competitive exclusion would occur. In this way, predation acts very much like disturbance.

12. Species living in the same area have sympatric ranges. Species that are geographically isolated have allopatric ranges. Closely related sympatric species may diverge in morphology or behavior, a phenomenon called character displacement. Some have suggested that competitive interactions lead to character displacement.

EXERCISES

1. Make a list of four situations in the natural environment near where you live where interspecific competition may be in effect. Design an experiment to test the presence of competition on one of the situations.

2. If competitive exclusion has occurred in a particular community, then a species that was once present in the community is no longer present there. How would an ecologist know that this has happened?

3. Review the experiments of Resetarits described in Section 22.6. State an appropriate null and alternative hypothesis for the experiments and describe the dependent variable.

4. Consider Figure 22-11 and the material presented in Section 14.2. Explain how the Poisson distribution can be used to model the frequency of different growth rates in a population of competing plants. Assume that there is a zone of depletion around each plant (the circles in Figure 22-13) from which it takes resources. The number of other plants within a plant's zone of depletion is the number of competing neighbors that the plant has. The growth rate of an individual plant is decreased by some amount for every competing neighbor in its zone of depletion.

5. Explain how predation and disturbance can affect competitive interactions.

CHAPTER 23

Predation

GUIDING QUESTIONS

- How can predators limit the populations of their prey?
- How might the interactions of predator and prey result in oscillations of the population sizes of both?
- How do the parasitoid-host models of Nicholson and Bailey differ from the Lotka-Volterra predator-prey model?
- What are the three types of predator functional responses, and how do they affect prey populations?
- What biological conditions might lead to stability in predator-prey interactions?
- What are the implications of a predator-prey system having two equilibrium points for agricultural pest control?
- How does prey behavior affect the outcome of predator-prey interactions?

In Chapter 7 we described the adaptations of predators and prey to show how evolution works to shape both the body forms and behaviors of organisms. Most of that discussion focused on the interactions between individual predators and prey and on the immediate outcomes of those interactions: death or survival in the case of the prey and successful or unsuccessful capture in the case of the predator. In this chapter, we expand our view of predator-prey interactions by considering how the outcomes of the many encounters between individual predators and individual prey in an area collectively affect the population dynamics of both predator and prey. What factors influence the size and stability of populations? This is one of the most fundamental questions of ecology. We saw in Chapters 16 and 18 that density-dependent factors and time delays modify the responses of birth and death rates to population density and result in population regulation. Species are often both consumers and resources for other consumers, and populations of organisms might be limited both by what they eat and by what eats them.

The study of predator-prey interactions attempts to answer at least two important questions. First, do predators reduce the size of their prey populations below their carrying capacity? Second, do the dynamics of predator-prey interactions cause populations to oscillate? The first question is of great practical concern both to those interested in the management of crop pests, in which predators may be of great benefit in pest control, and to those who manage populations of game and endangered species, on which predators may have a strong negative effect. It also has far-reaching implications for our understanding of the interactions among species that share resources and, therefore, for our understanding of the structure of biological communities. The second question is motivated by observations of predator-prey cycles in nature (see Chapter 18) and directly addresses the issue of stability in natural systems. Ecologists have tried to answer these questions with a combination of observation, theory, and experimentation.

In this chapter, we shall consider both predator-prey and parasitoid-host systems. Although predation and parasitoidism are distinct processes, many of their characteristics can be modeled using the same assumptions. In the first part of this chapter we shall describe some empirical studies demonstrating that predators may limit prey populations, and that some predator and prey populations vary in what appear to be linked cycles or oscillations. Using these empirical studies as background, we shall develop a body of simple theory predicting the oscillations of predator and prey populations, and in the process introduce some mathematical and graphic techniques that will be used in later chapters. We will expand and elaborate on these models in the remainder of the chapter to explore how predator behavior, habitat heterogeneity, and the spatial arrangement of predator and prey populations affect population interactions.

Predation differs from competition in that it is always antagonistic at the level of the individual participants. The act of predation always results in death for the prey. Competition and predation—as well as herbivory and parasitism, which will be covered in the next chapter—are consumer-resource interactions. However, predation differs from competition in that the resource involved in the interaction is one of the participants in that interaction. Because of this, the population effects of predator-prey interactions can be represented in relatively straightforward ways with mathematical models that account for the loss and gain of individuals in the interaction, as we shall see.

23.1 Predators and parasitoids may effectively limit prey populations.

A good place to begin in our discussion of predator-prey interactions is to ask what evidence there is that predators affect the populations of their prey. This is a particularly important question in agriculture, in which attempts have been made since the early 1900s to use predators, parasites, and parasitoids to control populations of pest insects that eat crops. A number of examples will demonstrate that, in some situations, predators may effectively limit prey populations.

Mite-Mite Interactions on Strawberries

The cyclamen mite (*Tarsonemus pallidus*) is a pest of strawberry crops in California. Populations of these mites are usually kept under control by a species of predatory mite of the genus *Typhlodromus.* Cyclamen mites typically invade a strawberry crop shortly after it is planted, but their populations do not reach damaging levels until the second year. Predatory mites usually invade fields during the second year, rapidly subdue the cyclamen mite populations, and keep them from reaching damaging levels a second time.

Greenhouse experiments have demonstrated the role of predation in keeping the cyclamen mites in check (Huffaker and Kennett 1956). One group of strawberry plants was stocked with both predator and prey mites; a second group was kept predator-free by regular applications of parathion, an insecticide that kills the predatory species but does not affect the cyclamen mite. Throughout the study, populations of cyclamen mites remained low in plots shared with *Typhlodromus,* but their infestations attained damaging proportions on predator-free plants (Figure 23-1). In field plantings of strawberries, the cyclamen mites also reached damaging levels where predators were eliminated by parathion, but they were effectively controlled in untreated plots (a good example of an insecticide having the wrong effect). When cyclamen mite populations began to increase in an untreated planting, the predator populations quickly responded to reduce the outbreak. On average, cyclamen mites were about twenty-five times more abundant in the absence of predators than in their presence.

Typhlodromus owes its effectiveness as a predator to several factors in addition to its voracious appetite (Huffaker and Kennett 1969). Its population can increase as rapidly as that of its prey. Both species reproduce parthenogenetically. Female cyclamen mites lay

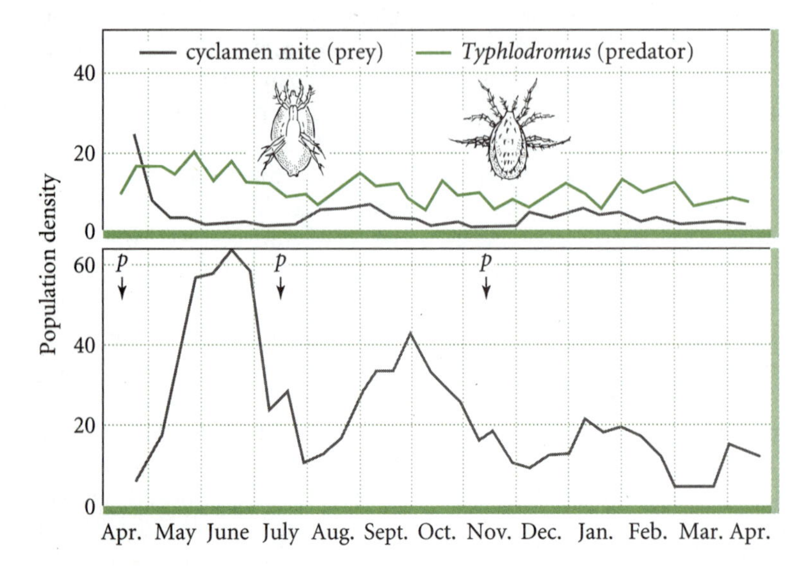

FIGURE 23-1 Infestation of strawberry plots by cyclamen mites (*Tarsonemus pallidus*) in the presence of the predatory mite *Typhlodromus* (a) and in its absence (b). Prey populations are expressed as numbers of mites per leaf; predator levels are the number of leaflets in 36 on which one or more *Typhlodromus* was found. Parathion treatments are indicated by the arrows. The figure shows that the predatory mite exerted better control of the pest mite population than the use of pesticides. (*After Huffaker and Kennett 1956.*)

three eggs per day over the 4 or 5 days of their reproductive life span; female *Typhlodromus* lay two or three eggs per day for 8 to 10 days. Seasonal synchrony of *Typhlodromus* reproduction with the growth of prey populations, ability to survive at low prey densities, and strong dispersal ability also contribute to the predatory efficiency of *Typhlodromus.* During winter, when cyclamen mite populations dwindle to a few individuals hidden in the crevices and folds of leaves in the crown of the strawberry plants, the predatory mites subsist on the honeydew produced by aphids and whiteflies. They do not reproduce, however, except when they are feeding on other mites. Whenever predators appear to control prey populations, one usually finds that the predators have a high reproductive capacity compared with that of the prey. Effective predators also have strong dispersal powers and the ability to switch to alternative food resources when their primary prey are unavailable.

Avian Control of the Spruce Budworm

Torgerson and Campbell (1982) studied the effect of avian predation on the western spruce budworm (*Choristoneura occidentalis*), a moth whose caterpillar larvae (called budworms) cause extensive damage to spruce trees during outbreak years. Birds eat western spruce budworm caterpillars. The researchers constructed wire cages around branches infested by caterpillars in order to exclude bird predators. Inside the cages, the caterpillars grew and eventually formed pupae from which adult moths would emerge. Torgerson and Campbell compared the numbers of pupating moth larvae on the caged branches with those on uncaged control branches a month later. At two out of three experimental sites, densities of pupae on caged branches were about three times those on control branches (5.1 versus 1.6 per 100 current-season shoots). These results show that bird predators may have a dramatic effect on the population of their prey, the western spruce budworm, by decreasing the number of larvae that survive to pupate.

23.2 Predation may establish coupled oscillations of predator and prey populations.

Populations of some predators and prey vary in what appear to be closely linked cycles, as we have seen in the case of the snowshoe hare and its predator, the lynx (see Chapter 18). Because such cycles persist, they appear to represent a stable interaction between predator and prey. One of the earliest goals of population biologists was to establish such cycles in experimental populations so that the dynamics of the relationship could be examined. We can examine the dynamics of such cycles by looking at a parasitoid-host system. As we saw in Chapter 7, parasitoids place their eggs in a living host. As the larvae develop, they consume the host from the inside out, thereby killing it. Thus, with respect to the prey population, the effect of a parasitoid-host interaction is the same as that of a predator-prey interaction: individual prey are removed from the population.

When azuki bean weevils (*Callosobruchus chinensis*) are maintained in laboratory cultures with *Heterospilus,* a parasitoid braconid wasp, the populations of parasitoid and host fluctuate out of phase with each other in regular cycles (Figure 23-2). Introduced to a population of weevils ultimately limited by a constant ration of seeds, the wasps rapidly increase in number. As their population grows, parasitoidism becomes a major source of mortality for the weevils. When deaths resulting from the activities of the parasitoid exceed the reproductive capacity of the weevil population, the number of weevils begins to decline, but because *Heterospilus* is an efficient parasitoid, it continues to prey heavily on the weevils, even as their population is reduced. Eventually, the weevils are nearly exterminated, and then the parasitoid population, lacking adequate food, decreases rapidly. The wasp is not so efficient, however, that all weevil larvae are attacked; hence a small but persistent reserve of weevils always remains to initiate a new cycle of host population growth after the parasitoid wasps have become scarce.

Extremely efficient predators often eat their prey populations to extinction, and then follow suit. This hopeless situation can be stabilized, however, if some of the prey—or, as in the case above, hosts—can find refuge from the predators or parasitoids. Gause (1934) demonstrated this principle in some of the earliest experimental studies on predator-prey systems. He employed *Paramecium* as prey and another ciliated protozoan, *Didinium,* as predator. In one experiment, predator and prey individuals were introduced to a nutritive medium in a plain test tube. By creating such

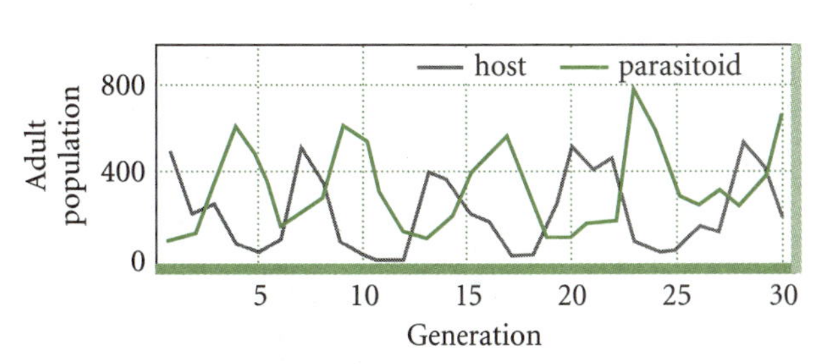

FIGURE 23-2 Population fluctuations of the azuki bean weevil (*Callosobruchus chinensis*) (host; gray line) and its braconid wasp parasitoid (green line), *Heterospilus,* in a laboratory culture. The two populations fluctuate slightly out of phase. (*After Utida 1957.*)

a simple environment, Gause had stacked the deck against the prey; the predators readily found all of them. When the last *Paramecium* had been consumed, the predators starved. In a second experiment, Gause added some structure to the environment by placing glass wool, in which the *Paramecium* could escape predation, at the bottom of the test tube. The tables having thus been turned, the *Didinium* population starved after consuming all the readily available prey, but the *Paramecium* population was restored by individuals concealed in the glass wool. In both these experiments, no cycle of predator and prey populations was achieved. In the first case, both predator and prey disappeared. In the second, because the presence of refuge allowed some of the *Paramecium* to avoid predation, the prey population persisted after all of the predators had starved. In order to achieve a predator-prey cycle in this simple system, Gause had to periodically restock the cultures with small numbers of predators.

Gause's experiment provides a number of clues about the dynamics of predator-prey cycles in nature. First, the experiments demonstrate the potential for predators to reduce their prey populations to extinction. (Here, the term *extinction* is used to refer to the demise of a local population. In Chapter 19 we described different meanings of the term.) Second, habitat structure may alter the outcome of predator-prey interactions. In the case of the *Paramecium-Didinium* system, the presence of a refuge prevented the predator (*Paramecium*) from completely eliminating the prey (*Didinium*). In that case, it was the predator population that was eliminated from the system. We shall see below that the spatial arrangement of habitat may also affect the outcome of predator-prey interactions. You have probably already considered the possibility that in more complex systems—say, with one predator and two acceptable prey species—the predator might survive the decline of one population of prey by switching to the other. We shall discuss that situation shortly. Third, the maintenance of predator-prey cycles may require interactions on a landscape level. Gause achieved a predator-prey cycle by periodically introducing small numbers of predators from outside the system. Presently, we shall show how we can use the metapopulation paradigm to conceptualize such a natural system.

23.3 Simple predator and prey models predict oscillations in population size.

The examples presented above show that under some conditions, predators can limit the population size of their prey, and that the interaction of predator and prey may result in oscillations in population size for both populations. Empirical studies such as these provide essential information regarding the interactions of predators and prey. However, because each predator-prey system is unique, experimental results are often not generalizable, and thus, mathematical models are used. Mathematical models of predator-prey interactions have been well developed. We begin our discussion with the simple models first presented by Lotka (1925) and Volterra (1926). These models provide a foundation for understanding how predator and prey populations sometimes come to oscillate with respect to one another.

It is common practice in predator-prey models to designate the predator population size with the uppercase letter P and the prey population size with the letter H. (If you think of the predator-prey interaction taking place low in the food web, you can think of H as standing for herbivore.) Lotka and Volterra expressed the rate of growth of both predator and prey populations by differential equations having the form $dH/dt = f(H,P)$ for the prey and $dP/dt = g(H,P)$ for the predator, where f and g designate arbitrary functions of the variables H and P. Thus, the rate of growth of the prey population, dH/dt, is some joint function, f, of the size of the prey population (H) and the size of the predator population (P). You will recognize that the general form of these models is the same as that of the population growth models that we discussed in Chapters 15 and 16, in which we represented the change in population size, dN/dt, as some function, f, of N, $f(N)$. The Lotka-Volterra predator-prey models presented here are a straightforward elaboration of this idea. What we will include in the functions $f(H,P)$ and $g(H,P)$ are terms that account for how consuming and being consumed add and remove individuals from the populations of predator and prey.

We wish to emphasize again that because these models represent changes in populations in units of individual organisms, they may not be appropriate for population interactions other than predator-prey interactions. The interactions of herbivores and parasites with their hosts, in which individual hosts may not die from the interaction, require different modeling approaches, as we shall see.

Prey Population Growth Rate

Let us begin by describing algebraically and graphically how a prey population might grow in the presence of a predator. We know from Chapter 15 that if the prey population experienced no mortality (either from predation or from other causes, such as disease or old age), it would grow at an exponential rate, which can be represented as rH, where r is the intrinsic rate of increase of the prey population. Mortality will curb this exponential growth, as we showed in Chapter 16. For the moment, let us assume that the prey population experiences only one kind of mortality, that inflicted by the predator (that is, we assume that a prey organism lives forever if it is not eaten). Using these simple assumptions, we can write an equation for the instantaneous growth rate of the prey:

$$\frac{dH}{dt} = rH - \text{(mortality inflicted by predator)}.$$

What is a reasonable algebraic representation of that mortality?

Lotka and Volterra both assumed that the consumer (predator) and resource (prey) populations interact with each other by the law of mass action—that is, that predation varies in direct proportion to the product of the prey and predator populations, HP. The idea is that as the numbers of predators and prey increase, so does the likelihood that a predator will encounter a prey. Some fraction of those encounters will result in a kill by the predator and the removal of an individual from the prey population. If we let p be the proportion of the encounters that result in a kill ($0 < p < 1$)—the predator efficiency—we can complete the equation:

$$\frac{dH}{dt} = rH - pHP. \quad (23\text{-}1)$$

The term HP gets larger whenever the population size of the prey or the predator increases, assuming no commensurate decrease in the other population.

The dynamics of this equation are shown in Figure 23-3a, which depicts the instantaneous rate of change of the prey population, dH/dt, as the predator population increases for three different prey population levels, $H = 10, 30, 50$, and parameter values $p = 0.01$ and $r = 1.2$. (The figure does not show population trends, only prey growth rates under different levels of P and H.) The intersection of each line with the vertical axis ($P = 0$) gives the growth rate of the prey population in the absence of the predator. For example, for the prey population size $H = 10$, the growth rate is $rH = 1.2(10) = 12$, which is point A in Figure 23-3a. When $H = 50$, the growth rate will be 60, which is point C.

In the presence of one or more predators, the prey growth rate declines with rate $-pH$ as the predator population increases. The graph shows that the number of predators, P, for which the prey population growth rate $dH/dt = 0$—the equilibrium value of $\hat{P}$—is the same for all prey population sizes. This can be shown analytically by setting dH/dt equal to zero and solving for P, as we have done in the past. So, $dH/dt = 0 = rH - pHP$ is solved for P to obtain

$$\hat{P} = \frac{r}{p}.$$

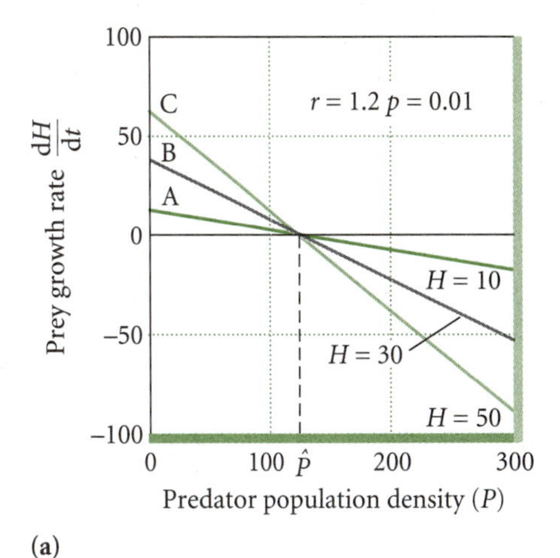

(a)

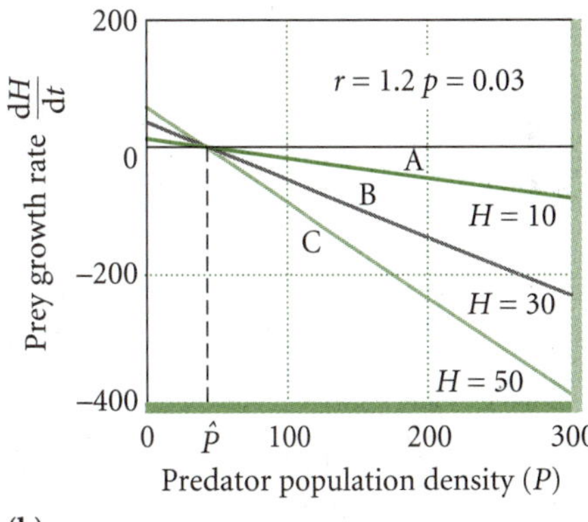

(b)

FIGURE 23-3 (a) Instantaneous rate of change of a prey population, dH/dt (vertical axis), with increasing predator density (horizontal axis) for three different prey densities (dark green line, $H = 10$; gray line, $H = 30$; light green line, $H = 50$) and parameter values $p = 0.01$, $r = 1.2$. The intersection of each line with the vertical axis denotes the growth rate of the prey in the absence of predators, rH (A = 12; B = 36; C = 60). $\hat{P} = r/p = 1.2/0.01 = 120$. (b) Same relationship as in (a), except $p = 0.03$. Note that increasing the predator efficiency, p, decreases the equilibrium value, $\hat{P}$.

As before, we use the "hat" to denote the equilibrium value. This formula verifies what the graph shows, that $\hat{P}$ is a constant. For any value of H in the current example, $\hat{P} = 1.2/.01 = 120$. The effect of increasing the predator efficiency is shown in Figure 23-3b, where $p = 0.03$. The increase in efficiency decreases the level of the predator population at which the prey growth rate equals zero from 120 to $\hat{P} = 1.2/0.03 = 40$.

One aspect of this model of prey population growth bears emphasis. You will notice that the higher the prey population, the greater the rate of decrease in the population as the number of predators increases (compare the line for $H = 50$ [greatest negative slope] with the line for $H = 10$ [smallest negative slope]). This pattern is an artifact of the principal of mass action in the term pHP. For a given prey population size, pH is a constant set by the size of that population and the efficiency of predation. So as prey become more numerous, the entire term gets larger, and thus the relationship $rH - pHP$ gets smaller. This is, of course, obvious from an algebraic standpoint, but what are the biological implications of the model, and are they realistic? Two issues come to mind. First, the model shows that the size of the prey population is important in the predator-prey interaction. Compare what happens to the three populations represented by the lines in Figure 23-3a when $P = 50$. At that predator level the smallest prey population has the smallest growth rate and the largest population the largest growth rate. Thus, in this model, two prey populations of different sizes that are subjected to the same predation pressure (predator population size) will both have their growth rates reduced, but the larger population will retain a higher growth rate. This is, of course, not true when the predator population size is greater than the equilibrium size ($P > \hat{P}$).

The second issue has to do with the biological implications of greater predator population size. The model says that increasing the number of predators has a proportional discounting effect on prey growth rate—that is, that nothing happens when more predators are added other than consumption of more prey. You can appreciate the weakness of this assumption by thinking about what happens when you add another cat to the yard where your cat is already hunting. The model tells us that the two cats should go happily about eating mice and, because there are two of them now gobbling up mice, the rate of growth of the mouse population will decline. But your understanding of how cats behave no doubt suggests that this assumption is far too simple. When a second cat is added to the yard, it is far more likely that neither cat will eat mice because they will be preoccupied with each other. In Section 23.4 we shall see how two ecologists developed a predator-prey model that addressed this issue in part.

Predator Population Growth Rate

Let us turn our attention to the predator half of the predator-prey interaction. How does the predator population grow in the presence of prey? We know that predators hunt to obtain nourishment for growth and reproduction. So, let us assume at the outset that the birth rate of the predator is related positively to the number of prey captured. Recall from the discussion above that the number of prey captured may be represented as pHP, where p is the efficiency with which prey are captured. The physiological mechanisms of the predator for converting the energy obtained from eating prey into reproduction are not 100% efficient. So, if $p = 0.03$ and there are $H = 1{,}000$ prey and $P = 10$ predators, the predator population will take $(0.03)(1{,}000)(10) = 300$ prey according to the model. But the predators will not be able to convert these 300 prey into 300 baby predators. We expect that some fraction, a, of the energy contained in those prey will be converted to young ($0 < a < 1$) and, thus, that the predator population will increase by $apHP$. We can now write the following equation for the instantaneous rate of change of the predator population:

$$\frac{dP}{dt} = apHP - (\text{mortality of predators}).$$

For simplicity, let us assume that predators die at a constant rate, dP, that is independent of the number of prey in the population. Such mortality might be the result of weather, for example. This gives us a complete equation for the growth rate of a predator population in terms of a prey population,

$$\frac{dP}{dt} = apHP - dP. \qquad (23\text{-}2)$$

Graphs of this equation bear a striking resemblance to graphs of the prey population dynamic (equation [23-1]), except that the lines trend upward rather than downward (Figure 23-4). The reason for the change in direction is that growth rate of the predator population is hindered by the *absence* of prey, whereas prey populations are negatively affected by the *presence* of predators. Thus, in the prey graphs (see Figure 23-3), negative growth rates occur when ($P > \hat{P}$), at high predator populations, but in the predator graphs, negative growth rates occur when $H < \hat{H}$, or in situations of low prey density. As with the prey graphs, the equilibrium point in the predator equation is a constant,

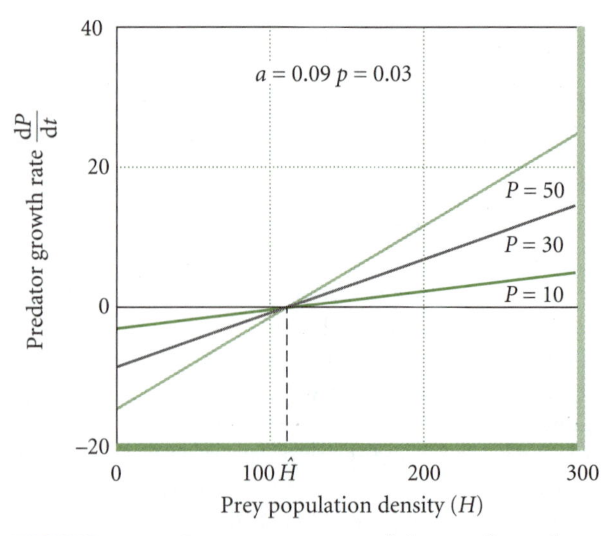

FIGURE 23-4 Instantaneous rate of change of a predator population, dP/dt, with increasing prey density for three different predator densities (dark green line, $P = 10$; gray line, $P = 30$; light green line, $P = 50$) and parameter values $a = 0.09, p = 0.03, d = 0.3$. $\hat{H} = d/ap = 111$.

which we obtain by setting equation (23-2) equal to zero and solving for H to obtain

$$\hat{H} = \frac{d}{ap}.$$

In Figure 23-4, where $a = 0.09$, $p = 0.03$, and $d = 0.3$, the level of prey at which the predator population growth rate equals zero is approximately $\hat{H} = 111$. So long as the prey population is greater than $\hat{H}$, the predator population will have a positive growth rate.

Joint Predator-Prey Population Growth

In the graphs presented thus far, we placed the rate of change of the population, either dH/dt or dP/dt, on the vertical axis and the population density of H or P on the horizontal axis. This approach gives us a two-dimensional way of examining how the rate of change behaves under certain conditions of predator or prey population size. But, the equations for dH/dt and dP/dt each have both H and P in them. Thus, what we would like to visualize is how the predator growth rate, dP/dt, behaves for all combinations of H and P, and likewise, how dH/dt changes for all values of predator and prey population size. A generalized picture of what we want is given in Figure 23-5a. The figure is composed of a "floor" that is defined by axes of predator and prey population size. Projecting through the floor is an axis representing either the predator or the prey population growth rate. For now, we need not bother with the actual value of the growth rate. Instead, we will focus on whether the rate is positive, lying in the space above the floor of the graph, or negative, in the volume below the floor. We can refer to Figures 23-3 and 23-4 to determine what happens at each of four points (A, B, C, D) on the floor of the figure. At point A, both the prey population and the predator population values are below the equilibrium values. That is, $H < \hat{H}$ and $P < \hat{P}$. We know from Figure 23-3 that when the number of predators is less than $\hat{P}$, the prey population will have a positive growth rate, which is shown as an upward-directed arrow from point A parallel to the vertical axis. We know from Figure 23-4 that when the number of prey is below H, the predator population will have a negative growth rate, hence the arrow directed into the negative space below the floor of the graph at point A. The directions of the prey and predator growth rates are reversed at point C because at that point there are more than $\hat{H}$ prey, thereby giving the predator population a positive growth rate, and more than $\hat{P}$ predators, giving the prey population a negative growth rate. Using the same reasoning, we can see why both the predator and prey populations have positive growth rates at point B and both have negative growth rates at point D.

When the growth rate of the population is positive (upward-pointing arrow), individuals are being added to the population. Thus, if a prey population is greater than $\hat{H}$, the predator population will be increasing. Likewise, if the predator population is

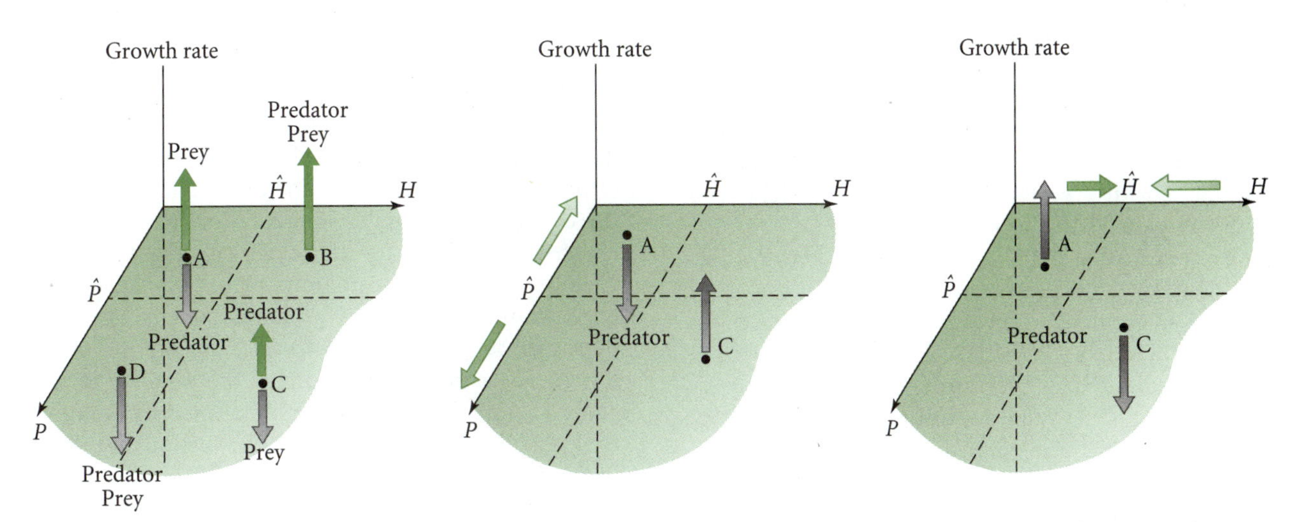

FIGURE 23-5 Three-dimensional representations of predator or prey growth rate (vertical axis) for combinations of predator population size (P) and prey population size (H). Arrows pointing above the plane defined by the P and H axes denote conditions of predator and prey population size for which growth rate is positive. Arrows pointing below the plane denote negative growth rate. (a) At point A, $H < \hat{H}$, so the predator population has a negative growth rate (downward-directed arrow), and $P < \hat{P}$, so the prey population has a positive growth rate (upward-directed arrow). The directions of the arrows at points B, C, and D reflect the relationship of the population size to the equilibrium value. (b) Similar graphs showing the trend in predator population growth when growth rate is negative (point A) and positive (point C). A positive growth rate implies that the individuals are being added to the population (dark green arrow parallel to the P-axis). A negative growth rate implies that individuals are being lost from the population (light green arrow parallel to the P-axis). (c) Same as (b), but showing the trends for the prey population.

less than $\hat{P}$, the prey population will have a positive growth rate and will be increasing. In Figure 23-5b, this means that when the predator arrow points upward (point C), the predator population trends along the P-axis in a direction away from the origin of the graph, as shown by the dark green arrow parallel to the P-axis. When the predator arrow is pointing downward (negative growth rate; point A), the predator population will be losing individuals, and the population will tend to move along the P-axis toward the origin, as shown by the light green arrow. A similar graph is shown in Figure 23-5c for the prey population. Thus, we do not need to know precisely what the value of the growth rate is in order to understand the dynamics of this simple predator-prey model. All we need to know are the sizes of the predator and prey populations relative to $\hat{H}$ and $\hat{P}$. Graphically, this means that we have only to look at the floor of Figure 23-5, the two-dimensional space formed by the H and P axes, which we show in Figure 23-6. The equilibrium population values of the predator ($\hat{P}$) and prey ($\hat{H}$) partition the graph into four regions. The line $\hat{P} = r/p$, representing the condition $dH/dt = 0$, is called the **equilibrium isocline of the prey.** For any combination of predator and prey numbers that lies in the region below the line, the prey increase because there are few predators to eat them. The **equilibrium isocline of the predator** ($dP/dt = 0$) defines the minimum level of prey ($\hat{H} = d/ap$) that can sustain the growth of the predator population. When H is to the left of this line, the predator population will decline.

At point A in Figure 23-6, the prey growth rate is positive (because there are fewer predators than $\hat{P}$). That is, it has an upward projecting arrow in Figure 23-5. Because of this, the prey population is moving along the horizontal axis from left to right, which we represent with a vector projecting to the right from point A parallel to the H-axis. The predator growth rate is negative at point A (too few prey), and is thus represented by a downward-projecting arrow from point A in Figure 23-5. Because of this, in Figure 23-6 we draw a vector projecting downward from A and parallel to the P-axis. The result of the negative predator growth rate and the positive prey growth rate is a general movement downward and to the right, which is shown by the black resultant vector. Similar reasoning reveals the general trends in each of the other three quadrants in the space.

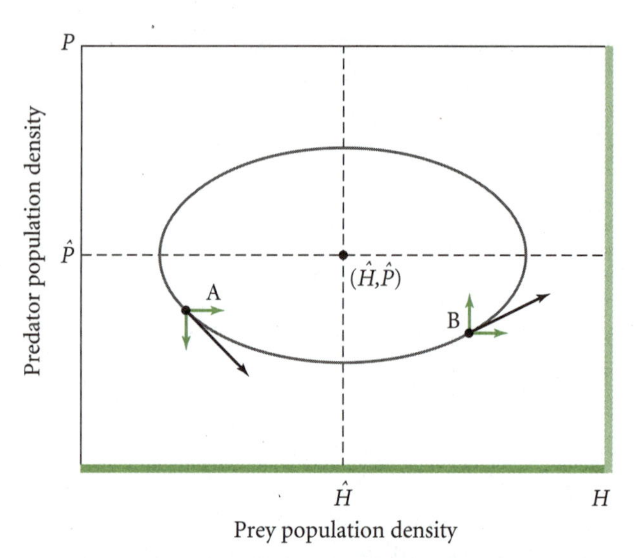

FIGURE 23-6 Space defined by the population sizes of the predator and prey. At points A and B, one vector (parallel to H-axis) represents the growth trend in the prey and another (parallel to P-axis) represents the trend in the predator population. The resultant vector (black) gives the general direction of the joint predator-prey system. For example, at point A, $H < \hat{H}$, indicating a negative growth rate for the predator. Thus the predator vector at point A points downward. $P < \hat{P}$ at point A, so the prey vector points to the right, indicating a growing population. The resultant vector indicates a general decrease in the predator population and an increase in the prey population. The directions of the resultant vectors in the graph indicate a counterclockwise trajectory of the joint predator and prey population dynamics.

Predator-Prey Population Oscillations

An inspection of the resultant vectors (the black vectors) in Figure 23-6 shows a counterclockwise trend in the space defined by P and H. We call these trends **trajectories.** Lotka and Volterra extended their analysis of equations (23-1) and (23-2) to show that when the growth rates of the predator and prey populations are compared directly (in mathematical terms, this means dividing one rate by the other) near the point (H,P), the result is the formula for an ellipse. We omit the details of that analysis here, however. A family of ellipse-like trajectories is shown in Figure 23-7a. Changes in the predator and prey population size move around the point $(\hat{H},\hat{P})$ along the curve defined by the ellipse. The resultant vector for any point on the curve is in the direction of the tangent to the curve (Figure 23-7b). Thus, moving through points a through e in Figure 23-7b results in vectors that point increasingly upward. At point e $(H_2,\hat{P})$, the vector is parallel to the P-axis. Such a system creates oscillations in the predator and prey populations. To see this, think of points 1 through 8 in Figure 23-7b as being eight points in time, beginning with 1. A plot of the predator and prey population size at each point shows the oscillation (Figure 23-7c). Initially (time = 1), the predator population lies below $\hat{P}$ and the prey population above $\hat{H}$. Between time 1 and time 4, the predator population continues to increase. The prey

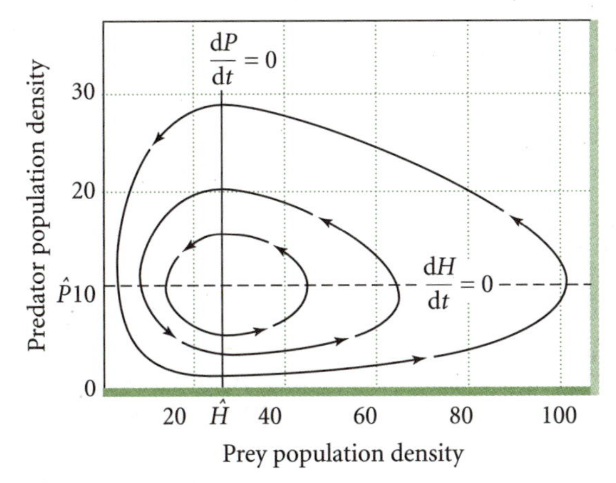

(a)

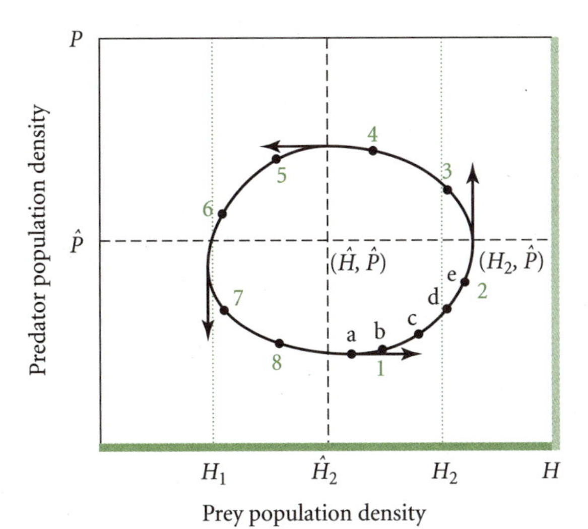

(b)

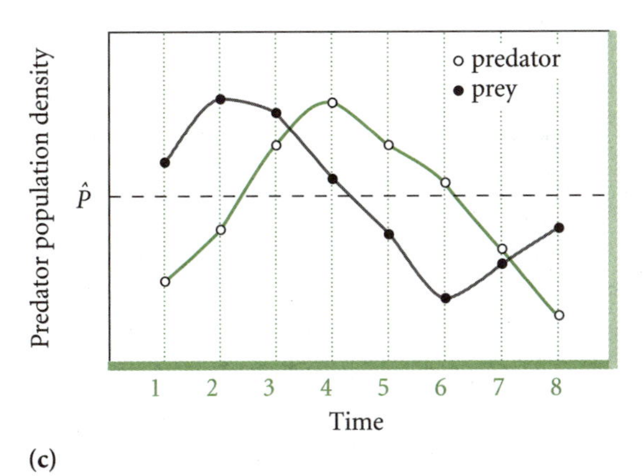

(c)

FIGURE 23-7 (a) Simulated population trajectories of predator and prey according to a Lotka-Volterra model. The degree of oscillation about the equilibrium point reflects the initial population sizes in the simulation. (b) A single elliptical trajectory showing resultant vectors at various points. Trajectories are tangent to the ellipse. Resultant vectors point increasingly upward as one moves from point a through point e because the curve trends upward. Whenever the curve crosses one of the equilibrium isoclines, the resultant vector will be parallel to one of the axes. For example, at point $(H_2,\hat{P})$, the resultant vector points upward parallel to the P-axis because at that point the prey growth rate is $dH/dt = 0$. (c) If the predator and prey population sizes are plotted for times 1 through 8 shown in (b), the predator-prey oscillation is revealed. (*a from Elseth and Baumgardner 1981.*)

population also increases, but only until time 3, when it peaks and then begins to decline. Thus, we see that the predator and prey populations cycle in abundance slightly out of phase with each other.

There are a number of important caveats regarding this model. Lotka and Volterra discovered in their analysis of equations (23-1) and (23-2) that the system is not stable at the equilibrium point $(\hat{H},\hat{P})$. Indeed, for some conditions of $\hat{H}$ and $\hat{P}$, there is no solution for their model. This means that a predator-prey system operating under the constraints of this model would never settle to a stable equilibrium. Instead, the system would continue to cycle about $(\hat{H},\hat{P})$. Systems that behave this way are said to have a **neutral equilibrium,** which means that no internal forces act to restore the populations to the intersection of the predator and prey isoclines. Lotka (1925) pointed out that adding higher-order terms to the differential equations for predators and prey would tend to create an inward spiraling of the population trajectory toward the joint equilibrium. For example, the expression $dH/dt = rH - pHP - cH^2$ leads to a damped oscillation and a stable equilibrium. Density dependence in the predator population has the same effect. Other biologically realistic modifications include time lags, which tend to make a predator-prey system unstable and lead to ever greater oscillations, and refuges for prey, which increase the likelihood of persistence.

Another constraint on the model is that it works only for values of H and P that are relatively close to $(\hat{H},\hat{P})$. The implications of this can be shown graphically by examining what happens in a situation in which a random perturbation, such as a sudden increase in mortality because of a natural disaster, changes the trajectory of the predator-prey system. Suppose, for example, that the predator population experiences a dramatic decrease in population size, moving from point A to point B in Figure 23-8. The decrease in predator population size puts the system on a new, larger, trajectory. Because the system has a neutral equilibrium and thus will not move toward $(\hat{H},\hat{P})$, in the larger trajectory, the predator population will eventually become so large that it simply outstrips its prey resource, and at point A the prey population will becomes locally extinct.

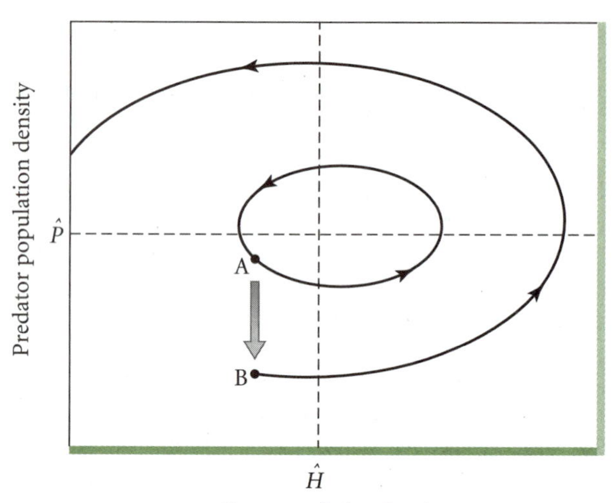

FIGURE 23-8 Implications of large perturbations of population size in the Lotka-Volterra model. The predator population is depressed from point A to point B because of a natural disaster. Since the model has no internal features that tend to return it to an equilibrium, the system will continue on the new larger trajectory until the prey population goes to zero and the system loses integrity.

23.4 Nicholson and Bailey proposed an alternative to the Lotka-Volterra model.

The *pHP* term in the Lotka-Volterra models is sometimes called the **predation term.** Implicit in this term is the assumption that all that matters in the predator-prey system is the number of predators and the number of prey. That is, at a particular prey population size, *H*, and predator efficiency, *p*, more predators should take more prey. We have already mentioned an important weakness of this assumption: it ignores possible interactions among predators that might decrease the number of prey taken. To remedy this situation, A. J. Nicholson and V. A. Bailey (1935) developed an alternative to the Lotka-Volterra models for parasitoids and their hosts.

Features of Parasitoids

The work of Nicholson and Bailey is a good example of how careful observations of natural history drive the questions that ecologists ask. They were interested in the ecology of parasitoids, insects whose larvae live in and consume a host, usually the larva or pupa of another insect. There are similarities between predators and parasitoids, the most notable of which is that, as a result of their activities, the prey or host is killed. This distinguishes these two types of consumers from herbivores and parasites, whose hosts may not be, and indeed usually are not, killed. However, there are some important differences between predation and parasitoidism. The first act of parasitoidism is the deposition of an egg by an adult female into or on a host. This may be thought of as an attack by the parasitoid on the host, but, unlike the attack of a predator on prey, it does not result in the immediate death of the host, and the attacker—the adult parasitoid—does not consume the host. Consumption and death come later, when the parasitoid egg develops into a larva and eats its way out of the host. Because the attack does not cause the immediate death of the host, and thus, does not remove the host from the host population—it is possible for a single host to be attacked sequentially by more than one parasitoid. The number of parasitoids that can be reared on a single host is limited, however, usually to just a few. Because of this, the dynamics of the parasitoid and its host depend not on the number of encounters, as indicated in the predation term *pHP* in the Lotka-Volterra model, but rather on the number of attacks.

Another characteristic of parasitoid-host systems is that generations are discrete and do not overlap. Thus, Nicholson and Bailey used difference equations (see Chapter 18) to develop their models, rather than the continuous time models used by Lotka and Volterra. It was with these features of parasitoid-host systems in mind that Nicholson and Bailey developed their model.

Because *P* and *H* may stand for "parasitoid" and "host" as well as for "predator" and "herbivore," we will continue to use this notation to denote the size of the consumer (parasitoid) and resource (host) populations. In keeping with previous notation related to difference equations, we will denote time, *t*, in parentheses.

Parasitoid-Host Population Dynamics

In each generation there will be $H(t)$ hosts available, of which H_a will be attacked by parasitoids and, by the feeding activity of the parasitoid larvae, removed from the host population. The number of hosts in the next generation $(t + 1)$ is simply the number of hosts not attacked $[H(t) - H_a]$ in the previous generation (t) times the per capita birth rate of hosts (b). Writing this as a difference equation yields

$$H(t + 1) = b[H(t) - H_a]. \tag{23-3}$$

Each of the H_a attacked hosts yields a certain number of parasitoid offspring, c. Thus, the number of parasitoids in the next generation may be represented as

$$P(t + 1) = cH_a. \tag{23-4}$$

You will notice that neither of these equations takes into consideration the density of the parasitoid population. To develop a meaningful density relationship,

Nicholson and Bailey began by assuming that parasitoids search for hosts at random with some search efficiency a. The parameter a is sometimes called the **area of discovery** because, as the parasitoid density increases, individual parasitoids often have to search over a greater area to find unattacked hosts. We showed in Chapter 14, in our discussion of dispersion, how random processes can be represented with the Poisson probability distribution, $p(X) = \mu^X e^{-\mu}/X!$, where μ is the mean value of the random variable X, e is the base of the natural logarithm (e = 2.78), and $X!$ is the factorial of X ($4! = 4 \times 3 \times 2 \times 1$; $0! = 1$). Nicholson and Bailey used the Poisson distribution to model the probability distribution of the number of attacks per host. Thus, the probability that a host will be attacked three times is given as $p(3) = \mu^3 e^{-\mu}/3!$ For an average attack rate of $\mu = 0.04$ attacks/host, $p(3) = 0.00001$.

The Poisson is a probability distribution; thus the sum of the probabilities for all X is 1. Because of this, we can think of the Poisson probabilities as proportions. The proportion of hosts that are not attacked is the same as the probability of not being attacked, $P(0)$, which is $P(0) = \mu^0 e^{-\mu}/0! = e^{-\mu}$ (since μ^0 and 0! are by both by definition 1). If the proportion of hosts not attacked is $e^{-\mu}$, then the proportion of hosts attacked (at least once) is $1 - e^{-\mu}$. Nicholson and Bailey assumed that the average number of parasitoid eggs laid per host in a particular area and time was $aH(t)$ and, consequently, the total number of eggs laid in the area would be $aP(t)H(t)$. The average number of eggs per host may then be given as $\mu = aP(t)$. So, the number of hosts attacked is $H_a = 1 - e^{-aP(t)}$. A plot of this relationship shows that as the density of the parasitoid increases, the proportion of hosts attacked also increases, but at a *decreasing rate* (Figure 23-9). For example, letting $a = 0.068$, an increase in parasitoid density from 10 to 15 results in an increase of 0.146 in the proportion of hosts attacked (from $1 - e^{-0.068(10)} = 0.493$ to $1 - e^{-0.068(15)} = 0.639$), whereas an increase in parasitoid density from 50 to 55 results in only a 0.01 change in the proportion of hosts attacked (from 0.966 to 0.976). The reason for this decelerating curve is that, at high densities, parasitoids have a hard time finding a host that has not been attacked already. Contrast this with the situation depicted in the Lotka-Volterra models, in which increasing predator density simply means that more prey will be eaten.

The quantity $H_a = 1 - e^{-aP(t)}$ can be substituted into equations (23-3) and (23-4) to yield the basic Nicholson-Bailey model:

$$H(t + 1) = bH(t)[e^{-aP(t)}] \tag{23-5}$$

and

$$P(t + 1) = cH(t)[1 - e^{-aP(t)}]. \tag{23-6}$$

Figure 23-10a shows a simulation of these equations with parameter values $b = 2$, $a = 1$, and $c = 1$. The figure reveals an interesting, and quite distressing, feature of the Nicholson-Bailey models: the equations give rise to ever-increasing oscillations. Figure 23-10a shows two such oscillations for both the parasitoid and the host. In the end, the model predicts that the two will not be able to coexist, since it is possible for either the host or the parasitoid to reach very low numbers (note the two places where the density curves of the host and the parasitoid touch the horizontal axis). If we draw the model in a space defined by the abundance of the parasitoid (vertical axis) and the host (horizontal axis), the trajectory of the population is seen to spiral outward from an equilibrium point (Figure 23-10b). A simulated time course of this trajectory for points 1 through 9 shows the increasing oscillations more clearly (Figure 23-10c).

If we reconsider a number of the features of the Nicholson-Bailey model, we can discover what factors work to stabilize a parasitoid-host system. At least two factors may lead to coexistence of parasitoid and host. First, it can be shown that if the birth rate, b, of the parasitoid is made to decrease as the density of the parasitoid increases [in the current model, b is a constant, independent of the parasitoid density $P(t)$], the expanding oscillations of the model will changed to damped oscillations. A second way to achieve a stable outcome in the Nicholson-Bailey model is to recognize that as the parasitoid population density increases, the efficiency with which each parasitoid finds unattacked hosts, a, will decrease. If either of these

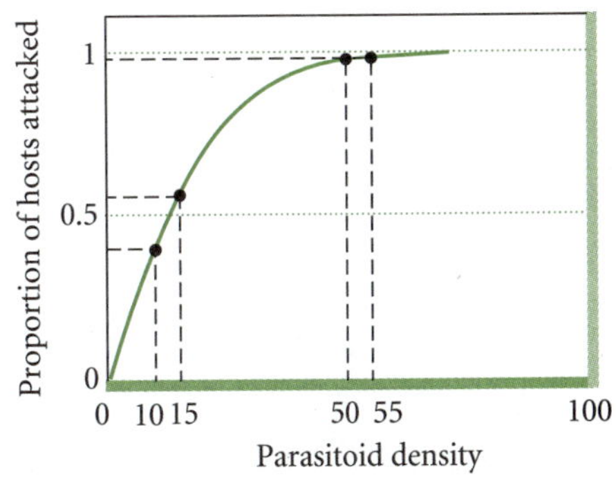

FIGURE 23-9 Relationship between the density of parasitoids, P, and the proportion of hosts attacked ($1 - e^{-aPt}$), with $a = 0.068$. The proportion of hosts attacked increases at a decreasing rate as P increases. At a low parasitoid density, a change in the number of parasitoids from 10 to 15 results in an increase of 0.146 in the proportion of hosts attacked, whereas at a high parasitoid density, the same increase (from $P = 50$ to $P = 55$) results in only a 0.01 increase in the proportion of hosts attacked. The reason for this that at high parasitoid densities, nearly all the hosts have already been attacked.

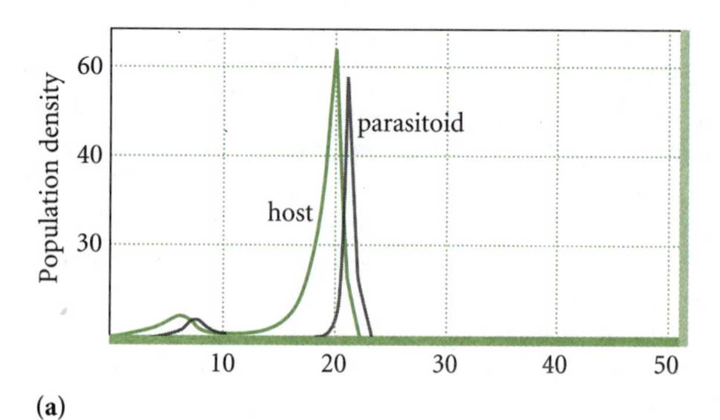

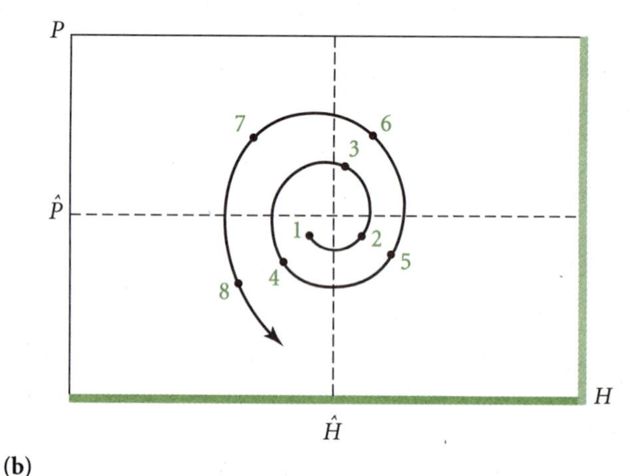

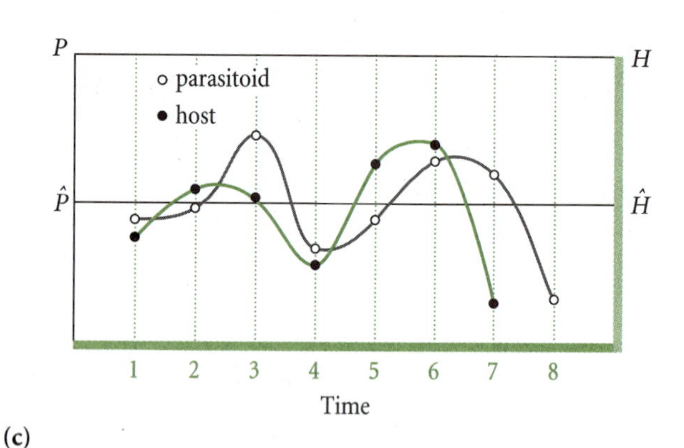

FIGURE 23-10 A simulation using the Nicholson-Bailey parasitoid-host model. (a) The simple Nicholson-Bailey model with $b = 2, a = 1, c = 1$. The graph shows two oscillations, the second bigger than the first. The basic Nicholson-Bailey model results in ever-increasing oscillations and is unstable. (b) Hypothetical Nicholson-Bailey trajectory in *H-P* space. (c) The time course of the trajectory. The numbers in (b) correspond to the numbers in the time axis in (c). (*a from Bulmer 1994.*)

enhancements is included in the model, a stable interaction is predicted.

The Nicholson-Bailey model presented in equations (23-5) and (23-6) has three parameters (*a, b,* and *c*), all of these parameters can be determined experimentally and then applied to data on

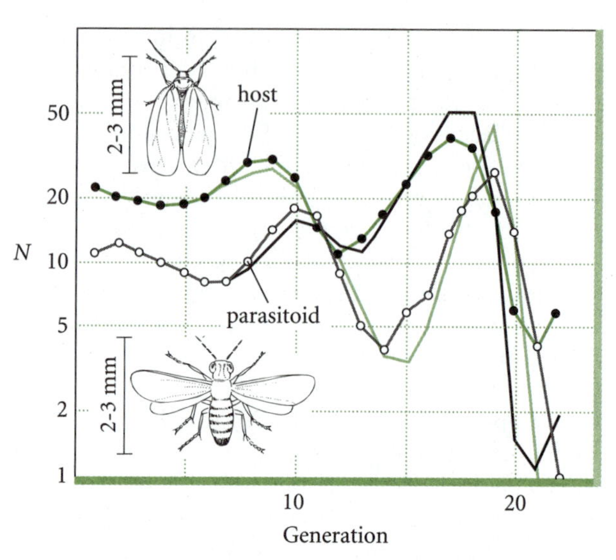

FIGURE 23-11 Population fluctuations of the whitefly *Trialeurodes vaporariorum* (solid symbols) and its chalcid parasitoid *Encarsia formosa* (open symbols). The black and light green lines are trajectories predicted by the Nicholson-Bailey model. (*From Begon and Mortimer 1986; after Hassell 1978.*)

parasitoid-host populations. For example, Burnett (1958) modeled the interaction between the greenhouse whitefly *Trialeurodes vaporariorum* and its chalcid wasp parasitoid *Encarsia formosa* (Figure 23-11). The values $a = 0.068$, $b = 2$ (experimentally imposed), and $c = 1$ gave a reasonable fit to the data over several population cycles.

23.5 The response of predators to prey density is not linear.

Nicholson and Bailey criticized the Lotka-Volterra model because its linear relationship between attack rate and predator (or parasitoid) density seemed unrealistic. The Canadian entomologist C. S. Holling (1959) voiced a similar complaint about the linear relationship between the number of prey consumed per predator and the density of prey. In the Lotka-Volterra model, the rate at which individuals are removed from the prey population is described by the term pHP. Thus, for a given density of predators (P), the rate of exploitation increases in direct proportion to the density of prey (H). Many biological factors ought to alter the form of this relationship and perhaps thereby alter the dynamics of the predator and prey populations.

The relationship of prey density to an individual predator's rate of food consumption was labeled the **functional response** by Holling (1959). There are three general types of functional responses (Figure 23-12). Type I is the linear relationship of the Lotka-Volterra model, in which the number of prey taken, pPH, increases linearly as H increases. The

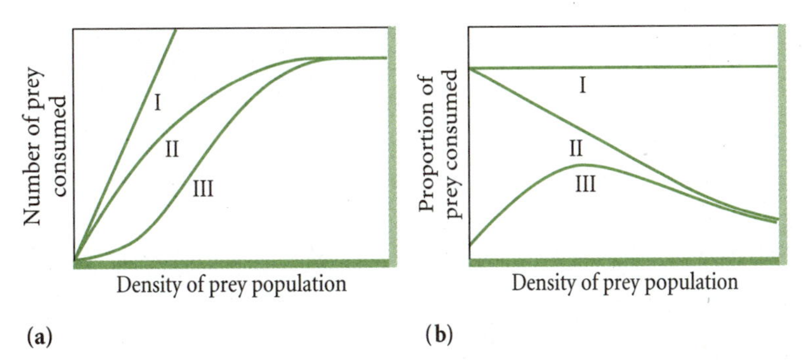

FIGURE 23-12
Three types of functional responses of predators to increasing prey density: (a) the functional response in terms of the number of prey consumed; (b) the functional response in terms of the proportion of prey consumed. Type I: the predator consumes a constant number or proportion of the prey regardless of the prey density. Type II: Predation rate decreases as predator satiation sets an upper limit on food consumption. Type III: Predator response is depressed at low prey density because of low hunting efficiency or absence of a search image.

linear type I functional response curve predicts that there will be no upper limit to the prey consumption rate of the predator, a prediction that is too simple to reflect nature. However, over some ranges of prey densities, predators may have linear functional responses (e.g., Korpimäki and Norrdahl 1991). For some filter-feeding organisms, the curve may be fairly representative of the relationship between consumption rate and prey density.

Typically, the number of prey consumed per predator levels off at some prey density, and the type II and type III functional responses reflect this (see Figure 23-12). Two factors dictate that the functional response should reach a plateau. First, predators may become satiated, at which point their rate of feeding is limited by the rate at which they can digest and assimilate food, rather than by the rate at which they can catch it. Second, as a predator captures more prey, the time spent handling and eating the prey cuts into searching time. Eventually, the two reach a balance, and prey capture rate levels off.

Type II Functional Response

The type II functional response describes a situation in which the number of prey consumed per predator initially rises quickly as the density of prey increases, but then levels off with further increases in prey density. The dynamics of the type II functional response were elucidated by Holling (1959), who considered how the time the predator spends handling and searching for prey affects consumption rate. Holling described the handling time by a simple expression, known as the **disc equation** because of its application to experiments in which blindfolded human subjects were required to discover and pick up small discs of paper on a flat surface. Any such task, including the subduing and eating of prey, requires a certain **handling time**, T_h. The total handling time is therefore the handling time per item (T_h) times the number of encounters (E), or T_hE. The time left over for searching—the search time, T_s—is the total time, T_{total}, minus the total handling time; that is, $T_s = T_{total} - T_hE$. The number of encounters can, itself, be defined as the product of search time, prey density (H), and a constant (a) for the efficiency of searching (number of prey detected per unit of time); that is, $E = aT_sH = a(T - T_hE)H$. If $E = a(T - T_hE)H$ is substituted into the equation for T_s, we may solve for E to obtain

$$E = \frac{aHT}{1 + aHT_h}. \qquad (23\text{-}7)$$

In this equation, the quantity aH is often called the **encounter rate.**

Equation (23-7) is plotted in Figure 23-13a for three different handling times and $a = 1$, and in Figure 23-13b for three values of a when $T_h = 0.1$. When prey are scarce, the denominator term aHT_h is small compared with 1, and the number of prey encountered approaches aHT. When prey are numerous, aHT_h is large compared with 1, and E approaches the ratio T/T_h, which is a constant value defining the maximum number of prey that can be captured in time T. For example, in Figure 23-13a, if $H = 30$ and $T_h = 0.1$, $aHT_h = 1(30)(0.1) = 3$, and the number of encounters, E, is $30/[1 + 3] = 7.5$, which approaches $T/T_h = 1/0.1 = 10$. Thus at a high prey density, search time drops to near zero, and the number of prey captured is limited only by how long the predator requires to handle each one;

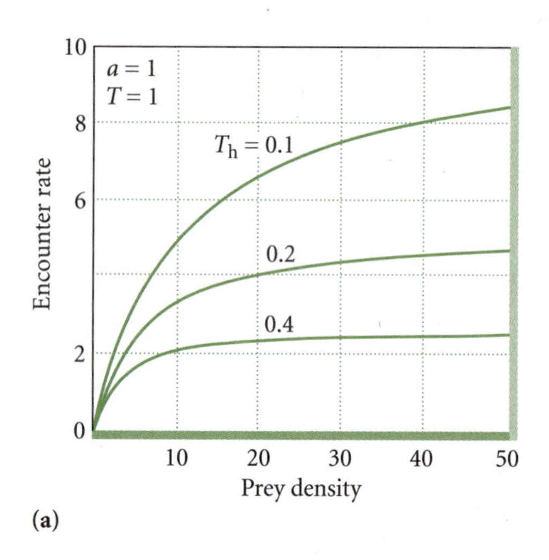

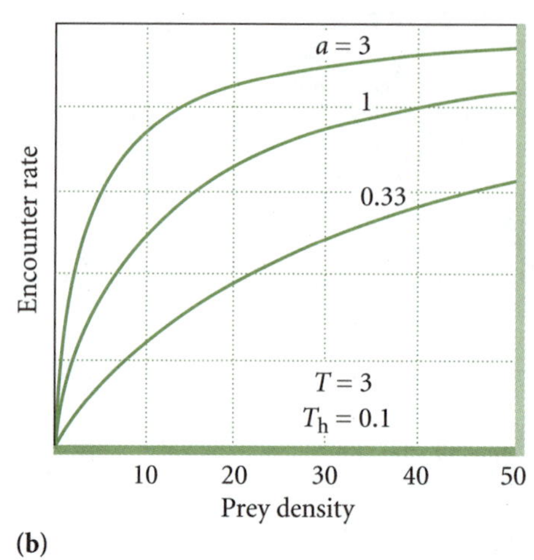

FIGURE 23-13 The relationship between encounter rate (E) and prey density (P), according to Holling's disc equation, at (a) different handling times (T_h) per prey and (b) different levels of hunting efficiency (a).

the shorter the handling time, the more prey can be captured.

Hassell (1978) fitted equation (23-7) to the relationship between attack rate and prey density observed in several laboratory systems (Table 23-1). As one would expect, both handling time and search efficiency varied widely, depending on the characteristics of the predator and prey and the structure of the laboratory environment. The very high handling time of the parasitoid *Nasonia* on housefly (*Musca domestica*) hosts reflects the fact that many pupae were rejected because they had already been parasitized (encountered). In that study, the number of encounters referred only to those not previously attacked, but handling time included time spent rejecting any pupa encountered, parasitized or not.

Dale et al. (1994) studied the predatory behavior of wolves (*Canis lupus*) in Gates of the Arctic National Park and Preserve in the central Brooks Range of Alaska, a wilderness area inside the Arctic Circle. Wolves prey primarily on caribou (*Rangifer tarandus*), although they will also take Dall sheep (*Ovis dalli*)

TABLE 23-1 Estimated values of handling time in consumer-resource equations for a selection of parasitoids and arthropod predators

Parasitoid or predator	Host	Handling time (h) T_h	T_h/T
Parasitoids			
Nemeritis canescens	*Ephestia cautella*	0.007	<0.0001
Chelonus texanus	*Ephestia kuhniella*	0.12	<0.001
Dahlbominus fuscipennis	*Neodiprion lecontei*	0.24	<0.003
Pleolophus basizonus	*Neodiprion sertifer*	0.72	<0.02
Dahlbominus fuscipennis	*Neodiprion sertifer*	0.96	<0.01
Cryptus inornatus	*Loxostege sticticalis*	1.44	>0.02
Nasonia vitripennis	*Musca domestica*	12.00*	<0.1
Predators			
Anthocoris confusus	*Aulacorthum circumflexus*	0.38	<0.001
Notonecta clauca	*Daphnia magna*	0.76	<0.005
Ischnura elegans	*Daphnia magna*	0.82	<0.002
Harmonia axyridis	*Aphis craccivora*	1.61	<0.002
Phytoseiulius persimilis	*Tetranychus urticae*	1.87	<0.005

**This figure is the handling time for each host, in which a female lays many eggs. The handling time per egg is roughly 0.4 h. (From Hassell 1978.)*

and moose (*Alces alces*). Dale and coworkers estimated the functional response of wolves to caribou abundance, which varies substantially, by observing four wolf packs for a period of 30 days in the winter of 1990 and counting the number of kills per wolf and the number of caribou available on each day. When the number of kills per wolf per day was plotted (vertical axis) against the estimated number of caribou per km^2, the data fit the Holling type II functional response curve (Figure 23-14).

The disc equation gives the number of encounters for a particular level of prey availability and a constant search efficiency, a, and handling time, T_h. The equation does not distinguish among predators with different search strategies. So, for example, over the same period of time, a predator that detects prey at regular intervals and one that sleeps most of the time and then, in a burst of foraging activity, rapidly detects a large number of prey could have the same search efficiency (prey detected/time) (Getty and Pulliam 1993).

Type III Functional Response

Both the type II and the type III functional response curves show that consumption increases more slowly as prey density increases. Eventually, in both functions, consumption will reach a constant value because of satiation. However, the type III functional response curve behaves differently than the type II curve when the prey density is low. At the very lowest prey densities, predators with type III functional responses consume very low numbers of prey. The highest consumption for such predators occurs at intermediate prey densities (see Figure 23-12a). There are a number of important consequences of such a functional response. One of the most important is that when their population is at a low density, prey are relieved of predation pressure.

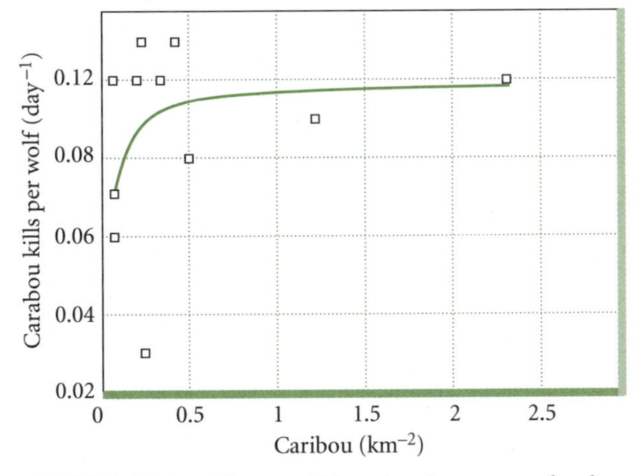

FIGURE 23-14 The type II functional response of wolves (*Canis lupus*) feeding on caribou (*Rangifer tarandus*) in winter in Gates of the Arctic National Park and Preserve. (*From Dale et al. 1994.*)

Several factors may lead to a decreased predator response at lower prey densities: (1) a heterogeneous habitat may afford a limited number of safe hiding places, which protect a larger proportion of the prey at lower densities than at higher densities; (2) lack of reinforcement of learned searching behavior owing to low rates of prey encounter may reduce hunting efficiency at low prey densities; and (3) switching to alternative sources of food when prey are scarce reduces hunting pressure on the prey. The relationship of prey consumption to prey density depends on a change in average prey vulnerability in the case of (1), on search and capture efficiency in the case of (2), and on motivation to hunt or searching time in the case of (3).

Learning influences the searching behavior of many predators, especially vertebrates with complex nervous systems, but also invertebrates (Papaj and Lewis 1993). An object is always easier to find if one has a preconception of what it looks like and where it occurs. Such **search images** (Tinbergen 1969) are acquired by experience—that is, by learning. When the prey population is dense, predators encounter prey frequently; thus search efficiency increases and handling time decreases as a result of experience. At low prey densities, neither are so well developed, and the capture rate is therefore depressed. The concept of search image may serve as a link between individual behavior and predator-prey dynamics because it encompasses mechanisms underlying how individual predators obtain prey and, thus, how individual behavior affects prey populations. Few comprehensive models of the effects of search image on predator-prey dynamics exist, however (Morgan and Brown 1996).

Regardless of the mechanisms underlying the type III response, predators often switch to a second prey species as the density of the first is reduced (Murdoch 1969). When Lawton et al. (1974) presented the predatory water bug *Notonecta glauca* with two types of prey—an isopod, *Ascellus aquaticus*, and larvae of the mayfly *Cloeon dipterum*—they found that the predators consumed the more abundant prey species in a greater proportion than its percentage of occurrence (Figure 23-15). This **switching** depended to some degree on variation in attack success on isopods as a function of their relative density. When water bugs encountered isopods infrequently, fewer than 10% of attacks were successful. At higher densities, and therefore higher encounter rates, attack success rose to almost 30% (also see Cornell and Pimentel 1978).

The consequences of switching for predator-prey stability have been explored in some detail by Murdoch and Oaten (1975; Oaten and Murdoch 1975). Using continuous time models, they found that

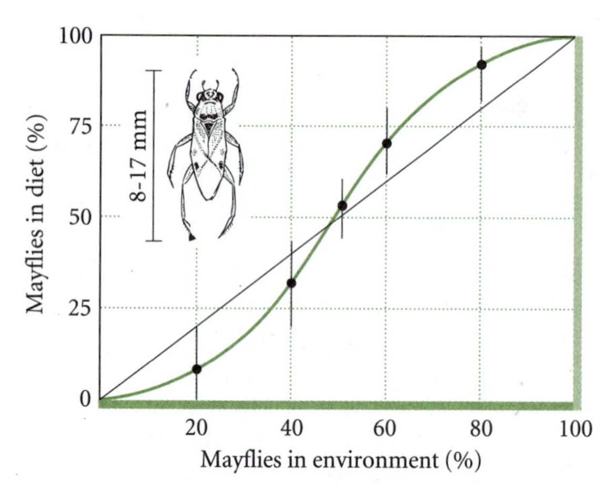

FIGURE 23-15 The proportion of mayfly larvae in the diet of the water bug *Notonecta* as a function of their relative abundance among available prey. The straight line indicates the situation if the predator took prey in proportion to their availability in the environment (no preference). The curved line shows the proportions of mayflies in the diet at different relative prey densities (solid circles show the means and vertical bars the ranges over several trials at each density). The diet of the water bugs contained fewer mayflies than represented in the environment when mayflies were less than one-half of the available prey in the environment. The bugs' diets contained a higher proportion than found in the environment when mayflies made up over one-half of the available prey. (*From Begon and Mortimer 1986; after Lawton et al. 1974.*)

the type III response characteristic of switching could stabilize predator and prey populations. Hassell and Comins (1978) pointed out, however, that switching does not produce the same effect in discrete time-difference equation models because of the overwhelming time delay in the response of predators to prey population density.

Temporal Scale of the Functional Response

The functional response models that we have just presented are called **differential functional response** models because they reflect the instantaneous predator consumption rate as a function of prey density (the word *differential* is used in the sense of calculus to reflect instantaneous rate). But feeding by the predator does not take place instantaneously. Rather, a feeding bout may last for hours or even days. If the functional response of the predator is measured over an entire foraging bout, it is referred to as the **integrated functional response.** In terms of calculus, this would be the integral of the differential response over the time period of the foraging event. Each predator is faced with decisions about how much time it should spend foraging and how much time it should spend doing other things, such as sleeping or looking for a mate. The **optimal functional response** is the differential functional response integrated over the time spent foraging and the time that the predator spends doing other things (Mitchell and Brown 1990, Morgan et al. 1997). The integrated and optimal functional responses incorporate more biological realism than the simpler differential functional response, and thus may reflect nature more closely.

23.6 Predator populations can respond to an increase in prey density by growth and immigration.

As we saw in the last section, individual predators can increase their consumption of prey only to the point of satiation. However, predators can respond to abundant prey by increasing their numbers, either by immigration or by population growth, which together constitute the **numerical response.** Populations of most predators grow slowly, especially when the reproductive potential of the predator is much less than that of the prey, and its life span longer. Immigration from surrounding areas contributes importantly to the numerical response of mobile predators, which may opportunistically congregate where resources become abundant. The bay-breasted warbler, a small insectivorous bird of eastern North America, exhibits such behavior during periodic outbreaks of the spruce budworm. During years of outbreak in a particular area, the density of warblers may reach 120 pairs per 100 acres, compared with about 10 pairs per 100 acres during non-outbreak years (Morris et al. 1958). This population behavior clearly shows how a predator may take advantage of a shifting mosaic of prey abundance.

In a study of the larch sawfly in tamarack swamps in Manitoba, Buckner and Turnock (1965) found that while avian predators were more abundant in areas of high sawfly density, they consumed a smaller proportion of the sawflies there than in areas with low sawfly populations. The sawflies varied in number over three orders of magnitude, while the numerical response of the birds resulted in only a doubling of population size. So while 6% of larvae and 65% of adult sawflies were eaten in one area of low sawfly population density, only 0.5% and 6% were consumed where sawflies were 50 times more abundant (see also Crawford and Jennings 1989).

Owls and raptors (birds of prey) often show strong numerical responses to cycles in their prey (Korpimäki and Norrdahl 1991, Korpimäki 1992, Rohner 1996). Three predatory birds, the pomarine jaeger (*Stercorarius pomarinus*), the snowy owl (*Nyctea scandiaca*), and the short-eared owl (*Asio flammeus*), each respond in a different manner to

TABLE 23-2 Response of predatory birds to different densities of the brown lemming near Barrow, Alaska

	1951	1952	1953
Brown lemming (individuals per acre)	1–5	15–20	70–80
Pomarine jaeger	Uncommon, no breeding	Breeding pairs 4 mi^{-2}	Breeding pairs 18 mi^{-2}
Snowy owl	Scarce, no breeding	Breeding pairs 0.2–0.5 mi^{-2}, many nonbreeders	Breeding pairs 0.2–0.5 mi^{-2}, few nonbreeders
Short-eared owl	Absent	One record 3–4 mi^{-2}	Breeding pairs

(From Pitelka et al. 1955.)

varying densities of lemmings on the arctic tundra (Table 23-2). Lemming populations exhibit great fluctuations; high and low points in a population cycle may differ by a factor of 100. At Barrow, Alaska, during the summer of 1951, when lemmings were scarce, none of the predatory birds bred; short-eared owls did not even appear in the area. During the following summer, one of moderate lemming density, both the jaeger and snowy owl bred, but short-eared owls again were absent. In 1953, a peak year for lemmings, all three species of avian predators bred. Jaegers were four times more abundant in 1953 than in 1952, showing a strong numerical response. In contrast, the density of snowy owls did not increase, but each pair of birds instead reared more young. Whereas most snowy owl nests contained 2 to 4 eggs during the year of moderate lemming abundance, clutches of up to 12 eggs were laid during the peak year of 1953. A recently completed study of the numerical response of great horned owls (*Bubo virginianus*) to the 10-year population cycle of its favorite prey, the snowshoe hare (*Lepus americanus*), in Canada revealed increases both in the number of juvenile owls and the ratio of juvenile to adult owls in years of high snowshoe hare populations (Rohner 1996) (Figure 23-16).

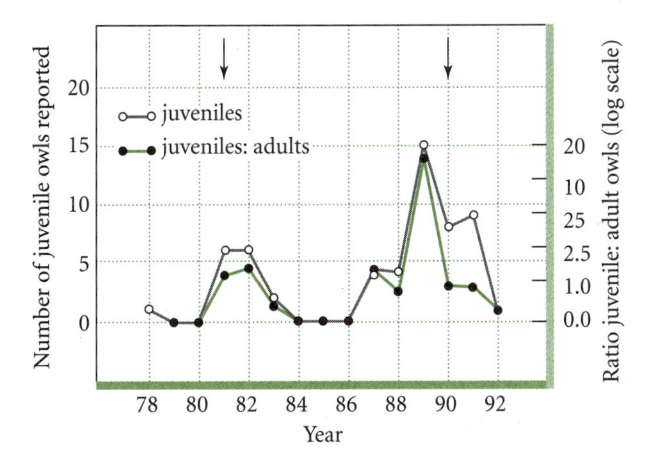

FIGURE 23-16 Numerical response of great horned owls (*Bubo virginianus*) in Canada feeding on snowshoe hares (*Lepus americanus*). The snowshoe hare population peaked (arrows) twice during the time of the study. Each peak resulted in more juvenile owls and a higher juvenile:adult ratio in the owl population. (*From Rohner 1996.*)

23.7 Graphic analyses demonstrate the conditions for stability in predator-prey systems.

Studies of the functional and numerical responses of predators, as well as the population dynamics of prey populations, have revealed the biological limitations of simple predator-prey models such as the Lotka-Volterra equations. Such limitations become especially evident when accurate predictions are needed, as when predator or prey populations are being managed. Many complications of biological reality have been incorporated into models and their contributions to the stability of predator-prey interactions studied (May 1975b, Hassell 1978, Arditi and Saiah 1992, Berryman 1992). But we can accomplish the same goal and avoid much of the difficult mathematics involved by using two graphic approaches to the issue of stability in predator-prey systems. One is based on modifications of the predator and prey isoclines drawn on a graph of their joint abundances (for example, see Figure 23-7). The other, which we shall take up first, builds upon the concept of functional response.

Recall our graphic representation of the Lotka-Volterra model of predator-prey dynamics, which we represented in a space defined by the predator (P) and prey (H) population sizes (see Figures 23-5 and 23-6). In Figure 23-5, the predator isocline was represented as a vertical line extending upward from the H-axis and the prey isocline as a horizontal line extending outward from the P-axis. The nature of these lines derives from the fact that both $\hat{P}$ and $\hat{H}$ are constants in the Lotka-Volterra formulations. These isoclines result in a neutral equilibrium. We have already hinted at the possibility that these lines might be far too simple to represent the situation in the natural world, and below, we will discuss some conditions that might lead to isoclines of different shape and orientation. But first, let us reexamine the basic Lotka-Volterra representation in order to gain some appreciation for the implications of rearranging the isoclines.

Consider Figure 23-17, which has four panels. Panel (b) is a reproduction of Figure 23-6, showing the vertical and horizontal isoclines and the resultant vectors for four points. In this case, the vectors show a counterclockwise trajectory and a neutral equilibrium. How would a change in the orientation of the isoclines affect the stability of the system? When we rotate the isoclines about the point $(\hat{H},\hat{P})$ while keeping points A, B, C, and D in the same positions relative to the isoclines, notice what happens to the resultant vectors at those points. Panel (a) shows a one-quarter rotation of the vectors in the counterclockwise direction and the repositioned points. Compare what happens at point A before and after the rotation. Before the rotation panel (b), point A lies in a position at which the predator growth rate is negative $(H < \hat{H})$ and the prey growth rate is positive $(P < \hat{P})$. The direction of change in predator and prey population size is given by the green vectors parallel to the P and H axes. The resultant vector (black) points downward and to the right, indicating a general decline in the predator population and an increase in the prey population. In panel (a), the relative position of point A with respect to the predator and prey isoclines remains the same, and thus, so do the green vectors indicating the growth trends. But in the new configuration, the resultant vectors indicate that the trajectories will collide with one of the axes and thus destabilize the system. The trick to understanding the effect of rotating the axes is to appreciate that as long as point A stays in the same relative position with respect to the isoclines, the green vectors will remain parallel to the H and P axes. Thus, counterclockwise rotation of the axes destabilizes the system.

When the axes are rotated in the clockwise direction, stability is achieved. Panel (c) shows a quarter-turn rotation in the clockwise direction. The resultant vectors now point directly toward the isoclines, indicating that the trajectory will remain stable. A half-turn clockwise rotation panel (d) yields a situation leading to the most stable condition, in which the trajectories lead directly to the point $(\hat{H},\hat{P})$. The point of this exercise is to emphasize that in models such as the Lotka-Volterra equations, the orientation of the isoclines with respect to the predator and prey abundances will determine the degree of stability of the system. Now, let us discuss some biological conditions that might lead to isoclines that differ from the simple vertical and horizontal lines of the Lotka-Volterra model.

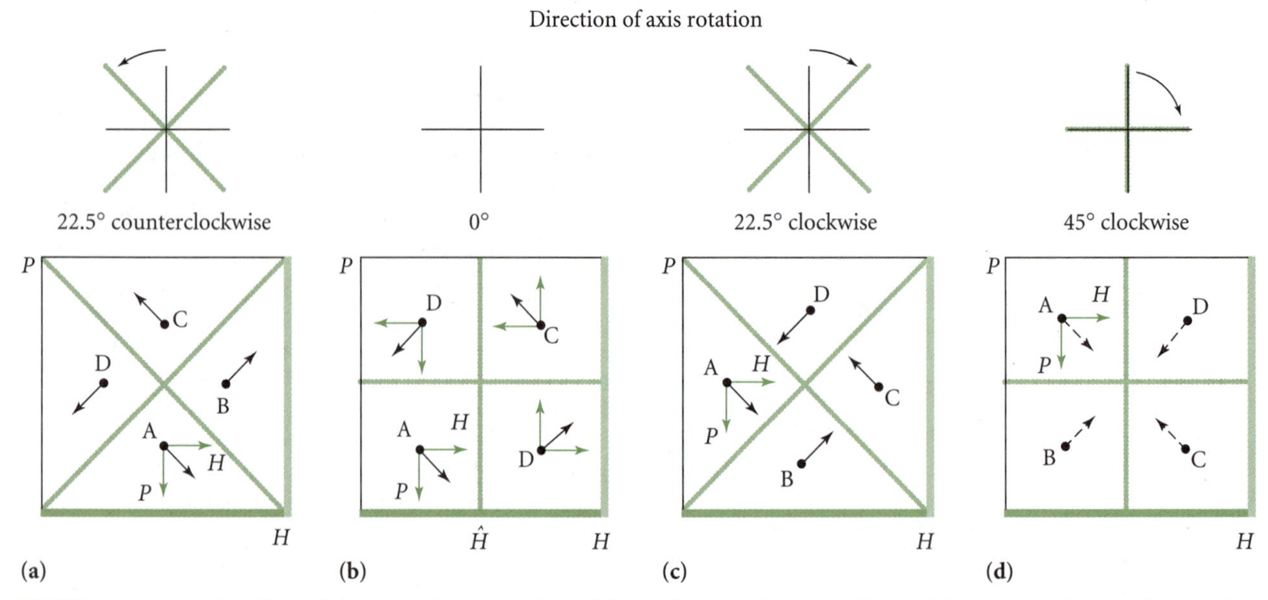

FIGURE 23-17 The effect of changing the orientation of the predator and prey isoclines with respect to the P and H axes in the Lotka-Volterra model. (a) The isoclines are rotated one quarter-turn in the counterclockwise direction, leading to a destabilization of the system. (b) The basic conditions of the model, which result in neutral equilibrium. (c) A quarter-turn clockwise rotation. (d) A half-turn clockwise rotation. Both clockwise rotations lead to stability of the system. The points A, B, C, and D are held in the same position relative to the isoclines during a rotation. The resultant vector may change with rotation. (See Figure 23-6.)

Effect of Logistic Self-Limitation

A number of ecologists have suggested that one way to add biological realism to the Lotka-Volterra models is to introduce a density-dependent term that limits the growth of the predator or prey population (Berryman 1992). For example, in the prey equation (23-1), we could multiply the term $(1 - H/K)$, where K is the carrying capacity of the prey, by the growth rate of the prey (rH) to obtain the following equation,

$$\frac{dH}{dt} = rH\left(1 - \frac{H}{K}\right) - pHP. \tag{23-8}$$

Setting this equation equal to zero and solving for the equilibrium value of P (see section 23.2), we find that $\hat{P} = r/p(1 - H/K)$. In this formulation, the equilibrium value $\hat{P}$ is a decreasing function of the prey population density H. This situation is depicted in Figure 23-18a. Self-limitation in the prey population yields a stable system.

Logistic self-limitation can also be incorporated into the predator equation. Leslie (1948) proposed the following form for the predator equation,

$$\frac{dP}{dt} = apP\left(1 - \frac{cP}{H}\right), \tag{23-9}$$

where c is the number of prey required for one predator to maintain itself and produce one offspring. This equation has the unique feature of employing the ratio of the predator and prey (P/H) instead of the product (PH). Such models are called **ratio-dependent models** (Berryman 1992, Ginzburg and Akçakaya 1992, Arditi and Saiah 1992, Gutierrez 1992, Slobodkin 1992). Solving for H in the equation, we obtain $\hat{H} = cP$, indicating that the equilibrium of the prey increases linearly with the density of the predator. When this predator equation is employed with the horizontal prey isocline, the system is unstable. However, when both of the above equations are included, conditions of stability result (Figure 23-18b).

The logistic self-limitation term in the prey equation (23-8) allows the prey isocline to decline with prey population size; that is, as the number of prey increases, the equilibrium number of prey, $\hat{H}$, decreases. The inclusion of the logistic term in the predator equation (23-9) likewise yields a slanted predator isocline representing a positive response in the predator

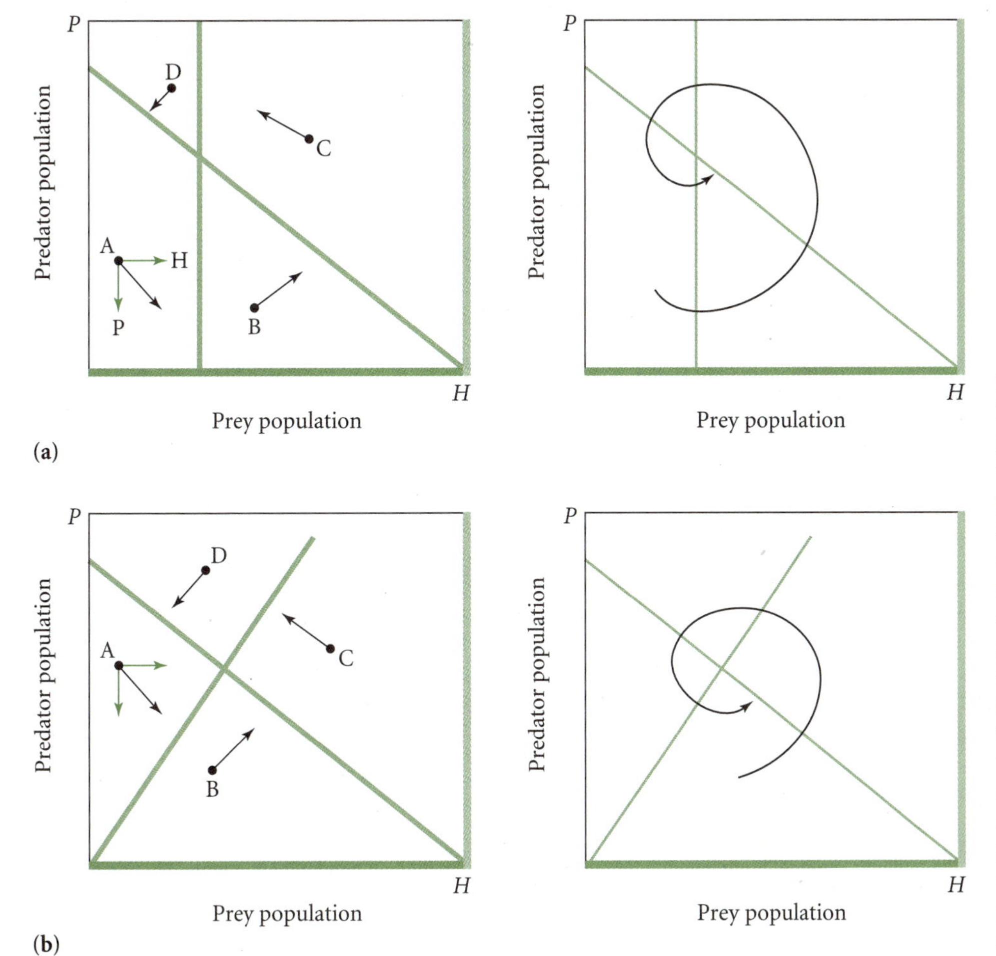

FIGURE 23-18 The Lotka-Volterra predator-prey model. The resultant vectors for four different points are shown at the left, and a possible trajectory is shown at the right. (a) A model in which there is self-limitation in the prey growth rate. This results in a prey isocline that slants from left (low prey density) to right (high prey density). The system is stable as long as H is relatively small. (b) A model in which there is self-limitation in both the predator and prey populations. The system is stable.

population to an increase in prey. The slanting predator and prey isoclines are more biologically reasonable. Moreover, the model yields a stable equilibrium, as shown above.

Nonlinear Isoclines

Michael L. Rosenzweig and Robert H. MacArthur (1963) showed that certain biological considerations may alter not only the orientation of the isoclines with respect to the axes, as we saw above, but their shapes as well. Altering the shapes of the isoclines may change the stability of the system. Consider first the biology of the prey. In the absence of predators, the prey population is limited by the carrying capacity of its environment (K), which is determined by the availability of food or other resources. As a result, the prey isocline bends downward toward the H-axis as the number of prey increases, intersecting the axis at K. This adjustment is shown in Figure 23-19b. This model follows the pattern of logistic population growth. It is equivalent to a clockwise rotation of the

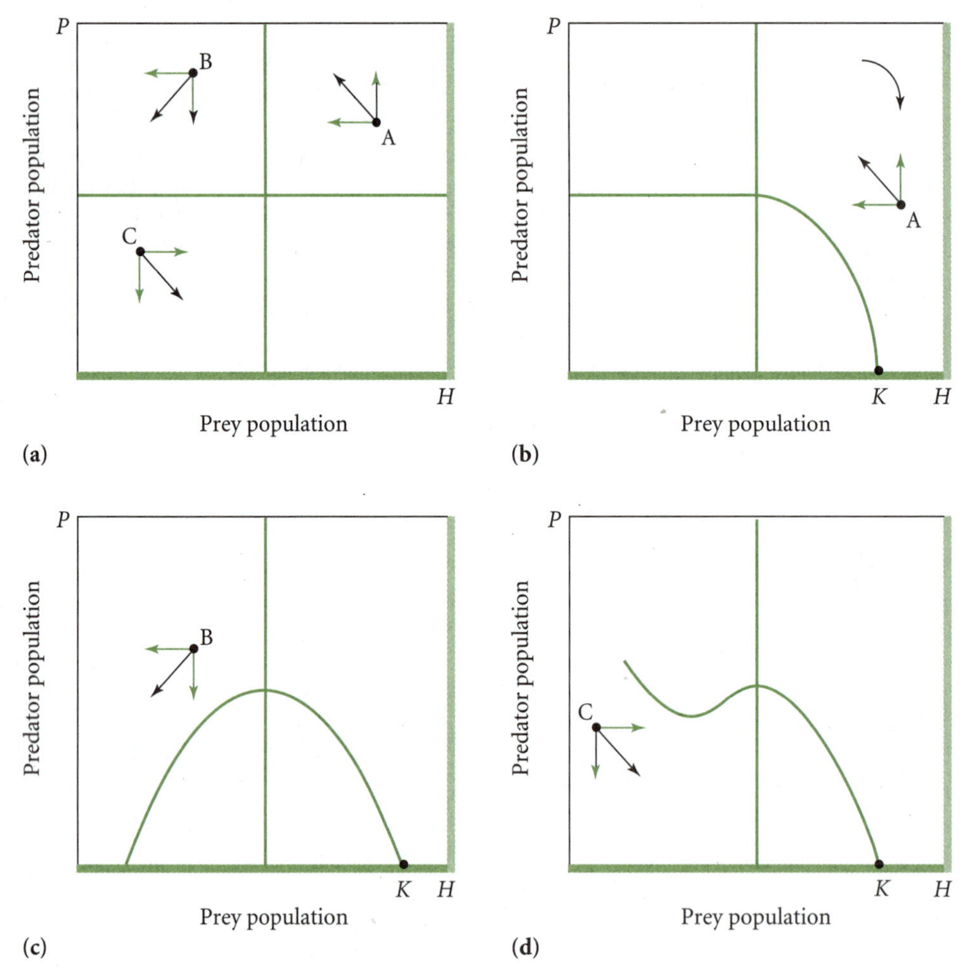

FIGURE 23-19 Possible reconfigurations of the prey isocline to incorporate more biological realism. (a) The basic Lotka-Volterra model with vertical predator and horizontal prey isoclines. (b) At low predator densities, the prey population may be limited by its own density, resulting in a bending of the isocline toward the H-axis, meeting it there at the carrying capacity (K). This is equivalent to a clockwise rotation of the isocline, and thus has a stabilizing effect. (c) Prey population may be limited at low prey densities because of limited recruitment of new individuals into the population, which is represented by a bending of the prey isocline toward the H-axis at low prey densities, equivalent to a counterclockwise rotation. (d) Predators may have a difficult time finding prey when prey are at low densities, and thus may reach high levels without affecting the prey population size. This is represented by an upswing in the prey isocline at low prey densities, equivalent to a clockwise (stabilizing) rotation of the isocline.

prey axis, as shown by the reorientation of point A, and therefore has a stabilizing effect.

Two opposing factors influence the shape of the prey isocline at low densities. On the one hand, small prey populations can support fewer predators than large populations because the recruitment rate of smaller populations is lower. As pointed out by Rosenzweig (1969), this gives the prey isocline a characteristic humped shape, with the line bending downward toward the prey axis at low densities as well as at high densities. Bending the prey isocline toward the *H*-axis at low prey densities is equivalent to a counterclockwise rotation, and thus has a destabilizing effect (note the change of position of point B in Figure 23-19c). On the other hand, when scarce prey are more difficult to locate than abundant prey because, for example, a larger proportion can find good hiding places, predation is reduced, and the prey population can persist in the presence of denser predator populations. Graphically, this corresponds to an upward swing (clockwise rotation) in the prey isocline at the left (Figure 23-19d). Whether the isocline swings downward or upward at low prey densities depends on the availability of alternative prey and on the heterogeneity of the habitat, the latter of which we discuss in more detail below.

The predator isocline also bends under the weight of biological reality (Figure 23-20). As a population of predators increases, its food requirement increases, and predators must capture more prey to maintain their population at a constant level. The predator isocline, therefore, should be rotated clockwise (Figure 23-20b). Furthermore, greater social interference among predators (territoriality, for example) would reduce the efficiency with which they utilize prey resources and would bend the predator isocline further to the right. The availability of suitable breeding sites or of alternative food resources also could limit the size of the predator population independently of the abundance of the prey, causing the predator isocline to become horizontal at high prey population densities.

As the graphic approach also shows, extrinsic limits to the growth of the predator population, the availability of hiding places or refuges for the prey, and under some circumstances, the availability of alternative prey all act to stabilize the predator-prey interaction. The use of alternative prey allows predator populations to increase in the absence of the prey species of concern; hence the predator isocline rotates clockwise toward a horizontal position, at which extreme its growth would be completely independent of the density of the one type of prey (Figure 23-20c).

Biological considerations alter the stability of predator-prey systems in many ways. For a humped prey isocline, the system is stable when the predator isocline lies to the right of the hump and unstable when it lies to the left (Figure 23-21). That is to say, increasing predator efficiency reduces the stability of the system, all else being equal. When a vertical predator isocline lies close to the center of the hump, the system may exhibit internal stabilizing properties that lead to stable limit cycles (May 1972a, 1975b, Gilpin 1974).

Graphic analyses suggest that under most biologically realistic conditions, we should expect predator and prey interactions to be highly stable. Population cycles might even be the exception. How, then, can we account for the widespread existence of population cycles in nature? One possibility is that they represent

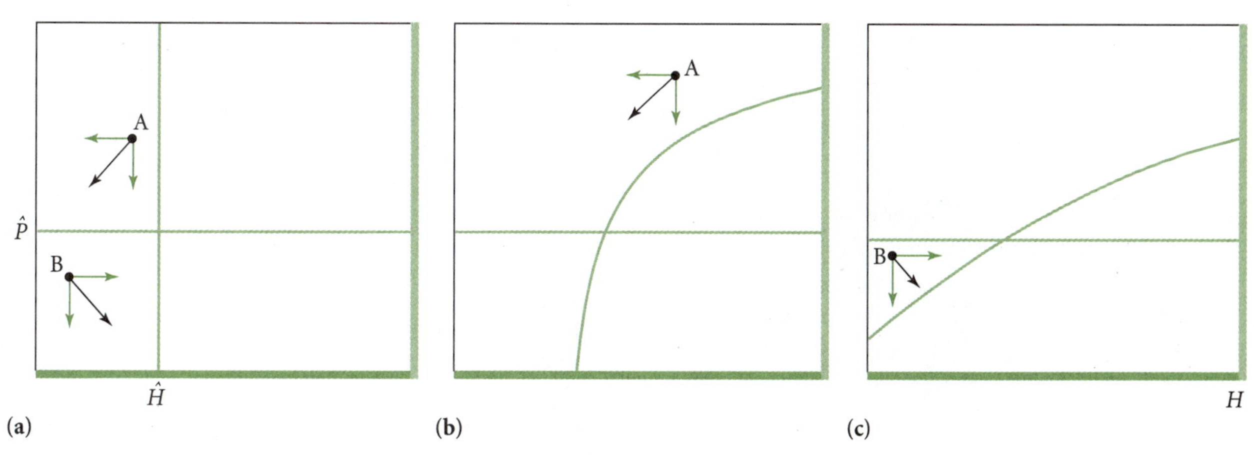

FIGURE 23-20 Possible reconfigurations of the predator isocline to incorporate more biological realism. (a) The basic Lotka-Volterra model with vertical predator and horizontal prey isoclines. (b) The predator isocline may bend to the right at high predator densities because of social interference among predators, limited breeding sites, or increased food requirements of a large number of predators. (c) If the predator switches prey at low prey densities, the predator isocline may become nearly horizontal.

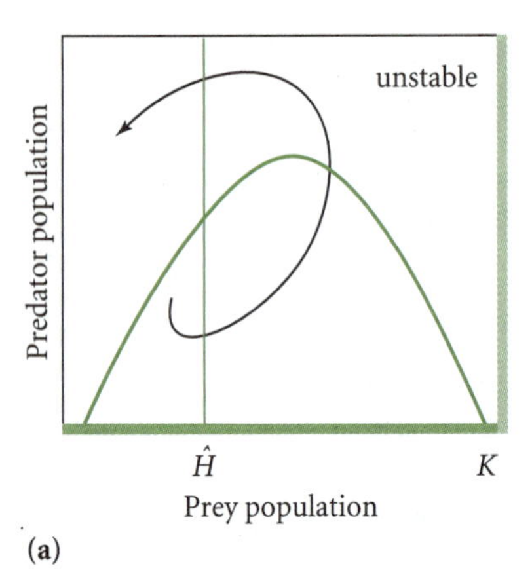

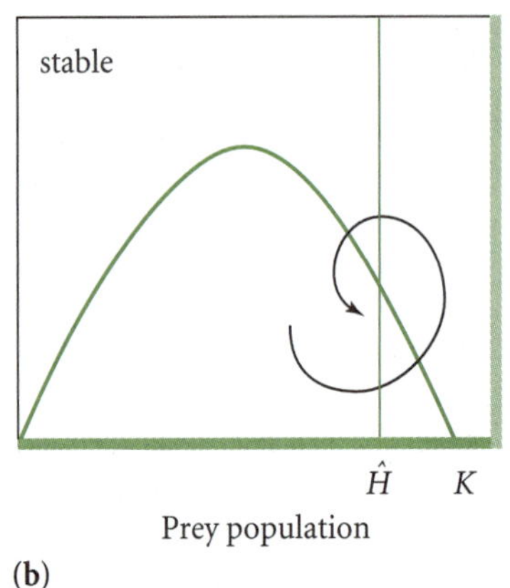

FIGURE 23-21 (a) When the predator isocline is positioned to the left of the peak of a hump-shaped prey isocline, the system is unstable. (b) When the predator isocline is to the right of and close to the peak, the system is stable.

stable limit cycles (stable oscillations; see Chapter 18). But the conditions required to maintain such stable oscillations are narrowly restricted and depend on a very special geometry of the predator and prey isoclines. For example, when a predator is efficient but its prey has suitable refuges or hiding places that protect it at very low population densities, the geometry of the predator and prey isoclines can create a stable cycle (Figure 23-22). Whereas the intersection of the predator and prey isoclines may result in an outward-spiraling population trajectory, this spiral is finally arrested by the presence of refuges or, equivalently, of some constant rate of prey immigration. Upon hitting the upturned prey isocline, the predator population trajectory would descend along it as if it were a "safe" corridor until predators became scarce enough that the prey could increase again. This would result in a stable oscillation because no matter where the predator population trajectory first struck the safe zone, it would always continue to the same point (A) before embarking on a new cycle.

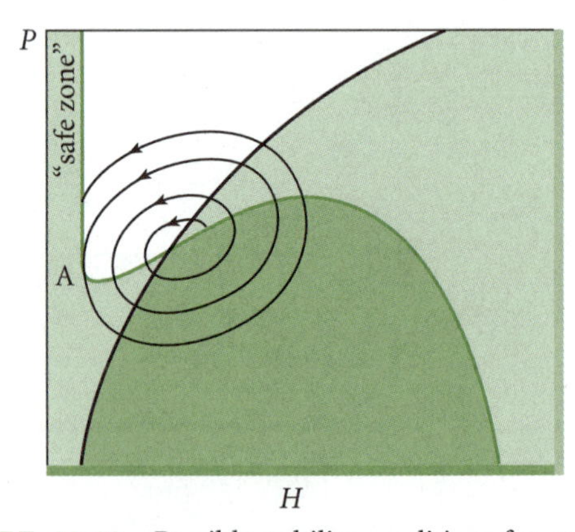

FIGURE 23-22 Possible stability conditions for an oscillating predator-prey system. The cycle is maintained by the balance between destabilizing influences at the intersection of the predator (black) and the prey isoclines (green) and the stabilizing influence of a "safe" zone.

23.8 Predator-prey systems may have two stable equilibria.

When the equilibrium isocline of the prey is humped and the predator can switch to alternative prey, the predator and prey equilibrium lines may cross in two places (Figure 23-23a). In this case, the equilibrium point on the right is stable (see Figure 23-21). The one the left, however, is not. In the proximity of the left-hand equilibrium point, combinations of predator and prey populations within the region of prey increase will tend to move toward the higher, stable equilibrium; those combinations outside that region will tend to move toward the predator axis, and hence to the elimination of the prey. Because populations tend to move directly away from the left-hand equilibrium point when displaced from it, it is referred to as unstable.

When prey have a refuge at low densities, the isocline of a predator with alternative prey may cross that of the prey at three points, two of which are stable equilibria (Figure 23-23b). In such cases, the predator-prey system may be regulated at one of two points, and certain conditions of the environment may cause it to switch from one point to the other. For example, if a predator-prey system were at equilibrium at point A in Figure 23-23b, a large reproductive output by prey that moved the prey population from H_1 to H_2 would bring the system under the influence of the equilibrium point at C. Recall that predator switching is one reason that the prey isocline is nearly horizontal. That is, the predator population density may be high at low prey densities because the predator simply switches to another prey type. Because of the potential role of predator switching in creating this isocline geometry, predator-prey interactions have been subjected to a second kind of graphic analysis, based upon the functional response curve, which we discuss next.

Let us turn for a moment from thinking about predator-prey interactions within the space of the

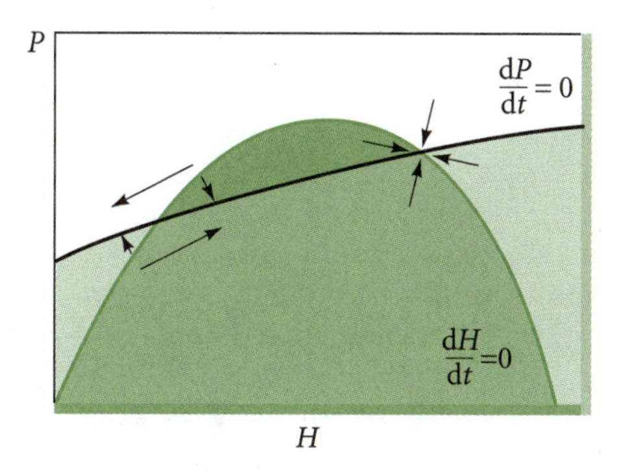

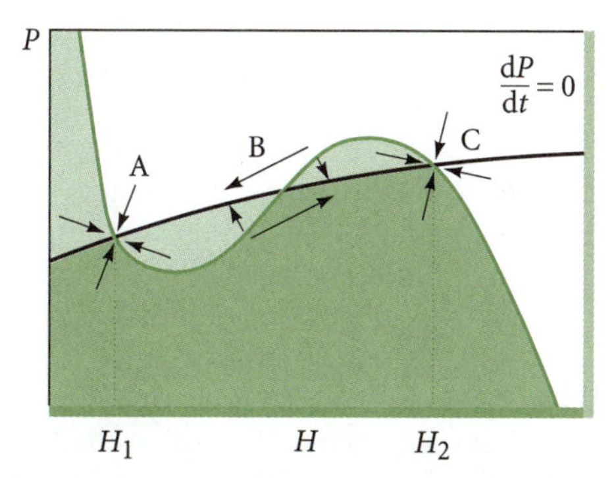

FIGURE 23-23 (a) An example of an interaction between predator (black) and prey (green) populations having two equilibrium points, the higher one stable and the lower one unstable. (b) An example of an interaction between predator (black) and prey (green) populations in which there are upper and lower stable equilibria (A and C) and an intermediate unstable point (B). The system may jump from one equilibrium to another if an event causes a large change in the population size of the prey, as shown by the black line, which indicates an increase in prey from H_1 to H_2. At point H_2, the system comes under the influence of the equilibrium C.

predator and prey population sizes (the space defined by the *H* and *P* axes) and consider the relationship between the rate at which prey add individuals to their population, the prey recruitment rate, and the rate at which predators take them out of the population, the functional response (Figure 23-24). Both recruitment rate and functional response are dependent on the prey population size, so we retain the *H*-axis. Let the vertical axis represent some proportion of the prey population. The **predation curve** (green line) is the sum of the functional and numerical responses of the predator population. The predation rate may be low at low prey densities owing to switching or the difficulty of locating scarce prey; it also tails off at high prey densities because of predator satiation and extrinsic limits to predator populations. The predation curve in Figure 23-24 thus represents a type III functional response. The recruitment curve (gray line) is the net contribution of prey births and deaths in the absence of predators. Hence the per capita recruitment rate is high when the prey population is small and goes to zero as the population approaches its carrying capacity.

The recruitment and predation curves in Figure 23-24 were drawn to produce three equilibrium points, which are homologous to the equilibria in Figure 23-23b. The highest and lowest points represent stable equilibria around which populations are regulated; the middle equilibrium is unstable. The lower equilibrium point (A) corresponds to the situation in which predators regulate a prey population substantially below its carrying capacity (*K*). The upper equilibrium (C) corresponds to the situation in which a prey population is regulated primarily by availability of food and other resources; predation exerts a minor depressing influence on population size.

Suppose the prey in the system shown in Figure 23-24 is a pest of an important agricultural crop, and the predator is a biological control agent. From the point of view of the farmer, the fewer prey in the field, the better, even though a good crop might be harvested even if there is some pest damage, as long as the damage is not extensive. Thus, levels of prey below H_1 might be acceptable, in which case the switching behavior of the predator would not cause a problem. However, if even small numbers of prey can cause significant economic damage, the predator is probably not an effective control agent, since at low prey densities, prey increase. High prey levels, however, such as between H_2 and H_3, would be unacceptable to the farmer, and therein lies a problem with this particular system. Predators maintain a shaky hold on prey populations at point A. If a heavy frost or an introduced disease reduced the predator population long enough

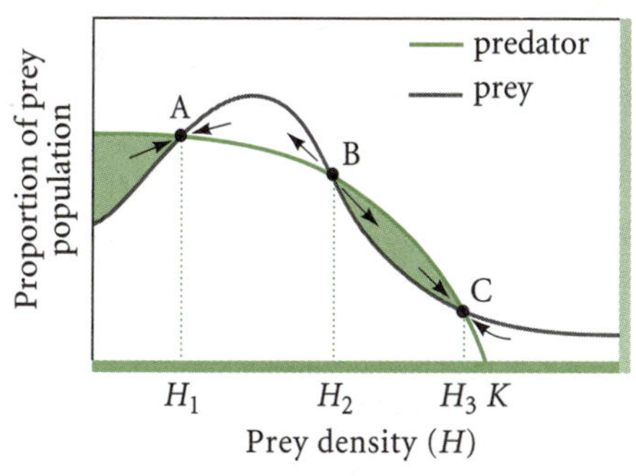

FIGURE 23-24 Predation and recruitment rates in a hypothetical predator-prey system. The green line gives the combined functional and numerical response of the predator, which shows a low prey consumption rate (measured as proportion of the prey population consumed) at low and high prey densities. The highest consumption is at intermediate prey densities (between H_1 and H_2). The prey recruitment rate in the absence of predators is given by the gray line. When predation exceeds recruitment, prey populations decrease. This occurs whenever the prey population is less than H_1 or between H_2 and H_3. Points A and C are stable equilibria for the prey population.

to allow the prey population to slip above point B, the prey would continue to increase to the higher stable equilibrium point (C), regardless whether the predator population recovered. To the farmer, this means that a crop pest, normally controlled at harmless levels by predators and parasites, suddenly becomes a menacing epidemic. After such an outbreak, predators could exert little control over the pest population until some quirk of the environment brought its numbers below point B, back within the realm of predator control.

As an example of such a scenario, outbreaks of tent caterpillars in the prairie provinces of Canada are generally preceded 2 to 4 years earlier by a year in which the winter is abnormally cold and the spring unusually warm (Ives 1973). These conditions presumably upset the normal balance between tent caterpillars and their predators and parasites. Infestations are subsequently brought under control by several cold winters that kill most of the tent caterpillar eggs (Witter et al. 1975).

Using the predation-recruitment diagram in Figure 23-24, we can examine the consequences of different levels of predation for prey population control (Figure 23-25). Inefficient predators cannot regulate prey populations at low densities; they depress prey numbers slightly, but the prey population remains near the equilibrium level set by resources (Figure 23-25a, point C). Increased predator efficiency at low prey densities can result in predator control at point A (Figure 23-25b).

When functional and numerical responses are sufficient to maintain high densities of predators, predation may effectively limit prey growth under all circumstances, and equilibrium point C disappears (Figure 23-25c). Finally, predation may be so intense at all prey densities that the prey are eaten to extinction (Figure 23-25d; no equilibrium point). We might expect this situation only in simple laboratory systems or when predator populations are maintained at high densities by the availability of some alternative, but less preferred, prey (hence no switching). Indeed, many ecologists have advocated providing parasites and predators of a pest with innocuous alternative prey to enhance biological control. At the very least, the curves suggest that the position of a predator-prey equilibrium, whether it is at very low levels of prey or close to their carrying capacity, may shift between extremes with small changes in closely matched predation and recruitment curves. Such considerations would appear to make equilibria at intermediate prey densities—that is, close to the hump of the prey equilibrium isocline—very unlikely.

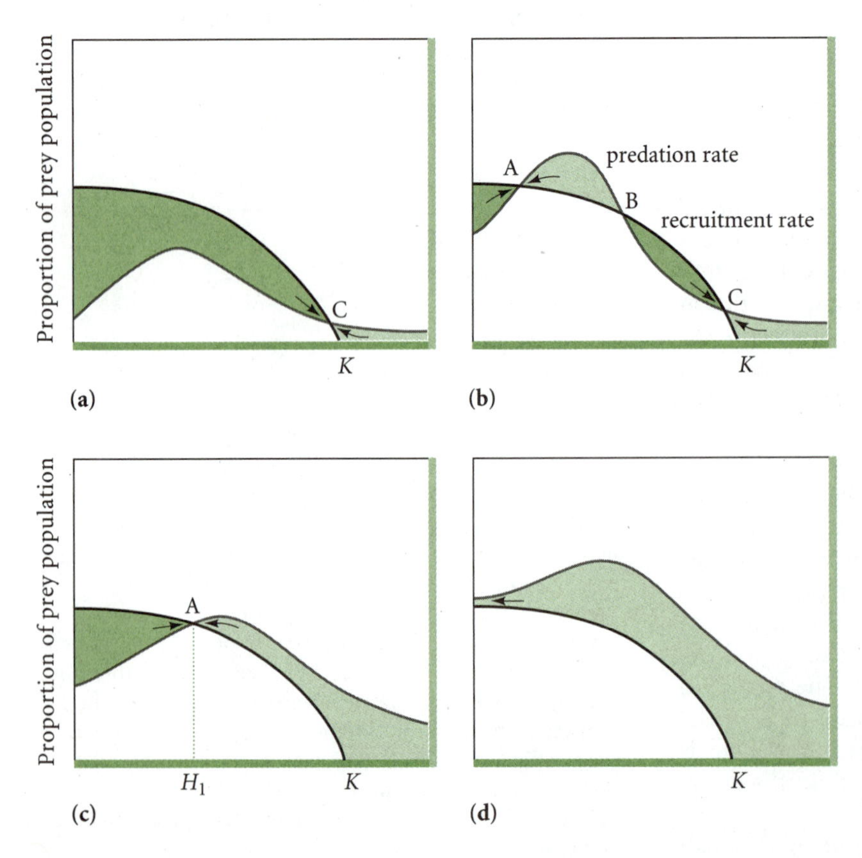

FIGURE 23-25 Predation and recruitment curves at different intensities of predation, showing the effect on the number of equilibria. The predation rate (gray line) is the result of the combined functional and numerical response of the predator. The prey recruitment curve in the absence of the predator is represented by the black line. The shaded areas show conditions under which the prey population increases. (a) The effect of predation does not influence the prey population as long as it remains below its carrying capacity. (b) The situation shown in Figure 23-24, in which the prey consumption rate is low at both low and high prey densities and high at intermediate densities. (c) The prey population experiences growth so long as it is below point H_1, which is about one-half its carrying capacity. (d) Predation pressure is strong at all prey population densities. Such a situation would lead to the elimination of the prey from the system.

Grazing resists analysis as a predator-prey system because population effects resulting from the consumption of vegetation are difficult to determine. Noy-Meir (1975) showed, however, that one may treat plant biomass increase in the same manner as population increase and thus determine the stable points of herbivore-plant interactions by analyzing graphic analogues of predator-prey equations, as we have done earlier. Hence the foregoing consideration of multiple stable points in predator-prey interactions applies equally well to herbivore-plant systems.

23.9 Predator-prey systems achieve characteristic population ratios.

The ability of a prey population to support predators varies with its density. A small prey population can support few predators because, while each prey individual's reproductive potential may be high, the total recruitment rate of a small population is low. Prey populations near their carrying capacities also are unproductive because, although prey are numerous, each individual's reproductive potential is severely limited by the effects of crowding.

At some intermediate density, the overall recruitment rate of the prey population reaches a maximum (Ricker 1954, Beverton and Holt 1957, Watt 1968). Because predators can remove a number of individual prey equivalent to the annual recruitment rate without reducing the size of the prey population, the prey population density that yields the maximum recruitment generally will support the greatest number of predators. This point corresponds to the peak of the hump of the prey isocline in a predator-prey graph. The rate of recruitment at this point is known as **maximum sustainable yield** (see May et al. 1979).

Ranchers and game managers are clearly concerned with maintaining populations of beef cattle, deer, and geese at their most productive levels to maximize man's ability to harvest these species without reducing their populations. We might ask whether predators also prudently manage their prey populations to maximize the productivity of their own populations. If so, how could such behavior evolve?

Territorial animals, which exclude competitors from their feeding areas, could indeed space themselves with respect to their prey so to achieve maximum yields. When the feeding areas of predators overlap, however, intraspecific competition dictates that each predator maximize its immediate harvest at the expense of long-term yields. Humans behave no differently. Intelligently managed ranches, with fences to exclude competing livestock, can achieve maximum sustainable yields. Alas, in highly competitive situations—fishing in international waters, to name one—humanity has proven to be pathetically short-sighted and imprudent, and stocks of many species of fish and other seafoods have declined dramatically (Beverton and Holt 1957, Cushing 1975). Overexploitation of whale populations has similarly led to the near extinction of some species, and has doomed the whaling industry (Laws 1962, 1977, McVay 1966).

In most cases, the level of exploitation of a prey population is determined by the ability of the predator to capture prey compared with the ability of the prey to avoid being captured. Both abilities are evolved characteristics. Regardless whether predators act prudently to manage prey populations for maximum sustainable yield, they often achieve a characteristic equilibrium with their prey populations. The relationship between wolves and various prey populations in several areas demonstrates this equilibrium rather well (Table 23-3). Population ratios and, in particular, the biomass ratios (1 pound of wolf for each 150 to 300 pounds of prey) are relatively constant despite the fact that both the the species and densities of the wolf's principal prey vary considerably with locality.

Different predator-prey systems may achieve different equilibria. The population ratio of mountain lions to deer in California is 1:500–600, which is equivalent to a biomass ratio of about 1:900, and the exploitation rate is only 6%, compared with values of 18% and 37% for wolves in two other areas. A mountain lion-elk-mule deer system in Idaho has a biomass ratio of 1:524, and an exploitation rate of 5% for the elk population and 3% for the mule deer population. Evidently, wolves exploit their prey more efficiently than mountain lions, perhaps because of their social hunting habits.

Where predators feed on more abundant populations of prey, as in savanna, grassland, and tundra habitats, predators not only are more numerous, but they achieve higher biomass ratios (1:50–1:150, see Table 23-3). Such conditions seem to enhance both prey productivity and predator efficiency. Large cats—lions and cheetahs—remove 16% of prey biomass in Nairobi Park, Kenya, where their biomass ratio is 1:140. Indeed, as we shall see in the next chapter, the high productivity of East African grasslands is of intense interest to ecologists.

We shall revisit the topic of predator-prey ratios again when we discuss community ecology (see Chapter 22). There is some controversy among community ecologists about whether such ratios represent an important measure of community structure, as some have claimed (Wilson 1996).

TABLE 23-3 The relationship between populations of predators and their prey in several localities

Locality	Predator	Principal prey	Density of predators (individuals 100 mi^{-2})	Ratio between predator and prey populations	
				Numbers	Biomass
Jasper National Park[1]	Wolf	Elk, mule deer	1	1:100	1:250
Wisconsin[2]	Wolf	White-tailed deer	3	1:300	1:300
Isle Royale[3]	Wolf	Moose	10	1:30	1:175
Algonquin Park[4]	Wolf	White-tailed deer	10	1:150	1:150
Canadian Arctic[5]	Wolf	Caribou	1.7	1:84	1:186
Utah[6]	Coyote	Jackrabbits	28	1:1,000	1:100
Idaho Primitive Area[7]	Mountain lion	Elk, mule deer	7.5	1:116	1:524
Ngorongoro Crater, Tanzania[8]	Hyena	Ungulates	440	1:135	1:46
Nairobi Park, Kenya[9]	Felids	Ungulates	96	1:97	1:140
Alaska[10]	Pomarine jaeger	Lemmings		1:1,263	1:90

References: 1, Cowan 1947; 2, Thompson 1952; 3, Mech 1966; 4, Pimlott 1967; 5, Kelsall 1968; 6, Clark 1972; Wagner and Stoddart 1972; 7, Hornocker 1970; 8, Kruuk 1969; 9, Foster and Coe 1968; 10, Maher 1970.

23.10 Risky prey behavior has population consequences.

Thus far we have focused principally on how predators affect the populations of their prey. This emphasis reflects both the historical development of predator-prey theory and the keen, and growing, interest in biological pest control. But what can we learn by taking the perspective of the prey, the putative loser in a predator-prey encounter?

Consider the predicament of a prey population in an environment with diminishing resources. As food becomes more and more scarce, each prey individual must spend more time searching for food, and this may mean a longer exposure to predators and, thus, a greater risk of being eaten. In such a situation, the regulatory effect of a predator on the prey population might be enhanced as a consequence of the increase in risky prey behavior. This situation is depicted in Figure 23-26, in which the prey recruitment rate and the prey population density are shown along with the total response curve of the predator. (The **total response** of the predator is the product of the functional and numerical responses and is plotted as per capita mortality of prey.) In Figure 23-26a, the total response of the predator is strongly dependent on prey density, and the prey population is held at a level far below its carrying capacity. In this situation, the prey population never reaches a level at which food limitation is critical; therefore, risky food-gathering behavior that would expose some prey to greater risk of predation is not present, and all prey have roughly the same probability of being eaten. When the total response of the predator takes the form shown in Figure 23-26b, the prey population is held at a level near its carrying capacity, K, at which intraspecific competition for food among prey is important. In this situation, predation events are more likely to be influenced by risky behavior on the part of the prey. This model of predator-prey interaction is called the **predation-sensitive food hypothesis** (Sih 1982, Sinclair et al. 1985, McNamara and Houston 1987, Sih and Moore 1990, Abrams 1991).

The predation-sensitive food hypothesis recognizes the trade-off between the need to avoid the tooth and claw of the resident predator and the need to obtain sufficient nourishment to survive in a limited-food environment. The hypothesis emphasizes the idea that predation may not be the only factor operating to limit prey population size; intraspecific competition during times of food limitation may also play an important role (Sinclair and Arcese 1995).

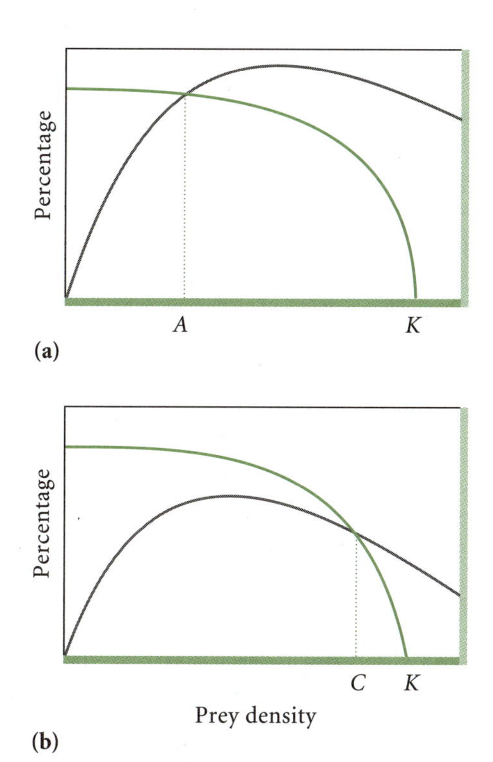

FIGURE 23-26 Effect of total response of a predator (gray) on the equilibrium density of prey in a theoretical predator-prey system. Both total predator response and prey recruitment are density dependent (green). (a) The predator population is strongly dependent on prey levels, and thus prey are held well below *K* (point A). (b) The predator population is less strongly dependent on prey levels, and prey populations stabilize near *K* (point C).

The predation-sensitive food hypothesis can be contrasted with two other notions about prey population regulation. First, as we indicated, intraspecific competition for food will not be a major factor in prey populations if predator pressure is sufficiently strong to keep the prey population below the level at which food becomes limiting (see Figure 23-26a). In such cases, the prey is said to be under **simple predator regulation.** This type of interaction is a feature of most of the models considered thus far in this chapter. Competition among prey may also be limited if predators take predominantly those prey individuals that are weakest or which, for whatever reason, stray into exposed or marginal habitats. This idea is sometimes called the **surplus hypothesis.** Note the relationship between these ideas and the nonequilibrium models of competition that we discussed in Chapter 21.

One may determine whether the predation-sensitive food hypothesis, simple predator regulation, or the surplus hypothesis applies to a particular system by comparing the condition of prey killed by predators with that of live animals, and of those that die of causes other than predation, such as starvation (Clark 1994, Sinclair and Arcese 1995). If the prey population is under simple predator regulation, then there should be little difference in the condition of living and dead prey, since the prey population will be at a level below which starvation is important. If predators are feeding on "surplus" prey (those starving or in marginal habitats), then the condition of all dead prey should be poorer than that of live prey, since, presumably, those prey animals forced into exposed areas are the ones with the greatest nutritional needs. If the predator-sensitive food hypothesis is relevant, than we should expect to see a hierarchy of conditions, with live animals being in the best shape, prey dying of natural causes or of starvation in the worst shape, and prey animals caught by predators somewhere in between (Clark 1994, Sinclair and Arcese 1995).

A. R. E. Sinclair of the University of British Columbia and P. Arcese of the University of Wisconsin undertook such a comparison for a wildebeest herd of the Serengeti by examining the body fat content of live animals, those killed by predators, and those dying of other causes during the periods 1968–1973, when wildebeest food was abundant, and 1977–1991, when food was limiting (Sinclair and Arcese 1995). Body condition was measured by visually determining the condition of the fat of the bone marrow of the femur. Marrow was classified as either solid white fatty, opaque gelatinous, or translucent gelatinous. Animals in the first or second categories were considered to be in good condition, whereas those with translucent gelatinous fat were considered to be in poor condition. The results of this study support the predation-sensitive hypothesis, though not unequivocally (Figure 23-27). Animals dying of predation or of some nonpredation causes were in poorer shape than live animals in both food-limited and non-food-limited years. When food was limiting, animals killed by predation were generally in better condition than those dying of other causes.

23.11 The spatial arrangement of predator and prey or parasitoid and host may affect the stability of the interaction.

Our treatment of predator-prey systems so far has been based on the population dynamics of the predator and the prey. The growth rates of predator and prey have been shown to be dependent on the relative sizes of their populations with respect to equilibrium densities, $\hat{H}$ and $\hat{P}$. But the interaction of population dynamics with environmental factors may affect the outcome of predator-prey interactions. In particular, the spatial distributions of the predator and the prey may have a dramatic effect on whether the two populations can coexist. Our interest in the spatial

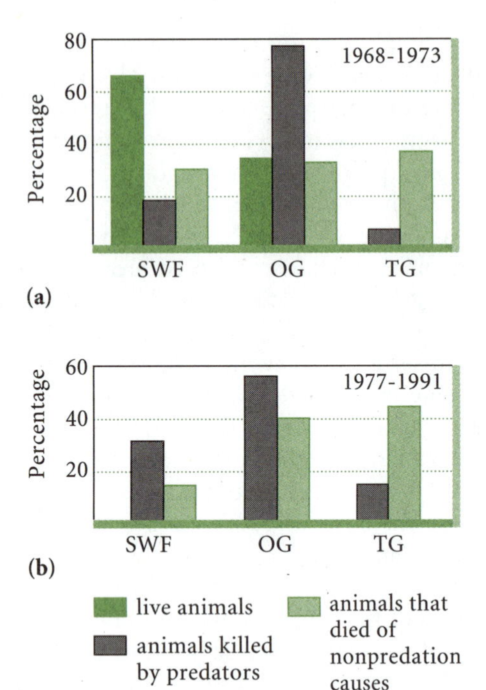

FIGURE 23-27 Frequency distribution of three categories of fat condition of the bone marrow of wildebeests (solid white fatty [SFW], opaque gelatinous [OG], translucent gelatinous [TG]) sampled during (a) high food years (1968–1973) and (b) low food years (1977–1991). Live animals, animals killed by predators, and animals that died of nonpredation causes were sampled. Fat scores of SWF or OG indicate good condition, whereas animals with scores of TG were in poor condition. These results support the predation-sensitive hypothesis. (*From Sinclair and Arcese 1995.*)

arrangement of predators, parasitoids and their prey and hosts derives from more than an intellectual interest. Many predator-prey and parasitoid-host systems that are of great importance to humans, particularly those of agricultural ecosystems, operate within spatial structures that are controlled by humans. The way in which adjacent crop fields are planted in accordance with crop rotation plans, for example, may juxtapose different crops, and the predators and prey associated with them, is different ways in different seasons. Human land management practices make structural changes in natural populations as well, which may result in changing population levels in predators and their prey.

C. B. Huffaker, a University of California biologist who pioneered the biological control of crop pests, attempted to produce a mosaic environment in the laboratory that would allow predator and prey to persist without restocking either population (Huffaker 1958). The six-spotted mite (*Eotetranychus sexmaculatus*) was the prey; another mite, *Typhlodromus occidentalis,* was the predator. Oranges provided the prey's food. Huffaker established experimental populations on trays within which he could vary the number, exposed surface area, and dispersion of the oranges (Figure 23-28). Each tray had forty positions arranged in four rows of ten each; where oranges were not placed, rubber balls of about the same size were substituted. The exposed surface area of the oranges was varied by covering the oranges with different amounts of paper, the edges of which were sealed in wax to keep the mites from crawling underneath. The exposed area was divided into numbered sections of equal area to facilitate counting the mites. In most experiments, Huffaker first established the prey population with twenty females per tray, then introduced two female predators 11 days later. Both species reproduce parthenogenetically; males were not required.

When six-spotted mites were introduced to the trays alone, their populations leveled off at between 5,500 and 8,000 mites per orange area. When predators were added, they rapidly increased in number and soon wiped out the prey population. Their own extinction followed shortly. Although predators always

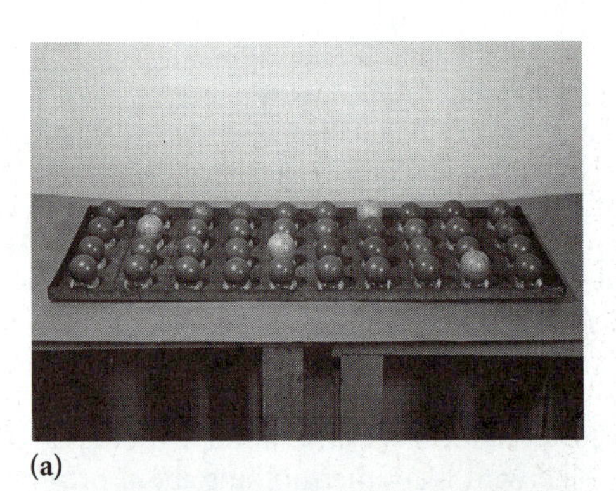

(a)

(b)

FIGURE 23-28 (a) One of Huffaker's experimental trays with four oranges, half exposed, distributed at random among the forty positions in the tray. The other positions are occupied by rubber balls. (b) Each orange was wrapped with paper and its edges sealed with wax. The exposed area was divided into numbered sections to facilitate counting the mites. (*From Huffaker 1958; courtesy of C. B. Huffaker.*)

A WALK THROUGH THE RAIN FOREST

A field worker searches for health-giving plants in the rain forest of Costa Rica. The medicinal properties of rain forest plants are numerous, and are still being discovered. Increasingly, pharmacologists are turning their attention to the healing practices of aboriginal forest peoples—one out of every four prescription drugs used today was discovered from studies of plants used medicinally by indigenous populations. *(Stephen Ferry, Liason.)*

"Delight . . . is a weak term to express the feelings of a naturalist, who for the first time, has wandered by himself in a Brazilian forest."
—Charles Darwin, 1832

Leaves in the rain forest are typically larger and more plentiful than in other forests. Up to 3 feet long, the leaves trap both rain and sunlight as they filter through the forest canopy. *(Courtesy of E. Hebets.)*

The iridescent wings of the blue morpho butterfly stand out against the browns and greens of the inner rain forest. While their brilliant wings assist in attracting mates, they also make them more noticeable to predators. The morpho can, however, mask its signature sheen simply by closing its wings: the undersides are the color of leaves and bark. *(Courtesy of Dr. Phil Schappert, University of Texas.)*

With the high productivity of rain forest plants, herbivores are abundant and varied. The green iguana, here camouflaged against the forest foliage, is a major consumer of leaves. *(Photograph by M. A. Guerra, courtesy of Smithsonian Tropical Research Institute.)*

Howler monkeys (*Alouatta pigra*) are so called for their loud, throaty cries that echo through the rain forest at sunrise and sunset. Howlers use their voices to mark their territory; through their calls, they communicate the location of their territory to other troops. *(W. M. Grenfell/Visuals Unlimited.)*

The agile coatimundi (*Nasua narica*) lives both in the trees of the rain forest and on the ground, where it hunts and forages for food. Naturally a diurnal animal, the coati is becoming increasingly nocturnal to avoid the human predators who hunt it for food. *(A. Kerstitch/Visuals Unlimited)*

The two-toed sloth is almost completely tree-dwelling. Hanging upside down, the sloth sleeps soundly for about 18 hours each day. *(Photograph by M. A. Guerra, courtesy of Smithsonian Tropical Research Institute.)*

The three-toed sloth—this one photographed at La Selva Biological Station, Costa Rica—moves at a rate of about 6 inches per second. *(Courtesy of E. Hebets.)*

Falling trees cause breaks in the forest canopy allowing sunlight to penetrate to the forest floor sustaining plant growth and promoting biodiversity. *(Courtesy of E. Hebets.)*

A black hawk (*Buteogallus anthracinus*) soaring above the Costa Rican rain forest. *(Daniel W. Gotshall/Visuals Unlimited.)*

The predominantly green color of this tree frog is cryptic, rendering it "hidden" from its predators, but the bright colors of its legs and ventral surfaces may be displayed during behavioral encounters. *(Photograph by M. A. Guerra, courtesy of Smithsonian Research Institute.)*

A night photograph of one of the many katydids at La Selva Biological Station, Costa Rica, attacking a leaf with a strong bite. Although katydids are common at La Selva, this type is rarely seen because of its nocturnal habits. *(Courtesy of E. Hebets.)*

Many rain forest creatures are ingeniously camouflaged, blending into their surroundings to avoid the notice of predators. Others are brightly colored. Vibrant colors can also be protective, advising predators that some prey is unpalatable, even poisonous. This use of color to warn is called aposematic coloration.

The brilliant colors of the green-eyed tree frog actually protect it from its predators, warning them of alkaloyd toxins in its body. *(Courtesy of E. Hebets.)*

Conspicuous banding on a caterpillar's body often indicates the presence of dangerous chemicals within. *(Photograph by J. Burgett, courtesy of the Smithsonian Tropical Research Institute.)*

Distinctly marked noxious insects, such as these true bugs (hemipterans), congregate to emphasize their coloration, making it a greater deterrent to predators. *(Photograph by C. C. Hansen, courtesy of the Smithsonian Tropical Research Institute.)*

The coloration and the antenna-like tails on the wings of this tropical butterfly trick predators into confusing its hind end and its head. *(Photograph by C. C. Hansen, courtesy of Smithsonian Tropical Research Institute.)*

ECOLOGISTS AT WORK

Mathematical relationships form the basis of ecological hypotheses.

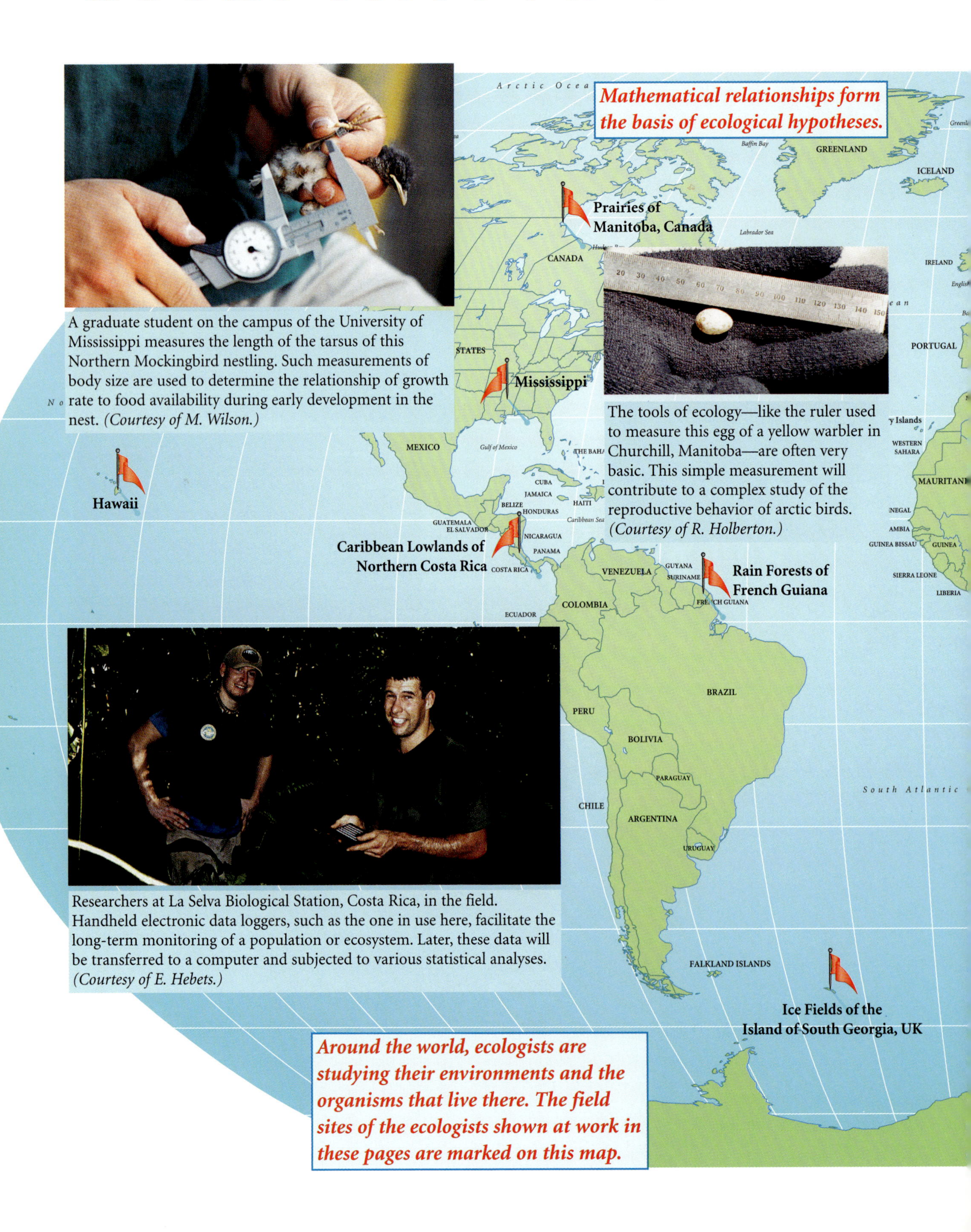

A graduate student on the campus of the University of Mississippi measures the length of the tarsus of this Northern Mockingbird nestling. Such measurements of body size are used to determine the relationship of growth rate to food availability during early development in the nest. *(Courtesy of M. Wilson.)*

The tools of ecology—like the ruler used to measure this egg of a yellow warbler in Churchill, Manitoba—are often very basic. This simple measurement will contribute to a complex study of the reproductive behavior of arctic birds. *(Courtesy of R. Holberton.)*

Researchers at La Selva Biological Station, Costa Rica, in the field. Handheld electronic data loggers, such as the one in use here, facilitate the long-term monitoring of a population or ecosystem. Later, these data will be transferred to a computer and subjected to various statistical analyses. *(Courtesy of E. Hebets.)*

Around the world, ecologists are studying their environments and the organisms that live there. The field sites of the ecologists shown at work in these pages are marked on this map.

Careful observation and systematic study reward ecologists with insights into the way nature works.

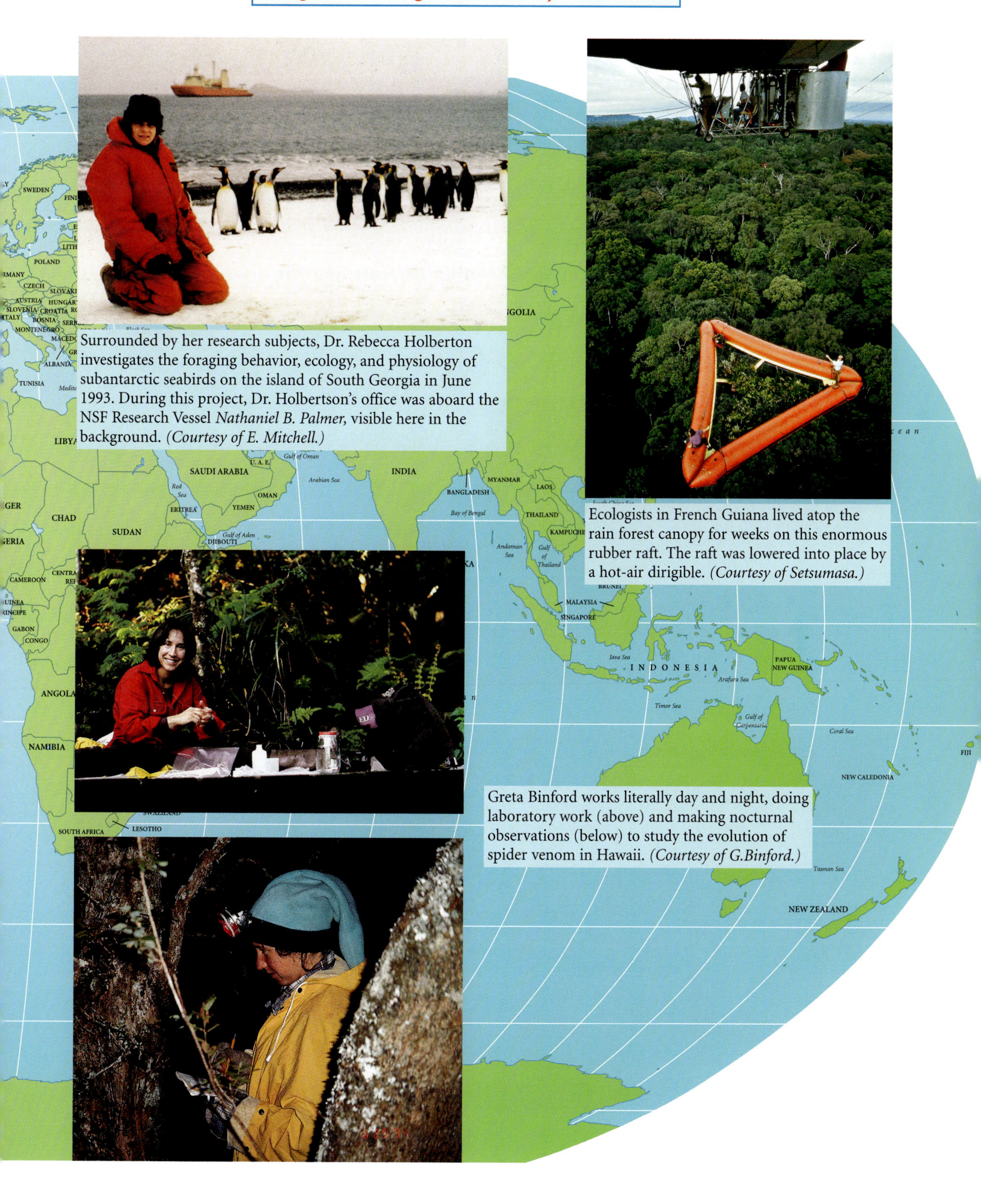

Surrounded by her research subjects, Dr. Rebecca Holberton investigates the foraging behavior, ecology, and physiology of subantarctic seabirds on the island of South Georgia in June 1993. During this project, Dr. Holbertson's office was aboard the NSF Research Vessel *Nathaniel B. Palmer*, visible here in the background. *(Courtesy of E. Mitchell.)*

Ecologists in French Guiana lived atop the rain forest canopy for weeks on this enormous rubber raft. The raft was lowered into place by a hot-air dirigible. *(Courtesy of Setsumasa.)*

Greta Binford works literally day and night, doing laboratory work (above) and making nocturnal observations (below) to study the evolution of spider venom in Hawaii. *(Courtesy of G.Binford.)*

The process of ecological inquiry begins with the generation of questions through exploration and observation. Here, researchers take measurements of trees (left) and collect samples from a marsh (right). Their data will eventually lead them to inferences about the world around them. *(Photographs by M. A. Guerra, courtesy of the Smithsonian Tropical Research Institute.)*

Florida Keys

These ecology students use a net called a seine to catch small fish in a stream. The seine, weighted along the lower edge so it drags along the bottom of the stream, is pulled through the water by two people. Every few yards or so, the students pull it out of the water and examine their catch. *(Courtesy of C. Britson.)*

Amazon Forest of Peru

Eileen Hebets photographs the Amazon forest of Peru. *(Courtesy of E. Hebets.)*

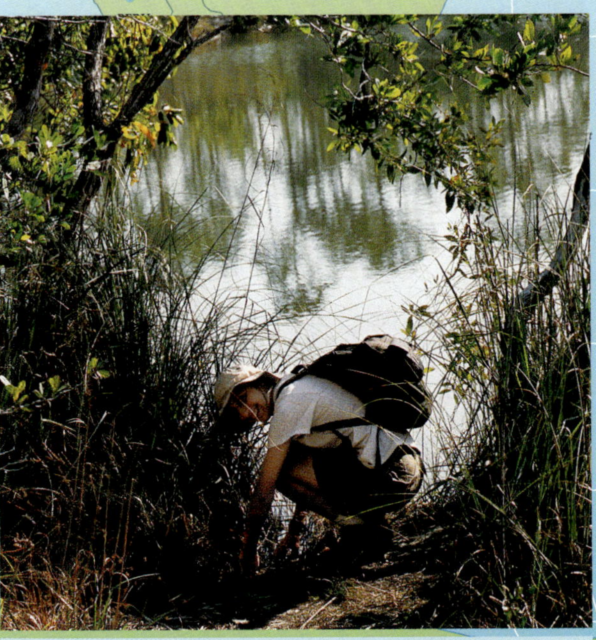

Karen Ober collecting carabid beetles in the Florida Keys for phylogenetic classification. *(Courtesy of E. Hebets.)*

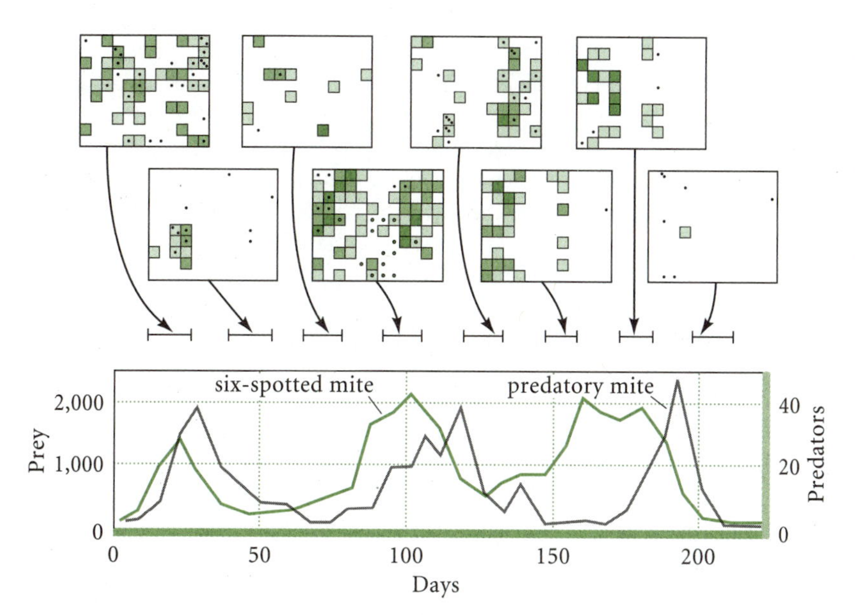

FIGURE 23-29 Population cycles of the six-spotted mite and the predatory mite *Typhlodromus* in a laboratory situation. The boxes show the relative densities and positions of the mites in the trays. The shading indicates the relative density of six-spotted mites; circles indicate the presence of predatory mites. (*After Huffaker 1958.*)

eliminated the six-spotted mites, the position of the exposed areas of oranges influenced the course of extinction. When the orange areas were in adjacent positions, minimizing dispersal distance between food sources, the prey reached maximum populations of only 113 to 650 individuals, and were driven to extinction within 23 to 32 days after the beginning of the experiment. The same area of exposed oranges randomly dispersed throughout the forty-position tray supported prey populations of up to 2,000 to 4,000 individuals that persisted for 36 days. Thus survival of the prey population could be prolonged by providing remote areas of suitable habitat to which predators dispersed slowly.

Huffaker reasoned that if predator dispersal could be further retarded, the two species might coexist. To accomplish this, he increased the complexity of the environment and introduced barriers to dispersal. The number of possible food positions was increased to 120, and the equivalent area of six oranges was dispersed over all 120 positions. A mazelike pattern of Vaseline barriers was placed among the food positions to slow the dispersal of the predators; *Typhlodromus* must walk to get where it is going, but the six-spotted mite spins a silk line that it can use like a parachute to float on wind currents. To take advantage of these behaviors, Huffaker placed vertical wooden pegs throughout the trays, which the mites used as jumping-off points in their wanderings. This arrangement finally produced a series of three population cycles over 8 months (Figure 23-29). The distributions of the predators and prey continually shifted as the prey, exterminated in one feeding area, recolonized the next a jump ahead of their predators.

Although the predator-prey cycle achieved by Huffaker was tenuous, we see that a spatial mosaic of suitable habitats did allow the stable coexistence of predator and prey populations. But, as we saw in Gause's experiments with protozoa (see Chapter 21), predator and prey also may coexist locally if some prey can take refuge in hiding places. And when the environment is so complex that predators cannot easily find scarce prey, stability can be achieved.

23.12 The dynamics of predator-prey systems in metapopulations are influenced by spatial relationships and movement patterns.

Spatially sensitive predator-prey models incorporate space either as a continuum or as a collection of patches of suitable habitat (Levin 1978, McLaughlin and Roughgarden 1993). It is the dynamics of predator and prey populations arranged in patches that most interest us here. An appreciation of what stabilizes or destabilizes predator-prey systems in patches requires some refinement of our ideas about stability. Recall our earlier discussion of metapopulations, in which we emphasized that the dynamics of single populations arranged in patches occur on two spatial scales: local and regional (see Chapter 17). The persistence of that portion of the metapopulation occupying a particular patch is dependent on the demographic characteristics of that population (birth, death, immigration, emigration) and on local environmental conditions, which may be different from the conditions in other patches. Populations in patches may go extinct and be recolonized by dispersal from other patches. The population as a whole, then, may be stable even though bits and pieces of it come and go. Thus, we may think of the population as having both local and regional (or global) stability properties (McLaughlin and Roughgarden 1993).

Patchy predator-prey systems also display properties of local and regional stability, but the situation is a bit more complicated than in the case of a single population owing to the complex dynamics of predator and prey behavior. Mite predator-prey interactions provide an excellent example of the complexity of such systems. Phytophagous mites such as spider mites are common plant pests that occur in patches, both because the plants on which they feed are usually distributed heterogeneously and because of their specialized feeding habits, which result in clumping on certain parts of individual plants (Sabelis et al. 1991). Dispersal from one plant to another is usually passive, highly risky, and restricted to certain life stages of the mite. Predatory mites generally have greater dispersal capabilities, and thus move to patches of plant-eating mites, where their populations expand as they feed on their prey. Thus, three types of patches may be envisioned: empty patches (not yet invaded by phytophagous mites), prey patches (phytophagous mites only), and predator-prey patches (phytophagous mites and their mite predators). There is, of course, a fourth type of patch—that containing only predators—which we will not consider here, because it is assumed that unless the prey migration rate is extremely high (and it usually is not in the case of phytophagous mites; see below), thereby providing the predators in the predator-only patches something to eat, such patches will be transient.

A number of interacting dynamics underlie the local and regional stability of such systems. Phytophagous mites have a tendency to overexploit their food resources in the absence of predator mites, and thus prey patches may be ephemeral. This situation may be exacerbated in the presence of high local variation in environmental conditions. The dispersal capabilities of many phytophagous mites are limited, and thus the rate of patch colonization may be low. Predatory mites must locate patches containing prey (there is some evidence that the plant aids in this process through chemical responses to the phytophagous mites) and, once those patches are exploited, move to another prey patch. Local extinction rates may be high because of these factors.

A Simple Lotka-Volterra Model of Predator-Prey Patch Dynamics

The conditions under which regional stability may occur can be revealed by the application of Lotka-Volterra models of predator-prey patch dynamics. If we ignore the dynamics of predator and prey populations within patches, we can construct a simple predator-prey patch model that depends on the predator and prey population sizes, rates of colonization of new patches by prey, and the extent to which predators respond to new prey colonies by moving to invade them. Let N be the population size of the prey and M be the size of predator population. The rates of change in these two populations living in patches may be represented as

$$\frac{dN}{dt} = aN - bNM - cN \qquad (23\text{-}10)$$

and

$$\frac{dM}{dt} = bNM - dM, \qquad (23\text{-}11)$$

where a is a coefficient representing the rate at which dispersing prey found new colonies, b is the rate at which predators invade prey patches, and c and d are the rates at which arbitrary patches of prey and predator go extinct (Sabelis et al. 1991). Using the same approach that we employed in Section 23.3, we may set $dN/dt = 0$ and $dM/dt = 0$ and solve for the equilibrium population sizes $\hat{N}$ and $\hat{M}$, which are found to be, respectively, $\hat{N} = d/b$ and $\hat{M} = (c - a)/b$. Were we to examine this system further, we would discover that the predator and prey populations oscillate in a continuous cycle about a neutral equilibrium.

To be sure, the model presented above overly simplifies the complex dynamics of predator-prey populations. On the scale of individual patches (local scale), encounters of predator or parasitoid with prey or host more often than not lead to disaster for the prey (local extinction), as we saw in Huffaker's experiments. Regionally, though, predator-prey systems persist, most likely because dispersal patterns of predator and prey against a background of habitat heterogeneity make for a mosaic of extinction and recolonization among patches (Mclaughlin and Roughgarden 1993), as we shall see next.

In this chapter we have explored the dynamics of populations of predators and parasitoids and their prey and hosts. Unlike competition, predation is always antagonistic, and the resource involved in the interaction is one of the participants—the prey or host. In the next chapter, we will discuss another set of interactions between consumer and host, herbivory and parasitism. Those interactions are similar to predation in that one of the participants represents the resource, and the relationship between individual participants is antagonistic. However, we shall see that they differ in some very important ways, not the least of which is that the consumer (the herbivore or the parasite) may gain from the interaction without killing the resource (the host plant or host).

SUMMARY

1. Studies of mite-mite interactions in strawberries and of avian predation on spruce budworms show that predators may limit the size of their prey populations. Studies of azuki bean weevils and of protozoans show that the interactions of predator and prey may lead to population oscillations.

2. A. J. Lotka and V. Volterra, in the 1920s, devised a simple model of predator and prey dynamics that predicted population cycles. This model used differential equations in which the rate of prey removal was directly proportional to the product of the predator and prey populations. The model predicts that predator and prey populations will oscillate. Because the model includes no factors that tend to move the predator and prey populations toward their joint equilibrium $(\hat{H}, \hat{P})$, it is said to have a neutral equilibrium.

3. In 1933 and 1935, Nicholson and Bailey proposed alternative models in which the rate of prey removal was an asymptotic function of predator density. These models were based explicitly on the relationship between parasitoids and their hosts. The simplest form of the models predicts that parasitoid-host interactions will be unstable because they will undergo ever-increasing oscillations.

4. In 1959, C. S. Holling introduced the concept of the functional response, which described the asymptotic relationship between prey removal rate per predator and the density of the prey. Three general types of functional responses are recognized: type I, in which the number of prey consumed per predator increases linearly with the prey population density; type II, in which the number of prey consumed increases at a decreasing rate until some satiation level is reached; and type III, in which the number of prey consumed increases slowly at low prey densities, rapidly at intermediate densities, and eventually levels off to a satiation level at high prey densities. The type III functional response curve may result in stable regulation of the prey population at low densities.

5. The numerical response describes the response of a predator population to increasing prey density by population growth and immigration.

6. Graphic analyses of predator-prey interactions based on simple models, but with such features as density dependence, prey refuges, and alternative prey added to give realism, demonstrate the conditions for stability in predator-prey interactions. In general, stability is promoted by density dependence of either the predator or prey, refuges or hiding places in which some prey can escape predation, reduced predator efficiency, and, in some circumstances, availability of alternative prey. Stable population cycles in nature apparently express the balance of these stabilizing factors and the destabilizing influence of time delays in population responses.

7. Models of consumer populations suggest that systems can have two stably regulated points between which populations may move, depending on environmental conditions. The lower equilibrium is determined by the strong depressing influence of predators on prey populations; the upper equilibrium is close to the carrying capacity of the prey in the absence of predation.

8. Each prey population has a density, at which recruitment is greatest, which will support the greatest number of predators. The recruitment rate at this point is called the maximum sustainable yield. When a single predator can completely control its prey, as humans can do in the cases of many domestic and game species, maximum sustainable yields can be achieved. When predators compete for the same resources, maximization of short-term yields generally precludes the achievement of maximum sustainable yield.

9. In order to obtain nourishment in food-limited situations, prey animals may adopt behaviors that put them at greater risk of predation and thus effectively augment predation pressure, an idea called the predation-sensitive food hypothesis. Prey populations may also be regulated simply by predation pressure or by a reduction in intraspecific prey competition because of consumption of surplus prey by predators.

10. Experimental studies have demonstrated that predator and prey populations can be made to oscillate in the laboratory. C. B. Huffaker's experiments showed that maintenance of population cycles usually requires a complex environment in which prey are able to establish themselves in refuges.

11. The stability of predator-prey systems in patchy environments is dependent on local patch dynamics and on the dispersal patterns of both predator and prey. Stability of parasitoid-host metapopulations depends on diffusive dispersal.

EXERCISES

1. Determine whether there are predators or parasites used to control pests in the agricultural areas near where you live.

2. The mortality term of the Lotka-Volterra prey population growth equation (the term pHP in equation [23-1]) assumes that the greater the predator (P) and prey (H) populations, the more consumption will occur (law of mass action). Can you think of situations in which this will not be true? (Hint: Consider specific behaviors and spatial arrangements of predator and prey such as those discussed in Section 23.4.)

3. The dynamics of predator and prey populations as described by the Lotka-Volterra models (equations [23-1] and [23-2]) assume that the efficiency with which the predator takes prey, p, and the efficiency with which the predator converts the energy of the prey into young, a, are both constant. Using what you have learned about the responses of organisms to the physical and biotic environment (Chapters 3–8) and your general knowledge about variation in the natural world, prepare a short essay discussing (a) how these assumptions are violated in the natural world and (b) how the predictions of the models might change if the assumptions are relaxed.

4. If the equilibrium prey population density for a particular predator-prey interaction is $\hat{H} = 100$ and $d = 0.2$, $a = 0.3$, and $r = 0.7$, what is the equilibrium predator population $\hat{P}$?

5. The Lotka-Volterra models (equations [23-1] and [23-2]) are equilibrium models. In light of our discussion in Chapter 21 about the assumptions and weaknesses of equilibrium models, discuss how factors such as disturbance might affect the outcome of predator-prey interactions.

6. The Nicholson-Bailey model of parasitoid-host interactions relies on the Poisson distribution, which is used in other ecological models. Explain the various uses of the Poisson distribution in ecology.

7. The Holling disc equation models a situation in which a predator moves about looking for prey, as in the exercise in which blindfolded humans are required to discover small discs of paper, the activity for which the equation is named. The rate at which predators encounter prey is determined by the prey population density and the amount of time that the predator takes to handle each prey. Suppose that you were asked to develop a model for an animal that forages by undertaking sorties from a certain location. That is, the predator leaves its position—say, a perch—goes to get a prey item, and then brings it back to the perch before eating it. Would the disc equation fairly represent this situation? What other factors would have to be considered in developing a model for this predator? (Hint: See Chapter 31.)

8. As we have seen, predators may limit the size of prey populations and thus may be useful in pest control in agriculture. If you were choosing a predator to control a particular agricultural pest, what features of functional and numerical response would you be hoping to see in the predator? How would the number of equilibria in the predator-prey system influence the effectiveness of the predator as a control agent?

CHAPTER 24

Herbivory and Parasitism

GUIDING QUESTIONS

- What characteristics distinguish herbivory and parasitism from parasitoidism and predation?
- How do plants compensate for the effects of herbivory?
- How can herbivores change the architecture of a plant?
- What is overcompensation, and why is it a controversial idea?
- What is the nature of the feedback relationships between herbivores and plant production?
- How can multiple consumers affect plant performance and population dynamics?
- How does the timing of herbivory with respect to developmental stage affect plants?
- How does plant distribution affect herbivory by insects?
- How does vegetation texture affect plant susceptibility to herbivory?
- How does the herbivore functional response differ from that of predators?
- What is the difference between direct and indirect transmission of parasites?
- What proportion of a population must be immunized to stop the spread of a disease?
- What are the mechanisms that promote periodic outbreaks of disease?
- How does host mobility affect the evolution of virulence?

In the previous chapter, we portrayed the interactions of predators and their prey and parasitoids and their hosts using models that accounted for changes in the densities of the interacting species. We now turn our attention to two other groups of consumers, herbivores and parasites. Herbivory and parasitism are qualitatively distinct from predation and parasitoidism because individual interactions do not always result in the death of one of the participants. Thus, we will focus on the physiological and morphological responses of individuals rather than on population densities. In the case of herbivores, we will discuss how plants compensate for partial consumption and how herbivores respond to changing availability of plants. In the case of parasites, we will focus on the dynamics of disease transmission, paying particular attention to the problem of immunization. To be sure, the interactions of herbivores and parasites with their hosts do have population consequences for all participants, and these will be discussed as well.

24.1 Herbivory and parasitism are complex processes that differ from predation and parasitoidism.

The term **herbivory** refers to many types of consumer-resource interactions involving the consumption of plants and plant parts. In cases in which the herbivore is smaller than the plant, the term **host plant** is used to refer to the plant (for efficiency, when discussing herbivory and parasitism together, we may use the term *host* to refer both to host plants and to parasite hosts). Predation and parasitoidism can be compared with parasitism and the various types of herbivory along two continua (Figure 24-1). The terms **grazing** (eating grass or forbs) and **browsing** (eating leaves or young shoots) are used most often to refer to the activities of relatively large herbivores such as deer, moose, or wildebeests, or in reference to algae-eating marine organisms of intertidal and subtidal areas, but grasshoppers and locusts might also be properly considered grazers (Pollard 1992). Grazing rarely results in the death of the grazed plant. Grazers generally engage in only a brief interaction with each individual plant, cropping off a leaf or a blade of grass and moving along, leaving the stems and roots of the plant behind. Like grazing, insect herbivory rarely results in the demise of the plant host, but, as with parasites, the relationship of an individual insect herbivore with its host plant is often lengthy. The distinction between herbivory and predation becomes somewhat blurred in cases in which the herbivore consumes the entire plant, resulting in its death. The consumption of small aquatic plants, such as duckweed; plantlike planktonic organisms, such as diatoms or dinoflagellates, by aquatic herbivores; the eating of small saplings roots and all; and the consumption of seeds are all examples of such herbivory. These types of interactions may be quite properly viewed as predation events. Indeed, the consumption of seeds is often referred to as **seed predation** or, more generally, **granivory.** The relationship between a parasitoid and its host is, like that of a parasite or an herbivorous insect and its host, generally very intimate, and, like that of a predator and its prey, nearly always results in the death of the host.

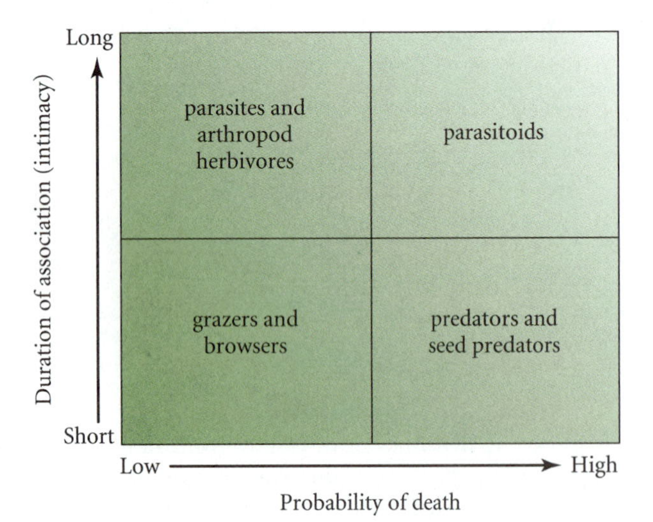

FIGURE 24-1 Consumer-resource interactions may be categorized by the length or intimacy of the interaction between an individual consumer and resource and by the probability that the interaction will result in the death of the resource. Parasites and arthropod herbivores engage in long-term interactions that rarely result in the death of the host. Parasitoids usually interact with their hosts for a relatively long period of time, and the interaction results in the death of the host. Grazing and browsing involve a brief encounter in which a bit of a plant or bush is removed. Predation is a relatively short-lived event resulting in the death of the prey. (*After Pollard 1992.*)

As we saw in the last chapter, the dynamics of predator-prey population interactions are measured in units of the individual: individual predators and prey are added to the population by birth (in population models, N is increased) and removed by death (N is decreased). The outcome of successful predation is unequivocal: death occurs, and an individual is removed from the prey population. The accounting of births and deaths, or of rates of births and deaths, forms the basis of predator-prey population models. Populations of host plants and parasite hosts are not so clearly responsive to the actions of herbivores and parasites. Because herbivory and parasitism may not result in the outright death of the host, the size of the host population may not change as a result of herbivory or parasitism. Rather, the effect of consumption is manifested in changes in the physiology and morphology of individual hosts, changes that may or may not alter rates of mortality or natality within the host population. Indeed, as we shall see, in some cases, the effect of grazing and browsing is almost like knocking the plant back to a juvenile stage at which growth is more vigorous. These features of plant-herbivore and host-parasite interactions, coupled with the persistence of these interactions through time, make them considerably more intimate than the interactions between predators and prey. An evolutionary consequence of this intimacy is the astounding variety of specializations in both the adaptations of herbivores and parasites for selecting and consuming plants, plant parts, or hosts, and the defensive reactions of plants and hosts.

In previous chapters we have emphasized that spatial dynamics are an important force in ecological processes (see Chapters 14, 17, 19, and 23). Like predators and parasitoids, herbivores and parasites are influenced by the distributions of their hosts. Plants are rooted in the soil, or, as in the case of mistletoe and other epiphytic plants, affixed to another plant. Thus, herbivores often exist in metapopulations, in which patches of suitable habitat are interspersed with areas where the preferred host does not exist. The larvae of butterflies and moths are among the most common herbivores (see Figures 1-9 and 24-6b). Butterflies, which have been particularly well studied owing to the alarming decline in their numbers worldwide, demonstrate the evolutionary tightrope created by spatial dynamics. Our vision of flocks of migrating monarch butterflies notwithstanding, most butterflies do not range far from the patch of vegetation where they were placed as eggs by their mothers (Singer and Ehrlich 1979, Ehrlich 1984). Small local populations often go extinct because of insufficient immigration, but some butterflies do move about, and in undisturbed areas, a balance between extinction and colonization is reached. At the metapopulation scale, the population persists. However, when habitat fragmentation by humans or natural disturbance changes the spatial arrangement of the habitat patches, the butterfly population may be imperiled (Thomas and Hanski 1997).

The complex life cycles of most parasites and disease organisms may be attributed in part to the complex spatial dynamics of their hosts. The parasite environment is decidedly patchy; each individual host serves as a favorable patch, and the area between hosts is highly unfavorable. Many parasites infect several hosts during their life cycle. Each host will have different genetic characteristics, physiology, behavior, and morphological structure. Moreover, hosts may be highly mobile. The parasite's adaptations for survival within a host may not be consistent with its need to move from one host to the next. You need only think of a tapeworm to understand that the morphological characteristics required for the worm to hold its position inside the digestive system of a vertebrate host provide no mechanism for finding, moving to, and infecting a new host should the worm find itself outside the vertebrate body. To be sure, other adaptations, particularly those related to reproduction and development, arise in parasites to overcome this problem. The point is that, if a parasite population is to persist, these problems must be overcome. Problems of locating hosts (contagion) figure importantly in models designed to represent parasite populations and, particularly, the dynamics of diseases caused by parasites. In the case of endoparasites, population growth may occur within only one of the several hosts required in the parasite life cycle, resulting in a subdivision of the parasite population and fostering population interactions among the descendants of infecting individuals.

24.2 Plant-herbivore interactions are taxonomically and ecologically diverse.

Herbivorous organisms are found in at least ten phyla of animals, representing body plans from simple single-celled organisms such as amoebas (protozoa) that eat plantlike single-celled organisms such as diatoms, to vertebrates (Table 24-1) (Crawley 1983). Numerous species of nematodes (roundworms) consume belowground plant parts. Species of echinoderms and mollusks such as sea urchins, marine slugs, and snails are prominent consumers of marine algae. The remarkable water bears (tardigrades), tiny organisms that are restricted to patches of moss, are all herbivorous, obtaining food by sucking the contents from individual moss cells. A vast number of arthropods—the most diverse animal phylum—are herbivorous. All of us have witnessed grasshoppers and caterpillars chewing on leaves, or heard complaints about various thrips, aphids, scale insects, and beetles from gardener friends (Figure 24-2). Among the crustaceans, which include the familiar crabs and lobsters as well as a host of small planktonic forms such as water fleas (*Daphnia*), copepods, and the like, are species that eat seaweed, phytoplankton, and plant fruits. A few species of forest floor millipedes are sap feeders. The greatest diversity of arthropod herbivores occurs among the insects. Eight of the twenty-six orders of insects, representing about 25% of all insects, include herbivorous species (Figure 24-3; Weis and Berenbaum 1989). Although most arachnids (spiders, mites, etc.) are carnivorous or omnivorous, many species of mites (order Acarina) make a living by sucking plant juices. The effects of mite herbivory are of considerable importance in agricultural ecosystems.

Herbivorous species are found within every major class of vertebrates. Carnivorous feeding habits are more common than herbivory among fishes, but many marine and freshwater species are specialized to eat phytoplankton, algae, or other plant material (Moyle and Cech 1988). Some of the most spectacular

FIGURE 24-2 Aphids are insect herbivores of considerable economic importance owing to the damage they cause to crop plants. (*Courtesy of R. B. Suter.*)

TABLE 24-1 Various types of plant tissue and their herbivores

Tissue	Mode of feeding	Examples of feeders
Leaves	Clipping	Ungulates, slugs, sawflies, butterflies
	Skeletonizing	Beetles, sawflies, capsid bugs
	Holing	Moths, weevils, pigeons, slugs
	Rolling	Microlepidoptera, aphids
	Spinning	Lepidoptera, sawflies
	Mining	Microlepidoptera, Diptera
	Rasping	Slugs, snails
	Sucking	Aphids, psyllids, hoppers, whiteflies, mites
Buds	Removal	Finches, browsing ungulates
	Boring	Hymenoptera, Lepidoptera, Diptera
	Deforming	Aphids, moths
Herbaceous stems	Removal	Ungulates, sawflies
	Boring	Weevils, flies, moths
	Sucking	Aphids, scales, cochineals, bugs
Bark	Tunneling	Beetles, wasps
	Stripping	Squirrels, deer, goats, voles
	Sucking	Scales, bark lice
Wood	Felling	Beavers, large ungulates
	Tunneling	Beetles, wasps
	Chewing	Termites
Flowers	Nectar drinking	Bats, hummingbirds, butterflies
	Pollen eating	Bees, butterflies, mice
	Receptacle eating	Diptera, microlepidoptera, thrips
	Spinning	Microlepidoptera
Fruits	Beneficial	Monkeys, thrushes, ungulates, elephants
	Destructive	Wasps, moths, rodents, finches, flies
Seeds	Predation	Deer, squirrels, mice, finches, pigeons
	Boring	Weevils, moths, bruchids
	Sucking	Lygaeid bugs
Sap	Phloem	Aphids, whiteflies, hoppers
	Xylem	Spittlebugs, cicadas
	Cell contents	Bugs, hoppers, mites, tardigrades
Roots	Clipping	Beetles, flies, rodents, ungulates
	Tunneling	Nematodes, flies
	Sucking	Aphids, cicadas, nematodes
Galls	Leaves	Hymenoptera, Diptera, aphids, mites
	Fruits	Hymenoptera
	Stems	Hymenoptera, Diptera
	Roots	Aphids, weevils, Hymenoptera

(From Crawley 1983.)

of these are the herbivorous fishes of coral reef ecosystems, many of which, like the highly territorial damselfishes, are brightly colored and leave the human visitor with the impression that they dominate the coral reef fish fauna. Carp are among the most common freshwater fish herbivores. The diets of many turtles and land tortoises are composed exclusively or predominantly of plant material. The giant land tortoises of the Galápagos Islands are some of the best-known turtle herbivores. The marine iguana of the Galápagos Islands eats seaweeds (Figure 24-4). Other iguanas consume the leaves of terrestrial plants as well as some

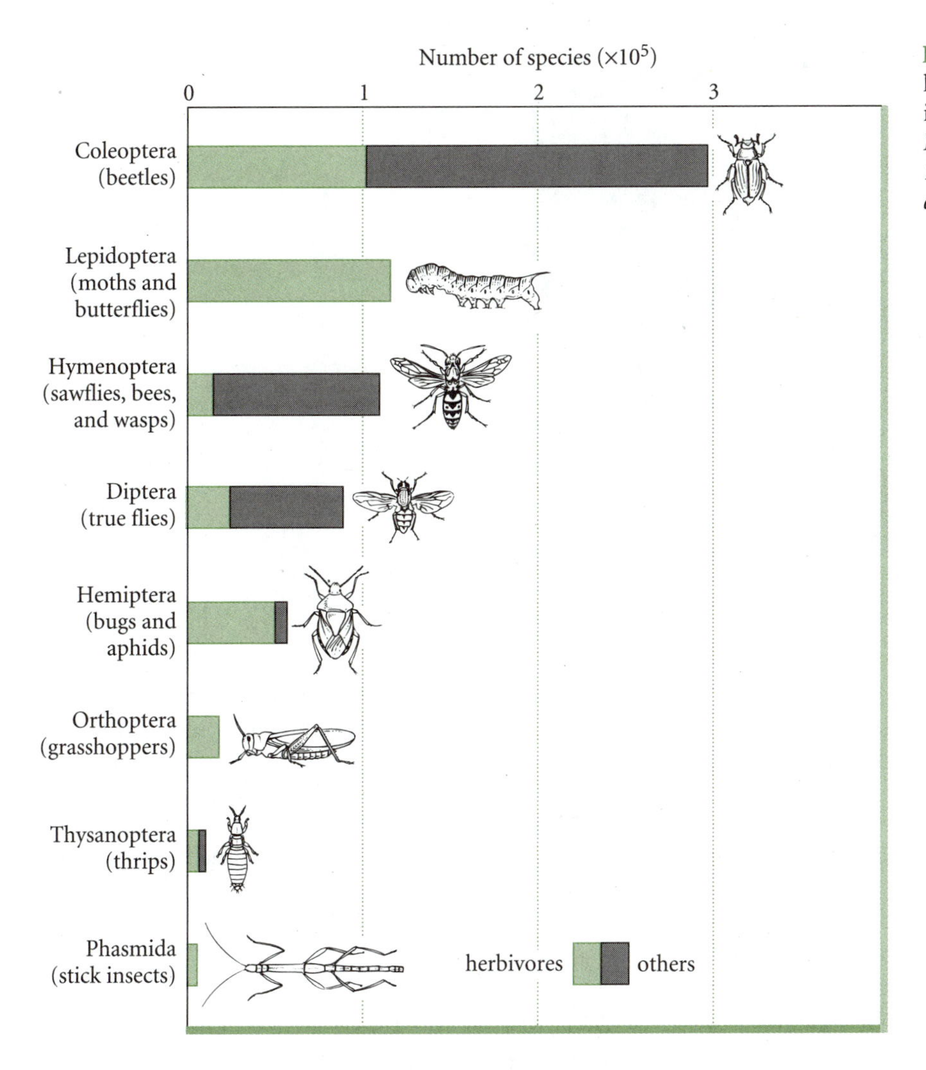

FIGURE 24-3 The numbers of herbivorous species in selected insect orders. (*After Weis and Berenbaum 1989; data from Lewis 1973, Borrer et al. 1976, Richards and Davies 1977, and Price 1977.*)

root material. Among the most prominent plant-eating birds are the ducks and geese, which subsist by grazing, the fruit- and seed-eating parrots, and seed-eating finches. Hummingbirds are highly specialized plant feeders, subsisting on the nectar of flowers. Various species of birds specialize on buds, fruits, or seeds, a type of herbivory that may benefit the plant through pollination or dispersal of seeds.

Nearly one-half of all mammals are plant eaters, including many species that specialize on fruits or seeds (Southwood 1985, Lindroth 1989). Among the most familiar of the mammalian herbivores are the grazers of the African grasslands, which, thanks to the proliferation of television nature shows, we have come to envision as large herds of wildebeests and gazelles nibbling at the grass under the seemingly indifferent gaze of a sleepy lion. But mammalian herbivores include such disparate types as the fruit-eating bats, which employ vision rather than echolocation to forage during moonlit nights, the tree sloth of South America, which spends nearly all of its time in the treetops, coming to the ground only to defecate, and even a number of mammals from "carnivorous" taxa, such as the giant panda, which eats only bamboo. Most primates consume primarily plant material (even some humans are strict herbivores).

The consumption of plant material is an essential process of energy flow in every ecosystem—herbivory is everywhere. But the nature of the interactions among plants and their herbivores, and the extent to which herbivory drives ecosystem processes, differ among types of ecosystems. Herbivory in marine ecosystems usually involves the consumption of macroscopic algae by organisms such as limpets, urchins, crabs, or fish. In intertidal and subtidal areas, the processes of herbivory, predation on herbivores, and patterns of environmental change—in particular, the physical affects of wave action—interact to determine the distribution of algae.

Several types of plant-herbivore interactions may be observed in freshwater lakes and streams. The open waters of lakes often contain both **macrograzers,** such as large crustaceans that feed on aquatic vegetation, and **micrograzers,** such as small crustaceans, rotifers, and protozoans, which, along with the macrograzers, feed on phytoplankton of various sizes. Submerged and

FIGURE 24-4 Marine iguanids of the Galápagos Islands. (*Courtesy of R. B. Suter.*)

emergent macrophytes populate the shorelines of most shallow lakes, and these plants are subject to herbivory by a variety of organisms, such as snails, birds, fish, and some zooplankton (Schutten et al. 1994). Organisms that make a living scraping or rasping periphyton from rocks and other hard substrates are extremely common in forest streams (Allen 1995). Extremely small and highly specialized stream insects of the family Hydroptilidae (microcaddisflies) pierce and suck the fluids from individual algal cells (Cummins and Klug 1979).

In terrestrial ecosystems such as forests and meadows, insect herbivores have been estimated to consume about 10% of the leaf area per year (Crawley 1989). Chewing caterpillars may take a heavy toll on flowers; a number of species of beetles specialize on eating seeds. The dynamics of insect herbivory are of considerable interest to agricultural ecologists. Insect pests of crop plants represent enormous economic losses each year, and herbivores of important weed species offer the potential for biological control of those weeds.

A substantial number of fossorial (burrowing) organisms have adapted to consuming the belowground parts of plants. Nematodes, which are often categorized as plant parasites, are extremely common and of considerable economic importance in agricultural ecosystems. The immature stages of species in four orders of insects (Coleoptera, Lepidoptera, Diptera, and Homoptera) consume belowground plant material. A number of species of burrowing rodents eat roots and belowground stems (Andersen 1987). Among the most interesting of such herbivores are the periodical cicadas (*Magicicada*), some species of which live 13 years and others for 17 years in nymphal form beneath the ground, where they suck sap from tree roots. Once every 13 or 17 years, depending on the species, all the cicadas in a particular area emerge at once in remarkable synchrony. Thousands of nymphs dig out of the soil and climb the nearest tree, post, or structure, there to emerge as winged adults that live a short 5 or 6 weeks, during which time they mate. Eggs are laid in small twigs, which eventually drop to the ground, where nymphs hatch out and dig into the ground to live for another 13 or 17 years.

24.3 Herbivory may induce compensatory responses by plants.

Predation is a one-sided interaction: the prey is eaten and the predator satisfied. Herbivory is less straightforward. A caterpillar's munching on leaves or a nematode's sucking of juices from roots does not kill a plant; rather, it invokes a response, which in some cases compensates for the damage done by the herbivore. It is this ability of individual plants to respond to loss of tissue that makes herbivory so different from predation and parasitoidism. The interaction is characterized by feedback between plant and herbivore: the herbivore consumes a portion of the plant, the plant responds by adjusting the distribution of resources within its body or by changing the pattern of its growth, the herbivore adjusts to this change, and so on. Proximally, this feedback results in patterns of physiological and morphological change in the plant, adjustments in the behavior of the herbivore, and, sometimes, changes in the population dynamics of either or both. Evolutionarily, these interactions result in specialization of plant defenses and herbivore food preferences.

The effects of herbivory on plants may be manifested in one of three ways (Jefferies 1988, Trumble et

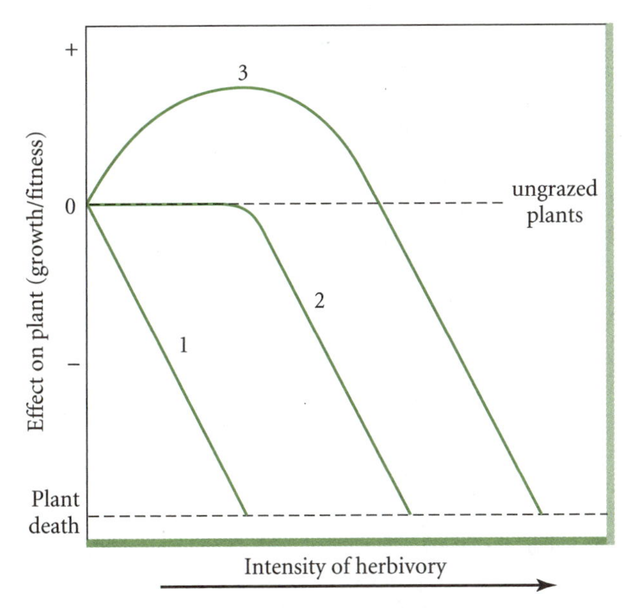

FIGURE 24-5 Three ways in which herbivory may affect plant growth and performance. Performance may decline in direct proportion to the intensity of herbivory (1), be maintained by compensation through some range of increasing herbivore pressure and then decrease (2), or increase through some range of herbivore pressure before decreasing (3). (*From Jefferies 1988; after Dyer 1975, McNaughton 1979, and Dyer et al. 1982.*)

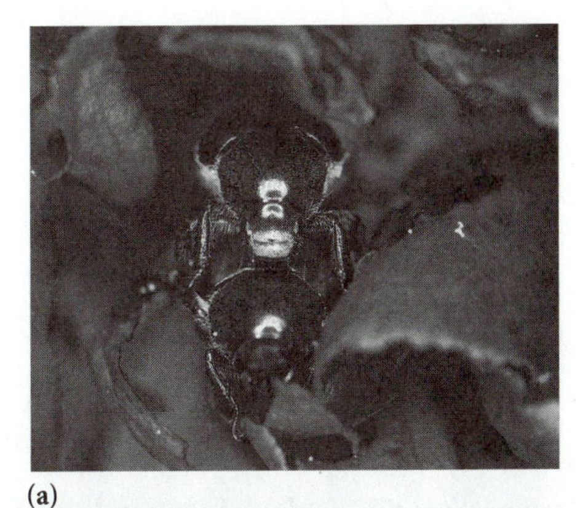

(a)

(b)

FIGURE 24-6 Some insect herbivores that cause damage to ornamental or agricultural plants: (a) Japanese beetle; (b) tomato hornworm. (*Courtesy of R. B. Suter.*)

al. 1993) (Figure 24-5). First, plant performance or reproductive output may be negatively affected by herbivory regardless of its intensity. Second, in some cases, the plant may sustain low to moderately high levels of herbivory through the process of **compensation,** in which morphological and physiological adjustments counter losses up to a specific threshold of intensity of herbivory, after which compensation can no longer overcome its effects. Third, some ecologists have suggested that, in some instances, plants respond to herbivory with **overcompensation,** an increase in performance, growth, or fitness in the face of herbivory. This idea has been vigorously debated, as we shall see below.

Negative Effects of Herbivory

Herbivory may negatively affect plant growth, survival, and fecundity. Outbreaks of pests provide the most dramatic demonstrations of the potential of herbivores to have strong negative effects on plants. Defoliation of oak trees by the gypsy moth may result in large stands of brown and dying trees. If you live in the eastern United States, you may have observed swarms of Japanese beetles devouring the plants in your backyard (Figure 24-6a). A single tomato hornworm can easily destroy your beautiful tomato plant in a day or two (Figure 24-6b). Agricultural crops may suffer extensive damage from pests that defoliate plants and suck juices from the stems and roots.

Plants subjected to even moderate levels of herbivory may suffer reductions in growth or seed production. Bach (1994) found that the sand-dune willow (*Salix cordata*) showed reduced growth (measured as change in height) in response to three different levels of herbivory by flea beetles (*Altica sublicata*) (Figure 24-7). Such effects are often difficult to detect because plants may compensate for herbivory in a number of ways, which we discuss below. Even when herbivory causes a negative effect in one aspect of plant population dynamics, the overall effect may be negligible. For example, a study of the effects of three levels of sheep grazing (no grazing, moderate grazing, and unrestricted grazing) on the demography of the legume *Anthyllis vulneraria* in France revealed that although moderate and unrestricted grazing lowered the seed output of the plants, it had no effect on population density because of a compensating increase in the survival of juveniles and vegetative adults (Bastrenta 1991).

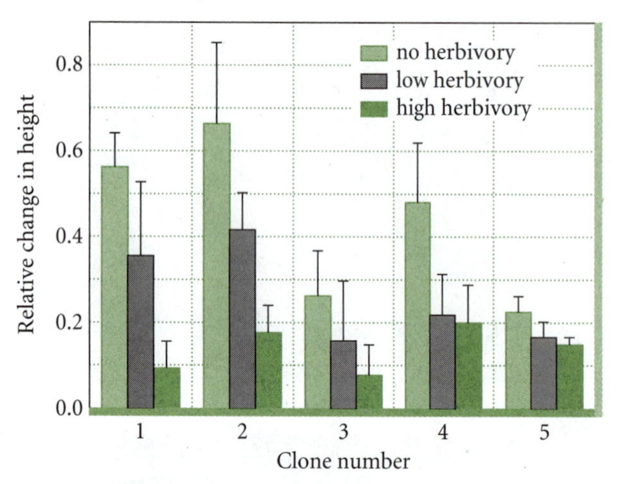

FIGURE 24-7 Relative change in height of the sand-dune willow (*Salix cordata*) when subjected to three levels of herbivory by the flea beetle, *Altica sublicata.* The numbers on the horizontal axis represent different clones of the willow. (*After Bach 1994.*)

Compensation

Plants may compensate for the effects of herbivory in a number of ways (Trumble et al. 1993). Compensation for the loss of leaves and other photosynthetic tissues may involve internal reallocation of materials. Photosynthesis takes place in leaves and other green parts of the plant. The products of photosynthesis, which are required by living tissues throughout the plant, are translocated in the form of sucrose from leaves to other, nonphotosynthetic, tissues in the plant body, where they are used for respiration or biosynthesis, or converted to starch for long-term storage. Leaves and other photosynthesizing plant parts are sources of the products of photosynthesis, called **photosynthate,** and the nonphotosynthesizing organs and tissues, such as roots, are sinks. (The mechanisms of translocation are complicated and will not be discussed here.) To some extent, the rate of photosynthesis is controlled by the availability of sinks to which photosynthate may be translocated. The rate of photosynthesis may decline if there is not a demand for photosynthate (low sink strength). Likewise, an increase in sink strength may result in increased photosynthetic activity. In the context of plant sources and sinks, then, it has been suggested that plants may respond to herbivory with either an increase or a decrease in photosynthetic rate, depending on whether a source or a sink is eaten. Herbivores such as sucking insects may reduce the sink strength of plants by destroying rapidly growing plant tissues such as stems and fruits, causing a reduction in photosynthetic rate. Defoliation, on the other hand, reduces photosynthate production relative to demand (increases relative sink strength), and may lead to an increase in photosynthesis in the remaining tissues (Crawley 1983, Trumble et al. 1993).

Much of the photosynthesis in wheat takes place in the long flag leaf, at the base of which lies the wheat ear, which is a sink for photosynthate. King et al. (1967) removed the wheat ears from wheat plants and observed that the rate of photosynthesis dropped dramatically during the following 24-hour period, presumably as a result of reduction of sink strength. When artificial sinks were created by shading lower leaves on the plant, thereby reducing their ability to photosynthesize and simultaneously increasing their demand for photosynthate (increasing sink strength), photosynthesis in the flag leaf increased (Figure 24-8).

One of the most characteristic and overlooked dynamics of terrestrial plants is the interaction between the aboveground stems and the system of belowground roots and stems. Photosynthesis takes place in the aboveground portion of the plant, which is also the site of much water loss. Thus, the shoots demand high levels of water and nutrients, materials that are supplied by the root system. If either water or nutrients are limiting, the root system must grow to support a given demand by the shoots. This root growth is subsidized by carbohydrates that are produced in the shoots, creating a complex feedback system that determines the ratio of belowground to aboveground plant biomass, called the **root/shoot ratio.** Where water or nutrients are limited, plants tend to have extensive root systems relative to their aboveground parts; that is, high root/shoot ratios. This pattern is often observed

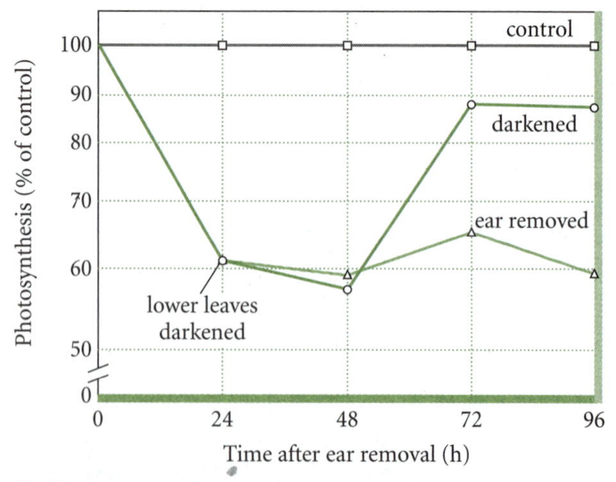

FIGURE 24-8 Results of experimentally increasing or decreasing the photosynthate sink strength in wheat. Photosynthesis takes place in the long flag leaf, and the wheat ear is a sink for the products of photosynthesis. Removal of the ear (reducing sink strength) resulted in a reduction in photosynthesis in the flag leaf. Shading the lower leaves (increasing sink strength) resulted in an increase in photosynthesis in the flag leaf. (*From Crawley 1986; after King et al. 1967.*)

in deserts where water is very limited. Species inhabiting areas with more abundant water and nutrients have lower root/shoot ratios. The productivity of the root system is also constrained by resources, with plants living in resource-limited environments having lower root productivity overall (Barbour et al. 1987). The belowground parts of many plants serve as storage reservoirs for minerals and other assimilated materials, the nutritional value of which may be extremely high (Andersen 1987). It is not surprising, therefore, that a large number of herbivores are adapted for feeding on the belowground parts of plants.

To be sure, the rate of photosynthesis is influenced by a variety of mechanisms other than the relative differences in the demand for photosynthate and the rate at which it is being produced. As we have seen in previous chapters, environmental factors such as temperature, light intensity, and nutrient availability affect the rate of plant production. Developmental processes related to the normal growth of plants also affect the availability of sources and the strength of sinks. For example, leaves, the principal source of photosynthetic products in most plants, represent strong sinks during their early development. The effects of herbivores interact with all of these factors to play a role in the physiology of the plant.

Plant Architecture

Plant architecture is the set of morphological features, such as height, root/shoot ratio, branching pattern, and leaf area, that define the growth form of a plant. Plant architecture is both affected by herbivory and affects it. Herbivores may alter growth in plants by attacking apical meristems, where stem growth normally occurs. Generally, intact apical buds prevent growth at lateral buds on the same stem, a phenomenon called **apical dominance.** When the apical bud is removed by an herbivore, lateral growth may occur, thereby replacing linear growth of the stem with the development of lateral branches, a change in plant architecture. The same thing happens when you prune the shrubs in your yard. The removal of the apical buds promotes lateral growth and makes the shrubs bushier. Grazing by greater snow geese (*Chen caerulescens atlantica*) was found to reduce the overall height of individuals of two species of high arctic plants (*Dupontia fisheria* and *Eriophorum scheuchzeri*) on Bylot Island, Canada (Beaulieu et al. 1996). Sea urchins (*Lytechinus variegatus*) have been shown to affect the aboveground biomass of seagrass so dramatically that the form of individual seagrass plants and the seagrass density in the area where they forage is changed, and the sea urchins become more vulnerable to predation (Heck and Valentine 1995).

Because of the modular construction of plants, the location of an herbivore attack will determine whether there is a change in plant architecture and what form that change will take (Marquis 1996). For example, defoliation by hungry caterpillars will result in decreased growth of the affected branches because plants are not designed to swiftly and efficiently replenish the photosynthate lost to herbivory at a specific location on the plant. A grazing or browsing animal, or a gall-forming insect that attacks buds, may cause a qualitatively different compensatory response in plant architecture. In that case, a general regrowth of lateral stems will result.

Plant architecture may also affect the number of herbivore species (herbivore diversity) that inhabit a particular variety of plant. Leafhoppers are common sucking insects that feed on a wide variety of different types of plants. In an experiment in which numbers and species of leafhoppers were monitored in grasslands subjected to different intensities of sheep grazing, leafhopper diversity was influenced more by the intensity of grazing than by any other variable measured (Brown et al. 1992).

Overcompensation

A number of empirical studies have shown that, in some instances, herbivory appears to enhance the performance, growth, or fitness of plants, a phenomenon called overcompensation (see Figure 24-5) (McNaughton 1983, Paige and Whitham 1987, Paige 1992, Alward and Joern 1993, Hjaelten and Danell 1993, Dangerfield and Modukanele 1996). Paige and Whitham (1987) discovered that seed production increased in the biennial herb *Ipomopsis aggregata*, a common plant in the mountain regions of the western United States, when the plants were browsed by mule deer and elk. The response of the plants resulted from changes in plant architecture. When the single inflorescence of the plant was removed by a browsing deer or elk, the plant produced a large number of flowering stalks, thereby increasing fruit production. The average number of seeds in each fruit was the same for browsed and unbrowsed plants, and there appeared to be no difference in the seed germination rate between the two groups. Thus, the increase in fitness was the result principally of a larger number of fruits (Paige 1992). These findings have been questioned, however, because some populations do not show the same response found by Paige and Whitham (Bergelson et al. 1996).

Some ecologists have suggested that overcompensation in response to herbivory results in an increase in plant fitness (e.g., McNaughton 1979). This idea has been proposed as an explanation for the success of many plant species of the grasslands of North America and Africa, where grazing by vertebrates may be more or less continuous (Figure 24-9). Whether these

FIGURE 24-9 Grazing animals on the Serengeti plain of Africa.

changes provide some benefit to the individual plant, as some have suggested, is less certain. Bergelson et al. (1996) contend that there is little evidence that individual plants benefit from overcompensation. Belsky et al. (1993) point out that it is difficult to tell whether the rapid regrowth after herbivory that is observed in some plants is a direct response to the actions of the herbivores or part of a broader response that has evolved to reduce the effect of any form of damage to plant tissue, such as that resulting from fire, wind, or trampling, as well as herbivory. Moreover, as we saw above in the example of the perennial grass *Ipomopsis aggregata,* although individual plants produced more fruit as a result of browsing, the number of seeds per fruit did not change, nor did seedling survival.

The issue of whether plants benefit individually by herbivory has important practical implications. An uncritical acceptance of the idea that herbivory has a beneficial effect on plants has led some land managers to conclude that heavy livestock grazing is a beneficial practice in grasslands, creating a situation that is having important effects on rangeland areas of the western United States (Painter and Belsky 1993).

Inducible Plant Defenses

Some of the morphological features of plants provide them with defenses against herbivores. Most of us have been stuck by the spines of a rose bush. Features of the life histories of plants also provide protection of a kind against herbivores. Some plants have extensive underground parts (roots or underground stems) that are protected from aboveground herbivores. In some cases, as in grasses, the meristem areas are positioned near ground level where they cannot be cropped off by an herbivore, thereby allowing rapid regrowth of aboveground parts. Some defensive mechanisms of plants are triggered by the activities of an herbivore. These mechanisms are called **inducible defenses,** and many of them involve the release of noxious and sometimes toxic chemicals produced by the plant.

Plants have many inducible defenses against the attacks of herbivores, microbes, and viruses. In response to wounding, a plant may produce toxic, noxious, or nutrition-reducing compounds—either in the area of the wound or systemically throughout the plant—that reduce subsequent herbivory. In some cases, these responses may take only minutes or hours; in others, they require a new season of growth.

When shoots of aspen, poplar, birch, and alder are heavily browsed by snowshoe hares, shoots produced during the following year have exceptionally high concentrations of terpene and phenolic resins, which are extremely unpalatable to the hares (Bryant 1981). By extracting the resins with ether and placing them in varying concentrations on shoots of unbrowsed trees, J. Bryant demonstrated experimentally that hares will not touch shoots containing 80 mg or more of resin per gram dry weight, regardless of the amount of other food available. The inducible response of browsed plants could result in a considerable time lag in the effect of hare density on subsequent hare survival and reproduction, particularly if second-year regrowth were also protected by resins. The snowshoe hare is the classic example of an oscillating population (see Figure 18-10), and Bryant's data suggest that the hare-plant system, rather than the lynx-hare system, may underlie the predator-prey oscillation (but see Lindroth and Batzli 1986).

24.4 Grazing may alter plant growth and affect primary production.

Overcompensation often involves the growth of lateral buds and the subsequent development of more stems, more fruit, and overall greater plant biomass. Some ecologists have suggested that, in some instances, these responses also lead to an increase in overall productivity among plants suffering damage from herbivory. McNaughton (1985) found an increase in aboveground net primary productivity resulting from grazing in plants of the Serengeti of Africa, one of the most diverse grassland ecosystems in the world (see Figure 24-9). Bazely and Jefferies (1989) found that changes in the number and size distribution of leaves and shoots of the salt marsh plant *Carex subspathacea* when grazed by lesser snow geese resulted in greater net primary production because of increased growth of existing shoots.

These results have been questioned, however (even by the those who reported them), because of uncertainty about how plant competition, environmental variation, and the form and function of the plant itself interact to alter the rate of primary production under grazing pressure (Jefferies 1988). Primary production in the Serengeti is extremely variable, even over relatively small spatial scales, and is correlated with rainfall (McNaughton 1985). Moreover, grazers do not operate like simple lawn mowers, chopping off grass indiscriminately. Rather, they are highly selective, both of the types of plants that they eat and the parts of those plants that they consume. Further, the process of grazing is not unidirectional, with herbivores only taking and grasses only giving. The grazers return nutrients to the grass ecosystem when they defecate, and the physical processes of the soil are changed by the tilling that results from the movement of heavy animals through the grass.

In view of these considerations, Jefferies (1988) proposed a model of grazing and plant productivity that incorporates feedback among various parts of the grazing ecosystem (Figure 24-10). In Jefferies's model, which is based on his studies of snow geese grazing in salt marsh ecosystems, a number of positive inputs into the ecosystem result from grazing. Nutrients released by herbivores through defecation are available for uptake by marsh grasses, and also promote production and subsequent nutrient release by marsh algae. Together, these nutrient inputs may increase plant production.

Of course, an increase in productivity in a grassland community resulting from interactions with grazers need not result in a change in grass abundance or biomass. The number of plants may remain the same under varying degrees of grazing pressure. But grazing may alter the form and, consequently, the demography of plant parts. For example, Jefferies (1988) showed that the average rates of leaf birth and leaf death differed when the plant *Puccinellia phryganodes* was grown under four different grazing intensities. Leaf births on the main and axillary shoots were greatest in areas that were rarely grazed and lowest in more heavily grazed areas. Leaf deaths were generally higher in the more intensively grazed areas. Alterations in the number of flowers or seeds resulting from herbivory might negatively affect recruitment and alter population size.

The response of plant populations to grazing cannot be decoupled from competitive interactions among individual plants of the same species and among plants of different species. Selective grazing may change the composition of the plant community, and as a consequence, the nature of the competitive interactions among plants in that community (Alward and Joern 1993). In sparse Australian grasslands, rabbit grazing can result in a reduction in abundance of certain grasses and an increase in weedy species, which may then outcompete grasses (Crawley 1983, Myers and Poole 1963). A discussion of the influence of competition in natural communities is presented in the next part of the book. It is sufficient to say at this point that a plant's response to grazing or other forms of herbivory is complicated by its interaction with other plants in its neighborhood.

24.5 The effect of herbivory on plant populations is complicated by multiple herbivores, plant development, and population age structure.

While herbivores affect the architecture and performance of individual plants in many ways, the extent to which their activities influence responses at the level of the plant population is uncertain. One reason for this uncertainty is the enormous variation in individual plant responses to herbivory, which we have emphasized in the preceding sections. Generalizations about the strength of the correlation between, say, flower consumption by insects and the reproductive output of the plant whose flowers are consumed are difficult to make. Variation in the effects of different types of herbivores also complicates the issue, particularly inasmuch as herbivores with several different feeding strategies may affect a single plant population. For example, grazers generally remove more biomass from the host plant than do plant-sucking insects. The effect of seed predators on plant recruitment is dependent to some extent on whether the plant population is limited

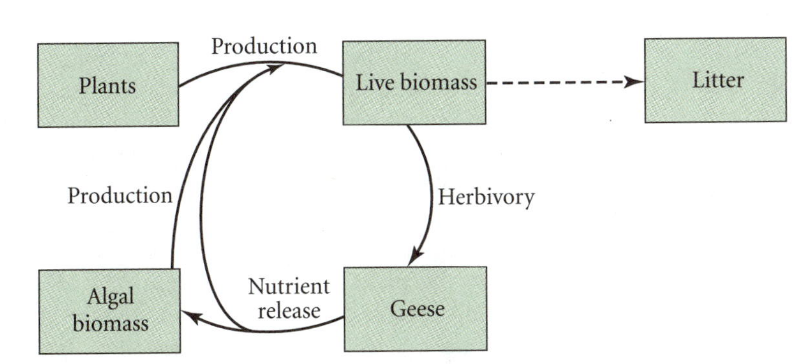

FIGURE 24-10 A positive feedback model of the interrelationship between goose grazers and their forage species. Nutrients released into the ecosystem by geese through defecation enhance algal and plant production and result in increases in biomass. (*After Jefferies 1988.*)

by the number of seeds produced or by the quality of the habitat where the seeds land.

Another reason for our poor understanding of the effects of herbivores on plant populations is the uncertainty about what mechanisms operate to regulate plant populations. Because plants are generally sessile, local competition for resources may be intense, and may ultimately be more important than herbivory in limiting growth and reproduction, particularly where herbivore densities are relatively low, as they are in most natural systems (Crawley 1989). Moreover, the effects of herbivory interact strongly with environmental conditions and with plant growth and reproductive patterns. Thus, herbivores may have a dramatic effect on plant abundance in one part of the year and little or no effect in another season.

Despite these difficulties, there are situations in which herbivores have a clear effect on the distribution and abundance of their plant hosts. Marine grazers may exert a strong influence on the local distributions of algae (John et al. 1992). Some natural populations of forest insects may increase mortality rates in the tree species on which they feed during outbreak years. Among the most important of these herbivores are bark beetles, the spruce budworm, and the gypsy moth. Evidence from attempts to introduce insect herbivores to control weeds and from experimental investigations involving removal of insect herbivores suggest that, under some circumstances, insect herbivores possess substantial capabilities to regulate plant populations (Crawley 1989). The weevil *Rhinocyllus conicus,* a seed head feeder, was introduced into Canada in 1969 in an attempt to control the nodding thistle *Carduus nylans,* which is a pest of grazing land in Canada and the United States (Harris 1984, Crawley 1983). Control of the thistle was achieved after about 10 years (Figure 24-11). The mechanism of control was probably not the straightforward increase in plant consumption by the expanding weevil population. Herbivory caused changes in both the age structure and the spatial distribution of the plant population, which no doubt contributed to the overall reduction in plant abundance (Harris 1984).

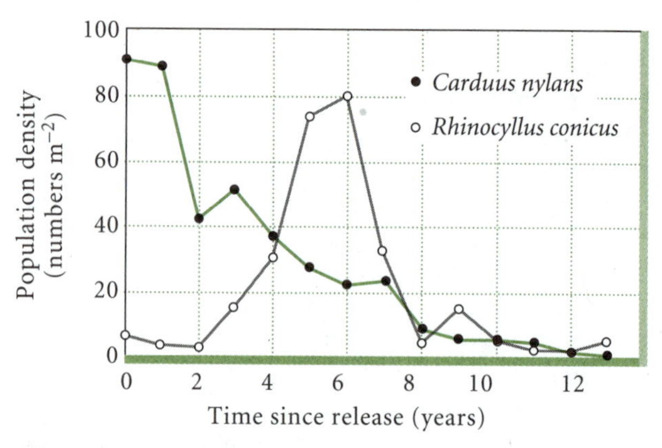

FIGURE 24-11 Biological control of the pest nodding thistle *Carduus nylans* by the seed head-feeding weevil *Rhinocyllus conicus.* Control was achieved in 10 years, probably through the effects of weevil herbivory on plant population age structure and spatial distribution rather than direct reduction by consumption. (*From Crawley 1986; data from Harris 1984.*)

In building models to represent the effects of herbivores on the populations of their host plants, the total biomass of the plant population may be more important than the number of individual plants in the population. Heavy grazing, for example, might not decrease the number of individual plants in the population, but it might dramatically decrease the amount of aboveground plant tissue in the area.

Plant Development and Herbivore Pressure

During growth, the size, shape, and distribution of biomass of a plant change. The proportion of the plant's biomass contained in vegetative structures, reproductive structures, and seeds changes. The life history patterns and population dynamics of many specialized herbivores have evolved to coincide with the development of specific plant parts. In the course of a plant's growth, then, the intensity of herbivory by a particular herbivore may change. Planthoppers (*Prokelisia marginata*), for example, are insects that suck the juices from the leaves of the common salt marsh grass *Spartina alterniflora.* Both the pattern of planthopper distribution on individual plants and their abundance on those plants change during the development of the plant (Figure 24-12).

Marquis (1992) examined the effects of **folivory** (consumption of leaves) on the shrub *Piper arieianum* of the wet tropical forests of Costa Rica. Marquis wanted to know whether the location or the timing of leaf damage resulted in different plant responses. He also wanted to know to what extent the plant's response to loss of leaf tissue was distributed throughout the plant. The number of seeds produced was selected as the dependent variable in the experiment (see Chapter 2). Three folivory treatments were applied. Folivory was simulated by removing 10% of the plant's leaf area, either from a single reproductive branch (referred to as the focal branch) or from various places spread throughout the canopy of the shrub. In the controls, no leaf area was removed. Marquis applied these treatments to two groups of plants: those yet to undergo flowering (preflowering group) and those at the height of flowering (flowering group) Marquis then counted the seeds produced on the focal branch and on the plant as a whole (Figure 24-13). In preflowering plants, he found that damaged branches produced fewer seeds than branches that

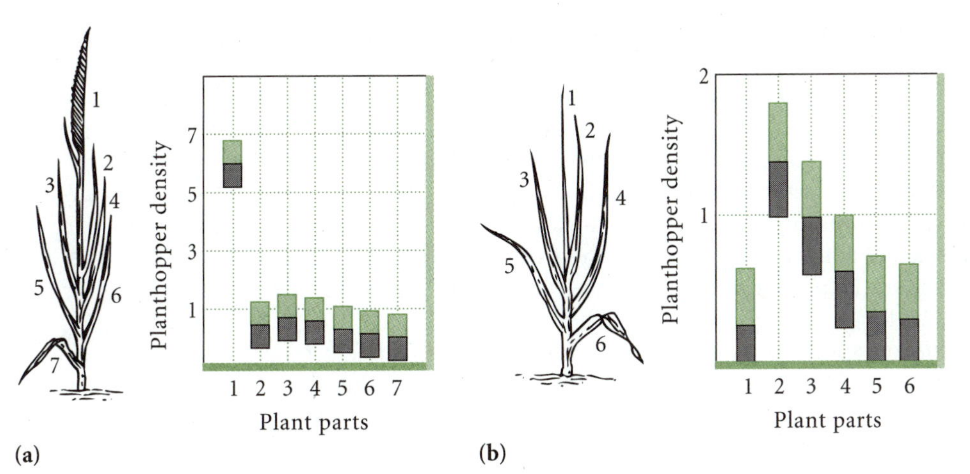

FIGURE 24-12 Number of adult planthoppers (*Prokelisia marginata*) occurring on the marsh plant *Spartina alterniflora.* The numbers on the horizontal axis refer to the positions on the plant designated in the drawing. (a) Flowering plant with seed head (1) and basal leaf (7). (b) Nonflowering plant with terminal leaf (1) and basal leaf (6). (*From Denno 1980, 1983.*)

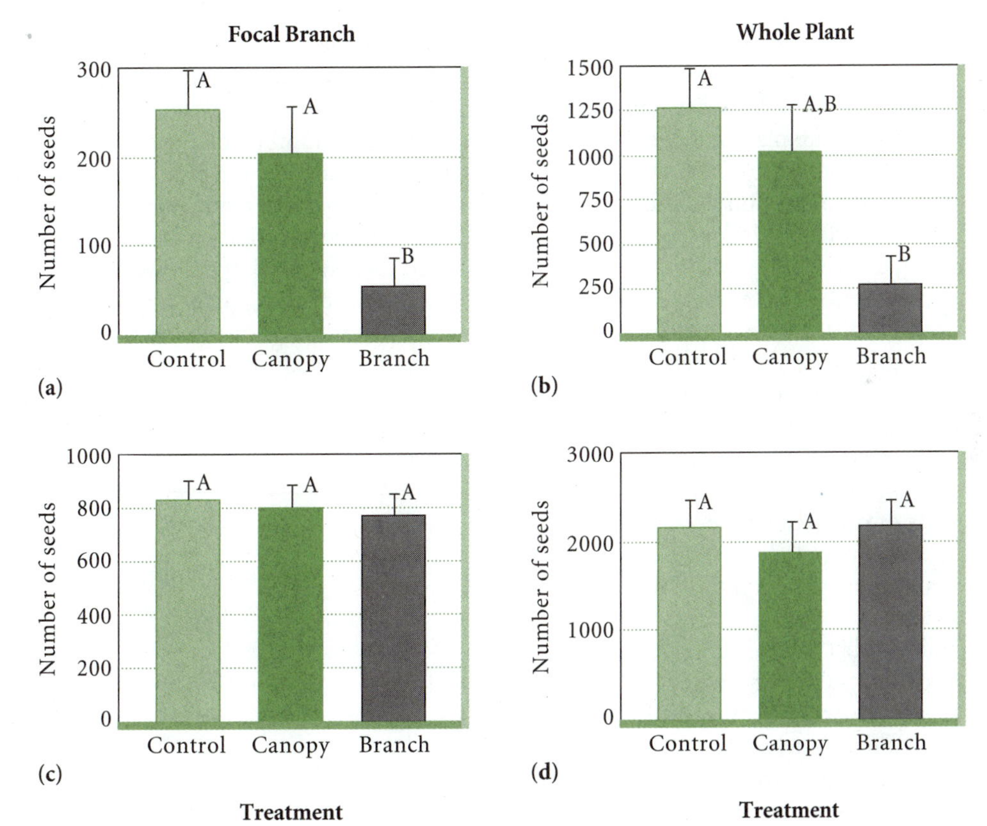

FIGURE 24-13 Marquis's experiment on the extent and timing of herbivory. Folivory (leaf eating) was simulated by removing 10% of the shrub leaf area of *Piper arieianum,* either from single branches (focal branch) or throughout the canopy of the shrub. A control treatment, in which no leaves were removed, was also applied. The effect of folivory was measured by counting the number of seeds produced on the focal branch (a and c) and on the plant as a whole (b and d) by plants that had not yet flowered, that is, preflowering plants (a) and (b), and by plants that were at the height of flowering (c) and (d). In preflowering plants, seed production was reduced by loss of leaf tissue, both on focal branches and on the plant as a whole. (Bars with different letters indicate that the null hypothesis that the treatments are the same is rejected; see Chapter 2.) The reduction was greatest on focal branches when leaves were removed from the focal branch only. In flowering plants, the number of seeds produced did not differ among the treatments (notice that in panels c and d, the same letter is assigned to each of the bars, indicating that the hypothesis that the number of seeds produced is the same among the treatments cannot be rejected). (*From Marquis 1992.*)

were undamaged or branches on shrubs where the leaf damage was distributed throughout the canopy (Figure 24-13a). Moreover, a shrub with damaged branches produced fewer seeds on all its branches than did control plants or those undergoing canopy damage (Figure 24-13b). When the simulated leaf damage was conducted during the peak of flowering, no difference in seed production among the three treatments was detected either on focal branches or on the entire plant (Figure 24-13c,d). Marquis's results indicate that in *Piper arieianum,* folivory occurring early in the reproductive cycle of the plant has a greater effect on seed production than herbivory occurring later.

Effects of Plant Age Structure and Multiple Herbivores

Individual plants are often host to many different types of herbivores, particularly the many groups of plant-feeding arthropods. The leaves, stems, buds, flowers, and roots of a mature plant may be subject to consumption simultaneously by many different species, each specialized to feed on a different plant part. Plants may also suffer a kind of serial assault during their growth, as herbivores with different specialized feeding preferences move to the plant and consume specific parts of the plant as those parts develop. Tender young shoots, the preferred food of some grazers, are not subject to consumption by flower or bud feeders. Seed predators do not affect the plant body at all, focusing their attention on the propagules of the plant. In age-structured plant populations (with plants of different ages occurring simultaneously in the same population; see Chapter 15), such serial herbivory may have an important effect on population dynamics.

The grassland shrub *Gutierrezia microcephala* of central New Mexico is subject to attack by both a grasshopper (*Hesperotettix viridis*), which eats the leaves of the plant, and a root-boring beetle (*Crossidius pulchellus*), which tunnels through the center of the taproot of the plant, consuming vital tissues as it goes. Parker (1985) examined how the consumption of different plant parts by these two herbivores affects mortality and senescence in an age-structured population of the plant. Populations of *G. microcephala* are composed of seedlings, juveniles, and adult plants of various ages. In the spring, photosynthetic stems and small leaves emerge on the shrubs, forming a small canopy that persists until late summer or early fall, when flowers are formed and wind-dispersed seeds are produced. Single-stemmed seedlings emerge the following spring. Both the grasshopper and the beetle appear to specialize on the shrub in the region where Parker studied them. Eggs are laid by adult grasshoppers in the fall, and nymphs emerge in early spring to begin feeding on the stems and foliage of the plant. Individual grasshopper growth is timed to coincide with the development of the stems and leaves of the plant, both peaking in mid-July. Adults of the beetle emerge from the ground and mate in late summer, during which time they may feed on various parts of the shrub. Beetle eggs are deposited at that time, and the developing larvae tunnel into the roots of the shrub.

The effects of these two herbivores are dependent both on the age of the plant and on the density of the plant population. Beetle larvae consistently preferred large plants (Figure 24-14), while grasshoppers showed no such preference. To determine the effects of grasshopper foraging on plant reproduction and mortality, Parker constructed cages around samples of plants of different ages to exclude grasshoppers, then compared the protected plants with those that were exposed to natural herbivory by the insect. The effect of grasshopper herbivory on the number of plants dying and the number of flower heads per plant produced was different for plants of different ages (Table 24-2). Both mature plants and first-year seedlings enjoyed reduced mortality and greater flower head production when grasshoppers were excluded. But young plants exposed to grasshopper herbivory suffered a much higher rate of mortality than did mature plants. Parker suggested that *G. microcephala* populations persist in spite of high grasshopper herbivory because of the vast number of seeds they produce.

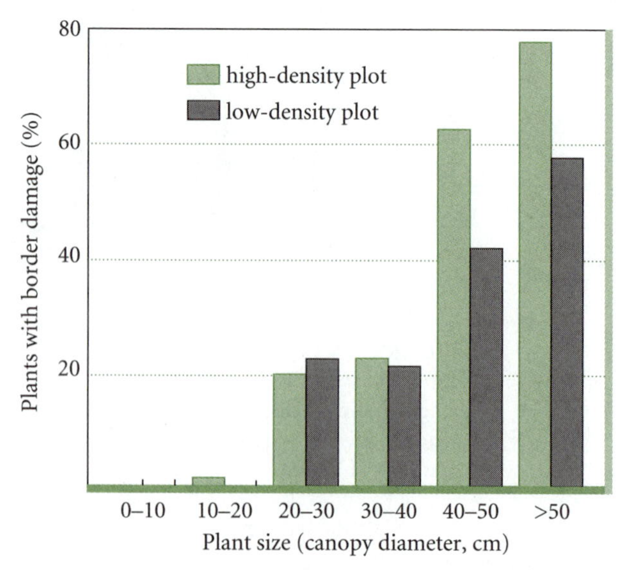

FIGURE 24-14 Percentage of individuals of different sizes of the shrub *Gutierrezia microcephala* damaged by the larvae of the root-boring beetle *Crossidius pulchellus* in experimental plots of high and low density in central New Mexico. Beetle larvae prefer larger plants regardless of density. (*From Parker 1985.*)

TABLE 24-2 The effect of herbivory by the grasshopper *Hesperotettix viridis* on reproduction and mortality in the grassland shrub *Gutierrezia microcephala* of central New Mexico

	Reproduction and mortality			
	Protected plants		Unprotected plants	
	Plants dying by September 1980*			
	(%)	(Number per group)	(%)	(Number per group)
Mature plants	0	29	4.9	81
Yearlings	6.9	29	33.3	210
	Flower heads in September 1980†			
	(Number per plant)		(Number per plant)	
Mature plants				
Mean ± 1 SD	1,310 ± 159		3 ± 1	
Range	352–2,713		0–80	

**The mortality rate of unprotected yearlings was significantly higher than that of all other groups (log-likelihood ratio test, $P < 0.005$).*

†No yearlings flowered in 1980.

(From Parker 1985.)

Insect Herbivory and Plant Distributions

Are plant distributions influenced by persistent herbivory? Recent experimental investigations suggest that, in some cases, insect herbivory may explain, in part, the ranges of plant species by differentially affecting abundance under some environmental conditions but not under others. Louda and Rodman (1996) investigated whether bittercress (*Cardamine cordifolia*), a perennial plant of the Rocky Mountains, is affected differently by insect herbivores in the sun and in the shade. Bittercress is found primarily in the shade of willows, where it flowers in July and bears fruit in late summer. From previous studies, Louda and Rodman knew that insect herbivores generally prefer the sun to the shade, and they wondered whether the restriction of bittercress to shady areas was the result of chronic herbivory. They believed that two complementary lines of evidence would be required to support this idea. First, in order to demonstrate a difference in herbivory between shady and sunny areas, it would be necessary to show that the intensity of insect herbivory increases on plants that have been subjected to a change in habitat from shade to sun. Second, if higher insect herbivory in sunny areas restricts bittercress to the shade, then plants grown in the sun and protected from herbivores should show similar or better performance than plants grown in the shade.

Louda and Rodman set up two experiments to test their idea. In the first, sixteen 50 cm × 50 cm plots were established beneath willow trees. In half the plots (treatment plots), the overhanging willow branches were pruned, thereby exposing the plots to sun for most of each day; the remaining plots were left in the shade (control plots). Measurements of insect damage and plant performance were obtained periodically over several years. In the second experiment, Louda and Rodman transplanted clones of bittercress (obtained by dividing one plant into four parts) into one of four experimental situations: (1) exposed to sun and protected from herbivores with pesticide applications, (2) exposed and unprotected, (3) shaded and protected, and (4) shaded and unprotected.

The average amount of damage per leaf for plants exposed to the sun increased shortly after exposure and remained higher than that of the shaded plants for 3 years after the beginning of the experiment (Figure 24-15a). Protection from insects yielded greater transplant survival, stem height, and fruit maturation in both exposed and shaded treatments (Figure 24-15b). These results suggest that insect herbivory affects the

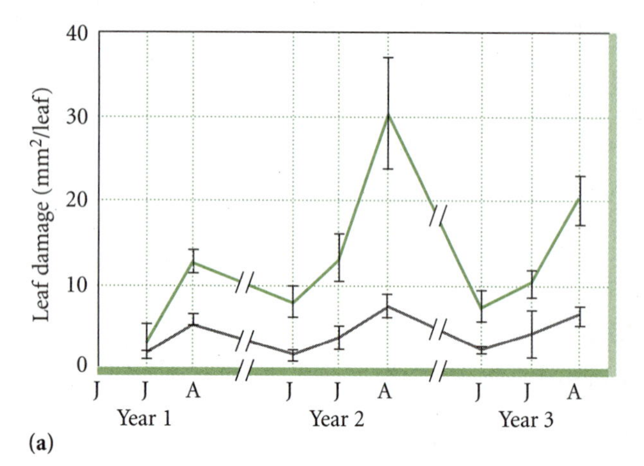

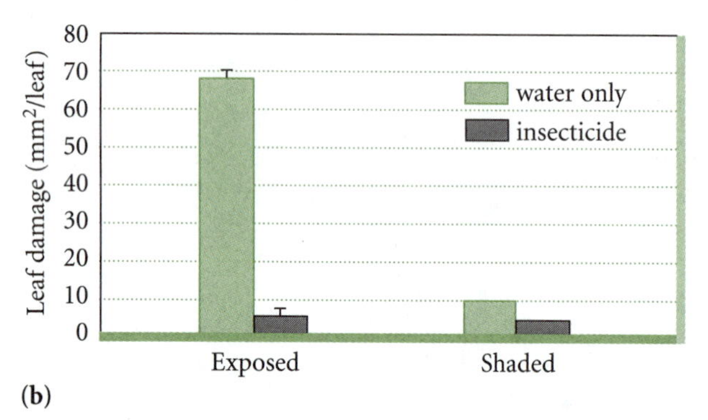

FIGURE 24-15 (a) The amount of leaf damage resulting from herbivory per leaf (mm²/leaf) for bittercress (*Cardamine cordifolia*) living in shaded areas (gray line) and sunny areas (green line). (b) Response of bittercress to insect herbivory when shaded and protected from insects by the application of insecticides. (*From Louda and Rodman 1996.*)

local distribution of bittercress. The plants are found more often in shady habitats, where damage resulting from herbivory is not as extensive as in habitats receiving full sunlight.

24.6 The susceptibility of plant populations to herbivory may be influenced by the spatial dynamics of the plant community.

The arrangement of plants in space, and the manner in which variation in plant form and function is mapped onto that spatial pattern, will influence the interaction of individual plants with herbivores, and thus, the response of plant populations to herbivory. Overall plant density, the dispersion of the individuals of plant species that are targets of herbivores, the type of vegetation occurring between those plants, and the number, density, and arrangement of plant patches—characteristics that are collectively referred to as **vegetation texture**—are known to affect the movement patterns and feeding intensity of herbivores, particularly insects (Kareiva 1983). Spatial variation in the nutritional quality or defensive responses of plants will influence local abundances of herbivores (Denno 1983).

The observation that polycultures of agricultural crops often support fewer types of insect pests at lower densities than monocultures has led plant ecologists to suggest that plants living in diverse communities may be less susceptible to herbivory than those found in communities with less diversity, a phenomenon sometimes referred to as **associational resistance** (Tahvanainen and Root 1972, Risch 1981). R. B. Root of Cornell University (1973) suggested two mechanisms by which such resistance might arise in diverse communities. The first idea, which is known as the **enemies hypothesis,** holds that, since diverse plant assemblages have a greater variety of different microhabitats, owing to the presence of plants with a variety of growth forms, they support a more diverse array of predators (enemies of herbivores), thereby keeping herbivore populations low. Root's second idea, which he termed the **resource concentration hypothesis,** suggests that host plants in diverse communities may simply be harder for herbivores to find. This idea also suggests that herbivores with movement capabilities, such as many insects, might not stay as long in areas of relatively low resource concentration (host plant density) as in areas where their host plants are abundant.

S. J. Risch, also of Cornell University, conducted an experimental test of these two ideas in a corn-bean-squash agroecosystem in Costa Rica (Risch 1981). Risch planted monocultures of corn, beans, and squash, polycultures of the three possible pairwise combinations of the crops (corn-bean, corn-squash, squash-bean), and a polyculture containing all three crops in each of three planting seasons between 1976 and 1978 (Figure 24-16). The abundances of six beetle species of the family Chrysomelidae, all of which are either specialist or generalist herbivores of the three crops, were monitored in the plots. The movement of beetles from each plot was measured by using special directional traps placed near the edges of the plots. In one experiment, vertical obstructions in the form of tall cornstalks were placed in plots to determine the effect of increasing architectural complexity on beetle movement patterns. Risch found that the number of beetles per host plant was substantially lower in polycultures containing at least one nonhost species (Figure 24-17). The movement of beetles from plots planted with one or more nonhost species was greater than the movement of beetles from pure plots of their host. The addition of vertical obstructions tended to reduce beetle movement. The results of Risch's experiment support the resource concentration hypothesis.

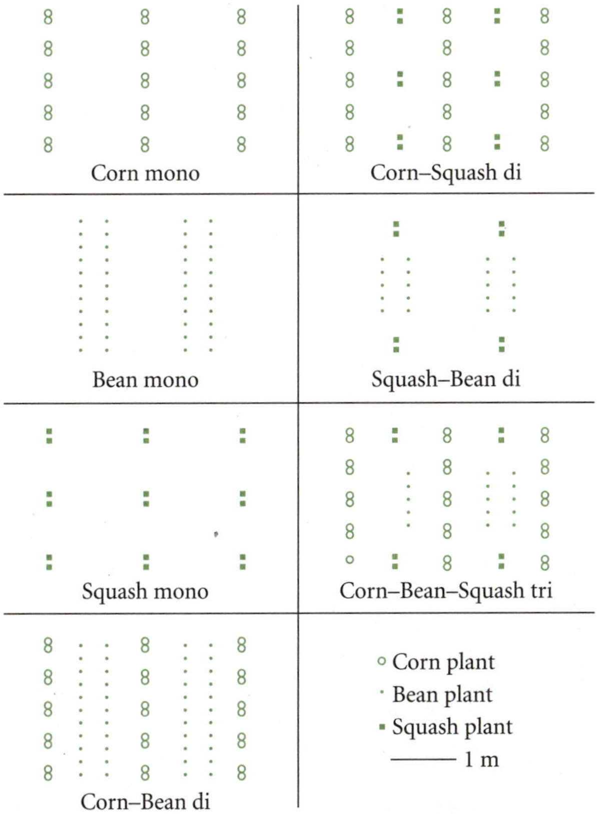

Corn mono

Corn–Squash di

Bean mono

Squash–Bean di

Squash mono

Corn–Bean–Squash tri

Corn–Bean di

◦ Corn plant
· Bean plant
▪ Squash plant
——— 1 m

FIGURE 24-16 Spatial arrangement of experimental plots in S. J. Risch's experiment on the effects of plant dispersion on herbivore damage in agricultural monocultures and polycultures. (*From Risch 1981.*)

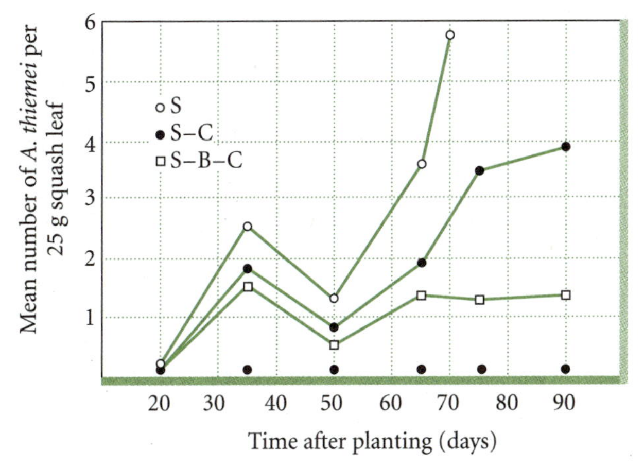

FIGURE 24-17 Average number of larvae of the chrysomelid beetle *Acalymma thiemei* on squash plants (S) in monoculture and in communities with one and two additional nonhost species. *A. thiemei* specializes on squash, so in this experiment, beans (B) and corn (C) are nonhost plant species. (*From Risch 1981.*)

24.7 Herbivores show a functional response to plant availability.

As discussed in Chapter 23, the functional response is defined as the relationship between the rate of exploitation of prey or hosts by predators or parasitoids and prey or host abundance. Holling's disc equation incorporates the important aspects of predator functional response. Do herbivores have functional responses? The answer is yes, but the herbivore functional response is different from that of predators in a number of important ways.

For predators, searching for prey and consuming them are more or less separate events. Predators that are in the process of consuming prey are usually not actively searching for the next prey. We may think of predators, then, as foraging among patches, each prey representing a patch (Spalinger and Hobbs 1992). Most predators consume all of a prey (patch) before moving on to the next prey, although instances of partial consumption are known. For the predator, search time and handling time do not overlap. For herbivores, however, the resource is usually not a single plant. Rather, it is a patch of plants or a particular part of an individual plant. The herbivore may not consume the entire patch and, importantly, it may continue to process plant material (e.g., chew) as it searches for new food, giving rise to an overlap of search time and handling time. Moreover, the density of plants within a patch and the size of the plant material relative to the herbivore's mouth size affect the rate of consumption of plant material by the herbivore.

We can illustrate the general features of the herbivore functional response by considering mammalian grazers. Mammalian herbivores forage by three different processes that are determined by different conditions of plant density and plant apparency (Table 24-3). **Plant apparency** refers to how easy or difficult it is for the herbivore to locate plants. Herbivores that feed on plants that are dispersed in space (low density) and difficult to locate (low apparency), such as small shoots hidden beneath taller grass, forage by process 1. Process 2 foraging occurs when easily located plants occur at low densities. Animals move directly from one plant to the next because they can easily detect the next plant from the position of the one that they are eating. Process 3 foraging occurs when plants are highly concentrated and apparent.

The number of bites (sometimes called the bite density) and the size of the bites that the herbivore takes while foraging interact with the density and apparency of the plants to regulate intake rate and to determine the herbivore functional response. An herbivore foraging by process 1 encounters a hidden plant and takes one or more bites from it. The size of the

TABLE 24-3 Characteristics that define three processes of foraging in mammalian herbivores

Process	Plant density	Plant apparency
1	Dispersed	Hidden
2	Dispersed	Apparent
3	Concentrated	Apparent

(From Spalinger and Hobbs 1992.)

bites is determined by the relationship between the plant morphology and the size of the herbivore's mouth. The number of bites depends on how dense the plants are in a particular area. For example, more bites are possible in a square meter containing ten individual hidden plants than in one containing five plants. Generally, it is bite density that regulates intake in process 1 and process 2 foraging (Figure 24-18). In both cases, as bite density increases, the intake rate increases in a type II functional response. The intake rate is somewhat higher at low bite densities for process 2 because plants are more apparent, making it easier for the herbivore to spot and move to the next plant and resulting in greater intake per minute. When the herbivore forages by process 3 (on concentrated or clumped plants that are apparent), the bite size, rather than the bite density, regulates intake, as shown in Figure 24-18. The shape of the functional response is the same as for the other two foraging processes.

For an herbivore that feeds on more than one type of plant species (a generalist), the diversity of the plant community may affect the intake rate because some plant species are more easily eaten than others. Beavers (*Castor canadensis*) have a higher food intake rate when foraging in pure stands of aspen (*Populus tremuloides*) seedlings than when they forage in stands that include aspen mixed with alder (*Alnus rugosa*) and red maple (*Acer rubrum*). The size of the plants may also affect the functional response. Beavers have a lower intake rate when foraging in stands of aspen of mixed size than when they feed in even-aged stands (Figure 24-19).

24.8 Plants and animals harbor a wide variety of parasites.

Parasites may live on the inside of their host's body, in which case they are referred to as endoparasites, or they may attach themselves to the outside of the host, in which case they are called ectoparasites (see Chapter 7). Parasites may also be categorized generally by their size. Microparasites are very small parasites such as viruses, bacteria, and protozoans. Most microparasites are also endoparasites. Microparasites may cause profound physiological and behavioral changes in their host, which manifest themselves as disease. Parasites that cause disease are called **disease organisms** or **pathogens.** The most important of these are bacteria and viruses, such as those that cause infection, colds, and flu. Animals may also serve as hosts for various types of protozoan parasites, such as those that cause sleeping sickness (trypanosomes) and the species of *Plasmodium* that causes malaria. Both of these parasites are of great economic importance because they cause sickness in human populations in many areas of the world and restrict human activities in some areas. Plants, too, suffer from parasitism by bacteria and viruses.

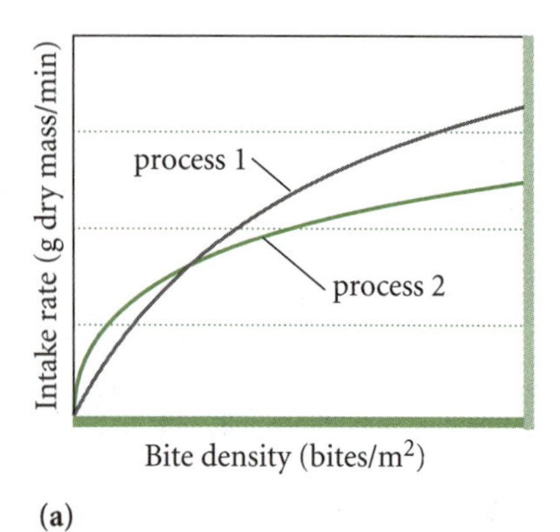

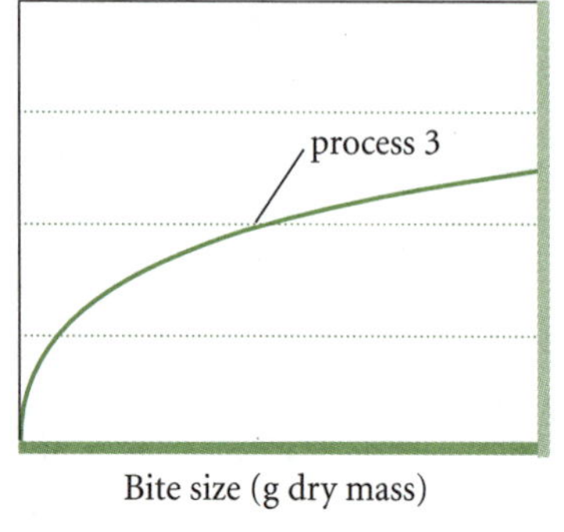

FIGURE 24-18 The intake rate (g dry mass/min) of a mammalian grazer increases in a pattern similar to the Holling type II functional response of predators (see Figure 23-12) as either the bite density (number of bites per unit of area) or the bite size increases. (*After Spalinger and Hobbs 1992.*)

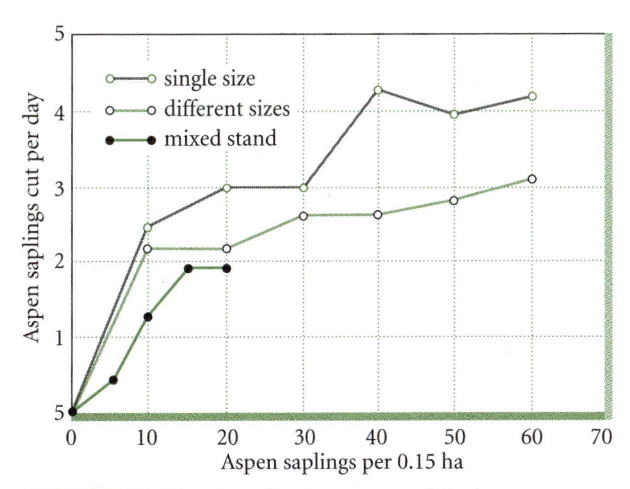

FIGURE 24-19 Functional response of the beaver (*Castor canadensis*) when feeding in plant communities having different size distributions. The rate of increase in aspen (*Populus tremuloides*) seedlings taken increases with the number of seedlings in pure stands of aspen seedlings of similar size. The rate of increase is lower in aspen stands of mixed size, and even lower when aspen is mixed with saplings of alder (*Alnus rugosa*) and red maple (*Acer rubrum*). (*After Fryxell and Doucet 1993.*)

Larger parasites, such as parasitic roundworms and tapeworms, which are endoparasites, and creatures such as ticks, fleas, mosquitoes, and mites, which are ectoparasites, are referred to as macroparasites. Macroparasites often produce eggs or other life cycle stages that leave one host and are carried to another. Endoparasitic tapeworms (platyhelminths) and roundworms (nematodes) live as adults in the body cavities or intestinal tracts of animals, where they feed by absorbing body fluids. Some ectoparasitic worms attach to the outside of their animal host. A number of flatworms feed on the skin of fish, reptiles, and amphibians and other marine and freshwater animals. Fungi may also attack animals; the common athlete's foot is a good example. Fungi are the most common macroparasites of plants, causing what plant pathologists call rust or smut. Gall-forming insects that attack the flowers or stems of plants may also be considered macroparasites.

Transmission

The complex life histories of parasites involve a variety of interactions with hosts, and different sets of factors affect each stage of the life cycle. The problem of locating hosts (**contagion** or dispersal) is crucial in parasite and, particularly, pathogen populations. In the case of endoparasites, population growth may occur within a single host, resulting in the subdivision of a parasite population and fostering population interactions among the descendants of infecting individuals.

The process of **transmission** is the movement of a parasite from one host to the next. Most microparasites do not have specialized structures for transmission, relying instead on **direct transmission** between hosts. The transmission of flu and cold viruses, for example, occurs when an infected individual comes close enough to an uninfected individual that the virus may be transmitted by direct contact or through the air in a sneeze. The virus is transmitted even more commonly when the hands of an uninfected person become contaminated by contact with an infected person and then are placed in the mouth or nostrils. Some microparasites may leave their host and reside in an inactive state in water or food, becoming active when the water or food is consumed by a host. This is the way the protozoan *Entamoeba histolytica,* which causes amoebic dysentery, is transmitted. The transmission of pathogens often involves **vectors,** which are organisms that transmit pathogens from one host to the next. For example, the trypanosomes that cause sleeping sickness are transmitted between hosts by the tsetse fly (*Glossina*), and the malaria-causing *Plasmodium* is transmitted by mosquitoes (Figure 24-20).

The transmission of macroparasites may be accomplished by direct movement of the adults, a process that usually requires hosts to be close to one another. You may have observed this process if you have seen fleas hopping from your cat or dog onto your skin. Many of the roundworms (nematodes) that infect humans are transmitted directly from one host to the next. Fungi, too, may often be transmitted directly from one host to the next. **Indirect transmission** occurs when the adult parasite, which resides in a primary host, produces immature stages that infect one or more intermediate hosts before infecting another primary host. The important human parasites *Schistosoma* and *Plasmodium* have this kind of life cycle (see Chapter 7).

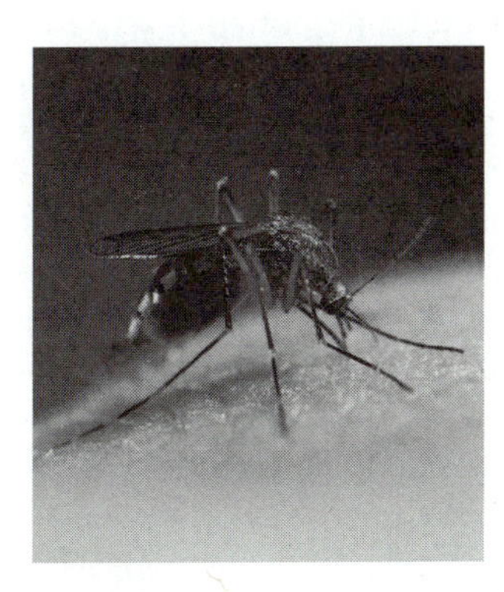

FIGURE 24-20 The mosquito is a vector of considerable importance to humans because it transmits microparasites such as those that cause malaria.

Parasites of Wildlife

Wild animals whose populations are monitored and managed for hunting and fishing are often collectively called **wildlife.** They include animals such as white-tailed deer, mule deer, wild turkeys and other wild birds, and a wide variety of fish, many of which are in the salmonid (trout and salmon) family. Wildlife managers are keenly interested in parasites and diseases of these animals because of the effect that diseases may have on mortality and, thus, the size of hunting and fishing stocks. A number of parasites affect wildlife, and many of these parasites have the potential to cause significant damage to wildlife populations (Scalet et al. 1996). The bacterial disease brucellosis infects the bison and elk herds of Yellowstone National Park. Some have suggested that this disease is transmitted from these wild animals to domesticated cattle in the area of these herds. Viruses infect white-tailed deer, causing outbreaks of epizootic hemorrhagic disease, for which there is no effective control. Even tuberculosis is a problem in some wildlife populations. Various types of bacterial diseases affect wild game birds. For example, botulism, caused by the bacterium *Clostridium botulinum,* can quickly spread in some bird populations. Fish fall victim to bacterial and fungal diseases in crowded conditions, such as in hatcheries.

24.9 Microparasite population dynamics may be modeled as infection.

Many microparasites cause diseases of humans, other animals, and plants. In most cases, the transmission of the disease from an individual host that harbors the parasite—an infected individual—to one that does not—a susceptible individual, occurs through direct contact. Sometimes, as in the case of human cold and flu viruses, transmission occurs between two individuals of the same species. In other cases, transmission occurs between hosts of different species. For example, Rocky Mountain spotted fever, a disease of humans caused by the bacterium *Rickettsia rickettsia,* is transmitted to humans by direct contact with a tick, an intermediate host.

Since pathogens depend on their hosts for their survival, pathogen population dynamics mirror the population dynamics of the hosts. At any given time, the host population will contain three types of individuals: those that are susceptible to infection by the disease, those that are infected, and those that have been infected but have now recovered. Recovered individuals may be immune to reinfection. A model that accounts for the dynamics of these three host conditions can represent the dynamics of the pathogen in the population of its hosts. Such a model is called an **infection model,** and involves the development of equations that track the dynamics of these three types of individuals in the host population, as we shall see below.

An Infection Model of Pathogen Population Dynamics

We shall follow the presentation of Bulmer (1994) in developing an infection model of pathogen population dynamics. Let $x =$ the number of susceptible individuals in the population, $y =$ the number of individuals who are infected, and $z =$ the number of individuals who have recovered. For simplicity, we shall assume that once an individual has recovered from the disease it retains an immunity against it, but that the immunity is not passed on to its offspring. Thus, recovered individuals are not susceptible, but their offspring are. This assumption means that susceptible individuals may enter the population only by birth. Susceptible individuals could also enter the population by immigration, but that possibility is not considered here. If the instantaneous birth rate, b, is the same for all three types of individuals in the population, then we may represent the number of new susceptible individuals added to the population at any instant as $b(x + y + z)$.

The number of susceptible individuals in the population may be reduced in two ways: by infection or by death. The transmission of a microparasite from an infected individual to one that is not yet infected but susceptible usually requires close contact between the individuals. Of course, not every contact between an infected and a noninfected individual will result in the transmission of the disease to the uninfected individual. The rate at which individuals are infected (and removed from the susceptible portion of the population) is thus best represented as the mass action term βxy, where β is a rate constant called the **transmission coefficient.** We expect that $0 < \beta < 1$. Pause for a moment to compare this term to the term aPH that appeared in the predator-prey models that we presented in the previous chapter. That term represents the same kind of phenomenon. In order for the predator to eat a prey, they have to come into close contact, and, of course, not every encounter results in a successful kill—thus the constant a.

Another way that individuals may be removed from the susceptible group is by mortality, which we may represent as a constant, d, for simplicity. Putting these terms together, we can write an equation for the instantaneous rate of change of the susceptible portion of the population,

$$\frac{\mathrm{d}x}{\mathrm{d}t} = b(x + y + z) - \beta xy - dx. \qquad (24\text{-}1)$$

In words, this equation says that the rate of change in the number of susceptible individuals ($\mathrm{d}x/\mathrm{d}t$) in the population is equal to the total number of individuals added to the population by birth in all three of the host conditions [infected, susceptible, recovered; $b(x + y + z)$] decreased by the number of susceptible individuals infected (βxy) and the number of susceptible individuals dying (dx).

The number of infected individuals in the population will increase by βxy, as we have just shown. The number of infected individuals decreases whenever an infected individual recovers from infection or dies. It seems reasonable to assume that the infected portion of the population will have a higher death rate than the susceptible or recovered parts of the population. To account for this, we will assign a death rate of D to the infected group, where $D > d$. Let the rate at which infected individuals recover be γ. Since both death and recovery remove individuals from the infected group, we combine them in the term $(D + \gamma)y$. Thus, we may write the equation for the infected part of the population as

$$\frac{dy}{dt} = \beta xy - (D + \gamma)y. \qquad (24\text{-}2)$$

This equation says that the instantaneous rate of change of the number of infected individuals in the population (dy/dt) is increased by the rate of infection (βxy) and decreased by death and recovery ($(D + \gamma)y$).

The rate of change in the number of recovered individuals is simply the rate of recovery, γy, which adds individuals to the recovered group, minus the rate of death of recovered individuals, dz, or

$$\frac{dz}{dt} = \gamma y - dz. \qquad (24\text{-}3)$$

A schematic representation of the relationships among these equations is presented in Figure 24-21.

Let us assume that a population harboring a disease will contain $N = x + y + z$ individuals. Thus, if we know how many susceptible and infected individuals reside in the population, and we know the population size, we can determine the number of recovered individuals; for example, $z = N - x - y$. This is wishful thinking, of course, since we rarely know the exact size of a natural population or the exact numbers of susceptible and infected individuals. Suppose further that the number of births and the number of deaths are the same at population size N. That is, $b(x + y + z) = d(x + z) + Dy$. Substituting this result into equation (24-1) for dx/dt, we obtain

$$\frac{dx}{dt} = d(x + z) + Dy - \beta xy - dx.$$

Substituting the relationship $z = N - x - y$ into this intermediate result gives us

$$\frac{dx}{dt} = d(x + N - x - y) + Dy - \beta xy - dx.$$

Multiplying through by d in the first term, grouping terms, and eliminating give us a new equation for the rate of change in the susceptible portion of the population,

$$\frac{dx}{dt} = d(N - x) + (D - d)y - \beta xy. \qquad (24\text{-}4)$$

This new equation eliminates the need for equation (24-3), the rate of change in the number of individuals who recover from the infection, so we need only

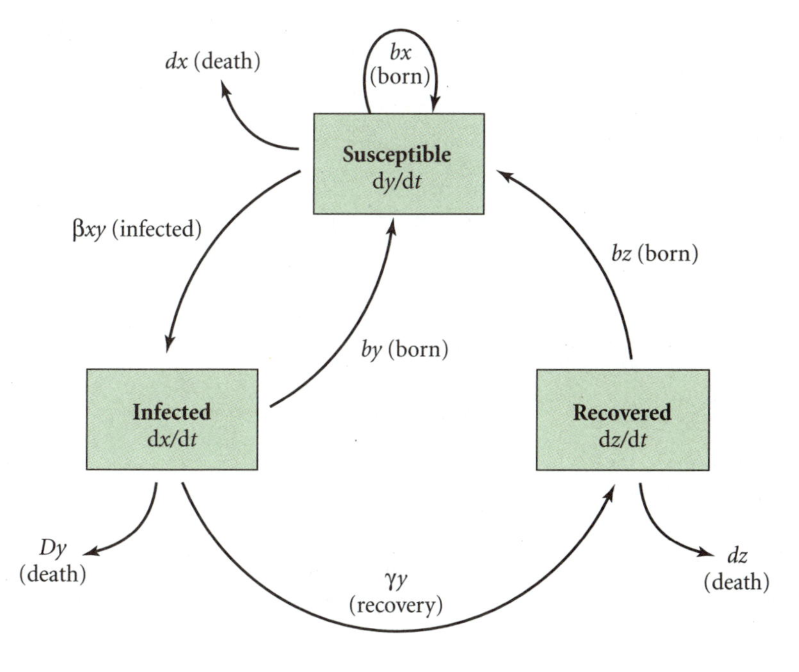

FIGURE 24-21 Schematic representation of the relationships between susceptible, infected, and recovered individuals in an infection model.

consider the new equation (24-4) and equation (24-2): $dy/dt = \beta xy - (D + \gamma)y$.

These two equations present conditions under which the infection is expected to persist in the population. To see this, consider the situation in which $dx/dt = 0$ and $dy/dt = 0$. Clearly, if $y = 0$, there will be no infection, because there will be no infected individuals in the population to infect other individuals. That is, the parasite does not exist in the population. When $y > 0$, there is a chance that a susceptible individual will be infected, and the mortality factors related to infected individuals will be in force. It can be shown that when $y > 0$, $dy/dt = 0$ when $x = (D + \gamma)/b$. This value is called the **threshold value** of the infection.

Reproductive Rate of Infection

To understand the implications of the threshold value, and to appreciate how the infection model works, we must consider the **reproductive rate of infection,** denoted $R(x)$. An infected individual will infect one or more susceptible individuals, and each of those newly infected individuals will infect more individuals. The reproductive rate of infection of an individual is the total number of individuals that are infected by it, either primarily or secondarily by individuals that it infected, during its lifetime. The reproductive rate of infection for the population, $R(x)$, is the average of the reproductive rates of the individuals in the population. The reproductive rate is dependent on the lifetime of an infected host. The longer an infected individual lives, the more other individuals it will infect. From the equations above, we see that infected individuals are removed from the population by recovery or death at a rate of $(D + \gamma)$. It is reasonable to say that the average lifetime of an infected host is the reciprocal of that rate, or $1/(D + \gamma)$ (Bulmer 1994). Susceptible individuals become infected at rate βx. The reproductive rate of infection is the product of the infection rate and the average lifetime of the infected individuals,

$$R(x) = \frac{\beta x}{(D + \gamma)}. \qquad (24\text{-}5)$$

The reproductive rate is given in terms of infected individuals, thus, if $R(x) < 1$, then on average, infected individuals are not infecting even a replacement for themselves, and the disease will not persist in the population. If, on the other hand, $R(x) > 1$, infected individuals are infecting more individuals than are required to replace themselves in the infected group, and the disease spreads. As long as the infection rate is greater than the average lifetime of infected hosts, the disease will persist. Recall that the average lifetime is the reciprocal of the rate of removal of infected individuals from the population by death or recovery. A long lifetime in infected individuals implies that rates of death and recovery from the disease are low. The value of x, the susceptible population, at which $R(x) = 1$ is $(D + \gamma)/\beta$, which, as we noted above, is the equilibrium value of x in the equations above (threshold value). This means that the infection cannot spread if there are fewer than $(D + \gamma)/\beta$ susceptible individuals in the population.

Immunization Against Disease

The results of the infection model presented above have important implications for public health. Any program that seeks to immunize a population against a disease must immunize enough individuals to reduce the susceptible portion of the population below the threshold value, at which point the disease will not persist. When this point is reached, the immunization program is said to provide **herd immunity.**

Let us examine how we obtain herd immunity. Before the immunization program begins, the reproductive rate of the disease is $R(x) > 1$. Rearranging the relationship $R(x) = \beta x/(D + \gamma)$ to solve for x gives $x = R(x)(D + \gamma)/\beta$. Designate this x as x_0. If the immunization program is successful, then $R(x) = 1$, and the equation becomes $x = (D + \gamma)/\beta$. Let this x be called x_1. The proportion of the population that must be vaccinated in order to halt the infection, p_v, is thus $p_v = 1 - x_1/x_0$, or $1 - 1/R(x)$. Figure 24-22 shows the proportion of the population that must be vaccinated to halt a number of common human diseases (Anderson and May 1991).

Suppose a disease breaks out in a population of 1 million people. The disease has the following characteristics: $D = 0.7$, $\gamma = 0.1$, $\beta = 0.9$. That is, the death rate related to the disease is high, the recovery rate low, and transmission between individuals

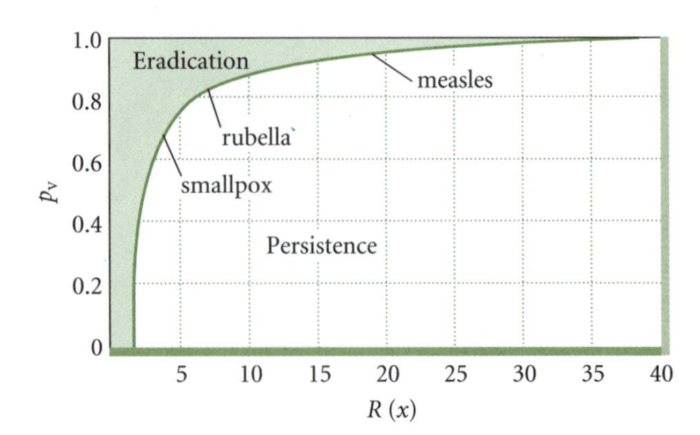

FIGURE 24-22 The proportion of a susceptible population, p_v, that must be vaccinated in order to eradicate a disease from a population for some common human diseases. $R(x)$ is the reproductive rate of the infection. (*After Anderson and May 1991.*)

extremely likely. If the reproductive rate of the infection is $R(x) = 2$, then the proportion of the population that must be vaccinated in order to provide herd immunity is

$$p_v = 1 - \frac{x_1}{x_0}$$
$$= 1 - \left[\frac{0.7 + 0.1}{0.9}\right] \Big/ \left[\frac{2(0.7 + 0.1)}{0.9}\right]$$
$$= 0.50$$

Thus one-half of the population (500,000) will have to be vaccinated against the disease to curtail its expansion in the population. If the reproductive rate of the disease is $R(x) = 10$—that is, if on average, every infected individual is responsible for 10 more infections during its lifetime—then $p_v = 0.90$, 90% of the population, and 900,000 people will have to receive immunization against the disease.

Immunity and Disease Cycles

Organisms often recover from a disease, and in response to the disease, develop some short-term or long-term immunity to it. Humans, for example, develop a lifetime immunity to measles once they are infected. In a population to which no new susceptible individuals are being added by birth or by immigration, and where immunity arises, the pathogen will eventually infect everyone in the population, everyone will become immune, and the pathogen will die out for lack of susceptible hosts. If the disease spreads more quickly than new susceptible individuals are added to the population—a situation that is typical for virulent diseases like measles—then the disease may be characterized by periodic cycles or outbreaks. Measles spreads quickly in a population, and immunity develops in infected individuals on a scale much faster than new susceptible individuals are added to the population (Figure 24-23). Immunization programs often focus on young individuals who are susceptible at birth, thereby preventing the resurgence of the disease by keeping the susceptible population low.

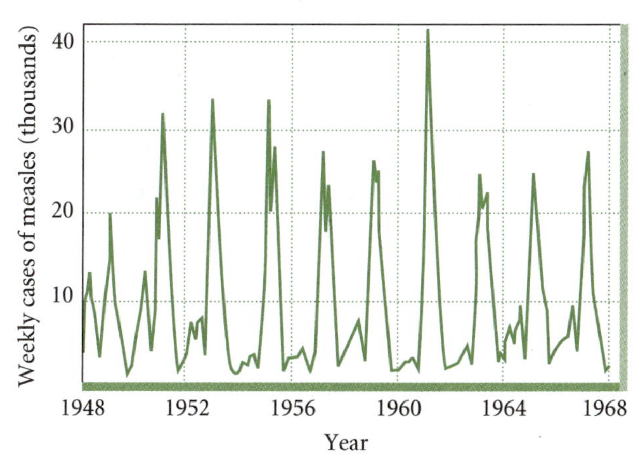

FIGURE 24-23 Weekly cases of measles in England and Wales between 1948 and 1968 prior to mass vaccination. (*From Anderson and May 1991.*)

24.10 The virulence of parasites depends on transmission properties and host immune responses.

The **virulence** of a parasite or pathogen is the extent to which it causes harm or mortality to its host. It has long been observed that whereas some parasites are extremely virulent, other, closely related parasites are only mildly harmful to the host (Ewald 1995, Lenski and May 1994, Antia et al. 1994). Often, one or two mutations in a relatively harmless form of a bacterium or virus will transform it into a highly virulent form (Rosqvist et al. 1988), as is the case with a number of bacteria related to the bacterium that causes bubonic plague. The question of why some parasites are more virulent than others, and the related question of how virulence evolves, have come under intense scrutiny in recent years.

Until relatively recently, most parasitologists accepted the view that parasites and their hosts evolve toward a more or less stable coexistence that is characterized by mild virulence. That is, parasites tend to lose virulence over a long evolutionary association with a host. This idea is grounded in the assumption that in order for the parasite to survive, it must avoid overexploitation and possible extinction of the host. Recent theoretical evaluations of this idea suggest that, indeed, under some circumstances, parasites and pathogens can be expected to become less virulent over time (Lenski and May 1994). The idea has been questioned, however, on the basis of flaws in its precepts, most particularly misconceptions about the way natural selection works, and the mechanisms by which virulence evolves. Let us take a look at one aspect of this debate, the importance of host mobility and transmission.

One of the problems with the idea that the most adapted parasites are benign parasites has to do with the assumption that selection favors parasites that do not overexploit their hosts. Is this really true? Consider a parasite that displays some variation in the way in which it exploits its host. If, by increasing its exploitation of the host, a particular variant of the parasite increases its own fitness (greater reproduction), it will be favored by natural selection. As long as increasing the exploitation of the host has no harmful effects on the host, and as long as greater exploitation means greater reproductive output for the parasite, selection will favor increased exploitation. What if the parasite increases exploitation of the host to the point at which it begins to harm the host? Selection should still favor exploitation as long as the benefit to the parasite is greater than the cost to the parasite incurred by the increased harm to the host. Harm to the parasite occurs when exploitation negatively affects transmission of the parasite. Thus, natural selection should favor greater exploitation of the host as long as transmission is not reduced (Ewald 1995). To be sure, evolutionarily speaking, the host has something to say about all of this. As we discussed in Chapter 7, a change in the intensity of exploitation by the parasite constitutes a selective pressure on the host.

The mode of transmission, then, is an important factor in the evolution of virulence in parasites. In thinking about parasite transmission, we often focus on the mobility of the host because the activities of the parasite—the exploitation of the host by the parasite—often affect the host's mobility, and diminished mobility often signifies severe disease symptoms. The more dependent the parasite is on the mobility of the host for its transmission, the more likely it is that any exploitation of the host that diminishes host mobility will exact a high cost on the parasite. Conversely, if transmission is independent of host mobility, exploitation should have little effect on transmission. Parasites that rely on arthropod vectors, such as mosquitoes, or that are carried in water do not suffer from host immobility. We may expect such parasites to evolve high virulence. Diarrheal parasites are a good example (Ewald 1991).

In this chapter, we have discussed the interactions of herbivores and their host plants and parasites and their hosts. These interactions have effects on the individuals involved, but mortality of the host is usually not one of them. The performance of individual plants may be affected by the activities of herbivores, and this may lead to lower reproductive output and thus a change in population size or total plant density within a population. The architecture, spatial distributions, and defensive mechanisms of plants affect the ability of herbivores to consume them, and thus have an effect on the herbivore population. Parasites, like herbivores, rarely cause death in their hosts. The population of each may be affected, however, by the interaction. Some parasites cause diseases that affect the host population size. Understanding the dynamics of these diseases is important in order to reduce their effects in humans and in domesticated and wild populations of plants and animals.

SUMMARY

1. Herbivory refers to all types of consumer-resource interactions involving the consumption of plants and plant parts. Herbivores display grazing, browsing, seed predation, and a variety of other plant-eating strategies. Unlike predator-prey interactions, herbivore-plant interactions are not easily characterized by a simple accounting of deaths and births because herbivory often does not cause death.

2. A vast number of different species representing nearly every major animal group specialize in eating plants or plant parts. The arthropods, and in particular, the insects include the greatest abundance and diversity of herbivorous forms. Herbivory is an important part of every ecosystem.

3. Herbivory may reduce plant growth, survival, and fecundity, a result most obvious when there is an outbreak of an insect herbivore, such as the Japanese beetle or gypsy moth. In many cases, herbivory induces compensatory responses in plants, such as a reallocation of resources within the plant body or a change in plant architecture. Some have suggested that some plants overcompensate for damage inflicted by herbivory to the extent that their fitness is actually increased, an idea that is controversial.

4. Grazing by vertebrate animals may influence the community structure and productivity of the plant community through feedback processes involving nutrient recycling.

5. Many plants are subject to simultaneous or successive attack by a number of different herbivores that specialize on specific plant parts. Thus, plants suffer different levels of herbivore pressure through their development, and herbivore damage is influenced by the age structure of the plant population. Evidence for direct population regulation of plants by herbivores is not extensive; however, recent studies demonstrating that insect herbivory may restrict plant ranges support the notion that such regulation is possible.

6. The susceptibility of plants to herbivory and the responses of plant populations to herbivore pressure are influenced by the spatial dynamics of the plant population. Specialized herbivores may have difficulty finding suitable host plants in diverse plant communities. Vegetation texture may restrict the movement of herbivores from one host to another.

7. The functional response of mammalian grazers is dependent on the density and the apparency of plant material and on the size and frequency of the bites that the grazer takes while foraging.

8. Microparasites often cause physiological changes in their hosts that we recognize as disease. Disease organisms, called pathogens, may be directly or indirectly transmitted from an infected individual to an uninfected individual.

9. The dynamics of infection provide an excellent model for microparasite populations. Infection models track the number of susceptible, infected, and recovered hosts in the population. The reproductive rate of the infection, $R(x)$, is the average number of infections caused by each infected individual during its lifetime. When the number of susceptible individuals, x, is reduced to the point that $R(x) < 1$, the disease will not persist in the population. The purpose of immunization programs is to achieve this threshold value of susceptible individuals.

10. The evolution of parasite virulence is related to the cost of host immobilization. Parasites that do not depend on their hosts for transmission are expected to become more virulent than those that rely on host mobility for movement from infected to susceptible host individuals.

EXERCISES

1. Deserts are places where there are few leafy green plants for herbivores to eat. Using this book and other sources, describe the natural history of a desert herbivore.

2. Herbivory may affect the distribution of plants on a local scale, as was demonstrated in the experiments of Louda and Rodman using bittercress. How might the effects of herbivory be manifested on a regional or landscape scale?

3. The models of predator-prey interactions focus on changes in the size of the predator and prey populations resulting from the interaction. The prey population is reduced by predation and other sources of mortality. The number of predators is reduced by mortality or the action of another predator. Both predator and prey populations are increased by reproduction. The consequences of herbivory on the populations of the herbivore and the host plant are not so straightforward. Discuss why this is the case.

4. A population of $N = 1{,}000$ individuals contains 10 individuals that are infected with a disease caused by a microparasite. What is the threshold value of the infection in the population?

5. Calculate p_v for a disease for which the death rate of the infected group is 0.15, the rate at which infected individuals recover is 0.85, the rate at which the disease is transmitted is 0.01, and the reproductive rate of the disease is 3.

CHAPTER 25

Coevolution and Mutualism

GUIDING QUESTIONS

- What are the processes of coevolution, specialization, and mutualism?
- What are the differences between population and species coevolution?
- What observations support the geographic mosaic theory of coevolution?
- How does the elicitor-receptor model explain the mechanism of coevolution?
- What are the three types of mutualism?
- How have mutualist relationships evolved in ants?
- What is the coevolutionary relationship between butterflies and umbelliferous host plants?
- How has coevolution shaped the relationship between yucca plants and their moth pollinators?

When two populations interact, each may evolve in response to characteristics of the other that affect its well-being—that is to say, its evolutionary fitness. Such reciprocal evolutionary responses of interacting species or populations are referred to as **coevolution** (Thompson 1982, 1994, Gilbert and Raven 1975, Berenbaum 1983, Nitecki 1983, Futuyma and Slatkin 1983, Boucher 1985, Armbruster 1992, Thompson 1994). Direct evidence of reciprocal evolutionary responses is often difficult to obtain, leading some ecologists to suggest that coevolution is rare in nature. Much of the debate about the prevalence of coevolution has centered on the phenomenon of **specialization,** in which an organism interacts with a limited number of other species (Thompson 1994). Specialization may result from reciprocal interactions among populations, but, as we shall see, this does not always imply that coevolution has occurred. One type of population interaction that may involve coevolution is **mutualism,** in which both participants benefit from the association. Mutualistic interactions often reflect strong coevolution and result in a high degree of specialization. Our discussion of coevolution, specialization, and mutualism in this chapter provides an opportunity to place population interactions in an evolutionary perspective.

25.1 Evolutionary relationships between antagonists often demonstrate coevolution.

The interactions of parasites and parasitoids with their hosts are among the most specialized of all population interactions (Price 1980, Thompson 1994), and thus it is to those groups that we turn first for examples of coevolution. One of the first population interactions to be analyzed in the context of coevolution was the relationship between the European rabbit that was introduced into Victoria, Australia, in 1959 and the myxoma virus (a virus related to smallpox) that was eventually employed to control the rabbit.

Shortly after the release of a few pairs in 1859, the rabbit became a major pest in Australia. To give you an idea of how fast the rabbit population increased, within a few years of the introduction, local ranchers

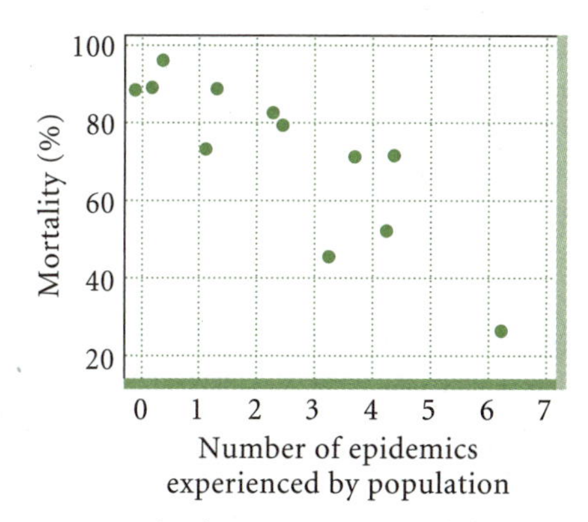

FIGURE 25-1 The decrease in susceptibility of wild rabbits to a myxoma virus with virulence causing 90% mortality in genetically unselected wild rabbits. (*From Fenner and Ratcliffe 1965.*)

were erecting rabbit fences and organizing rabbit brigades—shooting parties—in vain attempts to keep their numbers under control. Eventually, hundreds of millions of rabbits were distributed throughout most of the continent, where they destroyed range and pasturelands and threatened wool production. The Australian government tried poisons, predators, and other potential controls, all without success. After much investigation, the answer to the rabbit problem seemed to be a myxoma virus discovered in populations of a related South American rabbit. The myxoma virus produced a small, localized fibroma (a fibrous cancer of the skin). Its effect on South American rabbits was not severe, but European rabbits infected by the virus died quickly of myxomatosis.

In 1950, the myxoma virus was introduced locally in Victoria. An epidemic of myxomatosis broke out among the rabbits and spread rapidly. The virus was transmitted primarily by mosquitos, which bite infected areas of the skin and carry the virus on their mouthparts. The first epidemic killed 99.8% of the infected rabbits, reducing their populations to very low levels. But during the following myxomatosis season (which coincides with the mosquito season), only 90% of the remaining population was killed. During the third outbreak, only 40–60% of infected rabbits succumbed, and their population began to grow again.

The decline in the lethality of myxomatosis in the Australian rabbits resulted from evolutionary responses in both the rabbit and the virus populations. Before the introduction of the myxoma virus, some rabbits had genetic factors that conferred resistance to the disease. Although nothing had spurred an increase in these factors before, these individuals were strongly selected by the myxomatosis epidemic, until most of the surviving rabbit population consisted of resistant animals (Figure 25-1). At the same time, virus strains with less virulence increased because reduced virulence lengthened the survival time of infected rabbits and thus increased the mosquito-borne dispersal of the virus (mosquitoes bite only living rabbits). A virus organism that kills its host quickly has less chance of being carried by mosquitoes to other hosts.

Left on its own, the Australian rabbit-virus system would probably evolve to an equilibrial state of benign, endemic disease, as it had in the population of South American rabbits from which the myxoma virus was isolated. Pest management specialists keep the system out of equilibrium and maintain the effectiveness of myxoma as a control agent by finding new strains of the virus to which the rabbits have yet to evolve immunity.

The decreasing virulence of the myxoma virus results from the fact that mosquitoes disperse the virus to new hosts more efficiently when the infected host has a longer life span. Thus, less virulent strains of myxoma have a higher rate of growth in populations of hosts as a whole, if not within individual hosts. Highly contagious diseases that are spread directly through the atmosphere or water have no such constraint on dispersal and often exhibit high levels of virulence, with debilitating and even fatal consequences for their hosts. Similarly, most predators do not rely on a third party to find prey, so rather than evolving toward a benign equilibrium of restraint and tolerance, predator and prey tend to become locked in an evolutionary battle of persistent intensity. The outcome of the battle depends on which population gets the evolutionary upper hand.

In a series of classic experiments, David Pimentel and colleagues (Pimentel et al. 1968) explored the evolution of host-parasitoid relationships using the housefly and a wasp parasitoid of the pupal stage of the fly, *Nasonia vitripennis* (Figure 25-2). In one population cage, *Nasonia* was allowed to parasitize a fly population that was kept at a constant level by replenishment from a stock that had not been exposed to the wasp. Any flies that escaped attack by wasp parasitoids were removed from the population cage, so

FIGURE 25-2 The wasp *Nasonia* parasitizing a pupa of the housefly. (*Courtesy of D. Pimentel; from Pimentel 1968.*)

that the wasps were provided only with evolutionarily "naive" hosts.

In a second population cage, fly hosts were kept at the same constant number, but because emerging flies were allowed to remain, the population could evolve resistance to the wasps. The population cages were maintained for about 3 years, long enough for evolutionary change to occur. Over the course of this experiment, the reproductive rate of wasps in the cage that permitted evolution dropped from 134 to 39 progeny per female, and their longevity decreased from 7 to 4 days. The average size of the parasitoid population also decreased (1,900 adult wasps versus 3,700 in the nonevolving system), and population size was more constant than in the nonevolving cage. These results suggest that the flies evolved additional defenses when subjected to intense parasitism.

Next, Pimentel and colleagues established thirty-compartment population cages in which the numbers of flies were allowed to vary freely. One such experiment was started with flies and wasps that had had no previous contact with each other, and a second was established with animals from the evolving population described above. In the first cage, the wasps were efficient parasitoids, and the system underwent severe oscillations. In the second cage, however, the wasp population remained low, and the flies attained a high and relatively constant population level (Figure 25-3). This result strongly reinforced the conclusion, drawn from the earlier experiments, that the flies had evolved resistance to the wasp parasitoids.

When the coevolutionary relationship between two species is antagonistic, as in the above examples, the species can become locked in an evolutionary battle to increase their own fitness, each at the expense of the other. Such a struggle may lead to an evolutionary stalemate in which both antagonists continually evolve in response to each other, or they may simply run out of the genetic variability needed to fuel further evolutionary change. In either event, the net outcome of their interaction may be a steady state. Alternatively, when one of the antagonists cannot evolve fast enough, it may be driven to extinction. Coevolution between organisms that mutually depend on each other may also lead to stable arrangements of complementary adaptations.

In the two examples of coevolution above, each of the two populations—the European rabbit and the myxoma virus in the first case and the housefly and its parasitoid *Nasonia vitripennis* in the second—evolves in response to characteristics of the other that affect its evolutionary fitness. This reciprocal evolutionary response is coevolution. Ecologists often apply the term *coevolution* in a narrower sense to situations in which one species evolves an adaptation specifically in response to an adaptation in another species that affects their interaction. The following example is a bit of a stretch, but makes the point. Hyenas have jaws and associated muscles that are strong enough to crack the bones of their prey. These modifications clearly are adaptations for eating selected by attributes of the prey. They cannot, however, be considered an example of coevolution because the bones of gazelles have not evolved to resist being eaten by hyenas, or any other predator. By the time a hyena has reached that part of its meal, bone structure has no consequence for prey survival. That is, there is no reciprocal evolutionary response. In contrast, when an herbivore evolves the ability to detoxify substances that a plant has produced specifically to deter herbivory, the requirements of the narrow definition of coevolution are more likely to be met. When this restrictive definition is applied, ecologists sometimes have difficulty identifying unambiguous cases of coevolution. This restrictive definition is applied in the discussion of mutualism in the latter part of this chapter.

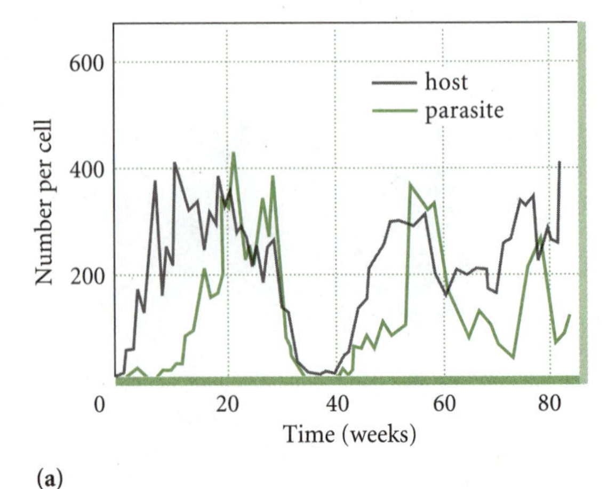

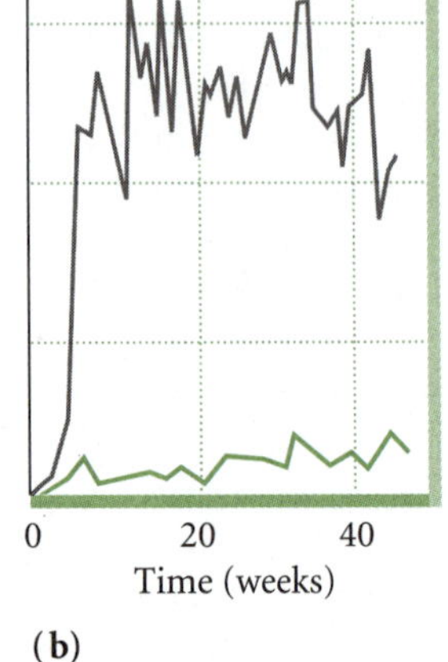

FIGURE 25-3 Populations of houseflies and the wasp parasitoid *Nasonia vitripennis* in thirty-cell laboratory cages. (a) Control: flies had no previous experience with the wasp. (b) Experimental: flies had been exposed to wasp parasitism for more than 1,000 days. (*After Pimentel 1968.*)

25.2 An appreciation of temporal and spatial scale are important in understanding coevolution.

At a number of places in this book, we have pointed out that our perception of how a particular natural process works is, to a large extent, dependent on the scale of resolution of time or space with which we observe that process. This is true for the process of coevolution as well, and this fact has provided an opportunity for considerable discussion among ecologists about whether, and under what conditions, coevolution occurs (Thompson 1994).

Most ecologists believe that coevolution involves a lengthy association between interacting populations. This view arises primarily from the perception that high degrees of specialization require considerable evolutionary fine-tuning over long periods of time. We shall encounter this issue when we discuss mutualisms in the last part of this chapter. Mutualisms often involve rather extraordinary complementarity in morphology, behavior, and physiology among coevolved populations. However, specialization need not require vast amounts of time for its evolution. Moreover, the presence of specialization itself does not necessarily imply that coevolution has taken place.

Specialization and Coevolution

It has been suggested that most species show substantial specialization, interacting with only a limited number of other species (Thompson 1994). Indeed, it is widespread specialization that gives rise to much of the world's biodiversity. Phytophagous (plant-eating or plant-sucking) insects, fungi, nematodes, and mites, among others, display high levels of specialization, and these groups make up large proportions of the fauna of many ecosystems (Hawksworth 1991, Thompson 1984, 1994).

The ubiquity of specialization in nature, particularly among parasites, fungi, and herbivores, argues for the pervasiveness of coevolution. However, more careful consideration reveals a number of weaknesses in this conclusion. How do we know whether specialization is the result of coevolutionary processes, or whether it is the result of a fortuitous pairing of species that occurs as part of the turnover (introduction and loss) of species in natural communities? For example, suppose that a community contains a plant species upon which two highly specialized herbivores feed. One is a nematode that feeds on the roots of the plant, and the other is a tiny leaf-boring insect. If a species of stem-boring insect capable of utilizing the plant were suddenly to appear in the community, it might immediately begin boring into stems of the plant. When we observe the stem-boring species, we might marvel at how specialized it appears to be, and in fact, our excitement would be justified, since it is likely that the stem borer is capable of using only a small number of different species of plants. But the mere presence of the stem borer does not imply that coevolution has occurred between it and its newly adopted host plant, as no reciprocal evolutionary responses between the plant and the stem borer have occurred. The stem borer has simply been thrust into an environment where there happens to have been a suitable host plant. Ecological opportunities such as the one enjoyed by the stem-boring insect are probably quite common in nature. Human activities, which disrupt ecosystems and move species from one place to another, probably contribute substantially to the number of such opportunities.

Coevolution and Spatial Scale

Investigation of a coevolved relationship requires painstaking observations of all the players, so ecologists generally focus their attention on local populations. The discovery of local specializations between mutualists, so-called **population coevolution,** leads us to ask whether such specializations occur in the same form and with the same intensity everywhere in the geographic range of the interacting species, a phenomenon referred to as **species coevolution.** The answer to this question appears to be no. Two species may interact differently in different populations because of variations in the biotic and abiotic environment. Thus, a particular interaction between two species may evolve in one population but not another, or there may be interpopulation differences in the degree of specialization in the interaction observed. There may also be evolutionary responses to a generalized type of organism rather than to a particular species, a situation referred to as **diffuse coevolution.** Such responses lead to less stringent specialization and changes in coevolutionary partners in different parts of a species' range.

Thompson has suggested a way to organize our thinking about coevolution and spatial scale based on five observations of coevolution and specialization. First, the outcome of species interactions is unlikely to be the same for each interacting species. Asymmetrical competition (see Chapter 22) is one way in which such differences may arise as part of the biotic interaction, but populations may also show varying responses because of their different responses to the physical environment.

Second, because these differences exist, a particular species interaction may result in coevolution in one local population and not in another. For example, suppose that a particular species of plant is known to fall victim to a species of root-eating nematode, and that both the plant and the nematode are known to occur together in three different communities in Mississippi. Because the soil conditions, moisture regimes, plant community structure, and population structure of both the plant and the nematode most likely differ among the communities, the outcome of the interaction between the plant and the nematode may also differ. Indeed, coevolution may occur in one of the communities but not in the other two. It is possible that the nematode has only a negligible effect on the plant in one of the communities because that community contains other closely related plants on which the nematode may feed.

Third, in addition to interpopulation variation in the outcome of species interactions, there is within-species variation in the degree to which specialization occurs. For example, a certain phytophagous insect might specialize on a single species of plant in one area, but the same species of insect may utilize two species of plants in another area. It is important to appreciate as well that specialization alone does not necessarily imply that coevolution has occurred. Thus, variation in the degree of specialization of species adds further texture to the coevolutionary landscape.

Fourth, the variation in response discussed above will create a "geographic mosaic of coevolution" (Thompson 1994). In other words, if we take a landscape view of an interaction such as the one described in the hypothetical example above, we will see places in the landscape where the interaction results in coevolution and places where it does not. You have probably already surmised that the existence of such geographic variation in the outcome of a population interaction could result in a situation in which, at the landscape level, a species coevolves with more than one other species.

Finally, spatial variation in coevolution and specialization is not static. Changes in population interactions and in the physical environment may result in local extinctions and introductions, creating new ecological opportunities and thus new chances for coevolution and specialization. Thus, the shifting mosaic of coevolutionary responses has a temporal component.

Thompson's model suggests that complex coevolved relationships can arise more readily in a spatially subdivided population because various components of the interaction may arise locally in one population or another, and these may be built up into more complex adaptive complexes by gene flow between populations. In a way, this is like building a complex structure from simple individual steps, which Thompson sees as taking place in different parts of the population independently before being brought together.

25.3 A gene-for-gene genetic process has been proposed as a mechanism of coevolution.

The genetic basis of coevolution would, at first glance, appear to be relatively straightforward. For a gene for a particular trait in one of the coevolving species, there exists a corresponding but antagonistic gene in the other species. Thus, for example, if European rabbits have a genetic factor that conveys resistance to the myxoma virus, it is reasonable to presume that viral mutations that restore virulence may arise and be selected in direct response. How might such a gene-for-gene genetic mechanism work in coevolution?

An early mathematical model of a single-allele complementary system was presented by Mode (1958). Mode built his model around a crop-rust system. Interactions between plants and their fungal parasites still provide the strongest evidence of gene-for-gene coevolution today (Thompson 1994). Mode assumed a long evolutionary relationship between the plant and the parasite that involved two opposing selection pressures: one exerted by the host on the parasite and the other exerted by the parasite on the host. This idea is essentially the same as the one we invoked in Chapter 7 in our discussion of the responses of organisms to their biotic environment, except that there we did not assume that the response involved a single allele. Mode suggested that the interaction between host and parasite would result eventually in an equilibrium between the two. Of course, such an equilibrium could be upset by the appearance of a new virulence gene in the pathogen or a new resistance gene in the host. Hence, stability in the coevolved systems depends upon the constancy of genotypes within each population.

A more recent conceptualization of gene-for-gene coevolution is the **elicitor-receptor model** (Keen et al. 1990, de Wit 1992, Thompson and Burdon 1992). Like the model of Mode, this model was developed in the context of a plant and its parasite. Envision a plant-parasite system based on Mendelian genetics in which a single locus in the plant may have a dominant allele, *R*, conferring resistance to the parasite, or a recessive allele, *r*, which is susceptible. The parasite has a corresponding locus with a dominant allele, *V*, conferring avirulence, and a recessive allele, *v*, conferring virulence (Table 25-1). Parasites that have the *V* allele (genotype *VV* or *Vv*) produce a product called an **elicitor,** which can be detected by those plants that possess the *R* allele (genotype *RR* or *Rr*). In order for resistance to occur, *both* the *R* gene and the *V* gene must be involved in the interaction. Thus, parasites with genotype *vv* would be virulent in any genotype of the plant (*RR*, *Rr*, or *rr*) because of the absence of the *V* allele. The *V* allele persists in the population

TABLE 25-1 Compatibility between genotypes in a single-locus gene-for-gene interaction

	Host genotype	
Pathogen genotype	*RR*	*rr*
VV	Resistant	Compatible
vv	Compatible	Compatible

(*From Thompson 1994.*)

owing to the fact that virulence and resistance are conditional on the genetic factors of the host. The best evidence for this model comes from studies of the parasites of agricultural crops, but it is still uncertain whether such mechanisms occur in other types of organisms (Thompson 1994).

25.4 Organisms often form mutualistic relationships.

Much of our interest in coevolution has focused on mutualism, the situation in which two species interact in such a way that both species benefit. Three types of mutualism are recognized: **trophic mutualism, defensive mutualism,** and **dispersive mutualism.** Sometimes the relationship between mutualists has evolved to the point where the species are fully dependent on each other, a situation that we refer to as **obligate mutualism.** Other mutualisms are considered **facultative mutualisms** because the partners could do without each other if necessary. Obligate mutualisms provide some of the best examples of coevolution (Boucher et al. 1982, Janzen 1985), and we shall discuss a number of those relationships in more detail in subsequent sections. First, let us consider a few examples of the three types of mutualistic relationships.

Trophic Mutualism

Trophic mutualism usually involves partners specialized in complementary ways to obtain energy and nutrients from each other; hence the term *trophic.* We have seen trophic mutualisms in the symbiotic associations of algae and fungi to form lichens (Chapter 1), of fungi and plant roots to form mycorrhizae (Chapter 12), and of *Rhizobium* bacteria and plant roots to form nitrogen-fixing root nodules (Chapter 11). In these cases, each of the partners supplies a limited nutrient or energy that the other cannot obtain by itself. *Rhizobium* can assimilate molecular nitrogen (N_2) from the soil, but requires carbohydrates supplied by a plant for the energy needed to do this. Bacteria in the rumens of cows and other ungulates can digest the cellulose in plant fibers, which a cow's own digestive enzymes cannot do. The cows benefit because they assimilate some of the by-products of bacterial digestion and metabolism for their own use (and they also digest some of the bacteria themselves). The bacteria benefit by having a steady supply of food in a warm, chemically regulated environment that is optimal for their own growth.

Ants belonging to the tropical group Attinae harvest leaves and bring them to their underground nests, where they use them to cultivate a highly specialized species of fungus. These leaf-cutter ants consume the fungus; in fact, it is their only source of food. They also provide a living environment for the fungus, which can live nowhere else in nature. Thus, the organisms are totally dependent on each other. Such mutualistic relationships are extremely stable, especially compared with consumer-resource interactions (see Chapter 20), because both partners cooperate and are mutually evolved to each other's benefit as well as to their own. Genetic studies indicate that some of these relationships go back more than 20 million years (Armbruster 1992).

Defensive Mutualism

Defensive mutualisms involve species that receive food or shelter from their mutualistic partners in return for defending those partners against herbivores, predators, or parasites. For example, in marine systems, specialized fishes and shrimps clean parasites from the skin and gills of other species of fish. These cleaners benefit from the food value of parasites they remove, and the groomed fish are unburdened of some of their parasites. Such relationships, often referred to as **cleaning symbioses,** are most highly developed in clear, warm tropical waters, where many cleaners display their striking colors at locations, called cleaning stations, to which other fish come to be groomed. As might be expected, a few species of predatory fish mimic the cleaners; when other fish come and expose their gills to be groomed, they get a bite taken out of them instead.

Dispersive Mutualism

Dispersive mutualism generally involves animals that transport pollen between flowers in return for rewards such as nectar, or that disperse seeds to suitable habitats as they eat the nutritional fruits that contain the seeds. Dispersive mutualisms rarely involve close living arrangements between partners. Seed dispersal mutualisms are not usually highly specialized; a single bird species may eat many kinds of fruit, for example, and each kind of fruit may be eaten by many kinds of birds. Plant-pollinator relationships tend to be more restrictive because it is in a plant's interest that a flower visitor carry pollen to another plant of the same species. Specializations related to plant pollination have been extensively studied; we shall provide more detail on those in Section 25.7.

25.5 Many mutualistic relationships have developed among ants and other organisms.

Some of the most interesting mutualistic relationships are those formed between ants and other organisms. We have already mentioned the trophic mutualism of the leaf-cutter ants and their fungus. Here we examine ant mutualisms in more detail, focusing on a few well known examples.

Ants and Defensive Mutualism

Daniel Janzen's study (1966, 1967) of interdependence between certain kinds of ants and swollen-thorn acacias in Central America provides a wonderful example of the outcome of coevolution. The acacia plant provides food and nesting sites for ants, which protect the plant from herbivorous insect pests. The bull's-horn acacia (*Acacia cornigera*) has large hornlike thorns with a tough woody covering and a soft pithy interior (Figure 25-4). To start a colony in the acacia, a queen ant of the species *Pseudomyrmex ferruginea* bores a hole in the base of one of the enlarged thorns and clears out some of the soft material inside to make room for her brood. In addition to housing the ants, the acacias provide carbohydrate-rich food for the ants in nectaries at the bases of their leaves, as well as fats and proteins in the form of nodules, called Beltian bodies, at the tips of some leaves (Figure 25-5). As the colony grows, more and more of the thorns on the plant are filled. A colony may grow to more than a thousand workers within a year, and eventually may have tens of thousands of workers. At any time, about a quarter of the ants are outside the nest actively gathering food and defending the acacia against herbivorous insects. The relationship between *Pseudomyrmex* and *Acacia* is obligatory: neither the ant nor the plant can survive without the other. Other ant-acacia associations are facultative; that is, the ant and the acacia can co-occur to their mutual benefit, but they can both exist independently as well. Species of acacia that lack the protection of ants altogether frequently produce toxic compounds that defend their leaves against herbivores (Rehr et al. 1973).

FIGURE 25-4 The thorns of *Acacia hindsii*, like those of *A. cornigera*, are greatly enlarged and are filled with a soft pith that ants excavate for nests. Thorns from a non-ant acacia are shown at the left for comparison. (*Courtesy of D. H. Janzen.*)

To test the influence of ants on the growth and survival of acacia plants, Janzen kept ants off new acacia shoots and compared their growth with that of shoots that had ants. After 10 months of one such experiment conducted in southern Mexico, the weight of the shoots lacking ants was less than one-tenth that of shoots with intact ant colonies, and they produced fewer than half the number of leaves and a third the number of swollen thorns.

The mutualism between ants and acacias has been accompanied by adaptations of both species to increase the effectiveness of their association. For example, *Pseudomyrmex* is active both night and day, an unusual trait for ants, and thereby provides continuous protection for the acacia. Also, these ants have a true sting like their wasp relatives and will swarm vertebrate herbivores that attempt to feed on their host plant. The ants also clear away potential plant competitors by attacking seedlings near their host plant's base as well as vines or overhanging branches of other plants. In a similar adaptive gesture, the acacia retains its leaves throughout the year, and thereby provides a continuous source of food for the ants. Most related species lose their leaves during the dry season. The mutual benefits of the ant-acacia relationship and the highly specialized adaptations of both the acacia and the ants provide a strong case for coevolution based upon a long evolutionary association between the two species.

A similar situation in which ants protect aphids and leafhoppers from predators and harvest the nutritious "honeydew" that they excrete is more difficult to interpret (Way 1963, Buckley 1987). One such system involving aphids and leafhoppers on ironweed (*Veronia noveboracensis*) in New York State has been studied experimentally by C. M. Bristow (1984). Aphids (*Aphis*) are small, sedentary, and form dense colonies

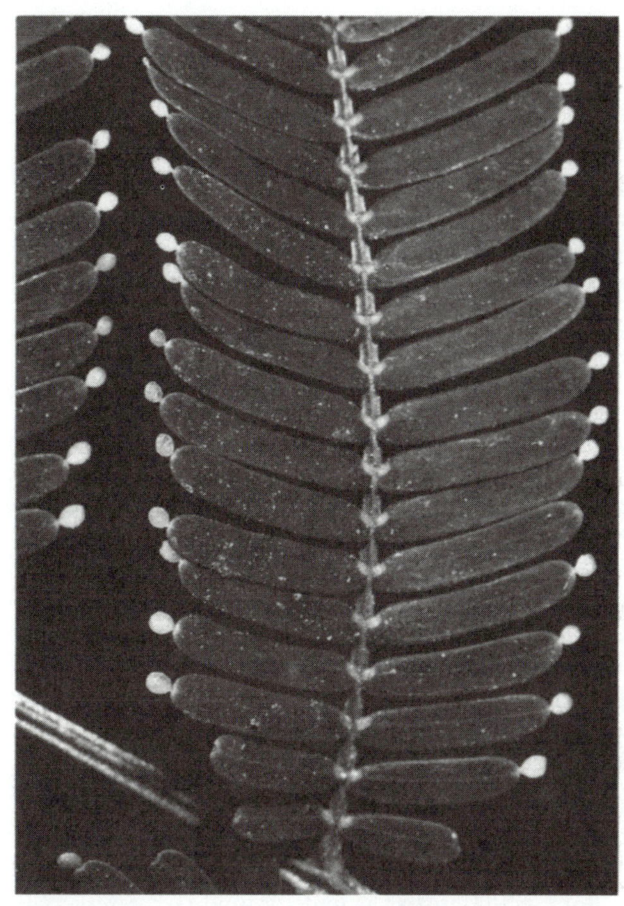

FIGURE 25-5 The leaves of *Acacia collinsii*, like those of *A. cornigera*, provide ants with food in the form of Beltian bodies at the tips of leaflets (a) and nectaries at the leaf base (b). (*Courtesy of D. H. Janzen; from Janzen 1966.*)

on the inflorescences of the plant (see Figure 24-2). Also occurring on ironweed is the larger leafhopper (Membracidae) *Publilia*, which sucks plant juices from the leaves. These insects are tended by three species of ants. One, in the genus *Tapinoma*, is tiny (2–3 mm) but abundant. The other two (*Myrmica*) are larger (4–6 mm) and more aggressive, but less common. The two genera of ants rarely co-occur on the same plant.

The presence of *Tapinoma* greatly enhances the survival of aphid colonies, but has less effect on the survival of leafhoppers. The larger *Myrmica* offers substantial protection to leafhoppers, but is less effective in warding off predators of aphids. Where Bristow excluded both species of ants, predators were more numerous. Where she added the predatory larvae of ladybird beetles, *Myrmica* and, to a lesser extent, *Tapinoma* effectively reduced their predation on leafhoppers.

This system has all the elements expected of coevolution, but it is not clear that the adaptations of the ant and homopteran participants evolved in response to each other. Most insects that suck plant juices produce large volumes of excreta from which they either do not or cannot extract all the nutrients. Thus honeydew production may reflect diet rather than being an adaptation to encourage protection by ants. Ants are voracious generalists, likely to attack any insect they encounter. Hence no special adaptation may be required to confer the benefit of defense on the aphids and leafhoppers upon whose excreta they also feed. The fact that different genera of ants more effectively protect different honeydew sources may simply reflect their different sizes and levels of aggression, which probably evolved in response to unrelated environmental factors.

Why don't ants eat the aphids and leafhoppers they tend? Perhaps this restraint is an evolved character of ants that facilitates the ant–homopteran mutualism. It may even have arisen as an extension of the common ant behavior of defending plant structures, such as flowers or specialized nectaries, from which they obtain food (see Figure 25-5). So, although this system has apparent specificity of interactions, that is not sufficient to prove coevolution.

Another example of a defensive ant–plant mutualism involves the plant genus *Lobelia*, which is widely distributed in both tropical and temperate regions. Some species of *Lobelia* produce **extrafloral nectaries**, nectar-producing structures located outside the flower. These nectaries attract foraging ants that feed on the sugar-rich nectar, and in many cases, the ants provide the plant with protection against herbivores.

Ants and Dispersal Mutualism

An ant–plant mutualism in the Cape region of South Africa highlights the importance of particular adaptations of ants, whether coevolved or not, in maintaining these mutualisms. There, many species of plants in the family Proteaceae have seeds with fleshy, edible, attached structures called elaiosomes, as illustrated for some Australian plants in Figure 25-6. Foraging ants pick up the seeds and transport them to their underground nests, where the elaiosomes are eaten. The seeds themselves, which the ants cannot eat, are then discarded, either in underground chambers or in refuse heaps on the surface. In many regions, these disposal sites are suitable for seed germination and seedling establishment (see Berg 1975, Culver and Beattie 1978). But in the fynbos (brushy, chaparral-like habitat), germination of many species of plants occurs only after fires have swept the habitat, and the only seeds that germinate are those stored by ants in underground nest chambers.

The Argentine ant *Iridomyrmex humilis* has invaded areas of fynbos shrublands and displaced many of the less aggressive native ants. *Iridomyrmex* does not store seeds within its nests; it removes the elaiosomes on the surface, where the seeds are dropped. Bond and Slingsby (1984) found that the germination of one species of *Mimetes* following fire was drastically reduced in areas invaded by *Iridomyrmex humilis* (Figure 25-7). With the continued persistence of the Argentine ant, it is likely that much of the native Cape flora will disappear as underground seed reserves are depleted. In a similar case, S. Temple (1977) has suggested that the virtual extinction of the tree *Calvaria major* on the island of Mauritius followed upon the extinction more than 300 years ago of the dodo, the only native species capable of effectively dispersing *Calvaria*.

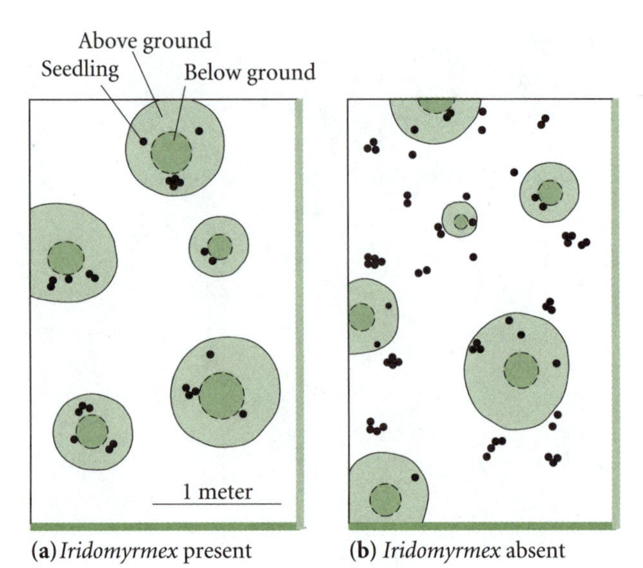

FIGURE 25-7 Seedling dispersion in *Mimetes cucullatus* populations after a burn, (a) in the presence of *Iridomyrmex* and (b) in its absence. The extent of aboveground and belowground parts of mature plants are indicated by circles and shading. In the absence of *Iridomyrmex,* native ants distribute the seeds of *Mimetes* widely throughout the habitat. (*From Bond and Slingsby 1984.*)

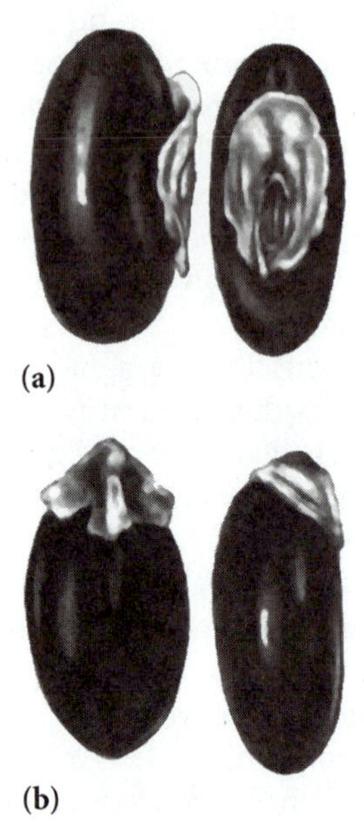

FIGURE 25-6 Ant-dispersed seeds of two Australian plants (a) *Kennedia rubicunda* and (b) *Beyeria viscosa,* showing the edible, light-colored appendage (elaiosome) that attracts ants. (*After Berg 1975.*)

Clearly, particular adaptations of ants influence their effectiveness as seed dispersers. But these adaptations may be evolutionarily independent of the plants themselves. The ant dispersers are diet generalists, and their food-caching behavior probably evolved for reasons unrelated to their role as seed dispersers of *Mimetes* and its relatives. The plants may have merely evolved to take advantage of this fortuitous element in their environment without any reciprocal evolution on the part of the ants. Whereas some mutualisms, such as that between ant and acacia described above, clearly involve specialized adaptations of both parties and seem to represent cases of coevolution, many mutualisms may involve more serendipitous arrangements.

25.6 Variation in plant defensive chemistry has a genetic basis.

Differences in the defensive chemicals of plants can be related to genetic changes, particularly when the pathways of biochemical synthesis and the responsible enzymes are known. M. Berenbaum (1978, 1981, 1983) has placed elements of the relationship between certain butterflies and their umbelliferous host plants in the context of coevolution. Umbellifers produce many noxious chemicals, among the most prominent of which are the furanocoumarins. The biosynthetic pathway of these chemicals (Figure 25-8) leads from para-coumaric acid—which, being a precursor of lignin, is found in virtually all plants—to hydroxycoumarins, such as umbelliferone, and then to furanocoumarins. These last include both linear and angular forms, which are produced directly from hydroxycoumarins by different enzyme reactions. As one proceeds down the biosynthetic pathway from para-coumaric acid to hydroxycoumarins to linear or angular furanocoumarins, toxicity increases and frequency of occurrence among plant families decreases. Hydroxycoumarins possess some biocidal properties; linear furanocoumarins bind with pyrimidine bases and interfere with DNA replication in the presence of ultraviolet light; angular furanocoumarins interfere with growth and reproduction quite generally, although their mechanisms of action have not been detailed. Para-coumaric acid is widespread among plants, occurring in at least a hundred families. Berenbaum (1983) lists only thirty-one families in which hydroxycoumarins have been found. Linear furanocoumarins (LFCs) are restricted to eight plant families and are widely distributed only in two—Umbelliferae (parsley family) and Rutaceae (citrus family). Angular furanocoumarins (AFCs) are known only from two genera of Leguminosae (pea family) and ten genera of Umbelliferae.

Berenbaum (1981) studied herbaceous umbellifers growing in New York State. Some of these plants (especially those growing in woodland sites with low levels of UV light) lack furanocoumarins, others contain linear furanocoumarins only, and some contain both linear and angular furanocoumarins. From a survey of the herbivorous insects collected from these plants, Berenbaum concluded that host plants containing both angular and linear furanocoumarins were attacked by more species of insects than plants with only linear furanocoumarins, or none; that the herbivores on AFC/LFC plants tended to be extreme diet specialists, most having been found on no more than three genera of plants; and that these specialists tended to be abundant compared with the numbers of the few generalists found on AFC/LFC plants, and compared with the numbers of any herbivores found on either LFC plants or umbellifers lacking furanocoumarins.

Although linear and, especially, angular furanocoumarins are extremely effective deterrents to most species of herbivorous insects, some genera that have evolved to tolerate these chemicals have become successful specialists. Berenbaum (1983) makes a strong case for coevolution here in the restricted sense. The taxonomic distribution of hydroxycoumarins, linear furanocoumarins, and angular furanocoumarins across host plants suggests that plants containing LFCs are a subset of those containing hydroxycoumarins, and that those containing AFCs are an even smaller subset of those containing LFCs. This conclusion is consistent with an evolutionary sequence of plant defenses progressing from hydroxycoumarins to LFCs and AFCs. Furthermore, insects specialized on plants containing LFCs belong to groups that characteristically feed on plants containing hydroxycoumarins, and those specialized on AFCs have close relatives that feed on plants containing LFCs. Although insect phylogenetic relationships and their history of host plant utilization have not been matched, the taxonomic distributions of insects across host plants produce patterns that would be expected of a coevolved system.

25.7 Pollination is a common form of plant-animal mutualism.

Some plants are wind-pollinated, and their pollen grains land on the receptive flowers of conspecific individuals just by chance. Where many species live together within an area and where great distances separate individuals, wind pollination is relatively inefficient. Various kinds of animals—including insects, birds, and bats and other mammals—visit flowers to feed on highly nutritious pollen and nectar, to gather other substances such as oils or fragrances, to lay their eggs, or to mate. In doing so, they also transport pollen from flower to flower with a relatively high efficiency.

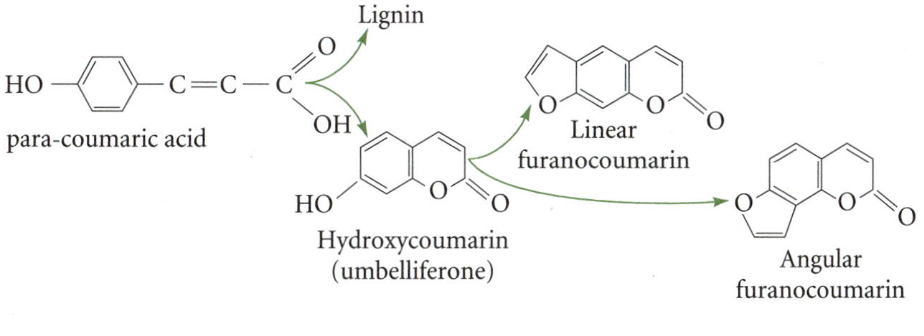

FIGURE 25-8 The biosynthetic pathways in the synthesis of furanocoumarins. Each step is controlled by a different enzyme.

Plant-pollinator relationships may have originated as purely consumer-resource interactions: pollen is an excellent food, and the ovaries of flowers, where seeds develop, are excellent brood sites for insect larvae. Even pure acts of consumption result in some pollen being transferred fortuitously between plants.

Since this pattern began, floral structures have been modified through evolution to increase the efficiency of pollen transfer. Many of these modifications involve offering accessible rewards, such as nectar, that are relatively economical for a plant to produce, and arranging flower parts in such a way that pollen is transferred to the bodies of particular animal visitors (Figure 25-9). As flower structure becomes more highly specialized, fewer and fewer types of animals fit a flower in such a way that they contact the anthers and transfer pollen efficiently to the stigmas of other flowers. Thus, flower morphology can exclude certain types of flower visitors and increase the efficiency of pollen transfer.

Bee Pollination in Orchids

Plant-pollinator relationships are highly developed in the orchid family, with its variety of flower shapes, colors, and smells. The intricate tie between flower and pollinator is exemplified by the orchid *Stanhopea grandiflora* and the tropical bee *Eulaema meriana.* This obligate mutualism is unusual in that *Stanhopea* flowers produce no nectar and only male *Eulaema* bees visit them. The flowers are extremely fragrant, and each species of *Stanhopea* orchid has its own unique combination of odors so that its specialist pollinator can find it without confusion. Each type of orchid tends to attract a single type of bee.

When a male *Eulaema* bee visits an orchid, it collects a perfume that it uses to attract female bees. Each bee species uses a slightly different scent to attract mates. The male bee brushes part of the flower with specially modified forelegs and then transfers the collected substance to the tibia of its hind leg, which is enlarged and has a storage cavity. In *Stanhopea,* a bee enters a flower from the side and brushes at a saclike structure on the lip of the flower (Figure 25-10). The surface of the lip is very smooth, and bees often slip when they withdraw from the flower. (The orchid fragrances may also intoxicate bees and cause them to lose their footing.). When a bee slips, it may brush against the column of the orchid flower, where pollinaria (saclike structures filled with pollen) are precisely placed so as to stick to the hindmost part of the thorax of the bee. If a bee with an attached pollinarium slips and falls out of another flower, the pollinarium catches on the stigma and pollinates the flower. Thus, flower structure and bee behavior are mutually adapted to increase the efficiency of pollen transfer.

Bird Pollination in *Lobelia*

Some of the members of the genus *Lobelia,* particularly those in the subgenus *Centropogon,* are pollinated by hummingbirds. In some highly specialized species of *Centropogon,* an exclusive relationship exists with the sicklebill hummingbirds (*Eutoxeres*), which possess long sickle-shaped bills. The bills of these hummingbirds represent the only structure capable of obtaining nectar from the long, curved flowers of the plants, and thus these hummingbirds are their exclusive pollinators. Unlike other hummingbirds, sicklebills must perch in order to feed. Thus, in addition to the curved flower shape, species of *Centropogon* that are specialized for pollination by sicklebills have flowers that are more compact and sturdy than close relatives, thus providing a place for the birds to perch (Stein 1992).

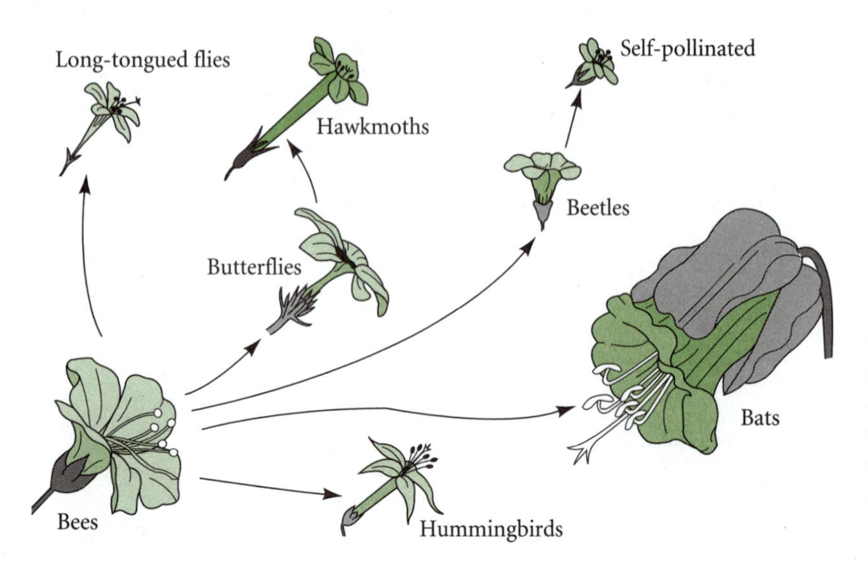

FIGURE 25-9 Diversity of flower types in the plant family Polemoniaceae, with their major pollinators indicated. The arrows suggest possible pathways in the evolution of the various pollination syndromes. (*After Ehrlich and Holm 1963; based on Verne Grant.*)

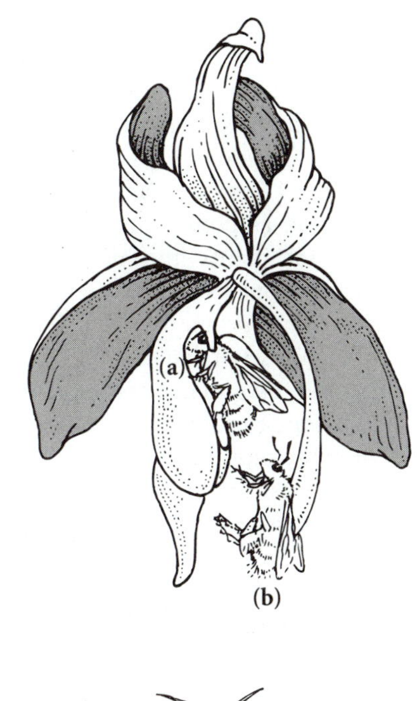

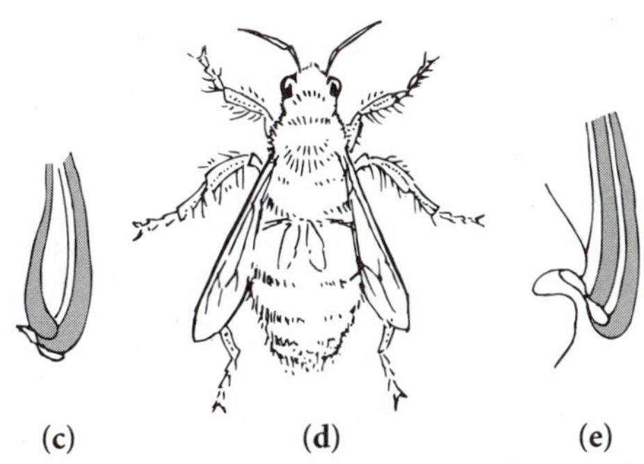

FIGURE 25-10 Pollination of the orchid *Stanhopea grandiflora* by *Eulaema meriana.* A bee enters from the side and brushes at the base of the orchid flower's lip (a). If it slips (b), the bee may fall against the pollinarium, which is located on the end of a column (c), in which case the pollinarium becomes stuck to the hind end of the bee's thorax (d). If a bee with an attached pollinarium later falls out of another flower, the pollinarium may catch on the stigma (e), resulting in pollination. (*After Dressler 1968.*)

25.8 The yucca moth is a coevolved pollinator of the yucca.

One plant-pollinator mutualism merits special attention here, both because it has long been recognized as a highly coevolved system and because it has been the focus of a number of excellent recent studies on phylogenetic patterns of coevolution and the evolutionary conflict between participants in mutualistic relationships (Pellmyr and Thompson 1992, Pellmyr and Huth 1994, Pellmyr et al. 1996). The curious pollination relationship between species of yucca plants (*Yucca,* in the lily family) and moths of the genus *Tegeticula* (Figure 25-11) was first described by C. V. Riley more than a century ago (1892) and has been considerably elucidated since (Powell and Mackie 1966). A moth enters a yucca flower and deposits one to five eggs on the ovary. After the moth has laid her eggs, she scrapes pollen off the anthers in the flower and rolls it into a small ball, which she grasps with specially modified mouthparts. She then flies to another plant, enters a flower, and proceeds to place the pollen ball onto the stigma of the flower before laying another batch of eggs. When the eggs hatch, the larvae burrow into the ovary, where they feed on the developing seeds.

The relationship between the moth and the yucca is obligatory. *Tegeticula* larvae can grow nowhere else; *Yucca* has no other pollinator. In return for the pollination of its flowers, the yucca seemingly tolerates the moth larvae feeding on its seeds, but the extent of this loss of potential reproduction is small, rarely exceeding 30%, and more nearly half that value on average, in *Yucca whipplei* (Powell and Mackie 1966).

Yucca and *Tegeticula* are specialized with respect to each other. On the yucca's part, its pollen is sticky and can easily be formed into a ball that the moth can

FIGURE 25-11 The mohave yucca (*Yucca shidigera*) and a yucca moth of the genus *Tegeticula.* (*After Powell and Mackie 1966.*)

carry, and the stigma is specially modified as a receptacle to receive pollen. On the moth's part, individuals visit flowers of only one species of yucca, mate within the flowers, lay their eggs in the ovary within the flowers, exhibit restraint in the number of eggs laid per flower, and have specially modified mouthparts and behaviors to obtain and carry pollen. Because the mutualism of *Tegeticula* and *Yucca* is so tight, one might expect all these characteristics to have evolved as a result of coevolution between the two.

In fact, however, many aspects of this mutualism are present in the larger lineage of nonmutualistic moths (Prodoxidae) within which *Tegeticula* evolved (Pellmyr and Thompson 1992). Examination of a phylogenetic tree of Prodoxidae (Figure 25-12) shows that several of the highly specialized characters of *Tegeticula* are found in other members of its family. Indeed, host specialization and mating on the host plant are basal (primitive) features of the family found in all its members. The trait of ovipositing in flowers has evolved independently at least three times in the family and has reversed (reverted to the ancestral state) at least twice, in *Parategeticula* and *Agavenema.* Of the species that oviposit in flowers, only *Tegeticula* and one species of *Greya* actually function as pollinators; the others are strictly parasites of the plants in which their larvae grow. It should be mentioned that *Greya politella* pollinates *Lithophragma parviflorum,* in the saxifrage family, which is not even closely related to the yuccas. We see from this example that many of the adaptations that are part of the yucca-moth mutualism appear to have been present in the moth lineage before the establishment of the mutualism itself. Such traits are often referred to as **preadaptations.**

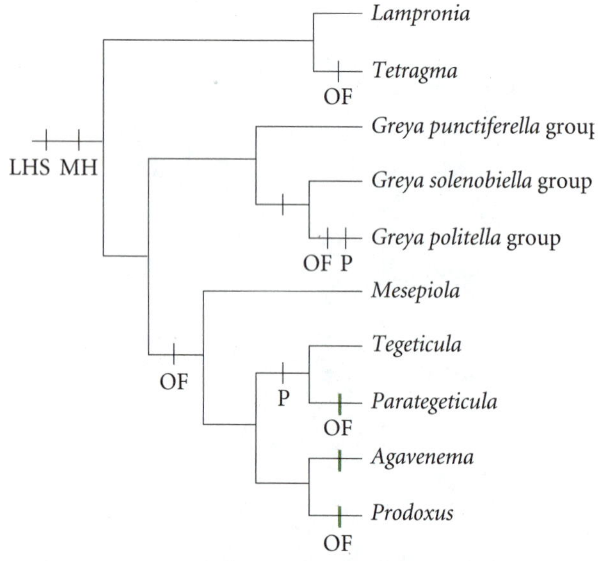

FIGURE 25-12 Phylogenetic tree of the moth family Prodoxidae, showing the sites of the evolution of traits critical to the yucca moth-yucca mutualism in moths of the genus *Tegeticula.* The characters are: LHS, local host specificity; MH, mating on host; OF, oviposition in flower; P, pollinator. Black bars indicate the gain of a character shared by descendants of the lineage; green bars indicate loss of a character. (*From Pellmyr and Thompson 1992.*)

25.9 Mutualism involves an inherent conflict among participants that may lead to cheating.

At first glance, it may seem that the participants in a mutualistic relationship have struck a good deal. Each gives a little, and in return, reaps a benefit. But if we examine mutualism a bit more closely, we discover that mutualism is not all that it seems to be. Indeed, there is an inherent conflict between the participants in a mutualistic relationship that stems from the dependency of the relationship. In order for one of the participants to increase its fitness (the genetic contribution of an individual's descendants to future generations; see Part 7), it must do so at the expense of the other participant. For example, in the yucca-moth mutualism just discussed, in order for the moth to increase its fitness, it must lay more eggs per flower. As we mentioned in the previous section, moths deposit eggs in only about 30% of the yucca seeds. The moth's restraint in this regard suggests the presence of some constraint on how much the partners in a mutualistic relationship may exploit each other to increase their own fitness. Indeed, such mechanisms do exist in the yucca-moth system, and it is the yucca that regulates the number of the eggs laid per flower (Pellmyr and Huth 1994).

When too many eggs are laid in the ovary of a particular flower—too many being enough to eat a majority of the developing seeds—the flower is aborted, and the moth larvae die. While this strategy would also seem to reduce the seed production of the yucca, resources that would have supported the production of seeds in the now aborted flower are diverted to other flowers. Yuccas, like many plants, produce far more flowers than they need. Flowers are relatively cheap to produce, and seeds are expensive, so it is almost always the case among plants that some flowers, even those that are pollinated, are aborted before setting seed. Selective abortion of insect-damaged fruit occurs widely among plants (Janzen 1971, Stephenson 1981), and yuccas use this mechanism to keep their moth pollinators in line (Udovic and Aker 1981, Aker 1982).

The natural tension between partners in a mutualism can lead to cheating. For example, the mutualistic yucca moths belong to a complex of very similar species, all of which oviposit their eggs in the ovaries of yuccas. However, some of the members of this species group do not pollinate the yucca; that is, they take, but they do not give. Pellmyr and coworkers (Pellmyr et al. 1996) analyzed the phylogenetic relationships among these moths, and discovered that cheaters have arisen from obligate pollinators a number of times. It is interesting to point out that because cheaters are just as specialized for ovipositing in yuccas, they are dependent on noncheaters to keep the system stable.

In the six chapters of this part of the book, we have discussed the details of the interactions of populations of organisms. In many cases, populations interact by competing for resources either directly or indirectly. Competition may limit the population sizes of the competitors. Some interactions, such as predation, parasitoidism, parasitism, and herbivory—all strong antagonistic interactions—involve the consumption of all or part of individuals of one population by individuals of another population. And, as we have seen in this chapter, there are other interactions, called mutualisms, in which both of the participants in the interaction benefit. In the next part of the book we will explore how these interactions work to shape communities of organisms.

SUMMARY

1. Coevolution involves reciprocal evolutionary responses of interacting species or populations. Specialization is a situation in which a species interacts with only a limited number of other species. Mutualism is a type of population interaction in which both participants benefit from the association.

2. The interaction between the European rabbit introduced into Australia and the myxoma virus, and the interaction between houseflies and the wasp parasitoid *Nasonia vitripennis,* are examples of relationships between antagonists, which may result in coevolution.

3. Specialization may occur rapidly, and its presence does not necessarily suggest that coevolution has occurred. Local specialization between mutualists is called population coevolution, whereas coevolution throughout the species' range is referred to as species coevolution. The geographic mosaic model of coevolution recognizes that the outcome of population interactions varies spatially and temporally, and thus a species may be subjected to different intensities of coevolution and specialization throughout its range.

4. The elicitor-receptor model of gene-for-gene coevolution holds that for coevolution to occur, complementary genotypes must be present in the interacting populations. The strongest evidence for gene-for-gene coevolution is in crop-parasite systems.

5. Organisms may form a number of different types of mutualistic relationships, including trophic mutualisms, defensive mutualisms, and dispersive mutualisms. Some mutualistic relationships are obligatory in that each partner in the relationship relies on the other partner for its survival. Other mutualisms are facultative, meaning that partners can do without each other if necessary.

6. The most convincing demonstrations of coevolution involve cases of mutualism, such as the obligate interdependence of *Pseudomyrmex* ants and *Acacia* plants. The ant keeps the plant free of herbivores while the plant provides the ant with food and housing; both have adaptations of structure and behavior or phenology that promote the relationship.

7. Analysis of metabolic pathways shows the genetic steps in the development of toxic chemicals in plants. When variations in these pathways (and in the abilities of insects to detoxify the chemicals) are overlaid upon taxonomic relationships within each group, one can infer the evolutionary history of the plant-insect interaction.

8. The relationships between plants and their pollinators are well-studied mutualisms. Pollination relationships with bees are highly developed in orchids. High degrees of specialization occur between some flowers and their hummingbird pollinators.

9. The interaction between yucca moths and yuccas is an obligate mutualism in which the moth pollinates the plant, but its larvae consume developing seeds. Both the moths and the yuccas have specializations that promote this relationship, but phylogenetic analysis shows that some of the adaptations of the moths are present in close relatives that are not mutualists of yuccas. Such traits are called preadaptations.

10. Phylogenetic analysis also shows that, within the yucca moth species complex, there are species that cheat the yucca plant in that they oviposit in the plant's ovaries, but do not pollinate the plant. Such cheating is an expected result of the natural conflict that exists between mutualists.

EXERCISES

1. Using diagrams, explain the geographic mosaic model of coevolution.

2. Suppose that a certain species of plant has two distinct genotypes, one resistant to a particular fungus, the other not resistant to the fungus. Both forms may occur in the same population. Explain the genetic basis of resistance in the plant.

3. You notice that ants are always associated with a particular plant in your garden, and that plant also harbors several types of insect herbivores. You wonder if there is a mutualistic relationship present among the ants and the sucking insects. Provide a detailed description of the natural history observations that you would make and the experiments that you would conduct to determine whether a mutualism exists. Include in your description such details as the time of day (or night) and the season that observations would be made, the questions to be asked and the hypothesis to be tested, and the materials that would be required to conduct the experiments.

PART 6
COMMUNITY ECOLOGY

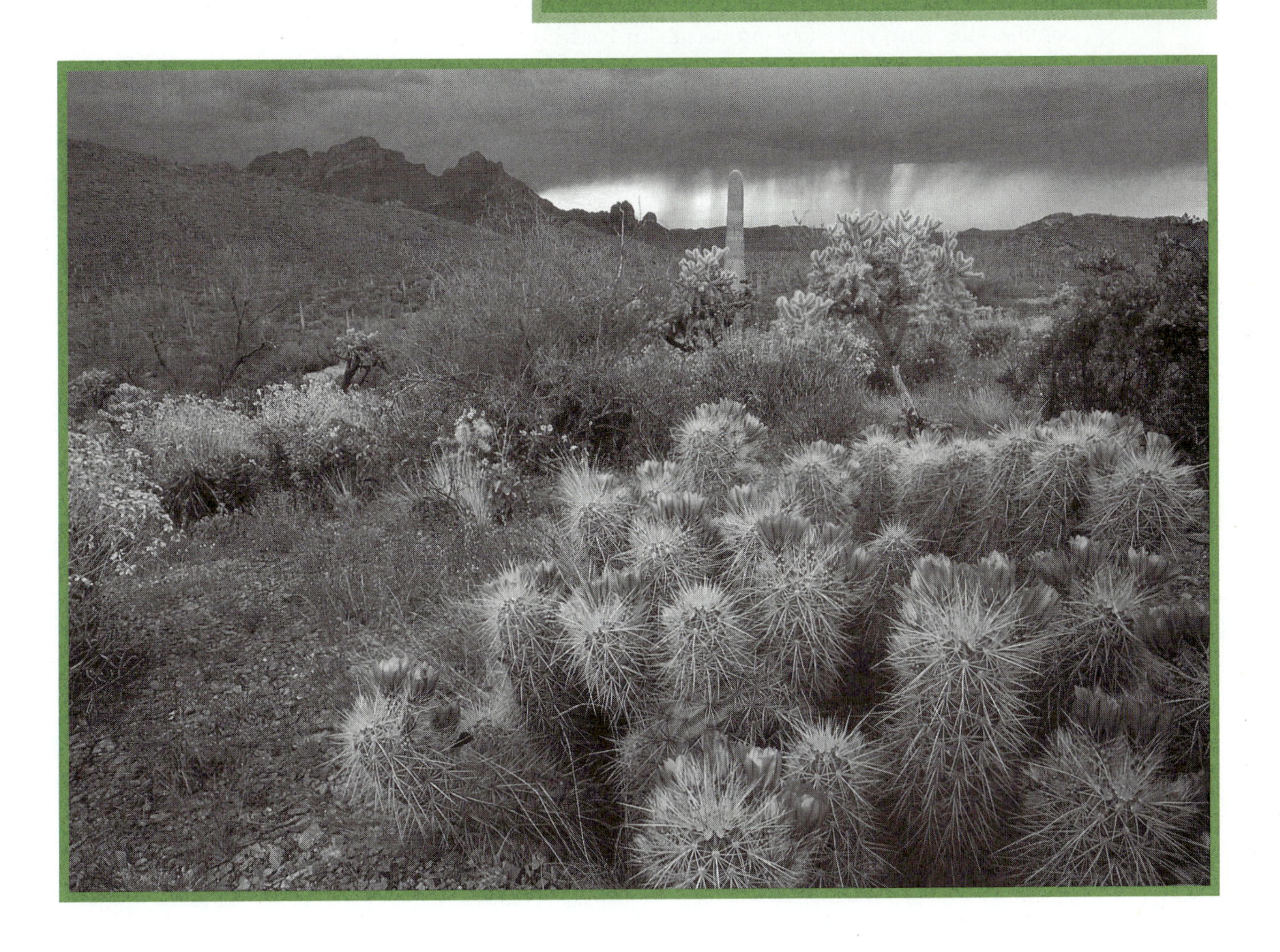

Most places on earth contain a number of different types of living things. We easily recognize a farm field, a forest, and a desert by the dominant plants and animals in each. These associations of populations of different organisms are referred to as ecological communities, the dynamics of which we take up in Part 6.

In Chapter 26 we will trace the development of the modern concept of the ecological community. We will see that the history of the ideas about natural communities closely parallels thinking about the nature of the ecosystem. Like populations, communities have structural properties such as the number and types of species found in the community and the pattern of trophic interactions. These will be discussed in Chapter 27. If we observe an abandoned farm field over a period of years we will see obvious changes in the number and type of plants in the field. The characteristics of such community change will be discussed in Chapter 28.

The number of different species in a community is an extremely important component of biodiversity. The mechanisms that determine the species diversity will be discussed in Chapter 29. We will also show in this chapter how species diversity is affected by immigration and extinction, competitive interactions, and disturbance emphasizing the different roles of regional and local processes.

CHAPTER 26

The Concept of the Community

GUIDING QUESTIONS

- What is the difference between an ecosystem, a community, a guild, and an assemblage?
- What is meant by community structure and community function?
- How did the views of F. E. Clements and H. A. Gleason regarding the nature of the biological community differ?
- What is the relationship between ecotones and habitat fragmentation?
- What is the continuum concept of community organization, and how can it be used to describe communities?
- Why do contemporary ecologists think of communities as being open in time and space?
- How does community organization reflect phylogenetic and biogeographic processes?
- How can phylogenetic analysis be used to distinguish among alternative explanations for contemporary community structure?
- How can coevolution affect community structure?

The individuals of a species that occupy a particular place represent a population. Every place on earth—each meadow, each pond, each rock at the edge of the sea—is shared by many coexisting populations. An association of such populations is called a **community.** Communities are usually defined by the nature of the interactions among the populations in the association or by the place in which the association occurs.

We have discussed ecosystem processes, population dynamics, and the mechanisms of interaction between pairs of populations, such as predator and prey or herbivore and host, in earlier chapters. Observations and experiments have clarified the mechanisms underlying these processes, which are now understood in broad outline. However, several related issues that have challenged ecologists for decades and which emerge in the context of the ecological community have not yet been addressed in this book. Among them are questions that focus on the mechanisms that determine the numbers and kinds of plants and animals in a particular place, and on whether patterns of trophic interactions correspond to particular types of communities. These are the issues that we shall discuss in this part of the book.

26.1 The community is an association of populations.

The term *community* has been given a variety of meanings by ecologists (Schoener 1986, Fauth et al. 1996). Historically, the term has often denoted associations of plants and animals occurring in a particular locality and dominated by one or more prominent species or by some physical characteristic (Slobodkin 1961, Shimwell 1971, Daubenmire 1968b). We speak of an oak community, a sagebrush community, or a pond community, meaning all the plants and animals found in a particular place dominated by the community's namesake. Used in this way, the term is unambiguous: a community is spatially defined and includes all the populations within its boundaries.

Communities defy delineation, however, when populations extend beyond arbitrary spatial boundaries. Migrations of birds between temperate and tropical regions link communities in each area. Within some tropical localities, as many as half the birds present during the northern winter are migrants. Salamanders, which complete their larval development in streams and ponds but pursue their adult existence in the surrounding woods, tie together aquatic and terrestrial communities, just as trees do when they shed their leaves into streams and thereby support aquatic detritus-based food chains.

We may add precision to our definition of community by thinking about how ecologists select sets of populations for study. Some studies require an examination of sets of populations, called taxa, that are phylogenetically related (set A in Figure 26-1). For example, a researcher might choose to study the various ways in which different plant species in a particular genus or family are pollinated. An **assemblage** is a taxonomically related group that occurs in the same geographic area. Sometimes the focus lies on a group of populations that uses a suite of resources in the same manner. Such groups are called **guilds** (set C in Figure 26-1). The members of a guild may occur in the same area, in which case the group is referred to as a **local guild.** Groups of populations that are delineated primarily by geography are considered communities proper (set B in Figure 26-1). Communities, then, are not defined by a particular type of interaction or by the evolutionary relationships among the members, although those relationships are reflected in the community. It is essential only that the members reside in the same place.

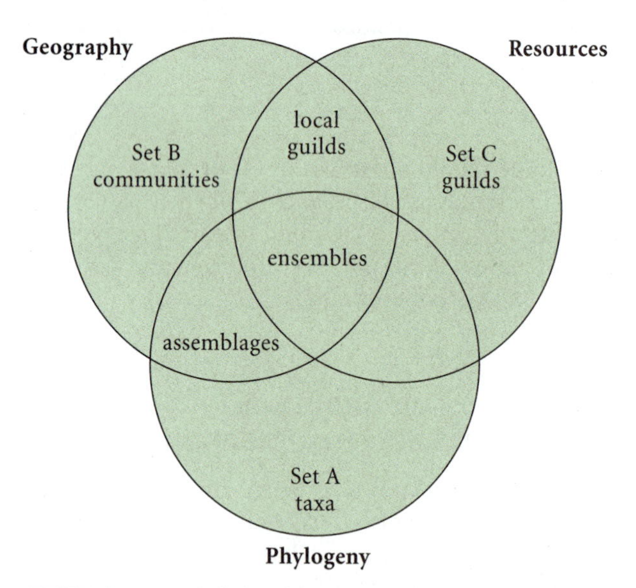

FIGURE 26-1 Relationships among the various terms used to describe groups of organisms. Taxa are sets of organisms that are phylogenetically related. Taxa located in the same geographic region are referred to as assemblages. Communities are defined by geography, and guilds by resource use patterns. (*From Fauth et al. 1996.*)

It is important to appreciate the difference between an ecosystem and a community. Ecosystems include all the interacting physical and biological components of an area. Communities include only the organisms. Unfortunately, the two terms are sometimes used synonymously.

Like populations, communities are characterized by a number of unique properties, which we refer to as **community structure** and **community function.** These terms are used to set apart suites of interacting processes in communities rather than to delineate single community processes or features. The number of species, called **species richness,** the types of species present and their relative abundances, the physical characteristics of the vegetation, and the trophic relationships among the interacting populations in the community are attributes of community structure. Rates of energy flow, properties of community resilience to perturbation, and productivity are examples of community function. To be sure, community structure and function are interrelated. We shall discuss these in detail in Chapter 27.

Community structure and function are manifestations of a complex array of interactions, directly or indirectly tying all members of a community together into an intricate web (Wootton 1994, Miller and Travis 1996). The influence of a population extends to ecologically distant parts of the community through its competitors, predators, and prey. Insectivorous birds do not eat trees, but they do prey on many of the insects that feed on foliage or pollinate flowers. By preying upon pollinators, birds indirectly affect the number of fruits produced, the amount of food available to animals that feed upon fruits and seedlings, and the predators and parasites of those animals. Examples of the complex web of interactions in two different ecosystems—a midwestern lake and a rocky intertidal ecosystem—are presented in Figure 26-2.

26.2 Is there a natural unit at the community level of ecological organization?

One of the issues that perplexed and polarized early community ecologists was the issue of unity of the community. At issue was whether or not the populations that make up a community interact with one another in the same complex and mutually dependent

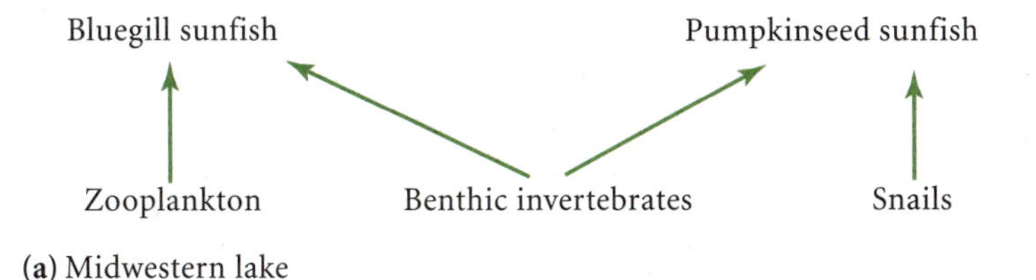

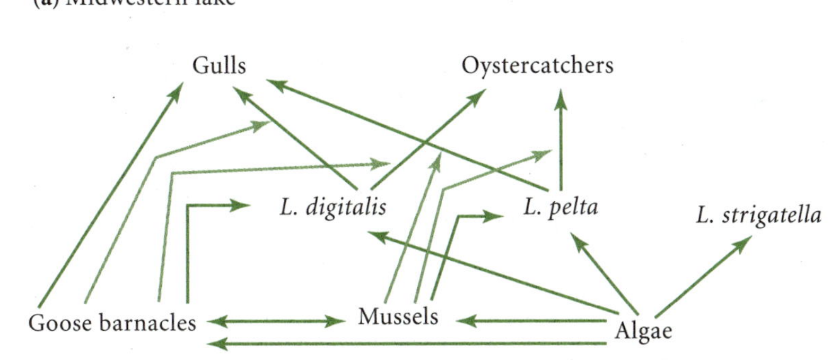

FIGURE 26-2 Interactions in two communities: (a) a midwestern lake and (b) a rocky intertidal community. The arrows indicate the directions of energy flow. (*After Wootton 1994.*)

way that the organs of your body interact with one another. That is, is the community a superorganism whose function and organization can be appreciated only by considering its place in nature as a whole, or does community structure and function simply express the interactions of the individual species constituting the local association without organization, purposeful or otherwise, above the species level? You will recognize this issue from our discussion of the development of the ecosystem concept, in which the idea of the superorganism was also raised (see Chapter 9). Most ecologists today do not accept to the superorganism concept of the community. But it is instructive to briefly examine the history of the idea because subsumed in the debate are useful ideas about community boundaries.

Certainly the most influential person to espouse the organismal viewpoint was the American plant ecologist F. E. Clements (1916, 1936), who perceived communities as discrete units with sharp boundaries, each with a unique organization. Clements's view was reinforced by the conspicuousness of many dominant vegetation types. A forest of ponderosa pines, for example, appears distinct from the fir forests that grow in moister habitats and from the shrubs and grasses typical of drier sites. The boundaries between these community types are sometimes so sharp that we may cross them within a few meters along a gradient of climatic conditions. Some community boundaries, such as that between deciduous forest and prairie in the midwestern United States, or between broad-leaved and needle-leaved forest in southern Canada, are respected by most species of plants and animals.

An opposing view of community organization was held by H. A. Gleason (1926, 1939), who suggested that the community, far from being a distinct unit like an organism, was merely a fortuitous association of organisms whose adaptations enabled them to live together under the particular physical and biological conditions that characterized a particular place. A plant association, he said, was "not an organism, scarcely even a vegetational unit, but merely a coincidence."

Clements's and Gleason's concepts of community organization predict different patterns in the distribution of species over ecological and geographic gradients. On the one hand, Clements believed that the species belonging to a community were closely associated with one another, and that the ecological limits of distribution of each species coincided with the distribution of the community as a whole. This concept of community organization is commonly referred to as a **closed community.** On the other hand, Gleason believed that each species was distributed independently of others that co-occurred with it in a particular association, a concept referred to as an **open community.** Open communities have no natural boundaries; therefore, their limits are arbitrary with respect to the geographic and ecological distributions of their component species, which may extend their ranges independently into other associations.

26.3 Ecotones occur at sharp physical boundaries or where habitat-dominating growth forms change.

The structure of closed and open communities is depicted schematically in Figure 26-3. In Figure 26-3a, the distributions of the species in each community are clustered together along a gradient of environmental conditions—for example, from dry to moist. Closed

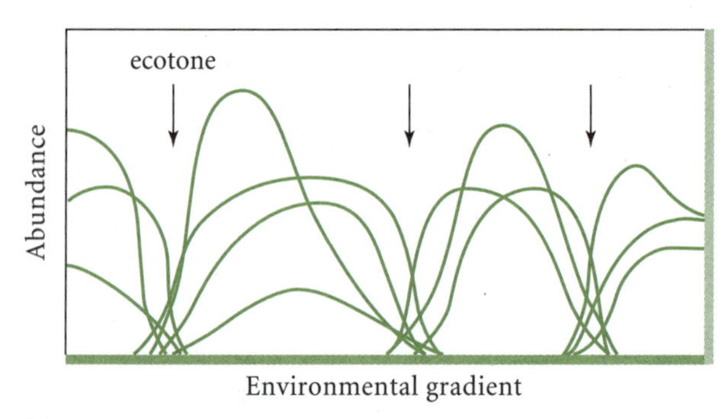

(a)

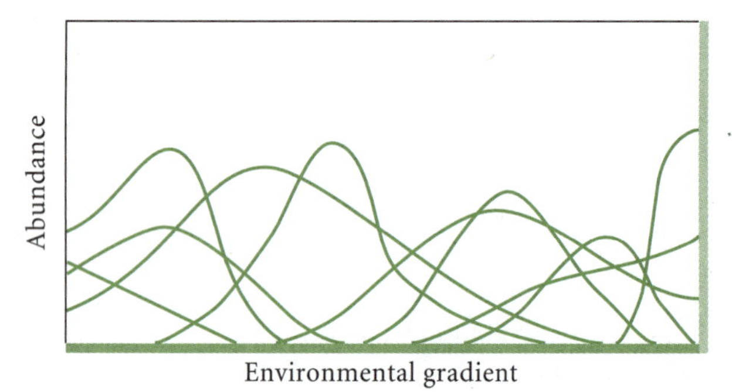

(b)

FIGURE 26-3 Hypothetical distributions of species according to two concepts of communities. (a) Species are organized into distinct assemblages, referred to as closed communities. The communities are separated by ecotones (indicated by arrows). (b) Species are distributed at random along a gradient of environmental conditions, indicating open communities.

communities are natural ecological units with distinct boundaries. The zones of transition between such communities, called **ecotones,** are regions of rapid replacement of species along the gradient. In Figure 26-3b, species are distributed at random with respect to one another, giving an open structure. We may arbitrarily delimit a "community" at some point—perhaps a dry forest community near the left-hand end of the moisture gradient—while recognizing that some of the species included are more characteristic of drier portions of the gradient, while others reach their greatest abundance in wetter sites.

The concepts of open and closed communities both have some validity in nature. We observe distinct ecotones between associations in two circumstances. First, when the physical environment changes abruptly—for example, at the transition between aquatic and terrestrial communities, between distinct soil types, or between north-facing and south-facing slopes of mountains—the transition between communities is abrupt and obvious. Second, when one species or life form so dominates the environment that the edge of its range signals the distributional limits of many other species, an ecotone is present.

Sharp physical boundaries create well-defined ecotones. Such boundaries occur at the interface between most terrestrial and aquatic (especially marine) communities (Johannesson 1989; Figure 26-4) or where underlying geologic formations cause the mineral content of soil to change abruptly. An ecotone between plant associations on serpentine and nonserpentine soils in southwestern Oregon is shown in Figure 26-5. Levels of nickel, chromium, iron, and magnesium increase across the boundary into serpentine soils; the copper and calcium levels of the soil drop off. The edge of the serpentine soil marks the range boundaries of many species that are either excluded from, or restricted to, serpentine outcrops. A few species are found only within the narrow zone of transition; others, seemingly unresponsive to variation in soil minerals, extend across the ecotone.

A change in soil acidity often accompanies a transition between broad-leaved and coniferous needle-leaved forests, an example of the second type of ecotone. The decomposition of conifer needles produces organic acids more abundantly than the breakdown of leaves of flowering plants. Furthermore, because needles tend to decompose slowly, a thick layer of partly

FIGURE 26-4 A sharp community boundary (ecotone) in the Bay of Fundy, New Brunswick, Canada, associated with an abrupt change in the physical properties of adjacent habitats. Seaweeds extend only to the high tide mark. Between the high tide mark and the spruce forest, waves wash soil from the rocks and salt spray kills pioneering land plants, leaving the area devoid of vegetation.

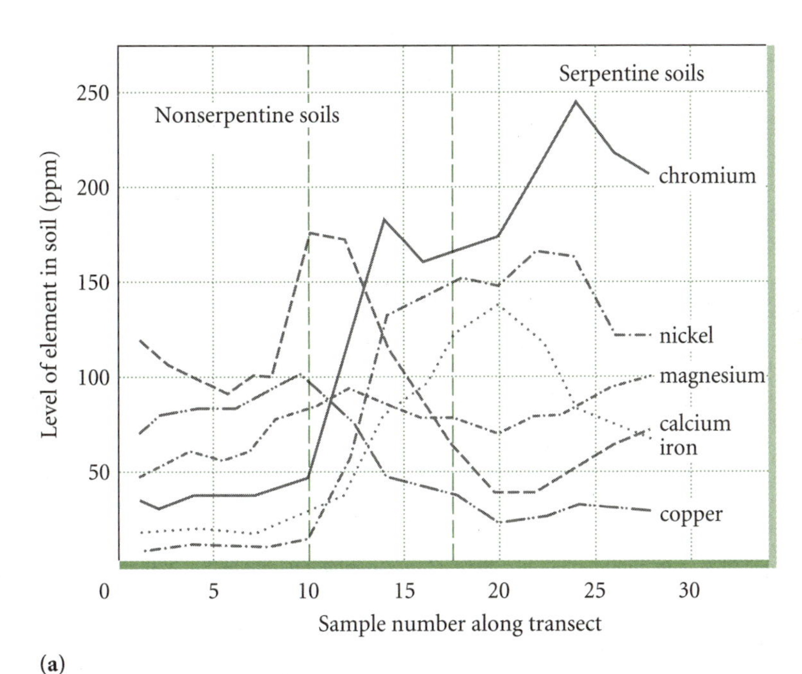

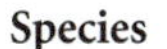

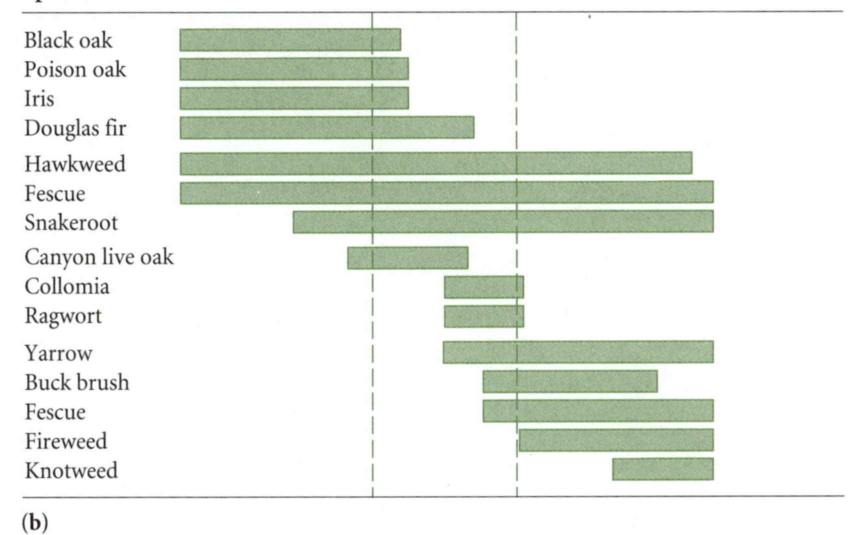

FIGURE 26-5 An ecotone resulting from soil conditions: (a) changes in the concentration of elements in the soil and (b) replacement of plant species across the boundary between nonserpentine (sample numbers 1 to 10) and serpentine (sample numbers 18 to 28) soils in southwestern Oregon. The transect diagrammed here is somewhat atypical in that magnesium does not increase as abruptly as usual across the serpentine ecotone. (*After* White 1971.)

decayed organic material accumulates at the soil surface. These conditions prevent the growth of broad-leaved species in the understory of the coniferous forest. This dramatic shift in environment between broad-leaved and needle-leaved forests marks the edges of distributions of many understory species within each forest type. Similarly, at boundaries between grassland and shrubland, or between grassland and forest, sharp changes in surface temperature, soil moisture, light intensity, and fire frequency result in many species replacements. Boundaries between grasslands and shrublands are often sharp because when one or the other vegetation type holds a slight competitive advantage, it dominates the community (Schultz et al. 1955). Grasses prevent the growth of shrub seedlings by reducing the moisture content of surface layers of soil. Shrubs depress the growth of grass seedlings by shading them. Fire evidently maintains a sharp edge between prairies and forests in the midwestern United States (Borchert 1950). Perennial grasses resist fire damage that kills tree seedlings outright, but fires do not penetrate deeply into the moister forest habitats.

It is important to appreciate that the concept of an ecotone is not restricted to the interaction among communities, nor to a transition in the number of species. Any interface between populations or between ecosystems, as well as between communities, might properly be viewed as an ecotone. And ecotone transitions might include fluxes of materials as well as transitions in assemblage structure (Holland 1988, di Castri and Hansen 1992, Wiens 1992).

Habitat Fragmentation and Ecotone Dynamics

We will make the assertion that, because of the underlying continua of environmental conditions and biotic interactions, most natural communities have predominately open structures characterized by gradients of change and smooth transitions among distinct community types. However, in areas of high human activity, this generalization often does not hold. Habitat fragmentation by agriculture, forestry, and other human activities disrupts ecological continua and increases the extent of sharp ecotones. In short, habitat fragmentation moves communities from an open to a closed structure (Merriam and Wegner 1992).

Figure 26-6 shows how ecotones are expanded by the fragmentation of a hypothetical rectangularly shaped forest community having a total area of 200 km^2 (10 km × 20 km). The unfragmented core forest has a total edge of 2(10 + 20) = 60 km, and the ratio of the amount of edge to the amount of area (edge/area) is 0.30 (Figure 26-6a). Suppose that a 4-km wide swath of the forest is taken for a large highway construction project, thus dividing the forest into two smaller forests, each having an area of 8 × 10 = 80 km^2. The total area of the forest is the sum of the two smaller forests, or 160 km^2, 40 km^2 less than the original forest. The total amount of edge in the divided forest is 2(16 + 20) = 72 km, or 12 km more than the edge of the original forest. The edge/area ratio is 0.45, higher than that of the undisturbed forest (Figure 26-6b). Further subdivision of each of the two forests to produce four smaller forest areas results in a loss of 32 km^2 and an edge/area ratio of 0.75 (Figure 26-6c). The figure shows clearly that continual fragmentation of habitat results in increasingly less core forest area relative to forest edge. The four small forests contain a total of 128 km^2 of forest, representing a loss of only about one-third of the original forest area. But the amount of edge in the system has increased by almost two-thirds.

The shape of the forest also affects the edge/area ratio. Compare Figures 26-6a and 26-6d, which show two different forest shapes having the same total area. The long, narrow forest has a total of 2(40 + 5) = 90 km of edge (edge/area = 0.45), more than the forest shown in Figure 26-6a. Fragmentation of the narrow forest will lead to a situation in which there is little core forest.

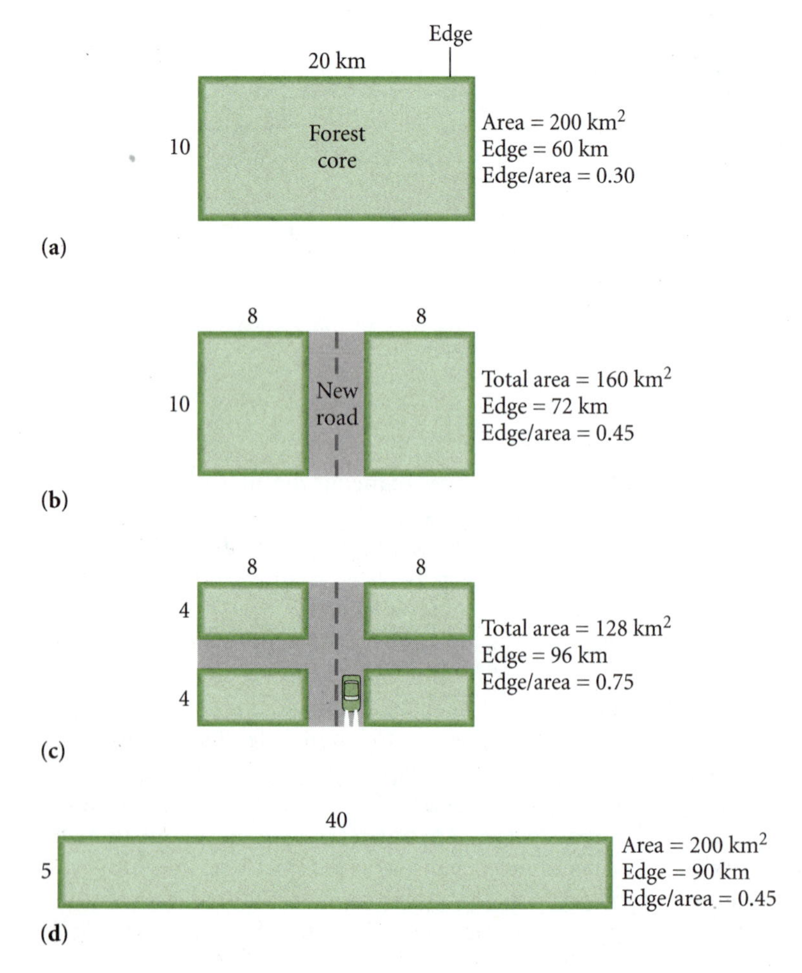

FIGURE 26-6 Schematic representation of how the amount of edge and interior space changes as a hypothetical forest becomes more and more fragmented. The original forest (a) has an area of 200 km^2 and a total edge of 60 km, with an edge/area ratio of 0.30. A 4-km wide swath is cleared through the middle of the forest to construct a highway (b), resulting in a fragmented forest having two areas of 80 km^2 each (160 km^2 total forest area). The total amount of edge in the fragmented forest is 72 km, with edge/area = 0.45. A 2-km path then bisects the forest at right angles to the highway, resulting in four equal-sized forest fragments (c), each 32 km^2 in area (128 km^2 total area). This fragmented forest has a total of 96 km of edge, with edge/area = 0.75. A long, narrow forest (d) having the same total area as the forest shown in panel a will have a third more edge (total edge 90 km), with edge/area = 0.45.

The fragmented habitat shown in Figure 26-6c is reminiscent of the metapopulation structure that we discussed in Chapter 17 (see Figure 17-1). Indeed, for a particular species in the core forest, the continual fragmentation of the forest creates a situation in which its persistence is dependent on the pattern of local extinction and recolonization among the forest fragments. Are the communities in such small patches subject to the same kinds of dynamics? We may define a **metacommunity** as a set of local communities in different locations that are connected to one another via the dispersal of one or more members of the community (Gilpin and Hanski 1991, Holt 1997). The extent to which the various local communities persist and remain similar in structure to one another depends on the suitability of the habitat patches for all the species in the local community. The creation of relatively more edge as a result of fragmentation may change the species composition of local communities. For example, Robbins et al. (1989) found that certain species of birds, such as gray catbirds and robins, prefer edge habitats, while others, such as worm-eating warblers and ovenbirds, live primarily in the interior portions of forests. The size of the area, too, will affect the community composition. Most communities consist primarily of relatively rare species (Gaston 1994), which typically disappear as habitat area decreases. In the example shown in Figure 26-6, the fragmentation has created habitat patches of equal size and shape. If, however, the fragmentation had resulted in habitat patches with a variety of total areas and edge/area ratios, we would expect that the local communities in those patches would become dissimilar. We will discuss a model of such change in Chapter 27 when we talk about community structure.

26.4 The structure of natural communities may be described in relation to ecological continua.

The deciduous forests of eastern North America are bounded to the north by cold-tolerant needle-leaved forests, to the west by fire- and drought-resistant grasslands, and to the southeast by fire-resistant pine forests. Within the region of their distribution, deciduous forests present a physiognomically uniform appearance. As a result of early botanical explorations, ecologists were aware that different species of trees and other plants occurred in different areas within the forest biome. For example, upstate New York had beech, sugar maple, and sweet birch, whereas forest areas in West Virginia included mainly red oak, white oak, and hickory. According to Clements's closed-community viewpoint, the distinctive vegetation of each area represented a distinct community separated by sharp vegetational transitions from other communities. But as plant distributions were described in more detail, their associations were found to fit the closed-community concept less and less well; classifications of plant communities became more and more finely split until absurd levels of distinction were reached.

Out of this mounting chaos arose the **continuum** concept of community organization. Within broadly defined habitats, such as forest, grassland, or estuary, populations of plants and animals gradually replace each other along gradients of physical conditions. The forest environments of the eastern United States form a continuum, with a north-south temperature gradient and an east-west rainfall gradient. The species of trees found in any one region—for example, those native to eastern Kentucky—have different geographic ranges, suggesting a variety of evolutionary backgrounds (Figure 26-7). Some of the species reach their northern limits in Kentucky; some their southern limits. Because few species have broadly overlapping geographic ranges, associations of plant species found in eastern Kentucky do not represent closed communities. Each species has a unique evolutionary history and present-day ecological position, with a variable degree of association with other species in the local community.

A more detailed view of eastern Kentucky forests would reveal that many of the tree species are segregated along local gradients of conditions. Some are found along ridge tops; others along moist river bottoms; some on poorly developed, rocky soils; others on rich organic soils. The species represented in each of these more narrowly defined associations might exhibit correspondingly closer ecological distributions, but the open community concept would still dominate our thinking about these associations.

Ecologists have developed a number of techniques to describe community structure and function within the community continuum concept. Let us take a look at two of these techniques.

Gradient Analysis and Ordination

One way to portray change in community structure is to plot the abundances of species along some continuous gradient of ecological conditions (Loucks 1962), a procedure referred to as **gradient analysis** (Whittaker 1967). The gradient might be based on any number of physical variables, such as moisture, temperature, salinity, exposure, or light level (Figure 26-8). The ecologist examines different localities along the gradient, taking note both of the exact conditions at each position and the abundances of each species there. The species abundances are then plotted as a function

FIGURE 26-7 The geographic range limits of three species of trees found in plant associations in eastern Kentucky (indicated by the star). (*After Fowells 1965.*)

of the gradient conditions. Closed community organization can be identified by the presence of sharp ecotones in species distributions (see Figure 26-3a). Figure 26-9 shows the distribution of six species of grasses along a gradient of soil pH.

The gradient studies conducted by Cornell University ecologist Robert Whittaker (1960) and others have been influential in putting to rest the extreme Clementsian view of the closed community. Most of Whittaker's work was conducted in mountainous areas, where moisture and temperature vary over short distances according to elevation, slope, and exposure, which in turn determine light, temperature, and moisture levels at a particular site. When Whittaker plotted the abundances of each plant species found at sites at the same elevation distributed along a continuum of soil moisture, he found that the species occupied unique ranges, with peaks of abundance scattered along the environmental gradient (Figure 26-10). Compared with communities in the mountains of southeastern Arizona, fewer species of plants occur in the mountains of Oregon, but each species has a wider ecological distribution, on average.

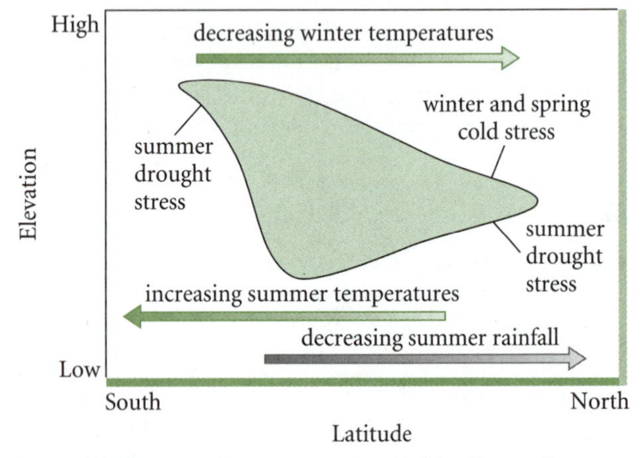

FIGURE 26-8 Schematic portrayal of gradients of temperature and moisture within elevational and latitudinal space. The shaded area represents suitable conditions for a hypothetical plant community. (*From Neilson et al. 1992; data from Neilson and Wullstein 1983, Neilson 1987.*)

In the Great Smoky Mountains of Tennessee, the dominant tree species are widely distributed outside the plant associations that bear their names (Figure 26-11). For example, red oak is most abundant in relatively dry sites at high elevations, but its distribution extends into forests dominated by beech, white oak, chestnut, and even hemlock (an evergreen, coniferous species), and extends throughout the entire range of elevation in the Smoky Mountains. Beech prefers moister situations than red oak, and white oak reaches its greatest abundance in drier situations, but the distributions of all three species in the same area were also independent of one another (Whittaker 1952).

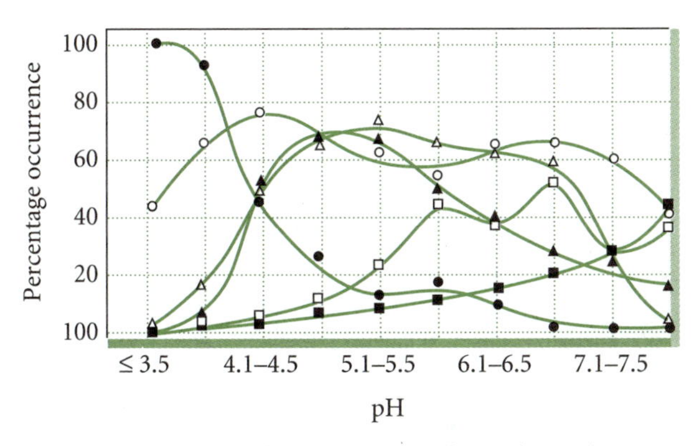

FIGURE 26-9 The distribution of six species of grasses along a gradient of pH. *Deschampsia flexuosa* and *Festuca ovina* can withstand relatively acidic conditions, whereas most of the other species are more likely to be found under more basic conditions. (*After Grime and Lloyd 1973.*)

A second method of describing community structure within the continuum concept is to place communities along one or more artificial axes based on data obtained from the communities themselves. Such a procedure results in a graph in which each community is represented as a point, with similar communities occurring near one another on the graph. The statistical procedures used to accomplish this are together referred to as **ordination** (Curtis and McIntosh 1951, Bray and Curtis 1957, Loucks 1962, Gauch 1982, Ter Braak and Prentice 1988). To obtain data for an ordination, an ecologist visits a large number of different local

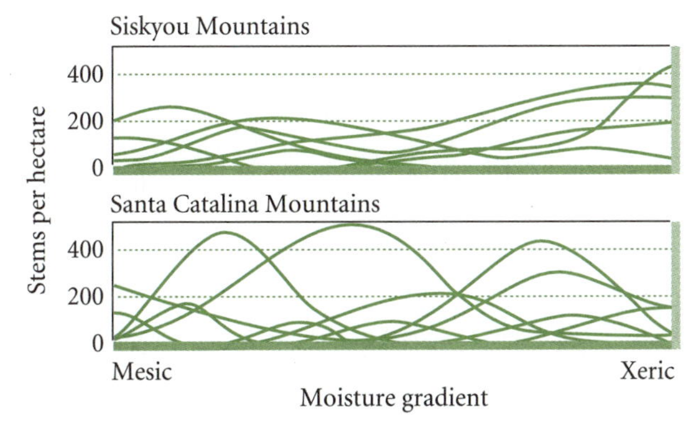

FIGURE 26-10 Distribution of plant species along moisture gradients at 460–470 meters elevation in the Siskyou Mountains of Oregon and at 1,830–2,140 meters elevation in the Santa Catalina Mountains of southeastern Arizona. Species in the more diverse Arizona flora occupy narrower ecological ranges; thus, in spite of the greater total number of species in the flora of the Santa Catalina Mountains, they and the Siskyou Mountains have a similar number of species at each sampling locality. (*After Whittaker 1960, Whittaker and Niering 1965.*)

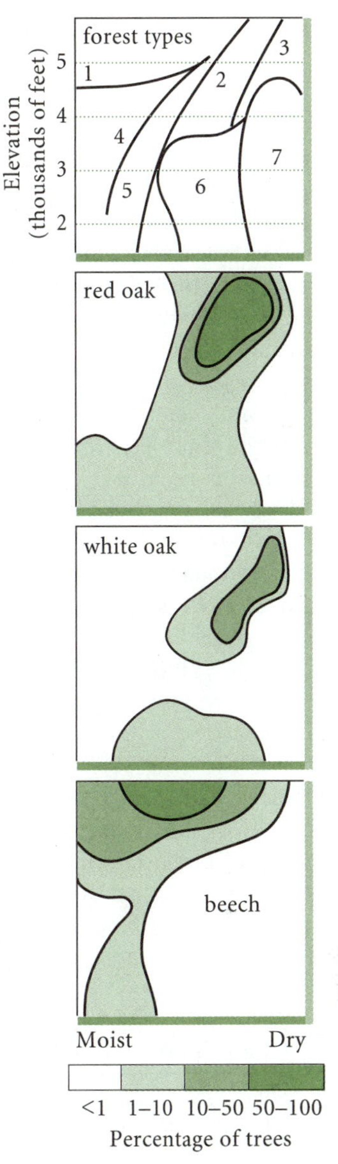

FIGURE 26-11 Distributions of red oak, white oak, and beech with respect to altitude and soil moisture in the Great Smoky Mountains of Tennessee. The approximate boundaries of the major forest associations in this area are shown in the top diagram: 1, beech; 2, red oak-chestnut; 3, white oak-chestnut; 4, cove; 5, hemlock; 6, chestnut oak-chestnut; 7, pine. Relative abundance, represented by the degree of shading, corresponds to the percentage of tree stems of the species more than 1 centimeter in diameter in samples of approximately 1,000 stems. (*After Whittaker 1956.*)

communities and obtains measurements of important environmental variables such as pH, moisture, and temperature, and determines abundances of the species in each community. Once the data are collected, statistical procedures (not discussed here) are used to construct a space based on the environmental data in which the communities can be arranged. Typically, a two-dimensional space is created in which each of the axes that define the space is composed of some combination of the environmental variables. For example, one axis may be composed of variables that affect soil moisture and the other of factors related to temperature (Figure 26-12). The distribution of the communities plotted within such a space reveals the degree to which there are distinct boundaries among them.

Modern ordination techniques rely on statistical procedures such as principal components analysis, reciprocal averaging (correspondence analysis), or detrended correspondence analysis (Gauch 1982) to determine the principal axes for ordination. These procedures use sample data on species abundance, distribution, or phylogeny, rather than on gradients of physical conditions, to determine similarities among communities. Such data are shown in Table 26-1, which is a species-by-sample (stand) incidence matrix, in which the abundance (given as the percentage of the total number of trees in the stand) of each tree species in each sample locality (A through J) is recorded. The ordering of the samples across columns and of the species down rows may be taxonomic, geographic, chronological, or arbitrary, but not ecological. Each column in the matrix represents the composition of one forest stand. Stand A, for example, is composed primarily of three species: *Acer saccharum*, *Fagus grandifolia*, and *Liriodendron tulipifera*. Stand A is very similar to stand B, but quite different from stand C. Such matrices may be used to develop axes for ordination.

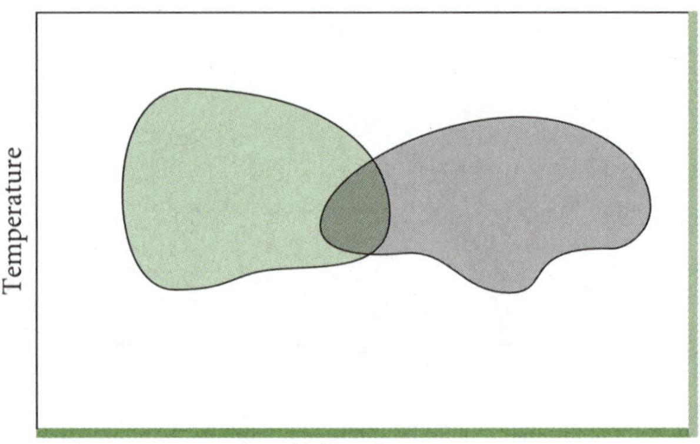

FIGURE 26-12 Representation of a hypothetical ordination analysis. Data from measurements of the physical conditions of many communities are subjected to statistical analysis to produce two ordination axes. Here, one axis is formed from combinations of measurements having to do with soil moisture, and the other from measurements that are correlated with temperature. Each community is then placed in the space defined by the axes according to the conditions in that community. In this analysis, two groups of communities are observed. The groups are similar with respect to temperature, but differ in regard to soil moisture.

A shortcoming of the gradient and ordination approaches is that they rely entirely on proximate characteristics of communities and ignore the phylogenetic history of the species that make up communities. Species bring to the present the constraints and

TABLE 26-1 **Composition of forest stands in the beech-maple forest region of the midwestern United States**

	Percentage of trees in stand									
Species	A	B	C	D	E	F	G	H	I	J
Acer rubrum						8	19		9	
Acer saccharum	17	13		14	7	28	4	6		49
Carya ovata	6	6	7	5			3		6	
Fagus grandifolia	33	21	5	17	72	40	7			
Fraxinus americanus	3	2		7	5	1	8	7	5	4
Juglans nigra		1		10				4		
Liriodendron tulipifera	21	15	2	5	10	1	1			
Nyssa sylvatica	4				2	6	1			
Quercus alba	8	1	63	7	15	46	3	13	8	
Quercus borealis	5	2	18	2			8	7	21	19
Quercus macrocarpa								4	1	
Tilia americana		13		2				31	19	16
Ulmus americana		1		9			3	36	25	1

Note: Not all columns sum to 100% owing to minor species excluded from the table. Locations are: A–D, Turkey Run State Park, Indiana; E, Heston's Woods, Oxford, Ohio; F, Canfield, Ohio; G, Graber Woods, Wayne County, Ohio; H–J, Harms Woods, Evanston, Illinois. (From Braun 1950.)

capabilities of a long evolutionary history. Whether community boundaries are sharp (closed community) or blurred (open community) will be influenced to some extent by the evolutionary history of the species in the community. It is also important to appreciate that the procedure of matching species distribution and abundance data to actual environmental gradients (gradient analysis), or of assembling species in communities based on distribution and abundance data alone by use of a statistical procedure (ordination), presupposes that local environmental conditions are chiefly responsible for species associations. Regional or landscape-scale interactions are not accounted for in such procedures, nor are biological interactions such as predation and herbivory. The composition of a community may be dependent on regional processes (Ricklefs 1987, Ricklefs and Schluter 1993a,b) and phylogenetic history (Losos 1996), as we shall see below.

26.5 The historical record reveals both change and continuity in communities.

Both the phylogenetic origins of taxa and the biogeographic processes that gave rise to species are reflected to some extent in the contemporary communities in which the species occur (Ricklefs and Schluter 1993a, b). The historical record preserved in fossils, fragmentary as it is concerning the organization of communities, equally supports the opposing perspectives of change and of stable association. New species arise and others disappear from the fossil record, to be sure, but their individual spans usually are long enough for much evolutionary adjustment to their cohabitants. Whereas associations among major codominants in systems appear ephemeral in the long history of the fossil record, associations between some of these dominant species and the smaller organisms that live on or within them may have arisen many millions of years ago.

During the Pleistocene (between 2 and 10 million years ago), massive glaciers periodically covered much of North America and the Eurasian continent (see Figure 28-10). Fluctuating global climatic conditions during that time resulted in periods of low temperatures, during which glaciers extended southward, and periods of mild temperatures, when glaciers retreated northward. The latter times are referred to as **interglacial periods.** During interglacial periods, plant communities developed in the areas once covered by glaciers. We can learn a great deal about the history of these communities by examining pollen grains that were deposited in the lakes and bogs formed by retreating glaciers in the northeastern United States and northern Europe. Such studies show that the composition of plant associations changes over time through range extensions and contractions and through integration of new species. For example, Wright (1964, 1968) showed that the sequence of reforestation following the most recent major glaciation in North America contained intermediate forest associations that are not found in the area today. The general pattern of reforestation began with spruce forest, which dominated the area until about 10,000 years ago, followed by extensive associations of pine and birch, which were later replaced by more temperate elm and oak forests.

During periods of glaciation, plants and animals survived south of the southern edge of the glaciers in areas called **refugia.** When glaciers retreated during warmer periods, many species extended their ranges to the north. The migration of tree species from their southern refugia since the height of the most recent glaciation (10,000–15,000 years ago) has been summarized by Davis (1965, 1976) and is illustrated for representative species in Figure 26-13. For hemlock and hickory, postglacial migration involved northerly range extensions from southern regions across most of the eastern United States. In contrast, white pine and chestnut appear to have emerged from refugia in

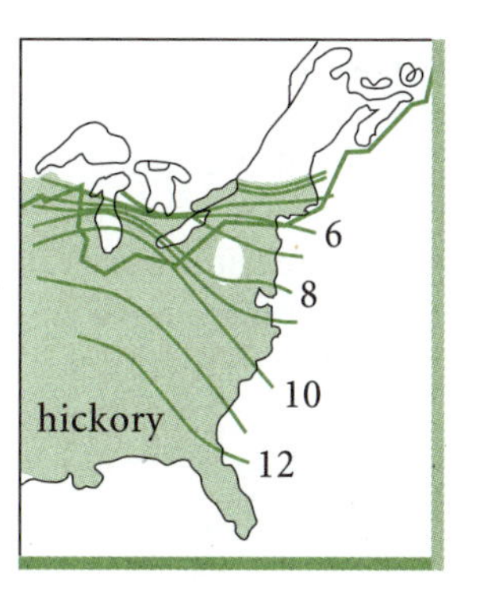

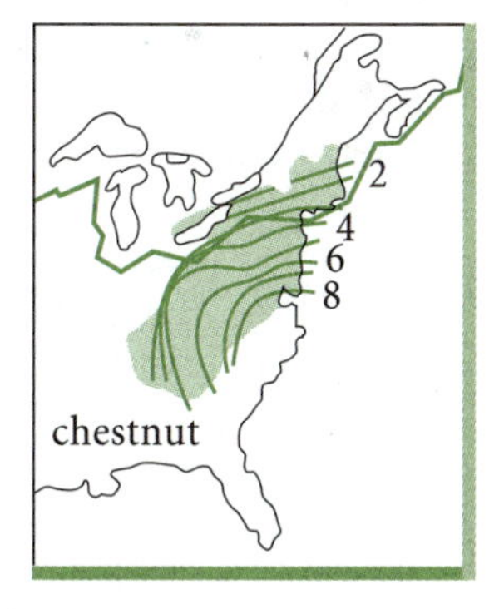

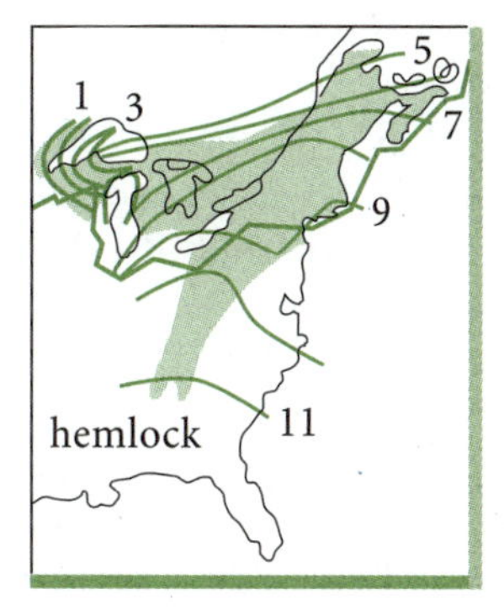

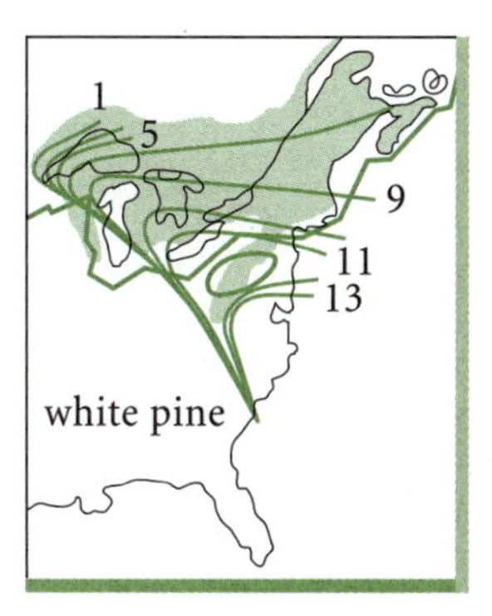

FIGURE 26-13 Migration of four tree species in eastern North America from Pleistocene refuges to their present distributions following the retreat of the glaciers. Numbers associated with distribution lines indicate thousands of years before the present. (*From Davis 1976.*)

the Carolinas and expanded their ranges to the west as much as to the north.

Glaciation per se was not the only agent of change in the history of the earth. A recent analysis of fossil deposits of reef-building communities by Kauffman and Fagerstrom (1993) showed extensive changes in species diversity through geologic time. Mass extinctions, which eliminated all but a few species, were common. Kauffman and Fagerstrom suggest that fluctuations in environmental conditions are chiefly responsible for this great variation in community organization. This evidence from fossil reef-building communities reveals patterns of change substantially greater than that reported by Davis for North American forests. The important point here is that the communities that we see today are not the ones that existed in the same locations in the past.

In contrast to the seeming flux in species composition of trees in broad-leaved temperate zone forests and of reef-building communities is the stability of many associations involving plants and animals. Historical information concerning these associations is often indirect. Consistencies between taxonomic relationships among hosts and taxonomic relationships among their herbivores or disease organisms suggest a long evolutionary history of association (Linsley 1961, Brooks 1985). In some cases, fossils reveal evidence of consumer activities indistinguishable from traces left by present-day consumers. For example, Opler (1973) reported the characteristic tracks of present-day leaf-mining moths in oak leaves fossilized more than 20 million years ago. Smith et al. (1985) found drill holes of gastropod predators in Devonian period fossils of marine brachiopods indistinguishable from those occurring more than 100 million years later in other fossils or those made by living gastropods. Mycorrhizal associations appear in fossils more than 100 million years old (Stubblefield et al. 1987). While the species in these interactions undoubtedly were not the same over that span of time, the evidence reveals long associations between groups of interacting organisms over their mutual evolutionary history.

Biogeographic evidence can also reveal prolonged associations, particularly when groups of species share distributions (Latham and Ricklefs 1993). During the early part of the Cenozoic period, about 60 million years ago, vast areas of North America, Asia, and Europe were covered by temperate forest more or less continuously distributed across the land areas at high latitudes. As the climate of the earth cooled and temperate vegetation retreated southward, the Asian and American remnants of this forest were separated, until only relict groups of species remained in isolated areas (Graham 1972, Thorne 1972). A part of this vast flora remains today as sets of closely related species, including magnolias, rhododendrons, tulip trees, and gums, now found only in parts of the United States and in southeastern Asia. The affinities of the floras of these two areas have been recognized for more than a century (Gray 1860, Fernald 1929, Li 1952, Wolfe 1975, 1981). At present, the components of this flora constitute a sizable fraction of the perennial plants in the woodlands and wetlands of these areas, suggesting that these plant communities and the animals associated with them might have had a long history of coevolution. Other components of the woodland flora in these areas have had different biogeographic histories, and the floras of fields and disturbed habitats clearly have different origins, yet the mixing of floras that makes up the present-day communities may not have occurred so rapidly as to prevent evolutionary accommodations among their species.

26.6 Evolutionary history may leave a distinctive imprint on community organization.

History affects community processes in two important ways. First, natural selection can act only on the traits that species bring to the community, and those traits reflect the evolutionary history of the species. For example, it is reasonable to hypothesize that selection working through competitive interactions accounts for divergence in the sizes and shapes of the wings and bills of tropical hummingbirds, since these morphological features are characteristic of hummingbirds. But the hummingbird community will not be characterized by a diversity of running speeds because hummingbirds do not run. Thus, the nature of a community is constrained by the features of the constituents of the community.

The fact that the distribution of life forms over the face of the earth is by no means uniform reveals a second consequence of evolutionary history: some regions lack groups that are abundantly represented elsewhere. Many irregularities in distribution patterns are linked to major climate patterns; for example, snakes and lizards cannot tolerate the cold of arctic environments. Historical accidents of distribution, caused by geographic barriers to dispersal, have also played important roles in the distributions of major groups of animals and plants. Anomalies of distribution are most obvious on islands (though the major continental landmasses have also been sufficiently isolated to reduce the exchange of forms evolved in each area). Australia lacks most groups of mammals except for marsupials and bats, which are highly diversified there. Few species of any kind reach small, remote islands.

Phylogenetic Analysis of Community Structure

The evolutionary relationship among species or other taxa is referred to as **phylogeny**, and the analysis of phylogeny is called **phylogenetic analysis** (Harvey and Pagel 1991). The goal of phylogenetic analysis is to understand the relationships among species or taxa that have descended from a common ancestor. These relationships are displayed in a diagram called a cladogram (see Figure 2-11), which resembles a simple tree or bush with the common ancestor at the base and the extant taxa represented by the branches. Wherever there is a bifurcation of a branch, an evolutionary event has given rise to a new species or taxon. Where the line extending from the ancestor to a modern species is unmarked, it is assumed that the modern species resembles the ancestor. Where the line is marked in some way, usually with a short line or a small box, it has been determined that the extant species has undergone a change. We shall not discuss here the details of how to construct a cladogram; rather, we will use cladograms to demonstrate how phylogeny may affect community structure.

In recent years, ecologists have used phylogenetic methodologies to study the effects of historical processes on species diversity and community structure (reviewed by Losos 1996). In applying phylogenetic analysis to questions about community patterns, the goal is to decouple the effects of relatively recent interactions, such as competition and predation, from the effects of lineage and historical accident on the distributions of species. Phylogenetic methods also provide ecologists with a means of understanding the importance of environmental factors in the convergence of community structure. Methodologies for the application of phylogenetic analysis to the study of community structure and function have been advanced by Brooks and McLennan (1991, 1993) and by Losos (1992, 1996), and the subject has been recently reviewed by Miles and Dunham (1993).

Figure 26-14 shows how phylogenetic analysis might be used to compare simple communities in three different regions. Each hypothetical community (one in California, one in Virginia, and one in Florida) contains three species: a plant, an herbivore, and a carnivore. Four different situations are shown. In Figure 26-14a, we can see that the trophic structure of the three hypothetical communities is ancestral to the formation of the three communities. Losos (1996) has described a situation like this for *Anolis* lizards of small islands of the western Caribbean. The species *A. sagrei* and *A. maynardi* co-occur on many islands in the region, and wherever they live, they occupy different habitats. While such habitat partitioning could have arisen repeatedly through competitive interactions, Losos suggests that the more likely scenario is that differences in habitat preference arose very early and were retained as the two species expanded their ranges into other islands.

Figure 26-14b depicts a situation in which a new ecological association—the addition of a parasite of the predator—arises in the Florida community. The parasite species is not found in either the California or Virginia communities, but is known from communities in Mexico and Alabama. Thus, the new ecological association in the Florida community is a result of the introduction of new component to the community. New associations can also arise when speciation occurs, as shown in Figure 26-14c. In this situation, speciation in California has given rise to another community member, one with a different foraging strategy (omnivory). The new association is not observed in the Virginia and Florida communities. Figure 26-14d shows a situation in which a difference in ecological relationships arises because of the invasion of a second, bigger predator at a time before the Virginia and Florida populations became established. Thus, the ancestral trophic structure, having both the predator and the bigger predator, persists in the contemporary communities in Virginia and Florida.

This simple analysis shows some of the ways in which speciation events and historical changes in ecological associations shape the structure of contemporary populations. It also demonstrates how an ecologist might focus her studies of community patterns. For example, a phylogenetic analysis might suggest that the similarity in community structures shown in Figure 26-14b is more a result of phylogenetic history than of responses to similarities in local environmental conditions. The absence of the parasite in the California and Virginia communities is probably best explained by the geographic overlap of the parasite's range with the Florida community, rather than by divergence in ecological associations among the California, Virginia, and Florida communities. The explanation of why the Virginia and Florida communities have a second predator is found by focusing on what happened before those two communities were established, rather than by comparing their ecological relationships with those in the California community.

Convergence and Community Structure

The different evolutionary histories and taxonomic affinities of the plants and animals of the earth's regions are obscured by convergence in form and function. Where woodpeckers are missing from a fauna, other species may adapt to fill their role (Figure 26-15). Rain forests in Africa and in South America are inhabited by plants and animals with different evolutionary origins but having remarkably similar

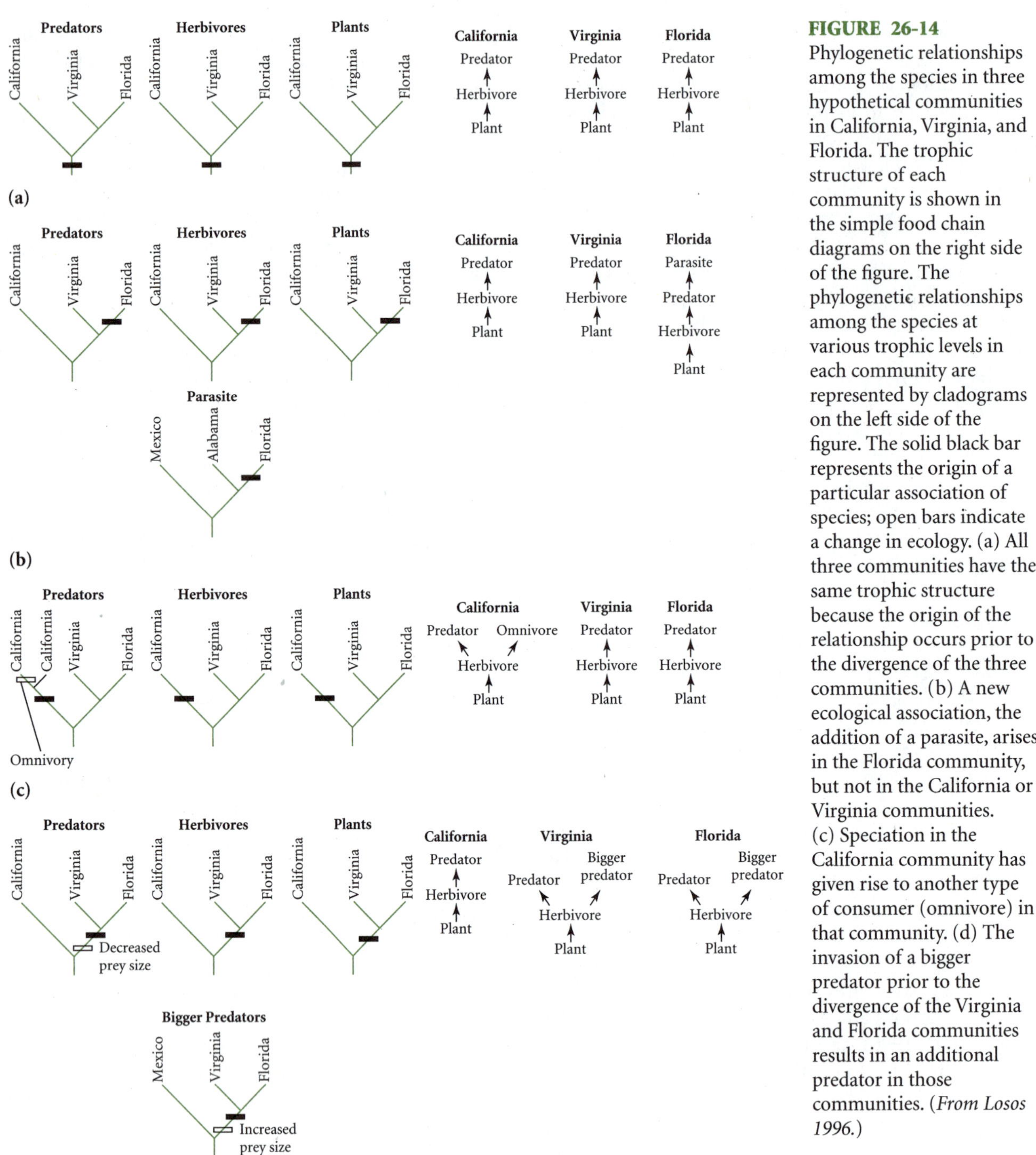

FIGURE 26-14 Phylogenetic relationships among the species in three hypothetical communities in California, Virginia, and Florida. The trophic structure of each community is shown in the simple food chain diagrams on the right side of the figure. The phylogenetic relationships among the species at various trophic levels in each community are represented by cladograms on the left side of the figure. The solid black bar represents the origin of a particular association of species; open bars indicate a change in ecology. (a) All three communities have the same trophic structure because the origin of the relationship occurs prior to the divergence of the three communities. (b) A new ecological association, the addition of a parasite, arises in the Florida community, but not in the California or Virginia communities. (c) Speciation in the California community has given rise to another type of consumer (omnivore) in that community. (d) The invasion of a bigger predator prior to the divergence of the Virginia and Florida communities results in an additional predator in those communities. (*From Losos 1996.*)

appearances (Keast 1972, Bourlière 1973; Figure 26-16). Plants and animals of North and South American deserts are more similar in morphological characteristics than one would expect from their different phylogenetic origins (Mares 1976, Orians and Solbrig 1977). Similarities also have been noted in the behavior and ecology of Australian and North American lizards, despite the fact that they belong to different families and have been separated for perhaps 100 million years (Pianka 1971). Plants in the Mediterranean climate zones of southern Europe, South Africa, California, Chile, and Australia are remarkably similar in their morphological and physiological adaptations to their winter-rainfall, summer-drought environments (Mooney and Dunn 1970, Mooney 1977).

Wherever one looks, one finds convergence, and convergence reinforces the idea that community organization might depend more on local conditions of the environment than on the evolutionary origins of the species that constitute the community. In many

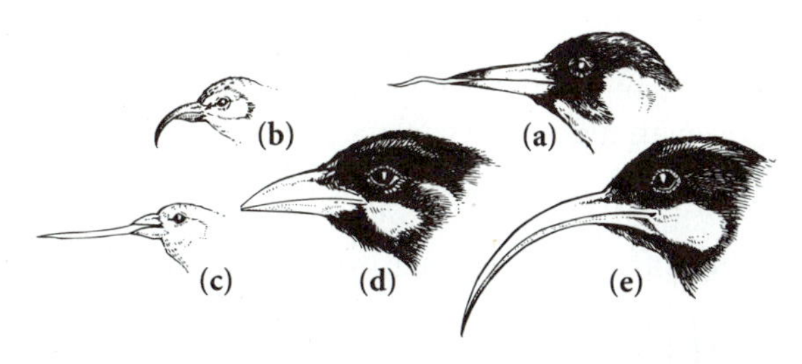

FIGURE 26-15 Unrelated birds that have become adapted to extract insects from wood: (a) the European green woodpecker, which excavates with its beak and probes with its long tongue; (b) the Hawaiian honeycreeper (*Heterorhynchus*), which taps with its short lower mandible and probes with its long upper mandible; (c) the Galápagos woodpecker-finch, which trenches with its beak and probes with a cactus spine; and the New Zealand huia (now extinct), which divided foraging roles on the basis of sex—the male (d) excavated with his short beak and the female (e) probed with her long beak. (*After Lack 1947.*)

instances, species-for-species matches have been made (Cody 1974, Fuentes 1976, Mares 1976), suggesting that environments may closely specify the particular characteristics of species that inhabit them, and that these specifications depend only on climate and other physical factors; plants and animals have little additional modifying influence on one another by reason of their different phylogenetic backgrounds. The situation might be viewed as similar to that of a basketball team, in which different individuals may take on the roles of center, forward, or guard, but the structure of the team (five players on the floor at one time) remains constant, as determined by the rules of the game.

This view is much too simple, however. Cases of species-for-species matching have not stood up under close scrutiny (Ricklefs and Travis 1980). In fact, detailed studies of convergence are as likely to turn up remarkable differences between the plants and

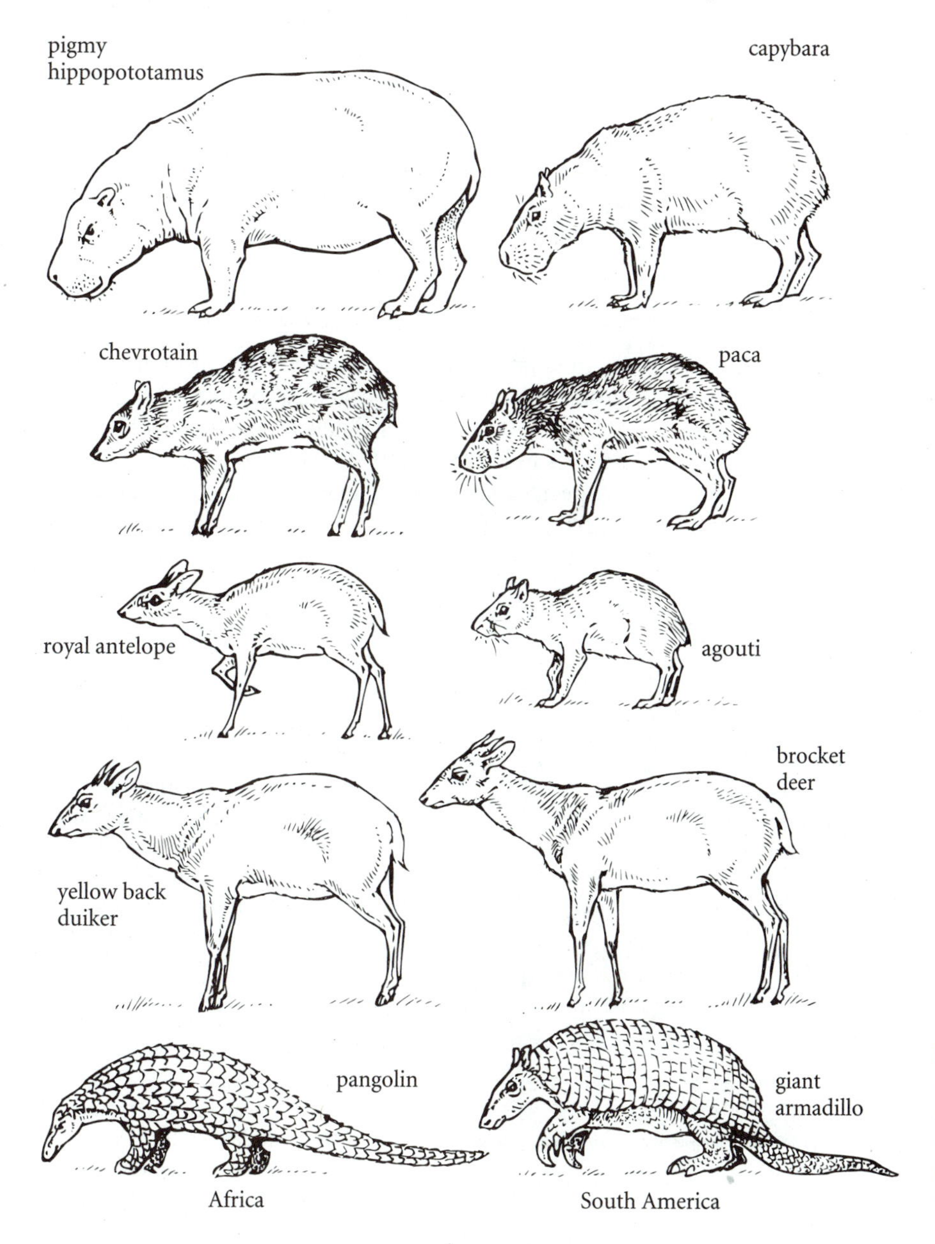

FIGURE 26-16 Morphological convergence among unrelated African (*left*) and Neotropical (*right*) rain forest mammals. Each pair is drawn to the same scale. (*After Bourlière 1973.*)

FIGURE 26-17 Mangrove vegetation in an estuary on the Pacific coast of Costa Rica. Note the prop roots of *Rhizophora* trees at left and the buttressed trunks of *Pelliciera* at right. These trees are established on mucky substrate within the tidal zone; hence the soil is flooded periodically with salt water.

animals in superficially similar environments. In spite of striking convergences, Mares (1976) noted that the ancient Monte Desert of South America is the only desert region of the world lacking seed-eating, bipedal, water-independent rodents like the kangaroo rats of North America and the gerbils of Asia. Among frogs and toads, however, several South American forms have taken adaptation to desert environments a step further than their North American counterparts: they construct unique nests of foam in which their eggs are kept from drying (Blair 1975). Differences between the Australian agamid lizard *Amphibolurus inermis* and its North American iguanid analogue, *Dipsosaurus dorsalis,* include diet, optimum temperature for activity, burrowing behavior, and annual cycle, even though at first glance the species are dead ringers for each other (Pianka 1971).

A study by Ricklefs and Latham (1993) revealed striking differences in mangrove communities in similar environments around the world. Mangroves are tropical forests that occur within tidal zones along coastlines and river deltas (Figure 26-17). Mangrove trees tolerate high salt concentrations and anaerobic conditions in the water-saturated sediments in which they take root. Fifteen lineages of terrestrial trees have independently colonized mangrove habitat, and several of these have subsequently diversified there. We cannot explain the much greater diversity of Indo-West Pacific mangroves on the basis of habitat, because both regions have roughly equal areas of mangrove habitat (Figure 26-18). Within the Indo-Pacific region, the small areal extent of suitable mangrove habitat in East Africa and Madagascar (region D) may explain low mangrove diversity there. The large variation in the number of mangrove species appears to have resulted from plant taxa invading mangrove habitat more frequently in the Indo-West Pacific than in the Caribbean region, although the reasons for this are not clear.

Coevolution and Community Structure

To understand the importance of evolutionary history in shaping community organization, we must appreciate the interdependent nature of the evolutionary process (Thompson 1994). The evolutionary fortunes of species may be tied inextricably and mutualistically with those of other species through the process of coevolution, as we saw in Chapter 25. The intricate (and

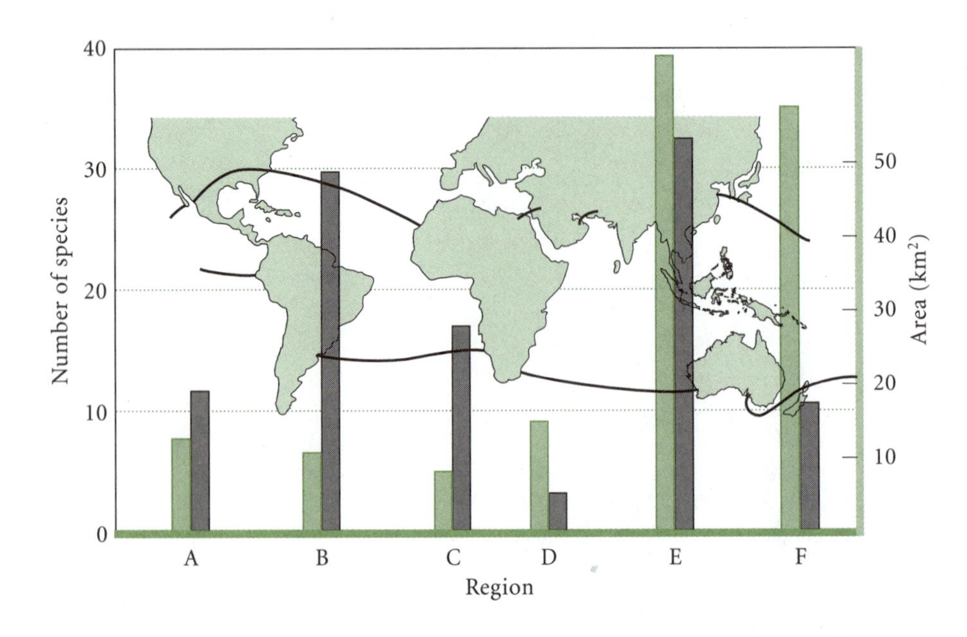

FIGURE 26-18 Limits of the distribution of mangrove vegetation (black lines) along coasts of continents and islands of the world. Bars show the areal extent of mangrove habitat (gray) and numbers of species of mangrove trees and shrubs (green) in each of six regions: A, eastern Pacific Ocean; B, Caribbean Sea and western Atlantic Ocean; C, eastern Atlantic Ocean; D, western Indian Ocean; E, eastern Indian Ocean; and F, western Pacific Ocean. (*After Chapman 1970, Saenger et al. 1983.*)

beautiful) morphological and behavioral coadaptations among plants and their herbivores and pollinators, and among parasites and their hosts, leave little doubt that community organization will somehow reflect these intimate relationships. But how does coevolution shape community organization?

The answer to this question is still emerging, but we may fruitfully address it by invoking the geographic mosaic model of coevolution (Thompson 1994). Suites of coevolved species constitute prominent components of many local communities. For example, distinctive assemblages of nectar-producing flowers and their hummingbird pollinators are features of Central and South American communities (Feinsinger 1978, Stiles 1985). Recent experimental and field studies of the local interactions among the constituents of these communities have revealed patterns of competition among pollinators and their flower hosts that suggest mechanisms for local community organization (Feinsinger and Tiebout 1991, Feinsinger et al. 1991). But the interrelationships among flowers and their bird pollinators vary, sometimes dramatically, on a regional scale. Bird species that specialize on one flower species in one location may prefer a different, often taxonomically distant, species at another location. Just as studies of local competition and predation between pairs of species cannot reveal completely the importance of those interactions in shaping community structure unless one considers the geographic and historical influences on the community as well, neither can observation of the interactions of local coadapted species provide an explanation of the importance of coevolution in community structure.

26.7 The characteristics of the community emerge from a hierarchy of processes over scales of time and space.

There is little agreement among ecologists about what a community is or how its structure is regulated. It is clear, however, that the concept of community cannot usefully be applied to a contemporary, physically circumscribed place, because its characteristics are molded to some extent by historical, regional, and even global processes. Rather than limiting our perception to a delineated area, we might think of a community as a single point of reference in time and space from which population and evolutionary influence emanates, with a force that diminishes over time and distance. Just as a point emits such influence, it also receives similar influence from other points, the strength of which decreases with distance. The characteristics of a community at a given point express all these influences; a community, therefore, has no fixed boundary, but is more like a sphere of influence with a density that is great at the center and diminishes with increasing radius. Thus the area of interest to the community ecologist must encompass the scales of processes important in shaping community structure at a given point.

Several kinds of processes are important, each with a different characteristic scale of time and space. Scale in space varies between the activity range of the individual, the geographic and ecological dispersal of individuals within populations, and the expansion and contraction of the geographic ranges of populations. Scale in time varies according to the rates of individual and population movements, the dynamics of population interactions, and the selective replacement of genotypes within populations. The relative predominance of local, contemporary processes versus regional or historical processes in shaping community attributes depends on the relative spatial and temporal scales of these processes. The fate of a population at a particular point depends on the balance between the tendency of intolerable physical conditions, interspecific competition, and predation to exclude that population locally and the rate of dispersal of individuals to the point from surrounding areas of population surplus (Shmida and Wilson 1982, Pulliam 1988). The diversity of species at a point depends on the balance between local rates of extinction—resulting from predators, disease, competitive exclusion, change in the physical environment—and regional rates of species production and immigration (MacArthur 1969). Every point on earth provides limited access via dispersal to sources of colonizing species. Local diversification depends not only on the capacity of the environment to support a variety of species, but also on the accessibility of a region to colonists, the capacity of a region to generate new forms through speciation, and the ability of a region to sustain varieties in the face of varying environments. Although ecology has traditionally focused upon local, contemporary systems, it clearly must expand its concept to embrace global and historical processes.

SUMMARY

1. A community is a group of organisms in a particular location. Communities are characterized by their structure, which includes features such as the number, types, and relative abundances of species in the community, the physical characteristics of the vegetation, and the trophic relationships among the species. Communities also have features of function, which include rates of energy flow, productivity, and resilience to disturbance.

2. Groups of organisms that use a suite of resources in a similar manner are referred to as guilds. Organisms that are phylogenetically related are called assemblages. An ecosystem includes all the interacting parts of the physical and biological worlds, whereas a community includes only the organisms in a particular place.

3. The early 1900s saw two opposing views of the nature of the community. The influential botanist F. E. Clements held the view that communities were superorganisms whose components interacted in much the same way as organs in a body. Thus, the ranges of the species in the community were inextricably tied to the community as a whole, and, thus, the community was closed. H. A. Gleason suggested that communities were merely fortuitous associations of organisms whose distributions corresponded to individual adaptations, and that, thus, the community was open.

4. Generally speaking, communities are not discrete units separated by abrupt transitions in species composition (closed communities), as Clements suggested. Species tend to be distributed over ecological gradients of conditions independently of the distributions of other species (open communities), as Gleason believed.

5. Discontinuities between associations of plants and animals, called ecotones, sometimes occur at sharp physical boundaries or accompany change in habitat-dominating growth forms. The aquatic-terrestrial transition is an example of the first kind of ecotone; the prairie-forest transition is an example of the second. Habitat fragmentation affects the extent of ecotones in a particular area. The more fragmented a habitat, the more area lies at the edge of the habitat relative to the amount that is in interior spaces. The term metacommunity refers to a set of local communities occurring in separate habitat patches connected by dispersal of organisms in the communities.

6. To analyze distributions of species with respect to environmental conditions and with respect to the distributions of other species, ecologists have devised various types of gradient analysis, in which they position sample localities on scales of physical conditions. The distributions of species along these environmental gradients emphasize the open structure of communities.

7. Fossil records and biogeographic evidence indicate that communities may change through time with the differential migration and dispersal of their component populations.

8. Phylogenetic analysis is used to study the evolutionary relationships among species, which are displayed in a cladogram. Evaluation of the cladogram can reveal clues about the historical development of community structure.

9. Studies of communities developing under similar ecological conditions in widely separated parts of the world suggest that, in spite of the convergence of many adaptations, history plays an important role in shaping the attributes of local communities.

10. The characteristics of the community emerge from the interaction of a hierarchy of processes acting over different scales of time and space. Ecological investigation of communities must encompass the study of regional, historical processes as well as local, contemporary processes.

EXERCISES

1. Write a brief explanation of the difference between an ecological community and an ecosystem that will be understood by someone who has never had a class in ecology.

2. Write a historical comparison of the emergences of the ideas of the ecological community and the ecosystem.

3. Explain how the geographic mosaic notion of coevolution may be applied to community structure.

CHAPTER 27

Structure of the Community

GUIDING QUESTIONS

- How do local and regional factors interact to structure communities?
- What is niche preemption, and how is it used in modeling species abundance?
- What is the difference between the random niche and geometric series models of species abundance?
- What is the significance of the lognormal distribution in community ecology?
- How are the Simpson and the Shannon-Weaver indices used to measure species diversity?
- What is the difference between taxonomic and ecological diversity?
- Why is rarefaction useful in the study of diversity?
- What are some of the hypotheses proposed to account for species-area relationships?
- What are the differences between topological, energy flow, and interaction food webs?
- What are the major generalizations arising from analyses of topological food webs?
- What was Robert May's idea about the relationship between food web species number and connectance and species interaction?
- What are the five types of indirect effects in food webs?
- What is a community matrix?
- What is the difference between pulse and press experiments?

Communities are associations of populations whose characteristics are greatly influenced by the physical environment and by the interactions of the populations within the community, and whose ultimate features are shaped by the evolutionary histories of their constituent species. The patterns of species abundance and population interactions within a community are referred to as community structure. The study of community structure involves both the description of more or less static features of communities and experimental analysis of population interactions. An appreciation of the complexities of population interactions is essential if we are to gain an understanding of community stability and resilience, features of communities that determine their integrity in the face of human disturbance. For this reason, the dynamics of community structure is a very active field of ecological study.

27.1 Understanding community structure requires that we adopt multiple perspectives.

We may examine community structure from different spatial and temporal perspectives. First, we may address patterns within a small area of relatively uniform habitat. Such local areas may be thought of as ecologically homogeneous patches encompassing the daily activities of individuals. Thus, descriptions of local community structure represent the ecological interrelationships of individuals co-occurring at a single point: their adaptations and responses to the physical environment and their interactions with one another, including direct conflict over resources, predation, disease, chemical interference, and so on. Within local areas, investigators characterize the community by the number of species present (species richness), the relative abundances of those species, their feeding relationships (food webs), and the way in which they partition the spectrum of ecological resources.

These local observations, however, give us a very limited view of the community structure and provide only a hint of the broader processes that give rise to that structure, because individual adaptation reflects a much larger gene pool than the local community, one that reflects a long history of selection. Furthermore, the presence of individuals in a local community may reflect population processes over wide areas linked together by dispersal, and thus have little to do with the interactions that we observe in the local community (Shmida and Wilson 1985, Pulliam 1988, Ricklefs and Schluter 1993a,b).

A second perspective on community structure addresses patterns of distribution of species over large areas containing a variety of habitats. Individuals of a particular species occur within particular habitats in such landscape mosaics (Cody 1985, Ricklefs and Schluter 1993a). This view may be thought of as a regional perspective. Each habitat has a distinctive assemblage of plants, animals, and microorganisms, and each population is specialized with respect to habitat. The presence of species at particular places depends both on the adaptations of individuals to local conditions and on the interactions among populations at those places (Rosenzweig 1981, 1987). Thus, as we broaden our perspective from the local, more or less uniform, habitat to the larger region encompassing many habitats, we witness, first, the realization of the adapted attributes of, and direct interactions between, individuals that co-occur locally, and second, the expression of dynamic interactions between populations, which we discussed in detail in Part 5.

Within the local community, structure is defined by the abundances and activities of individuals of each species: where they live, what they eat, and so on. Within the region, structure is defined by patterns of habitat selection.

We may also view communities from two temporal perspectives. The first encompasses the processes responsible for the production of new species, referred to as **speciation.** Speciation results from the geographic, ecological, and behavioral splitting of species into independently evolving lineages. How speciation occurs depends on the genetic structure of populations, including mating system and dispersal characteristics, and on the ecological circumstances of populations. The rate of speciation affects **species richness** (number of species) in the long term. The second perspective includes historical events that may affect the structure of a community. Accordingly, the presence or absence of a species may reflect routes of dispersal between ecologically suitable regions, or unlikely biological and physical circumstances, such as an extended drought, local volcanism, or a new, virulent disease. Stochastic fluctuations may also exert an influence on community composition. Together, such factors can result in unique assemblages of species at each place on earth, depending on its history of climate and geography.

27.2 Lists of species provided the first descriptions of biological communities.

During the latter part of the nineteenth century, European naturalists turned their attention from describing new species to characterizing local floras according to their species composition (Shimwell 1971, Mueller-Dombois and Ellenberg 1974). This avenue of study, sometimes called **floristic analysis** or **phytosociology** (Braun-Blanquet 1932, 1965), led directly to the functional concepts of the community explored during the early part of the twentieth century by Frederic Clements and others. But the initial concern of floristic analysis was one of classification, in which species composition was used to determine the relationships among plant associations, just as morphological characters indicate relationships among species. But whereas species have, on the whole, proved to be discrete types more or less distinguishable from one another, associations, as a rule, merge without sharp boundaries. Thus, delineating different communities was often not straightforward. As floristic data became more complete, this difficulty became more insistent, and the original premise of the endeavor began to unravel.

Ecologists soon recognized that floristic studies alone could not describe communities adequately. Studies that sought additional information about communities yielded new insights. These studies at first addressed quantitative patterns of relative abundance among species, but then proceeded to the description of functional relationships among species within the community—the food web.

27.3 The relative abundance of species is a measure of community structure.

The Danish botanist Christen Raunkiaer (1918) was one of the first investigators to notice that the abundances of species within local communities assumed a regular distribution. When he plotted the numbers of species in each of several abundance classes, the points followed a reversed J-shape (Figure 27-1). This pattern suggests that within a particular community, a few species attain high abundance—they are the **dominants** in the community—whereas most other species are relatively rare. Raunkiaer did not use a mathematical expression to describe his "law" of frequency, but its generality and pervasiveness among different communities have tempted others to mathematical description ever since.

A number of different theoretical distributions of relative species abundance have been proposed, three of which are discussed below. These mathematical representations can serve two purposes. One is to describe many data sets of species abundance with simple equations whose parameters may be used to make comparisons among different samples of species. A second is to use the logic of the mathematical model to infer the processes that produce the distributions observed. Each of the models described below was proposed specifically to test predictions following upon certain community processes.

Most ecologists now question the utility of these models because the mathematical assumptions are often difficult to link to real biological processes, and because the models are often highly sensitive to sampling and random effects (e.g., Tokeshi 1990, Wilson 1991). Nevertheless, we believe it is of value to discuss them here, because they represent the fruits of much early work regarding the nature of community structure, and because they still serve to stimulate thinking about community dynamics. Moreover, the development of these models provides good examples of how hypotheses about the natural world arise from theoretical considerations.

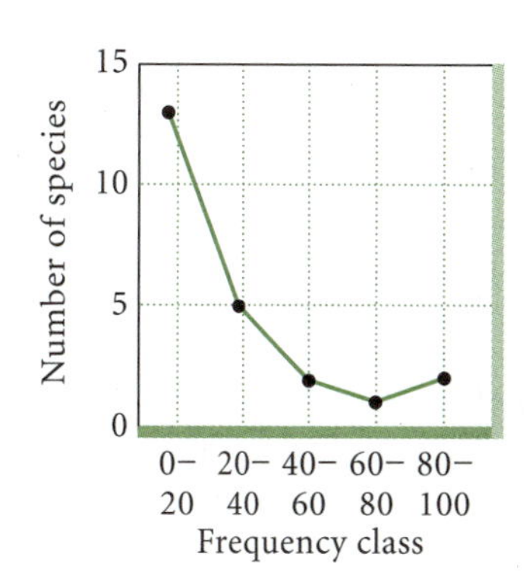

FIGURE 27-1 Number of species of plants in a peat bog near Kalamazoo, Michigan, in each of five frequency classes, based on the percentages of twenty-five 0.1 m^2 sampling areas occupied. (*Data from Kenoyer 1927.*)

The **geometric series** model of species abundance derives from the concept of **niche preemption,** in which species colonize an area in succession and each manages to preempt a constant fraction (k) of the remaining resources. The **niche** of a species is its role in the environment as defined by the ranges of conditions and resource qualities within which it may survive (see Chapter 29). Think of an unoccupied area having a total level of resources equal to 1. Suppose that each time a new species colonizes the area, it takes for itself $k = 0.5$ of the resources that are not being utilized by the species already in the area. After the first species arrives, $1 - 0.5 = 0.5$ resources remain unused and available for the second species to arrive in the area (Table 27-1). The second species, then, will co-opt half of the one-half of the total resources available, or $0.5(1 - 0.5) = 0.25$, leaving 0.25 available for the third species. Table 27-1 shows how the third and fourth species to arrive in the area use the remaining resources. In general, if each new species takes k proportion of the resources, then the first species will take k, the second $k(1 - k)$, the third $k(1 - k)^2$, and the ith

TABLE 27-1 Calculation of resource use according to the geometric series (niche preemption) model

Arrival	Resources available	Amount used	Amount left
First	10	$(0.5)(10) = 5$	5
Second	5	$(0.5)(5) = 2.5$	2.5
Third	2.5	$(0.5)(2.5) = 1.25$	1.25
Fourth	1.25	$(0.5)(1.25) = 0.625$	0.625

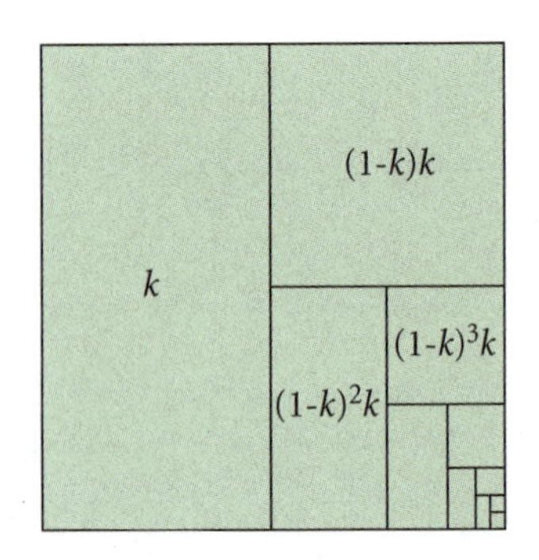

FIGURE 27-2 Geometric representation of the niche preemption model of relative abundance, in which each successive species utilizes a constant fraction (k) of the remaining resources. The ith species to arrive will take $k(1 - k)^{i-1}$ of the resources.

species $k(1 - k)^{i-1}$ (Figure 27-2). Assuming abundance to be directly proportional to resources, the abundance of the first species would be proportional to k, that of the second to $k(1 - k)$, and so on.

In 1957, Robert MacArthur proposed that the abundance of each species within a community was determined by processes that resembled the random partitioning of resources distributed along a continuum of resource types. This model is called the **random niche model.** For the purposes of his model, MacArthur envisioned resources as if they were distributed evenly along the length of a stick. To predict the relative abundances of N species according to this caricature, $N - 1$ points are chosen at random along the length of the stick, and the stick is broken at these points. (It is the imaginary breaking of the stick that led to the term "broken stick model" that is sometimes used for this idea.) The length of each segment of the stick corresponds to the abundance of a species. When the segments are arranged on a logarithmic scale of decreasing rank, the expected distribution of stick lengths decreases approximately linearly (Figure 27-3). When species rank (higher rank means higher abundance) is plotted on an arithmetic scale against log abundance, the broken stick model yields an S-shaped curve (Figure 27-4). An example of how species abundance curves may be generated using the random niche model is presented in Figure 27-5. Birds, fish, ophiuroid worms, and predatory gastropods were found to fit the model well (MacArthur 1957, 1960, King 1964), whereas the abundances of many small-bodied, short-lived organisms, such as soil arthropods, nematodes, protozoans, and phytoplankton, were found to have less equitably distributed abundances, with fewer very common species and many more rare ones according to abundance rank (Hairston 1959, King 1964, Batzli 1969, Whittaker 1965, 1972).

The random niche model can be viewed in at least two biological contexts. First, as MacArthur intended, one may envision a group of species simultaneously partitioning a single resource at random. The abundance of the various species falls out as the distribution of broken sticks. We might also envision a situation in which species are invading a particular area sequentially in the manner discussed for the geometric series model, except that as each species colonizes, it co-opts a random portion of the resources available, rather than a fixed amount (k), as in the geometric model (Gotelli and Graves 1996).

One of the difficulties with the random niche model is that all distributions of species abundance are equally probable, since the distribution is formed by a random process (Pielou 1975). The distribution of a single community, therefore, does not reveal whether the broken stick model, or some other model, is in force. In order to evaluate which model applies, one must take samples of the same size from many

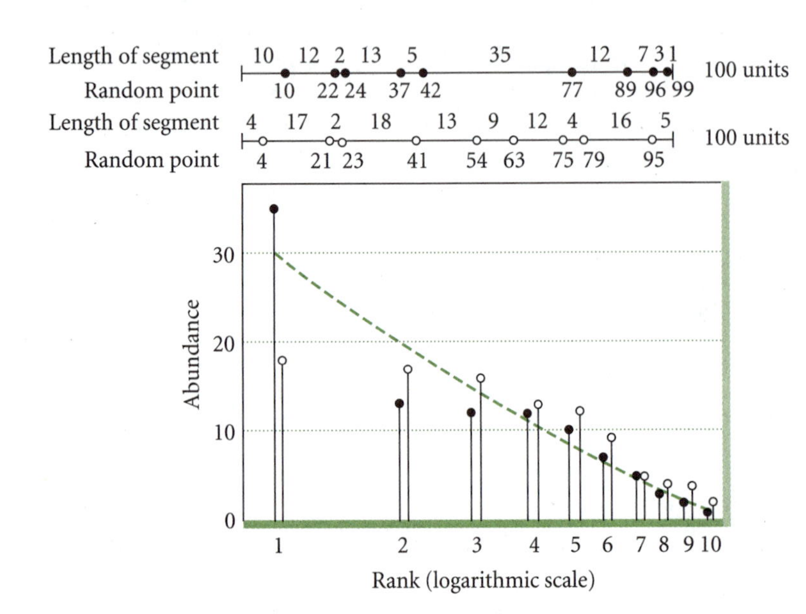

FIGURE 27-3 Two simulations of relative abundance of species in a community according to a random niche (broken stick) process. The abundances of N species are determined by $N - 1$ random points (the numerals below each line), which divide the line into N segments, whose lengths are indicated by numerals above the line. The segments are ranked by abundance and arrayed on a logarithmic scale of decreasing rank. The expected (average) distribution of abundances according to the broken stick model for a sample of 10 species and 100 individuals is indicated by the dashed line. Since the segments are generated by a random process, any number of distributions of segment length are possible.

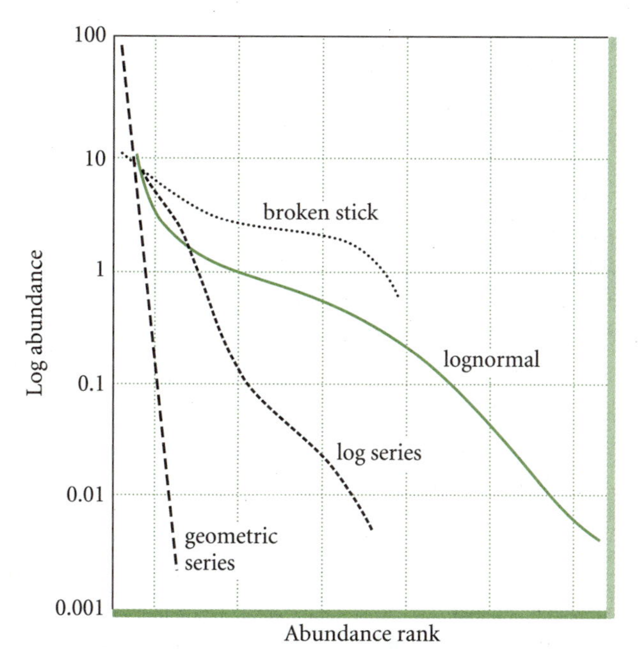

FIGURE 27-4 Relationship between species abundance (log scale) and abundance rank (greater abundance means greater rank) for the geometric series, random niche (broken stick), and lognormal models of species abundance. (*From Gotelli and Graves 1993, Magurran 1988.*)

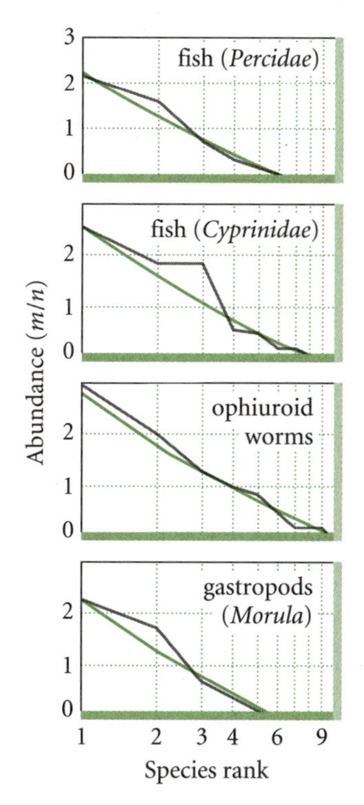

FIGURE 27-5 Abundances, in units of *m* individuals divided by *n* species, of species ordered according to abundance rank; the ranks are portrayed on a logarithmic scale. The green line is the prediction of the random niche (broken stick) model. Compare this figure with Figure 27-6 to see how changing the scale of the rank axis to a logarithmic scale changes the shape of the random niche model. (*From King 1964.*)

independent communities and evaluate both the average distribution of the ranked abundances and their variation among samples, an activity that is nearly impossible because replicates of natural communities are not likely. Recent evaluations of the niche preemption model suggest that it is most representative of small communities (<20 species) of animals, but has little utility in plant communities (Wilson 1991).

The geometric series and random niche models represent very simple processes of dividing up resources among relatively few species. Newly arriving species take either a set amount (k) or a random amount of the resources available. In these models, the abundance of the species is proportional to the amount of resources that it obtains. Such models are likely to represent only very simple communities because the abundance of a particular species reflects the balance between a large number of factors and processes, variations in each of which may result in small increments or decrements in abundance (May 1975a). A simple way to think about how many factors can affect the size of one population is to think of the sum of the effects of all the factors that are important to the species. These factors may be such things as the amount of various types of food, the conditions of the soil, the presence of predators, and so on. Statisticians have shown that the sums of many independent factors with small effects tend to assume a normal distribution, the familiar bell-shaped curve (see Chapter 2). If the abundance of each species is related to the sum of the effects of many factors on it, then we might expect that species abundances would also have a bell-shaped distribution.

In 1948, Frank Preston published a seminal paper entitled "The Commonness, and Rarity, of Species," in which he characterized the distribution of species abundances by a lognormal curve. Preston assigned species to classes of abundance based on the following scale of numbers of individuals per species: 1–2 individuals, 2–4 individuals, 4–8 individuals, 8–16 individuals, and so on. Each class was twice as big as the one before it, so Preston called the classes "octaves." (In the musical scale, the vibration frequency of each note is twice that of the note one octave lower.) This procedure results in a logarithmic scale. To see this, consider the upper numbers of Preston's classes (2, 4, 8, 16). These numbers represent a series in which each member of the series is 2^y, where y is the position in the series. Thus, $2 = 2^1$, $4 = 2^2$, $8 = 2^3$, $16 = 2^4$, and so on. Recall from algebra that if $x = a^y$, then $y = \log_a x$. Using this relationship on the series of upper class boundaries, we obtain $1 = \log_2 2$, $2 = \log_2 4$, $3 = \log_2 8$, and $4 = \log_2 16$. The difference between each step in the log-transformed series is a constant, 1. Thus, on the $\log_2$ scale, we may plot the classes evenly spaced on a line. A large sample of individuals from a community will contain individuals from a number of different species. When the number of individuals in each species, the frequency, is plotted against the logarithmic abundance classes, a normal distribution results. This distribution is called the **lognormal distribution.**

Figure 27-6 shows the relative abundances of species of moths attracted to light traps near Orono, Maine. (You know from being around porch lights or street lights at night that moths and other insects are attracted to the light. A light trap is simply a device that uses light to attract moths to an apparatus that traps them.) Four different scales (A through D) are shown on the horizontal axis: A, the number of individuals of each species; B, $\log_2$ of that number; C, the octave number, shown as the midpoint between the lower and upper boundary of the octave, and D, a rescaled octave number (see below). Thus, the third octave corresponds to between 4 and 8 individuals, or $\log_2 4 = 2$ and $\log_2 8 = 3$ individuals on a log scale, and the eighth octave corresponds to between 128 and 256 individuals, or $\log_2 128 = 7$ and $\log_2 256 = 8$ individuals on a log scale. The vertical axis shows the number of species with the corresponding abundance. The data show that there are about 50 species that have between 4 and 8 individuals (octave 3), about 25 that have between 64 and 128 individuals (octave 7), and very few species that have more than 1,000 individuals (octave > 11).

We may generalize this pattern by standardizing the abundance values in the following way. Let R be the values of a new octave scale on which the most common, or modal, abundance class—in the case of the moth data, octave 3 (between 4 and 8 individuals)—is assigned the value $R = 0$. Rescale the octave numbers to correspond to the modal octave, thus creating scale D of Figure 27-6. Octave 2 becomes $R = -1$ on the new scale because it is the octave to the immediate left of $R = 0$ (formerly octave 3). Octave 4 becomes $R = 1$ because it is to the immediate right of $R = 0$. Using this scale, the general frequency pattern is given by the equation for the normal distribution,

$$n_R = n_0 e^{-(1/2)(R/s)^2}$$

in which n_R is the number of species whose abundance is R octaves greater or less than the modal abundance of species within the community, n_0 is the modal number of species (that is, the number in the most abundant class), and s (standard deviation) is a measure of dispersion or breadth of the normal curve (Figure 27-7). The scale shown at the top of Figure 27-7 is the scale of R, with the octaves adjusted to the modal octave (scale D in Figure 27-6), and the scale at the bottom of the figure is the number of individuals per species (scale A in Figure 27-6). Since either of these can be represented as a log scale (scale B in Figure 27-6), this distribution is a lognormal distribution.

You will notice that in the moth abundance data shown in Figure 27-6, there are no species with fewer than one individual in the sample. Some species are too rare to be represented by one or more individuals in a sample of a particular size. These species fall below the **veil line** of the distribution, and only

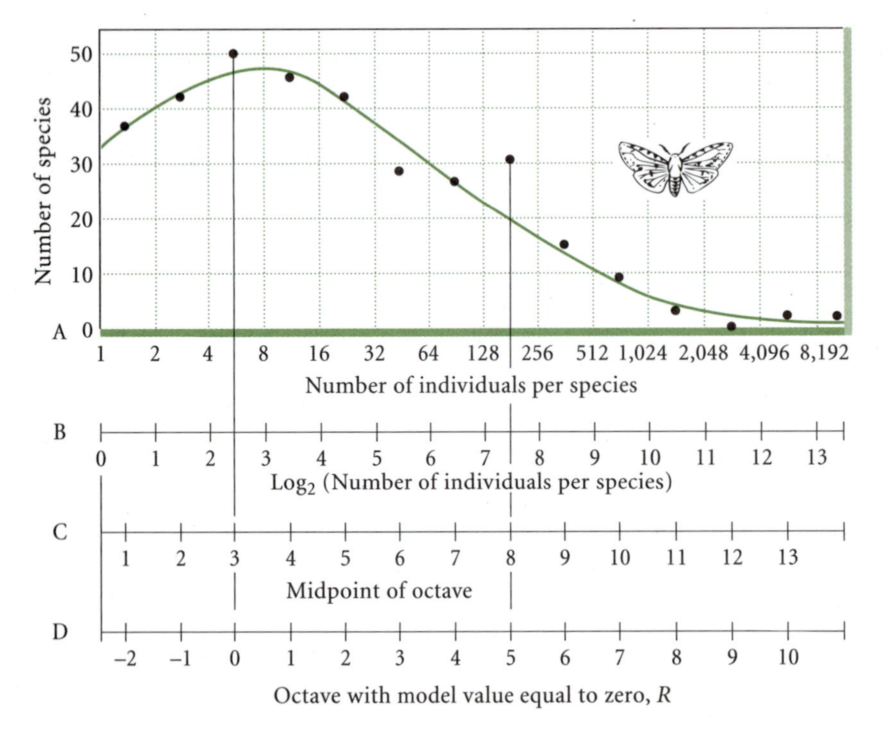

FIGURE 27-6 Relative abundances of species of moths attracted to light traps near Orono, Maine, plotted against (A) the number of individuals of each species (from 1 to >8,192), (B) $\log_2$ of the number of individuals (from 0 to 13), (C) the octave number, shown as the midpoint between the lower and upper boundary of the octave, and (D) an octave scale with the modal value as zero. (*After Preston 1948.*)

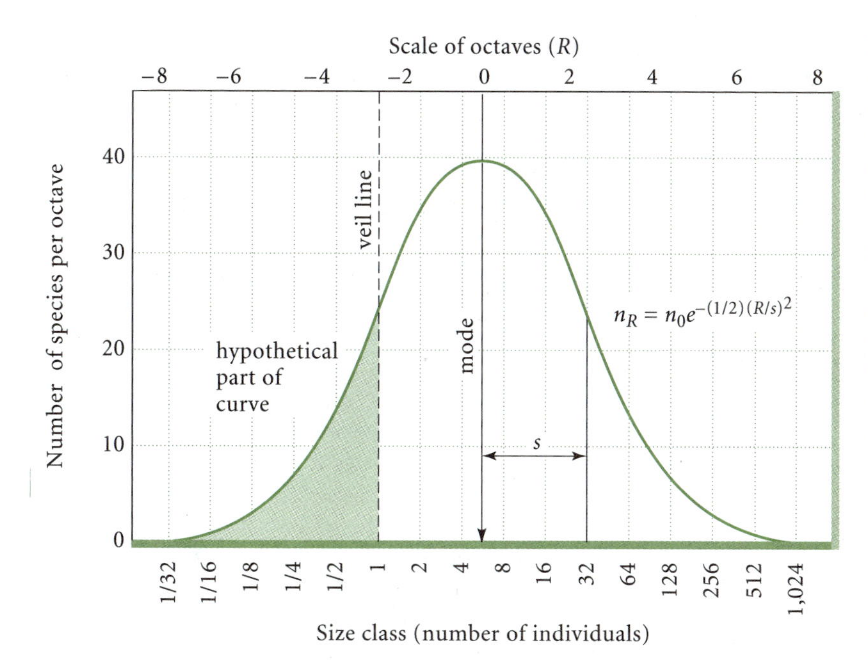

FIGURE 27-7 Lognormal distribution of species abundance. The top scale corresponds to scale D in Figure 27-6, with octaves adjusted to the modal octave, and the bottom scale corresponds to scale A in Figure 27-6, the number of individuals. The part of the curve to the left of the "veil line" (shaded green), which corresponds to species with less than one individual in the sample, and thus not represented, is hypothetical. One standard deviation (s) on both sides of the mode includes about two-thirds of all the species in the sample. (*After Preston 1948.*)

increasing the sample size can reveal their presence. The horizontal axis of Figure 27-7 has been extended to include size classes below the veil line in order to show the symmetry of the lognormal distribution.

A useful feature of Preston's lognormal model is that it takes sample sizes into account. One can predict the total number of species (N) in a community, including those not represented in the sample, knowing only the number of species in the modal abundance class (n_0) and the breadth of the lognormal distribution (s), by the equation

$$N = n_0\sqrt{2\pi s^2} = 2.5sn_0.$$

For the sample of moths graphed in Figure 27-6, $n_0 = 48$ and $s = 3.4$ octaves. Therefore $N = 2.5(3.4)(48) = 408$ species. The actual sample of over 50,000 specimens contained only 349 species, 86% of the number theoretically present in the sample area if abundances were distributed lognormally. In practice, confidence limits on estimates of N are so broad that the estimate is of limited value. When the sample is small enough, the veil line may fall close to, or even to the right of, the mode, and the revealed lognormal distribution then resembles the geometric series as well as the random niche model. In a sense, the lognormal model can be considered a more general statistical description of species abundances.

The dispersion (s) of lognormal curves fitted to large samples is remarkably similar for various groups of organisms. Preston (1948) obtained values of 2.3 for birds and 3.1–4.7 for moths. Patrick et al. (1954) obtained values of 2.8–4.7 for diatoms. May (1975) suggested that this range of values is to be expected from statistical sampling properties and is independent of the influence of the environment on the community. But, as MacArthur (1969) pointed out, dispersion values do vary with environment. Working with censuses of forest birds, MacArthur obtained values of $s = 0.98$ for lowland tropical localities, 1.36 for temperate localities, and 1.97 for islands. MacArthur's data indicate that there are greater discrepancies between abundances of species on islands than in similarly sized areas in temperate and, especially, tropical mainland areas. Patrick (1963) found similar variations in the dispersion of the lognormal curve fitted to diatom samples from streams with different water conditions in the eastern United States.

The models presented above were developed from mathematical theory and then applied to communities for comparative purposes. Another way to examine community structure is to develop mathematical expressions that reflect actual measurements of features of real communities. Such expressions are often called indices. Ecologists have developed a number of indices of community structure, as we shall see in the next section.

27.4 Diversity indices incorporate species richness and species abundance.

Communities differ in their numbers of species (species richness) and in the relative abundances of those species, a feature that is referred to as **species diversity.** Ecologists have sought mathematical expressions to measure species diversity, thus providing a means of objective comparison of community

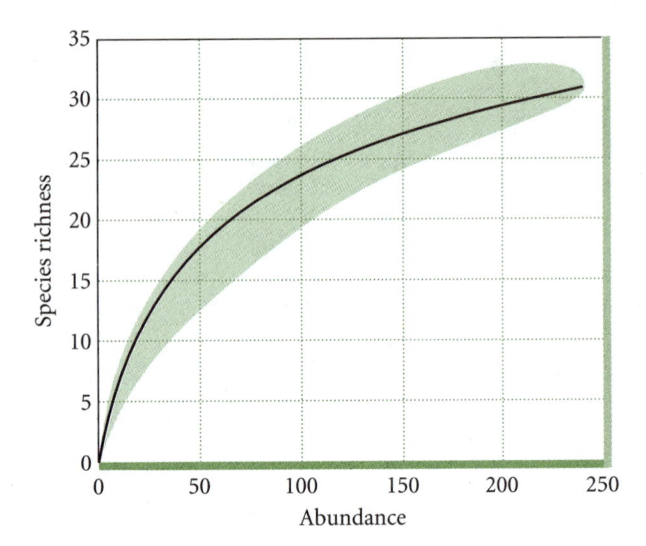

FIGURE 27-8 A rarefaction curve for a carabid beetle assemblage of young pine habitat. (*After Gotelli and Graves 1993; data from Niemelä et al. 1988.*)

structures. However, in so doing, they are faced with two practical problems. First, when an ecologist samples a community to determine the number of species present, the total number of species included in the sample varies with sample size, because as more individuals are sampled, the probability of encountering very rare species increases (see the discussion of the lognormal distribution above). Suppose, for example, that an ecologist samples spiders in a pine forest and a hardwood forest. She collects twice as many individual spiders in the hardwood forest as in the pine forest, and discovers that more species are represented in the hardwood forest. Is the higher species richness in the hardwood forest a result of the larger number of individuals collected, or is there a real difference in species richness between the two habitats? One way to overcome this problem is to transform counts of numbers of species into the rate at which species are added as the sample size is increased. Various methods by which this transformation may be accomplished have been presented elsewhere (Magurran 1988).

Another way to overcome the problem is through the application of sampling techniques that use a procedure called **rarefaction** (Sanders 1968, Simberloff 1976). Rarefaction works in the following way. Suppose a large sample of N individuals representing S species is obtained from a particular habitat. To perform a rarefaction, one extracts subsamples from the large sample and calculates the expected species richness of the subsamples based on the abundance distribution of the sample. The process is repeated for subsamples of different sizes. The smallest subsample possible, of course, is one individual, and that subsample will represent a single species. The largest "subsample" is the total sample, which contains N individuals and S species. A rarefaction curve shows the change in the expected value of species richness with subsample size (Figure 27-8).

A schematic representation of the steps in comparing the species richness of two communities through rarefaction is presented in Figure 27-9. Species abundance distributions are obtained from each community through sampling. Subsamples are obtained from these distributions to produce the rarefaction curves. The figure illustrates the utility of this technique. It is difficult to compare the species abundance distributions directly because they are of different shapes. The rarefaction curves, however, may be compared on a point-by-point basis. Among the assumptions of the technique are that the subsamples are not too small, that the spatial distribution of the

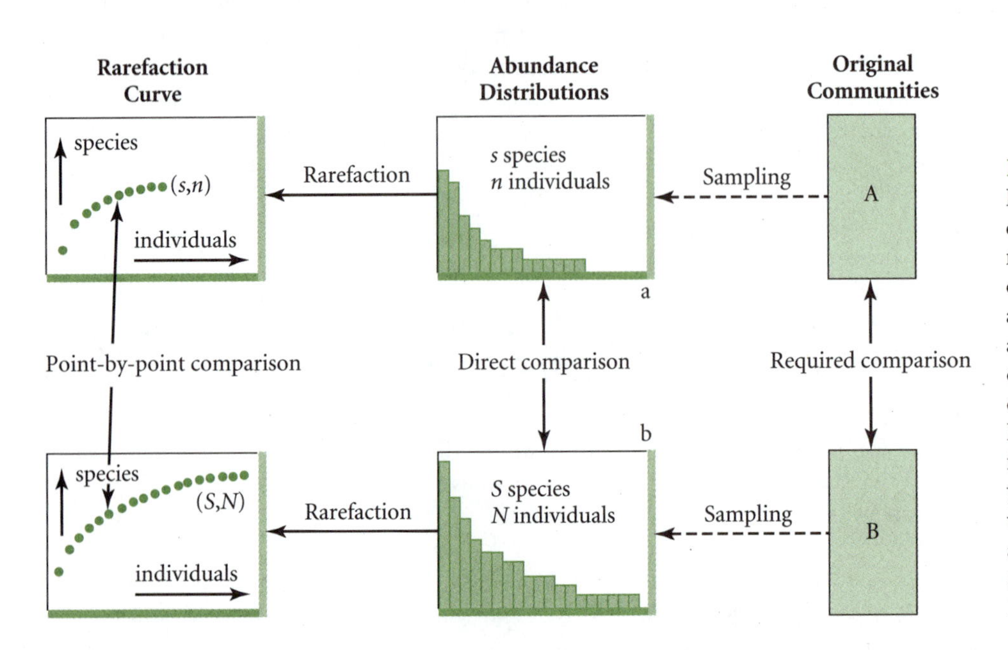

FIGURE 27-9 Rarefaction is used to compare the species richness of two communities. Species abundance distributions are obtained from community A and community B by sampling. Repeated subsampling from the two distributions yields rarefaction curves, which may be compared on a point-by-point basis. (*After Gotelli and Graves 1993; from Tipper 1979.*)

organisms in the community is not clumped, and that the same sampling technique is used for all subsamples. If these assumptions are met, and an adequate range of subsample sizes is obtained, the species richness of the two communities can be compared directly for a particular subsample size. In the figure, community B appears to have higher species richness than community A. The rarefaction curves of bird species in five different forest habitats are presented in Figure 27-10 as an example of the use of this method.

A second difficulty in the development of species diversity indices is that not all species should count equally toward an estimate of species diversity because their functional roles in the community vary, to some degree, in proportion to their overall abundances. Ecologists have tackled the second problem by formulating diversity indices to which the contribution of each species is in some way weighted by its relative abundance (Pielou 1966, 1977, Whittaker 1972, May 1975). Two such indices have been widely used in ecology: **Simpson's index** (Simpson 1949), and the **Shannon-Weaver index** (Shannon and Weaver 1949), made popular by Margalef (1958) and MacArthur (1955, 1957). In both cases, the indices are calculated from the proportions (p_i) of the species (i) in the total sample of individuals. Simpson's index is

$$D = 1/\Sigma p_i^2.$$

For any particular number of species in a sample (S), the value of D can vary from 1 to S, depending on the evenness of species abundances. If a sample contains, for example, five species that are equally abundant, each p_i is 0.20. Therefore, each $p_i^2 = 0.04$, the sum of the p_i^2s is 0.20, and the reciprocal of the sum is

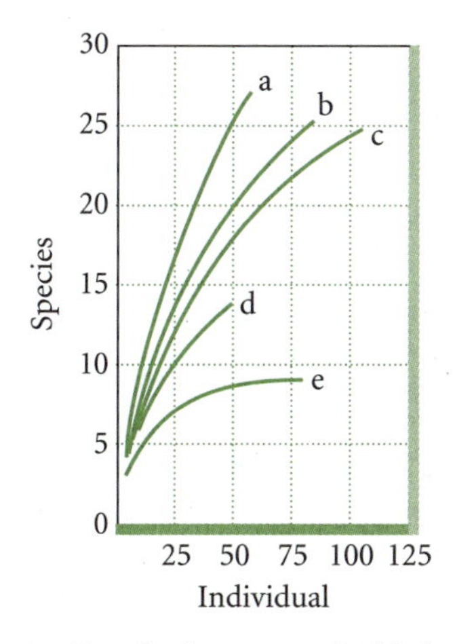

FIGURE 27-10 Rarefaction curves for bird communities from five different habitats: (a) maple-pine-oak second-growth forest, (b) tulip tree-maple-oak forest, (c) cottonwood floodplain forest, (d) jack pine, and (e) birch forest. (*From Gotelli and Graves 1996; data from Niemelä et al. 1988.*)

5, the number of species in the sample. Similar calculations for some hypothetical communities are presented in Table 27-2, in which you can see that rarer species contribute less to the value of the diversity index than do common species.

The Shannon-Weaver index, developed from information theory, is calculated by the equation

$$H = -\Sigma p_i \log_e p_i,$$

and, like Simpson's index, gives less weight to rare species than to common ones. Because H is roughly proportional to the logarithm of the number of species, it is sometimes preferable to express the index as e^H, which is proportional to the number of species. The values in Table 27-2 are presented in this way so that they may be compared with Simpson's index.

The results of most studies are relative insensitive to which index of diversity is applied, or to whether an index of any kind is used in place of a simple count of the species present. This may be attributed to the fact that communities have characteristic patterns of relative abundances among their member species. As a result, each diversity index tends to bear a consistent relationship to all other indices as well as to the number of species in the community. Moreover, the species richness is typically related to the various diversity indices in some predicable way. Thus, if species richness is normalized from sample size, it alone is a good measure for comparisons among communities (Schluter and Ricklefs 1993).

The diversity indices discussed above are based on taxonomy; that is, the species, or some other higher taxonomic unit, is the basic unit of comparison. Thus, these indices measure **taxonomic diversity.** When taxonomic diversity indices are employed with higher-level taxonomic categories, such as genera or families, we may learn something about the origin of diversity in the community (Ricklefs 1987, Schluter and Ricklefs 1993). But species play different roles within the community that are reflected in differences in diet, habitat preference, activity period, and a myriad of other aspects of behavior and ecology. And, so, there is within the community not only taxonomic diversity, but **ecological diversity** as well. Studies of birds and mammals have shown high levels of concordance between the two kinds of diversity (Ricklefs and Miles 1993), but this is not always the case (Cousins 1991). For example, in their study of the insect herbivores of bracken ferns (*Pteridium aquilinum*), Lawton et al. (1993) found that species diversity was more a result of random colonizations over evolutionary time than of contemporary ecological interactions within the community. In the case of the herbivores of bracken ferns, ecological diversity and taxonomic diversity are little related.

TABLE 27-2 Comparison of the Simpson and Shannon-Weaver diversity indices for three communities having different relative abundance and species richness values

	n	p_i	p_i^2	$-p_i \ln p_i$
Community A (species richness = 5)				
Species 1	10	0.50	0.25	0.35
Species 2	10	0.50	0.25	0.35
Species 3	0	0.00	0.00	0.00
Species 4	0	0.00	0.00	0.00
Species 5	0	0.00	0.00	0.00
Simpson's index, D	2			
Shannon-Weaver index, H	0.69			
Community B (species richness = 5, high evenness)				
Species 1	4	0.20	0.04	0.32
Species 2	4	0.20	0.04	0.32
Species 3	4	0.20	0.04	0.32
Species 4	4	0.20	0.04	0.32
Species 5	4	0.20	0.04	0.32
Simpson's index, D	5			
Shannon-Weaver index, H	1.61			
Community C (species richness = 5, low evenness)				
Species 1	18	0.80	0.64	0.18
Species 2	1	0.05	0.00	0.15
Species 3	1	0.05	0.00	0.15
Species 4	1	0.05	0.00	0.15
Species 5	1	0.05	0.00	0.15
Simpson's index, D	1.54			
Shannon-Weaver index, H	0.78			

If we sample more intensively in an area, our sample may include more species, since we have a better chance of finding individuals of rare species. The number of species in our sample will also change if we sample a larger area, a property of communities that we take up next.

27.5 The number of species encountered increases in direct proportion to the area sampled.

As a rule, more species occur within large areas than within small areas. The botanist Olaf Arrhenius (1921) first formalized this **species-area relationship.** Since then, it has been common practice to portray the relationship between species richness (S) and area (A) with a power function of the form

$$S = cA^z,$$

where c and z are constants fitted to the data (Preston 1960, 1962, MacArthur and Wilson 1967; Figure 27-11). Graphic portrayals of species-area relationships plot the logarithm of species number against the logarithm of area (Figure 27-12). After log transformation, the species-area relationship becomes

$$\log S = \log c + z \log A,$$

which is the equation for a straight line with slope z. (It should be noted that the practice in some studies, particularly in studies of plants, is to relate the number of species to the logarithm of area; that is, $S = \log A$ [for example, Gleason 1922, Hopkins 1955]; empirical support can be found for both relationships [Schoener 1974].)

Analysis of species-area relationships among many groups of organisms during the 1960s and 1970s revealed that most values of z fell within the range 0.20–0.35 (May 1975, Connor and McCoy 1979). The persistent consistency of z suggested several

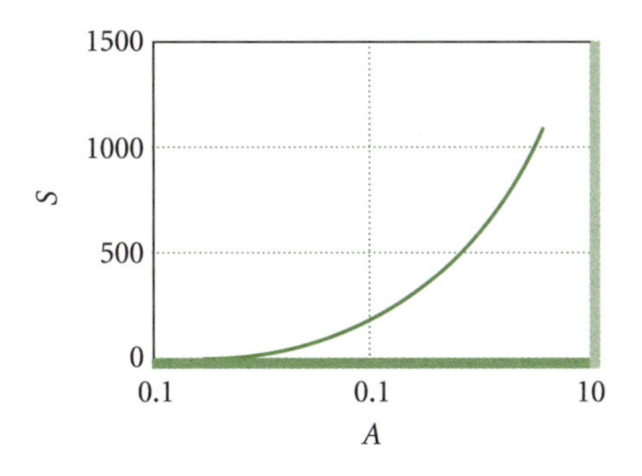

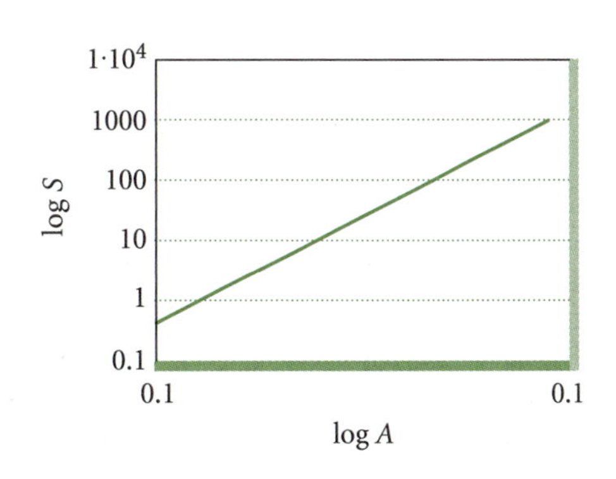

FIGURE 27-11 The power curve $S = cA^z$, where $c = 1$ and $z = 1.2$, and the linear transformation, $\log S = \log c + z \log A$.

possibilities. First, observed z values might be a simple statistical consequence of the lognormal distribution of species abundances. As one increases the area of a sample, one usually increases the number of individuals included within it, and as the sample size increases, the veil line is displaced to the left, exposing more and more species (May 1975). It is also important to appreciate that the parameter z has different interpretations depending on whether the power function or the linear function (log transformation of the power function; see Figure 27-11) is used to fit the data (Loehle 1990). Sugihara (1980) pointed out, however, that a wide range of values of z is possible within the context of a lognormal sampling scheme, and suggested that values of z provide information about the processes responsible for the generation of community structure (see Wright 1988). Empirical studies have demonstrated z values differing greatly from 0.25 in association with differences in the biological attributes of samples (Ricklefs and Cox 1972). Indeed, z values obtained for continental areas of different sizes tend to be lower than those for series of islands within a comparable size range.

Mechanisms of Species-Area Relationships

Owing to the historical preeminence of MacArthur and Wilson's equilibrium model of species-area relationships on islands (see Chapter 29), the parameter z of the species-area equation has often been interpreted as a measure of the degree of isolation of island communities. In this conception, species richness is viewed as a balance between colonization from a mainland source community and extinction of island species. High values of z correspond to a high degree of isolation (Figure 27-13). This idea is often referred to as the **equilibrium hypothesis** of species-area relationships.

Species-area relationships may arise by mechanisms other than an equilibrium of colonization and extinction. At least three other hypotheses have been suggested (Connor and McCoy 1979, Gotelli and

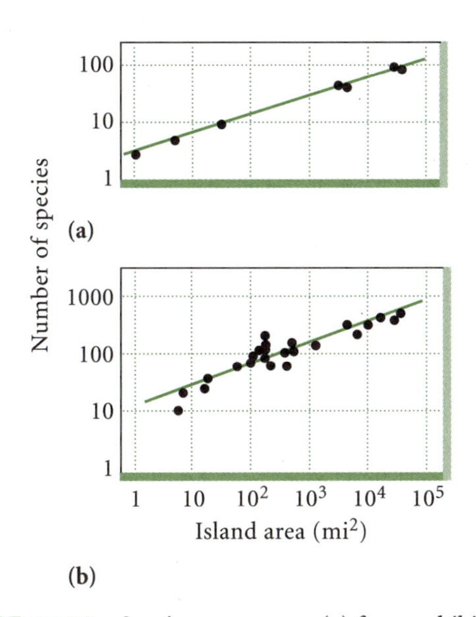

FIGURE 27-12 Species-area curves (a) for amphibians and reptiles in the West Indies and (b) for birds in the Sunda Islands, Malaysia. (*From MacArthur and Wilson 1963, 1967.*)

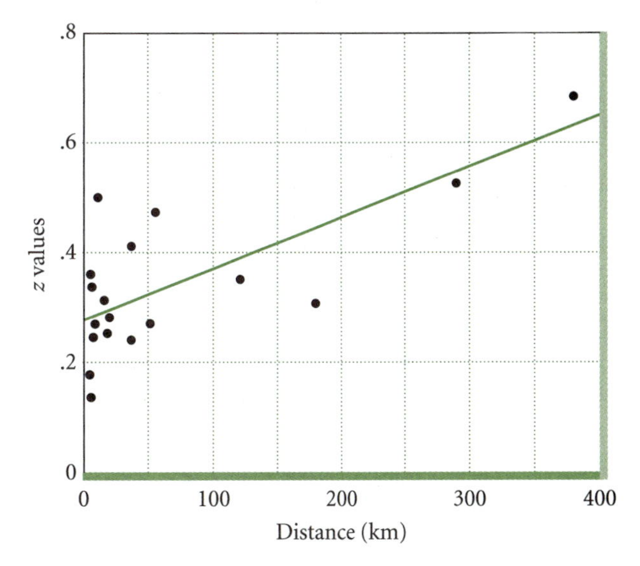

FIGURE 27-13 Higher values of z are associated with greater isolation (distance from a source of colonists) for mammals in a number of habitat types. (*From Lomolino 1984.*)

Graves 1996). The **disturbance hypothesis** holds that populations on small islands (or small, isolated inland areas) are more susceptible to disturbance. If this hypothesis is correct, then one would expect the number of species to decrease as area decreases. A second idea, referred to as the **habitat diversity hypothesis,** recognizes that, in general, large areas possess a more diverse array of habitat types than smaller areas, and proposes that, assuming that habitat specialization is important, larger areas should therefore contain more species (Williams 1943). Finally, larger areas may contain more species simply because they represent bigger "targets" for immigration, an idea referred to as the **passive sampling hypothesis** (Colman et al. 1982). In addition to these mechanisms, the geographic distribution of species plays a role in species-area relationships.

Where a flora or fauna is perfectly known (that is, all species have been sampled), the sampling properties of the lognormal distribution cannot be held accountable for any relationship between species and area; no species are hiding behind the veil line. In spite of the fact that a new bird species was discovered on the island of Puerto Rico during the 1960s, we possess near-perfect knowledge of the land bird fauna of the West Indies, among which there is a pronounced species-area relationship, with a slope of about $z = 0.24$ (Figure 27-14). In this case, differences in species diversity between large and small islands must express differences in their intrinsic qualities. Likely candidates are habitat heterogeneity, which undoubtedly increases with the size and consequent topographic heterogeneity of the island, and size per se (Williamson 1981). Larger islands are probably better "targets" for potential immigrants from mainland sources of colonization. In addition, the larger populations on larger islands probably persist longer, owing to greater genetic diversity, broader distributions over area and habitat, and numbers large enough to avoid chance extinction.

Larger islands usually are more heterogeneous than small ones. High mountains create elevational

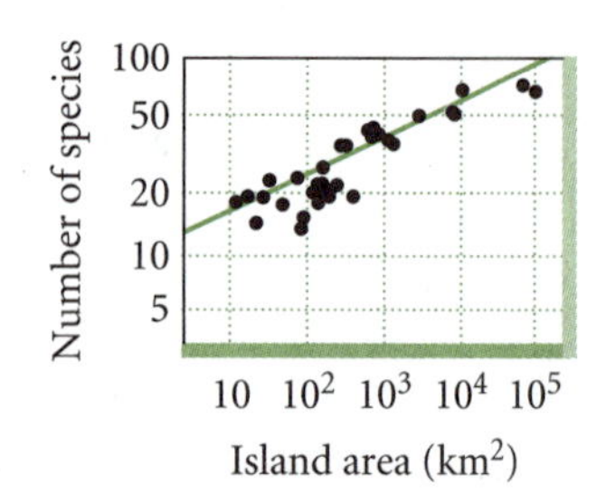

FIGURE 27-14 Species-area curve for land birds of the West Indies, including both the Greater and Lesser Antilles. The slope of the line is about $z = 0.24$. (*After Ricklefs and Cox 1972.*)

gradients associated with temperature and moisture. In the West Indies, large islands such as Puerto Rico and Guadaloupe have rain forest, desert, and the whole range of habitats in between. But there are also sufficient differences among the islands in elevation, habitat diversity, and distance from the mainland and nearby islands to statistically separate the contributions of each of these factors to the species-area relationship. Hamilton et al. (1963) were the first to attempt such an analysis, using data on the numbers of plant species on islands in the Galápagos archipelago off Ecuador. They found that elevation and degree of isolation predicted species richness better than did island area. Hamilton's studies and those of other researchers (e.g., Simpson 1974, Connor and Simberloff 1978) employed multiple regression techniques in an attempt to tease apart the effects of the various factors on the species-area relationship. In this approach, a linear combination of various factors that might affect species abundance, such as area, habitat heterogeneity, elevation, distance to source communities, and the like (independent variables), is constructed as a predictor of the number of plant or animal species (dependent variable). Statistical procedures are then employed to determine the relative strengths of the effects of each of the terms of the linear model.

The various islands of the Lesser Antilles host from eleven to forty-two species of resident land birds (see Figure 27-14). A statistical analysis comparing number of species with island area, altitude, distance from the mainland, and distance to the nearest island showed that each of these independent variables makes a unique contribution to variation in species richness. As one would expect, species richness increases with area and elevation and decreases with degree of isolation. Barbados is larger than St. Vincent, but, being low and isolated, harbors only half as many species of birds. Such analyses suggest that area per se may be an important determinant of diversity, but it is still possible that island size is associated with variation in habitat diversity over and above that attributable to elevation.

A positive relationship between the geographic range of a species and its local or regional abundance is often observed (Hanski et al. 1993). Does this observation help explain species-area relationships? Is the manner in which species are accumulated with increasing area dependent, in part, on the distributions of those species? Clearly, species distributions are restricted with respect to habitat (Cody 1985), and as larger areas include more habitats, more species will be included in the total sample. But larger areas of uniform habitat tend to have more species than smaller areas of different, but equally uniform, habitat.

A classic example of this relationship is found in the numbers of insects collected from species of trees with different geographic ranges and abundances within those ranges. Although Karban and Ricklefs (1983) showed that the species richness of insect herbivore faunas was independent of the local abundance of the host, Southwood and his colleagues (Southwood 1961, Southwood et al. 1982) have demonstrated that insect herbivore faunas in Britain increase in diversity with increasing geographic range of the host. This relationship could result from three causes. First, more complete sampling from a larger area may expose more of the lognormal curve of species abundances (e.g., Kuris et al. 1980, Lawton et al. 1981, Rey et al. 1981). Second, the mode (n_0) or breadth (s), or both, of the lognormal curve may be larger for hosts with a greater geographic extent. Perhaps hosts with larger ranges accumulate more herbivore species because they are better targets for ecological and evolutionary colonization (Janzen 1968, 1973), or perhaps the larger herbivore populations on more widespread hosts are demographically more stable and have lower probabilities of extinction. Third, larger areas contain larger numbers of habitat subdivisions, each with its own, partially independent, lognormal distribution of species abundance.

Southwood et al. (1982) and Strong et al. (1984) suggested that the British insect fauna of trees has been so thoroughly sampled that it is essentially perfectly known. If this is true, we can rule out the first factor as a cause of the species-area relationship. Distinguishing the second and third possibilities requires systematic local sampling from widely distributed and narrowed distributed tree species. Stevens (1983) accomplished this in a study of wood-boring beetles (Scolytidae) in the eastern deciduous forests of the United States. He set out measured lengths of cut wood from the tree species of interest in five localities, placing them so that each tree species was sampled to exactly the same degree in each of the forest areas in which it was present. He found that, although the number of wood-boring beetle species attracted to each tree species in all five localities increased with the geographic distribution of the host species, and hence the number of localities in which it was sampled, local diversity was independent of the host range. That is, the number of beetle species collected from logs of a particular tree species in a single locality—say, in North Carolina—was unrelated to the geographic range of the host. This particular result favors the third alternative, and indicates that there is a geographic turnover of herbivore species within the range of a single host, perhaps associated with variation in climate, parasites, or the resource quality of the host.

Models of relative abundance, species diversity, and species-area relationships provide valuable information about community structure. In particular, they are important tools for distinguishing between local and regional patterns of community composition and for identifying large-scale patterns of community change. However, for the most part, they do not provide much information about the mechanisms that give rise to local community structure, or about the dynamics of that structure, which are shaped by population interactions and phylogenetic relationships. To obtain this type of information about communities, ecologists often turn to the analysis of food webs.

27.6 Food web analysis is used to reveal community structure.

A **food web** represents the various ways in which energy passes through populations in communities. In effect, it shows who eats whom. Food webs are composed of **food chains,** which represent the passage of energy from a primary producer through a series of consumers at progressively higher trophic levels. We first introduced the concept of the food web in Chapter 10, where we explained that the term *food web* arose from Elton's "food-cycles." We used the concept in Chapter 12 in our discussion of the movement of materials in terrestrial and aquatic ecosystems. Our representation of trophic cascades in Chapter 13 utilized a simple food web diagram (see Figure 13-15). We also presented simple food web diagrams in Chapter 20 (see Figures 20-1, 20-21, and 20-22) in order to illustrate some of the properties of consumer-resource dynamics. The analysis of food webs is an important means of understanding the patterns and dynamics of communities, and is a very active field of contemporary ecological study (Briand 1983, Briand and Cohen 1984, 1987, Cohen 1989, Cohen et al. 1990, Pimm 1991, Polis 1991, Polis and Winemiller 1996, Schoenly et al. 1991, Schmitz 1997).

The first food web analyses were quite descriptive. In the early twentieth century, ecologists began portraying food webs with diagrams in which arrows connected species in a community according to their feeding relationships. These diagrams were often complex and difficult to compare quantitatively between communities. To some ecologists, such diagrams merely emphasized the overwhelming complexity of natural systems and the need to simplify their structure by dividing them into trophic groups. Description gave way to a more analytical approach in the mid-1950s with the suggestion that increasing complexity of community organization leads to increasing dynamic stability (MacArthur 1955). The reasoning was simple: when predators have alternative prey, their own numbers depend less on fluctuations in the numbers of a particular species. Where energy can take multiple routes through a system, disruption of one pathway merely shunts more energy through another, and the overall flow continues uninterrupted.

The analytical approach linked community stability to species diversity and food web complexity, and it stimulated a flurry of theoretical, comparative, and experimental work. This work has taken two approaches to the use of food webs in the study of community structure. The first approach involves the study of the properties of food web diagrams with the goal of uncovering general patterns that suggest mechanisms of community stability. This is done both by the comparison of food webs from natural communities and by the use of simulation and mathematical modeling to study hypothetical food webs. This research has yielded much of the terminology now associated with food webs and has generated a body of food web theory that includes a number of hypotheses about community structure.

The second approach, which grew from early theoretical and experimental community studies, involves the dynamic analysis of food webs to determine not only the pattern of interactions among the populations in the community, but also the relative strengths of those interactions. Dynamic food web analysis also seeks to uncover interactions that are not obvious from simple food web diagrams, so-called indirect interactions (see below). It requires the careful merging of experimental and theoretical approaches. We shall consider these two approaches separately in the remaining sections of this chapter. But first, let us distinguish a number of different types of food webs.

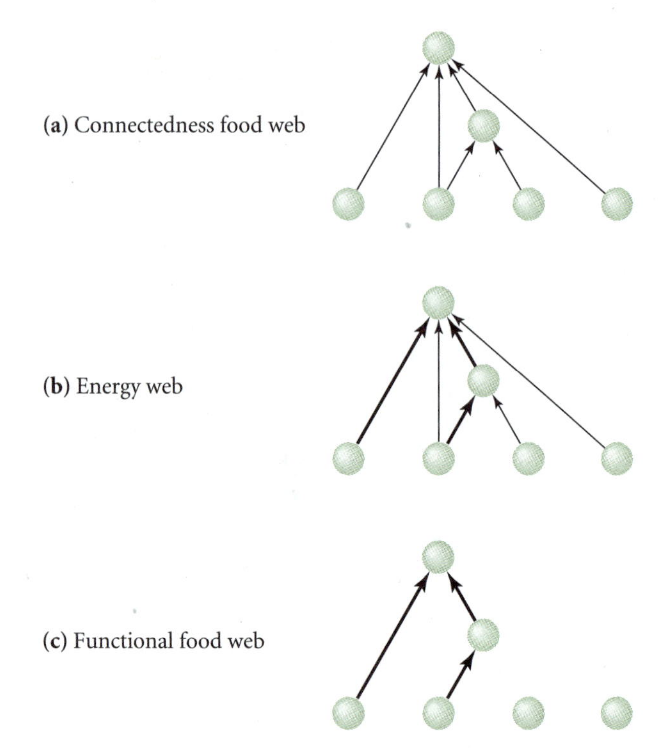

FIGURE 27-15 Three types of food webs. (a) Connectedness food webs display all of the trophic interactions in the community without reference to their strengths. (b) Energy flow webs quantify the connections between populations by the flux of energy between consumer and resource. The heavier the arrow, the greater the energy flux. (c) Functional food webs show those interactions most important to the structure of the community.

Types of Food Webs

Paine (1980) distinguished three different concepts of food webs (Figure 27-15). **Connectedness webs,** which are now more often called **topological food webs,** emphasize feeding relationships among organisms, portrayed as links in the web. Such webs depict only the presence or absence of a trophic interaction, not the strength of the interaction, nor do they show any change in trophic relationships that may occur because of the growth of individuals or changes in the age structure of the consumer or resource populations. For these reasons, topological food webs are sometimes referred to as **static food webs.** It is the analysis of topological food webs that is the focus of the first approach to food web analysis mentioned above. **Energy flow webs** (sometimes referred to simply as **flow webs** or, more descriptively, as **bioenergetic webs**) represent an ecosystem viewpoint, in which connections between populations are quantified by the flux of energy between a resource and its consumer. These are the types of webs that we discussed in Part 3 of this book, and they will not be considered further here. **Functional** or **interaction food webs** are those that identify the feeding relationships within the topological food web that are most important to community structure. These food webs depict the importance of each population in maintaining the integrity of a community as reflected in its influence on the growth rates of other populations. Such controlling roles, which only experiments can reveal, need not correspond to the amount of energy flowing through a particular link in the food web of an intact community. It is the development and analysis of interaction food webs that are the focus of dynamic food web studies.

27.7 Ideas about the influence of food web structure on community stability were developed from the analysis of topological food webs.

An important question about community structure is how differences in the structure of food webs affect the dynamics, stability, and persistence of communities. The issue of community stability and persistence is particularly important in conservation biology, in which the challenge is often to preserve whole communities in the face of human disturbance. Is a particular arrangement of feeding relationships among species intrinsically more stable than a different arrangement among the same number of species? How important is food web stability in shaping the structure of natural communities?

The comparison of food webs from many different communities and the theoretical analysis of the properties of topological food webs are parts of one approach that ecologists have used to address these questions. A number of ecologists have argued that some attributes of topological food web design are consistent with and dependent upon qualities that enhance the intrinsic dynamic stability of the community (Pimm and Lawton 1978, Pimm 1980, 1991, Lawton 1989, Cohen et al. 1990, Hall and Raffaelli 1993). Studies of the properties of topological food webs have resulted in a number of specific hypotheses about the nature of community structure and community stability. Understanding these generalizations requires knowledge of food web terminology.

Food Web Terminology

Food webs are composed of a number of **nodes,** each of which represents a single species or a group of species thought to be trophically similar—that is, to have similar prey and similar predators. The latter are sometimes called **trophospecies.** It is common for ecologists to use the word *species* synonymously with the word *node* regardless of whether there is just one species or more than one species involved. We shall follow that convention. The number of species (or nodes) is denoted *S*. Species occupying the lowest trophic level in the food web are called **basal species** (or basal nodes). Top predators and intermediate species are also often distinguished. Interactions between species are represented as **links,** which in food web diagrams are lines drawn from one node to another (see Figure 27-15). As we discussed in Chapter 20, interactions may negatively affect both interacting species (competition), positively affect one species and negatively affect the other (predation, parasitism), or positively affect both (mutualism). Links in food web diagrams are often labeled with a plus or a minus sign near a node to denote the nature of the effect. The average number of links per species is called the **linkage density** of the food web. The total number of links in the food web divided by the number of possible links is called the **connectance** of the food web, which is denoted *C*. The number of links from the top predator to a basal species is called a **chain length.** The maximum chain length is a measure of the number of trophic levels in the food web.

Food webs may reflect a number of special interspecific and intraspecific interactions (Figure 27-16). A **cycle** is a situation in which there is a reciprocal predatory relationship between two species; for example, species A eats species B, and species B eats species A. **Cannibalism** is a cycle in which a species preys upon itself. Cycles and cannibalism are probably fairly common in some communities, as, for example, in forest floor spider communities. Some species may feed on species that occupy more than one trophic level, a situation called **omnivory.** Omnivory is represented by the dashed line in Figure 27-16.

The ratio of the number of predator species to the number of prey species in the food web, and the proportion of species occupying lower trophic levels in comparison to the proportion at higher levels, are characteristics of food webs that are often explored. Additionally, **compartmentalization** in a food web is of interest. A compartmentalized food web is one in which there are several groups of species within which interactions are stronger than interactions between those groups.

Generalizations from Topological Food Web Studies

At least six important generalizations have emerged from the analysis of topological food webs (Pimm

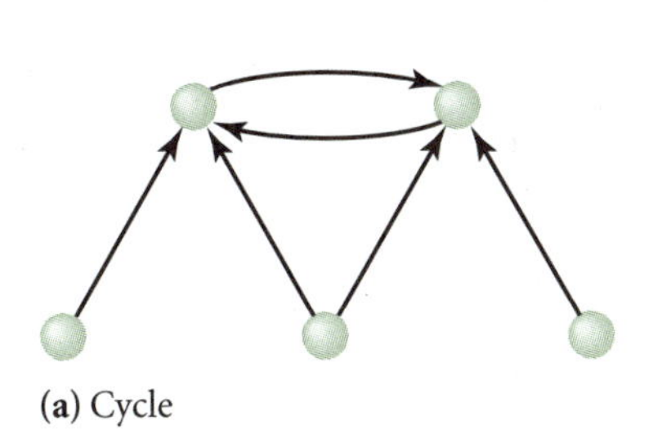
(a) Cycle

(b) Cannibalism

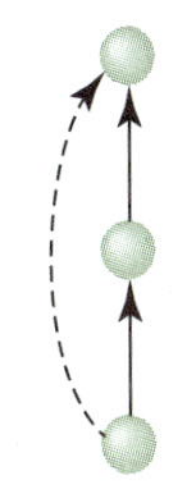
(c) Omnivory

FIGURE 27-16 Features of food webs. (a) A cycle occurs when there is a reciprocal predatory relationship between two species. (b) Cannibalism is a cycle in which individuals of a species prey upon other individuals of the same species. (c) Omnivory involves feeding at more than one trophic level.

et al. 1991, reviewed in Polis and Winemiller 1996). Most of these generalizations focus on changes in the structure of food webs as the number of species in them increases.

First, the number of trophic levels in food webs appears to be relatively small. Cohen et al. (1990) examined 113 food webs from marine, terrestrial, and freshwater communities and found that nearly all had a maximum chain length of five or fewer links and the most common maximum chain length was four. One suggestion about why the number of trophic levels may be limited in nature has to do with inefficiencies in energy transformations. The metabolic costs of biosynthesis and tissue maintenance by primary producers represent energy lost to respiration and thus unavailable to higher trophic levels (see Figure 10-1). Energy is lost between all subsequent trophic levels because ecological efficiencies are less than 100% (see Figure 10-8). The idea that energetic constraints limit the number of trophic levels in food webs is called the **energy flow hypothesis** (Hutchinson 1959). The energy flow hypothesis suggests that the rate of primary production will influence the number of trophic levels in the food web. Studies of food webs with vastly different levels of primary production have failed to show this (Briand and Cohen 1987, Schoenly et al. 1990, Pimm 1991). Thus, most ecologists reject the notion that the number of trophic levels is controlled primarily by energetic constraints.

Another idea, sometimes called the **dynamic stability hypothesis,** holds that more complex food webs—those with a relatively large number of trophic levels and greater species diversity—will not recover from disturbance as readily as food webs that are relatively short (few trophic levels), and thus will not be as stable as less complex webs (May 1972b, Pimm and Lawton 1977, Pimm 1991). But the response of communities to disturbance is dependent on factors other than diversity, such as the level of intensity of the disturbance and the growth rates of the populations in the community (DeAngelis 1975, Huston 1994), as we shall see in the next chapter.

A second generalization emerging from the analysis of topological food webs—a generalization that has come under some scrutiny—holds that as the number of species in the food web increases, the linkage density of the web remains relatively constant. Indeed, some food web analyses suggest that the number of interactions per species is about two, regardless of the number of species in the community (Cohen and Briand 1984, Cohen et al. 1990, Pimm 1991). This idea is sometimes called the **link-species scaling hypothesis.** It means that as the number of species increases, the connectance of the food web decreases, suggesting that there are relatively fewer interactions in larger food webs than in smaller food webs.

Martinez (1991a,b) questioned the idea of constant linkage density by pointing out that, among other things, it requires unreasonable assumptions about diet breadth (number of different types of diet items taken). He offered the following example: Suppose an insectivorous bird species occupies two different communities. Community A has twice as many insect species in it as community B. Under the assumption of constant linkage density, the bird would eat the same number of insect species in both communities. But most insectivorous birds are generalists that eat many different species of insects (large diet breadth). Martinez suggested that it is more reasonable to assume that the bird will eat more different kinds of insects in community A than in community B. His own analysis of 175 different food webs suggested that instead of decreasing as the number of species in the community increases, connectance remains constant. This idea is called the **constant connectance hypothesis.** It holds that the number of links increases approximately as the square of the number of species in the web. The insectivorous bird example, however, belies the complexity of diet adaptations in organisms. Highly specialized feeders may in fact show invariance in links among communities having different numbers of species. A productive avenue of investigation regarding this generalization about food webs would focus on physiological and evolutionary constraints on feeding (Schoener 1989, Martinez 1991).

The proportion of organisms in the top, intermediate, and basal trophic levels appears to be relatively constant among food webs (Briand and Cohen 1984, Jeffries and Lawton 1985, Cohen et al. 1996, Hall and Raffaelli 1993). This third generalization is shown in Figure 27-17 for 92 freshwater invertebrate food webs. Regardless of the number of species in the food web, the number of prey per predator species is roughly between two and three.

Figure 27-18 shows food webs with different levels of omnivory. Early analyses of food webs led to the generalization that omnivory (feeding at more than one trophic level) was uncommon in natural communities (Pimm 1982). Subsequent detailed studies of communities in a variety of aquatic and terrestrial habitats have suggested that this generalization does not hold. Polis (1991) found that about 78% of the species in a complex desert food web were omnivorous. Martinez (1991) found similar patterns for a number of aquatic food webs. Omnivory may be overlooked if food webs do not include detailed information about the feeding habits and life histories of the species in the community, or if there is substantial lumping of species into trophospecies. For example, because many species undergo dietary changes during their development, over their entire life span they are omnivores. This type of omnivory is indicated by **diet loops** in the food web, in

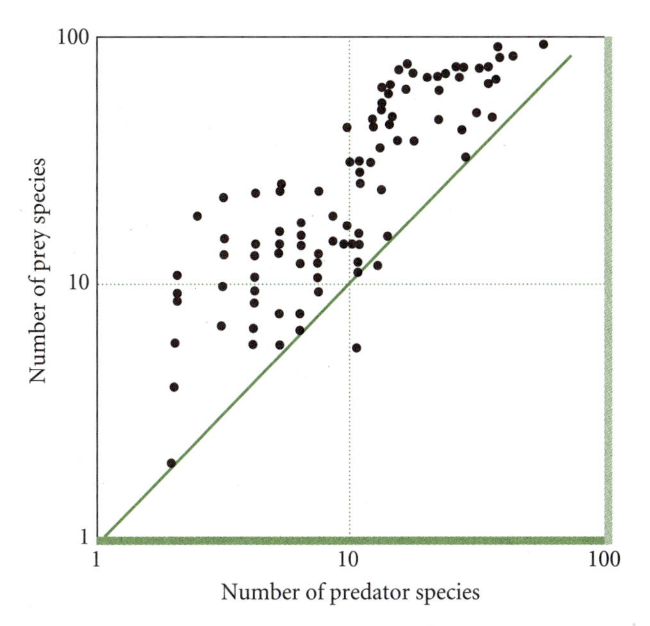

FIGURE 27-17 The relationship between the number of predator species and the number of prey species in 92 freshwater invertebrate food webs. As the number of species in the web (predator species + prey species) increases, the proportion of predators to prey remains relatively constant at 2:1 prey:predator in the smaller webs and 3.5:1 prey:predator in the larger webs. (*Data from Jeffries and Lawton 1985.*)

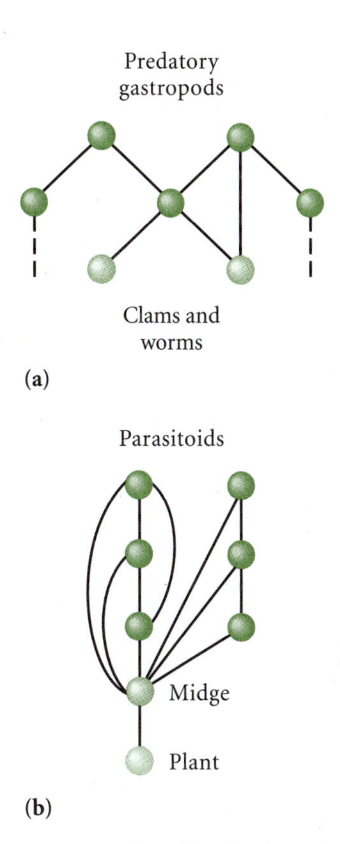

FIGURE 27-18 Examples of food webs with little and frequent omnivory. (a) A mudflat community containing intertidal gastropods, bivalves, and their prey. This food web has seven species and eight links. (b) The plant *Bacharis,* its insect herbivores, and their parasitoids. This food web has eight species and twelve links. Not all prey species are depicted. (*After Pimm 1982.*)

which adults of species A eat juveniles of species B and adults of species B eat juveniles of species A. Such loops are difficult to detect without careful study of the food web over an extended period of time (Pimm and Rice 1987, Polis 1991).

Compartmentalization is thought to be rare in food webs (Pimm and Lawton 1980). At the extreme, each compartment could be considered a separate community. Compartments might be established in association with habitat patches or distinct species within habitats. For example, in southern Canada, broad-leaved trees and needle-leaved trees each have distinct associations of lepidopteran herbivores, with few species or genera of moths and sawflies feeding on plants in both groups. In mixed forests, therefore, one might consider the faunas of broad-leaved trees and needle-leaved trees as separate communities. By way of contrast, moths feed widely among different species of broad-leaved trees, and although most are specialized to a greater or lesser degree, they seem not to fall into discrete subsets. Pimm (1991) has pointed out that resistance to interaction across compartments is weak. Consider what happens when a species requiring both plant and animal food (an omnivore) invades a compartmentalized community (Figure 27-19). The new species may find its plant and its animal food within a single food chain, where it will be both a competitor and a predator. Or, it may seek its plant food in another compartment, thereby assuming only a predatory role in one food chain and only the role of an herbivore in the other, and linking the two compartments. Pimm believes that the latter strategy is the more likely outcome of such an invasion, and thus, that compartments should be rare.

Problems with Generalizations from Topological Food Webs

Many ecologists have suggested that generalizations derived from the study of topological food webs are weak on methodological grounds (Polis 1991, Polis and Winemiller 1996). One of the chief criticisms is that many of the generalizations are drawn from comparative studies of published food webs (e.g., Cohen 1978, Pimm 1980, 1982, Briand 1983, Cohen et al. 1990), which represent studies that were conducted with different objectives, using different techniques, and which consequently have different levels of resolution. The number of trophic levels in a food web is dependent on how that food web is constructed. Webs from studies in which there is little lumping (few

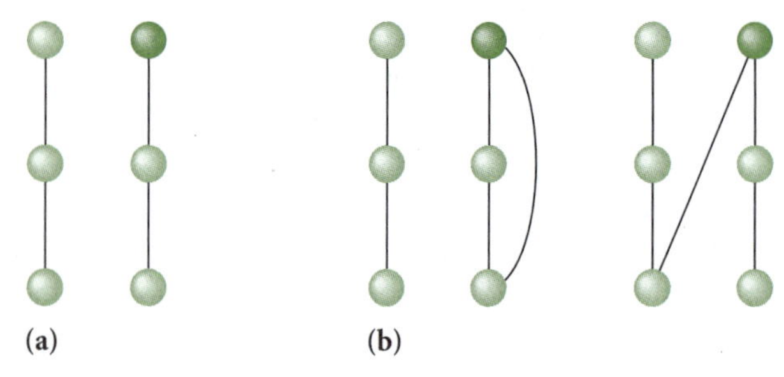

FIGURE 27-19 Why compartmentalization may be rare. (a) A compartmentalized food web, in which species in the left-hand food chain do not interact with species in the right-hand food chain. (b) The dark green circle represents an omnivore that has invaded a compartmentalized food web. The species may feed within a single food chain (*left*), thereby becoming both competitor and predator in that food chain. Alternatively, it may feed across compartments (*right*), thereby assuming a single role in each food chain. The latter outcome is thought to be more likely. (*After Pimm 1991.*)

trophospecies) may have a greater number of trophic levels than webs in which there are many trophospecies (Hall and Raffaelli 1993, Martinez 1993). Lumping is particularly characteristic of early food web studies, which are thought to contain, in general, too few species, too few feeding links, and an underrepresentation of omnivory and cannibalism (Winemiller 1990, Polis 1991).

27.8 The analysis of interaction food webs has roots in theoretical and experimental ecology.

A promising contemporary avenue for the study of community structure is the dynamic analysis of interaction food webs. The theoretical roots of this approach may be found in Robert May's linking of the metrics of topological food webs (number of species, S, and connectance, C) with a measure of the interaction of populations in the community. The empirical roots of the approach may be found in Robert Paine's classic field experiments on intertidal community dynamics.

Food Web Topology and Community Stability

May (1972, 1975) proposed a theoretical model for predicting community stability, derived from Lotka-Volterra models of population interactions. He suggested that communities are stable when

$$b(SC)^{1/2} < 1,$$

where b is the average magnitude of species interactions and S and C are the number of species and the connectance, respectively. The key to this formulation is the parameter b, a measure of the average intensity of the interactions within a community. Rearranging the equation above, we may obtain the following stability criterion:

$$SC < b^{-2}.$$

Thus, community stability depends on the strength of the interactions in the community.

To illustrate, let the average intensity of the interactions among the populations in the community be $b = 0.5$. Communities of $S = 10$ species would be stable so long as the connectance, C, was less than 0.4 ($C = 0.5^{-2}/10 = 4/10 = 0.4$). However, for communities of $S = 20$ species, stability is achieved when $C = 0.2$, and for $S = 50$, when $C = 0.08$. Without derivation, we say that the number of interacting species pairs in the community is $CS(S - 1)$. Thus, when $S = 10$ and $b = 0.5$, $C = 0.4$, and there are $CS(S - 1) = (0.4)(10)(9) = 36$ interacting pairs of species, and 76 and 196 pairs for the other two sets of conditions respectively.

May's results indicate that no matter how large the system, the number of interactions per species has an upper limit determined by the average intensity of interaction. Hence it is unlikely that more diverse systems are more complex, because more complex systems are not stable. Methodological objections to May's theory have been raised (Haydon 1994, Winemiller and Polis 1996), but it has succeeded in focusing attention on species interactions in the community, rather than simply on the topology (C and S) of the food web. As we shall see in the next section, many ecologists believe that it is in the careful delineation of these interactions that a greater understanding of community structure will arise.

The Role of Consumers in Community Structure

Paine's work on the role of predators in rocky shore communities (1966) is a classic experimental ecological study. He studied food webs in the Gulf of California and on the coast of Washington State, both of which are dominated by imposing predators, the starfishes *Pisaster* and *Heliaster* (Figure 27-20). By removing starfish from experimental areas on the coast of Washington, Paine demonstrated the crucial role of those predators in maintaining the structure of the community. Released from predation by this manipulation, mussels (*Mytilus*) spread very rapidly, crowding other organisms out of the experimental plots and reducing the diversity and complexity of local food webs. Removal of the urchin *Strongylocentrotus*, an herbivore, similarly allowed a small number of competitively superior algae to dominate an area, crowding out many ephemeral or grazing-resistant

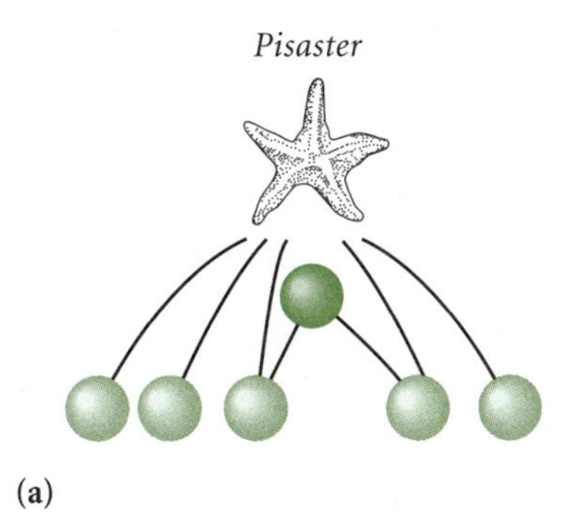

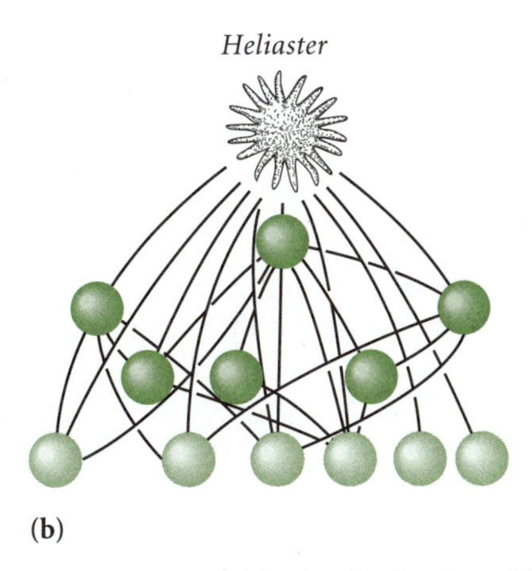

FIGURE 27-20 Intertidal food webs dominated by keystone predators: the starfishes (a) *Pisaster,* on the coast of Washington, and (b) *Heliaster,* in the northern Gulf of California. The lowest trophic levels of these food webs include herbivores such as chitons, limpets, herbivorous gastropods, and barnacles (light green circles). (*After Paine 1966.*)

species. Paine showed that predators and herbivores can manipulate competitive relationships among species at lower trophic levels and thereby control the structure of the community. Such species are called **keystone predators** because when they are removed, the edifice of the community tumbles. To be sure, species other than predators may function in a keystone role in communities (Hunter and Price 1992). The keystone concept is probably too simple to explain the importance of species in all but a few situations.

Paine's (1966) food webs of rocky shore intertidal communities are, in essence, interaction webs. The analysis of interaction food webs requires not only that we evaluate the relative strengths of the direct interactions in the web, but that we determine the extent of indirect interactions as well. Let us examine the nature of indirect interactions in communities, and then look closely at an example of the experimental analysis of an interaction food web.

27.9 Indirect interactions are important features of community structure.

Topological food webs provide a tabulation of the trophic interactions within a community, but they provide no weighting of the intensity or importance of those interactions. Ultimately, in order to understand the structure of the community, we must determine which interactions of the possible set are the most important (Paine 1992). One of the greatest challenges is in determining the number and intensity of indirect interactions.

When the interaction of one species with another involves direct physical contact, as in predation, herbivory, and parasitism, or when one species physically interferes with or somehow benefits (by, say, protection) another, the interaction is said to be a **direct interaction.** When one species affects the abundance of another through its interaction with a third species, called a **intermediary species,** the interaction is called an **indirect interaction** (Figure 27-21). Two mechanisms resulting in indirect interactions may be distinguished depending on the role of the intermediary species. An indirect effect may result from a chain of direct effects, known as an **interaction chain.** When a species affects the *interaction* between two other species, an **interaction modification** has taken place (Wootton 1993, 1994). Interaction chains are hypothesized by topological food webs, but interaction modifications are difficult to uncover without careful experimental analysis.

We may identify five simple types of indirect effects, three of which we have already encountered (Figure 27-22). Both interspecific competition and apparent competition (Chapters 21 and 22) may be considered indirect interactions. In interspecific competition, two consumers, A and C, utilize a single resource species, B. If the activities of one of the

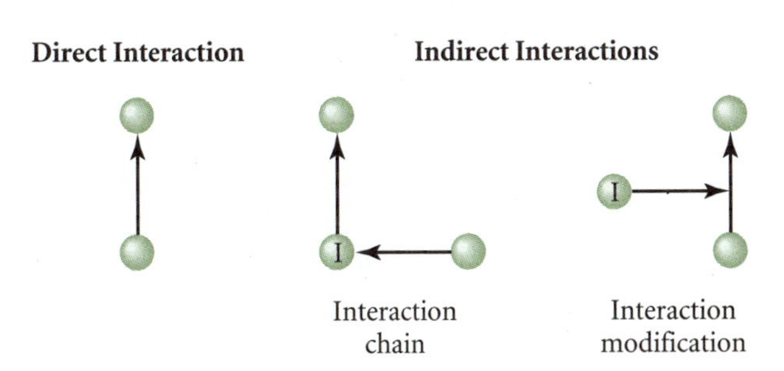

FIGURE 27-21 Direct interactions are those involving direct physical contact (e.g., predation, interference competition). Indirect interactions are those that involve an intermediary species (I). The intermediary species may be part of an interaction chain, or it may modify a direct interaction. (*After Wootton 1994.*)

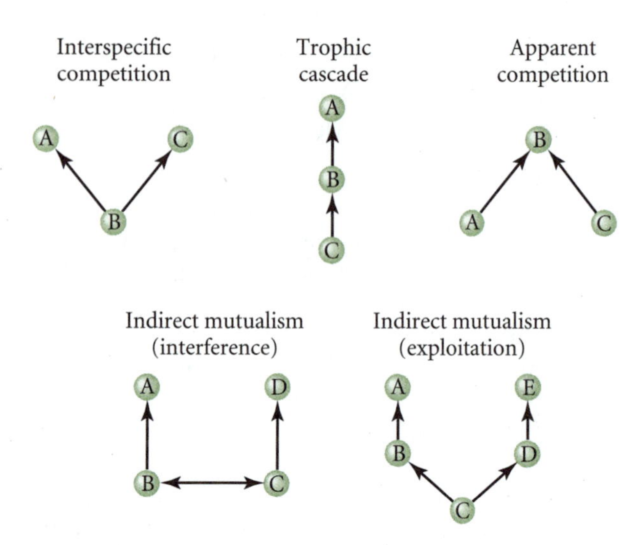

FIGURE 27-22 Five simple types of indirect effects are recognized. In interspecific competition, two consumer species, A and C, may indirectly affect each other through the intermediary resource species, B. A trophic cascade involves an interaction chain (species B is the intermediary species) across three or more trophic levels. In apparent competition, two resource species may affect each other through an intermediary consumer species (species B). Indirect mutualism may arise either from interference or exploitation competition. (*After Wootton 1994.*)

consumers—say, A—reduce the abundance of resource species B, then the abundance of consumer species C is affected indirectly. In this case, the resource species, B, is the intermediary species. In apparent competition, a single consumer, B, utilizes two resource species, A and C. An increase in the abundance of one of the resource species—for example, A—may cause an increase in the abundance of consumer species B, resulting in a decrease in resource species C. Thus, one resource species affects the other through the intermediary consumer species. Trophic cascades represent indirect interaction chains involving three or more trophic levels (see Chapter 13).

Indirect mutualisms are indirect interactions involving positive effects of one species on another (Dethier and Duggins 1984, Schoener 1993, Wootton 1994). Indirect mutualisms may arise when interference or exploitative competition is linked with a direct effect. For example, suppose that there are two species, B and C, that compete directly for space (interference competition), and suppose further that species B is a superior competitor. Species B is consumed by species A, and species C is consumed by species D (see Figure 27-22). If species A reduces the abundance of superior competitor species B, the abundance of species C may increase, and so may the abundance of species D. Such an interaction has been observed in the southwestern deserts of the United States, where kangaroo rats prefer to eat plants that produce large seeds over plants that produce small seeds. The reduction in plants that produce large seeds relative to those that produce small seeds favors ants, which prefer small seeds (Davidson et al. 1984).

To be sure, indirect effects are often more complex than the simple interactions shown in Figure 27-22. Moreover, indirect effects, even simple ones, are difficult to detect without extensive experimentation. The use of experimental approaches to develop interaction food webs that show the major direct and indirect effects is growing rapidly in ecology. Let us examine one such experiment on a food web in an old field.

27.10 Analysis of interaction food webs requires experimentation based on theory.

To fully understand community structure, one must understand the strengths of the various possible interactions in the community food web and the extent to which indirect effects occur. This requires experimentation (Bender et al. 1984, Schoener 1993, Paine 1992, Yodzis 1995, Schmidt 1997). Recall that an experiment involves the application of some treatment or perturbation to a system and the subsequent comparison of that system with a control system that did not receive the treatment, but that was handled the same way in all other respects. Natural communities are perturbed when there is an increase or decrease in the abundance of one or more of the constituent species, whether such changes are the result of biotic events or natural disturbance. Changes in the abundances of one or more species have effects on other species, and it is the pattern of these effects that constitutes the result of the perturbation. Overall, the community may respond with a variety of direct and indirect effects. This complexity of response presents a major challenge to community ecologists. Related to this challenge is the daunting challenge of determining what, exactly, the perturbed community should be compared with, since it is unlikely that all the possible direct and indirect routes of change within the community are known. Overcoming these challenges in the analysis of food webs is a major focus of contemporary community ecology.

The Theoretical Basis

There are two major experimental approaches to the analysis of interaction food webs, both of which are closely wedded to classic community theory. To develop the foundation of this theory and to introduce the experimental approaches, let us consider a community having three species, denoted 1, 2, and 3, in which species 1 and 2 are in competition with each other and species 3 preys upon both species 1 and 2. We may write a set of dynamic equations to describe the system:

$$dN_1/dt = f_1(N_1,N_2,N_3)$$
$$dN_2/dt = f_2(N_1,N_2,N_3)$$
$$dN_3/dt = f_3(N_1,N_2,N_3)$$

where N_1, N_2, and N_3 are the densities of populations 1, 2, and 3 respectively. For our purposes here, it is not important to know the mechanism of the competition between species 1 and 2. We need only know that the two hinder each other in some way. At this point, we encourage you to review Section 21.5, where we develop a model of competition between two species. There, we give specific functions for f_i based on logistic growth (see equations [21-3] and [21-4]).

Such a system of dynamic equations has an equilibrium, $\mathbf{N_e}$, which is the vector of population densities ($\hat{N}_1$, $\hat{N}_2$, $\hat{N}_3$) for which $dN_1/dt = dN_2/dt = dN_3/dt = 0$. At equilibrium, each of the species has a per capita effect on the growth rate of the other two species. Let us denote the effect of one species, say, *i*, on another species, *j*, as A_{ij}. Thus, for example, the effect of species 2 on species 1 is given as A_{12}. The matrix of these effects is called the **community matrix,** as shown below:

$$\begin{bmatrix} A_{11} & A_{12} & A_{13} \\ A_{21} & A_{22} & A_{23} \\ A_{31} & A_{32} & A_{33} \end{bmatrix}$$

The diagonal elements of the community matrix (A_{11}, A_{22}, A_{33}) represent the effect that a population has on its own growth through intraspecific competition and thus have negative values (Yodzis 1988).

An increase or decrease in N_1, N_2 or N_3 constitutes a perturbation of the system, the result of which should be a change in dN_1/dt, dN_2/dt, and dN_3/dt. As we saw in Chapter 22, an ecologist could experimentally perturb the system by adding or removing individuals of one or more of the three species. Such additions or removals could be carried out in one of two ways. First, individuals could be added or removed as a single group over a short period of time. For example, a community or portion of a community could be enclosed by a fence or some other barrier, and all or most of the individuals of one species removed from inside the enclosure, thereby decreasing the abundance of that population in the enclosure. Or, individuals could be added to the enclosure to increase the abundance of that population. Such an experiment is called a **pulse experiment** because the perturbation is applied more or less instantaneously and no other changes are made (Bender et al. 1984). In order to study the results of such an experiment, one must follow the community from its initial state through its transition owing to the application of the pulse to its return to its final state.

Another way to perturb the system is to continue to add or remove individuals in such a way that the perturbation is sustained. This type of experiment is called a **press experiment** (Bender et al. 1984). The effect of a press experiment is to move the system from its original state to a new state, where it is held. In effect, the press experiment moves the system from its original equilibrium, N_e, to a new equilibrium. Although it is often much more logistically difficult to perform a press experiment because it requires that the treatment be continually applied, the results of such experiments are often more easily observed, since there is no need to study a transient condition of the community. One can instead compare the initial state with the new state created by the perturbation.

Let us return to the hypothetical three-species community above to demonstrate an important property of press experiments. We may represent a press addition of species 1 by rewriting the dynamic equations above as follows:

$$\frac{dN_1}{dt} = f_1(N_1,N_2,N_3) + I_1$$
$$\frac{dN_2}{dt} = f_2(N_1,N_2,N_3)$$
$$\frac{dN_3}{dt} = f_3(N_1,N_2,N_3)$$

where I_1 is the addition of individuals of species 1. The press addition of individuals of species 1 will change the equilibrium densities of the other species. It can be shown that if the stability of such a system is evaluated mathematically, the negative inverse of the community matrix provides a summary of all possible outcomes of press experiments (Yodzis 1988). The elements of the negative inverse of the community matrix may be represented in general as $-[A^{-1}]_{ij}$ for all *i* and *j*. The sign of each element denotes whether the interaction has a positive or negative effect on the per capita growth rate. If $-[A^{-1}]_{ij} < 0$, then the press addition of individuals of species *j* causes a decrease in the equilibrium density of species *i*. A very important property of these values is that they contain the *total* effect of one species on another in the community over *all* direct and indirect pathways in the community.

The significance of this property of the inverse community matrix is revealed in a simple example involving apparent competition. Suppose that in the three-species community that we have been considering, species 1 and species 2 do not compete. That is, there is no interaction between the two species. A press experiment that increases the number of individuals of species 2 will no doubt result in an increase in species

3, which preys upon it. But it may also have a negative effect on species 1 because the greater population density of the predator depresses the population of both of its prey (apparent competition). The inverse community matrix of such an analysis would yield a value of $-[A^{-1}]_{21} < 0$, revealing the indirect interaction.

The values of the negative inverse of the matrix provide a set of predictions of the outcome of press experiments. But the use of such an approach presents an enormous practical challenge to ecologists for a number of reasons. First, as we mentioned above, the resolution of the topological food web upon which the analysis is based will affect the results. Interactions in webs having large numbers of trophospecies are likely to be misleading. Second, the community matrix and its inverse are based on the mathematical evaluation of the stability of a dynamic system of equations such as those presented above, an analysis that we did not present here because of its complexity. In order for this to be accomplished, the form of f_1, f_2, and f_3 must be known, or at least hypothesized. And, finally, once the equations for food web interactions are obtained, values for their parameters must be obtained via experimentation. A number of recent studies using press experiments have grappled with these problems. Let us examine one of them.

Experimental Analysis of an Old Field Community

O. J. Schmitz used theoretical and experimental approaches to investigate the interaction food web of an old field in Ontario (Schmitz 1993, 1994, 1997). The study site was a farm field that had been abandoned in 1954. Once it was abandoned, the plant and animal community in the field began to change. We shall discuss the nature of such changes, called succession, in great detail in the next chapter. At the time of Schmitz's study, the vegetation in the field included a number of perennial dicots and grasses. Four plant species dominated the field: *Hieracium aurantiacum, Solidago altissima, Phleum pratense,* and *Fragaria canadensis.* In addition to the plants, the community contained a number of species of phytophagous insects, including the grasshopper *Melanoplus borealis,* which dominated the animal portion of the community.

Considerable concordance in life history patterns existed among the plants and animals in the community. The perennial plant species dominated the field between late May, after the last frost, and late July. The nymphs of the grasshopper *M. borealis* emerged in early June, developed into adults in early July, and began to die off in late July, at about the same time as the plants senesced.

Schmitz's investigation of this old field community included three approaches. First, he conducted a number of field and laboratory experiments to develop a set of hypotheses about the interactions in the community. Essentially, this part of the work involved the development of a topological food web for the community. Based on the topological food web, and using his knowledge of the natural history of the system and the theoretical and empirical results of other studies, he developed a set of dynamic equations like the ones above for the major components of the community. Using these equations with parameter values from his field studies, he calculated the negative inverse community matrix and refined the values using simulation techniques. These values represented predictions, which he then compared with the results of field press experiments.

Figure 27-23 shows the major players in Schmitz's community and the hypothesized direct interactions among them based on field experiments. The community contains three components: nitrogen, which is limiting in the community, primary producers, and grasshopper herbivores (predominately *M. borealis*). Schmitz developed the following set of equations to describe the dynamics of the community:

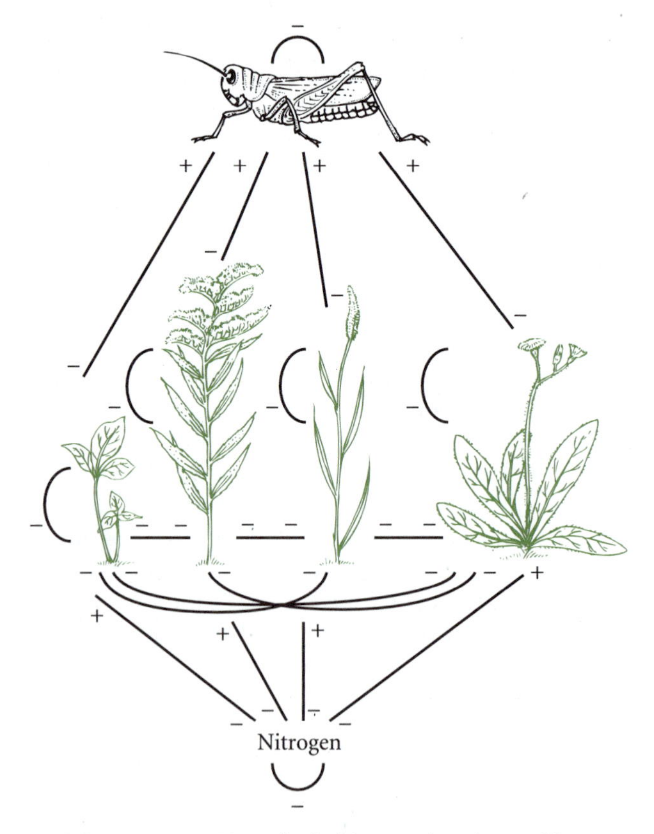

FIGURE 27-23 Hypothetical interactions in an old field community. The interactions were surmised by the evaluation of the inverse of the community matrix based on a set of equations that hypothesized the rates of change of the three major components of the web: nitrogen, plants, grasshopper. The negative inverse of the community matrix was then derived for a series of press experiments and the predictions of the matrix tested with actual experiments (see Table 27-4).

$$\frac{dN}{dt} = S_N - N\sum \mu_i V_i$$

$$\frac{dV_i}{dt} = V_i\left[a_i\mu_i N - \beta_i V_i - \sum \alpha_{ij}V_j - f_i H\right]$$

$$\frac{dH}{dt} = H\left[\sum e_i f_i V_i - \beta_H H\right]$$

where N represents the amount of nitrogen, V_i the biomass of plants of species i, and H the biomass of the herbivore. Biomass was used instead of density for practical reasons. The parameters of the equations are summarized and explained in Table 27-3.

These equations reveal the pattern of interactions in the community. The amount of nitrogen is dependent on the rate at which it is supplied to the community (S_N) and the rate at which it is taken up by the plant species ($N\Sigma\mu_i V_i$). The rate of change in the biomass of each plant species is dependent on the density of that species (V_i), the rate at which it converts nitrogen into biomass ($a_i\mu_i N$), which is dependent both on the amount of nitrogen available and the rate at which it is taken up by the plant, the loss of biomass due to intraspecific interference competition ($\beta_i V_i$), the loss of biomass due to interspecific competition ($\Sigma\alpha_{ij}V_j$), and the loss of biomass due to the feeding activities of the herbivore ($f_i H$). The rate of change in the biomass of the herbivore is related to the biomass of the herbivore (H), the rate at which plant material is converted into herbivore biomass ($\Sigma e_{ij} f_i V_i$), and the loss of biomass related to herbivore density ($\beta_H H$). Using field experiments, Schmitz arrived at values for the parameters of these equations and from these, calculated the negative inverse of the community matrix.

Using this system of equations, Schmitz predicted the outcome of nitrogen and herbivore press experiments. The predictions, presented in Table 27-4, suggest that an increase in nitrogen should cause an

TABLE 27-3 Summary of parameters for the dynamic equations of a simple old field food web

Parameter	Description
S_N	Supply rate of soil nitrogen (grams of N per day)
μ_i	Per capita uptake rate of nitrogen by plant species *i* (fraction of N taken up per gram of plant species *i* per day)
a_i	Conversion of nitrogen into plant biomass (fraction of plant biomass produced per g N taken up)
β_i	Per capita loss rate of plant biomass due to intraspecific interference competition (fraction of biomass per day)
α_{ij}	Per capita loss rate of plant biomass due to interspecific interference competition with members of plant species *j* (fraction of biomass per plant species *j* per day)
f_i	Per capita loss rate of plant biomass due to herbivory (fraction of plant biomass lost per time per herbivore)
e_i	Conversion of plant biomass into herbivore biomass (grams of herbivore biomass produced per gram of plant biomass consumed)
β_H	Per capita loss rate of herbivores due to herbivore density (fraction of herbivore production lost per day per herbivore)

(From Schmitz 1997.)

TABLE 27-4 Predictions and outcomes of press experiments on a simple old field food web

		FREQUENCY	
Experiment	Species	Predicted*	Observed
Nitrogen press	*Hieracium*	0.93–0.83	0.63
	Solidago	0.77–0.54	0.94
	Phleum	0.79–0.65	0.67
	Fragaria	0.92–0.69	0.63
Herbivore press	*Hieracium*	0.53–0.34	0.50
	Solidago	0.73–0.60	0.73
	Phleum	0.63–0.44	0.63
	Fragaria	0.64–0.53	0.63

**95% confidence intervals.*

increase in all four of the plant species, whereas a press of herbivores on the communities would result in a negative effect on *Hieracium, Solidago,* and *Phleum* and a positive effect on *Fragaria.* Subsequent press experiments verified the direction (sign) of these predictions, even though their magnitude was sometimes not in accordance with the predictions. Such analyses of interaction food webs are likely to be of great practical value in understanding the effects of human perturbations on natural ecosystems, determining the important factors driving agricultural food webs, and detecting ecosystem stress (Crowder et al. 1996).

The study of the structure of the community usually focuses on the state of the community over a relatively short period of time. But communities are always changing. We shall discuss the dynamics of that change in the next chapter.

SUMMARY

1. Ecologists characterize communities by the numbers of species present, their relative abundances, and their feeding and other ecological relationships. These features are part of the community structure. Community structure also includes historical patterns of community change and the process of species development (speciation).

2. Historically, much of community ecology grew out of attempts to classify plant associations on the basis of their species composition. Floristic analysis provided data that forced ecologists to abandon the concept of communities as discrete entities.

3. An important feature of community structure is the relative abundances of the species in the community. Three models of abundance distributions serve as hypotheses of specific ecological processes. The geometric series model hypothesizes the process of niche preemption, in which successive new species in a community take a set proportion of the remaining resources. The random niche model assumes that species utilize resources in random fashion. The lognormal distribution assumes that the effects of many factors sum together to affect species abundances.

4. The lognormal distribution of species abundances has proved to be a useful empirical device, characterizing frequency distributions by a modal abundance class (n_0) and the dispersion of abundances about the mode (s). The term n_0 depends on the size of the sample, but the value of s is an intrinsic property of the community. Species with predicted abundances of less than one individual—those beyond the veil line—are not detected unless the size of the sample is increased. The lognormal model stresses the dependence of diversity estimates upon sample size.

5. Various indices of diversity, most notably Simpson's index and the Shannon-Weaver index, have been devised to take into account variation in abundance distributions when making comparisons between samples. These indices account for both species richness and species abundance.

6. Communities possess both taxonomic diversity (diversity of species) and ecological diversity. The latter reflects the variation in ecological roles of species in the community.

7. Species abundance distributions are likely to differ in shape between communities and, thus, may not be possible to compare directly. Rarefaction is a process whereby the diversities of two communities may be compared directly by obtaining curves of expected species richness for a range of subsample sizes.

8. The number of species increases in direct relation to the area sampled. In part, this pattern results from larger areas giving rise to large total samples. But studies of well-known faunas and floras also indicate that larger areas are more heterogeneous ecologically, providing opportunities to sample more kinds of habitats, and additionally, that larger islands have more species because they are better targets for colonization and because larger populations better resist extinction. Differences in the effects of disturbance on islands of different sizes may also influence species-area relationships.

9. Different concepts of the food web emphasize different ways in which populations influence one another. Feeding relationships are portrayed in connectedness or topological webs, energy flux is shown in energy flow webs, and the influences of populations on one another's growth rates are depicted in interaction or functional webs.

10. A number of generalizations have arisen from the analysis of topological food webs. Food web characteristics such as the length of food chains, the amount of omnivory, the extent of compartmentalization, and the ratio of predators to prey provide clues to community persistence and the ability of the community to withstand disturbance.

11. The analysis of interaction food webs involves the application of theory and experimentation. May's theoretical studies linked food web metrics to the interactions of populations. Experimental studies of food webs, such as those conducted by Paine in intertidal communities, revealed the roles of individual species in the maintenance of community structure.

12. Experimental analysis of interaction food webs involves the application of pulse or press experiments. The outcomes of press experiments are predicted by the inverse community matrix, the elements of which show the per capita effect of one species on another via all the direct and indirect interactions in the food web.

EXERCISES

1. Construct a table like Table 27-1 for the situation in which k, the proportion of community resources used by arriving species, is 0.75.

2. An ecologist has collected over 100,000 spider specimens from an extensive sampling study in a hardwood forest. Another study of forest spiders yielded a lognormal distribution of spider abundance with breadth $s =$ 4.5 octaves. The modal abundance class for the current study is 35. Roughly how many species are included in the 100,000 spiders collected?

3. Calculate the Shannon-Weaver diversity index for the following data:

Species	Number in sample
1	100
2	35
3	6

4. Letting the intensity of interactions within a food web be $b = 0.3$, construct a graph with S on the horizontal axis and C on the vertical axis that shows community stability.

5. Examine Schmitz's equations for his simple old field community and describe how you would go about experimentally determining the various parameters in the equations.

CHAPTER 28

Community Development

GUIDING QUESTIONS

- What is the difference between primary and secondary succession?
- What is the difference between autogenic and allogenic succession?
- How does the intensity of disturbance affect succession?
- What life history characteristics make organisms good colonizers?
- What are the three ways in which one species may affect the probability of establishment of another?
- How can Markovian dynamics be used to make predictions about community development?
- What kinds of theories about the nature of the climax community have been proposed?

When viewed for an instant, or even over the course of several days or weeks, most communities appear to be stable, unchanging features of the landscape. However, when viewed closely for an extended period, communities prove to exist in a continual state of flux. Organisms die, and other organisms are born and take their places. Energy and nutrients pass through the community. If they are undisturbed, the change in appearance and composition of most communities occurs very slowly. Oaks replace oaks, squirrels replace squirrels, and so on, in continual self-perpetuation. But when a habitat is disturbed—a forest cleared, a prairie burned, a coral reef obliterated by a hurricane—the community rebuilds. Pioneering species adapted to the disturbed habitat are successively replaced by others until another community, often with a structure and composition similar to the former community, arises (Figure 28-1).

The sequence of changes initiated by disturbance is called **succession.** This natural process caught the attention of early ecologists such as Henry Cowles and Frederic Clements (Drury and Nisbet 1973, McIntosh 1974, 1985). By 1916, Clements had outlined the basic features of succession, supporting his conclusions with detailed studies of change in plant communities in a variety of environments. Since then, the study of community development has grown to include the processes that underlie successional change, the adaptations of organisms to the different conditions of early and late succession, and interactions between early colonists and the species that replace them. Succession has important practical implications, particularly for landscape architects, range managers, farmers, utility mangers, and even golf course designers, because it represents a force of change in such designed and managed communities. In this chapter, we shall examine the course and causes of succession.

28.1 Succession follows an orderly pattern of species replacements.

The creation of any new habitat—a plowed field, a sand dune at the edge of a lake, an elephant's dung, a temporary pond left by a heavy rain, even the carcass of a dead animal—invites a host of species that are particularly adapted to be good invaders. These first

FIGURE 28-1 Stages of succession leading to an oak-hornbeam forest in southern Poland. From (a) to (f), the time since clear-cutting progresses from 0 to 7, 15, 30, 95, and 150 years. (*Photographs by Z. Glowacinski, courtesy of O. Jarvinen; from Glowacinski and Jarvinen 1975.*)

colonists are followed by others that are slower to take advantage of the new habitat, but eventually will be more successful than the pioneering species in maintaining a viable population within the site. In turn, these species change the environment, and in this way, the character of the community changes with time. Community change is characterized by the loss and gain of different types of organisms in the community.

We may observe succession in a number of ways. In some cases, a dramatic disturbance essentially removes all of the living things from an area. We can then watch as life regains a foothold. Such a situation occurred after the volcanic eruption of Mount St. Helens in the spring of 1980, which created vast areas denuded of vegetation and animal life. In some cases, we can examine sites having similar conditions that were disturbed at different times. The opportunity to observe succession is almost always at hand in abandoned fields of various ages (Figure 28-2). In the Piedmont Region of North Carolina, bare fields are quickly covered by a variety of annual plants. Within a few years, most of these annuals are replaced by herbaceous perennials and shrubs. The shrubs are followed by pines, which eventually crowd out the earlier successional species; pine forests are in turn invaded and then replaced by a variety of hardwood species (Oosting 1942). Change is rapid at first. Crabgrass quickly enters an abandoned field, hardly allowing time for the plow's furrows to smooth over. Horseweed and ragweed dominate the field in the first summer after abandonment, aster in the second, and broomsedge in the third. The pace of succession falls off as slower-growing plants appear. The transition from a bare field to pine forest requires 25 years. Another century must pass before the developing hardwood forest begins to resemble the natural climax vegetation of the area. And the sequence may be disrupted at any time by a fire, a hurricane, or a new round of development or agriculture.

The transition from abandoned field to mature forest is only one of several successional sequences leading to the same community type. In the eastern United States and Canada, forests are the end point of several successional series, or **seres,** each having a different beginning (Christensen and Peet 1984). The sequence of species on newly formed sand dunes at the southern end of Lake Michigan differs from the sere that develops on abandoned fields a few miles away (Cowles 1899, Olson 1958). The sand dunes are first invaded by marram and bluestem grasses. Plants of

FIGURE 28-2 An old field on the Piedmont of North Carolina. Such habitats develop after abandonment of agricultural land.

these species established in soils at the edge of a dune send out rhizomes (runners) under the surface of the sand, from which new shoots sprout. These grasses stabilize the dune surface and add organic detritus to the sand. Numerous annuals follow the perennial grasses onto the dunes, further enriching and stabilizing them and gradually creating conditions suitable for the establishment of shrub species. Sand cherry, dune willow, bearberry, and juniper form shrub layers before pines become established. As in the abandoned fields of North Carolina, pines persist for only one or two generations, with little reseeding after initial establishment, giving way in the end to the beech-oak-maple-hemlock forest characteristic of the region. Succession follows a similar course on Atlantic coastal dunes, where beach grass initially stabilizes the dune surface, followed by bayberry, beach plum, and other shrubs (Oosting 1954). Shrubs act like the snow fencing often used to keep dunes from blowing out; they are called dune builders because they intercept blowing sand and cause it to pile up around their bases (Figure 28-3).

Ecologists classify seres into two groups according to their origin. The establishment and development of plant communities in newly formed habitats previously without plants, such as sand dunes, lava flows, or rock bared by erosion or exposed by a receding glacier (see Figure 28-10), is called **primary succession.** The return of an area to its natural vegetation following a disturbance is called **secondary succession.** The distinction between the two is blurred because disturbances vary in the degree to which they destroy the fabric of the community and its physical support systems. A tornado that levels a large area of forest usually leaves the soil's bank of nutrients, seeds, and sproutable roots intact. In contrast, a severe fire may burn through the organic layers of the soil, destroying hundreds or thousands of years of biologically mediated development.

Ecologists distinguish between succession resulting from changes brought about by the organisms themselves and succession that results from factors external to the community. The first type of succession is called **autogenic succession.** The transition of an abandoned agricultural field to a mature forest described above is

(a)

(b)

FIGURE 28-3 Initial stages of plant succession on sand dunes along the coast of Maryland. (a) Beach grass on the frontal side of a dune. This grass is used widely to stabilize dune surfaces. (b) Invasion of back dune areas by bayberry and beach plum. (*Courtesy of the U.S. Soil Conservation Service.*)

an example of autogenic succession. When organisms change the environment in which they live, other species in the community may be harmed or benefited. Harm may also befall the organisms making the change, resulting in their loss from the community. Community change also creates new ecological opportunities for species that are not in the community but which may enter it through dispersal. Autogenic succession occurs on time scales commensurate with the life span of the organisms in the community.

In autogenic succession, the principal force of change comes from within the community. Succession driven by forces primarily outside the community is called **allogenic succession.** Such external forces may include massive disturbance, such as that occurring on Mount St. Helens, climate change, or daily or seasonal changes in temperature and other environmental factors. Allogenic succession operates on a time scale commensurate with the time scale of the disturbance. Communities of phytoplankton in freshwater lakes may undergo succession each year resulting in part from seasonal changes in temperature and light intensity. Allogenic succession resulting from climate change may occur over many thousands of years.

Autogenic and allogenic succession are not mutually exclusive processes. The more or less proximate responses of organisms to autogenic forces occur even as slower allogenic succession proceeds. The relative importance of the two depends on the time scale of the allogenic changes. When allogenic change occurs over a long period of time in response to gradual changes in the environment, as might occur in the case of climate change, then autogenic succession is likely to be the most important factor determining the community structure at any particular time. When disturbance occurs frequently, allogenic succession is likely to be a more important force of community change (Overpeck et al. 1990, Huston 1994).

Succession has traditionally been viewed as leading inexorably toward an ultimate expression of community development, the **climax** community (Clements 1936, Shimwell 1971). Early studies of succession suggested that the many seres found within a region, each developing under a particular set of local environmental circumstances, progress toward the same climax (Cooper 1913, Oosting 1956). The nature of the local climax was thought to be determined solely by climate. Aberrations in community composition caused by soils, topography, fire, or animals (especially grazing) were thought to represent interrupted stages in the transition toward the local climax—immature communities. Clements recognized fourteen climaxes in the terrestrial vegetation of North America, including two types of grassland (prairie and tundra), three types of scrub (sagebrush, desert scrub, and chaparral), and nine types of forest, ranging from pine-juniper woodland to beech-oak forest.

In recent years, the concept of the climax has been greatly modified—to the point of outright rejection by many ecologists—with the recognition of communities as open systems whose composition varies continuously over environmental gradients. Whereas in 1930, for example, plant ecologists described the climax vegetation of much of Wisconsin as a sugar maple-basswood forest, by 1950 ecologists placed this forest type on an open continuum of climax communities extending over both broad, climatically defined regions and local, topographically defined areas (Whittaker 1953, McIntosh 1967, Peet and Loucks 1977). To the south, beech increased in prominence; to the north, birch, spruce, and hemlock were added to the climax community; in drier regions bordering prairies to the west, oaks became prominent. Locally, quaking aspen, black oak, and shagbark hickory, long recognized as successional species on moist, well-drained soils, came to be accepted as climax species on drier upland sites.

Mature stands of forest in Wisconsin, representing the end points of local seres, were ordered by Curtis and McIntosh (1951) along a **continuum index.** Forest stands were categorized according to species composition and arranged along a gradient of moisture conditions, ranging from dry sites dominated by oak and aspen to moist sites dominated by sugar maple, ironwood, and basswood. The forest types were assigned arbitrary values, ranging between extremes of 300 for a pure stand of bur oak to 3,000 for a pure stand of sugar maple. Although increasing values of the index correspond to seral stages leading to the sugar maple climax, they also represent local climax communities determined by topographic or soil conditions. Thus, the so-called climax vegetation of southern Wisconsin actually represents a continuum of forest (and, in some areas, prairie) types (Figure 28-4).

Historically, ideas about succession arose from studies of terrestrial plant communities, and much contemporary work continues to focus on such communities. However, succession occurs in freshwater and marine aquatic communities as well. We mentioned above that allogenic factors may cause seasonal successional changes in freshwater plankton communities (Wetzel 1983). Succession in the freshwater ponds associated with the sand dunes of Lake Michigan is thought to have resulted primarily from allogenic forces (Jackson et al. 1988). Succession also occurs in soil communities, and, as we shall see below, the process even occurs in the community of organisms associated with the decomposing bodies of animals.

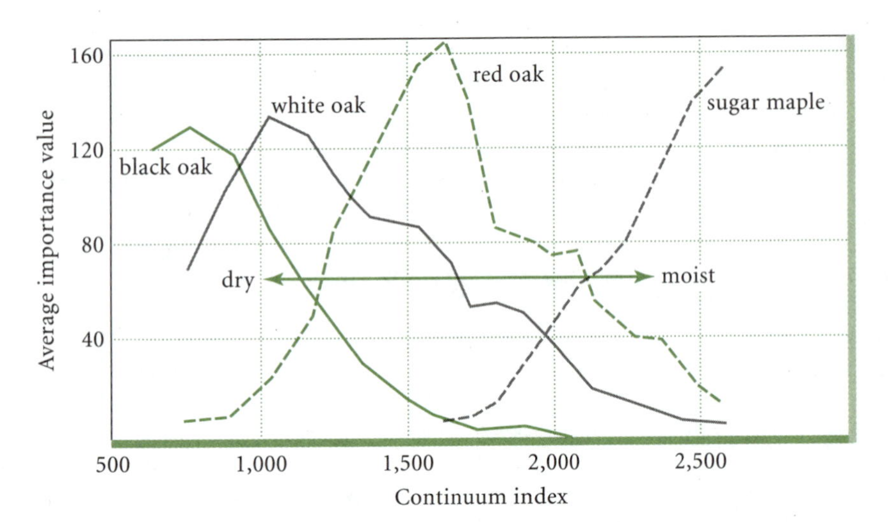

FIGURE 28-4 Relative importance (a measure of abundance) of several tree species in forest communities of southwestern Wisconsin, arranged along a continuum index. Soil moisture, exchangeable calcium, and pH increase toward the right of the continuum index. (*After Curtis and McIntosh 1951.*)

28.2 Primary succession develops in habitats newly exposed to colonization by plants and animals.

Areas that are subjected to extreme disturbance, such as the deposits left by a receding glacier or the spaces covered by the lava flow from a volcano, and areas where there are few essential nutrients, such as rock outcrops and sand dunes, are devoid of life. But in many cases organisms slowly take hold in these areas, changing the conditions in their immediate vicinity, and thereby creating conditions suitable for other species to colonize.

Species colonizing the thin deposits of clay left by receding glaciers in the Glacier Bay region of southern Alaska (see Figure 28-10) must cope with deficiencies of nutrients, particularly nitrogen, and with stressful wind and cold. Here the sere begins with mat-forming mosses and sedges, and then progresses through prostrate willows, shrubby willows, alder thicket, sitka spruce, and finally, to spruce-hemlock forest. Succession is rapid, reaching the alder thicket stage within 10 to 20 years and tall spruce forest within 100 years (Crocker and Major 1955, Lawrence et al. 1967, Reiners et al. 1971, Noble et al. 1984, Chapin et al. 1994).

An extraordinary opportunity to study terrestrial primary succession arose on May 18, 1980, when Mount St. Helens in Washington State exploded in a violent volcanic eruption. The blast blew off the top of the mountain, and the ensuing mud and lava flows created vast lifeless areas. Extensive studies of primary succession in these areas have been conducted (Del Moral and Wood 1988a,b, 1993, Wood and del Moral 1987, 1988, del Moral 1993). The establishment of vegetation in the barren areas around Mount St. Helens is dependent on the dispersal of seeds into the area from surrounding undisturbed regions. The first plant to become established was a lupine (*Lupinus lepidus*), a species that can fix nitrogen and thus contribute to the nutrient enrichment of soil. Other species, such as *Aster ledophyllus*, became established near lupine plants (del Moral 1993). Such early arrivals are referred to as **pioneer species.** The survival of the lupines and other plants was tenuous, however. In order for vegetation to become established in vast barren areas like those of Mount St. Helens, the pioneering species must have a high level of tolerance for the extreme local conditions and a high dispersal ability. Many of the early successional species on Mount St. Helens did not possess these qualities. Lupines lived only a couple of years in the lava flows, and the *Aster* species that colonized around the lupines, while possessing wind-dispersed seeds and, thus, high dispersal capabilities, proved to be unable to withstand the harsh conditions of the area. Thus primary succession in the area has been very slow. In the decade following the eruption, only eleven species of plants became established in the area.

The time required for succession to proceed from a cleared habitat to a climax community varies with the nature of the climax and the initial quality of the soil. Clearly, succession is slower to gain momentum when starting on bare rock, as on Mount St. Helens, than on a recently cleared field. A mature oak-hickory forest climax develops within 150 years on cleared fields in North Carolina (Oosting 1942). Climax stages of western grasslands are reached in 20 to 40 years of secondary succession (Schantz 1917). On the basis of radiocarbon dating methods, Olson (1958) suggested that complete primary succession to a beech-maple climax forest on Michigan sand dunes requires up to 1,000 years. In the humid Tropics, forest communities regain most of their climax elements within 100 years after clear-cutting, provided that the soil is not abused by farming or prolonged exposure to sun and rain (Budowski 1965). But the development of a truly mature tropical forest devoid of any remnants of successional species requires many centuries.

28.3 The intensity and extent of disturbance influence the pattern of secondary succession.

Breaks in the forest canopy caused by tree death or blowdown, the opening of patches by wave action in intertidal environments, and the denuding of areas by fire are examples of **physical disturbance** that set in motion successional processes. Behaviors of animals such as digging, grazing, predation, and herbivory may sometimes be so disruptive as to constitute a **biological disturbance.**

Disturbed sites are colonized from three sources. The first source is, of course, the area surrounding the disturbance. Pools of dispersants in areas some distance away from the disturbed area represent a second source of colonists. For example, the pelagic larvae of epifaunal (surface-growing) marine invertebrates or wind-dispersed seeds from surrounding plant communities may colonize the disturbed area. Finally, the disturbance may fail to eliminate buried seeds or eggs, and these may germinate and develop in situ, thereby providing a third source of organisms to the emerging community.

But disturbance does not always result in total destruction of a habitat, thereby requiring total renewal. Periodic disturbance is a fact of life in most communities, and may represent a strong selective force. Plants adapted to fire or grazing survive those disturbances, even though the community in which they find themselves is greatly altered. Also, succession in communities where disturbance is common may be more rapid than in communities where disturbance is uncommon. The slow successional development on the lava flows of Mount St. Helens demonstrates the challenges presented by uncharacteristic and widespread disturbance. Community development in such cases may not follow classic successional patterns (Sousa 1984, Whelan 1995). The pattern of community development after a disturbance depends on a number of interacting factors, including the intensity of the disturbance and the size of the disturbed area, the microclimate in the disturbed area, the characteristics of growth and reproduction of the organisms that survived the disturbance, the competitive abilities of colonizing species, and the responses of predators and herbivores to the new opportunities created by the disturbance.

For the most part, ecologists have focused their attention on how disturbance affects communities of sessile organisms such as plants and marine invertebrates (Sousa 1979, 1984), and that is the approach that we shall take here. It is worth noting, however, that extreme physical disturbance may eliminate or significantly alter the structure of populations and communities of mobile animals as well (Sousa 1984).

Succession in Marine Epifaunal Habitats

The influence of gap size on succession has been investigated in several marine habitats, where disturbance and recovery frequently follow upon each other. Working in southern Australia, Keough (1984) investigated the colonization of artificially created patches, ranging in size from 25 to 2,500 cm^2 (5 to 50 cm on a side), by various subtidal encrusting invertebrates that grow on hard surfaces. The major epifaunal taxa vary considerably in their colonizing abilities and competitive abilities, which are generally inversely related (Table 28-1). When Keough created bare patches of different sizes within larger areas of rock occupied by encrusting invertebrates, the exposed areas were quickly occupied by such highly successful competitors as tunicates and sponges, which grew in from surrounding areas. In this

TABLE 28-1 Summary of life history attributes of the major epifaunal taxa at Edithburg, Australia

Taxon	Growth form	Colonizing ability	Competitive ability	Capacity for vegetative growth
Tunicates	Colonial	Poor	Very good	Very extensive; up to 1 m^2
Sponges	Colonial	Very poor	Good	Very extensive; up to 1 m^2
Bryozoans	Colonial	Good	Poor	Poor, up to 50 cm^2
Serpulid polychaetes	Solitary	Very good	Very poor	Very poor; up to 0.1 cm^2

(From Keough 1984.)

case, patch size had little influence on community development, because the distances from the edges to the centers of the patches (less than 25 cm) were easily spanned by growth. Bryozoan and polychaete larvae are good colonizers, and many attempted to colonize the patches. However, these were quickly overgrown by the tunicates and sponges, which enjoy a competitive advantage.

Keough also created isolated patches by placing hard substrates in sand, mimicking the shells of *Pinna* clams. In these patches, size was very important. Just by chance, few of the small patches were colonized by tunicates and sponges, which produce relatively few propagules. Their absence allowed bryozoans and polychaetes to obtain a foothold. Because they were bigger targets, many of the large patches were settled by a few larvae of tunicates and sponges, which then spread rapidly and eliminated other types of species that had colonized along with them. As a result, tunicates and sponges predominated in the larger isolated patches, but bryozoans and polychaetes, which, once established, can deter the colonization of tunicate and sponge larvae, were able to dominate many of the smaller patches. In this system, bryozoans and polychaetes are disturbance-adapted species—what botanists call **weeds.** They get into open patches quickly, mature and produce offspring at an early age, and then often are eliminated by more slowly colonizing but superior competitors. Such weedy species require frequent disturbances to stay in the system.

The size of a patch partly determines whether predators and herbivores will be active there. These consumers can affect the course of succession. Some consumers may select large patches for feeding because they are easy to find and require less travel time between patches. Other consumers that are themselves vulnerable to predators may require the cover of intact habitat, from whose edges they venture to feed in newly exposed areas. In this case, small patches are likely to be grazed more intensively than the centers of large patches.

Succession in Rocky Intertidal Algal Communities

Rabbits rarely feed far from the cover of brush or trees to avoid being seen by predators far from safety. Limpets (grazing mollusks) similarly do not venture far from the safety of mussel beds to feed on algae. Sousa (1984) demonstrated this point forcefully in an intertidal rocky shore habitat in central California, where he cleared patches of either 625 or 2,500 cm^2 in mussel beds, and excluded limpets from half the patches in each of these sets by applying a barrier of copper paint along their edges. He then monitored the colonization of the cleared patches over the following 3 years. Limpets live in the crevices between mussels when they are not feeding, so as to avoid predation and desiccation. Because this behavior limits their foraging range (Figure 28-5), densities of limpets in the small patches (surrounded by more edge compared with area) exceeded those in the large patches. Not surprisingly, throughout the course of the experiment, algae grew more densely in the larger patches. Where limpet grazing was prevented, total cover by all species of algae was high and did not differ between patches of different size.

As one would expect, limpet grazing depressed the establishment and growth of most species of algae, but it favored three relatively rare species: the brown alga *Analipus,* the green *Cladophora,* and the red *Endocladia.* All of these have a low-lying, crustose or turf growth form that makes them less vulnerable to grazers, but more vulnerable to shading and overgrowth by other species. In addition, establishment of *Endocladia* was sensitive to patch size, generally being more common in larger patches regardless of limpet grazing.

Patches from which limpets were excluded by copper barriers and large patches where limpet densities were relatively low were colonized by a mussel species that is an inferior competitor to the mussels that form the beds in Sousa's study area. This mussel establishes itself best in the short algal turfs that develop only where limpet densities are low. The interaction between patch size and grazing explained the abundances of algae and of this second mussel species. Large patches having low limpet densities or patches from which limpets were excluded favored the formation of algal turfs and subsequent colonization by the second mussel species.

FIGURE 28-5 A natural cleared patch in a bed of mussels (*Mytilus californianus*) on the central coast of California. The patch is about 1 m across and has been colonized by a heavy growth of the green alga *Ulva.* Note the distinct browse zone around the perimeter of the patch. It is created by limpets, which feed only short distances away from refuge in the mussel bed. (*Courtesy of W. P. Sousa; from Sousa 1984.*)

Succession After Disturbance by Fire

The magnificent violence of fire predisposes us to think of it as a catastrophic event (see Figure 1-12). When fires occur in areas where they are rare, or when particularly intense or out-of-season fires occur in habitats where fire is common, the result may indeed be a catastrophe. However, in many ecosystems, fire is common, and our awe of its fierce nature notwithstanding, its effects are not devastating. Rather, communities respond to it with more or less predictable patterns of regrowth and regeneration. Indeed, the structure of some grassland, shrubland, and forest habitats appears to be maintained by periodic fires (see Section 28.7).

The intensity of a fire is determined by a number of factors, including the type, amount, and location of available fuel, moisture conditions, and the direction and strength of winds. Some fires burn close to the ground, sparing large trees and having little effect on the subsurface parts of the low vegetation. Ground fires may become quite intense, however, causing damage to organic material in the soil as well as destroying the aboveground vegetation. Intense fires may reach to the canopy of trees, causing great damage to the foliage there.

A classic pattern of successional change resulting from fire may be observed in high-latitude forest and forest-tundra regions, where fire often destroys most of the plants in an area (Whalen 1995). After the passage of the fire, opportunistic species having light seeds and rapid growth strategies invade (or reinvade) and establish communities. Community development is driven by a continuum of successive changes through time, brought about by successive interactions of communities with the environment. These interactions change the environment and give rise to a series of different communities.

But the classic view of secondary succession, in which post-disturbance communities develop from scratch in an area practically denuded of living forms via colonization from remote areas, is probably too simplistic a framework within which to understand the dynamics of fire disturbance in most communities. Many plants, such as pitch pine and some oaks, are adapted to withstand the heat of fire and, thus, live to see another day. In the case of jack pine, the seeds are held in cones for many years. A fire in the crowns of stands of these trees opens the cones, releasing the seeds, a phenomenon called **serotiny.** In many cases, fire does not affect the belowground component of the plant community or the seeds, so "colonizing" plants may arise from the undisturbed seed bank or from belowground stems and roots, giving rise to a post-disturbance community with a species composition similar to that of the pre-disturbance community. Moreover, the intensity of the fire disturbance itself is, to a large extent, dependent on the nature of the community in which the fire occurs. The amount of fuel in the community, for example, is related to the ratio of living to dead biomass and to population densities in the community, among other things. Different flammabilities of the components of the community will also affect the intensity of the fire. The unique features of fire as a physical disturbance in both natural and managed ecosystems have been reviewed by Whelan (1995). We shall discuss its role in late successional stages in Section 28.7.

28.4 Succession results from variation in the ability of organisms to colonize disturbed areas and from changes in the environment following the establishment of new species.

Two factors interact to determine the position of a species in a sere: the life history characteristics of the species and the nature of the changes that occur through the course of succession.

Life History Features and Succession

In general, the properties of organisms that make them good colonizers of disturbed areas are not the same properties that provide advantages in developed communities. For example, organisms that reproduce very early in life, produce lots of small propagules that are easily dispersed, and grow fast will not compete well in a community near climax where disturbance is moderate or low. Species with good colonizing abilities usually are not particularly strong competitors. We saw this in Section 28.3 in Keough's study of succession in epifaunal communities in Australia, where bryozoans and polychaetes colonized disturbed areas well, but were competitively inferior to tunicates and sponges.

Many life history and physiological characteristics are known to differ between early and late successional plants (Budowski 1965, 1970, Grubb 1977, Bazzaz 1979, Huston 1994). A number of these are listed in Table 28-2. Early successional species often produce many small seeds that usually are wind dispersed or carried long distances by birds or bats, making them good colonizers. The seeds of many early successional species are long-lived and can remain dormant in the soils of forests and shrub habitats for years until fires or treefalls create the bare-soil conditions required for germination and growth (Harper 1977). The seeds of late successional species, which are relatively large, provide their seedlings with ample

TABLE 28-2 General characteristics of early and late successional plants

Characteristic	Early succession	Late succession
Photosynthesis		
Light saturation intensity	High	Low
Light compensation point	High	Low
Efficiency at low light	Low	High
Photosynthetic rates	High	Low
Dark respiration rates	High	Low
Water use efficiency		
Transpiration rates	High	Low
Mesophyll resistance	Low	High
Seeds		
Number	Many	Few
Size	Small	Large
Dispersal distance	Large	Small
Dispersal mechanism	Wind, birds, bats	Gravity, mammals
Viability	Long	Short
Induced dormancy	Common	Uncommon?
Resource acquisition rates	High	Low?
Recovery from nutrient stress	Fast	Slow
Root/shoot ratio	Low	High
Mature size	Small	Large
Structural strength	Low	High
Growth rate	Rapid	Slow
Maximum life span	Short	Long

(From Huston 1994; data from Budowski 1965, 1970, Pianka 1970, Ricklefs 1973, Bazzaz 1979.)

nutrients to get started in the highly competitive environment of the forest floor (Salisbury 1942).

The survival of seedlings in shade is directly related to seed weight (Figure 28-6). The ability of seedlings to survive the shady conditions of climax habitats is inversely related to their growth rate in the direct sunlight of early successional habitats (Grime and Jeffrey 1965). When placed in full sunlight, early successional herbaceous species grew ten times more rapidly than shade-tolerant trees. Shade-intolerant trees, such as birch and red maple, had intermediate growth rates.

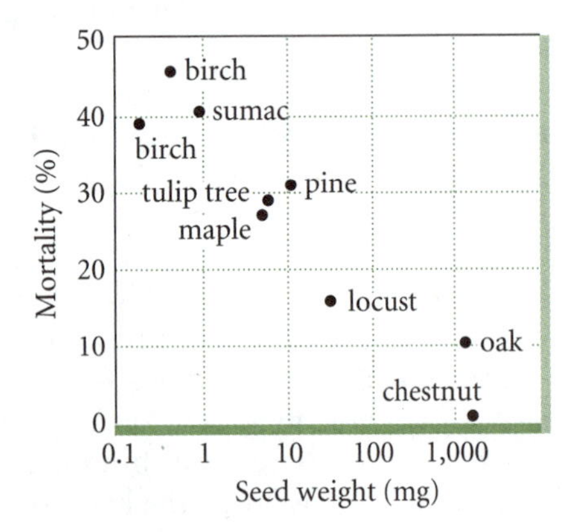

FIGURE 28-6 Relationship between seed weight (log scale) and mortality of seedlings after 3 months under shaded conditions. *(After Grime and Jeffrey 1965.)*

Shade tolerance and growth rate are adaptations that are balanced evolutionarily in response to the environment. The rapid growth of early successional species is due partly to the relatively large proportion of seedling biomass allocated to leaves (Abrahamson and Gadgil 1973). Leaves carry on photosynthesis, and their productivity determines the net accumulation of plant tissue during growth. Hence the growth rate of a plant is influenced by the allocation of tissue to the roots and to the aboveground parts (shoot). In the seedlings of annual herbaceous plants, the shoot typically constitutes 80–90% of the entire plant; in biennials, 70–80%; in herbaceous perennials, 60–70%; and in woody perennials, 20–60% (Monk 1966). The allocation of a large proportion of production to shoot biomass in early successional plants leads to rapid growth and production of large crops of seeds.

Because annual plants must produce seeds quickly and copiously, they never attain a large size. Climax species allocate a larger proportion of their production to root and stem tissue and to the costs of its maintenance, including resistance to stress and herbivory, so as to increase their competitive ability. This reallocation means a slower growth rate. The progression of successional species is therefore accompanied by a shift in the balance between adaptations promoting dispersal and adaptations enhancing competitive ability.

Effects of Species on One Another

Joseph Connell and R. O. Slatyer (1977) suggested three mechanisms by which the presence of one species affects the probability of establishment of a second: facilitation, inhibition, and tolerance. **Facilitation** embodies Clements's view of succession as a developmental sequence in which each stage paves the way for the next, just as structure follows structure during an organism's development and during the building of a house. Colonizing plants enable climax species to invade, just as wooden forms are essential to the pouring of a concrete wall, but have no place in the finished building. Early stages facilitate the development of later stages by contributing to the nutrient and water levels of the soil and by modifying the microenvironment of the soil surface. Facilitation is thought to be particularly important where primary succession occurs in rather severe environments, where the disturbance leaves an area severely deficient in nutrients. In such cases, the activities of pioneering species augment the poor soil sufficiently to give other species a foothold in the area. We have already mentioned how the nitrogen-fixing ability of lupines augmented the barren soil of the Mount St. Helens volcano fields, facilitating later colonization by other plants. Black locust, which harbors nitrogen-fixing bacteria in its roots, appears to contribute nitrogen to soils in the same way during early succession in the southern Appalachian region of the United States (Boring and Swank 1984).

Soils do not develop in marine systems, but facilitation is often encountered when one species enhances the quality of settling and establishment sites for another. Working with experimental panels placed subtidally in Delaware Bay, Dean and Hurd (1980) found that the presence of hydroids enhanced the settlement of tunicates, and both facilitated the settlement of mussels. In southern California, early-arriving, fast-growing algal stands provide dense protective cover for the reestablishment of kelp plants following their removal by winter storms. When Harris et al. (1984) kept areas clear of early successional species of algae, kelp sporophytes that settled there were quickly removed by grazing fish. Turner (1983) found that the establishment of the surfgrass *Phyllospadix scouleri* in rocky intertidal communities depends on the presence of certain early successional algae, to which its seeds cling and then germinate (Figure 28-7). In the absence of these algae, the seagrass cannot invade the community.

Inhibition results when one species is prohibited from becoming established by the presence of another species. This process is intimately connected with species replacement. A newly disturbed area may be colonized by a number of different species. One of those species may inhibit another by eating it, by reducing resources below the level the second species can subsist upon, or by direct conflict. With respect to succession, climax species, by definition, inhibit species characteristic of earlier stages: the latter cannot invade the climax community except following disturbance.

Sousa (1979) provided an excellent example of the role of inhibition in his study of succession in an intertidal algal community of the rocky shore of southern California. In this area of the southern California coast, there are fields of sandstone boulders that have been washed into the ocean from nearby creeks. Various algae and sessile invertebrates grow on these boulders, forming communities there. In Sousa's study area near Ellwood Beach, California, winter storms often produce waves strong enough to overturn boulders, killing the algae and invertebrates that grow on the top of the boulder (death may occur via anoxia or grazing by sea urchins) and exposing its uncolonized

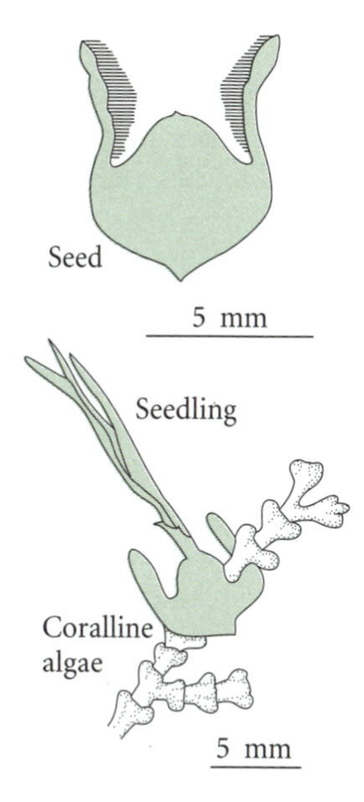

FIGURE 28-7 Seeds of the surfgrass *Phyllospadix* have barbs that allow them to become attached to certain types of erect algae, to which they cling while germinating. (*After Turner 1983.*)

underside. The successional mechanisms involved in the recolonization of these exposed boulder surfaces (and of experimental surfaces placed in the water) were the focus of Sousa's study.

Sousa found that within a month, exposed areas were colonized by mats of the green alga *Ulva*. *Ulva* becomes established on exposed areas quickly and grows rapidly. Because it reproduces year-round, it is able to quickly recolonize open areas. Several species of red algae that occur in the area are also capable of colonizing exposed surfaces. But these species grow relatively slowly and reproduce only seasonally. Thus, they are inferior to *Ulva* in competition for exposed rock surfaces. Sousa found that so long as the *Ulva* mats remained undamaged, they inhibited the establishment of other species of algae. Several factors worked to eventually diminish the inhibitory effect of *Ulva*, however, leading to the establishment of a variety of red algae species and, in most cases, the eventual dominance of one of those species (*Gigartina canaliculata*). The most important factor disrupting the inhibitory effect of *Ulva* was grazing by the crab *Pachygrapsus crassipes,* which prefers *Ulva* over other forms of algae. The study showed that *Ulva* inhibits successional change for a time by preventing the colonization of other algae, but that its inhibitory effect is eventually disrupted by disturbance, and succession proceeds.

Inhibition can give rise to an interesting situation when the outcome of an interaction between two species depends upon which becomes established first. Colonizing propagules often are the most sensitive stage of the life history, and sometimes neither species of a pair can become established in the presence of competitively superior adults of the other. In such cases, the course of succession depends upon precedence. Precedence, in turn, may be strictly random, depending on which species reaches a disturbed site first, or it may follow upon certain properties of the disturbed site—its size, location, the season, and so on. We have seen such a case in the subtidal zone of southern Australia, where bryozoans can prevent the establishment of tunicates and sponges when they become established first. Because of their stronger powers of dispersal, this is more likely to happen on small, isolated substrates.

Tolerance refers to a situation in which species are equally capable of invading newly exposed habitat and becoming established. The ensuing sere is then determined by the competitive abilities of the colonists. Early stages will be dominated by poor competitors that have short life spans but become established quickly; climax species will be superior competitors, but may grow more slowly and not express their dominance in the sere until other species have grown up and reproduced. Whereas facilitation by early successional species will have a positive effect on subsequent colonists, and inhibition will have a negative effect, tolerance in early-arriving species will have little effect on the growth and recruitment of later-arriving species.

The replacement of one species by another during succession is an example of competitive exclusion (see Chapter 21). Tilman's resource ratio hypothesis (1984, 1985, 1986, 1990; see Section 21.9) may explain how competitive interactions lead to successional change. According to this model, plants are viewed as competing simultaneously for light and some soil nutrient such as nitrogen. The outcome of competition for these two resources determines which species of plants will predominate at a particular time. The amounts of the resources available, which may be represented as the ratio of the two resources (resource ratio), will change gradually through time because of the activities of the plants in the community (as, for example, through facilitation). The changes in resource levels will alter the outcome of competitive interactions and lead to community change. For example, during early succession, when there are few plants established in a community, there will be little shade and poorly developed soils containing few nutrients. Species having very low nutrient requirements and a tolerance of high light levels will appear early in the sere, outcompeting other species requiring higher levels of nutrients and lower levels of light. The growth of early successional species will alter the environment in two ways: by increasing soil nutrients through litter input and by shading the soil surface. As these changes occur, species with lower light and higher nutrient requirements will outcompete the early successional plants and, eventually, dominate the community. These plants, in turn, will cause changes in the environment that will eventually render them the less competitive species. Similar competition models have been proposed by Huston and Smith (1987).

28.5 Succession in old fields and glacial areas illustrates the development of the sere.

Clearly all three of Connell and Slatyer's mechanisms—facilitation of establishment, inhibition of establishment, and competitive exclusion (replacement of established species)—together with the life history characteristics of successional species, are important in every sere. The importance of these mechanisms is demonstrated by succession in old fields of North Carolina and in the barren areas left by glacial retreat in Alaska.

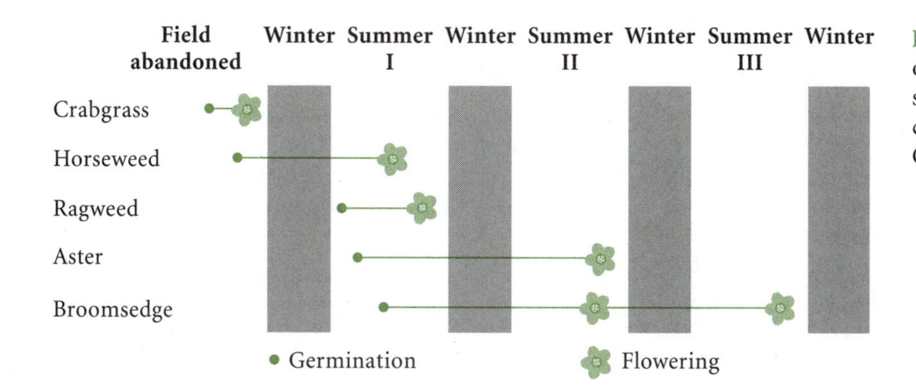

FIGURE 28-8 Schematic summary of the life histories of five early successional species of plants that colonize abandoned fields in North Carolina.

Succession in an Old Field

Early stages of plant succession on old fields in the Piedmont Region of North Carolina (see Figure 28-2) demonstrate how the mechanisms of succession combine in a particular sere (Oosting 1942, Keever 1950, Monk and Gabrielson 1985). The first 3 to 4 years of old-field succession are dominated by a small number of species that replace each other in rapid sequence: crabgrass, horseweed, ragweed, aster, and broomsedge. The life history of each species partly determines its place in the succession (Figure 28-8). Crabgrass, a rapidly growing annual, is usually the most conspicuous plant in a cleared field during the year in which the field is abandoned. Horseweed is a winter annual, whose seeds germinate in the fall. Through the winter, the plant exists as a small rosette of leaves; it blooms by the following midsummer. Because horseweed disperses well and develops rapidly, it usually dominates 1-year-old fields. But because its seedlings require full sunlight, horseweed is quickly replaced by shade-tolerant species. Ragweed is a summer annual; its seeds germinate early in the spring, and the plants flower by late summer. Ragweed dominates the first summer of succession in fields that are plowed under in the late fall, after horseweed normally germinates. Aster and broomsedge are biennials that germinate in the spring and early summer, exist through the winter as small plants, and bloom for the first time in their second autumn. Broomsedge persists and flowers during the following autumn as well.

Horseweed and ragweed both disperse their seeds efficiently and, as young plants, tolerate desiccation. These abilities allow them to invade cleared fields rapidly and produce seed before competitors become established. Decaying horseweed roots stunt the growth of horseweed seedlings; this self-inhibiting effect, whose function and origin are not understood, cuts short the life of horseweed in the sere. These growth inhibitors presumably are the by-products of other adaptations that increase the fitness of horseweed during the first year of succession. One might postulate that if horseweed plants had little chance of persisting during the second year, owing to invasion of the sere by superior competitors, self-inhibition would have little negative selection value. At any rate, self-inhibition is fairly common in early stages of succession (Rice 1984).

Aster successfully colonizes recently cleared fields, but it grows slowly and does not dominate the habitat until the second year. The first aster plants to colonize a field thrive in the full sunlight; the seedlings, however, are not shade-tolerant, and adult plants shade their progeny out of existence. Furthermore, asters do not compete effectively with broomsedge for soil moisture. Keever (1950) observed this when she cleared a circular area, 1 meter in radius, around several broomsedge plants and planted aster seedlings at various distances from them. After 2 months, the dry weight of asters planted 13, 38, and 63 cm from the bases of the broomsedge plants averaged 0.06, 0.20, and 0.46 gram; available soil water at these distances was 1.7, 3.5, and 6.4 grams per 100 grams of soil (Figure 28-9).

The pattern of early succession on the Piedmont suggests the importance of inhibition of seedling establishment and replacement through competitive exclusion. To demonstrate facilitation, one would have to show that late successional species cannot become

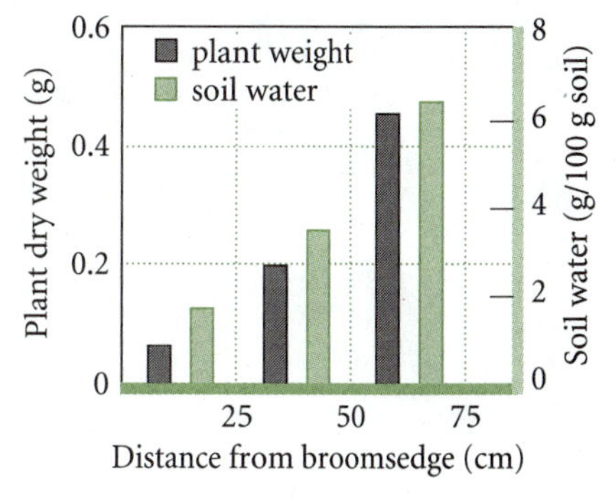

FIGURE 28-9 Growth response of aster (dry weight) and soil water content as a function of distance from broomsedge plants in an old field. (*From Keever 1950.*)

established unless they are preceded by earlier colonists. On experimental old-field plots in southeastern Pennsylvania, J. McCormick painstakingly picked seedlings of early successional species, and discovered that the later successional species still invaded the plots and became established. Probably, agricultural clearing does not disturb the soil enough to reduce its nutrient status and water-holding ability below levels sufficient to support the establishment of most species in the sere. Facilitation is undoubtedly more conspicuous in primary succession.

Succession in an Area of Glacial Retreat

A rapid glacial retreat in Glacier Bay, Alaska (Figure 28-10) exposed extensive areas of barren glacial till that underwent primary succession. A recent study of succession in this area provided a direct experimental analysis of the interaction of life history characteristics with inhibition and facilitation.

Four different successional stages have been identified in the Glacier Bay area:

1. A pioneer stage, consisting of blue-green algal mats, lichens, liverworts, some forbs, and a scattering of willows, cottonwoods, and spruce, along with a mat-forming dwarf shrub (*Dryas drummondii*) that is capable of fixing nitrogen.
2. A second stage characterized by a thick mat of *Dryas,* interspersed with a few willows, cottonwoods, alder, and spruce, that emerges about 30 years after deglaciation. This stage is referred to as the *Dryas* stage.
3. A third stage, featuring alder (*Alnus sinuata*), that appears after about 50 years, called the alder stage.
4. A spruce climax stage, appearing after about 100 years.

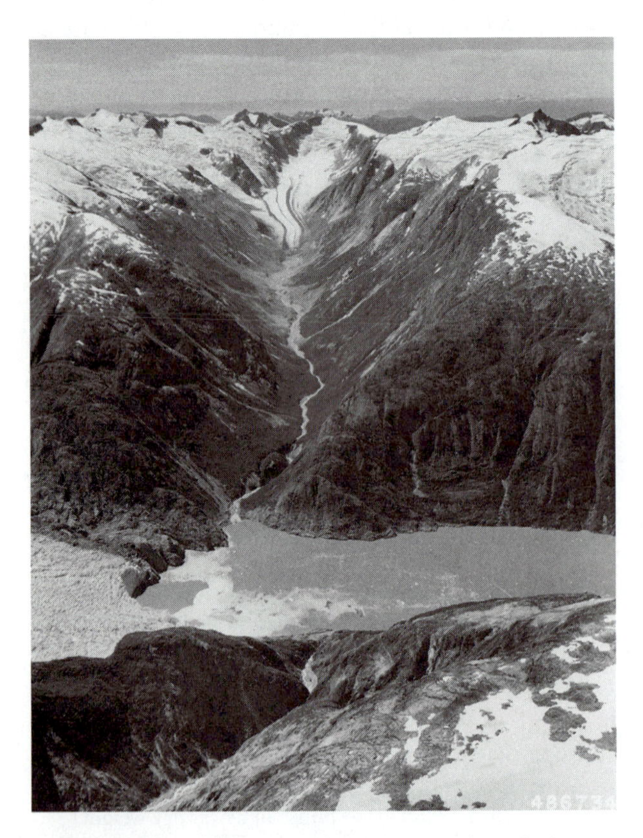

FIGURE 28-10 A valley exposed by a receding glacier, visible at the top center, in North Tongass National Forest, Alaska. (*Courtesy of the U.S. Forest Service.*)

Spruce, which are slow-growing, occur in all successional stages, but do not achieve overgrowth until the alder stage.

It had long been thought that the emergence of the spruce community from the pioneering stage at Glacier Bay results primarily from the accumulation of nitrogen in the soils via the early establishment of *Dryas,* which fixes nitrogen, thereby facilitating the later stages. The transition from one dominant species to the next was thought to result primarily from competitive replacement, which occurs because of both the large sizes and the longer life spans of successive species. Alder lives longer and grows taller than *Dryas* and the other species associated with that stage. Thus, once established, it grows above the *Dryas* community, creating shade and making the environment less favorable for it. Likewise, spruce live longer and grow taller than alder, and once established in the alder community, eventually overgrow the alder and dominate the community. F. S. Chapin III and coworkers (1994) set out to test these ideas using field and laboratory experiments between 1987 and 1989. In particular, they asked whether the establishment of spruce was the result of the presence of nitrogen-fixing *Dryas* in earlier successional stages, and whether other causes of succession, such as life history characteristics, were important in primary succession at Glacier Bay.

Chapin and colleagues found that a complicated interaction between facilitative and inhibitory effects and the life history characteristics of the successional species explains succession at Glacier Bay (Figure 28-11). Two factors tend to work against the transition from pioneer to *Dryas* to alder to spruce. First, neither alder nor spruce has high dispersal capabilities, both because of their heavy seeds and because pioneer and *Dryas* areas are located near the edge of the retreating glacier, quite a distance from large concentrations of alder and spruce. (Chapin and colleagues estimated the average seed mass [± standard error] of alder to be 494 ± 23 μg/seed and that of spruce to be 2,694 ± 26 μg/seed—extremely heavy compared with an average mass of 97 ± 18 μg/seed for *Dryas.*) The researchers placed framed cloth squares 1.5 m^2 in size in pioneer and *Dryas* areas to

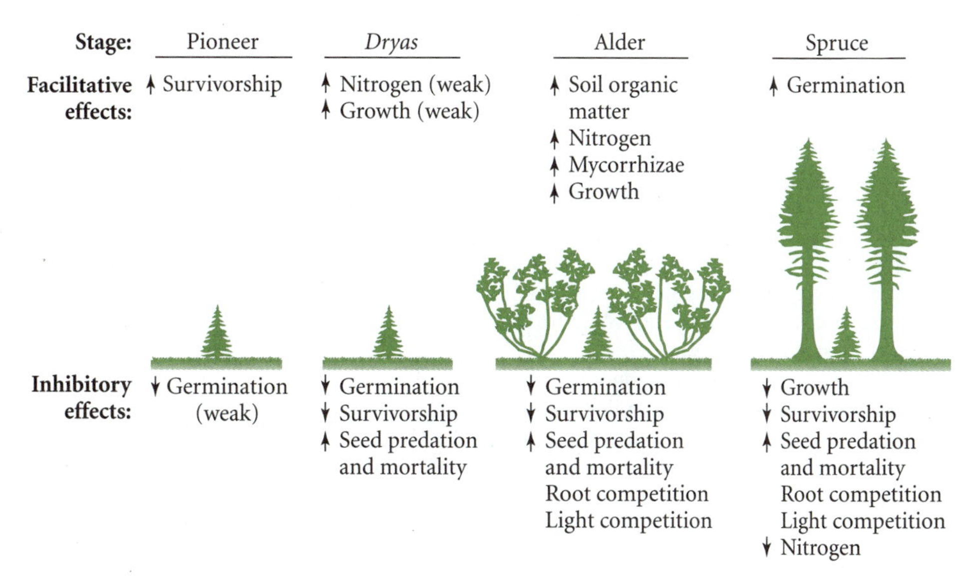

FIGURE 28-11 Diagram showing the complex influences of early successional stages on the establishment of spruce seedlings at Glacier Bay, Alaska. Germination of spruce seeds is inhibited during all three early successional stages, but growth and survival of established seedlings is facilitated. Few spruce seedlings become established in the early stages, but those that do have a high probability of surviving. Nitrogen is increased in the soil by the activities of *Dryas* and alder. (*From Chapin et al. 1994.*)

collect seeds blown in by the wind. They also placed plastic trays (0.134 m^2) beneath the vegetation in alder and spruce areas to catch seeds dropping from those plants. This seed trap experiment, which was conducted in 1987 and 1988, revealed that no spruce seeds, and just a few alder seeds, arrived in the pioneer and *Dryas* areas during the study. Nevertheless, a scattering of spruce seedlings did occur in the pioneer stage, and survivorship of those seedlings was high.

The second factor working against the successional transition at Glacier Bay was the strong inhibition in pioneer and *Dryas* areas of the establishment of both alder and spruce. Soil analysis showed that the availability of nitrogen and the amount of soil organic matter and moisture increased through the four successional stages. Experimental studies revealed that growth of alder and spruce seedlings was supported in lower successional stages. But pioneer and *Dryas* areas exerted inhibitory effects on spruce germination and survivorship, and tended to increase seed predation and seedling mortality in both alder and spruce, thus creating a net inhibitory effect on those species. In addition, in the alder stage, alder and spruce compete for light and soil nutrients (root and shoot competition; see Chapter 22). However, because of the better soil conditions in the alder stage, spruce overgrows alder. Thus, were it not for the ability of the odd spruce to establish itself in early successional stages, inhibition would prevent the transition to a spruce forest.

28.6 Analytical models of succession are based on transitions from one successional stage to the next.

Models of element transfer among ecosystem compartments form the basis of theory of ecosystem regulation (see Chapter 13). Our understanding of population growth and regulation is enhanced by mathematical formulas that allow us to study the dynamics of idealized populations. Likewise, succession may be understood by comparing natural successional processes with analytical models of succession. In most cases, such models represent the probabilities of transition between one community and the next during succession. Let us examine the characteristics and application of one such modeling approach.

During the course of succession, individuals of various species become established, grow, reproduce, and die. Those that die are replaced by the growth of neighbors or the establishment of new individuals of the same or different species. Each change in the system over time may be represented as a transition from one state to another; each transition occurs with a certain probability. Suppose we designate open space (or a recent disturbance) state O, and individuals of the several species that may be present in the system as states A, B, . . . and so on. For sessile organisms like plants,

N_O
N_A
N_B

State vector (**N**)

Future state	Present state		
	O	A	B
O	P_{OO}	P_{AO}	P_{BO}
A	P_{OA}	P_{AA}	P_{BA}
B	P_{OB}	P_{AB}	P_{BB}

Transition probabilities (**P**)

FIGURE 28-12 A state vector, **N**, and matrix transition probabilities, **P**, in a system in which two species (A and B) colonize open space (O) or replace each other. Open space is generated by the death of individuals without replacement. Entries in the state vector represent the relative proportions of each state at time t; entries in the transition matrix represent the probability of change from one state to another.

you may think of particular sites as being either unoccupied (O) or occupied by species A, B. . . . The death of an individual of species A would be the transition A → O; the replacement of an individual of species B by one of A would be B → A. Each of these transitions has a certain probability of occurring (P_{AO}, P_{BA}). Such a system could be represented by a matrix of such transition probabilities, which we may designate **P** (Figure 28-12). (The probabilities P_{OO}, P_{AA}, and P_{BB} represent the probability that the state does not change from one time period to the next.)

This model represents succession as a **Markov process** (named after the Russian mathematician Andrei Andreevich Markov, who developed the theory in the early 1920s), which is a process in which the probability of occurrence of a particular state is dependent only on the present state of the system and not on the path by which the system arrived at its present state. The state of the system at time t is represented by a column vector, **N**, referred to as the **state vector,** of the relative proportions of the various states, O, A, B . . . , each proportion being designated $\mathbf{N}_O$, $\mathbf{N}_A$, . . . (see Figure 28-12). If we assume that the transition probabilities do not change from one time period to the next, then the state of the system at time $t + 1$ may be obtained by

$$\mathbf{PN}(t) = \mathbf{N}(t + 1).$$

The new state, $\mathbf{N}(t + 1)$, can then be multiplied by the transition matrix to obtain the state of the system at $t + 2$ [$\mathbf{PN}(t + 1) = \mathbf{N}(t + 2)$], which is two time periods from the starting point. These matrix operations are the same as those used to project population growth using the Lewis-Leslie matrix (see Chapter 15). Under some circumstances, this iterative process will yield a vector called the **steady-state vector** (also referred to as the vector of stationary distribution of states) that does not change with further multiplications by the transition matrix **P**.

In Figure 28-13 we provide four examples of community transition matrices. The first three panels represent Connell and Slatyer's facilitation, inhibition, and tolerance succession models, which we discussed above. The fourth panel represents a situation that results in cyclic succession, which will be discussed shortly (Section 28.7). We have not given actual transition probabilities, since our goal is to show the general pattern of community change over time. A plus sign in the matrix indicates that the particular transition is possible; a zero indicates that it is not (that is, it would have a transition probability of 0.00). The equilibrium frequencies of the states will be determined by the particular transition probabilities and may be calculated analytically (Horn 1975).

The transition matrix simplifies nature in a number of important ways. Systems are divided into dis-

Facilitation model

Future state	Present state		
	O	A	B
O	+	+	+
A	+	+	0
B	0	+	+

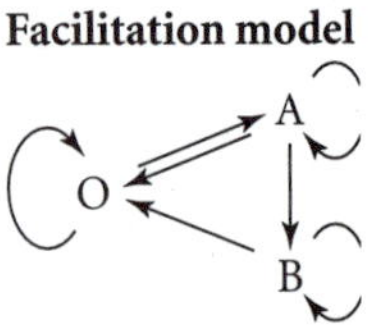

(a)

Inhibition model

Future state	Present state		
	O	A	B
O	+	+	+
A	+	+	0
B	+	0	+

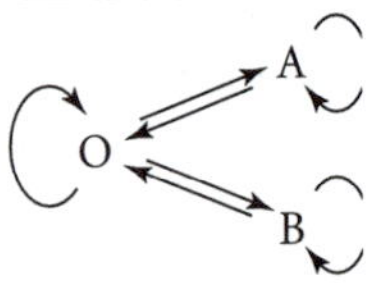

(b)

Tolerance model

Future state	Present state		
	O	A	B
O	+	+	+
A	+	+	0
B	+	+	+

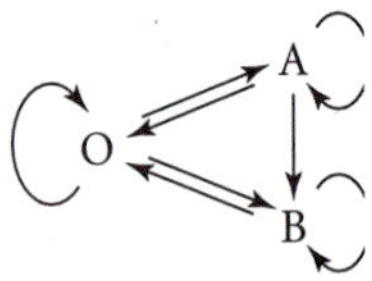

(c)

Cyclic model

Future state	Present state		
	O	A	B
O	+	+	+
A	+	0	+
B	+	+	0

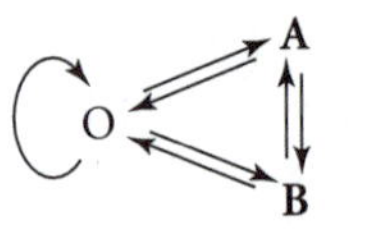

(d)

FIGURE 28-13 Transition probabilities for successional systems behaving according to facilitation, inhibition, tolerance, and cyclic models of species interaction. A plus sign indicates that the transition is possible; a zero indicates that it is not possible. (a) Only species A can colonize an empty area. Species A facilitates species B. If species B is present, the next state will be O or B. (b) Both A and B can colonize empty areas. When A is present it inhibits B, and when B is present it inhibits A. (c) Both A and B can colonize, and the two species can live together. (d) Both A and B can colonize. If A colonizes first, it facilitates B, but B inhibits itself. If B colonizes first, it facilitates A while inhibiting itself.

crete subunits that correspond only loosely to individuals or areas of habitat; in nature, transition matrices undoubtedly vary with season, age of stand, and other factors (Lippe et al. 1985). Transition probabilities may in fact be influenced by prior transitions. Nevertheless, the technique has been employed to great advantage in a number of community studies. Let us take a look at how the Markov model has been applied to a simple desert community and a more complex forest community.

Markovian Dynamics in a Desert Plant Community

J. R. McAuliffe applied a Markov modeling approach to his study of two relatively simple desert plant communities near Dateland and San Luis in Yuma County, Arizona (McAuliffe 1988). Both areas are predominantly open desert (<12% canopy cover at Dateland and <6% cover at San Luis). Two desert shrub species, *Larrea tridentata* and *Ambrosia dumosa,* are the predominant vegetation in both areas. McAuliffe found that the distribution of *Larrea* was not independent of that of *Ambrosia.* Specifically, he found that at both sites, nearly all young Larrea were rooted under the canopy of the larger *Ambrosia.* Moreover, a high proportion of young *Larrea* were rooted next to dead *Ambrosia,* indicating the possibility of facilitation. Figure 28-14 depicts the hypothesized dynamics between open sites, which predominate in McAuliffe's system, and sites occupied by *Larrea* and *Ambrosia.*

To understand the dynamics of this simple community, McAuliffe used a variety of information about plant densities, mortality rates, wood decay rates, and recruitment to calculate transition probabilities for a simple Markov transition matrix (Table 28-3). As we pointed out earlier, the calculation of transition probabilities results in a simplification of the system. For example, in the case of McAuliffe's communities, the replacement of *Ambrosia* by the *Larrea* growing beneath its canopy may take decades, and thus, year-to-year transition probabilities would be unrepresentative of the system. Because the mortality rate of *Larrea* is so low after plants reach 1 or 2 years of age, McAuliffe assumed that situations in which *Larrea* occurred beneath *Ambrosia* would lead inevitably to the replacement of *Ambrosia.* Other assumptions such as this, based on McAuliffe's knowledge of the natural history of the system, allowed the construction of a year-to-year transition matrix for each site.

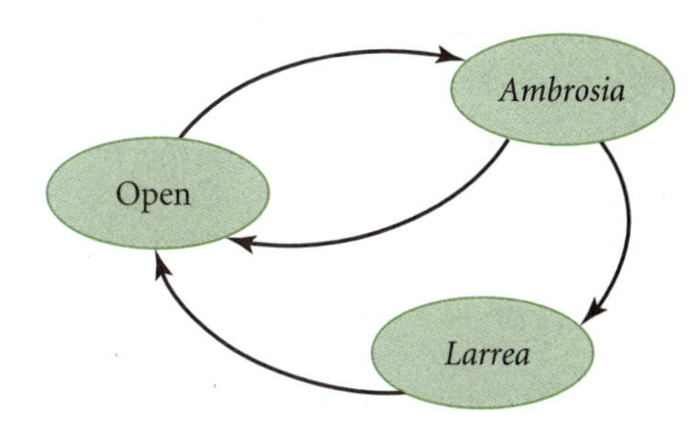

FIGURE 28-14 Diagram of the relationships between the three states of the *Larrea-Ambrosia* communities of the deserts of Arizona. Open sites may be occupied by *Ambrosia* but not by *Larrea.* Compare this figure with panel (a) of Figure 28-13 to appreciate the possible importance of facilitation in this community. (*From McAuliffe 1988.*)

The transition probabilities reveal much about the dynamics of this community. In both the Dateland and San Luis study sites, the diagonals of the transition matrix have very large probabilities (exceeding 0.90 in each case). This indicates that the rate of change in these communities is very slow. The most likely transition from one time period to the next is that nothing will happen; that is, colonization of open sites or replacement of one species by another is unlikely. The probability of transition from an open site to *Ambrosia* is greater than the probability of transition from an open site to *Larrea* because of the apparent facilitation of *Larrea* by *Ambrosia.* The probability of a transition from *Ambrosia* to *Larrea* is estimated to be negligible.

McAuliffe calculated the steady-state vector for his system, which serves as a predictor of the relative frequencies of the states of the system at equilibrium. A

TABLE 28-3 Transition matrices for two desert communities in Arizona

	Present state					
	Dateland			San Luis		
Future state	Open	*Ambrosia*	*Larrea*	Open	*Ambrosia*	*Larrea*
Open	0.998	0.028	0.00012	0.998	0.031	0.0016
Ambrosia	0.0016	0.917	0.00030	0.0013	0.968	0.00
Larrea	0.00012	0.00	0.999	0.00016	0.00058	0.984

(*From McAuliffe 1988.*)

TABLE 28-4 A comparison of the observed relative frequencies of occurrence of three states (open site, occupied by *Ambrosia*, occupied by *Larrea*) with the frequencies predicted by Markov analysis for two desert communities in Arizona)

	Dateland		San Luis	
Cover type	Observed	Predicted	Observed	Predicted
Open	0.883	0.814	0.947	0.865
Ambrosia	0.057	0.046	0.025	0.036
Larrea	0.059	0.139	0.028	0.099

Note: *Proportions do not add to 1 because of rounding errors. (From McAuliffe 1988.)*

comparison of the observed and predicted state frequencies for both study sites is presented in Table 28-4. The results of this analysis show considerable concordance between the predictions of the Markov model and the communities observed. The Markov projection predicts that most sites will be unoccupied, and that both *Larrea* and *Ambrosia* will co-occur in the system.

Markovian Dynamics in a Forest Community

Markovian models have also been applied to forest communities. As with the desert system above, assumptions about recruitment, mortality, life span, and a myriad other factors must be made in order to estimate transition probabilities. Nevertheless, such modeling approaches can reveal a great deal about forest community dynamics. A good illustration of the use of this approach to understanding forest dynamics is the work of Henry Horn (1975) in a forest near Princeton, New Jersey.

Horn counted the number of saplings of the five principal species of trees in the Princeton forest and noted how they were distributed beneath the canopies of adults of those species (Table 28-5). He took these proportions to be transition probabilities. That is, he assumed that the proportion of the saplings of one species that occurred beneath the canopy of another species represented the replacement rate of the canopy species by the sapling species. For example, he found that 10% of the saplings beneath red oak trees were sweet gum. Thus, he assumed that the probability of a transition from red oak to sweet gum in the canopy was $P_{\text{red oak, sweet gum}} = 0.10$. To be sure, these probabilities are not strictly transition probabilities, since much can happen between the opening of a gap in the canopy and its being filled by another individual, and since no consideration is given to variable mortality rates, the distributions of the various species in the canopy—which influence the number of saplings of each species found below the canopy—and other factors, such as the rate at which canopy gaps open. But if one assumes that the sapling distributions on the forest floor are adequate estimates of transition probabilities, hypotheses about forest transitions may be developed using a Markov projection.

TABLE 28-5 The proportions of saplings of five tree species occurring under mature individuals of each of those species in a forest near Princeton, New Jersey

	Canopy				
Saplings	Sweet gum	Red oak	Hickory	Red maple	Beech
Sweet gum	39	10	1	13	1
Red oak	9	11	16	11	1
Hickory	9	11	5	26	1
Red maple	34	45	58	18	7
Beech	9	23	20	32	90
Total	100	100	100	100	100

(From Horn 1975.)

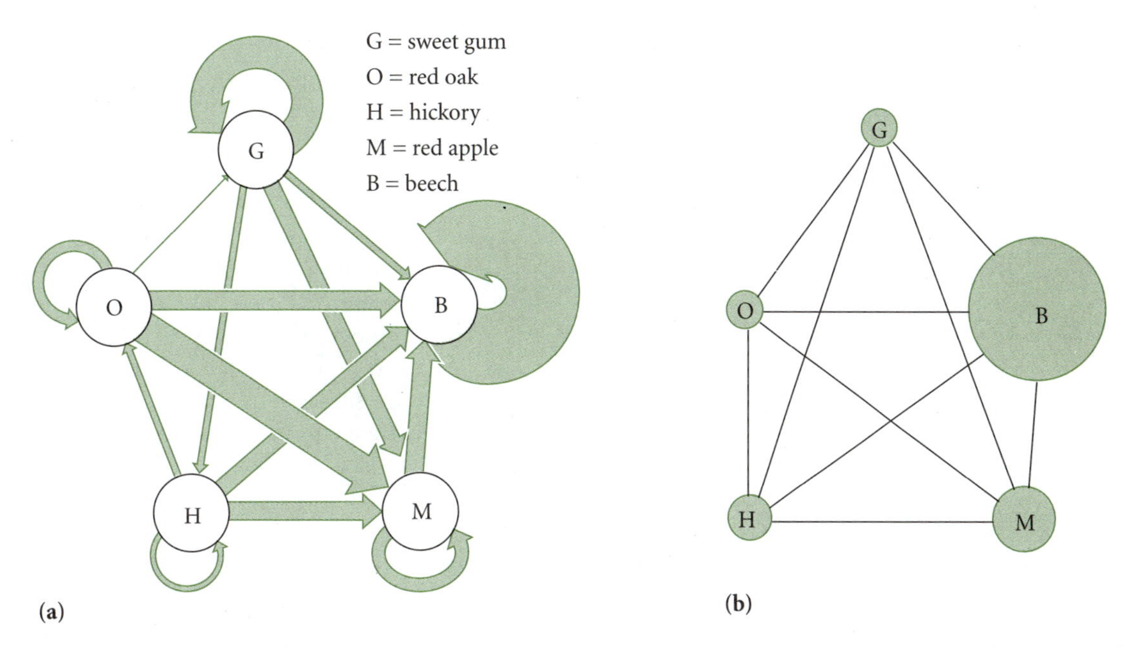

FIGURE 28-15 Predicted transitions in a deciduous forest community near Princeton, New Jersey, based on the probabilities shown in Table 28.5. (a) The thicknesses of the arrows are proportional to the transition probabilities. (b) The sizes of the circles represent the relative abundances of each species after the system has come into equilibrium.

Figure 28-15 shows a transition in the forest based on these probabilities. The projection suggests that beech trees should come to dominate the forest canopy both because of the high proportion of beech saplings beneath beech trees ($P_{\text{beech, beech}} = 0.39$) and the high transition probabilities from oak to beech, hickory to beech, and maple to beech. The high transition probabilities of oak to maple, and hickory to maple, suggest that the presence of maple will increase in the canopy. Similar but more complex models have been constructed to predict the course of succession in particular forest stands (Botkin et al. 1972, Shugart 1984).

28.7 The character of the climax is determined by local conditions.

In the absence of strong disturbance, succession eventually leads to a situation in which environmental conditions change slowly and newly invading species are not able to replace existing species at a site. For example, once forest vegetation establishes itself, patterns of light intensity and soil moisture do not change, except in the smallest details, with the introduction of new species of trees. Beech and maple replace oak and hickory in northern hardwood forests because their seedlings are better competitors in the shade of the forest floor environment, but beech and maple seedlings probably develop as well under their own parents as they do under the oak and hickory trees they replace. At this point, succession reaches a climax; the community has come into equilibrium with its physical environment (Leak 1970; Waggoner and Stephens 1970).

To be sure, subtle changes in species composition usually follow the attainment of the climax growth form of a sere. For example, a site near Washington, D.C., left undisturbed for nearly 70 years developed a tall forest community dominated by oak and beech. When saplings and large trees in this stand were sampled (Figure 28-16), it became clear that the community had not yet reached an equilibrium because the saplings in the forest understory—which eventually replace the existing trees—included neither white or black oak (Dix 1957). In another century, the forest will probably be dominated by the species showing the most vigorous reproduction, namely, red maple, sugar maple, and beech.

The end of successional change does not mean the end of community development. Climax communities undergo changes in structure as a result of birth, death, and growth processes in the community. To reemphasize, these changes are less dramatic than the community transformations observed during succession. Nevertheless, modifications of the structure of climax communities are measurable. In a comparison of twenty-five old-growth hemlock-hardwood forests in northern Wisconsin and Michigan, Tyrrell and Crow (1994) found that stands of different ages sometimes differed in the distribution of tree size classes, tree density, and the volume of dead logs in the forest. Differences in the number of canopy gaps

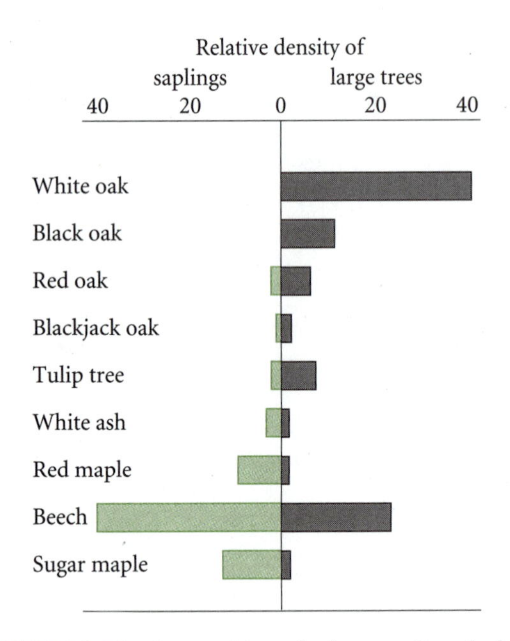

FIGURE 28-16 Composition of a forest undisturbed for 67 years near Washington, D.C. The relative predominance of beech and maple saplings in the understory foretells a gradual successional change in the community beyond the present oak-beech stage. (*After Dix 1957.*)

among different-aged stands were also observed. Such results demonstrate the dynamic nature of climax communities.

The biological properties of a developing community change as species enter and leave the sere. As a community matures, the ratio of biomass to productivity increases; the maintenance requirements of the community also increase until production no longer can meet the demand, at which point the net accumulation of biomass in the community stops (Odum 1969, Whittaker 1975, Peet 1981). The end of biomass accumulation does not necessarily signal the attainment of climax; species may continue to invade the community and replace others. But the attainment of a steady-state biomass does mark the end of major structural change in the community; further changes are limited to the adjustment of details.

As plant size increases with succession, a greater proportion of the nutrients available to the community are tied up in organic materials. Furthermore, because the vegetation of mature communities has more supportive tissue, which is less digestible than photosynthetic tissue, a larger proportion of their productivity enters the detritus food chain rather than the consumer food chain. Other aspects of the community change as well (see, for example, Vitousek and White 1981). Soil nutrients are held more tightly in the ecosystem because they are not exposed to erosion; minerals are taken up more rapidly and stored to a greater degree by the well-developed root systems of forests; the environment near the ground is protected by the canopy of the forest; conditions in the litter are more favorable to detritus-feeding organisms.

Nature of the Climax

Clements believed that succession resulted in a single true climax community that was determined primarily by the climate of the region. This view, which is called the **monoclimax theory** of succession, holds that the many different vegetation communities found in a region are successional stages of the true climax community. Such communities were often called subclimax, preclimax, or postclimax communities. The monoclimax theory suggests that, given sufficient time, the differences in local conditions of soil moisture, temperature, nutrient availability, hydrology, and so forth that give rise to different vegetation types would be overcome and a homogeneous true climax would emerge. This view naturally gave way to the **polyclimax theory** of succession, which recognized the validity of many different types of vegetation as climaxes, depending on local conditions. More recently, the development of the continuum index and gradient analysis fostered the broader **pattern-climax theory** of Robert Whittaker (1953), which recognizes a regional pattern of open climax communities whose composition at any one locality depends on the particular environmental conditions at that point.

Many factors determine the nature of the climax community, among them soil nutrients, moisture, slope, and exposure. Fire is an important feature of many climax communities, favoring fire-resistant species and excluding others that otherwise would dominate (Cooper 1961, Kozlowski and Ahlgren 1974, Gill 1975, Christensen 1985, Riggan et al. 1988, Whelan 1995). The vast southern pine forests of the Gulf Coast and southern Atlantic Coast states are maintained by periodic fires. The pines are adapted to withstand scorching that destroys oaks and other broad-leaved species (Figure 28-17). Some species of pines do not even shed their seeds unless their release is triggered by the heat of a fire passing through the understory below. After a fire, pine seedlings grow rapidly in the absence of competition from other understory species.

Any habitat that is occasionally dry enough to create a fire hazard but normally wet enough to produce and accumulate a thick layer of plant detritus is likely to be influenced by fire. The chaparral vegetation of seasonally dry habitats in California is a fire-maintained climax that is replaced by oak woodland when fire is prevented. The forest-prairie edge in the midwestern United States separates climatic climax and fire climax communities (Borchert 1950). The forest-prairie edge occasionally shifts back and forth

(a)

(b)

(c)

FIGURE 28-17 (a) A stand of longleaf pine in North Carolina shortly after a fire. Although the seedlings are badly burned (b), their growing shoots are protected by the dense, long needles—shown on an unburned individual, (c)—and often survive. In addition, the slow-growing seedlings have extensive roots that store nutrients to support the plant following fire damage.

across the landscape, depending on the intensity of recent drought and the extent of recent fires. After prolonged wet periods, the forest edge may advance out onto the prairie as tree seedlings grow up and begin to shade out the grasses. Prolonged drought followed by intense fire can destroy tall forest and allow rapidly spreading prairie grasses to gain a foothold. Once prairie vegetation is established, however, fires become more frequent owing to the rapid buildup of flammable litter. Reinvasion by forest species then becomes more difficult. Frequent burning eliminates seedlings of hardwood trees, but the perennial grasses sprout from their roots after a fire (Daubenmire 1968a). By the same token, mature forests resist fire and rarely become damaged enough to allow the encroachment of prairie grasses. Hence the forest-prairie boundary remains generally stable.

Grazing pressure also can modify the nature of the climax community (Harper 1969). Grassland can be turned into shrubland by intense grazing. Herbivores may kill or severely damage perennial grasses and allow shrubs and cacti unsuitable for forage to establish themselves. Most herbivores graze selectively, suppressing favored species of plants and bolstering competitors that are less desirable as food. On the African plains, grazing ungulates move through an area in a regular succession of species, each using different types of forage (Vesey-Fitzgerald 1960, Gwynne and Bell 1968, Jarman and Sinclair 1979, McNaughton 1979, Walker 1981). By excluding wildebeests, the first of the successional species, from large fenced-off areas, McNaughton (1976) was able to show that the subsequent wave of Thompson's gazelles preferred to feed in areas previously used by wildebeests or other

FIGURE 28-18 Zebras and Thompson's gazelles feed side by side in the Serengeti ecosystem of East Africa, but utilize different food plants.

FIGURE 28-19 Vultures feeding on a wildebeest carcass in Masai Mara Park, Kenya.

large herbivores (Figure 28-18). Apparently, heavy grazing by wildebeests stimulates the growth of the gazelles' preferred food plants and reduces cover within which predators of the smaller gazelles could conceal themselves. In western North America, grazing allows the invasion of the alien cheatgrass (*Bromus tectorum*), which promotes fire and may lead succession to an alternative stable state.

Transient and Cyclic Climaxes

We usually view succession as a series of changes leading to a climax determined by, and in equilibrium with, the local environment. Once established, the beech-maple forest is self-perpetuating, and its general appearance does not change in spite of the constant replacement of individuals within the community. Yet not all climaxes are persistent. Simple cases of **transient climaxes** include the development of animal and plant communities in seasonal ponds—small bodies of water that either dry up in the summer or freeze solid in the winter and thereby regularly destroy the communities that become established each year during the growing season. Each spring the ponds are restocked either from larger, permanent bodies of water or from spores and resting stages left by plants, animals, and microorganisms before the habitat disappeared the previous year.

Succession recurs whenever a new environmental opportunity appears. For example, excreta (Mohr 1943) and dead organisms are a resource for a wide variety of scavengers and detritus feeders. On African savannas, carcasses of large mammals are fed upon by a succession of vultures (Figure 28-19), beginning with large, aggressive species that devour the largest masses of flesh, followed by smaller species that glean smaller bits of meat from the bones, and finally by a kind of vulture that cracks open the bones to feed on the marrow (Kruuk 1967, Houston 1979). Scavenging mammals, maggots, and microorganisms enter the sere at different points and ensure that nothing edible remains. This succession has no climax because all the scavengers disperse when the feast is concluded. We may, however, consider all the scavengers a part of a climax, which is the entire savanna community.

In simple communities, particular life history characteristics in a few dominant species can create a **cyclic climax.** Suppose, for example, that species A can germinate only under species B, B can germinate only under species C, and C only under A. This situation would create a regular cycle of species dominance in the order A, C, B, A, C, B, A, . . . , with the length of each stage determined by the life span of the dominant species.

Stable cyclic climaxes, which are known from a variety of localities, usually follow this pattern, often with one of the stages being bare substrate (Watt 1947, Forcier 1975, Sprugel 1976). Wind or frost heaving sometimes drives the cycle. When heaths and other vegetation forms suffer extreme wind damage, shredded foliage and broken twigs create an opening for further damage, and the process becomes self-accelerating. Soon a wide swath is opened in the vegetation. Regeneration occurs on the protected side of the damaged area while wind damage further encroaches upon the exposed vegetation. As a result, waves of damage and regeneration move through the community in the direction of the wind (Figure 28-20). If we watched the sequence of events at any one location, we would witness a healthy heath being reduced to bare earth by wind damage and then regenerating in repeated cycles (Figure 28-21). Similar cycles occur in windy regions where hummocks of earth form around the bases of clumps of grasses. As the hummocks grow, the soil

FIGURE 28-20 Waves of regeneration in balsam fir forests on the slopes of Mt. Katahdin, Maine. (*Courtesy of D. G. Sprugel; from Sprugel and Bormann 1981.*)

becomes more exposed and better drained. With these changes in soil quality, shrubby lichens take over the hummock and exclude the grasses around which the hummock formed. The shrubby lichens are worn down by wind erosion and eventually are replaced by prostrate lichens, which resist wind erosion but, lacking roots, cannot hold the soil. Eventually the hummocks are completely worn down, and grasses once more become established and renew the cycle.

Mosaic Patterns of Successional Stages

Mosaic patterns of vegetation types are common in any climax community where the death of individuals alters the environment. Treefalls open the forest canopy and create patches of habitat that are dry, hot, and sunlit compared with the forest floor under the unbroken canopy. These openings are often invaded by early seral forms, which persist until the canopy closes (Aubréville 1938, Forcier 1975, Williamson 1975, Brokaw 1985). Treefalls thus create a mosaic of successional stages within an otherwise uniform community. Indeed, adaptation by some species to grow in the conditions created by different-sized openings in the canopy could enhance the overall diversity of the climax community (Denslow 1980, 1985, Orians 1982, Pickett 1983). A similar pattern is found in intertidal regions of rocky coasts, where wave damage and intense predation continually open new patches of habitat (Dayton 1971, Levin and Paine 1974, Connell 1978, Sousa 1985, Connell and Keough 1985).

Cyclic patterns of change and mosaic patterns of distribution must be incorporated into the concept of the climax community. The climax is a dynamic, self-perpetuating state, even if it perpetuates itself through regular cycles of change. Persistence is the key to the climax. If a cycle persists, it is inherently as much of a climax as an unchanging steady state.

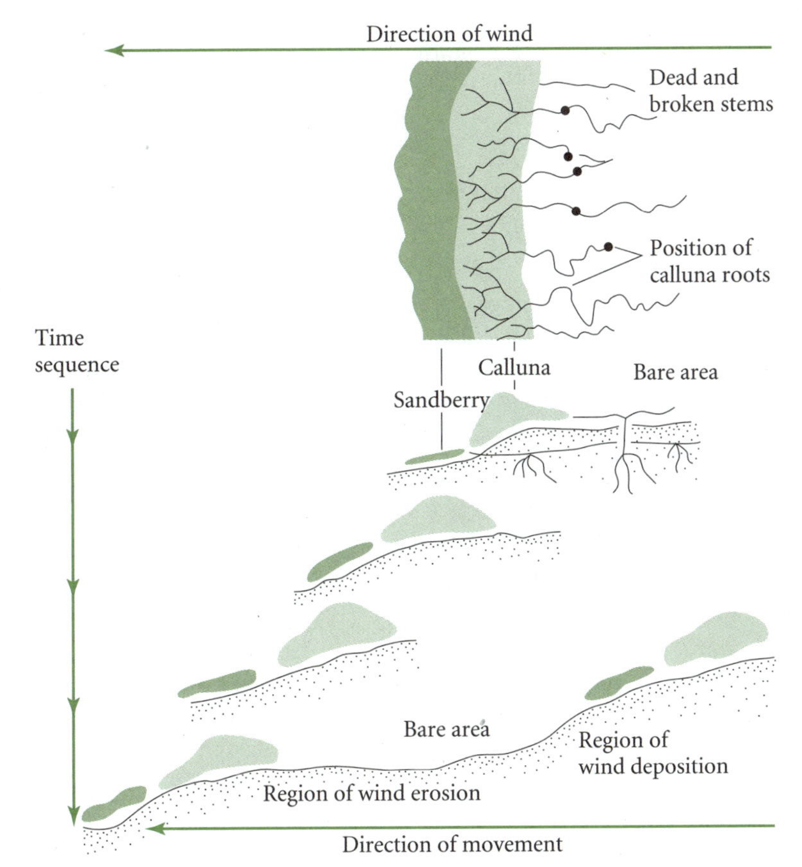

FIGURE 28-21 Sequence of wind damage and regeneration in the dwarf heaths of northern Scotland. (*After Watt 1947.*)

Succession and Species Diversity

Historically, many ecologists have held that communities become more diverse and complex as succession progresses (Margalef 1968, Odum 1969). However, Whittaker (1975) suggested that, in some seres, intermediate stages of succession may be more diverse because they contain elements of early seral stages as well as elements of the climax community. Indeed, such a situation occurred in Sousa's study (1979) of succession in intertidal algal communities, where species diversity increased initially and then declined in later successional stages as one algal species came to dominate the sites. Higher levels of species diversity in the intermediate stages of succession have been observed in many communities (Shugart and Hett 1973, Sousa 1980, Huston 1994).

The rapid increase in diversity early in succession is probably the result both of the arrival of species in an area where few species exist and of augmentation of the soil by early colonists facilitating the survival of species arriving later (Huston 1994). The improvement of soil by early colonists has long been suggested as the major force of successional change in primary succession in areas of glacial retreat (Crocker and Major 1955), though the recent studies by Chapin et al. (1994) show that other factors are also important.

Succession emphasizes the dynamic nature of the biological community. By upsetting the natural balance, disturbance reveals to us the forces that determine the presence or absence of species within a community and the processes responsible for the regulation of community structure. Succession also emphasizes that the structure of the community comprises a patchwork mosaic of successional stages, and that community studies must consider disturbance cycles on many scales of time and space. The next three chapters address the structure of the community and its regulation. The dynamics and structural consequences of disturbance and succession will never be far from the surface of our discussions.

SUMMARY

1. Succession refers to change in a community following either disturbance or the colonization of newly exposed substrate. The particular sequence of communities at a given location is referred to as a sere, and the ultimate stable association of plants and animals is the climax.

2. Succession on newly formed substrates, such as sand dunes, landslides, and lava flows—referred to as primary succession—involves substantial modification of the environment by early colonists. Moderate disturbances, which leave much of the physical structure of the ecosystem intact, are followed by secondary succession. Autogenic succession refers to succession brought about by forces internal to the community; allogenic succession is succession initiated by factors outside the community.

3. Physical or biological disturbance may initiate the process of succession. Succession in marine habitats shows that the initial stages of the sere depend upon the intensity and extent of the disturbance, and that the responses of predators and herbivores to successional changes can influence the pattern of community development. Studies of succession following fire show how the structure of the community influences the intensity of the disturbance in some cases.

4. The presence and persistence of a species in a sere, especially in secondary succession, depend upon its colonizing and competitive abilities. Members of early seral stages tend to disperse well and grow rapidly; those of later stages tend to tolerate low resource levels or dominate direct interactions with other species.

5. Connell and Slatyer categorized the processes by which one established species affects the probability of colonization by a second potential invader as facilitation, inhibition, and tolerance. Facilitation is most prominent in early stages of primary succession. Inhibition is a more common feature of secondary succession, and may be expressed in priority effects, conferring competitive dominance on the first arrival.

6. Patterns of succession in North Carolina old fields and in an area of glacial retreat in Alaska demonstrate the complexity of succession.

7. The interactions that drive succession may be described by a transition matrix of probabilities of replacement of species A by species B. The dynamics of succession and the composition of the local climax may be determined from the properties of this matrix.

8. Succession continues until the community is dominated by species that are capable of becoming established in their own and one another's presence. At this point the community becomes self-perpetuating, but changes in the structure and development of the community do not cease.

9. Climax communities are continua of community types determined primarily by climate and topography. Within a region, seres tend to converge to a single type. The character of the climax may be influenced profoundly by local conditions, such as fire and grazing, that alter interactions between seral species.

10. Transient climaxes develop on ephemeral resources and habitats, such as temporary ponds and the carcasses of individual animals. In such cases, the regional climax may be thought of as including transient seres.

11. Cyclic local climaxes may develop where each species can become established only in association with some other species. Cyclic climaxes are often driven by harsh physical conditions, such as frost and strong winds. The regional climax may be thought of as including the local cyclic sere.

EXERCISES

1. Find an area close to where you live where there has been a relatively recent disturbance. Describe the features of the vegetation that you find growing in that area.

2. Describe a situation in which you might observe succession in an animal community.

3. Explain how succession might affect the movement of materials and energy in ecosystems.

4. A simple community contains three species: A, B, and C. At a particular time, there are 10 individuals of species A, 20 individuals of species B, and 30 individuals of species C present in the community. The presence of species A inhibits species B and facilitates species C. Species B and C tolerate each other, and both of these species inhibit species A. Using arbitrary values, construct a transition matrix of successional change in the community and project it through three time periods. Set the diagonals of the matrix—the probabilities that there is no change from the present condition—to 0.5. (Hints: 1. Review the material in Chapter 15 regarding matrix operations. 2. The state vector, **N**, is the number of individuals of each species in the community at the present time.)

CHAPTER 29

Biodiversity

GUIDING QUESTIONS

- What are the various types of diversity?
- What general patterns of species diversity have ecologists observed?
- What is the difference between regional/historical and local/deterministic thinking with respect to diversity?
- What are the relationships among alpha, beta, and gamma diversity?
- What is the relationship between species diversity and immigration and extinction on islands?
- What information may be gained from species/genus ratios?
- What is the time hypothesis of diversity?
- What is the relationship between species richness and niche dynamics?
- What is the paradox of enrichment?
- How do predators and herbivores affect local species diversity?
- What are the mechanisms by which disturbance affects species diversity?
- How do ecologists use null models?
- What is community character displacement?

Throughout this book we have emphasized the great variation in the natural world. The earth holds a vast diversity of living things. Although taxonomists so far have catalogued fewer than 2 million species, by extrapolating rates of discovery of new insects and other life forms, some biologists have estimated that as many as 30 million species of animals and plants may inhabit the earth, most of them small insects in tropical forests (Wilson 1992). Recent estimates suggest that the number of bacterial species is probably two hundred times higher than the number described. In addition to this great number of different kinds of plants, animals, and microorganisms, the earth holds an immense variety of habitats and ecosystems, some of which were described in Chapter 8. Taken together, all of this variation in species and habitats is referred to as **biodiversity.** Many of the mechanisms that promote and maintain this biodiversity can be understood in the context of community dynamics.

It is important to emphasize that biodiversity includes a number of different levels of variation in the natural world (e.g., see Noss 1990). Genetic diversity, of course, fuels the engine of evolution. We have already discussed how the loss of genetic diversity in populations of highly endangered plants and animals can spell doom for those species (see Chapter 19). We shall discuss genetic variation in more detail in Part 7. There is also diversity related to the numbers and relative abundances of species within a community. In Chapter 27, we developed diversity indices that combined these two attributes of community structure, which are referred to together as **species diversity.** We noted that the most important of these attributes is species richness. We may also think of the richness of higher taxa such as genera and families. The diversity of genera and families provides clues about the distinctiveness of various lineages of organisms. Each lineage harbors unique genetic variation, which is important to preserve. Biodiversity may also be viewed at the ecosystem level—so-called **ecosystem diversity**—which encompasses the great variety of habitat types and biomes that we discussed in Part 2. These different levels of diversity describe a hierarchy from the individual and population levels of genetic variation, through community levels of species diversity and diversity of higher taxa, to the ecosystem level.

In this chapter we will focus primarily on species diversity (which we shall sometimes refer to simply as *diversity*). Our goal is to explore the ways in which species richness arises and is maintained in communities. We will first briefly describe a number of general patterns of species diversity that ecologists have observed in time and space. Then we will distinguish between regional and local views of species diversity and discuss the dynamics of species richness in island systems as an example of the application of a regional perspective. In the last part of the chapter we shall turn to ideas about what limits species diversity in local communities.

29.1 A number of general patterns of species diversity have been observed.

Ecologists have observed a number of different patterns of species richness. These patterns serve as the basis for observational and experimental studies of the mechanisms of diversity.

Diversity in Geologic Time

The number of species on earth has not been constant since the emergence of cellular life some 3.2 billion years ago. Indeed, it is likely that the present time is not the most diverse in the earth's history (Gould 1989). Even though the fossil record is incomplete, particularly for periods of time close to the beginning of life on earth, the preservation of organisms with hard body parts, such as marine invertebrates, provides a glimpse of changes in diversity through geologic time. In Chapter 19 we discussed the geologic history of mass extinctions on earth. Between these great extinctions were periods of great evolutionary radiation and increase in the richness of families and higher taxa. Figure 29-1 shows how the number of families of marine invertebrates has changed over geologic time. There appear to have been at least three periods of lineage proliferation beginning in the Cambrian, each achieving a higher level of family diversity. Following the initial diversification of complex life forms, there was a period of pervasive extinction at the end of the Cambrian. Diversity subsequently rose to a new peak in the Paleozoic, only to be decimated once more by mass extinction at the end of Permian.

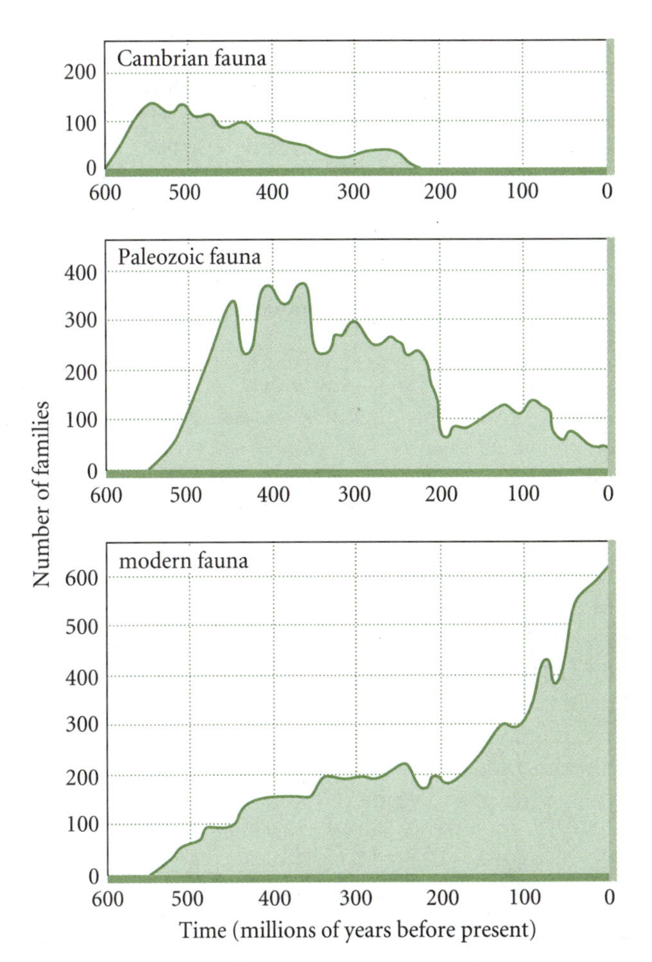

FIGURE 29-1 Number of families of organisms that arose during three geologic periods. (*From Meffe and Carroll 1997b; after Sepkoski 1984.*)

The most recent rapid rise in the number of families has contributed to the development of the modern fauna (Sepkoski 1984).

Latitudinal Gradients of Diversity

The great naturalist-explorers of the nineteenth century—Charles Darwin, Henry Bates, Alfred Russel Wallace, and others—recognized that the Tropics held a great store of undescribed species, many with bizarre forms and habits. Within most large taxonomic groups of organisms—plants, animals, and perhaps microbes—the number of species increases markedly, with a few exceptions, toward the equator (Fischer 1960, Stehli 1968, Stehli et al. 1969, Stevens 1989). For example, within a small region at 60° north latitude one might find 10 species of ants; at 40°, there may be between 50 and 100 species; and in a similar sampling area within 20° of the equator, between 100 and 200 species. Greenland is home to 56 species of breeding birds, New York boasts 105, Guatemala 469, and

Colombia 1,395 (Dobzhansky 1950). Bird diversities in some places in the Amazon exceed 100 species in a single small sampling location (Terborgh et al. 1990). Species, genus, and family diversity in marine environments follow a similar trend. Terrestrial birds, trees, and mammal species all show greater species richness toward the Tropics (Figure 29-2; Briggs 1995).

Habitat and Physical Conditions

Within a given belt of latitude around the globe, the number of species may vary widely among habitats according to their productivity, degree of structural heterogeneity, and suitability of physical conditions (see, for example, Kotler and Brown 1988, Gentry 1988, Rosenzweig and Abramsky 1993). The relationship between structure and diversity is apparent to birdwatchers and other naturalists, but MacArthur and MacArthur (1961) were the first to set it forth in plain graphic form for ecologists by plotting the diversity of birds observed in different habitats according to foliage height diversity—a measure of the structural complexity of vegetation (Figure 29-3). Greenstone (1984) has shown a similar pattern for web-building spiders, noting that the diversity of species varies in direct relationship to heterogeneity in the heights of the tips of the vegetation to which such spiders attach their webs. Pianka (1967) found a close relationship between the number of species of lizards and the total volume of vegetation per unit of area in desert habitats of the southwestern United States.

On a regional basis, the number of species is sensitive to the suitability of physical conditions, the heterogeneity of habitats, and isolation from centers of dispersal. In North America, the number of species in most groups of animals and plants increases from north to south, but the influence of geographic heterogeneity and the isolation of peninsulas is apparent.

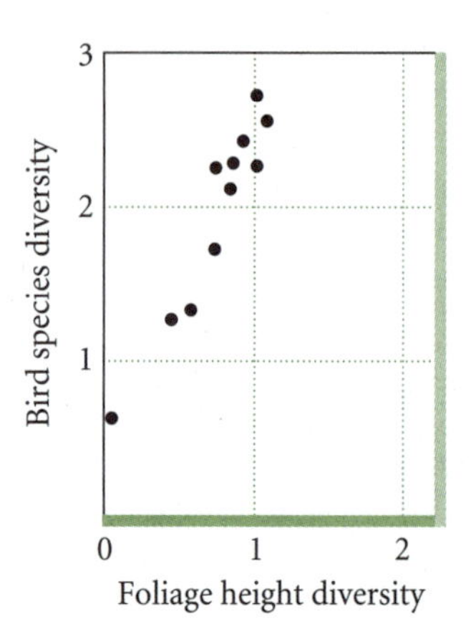

FIGURE 29-3 Relationship between bird species diversity and foliage height diversity in areas of deciduous forest in eastern North America. (*From MacArthur and MacArthur 1961.*)

Simpson (1964) tabulated the numbers of species of mammals in 150-square-mile blocks distributed over the entire area of North America to the Isthmus of Panama. The number of species per block increased from 15 in northern Canada to more than 150 in Central America (Figure 29-4). Across the same latitude in the middle of the United States, more species of mammals were found in the topographically heterogeneous western mountains (90–120 species per block) than in the more uniform environments of the East (50–75 species per block). The number of species of breeding land birds follows a similar pattern (MacArthur and Wilson 1967, Cook 1969), but reptile and amphibian faunas do not (Kiester 1971). Reptiles are more diverse in the eastern half of the United States than in the mountainous western regions; amphibians are strikingly underrepresented in the deserts

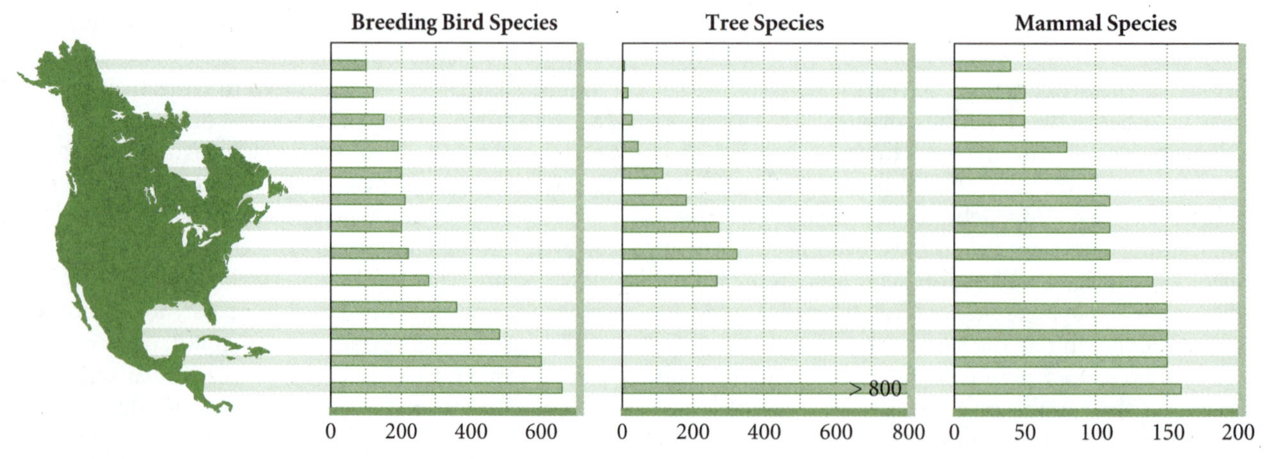

FIGURE 29-2 Numbers of species of breeding birds, trees, and mammals by latitude. The bars correspond to positions on the map to the left. (*From Meffe and Carroll 1997; after Briggs 1995.*)

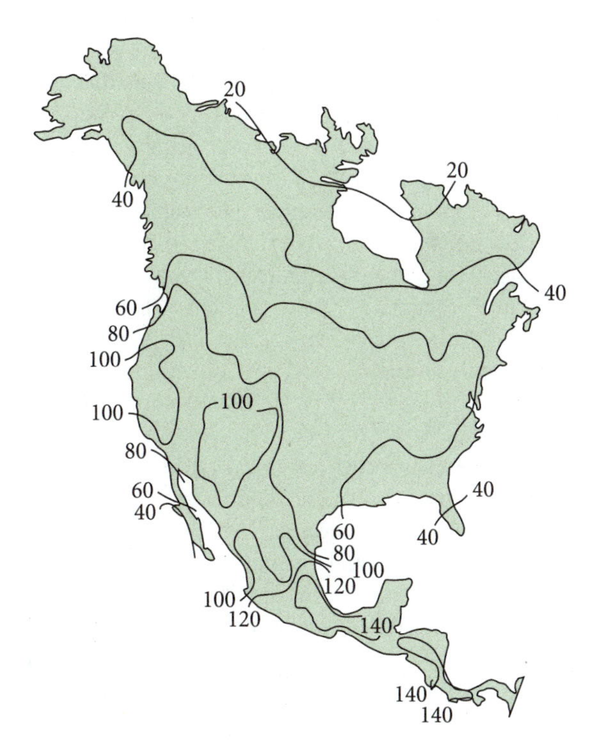

FIGURE 29-4 Species diversity contours for mammals in 150-square-mile blocks in continental North America. (*From Simpson 1964.*)

of the Southwest, owing to the requirement of most species for abundant water.

29.2 Contemporary thinking about community organization reconciles the regional/historical and local/deterministic views of regulation of diversity.

Until the late 1950s, ecologists viewed species diversity as a regional phenomenon representing the outcome of historical events. We shall refer to this view as the **regional/historical** view of species richness. Subsequently, ecologists began to ask questions about how population interactions such as predation and competition affect species diversity. Fundamental to this **local/deterministic** view is the idea that local interactions, which tend to reduce diversity through competitive exclusion and extinction, somehow balance regional processes that increase diversity through speciation and migration, maintaining a kind of equilibrium. Equilibrium thinking has dominated theories of diversity for nearly 30 years. The emerging appreciation of the interaction between regional and local processes, along with recognition of the importance of factors such as disturbance and evolutionary history in community structure—the latter development being aided by advances in the study of phylogenetic relationships—has revitalized the regional/historical perspective. The goal of ecologists today is to understand how factors operating in hierarchies of scales of time and space shape community organization (Ricklefs and Schluter 1993b and articles therein; Rosenzweig 1995). To understand how this contemporary view emerged, let us examine the local/deterministic and regional/historical views in more detail.

The Local/Deterministic View of Diversity

Local determinism proposes that community diversity is a result principally of competition, predation, and other interspecific interactions at the local community level. Ecologists were attracted to this view because it placed the problem of species diversity within their conceptual domain of present-day processes taking place within small areas. The earlier idea that diversity was a reflection of history left ecologists with little to say about the matter. The idea of local determinism became so popular, however, that ecologists embraced it nearly to the exclusion of any consideration of the influence of regional/historical factors (Ricklefs 1987, Ricklefs and Schluter 1993a). Instead of viewing diversity as a balance between regional production of species and local extinction, ecologists began to believe that levels of diversity were determined largely by local conditions. Hence their investigations concentrated almost exclusively on local interactions and niche relationships among populations.

One can trace a logical progression of ideas leading to the hypothesis of local determinism. The transition from the regional/historical viewpoint to the local/deterministic one of the 1960s and 1970s began nearly a half century earlier with Lotka's and Volterra's mathematical representations of species interactions. Gause then demonstrated experimentally what had been predicted by the theory, that species could not coexist on a single limiting resource—the so-called competitive exclusion principle (see Chapter 21). In the 1940s, Lack applied the concept of competitive interaction to the problems of ecological diversification, culminating in his landmark treatise on Darwin's finches in the Galápagos Islands (Lack 1947). Then, in the 1950s, Hutchinson (1957, 1959) developed the concept of species packing in multidimensional niche space (see Section 29.6) and planted the seed of the idea of limiting similarity—that there is an upper limit to the ecological similarity of species set by their competitive interactions. During the 1960s, the consequences of community interactions were formalized by Robert MacArthur, Richard Levins, and Robert May, who explored the deterministic, equilibrium properties of systems characterized by matrices of interactions among their component species (MacArthur and Levins 1967, Vandermeer 1972, Roughgarden 1974, May 1973a, b, c).

The Regional/Historical Perspective on Species Richness

While emphasizing the importance of local population interactions in determining local diversity, ecologists were still faced with the reality of regional processes. Indeed, the resurgence of the regional/historical perspective on community diversity grew from a kind of rationalization that ecologists had come to apply to this difficulty. Almost subconsciously, many ecologists adapted their thinking in such a way that they could divorce the problems of local and regional diversity and attribute each to different causes having their own scales in space and time. The line of reasoning goes like this: Local interactions, taking place within a milieu of local conditions, determine the number of species that can coexist in the local community. This number is the saturation point, beyond which no new species can be added to the community. Regional processes, such as species production, migration, and historical accidents of geographic location, determine regional diversity. The difference between the two is accommodated by differences in the degree of habitat specialization, or beta diversity, which is adjusted to maintain the number of species locally in accordance with local conditions while the number of species in the region may vary.

We can reconcile the regional/historical and local/deterministic views by adopting a hierarchical perspective on the processes that influence species diversity in time and space (Figure 29-5). Interactions occur at progressively broader spatial scales, beginning with the local population and moving outward to encompass patches, landscapes, regions, and so on, the interactions at each scale affecting all the processes at lower scales. Likewise, present-day interactions are shaped by events of the past in a temporal hierarchy (Ricklefs and Schluter 1993b).

Local and Regional Components of Diversity

Local and regional factors are expressed in different components of species diversity, two of which are **alpha** (or **local**) **diversity** and **gamma** (or **regional**) **diversity** (Whittaker 1972). Local diversity is the number of species in a small area of more or less uniform habitat. Clearly, local diversity is sensitive to definition of habitat, area, and intensity of sampling effort. Regional diversity is the total number of species observed in all habitats within a region. By *region,* ecologists generally mean a geographic area that includes no significant barriers to dispersal of organisms. Thus, the boundaries of a region depend on which organisms we consider. The important point is that within a region, distributions of species should reflect their selection of suitable habitats rather than their inability to disperse to a particular locality.

When each species occurs in all habitats within a region, local and regional diversities are the same (Figures 29-6a and d). When each habitat has a unique flora and fauna, regional diversity equals the average local diversity times the number of habitats in the region (Figure 29-6b). Ecologists refer to the difference in species from one habitat to the next as **beta diversity.** The greater the difference, or turnover, of species between habitats, the greater the beta diversity. There are many different ways of quantifying beta diversity, but a useful one is the number of unique habitats recognized by species within a region. When all species are habitat generalists, there is effectively only a single habitat within the region, and beta diversity is equal to 1 (Figures 29-6a and d). As habitat specialization increases, more habitats are recognized. Accordingly, gamma diversity equals alpha diversity × beta diversity. It is not practical to measure beta diversity directly because the habitat distributions of species overlap. But we can calculate the number of unique habitats recognized by species within a region from the following relationship: beta diversity = gamma diversity/alpha diversity.

Where many species coexist within a region, each occurs in relatively few kinds of habitats (MacArthur et al. 1966). Changes in gamma diversity generally result from parallel changes in both alpha and beta diversity. This relationship has been most carefully noted in comparisons of islands and mainland regions, in which one may examine a range of species diversities (resulting from different degrees of geographic isolation; see Section 29.3) within similar ranges of physical conditions. Islands usually have

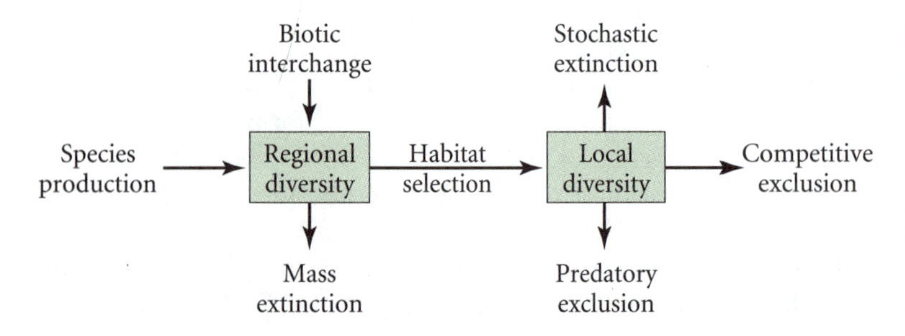

FIGURE 29-5 Factors affecting regional and local species diversity. Numbers of species are increased at the regional level by speciation and immigration. Ecological interactions influence diversity at local levels. Each local community contains a small sample of the total regional diversity because species are habitat specialists. Thus habitat selection connects regional and local diversity.

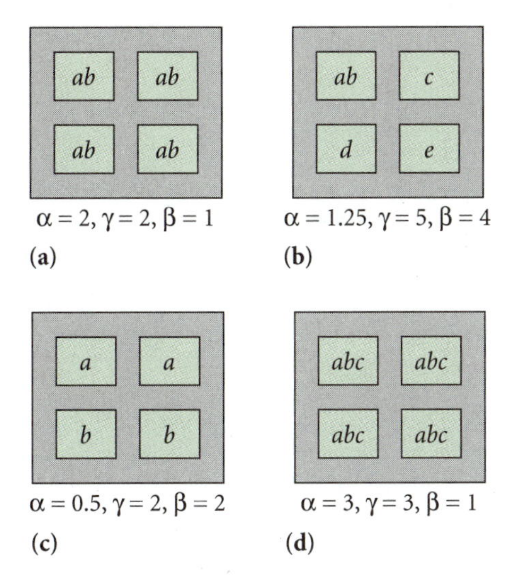

FIGURE 29-6 Relationship between alpha (local), gamma (regional), and beta (turnover) diversity. Each panel shows a region (large gray box) containing four habitats (small green boxes). (a) The diversity in each habitat (alpha diversity) is the same for all four habitats (each contains species *a* and *b*, species richness of 2). The regional diversity (gamma) is 2. The beta (turnover) diversity is gamma/alpha = 2/2 = 1. (b) Alpha diversity is 2 for one habitat (species *a* and *b*) and 1 for the other three (species *c*, *d*, and *e* occur alone in a habitat), yielding an average alpha diversity of 1.25. Gamma diversity is 5 (number of habitats × average alpha diversity), so beta diversity is gamma/alpha = 5/1.25 = 4. (c) Average alpha diversity = 0.5, gamma diversity = 2, beta diversity = 2/1 = 2. (d) Alpha diversity = 3, gamma diversity = 3, beta diversity = 1. Regions (a) and (d) have different gamma diversities, but the same beta diversity, indicating little species turnover in those areas.

fewer species than comparable mainland areas. Island species often attain greater densities than their mainland counterparts, a phenomenon called **density compensation** (Crowell 1962). Also, they expand into habitats that would normally be filled by other species on the mainland, a phenomenon called **habitat expansion** (MacArthur et al. 1972, Wright 1980). Collectively, these phenomena are referred to as **ecological release.** On the island of Puerto Rico, MacArthur and his coworkers found that many bird species occupied most of the habitats on the island. In Panama, which has a similar variety of tropical habitats, each species occupied fewer habitats, often a single type.

Ecological release has been demonstrated in surveys of bird communities in seven tropical regions and islands within the Caribbean basin. These areas range in size from mainland Panama to St. Kitts, a small island in the Lesser Antilles. These surveys show that where fewer species occur, each is likely to be more abundant and to live in more habitats (Table 29-1). Similar numbers of individuals of all species added together were seen in each of the seven localities, although the total number of species (regional diversity) differed by a factor of almost 7 between Panama and St. Kitts. In each habitat in Panama (mainland), about three times as many species (alpha diversity) were recorded, and populations of each species were about half as dense, as in corresponding habitats on St. Kitts (the smallest island). Beta diversity (the number of habitats recognized by species) increased by a factor of almost 3 between St. Kitts and Panama.

TABLE 29-1 **Relative abundances and habitat distributions of resident land birds in seven tropical localities within the Caribbean basin**

Locality	Number of species observed (regional diversity)	Average number of species per habitat (local diversity)	Habitats per species	Relative abundance per species per habitat (density)	Relative abundance per species	Relative abundance of all species
Panama	135	30.2	2.01	2.95	5.93	800
Trinidad	106	28.2	2.35	3.31	7.78	840
Jamaica	56	21.4	3.43	4.97	17.05	955
Tobago	53	21.4	3.63	4.71	17.10	906
St. Lucia	33	15.2	4.15	5.77	23.95	790
Grenada	30	15.5	4.63	5.36	24.82	745
St. Kitts	20	11.9	5.35	5.88	31.45	629

Note: *Based on ten counting periods in each of nine habitats in each locality. The relative abundance of each species in each habitat is the number of counting periods in which the species was seen (maximum 10); this times the number of habitats gives the relative abundance per species; this times the number of species gives the relative abundance of all species together. (From Cox and Ricklefs 1977, Wunderle 1985.)*

29.3 The number of species on islands depends on immigration and extinction rates.

One of the most influential ideas regarding species diversity was the **theory of island biogeography** developed by MacArthur and Wilson (1963, 1967). This theory, which adopts the regional/historical perspective, states that the number of species on islands balances regional processes governing immigration against local processes governing extinction. On islands too small to support speciation through geographic isolation of populations, the number of species increases solely by immigration from other islands or from the mainland. Whereas we know little about rates of speciation within continents, we may reasonably assume that fewer species immigrate to islands that are farther from the mainland. Hence the advantage of the island model.

Consider an offshore island. The flora and fauna of the adjacent mainland constitute the **species pool** of potential colonists of the island. The rate of immigration of new species to the island decreases as the number of species on the island increases; that is, as more and more of the potential mainland colonists have already colonized the island, fewer immigrants belong to new species (Figure 29-7). When all the mainland species occur on the island, the immigration rate of new species will be zero. If species on the island go extinct at random, the extinction rate will increase with the number of species present on the island. Where the immigration and extinction curves cross, the corresponding number of species on the island is at an equilibrium ($\hat{S}$).

Immigration and extinction rates probably do not vary in strict proportion to the number of potential colonists and the number of species established on the island. Some species undoubtedly are better colonizers than others and reach the island first. Thus the rate of immigration to the island initially decreases more rapidly than it would if all mainland species had equal potential for dispersal; as a result, the relationship between immigration rate and island diversity follows a curved line (Figure 29-8). At least two factors may underlie the curved relationship between extinction rate and species diversity. Competition between species on islands may increase extinction, so the extinction curve rises progressively more rapidly as species diversity increases. Such a relationship may also arise if the extinction probabilities of the species immigrating to the island are unequal, which is quite likely.

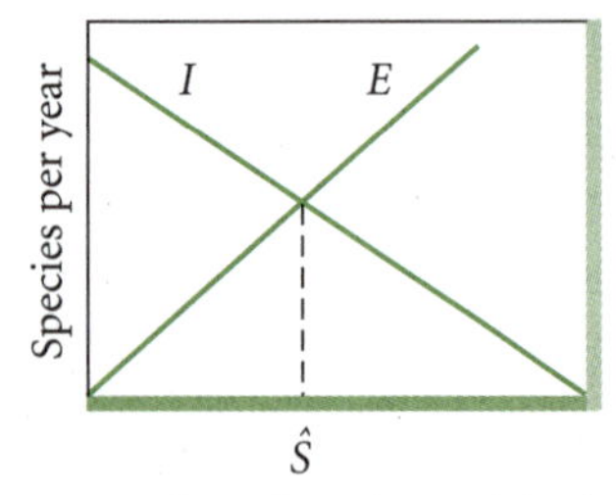

FIGURE 29-7 Equilibrium model of the number of species on islands. The equilibrium number of species ($\hat{S}$) is determined by the intersection of the immigration (I) and extinction (E) curves. (*After MacArthur and Wilson 1963, 1967.*)

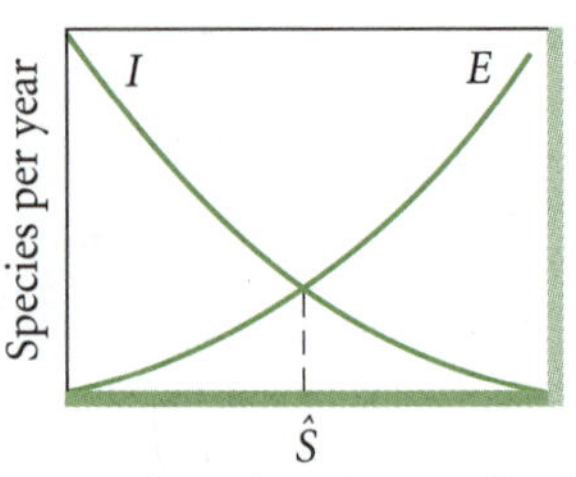

FIGURE 29-8 Several biological considerations influence the shape of the immigration and extinction curves. The immigration rate initially drops off rapidly as the best colonists become established on the island. The extinction rate increases more rapidly at high species numbers because of increased competition between species. (*After MacArthur and Wilson 1963, 1967.*)

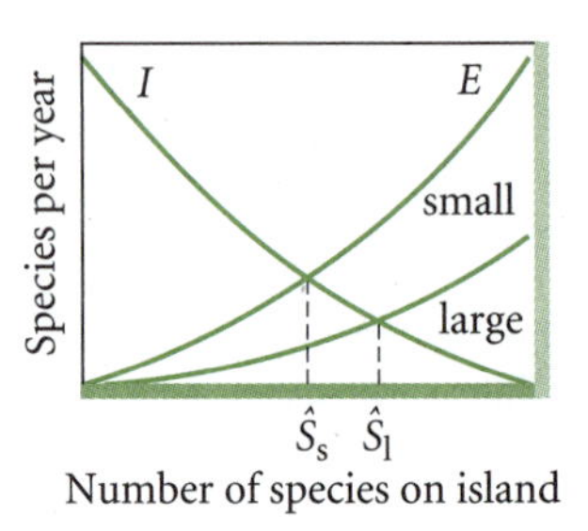

FIGURE 29-9 According to the MacArthur-Wilson equilibrium model, small islands support fewer species owing to higher extinction rates.

One can assume that smaller populations are more prone to extinction than larger populations. Thus, if the probability of extinction increased as absolute population size decreased, extinction curves for species on small islands would be higher than those for species on larger islands. Therefore, small islands would support fewer species than large islands (Figure 29-9). If the rate of immigration to islands decreased with increasing distance from mainland sources of colonists, the immigration curve would be lower for far islands than for near islands. As a result, the equilibrium number of species for distant islands should lie to the left of the equilibrium for islands close to the mainland (Figure 29-10). These predictions have been

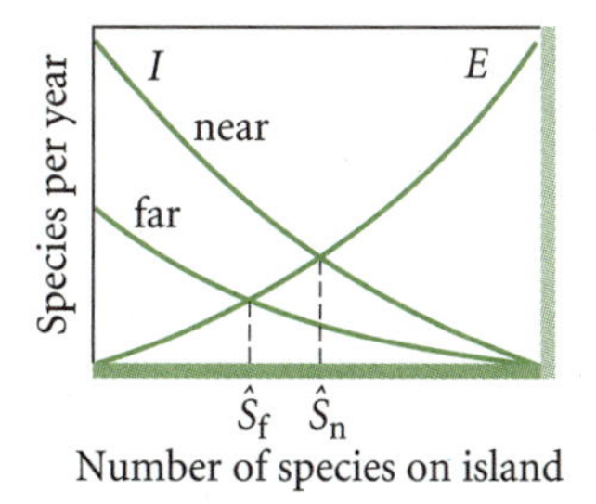

FIGURE 29-10 According to the MacArthur-Wilson equilibrium model, islands close to the mainland support more species owing to higher immigration rates.

verified for islands throughout the world (MacArthur and Wilson 1967).

If the number of species on an island behaves according to the equilibrium model, a change in the number of species should lead to a response tending to restore the equilibrium diversity. A natural test of this prediction was begun quite spectacularly in 1883 when the island of Krakatoa, located between Sumatra and Java in the East Indies, blew up after a long period of repeated volcanic eruptions. At least half the island disappeared beneath the sea, and hot pumice and ash covered its remaining area. The entire flora and fauna of the island were obliterated, as was apparent to the first visitors after the explosion. During the years that followed, plants and animals recolonized Krakatoa at a surprisingly high rate: within 25 years, more than 100 species of plants and 13 species of land and freshwater birds were found there. During the ensuing 13 years, 2 species of birds disappeared and 16 were gained, bringing the total to 27. During the next 14-year period between exploring expeditions, the number of species on the island did not change, but 5 species disappeared and 5 new ones arrived, suggesting that the number of species had reached an equilibrium, at a level that would be expected for an island the size of Krakatoa in the East Indies. (No one knew how many species were present before the explosion.) The turnover rate during this period was 1.3% of species per year [5 species/(27 species × 14 years)]. Experimental studies involving the colonization of glass slides by diatoms (Patrick 1967), sponges by protozoans (Cairns et al. 1967), water-filled vials by microorganisms (Maguire 1963), and mangrove islands by arthropods (Simberloff 1969, Simberloff and Wilson 1969; Figure 29-11) have repeated the pattern of colonization and attainment of an equilibrium seen on Krakatoa.

In a more recent study, Becker (1992) demonstrated that the species diversity patterns of herbivorous beetles of the Canary Islands were dependent to a larger extent on the availability of host plants on an island than on island size per se. This study underscores what most ecologists have come to appreciate: that island size and distance from the mainland may not be the only factors that influence species diversity patterns on islands. Habitat structure and local population interactions no doubt influence colonization and extinction rates on islands.

The equilibrium view of diversity can be applied to mainland assemblages of species as well as to those on oceanic islands. The major difference is that regional production of species (speciation) augments the addition of species from outside the system by immigration (MacArthur 1969, Rosenzweig 1975). In a large region isolated from others by barriers to dispersal (an island continent, for example), new species must originate primarily by speciation events within the region. Curves relating rates of species production and extinction to regional diversity might look like those drawn in Figure 29-12. The curvature of the lines would vary depending on the processes that produced them. Probability of extinction per species

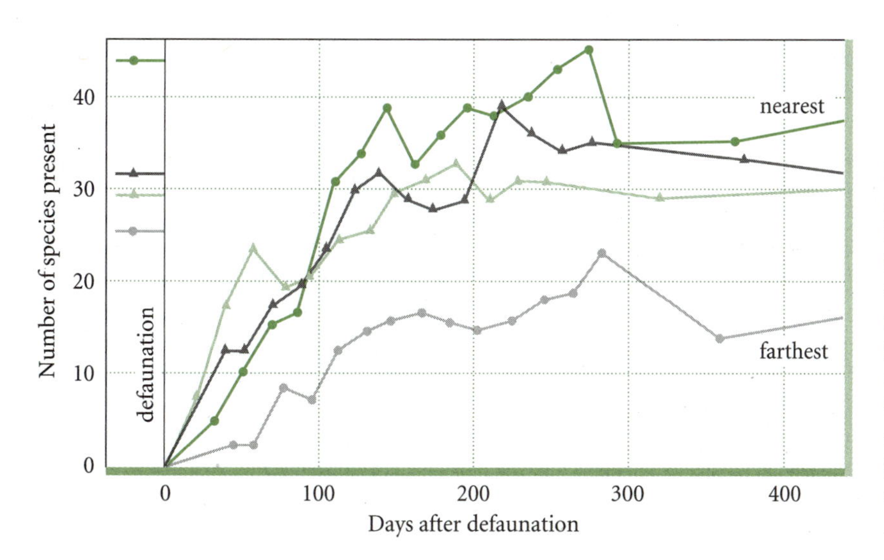

FIGURE 29-11 Recolonization curves for four small mangrove islands in the lower Florida Keys whose entire faunas, consisting almost solely of arthropods, were exterminated by methyl bromide fumigation. Estimated numbers of species present before defaunation are indicated at left. Species accumulated more slowly and achieved lower equilibrium numbers on islands more distant from sources of colonists. (*From Simberloff and Wilson 1969.*)

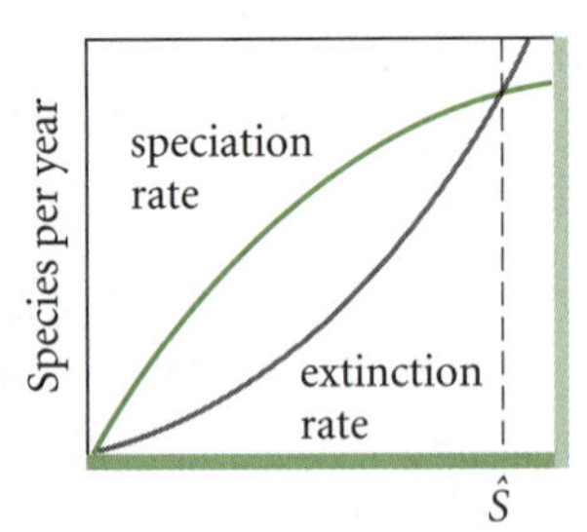

FIGURE 29-12 Equilibrium model of the number of species in a mainland region with a large area; new species are generated by the evolutionary process of speciation rather than immigration from elsewhere. (*After MacArthur 1969.*)

could increase if competitive exclusion increased with diversity, whereas it could decrease if mutualisms and alternative paths to energy flow buffered diverse communities from external perturbations. The rate of speciation per species could level off if opportunities for further diversification were restricted by increasing diversity, whereas the rate could increase if diversity led to greater specialization and higher probability of reproductive isolation of subpopulations.

Regardless of the particular shape of the immigration, speciation, and extinction curves, most biologically reasonable models define an equilibrium level of diversity. Therefore, although such models provide a valuable perspective, they do not necessarily explain variation in diversity; most causes of variation can be incorporated into an equilibrium model. Alternatively, local communities might be saturated with species while diversity continued to increase regionally, the growing difference being made up by increasing beta diversity (MacArthur 1969). If diversity were at equilibrium, the positions of species production and extinction curves would be affected by a variety of factors (Connell and Orias 1964, Pianka 1966, Buzas 1972), each of which could shift the equilibrium.

29.4 Are species produced more rapidly in the Tropics than at higher latitudes?

We have emphasized the interplay between local and regional processes and between current and historical events in the regulation of community structure. Regional processes, however, are difficult to contemplate, much less observe directly, and correspondingly little has been written about their role in determining local and regional diversity. Many systematists and biogeographers have suggested that the rate of species production and the climatic history of an environment contribute to regional diversity, and that these factors are responsible for the latitudinal gradient in diversity (Willis 1922, Haffer 1969, Simpson and Haffer 1978, Prance 1982). Because thousands of generations may be needed for isolated populations of a species to develop barriers to reproduction and themselves become distinct species, direct observation of or experimentation on this process is not possible. However, a number of different types of analysis can be applied to the problem, two of which we discuss below.

Species/Genus Ratios

Haffer (1969, 1974) and Prance (1982) have suggested that fragmentation of tropical forests during the periodic dry periods of the recent Ice Age (Figure 29-13) provided opportunities for allopatric speciation in the Tropics at a time when harsh conditions and restriction of habitats in temperate and arctic zones may have caused an increase in extinction rates there. If differences in diversity between temperate and tropical forests express recent high rates of species production in the Tropics, one would expect to find more species per genus in tropical forests than in their temperate zone counterparts. But, in fact, tropical forests present their tremendous diversity to us as much at the family and genus levels as at the species level. Samples of a similar number of forest tree species from small areas in the Republic of Panama (46 species) and Ontario, Canada (33 species), for example, reveal many more species per genus (including four species of *Populus*) in the temperate forest, but similar numbers of genera per family in both localities (Table

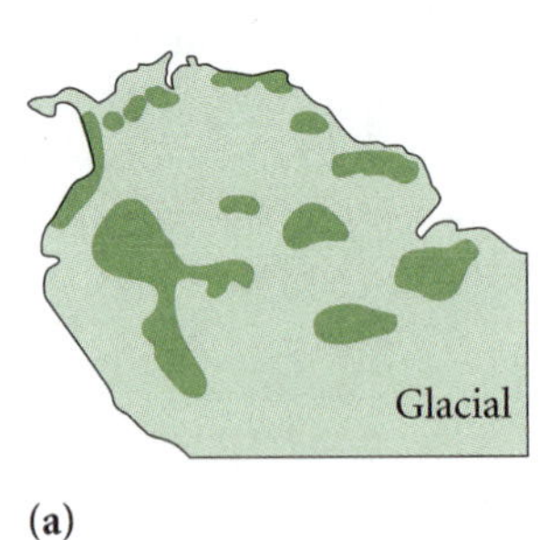

(a)

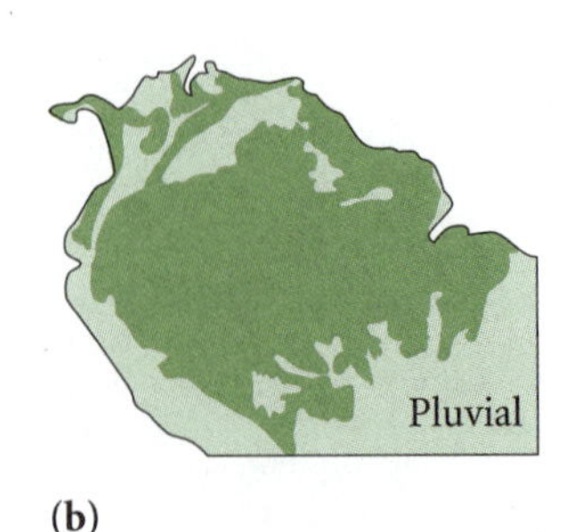

(b)

FIGURE 29-13 Approximate distribution of lowland rain forest in South America (a) during the height of glacial periods in the Northern Hemisphere and (b) at present.

TABLE 29-2 Taxonomic levels of diversity among forest trees in several regions

Taxa	Number of taxa: Europe	Eastern North America	Eastern Asia
Orders	16	26	37
Families	21	46	67
Genera	43	90	177
Species	124	253	729
Percentage of genera predominantly tropical	5	14	32
Number of genera in fossil record	130	60	122

(From Latham and Ricklefs 1993.)

29-2). The difference in diversity between the species-rich tropical site (with a total of more than 300 species of trees: Knight 1975, Croat 1978, Gentry 1988) and the temperate zone site, having fewer than 50 species (for example, Braun 1950, Beschel et al. 1962), resides primarily at the family level. In fact, the tropical forests are decidedly poor in closely related species.

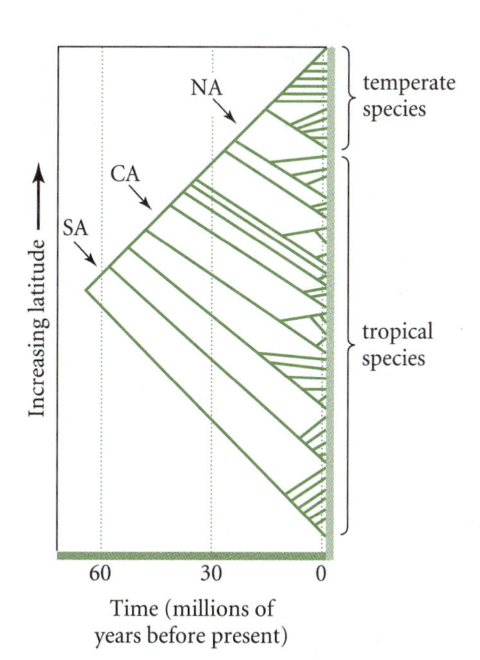

FIGURE 29-14 Hypothetical primitively tropical clade having both temperate and tropical species. Species occur in South America (SA), Central America (CA), and North America (NA). The tropical portion of the clade has more species because it is older. The rate of species diversification is actually higher in the temperate portion of the clade. *(From Farrell and Mitter 1993.)*

Their great number of higher taxa reveals the ancient roots of diversity there.

Phylogenetic Analysis

Another way of comparing the rates of species diversification in temperate and tropical areas is through phylogenetic analysis. Figure 29-14 depicts a hypothetical clade that is tropical in origin (the figure shows the origins of the clade in South America), but with contemporary species in both tropical and temperate areas. The clade contains many more tropical species, but on close examination, it can be seen that the rate of species differentiation is greater in the temperate region than in the Tropics. The greater diversification in the Tropics is shown to be the result of the clade's having occupied that region for a longer time. The figure also shows that the greater diversity of the tropical fauna lies primarily at taxonomic levels above the species.

29.5 The time hypothesis suggests that older habitats are more diverse.

In 1878, the English naturalist Alfred Russel Wallace suggested that diversity in the Tropics was greater than in temperate regions because tropical conditions appeared on the earth's surface earlier than more polar environments. Furthermore, temperate regions have been subjected to more frequent and widespread disturbance—for example, glaciation—than tropical regions during the history of the earth. Thus, tropical regions have enjoyed longer periods of stability and, thus, have had more time for species differentiation.

This idea is now sometimes referred to as the **time hypothesis** of species diversity.

The earth's climate has undergone several cycles of warming and cooling. Sediment deposits and fossils have left a record of the influence of those cycles on vegetation and ocean temperatures. As the climate of the earth warmed, as it last did during the Oligocene epoch, perhaps 30 million years ago, the area of tropical and subtropical climates expanded, reaching what is now the United States and southern Canada. As a result, the temperate and arctic zones were squeezed into smaller areas closer to the poles. During the last 25 million years, the climate of the earth has become cooler and drier, and the Tropics have contracted.

Both high and low latitudes have experienced drastic fluctuations in climate during the Ice Age of the last 2 million years. Temperate and arctic areas witnessed the expansion and retreat of glaciers, causing major habitat zones to be displaced geographically and, possibly, to disappear. Periods of glacial expansion were coupled with high rainfall in the Tropics (pluvial periods). The Amazonian rain forest, which today covers vast regions of the Amazon River's drainage basin, was repeatedly restricted to small, isolated refuges during periods of drought (Haffer 1969, Prance 1982). Restriction and fragmentation of the rain forest habitat could have caused the extinction of many species; conversely, the isolation of populations in patches of rain forest could have facilitated the formation of new species.

A number of types of evidence may be used to examine the time hypothesis. First, if tropical climates predate others, then the fauna and flora of cooler regions should be derived from those of the Tropics. Careful studies of phylogenetic relationships have shown this to be the case for phytophagous insects (Farrell and Mitter 1993). But the pattern may not be applicable to all groups of organisms. For example, coniferous trees, which predominate in the forests of cold regions, generally predate the flowering trees that have replaced them in the Tropics. Conifers did not necessarily evolve in temperate zones, but cold-hardy spruces and firs almost certainly have occurred in habitats similar to those in which they are found today for many tens of millions of years. Other distinct groups of temperate zone plants, such as the oaks and maples, have fossil histories that carry their origins back into the Mesozoic, over 60 million years ago. It is difficult to argue, therefore, that the relatively depauperate temperate and arctic communities to which these species belong have had less time to develop than more diverse tropical communities. To be sure, the areas of temperate and arctic habitats may have been so restricted in the past that few of their representatives survived, but the habitats themselves and the lineages of the plants and animals that live there appear to be ancient.

Evidence from the fossil record that diversity has increased over time would lend support to the time hypothesis. The fossil record is so fragmentary that this test can be applied to only a few taxa and is restricted to certain types of habitats, particularly marine habitats (for example, see Stehli et al. 1969, Hallam 1977). Figure 29-15 shows how worldwide diversity has increased since the beginning of the Paleozoic era, the earliest time for which fossilized remains of organisms are abundant, particularly as plants and animals have invaded new adaptive zones (Simpson 1969, Sepkoski

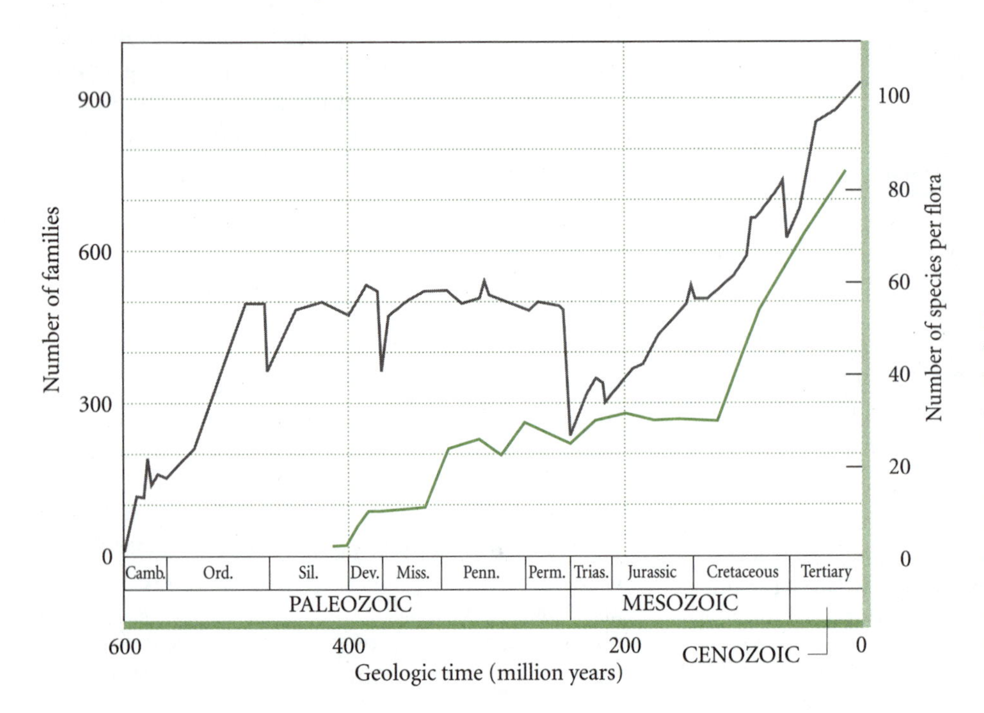

FIGURE 29-15 Changes in the global total number of families of marine animals since the beginning of the Paleozoic era (gray line), and changes in the average number of species represented in local fossil floras of terrestrial plants since the mid-Paleozoic (green line; scale 10×). *(From Sepkoski 1984; after Knoll 1986.)*

1978, Sepkoski et al. 1981). Less information is available for particular habitats or communities, or for established groups of organisms. Buzas (1972) and Gibson and Buzas (1973) have reported that the diversity of temperate zone communities of Foraminifera (marine shelled protozoa) has not changed during the last 15 million years. But the species themselves also have changed little, and so Foraminifera may not be representative of other groups of organisms in which evolution continues to proceed more rapidly.

To most ecologists, the idea that time alone accounts for latitudinal differences in species diversity is too simplistic to provide a full explanation of tropical diversity. No doubt many other factors are involved; among them may be the greater area encompassed by tropical regions as well as differences in solar radiation, the extent of habitat heterogeneity, and the frequency and intensity of local and regional disturbances (Wilson 1992). Moreover, even if temperate species are relatively recently derived from tropical lineages, the constraints on their diversification in the temperate zone may have more to do with coevolutionary relationships than simply with the amount of time available to them for species diversification. For example, phylogenetic studies of tropical and temperate leaf-eating beetles by Farrell and Mitter (1993) revealed a higher species diversity in the Tropics. The temperate beetle fauna was clearly derived phylogenetically from lineages of tropical beetles. However, their studies also revealed that the rate of speciation—the development of new species—showed no latitudinal gradient. They suggest that the diversification of temperate beetle species has been constrained by the close coevolutionary relationship between phytophagous beetles and their host plants. These temperate host plants have their origins in the Tropics. It is constraints on their diversification, rather than on that of the beetles, that explain the temperate-tropical difference in beetle diversity in this case.

One variation of the time hypothesis suggests that, because of more frequent and more severe disturbances in the temperate zone, community diversity in those regions is less likely to reach equilibrium. That is to say, such communities are simply not saturated with species because they have had insufficient time to reach the equilibrium level of species richness. Most ecologists do not support this idea (Rosenzweig and Abramsky 1993).

We shall now turn our attention to local/deterministic factors that may affect species diversity. These factors include primary production, the structural features of the habitat, the action of predators and herbivores, disturbance, and competition. First, we will introduce a body of theory that will help organize our thinking about the relationship between community structure and species diversity.

29.6 Niche theory provides the framework for the theory of regulation of species diversity.

Ecologists use the term **niche** to express the relationship of individuals or populations to all aspects of their environments—and hence their ecological roles within communities. A niche represents the range of conditions and resource qualities within which an individual or species can survive and reproduce. Thus, for example, the boundaries of a particular species' niche might extend between temperatures of 10°C and 30°C, prey sizes of 4 and 12 mm, and the hours from dawn to dusk. Of course, the niche of any species would include many more variables than these three, and ecologists often cite the multidimensional nature of the niche (see below) to acknowledge the complexity of species-environment relationships.

The Niche Concept

Hutchinson (1957) first defined the niche concept formally. Ideally, he said, one could describe the activity range of any species along every dimension of the environment, including such physical and chemical factors as temperature, humidity, salinity, and oxygen concentration and such biological factors as prey species and resting backgrounds against which an individual may escape detection by predators. Each of these dimensions may be thought of as a dimension in space. If there were n dimensions, then the niche would be described in n-dimensional space. Of course, we cannot visualize a space with more than three dimensions; the concept of the n-dimensional niche is an abstraction. But we can deal with multidimensional concepts mathematically and statistically, and depict their essence by physical or graphic representations in three or fewer dimensions. For example, a graph relating biological activity to a single environmental gradient would represent the distribution of a species' activity along one niche dimension (Figure 29-16a). The level of activity, whether oxygen metabolism as a function of temperature or consumption rate as a function of prey size, conveys the ability of an individual to exploit resources in a particular part of the niche space. Conversely, it conveys the degree to which the environment can support that species. In two dimensions, the species' niche may be depicted as a hill, with contours representing the various levels of biological activity (Figure 29-16b). In three dimensions, we must think of a cloud in space whose density conveys niche utilization (Figure 29-16c). Beyond three dimensions, the mind boggles.

The niche of each species occupies a part of the n-dimensional volume that represents the total resource space, or niche space, available to the

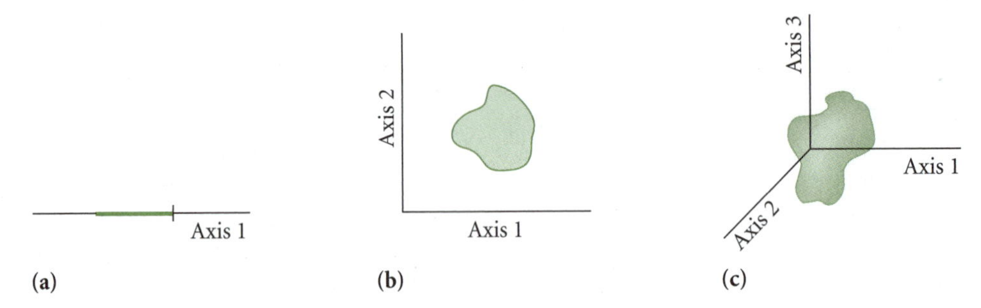

FIGURE 29-16 Portrayal of an ecological niche with a single axis (a), two axes (b), and three axes (c). The shaded area represents the utilization of resources or the tolerance of conditions of the species. The pictorial representation of more than three niche dimensions is not possible.

community. We may think of the total niche space of a community as a volume into which the niches of all the species fit, as do balls of various sizes packed into a box. The number of species in the community, therefore, depends upon the total amount of niche space and the average size of each species' niche. Community ecologists are interested in the factors that determine both quantities because they are among the processes that underlie patterns of biodiversity.

It is very difficult to identify and measure all the *n* dimensions of a species' niche. We may nonetheless begin to characterize niche relationships within biological communities by observing the patterns of resource utilization and microhabitat preferences of community members on one or a few niche dimensions. Imagine that there is a single resource or condition along which all the members of a particular community may be arranged according to their utilization of that resource or their tolerance of that condition. Such a niche dimension might represent prey size (a resource) or soil moisture (a condition of the environment). When we plot the preferences of individuals on a line representing the resource or condition, a frequency distribution of utilization or tolerance is obtained for each species (Figure 29-17). In the figure, the average utilization of individuals of species *i* is noted as point *i*. Some individuals of species *i* use much more or less of the resource, or are tolerant of conditions above and below those withstood by most individuals. The average utilization of the resource by species *j* is given by point *j*. Each distribution can be characterized by the location of its peak (which corresponds to the points *i* and *j* on the niche dimension), its breadth, and its height. The **niche breadth,** represented by the width of each distribution, represents the extent of the variety of resources used or the range of conditions tolerated by the individuals in the population. (Ecologists sometimes use the term *niche size,* where "size" refers to the variety of forms or conditions, not to the amount.) It should be mentioned that the term **niche width,** which is sometimes used synonymously (incorrectly) with niche breadth, is defined more formally as the standard deviation of the distribution of the resource use. Individuals of species *i* that show a higher utilization of the resource than the average for their population overlap in their utilization with individuals of species *j* having a lower than average utilization (shaded region in Figure 29-17). The extent of this similarity of resource use or tolerance of conditions is called the **niche overlap.**

Measures of niche overlap may indicate the intensity of competition between species in the community for a resource (Schoener 1986). If we move the peaks of the distributions of species *i* and *j* in Figure 29-17

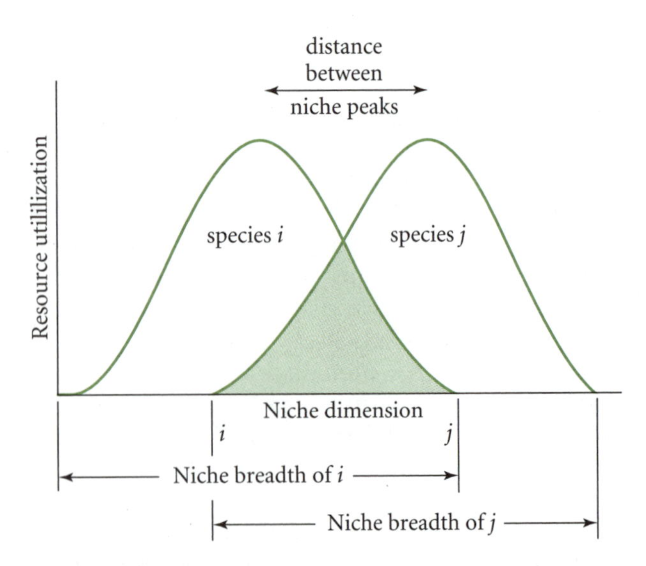

FIGURE 29-17 Positions of two species *i* and *j* along a single resource dimension. The niche breadth is the range of utilization of the resource. The niche overlap (shaded area) represents the portion of the resource that is used by both species. Resource utilization is the frequency with which a certain portion of the resource is used.

closer together, the niche overlap will increase. When the distributions are arranged so that $i = j$, the utilization of the resource by the two species is identical. If the single niche axis represented the only resource used by the two species, we would expect that competitive exclusion (see Chapter 21) would eliminate one of the species from the system. The extent to which species may be similar in their use of a resource and still coexist is called **limiting similarity** (MacArthur and Levins 1967, May 1973, Abrams 1983). Nearly all coexisting species use more than a single niche axis. Often, when species are too similar to coexist on one axis, they differ substantially in their utilization of other resources. This situation is referred to as **niche complementarity.**

We have cast the discussion above in terms of a population, so that the points along the niche axis represent different individuals in the population. This requires some elaboration. Just as there will be a distribution of resource use in the population (not every individual will use the resources in the same way), there will also be a distribution of resource use for each individual in the population. That is, we do not expect that an individual will use resources in the same way all the time. Each distribution of individual resource use will have a mean and a standard deviation. These distributions will have a smaller standard deviation than the distribution of resource use for the entire population, and the sum of these distributions will give the population resource use distribution (Taper and Case 1985; Figure 29-18).

Niche and Species Diversity

By expanding the concepts discussed above to include more than two species, we can uncover some fundamental ideas about the mechanisms regulating diversity in communities. Adding species to or removing species from the single-axis niche space has certain geometric consequences. The possibilities are illustrated in Figure 29-19, which portrays distributions of species niches along a single continuous axis. The species richness of the community portrayed in Figure 29-19a is five. How might this community accommodate three additional species, increasing the species richness to eight? There are three possibilities. First, the five original species and the three additional ones could maintain the same niche widths and overlaps, in which case the total niche space would have to be increased. In the figure, this situation is depicted by extending the niche axis line, signifying the addition of new resource types. Thus, biodiversity may be increased by increased resource diversity. To be sure, the increased species richness depends on the fact that the average utilization curves of the eight species (1, 2, 3, . . . , 8) are spaced evenly along the niche axis, which is a simplification.

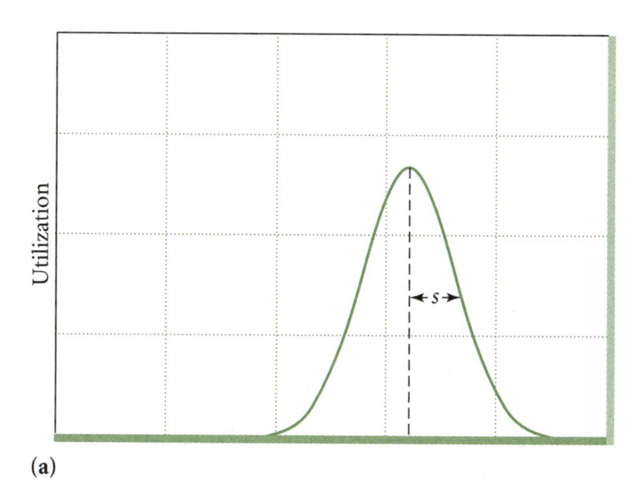

(a)

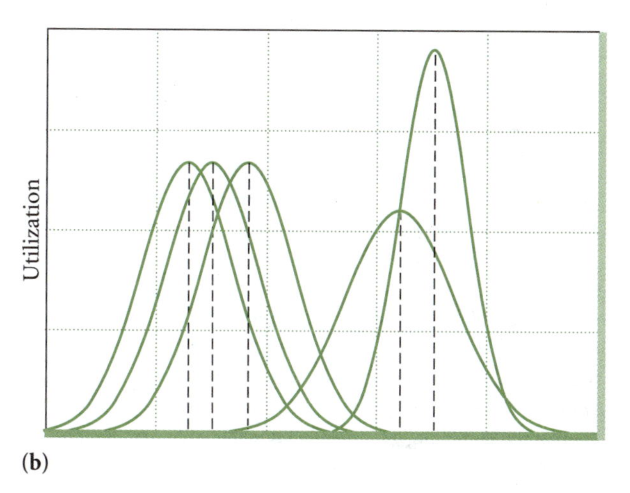

(b)

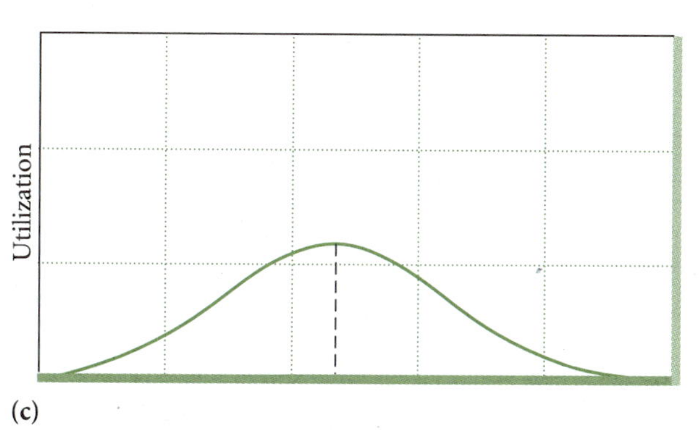

(c)

FIGURE 29-18 (a) Resource use by a single individual has a normal distribution with standard deviation s. (b) In a population, there are many different individuals, each having a distribution of resource use. (c) The sum of these distributions yields the population resource utilization curve. The resource use curves of individuals have smaller standard deviations than the sum of those curves. (*From Taper and Case 1985.*)

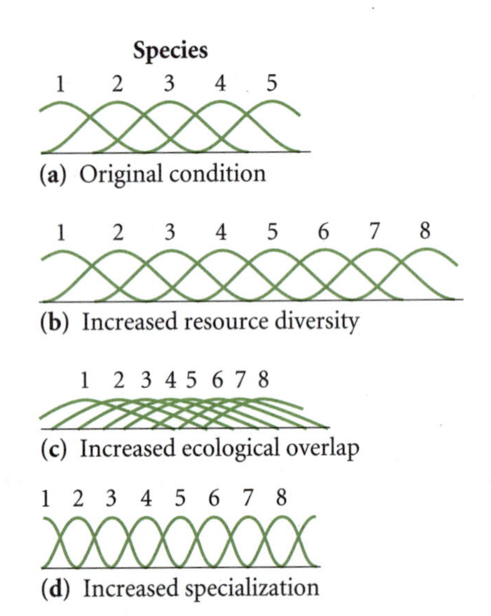

FIGURE 29-19 Schematic diagram showing how resource utilization along a single niche axis can be altered to accommodate more species. (a) The original community has a species richness of five. Each species in the community has the same niche breadth (see Figure 29-8). The species richness may be increased from five to eight if (b) additional resource types are added (shown by extending the axis), (c) the niche overlap among species increases, or (d) the niche breadth of the species decreases. Both (c) and (d) are forms of species packing, an increase in species richness without an increase in resource types. Species packing may result in an overall reduction in the productivity of the species because each will have fewer resources available to it.

Second, without a change in resource diversity, increased diversity could be accommodated by increased niche overlap (Figure 29-19b). In this case, the average productivity of each species would decline as a consequence of increased sharing of resources, all else being equal. Finally, without an increase in niche overlap, increased specialization could accommodate additional species within a community's niche space (Figure 29-19c). Here, too, average productivity would decline because each species would have access to a narrower range of resources. An increase in species richness without a change in resource diversity (Figures 29-19c and d) is sometimes called **species packing.**

Most ecologists agree that the high diversity in the Tropics results at least in part from there being a greater variety of ecological roles there (see Figure 29-19b). That is, the total community niche occupies a greater volume near the equator, where there are many species, than it does toward the poles, where there are few. For example, part of the increase in the number of birds species toward the Tropics is related to an increase in fruit- and nectar-feeding species and in insectivorous species that hunt by searching for their prey while quietly sitting on perches—a behavior uncommon among birds in temperate regions (Orians 1969). Among mammals, the Tropics are species-rich primarily because of the many species of bats in tropical communities (Wilson 1974). Nonflying mammals are no more diverse at the equator than they are in the United States and other temperate regions at similar latitudes, although their variety does decrease as one goes farther north.

In streams and rivers, the number of species in most taxonomic groups increases from the headwaters to the mouth of the river. One presumes that as a river increases in size, it presents a greater variety of ecological opportunities, and its physical conditions become more stable and therefore more reliable (Allen 1995). Local communities reflect these changes. For example, a headwater spring in the Rio Tamesi drainage of east central Mexico supports only one species of fish, a detritus-feeding platyfish (Figure 29-20). Farther

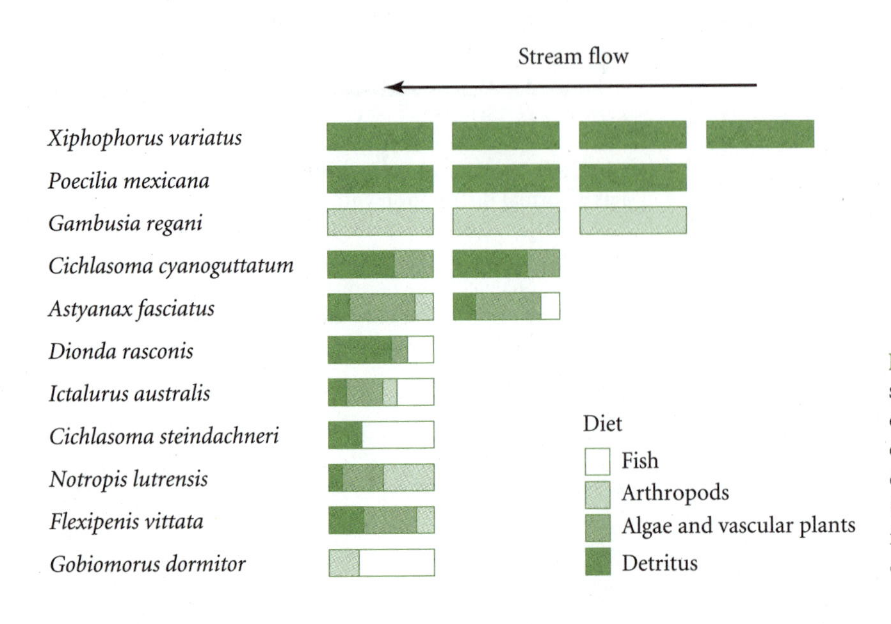

FIGURE 29-20 Food habits of fish species in four communities (vertical columns) from a headwater spring with one species (*right*) to downstream communities with up to eleven species (*left*). The communities sampled were in the Rio Tamesi drainage of east central Mexico. (*From Darnell 1970.*)

downstream, three species occur: the platyfish, a detritus-feeding molly (*Poecilia*) that prefers slightly deeper water than the platyfish, and a mosquito fish (*Gambusia*) that eats mostly insect larvae and small crustaceans. Species that appear in the community farther downstream include additional carnivores—among them, fish eaters—and other fish that feed primarily on filamentous algae and vascular plants. None of the species drops out of the community downstream from any of the sampling localities. Thus diversity increases as the stream becomes larger and presents more kinds of habitats and a greater variety of food items.

Although these and other examples suggest an increase in total niche space with increasing diversity, two considerations must be borne in mind. First, a decrease in niche breadth (specialization) and niche overlap may also accompany an increase in diversity, although these are more difficult to measure. Second, increased community niche space may be in part a consequence of increased diversity rather than an underlying cause or permissive factor. The mere presence of more species makes for more possible roles—a greater variety of interactions. The larger proportion of specialized, parasitic species in collections of insects from sites with high diversity suggests enhancement of diversity through biotic interaction (Janzen and Schoener 1968).

Escape Space and Aspect Diversity

Niche dimensions are not limited to resources and physical conditions. Avoidance of predation is equally important to population processes, and avenues of predator escape constitute dimensions of the niche along which species may diversify (Lawton and Strong 1981). That part of the niche space that is defined by adaptations (including behaviors) of prey organisms that help them avoid predation is referred to as **escape space** (Blest 1963, Rand 1967). We would expect predators to be most efficient when they focus their attention on portions of the niche space most densely occupied by prey species (Martin 1988). Where many prey species use the same mechanisms to escape predation, predators having adaptations or learned behaviors that enable them to exploit such prey should be favored. Thus these prey populations should suffer increased mortality. Conversely, prey having unusual adaptations for predator escape (see Chapter 7) should be strongly selected—a variant of the odd-man-out idea. As a result, predation pressure should diversify prey with respect to escape mechanisms, and prey species should therefore tend to become uniformly distributed within available escape space. The quality of a particular location within escape space depends upon predator attributes (hunting methods, body size, visual acuity and color perception, and so on) and on the attributes of the prey that help them to escape (color and pattern of resting background for cryptic prey, availability of hiding places and other refuges, structure of the vegetation for those relying on fleeing to escape, and so on). But compared with trophic niche relationships, the positions of species with respect to escape space have received little attention outside papers on cryptic insects (Sargent 1966, Ricklefs and O'Rourke 1975, Otte and Joern 1977, Endler 1978, 1984, Pearson 1985).

One study used morphology to estimate the packing of moth species in escape space in different habitats (Ricklefs and O'Rourke 1975). In this case, escape space corresponded to the variety of backgrounds against which day-resting moths conceal themselves to avoid detection by diurnal visual predators. Among cryptic species, appearance has evolved to match the background against which the species rests. Hence we assume that the morphological appearance, or **aspect,** reflects characteristics of the resting place and the searching techniques of the predators to be avoided (Figure 29-21). A. Stanley Rand (1967) has referred to the variety of cryptic patterns as **aspect diversity.**

FIGURE 29-21 Representative species of moths from Panama photographed against the window screens to which they were attracted by ultraviolet lights. These moths show the variety of appearances in the community, which reflect the characteristics of their resting places and the searching techniques of their predators. The variety of different cryptic patterns is called aspect diversity. (*From Ricklefs and O'Rourke 1975.*)

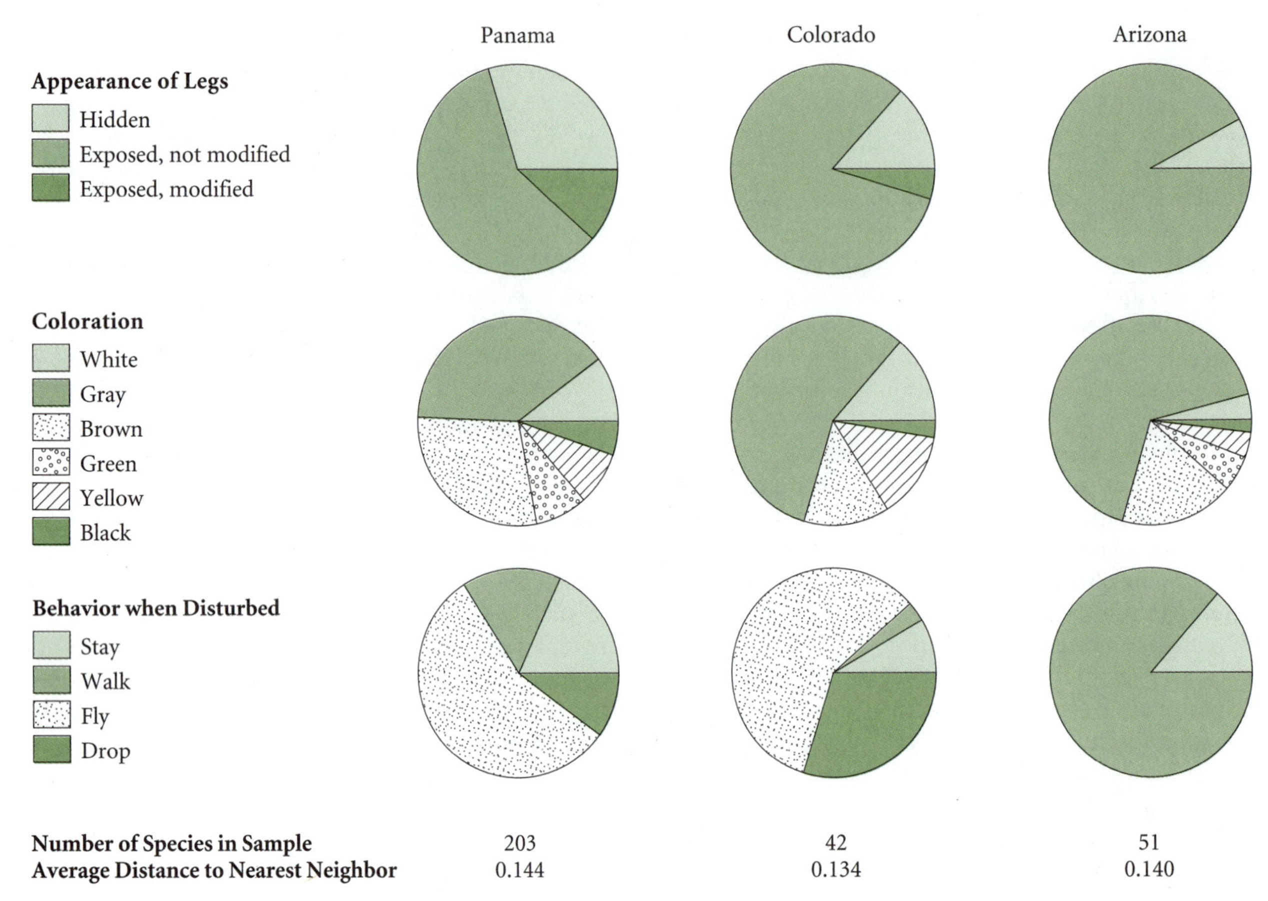

FIGURE 29-22 Diversity in the appearance and behavior of moths from three localities exemplified by three of the twelve characters utilized by Ricklefs and O'Rourke. Aspect diversity is greatest in the most diverse sample, that from Panama. (*From Ricklefs and O'Rourke 1975.*)

Ricklefs and O'Rourke described the appearances of moths using twelve characters, including morphology and position of the legs, coloration, and reaction to disturbance (Figure 29-22). Samples of moths were analyzed from three areas: a spruce-aspen forest in Colorado (43 species), a Sonoran desert habitat in Arizona (51 species), and a lowland rain forest habitat in Panama (203 species). The numbers of moth species sampled probably reflect the differences in diversity between the areas. The variety of morphological characters represented in a sample was greatest in Panama. A full analysis of their data led Ricklefs and O'Rourke to conclude that species were added to the communities by expansion of the niche space utilized (see Figure 29-19b) rather than by denser packing of species in the same space. Variation in the amount of escape space used could arise from a number of factors: variation in the diversity of escape possibilities presented by the environment, greater predation pressure, greater pressure for diversification through competition for escape space, and more opportunities for diversification owing to a greater variety of independently evolving populations.

29.7 Species diversity increases with primary production in some cases.

Recall that primary production is the accumulation of energy by green plants and other autotrophs. Primary production sets a limit on the amount of energy available for use by species within a community. Thus, we might expect that species richness would be limited by the productivity of the environment. Connell and Orias (1964) suggested that species richness should be greatest in relatively stable environments having high rates of productivity. This idea, which was first proposed by Whittaker and Niering (1965) and given theoretical grounding by Tilman (1982), is called the **productivity-stability hypothesis** of species diversity. On a regional scale—that is, over areas of about 10^6 km^2 (Rosenzweig and Abramsky 1993)—species diversity has been found to be positively related to productivity in some cases (e.g., Currie and Paquin 1987, Aronson and Shmida 1990, Currie 1991, Rosenzweig and Abramsky 1993, Tilman and Pacala 1993, Wright et al. 1993). However, it is unclear how strong the relationship between productivity and species richness is, or what the possible mechanisms for such a relationship might be. Tilman and Pacala (1993) suggest that diversity does not increase monotonically with productivity for any group of species, but that species richness varies depending on what environmental factor is used as a measure of productivity and which species are being considered. Currie (1991) found that bird, mammal, amphibian, and reptile species richness increased with increasing potential evapotranspiration (PET) (mm yr^{-1}) (Figure 29-23). For amphibians and reptiles, the increase was monotonic; for birds and mammals, there was a slight but insignificant decrease in richness at the higher PET levels.

Rosenzweig and Abramsky (1993) have recently evaluated other ideas regarding the relationship between productivity and species richness. One possibility is that productivity is simply correlated with species richness, rather than a determinant of it. That is to say, some other factor, such as disturbance, the spatial distribution of habitats, or some other as yet unidentified variable that is correlated with productivity may be at play. For example, some have suggested that predator-prey ratios increase as productivity increases, and thus, at high productivity levels, predators consume a disproportionate share of the available production, thereby causing a reduction in community diversity. Analyses of topological food webs have suggested that the predator-prey ratio is stable across webs having different numbers of species (S), a result that works against this argument (Pimm 1991). However, as we mentioned in Chapter 27, the utility of generalizations from topological food webs is in question.

Rosenzweig and Abramsky (1993) have suggested that two hypotheses are most worthy of further consideration. The first is what they call the **intertaxon competition hypothesis.** This idea holds that the peaks of species diversity for different multispecies taxa (such as orders or classes) should occur in areas having different productivity levels. For example, they point out that among small mammals in the southwestern deserts of the United States, rodent diversity peaks in areas of low productivity west of the El Paso River, whereas carnivore species richness is highest in East Texas in areas of much higher productivity. Rosenzweig and Abramsky suggest that, as a whole, a taxon will compete better at a certain productivity level than at others. Consequently, where that productivity level exists, that taxon will have an advantage over another taxon whose richness peaks at another productivity level (which presumably occurs in some other location), thereby reducing overall species richness through intertaxon competition. Rosenzweig and Abramsky (1993) hasten to point out that this idea is virtually untested. Another hypothesis worthy of consideration is that of Tilman (1982) and others, who have suggested that habitat heterogeneity increases with productivity to a certain point, after which it decreases. The relationship of habitat heterogeneity to species richness is explained in some detail in Section 29.8.

The observation of a positive relationship between species richness and productivity is by no means a universal phenomenon. For example, in censuses of birds in small areas (usually 5–20 ha) of relatively uniform habitat, both the lowest species richness (six species) and the highest (twenty-four

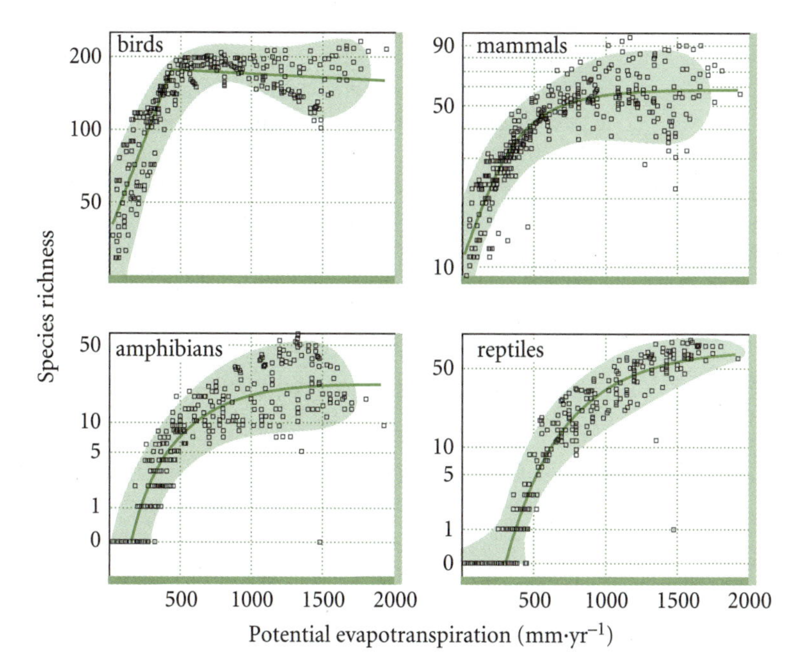

FIGURE 29-23 Relationship of potential evapotranspiration (PET) to species richness for birds, mammals, amphibians, and reptiles. The species richness axes (vertical) are given on a log scale. (*From Currie 1991.*)

TABLE 29-3 Plant productivity and the number of bird species in representative temperate zone habitats

Habitat	Approximate productivity ($g\ m^{-2}\ yr^{-1}$)	Average number of bird species
Marsh	2,000	6
Grassland	500	6
Shrubland	600	14
Desert	70	14
Coniferous forest	800	17
Upland deciduous forest	1,000	21
Floodplain deciduous forest	2,000	24

(From Tramer 1969; productivity data from Whittaker 1975.)

(a)

(b)

FIGURE 29-24 The paradox of enrichment: (a) Sonoran desert of Baja California; (b) Marsh at Malheur Refuge, Oregon. Deserts are less productive than marshes, but they are more diverse. In general, habitat structure overrides productivity in determining species diversity. (*Courtesy of U.S. Department of the Interior.*)

species) occurred in the habitats with the highest productivities; that is, in marshes and in floodplain deciduous forests respectively (Table 29-3). It is clear from such data that species richness is not determined by productivity alone. Indeed, there are many cases in which highly productive environments have relatively low species diversities. This observation, coupled with the knowledge that experimental additions of nutrients or other resources that increase productivity often lead to reductions in diversity, led to the use of the term **paradox of enrichment** (Rosenzweig 1975, Riebesell 1974) to describe a negative relationship between species diversity and production. Most ecologists agree that habitat structure generally overrides productivity in determining species diversity. Marshes are productive but structurally uniform, and they contain relatively few species. By contrast, desert vegetation is less productive, but its great variety of structure apparently makes room for more kinds of habitats (Figure 29-24). Huston (1994) has suggested that the extent to which diversity is related to productivity is dependent, in part, on the scale on which one considers the phenomenon. For example, increasing productivity may lead to decreased diversity within groups of competing primary producers, but may increase diversity in the community as a whole.

29.8 Environmental and life history variation may affect species diversity.

Variation in soil type, moisture, temperature, and other environmental factors may affect species diversity among plants. Abundant evidence suggests that tropical forest trees may be specialized to certain soil and climate conditions. Could greater variation in the physical environment in the Tropics possibly account for the tenfold or even greater diversity of plants compared with temperate zone forests? Tilman (1982) proposed a possible answer to this question based on his model of competition among consumers (see Chapter 21).

Consider a graph whose axes are the levels of two essential resources (Figure 29-25). The population of a species can increase when each of the two resources exceeds some specified level; hence it can persist in the absence of competition within a specified area (the shaded portion of the graph). When competitors are present, the area of the graph within which each species can persist is reduced to a narrow segment (Figure 29-26). Within each geographic region, there will be a natural heterogeneity of conditions (resource levels). Species whose boundaries of persistence fall within those areas may coexist (Figure 29-27). Now, notice that at low resource levels, a given range of environmental heterogeneity intersects the persistence areas of more species than a similar range of heterogeneity at high resource levels. An inverse relationship between resource level and diversity has some empirical and experimental support (Tilman 1982). Possibly, therefore, the low resource levels of soils under many tropical forests may promote their high diversity. The spatial scale of Tilman's model must be carefully defined, since variation in environmental conditions will increase with the size of the area considered, and, as we showed in the last chapter, so will the number of species.

Unpredictable temporal heterogeneity generally is not thought to promote diversity (May 1974). However, Chesson and Warner (1981) have suggested that year-to-year variation in reproductive rates, such that each species is favored in some years, may lead to coexistence. This mechanism relies on a kind of frequency dependence. Suppose that a certain fraction of individual trees dies each year, and each tree is replaced by an individual of the same or a different species in direct proportion to the production of seeds by each of the species. When a species is very rare, a bad year for seed production has little effect on the probability of seedling establishment, but a good year can bring a comparative bonanza—a little less of almost nothing is almost nothing, but a little more represents a huge relative increase! Although this mechanism may work in theory, it rarely has a significant effect. Thus any competitive inferiority will push a species into the very rare category. Moreover, because each species must experience some years during which it produces more seeds than all other

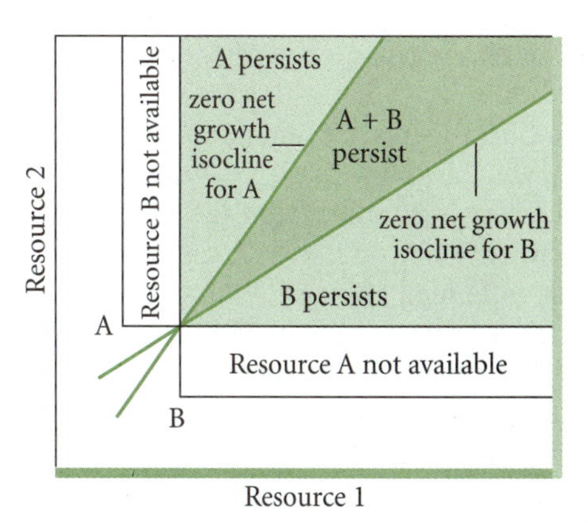

FIGURE 29-26 Conditions for the coexistence of two species according to Tilman's resource model. Net growth is zero on the isoclines. (See Figure 21-12).

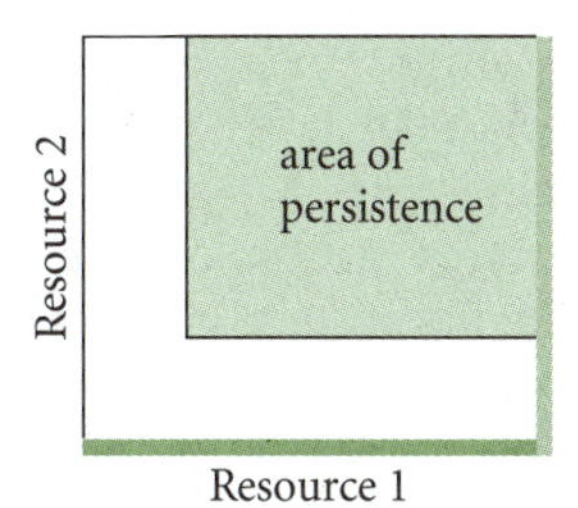

FIGURE 29-25 The persistence of a species requires minimum critical levels of two resources (shaded area) that may independently limit the population.

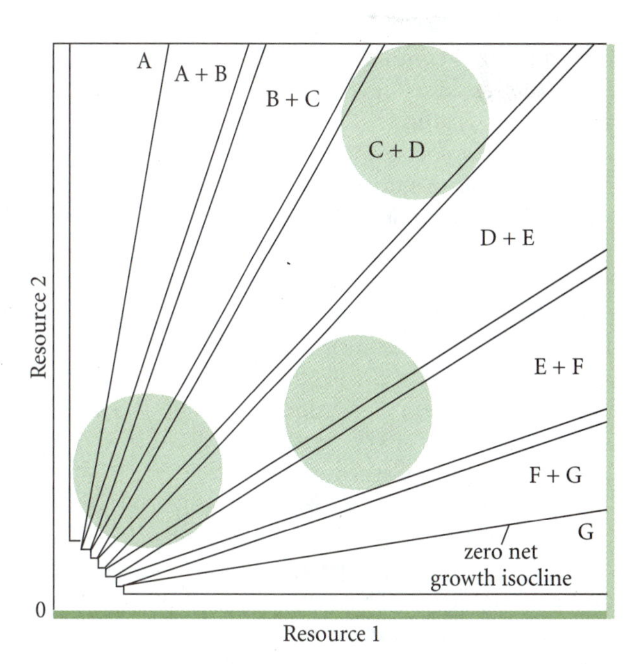

FIGURE 29-27 Seven-species competition for two essential resources, showing the regions of coexistence (never more than two species). The shaded areas represent the ranges of resource conditions available in different habitats. Habitats with lower overall resource abundance intersect the persistence zones of more species, provided the absolute range is comparable between habitats. (*From Tilman 1982.*)

species, relatively few species can coexist solely by this mechanism.

The Chesson-Warner model was inspired by coral reef fish, which exhibit diversity comparable to that of tropical forest trees and exhibit little niche diversification (Ehrlich 1975, Sale 1980). Considerable controversy has attended the issue of competition among these species (Anderson et al. 1981, Sale and Williams 1982. Sale (1977, 1978) suggested that juvenile fish colonize coral heads at random. Subsequently, individuals of all species have equal opportunity to take the place of adults that die or otherwise leave their territories in the reef. This idea became known as the **lottery hypothesis.** Colonization by lottery, which produces random variation in time, reduces competitive exclusion to chance extinction and may contribute to the coexistence of large numbers of fish species in tropical reef communities. But the lottery model cannot explain the high diversity of larval fish in the plankton from which coral residents come. Nor can it explain the difference in fish diversity between tropical and temperate oceans.

29.9 The activities of predators and herbivores may affect species diversity.

Predators and herbivores may affect the diversity of prey and host communities. When predators reduce populations of prey species below the carrying capacity of their resources, they may reduce competition and promote coexistence (Parrish and Saila 1970). Moreover, selective predation or herbivory on superior competitors may allow competitively inferior species to persist in a system (for example, Paine 1966, Harper 1967). The effects of predation on diversity have been particularly well documented in aquatic systems, in which the introduction of a predatory starfish, salamander, or fish can greatly change the community of primary consumers and producers (Paine 1966, 1974, Zaret and Paine 1973, Porter 1972b, Zaret 1980, Wilbur et al. 1983, Carpenter et al. 1987, 1988).

Let us look first at some aspects of predator-prey interactions that suggest mechanisms for the effects of predation on prey species diversity at the local and regional levels. Then, we will turn our attention to how the consumption of plants, plant parts, or plant propagules may affect plant diversity in the Tropics.

Predator-Prey Interactions

The activities of predators may promote species diversity in at least two ways. The first is related to prey switching, whereby a predator, faced with the decline of a favored prey, turns its attention to another prey species that is more abundant. This behavior releases the first prey from predation pressure and provides an opportunity for its population to expand more rapidly. A reduction in competitive interactions among prey brought about by such frequency-dependent predation could promote coexistence among prey species. Another mechanism by which predators can promote species diversity is through feeding preferences. If a predator prefers the best competitor from among a suite of competing prey, its activities may relieve competitive pressure on other prey, thereby promoting coexistence. Neither of these mechanisms takes into consideration the spatial distributions of predator or prey.

Models of predator-prey interactions that include spatial dynamics provide clues about how predators affect prey species diversity at the regional level. For example, in a predator-prey system arranged in patches, prey species may persist on a regional scale—that is, exist in one or more patches—while suffering local extinction due to predation in a number of individual patches. The balance between regional and local diversity in such a case will depend on the spatial dynamics and the relative mobilities of the predator and prey species (McLaughlin and Roughgarden 1993). Theoretical models of predator-prey interactions that include information on mobility and spatial dynamics predict that, in general, if a local population is subject to extinction because of predation, then at least a moderate amount of mobility of both predator and prey is required in order for the prey to persist at the regional scale (McLaughlin and Roughgarden 1993).

Spatial models need not involve patch dynamics. McLaughlin and Roughgarden (1991a,b) have suggested that prey persistence in the face of predation in continuous (nonpatchy) environments may be understood by envisioning a system in which a predator moves faster and on a larger scale relative to its prey and in which there is a heterogeneous pattern of prey population growth. Think of a predator-prey system operating in a linear space where the prey population grows faster at location *a* than at other locations in the space (Figure 29-28). If the difference between the rate and extent of movement of predator and prey is large, the prey population will grow only in areas that support the highest rate of growth (patch A in Figure 29-28). In other areas, where high rates of prey population growth are not sustained, prey will be picked off by the widely ranging predator before the prey population can take hold. If the relative difference in movement abilities of predator and prey is small, the prey population will persist in more of the area. Thus, on a regional scale, prey species diversity is maintained via divergent scales of operation of predator and prey.

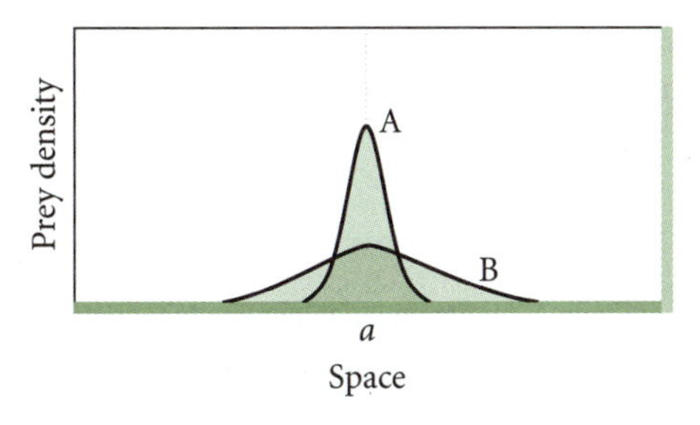

FIGURE 29-28 Effect of the extent and rate of predator and prey movement on prey density. The highest rate of prey population growth occurs at point *a* in the space. The prey density distribution is more peaked when the predator moves fast relative to the prey, thereby preventing the prey from colonizing areas of the habitat where its growth rate is low (curve A). When there is little difference in movement between predator and prey, the prey population can exist in more of the space (curve B). (*From McLaughlin and Roughgarden 1993.*)

Effects of Herbivory on Plant Species Diversity

Since Darwin's time, at least, naturalists have thought that both selective and nonselective herbivory may influence the diversity of plant species (Harper 1967). In particular, several authors, including Doutt (1960), Gillette (1962), and Janzen (1970), have suggested that herbivory could promote the high diversity of tropical forests. Janzen argued that herbivores feed upon the buds, seeds, and seedlings of abundant species so efficiently that their densities are reduced. This allows other, less common species to grow. The key to this idea is that abundance per se, rather than the intrinsic quality of individuals as food items, makes a species vulnerable to consumers. Consumers locate abundant species easily, and their own populations grow to high levels. This idea became known as the **pest pressure hypothesis.**

Several lines of evidence support the pest pressure hypothesis. For example, attempts to establish crop plants in monoculture frequently are doomed by infestations of herbivores. Dense plantations of rubber trees in their native habitats of the Amazon basin, where many species of herbivores have evolved to exploit them, have met with singular lack of success. However, rubber tree plantations thrive in Malaya, where specialist herbivores are not (yet) present. Attempts to grow many other commercially valuable crops in single-species stands in the Tropics have met the same disastrous end that befell the rubber plantations. Cacao, the South American plant that is the source of chocolate, provides a conspicuous exception. Insect pests in cacao plantations are no more numerous or diverse within the plant's native range than in plantations in Africa and Asia (Strong 1974).

The difficulties of monoculture extend beyond the Tropics. Temperate zone trees also have their herbivores; few acorns escape predation by squirrels and weevils, and seedlings are attacked by herbivores and pathogens just as they are in the Tropics. If pest pressure does promote greater diversity in the Tropics, it must operate differently in different latitude belts. In particular, tropical herbivores and plant pathogens must either be more specialized with respect to host plant species, or their populations must be more sensitive to the density and dispersion of host populations.

One prediction of the pest pressure hypothesis is that the establishment of seedlings should be more difficult close to adults of the same species than at a distance from them (Janzen 1970). Adult individuals may harbor populations of specialized herbivores and pathogens that readily infest their nearby progeny; furthermore, because most seeds germinate close to a parent, herbivores may be attracted to the abundance of seedlings there while overlooking the few that germinate at some distance (Figure 29-29). A number of studies have tested

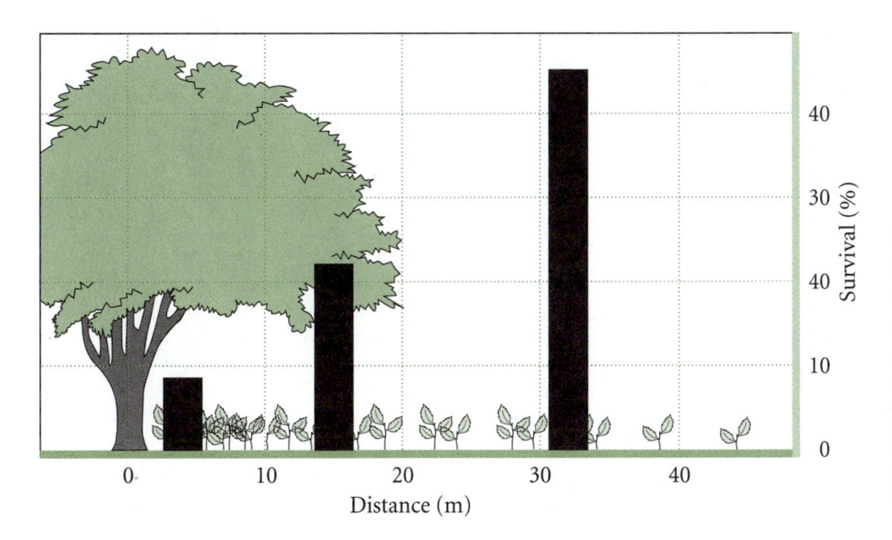

FIGURE 29-29 The pest pressure hypothesis. This hypothesis suggests that seedlings are most dense close to the parent tree, but that survival of seedlings is highest at a distance from the parent, because herbivores will be more common among the dense seedlings near the parent tree. (*Data from Janzen 1970, Clark and Clark 1984.*)

the prediction of distance-dependent germination and establishment success, with varied outcomes, but generally supportive results (Augspurger 1983, Clark and Clark 1984, Howe et al. 1985).

If seedlings establish themselves more readily at greater distances from the parent or other conspecifics, then trees of the same species should be widely dispersed within a forest rather than randomly associated or clumped. This prediction may be tested by mapping individuals of each species and applying one of several statistics to test the degree of dispersion. Most studies have shown that many species of trees, especially the more common ones, have significantly clumped distributions (for example, Hubbell 1979, Hubbell and Foster 1983). This result weighs against a major role for pest pressure in forest community structure, but degree of clumping is relative, and herbivores and pathogens may cause greater dispersion of mature individuals than would occur otherwise.

29.10 Can reduced competition explain high diversity?

Both theory and experiment have shown that intense competition results in exclusion of species from a community. Many ecologists have argued that less severe competition for shared resources should allow more species to coexist. How might interspecific competition be reduced? Greater ecological specialization, greater resource availability, reduced resource demand, and intensified predation have all been cited as possibilities. In every case, however, intraspecific competition, as well as interspecific competition, influences how populations adjust to these factors. We find it difficult to imagine reduced competition as an intrinsic property of any population. Density-dependent selection tends to increase population size. Therefore, if variation in competition were to explain variation in diversity, then the relative strengths of intraspecific competition and interspecific competition would have to vary in their influence on the gene pool of the population.

Perhaps differences in these relative strengths may be envisioned in terms of the geometry of niche relationships. In a niche space having few dimensions, each population has relatively few neighbors. As the number of dimensions increases, the same number of species may pack themselves differently into the niche space, increasing the number of neighbors but reducing the asymmetry of competition. This variety may prevent each individual population from adapting to compete effectively with its neighbors in communities of high niche dimensionality, preventing the emergence of competitive dominants. High dimensionality may occur in physically undemanding environments, such as the wet Tropics or the abyssal depths of the oceans. In harsher environments, a few physical factors may dominate the character of the niche space by establishing a small number of critically important dimensions. Furthermore, because biological factors generate many niche dimensions, the dimensionality of niche space may increase as biological communities build over time, further strengthening their capacity to support many species against the evolutionary tendency of a small number of species to dominate.

29.11 Disturbance may affect species diversity.

Periodic physical disturbance is a feature of many communities. Human activities that alter habitat characteristics dramatize the relationship between habitat structure and local species richness. The effects of timber harvesting in riparian zones (areas along the banks of rivers and streams), for example, may change stream flow, increase sediment loads, alter the distribution and availability of microhabitats, and increase local productivity in streams (Allen 1995). The detrimental effects of such changes on the diversity of vertebrate communities, particularly fish, are well documented (Bisson et al. 1992). An example of the effects of timber harvesting on amphibian species richness is shown in Figure 29-30. The proportion of streams containing two or fewer amphibian species was dramatically higher in logged forests. This difference in amphibian

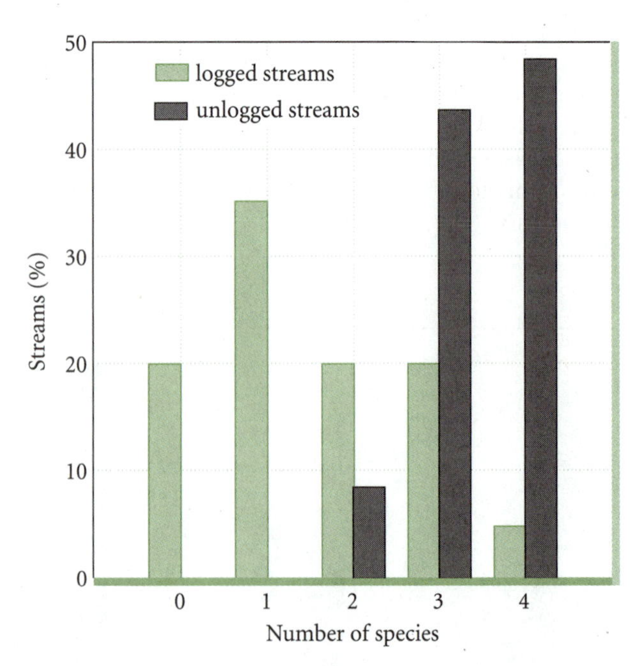

FIGURE 29-30 Amphibian species diversity in streams in unlogged (gray bars; 23 streams) and logged (green bars; 20 streams) areas in western Oregon. Logging reduces the availability of habitats, leading to a reduction in the number of species. (*From Allen 1995; after Corn and Bury 1989.*)

abundance resulted in part from a redistribution of habitat types in the streams. The relative availability of larger cobble and boulder habitats decreased in the logged streams (Corn and Bury 1989).

A number of mechanisms have been proposed to explain the effects of disturbance on species diversity. Consideration of tropical rain forests and coral reefs led Connell (1978) to relate high diversity to intermediate levels of disturbance, an idea referred to as the **intermediate disturbance hypothesis.** Disturbances caused by physical conditions, predators, or other factors open space for colonization and initiate a cycle of succession by species adapted to colonize disturbed sites. With a moderate level of disturbance, the community becomes a mosaic of patches of habitat at different stages of regeneration; together these patches contain the full variety of species characteristic of the successional sere. For this hypothesis to account satisfactorily for differences in diversity between regions, especially on the magnitude of the latitudinal difference in tree species diversity, there must be comparable differences in levels of disturbance. Rates of turnover of individual forest trees (that is, the inverse of average life span) do not differ systematically between temperate and tropical areas (Table 29-4). Nor is it likely that major disturbances such as storms and fires are more frequent in the Tropics. Thus, while disturbance may promote diversity, it seems unlikely to account for much of the observed variation in diversity among forests or, indeed, other types of communities.

Ricklefs (1977a) proposed a somewhat different mechanism by which forest gap formation might generate diversity, based upon the idea that disturbances create a range of conditions for seed germination and seedling establishment within which different species of trees may specialize (Denslow 1980, 1987, Orians 1982, Pickett 1983). Disturbance creates transient environmental heterogeneity during a critical life stage of plants and other organisms. Furthermore, the interaction of structural gaps in the forest canopy with physical conditions of the environment may create greater heterogeneity in the Tropics than in temperate regions. Gaps in forest canopies created by treefalls admit light to the forest floor, thereby changing the physical conditions for seedling establishment and the decomposition of organic detritus. Furthermore, with fewer living tree roots in the forest gap, nutrients are more readily leached out of the soil. These processes occur more intensely in the Tropics: sunlight enters gaps from directly overhead rather than at the shallower angles of higher latitudes (Figure 29-31); more rain falls in the Tropics, so decomposition and leaching proceed more rapidly. Furthermore, conditions under the intact canopy are more protected in tropical forests than in temperate zones owing to the denser canopy and the year-round persistence of leaves. As a result, the forest-environment interaction creates a greater variety of physical conditions available for seedling germination in the Tropics than at higher latitudes. If trees specialize along this gradient of conditions, then perhaps their coexistence becomes easier in the Tropics. It may be relevant that the latitudinal gradient in diversity that characterizes trees does not appear among shrubs, herbs, and grasses, which

TABLE 29-4 Turnover of canopy trees in primary forests in tropical and temperate localities

Locality	Turnover time (years)*	Turnover rate (% per year)
Tropical		
Panama	62–114	0.9–1.6
Costa Rica	80–135	0.7–1.3
Venezuela	104	1.0
Gabon	60	1.7
Malaysia	32–101	1.0–3.1
Temperate		
Great Smoky Mountains	49–211	0.5–2.0
Tionesta, Pennsylvania	107	0.9
Hueston Woods, Ohio	78	1.3

**Turnover time does not include the time required to grow into the canopy, which is estimated to be 54–185 years for various temperate zone species.*
(Data from Runkle 1985, Putz and Milton 1982, Brokaw 1985.)

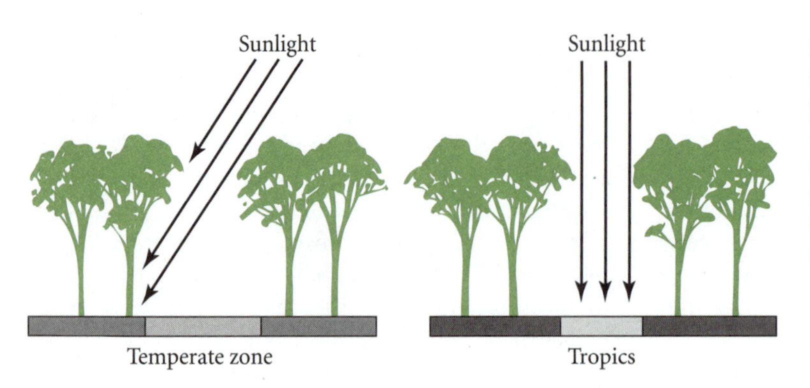

FIGURE 29-31 Light admitted to the forest floor through treefall gaps changes the physical conditions for seedling establishment and decomposition on the forest floor. These changes are more intense in the Tropics, where sunlight enters the gaps from directly overhead, thereby providing more light.

lack the distinctive physical structure of the forest canopy.

Huston (1979, 1994) proposed that the existence of highly diverse communities of species having seemingly identical ecological functions might be understood in terms of the interaction between the extent of disturbance in the community and the rates of population growth of the species in the community. His model, called the **dynamic equilibrium model,** is an elaboration of Connell's intermediate disturbance hypothesis (Connell 1978). It is represented graphically in Figure 29-32.

The figure is constructed in a manner similar to Figure 21-4. The figure represents three dimensions. The vertical dimension, which projects out of the page and thus cannot be seen in the figure, represents the species diversity of a community. On the page, the horizontal axis is the rate of population growth, and the vertical axis represents the level or intensity of disturbance in the area. Competitive exclusion reduces diversity and occurs when population growth rates are high. Disturbance may also reduce diversity, but only if populations have low growth rates and cannot recover from the disturbance. The figure shows how disturbance interacts with competition to affect biodiversity. At point A, population growth rates are low, and thus high levels of disturbance result in the local extinction of some populations and a reduction in diversity. At Point B, the level of disturbance is the same as at point A, but because population growth rates are high, species recover from disturbance and persist in the community. The maximum diversity occurs when there is a balance between competitive exclusion and disturbance (the 45° line). Thus a disturbance-free community with slow-growing populations may have a diversity similar to that of a community containing rapidly growing populations located in an area of high disturbance.

As the dynamic equilibrium model demonstrates, the effects of the various factors we discussed above (productivity, predation, competition, disturbance, etc.) are probably not mutually exclusive in all cases. Species diversity is likely to be a result of the simultaneous effects of a number of these factors mediated by the conditions of the environment and operating at a number of different spatial scales.

The processes discussed in the last few sections suggest mechanisms by which species richness may arise and be maintained. What can we learn from communities themselves about these mechanisms?

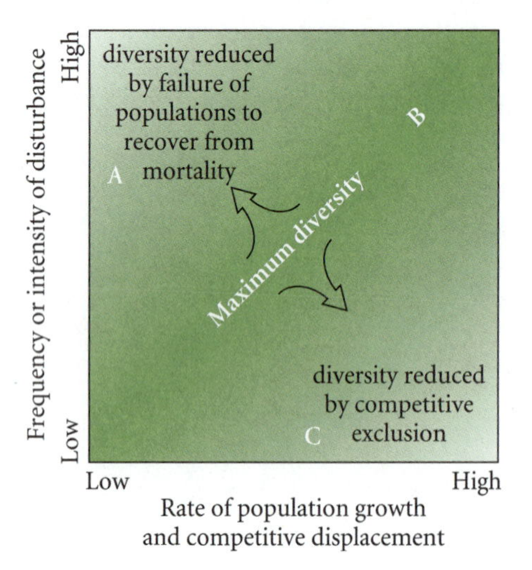

FIGURE 29-32 Relationship between the effects of disturbance and of population growth on local diversity. At low population growth rates, high rates of disturbance reduce the number of species in the community because many slow-growing populations cannot recover from disturbance (point A). In areas with high population growth rates, high rates of disturbance may prevent competitive exclusion, maintaining high diversity (point B). High diversity may also occur in low-disturbance areas where populations have low growth rates (point C) because populations do not grow rapidly enough to outcompete other populations. (*After Huston 1979, 1994.*)

29.12 Do communities reveal evidence of competition between species?

The role of species interactions, particularly competition, in molding the structure of communities has received considerable attention. Many ecologists have suggested that if competition is an important factor in the diversity of communities, then its effect should be evident in niche relationships among coexisting species. In particular, competition should lead to regular spacing among the positions of species within niche space. A simple example could involve the positions of species along a one-dimensional niche space. If the position of each species is sensitive to the positions of others—that is, if the coexistence of species depends on somehow minimizing competition between them—then the species should be regularly spaced along the niche dimension. Conversely, if the positions of the species are not influenced by interactions among them, then their positions along the niche dimension should be independently, or randomly, spaced.

Consideration of whether community features reflect competitive interactions has been somewhat limited by the methodologies available for determining the nature of random distributions in niche space. Randomness of distribution can be tested statistically through the use of null models, which ecologists began to apply in the late 1970s. A **null model** generates a pattern by randomizing ecological data by randomly sampling from some known or contrived distribution of ecological measurements (Gotelli and Graves 1996). Patterns obtained from field observations are then compared with the results of the null model to determine whether there is evidence of the presence of some underlying biological interaction, such as competition. Many authors of such studies have failed to find pervasive evidence of interaction (nonrandomness), but an equal number of authors have raised objections to their results, questioning either their analytical procedures or their interpretations. Null model techniques continue to be refined and used in the analysis of community patterns.

Analysis of Geographic Distributions

One way to get at the importance of competition in determining species diversity is to compare the geographic distributions of species on islands. Some have suggested that significant nonoverlap of distributions among close relatives would be consistent with the notion that competitive interaction has occurred between those species. Diamond (1975) examined the distributions of closely related (congeneric) birds in the Bismarck Archipelago, near New Guinea (Figure 29-33). He found that two species of cuckoo-doves (*Macropygia*) occupy six and fourteen islands respectively; neither occurs on thirteen islands in the area, but most significantly, no island is inhabited by both species. The probability of such a distribution occurring by chance, if the two species inhabited islands independently of each other (that is, were randomly distributed), is less than 1 in 40 ($P < 0.05$). Hence Diamond concluded that interaction plays an important role in distribution and local community diversity.

Connor and Simberloff (1979) questioned the validity of Diamond's analysis on methodological grounds. They reasoned that if the distribution of *Macropygia* species might have occurred by chance 1 time in 40, then for every 40 pairs of species one examined, one would be likely to observe a pair with a distribution as nonrandom in appearance as that of the two cuckoo-doves. To test for interaction adequately, reasoned Connor and Simberloff, one would have to examine all possible pairs of species in an area. They then did this for birds in the West Indies. In their analysis, they accepted the number of species per island and the number of islands occupied per species as fixed properties of the system. Within these constraints, they then randomized the distributions of the 211 species in the sample several times and examined the results. Of the 22,155 possible pairs of species [all combinations of

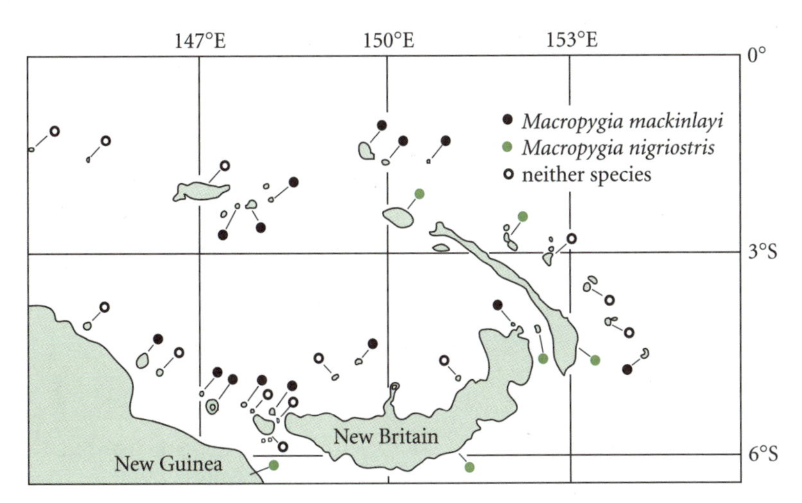

FIGURE 29-33 Distribution of cuckoo-doves of the genus *Macropygia* in the Bismarck Archipelago. Most islands have one of the two species, no island has both, and some have neither. (*From Diamond 1975.*)

n species equals $n(n - 1)/2$], an average of 12,448 had exclusive distributions (no co-occurrence on any island) in the randomized set of species. The exclusive distributions among pairs of species in the actual avifauna of the West Indies number 12,757, so close that one must accept general agreement with the randomly generated pattern. Connor and Simberloff found similar agreement between observed and randomized distributions in a number of other tests, and concluded that interaction between species was not an important determinant of their geographic distributions.

Diamond and Gilpin (1982) responded to the Connor-Simberloff analysis, protesting that by looking at all species pairs, one reduced the chances of recognizing interaction, which one would expect only among ecologically similar species. They also objected that the method used by Connor and Simberloff to produce their random distributions was flawed in such a way as to nearly guarantee a resemblance between randomized communities and the observed communities from which they were derived. The problem here is the procedure for producing the random community—the null model (Harvey et al. 1983, Quinn and Dunham 1983, Simberloff 1983, Gotelli and Graves 1996). These disagreements about null model methodology persist. Recent evaluations of the problem by Stone and Roberts (1990, 1992) suggest that both approaches are weak, and that while random colonization by island species is unlikely, forces other can competition may be at play.

Community Character Displacement

The divergence of the characteristics of two otherwise similar species where their ranges overlap is referred to as **character displacement** (see Chapter 22). Regular differences in morphological features such as body length, size of appendages, or size of feeding structures among species within a community have often been interpreted as indirect evidence of the importance of competition in the community. Strong et al. (1979) used the term **community character displacement** to describe the situation in which, when the species of a community are ranked according to the size of a particular morphological characteristic, the ratios of adjacent species are more or less equal.

Strong et al. (1979) investigated the size ratios of confamilial species of birds inhabiting the Tres Marías Islands off the western coast of Mexico. In an earlier study of those communities, Grant (1966) had suggested that size ratios exceeded those found in mainland communities, indicating that competition and ecological divergence had exerted important influences on the island communities. Strong and his colleagues tested this hypothesis by drawing sets of species at random from the "pool" of species inhabiting the nearby mainland of Mexico. They found that the size ratios of related birds on the Tres Marías Islands were no greater than among those drawn at random from the mainland, and thus rejected Grant's claim. However, while Strong and colleagues tested the hypothesis of interaction among species, their null model contained the assumption that all mainland species have equal probability of colonizing offshore islands. This assumption can be rejected easily; ecologists well know that colonizing ability varies with habitat, abundance, and behavior. Some species are better colonists than others. One can say with certainty, therefore, that the null model in this study is unrealistic, but devising a better procedure for constructing random communities poses a great challenge.

Simberloff and Boecklen (1981) partly avoided the problem of a source pool of species by assuming a certain statistical distribution of, for example, sizes of species from which to draw. For the purposes of their analysis, they assumed that species occupied a uniform probability distribution of sizes. Three points (species) drawn at random from such a distribution defined two intervals, that between the smallest and the medium-sized species and that between the medium and the largest (Figure 29-34). The ratio (a) between the smallest and largest intervals in randomly drawn trios of species itself has an expected probability distribution between 0 and 1, against which one may test the size of intervals separating trios of actual species. The probability of observing a ratio (r) less than some value a in a randomly drawn community is $1/[(1/2) + (1/2a)]$. Thus the probability of $r < 1$ is 1.0, while the probability of $r < 0.5$ is 0.67, and that of $r = 0.33$ is 0.5; one would expect half the ratios to exceed 0.33 and half to be less than that value. Simberloff and Boecklen worked out a similar distribution for the expected size of the smallest of $n - 1$ segments produced by drawing n points at random. If similarity between species limits coexistence, one should find fewer small ratios than expected at random. If species were spaced more evenly than

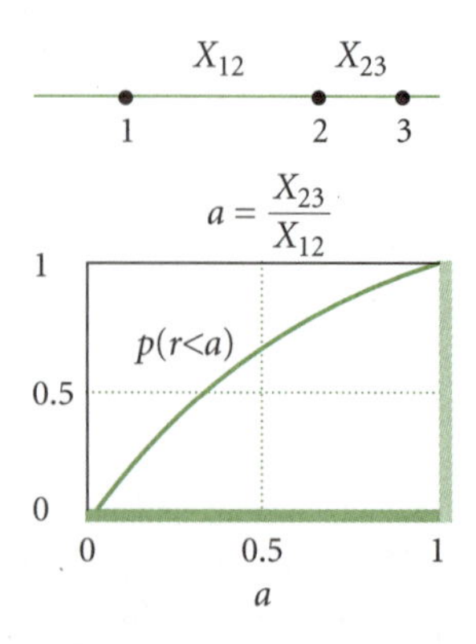

FIGURE 29-34 The probability that the ratio between the smaller and the larger of two randomly determined intervals will be less than the value of a.

expected based on the random distributions, the distribution of r would be shifted from that of a toward 1. After making such comparisons, Simberloff and Boecklen found little support for nonrandomness in the data for which claims of spacing had been made.

Ricklefs and Travis (1980) took a similar null model approach to assess regularity of spacing of bird species in the scrub habitats described by Cody (1974). They constructed random communities of 5, 9, 13, and 17 species, either by drawing species at random from the total species pool or by generating points (synthetic species) at random from the morphological space occupied by the species. If interaction influenced community structure, one would expect species to be spaced more evenly in the natural communities than in the null communities. The criterion for even spacing was the standard deviation of nearest-neighbor distances, which decreases to zero as spacing becomes perfectly even. As one can see in Figure 29-35, natural communities do not differ from randomly generated ones; thus, morphological spacing does not reveal species interaction.

The absence of nonrandom spacing need not imply that competition is absent (experiments demonstrate that it is present in many cases!) or unimportant in regulating diversity. We may conclude only that the evolutionary adjustments of species or their selective establishment and extinction locally do not produce nonrandom patterns. Because species extend geographically and ecologically over large areas, we might not expect their evolved properties to reflect local ecological conditions. Furthermore, because each species may compete with many others in the system over a large number of ecological dimensions, small variations in proximity between nearest neighbors along a single dimension of niche space may have a negligible influence on the demography of any particular population. That is, selection for even spacing may be very weak.

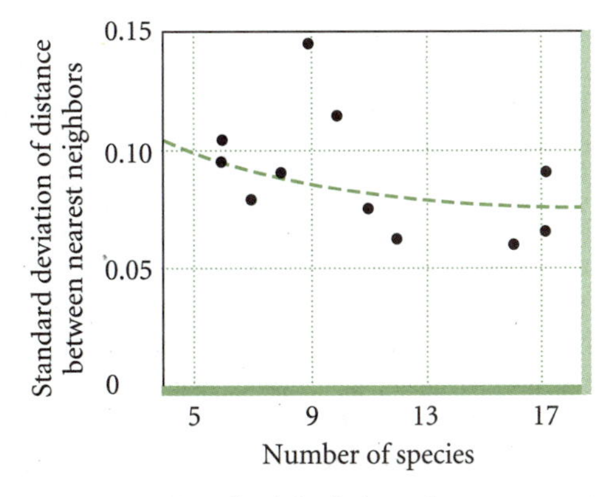

FIGURE 29-35 Standard deviation of average nearest-neighbor distance in morphological space among passerine birds inhabiting scrub communities. The dashed line is the predicted value from randomly generated communities. (*After Ricklefs and Travis 1980.*)

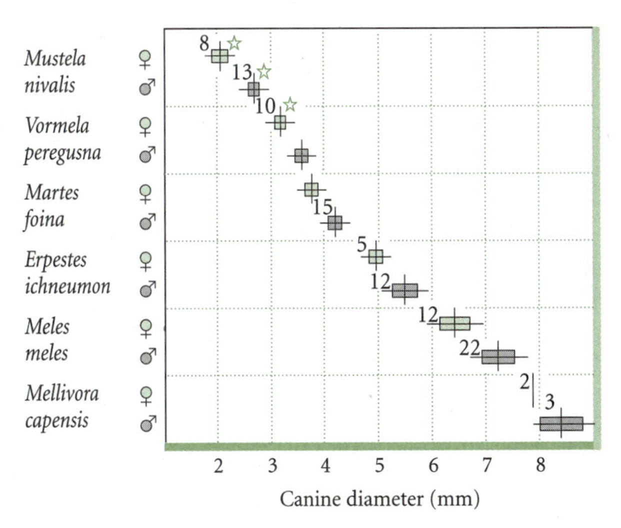

FIGURE 29-36 Community character displacement in canine diameter in species of mustelids and viverrids in Israel. The species *Mustela nivalis* is extinct. Vertical lines represent means (based on the sample size indicated beside each box) with standard deviation (horizontal bars) and standard error (boxes). Males and females are presented separately because of pronounced sexual size dimorphism in the species. The species are ranked along the vertical axis from smallest canine diameter to largest. The ratio of canine diameter between adjacent species is remarkably constant, indicating community character displacement. (*From Dayan et al. 1989.*)

T. Dayan and coworkers have presented perhaps the most compelling evidence for the existence of community character displacement (Dayan et al. 1989, 1990, Dayan and Simberloff 1994). Dayan et al. (1989) measured the maximum diameter of the upper canine tooth and a number of other characters of males and females of five extant and one extinct species of mammals of the families Mustelidae (weasels) and Viverridae (ferrets) that occur (or, in the case of the extinct species, occurred) together in various communities in Israel. Their results show a remarkable community character displacement (Figure 29-36). Dayan and colleagues found similar results for mustelids in North America, Britain, and Ireland and for small wild cats (Felidae) of Israel.

29.13 Lack of strong evidence of local species saturation and community convergence suggests that regional/historical factors play a major role in community diversity.

In determining the relative effects of local and regional processes on species diversity, it would be helpful to know whether local communities are saturated with species. That is, do species fill ecological space in communities up to some fixed capacity determined by their local interactions? If so, we would expect the local diversity of similar habitats to be similar irrespective of regional diversity, the discrepancy being made up by beta diversity (turnover of species among habitats). We would also like to know whether communities converge in their structure. Local process theories of species diversity predict such convergence. That is, communities occurring under similar physical conditions should have similar numbers of species, regardless of geographic location or history. Evidence supporting local species saturation and community convergence would favor local/deterministic concepts of species diversity over regional/historical theories. Let us examine some of the studies that have addressed these questions.

Saturation of Local Communities

The idea of species saturation can be tested by plotting the relationship between local and regional diversity for many areas with similar environments. Using this technique, Terborgh and Faaborg (1980) found evidence for saturation in bird communities of the West Indies; Cox and Ricklefs (1977), working with slightly different samples from the same system, did not (Figure 29-37). In two other tests of this hypothesis, Cornell (1985) failed to find evidence for saturation in the local diversity of gall-forming wasps on oak trees in California (Figure 29-38), whereas the results of Stevens (1986) supported the hypothesis of saturation for wood-boring scolytid beetles in hardwood stands of eastern North America. Clearly, we cannot reach definitive conclusions about ecological saturation at this point. To the extent that supporting evidence surfaces, it will weigh in favor of local processes and against regional/historical processes in regulating local species diversity.

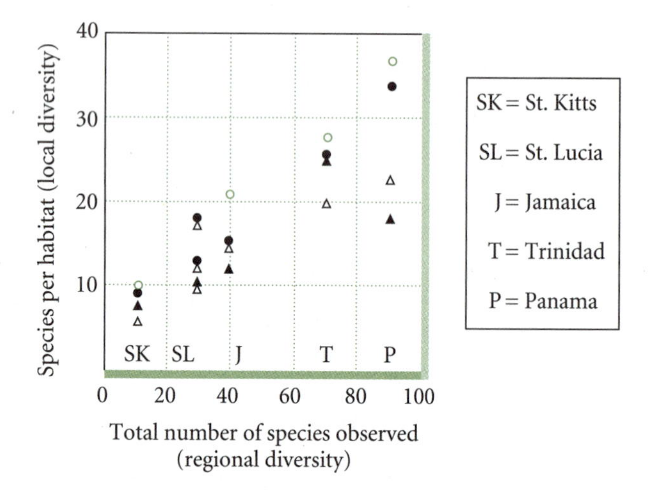

FIGURE 29-37 Local species richness of birds in the Caribbean region is sensitive to regional diversity, which is determined primarily by biogeographic considerations. Within each area, standardized counts were made of songbird species within small, homogeneous areas of habitat, each type indicated by a different symbol. (*From Ricklefs 1987.*)

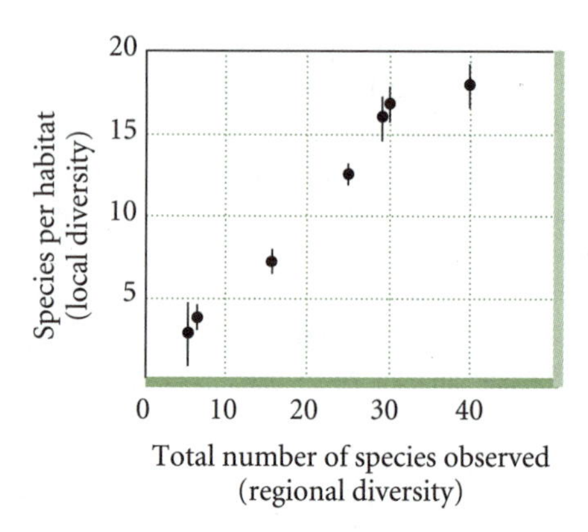

FIGURE 29-38 Local species richness of cynipine wasps on oaks is directly related to the total number of species of cynipines recorded from throughout the range of each oak species (represented by a single data point). The solid line indicates local diversity equal to regional diversity. (*From Cornell 1985.*)

Community Convergence

Plants and animals of similar habitats in different geographic locations often reveal convergence of form and function. It has been suggested that properties of communities also may exhibit convergence (Cody and Mooney 1978, Orians and Paine 1983). Community convergence occurs when assemblages of species become more similar to one another than their ancestors who lived in a similar habitat. Evidence bearing upon community convergence is growing (Schluter and Ricklefs 1993). Here, we mention just two examples: temperate deciduous forests and tropical mangrove ecosystems.

The temperate deciduous forests of eastern North America include 253 species of trees, more than twice the number (124) found in similar habitats in Europe (see Table 29-2). Temperate eastern Asia, whose climate is also similar to that of eastern North America, has 729 species of trees. Thus, although the climates of the three regions are similar and their forests have similar structures—they are all dominated by deciduous, broad-leaved trees—species diversity varies by a factor of nearly 6 among the different regions. These figures represent the total diversity of each region, but local diversity within small areas of uniform habitat exhibits parallel differences. Thus regional diversity and local diversity appear to be closely related.

A similar difference exists between the species diversities of mangrove forests in the Caribbean region and the Indo-West Pacific region (Ricklefs and Latham

1993). In the Tropics of the New World and western Africa, mangrove communities consist of the same three or four species of specialized trees distributed throughout the region (Davis 1940; see Figure 26-17). In physically similar environments in Malaysia, mangrove vegetation comprises seventeen "principal" and twenty-three "subsidiary" species (Watson 1928), representing an order of magnitude greater diversity than in the New World.

Because the environments are similar but community diversity differs, these findings suggest that factors other than the outcome of ecological interactions within local communities must have an influence on species richness both within regions and in local communities. Thus, spatial patterns of diversity reflect both ecological processes occurring on small scales of time and space and large-scale processes such as those responsible for species production and accidents of history.

Our discussion here has focused on how species diversity arises and is maintained in nature. The services that this great diversity provides are manifested in the energy transformations that sustain life (see Part 3). Human activities tend to reduce diversity and put these services at risk. The way to preserve biodiversity is to set aside large areas of the variety of natural habitats found on earth and maintain their capacity for supporting species. Basically, this means minimizing human impacts of all kinds over representative areas of the earth's surface. As the human population grows, however, this goal recedes farther into the distance and becomes, in the eyes of most people, less pressing than the problem of maintaining basic life-supporting systems for humans. Resolution of the conflict between natural and human values will ultimately depend on erasing the distinction between them and somehow making them compatible.

SUMMARY

1. Some communities contain more species (greater species richness) than others. Patterns of species richness in geologic time, along latitudinal gradients, related to habitat structure and disturbance, and associated with differences in productivity have been observed by ecologists.

2. Two views of species diversity have prevailed in ecology. The local/deterministic view holds that diversity is determined principally by biological interactions such as competition and predation occurring at the local level. The regional/historical view emphasizes the importance of species differentiation and movement at the regional level and the interaction between local and regional processes. Contemporary ecology incorporates both of these views.

3. Alpha diversity refers to the number of species in a small area of relatively homogeneous habitat. Gamma diversity is the number of species in a region. A measure of the difference in diversity from one habitat to the next (species turnover) is beta diversity, which is the ratio of gamma diversity to alpha diversity. Changes in alpha (local) and beta (turnover) diversity usually result in changes in gamma (regional) diversity. Greater gamma diversity usually means greater habitat specialization.

4. Differences in the numbers of species on islands emphasize the importance of regional processes—immigration from the mainland or other islands, in this case—to the maintenance of species diversity. On continents, immigration of species to local areas reflects, in part, the rate of production of new species, which is also a regional process.

5. The species/genus ratio is one way in which to compare rates of species development between regions, with high ratios indicating recent diversification. Tropical areas have low species/genus ratios, indicating that diversity there emerged over a long period of time. Phylogenetic analysis may also be used to determine speciation rates.

6. The time hypothesis holds that areas, such as the Tropics, that have enjoyed long periods of stability should have more species than areas with more frequent and widespread disturbance. Fossil evidence and the results of phylogenetic studies suggest that the hypothesis should be accepted with caution.

7. The niche concept expresses the relationship of the individual to physical and biological aspects of its environment. Each factor may be thought of as an axis of a multidimensional niche space, within which organisms and populations occupy characteristic spaces. The niche of a species is defined by the range of values on each niche axis within which the population can persist.

8. The ecological relationship between two species can be described by the degree to which their niches overlap, which is a function of niche breadth and niche separation.

9. The community may be conceptualized as an assemblage of species each occupying a niche defined by axes of resource quality and ecological conditions. Additions of species to the community may increase the total niche space, decrease niche overlap, or decrease niche breadth (increase specialization). Opportunities for escaping predators are an important component of the niche and can be considered a resource for which prey compete.

10. Ecologists have noticed that in some cases, species diversity increases as the productivity of the environment

increases. At least two ideas have been put forth to explain this pattern: the productivity-stability hypothesis and the intertaxon competition hypothesis. In some cases, high productivity corresponds to low diversity, a situation referred to as the paradox of enrichment.

11. Variation in soil type, moisture, and temperature may affect the species diversity of plants and thus the animals living on and among the plants. Fluctuations in environmental conditions may promote species diversity in some situations.

12. Predators are thought to enhance diversity among their prey by reducing prey populations (and hence competition for resources), thereby easing conditions for coexistence. Evidence that predators and diseases may act in a density-dependent manner also favors this hypothesis. Spatial models of predator-prey interactions show how regional and local processes interact to determine predator and prey diversity.

13. Disturbance may affect species diversity. The intermediate disturbance hypothesis holds that diversity is highest in areas where there is an intermediate amount of disturbance. This idea does not account for regional patterns of diversity. The dynamic equilibrium model suggests that the effects of disturbance depend on the growth rates of the populations in the community through a balance between the intensity of disturbance and competitive exclusion.

14. Ecologists believe that if competition is an important determinant of species diversity, its effect should be reflected in the niche relationships among coexisting species. Null models, which are used to compare the observed relationships of a natural community with those of a statistically derived community, have been widely used to investigate this idea, but with mixed results. One indication of the importance of competition is community character displacement, which has been observed in a number of communities.

EXERCISES

1. Most of this chapter dealt with species diversity. Using the information in Chapter 8 and readings from other sources, discuss the concept of ecosystem diversity.

2. Prepare a short essay that compares how our approach to the preservation of endangered species would be shaped differently by the regional/historical and the local/deterministic views of the regulation of diversity.

3. Prepare a written explanation of niche theory for a group of people who have never had an ecology course. Use your explanation to introduce the group to the concept of species diversity.

4. Are there groups of animals that are more likely to increase in diversity in environments with high productivity?

5. E. O. Wilson (1984) suggested that our descendants will be more likely to forgive us for economic collapse and nuclear war than for the loss of global biodiversity. Explain this sentiment in terms of ecological theory.

PART 7

EVOLUTIONARY ECOLOGY

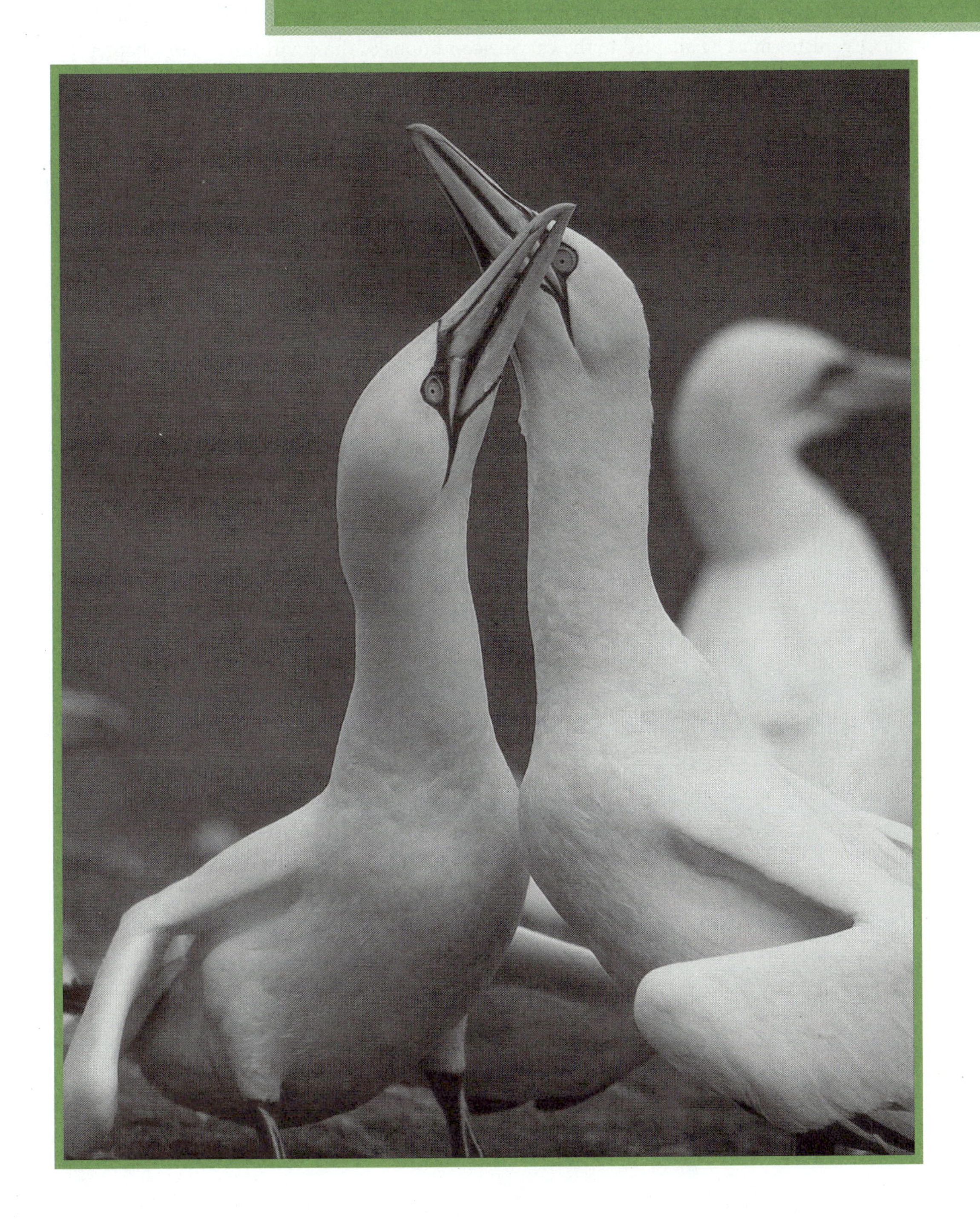

Patterns of reproduction, foraging, social interaction, growth, and senescence are shaped by natural selection through the interactions of organisms and their environment. Those behavioral, physiological, or developmental responses that allow an organism to accommodate or acclimate to the current conditions are called evolutionary adaptations. In this part of the book we shall examine the mechanisms and evolutionary basis of such patterns.

In Chapter 30 we will discuss the nature of adaptations, emphasizing the genetic basis of natural selection and the interactions of the phenotype and the environment. Organisms must be adapted to the varying conditions of the environment. In Chapter 31 we will emphasize how predators respond to varying conditions of prey. Natural selection adjusts the allocation of time and resources among competing demands such as reproduction and growth. The consequences of these trade offs will be examined in Chapter 32. Adaptations related to sexual reproduction will be discussed in Chapter 33. There we shall explore the dynamics of mate selection and parental care and show how selection shapes these aspects of the ecology of sexually reproducing organisms. A number of organisms have adopted social systems whereby some individuals give up or postpone reproduction in order to assist with the rearing of young. The evolution of these systems will be discussed in Chapter 34.

CHAPTER 30

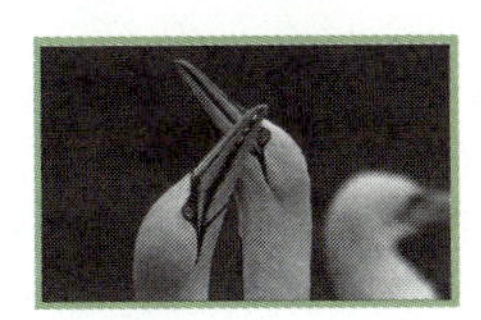

Evolution and Adaptation

GUIDING QUESTIONS

- What are the five programs of inquiry in evolutionary ecology?
- How does natural selection change genotype frequencies in populations?
- What factors affect the rate of evolution?
- What are the various components of phenotypic variance?
- What is the relationship among heritability, response to selection, and selection differential?
- How do correlated responses to selection limit evolutionary response?
- What are the costs and benefits of sexual reproduction?
- How do inbreeding, outcrossing, and selective abortion in plants affect genetic variation?
- What is the difference between the phenotypic optimization and evolutionary stable strategy approaches to understanding adaptation?
- What are Gould and Lewontin's criticisms of the adaptationist program?

From its beginning, ecology has formed links with three related biological disciplines: genetics, which studies the mechanisms of heredity; evolution, which includes the study of change in the genetic makeup of populations; and development, which represents the realization of the genetic blueprint in the form of the individual. A distinct subdiscipline, called **evolutionary ecology,** concentrates especially on the interpretation of form and function in organisms as adaptations to their environments. In this chapter we shall explore the nature of these adaptations in detail.

30.1 Adaptations have a genetic basis.

The ecology of plants, animals, and microorganisms reflects interactions between gene pools of populations and selective factors in the environment. The environment embodies the selective pressures that establish fitness differences between individuals with different genotypes. Evolutionary responses of the phenotype to those pressures depend on functional interrelationships that limit form and function to combinations of traits that work, in the sense that they obey physical laws. Evolutionary responses also depend upon the availability of genetic variation in the population, upon which selection acts. Moreover, genetic variation itself makes advantageous certain types of adaptations, principally of the breeding structure of the population, that optimally manage the expression of genetic variation in the progeny of each individual.

30.2 There are five research approaches to evolutionary ecology.

Evolutionary ecologists employ one or more of five approaches to their work. The first approach involves understanding the mechanisms of evolutionary change from a genetic standpoint, including the limits imposed by the availability of genetic variation. This approach is discussed in detail in texts on population genetics and evolution (e.g., Hartl 1980, Milkman 1982, Futuyma 1986, Hartl and Clark 1989), and will be touched upon only briefly here. A second approach focuses on understanding how genetic variation in populations confronts individuals with the problem of choosing the genotypes of their mates so as to optimize the genotypes of their offspring. This approach involves the study of breeding systems and patterns of mate choice, topics that will be discussed in detail in Chapters 33 and 34.

The third research approach, to which we shall devote most of our attention in this chapter, is the interpretation of form and function in the context of adaptation. This approach, often called the **adaptationist program** (Gould and Lewontin 1979) or "selection thinking" (Charnov 1982), presumes that evolution can achieve that combination of traits best suited to any particular environment and permissible within the bounds set by physical laws. The major goals of this approach are to learn how physical limits on form and function prescribe the possible phenotypes among which the environment selects; to determine appropriate measures of fitness, especially with respect to traits that govern the interactions of individuals within a population; and to understand where the adaptationist paradigm can be applied validly and where it cannot.

The fourth approach elaborates the last point: it seeks to determine the degree and mechanisms of matching of phenotypes to environments. Matching can occur both through selection of the phenotype by the environment and selection of the environment by the phenotype. At issue is the degree to which response to the environment is a property of the individual or of the gene pool of the population. Furthermore, organisms retain tangible evidence of their evolutionary history in the form of traits that are shared with related species regardless of the environment. For example, plants in the rose family have flowers with five-part symmetry; those in the lily family are endowed with a three-part or six-part symmetry. An understanding of the evolutionary significance of flower symmetry thus requires an exploration both of its proximate ecological utility and of its historical development.

Finally, the fifth program of evolutionary ecology seeks to determine the extent to which properties of larger ecological systems—communities and ecosystems—depend upon evolutionary relationships among their parts.

An important concept throughout the chapters in this part of the book is the concept of evolutionary fitness. We define **fitness** as the genetic contribution of an individual's descendants to future generations of a population. Several aspects of the concept of evolutionary fitness must be emphasized. First, it is important to appreciate that it is not the *number* of offspring produced by an individual, but rather, the *contribution* of those offspring to future generations that counts. Second, fitness is measured over the lifetime of the individual, not just for one breeding season. The term *fitness* is not synonymous with the term *fecundity*, which refers to the rate at which an individual produces offspring. Finally, when making comparisons of fitness, we are usually interested in the fitness of an individual relative to others in its population, rather than its absolute fitness. So, an individual that leaves in its lifetime four descendants that reproduce would have high relative fitness in a population where most individuals leave a single reproducing offspring, but low relative fitness in a population where individuals leave on average ten reproducing offspring in their lifetime.

30.3 Evolutionary response to selection proceeds by the substitution of genes within populations.

In a natural population we observe variation in form and function among the individuals. Not all the Nile perch in Lake Victoria (see Chapter 1) are identical in size, maximum swimming speed, rate of digestion, and so forth. Not everyone in your ecology class is the same height or has the same skin color. This variation has a genetic basis. In some cases, the differences between individuals express variation at a single genetic locus. The principal color groups of eyes (blue or brown) in humans and typical and melanic coloration in the peppered moth result from alternative forms of the same gene, called alleles. In other cases, traits come under the influence of many loci, whose effects may be additive, complementary, or modifying.

A number of factors can affect variation in a population. The ultimate source of natural variation is mutations. Mutations are thought to exert a relatively weak force on variation in nature. Random variations in fecundity or mortality may also affect variation in populations, particularly small ones—a phenomenon called genetic drift (see Chapter 14). A third factor affecting variation is immigration and emigration. **Assortative mating** may also change allele frequencies in populations. Assortative mating simply means that mates are chosen nonrandomly with respect to genotypes. Like mating with like, which could include inbreeding, is referred to as **positive assortative mating;** a tendency for mates to differ genetically is referred to as **negative assortative mating.** Positive assortative mating (for example, $A_1A_1 \times A_1A_1$) tends to reduce the proportion of heterozygotes in the population, whereas the opposite results from negative assortative mating ($A_1A_1 \times A_2A_2$). Thus, the expression of recessive alleles, including rare and harmful genes, increases with positive assortative mating. Finally, and most importantly, natural selection may affect variation.

It is the action of natural selection that will command most of our attention in this part of the book. When selection is applied to some traits, such as the amount of black pigment in the wing scales of peppered moths, a single gene locus or a few gene loci are affected. Selection upon other traits, such as body shape or patterns of social behavior, may affect numerous genes responsible for producing the trait, many of which in turn influence other characteristics of the phenotype.

The Hardy-Weinberg Model

The evolutionary mechanics of selection and genetic responses are the subject of **population genetics** (Crow and Kimura 1970, Hartl and Clark 1989, Tamarin 1993). A primary task of population geneticists since the late 1920s has been to develop quantitative predictions of changes in gene frequencies in response to selection. While population genetic models are often extremely complex, the essence of natural selection may be understood using a very simple genetic model, which was discovered independently in 1908 by G. H. Hardy and W. Weinberg. The purpose of the model was to show how genetic variation is retained in Mendelian inheritance. The model, called the **Hardy-Weinberg law,** demonstrates that the frequencies of alleles and genotypes remain constant from generation to generation in large populations in which there is random mating (each zygote is formed from the random combination of any two gametes), no selection, no mutation, and no migration to or from the population. This model may be used as a point of comparison for studying the effects of natural selection. If selection is applied to a reasonably large closed population having random mating and no mutation, then changes in the frequencies of the alleles in that population over several generations may be attributed to selection. Let us examine the Hardy-Weinberg law in more detail and show how it can be used to demonstrate the effects of selection.

When a population exists at Hardy-Weinberg equilibrium, the proportions of homozygotes and heterozygotes take on equilibrium values, which we can calculate from the proportions of each allele in the population. Thus, two alleles A_1 and A_2 of the locus A might occur in the population with frequencies p and q respectively ($p + q = 1$, and therefore $q = 1 - p$). The three possible genotypes in the population are the two homozygotes A_1A_1 and A_2A_2 and the heterozygote A_1A_2, which, in a population at Hardy-Weinberg equilibrium, will occur with frequencies p^2, q^2, and $2pq$, respectively. Because of random mating, $p^2 + 2pq + q^2 = 1$. If alleles A_1 and A_2 occur with frequencies 0.7 and 0.3 respectively in the population, than the proportion of A_1A_1 genotypes in the population is $0.7^2 = 0.49$ (49%), if all the assumptions of the Hardy-Weinberg law are met. In the same manner, we see that 42% of the genotypes will be A_2A_2 and 9% will be the heterozygote A_1A_2.

We may use the Hardy-Weinberg law to evaluate relative fitness. Suppose genotype A_1A_1 has a fitness of 1. If A_1 is dominant over A_2, then the genotype A_1A_2 will also have a fitness of 1. Let genotype A_2A_2 have a lower fitness by some fraction s, which we denote $1 - s$. Prior to selection, the frequencies of the genotypes A_1A_1, A_1A_2, and A_2A_2 are p^2, $2pq$, and q^2 respectively, according to the Hardy-Weinberg law (Table 30-1). How much will the frequency of each allele change in the course of one generation of selection? For each generation, the frequencies of the genotypes are multiplied by their respective fitness values to obtain the relative number of their descendants in the next generation. If the fitnesses of the A_1A_1 and A_2A_2 phenotypes are 1 and $1 - s$ respectively, the relative numbers of progeny of each of the genotypes are p^2, $2pq$, and $(1 - s)q^2$. By counting up the relative numbers of A_1 and A_2 alleles in the progeny, we can calculate the change in frequency of the A_2 allele caused by selection against homozygous genotypes (Table 30-1).

From Table 30-1, we see that the relative proportion of the A_2 allele in the descendant population (q') is the ratio of the A_2 alleles to the total, or

$$q' = \frac{pq + (1 - s)q^2}{p^2 + pq + (1 - s)q^2}, \qquad (30\text{-}1)$$

TABLE 30-1 Rate of change of allele frequencies under selection

	GENOTYPE		
	A_1A_1	A_1A_2	A_2A_2
Initial genotype frequency	p^2	$2pq$	q^2
Reproductive success (fitness)	1	1	$(1 - s)$
Relative proportion of descendants	p^2	$2pq$	$(1 - s)q^2$
Relative proportion of A_1 alleles in the descendant population	p^2	pq	
Relative proportion of A_2 alleles in the descendant population		pq	$(1 - s)q^2$

which may be simplified to

$$q' = \frac{q(1 - sq)}{1 - sq^2}. \quad (30\text{-}2)$$

The change in allele frequency from one generation to the next, Δq, is $q' - q$, which, with a little algebra, can be rearranged to give

$$\Delta q = \frac{-sq^2(1 - q)}{1 - sq^2}. \quad (30\text{-}3)$$

When the recessive homozygote is lethal ($s = 1$), the equation above simplifies to $\Delta q = -q^2/(1 + q)$. That is, the change in the frequency of allele A_2 is entirely dependent on the proportion of A_2 in the population. When selection is very weak (s is very small, perhaps less than 0.01, or 1%), the equation becomes approximately $\Delta q = -sq^2(1 - q)$.

These equations show that selection against the A_2A_2 genotype always causes a decrease in the frequency of the A_2 allele. Additionally, the rate of change in q depends on both the selective pressure on a population and the frequency of the A_2 allele. For example, change in q is fastest when q is relatively large because a larger proportion of the A_2 alleles are exposed in homozygous form (Figure 30-1). Also,

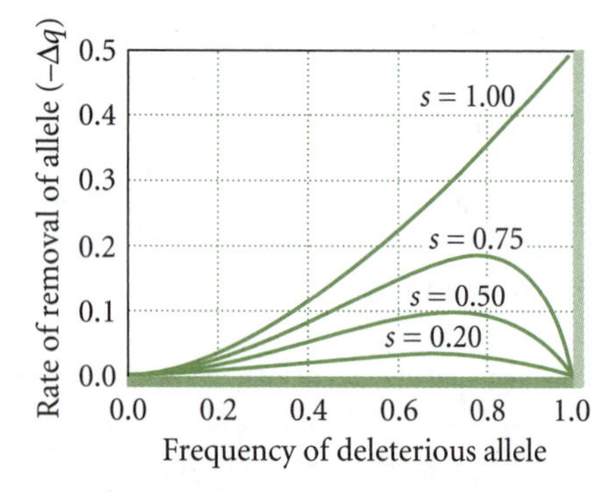

FIGURE 30-1 Rate of change in the frequency of a recessive, harmful allele as a function of allele frequency and strength of selection.

evolution stops ($\Delta q = 0$) only when q is equal to either 0 or 1, in which case either the A_1 or the A_2 allele is fixed in the population, and there is no longer any genetic variation for selection act upon.

Rates of Evolution

The models above predict the change in a single gene with one allele dominant over the other. Equations that predict the change in allele frequencies resulting from one generation of selection can be used to show how a population evolves over many generations of continued selection and to predict how rapidly a population can respond genetically to a change in its environment.

The time required for a dominant allele to replace a recessive allele depends on its initial frequency in the population and on the strength of selection. To illustrate this, let us take another look at the peppered moth (*Biston betularia*) that inhabits woodland areas of England, where it rests on lichen-covered trees during the daytime. Recall from Chapter 1 that early in the nineteenth century, occasional dark or melanistic specimens of this common moth were collected, and subsequently, over period of about 100 years, the dark form, referred to as *carbonaria,* became increasingly common in forests near heavily industrialized regions of England. It appeared that environmental conditions had somehow been altered to give the dark forms a survival advantage over the light forms, and natural selection led to the replacement of typical light individuals with *carbonaria* individuals. To test this hypothesis, the English biologist H. B. D. Kettlewell measured the relative fitnesses of the two forms independently of the fact that the frequency of one had increased over that of the other.

From the results of Kettlewell's experiments, we can estimate that the fitness of the recessive homozygous genotype for typical (light) coloration was only 47% that of the *carbonaria* genotype in woods affected by industrial pollution; hence the fitness differential, or strength of selection against the typical

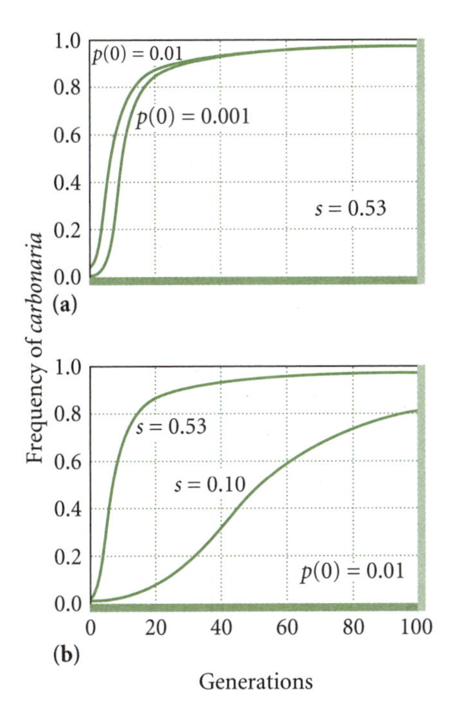

FIGURE 30-2 Simulation of the change in frequency of the *carbonaria* allele in a population of peppered moths in a region of industrial pollution, assuming selection differentials against homozygous genotypes with typical coloration of $s = 0.53$, as measured in Kettlewell's experiments, and $s = 0.1$. (a) Comparison of different initial allele frequencies; (b) comparison of different strengths of selection.

form, was $s = 0.53$. Using this information and equation (30-3) above in an iterative manner, and assuming selection against homozygous genotypes having typical coloration, the change in the frequency of the *carbonaria* allele through time can be predicted (Figure 30-2). Because *carbonaria* is a dominant allele, it is exposed to selection even at low frequencies, and the initial frequency of the allele has little effect on the rate of evolutionary change (Figure 30-2a). Simulation of the selection process predicts that the population will consist mostly of melanistic forms in about 50 generations. If selection is weaker, however, the transition takes longer (Figure 30-2b).

30.4 Many traits of ecological interest have a polygenic basis.

The relationships between organisms and their environments often depend upon modifications of continuously varying traits, such as lengths of appendages, body sizes and shapes, thickness of hair or cuticles, and continuous gradations of behavior. Because variation in these characters depends on the contributions of many genetic loci, their response to selection cannot be analyzed by the simple population genetic models that have been applied to single-gene traits. Animal and plant breeders interested in such traits as milk production, oil content of seeds, and rate of egg production have developed a mathematical treatment of continuously varying traits and their responses to selection, known as **quantitative genetics** (Bulmer 1980, Falconer 1981). The theory of quantitative genetics rests on the assumption that variation within a population results from the additive contributions of many genes with similar effects. Thus, the length of an appendage may come under the influence of a dozen genetic loci, each of which may cause a small increase or decrease in length relative to the population average, depending on the allele. Individuals with a net excess of length-increasing alleles at these twelve loci would have appendages longer than the average.

Quantitative genetics greatly simplifies our concept of gene action (Thompson and Thoday 1979), and has proved successful in predicting the results of selective breeding programs. To the extent that artificial selection mimics selection in nature, quantitative genetics can help us to understand the evolutionary responses of natural populations.

Variance of a Trait

The heart of quantitative genetics is the variance of a trait within a population. Variance is a statistical measure of variation—it is the average of the squared deviations of individuals from the mean of the population. Recall from Chapter 2 that variance s^2 in a variable X (a measurement) in a population of n individuals is $s^2 = \Sigma(X_i - \overline{X})^2/(1 - n)$ where X_i is the value for each individual i ($i = 1$ to n), and $\overline{X}$ is the mean value of X in the population. In genetics, X is the measurement of some character or trait. In population genetics, V is used to denote variance instead of s^2. We shall follow that convention here.

Continuously variable traits often exhibit a bell-shaped distribution of values in natural populations, with most individuals clustered near the mean and with frequencies diminishing at extreme values (Figure 30-3; see also Figure 2-3). Each individual's value of a particular trait—the **phenotypic value**—is determined by genetic and environmental influences. Because both sources of deviation from the population mean enter into the calculation of variance for all values in the population, we may speak of **phenotypic variance** (V_P) as having two components, one attributable to genetic constitution (V_G), the **genotypic variance,** and one resulting from environmental factors (V_E), the **environmental variance.** The two components added together equal the total phenotypic variance, or $V_P = V_G + V_E$.

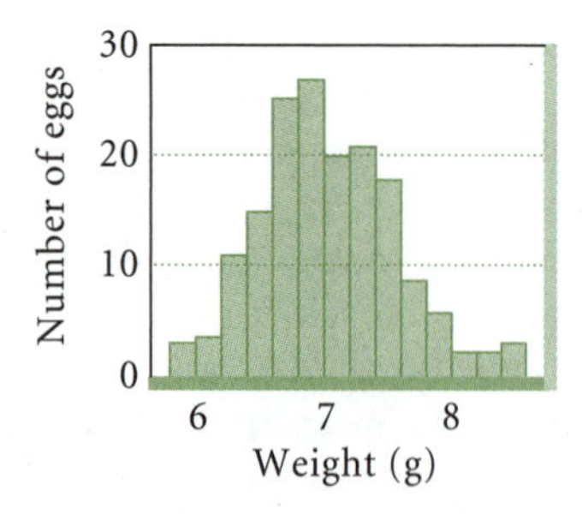

FIGURE 30-3 The frequency distribution of the weights of eggs of the European starling (*Sturnus vulgaris*) sampled near Philadelphia, Pennsylvania, like many distributions of continuously varying traits found in natural populations, forms a bell-shaped curve (see also Figure 2-3).

The genotypic variance can be subdivided further: V_A is the **additive variance** determined by the expression of alleles in homozygous form; V_D is the **dominance variance** determined by the interaction of alleles in heterozygous form; and V_I is the **interaction variance,** comprising the influences of different genes on the expression of alleles at a particular locus. In general, $V_G = V_A + V_D + V_I$, although variance may be further incremented by the correlation between particular genotypes and particular environments and other, usually minor, factors.

The principal task of quantitative genetics is to estimate the magnitude of the several components of phenotypic variance. This is made necessary by the fact that response to selection derives only from the additive genetic component of variance, because only V_A reflects the genetic diversity of the population—that is, the different alleles that replace, and are replaced by, others during evolutionary change. Phenotypic variance can be partitioned into its several components by statistical analyses of the results of breeding programs designed for the purpose (Falconer 1981). These analyses utilize correlations of phenotypic values between close relatives, usually between parents and their offspring or between siblings.

Heritability

The proportion of the phenotypic variance that is due to additive genetic factors is often expressed as a ratio, called the **heritability** (h^2) of a trait: $h^2 = V_A/V_P$. Many studies of heritability involve traits of commercial value in livestock, poultry, and crops, for which representative values of h^2 appear in Table 30-2. The data indicate that sizes have higher heritabilities (0.50–0.70), hence less environmental influence, than weights (0.20–0.35). Among traits related to production and fecundity, those creating the greater drain on energy and nutrients have the lower heritabilities. Thus the percentage of butterfat in milk is under strong genetic control ($h^2 = 0.60$), while total milk production has a low heritability (0.30); variation in egg size in chickens has a large additive genetic component (0.60), while rate of egg production has a lower heritability (0.30). Any character that requires a large commitment of resources must be sensitive to environmental variation in those resources. The heritabilities of fecundity and life history characteristics are generally low (0.05–0.50). The heritability estimates that are now accumulating for traits of individuals in wild populations resemble those of domestic animals and crops (Boag 1983, Price et al. 1984, Noordwijk 1984, Cheverud and Dittus 1992, Gu and Danthanarayana 1992, Hakkarainen et al. 1996).

Heritability may be estimated by comparing average parent and offspring phenotypes. The resemblance between parent and offspring is almost entirely the result of additive genetic effects. The average phenotype of an individual's parents is called the midparent value. The relationship between parent and offspring values may be visualized graphically by plotting the **midparent values** on the x-axis against the offspring phenotypes on the y-axis (Figure 30-4). The slope of the regression line through these points is an estimate of h^2. Environmental interactions and genetic processes such as epistasis (in which alleles of one locus alter the expression of alleles of another locus) may lower the correlation between the parent and offspring values and affect the heritability of the trait.

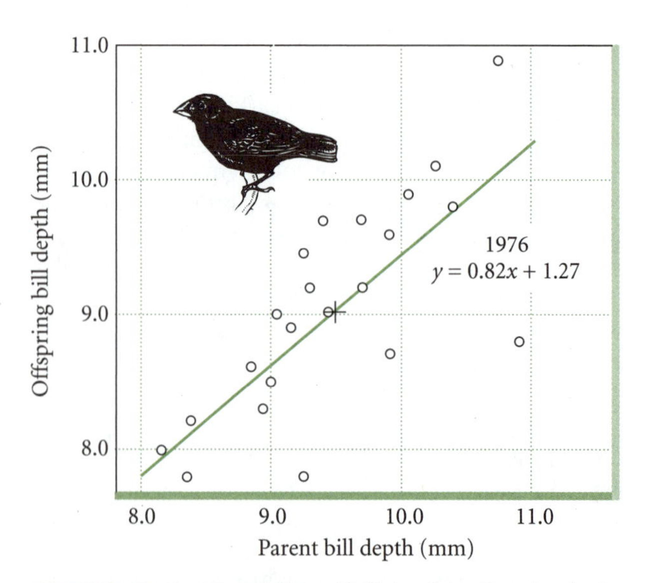

FIGURE 30-4 Heritability of bill depth in *Geospiza fortis*, one of Darwin's finches of the Galápagos Islands. The horizontal axis is the midparent value. The vertical axis is the offspring bill depth. The slope of the line is about 0.82, suggesting a strong heritability of bill depth. (*After Boag 1983.*)

TABLE 30-2 Heritabilities of several traits in domesticated and laboratory animals and plants

Trait	Organism	Heritability
Size or length		
Plant height	Corn	0.70
Root length	Radishes	0.65
Tail length	Mice	0.60
Length of wool	Sheep	0.55
Body length	Pigs	0.50
Weight		
Body weight	Sheep	0.35
Body weight	Pigs	0.30
Body weight	Chickens	0.20
Production		
Butterfat content of milk	Cattle	0.60
Egg weight	Chickens	0.60
Thickness of back fat	Pigs	0.55
Weight of fleece	Sheep	0.40
Milk yield	Cattle	0.30
Fecundity		
Egg production	Chickens	0.30
Yield	Corn	0.25
Litter size	Pigs	0.15
Litter size	Mice	0.15
Conception rate	Cattle	0.05
Life history		
Age at onset of laying	Chickens	0.50
Age at puberty	Rats	0.15
Viability	Chickens	0.10

(From Falconer 1981.)

30.5 Artificial selection of quantitative traits illustrates some characteristics of evolution in natural populations.

Animal and plant breeders apply selection artificially by determining which individuals will breed. Thus, breeders of sheep who want to produce wool will pair individual ewes and rams having the desired coat characteristics. Just as in nature, artificial selection relies on variation; after all, the sheep breeder must have choices from which to select pairs for mating. Artificial selection is usually more intense than selection that occurs in natural populations. Indeed, breeders often select individuals that would have no chance of surviving in the wild, but have characteristics that are of some economic value (domestic turkeys are a good example). Despite these differences, selective breeding programs suggest the kinds of evolutionary responses that one might observe under natural conditions.

Response to Selection

The change in a quantitative trait (a trait under polygenic control) resulting from a single generation of selection (R, for response) depends on the difference of selected individuals from the mean value of the population, referred to as the **selection differential** (denoted S), and the heritability of the trait, according to the relationship

$$R = h^2S. \qquad (30\text{-}4)$$

For example, if h^2 were 0.5 and males and females 10 size units larger than the population average

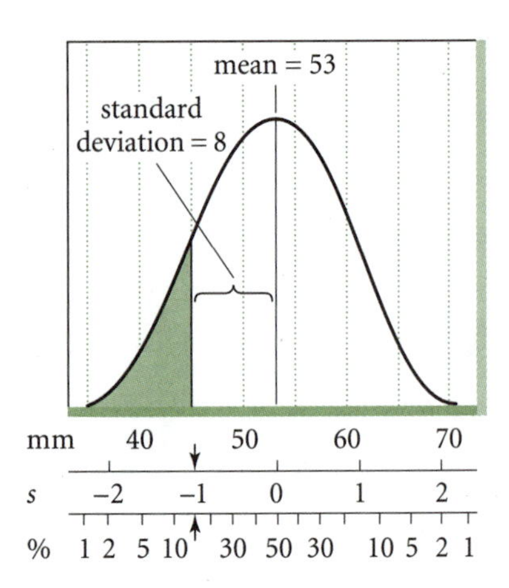

FIGURE 30-5 Schematic diagram of variation in a hypothetical trait, showing the relationship between the measurement scale, the standard deviation, and the proportions of the population with phenotypic values more extreme than a particular value. For example, 16% of the population have phenotypic values greater than one standard deviation above or below the mean (green shaded area).

(S = 10) were bred together, their progeny would be 5 size units overall above the average of the unselected population. The greater the heritability of the trait, the more rapidly it can respond to selection.

Values of R and S are conveniently expressed as multiples of the standard deviation of measurements within the population, the standard deviation (s) being the square root of the sample variance. Each value expressed in standard deviation units corresponds to a particular percentile rank in the population. Zero s units is the population mean; when values are symmetrically distributed about the mean, half the individuals lie above that value and half below. When values have a normal distribution, as in Figure 30-5, 31% of individuals lie above +0.5 s and 31% lie below −0.5 s from the mean; 16% have values more extreme than +1.0 s or −1 s; 7% exceed 1.5 s; and only 2.3% exceed 2.0 s in each direction from the mean. As a result, the more intense is selection (the larger of the value of S), the smaller the number of individuals selected, and the smaller the number of resulting progeny for the next generation of selection. When selection is too strong, the population dwindles, eventually to extinction. Even in artificial selection programs, the strength of selection is limited by the size of the stock population and the reproductive rate of selected individuals, which must be at least enough to replace the individuals eliminated by selection each generation.

The relationship between selection intensity (percentage of individuals selected), selection differential (S), and response to selection (R) may be illustrated by a program of selection for rate of egg laying in the flour beetle *Tribolium castaneum* (Ruano et al. 1975). The goal of the study was to select for greater egg production. In the stock population, the number of eggs laid from 7 to 11 days after adult emergence had a mean of 19.0, a standard deviation of 11.8, and a heritability (h^2) of 0.30. One control line and five lines subjected to different intensities of selection (ranging from 50% to 5%) were established (Table 30-3). Knowing the variability, heritability, and selection differential, one can estimate the initial response of the population to selection. For example, in the C line, a selection intensity of 80% corresponds to a selection differential of 1.4 s (Falconer 1981), or 16.5 eggs (1.4 × 11.8). With a heritability of 0.30, the response to selection should be about 5.0 eggs per generation ($R = h^2S = 0.30 \times 16.5$). The observed response fell somewhat short of this prediction (about 3.0 eggs per generation), probably because the estimate of heritability included maternal and dominance effects as well as additive genetic variation. But the beetle population did behave as predicted in that the rate of

TABLE 30-3 Selection procedure in six lines of *Tribolium castaneum* under selection for fecundity

Line	Number of families scored per generation	Number of females scored per family	Total scored	Total selected	Selection intensity (percent removed)	Selection differential (S)*
A	10	20	200	10	95	2.0
B	20	10	200	20	90	1.8
C	40	5	200	40	80	1.4
D	66	3	198	66	67	1.1
E	100	2	200	100	50	0.8
F	200	1	200	200	0	0.0

** Standard deviation units.*
(From Ruano et al. 1975.)

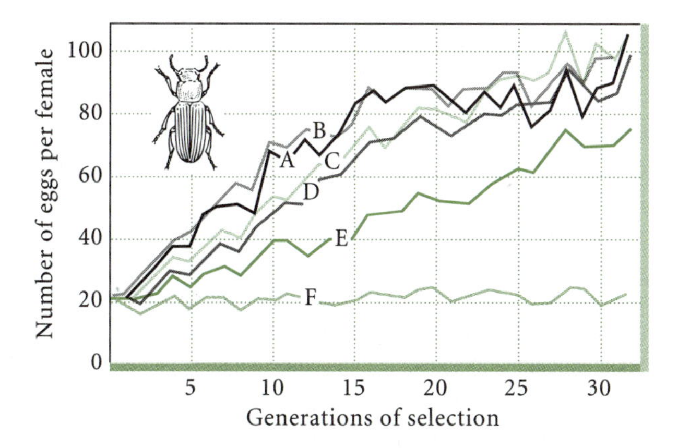

FIGURE 30-6 Change in rate of egg laying in *Tribolium castaneum* lines exposed to different intensities of selection, increasing from F to A (see Table 30-3). (*From Ruano et al. 1975.*)

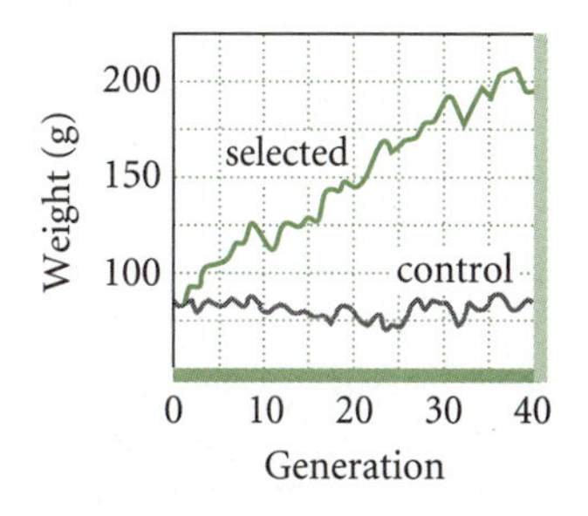

FIGURE 30-7 Body weights of Japanese quail at four weeks of age in control lines and in lines selected for high body weight. (*After Marks 1978.*)

response varied in direct proportion to the intensity of selection (Figure 30-6).

With continued selection pressure on experimental populations, the response to selection eventually stops, as seen in Figure 30-6. The slowed response is the result of two factors: erosion of genetic variation by selection and the application of opposing selection by correlated changes in other traits. Selection works only when individuals vary genetically within a population. When all unfit alleles are removed by selection, evolution pauses until new mutations or gene combinations appear.

Maximum Response to Selection

What is the maximum long-term response to selection? Somewhat surprisingly, laboratory studies have shown that selection can produce phenotypes well beyond the extreme value observed in an unselected population. Mature body weights of unselected Japanese quail average about 91 g, with a standard deviation of 8 g. After 40 generations of selection for high body weight, the population average in one study increased to 200 g, or almost 14 s units above the mean of the unselected population (Figure 30-7; Marks 1978). Realized heritabilities ($h^2 = R/S$) decreased from values of 0.30–0.45 in generations 1 to 10, to 0.15–0.20 in generations 11 to 30, to 0.05–0.10 in generations 31 to 40.

We may obtain estimates of the long-term response to selection in the absence of new genetic variation from properties of the probability distributions governing quantitative traits (Robertson 1970, Dudley 1977). Suppose that some measure of body size or length is influenced equally by n genetic loci, each having two alleles: (−) and (+). Individuals that are homozygous for (−) alleles at each of the n loci exhibit the minimum measurement. Each (+) allele in the genotype adds a single unit increment to the measurement, so that the maximum possible measurement is $2n$ units greater than the minimum (remember that there are two copies of each gene in diploid organisms). Suppose that the frequency in the population of the (+) allele is the same at each locus, and is p; then the mean value of the measurement in the population is $2np$ greater than the minimum possible. If the alleles segregate randomly, probability theory tells us that the variance in the number of (+) alleles among individuals should be $2npq$, hence the standard deviation would be $\sqrt{2npq}$.

How much of a selection response can be expected when the (+) alleles are strongly selected? Without the addition of new genetic variation, the maximum possible measurement is $2n$ above the minimum. Thus the maximum response (maximum phenotype − average phenotype) is $2n - 2np$, which may be expressed as $2n(1 - p)$ or $2nq$. In terms of standard deviations, the magnitude of this response is $2nq/\sqrt{2nqp}$, which may be rearranged to give $(\sqrt{2nq})/p$. For example, when a trait is controlled by eight loci at which the frequency of (+) alleles is 0.5, the maximum phenotypic response to selection is $[(2 \times 8 \times 0.5)/0.5] = 4$ standard deviations above the mean. Probability theory reminds us that the proportion of individuals receiving sixteen (+) alleles at eight loci just by chance, when the frequency of (+) is 0.5, is less than 1 in 66,000 (that, is 0.5^{16}); twelve or more out of sixteen (+) alleles (1.4 s units above the mean) occur in less than 4% of individuals. So we see that the short-term evolutionary potential of quantitatively varying traits may extend far beyond the existing distribution of phenotypes in a population, providing that the multilocus additive model of quantitative variation is reasonable.

What is often more difficult to explain than the response of a trait to selection is the leveling off of the response, often after only modest change. This leveling off usually is not due to exhaustion of genetic variation for the trait, because reverse selection (back toward the mean of the unselected population) typically produces an immediate response, which can result only from genetic variation remaining in the population. Furthermore, when selection is merely relaxed, so that selection coefficients are zero, a selected trait will sometimes return toward the preselection measurement, apparently by itself. The most reasonable explanation for these results is that selection applied to one trait causes changes in other traits that affect the fitness of the organism. Increases in the rate of egg laying, for example, may cause physiological or morphological changes that reduce viability and thus oppose the artificial selection regime.

30.6 Correlated responses to selection limit evolutionary response.

Evolutionary responses often include characters other than the one selected. Both development and function integrate the parts of organisms, bringing about an interdependence of phenotypic traits, particularly among those involving size or rate of growth and production. Animal and plant breeders have estimated genetic correlations between traits in many domestic species. In poultry, for example, the genetic correlation between body weight and egg weight is 0.50; between body weight and egg production it is -0.16. Therefore, one cannot apply selection to body weight without also obtaining a relatively rapid increase in egg size and a slower, but steady, decrease in rate of laying. Simultaneous selection of large egg size and small body size goes against the grain of genetic correlation, and usually is unsuccessful (Nordskog 1977).

Many characters, such as body size and egg size, are linked in developmentally, genetically, or functionally related groups that tend to respond to selection in concert. For example, Leamy (1977) determined genetic correlations among skeletal measurements of mice and identified four clusters of traits that were highly integrated genetically among themselves, but were relatively independent of one another: (1) skull length; (2) skull width, body weight, and tail length; (3) skull width (providing a link to group 2), scapula length, and other measurements associated with the pectoral girdle; and (4) limb bones and total body length. In mice, therefore, selection for body weight will produce responses in tail length and in the proportions of the skull, as well as in body weight itself.

When Dawson (1966) selected for fast and slow larval development in *Tribolium castaneum* and *T. confusum,* he observed a number of correlated responses that decreased the beetles' fitness regardless of the direction of selection. Selection for rapid development resulted in decreased size and, in *T. castaneum,* decreased larval survival and an increase in the incidence of adult abnormalities. Selection for slow development resulted in increased adult weight, increased frequency of adult abnormalities, and decreased fertility in females.

Dawson (1967) assessed the fitness of each of his selected strains of flour beetles by placing them in competition with unselected stocks of the other beetle species (see Chapter 21). Both the "fast" and "slow" selected lines suffered decreased fitness at the early stages of selection, suggesting that flour beetle development rate is optimized with respect to its effect on fitness. (As Dawson continued his selection experiments, the fitness of one of the selected lines began to improve, apparently because of increased cannibalism by the selected larvae. Evolution frequently springs such surprises on us.)

Correlated responses to selection tend to work against an artificial selection program because they produce counteracting selective pressures. For example, correlated responses in reproductive characters have appeared in selected lines of chickens. Selection for both increases and decreases in body weight and egg weight has resulted in declines in reproductive fitness, as indicated by rates of egg production, hatch rates, and survival of chicks. Regardless of the character or direction of selection, fitness in the selected lines varied from 54% to 85% of that in the unselected line. These results of artificial selection programs suggest that populations are balanced genetically and that their adaptations are both well tuned to the environment and finely adjusted to one another.

30.7 Population genetics offers a number of important messages for ecologists.

The basic concepts of population genetics discussed in the previous several sections are important to ecologists in at least three ways. First, every population harbors some genetic variation that influences fitness. This means that evolution is a continuous process in all populations. It also means that individual organisms should be expected to have adaptations that help them to reduce the harmful effects of deleterious mutations on themselves and their offspring. Adaptations to avoid inbreeding (see below) are one such mechanism by which organisms manage the ubiquitous genetic variability in populations.

Second, changes in selective factors in the environment of a population will almost always be met by evolutionary responses that lead to shifts in phenotypes within the population. The response itself is not always predictable and depends on the particular genetic variation present in the population at a given time. Most quantitative traits have enough genetic variation to respond to selection, but the range and extent of a response may be limited by correlated responses of other traits that have negative fitness consequences. Given enough time, populations may reach some sort of evolutionary optimum and become stabilized, but we have little idea of how much time is required.

Third, intense selection pressures brought about by human-caused changes in the environment, the introduction of predator or disease organisms, or the appearance of genetic novelties in those enemies will often exceed the capacity of a population to respond by evolution. In these circumstances, the decline of populations toward extinction is a distinct possibility.

Let us now turn our attention to ideas about how the behavior of organisms can influence genetic variation. We will begin with a discussion of sex, which tends to increase genetic variation.

30.8 Sex is thought to benefit individuals by increasing genetic variation among their progeny.

Natural selection tends to reduce genetic variation in populations. Several other processes increase genetic variation. Most genetic variation in a local population arises from mutation, immigration of individuals from other areas having different selective environments, temporally varying selective factors, and changes in gene frequencies from generation to generation arising purely by chance in small populations (genetic drift; see above). Another process that increases genetic variation is sex. The origin and maintenance of sexual reproduction continues to be vigorously discussed among ecologists, and a number of excellent reviews and collections of papers on this subject are available (for example, Bell 1982, Stearns 1987, Stevens and Bellig 1988, Slater and Halliday 1994, Barrett and Harder 1996). Much of the discussion turns on the question of the relative advantages of sexual and asexual reproduction under various environmental conditions. Related to this question are questions about who mates with whom, how mates are chosen, and what proportion of the population is represented by one sex or the other. We will defer much of our consideration of sexual reproduction until Chapter 33, but we will present a few aspects of it here in order to complete our discussion of the nature of variation.

Sexual reproduction mixes the genetic material of two individuals, resulting in new combinations of genes in their offspring. As a result of this **recombination,** siblings differ from one another genetically. Thus, in variable environments, at least some of the offspring of a sexual union are likely to have a genetic constitution that enables them to survive and reproduce, regardless of the particular conditions. Sexual reproduction may also produce new genotypes previously absent from a population and reduce the effects of deleterious mutations. In contrast, individuals reproducing asexually produce genetically exact copies of themselves (mutations notwithstanding). Both asexual and sexual reproduction are viable alternative strategies. Asexual reproduction is widespread among plants and is found in all major groups of animals, with the exception of birds and mammals. Sexual reproduction evolved very early in the history of life and is also widespread among plants and animals.

Costs of Sexual Reproduction

The fact that sexual reproduction is evolutionarily old and phylogenetically widespread suggests that selection favors such a strategy under a wide range of conditions. Yet, on close examination, there is a tremendous fitness incentive for the individual to give up sex. That is to say, sex, in evolutionary terms, comes with high costs. When the sexes are separate—that is, when individuals are either male or female—sexual reproduction has a much higher cost arising from the fact that only half the genetic material of each individual offspring comes from a given parent. Thus, compared with asexually produced offspring, which contain only the genes of their single parent, the progeny of a sexual union contribute only half as much to the evolutionary fitness of each parent. This 50% cost of sexual reproduction to the individual parent is sometimes referred to as the **cost of meiosis.**

Other costs of sexual reproduction have been identified (Lewis 1989). The act of finding a mate and effecting fertilization exposes organisms to risks of predation and disease prior to the production of young. Reproduction may be delayed or prevented in populations with extremely low densities because of difficulties in finding a mate. Inefficiencies in the actual transfer of gametes from one sex to the other may reduce the number of offspring produced. Recent studies that examined the DNA sequences of males and females have shown that most mutations occur in the male germ line in humans, primates, and rodents. Using computer simulations, Redfield (1994) has demonstrated that this presents an additional high cost to the female. There is also the problem of conflicts between sexes, which may divert energy away from reproduction or reduce opportunities to mate (see Chapter 33).

Role of Sexual Reproduction in the Maintenance of Genetic Variation

Most attempts to explain how sexual reproduction can be maintained in the face of the overwhelming short-term advantages of asexual reproduction have proposed that the compensating advantage of sex resides in either an increase in the genetic variation among progeny or a reduction of deleterious genetic variation. Variable offspring may confer a fitness advantage in a highly unpredictable environment, where variation assures that at least some progeny will be well suited to whatever environmental conditions should come their way. In strongly competitive environments, variation among progeny increases the chance that at least some will have the extremely high fitness necessary to persist (Williams and Mitton 1973, Williams 1975).

Alternative explanations for sex argue that the genetic recombination made possible by sexual reproduction is necessary to eliminate deleterious mutations from the germ line (Muller 1964, Felsenstein 1974). In an asexual clone, deleterious mutations can be eliminated only by selection among progeny. When the mutation rate per genome exceeds the number of selective deaths per generation, mutations will accumulate and reduce the vitality of the clone. In contrast, individuals in sexual populations continually exchange genetic material between family lines. Because the entire population is joined into a common gene pool, selection picks and chooses among continually reshuffled genotypes. A deleterious mutation that crops up in one line can be exchanged for a beneficial allele through recombination.

Whatever the raison d'être of sex, it provides three potential benefits: long-term evolutionary flexibility of the population, medium-term elimination of deleterious mutations from the family line, and short-term production of variation among the progeny of the individual.

30.9 Breeding systems manage genetic variation in sexual populations.

The advantage to the individual of genetic variation is that it may allow the production of variable offspring, at least some of which will be adapted to particular patches or to changed conditions. But most genetic variation is negative, in the sense that mutations usually make the individual less suited to the environment in which it lives. Genetic variation is maintained in populations primarily by mutation and by gene flow from other localities in which different genes have a selective advantage. However, organisms have a variety of mechanisms for reducing the negative effects of genetic variation. Chief among these are sexual strategies that determine the degree to which inbreeding occurs (Willson and Burley 1983, Wyatt 1983, Stephenson and Bertin 1983, Schoen 1982, 1983, Pusey and Wolf 1996).

Inbreeding

Most of us think of **inbreeding**—mating among close relatives—as bad (Ralls et al. 1988). Brother-sister matings and, where possible (especially in plants), **selfing** (self-mating) may result in the expression of deleterious recessive genes. The mechanism of this effect is simple: Suppose an individual is heterozygous for a rare, deleterious, recessive allele (the gene pool is full of them, and most individuals have some). If it were to mate with an individual from the general population, which probably would not have the same rare allele, half of their progeny would be heterozygous, like the one parent, and half would be homozygous for the common form of the gene, like the other parent. None of the progeny would be disadvantaged by the union. If, however, the individual selfed, one-quarter of its offspring would be homozygous for the deleterious allele and would suffer loss of fitness as a result. Matings between close relatives produce the same result, only less frequently.

Most species employ mechanisms, including dispersal of progeny, recognition of close relatives, and negative assortative mating, to reduce the occurrence of inbreeding. Hermaphroditic species of plants, in which individuals bear both male and female sexual organs, have additional mechanisms to prevent selfing, including self-incompatibility, temporal separation of male and female functions, and elaborate flower structures designed to make self-fertilization difficult (Willson 1983).

While close inbreeding generally creates problems, it may confer benefits as well (Lloyd 1979, 1980, Wells 1979). In particular, selfers can guarantee fertilization of their flowers in habitats lacking suitable pollinators or where individuals are widely spaced (Baker 1965, Ghiselin 1969). Many weedy species that colonize isolated patches of disturbed habitat (for example, dandelions) are selfers (Jain 1976). One assumes that most deleterious variation was weeded out of such populations as it was exposed in homozygous individuals during the evolutionary transition between outcrossing and selfing (Ritland and Ganders 1987).

Outcrossing

Outcrossing at great distance may also reduce fitness. Some populations of plants include spatially defined ecotypic variation over small scales of distance, particularly in complex, heterogeneous environments (see Chapter 6). In such cases, local adaptation to particular habitat patches enhances fitness, and receiving pollen from individuals adapted to different habitat conditions may reduce the fitness of progeny that become established near the female parent.

Several studies have reported an **optimal outcrossing distance** in populations of plants (Levin 1984). Nearby individuals are likely to be close relatives, which raises the specter of inbreeding. Distant individuals are likely to be adapted to different conditions. Price and Waser (1979) fertilized flowers of the larkspur *Delphinium nelsoni* in central Colorado with pollen obtained from the same individual and from individuals located at distances of 1, 10, 100, and 1,000 meters. Their results showed that the number of seeds set per flower was greatest when the pollen source came from a distance of 10 meters, and was least for selfed pollen and that obtained 1,000 meters distant. Furthermore, when these seeds were planted, survivorship to 1 and 2 years greatly favored matings across the intermediate distance of 10 meters.

Selective Fertilization and Abortion in Plants

Although Price and Waser's study indicated an optimum outcrossing distance, larkspurs cannot exploit this advantage. The plants are pollinated primarily by bees and hummingbirds, which tend to visit nearby flowers in succession. Colored dye particles placed along with natural pollen on the male anthers of larkspurs were recovered primarily on flowers within 2 meters of their source. Perhaps no adaptation of flower structure could modify the behavior of pollinators (which have little interest in the plant's fitness) to achieve the optimal outcrossing distance.

Plants may not be able to control the distances that their pollinators travel between flowers, but they can manage genetic variation by establishing competition between pollen grains for opportunities to fertilize ovules and by selectively aborting developing ovules on the basis of the genotype of the embryo (Willson and Burley 1983). Most plants produce many more flowers than they can mature as fruits; flowers are relatively cheap, fruits expensive. The excess of fertilized ovules is reduced in part by predation and other extrinsic damage and in part by programmed abortion (Casper 1984). Abscission of flowers and fruits can be highly selective with respect to pollen loads, numbers of ovules fertilized per flower, and the genotypes of developing embryos.

An example of such a mechanism is found in *Banksia spinulosa*, a partially self-compatible, but normally outcrossing, Australian shrub that is pollinated by small nectar-feeding birds. Each inflorescence has about 800 flowers, but produces fewer than 50 fruits. Fruit production appears to be resource limited rather than pollen limited, because removal of one-third of the flowers from either the base or the top of an inflorescence does not significantly depress fruit set. To determine whether these plants can distinguish the quality of pollen, Australian botanists G. Vaughton and

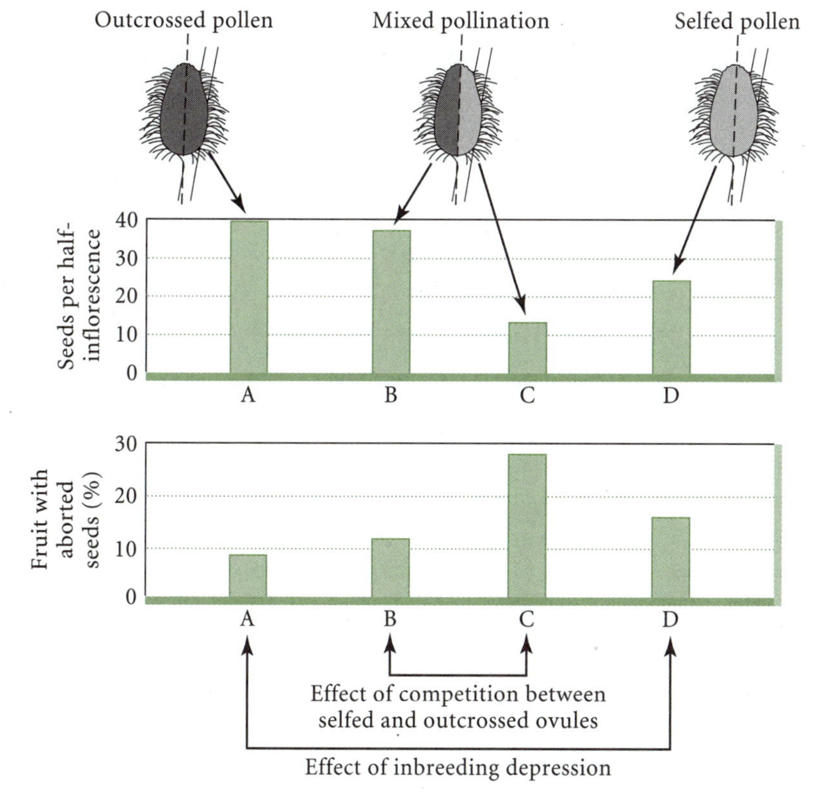

FIGURE 30-8 Results of a pollination experiment with the Australian shrub *Banksia spinulosa*, in which the two halves of an individual inflorescence were fertilized with pollen either from the same plant or from other plants. The results show the negative effects of inbreeding (D compared with A) and the further discrimination against selfed ovules in competition with outcrossed ovules growing in the same inflorescence (C compared with B). (*Data from Vaughton and Carthew 1993.*)

S. Carthew (1993) hand-pollinated *Banksia* inflorescences with pollen obtained either from the same plant (selfed pollen) or from neighboring plants (outcrossed pollen). In some plants, they pollinated half the inflorescence with selfed pollen and half with outcrossed pollen (mixed pollination). After fruits had developed, the numbers of fruits and seeds (no more than one seed per fruit) were counted on each side of the inflorescence. Compared with cross-pollination, selfing reduced seed set by 38% (24 versus 39 seeds per half-inflorescence), and fruits with aborted seeds increased from 8% in outcrossed to 16% in selfed inflorescences (Figure 30-8). These results clearly indicated **inbreeding depression,** or reduction of fitness caused by inbreeding. When one-half of an inflorescence was cross-pollinated and the other half selfed, seed set in the selfed half dropped further to 14%, and 28% of the fruits aborted their seeds. These results show that self-pollinated ovules did not fare well in competition with cross-pollinated ovules. Thus, plants are capable of making distinctions among developing embryos on the basis of their genotypes. The extent to which this mechanism is used to manage the transmission of genetic variation to offspring in natural populations has yet to be determined.

30.10 Evolutionary ecologists interpret form and function as adaptations to the environment.

Organisms are well adapted to their environments. This assumption has provided a powerful tool for conceptualizing the organism-environment interaction and understanding the design limitations placed upon the responses of organisms to environmental change (Wake 1982). The adaptationist research program also has led to the discovery of new components of the fitness of organisms, and has emphasized the important roles of interactions between individuals—predators, competitors, mutualists, society members, mates, parents, and offspring—in directing the course of adaptation. Therefore, even if the assumption of adaptation is not fully correct, it has expanded our ecological concept tremendously.

The primary question addressed by evolutionary ecology is: How do the adaptations of organisms reflect their environment? Answering this question requires an understanding of the selective factors in the environment and the evolutionary responsiveness of the phenotype. To have scientific validity, any idea concerning the phenotype-environment relationship must include a suitable criterion for the fitness of phenotypes, the genetic basis of phenotypic variation, and a model to link aspects of form and function that determine fitness to one another and to conditions of the environment (Arnold 1983).

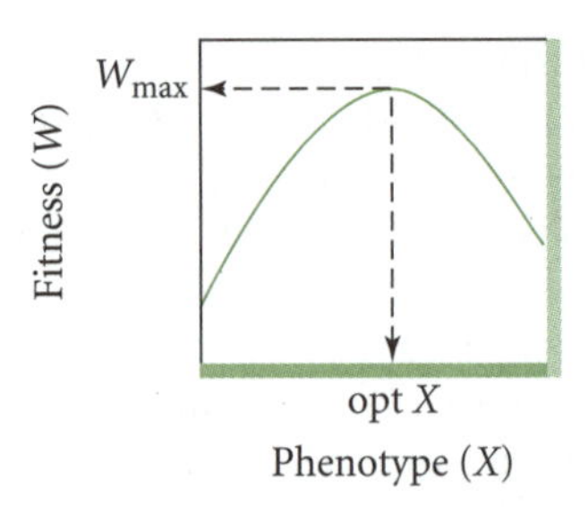

FIGURE 30-9 Relationship between fitness and phenotypic value when there is a single optimum phenotype.

One approach to understanding adaptation, known as **phenotypic optimization,** does not incorporate genetic variation explicitly, but rather assumes that phenotypic variation has a parallel genetic basis and that the selection of optimum phenotypes brings about appropriate genetic change. For example, particular conditions may dictate an optimal allocation of resources between producing reproductive structures and continuing to grow; a certain genotype presumably produces that optimum phenotype, and it will be selected irrespective of the particular genetic basis of the trait. The core of an optimization model is the relationship between phenotype and fitness (Figure 30-9). Each such relationship is unique for a particular environment. To visualize the relationship, one must either measure the fitnesses of a range of phenotypes directly or devise a realistic model that predicts the fitnesses of nonexistent phenotypes. The degree to which observed phenotypes match the predictions of the model measures the validity of the mechanics embodied in that model, although we must keep in mind that alternative models may make similar predictions.

30.11 An evolutionarily stable strategy (ESS) resists invasion by all other phenotypes.

Another approach to understanding adaptation, particularly useful in the case of discretely varying phenotypes and when phenotypes interact with one anther, is that of determining the **evolutionarily stable strategy** (Smith and Price 1973, Smith 1974, 1982, Parker 1984). An evolutionarily stable strategy, or **ESS,** is that phenotype or combination of phenotypes that, when constituting a population, makes it impossible for individuals with alternative phenotypes to invade the population. As John Maynard Smith (1982) puts it, as ESS is "strategy such that, if all the members of a population adopt, no mutant strategy can invade." Referring back to the relationship between phenotype and fitness in Figure

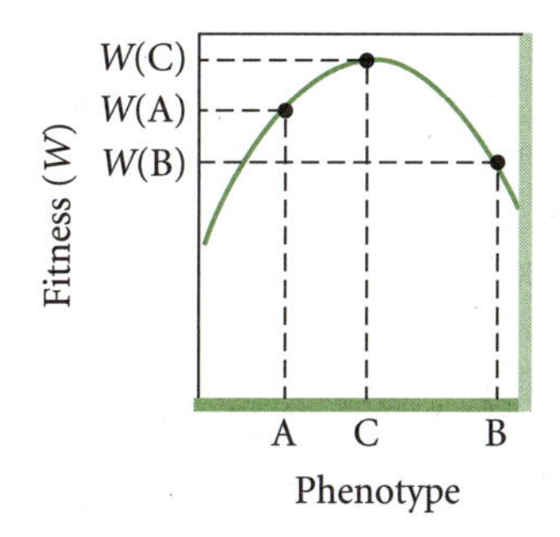

FIGURE 30-10 Phenotype C is an evolutionarily stable strategy because no other phenotype can invade a population of C individuals.

30-9, let us suppose that only three phenotypes, A, B, and C, occur in a species (Figure 30-10). A population consisting only of B individuals can be invaded by A individuals—that is, the A phenotype will increase when rare owing to its superior fitness. A population consisting only of A individuals resists invasion by the B phenotype; hence A is the ESS. Considering the range of all possible phenotypes, it is clear that only C, which confers the greatest fitness, will resist invasion by all other phenotypes. Therefore, a phenotype identified by the maximum fitness criterion is also an evolutionarily stable strategy.

The utility of ESS thinking becomes more apparent when phenotypes interact with one another and the fitness of each depends on the proportions of other phenotypes in the population (Smith 1982, Parker 1984). Hence the concept of the evolutionarily stable strategy has found broad application in the study of social behavior and mating systems, as we shall see a few chapters hence.

30.12 The influence of the phenotype-environment interaction on fitness is the key to understanding adaptation.

Regardless of the fitness criterion adopted, and irrespective of whether genotypic or phenotypic models are employed, the principal challenge to the evolutionary ecologist is to understand how small changes in phenotype affect fitness. To accomplish this goal, ecologists must understand the interrelationships of the many characteristics of the organism and the fullness of its interaction with all aspects of its environment. At any given time, our maturing concept of the organism-environment complex can be summarized in the form of models, whose consistency with observations on natural systems is continually scrutinized by research, followed by reformulation of ideas when inconsistencies arise.

We shall illustrate this process with an example from the work of Ricklefs on the life histories of birds. In 1968, David Lack, the prominent English ecologist, suggested that the growth rates of birds optimally balance two environmental factors: predation and other sources of mortality, which select rapidly growing individuals that pass through vulnerable developmental stages quickly; and food supply, which selects more

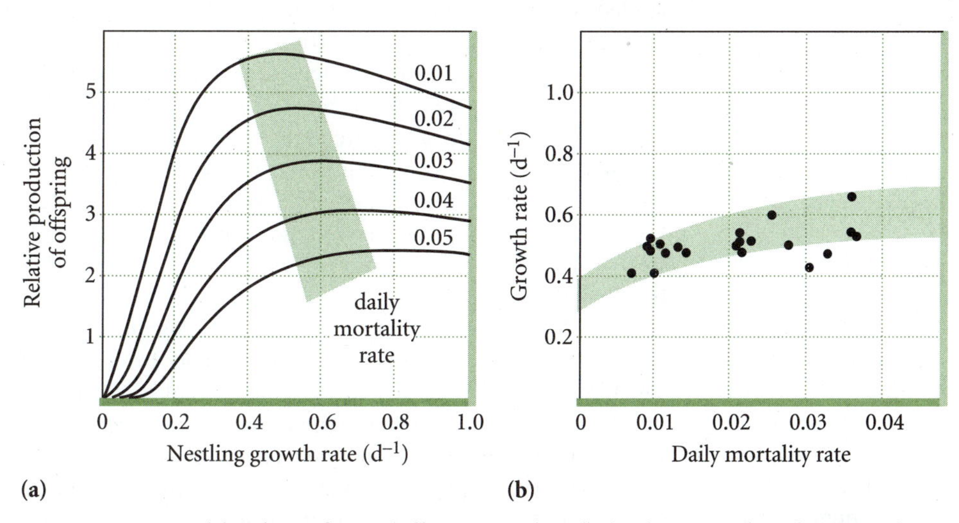

FIGURE 30-11 (a) Relative fitness (offspring produced) as a function of nestling growth rate, as predicted by a model relating the two through the influence of growth rate on brood size, number of broods per season, and brood survival, for brood mortality rates of 0.01–0.05 d^{-1}. The shaded area indicates the optimum growth rate for each level of mortality. (b) Observed growth rates of various species as a function of daily mortality rate. The shaded area corresponds to the optimum determined by the model. (*After Ricklefs 1984.*)

slowly growing individuals that demand less food and thereby allow their parents to provide for more offspring. From this simple idea, Lack suggested that growth rate should vary in direct proportion to the vulnerability of nestlings to mortality factors. But observations on birds revealed, to the contrary, that growth rate is relatively insensitive to variation in mortality rate, suggesting that Lack's hypothesis was not sufficient (Ricklefs 1969).

In an attempt to resolve this problem, Ricklefs developed a model to describe the influence of growth rate on fitness through the consequences of growth rate for the number of chicks produced per brood and the number of surviving broods per season (Ricklefs 1984). Working out the details of such a model requires knowledge of the food requirements of growing chicks and of the frequency of initiation of new nesting attempts following both success and failure of the previous attempt. A set of expressions for each of these effects resulted in equations relating an index of fitness (number of young reared per season) to growth rate. The curves defined by these equations have maxima at intermediate growth rates close to rates measured in nature (Figure 30-11). Furthermore, when nestling mortality rate is varied over a wide range in the model, the position of the optimum growth rate changes little, contrary to Lack's expectation but consistent with observations. This finding is related to the fact that birds rapidly replace destroyed clutches, and so the effect of mortality on the number of young reared per season is correspondingly diminished. The consistency of the model's predictions with observations suggests that Lack's concept of how growth rate influences fitness is essentially correct, and that factors other than the ones included in the model are not important to the evolutionary optimization of growth rate.

30.13 The adaptationist program has many difficulties.

Confidence in our understanding of growth rates in birds rests firmly on the assumption that the organism-environment relationship reflects solely the fitness relationships of phenotypes with selective factors in their surroundings. This general view has been challenged on a number of grounds. One of these is the many processes that oppose the perfection of adaptation: lack of suitable genetic variation for response to the environment; continuous production of less fit phenotypes due to mutation and immigration; changes in the environment that leave previously well-adapted phenotypes behind. In addition, some phenotypes of potentially great fitness are not possible because of limits imposed by physical laws. Some things, such as rates of physical diffusion, are beyond the realm of biological influence. Evolution also drags along the baggage of past adaptation, which may contain no particular relevance for the present. The body plans of major taxa have different numbers of limbs: four in terrestrial vertebrates, six in insects, eight in spiders, ten in crabs, and so on. The number can be changed by the reduction of appendages (snakes have zero limbs) or the modification of other parts, such as antennae or mouthparts, to function partly as limbs, but the basic number of limbs characteristic of each major group, even though it has no particular meaning, other than historical, may restrict future evolutionary potential. Finally, some adaptations have fortuitous consequences for developmentally or genetically linked features of the phenotype. Lande's observation that selection for large body size alone automatically results in an increase in brain size, owing to the genetic intercorrelation of these characters, makes this point.

Some difficulties with adaptationist thinking were eloquently pointed out by Harvard evolutionary biologists Stephen J. Gould and Richard Lewontin (1979) in a paper entitled "The spandrels of San Marco and the Panglossian paradigm: A critique of the adaptationist programme." Spandrel refers to the space, generally triangular, between an arch supporting a ceiling or other horizontal structure and the ceiling itself. In many cathedrals, these were decorated with painted scenes. Gould and Lewontin's point was that the spandrel arises as a consequence of architecture, but is not a key feature of building design, as are arches and the horizontal members they support. The analogy to organism architecture cautions us that some structures may be fortuitous or secondary consequences of other, strongly selected adaptations. As Gould and Lewontin put it, "One must not confuse the fact that a structure is used in some way . . . with the primary evolutionary reason for its existence." They summarized their reservations about the adaptationist program by questioning the ubiquity of its basic assumption:

> This [adaptationist] programme regards natural selection as so powerful and the constraints upon it so few that direct production of adaptation through its operation becomes the primary cause of nearly all organic form, function, and behavior. . . . We would not object so strenuously to the adaptationist programme if its invocation, in any particular case, could lead in principle to its rejection for want of evidence.

In this passage, Gould and Lewontin suggest that evolutionary ecologists should test the basic assumption of adaptation as well as determine the validity of adaptationist explanations for patterns in nature (for one approach, see Clutton-Brock and Harvey 1979, 1984).

30.14 Taxonomically useful characters illustrate that, once established, some adaptations resist further change.

Evolutionists divide the characteristics of organisms into one set that reveals the phylogenetic history of a group and another set that responds easily to selection and reflects the contemporary environment. Characters do not, of course, fall discretely into taxonomically revealing and ecologically revealing sets. Rather, they are arranged along a continuum from evolutionarily stable to labile. Different characters are useful in distinguishing different taxonomic levels (Mayr et al. 1953, Crowson 1970). Among birds, for example, orders are distinguished primarily by skeletal features, such as the structure of the palate and the arrangement of bones in the skull, whereas families are distinguished by variations in the beak, patterns of scales on the tarsus (lower leg), and numbers and relative lengths of the primary (flight) feathers. The lower taxonomic categories—genera and species—are often based upon small differences in measurements and ratios of measurements, as well as plumage coloration and song.

The differences that distinguish mammals from reptiles—reflecting a truly drastic reorganization of certain parts of the body plan, physiological processes, and patterns of reproduction—evolved over tens of millions of years through intermediate stages that enjoyed varying levels of success, judging from their abundance in the fossil record. Eventually, however, evolution arrived at a particularly providential set of characteristics—those shared by all modern mammals—and the group underwent an explosion of evolutionary diversification between 70 and 60 million years ago. Regardless of the subsequent modifications that now distinguish bears, rabbits, mice, seals, and wildebeests, all mammals retain the fundamental class characteristics, including warm-bloodedness and nursing of the young. At each lower taxonomic level, other traits have been similarly set aside by evolution.

How do characters at each level of this hierarchy match up with the ecological distributions of organisms? Which ones may be thought of as ecologically revealing characters in the sense of the adaptationist program, and which are too remote to infer the ecological mold in which they were cast? The matching of organisms to their environments comes about in two ways: by the organism's choosing its environment to match its evolved features, and by selective modification of the gene pool of a population. Larger taxonomic groups appear to be widespread with respect to climate and other aspects of the physical environment, but are often specialized with respect to diet. Thus the insect order Homoptera (leafhoppers, cicadas) may be found wherever vascular plants occur, but because their mouthparts are specialized for sucking plant juices, their local ecological roles are narrowly prescribed. All the evolutionary modifications of this group have taken place within an ecological context established by the ordinal characters. As Gould (1982) has put it, "current utility permits no necessary conclusion about historical origin. Structures now indispensable for survival may have arisen for other reasons and been 'coopted' by functional shift for their new role."

Smaller taxonomic groups are often distinguished by differences in body size, habitat, microhabitat, and selection of diet according to prey or host species, rather than by manner of feeding. We may presume that the modifications responsible for such diversification are evolutionarily more malleable and, therefore, more amenable to study by evolutionary ecologists. As a general rule of thumb, adaptationist thinking may provide a useful tool for understanding aspects of the organism-environment interaction based on characters that distinguish close relatives.

30.15 Do large systems have uniquely evolved properties?

The functioning of biological communities and ecosystems is determined by the collective adaptations of all their constituent species. To a large degree, these species constitute important aspects of one another's selective environments, and so each has evolved with respect to the evolution of others in the system. This mutual accommodation of species to one another is broadly referred to as coevolution, a term that has also been given some specialized meanings, as we saw in Chapter 25.

Ecologists have often wondered about the importance of coevolution to the maintenance of community function, and whether there is a criterion for fitness at the level of the system as well as at the level of the individual genotype. One approach to this question would be to perform an experiment—impossible to pull off other than with laboratory microcosms—in which ecosystems are constituted with species appropriate to a natural system—that is, representing all the ecological roles found in a natural system—but obtained from different places and thereby having independent evolutionary histories. If such artificially concocted systems functioned less well than corresponding natural systems, one could conclude that "coevolved" properties conferred unique qualities on large ecological systems.

The degree to which communities and ecosystems have an evolved genetic integrity has not been resolved by ecologists, and may not be for decades, if ever. Smaller parts of this larger question, concerning the role of adaptation in suiting the organism to its environment and the mutual coevolution of populations, have, however, contributed importantly to our concepts and understanding of ecology, as we shall see in the chapters that follow.

SUMMARY

1. The study of evolutionary ecology is based upon the assumption that differences between the adaptations of organisms can be interpreted as evolutionary responses to different selective pressures in the environment. An important consideration for ecologists is the degree to which adaptation to local environments is determined by the availability of suitable genetic variation. Another concern is the extent to which organisms have evolved to manage genetic variation within the population. Evolutionary fitness is the number of descendants that an individual leaves.

2. Population genetic models demonstrate how evolution can proceed by the substitution of alleles according to their relative fitnesses.

3. Many adaptations of ecological interest involve modifications of continuously varying traits. The variation in a trait within a population is described by its variance, which has environmental and genetic components. The science of quantitative genetics has developed statistical analyses to tease apart these components from the results of certain breeding programs.

4. Heritability (h^2) is the ratio of additive genetic variance to phenotypic variance; its value ranges between 0 and 1. Heritabilities of traits in natural populations are on the order of 0.5–0.7 for many size-related traits, but often lower for production-related traits.

5. The response of a trait (R) to selection is equal to the heritability times the selection differential (S). In animal and plant breeding, the stronger the selection, the faster the response, so long as enough offspring are produced to replace individuals selectively removed.

6. Response to selection levels off when genetic variation is exhausted or, more frequently, when correlated responses in other traits reduce the fitness of selected individuals.

7. The response of one trait to selection applied on a second depends on the genetic correlation between them. The phenotype often consists of groups of genetically intercorrelated characters that tend to respond to selection in concert. Such correlations may inhibit the independent evolutionary response of two traits to opposing selective pressures.

8. Genetic variation maintains the long-term evolutionary potential of a population, and it has been argued that a primary function of sexual reproduction is to increase genetic variation through recombination of genotypes.

9. Genetic variation may also impair the ability of individuals to adapt to local variation in the environment. To some degree, individuals can minimize the negative effects of genetic variation on the genotypes of their offspring by selective mating to control level of inbreeding and outcrossing distance.

10. The outcome of evolution may be interpreted as a phenotypic optimum or an evolutionarily stable strategy (ESS). Phenotypic optimization is a mathematical technique of finding the point on a phenotypic continuum having the greatest fitness. An ESS is a trait that, if adopted by all members of a population, is superior to all other possible alternative traits. ESS analysis is particularly useful for adaptations involving interactions between members of a population.

11. Adaptationist thinking should be applied with caution to the extent that evolutionary response cannot keep up with environmental change, or that adaptation cannot be perfected owing to mutation and immigration of locally less fit alleles. Furthermore, some imaginable phenotypes are not possible owing to limits imposed by physical laws beyond the reach of evolution. Stephen J. Gould and Richard Lewontin have emphasized that many attributes of form and function are fortuitous consequences of selection on other traits, and that adaptations may have evolved in response to environmental conditions of the past that are no longer relevant to their present-day utility.

12. Characteristics that distinguish higher taxonomic groups (orders, families) usually are thought of as being less responsive to selection than those that vary between lower taxonomic categories (genera, species). Whether the characteristic traits of higher orders lack appropriate genetic variation, are too highly integrated through genetic correlation to have evolutionary latitude, are protected from selection pressures in the environment, or place restrictions on habitat use by their bearers is poorly understood.

EXERCISES

1. Consider a population with two alleles for a particular gene, A_1 and A_2, where A_1 is dominant over A_2. The initial frequency p of allele A_1 is 0.6. If selection against the homozygous genotype A_2A_2 reduces the fitness of that genotype by $s = 0.2$, what is the change in the frequency of the allele A_2 after one generation of selection?

2. Review the material presented in Chapter 19 regarding the restoration of highly threatened species. Using that information and what you have learned in this chapter, discuss the problems involved with using artificial breeding programs to save such endangered species.

3. What are the implications of loss of species diversity for the operation of evolution by natural selection?

4. Suppose that the heritability of male pedipalp length in a particular species of spider is 0.6 mm. (The pedipalp of a spider is a modified appendage used as by males to transfer sperm to females.) If males having pedipalps 0.5 mm longer than the population average bred with females, how much larger than the average of the unselected population will the pedipalps of their progeny be?

CHAPTER 31

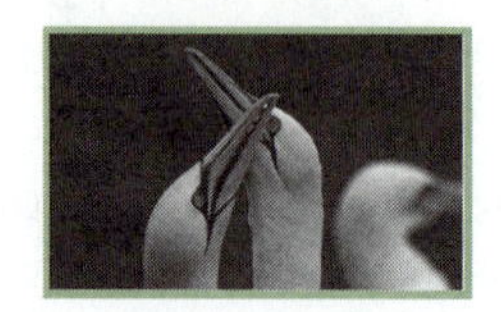

Adaptations to Heterogeneous Environments

GUIDING QUESTIONS

- What is an adaptive landscape and what are the processes that might change it?
- What are fitness sets?
- What is the difference between negative and positive frequency-dependent selection?
- What is phenotypic plasticity, and how can it be studied experimentally?
- What are the three important elements of optimality models?
- What are the differences between prey and patch models of foraging behavior?
- What are the predictions of the classic model of optimal foraging?
- What is an ideal free distribution?
- What is the difference between risk-prone and risk-averse foraging?
- How does foraging behavior change under the risk of predation?
- What does the stochastic dynamic programming approach attempt to predict?

The world is anything but constant and uniform. No single genotype can be best suited to all the conditions encountered by each individual. These facts pose a fundamental and difficult challenge to understanding adaptation, raising several related issues. One is the definition of fitness under variable conditions; that is, how does one best integrate the varied performance of a genotype under different conditions into a single measure of fitness? A second issue is whether environmental heterogeneity facilitates the coexistence of more than one genotype in a population. Third, heterogeneous environments present individuals with choices concerning habitat use, prey selection, mate selection, and so on. The behaviors that govern these choices are, to a large extent, under genetic control and therefore are evolved adaptations. To comprehend such adaptations, one must understand the consequences of rules adopted for making choices. These are the issues that we take up in this chapter.

31.1 Populations may be viewed metaphorically as situated atop peaks within an adaptive landscape.

In the previous chapter we portrayed evolutionary change simply as change in allele frequencies resulting from mutation, genetic drift, or natural selection. Such a view is too simplistic when the environment is heterogeneous because fitness is determined by the interaction of the genotype with its surroundings. The evolutionist Sewall Wright provided a useful metaphor for thinking about populations in changing environments, which we discuss here (Wright 1932, 1982).

The genetic composition of a particular population—measured by the frequencies of the various alleles of each gene—most often represents a subset of a larger number of possible genotypes, which we may refer to as the **genetic potential** or the **genotype space.** Under a particular set of environmental conditions, some genotypes will have a higher average fitness than others will, and selection will therefore favor those genotypes. Inasmuch as natural selection can only increase fitness, evolutionary change in populations may be viewed as movement upward toward some fitness peak, which ecologists sometimes refer to as an **adaptive peak.** Think about this adaptive peak as being situated within a landscape, an **adaptive landscape** if you will, of possible adaptive peaks of different heights, each representing a different genetic configuration of the population (Figure 31-1). The peaks are separated by valleys, which are areas in the adaptive landscape where average fitness is low. Put another way, the peaks of the adaptive landscape represent favorable genotypes, and the valleys unfavorable genotypes, for the prevailing environmental conditions. Were the environment to change, so would the adaptive landscape, because previously unfavorable genotypes—those in a valley of the adaptive landscape—might be favored under the new environmental conditions, and selection would move the population uphill toward a new adaptive peak.

Armbruster (1985, 1990, 1991) worked out an approximation of the adaptive landscape of pollination for species of tropical plants of the genus *Dalechampia.* The inflorescence of *Dalechampia* contains both male and female flowers merged together into a single functional unit (Figure 31-2). The male flowers contain a resin gland that varies in size from flower to flower. Although *Dalechampia* can self-fertilize, it may also be pollinated by bees that visit the inflorescence to collect resin from the gland for use in their nests. Cross-pollination is thought to convey a fitness advantage compared with selfing in these plants. Whether or not visiting bees pollinate a flower depends on the size of the bee, the size of the resin gland, and the distances between the various parts of the inflorescence. Large bees are attracted to large resin glands. Bees that visit inflorescences land with their heads near the resin gland of an individual flower. Large bees may also touch an anther and stigma and effect pollination, something that small bees cannot do. But whether a bee pollinates a flower or not is also influenced by the distance between the gland and the stigma and the distance between the gland and the anther. If these distances are too large, even big bees are not able to pollinate.

Armbruster concluded from his studies that the morphology of *Dalechampia* constitutes an adaptive landscape (Figure 31-3). Low pollination rates occur when the resin gland is small compared with the

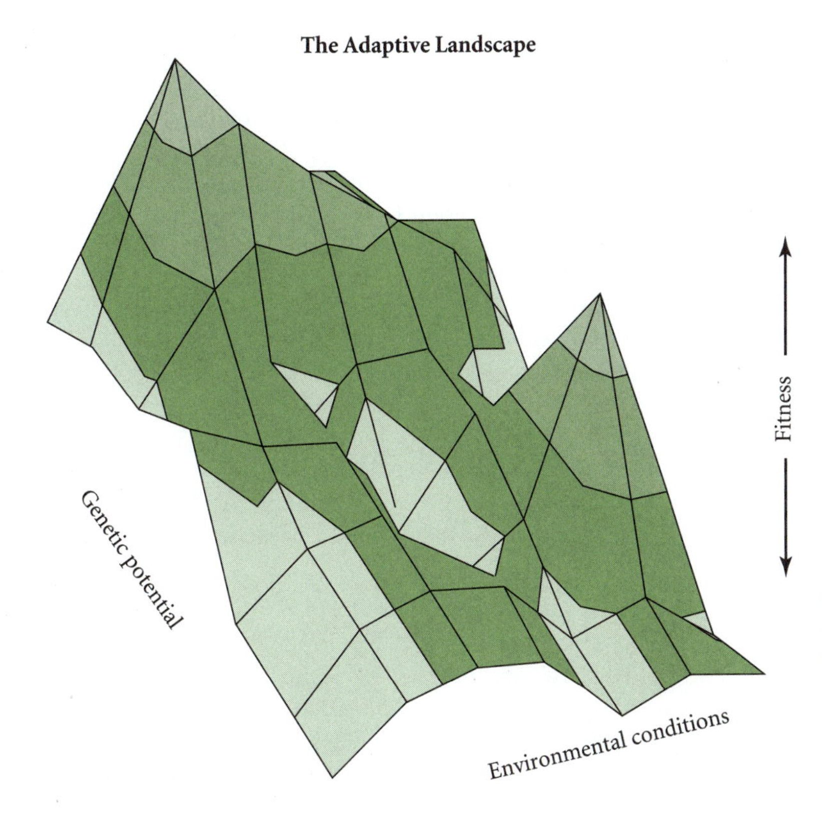

FIGURE 31-1 The adaptive landscape is conceptualized as a series of adaptive peaks occurring in a space defined by genetic potential and environmental conditions. Natural selection moves populations up a peak. Changes in peaks occur with shifts in environmental conditions.

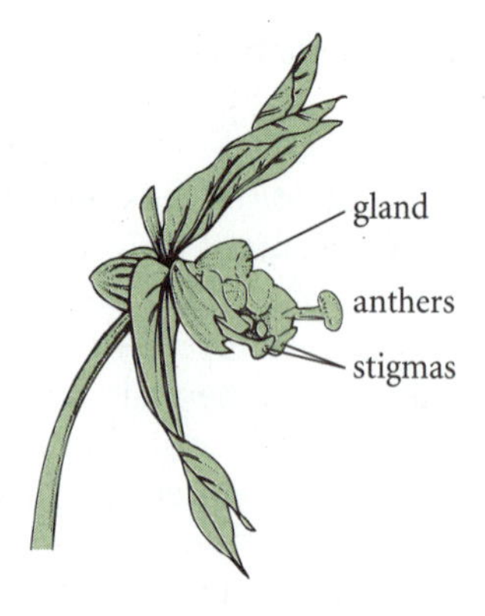

FIGURE 31-2 Inflorescence of *Dalechampia scandens* showing the resin gland, anthers, and stigmas. (*After Armbruster 1985.*)

distance between the gland and the flowers. Small resin glands attract only small bees, which are unable to bridge a large distance between the resin gland and the flowers. Armbruster also suggested that the costs of the production of resin in very large glands would exceed the benefits of attracting bees large enough to pollinate the flowers. A ridge of adaptive peaks separates these two adaptive valleys (Figure 31-3).

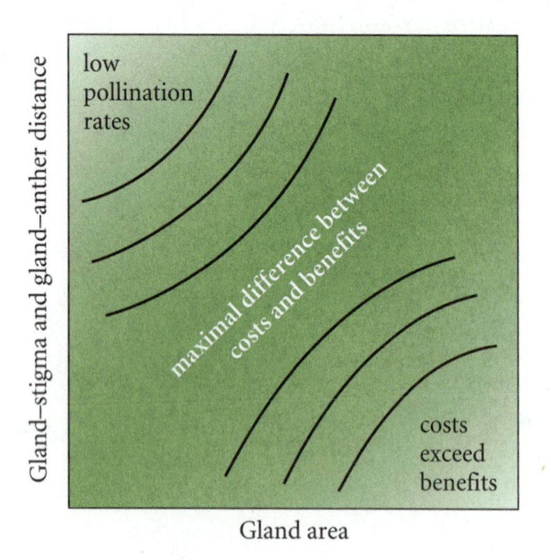

FIGURE 31-3 Adaptive landscape of the inflorescence of *Dalechampia*. Increasing gland-stigma and gland-anther distances are shown along the vertical axis. Gland size is shown along the horizontal axis. A ridge of adaptive peaks runs from the lower left corner of the graph (low gland-stigma and gland-anther distance; low gland area) to the upper right corner (high gland-stigma and gland-anther distance; high gland area), with adaptive valleys on either side. One valley occurs where gland area is small and distances between the gland and the flower parts are large because such an arrangement attracts small bees that are unable to pollinate. A second adaptive valley occurs where gland area is extremely large because the high cost to the plant of producing resin outweighs any benefit from pollination. (*After Armbruster 1990.*)

A problem with the adaptive landscape metaphor is the difficulty of explaining how a population moves from one adaptive peak to another across a valley of unfavorable genotypes. Also, changes in the adaptive landscape may occur in the absence of environmental change. This may occur in small populations if genetic drift overcomes the strength of selection; that is, drift may move the population down an adaptive peak. Peak shifts may also occur as a result of correlated responses to selection (Price et al. 1993). Recent conceptual modifications of the adaptive landscape idea have suggested that if the adaptive landscape is viewed in a multidimensional space rather than in three dimensions, peak shifts are more easily explained (Gavrilets 1997).

31.2 Polymorphisms may be maintained by evolution in heterogeneous environments.

Polymorphism is the coexistence of more than one distinct genotype or phenotype in a population. Until the mid-1960s, when biochemical techniques uncovered vast amounts of previously unknown genetic variation, observed polymorphisms were thought to arise from the superior fitness of heterozygotes, referred to as **heterosis.** The recessive allele that causes sickle-cell anemia, for example, is maintained in human populations because heterozygotes are resistant to malaria (Allison 1956, Cavalli-Sforza and Bodmer 1971). But experiments with flies and other organisms had also shown that constant laboratory conditions revealed no such heterosis; rarely did two or more phenotypes or genotypes coexist in population cages. Consideration of how polymorphisms are maintained in populations has focused on the dynamics of evolution in temporally or spatially varying environments. This work emphasizes the implications of the relative fitnesses of different forms of a polymorphic trait under different conditions. In this section we first discuss briefly the theoretical work of Richard Levins. Although Levins failed to provide a compelling explanation for the maintenance of genetic polymorphisms, his ideas leapt beyond contemporary bounds of empirical data and experimental results and provided a basis for discussion of the importance of environmental heterogeneity in evolution. Levins's work leads naturally to the concept of frequency-dependent selection, which we will also discuss briefly.

Fitness Sets

A. B. da Cunha and Theodosius Dobzhansky (1954), Richard Lewontin (1958), and others had suggested that polymorphism persisted only when different genotypes were selectively favored in different parts of the environment or at different times. Levins (1962)

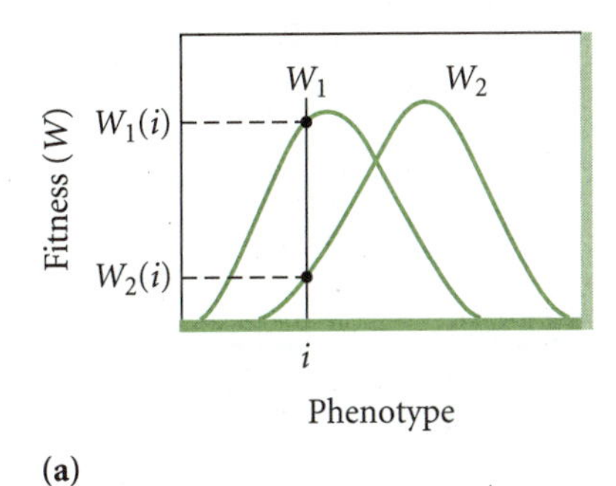

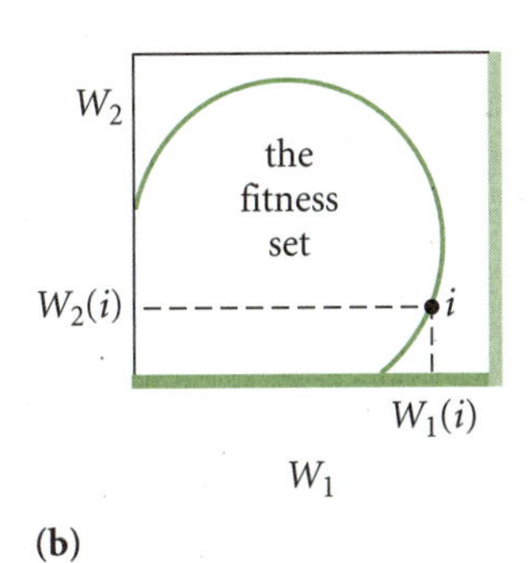

FIGURE 31-4 (a) Relationship between phenotype and fitness in two different environments (1 and 2). The fitnesses of a single phenotype (i) in the two environments (W_1 and W_2) are indicated. (b) The fitness set portrays the relationships between the fitnesses of all possible phenotypes in environments 1 and 2. Each point on the perimeter of the set represents a single phenotype (i) whose position on the graph is determined by its fitnesses $W_1(i)$ and $W_2(i)$.

developed a theoretical approach to this issue that allowed researchers to predict the outcome of evolution in heterogeneous environments.

The heart of Levins's theory was fitness set analysis. A **fitness set** portrays phenotypes on a graph whose axes are components of fitness. One then superimposes a fitness criterion upon this graph to identify that phenotype having the greatest fitness. For simplicity, Levins restricted his analysis to the case of a population exposed to two different environments (1 and 2). Figure 31-4a shows the distributions of the fitnesses of individuals in a population in two different environments. The uppercase letter W is used to denote fitness in evolutionary ecology. The figure shows that an individual with phenotype i has a higher fitness in environment 1 [$W_1(i)$] than it does in environment 2 [$W_2(i)$]. When all possible phenotypes (all points along the horizontal axis of Figure 31-4a) are positioned on a graph according to their fitnesses in environment 1 (W_1) and environment 2 (W_2), the solid figure outlined is the fitness set (Figure 31-4b). Only those phenotypes lying on the periphery of the figure (the curved line) are of interest because at least one of these will have a fitness superior to any phenotype within. But which point on the periphery represents the greatest fitness? When conditions are constant, and individuals experience only environment 1 or only environment 2, the selected phenotype will be the one having the greatest fitness in that environment.

To determine the most fit phenotype in a heterogeneous environment, Levins devised the **adaptive function,** a mathematical formulation combining fitnesses under the conditions of each of the environments 1 and 2 into a measure of the overall fitness of a phenotype in a heterogeneous environment. When the environment varies spatially and organisms encounter habitat patches of type i in direct proportion to their frequency (p_i), the overall fitness of a phenotype (W) is the average of its fitnesses in each habitat type weighted by their frequency. Thus, $W = p_1W_1 + p_2W_2$. By rearranging this expression to solve for W_2, we obtain

$$W_2 = \frac{W}{p_2} - \left(\frac{p_1}{p_2}\right)W_1,$$

which is the equation for a straight line on a graph whose axes are W_1 and W_2 (Figure 31-5), the same axes used to define the fitness set. This equation defines a family of lines having a slope $-p_1/p_2$ and an intercept, or distance from the origin of the graph, determined by the value W—the overall fitness. Each line represents the combinations of W_1 and W_2 resulting in the same overall fitness. Levins called these lines of equal fitness adaptive functions. Superimposed upon the fitness set, these adaptive functions identify the phenotype with the highest overall fitness. This is the point on the periphery of the fitness set that is touched by the adaptive function lying farthest from the origin of the graph, hence representing the highest value of W among all phenotypes (Figure 31-6). This adaptive function is a tangent to the fitness set.

As the ratio of habitat patches varies from place to place, the slope of the adaptive function (p_1/p_2) varies, and the tangent adaptive function touches the fitness set at different points (Figure 31-7). From this result, Levins suggested that along a cline of frequency in habitat patches, one should expect to see a cline in selected phenotypes.

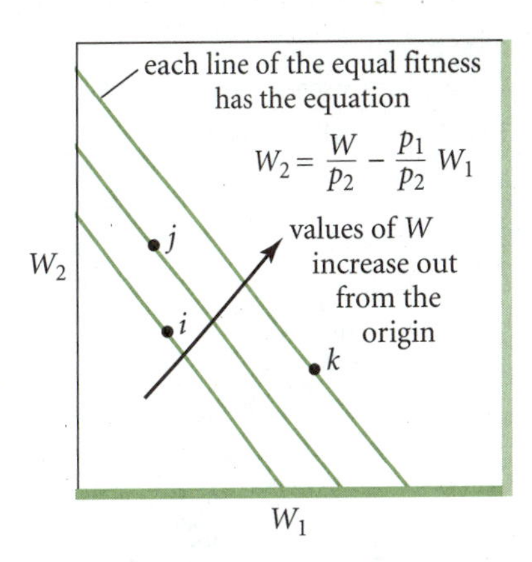

FIGURE 31-5 Adaptive functions are lines of equal fitness (W) whose positions on the graph are determined by the relative frequencies of environments 1 and 2. The fitness values of points i, j, and k represent increasing values of W with distance from the origin of the graph.

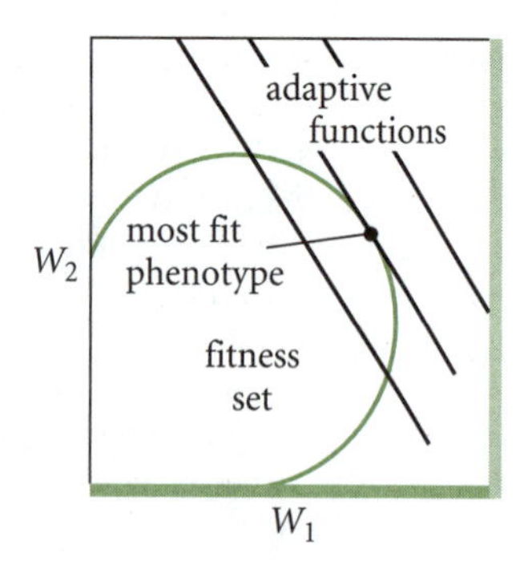

FIGURE 31-6 The superimposition of adaptive functions on the fitness set reveals the most fit phenotype as that touched by the adaptive function of highest value (the one that is tangent to the fitness set).

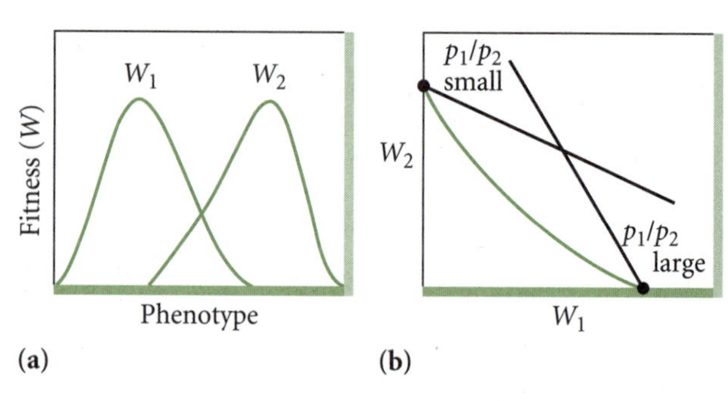

FIGURE 31-8 When environments 1 and 2 differ greatly (a), the fitness set is concave (b). Changes in the frequencies of the environments cause a stepwise shift in the optimum phenotype because the phenotypes lying on the inward-bulging curve have lower fitnesses than those at either end point of the curve.

The fitness set portrayed in Figures 31-4, 31-6, and 31-7 is based on habitat types sufficiently similar that the fitnesses of each phenotype in the two habitats do not differ greatly. Levins also explored the situation in which the habitat patches present markedly different selective conditions. In such cases, the fitness set has a depression in its center, which results in a different outcome of selection (Figure 31-8). Linear adaptive functions superimposed on this fitness set show that phenotypes lying along the inward-bulging part of the fitness set's periphery cannot be selected. As the ratio of habitat patches changes, the phenotype with maximum fitness shifts abruptly from one lobe of the fitness set to the other, skipping over intermediates. Intuitively, we can see that when habitats differ greatly, phenotypes intermediate between those that are superior in each type of habitat will not be well suited to either. Levin's result gave the adage "jack of all trades, master of none" considerable currency in ecological thinking. It also led to an interesting prediction: along a gradual cline of frequency of habitat patches, one could expect to see an abrupt change in phenotype (see Endler 1977).

When the environment varies temporally, so that habitat types are encountered in a time sequence, the adaptive function may be more appropriately calculated as a geometric mean, rather than an arithmetic mean. This is because fitnesses represent factorial increases in the numbers of individuals with each phenotype from generation to generation. When the population growth rate varies over generations, the long-term expectation of population size is directly proportional to the product of the population growth rates experienced during each time segment. Therefore, in a temporally varying environment, the average fitness of a phenotype is $W = W_1 p_1 W_2 p_2$, which can be rearranged to give the adaptive function

$$W_2 = \left(\frac{W}{W_1 p_1}\right)^{1/p_2},$$

which is a hyperbola. This equation is portrayed graphically in Figure 31-9.

Levins extended his ideas by asserting the existence of an extended fitness set (F'), which would include all possible mixtures of phenotypes in the environment (Figure 31-10). According to Levins, selection favored a mixture of genotypes represented by the tangent of the hyperbolic adaptive function on

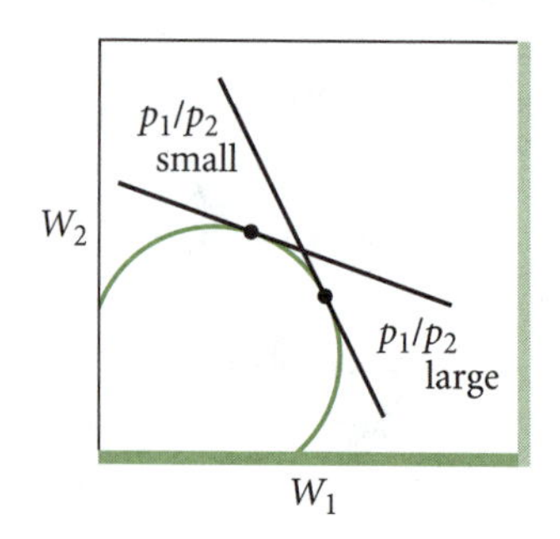

FIGURE 31-7 As the slope of the adaptive function changes with changes in the frequencies of environments 1 and 2, the most fit phenotype also changes.

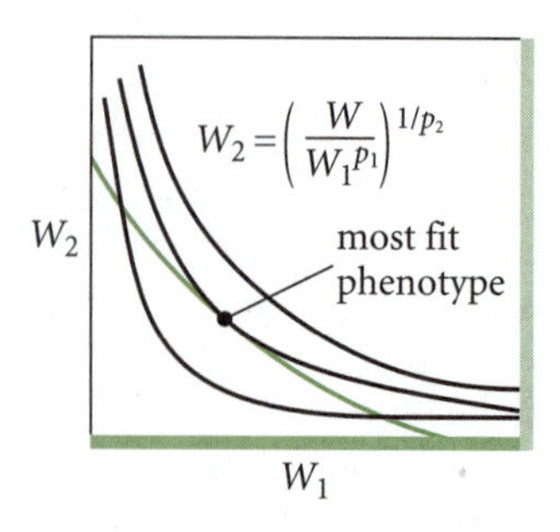

FIGURE 31-9 When environments are encountered in sequence, the adaptive functions are hyperbolas and may favor an intermediate phenotype, even for a concave fitness set.

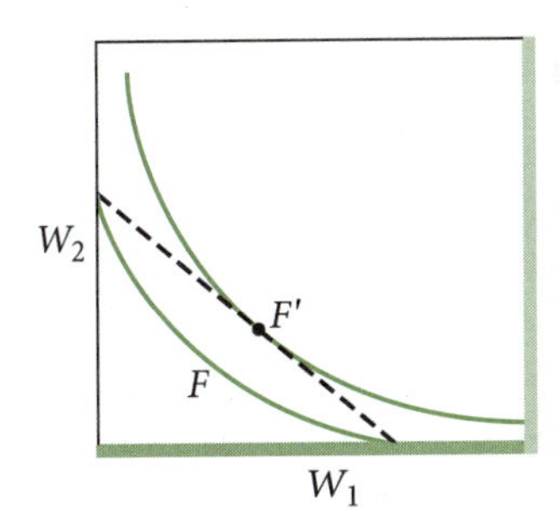

FIGURE 31-10 Levins's concept of the extended fitness set (F') as a mixture of different proportions of extreme phenotypes, which he thought could explain the maintenance of polymorphisms in populations.

the extended fitness set. Thus he believed that he had achieved an explanation for polymorphism in populations, and further predicted that along a cline of frequency of temporally heterogeneous conditions, one should observe a cline in the frequency of two phenotypes (a morph-ratio cline).

Levins erred in thinking that, in a temporally varying environment, mixtures of two phenotypes assume the arithmetic mean of their component fitnesses. In fact, the "fitness" (that is, the population growth rate) of a mixture of phenotypes is the average of the geometric mean of the component fitnesses of each phenotype. Hence Levins's extended fitness set does not exist. The absurdity of the concept can be appreciated most readily when the fitness set bows inward from one axis to the other. A mixture of two phenotypes, each well suited to one of two conditions but having zero fitness in the alternate condition, has zero fitness overall: neither of the phenotypes can persist in a temporally varying environment. Levins made this mistake because he ignored the mechanics of selection in his model.

Frequency-Dependent Selection

Polymorphisms may be maintained in a population by the process of **frequency-dependent selection.** Suppose that each of two genotypes in a polymorphic population has fitness that is dependent on its frequency in the population with respect to the other genotype. That is to say, the fitness of one form, say, form A, changes as it becomes more common in the population with respect to form B. A change in the frequency of form A will result in a change in its selective advantage. Thus, selection on form A is frequency-dependent. Frequency-dependent selection may be either negative or positive. **Negative frequency-dependent selection** means that the fitness of the trait decreases as the frequency of the trait increases. An increase in fitness with frequency is referred to as **positive frequency-dependent selection.** Let us examine an example to show how this process might maintain a polymorphism in a natural population.

Among the many species of African cichlid fishes (see Chapter 1) are seven species that specialize in eating the scales of other fish. In order to obtain this rather unusual food, scale-eating fishes approach their prey—perhaps more properly called a host, since it usually does not die from the attack—from the rear and then quickly dart up to snatch a scale or two from the flank of the prey. A rather unusual polymorphism exists in these fishes: the mouth is asymmetrical. That is, in some individuals of a species, the mouth opens to the left (sinistral) and in other individuals of the same species, the mouth opens to the right (dextral) (Figure 31-11). It is quite proper to say that there is "handedness" to the direction of the mouth opening in these fishes. Individuals with a dextral mouth opening always attack the prey's left flank, and those with a sinistral mouth opening attack the right flank (Hori 1993).

Studies of one of these species, *Perissodus microlepis,* which occurs in Lake Tanganyika, suggest that the left- or right-mouthed condition is inherited as a simple Mendelian system. Prey species actively guard against attack from the scale-eating fish. When one or the other type—left-mouthed or right-mouthed—of *P. microlepis* is more abundant, prey guard more intensively against that type, thereby reducing the success of that type at obtaining scales relative to the other type. Hori (1993) found that such preferential guarding acted as a form of frequency-dependent selection on the scale eater and maintained the polymorphism in mouth direction in the population. We shall see below that the concept of frequency-dependent selection is also useful in considering the

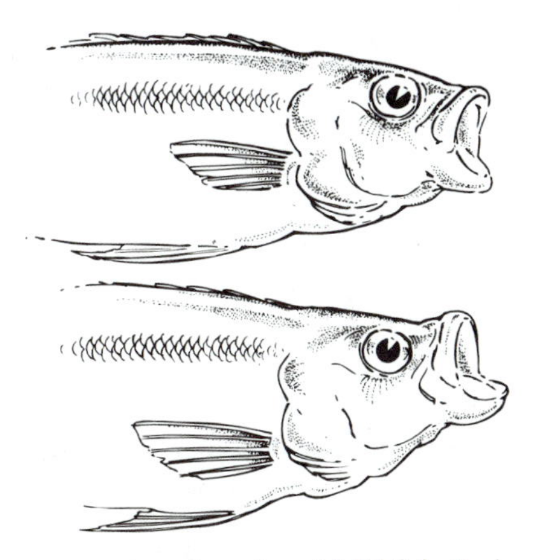

FIGURE 31-11 A scale-eating cichlid fish, *Perissodus microlepis,* of Lake Tanganyika, Africa. The moth opens to the right (*top*) in some individuals and to the left (*bottom*) in others. The polymorphism in mouth direction is maintained in the population by frequency-dependent selection. (*After Hori 1993.*)

balance among alternative foraging or mating strategies of animals in populations.

31.3 The relationship between phenotype and environment is a property of individual organisms.

How can we distinguish between evolutionary adaptations of populations and responses of individuals to the range of environmental conditions they normally encounter? Individual responses are nongenetic, in the sense that an individual has one set of genetic material that does not change during its lifetime, but the ways in which an individual can respond to its environment may be under genetic control and thus subject to evolutionary change by natural selection.

Virtually all attributes of an individual are affected by environmental conditions, at least to some extent. The relationship between the phenotype of an individual and the environment is referred to as a **reaction norm** (Figure 31-12). Many of the responses discussed in Chapter 6 are examples of reaction norms. For example, the way your body responds to changes in temperature represents the relationship between your phenotype—in this case, the physiological processes related to temperature regulation—and the environment. Your response may not be exactly like everyone else's. The general responsiveness of phenotypes to environmental conditions, which is called **phenotypic plasticity,** has received considerable recent attention (e.g., Schlichting 1986, Trexler and Travis 1990a,b, Berrigan and Koella 1994, Gomulkiewicz and Kirkpatrick 1992, Houston and McNamara 1992, Kawecki and Stearns 1993, Scheiner 1993, Weeks and Meffe 1996).

Some reaction norms are a simple consequence of the influence of the physical environment on life processes. Heat energy accelerates most life processes. Therefore, we should not be surprised that caterpillars of the swallowtail butterfly *Papilio canadensis* grow faster at higher temperatures. However, the fact that individuals of that species from Michigan and from Alaska exhibit different relationships between growth rate and temperature indicates that reaction norms may be modified by evolution. In one experiment, larvae from Alaskan populations grew more rapidly at lower temperatures and larvae from Michigan grew more rapidly at higher temperatures, as one might have predicted from the range of temperatures found in each location during the growing season (Ayres and Scriber 1994; Figure 31-13).

As swallowtail growth rates show, genotype and environment interact to determine phenotypic traits. **Genotype-environment interaction** means simply that the slope of the reaction norm differs between genotypes. When two reaction norms cross, as they do in the case of the swallowtail butterfly, then each genotype (or population) performs better in one environment and worse in another. The effects of such interactions between genetic factors and environmental factors on performance are the basis for the evolution of specialization. Over time, when two populations are exposed to different ranges of environmental con-

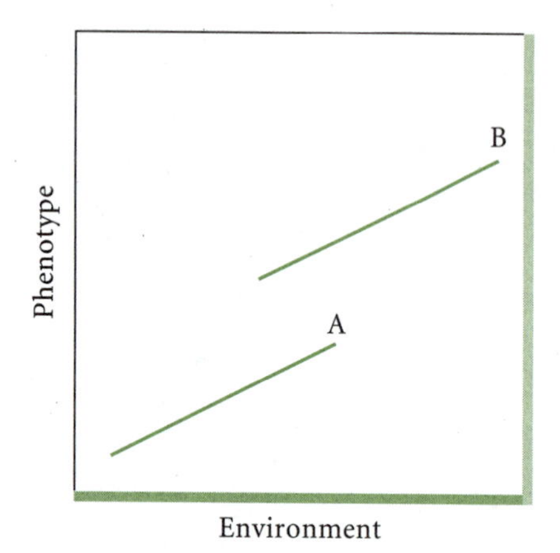

FIGURE 31-12 The relationship between phenotype and environment, called the reaction norm, is a property of the individual organism. The responsiveness of individuals to their environment is under genetic control and may be modified by evolution, as indicated by a shift in the reaction norm of species A as compared with that of related species B.

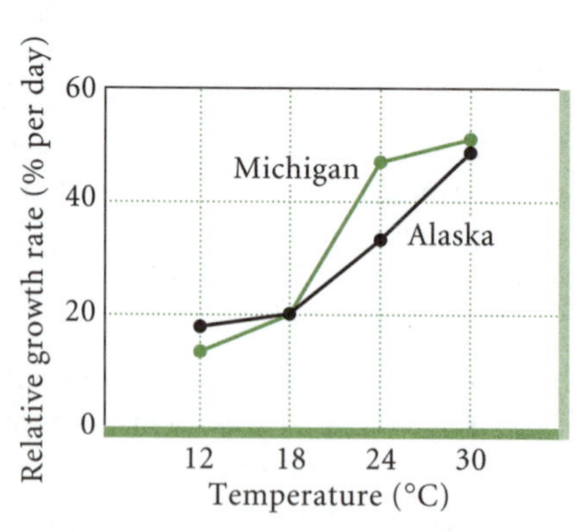

FIGURE 31-13 Growth rates of fourth-instar larvae of the swallowtail butterfly *Papilio canadensis* as a function of temperature. Each of the two relationships shown is a reaction norm for a population. Larvae were obtained from populations in Alaska and Michigan and reared on balsam poplar (*Populus balsamifera*). Note that Alaskan butterflies grew more rapidly than Michigan individuals at colder temperatures. The reverse was true at warmer temperatures, even though individuals from both populations grew faster as temperatures increased. These results are fairly typical of genotype-environment interactions in that differences between populations are small. (*After Ayres and Scriber 1994.*)

ditions, genotype-environment interactions will cause different genotypes to predominate in each population. The populations will therefore become differentiated and will have different reaction norms, each of which enables individual organisms to perform better in their own environments.

Whether differences between populations are due to evolutionary differentiation or to phenotypic responses of individuals to different environments often can be revealed by **reciprocal transplant experiments.** Transplant studies compare the phenotypes of individuals kept in their native environment with those of individuals transplanted to a different environment. Reciprocal transplants involve the switching of individuals between two localities. The traits of interest are assumed to be genetically determined when the phenotypic values of native and transplanted individuals do not vary between the two environments. When trait values reflect where an individual is living (environment) rather than where it came from (genotype), then the results of the experiment are consistent with phenotypic plasticity as a cause of differences between populations. Of course, intermediate results are possible, in which case one might conclude that the reaction norm has been subject to evolutionary modification.

In one reciprocal transplant experiment, fence lizards (*Sceloporus undulatus*) from New Jersey and Nebraska were switched between those locations (Niewiarowski and Roosenburg 1993). The effect of the transplants on growth rate revealed both genetic determination and phenotypic plasticity (Figure 31-14). The growth rates of Nebraska lizards, about twice those of New Jersey lizards in their native environments, decreased by half—to the New Jersey level—in individuals transplanted to New Jersey. In contrast, New Jersey lizards did not grow faster in Nebraska. A simple interpretation of these results is that resources available for growth are consistently fewer in New Jersey than in Nebraska and that Nebraska lizards transplanted to New Jersey cannot gather resources fast enough to support their natural growth rates. Apparently, New Jersey lizards have a genetically regulated growth rate that is adjusted to a low resource level. That is, they have lost the ability to modify individual growth rates in response to higher resource levels—levels that they probably experience rarely, if ever.

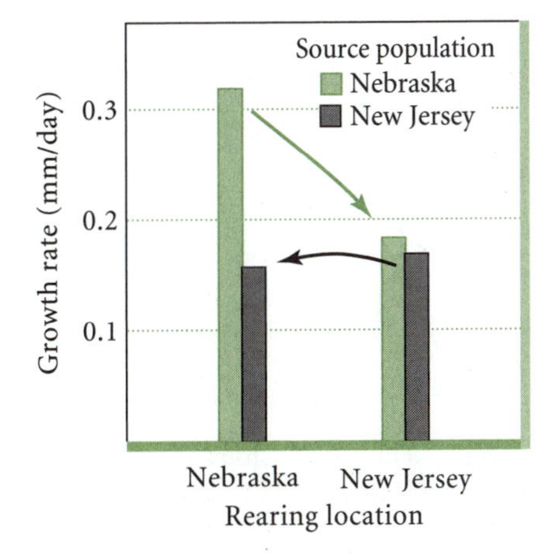

FIGURE 31-14 Growth rates of juvenile eastern fence lizards (*Sceloporus undulatus*) from populations in Nebraska and New Jersey exchanged in a reciprocal transplant experiment. Arrows indicate transplanted populations. (*Data from Niewiarowski and Roosenburg 1993.*)

It is a fair question whether the slower growth rate of Nebraska lizards transplanted to New Jersey is adaptive or merely a consequence of reduced resources. If phenotypic plasticity is adaptive, then a change in the form or function of an organism should reduce the negative effects of environmental change on fitness. A simple experiment conducted by Schew (1991) illustrates the difference between adaptive and nonadaptive phenotypic plasticity. Schew restricted the food supplies of hand-reared chicks of Japanese quail and European starlings, for periods of 10 and 3 days respectively, to a level that was just sufficient to maintain a constant weight. Thus, the experiment simulated the effects of poor feeding conditions during the period of rapid chick growth.

Although chicks of neither species could gain body weight, the responses of other aspects of growth and development to food restriction differed markedly between the two species. The quail chicks quickly decreased their body temperatures and metabolic rates (thereby saving energy). Their feathers and extremities, particularly the long bones of the legs and wings, ceased growing almost immediately. In effect, the quail stopped their developmental clocks and remained at a physiological age equivalent to that at the beginning of the experiment. When food restrictions were lifted, normal growth and development resumed, and the quail subsequently reached normal adult size.

The starlings showed none of these compensating responses. Their metabolism and body temperatures remained high, and their bones and feathers continued to grow throughout the restriction period, at the expense of the size of their internal organs. As a result, physiological age kept pace with chronological age, and the chicks became progressively more undernourished. When food restrictions were lifted, growth resumed at a rapid rate, but the chicks did not reach adult size before their developmental program told them to stop growing. As a result, the restricted chicks became stunted adults, with bones and feathers significantly shorter than those of adults that were well nourished during the growth period.

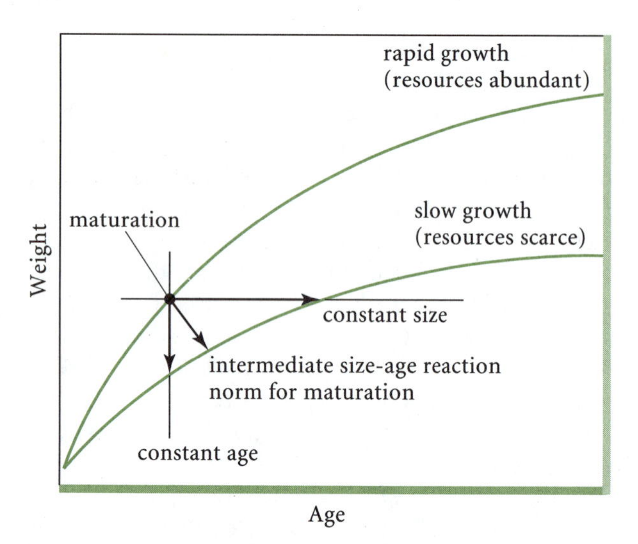

FIGURE 31-15 Possible relationships between age and size at maturation or metamorphosis when growth rates differ. Individuals may switch at a constant age, a constant size, or some intermediate between the two. (*From Stearns 1983, Stearns and Koella 1986.*)

The story of the quail and starlings shows that although organisms may have little control over their rate of growth, other aspects of their life histories can be modified in response to growth performance. Many types of organisms undergo dramatic changes during the course of their growth; metamorphosis from larval to adult forms and sexual maturation are the most prominent of these changes. The optimal timing of such developmental events depends on resources and natural enemies, and is made more complicated by variations in the rate of growth due to food supply, temperature, and other environmental factors.

Imagine two growth curves resulting from two levels of food supply (Figure 31-15). Let us suppose that under a good nutritional regime resulting in rapid growth, an individual matures at a given weight and age. Poorly nourished individuals clearly cannot reach the same weight at a given age, and therefore must use a different transition point for maturation. Faced with such environmental variation, an individual may adopt one of two rules for the switch, or some intermediate between them. First, the individual may mature at a predetermined weight; with poor nourishment, this will take longer to achieve and will therefore expose the individual to a longer period of risk prior to reproduction. Alternatively, the individual may mature at a predetermined age; with poor nourishment, this will result in smaller size at maturity and perhaps a reduced reproductive rate as an adult. The optimum solution is usually somewhere in between, depending in part on the risk of death as a juvenile (high risk favors earlier maturation at a smaller size) and the slope of the relationship between fecundity and size at maturity (higher values favor delayed maturation at a larger size because the fecundity payoff is greater).

Tadpoles raised under high and low food availabilities exhibit different growth rates, as one would expect. In one experiment, the tadpoles given a poorer diet metamorphosed into adult frogs at a smaller size, but at a later age, than those reared with abundant food (Figure 31-16). This finding supports the theoretical conclusion that timing of metamorphosis should be sensitive to both age and size: poor nutrition slows the developmental program in frogs, but does not stop it altogether, as it seems to in Japanese quail. The relationship between age and size at metamorphosis under different feeding regimes is the reaction norm for metamorphosis with respect to age and size. To illustrate the generality of this pattern, we note a similar reaction norm for maturity in human females in the United States and other developing countries, who are, on average, much better nourished now than they were a century ago. In response to a shifted weight-age relationship, the age at maturity of young women has decreased by about 4 years, and weight at maturity has increased by about 2 kilograms, since the beginning of the twentieth century.

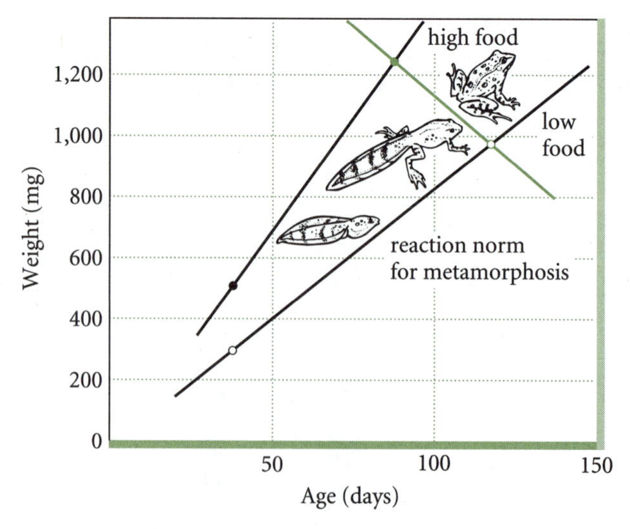

FIGURE 31-16 The relationship between age and size at metamorphosis in samples of frogs grown at high and low food availabilities. The metamorphosis reaction norm (green line) lies between constant size and constant age extremes. Circles represent weights of tadpoles at 40 days and at metamorphosis, which were used to obtain the growth relationship (black lines) of each treatment group. (*Data from Travis 1984.*)

31.4 Optimal foraging theory addresses the problem of choice among resources or habitats.

When a predator moves through a habitat in search of food, it sequentially encounters potential prey. With each encounter, the predator is confronted with the choice of pursuing and eating the prey, which requires time and the expenditure of energy, or passing up the potential prey in favor of continuing to search. During foraging, each choice is associated with potential benefits and potential costs. Intuitively, we accept that animals will make choices that minimize costs relative to benefits. That is, animals should behave in a way that leads to the intake of as much food as possible for a given amount of energy spent in obtaining that food. The body of ecological theory that focuses on the analysis of costs and benefits is referred to as **optimality theory. Optimal foraging theory** is based on the idea that natural selection molds the foraging behavior of animals to maximize fitness (Stephens and Krebs 1986, Mangel and Clark 1988, Krebs and Kacelnik 1991, Bulmer 1994). Fitness is usually not measured directly. Rather, some surrogate measurement, such as the rate of average net energy intake, is obtained.

Elements of Optimality Modeling

Optimality modeling rests on the proposition that animals are decision makers who choose from a number of alternative behaviors so as to maximize their fitness. Ecologists do not believe that animals make conscious choices during foraging. Rather, animals choose to take a particular prey type or not based on their recent experience, physiological state, morphological ability to capture and hold the prey, and other more or less proximate feedbacks.

Optimality models include three basic elements (Krebs and Kacelnik 1991): a currency, constraints, and a decision variable. In the business world, behavior is optimized to reduce costs relative to profits, which are measured at the bottom line in currencies such as American dollars, German marks, or Japanese yen. The currency of a foraging animal is energy, which may be estimated by the amount of time spent in various activities. The risk of being killed is also an important currency. Animals forage under constraints related to their environment and their own morphological and physiological limits. The availability of various prey types and the escape behaviors of those prey types are environmental factors that constrain the time that it takes to find, capture, and consume prey. The size of the feeding structures of the predator relative to the size of the prey is an example of a morphological constraint on foraging.

Animals foraging within the constraints of their structure and environment will make choices that optimize the currency. To represent such decision making in an analytical model that is capable of generating hypotheses about foraging behavior, it is necessary to specify clearly the **decision variable** of the model. That is, we must know what is being decided at each time in the foraging model. One important decision variable is whether to eat a particular prey once that prey is encountered. Another is whether to continue foraging in a patch of food or to leave that patch to seek another patch.

Criticisms of Optimality Modeling

Optimality modeling has come under some criticism (Krebs and Kacelnik 1991). Indeed, some of the criticism is quite passionate, as evidenced by the title of a paper by Pierce and Ollason (1987), "Eight reasons why optimal foraging theory is a complete waste of time." Some of the difficulty stems from the fact that such approaches are rooted in the adaptationist program, which assumes that natural selection will lead inevitably to phenotypes that optimize cost-benefit ratios (Gould and Lewontin 1979) (see Chapter 30). Other criticisms turn on whether optimality models adequately represent the behaviors of animals in the natural world (Gray 1987). These criticisms, and others that suggest that careful descriptive studies will move our understanding of the evolution of behavior along much faster than theoretical studies, are commonly leveled at analytical models in which one-for-one predictions are made based on highly specific and, admittedly, sometimes biologically unrealistic constraints. However, as we emphasized in Chapter 2, the utility of analytical models is to represent the general shape and design of a natural process and to generate hypotheses for experimental evaluation, not to mirror the details of the process. Optimal foraging theory is thus a valuable tool in understanding the patterns of behavior of animals in nature.

31.5 Predators may optimize the number of different types of prey in their diet.

We may consider how a foraging animal might divide its time so as to maximize energy intake by employing a **prey model** of optimal foraging. In a prey model, the decision variable is whether to eat a particular prey or not. Energy intake is optimized by managing the **diet breadth,** the number of different types of prey that are included in the diet. If we stipulate that all prey have the same nutritional value to the predator, then the abundances of prey in the environment and their ability to escape predation distinguish prey types. To the predator, less abundant prey or prey with well-developed abilities of escape are of lower quality because they cost more to obtain. Here, we measure cost in the currency of time. Time spent searching for and pursuing prey is a cost because it is time that is not spent eating. A predator has a finite amount of time to spend foraging. Under these conditions, what determines the number of prey types to be eaten? To address this question, MacArthur and Pianka (1966) developed a simple graphic model of a predator's behavior under these conditions.

One might think that the best strategy for a predator is simply to eat everything that it encounters. Increasing the diet breadth would seem to be a good thing. However, adding new kinds of prey to the diet may affect the rate of prey consumption both positively and negatively. A broader diet presents a predator with more potential prey individuals and thus may increase prey consumption. However, different prey types differ in their abundances and in their ability to escape predation. As more prey types are added to the diet, the average time that it takes the predator to pursue, capture, and consume prey may increase, provided that the predator always includes prey species in its diet in descending order of quality.

In MacArthur and Pianka's model (Figure 31-17), the average search time (T_s) and pursuit and handling time (T_p) per prey item are graphed as a function of diet breadth when prey are ranked according to their suitability. The optimum diet breadth is that which results in the lowest sum of search and pursuit time ($T_s + T_p$). Remember, the time a predator spends searching for and pursuing prey is a cost, and optimization means minimizing this cost. The slope of the search time curve becomes less steep with increasing diet breadth because each new prey species adds proportionately less to the total diet. The slope of the pursuit time curve increases because as the suitability of the prey decreases, the capture and handling time increase more rapidly. Because of the shapes of the T_s and T_p curves, the sum of the two usually assumes a U-shaped curve, the lowest point of which defines the optimum diet breadth.

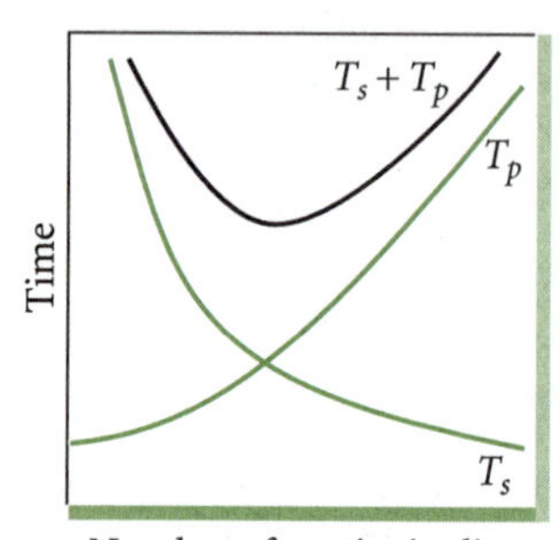

FIGURE 31-17 A graphic model of the optimization of diet breadth by a predator. Search time (T_s) and pursuit and handling time (T_p) are plotted as a function of diet breadth. The sum of the two ($T_s + T_p$) is inversely related to the rate of prey consumption. Hence optimum diet breadth corresponds to the minimum of the ($T_s + T_p$) curve.

A change in the overall abundance of prey, resulting from a change in the productivity of the habitat or the number of competing species, will change search time, but not pursuit and handling time; only the quantity of prey will vary, not their quality. When productivity increases, the T_s curve decreases, and the optimum diet breadth shifts to the left. Thus, increased productivity favors specialization. When competition increases, prey become more scarce, and the T_s curve increases, shifting the optimum to the right and favoring increased generalization (Figure 31-18).

How does prey diversity affect diet breadth? When the number of potential prey species increases without an increase in their overall variety (as assessed by their predator escape characteristics) or total abundance, the T_s curve increases because each species becomes proportionately less abundant, the T_p curve decreases somewhat because the number of easily caught prey species increases, and optimum diet breadth shifts to the right. But although the diet includes more species of prey, the proportion of available prey eaten and the variety of prey do not change. When there is an increase in the variety of prey species (adding species more difficult to capture), only the search time curve increases (each prey species becomes less abundant), and although diet breadth increases slightly, the proportion of available prey eaten, as well as the rate of prey capture, decreases. In general, then, increased production leads to diet specialization and increased prey consumption. Increased competition is equivalent to reduced production and has an opposite effect. Increased variety of prey species may lead to a somewhat broader food niche,

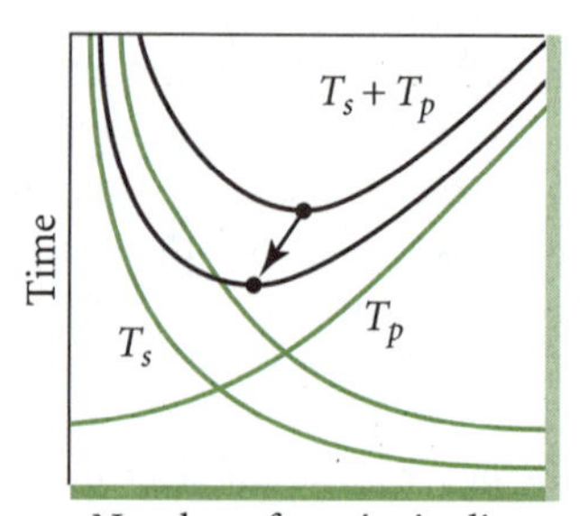

FIGURE 31-18 When availability of prey increases, either because of increased habitat production or reduced competition, search time (T_s) decreases, and the optimum diet breadth narrows.

but a reduced rate of prey consumption, all other things being equal.

31.6 The "classic" model of optimal foraging incorporates the concepts of encounter rate and profitability of individual prey.

Optimization of choice by predators and parasitoids has been explored theoretically by a number of authors (Schoener 1969a, 1971, Rapport 1971, Cody 1974, Orians and Pearson 1979, Townsend and Hughes 1981, Sih 1984, Charnov and Stephens 1988, Krebs and Kacelnik 1991; for a general treatment, see Stephens and Krebs 1986; for a collection of papers regarding foraging studies, see Kamil et al. 1987). These treatments agree with MacArthur and Pianka's prediction (1966) that increased production of resources should favor increased diet specialization. Mathematically, the problem of optimal diet has been characterized by the "classic" model of prey choice. The exposition here follows that of Krebs and Davies (1987), who, for the sake of simplicity, considered a predator hunting two types of prey (1 and 2). The pertinent characteristics of the prey are encounter rate (λ), reward (E), handling time (h), and profitability (E/h). A predator searches until it encounters the first prey (i). It then pursues, subdues, and consumes that prey with handling time h_i. During this handling period, the predator cannot search, and therefore passes up any other prey that it might have encountered had it continued searching. When prey i is consumed, the predator once again starts searching.

When two types of prey are sought, the total encounter rate of prey is the sum of the encounter rates of each individually ($\lambda_1 + \lambda_2$). The total number of prey encountered in a series of search intervals of total time T_s is equal to $T_s(\lambda_1 + \lambda_2)$, and the expected reward obtained from these prey is

$$E = T_s(\lambda_1 E_1 + \lambda_2 E_2).$$

The total handling time associated with search time T_s is $T_s(\lambda_1 h_1 + \lambda_2 h_2)$, and therefore the total time required to obtain reward E is

$$T = T_s + T_s(\lambda_1 h_1 + \lambda_2 h_2),$$

the sum of the search and handling times. The amount of reward received per unit of time when the diet includes prey items 1 and 2 is therefore

$$\frac{E}{T} = \frac{\lambda_1 E_1 + \lambda_2 E_2}{1 + \lambda_1 h_1 + \lambda_2 h_2}.$$

(The elimination of the total search time, T_s, is an example of the kind of unexpected result that we sometimes obtain from manipulating ecological models; see Chapter 2.) The reward per unit of time for selecting only one of the two prey types is, similarly,

$$\frac{E_1}{T_1} = \frac{\lambda_1 E_1}{1 + \lambda_1 h_1}.$$

Supposing that prey type 1 provides more reward per unit of time than type 2—that is, $E_1/T_1 > E_2/T_2$—the predator should specialize on prey type 1 rather than including both 1 and 2 in its diet when $E_1/T_1 > E_{1,2}/T_{1,2}$, or

$$\frac{1}{\lambda_1} < \frac{E_1}{E_2}(h_2 - h_1).$$

In words, the more specialized diet results in a higher feeding rate when the time to encounter the next item of type 1 ($1/\lambda_1$) is generally less than the ratio of the food value (E_1/E_2) times the difference in handling times ($h_2 - h_1$). High encounter rates, high food values, and low handling times therefore favor specialization. Notice that regardless of the abundance or food value of prey type 1, if it takes longer to handle than prey type 2, the predator should always include both types in its diet.

Krebs et al. (1977) attempted to test the "classic" foraging model by putting great tits (*Parus major*) in an ingenious prey choice situation. Pieces of mealworms of two sizes were placed on a conveyor belt that passed under an opening, allowing the caged birds access to the "prey." Reward (E) was related to the size of the prey, which was under the experimenters' control; handling time (h) was measured; and encounter rate (λ_1) could be varied. According to the theory, when the benefit of selecting the single most profitable prey type ($E_1/T_1 - E_{1,2}/T_{1,2}$) exceeded zero, the subjects should have switched from picking prey items off the belt at random to selecting only the larger pieces. The results of the experiment conformed generally to this prediction (Figure 31-19), but, contrary to expectation, prey type 1 was never selected to the exclusion of prey type 2.

Like other ecological models, the theory described above oversimplifies both the food resource and the behavioral response of an astute predator on that resource. Certainly factors such as errors in discriminating prey types, long-term learning of preferences for one type or another, short-term learning during runs of one or the other prey type, simultaneous encounters with more than one prey, and nonsubstitutability of prey because each provides a different essential requirement add complications that are difficult to incorporate into simple models. Risks associated with variable prey availability and with foraging under the threat of predation must also be taken into account, as we shall see below.

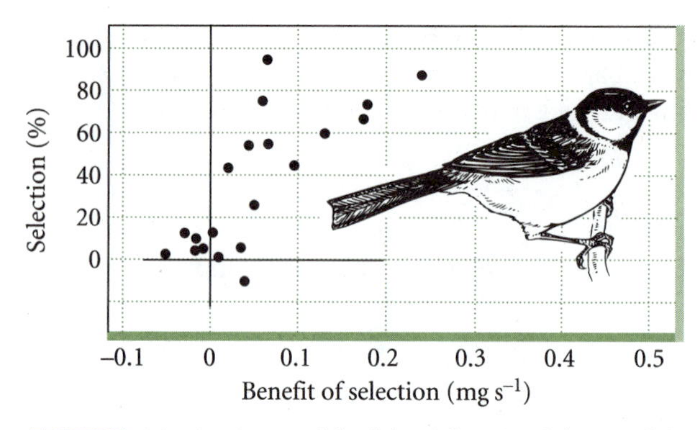

FIGURE 31-19 A test of the "classic" optimal diet model with great tits (*Parus major*), showing the relationship between preference for a more profitable prey and the benefit of specializing on that prey. The model predicts an abrupt switch when the benefit of specialization exceeds zero, a prediction to which the subjects only partially conformed. (*After Krebs and Davies 1981.*)

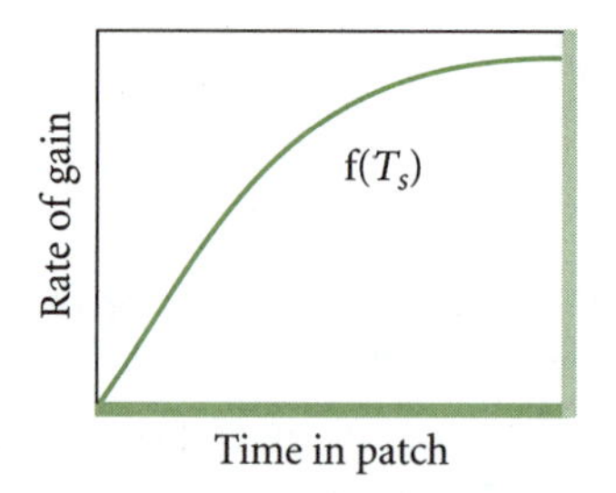

FIGURE 31-20 Gain curve, f(T_s), of an animal foraging in a patch. The rate of gain decreases with time, eventually leveling off.

31.7 Optimal patch use depends on patch quality and traveling time between patches.

In the prey models presented above, the arrangement of prey in space is not considered. In nature, most prey organisms occur in patches of high abundance separated by unsuitable habitat. Since predators travel among patches, they are confronted with choices about when to stop foraging in one patch and move on to another. Models predicting how foraging animals should choose among patches are called **patch use models.** Like optimal prey choice models, patch use models have received considerable attention from theoreticians (Charnov 1976, McNair 1982) as well as experimentalists (e.g., Cuthill et al. 1990). The formulations of patch use models may also be applied to animals foraging from a central location, such as a nest—a situation referred to as **central-place foraging.**

Consider a situation in which a predator can forage in one of two patches that are identical in the amount and quality of the prey available. While feeding in a patch, the animal gains food (energy) at a rate that is some function [f(T_s)] of the amount of time that it spends in the patch (T_s). As the predator depletes the resource, its reward, or gain, per unit of time decreases, and, eventually, the rate of gain from the patch begins to level off. This may be represented as a **gain curve** (Figure 31-20).

The time that it takes for the predator to move from one patch to the other is referred to as the **travel time,** and is denoted (T_t). In Figure 31-21, travel time is measured along that portion of the horizontal axis to the left of 0; the greater the distance from 0, the greater the travel time. As the predator initially searches for food within a patch, it encounters and consumes food at a high rate because food is plentiful and easy to find. Time spent in the patch is that part of the horizontal axis of Figure 31-21 to the right of 0. (Unfortunately, the term "search time" is sometimes used in the literature to denote both the time it takes to search for a new patch and the time spent searching for food inside a patch. We adopt the latter usage here and, thus, have used the subscript *s* in the notation T_s.) Gain may be measured in any of several currencies: units of energy, access to hosts for parasitoids, the amount of food that a central-place forager obtains to carry back to the nest (see below), or other correlates of fitness. The gain [f(T_s)] decreases as the search time increases because prey become more difficult to find as they become less abundant. The rate of gain decreases to zero as the resource is exhausted, but prey may become difficult to find well before that time. When this happens, the predator will receive a greater reward if it leaves the patch and moves to another patch containing a greater abundance of prey. The amount of time that the predator stays in a patch before leaving to find another patch is referred to as the **giving-up time.** In choosing when to give up, the predator must weigh the cost of traveling to the next patch. Intuition suggests that if there is a new patch with high levels of prey nearby—that is, T_t is small—the predator should leave the current patch, where returns are diminishing, sooner. Examination of Figure 31-21 leads us to this prediction.

Suppose that the travel time to the next patch is T_t = A (see Figure 31-21). The optimal amount of time the predator should spend in the current patch before leaving for the next patch may be found by drawing a line from the point A to the point tangent to f(T_s)—in this case, *a*. The point *a* is the optimal giving-up time. If the predator rushes off to the next patch too quickly, it will never reach the maximum gain in the current patch. If it stays too long, its efforts in that patch will be poorly rewarded. The major prediction of the model can be seen by comparing *a* to the optimum giving-up time if the next patch is farther away, T_t = B. In that case, the predator should

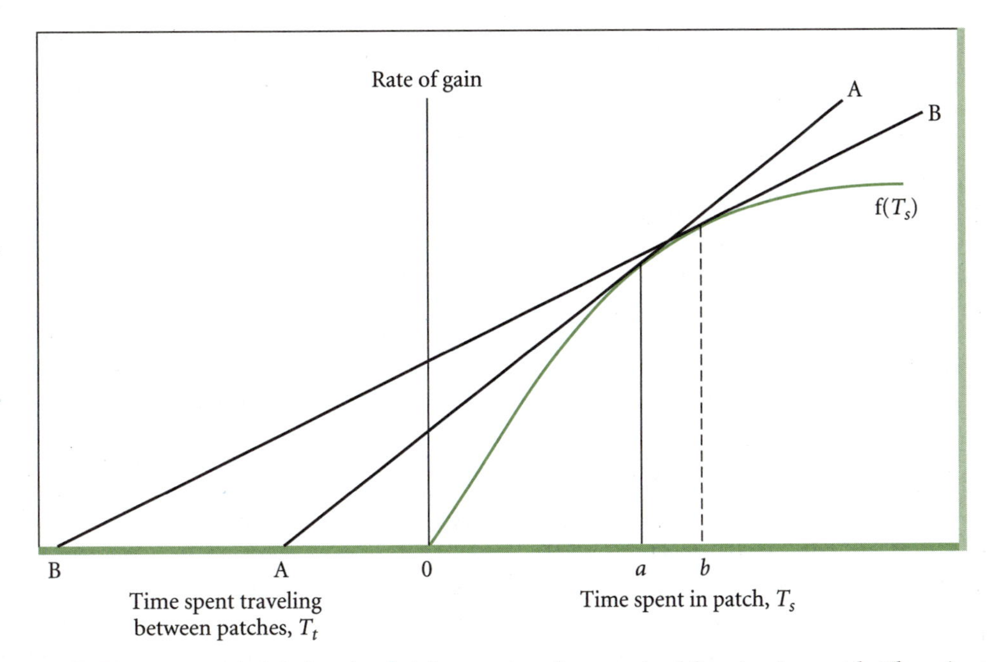

FIGURE 31-21 Model of optimal giving-up time for an animal foraging in a patch. The gain curve, f(T_s), for the current patch is given on the right side of the graph (green line). The portion of the horizontal axis to the right of zero represents time spent in the patch (T_s). The portion of the horizontal axis to the left of zero represents the time required to travel from the current patch to the next patch (T_t). The travel times, A and B, to two alternative patches are shown. If the travel time to the next patch is A, the model predicts that the forager should give up in the current patch after time *a*. If the travel time to the next patch is greater than A—say, B—then the forager should stay in the current patch longer (until time *b*), even though the rate of return is decreasing.

stay longer in the patch, as indicated by the greater optimum giving-up time *b*.

It is unlikely that different patches will be identical in the quality and quantity of prey they contain. The situation in which two patches have different gain curves is represented in Figure 31-22. If the travel time between the two patches is the same, T_t = A, then the predator will spend more time in the patch with the lower rate of gain than in the patch with the higher gain curve before moving to another patch.

Central-Place Foraging

Central-place foraging involves traveling from a central place to a patch containing food, loading up with food, and returning to the central place. One of the most studied of such foragers is the starling, *Sturnus vulgaris,* which can be easily trained and handled in a laboratory setting (Kacelnik 1984, Cuthill and Kacelnik 1990, Cuthill et al. 1990). When feeding their young, starlings fly from their nests, which are most often located in tree holes or nest boxes, to patches of grass, where they load their beaks with insects. The number of insects that a starling accumulates is related to the time that it spends searching for prey. In the graphic model above (see Figure 31-21), the vertical axis could be considered the load curve, which is a function of the bird's search time. The problem facing the starling is how big a load to take back to the nest in order to optimize the rate at which the young are fed. Application of the optimal patch use model above suggests that, as with the predator foraging in patches, if all patches have the same load curve, the optimum time

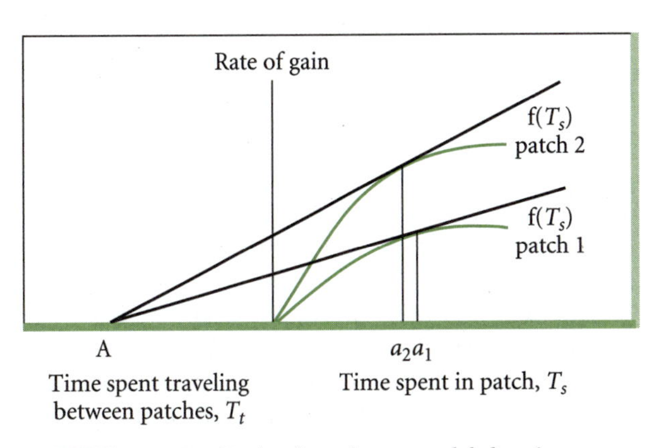

FIGURE 31-22 Optimal patch use model showing two patches with different gain curves. If the travel time to the next patch is A regardless whether the forager is in current patch 1 or 2, then the model predicts that it will stay in patch 1 longer than it will stay in patch 2 before moving to the next patch.

spent in the patch is determined by how far the patch is from the nest. The farther the bird has to fly to get back to the nest, the greater the load.

31.8 The ideal free distribution equalizes gains among individuals in a population.

The models of optimal choice that we have discussed to this point address the behavior of a single individual confronted with environmental heterogeneity. Frequently, however, many individuals face the same options, and optimal behavior may be influenced by the decisions of other members of the population. For example, a patch becomes less attractive to newcomers as the number of predators exploiting it increases. A patch containing other predators probably has fewer remaining resources than an undiscovered patch; furthermore, competing individuals may precipitate costly behavioral conflicts. Presented with many patches of varying intrinsic quality, consumers having complete knowledge and freedom of choice should occupy or exploit patches in direct proportion to their quality. Those with the highest levels of resources should attract the most consumers; those with the lowest levels, the fewest. Each consumer bases its choice of a patch on criteria that maximize its own rate of gain of resources.

Imagine two patches, one having more resources than the second. At first, consumers will choose the intrinsically better patch. But as the population of consumers builds up in that patch, its apparent quality decreases, owing to depletion of resources and antagonistic interactions, until the second patch becomes the better choice. At this point, additional consumers choose the second and first patches alternately as the quality of both continues to decrease. As a result, each individual in the population will exploit a patch of equal realized quality, regardless of the variation in intrinsic patch quality in the absence of consumers. This pattern is called the **ideal free distribution,** a term used by Stephen Fretwell and H. L. Lucas (1970) to describe the filling of different habitat patches by territorial species of birds.

The concept of the ideal free distribution has been applied to several phenomena, perhaps most successfully by G. A. Parker (1974, 1978), who examined the numbers of male dungflies (*Scatophaga*) that compete for mates on cowpats (also known as meadow muffins and cow pies) of different sizes, conferring different resource quality and attractiveness to females. The distribution of males among these patches of "habitat" approximated an ideal free distribution, in which males gained similar numbers of matings regardless of patch quality.

The achievement of an ideal free distribution has also been investigated in the laboratory, where the quality of patches can be controlled. For example, Milinski (1979) conducted experiments in which six stickleback fish were provided with food (water fleas) at different rates at opposite ends of an aquarium. Each end was regarded as a patch, and the system had the following conditions conducive to establishing an ideal free distribution: (1) the two patches differed in profitability; (2) profitability decreased as the number of fish using a patch increased; (3) the fish were free to move between patches, hence they had free choice. Hungry fish were placed in the aquarium about 3 hours before the start of the experiment. During trials, the numbers of fish in each half of the tank were recorded at 20-second intervals. Before any food was added to the tank, the fish were distributed equally between the two halves. In one experiment, water fleas were added at a rate of thirty per minute to one end of the tank and six per minute to the other, a ratio of 5 to 1. Within 5 minutes, the fish had distributed themselves between the two halves in the same ratio as predicted for an ideal free distribution. In a second experiment, water fleas were provided at rates of 30 and 15 per minute, a 2 to 1 ratio. Again the distribution of fish followed suit, and when the better and poorer patches were reversed in the tank, the fish reversed their distribution within about 5 minutes (Figure 31-23). The behavioral mechanisms the fish used to achieve an ideal free distribution were not

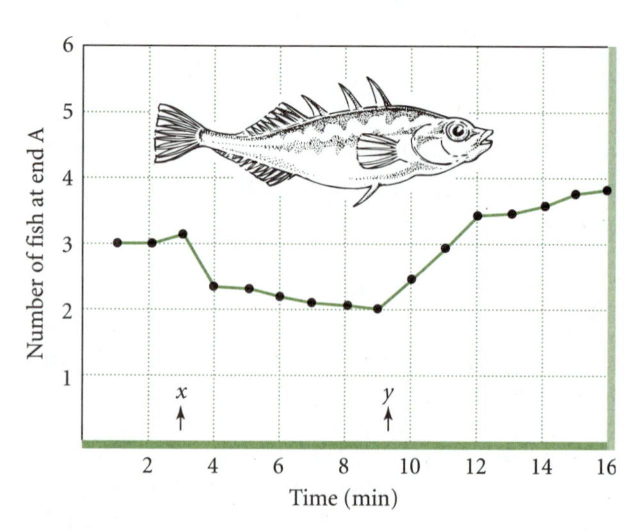

FIGURE 31-23 The number of sticklebacks at one end of an aquarium over time. At 3 minutes (point *x*), the amount of food in the other end of the aquarium was twice the amount of food in the end shown in the graph. At about 9 minutes (point *y*), the food abundance was reversed. The graph shows that fish moved to the side where there was more food. Each dot on the graph represents the mean of several experiments. (*From Krebs and Davies 1993; after Milinski 1979.*)

determined, but cues for behavioral choices must incorporate both the rate of food provisioning and the number of competitors within the patch.

Field studies have also demonstrated results consistent with the ideal free distribution hypothesis. Wahlström and Kjellander (1995) studied the distributions of female roe deer (*Capreolus capreolus*) in two habitats in Sweden: fields and forests. The amount of nitrogen in feces collected in the two areas was used as a measure of habitat quality (the higher the nitrogen content, the greater the quality). Telemetry was used to measure summer home range sizes of male deer, which estimated resource abundance (the smaller the home range, the greater the resource abundance). Fields were found to have resources of higher quality and abundance than forest areas. Deer could freely move between the two areas. Wahlström and Kjellander found higher population densities in the fields, consistent with the ideal free distribution hypothesis. Field observations such as this one and experimental results such as Milinski's demonstrate considerable sensitivity of organisms to conditions of their environment as well as behavioral flexibility in making choices.

31.9 Risk-sensitive foraging models focus on the minimization of the risk of starvation.

Consider the situation in which an animal may choose between two foraging strategies. It may choose to ensure that it will receive 10 grams of food from each foraging bout. On the other hand, it may adopt a strategy that yields 5 grams of food on one-half of the foraging bouts and 20 grams of food on the other half of the foraging bouts. In the first strategy, there is no variability (variance = 0) in the amount of food that the animal obtains. It is always 10 grams. In the second strategy, there is a variance in the amount of food that is obtained, but the average amount of food obtained over the long term will be greater than the 10 grams that can be expected if the animal adopts the first strategy. The first strategy is referred to as a **risk-averse** foraging strategy because the forager is willing to accept a lower mean gain in order to reduce the variation in the gain. We call the second strategy a **risk-prone** strategy because the forager is willing to accept higher variability in food intake in order to obtain a higher average gain. A body of theory called **risk-sensitive foraging theory** has arisen that considers the implications of these types of foraging choices (Caraco et al. 1980, Caraco et al. 1990, and see reviews by McNamara and Houston 1992, Krebs and Davies 1993, and Bulmer 1994).

Which of these two strategies should a foraging animal adopt? The answer clearly depends on what the benefit—sometimes referred to as utility—of each of the two strategies is in the context of the life of the forager. For example, suppose an animal has reached the end of a long day during which it has been unsuccessful in finding food. In order for the hungry animal to survive through the night without feeding, it must obtain 11 grams of food during the day. If the animal adopts the risk-averse strategy with the intake rates used above, it will be sure of obtaining food (10 grams to be exact), but the utility of that food will be zero, since the animal will die during the night. Risk-sensitive theory predicts that the animal will adopt the more risky strategy, which is the only one of the two that offers a chance of survival. In risk-sensitive foraging theory, the currency of interest is the risk of death.

31.10 Foraging animals may be at risk of predation.

So far we have suggested that the decisions that foraging animals make have to do with the rate of energy gain, the proximity of the next foraging location, and the mean and variance of the rate of intake. However, animals must do more than just forage. They must court, mate, and reproduce. They must build nests and provide food for their young. Many animals migrate long distances. And, of course, many animals must invest energy to ensure that they do not become prey themselves. All these concerns present animals with important trade-offs with respect to foraging. Should the animal forage or guard the nest? Should the animal forage or look for a mate? Should the animal forage or look about for an approaching predator?

Consider the risk of foraging under the specter of being eaten. Considerable empirical evidence demonstrates that foraging animals do not behave as predicted by the models that we have discussed when there is some threat that they may become prey themselves (Krebs and Kacelnik 1991). Gray squirrels (*Sciurus carolinensis*) distributed themselves according to the ideal free distribution when given a choice between a large tray of food and a small tray of food placed nearby. However, when the larger tray was moved into an open area, where, presumably, the squirrels were exposed to possible attack by predators, fewer squirrels foraged in that tray than predicted (Newman and Caraco 1987). Dog whelks (*Nucella lapillus*) that are allowed to forage in a tank with flowing water become inactive when they sense the scent of a crab, a potential predator, or the odor of a damaged dog whelk, presumably signaling that a predation event has taken place (Vadas et al. 1994). When given a choice between low- and high-quality seeds in field experiments, seed-eating field mice (*Peromyscus polionotus*) chose to eat lower-quality seeds in a safer location when in the presence of a predator (Phelan and Baker 1992).

A number of factors may affect the way in which foraging animals respond to the threat of predation. One of the most important of these is the level of hunger of the forager. Intuition tells us that a starving animal may be willing to take more risks to obtain food than one that has just eaten. Indeed, in the study of dog whelks mentioned above, Vadas et al. (1994) found that starved animals were less likely to become inactive in the presence of the odor of the crab. Many other studies have demonstrated that foraging animals may throw caution to the wind when in the grip of hunger (e.g., Milinski and Heller 1978, Krebs 1980, Dill and Fraser 1984).

31.11 Stochastic dynamic programming is a modeling approach that evaluates how short-term decisions contribute to lifetime fitness.

One of the important lessons from our discussion of risk-sensitive foraging and foraging under risk of predation is that the particular condition of a foraging animal and the variability of the environment have a great deal to do with what foraging decision the animal will make at a particular time. As an example, think of your own foraging behavior. You may not feel like eating a dozen doughnuts if you have just downed the double cheese special at the local burger barn. However, hours later, those doughnuts might look pretty good to you. And, of course, there is always uncertainty about the future. If you were offered the doughnuts right after lunch and you were pretty sure that you would not have time to eat dinner, you might make room for them. Similar changes of state affect the foraging behavior of animals. The relatively new technique of **stochastic dynamic programming** (SDP) is now commonly employed to analyze the effects of short-term behavioral decisions, which are influenced by the state of the individual, on the long-term fitness of the individual. Reviews by Houston et al. (1988) and Mangel and Clark (1988) should be consulted for a complete discussion of this approach. Our description of the technique below follows closely that presented by Mangel and Clark.

Think of the various activities of an organism as "patches." In this context, we use the concept of the patch as an abstraction denoting various choices of activities that an animal has, rather than a specific foraging location. Now, consider what may happen whenever an animal makes a choice. It may get some reward, such as food or the opportunity to mate. The activity no doubt will have some cost to the individual in the form of energy. Also, there may be some risk involved in the activity. Here we think of risk as the probability of being killed. To summarize, each choice has a potential gain, a cost, and a potential risk associated with it. As an animal goes through its life, it will make choice after choice, each choice having some benefits, some costs, and some risk. And the results of each choice affect the animal's state, which in turn affects the next choice it makes. Dynamic programming models address the way in which all of an animal's choices affect its lifetime fitness.

Consider the simple example given by Mangel and Clark (1988) of a nonreproductive animal that can forage in any of three patches (Figure 31-24). Suppose that we can determine the state (energy reserves) of the animal at time t. Since we expect the state of the animal to change over time, we may think of it as a function, $X(t)$, of time, t. Over some arbitrarily long period of time, T, we may think of $X(t)$ as changing with each discrete interval of time. The change in state from one time interval to the next—that is, the difference between $X(t)$ and $X(t + 1)$—is determined by the relative balance of benefit, cost, and risk in each time interval. For each choice in Figure 31-24 there is some probability of finding food, some cost of foraging in the patch, and some probability that the animal will be eaten or die by other means in the patch. The overall probability that the animal will survive from time t to the end of the arbitrarily long period T is given by what is referred to as the **lifetime fitness function.** The formulation and application of the lifetime fitness function is complex, and will not be covered here. The idea of stochastic dynamic programming is to understand how individual choices affect the lifetime fitness of an individual.

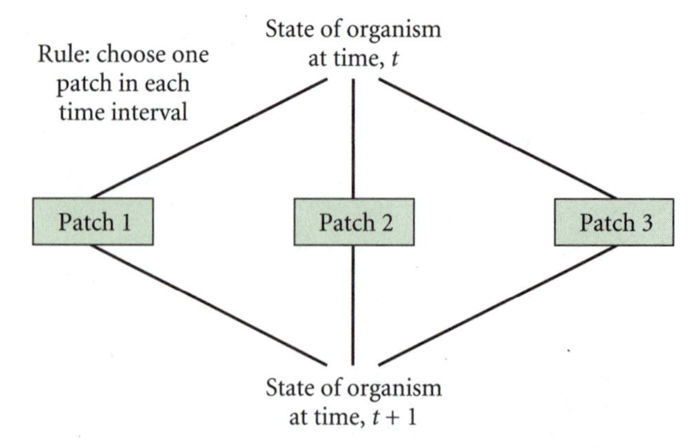

FIGURE 31-24 Schematic representation of the essential approach to dynamic programming. An organism at time t has a particular state, which changes depending on which of three choices it makes. Dynamic programming models determine how such choices, when made over a long period of time, affect the fitness of the individual. (*From Mangel and Clark 1988.*)

SUMMARY

1. The environment of an individual varies both temporally and spatially, and includes a diverse array of resources. Analyses based on the performance of organisms in different environments and upon different resources provide insights into the rules governing decisions about diet and habitat selection.

2. The adaptive landscape is a metaphor for the genetic composition of a population. Adaptive peaks are genotypes that are favored within the current environment. Natural selection moves populations upward on a peak. Changes in the environment may alter the adaptive landscape by changing which genotypes are favored.

3. Fitness set analysis, devised by Richard Levins, portrays phenotypes in a fitness space whose axes are component fitnesses in different habitat or resource conditions. The adaptive function provides a criterion by which one can identify the phenotypes in the fitness set having the highest fitness.

4. Polymorphisms may be maintained in populations by the process of frequency-dependent selection, whereby the fitness of each genotype is dependent on the frequency with which it occurs in the population.

5. The relationship between the phenotype of an organism and its environment is called the reaction norm. The general responsiveness of a phenotype to environmental variation is referred to as phenotypic plasticity. Reciprocal transplant experiments provide a means of determining whether differences between populations are due to evolutionary differentiation or to phenotypic plasticity.

6. Optimality theory focuses on the analysis of costs and benefits. Optimality models include three basic elements: a currency, constraints, and a decision variable. Optimal foraging theory predicts that organisms will behave so as to minimize the costs and maximize the benefits of foraging.

7. Optimal foraging models that predict how animals will manage their diet breadth are called prey models. Simple prey models predict that animals will specialize—that is, that they will have a lower diet breadth—when the productivity of the environment increases.

8. Simple models of prey choice based upon considerations of search time, handling time, and reward—so-called classic foraging models—predict that predators should specialize more when prey are abundant than when prey are scarce.

9. Models that predict how long a foraging animal should remain in a patch of food before moving to another patch are called patch use foraging models. The optimum time that a consumer should forage in a patch of resources before moving on to another patch is the optimal giving-up time. According to simple models, giving-up time increases as the initial level of resources within the patches and the traveling time between patches increase. A type of patch use model in which the forager sorties from a central location to obtain food is referred to as a central-place foraging model.

10. When many individuals choose among patches, assessment of patch quality is influenced by the presence of other, competing individuals. In theory, individuals should distribute themselves among habitat patches so as to equalize their fitnesses, resulting in what is called an ideal free distribution.

11. Risk-averse foragers accept a lower average gain in food in order to reduce variation in their food intake. Risk-prone foragers are those that accept high variation in their rate of food intake in order to have a larger average intake. Whether an animal is risk-averse or risk-prone depends in part on its current state.

12. Animals that forage under the risk of predation behave differently than predicted by foraging models. Starving animals are more likely to take risks to obtain food than well-fed animals.

13. The technique of stochastic dynamic programming attempts to analyze the effects of short-term behavioral decisions on the long-term fitness of the individual.

EXERCISES

1. The application of optimality theory to nonhuman organisms rests on the assumption that behaviors are selected to maximize expected fitness. What are the problems with applying such theory to human activities such as logging or fishing? (Hint: See Bulmer 1994 for a simple review.)

2. The profitability (E/T) for a predator feeding on two prey species, where λ_1 and λ_2 are the encounter rates of prey species 1 and 2 respectively, E_1 and E_2 represent the amount of energy obtained from each, and h_1 and h_2 are the respective prey handling times, is $E/T = (\lambda_1 E_1 + \lambda_2 E_2)/(1 + \lambda_1 h_1 + \lambda_2 h_2)$. It can be shown that the profitability of eating only prey type 1 is $E_1/T = \lambda_1 E_1/(1 + \lambda_1 h_1)$, and that of eating only prey type 2 is $E_2/T = \lambda_2 E_2/(1 + \lambda_2 h_2)$. If $\lambda_1 = 1$/min and $\lambda_2 = 3$/min; $E_1 = 1$ cal and $E_2 = 2$ cal; $h_1 = 1.5$ and $h_2 = 0.5$, should the predator eat both prey or specialize on one or the other?

CHAPTER 32

Evolution of Life Histories

GUIDING QUESTIONS

- What are the important components of life histories?
- What are some of the important trade-offs that are part of life history evolution?
- What are the trade-offs involved in the determination of age at first reproduction?
- What factors favor annual and perennial reproduction?
- What is the relationship between optimal reproductive effort and adult survival?
- What are the implications of different selective pressures on fecundity and survival rates at different ages?
- What is bet hedging?
- What is semelparity, and under what conditions might it evolve?
- What theories have been put forward to explain the evolution of senescence?
- What is the *r*- and *K*-selection spectrum?

Organisms are generally well suited to the conditions of their environments. Form and function vary in parallel with the ranges of temperature, water availability, salinity, oxygen, and other factors encountered by each species. We have seen how homeostatic mechanisms enable individuals to respond to temporal and spatial variation in their environments (Chapter 6). Whether the differences we observe among populations and species are evolved or result from individual responses to different environments (phenotypic plasticity; see Chapter 31), we presume that modification of form and function improves either survival or reproduction or both. That is, we assume that the evolutionary and individual modifications are adaptive and that they increase the fitness of individuals. When structure and function respond to prominent physical conditions in the environment, simple engineering principles often suggest how these modifications could improve fitness. From an engineering perspective, we can see why desert plants have small leaves with thick cuticles to reduce water loss, or why swift runners have long legs. Many other adaptations are equally straightforward. The close color matching of grasshoppers to their backgrounds, for example, makes sense when one understands that they are eaten by visually hunting predators.

Organisms have limited time, energy, and nutrients at their disposal. Adaptive modifications of form and function serve two purposes in this regard. One is to increase the resources available to individuals. The other is to use those resources to their best advantage, that is, in a manner that maximizes the survival and reproduction of individuals in their particular environmental settings. Every modification involves a **trade-off,** meaning that an increase in any one thing implies a decrease in another. If resources are limited, then the time, energy, or materials devoted to one structure or function cannot be allotted to another. Therefore, each individual is faced with the problem of **allocation:** given that resources are limited, how can the organism best use its time and resources?

Practical solutions to the allocation problem depend on how changes in any given structure or function affect fitness. When modification of a trait influences several components of survival and reproduction, as is often the case, the evolution of that trait can be understood only by considering the entire life strategy. For example, an increase in the number of seeds produced by a plant may contribute to fitness by increasing fecundity (number of offspring), but it may also reduce the survival of seedlings (if seed size is reduced to make more of them), the survival of adult plants (if resources are shifted from root growth to support increased seed production), or subsequent fecundity (if seed production in one year reduces the growth of a plant and therefore its size in subsequent years).

In this chapter, we shall consider some general rules governing the allocation of time and resources in the life strategies of plants and animals. From an evolutionary point of view, the object of life is to produce successful progeny—as many as possible. Reproduction involves choosing among options: when to begin to breed, how many offspring to have at one time, how much care to bestow upon them. The set of rules and choices pertaining to an individual's schedule of reproduction is referred to as its **life history** (Lessells 1991, Roff 1992, Stearns 1992). Each life history has many components, the most important of which concern **maturity** (age at first reproduction), **parity** (number of episodes of reproduction), **fecundity** (number of offspring produced per reproductive episode), and termination of life (senescence and programmed death). Each of these components affects other aspects of an individual's life. Because breeding takes time and resources from other activities and entails risks, investment in offspring generally diminishes the survival of parents. In many cases, rearing offspring drains the parent's resources so much that fewer offspring are produced later. Thus, an optimized life history represents a resolution of conflicts between the competing demands of survival and reproduction to the best advantage of the individual in terms of perpetuating its lineage.

32.1 Interest in life history adaptations has been stimulated by their variation among species.

During the early 1940s Reginald Moreau, a British ornithologist who had worked in Africa for many years, called attention to the fact that songbirds in the Tropics lay fewer eggs (two or three, on average) than their counterparts at higher latitudes (generally four to ten, depending on the species) (Moreau 1944). The study of life histories is mostly a legacy of one of Moreau's colleagues, David Lack, whose influence is ubiquitous in population biology and evolutionary ecology. Lack recognized that an increase in clutch size (number of eggs per nest) would increase the overall reproductive success of the parents, unless something reduced the survival of offspring in large broods (Lack 1944). He presumed that the ability of adults to gather food for their young was limited, and accordingly, that in broods having too many mouths to feed relative to the availability of food, chicks would be undernourished and survive poorly. Lack further suggested that because of the longer day length at higher latitudes in the season when offspring are reared, birds living at temperate and arctic latitudes could gather more food, and therefore rear more offspring, than birds breeding in the Tropics, where day length remains close to 12 hours year-round.

Lack made three important points: First, he related life history traits, such as fecundity, to reproductive success and thus to evolutionary fitness. Second, he demonstrated that life histories vary consistently with respect to factors in the environment, which suggested the possibility that life history traits are molded by natural selection. Third, he proposed a hypothesis that could be subject to experimentation. In the particular case of clutch size, Lack suggested that food supply limits the number of offspring that parents can rear. To test this idea, one could add eggs to nests to create enlarged clutches and broods. According to Lack's hypothesis, parents should be unable to rear added chicks because they cannot gather the additional food required. We shall return to this hypothesis below. Lack also noted that clutch size and day length are positively correlated; additional explanations for the increase in clutch size with latitude have been based on average temperatures, seasonal variation in temperatures and food supplies, and rates of predation on bird nests, all of which vary consistently with latitude.

Variation in Life Histories

Some life history traits, such as fecundity and mortality, tend to vary in close association (Figure 32-1). At one extreme, elephants, albatrosses, giant tortoises, and oak trees exhibit long life, slow development, delayed maturity, high parental investments, and low reproductive rate; at the other extreme, one finds mice, fruit flies, and weedy plants (Pianka 1970). In broad comparisons within the plant and animal kingdoms, such associations of traits vary in close relation to body size and undoubtedly reflect the relative slowness of all life processes in large organisms (Peters 1983, Calder 1984, Schmidt-Nielsen 1990). But even among organisms of similar size and body plan, different environments produce widely divergent life

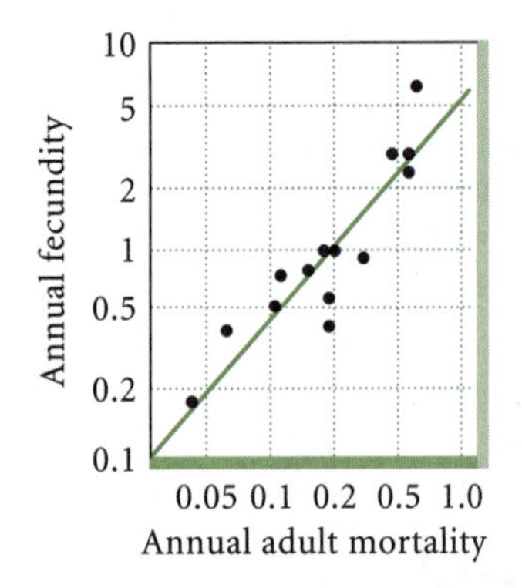

FIGURE 32-1 Relationship between annual fecundity and adult mortality in several populations of birds ranging from albatrosses (low values) to sparrows (high values). (*Data from Ricklefs 1977b.*)

histories. Storm petrels, which are seabirds the size of thrushes, rear at most a single chick each year, do not begin to reproduce until 4 or 5 years of age, and may live to 30 or 40 years. Thrushes themselves may produce several broods of three or four young each year beginning with their first birthday, but rarely live beyond 3 or 4 years. Similarly varied life histories may be found even among different populations of the same species, as illustrated by fence lizards (Table 32-1) and other organisms (e.g., Tinkle and Ballinger 1972, Schaffer and Elson 1975, Leggett and Carscadden 1978, Jerling 1988, Fleming and Gross 1989, Sand 1996).

Beyond the strong associations of life history traits with environmental conditions, many taxonomic groups also exhibit characteristic values of life history adaptations. Ducks usually lay eight to ten eggs per clutch, shorebirds four, hummingbirds two, and petrels one. These differences probably reflect ways in which taxonomically conservative traits affect the selection of habitats by organisms and their particular interactions with the environment. Ducklings can feed themselves, and so their number is not limited by the ability of parents to gather food for them (what does limit their number remains a mystery; Rohwer 1985, Lessells et al. 1989). Shorebirds typically lay very large eggs whose number may be limited by the ability of parents to incubate them successfully. The taxonomic affinity of a trait, however, does not reveal, for example, the significance of large egg size to shorebirds (why don't other birds lay large eggs?) or why shorebirds do not modify nest structure (usually a shallow depression in the ground) to accommodate more eggs (ducks incubate much larger egg masses than shorebirds).

Life History Variation in Plants

The English plant ecologist J. P. Grime emphasized the relationship between the life history traits of plants and certain conditions of the environment. He envisioned variation in life history traits between three extreme apexes, like points of a triangle, and called plants with life histories at these extremes **stress tolerators, ruderals,** and **competitors.** As the name implies, stress tolerators grow under extreme environmental conditions, growing slowly and conserving resources. Because seedling establishment is difficult in stressful environments, vegetative spread is emphasized. Where conditions for plant growth are more favorable, ruderals and competitors occupy opposite ends of a spectrum of disturbance. Ruderals colonize disturbed patches of habitat, where they exhibit rapid growth, early maturation, high reproductive rates, and easily dispersed seeds, which enable them to reproduce quickly and disperse their progeny to other disturbed sites before being overgrown by superior competitors. Plants with the competitor suite of life

TABLE 32-1 Life history traits of several populations of the fence lizard *Sceloporus undulatus*

	Location							
Trait	Arizona	Utah	Colorado	New Mexico	Kansas	Texas	Ohio	South Carolina
Clutch size	8.3	6.3	7.9	9.9	7.0	9.5	11.8	7.4
Clutches per year	3	3	2	4	1–2	3	2	3
Egg weight (g)	0.29	0.36	0.42	0.24	0.26	0.22	0.35	0.33
Relative clutch mass	0.22	0.21	0.23	0.21	0.28	0.27	0.25	0.23
Age at maturity (months)	12	23	21	12	12	12	20	12
Survival to breeding	0.07	0.05	0.11	0.03	0.10	0.06	0.03	0.11
Annual adult survival	0.24	0.48	0.37	0.20	0.27	0.11	0.44	0.49

(From Tinkle and Dunham 1986.)

TABLE 32-2 Typical life histories of plants in environments with different selective factors

Competitors	Ruderals	Stress tolerators
Herbs, shrubs, or trees	Herbs, usually annuals	Lichens, herbs, shrubs, or trees; usually evergreen
Large, with a fast potential growth rate	High potential growth rate	Potential growth rate slow
Reproduction at a relatively early age	Reproduction at an early age	Reproduction at a relatively late age
Small proportion of production to seeds	Large proportion of production to seeds	Small proportion of production to seeds
Seed bank sometimes, vegetative spread often, important	Seed bank and/or highly vagile seeds	

(From Grime 1979.)

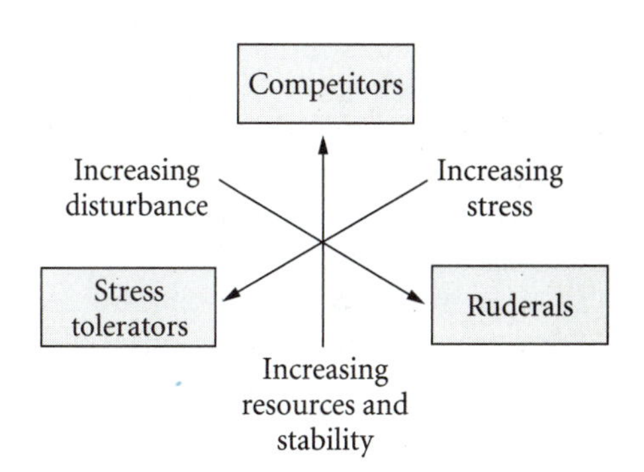

history traits tend to grow to large stature, mature at a large size, and exhibit long life spans. The competitor life history therefore requires stable, favorable conditions for its success. Some typical life histories of plants in environments with different selective factors are presented in Table 32-2.

32.2 Life history theory developed rapidly during the 1960s.

The history of evolutionary ecology begins with Charles Darwin. However, with regard to the types of life history traits discussed in this chapter, Darwin was uncharacteristically silent. He recognized the great reproductive capacity of all living beings, but regarded the environment as an agent of selective mortality, undiscriminating with regard to birth rate.

Darwin did not have the benefit of life table analysis to guide his thinking. Even its development by A. J. Lotka and others, and its application to evolutionary phenomena by population geneticist Ronald Fisher in his classic book *The Genetical Theory of Natural Selection* (1930), did not generate enough interest to launch a new field of inquiry. In fact, the next major step toward the establishment of life history as an important focus for ecologists was the publication of Moreau's and Lack's papers on clutch size in birds during the mid-1940s. Although these papers attracted considerable attention, they still did not spark the flame that was to flare two decades later. In part, Lack's theory was too simplistic; it isolated clutch size (or reproductive rate more generally) as a single adaptation unrelated to other aspects of the phenotype. The ability of adults to deliver food to their offspring was accepted as determined by food availability in the environment. Neither the effort devoted to gathering food nor the time and effort devoted to caring for the young entered into the equation for fitness. Indeed, Lack's approach was decidedly nonquantitative. In addition, population biology during the late 1940s and most of the 1950s was embroiled in a controversy over the role of density dependence in the regulation of population size (see Chapter 16). Absorbed as ecologists were by the "balance of nature" (Egerton 1973), their attention was diverted from inquiry into life histories.

The early 1960s marked a turning point in population studies and saw the birth of modern evolutionary ecology. It is sometimes difficult to know what forces urge a discipline in one direction or another. The year 1959 was the centennial of the publication of *On the Origin of Species.* Dover Publications reprinted Fisher's *The Genetical Theory of Natural Selection* in 1958. At that time, George Williams, at the State University of New York at Stony Brook, was pondering the adaptive bases for the evolution of senescence (Williams 1957) and of insect societies (Williams and Williams 1957). In 1960, papers were published by A. W. F. Edwards, W. A. Kolman, and H. Kalmus and by C. A. B. Smith on the adaptive significance of the 1:1 ratio of males to females in most populations, a topic not touched since Fisher's treatment in 1932. This period saw the reunification of ecology and evolution.

Life history study burst into this arena in 1966 with the publication of papers by Martin Cody and George C. Williams. Cody made two points of lasting significance. First, he applied Levins's ideas about fitness sets (see Chapter 31) to life history evolution, calling attention to the fact that different components of fitness may be under conflicting selective pressures. Adaptation, he said, is largely the making of compromises in the allocation of time and energy to competing demands. Second, Cody introduced the idea that different life history adaptations are favored under conditions of high and low population densities relative to the carrying capacity of the environment. At high densities, selection favors adaptations that enable individuals to survive and reproduce with few resources; hence efficiency carries a premium. At low densities, adaptations promoting rapid population increase are selected; hence high rates of productivity, regardless of efficiency, increase fitness. These contrasting strategies were referred to as **K-selected** and **r-selected** traits, respectively, after the variables of the logistic equation for population growth (Boyce 1984).

Williams's (1966) paper explored the demographic coupling among life history traits. He pointed out that each increment of reproductive effort influenced both contemporary fecundity and survival to reproduce in the future. Williams quantified present and future components of fitness in the uniform currency of reproductive value, based on life table calculations. He showed how the conflict between the effects of a life history modification on present and on future reproduction was resolved according to the relative values of present offspring and the expectation of future offspring. In a population in which individuals have a high probability of survival between breeding seasons, the expectation of future reproduction has a high value, and selection tends to diminish reproductive investment in favor of protecting an otherwise high survival rate. In a population in which most individuals die before the next breeding season, selection favors a high investment in the current crop of offspring.

W. D. Hamilton (1966) provided a theoretical contribution showing explicitly how variation in each of the life table entries—that is, fecundity and survival at each age—results in variation in fitness. This result and other related derivations formed the mathematical basis for the subsequent development of most life history theory. Papers by M. Gadgil and W. H. Rossert (1970), showing that the outcome of selection depended on the form of the relationship among life table values determined by particular life history adaptations, and by G. Murphy (1968) and W. M. Schaffer (1974), raising the issue of life history evolution in a variable environment, essentially completed the groundwork.

32.3 Natural selection adjusts the allocation of limited time and resources among competing demands.

Ecologists believe that observed life histories represent the best resolutions of conflicting demands on organisms. Therefore, a critical effort in the study of life histories has been to understand the allocation of limited time and resources to competing functions. An in-

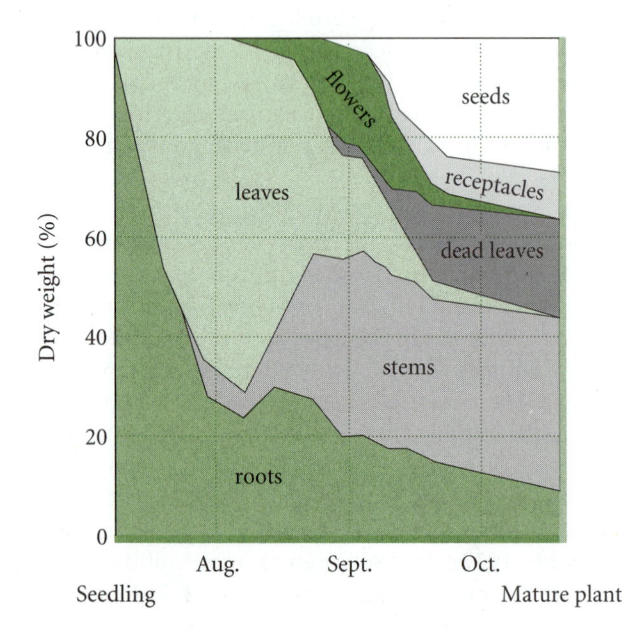

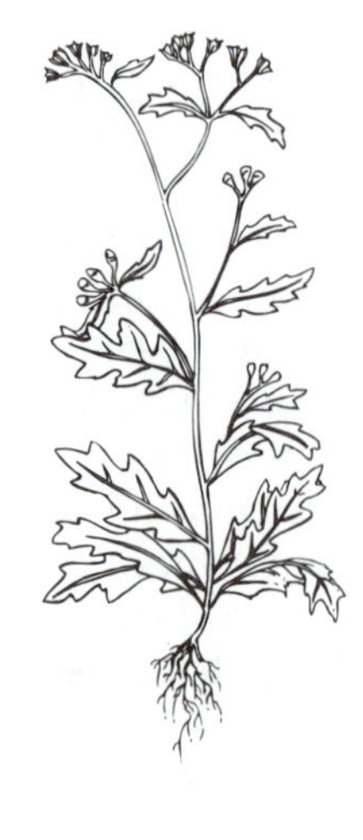

FIGURE 32-2 Proportional distribution of dry weight among different plant parts of the groundsel, *Senecio vulgaris* (Compositae), during its life cycle. Note the development of reproductive parts at the expense of leaves and roots toward the end of the growing season. (*After Harper 1967.*)

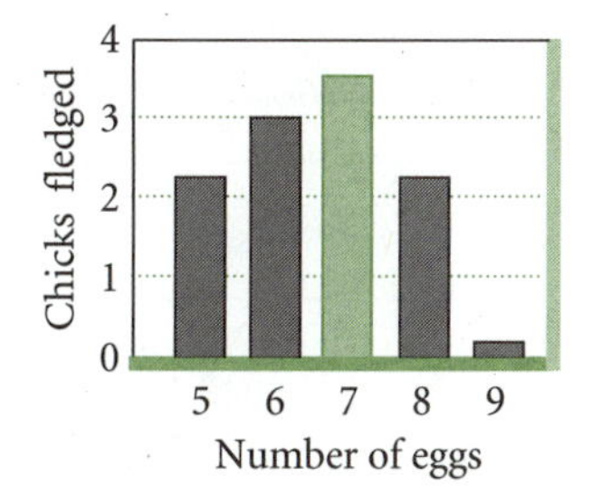

FIGURE 32-3 Number of chicks fledged from nests of European magpies in which seven eggs were laid, but the experimenter added or removed eggs to make up manipulated clutches of between five and nine eggs. The most productive clutch size was seven eggs. (*After Hogstedt 1980.*)

crease in any one function requires a shift in the allocation of time, energy, and nutrients. Because these are limited in supply, any increase in one component of demand, such as that created by reproduction, results in a decrease in allocation to some other component of demand. Time spent searching for food, for example, cannot be applied to caring for offspring or watching for predators.

Energy and nutrients used for growth cannot be earmarked for reproduction. In many species, eggs are produced in direct proportion to the size of the gonads, seeds in direct proportion to flower number. Therefore, when growth is shifted from reproductive structures to other parts of the body, fecundity also decreases. Photosynthetic rate depends in part on how much of its production a plant has allocated to photosynthetic tissues at the expense of root and support tissues (Figure 32-2). Plants also are built upon a modular body plan. In some species, nodes at points of leaf attachment may produce either lateral branches or flowers, but not both, thus trading off growth against reproduction.

In spite of the presumed pervasiveness of trade-offs in life histories, demonstrating their existence has proved difficult because this requires controlled experimental manipulation of individual components of the phenotype. Adding and subtracting eggs in the nests of birds has, in many cases, revealed an inverse relationship between the number of chicks in a nest and their survival. As a result, production of offspring is often greatest from clutches of intermediate size, as predicted by David Lack. For example, Hogstedt (1980) showed that the clutch laid by an individual female magpie produced the maximum number of chicks that she could nourish. Either adding or subtracting eggs resulted in fewer offspring fledged (Figure 32-3). In some species, however, the most productive clutch size is larger than the commonest clutch size observed in a population (Boyce and Perrins 1987; Figure 32-4). To explain such results, we may hypothesize that rearing a larger clutch has negative effects on the parents. There is also a possible trade-off between the size of eggs and the number of eggs produced. Producing a larger egg is simply a way for a female to allocate more resources to her offspring prior to hatching. A few well-provisioned eggs may, in some circumstances, result in higher offspring survival than producing many, smaller, eggs. This trade-off has been difficult to demonstrate, however (Lessells et al. 1989).

In parasitoids, the trade-off between the number and fitness of offspring is made prior to the development of the young when the female chooses a host and decides how many eggs to lay in it. The size and

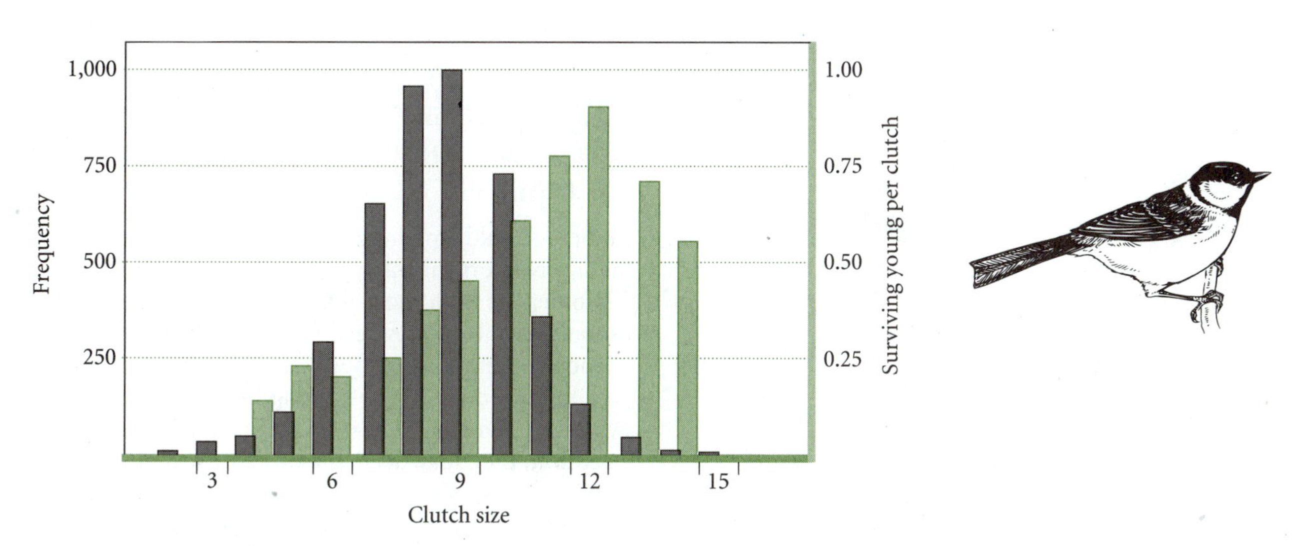

FIGURE 32-4 The frequencies of clutch sizes (gray bars, left scale) among 4,489 clutches of the great tit (*Parus major*) near Oxford, England, between 1960 and 1982, and number of young per clutch surviving at least to the next season (green bars, right scale) as a function of clutch size. Note that the most common clutch size is not the most productive. (*From Boyce and Perrins 1987.*)

age of hosts are known to affect the survival of parasitoid larvae to the adult stage, the size and fecundity of the parasitoid, and the time that it takes the parasitoid to develop (Godfray 1994). Since the host represents a limited pool of resources, the number of eggs laid in or on the host will also affect offspring fitness. The bruchid beetle *Callosobruchus maculatus* lays eggs in black-eyed beans, which provide resources for the development of the larvae (here the bean seed is the host of the parasitoid). The survival of larvae to adulthood decreases as the number of larvae per bean increases (Figure 32-5). In this species, lifetime egg production is related to the weight of the adult at emergence, which has been shown to decrease as the number of larvae per bean increases (Wilson 1989, Lessells 1991).

Other manipulations have failed to demonstrate trade-offs between life history traits. For example, in a study of guppies by Reznick (1983), the resources allocated to reproduction were experimentally reduced by preventing females from mating with males. If growth and reproduction compete for allocation of assimilated resources, the experimental fish should have attained larger sizes by the end of the study period. In fact, little of the difference between mated and unmated females in reproductive tissue accumulated was converted to growth (Figure 32-6). These results suggest several interpretations. Possibly, food availability does not limit reproduction in guppies. Alternatively, genetic factors may determine the growth of females at a rate that corresponds to the expected availability of resources (guppies normally do not abstain from mating).

Most issues concerning life histories can be phrased in terms of three questions: When should an individual begin to produce offspring? How often should it breed? How many offspring should it attempt to produce in each breeding episode? The ways in which different species have answered these questions express in different ways the fundamental trade-off between fecundity and adult growth and survival—that is, between present and future reproduction. We shall take up these questions in the next several sections.

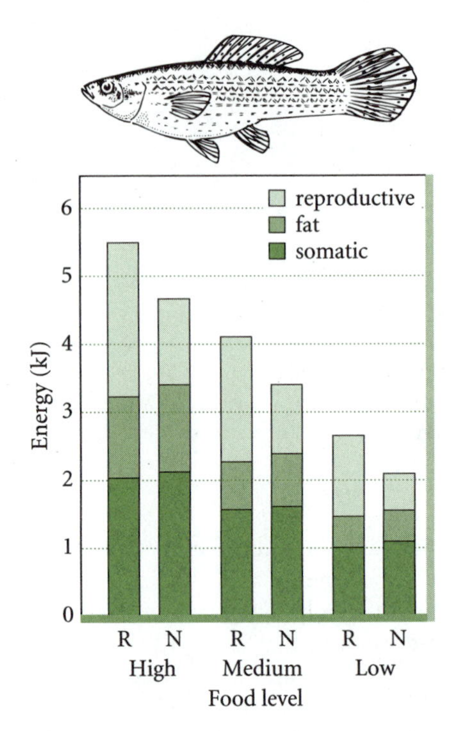

FIGURE 32-6 Content of energy in somatic tissues, fat deposits, and reproductive tissues (including eggs) in female guppies raised at three food levels and which were either permitted to be (R), or prevented from being (N), courted and inseminated by males. (*From Reznick 1983.*)

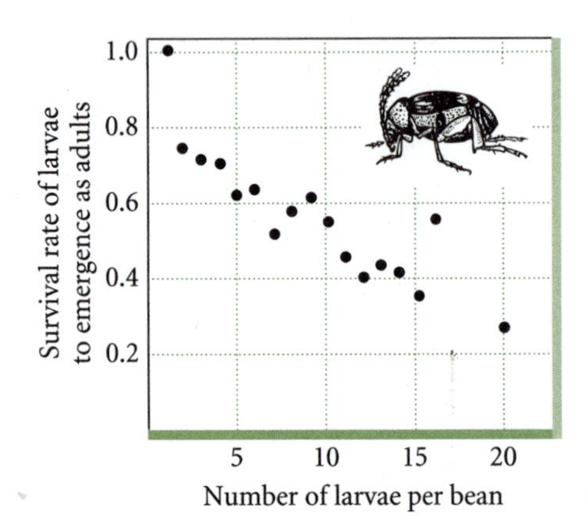

FIGURE 32-5 Decrease in the rate of larval survival to emergence with increasing numbers of larvae deposited per bean in the bruchid beetle *Callosobruchus maculatus.* (*From Lessells 1991; data from Wilson 1989.*)

32.4 Age at first reproduction generally increases in direct relation to adult life span.

When should an animal or plant begin to breed? Long-lived organisms typically begin to reproduce at an older age than short-lived ones (Figure 32-7). Why should this be so? At every age, an individual must choose, whether consciously or not, between attempting to reproduce and abstaining from breeding. When young individuals resolve this choice in favor of abstention, they may delay the onset of sexual maturity. Thus we may understand age at first reproduction in terms of the benefits and costs of breeding at a particular age. The benefit appears as an increase in fecundity at that age. Costs may appear as reduced survival to older ages or reduced fecundity at older ages, or both.

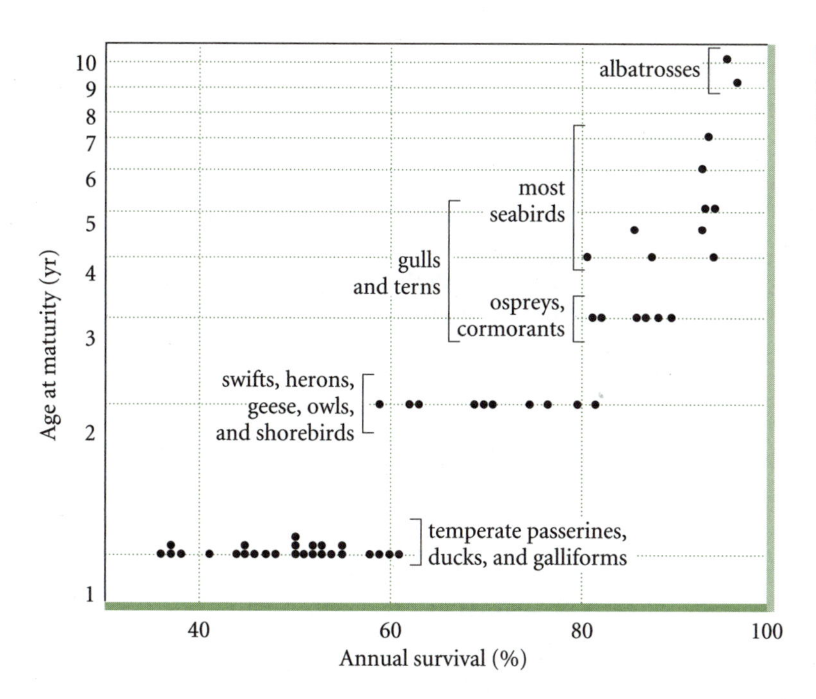

FIGURE 32-7 Relationship between age at maturity and annual adult survival rate, which is directly proportional to life span, in a variety of birds. (*From Ricklefs and Farner 1973.*)

Consider the following hypothetical example. A type of lizard continues to grow only until it reaches sexual maturity. Its fecundity is directly proportional to body size at maturity. Suppose that number of eggs laid per year increases by ten for each year that an individual delays reproduction, so that individuals breeding in their first year produce ten eggs that year and the same number each year thereafter. Individuals first breeding in their second year produce twenty eggs per year, individuals first breeding in their third year produce thirty eggs per year, and so on. Comparing the cumulative egg production of early- and late-maturing individuals (Table 32-3) reveals that the age at maturity that maximizes lifetime reproduction varies in direct proportion to the life span. For a life span of 3 years, maturing in the first year results in the production of a total of thirty eggs during the lifetime of the individual (ten in each of the 3 years). Maturing in the third year also results in the production of thirty eggs (with all thirty produced in one year). However, maturing in the second year results in a total of forty eggs (twenty in the second year and twenty in the third year). When the life span is 7 years, 4 years is the best age to mature.

For organisms that do not grow after their first year (most birds, for example), the choice between breeding or not depends on balancing current reproduction against survival. Nonbreeding individuals avoid the risks of courtship, nest building, or migration to breeding areas. Life experience gained with age may also reduce the risks of breeding, or increase the realized fecundity of a certain level of parental investment, favoring delayed reproduction. Among birds, age at maturity varies directly with the annual survival

TABLE 32-3 **Cumulative egg production by individuals in a hypothetical population as a function of life span and age at first reproduction**

Age at first reproduction (years)	Life span (years)							
	1	2	3	4	5	6	7	8
1	**10***	20	30	40	50	60	70	80
2	0	**20**	**40**	**60**	80	100	120	140
3	0	0	30	**60**	**90**	**120**	150	180
4	0	0	0	40	80	**120**	**160**	**200**
5	0	0	0	0	50	100	150	**200**
6	0	0	0	0	0	60	120	180

**Bold type indicates the most productive ages at first reproduction for a given life span.*

rate of adults, up to about 10 years in certain long-lived seabirds. Tending to offset the advantages of delayed reproduction are many factors that reduce the expectation of future reproduction. These factors include high predation rates, encroaching senescence at old age, and, for organisms that live a single year or less in seasonal environments, the end of the reproductive season.

32.5 Perennial life histories are favored by high and relatively constant adult survival.

Organisms either reproduce during a single season and die, a habit referred to as **annual** reproduction, or have the potential to reproduce over a span of many seasons, called **perennial** reproduction. Population biologists have pondered the relative advantages of each habit in terms of the trade-off between survival probability and fecundity. To survive the nonreproductive winter period, a perennial plant must allocate resources to the storage of materials in roots and the formation of freeze-resistant or drought-resistant buds, presumably at the expense of reproduction. The advantages of the perennial habit must therefore outweigh the costs of reduced fecundity relative to that of the annual habit. The pertinent consideration is the ratio of survival of adult plants to that of immature plants. When few individuals survive from one breeding season to the next, or when individuals produce relatively few seeds, the annual habit is favored. When individuals have a high probability of survival once they are established, but seedlings survive poorly (a high ratio of adult to immature survival), an annual must have many seeds in order to outproduce a perennial. Thus, annuals tend to dominate the floras of deserts, where few adult plants can survive long drought periods, and perennials tend to dominate tropical floras, where competition and predator pressure make establishment of seedlings difficult.

Following the earlier lead of Lamont Cole (1954), Charnov and Schaffer (1973) compared annuals and perennials in an algebraic model. Suppose a population of plants contains some individuals that produce a large number of seeds at the end of their first growing season and then die (annuals), and others that produce fewer seeds, but survive through the winter to reproduce in subsequent growing seasons (perennials). Which has the greater fitness? For the purposes of their model, Charnov and Schaffer assumed that annual and perennial plants have the same probability of survival during their first (in the annual's case, only) growing season (S_0), and that perennials have a constant probability of survival thereafter (S_p).

The factor by which a population of an annual plant grows (λ_a) equals the number of seeds each individual produces (B_a) times their survival to reproductive age (S_0), or

$$\lambda_a = B_a S_0.$$

The increase of a population of a perennial plant equals the number of seeds (B_p) times their survival (S_0), plus the probability of survival of the parent (S_p), hence

$$\lambda_p = B_p S_0 + S_p.$$

The population growth rate of the annual exceeds that of the perennial ($\lambda_a > \lambda_p$) when $B_a S_0 > B_p S_0 + S_p$. Dividing both sides of the inequality by S_0, we obtain $B_a > B_p + S_p/S_0$, or, by rearranging,

$$B_a - B_p > \frac{S_p}{S_0}.$$

Accordingly, an annual life history is favored when the number of seeds produced by the annual exceeds the fecundity of the perennial by the ratio S_p/S_0. When few perennials survive from one breeding season to the next, or when perennials produce relatively few seeds, the annual habit is favored. When individuals survive well once established, but seedlings survive poorly (high S_p/S_0), the annual habit must result in extremely high fecundity to be favored. Adding complexity to the model by incorporating growth from year to year and annual variation in survival probabilities does not destroy the basic qualitative conclusion that the life history strategy is determined primarily by the ratio of adult to juvenile survival (Stearns 1976, Bulmer 1985).

32.6 Optimal reproductive effort varies inversely with adult survival.

For annual plants, expectation of life beyond the first breeding season is so small that all resources are devoted to current reproduction. Perennials, however, must allocate resources between current reproduction and adaptations that prolong life. When particular adaptations affect both fecundity and survival, the trade-off between the two must be optimized. Intuitively, when life span is short, regardless of the consequences of reproduction, the balance of allocation should tip in favor of current fecundity. When potential life span is great, current fecundity should not unduly jeopardize future reproduction. This can be shown algebraically quite simply. Let the rate of population growth for perennials be $\lambda_p = B_p S_0 + S_p$, as above. If we partition adult survival into two components, one directly related to reproduction (S_R) and the other independent of reproduction (S), the rate of growth may be expressed in the following way:

$$\lambda_P = B_P S_0 + S_P S_R.$$

Certain reproductive traits that cause small changes in the values of survival (ΔS_R) and fecundity (ΔB) will influence the number of descendants as follows:

$$\Delta\lambda_P = S_0 \Delta B_P - S_P \Delta S_R.$$

When changes that enhance fecundity (ΔB_P positive) also reduce survival (ΔS_R negative), their effects on $\Delta\lambda_P$ depend on the relative values of S_P and S_0. In general, when S_P is large compared with S_0, selection favors adaptations that increase adult survival at the expense of fecundity, and vice versa. Thus one expects parental investment in offspring to decrease with increasing adult life span.

The problem of finding a balance between growth and reproduction is not exclusive to perennials, nor is it restricted to plants. Consider the paper wasps that commonly build nests on the eaves of your house or, annoyingly, the ceiling of the front porch or the garage closet. These nests are annual colonies founded in the spring by a single queen, who overwintered fertilized, and from which a number of winged reproductive males and females exit some months later. During the spring and summer, the queen produces nonreproductive workers that assist her in developing the colony. A critical choice for the queen is when to switch from making nonreproductive workers to making reproductives in order to maximize the number of reproductives produced. Stochastic dynamic programming approaches (see Chapter 31) have suggested that, when the end of the season is predictable, an abrupt switch is favored over a more graded change (Macevicz and Oster 1976, Alexander 1982, Bulmer 1994).

Many plants and invertebrates, as well as some fishes, reptiles, and amphibians, do not have a characteristic adult size. They grow, at a continually decreasing rate, throughout their adult lives, a phenomenon referred to as **indeterminate growth.** Fecundity is directly related to body size in most species with indeterminate growth. Because egg production and growth draw upon the same resources of assimilated energy and nutrients, increased fecundity during one year must be weighted against reduced expectation of fecundity in subsequent years. For organisms having longer life expectancies, growth should be favored over fecundity during each year. For organisms with less chance of living to reproduce in future years, resources allocated to growth instead of to eggs are largely wasted.

Consider two hypothetical fish species, each weighing 10 grams at sexual maturity, but which allocate resources to growth and reproduction differently. Both gather enough food each year to reproduce their weight in new tissue or eggs. Fish A allocates two-tenths of its production to growth and eight-tenths to eggs, whereas fish B allocates one-half each to growth and eggs. Calculations of growth, fecundity, and cumulative fecundity (Table 32-4) show that for fish living 4 or fewer years, on average, high fecundity and slow growth give the greater overall productivity, whereas

TABLE 32-4 Numerical comparison of the strategies of slow growth/high fecundity and rapid growth/low fecundity in two hypothetical fish species

	Year					
Characteristic	1	2	3	4	5	6
Slow growth/high fecundity						
Body weight	10	12	14.4	17.3	20.8	25.0
Growth increment	2	2.4	2.9	3.5	4.2	5.0
Weight of eggs	8	9.6	11.5	13.8	16.6	20.0
Cumulative weight of eggs	8	17.6	29.1	42.9	59.5	79.5
Rapid growth/low fecundity						
Body weight	10	15	22.5	33.8	50.7	76.1
Growth increment	5	7.5	11.3	16.9	25.4	38.1
Weight of eggs	5	7.5	11.3	16.9	25.4	38.1
Cumulative weight of eggs	5	12.5	23.8	40.7	66.1	104.2

Note: *All weights in grams. Body weight + growth increment = next year's body weight. Cumulative weight of eggs to last year + weight of eggs = cumulative weight of eggs to this year. Growth increment and weight of eggs in each year are equal to the body weight.*

for fish living longer than 4 years, more rapid growth and lower fecundity is the superior strategy. Adult mortality, therefore, determines the optimum allocation of resources between growth and reproduction.

32.7 When survival and fecundity vary with age, models of life history evolution must be based on the life table.

In the preceding models, we assumed that fecundity and adult survival were constant values, unvarying over time. In reality, however, among animals and plants that reproduce repeatedly, rates of survival and fecundity vary with age. Survival and fecundity are influenced both by the environment and by genetic factors. Thus, to some extent, natural selection acts on fecundity and survival with different intensities at different ages. We can use life tables to examine these differences.

Recall from Chapter 14 that life tables provide age-specific values for fecundity (b_x), survival rate (s_x), the proportion of individuals that survive to a particular age x, called survivorship (l_x), and other characteristics such as life expectancy (see Table 14-4). We may calculate the rate of growth of a population from the life table (assuming a stable age distribution) by using the characteristic equations (15-3) and (15-4) (see Chapter 15). Theoretical derivations from these equations allow ecologists to see how small changes in survival rate, s_x, or fecundity, b_x, affect population growth rates (Hamilton 1966, Emlen 1970). Those individuals of a particular age in the population that have high survival or fecundity will have greater reproductive success and will thus be favored by selection. The presence of such individuals will also increase the population growth rate. However, increases in survivorship or fecundity in individuals that are older will not be selected as strongly. For example, if 50% of individuals survive to age 1 and only 25% to age 2, then selection on change in fecundity at age 1 would be twice as strong as that on the same change in fecundity at age 2. A change in survivorship or fecundity at a younger age has a greater effect on reproductive success because more reproductive potential is at stake.

In general, the strength of selection on fecundity at a given age is proportional to the number of individuals surviving to that age (l_x). Thus, selection will be stronger at younger ages at which survivorship is high. The strength of selection on a change in survival rate also decreases with age. We may see the implications of this by examining a life table for a hypothetical population in which $s_x = 0.5$ and $b_x = 1.0$ for all x (note that $b_0 = 0$ is an exception) (Table 32-5). The exponential growth rate of the population, r, calculated from the characteristic equation (see Chapter 15), is $r = 0.103$, indicating about a 10% increase per year.

Now consider three cases in which either survival rate or fecundity is changed. In case 1, a mutation increases the fecundity of individuals in age classes 1 and 2 from 1.0 to 1.1. This change increases the popu-

TABLE 32-5 Life table for a hypothetical population without and with senescence

	Nonsenescing population					Case 1		Case 2		Case 3	
x	s_x	l_x	b_x	l_xb_x	xl_xb_x	s_x	b_x	s_x	b_x	s_x	b_x
0	0.5	1.000	0.0	0.000	0.000	0.5	0.0	0.5	0.0	0.5	0.0
1	0.5	0.500	1.0	0.500	0.500	0.5	1.1	0.5	1.1	0.5	1.1
2	0.5	0.250	1.0	0.250	0.500	0.5	1.1	0.5	1.1	0.5	1.1
3	0.5	0.125	1.0	0.125	0.375	0.5	1.0	0.5	1.0	0.5	1.1
4	0.5	0.063	1.0	0.063	0.252	0.5	1.0	0.5	1.0	0.4	1.1
5	0.5	0.031	1.0	0.031	0.155	0.5	1.0	0.5	0.9	0.4	1.1
6	0.5	0.016	1.0	0.016	0.090	0.5	1.0	0.5	0.9	0.4	1.1
7	0.5	0.008	1.0	0.008	0.056	0.5	1.0	0.5	0.9	0.3	1.1
8	0.5	0.004	1.0	0.004	0.032	0.5	1.0	0.5	0.9	0.3	1.1
9	0.5	0.002	1.0	0.002	0.018	0.5	1.0	0.5	0.9	0.3	1.1
10	0.5	0.001	1.0	0.001	0.010	0.5	1.0	0.5	0.9	0.3	1.1
	Exponential growth rate (r)			0.103		0.123		0.118		0.107	

(From Ricklefs and Finch 1995.)

lation growth rate from 0.103 to 0.123. The mutation thus increases in frequency in the population. In case 2, two changes occur. The fecundity of 1- and 2-year-olds is increased as in case 1. This change is accompanied by a decrease in the fecundity of individuals in age class 5 and older from 1.0 to 0.9. The effect is to lower the fecundity of older age classes by the same amount that fecundity is raised in younger age classes. The result of such a genetic change is an increase in the population growth rate to 0.118, an increase not as large as that seen in case 1. Older members of the population lose some of their breeding potential in this scenario, and this suggests an evolutionary acceleration of aging in the population. In case 3, consider a situation in which an increase in fecundity at all age classes (from 1.0 to 1.1) is accompanied by a decrease in survival from 0.5 to 0.4 in age classes 4, 5, and 6, and from 0.5 to 0.3 in the remaining age classes. In this situation, the population growth rate increases only slightly over the original rate. As in case 2, an acceleration of aging will occur because older individuals lose some of their reproductive potential (Ricklefs and Finch 1995).

32.8 Bet hedging minimizes reproductive failure in an unpredictable environment.

When the environment varies unpredictably over the life span of the individual, selection may favor spreading reproduction over many seasons or concentrating it early in life, depending on the circumstances (Goodman 1979, Hastings and Caswell 1979, Murphy 1968, Schaffer 1974, Yoshimura and Clark 1993). When recruitment of offspring is unpredictable from year to year, selection favors adult survival at the expense of present fecundity, a strategy referred to as **bet hedging** (Stearns 1976). The logic of this strategy is best appreciated by considering the extreme case, breeding only once. If conditions fluctuated such that in some years breeding success was zero, one-time breeders would occasionally fail to reproduce, and their lines would die out. Spreading reproduction over several years, even at the expense of annual fecundity, would be favored under such conditions.

Mosquitofish (*Gambusia affinis*) were introduced into the Hawaiian Islands early in the twentieth century, and they have maintained populations in dozens of reservoirs ever since. In some of these reservoirs, water depth is kept at stable levels. In others, water levels fluctuate markedly in response to rainfall and demand for irrigation water. Mosquitofish from stable and from fluctuating reservoirs differ consistently in their life history traits (Figure 32-8). Mature females from fluctuating reservoirs tend to be smaller, allocate more of their body mass to reproduction, and produce larger numbers of offspring, each with a smaller size, than females from stable reservoirs. Although variations in the mortality rates of adults and juveniles have not been measured directly in stable and fluctuating reservoirs, the differences in reproductive allocation are consistent with more variable adult survival, and perhaps lower average adult survival, in the fluctuating reservoirs. Smaller offspring size also indicates a reduced commitment to each individual offspring, perhaps because juveniles in fluctuating reservoirs also have highly variable survival rates (Scribner et al. 1992).

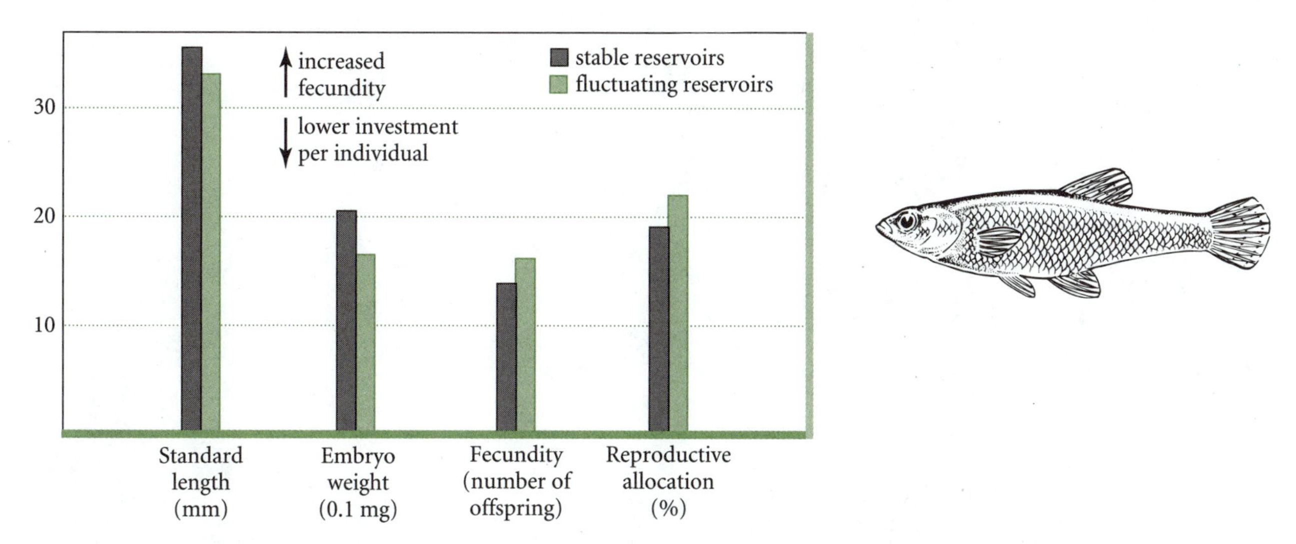

FIGURE 32-8 Life history characteristics of populations of mosquitofish (*Gambusia affinis*) introduced into reservoirs in Hawaii indicate greater total reproductive investment, but lower investment in individual offspring, in fluctuating environments than in stable environments. Reproductive allocation is the proportion of the dry mass constituting embryos. (*Data from Scribner et al. 1992.*)

32.9 Extensive preparation for breeding and uncertain or ephemeral environmental conditions may favor a single, all-consuming reproductive episode.

Unlike mosquitofish, which breed repeatedly, some species of salmon grow rapidly for several years, then undertake a single episode of breeding. During this one burst of reproduction, females convert a large portion of their body tissues into eggs, then die shortly after spawning. Gadgil and Rossert (1970) reasoned that because salmon make so great an effort to migrate upriver just to reach their spawning grounds, it may be to their advantage to make the trip just once, at which time they should produce as many eggs as possible, even if this supreme reproductive effort requires the conversion of their muscle and digestive tissue to eggs and ensures death. This pattern is called **programmed death.**

The salmon life history pattern is sometimes referred to as "big-bang" reproduction, but more properly as **semelparity.** This term comes from the Latin *semel,* "once," and *pario,* "to beget"; it is contrasted with **iteroparity,** from *itero,* "to repeat." It is important to distinguish semelparity from annual reproduction. Annuals may experience more than one reproductive bout, or in some cases, prolonged, continuous reproduction, during their 1-year life cycle. And, like perennials, semelparous individuals must survive at least one nonreproductive season before becoming sexually mature. Semelparity is rarely encountered among animals and plants that live for more than 1 or 2 years. Usually, the effort and the allocation of resources required to survive between growing seasons are so much greater than those needed to prepare for breeding that, once a perennial life habit has been adopted, reproduction every year seems the most productive pattern.

The best-known cases of semelparous reproduction in plants occur in the bamboos (Janzen 1976) and the agaves (century plants; Schaffer and Schaffer 1977), two distinctly different groups. Most bamboos are tropical or warm temperate zone plants that form dense stands in disturbed habitats. Reproduction in

(a)

(b)

FIGURE 32-9 Stages in the life cycle of the Kaibab agave (*Agave kaibabensis*) in the Grand Canyon of Arizona. An individual plant grows as a rosette of thick, fleshy leaves (a) for up to 15 years. Then it rapidly sends up its flowering stalk (b) and sets fruit, after which the entire plant dies.

bamboos does not appear to require substantial preparation or resources, but opportunities for successful seed germination probably are rare. Once established, a bamboo plant increases by asexual reproduction, continually sending up new stalks, until the habitat in which it germinated is packed with bamboo. Only at this point, when vegetative growth becomes severely limited, do plants benefit from producing seeds, which may colonize disturbed sites.

Most species of agaves inhabit climates with sparse and erratic rainfall. These plants grow vegetatively for several years, the number varying from species to species, then send up a gigantic flowering stalk. After producing its seeds, the agave dies (Figure 32-9). One curious fact about agaves is that they frequently live side by side with yuccas, a group of plants with a similar growth form, but which flower year after year. The root systems of agaves do, however, differ from those of yuccas; yucca roots descend deeply to tap persistent sources of groundwater. Agaves have shallow, fibrous roots that catch water percolating through the surface layers of desert soils after rain showers, but are left high and dry during drought periods. The erratic water supply of the agave may prevent successful seed production or seedling establishment every year, and the period between suitable years may be very long. Under these conditions, it may be most advantageous for the agave to grow and store nutrients until an unusually wet year comes—perhaps 1 year in 10 or even 1 in 100—and then to put all its resources into reproduction.

Several explanations have been proposed for the occurrence of semelparous and iteroparous reproduction in plants (Young 1990, Young and Augspurger 1991, Silvertown 1996). First, variable environments might favor iteroparity, which would reduce the variation in lifetime reproductive success by spreading reproduction over both good and bad years. This hypothesis can be rejected because semelparous plants tend to occur in more variable (usually drier) environments than their iteroparous relatives. Second, variable environments might favor semelparity when a plant can time its reproduction to occur during a very favorable year. In these circumstances, storing resources for the big event makes sense, just as not holding back resources for an uncertain episode of future reproduction also makes sense. *Carpe diem:* "seize the day." This tactic is particularly favored when adult survival is relatively low and the interval between good years is long. Finally, attraction of pollinators to massive floral displays might favor plants that put all of their effort into one reproductive episode. The few observations on this point are equivocal to mildly supportive. For example, in the semelparous rosette plant *Lobelia telekii,* which grows high on the slopes of Mount Kenya in Africa, a doubling of inflorescence size was seen to result in a fourfold increase in seed production (Young 1990). Comparison of *Lobelia telekii* with its iteroparous relative *Lobelia keniensis* (Table 32-6) suggests that semelparity is associated with dry habitats that are highly variable in both space and time. Presumably, infrequent conditions that are highly favorable for the establishment of seedlings trigger the massive flowering episodes in *L. telekii.*

TABLE 32-6 Ecological, life history, demographic, and reproductive traits of *Lobelia telekii* and *Lobelia keniensis* on Mount Kenya

Trait	*Lobelia telekii*	*Lobelia keniensis*
Life history	Semelparous	Iteroparous
Habitat	Dry rocky slopes	Moist valley bottoms
Growth form	Unbranched	Branched
Reproductive output	Larger inflorescences, more seeds	Smaller inflorescences, fewer seeds
Variation in inflorescence size	Highly variable, increases with soil moisture	Relatively invariable, independent of soil moisture
Demography	Virtually no adult survivorship	Populations in drier sites have lower adult survivorship and less frequent reproduction
Variation in number of seeds per pod	Strongly positively correlated with inflorescence size	Independent of inflorescence size, positively correlated with number of rosettes
Effects of pollinators	Increased seed quality, but not seed quantity	Increased seed quality, but not seed quantity

(From Young 1990.)

32.10 Senescence evolves because of the reduced strength of selection in old age.

While few organisms exhibit programmed death associated with reproduction, most do experience the gradual increase in mortality and decline in fecundity resulting from deterioration of physiological function known as **senescence** (Kohn 1971, Rockstein 1974, Lamb 1977, Calow 1978, Rose 1991, Kirkland 1992, Partridge and Barton 1993, Ricklefs and Finch 1995). Senescence, reproductive decline, and death in old age do not result from abrupt physiological changes. Rather, the demographic consequences of senescence result from a gradual decrease in physiological function with age. For example, the rates of most physiological functions in humans decrease in a roughly linear fashion between the ages of 30 and 85 years, to 80–85% of the value in 30-year-old individuals for nerve conduction and basal metabolism, 40–45% of the volume of blood circulated through the kidneys, and 37% of maximum breathing capacity (Mildvan and Strehler 1960). Many birth defects related to chromosomal abnormalities occur with increasing prevalence in women progressively older than 30 years (Figure 32-10). Such changes are found throughout the animal kingdom (Comfort 1956, Strehler 1960, Rose 1991).

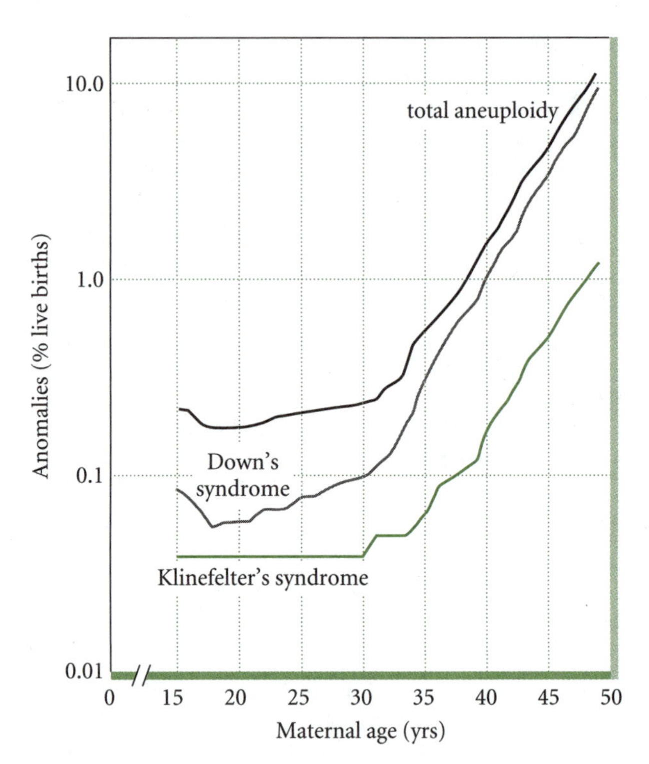

FIGURE 32-10 In humans, the risk of abnormal numbers of chromosomes (aneuploidy) in offspring increases with the mother's age. Down's syndrome, the most common chromosomal abnormality, is due to an extra copy of chromosome 21. Klinefelter's syndrome is a condition in males resulting from an extra X chromosome, which causes sterility and the development of breasts resembling those of women. (*From Ricklefs and Finch 1995.*)

How can senescence evolve? Why is it not eliminated by selection when survival presumably is advantageous to an individual at any age? The answer to these questions is generally thought to originate in the declining strength of selection on genes expressed at progressively greater ages, owing to the fact that fewer individuals bearing those genes survive to express them (Williams 1957, Hamilton 1966, Rose 1992). Given this age dependence of selection, senescence can arise in a number of ways.

Senescence may reflect the accumulation of molecular defects that fail to be repaired, just as an automobile eventually deteriorates and has to be junked. Ionizing radiation and highly reactive free radicals break chemical bonds; macromolecules become cross-linked; DNA accumulates mutations. We refer to this as the **mutation-accumulation theory** of senescence. Recent work by Hughes and Charlesworth (1994) has shown that the variance in the rate of mortality in the fruit fly (*Drosophila*) increases with age, a finding that lends support the mutation-accumulation theory.

This wear and tear cannot be the entire explanation for patterns of aging, however, because maximum longevity varies widely even among species of similar size and physiology. Many small insectivorous bats achieve ages in captivity of 10–20 years, whereas mice of similar size rarely live beyond 3–5 years. In addition, cellular mechanisms to repair damaged DNA and protein molecules appear to be better developed in long-lived animals than in their short-lived relatives. These observations suggest that rates of senescence may be under the influence of natural selection and evolutionary modification. Like any other life history trait, postponement of senescence may exact costs in terms of reduced reproduction at younger ages. For example, if repair processes require both energy and resources, and if mortality is so high that few individuals live to old age, it may be more productive to allocate resources to early reproduction and let the body fall apart eventually.

In a population without senescence, accidental causes of death would affect young and old equally. Even so, progressively fewer individuals would survive as age increased. With few individuals living to old age, progressive physiological deterioration has little effect, on average, on lifetime reproduction. As we have seen, the strength of selection on changes in survival or fecundity declines as those changes are confined to older ages, so selection tends to favor improvements in reproductive success at a young age over those later in life. The lower the survival rate of adults, on the whole, the weaker is selection for improvement in reproductive success at older ages, and the faster senescence should progress. This relationship can be tested by comparing the rate of senescence with the "baseline" mortality rate experienced by young adults in a population prior to the onset of aging. As you can see in Figure 32-11, data for birds and mammals seem to bear out this prediction.

Some alleles may act pleiotropically (affect many different characteristics) to enhance fitness at early ages but reduce fitness in later life (Williams 1957, Rose and Charlesworth 1981, Rose 1982). Such alleles will tend to remain in the gene pool because effects expressed at young ages contribute more to fitness than do those expressed at old ages. This idea is of senescence is called **antagonistic pleiotropy.** Evidence for this notion comes from studies of *Drosophila* in which selection for later egg laying also increased longevity (Table 32-7). In another study, fruit flies selected for increased early survival had reduced survival rates later in life (Rose and Charlesworth 1981).

Kirkwood and Rose (1991) have suggested that the most important trade-off involved in senescence is between allocating energy to the repair and maintenance of germ cell DNA and somatic DNA (DNA not in germ cells). Their thinking is that, since it is extremely expensive to make error-free DNA, allocations of energy toward that process should be restricted to

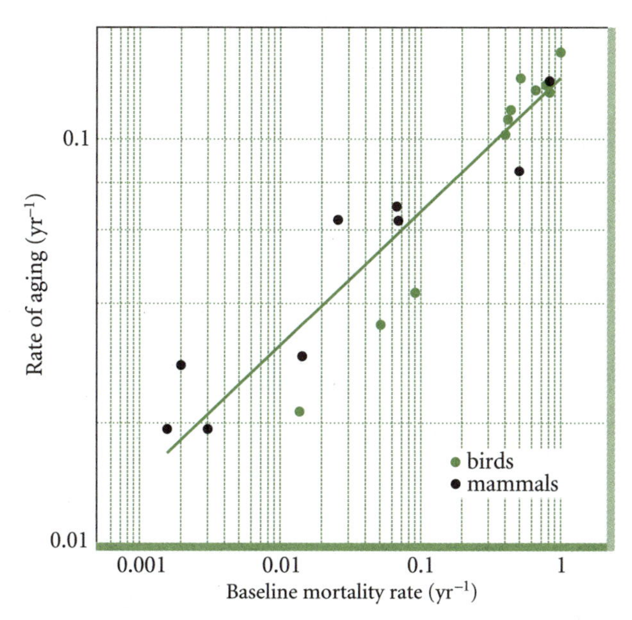

FIGURE 32-11 Species of birds and mammals with a high baseline mortality rate appear to have a high rate of aging as well. The rate of aging is the inverse of the number of years that would be required for 90% of adult individuals to die from causes of aging only. The baseline mortality rate is the rate for young adults. (*From Ricklefs and Finch 1995.*)

TABLE 32-7 Results of selection for late reproductive output in the fruit fly (*Drosophila*)

Days	Eggs laid (5-day total)	
	Selected line*	Control line
1–5	422	551
11–15	392	323
16–20	287	239
21–25	183	137
Longevity (days)	30.2	26.8

**The selected line is from adults that survived at least 21 days.*
(After Rose and Charlesworth 1981.)

cells from which gametes are to be produced, at the expense of allowing errors to accumulate in somatic DNA. This idea, known as the **disposable-soma theory** of senescence, is reviewed by Bulmer (1994).

32.11 Life history patterns vary according to the growth rate of the population.

The relative strength of selection on life history traits expressed at different ages depends on the growth rate of the population. This fact has had important consequences for thinking about life history evolution. For example, reproductive rates have been linked to the growth rates of populations to explain latitudinal variations in fecundity (Skutch 1949, Cody 1966, MacArthur and Wilson 1967). The argument runs as follows: In temperate and arctic regions, populations are periodically reduced by catastrophic weather conditions, and individuals die with little regard to their genotypes. These population crashes are followed by longer periods of population increase, during which adaptations that increase the intrinsic population growth rate (r)—including increased fecundity and earlier maturity—are selected. In "constant" tropical environments, where populations fluctuate little, populations remain near the limit imposed by resources (K), and adaptations that improve competitive ability and efficiency of resource utilization are selected.

The distinction between temperate and tropical patterns has been described as the r- and K-selection spectrum (Pianka 1970). The term r refers to the growth capacity (exponential growth rate) of the population, and K denotes the carrying capacity of the environment for the population—the upper resource limit to population size. The naming of this concept set off a minor semantic battle among population ecologists (Hairston et al. 1970, Pianka 1972, Wilbur et al. 1974, Boyce 1984, Roff 1992 and references cited therein).

Eric Pianka (1970) listed a variety of traits that could be considered either *r*-selected or *K*-selected (Table 32-8). Selection favoring *r*-selected traits under conditions of population growth could arise in two ways. First, individuals in populations reduced below their carrying capacities, and therefore presented with abundant resources, should be able to grow more rapidly, reproduce at an earlier age, and produce more progeny than can individuals in populations at the carrying capacity. This hypothesis provides a resource-based explanation for the presence of *r*- and *K*-selected traits in different populations. In populations regulated by density-dependent processes, all modifications of the phenotype influence the relationship between population growth rate and density. Modifications that enhance growth rate at low population densities but reduce growth rate at high densities would be favored only when population densities are low (and presumably growing); hence these are distinctly *r*-selected traits. Conversely, modifications that enhance population growth rate at high densities, even at the expense of growth rate at low densities, are *K*-selected traits.

A second mechanism for generating divergent *r*- and *K*-selected traits derives from the dependence of strength of selection at different ages on the rate of population growth. As we have seen, in a growing population, modifications of traits expressed at later ages are more weakly selected than those expressed at earlier ages. As a result, in a growing population, selection favors early reproduction at the expense of longevity and continued fecundity. Early reproduction and high reproductive rates are traits listed by Pianka (1970) as *r*-selected.

Although the theory is plausible, a direct relationship between population growth rates or population fluctuations and life history characteristics has not been established. Pianka (1970) placed insects at the *r*-selected end of the spectrum and mammals at the *K*-selected end, reasoning that insect populations fluctuate more than mammal populations. But the differences in life history traits between the two groups could be attributable to differences in body size, over which differences arise in the time and power scale of all physiological processes. Small organisms move more rapidly relative to body length, use more energy relative to body weight, and have more rapid development and shorter generations than large animals. These traits may be inherently correlated with size through physical and physiological relationships—just as a pendulum swings at a rate inversely related to its length—and thus largely insensitive to environmental influences. The importance of *r*- and *K*-selection theory, relative to other sources of variation in life histories, depends on demonstrating a direct link between differences in population fluctuations and life history traits in pairs of otherwise similar organisms. A further difficulty with the interpretation of adaptations according to *r*- and *K*-selection theory is that the traits attributed to different levels of population fluctuation are similar to those predicted for different levels of adult mortality and population turnover, even among populations with constant sizes.

TABLE 32-8 Some attributes of *r*- and *K*-selected species

	r-selection	*K*-selection
Mortality	Variable and unpredictable	More constant and predictable
Population size	Variable, below carrying capacity	Constant, close to carrying capacity
Intraspecific and interspecific competition	Variable, often weak	Usually strong
Selection favors	Rapid development	Slow development
	High r_m	Low resource thresholds
	Early reproduction	Delayed reproduction
	Small body size	Large body size
	Semelparity	Iteroparity
Length of life	Usually shorter	Usually longer
Leads to	High productivity	High efficiency

(From Pianka 1970.)

Several investigators have attempted to contrast genetic responses to r-selected and K-selected regimes in laboratory populations. Francisco Ayala (1965) found that when populations of *Drosophila* were maintained for long periods under crowded conditions, the numbers of adults per cage gradually increased, presumably owing to selection of traits that improved fecundity and survival at high densities. Further experiments in which *Drosophila* populations were kept considerably below carrying capacity by removing adults (Mueller and Ayala 1981, Taylor and Condra 1980) confounded the selective effects of low density with those of high mortality (Reznick 1985). Similar experiments on laboratory populations of bacteria (Luckinbill 1978, 1984) and protozoans (Luckinbill 1979) have also produced ambiguous or negative results.

The idea of r- and K-selection occupies an important place in the development of theory of life history patterns (Stearns 1976, Boyce 1985), but many ecologists believe that these ideas have been applied uncritically (Mueller 1988, Roff 1992). In his recent review of life history theory, Roff (1992) wrote: "To summarize, the concept of r- and K-selection has been useful in helping to formalize the definition of fitness in density-regulated populations, but attempts to transfer the concept to actual populations without regard to the realities of the complexities in life history have probably been detrimental rather than helpful."

SUMMARY

1. The life history traits of organisms are those associated with reproductive rate, age at first reproduction, and life span. Life history evolution involves compromises between competing demands on time, energy, and resources, referred to as trade-offs.

2. Variation in reproduction among songbirds stimulated the study of life history variation in the 1940s, which led to extensive studies in the 1960s.

3. Life history characteristics represent trade-offs between conflicting demands on organisms. Demonstrating such trade-offs is an important challenge in the study of life histories. Among the most important trade-offs are those between the number and size of young and that between growth and reproduction.

4. The age at which an organism first reproduces may be seen as a trade-off between reproduction and survival. By delaying reproduction, organisms risk dying before they can produce young. By reproducing early, they may compromise survival or future reproduction.

5. When survival between breeding seasons is low or requires large sacrifices in fecundity, annual life histories are favored relative to perennial life histories.

6. Life table analysis may be used to model life history evolution when fecundity and survival vary with age.

7. When recruitment of young is unpredictable from year to year, selection favors a strategy of high adult survival at the expense of reproduction, called bet hedging.

8. When reproduction requires costly preparation, selection may favor a single all-consuming reproductive event followed by death, as in salmon, agaves, and bamboos. This strategy is called semelparity, which is contrasted with a strategy of repeated reproductive events, called iteroparity.

9. Senescence arises owing to the declining strength of selection upon traits expressed at progressively older ages. The strength of selection wanes primarily because progressively fewer individuals survive to older ages to prosper or suffer because of genes expressed at those ages.

10. Life history patterns vary according to population growth rate. Some have suggested that, in unpredictable environments, such as temperate regions, adaptations that increase population growth rates are selected (r-selection), whereas in more predictable environments, adaptations that improve competitive ability and efficiency of resource use are favored (K-selection).

EXERCISES

1. Compared with most other mammals, humans delay maturation and reproduction for an inordinately long time. Speculate on why this is the case.

2. Use the Hardy-Weinberg law discussed in Chapter 30 to show how rare deleterious alleles can accumulate in a population if selection against such alleles diminishes with age. [Hint: Think of the counterbalance between selection, which works to diminish the frequency of deleterious alleles, and mutation rate, which works to increase their frequency. To get started, let q = frequency of a deleterious allele and u = the mutation rate. The frequency of allele q in the next generation may be represented as $(1 - s)q$.]

3. In the context of this chapter, the term *bet hedging* is used to distinguish an evolutionary strategy whereby adult survival is favored at the expense of present fecundity in the face of unpredictable recruitment of offspring. Can you think of ways that you bet-hedge in a more proximate sense in your everyday life?

CHAPTER 33

Sex

GUIDING QUESTIONS

- Under what conditions might hermaphroditism be favored over the gonochoristic arrangement of sexes?
- What is the relationship between age and size and the evolution of hermaphroditism?
- What is Fisher's theory of sex ratio?
- Under what circumstances do haplodiploid females vary the sex ratios of their offspring?
- How do local mate competition and local resource competition affect sex ratios?
- What are the different mating systems found among animals?
- What is the relationship between parental investment and mating system?
- What determines an evolutionarily stable strategy for parental investment?
- In what circumstances is polygyny likely to evolve, according to the polygyny threshold model?
- What is a lek?
- What environmental factors determine mating systems in plants?
- What is the difference between intrasexual and intersexual selection?
- What are the various hypotheses regarding the origin of female choice?

We measure fitness in terms of numbers of descendants that survive to reproduce. In most plants and animals, the production of the next generation is accomplished sexually, by the joining together of male gametes (sperm or pollen) and female gametes (eggs or ovules) in the process of fertilization. Individuals are either male or female—or both. In any case, organisms must divide their resources between male and female functions, a process referred to as **sex allocation.** Some of the most important and fascinating attributes of life concern sex allocation. In this chapter, we shall consider how sex influences the evolutionary modification of the morphology and behavior of organisms.

Sex is the union of two haploid gametes to form a diploid cell. You will recall from your general biology course that gametes are the result of meiosis. As we saw in Chapter 30, sexual reproduction mixes the genetic material of two individuals, resulting in new combinations of genes in their offspring. In the view of many biologists, sexual reproduction evolved very early in the history of life as a means of generating the genetic diversity necessary to cope with a varied and changing environment.

The alternative to sex is **asexual reproduction,** by which an individual reproduces genetically exact copies of itself (except for the occasional mutation), either by vegetative means or by producing a diploid, egglike propagule that develops directly without fertilization. Vegetative reproduction occurs in many plants, most of whose cells retain the ability to produce an entire new individual. Thus, shoots may sprout from roots or even from the margins of leaves, and then separate from the parent plant to become new individuals (Figure 33-1). Many simple animals, such as hydras, corals, and their relatives, can form buds in their body walls that develop into new individuals. When these remain attached to the parent individual, a colony forms, as in the case of hydroids, corals, bryozoans, and many other aquatic animals; when buds detach, independent new individuals are formed. Progeny produced by asexual reproduction

FIGURE 33-1 The walking fern propagates asexually by sprouting a fully formed plant from the tip of a leaf. (*After Greulach and Adams 1962.*)

are identical to one another and to their parent, and thus none of them is likely to be well adapted to novel conditions.

In this chapter, we will discuss the conditions that favor having male and female functions combined into one individual and those that favor separate sexes; how male and female function should be allocated; the number of mates one chooses and how these arrangements are organized and managed; and how choices exercised by one sex may serve as selection pressures on traits of the other sex through the process of sexual selection.

33.1 Separation of sexes is favored when fixed costs of sex are high and the sexual functions compete strongly for resources.

As with many issues, botanists and zoologists have divergent views of sex. In most species of plants, individual plants have both male and female function, leading plant ecologists to ask: What conditions favor separation of the sexes on different plants? In most animal species, male and female functions are separated between individuals, and animal ecologists ask: What conditions favor the joining of both sexual functions in a single individual? These questions are, of course, opposite sides of the same coin. Before attempting to answer them, however, we shall present some definitions.

Sex and Sex Allocation

The existence of separate male and female individuals is referred to as **gonochorism** when applied to animals, from the Greek *gonos,* "offspring" (as in gonad), and *choris,* "apart or separately." Botanists refer to the same condition as **dioecy,** from the Greek *di-,* "two," and *oikos,* "dwelling." (The term *gender* is most properly restricted to human maleness or femaleness inasmuch as the word also has important social and cultural meanings. The word *sex* is the appropriate biological term for the distinction between males and females.)

When structures for both sexual functions occur in the same individual, the individual is referred to as a **hermaphrodite,** after Hermaphroditos, son of Hermes and Aphrodite, who while bathing became joined in one body with a nymph. (The circumstances of this happy union are not important here.) Hermaphrodites may be divided into two broad types based on sex allocation. Hermaphrodites may be **simultaneous,** as in the case of many snails and most worms, or they may be **sequential,** in which case first one sexual function appears, then the other. When male function is followed by female function, as in some mollusks and echinoderms, the individual is said to be **protandrous** (Greek *andros,* "male"). When female function comes first, as in some fishes, the individual is **protogynous** (Greek *gyne,* "woman"). The distinction between simultaneous and sequential hermaphrodites belies the extensive variation among hermaphrodites (St. Mary 1994). Some fishes undergo multiple sex changes (Kobayashi and Suzuki 1992). In some animals, sex allocation varies with size or age (Petersen 1987, 1990, Cole 1990, Dolgov 1991, St. Mary 1994).

Botanists apply an additional term, **monoecious,** to an individual plant bearing separate male and female flowers, rather than *perfect* flowers with both male and female parts. While perfect-flowered hermaphrodites are the rule among plants (72% of species, by one estimate), as with animals, almost all imaginable combinations of sexual functions are known, including populations with hermaphrodites and either male or female plants, populations with male, female, and monoecious plants, and hermaphrodite individuals with perfect flowers that also bear either male or female flowers (Yampolsky and Yampolsky 1922, Charnov 1982, 1993). Multiple sex changes and variation in sex allocation with size are also known to occur in plants (Freeman et al. 1980, Ackerly and Jasienski 1990).

Evolution of Hermaphroditism

When is hermaphroditism favored over having sexual function allocated to separate individuals? We may use Levins's fitness set approach to gain an intuitive understanding of this question (see Chapter 31). Recall that we developed fitness sets by thinking of a population exposed to two different environments. A two-dimensional graph whose axes represent the components of fitness in each of the two environments constitutes a simple fitness set. When the fitnesses of each phenotype do not vary much between the two environments, the fitness set is curved outward (convex) from the axis. When there is a great difference in fitness between the habitats, the set curves inward (concave). In the case of sex allocation, we may think of the two environments as being male and female function. Thus, the axes of the fitness set represent the fitnesses of those two sexual functions. Let us denote fitness due to the male and female functions as W_m and W_f respectively. Further, let $W_m(m)$ be the fitness of separate males and $W_f(f)$ be the fitness of separate females. $W_m(h)$ and $W_f(h)$ represent the fitnesses associated with the male and female portions of the hermaphrodite $[W(h) = W_m(h) + W_f(h)]$ (Figure 33-2).

It can be shown that hermaphroditism will represent an evolutionarily stable strategy (ESS)—that is, will invade a system having separate male and female sexual function—when the sum of the combined male and female contributions $[W_m(h) + W_f(h)]$ exceeds the sum of the contributions from each sex separately (Figure 33-2a). When a female can achieve a certain amount of male function by giving up a smaller amount of her female function, she (subsequently "it" or "they") should do so. Similarly, males should add female function when doing so does not cut deeply into their male productivity. For the flowers of female plants to produce a little bit of pollen, or increase the rate of pollen transfer, seemingly few resources would have to be transferred from female to male function (Bell 1985, Stanton et al. 1986), and so the hermaphrodite strategy should be adopted frequently, as it is among plants.

For hermaphrodites to be excluded from a population, the fitness set must bulge inward so that gains in adding the function of one sex are more than offset by losses in function of the other (Figure 33-2b). This may occur when the establishment of new sexual function in an individual requires a substantial fixed cost before *any* gametes can be produced. Sexual function may require gonads, ducts, and other structures for transmitting gametes, and secondary sexual characteristics for attracting and competing for mates. In many animals, in which maleness requires specializations for mate attraction and antagonistic interaction with other males, or femaleness requires specializations for egg production or brood care, fixed costs may be high and hermaphroditism disadvantageous compared with sexual specialization. In fact, hermaphroditism is rare among active animal species and those that practice brood care. It is particularly common among sedentary aquatic species that shed ga-

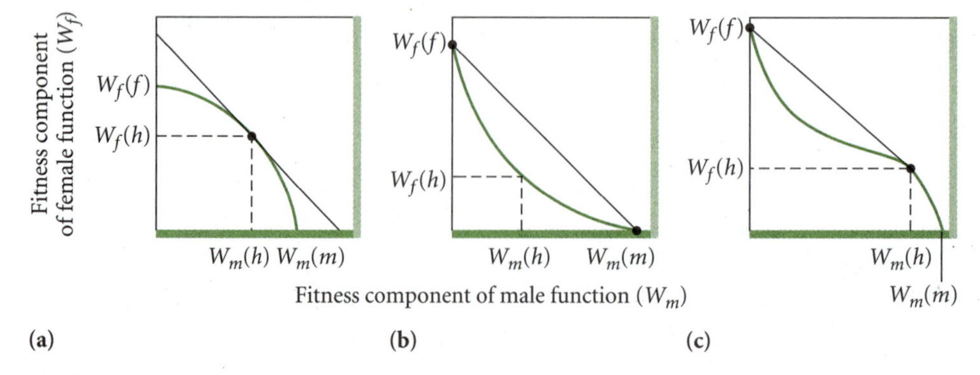

FIGURE 33-2 Fitness set for allocation to male function between 0 and 1. (a) The adaptive function of a convex fitness set indicates that the evolutionarily stable strategy (ESS) is a hermaphrodite with intermediate allocation between male and female sexual functions. (b) When the fitness set is concave, extreme phenotypes (males and females) are selected, and hermaphrodites are excluded from the population. (c) Fitness set representation of the conditions required to maintain a mixed population of females and hermaphrodites. (*After Charnov 1982.*)

metes into the water, such as bivalves (Dologov 1991). Some populations include both hermaphrodites and individuals of one sex, usually females. Fitness set analysis shows that this situation may arise when female function is gained at a large cost in male function (Figure 33-2c)(Charnov 1982).

Sequential hermaphroditism reflects the changing costs and benefits of male and female sexual function as an organism grows. In some marine gastropods practicing internal fertilization, such as the slipper shell *Crepidula*, insemination requires the production of only small amounts of sperm. Hence male function requires few resources and has little effect on somatic growth. As a consequence, many such species are protandrous hermaphrodites, male when small and female when they are large and can produce correspondingly large clutches of eggs (Hoagland 1978). When male-male competition for mates is important, large size can be advantageous for males. This situation apparently has led to protogynous hermaphroditism in some fish species that inhabit reefs, where males compete among themselves for breeding territories (Warner 1975, Charnov 1979).

Variation in Sex Allocation Related to Size or Age

An important prediction of sex allocation theory is that sequential hermaphroditism should evolve in situations in which the reproductive success of both males and females increases with age or size, but the rate of such increase is different between the sexes (Ghiselin 1969, Charnov 1982, Parker 1984). This idea is sometimes referred to as the **size advantage** model of the evolution of hermaphroditism. Milinski and Parker (1991) developed a simple graphic presentation of the essentials of this model, which we summarize below.

Which sex, male or female, should appear first? Differences in size-related reproductive success suggest that there are fitness consequences to this choice. At what size (or age) should an individual switch from one sex to the other? In terms of an ESS, how should phenotypes (sizes) be divided between male function and female function so as to avoid invasion of other distributions? Milinski and Parker (1991) applied competition theory to these questions (see Chapters 21 and 22). Consider a situation in which the reproductive success of males increases linearly faster than that of females, which also increases linearly. A sequential hermaphrodite will be one sex through some lower range of size or age and then switch to the other sex at a larger size or greater age. The point at which that switch should be made to result in an ESS is referred to as the **phenotypic boundary,** S^*.

Milinski and Parker's graphic resolution of the questions posed above is shown in Figure 33-3. Figure 33-3a shows the ESS phenotypic boundary for the switch between the smaller males and the larger females. By switching to female at point S^*, the individual enjoys a higher rate of increase in fitness with size (compare the dashed portion of the male line

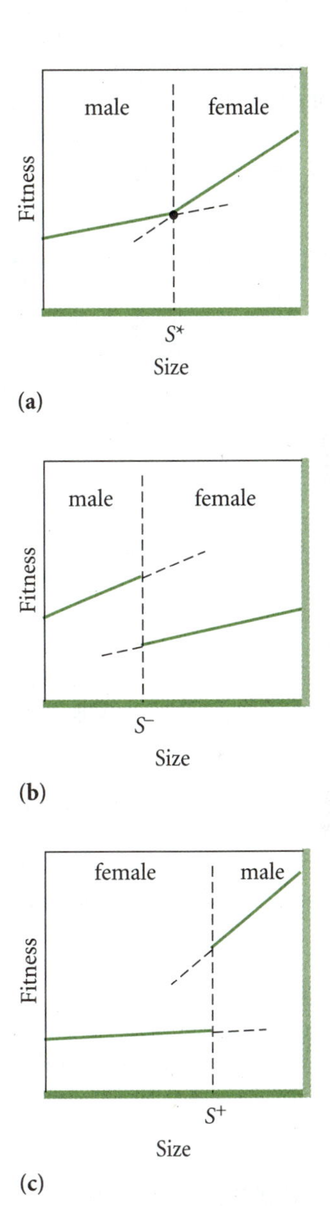

FIGURE 33-3 The size advantage model of the evolution of hermaphroditism. (a) Phenotypic boundary, S^*, between small males and large females. Both males and females increase in fitness as they grow, but the rate of the increase is greater for females. (b) Males range in size $< S-$ and females are $> S-$. For both sexes, fitness increases with size, but for females, it increases more slowly. Thus, if the switch from male function to female function is made at size $S-$, which is below the phenotypic boundary, individuals experience a lower reproductive rate. (c) Here females are the smaller sex (represented by the line below S^*). As in (a), females have a lower rate of increase in reproductive success with size. Increased reproductive success can occur only if the individual switches from female to male (at S^*).

extending past the point S^* with the female line, which has a greater slope). Figure 33-3b shows a situation in which the switch from male to female is made at some point $S-$ below the ESS phenotypic boundary S^*. In this situation, the switch from male to female is made at a smaller size than that predicted by the ESS. Switching sexual function at this point results in a lower rate of increase in fitness with size (compare the dashed portion of the male line, the portion to the right of $S-$, with the female line). This shows that an individual making the switch at $S-$ should experience lower reproductive success. Indeed, in this situation, all individuals should remain males.

Figure 33-3c shows a situation in which females precede males, and the changeover is made at a size greater than the ESS phenotypic boundary, S^*. Here, there is no advantage to remaining female because the phenotypic boundary occurs at a point at which growth leads to increased reproductive success only if there is a switch in sex. Individuals that continue to grow as females (move along the dashed portion of the female line to the right of S^*) will experience lower reproductive success than those that make the switch.

For the most part, theories of sex allocation in simultaneous hermaphrodites have focused on the fitness differences among organisms with different proportions of sexual function allocated to male and female functions without consideration of size (Charnov et al. 1976, Fischer 1984, St. Mary 1994). However, many simultaneous hermaphrodites do exhibit a sequential change in sex allocation, changing from predominately female to predominately male, for example, and in some species such changes are related to body size (Petersen 1990, St. Mary 1994). For example, female-biased individuals of the gobiid fish *Lythrypnus dalli,* which is native to the waters off southern California and Baja California, Mexico, reallocate sex to the male function as they grow larger (St. Mary 1994).

33.2 An optimal progeny sex ratio balances contributions to fitness through male and female function.

When sexes are separate, one may define a **sex ratio** among the progeny of an individual, or within a population, as the number of males relative to the number of females. The ratio of males to females among zygotes is referred to as the **primary sex ratio.** The relative proportions of males and females available at the time of mating, the **operational sex ratio,** may be different from the primary sex ratio owing to differences in mortality between the sexes. A considerable literature has grown around questions related to the primary and operational sex ratios (see reviews by Charnov 1982, Godfray 1994, Bulmer 1994).

Two facts are remarkable about sex ratio: first, that so many populations have very nearly equal numbers of males and females, and second, that there are so many exceptions to this rule. Because of its prevalence, symmetry, and application to the human population, the even, or 1:1, sex ratio is considered the standard condition, deviations being special cases.

Fisher (1930) summed up the theoretical basis of sex ratio theory when he observed that every product of a sexual union has exactly one mother and one father. One consequence of this truism is that individuals of the rarer sex in the population will enjoy greater fitness because they compete with fewer others of the same sex for matings. For example, if a population of 5 males and 10 females produces 100 offspring, each male contributes 20 sets of genes, but each female contributes only 10 sets of genes. With such a skewed sex ratio, any parental genotype giving rise to a larger proportion of male offspring would be favored, and the frequency of males in the population would increase. Similarly, if females were the rarer sex, genotypes that increased the proportion of female progeny would be favored, and the frequency of females would increase in the population. When males and females occur with equal frequency, fitnesses are balanced, and there is no selective pressure to alter the sex ratio. In this case, individuals of both sexes contribute equally to future generations, and the frequencies of males and females among the progeny of an individual are of no consequence to its fitness.

With respect to selection on the primary sex ratio, the relevant measure of fitness is not the number of progeny, but the number of grandprogeny. This is because a gene that influences sex ratio expresses itself in the proportion of males or females among the progeny, not the total number of progeny, of the individual bearing that gene. Its consequences are not felt until the progeny have passed their genes on to the following generation.

Williams (1979) considered the 1:1 sex ratio to be a trivial consequence of the most common sex-determining mechanism. In most vertebrates, one pair of chromosomes (the sex chromosomes) for which the population is polymorphic (for example, X and Y) determines an individual's sex. When an individual is homozygous for one of the chromosomes, it is one sex; when heterozygous, it is the other sex. Thus, in humans, XX homozygotes are female and XY heterozygotes are male. Because matings occur between XX and XY individuals, the genotypes of the offspring are half XX and half XY—half male and half female. The heterozygous sex may be either male (mammals) or female (birds and butterflies), but the sex-determining mechanism will produce equal numbers of males and females in either case.

Charnov (1982) objected to Williams's "nonselection" thinking on two grounds. First, the occurrence of sex chromosomes may signal the predominant selective advantage of an even sex ratio. That is, rather than sex ratio being the inevitable consequence of a particular sex-determining mechanism, the mechanism itself may have evolved to its prevalence because it produced the most fit ratio of the sexes among progeny. After all, animals employ many other sex-determining mechanisms (Bull 1983), including, in many reptiles, temperature-sensitive sex ratios (Bull 1980) and other, undetermined, mechanisms in mammals (Clutton-Brock and Iason 1986).

Charnov's second objection to Williams's thinking was that the existence of sex chromosomes may make it difficult to evolve toward some alternative system of sex determination. Thus, even when selection favors an unbalanced sex ratio, there may be no genetic variation in sex ratio upon which natural selection can work. This does not invalidate selection thinking about sex ratio, but rather illustrates how the genetic system might limit evolution.

We shall see shortly that in some organisms, the sex ratio of the offspring is not so constrained by chromosomal arrangement as in birds and mammals. In some cases, the parent can select the sex of each offspring and thus exercise considerable control over the primary sex ratio, and of course, her fitness.

33.3 In certain situations, mothers should vary the sex ratio of their offspring in relation to their own breeding condition.

In many species, competition among individuals of one sex (usually males) for matings leads to tremendous variance in reproductive success among individuals (Arnold and Wade 1984, Trail 1985, Clutton-Brock 1988). Presumably, where competition is keen, some males achieve many matings, and others none. In harem-forming species, for example, a few males control most of the females, and others lower in social dominance have little access to mates. Generally, contests among males are won by the largest individuals. Trivers and Willard (1973) have suggested that if the condition of progeny is directly influenced by the condition of the mother, then females in good condition should produce male offspring, which will grow large and fare well in male-male competition for mates. Females in poor condition should place their investment in female offspring, which will mate successfully regardless of the parental care they are given. Certainly in mammals, in which the female cares directly for her offspring through the periods of gestation and lactation, the condition of the mother will probably influence that of her offspring.

Experimental confirmation of Trivers and Willard's idea has emerged from a laboratory study on wood rats (*Neotoma floridana*) (McClure 1981). In this species, as in most mammals, females normally invest equally in male and female offspring. But when McClure restricted food drastically during the first 3 weeks of the lactation period to below the maintenance level of a nonreproductive female, male offspring were selectively starved (mothers actively rejected their attempts to nurse), and the sex ratio at 3 weeks was altered to about 0.5 males for each female (also see Gosling 1986). Observations of red deer support these findings. Red deer are harem-forming species in which females are organized in a dominance hierarchy. Dominant females have more access to resources during the time when they are lactating than the less dominant members of the harem. Clutton-Brock and coworkers (1982) found that dominant females tended to produce sons while subordinate females produced daughters.

33.4 In hymenopterans and other haplodiploid invertebrates, the sex ratio of offspring is controlled facultatively in response to local mate competition.

Males compete with one another for matings. An assumption of the 1:1 sex ratio model is that this competition occurs throughout the population. That is, it is assumed that nothing constrains the dispersal of males in a population and, thus, their opportunity to find mates. However, very often dispersal is limited in some way, and this assumption is violated. When individuals disperse only short distances or not at all, or when they mate prior to dispersal, mating often takes place among close relatives, sometimes among the progeny of an individual parent, called **sibling mating.** This situation is referred to as **local mate competition,** and it has important consequences for brood and population sex ratios.

Local mate competition influences the fitness of mothers who produce sons. Consider the most extreme case of complete inbreeding, in which brothers mate with sisters (sibling mating). This situation might arise in the case of a parasitoid in which a mother places all her eggs in one host, such as a seed or an insect larva, and that host is not utilized by any other female in the population. In this case, all the mother's daughters will be fertilized by her sons (assuming that mating occurs on the host before females disperse; see below). Suppose, for example, that three eggs are laid, two sons and one daughter, and further, that the daughter accepts only one mating. Clearly, from the perspective of the mother, one of the sons is "wasted"; her fitness would have been enhanced by producing a daughter instead. Our prediction is that selection should favor a female-biased sex ratio in the face of local mate competition. (Godfray [1994] provides an extensive review of local mate competition in parasitoids.)

Much of our understanding of local mate competition comes from studies of insects in the order Hymenoptera (bees, ants, and wasps), which have an unusual sex-determining mechanism by which fertilized eggs produce females and unfertilized eggs produce males. Sibling mating is common in these insects, so we will examine their natural history as a way of exploring further the importance of local mate competition.

Haplodiploidy

In the bees, ants, and wasps, females are diploid (developing from fertilized eggs) and males are haploid (developing from unfertilized eggs), giving rise to the term **haplodiploidy.** Reproductive females can control the sex of their offspring simply by fertilizing eggs or not. Thus, the sex ratio is under direct control of the female.

Haplodiploidy has evolved a number of times from more usual diploid sex-determining systems. Two steps are required. First, genes must arise in the maternal genome that enable the female to suppress the fertilization of eggs; these eggs also must spontaneously initiate development without being fertilized and develop into males. Second, genes must arise that suppress the development of maleness in diploid individuals. The first step confers a tremendous advantage on the genes responsible for it. The haploid male offspring contain only genes from their mother, including the genes responsible for haploidy. These genes are transmitted to the next generation at twice the frequency of maternal genes transmitted through sexually produced offspring, in which half of the genome comes from the father. Therefore, if uniparental males make at least half the fitness contribution to the next generation that biparental males do, genes favoring haplodiploidy can invade a population (Bull 1983). Why isn't the uniparental sex female? Presumably this would lead to strict parthenogenesis with haploid clones of asexually reproducing females. In spite of the initial advantages of this system in terms of the transmission of genes to future generations, such clones might die out owing to the lack of sexual reproduction and the reduced potential for genetic recombination in haploids (heterozygotes are not possible in haploids).

Haplodiploidy and Local Mate Competition

Many wasps are parasitoids on other insects or complete their larval development within the fruits of certain plants. For some of these species, hosts are so scarce, and mates so difficult to find, that females mate on the host on which they grew up before dispersing to find new hosts on which to lay their own eggs. When a host is parasitized by a single female wasp, females within each brood are limited to mating with their brothers. In these circumstances, because the number of grandprogeny of a female wasp is directly proportional to the number of female offspring that disperse to search out new hosts, male offspring have reduced fitness value. One might therefore expect a reduced proportion of males in each brood. In other words, when males compete only with their brothers for the opportunity to mate, one male offspring is as good as several from the standpoint of their mother. Thus it is not surprising that most species of parasitoid wasps have sex ratios skewed greatly in favor of females. Several produce only one male per brood. Such males are often wingless. In extreme cases, males fertilize females as larvae within the host, or, as in viviparous pyemotid mites, within the mother's body (Hamilton 1967, 1979). In the last case, males become sexually functional as larvae and never develop into adult forms. These observations confirm the idea that when local mate competition occurs only among siblings, the fitness value of males is greatly reduced, and where possible, mothers limit the number of males among their progeny (Herre 1986). In normally outcrossing haplodiploid species, a primary sex ratio closer to 1 : 1 is the rule.

In many parasitoid species, hosts are **superparasitized,** meaning that more than one female may lay her eggs in the same host. Also, in most cases, ovipositing wasps can use chemical cues to determine whether eggs have been previously laid in a host by another female or not. When a female attacks a "virgin" host, and when few hosts are superparasitized, her daughters will probably engage in sibling mating because their brothers will be the only males around. When a female attacks a host that has already been parasitized, her sons can compete to mate with the daughters of the first parasitoid female, and their value will be considerably enhanced. One might expect superparasitic females to produce more male offspring per brood than the first female to lay on a particular host. And this is, in fact, frequently the case. Moreover, the proportion of sons in the progeny of the second female should increase as the ratio of the number of offspring of the first female to the number of her own offspring increases, because opportunities for outcrossing increase in direct proportion to this ratio (Hamilton 1967).

Werren (1980) established conditions for testing this idea in laboratory experiments with the wasp *Nasonia,* which parasitizes larvae of the fly *Sarcophaga.* The results conformed to the predictions of the ideas discussed above (Figure 33-4). The progeny of primary parasites were 9% male. The proportion of males produced by secondarily parasitic females varied from a similarly low level when their own offspring predominated in the host to 100% males when their offspring were few compared with those of the primary parasite.

Haplodiploid wasps also provide an opportunity to test Trivers and Willard's (1973) idea about mothers controlling the sex of offspring in relation to their expected fitness contributions in a variable environment (King 1988). In many parasitoids, larger hosts result in the production of larger female offspring with greater fecundity than the smaller females that emerge from smaller hosts. Males presumably are less disadvantaged by developing to small size within a small host because male sexual function is less dependent on body size than is female sexual function. In his 1979 paper on sexual selection, Charnov quoted the results of an experiment by the Russian Ivan Chewyreuv, published in 1913, on the ichneumonid wasp *Pimpla instigator,* which parasitizes the pupae of moths and butterflies, laying a single egg on each pupa. When Chewyreuv presented females with large hosts, all of 23 offspring produced were female; when he presented them with small hosts, 38 out of 47 offspring were male. Large and small hosts could be alternated, in which case the female would either fertilize the egg or not as appropriate. In these laboratory experiments, pupae of different species of moths were used to present hosts of different sizes. But Chewyreuv also showed that when the wasp *Exenterus* parasitized cocoons of the moth *Lophyrus* in nature, larger host pupae tended to produce a preponderance of females. In *Lophyrus,* males are only half the size of females. Of the wasps emerging from the larger, female pupae, 79% were females, compared with only 47% of those emerging from the smaller, male pupae.

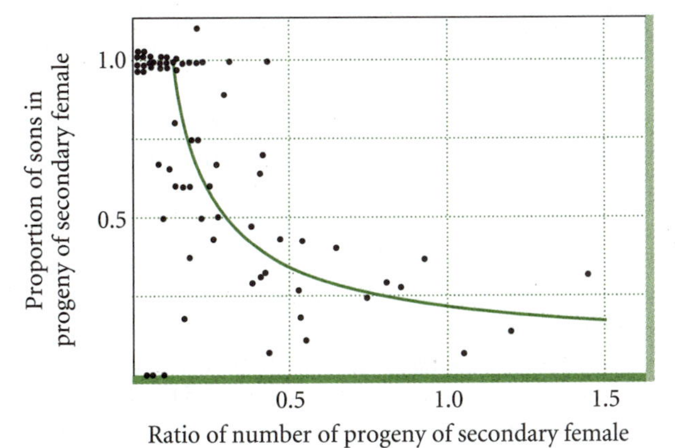

FIGURE 33-4 Relationship between the sex ratio in progeny of secondary females of the parasitoid wasp *Nasonia* and the probability that their offspring will mate with the offspring of other females. (*From Werren 1980.*)

Werren's and Chewyreuv's results show that when females can control the sex of their progeny, they do produce more of the sex with the greater fitness contribution. This contribution depends on the expected fecundity of progeny of each sex, which depends in turn on resources for egg production in female wasps and the availability of potential outcrossed matings for males.

33.5 Local resource competition may lead to male-biased sex ratios.

We have just seen how competition among males for mates (local mate competition) provides selective pressure for female-biased sex ratios. Competition for resources, both among siblings and between offspring and parent, may influence sex ratios in the other direction, toward more males. In mammals and a few birds, males disperse farther from their place of birth—a phenomenon called **natal dispersal**—than do females (Miller and Carroll 1989). The tendency to stay near the place of birth is called **philopatry.** Daughters remain near the home of their mother (their birthplace), where they compete among one another and with their mother (if she is alive) for the resources necessary for reproduction. From the point of view of the daughter, there is some advantage in inheriting her birthplace from her mother because its quality has been proved by the mother's reproductive success there. However, from the point of view of the mother, there seems little point in making lots of daughters, since most of them will fail to inherit the home site and will have to remain nearby and compete with other philopatric (home-loving) daughters, where, presumably, they will have little chance for success. We use the term **local resource competition** to refer to such competition among offspring. Where local resource competition is intense, mothers will benefit by producing more sons than daughters (Clark 1978, Craig 1980, Charnov 1982, Bulmer 1983, Johnson 1988).

The phenomenon of local resource competition was first described by Clark (1978), who observed that South African bush babies, a prosimian primate, routinely produced more sons than daughters. She observed that young males dispersed some distance from

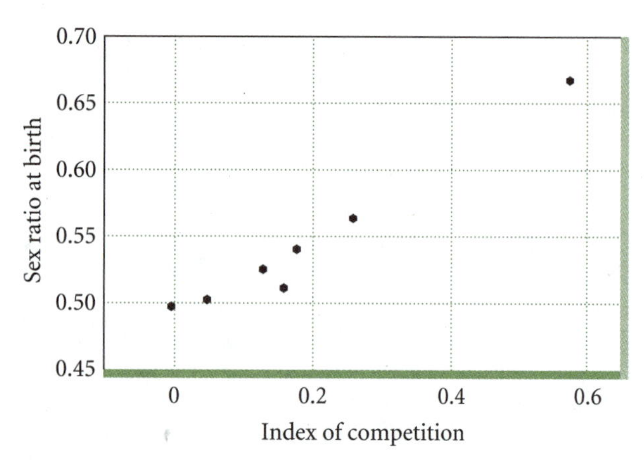

FIGURE 33-5 Relationship between the primary sex ratio and an index of the intensity of competition among females in several genera of primates. The index is the slope of the regression of reproductive success and group size. (*From Bulmer 1994; after Johnson 1988.*)

the place of their birth, but that young females remained close to home, where they competed with their sisters and their mother for resources. Johnson (1988) made similar observations in species of primates of several genera. Johnson also showed that, as with local mate competition, the greater the intensity of resource competition among philopatric females, the more biased is the sex ratio (Figure 33-5).

Local resource competition has been invoked to explain the male-biased sex ratios in some social insects that swarm to establish new colonies (e.g., honeybees and army ants). In these species, one or a few new queens leave the natal colony in a swarm of males to form a new colony (Craig 1980, Bulmer 1983). Because queens have no value without a swarm of males, and because males may mate with queens from other colonies, there is little advantage to making large numbers of females.

We hasten to point out that philopatry alone does not necessarily result in a skewed sex ratio. In some birds, such as the red-cockaded woodpecker and the Florida scrub jay, philopatric males assist in the rearing of their parents' additional broods. That is, they provide assistance to young that are not their own. While on its face this would seem to be wholly nonadaptive, we shall see in Chapter 34 that such behavior may in fact enhance the fitness of the helper. Also, it is important to appreciate that local resource competition will have an effect only if reproductive success is related to resource gain in some way, as it has been shown to be in many primates. The grouping of juvenile males and females near the home site may have little effect on reproductive success if there is no competition among those juveniles for resources or mates.

33.6 Mating systems depend on the degree to which individuals of one sex can monopolize resources.

The **mating system** of a population is the pattern of matings between males and females—specifically, the number of simultaneous and sequential mates and the permanence of the pair bond. Like the sex ratio, the mating system of a population is based on behavioral tendencies of individuals that are subject to natural selection and evolutionary modification. As a consequence, mating systems are usually correlated with attributes of the species and the environment.

The theory of mating systems has as its foundation a basic asymmetry of life: the reproductive success of a female is limited by her ability to make eggs and otherwise provide for her offspring, while the reproductive success of a male is usually limited by the number of matings that he can procure. Because the female gamete is larger than the male gamete, it requires more resources to produce, and a female's fecundity is likely to be limited by her ability to gather resources to make eggs. When males contribute no resources to their mates, or when males do not care for their young, a male can increase his own fecundity only by mating with additional females.

It would seem that ideally males should mate with as many females as possible, in each case leaving the female to care for the young. For females, the ideal strategy might be to mate and then leave the male to care for the young while she goes off to gather resources for the production of more eggs. Of course, things are not that simple. The costs and benefits of parental care are different for males and females, affected by ecological and physiological factors, and influenced by the way in which resources are distributed in nature. Nevertheless, much insight about reproductive behavior and mating systems can be gained by appreciating how the fundamental interests of males and females are adjusted to ecological conditions (Davies 1991, Krebs and Davies 1993).

When a male mates with as many females as he can locate and properly persuade, and provides his offspring nothing more than a set of genes, he is said to be **promiscuous.** Promiscuity generally precludes a lasting pair bond. When adopted by a population, it also tends to increase variance in mating success among males, with some individuals obtaining perhaps dozens of matings while others get none. Among animal taxa as a whole, promiscuous mating is by far the commonest system, and it is universal among outcrossing plants.

When males can contribute to the fecundity of their mates by procuring resources for the female or by caring for the offspring directly, pair bonds may occur. As males increasingly contribute to the realized fecundity of a single mating, mating systems progress from promiscuity (least contribution), through polygamy and serial polygamy, to strict monogamy (greatest contribution, relative to the female's). **Polygamy** describes the situation in which a single individual of one sex forms long-term pair bonds with more than one individual of the opposite sex. (The length of a pair bond is dependent on the life history of the organism. "Long-term" may mean a single season in some species and more than one season in others.) Normally, males are mated to more than one female, in which case the system is referred to as **polygyny** (literally, "many females"). Polygyny may be expressed through the defense of a group of females, called a **harem,** against the mating attempts of other males, or the defense of a territory or nesting site to which more than one female is attracted to breed as well as mate. Thus polygyny may arise through the ability of a male to control matings by defending females, in which case the contribution of the male to his progeny may be primarily genetic, or through the ability of a male to control or provide resources necessary for the female to reproduce. **Polyandry** is the situation in which a single female enters into a durable pair bond with more than one male.

In **serial polygamy,** an individual of one sex forms a bond with one individual of the opposite sex, but eventually abandons its mate, leaving it to care for their offspring, while it goes off to seek a new mate. There is never more than one pair bond formed at a time in such systems, but one sex always takes the primary responsibility for caring for the offspring of a mating. Most commonly it is the male that abandons the female to seek other mates (serial polygyny), but the reverse situation (serial polyandry) is also known.

Monogamy refers to the formation of a pair bond between one male and one female that often persists through the period required to rear the offspring of a mating, and which may last until one of the pair dies. Monogamy arises primarily in situations in which males can make a large contribution to the number and survival of offspring. Hence it is most common in species with prolonged dependence of offspring and in which both sexes can provide for the young. Monogamy is not common in mammals because providing milk is a specialized task of the female (Eisenberg 1981). However, it is common among birds, especially those in which parents feed their offspring, a task that both sexes are equally capable of performing.

33.7 Mating systems are associated with habitat and diet.

An early and crucial step toward interpreting mating systems as evolved adaptations was recognizing associations between the mating systems of populations and their habitats, habits, and diets. The English behaviorist John Crook (1964, 1965) was among the first to recognize the relationship between mating systems and ecology. Surveying species of African weaverbirds, he noted that monogamy predominated among insectivorous species inhabiting forests and savannas (see Chapter 8), whereas polygyny was the rule among seed-eating species of open savannas and grasslands (Table 33-1). Furthermore, most of the monogamous species bred as solitary pairs on widely dispersed territories, while polygynous species were invariably

TABLE 33-1 Relationship of habitat and diet to mating system in African weaverbird species

Habitat			Food		Pair bond		Sociality		
Forest	Savanna	Grassland	Insects	Seeds	Monogamous	Polygamous	Solitary	Grouped territories	Colonial
+			+		17	0	15	0	1
	+		+		5	0	4	0	2
+			+	+	3	0	2	0	0
	+		+	+	1	4	1	0	4
		+	+	+	1	1	1	0	0
	+			+	2	10	0	1	16
		+		+	0	15	0	13	3

(Data from Clutton-Brock and Harvey 1984.)

colonial (nesting in large groups). In a broader study of North American songbirds, Verner and Willson (1966, 1969) found a similarly strong association of polygyny with grassland, prairie, and marsh habitats. But such simple correlations between variables cannot explain how these associations arose. Empirical patterns may, however, suggest hypotheses or models that embody a plausible mechanism relating selective factors in the environment to evolved behaviors.

Verner (1964), with later elaboration by Verner and Willson (1966), proposed a model for the origin of polygynous mating systems. As long as his territory holds sufficient resources, a male gains by increasing the number of his mates. A female benefits from choosing a territory or a mate of high quality. Thus, polygyny arises when a female can obtain greater reproductive success by sharing a male with one or more other females than she can by forming a monogamous relationship. Orians (1969) devised a simple graphic representation of the process of mate selection according to this Verner-Willson **polygyny threshold** model (Figure 33-6). The model contrasts the fitness of a female mated to a monogamous male with that of a female mated to a polygynous male as a function of the quality of the male's territory. Polygyny will occur only when the quality of territories varies so much that some females will have higher fitness mated to a polygynous male on a territory of high quality than they would mated monogamously to a male (and receiving his undivided attention) on a territory of poor quality. This point is called the polygyny threshold of territory quality. If females can assess the quality of territories, the first individuals to pair should mate monogamously, with latecomers choosing mates with territories of progressively lower quality until the polygyny threshold is reached. At this point, females should be ambivalent about pairing with unmated or with previously mated males (Alatalo et al. 1982).

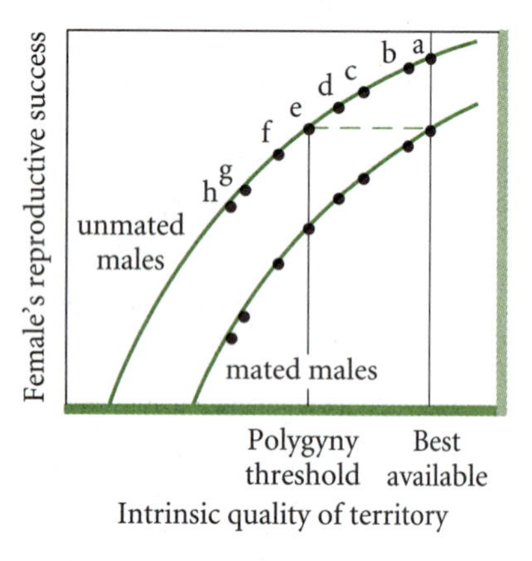

FIGURE 33-6 The polygyny threshold model. Female fitness (reproductive success) varies with the intrinsic quality of a male's territory. The presence of a mated female reduces that quality to subsequent females. Females should select unmated males (a through e) until the quality of the territory of any remaining unmated males (f, g, and h) drops below the "net" quality (with the presence of another female "factored in") of the best territory controlled by a mated male (a). At this point, called the polygyny threshold, female choice will alternate between mated males (a through e, which then become polygamists) and unmated males (f through h). (*After Orians 1969.*)

FIGURE 33-7 Male red-winged blackbird. (*Photograph by A. Morris, courtesy of VIREO, Academy of Natural Sciences, Philadelphia.*)

In cattail marshes throughout North America, male red-winged blackbirds establish territories in early spring (Figure 33-7). Marsh habitat is heterogeneous with respect to vegetation cover and water depth, which affect food supply and the safety of nests, and therefore territories vary greatly in their intrinsic quality. Females return to the breeding grounds after the males, at which time they settle in and pair with established males. Female blackbirds appear to assess the quality of male territories, and the first individuals to arrive pair monogamously with the best males—that is, those holding the best territories. Latecomers are faced with the choice between pairing monogamously with a low-quality male or pairing polygynously with a high-quality male, but sharing his territory's resources with one or more other females (Lenington 1980, Orians 1980, Nero 1984; Figure 33-8). In contrast to blackbirds, birds of forests live in habitats that are more homogeneous than marshes. In forests, bird territories vary less in quality, and most species of birds are primarily monogamous; few territories drop below the polygyny threshold.

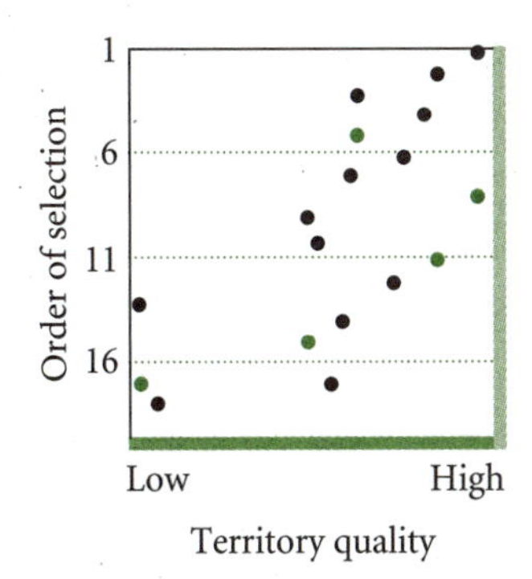

FIGURE 33-8 Observed order of selection of males as a function of territory quality by primary females (black dots) and secondary females (green dots) in a marsh-nesting population of red-winged blackbirds in New Jersey. (*After Lenington 1980.*)

33.8 The configuration of mating systems depends to a large extent on the arrangement of parental care.

To reproduce successfully, individuals must make an investment in gametes and in the activities necessary to obtain one or more mates. In some cases, this initial investment may be enhanced by a further expenditure of time and energy in the care or protection of young, an activity ecologists call **parental investment.** If parental investment increases the probability of survival of individual offspring, and the cost of providing for young is not too high, parental investment is an advantage. In large measure, the configuration of mating systems depends on which parent, if any, gives up other matings to stay with and care for the young.

In birds, it is common for both males and females to engage in parental care. In most birds, the survival of young, and thus the reproductive success of the parents, is highly dependent on the rate at which food is delivered to the nest (Krebs and Davies 1993). The feeding rate is roughly doubled if both parents engage in food gathering for the young. The advantage of such dual parental investment is probably the reason that monogamy is so common among birds.

Because promiscuous mating is so rare among birds—being predominant only among pheasants and grouse and a few groups of tropical frugivorous (fruit-eating) species such as manakins, cotingas, and birds of paradise—ecologists have paid considerable attention to those cases in which promiscuity has surfaced on the calm sea of avian monogamy. The key to the evolution of promiscuity in birds appears to be the emancipation of males from parental investment without substantially cutting into the reproductive success of their mates. This shift can happen in several situations. First, because grouse chicks can feed themselves, males can do relatively little to increase the fecundity of their mates. Females of such species can rear nearly as many young by themselves as they could with the help of males. The same is true of mammals, in which the female alone nurses the young, and potential male contributions (defending resources, protecting the litter, and providing food directly to the female) pale by comparison. Second, the resources of some fruit-eating birds may be conspicuous and of fixed quantity, in which case a female alone can gather as much as a male and a female working together. Clearly, when males can desert their mates without diminishing the number of their progeny, they can increase their reproductive success by attempting to attract additional mates.

In mammals, parental care is almost exclusively the job of the female. Physiological and anatomic differences between males and females make it difficult for males to provide parental care in most cases. Many mammals have long gestation periods during which the male can do nothing to nurture the young. Moreover, postpartum young are nourished by their mother's milk, and, of course, male mammals do not lactate. Very little parental care occurs among fishes (Gross and Sargent 1985), and when it is observed, it is usually the male that provides it. In most fishes, both eggs and sperm are extruded into the water, where fertilization takes place (external fertilization). In these cases, male parental care is most common. When female parental care occurs, it is usually in species with internal fertilization, although this is by no means the rule.

An ESS Model of Parental Investment

The decision of a male or a female to provide parental care or not depends on the costs and benefits of that choice as they relate to that individual, not as they relate to its mate. Males, for example, will desert the nest or not according to their own best interest, as measured by their reproductive fitness. Although the choice between caring for the young and deserting them resides with each individual parent, its consequences depend in large part on the choice made by its mate. Frequently one parent may abstain from caring for its offspring with little reduction in fitness as long as the other parent remains faithful to the clutch or brood; however, when both parents abandon their progeny, few or none will survive. It is generally assumed that by reducing care for one set of offspring, a parent can go on to lay more eggs or obtain more matings.

The evolutionarily stable strategy for the two parents can be investigated by game theory analysis. Our

TABLE 33-2 **Payoffs to the male and female parent for guarding or desertion on the part of either one**

		FEMALE	
		Guards (v eggs)	Deserts (V eggs)
MALE	Guards ($1 + p$ matings)	vS_2 / $vS_2(1 + p)$	VS_1 / $VS_1(1 + p)$
	Deserts ($1 + P$ matings)	vS_1 / $vS_1(1 + p)$	VS_0 / $VS_0(1 + p)$

Note: *Payoffs for females are indicated in the upper right half of each box; payoffs for males in the lower left.*

discussion here follows closely upon the treatment of John Maynard Smith (1982). Each parent may either care for the offspring of a mating, which Maynard Smith refers to as guarding, or desert them. Let us assume that the male may obtain an additional mating with probability p if he guards the brood and P if he deserts. It is assumed that $P > p$. The number of eggs that a female can produce is V if she deserts the clutch and v if she remains to guard. It is presumed that a female that deserts a clutch will produce another ($V > v$). The probability of survival of the eggs is S_0 when neither parent guards, S_1 when one parent guards, and S_2 when both parents guard. The number of offspring produced by each parent under the four possible combinations of male and female guarding and deserting—the so-called **payoff matrix**—is shown in Table 33-2.

Guarding by both parents is the evolutionarily stable strategy when the strategy of desertion by neither males nor females can invade the population. That is, biparental care is an ESS when $vS_2 > VS_1$ for females and $S_2(1 + p) > S_1(1 + P)$ for males. In words, these conditions require that remaining to care for the offspring increases fitness more than abandoning them increases the number of matings for both parents; that is, $S_2/S_1 > V/v$ and $(1 + P)/(1 + p)$. The ESS conditions for all four combinations of behavior (Table 33-3) reveal that when biparental care is an ESS, uniparental care cannot be. Also, when biparental abandonment is an ESS, uniparental abandonment (= uniparental care) cannot be. Both biparental care and biparental abandonment may be ESSs under the special conditions $S_2/S_1 > V/v > S_1/S_0$ and $S_2/S_1 > (1 + P)/(1 + p) > S_1/S_0$. Because neither type of uniparental care is an ESS under these conditions, the evolutionary step between biparental and uniparental care may be impossible to achieve, even if the alternative strategy would confer greater fitness on both males and females. When these inequalities are reversed, both female and male uniparental care can be evolutionarily stable strategies, and the outcome of evolution is ambivalent.

Unambiguous female uniparental care occurs when the ESS for the female is to guard no matter what the male does—that is, $vS_2 > VS_1$ and $vS_1 > VS_0$—and that for the $VS_1 > vS_2$; male is to desert as long as the female cares, $S(1 + P) > S_2(1 + p)$. These inequalities can be rearranged to give the conditions S_2/S_1 and S_1/S_0 for females and $S_2/S_1 < (1 + P)/(1 + p)$ for males. Quite simply, the female's gain must be greater through caring for offspring than through leaving eggs either unattended or in the care of a male; the male must be better off deserting when his mate remains with the brood. The conditions for male uniparental care are the reverse.

In birds, the ratio S_1/S_0 is almost always very large; in no species do both parents abandon the eggs,

TABLE 33-3 **ESS conditions for different combinations of male and female parental care**

1. Both parents care for offspring; female $vS_2 > VS_1$; male $vS_2(1 + p) > vS_1(1 + P)$
2. Female cares, male deserts; female $vS_1 > VS_0$; male $vS_1(1 + P) > vS_2(1 + p)$
3. Male cares, female deserts; female $VS_1 > vS_2$; male $vS_1(1 + p) > VS_0(1 + P)$
4. Neither cares for offspring; female $VS_0 > vS_1$; male $VS_0(1 + P) > VS_1(1 + p)$

although the newly hatched chicks of megapodes are free of all parental care. Female uniparental care is quite common, being the situation in species with promiscuous mating systems, but male uniparental care is known only in rheas, the mallee fowl, and a few polyandrous shorebirds, in which the female lays a set of eggs that are incubated by her first mate while she lays a second set of eggs fertilized by a second male. The predominance of female uniparental care is undoubtedly due to the fact that $(1 + P)/(1 + p)$ is usually much larger than V/v; when either sex abandons the clutch, males can gain additional matings more rapidly than females can produce more eggs.

Under the ambivalent conditions in which either parent's abandoning the clutch is an ESS, which one stays and which leaves depends on the relationship between fertilization and egg laying (Dawkins and Carlisle 1976, Ridley 1978). When fertilization is internal, the male's mating function is completed before the eggs are laid, and the male is free to leave the female—holding her bag of eggs, so to speak. When fertilization is external, the eggs are laid first, and the male must subsequently fertilize them. Smith (1978a) summarized data of Breder and Rosen (1966) showing a strong relationship between external fertilization and male uniparental care in fishes and amphibians (Table 33-4).

A final consideration pertinent to male uniparental care is the certainty of paternity (Trivers 1972). When fertilization is external and eggs are laid in a nest prepared and guarded by the male, the male can be certain that the offspring are his own progeny and that subsequent care is likely to enhance the survival of his own genes. When fertilization is internal and females mate at random within the population, males cannot know for certain that they have fertilized the eggs laid by a particular female, unless they guard the female throughout her receptive period (see, for example, Burton 1985).

33.9 Populations may include individuals of the same sex having different reproductive strategies.

In many fishes with promiscuous mating systems, males defend territories to which they attract females to spawn with them. The males then take care of the eggs until they have hatched. Indeed, fertilized eggs do not survive without the constant attention and protection of a territorial male. The bluegill sunfish (*Lepomis macrochirus*) is typical of such species (Gross and Charnov 1980). Because males must compete vigorously for good spawning sites, they usually do not mature until they are at least 7 years old and large enough to fend off other males. However, a small number of males adopt a different tactic: they mature sexually at 2 years and a small body size, and they sneak matings with females attracted to territorial males. Sneaker males are reproductive parasites in that the survival of their offspring depends on the care of territorial males. Whether a male becomes a territory holder or a sneaker is probably genetically based.

The mating system of the bluegill sunfish is a good example of a **mixed reproductive strategy**, in which two or more phenotypes are stably maintained in a population because each has a higher fitness when it is rare. We have already seen this principle of frequency-dependent selection operating on the proportion of males and females among progeny to maintain a balanced 1:1 sex ratio. In the case of the bluegill sunfish, opportunities for sneak matings are limited by vigilance of territory holders, so as more

TABLE 33-4 Relationship of location of fertilization to parental care in fish and amphibian species

	Care delivered by			
Fertilization	Both parents	Male	Female	Neither
Fishes				
External	8	28	6	191
Internal	0	2	10	
Amphibians				
External	0	14	8	10
Internal	0	2	11	0

(From Smith 1978; after Breder and Rosen 1966, Gross and Shine 1981.)

(a)

(b)

FIGURE 33-9 (a) Two male sharp-tailed grouse displaying on a communal courting area (lek) in southern Michigan. (b) The female has a dull plumage compared with that of the male. (*Courtesy of U.S. Soil Conservation Service.*)

and more sneakers compete for these opportunities, their reproductive success goes down.

A conspicuous feature of the mating systems of many promiscuous birds, as well as other promiscuous animals such as frogs, is the existence of communal mating areas, called **leks,** where males congregate to perform elaborate displays to attract females to mate with them (Clutton-Brock et al. 1988, Oring 1982, Wells 1977, Thornhill and Alcock 1983, Davies 1991; Figure 33-9). Leks appear to form because several males displaying together attract more females than do single males displaying alone.

The ruff (*Philomachus pugnax,* literally, "quarrelsome lover of combat") derives its English name from the distinctive collar of elongated feathers ringing the neck of the male; it derives its scientific name from the intense contests that take place among males on their leks. This system of mating behavior is unusual among the group of wading birds to which the ruff belongs, and is further distinguished by polymorphism in appearance and behavior among the males (Hogan-Warburg 1966, Widemo and Owens 1995). Detailed observations on ruffs in Sweden (Widemo and Owens 1995) showed that the overall frequency of matings increased with lek size up to a mean lek size of five males (Figure 33-10a). The maximum copulation rate on a lek was about three per hour. Most of these matings were gained by the highest-ranking male within the lek. Only after lek size had increased to more than five males did the second- and third-ranking males gain a significant share of the matings, at the expense of the highest-ranking male (Figure 33-10b). Because lower-ranking males fare so poorly, one wonders why they join leks at all. Of course, we cannot think like ruffs, but we can easily observe that males on a lek engage in a variety of behaviors—wing flapping, jumping up and down, and hovering over the lek—to entice other males to join them. Not surprisingly, high-ranking males lose interest in enticing other males when the leks reach a size of four or five males. Low-ranking birds appear to benefit from the commotion on a crowded lek to sneak matings from time to time, and so they solicit other males to join even when a lek already has more than five males (Figure 33-10c).

33.10 Three mating systems predominate in plants.

Almost three-quarters of all species of flowering plants are hermaphroditic, meaning that an individual plant has both male and female sexual functions. Most of these species have mechanisms that prevent self-fertilization, resulting in a fully outcrossed population. In the terms applied to animal populations, such plants would be referred to as promiscuous in their mating behavior. In fact, individual plants have relatively little control over where their pollen will land or from whom they will receive pollen. Self-compatibility and inbreeding (selfing) are apparently favored, and do occur, in habitats where pollinators are scarce (windswept coastal landscapes, for example), and in species that are good colonizers of remote locations. Often only a single colonizing individual gains an initial foothold, so that selfing is the only option that can lead to a new population. The other one-quarter of plant species have adopted a variety of mating systems. Two of the most common of these are **gynodioecy,** in which populations contain a mixture of

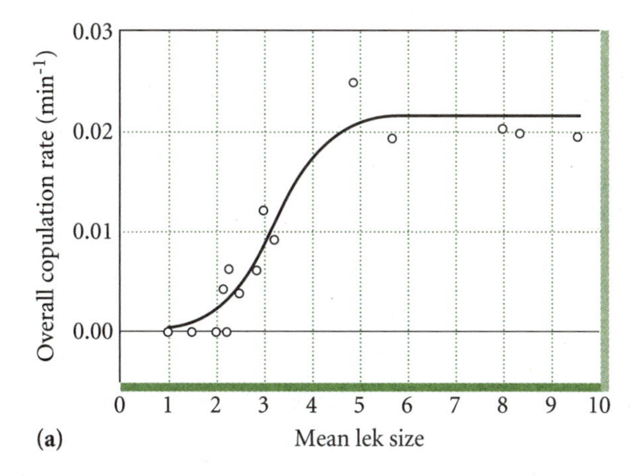

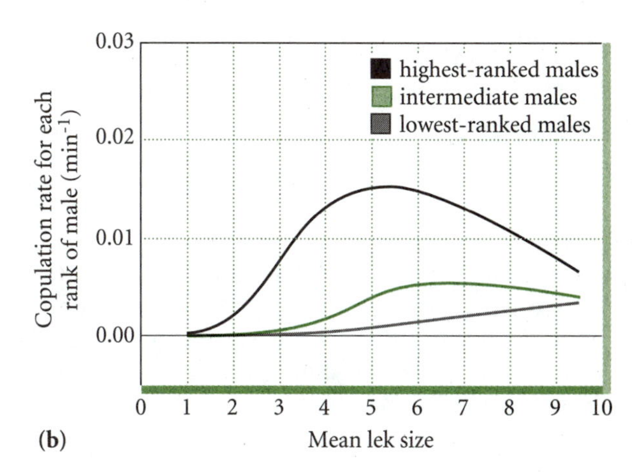

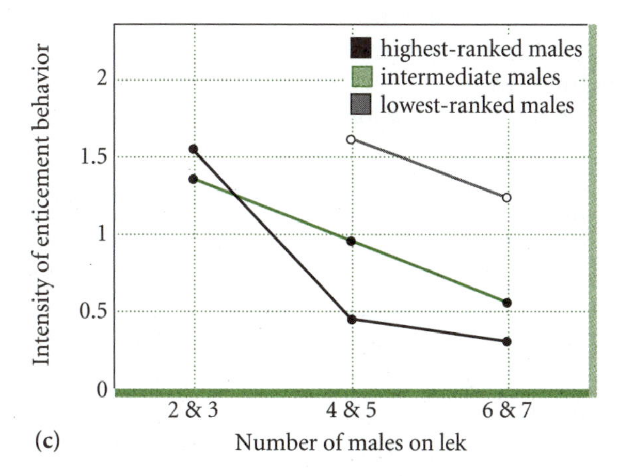

FIGURE 33-10 The influence of lek size on male mating success and enticement behavior in ruffs (*Philomachus pugnax*). (a) The rate of copulation on a lek increases with lek size up to a size of about five males. (b) Most of the increase is enjoyed by the highest-ranking male, but subordinate males benefit by obtaining matings when leks exceed five individuals. (c) High-ranking males try to entice other males to join the lek only when it is small, but subordinate males maintain enticement behaviors even after a lek has grown to a large size. (*After Widemo and Owens 1995.*)

female and hermaphroditic individuals, and dioecy, in which populations are made up of separate male and female individuals.

The ecological conditions that favor one or another mating system in plants are poorly understood (Barrett and Harder 1996). Several possible factors have been suggested, including specialization of male and female individuals for different ecological roles, and interference between male and female function when these occur on the same plant. There is little evidence for ecological segregation between the sexes in most dioecious species, so the first hypothesis has little empirical support. However, male and female sexual function may interfere with each other in several ways. Wind-pollinated species produce prodigious amounts of pollen because their pollen is scattered indiscriminately, rather than being delivered more precisely by an animal pollinator. High pollen production could result in the female parts of hermaphroditic flowers becoming clogged with pollen from the same plant, which could reduce the success of the smaller amount of pollen that arrives from other individuals. It is perhaps a consequence of this problem that dioecy is particularly common in wind-pollinated species.

Individual plants may lack male function due to their allocation of energy and nutrients to female function. Because male function depends on attracting pollinators, it requires the use of resources to produce showy displays of flowers, typically many more than are needed to develop into fruits. In general, female function is more severely limited than male function by the availability of resources for seed and fruit production. Of course, dioecy also guarantees outcrossing, but so many other mechanisms exist to prevent selfing that it is unlikely that this factor is responsible for the evolution of dioecy. (Barrett and Harder 1996 presents an excellent review of plant mating systems.)

33.11 Sexual selection has led to the elaboration of courtship behavior.

Sexual selection is the situation in which one sex, usually the female, determines the fitness of traits expressed in the other sex by exercising choice in mating (Harvey and Arnold 1982, Partridge and Halliday 1984, Bradbury and Andersson 1987, Maynard Smith 1991, Andersson 1994). The usual result of sexual selection is strong **sexual dimorphism,** a difference in the phenotypes of male and female individuals of the same species, especially in ornamentation, coloration, and courtship behavior. Such traits, which distinguish sex over and above the primary sexual organs, are known as **secondary sexual characteristics.** (The considerable controversy and discussion about this and other aspects of sexual selection theory are covered in excellent recent reviews by Maynard Smith [1991] and Andersson [1994].)

FIGURE 33-11 Extreme sexual dimorphism in size in the garden spider *Argiope argentata.* The male is much smaller than the female, which is portrayed in her normal resting position at the hub of her web.

Charles Darwin, in his book *The Descent of Man and Selection in Relation to Sex* (1871), was the first to propose that sexual dimorphism could be explained by selection applied uniquely to one sex. Sexual dimorphism can arise in three ways. First, the dissimilar sexual functions of males and females emphasize different considerations in the evolution of their life histories and ecological relationships. For example, because females produce large gametes, fecundity is often directly related to body size. This may provide a basis for the larger size of females in many species (Figure 33-11). Furthermore, females must acquire additional nutrients to make eggs, and the duties of protecting the eggs and young usually fall upon the female. These special requirements may lead females to use the environment differently than males. Simply because they have to find suitable nest sites, females may exploit different habitats than do males during the nesting season.

Second, sexual dimorphism may result from contests between males, which may favor the evolution of elaborate weapons for combat, such as the antlers of deer (Figure 33-12) and the horns of mountain sheep. Males that win such contests are more likely to gain

FIGURE 33-12 Elk have immense antlers that are used during contests between males to establish control over harems of females. (*Courtesy of U.S. Department of the Interior.*)

access to females. Selection based on competition among individuals of the same sex is referred to as **intrasexual selection.**

Third, sexual dimorphism may arise through the direct exercise of choice by individuals of the opposite sex. With few exceptions, females do the choosing, and males attempt to influence their choices with magnificent courtship displays. That females choose, and males compete among themselves for the opportunity to mate, is a consequence of the asymmetry of parental investment that defines the male and female conditions (Bateman 1948, Trivers 1972). Males enhance their fecundity in direct proportion to the number of matings they obtain; females are limited in their number of offspring by the number of eggs they can produce, but they stand to improve the quality of their offspring by choosing to mate with males that have superior genotypes. When one sex chooses among individuals of the opposite sex based upon their appearance and behavior, **intersexual selection** may occur.

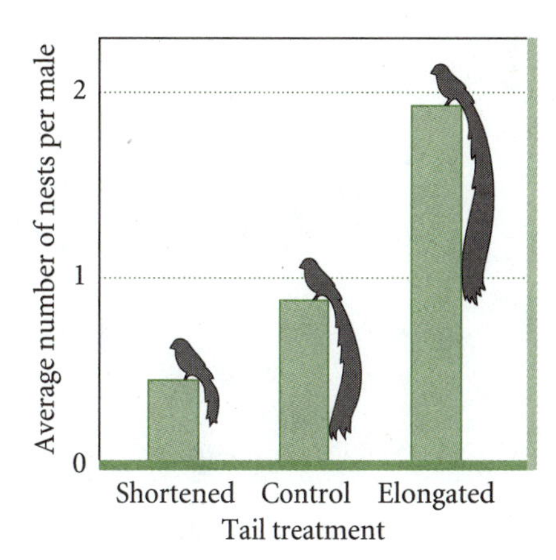

FIGURE 33-13 Relative reproductive success of male long-tailed widowbirds (*Euplectes progne*) with artificially shortened and artificially elongated tails. (*After Andersson 1982.*)

Female Choice

Female choice is experienced at some level by most males. One of the first demonstrations of female choice came from an experimental study of tail length in male long-tailed widowbirds (*Euplectes progne*) (Andersson 1982). This polygynous species inhabits open grasslands of central Africa. The females, which are about the size of sparrows, are mottled brown, short-tailed, and altogether ordinary in appearance. During the breeding season, the males are jet black, with a red shoulder patch, and sport a half-meter-long tail that is conspicuously displayed in courtship flights. Males may attract up to half a dozen females to nest in their territories, but they provide no care for their offspring. The tremendous variation in male reproductive success in this species provides classic conditions for sexual selection. In a simple yet elegant experiment, Andersson (1982) cut the tail feathers of some males to shorten them, and then glued the clipped feather ends onto other males's tails to lengthen them. Length of tail had no effect on a male's ability to maintain a territory, but males with experimentally elongated tails attracted significantly more mates than those with shortened or unaltered tails (Figure 33-13). This result strongly suggests female choice of mates on the basis of tail length.

Many subsequent studies have demonstrated that females choose their mates on the basis of conspicuous differences among males. Indeed, sexual selection has been studied experimentally in over two hundred species (Andersson 1994). There are nonetheless many issues regarding male traits and female preferences that have not been fully resolved. Which came first, female choice or male traits that indicated intrinsic quality? How are the various ornaments of males related to fitness attributes? Why don't low-quality males cheat by taking on a high-quality appearance?

Origin of Female Choice

There are two principal hypotheses for the origin of female choice. These hypotheses differ regarding whether choice arises first and drives evolution of male traits, or whether choice evolves secondarily in response to manifest variation in male quality. The first, called the **sensory exploitation hypothesis,** assumes that females have intrinsic preferences for certain appearances or behaviors because of the ways in which their sensory systems (which are designed for many things besides mate choice) receive and process sensory information (Ryan 1990, Ryan and Keddy-Hector 1992). (Reviews of the ways in which the evolution of signals and communication is constrained by sensory systems may be found in Endler 1992 and Reeve and Sherman 1993.) Two groups of animals have provided support for this hypothesis.

Swordtails are small fish of the genus *Xiphophorus* in which the lower part of the male's tail fin is greatly elongated. It has been shown that female swordtails prefer males with longer tail fins (Figure 33-14). It has been further demonstrated, however, that the females of close relatives of swordtails that lack the long tail fin prefer experimentally produced males that have a swordlike tail (Ryan and Wagner 1987, Basolo 1990a,b). This finding supports the argument that the female preference for long tail fins was present in the ancestors of swordtails and that males having genetic factors that produced elongated tail fins were greatly favored. The absence of the long tail in the relatives of swordtails could be attributed either to selection

FIGURE 33-14 This photograph of a male green swordtail (*Xiphophorus helleri*), shows the greatly elongated lower portion of the tail fin. (*Courtesy of A. Basolo.*)

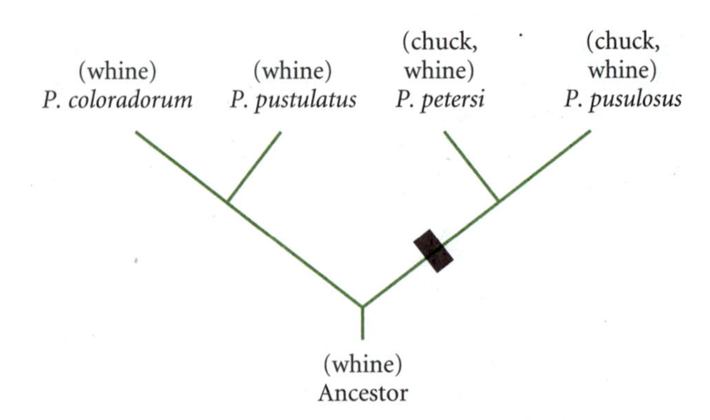

FIGURE 33-15 Cladogram showing hypothetical phylogenetic relationship among four species of frogs of the genus *Physalaemus*. The males of two species, *P. coloradorum* and *P. pustulatus*, give only the *whine* call, whereas the males of two other species, *P. petersi* and *P. pustulosus*, give both the *whine* and the *chuck* call. Females of all species prefer males that give the *chuck* call. The black bar designates the origin of the *chuck* call. The cladogram suggests that the sensory capability of detecting the *chuck* call existed prior to the time that the call evolved. (*After Kirkpatrick and Ryan 1991.*)

against the trait by other factors, such as predators, or to lack of genetic variability for tail length in the population. Why long tails tickle the fancy of female swordtails is not known.

Another group of species that appear to support the sensory exploitation hypothesis are tropical frogs of the genus *Physalaemus*, in which males court females with vocalizations. These frogs have been the subject of some of the most comprehensive studies of sexual selection (Ryan 1980, 1985, Ryan et al. 1982, 1990, Green 1990, Kirkpatrick and Ryan 1991). One can experiment with this system efficiently because artificial courtship calls can be constructed using sound synthesizers, and females will respond to those calls by hopping toward speakers that play them. One can also measure the neural sensitivity of females to sounds of different frequencies. The male vocalization consists of a *whine* call and, in some species, a *chuck* call. Experiments have shown that females of *P. pustulosus* prefer *chuck* calls with sound frequencies closer to their maximum auditory sensitivity than the *chuck* calls that males of their species actually give. This finding suggests that there is ongoing directional selection for deeper-voiced frogs in this species. Moreover, females of other *Physalaemus* species in which males do not make the *chuck* call prefer the *chuck* call to the calls of males of their own species.

How might such a preference evolve? One possibility is shown in Figure 33-15, which depicts the hypothetical evolutionary relationship among four species of *Physalaemus* (*P. coloradorum, P. pustulatus, P. petersi,* and *P. pustulosus*). The females of all four of these species show a preference for the *chuck* call, but the males of only *petersi* and *pustulosus* actually give the call, suggesting that the *chuck* call arose from the common ancestor of those two species. If the phylogenetic relationship between the four species as shown in the cladogram is correct, this would imply that the ancestor of the four species possessed a sensitivity for the *chuck* call, which was later exploited (Kirkpatrick and Ryan 1991). Of course, another possibility is that the ancestor of the four species possessed both the sensitivity for the call and the call itself, and that the current situation arose from a loss of the *chuck* call in the line leading to *coloradorum* and *pustulatus* (Gardner 1990). This example demonstrates the importance of understanding the phylogenetic relationships among organisms. It also suggests that more work is needed to resolve the issue of the evolution of sensory exploitation.

The second hypothesis proposed to explain sexual selection is that female preferences are based on perceptible variations in the quality of male genotypes; that is, that females choose to mate with males that will confer higher fitness on their offspring. In this scenario, female preference evolves because those females that make the best choices of mates leave the most descendants. Any genetic trait that influences preference by females thereby comes under strong selection.

Male Secondary Sexual Characteristics

Regardless of whether female preference arises by sensory bias or in response to variation in male quality, once female choice is established in a population, it exaggerates fitness differences among males, and may create what is known as **runaway sexual selection.** R. A. Fisher (1930) was the first to provide a detailed explanation for the evolution of male adornments by runaway selection (a term that Fisher coined). The elements of Fisher's model progressed logically as follows: (1) variation among males in a fitness-related trait made some individuals more desirable as mates than others; (2) females that perceived this difference among males and selected mates accordingly had higher fitness than nonselective females; (3) persistent female choice of the male trait under selection led to continued male response and eventually to extreme values of the trait, as in the bizarre courtship antics of the birds of paradise, to pick a conspicuous example. Fisher's model sees both female choice and male sexually selected traits as adaptive modifications. Also, the "runaway" behavior of the model can be achieved only if female choice is based on comparison among males, rather than upon some absolute ideal. In the latter case, selection would stop when male adaptations coincided with the ideal. In the former case, sources of new variation would always provide a superior male by comparison with others in the population, just as, in artificial selection programs, breeders continually up the ante even as the population pays through selective deaths to stay in the game.

If sensory exploitation is the basis for female preference, then sexually selected male traits need not have any direct connection to the quality of males; "quality" is determined solely by what females prefer. If selected traits indicate—at least initially, before runaway selection takes hold—intrinsic attributes of male quality, then we are then faced with a paradox. Presumably, such outlandish traits as the tail of the long-tailed widowbird burden males by making them more conspicuous to predators and by requiring energy and resources to maintain. How can such traits indicate, let alone contribute to, male quality?

One intriguing possibility, suggested and elaborated by Zahavi (1975, 1977, 1987, 1991), is that male secondary sexual characteristics act as handicaps. That a male can survive while bearing such a handicap indicates to a female that he has an otherwise superior genotype. This idea is known as the **handicap principle.** It may sound crazy, but if you wanted to demonstrate your strength to someone, you might make your point by carrying around a large set of weights. A weaker individual couldn't do it, and thus could not falsely advertise strength. Accordingly, the greater the handicap borne, the greater the ability of the individual to offset the handicap by other virtues—and to pass genes for those virtues on to his offspring. One small European songbird, the wheatear, takes the iron-pumping analogy literally and festoons its nesting ledge with up to 2 kilograms of small stones carried from a distance in its beak.

One virtue that males might possess, and which might be demonstrated by producing a showy plumage, is resistance to parasites and other disease-causing organisms (Hamilton and Zuk 1982). Only individuals having genetic factors for resistance to parasite infection could produce or maintain a bright and showy plumage. Thus, an elaborate and well-maintained sexual display may provide a convincing demonstration of high male fitness, even if the display itself is an encumbrance (Figure 33-16). Indeed, the fact that a male survives with such an encumbrance may be evidence enough of his superior constitution. The importance of parasites to this theory is that they evolve rapidly and thereby continually apply selection for genetic resistance factors.

The **Hamilton-Zuk hypothesis,** along with its subsequent modifications, comes under the general heading of **parasite-mediated sexual selection** (general reviews of these ideas may be found in Read 1990, Atkinson 1991, Clayton et al. 1992, Folstad and Karter 1992, and Andersson 1994). Its general assumptions—that parasites reduce host fitness, that parasites alter male showiness, that parasite resistance is inherited, and that females choose less parasitized males—are generally supported by experiments and field observations. For example, feather lice produce obvious damage by eating the downy portions of feathers and the barbules of feather vanes (Figure 33-17). In feral rock doves, females preferred clean to

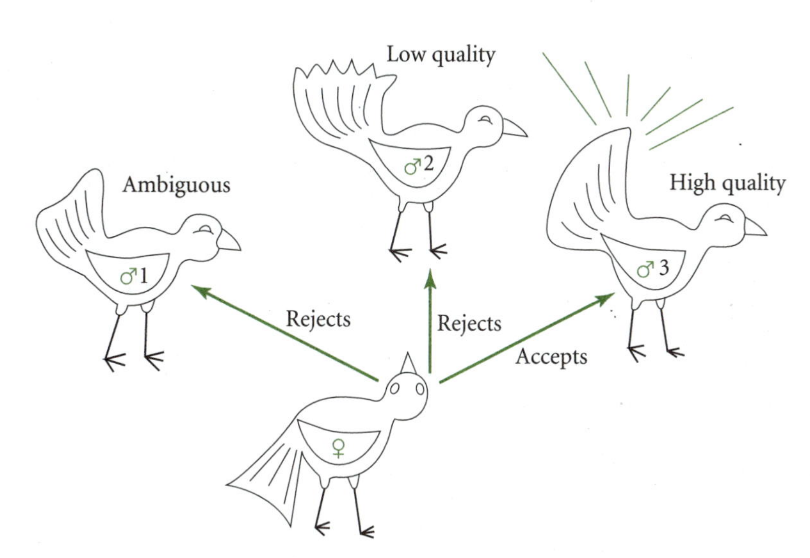

FIGURE 33-16 Parasite-mediated mate choice. Given a choice of three males, a female rejects the one with a short tail, which is too inconspicuous to reveal his parasite load. She also rejects the male whose long tail is obviously damaged by parasites, and chooses the male whose well-kept long tail reveals that he is parasite-free. (*After Clayton 1991.*)

(a)

(b)

FIGURE 33-17 (a) Scanning electron microscopic view of a louse on a host's feather. The louse is about 1 mm long, and is seen in dorsal view. (b) Average (*center*) and heavy (*right*) damage to abdominal contour feathers by feather lice. A normal feather is at the left. (*Courtesy of D. H. Clayton; from Clayton 1990.*)

lousy males by a ratio of three to one; furthermore, highly infested males had higher metabolic requirements in cold weather because of the reduced insulative quality of their plumage, and they were lighter in weight (Clayton 1990).

33.12 Lack of choices for females may lead to extra-pair copulations.

The application of contemporary genetic analysis that allows for the determination of paternity has revealed that, in many species of birds, females routinely copulate with males who are not their identified social mates (Birkhead and Møller 1992, Kempenaers et al. 1992, Graves et al. 1993, Lubjuhn et al. 1993, Hill et al. 1994, Weatherhead et al. 1994, Freeman-Gallant 1996). In some species, as many as one-third of the broods contain one or more offspring sired by these so-called **extra-pair copulations,** or **EPCs.** Most EPCs involve males on neighboring territories, indicating considerable opportunism and infidelity in natural populations. The fitness of the neighboring male is probably increased substantially and at little cost with this strategy. However, recent studies suggest that the cost to the female may be considerable (Weatherhead et al. 1994). The constant threat of EPCs also has selected strongly for mate-guarding behaviors on the part of male birds during their mates' periods of fertility. It is also reasonable to suppose that the intensity of male parental care should decrease with increasing uncertainty of paternity, and so frequent EPCs might even lead to the breakdown of a monogamous mating system.

If females are choosy when it comes to mates, what would lead a female who has chosen a mate, presumably wisely, to engage in extra-pair copulations? The explanation has to do with variation in male quality and in the availability of males to the female (Møller 1992). Females are often not afforded a completely free choice. For example, in monogamous, territorial animals, such as many birds, males are taken out of circulation once they form a pair bond with a female. As high-quality males are paired up, only less desirable males remain for females who have not yet chosen. Females that end up with less desirable males might find it advantageous to engage in extra-pair copulations with more desirable, albeit mated, males in nearby territories if possible. Studies of extra-pair

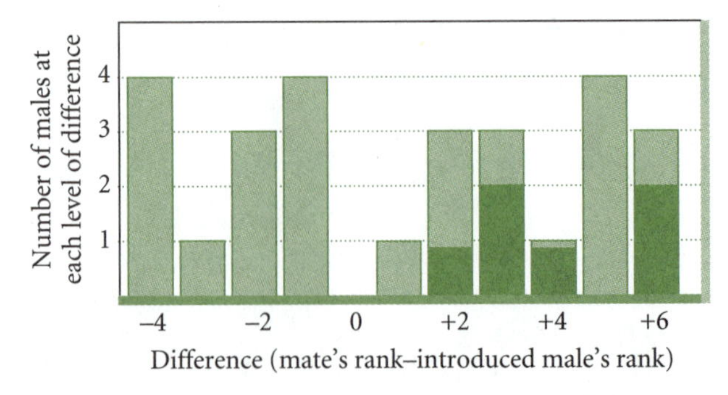

FIGURE 33-18 The number of extra-pair copulations by mated female zebra finches (*Taeniopygia guttata*) (dark green portions of the bars) in relation to how attractive (measured by song rate) those males were when compared with the female's mate. A (+) rank means that the introduced male was more attractive than the female's mate. Females engaged in extra-pair copulations only with more attractive males. (*From Houtman 1992.*)

copulations in zebra finches (*Taeniopygia guttata*) (Birkhead et al. 1990, Houtman 1992) provide supporting evidence for this idea. In that species, when mated females were given a choice of engaging in an extra-pair copulation with males of greater and lesser quality than their mate (quality was measured by the male song rate), they chose only males of higher quality (Figure 33-18).

The advantage to the female of extra-pair copulations is that she may obtain sperm and produce young from a more desirable male than the one she is mated with. This situation may be thought of as a kind of male-male competition, referred to as **sperm competition,** in which males vie to have their sperm fertilize a female's eggs. Very elaborate sperm competition exists in some invertebrate animals.

SUMMARY

1. In most plants and animals, reproduction is accomplished sexually. Sexual function influences the adaptations of organisms.

2. Sexual reproduction results in genetic recombination, yielding offspring that differ genetically. Genetic variation among offspring may be an advantage in variable environments. Asexual reproduction results in offspring that are genetically identical to the parent.

3. Separation of male and female sexual function between individuals is relatively rare among plants, but is common among animals. It is favored when the fixed costs of sexual function are high and the two functions compete strongly. Hermaphroditism is the condition in which a single individual has both sexual functions. Hermaphroditism is favored when the sum of the fitness contribution of the male and female sexual functions in the hermaphrodite is greater than the fitness contribution of separate sexes.

4. The optimal sex ratio in a population balances fitness contributions to progeny through male and female function. In general, because the rarer sex is favored, frequency-dependent selection produces an even sex ratio in most populations. Sex ratio may, however, be influenced by differential costs and benefits of producing males and females, differential mortality, and variation in the condition of the female parent.

5. Females may vary the sex ratio of their offspring in relation to their own breeding condition. Observations and experiments have shown that, in species in which males compete with one another for access to females, females will produce sons when they are in good condition and daughters when they are in poor condition.

6. In some parasitic wasps, males compete with siblings for matings, a situation referred to as local mate competition, and the sex ratio is shifted in favor of females.

7. Competition for resources usually moves the sex ratio toward more males.

8. Mating systems may be monogamous (a lasting bond formed between one male and one female), polygynous (with more than one mate, usually female, per individual), or promiscuous (mating at large within the population, without lasting pair bonds).

9. Polygamy arises when individuals of one sex can monopolize either resources or mates through intrasexual competition. In birds, polygyny is associated with heterogeneous habitats, such as grasslands and marshes, in which the quality of breeding sites varies greatly. Accordingly, some females gain greater fitness by joining a mated male that holds a superior territory than by joining an unmated male on an inferior territory.

10. The configuration of a mating system depends in part on the relative amounts of male and female parental investment. An ESS model of parental investment shows the conditions under which biparental, uniparental, and no parental care evolve. When either parent can care for the offspring, but two are little better than one, the sex of the caring parent may depend on whether fertilization is internal, in which case the female remains with the eggs she lays, or external, in which case the male remains with the eggs he fertilizes.

11. In a few species, individuals of one sex, usually males, may adopt different strategies for gaining matings. These usually contrast dispersing versus nondispersing forms, or territory-holding versus nonterritorial, opportunistic forms. Communal mating areas called leks are a conspicuous feature of some promiscuous birds and other animals. Leks often include males having different courtship behaviors.

12. Three mating systems predominate in plants. Most flowering plants are hermaphroditic. Gynodioecy is the situation in which the population contains a mixture of female and hermaphroditic individuals. Dioecy is the situation in which the population is composed of separate male and female individuals.

13. When males compete with one another for mates, females are able to choose among them. This leads to sexual selection of traits in males that are believed to indicate male fitness. These traits may be in the form of "handicaps" that only the more fit males can bear without encumbrance. Strong female preferences established within a population are believed to result in "runaway" selection of outlandish traits that serve no purpose to the male other than attracting mates, and may be otherwise detrimental.

EXERCISES

1. Are there aspects of human sexual behavior that may be explained in part by the different evolutionary imperatives of males and females?

2. A lek is a communal mating area where males congregate to perform displays to attract females. Are there places in your town that you would consider human leks?

3. Show that under the following conditions, the ESS is for both parents to desert the eggs: $S_0 = S_1 = S_2 = 0.2$ (the same regardless of what combination of parents care for the eggs); $V = 20$ and $v = 10$; $P = 0.6$ and $p = 0.4$.

CHAPTER 34

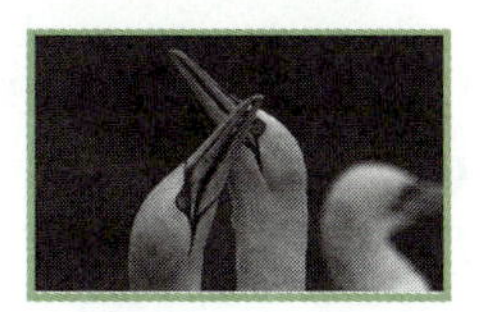

Evolution and Social Behavior

GUIDING QUESTIONS

- What is the relationship between territoriality and dominance hierarchies?
- What is meant by ritualized behavior, and what is its importance in social interaction?
- What are the costs and benefits of flocking behavior?
- What are information centers?
- What are the four types of social interaction?
- What is kin selection, and how does it depend on the concepts of coefficient of relationship and inclusive fitness?
- What are the costs and benefits of altruistic behavior?
- What does the hawk-dove game model tell us about the evolution of cooperative behavior?
- What is reciprocal altruism, and what is required for it to evolve?
- What are the competing evolutionary interests of parent and offspring, and how are they balanced?
- What are the various types of insect social systems?
- How are ant societies organized?

During the course of its life, each individual interacts with many others of the same species: mates, offspring, nondescendant relatives, and unrelated individuals. Each interaction requires the individual to perceive the behavior of others and make appropriate responses. All interactions between conspecifics delicately balance conflicting tendencies of cooperation and competition, altruism and selfishness. In general, interactions between close relatives are more likely to be mutually supportive than interactions between unrelated individuals because of the shared genetic factors that relatives have inherited from their ancestors. In this chapter, we shall explore some of the consequences for individuals of interactions within social and family groups, and we shall describe various ways in which these relationships are managed behaviorally.

Humans are the most social of all animals. Our societies are sustained by role specialization and the cooperation that such interdependence requires. Yet humans also are competitive—to the point of violence—within this mutually supportive structure. Our social life acts to balance these contrasting tendencies toward mutual assistance and conflict. Some animal populations also exhibit much of the complexity of human societies. The social insects—ants, bees, wasps, and termites—are remarkable both for their division of labor and for their behavioral integration within the hive or nest. Similar subtlety of social interaction, including role specialization and altruistic behavior, is being discovered increasingly among other animals, especially among mammals and birds.

34.1 Territoriality and dominance hierarchies organize social interaction within populations.

Social behavior includes all types of interactions between individuals in a population, from cooperation to antagonism. Sometimes social behavior provides a means of organizing and ritualizing the expression of competition within populations and social groups. For example, outright conflict is averted when it is channeled into the posturing of males for social rank or access to mates. The defense of territories and the establishment of dominance hierarchies also may serve this purpose with a minimum of social strife.

Territoriality

Any area defended against the intrusion of others may be regarded as a **territory.** Territories may be transient or more or less permanent, depending on the stability of resources and an individual's need of them. Territoriality is most conspicuously displayed in birds, which may actively defend areas throughout the year or only during the breeding season. Many migratory species establish territories on both the breeding and wintering grounds; shorebirds defend feeding areas for a few hours or days on stopover points on their long migrations. Hummingbirds defend individual flowering bushes and abandon them when their flowering periods are over. Male ruffs and grouse defend a few square meters of space on a communal lek. During the egg-laying period, males of many species accompany their mates and chase off would-be cuckolders. Of course, other animals besides birds defend territories at various times during their life cycles. Among them are mammals, lizards, fishes, and insects. In some animals, territories are defended by groups of individuals rather than by a single individual. Wolf packs are a good example (e.g., White et al. 1996), as are the groups that are formed by many primate species (e.g., Garber et al. 1993). Lions have received considerable attention in this regard (e.g., Grinnell et al. 1995, Heinsohn and Packer 1995, Packer et al. 1990, Heinsohn et al. 1996).

Historically, ecologists have asked two important questions about territoriality: First, under what conditions is territoriality favored over a nonterritorial social organization, and second, when territories are established, what determines their optimal size? Let us briefly explore each of these questions.

Intuition tells us that it takes energy to establish and defend a territory. This intuition, coupled with what we have learned in the previous chapters in this section about optimality modeling, may lead us to reformulate the first question above: When is the benefit of defending a territory greater than the cost? The idea that territoriality should be favored when territories are economically defensible—that is, when the gain from defending the territory outweighs the cost of the defense—was first proposed by Brown (1964). The economic defensibility of a territory is related to the abundance of the resource that is being defended, the intensity with which others compete for the resource, and the spatial distribution of the resource. If resources are scarce, widely dispersed, or if the abundance of the resource changes temporally, as the availability of food might change with season, it may simply be too costly to an individual to defend the resource against others who need it. In that case, it may be more advantageous to live in nonterritorial groups. It is also possible that a resource may be so plentiful that mounting a defense of a particular portion of the resource is a waste of energy. These factors also interact to determine the optimal territory size. Carpenter et al. (1983) found that on stopovers during migration, rufous hummingbirds (*Selasphorus rufus*) defended patches of Indian paintbrush (*Castilleja linariaefolia*) that were just large enough to maximize their weight gain (Figure 34-1).

Dominance Hierarchies

In some situations, dispersion of individuals on territories may not be practical because of the pressures of high population density, the transience of critical resources, or the overriding benefits of living in groups. In such circumstances, conflict among individuals may be resolved by contest, but social rank, rather than space, is the winner's prize. The result of such

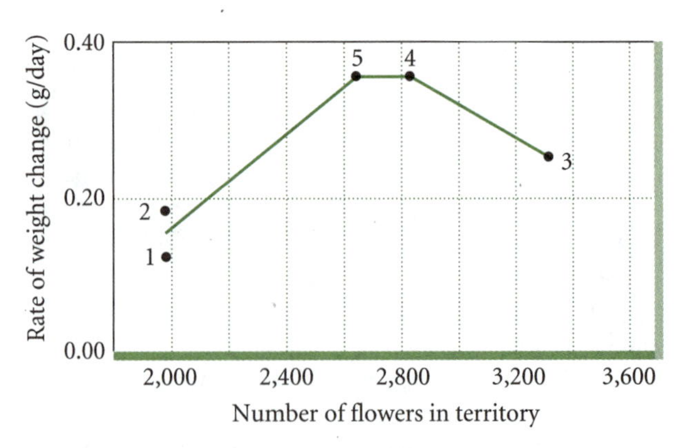

FIGURE 34-1 Rate of weight gain (g d^{-1}) of a rufous hummingbird (*Selasphorus rufus*) on each of five successive days (represented as numbers 1 through 5) on each of which it occupied a territory of different size (as measured by the number of flowers in the territory). The greatest weight gain occurred at intermediate territory size. *(From Krebs and Davies 1993; data from Carpenter et al. 1983.)*

contests is a **dominance hierarchy,** or ranking by social status. Female red deer organize themselves in such a way within the harem, as we mentioned in the last chapter. Once individuals order themselves into a dominance hierarchy, subsequent contests between them are resolved quickly in favor of higher-ranking individuals. When a social hierarchy is linearly ordered, the first-ranked member of a group dominates all others, the second-ranked dominates all but the first-ranked, and so on down the line to the last-ranked individual, who dominates none.

Dominance hierarchies form as a way of organizing access to food, to mates, or to space that may contain food or mates. Recent studies have shown that in a number of types of animals, whether or not an individual obtains access to the best feeding sites is dependent on its position within a dominance hierarchy (Hughes 1992, Grand and Grant 1994, Kaiser and Mushinsky 1994, Mueller et al. 1994, Nakano and Furukawa-Tanaka 1994, Iwasaki 1995, Nakano 1995). Access to mates or number of copulations is a consequence of social rank in some insects (e.g., Choe 1994), mammals (Thompson 1993, Haley et al. 1994), and fishes (Kodric-Brown 1993), to name a few. In some ants, dominance hierarchies of females determine which female in a colony will be the egg layer (Ortius and Heinze 1995). Dominance hierarchies may not be restricted to animals that occur in social groups. For example, Cigliano (1993) demonstrated that normally solitary octopuses (*Octopus bimaculoides*) form relatively stable dominance hierarchies when confronted experimentally with a limited number of den sites. Bigger octopuses (measured by wet weight and length of the mantle) were most dominant, and as a consequence, obtained a den in contests with lighter and shorter individuals. We should also point out that, although we think of dominance hierarchies as involving primarily individuals of the same species, interspecific dominance hierarchies may also arise in nature. For example, Nakano and coworkers observed interspecific dominance hierarchies between white-spotted char (*Salvelinus leucomaenis*) and Dolly Varden (*Salvelinus malma*), two species of fish with similar foraging strategies that occupy streams in Japan (Nakano and Furukawa-Tanaka 1994).

Relationship Between Territoriality and Dominance Hierarchies

Territoriality and dominance hierarchies are alternative expressions of the same social tendencies. We see this most clearly when a population switches from one to the other as circumstances change. For example, the dragonfly *Leucorrhinia rubicunda* switches from a territorial system with strong site fidelity at low densities to one of broadly overlapping feeding areas at high densities (Pajunen 1966). In one area of Finland, dragonflies in a sparse population were found to be spaced 3 to 7 meters apart along the edge of a pool; in a dense population, individuals were a half meter apart on average. Pajunen (1966) scored the intensity of interactions between males (only males are territorial) from low values of 1 for no interaction, through intermediate values corresponding to varying degrees of threat and pursuit, to 5 for threats followed by fighting. Where the dragonflies were sparse, the frequency and intensity of aggressive interactions were high (average score 4.2) compared with those in densely inhabited areas (3.1) because of frequent territorial defense in the first area. In dense populations, individual dragonflies rarely returned to specific resting sites, which are usually stems of emergent vegetation. Instead, they flew over larger areas of the ponds and alighted to rest less frequently than did dragonflies in sparser populations (Figure 34-2). Both the level of territorial defense and the level of site fidelity decreased as the density of dragonflies in an area increased. Territoriality also is more likely to be manifested as resources become more defensible and valuable. In an experimental study, Anna's hummingbirds (*Calypte anna*) defended artificial feeders more aggressively as their provision of sugar water was increased to a substantial proportion of the birds' daily energy requirement (Ewald and Carpenter 1978, Ewald and Orians 1983).

The position of an individual in a dominance hierarchy is sometimes reflected by its spatial position within a group. In large foraging flocks of wood pigeons, individuals low in the dominance hierarchy tend to be at the periphery, where they are more exposed to predators than the dominant individuals occupying the center of the flock. Peripheral birds

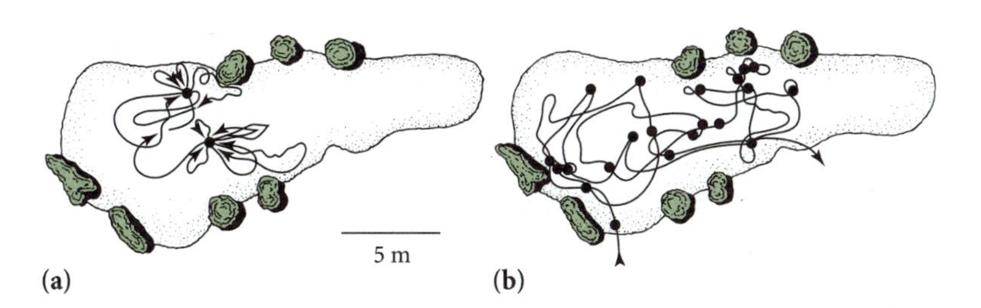

FIGURE 34-2 Examples of male flight activity in the dragonfly *Leucorrhinia rubicunda* showing (a) well-developed site fidelity and (b) weak site fidelity. The outline of a pool and several bushes are indicated; periods of observation were 1 and 13 minutes, respectively. (*After Pajunen 1966.*)

appear to be nervous, and because they spend much of their time looking up from feeding, they are often undernourished. Birds in the center of the flock are generally calmer and feed more because they are protected from the surprise attack of a predator by the vigilance of the individuals at the periphery (Murton 1970).

34.2 The communication of social dominance usually is ritualized.

Most conflict associated with social rank or territorial defense involves **ritualized behaviors** that rarely lead to risky physical struggle. These behaviors allow individuals to ascertain their chances of winning a contest, and presumably they have evolved because they allow contests to be resolved without bloodshed. Certain appearances or behaviors appear to signal higher status than others. Why, then, don't less dominant individuals assume the behavior of their betters? That is, why don't they cheat the system and deceive other members of the population into thinking that they are more aggressive than they really are? Part of the answer is that some ritualized behaviors allow contestants to judge each other's size, which is difficult to fake and which often determines the outcome of earnest combat. Another part of the answer is that signaling status and being able to back it up must go together. This has been demonstrated by a remarkable set of experiments by Sievert Rohwer (Rohwer 1977, 1982, Rohwer and Rohwer 1978, Rohwer and Ewald 1981) on Harris's sparrows (*Zonotrichia querula*).

Harris's sparrows, which are related to the more familiar white-throated and white-crowned sparrows, breed in the Canadian Arctic and winter in small flocks in the central United States. Social status within flocks is highly correlated with the amount of dark coloration in the plumage of the throat and upper breast (Figure 34-3). Upon first encounters, lighter birds generally avoid darker individuals, and dark coloration meaningfully signals status. When Rohwer dyed the plumage of light individuals dark, they were mercilessly attacked by others who easily saw through the ruse. However, dark individuals bleached to a lighter color found themselves constantly having to attack naturally light birds to regain their status. The system works because plumage and aggression go together—there are no cheaters. When Rohwer implanted testosterone into lighter birds to raise their level of aggression, implanted individuals that also were dyed darker rose in the dominance hierarchy. Those left with their light plumage were persecuted by naturally dark birds and could not rise in status in spite of their raised level of aggression. Behavior and plumage probably are affected by common physiological conditions associated with hormone levels. When the two do not match, birds get into trouble, either because they do not get their due respect or because they are treated as impostors.

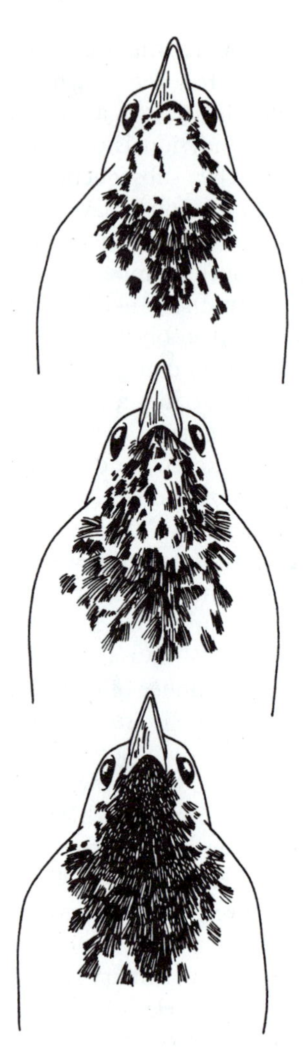

FIGURE 34-3 Examples of variation in the throat plumage of Harris's sparrows. Individuals having naturally dark plumage enjoy a higher status in the dominance hierarchy. (*After Morse 1980.*)

Rohwer's experiments raise a difficult question: If social status depends on the level of a hormone whose production imposes little physiological cost, why don't all members of the population have the highest level of aggression? If a sparrow's social rank can be elevated with a nickel's worth of hormone and dye, what's to stop any individual from assuming the trappings of high rank? Similar hormone implants have turned monogamous song sparrows into bigamists (Wingfield 1984) and nonterritorial red grouse into landowners (Watson and Parr 1981). At this point, we can only presume either that status is linked to age and experience, according to which hormone-mediated levels of behavior are adjusted to serve other purposes, or that variation in rank is a mixed evolutionarily stable strategy in which dominance has its costs and all members of the hierarchy have roughly equivalent fitness.

Given that status truly represents an individual's capabilities in behavioral encounters, one might also ask: Why do low-ranking individuals, which are excluded from food and mates and exposed to higher risks, remain with the flock? The answer must simply be that it is better to be a low-ranking individual in a flock, perhaps rising in rank with age, than to be a loner. Both the establishment of flocks and the sizes of flocks must balance costs and benefits to their members. One presumes that individuals will not associate with others unless they gain some personal advantage from it.

Many conflicts between individuals occur over territories or mates. In the case of territories, usually one of the contestants holds the territory over which the contest ensues. In such cases, if the intruder could somehow gauge the ability of the territory owner to defend the territory relative to the intruder's own ability to overpower the owner, then it would have useful information with which to decide whether to continue the engagement. An organism's ability to defend a territory or resource is referred to as its **resource holding potential** (RHP). Body size and the availability of weapons such as horns or antlers, among other things, determine an individual's RHP. In some cases, prior ownership of a resource may enhance an individual's RHP. In general, animals can reduce the chance of injury or failure in a contest if they can assess the RHP of their opponent (Parker 1974, Maynard Smith 1982).

34.3 Group living confers advantages and disadvantages.

Animals get together for a variety of reasons. Sometimes they are independently attracted to suitable habitat or resources and fortuitously form aggregations, like those of vultures around a carcass or dungflies on a cowpat. Within such groups, individuals may interact, usually to contest space, resources, or mates. In other cases, progeny remain with their parents to form family groups. In this case, aggregation results from a failure to disperse. True social groups arise through the attraction of individuals to one another—that is, through a purposeful joining together. The evolutionary motivation for such behavior presumably resides in increased individual fitness through facilitated feeding, protection from predators, or increased access to mates.

Vigilance

In groups, individuals tend to spend more time feeding and less time looking out for predators, a behavior referred to as **vigilance.** Consider the data presented in Figure 34-4 for the European goldfinch (*Carduelis carduelis*), which feeds on seed heads of plants in open fields and hedgerows. Two factors control optimal flock size in these birds. As flock size increases, each individual spends less time looking out for predators. If you watch closely as birds feed, you will notice that they raise their heads and look around from time to time. In a larger group, an individual goldfinch can spend more time going about the business of eating,

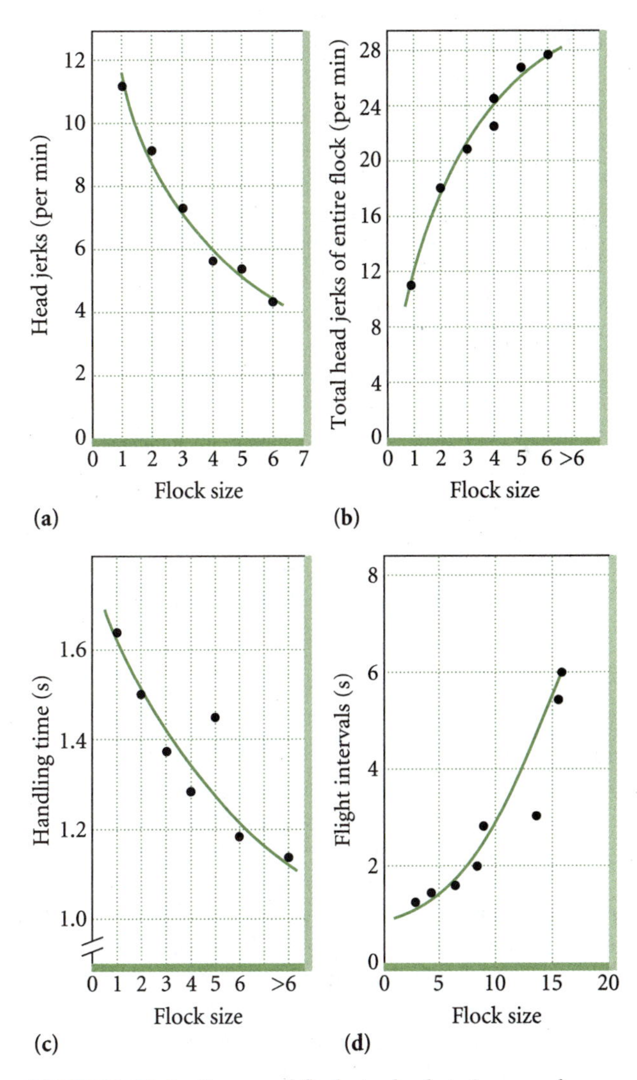

FIGURE 34-4 Increased flock size leads to increased security from predation, but reduced efficiency of foraging, in the European goldfinch (*Carduelis carduelis*). (a) Mean rates of looking up from foraging per individual; (b) total vigilance rate for the entire flock; (c) time required to husk each seed of sorrel (*Rumex acetosa*); (d) time required to move from one plant to the next for foraging individuals, at different flock sizes. (*After Glück 1987.*)

and can gather and husk seeds more rapidly, because the total vigilance of the flock is higher (Glück 1987). Balancing this advantage of reduced individual vigilance time, a larger flock depresses a local food supply faster, and individuals are forced to fly farther between suitable foraging patches, using valuable feeding time and energy and perhaps increasing their vulnerability to predators. Thus, joining a flock is a good choice for an individual as long as the flock is not too large. This situation is reminiscent of the problem of optimal lek size (see Chapter 33).

The idea that animals may flock together because many eyes are better than one has been considered in detail (Lima 1995, Lima and Zollner 1996). It has been presumed that such a system works because the perception of danger by one animal is communicated to all the other animals in the flock, a behavior referred to as **collective detection.** It is also assumed that each animal must be able to gauge the level of vigilance of other animals in the flock in order to know how to adjust its own vigilant behavior. That is, if an individual is to benefit from being in the flock, it must know how much others are assisting in watching out for predators in order to know how much less vigilance is required on its part. Recent studies of flocks of dark-eyed juncos, *Junco hyemalis,* and American tree sparrows, *Spizella arborea,* provide only limited support for the ideas of collective detection and individual monitoring of vigilance. Lima (1995) has suggested that the advantage of flocks is most likely in simultaneous detection of predators by multiple individuals, rather than the communication of danger from one individual to the rest of the flock. Clearly, this is an area of behavioral ecology that merits more study.

Information Centers

Individuals in flocks may learn the location of food from others, cooperate in obtaining food, or obtain information about their environment from one another. Ward and Zahavi (1973) suggested that breeding colonies, roosting congregations, and other group formations may serve as **information centers,** meaning that individuals can learn about places to feed, for example, from the behavior of others, perhaps by following them to a food source. Certainly animals are capable of learning from the behavior of others (Klopfer 1959, Alcock 1969, Mason and Reidinger 1981, Langen 1996), and both field observations (Krebs 1974, Pitcher et al. 1982, Brown 1986, Heinrich and Marzluff 1995) and experiments (de Groot 1980, Galef and Wigmore 1983, Templeton and Giraldeau 1995) suggest that individuals can utilize the behavior of others—so-called public information—to locate resources.

Some colonially nesting seabirds appear to gain information about the quality of a nesting location from the reproductive success of their neighbors. Kittiwakes are small seabirds, related to gulls, that nest on cliffs (Figure 34-5). Nesting areas vary in quality due to variation in the abundance of ticks, which infest chicks. When a pair of kittiwakes successfully rears a family in one year, it usually returns to the same cliff to breed in the next. When a pair is unsuccessful, its decision to return to the same cliff or move elsewhere depends on the success of others breeding in the same area. Danchin (1987) observed pairs on the coast of Brittany, in France, that failed in the egg stage of the nesting cycle. Nearly all pairs nonetheless returned to the same site the following year when the failure rate of others around them was less than 25%, but more than one-third abandoned the cliff when the failure rate of their neighbors exceeded 50%. The success or failure of an individual pair in any given year is largely a matter of chance. By observing what is happening around them, individuals can gain a better estimate of their own probability of success in the future.

Social groups open the door to cooperative behavior. Cooperation involves the coordination of individual behavior toward a common goal (Packer and Ruttan 1988), as in defense of the herd by male musk oxen; pack hunting by wolves, killer whales, and other mammals (Mech 1970, Kruuk 1972, Schaller 1972) and by some birds (Bednarz 1988); or the coordinated behavior of aquatic birds that corral fish into a small area where they can be fed upon easily (Bartholomew

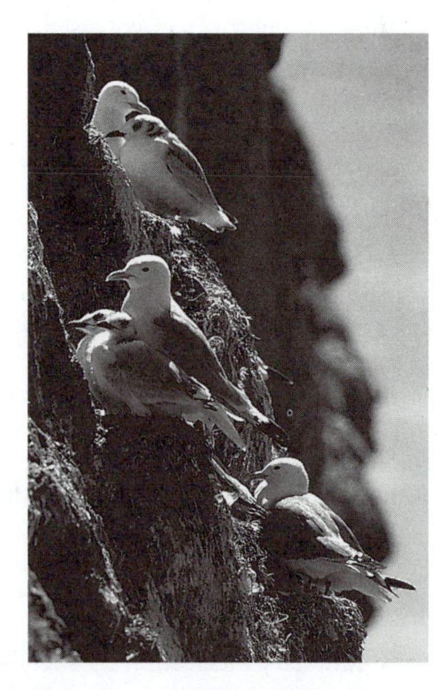

FIGURE 34-5 Black-legged kittiwakes (*Rissa tridactyla*) on a nesting ledge on the coast of Newfoundland, Canada. The chicks have a black collar marking.

1942, Emlen and Ambrose 1970). Such instances of cooperation are not common, but where they do occur, social behavior serves more to foster cooperation among individuals than to organize competition among them: individual behavior becomes subservient to the group, and groups may assume a characteristic behavior of their own. Considering our own position as the most social of all species, it is natural that humans have devoted considerable effort to learning about the evolution and organization of animal societies.

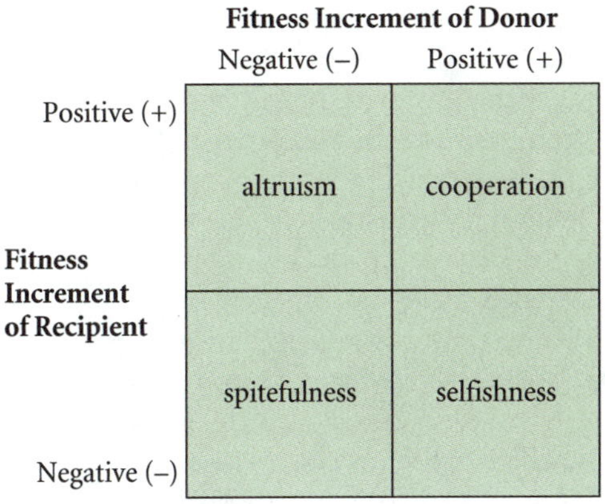

34.4 The evolutionary modification of social interaction balances the costs and benefits of social behavior.

Any social interaction other than mutual display can be dissected into a series of behavioral acts by one individual, referred to as the **donor** of the behavior, directed toward another, called the **recipient.** One individual delivers food, the other receives it; one threatens, the other is threatened. When one individual attacks another, the attacker may be thought of as the donor of a behavior. The attacked individual (the recipient in this case) may respond by standing its ground or by fleeing. In either case, it thereby becomes the donor of a subsequent behavior. The donor-recipient distinction is useful because each act has the potential to affect the reproductive success of both the donor and the recipient of the behavior. These incremental effects on reproductive success, or fitness, may be positive or negative, depending on the interaction.

Four combinations of cost and benefit to donor and recipient can be used to organize social interactions into four categories (Figure 34-6). **Cooperation** and **selfishness** both benefit the donor of the behavior, and therefore occur frequently. **Spitefulness**—behavior that reduces the fitness of both donor and recipient—cannot be favored by natural selection under any circumstances, and presumably does not occur in natural populations. The fourth type of behavior, **altruism,** benefits the recipient at a cost to the donor. Altruism presents a difficult problem because it requires the evolution of behaviors that reduce the fitness of the individuals performing them. We would expect selfish behaviors to prevail to the exclusion of altruism because they increase the reproductive success of the donor. However, altruism appears to have arisen in several groups of animals. For example, in colonies of social insects, workers forgo personal reproduction to rear the offspring of the queen, their mother. A similar situation exists in the burrow-dwelling naked mole rat (*Heterocephalus glaber*), in which nonreproductive individuals tend the nest and rear the young of a single female who breeds with a few males. We humans also like to think that we are not only capable of altruistic behavior, but that such interactions hold together the fabric of our society.

Fitness Increment of Recipient	Fitness Increment of Donor: Negative (–)	Fitness Increment of Donor: Positive (+)
Positive (+)	altruism	cooperation
Negative (–)	spitefulness	selfishness

FIGURE 34-6 The four types of social behavior classified according to their effects on the fitness of donors and recipients of the actions.

34.5 Group selection, kin selection, and reciprocal altruism have been proposed to explain the occurrence of altruistic behavior.

The evolutionary problem posed by altruism has been appreciated for a long time. Darwin recognized that the presence of nonreproductive individuals in colonies of social insects represented a possible difficulty for his notion of evolution by natural selection. The distinctive feature of these sterile castes of insects is that they forgo personal reproduction and devote themselves to the fertility of the colony as a whole. Darwin explained this apparent inconsistency by suggesting that selection might apply to colonies as well as to individuals. Darwin thus considered insect colonies to be family groups whose fitness was measured in competition with other colonies. The adaptations of the sterile castes—and by extension, the existence of sterile castes at all—were selected according to their contribution to the productivity of the colony as a whole.

This concept of intergroup selection, originally applied to insect societies, eventually was adopted more widely to explain the evolution of social behavior in all kinds of animal groups (Wright 1931, 1945). It was accepted with little interest and more or less uncritically for many years. Then, in the early 1960s, the Scottish zoologist V. C. Wynne-Edwards (1962, 1963) advocated **group selection** so forcefully, and saw its magnificent consequences for social behavior so universally, that evolutionary ecologists were forced to consider the argument more carefully. Wynne-Edwards believed that individuals restrained themselves from reproduction in order to keep their population from overtaxing its resources. Restraint from reproduction is altruistic behavior toward selfish individuals, which attempt to monopolize resources and enhance their personal fitnesses within the population. Wynne-Edwards argued that selection of groups that altruistically nurtured their resources predominated over selection of selfish individuals within groups, which led to overexploitation.

Opponents of Wynne-Edwards's views offered two counterarguments: first, few populations have the kind of group structure required for intergroup selection, and second, even where such selection could exist, it was bound to be weaker than individual selection because groups displace one another within populations more slowly than individuals displace one another within groups. By the mid-1960s, individual selectionists had clearly won the day (Lack 1966, Williams 1966a, b), and group selection theory faded from view. But without group selection, how was one to explain cases of altruistic behavior? One solution was to argue that such behavior did not exist; another was an idea proposed by John Maynard Smith (1964) and William D. Hamilton (1964), which they called kin selection.

34.6 Kin selection may favor altruistic acts between related individuals.

The evolutionary dilemma posed by the apparent altruism of social insects is resolved when one realizes that their colonies are discrete family units, containing mostly the offspring of a single female (the queen). Therefore, behavioral interactions within an ant colony or beehive occur between close relatives—in this case, siblings. When an individual directs a behavior toward a sibling or other close relative, it influences the reproductive success of an individual with which it shares more of its own genetic makeup than it does with an individual drawn at random from the population. This special outcome of social behavior among close relatives is referred to as **kin selection.**

Close relatives have a certain probability of inheriting copies of the same gene from a particular ancestor. The likelihood that two individuals share copies of any particular gene is the probability of **identity by descent,** the value of which varies with degree of relationship (Table 34-1). For example, two siblings have a 50% probability of inheriting copies of the same gene from one parent. This probability is also called their **coefficient of relationship.** Two cousins have a probability of one in eight (0.125) of inheriting copies of the same gene from one of their grandparents, which are their closest shared ancestors. The coefficient of relationship between an individual and one of its parents is 0.50, since it must have received the gene in question from one parent or the other. Reciprocally, because each parent contributes half of its genes to each of its offspring, the probability that a son or daughter will inherit a particular gene is also 0.50. The genetic relationship between full siblings is 0.50, because the gene possessed by one individual was inherited from either its mother or its father, which, in either case, would pass the gene on to a sibling with a probability of 0.50. A schematic representation of these probabilities is provided in Figure 34-7.

TABLE 34-1 Probabilities of identity by descent between one individual and others having various degrees of relationship

Relationship	Probability of identity by descent
Parent	0.50
Offspring	0.50
Full sibling	0.50
Half-sibling	0.25
Grandparent	0.25
Grandchild	0.25
Uncle or aunt	0.25
Nephew or niece	0.25
First cousin	0.125

From this pattern we can see that when an individual behaves in a particular way toward a close relative, that act influences not only its own personal fitness, but also the fitness of an individual that shares a portion of its genes. Suppose that an act of altruism is caused by a genetic factor that was inherited from one parent. When this act is directed toward a sibling, the probability that the recipient of the behavior will also have a copy of that gene is 50%. Therefore, the occurrence of the gene within the population as a whole will be determined both by its influence on the fitness of the donor of the behavior and by its influence on the fitness of the recipient, discounted by the coefficient of relationship between them.

The fitness of an individual plus the fitness of its relatives, weighted according to the coefficient of relationship, is called the **inclusive fitness** of that individual. The inclusive fitness of a gene responsible for a particular behavior equals the contribution to the

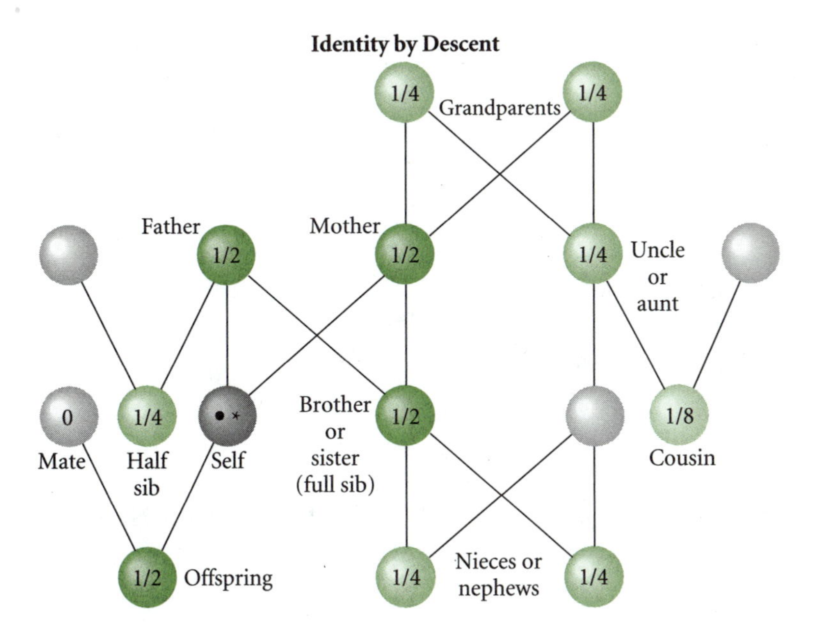

FIGURE 34-7 Degrees of genetic relationship (identity by descent: probability of occurrence of a copy of the same gene) among relatives.

reproductive success of the donor resulting from its own behavior plus the product arrived at by multiplying the contribution of the behavior to the reproductive success of the recipient times the probability that it carries a copy of the same gene. Therefore, the inclusive fitness of an altruistic gene would exceed that of its selfish alternative as long as the cost to the altruist was less than the benefit to the recipient. Thus, a genetic factor for altruism will have a positive fitness and increase in the population when the cost (C) of a single altruistic act to the donor is less than the benefit (B) to the recipient times the coefficient of relationship (r) between donor and recipient; that is, when $C < Br$. When this equation is rearranged, the condition for the evolution of altruism becomes $C/B < r$; that is, the cost-benefit ratio, which is a measure of how altruistic the behavior is, must be less than the average coefficient of relationship of the recipient of the altruistic behavior to the donor.

Just as inclusive fitness makes possible the evolution of altruism among close relatives, the same considerations constrain the evolution of selfish behavior. With B now representing the benefit to the donor of a behavior and C the cost to the recipient, selfish behavior among close relatives can evolve only when $B > Cr$, or $C/B < 1/r$. The cost-benefit ratio (C/B) is, in this case, a measure of the selfishness of the behavior. The higher the coefficient of relationship (r) between donor and recipient, the lower the level of selfishness that can evolve.

Hamilton (1964) devised a formula for this cost-benefit relationship that can be applied either to altruistic or to selfish behavior. Suppose that the fitness contributions of a single behavior were w_s to the donor (self) and w_i to the recipient. The net fitness contribution of the act would be $w_s + r_i w_i$, which is the inclusive fitness because it includes components (which may be gains or losses) realized through the personal fitness of both the donor and the recipient. Summed over all the interactions of the donor, the total fitness contribution of the behavior is

$$W = w_s + \sum f_i r_i w_i,$$

where f_i is the proportion of interactions with individuals having the coefficient of relationship r_i. (This equation applies strictly only when genes for the behavior in question are rare; otherwise, distantly related individuals may bear the same gene with high probability. It is therefore most useful in evaluating behaviors as evolutionarily stable strategies.)

For behaviors directed toward individuals having a coefficient of relationship i, $W > 0$ when $w_s > -r_i w_i$, or $w_i > -w_s/r_i$. This relationship between w_i and w_s, which increases inclusive fitness, is shown graphically in Figure 34-8. As with Figure 34-6, the graph shows both negative and positive values for both w_s and w_i. The line $w_s = -0.5_i w_i$ is shown, indicating the situation in which the recipient is a full sibling of the donor. The net fitness $W = 0$ when $w_s = 0$, so the line passes through the origin. Points to the right of the line correspond to behaviors for which the fitness contribution to the donor is greater than half that to the recipient ($w_s > 0.5w_i$), and points to the left of the line correspond to behaviors for which the fitness contribution to the donor is less than half that to the recipient ($w_s < 0.5w_i$). The diagram shows how the concept of inclusive fitness applies to both altruistic and selfish behavior. When siblings interact, the evolutionarily sound limits of generosity are

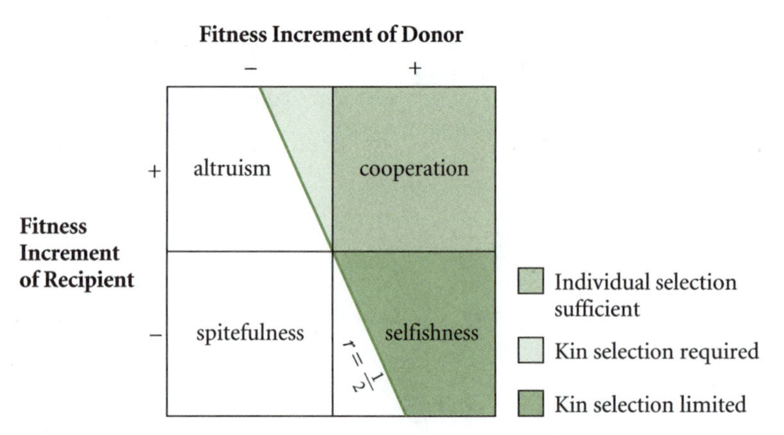

FIGURE 34-8 Combinations of fitness increments to donors and recipients of behaviors having positive inclusive fitnesses when the coefficient of relationship between donor and recipient is $\frac{1}{2}$.

$-w_s < 0.5w_i$, and the evolutionarily tolerable limits of selfishness are $w_s < -2w_i$. As the coefficient of relationship decreases below 0.50, altruistic behavior becomes less likely to increase inclusive fitness ($-w_s < r_i w_i$), and selfish behavior more likely ($w_s < -w_i r_i$), as shown in Figure 34-9.

The maintenance of altruistic behavior by kin selection requires that such behavior be restricted to close relatives. Certainly individuals of many species tend to associate in family groups. When dispersal is limited, the nearby individuals with which interactions are most frequent are likely to be close relatives. Moreover, individuals seem to be sensitive to their degree of relationship to others, even when they have had no family experience that would help them learn who their relatives are (Hoogland 1982, Bateson 1983, Carlin and Hölldobler 1983, Lacy and Sherman 1983, Klahn and Gamboa 1983, Holmes and Sherman 1985, Fletcher and Michener 1987, Waldman 1988, Pfennig et al. 1993). The cues that indicate relationship must involve some subtle matching of chemical, acoustic, or visual signals that are under genotypic control and are extremely variable within populations.

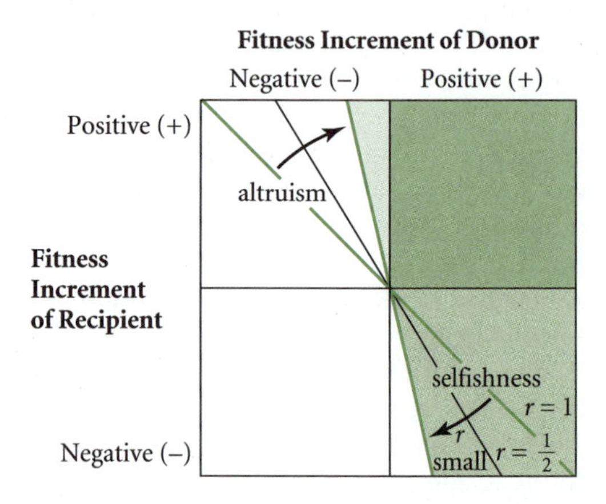

FIGURE 34-9 Reducing the coefficient of relationship between donor and recipient increases the inclusive fitness of selfish behaviors and reduces that of altruistic behaviors.

34.7 Several behavioral systems suggest the operation of kin selection.

We find evidence of kin selection in a number of the behavioral systems of animals: alarm calling, and two aspects of the phenomenon of the extended family, in which offspring stay for some time with their parents.

Alarm Calling

Alarm calls are often given by animals when they see or are captured by a predator. Paul Sherman's observations (1977) on the behavior of Belding's ground squirrels (*Citellus beldingi*) in the Sierra Nevada of California offered strong support for the hypothesis that the alarm trills given in the presence of terrestrial predators are maintained by kin selection. (Alarm whistles to aerial predators appear not to be so maintained [Sherman 1985].)

Because ground squirrels are social creatures and are exposed to view in their montane meadow habitat (Figure 34-10), the threat of predation is great. Reducing this threat with alarm calls is an important factor in their lives. Sherman found that alarm callers to terrestrial predators were adult and yearling females. Males and juvenile females did not exhibit the behavior (Figure 34-11). He also found that males disperse widely from their place of birth, whereas females are philopatric. Moreover, females called more frequently when they had occupied the same territory for many years. Thus, those most likely to give alarm calls (females) resided in areas where relatives were likely to be nearby.

Sherman also determined that alarm calling to terrestrial predators has a cost. When predators approached groups of ground squirrels, at least one of whom gave an alarm call, 14 of 107 calling individuals were attacked (13%) compared with 8 of 168 silent individuals (5%). These observations are consistent with kin selection playing a role in the evolution of alarm calling.

FIGURE 34-10 Belding's ground squirrel (*Citellus beldingi*) at Tioga Pass, California. (*Courtesy of P. W. Sherman; from Sherman 1977.*)

Extended Families

Extended families in humans include the nuclear family of a mated pair and their young progeny as well as, to varying degrees, older siblings, grandparents, uncles and aunts, cousins, nephews and nieces, and sometimes individuals of uncertain relationship to the rest. These are complex social units within which occur a tremendous variety of social interactions, most of them cooperative but many competitive enough to stress the bonds that hold a family unit together. Rarely do human extended families include more than one child-producing pair, and at least a portion of the behavior of the non-nuclear members of the family is directed toward supporting the well-being and upbringing of the children.

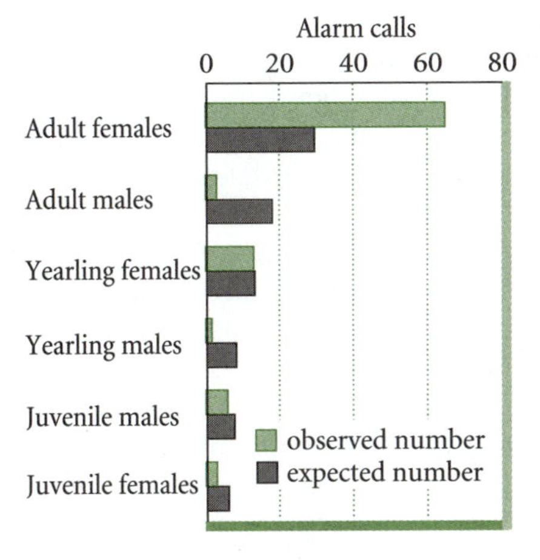

FIGURE 34-11 Expected and observed frequencies of alarm calling in response to the presence of a predatory mammal by sex and age classes of Belding's ground squirrels. Expected values were computed by assuming that individuals call in direct proportion to the number of times they are present when a predatory mammal appears. (*From Sherman 1977.*)

For the most part, family life in animals—and in plants, for that matter—is not so complex. In most animal groups, offspring receive no support from their parents or their relatives beyond nutrients placed in the egg to nourish the embryo. Even when parents care for their young, as is the case in all mammals, in most birds, and sporadically in other groups, the family is a transient phenomenon limited to the period of growth and early development, after which offspring disperse to begin their own independent lives.

Occasionally, however, offspring stay with their parents for extended periods, and may even remain permanently associated with them in extended families. This is the case for several hundred species of birds, including some species, such the white-winged choughs (*Corcorax melanorhamphos*) of Australia, that require helpers in order to successfully produce young (Heinsohn 1995). One of the most studied of these species is the Florida scrub jay (*Aphelocoma coerulescens*), in which offspring remain with their parents for a year or more (Woolfenden and Fitzpatrick 1984, Mumme 1992, Hailman et al. 1994, Schoech 1996, Schoech et al. 1996). Detailed observations of this species have shown that the stay-at-home jays do not themselves breed, but rather help to care for the offspring produced by their parents during the next breeding season, which are their younger brothers and sisters. Such behavior, now documented in many species, prompted many researchers to include this "helping at the nest" in the category of altruistic behaviors. In many such cases, including that of the scrub jay, it has been shown that helpers increase the reproductive success of their parents. That is reason enough for parents to tolerate their offspring staying at home.

Whether helping is a purely altruistic behavior, however, depends on the alternatives available. Scrub jay habitat is densely occupied, and young birds have little chance of setting up territories of their own. One-year-olds can increase their inclusive fitness by helping their parents while gaining experience that may later improve their own personal fitness as independent breeders. Staying at home provides another immense benefit: the possibility of inheriting the family territory when the father dies.

Studies of the white-fronted bee-eater, a bird species that lives in East Africa, have revealed complex extended families (Emlen et al. 1995). These families are typically multigenerational groups of three to seventeen individuals, often including two or three mated pairs plus assorted single birds—unpaired young and widowed older individuals. Careful observations of

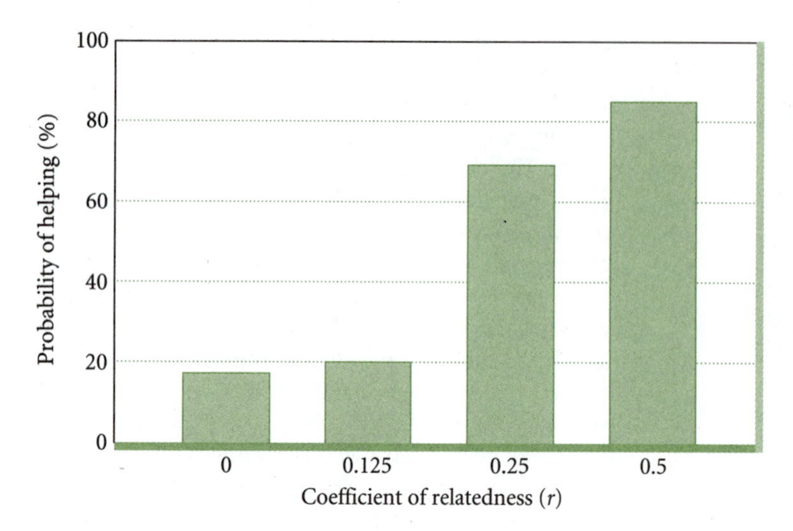

FIGURE 34-12 Distribution of altruistic behaviors as a function of degree of relationship within family groups in the white-fronted bee-eater. (*From Emlen et al. 1995.*)

individually marked birds over several years have shown that these social groups are truly extended families, comprising only related individuals and their mates, which normally come from other families. Although relationships within extended families tend to be cooperative, bee-eater family groups are hardly models of harmonious behavior. One sees the usual squabbling over food, nest sites, and who mates with whom typical of society at large. Remarkably, however, selfless and selfish acts appear to be directed toward other individuals very much in accordance with their degree of relationship. Brothers and sisters are treated better than half-siblings and uncles, for example, and cousins fare almost as badly as nonrelatives outside the family group (Figure 34-12). Through their behavior, bee-eaters tell us that individuals know who their relatives are and can distinguish subtle differences in degree of relationship. We can also conclude from the distribution of helpful and harmful behaviors in this species that inclusive fitness is the appropriate measure of selection on social behavior. Altruistic behaviors can indeed evolve among close relatives by kin selection.

34.8 Kin selection has been implicated in the evolution of warning coloration.

Can cooperative or altruistic behavior evolve among nonrelatives within a society? Clearly, social groups can form out of the self-interest of group members seeking protection from predators or perhaps some efficiency gained by foraging or hunting in groups. Whether groups of unrelated individuals can take the next step toward true cooperation is a fundamental issue in the evolution of social behavior.

Warning coloration signals predators that potential prey items are unpalatable, poisonous, or both. Conspicuous advertising of these qualities benefits the vulnerable prey if it dissuades would-be captors. It also benefits predators who recognize the meaning of the display and avoid unwise food choices. Adopted by all members of both prey and predator populations, these conventions of advertisement and appropriate response are unquestionably evolutionarily stable strategies. But how does the system get started? Can a genetic mutation leading to warning coloration invade a population of unpalatable, cryptic prey hunted mercilessly by visually oriented predators that cannot distinguish them from other, palatable species? A single individual with warning coloration would be rendered conspicuous and vulnerable in a population of otherwise cryptic organisms, and would probably be nailed by an unsuspecting predator. Noxiousness and warning coloration can confer fitness only if the potential predator has opportunities to learn their meaning—by eating another individual.

In the case of a novel mutation, the most likely "other individual" in a population with the same rare gene would be a parent or sibling. Half of the individuals in a single brood are likely to have the same unique gene inherited from one parent. If the conditioning stimulus is strong enough, one attempt by a predator to eat a noxious, aposematic individual may save the lives of several of its brothers and sisters. With its inclusive fitness thus enhanced, the mutation could spread rapidly through the population.

This scenario requires that members of a brood remain in close proximity so that the experience of a single predator is translated into the rejection of a sibling. Among butterflies and moths, warning coloration and aggregation of broods of larvae are strongly associated (Sillén-Tullberg and Leimar 1988), which is consistent with the idea that kin selection is important to the evolution of warning coloration (Figure 34-13). Among species that disperse their eggs

	Dispersion of larvae	
Appearance	aggregated	solitary
Aposematic	9 species	11
Cryptic	0	44

FIGURE 34-13 Relationship between the appearance of larvae of various lepidopteran species and their degree of aggregation. (*From Jarvi et al. 1981.*)

widely over many food plants, warning coloration is less frequent.

34.9 A game theory model indicates how individuals should interact socially within a large population.

Self-interest rules behavior among distantly related individuals. A paradox of selfish behavior in a social setting is that conflict can reduce the reproductive success of selfish individuals below the likely success of cooperative individuals. Because natural selection favors increased reproductive success, it should be possible for cooperation to evolve within societies at large. The problem with this reasoning is that when most of a society consists of cooperative members, a selfish individual can greatly increase its personal reproductive success by "cheating"; thus, selfish behavior will always be favored by natural selection, which will prevent groups from crossing the threshold of cooperative behavior to become true societies.

The Hawk-Dove Game

The logic of this somewhat pessimistic argument can be shown by a simple game theory analysis (Smith 1982, Maynard Smith 1982). The analysis used is the **hawk-dove game.** Let us assume that there are two types of individuals in a population. One type always behaves selfishly in conflict situations, always being willing to fight over contested resources and taking all the reward when it wins: this is hawk behavior (H). In contrast, doves (D) never contest a potential resource, but share it evenly with other doves. Each contest between two individuals has a potential reward, or benefit (B), and has a cost (C) when a contest results in physical conflict. The payoff, both to hawks and to doves, depends on the behavior of the second contestant—that is, on whether it is a hawk or a dove (Table 34-2). For example, two hawks always fight, and on average, get half the resource, so that the payoff is $\frac{1}{2}B - C$. When a hawk confronts a dove, the hawk gains the entire uncontested resource without a cost; thus, the payoff is B. When two doves interact, the payoff to each is $\frac{1}{2}B$; there is no cost.

The average payoff (fitness contribution) to hawks and doves depends on the relative proportions of the two kinds of individuals in a population. Let p be the proportion of hawks and $(1 - p)$ the proportion of doves. The payoffs are now as follows: hawks receive $p(\frac{1}{2}B - C) + (1 - p)B$, and doves receive $\frac{1}{2}(1 - p)B$. One can see that a population consisting only of hawks ($p = 1$) has an average payoff of $\frac{1}{2}B - C$, which is less than the average payoff of $\frac{1}{2}B$ in a population consisting only of doves ($p = 0$). Clearly the dove strategy would be better from a social point of view.

The problem is that dove behavior is not an evolutionarily stable strategy. It cannot resist evolutionary invasion by an alternative strategy (a genetic mutation, if you like)—namely, hawk behavior. A single hawk in a population of doves (p close to 0) receives twice the average payoff that doves do (B versus $\frac{1}{2}B$) because it never encounters another hawk, and conflicts are never contested. Thus, in a world of doves, the hawk strategy increases rapidly. Not only can hawk behavior invade a dove population, but a pure hawk population is also resistant to invasion by doves, except when the cost of conflict is very high relative to the benefit. In that case, doves can survive in a hawk population because the hawks fight so much among

TABLE 34-2 Costs and benefits in the hawk-dove game

Behavior of first player	Response of second player: Hawk	Response of second player: Dove
Hawk	Shares both the benefit and the cost of conflict $\frac{1}{2}B - C$	Gains entire benefit B
Dove	Neither gains benefit nor assumes cost 0	Shares benefit without cost of conflict $\frac{1}{2}B$

themselves and incur the costs of doing so. When p is close to 1 (a pure hawk population), the payoff to hawks is $\frac{1}{2}B - C$, and that to doves is 0. Thus, hawkish behavior is an evolutionarily stable strategy as long as $B > 2C$. When the benefit is less than twice the cost of conflict, doves can invade the hawk population, and the eventual outcome is a mixed population of hawks and doves with the proportion of hawks (p) equal to $\frac{1}{2}B/C$. The hawk-dove game, as well as more complex game theory analyses, demonstrates how difficult it is for cooperative behavior to evolve among unrelated individuals.

Reciprocal Altruism

One way around these constraints on the evolution of altruism is the strategy of **reciprocal altruism,** in which an individual is cooperative (altruistic) toward doves, but fights back against hawks (Trivers 1971, 1985). In this case, the behavior of the individual is contingent on the behavior of others in the population. This strategy is also referred to as **tit-for-tat,** that is to say, giving back in kind. The average payoff to a reciprocal altruist in the hawk-dove game is $p(\frac{1}{2}B - C) - \frac{1}{2}(1 - p)B$, or $\frac{1}{2}B - pC$. This payoff is always as good or better than that of a dove, and never worse than that of a hawk; thus, reciprocal altruism is a superior strategy that could invade and predominate in any mixed population of doves and hawks.

How do reciprocal altruists behave toward each other? In a tit-for-tat world, the first encounter colors all future interactions between two individuals. Thus, a benefit-of-the-doubt strategy—cooperate initially, until one finds out that another individual's intentions are not altruistic—is the only one that will lead to a cooperative society. Under this scenario, reciprocal altruism will lead to cooperation generally within the population—reciprocal altruists are nice guys who are willing to fight against cheaters.

Although reciprocal altruism provides a model for the evolution of altruistic behavior, it relies on long-term associations among individuals during which they can learn individual behavior patterns and act toward others in their group accordingly. Individual recognition and a high probability of return altruistic acts in the future are both necessary ingredients for the evolution of reciprocal altruism. These conditions appear to have been met in some bird and mammal species in which individuals feed unrelated individuals, a behavior referred to as **alloparental care.** It has been hypothesized that this behavior may be adaptive if the recipients of such care learn to perceive the donor as a relative and later, based on the perceived kinship, lend a hand to the donor. This process is termed **kinship deceit** (e.g., Connor and Curry 1995).

Perhaps the best-documented case of reciprocal altruism is that of vampire bats, in which successfully foraging individuals share their blood meals with others that fail to feed on a particular night (Wilkinson 1984). Vampires roost in small groups whose composition may be stable for many months or years, so individuals get to know one another well and could easily refuse others who do not cooperate. Because fewer than 10% of bats are unsuccessful on a particular night, sharing meals is not overly costly (less than 5% of total food intake if sharing is equally spread), but the survival benefits of receiving part of a meal may be substantial.

A further difficulty with reciprocal altruism is that because it requires a reciprocating partner in order to work, it can confer little fitness when reciprocators are rare. In fact, an altruistic gene cannot easily invade a nonreciprocating population. However, reciprocal altruism can be an evolutionarily stable strategy, as Axelrod and Hamilton (1981), Smith (1982), and Axelrod and Dion (1988) showed using game theory analysis. Suppose that a population consists of three genotypes: selfish individuals, which always take what is offered them, with a gain of two units of fitness (the accept strategy, A); altruistic individuals, which always give help, at a cost of one unit (offer, O); and reciprocators (R), which always do what an individual last did to them, except that, being ever hopeful, their first act is to offer. The payoff matrix for repeated interactions between two individuals in such a population is shown in Figure 34-14. In a population consisting of all three types, fitnesses are ranked in the order $W_A > W_R > W_O$. As the altruistic offerers are eliminated from the population, the reduced payoff matrix in the absence of offerers favors reciprocators over acceptors, and reciprocal altruism prevails. But one can also see that in a pure population of acceptors, the reciprocator has no net fitness advantage. Hence the evolution of altruism through reciprocity requires some mechanism to increase the proportion of reciprocators initially to the point that they can do each other some good. This may be particularly difficult because reciprocation is a complicated strategy involving responses conditioned on prior experience.

		Strategy of recipient		
		accept	offer	reciprocate
Strategy of donor	accept	0	2	0
	offer	–1	1	1
	reciprocate	0	1	1

FIGURE 34-14 Payoffs in a game theory analysis of reciprocal altruism. (*After Smith 1982.*)

Clearly, mechanisms exist for the evolution of cooperative and altruistic behavior in small family groups and in social groups. Whether these behaviors can balance the universal self-interest of the individual in such large groups as human societies remains to be seen. Mechanisms of reciprocal altruism and cooperation are firmly embedded in our culture and social institutions, but cheating and selfish behavior are also widespread. Regardless of how this issue is resolved, the study of social behavior emphasizes the conflicting values of self-interest and group interest. Nowhere is this conflict stronger (and perhaps more personally felt) than in relations between parent and offspring.

34.10 The optimum level of parental investment can differ for parents and their offspring.

Rather than passively accepting whatever their parents offer, most offspring actively solicit care. Young animals beg for food and solicit brooding; eggs actively take up nutrients from the ovarian tissues or bloodstream of the mother. For the most part, the fitness interests of parent and offspring are compatible; when progeny thrive, so do their parents' genes. But when the selfish accumulation of resources by one offspring reduces the overall reproductive success of the parents, parent and offspring can come into **parent-offspring conflict.** Trivers (1974) was the first to give parent-offspring conflict a theoretical basis. Each act of parental care benefits the offspring by enhancing its survival, but costs the parent by decreasing the number of other contemporary or future offspring. Resources allocated to one offspring cannot be delivered to others. Prolonged parental care delays the birth of subsequent offspring. The risks of caring for today's offspring decrease the probability that parents will survive to rear tomorrow's. Thus, there is always a conflict between present and future reproductive success. Offspring try to resolve that conflict in favor of present reproductive success (that is, themselves); parents benefit from a more balanced distribution of their investment.

From the standpoint of the parent, its offspring are genetically equivalent, and it should not exhibit preferences among them in delivering care. From the standpoint of an individual offspring, however, the self has twice the genetic value of a sibling because a sibling shares only half of an individual's genes. Therefore, when an individual possesses a genetic factor that increases the care it receives from its parents, that trait is favored as long as the cost to the parents, in terms of number of siblings, is less than twice the benefit to the individual. This is the limit to selfish behavior under kin selection that we discussed above.

As the offspring develop and become more self-sufficient, the ratio of benefit to cost of continuing to care for them decreases (Figure 34-15). The cost of a particular parental act may change little over the age of the offspring, but as the young mature and are better able to care for themselves, the benefits of parental care dwindle. When the benefit-cost ratio drops below 1, the parent should cease to provide care to that offspring in favor of producing additional ones. Suppose, however, that an offspring has a gene that increases its solicitation of parental care. Because the inclusive fitness of the offspring discounts the cost of care to the parent by one-half the cost of not delivering care to siblings, it "prefers" that parental care continue until the benefit-cost ratio is 0.50 if its parents' future offspring are full siblings, and less if they are not. Thus, the period of time between the age at which B/C is 1 and the age at which it is 0.50 is one of conflict between parent and offspring.

Postadolescent humans may encounter this region of conflict if they express a wish to live with their parents after finishing college (unless they also engage in helping behavior). Biologists believe that similar conflicts over time of weaning can be seen in many species of mammals and birds that provide extensive postnatal care. Young, often perfectly capable of taking care of themselves, sometimes hound their parents mercilessly for food. One would think that parents would have the upper hand in any conflict with their offspring, but one must remember that parents are adapted to respond positively to solicitations by their

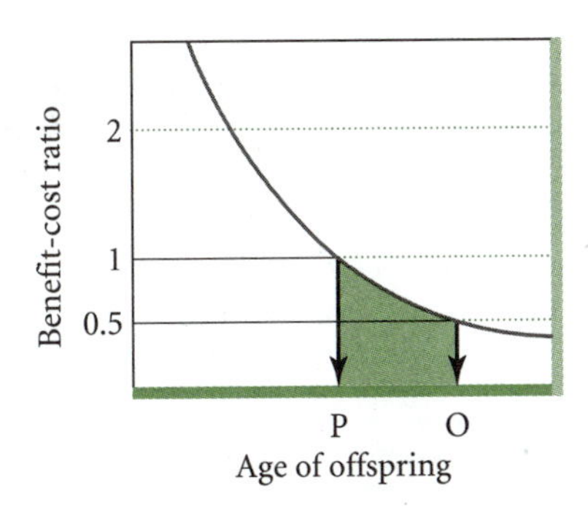

FIGURE 34-15 The ratio of benefit to cost of an act of parental care decreases with the offspring's age as it grows and becomes more self-sufficient. Because all offspring are of equal genetic value to an individual parent, parents should shift care to succeeding offspring when the benefit-cost ratio falls below 1 (age P). Because siblings have a coefficient of relationship of 0.50, however, inclusive fitness arguments dictate that offspring should solicit parental care until the benefit-cost ratio equals 0.50 (age O). This difference establishes a region of parent-offspring conflict (shaded). (*After Trivers 1972.*)

offspring while they are growing and not yet independent. By prolonging juvenile appearance and dependent behavior, offspring may be able to take advantage of their parents' responsiveness and prolong parental care.

34.11 Eusocial insect societies are based upon sibling altruism and parental despotism.

The complex societies of the termites, ants, bees, and wasps have presented a formidable challenge to evolutionary ecologists, primarily because of the existence of nonreproductive castes (Brockman 1984, West-Eberhard 1981, Choe and Crespi 1997). There is some disagreement among behavioral ecologists regarding the terminology to be applied to social systems in insects and other arthropods (Wilson 1971, Sherman et al. 1995, Crespi and Choe 1997). One scheme proposed by Crespi and Yanega (1995) identifies five different social systems that represent progressively more complex steps in the evolution of social behavior (Table 34-3). **Subsocial** systems are those that are characterized simply by the presence of parental or biparental care. In **colonial** societies, multiple females share breeding sites where they each provide care separately to their young. One of the best understood of such systems is that of the colonial web-building spider *Metepeira incrassata,* which has been the subject of research by G. Uetz and colleagues for over a decade (see review by Uetz and Hieber 1997). *Metepeira* spiders occur both solitarily and in colonies, in which their orb webs share a common silk framework. When compared with solitary individuals, spiders in colonies were found to capture more prey and to experience less variation in prey intake (Rypstra 1989, Uetz 1988, Spiller 1992). Colonies of *Metepeira* attract predators and parasitoids (Spiller and Schoener 1989, Hieber and Uetz 1990, Uetz and Hieber 1994), which represents a cost of colonial living, but this cost may be balanced to some extent because spiders in the interior of colonies are subject to less predation (Rayor and Uetz 1990, 1993).

Communal, cooperatively breeding, and **eusocial** (true social) systems lie along continua of increasing cooperation among different individuals in parental care, increasing division of labor in that care, and increasing developmental specialization for care of young. Individuals of communal species share in the care of young, and such care involves some division of labor in cooperatively breeding species. In eusocial systems, individuals are present that are irreversibly behaviorally specialized (Crespi and Yanega 1995). Eusociality is limited among insects to the termites (Isoptera) and the ants, bees, and wasps (Hymenoptera), although the elements of eusociality are present in some spider species and, as we mentioned earlier, one mammal, the African naked mole rat.

The complex organization of the eusocial insect society is dominated by one or a few egg-laying females referred to as **queens.** Nonreproductive progeny of the queen gather food and care for developing brothers and sisters, some of which become sexually mature, leave the colony to mate, and establish new colonies. Most insect societies are huge extended families.

Biologists believe that eusociality evolved independently many times in bees, wasps, and ants. It is less clear what route was followed to get there. The most widely accepted sequence of evolutionary steps includes a lengthened period of parental care of the developing brood, with parents either guarding their nests or continuously provisioning their larvae in a manner similar to birds feeding their young. If parents lived and continued to produce eggs after their first progeny emerged as adults, then their offspring would be in a position to help raise subsequent broods, consisting of their younger siblings. This overlapping of generations, which is not common among insects, and extensive parental care are necessary ingredients in the recipe for eusociality. Once progeny remain with their mother after they attain adulthood, the way is open to relinquishing their own reproductive function solely to support hers. (Seger [1991] provides a concise

TABLE 34-3 Types of arthropod social systems

Type of society	Brood care	Shared breeding site	Cooperation in brood care	Alloparental brood care	Castes
Subsocial	+				
Colonial	–	+			
Communal	+	+	+		
Cooperatively breeding	+	+	+	+	
Eusocial	+	+	+	+	+

(From Crespi and Choe 1997.)

overview of the ideas regarding the evolution of social organization.)

Bee societies are simply organized: females are divided between a sterile worker caste and a reproductive caste that is produced seasonally. Whether an individual will become a sterile worker or a fertile reproductive is controlled by the quality of nutrition given the developing larva (Light 1942, 1943). In general, differentiation of sterile castes is stimulated by environmental, usually nutritional, factors (Brian 1979, 1980). The development of sexual forms can also be inhibited by substances produced by the queen and fed to the larvae. In bees, the worker caste represents an arrested stage in the development of the reproductive female, stopped short of sexual maturity.

Ant and termite colonies often have a continuous gradation of worker castes, ranging from very small individuals that are primarily responsible for the nutrition of the colony to larger individuals that are specialized morphologically to defend against intruders (Oster and Wilson 1978; Figure 34-16). In the leaf-cutter ant (*Atta*), small workers sometimes ride on sections of leaves carried by large workers from which they ward off parasitic flies, like the sidekick riding shotgun on a stagecoach (Figure 34-17). In most insect societies, workers assume specialized roles at any given time, although these may change with the age of a worker or conditions in the colony. Colonies of harvester ants in the southwestern deserts of the United States have five distinct worker roles, each with a characteristic pattern of behavior: nest interior worker (brood care, nest construction, seed storage), midden (refuse pile) worker, forager, patrolling ant, and nest surface maintenance worker (Figure 34-18a). Changes in conditions at a nest cause certain ants to switch to different tasks, and once a switch is made, it is permanent. For example, increased food availability causes midden workers, nest surface maintenance workers, and patrolling ants to switch to foraging behavior; threats of intrusion by foreign ants cause nest surface maintenance workers to become patrolling ants, which defend the colony (Figure 34-18b.)

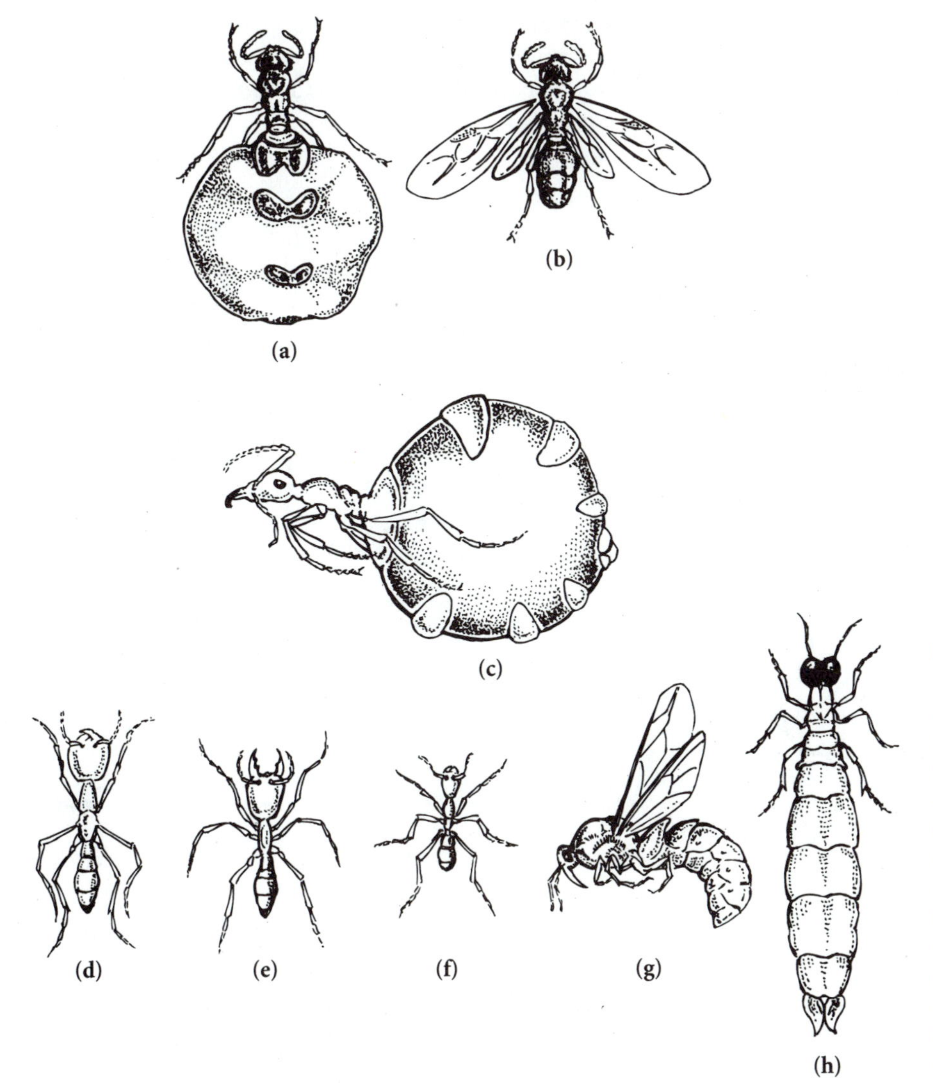

FIGURE 34-16 Variation among castes of several species of ants: (a) a virgin queen and (b) an old egg-laying queen of the workerless social parasite *Anergates atralulus* of Europe; (c) a so-called replete worker of the honey ant *Myrmecocytus melliger* from Mexico; (d–h) three sizes of blind workers, a winged male, and a queen, blind and wingless, of the African visiting ant *Dorylus nigricans*. (*After Wheeler 1923, Grasse 1951.*)

FIGURE 34-17 Leaf-cutter ants returning to their nest. Note the small worker riding on one of the leaf cutouts.

Unlike ant, bee, and wasp societies, termite colonies are headed by a mated pair—the king and queen—which produce all the workers by sexual reproduction. Workers are both male and female, but neither sex matures sexually unless either the king or queen dies. The queens that produce colonies of ants, bees, and wasps mate only once during their lives and store enough sperm to produce all their offspring—up to a million or more over 10–15 years in some army ants. As we saw in the last chapter, hymenopterans have a haplodiploid sex-determining mechanism: workers are all females produced from fertilized eggs.

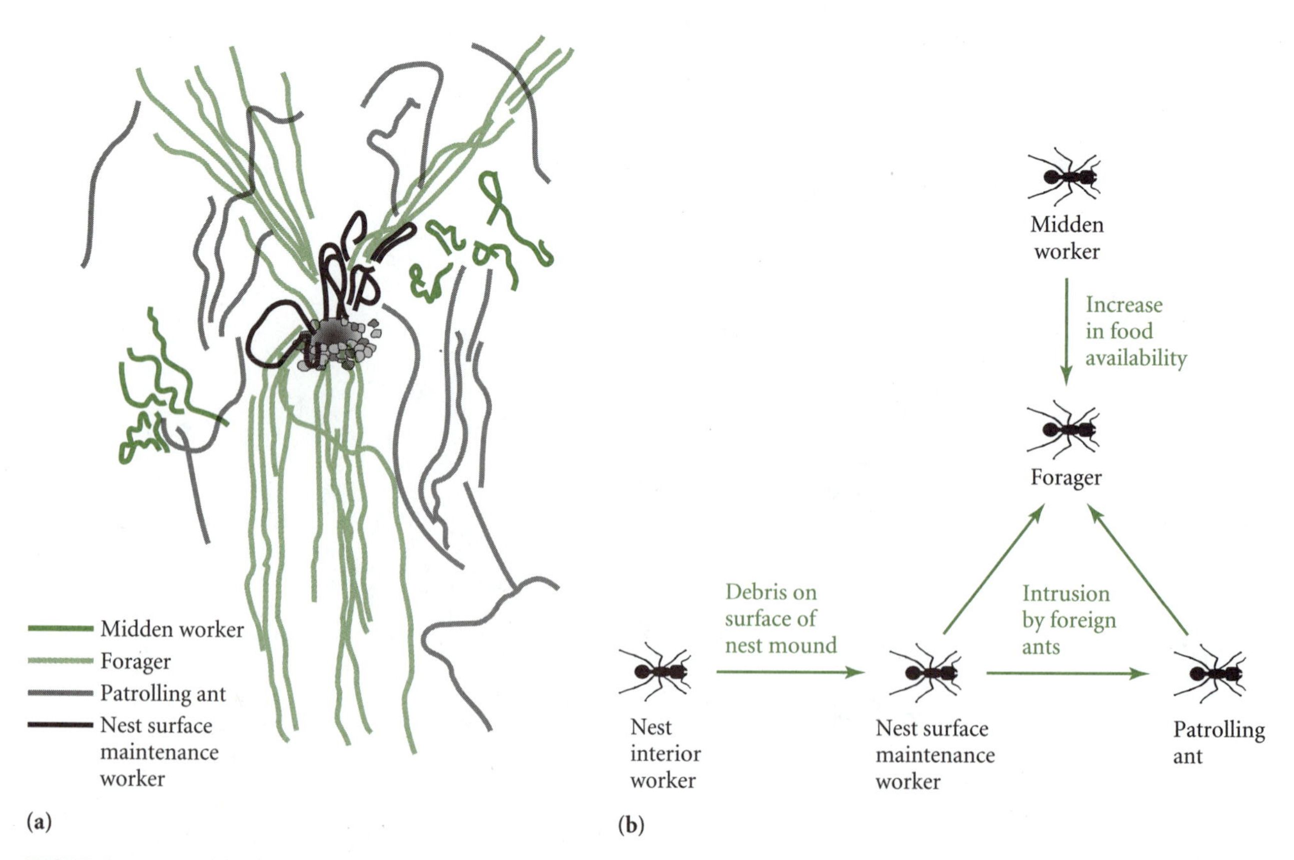

FIGURE 34-18 (a) Characteristic patterns of paths taken by four types of harvester ant (*Pogonomyrmex barbatus*) workers outside the nest, and (b) the directions of switching between these roles when conditions at the nest change. (*After Gordon 1995.*)

TABLE 34-4 Probabilities of identity by descent between male and female individuals and their relatives in eusocial hymenopterans

Probability of identity by descent with	Male	Female
Mother	0.50	0.50
Father	0.00	0.50
Brother	0.50	0.25
Sister	0.25	0.75
Son	0.00	0.50
Daughter	1.00	0.50

Males, which develop from unfertilized eggs, appear in colonies only as reproductives (drones) that leave to seek mates.

As mentioned above, the caste structure of insect societies is maintained primarily by nutrition and by means of substances produced by the queen that arrest sexual development in the worker castes. In the hymenopterans, workers acquiesce to this system—that is, they respond appropriately to chemical signals and nutrition—partly because they would be killed by the queen or by other workers if they embarked on sexual development, and partly because their inclusive fitness differs little whether they rear siblings or their own offspring. Haplodiploidy creates strong asymmetries in coefficients of genetic relationship within hymenopteran societies (Table 34-4). In particular, a female worker's coefficient of relationship to a female sibling is 0.75, whereas to a male sibling it is 0.25. The queen herself has the same genetic relatedness to sons and to daughters (0.50), so she can be relatively ambivalent about the sex of her offspring, especially when the sex ratio among reproductives in the population as a whole is near equality. The skewed genetic relatedness among siblings means that cooperation is likely to be greater among all-female castes than among male castes or, especially, among mixed castes. This may explain why workers in hymenopteran societies are all female, and why broods of reproductives usually favor females, by about 3:1 on a weight basis. Furthermore, when a female worker can help to produce more female than male reproductives, her own inclusive fitness may actually be higher than it would be if she raised a brood of her own consisting of an equal number of males and females. Under this bizarre circumstance, it is not surprising that sterile castes might have evolved.

In many species of social insects, colonies have multiple queens, which may not even be close relatives. In this situation, the degree of genetic relationship among workers is greatly reduced on average, and selection of altruistic traits by means of high inclusive fitness may not be strong enough to maintain the integrity of the colony. The fact that such multiqueen social units exist suggests that queens themselves exercise decisive control over caste formation. The balance must be very tenuous, however, as shown by the following example (Keller and Ross 1993). The fire ant *Solenopsis invicta,* an introduced pest in the southern United States, has two kinds of colonies: in one, there is a single queen; in the other, up to hundreds of queens may be sexually active, each laying relatively small numbers of eggs. The multiqueen colonies produce smaller workers and a lower proportion of reproductives, many of which remain within the colony to breed; colonies are founded by multiple, unrelated queens, and new queens may also be adopted by a nest from outside the colony. The *Solenopsis* population in the United States is polymorphic for an enzyme-encoding genetic locus known as *Pgm-3.* Allele *a* of this gene, when present in homozygous form, causes queens in multiqueen colonies to have a higher reproductive rate than *a*/*b* heterozygotes and *b*/*b* homozygotes. Apparently, the *a*/*a* genotype upsets the delicate balance of relationships within a colony, and *a*/*a* queens are quickly killed by workers. How they recognize *a*/*a* queens is not known, but the process probably involves a chemical cue.

Behavioral relationships among the social insects represent one extreme along a continuum of social organization from animals that live alone except to breed to those that aggregate in large groups organized by complex behavior. Regardless of its complexity, however, all behavior balances costs and benefits to the individual and to close relatives affected by the behavior. Like morphology and physiology, behavior is strongly influenced by genetic factors and thus is subject to evolutionary modification by natural selection. The evolution of behavior becomes complicated when individuals interact within a social setting, where the interests of individuals within a population may either coincide or conflict. Understanding the evolutionary resolution of social conflict in animal societies continues to be one of the most challenging and important concerns of biology.

SUMMARY

1. Selection imposed by behavioral interactions with members of one's family and with unrelated individuals within one's population provides the evolutionary basis for social behavior.

2. Territoriality is the defense of an object or area from intrusion by other individuals. Territories are maintained when the resources thus gained are rewarding and defensible.

3. Dominance hierarchies order individuals in social groups by rank, which is established by direct confrontation. Because rank is generally respected, dominance relationships reduce conflict within the group. Contestants over resources often have ways of assessing their opponent's resource holding potential—its ability to defend a resource against a competitor. Depending on resource dispersion and population density, populations may switch between territorial and dominance organization.

4. Living in large social groups may benefit individuals by enabling them to better detect and defend against predators or to obtain food more efficiently. Groups form to the extent that these advantages outweigh the negative effects of competition among group members.

5. Individual acts of social behavior involve a donor and recipient. When both benefit, the behavior is termed cooperation or mutualism; when the donor benefits at a cost to the recipient, the behavior is selfish; when the recipient benefits at a cost to the donor, the behavior is altruistic.

6. The presence of altruistic behavior in populations has been explained in terms of kin selection. Kin selection arises because when an individual interacts with a relative, it affects the fitness of that portion of its own genotype that is also inherited by the relative directly from a common ancestor.

7. Inclusive fitness expresses the benefit (or cost) of a behavior to the donor plus the benefit (or cost) to the recipient, adjusted by the coefficient of their relationship. In the case of interactions between siblings, which have a coefficient of relationship of 0.50, selection will favor any altruistic behavior whose cost to the donor is less than one-half the benefit it confers on the recipient.

8. Inclusive fitness arguments have been invoked to explain such behaviors as alarm calling and delaying one's own sexual maturation to help one's parents rear younger siblings. In general, the distribution of cooperation and altruism within social groups is sensitive to degree of genetic relatedness between individuals.

9. Game theory analyses, such as the hawk-dove game, indicate that cooperative behavior cannot evolve among nonrelatives even though the average benefit to individuals in a purely cooperative social group exceeds that gained by individuals through confrontation and conflict. The reason is that cooperative behavior is not an evolutionarily stable strategy, but can be invaded by selfish cheaters.

10. When individuals live in close association over long periods, the choice of cooperation or conflict by one individual may be made contingent on the experienced behavior of others in the group. In this way, cooperation may pervade through the mechanism of reciprocal altruism, in which members of the social group withhold cooperation from cheaters.

11. Conflict may arise between parents and offspring over the optimal level of parental investment. All siblings are genetically equal in the eyes of their parents, but siblings are genetically related to one another by only 0.50. Therefore, individual offspring should prefer unequal parental investment in themselves, even when parental fitness is reduced as a result.

12. Social insects (termites, ants, wasps, and bees) live in extended family groups in which most offspring are retained in the colony as sterile workers, increasing their mother's fitness by rearing reproductive siblings.

13. The haplodiploid sex-determining mechanism of the Hymenoptera results in females having a coefficient of relationship of 0.75 to sisters, but only 0.25 to brothers. This skew probably has contributed to the facts that workers in ant, bee, and wasp colonies are all female and that more female reproductives are produced than males.

EXERCISES

1. Consider a bird that feeds on the nectar of flowers. The amount of nectar available per flower depends on the rate of nectar production and the rate at which it is extracted by nectar feeders. The amount of time that the bird must forage each day to obtain enough nourishment to survive is dependent on the average amount of nectar contained in the flowers in the area in which it forages. The bird could increase the average amount of nectar in the flowers by defending a feeding territory, thereby excluding other birds that would deplete the nectar supply. However, there is a cost to the territorial defense. Using what you learned in Chapter 31, develop a simple verbal model to explain the behavior of the bird if it does only two things: forage and defend the territory.

2. In our discussion of kin selection theory we emphasized the importance of the degree of relationship between donor and recipient. Another important aspect of kin selection is kin recognition: the ability of individuals to distinguish those who are related to them from those who are not. Speculate on the mechanisms of kin recognition. How would the spatial distribution of the organisms come into play?

3. Are there ritualized behaviors in humans?

4. Spitefulness—behavior that reduces the fitness of both donor and recipient—cannot be favored by natural selection under any circumstances. Yet, the behavior occurs in human populations. How might this be explained?

5. To what extent do the principles discussed in this chapter apply to human social systems?

GLOSSARY

Abscission. The process by which a part of an organism separates from the rest.

Acclimation. A reversible change in the morphology or physiology of an organism in response to environmental change; also called acclimatization.

Active flux. *See* Active transport.

Active transport. Movement of ions or other substances across a membrane against a concentration gradient, requiring the expenditure of energy. Also referred to as active flux.

Activity space. The range of environmental conditions suitable for the activity of an organism.

Adaptation. A genetically determined characteristic that enhances the ability of an individual to cope with its environment; an evolutionary process by which organisms become better suited to their environments.

Adaptationist program. The interpretation of form and function strictly in the context of adaptation.

Adaptive allometry. The adaptive significance of allometric relationships.

Adaptive function. In fitness set analysis, a mathematical expression combining the fitnesses of a phenotype in each of several environments into a measure of overall fitness in a heterogeneous environment.

Adaptive landscape. A metaphor for evolutionary change that holds that populations occupy positions in a landscape of allele frequencies that represent possible levels of fitness.

Adaptive peak. The combination of allele frequencies at which a population has a local maximum fitness.

Adaptive radiation. The process of diversification through evolution.

Additive variance (V_A). Variation in a phenotypic value within a population due to the difference in expression of alleles in the homozygous state.

Adhesion. The attraction of water molecules to surfaces having a charge.

Adiabatic cooling. Decrease in temperature caused by the expansion of air with lower atmospheric pressures at higher altitudes.

Aerenchyma. Specialized vascular tissue in some plants that conducts air directly from the atmosphere to the roots.

Age class. The individuals in a population of a particular age.

Age structure. The relative proportions of a population in different age classes.

Aggregation. *See* Clumped distribution.

Aggressive mimicry. A situation in which one organism mimics the morphology or behavior of another in order to lure it close enough to eat.

Alarm pheromone. A chemical exuded by an organism that is injured that warns conspecifics of possible danger.

Alary polymorphism. A situation in which individuals may have one of several wing lengths depending on the environmental conditions in which they live.

Alkaloids. Nitrogen-containing compounds, such as morphine and nicotine, that are produced by plants and are toxic to many herbivores.

Allele. One of several alternative forms of a gene.

Allelopathy. Direct inhibition of one species by another using noxious or toxic chemicals.

Allochthonous. Originating outside of a system, such as minerals and organic matter transported into streams and lakes. *Compare with* Autochthonous.

Allometric constant. Slope of the relationship between the logarithm of one measurement of an organism and the logarithm of another, usually its overall size.

Allometric equation. An equation representing the relationships between rates of processes and dimensions of objects: $Y = aX^b$, where a is the constant of proportionality, b is the allometric constant, and X and Y are rates of change of size and physiological process respectively.

Allometry. A relative increase in some part of an organism or some measure of its physiology or behavior in relation to some other measure, usually its overall size.

Alloparental care. Care, such as feeding, that is provided to unrelated individuals.

Allopatric. Occurring in different places; usually referring to geographic separation of populations.

Allozyme. One of several slightly different forms of a protein. Allozymes often have different electric charges and thus move different distances on electrophoresis gels. *See also* Electrophoresis.

Alluvial. Referring to the deposition of sediment by running water.

Alpha diversity. The variety of organisms occurring in a particular place or habitat; often called local diversity.

Alternative hypothesis. One of several possible mechanisms different from the one that is identified for immediate investigation (null hypothesis). If support cannot be found for a null hypothesis, then one of the alternatives is likely.

Altruism. A social interaction that enhances the fitness of an unrelated recipient while reducing the fitness of the donor.

Ambient. Referring to conditions of the environment surrounding the organism.

Amino acid. One of about thirty organic acids containing the group NH_2, which are the building blocks of proteins.

Ammonification. Metabolic breakdown of proteins and amino acids with ammonia as an excreted by-product.

Anaerobic. Without oxygen.

Anhydrobiosis. A condition in which an organism becomes nearly completely dehydrated and ceases metabolic activity in order to withstand harsh conditions.

Anion. A part of a dissociated molecule carrying a negative electric charge.

Annual. Referring to an organism that completes its life cycle from birth or germination to death within a year.

Annual aboveground net productivity (AANP). The amount of production in the aboveground parts of plants in a year; measured by harvesting.

Anoxic. Lacking oxygen; anaerobic.

Antagonistic resource. A resource that interacts with another resource such that a consumer requires more of a mixture of those resources than of either one alone. *Compare with* Complementary resource.

Anthropogenic. Caused by humans.

Apical dominance. The prevention of growth in the lateral buds of a plant by the presence of an apical bud.

Aposematism. *See* Warning coloration.

Apostatic prey selection. Preferential consumption of the most common or most obvious prey types.

Apparent competition. A situation in which two or more species negatively affect one another indirectly through their interaction with a common predator.

Area of discovery. The area over which parasitoids must search for unattacked hosts. As parasitoid density increases, the area of discovery increases because more hosts have been attacked.

Arena. *See* Lek.

Arithmetic mean. The sum of a series of values divided by the number of values; a measure of the central tendency of a sample. *Compare with* Geometric mean.

Artificial selection. Manipulation by humans of the fitnesses of individuals in a population to produce a desired evolutionary response.

Asexual reproduction. Reproduction by which an individual produces genetically exact copies of itself by vegetative means or by producing diploid, egglike propagules that develop directly without fertilization.

Aspect. Morphological appearance.

Aspect diversity. The variety of outward appearances of species, as in the case of moths that live in the same habitat and are eaten by visually hunting predators.

Assimilation. Incorporation of any material into the tissues, cells, and fluids of an organism.

Assimilation efficiency. A percentage expressing the proportion of ingested energy that is absorbed into the bloodstream.

Assimilatory. Referring to a biochemical transformation that results in the reduction of an element to an organic form, hence its gain by the biological compartment of the ecosystem.

Association. A group of species living in the same place.

Associational resistance. Resistance of plants to insect pests that comes from living in association with other species of plants. *See also* Enemies hypothesis; Resource concentration hypothesis.

Assortative mating. A situation in which mates are chosen nonrandomly with respect to genotype. Positive assortative mating occurs when like mates with like; negative assortative mating refers to the tendency for mates to differ genetically.

Asymmetrical competition. An interaction between species in which one exploits a particular resource more efficiently than the other; the second may persist by better avoiding predation or by subsisting on an exclusive resource.

Atoll. A coral reef surrounding a lagoon, often built upon a sunken volcano.

Autochthonous. Originating within a system, such as organic matter produced, and minerals cycled, within streams and lakes. *Compare with* Allochthonous.

Autotroph. An organism that assimilates energy from either sunlight (green plants) or inorganic compounds (sulfur bacteria). *Compare with* Heterotroph.

Background extinction. The slow natural replacement of species that occurs as ecosystems change.

Ballistic seed dispersal. The dispersal of seeds by their forceful expulsion from the seed pod.

Barren. An area with sparse vegetation owing to some physical or chemical property of the soil.

Basal metabolic rate (BMR). The energy expenditure of an organism that is at rest, fasting, and in a thermally neutral environment.

Basal species. Species occupying the lowest trophic level in a food web; basal node.

Batesian mimicry. Resemblance of an edible species (mimic) to an unpalatable species (model) to deceive predators.

Bedrock. Unweathered solid rock underlying soil.

Benthic. Referring to the bottom of rivers, lakes, and oceans.

Bet hedging. Spreading of risk; reducing the risk of catastrophic failure or death. In life history evolution, bet hedging leads to more frequent or prolonged, but less intense, episodes of reproduction.

Beta diversity. The variety of organisms within a region arising from turnover of species among habitats.

Beverton-Holt model. A model in which recruitment increases monotonically from zero. (*See* Figure 18-11.)

Biennial. Requiring 2 years to complete the life cycle.

Bioassay experiment. An experiment in which the dependent variable is the response of a focal organism or group of organisms to an experimental treatment.

Biodiversity. The variety of types of organisms, habitats, and ecosystems on earth or in a particular place.

Bioenergetic food web. *See* Energy flow food web.

Biological community. *See* Community.

Biological oxygen demand (BOD). The amount of oxygen required to oxidize organic material in water samples; high values in aquatic habitats often indicate pollution by sewage and other sources of organic wastes, or the overproduction of plant material resulting from overenrichment by mineral nutrients.

Biomass. Weight of living material, usually expressed as a dry weight, in all or part of an organism, population, or community. Commonly presented as weight per unit of area, or biomass density.

Biomass accumulation ratio. The ratio of biomass to annual production.

Biome. A major category of ecological communities (e.g., grassland biome.).

Biota. Fauna and flora.

Birth rate (b_x). The average number of offspring produced per individual per unit of time, often expressed as a function of age (x).

Bomb calorimeter. A device that measures the energetic content of organic material by burning the material in a small chamber. Heat absorbed by water in a jacket surrounding the chamber estimates the heat energy released by the combusted material.

Boreal. Northern; often refers to the coniferous forest regions that stretch across Canada, northern Europe, and Asia.

Bottom-up control. Control of ecosystem function by nutrient flux and the conditions of the physical environment. *Compare with* Top-down control.

Boundary layer. A layer of still or slow-moving water or air close to the surface of an object.

Breakpoint. In logistic models of population regulation, a point that separates the domains of attraction of two equilibrium points. (*See* Figure 16-6.)

Breeding system. *See* Mating system.

Broken stick model. *See* Random niche model.

Brood parasite. An organism that lays its eggs in the nest of another species or that of another individual of the same species.

Browsers. Organisms that consume leaves from woody plants or, in the case of fish or aquatic invertebrates, that feed on pieces of macroalgae or other material that they bite off in pieces.

Browsing. A type of herbivory involving the consumption of leaves and young shoots by herbivores that are large relative to the plant parts that they consume. *Compare with* Grazing.

Bundle sheath cells. Specialized cells surrounding the leaf veins in plants where the Calvin-Benson cycle occurs.

C_3 photosynthesis. Photosynthetic pathway in which carbon dioxide is initially assimilated into a three-carbon compound, phosphoglyceraldehyde (PGA), in the Calvin-Benson cycle.

C_4 photosynthesis. Photosynthetic pathway in which carbon dioxide is initially assimilated into a four-carbon compound, such as oxaloacetic acid (OAA) or malate.

Calvin-Benson cycle. The basic assimilatory sequence of photosynthesis during which an atom of carbon is added to the 5-carbon ribulose bisphosphate (RuBP) molecule to produce phosphoglyceraldehyde (PGA) and then glucose.

CAM photosynthesis. Photosynthetic pathway in which the initial assimilation of carbon dioxide into a four-carbon compound occurs at night; found in some succulent plants in arid habitats.

Cannibalism. A intraspecific consumer-resource interaction in which individuals consume other individuals of the same species.

Canopy gap. An opening in the canopy of a forest created by a treefall.

Capillary action. The movement of water in a small tube against the force of gravity because of the forces of adhesion and cohesion.

Carbonate ion. An anion (CO_3^{2-}) formed by the dissociation of carbonic acid or one of its salts.

Carbonic acid. A weak acid (H_2CO_3) formed when carbon dioxide dissolves in water.

Carnivore. An organism that consumes mostly flesh.

Carotenoids. Yellow pigments found in plant tissue that absorb light that is used in photosynthesis.

Carrying capacity. The number of individuals in a population that the resources of a habitat can support; the asymptote, or plateau, of the logistic and other sigmoid equations for population growth.

Caste. A subgroup of individuals within a social group sharing a specialized form or behavior.

Cation. A part of a dissociated molecule carrying a positive electric charge.

Cation exchange capacity. The ability of soil particles to hold positively charged ions, such as hydrogen (H^+) and calcium (Ca^{2+}).

Cellulose. A long-chain molecule made up of glucose subunits, found in the cell walls and fibrous structures of plants.

Central-place foraging. Foraging that involves sorties from a central location, such as a nest.

Chain length. The number of links in a food web from the top predator to a basal species.

Chamaephytes. Small shrubs and herbs that grow close to the ground; one of five life forms identified by Christen Raunkiaer. (*See* Figure 8-25.)

Chaos. Erratic change in the size of a population governed by difference equations and having a high intrinsic rate of growth.

Character displacement. Divergence in the characteristics of two otherwise similar species where their ranges overlap, caused by the selective effects of competition between the species in the area of overlap.

Characteristic equation. An equation relating the exponential growth rate of a population to age-specific survivorship and fecundity under stable age conditions; also called Euler's equation.

Characteristic return time. The inverse of the exponential growth rate; an index to the rate at which a population regains its equilibrium value after perturbation.

Chemical competition. Competition that involves the production of a toxin that diffuses through the environment.

Chemical weathering. Weathering that occurs when water dissolves soluble constituents of rock.

Chemoautotroph. An organism that oxidizes inorganic compounds (often hydrogen sulfide) to obtain energy for the synthesis of organic compounds; for example, sulfur bacteria.

Chemostat. A culture vessel that provides a constant flux of nutrient medium.

Chlorophylls. Green pigments found in leaves that absorb light that is used in photosynthesis.

Chromosomes. Rodlike structures in eukaryotic cells on which genes reside.

Cladogram. A representation of the relationships among a monophyletic group of species or other taxa based on cladistic analysis.

Clay. A fine-grained component of soil, formed by the weathering of granitic rock, composed primarily of hydrous aluminum silicates.

Cleaning symbiosis. A type of defensive mutualism involving a species that is specialized for removing parasites from another species.

Climax. The end point of a successional sequence, or sere; a community that has reached a steady state under a particular set of environmental conditions.

Cline. A gradual change in population characteristics or adaptations over a geographic area.

Clone. A population of genetically identical individuals.

Closed community. The concept of a community, popularized by F. C. Clements, as a distinct association of highly interdependent species. *Compare with* Open community.

Clumped distribution. A spatial arrangement of organisms in which individuals are found in groups. *Compare with* spaced distribution; random distribution.

Clutch size. The number of eggs laid at one time; usually with reference to the nests of birds.

Coadaptation. Evolution of characteristics of two or more species in response to changes in one another, often to their mutual advantage.

Coarse-grained. Referring to qualities of the environment that occur in large patches with respect to the activity patterns of an organism, and among which the organism can select. *Compare with* fine-grained.

Coarse particulate organic matter (CPOM). Organic material with particle sizes greater than 1 mm in linear dimension. *Compare with* Fine particulate organic matter (FPOM).

Coefficient of competition (a_{ij}). A measure of the degree to which one consumer, *j*, utilizes the resources of another, *i*, expressed in terms of the population consequences of the interaction.

Coefficient of relationship. The probability that two individuals share copies of any particular gene; two siblings, for example, have a 50% probability of inheriting copies of the same gene from one parent.

Coevolution. The evolution of traits in two or more species selected by the mutual interactions controlled by those traits. *See also* Population coevolution; Species coevolution.

Coexistence. Occurrence of two or more species in the same habitat; usually applied to potentially competing species.

Cohesion. The attraction of like molecules to one another. Cohesion of water molecules causes surface tension.

Cohort. A group of individuals of the same age recruited into a population at the same time.

Cohort life table. *See* Dynamic life table.

Collective detection. The communication of the perception of danger by one animal in a group to all of the other animals in the group. *See also* Vigilance; Information center.

Colloidal phosphate. A soluble form of inorganic phosphate that is not available to organisms. *Compare with* Orthophosphate.

Community. An association of interacting populations, usually defined by the nature of their interaction or by the place in which they live.

Community character displacement. A situation in which, when the species of a community are ranked according to a particular morphological characteristic, the ratios between adjacent species are more or less equal. (*See* Figure 29-36.)

Community ecology. A subfield of ecology involving the study of ecological communities.

Community matrix. A square matrix, the entries of which are the pairwise effects of species within a community.

Comparative studies. Studies that make broad comparisons across time or spatial scales, or among a number of different taxonomic groups.

Compartment model. A representation of a system in which the various parts are portrayed as units (compartments) that receive inputs from, and provide outputs to, other such units.

Compartmentalization. The subdivision of a food web into groups of strongly interacting species that are somewhat isolated from other such groups.

Compatible solutes. Compounds such as amino acids, some polyols, and some methylated ammonium compounds that reside in the cytoplasm in high concentrations, thereby increasing its osmotic pressure without harmful effects on the cell.

Compensation. The morphological and physiological adjustments that plants make to cope with losses resulting from herbivory. *See also* Overcompensation.

Compensation point. The depth of water or level of light at which respiration and photosynthesis balance each other; the lower limit of the euphotic zone.

Competition. Use or defense of a resource by one individual that reduces the availability of that resource to other individuals, whether of the same species (intraspecific competition) or of other species (interspecific competition).

Competition coefficient. *See* Coefficient of competition.

Competitive exclusion principle. The hypothesis that two or more species cannot coexist on a single resource that is scarce relative to the demand for it.

Complementary resource. A resource that interacts with another resource synergistically, such that a consumer requires less of a mixture of those resources than of either one alone.

Condition. A physical or chemical attribute of the environment that, while not being consumed, influences biological processes and population growth; for example, temperature, salinity, acidity. *Compare with* Resource.

Conductance. The capacity of heat, electricity, or a substance to pass through a particular material.

Conduction. The ability of heat to pass through a substance.

Conformer. An organism that allows its internal environment to vary with external conditions. *Compare with* Regulator.

Congeneric. Belonging to the same genus.

Connectance (*C*). The total number of links in a food web divided by the number of possible links.

Connectedness food web. *See* Topological food web.

Conservation biology. A field of study involving the application of genetics and population and community ecology to problems of biodiversity loss.

Conservationism. A social movement aimed at preserving the natural world against human industrialization. *See also* Environmentalism.

Conspecific. Belonging to the same species.

Constant connectance hypothesis. A hypothesis of food web complexity that holds that connectance remains constant as the number of species in the community increases.

Constant of proportionality. *See* Allometric equation.

Consumer. An individual or population that utilizes a particular resource.

Consumer-resource interactions. Interactions in which individuals of one species consume individuals of another. Consumer-resource interactions affect the consumer positively and the resource negatively.

Consumptive competition. Competition for a renewable resource.

Contagion. The transmission of a disease by direct or indirect contact.

Continental shelves. Areas of relatively shallow seas surrounding continents and lying on the edges of the continental plates.

Continuum. A gradient of environmental characteristics or of change in the composition of communities.

Continuum index. A scale of an environmental gradient based upon changes in physical characteristics or community composition along that gradient.

Convection. Transfer of heat by the movement of a fluid (such as air or water).

Cooperation. A social interaction that benefits both the donor and the recipient.

Correlated response. Response of the phenotypic value of one trait to selection upon another trait.

Correlation analysis. A study in which the correlations between variables are calculated.

Cost of meiosis. The loss of fitness resulting from sexual reproduction. The progeny of a sexual union contribute only half as much to the evolutionary fitness of each parent as asexually produced offspring.

Counteradaptation. Evolution of characteristics of two or more species based upon their mutual antagonism, as in competitive and consumer-resource interactions.

Countercurrent circulation. Movement of fluids in opposite directions on either side of a barrier through which heat or dissolved substances may pass.

Covalent bond. A bond between two atoms that involves the sharing of electrons.

Covariance. The average product of the deviations of two variables from their respective means.

Convergence. Resemblance among organisms of different taxonomic groups resulting from adaptation to similar environments.

Correlated extinction. The extinction of multiple local populations in a metapopulation that respond to environmental fluctuation in the same way.

Crassulacean acid metabolism. *See* CAM photosynthesis.

Critical resource level (R^*). The level of a resource at which mortality balances growth in a population.

Cross-resistance. Resistance or immunity to one disease organism resulting from infection by another, usually closely related, organism.

Cryptophytes. Plants that are protected from freezing and desiccation by having buds that are completely buried beneath the soil; one of five life forms identified by Christen Raunkiaer. (*See* Figure 8-25.)

Crypsis. An aspect of the appearance of an organism whereby it avoids detection by others.

Cycle. Recurrent variation in a system that periodically returns to its starting point. In food webs, a situation in which there is a reciprocal predatory relationship between two species (species A eats species B, and species B eats species A).

Cyclic climax. A steady-state, cyclic sequence of seral stages, none of which by itself is stable.

Damped oscillation. Cycling with progressively smaller amplitude, as in some populations approaching their equilibria.

Death rate (d_x). The percentage of newborns dying during a specified interval, often expressed as a function of age (x). *Compare with* Mortality.

Decision variable. In optimality models of behavior, the variable upon which decisions about changes in behavior are based.

Defensive mutualism. A mutualistic relationship in which one species receives food or shelter from another species in return for defending its partner against herbivores, predators, or parasites.

Deme. A local population within which mating occurs among individuals more or less at random.

Demographic stochasticity. Random variation in birth and death rates in a population.

Demography. The study of population structure and growth.

Denitrification. The biochemical reduction, primarily by microorganisms, of nitrogen from nitrate (NO_3^-), eventually to molecular nitrogen (N_2).

Density. Referring to a population, the number of individuals per unit of area or volume; referring to a substance, the weight per unit of volume.

Density compensation. An increase in population size in response to a reduction in the number of competing populations; often observed on islands.

Density dependent. Having an influence on individuals in a population that varies with the density of that population.

Density independent. Having an influence on individuals in a population that does not vary with the density of that population.

Density vagueness. A situation in which large demographic or environmental stochasticity obscures density-dependent processes.

Dependent variable. In an experiment, the variable in which a change resulting from the experimental procedure is observed; also known as the response variable. *Compare with* Independent variable.

Derived trait. An evolved characteristic of an organism that is changed from the ancestral condition.

Deterministic. Having an outcome that is not subject to stochastic (random) variation.

Detoxification. The biochemical neutralization of a toxic compound; an adaptation of some herbivores to counter toxic secondary compounds in plants.

Detritivore. An organism that feeds on detritus.

Detritivore-detritus interactions. Interactions in which individuals consume detritus. Detritivore-detritus interactions affect the consumer positively and do not affect the detritus.

Detritus. Freshly dead or partially decomposed organic matter.

Detritus-based food chain. The food chain involving soil organic matter and decomposition.

Developmental response. The acquisition of one of several alternative forms by an organism depending on the environmental conditions under which it grows.

Diapause. A temporary interruption in the development of insect eggs or larvae, usually associated with a dormant period.

Diet breadth. The number of different types of prey that are included in the diet.

Diet loop. A type of omnivory in which adults of species A eat juveniles of species B, and adults of species B eat juveniles of species A; shown as a loop in a food web.

Difference equation. An equation describing the change in a quantity over an interval of discrete time, as, for example, the change in the number of individuals in a population between time t and time $t + 1$.

Differential equation. Equation describing the instantaneous rate of change in a quantity, as, for example, dN/dt describes the change in the number of individuals in a population (N) over time (t).

Differential functional response. A functional response that reflects the instantaneous prey consumption rate as a function of prey density. *Compare with* integrated functional response; optimal functional response.

Diffuse coevolution. The evolution of traits influencing species interactions that are subject to selection by a wide variety of species interacting with different intensities.

Diffuse competition. The sum of weak competitive interactions among species that are ecologically distantly allied.

Diffusion. Movement of gas or liquid particles from regions of high to low concentration by means of their own spontaneous motion.

Dimorphism. The occurrence of two forms of individuals within a population.

Dioecy. In plants, the occurrence of reproductive organs of the male and female sex on different individuals. *Compare with* Monoecy.

Diploid. Having two sets of chromosomes. *Compare with* Haploid.

Direct competition. Exclusion of other individuals from resources by means of aggressive behavior or use of toxins. *Compare with* Indirect competition.

Direct interaction. An interaction between organisms that involves direct physical contact between the interactors (e.g., predation and herbivory). *Compare with* Indirect interaction.

Direct transmission. Transmission of a parasite that occurs when two hosts come into contact with each another. *Compare with* Indirect transmission.

Disc equation. An equation relating the rate of encounter of prey by a predator to searching efficiency, prey abundance, and handling time.

Disease vector. An organism that transmits pathogens from one host to the next. *See also* Transmission.

Dispersal. Movement of organisms away from their place of birth or from centers of population density.

Dispersion. The spatial pattern of distribution of individuals within populations.

Dispersive mutualism. A mutualism, usually between plants and animals, involving the transport of pollen between flowers in return for rewards such as nectar, or the dispersal of seeds to suitable habitats as animals consume the fruit.

Dissimilatory. Referring to a biochemical transformation that results in the oxidation of the organic form of an element, and hence its loss from the biological compartment of the ecosystem.

Dissociation. The breaking up of a compound into its component parts; especially the formation of ions in an aqueous solution.

Dissolution. The entry of a substance into solution with water.

Distribution. The geographic extent of a population or other ecological unit.

Disturbance hypothesis. A hypothesis of species-area relationships that holds that populations on small islands are more susceptible to disturbance than those on large islands, and thus that small islands will have fewer species than large islands.

Diversity. The number of species in a local area (alpha diversity) or region (gamma diversity). Also, a measure of the variety of species in a community that takes into account the relative abundance of each species.

DNA fingerprinting. A technique used to detect variation in short segments of DNA having repeating sequences of nucleotides called variable number tandem repeats (VNTR).

Domain of attraction. In logistic models of population regulation, the range of population densities around an equilibrium point at which the population density tends toward the equilibrium point. (*See* Figure 16-6.)

Dominance hierarchy. An orderly ranking of the individuals in a social group, based on the outcome of aggressive encounters.

Dominance variance (V_D). Variation in a phenotypic value within a population due to the unequal expression of alleles in the heterozygous state.

Dominants. The few species that attain high abundances in a community.

Donor. In social interactions, the individual performing a particular behavior that is directed toward another individual. *Compare with* Recipient.

Dormancy. An inactive state, such as hibernation, diapause, or seed dormancy, usually assumed during a period of inhospitable conditions.

Downstream drift. The movement of materials, including animals and plants, down streams by the flow of the water.

Dynamic equilibrium model. A model of species diversity whereby the interaction between the extent of disturbance in a community and the rates of population growth of species in the community determines the community's diversity. (*See* Figure 29-32.)

Dynamic life table. The age-specific survival and fecundity of a cohort followed from birth to the death of the last individual; also called a cohort life table.

Dynamic stability hypothesis. A hypothesis of food web complexity that holds that more complex food webs (i.e., with large number of trophic levels) will not recover from disturbance as readily as food webs that are relatively simple.

Dynamic steady state. A condition in which fluxes of energy or materials into and out of a system are balanced.

Ecdysis. Periodic shedding of the outer skin, as in reptiles.

Ecological diversity. Diversity of ecological function or types. *Compare with* Taxonomic diversity.

Ecological efficiency. The percentage of energy in the biomass produced by one trophic level that is incorporated into the biomass produced by the next higher trophic level.

Ecological isolation. Avoidance of competition between two species by differences in food, habitat, activity period, or geographic range.

Ecological range. The set of conditions under which individuals of a species can survive.

Ecological release. Expansion of habitat and resource utilization by populations in regions of low species diversity resulting from reduced interspecific competition.

Ecology. The study of the natural environment and of the relations of organisms to one another and to their surroundings.

Ecophysiology. *See* Physiological ecology.

Ecosystem. All of the interacting parts of the physical and biological worlds.

Ecosystem diversity. The variety of different ecosystems.

Ecosystems ecology. A subfield of ecology involved with the study of ecosystems.

Ecotone. A habitat created by the juxtaposition of two or more distinctly different habitats; an edge habitat; a zone of transition between habitat types.

Ecotoxicology. The study of the fate and action of anthropogenic chemicals in the environment.

Ecotype. A genetically differentiated subpopulation that is restricted to a specific habitat.

Ectomycorrhizae. Mutualistic associations between fungi and the roots of plants in which the fungus forms a sheath around the outside of the root. *Compare with* Endomycorrhizae.

Ectoparasite. A parasite that lives on, or attached to, the surface of its host. *Compare with* Endoparasite.

Ectoparasitoid. A parasitoid in which the female deposits eggs on the outside of the host. *Compare with* Endoparasitoid.

Ectothermy. The capacity to maintain body temperature by gaining heat from the environment, either by conduction or by absorbing radiation.

Edaphic. Pertaining to, or influenced by, the soil.

Effective population size (N_e). The average size of a population expressed in terms of the number of individuals assumed to contribute genes equally to the next generation; generally smaller than the actual size of the population, depending on the variation in reproductive success among individuals.

Egested energy. The energy contained in animal food that is resistant to digestion and assimilation (e.g., hair, feathers, insect exoskeletons, cartilage, cellulose, and lignin).

Egestion. Elimination of undigested food material.

Electric potential (Eh). The relative capacity, measured in volts, of one substance to oxidize another.

Electron acceptor. A substance that readily accepts electrons, hence, one that is capable of oxidizing another substance.

Electrophoresis. A technique for separating and visualizing proteins or nucleic acids whereby the substances move via an electric current in a polyacrylamide or starch gel. Different substances move different distances owing to differences in electric charge.

Elicitor-receptor model. A genetic model of gene-for-gene coevolution.

El Niño. A warm current from the Tropics that intrudes each winter along the west coast of northern South America. *See also* ENSO.

Eluviation. The downward movement of dissolved soil materials from the topmost (A) horizon, carried by percolating water.

Emergent property. A feature of a system not deducible from lower-order processes.

Emergent trees. Trees that rise above the tall canopy of tropical forests, often to heights exceeding 50 meters.

Emigration. Movement of individuals out of a population. *Compare with* Immigration.

Encounter competition. Competition that involves a transient interaction over a resource; may result in physical harm, loss of energy, or theft of food.

Encounter rate. The rate at which foraging animals encounter prey; portrayed in foraging models as the product of prey density and a constant representing the efficiency of searching.

Endemic. Confined to a certain region.

Endomycorrhizae. Mutualistic associations between fungi and the roots of plants in which part of the fungus resides within the root tissues. *Compare with* Ectomycorrhizae.

Endoparasite. A parasite that lives within the tissues or bloodstream of its host. *Compare with* Ectoparasite.

Endoparasitoid. A parasitoid in which the female lays eggs on the inside of the host with the aid of a long ovipositor. *Compare with* Ectoparasitoid.

Endophytophagy. The consumption of plant material by feeding within the tissue. A number of insect larvae adopt this feeding strategy.

Endothermy. The capacity to maintain body temperature by the metabolic generation of heat.

Enemies hypothesis. The proposition that diverse plant communities have a greater diversity of potential predators of pests because of the greater number of microhabitats in the community. *See also* Associational resistance; Resource concentration hypothesis.

Enemy-free space. *See* shared predation.

Energetic efficiency. The ratio of useful work or energy storage to energy intake.

Energy. Capacity for doing work.

Energy flow food web. A food web that quantifies the flux of energy between resources and consumers; also called a flow web or bioenergetic web.

Energy flow hypothesis. The idea that energetic constraints limit the number of trophic levels in a food web.

ENSO. El Niño/Southern Oscillation; an occasional shift in winds and ocean currents in the South Pacific region from a westerly direction to an easterly direction, with worldwide consequences for climate and biological systems.

Environment. The surroundings of an organism, including the plants, animals, and microbes with which it interacts.

Environmentalism. A social movement aimed at understanding human interactions with the environment. *See also* Conservationism; Preservationism.

Environmental science. A subfield of ecology that focuses on the specific effects of humans on the environment.

Environmental stochasticity. Random variation in the abiotic environment.

Environmental variance (V_E). Variation in a phenotypic value within a population due to the influence of environmental factors.

Enzyme. An organic compound in a living cell, or secreted by it, that accelerates a specific biochemical transformation without itself being affected.

Epidemiology. The study of factors influencing the spread of disease through a population.

Epifauna. Animals living on the surface of a substrate.

Epilimnion. The warm, oxygen-rich surface layers of a lake or other body of water. *Compare with* Hypolimnion.

Epiphytes. Plants that grow on the branches of other plants and are not rooted in the soil.

Equilibrium. A state of balance between opposing forces.

Equilibrium hypothesis. A hypothesis of species-area relationships applied to island communities that holds that the species richness declines as the degree of isolation of the island increases.

Equilibrium isocline. A line on a population graph designating combinations of competing populations, or predator and prey populations, for which the growth rate of one of the populations is zero.

Escape space. Refuge from predators and parasites; often reflected in the adaptations of prey organisms to fight, flee, or escape detection.

Essential resource. A resource that limits a consumer population independently of other resources; that is, without synergism or the possibility of substitution.

Estuary. A semi-enclosed coastal water, often at the mouth of a river, having a high input of fresh water and great fluctuation in salinity.

Euler's equation. *See* Characteristic equation.

Euphotic zone. The surface layer of water to the depth of light penetration at which photosynthesis balances respiration. *See also* Compensation point.

Eurytypic. Having a wide range of tolerances for environmental conditions. *Compare with* Stenotypic.

Eusociality. The complex social organization of termites, ants, and many wasps and bees, dominated by an egg-laying queen that is tended by nonreproductive offspring.

Eutrophic. Rich in the mineral nutrients required by green plants; pertaining to an aquatic habitat with high productivity.

Eutrophication. Enrichment of water by nutrients required for plant growth; often refers to overenrichment caused by sewage and runoff from fertilized agricultural lands and resulting in excessive bacterial growth and oxygen depletion.

Evaporation. The conversion of a liquid to a vapor requiring heat; a cooling process.

Evapotranspiration. The sum of transpiration by plants and evaporation from the soil. *See also* Potential evapotranspiration.

Evolutionarily stable strategy (ESS). A strategy such that, if all members of a population adopt it, no alternative strategy can invade the population.

Evolutionary ecology. The integrated science of evolution, genetics, adaptation, and ecology; interpretation of the structure and function of organisms, communities, and ecosystems in the context of evolutionary theory.

Excreted energy. The portion of the energy assimilated by animals that is eliminated in the form of nitrogen-containing organic wastes.

Excretion. Elimination from the body, by way of the kidneys, gills, and dermal glands, of excess salts, nitrogenous waste products, and other substances.

Expectation of further life (e_x). The average remaining lifetime of an individual of age x.

Experiment. An activity whereby natural processes are allowed to proceed under conditions that are controlled by the experimenter.

Exploitation. Removal of individuals or biomass from a population by consumers.

Exploitation efficiency. The efficiency with which the biological production of an entire trophic level is consumed.

Exploitative competition. Competition between individuals by way of their reduction of shared resources.

Exponential growth. Population growth according to the equation $N_t = N_0e^{rt}$, where e is the base of the natural logarithm (about 2.72), r is the exponential growth rate, N_0 is the population density at some arbitrary time $t = 0$, and N_t is the density at time t.

Exponential growth rate (r). The rate at which a population is growing at a particular time, expressed as a proportional increase per unit of time. *Compare with* Geometric growth rate.

Exponential mortality rate (k_x). Negative logarithm of the probability of survival between ages of x and $x + 1$. *See also* Killing power.

External forcing function. In systems modeling, a material input from outside the system or a condition of the environment of the system that influences its structure and function.

Extinction. The disappearance of a species or other taxon from a region or biota.

Extirpation. Extinction of a local population.

Extrafloral nectaries. In plants, nectar-producing structures located outside the flower.

Extra-pair copulation (EPC). Copulation by an individual with another individual who is not identified as its partner; best known in birds.

Facilitation. Enhancement of conditions for a population of one species by the activities of another, particularly during early succession.

Facultative. Available under certain conditions or circumstances; optional for the organism. *Compare with* Obligate.

Facultative endothermy. The ability in an ectothermic animal to increase its body temperature by means of some physiological process.

Facultative mutualism. *See* mutualism.

Fall bloom. The rapid growth of algae in temperate lakes following fall overturn.

Fall overturn. Vertical mixing of water layers in temperate lakes in the fall following the breakdown of thermal stratification.

Fecundity. The rate at which an individual produces offspring.

Fermentation. Anaerobic, energy-releasing transformation of organic substances; often involving the transformation of pyruvate to ethanol or lactic acid.

Field capacity. The amount of water that soil can hold against the pull of gravity.

Filter feeder. An organism that strains tiny food particles from its aqueous environment by means of sievelike structures; for example, clams and baleen whales.

Fine-grained. Referring to qualities of the environment that occur in small patches with respect to the activity patterns of an organism, and among which the organism cannot usefully distinguish. *Compare with* Coarse-grained.

Fine particulate organic matter (FPOM). Organic material with particle sizes ranging from 1 mm to 0.5 mm. *Compare with* Coarse particulate organic matter (CPOM).

Fitness. The genetic contribution by an individual's descendants to future generations of a population.

Fitness set. A graphic portrayal of phenotypes within a population according to their values for various components of fitness.

Fixation. In genetics, the situation in which selection causes the disappearance of one (or more) alleles, leaving one (fixed) in the population. In ecosystems ecology, the assimilation of an element such as carbon or nitrogen.

Floristic. Referring to the species composition of plant communities.

Floristic analysis. An analysis of a community that consists primarily of identifying the different species of plants present in the community.

Fluvial. Referring to rivers or moving water.

Flux. Movement of energy or materials into or out of a system.

Food chain. A representation of the passage of energy through populations in a community.

Food chain efficiency. *See* Ecological efficiency.

Food web. A representation of the various paths of energy flow through populations in a community.

Forbs. Herbaceous, broad-leaved vegetation (i.e., other than grasses) consumed by grazers.

Founder event. A single episode of small population size, which may occur when a new area is colonized by a few individuals from a larger population.

Frequency-dependent selection. Selection in which the intensity of the selective force varies with the relative proportions of phenotypes in the population. Negative frequency-dependent selection means that the fitness of the trait decreases as the frequency of the trait increases. Positive frequency-dependent selection refers to an increase in fitness with frequency.

Front. A meeting of two air or water masses having different characteristics.

Functional food web. A food web that identifies only the most important feeding relationships within the community; also called an interaction food web.

Functional response. A change in the rate of exploitation of prey by an individual predator as a result of a change in prey density. *Compare with* Numerical response.

Gain curve. A graph showing the reward, or gain, per unit of time of foraging. (*See* Figure 31-20.)

Game theory. A method of analyzing the outcomes of social interactions between individuals with different behavior programs.

Gamete. A haploid cell that fuses with another haploid cell of the opposite sex during fertilization to form the zygote. In animals, the male gamete is called the sperm and the female gamete, the egg or ovum.

Gamma diversity. The inclusive diversity of all the habitat types within an area.

Gene. Generally, a unit of genetic inheritance. In biochemistry, the part of the DNA molecule that encodes a single enzyme or structural protein.

Gene flow. Exchange of genetic traits between populations by movement of individuals, gametes, or spores.

Gene frequency. The proportion of a particular allele of a gene in the gene pool of a population.

Gene locus. *See* Locus.

Generalist. A species with broad food or habitat preferences. *Compare with* Specialist.

Generation time. The average age at which a female gives birth to her offspring, or the average time for a population to increase by a factor equal to the net reproductive rate.

Genet. A group of individuals derived from the same zygote. *Compare with* Ramet.

Genetic drift. Changes in gene frequency due to random variations in fecundity and mortality in a population.

Genetic potential. The number of possible genotypes of which a particular population could be composed.

Genetic structure. The characteristics of genetic variation among the individuals of a population or a community.

Genetic variance (V_G). Variation in a phenotypic value within a population due to the expression of genetic factors.

Genotype. All of the genetic characteristics that determine the structure and functioning of an organism; often applied to a single gene locus to distinguish one allele, or combination of alleles, from another.

Genotype-environment interaction. A situation in which the slope of the reaction norm differs between genotypes.

Genotype space. *See* Genetic potential.

Genotypic variance. That portion of phenotypic variance that is attributable to the genetic constitution.

Geographic distribution. The geographic area in which a species occurs. Also referred to as the geographic range, or simply the range.

Geographic range. *See* Geographic distribution.

Geometric growth. Population growth in which the rate of increase is proportional to the number of individuals in the population at the beginning of the breeding season.

Geometric growth model. A population growth model based on geometric growth.

Geometric growth rate (λ). In populations that are augmented during discrete reproductive periods, the ratio of the population density in one year to that in the previous year $[N(1)/N(0)]$. *Compare with* Exponential growth rate.

Geometric mean. The *n*th root of the product of *n* values; alternatively, the antilogarithm of the mean of the logarithms of a series of values. *Compare with* Arithmetic mean.

Geometric series. A series in which each element is formed by multiplying the previous one by some factor. A characterization of the relative abundances of species in which the logarithm of abundance decreases linearly with the abundance rank of the species.

Giving-up time. In foraging models, the amount of time that a foraging animal stays in a patch before leaving to find another patch.

Gonochorism. In animals, the condition of separate male and female individuals; in plants, called dioecy.

Gradient. A difference in a condition or concentration of a substance between two points or across a boundary.

Gradient analysis. Portrayal and interpretation of the abundances of species along gradients of physical conditions.

Grain. The scale of heterogeneity of habitats in relation to the activities of organisms.

Granivory. *See* Seed predation.

Grazers. Organisms that eat grasses or other herbs or, in the case of fish and aquatic invertebrates, scrape or suck plant material from some surface.

Grazing. A type of herbivory involving the consumption of grasses or forbs by herbivores that are larger than the plants and plant parts that they consume. *Compare with* Browsing.

Greenhouse effect. Warming of the earth's climate owing to the increased concentration of carbon dioxide and certain other pollutants in the atmosphere.

Greenhouse gases. Gases such as carbon dioxide that trap infrared radiation in the earth's atmosphere, creating the greenhouse effect.

Gross primary production. The total energy or nutrients assimilated by an organism, a population, or a community. *Compare with* Net primary production.

Gross production efficiency. The percentage of ingested food utilized for growth and reproduction by an organism. *Compare with* Net production efficiency.

Groundwater. Water that percolates through the soil and through cracks and interstices in bedrock.

Group selection. Elimination of groups of individuals with a detrimental genetic trait caused by competition with other groups lacking the trait; often called intergroup selection.

Growing season. The frost-free season.

Gynodioecy. A situation in which a population contains a mixture of female and hermaphroditic individuals.

Habit. In plants, the general life form of an individual (for example, erect, prostrate, woody).

Habitat. The place where an animal or plant normally lives, often characterized by a dominant plant form or physical characteristic (that is, the stream habitat, the forest habitat).

Habitat diversity hypothesis. A hypothesis about species-area relationships that holds that larger areas will have greater species diversity because they will have a greater variety of habitats.

Habitat expansion. An increase in the average breadth of habitat distribution of species in depauperate biotas, especially on islands, compared with that of species in more diverse biotas.

Habitat patch. An area of habitat that contains the necessary resources and conditions for a population to persist.

Halophilic. Having a tolerance to high salt concentrations.

Halophytes. Plants that possess adaptations to withstand high salt concentrations.

Hamilton-Zuk hypothesis. *See* Parasite-mediated sexual selection.

Handicap principle. The idea that elaborate, sexually selected displays and adornments act as handicaps that demonstrate the generally high fitness of the bearer.

Haplodiploidy. A sex-determining mechanism by which females develop from fertilized eggs and males from unfertilized eggs.

Haploid. Containing only one set of chromosomes. *Compare with* Diploid.

Hardy-Weinberg law. Holds that the frequency of alleles and genotypes remains constant from generation to generation in large populations in which there is random mating, no selection, no mutation, and no migration to or from the population.

Harem. A group of females that is guarded by one or more males against the mating attempts of other males.

Harvesting. Removing animals or plants from a population.

Hawk-dove game. A game theory analysis of the interactions of two types of individuals in a population in conflicts over a resource: one that always contests the resource (hawk) and one that never contests the resource (dove).

Heat. A measure of the kinetic energy of the atoms or molecules in a substance.

Heat of melting. Amount of heat energy that must be added to a substance to make it melt.

Heat of vaporization. Amount of heat energy that must be added to a substance to make it vaporize.

Hemicryptophytes. Plant life forms that persist through extreme conditions by dying back to ground level, where the regenerating bud is protected by soil and withered leaves; one of five life forms identified by Christen Raunkiaer. (*See* Figure 8-25.)

Hemiessential resource. Either of two resources, one essential and the other nonessential, that interact in such a way that, as the amount of the nonessential resource declines, more of the essential resource is required.

Herbivore. An organism that consumes living plants or their parts.

Herbivory. A consumer-resource interaction involving the consumption of plants or plant parts.

Herd immunity. A situation in which sufficient individuals in a population (often less than 100% of the population) are immunized against a disease so that it cannot persist in that population.

Heritability (h^2). The proportion of variance in a phenotypic trait that is due to the effects of additive genetic factors.

Hermaphrodite. An organism that has the reproductive organs of both sexes. The male and female reproductive organs may occur simultaneously or sequentially in an individual.

Heterosis. A situation in which the heterozygous genotype is more fit than either homozygote.

Heterothermy. Facultative reduction in body temperature by an endothermic animal.

Heterotroph. An organism that utilizes organic materials as a source of energy and nutrients. *Compare with* Autotroph.

Heterozygous. Having two different alleles of a gene, one derived from each parent. *Compare with* Homozygous.

Hibernation. A state of winter dormancy associated with lowered body temperature and metabolism.

Higher-order interaction. The influence of one species on the interaction between two others.

Homeostasis. The maintenance of constant internal conditions in the face of a varying external environment.

Homeothermy. The ability to maintain a constant body temperature in the face of fluctuating environmental temperatures; warm-bloodedness.

Homozygous. Having two identical alleles of a gene. *Compare with* Heterozygous.

Horizon. A layer of soil distinguished by its physical and chemical properties.

Host plant. A plant that is subject to consumption by small herbivores, such as insects.

Humus. Fine particles of organic detritus in soil.

Hydrogen bond. The weak attraction of polar molecules.

Hydrological cycle. The movement of water throughout the ecosystem.

Hydrolysis. A biochemical process by which a molecule is split into parts by the addition of the parts of water molecules.

Hyperdispersion. A pattern of distribution in which distances between individuals are more even than expected from random placement.

Hyperosmotic. Having an osmotic potential (generally, a salt concentration) greater than that of the surrounding medium. *Compare with* Hypo-osmotic.

Hyphae. Threadlike cells of fungi that are connected end-to-end to form a network called the mycelium.

Hypolimnion. The cold, oxygen-depleted part of a lake or other body of water that lies below the zone of rapid change in water temperature (thermocline). *Compare with* Epilimnion.

Hypo-osmotic. Having an osmotic potential (generally, salt concentration) less than that of the surrounding medium. *Compare with* Hyperosmotic.

Hypothesis. A specific idea about a natural phenomenon that is identified for study. Hypotheses are declarative statements that assert a mechanism. Experiments are used to falsify hypotheses. *See also* Null hypothesis; Alternative hypothesis.

Ideal free distribution. The distribution of individuals across resource patches of different intrinsic quality that equalizes the net rate of gain of each when competition is taken into account.

Identity by descent. The situation in which copies of the same gene are inherited by different individuals from a common ancestor.

Igneous rock. Rock formed when the hot magma of the earth's mantle rises to near the surface and cools.

Illuviation. The accumulation of dissolved substances within a soil layer, usually the middle (B) horizon.

Immigration. Movement of individuals into a population. *Compare with* Emigration.

Inbreeding. Mating among related individuals.

Inbreeding depression. Reduction in fitness caused by inbreeding.

Incidence matrix. A table of occurrences of species in samples.

Inclusive fitness. The sum of the fitness of an individual and the fitnesses of its relatives, the latter weighted according to the coefficient of relationship; usually applied to the consequences of social interaction between relatives.

Independent variable. In an experiment, a variable that is under the control of the experimenter. *Compare with* Dependent variable.

Indirect competition. Exploitation of a resource by one individual that reduces the availability of that resource to others. *Compare with* Direct competition.

Indirect interaction. An interaction between two individuals that involves one or more intermediary species. *Compare with* Direct interaction.

Indirect mutualism. An indirect interaction having positive effects on both interacting species.

Indirect transmission. Transmission that occurs when an adult parasite residing in a primary host produces immature stages that infect one or more intermediate hosts. *Compare with* Direct transmission.

Individual-based models. Ecosystems models in which organism-level processes, such as growth, reproduction, or nutrient uptake, are averaged.

Individual distance. The distance within which one individual does not tolerate the presence of another.

Inducible response. Any change in the state of an organism caused by an external factor; usually reserved for the response of organisms to parasitism and herbivory.

Industrial melanism. The evolution of dark coloration by cryptic organisms in response to industrial pollution, especially by soot, of their environments.

Information center. An aggregation of individuals within which information about food or other resources is obtained, generally by observing the return and departure of foraging individuals. *See also* Collective detection.

Infrared radiation. Electromagnetic radiation having a wavelength longer than about 700 mm.

Inhibition. The suppression of a colonizing population by another that is already established, especially during successional sequences.

Instantaneous rate of increase. The rate of increase of a population undergoing exponential growth at a particular time.

Integrated functional response. The functional response of a predator measured over an entire foraging bout. *Compare with* differential functional response; optimal functional response.

Interaction food web. *See* Functional food web.

Interaction modification. A situation in which a species affects the interaction between two other species.

Interaction variance (V_I). The portion of the genotypic variance comprising the influences of different genes on the expression of alleles at a particular locus.

Interactive essential resources. Essential resources that interact in such a way that a reduction in one requires an organism to use more of the second.

Interactive traits. Adaptations that arise from reciprocal interactions between organisms.

Interference competition. Competition in which individuals actively interfere with one another's access to resources.

Intergroup selection. *See* Group selection.

Intermediary species. A species through which two or more other species interact indirectly. *See also* Indirect interaction.

Intermediate disturbance hypothesis. The idea that species diversity is greatest in habitats with moderate amounts of physical disturbance, owing to the coexistence of early and late successional species.

Intermediate host. A host that harbors an asexual stage of the life cycle of a parasite or disease organism.

Intersexual selection. Selection based on the choice of mates based upon their appearance and behavior.

Interspecific competition. Competition between individuals of different species.

Intertaxon competition hypothesis. A hypothesis of species diversity that holds that peaks of species diversity for different multispecies taxa (e.g., orders and classes) should occur in areas having different productivity levels. *Compare with* Productivity-stability hypothesis.

Intrasexual selection. Selection based on competition among individuals of the same sex.

Intraspecific competition. Competition between individuals of the same species.

Intrinsic rate of increase (r). The exponential growth rate of a population with a stable age distribution; that is, under constant conditions.

Ion. A dissociated part of a molecule that carries an electric charge, either positive (cation) or negative (anion).

Isocline. In a population graph, a line designating combinations of competing populations, or predator and prey populations, for which the growth rate of one of the populations is zero.

Isoenzymes. Slightly different forms of an enzyme resulting from a substitution of one or more amino acids; also called isozymes.

Iteroparity. The condition of reproducing repeatedly during the lifetime. *Compare with* Semelparity.

***K*-selection.** Selection on traits that determine fecundity and survival to favor competitive ability at densities near the carrying capacity (K). *Compare with* r-selection.

Keystone species. A species, often a predator, having a dominating influence on the composition of a community, which may be revealed when the keystone species is removed.

Killing power (k). Logarithm of the probability of survival from one age class (x) to the next; k-value. *See also* Exponential mortality rate.

Kin selection. Differential reproduction among lineages of closely related individuals based upon genetic variation in their social behavior.

Kinetic energy. Energy associated with motion.

Kinship deceit. A situation in which the recipient of alloparental care learns to perceive the donor of the care as a relative and later, based on this perception, assists the donor in some way.

Landscape. A large area that includes one or more ecosystems.

Landscape ecology. A subfield of ecology that focuses on how ecological processes operate on large spatial scales.

Larva. An immature stage of an invertebrate that is fundamentally unlike the adult and must undergo some kind of metamorphosis.

Laterite. A hard substance rich in oxides of iron and aluminum, frequently formed when tropical soils weather under alkaline conditions.

Laterization. Leaching of silica from soil, usually in warm, moist regions, which results in alkaline soil.

Law of constant final yield. The idea that beyond some plant population density, total yield will not increase (*See* Figure 14-12.)

Law of mass action. In chemistry, the principle that the rate of a reaction is directly proportional to the concentration of the reacting substances. In consumer-resource models, the principle that the rate of growth of the resource is proportional to the "concentration" of the consumer and the resource.

Leaching. Removal of soluble compounds from leaf litter or soil by water.

Legume. Any of a number of plant species belonging to the pea family (Leguminosae).

Lek. A communal courtship area on which several males hold courtship territories to attract and mate with females; sometimes called an arena.

Lewis-Leslie matrix. A matrix of values of age-specific fecundity and survivorship used to project the size and age structure of a population through time; a population matrix.

Liana. A climbing plant of tropical rain forests, usually woody, that is rooted in the ground.

Liebig's law of the minimum. The idea that the growth of an individual or population is limited by the essential nutrient present in the lowest amount relative to the demand for it.

Life cycle graph. A graph that depicts the transitions among the various stages of a stage-classified population. (*See* Figure 15-6.)

Life history. The set of adaptations of an organism that influence the life table values of age-specific survival and fecundity, such as reproductive rate or age at first reproduction.

Life table. A summary by age of the survivorship and fecundity of individuals in a population.

Life zone. A more or less distinct belt of vegetation occurring within, and characteristic of, a particular latitude or range of elevation.

Lifetime fitness function. In stochastic dynamic programming, the probability that an animal will survive from time t to the end of an arbitrarily long period T.

Lignin. A long-chain, nitrogen-containing molecule made up of phenolic subunits, occurring in woody structures of plants, that is highly resistant to digestion by herbivores.

Limestone. A rock formed chiefly by the sedimentation of shells and the precipitation and sedimentation of calcium carbonate ($CaCO_3$) in marine systems.

Limit cycle. The oscillation of predator and prey populations that occurs when the stabilizing and destabilizing tendencies of their interaction balance.

Limiting resource. A resource that is scarce relative to demand for it.

Limiting similarity. The minimum degree of ecological similarity compatible with the coexistence of two or more populations.

Limnetic. Pertaining to open water.

Limnology. The study of freshwater habitats and communities, particularly lakes, ponds, and other standing waters.

Link. An interaction between species in a food web; represented as a line drawn between nodes.

Link-species scaling hypothesis. A hypothesis of food web complexity that holds that the number of species in a community decreases as food web connectance decreases.

Linkage density. The average number of links per species in a food web.

Littoral. Pertaining to the shore.

Loam. Soil that is a mixture of coarse sand particles, fine silt, clay particles, and organic matter.

Local/deterministic view of species richness. The view that the number of species in a community is primarily the result of local processes, such as competition and predation. *Compare with* Regional/historical view of species richness.

Local diversity. *See* Alpha diversity.

Local mate competition. A situation in which males compete for mates at or near their place of birth, and hence frequently with close relatives.

Local population. The individuals of a species that live in a habitat patch; also called a subpopulation.

Local resource competition. Competition among siblings for resources.

Locus. The position on a chromosome where a particular gene is located.

Logistic equation. The mathematical expression for a particular sigmoid growth curve in which the percentage rate of increase decreases in linear fashion as population size increases.

Lognormal distribution. A characterization of the number of species in a community in logarithmically scaled abundance classes, according to which most species have moderate abundances and fewer have either extremely high or low abundances.

Long-term monitoring studies. Ecological studies that involve taking measurements of variables over periods of years in order to gain an understanding of processes that take place at large temporal scales.

Lottery hypothesis. The idea that colonizing juveniles of all species in a community have an equal opportunity to take the place of adults that die or otherwise leave their territories, and thus colonize those territories at random.

Lower critical ecological temperature. The ambient temperature below which homeotherms are unable to produce enough heat both to gather food and to maintain body temperature.

Lower critical physiological temperature. The ambient temperature below which homeotherms cannot sustain metabolic energy production for body temperature regulation.

Lower critical temperature (T_{lc}). The ambient temperature below which homeotherms must generate heat to maintain their body temperature.

Luxury consumption. Consumption and storage of a nutrient beyond the amount needed for growth as a way of taking advantage of short-term abundance of the nutrient.

Macrograzers. Those relatively large grazers (e.g., crustaceans) occurring in open water that feed on aquatic vegetation. *Compare with* Micrograzers.

Macroparasites. Large, non-disease-causing parasites. *Compare with* Microparasites.

Mainland-island metapopulation. A metapopulation consisting of patches or islands situated near another, larger, patch, the mainland, from which dispersers can reach all the islands.

Mark-recapture method. A way of estimating the size of a population by the recapture rate of marked individuals.

Markov process. A stochastic process in which the future state of a system depends only upon its present state.

Mass balance. The balance between the amounts of nutrients and other materials entering and leaving a large ecosystem.

Mass extinction. The extinction of large numbers of species as a result of a natural catastrophe.

Mating system. The pattern of matings between individuals in a population, including number of simultaneous mates, permanence of pair bond, and degree of inbreeding.

Matric potential. The average strength with which the least tightly held water molecules are held by soil.

Matrix. A rectangular array (rows and columns) of mathematical elements (such as the coefficients of simultaneous equations) that is subject to special mathematical manipulations.

Maximum sustainable yield (MSY). The greatest rate at which individuals may be harvested from a population without reducing the size of the population; that is, at which recruitment equals or exceeds harvesting.

Mechanical weathering. Weathering caused by scouring by wind, repeated freezing and thawing of water in rock crevices, and the physical actions of roots.

Mediterranean climate. A climate pattern found in middle latitudes characterized by cool, wet winters and warm, dry summers.

Meiofauna. Animals that live within a substrate, such as soil or aquatic sediment.

Meiosis. A series of two divisions by cells destined to produce gametes, involving pairing and segregation of homologous chromosomes, and a reduction of chromosome number from diploid to haploid.

Melanism. Occurrence of black pigment, usually melanin.

Mesic. Referring to habitats with plentiful rainfall and well-drained soils.

Mesophyte. A plant that requires moderate amounts of moisture.

Mesoplankton. Plankton whose cells are greater than 200 mm in size.

Metabolism. Biochemical transformations responsible for the building up and breaking down of tissues and the release of energy by the organism.

Metamorphic rock. Rock formed when either igneous or sedimentary rock is subjected to the intense heat and pressure of the earth's core.

Metamorphosis. An abrupt change in form during development that fundamentally alters the function of the organism.

Metapopulation. A set of local populations occupying an array of habitat patches and connected to one another by the movement of individuals among them.

Metapopulation persistence time. The length of time until all of the local populations in a metapopulation become extinct.

Metapopulation structure. The characteristics of spatial distribution and demographics of metapopulations.

Micelle. A complex soil particle resulting from the association of humus and clay particles, with negative electric charges at its surface.

Michaelis-Menton constant (K_m). The concentration of a substrate at which the velocity of an enzyme-catalyzed reaction proceeds at half its maximum rate.

Micrograzers. Those relatively small grazers (e.g., protozoans) occurring in open water that feed on phytoplankton. *Compare with* Macrograzers.

Microhabitat. The particular parts of the habitat that an individual encounters in the course of its activities; also called microenvironment.

Micronutrients. Elements such as boron, selenium, zinc, and copper that are required in very small quantities by plants.

Microparasites. Endoparasites that cause diseases, such as dysentery or malaria. *Compare with* Macroparasites.

Microplankton. Plankton whose cells are between 20 and 200 mm in size.

Midparent value. The average phenotype of an individual's parents.

Mimic. An organism that resembles some other organism or object in the environment, thereby deceiving predators or prey into confusing it with that which it mimics.

Mineral nutrients. Elements such as nitrogen, phosphorus, sulfur, potassium, calcium, and others that are necessary for the growth and development of plants.

Mineralization. Transformation of elements from organic to inorganic forms, often by dissimilatory oxidations.

Minimum viable population (MVP). The smallest population that can persist for some arbitrarily long time, usually 1,000 years.

Mixed evolutionarily stable strategy. An ESS comprising more than one phenotype within a population; generally the outcome of frequency-dependent fitnesses of the phenotypes.

Model. An organism, usually distasteful or otherwise noxious, upon which a mimic is patterned.

Modular construction. The features of plants such as the occurrence of growth in specific areas called meristems and the more or less independent nature of plant parts.

Monod equation. An equation describing the growth rate of a population as a monotonic, asymptotic function of the concentration of resources.

Monoecy. In plants, the occurrence of male and female reproductive organs in different flowers on the same individual. *Compare with* Dioecy.

Monogamy. A mating system in which each individual mates with only one individual of the opposite sex, generally involving a strong and lasting pair bond. *Compare with* Polygamy.

Morph. A specific form, shape, or structure.

Mortality (m_x). Ratio of the number of deaths to individuals at risk, often described as a function of age (x). *Compare with* Death rate.

Müllerian mimicry. Mutual resemblance of two or more conspicuously marked, unpalatable species to enhance predator avoidance.

Multivariate ENSO index (MEI). An index composed of information on atmospheric pressure at sea level, direction of surface winds, sea surface temperature, and air temperature used to compare ENSO events.

Multivariate statistics. Statistical procedures that simultaneously analyze groups of variables, usually based on the intercorrelations among the variables, in order to find general patterns among the variables.

Mutation. Any change in the genotype of an organism occurring at the gene, chromosome, or genome level; usually applied to changes in genes to new allelic forms.

Mutualism. A relationship between two species that benefits both. A mutualism is obligate when the participating species are fully dependent on one another and facultative when either partner can live alone if necessary.

Mycelium. The network of hyphae that makes up the body of a fungus.

Mycorrhizae. Close associations of fungi and tree roots in the soil that facilitate the uptake of minerals by trees.

Myxomatosis. A viral disease of some mammals, transmitted by mosquitoes, that causes a fibrous cancer of the skin in susceptible individuals.

Natal dispersal. Dispersal of young animals from their place of birth.

Natural selection. Change in the frequency of genetic traits in a population through differential survival and reproduction of individuals bearing those traits.

Negative assortative mating. *See* Assortative mating.

Negative feedback. The tendency of a system to counteract externally imposed change and return to a stable state.

Negative frequency-dependent selection. *See* frequency-dependent selection.

Neighborhood size. The number of individuals in a population included within the dispersal distance of a single individual.

Neritic zone. That portion of the oceans extending to depths of about 200 meters, which corresponds to the edge of the continental shelf.

Net ecosystem production (NEP). The difference between gross primary production and the energetic costs of both the plants and the soil organisms (NEP = gross primary production − respiration of plants − respiration of soil organisms).

Net primary production (NPP). The total energy or nutrients accumulated by an organism, a population, or a community by growth and reproduction (NPP = gross primary production − respiration).

Net production efficiency. The percentage of assimilated food utilized for growth and reproduction by an organism. *Compare with* Gross production efficiency.

Net reproductive rate (R_0). The expected number of offspring produced by a female during her lifetime.

Neutral equilibrium. The state of a system that has no forces acting upon it; in Lotka-Volterra predator-prey models, the situation in which no internal forces act to restore the predator and prey populations to an equilibrium point.

Newton's law of cooling. The principle that heat loss varies in direct proportion to the gradient between body and ambient temperature.

Niche. The ecological role of a species in the community; the many ranges of conditions and resource qualities within which the organism or species can persist, often conceived as a multidimensional space.

Niche breadth. The variety of resources utilized and range of conditions tolerated by an individual, population, or species. *Compare with* Niche width.

Niche complementarity. A situation in which species that overlap extensively in their use of one resource differ substantially in their use of another resource.

Niche overlap. The sharing of niche space by two or more species; similarity of resource requirements and tolerance of ecological conditions.

Niche preemption. A model in which species successively procure a proportion of the available resources in an area, leaving less for the next colonizing species.

Niche width. The standard deviation of the distribution of resource use. *Compare with* Niche breadth.

Nitrification. The breakdown of nitrogen-containing organic compounds by microorganisms, yielding nitrates and nitrites.

Nitrogen fixation. Biological assimilation of atmospheric nitrogen to form organic nitrogen-containing compounds.

Node. A position in a food web representing a single species or a group of species thought to be trophically similar. *See also* Trophospecies.

Nodules. Swellings on the roots of some legumes and other plants within which nitrogen fixation is carried out by symbiotic bacteria.

Nonequilibrium system. A system, such as a system in which two species compete, in which disturbance and environmental fluctuation prevent populations from growing to the point at which the stable equilibrium predicted by logistic models occurs.

Nonlinearity. The dependence of interaction coefficients on population density.

Nonrenewable resource. A resource present in a fixed quantity, such as space, that can be utilized completely by consumers.

Normal distribution. A bell-shaped statistical distribution in which the probability density varies in proportion to $\frac{1}{\sigma\sqrt{2\pi}} e^{\frac{-1}{2}[(x-\mu)/\sigma]^2}$, where μ is the mean and σ is a distance (the standard deviation) from the mean.

Null hypothesis. A specific idea identified for consideration about a mechanism or phenomenon that is investigated via experimentation given as a declarative statement. *Compare with* Alternative hypothesis.

Null model. A set of rules for generating community patterns, presupposing no interaction between species, with which observed community patterns can be compared statistically.

Numerical response. A change in the population size of a predator species as a result of a change in the density of its prey. *Compare with* Functional response.

Nutrient. Any substance required by organisms for normal growth and maintenance.

Nutrient addition experiment. An experiment involving the measurement of the response of an ecosystem to the addition to a nutrient such as nitrogen or phosphorus; often carried out in lakes.

Nutrient cycle. The path of an element through the ecosystem, including its assimilation by organisms and its regeneration in a reusable inorganic form.

Nutrient spiral. The spiraling movement of nutrients in streams resulting from stream flow. (*See* Figure 12-19.)

Nutrient use efficiency (NUE). Ratio of production to uptake of a required nutrient.

Nymph. Adult-like immature stage of an arthropod, especially an insect, with direct development.

Obligate. Essential; without alternatives; not optional. *Compare with* Facultative.

Obligate mutualism. *See* mutualism.

Oceanic zone. The great depths of the ocean thousands of meters below the surface.

Oligotrophic. Poor in the mineral nutrients required by green plants; pertaining to an aquatic habitat with low productivity.

Omnivore. An organism whose diet is broad, including both plant and animal foods; specifically, an organism that feeds at more than one trophic level.

Omnivory. In food web analysis, feeding at more than one trophic level.

Open community. The concept of a community, advocated by H. A. Gleason and R. H. Whittaker, as a local expression of the independent geographic distributions of species. *Compare with* Closed community.

Operational sex ratio. The male/female ratio among mating individuals in a population. *Compare with* Primary sex ratio.

Optimal foraging theory. A body of theory according to which organisms maximize their food intake per unit of time or minimize the time required to meet their food requirements; risk of predation may also enter the equation for optimal foraging.

Optimal functional response. The functional response of a predator integrated over the time spent foraging and the time spent doing other things. *Compare with* differential functional response; integrated functional response.

Optimal outcrossing distance. The distance between populations whereby the negative effects of inbreeding are not revealed.

Optimality theory. A body of theory focusing on the analysis of the costs and benefits of behaviors.

Optimum. In ecology, the most efficient or best set of conditions or adaptations.

Orbital. The position of an electron, represented as a volume surrounding the nucleus of an atom within which an electron may be found 90% of the time.

Order of magnitude. A factor of 10; for example, three orders of magnitude is a factor of 1,000.

Ordination. A set of mathematical methods by which communities are ordered along physical gradients or along derived axes over which distance is related to dissimilarity in species composition.

Organic detritus. *See* Detritus.

Organismal viewpoint. The idea that a community is a discrete, highly integrated association of species within which the function of each species is subservient to the well-being of the whole.

Orthophosphate. An inorganic form of phosphate (PO_4^{3-}) that is readily assimilated by organisms. *Compare with* Colloidal phosphate.

Oscillation. Cyclic fluctuation above and below some mean value.

Osmoregulation. Regulation of the solute concentration of cells and body fluids.

Osmosis. Diffusion of substances in aqueous solution across the membrane of a cell.

Osmotic component of water potential. The component of water potential resulting from solute concentration.

Osmotic desert. An environment where there is plenty of water but where the salt concentration of the water is extremely high, as in mangrove habitats.

Osmotic pressure. The attraction of water to an aqueous solution owing to the concentration of ions and other small molecules.

Outcrossing. Mating with unrelated individuals within a population.

Overcompensation. An increase in the performance, growth, or fitness of a plant in response to herbivory. *See also* Compensation.

Overgrowth competition. Competition that occurs when one individual grows upon or over another.

Overlapping generations. The co-occurrence of parents and offspring in the same population as reproducing adults.

Oxic. Having oxygen.

Oxidation. The removal of one or more electrons from an atom, ion, or molecule.

Oxidation-reduction reaction. *See* Redox reaction.

Paradox of enrichment. The observation that experimental additions of nutrients or other resources that increase productivity often lead to reductions in diversity.

Parameter. In statistics, an unknown true characteristic of a statistical population. It is usually impossible to know the value of a parameter. A statistic estimates a parameter.

Parasite. An organism that consumes part of the blood or tissues of its host, usually without killing the host.

Parasite-mediated sexual selection. A theory of sexual selection that holds that females choose less parasitized males.

Parasitoid. Any of a number of insect species whose larvae live within and consume their host, usually another insect.

Parent material. Unweathered rock from which soil is derived.

Parent-offspring conflict. The situation arising when the optimum level of parental investment in a particular offspring from the viewpoint of the parent and from that of the offspring differs.

Parental investment. An act of parental care that enhances the survival of individual offspring or increases their number.

Parthenogenesis. Reproduction without fertilization by male gametes, usually involving the formation of diploid eggs whose development is initiated spontaneously.

Partial pressure. The proportional contribution of a particular gas to the total pressure of a mixture.

Partial regulation. A strategy in which organisms regulate their internal environment over moderate ranges of external conditions, but conform to the external environment under extreme conditions.

Particulate organic phosphate. That phosphorus incorporated into the cells of living organisms or as part of dead organic material.

Passive flux. The natural thermodynamic tendency of materials or energy to move from areas of high concentration to areas of low concentration.

Passive sampling hypothesis. A hypothesis of species-area relationships that holds that large islands will have greater species diversity because they present bigger targets for immigration.

Patch. *See* Habitat patch.

Patch metapopulation model. A simple metapopulation model in which the population dynamics of the local populations are ignored.

Patch population. A population that displays spatial patchiness, but in which individuals are not segregated to any great degree by the spatial structure of the population.

Patch use model. An optimal foraging model that predicts how foraging animals should choose among patches containing prey.

Pathogen. A parasite that causes disease in its host.

Payoff matrix. A matrix showing the costs and benefits of various combinations of behaviors. (*See* Table 33-2).

Pelagic. Pertaining to the open sea.

Per capita. Expressed on a per-individual basis.

Perennial. Referring to an organism that lives for more than one year; lasting throughout the year.

Perfect flower. A flower having both male and female sex organs (anthers and pistils).

Periphyton. The community of producers such as cyanobacteria, diatoms, and algae that covers surfaces in aquatic systems.

Permafrost. A layer of permanently frozen ground in very cold regions, especially the Arctic and Antarctic.

Permanent charge. Electric charges in the soil resulting from the substitution of metal atoms in clay particles.

Permeability. The capacity of a material, such as a biological membrane, to allow substances to pass through it.

Persistence time. *See* Time to extinction.

Pest pressure hypothesis. An idea regarding the effects of herbivory on plant species diversity that holds that less abundant plant species may thrive when herbivores forage differentially on more abundant plants.

pH. A scale of acidity or alkalinity; the logarithm of the concentration of hydrogen ions.

pH-dependent charge. Electric charge in the soil resulting from the dissociation of functional groups.

Phagoplankton. Planktonic organisms that assimilate dissolved organic material from the water.

Phanerophytes. Plants that carry their buds on the tips of branches, exposed to extremes of climate; one of five life forms identified by Christen Raunkiaer. (*See* Figure 8-25.)

Phenolic. Pertaining to compounds, such as lignin, based on the phenol chemical structure, a hydroxylated six-carbon ring (C_6H_5OH).

Phenolics. Aromatic hydrocarbons produced by plants, many of which exhibit antimicrobial properties.

Phenotype. The physical expression in the organism of the interaction between the genotype and the environment; the outward appearance and behavior of the organism.

Phenotypic optimization. The idea that genetic change occurs as a result of selection of phenotypes best suited to the environment.

Phenotypic plasticity. The general responsiveness of phenotypes to environmental conditions.

Phenotypic value. The measure of a particular phenotypic trait in a particular organism.

Phenotypic variance (V_P). A statistical measure of the variation in a phenotypic value among individuals in a population.

Pheromones. Chemical substances used for communication between individuals.

Philopatry. The tendency to stay near the place of birth.

Photic. Pertaining to surface waters to the depth of light penetration.

Photoautotroph. An organism that utilizes sunlight as its primary energy source for the synthesis of organic compounds.

Photomorphogenesis. The influence of light on seed germination, seedling growth, plant development, and flowering.

Photoperiod. Length of the daylight period each day.

Photorespiration. Oxidation of carbohydrates to carbon dioxide and water by the enzyme responsible for CO_2 assimilation in the presence of bright light.

Photosynthate. The organic products of photosynthesis; that is, simple carbohydrates.

Photosynthesis. Utilization of the energy of light to combine carbon dioxide and water into simple sugars.

Photosynthetic efficiency. The percentage of light energy assimilated by plants, based either on net production (net photosynthetic efficiency) or on gross production (gross photosynthetic efficiency).

Phylogeny. The pattern of evolutionary relationships among species or other taxonomic groups. *See also* Cladogram.

Physiological ecology. A subfield of ecology that involves the study of the ways in which the physiological characteristics of organisms respond to the environment.

Phytophagous. Consuming plants and plant material; usually refers to plant-eating insects.

Phytoplankton. Microscopic floating aquatic plants.

Picoplankton. Plankton with cells less than 2 mm in size.

Pit organ. A recessed organ in the head of pit vipers for detecting infrared radiation (heat) given off by prey organisms.

Plankton. Microscopic floating aquatic plants (phytoplankton) and animals (zooplankton).

Plant apparency. The ease or difficulty with which herbivores can locate their preferred plant.

Plant architecture. The set of morphological features, such as height, root/shoot ratio, branching pattern, and leaf area, that the define the growth form of a plant.

Plant defense guilds. Associations of plants and herbivores in groups based on plant chemistry and structure.

Pleiotropy. The influence of one gene on the expression of more than one trait in the phenotype.

Plotless sampling. A method of estimating the density and dispersion of populations from the distribution of distances between individuals.

Podsolization. The breakdown and removal of clay particles from the acidic soils of cold, moist regions.

Poikilothermy. Inability to regulate body temperature; cold-bloodedness.

Poisson distribution. A statistical description of the random distribution of items among categories, often applied to the distribution of individuals among sampling plots.

Polar covalent bond. A covalent bond in which the shared electrons spend more time around one of the atoms in the bond than the other, as in the water molecule. Such bonds give the molecule polarity.

Polyandry. A mating system in which a female mates with more than one male at the same time or in quick succession.

Polygamy. A mating system in which a male pairs with more than one female at a time (polygyny) or a female pairs with more than one male at a time (polyandry). *Compare with* Monogamy.

Polygenic. Determined by the expression of more than one gene.

Polygyny. A mating system in which a male mates with more than one female at the same time or in quick succession.

Polygyny threshold. The level of variation among the intrinsic values of territories or males at which the realized values of an unmated male on a poorer territory and a mated male on a better territory are equal in the eyes of an unmated female.

Polymorphism. The occurrence of more than one distinct phenotype or genotype in a population.

Polysaccharide. A long-chain carbohydrate, such as cellulose, made up of many monosaccharides (simple sugars).

Pool. A deep stretch of slowly moving water.

Population. Those organisms of the same species living in the same place and time.

Population bottleneck. The existence of a population at a small size for some period of time, which may result in a loss of genetic variation owing to genetic drift and inbreeding.

Population coevolution. Coevolution that results in local specializations in mutualists. *Compare with* Species coevolution.

Population cycle. Recurrent variation in population size in which numbers fluctuate between extremes over a regular period.

Population genetics. The study of the evolutionary mechanics of selection and genetic responses.

Population matrix. *See* Lewis-Leslie matrix.

Population structure. All of the features of a population, such as density, spacing, movement of individuals, genetic variation, age classes, and arrangement and size of habitats, that characterize the population. *See also* Age structure; Genetic structure; Spatial structure.

Population trajectory. The sum of the vectors of two populations on a graph portraying the abundance of one as a function of the abundance of the other; a representation of the simultaneous change in the sizes of two, usually interacting, populations.

Population viability analysis (PVA). The strategic analysis of the ecological, economic, and political issues and challenges related to the conservation of an endangered species, community, or ecosystem.

Positive assortative mating. *See* Assortative mating.

Positive frequency-dependent selection. *See* Frequency-dependent selection.

Potential energy. The energy of a substance or object deriving from its position, particularly with respect to gravity.

Potential evapotranspiration (PET). The amount of transpiration by plants and evaporation from the soil that would occur, given the local temperature and humidity, if water were not limited.

Prairie. An extensive area of level or rolling, almost treeless grassland in central North America.

Predation curve. The sum of the functional and numerical responses of a predator population. *Compare with* Total response of predator.

Predation-sensitive food hypothesis. The idea that prey should engage in behavior that puts them at greater risk of predation when intraspecific competition for food among the prey is intense, such as in times of food limitation.

Predation term. The algebraic term in Lotka-Volterra predator-prey models that portrays the density-dependent relationship between the predator and prey populations.

Predator. An animal (rarely a plant) that kills and eats animals.

Preemptive competition. Competition by means of the occupation of open space. *Compare with* Territorial competition.

Preservationism. A social movement that aims to preserve the natural world against human industrialization. *See also* Environmentalism.

Press experiment. An experiment in which a system is perturbed in such a way that the perturbation is sustained.

Pressure component of water potential. The component of water potential resulting from pressure.

Prey model. An optimal foraging model in which the decision variable is whether to eat a particular prey or not. *Compare with* Patch use model.

Primary consumer. An herbivore, the lowermost consumer on the food chain.

Primary host. A host that harbors the sexual stage of the life cycle of a parasite or disease organism.

Primary minerals. Products of the original rock formation, such as olivine and plagioclase. *Compare with* Secondary minerals.

Primary producer. A green plant or other autotroph that assimilates the energy of light to synthesize organic compounds.

Primary production. Assimilation (gross primary production) or accumulation (net primary production) of energy and nutrients by green plants and other autotrophs.

Primary sex ratio. The male/female ratio among zygotes in a population. *Compare with* Operational sex ratio.

Primary succession. The sequence of communities developing in a newly exposed habitat devoid of life. *Compare with* Secondary succession.

Probability of death (m_x). The probability that an individual will die between ages x and $x + 1$.

Probability of survival (s_x). The probability that an individual will survive from age x to age $x + 1$. *Compare with* Probability of death.

Production. Accumulation of energy or biomass.

Productivity. The rate at which energy is accumulated.

Productivity-stability hypothesis. A hypothesis of species diversity that suggests that diversity is limited by primary production. *Compare with* Intertaxon competition hypothesis.

Profundal. Pertaining to the very deepest parts of oceans or deep lakes.

Promiscuous. Mating with many individuals within a population, without the formation of strong or lasting pair bonds.

Protandry. Sequential hermaphroditism in which the male function is followed by the female function.

Protogyny. Sequential hermaphroditism in which the female function is followed by the male function.

Proximate factor. An aspect of the environment that an organism uses as a cue for behavior, such as day length, which often is not directly important to the organism's well-being. *Compare with* Ultimate factor.

Pseudoextinction. A situation in which a species evolves sufficiently that individuals are no longer recognized as belonging to the same taxon as their ancestors.

Pulse experiment. An experiment in which a treatment is applied more or less instantaneously and no other applications of the treatment are made.

Pupa. In insects with indirect development (complete metamorphosis), the transition stage between larva and adult.

Pyramid of energy. The concept that the energy flux through a given link in the food chain decreases at progressively higher trophic levels.

Pyramid of numbers. Charles Elton's idea that the sizes of populations decrease at progressively higher trophic levels; that is, as one progresses along the food chain.

Quantitative genetics. The study of the inheritance and response to selection of continuously varying traits having polygenic inheritance.

Quantitative trait. A trait having continuous variability within a population and revealing the expression of many gene loci.

***r*-selection.** Selection on traits that determine fecundity and survival to favor rapid population growth (r) at low population densities. *Compare with* K-selection.

Radiation. Energy emitted in the form of waves.

Rain shadow. A dry area on the leeward side of a mountain range.

Ramet. An individual within a clone, arising from an asexual process. *Compare with* Genet.

Random distribution. A distribution in which the position or value of one observation is not related to the positions or values of other observations. In a random spatial distribution, individuals are distributed throughout an area without regard to the other individuals in the area. *Compare with* Clumped distribution; Spaced distribution.

Random niche model. A model of species diversity suggesting that species abundance is determined by processes that randomly partition resources among species; also called the broken stick model.

Range. *See* Geographic distribution.

Rarefaction. A method of determining the relationship between species diversity and sample size by randomly deleting individuals from a sample.

Ratio-dependent model. A model in which the ratio of two populations (e.g., predator-prey, two competitors) is explicitly incorporated.

Reaction norm. The relationship between the phenotype of an individual and the environment. (*See* Figure 31-12.)

Recipient. In social interactions, the individual toward whom a particular behavior by another individual is directed. *Compare with* Donor.

Reciprocal altruism. The exchange of altruistic acts between individuals. Also referred to as tit-for-tat.

Reciprocal transplant experiment. An experimental procedure used to compare the genotype-environment interactions of individuals from different environments, involving the switching of individuals between two localities.

Recombinant DNA technology. Techniques used to examine genes and gene products.

Recombination. The mixing of genetic material via sexual reproduction.

Recruitment. Addition of new individuals to a population by reproduction; often restricted to the addition of breeding individuals.

Recruitment function. A function that depicts the size of a population at time $t + 1$ given its current size; of the form $N(t + 1) = \mathrm{F}[N(t)]$, where F is the recruitment function. (*See* Figures 18-11 and 18-12.)

Redfield ratio. The ratio of the atomic weights of carbon, nitrogen, and phosphorous atoms in an ecosystem.

Redox potential (Eh). The relative capacity of an atom or compound to donate or accept electrons, expressed in volts (electric potential); higher values indicate more powerful oxidizers.

Redox reaction. A reaction involving the oxidation of one constituent and the concomitant reduction of another.

Reduction. The addition of one or more electrons to an atom, ion, or molecule.

Regional/historical view of species richness. The view that the number of species in a community is primarily the result of historical events operating on a regional scale. *Compare with* Local/deterministic view of species richness.

Regulator. An organism that maintains an internal environment different from external conditions. *Compare with* Conformer.

Regulatory response. A rapid, reversible physiological or behavioral response by an organism to change in its environment. *Compare with* Acclimation.

Relative abundance. The proportional representation of a species in a sample or a community.

Renewable resource. A resource that is continually supplied to the system so that it cannot be fully depleted by consumers.

Replacement series. An experimental investigation of competition between two species in which the ratio of the species is varied while the total density of organisms is held constant. Also referred to as a substitution experiment.

Replacement series diagram. A diagrammatic representation of the outcome of a replacement series experiment in which the response of each species is plotted separately as a function of the planting ratio or density. (*See* Figure 22-14.)

Reproductive effort. Allocation of time or resources, or assumption of risk, in order to increase fecundity.

Reproductive rate of infection. The total number of individuals infected directly or indirectly by an individual having an infectious disease.

Rescue effect. Prevention of the extinction of a local population by immigration of individuals from elsewhere, often from a more productive habitat.

Residence time. The ratio of the size of a compartment to the flux through it, expressed in units of time; thus, the average time spent by energy or a substance in the compartment.

Resilience. The rate at which a population returns to equilibrium after a disturbance.

Resource. A substance or object required by an organism for normal maintenance, growth, and reproduction. *See also* limiting resource; renewable resource; nonrenewable resource.

Resource concentration hypothesis. The proposition that plant pests have a more difficult time finding their preferred plant host when the plant occurs in a diverse community. *See also* Associational resistance; Enemies hypothesis.

Resource holding potential (RHP). An organism's ability to defend a territory or resource.

Respiration. The use of oxygen to break down organic compounds metabolically for the purpose of releasing chemical energy.

Respired energy. The portion of the energy assimilated by an animal used to fulfill its metabolic needs.

Response to selection (*R*). The difference between the mean phenotypic value of the offspring of selected individuals and that of the population from which their parents were drawn. *See also* Selection differential.

Response variable. *See* Dependent variable.

Resting metabolic rate (RMR). The metabolic rate of an organism in an inactive, fasting, thermoneutral state.

Restoration ecology. A subfield of ecology focusing on restoring species and ecosystems after human perturbation.

Restriction enzymes. Enzymes that recognize specific base sequences on the DNA molecule and cut both strands at the site of those sequences, resulting in a restriction fragment.

Restriction fragment. *See* Restriction enzymes.

Restriction fragment length polymorphism (RFLP). The existence of restriction fragments of different lengths resulting from the application of restriction enzymes. RFLPs are heritable and thus serve as a basis for determining genetic variation.

Rhizome. An underground, usually horizontal stem of a plant that produces both roots and aboveground shoots, and which may be modified to store carbohydrate nutrient reserves.

Ricker model. A model in which recruitment is highest at intermediate population densities. (*See* Figure 18-12.)

Riffle. An area in a small stream where water runs rapidly over a rocky substrate.

Riparian. Referring to the banks of a river or lake.

Risk-averse foraging strategy. A foraging strategy in which the forager is willing to accept a lower mean gain in order to reduce the variation in that gain. *Compare with* Risk-prone foraging strategy.

Risk-prone foraging strategy. A foraging strategy in which the forager is willing to accept higher variation in food intake in order to obtain a higher average gain. *Compare with* Risk-averse foraging strategy.

Ritualized behavior. A predictable behavior pattern, often used to ascertain the chances of winning a contest without combat.

River continuum. The concept of a river as a single ecosystem having gradients of structure and function.

Root competition. Competition for water and nutrients among the belowground parts of plants in an area. *Compare with* Shoot competition.

RuBP. Ribulose bisphosphate, a five-carbon carbohydrate to which a carbon atom is attached during the assimilatory step of the Calvin-Benson cycle in photosynthesis.

RuBP carboxylase. An enzyme responsible for the reaction of ribulose bisphosphate and carbon dioxide to form two molecules of phosphoglyceraldehyde in the Calvin-Benson cycle of photosynthesis.

Ruderal. Pertaining to or inhabiting highly disturbed sites. *See also* Weed.

Rumen. An elaboration of the forepart of the stomach of certain ungulate mammals within which cellulose is broken down by symbiotic bacteria.

Runaway sexual selection. The situation in which females persistently choose the most extreme male phenotypes in a population, leading to continuous elaboration of secondary sexual characteristics.

Salt gland. An organ in birds and reptiles used to excrete sodium chloride in high concentration.

Sample. A collection of measurements of a natural phenomenon from which statistics are calculated.

Sampling study. An ecological study that involves systematic sampling from a natural system, in which statistical techniques are used to interpret the results.

Saturation. In reference to biological communities, an upper limit to the number of species that can coexist locally, set by their competitive interactions.

Saturation point. In reference to primary production, the amount of light that causes photosynthesis to attain its maximum rate.

Search image. A behavioral prey selection mechanism that enables predators to increase searching efficiency for prey that are abundant and worth capturing.

Secondary host. A host that harbors an asexual stage of the life cycle of a parasite or disease organism.

Secondary minerals. Minerals that are the products of weathering. *Compare with* Primary minerals.

Secondary plant compounds. Chemical products of plant metabolism specifically for the purpose of defense against herbivores and disease organisms.

Secondary sexual characteristics. Traits that distinguish sex over and above the primary sexual organs.

Secondary succession. The sequence of communities developing in habitats in which the climax community has been disturbed or removed. *Compare with* Primary succession.

Sedimentary rock. Rock formed when deposits of material in lakes and oceans accumulate and merge over thousands of years.

Seed pool. The repository of seeds in the soil.

Seed predation. A type of herbivory involving the consumption of seeds; also referred to as granivory.

Selection. Differential survival or reproduction of individuals in a population owing to phenotypic differences among them.

Selection differential (*S*). The difference between the mean phenotypic value of selected individuals and that of the population from which they were drawn. *See also* Response to selection.

Self-incompatible. Unable to mate with oneself; in the case of a hermaphrodite, owing to structural or biochemical factors that prevent self-fertilization.

Self-thinning curve. In populations of plants limited by space or other resources, the characteristic relationship between the logarithms of density and biomass.

Selfing. Mating with oneself; applicable, of course, only to individuals (usually plants) having both male and female sexual organs.

Selfishness. A social interaction that benefits the donor, but not the recipient, of the behavior.

Semelparity. The condition of having only one reproductive episode during the lifetime. *Compare with* Iteroparity.

Senescence. Gradual deterioration of function in an organism with age, leading to increased probability of death; aging.

Sensory exploitation hypothesis. A hypothesis regarding the origin of female choice that assumes that females have intrinsic preferences for certain appearances or behaviors based on the sensitivity of their sensory systems.

Sequential hermaphrodite. *See* Hermaphrodite.

Sere. A series of stages of community change in a particular area leading toward a stable state, or climax.

Serial polygamy. A mating system in which an individual mates with several individuals of the opposite sex in succession; the broods of each mating are generally cared for, in part, simultaneously.

Serpentine. An igneous rock rich in magnesium that forms soils toxic to many plants.

Set point. In a negative feedback system, the desired condition with which the existing condition is compared.

Sex ratio. Ratio of the number of individuals of one sex to that of the other sex in a population.

Sexual dimorphism. A difference in the phenotypes of male and female individuals of the same species, especially in ornamentation, coloration, and courtship behavior.

Sexual selection. Selection by one sex for specific characteristics in individuals of the opposite sex, usually exercised through courtship behavior. *See also* Intersexual selection; Intrasexual selection.

Shannon-Weaver index (*H*). A logarithmic measure of the diversity of species weighted by the relative abundance of each.

Shared predation. A type of apparent competition in which two species fall victim to a single predator and may compete for enemy-free space in which to avoid the predator.

Shoot competition. Competition, mainly for light, among the aboveground parts of adjacent plants. *Compare with* Root competition.

Sibling mating. Mating among the progeny of an individual parent.

Simple predator regulation. A situation in which predation pressure is sufficiently strong to keep the prey population at a level below which intraspecific competition among prey is important.

Simpson's index (*D*). A measure of the diversity of species weighted by the relative abundance of each.

Simultaneous hermaphrodite. *See* Hermaphrodite.

Sink. An ecosystem, habitat, population, or community that receives input of materials or individual organisms. *Compare with* Source.

Sink strength. In general, the rate of input to a sink, in reference to the translocation of photosynthate in the plant body, the demand for photosynthate in those plant parts, such as the roots, that do not photosynthesize.

Size advantage model. A model that holds that sequential hermaphroditism should evolve when the reproductive success of both males and females increases with age or size, but the rate of such increase is different between the sexes. (*See* Figure 33-3.)

Social behavior. Any direct interaction among distantly related individuals of the same species; usually does not include courtship, mating, parent-offspring, and sibling interactions.

Social dominance. Physical domination of one individual by another, initiated and sustained by aggression within a population.

Sociobiology. The study of the biological basis of social behavior.

Soil. The solid substrate of terrestrial communities resulting from the interaction of weather and biological activities with the underlying geologic formation.

Soil horizon. A layer of soil formed at a characteristic depth and distinguished by its physical and chemical properties.

Soil profile. A characterization of the structure of soil vertically through its various horizons, or layers.

Soil skeleton. The physical structure of mineral soil, referring principally to sand grains and silt particles.

Solar equator. The parallel of latitude that lies directly under the sun at any given season.

Source. An ecosystem, habitat, population, or community from which materials or organisms move. *Compare with* Sink.

Source-sink metapopulation. A metapopulation in which some local populations (sources) have a positive growth rate at low densities and others (sinks) have a negative growth rate in the absence of immigration.

Southern oscillation (SO). A fluctuation in atmospheric pressure between the western and eastern Pacific in which high pressure in one region corresponds to low pressure in the other.

Spaced distribution. A spatial arrangement in which each individual maintains a minimum distance between itself and its neighbors. *Compare with* Clumped distribution; Random distribution.

Spatial structure. The arrangement of the individuals of a population in space.

Spatially explicit metapopulation model. A model of a metapopulation that relates the extent of migration among local populations to the distance between them.

Spatially implicit metapopulation model. A model of a metapopulation in which it is assumed that all local populations are equally connected by migration.

Specialist. An organism with restricted use of habitats or resources. *Compare with* Generalist.

Specialization. A situation in which an organism interacts with a limited number of other organisms.

Speciation. The production of new species.

Species. A group of actually or potentially interbreeding populations that are reproductively isolated from all other kinds of organisms.

Species-area relationship. The number of species in relation to area.

Species coevolution. Coevolution leading to specializations that occur in the same form and with the same intensity everywhere in the geographic range of the interacting species. *Compare with* Population coevolution.

Species diversity. The number and/or relative abundances of species within a community.

Species packing. An increase in species richness without a change in resource diversity. (*See* Figure 29-19.)

Species pool. The flora and fauna of the mainland adjacent to offshore islands, or of a large area from which species migrate to populate smaller areas.

Species richness. The number of species in a community.

Specific heat. The amount of energy that must be added or removed to change the temperature of a substance by a specific amount. By definition, 1 calorie of energy is required to raise the temperature of 1 gram of water by 1 degree Celsius.

Sperm competition. A type of male-male competition in which males vie to have their sperm fertilize a female's egg.

Spitefulness. A social interaction in which the donor of a behavior incurs a cost in order to reduce the fitness of a recipient.

Spring bloom. An increase in phytoplankton growth during early spring in temperate lakes associated with vertical mixing of the water column.

Spring overturn. Vertical mixing of water layers in temperate lakes in spring as surface ice disappears.

Stable age distribution. The proportions of individuals in various age classes in a population that has been growing at a constant rate.

Stable equilibrium. The state to which a system returns if displaced by an outside force.

Stage-classified population. A population containing individuals of different developmental stages (e.g., adults and larvae) in the same or different habitats. *See also* Life cycle graph.

Standard deviation (s). A measure of the variability among items in a sample, such as individuals in a population; the square root of the variance, hence the square root of the average squared deviation from the mean.

Standard error (SE). A measure of the precision of repeated measurements of a variable. The standard error of the normal distribution is $s/\sqrt{n}$.

Static food web. A food web that shows only feeding relationships without reference to the strength of interactions or to changes in feeding relationships related to growth and population age structure.

Static life table. The age-specific survival and fecundity of individuals of different ages within a population at a given time; also called a time-specific life table.

Statistic. A value or index based on a sample of measurements that estimates a parameter.

Statistical population. The group of all possible measurements of a variable.

Statistics. The study and analysis of quantitative data based on probability theory.

Steady state. The condition of a system in which opposing forces or fluxes are balanced.

Stenotypic. Having a narrow range of tolerances for environmental conditions. *Compare with* Eurytypic.

Steppe. A usually treeless plain, especially in southeastern Europe and Asia in regions of extreme temperatures and sandy soils.

Stochastic. Resulting from random effects.

Stochastic dynamic programming. An ecological modeling approach that analyzes the long-term fitness of behavioral decisions based on the state of the individual and the pattern of short-term behavioral decisions.

Stochastic fluctuations. Changes in populations owing to chance events.

Stratification. The establishment of distinct layers of temperature or salinity in bodies of water owing to the different densities of warm and cold water or saline and fresh water.

Stream. A body of moving water that forms where precipitation exceeds evaporation and excess water drains from the surface of the land.

Subpopulation. *See* Local population.

Substitutable resource. A resource that can satisfy a particular requirement in the place of some other resource; that is, interchangeably.

Substitution experiment. *See* Replacement series.

Succession. Replacement of populations in a habitat through a regular sequence to a stable state.

Supercooling. A situation in which a fluid falls below the freezing point without ice crystals forming.

Superorganism. An association of individuals in which the function of each promotes the well-being of the entire system.

Superparasitism. The parasitizing of a single host by more than one parasitoid individual.

Surface tension. The attraction of water molecules to one another (cohesion) at the surface of a body of water.

Surplus hypothesis. A hypothesis of predator-prey interaction that holds that competition among prey is limited because predators take prey individuals that are weak or that stray into exposed or marginal habitats.

Survival (l_x). Proportion of newborn individuals alive at age x; also called survivorship.

Survivorship. *See* Survival.

Survivorship curve. Curve showing the number of individuals surviving to age x (log scale) plotted against age. (*See* Figure 14-18.)

Switching. A change in diet to favor food items of increasing suitability or abundance.

Symbiosis. An intimate, and often obligatory, association of two species, usually involving coevolution. Symbiotic relationships can be parasitic or mutualistic.

Sympatric. Occurring in the same place; usually refers to areas of overlap in species distributions.

Synergism. The interaction of two causes such that the total effect is greater than the sum of two acting independently.

Systematics. The classification of organisms into a hierarchical set of categories (taxa) emphasizing their evolutionary interrelationships.

Systems ecology. The study of an ecological system as a set of components linked by fluxes of energy and nutrients or by population interactions.

Systems modeling. Portrayal of a system's function by mathematical functions describing the interactions among its components.

Tannins. Polyphenolic compounds, produced by most plants, that bind proteins, thereby impairing digestion by herbivores and inhibiting microbes.

Taxon cycle. A cycle of expansion and contraction of the geographic range and population density of a species or higher taxon.

Taxonomic diversity. Diversity based on taxonomic categorizations of organisms. *Compare with* Ecological diversity.

Taxonomy. The description, naming, and classification of organisms.

Temperature profile. The relationship of temperature to depth below the surface of water or the soil, or height above the ground.

Territorial competition. Competition involving the defense of occupied space. *Compare with* Preemptive competition.

Territoriality. A situation in which individuals defend exclusive spaces, or territories.

Territory. Any area defended by one or more individuals against intrusion by others of the same or different species.

Theory of island biogeography. A theory of the relationship between species richness (number of species), island area, and the distance of islands from the mainland.

Thermal conductance. The rate at which heat passes through a substance.

Thermal stratification. A sharp delineation of layers of water by temperature, with the warmer layer generally lying over the colder layer.

Thermocline. The zone of water depth within which the temperature changes rapidly between the upper warm water layer (epilimnion) and the lower cold water layer (hypolimnion).

Thermodynamic. Relating to heat and motion.

Thermodynamic steady state. A situation in which a system loses heat at the same rate as it gains it.

Thermohaline circulation. Patterns of downwelling and upwelling driven by temperature and salinity differences in surface waters.

Thermophilic bacteria. Bacteria adapted to live in extremely hot environments.

Therophytes. Plants that die during the unfavorable season of the year and do not have persistent buds; one of five life forms identified by Christen Raunkiaer. (*See* Figure 8-25.)

Theoretical ecological model. A verbal, algebraic, or graphic construction that hypothesizes the relationships among variables of a natural system.

Three-halves power law. A generalization proposing that the relationship between the logarithms of biomass and density of a population of plants has a slope of $-3/2$.

Time delay. A delay in the response of a population or other system to conditions of the environment; also called a time lag.

Time hypothesis. The idea that areas that have enjoyed long periods of stability have greater biodiversity than areas subject to frequent and widespread disturbance.

Time lag. *See* Time delay.

Time-specific life table. *See* Static life table.

Time to extinction. The time that elapses between colonization of a site and extinction.

Tit-for-tat. *See* Reciprocal altruism.

Tolerance. The indifference of establishment of one species to the presence of others in a successional sequence.

Top-down control. Regulation of ecosystem function via trophic interactions. *Compare with* Bottom-up control.

Topological food web. A food web that emphasizes feeding relationships among organisms; also called a connectedness food web.

Torpor. Loss of the power of motion and feeling, usually accompanied by a greatly reduced rate of respiration.

Total response of predator. The product of the functional and numerical responses plotted as per capita mortality of prey. *Compare with* Predation curve.

Tracheids. In plants, one of two types of xylem cells used for the transport of water.

Trade winds. The westward deflection of surface air flows in the Tropics resulting from the earth's rotation. *Compare with* Westerlies.

Trajectory. In ecological models that portray the joint behavior of two populations (e.g., two competitors or predator and prey), the curve of all of the joint densities of the populations.

Transfer function. In systems modeling, the equation describing the flux of energy or a substance between one component and another.

Transition matrix. A matrix of probabilities of changes from one state to another.

Transit time. The average time that a substance or energy remains in the biological realm or in any compartment of a system; the ratio of biomass to productivity.

Transmission. The movement of a parasite from one host to the next. *See also* Disease vector.

Transmission coefficient. In models of disease transmission, a parameter that represents the rate at which transmission of the disease from one host to the next occurs on average.

Transpiration. Evaporation of water from leaves and other parts of plants.

Transpiration efficiency. The ratio of net primary production to transpiration of water by a plant, usually expressed as grams per kilogram of water; also called water use efficiency.

Transpirational pull. Tension placed on the water in xylem tissue as a result of transpirational water loss in the leaves of a plant.

Travel time. The time that it takes for a predator to move from one patch to another while foraging.

Trophic. Pertaining to food or nutrition.

Trophic cascade. A food chain with top-down regulation. (*See* Figure 13-16.)

Trophic level. Position in the food chain, determined by the number of energy transfer steps to that level.

Trophic mutualism. A mutualism involving interacting species specialized in complementary ways to obtain energy and nutrients from each other.

Trophic structure. A community structure determined by feeding relationships.

Trophospecies. A group of species with similar trophic ecology; often lumped together in food webs.

Turgor. The rigidity of plants resulting from osmotic pressure in cells causing them to take in water, thus pushing the cell membrane against the cell wall.

Type I error. The rejection of a true null hypothesis.

Type II error. The acceptance of a false null hypothesis.

Ultimate factor. An aspect of the environment that is directly important to the well-being of an organism (for example, food availability). *Compare with* Proximate factor.

Ultraviolet (UV) radiation. Electromagnetic radiation having a wavelength shorter than about 400 mm.

Unstable equilibrium. The state of a system at which forces are precisely balanced, but away from which the system moves when displaced.

Upwelling. Vertical movement of water, usually near coasts and driven by offshore winds, that brings nutrients from the depths of the ocean to surface layers.

Urea. A nitrogenous waste product excreted by mammals and some other organisms; also employed in the bloodstream of sharks for osmoregulation.

Vapor pressure. The pressure exerted by a vapor that is in equilibrium with its liquid form.

Variable. A characteristic or measure of the natural world that may take on any of a number of different values.

Variable number tandem repeats. *See* DNA fingerprinting.

Variance (s^2). A statistical measure of the dispersion of a set of values about its mean.

Vector. In matrix algebra, a row or column of mathematical elements that is subject to special mathematical manipulations. *See also* Disease vector.

Vegetation texture. The density, dispersion pattern, and species composition of plants in a patch of habitat or the number, density, and arrangement of patches of plants.

Veil line. The point on a lognormal distribution of species abundances below which one expects fewer than one individual per species in a sample; hence the veil line separates observed from potentially observed species.

Vertical mixing. Exchange of water between deep and surface layers.

Vesicular-arbuscular mycorrhizae. Endomycorrhizal associations between fungi and the roots of plants distinguished by the branching pattern of growth of the fungus within the root tissues.

Vessel elements. In plants, one of two types of xylem cells used for the transport of water.

Viscosity. The quality of a fluid that resists internal flow.

Vigilance. Behaviors associated with looking out for predators. *See also* Collective detection; Information center.

Warm event. *See* ENSO.

Warning coloration. Conspicuous patterns or colors adopted by noxious organisms to advertise their unpalatability or dangerousness to potential predators; also called aposematism.

Water potential. The force by which water is held in the soil by capillary and hygroscopic attraction; the free energy of water in a system.

Watershed. The drainage area of a stream or river.

Water use efficiency. *See* Transpiration efficiency.

Weathering. The physical and chemical breakdown of rock and its component minerals at the base of the soil.

Weed. A plant or animal, generally having high powers of dispersal, capable of living in highly disturbed habitats.

Westerlies. The eastward deflection of surface air flows in the middle latitudes resulting from the earth's rotation. *Compare with* Trade winds.

Wildlife. Wild animals whose populations are monitored and managed for hunting and fishing, such as white-tailed deer and turkeys.

Wilting. A loss of turgor in plant cells that occurs when water is in short supply, causing the plant to droop.

Wilting coefficient. The minimum water content of the soil at which plants can obtain water.

Xeric. Referring to habitats in which plant production is limited by availability of water.

Xerophyte. A plant that tolerates dry (xeric) conditions.

Yield. The amount of plant or animal biomass. *See also* Law of constant final yield.

Zero net growth isocline (ZNGI). An equilibrium isocline representing the lowest level of a resource, or combination of resources, for which population growth is positive.

Zooplankton. Tiny floating aquatic animals.

Zygote. A diploid cell formed by the union of male and female gametes during fertilization.

BIBLIOGRAPHY

Aber, J. D., and J. M. Melillo. 1991. Terrestrial Ecosystems. Saunders College Publishing, Philadelphia.

Aber, J. D., J. M. Melillo, and C. A. Federer. 1982. Predicting the effects of rotation length, harvesting intensity, and fertilization on fiber yield from northern hardwood forests in New England. For. Sci. 28:31–45.

Abrahamson, W. G., and M. Gadgil. 1973. Growth form and reproductive effort in goldenrods (*Solidago,* Compositae). Am. Nat. 107:651–661.

Abrams, P. A. 1983. The theory of limiting similarity. Annu. Rev. Ecol. Syst. 14:359–376.

Abrams, P. A. 1991. The relationship between food availability and foraging effort: Effects of life history and time scale. Ecology 72:1242–1252.

Abrams, P. A., and H. Matsuda. 1996. Positive indirect effects between prey species that share predators. Ecology 77:610–616.

Adams, R. M., C. Rosenzweig, P. M. Peart, J. T. Ritchie, B. A. McCarl, J. D. Glyer, R. B. Curry, J. W. Jones, K. J. Boote, and L. H. Allen Jr. 1990. Global climate change and U.S. agriculture. Nature 345:219–224.

Adolph, S. C., and W. P. Porter. 1993. Temperature, activity, and lizard life histories. Am. Nat. 142:273–295.

Aker, C. L. 1982. Regulation of flower, fruit and seed production by a monocarpic perennial, *Yucca whipplei.* J. Ecol. 70:357–372.

Alatalo, R. V., A. Lundberg, and K. Stahlbrandt. 1982. Why do pied flycatcher females mate with already-mated males? Anim. Behav. 30:585–593.

Alcock, J. 1969. Observational learning in three species of birds. Ibis 111:308–321.

Alerstam, T. 1990. Bird Migration. Cambridge University Press, Cambridge.

Alerstam, T. 1991. Bird flight and optimal migration. Trends Ecol. Evol. 6:210–215.

Alexander, R. M. 1982. Optima for animals. Arnold, London.

Allee, W. C., O. Park, A. E. Emerson, T. Park, and K. P. Schmidt. 1949. Principles of Animal Ecology. Saunders, Philadelphia.

Allen, J. D. 1995. Stream Ecology: Structure and Function of Running Waters. Chapman & Hall, New York.

Allison, A. C. 1956. Sickle cells and evolution. Sci. Am. 195:87–94.

Alward, R. D., and A. Joern. 1993. Plasticity and overcompensation in grass responses to herbivory. Oecologia 95:358–364.

Amadon, D. 1950. The Hawaiian honeycreepers (Aves, Drepanidae). Bull. Am. Mus. Nat. Hist. 95:151–262.

Amarasekare, P. 1994. Spatial population structure in the banner-tailed kangaroo rat, *Dipodomys spectabilis.* Oecologia 100:166–176.

Amos, W. H. 1959. The life of a sand dune. Sci. Am. 201:91–99.

Andersen, D. C. 1987. Below-ground herbivory in natural communities: A review emphasizing fossorial animals. Q. Rev. Biol. 62:261–286.

Anderson, G. R. V., A. H. Ehrlich, P. R. Ehrlich, J. D. Roughgarden, B. C. Russell, and F. H. Talbot. 1981. The community structure of coral reef fishes. Am. Nat. 117:476–495.

Anderson, J. M., W. S. Chow, and Y.-I. Park. 1995. The grand design of photosynthesis: Acclimation of the photosynthetic apparatus to environmental cues. Photosyn. Res. 46:129–139.

Anderson, N. H., and J. R. Sedell. 1979. Detritus processing by macroinvertebrates in stream ecosystems. Annu. Rev. Entomol. 24:351–377.

Anderson, R. M., and R. M. May. 1991. Infectious Diseases of Humans: Dynamics and Control. Oxford University Press, Oxford.

Andersson, M. 1982. Female choice selects for extreme tail length in a widowbird. Nature 299:818–820.

Andersson, M. 1994. Sexual Selection. Princeton University Press, Princeton, NJ.

Andersson, M., and S. Erlinge. 1977. Influence of predation on rodent populations. Oikos 29:591–597.

Andersson, M., and J. Krebs. 1978. On the evolution of hoarding behavior. Anim. Behav. 26:707–711.

Andrewartha, H. G., and L. C. Birch. 1954. The Distribution and Abundance of Animals. University of Chicago Press, Chicago.

Antia, R., B. R. Levin, and R. M. May. 1994. Within-host population dynamics and the evolution and maintenance of microparasite virulence. Am. Nat. 144:457–472.

Antonovics, J. 1971. The effects of a heterogeneous environment on the genetics of natural populations. Am. Sci. 59:593–599.

Applebaum, S. W. 1964. Physiological aspects of host specificity in the Bruchidae. I. General considerations of developmental compatibility. J. Insect Physiol. 10:783–788.

Applebaum, S. W., B. Gestetner, and Y. Birk. 1965. Physiological aspects of host specificity in the Bruchidae. IV. Developmental incompatibility of soybeans for *Callosobruchus.* J. Insect Physiol. 11:611–616.

Arcese, P., and J. N. M. Smith. 1988. Effects of population density and supplemental food on reproduction in song sparrows. J. Anim. Ecol. 57:119–136.

Arditi, R., and H. Saiah. 1992. Empirical evidence of the role of heterogeneity in ratio-dependent consumption. Ecology 73:1544–1551.

Armbruster, W. S. 1985. Patterns of character divergence and the evolution of reproductive ecotypes of *Dalechampia scandens.* Evolution 39:733–752.

Armbruster, W. S. 1990. Estimating and testing the shapes of adaptive surfaces: The morphology and pollination of *Dalechampia* blossoms. Am. Nat. 135:14–31.

Armbruster, W. S. 1991. Multilevel analysis of morphometric data from natural populations: Insights into ontogenetic, genetic, and selective correlations in *Dalechampia scandens.* Evolution 45:1229–1244.

Armbruster, W. S. 1992. Phylogeny and the evolution of plant-animal interaction. BioScience 42:12–20.

Arnold, S. J. 1983. Morphology, performance and fitness. Am. Zool. 23:347–361.

Arnold, S. J. 1993. Foraging theory and prey-size–predator-size relations in snakes. In R. A. Seigel and J. T. Collins (eds.), Snakes: Ecology and Behavior, 87–115. McGraw-Hill, Inc., New York.

Arnold, S. J., and M. J. Wade. 1984. On the measurement of natural and sexual selection: Theory. Evolution 38:709–719.

Aronson and Shmida 1990, chapter 29.

Arrhenius, O. 1921. Species and area. J. Ecol. 9:95–99.

Atkinson, D. 1991. Sexual showiness and parasite load: Correlations without parasite coevolution cycles. J. Theor. Biol. 150:251–260.

Atsatt, P. R., and D. J. O'Dowd. 1976. Plant defense guilds. Science 193:24–29.

Aubréville, A. 1938. La fôret coloniale: Les fôrets de l'Afrique occidentale française. Annals of the Academic Society of Colon, Paris 9:1–245.

Augner, M. 1995. Low nutritive quality as a plant defence: Effects of herbivore-mediated interactions. Evol. Ecol. 9:605–616.

Augspurger, C. K. 1983. Seed dispersal of the tropical tree, *Platypodium elegans,* and the escape of its seedlings from fungal infection. J. Ecol. 71:759–771.

Austin, G. T. 1974. Nesting success of the cactus wren in relation to nest orientation. Condor 76:216–217.

Axelrod, R., and D. Dion. 1988. The further evolution of cooperation. Science 242:1385–1390.

Axelrod, R., and W. D. Hamilton. 1981. The evolution of cooperation. Science 211:1390–1396.

Ayala, F. J. 1965. Evolution of fitness in experimental populations of *Drosophila serrata.* Science 150:903–905.

Ayala, F. J. 1970. Competition, coexistence, and evolution. In M. K. Hecht and W. C. Steere (eds.), Essays in Evolution and Genetics, 121–158. Appleton-Century Crofts, New York.

Ayala, F. J. 1971. Competition between species: frequency dependence. Science 171:820–824.

Azam, F. T. Fenchel, J. D. Field, L. A. Meyer-Reil, and F. Thingstad. 1983. The ecological rate of water-column microbes in the sea. Mar. Ecol. Prog. Ser. 10:257–263.

Bach, C. E. 1994. Effects of herbivory and genotype on growth and survivorship of sand-dune willow (*Salix cordata*). Ecol. Entomol. 9:303–309.

Baker, H. G. 1965. Characteristics and modes of origin of weeds. In H. G. Baker and G. L. Stebbins (eds.), The Genetics of Colonizing Species, 147–172. Academic Press, New York.

Baker, J. R. 1938. The evolution of breeding seasons. In G. R. de Beer (ed.), Evolution: Essays on Aspects of Evolutionary Biology, 161–177. Oxford University Press, London and New York.

Baker, R. R. 1978. The Evolutionary Ecology of Animal Migration. Holmes and Meier, New York.

Barbour, M. G., J. H. Burk, and W. D. Pitts. 1980. Terrestrial Plant Ecology. Benjamin/Cummings, Menlo Park, CA.

Barbour, M. G., J. H. Burk, and W. D. Pitts. 1987. Terrestrial Plant Ecology. Benjamin/Cummings, Menlo Park, CA.

Barel, C. D. N., R. Dorit, P. H. Greenwood, G. Fryer, N. Hughes, P. B. N. Jackson, H. Kawanabe, R. H. Lowe-McConnell, M. Nagoshi, A. J. Ribbink, E. Trewavas, F. Witte, and K. Yamaoka. 1985. Destruction of fisheries in Africa's lakes. Nature 315:19–20.

Barko, J. W., and R. M. Smart. 1981. Sediment-based nutrition of submersed macrophytes. Aquat. Bot. 10:339–352.

Barnes, R. K., and K. H. Mann (eds.). 1980. Fundamentals of Aquatic Ecosystems. Blackwell, Oxford.

Barrett, S. C. H., and L. D. Harder. 1996. Ecology and evolution of plant mating. Trends Ecol. Evol. 11:73–79.

Barrios, E., and R. Herrera. 1994. Nitrogen cycling in a Venezuelan tropical seasonally flooded forest: Soil nitrogen mineralization and nitrification. J. Trop. Ecol. 10:399–416.

Bartholomew, B. 1970. Bare zone between California shrub and grassland communities: The role of animals. Science 170:1210–1212.

Bartholomew, G. A. 1942. The fishing activities of double-crested cormorants on San Francisco Bay. Condor 44:13–21.

Baskin, C. C., and J. M. Baskin. 1988. Germination ecophysiology of herbaceous plant species in a temperate region. Am. J. Bot. 75:286–305.

Basolo, A. 1990a. Female preference for male sword length in the green swordtail, *Xiphophorus helleri* (Pisces: Poeciliidae). Anim. Behav. 40:332–338.

Basolo, A. 1990b. Female preference predates the evolution of the sword in swordtail fish. Science 250:808–810.

Bastrenta, B. 1991. Effect of sheep grazing on the demography of *Anthyllis vulneraria* in southern France. J. Ecol. 72:275–284.

Bateman, A. J. 1948. Intra-sexual selection in *Drosophila*. Heredity 2:349–368.

Bateson, P. (ed.). 1983. Mate Choice. Cambridge Univ. Press, Cambridge.

Batzli, G. O. 1969. Distribution of biomass in rocky intertidal communities on the Pacific coast of the United States. J. Anim. Ecol. 38:531–546.

Baust, J. G. 1973. Mechanisms of cryoprotection in freezing tolerant animal systems. Cryobiology 10:197–205.

Baust, J. G., and R. E. Morrissey. 1975. Supercooling phenomenon and water content independence in the overwintering beetle, *Coleomegilla maculata*. J. Insect Physiol. 21:1751–1754.

Bayram, A., and M. L. Luff. 1993. Cold-hardiness of wolf-spiders (Lycosidae, Araneae) with particular reference to *Pardosa pullata* (Clerck). J. Therm. Biol. 18:263–268.

Bazely, D. R., and J. L. Jefferies. 1989. Leaf and shoot demography of an arctic stoloniferous grass, *Puccinellia phryganodes*, in response to grazing. J. Ecol. 77:811–822.

Bazzaz, F. A. 1979. The physiological ecology of plant succession. Annu. Rev. Ecol. Syst. 10:351–357.

Bazzaz, F. A. 1990. The response of natural ecosystems to the rising global CO_2 levels. Annu. Rev. Ecol. Syst. 21:167–196.

Bazzaz, F. A., S. L. Miao, and P. M. Wayne. 1993. CO_2-induced growth enhancements of co-occurring tree species decline at different rates. Oecologia 96:478–482.

Beals, E. W., and J. B. Cope. 1964. Vegetation and soils in eastern Indiana woods. Ecology 45:777–792.

Beaulieu, J., G. Gauthier, and L. Rochefort. 1996. The growth response of graminoid plants to goose grazing in a high arctic environment. J. Ecol. 84:905–914.

Beck, S. D. 1965. Resistance of plants to insects. Annu. Rev. Entomol. 10:207–232.

Beck, S. D. 1980. Insect Photoperiodism. 2nd ed. Academic Press, New York.

Becking, J.-H. 1992. The *Rhizobium* symbiosis of the nonlegume *Parasponia*. In G. Stacey, R. H. Burris, and H. J. Evans (eds.), Biological Nitrogen Fixation, 497–559. Chapman & Hall, New York.

Beckman, W. A., J. W. Mitchell, and W. P. Porter. 1973. Thermal model for prediction of a desert iguana's daily and seasonal behavior. J. Heat Transfer, May 1973:257–262.

Bednarz, J. C. 1988. Cooperative hunting in Harris' hawks (*Parabuteo unicinctus*). Science 239:1525–1527.

Beer, T., and M. D. Swaine. 1977. On the theory of explosively dispersed seeds. New Phytol. 78:681–694.

Begon, M., and M. Mortimer. 1986. Population Ecology. 2nd ed. Blackwell, Oxford.

Bell, G. 1982. The Masterpiece of Nature: The Evolution and Genetics of Sexuality. University of California Press, Berkeley.

Bell, G. 1985. On the function of flowers. Proc. R. Soc. Lond. B 224:223–265.

Bell, W. J., and Cardé, R. T. 1984. Chemical Ecology of Insects. Sinauer Associates, Sunderland, MA.

Belovsky, G. E. 1984. Moose and snowshoe hare competition and a mechanistic explanation from foraging theory. Oecologia 61:150–159.

Belsky, A. J., W. P. Carson, C. Jensen, and G. A. Fox. 1993. Overcompensation by plants: Herbivore optimization or red herring? Evol. Ecol. 7:109–121.

Bender, E. A., T. J. Case, and M. E. Gilpin. 1984. Perturbation experiments in community ecology: Theory and practice. Ecology 65:1–13.

Bengtsson, J. 1991. Interspecific competition in metapopulations. In M. E. Gilpin and I. Hanski (eds.), Metapopulation Dynamics: Empirical and Theoretical Investigations, 219–237. Academic Press, London.

Berenbaum, M. R. 1978. Toxicity of a furanocoumarin to armyworms: A case of biosynthetic escape from insect herbivores. Science 201:532–534.

Berenbaum, M. R. 1981. Patterns of furanocoumarin distribution and insect herbivory in the Umbelliferae: Plant chemistry and community structure. Ecology 62:1254–1266.

Berenbaum, M. R. 1983. Coumarins and caterpillars: A case for coevolution. Evolution 37:163–179.

Berg, R. Y. 1975. Myrmecochorous plants in Australia and their dispersal by ants. Aust. J. Bot. 23:475–508.

Bergelson, J., T. Juenger, and M. Crawley. 1996. Regrowth following herbivory in *Ipomopsis aggregata:* Compensation but not overcompensation. Am. Nat. 148:744–755.

Berger, J. 1990. Persistence of different-sized populations: An empirical assessment of rapid extinctions in bighorn sheep. Conserv. Biol. 4:91–98.

Bergerud, A. T. 1974. Decline of caribou in North America following settlement. J. Wildl. Mgmt. 38:757–770.

Bergerud, A. T. 1980. A review of the population dynamics of caribou and wild reindeer in North America. In E. Reimers, E. Gaare, and S. Skenneberg (eds.), Proceedings of the Second International Reindeer/Caribou Symposium, 556–581. Direktoratet for vilt og ferkvannsfisk, Trondheim, Norway.

Bergerud, A. T., and W. E. Mercer. 1989. Caribou introductions in eastern North America. Wildl. Soc. Bull. 17:111–120.

Bernays, E. A., and R. F. Chapman. 1970a. Food selection by *Chorthippus parallelus* (Zetterstedt) (Orthoptera: Acrididae) in the field. J. Anim. Ecol. 39:383–394.

Bernays, E. A., and R. F. Chapman. 1970b. Experiments to determine the basis of food selection by *Chorthippus parallelus* (Orthoptera: Acrididae) in the field. J. Anim. Ecol. 39:761–776.

Bernays, E. A., and R. F. Chapman. 1994. Host-plant selection by phytophagous insects. Chapman & Hall, New York.

Berrigan, D., and J. C. Koella. 1994. The evolution of reaction norms: Simple models for age and size at maturity. J. Evol. Biol. 7:549–566.

Berry, J. A. 1975. Adaptation of photosynthetic processes to stress. Science 188:644–650.

Berry, J. A., and O. Bjorkman. 1980. Photosynthetic response and adaptation to temperature in higher plants. Annu. Rev. Plant Physiol. 31:491–543.

Berryman, A. A. 1981. Population Systems: A General Introduction. Plenum Press, New York.

Berryman, A. A. 1992. The origins and evolution of predator-prey theory. Ecology 73:1530–1535.

Berthold, P. 1993. Bird Migration: A General Survey. Oxford University Press, Oxford.

Beschel, R. E., P. J. Webber, and R. Tippett. 1962. Woodland transects of the Frontenac Axis Region, Ontario. Ecology 43:386–396.

Bess, H. A., R. van den Bosch, and F. A. Haramoto. 1961. Fruit fly parasites and their activities in Hawaii. Proc. Haw. Entomol. Soc. 17:367–378.

Beverton, R. J. H., and S. J. Holt. 1957. On the dynamics of exploited fish populations. Fish. Invest. 19:1–533.

Billings, W. D., and H. A. Mooney. 1968. The ecology of arctic and alpine plants. Biol. Rev. 43:481–529.

Birch, L. C. 1953. Experimental background to the study of the distribution and abundance of insects. 1. The influence of temperature, moisture and food on the innate capacity for the increase of three grain beetles. Ecology 34:698–711.

Birch, L. C., and D. P. Clark. 1953. Forest soil as an ecological community with special reference to the fauna. Q. Rev. Biol. 28:13–36.

Birkhead, T. R., T. Burke, R. Zann, F. M. Hunter, and A. P. Krupa. 1990, Extra-pair paternity and intra-specific brood parasitism in wild zebra finches *Taeniopygia guttata* revealed by DNA fingerprinting. Behav. Ecol. Sociobiol. 27:315–324.

Birkhead, T. R., and A. P. Møller. 1992. Sperm Competition in Birds. Academic Press, London.

Bisson, P. A., T. P. Quinn, G. H. Reeves, and S. V. Gregory. 1992. Best management practices, cumulative effects and long-term trends in fish abundance in Pacific Northwest river systems. In R. J. Naiman (ed.), Watershed Management: Balancing sustainability and Environmental Change, 189–233. Springer-Verlag, New York.

Bjorkman, O. 1968. Further studies on differentiation of photosynthetic properties of sun and shade ecotypes of *Solidago virgaurea*. Physiol. Plant. 21:84–99.

Bjorkman, O., and J. Berry. 1973. High efficiency photosynthesis. Sci. Am. 229:80–93.

Black, H. S., and J. T. Chan. 1977. Experimental ultraviolet light carcinogenesis. Photochemical Photobiol. 26:183–199.

Blair, W. F. 1960. The Rusty Lizard, A Population Study. University of Texas Press, Austin.

Blair, W. F. 1975. Adaptation of anurans to equivalent desert scrub of North and South America. In D. W. Goodall (ed.), Evolution of Desert Biota, 197–222. University of Texas Press, Austin.

Blaustein, A. R., B. Edmond, J. M. Kiesecker, J. J. Beatty, and D. G. Hokit. 1995. Ambient ultraviolet radiation causes mortality in salamander eggs. Ecol. Appl. 5:740–743.

Blaustein, A. R., P. D. Hoffman, D. G. Hokit, J. M. Kiesecker, S. C. Walls, and J. B. Hays. 1994. UV repair and resistance to solar UV-B in amphibian eggs: A link to population declines? Proc. Natl. Acad. Sci. USA 91:1791–1795.

Blaustein, A. R., J. M. Kiesecker, D. G. Hokit, and S. C. Walls. 1995. Amphibian declines and UV radiation. BioScience 45:514–515.

Blaustein, A. R., and D. B. Wake. 1990. Declining amphibian populations: A global phenomenon? Trends Ecol. Evol. 5:203.

Blest, A. D. 1957. The function of eye-spot patterns in Lepidoptera. Behaviour 11:209–256.

Blest, A. D. 1963. Relations between moths and predators. Nature 197:1046–1047.

Blest, A. D. 1964. Protective display and sound production in some New World arctiid and ctenuchid moths. Zoologica 49:161–181.

Block, B. A. 1987. Strategies for regulating brain and eye temperatures: A thermogenic tissue in fish. In P. Dejours, L. Bolis, C. R. Taylor, and E. R. Weibel (eds.), Comparative Physiology: Life in Water and on Land, 401–420. Fidia Research Series, IX. Liviana Press, Padova, Italy.

Blondel, J., P. C. Dias, M. Maistre, and P. Perret. 1993. Habitat heterogeneity and life-history variation of Mediterranean blue tits. Auk 110:511–520.

Bloom, B. R. 1979. Games parasites play: How parasites evade immune surveillance. Nature 279:21–26.

Boag, P. T. 1983. The heritability of external morphology in Darwin's finches (*Geospiza*) on Isla Daphne Major, Galapagos. Evolution 37:877–894.

Bock, C. E., and R. E. Ricklefs. 1983. Range size and local abundance of some North American songbirds: A positive correlation. Am. Nat. 122:295–299.

Boesch, C. 1994. Cooperative hunting in wild chimpanzees. Anim. Behav. 48:653–667.

Bogert, C. M. 1949. Thermoregulation in reptiles, a factor in evolution. Evolution 3:195–211.

Bohn, H., B. McNeal, and G. O'Connor. 1979. Soil Chemistry. Wiley, New York.

Boje, R., and M. Tomczak (eds.). 1978. Upwelling Ecosystems. Springer-Verlag, New York.

Bonaccorso, F. J., A. Arends, M. Genoud, D. Cantoni, and T. Morton. 1992. Thermal ecology of moustached and ghost-faced bats (Mormoopidae) in Venezuela. J. Mammal. 73:365–378.

Bond, W., and P. Slingsby. 1984. Collapse of an ant-plant mutualism: The Argentine ant (*Iridomyrmex humilis*) and myrmecochorous Proteaceae. Ecology 65:1031–1037.

Boonstra, R., J. M. Eadie, C. J. Krebs, and S. Boutin. 1995. Limitations of far infrared thermal imaging in locating birds. J. Field Ornithol. 66:192–198.

Borchert, J. R. 1950. The climate of the central North American grassland. Ann. Assoc. Am. Geogr. 40:1–39.

Boring, L. R., and W. T. Swank. 1984. The role of black locust (*Robinia pseudoacacia*) in forest succession. J. Ecol. 72:749–766.

Bormann, F. H. 1958. The relationships of ontogenetic development and environmental modification of photosynthesis in *Pinus taeda* seedlings. In K. V. Thimann (ed.), The Physiology of Forest Trees, 197–215. Ronald Press, New York.

Bormann, F. H., and G. E. Likens. 1967. Nutrient cycling. Science 155:424–429.

Bormann, F. H., G. E. Likens, and J. M. Melillo. 1977. Nitrogen budget for an aggrading northern hardwood forest ecosystem. Science 196:981–983.

Borrer, D. J., D. M. Delong, and C. A. Triplehorn. 1976. An introduction to the study of insects. 4th ed. Holt, Reinhart and Winston, New York.

Botkin, D. B. 1990. Discordant Harmonies: A New Ecology for the Twenty-First Century. Oxford University Press, New York.

Botkin, D. B., J. F. Janak, and J. R. Wallis. 1972. Some ecological consequences of a computer model of forest growth. J. Ecol. 60:849–872.

Botkin, D. B., G. M. Woodwell, and N. Tempel. 1970. Forest productivity estimated from carbon dioxide uptake. Ecology 51:1057–1066.

Bottomley, P. J. 1992. The ecology of *Bradyrhizobium* and *Rhizobium.* In G. Stacey, R. H. Burris, and H. J. Evans (eds.), Biological Nitrogen Fixation, 293–348. Chapman & Hall, New York.

Boucher, D. H. (ed.). 1985. The Biology of Mutualism. Croom Helm, London.

Bourlière, F. 1973. The comparative ecology of rainforest mammals in Africa and tropical America: Some introductory remarks. In B. J. Meggars, E. S. Ayensu, and W. D. Duckworth (eds.), Tropical Forest Ecosystems in Africa and South America: A Comparative Review, 279–292. Smithsonian Institution Press, Washington, D.C.

Bowen, G. D. 1973. Mineral nutrition of mycorrhizas. In G. C. Marks and T. T. Kozlowsky (eds.), Ectomycorrhizas, 151–201. Academic Press, New York and London.

Bowers, M. D. 1993. Aposematic caterpillars: Life-styles of the warningly colored and unpalatable. In N. E. Stamp and T. M. Casey (eds.), Caterpillars: Ecological and Evolutionary Constraints on Foraging, 331–371. Chapman & Hall, New York.

Bowles, M. L., and C. J. Whelan. 1994. Restoration of Endangered Species. Cambridge University Press, Cambridge.

Boyce, M. S. 1984. Restitution of *r*- and *K*-selection as a model of density-dependent natural selection. Annu. Rev. Ecol. Syst. 15:427–447.

Boyce, M. S., and C. M. Perrins. 1987. Optimizing great tit clutch size in a fluctuating environment. Ecology 68: 142–153.

Boynton, W. R., W. M. Kemp, and C. W. Keffe. 1982. A comparative analysis of nutrients and other factors influencing estuarine phytoplankton production. In V. S. Kennedy (ed.), Estuarine Comparisons, 69–90. Academic Press, New York.

Bradbury, J. W., and M. B. Andersson (eds.). 1987. Sexual Selection: Testing the Alternatives. Wiley-Interscience, New York.

Bradshaw, A. D. 1987. Comparison: Its scope and limits. In I. H. Rorison, J. P. Grime, R. Hunt, G. A. F. Hendry, and D. H. Lewis (eds.), Frontiers of Comparative Plant Ecology, 3–21. Academic Press, London.

Bradshaw, J. W. S., and P. E. Howse. 1984. Sociochemicals of ants. In W. J. Bell and R. T. Cardé (eds.), Chemical Ecology of Insects, 429–473. Sinauer Associates, Sunderland, MA.

Brady, N. C. 1974. Nature and Property of Soils. 8th ed. Macmillan, New York.

Brattstrom, B. H. 1962. Thermal control of aggregation behavior in tadpoles. Herpetologica 18:38–46.

Braun, E. L. 1950. Deciduous Forests of Eastern North America. Free Press, New York. Reprinted 1974, Hafner, New York.

Braun-Blanquet, J. 1932. Plant Sociology: The Study of Plant Communities. Trans. G. D. Fuller and H. S. Conard. McGraw-Hill, New York.

Braun-Blanquet, J. 1965. Plant Sociology: The Study of Plant Communities. Revised translation. Hafner, New York.

Bray, J. R., and J. T. Curtis. 1957. An ordination of the upland forest communities of southern Wisconsin. Ecol. Monogr. 27:325–349.

Breder, C. N., and D. E. Rosen. 1966. Modes of Reproduction in Fishes. Natural History Press, New York.

Brewer, R. 1994. The Science of Ecology. 2nd ed. Saunders, New York.

Brezonik, P. L., and G. F. Lee. 1968. N Denitrification as a nitrogen sink in Lake Mendota, Wisconsin. Environ. Sci. Technol. 2:120–125.

Brian, M. V. 1979. Caste differentiation and division of labor. In H. R. Hermann (ed.), Social Insects, Vol I, 121–222. Academic Press, New York.

Brian, M. V. 1980. Social control over sex and caste in bees, wasps, and ants. Biol. Rev. 55:379–415.

Briand, F. 1983. Environmental control of food web structure. Ecology 64:253–263.

Briand, F., and J. E. Cohen. 1984. Community food webs have scale-invariant structure. Nature 307:264–267.

Briand, F., and J. E. Cohen. 1987. Environmental correlates of food chain length. Science 238:956–960.

Briggs, J. C. 1995. Global Biogeography. Elsevier, Amsterdam.

Brinkhurst, R. O. 1959. Alary polymorphism in the Gerroidae (Hemiptera-Heteroptera). J. Anim. Ecol. 28:211–230.

Bristow, C. M. 1984. Differential benefits from ant attendance to two species of homoptera on New York ironweed. J. Anim. Ecol. 53:715–726.

Brock, T. D. 1970. High temperature systems. Annu. Rev. Ecol. Syst. 1:191–220.

Brock, T. D. 1985. Life at high temperatures. Science 230:132–138.

Brock, T. D., and G. K. Darland. 1970. Limits of microbial existence: Temperature and pH. Science 169:1316–1318.

Brockmann, H. J. 1984. The evolution of social behaviour in insects. In J. R. Krebs and N. B. Davies (eds.), Behavioural Ecology: An Evolutionary Approach, 2nd ed., 340–361. Blackwell Scientific Publications, Oxford.

Brodie, C., G. Houle, and M. Fortin. 1995. Development of *Populus balsamifera* clone in subarctic Quebec reconstructed from spatial analysis. J. Ecol. 83:309–320.

Brokaw, N. V. L. 1985. Treefalls, regrowth, and community structure in tropical forests. In S. T. A. Pickett and P. S. White (eds.), The Ecology of Natural Disturbance and Patch Dynamics, 53–69. Academic Press, Orlando, FL.

Brook, G. A., M. E. Folkoff, and E. O. Box. 1983. A world model of soil carbon dioxide. Earth Surface Processes and Landforms 8:79–88.

Brooks, D. R. 1985. Historical ecology: A new approach to studying the evolution of ecological associations. Ann. Mo. Bot. Gard. 72:660–680.

Brooks, D. R., and D. A. McLennan. 1991. Phylogeny, Ecology, and Behavior: A Research Program in Comparative Biology. University of Chicago Press, Chicago.

Brooks, D. R., and D. A. McLennan. 1993. Historical ecology: Examining phylogenetic components of community evolution. In R. E. Ricklefs and D. Schluter (eds.), Species Diversity in Ecological Communities: Historical and geographical perspectives, 267–280. University of Chicago Press, Chicago.

Broughton, W. J. (ed.). 1983. Nitrogen Fixation. Vols. 1–3. Oxford University Press, New York.

Brower, J. V. Z. 1958. Experimental studies of mimicry in some North American butterflies. Part I, The monarch, *Danaus plexippus* and viceroy, *Limenitis archippus.* Part II, *Battus philenor* and *Papilio troilus, P. polyxenes,* and *P. glaucus.* Part III, *Danaus gilippus berenice* and *Limenitis archippus floridensis.* Evolution 12:32–47, 123–136, 273–285.

Brower, J. V. Z., and L. P. Brower. 1962. Experimental studies of mimicry. Part 6, The reaction of toads (*Bufo terrestris*) to honeybees (*Apis mellifera*) and their dronefly mimics (*Eristalis vinetorum*). Am. Nat. 96:297–308.

Brower, L. P. 1969. Ecological chemistry. Sci. Am. 220:22–29.

Brown, C. M., D. S. McDonald-Brown, and J. L. Meers. 1974. Physiological aspects of inorganic nitrogen metabolism. Adv. Microbial Physiol. 11:1–52.

Brown, C. R. 1986. Cliff swallow colonies as information centers. Science 234:83–85.

Brown, G. E., D. P. Chivers, and R. J. F. Smith. 1996. Effects of diet on localized defecation by northern pike, *Esox lucius.* J. Chem. Ecol. 22:467–475.

Brown, J. H. 1995. Macroecology. University of Chicago Press, Chicago.

Brown, J. H., and D. W. Davidson. 1977. Competition between seed-eating rodents and ants in desert ecosystems. Science 196:880–882.

Brown, J. H., D. W. Davidson, and O. J. Reichman. 1979. An experimental study of competition between seed-eating desert rodents and ants. Am. Zool. 19:1129–1143.

Brown, J. H., and B. A. Maurer. 1989. Macroecology: The division of food and space among species on continents. Science 243:1145–1150.

Brown, J. L. 1964. The evolution of diversity in avian territorial systems. Wilson Bull. 76:160–169.

Brown, J. S. 1994. Restoration ecology: Living with the prime directive. In M. L. Bowles and C. J. Whelan (eds.), Restoration of Endangered Species, 355–380. Cambridge University Press, Cambridge.

Brown, V. K., C. W. D. Gibson, and J. Kathirithamby. 1992. Community organisation in leaf hoppers. Oikos 65:97–106.

Brussard, P. F., and A. T. Vawter. 1975. Population structure, gene flow, and natural selection in populations of *Euphydryas phaeton.* Heredity 34:407–415.

Bryant, J. P. 1981. Phytochemical deterrence of snowshoe hare browsing by adventitious shoots of four Alaskan trees. Science 213:889–890.

Bryant, J. P., and P. J. Kuropat. 1980. Selection of winter forage by subarctic browsing vertebrates: The role of plant chemistry. Annu. Rev. Ecol. Syst. 11:261–285.

Buckley, R. C. 1987. Interactions involving plants, homoptera, and ants. Ann. Rev. Ecol. Syst. 18:111–135.

Buckner, C. H., and W. J. Turnock. 1965. Avian predation on the larch sawfly, *Pristiphora erichsonii* (Htg.) (Hymenoptera: Tenthredinidae). Ecology 46:223–236.

Buchsbaum, R. 1948. Animals without backbones (2nd ed.) Univ. Chicago Press, Chicago.

Budowski, G. 1965. Distribution of tropical American rain forest species in the light of successional processes. Turrialba 15:40–42.

Budowski, G. 1970. The distinction between old secondyear and climax species in tropical Central American lowland forests. Trop. Ecol. 11:44–48.

Bull, J. J. 1980. Sex determination in reptiles. Quart. Rev. Biol. 55:3–21.

Bull, J. J. 1983. Evolution of Sex Determining Mechanisms. Benjamin/Cummings, Menlo Park, CA.

Bulmer, M. G. 1980. The Mathematical Theory of Quantitative Genetics. Oxford University Press, Oxford.

Bulmer, M. G. 1983. Sex ratio theory in social insects with swarming. J. Theor. Biol. 100:329–339.

Bulmer, M. G. 1985. Selection for iteroparity in a variable environment. Am. Nat. 126:63–71.

Bulmer, M. G. 1994. Theoretical Evolutionary Ecology. Sinauer Associates, Sunderland, MA.

Bunning, E. 1967. The Physiological Clock. 2nd rev. ed. Springer-Verlag, New York.

Bunt, J. S. 1973. Primary production: Marine ecosystems. Human Ecol. 1:333–345.

Bunting, B. T. 1967. The Geography of Soil. Rev. ed. Aldine, Chicago.

Buol, S. W., F. D. Hole, and R. J. McCracken. 1973. Soil Genesis and Classification. Iowa State University Press, Ames.

Burke, M. J., L. V. Gusta, H. A. Quamme, C. J. Weiser, and P. H. Li. 1976. Freezing and injury in plants. Annu. Rev. Plant Physiol. 27:507–528.

Burnett, T. 1958. Dispersal of an insect parasite over a small plot. Can. Entomol. 90:279–283.

Burton, R. S. 1985. Mating system of the intertidal copepod *Tigriopus californicus.* Mar. Biol. 86:247–252.

Buzas, M. A. 1972. Patterns of species diversity and their explanation. Taxon 21:275–286.

Cable, D. R. 1975. Influence of precipitation on perennial grass production in the semidesert Southwest. Ecology 56:981–986.

Cairns, J., M. L. Dahlberg, K. L. Dickson, N. Smith, and W. T. Waller. 1969. The relationship of fresh-water protozoan communities to the MacArthur-Wilson equilibrium model. Am. Nat. 103:439–454.

Calder, W. A. 1968. Nest sanitation: A possible factor in the water economy of the roadrunner. Condor 70:279.

Calder, W. A. III. 1984. Size, Function, and Life History. Harvard University Press, Cambridge, MA.

Calder, W. A., and J. R. King. 1974. Thermal and caloric relations of birds. In D. S. Farner and J. R. King (eds.), Avian Biology, vol. 4, 259–413. Academic Press, New York.

Calow, P. 1978. Life Cycles: An Evolutionary Approach to the Physiology of Growth, Reproduction, and Ageing. Chapman & Hall, London.

Caraco, N. F. 1993. Disturbance of the phosphorus cycle: A case of indirect effects of human activity. Trends Ecol. Evol. 8:51–54.

Caraco, N. F., J. J. Cole, and G. E. Likens. 1989. Evidence for sulphate-controlled phosphorus release from sediments of aquatic systems. Nature 341:316–318.

Caraco, T. S., W. U. Blanckenhorn, G. M. Gregory, J. A. Newman, G. M. Recer, and S. M. Zwicker. 1990. Risk-sensitivity: Ambient temperature affects foraging choice. Anim. Behav. 39:338–345.

Caraco, T. S., S. Martindale, and T. W. Whitham. 1980. An empirical demonstration of risk-sensitive foraging preferences. Anim. Behav. 28:820–830.

Carbyn, L. N., H. J. Armbruster, and C. Mamo. 1994. The swift fox reintroduction program in Canada from 1983 to 1992. In M. L. Bowles and C. J. Whelan (eds.), Restoration of Endangered Species, 247–271. Cambridge University Press, Cambridge.

Carey, F. G. 1982. A brain heater in the swordfish. Science 216:1327–1329.

Carey, F. G., J. M. Teal, J. W. Kanwisher, and K. D. Lawson. 1971. Warm-bodied fish. Am. Zool. 11:137–145.

Carlin, N. F., and B. Hölldobler. 1983. Nestmate and kin recognition in interspecific mixed colonies of ants. Science 222:1027–1029.

Carpenter, E. J., and J. L. Culliney. 1975. Nitrogen fixation in marine shipworms. Science 187:551–552.

Carpenter, F. L., D. C. Paton, and M. A. Hixon. 1983. Weight gain and adjustment of feeding territory size in migrant hummingbirds. Proc. Natl. Acad. Sci. USA 80:7259–7263.

Carpenter, S. R. (ed.). 1988. Complex Interactions in Lake Communities. Springer-Verlag, New York.

Carpenter, S. R., and J. F. Kitchell. 1993. The trophic Cascade in Lakes. Cambridge University Press, Cambridge.

Carpenter, S. R., J. F. Kitchell, J. R. Hodgson, P. A. Cochran, J. J. Elser, M. M. Elser, D. M. Lodge, D. Kretchmer, X. He, and C. N. VonEnde. 1987. Regulation of lake primary productivity by food web structure. Ecology 68:1863–1876.

Carpenter, S. R., P. R. Leavitt, J. J. Elser, and M. M. Elser. 1988. Chlorophyll budgets: Response to food web manipulation. Biogeochemistry 6:79–90.

Carroll, G. C., and D. T. Wicklow (eds.). 1992. The fungal community: Its organization and role in the ecosystem. 2nd ed. Mycology Series, vol. 9. Marcel Dekker, New York.

Cartwright, F. F. and M. D. Biddiss. 1972. Disease and History. Crowell, New York.

Casaretto, J. A., and L. J. Corcuera. 1995. Plant proteinase inhibitors: A defensive response against insects. Biol. Res. 28:239–249.

Casper, B. B. 1984. On the evolution of embryo abortion in the herbaceous perennial *Cryptantha flava.* Evolution 38:1337–1349.

Caswell, H. 1989. Matrix Population Models. Sinauer Associates, Inc., Massachusetts.

Caswell, H., and R. J. Etter. 1993. Ecological interactions in patchy environments: From patch-occupancy models to cellular automata. In S. A. Levin, T. M. Powell, and J. H. Steele (eds.), Patch Dynamics, 93–109. Springer-Verlag, Berlin.

Catley, K. M. 1992. Supercooling and its ecological implication in *Coelotes atropos* (Arraneae, Agelenidae). J. Arachnol. 20:58–63.

Caughley, G. 1977. Analysis of Vertebrate Populations. Wiley, New York and London.

Caughley, G., H. Dublin, and I. Parker. 1990. Projected decline of the African elephant. Biol. Conserv. 54:157–164.

Cavalli-Sforza, L. L., and W. F. Bodmer. 1971. The Genetics of Human Populations. W. H. Freeman, San Francisco.

Center, T. D., and C. D. Johnson. 1974. Coevolution of some seed beetles (Coleoptera: Bruchidae) and their hosts. Ecology 55:1096–1103.

Chaetum, E. L., and C. W. Severinghaus. 1950. Variations in fertility of white-tailed deer related to range conditions. Trans. N. Am. Wildl. Conf. 15:170–189.

Chapin, F. S. III, P. C. Miller, W. D. Billings, and P. I. Coyne. 1980. Carbon and nutrient budgets and their control in costal tundra. In J. Brown, P. C. Miller, L. Tieszen and F. L. Bunnel (eds.), An Arctic Ecosystem: The Coastal Tundra at Barrow, Alaska, 458–484. Dowden, Hutchinson, and Ross, Stroudsburg, PA.

Chapin, F. S. III, L. R. Walker, C. L. Fastie, and L. C. Sharman. 1994. Mechanisms of primary succession following deglaciation at Glacier Bay, Alaska. Ecol. Monogr. 64:149–175.

Chapman, R. 1928. The quantitative analysis of environmental factors. Ecology 9:111–122.

Charles-Dominique, P. 1993. Speciation and coevolution: An interpretation of frugivory phenomena. In T. H. Fleming and A. Estrada (eds.), Frugivory and Seed Dispersal: Ecological and Evolutionary Aspects, 75–84. Kluwer Academic Publishers, Dordrecht.

Charlesworth, B. 1994. Evolution of age-structured populations, second edition. Cambridge University Press, Cambridge.

Charnov, E. L. 1976. Optimal foraging, the marginal value theorem. Theor. Pop. Biol. 9:129–136.

Charnov, E. L. 1979. Natural selection and sex change in pandalid shrimp: Test of a life history theory. Am. Nat. 113:715–734.

Charnov, E. L. 1982. The Theory of Sex Allocation. Princeton University Press, Princeton, NJ.

Charnov, E. L. 1993. Life History Invariants. Oxford University Press, Oxford.

Charnov, E. L., and W. M. Schaffer. 1973. Life history consequences of natural selection: Cole's result revisited. Am. Nat. 107:791–793.

Charnov, E. L., J. M. Smith, and J. Bull. 1976. Why be an hermaphrodite? Nature 263:125–126.

Charnov, E. L., and D. W. Stephens. 1988. On the evolution of host selection in solitary parasitoids. Am. Nat. 132: 707–722.

Chesson, P. L., and R. R. Warner. 1981. Environmental variability promotes coexistence in lottery competitive systems. Am. Nat. 117:923–943.

Cheverud, J. M., and W. P. J. Dittus. 1992. Primate population studies at Polonnaruwa. Heritability of body measurements in a natural population of Toque Macaques (*Macaca sinica*). Am. J. Primatol. 27:145–156.

Chew, R. M., and A. E. Chew. 1970. Energy relationships of the mammals of a desert scrub. Ecol. Monogr. 40:1–21.

Choe, J. C. 1994. Sexual selection and mating system in *Zorotypus gurneyi* Choe (Insecta: Zoraptera). I. Dominance hierarchy and mating success. Behav. Ecol. Sociobiol. 34:87–93.

Choe, J. C., and B. J. Crispi. 1997. The Evolution of Social Behavior in Insects and Arachnids. Cambridge University Press, Cambridge.

Christensen, N. L. 1985. Shrubland fire regimes and their evolutionary consequences. In S. T. A. Pickett and P. S. White (eds.), The Ecology of Natural Disturbance and Patch Dynamics, 85–100. Academic Press, Orlando, FL.

Christensen, N. L., and R. K. Peet. 1984. Convergence during secondary forest succession. J. Ecol. 72:25–36.

Chróst, R. J. 1990. Microbial ectoenzymes in aquatic environments. In J. Overbeck and R. J. Chróst, Aquatic Microbial Ecology, 47–78. Springer-Verlag, New York.

Cigliano, J. A. 1993. Dominance and den use in *Octopus bimaculoides*. Anim. Behav. 46:677–684.

Clark, A. B. 1978. Sex ratio and local resource competition in a prosimian primate. Science 201:163–165.

Clark, C. W. 1994. Antipredator behavior and the asset-protection principle. Behav. Ecol. 5:159–170.

Clark, D. A., and D. B. Clark. 1984. Spacing dynamics of a tropical rain forest tree: Evaluation of the Janzen-Connell model. Am. Nat. 124:769–788.

Clark, F. W. 1972. Influence of jackrabbit density on coyote population change. J. Wildl. Mgmt. 36:343–356.

Clark, P. J., and F. C. Evans. 1954. Distance to nearest neighbor as a measure of spatial relationships in populations. Ecology 35:445–453.

Clark, T. W., and J. R. Cragun. 1994. Organizational and managerial guidelines for endangered species restoration programs and recovery teams. In M. L. Bowles and C. J. Whelan (eds.), Restoration of Endangered Species, 9–33. Cambridge University Press, Cambridge.

Clarke, C. A., G. S. Mani, and G. Wynne. 1986. Evolution in reverse: clean air and the peppered moth. Biol. J. Linn. Soc. 26:189–199.

Clarkson, D. T., and J. B. Hanson. 1980. The mineral nutrition of higher plants. Annu. Rev. Plant. Physiol. 31: 239–298.

Clausen, J., D. D. Keck, and W. M. Hiesey. 1948. Experimental studies on the nature of species. III: Environmental responses of climatic races of *Achillea*. Carnegie Inst. Wash. Publ. 581:1–129.

Clayton, D. H. 1990. Mate choice in experimentally parasitized rock doves: Lousy males lose. Am. Zool. 30:251–262.

Clayton, D. H. 1991. The influence of parasites on host sexual selection. Parasitology Today 71:329–334.

Clayton, D. H., S. G. Pruett-Jones, and R. Lande. 1992. Reappraisal of the interspecific prediction of parasite-mediated sexual selection: Opportunity knocks. J. Theor. Biol. 157:95–108.

Clayton, H. H. 1944. World weather records. Smithsonian Misc. Coll. 79:1–1199.

Clayton, H. H., and F. L. Clayton. 1947. World weather records 1931–1940. Smithsonian Misc. Coll. 105:1646.

Clements, F. E. 1916. Plant succession: Analysis of the development of vegetation. Carnegie Inst. Wash. Publ. 242: 1–512.

Clements, F. E. 1936. Nature and structure of the climax. J. Ecol. 24:252–284.

Cloudsley-Thompson, J. L. 1991. Ecophysiology of desert arthropods and reptiles. Springer-Verlag, Berlin.

Clutter, M. E. (ed.). 1978. Dormancy and Developmental Arrest: Experimental Analysis in Plants and Animals. Academic Press, New York.

Clutton-Brock, T. H. (ed.). 1988. Reproductive Success: Studies of Individual Variation in Contrasting Breeding Systems. University of Chicago Press, Chicago.

Clutton-Brock, T. H., D. Green, M. Hiraiwa-Hasegawa, and S. D. Albon. 1988. Passing the buck: Resource defence, lekking and mate choice in fallow deer. Behav. Ecol. Sociobiol. 23:281–296.

Clutton-Brock, T. H., F. E. Guinness, and S. D. Albon. 1982. Red Deer: Behavior and Ecology of Two Sexes. University of Chicago Press, Chicago.

Clutton-Brock, T. H., and P. H. Harvey. 1979. Comparison and adaptation. Proc. R. Soc. Lond. B 205:547–565.

Clutton-Brock, T. H., and P. H. Harvey. 1984. Comparative approaches to investigating adaptation. In J. R. Krebs and N. B. Davies (eds.), Behavioural Ecology: An Evolutionary Approach, 2nd ed., 7–29. Blackwell Scientific Publications, Oxford.

Clutton-Brock, T. H., and G. R. Iason. 1986. Sex ratio variation in mammals. Q. Rev. Biol. 61:339–374.

Clutton-Brock, T. H., O. F. Price, S. D. Albon, A. Robertson, and P. A. Jewell. 1991. Population regulation in Soay sheep. J. Anim. Ecol. 60:593–608.

Cody, M. L. 1966. A general theory of clutch size. Evolution 20:174–184.

Cody, M. L. 1974. Optimization in ecology. Science 183:1156–1164.

Cody, M. L. (ed.). 1985. Habitat Selection in Birds. Academic Press, New York.

Cody, M. L., and H. A. Mooney. 1978. Convergence versus nonconvergence in Mediterranean-climate ecosystems. Annu. Rev. Ecol. Syst. 9:265–321.

Cogger, H. G. 1974. Thermal relations of the mallee dragon *Amphibolurus fordi* (Lacertilia: Agamidae). Aust. J. Zool. 22:319–339.

Cohen, J. E. 1973. Heterologous immunity in human malaria. Q. Rev. Biol. 48:467–489.

Cohen, J. E. 1989. Food webs and community structure. In J. Roughgarden, R. M. May, and S. A. Levin (eds.), Perspectives in Ecological Theory, 181–202. Princeton University Press, Princeton, NJ.

Cohen, J. E., and F. Briand. 1984. Trophic links of community food webs. Proc. Natl. Acad. Sci. USA 81:4105–4109.

Cohen, J. E., F. Briand, and C. M. Newman. 1990. Community food webs: Data and theory. Springer-Verlag, New York.

Cole, J. J., S. Findlay, and M. L. Pace. 1988. Bacterial production in fresh and saltwater ecosystems: a cross-system overview. Mar. Ecol. Prog. Ser. 43:1–10.

Cole, J., G. Lovett, and S. Findlay. 1991. Comparative Analyses of Ecosystems: Patterns, Mechanisms, and Theories. Springer-Verlag, New York.

Cole, J. J., N. F. Caraco, G. W. Kling, and T. K. Kratz. 1994. Carbon dioxide supersaturation in the surface waters of lakes. Science 265:1568–1570.

Cole, K. S. 1990. Patterns of gonad structure in hermaphroditic gobies (Teleostei: Gobiidae). Environ. Biol. Fishes 28:125–142.

Cole, L. C. 1951. Population cycles and random oscillations. J. Wildl. Mgmt. 15:233–252.

Cole, L. C. 1954. The population consequences of life history phenomena. Q. Rev. Biol. 29:103–137.

Coleman, B. D., M. A. Mares, M. R. Willig, and Y.-H. Hsieh. 1982. Randomness, area, and species richness. Ecology 63:1121–1133.

Coleman, D. C., C. P. P. Reid, and C. V. Cole. 1983. Biological strategies of nutrient cycling in soil systems. Adv. Ecol. Res. 13:1–55.

Coley, P. D. 1983. Herbivory and defensive characteristics of tree species in a lowland tropical forest. Ecol. Monogr. 53:209–233.

Comfort, A. 1956. The Biology of Senescence. Rinehart, New York.

Comins, H. N., M. P. Hassell, and R. M. May. 1992. The spatial dynamics of host-parasitoid systems. J. Anim. Ecol. 61:735–748.

Connell, J. H. 1961. The influence of interspecific competition and other factors on the distribution of the barnacle *Chthamalus stellatus*. Ecology 42:710–723.

Connell, J. H. 1978. Diversity in tropical rain forests and coral reefs. Science 199:1302–1310.

Connell, J. H. 1983. On the prevalence and relative importance of interspecific competition: Evidence from field experiments. Am. Nat. 122:661–696.

Connell, J. H. 1990. Apparent versus "real" competition in plants. In J. B. Grace, and D. Tilman (eds), Perspectives on Plant Competition. Academic Press, New York.

Connell, J. H., and M. J. Keough. 1985. Disturbance and patch dynamics of subtidal marine animals on hard substrata. In S. T. A. Pickett and P. S. White (eds.), The Ecology of Natural Disturbance and Patch Dynamics, 125–151. Academic Press, Orlando, FL.

Connell, J. H., and E. Orias. 1964. The ecological regulation of species diversity. Am. Nat. 98:399–414.

Connell, J. H., and R. O. Slatyer. 1977. Mechanisms of succession in natural communities and their role in community stability and organization. Am. Nat. 111:1119–1144.

Connor, R. C., and R. L. Curry. 1995. Helping non-relatives: a role for deceit? Anim. Behav. 49:389–394.

Connor, E. F., and E. D. McCoy. 1979. The statistics and biology of the species-area relationship. Am. Nat. 113: 791–833.

Connor, E. F., and D. S. Simberloff. 1978. Species number and compositional similarity of the Galápagos flora and avifauna. Ecol. Monogr. 48:219–248.

Connor, E. F., and D. S. Simberloff. 1979. The assembly of species communities: Chance or competition? Am. Nat. 113:791–833.

Cook, R. E. 1969. Variation in species density in North American birds. Syst. Zool. 18:63–84.

Cook, R. E. 1977. Raymond Lindeman and the trophic-dynamic concept in ecology. Science 198:22–26.

Cooper, C. F. 1961. The ecology of fire. Sci. Am. 204: 150–160.

Cooper, W. S. 1913. The climax forest of Isle Royale, Lake Superior, and its development. Bot. Gaz. 55:1–44, 115–140, 189–235.

Cormack, R. M. 1979. Models for capture-recapture. In R. M. Cormack, G. P. Patil, and D. S. Robson (eds.), Sampling Biological Populations, 217–255. International Cooperative Publishing House, Fairland, MD.

Cormack, R. M. 1981. Loglinear models for capture-recapture experiments on open populations. In R. W. Hiorns and D. Cooke (eds.), The Mathematical Theory of the Dynamics of Biological Populations, 197–215. Academic Press, London and New York.

Corn, P. S. 1998. Effects of ultraviolet radiation on boreal toads in Colorado. Ecol. Monogr. 8:18–26.

Corn, P. S., and R. B. Bury. 1989. Logging in Western Oregon: Responses of headwater habitats and stream amphibians. Forest Ecol. Manage. 29:39–57.

Corn, P. S., and J. C. Fogleman. 1984. Extinction of montane populations of the northern leopard frog (*Rana pipiens*) in Colorado. J. Herpetol. 18:147–152.

Cornell, H. V. 1985. Species assemblages of cynipid gall wasps are not saturated. Am. Nat. 126:565–569.

Cornell, H. V., and D. Pimentel. 1978. Switching in the parasitoid *Nasonia vitripennis* and its effect on host competition. Ecology 59:297–308.

Coulsen, J. C. 1968. The influence of the pair-bond and age on the breeding biology of the kittiwake gull *Rissa tridactyla.* J. Anim. Ecol. 35:269–279.

Cousins, S. H. 1991. Species diversity measurements: Choosing the right index. Trends Ecol. Evol. 6:190–192.

Cowan, I. M. 1947. The timber wolf in the Rocky Mountain National Parks of Canada. Can. J. Res. 25:139–174.

Cowles, H. C. 1899. The ecological relations of the vegetation on the sand dunes of Lake Michigan. Bot. Gaz. 27:95–117, 167–202, 281–308, 361–391.

Cowles, R. B., and C. M. Bogert. 1944. A preliminary study of the thermal requirements of desert reptiles. Bull. Am. Mus. Nat. Hist. 83:265–296.

Cox, G. W., and R. E. Ricklefs. 1977. Species diversity, ecological release, and community structuring in Caribbean land bird faunas. Oikos 29:60–66.

Coyne, J. A., J. Bundgaard, and T. Prout. 1983. Geographic variation of tolerance to environmental stress in *Drosophila pseudoobscura.* Am. Nat. 122:474–488.

Craig, R. 1980. Sex investment ratios in social Hymenoptera. Am. Nat. 116:311–323.

Crawford, D. L., and R. L. Crawford. 1980. Microbial degradation of lignin. Enzyme and Microbial Technol. 2:11–22.

Crawford, H. S., and D. T. Jennings. 1989. Predation by birds on spruce budworm *Choristoneura fumiferana:* Functional, numerical, and total responses. Ecology 70:152–163.

Crawley, M. J. 1983. Herbivory: The Dynamics of Animal-Plant Interactions. University of California Press, Berkeley.

Crawley, M. J. 1989. Insect herbivores and plant population dynamics. Annu. Rev. Entomol. 34:531–564.

Crespi, B. J., and J. C. Choe. 1997. Introduction. In J. C. Choe and B. J. Crespi (eds.), The Evolution of Social Behavior in Insects and Arachnids, 1–7. Cambridge University Press, Cambridge.

Crespi, G. J., and D. Yanega. 1995. The definition of eusociality. Behav. Ecol. 6:109–115.

Croat, T. B. 1978. Flora of Barro Colorado Island. Stanford University Press, Stanford, CA.

Crocker, R. L., and J. Major. 1955. Soil development in relation to vegetation and surface age at Glacier Bay, Alaska. J. Ecol. 43:427–448.

Croghan, P. C. 1958. The osmotic and ionic regulation of *Artemia salina* (L). J. Exp. Biol. 35:219–233.

Crombie, A. C. 1944. On intraspecific and interspecific competition in larvae of graminivorous insects. J. Exp. Biol. 20:135–151.

Crombie, A. C. 1945. On competition between different species of graminivorous insects. Proc. R. Soc. Lond. B 132:362–395.

Crombie, A. C. 1947. Interspecific competition. J. Anim. Ecol. 16:44–73.

Crook, J. H. 1964. The evolution of social organization and visual communication in the weaver birds (Ploceinae). Behavior, Suppl. 10:1–178.

Crook, J. H. 1965. The adaptive significance of avian social organizations. Symp. Zool. Soc. Lond. 14:181–218.

Crow, J. F., and M. Kimura. 1970. An Introduction to Population Genetics Theory. Harper & Row, New York.

Crowder, L. B., D. P. Reagan, and D. W. Freckman. 1996. Food web dynamics and applied problems. In G. A. Polis and K. O. Winemiller (eds.), Food Webs: Integration of Patterns and Dynamics, 327–336. Chapman & Hall, New York.

Crowell, K. L. 1962. Reduced interspecific competition among the birds of Bermuda. Ecology 43:75–88.

Crowson, R. A. 1970. Classification and Biology. Aldine, Chicago.

Crump, M. L., F. R. Hensley, and K. L. Clark. 1992. Apparent decline of the gold toad: Underground or extinct? Copeia 1992:413–420.

Crumpacker, D. W., and J. S. Williams. 1973. Density, dispersion, and population structure in *Drosophila pseudoobscura.* Ecol. Monogr. 43:499–538.

Cui, M., and P. S. Nobel. 1992. Nutrient status, water uptake and gas exchange for three desert succulents infected with mycorrhizal fungi. New Phytol. 122:643–649.

Culver, D. C., and A. J. Beattie. 1978. Myrmecochory in *Viola:* Dynamics of seed-ant interactions in some West Virginia species. J. Ecol. 66:53–72.

Cummins, K. W., and M. J. Klug. 1979. Feeding ecology of stream invertebrates. Annu. Rev. Ecol. Syst. 10:147–172.

Currie, D. J. 1991. Energy and large-scale patterns of animal- and plant-species richness. Am. Nat. 137:27–49.

Currie, D. J., and J. Kalff. 1984a. A comparison of the abilities of freshwater algae and bacteria to acquire and retain phosphorus. Limnol. Oceanogr. 29:298–310.

Currie, D. J., and J. Kalff. 1984b. The relative importance of bacterioplankton and phytoplankton in phosphorus uptake in freshwater. Limnol. Oceanogr. 29:311–321.

Currie, D. J., and V. Paquin. 1987. Large-scale biogeographical patterns of species richness of trees. Nature 329: 326–327.

Curtis, J. T., and R. P. McIntosh. 1951. An upland forest continuum in the prairie-forest border region of Wisconsin. Ecology 32:476–496.

Cushing, D. H. 1975. Marine Ecology and Fisheries. Cambridge University Press, Cambridge.

Cuthill, I. C., and A. Kacelnik. 1990. Central place foraging: A re-appraisal of the "loading effect." Anim. Behav. 40:1087–1101.

Cuthill, I. C., A. Kacelnik, J. R. Krebs, P. Haccou, and Y. Iwasa. 1990. Patch use by starlings: The effect of recent experience on foraging decisions. Anim. Behav. 40: 625–640.

Da Cunha, A. B., and T. Dobzhansky. 1954. A further study of chromosomal polymorphism in *Drosophila willistoni* in relation to its environment. Evolution 8:119–134.

Dale, B. W., L. G. Adams, and R. T. Bowyer. 1994. Functional response of wolves preying on barren-ground caribou in a multiple-prey ecosystem. J. Anim. Ecol. 63:644–652.

Damian, R. T. 1964. Molecular mimicry: Antigen sharing by parasite and host and its consequences. Am. Nat. 98: 129–147.

Dangerfield, J. M., and B. Modukanele. 1996. Overcompensation by *Acacia erubescens* in response to simulated browsing. J. Trop. Ecol. 12:905–908.

Danks, H. V., O. Kukal, and R. A. Ring. 1994. Insect cold-hardiness: Insights from the Arctic. Arctic 47:391–404.

Darnell, R. M. 1970. Evolution of the ecosystem. Am. Zoologist 10:9–15.

Darwin, C. 1859. On the Origin of Species. Murray, London.

Darwin, C. 1871. The Descent of Man and Selection in Relation to Sex. Murray, London.

Darwin, C. 1872. On the Origin of Species. 6th ed. Murray, London.

Daubenmire, R. 1968a. Ecology of fire in grasslands. Adv. Ecol. Res. 5:209–266.

Daubenmire, R. 1968b. Plant Communities: A Textbook of Plant Synecology. Harper & Row, New York.

Daubenmire, R., and D. C. Prusso. 1963. Studies of the decomposition rates of tree litter. Ecology 44:589–592.

Davidson, D. W., R. S. Inouye, and J. H. Brown. 1984. Granivory in a desert ecosystem: Experimental evidence for indirect facilitation of ants by rodents. Ecology 65: 1780–1786.

Davidson, J. 1938. On the growth of the sheep population in Tasmania. Trans. R. Soc. S. Aust. 62:342–346.

Davidson, J., and H. G. Andrewartha. 1948a. Annual trends in a natural population of *Thrips imaginis* (Thysanoptera). J. Anim. Ecol. 17:193–199.

Davidson, J., and H. G. Andrewartha. 1948b. The influence of rainfall, evaporation and atmospheric temperature on fluctuations in the size of a natural population of *Thrips imaginis* (Thysanoptera). J. Anim. Ecol. 17:200–222.

Davies, N. B. 1991. Mating systems. In J. R. Krebs and N. B. Davies, Behavioural Ecology: An Evolutionary Approach, 3rd ed., 362–294. Blackwell Scientific Publications, Oxford.

Davis, C. C. 1964. Evidence for the eutrophication of Lake Erie from phytoplankton records. Limnol. Oceanogr. 9:275–283.

Davis, J. H. 1940. The ecology and geologic role of mangroves in Florida. Publ. Carnegie Inst. 517:303–412.

Davis, M. B. 1965. Phytogeography and palynology of northeastern United States. In H. E. Wright Jr. and D. G. Frey (eds.), Quaternary of the United States, 377–401. Princeton University Press, Princeton, NJ.

Davis, M. B. 1976. Pleistocene biogeography of temperate deciduous forests. Geoscience and Man 13:13–26.

Dawkins, R., and T. R. Carlisle. 1976. Parental investment, mate desertion and a fallacy. Nature 262:131–133.

Dawson, P. S. 1966. Correlated responses to selection for developmental rate in *Tribolium.* Genetica 37:63–77.

Dawson, P. S. 1967. Developmental rate and competitive ability in *Tribolium.* II. Changes in competitive ability following further selection for developmental rate. Evolution 21:292–298.

Dayan, T., and D. Simberloff. 1994. Character displacement, sexual dimorphism, and morphological variation among British and Irish mustelids. Ecology 75:1063–1073.

Dayan, T., D. Simberloff, E. Tchernov, and Y. Yom-Tov. 1989. Inter-and intraspecific character displacement in mustelids. Ecology 70:1526–1539.

Dayan, T., D. Simberloff, E. Tchernov, and Y. Yom-Tov. 1990. Feline canines: Community-wide character displacement among the small cats of Israel. Am. Nat. 136:39–60.

Dayton, P. K. 1971. Competition, disturbance, and community organization: The provision and subsequent utilization of space in a rocky intertidal community. Ecol. Monogr. 41:351–389.

Dean, T. A., and L. E. Hurd. 1980. Development in an estuarine fouling community: The influence of early colonists on later arrivals. Oecologia 46:295–301.

DeAngelis, D. L. 1975. Stability and connectance in food web models. Ecology 56:238–243.

DeAngelis, D. L., R. H. Gardner, and H. H. Shugart. 1981. Productivity of forest ecosystems studied during the IBP:The woodlands data set. In D. E. Reichle (ed.), Dynamic Properties of Forest Ecosystems, 567–672. Cambridge University Press, New York.

DeBach, P. 1966. The competitive displacement and coexistence principles. Annu. Rev. Entomol. 11:183–212.

DeBach, P. 1974. Biological Control by Natural Enemies. Cambridge University Press, London.

DeBach, P., R. M. Hendrickson Jr., and M. Rose. 1978. Competitive displacement: Extinction of the yellow scale, *Aonidiella* (Coq.) (Homoptera: Diaspididae), by its ecological homologue, the California red scale *Aonidiella aurantii* (Mask.) in southern California. Hilgardia 46:1–35.

DeBach, P., and R. A. Sundby. 1963. Competitive displacement between ecological homologues. Hilgardia 34: 105–166.

Debinski, D. M. 1994. Genetic diversity assessment in a metapopulation of the butterfly *Euphydryas gillettii.* Biol. Conserv. 70:25–31.

Deevey, E. S. Jr. 1947. Life tables for natural populations of animals. Q. Rev. Biol. 22:283–314.

DeGroot, P. 1980. Information transfer in a socially roosting weaver bird (*Quelea quelea:* Ploceinae): An experimental study. Anim. Behav. 28:1249–1254.

Del Giorgio, P. A., and R. H. Peters. 1994. Patterns in planktonic P:R ratios in lakes: Influence of lake trophy and dissolved organic carbon. Limnol. Oceanogr. 39:772–787.

Del Moral, R. 1993. Mechanisms of primary succession on volcanos: The view from Mount St. Helens. In J. Miles (ed.), Primary Succession on Land, 79–100. Blackwell, Oxford.

Del Moral, R., and D. M. Wood. 1988a. Dynamics of herbaceous vegetation recovery on Mount St. Helens, Washington, USA, after a volcanic eruption. Vegetatio 74:11–27.

Del Moral, R., and D. M. Wood. 1988b. The high elevation flora of Mount St. Helens, Washington. Madroño 35: 309–319.

Del Moral, R., and D. M. Wood. 1993. Early primary succession on the volcano Mount St. Helens. J. Vegetation Sci. 4:223–234.

Delwiche, C. C., and B. A. Bryan. 1976. Denitrification. Annu. Rev. Microbiol. 30:241–262.

Demauro, M. M. 1994. Development and implementation of a recovery program for the federal threatened lakeside daisy (*Hymenoxys acaulis* var. *glabra*). In M. L. Bowles and C. J. Whelan (eds.), Restoration of Endangered Species, 298–321. Cambridge University Press, Cambridge.

Deming, J. W., and J. A. Baross. 1993. Deep-sea smokers: Windows to a subsurface biosphere? Geochim. Cosmochim. Acta 57:3219–3230.

Denno, R. F. 1983. Tracking variable host plants in space and time. In R. F. Denno and M. S. McClure (eds.), Variable Plants and Herbivores in Natural and Managed Systems, 291–341. Academic Press, New York.

Denslow, J. S. 1980. Gap partitioning among tropical rainforest trees. Biotropica 12 (Suppl.):47–55.

Denslow, J. S. 1985. Disturbance-mediated coexistence of species. In S. T. Pickett and P. S. White (eds.), The Ecology of Natural Disturbance and Patch Dynamics, 307–323. Academic Press, Orlando, FL.

Denslow, J. S. 1987. Tropical rainforest gaps and tree species diversity. Annu. Rev. Ecol. Syst. 18:431–451.

Dethier, M. N., and D. O. Duggins. 1984. An "indirect commensalism" between marine herbivores and the importance of competitive hierarchies. Am. Nat. 124:205–219.

DeVries, A. L. 1980. Biological antifreezes and survival in freezing environments. In R. Gilles (ed.), Animals and Environmental Fitness, 583–607. Pergamon Press, New York.

DeVries, A. L. 1982. Biological antifreeze agents in coldwater fishes. Comp. Biochem. Physiol. 73a: 627–640.

deWit, P. J. G. M. 1992. Molecular characterization for gene-for-gene systems in plant-fungus interactions and the application of avirulence genes in control of plant pathogens. Annu. Rev. Phytopathol. 30:391–418.

DeWitt, C. B. 1967. Precision of thermoregulation and its relation to environmental factors in the desert iguana, *Dipsosaurus dorsalis.* Physiol. Zool. 40:49–66.

Dhillion, S. S., and R. C. Anderson. 1993. Growth dynamics associated with mycorrhizal fungi of little bluestem grass (*Schizachyrium scoparium* (Michx) Nash) on burned and unburned sand prairies. New Phyt. 123:77–91.

Diamond, J. M. 1975. Assembly of species communities. In M. L. Cody and J. M. Diamond (eds.), Ecology and Evolution of Communities, 342–444. Harvard University Press, Cambridge, MA.

Diamond, J. M. 1984. "Normal" extinctions of isolated populations. In M. H. Nitecki (ed.), Extinctions, 191–246. University of Chicago Press, Chicago.

Diamond, J. M., and M. E. Gilpin. 1982. Examination of the "null" model of Connor and Simberloff for species co-occurrences on islands. Oecologia 52:64–74.

Diamond, J. M., and R. M. May. 1977. Species turnover on islands: Dependence on census interval. Science 197:266–270.

Di Castri, F., and A. J. Hansen. 1992. The environment and development crises as determinants of landscape dynamics. In A. J. Hansen and F. di Castri (eds.), Landscape Boundaries, 3–18. Springer-Verlag, New York.

Dickinson, C. H., and G. J. F. Pugh (eds.). 1974. Biology of Plant Litter Decomposition. Vols. I and II. Academic Press, London and New York.

Diggle, P. J. 1983. Statistical analysis of spatial point patterns. Academic Press, New York.

Dill, L. M., and A. H. G. Fraser. 1984. Risk of predation and feeding behaviour of juvenile coho salmon (*Oncorhynchus kisutch*). Behav. Ecol. Sociobiol. 16:65–71.

Dingle, H. (ed.). 1978. Evolution of Insect Migration and Diapause. Springer-Verlag, New York.

Dingle, H. 1980. Ecology and evolution of migration. In S. A. Gauthreaux Jr. (ed.), Animal Migration, Orientation, and Navigation, 2–100. Academic Press, New York.

Dix, R. L. 1957. Sugar maple in forest succession at Washington, D.C. Ecology 30:663–665.

Dixon, A. F. G. 1959. An experimental study of the searching behaviour of the predatory coccinellid beetle *Adalia decempunctata* (L.). J. Anim. Ecol. 28:259–281.

Dobzhansky, T. 1950. Evolution in the Tropics. Am. Sci. 38:209–221.

Dobzhansky, T., and S. Wright. 1943. Genetics of natural populations. X. Dispersion rates in *Drosophila pseudoobscura.* Genetics 28:304–340.

Dobzhansky, T., and S. Wright. 1947. Genetics of natural populations. XV. Rates of diffusion of a mutant gene through a population of *Drosophila pseudoobscura.* Genetics 32:303–324.

Dolinger, P. M., P. R. Ehrlich, W. L. Fitch, and D. E. Breedlove. 1973. Alkaloids and predation patterns in Colorado lupine populations. Oecologia 13:191–204.

Dologov, L. V. 1991. Sexual structure of a *Tridacna squamosa* population: Relative advantages of sequential and simultaneous hermaphroditism. J. Moll. Stud. 58:21–27.

Doutt, R. L. 1960. Natural enemies and insect speciation. Pan-Pacific Entomol. 36:1–13.

Downes, R. W., and J. D. Hesketh. 1968. Enhanced photosynthesis at low O_2 concentrations: Differential response of temperate and tropical grasses. Planta 78:79–84.

Drent, R. H., and S. Daan. 1980. The prudent parent: Energetic adjustments in avian breeding. Ardea 68:225–252.

Dressler, R. L. 1968. Pollination by euglossine bees. Evolution 22:202–210.

Drury, W. H., and I. C. T. Nisbet. 1973. Succession. J. Arnold Arboretum 54:331–368.

Dudley, J. W. 1977. 76 generations of selection for oil and protein percentage in maize. In E. Pollack, O. Kempthorne, and T. Bailey (eds.), Proc. Intern. Conf. Quant. Genet., 1976, 459–473. Iowa State University Press, Ames.

Dunham, A. E. 1980. An experimental study of interspecific competition between the iguanid lizards *Sceloporus merriami* and *Urosaurus ornatus.* Ecol. Monogr. 50: 309–330.

Dunson, W. A. 1976. Salt glands in reptiles. In C. Gans and W. R. Dawson (eds.), Biology of the Reptilia, Vol. 5, 413–445. Academic Press, New York.

Dusenbery, D. B. 1992. Sensory Ecology. W. H. Freeman, New York.

Dussourd, D. E. 1993. Foraging with finesse: Caterpillar adaptations for circumventing plant defenses. In N. E. Stamp and T. M. Casey (eds.), Caterpillars: Ecological and Evolutionary Constraints on Foraging, 92–131. Chapman & Hall, New York.

Duvigneaud, P., and S. Denayer-de-Smet. 1970. Biological cycling of minerals in temperate deciduous forests. In. D. E. Reichle (ed.), 199–225. Analysis of Temperate Forest Ecosystem. Springer-Verlag, New York.

Dyer, A. R., and K. J. Rice. 1997. Intraspecific and diffuse competition: The response of *Nessella pulchra* in a California grassland. Ecol. Appl. 7:484–492.

Dyer, M. I. 1975. The effects of Red-winged Blackbirds (*Agelaius phoeniceus* L.) on biomass production of corn grains (*Zea mays* L.). J. Appl. Ecol. 12:719–726.

Dyer, M. I., D. C. Detling, D. C. Coleman, and D. W. Hilbert. 1982. The role of herbivores in grasslands. In J. R. Estes, R. J. Tyre, and J. N. Brunken (eds.), Grasses and Grasslands: Systematics and Ecology, 255–295. University of Oklahoma Press, Norman.

Eberhard, W. G. 1977. Aggressive mimicry by a Bolas Spider. Science 198:1173–1175.

Edmondson, W. T. 1970. Phosphorus, nitrogen, and algae in Lake Washington after diversion of sewage. Science 169:690–691.

Edmunds, M. 1974. Defense in Animals: A Survey of Anti-Predator Defenses. Longmans, London.

Edwards, C. A., and G. W. Heath. 1963. The role of soil animals in breakdown of leaf material. In J. Doeksen and J. Van Der Drift (eds.), Soil Organisms, 76–84. North-Holland, Amsterdam.

Edwards, G., and D. Walker. 1983. C_3, C_4: Mechanisms, and Cellular and Environmental Regulation, of Photosynthesis. University of California Press, Berkeley.

Edwards, R. T., and J. L. Meyer. 1987. Metabolism of a subtropical low gradient blackwater river. Freshwater Biol. 17:251–263.

Edwards, R. T., J. L. Meyer, and S. E. G. Findlay. 1990. The relative contribution of benthic and suspended bacteria to system biomass, production, and metabolism in a low-gradient blackwater river. J. N. Am. Benthol. Soc. 9: 216–228.

Egerton, F. N. 1973. Changing concepts in the balance of nature. Q. Rev. Biol. 48:322–350.

Ehleringer, J. R. 1984. Ecology and ecophysiology of leaf pubescence in North American desert plants. In E. Rodriguez, P. Healey, and I. Mehta (eds.), Biology and Chemistry of Plant Trichomes, 113–132. Plenum Press, New York.

Ehleringer, J. R., T. E. Cerling, and B. R. Helliker. 1997. C_4 photosynthesis, atmospheric CO_2, and climate. Oecologia 112:285–299.

Ehleringer, J. R., and I. Forseth. 1980. Solar tracking by plants. Science 210:1094–1098.

Ehleringer, J. R., R. F. Sage, L. B. Flanagan, and R. W. Pearcy. 1991. Climate change and the evolution of C_4 photosynthesis. Trends Ecol. Evol. 6:95–97.

Ehrlich, P. R. 1975. The population biology of coral reef fishes. Annu. Rev. Ecol. Syst. 6:211–247.

Ehrlich, P. R. 1984. The structure and dynamics of butterfly populations. In R. I. Vane-Wright and P. R. Ackery (eds.), The Biology of Butterflies, 25–40. Princeton University Press, Princeton, NJ.

Ehrlich, P. R., and L. C. Birch. 1967. The "balance of nature" and "population central." Amer. Nat. 101:97–107.

Ehrlich, P. R., and A. H. Ehrlich. 1981. Extinction: The Causes and Consequences of the Disappearance of Species. Random House, New York.

Ehrlich, P. R., and R. W. Holm. 1963. The Process of Evolution. McGraw-Hill, New York.

Eickmeier, W., M. Adams, and D. Lester. 1975. Two physiological races of *Tsuga canadensis*. Can. J. Bot. 53:940–951.

Einarsen, A. S. 1942. Specific results from ring-necked pheasant studies in the Pacific Northwest. Trans. N. Am. Wildl. Conf. 7:130–145.

Einarsen, A. S. 1945. Some factors affecting ring-necked pheasant population density. Murrelet 26:39–44.

Eisenberg, J. F. 1981. The Mammalian Radiations. University of Chicago Press, Chicago.

Eisner, T., and J. Meinwald. 1966. Defensive secretions of arthropods. Science 153:1341–1350.

Elser, J. J., E. R. Marzolf, and C. R. Goldman. 1990. Phosphorus and nitrogen limitation of phytoplankton growth in the freshwaters of North America: A review and critique of experimental enrichments. Can. J. Fish. Aquat. Sci. 47: 1468–1477.

Elseth, G. D., and K. D. Baumgardner. 1981. Population Biology. Van Nostrand, New York.

Elton, C. 1924. Periodic fluctuations in the numbers of animals: Their causes and effects. Brit. J. Exp. Biol. 2:119–163.

Elton, C. 1927. Animal Ecology. Macmillan, New York.

Emerson, R., and C. M. Lewis. 1942. The photosynthetic efficiency of phycocyanin in *Chroococcus*, and the problem of carotenoid participation in photosynthesis. J. Gen. Physiol. 25:579–595.

Emlen, J. M. 1970. Age specificity and ecological theory. Ecology 51:588–601.

Emlen, J. M. 1984. Population Biology: The Coevolution of Population Dynamics and Behavior. Macmillan, New York, and Collier Macmillan, London.

Emlen, J. T. Jr. 1940. Sex and age ratios in the survival of California quail. J. Wildl. Mgmt. 4:2–99.

Emlen, S. T., and H. W. Ambrose III. 1970. Feeding interactions of snowy egrets and red-breasted mergansers. Auk 87:164–165.

Emlen, S. T., P. H. Wrege, and H. J. Demong. 1995. Making decisions in the family: an evolutionary perspective. Am. Sci. 83:148–157.

Endler, J. A. 1977. Geographic Variation, Speciation, and Clines. Princeton University Press, Princeton, NJ.

Endler, J. A. 1978. A predator's view of animal color patterns. Evol. Biol. 11:319–364.

Endler, J. A. 1984. Progressive background matching in moths, and a quantitative measure of crypsis. Biol. J. Linn. Soc. 22:187–231.

Endler, J. A. 1991. Interactions between predators and prey. In J. R. Krebs and N. B. Davies (eds.), Behavioural Ecology: An Evolutionary Approach, 3rd ed, 169–202. Blackwell Scientific Publications, Oxford.

Endler, J. A. 1992. Signals, signal conditions, and the direction of evolution. Am. Nat. 139:S125.

Epling, C. H., and T. Dobzhansky. 1942. Genetics of natural populations. VI. Microgeographic races in *Linanthus parryae*. Genetics 27:317–332.

Erickson, R. C. 1945. The *Clematis fremontii* var. *riehlii* population in the Ozarks. Ann. Mo. Bot. Gard. 32:413–460.

Espenshade, E. B. Jr. (ed.). 1971. Goode's World Atlas. 13th ed. Rand-McNally, Chicago.

Evans, F. C. 1956. Ecosystem as the basic unit in ecology. Science 123:1127–1128.

Ewald, P. W. 1983. Host-parasite relations, vectors, and the evolution of disease severity. Annu. Rev. Ecol. Syst. 14:465–485.

Ewald, P. W. 1991. Waterborne transmission and the evolution of virulence among gastrointestinal bacteria. Epidemiology and Infection 106:83–119.

Ewald, P. W. 1995. The evolution of virulence: A unifying link between parasitology and ecology. J. Parasitol. 81: 659–669.

Ewald, P. W., and F. L. Carpenter. 1978. Territorial responses to energy manipulations in the Anna hummingbird. Oecologia 31:277–292.

Ewald, P. W., and G. H. Orians. 1983. Effects of resource depression on use of inexpensive and escalated aggressive behavior: Experimental tests using Anna hummingbirds. Behav. Ecol. Sociobiol. 12:95–101.

Eyre, S. R. 1968. Vegetation and Soils: A World Picture. 2nd ed. Aldine, Chicago.

Fahrig, L., and G. Merriam. 1994. Conservation of fragmented populations. Conserv. Biol. 8:50–59.

Falconer, D. S. 1981. Introduction to Quantitative Genetics. 2nd ed. Ronald Press, New York.

Farrell, B. D., and C. Mitter. 1993. Phylogenetic determinants of insect/plant community diversity. In. R. E. Ricklefs and D. Schluter (eds.), Species Diversity in Ecological Communities: Historical and Geographical Perspectives, 253–266. University of Chicago Press, Chicago.

Fauth, J. E., J. Bernardo, M. Camara, W. J. Resetarits Jr., J. Van Buskirk, and S. A. McCollum. 1996. Simplifying the jargon of community ecology: A conceptual approach. Am. Nat. 147:282–286.

Feder, M. E., A. G. Gibbs, G. A. Griffith, and J. Tsuji. 1984. Thermal acclimation of metabolism in salamanders: Fact or artefact? J. Therm. Biol. 9:255–260.

Feeny, P. P. 1968. Effect of oak leaf tannins on larval growth of the winter moth *Operophtera brumata*. J. Insect Physiol. 14:805–817.

Feeny, P. P. 1969. Inhibitory effect of oak leaf tannins on the hydrolysis of proteins by trypsin. Phytochemistry 8: 2119–2126.

Feinsinger, P. 1976. Organization of a tropical guild of nectarivorous birds. Ecol. Monogr. 46:257–291.

Feinsinger, P. 1978. Ecological interactions between plants and hummingbirds in a successional tropical community. Ecol. Monogr. 48:269–287.

Feinsinger, P., and H. M. Tiebout III. 1991. Competition among plants sharing hummingbird pollinators: Laboratory experiments on a mechanism. Ecology 72:1946–1952.

Feinsinger, P., H. M. Tiebout III, and B. E. Young. 1991. Do tropical bird-pollinated plants exhibit density-dependent interactions? Field experiments. Ecology 72:1953–1963.

Feldman, G. C. 1984. Satellites, seabirds, and seals. Trop. Ocean-Atmos. Newsl. 28:4–5.

Felsenstein, J. 1974. The evolutionary advantage of recombination. Genetics 78:737–756.

Fenchel, T. 1988. Marine plankton food chains. Annu. Rev. Ecol. Syst. 19:19–38.

Fenchel, T., and B. J. Finlay. 1995. Ecology and evolution in anoxic worlds. Oxford University Press, Oxford.

Fenner, F., and F. N. Ratcliffe. 1965. Myxomatosis. Cambridge University Press, London and New York.

Fenster, C. B., and M. R. Dudash. 1994. Genetic considerations for plant population restoration and conservation. In M. L. Bowles and C. J. Whelan (eds.), Restoration of Endangered Species, 34–62. Cambridge University Press, Cambridge.

Fernald, M. L. 1929. Some relationships of the floras of the Northern Hemisphere. Proc. Intl. Congr. Plant Sci., Ithaca, 2:1487–1507.

Fischer, A. G. 1960. Latitudinal variation in organic diversity. Evolution 14:64–81.

Fischer, E. A. 1984. Local mate competition and sex allocation in simultaneous hermaphrodites. Am. Nat 124: 590–596.

Fisher, R. A. 1930. The Genetical Theory of Natural Selection. Clarendon Press, Oxford.

Fisher, R. A., and N. C. Turner. 1978. Plant productivity in the arid and semiarid zones. Annu. Rev. Plant Physiol. 29:277–317.

Fisher, S. G., L. J. Gray, N. B. Brimm, and D. E. Busch. 1982. Temporal succession in a desert stream ecosystem following flash flooding. Ecol. Monogr. 52:93–110.

Fleming, I. A., and M. R. Gross. 1989. Evolution of adult female life history and morphology in a Pacific salmon (Coho: *Oncorhynchus kisutch*). Evolution 43:141–157.

Fletcher, D. J. C., and C. D. Michener (eds.). 1987. Kin Recognition in Animals. Wiley-Interscience, New York.

Fochte, D. D., and W. Verstraete. 1977. Biochemical ecology of nitrification and denitrification. Adv. Microbial Ecol. 1:135–214.

Foley, P. 1997. Extinction models for local populations. In I. A. Hanski and M. E. Gilpin (eds.), Metapopulation Biology: Ecology, Genetics, and Evolution, 215–246. Academic Press, San Diego.

Folstad, I. and A. J. Karter. 1992. Parasites, bright males, and the immunocompetence handicap. Am. Nat. 139:603–622.

Foose, T. J., and J. D. Ballou. 1988. Population management: Theory and practice. Intl. Zoo Yrbk. 27:26–41.

Forcier, L. K. 1975. Reproductive strategies and the cooccurrence of climax tree species. Science 189:808–809.

Forman, R. T. T. (ed.). 1979. Pine Barrens: Ecosystem and Landscape. Academic Press, New York.

Forman, R. T. T., and M. Godron. 1989. Landscape Ecology. Wiley, New York.

Forys, E. S., and S. R. Humphreys. 1996. Home range and movements of the Lower Keys marsh rabbit in a highly fragmented habitat. J. Mammal. 77:1042–1048.

Fossing, H., V. A. Gallardo, and J Kuver. 1995. Concentration and transport of nitrate by mat-forming sulphur bacterium *Thioploca*. Nature 374:713–715.

Foster, J. B., and M. J. Coe. 1968. The biomass of game animals in Nairobi National Park, 1960–1966. J. Zool. Lond. 155:413–425.

Fowells, H. A. 1965. Silvics of Forest Trees of the United States. Agriculture Handbook no. 271. U.S. Dept. of Agriculture, Washington, D.C.

Fowler, S. V., and M. MacGarvin. 1985. The impact of hairy wood ants, *Formica lugubris*, on the guild structure of herbivorous insects on birch, *Betula pubescens*. J. Anim. Ecol. 54:847–855.

Fowler, S. V., and M. MacGarvin. 1986. The effects of leaf damage on the performance of insect herbivores on birch, *Betula pubescens*. J. Anim. Ecol. 55:565–573.

Fox, R., S. W. Lehmkuhle, and D. H. Westendorf. 1976. Falcon visual acuity. Science 192:263–266.

Fraenkel, G. S. 1959. The raison d'etre of secondary plant substances. Science 129:1466–1470.

Fraenkel, G. S. 1969. Evaluation of our thoughts on secondary plant substances. Entomol. Exp. Appl. 12:474–486.

Frank, P. W., C. D. Boll, and R. W. Kelly. 1957. Vital statistics of laboratory cultures of *Daphnia pulex* De Geer as related to density. Physiol. Zool. 30:287–305.

Frankie, G. W., H. G. Baker, and P. A. Opler. 1974. Comparative phenological studies of trees in tropical lowland wet and dry forest sites of Costa Rica. J. Ecol. 62:881–919.

Fraser Rowell, C. H. 1970. Environmental control of coloration in an acridid, *Gastrimargus africanus* (Saussure). Anti-Locust Bull. 47:1–48.

Freeman, C. C., K. T. Harper, and E. L. Charnov. 1980. Sex change in plants: Old and new observations and new hypotheses. Oecologia 47:222–232.

Freeman-Gallant, C. R. 1996. DNA fingerprinting reveals female preference for male parental care in Savannah Sparrows. Proc. R. Soc. Lond. B 263:157–160.

Fretwell, S. D., and H. L. Lucas. 1970. On territorial behaviour and other factors influencing habitat distribution in birds. Acta Biotheor. 19:16–36.

Fry, F. E. J., and J. S. Hart. 1948. Cruising speed of goldfish in relation to water temperature. J. Fish. Res. Bd. Can. 7:169–175.

Fryxell, J. M., and C. M. Doucet. 1993. Diet choice and functional response of beavers. Ecology 74:1297–1306.

Fuentes, E. R. 1976. Ecological convergence of lizard communities in Chile and California. Ecology 57:3–18.

Fullick, T. G., and J. J. D. Greenwood. 1979. Frequency dependent food selection in relation to two models. Am Nat. 113:762–765.

Futuyma, D. J. 1986. Evolutionary Biology. 2nd ed. Sinauer Associates, Sunderland, MA.

Futuyma, D. J., and M. Slatkin (eds.). 1983. Coevolution. Sinauer, Sunderland, Massachusetts.

Gadgil, M., and W. H. Rossert. 1970. Life historical consequences of natural selection. Am. Nat. 104:1–24.

Galef, B. G., and S. W. Wigmore. 1983. Transfer of information concerning distant foods: A laboratory investigation of the "information centre hypothesis." Anim. Behav. 31:748.

Galloway, J. N., H. Levy II, and P. S. Kasibhatla. 1994. Year 2020:Consequences of population growth and development on deposition of oxidized nitrogen. Ambio 23:120–123.

Gans, C., and F. H. Pough (eds.). 1982. Biology of the Reptilia. Vol. 12. Physiology C. Part 1. Temperature Regulation and Thermal Relations. Academic Press, New York.

Garber, P. A., J. D. Pruetz, and J. Isaacson. 1993. Patterns of range use, range defense, and intergroup spacing in moustached tamarin monkeys (*Saguinus mystax*). Primates 34:11–25.

Gardner, R. 1990. Mating calls. Nature 344:495–496.

Gaston, K. J. 1994. Rarity. Chapman & Hall, London.

Gates, D. M. 1980. Biophysical Ecology. Springer-Verlag, New York.

Gathreaux, S. A. Jr. (ed.). 1981. Animal Migration, Orientation and Navigation. Academic Press, New York.

Gauch, H. G. Jr. 1982. Multivariate Analysis in Community Ecology. Cambridge University Press, Cambridge.

Gause, G. F. 1932. Experimental studies on the struggle for existence. I. Mixed population of two species of yeast. J. Exp. Biol. 9:389–402.

Gause, G. F. 1934. The Struggle for Existence. Williams and Wilkins, Baltimore.

Gavrilets, S. 1997. Evolution and speciation on holey adaptive landscapes. Trends Ecol. Evol. 12:307–312.

Gentry, A. H. 1988. Changes in plant community diversity and floristic composition on environmental and geographical gradients. Ann. Mo. Bot. Gard. 75:1–34.

George, M. F., M. R. Becwar, and M. J. Burke. 1982. Freezing avoidance by deep undercooling of tissue water in winter-hardy plants. Cryobiology 19:629–639.

Getty, T., and H. R. Pulliam. 1993. Search and prey detection by foraging sparrows. Ecology 74:734–742.

Ghiselin, M. T. 1969. The evolution of hermaphroditism among animals. Q. Rev. Biol. 44:189–208.

Gibson, A. R., D. A. Smucny, and J. Kollar. 1989. The effect of feeding and ecolysis on temperature selection by young garter snakes in a simple thermal mosaic. Can. J. Zool. 67:19–23.

Gibson, T. G., and M. A. Buzas. 1973. Species diversity: Patterns in modern and Miocene foraminifera of the eastern margin of North America. Geol. Soc. Am. Bull. 84:217–238.

Gilbert, L. E., and P. H. Raven (eds.). 1975. Coevolution of Animals and Plants. University of Texas Press, Austin.

Gillette, J. B. 1962. Pest pressure, an underestimated factor in evolution. Syst. Assoc. Publ. 4:37–46.

Gill, A. M. 1975. Fire and the Australian flora. Aust. For. 38:4–25.

Gilpin, M. E. 1974. A model of the predator-prey relationship. Theor. Pop. Biol. 5:333–344.

Gilpin, M. E. 1991. Extinction of finite metapopulations in correlated environments. In B. Shorrocks and I. R. Swingland (eds.), Living in a Patchy Environment, 177–186. Oxford University Press, Oxford.

Gilpin, M. E., and F. J. Ayala. 1973. Global models of growth and competition. Proc. Natl. Acad. Sci. USA 70:3590–3593.

Gilpin, M. E., and I. Hanski (eds.). 1991. Metapopulation Dynamics: Empirical and Theoretical Investigations. Academic Press, London.

Gilpin, M. E., and K. E. Justice. 1972. Reinterpretation of the invalidation of the principle of competitive exclusion. Nature 236:273–274, 299–301.

Gilpin, M. E., and K. E. Justice. 1973. A note on nonlinear competition models. Math. Biosci. 17:57–63.

Gilpin, M. E., and M. E. Soulé. 1987. Minimum viable populations: Processes of species extinction. In M. E. Soulé (ed.), Conservation Biology: The Science of Scarcity and Diversity, 19–34. Sinauer Associates, Sunderland, MA.

Ginzburg, L. R., and L. R. Akçakaya. 1992. Consequences of ratio-dependent predation for steady-state properties of ecosystems. Ecology 73:1536–1543.

Gleason, H. A. 1922. On the relation between species and area. Ecology 3:158–162.

Gleason, H. A. 1926. The individualistic concept of the plant association. Torrey Bot. Club Bull. 53:7–26.

Gleason, H. A. 1939. The individualistic concept of the plant association. Am. Midl. Nat. 21:92–110.

Gleeson, S. K., and D. S. Wilson. 1986. Equilibrium diet: Optimal foraging and prey coexistence. Oikos 46:139–144.

Glowacinski, Z., and O. Jarvinen. 1975. Rate of secondary succession in forest bird communities. Ornis Scand. 6:33–40.

Glück, E. 1987. Benefits and costs of social foraging and optimal flock size in goldfinches (*Carduelis carduelis*). Ethology 74:65–79.

Glynn, P. W. 1988. El Niño-Southern Oscillation 1982–1983:Nearshore population, community, and ecosystem responses. Annu. Rev. Ecol. Syst. 309–345.

Godfray, H. C. J. 1994. Parasitoids: Behavioral and Evolutionary Ecology. Princeton University Press, Princeton, NJ.

Godfray, H. C. J., and B. T. Grenfell. 1993. The continuing quest for chaos. Trends Ecol. Evol. 8:43–44.

Gogan, P. J. P., and J. F. Cochrane. 1994. Restoration of woodland caribou in the Lake Superior region. In M. L. Bowles and C. J. Whelan (eds.), Restoration of Endangered Species, 219–242. Cambridge University Press, Cambridge.

Gogan, P. J. P., P. A. Jordan, and J. L. Nelson. 1990. Planning to reintroduce woodland caribou to Minnesota. Trans. N. Am. Wildl. Nat. Res. Conf. 55:599–608.

Goldschmidt, T. T., F. Witte, and J. Wanink. 1993. Cascading effects of the introduced Nile perch on the detritivorous/phytoplanktivorous species of the sublittoral areas of Lake Victoria. Conserv. Ecol. 7:686–700.

Gomulkiewicz, R., and M. Kirkpatrick. 1992. Quantitative genetics and the evolution of reaction norms. Evolution 46:390–411.

Gonzalez-Prieto, S. J., and T. Carballas. 1995. N biochemical diversity as a factor of soil diversity. Soil Biol. Biochem. 27:205–210.

Goodman, D. 1979. Regulating reproductive effort in a changing environment. Am. Nat. 113:735–748.

Gopal, B., and U. Goel. 1993. Competition and allelopathy in aquatic plant communities. Bot. Rev. 59:155–210.

Gordon, D. M. 1995. The development of organization in an ant colony. Am. Sci. 3:50–57.

Gordon, M. S. 1968. Animal Function: Principles and Adaptations. Macmillan, New York.

Gosling, L. M. 1986. Selective abortion of entire litters in the coypu: Adaptive control of offspring production in relation to quality and sex. Am. Nat. 127:772–795.

Gotelli, N. J. 1995. A Primer of Ecology. Sinauer Associates, Sunderland, MA.

Gotelli, J. J., and G. R. Graves. 1996. Null Models in Ecology. Smithsonian Institution, Washington, D.C.

Gould, S. J. 1982. Darwinism and the expansion of evolutionary theory. Science 216:380–387.

Gould, S. J. 1989. Wonderful Life: The Burgess Shale and the Nature of History. W. W. Norton, New York.

Gould, S. J., and R. C. Lewontin. 1979. The spandrels of San Marco and the Panglossian paradigm: A critique of the adaptationist programme. Proc. R. Soc. Lond. B 205:581–598.

Goulden, C. E., L. L. Henry, and A. J. Tessier. 1982. Body size, energy reserves, and competitive ability in three species of Cladocera. Ecology 63:1780–1789.

Goulden, C. E., and L. L. Hornig. 1980. Population oscillations and energy reserves in planktonic Cladocera and their consequences to competition. Proc. Natl. Acad. Sci. USA 77:1716–1720.

Graedel, T. E., and P. J. Crutzen. 1989. The changing atmosphere. In Managing Planet Earth: Readings from Scientific American, 13–23. W. H. Freeman, New York.

Graham, A. (ed.). 1972. Floristics and Paleofloristics of Asia and Eastern North America. Elsevier, New York.

Graham, W. F., and R. A. Duce. 1979. Atmospheric pathways of the phosphorus cycle. Geochim. Cosmochim. Acta 43:1195–1208.

Grand, T. C., and J. W. A. Grant. 1994. Spatial predictability of resources and the ideal free distribution in convict cichlids, *Cichlasoma nigrofasciatum.* Anim. Behav. 48: 909–919.

Grant, P. R. 1966. Ecological incompatibility of bird species on islands. Am. Nat. 100:451–462.

Grant, P. R. 1972a. Convergent and divergent character displacement. Biol. J. Linn. Soc. 4:39–68.

Grant, P. R. 1972b. Interspecific competition among rodents. Annu. Rev. Ecol. Syst. 3:79–106.

Grant, P. R. 1986. Ecology and Evolution of Darwin's Finches. Princeton University Press, Princeton, NJ.

Grant, P. R., and B. R. Grant. 1992. Demography and the genetically effective sizes of two populations of Darwin's finches. Ecology 73:766–784.

Grant, W. D., and P. E. Long. 1981. Environmental Microbiology. Wiley (Halsted Press), New York.

Grasse, P.–P. 1951. Traité de Zoologie. Vol. X. Insects Superieurs et Hemipteroides. Part II. Masson, Paris.

Grassle, J. F. 1985. Hydrothermal vent animals: Distribution and biology. Science 229:713–717.

Grassle, J. F. 1996. The ecology of deep sea hydrothermal vent communities. Adv. Mar. Biol. Ecol. 23:301–362.

Graves, J., J. Ortega-Rauno, and P. J. B. Slater. 1993. Extrapair copulations and paternity in shags: Do females choose better males? Proc. R. Soc. Lond. B 253:3–7.

Graves, J. E., R. H. Rosenblatt, and G. N. Somero. 1983. Kinetic and electrophoretic differentiation of lactate dehydrogenases of teleost species-pairs from the Atlantic and Pacific coasts of Panama. Evolution 37:30–37.

Gray, A. 1860. Illustrations of botany of Japan and its relation to that of central and northern Asia, Europe, and North America. Proc. Am. Acad. Arts Sci. 4:131–135.

Gray, R. D. 1987. Faith and foraging: A critique of the "paradigm argument from design." In A. C. Kamil, J. R. Krebs, and H. R. Pulliam (eds.), Foraging Behavior, 69–140. Plenum Press, New York.

Green, A. J. 1990. Determinants of chorus participation and the effects of size, weight and competition on advertisement calling in the tungara frog, *Physalaemus pustulosus* (Leptodactulidae). Anim. Behav. 39:620–638.

Greene, E., L. J. Orsak, and D. W. Whitman. 1987. A tephritid fly mimics the territorial displays of its jumping spider predators. Science 236:310–312.

Greenland, D. J., and J. M. L. Kowal. 1960. Nutrient content of a moist tropical forest of Ghana. Plant Soil 12:154–174.

Greenslade, P. J. M. 1968. Island patterns in the Solomon Islands bird fauna. Evolution 22:751–761.

Greenslade, P. J. M. 1969. Land fauna: Insect distribution patterns in the Solomon Islands. Phil. Trans. R. Soc. B 255:271–284.

Greenstone, M. H. 1984. Determinants of web spider species diversity: Vegetation structural diversity vs. prey availability. Oecologia 62:299–304.

Greenwood, J. J. D. 1974. Effective population numbers in the snail *Cepaea nemoralis.* Evolution 28:513–526.

Greenwood, J. D., and S. R. Ballie. 1991. Effects of density-dependence and weather on population changes of English passerines using a non-experimental paradigm. Ibis 133: 121–133.

Gregory, P. T. 1982. Reptilian hibernation. In C. Gans and F. H. Pough (eds.), Biology of the Reptilia, vol. 13, 53–154. Academic Press, New York.

Griffin, D. H. 1981. Fungal Physiology. Wiley, New York.

Griffin, D. M. 1972. Ecology of Soil Fungi. Chapman & Hall, London.

Griffith, D. M., and T. L. Poulson. 1993. Mechanism and consequences of intraspecific competition in a carabid cave beetle. Ecology 74:1373–1383.

Grime, J. P. 1977. Evidence for the existence of three primary strategies in plants and its relevance to ecological and evolutionary theory. Am. Nat. 111:1169–1194.

Grime, J. P. 1979. Plant Strategies and Vegetation Processes. Wiley, New York.

Grime, J. P., and D. W. Jeffrey. 1965. Seedling establishment in vertical gradients of sunlight. J. Ecol. 53:621–642.

Grime, J. P., and P. S. Lloyd. 1973. An ecological atlas of grassland plants. Edward Arnold, London.

Grimm, N. B., and S. G. Fisher. 1986. Nitrogen limitation in a Sonoran Desert stream. J. N. Am. Benthol. Soc. 5:2–15.

Grinnell, A. D. 1968. Sensory physiology. In M. S. Gordon (ed.), Animal Function: Principles and Adaptations, 396–460. Macmillan, New York.

Grinnell, J., C. Packer, and A. E. Pusey. 1995. Cooperation in male lions: Kinship, reciprocity or mutualism? Anim. Behav. 49:95–105.

Grodzinski, B. 1992. Plant nutrition and growth regulation by CO_2 enrichment. BioScience 42:517–525.

Grodzinski, W., and B. A. Wunder. 1975. Ecological energetics of small mammals. In F. B. Golley, K. Petrusewicz, and L. Ryszkowski (eds.), Small Mammals: Their Productivity and Population Dynamics, 173–204. Cambridge University Press, Cambridge.

Groom, M. J., and N. Schumaker. 1993. Evaluating landscape change: Patterns of worldwide deforestation and local fragmentation. in P. M. Kareiva, J. G. Kingsolver, and R. B. Huey (eds.), Biotic Interactions and Global Change, 24–44. Sinauer Associates, Sunderland, MA.

Groot, C., and T. P. Quinn. 1987. Heming migration of sockeye salmon, *Oncorhynchus nerka* to the Fraser River. Fisher Bulletin 85:455–469.

Gross, A. D. 1947. Cyclic invasions of the snowy owl and the migration of 1945–1946. Auk 64:584–601.

Gross, M. R., and E. L. Charnov. 1980. Alternative male life histories in bluegill sunfish. Proc. Natl. Acad. Sci. USA 77:6937–6940.

Gross, M. R., R. M. Coleman, and R. M. McDowall. 1988. Aquatic productivity and the evolution of diadromous fish migration. Science 239:1291–1293.

Gross, M. R., and R. C. Sargent. 1985. The evolution of male and female parental care in fishes. Am. Zool. 25:775–793.

Gross, M. R., and R. Shine. 1981. Parental care and mode of fertilization in ectothermic vertebrates. Evolution 35:775–793.

Grubb, P. J. 1977. The maintenance of species diversity in plant communities: The importance of the regeneration niche. Biol. Rev. 52:107–145.

Grubb, P. J. 1995. Mineral nutrition and soil fertility in tropical rain forests. In A. E. Lugo and C. Lowe (eds.), Tropical Forests: Management and Ecology, 308–330. Springer-Verlag, New York.

Gu, H., and W. Danthanarayana. 1992. Quantitative genetic analysis of dispersal in *Epiphyas postvittana.* I. Genetic variation in flight capacity. Heredity 68:53–60.

Guilford, T. 1990. The evolution of aposematism. In D. L. Evans and J. O. Schmidt (eds.), Insect Defense: Adaptive Mechanisms and Strategies of Prey and Predator, 23–61. State University of New York Press, New York.

Gunn, D. L. 1960. The biological background of locust control. Annu. Rev. Entomol. 5:279–300.

Gutierrez, A. P. 1992. Physiological basis of ratio-dependent predator-prey theory: The metabolic pool model as a paradigm. Ecology 73:1552–1563.

Gutiérrez, R. J., and S. Harrison. 1996. Applying metapopulation theory to spotted owl management: A history and critique. In D. R. McCullough (ed.), Metapopulations and Wildlife Conservation, 167–185. Island Press, Washington, D.C.

Gutschick, V. P. 1981. Evolved strategies in nitrogen acquisition by plants. Am. Nat. 118:607–637.

Gwinner, E. (ed.). 1990. Bird Migration: The Physiology and Ecophysiology. Springer-Verlag, Berlin.

Gwynne, M. O., and R. H. V. Bell. 1968. Selection of vegetation components by grazing ungulates in the Serengeti National Park. Nature 220:390–393.

Gyllenberg, M., I. Hanski, and A. Hastings. 1997. Structured metapopulation models. In I. Hanski and M. E. Gilpin (eds.), Metapopulation Biology: Ecology, Genetics, and Evolution, 93–122. Academic Press, San Diego.

Hadley, N. F. 1970. Desert species and adaptation. Am. Sci. 60:338–347.

Hadley, N. F. 1994. Water relations of terrestrial arthropods. Academic Press, San Diego.

Haffer, J. 1969. Speciation in Amazonian forest birds. Science 165:131–137.

Haffer, J. 1974. Avian speciation in tropical South America. Publication of the Nuttall Ornithological Club, no. 14.

Hagerman, A. E., and L. Butler. 1991. Tannins and lignins. In G. A. Rosenthal and M. R. Berenbaum (eds.), Herbivores: Their Interactions with Secondary Plant Metabolites, 2nd ed., vol. 1, The chemical participants, 355–388. Academic Press, New York.

Hahn, W. E., and D. W. Tinkle. 1965. Fat body cycling and experimental evidence for its adaptive significance to ovarian follicle development in the lizard *Uta stansburiana.* J. Exp. Zool. 158:79–86.

Hailman, J. P., K. J. McGowan, and G. E. Woolfenden. 1994. Role of helpers in the sentinel behaviour of the Florida scrub jay (*Aphelocoma c. coerulescens*). Ethology 97:119–140.

Hainsworth, F. R. 1995. Optimal body temperatures with shuttling: Desert antelope ground squirrels. Anim. Behav. 49:107–116.

Hainsworth, F. R., B. G. Collins, and L. L. Wolf. 1977. The function of torpor in hummingbirds. Physiol. Zool. 50:214–222.

Hainsworth, F. R., and L. L. Wolf. 1970. Regulation of oxygen consumption and body temperature during torpor in a hummingbird, *Eulampis jugularis.* Science 168:368–369.

Hairston, N. G. 1959. Species abundance and community organization. Ecology 40:404–416.

Hairston, N. G. 1965. On the mathematical analysis of schistosome populations. Bull. World Health Org. 33:45–62.

Hairston, N. G. 1973. Ecology, selection, and systematics. Breviora 414:1–21.

Hairston, N. G. 1980. The experimental test of an analysis of field distributions: Competition in terrestrial salamanders. Ecology 61:817–826.

Hairston, N. G. 1983. Alpha selection in competing salamanders: Experimental verification of an a priori hypothesis. Am. Nat. 122:105–113.

Hairston, N. G., J. D. Allan, R. K. Colwell, D. J. Futuyma, J. Howell, M. D. Lubin, J. Mathias, and J. H. Vandermeer. 1968. The relationship between species diversity and stability: An experimental approach with protozoa and bacteria. Ecology 49:1091–1101.

Hairston, N. G., F. E. Smith, and L. B. Slobodkin. 1960. Community structure, population control, and competition. Am. Nat. 94:421–425.

Hairston, N. G., D. W. Tinkle, and H. M. Wilbur. 1970. Natural selection and the parameters of population growth. J. Wildl. Mgmt. 34:681–690.

Hakkarainen, H., E. Korpimäki, J. Ryssy, and S. Vikstrom. 1996. Low heritability in morphological characters of Tengmalm's owls: The role of cyclic food and laying date. Evol. Ecol. 10:207–219.

Haldane, J. B. S. 1932. The Causes of Evolution. Longmans, Green, New York, London, and Toronto.

Haldane, J. B. S. 1949a. Disease and evolution. Symposium sui fattori ecologici e genetici della speciazone negli animali. Ric. Sci. 19 (suppl.):3–11.

Haldane, J. B. S. 1949b. Suggestions as to quantitative measurement of rates of evolution. Evolution 3:51–56.

Haley, M. P., C. J. Deutsch, and B. J. LeBoeuf. 1994. Size, dominance and copulatory success in male northern elephant seals, *Mirounga angustirostris.* Anim. Behav. 48: 1249–1260.

Hall, S. J., and D. Raffaelli. 1993. Food webs: Theory and Reality. In M. Begon and A. H. Fitter (eds.), Advances in Ecological Research, 187–239. Academic Press, London.

Hallam, A. (ed.). 1977. Patterns of Evolution as Illustrated by the Fossil Record. Elsevier, Amsterdam and New York.

Halle, F. 1990. A raft atop the rain forest. National Geographic 178:128–138.

Hamilton, T. H., I. Rubinoff, R. H. Barth Jr., and G. L. Bush. 1963. Species abundance: Natural regulation of insular variation. Science 142:1575–1577.

Hamilton, W. D. 1964. The genetical evolution of social behaviour. J. Theor. Biol. 7:1–52.

Hamilton, W. D. 1966. The moulding of senescence by natural selection. J. Theor. Biol. 12:12–45.

Hamilton, W. D. 1967. Extraordinary sex ratios. Science 156:477–488.

Hamilton, W. D. 1979. Wingless and fighting males in fig wasps and other insects. In M. S. Blum and N. A. Blum (eds.), Sexual Selection and Reproductive Competition in Insects, 167–220. Academic Press, New York.

Hamilton, W. D., and M. Zuk. 1982. Heritable true fitness and bright birds: A role for parasites? Science 218:384–387.

Hammel, H. T. 1968. Regulation of internal body temperature. Annu. Rev. Physiol. 30:641–710.

Hammel, H. T., F. T. Caldwell, and R. M. Abrams. 1967. Regulation of body temperature in the blue-tongued lizard. Science 156:1260–1262.

Hansen, A. J., and F. di Castri. 1992. Landscape boundaries: Consequences of biotic diversity and ecological flows. Springer-Verlag, New York.

Hansen, T. A. 1978. Larval dispersal and species longevity in Lower Tertiary gastropods. Science 199:885–886.

Hansen, T. A. 1980. Influence of larval dispersal and geographic distribution on species longevity in neogastropods. Paleobiology 6:193–207.

Hanski, I. 1991. Single-species metapopulation dynamics: Concepts, models and observations. Biol. J. Linn. Soc. 42:17–38.

Hanski, I. 1997. Metapopulation dynamics: From concepts and observations to predictive models. In I. A. Hanski and M. E. Gilpin (eds.), Metapopulation Biology: Ecology, Genetics, and Evolution, 69–91. Academic Press, San Diego.

Hanski, I., and M. Gilpin. 1997. Metapopulation Biology: Ecology, Genetics, and Evolution. Academic Press, San Diego.

Hanski, I., L. Hansson, and H. Henttonen. 1991. Specialist predators, generalist predators, and the microtine rodent cycle. J. Anim. Ecol. 60:353–367.

Hanski, I. L., and H. Henttonen. 1996. Predation on competing rodent species: A simple explanation of complex patterns. J. Anim. Ecol. 65:220–232.

Hanski, I., J. Kouki, and A. Halkka. 1993. Three explanations of the positive relationship between distribution and abundance of species. In R. E. Ricklefs and D. Schluter (eds.), Species Diversity in Ecological Communities: Historical and Geographical Perspectives, 108–116. University of Chicago Press, Chicago.

Hanski, I., M. Kuussaari, and M. Nieminen. 1994. Metapopulation structure and migration in the butterfly *Melitaea cinxia.* Ecology 75:747–762.

Hanski, I., T. Pakkala, M. Kuussaari, and G. Lei. 1995. Metapopulation persistence of an endangered butterfly in a fragmented landscape. Oikos 72:21–28.

Hanski, I., and D. Simberloff. 1997. The metapopulation approach, its history, conceptual domain, and application to conservation. In I. Hanski and M. Gilpin (eds.), Metapopulation Biology: Ecology, Genetics, and Evolution, 5–26. Academic Press, San Diego.

Harborne, J. B. 1982. Introduction to Ecological Biochemistry. 2nd ed. Academic Press, London and New York.

Hardin, G. 1960. The competitive exclusion principle. Science 131:1292–1297.

Harley, J. L. 1972. Fungi in ecosystems. J. Anim. Ecol. 41:1–16.

Harley, J. L., and S. E. Smith. 1983. Mycorrhizal Symbiosis. Academic Press, London.

Harmon, M. E., J. F. Franklin, F. J. Swanson, P. Sollins, S. V. Gregory, J. D. Lattin, N. H. Anderson, S. P. Cline, N. G. Aumen, J. R. Sedell, G. W. Lienkaemper, K. Cromack Jr., and K. W. Cummins. 1986. Ecology of coarse woody debris in temperate ecosystems. Adv. Ecol. Res. 15:133–302.

Harper, J. L. 1967. A Darwinian approach to plant ecology. J. Ecol. 55:247–270.

Harper, J. L. 1969. The role of predation is vegetational diversity. Brookhaven Symp. Biol. 22:48–62.

Harper, J. L. 1977. Population Biology of Plants. Academic Press, New York and London.

Harper, J. L., J. T. Williams, and G. R. Sagar. 1965. The behaviour of seeds in soil. J. Ecol. 51:273–286.

Harris, L. G., A. W. Ebeling, D. R. Laur, and R. J. Rowley. 1984. Community recovery after storm damage: A case of facilitation in primary succession. Science 224:1336–1338.

Harris, P. 1984. *Carduus nutans* L., nodding thistle and *C. acanthoides* L., plumeless thistle (Compositae). In J. S. Kellener and M. A. Hulme (eds.), Biological Control Programmes against Insects and Weeds in Canada 1969–1980, 114–126. Commonwealth Agricultural Bureaux, London.

Harrison, A. T., E. Small, and H. A. Mooney. 1971. Drought relationships and distribution of two Mediterranean climate California plant communities. Ecology 52:869–875.

Harrison, J. J., and M. L. Van Buren. 1995. A phosphate transporter from the mycorrhizal fungus *Glomus versiforme.* Nature 378:626–629.

Harrison, S. 1991. Local extinction in a metapopulation context: An empirical evaluation. Biol. J. Linn. Soc. 42: 73–88.

Harrison, S. 1994. Metapopulations and conservation. In P. J. Edwards, R. M. May, and N. Webb (eds.), Large Scale Ecology and Conservation Biology, 111–128. Blackwell, Oxford.

Harrison, S., and J. F. Quinn. 1989. Correlated environments and the persistence of metapopulations. Oikos 56: 293–298.

Harrison, S., and A. D. Taylor. 1997. Empirical evidence for metapopulation dynamics. In I. A. Hanski and M. E. Gilpin (eds.), Metapopulation Biology: Ecology, Genetics, and Evolution, 27–42. Academic Press, San Diego.

Hartenstein, R. 1986. Earthworm biotechnology and global biogeochemistry. Adv. Ecol. Res. 15:379–409.

Hartl, D. L. 1980. Principles of Population Genetics. Sinauer Associates, Sunderland, MA.

Hartl, D. L., and A. B. Clark. 1989. Principles of Population Genetics. 2nd ed. Sinauer Associates, Sunderland, MA.

Harvey, P. H., and S. J. Arnold. 1982. Female mate choice and runaway sexual selection. Nature 297:533–534.

Harvey, P. H., R. K. Colwell, J. W. Silvertown, and R. M. May. 1983. Null models in ecology. Annu. Rev. Ecol. Syst. 14: 189–211.

Harvey, P. H., and M. D. Pagel. 1991. The Comparative Method in Evolutionary Biology. Oxford Series in Ecology and Evolution. Oxford University Press, Oxford.

Hassell, M. P. 1978. The Dynamics of Arthropod Predator-Prey Systems. Princeton University Press, Princeton, NJ.

Hassell, M. P., and H. N. Comins. 1978. Sigmoid functional responses and population stability. Theor. Pop. Biol. 14: 62–67.

Hassell, M. P., J. H. Lawton, and R. M. May. 1976. Patterns of dynamical behaviour in single-species populations. J. Anim. Ecol. 45:471–486.

Hassell, M. P., and R. M. May. 1973. Stability in insect host-parasite models. J. Anim. Ecol. 42:693–736.

Hassell, M. P., and R. M. May. 1989. The population biology of host-parasite and host-parasitoid associations. In J. Roughgarden, R. M. May, and S. A. Levin (eds.), Perspectives in Ecological Theory, 319–347. Princeton University Press, Princeton, NJ.

Hastings, A., and H. Caswell. 1979. Role of environmental variability in the evolution of life history strategies. Proc. Natl. Acad. Sci. USA 76:4700–4703.

Hastings, A., C. L. Hom, S. Ellner, P. Turchin, and H. C. J. Godfray. 1993. Chaos in ecology: Is mother nature a strange attractor? Annu. Rev. Ecol. Syst. 24:1–33.

Hatch, M. D., and C. R. Slack. 1966. Photosynthesis by sugar-cane leaves: A new carboxylation reaction and the pathway of sugar formation. Biochem. J. 101:103–111.

Haukioja, E. 1980. On the role of plant defenses in the fluctuation of herbivore populations. Oikos 35:202–213.

Hawksworth, D. L. 1991. The fungal dimension of biodiversity: Magnitude, significance, and conservation. Mycol. Res. 95:641–655.

Haxo, F. T., and L. R. Blinks. 1950. Photosynthetic action spectra of marine algae. J. Gen. Physiol. 33:389–422.

Hay, M. E. 1981a. The functional morphology of turf-forming seaweeds: Persistence in stressful marine habitats. Ecology 62:739–750.

Hay, M. E. 1981b. Herbivory, algal distribution, and the maintenance of between-habitat diversity on a tropical fringing reef. Am. Nat. 118:520–540.

Hay, M. E. 1986. Associational plant defenses and the maintenance of species diversity: Turning competitors into accomplices. Am. Nat. 128:617–641.

Hay, M. E., T. Colburn, and D. Downing. 1983. Spatial and temporal patterns in herbivory on a Caribbean fringing reef: The effects on plant distribution. Oecologia 58: 299–308.

Hay, M. E., and P. R. Taylor. 1985. Competition between herbivorous fishes and urchins on Caribbean reefs. Oecologia 65:591–598.

Haydon, D. 1994. Pivotal assumptions determining the relationship between stability and complexity: An analytical synthesis of the stability-complexity debate. Am. Nat. 144:14–29.

Hayes, A. J. 1979. The microbiology of plant litter decomposition. Sci. Progress 66:25–42.

Heal, O. W., and J. P. Grime. 1991. Comparative analysis of ecosystems: Past lessons and future directions. In J. Cole, G. Lovett, and S. Findlay (eds.), Comparative Analysis of Ecosystems, 7–23. Springer-Verlag, New York.

Heath, J. E. 1965. Temperature regulation and diurnal activity in horned lizards. University of California Publications in Zoology 64:97–136.

Heatwole, H. 1970. Thermal ecology of the desert dragon *Amphibolurus inermis*. Ecol. Monogr. 40:425–457.

Heatwole, H. 1976. Reptile Ecology. University of Queensland Press, St. Lucia, Queensland.

Heck, K. L., and J. F. Valentine. 1995. Sea urchin herbivory: Evidence for long-lasting effects in subtropical seagrass meadows. J. Exp. Mar. Biol. Ecol. 189:205–217.

Hedrick, P. W., and M. E. Gilpin. 1997. Genetic effective size of a metapopulation. In I. A. Hanski and M. E. Gilpin (eds.), Metapopulation Biology: Ecology, Genetics, and Evolution, 166–181. Academic Press, San Diego.

Heinrich, B. 1979. Bumblebee Economics. Harvard University Press, Cambridge, MA.

Heinrich, B. 1993. How avian predators constrain caterpillar foraging. In N. E. Stamp and T. M. Casey (eds.), Caterpillars: Ecological and Evolutionary Constraints on Foraging, 224–227. Chapman & Hall, New York.

Heinrich, B., and G. A. Bartholomew. 1971. An analysis of pre-flight warm-up in the sphinx moth, *Manduca sexta*. J. Exp. Biol. 55:223–239.

Heinrich, B., and S. L. Collins. 1983. Caterpillar leaf damage and the game of hide-and-seek with birds. Ecology 64:592–602.

Heinrich, B., and J. Marzluff. 1995. Why ravens share. Am. Sci 83:342–349.

Heinsohn, R. 1995. Hatching asynchrony and brood reduction in cooperatively breeding white-winged choughs *Corcorax melanorhamphos*. Emu 95:252–258.

Heinsohn, R., and C. Packer. 1995. Complex cooperative strategies in group-territorial African lions. Science 269:1260–1262.

Heinsohn, R., C. Packer, and A. E. Pusey. 1996. Development of cooperative territoriality in juvenile lions. Proc. R. Soc. Lond. B 263:475–479.

Heller, H. C. 1971. Altitudinal zonation of chipmunks (*Eutamias*): Interspecific aggression. Ecology 52:312–319.

Heller, H. C., L. I. Crawshaw, and H. T. Hammel. 1978. The thermostat of vertebrate animals. Sci. Am. 239:102–113.

Hellmers, H., J. S. Horton, G. Juhren, and J. O'Keefe. 1955. Root systems of some chaparral plants in southern California. Ecology 36:667–678.

Herczeg, A. L. 1987. A stable carbon isotope study of dissolved inorganic carbon cycling in a softwater lake. Biogeochemistry 4:231–263.

Herre, E. A. 1985. Sex ratio adjustment in fig wasps. Science 228:896–898.

Hertz, P. E. 1992. Temperature regulation in Puerto Rican *Anolis* lizards: A field test using null hypotheses. Ecology 73:1405–1417.

Hieber, C. S., and G. W. Uetz. 1990. Colony size and parasitoid load in two species of colonial *Metepeira* spiders from Mexico (Araneae: Araneidae). Oecologia 82:145–150.

Hiesey, W. M., and H. W. Milner. 1965. Physiology of ecological races and species. Annu. Rev. Plant Physiol. 16: 203–216.

Hill, G. E., R. Montgomerie, C. Roeder, and P. Boag. 1994. Sexual selection and cuckoldry in a monogamous songbird: Implications for sexual selection theory. Behav. Ecol. Sociobiol. 35:193–199.

Hjaelten, J., and K. Danell. 1993. Effects of simulated herbivory and intraspecific competition on the compensatory ability of birches. Ecology 74:1136–1142.

Hoagland, K. E. 1978. Protandry and the evolution of environmentally-mediated sex change: A study of the Mollusca. Malacologia 17:365–391.

Hochachka, P. W., and G. N. Somero. 1973. Strategies of Biochemical Adaptation. Saunders, Philadelphia.

Hochachka, P. W., and G. N. Somero. 1984. Biochemical Adaptation. Princeton University Press, Princeton, NJ.

Höfle, M. G. 1990. RNA chemotaxonomy of bacterial isolates and natural microbial communities. In J. Overbeck and R. J. Chróst, Aquatic Microbial Ecology, 129–159. Springer-Verlag, New York.

Hogan-Warburg, A. J. 1966. Social behavior of the ruff, *Philomachus pugnax* (L.). Ardea 54:109–225.

Hogsted, G. 1980. Evolution of clutch size in birds: Adaptive variation in relation to territory quality. Science 210: 1148–1150.

Holdren, G. C., and D. E. Armstrong. 1980. Factors affecting phosphorus release from intact lake sediment cores. Environ. Sci. Technol. 14:79–87.

Holdridge, L. 1967. Life Zone Ecology. Tropical Science Center, San Jose, Costa Rica.

Holland, K. N., and J. R. Sibert. 1994. Physiological thermoregulation in bigeye tuna, *Thunnus obesus*. Environ. Biol. Fishes 40:319–327.

Holland, M. M. 1988. SCOPE/MAB technical consultations on landscape boundaries. Biol. Intl. 17:47–106.

Holling, C. S. 1959. The components of predation as revealed by a study of small mammal predation of the European pine sawfly. Can. Entomol. 91:293–320.

Holmes, W. G., and P. W. Sherman. 1983. Kin recognition in animals. Am. Sci. 71:46–55.

Holt, R. D. 1977. Predation, apparent competition, and the structure of prey communities. Theor. Pop. Biol. 12: 197–229.

Holt, R. D. 1984. Spatial heterogeneity, indirect interactions, and the coexistence of prey species. Am. Nat. 124:377–406.

Holt, R. D. 1997. From metapopulation dynamics to community structure. In I. Hanski and M. E. Gilpin (eds.), Metapopulation Biology: Ecology, Genetics and Evolution, 149–164. Academic Press, San Diego.

Holt, R. D., and J. H. Lawton. 1994. The ecological consequences of shared natural enemies. Annu. Rev. Ecol. Syst. 25:495–520.

Hoogland, J. L. 1982. Prairie dogs avoid extreme inbreeding. Science 215:1639–1641.

Hopkins, B. 1955. The species-area relations of plant communities. J. Ecol. 43:409–426.

Hopkins, W. G. 1995. Introduction to Plant Physiology. John Wiley & Sons, New York.

Hori, M. 1993. Frequency-dependent natural selection in the handedness of scale-eating cichlid fish. Science 260:216–219.

Horn, H. S. 1975. Markovian properties of forest succession. In M. L. Cody and J. M. Diamond (eds.), Ecology and Evolution of Communities, 196–211. Harvard University Press, Cambridge, MA.

Hornocker, M. G. 1970. An analysis of mountain lion predation upon mule deer and elk in the Idaho Primitive Area. Wildl. Monogr. 21:3–39.

Hough, A. F., and R. D. Forbes. 1943. The ecology and silvics of forests in the high plateaus of Pennsylvania. Ecol. Monogr. 13:299–320.

Houston, A. I., C. W. Clarke, J. M. McNamara, and M. Mangel. 1988. Dynamic models in behavioural and evolutionary ecology. Nature 332:29–34.

Houston, A. I., and J. M. McNamara. 1992. Phenotypic plasticity as a state-dependent life-history decision. Evol. Ecol. 6:243–253.

Houston, D. C. 1979. The adaptations of scavengers. In A. R. E. Sinclair and M. Norton-Griffiths (eds.), Serengeti: Dynamics of an Ecosystem, 263–286. University of Chicago Press, Chicago.

Houtman, A. M. 1992. Female zebra finches choose attractive partners for extra-pair copulations. Proc. R. Soc. Lond. B. 249:3–6.

Houvenaghel, G. T. 1984. Oceanographic setting of the Galápagos Islands. Pp 43–54 in R. Perry (ed.), Galapagos (Key Environment Series), Pergamon Press, New York.

Howard, R. D. 1979. Estimating reproductive success in natural populations. Am. Nat. 114:221–231.

Howarth, R. W. 1984. The ecological significance of sulfur in the energy dynamics of salt marsh and coastal marine sediments. Biogeochemistry 1:5–27.

Howarth, R. W. 1988. Nutrient limitation of net primary production in marine ecosystems. Annu. Rev. Ecol. Syst. 19:89–110.

Howe, H. F., E. W. Schupp, and L. C. Westley. 1985. Early consequences of seed dispersal for a Neotropical tree (*Virola surinamensis*). Ecology 66:781–791.

Howse, P. E. 1984, Sociochemicals of Termites. In W. J. Bell and R. T. Cardé (eds.), Chemical Ecology of Insects, 475–519. Sinauer Associates, Inc., Sunderland, Massachusetts.

Hubbell, S. P. 1979. Tree dispersion, abundance, and diversity in a tropical dry forest. Science 203:1299–1039.

Hubbell, S. P., and R. B. Foster. 1983. Diversity of canopy trees in a Neotropical forest and implications for conservation. In S. Sutton, T. C. Whitmore, and A. Chadwick (eds.), Tropical Rain Forest: Ecology and Management, 25–41. Blackwell, Oxford.

Huey, R. B. 1974. Behavioral thermoregulation in lizards: Importance of associated costs. Science 184:1001–1003.

Huey, R. B., C. R. Peterson, S. J. Arnold, and W. P. Porter. 1989. Hot rocks and not-so-hot rocks: Retreat-site selection by garter snakes and its thermal consequences. Ecology 70:931–944.

Huffaker, C. B. 1958. Experimental studies on predation: Dispersion factors and predator-prey oscillations. Hilgardia 27:343–383.

Huffaker, C. B., and C. E. Kennett. 1956. Experimental studies on predation: Predation and cyclamen-mite populations on strawberries in California. Hilgardia 26: 191–222.

Huffaker, C. B., and C. E. Kennett. 1969. Some aspects of assessing efficiency of natural enemies. Can. Entomol. 101: 425–447.

Hughes, K. A., and B. Charlesworth. 1994. A genetic analysis of senescence in *Drosophila.* Nature 367:64–66.

Hughes, N. F. 1992. Selection of positions by drift-feeding salmonids in dominance hierarchies: Model and test for Arctic grayling (*Thymallus arcticus*) in subarctic mountain streams, interior Alaska. Can. J. Fish. Aquat. Sci. 49: 1999–2008.

Hunter, M. D., and P. W. Price. 1992. Playing chutes and ladders: Heterogeneity and the relative roles of bottom-up and top-down forces in natural communities. Ecology 73: 724–732.

Huntley, M., and C. Boyd. 1984. Food-limited growth of marine zooplankton. Am. Nat. 124:455–478.

Huston, M. A. 1979. A general hypothesis of species diversity. Am. Nat. 113:81–101.

Huston, M. A. 1986. Size bimodality in plant populations: An alternative hypothesis. Ecology 67:265–269.

Huston, M. A. 1994. Biological Diversity: The Coexistence of Species on Changing Landscapes. Cambridge University Press, Cambridge.

Huston, M. A., and D. L. DeAngelis. 1994. Competition and coexistence: The effects of resource transport and supply rate. Am. Nat. 144:954–977.

Huston, M. A., D. DeAngelis, and W. Post. 1988. New computer models unify ecological theory. BioScience 38: 682–691.

Huston, M. A., and T. M. Smith. 1987. Plant succession: Life history and competition. Am. Nat. 130:168–198.

Hutchinson, G. E. 1948. Circular causal systems in ecology. Ann. N.Y. Acad. Sci. 50:221–246.

Hutchinson, G. E. 1957a. Concluding remarks. Cold Spring Harbor Symp. Quant. Biol. 22:415–427.

Hutchinson, G. E. 1957b. A Treatise on Limnology. Vol. 1: Geography, Physics, and Chemistry. Wiley, New York.

Hutchinson, G. E. 1959. Homage to Santa Rosalia, or Why are there so many kinds of animals? Am. Nat. 93: 145–159.

Hutchinson, G. E. 1961. The paradox of the plankton. Am. Nat. 95:137–145.

Hutchison, V. H., H. G. Dowling, and A. Vinegar. 1966. Thermoregulation in a brooding female Indian python, *Python molurus bivittatus.* Science 151:694–696.

Hutchison, V. H., and R. K. Dupré. 1992. Thermoregulation. In M. E. Feder and W. W. Burggren (eds.), Environmental Physiology of the Amphibians, 206–249. University of Chicago Press, Chicago.

Immelmann, K. 1971. Ecological aspects of periodic reproduction. In D. S. Farrier and J. R. King (eds.), Avian Biology, vol. 1, 341–389. Academic Press, New York.

Irlandi, E. A., W. G. Ambrose, and B. A. Orando. 1995. Landscape ecology and the marine environment: How spatial configuration of seagrass habitat influences growth and survival of the bay scallop. Oikos 72:307–313.

Irving, L. 1966. Adaptations to cold. Sci. Am. 214:94–101.

Ives, W. G. H. 1973. Heat units and outbreaks of the forest tent caterpillar, *Malacosoma disstria* (Lepidoptera: Lasiocampidae). Can. Entomol. 105:529–543.

Iwasaki, K. 1995. Dominance order and resting site fidelity in the intertidal pulmonate limpet *Siphonaria sirius* (Pilsbry). Ecol. Res. 10:105–115.

Jablonski, D. 1986. Background and mass extinctions: The alteration of macroevolutionary regimes. Science 231: 129–133.

Jablonski, D., and R. A. Lutz. 1980 Larval shell morphology: Ecology and paleoecological applications. In D. C. Rhoads and R. A. Lutz (eds.), Skeletal Growth of Aquatic Organisms, 323–377. Plenum Press, New York.

Jablonski, D., and J. W. Valentine. 1990. From regional to total geographic ranges: Testing the relationship in recent bivalves. Paleobiology 16:126–142.

Jackson, G. J., R. Herman, and I. Singer (eds.). 1969–1970. Immunity to Parasitic Animals. Vol. I and II. Appleton-Century-Crofts, New York.

Jackson, J. B. C., and L. Buss. 1975. Allelopathy and spatial competition among coral reef invertebrates. Proc. Natl. Acad. Sci. USA 72:5160–5163.

Jackson, S. T., R. P. Futuyma, and D. A. Wilcox. 1988. A paleoecological test of a classical hydrosere in the Lake Michigan dunes. Ecology 69:928–936.

Jagger, J. 1985. Solar-UV Actions on Living Cells. Praeger, New York.

Jain, S. K. 1976. The evolution of inbreeding in plants. Annu. Rev. Ecol. Syst. 7:469–495.

Janinasch, H. W., and M. J. Mottl. 1985. Geomicrobiology of deep-sea hydrothermal vents. Science 229:717–725.

Janos, D. P. 1980. Vesicular-arbuscular mycorrhizae affect lowland tropical rain forest plant growth. Ecology 61:151–162.

Janzen, D. H. 1966. Coevolution of mutualism between ants and acacias in Central America. Evolution 20:249–275.

Janzen, D. H. 1967. Interaction of the bull's-horn acacia (*Acacia cornigera* L.) with an ant inhabitant (*Pseudomyrmex ferruginea* F. Smith) in eastern Mexico. Univ. Kans. Sci. Bull. 47:315–558.

Janzen, D. H. 1968. Host plants as islands in evolutionary and contemporary time. Am. Nat. 102:592–595.

Janzen, D. H. 1969. Seed-eaters versus seed size, number, toxicity and dispersal. Evolution 23:1–27.

Janzen, D. H. 1970. Herbivores and the number of tree species in tropical forests. Am. Nat. 104:501–528.

Janzen, D. H. 1971. Seed predation by animals. Annu. Rev. Ecol. Syst. 2:465–492.

Janzen, D. H. 1973. Host plants as islands. II. Competition in evolutionary and contemporary time. Am. Nat. 107: 786–790.

Janzen, D. H. 1976. Why bamboos wait so long to flower. Annu. Rev. Ecol. Syst. 7:347–391.

Janzen, D. H. 1985. The natural history of mutualisms. In D. H. Boucher (ed.), The Biology of Mutualism, 40–99. Croom Helm, London.

Janzen, D. H. 1988. Tropical ecological and biocultural restoration. Science 239:243–244.

Janzen, D. H., and T. W. Schoener. 1968. Differences in insect abundance and diversity between wetter and drier sites during a tropical dry season. Ecology 49:96–110.

Jarman, P. J., and A. R. E. Sinclair. 1979. Feeding strategy and the pattern of resource partitioning in ungulates. In A. R. E. Sinclair and M. Norton-Griffiths (eds.), Serengeti: Dynamics of an Ecosystem, 130–163. University of Chicago Press, Chicago.

Jarvi, T., B. Sillén-Tulberg, and C. Wiklund. 1981. The cost of being aposematic: An experimental study of predation of larvae of *Papitio machaon* by the great tit, *Parus major.* Oikos 36:267–272.

Jefferies, R. L. 1988. Vegetational mosaics, plant-animal interactions and resources for plant growth. In L. D. Gottlieb and S. K. Jain (eds.), Plant Evolutionary Biology, 341–369. Chapman & Hall, London.

Jeffries, M. J., and J. H. Lawton. 1985. Predator-prey ratios in communities of freshwater invertebrates: The role of enemy free space. Freshwater Biol. 15:105–112.

Jenny, H. 1941. Factors in Soil Formation. McGraw-Hill, New York.

Jenny, H. 1980. The Soil Resource: Origin and Behavior. Springer-Verlag, New York.

Jerling, L. 1988. Genetic differentiation in fitness related characters in *Plantago maritima* along a distributional gradient. Oikos 53:341–350.

Johannesson, K. 1989. The bare zone of Swedish rocky shores: Why is it there? Oikos 54:77–86.

John, D. M., S. J. Hawkins, and J. H. Price. 1992. Plant-animal interactions in the marine benthos. Clarendon Press, Oxford.

Johnson, C. N. 1988. Dispersal and the sex ratio at birth in primates. Nature 332:726–728.

Johnson, D. W., D. W. Cole, S. P. Gessel, M. J. Singer, and R. V. Minden. 1977. Carbonic acid leaching in a tropical, temperate, subalpine, and northern forest soil. Arctic and Alpine Res. 9:329–343.

Johnson, D. W., and G. S. Henderson. 1989. Terrestrial nutrient cycling. In D. W. Johnson and R. I. Van Hook (eds.), Analysis of Biogeochemical Cycling Processes in Walker Branch Watershed, 233–300. Springer-Verlag, New York.

Johnson, D. W., G. S. Henderson, and D. E. Todd. 1988. Changes in nutrient distribution in forests and soils of Walker Branch watershed, Tennessee, over an eleven-year period. Biogeochemistry 5:275–293.

Johnson, D. W., and R. I. Van Hook. 1989. Analysis of Biogeochemcial Cycling Processes in Walker Branch Watershed. Springer-Verlag, New York.

Jolly, G. M. 1965. Explicit estimates from capture-recapture data with low death and immigration stochastic model. Biometrika 52:315–337.

Jolly, G. M. 1979. Sampling of large objects. In R. M. Cormack, G. P. Patil, and D. S. Robson (eds.), Sampling Biological Populations, 193–201. International Co-operative Publishing House, Fairland, MD.

Jones, F. R. H. 1968. Fish Migration. Edward Arnold, London.

Jones, M. L., S. L. Swartz, and S. Leatherwood (eds.). 1984. The Gray Whale *Eschrichtius robustus.* Academic Press, Orlando, FL.

Jordan, C. F. 1985. Nutrient Cycling in Tropical Forest Ecosystems. Wiley, New York.

Jordan, C. F., and R. Herrera. 1981. Tropical rain forests: Are nutrients really critical? Am. Nat. 117:167–180.

Jordan, P., and G. Webbe. 1969. Human Schistosomiasis. Heinemann Medical Books, London.

Jordan, W. R. III, M. E. Gilpin, and J. D. Aber. 1987. Restoration Ecology: A Synthetic Approach to Ecological Research. Cambridge University Press, Cambridge.

Julliot, C. 1997. Impact of seed dispersal by red howler monkeys *Alouatta seniculus* on the seedling population in the understory of tropical rain forest. J. Ecol. 85:431–440.

Juniper, S. K., V. Tunnicliffe, and E. C. Southward. 1992. Hydrothermal vents in turbidite sediments on a Northeast Pacific spreading centre: Organisms and substratum at an ocean drilling site. Can. J. Zool. 70:1792–1823.

Junk, W. J., P. B. Bayley, and R. E. Sparks. 1989. The flood pulse concept in river-floodplain systems. In D. P. Dodge (ed.), Proceedings of the International Large River Symposium, 110–127. Can. Spec. Publ. Fish. Aquat. Sci. 106.

Kacelnik, A. 1984. Central place foraging in starlings (*Sturnus vulgaris*). 1. Patch residence time. J. Anim. Ecol. 53: 283–299.

Kaiser, B. W., and H. R. Mushinsky. 1994. Tail loss and dominance in captive adult male *Anolis sagrei.* J. Herpetol. 28:342–346.

Kamil, A. C., J. R. Krebs, and H. R. Pulliam (eds.). 1987. Foraging Behaviour. Plenum Press, New York.

Karban, R., and J. R. Carey. 1984. Induced resistance of cotton seedlings to mites. Science 225:53–54.

Karban, R., and R. E. Ricklefs. 1983. Host characteristics, sampling intensity, and species richness of Lepidoptera larvae on broad-leaved trees in southern Ontario. Ecology 64:636–641.

Kareiva, P. M. 1983. Influence of vegetation texture on herbivore populations: Resource concentration and herbivore movement. In R. F. Denno and M. S. McClure (eds.), Variable Plants and Herbivores in Natural and Managed Systems, 259–289. Academic Press, New York.

Kareiva, P. M. 1987. Habitat fragmentation and the stability of predator-prey interactions. Nature 326:388–390.

Kauffman, E. G., and J. A. Fagerstrom. 1993. The Phanerozoic evolution of reef diversity. In R. E. Ricklefs and D. Schluter (eds.), Species Diversity in Ecological Communities: Historical and Geographical Perspectives, 315–329. University of Chicago Press, Chicago.

Kaufman, L. 1992. Catastrophic change in species-rich freshwater ecosystems: The lessons of Lake Victoria. BioScience 42:846–858.

Kawecki, T. J., and S. C. Stearns. 1993. The evolution of life-histories in spatially heterogeneous environments: Optimal reaction norms revisited. Evol. Ecol. 74:673–684.

Kazacos, K. R., and R. E. Thorson. 1975. Cross-resistance between *Nippostrongylus brasiliensis* and *Strongyloides ratti* in rats. J. Parasitol. 61:525–529.

Keast, A. 1972. Ecological opportunities and dominant families, as illustrated by the Neotropical Tyrarmidae (Aves). Evol. Biol. 5:229–277.

Keast, A., and E. S. Morton. 1980. Migrant Birds in the Neotropics: Ecology, Behavior, Distribution, and Conservation. Smithsonian Institution Press, Washington, D.C.

Keen, N. T., S. Tamaki, D. Kobayashi, D. Gerhold, M. Stayton, H. Shen, S. Gold, J. Lorang, H. Thordal-Christensen, D. Dahlbeck, and D. Staskawicz. 1990. Bacteria expressing avirulence gene D produce a specific elicitor of the soybean hypersensitive reaction. Molecular Plant-Microbe Interactions 3:112–121.

Keever, C. 1950. Causes of succession on old fields of the Piedmont, North Carolina. Ecol. Monogr. 20:230–250.

Keller, L., and K. G. Ross. 1993. Phenotypic plasticity and "cultural" transmission of alternative social organizations in the fire ant *Solenopsis invicta.* Behav. Ecol. Sociobiol. 33:121–129.

Kelsall, J. P. 1968. The Migratory Barren-ground Caribou of Canada. Canadian Wildlife Service, Ottawa.

Kempenaers, B., G. R. Verheyen, M. Vanden Broeck, T. Burke, C. Van Broeckhoven, and A. A. Dhondt. 1992. Extra-pair paternity results from female preference for high quality males in the blue tit. Nature 357:494–496.

Kenoyer, L. A. 1927. A study of Raunkaier's law of frequence. Ecology 8:341–349.

Keough, M. J. 1984. Effects of patch size on the abundance of sessile marine invertebrates. Ecology 65:423–437.

Kerster, H. W. 1964. Neighborhood size in the rusty lizard, *Sceloporus olivaceus.* Evolution 18:445–457.

Kessel, B. 1953. Distribution and migration of the European starling in North America. Condor 55:49–67.

Kettlewell, H. B. D. 1955. Selection experiments on industrial melanism in the Lepidoptera. Heredity 10:287–301.

Kettlewell, H. B. D. 1956. Further selection experiments on industrial melanism in the Lepidoptera. Heredity 10:287–301.

Kettlewell, H. B. D. 1959. Darwin's missing evidence. Sci. Am. 200:48–53.

Keynes, R. D., and H. Martins-Ferreira. 1953. Membrane potentials in the electroplates of the electric eel. J. Physiol. 119:315–351.

Kiester, A. R. 1971. Species density of North American amphibians and reptiles. Syst. Zool. 20:127–137.

Kimura, S., K. Tsukamoto, and T. Sugimoto. 1994. A model for the larval migration of the Japanese eel: Roles of the trade winds and salinity front. Mar. Biol. 119:185–190.

King, B. H. 1988. Sex-ratio manipulation in response to host size by the parasitoid wasp *Spalangia cameroni:* A laboratory study. Evolution 42:716–727.

King, C. E. 1964. Relative abundance of species and MacArthur's model. Ecology 45:716–727.

King, J. R. 1974. Seasonal allocation of time and energy resources in birds. In R. A. Paynter Jr. (ed.), Avian Energetics, 4–70. Nuttall Ornithol. Club, Cambridge, MA.

King, J. R., and D. S. Farner. 1961. Energy metabolism, thermoregulation and body temperature. In A. J. Marshall (ed.), Biology and Comparative Physiology of Birds, vol. II, 215–288. Academic Press, New York.

King, R. W., I. F. Wardlaw, and L. T. Evans. 1967. Effects of assimilate utilization and photosynthetic rate in wheat. Planta 77:261–276.

Kingsland, S. E. 1985. Modeling Nature: Episodes in the History of Population Ecology. University of Chicago Press, Chicago.

Kira, T., and T. Shidel. 1967. Primary production and turnover of organic matter in different forest ecosystems of the western Pacific. Jap. J. Ecol. 17:70–87.

Kirk, T. K., W. J. Connors, and J. G. Zeikus. 1977. Advances in understanding the microbiological degradation of lignin. Rec. Adv. Phytochem. 11:369–394.

Kirkland, J. L. 1992. The biochemistry of mammalian senescence. Clin. Biochem. 25:61–75.

Kirkpatrick, M and R. J. Ryan. 1991. The evolution of mating preferences and the paradox of the lek. Nature 350: 33–38.

Kirkwood, T. B. L., and M. R. Rose. 1991. Evolution of senescence: Late survival sacrificed for reproduction. Phil. Trans. R. Soc. Lond. B 332:15–24.

Kitchell, J. F., and S. R. Carpenter. 1993. Cascading trophic interactions. In S. R. Carpenter and J. F. Kitchell (eds.), The Trophic Cascade in Lakes, 1–14. Cambridge University Press, Cambridge.

Klahn, J. E., and G. J. Gamboa. 1983. Social wasps: Discrimination between kin and nonkin brood. Science 221: 482–484.

Kleeberger, S. R. 1984. A test of competition in two sympatric populations of desmognathine salamanders. Ecology 65:1846–1856.

Kleiber, M. 1961. The Fire of Life. Wiley, New York.

Kling, G. W. 1994. Ecosystem-scale experiments: The use of stable isotopes in freshwater. In L. A. Baker (ed.), Environmental Chemistry of Lakes and Reservoirs, 91–120. ACS Advances in Chemistry Series, no. 237. American Chemical Society.

Kling, G. W., B. Fry, and W. J. O'Brien. 1992. Stable isotopes and planktonic trophic structure in arctic lakes. Ecology 73:561–566.

Kling, G. W., G. W. Kipphut, and M. C. Miller. 1991. Arctic lakes and streams as gas conduits to the atmosphere: Implications for tundra carbon budgets. Science. 251:298–301.

Kling, G., W. G. W. Kipphut, and M. C. Miller. 1992. The flux of CO_2 and CH_4 from lakes and rivers in arctic Alaska. Hydrobiologia 240:23–36.

Klopfer, P. 1959. Social interactions in discrimination learning with special reference to feeding behavior in birds. Behaviour 14:282–299.

Kluge, M., and I. P. Ting. 1978. Crassulacean Acid Metabolism. Springer-Verlag, Berlin.

Knauss, J. A. 1978. Introduction to Physical Oceanography. Prentice Hall, Englewood Cliffs, NJ.

Knight, D. H. 1975. A phytosociological analysis of species-rich tropical forest: Barro Colorado Island, Panama. Ecol. Monogr. 45:259–284.

Knoll, A. H. 1986. Patterns of change in plant communities through geological time. in J. Diamond and T. J. Case (eds.), Community Ecology, 126–141. Harper & Row, New York.

Knutson, R. M. 1974. Heat production and temperature regulation in eastern skunk cabbage. Science 186:746–748.

Kobayashi, K., and K. Suzuki. 1992. Hermaphroditism and sexual function in *Cirrhitichthys aureus* and the other Japanese hawkfishes (Cirrhitidae: Teleostei). Jap. J. Ichthyol. 38:397–410.

Kodric-Brown, A. 1993. Female choice of multiple male criteria in guppies: Interacting effects of dominance, coloration and courtship. Behav. Ecol. Sociobiol. 32:415–420.

Kohn, R. R. 1971. Principles of Mammalian Aging. Prentice-Hall, Englewood Cliffs, NJ.

Koller, D. 1969. The physiology of dormancy and survival of plants in desert environments. Symp. Soc. Exp. Biol. 23:449–469.

Koplin, J. R., and R. S. Hoffmann. 1968. Habitat overlap and competitive exclusion in voles (*Microtus*). Am. Midl. Nat. 80:494–507.

Korpimäki, E. 1992. Population dynamics of Fennoscandian owls in relation to wintering conditions and between-year fluctuations of food. In C. A. Galbraith, I. R. Taylor, and S. Percival (eds.), The Ecology and Conservation of European Owls, 1–10. Joint Natural Conservation Committee, Peterborough, United Kingdom.

Korpimäki, E., and K. Norrdahl. 1989. Avian predation on mustelids in Europe. 2. Impact on small mustelid and microtine dynamics—a hypothesis. Oikos 55:273–280.

Korpimäki, E., and K. Norrdahl. 1991. Numerical and functional responses of kestrels, short-eared owls, and long-eared owls to vole densities. Ecology 72:814–826.

Kotler, B. P., and J. S. Brown. 1988. Environmental heterogeneity and the coexistence of desert rodents. Annu. Rev. Ecol. Syst. 19:281–307.

Kozlovsky, D. G. 1968. A critical evaluation of the trophic level concept. 1. Ecological efficiencies. Ecology 49:48–59.

Kozlowski, T. T., and C. E. Ahlgren (eds.). 1974. Fire and Ecosystems. Academic Press, New York.

Kramer, P. J. 1958. Photosynthesis of trees as affected by their environment. In K. V. Thimann (ed.), The Physiology of Forest Trees, 157–186. Ronald, New York.

Kramer, P. J. 1969. Plant and Water Relationships: A Modern Synthesis. McGraw-Hill, New York.

Krebs, C. J. 1994. Ecology: The Experimental Analysis of Distribution and Abundance. 4th ed. Harper Collins, New York.

Krebs, J. R. 1974. Colonial nesting and social feeding as strategies for exploiting food resources in the great blue heron (*Ardea herodias*). Behaviour 51:99–134.

Krebs, J. R. 1980. Optimal foraging, predation risk and territory defence. Ardea 68:83–90.

Krebs, J. R., and N. B. Davies. 1981. An Introduction to Behavioural Ecology. Blackwell Scientific Publications, London.

Krebs, J. R., and N. B. Davies. 1987. An Introduction to Behavioural Ecology. 2nd ed. Blackwell Scientific Publications, London.

Krebs, J. R., and N. B. Davies (eds.). 1993. Behavioural Ecology: An Evolutionary Approach. 3rd ed. Blackwell Scientific Publications, London.

Krebs, J. R., J. T. Erichsen, M. I. Webber, and E. L. Charnov. 1977. Optimal prey selection in the great tit (*Parus major*). Anim. Behav. 25:30–38.

Krebs, J. R., and A. Kacelnik. 1991. Decision-making. In J. R. Krebs and N. B. Davies (eds.), Behavioural Ecology: An Evolutionary Approach, 105–136. Blackwell Scientific Publications, Oxford.

Kruckenberg, A. R. 1951. Intraspecific variability in the response of certain native plants to serpentine soil. Am. J. Bot. 38:408–419.

Krueger, O. 1997. Population density and intra- and interspecific competition of the African fish eagle *Haliaeetus vocifer* in Kyambura Game Reserve, southwest Uganda. Ibis 139:19–24.

Kruuk, H. 1967. Competition for food between vultures in East Africa. Ardea 55:171–193.

Kruuk, H. 1969. Interactions between populations of spotted hyenas *Crocuta crocuta* (Erxleben) and their prey species. In A. Watson (ed.). Animal Populations in Relation to Their Food Resources, 359–374. Blackwell, Oxford.

Kruuk, H. 1972. The Spotted Hyena: A study of predation and social behaviors. University of Chicago Press, Chicago.

Kuijt, J. 1969. The Biology of Flowering Parasitic Plants. University of California Press, Berkeley.

Kuris, A. M., A. R. Blaustein, and J. J. Alio. 1980. Hosts as islands. Am. Nat. 116:570–586.

Lack, D. 1944. Ecological aspects of species formation in passerine birds. Ibis 86:260–286.

Lack, D. 1947. Darwin's Finches. Cambridge University Press, Cambridge.

Lack, D. 1954. The Natural Regulation of Animal Numbers. Oxford University Press, London.

Lack, D. 1966. Population Studies of Birds. Clarendon Press, Oxford.

Lack, D. 1968. Ecological Adaptations for Breeding in Birds. Methuen, London.

Lacy, R. C. 1994. Managing genetic diversity in captive populations of animals. In M. L. Bowles and C. J. Whelan (eds.), Restoration of Endangered Species, 63–89. Cambridge University Press, Cambridge.

Lacy, R. C., and P. W. Sherman. 1983. Kin recognition by phenotype matching. Am. Nat. 121:489–512.

Lahaye, W. S., R. J. Gutiérrez, and H. R. Akçakaya. 1994. Spotted owl metapopulation dynamics in southern California. J. Anim. Ecol. 63:775–785.

Lahti, S., J. Tast, and H. Uotila. 1976. Pikkujyrsijöiden kannanvaihteluista kilpisjävellä vuosina 1950–1975. Luonnon Tutkija 80:97–107.

Lamb, M. J. 1977. Biology of Ageing. Blackie, Glasgow and London.

Lamb, R. J., and P. J. Pointing. 1972. Sexual morph determination in the aphid, *Acyrthosiphon pisum*. J. Insect Physiol. 18:2029–2042.

Lamont, B. 1983. The Biology of Mistletoes. Academic Press, Sydney, Australia.

Lande, R. 1993. Risks of population extinction from demographic and environmental stochasticity, and random catastrophes. Am. Nat. 142:911–927.

Lande, R., and G. F. Barrowclough. 1987. Effective population size, genetic variation, and their use in population management. In M. E. Soulé (ed.), Viable Populations for Conservation, 87–123. Cambridge University Press, Cambridge.

Langen, T. A. 1996. Skill acquisition and the timing of natal dispersal in the white-throated magpie-jay, *Calocitta formosa*. Anim. Behav. 51:575–588.

Larcher, W. 1980. Physiological Plant Ecology. 2nd ed. Springer-Verlag, New York.

Larsen, H. 1962. Halophilism. In I. C. Gunsalus and R. Y. Stanier (eds.), The Bacteria: A Treatise on Structure and Function, vol. 4, The Physiology of Growth, 297–342. Academic Press, New York.

Latham, M. C. 1975. Nutrition and infection in national development. Science 188:561–565.

Latham, R. E., and R. E. Ricklefs. 1993. Continental comparisons of temperate-zone tree species diversity. In R. E. Ricklefs and D. Schluter (eds.), Species Diversity in Ecological Communities: Historical and Geographical Perspectives, 294–314. University of Chicago Press, Chicago.

Laughlin, R. 1965. Capacity for increase: A useful population statistic. J. Anim. Ecol. 34:77–91.

Law, R., and A. R. Watkinson. 1987. Response-surface analysis of two-species competition: An experiment on *Phleum arenarium* and *Vulpia fasciculata.* J. Ecol. 75:871–886.

Lawler, G. H. 1965. Fluctuations in the success of year classes of whitefish populations with special reference to Lake Erie. J. Fish. Res. Bd. Can. 22:1197–1227.

Lawrence, D. B., R. E. Schoenike, A. Quispel, and G. Bond. 1967. The role of *Dryas drummondii* in vegetation development following ice recession at Glacier Bay, Alaska, with special reference to its nitrogen fixation by root nodules. J. Ecol. 55:793–813.

Laws, R. M. 1962. Some effects of whaling on the southern stocks of baleen whales. In E. D. LeCren and M. W. Holdgate (eds.), The Exploitation of Natural Animal Populations, 137–158. Wiley, New York.

Laws, R. M. 1977. The significance of vertebrates in the Antarctic marine ecosystem. In G. A. Llano (ed.), Adaptations within Antarctic Ecosystems, 411–438. Smithsonian Institution Press, Washington, D.C.

Lawton, J. H. 1989. Food webs. In J. M. Cherrett (ed.), Ecological Concepts: The Contribution of Ecology to An Understanding of the Natural World, 43–78. Blackwell Scientific, Oxford.

Lawton, J. H. 1994. What do species do in ecosystems? Oikos 71:367–374.

Lawton, J. H., H. Cornell, W. Dritschilo, and S. D. Hendrix. 1981. Species as islands: Comments on a paper by Kuris et al. Am. Nat. 117:623–627.

Lawton, J. H., J. R. Heddington, and R. Bonser. 1974. Switching in invertebrate predators. In M. B. Usher and M. H. Williamson (eds.), Ecological Stability, 141–158. Chapman & Hall, London.

Lawton, J. H., T. M. Lewinsohn, and S. G. Compton. 1993. Patterns of diversity for the insect herbivores on bracken. In R. E. Ricklefs and D. Schluter (eds.), Species Diversity in Ecological Communities: Historical and Geographical Perspectives, 178–184. University of Chicago Press, Chicago.

Lawton, J. H., and R. M. May. 1995. Extinction Rates. Oxford University Press, New York.

Lawton, J. H., and D. R. Strong Jr. 1981. Community patterns and competition in folivorous insects. Am. Nat. 118:317–338.

Leak, W. B. 1970. Successional change in northern hardwoods predicted by birth and death simulation. Ecology 51:794–801.

Leamy, L. 1977. Genetic and environmental correlations of morphometric traits in randombred house mice. Evolution 31:357–369.

Lee, R. E. Jr., C. Chen, and D. L. Denlinger. 1987. A rapid cold-hardening process in insects. Science 238:1415–1417.

Lee, R. E. Jr., J. P. Costanzo, P. E. Kaufman, M. R. Lee, and J. A. Wyman. 1994. Ice-nucleating active bacteria reduce the cold-hardiness of the freeze-intolerant Colorado potato beetle (Coleoptera: Chrysomelidae). J. Econ. Entomol. 87:377–381.

Lee, R. E. Jr., M. R. Lee, and J. M Strong-Gunderson. Insect cold-hardiness and ice nucleating active microorganisms including their potential use for biological control. J. Insect Physiol. 39:1–12.

Lees, A. D. 1966. The control of polymorphism in aphids. Adv. Insect Physiol. 3:207–277.

Le Gall, J., and J. R. Postgate. 1973. The physiology of sulphate-reducing bacteria. Adv. Microbial Physiol. 10:81–128.

Legaspi, J. C., and R. J. O'Neil. 1994. Developmental response of nymphs of *Podisus maculiventris* (Heteroptera: Pentatomidae) reared with low numbers of prey. Environ. Entomol. 23:374–380.

Leggett, W. C., and J. E. Carscadden. 1978. Latitudinal variation in reproductive characteristics of American shad (*Alosa sapidissima*): Evidence for population specific life history strategies in fish. J. Fish. Res. Bd. Can. 35:1469–1478.

Lenington, S. 1980. Female choice and polygyny in redwinged blackbirds. Anim. Behav. 28:347–361.

Lenski, R. E. 1984. Food limitation and competition: A field experiment with two *Carabus* species. J. Anim. Ecol. 53: 203–216.

Lenski, R. E., and R. M. May. 1994. The evolution of virulence in parasites and pathogens: Reconciliation between two competing hypotheses. J. Theor. Ecol. 169:253–265.

Leslie, P. H. 1945. On the use of matrices in certain population mathematics. Biometrika, 33:183–212.

Leslie, P. H. 1948. Some further notes on the use of matrices in population analysis. Biometrika 35:213–245.

Lessells, C. M. 1991. The evolution of life histories. In J. R. Krebs and N. B. Davies (eds.), Behavioural Ecology: An Evolutionary Approach, 3rd ed., 32–68. Blackwell Scientific Publications, Oxford.

Lessells, C. M., F. Cooke, and R. R. Rockwell. 1989. Is there a trade-off between egg weight and clutch size in wild lesser snow geese (*Anser c. caerulescens*)? J. Evol. Biol. 2:457–472.

Levin, D. A. 1976. The chemical defenses of plants to pathogens and herbivores. Annu. Rev. Ecol. Syst. 7:121–159.

Levin, D. A. 1984. Inbreeding depression and proximity-related crossing success in *Phlox drummondii.* Evolution 38:116–127.

Levin, S. A. 1978. Population models and community structure in heterogeneous environments. In S. A. Levin (ed.), Studies in Mathematical Biology, Part II, Populations and Communities, 439–479. Studies in Mathematics 16. Mathematical Association of America, Washington, D.C.

Levin, S. A., and R. T. Paine. 1974. Disturbance, patch formation, and community structure. Proc. Natl. Acad. Sci. USA 71:2744–2747.

Levins, R. 1962. Theory of fitness in a heterogeneous environment. 1. The fitness set and adaptive function. Am. Nat. 96:361–373.

Levins, R. 1969. Some demographic and genetic consequences of environmental heterogeneity for biological control. Bull. Entomol. Soc. Am. 15:237–240.

Levins, R. 1970. Extinction. In M. Gertenhaber (ed.), Some Mathematical Problems in Biology, 75–107. American Mathematical Society, Providence, RI.

Lewert, R. M. 1970. Schistosomes. In G. J. Jackson, R. Herman, and I. Singer (eds.), Immunity to Parasitic Animals, vol. 2, 981–1008. Appleton-Century-Crofts, New York.

Lewis, E. G. 1942. On the generation and growth of a population. Sankhya 6:93–96.

Lewis, T. 1973. Thrips: Their Biology, Ecology and Economic Importance. Academic Press, London.

Lewis, W. M. Jr. 1989. The cost of sex. In S. C. Stearns (ed.), The Evolution of Sex and Its Consequences, 33–57. Birkhäuser Verlag, Basel.

Lewontin, R. C. 1958. Studies on heterozygosity and homeostasis, II. Evolution 12:494–503.

Li, H.-L. 1952. Floristic relationships between eastern Asia and eastern North America. Trans. Am. Phil. Soc., New Ser. 42:371–429.

Liao, C. F., and D. R. S. Lean. 1978. Nitrogen transformations within the trophogenic zone of lakes. J. Fish. Res. Bd. Can. 35:1102–1108.

Licht, L. E., and K. P. Grant. 1997. The effects of ultraviolet radiation on the biology of amphibians. Am. Zool. 37: 137–145.

Liebig, J. 1840. Chemistry in Its Application to Agriculture and Physiology. Taylor and Walton, London.

Leith, H. 1973. Primary production: terrestrial ecosystem. Human Ecol. 1:303–332.

Lieth, H. 1975. Primary productivity of the major vegetation units of the world. In H. Lieth and R. H. Whittaker (eds.), Primary Productivity of the Biosphere, 201–215. Springer-Verlag, Berlin.

Light, S. F. 1942–43. The determination of costs in social insects. Quart. Rev. Biol. 17:312–326; 18:46–63.

Lijklema, L. 1994. Nutrient dynamics in shallow lakes: Effects of changes in loading and role of sediment-water interactions. Hydrobiologia 275/276:335–348.

Likens, G. E. 1985. An Ecosystem Approach to Aquatic Ecology: Mirror Lake and Its Environment. Springer-Verlag, New York.

Likens, G. E. 1992. The ecosystem approach: Its use and abuse. In O. Kinne (ed.), Excellence in Ecology, 3–166. Ecology Institute Nordbünte 23, Oldendorf, Germany

Likens, G. E., and F. H. Bormann. 1985. An ecosystem approach. In G. E. Likens (ed.), An Ecosystem Approach to Aquatic Ecology: Mirror Lake and Its Environment, 1–8. Springer-Verlag, New York.

Likens, G. E., and F. H. Bormann. 1995. Biochemistry of forested ecosystems, 2nd. ed. Springer-Verlag, New York.

Likens, C. E., F. H. Bormann, N. M. Johnson, and R. S. Pierce. 1967. The calcium, magnesium, potassium, and sodium budgets for a small forested ecosystem. Ecology 48:772–785.

Likens, G. E., F. H. Bormann, R. S. Pierce, J. S. Eaton, and N. M. Johnson. 1977. Biogeochemistry of a Forested Ecosystem. Springer-Verlag, New York.

Likens, G. E., C. T. Driscoll, and D. C. Busco. 1996. Long-term effects of acid rain: Response and recovery of a forest ecosystem. Science 272:244–245.

Lima, S. L. 1995. Back to the basics of anti-predatory vigilance: The group-size effect. Anim. Behav. 49:11–20.

Lima, S. L., and P. A. Zollner. 1996. Anti-predatory vigilance and the limits to collective detection: Visual and spatial separation between foragers. Behav. Ecol. Sociobiol. 38: 355–363.

Lindeman, R. 1942. The trophic-dynamic aspect of ecology. Ecology 23:399–418.

Lindroth, R. L. 1989. Mammalian herbivore-plant interactions. In W. B. Abrahamson, Plant-Animal Interactions, 163–206. McGraw-Hill, New York.

Lindroth, R. L., and G. O. Batzli. 1986. Inducible plant chemical defences: A cause of vole population cycles? J. Anim. Ecol. 55:431–449.

Lindsay, W. L., and E. C. Moreno. 1960. Phosphate equilibria in soils. Soil Sci. Soc. Am. Proc. 24:177–182.

Linsley, E. G. 1961. Bering Arc relationships of Cerambycidae and their host plants. In J. L. Gressitt (ed.), Pacific Basin Biogeography, 159–178. Bishop Museum Press, Honolulu.

Lippe, E., J. T. De Smidt, and D. C. Glenn-Lewin. 1985. Markov models and succession: A test from a heathland in the Netherlands. J. Ecol. 73:775–791.

List, R. J. 1966. Smithsonian Meteorological Tables. 6th rev. ed. Smithsonian Misc. Coll. 114:1–527.

Litvaitis, J. A., and R. Villafuerte. 1996. Factors affecting the persistence of New England cottontail metapopulations: The role of habitat management. Wildl. Soc. Bull. 24: 686–693.

Lloyd, D. G. 1979. Some reproductive factors affecting the selection of self-fertilization in plants. Am. Nat. 113:67–69.

Lloyd, D. G. 1980. Demographic factors and mating patterns in angiosperms. In O. T. Solbrig (ed.), Demography and Evolution of Plant Populations, 67–88. Blackwell, Oxford.

Lloyd, J. E. 1975. Aggressive mimicry in *Photuris* fireflies: Signal repertoires by femmes fatales. Science 187:452–453.

Lloyd, J. E. 1980. Firefly mate-rivals mimic their predators and vice versa. Nature 290:498–500.

Loach, K. 1967. Shade tolerance in tree seedlings. 1. Leaf photosynthesis and respiration in plants raised under artificial shade. New Phytol. 66:607–621.

Lodge, D. J., W. H. McDowell, and C. P. McSwiney. 1994. The importance of nutrient pulses in tropical forests. Trends Ecol. Evol. 9:384–387.

Lodge, D. M., J. W. Barko, D. Strayer, J. M. Melack, G. G. Mittelbach, R. W. Howarth, B. Menge, and J. E. Titus. 1988. Spatial heterogeneity and habitat interactions in lake communities. In S. R. Carpenter (ed.), Complex interactions in lake communities, 181–208. Springer-Verlag, New York.

Loehle, C. 1990. Proper statistical treatment of species-area data. Oikos 57:143–145.

Lofts, B. 1970. Animal Photoperiodism. Arnold, London.

Loik, M. E., and Nobel, P. S. 1993. Freezing tolerance and water relations of *Opuntia fragilis* from Canada and the United States. Ecology 74:1722–1732.

Lomolno, M. V. 1984. Mammalian island biogeography: effects of area, isolation, and vagility. Oecologia 61:376–382.

Losos, J. B. 1992. The evolution of convergent structure in Caribbean *Anolis* communities. Syst. Biol. 41:403–420.

Losos, J. B. 1996. Phylogenetic perspectives on community ecology. Ecology 77:1344–1354.

Lotka, A. J. 1907. Relation between birth rates and death rates. Science 26:21–22.

Lotka, A. J. 1922. The stability of the normal age distribution. Proc. Natl. Acad. Sci. USA 8:339–345.

Lotka, A. J. 1925. Elements of Physical Biology. Williams and Wilkins, Baltimore.

Lotka, A. J. 1932. The growth of mixed populations: Two species competing for a common food supply. J. Wash. Acad. Sci. 22:461–469.

Loucks, O. L. 1962. Ordinating forest communities by means of environmental scalars and phytosociological indices. Ecol. Monogr. 32:137–166.

Loucks, O. L. 1977. Emergence of research on agroecosystems. Annu. Rev. Ecol. Syst. 8:173–192.

Louda, S. M., and J. E. Rodman. 1996. Insect herbivory as a major factor in the shade distribution of a native crucifer (*Cardamine cordifolia,* A. Gray, bittercress). J. Ecol. 84:229–238.

Lubchenco, J. 1978. Plant species diversity in a marine intertidal community: Importance of herbivore food preference and algal competitive abilities. Am. Nat. 112:23–39.

Lubjuhn, T., E. Curio, S. Muth, J. Brun, and J. T. Epplen. 1993. Influence of extra-pair paternity on parental care in great tits (*Parus major*). In S. D. J. Pena, R. Chakraborty, J. T. Epplen, and A. J. Jeffreys (eds.), DNA Fingerprinting: State of the Science, 370–385. Birkhäuser Verlag, Basel.

Luckinbill, L. S. 1978. *r* and *K* selection in experimental populations of *Escherichia coli.* Science 202:1201–1203.

Luckinbill, L. S. 1979. Selection and the *r*/*K* continuum in experimental populations of protozoa. Am. Nat. 113: 427–437.

Luckinbill, L. S. 1984. An experimental analysis of a life history theory. Ecology 65:1170–1184.

Lugo, A. E., and S. Brown. 1991. Comparing tropical and temperate forests. In J. Cole, G. Lovett, and S. Findley (eds.), Comparative Analysis of Ecosystems: patterns, mechanisms, theories, 319–330. Springer-Verlag, New York.

Luse, R. A. 1970. The phosphorus cycle in a tropical rain forest. In H. T. Odum and R. F. Pigeon (eds.), A Tropical Rain Forest, H161–H166. U.S. Atomic Energy Commission, Washington, D.C.

Lutterschmidt, W. I., and H. K. Reinert 1990. The effect of ingested transmitters upon temperature preference of the northern water snake, *Nerodia s. sipedon.* Herpetologica 46:39–42.

Lyman, C. 1982. Hibernation and Torpor in Mammals and Birds. Academic Press, New York.

Lynch, M., and J. Shapiro. 1981. Predation, enrichment and phytoplankton community structure. Limnol. Oceanogr. 26:86–102.

Lythgoe, J. N. 1979. The Ecology of Vision. Clarendon Press, New York.

MacArthur, R. A., and A. P. Dyck. 1990. Aquatic thermoregulation of captive and free-ranging beavers (*Castor canadensis*). Can. J. Zool. 68:2409–2416.

MacArthur, R. H. 1955. Fluctuations of animal populations and a measure of community stability. Ecology 36:533–536.

MacArthur, R. H. 1957. On the relative abundance of bird species. Proc. Natl. Acad. Sci. USA 43:293–295.

MacArthur, R. H. 1960. On the relative abundance of species. Am. Nat. 94:25–36.

MacArthur, R. H. 1969. Patterns of communities in the Tropics. Biol. J. Linn. Soc. 1:19–30.

MacArthur, R. H., J. M. Diamond, and J. R. Karr. 1972. Density compensation in island faunas. Ecology 53:330–342.

MacArthur, R. H., and R. Levins. 1967. The limiting similarity, convergence, and divergence of coexisting species. Am. Nat. 101:377–385.

MacArthur, R. H., and J. MacArthur. 1961. On bird species diversity. Ecology 42:594–598.

MacArthur, R. H., and E. R. Pianka. 1966. On optimal use of patchy environment. Am. Nat. 100:603–609.

MacArthur, R. H., H. Recher, and M. Cody. 1966. On the relation between habitat selection and species diversity. Am. Nat. 100:319–332.

MacArthur, R. H., and E. O. Wilson. 1963. An equilibrium theory of insular zoogeography. Evolution 17:373–387.

MacArthur, R. H., and E. O. Wilson. 1967. The Theory of Island Biogeography. Princeton University Press, Princeton, NJ.

Mace, G. M. 1994. Classifying threatened species: Means and ends. Phil. Trans. R. Soc. Lond. B 344:91–97.

Macevicz, S., and G. Oster. 1976. Modeling social insect populations. II. Optimal reproductive strategies in annual eusocial insect colonies. Behav. Ecol. Sociobiol. 1:265–282.

Machin, K. E., and H. W. Lissmann. 1960. The mode of operation of the electric receptors in *Gymnarchus niloticus.* J. Exp. Biol. 37:801–811.

MacKinnon, J., and K. MacKinnon. 1986a. Review of the protected areas system in the Afrotropical realm. International Union for the Conservation of Nature and Natural Resources, Gland, Switzerland.

MacKinnon, J., and K. MacKinnon. 1986b. Review of the protected areas system in the Indo-Malayan realm. International Union for the Conservation of Nature and Natural Resources, Gland, Switzerland.

MacLulich, D. A. 1937. Fluctuations in the numbers of the varying hare (*Lepus americanus*). University Toronto Studies, Biol. Ser. no. 43.

MacMahon, J. A. 1997. Ecological restoration. In G. K. Meffe and C. R. Carroll (eds.), Principles of Conservation Biology, 2nd ed., 479–511. Sinauer Associates, Sunderland, MA.

MacMillen, R. E., and A. K. Lee. 1967. Australian desert mice: Independence of exogenous water. Science 158:383–385.

Maddock, L. 1979. The "migration" and grazing succession. In A. R. E. Sinclair and M. Norton-Griffiths (eds.), Serengeti: Dynamics of an Ecosystem, 104–129. University of Chicago Press, Chicago.

Maguire, B. 1963. The passive dispersal of small aquatic organisms and their colonization of isolated bodies of water. Ecol. Monogr. 33:161–185.

Magurran, A. E. 1988. Ecological Diversity and Its Measurement. Princeton University Press, Princeton, NJ.

Maher, W. J. 1970. The pomarine jaeger as a brown lemming predator in northern Alaska. Wilson Bull. 82:130–157.

Mangel, M., and C. W. Clark. 1988. Dynamic Modeling in Behavioral Ecology. Princeton University Press, Princeton, NJ.

Mani, G. S., and M. E. N. Majerus. 1993. Peppered moth revisited: Analysis of recent decreases in melanic frequency and prediction for the future. Biol. J. Linn. Soc. 48:157–165.

Mann, K. H. 1973. Seaweeds: Their productivity and strategy for growth. Science 182:975–981.

Mares, M. A. 1976. Convergent evolution of desert rodents: Multivariate analysis and zoogeographic implications. Paleobiology 2:39–63.

Margalef, D. R. 1958. Information theory in ecology. General Systems 3:36–71.

Margalef, R. 1963. On certain unifying principles in ecology. Am. Nat. 92:357–374.

Margalef, R. 1968. Perspectives in Ecological Theory. University of Chicago Press, Chicago.

Marks, H. L. 1978. Long term selection for four-week body weight in Japanese quail under different nutritional environments. Theor. Appl. Genet. 52:105–111.

Marquis, R. J. 1992. A bite is a bite is a bite? Constraints on response to folivory in *Piper arielianum* (Piperaceae). Ecology 73:143–152.

Marquis, R. J. 1996. Plant architecture, sectoriality and plant tolerance to herbivores. Vegetatio 127:85–97.

Marshall, A. J., and H. S. De S. Disney. 1957. Experimental induction of the breeding seasons in a xerophilous bird. Nature 180:647.

Marshall, D. R., and S. K. Jain. 1969. Interference in pure and mixed populations of *Avena fatua* and *A. barbata.* J. Ecol. 57:251–270.

Martin, J. H., and R. M. Gordon. 1988. Northeast Pacific iron distribution in relation to phytoplankton productivity. Deep Sea Res. 35:177–196.

Martin, J. H., R. M. Gordon, S. Fitzwater, and W. W. Broenkow. 1989. VERTEX: Phytoplankton studies in the Gulf of Alaska. Deep Sea Res. 36:649–680.

Martin, M. P. L. D., and R. J. Field. 1984. The nature of competition between perennial ryegrass and white clover. Grass For. Sci. 39:247–253.

Martin, T. E. 1988. On the advantage of being different: Nest predation and the coexistence of bird species. Proc. Natl. Acad. Sci. USA 85:2196–2199.

Martinez, N. D. 1991a. Artifacts or attributes: Effects of resolution on the Little Rock Lake food web. Ecol. Monogr. 61:367–392.

Martinez, N. D. 1991b. Constant connectance in community food webs. Am. Nat. 139:1208–1218.

Martinez, N. D. 1993. Effects of resolution on food web structure. Oikos 66:413–412.

Marzusch, K. 1952. Untersuchungen über di Temperaturabhängigkeit von lebensprozessen bei insekten unter besonderer Berücksichtigung winter-schlantender Kartoffelkäfer. Zeitschrift für Vergleicherde Physiologie 34:75–92.

Mason, J. R., and R. F. Reidinger. 1981. Effects of social facilitation and observational learning on feeding behavior of the red-winged blackbird (*Agelaius phoeniceus*). Auk 98: 778–784.

Massey, A. B. 1925. Antagonism of the walnuts (*Juglans nigra* L., and *J. cinerea* L.) in certain plant associations. Phytopathology 15:773–784.

Mather, M., and B. D. Roitberg. 1987. A sheep in wolf's clothing: Tephritid flies mimic spider predators. Science 236:308–310.

Matson, P. A., and P. M. Vitousek. 1990. Exosystem approach to a global nitrous oxide budget. BioScience 40:667–672.

Mattingly, P. F. 1969. The Biology of Mosquito-Borne Disease. American Elsevier, New York.

Maurer, B. A., and M-A. Villard. 1994. Population Density. National Geographic Research and Exploration. 10: 306–317.

May, J. M. 1961. Studies in Disease Ecology. Hafner, New York.

May, M. 1991. Aerial defense tactics of flying insects. Am. Sci. 79:316–328.

May, R. M. 1972a. Limit cycles in predator-prey communities. Science 177:900–902.

May, R. M. 1972b. Will a large complex system be stable? Nature 238:413–414.

May, R. M. 1973a. On relationships among various types of population models. Am. Nat. 107:46–57.

May, R. M. 1973b. Qualitative stability in model ecosystems. Ecology 54:638–641.

May, R. M. 1973c. Stability and Complexity in Model Ecosystems. Princeton University Press, Princeton, NJ.

May, R. M. 1974. Biological populations with nonoverlapping generations: Stable points, stable cycles, and chaos. Science 186:645–647.

May, R. M. 1975a. Patterns of species abundance and diversity. In M. L. Cody and J. M. Diamond (eds.), Ecology and Evolution of Communities, 81–120. Harvard University Press, Cambridge, MA.

May, R. M. 1975b. Stability and Complexity in Model Ecosystems. 2nd ed. Princeton University Press, Princeton, NJ.

May, R. M. 1976. Simple mathematical models with very complicated dynamics. Nature 261:459–467.

May, R. M. 1981. Theoretical Ecology. 2nd ed. Blackwell Scientific Publishers, Oxford.

May, R. M. 1987. Chaos and the dynamics of biological populations. Proc. R. Soc. Lond. A 413:27–44.

May, R. M., and R. M. Anderson. 1979. Population biology of infectious diseases: Part II. Nature 280:455–461.

May, R. M., J. R. Beddington, C. W. Clark, S. J. Holt, and R. M. Laws. 1979. Management of multispecies fisheries. Science 205:267–277.

May, R. M., and M. A. Nowak. 1994. Superinfection, metapopulation dynamics, and the evolution of diversity. J. Theor. Biol. 170:95–114.

Maybeck, M. 1982. Carbon, nitrogen, and phosphorus transport by world rivers. Am. J. Sci. 282:401–450.

Maynard Smith, J. 1964. Group selection and kin selection. Nature 201:1145–1147.

Maynard Smith, J. 1982. Evolution and the theory of games. Cambridge University Press, Cambridge.

Maynard Smith, J. 1991. Theories of sexual selection. Trends Ecol. Evol. 6:146–151.

Mayr, E., E. G. Linsley, and R. L. Usinger. 1953. Methods and Principles of Systematic Zoology. McGraw-Hill, New York.

McAuliffe, J. R. 1988. Markovian dynamics of simple and complex desert plant communities. Am. Nat. 131: 459–490.

McCauley, S., and J. Kaliff. 1981. Empirical relationships between phytoplankton and zooplankton biomass in lakes. Can. J. Fish. Aquat. Sci. 38:458–463.

McCleave, J. D., et al. (eds.). 1984. Mechanisms of Migration in Fishes. Plenum, New York.

McClelland, W. J. 1965. The production of cercariae by *S. mansoni* and *S. haematobium* and methods for estimating the numbers of cercariae in suspension. Bull. World Health Org. 33:270–275.

McClure, P. A. 1981. Sex-biased litter reduction in food restricted wood rats (*Neotoma floridana*). Science 211: 1058–1060.

McClusky, D. S. 1981. The Estuarine Ecosystem. Wiley, New York.

McCormick, J. 1970. The Pine Barrens: A Preliminary Ecological Inventory. New Jersey State Museum, Trenton.

McCullough, D. R. 1996. Metapopulations and wildlife conservation. Island Press, Washington, D.C.

McElroy, M. B., and S. C. Wofsy. 1986. Tropical forests: Interactions with the atmosphere. In G. T. Prance (ed.), Tropical Rain Forests and the World Atmosphere, 33–36. Westview Press, Boulder, CO.

McFalls, J. A. Jr. 1991. Population: A lively introduction. Pop. Bull. 46:1–43.

McGinnis, S. M., and L. L. Dickson. 1967. Thermoregulation in the desert iguana *Dipsosaurus dorsalis*. Science 156:1757–1759.

McIntosh, R. P. 1967. The continuum concept of vegetation. Bot. Rev. 33:130–187.

McIntosh, R. P. 1974. Plant ecology 1947–1972. Ann. Mo. Bot. Gard. 61:132–165.

McIntosh, R. P. 1985. The Background of Ecology: Concept and Theory. Cambridge University Press, Cambridge and New York.

McLaughlin, J. F., and J. Roughgarden. 1991a. Pattern and stability in predator-prey communities: How diffusion in spatially variable environments affects the Lotka-Volterra model. Theor. Pop. Biol. 40:148–172.

McLaughlin, J. F., and J. Roughgarden. 1991b. Predation across spatial scales in heterogeneous environments. Theor. Pop. Biol. 41:277–299.

McLaughlin, J. F., and J. Roughgarden. 1993. Species interactions in space. In R. E. Ricklefs and D. Schluter (eds.), Species Diversity in Ecological Communities: Historical and Geographical Perspectives, 89–98. University of Chicago Press, Chicago.

McLaughlin, S. B., M. G. Tjoelker, and W. K. Roy. 1993. Acid deposition alters red spruce physiology: Laboratory studies support field observations. Can. J. For. Res. 23:380–386.

McMaster, G. S., W. M. Jow, and J. Kummerow. 1982. Response of *Adenostoma fasciculatum* and *Ceanothus greggii* chaparral to nutrient additions. J. Ecol. 70:745–756.

McMillan, C. 1956. Edaphic restriction of *Cupressus* and *Pinus* in the coast ranges of central California. Ecol. Monogr. 26:177–212.

McNab, B. K. 1966. An analysis of the body temperatures of birds. Condor 68:47–55.

McNair, J. N. 1982. Optimal giving up times and the marginal value theorem. Am. Nat. 119:511–529.

McNamara, J. M., and A. I. Houston. 1987. Starvation and predation as factors limiting population size. Ecology 68:1515–1519.

McNamara, J. M., and A. I. Houston. 1992. Risk-sensitive foraging: A review of the theory. Bull. Math. Biol. 54: 355–378.

McNaughton, S. J. 1973. Comparative photosynthesis of Quebec and California ecotypes of *Typha latifolia*. Ecology 54:1260–1270.

McNaughton, S. J. 1976. Serengeti migratory wildebeest: Facilitation of energy flow by grazing. Science 191:92–94.

McNaughton, S. J. 1979. Grassland-herbivore dynamics. In A. R. E. Sinclair and M. Norton-Griffiths (eds.), Serengeti: Dynamics of an Ecosystem, 46–81. University of Chicago Press, Chicago.

McNaughton, S. J. 1983. Compensatory plant growth as a response to herbivory. Oikos 40:329–336.

McNaughton, S. J. 1985. Ecology of a grazing ecosystem: The Serengeti. Ecol. Monogr. 55:259–294.

McNaughton, S. J, M. Oessterheld, D. A. Frank, and K. J. Williams. 1989. Ecosystem-level patterns of primary productivity and herbivory in terrestrial habitats. Nature 341:142–144.

McPhee, J. 1968. The Pine Barrens. Farrar, Straus & Giroux, New York.

McVay, S. 1966. The last of the great whales. Sci. Amer. 215:13–21.

McQueen, D. J., M. R. S. Johannes, J. R. Post, T. J. Stewart, and D. R. S. Lean. 1989. Bottom-up and top-town impacts on freshwater pelagic community structure. Ecol. Monogr. 59:289–309.

Mech, L. D. 1966. The Wolves of Isle Royale. Fauna Series no. 7. U.S. National Park Service, Washington, D.C.

Mech, L. D. 1970. The Wolf: The Ecology and Behavior of an Endangered Species. Natural History Press, New York.

Meffe, G. K., and C. R. Carroll. 1997a. Conservation reserves and heterogeneous landscapes. In G. K. Meffe and C. R. Carroll, Principles of Conservation Biology, 2nd ed., 305–343. Sinauer Associates, Sunderland, MA.

Meffe, G. K., and C. R. Carroll. 1997b. Principles of Conservation Biology. 2nd ed. Sinauer Associates, Sunderland, MA.

Melillo, J. M., A. D. McGuire, D. W. Kicklighter, B. Moore III, C. J. Vorosmarty, and A. L. Schloss. 1993. Global climate change and terrestrial net primary production. Nature 363:234–240.

Melnikov, V. V., and A. V. Bobkov. 1994. Bowhead whale migration in the Chuckchee Sea. Russian J. Mar. Biol. 19: 180–185.

Menge, B. A. 1992. Community regulation: Under what conditions are bottom-up factors important on rocky shores? Ecology 73:755–765.

Meredith, D. H. 1977. Interspecific agonism in two parapatric species of chipmunks (*Eutamias*). Ecology 58:423–430.

Merriam, C. H. 1894. Laws of temperature control of the geographic distribution of terrestrial animals and plants. Nat. Geogr. 6:229–238.

Merriam, G., and J. Wegner. 1992. Local extinctions, habitat fragmentation, and ecotones. In A. J. Hansen and F. di Castri (eds.), Landscape Boundaries, 150–169. Springer-Verlag, New York.

Meyer, J. L. 1990. A blackwater perspective on riverine ecosystems. BioScience 40:643–651.

Milbank, J. W., and K. A. Kershaw. 1969. Nitrogen metabolism in lichens, 1:Nitrogen fixation in cephalodia of *Peltigera aphthosa.* New Phytol. 68:721–729.

Mildvan, A. S., and B. L. Strehler. 1960. A critique of theories of mortality. In B. L. Strehler (ed.), The Biology of Aging, 216–235. American Institute of Biological Sciences, Washington, D.C.

Miles, D. B., and A. E. Dunham. 1993. Historical perspectives in ecology and evolutionary biology: The use of phylogenetic comparative analysis. Annu. Rev. Ecol. Syst. 24: 587–619.

Milinski, M. 1979. An evolutionarily stable feeding strategy in sticklebacks. Z. Tierpsychol. 51:36–40.

Milinski, M., and R. Heller. 1978. Influence of a predator on the optimal foraging behaviour of sticklebacks (*Gasterosteus aculeatus*). Nature 275:642–644.

Milinski, M., and G. A. Parker. 1991. Competition for resources. In J. R. Krebs and N. B. Davies (eds.), Behavioural Ecology: An Evolutionary Approach, 3rd ed., 137–168. Blackwell Scientific Publications, Oxford.

Milkman, R. 1982. Perspectives on Evolution. Sinauer Associates, Sunderland, MA.

Miller, D. A. 1996. Allelopathy in forage crop systems. Agron. J. 88:854–859.

Miller, G. L., and B. W. Carroll. 1990. Modeling vertebrate dispersal distances: alternatives to the geometric distribution. Ecology 70:977–986.

Miller, L. A., and J. Olesen. 1979. Avoidance behavior in green lacewings. I. Behavior of free flying green lacewings to hunting bats and ultrasound. J. Comp. Physiol. 131: 113–120.

Miller, T. E., and J. Travis. 1996. The evolutionary role of indirect effects in communities. Ecology 77:1329–1335.

Milner, C., and R. E. Hughes. 1968. Methods for the Measurement of the Primary Production of Grassland. Blackwell, Oxford.

Minderman, G. 1968. Addition, decomposition and accumulation of organic matter in forests. J. Ecol. 56:355–362.

Minshall, G. W. 1978. Autotrophy in stream ecosystems. BioScience 28:767–771.

Mitchell, W. A., and J. S. Brown. 1990. Density-dependent harvest rates by optimal foragers. Oikos 57:180–190.

Mittler, R., and E. Lam. 1996. Sacrifice in the face of foes: Pathogen-induced programmed cell death in plants. Trends Microbiol. 4:10–15.

Mode, C. J. 1958. A mathematical model for the coevolution of obligate parasites and their hosts. Evolution 12:158–165.

Mohr, C. O. 1943. Cattle droppings as ecological units. Ecol. Monogr. 13:275–298.

Moiseff, A., B. S. Pollack, and R. R. Hoy. 1978. Steering responses of flying crickets to sound and ultrasound: Mate attraction and predator avoidance. Proc. Natl. Acad. Sci. USA Biological Sciences 75:4052–4056.

Molina, R. 1994. The role of mycorrhizal symbiosis in the health of giant redwood and other forest ecosystems. Proc. Symp. Giant Sequoias. 151:78–81.

Møller, A. P. 1994. Sexual relation and the Barn Swallow. Oxford University Press. Oxford.

Monk, C. D., and F. C. Gabrielson Jr. 1985. Effects of shade, litter and root competition on old-field vegetation in South Carolina. Bull. Torrey Bot. Club 112:383–392.

Monod, J. 1950. La technique de culture continue: Théorie et applications. Ann. Inst. Pasteur 79:390–410.

Mooney, H. A. (ed.). 1977. Convergent Evolution in Chile and California. Dowden, Hutchinson & Ross, Stroudsburg, PA.

Mooney, H. A. 1988. Lessons from Mediterranean-climate regions. In E. O. Wilson and F. M. Peter (eds.), Biodiversity, 157–165. National Academy Press, Washington, D.C.

Mooney, H. A., and W. D. Billings. 1961. Comparative physiological ecology of arctic and alpine populations of *Oxyria digyna.* Ecol. Monogr. 31:1–29.

Mooney, H. A., and E. L. Dunn. 1970. Convergent evolution of Mediterranean-climate evergreen schlerophyll shrubs. Evolution 24:292–303.

Moran, N., and W. D. Hamilton. 1980. Low nutritive quality as defense against herbivores. J. Theor. Biol. 86:247–254.

Moreau, R. E. 1944. Clutch-size: a comparative study, with special reference to African birds. Ibis 86:286–347.

Morgan, R. A., and J. S. Brown. 1996. Using giving up densities to detect search images. Am. Nat. 148:1059–1074.

Morgan, R. A., J. S. Brown, and J. M. Thorson. 1997. The effect of spatial scale on the functional response of fox squirrels. Ecology 78:1087–1077.

Morin, P. J. 1981. Predatory salamanders reverse the outcome of competition among three species of anuran tadpoles. Science 212:1284–1286.

Morris, R. F., W. F. Chesire, C. A. Miller, and D. G. Mott. 1958. Numerical responses of avian and mammalian predators during a gradation of the spruce budworm. Ecology 39:487–494.

Morse, D. 1980. Behavioral mechanisms in ecology. Harvard University Press, Cambridge, Massachusetts.

Mortimer, C. H. 1941–1942. The exchange of dissolved substances between mud and water in lakes. J. Ecol. 29:280–329; 30:147–201.

Moyle, P. B., and J. J. Cech Jr. 1982. Fishes: An Introduction to Ichthyology. 2nd ed. Prentice Hall, Englewood Cliffs, NJ.

Mrosovsky, N. 1976. Lipid programmes and life strategies in hibernators. Am. Zool. 16:685–697.

Muellar, K. W., G. D. Dennis, D. B. Eggleston, and R. I. Wicklund. 1994. Size-specific social interactions and foraging styles in a shallow water population of mutton snapper, *Lutjanus analis* (Pisces: Lutjanidae), in the central Bahamas. Environ. Biol. Fishes 40:175–188.

Mueller, L. D. 1988. Density-dependent population growth and natural selection in food-limited environments: The *Drosophila* model. Am. Nat. 132:786–809.

Mueller, L. D., and F. J. Ayala. 1981. Tradeoff between *r* selection and *K* selection in *Drosophila* populations. Proc. Natl. Acad. Sci. USA 78:1303–1305.

Mueller-Dombois, D., and H. Ellenberg. 1974. Aims and Methods of Vegetation Ecology. Wiley, New York.

Mulcahy, G. B., and D. L. Mulcahy. 1985. Pollen-pistil interactions. In G. B. Mulcahy, D. L. Mulcahy, and E. Ottaviano (eds.), Biotechnology and Ecology of Pollen: Proceedings of the International Conference on the Biotechnology and Ecology of Pollen, 173–178. Springer-Verlag, New York.

Mulholland, P. J., J. D Newbold, J. W. Elwood, L. A. Ferren, and J. R. Webster. 1985. Phosphorus spiralling in a woodland stream: Seasonal variations. Ecology 66:1012–1023.

Mulholland, P. J., and A. D. Rosemond. 1992. Periphyton response to longitudinal nutrient depletion in a woodland stream: Evidence of upstream-downstream linkage. J. N. Am. Benthol. Soc. 11:405–419.

Muller, C. H. 1966. The role of chemical inhibition (allelopathy) in vegetational composition. Bull. Torrey Bot. Club 93:332–351.

Muller, C. H. 1970. Phytotoxins as plant habitat variables. Rec. Adv. Phytochem. 3:106–121.

Muller, C. H., R. B. Hanawalt, and J. K. McPherson. 1968. Allelopathic control of herb growth in the fire cycle of California chaparral. Bull. Torrey Bot. Club 95:225–231.

Muller, C. H., W. H. Muller, and B. L. Haines. 1964. Volatile growth inhibitors produced by aromatic shrubs. Science 143:471–473.

Muller, H. J. 1964. The relation of recombination to mutational advance. Mutat. Res. 1:2–9.

Muller, W. H. 1965. Volatile materials produced by *Salvia leucophylla:* Effects on seedling growth and soil bacteria. Bot. Gaz. 126:195–200.

Mumme, R. L. 1992. Do helpers increase reproductive success? An experimental analysis in the Florida scrub jay. Behav. Ecol. Sociobiol. 31:319–328.

Murdoch, W. W. 1966. Population stability and life history phenomena. Amer. Nat. 100:5–11.

Murdoch, W. W. 1969. Switching in general predators: Experiments on predator specificity and stability of prey populations. Ecol. Monogr. 39:335–354.

Murdoch, W. W. 1994. Population regulation in theory and practice. Ecology 72:271–287.

Murdoch, W. W., J. Chesson, and P. Chesson. 1985. Biological control in theory and practice. Am. Nat. 125:344–366.

Murdoch, W. W., and A. Oaten. 1975. Predation and population stability. Adv. Ecol. Res. 9:2–131.

Murdoch, W. W., J. D. Reeve, C. B. Huffaker, and C. E. Kennett. 1984. Biological control of scale insects and ecological theory. Am. Nat. 123:371–392.

Murie, O. 1944. The Wolves of Mt. McKinley. Fauna Series no. 5. U.S. Dept. of the Interior, National Park Service, Washington, D.C.

Murphy, G. I. 1968. Patterns in life history and environment. Am. Nat. 102:390–404.

Murton, R. K. 1967. The significance of endocrine stress in population control. Ibis 109:622–623.

Murton, R. K. 1970. Why do some bird species feed in flocks? Ibis 113:534–535.

Murton, R. K., and N. J. Westwood. 1977. Avian Breeding Cycles. Clarendon Press, Oxford.

Mutch, W. R. 1970. Wildland fires and ecosystems: A hypothesis. Ecology 51:1046–1051.

Myers, K. 1970. The rabbit in Australia. In P. J. den Boer and G. R. Gradwell (eds.), Dynamics of Populations, 478–506. Centre Agric. Publ. Documentation, Wageningen, The Netherlands.

Myers, K., and W. E. Poole. 1963. A study of the biology of the wild rabbit, *Orytolagus cuniculus* (L.), in confined populations. IV. The effects of rabbit grazing on sown pastures. J. Ecol. 51:435–451.

Myers, N. 1997. Global biodiversity II: Losses and threats. In G. K. Meffe and C. R. Carroll (eds.), Principles of Conservation Biology, 2nd ed., 123–158. Sinauer Associates, Sunderland, MA.

Myers, R. A., J. A. Hutchings, and N. J. Barrowman. 1997. Why do fish stocks collapse? The example of cod in Atlantic Canada. Ecol. Appl. 7:91–106.

Nagy, K. A., D. K. Odell, and R. S. Seymour. 1972. Temperature regulation by the inflorescence of *Philodendron.* Science 178:1196–1197.

Nakano, S. 1995. Individual differences in resource use, growth and emigration under the influence of a dominance hierarchy in fluvial red-spotted masu salmon in a natural habitat. J. Anim. Ecol. 64:75–84.

Nakano, S. and T. Furukawa-Tanaka. 1994. Intra-and interspecific dominance hierarchies and variation in foraging tactics of two species of stream-dwelling chars. Ecol. Res. 9:9–20.

Nature Conservancy (UK). 1984. Nature Conservation in Great Britain. Nature Conservancy Council, Shrewsbury.

Nault, L. R., and P. L. Phelan. 1984. Alarm pheromones and sociality in pre-social insects. In W. J. Bell and R. T. Cardé (eds.), Chemical Ecology of Insects, 237–256. Sinauer Associates, Sunderland, MA.

Nee, S., and R. M. May. 1992. Dynamics of metapopulations: Habitat destruction and competitive coexistence. J. Anim. Ecol. 61:37–40.

Nee, S., R. M. May, and M. P. Hassell. 1997. Two-species metapopulation models. In I. L. Hanski and M. E. Gilpin (eds.), Metapopulation Biology: Ecology, Genetics, and Evolution, 123–147. Academic Press, San Diego.

Neill, W. E. 1974. The community matrix and the interdependence of the competition coefficients. Am. Nat. 108: 399–408.

Neilson, R. P. 1987. On the interface between current ecological studies and the paleobotany of pinyon-juniper woodlands. In R. L. Everett (ed.), Proceedings, Pinyon-Juniper Conference, 93–98. USDA Forest Service, General Technical Report INT-215, Ogden, Utah.

Neilson, R. P., G. A. King, R. L. Develice, and J. M. Lenihan. 1992. Regional and local vegetation patterns: The responses of vegetation diversity to subcontinental air masses. In A. J. Hansen and F. di Castri (eds.), Landscape Boundaries, 130–149. Springer-Verlag, New York.

Neilson, R. P., and L. H. Wullstein. 1986. Biogeography of two southwest American oaks in relation to atmospheric dynamics. J. Biogeogr. 10:275–297.

Nero, R. W. 1984. Redwings. Smithsonian Institution Press, Washington, D.C.

Nettancourt, N. de. 1977. Incompatibility in Angiosperms: Monographs on Theoretical and Applied Genetics. Springer-Verlag, New York.

Newbold, J. D., J. W. Elwood, R. V. O'Neill, and A. L. Sheldon. 1983. Phosphorus dynamics in a woodland stream ecosystem: A study of nutrient spiralling. Ecology 64: 1249–1265.

Newbold, J. D., R. V. O'Neill, J. W. Elwood, and W. Van Winkel. 1982. Nutrient spiralling in streams: Implications for nutrient limitation and invertebrate activity. Am. Nat. 120:628–652.

Newbould, P. J. 1967. Methods for Estimating the Primary Production of Forests. Blackwell, Oxford.

Newman, J. A., and T. Caraco. 1987. Foraging, predation hazard, and patch use in grey squirrels. Anim. Behav. 35:1804–1813.

Nicholson, A. J. 1958. The self-adjustment of populations to change. Cold Spring Harbor Symp. Quant. Biol. 22: 153–173.

Nicholson, A. J., and V. A. Bailey. 1935. The balance of animal populations. Proc. Zool. Soc. Lond. 3:551–598.

Nicol, J. A. C. 1967. The Biology of Marine Animals. 2nd ed. Wiley, New York.

Niemelä, J., Y. Haila, E. Halme, T. Lahti, T. Pajunen, and P. Punttila. 1988. The distribution of carabid beetles in fragments of old coniferous taiga and adjacent managed forest. Ann. Zool. Fenn. 25:107–119.

Nieminen, M. 1996. Migration of moth species in a network of small islands. Oecologia 108:643–651.

Niewiarowski, P. H., and W. Roosenburg. 1993. Reciprocal transplant reveals sources of variation in growth rates of the lizard *Sceloporus undulatus.* Ecology 74:1992–2002.

Nisbet, R. M., and W. S. C. Gurney. 1982. Modelling Fluctuating Populations. Wiley, New York.

Nitecki, M. H. (ed.). 1983. Coevolution. University of Chicago Press, Chicago.

Nixon, S. W. 1980. Between coastal marshes and coastal waters—a review of twenty years of speculation and research on the role of salt marshes in estuarine productivity and water chemistry. In P. Hamilton and K. B. MacDonald (eds.), Estuarine and Wetland Processes, 437–525. Plenum Press, New York.

Nobel, P. S. 1988. Environmental biology of Agaves and Cacti. Cambridge University Press, Cambridge.

Noble, M. G., D. B. Lawrence, and G. P. Streveler. 1984. *Sphagnum* invasion beneath an evergreen forest canopy in southeastern Alaska. Bryologist 87:119–127.

Noordwijk, A. J. van. 1984. Quantitative genetics in natural populations of birds illustrated with examples from the great tit, *Parus major.* In K. Wohrmann and V. Loeschcke (eds.), Population Biology and Evolution, 67–79. Springer-Verlag, New York.

Nordskog, A. W. 1977. Success and failure of quantitative genetic theory in poultry. In E. Pollack, O. Kempthorne, and T. B. Bailey (eds.), Proc. Intl. Conf. Quant. Genet., 1976, 569–586. Iowa State University Press, Ames.

Norris, K. S. 1953. The ecology of the desert iguana, *Dipsosaurus dorsalis.* Ecology 34:265–287.

Noss, R. F. 1990. Indicators for monitoring biodiversity: A hierarchical approach. Conserv. Biol. 4:355–364.

Noy-Meir, I. 1975. Stability of grazing systems: An application of predator-prey graphs. J. Ecol. 63:459–483.

Nunney, L., and K. A. Campbell. 1993. Assessing minimum viable population sizes: Demography meets population genetics. Trends Ecol. Evol. 8:234–239.

Oades, J. M. 1993. The role of biology in the formation, stabilization and degradation of soil structure. Geoderma 56:377–400.

Oaten, A., and W. W. Murdoch. 1975a. Functional response and stability in predator-prey systems. Am. Nat. 109: 289–298.

Oaten, A., and W. W. Murdoch. 1975b. Switching, functional response, and stability in predator-prey systems. Am. Nat. 109:299–318.

Odum, E. P. 1959. Fundamentals of Ecology. 2nd ed. Saunders, Philadelphia.

Odum, E. P. 1960. Organic production and turnover in old field succession. Ecology 41:34–49.

Odum, E. P. 1962. Relationships between structure and function in the ecosystem. Jap. J. Ecol. 12:108–118.

Odum, E. P. 1969. The strategy of ecosystem development. Science 164:262–270.

Odum, E. P. 1971. Fundamentals of Ecology. 3rd ed. Saunders, Philadelphia.

Odum, H. T. 1957. Trophic structure and productivity of Silver Springs, Florida. Ecol. Monogr. 27:55–112.

Odum, H. T. 1970. Rain forest structure and mineral-cycling homeostasis. In H. T. Odum and R. F. Pigeon (eds.), A Tropical Rain Forest, H-3–H-52. U.S. Atomic Energy Commission, Washington, D.C.

Odum, H. T. 1983. Systems Ecology: An Introduction. Wiley, New York.

Odum, H. T. 1988. Self-organization, transformity, and information. Science 242:1132–1139.

Ogren, W. L. 1984. Photorespiration: Pathways, regulation, and modification. Annu. Rev. Plant Physiol. 415–442.

Ohkawara, K., and S. Higashi. 1994. Relative importance of ballistic and ant dispersal in two diplochorous *Viola* species (Violaceae). Oecologia 100:135–140.

Olmsted, C. E. 1944. Growth and development in range grasses. IV. Photoperiodic responses in twelve geographic strains of side oats grama. Bot. Gaz. 106:4674.

Olson, J. S. 1958. Rates of succession and soil changes on southern Lake Michigan sand dunes. Bot. Gaz. 119: 125–170.

Olson, J. S. 1963. Energy storage and the balance of producers and decomposers in ecological systems. Ecology 44: 322–331.

O'Neill, R. V. 1968. Population energetics of the millipede, *Narceus americanus* (Beauvois). Ecology 49:803–809.

O'Neill, R. V., and D. L. DeAngelis. 1981. Comparative productivity and biomass relations of forest ecosystems. In D. E. Reichle (ed.), Dynamic Properties of Forest Ecosystems, 411–449. Cambridge University Press, Cambridge.

Oosting, H. J. 1942. An ecological analysis of the plant communities of Piedmont, North Carolina. Am. Midl. Nat. 28:1–126.

Oosting, H. J. 1954. Ecological processes and vegetation of the maritime strand in the southeastern United States. Bot. Rev. 20:226–262.

Oosting, H. J. 1956. The Study of Plant Communities. 2nd ed. Dover, New York.

Opdam, P. 1991. Metapopulation theory and habitat fragmentation: A review of holarctic breeding bird studies. Landscape Ecol. 5:93–106.

Opler, P. A. 1973. Fossil lepidopterous leaf mines demonstrate the age of some insect-plant relationships. Science 179:1321–1323.

Orians, G. H. 1969. The number of bird species in some tropical forests. Ecology 50:783–801.

Orians, G. H. 1980. Some Adaptations of Marsh-nesting Blackbirds. Princeton University Press, Princeton, NJ.

Orians, G. H. 1982. The influence of tree falls in tropical forests on tree species richness. Trop. Ecol. 23:255–279.

Orians, G. H., and R. T. Paine. 1983. Convergent evolution at the community level. In D. J. Futuyma and M. Slatkin (eds.), Coevolution, 431–458. Sinauer Associates, Sunderland, MA.

Orians, G. H., and N. E. Pearson. 1979. On the theory of central place foraging. In D. J. Horn, R. Mitchell, and G. R. Stair (eds.), Analysis of Ecological Systems, 155–177. Ohio State University Press, Columbus.

Orians, G. H., and O. T. Solbrig (eds.). 1977. Convergent Evolution in Warm Deserts. Dowden, Hutchinson & Ross, Stroudsburg, PA.

Oring, L. W. 1982. Avian mating systems. In D. S. Fayner, J. R. King, and K. C. Parkes, Avian Biology 6, 1–92. Academic Press, New York.

Ortius, D., and J. Heinze. 1995. Dynamics and consequences of hierarchy formation in the ant *Leptothorax* sp. A. Ethology 99:223–233.

Osmond, C. B. 1978. Crassulacean acid metabolism: A curiosity in context. Annu. Rev. Plant Physiol. 29:379–414.

Oster, G. F., and E. O. Wilson. 1978, Caste and Ecology in the Social Insects. Princeton University Press, Princeton, NJ.

Otte, D., and A. Joern. 1977. On feeding patterns in desert grasshoppers and the evolution of specialized diets. Proc. Acad. Nat. Sci. Phila. 128:89–126.

Overpeck, J. T., D. Rind, and R. Goldberg. 1990. Climate-induced changes in forest disturbance and vegetation. Nature 343:51–53.

Overton, J. M. 1994. Dispersal and infection in mistletoe metapopulations. J. Ecol. 82:711–723.

Overton, W. S. 1969. Estimating the numbers of animals in wildlife populations. In R. H. Giles Jr. (ed.), Wildlife Management Techniques, 3rd ed., 403–455. Wildlife Society, Washington, D.C.

Ovington, J. D. 1962. Quantitative ecology and the woodland ecosystem concept. Adv. Ecol. Res. 1:103–192.

Ovington, J. D. 1965. Organic production, turnover, and mineral cycling in woodlands. Biol. Rev. 40:295–336.

Packard, G. C., J. W. Lang, L. D. Lohmiller, and M. J. Packard. 1997. Cold tolerance in hatchling painted turtles (*Chrysemys picta*): Supercooling or tolerance for freezing? Physiol. Zool. 70:670–678.

Packer, C., and L. Ruttan. 1988. The evolution of cooperative hunting. Am. Nat. 132:159–198.

Packer, C., D. Scheel, and A. E. Pusey. 1990. Why lions form groups: Food is not enough. Am. Nat. 136:1–19.

Paige, K. N. 1992. Overcompensation in response to mammalian herbivory: From mutualistic to antagonistic interactions. Ecology 73:2076–2085.

Paige, K. N., and T. G. Whitham. 1987. Overcompensation in response to mammalian herbivory: The advantage of being eaten. Am. Nat. 129:407–416.

Paine, R. T. 1966. Food web complexity and species diversity. Am. Nat. 100:65–75.

Paine, R. T. 1971. The measurement and application of the calorie to ecological problems. Annu. Rev. Ecol. Syst. 2:145–164.

Paine, R. T. 1974. Intertidal community structure: Experimental studies on the relationship between a dominant competitor and its principal predator. Oecologia 15:93–120.

Paine, R. T. 1980. Food webs: Linkage, interaction strength and community infrastructure. J. Anim. Ecol. 49:667–685.

Paine, R. T. 1992. Food web analysis through field measurements of per capita interaction strength. Nature 355:73–75.

Paine, R. T., and R. Vadas. 1969. The effects of grazing by sea urchins, *Strongylocentrotus* spp., on benthic algal populations. Limnol. Oceanogr. 14:710–719.

Painter, E. L., and A. J. Belsky. 1993. Application of herbivore optimization theory to rangelands of the western United States. Ecol. Appl. 3:2–9.

Pajunen, V. I. 1966. The influence of population density on the territorial behavior of *Leucorrhinia rubicunda* L. (Odon., Libellulidae). Ann. Zool. Fenn. 3:40–52.

Peláez, J. and J. A. McGowan. 1986. Phytoplankton pigment patterns in the California Current as determined by satellite. Limnol. Oceanogr. 31:927–950.

Pandey, C. B., and J. S. Singh. 1992. Rainfall and grazing effects on net primary productivity in a tropical savanna, India. Ecology 73:2007–2021.

Papaj, D., and A. Lewis (eds.). 1993. Insect Learning. Chapman & Hall, New York.

Park, T. 1954. Experimental studies of interspecific competition, II: Temperature, humidity and competition in two species of *Tribolium.* Physiol. Zool. 27:177–238.

Park, T. 1962. Beetles, competition, and populations. Science 138:1369–1375.

Parker, G. A. 1974. The reproductive behavior and the nature of sexual selection in *Scatophaga stercoraria* L. (Diptera: Scatophagidae). IX. Spatial distribution of fertilization rates and evolution of male search strategy within the reproductive area. Evolution 28:93–108.

Parker, G. A. 1978. Evolution of competitive mate searching. Annu. Rev. Entomol. 23:173–196.

Parker, G. A. 1984. Evolutionarily stable strategies. In J. R. Krebs and N. B. Davies (eds.), Behavioural Ecology: An Evolutionary Approach, 2nd ed., 30–61. Blackwell Scientific Publications, Oxford.

Parker, M. A. 1985. Size-dependent herbivore attack and the demography of an arid grassland shrub. Ecology 66:850–860.

Parr, J. C., and R. Thurston. 1968. Toxicity of *Nicotiana* and *Petunia* species to larvae of the tobacco hornworm. J. Econ. Entomol. 61:1525–1531.

Parrish, J. D., and S. B. Saila. 1970. Interspecific competition, predation, and species diversity. J. Theor. Biol. 27: 207–220.

Parry, M. L. 1990. The potential impact on agriculture of the greenhouse effect. Land Use Policy 7:109–115.

Parry, M. L., J. H. Porter, and T. R. Carter. 1990. Climate change and its implications for agriculture. Outlook on Ag. 19:9–16.

Partridge, L., B. Barrie, K. Fowler, and V. French. 1994. Evolution and development of body size and cell size in *Drosophila melanogaster* in response to temperature. Evolution 48:1269–1276.

Partridge, L., and N. H. Barton. 1993. Optimality, mutation and the evolution of ageing. Nature 362:305–311.

Partridge, L., and T. Halliday. 1984. Mating patterns and mate choice. In J. R. Krebs and N. B. Davies (eds.), Behavioural Ecology: An Evolutionary Approach, 2nd ed., 222–250. Blackwell Scientific Publications, Oxford.

Pastor, J., and W. M. Post. 1986. Influence of climate, soil moisture, and succession on forest carbon nitrogen cycles. Biogeochemistry 2:3–27.

Patrick, R. 1963. The structure of diatom communities under varying ecological conditions. Ann. N.Y. Acad. Sci. 108:353–358.

Patrick, R. 1967. The effect of invasion rate, species pool, and size of area on the structure of the diatom community. Proc. Natl. Acad. Sci. USA 58:1335–1342.

Patrick, R., M. H. Hohn, and J. H. Wallace. 1954. A new method for determining the pattern of the diatom flora. Natulae Naturae, no. 259.

Peace, W. J. H., and P. J. Grubb. 1982. Interaction of light and mineral nutrient supply in the growth of *Impatiens parviflora.* New Phytol. 90:127–150.

Pearcy, R. E., and J. Ehleringer. 1984. Comparative ecophysiology of C_3 and C_4 plants. Plant Cell Environ. 7:1–13.

Pearcy, R. W., and D. A. Sims. 1994. Photosynthetic acclimation to changing light environments: Scaling from the leaf to the whole plant. In M. M. Caldwell and R. W. Pearcy (eds.), Exploitation of Environmental Heterogeneity in Plants, 145–174. Academic Press, San Diego.

Pearl, R. 1927. The growth of populations. Q. Rev. Biol. 2:532–548.

Pearl, R., and S. L. Parker. 1921. Experimental studies on the duration of life. 1. Introductory discussion of the duration of life in *Drosophila.* Am. Nat. 55:481–509.

Pearl, R., and L. J. Reed. 1920. On the rate of growth of the population of the United States since 1790 and its mathematical representation. Proc. Natl. Acad. Sci. USA 6:275–288.

Pearson, D. L. 1985. The function of multiple anti-predator mechanisms in adult tiger beetles (Coleoptera: Cicindelidae). Ecol. Entomol. 10:65–72.

Peet, R. K. 1981. Changes in biomass and production during secondary forest succession. In D. C. West, H. H. Shugart, and D. B. Botkin (eds.), Forest Succession: Concepts and Application, 324–338. Springer-Verlag, New York.

Peet, R. K., and O. L. Loucks. 1977. A gradient analysis of southern Wisconsin forests. Ecology 58:485–499.

Pehrson, A. 1983. Digestibility and retention of food components in caged mountain hares *Lepus timidus* during the winter. Holarctic Ecol. 6:395–403.

Pellmyr, O., and C. J. Huth. 1994. Evolutionary stability of mutualism between Yuccas and Yucca moths. Nature 372:257–260.

Pellmyr, O., J. Leebens-Mack, and C. J. Huth. 1996. Non-mutualistic yucca moths and their evolutionary consequences. Nature 380:155–156.

Pellmyr, O., and J. N. Thompson. 1992. Multiple occurrences of mutualism in the yucca moth lineage. Proc. Natl. Acad. Sci. USA: 2927–2929.

Peltonen, A., and I. Hanski. 1991. Patterns of island occupancy explained by colonization and extinction rates in three species of shrew. Ecology 72:1698–1708.

Penfound, W. T. 1956. Primary production of vascular aquatic plants. Limnol. Oceanogr. 1:92–101.

Perry, J. N., I. P. Woiwod, and I. Hanski. 1993. Using response-surface methodology to detect chaos in ecological time series. Oikos 68:329–339.

Peters, R. H. 1983. The Ecological Implications of Body Size. Cambridge University Press, Cambridge.

Peters, R. H., J. J. Armesto, B. Boeken, J. J. Cole, C. T. Driscoll, C. M. Duarte, T. M. Frost, J. P. Grime, J. Kolasa, E. Prepas, and W. G. Sprules. 1991. On the relevance of comparative ecology to the larger field of ecology. In J. Cole, G. Lovett, and S. Findlay (eds.), Comparative analysis of ecosystems, 46–63. Springer-Verlag, New York.

Petersen, C. W. 1987. Reproductive behavior and gender allocation in *Serranus fasciatus*, a hermaphroditic reef fish. Anim. Behav. 35:1601–1614.

Petersen, C. W. 1990. The relationships among population density, individual size, mating tactics, and reproductive success in a hermaphroditic fish, *Serranus fasciatus*. Behaviour 113:57–80.

Petersen, R. C., and K. W. Cummins. 1974. Leaf processing in a woodland stream. Freshwater Biol. 4:343–368.

Peterson, B. J., and B. Fry. 1987. Stable isotopes in ecosystems studies. Annu. Rev. Ecol. Syst. 18:293–320.

Peterson, B. J., L. Deegan, J. Helfrich, J. E. Hobbie, M. Hullar, B. Mollar, T. E. Ford, A. Hershey, A. Hiltner, G. Kipphut, M. A. Lock, D. M. Fiebig, V. McKinley, M. C. Miller, J. R. Vestal, R. Ventullo, and G. Volk. 1993. Biological responses of a tundra river to fertilization. Ecology 74:653–672.

Peterson, C. R., A. R. Gibson, and M. E. Dorcas. 1993. Snake thermal ecology: The causes and consequences of body-temperature variation. In R. A. Seigel and J. T. Collins (eds.), Snakes: Ecology and Behavior, 241–314. McGraw-Hill, New York.

Peterson, D. L., M. A. Spanner, S. W. Running, and K. T. Teuber. 1987. Relationship of thematic simulator data to leaf area index of temperate coniferous forest. Remote Sensing Environ. 22:323–341.

Pfennig, D. W., H. K. Reeve, and P. W. Sherman. 1993. Kin recognition and cannibalism in spadefoot toad tadpoles. Anim. Behav. 46:87–94.

Phelan, J. P., and R. H. Baker. 1992. Optimal foraging in *Peromyscus polionotus:* The influence of item-size and predation risk. Behaviour 121:95–109.

Pianka, E. R. 1966. Latitudinal gradients in species diversity: A review of concepts. Am. Nat. 100:33–46.

Pianka, E. R. 1967. On lizard species diversity: North American flatland deserts. Ecology 48:333–351.

Pianka, E. R. 1970. On *r* and *K* selection. Am. Nat. 104: 592–597.

Pianka, E. R. 1971. Comparative ecology of two lizards. Copeia 1971:129–138.

Pianka, E. R. 1972. *r* and *K* selection or *b* and *d* selection? Am. Nat. 106:581–588.

Pianka, E. R. 1988. Evolutionary Ecology, 4th ed. Harper & Row, New York.

Pickett, S. T. A. 1983. Differential adaptation of tropical species to canopy gaps and its role in community dynamics. Trop. Ecol. 24:68–84.

Pielou, E. C. 1966. Comment on a report by J. H. Vandermeer and R. H. MacArthur concerning the broken stick model of species abundance. Ecology 47:1073–1074.

Pielou, E. C. 1975. Ecological Diversity. Wiley, New York.

Pielou, E. C. 1977. Mathematical Ecology. Wiley, New York and London.

Pierce, G. J., and J. G. Ollason. 1987. Eight reasons why optimal foraging theory is a complete waste of time. Oikos 49:111–118.

Pimentel, D. 1968. Population regulation and genetic feedback. Science 159:1432–1437.

Pimentel, D. 1988. Herbivore population feeding pressure on plant hosts: feedback evolution and host conservation. Oikos 53:289–302.

Pimentel, D., E. H. Feinberg, P. W. Wood, and J. T. Hayes. 1965. Selection, spatial distribution, and the coexistence of competing fly species. Am. Nat. 99:97109.

Pimlott, D. H. 1967. Wolf predation and ungulate populations. Am. Zool. 7:267–278.

Pimm, S. L. 1980. Properties of food webs. Ecology 61:219–225.

Pimm, S. L. 1982. Food Webs. Chapman & Hall, London and New York.

Pimm, S. L. 1991. The balance of nature? University of Chicago Press, Chicago.

Pimm, S. L., H. L. Jones, and J. Diamond. 1988. On the risk of extinction. Am. Nat. 132:757–785.

Pimm, S. L., and J. H. Lawton. 1977. The number of trophic levels in ecological communities. Nature 268:329–331.

Pimm, S. L., and J. H. Lawton. 1978. On feeding on more than one trophic level. Nature 275:542–544.

Pimm, S. L., and J. H. Lawton. 1980. Are food webs compartmented? J. Anim. Ecol. 49:879–898.

Pimm, S. L., J. H. Lawton, and J. E. Cohen. 1991. Food web patterns and their consequences. Nature 350:669–674.

Pimm, S. L., and J. A. Rice. 1987. The dynamics of multispecies, multi-lifestage models of aquatic food webs. Theor. Pop. Biol. 32:303–325.

Pinder, A. W., K. B. Storey, and G. R. Ultsch. 1992. Estivation and hibernation. In M. E. Feder and W. W. Burggren (eds.), Environmental Physiology of the Amphibians, 250–274. University of Chicago Press, Chicago.

Pitcher, T. J., A. E. Magurran, and I. J. Winfield. 1982. Fish in larger shoals find food faster. Behav. Ecol. Sociobiol 10:149–151.

Pitelka, F. A., P. O. Tomich, and G. W. Treichel. 1955. Ecological relations of jaegers and owls as lemming predators near Barrow, Alaska. Ecol. Monogr. 25:85–117.

Pitelka, L. F., and D. J. Raynal. 1989. Forest decline and acidic deposition. Ecology 70:2–10.

Platt, T., D. U. Subba Roa, and B. Irwin. 1983. Photosynthesis of picoplankton in the oligotropic sea. Nature 301: 702–704.

Platt, W. J. 1975. The colonization and formation of equilibrium plant species associations on badger disturbances in tallgrass prairie. Ecol. Monogr. 45:285–305.

Platt, W. J., and I. M. Weiss. 1977. Resource partitioning and competition within a guild of fugitive prairie plants. Am. Nat. 111:479–513.

Polis, G. A. 1991. Complex trophic interactions in deserts: An empirical critique of food web theory. Am. Nat. 138: 123–155.

Polis, G. A., and K. O. Winemiller. 1996. Food Webs. Chapman & Hall, New York.

Pollard, A. J. 1992. The importance of deterrence: Responses of grazing animals to plant variation. In R. S. Fritz and E. L. Simms, Plant Resistance to Herbivores and Pathogens: Ecology, Evolution, and Genetics, 216–239. University of Chicago Press, Chicago.

Poole, R. W. 1974. An Introduction to Quantitative Ecology. McGraw-Hill, New York.

Porter, J. H., and J. L. Dooley Jr. 1993. Animal dispersal patterns: A reassessment of simple mathematical models. Ecology 74:2436–2443.

Porter, J. W. 1972a. Ecology and species diversity of coral reefs on opposite sides of the isthmus of Panama. Bull. Biol. Soc. Wash. 2:89–116.

Porter, J. W. 1972b. Predation by *Acanthaster* and its effect on coral species diversity. Am. Nat. 106:487–492.

Porter, J. W. 1974. Community structure of coral reefs on opposite sides of the Isthmus of Panama. Science 186: 343–345.

Porter, K. G., J. Gerritsen, and J. D. Orcutt Jr. 1982. The effect of food concentration on swimming patterns, feeding behavior, ingestion, assimilation, and respiration by *Daphnia.* Limnol. Oceanogr. 27:935–949.

Porter, K. G., J. D. Orcutt Jr., and J. Gerritsen. 1983. Functional response and fitness in a generalist filter feeder, *Daphnia magna* (Cladocera: Crustacea). Ecology 64: 735–742.

Porter, W. P., J. W. Mitchell, W. A. Beckman, and C. B. DeWitt. 1973. Behavioral implications of mechanistic ecology: Thermal and behavioral modeling of desert ectotherms and their microenvironment. Oecologia 13:1–54.

Post, W. M., W. R. Emanuel, P. J. Zinke, and A. G. Stangenberger. 1982. Soil carbon polls and world life zones. Nature 298:156–159.

Post, W. M., J. Pastor, P. J. Zinke, and A. G. Stangenberger. 1985. Global patterns of soil nitrogen storage. Nature 317:613–616.

Post, W. M., T.-H. Peng, W. R. Emanuel, A. W. King, V. H. Dale, and D. L. DeAngelis. 1990. The global carbon cycle. Am. Sci. 78:310–326.

Postgate, J. R., and S. Hill. 1979. Nitrogen fixation. In J. M. Lynch and N. J. Poole (eds.), Microbial Ecology: A Conceptual Approach, 191–213. Blackwell, Oxford.

Potter, M. A. 1990. Movement of North Island brown kiwi (*Apteryx australis mantelli*) between forest remnants. New Zealand J. Ecol. 14:17–24.

Pough, F. H. 1980. The advantages of ectothermy for tetrapods. Am. Nat. 115:92–112.

Powell, J. A., and R. A. Mackie. 1966. Biological interrelationships of moths and *Yucca whipplei.* U. Calif. Publ. Entomol. 42:1–59.

Power, M. E. 1992. Top-down and bottom-up forces in food webs: Do plants have primacy? Ecology 73:733–746.

Power, M. E., A. J. Steward, and W. J. Matthews. 1988. Grazer control of algae in an Ozark mountain stream: Effects of short-term exclusion. Ecology 69:1894–1898.

Prance, G. T. (ed.). 1982. The Biological Model of Diversification in the Tropics. Columbia University Press, New York.

Pratt, D. M. 1943. Analysis of population development in *Daphnia* at different temperatures. Biol. Bull. 85: 116–140.

Pregill, G. K., and S. L. Olson. 1981. Zoogeography of West Indian vertebrates in relation to Pleistocene climatic cycles. Annu. Rev. Ecol. Syst. 12:75–98.

Preston, F. W. 1948. The commonness, and rarity, of species. Ecology 29:254–283.

Preston, F. W. 1960. Time and space variation of species. Ecology 29:254–283.

Preston, F. W. 1962. The canonical distribution of commonness and rarity, parts 1 and 2. Ecology 43:185–215, 410–432.

Price, M. V., and N. M. Wasser. 1979. Pollen dispersal and optimal outcrossing in *Delphinium nelsoni.* Nature 277: 294–297.

Price, P. W. 1977. General concepts on the evolutionary biology of parasites. Evolution 31:405–420.

Price, P. W. 1980. Evolutionary Biology of Parasites. Princeton University Press, Princeton, NJ.

Price, T. D., P. R. Grant, and P. T. Boag. 1984. Genetic changes in the morphological differentiation of Darwin's ground finches. In K. Wohrmann and V. Loeschcke (eds.), Population Biology and Evolution, 49–66. Springer-Verlag, New York.

Price, T., M. Turelli, and M. Slatkin. 1993. Peak shifts produced by correlated response to selection. Evolution 47:280–290.

Prins, H. B. A., and J. T. M. Elzenga. Bicarbonate utilization: Function and mechanism. Aquat. Bot. 34:59–83.

Proctor, J., and S. R. J. Woodell. 1975. The ecology of serpentine soils. Adv. Ecol. Res. 9:255–366.

Prokopy, R. J. 1968. Visual responses of apple maggot flies, *Rhagoletis pomonella* (Diptera: Tephritidae):Orchard studies. Entomol. Exp. Appl. 11:403–422.

Prosser, C. L. 1973. Comparative Animal Physiology. 3rd ed. Saunders, Philadelphia.

Prosser, C. L., and F. A. Brown. 1961. Comparative Animal Physiology. 2nd ed. Saunders, Philadelphia.

Pulliam, H. R. 1988. Sources, sinks, and population regulation. Am. Nat. 132:652–661.

Pusey, A., and M. Wolf. 1996. Inbreeding avoidance in animals. Trends Ecol. Evol. 11:201–206.

Putnam, A. R., and C.-S. Tang (eds.). 1986. The Science of Allelopathy. Wiley, New York.

Putz, F. E., and K. Milton. 1982. Tree mortality rates on Barro Colorado Island. In E. G. Leigh Jr., A. S. Rand, and D. M. Windsor (eds.), The Ecology of a Tropical Forest, 95–100. Smithsonian Institution Press, Washington, D.C.

Quattrochi, D. A., and R. E. Pelletier. 1991. Remote sensing for analysis of landscapes: An introduction. In M. G. Turner and R. H. Gardner (eds.), Quantitative Methods in Landscape Ecology, 51–76. Springer-Verlag, New York.

Quinn, J. F., and A. E. Dunham. 1983. On hypothesis testing in ecology and evolution. Am. Nat. 122:602–617.

Quispel, A. (ed.). 1974. The Biology of Nitrogen Fixation. North-Holland, Amsterdam.

Rabinowitch, E., and Govindjee. 1969. Photosynthesis. Wiley, New York.

Raison, J. K., J. A. Berry, P. A. Armond, and C. S. Pike. 1980. Membrane properties in relation to the adaptation of plants to temperature stress. In N. C. Turner and P. J. Kramer (eds.), Adaptation of Plants to Water and High Temperature Stress, 261–273. Wiley, New York.

Ralls, K., J. D. Ballou, and A. Templeton. 1988. Estimates of the cost of inbreeding in mammals. Conserv. Biol. 2:185–193.

Rand, A. D. 1967. Predator-prey interactions and the evolution of aspect diversity. Atas do Simposio sobre a Biota Amazonica 5:73–83.

Randall, D. J. 1968. Fish physiology. Am. Zool. 8:179–189.

Rapport, D. J. 1971. An optimization model of food selection. Am. Nat. 105:575–587.

Raunkiaer, C. 1918. Recherches statistiques sur les formations végètales. K. Danske V.densk Selsk. Biol. Meddel. 1:1–47.

Raunkiaer, C. 1934. The Life Forms of Plants and Statistical Plant Geography. Clarendon Press, Oxford.

Raunkiaer, C. 1937. Plant Life Forms. Trans. H. Gilbert-Carter. Clarendon Press, Oxford.

Raup, D. M. 1991. Extinction: Bad Genes or Bad Luck. W. W. Norton, New York.

Raven, P. H. 1990. The politics of preserving biodiversity. BioScience 40:769–774.

Rayor, L. S., and G. W. Uetz. 1990. Trade-offs in foraging success and predation risk with spatial position in colonial spiders. Behav. Ecol. Sociobiol. 27:77–85.

Rayor, L. S., and G. W. Uetz. 1993. Ontogenetic shifts within the selfish herd: Predation risk and foraging trade-offs with age in colonial web-building spiders. Oecologica 95:1–8.

Read, A. F. 1990. Parasites and the evolution of host sexual behaviour. In C. J. Barnard and J. M. Behnke (eds.), Parasitism and Host Behaviour, 117–157. Taylor and Francis, London.

Reavey, D. 1993. Why body size matters to caterpillars. In N. E. Stamp and T. M. Casey (eds.), Caterpillars: Ecological and Evolutionary Constraints on Foraging, 248–279. Chapman & Hall, New York.

Redfield, A. C., B. H. Ketchum, and F. A. Richards. 1963. The influence of organisms on the composition of sea water. In M. N. Hill (ed.), The Sea, vol. 2, 26–77. Interscience, New York.

Redfield, R. J. 1994. Male mutation rates and the cost of sex for females. Nature 369:145–147.

Redford, K. H., A. Taber, and J. A. Simonetti. 1990. There is more to biodiversity than the tropical rain forests. Conserv. Biol. 4:328–330.

Reeve, H. K., and P. W. Sherman. 1993. Adaptation and the goals of evolutionary research. Q. Rev. Biol. 68:1–32.

Reeve, J. D. 1988. Environmental variability, migration, and persistence in host-parasitoid systems. Am. Nat. 132: 810–853.

Rehr, S. S., P. P. Feeny, and D. H. Janzen. 1973. Chemical defense in Central American non-ant acacias. J. Anim. Ecol. 42:405–416.

Reiners, W. A., I. A. Worley, and D. B. Lawrence. 1971. Plant diversity in a chronosequence at Glacier Bay, Alaska. Ecology 52:55–69.

Resetarits, W. J. Jr. 1997. Interspecific competition and qualitative competitive asymmetry between two benthic stream fish. Oikos 78:429–439.

Rettenmeyer, C. W. 1963. Behavioral studies of army ants. U. Kans. Sci. Bull. 44:281–465.

Rettenmeyer, C. W. 1970. Insect mimicry. Annu. Rev. Entomol. 15:43–74.

Reuss, J. O., and G. S. Innis. 1977. A grassland nitrogen flow simulation model. Ecology 58:379–388.

Reuss, J. O., and D. W. Johnson. 1986. Acid Deposition and the Acidification of Soils and Waters. Springer-Verlag, New York.

Rey, J. R., E. D. McCoy, and D. R. Strong Jr. 1981. Herbivore pests, habitat islands, and the species-area relationship. Am. Nat. 117:611–622.

Reznick, D. 1983. The structure of guppy life histories: The tradeoff between growth and reproduction. Ecology 64: 862–873.

Reznick, D. 1985. Costs of reproduction: An evaluation of the empirical evidence. Oikos 44:257–267.

Rheinheimer, G. 1980. Aquatic Microbiology, 2nd ed. Wiley, New York.

Rhoton, F. E., D. D. Tyler, and D. L. Lindbo. 1996. Fragipan soils in lower Mississippi River valley: Their distribution, characteristics, erodibility, productivity, and management. U. S. Department of Agriculture, ARS-137, 20 pp.

Rice, E. L. 1984. Allelopathy. 2nd ed. Academic Press, Orlando, FL.

Richards, O. W., and R. G. Davies. 1977. Imm's general textbook of entomology. Vol 2. Chapman & Hall, London.

Richey, J. E. 1983. The phosphorus cycle. In B. Bolin and R. B. Cook (eds.), The major biogeochemical cycles and their interactions, 51–56. Wiley, New York.

Ricker, W. E. 1954. Stock and recruitment. J. Fish. Res. Bd. Can. 11:559–623.

Ricklefs, R. E. 1969. Preliminary models for growth rates of altricial birds. Ecology 50:1031–1039.

Ricklefs, R. E. 1970. Stage of taxon cycle and distribution of birds on Jamaica, Greater Antilles. Evolution 24:475–477.

Ricklefs, R. E. 1973. Ecology. Chiron, Newton, MA.

Ricklefs, R. E. 1977a. Environmental heterogeneity and plant species diversity: An hypothesis. Am. Nat. 111: 376–381.

Ricklefs, R. E. 1977b. On the evolution of reproductive strategies in birds: Reproductive effort. Am. Nat. 111: 453–478.

Ricklefs, R. E. 1979. Ecology. 2nd ed. Chiron, New York.

Ricklefs, R. E. 1984. The optimization of growth rate in altricial birds. Ecology 65:1602–1616.

Ricklefs, R. E. 1987. Community diversity: Relative roles of local and regional processes. Science 235:167–171.

Ricklefs, R. E., and G. W. Cox. 1972. Taxon cycles in the West Indian avifauna. Am. Nat. 106:195–219.

Ricklefs, R. E., and G. W. Cox. 1978. Stage of taxon cycle, habitat distribution, and population density in the avifauna of the West Indies. Am. Nat. 122:875–895.

Ricklefs, R. E., and C. E. Finch. 1995. Aging: A natural history. Scientific American Library. W. H. Freeman, New York.

Ricklefs, R. E., and F. R. Hainsworth. 1968. Temperature dependent behavior of the cactus wren. Ecology 49:227–233.

Ricklefs, R. E., and F. R. Hainsworth. 1969. Temperature regulation in nestling cactus wrens: The nest environment. Condor 71:32–37.

Ricklefs, R. E., and R. E. Latham. 1993. Global patterns of diversity in mangrove floras. In R. E. Ricklefs and D. Schluter (eds.), Species Diversity in Ecological Communities: Historical and Geographical Perspectives, 215–240. University of Chicago Press, Chicago.

Ricklefs, R. E., and D. B. Miles. 1993. Ecological and evolutionary inferences from morphology: an ecological perspective. In P. C. Wainright and S. Reilly (eds.), Ecological Morphology: Integrative Organismal Biology, 13–41. University of Chicago Press, Chicago.

Ricklefs, R. E., and K. O'Rourke. 1975. Aspect diversity in moths: A temperate-tropical comparison. Evolution 29: 313–324.

Ricklefs, R. E., and D. Schluter. 1993a. Species diversity: Regional and historical influences. In R. E. Ricklefs and D. Schluter (eds.), Species Diversity in Ecological Communities: Historical and Geographical Perspectives, 350–363. University of Chicago Press, Chicago.

Ricklefs, R. E., and D. Schluter. 1993b. Species Diversity in Ecological Communities: Historical and Geographical Perspectives. University of Chicago Press, Chicago.

Ricklefs, R. E., and J. Travis. 1980. A morphological approach to the study of avian community organization. Auk 97:321–338.

Ridley, M. 1978. Paternal care. Anim. Behav. 26:904–932.

Riebesell, J. F. 1974. Paradox of enrichment in competitive systems. Ecology 55:183–187.

Riggan, P. J., S. Goode, P. M. Jacks, and R. N. Lockwood. 1988. Interaction of fire and community development in chaparral of southern California. Ecol. Monogr. 58: 155–176.

Riley, C. V. 1892. The yucca moth and yucca pollination. Third Annu. Rept. Mo. Bot. Garden: 181–226.

Riley, E. T., and E. E. Prepas. 1984. Role of internal phosphorus loading in shallow, productive lakes in Alberta, Canada. Can. J. Fish. Aquat. Sci. 41:845–855.

Risch, S. J. 1981. Insect herbivore abundance in tropical monocultures and polycultures: An experimental test of two hypotheses. Ecology 62:1325–1340.

Ritland, D. B. 1991. Unpalatability of viceroy butterflies (*Limenitis archippus*) and their purported mimicry models, Floida queens (*Danaus gilippus*). Oecologia 88:102–108.

Ritland, D. B. 1995. Comparative unpalatability of mimetic viceroy butterflies (*Limenitis archippus*) from four southeastern United States Populations. Oecologia 103:327–336.

Ritland, K., and F. R. Ganders. 1987. Covariation of selfing rates with parental gene fixation indices within populations of *Mimulus guttatus*. Evolution 41:760–771.

Ritter, W. E. 1938. The California Woodpecker and I: A Study in Comparative Zoology. University of California Press, Berkeley.

Robbins, C. S., D. K. Dawson, and B. A. Dowell. 1989. Habitat requirements of breeding forest birds of the Middle Atlantic states. Wildl. Monogr. 103.

Robbins, C. T., and A. N. Moen. 1975. Composition and digestibility of several deciduous browses in the Northeast. J. Wildl. Mgmt. 39:337–341.

Robert, D. 1989. The auditory behavior of flying locusts. J. Exp. Biol. 147:279–301.

Robert, H., and A. L. MacArthur. 1994. Beyond global warming: Ecology and global change. Ecology 75:1861–1876.

Robertson, A. 1970. A theory of limits in artificial selection with many linked loci. In K. Kojima (ed.), Mathematical Topics in Population Genetics, 246–288. Springer-Verlag, New York.

Robertson, I. C., B. D. Roitberg, I. Williamson, and S. E. Senger. 1995. Contextual chemical ecology: An evolutionary approach to the chemical ecology of insects. Am. Entomol. 41:237–239.

Robinson, M. H. 1969. Defenses against visually hunting predators. Evol. Biol. 3:225–259.

Rockstein, M. (ed.). 1974. Theoretical Aspects of Aging. Academic Press, New York.

Roff, D. A. 1992. The Evolution of Life Histories: Theory and Analysis. Chapman & Hall, New York.

Roff, D. A. 1994. Habitat persistence and the evolution of wing dimorphism in insects. Am. Nat. 144:227–233.

Rohner, C. 1996. The numerical response of great horned owls to the snowshoe hare cycle: Consequences of non-territorial "floaters" on demography. J. Anim. Ecol. 65: 359–370.

Rohwer, F. C. 1985. The adaptive significance of clutch size of prairie ducks. Auk 102:354–360.

Rohwer, S. 1977. Status signalling in Harris' sparrows: Some experiments in deception. Behaviour 61:107–129.

Rohwer, S. 1982. The evolution of reliable and unreliable badges of fighting ability. Am. Zool. 22:531–546.

Rohwer, S., and P. W. Ewald. 1981. The cost of dominance and advantage of subordination in a badge signalling system. Evolution 35:441–454.

Rohwer, S., and F. C. Rohwer. 1978. Status signalling in Harris' sparrows: Experimental deceptions achieved. Anim. Behav. 26:1012–1022.

Rome, L. C., E. D. Stevens, and H. B. John-Alder. 1992. The influence of temperature and thermal acclimation on physiological function. In M. E. Feder and W. W. Burggren (eds.), Environmental Physiology of the Amphibians, 183–205. University of Chicago Press, Chicago.

Root, R. B. 1973. Organization of a plant-arthropod association in simple and diverse habitats: The fauna of collards (*Brassica oleracea*). Ecol. Monogr. 43:95–125.

Roper, T. J., and S. Redston. 1987. Conspicuousness of distasteful prey affects the strength and durability of one-trial avoidance learning. Anim. Behav. 35:739–747.

Rose, M. R. 1982. Antagonistic pleiotropy, dominance, and genetic variation. Heredity 48:63–78.

Rose, M. R. 1991. Evolutionary Biology of Ageing. Chapman & Hall, London.

Rose, M. R., and B. Charlesworth. 1981. Genetics of life history in *Drosophila melanogaster.* II. Exploratory selection experiments. Genetics 97:187–196.

Rosemond, A. D., P. J. Mulholland, and J. Elwood. 1993. Top-down and bottom-up control of stream periphyton: Effects of nutrients and herbivores. Ecology 74:1264–1280.

Rosenthal, G. A., D. L. Dahlman, and D. H. Janzen. 1976. A novel means for dealing with L-canavanine, a toxic metabolite. Science 192:256–258.

Rosenthal, G. A., and D. H. Janzen (eds.). 1979. Herbivores: Their Interaction with Secondary Plant Metabolites. Academic Press, New York.

Rosenzweig, C. and M. L. Parry. 1994. Potential impact of climate change on world food supply. Nature 367:133–137.

Rosenzweig, M. L. 1969. Why the prey curve has a hump. Am. Nat. 103:81–87.

Rosenzweig, M. L. 1975. On continental steady states of species diversity. In M. L. Cody and J. M. Diamond (eds.), Ecology and Evolution of Communities, 121–140. Harvard University Press, Cambridge, MA.

Rosenzweig, M. L. 1981. A theory of habitat selection. Ecology 62:327–335.

Rosenzweig, M. L. 1987. Habitat selection as a source of biological diversity. Evol. Ecol. 1:315–330.

Rosenzweig, M. L. 1995. Species Diversity in Space and Time. Cambridge University Press, Cambridge.

Rosenzweig, M. L., and Z. Abramsky. 1993. How are diversity and productivity related? In R. E. Ricklefs and D. Schluter (eds.), Species Diversity in Ecological Communities: Historical and Geographical Perspectives, 52–65. University of Chicago Press, Chicago.

Rosenzweig, M. L., and C. W. Clark. 1994. Island extinction rates from regular censuses. Conserv. Biol. 8:491–494.

Rosenzweig, M. L., and R. H. MacArthur. 1963. Graphical representation and stability conditions of predator-prey interactions. Am. Nat. 97:209–223.

Rosqvist, R., M. Skurnik, and H. Wolf-Watz. 1988. Increased virulence of *Yersinia pseudotuberculosis* by independent mutations. Nature 334:522–525.

Roughgarden, J. 1974. Niche width: Biogeographic patterns among *Anolis* lizard populations. Am. Nat. 108:429–442.

Ruano, R. G., F. Orozco, and C. Lopez-Fanjul. 1975. The effect of different selection intensities on selection response in egg-laying of *Tribolium castaneum.* Genet.Res. 25:17–27.

Runkle, J. R. 1985. Disturbance regimes in temperate forests. In S. T. A. Pickett and P. S. White (eds.), The Ecology of Natural Disturbance and Patch Dynamics, 17–33. Academic Press, Orlando, FL.

Russell, E. W. 1961. Soil Conditions and Plant Growth. 9th ed. Wiley, New York.

Ryan, M. J. 1980. Female choice in a Neotropical frog. Science 209:523–525.

Ryan, M. J. 1985. The Tungara Frog: A Study in Sexual Selection and Communication. University of Chicago Press, Chicago.

Ryan, M. J. 1990. Sexual selection, sensory systems and sensory exploitation. Oxford Surv. Evol. Biol. 7:157–195.

Ryan, M. J., and A. Keddy-Hector. 1992. Directional patterns of female mate choice and the role of sensory biases. Am. Nat. 139:S4–S35.

Ryan, M. J., M. D. Tuttle, and A. S. Rand. 1982. Bat predation and sexual advertisement in a Neotropical anuran. Am. Nat. 119:136–139.

Ryan, M. J., and W. E. Wagner Jr. 1987. Asymmetries in mating preferences between species: Female swordtails prefer heterospecific males. Science 236:595–597.

Rypstra, A. L. 1989. Foraging success of solitary and aggregated spiders: Insights into flock formation. Anim. Behav. 37:274–281.

Ryther, J. H. 1956. Photosynthesis in the ocean as a function of light intensity. Limnol. Oceanogr. 1:61–70.

Ryther, J. H., and C. S. Yentsch. 1957. The estimation of phytoplankton production in the ocean from chlorophyll and light data. Limnol. Oceanogr. 2:281–286.

Sabelis, M. W., O. Diekmann, and V. A. A. Jansen. 1991. Metapopulation persistence despite local extinction: Predator-prey patch models of the Lotka-Volterra type. Biol. J. Linn. Soc. 42:267–283.

Sagers, C. L., and P. D. Coley. 1995. Benefits and costs of defense in a Neotropical shrub. Ecology 76:1835–1843.

Sale, P. F. 1977. Maintenance of high diversity in coral reef fish communities. Am. Nat. 111:337–359.

Sale, P. F. 1978. Coexistence of coral reef fishes: A lottery for living space. Environ. Biol. Fishes 3:85–102.

Sale, P. F. 1980. The ecology of fishes on coral reefs. Annu. Rev. Oceanogr. Mar. Biol. 18:367–421.

Sale, P. F., and D. M. Williams. 1982. Community structure of coral reef fishes: Are the patterns more than those expected by chance? Am. Nat. 120:121–127.

Salisbury, E. J. 1942. The Reproductive Capacity of Plants. Studies in Quantitative Biology. G. Bell and Sons, London.

Sand, H. 1996. Life history patterns in female moose (*Alces alces*): The relationship between age, body size, fecundity and environmental conditions. Oecologia 106:212–220.

Sanders, H. L. 1968. Marine benthic diversity: A comparative study. Am. Nat. 102:243–282.

Sanford, R. L. Jr. 1987. Apogeotropic roots in an Amazon rain forest. Science 235:1062–1064.

Sargent, T. D. 1966. Background selections of geometrid and noctuid moths. Science 154:1674–1675.

Scalet, C. G., L. D. Flake, and D. W. Willis. 1996. Introduction to Wildlife and Fisheries: An Integrated Approach. W. H. Freeman, New York.

Schad, G. A. 1966. Immunity, competition, and natural regulation of helminth populations. Am. Nat. 100:359–364.

Schaffer, W. M. 1974. Selection for optimal life histories: The effects of age structure. Ecology 53:291–303.

Schaffer, W. M., and P. F. Elson. 1975. The adaptive significance of variations in life history among local populations of Atlantic salmon in North America. Ecology 56:577–590.

Schaffer, W. M., and M. Kot. 1986. Chaos in ecological systems: The coals of Newcastle forgot. Trends Ecol. Evol. 1:58–63.

Schaffer, W. M., and M. V. Schaffer. 1977. The adaptive significance of variations in reproductive habit in the Agavaceae. In B. Stonehouse and C. Perrins (eds.), Evolutionary Ecology, 261–276. Macmillan, London.

Schaller, G. B. 1972. The Serengeti Lion: A Study of Predator-Prey Relations. University of Chicago Press, Chicago.

Scheiner, S. M. 1993. Genetics and evolution of phenotypic plasticity in plants. Annu. Rev. Ecol. Syst. 24:35–68.

Schemske, D. W. 1984. Population structure and local selection in *Impatiens pallida* (Balsaminaceae), a selfing annual. Evolution 38:817–832.

Schindler, D. W. 1974. Eutrophication and recovery in experimental lakes: Implications for lake management. Science 184:897–899.

Schindler, D. W. 1977. Evolution of phosphorus limitation in lakes. Science 195:260–262.

Schindler, D. W. 1978. Factors regulating phytoplankton production and standing crop in the world's freshwaters. Limnol. Oceanogr. 23:478–486.

Schindler, D. W. 1988. Effects of acid rain on freshwater ecosystems. Science 239:149–157.

Schlesinger, W. H. 1991. Biogeochemistry: An Analysis of Global Change. Academic Press, San Diego.

Schlichting, C. D. 1986. The evolution of phenotypic plasticity in plants. Annu. Rev. Ecol. Syst. 17:667–693.

Schmidt-Nielsen, K. 1983. Animal Physiology: Adaptations and Environment. 3rd ed. Cambridge University Press, London and New York.

Schmidt-Nielsen, K. 1990. Scaling: Why Is Animal Size So Important? 2nd ed. Cambridge University Press, Cambridge.

Schmidt-Nielsen, K., F. R. Hainsworth, and D. E. Murrish. 1970. Counter-current heat exchange in the respiratory passages: Effect on water and heat balance. Resp. Physiol. 9:263–276.

Schmidt-Nielsen, K., and B. Schmidt-Nielsen. 1952. Water metabolism of desert mammals. Physiol. Rev. 32:135–166.

Schmidt-Nielsen, K., and B. Schmidt-Nielsen. 1953. The desert rat. Sci. Am. 189:73–78.

Schmitz, O. J. 1993. Trophic exploitation in grassland food webs: Simple models and a field experiment. Oecologia 93:327–335.

Schmitz, O. J. 1994. Resource edibility and trophic exploitation in an old-field food web. Proc. Natl. Acad. Sci. USA 91:5364–5367.

Schmitz, O. J. 1997. Press perturbations and the predictability of ecological interactions in a food web. Ecology 78:55–69.

Schoech, S. J. 1996. The effect of supplemental food on body condition and the timing of reproduction in a cooperative breeder, the Florida scrub jay. Condor 98:234–244.

Schoech, S. J., R. L. Mumme, and J. C. Wingfield. 1996. Prolactin and helping behaviour in the cooperatively breeding Florida scrub-jay, *Aphelocoma c. coerulescens.* Anim. Behav. 52:445–456.

Schoen, D. J. 1982. Genetic variation in the breeding system of *Gilia achilleifolia.* Evolution 36:361–370.

Schoen, D. J. 1983. Relative fitnesses of selfed and outcrossed progeny in *Gilia achilleifolia* (Polemoniaceae). Evolution 37:292–301.

Schoener, T. W. 1969a. Models of optimal size for solitary predators. Am. Nat. 103:277–313.

Schoener, T. W. 1969b. Optimal size and specialization in constant and fluctuating environments: An energy time approach. Brookhaven Symp. Biol. 22:103–114.

Schoener, T. W. 1971. Theory of feeding strategies. Annu. Rev. Ecol. Syst. 2:369–404.

Schoener, T. W. 1974. The species-area relation within archipelagos: models and evidence from island land birds. Proc. 16th Intern. Ornithol. Congr., Canberra: 629–642.

Schoener, T. W. 1983. Field experiments on interspecific competition. Am. Nat. 122:240–285.

Schoener, T. W. 1986. Overview: Kinds of ecological communities—ecology becomes pluralistic. In J. Diamond and T. Case (eds.), Community Ecology, 467–479. Harper & Row, New York.

Schoener, T. W. 1989. Food webs from the small to the large: Probes and hypotheses. Ecology 70:1559–1589.

Schoener, T. W. 1991. Extinction and the nature of the metapopulation. Acta Oecol. 12:53–75.

Schoener, T. W. 1993. On the relative importance of direct versus indirect effects in ecological communities. In W. C. Kerfoot and A. Sih (eds.), Predation: Direct and Indirect Impacts on Aquatic Communities, 365–415. University Press of New England, Hanover, NH.

Schoener, T. W., and D. A. Spiller. 1987. Effect of lizards on spider populations: Manipulative reconstruction of a natural experiment. Science 236:949–952.

Schoener, T. W., and D. A. Spiller. 1992. Is extinction rate related to temporal variability in population size? An empirical answer for orb spiders. Am. Nat. 139:1176–1207.

Schoenly, K. R., R. A. Beaver, and T. A. Heumier. 1991. On the trophic relations of insects: A food web approach. Am. Nat. 137:597–638.

Scholander, P. F. 1955. Evolution of climatic adaptation in homeotherms. Evolution 9:15–26.

Scholander, P. F., and W. E. Schevill. 1955. Countercurrent vascular heat exchange in the fins of whales. J. Appl. Physiol. 8:279–282.

Schultz, A. M., J. L. Launchbaugh, and H. H. Biswell. 1955. Relationship between grass diversity and brush seedling survival. Ecology 36:226–238.

Schulz, J. P. 1960. Ecological Studies on Rain Forest in Northern Suriname. North-Holland, Amsterdam.

Schulze, E. D. 1989. Air pollution and forest decline in a spruce (*Picea abies*) forest. Science 244:776–783.

Schulze, E. D., R. H. Robichaux, J. Grace, P. W. Rundel, and J. R. Ehleringer. 1987. Plant water balance. BioScience 37: 30–37.

Schurmann, H., and J. S. Christiansen. 1994. Behavioral thermoregulation and swimming activity of two Arctic teleosts (subfamily Gadinae):The polar cod (*Boreogadus saida*) and the navaga (*Elginus navaga*). J. Therm. Biol. 19:207–212.

Schutten, J., J. A. Vander Velden, and H. Smit. 1994. Submerged macrophytes in the recently freshened lake system Volkerak-Zoom (The Netherlands), 1987–1991. Hydrobiologia 275/276:207–281.

Schwab, J. H. 1975. Suppression of the immune response by microorganisms. Bacteriol. Rev. 39:121–143.

Scriber, J. M. 1984. Host-plant suitability. In W. J. Bell and R. T. Cardé (eds.), Chemical Ecology of Insects, 159–202. Chapman & Hall, London.

Scriber, J. M., and F. Slansky. 1981. The nutritional ecology of immature insects. Annu. Rev. Entomol. 26:183–211.

Scribner, K. T., M. C. Wooten, and M. H. Smith. 1992. Variation in life history and genetic traits of Hawaiian mosquitofish populations. J. Evol. Biol. 5:267–288.

Seal, U. S., and R. C. Lacy. 1989. Florida Panther population viability analysis. Report to the U.S. Fish and Wildlife Service. Apple Valley: Captive Breeding Specialist Group, SSC, IUCN.

Seber, G. A. F. 1965. A note on the multiple recapture census. Biometrika 52:249–259.

Seber, G. A. F. 1973. Estimation of Animal Abundance. Hafner, New York.

Seger, J. 1991. Cooperation and conflict in social insects. In J. R. Krebs and N. B. Davies (eds.), Behavioural Ecology: An Evolutionary Approach, 3rd ed., 338–373. Blackwell Scientific Publications, London.

Seigler, D. S. 1996. Chemistry and mechanisms of allelopathic interactions. Agron. J. 88:876–885.

Seigler, D. S., and P. W. Price. 1976. Secondary compounds in plants: Primary functions. Am. Nat. 110:101–105.

Sepkoski, J. J. Jr. 1978. A kinetic model of Phanerozoic taxonomic diversity. I. Analysis of marine orders. Paleobiology 4:223–251.

Sepkoski, J. J. Jr. 1984. A kinetic model of Phanerozoic taxonomic diversity. III. Post-Paleozoic families and mass extinctions. Paleobiology 10:246–267.

Sepkoski, J. J. Jr., R. K. Bambach, D. M. Raup, and J. W. Valentine. 1981. Phanerozoic marine diversity and the fossil record. Nature 293:435–437.

Shaffer, M. L. 1981. Minimum population sizes for species conservation. BioScience 31:131–134.

Shaffer, M. L. 1990. Population viability analysis. Conserv. Biol. 4:39–40.

Shanmugam, K. T., F. O'Gara, K. Andersen, and R. C. Valentine. 1978. Biological nitrogen fixation. Annu. Rev. Plant Physiol. 29:263–276.

Shannon, C. E., and W. Weaver. 1949. The Mathematical Theory of Communication. University of Illinois Press, Urbana.

Shaver, G. R., W. D. Billings, F. S. Chapin III, A. E. Giblin, K. J. Nadelhoffer, W. C. Oechel, and E. B. Rastetter. 1992. Global change and the carbon balance of arctic ecosystems. BioScience 42:433–441.

Shaver, G. R., and F. S. Chapin III. 1980. Response to fertilization by various plant growth forms in an Alaskan tundra: Nutrient accumulation and growth. Ecology 61:662–675.

Shaver, G. R., K. J. Nadelhoffer, and A. E. Giblin. 1991. Biogeochemical diversity and element transport in a heterogeneous landscape, the North Slope of Alaska. In M. G. Turner and R. H. Gardner (eds.), Quantitative Methods in Landscape Ecology: The Analysis and Interpretation of Landscape Heterogeneity, 105–125. Springer-Verlag, New York.

Shelford, V. E., and W. P. Flint. 1943. Populations of the chinch bug in the Upper Mississippi Valley from 1823 to 1940. Ecology 24:435–455.

Sheppard, D. H. 1971. Competition between two chipmunk species (*Eutamias*). Ecology 52:320–329.

Sherman, P. W. 1977. Nepotism and the evolution of alarm calls. Science 197:1246–1253.

Sherman, P. W. 1985. Alarm calls of Belding's ground squirrels to aerial predators: Nepotism or self-preservation? Behav. Ecol. Sociobiol. 17:313–323.

Sherman, P. W., E. A. Lacey, H. K. Reeve, and L. Keller. 1995. The eusociality continuum. Behav. Ecol. 6:102–108.

Sherry, D. F. 1984. Food storage by black-capped chickadees: Memory for the location and contents of caches. Anim. Behav. 32:451–464.

Shimwell, D. W. 1971. Description and Classification of Vegetation. University of Washington Press, Seattle.

Shmida, A., and M. V. Wilson. 1985. Biological determinants of species diversity. J. Biogeogr. 12:1–20.

Short, H. L. 1971. Forage digestibility and diet of deer on southern upland range. J. Wildl. Mgmt. 35:698–706.

Shugart, H. H. 1984. A Theory of Forest Dynamics. The Ecological Implications of Forest Succession Models. Springer-Verlag, New York.

Shugart, H. H., and J. M. Hett. 1973. Succession: Similarities of turnover rates. Science 180:1379–1381.

Shugart, H. H., and D. C. West. 1977. Development of an Appalachian deciduous forest succession model and its application to assessment of the impact of the chestnut blight. J. Environ. Mgmt. 5:161–179.

Sieburth, J. M. 1979. Sea Microbes. Oxford University Press, New York.

Siegenthaler, U., and J. L. Sarmiento. 1993. Atmospheric carbon dioxide and the ocean. Nature 365:119–125.

Sih, A. 1982. Foraging strategies and the avoidance of predation by an aquatic insect, *Notonecta hoffmani.* Ecology 63:786–796.

Sih, A. 1984. Optimal behavior and density dependent predation. Am. Nat. 123:314–326.

Sih, A., and R. D. Moore. 1990. Interacting effects of predator and prey behavior in determining diets. In R. N. Hughes (ed.), Behavioural mechanisms of food selection, 771–796. Springer-Verlag, Berlin.

Silander, J. A. Jr. 1985. The genetic basis of the ecological amplitude of *Spartina patens.* II. Variance and correlation analysis. Evolution 39:1034–1052.

Sillén-Tullberg, B., and O. Leimar. 1988. The evolution of gregariousness in distasteful insects as a defense against predators. Am. Nat. 132:723–734.

Silvertown, J. W. 1982. Introduction to Plant Population Ecology. Longman, London.

Silvertown, J. W. 1996. Are sub-alpine firs evolving towards semelparity? Evol. Ecol. 10:77–80.

Silvertown, J. W., and J. L. Doust. 1993. Introduction to Plant Population Biology. Blackwell Scientific, Oxford.

Simberloff, D. S. 1969. Experimental zoogeography of islands: A model for insular colonization. Ecology 50:296–314.

Simberloff, D. S. 1976. Species turnover and equilibrium island biogeography. Science 194:572–578.

Simberloff, D. S. 1983. Competition theory, hypothesis testing, and other community ecological buzzwords. Am. Nat. 122:626–635.

Simberloff, D. S. 1994. Habitat fragmentation and population extinction of birds. Ibis 137:S105–S111.

Simberloff, D. S., and W. Boecklen. 1981. Santa Rosalia reconsidered: Size ratios and competition. Evolution 35:1206–1228.

Simberloff, D. S., and J. Cox. 1987. Consequences and costs of conservation corridors. Conserv. Biol. 1:63–71.

Simberloff, D. S., and E. O. Wilson. 1969. Experimental zoogeography of islands: The colonization of empty islands. Ecology 50:278–296.

Simpson, B. B. 1974. Glacial migrations of plants: Island biogeographical evidence. Science 185:698–700.

Simpson, B. B., and J. Haffer. 1978. Speciation patterns in the Amazonian forest biota. Annu. Rev. Ecol. Syst. 9: 497–518.

Simpson, E. R. 1949. Measurement of diversity. Nature 163:688.

Simpson, G. G. 1964. Species density of North American Recent mammals. Syst. Zool. 13:57–73.

Simpson, G. G. 1969. The first three billion years of community evolution. Brookhaven Symp. Biol. 22:162–177.

Sinclair, A. R. E., and P. Arcese. 1995. Population consequences of predation-sensitive foraging: The Serengeti wildebeest. Ecology 76:882–891.

Sinclair, A. R. E., H. Dublin, and M. Borner. 1985. Population regulation of the Serengeti wildebeest: A test of the food hypothesis. Oecologia 65:266–268.

Sinclair, A. R. E., and R. P. Pech. 1996. Density dependence, stochasticity, compensation and predator regulation. Oikos 75:164–173.

Sinervo, B., and K. D. Dunlap. Thyroxine affects behavioral thermoregulation but not growth rate among populations of the Western fence lizard (*Sceloporus occidentalis*). J. Comp. Physiol. B 164:509–517.

Singer, M. C., and P. R. Ehrlich. 1979. Population dynamics of the checkerspot butterfly *Euphydryas editha.* Forsch. Zool. 25:53–60.

Singleton, R. Jr., and R. E. Amelunxen. 1973. Proteins from thermophilic microorganisms. Bacteriol. Rev. 37: 320–342.

Skutch, A. F. 1949. Do tropical birds rear as many young as they can nourish? Ibis 91:430–455.

Slater, P. J. B., and T. R. Halliday (eds.). 1994. Behaviour and Evolution. Cambridge University Press, Cambridge.

Sleigh, M. 1973.The biology of Protozoa. American Elsevier, New York.

Slip, D. J., and R. Shine. 1988. Reptilian endothermy: A field study of thermoregulation by brooding diamond pythons. J. Zool. 216:367–378.

Slobodkin, L. B. 1961. Growth and Regulation of Animal Populations. Holt, Rinehart & Winston, New York.

Slobodkin, L. B. 1992. A summary of the special feature and comments on its theoretical context and importance. Ecology 73:1564–1566.

Smith, A. T. 1974a. The distribution and dispersal of pikas: Consequences of insular population structure. Ecology 55: 1112–1119.

Smith, A. T. 1974b. The distribution and dispersal of pikas: Influences of behavior and climate. Ecology 55:1368–1376.

Smith, A. T. 1978. Comparative demography of pikas (*Ochotoma*): Effects of spatial and temporal age-specific mortality. Ecology 59:133–139.

Smith, A. T. 1980. Temporal changes in insular populations of pika (*Ochotona princeps*). Ecology 60:8–13.

Smith, A. T., and M. E. Gilpin. 1997. Spatially correlated dynamics in a pika metapopulation. In I. A. Hanski and M. E. Gilpin (eds.), Metapopulation Biology: Ecology, Genetics, and Evolution, 407–428. Academic Press, San Diego.

Smith, F. E. 1961. Density dependence in the Australian thrips. Ecology 42:403–407.

Smith, J. M. 1974. The theory of games and the evolution of animal conflicts. J. Theor. Biol. 47:209–221.

Smith, J. M. 1978a. The Evolution of Sex. Cambridge University Press, Cambridge.

Smith, J. M. 1978b. The handicap principle: A comment. J. Theor. Biol. 70:251–252.

Smith, J. M. 1982. Evolution and the Theory of Games. Cambridge University Press, Cambridge.

Smith, J. M., and G. R. Price. 1973. The logic of animal conflict. Nature 246:15–18.

Smith, J. N. M., P. Arcese, and W. M. Hochachka. 1991. Social behavior and population regulation in insular bird populations: implications for conservation. In C. M. Perrins, J.-D. Lebreton, G. J. M. Hirons (eds.), Bird Population Studies: Relevance to Conservation and Management, 148–167. Oxford University Press, Oxford.

Smith, S. A., C. W. Thayer, and C. E. Brett. 1985. Predation in the Paleozoic: Gastropod-like drillholes in Devonian brachiopods. Science 230:1033–1035.

Smith, S. E. 1980. Mycorrhizas of autotrophic higher plants. Biol. Rev. 55:474–510.

Smith, S. M. 1975. Innate recognition of coral snake pattern by a possible avian predator. Science 187:759–760.

Smith, S. M. 1977, Coral-snake pattern recognition and stimulus generalisation by naive great kiskadees (Aves: Tyrarmidae). Nature 265:535–536.

Smith, W. G. 1975. Dynamics of pure and mixed populations of *Desmodium glutinosum* and *D. nudiflorum* in natural oak forest communities. Am. Midl. Nat. 94:99–107.

Smithers, S. R., R. J. Terry, and D. J. Hockley. 1969. Host antigens in schistosomiasis. Proc. R. Soc. Lond. B 171: 483–494.

Snucins, E. J., and J. M. Gunn. 1994. Coping with a warm environment: Behavioral thermoregulation by lake trout. Trans. Am. Fish. Soc. 124:118–123.

Snyder, L. L. 1947. The snowy owl migration of 1945–1946: Second report of the Snowy Owl Committee. Wilson Bull. 59:74–78.

Söderlund, R., and T. Rosswall. 1982. The nitrogen cycles. In O. Hutzinger (ed.), The Handbook of Environmental Chemistry, vol. 1, part B, The National Environment and the Biogeochemical Cycles, 61–81. Springer-Verlag, New York.

Solomon, M. E. 1957. Dynamics of insect populations. Annu. Rev. Entomol. 2:121–142.

Somero, G. N. 1978. Temperature adaptation of enzymes: Biological optimization through structure-function compromises. Annu. Rev. Ecol. Syst. 9:1–29.

Somme, L. 1964. Effects of glycerol in cold hardiness in insects. Can. J. Zool. 42:87–101.

Sonenshine, D. E. 1991. Biology of Ticks. Volume 1. Oxford University Press, Oxford.

Soulé, M. E. 1991. Conservation: Tactics for a constant crisis. Science 253:744–750.

Sousa, W. P. 1979. Experimental investigations of disturbance and ecological succession in a rocky intertidal algal community. Ecol. Monogr. 4:227–254.

Sousa, W. P. 1980. The response of a community to disturbance: The importance of successional age and species life histories. Oecologia 45:72–81.

Sousa, W. P. 1984. Intertidal mosaics: Patch size, propagule availability, and spatially variable patterns of succession. Ecology 65:1918–1935.

Sousa, W. P. 1985. Disturbance and patch dynamics on rocky intertidal shores. In S. T. A. Pickett and P. S. White (eds.), The Ecology of Natural Disturbance and Patch Dynamics, 101–124. Academic Press, Orlando, FL.

Southwood, T. R. E. 1961. The number of species of insect associated with various trees. J. Anim. Ecol. 30:1–8.

Southwood, T. R. E. 1985. Interactions of plants and animals: Patterns and processes. Oikos 44:5–11.

Southwood, T. R. E., V. C. Moran, and C. E. J. Kennedy. 1982. The richness, abundance and biomass of the arthropod communities on trees. J. Anim. Ecol. 51:635–649.

Spalinger, D. E., and N. T. Hobbs. 1992. Mechanisms of foraging in mammalian herbivores: New models of functional response. Am. Nat. 140:325–348.

Speidel, D. H., and A. F. Agnew. 1982. The Natural Geochemistry of Our Environment. Westview Press, Boulder, CO.

Spiller, D. A. 1992. Relationship between prey consumption and colony size in an orb spider. Oecologia 90:457–466.

Spiller, D. A., and T. S. Schoener. 1989. Effect of a major predator on grouping of an orb-weaving spider. J. Anim. Ecol. 58:509–523.

Sprugel, D. G. 1976. Dynamic structure of wave-regenerated *Abies balsamea* forests in the northeastern United States. J. Ecol. 64:889–911.

Sprugel, D. G., and F. H. Bormann. 1981. Natural disturbance and the steady-state in high-altitude balsam fir forests. Science 211:390–393.

Stacey, P. B., V. A. Johnson, and M. L. Taper. 1997. Migration within metapopulations: The impact upon local population dynamics. In I. A. Hanski and M. E. Gilpin, Metapopulation Biology, Ecology, Genetics, and Evolution, 267–291. Academic Press, San Diego.

Stacey, P. B., and M. Taper. 1992. Environmental variation and the persistence of small populations. Ecol. Appl. 2: 18–29.

Stacy, G., T. H. Burris, and H. J. Evans (eds.). 1992. Biological nitrogen fixation. Chapman & Hall, New York.

Stamp, N. E. and R. T. Wilkens. 1993. On the cryptic side of life: Being unapparent to enemies and the consequences for foraging and growth of caterpillars. In N. E. Stamp and T. M. Casey (eds.), Caterpillars: Ecological and Evolutionary Constraints on Foraging, 283–300. Chapman & Hall, New York.

Stanton, M. L., A. A. Snow, and S. N. Handel. 1986. Floral evolution: Attractiveness to pollinators increases male fitness. Science 232:1625–1627.

Stark, N., and C. F. Jordan. 1978. Nutrient retention by the root mat of the Amazonian rain forest. Ecology 59: 434–437.

Stearns, S. C. 1976. Life-history tactics: A review of the ideas. Q. Rev. Biol. 51:3–47.

Stearns, S. C. 1983. The evolution of life-history traits in mosquitofish since their introduction to Hawaii in 1905: Rates of evolution, heritabilities and developmental plasticity. Am. Zool. 23:65–76.

Stearns, S. C. 1992. The Evolution of Life Histories. Oxford University Press, Oxford.

Stearns and Koella. 1986. The evolution of phenotypic plasticity in life-history traits: predictions of reaction norms for age and size at maturity. Evol. 40:893–913.

Steele, J. H. 1962. Environmental control of photosynthesis in the sea. Limnol. Oceanogr. 6:137–150.

Steffan, W. A. 1973. Polymorphism in *Plastosciara perniciosa.* Science 182:1265–1266.

Stehli, F. G. 1968. Taxonomic gradients in pole location: The recent model. In E. T. Drake (ed.), Evolution and Environment, 163–227. Yale University Press, New Haven, CT.

Stehli, F. G., R. G. Douglas, and N. D. Newell. 1969. Generation and maintenance of gradients in taxonomic diversity. Science 164:947–949.

Stein, B. A. 1992. Sicklebill hummingbirds, ants, and flowers. BioScience 42:27–33.

Steneck, R. S. 1982. A limpet-coralline alga association: Adaptations and defenses between a selective herbivore and its prey. Ecology 63:507–522.

Stenseth, N. C., and W. Z. Lidicker (eds.). 1992. Animal Dispersal: Small Mammals as a Model. Chapman & Hall, London.

Stephens, D. W., and J. R. Krebs. 1986. Foraging Theory. Princeton University Press, Princeton, NJ.

Stephenson, A. G. 1981. Flower and fruit abortion: Proximate causes and ultimate functions. Annu. Rev. Ecol. Syst. 12:253–279.

Stephenson, A. G., and R. I. Bertin. 1963. Male competition, female choice, and sexual selection in plants. In L. Real (ed.), Pollination Biology, 110–149. Academic Press, New York.

Stevens, G. C. 1983. Patterns of plant use by wood-boring insects. Ph.D. thesis, University of Pennsylvania, Philadelphia.

Stevens, G. C. 1986. Dissection of the species-area relationship among wood-boring insects and their host plants. Am. Nat. 128:35–46.

Stevens, G. C. 1989. The latitudinal gradient in geographical range: How so many species coexist in the Tropics. Am. Nat. 133:240–256.

Stevens, G., and R. Bellig (eds.). 1988. The Evolution of Sex. Harper & Row, San Francisco.

Stevenson, R. D. 1985. The relative importance of behavioral and physiological adjustments controlling body temperature in terrestrial ecotherms. Am. Nat. 126:362–386.

Stewart, W. D. P. 1975. Nitrogen Fixation by Free-living Microorganisms. Cambridge University Press, Cambridge.

Stiles, F. G. 1985. Seasonal patterns and coevolution in the hummingbird-flower community of a Costa Rican subtropical forest. In P. A. Buckley, M. S. Foster, E. S. Morton, R. S. Ridgley, and F. G. Buckley (eds.), Neotropical Ornithology, 757–785. Ornithological Monographs, no. 36. American Ornithologists' Union, Washington, D.C.

St. Mary, C. M. 1994. Sex allocation in a simultaneous hermaphrodite, the blue-banded goby (*Lythrypnus dalli*): The effects of body size and behavioral gender and the consequences for reproduction. Behav. Ecol. 5:304–313.

Stockner, J. G., and K. G. Porter. 1987. Microbial food webs in freshwater planktonic ecosystems. In S. R. Carpenter (ed.), Complex Interactions in Lake Communities, 69–83. Springer-Verlag, New York.

Stoecker, R. E. 1972. Competitive relations between sympatric populations of voles (*Microtus montanus* and *M. pennsylvanicus*). J. Anim. Ecol. 41:311–329.

Stone, G. N. 1994. Patterns of evolution of warm-up rates and body temperatures in flight in solitary bees of the genus *Anthopohora.* Funct. Ecol. 8:324–335.

Stone, L., and A. Roberts. 1990. The checkerboard score and species distributions. Oecologia 85:74–79.

Stone, L., and A. Roberts. 1992. Competitive exclusion, or species aggregation? An aid in deciding. Oecologia 91:419–424.

Story, J. M., W. R. Good, and N. W. Callan. 1993. Supercooling capacity of *Urophora affinis* and *U. quadrifasciata* (Diptera: Tephritidae), two flies released on spotted knapweed in Montana. Environ. Entomol. 22:831–836.

Straube, W. L., J. W. Deming, C. C. Somerville, R. R. Colwell, and J. A. Baross. 1990. Particulate DNA in smoker fluids: Evidence for existence of microbial populations in hot hydrothermal systems. Appl. Environ. Microbiol. 56:1440–1447.

Strayer, D. L. 1991. Comparative ecology and undiscovered public knowledge. In J. Cole, G. Lovett, and S. Findlay (eds.), Comparative Analysis of Ecosystems: Patterns, mechanisms and theories, 1–6. Springer-Verlag, New York.

Strehler, B. L. (ed.). 1960. The Biology of Aging. American Institute of Biological Sciences, Washington, D.C.

Strickland, J. D. H. 1960. Measuring the production of marine phytoplankton. Bull. Fish. Res. Bd. Can. 12:1–172.

Strickland, J. D. H., and T. R. Parsons. 1968. A manual of sea water analysis. Bull. Fish. Res. Bd. Can. 125:1–311.

Strong, D. R. 1974. Rapid asymptotic species accumulation in phytophagous insect communities: The pests of cacao. Science 185:1064–1066.

Strong, D. R. 1986. Density-vague population change. Trends Ecol. Evol. 1:39–42.

Strong, D. R. 1992. Are trophic cascades all wet? Differentiation and donor-control in speciose ecosystems. Ecology 73:747–754.

Strong, D. R., J. H. Lawton, and T. R. E. Southwood. 1984. Insects on Plants: Community Patterns and Mechanisms. Harvard University Press, Cambridge, MA.

Strong, D. R. Jr., L. A. Szyska, and D. S. Simberloff. 1979. Tests of community-wide character displacement against null hypotheses. Evolution 33:897–913.

Stross, R. G. 1969. Photoperiod control of diapause in *Daphnia.* II. Induction of winter diapause in the Arctic. Biol. Bull. 136:264–273.

Stross, R. G., and V. C. Hill. 1965. Diapause induction in *Daphnia* requires two stimuli. Science 150:14621464.

Stubblefield, S. P., T. N. Taylor, and J. M. Trappe. 1987. Fossil mycorrhizae: A case for symbiosis. Science 237:59–60.

Stumm, W., and J. J. Morgan. 1981. Aquatic Chemistry. 2nd ed. Wiley, New York.

Subba Roa, N. S. (ed.). 1980. Recent Advances in Biological Nitrogen Fixation. Edward Arnold, London.

Sugihara, G. 1980. Minimal community structure: An explanation of species abundance patterns. Am. Nat. 116:770–787.

Sullivan, C. W., K. R. Arrigo, C. R. McClain. 1993. Distributions of phytoplankton blooms in the Southern Ocean. Science 262:1832–1886.

Sutcliffe, O. L., C. D. Thomas, and D. Peggie. 1997a. Area-dependent migration by ringlet butterflies generates a mixture of patchy population and metapopulation attributes. Oecologia 109:229–234.

Sutcliffe, O. L., C. D. Thomas, T. J. Yates, and J. J. Greatorex-Davies. 1997b. Correlated extinctions, colonizations and population fluctuations in a highly connected ringlet butterfly metapopulation. Oecologia 109:235–241.

Svardson, G. 1957. The "invasion" type of bird migration. Brit. Birds 50:314–343.

Swanberg, O. 1951. Food storage, territory and song in the thick-billed nutcracker. Proc. Tenth Intl. Congr. Ornithol., 545–554.

Swank, W. T., and D. A. Crossley Jr. 1988. Forest Hydrology and Ecology at Coweeta. Springer-Verlag, New York.

Swenson, M. J. (eds.). 1977. Dukes' Physiology of Domestic Animals. 9th ed. Cornell University Press, Ithaca, NY.

Tahvanainen, J. O., and R. B. Root. 1972. The influence of vegetational diversity on the population ecology of a specialized herbivore, *Phyllotreta cruciferae* (Coleoptera: Chrysomelidae). Oecologia 10:321–346.

Tamarin, R. H. 1993. Principles of Genetics. 4th ed. Wm. C. Brown, Dubuque, Iowa.

Tansley, A. G. 1917. On competition between *Galium saxatile* L. (*G. hercynicum* Weig.) and *Galium sylvestre* poll. (*G. asperum* Schreb.) on different types of soil. J. Ecol. 5:173–179.

Tansley, A. G. 1935. The use and abuse of vegetational concepts and terms. Ecology 16:204–307.

Tansley, A. G., and R. S. Adamson. 1925. Studies of the vegetation of the English chalk. III. The chalk grasslands of the Hampshire-Sussex border. J. Ecol. 13:177–223.

Taper, M. L., and T. J. Case. 1985. Quantitative genetic models for the coevolution of character displacement. Ecology 55:355–371.

Taylor, A. D. 1988. Parasitoid competition and the dynamics of host-parasitoid models. Am. Nat. 132:417–436.

Taylor, A. D. 1991. Studying metapopulation effects in predator-prey systems. In M. Gilpin and I. Hanski (eds.), Metapopulation Dynamics: Empirical and Theoretical Investigations, 305–323. Academic Press, London.

Taylor, C. R., and C. Condra. 1980. *r* and *K* selection in *Drosophila pseudoobscura.* Evolution 34:1183–1193.

Teal, J. M. 1962. Energy flow in the salt marsh ecosystem of Georgia. Ecology 43:614–624.

Teeri, J., and L. Stowe. 1976. Climatic patterns and the distribution of C_4 grasses in North America. Oecologia 23:1–12.

Temple, S. A. 1977. Plant-animal mutualism: Coevolution with dodo leads to near extinction of plant. Science 197:885–886.

Templeton, J. J., and L. A. Giraldeau. 1995. Patch assessment in foraging flocks of European starlings: Evidence for the use of public information. Behav. Ecol. 6:65–72.

Terashima, I, and K. Hikosaka. 1995. Comparative ecophysiology of leaf and canopy photosynthesis. Plant Cell Environ. 18:1111–1128.

Terborgh, J. W., and J. Faaborg. 1973. Turnover and ecological release in the avifauna of Mona Island, Puerto Rico. Auk 90:759–779.

Terborgh, J. W., S. K. Robinson, T. A. Parker III, C. A. Munn, and N. Pierpont. 1990. Structure and organization of an Amazonian forest bird community. Ecol. Monogr. 60:213–238.

Ter Braak, C. J. F., and I. C. Prentice. 1988. A theory of gradient analysis. Adv. Ecol. Res. 18:272–317.

Tessier, A. J., L. L. Henry, and C. E. Goulden. 1983. Starvation in *Daphnia:* Energy reserves and reproductive allocation. Limnol. Oceanogr. 28:667–676.

Tevis, L. Jr., and I. M. Newell. 1962. Studies on the biology and seasonal cycle of the giant red velvet mite, *Dinothrombium pandorae* (Acari, Thrombidiidae). Ecology 43:497–505.

Thomas, C. D., and I. Hanski. 1997. Butterfly metapopulations. In I. A. Hanski and M. E. Gilpin (eds.), Metapopulation Biology: Ecology, Genetics, and Evolution, 359–386. Academic Press, San Diego.

Thomas, C. D., M. C. Singer, and D. A. Boughton. 1996. Catastrophic extinction of population sources in a butterfly metapopulation. Am. Nat. 148:957–975.

Thompson, D. Q. 1952. Travel, range and food habits of timber wolves in Wisconsin. J. Mammal. 33:429–442.

Thompson, J. N. 1982. Interaction and coevolution. Wiley, New York.

Thompson, J. N. 1984. Insect diversity and trophic structure of communities. In C. B. Huffaker and R. L. Rabb (eds.), Ecological Entomology, 591–606. Wiley, New York.

Thompson, J. N. 1994. The Coevolutionary Process. University of Chicago Press, Chicago.

Thompson, J. N., and J. J. Burdon. 1992. Gene-for-gene coevolution between plants and parasites. Nature 360:121–125.

Thompson, J. N. Jr., and J. M. Thoday (eds.). 1979. Quantitative Genetic Variation. Academic Press, New York.

Thompson, K. V. 1993. Aggressive behavior and dominance hierarchies in female sable antelope, *Hippotragus niger:* Implications for captive management. Zool. Biol. 2:189–202.

Thorne, R. F. 1972. Major disjunctions in the geographic ranges of seed plants. Q. Rev. Biol. 47:365–411.

Thornhill, R., and J. Alcock. 1983. The Evolution of Insect Mating Systems. Harvard University Press, Cambridge, MA.

Tiebout, H. M. III. 1993. Mechanisms of competition in tropical hummingbirds: Metabolic costs for losers and winners. Ecology 74:405–418.

Tilman, D. 1982. Resource Competition and Community Structure. Princeton University Press, Princeton, NJ.

Tilman, D. 1985. The resource ratio hypothesis of succession. Am. Nat. 125:827–852.

Tilman, D. 1986. Resources, competition, and the dynamics of plant communities. In M. J. Crawley (ed.), Plant Ecology, 51–75. Blackwell, Oxford.

Tilman, D. 1990. Mechanism of plant competition for nutrients: The elements of a predictive theory of competition. In J. B. Grace, and D. Tilman (eds.), Perspectives in Plant Competition, 117–141. Academic Press, San Diego.

Tilman, D. 1994. Competition and biodiversity in spatially structured habitats. Ecology 75:2–16.

Tilman, D., M. Mattson, and S. Langer. 1981. Competition and nutrient kinetics along a temperature gradient: An experimental test of a mechanistic approach to niche theory. Limnol. Oceanogr. 26:1020–1033.

Tilman, D., and S. Pacala. 1993. The maintenance of species richness in plant communities. In R. E. Ricklefs and D. Schluter (eds.), Species Diversity in Ecological Communities Historical and Geographical Perspectives, 13–25. University of Chicago Press, Chicago.

Tilman, G. D. 1984. Plant dominance along an experimental nutrient gradient. Ecology 65:1445–1453.

Tinbergen, L. 1969. The natural control of insects in pinewoods. I. Factors influencing the intensity of predation by songbirds. Arch. Neerl. Zool. 13:266–336.

Tinkle, D. W, and R. E. Ballinger. 1972. *Sceloporus undulatus:* A study of the intraspecific comparative demography of a lizard. Ecology 53:570–584.

Tipper, J. C. 1979. Rarefaction and rarefiction: The use and abuse of a method in paleoecology. Paleobiology 5:423–434.

Tokeshi, M. 1990. Niche apportionment or random assortment: Species abundance patterns revisited. J. Anim. Ecol. 59:1129–1146.

Tomback, D. F. 1980. How nutcrackers find their seed stores. Condor 82:10–19.

Torgerson, T. R., and R. W. Campbell. 1982. Some effects of avian predators on the western spruce budworm in north central Washington. Environ. Entomol. 11:429–431.

Townsend, C. R., and R. N. Hughes. 1981. Maximizing net energy returns from foraging. In C. R. Townsend and P. Calow (eds.), Physiological Ecology: An Evolutionary Approach to Resource Use, 86–108. Blackwell, Oxford.

Trail, P. W. 1985. The intensity of selection: Intersexual and interspecific comparisons require consistent measures. Am. Nat. 126:434–439.

Tramer, E. J. 1969. Bird species diversity: Components of Shannon's formula. Ecology 50:927–929.

Transeau, E. N. 1926. The accumulation of energy by plants. Ohio J. Sci. 26:1–10.

Travis, 1984. Anuran size at metamorphosis: Experimental test of a model based on intraspecific competition. Ecology 65:1155–1160.

Trenberth, K. E., and T. J. Hoar. 1996. The 1990–1995 El Niño–Southern Oscillation event: Longest on record. Geophys. Res. Lett. 23:57–60.

Trexler, J. C., and J. Travis. 1990a. Phenotypic plasticity in the sailfin molly, *Poecilia latipinna* (Pisces: Poeciliidae). I. Field experiments. Evolution 44:157–167.

Trexler, J. C., and J. Travis. 1990b. Phenotypic plasticity in the sailfin molly, *Poecilia latipinna* (Pisces: Poeciliidae). II. Laboratory experiment. Evolution 44:157–167.

Trivers, R. L. 1971. The evolution of reciprocal altruism. Quart. Rev. Biol. 46:35–57.

Trivers, R. L. 1972. Parental investment and sexual selection. In B. Campbell (ed.), Sexual Selection and the Descent of Man 1871–1971, 136–179. Aldine, Chicago.

Trivers, R. L. 1974. Parent-offspring conflict. Am. Zool. 14:249–264.

Trivers, R. L. 1985. Social Evolution. Benjamin/Cummings, Menlo Park, CA.

Trivers, R. L., and D. E. Willard. 1973. Natural selection of parental ability to vary the sex ratio of offspring. Science 179:90–92.

Trumble, J. T., D. M. Kolodyn-Hirsch, and I. P. Ting. 1993. Plant compensation for arthropod herbivory. Annu. Rev. Entomol. 38:93–119.

Turchin, P., and A. D. Taylor. 1992. Complex dynamics in ecological time series. Ecology 73:289–305.

Turesson, G. 1922. The genotypic response of the plant species to the habitat. Hereditas 3:211–350.

Turner, J. R. G. 1984. Mimicry: The palatability spectrum and its consequences. In R. I. Vane-Wright and P. R. Ackery (eds.), The Biology of Butterflies, 141–161. Academic Press, New York.

Turner, M. G. 1989. Landscape ecology: The effect of pattern on process. Annu. Rev. Ecol. Syst. 20:171–197.

Turner, T. 1983. Facilitation as a successional mechanism in a rocky intertidal community. Am. Nat. 121:729–738.

Turner, T. F., J. C. Trexler, G. L. Miller, and K. E. Toyer. 1994. Temporal and spatial dynamics of larval and juvenile fish abundance in a temperate floodplain river. Copeia 1994: 174–185.

Tyrrell, L. E., and T. R. Crow. 1994. Structural characteristics of old-growth hemlock-hardwood forests in relation to age. Ecology 75:370–386.

Udovic, D., and C. Aker. 1981. Fruit abortion and the regulation of fruit number in *Yucca whipplei.* Oecologia 49: 245–248.

Uetz, G. A. 1988. Group foraging in colonial web-building spiders: Evidence of risk sensitivity. Behav. Ecol. Sociobiol. 22:265–270.

Uetz, G. W., and G. S. Hieber. 1994. Group size and predation risk in colonial web-building spiders: Analysis of attack abatement mechanisms. Behav. Ecol. 5:326–333.

Uetz, G. W., and G. S. Hieber. 1997. Colonial web-building spiders: Balancing the costs and benefits of group-living. In J. C. Choe and B. J. Crespi (eds.), The Evolution of Social Behavior in Insects and Arachnids, 458–475. Cambridge University Press, Cambridge.

Utida, S. 1957. Population fluctuation: An experimental and theoretical approach. Cold Spring Harbor Symp. Quant. Biol. 22:139–151.

Uvarov, B. P. 1961. Quantity and quality in insect populations. Proc. R. Entomol. Soc. Lond. C 25:52–59.

Vadas, R. L. Sr., M. T. Burrows, and R. N. Hughes. 1994. Foraging strategies of dogwhelks, *Nucella lapillus* (L.): Interacting effects of age, diet and chemical cues to the threat of predation. Oecologia 100:439–450.

Valiela, I., and J. M. Teal. 1979. The nitrogen budget of a salt marsh ecosystem. Nature 280:652–656.

Van Cleve, K., L. A. Viereck, and R. L. Schlentner. 1971. Accumulation of nitrogen in alder (*Alnus*) ecosystems near Fairbanks, Alaska. Arctic Alpine Res. 3:101–114.

Vandermeer, J. H. 1969. The competitive structure of communities: An experimental approach with protozoa. Ecology 50:362–371.

Vandermeer, J. H. 1972. Niche theory. Annu. Rev. Ecol. Syst. 3:107–132.

van der Meijden, E., and C. A. M. van der Veen-Van Wijk. 1997. Tritrophic metapopulation dynamics. In I. A. Hanski and M. E. Gilpin, Metapopulation Biology: Ecology, Genetics, and Evolution, 387–405. Academic Press, San Diego.

Van der Merwe, M., S. L. Chown, and V. R. Smith. 1997. Thermal tolerance limits in six weevil species (Coleoptera, Curculionidae) from sub-Antarctic Marion Island. Polar Biol. 18:331–336.

van der Pijl, L. 1969. Principles of Dispersal in Higher Plants. Springer-Verlag, Berlin.

Van Dyne, G. M. 1966. Ecosystems, systems ecology, and systems ecologists. U.S. Atomic Energy Commission, Oak Ridge Natl. Lab. Rept. ORNL 3957:1–31. Reprinted 1969 in G. W. Cox (ed.), Readings in Conservation Ecology. Appleton-Century-Crofts, New York.

Van Dyne, G. M. (ed.). 1969. The Ecosystem Concept in Natural Resource Management. Academic Press, New York and London.

Van Dyne, G. M., and J. C. Anway. 1976. A research program for and the process of building and testing grassland ecosystems models. J. Range Mgmt. 29:114–122.

Van Emden, H. F., and G. F. Williams. 1974. Insect stability and diversity in agroecosystems. Annu. Rev. Entomol. 19:455–475.

Vannote, R. L., G. W. Minshall, K. W. Cummins, J. R. Sedell, and C. E. Cushing. 1980. The river continuum concept. Can. J. Fish. Aquat. Sci. 37:130–137.

Van Riper III, C., S. G. Van Riper, M. L. Goff, and M. Laird. 1986. The epizootiology and ecological significance of malaria in Hawaiian land birds. Ecol. Monogr. 56:327–344.

Van Valen, L. 1971. The history and stability of atmospheric oxygen. Science 171:439–443.

Varley, G. C. 1949. Population changes in German forest pests. J. Anim. Ecol. 18:117–122.

Varley, G. C. 1963. The interpretation of change and stability in insect populations. Proc. R. Entomol. Soc. Lond. C 27:52–57.

Varley, G. C., and G. R. Gradwell. 1970. Recent advances in insect population dynamics. Annu. Rev. Entomol. 15:1–24.

Varley, G. C., G. R. Gradwell, and M. P. Hassell. 1975. Insect Population Ecology. Blackwell Scientific Publications, Oxford.

Vaughan, T. A. 1986. Mammalogy. 3rd ed. Saunders, Philadelphia.

Vaughton, G., and S. M. Carthew. 1993. Evidence for selective fruit abortion in *Banksia spirulosa* (Proteaceae). Biol. J. Linn. Soc. 50:35–46.

Vaurie, C. 1950. Notes on Asiatic nuthatches and creepers. Am. Mus. Novit. 1472:1–39.

Vegis, A. 1964. Dormancy in higher plants. Annu. Rev. Plant Physiol. 15:185–215.

Vepsalainen, K. 1971. The role of gradually changing day length in determination of wing length, alary polymorphism and diapause in a *Gerris odontogaster* (Zett.) population (Gerridae, Heteroptera) in South Finland. Ann. Acad. Sci. Fenn. Ser. A IV 183:1–25.

Vepsalainen, K. 1973. The distribution and habitats of *Gerris* Fabr. species (Heteroptera, Gerridae) in Finland. Ann. Zool. Fenn. 10:419–444.

Vepsalainen, K. 1974a. Determination of wing length and diapause in water-striders (*Gerris* Fabr., Heteroptera). Hereditas 77:163–176.

Vepsalainen, K. 1974b. The life cycles and wing lengths of Finnish *Gerris* Fabr. species (Heteroptera, Gerridae). Acta Zool. Fenn. 141:1–73.

Vepsalainen, K. 1974c. The winglengths, reproductive stages and habitats of Hungarian *Gerris* Fabr. species (Heteroptera, Gerridae). Ann. Acad. Sci. Fenn. Ser. A IV 202:1–18.

Verghese, M. W., and A. W. Nordskog. 1968. Correlated responses in reproductive fitness to selection in chickens. Genet. Res. 11:221–238.

Verhulst, P. F. 1838. Notice sur la loi que la population suit dans son accroissement. Corresp. Math. Phys. 10:113–121.

Verner, J. 1964. Evolution of polygamy in the long-billed marsh wren. Evolution 18:252–261.

Verner, J., and M. F. Willson. 1966. The influence of habitats on mating systems of North American passerine birds. Ecology 47:143–147.

Verner, J., and M. F. Willson. 1969. Mating systems, sexual dimorphism, and the role of male North American passerine birds in the nesting cycle. Ornithological Monographs, no. 9. American Ornithologists Union, Anchorage, KY.

Vesey-Fitzgerald, D. F. 1960. Grazing succession among East African game animals. J. Mammal. 41:161–172.

Via, S. 1984. The quantitative genetics of polyphagy in an insect herbivore. I. Genotype-environment interactions in larval performance on different host plant species. Evolution 38:881–895.

Vickery, W. L., and T. D. Nudds. 1991. Testing for density-dependent effects in sequential censuses. Oecologia 85:419–423.

Vietmeyer, N. D. 1986. Lesser-known plants of potential use in agriculture and forestry. Science 232:1379–1384.

Vince, S., and I. Valiela. 1973. The effects of ammonia and phosphate enrichments on chlorophyll *a*, a pigment ratio and species composition of phytoplankton of Vineyard Sound. Mar. Biol. 19:69–73.

Vincent, T. L, and J. S. Brown. 1988. The evolution of ESS theory. Annu. Rev. Ecol. Syst. 19:423–443.

Vince-Prue, D. 1975. Photoperiodism in Plants. McGraw-Hill, New York.

Vinegar, A., V. H. Hutchinson, and H. G. Dowling. 1970. Metabolism, energetics, and thermoregulation during brooding of snakes of the genus *Python* (Reptilia, Boidae). Zoologica 55:19–50.

Vitousek, P. M. 1982. Nutrient cycling and nutrient use efficiency. Am. Nat. 119:553–572.

Vitousek, P. M. 1984. Litterfall, nutrient cycling, and nutrient limitation in tropical forests. Ecology 65:285–298.

Vitousek, P. M. 1994. Beyond global warming: Ecology and global change. Ecology 75:1861–1876.

Vitousek, P. M., and P. A. Matson. 1991. Gradient analysis of ecosystems. In J. Cole, G. Lovett, and S. Findlay (eds.), Comparative Analysis of Ecosystems, 287–298. Springer-Verlag, New York.

Vitousek, P. M., and R. L. Stanford. 1986. Nutrient cycling in moist tropical forests. Annu. Rev. Ecol. Syst. 17:137–167.

Vitousek, P. M., and P. S. White. 1981. Process studies in succession. In D. C. West, H. H. Shugart, and D. B. Botkin (eds.), Forest Succession: Concepts and Application, 267–276. Springer-Verlag, New York.

Vogel, S. 1970. Convective cooling at low airspeeds and the shape of broad leaves. J. Exp. Bot. 21:91–101.

Vogel, S. 1981. Life in Moving Fluids: The Physical Biology of Flow. Princeton University Press, Princeton, NJ.

Vogel, S. 1988. Life's Devices. Princeton University Press, Princeton, NJ.

Vogl, R. J. 1973. Ecology of knobcone pine in the Santa Ana Mountains of California. Ecol. Monogr. 43:125–143.

Vogl, R. J., and P. K. Schorr. 1972. Fire and manzanita chaparral in the San Jacinto Mountains, California. Ecology 53:1179–1188.

Vogt, K. A., C. C. Grier, C. E. Meier, and R. L. Edmunds. 1982. Mycorrhizal role in net primary production and nutrient cycling in *Abies amabilis* ecosystems in western Washington. Ecology 63:370–380.

Vogt, K. A., C. C. Grier, and D. J. Vogt. 1986. Production, turnover, and nutrient dynamics of above- and belowground detritus of world forests. Adv. Ecol. Res. 15:303–337.

Volterra, V. 1926a. Variazioni e fluttuazioni del numero d'individui in specie animali conviventi. Mem. Acad. Lincei 2:31–113.

Volterra, V. 1926b. Variations and fluctuations of the numbers of individuals in animal species living together. Reprinted 1931 in R. N. Chapman, Animal Ecology. McGraw-Hill, New York.

Waage, J., and D. Greathead (eds.). 1986. Insect Parasitoids. Academic Press, London.

Wade, M. J. 1977. Experimental study of group selection. Evolution 31:134–153.

Wade, M. J. 1978. A critical review of the models of group selection. Q. Rev. Biol. 53:101–114.

Waggoner, P. E., and G. R. Stephens. 1970. Transition probabilities for a forest. Nature 255:1160–1161.

Wagner, F. H., and L. C. Stoddart. 1972. Influence of coyote predation on black-tailed jackrabbit populations in Utah. J. Wildl. Mgmt. 36:329–343.

Wahlström, L. K., and P. Kjellander. 1995. Ideal free distribution and natal dispersal in female roe deer. Oecologia 103:302–308.

Waisel, Y. 1972. Biology of Halophytes. Academic Press, New York.

Wake, D. B. 1982. Functional and evolutionary morphology. Perspect. Biol. Med. 25:603–620.

Waldman, B. 1988. The ecology of kin recognition. Annu. Rev. Ecol. Syst. 19:543–571.

Walker, B. H. 1981. Is succession a viable concept in African savanna ecosystems? In D. C. West, H. H. Shugart, and D. B. Botkin (eds.), Forest Succession: Concepts and Application, 430–447. Springer-Verlag, New York.

Walker, R. B. 1954. The ecology of serpentine soils. II. Factors affecting plant growth on serpentine soils. Ecology 35:259–266.

Wallace, A. R. 1876. The Geographical Distribution of Animals, Vol. 1 and 2. Reprinted 1962, Hafner, New York.

Wallace, A. R. 1878. Tropical Nature and Other Essays. Macmillan, New York and London.

Waloff, Z. 1966. The upsurges and recessions of the desert locust plague: An historical survey. Anti-Locust Mem. 8:1–111.

Walsh, J. 1986. Return of the locust: A cloud over Africa. Science 234:17–19.

Walters, R. G., and P. Horton. 1995. Acclimation of *Arabidopsis thaliana* to the light environment: Changes in composition of the photosynthetic apparatus. Planta 195: 248–256.

Wangensteen, O. D., H. Rahn, R. R. Burton, and A. H. Smith. 1974. Respiratory gas exchange of high altitude adapted chick embryos. Respir. Physiol. 21:61–70.

Ward, P. 1994. The End of Evolution. Bantam Books, New York.

Ward, P., and A. Zahavi. 1973. The importance of certain assemblages of birds as "information-centres" for food-finding. Ibis 115:517–534.

Ward, R. C. 1967. Principles of Hydrology. McGraw-Hill, New York and London.

Wardenaar, E. C. P., and J. Sevink. 1992. Comparative study of soil formation in primary stands of Scots pine (planted) and poplar (natural) on calcareous dune sands in the Netherlands. Plant-Soil 140:109–120.

Waring, R. H. 1983. Estimating forest growth and efficiency in relation to canopy leaf area. Adv. Ecol. Res. 13:327–354.

Waring, R. H., and J. Major. 1964. Some vegetation of the California coastal region in relation to gradients of moisture, nutrients, light, and temperature. Ecol. Monogr. 34:167–215.

Warner, R. R. 1975. The adaptive significance of sequential hermaphroditism in animals. Am. Nat. 109:61–82.

Warren, M. S. 1992. The conservation of British butterflies. In R. L. H. Dennis (ed.), The Ecology of Butterflies in Britain, 246–274. Oxford University Press, Oxford.

Warren, M. S. 1993. A review of butterfly conservation in central southern Britain: I. Protection, evaluation and extinction on prime sites. Biol. Conserv. 64:25–35.

Warren, W. G., and C. L. Batcheler. 1979. The density of spatial patterns: Robust estimation through distance methods. In R. M. Cormack and J. K. Ord (eds.), Spatial and Temporal Analysis in Ecology, 247–270. International Cooperative Publishing House, Fairland, MD.

Washburn, S. L., and I. DeVore. 1961. The social life of baboons. Sci. Am. 204:62–71.

Watkinson, A. R., and W. J. Sutherland. 1995. Sources, sinks and pseudo-sinks. J. Anim. Ecol. 64:126–130.

Watson, A., and D. Jenkins. 1968. Experiments on population control by territorial behaviour in red grouse. J. Anim. Ecol. 37:395–614.

Watson, A., and R. Moss. 1980. Advances in our understanding of the population dynamics of red grouse from a recent fluctuation in numbers. Ardea 68:103–111.

Watson, A., and R. Parr. 1981. Hormone implants affecting territory size and aggressive and sexual behaviour in red grouse. Ornis Scand. 12:55–61.

Watson, I. A. 1970. Changes in virulence and population shifts in plant pathogens. Annu. Rev. Phytopathol. 8:209–230.

Watson, J. G. 1928. The mangrove swamps of the Malay Peninsula. Malay. For. Rec. 6.

Watt, A. S. 1947. Pattern and process in the plant community. J. Ecol. 35:1–22.

Watt, K. E. F. 1968. Ecology and Resource Management. McGraw-Hill, New York.

Way, M. J. 1963. Mutualism between ants and honeydew producing Hornoptera. Annu. Rev. Entomol. 8:307–344.

Weatherhead, P. J., R. Montgomerie, G. Gibbs, and P. T. Boag. 1994. The cost of extra-pair fertilizations to female red-winged blackbirds.

Webb, W., S. Szarek, W. Lauenroth, R. Kinerson, and M. Smith. 1978. Primary productivity and water use in native forest, grassland, and desert ecosystems. Ecology 59: 1239–1247.

Webster, J. R., and B. C. Patton. 1979. Effects of watershed perturbation on stream potassium and calcium dynamics. Evol. Monogr. 49:51–72.

Wecker, S. C. 1963. The role of early experience in habitat selection by the prairie deer mouse, *Peromyscus maniculatus bairdii.* Ecol. Monogr. 33:307–325.

Wecker, S. C. 1964. Habitat selection. Sci. Am. 211:109–116.

Weeks, H. P. Jr., and C. M. Kirkpatrick. 1976. Adaptations of white-tailed deer to naturally occurring sodium deficiencies. J. Wildl. Mgmt. 40:610–625.

Weeks, S. C., and G. K. Meffe. 1996. Quantitative genetic and optimality analysis of life-history plasticity in the eastern mosquitofish. Evolution 50:1358–1365.

Weinberg, G. M. 1975. An Introduction to General Systems Thinking. Wiley, New York.

Weiner, J. 1988. Variation in the performance of individuals in plant populations. In A. J. Davy, M. J. Hutchings, and A. R. Watkinson (eds.), Plant Population Ecology, 59–81. Blackwell, Oxford.

Weiner, J. 1994. The Beak of the Finch. Alfred A. Knopf, New York.

Weis, A. E., and M. R. Berenbaum. 1989. Herbivorous insects and green plants. In W. G. Abrahamson, Plant-Animal Interactions, 123–162. McGraw-Hill, New York.

Weissburg, M. 1986. Risky business: On the ecological relevance of risk-sensitive foraging. Oikos 46:261–262.

Weisskopf, V. F. 1968. How light interacts with matter. Sci. Am. 219:60–71.

Welch, H. 1968. Relationships between assimilation efficiencies and growth efficiencies for aquatic consumers. Ecology 49:755–759.

Weller, D. E. 1987. A reevaluation of the −3/2 power rule of plant self-thinning. Ecol. Monogr. 57:23–43.

Weller, S. G. 1994. The relationship of rarity to plant reproductive biology. In M. L. Bowles and C. J. Whelan (eds.), Restoration of Endangered Species, 90–117. Cambridge University Press, Cambridge

Wellington, W. G. 1960. Qualitative changes in natural populations during changes in abundance. Can. J. Zool. 38: 289–314.

Wells, H. 1979. Self-fertilization: Advantageous or deleterious? Evolution 33:252–255.

Wells, K. D. 1977. The social behaviour of anuran amphibians. Anim. Behav. 25:666–693.

Werner, D. 1992. Physiology of nitrogen-fixing legume nodules: Compartments and functions. In G. Stacey, R. H. Burris, and H. J. Evans (eds.), Biological Nitrogen Fixation, 399–460. Chapman & Hall, New York.

Werner, E. E. 1977. Species packing and niche complementarity in three sunfishes. Am. Nat. 111:553–578.

Werner, E. E. 1986. Amphibian metamorphosis: Growth rate, predation risk, and the optimal size at transformation. Am. Nat. 128:319–341.

Werner, E. E., J. F. Gilliam, D. J. Hall, and G. G. Mittelbach. 1983. An experimental test of the effects of predation on habitat use in fish. Ecology 54:1540–1548.

Werner, E. E., and D. J. Hall. 1976. Niche shifts in sunfishes: Experimental evidence and significance. Science 191: 404–406.

Werner, E. E., and D. J. Hall. 1977. Competition and habitat shift in two sunfishes (Centrarchidae). Ecology 58:869–876.

Werner, E. E., and D. J. Hall. 1979. Foraging efficiency and habitat switching in competing sunfishes. Ecology 60: 256–264.

Werner, P. A., and W. J. Platt. 1976. Ecological relationships of co-occurring goldenrods (*Solidago:* Compositae). Am. Nat. 110:959–971.

Werren, J. H. 1980. Sex ratio adaptations to local mate competition in a parasitic wasp. Science 208:1157–1159.

West, G. C. 1972. Seasonal differences in resting metabolic rate of Alaskan ptarmigan. Comp. Biochem. Physiol. 42A: 867–876.

West Eberhard, M. J. 1975. The evolution of social behavior by kin selection. Q. Rev. Biol. 50:1–33.

West Eberhard, M. J. 1981. Intragroup selection and the evolution of insect societies. In R. D. Alexander and D. W. Tinkle (eds.), Natural Selection and Social Behavior, 3–17. Chiron, New York.

Westheimer, F. H. 1987. Why nature chose phosphates. Science 235:1173–1178.

Westoby, M. 1988. Comparing Australian ecosystems to those elsewhere. BioScience 38:549–556.

Wetzel, R. G. 1995. Death, detritus, and energy flow in aquatic ecosystems. Freshwater Ecol. 33:83–89.

Wetzel, R. G. 1983. Limnology. 2nd ed. Saunders College Publishing, Philadelphia.

Wheeler, W. M. 1923. Social Life among the Insects. Harcourt, Brace, New York.

Wheeler, W. M. 1928. The Social Insects: Their Origin and Evolution. Harcourt, Brace, New York.

Wheeler, W. M. 1933. Colony Founding among Ants. Harvard University Press, Cambridge, MA.

Whelan, R. J. 1995. The Ecology of Fire. Cambridge Studies in Ecology. Cambridge University Press, Cambridge.

White, C. D. 1971. Vegetation-Soil Chemistry Correlations in Serpentine Ecosystems. Ph.D. diss., University of Oregon, Eugene.

White, K. A. J., M. A. Lewis, and J. D. Murray. 1996. A model of wolf-pack territory formation and maintenance. J. Theor. Biol. 178:29–43.

Whitham, T. G., and S. Mopper. 1985. Chronic herbivory: Impacts on architecture and sex expression of pinyon pine. Science 228:1089–1091.

Whitlock, M. C. 1992. Nonequilibrium population structure in forked fungus beetles: Extinction, colonization, and the genetic variance among populations. Am. Nat. 139: 952–970.

Whitman, D. W. 1987. Thermoregulation and daily activity patterns in a black desert grasshopper, *Taeniopoda eques.* Anim. Behav. 35:1814–1826.

Whitman, D. W. 1988. Function and evolution of thermoregulation in the desert grasshopper *Taeniopoda eques.* J. Anim. Ecol. 57:811–825.

Whittaker, R. H. 1952. A study of summer foliage insect communities in the Great Smoky Mountains. Ecol. Monogr. 22:1–44.

Whittaker, R. H. 1953. A consideration of climax theory: The climax as a population and pattern. Ecol. Monogr. 23:41–78.

Whittaker, R. H. 1954. The ecology of serpentine soils. I. Introduction. Ecology 35:258–259.

Whittaker, R. H. 1956. Vegetation of the Great Smoky Mountains. Ecol. Monogr. 26:1–80.

Whittaker, R. H. 1960. Vegetation of the Siskiyou Mountains, Oregon and California. Ecol. Monogr. 30:279–338.

Whittaker, R. H. 1965. Dominance and diversity in land plant communities. Science 147:250–260.

Whittaker, R. H. 1967. Gradient analysis of vegetation. Biol. Rev. 42:207–264.

Whittaker, R. H. 1972. Evolution and measurement of species diversity. Taxon 21:213–251.

Whittaker, R. H. 1975. Communities and Ecosystems. 2nd ed. Macmillan, New York.

Whittaker, R. H., and P. P. Feeny. 1971. Allelochemics: Chemical interactions between species. Science 171:757–770.

Whittaker, R. H., and G. E. Likens. 1973a. Carbon in the biota. In G. M. Woodwell and E. V. Pecan (eds.). Carbon and the Biosphere, 281–302. Conf. 720510, National Technical Information Service, Washington, D.C.

Whittaker, R. H., and G. E. Likens. 1973b. Primary production: The biosphere and man. Human Ecol. 1:357–369.

Whittaker, R. H., and W. A. Niering. 1965. Vegetation of the Santa Catalina Mountains, Arizona: A gradient analysis of the south slope. Ecology 46:429–452.

Whittaker, R. H., and G. M. Woodwell. 1968. Dimension and production relations of trees and shrubs in the Brookhaven Forest, New York. J. Ecol. 56:1–25.

Whittaker, R. H., and G. M. Woodwell. 1969. Structure, production and diversity of the oak-pine forest at Brookhaven, New York. J. Ecol. 57:155–174.

Wickler, W. 1968. Mimicry in Plants and Animals. World University Library, London.

Widemo, F., and I. P. F. Owens. 1995. Lek size, male mating skew and the evaluation of lekking. Nature 373:148–150.

Wiebe, W. J. 1975. Nitrogen fixation in a coral reef community. Science 188:257–259.

Wiebes, J. T. 1979. Co-evolution of figs and their insect pollinators. Annu. Rev. Ecol. Syst. 10:1–12.

Wiegert, R. G. 1988. The past, present, and future of ecological energetics. In L. R. Pomeroy and J. J. Albert (eds.), Concepts of Ecosystem Ecology: A Comparative View, 29–55. Springer-Verlag, New York.

Wiegert, R. G., and D. F. Owen. 1971. Trophic structure, available resources and population density in terrestrial vs. aquatic ecosystems. J. Theor. Biol. 30:69–81.

Wiens, D. 1984. Ovule survivorship, brood size, life history, breeding systems, and reproductive success in plants. Oecologia 64:47–53.

Wiens, J. A. 1966. On group selection and Wynne-Edwards' hypothesis. Am. Sci. 273–287.

Wiens, J. A. 1973. Pattern and process in grassland bird communities. Ecol. Monogr. 43:237–270.

Wiens, J. A. 1977. On competition and variable environments. Am. Sci. 65:590–597.

Wiens, J. A. 1992. Ecological flows across landscape boundaries: A conceptual overview. In A. J. Hansen and F. di Castri (eds.), Landscape Boundaries, 217–235. Springer-Verlag, New York.

Wiens, J. A. 1997. Metapopulation dynamics and landscape ecology. In I. A. Hanski and M. E. Gilpin (eds.), Metapopulation Biology: Ecology, Genetics, and Evolution, 43–62. Academic Press, San Diego.

Wiklund, C. 1981. Generalist vs. specialist oviposition behaviour in *Papilio machaon* (Lepidoptera) and functional aspects of the hierarchy of oviposition preferences. Oikos 36:163–170.

Wilbur, H. M. 1972. Competition, predation, and the structure of the *Ambystoma-Rana sylvatica* community. Ecology 53:3–21.

Wilbur, H. M., and R. A. Alford. 1985. Priority effects in experimental pond communities: Responses of *Hyla* to *Bufo* and *Rana*. Ecology 66:1106–1114.

Wilbur, H. M., P. J. Morin, and R. N. Harms. 1983. Salamander predation and the structure of experimental communities: Anuran responses. Ecology 64:1423–1429.

Wilbur, H. M., D. W. Tinkle, and J. P. Collins. 1974. Environmental certainty, trophic level, and resource availability in life history evolution. Am. Nat. 108:805–817.

Wilde, S. A. 1968. Mycorrhizae and tree nutrition. BioScience 18:482–484.

Wiley, R. H. 1974. Evolution of social organization and life-history patterns among grouse. Q. Rev. Biol. 49:201–227.

Wilkinson, G. S. 1984. Reciprocal food sharing in the vampire bat. Nature 308:181–184.

Willey, R. W., and S. B. Heath. 1969. The quantitative relationships between plant population and crop yield. Adv. Agron. 21:281–321.

Williams, C. B. 1943. Area and the number of species. Nature 152:264–267.

Williams, C. B. 1964. Patterns in the Balance of Nature and Related Problems in Quantitative Ecology. Academic Press, New York.

Williams, G. C. 1957. Pleiotropy, natural selection, and the evolution of senescence. Evolution 11:398–411.

Williams, G. C. 1966a. Adaptation and Natural Selection. Princeton University Press, Princeton, NJ.

Williams, G. C. 1966b. Natural selection, the costs of reproduction, and a refinement of Lack's principle. Am. Nat. 100:687–690.

Williams, G. C. 1975. Sex and Evolution. Princeton University Press, Princeton, NJ.

Williams, G. C. 1979. The question of adaptive sex ratio in outcrossed vertebrates. Proc. R. Soc. Lond. B 205:567–580.

Williams, G. C., and J. B. Mitton. 1973. Why reproduce sexually? J. Theor. Biol. 39:545–554.

Williams, G. C., and D. C. Williams. 1957. Natural selection of individually harmful social adaptations among sibs with special reference to social insects. Evolution 11:32–39.

Williams, P. H. 1975. Genetics of resistance in plants. Genetics 79:409–419.

Williamson, G. B. 1975. Pattern and seral composition in an old-growth beech-maple forest. Ecology 56:727–731.

Williamson, M. 1981. Island Populations. Oxford University Press, Oxford.

Williamson, M. H. 1983. The land-bird community of Skokholm: Ordination and turnovers. Oikos 41:378–384.

Willis, J. C. 1922. Age and Area: A Study in Geographical Distribution and Origin in Species. Cambridge University Press, Cambridge.

Willson, M. F. 1983. Plant Reproductive Ecology. Wiley, New York.

Willson, M. F., and N. Burley. 1983. Mate Choice in Plants: Tactics, Mechanisms, and Consequences. Princeton University Press, Princeton, NJ.

Wilson, D. S. 1980. The Natural Selection of Populations and Communities. Benjamin/Cummings, Menlo Park, CA.

Wilson, D. S. 1983. The group selection controversy: History and current status. Annu. Rev. Ecol. Syst. 14:159–187.

Wilson, E. O. 1961. Nature of the taxon cycle in the Melanesian ant fauna. Am. Nat. 95:169–193.

Wilson, E. O. 1971. The Insect Societies. Belknap Press of Harvard University Press, Cambridge, MA.

Wilson, E. O. 1973. Group selection and its significance for ecology. BioScience 23:631–638.

Wilson, E. O. 1984. Biophilia. Harvard University Press, Cambridge.

Wilson, E. O. (ed.). 1988. Biodiversity. National Academy Press, Washington, D.C.

Wilson, E. O. 1992. The Diversity of Life. Belknap Press of Harvard University Press, Cambridge, MA.

Wilson, J. B. 1988. Shoot competition and root competition. J. Appl. Ecol. 25:279–296.

Wilson, J. B. 1991. Methods of fitting dominance/diversity curves. J. Vegetation Sci. 2:35–46.

Wilson, J. B. 1996. The myth of constant predator: prey ratios. Oecologia 106:272–276.

Wilson, J. W. 1974. Analytical zoogeography of North American mammals. Evolution 28:124–140.

Wilson, K. 1989. The evolution of oviposition behaviour in the bruchid *Callosobruchus maculatus.* Ph.D. thesis, University of Sheffield.

Wilson, S. D., and D. Tilman. 1991. Components of plant competition along an experimental gradient of nitrogen availability. Ecology 72:1050–1065.

Winemiller, K. O. 1990. Spatial and temporal variation in tropical fish trophic networks. Ecol. Monogr. 60:331–367.

Winemiller, K. O., and G. A. Polis. 1996. Food Webs: What do they tell us about the world? In G. A. Polis and K. O. Winemiller (eds.), Food Webs: Integration of Patterns and Dynamics, 1–24. Chapman & Hall, New York.

Wingfield, J. C. 1984. Androgens and mating systems: Testosterone-induced polygyny in normally monogamous birds. Auk 101:665–671.

Wit, C. T. de. 1960. On competition. Versl. Landbouwk. Onderz. 66:1–82.

Witkamp, M. 1966. Decomposition of leaf litter in relation to environment, microflora, and microbial respiration. Ecology 47:194–201.

Witkamp, M., and J. Van Der Drift. 1961. Breakdown of forest litter in relation to environmental factors. Plant Soil 15:295–311.

Witman, J. D. 1985. Refuges, biological disturbance, and rocky subtidal community structure in New England. Ecol. Monogr. 55:421–445.

Witman, J. D. 1987. Subtidal coexistence: Storms, grazing, mutualism, and the zonation of kelps and mussels. Ecol. Monogr. 57:167–187.

Wittenberger, J. F. 1978. The evolution of mating systems in grouse. Condor 80:126–137.

Witter, J. A., H. M. Kulman, and A. C. Hodson. 1975. Life tables for the forest tent caterpillar. Ann. Entomol. Soc. Am. 65:25–31.

Witztum, A., and K. Schulgasser. 1995. The mechanics of seed expulsion in Acanthaceae. J. Theor. Biol. 176:531–542.

Woiwod, I. P., and I. Hanski. 1992. Patterns of density dependence in moths and aphids. J. Anim. Ecol. 61: 619–629.

Wolanski, E., and W. M. Hammer. 1988. Topographically controlled fronts in the ocean and their biological significance. Science 241:177–181.

Wolcott, T. G. 1984. Uptake of interstitial water from soil: Mechanism and ecological significance in the Ghost Crab *Ocypode quadrata* and two gecarcinid land crabs. Physiol. Zool. 57:161–184.

Wolcott, T. G. 1992. Water and solute balance in the transition to land. Am. Zool. 32:428–437.

Wolda, H. 1989. The equilibrium concept and density dependence tests: What does it all mean? Oecologia 81: 430–432.

Wolfe, J. A. 1975. Some aspects of plant geography of the Northern Hemisphere during the late Cretaceous and Tertiary. Ann. Mo. Bot. Gard. 62:264–279.

Wolfe, J. A. 1979. Temperature parameters of humid to mesic forests of eastern Asia and relation to forests of other regions of the Northern Hemisphere and Australasia. Geol. Surv. Prof. Paper 1106:1–37.

Wolfe, J. A. 1981. Vicariance biogeography of angiosperms in relation to paleobotanical data. In G. Nelson and D. E. Rosen (eds.), Vicariance Biogeography: A Critique, 413–427. Columbia University Press, New York.

Wolin, M. J. 1979. The rumen fermentation: A model for microbial interactions in anaerobic systems. Adv. Microbial Ecol. 3:49–77.

Wolk, P. 1973. Physiology and cytological chemistry of blue-green algae. Bacteriol. Rev. 37:32–101.

Wolter, K. 1987. The Southern Oscillation in surface circulation and climate over the tropical Atlantic, Eastern Pacific, and Indian Oceans as captured by cluster analysis. J. Climate Appl. Meteor. 26:540–558.

Wolter, K, and M. S. Timlin. 1993. Monitoring ENSO in COADS with a seasonally adjusted principal component index. Proceedings of the 17th Climate Diagnostics Workshop, Normal, OK, NOAA/NMC/CAC, NSSL, Oklahoma Climate Survey, CIMMS and the School of Meteorology, University of Oklahoma, 52–57.

Wood, D. M., and R. Del Moral. 1987. Mechanisms of early primary succession in subalpine habitats on Mount St. Helens. Ecology 68:780–790.

Wood, D. M., and R. Del Moral. 1988. Colonizing plants on the pumice plains, Mount St. Helens, Washington. Am. J. Bot. 75:1228–1237.

Woodwell, G. M., J. E. Hobbie, R. A. Houghton, J. M. Melillo, B. Moore, B. J. Peterson, and G. R. Shaver. 1983. Global deforestation: Contribution to atmospheric carbon dioxide. Science 222:1081–1086.

Woolfenden, G. E., and J. W. Fitzpatrick. 1984. The Florida Scrub Jay: Demography of a Cooperative-Breeding Bird. Princeton University Press, Princeton, NJ.

Wootton, J. T. 1993. Indirect effects and habitat use in an intertidal community: Interaction chains and interaction modifications. Am. Nat. 141:71–89.

Wootton, J. T. 1994. The nature and consequences of indirect effects in ecological communities. Annu. Rev. Ecol. Syst. 25:443–466.

World Resources Institute. 1991. World Resources Report 1990–1991: A Guide to the Global Environment. Oxford University Press, New York.

Worthington, E. B. (ed.). 1975. The Evolution of IBP. Cambridge University Press, London.

Wright, D. H., D. J. Currie, and B. A. Maurer. 1993. Energy supply and patterns of species richness on local and regional scales. In R. E. Ricklefs and D. Schluter (eds.), Species Diversity in Ecological Communities: Historical and Geographical Perspectives, 66–74. University of Chicago Press, Chicago

Wright, H. E. Jr. 1964. Aspects of the early postglacial forest succession in the Great Lakes Region. Ecology 45:439–448.

Wright, H. E. Jr. 1968. The roles of pine and spruce in the forest history of Minnesota and adjacent areas. Ecology 49:937–955.

Wright, S. J. 1931. Evolution in Mendelian populations. Genetics 16:97–159.

Wright, S. J. 1932. The roles of mutation, inbreeding, crossbreeding and selection in evolution. Proc. VI Intl. Gen. Congr 1:356–366.

Wright, S. J. 1943. Isolation by distance. Genetics 28:114–138.

Wright, S. J. 1945. Tempo and mode in evolution: A critical review. Ecology 26:415–419.

Wright, S. J. 1946. Isolation by distance under diverse systems of mating. Genetics 31:39–59.

Wright, S. J. 1969. Evolution and the Genetics of Populations. Vol. 2. The Theory of Gene Frequencies. University of Chicago Press, Chicago.

Wright, S. J. 1980. Density compensation in island avifaunas. Oecologia 45:385–389.

Wright, S. J. 1982. The shifting balance theory and macroevolution. Annu. Rev. Genet. 16:1–19.

Wright, S. J. 1988. Patterns of abundance and the form of the species-area relation. Am. Nat. 131:401–411.

Wunderle, J. M., Jr. 1985. An ecological comparison of the avifaunas of Grenada and Tobago, West Indies. Wilson Bulletin 97:356–366.

Wunderle, J. M., M. S. Castro, and N. Fetcher. 1988. Risk averse foraging by bananaquits on negative energy budgets. Behav. Ecol. Sociobiol. 21:249–255.

Wyatt, R. 1983. Pollinator-plant interactions and the evolution of breeding systems. In L. Real (ed.), Pollination Biology, 51–95. Academic Press, New York.

Wynne-Edwards, V. C. 1962. Animal Dispersion in Relation to Social Behaviour. Oliver and Boyd, Edinburgh.

Wynne-Edwards, V. C. 1963. Intergroup selection in the evolution of social systems. Nature 200:623–628.

Wynne-Edwards, V. C. 1986. Evolution through Group Selection. Blackwell Scientific, Palo Alto, CA.

Yager, D. D., M. L. May, and B. M. Fenton. 1990a. Ultrasound-triggered, flight-gated evasive maneuvers in the flying praying mantis, *Paraphendale agrionina.* I. Free flight. J. Exp. Biol. 152:17–39.

Yager, D. D., M. L. May, and B. M. Fenton. 1990b. Ultrasound-triggered, flight-gated evasive maneuvers in the flying praying mantis, *Paraphendale agrionina.* II. Tethered flight. J. Exp. Biol. 152:41–58.

Yampolsky, E., and H. Yampolsky. 1922. Distribution of sex forms in phanerogamic flora. Bibl. Genet. 3:1–62.

Yancey, P. H., M. E. Clark, S. C. Hand, R. D. Bowlus, and G. N. Somero. 1982. Living with water stress: Evolution of osmolyte systems. Science 217:1214–1222.

Yancey, P. H., and G. N. Somero. 1980. Methylamine osmoregulatory solutes of elasmobranch fishes counteract urea inhibition of enzymes. J. Exp. Zool. 212:205–213.

Yang, R. S. H., and F. E. Guthrie. 1969. Physiological responses of insects to nicotine. Ann. Entomol. Soc. Am. 62:141–146.

Yeaton, R. L., and M. L. Cody. 1979. Distribution of cacti along environmental gradients in the Sonoran and Mojave deserts. J. Ecol. 67:529–541.

Yeaton, R. L, R. W. Yeaton, J. P. Waggoner III, and J. E. Horenstein. 1985. The ecology of *Yucca* (Agavaceae) over an environmental gradient in the Mohave Desert: Distribution and interspecific interactions. J. Arid Environ. 8:33–44.

Yoda, K., T. Kira, H. Ogawa, and K. Hozuml. 1963. Self thinning in overcrowded pure stands under cultivated and natural conditions. J. Biol., Osaka City University 14: 107–129.

Yodzis, P. 1988. The indeterminacy of ecological interactions as perceived through perturbation experiments. Ecology 69:508–515.

Yodzis, P. 1989. Introduction to Theoretical Ecology. Harper & Row, New York.

Yodzis, P. 1995. Food webs and perturbation experiments: Theory and practice. In G. Pollis and K. Winemiller (eds.), Food Webs: Integration of Patterns and Dynamics, 192–200. Chapman & Hall, New York.

Yoshimura, J., and C. W. Clark (eds.). 1993. Adaptation in Stochastic Environments. Lecture Notes in Biomathematics, vol. 98. Springer-Verlag, Berlin.

Young, E. C. 1965. Flight muscle polymorphism in British Corixidae: Ecological observations. J. Anim. Ecol. 34:353–390.

Young, J. P. W. 1992. Phylogenetic classification of nitrogen-fixing organisms. In G. Stacey, R. H. Burris, and H. J. Evans (eds.), Biological Nitrogen Fixation, 43–86. Chapman & Hall, New York.

Young, T. P. 1990. Evolution of semelparity in Mount Kenya lobelias. Evol. Ecol. 4:157–171.

Young, T. P., and C. K. Augspurger. 1991. Ecology and evolution of long-lived semelparous plants. Trends Ecol. Evol. 6:285–289.

Yule, G. U. 1924. A mathematical theory of evolution based on the conclusions of Dr. J. C. Willis F. R. S. Phil. Trans. R. Soc. Lond. B 213:21–87.

Zahavi, A. 1975. Mate selection: A selection for a handicap. J. Theor. Biol. 53:205–214.

Zahavi, A. 1977a. The cost of honesty (further remarks on the handicap principle). J. Theor. Biol. 67:603–605.

Zahavi, A. 1977b. Reliability in communication systems and the evolution of altruism. In B. Stonehouse and C. M. Perrins (eds.), Evolutionary Ecology, 253–259. Macmillan, London.

Zahavi, A. 1987. The theory of sexual selection and some of its implications. In V. P. Delfino (ed.), Proceedings of International Symposium on Biology and Evolution, 305–325. Adriatica Editrica, Bari.

Zahavi, A. 1991. On the definition of sexual selection, Fisher's model, and the evolution of waste and of signals in general. Anim. Behav. 42:501–503.

Zaret, T. M. 1980. Predation and Freshwater Communities. Yale University Press, New Haven, CT.

Zaret, T. M., and R. T. Paine. 1973. Species introduction in a tropical lake. Science 182:449–455.

Zehnder, A. J. B. (ed.). 1988. Biology of Anaerobic Microorganisms. John Wiley & Sons, New York.

ANSWERS TO SELECTED EXERCISES

CHAPTER 1

5. In preparing this answer consider that, in many cases, elected officials do not have an extensive background in science.

CHAPTER 2

2. To sketch out the rudiments of a simple mathematical model, you must understand the important processes and activities of the organism in which you are interested, an understanding that comes from observation. For example, photosynthesis in green plants requires light. A simple relationship may link the rate of photosynthesis, call that R, to the intensity of light, which you may call I. The question whether R increases linearly with I or whether there is a gradual increase in photosynthetic rate with increasing light intensity until some threshold of light intensity is reached, is the heuristic part of model development.

CHAPTER 3

2. Recall that your activities may have an indirect effect on the environment. For example, if you drove your car today, you contributed carbon dioxide to the atmosphere. If you purchased food at the grocery store, you probably bought with it a considerable amount of packaging material composed of plastic and natural products such as paper. In developing your list, think in both the short and long-term.

CHAPTER 7

4. A common misconception about evolution is that it is directed toward improvement in the sense of asthetic quality or utility in human terms. In this sense, organisms are not improved by evolution since evolutionary change is non-directional. In the sense that the evolutionary process adjusts behavior and morphology in the face of environmental change, organisms are improved.

CHAPTER 8

2. One possibility is a lake that occurs at a relatively high altitude in the temperate region. Such lakes might be stratified in all but the very warmest part of the year when water temperature increases enough for turnover to occur.

CHAPTER 9

4. Lindeman wrote: "Where λ_n is by definition positive and represents the rate of contribution of energy from Λ_{n+1} (the previous level) to Λ_n, while λ_n' is negative and represents the sum of the rate of energy dissipated from Λ_n and the rate of energy content handed on to the following level Λ_{n+1}." $\lambda_1 =$ "rate of contribution" $= (0.10)(100{,}000) = 10{,}000$. $\lambda_n' =$ "energy dissipated" $= 10{,}000 - 1000 = 9000$.

CHAPTER 10

1. One gram of CO_2 assimilated represents 0.614 g of carbohydrate produced. Thus, 9,201 g of $C_6H_{12}O_6$ would require $9{,}201/0.614 = 14{,}985$ g of CO_2.

2. Carbohydrates contain 17.6 kJ g^{-1} of energy, proteins contain 23.8 kJ g^{-1}, and fats contain 39.7 kJ g^{-1}. 85% of each gram of the plant tissue contains cellulose and lignan, which, cannot be used by consumers and thus have no energy content. 15% of a gram of plant tissue contains: $(0.15)[17.6 + 23.8 + 39.7] = 12.16$ kJ energy.

3. Partition the incident light (100 W m^{-2}) in the following way for graphing: photosynthetic efficiency $= (0.015)(100) = 1.5$; reflectance $= (0.26)(100) = 26$; transpiration $= 100 -$ (photosythetic efficiency $+$ reflectance) $= 100 - (26 + 1.5) = 72.5$.

5. The amount of energy stored in 2 years is $2(1000 \text{ kJ m}^{-2} \text{ y}^{-1}) = 2000 \text{ kJ m}^{-2}$. At this rate it will take 10 years to accumulate 10,000 kJ of energy.

CHAPTER 13

1. Consider the dynamics of each compartment separately. For compartment 1, $dX_1/dt = J_3 - J_1 = g_3X_3 - g_1X_1$. At equilibrium, $g_3X_3 - g_1X_1 = 0$ and, using the numbers given in the problem, $0 = g_3(200) - (2.3)(100)$. Thus, $g_3 = 1.15$. For compartment 2, $dX_2/dt = J_1 - J_2 = g_1X_1 - g_2X_2$ and, at equilibrium, $0 = (2.3)(100) - g_2(150)$ giving $g_2 = 1.53$.

CHAPTER 14

4. The average number of visits, M, to each feeding site is 1. The Poisson probability P(x) is $M^x e^{-M}/x!$. The probability that the bird will not visit a site is $P(0) = 1^0 e^{-1}/0! = 0.36$. The probability that the bird will visit a site twice is $P(2) = 1^2 e^{-1}/2! = 0.18$.

6. Calculate the number of breeding males (N_m) and females (N_f) in each population. In the first population the sex ratio is 1:1 so ½ of the 10,000 individuals are male. Half (0.50) are breeding adults, so $N_m = (0.50)(5000) = 2500$. Likewise, $N_f = 2500$. Similar calculations for population B yield $N_m = 833$ and $N_f = 1666$. Next calculate the separate male and female effective population sizes for each population. For population A the calculations are:

$$N_{males} = (N_m K_m - 1)/[(K_m + V_m/K_m) - 1$$
$$= [(2500)(1) - 1]/[(1 + 1) - 1]$$
$$= 2499$$
$$N_{females} = (N_f K_f - 1)/[(K_f + V_f/K_f) - 1$$
$$= [(2500(2) - 1]/[(1 + 1/2) - 1]$$
$$= 9998$$

Using these results, calculate the effective population size N_e as follows:

$$N_e = 4[1/N_{males} + 1/N_{females}]^{-1}$$
$$= 4[1/2499 + 1/9998]$$
$$= 7999.6$$

Similar calculations for population B give $N_e = 7617$.

CHAPTER 15

1. Ten times the original population size is $(10)(N_0)$ or $10 \times 10 = 100 = N_t$. Substituting all we know into the equation for exponential growth, $N_t = N_o e^{rt}$ we have $100 = (10)(2.718)^t$. Dividing through by 10 gives $10 = 2.718^t$ which is in the form $x = a^y$ where $x = 10$, $a = 2.718$, and $y = t$. When $x = a^y$ then $y = \log_a x$. Thus, $t = \log_{2.718} 10$ or $t = \ln 10 = 2.3$.

CHAPTER 16

1. Recognize that $r = r_0(1 - N/K)$ may be rewritten as $r = r_0 - r_0 N/K$ which is the equation for a straight line with slope r_0/K. For $r_0 = 1.5$ and $K = 50$, the line is $r = 1.5 - 0.03N$.

CHAPTER 20

1. Recall that the consumer population growth rate will become negative when $1 - aC/k_R < 0$. Convert $k_R = 5$ kg/hr to 5000 g/hr to give it the same units as $a = 5$ g/hr then calculate $a/k_R = 5/5000 = 0.001$. The growth rate is negative when $1 - 0.001C < 0$ or when $C < 1000$ individuals. Note that the information about the efficiency with which resources are converted to reproduction, b, is not needed to solve the problem.

CHAPTER 21

6. The rate of growth of population i is calculated using equation 21 − 3. $dN_i/dt = (1.2)(1000)[1 - 1000/1800 - (0.8)(1000)/350] = 1.2$.

7. The carrying capacity of population i is $K_i = 150$ and its competitive effect on population j is $a_{ji} = 0.8$. Patterns of disturbance lower the size of population i and it is suggested that this prevents it from competitively excluding population j. First, use equation 21 − 4 to show that population i excludes population j at high levels of population i. Arbitrarily choose $N_i = 130$, very close to the carrying capacity of 150 and let $N_j = 100$. Calculation of dN_j/dt reveals that it is negative, indicating that population j is declining. Now, dramatically decrease the size of population i to, say, $N_i = 50$ to simulate the effect of a disturbance. Calculation of dN_j/dt reveals a positive rate. If the size of population i is kept small by disturbance, population j will grow even though population i has the growth potential to out compete it.

8. To make the algebra a bit easier let $q = m_i/r_i$. Thus, the equation becomes $N_i = 0 = [K_1(1 - q) - a_{12}K_2]/[1 - a_{12}a_{21}]$. Solving for q we obtain $q = 1 - a_{12}K_2/K_1$. Using the carrying capacity and competition coefficient values given above we determine that $q = (1 - 0.5(100))/200 = 0.75$ for the example given in Figure 23-16. Thus, when $q = m_i/r_i = 0.75$, the population of species i will have been eaten into nonexistence. Substituting the ratio m_i/r_i for q in the result above gives us $m_i/r_i = 1 - a_{ij}K_j/K_i$ from which we may obtain the following equation $m_i = r_i(1 - a_{ij}K_j/K_i)$, which represents the level of mortality at equilibrium.

CHAPTER 22

4. The Poisson distribution is given as $P(x) = M^x e^{-M}/x!$, where M is the mean number of events occurring within some unit of space or time. Since growth rate is related negatively to the number of competing neighbors within the zone of depletion of a plant, the frequency distribution of growth rates may be represented as the frequency distribution of the number of neighbors. The Poisson distribution may be used to model the number of neighbors where $x = 0, 1, 3 \ldots$ competing neighbors.

CHAPTER 23

2. The law of mass action is a chemical principle that applies strictly to randomly moving particles. In nature, predators and their prey do not move about at random and, thus, an increase in P or H may not necessarily lead to an increase in the number of encounters between individual predators and prey. The answer to this question lies in the details of the natural history of the predator and prey, which, of course, cannot be included in a general model. You will recall that interest in such details motivated Nicholson and Bailey to develop a model for parasitoids and their hosts.

4. Solve for p using the relationship $\hat{H} = d/ap(100 = 0.2/0.3p;\ p = 0.0067)$. Substitute p into the relationship $\hat{p} = r/p$ to obtain $\hat{p} = 0.7/0.0067 = 104.4$.

CHAPTER 24

4. In a population of $N = 1000$ individuals having $y = 10$ infected individuals, there are $x = 990$ uninfected individuals. The rate of change in the number of infected individuals is given by equation (24 − 2) as $dy/dt = \beta xy - (D + \gamma)y$. Substituting N, y, and into this equation and evaluating at $dy/dt = 0$ gives a threshold value $(D + \gamma)/\beta = 990$.

5. The herd immunity is given by the relationship $p_v = 1 - x_1/x_0$ where $x_1 = (D + \gamma)/\beta$ and $x_0 = R(x)(D + \gamma)/\beta$. From the problem we have D = 0.15, $\gamma = 0.85$, $\beta = 0.01$, and R(x) = 3. Thus, $p_v = 1 - ([0.15 + 0.85]/0.01]/(3[0.15 + 0.85])/0.01 = .67$. About 67% of the population should be vaccinated in order to stop the expansion of the disease.

CHAPTER 27

2. The number of species, N, is determined using the equation $N = n_o\sqrt{2\pi s^2}$, where n_o is the modal number and s is the breadth of the lognormal distribution. For $n_o = 35$ and s = 4.5, N = 368.

4. According to May's result, stability is achieved only when $SC < b^{-2}$. Thus, points (S,C) below the line $S = .3^{-2}/C$ denote stability.

CHAPTER 28

2. One of the best places to observe succession in animal communities is on the carcass of a dead animal. As it decays suites of animals, usually insects, are replaced by other types. Forensic scientists make good use of this in determining the time since death of abandoned bodies.

4. The state vector, **N**, is [10, 20, 30]. Any number of transition matrices (**P**) are possible since the entries are arbitrary numbers. The important feature is the sign of the entry. Following is a possible transition matrix:

$$\begin{bmatrix} 0.5 & 0 & 0 \\ -0.3 & 0.5 & 0.5 \\ 0.5 & 0.5 & 0.5 \end{bmatrix}$$

Using this matrix and performing the operation **PN**(t) yields new values for the three species of A = 5, B = 22, C = 30 for the first transition.

CHAPTER 30

1. If $p = 0.6$ then q, the frequency of the A_2 allele is 0.4 since $p + q = 1$. Using equation (30 − 3) ($\Delta q = -sq^2(1 - q)/(1 - sq^2)$) we calculate $\Delta q = -0.0198$.

4. The response to selection, R, is 2.3. Therefore S, the selection differential, is $R/h^2 = 2.3/.6 = 3.8$.

CHAPTER 31

2. If the predator eats both prey type 1 and prey type 2, $E/T = (\lambda_1 E_1 + \lambda_2 E_2)/(1 + \lambda_1 h_1 + \lambda_2 h_2) = 1.75$. If the predator eats just prey type 1, $E_1/T = \lambda_1 E_1/(1 + \lambda_1 h_1) = 0.4$. If the predator eats just prey type 2, $E_2/T = \lambda_2 E_2/(1 + \lambda_2 h_2) = 4.0$. The predator should specialize on prey type 2.

CHAPTER 32

1. At least part of the reason that humans mature more slowly has to do with the amount of time that it takes to learn about the environment and social interactions. There is evidence that learning plays a part in the life history adaptations of other organisms as well. (See Ricklefs and Finch 1995.)

3. We may balance the forces of mutation and selection using the following equation that gives the allele frequency in the next generation is $q_{t+1} = (1 - s)q_t + u$. At equilibrium, q will remain constant from generation to generation ($q_{t+1} = q_t = q$) and, solving for q, we obtain $q = u/s$. From this we can see that if selection against deleterious alleles diminishes with age and the mutation rate is age dependent (a reasonable assumption in most cases), the frequency of the deleterious allele will increase with age (Bulmer 1994).

CHAPTER 33

3. Use Table 33-2 to calculate the inequalities for each of the four possibilities. You will find that the inequalities will not be true in any case except where both parents desert. For example, if both parents provide care then both $vS_2 > VS_1$ (female) and $vS_2(1 + p) > vS_1(1 + P)$ (male) must be true. However, neither is true using the numbers given in the problem: for the females $10(.2) \not> 20(.2)$ and for males $(10)(0.2)(1.4) \not> (10)(.2)(1.6)$. The conditions hold when both parents desert: for females $20(.2) > 10(.2)$ and for males $(20)(.2)(.16) > (20)(0.2)(1.2)$.

INDEX

Conversion Factors

Length

1 meter (m) = 39.4 inches (in)
1 meter = 3.28 feet (ft)
1 kilometer (km) = 3,281 feet
1 kilometer = 0.621 mile (mi)
1 micron (μ) = 10^{-6} meter
1 inch = 2.54 centimeters (cm)
1 foot = 30.5 centimeters
1 mile = 1,609 meters
1 Angstrom unit (Å) = 10^{-10} meter
1 millimicron (mμ) = 10^{-9} meter

Area

1 square centimeter (cm^2) = 0.155 square inch (in^2)
1 square meter (m^2) = 10.76 square feet (ft^2)
1 hectare (ha) = 2.47 acres (A)
1 hectare = 10,000 square meters
1 hectare = 0.01 square kilometer (km^2)
1 square kilometer = 0.386 square mile
1 square mile = 2.59 square kilometers
1 square inch = 6.45 square centimeters
1 square foot = 929 square centimeters
1 square yard (yd^2) = 0.836 square meter
1 acre = 0.407 hectare

Mass

1 gram (g) = 15.43 grains (gr)
1 kilogram (kg) = 35.3 ounces
1 kilogram = 2.205 pounds (lb)
1 metric ton (t) = 2,204.6 pounds
1 ounce (oz) = 28.35 grams
1 pound = 453.6 grams
1 short ton = 907 kilograms

Time

1 year (yr) = 8,760 hours (hr)
1 day = 86,400 seconds (s)

Volume

1 cubic centimeter (cc or cm^3) = 0.061 cubic inch (in^3)
1 cubic inch = 16.4 cubic centimeters
1 liter = 1,000 cubic centimeters
1 liter = 33.8 U.S. fluid ounces (oz)
1 liter = 1.057 U.S. quarts (qt)
1 liter = 0.264 U.S. gallon (gal)
1 U.S. gallon = 3.79 liters
1 British gallon = 4.55 liters
1 cubic foot (ft^3) = 28.3 liters (l)
1 milliliter (ml) = 1 cubic centimeter
1 U.S. fluid ounce = 29.57 milliliters
1 British fluid ounce = 28.4 milliliters
1 quart = 0.946 liter

Velocity

1 meter per second ($m\ s^{-1}$) = 2.24 miles per hour ($mi\ hr^{-1}$)
1 foot per second ($ft\ s^{-1}$) = 1.097 kilometers per hour
1 kilometer per hour = 0.278 meter per second
1 mile per hour = 0.447 meter per second
1 mile per hour = 1.467 feet per second

Energy

1 joule = 0.239 calorie (cal)
1 calorie = 4.184 joules
1 kilowatt-hour (kWh) = 860 kilocalories
1 kilowatt-hour = 3,600 kilojoules
1 British thermal unit (Btu) = 252.0 calories
1 British thermal unit = 1,054 joules
1 kilocalorie (kcal) = 1,000 calories

Power

1 kilowatt (kW) = 0.239 kilocalorie per second
1 kilowatt = 860 kilocalories per hour
1 horsepower (hp) = 746 watts
1 horsepower = 15,397 kilocalories per day
1 horsepower = 641.5 kilocalories per hour

Energy per unit area

1 calorie per square centimeter = 3.69 British thermal units per square foot
1 British thermal unit per square foot = 0.271 calorie per square centimeter
1 calorie per square centimeter = 10 kilocalories per square meter

Power per unit area

1 kilocalorie per square meter per minute = 52.56 kilocalories per hectare per year
1 footcandle (fc) = 1.30 calories per square foot per hour at 555 mμ wavelength
1 footcandle = 10.76 lux
1 lux (lx) = 1.30 calories per square meter per hour at 555 mμ wavelength

Metabolic energy equivalents

1 gram of carbohydrate = 4.2 kilocalories
1 gram of protein = 4.2 kilocalories
1 gram of fat = 9.5 kilocalories

Miscellaneous

1 gram per square meter = 0.1 kilogram per hectare
1 gram per square meter = 8.97 pounds per acre
1 kilogram per square meter = 4.485 short tons per acre
1 metric ton per hectare = 0.446 short tons per acre